Intermediate Algebra: Graphs and Models

THIRD EDITION

Marvin L. Bittinger
Indiana University Purdue University Indianapolis

David J. Ellenbogen
Community College of Vermont

Barbara L. Johnson
Indiana University Purdue University Indianapolis

PEARSON
Addison Wesley

Boston San Francisco New York
London Toronto Sydney Tokyo Singapore Madrid
Mexico City Munich Paris Cape Town Hong Kong Montreal

Publisher	Greg Tobin
Editor in Chief	Maureen O'Connor
Acquisitions Editor	Randy Welch
Project Editor	Katie Nopper
Editorial Assistant	Antonio Arvelo
Production Manager	Ron Hampton
Editorial and Production Services	Martha K. Morong/Quadrata, Inc.
Art Editor and Photo Researcher	The Davis Group, Inc.
Compositor	BeaconPMG
Senior Media Producer	Ceci Fleming
Software Development	Mary Dougherty and Marty Wright
Marketing Manager	Jay Jenkins
Marketing Coordinator	Alexandra Waibel
Prepress Supervisor	Caroline Fell
Manufacturing Manager	Evelyn Beaton
Senior Media Buyer	Ginny Michaud
Design Supervisor	Dennis Schaefer
Text Designer	The Davis Group, Inc.
Cover Designer	Leslie Haimes
Cover Photograph	© Getty Images/Jeremy Walker

Photo credits appear on page A-60.

Library of Congress Cataloging-in-Publication Data

Bittinger, Marvin L.
Intermediate algebra: graphs and models / Marvin L. Bittinger, David J. Ellenbogen, Barbara L. Johnson.—3rd ed.
p. cm.

ISBN-13: 978-0-321-41616-2/ISBN-10: 0-321-41616-3
1. Algebra. I. Ellenbogen, David. II. Johnson, Barbara L. (Barbara Loreen), 1962–III. Title.

QA154.3.B5824 2007
512.9—dc22 2006044688

1 2 3 4 5 6 7 8 9 10—DOW—10 09 08 07

Contents

Selected Keys of the Graphing Calculator

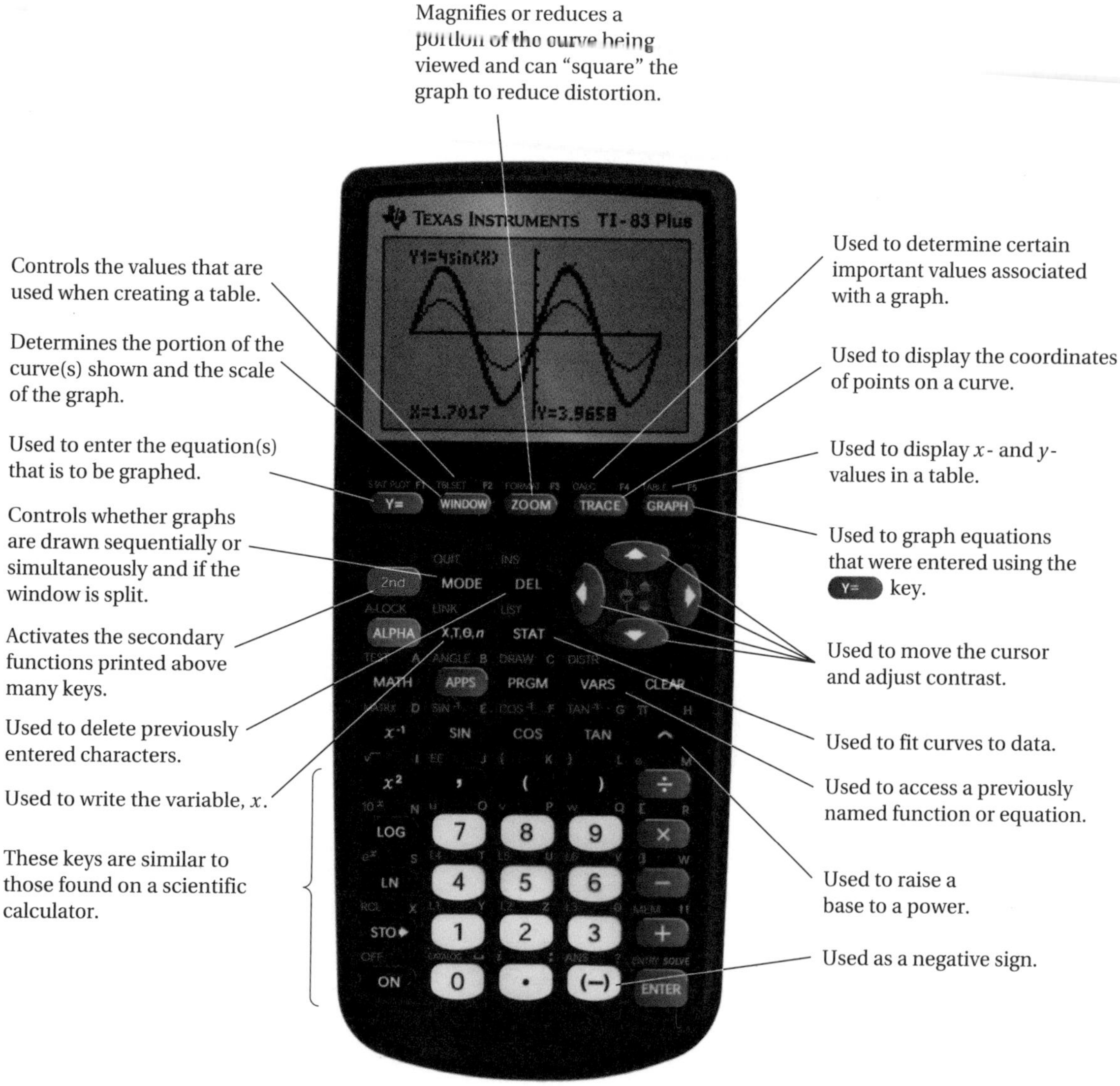
Magnifies or reduces a portion of the curve being viewed and can "square" the graph to reduce distortion.
Controls the values that are used when creating a table.
Determines the portion of the curve(s) shown and the scale of the graph.
Used to enter the equation(s) that is to be graphed.
Controls whether graphs are drawn sequentially or simultaneously and if the window is split.
Activates the secondary functions printed above many keys.
Used to delete previously entered characters.
Used to write the variable, x.
These keys are similar to those found on a scientific calculator.
Used to determine certain important values associated with a graph.
Used to display the coordinates of points on a curve.
Used to display x- and y-values in a table.
Used to graph equations that were entered using the Y= key.
Used to move the cursor and adjust contrast.
Used to fit curves to data.
Used to access a previously named function or equation.
Used to raise a base to a power.
Used as a negative sign.

9 Exponential and Logarithmic Functions 699

10 Conic Sections 791

11 Sequences, Series, and the Binomial Theorem 839

Preface

Appropriate for a one-term course in intermediate algebra, *Intermediate Algebra: Graphs and Models*, Third Edition, is intended for those students who have completed a first course in algebra. This text is more interactive than most other intermediate algebra texts. Our goal is to enhance the learning process by encouraging students to visualize the mathematics and by providing as much support as possible to help students in their study of algebra. This text is part of a series that includes the following texts:

- *Elementary Algebra: Graphs and Models*
- *Elementary and Intermediate Algebra: Graphs and Models,* Third Edition

Content Features

Problem Solving

One distinguishing feature of our approach is our treatment of and emphasis on problem solving. We use problem solving and applications to motivate the material wherever possible, and we include real-life applications and problem-solving techniques throughout the text. Problem solving not only encourages students to think about how mathematics can be used, it helps to prepare them for more advanced material in future courses.

In Chapter 1, we introduce the five-step process for solving problems: (1) *Familiarize*, (2) *Translate*, (3) *Carry out*, (4) *Check*, and (5) *State* the answer. These steps are then used consistently throughout the text whenever we encounter a problem-solving situation. Repeated use of this problem-solving strategy gives students a sense that they have a starting point for any type of problem they encounter, and frees them to focus on the mathematics necessary to successfully translate the problem situation. We often use estimation and carefully checked guesses to help with the *Familiarize* and *Check* steps (see pp. 72, 152–153, and 217).

Algebraic/Graphical Side-by-Sides

Algebraic/graphical side-by-sides give students a direct comparison between these two problem-solving approaches. They show the connection between algebraic and graphical or visual solutions and demonstrate that there is more than one way to obtain a result. This feature also illustrates the comparative efficiency and accuracy of the two methods. (See pp. 138, 223, and 419.) Instructors using this text have found that it works superbly both in courses where the graphing calculator is required and in courses where it is optional.

Applications

Interesting applications of mathematics help motivate both students and instructors. Solving applied problems gives students the opportunity to see their conceptual understanding put to use in a real way. In the third edition of *Intermediate Algebra: Graphs and Models*, the number of applications and source lines has been increased, and effort has been made to present the most current and relevant applications. As in the past, art is integrated into the applications and exercises to aid the student in visualizing the mathematics. (See pp. 89, 129, and 219.)

Interactive Discoveries

Interactive Discoveries invite students to develop analytical and reasoning skills while taking an active role in the learning process. These discoveries can be used as lecture launchers to introduce new topics at the beginning of a class and quickly guide students through a concept, or as out-of-class concept discoveries. (See pp. 117, 305, and 341.)

Pedagogical Features

New! **Concept Reinforcement Exercises.** This feature is designed to help students build their confidence and comprehension through true/false, matching, and fill-in-the-blank exercises at the beginning of most exercise sets. Whenever possible, special attention is devoted to increasing student understanding of the new vocabulary and notation developed in that section. (See pp. 170, 193, and 306.)

New! **Visualizing the Graph.** These matching exercises provide students with an opportunity to match an equation with its graph by focusing on the characteristics of the equation and the corresponding attributes of the graph. This feature occurs once in each chapter at the end of a related section. (See pp. 192, 330, and 423.)

New! **Student Notes.** These comments, strategically located in the margin within each section, are specific to the mathematics appearing on that page. Remarks are often more casual in format than the typical exposition and range from suggestions on how to avoid common mistakes to how to best read new mathematical notation. (See pp. 187, 201, and 312.)

New! **Tabbing for Success.** The new Tabbing for Success page, located before Chapter 1, provides students with forty color-coded, reusable tabs to quickly locate key examples, review important summaries, flag text topics, and highlight areas where they need help. Together, these features will help students better use their time and their textbooks to succeed in the course.

Connecting the Concepts. To help students understand the big picture, Connecting the Concepts subsections relate the concept at hand to previously learned and upcoming concepts. Because students occasionally lose sight of the forest because of the trees, this feature helps students keep their bearings as they encounter new material. (See pp. 150, 203, and 329.)

Study Tips. These remarks, located in the margin near the beginning of each section, provide suggestions for successful study habits that can be applied to both this and other college courses. Ranging from ideas for better time management to suggestions for test preparation, these comments can be useful even to experienced college students. (See pp. 133, 198, and 270.)

Skill Maintenance Exercises. Retention of skills is critical to a student's success in this and future courses. To this end, beginning in Section 1.2, every exercise set includes Skill Maintenance exercises that review skills and concepts from preceding sections of the text. Often, these exercises provide practice with specific skills needed for the next section of the text. **New to this edition,** some Skill Maintenance sections are titled **Focused Review.** These sections provide a set of mixed exercises **focused on reviewing connected skills,** such as factoring or solving equations, covered in separate sections or chapters. (See pp. 172, 230, and 332.)

Synthesis Exercises. Following the Skill Maintenance section, every exercise set ends with a group of Synthesis exercises that offers opportunities for students to synthesize skills and concepts from earlier sections with the present material, and often provide students with deeper insights into the current topic. Synthesis exercises are generally more challenging than those in the main body of the exercise set and occasionally include Aha! exercises (exercises that can be solved more quickly by intuition than by computation). (See pp. 116, 356, and 409.)

Thinking and Writing Exercises. Writing exercises have been found to aid in student comprehension, critical thinking, and conceptualization. Thus every set of exercises includes at least four thinking and writing exercises. Two of these appear just before the Skill Maintenance exercises. The other Thinking and Writing exercises are more challenging and appear as Synthesis exercises. All are marked with TW and require answers that are one or more complete sentences. Because some instructors may collect answers to writing exercises, and because more than one answer may be correct, answers to the Thinking and Writing exercises are listed at the back of the text only when they are within review exercises. (See pp. 131, 259, and 297.)

Collaborative Corners. Studies have shown that students who work together generally outperform those who do not. Throughout the text, we provide optional Collaborative Corner features that require students to work in groups to explore and solve problems. There is at least one Collaborative Corner per chapter, each one appearing after the appropriate exercise set. (See pp. 206, 260, and 281.)

Chapter Summary. Each chapter summary contains a list of key terms from the chapter, with definitions, as well as formulas from the chapter. The second part of the summary is a two-column table: An important concept is shown in the first column, with an example explaining that concept appearing in the second column. Page numbers or section references are provided so students can reference the corresponding exposition in the chapter. The Summary provides a terrific point from which to begin reviewing for a chapter test. (See pp. 174, 261, and 430.)

What's New in the Third Edition?

We have rewritten many key topics in response to user and reviewer feedback and have made significant improvements in design, art, pedagogy, and an expanded supplements package. Information about the content changes is available in the form of a conversion guide in the *Instructor and Adjunct Support Manual*. Following is a list of the major changes in this edition.

New Design

While incorporating a new layout, a fresh palette of colors, and new features, we have maintained the larger page dimension for an open look and a typeface that is easy to read. As always, it is our goal to make the text look mature without being intimidating. In addition, we continue to pay close attention to the pedagogical use of color to make sure that it is used to present concepts in the clearest possible manner.

Content Changes

A variety of content changes have been made throughout the text. Some of the more significant changes are listed below.

- Solving linear equations and applications is moved from Chapter 2 to Chapter 1, since these are topics that are covered thoroughly in Elementary Algebra.
- Chapter 1 now includes an improved discussion of the use of scientific notation and significant digits.
- Solving systems of equations using determinants and Cramer's rule is now included in Chapter 3.
- Solving equations and inequalities by graphing is discussed in a separate section after systems of equations are introduced. It is included in Chapter 4 to separate the idea from systems of equations.
- Chapter 6 now places greater emphasis on identifying the domain of a rational function.
- Conic sections have been moved from an appendix to Chapter 10, making the topic mainstream for those schools that require it. It will still be easy to omit for those schools that do not wish to cover it.
- An appendix covering unit conversion and dimension analysis is added for schools wishing to include this topic or for students who need a review of this topic.
- Throughout the text, there is an increased emphasis on students learning how to distinguish between equivalent expressions and equivalent equations.

Ancillaries

The following ancillaries are available to help both instructors and students use this text more effectively.

STUDENT SUPPLEMENTS

Student's Solutions Manual

(ISBN-13 978-0-321-42904-9)
(ISBN-10 0-321-42904-4)

- By James J. Ball and Rhea Meyerholtz, *Indiana State University*
- Contains completely worked-out solutions for all the odd-numbered exercises in the text, with the exception of the Thinking and Writing exercises, as well as completely worked-out solutions to all the exercises in the Chapter Reviews, Chapter Tests, and Cumulative Reviews.

Graphing Calculator Manual

(ISBN-13 978-0-321-42612-3)
(ISBN-10 0-321-42612-6)

- By Judith A. Penna, *Indiana University Purdue University Indianapolis*
- Uses actual examples and exercises from the text to help teach students to use the graphing calculator.
- Order of topics mirrors order in the text, providing a just-in-time mode of instruction.

Video Lectures on CD

(ISBN-13 978-0-321-42694-9)
(ISBN-10 0-321-42694-0)

- Complete set of digitized videos on CD-ROMs for student use at home or on campus.
- Presents a series of lectures correlated directly to the content of each section of the text.
- Features an engaging team of instructors including authors Barbara Johnson and David Ellenbogen who present material in a format that stresses student interaction, often using examples and exercises from the text.
- Ideal for distance learning or supplemental instruction.
- Includes an expandable window that shows text captioning. Captions can be turned on or off.

INSTRUCTOR SUPPLEMENTS

Annotated Instructor's Edition

(ISBN-13 978-0-321-42855-4)
(ISBN-10 0-321-42855-2

- Includes answers to all exercises printed in blue on the same page as those exercises.

Instructor's Solutions Manual

(ISBN-13 978-0-321-42534-8)
(ISBN-10 0-321-42534-0)

- By James J. Ball and Rhea Meyerholtz, *Indiana State University*
- Contains full, worked-out solutions to all the exercises in the exercise sets, including the Thinking and Writing exercises, and worked-out solutions to all the exercises in the Chapter Reviews, Chapter Tests, and Cumulative Reviews.

Printed Test Bank

(ISBN-13 978-0-321-42680-2)
(ISBN-10 0-321-42680-0)

- By Carrie Green
- Provides 8 revised test forms for every chapter and 8 revised test forms for the final exam.
- For the chapter tests, test forms are organized by topic order following the chapter tests in the text, and 2 test forms are multiple choice.

New! Instructor and Adjunct Support Manual

(ISBN-13 978-0-321-42856-1)
(ISBN-10 0-321-42856-0)

- Features resources and teaching tips designed to help both new and adjunct faculty with course preparation and classroom management.
- Resources include extra practice sheets, conversion guide, video index, and transparency masters.
- Also available electronically so course/adjunct coordinators can customize material specific to their schools.

STUDENT SUPPLEMENTS

Addison-Wesley Math Tutor Center
www.aw-bc.com/tutorcenter

- The Addison-Wesley Math Tutor Center is staffed by qualified mathematics instructors who provide students with tutoring on examples and odd-numbered exercises from the textbook. Tutoring is available via toll-free telephone, toll-free fax, e-mail, or the Internet. White Board technology allows tutors and students to actually see problems worked while they "talk" in real time over the Internet during tutoring sessions.

MathXL® Tutorials on CD

(ISBN-13 978-0-321-42783-0)
(ISBN-10 0-321-42783-1)

- Provides algorithmically generated practice exercises that correlate at the objective level to the content of the text.
- Includes an example and a guided solution to accompany every exercise and video clips for selected exercises.
- Recognizes student errors and provides feedback; generates printed summaries of students' progress.

INSTRUCTOR SUPPLEMENTS

TestGen with Quizmaster

(ISBN-13 978-0-321-42157-9)
(ISBN-10 0-321-42157-4)

- Enables instructors to build, edit, print, and administer tests.
- Features a computerized bank of questions developed to cover all text objectives.
- Algorithmically based content allows instructors to create multiple but equivalent versions of the same question or test with a click of a button.
- Instructors can also modify test-bank questions or add new questions by using the built-in question editor, which allows users to create graphs, input graphics, and insert math notation, variable numbers, or text.
- Tests can be printed or administered online via the Internet or another network. Quizmaster allows students to take tests on a local area network.
- Available on a dual-platform Windows/Macintosh CD-ROM.

MathXL® www.mathxl.com MathXL is a powerful online homework, tutorial, and assessment system that accompanies Addison-Wesley textbooks in mathematics or statistics. With MathXL, instructors can create, edit, and assign online homework and tests using algorithmically generated exercises correlated at the objective level to the textbook. They can also create and assign their own online exercises and import TestGen tests for added flexibility. All student work is tracked in MathXL's online gradebook. Students can take chapter tests in MathXL and receive personalized study plans based on their test results. The study plan diagnoses weaknesses and links students directly to tutorial exercises for the objectives they need to study and retest. Students can also access supplemental animations and video clips directly from selected exercises. MathXL is available to qualified adopters. For more information, visit our Web site at www.mathxl.com or contact your Addison-Wesley representative.

MyMathLab® www.mymathlab.com MyMathLab is a series of text-specific, easily customizable online courses for Addison-Wesley textbooks in mathematics and statistics. Powered by CourseCompass™ (Pearson Education's online teaching and learning environment) and MathXL® (our online homework, tutorial, and assessment system), MyMathLab gives instructors the tools they need to deliver all or a portion of their course online, whether students are in a lab setting or working from home. MyMathLab provides a rich and flexible set of course materials, featuring free-response exercises that are algorithmically generated for unlimited practice and mastery. Students can also use online tools, such as video lectures, animations, and a multimedia textbook, to independently improve their understanding and performance. Instructors can use MyMathLab's homework and test managers to select and assign online exercises correlated directly to the textbook, and they can also create and assign their own online exercises and import TestGen tests for added flexibility. MyMathLab's online gradebook—designed specifically for mathematics and statistics—automatically tracks students' homework and test results and gives the instructor control over how to calculate final grades. Instructors can also add offline (paper-and-pencil) grades to the gradebook. MyMathLab is available to qualified adopters. For more information, visit our Web site at www.mymathlab.com or contact your Addison-Wesley representative.

InterAct Math® Tutorial Web site www.interactmath.com Get practice and tutorial help online! This interactive tutorial Web site provides algorithmically generated practice exercises that correlate directly to the exercises in the textbook. Students can retry an exercise as many times as they like with new values each time for unlimited practice and mastery. Every exercise is accompanied by an interactive guided solution that provides helpful feedback for incorrect answers, and students can also view a worked-out sample problem that steps them through an exercise similar to the one they're working on.

Addison-Wesley Math Adjunct Support Center The Addison-Wesley Math Adjunct Support Center is staffed by qualified mathematics instructors with over 50 years of combined experience at both the community college and university level. Assistance is provided for faculty in the following areas:

- Suggested syllabus consultation
- Tips on using materials packed with your book
- Book-specific content assistance
- Teaching suggestions including advice on classroom strategies

For more information, visit www.aw-bc.com/tutorcenter/math-adjunct.html

Acknowledgments

No book can be produced without a team of professionals who take pride in their work and are willing to put in long hours. Thanks to James J. Ball and Rhea Meyerholtz, for authoring the *Instructor's Solutions Manual* and the *Student's Solutions Manual,* and to Carrie Green, for authoring the *Printed Test Bank.* Sybil MacBeth, Laurie A. Hurley, Holly Martinez, Paul Lorczak, and Jeremy Pletcher provided enormous help, often in the face of great time pressure, as accuracy checkers. We are also indebted to Scott Fallstrom for his help with applications research.

Martha Morong, of Quadrata, Inc., provided editorial and production services of the highest quality imaginable—she is simply a joy to work with. Geri Davis, of the Davis Group, Inc., performed superb work as designer, art editor, and photo researcher, and always with a disposition that can brighten an otherwise gray day. Network Graphics generated the graphs, charts, and many of the illustrations. Not only are the people at Network reliable, but they clearly take pride in their work. The many representational illustrations appear thanks to Bill Melvin, a gifted artist with true mathematical sensibilities.

Our team at Addison-Wesley deserves special thanks. Editorial Assistant Antonio Arvelo managed many of the day-to-day details—always in a pleasant and reliable manner. Project Editor Katie Nopper expertly provided information and a steadying influence along with gentle prodding at just the right moments. Acquisitions Editor Randy Welch provided many fine suggestions along with unflagging support. Production Manager Ron Hampton exhibited careful supervision and an eye for detail throughout production. Marketing Manager Jay Jenkins and Marketing Coordinator Alexandra Waibel skillfully kept us in touch with the needs of faculty; Senior Media Producer Ceci Fleming provided us with the technological guidance so necessary for our many supplements and our fine video series. To all of these people we owe a real debt of gratitude.

Reviewers

Marion Graziano, *Montgomery County Community College*
Michael Helinger, *Clinton Community College*
Celeste Hernandez, *Richland College*
Todd Hoff, *Wisconsin Indianhead Technical College*
Kandace Kling, *Portland Community College, Sylvania*
Julia Simms, *Southern Illinois University–Edwardsville*
Kay Willerton, *University of New Mexico–Los Alamos*

M.L.B.
D.J.E.
B.L.J.

Feature Walkthrough

What's New

New!

TABBING FOR SUCCESS

The new Tabbing for Success page before Chapter 1 provides students with forty color-coded, reusable tabs to quickly locate key examples, review important summaries, flag text topics, and highlight areas where they need help. Together, these features will help students better use their time and their textbooks to succeed in the course.

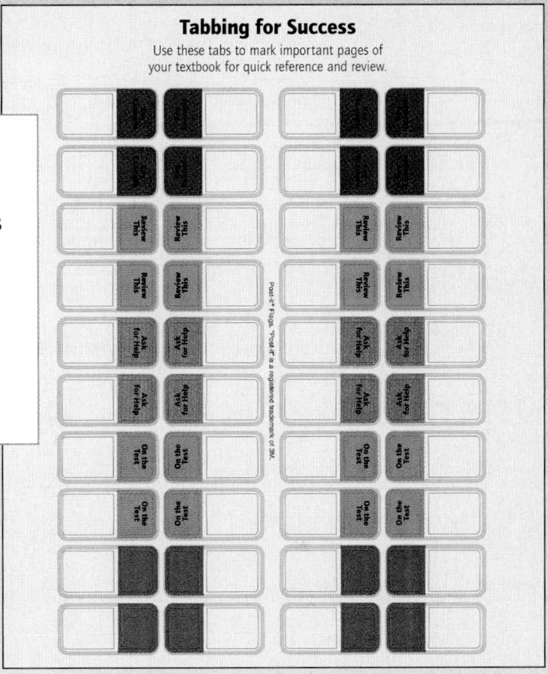
Tabbing for Success
Use these tabs to mark important pages of your textbook for quick reference and review.
Review This
Ask for Help
On the Test

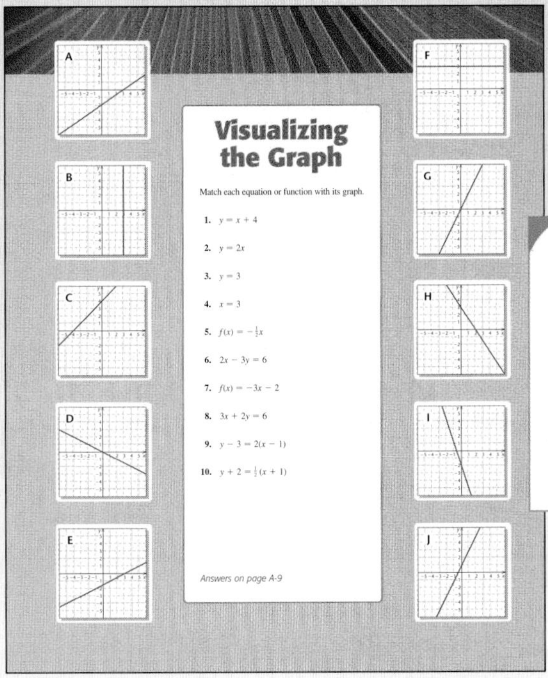
Visualizing the Graph
Match each equation or function with its graph.
1. $y = x + 4$
2. $y = 2x$
3. $y = 3$
4. $x = 3$
5. $f(x) = -\frac{1}{3}x$
6. $2x - 3y = 6$
7. $f(x) = -3x - 2$
8. $3x + 2y = 6$
9. $y - 3 = 2(x - 1)$
10. $y + 2 = \frac{1}{2}(x + 1)$
Answers on page A-9
A
B
C
D
E
F
G
H
I
J

New!

VISUALIZING THE GRAPH

Occurring once in each chapter, Visualizing the Graph matching exercises provide students with an opportunity to match an equation with its graph by focusing on the characteristics of the equation and the corresponding attributes of the graph.

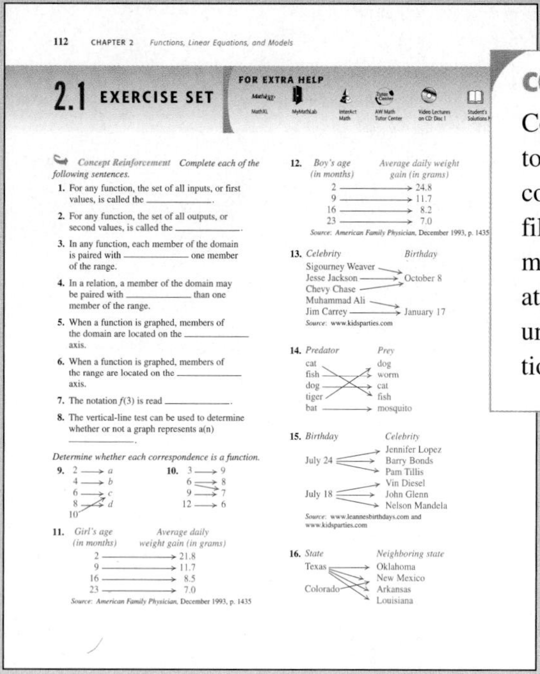
112 CHAPTER 2 Functions, Linear Equations, and Models

2.1 EXERCISE SET

FOR EXTRA HELP

MathXL MyMathLab InterAct Math AW Math Tutor Center Video Lectures on CD Disc 1 Student's Solutions Manual

Concept Reinforcement *Complete each of the following sentences.*

1. For any function, the set of all inputs, or first values, is called the ________.
2. For any function, the set of all outputs, or second values, is called the ________.
3. In any function, each member of the domain is paired with ________ one member of the range.
4. In a relation, a member of the domain may be paired with ________ than one member of the range.
5. When a function is graphed, members of the domain are located on the ________ axis.
6. When a function is graphed, members of the range are located on the ________ axis.
7. The notation $f(3)$ is read ________.
8. The vertical-line test can be used to determine whether or not a graph represents a(n) ________.

Determine whether each correspondence is a function.

9. 2 → a, 4 → b, 6 → c, 8 → d, 10 → d
10. 3 → 9, 6 → 8, 9 → 7, 12 → 6

11. *Girl's age (in months)* / *Average daily weight gain (in grams)*: 2 → 21.8, 9 → 11.7, 16 → 8.5, 23 → 7.0
Source: American Family Physician, December 1993, p. 1435

12. *Boy's age (in months)* / *Average daily weight gain (in grams)*: 2 → 24.8, 9 → 11.7, 16 → 8.2, 23 → 7.0
Source: American Family Physician, December 1993, p. 1435

13. *Celebrity* / *Birthday*: Sigourney Weaver, Jesse Jackson, Chevy Chase → October 8; Muhammad Ali, Jim Carrey → January 17
Source: www.kidsparties.com

14. *Predator* / *Prey*: cat, fish, dog, tiger, bat / dog, worm, cat, fish, mosquito

15. *Birthday* / *Celebrity*: July 24 → Jennifer Lopez, Barry Bonds, Pam Tillis; July 18 → Vin Diesel, John Glenn, Nelson Mandela
Source: www.leannesbirthdays.com and www.kidsparties.com

16. *State* / *Neighboring state*: Texas, Colorado / Oklahoma, New Mexico, Arkansas, Louisiana

CONCEPT REINFORCEMENT EXERCISES

New!

Concept Reinforcement Exercises are designed to help students build their confidence and comprehension through true/false, matching, and fill-in-the-blank exercises at the beginning of most exercise sets. Whenever possible, special attention is devoted to increasing student understanding of the new vocabulary and notation developed in that section.

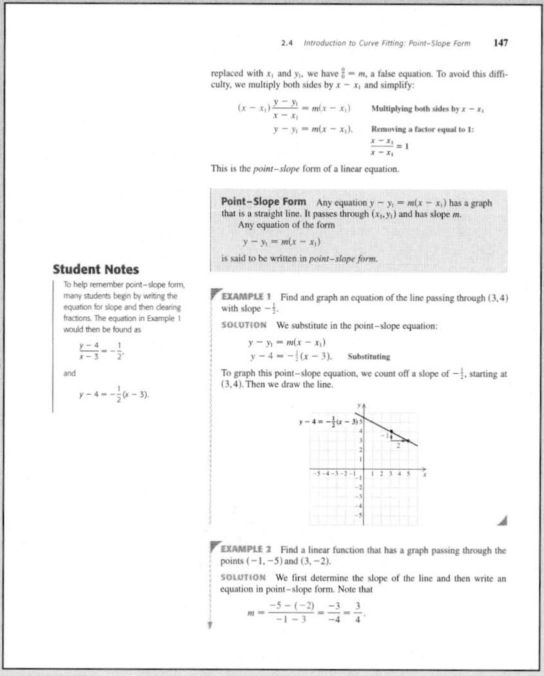
2.4 Introduction to Curve Fitting: Point–Slope Form 147

replaced with x_1 and y_1, we have $\frac{0}{0} = m$, a false equation. To avoid this difficulty, we multiply both sides by $x - x_1$ and simplify:

$$(x - x_1)\frac{y - y_1}{x - x_1} = m(x - x_1) \quad \text{Multiplying both sides by } x - x_1$$

$$y - y_1 = m(x - x_1). \quad \text{Removing a factor equal to 1: } \frac{x - x_1}{x - x_1} = 1$$

This is the *point–slope* form of a linear equation.

Point–Slope Form Any equation $y - y_1 = m(x - x_1)$ has a graph that is a straight line. It passes through (x_1, y_1) and has slope m. Any equation of the form $y - y_1 = m(x - x_1)$ is said to be written in *point–slope form.*

Student Notes
To help remember point–slope form, many students begin by writing the equation for slope and then clearing fractions. The equation in Example 1 would then be found as $\frac{y-4}{x-3} = -\frac{1}{2}$, and $y - 4 = -\frac{1}{2}(x - 3)$.

EXAMPLE 1 Find and graph an equation of the line passing through (3, 4) with slope $-\frac{1}{2}$.

SOLUTION We substitute in the point–slope equation:

$y - y_1 = m(x - x_1)$

$y - 4 = -\frac{1}{2}(x - 3)$. Substituting

To graph this point–slope equation, we count off a slope of $-\frac{1}{2}$, starting at (3, 4). Then we draw the line.

EXAMPLE 2 Find a linear function that has a graph passing through the points (−1, −5) and (3, −2).

SOLUTION We first determine the slope of the line and then write an equation in point–slope form. Note that

$$m = \frac{-5 - (-2)}{-1 - 3} = \frac{-3}{-4} = \frac{3}{4}.$$

STUDENT NOTES

New!

Student Notes are strategically located remarks in the margin within each section, and are specific to the mathematics appearing on that page. They range from suggestions on how to avoid common mistakes to how to best read new mathematical notation.

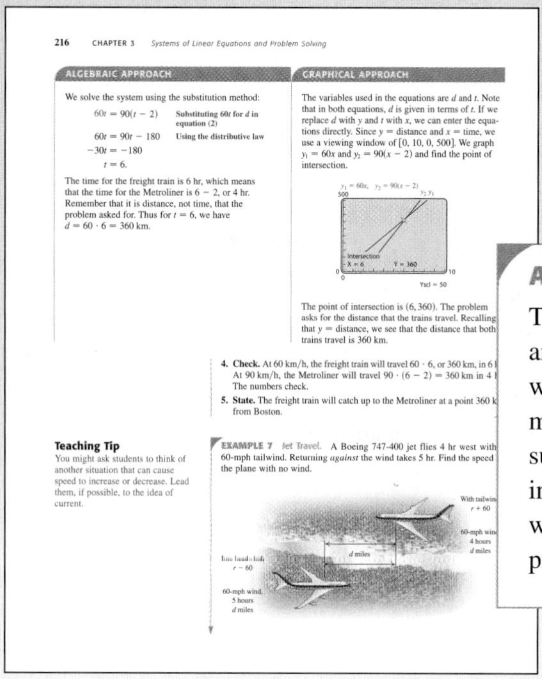
216 CHAPTER 3 Systems of Linear Equations and Problem Solving

ALGEBRAIC APPROACH

We solve the system using the substitution method:

$60t = 90(t - 2)$ Substituting $60t$ for d in equation (2)

$60t = 90t - 180$ Using the distributive law

$-30t = -180$

$t = 6.$

The time for the freight train is 6 hr, which means that the time for the Metroliner is 6 − 2, or 4 hr. Remember that it is distance, not time, that the problem asked for. Thus for $t = 6$, we have $d = 60 \cdot 6 = 360$ km.

GRAPHICAL APPROACH

The variables used in the equations are d and t. Note that in both equations, d is given in terms of t. If we replace d with y and t with x, we can enter the equations directly. Since y = distance and x = time, we use a viewing window of [0, 10, 0, 500]. We graph $y_1 = 60x$ and $y_2 = 90(x - 2)$ and find the point of intersection.

The point of intersection is (6, 360). The problem asks for the distance that the trains travel. Recalling that y = distance, we see that the distance that both trains travel is 360 km.

4. **Check.** At 60 km/h, the freight train will travel 60 · 6, or 360 km, in 6 hr. At 90 km/h, the Metroliner will travel 90 · (6 − 2) = 360 km in 4 hr. The numbers check.
5. **State.** The freight train will catch up to the Metroliner at a point 360 km from Boston.

Teaching Tip
You might ask students to think of another situation that can cause speed to increase or decrease. Lead them, if possible, to the idea of current.

EXAMPLE 7 *Jet Travel.* A Boeing 747-400 jet flies 4 hr west with a 60-mph tailwind. Returning *against* the wind takes 5 hr. Find the speed of the plane with no wind.

ANNOTATED INSTRUCTOR'S EDITION

The Annotated Instructor's Edition includes all the answers to the exercise sets, usually right on the page where the exercises appear, and Teaching Tips in the margins that give insights and classroom discussion suggestions that will be especially useful for new instructors. These handy answers and Teaching Tips will help both new and experienced instructors save preparation time.

CHAPTER OPENERS

Each chapter opens with a real-data application, including a data table (numerical representation) and a graphical representation of the data. Data tables and graphs are used frequently throughout the body of the text to show the relevance of the material as well as to appeal to the visual learners.

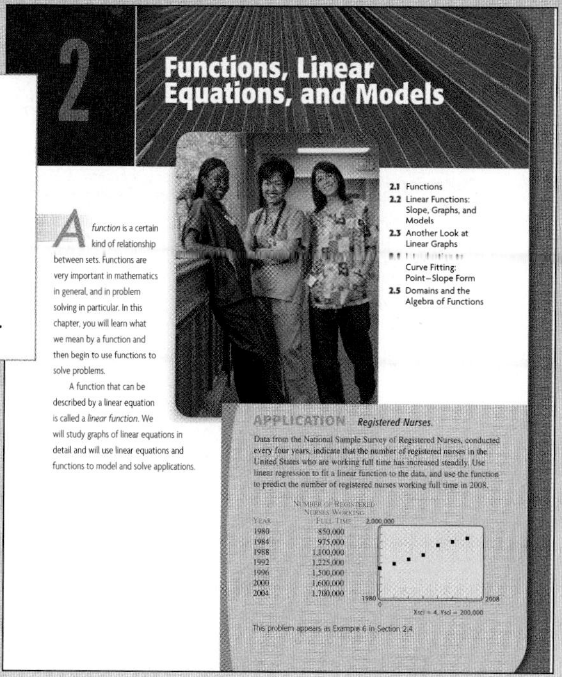
2

Functions, Linear Equations, and Models

A *function* is a certain kind of relationship between sets. Functions are very important in mathematics in general, and in problem solving in particular. In this chapter, you will learn what we mean by a function and then begin to use functions to solve problems.

A function that can be described by a linear equation is called a *linear function*. We will study graphs of linear equations in detail and will use linear equations and functions to model and solve applications.

APPLICATION *Registered Nurses.*

Data from the National Sample Survey of Registered Nurses, conducted every four years, indicate that the number of registered nurses in the United States who are working full time has increased steadily. Use linear regression to fit a linear function to the data, and use the function to predict the number of registered nurses working full time in 2008.

Year	Number of Registered Nurses Working Full Time
1980	850,000
1984	975,000
1988	1,100,000
1992	1,225,000
1996	1,500,000
2000	1,600,000
2004	1,700,000

This problem appears as Example 6 in Section 2.4.

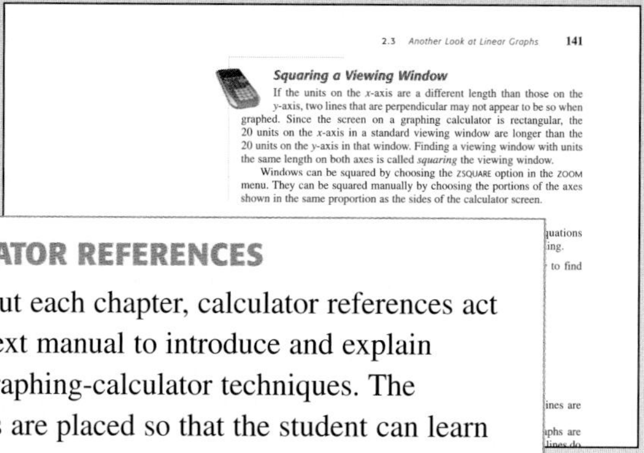

Squaring a Viewing Window

If the units on the x-axis are a different length than those on the y-axis, two lines that are perpendicular may not appear to be so when graphed. Since the screen on a graphing calculator is rectangular, the 20 units on the x-axis in a standard viewing window are longer than the 20 units on the y-axis in that window. Finding a viewing window with units the same length on both axes is called *squaring* the viewing window.

Windows can be squared by choosing the ZSQUARE option in the ZOOM menu. They can be squared manually by choosing the portions of the axes shown in the same proportion as the sides of the calculator screen.

CALCULATOR REFERENCES

Throughout each chapter, calculator references act as an in-text manual to introduce and explain various graphing-calculator techniques. The references are placed so that the student can learn the technique just prior to practicing it.

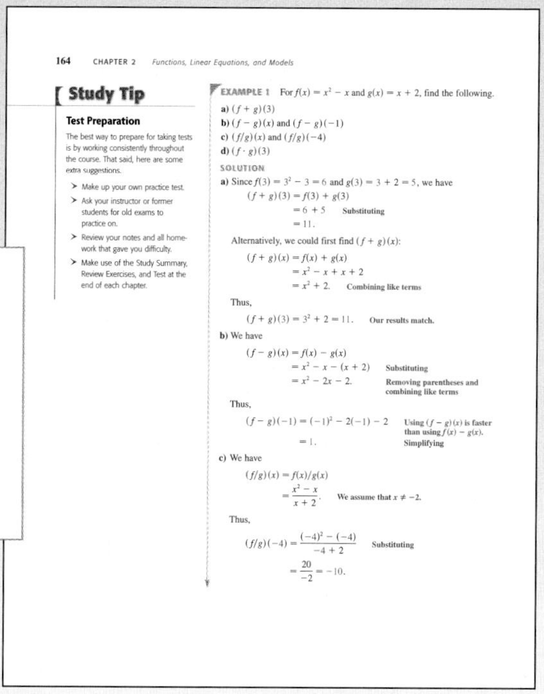

Study Tip

Test Preparation

The best way to prepare for taking tests is by working consistently throughout the course. That said, here are some extra suggestions.

- Make up your own practice test.
- Ask your instructor or former students for old exams to practice on.
- Review your notes and all homework that gave you difficulty.
- Make use of the Study Summary, Review Exercises, and Test at the end of each chapter.

EXAMPLE 1 For $f(x) = x^2 - x$ and $g(x) = x + 2$, find the following.

a) $(f + g)(3)$
b) $(f - g)(x)$ and $(f - g)(-1)$
c) $(f/g)(x)$ and $(f/g)(-4)$
d) $(f \cdot g)(3)$

SOLUTION

a) Since $f(3) = 3^2 - 3 = 6$ and $g(3) = 3 + 2 = 5$, we have

$(f + g)(3) = f(3) + g(3)$
$= 6 + 5$ Substituting
$= 11.$

Alternatively, we could first find $(f + g)(x)$:

$(f + g)(x) = f(x) + g(x)$
$= x^2 - x + x + 2$
$= x^2 + 2.$ Combining like terms

Thus,

$(f + g)(3) = 3^2 + 2 = 11.$ Our results match.

b) We have

$(f - g)(x) = f(x) - g(x)$
$= x^2 - x - (x + 2)$ Substituting
$= x^2 - 2x - 2.$ Removing parentheses and combining like terms

Thus,

$(f - g)(-1) = (-1)^2 - 2(-1) - 2$ Using $(f - g)(x)$ is faster than using $f(x) - g(x)$.
$= 1.$ Simplifying

c) We have

$(f/g)(x) = f(x)/g(x)$
$= \frac{x^2 - x}{x + 2}.$ We assume that $x \neq -2$.

Thus,

$(f/g)(-4) = \frac{(-4)^2 - (-4)}{-4 + 2}$ Substituting
$= \frac{20}{-2} = -10.$

STUDY TIPS

To help students develop good study habits throughout this course, Study Tips are placed near the beginning of each section in the margins. These tips range from how to approach assignments, to reminders of the various study aids that are available, to strategies for preparing for a final exam.

Examples

168 CHAPTER 2 *Functions, Linear Equations, and Models*

CAUTION! A graphing calculator is useful for checking whether a particular value of x is in the domain of a function. However, algebraic methods are needed in order to determine the entire domain, since it is impossible to check every real number with a calculator.

Interactive Discovery

Let $f(x) = \frac{5}{x}$ and $g(x) = \frac{2x - 6}{x + 1}$. Enter $y_1 = f(x)$, $y_2 = g(x)$, and $y_3 = y_1 + y_2$. Create a table of values for the functions with TblStart $= -3$, ΔTbl $= 1$, and Indpnt set to Auto.

1. What number is not in the domain of f?
2. What number is not in the domain of g?
3. What numbers are not in the domain of $f + g$?

Now enter $y_4 = y_1 - y_2$, $y_5 = y_1 \cdot y_2$, and $y_6 = y_1/y_2$. Create a table of values for the functions.

4. Does the domain of $f - g$ appear to be the same as the domain of $f + g$?
5. Does the domain of $f \cdot g$ appear to be the same as the domain of $f + g$?
6. Does the domain of f/g appear to be the same as the domain of $f + g$?

In the Interactive Discovery above, because division by 0 is not defined, we have

the domain of $f = \{x \mid x$ is a real number *and* $x \neq 0\}$

and

the domain of $g = \{x \mid x$ is a real number *and* $x \neq -1\}$.

In order to find $f(a) + g(a)$, $f(a) - g(a)$, or $f(a) \cdot g(a)$, we must know that a is in *both* of the above domains. Thus,

the domain of $f + g$ = the domain of $f - g$ = the domain of $f \cdot g$ $= \{x \mid x$ is a real number *and* $x \neq 0$ *and* $x \neq -1\}$.

The domain of f/g also excludes the number 3, because $g(3) = 0$:

the domain of $f/g = \{x \mid x$ is a real number *and* $x \neq 0$ *and* $x \neq -1$ *and* $x \neq 3\}$.

INTERACTIVE DISCOVERIES

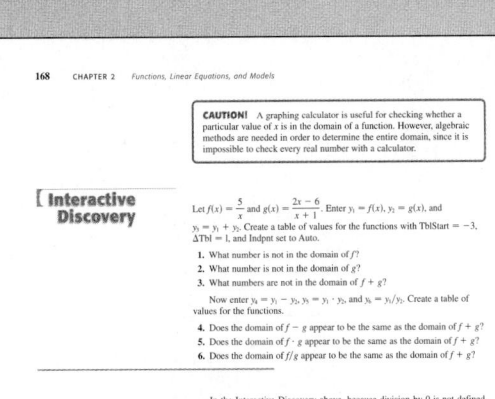

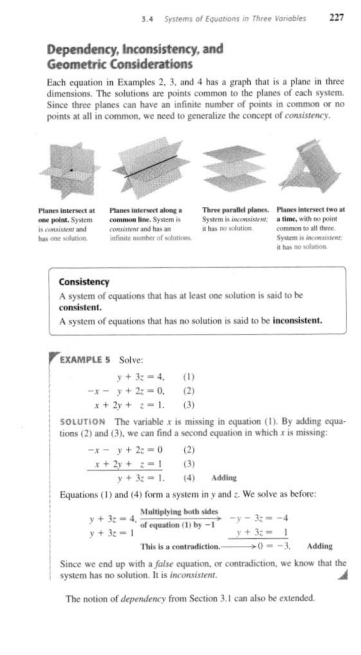

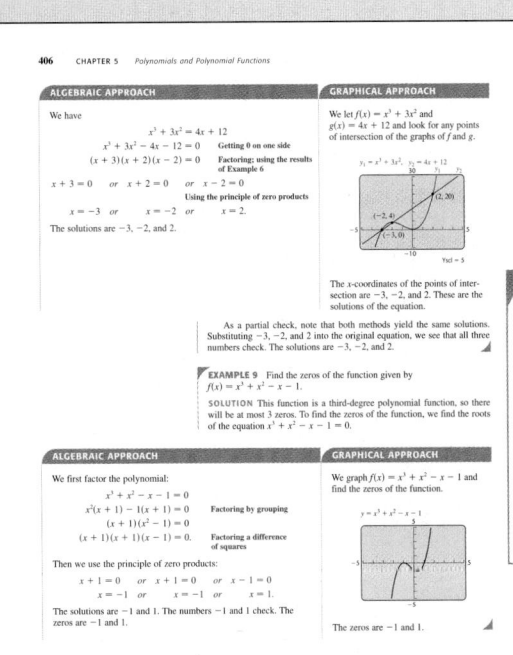

Interactive Discoveries invite students to develop analytical and reasoning skills while taking an active role in the learning process. These discoveries can be used as lecture launchers to introduce new topics at the start of a class and quickly guide students through a concept, or as out-of-class concept discoveries.

3.4 *Systems of Equations in Three Variables* 227

Dependency, Inconsistency, and Geometric Considerations

Each equation in Examples 2, 3, and 4 has a graph that is a plane in three dimensions. The solutions are points common to the planes of each system. Since three planes can have an infinite number of points in common or no points at all in common, we need to generalize the concept of *consistency*.

Planes intersect at one point. System is *consistent* and has one solution. **Planes intersect along a common line.** System is *consistent* and has an infinite number of solutions. **Three parallel planes.** System is *inconsistent;* it has no solution. **Planes intersect two at a time,** with no point common to all three. System is *inconsistent;* it has no solution.

Consistency

A system of equations that has at least one solution is said to be **consistent.**

A system of equations that has no solution is said to be **inconsistent.**

EXAMPLE 5 Solve:

$$\begin{aligned} y + 3z &= 4, && (1)\\ -x - y + 2z &= 0, && (2)\\ x + 2y + z &= 1. && (3) \end{aligned}$$

SOLUTION The variable x is missing in equation (1). By adding equations (2) and (3), we can find a second equation in which x is missing:

$$\begin{aligned} -x - y + 2z &= 0 && (2)\\ x + 2y + z &= 1 && (3)\\ \hline y + 3z &= 1. && (4) \quad \textbf{Adding} \end{aligned}$$

Equations (1) and (4) form a system in y and z. We solve as before:

$$\begin{aligned} y + 3z &= 4, \xrightarrow[\text{of equation (1) by } -1]{\text{Multiplying both sides}} & -y - 3z &= -4\\ y + 3z &= 1 & y + 3z &= 1\\ & \textbf{This is a contradiction.} \longrightarrow & 0 &= -3. \quad \textbf{Adding} \end{aligned}$$

Since we end up with a *false* equation, or contradiction, we know that the system has no solution. It is *inconsistent*.

The notion of *dependency* from Section 3.1 can also be extended.

ANNOTATED EXAMPLES

Learning is carefully guided with numerous color-coded art pieces and step-by-step annotations alongside examples. Substitutions and annotations are highlighted in red so students can see exactly what is happening in each step.

406 CHAPTER 5 *Polynomials and Polynomial Functions*

ALGEBRAIC APPROACH

We have

$$\begin{aligned} x^3 + 3x^2 &= 4x + 12\\ x^3 + 3x^2 - 4x - 12 &= 0 && \textbf{Getting 0 on one side}\\ (x + 3)(x + 2)(x - 2) &= 0 && \textbf{Factoring; using the results of Example 6} \end{aligned}$$

$x + 3 = 0$ *or* $x + 2 = 0$ *or* $x - 2 = 0$ **Using the principle of zero products**

$x = -3$ *or* $x = -2$ *or* $x = 2$.

The solutions are -3, -2, and 2.

GRAPHICAL APPROACH

We let $f(x) = x^3 + 3x^2$ and $g(x) = 4x + 12$ and look for any points of intersection of the graphs of f and g.

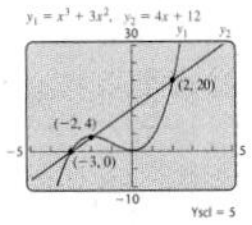

The x-coordinates of the points of intersection are -3, -2, and 2. These are the solutions of the equation.

As a partial check, note that both methods yield the same solutions. Substituting -3, -2, and 2 into the original equation, we see that all three numbers check. The solutions are -3, -2, and 2.

EXAMPLE 9 Find the zeros of the function given by $f(x) = x^3 + x^2 - x - 1$.

SOLUTION This function is a third-degree polynomial function, so there will be at most 3 zeros. To find the zeros of the function, we find the roots of the equation $x^3 + x^2 - x - 1 = 0$.

ALGEBRAIC APPROACH

We first factor the polynomial:

$$\begin{aligned} x^3 + x^2 - x - 1 &= 0\\ x^2(x + 1) - 1(x + 1) &= 0 && \textbf{Factoring by grouping}\\ (x + 1)(x^2 - 1) &= 0\\ (x + 1)(x + 1)(x - 1) &= 0. && \textbf{Factoring a difference of squares} \end{aligned}$$

Then we use the principle of zero products:

$x + 1 = 0$ *or* $x + 1 = 0$ *or* $x - 1 = 0$

$x = -1$ *or* $x = -1$ *or* $x = 1$.

The solutions are -1 and 1. The numbers -1 and 1 check. The zeros are -1 and 1.

GRAPHICAL APPROACH

We graph $f(x) = x^3 + x^2 - x - 1$ and find the zeros of the function.

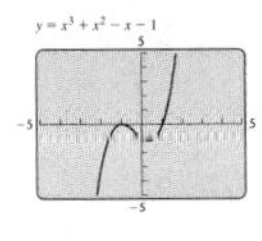

The zeros are -1 and 1.

ALGEBRAIC/GRAPHICAL SIDE-BY-SIDES

Algebraic/Graphical Side-by-Sides give students a direct comparison between these two problem-solving approaches. They show the connection between algebraic and graphical or visual solutions and demonstrate that there is more than one way to obtain a result. This feature also illustrates the comparative efficiency and accuracy of the two methods.

Student Guidance

FIVE-STEP PROBLEM-SOLVING PROCESS

Bittinger's five-step problem-solving process (Familiarize, Translate, Carry Out, Check, State) is introduced early in the text and used consistently throughout the text whenever students encounter an application problem. This provides students with a consistent framework for solving application problems.

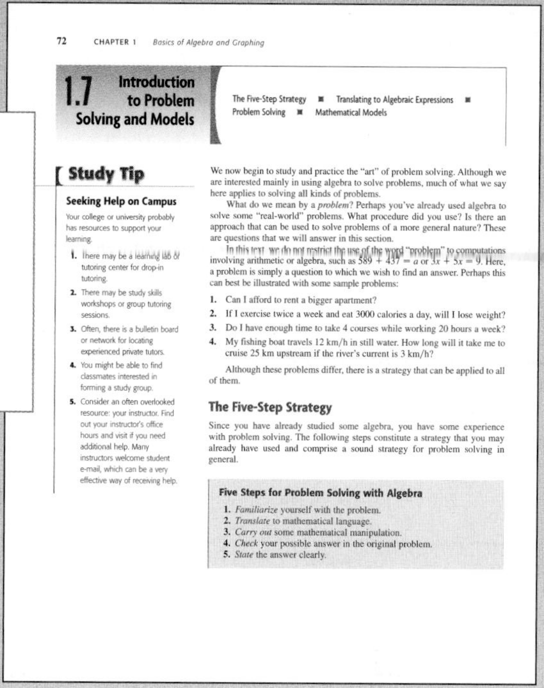
72 CHAPTER 1 *Basics of Algebra and Graphing*

1.7 Introduction to Problem Solving and Models

The Five-Step Strategy ■ Translating to Algebraic Expressions ■ Problem Solving ■ Mathematical Models

Study Tip

Seeking Help on Campus

Your college or university probably has resources to support your learning.

1. There may be a learning lab or tutoring center for drop-in tutoring.
2. There may be study skills workshops or group tutoring sessions.
3. Often, there is a bulletin board or network for locating experienced private tutors.
4. You might be able to find classmates interested in forming a study group.
5. Consider an often overlooked resource: your instructor. Find out your instructor's office hours and visit if you need additional help. Many instructors welcome student e-mail, which can be a very effective way of receiving help.

We now begin to study and practice the "art" of problem solving. Although we are interested mainly in using algebra to solve problems, much of what we say here applies to solving all kinds of problems.

What do we mean by a *problem*? Perhaps you've already used algebra to solve some "real-world" problems. What procedure did you use? Is there an approach that can be used to solve problems of a more general nature? These are questions that we will answer in this section.

In this text, we do not restrict the use of the word "problem" to computations involving arithmetic or algebra, such as $589 + 437 = a$ or $3x + 5x = 9$. Here, a problem is simply a question to which we wish to find an answer. Perhaps this can best be illustrated with some sample problems:

1. Can I afford to rent a bigger apartment?
2. If I exercise twice a week and eat 3000 calories a day, will I lose weight?
3. Do I have enough time to take 4 courses while working 20 hours a week?
4. My fishing boat travels 12 km/h in still water. How long will it take me to cruise 25 km upstream if the river's current is 3 km/h?

Although these problems differ, there is a strategy that can be applied to all of them.

The Five-Step Strategy

Since you have already studied some algebra, you have some experience with problem solving. The following steps constitute a strategy that you may already have used and comprise a sound strategy for problem solving in general.

Five Steps for Problem Solving with Algebra

1. *Familiarize* yourself with the problem.
2. *Translate* to mathematical language.
3. *Carry out* some mathematical manipulation.
4. *Check* your possible answer in the original problem.
5. *State* the answer clearly.

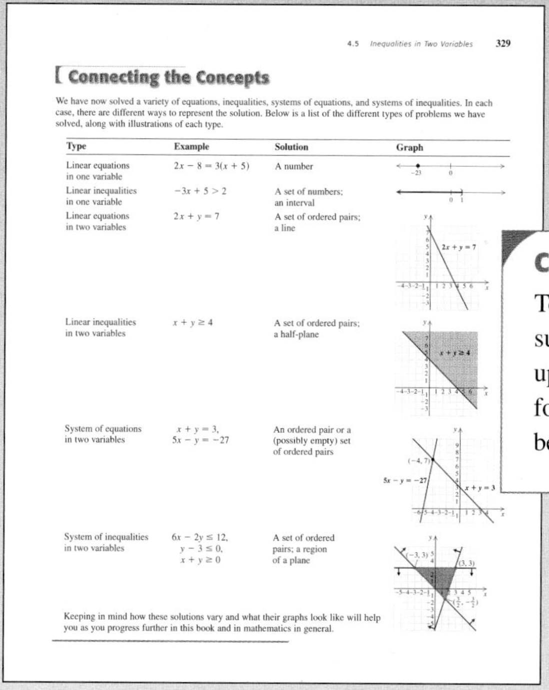
4.5 *Inequalities in Two Variables* 329

Connecting the Concepts

We have now solved a variety of equations, inequalities, systems of equations, and systems of inequalities. In each case, there are different ways to represent the solution. Below is a list of the different types of problems we have solved, along with illustrations of each type.

Type	Example	Solution	Graph
Linear equations in one variable	$2x - 8 = 3(x + 5)$	A number	
Linear inequalities in one variable	$-3x + 5 > 2$	A set of numbers; an interval	
Linear equations in two variables	$2x + y = 7$	A set of ordered pairs; a line	
Linear inequalities in two variables	$x + y \geq 4$	A set of ordered pairs; a half-plane	
System of equations in two variables	$x + y = 3,$ $5x - y = -27$	An ordered pair or a (possibly empty) set of ordered pairs	
System of inequalities in two variables	$6x - 2y \leq 12,$ $y - 3 \leq 0,$ $x + y \geq 0$	A set of ordered pairs; a region of a plane	

Keeping in mind how these solutions vary and what their graphs look like will help you as you progress further in this book and in mathematics in general.

CONNECTING THE CONCEPTS

To help students understand the big picture, Connecting the Concepts subsections relate the concept at hand to previously learned and upcoming concepts. Because students occasionally lose sight of the forest because of the trees, this feature helps students keep their bearings as they encounter new material.

COLLABORATIVE CORNERS

Collaborative Corner exercises appear one to three times per chapter, after the section exercise set. Designed for group work, these optional activities provide instructors and students with the opportunity to incorporate group learning and/or class discussions.

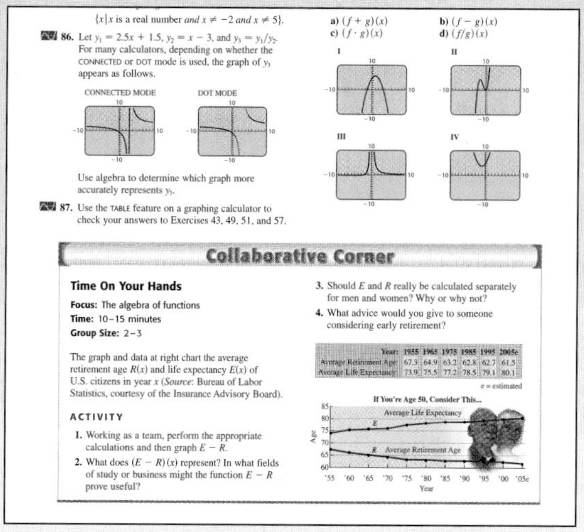
$\{x \mid x$ is a real number *and* $x \neq -2$ *and* $x \neq 5\}$.

86. Let $y_1 = 2.5x + 1.5$, $y_2 = x - 3$, and $y_3 = y_1/y_2$. For many calculators, depending on whether the CONNECTED or DOT mode is used, the graph of y_3 appears as follows.

Use algebra to determine which graph more accurately represents y_3.

87. Use the TABLE feature on a graphing calculator to check your answers to Exercises 43, 49, 51, and 57.

a) $(f + g)(x)$ b) $(f - g)(x)$
c) $(f \cdot g)(x)$ d) $(f/g)(x)$

Collaborative Corner

Time On Your Hands

Focus: The algebra of functions
Time: 10–15 minutes
Group Size: 2–3

The graph and data at right chart the average retirement age $R(x)$ and life expectancy $E(x)$ of U.S. citizens in year x (*Source:* Bureau of Labor Statistics, courtesy of the Insurance Advisory Board).

ACTIVITY

1. Working as a team, perform the appropriate calculations and then graph $E - R$.
2. What does $(E - R)(x)$ represent? In what fields of study or business might the function $E - R$ prove useful?
3. Should E and R really be calculated separately for men and women? Why or why not?
4. What advice would you give to someone considering early retirement?

Year:	1955	1965	1975	1985	1995	2005e
Average Retirement Age	67.3	64.9	63.2	62.8	62.7	61.5
Average Life Expectancy	73.9	75.5	77.2	78.5	79.1	80.1

e = estimated

If You're Age 50, Consider This...

Exercises

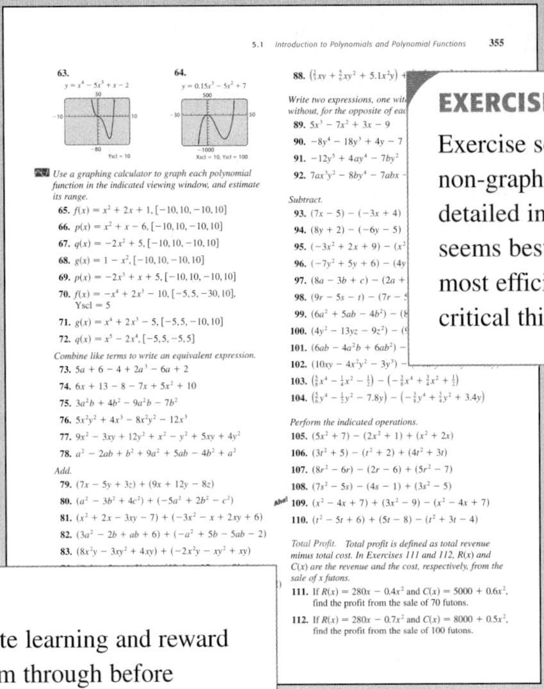

EXERCISE SETS

Exercise sets include both graphing calculator and non-graphing calculator exercises. In some cases, detailed instruction lines indicate the approach that seems best, and others let students choose the most efficient approach, thereby encouraging critical thinking.

AHA! EXERCISES

In an effort to discourage rote learning and reward students who think a problem through before solving it, we include exercises that can be solved quickly if the student has the proper insight. The *Aha!* designation is used the first time a new insight can be used on a particular type of exercise and indicates to the student that there is a way to complete the exercise with more concise computation. It's then up to the student to find the simpler approach and, in subsequent exercises, to determine if and when that particular insight might be used again.

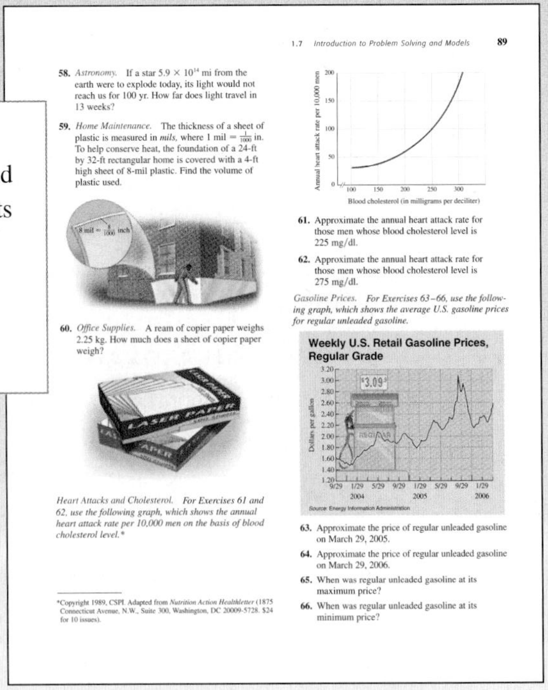

REAL-DATA APPLICATIONS

Real-data applications, often denoted in examples and exercises with a source line, serve to motivate students by showing them how the mathematics they are studying in their text relates to their everyday lives. The number of application problems has been increased significantly in this edition.

Determine whether each equation is linear. Find the slope of any nonvertical lines.

83. $5x - 3y = 15$

84. $3x + 5y + 15 = 0$

85. $16 + 4y = 10$

86. $3x - 12 = 0$

87. $3g(x) = 6x^2$

88. $2x + 4f(x) = 8$

89. $3y = 7(2x - 4)$

90. $2(5 - 3x) = 5y$

91. $g(x) - \frac{1}{x} = 0$

92. $f(x) + \frac{1}{x} = 0$

93. $\frac{f(x)}{5} = x^2$

94. $\frac{g(x)}{2} = 3 + x$

TW **95.** *Engineering.* Wind friction, or *air resistance,* increases with speed. Following are some measurements made in a wind tunnel. Plot the data and explain why a linear function does or does not give an approximate fit.

45	15.1
52	29.0

THINKING AND WRITING EXERCISES

Thinking and Writing exercises, denoted by TW, aid student comprehension and can also be used in class for discussion or group projects. Two appear right before the Skill Maintenance exercises, and two more appear in the Synthesis exercises.

TW **96.** *Meteorology.* Wind chill is a measure of how cold the wind makes you feel. Below are some measurements of wind chill for a 15-mph breeze. How can you tell from the data that a linear function will give an approximate fit?

SKILL MAINTENANCE EXERCISES

Skill Maintenance exercises help students review skills and concepts from preceding sections and prepare for upcoming topics. This feature is labeled **Focused Review** (*new to this edition*) when it provides mixed exercises that focus on connecting skills between topics, such as factoring and solving equations.

To the student and the instructor: Focused Review exercises such as the following replace Skill Maintenance exercises in selected exercise sets. These exercises focus on specific skills and often provide a mixed review connecting concepts of several sections or chapters.

Focused Review

Graph using the y-intercept and the slope. [2.2]

97. $y = \frac{1}{2}x + 3$

98. $f(x) = -x + 1$

Graph using intercepts. [2.3]

99. $3x - y = 3$

100. $2x + 3y = 6$

Graph each horizontal or vertical line. [2.3]

101. $f(x) = -2$

102. $x = 4$

Graph. [2.2], [2.3]

103. $5x + y = 5$

104. $y = \frac{1}{2}$

105. $f(x) = 2x - 3$

106. $x + y = 4$

Synthesis

TW **107.** Jim tries to avoid fractions as often as possible. Under what conditions will graphing using intercepts allow him to avoid fractions? Why?

TW **108.** Under what condition(s) will the x- and y-intercepts of a line coincide? What would the equation for such a line look like?

109. Give an equation, in standard form, for the line whose x-intercept is 5 and whose y-intercept is -4.

SYNTHESIS EXERCISES

Every exercise set ends with Synthesis exercises that help students synthesize skills and concepts from earlier sections with present material, providing deeper insight into the current topic.

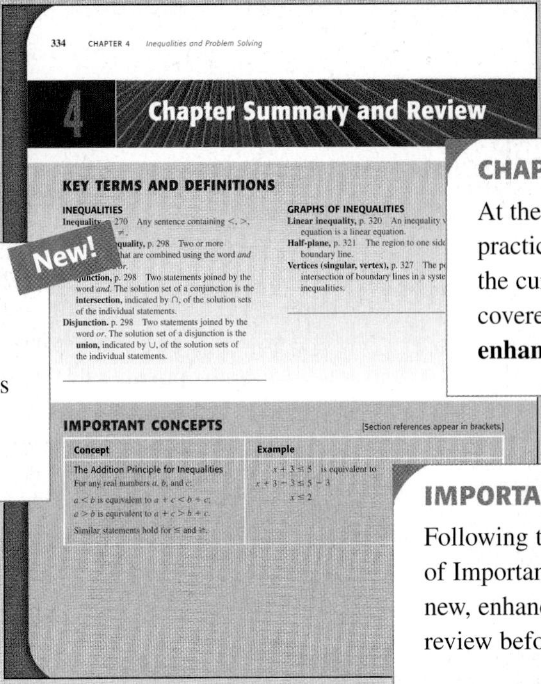

KEY TERMS AND DEFINITIONS

A list of Key Terms and Definitions from the chapter is provided along with the corresponding page numbers to serve as a reference point for students.

CHAPTER SUMMARY AND REVIEW

At the end of each chapter, students can practice all they have learned as well as tie the current chapter material to material covered in earlier chapters in the **enhanced** chapter summary and review.

IMPORTANT CONCEPTS

Following the Key Terms and Definitions, a list of Important Concepts with new examples in a new, enhanced format offers students a quick review before beginning the review exercises.

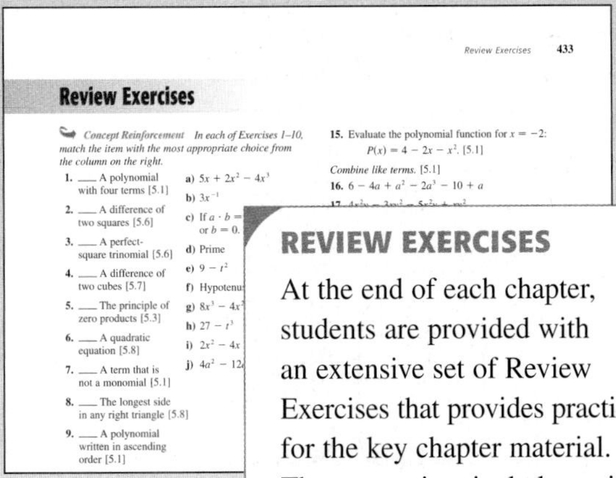

REVIEW EXERCISES

At the end of each chapter, students are provided with an extensive set of Review Exercises that provides practice for the key chapter material. These exercises include a wide variety of exercises, including concept reinforcement and synthesis exercises.

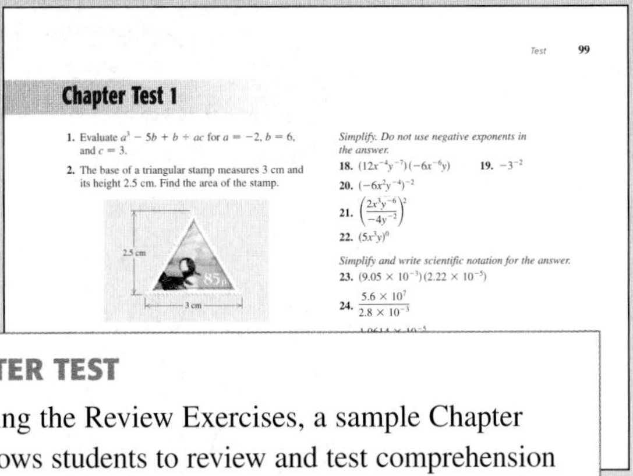

CHAPTER TEST

Following the Review Exercises, a sample Chapter Test allows students to review and test comprehension of chapter skills prior to taking an instructor's exam.

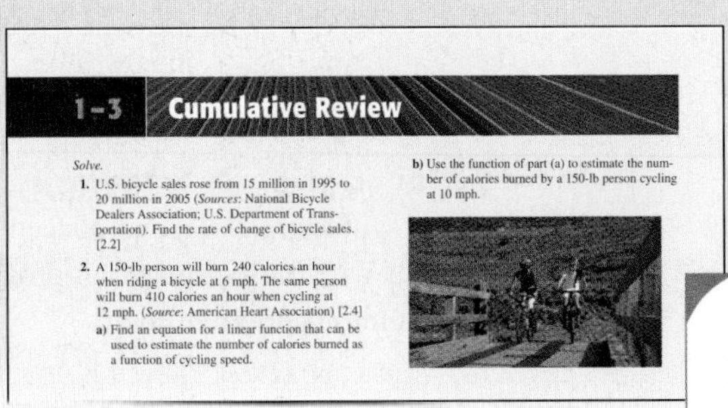

CUMULATIVE REVIEW

Following every three chapters, students will find a cumulative review, which covers skills and concepts from all preceding chapters of the text.

Tabbing for Success

Use these tabs to mark important pages of your textbook for quick reference and review.

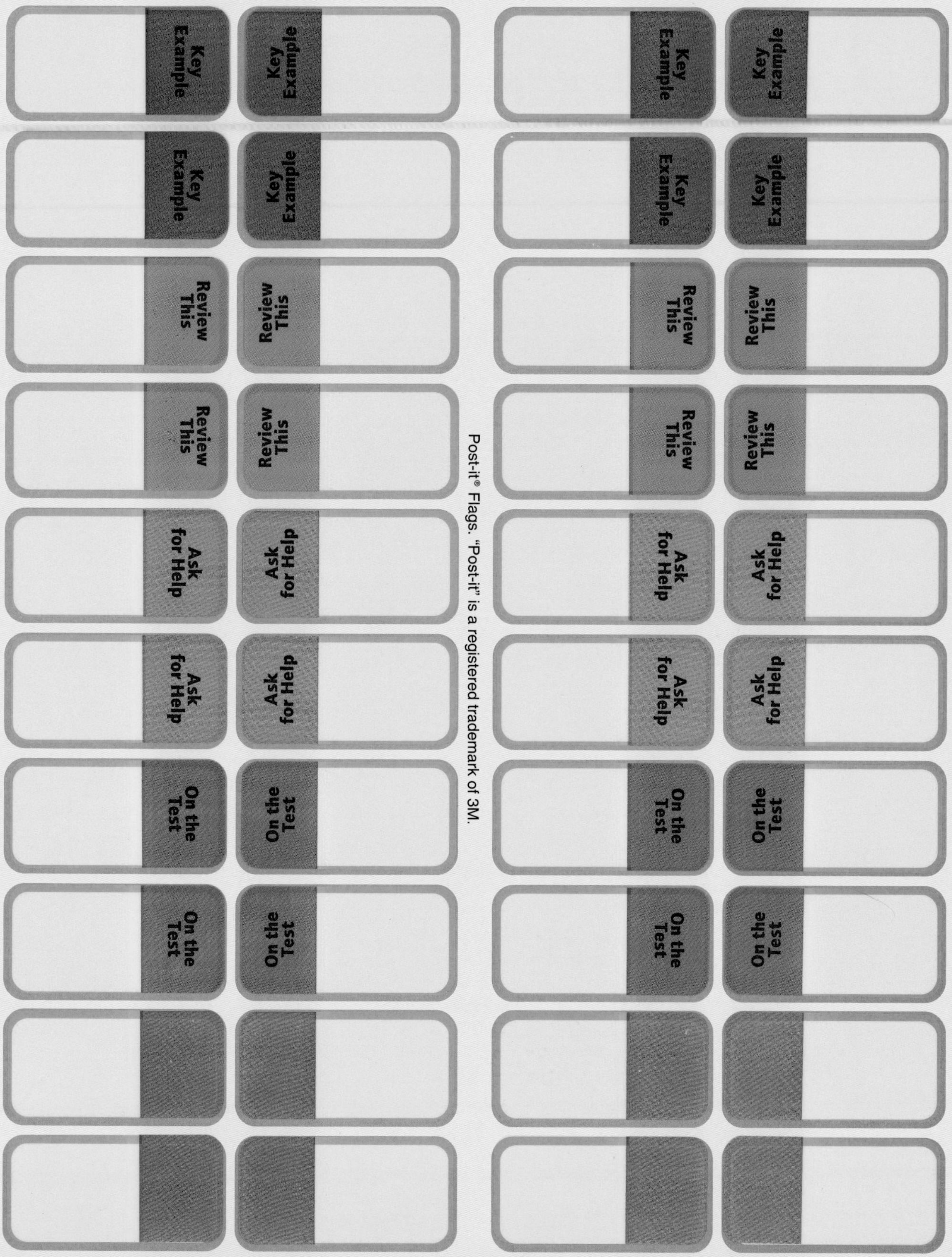

Using the Tabs

Customize Your Textbook and Make It Work for You!

These removable and reusable tabs offer you five ways to be successful in your math course by letting you bookmark pages with helpful reminders.

Key Example — Use these tabs to flag examples that will help you while doing your homework or preparing for your tests.

Review This — Mark important definitions, procedures, or key terms to review later.

Ask for Help — Not sure of something? Need more instruction? Place these tabs in your textbook to address any questions with your instructor during your next class meeting or with your tutor during your next tutoring session.

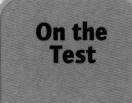

On the Test — If your instructor alerts you that something will be covered on a test, use these tabs to bookmark it.

Write your own notes or create more of the preceding tabs to help you succeed in your math course.

ISBN-13: 978-0-321-49680-5
ISBN-10: 0-321-49680-9

1 Basics of Algebra and Graphing

The principal theme of this text is problem solving in algebra. Symbolic algebra, numeric analysis, and graphs are used to develop models and solve applications. In this chapter, we introduce the basics of algebra, graphing, and the graphing calculator. An overall strategy for solving problems is presented in Section 1.7.

APPLICATION *Airline Consumer Complaints.*

The following table lists the number of complaints filed by consumers against airlines between 1995 and 2004 (*Source*: U.S. Department of Transportation, Aviation Consumer Protection Division). The number of consumer complaints can be modeled by graphing the data.

YEAR	NUMBER OF COMPLAINTS
1995	4,629
1996	5,782
1997	6,394
1998	7,980
1999	17,345
2000	20,564
2001	14,076
2002	7,697
2003	4,601
2004	5,863

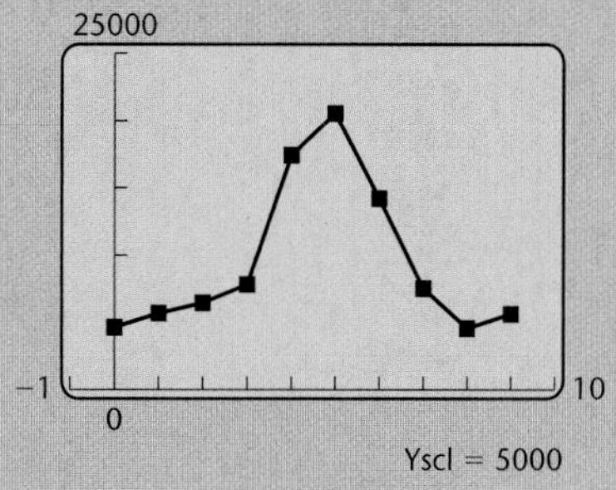

This problem appears as Example 7 in Section 1.7.

1.1 Some Basics of Algebra

Algebraic Expressions and Their Use ■ Evaluating Algebraic Expressions ■ Equations and Inequalities ■ Sets of Numbers ■ Introduction to the Graphing Calculator

The primary difference between algebra and arithmetic is the use of *variables.* In this section, we will see how variables can be used to represent various situations. We will also examine the different types of numbers that will be represented by variables throughout this text.

Algebraic Expressions and Their Use

We are all familiar with expressions like

$$95 + 21, \qquad 57 \times 34, \qquad 9 - 4, \quad \text{and} \quad \frac{35}{71}.$$

In algebra, we use these as well as expressions like

$$x + 21, \qquad l \cdot w, \qquad 9 - s, \quad \text{and} \quad \frac{d}{t}.$$

A letter that can be any one of various numbers is called a **variable.** If a letter represents a particular number that never changes, it is called a **constant.** Let $d =$ the number of hours it takes the moon to orbit the earth. Then d is a constant. If $a =$ the age of a baby chick, in minutes, then a is a variable since a changes as time passes.

An **algebraic expression** consists of variables, numbers, and operation signs. All of the expressions above are examples of algebraic expressions. When an equals sign is placed between two expressions, an **equation** is formed.

Algebraic expressions and equations arise frequently in problem-solving situations. Suppose, for example, that we want to determine by how much the sales of MP3 players increased from 2002 to 2005.

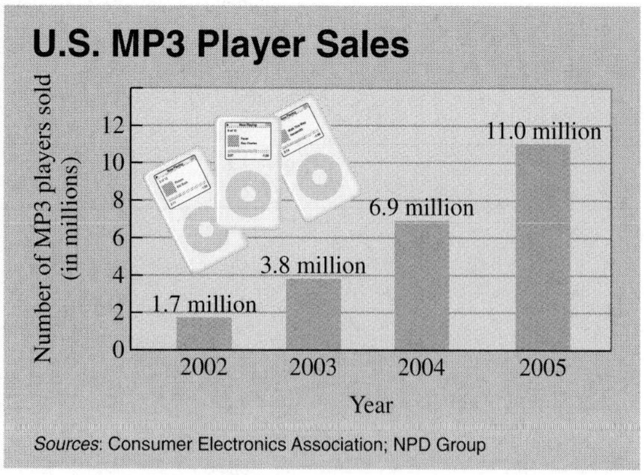

By using x to represent the increase in sales, we can form an equation:

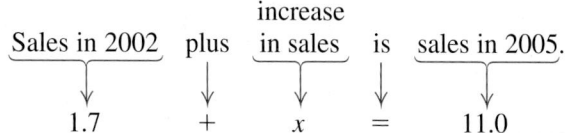

$$\underbrace{\text{Sales in 2002}}_{1.7} \quad \underbrace{\text{plus}}_{+} \quad \underbrace{\text{increase in sales}}_{x} \quad \underbrace{\text{is}}_{=} \quad \underbrace{\text{sales in 2005}}_{11.0}.$$

To find a **solution,** we can subtract 1.7 from both sides of the equation.

$$x = 11.0 - 1.7$$
$$x = 9.3$$

The sales increased by 9.3 million units from 2002 to 2005.

An expression like bh represents a product and can also be written as $b \cdot h$, $b \times h$, or $(b)(h)$. The multipliers b and h are called *factors.*

Some algebraic expressions contain *exponential notation.* Many different kinds of numbers can be used as *exponents.* Here we establish the meaning of a^n when n is a counting number, $1, 2, 3, \ldots$.

Exponential Notation The expression a^n, in which n is a counting number, means

$$\underbrace{a \cdot a \cdot a \cdot \cdots \cdot a \cdot a}_{n \text{ factors.}}$$

In a^n, a is called the *base* and n is the *exponent,* or *power.* When no exponent appears, it is assumed to be 1. Thus, $a^1 = a$.

The expression a^n is read "a raised to the nth power" or simply "a to the nth." We often read s^2 as "s-squared" and x^3 as "x-cubed." This terminology comes from the fact that the area of a square of side s is $s \cdot s = s^2$ and the volume of a cube of side x is $x \cdot x \cdot x = x^3$.

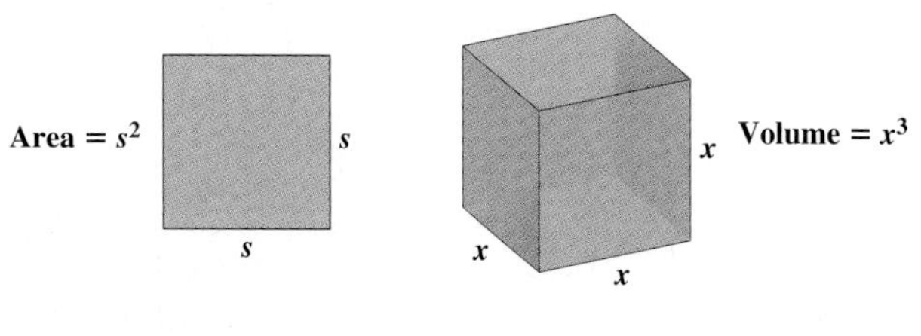

Evaluating Algebraic Expressions

When we replace a variable with a number, we say that we are **substituting** for the variable. The calculation that follows the substitution is called **evaluating the expression.**

Geometric formulas are often evaluated. In the following example, we use the formula for the area A of a triangle with a base of length b and a height of length h:

$$A = \tfrac{1}{2} \cdot b \cdot h.$$

This is an important formula that is worth remembering.

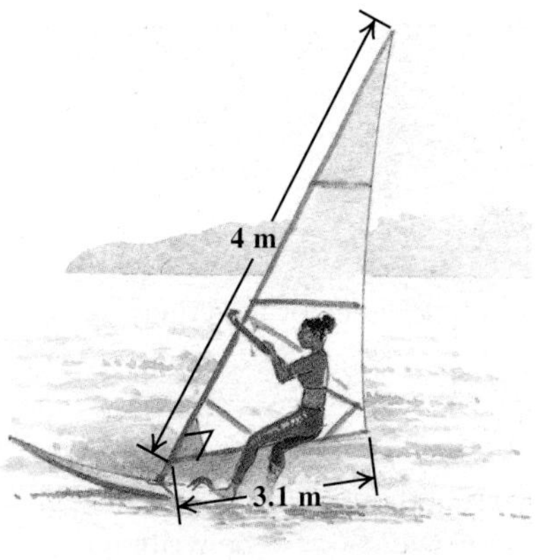

EXAMPLE 1 The base of a triangular sail is 3.1 m and the height is 4 m. Find the area of the sail.

SOLUTION We substitute 3.1 for b and 4 for h and multiply:

$$\frac{1}{2} \cdot b \cdot h = \frac{1}{2} \cdot 3.1 \cdot 4 \quad \text{We use color to highlight the substitution.}$$
$$= 6.2 \text{ square meters (sq m or m}^2\text{).}$$

Exponential notation tells us that 5^2 means $5 \cdot 5$, or 25, but what does $1 + 2 \cdot 5^2$ mean? If we add 1 and 2 and multiply by 25, we get 75. If we multiply 2 times 5^2, or 25, and add 1, we get 51. A third possibility is to square $2 \cdot 5$ to get 100 and then add 1 to get 101. The following convention indicates that only the second of these approaches is correct: We square 5, then multiply, and then add.

Student Notes

Note that rule (3) states that when division precedes multiplication, the division is performed first. Thus, $20 \div 5 \cdot 2$ represents $4 \cdot 2$, or 8. Similarly, $9 - 3 + 1$ represents $6 + 1$, or 7.

Rules for Order of Operations

1. Simplify within any grouping symbols, such as parentheses and brackets.
2. Simplify all exponential expressions.
3. Perform all multiplication and division, as either occurs, working from left to right.
4. Perform all addition and subtraction, as either occurs, working from left to right.

EXAMPLE 2 Evaluate $5 + 2(a - 1)^2$ for $a = 4$.

SOLUTION

$$\begin{aligned} 5 + 2(a - 1)^2 &= 5 + 2(4 - 1)^2 && \text{Substituting} \\ &= 5 + 2(3)^2 && \text{Working within parentheses first} \\ &= 5 + 2(9) && \text{Simplifying } 3^2 \\ &= 5 + 18 && \text{Multiplying} \\ &= 23 && \text{Adding} \end{aligned}$$

Step (3) in the rules for order of operations tells us to divide before we multiply when division appears first, reading left to right. This means that an expression like $6 \div 2x$ should be thought of as $(6 \div 2)x$.

EXAMPLE 3 Evaluate $9 - x^3 + 6 \div 2y^2$ for $x = 2$ and $y = 5$.

SOLUTION

$$\begin{aligned} 9 - x^3 + 6 \div 2y^2 &= 9 - 2^3 + 6 \div 2(5)^2 && \text{Substituting} \\ &= 9 - 8 + 6 \div 2 \cdot 25 && \text{Simplifying } 2^3 \text{ and } 5^2 \\ &= 9 - 8 + 3 \cdot 25 && \text{Dividing} \\ &= 9 - 8 + 75 && \text{Multiplying} \\ &= 1 + 75 && \text{Subtracting} \\ &= 76 && \text{Adding} \end{aligned}$$

Equations and Inequalities

An equation can be true or false. For example, the equation

$$13 + 3 = 16$$

is true, and the equation

$$4 + 25 = 30$$

is false. The equation

$$x + 12 = 35$$

is neither true nor false. However, when the algebraic expressions on each side of the equation are evaluated for a value for x, the equation becomes true or false, depending on the replacement for x. A replacement that makes the equation true is called a *solution* of the equation. This type of equation is called a *conditional equation.*

EXAMPLE 4 Tell whether each number is a solution of the equation $x + 12 = 35$: **(a)** 15; **(b)** 23.

SOLUTION

a) We replace x with 15 and evaluate the expressions on both sides of the equals sign.

$x + 12 = 35$		
$15 + 12$	35	
$27 \stackrel{?}{=} 35$		FALSE

Since $27 = 35$ is false, 15 *is not* a solution of the equation.

b) We replace x with 23 and evaluate the expressions on both sides of the equals sign.

$x + 12 = 35$		
$23 + 12$	35	
$35 \stackrel{?}{=} 35$		TRUE

Since $35 = 35$ is true, 23 *is* a solution of the equation.

The symbols $<$ (is less than), $>$ (is greater than), $\leq$ (is less than or equal to), $\geq$ (is greater than or equal to), and $\neq$ (is not equal to) are inequality symbols. An **inequality** is formed when an inequality symbol is placed between two algebraic expressions. Like equations, inequalities can be true, false, or neither true nor false. A *solution* of an inequality is a replacement that makes the inequality true.

EXAMPLE 5 Tell whether each number is a solution of the inequality $3y + 2 \leq 11$: **(a)** 0; **(b)** 5; **(c)** 3.

SOLUTION

a) We replace y with 0.

$$\begin{array}{r|l} 3y + 2 \le 11 & \\ \hline 3 \cdot 0 + 2 & 11 \\ 0 + 2 & \\ 2 \overset{?}{\le} 11 & \text{TRUE} \end{array}$$

Since the statement $2 \le 11$, read "2 is less than or equal to 11," is true, 2 *is* a solution of the inequality.

b) We replace y with 5.

$$\begin{array}{r|l} 3y + 2 \le 11 & \\ \hline 3 \cdot 5 + 2 & 11 \\ 15 + 2 & \\ 17 \overset{?}{\le} 11 & \text{FALSE} \end{array}$$

Since the statement "17 is less than or equal to 11" is false, 5 *is not* a solution of the inequality.

c) We replace y with 3.

$$\begin{array}{r|l} 3y + 2 \le 11 & \\ \hline 3 \cdot 3 + 2 & 11 \\ 9 + 2 & \\ 11 \overset{?}{\le} 11 & \text{TRUE} \end{array}$$

Since the statement "11 is less than or equal to 11" is true (11 is equal to 11), 3 *is* a solution of the inequality.

Sets of Numbers

When evaluating algebraic expressions, and in problem solving in general, we often must examine the *type* of numbers used. For example, if a formula is used to determine an optimal class size, any fraction results must be rounded up or down since it is impossible to have a fractional part of a student. Three frequently used sets of numbers are listed below.

Natural Numbers, Whole Numbers, and Integers

Natural Numbers (or Counting Numbers) Those numbers used for counting:

$$\{1, 2, 3, \ldots\}$$

Whole Numbers The set of natural numbers with 0 included:

$$\{0, 1, 2, 3, \ldots\}$$

Integers The set of all whole numbers and their opposites:

$$\{\ldots, -4, -3, -2, -1, 0, 1, 2, 3, 4, \ldots\}$$

The dots are called ellipses and indicate that the pattern continues without end.

The integers correspond to the points on a number line as follows:

To fill in the numbers between these points, we must describe two more sets of numbers. This requires us to first discuss set notation.

The set containing the numbers −2, 1, and 3 can be written $\{-2, 1, 3\}$. This way of writing a set is known as **roster notation.** Roster notation was used for the three sets listed above. A second type of set notation, **set-builder notation,** specifies conditions under which a number is in the set. The following example of set-builder notation is read as shown:

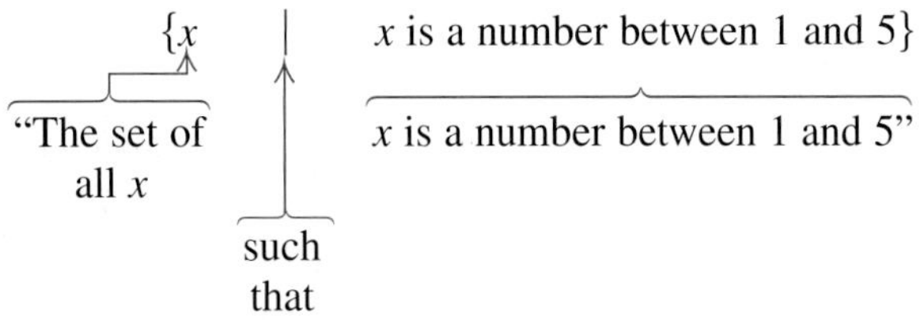

Set-builder notation is generally used when it is difficult to list a set using roster notation.

EXAMPLE 6 Using both roster notation and set-builder notation, represent the set consisting of the first 15 even natural numbers.

SOLUTION

Using roster notation: $\{2, 4, 6, 8, 10, 12, 14, 16, 18, 20, 22, 24, 26, 28, 30\}$

Using set-builder notation: $\{n \mid n \text{ is an even number between 1 and 31}\}$

The symbol $\in$ is used to indicate that an **element** or **member** belongs to a set. Thus if $A = \{2, 4, 6, 8\}$, we can write $4 \in A$ to indicate that 4 *is an element of A*. We can also write $5 \notin A$ to indicate that 5 *is not an element of A*.

EXAMPLE 7 Classify the statement $8 \in \{x \mid x \text{ is an integer}\}$ as true or false.

SOLUTION Since 8 *is* an integer, the statement is true. In other words, since 8 is an integer, it belongs to the set of all integers.

With set-builder notation, we can describe the set of all *rational numbers.*

Rational Numbers Numbers that can be expressed as an integer divided by a nonzero integer are called *rational numbers*:

$$\left\{ \frac{p}{q} \,\middle|\, p \text{ is an integer, } q \text{ is an integer, and } q \neq 0 \right\}.$$

Rational numbers can be written using fraction or decimal notation. *Fraction notation* uses symbolism like the following:

$$\frac{5}{8}, \quad \frac{12}{-7}, \quad \frac{-17}{15}, \quad -\frac{9}{7}, \quad \frac{39}{1}, \quad \frac{0}{6}.$$

In *decimal notation,* rational numbers either *terminate* (end) or *repeat* (have a repeating block of digits).

EXAMPLE 8 When written in decimal form, does each of the following numbers terminate or repeat? **(a)** $\frac{5}{8}$; **(b)** $\frac{6}{11}$.

SOLUTION

a) Since $\frac{5}{8}$ means $5 \div 8$, we perform long division to find that $\frac{5}{8} = 0.625$, a decimal that ends. Thus, $\frac{5}{8}$ can be written as a terminating decimal.

b) Using long division, we find that $6 \div 11 = 0.5454\ldots$, so we can write $\frac{6}{11}$ as a repeating decimal. Repeating decimal notation can be abbreviated by writing a bar over the repeating block of digits—in this case, $0.\overline{54}$.

Many numbers, like π, $\sqrt{2}$, and $-\sqrt{15}$, are not rational numbers. For example, $\sqrt{2}$ is the number for which $\sqrt{2} \cdot \sqrt{2} = 2$. A calculator's representation of $\sqrt{2}$ as 1.414213562 is an approximation since $(1.414213562)^2$ is not exactly 2.

To illustrate that $\sqrt{2}$ is a "real" point on the number line, it can be shown that when a right triangle has two legs of length 1, the remaining side has length $\sqrt{2}$. Thus we can "measure" $\sqrt{2}$ units and locate $\sqrt{2}$ on the number line.

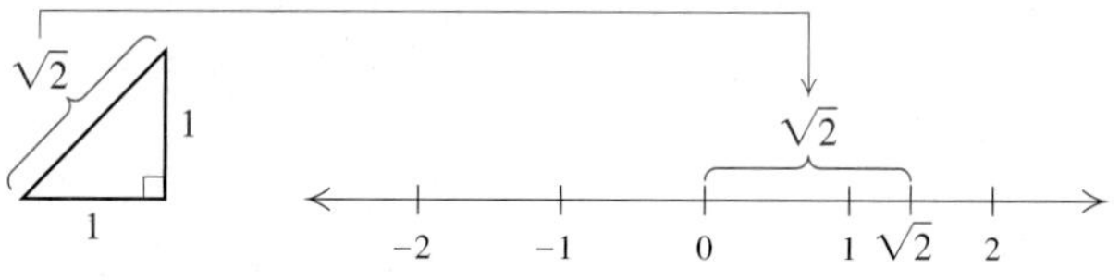

Numbers like π, $\sqrt{2}$, and $-\sqrt{15}$ are said to be **irrational.** Decimal notation for irrational numbers neither terminates nor repeats.

The set of all rational numbers, combined with the set of all irrational numbers, gives us the set of all **real numbers.**

Real Numbers Numbers that are either rational or irrational are called *real numbers.* The set of all real numbers is often represented as $\mathbb{R}$:

$$\mathbb{R} = \{x \mid x \text{ is rational or } x \text{ is irrational}\}.$$

Every point on the number line represents some real number and every real number is represented by some point on the number line.

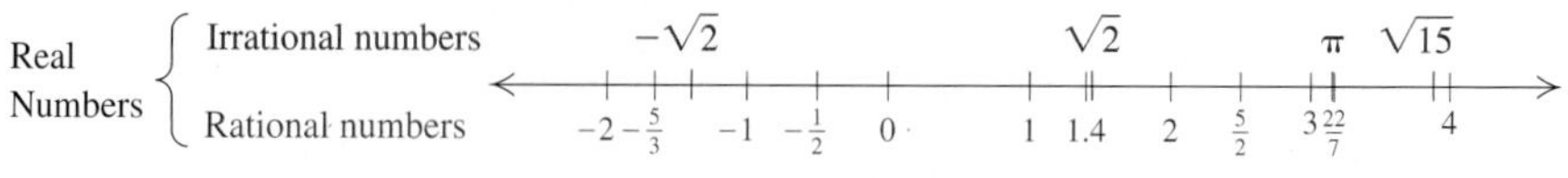

The following figure shows the relationships among various kinds of numbers, along with examples of how real numbers can be sorted.

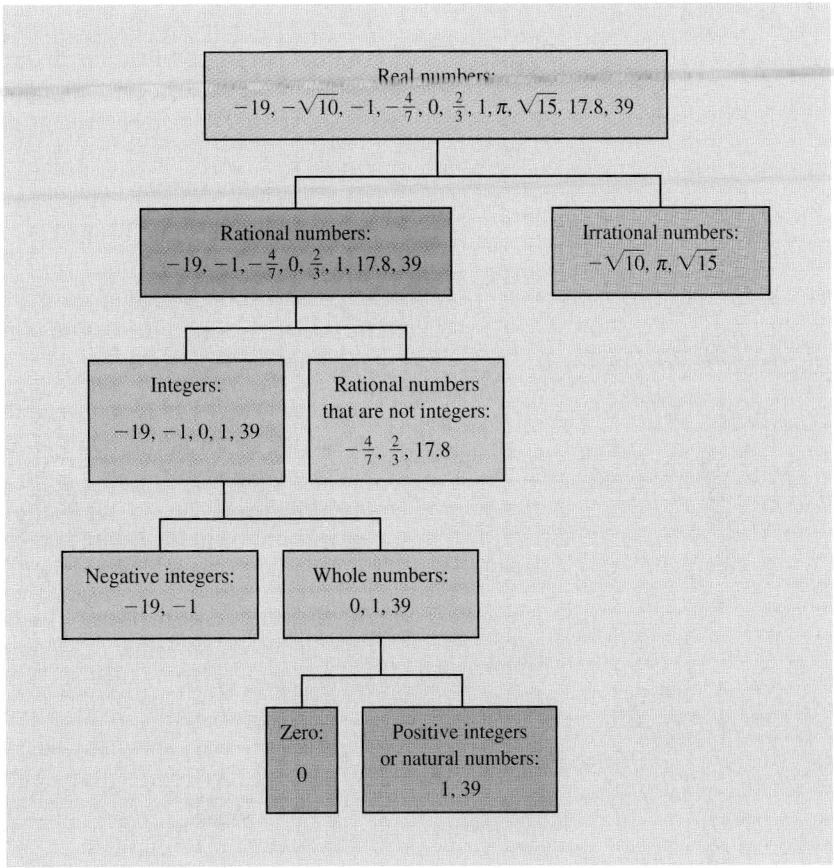

When all members of one set are found in a second set, the first set is a **subset** of the second set. Thus if $A = \{2, 4, 6\}$ and $B = \{1, 2, 4, 5, 6\}$, we write $A \subseteq B$ to indicate that *A is a subset of B*. Similarly, if $\mathbb{N}$ represents the set of all natural numbers and $\mathbb{Z}$ the set of all integers, we can write $\mathbb{N} \subseteq \mathbb{Z}$. Additional statements can be made using other sets in the diagram above.

Student Notes

Be sure that you understand and can use mathematical principles such as the rules for order of operations before you rely on a graphing calculator to do the operations. When you are learning the rules, a calculator can be useful to check your answers.

Introduction to the Graphing Calculator

Graphing calculators and computers equipped with graphing software can be valuable aids in understanding and applying algebra. Features and keystrokes vary among the many brands and models of graphing calculators available. In this text, we use features that are common to most graphing calculators. Specific keystrokes and instructions for certain calculators are included in the *Graphing Calculator Manual* that accompanies this book. For other procedures, you should consult your instructor or the user's manual for your particular calculator.

(continued)

There are two important things to keep in mind as you proceed through this text.

1. You will not learn to use a graphing calculator by simply reading about it; you must in fact *use* the calculator. Press the keys on your own calculator as you read the text, do the calculator exercises in the exercise set, and experiment with new options.
2. Your user's manual contains more information about your calculator than appears in this text. If you need additional explanation and examples, be sure to consult the manual.

Keypad A diagram of the keypad of a graphing calculator appears at the front of this text. The organization and labeling of the keys differ for different calculators. Note that there are options written above keys as well as on the keys. To access the options shown above the keys, press 2ND or ALPHA and then the key below the desired option.

Screen After you have turned the calculator on, you should see a blinking rectangle, or **cursor,** at the top left corner of the screen. If you do not see anything, try adjusting the **contrast.** On many calculators, this is done by pressing 2ND and the up or down arrow keys. To perform computations, you should be in the **home screen.** Pressing QUIT (often the 2nd option associated with the MODE key) will return you to the home screen.

Performing Operations To perform addition, subtraction, multiplication, and division using a graphing calculator, type in the expression as it is written. The entire expression will appear on the screen, and you can check your typing. A multiplication symbol will appear on the screen as *, and division is usually shown by the symbol /. Grouping symbols such as brackets or braces are entered as parentheses. If the expression appears to be correct, press ENTER. At that time, the calculator will evaluate the expression and display the result on the screen.

Catalog A graphing calculator's catalog lists all the functions of the calculator in alphabetical order. On many calculators, CATALOG is the 2nd option associated with the 0 key. To copy an item from the catalog to the screen, press CATALOG and then scroll through the list using the up and down arrow keys until the desired item is indicated. To move through the list more quickly, press the key associated with the first letter of the item. The indicator will move to the first item beginning with that letter. When the desired item is indicated, press ENTER.

Exponents On most graphing calculators, exponents are entered using the ^ key before the exponent. For the exponent 2, you can often use an x^2 key.

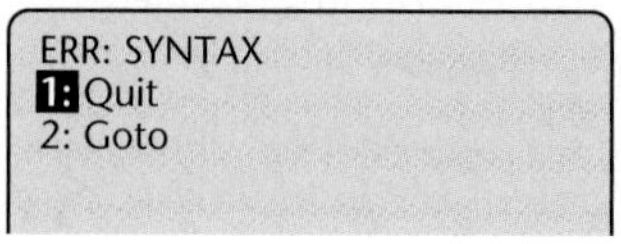

Error Messages When a calculator cannot complete an instruction, a message similar to the one shown in the figure at left appears on the screen. Press 1 to return to the home screen, and press 2 to go to the instruction that caused the error. Not all errors are operator errors; ERR:OVERFLOW indicates a result too large for the calculator to handle.

EXAMPLE 9 Evaluate $2(y - 3)^2 + 7$ for $y = 5$.

SOLUTION We replace y with 5 and enter the expression.

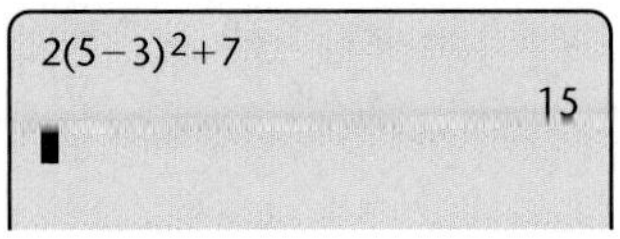

The result is 15.

Study Tip

It's Who You Know

Throughout this textbook, you will find a feature called *Study Tip*. These tips are intended to help improve your math study skills. On the first day of class, we recommend that you complete this chart.

Instructor:

Name ______________________

Office hours and location ______________________

Phone number ______________________

Fax number ______________________

E-mail address ______________________

Find the names of two students whom you could contact for information or study questions:

1. Name ______________________

Phone number ______________________

E-mail address ______________________

2. Name ______________________

Phone number ______________________

E-mail address ______________________

Math lab on campus:

Location ______________________

Hours ______________________

Phone ______________________

Tutoring:

Campus location ______________________

Hours ______________________

Addison-Wesley Tutor Center ______________________

To order, call ______________________

(See the preface for important information concerning this tutoring.)

Important supplements:

(See the preface for a complete list of available supplements.)

Supplements recommended by the instructor

____

____

1.1 EXERCISE SET

FOR EXTRA HELP

MathXL | MyMathLab | InterAct Math | AW Math Tutor Center | Video Lectures on CD: Disc 1 | Student's Solutions Manual

Concept Reinforcement In each of Exercises 1–10, fill in the blank with the appropriate word or words.

1. A letter representing a specific number that never changes is called a(n) ______________.

2. A letter that can be any one of a set of numbers is called a(n) ______________.

3. In the expression $7y$, the multipliers 7 and y are called ______________.

4. When all variables in a variable expression are replaced with numbers and a result is calculated, we say that we are ______________ the expression.

5. When no grouping symbols, exponents, division, or multiplication appear, we subtract before we add, provided the subtraction appears to the ______________ of any addition.

6. In a^b, a is called the ______________ and b is called the ______________.

7. A number that can be written in the form a/b, where a and b are integers (with $b \neq 0$), is said to be a(n) ______________ number.

8. A real number that cannot be written as a quotient of two integers is an example of a(n) ______________ number.

9. Division can be used to show that $\frac{7}{40}$ can be written as a(n) ______________ decimal.

10. Division can be used to show that $\frac{13}{7}$ can be written as a(n) ______________ decimal.

To the student and the instructor: Throughout this text, selected exercises are marked with the icon **Aha!**. *These "Aha!" exercises can be answered quite easily if the student pauses to inspect the exercise rather than proceeds mechanically. This is done to discourage rote memorization. Some "Aha!" exercises are left unmarked to encourage students to always pause before working a problem.*

Evaluate each expression for the values provided.

11. $7x + y$, for $x = 3$ and $y = 4$

12. $6a - b$, for $a = 5$ and $b = 3$

13. $2c \div 3b$, for $b = 2$ and $c = 6$

14. $3z \div 2y$, for $y = 1$ and $z = 6$

15. $25 + r^2 - s$, for $r = 3$ and $s = 7$

16. $n^3 + 2 - p$, for $n = 2$ and $p = 5$

Aha! **17.** $3n^2p - 3pn^2$, for $n = 5$ and $p = 9$

18. $2a^3b - 2b^2$, for $a = 3$ and $b = 7$

19. $5x \div (2 + x - y)$, for $x = 6$ and $y = 2$

20. $3(m + 2n) \div m$, for $m = 7$ and $n = 0$

21. $[10 - (a - b)]^2$, for $a = 7$ and $b = 2$

22. $[17 - (x + y)]^2$, for $x = 4$ and $y = 1$

23. $[5(r + s)]^2$, for $r = 1$ and $s = 2$

24. $[3(a - b)]^2$, for $a = 7$ and $b = 5$

25. $m^2 - [2(m - n)]^2$, for $m = 7$ and $n = 5$

26. $x^2 - [3(x - y)]^2$, for $x = 6$ and $y = 4$

27. $(r - s)^2 - 3(2r - s)$, for $r = 11$ and $s = 3$

28. $(m - 2n)^2 - 2(m + n)$, for $m = 8$ and $n = 1$

In Exercises 29–32, find the area of a triangular window with the given base and height.

29. Base = 5 ft, height = 7 ft

30. Base = 2.9 m, height = 2.1 m

31. Base = 7 m, height = 3.2 m

32. Base = 3.6 ft, height = 4 ft

Tell whether each number is a solution of the given equation or inequality.

33. $12 - y = 5$; **(a)** 7; **(b)** 5; **(c)** 12

34. $2x + 4 = 14$; **(a)** 3; **(b)** 7; **(c)** 5

35. $5 - x \leq 2$; **(a)** 0; **(b)** 4; **(c)** 3

36. $3y - 5 > 10$; **(a)** 3; **(b)** 5; **(c)** 7

37. $3m - 8 < 13$; **(a)** 6; **(b)** 7; **(c)** 9

38. $15 - 3n = 6$; **(a)** 3; **(b)** 2; **(c)** 5

Use roster notation to write each set.

39. The set of all vowels in the alphabet

40. The set of all days of the week

41. The set of all odd natural numbers

42. The set of all even natural numbers

43. The set of all natural numbers that are multiples of 5

44. The set of all natural numbers that are multiples of 10

Use set-builder notation to write each set.

45. The set of all odd numbers between 10 and 20

46. The set of all multiples of 4 between 22 and 35

47. $\{0, 1, 2, 3, 4\}$

48. $\{-3, -2, -1, 0, 1, 2\}$

49. The set of all multiples of 5 between 7 and 79

50. The set of all even numbers between 9 and 99

In Exercises 51–54, determine which numbers in the list provided are **(a)** *whole numbers?* **(b)** *integers?* **(c)** *rational numbers?* **(d)** *irrational numbers?* **(e)** *real numbers?*

51. $-8.7, -3, 0, \frac{2}{3}, \sqrt{7}, 6$

52. $-\frac{9}{2}, -4, -1.2, 0, \sqrt{5}, 3$

53. $-17, -4.13, 0, \frac{5}{4}, 3, \sqrt{77}$

54. $-9.1, -2, 0, 4, \sqrt{17}, \frac{99}{2}$

Classify each statement as true or false. The following sets are used:

$\mathbb{N}$ = the set of natural numbers;
$\mathbb{W}$ = the set of whole numbers;
$\mathbb{Z}$ = the set of integers;
$\mathbb{Q}$ = the set of rational numbers;
$\mathbb{H}$ = the set of irrational numbers;
$\mathbb{R}$ = the set of real numbers.

55. $5.1 \in \mathbb{N}$

56. $\mathbb{N} \subseteq \mathbb{W}$

57. $\mathbb{W} \subseteq \mathbb{Z}$

58. $\sqrt{8} \in \mathbb{Q}$

59. $\frac{2}{3} \in \mathbb{H}$

60. $\mathbb{H} \subseteq \mathbb{R}$

61. $\sqrt{10} \in \mathbb{R}$

62. $4.3 \notin \mathbb{Z}$

63. $\mathbb{Z} \not\subseteq \mathbb{N}$

64. $\mathbb{Q} \subseteq \mathbb{R}$

65. $\mathbb{Q} \subseteq \mathbb{Z}$

66. $9 \in \mathbb{N}$

To the student and the instructor: The symbol [graphing calculator icon] *indicates an exercise designed to be solved with a graphing calculator.*

Evaluate each of the following using a graphing calculator.

67. $13 - (y - 4)^3 + 10$, for $y = 6$

68. $(t + 4)^2 - 12 \div (19 - 17) + 68$, for $t = 5$

69. $3.86 + 2.7(2.1x + 1.7)$, for $x = 5.82$

70. $1.5(3.982 - a) + 2.3^2$, for $a = 2.19$

71. $3(m + 2n) \div m$, for $m = 1.6$ and $n = 5.9$

72. $1.5 + (2x - y)^2$, for $x = 3.25$ and $y = 1.7$

73. $\frac{1}{2}(x + 5/z)^2$, for $x = 141$ and $z = 0.2$

74. $a - \frac{3}{4}(2a - b)$, for $a = 116$ and $b = 207$

To the student and the instructor: The symbol TW *is used to denote thinking and writing exercises. These exercises are meant to be answered with one or more English sentences. Because many writing exercises have a variety of correct answers, these solutions are not listed in the answers at the back of the book.*

TW **75.** What is the difference between rational numbers and integers?

TW **76.** Devin insists that $15 - 4 + 1 \div 2 \cdot 3$ is 2. What error is he making?

Synthesis

To the student and the instructor: Synthesis exercises are designed to challenge students to extend the concepts or skills studied in each section. Many synthesis exercises require the assimilation of skills and concepts from several sections.

TW **77.** Is the following true or false, and why?

$$\{2, 4, 6\} \subseteq \{2, 4, 6\}$$

TW **78.** On a quiz, Francesca answers $6 \in \mathbb{Z}$ while Jacob writes $\{6\} \in \mathbb{Z}$. Jacob's answer does not receive full credit while Francesca's does. Why?

Use roster notation to write each set.

79. The set of all whole numbers that are not natural numbers

80. The set of all integers that are not whole numbers

81. $\{x \mid x = 5n, n \text{ is a natural number}\}$

82. $\{x \mid x = 3n, n \text{ is a natural number}\}$

83. $\{x \mid x = 2n + 1, n \text{ is a whole number}\}$

84. $\{x \mid x = 2n, n \text{ is an integer}\}$

85. Draw a right triangle that could be used to measure $\sqrt{13}$ units.

1.2 Operations with Real Numbers

Absolute Value and Opposites ■ Order ■ Addition, Subtraction, Multiplication, Division, and Reciprocals

In this section, we review addition, subtraction, multiplication, and division of real numbers. First, however, we must discuss absolute value.

Absolute Value

It is convenient to have a notation that represents a number's distance from zero on the number line. Note that distance is never negative.

> **Absolute Value** The notation $|a|$, read "the absolute value of a," represents the number of units that a is from zero on the number line.

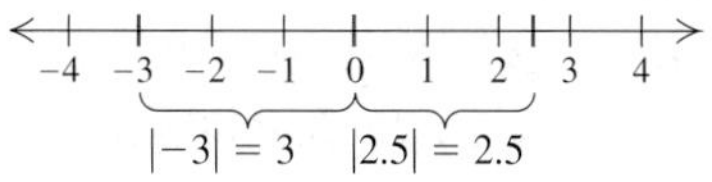

EXAMPLE 1 Find the absolute value: **(a)** $|-3|$; **(b)** $|2.5|$; **(c)** $|0|$.

SOLUTION

a) $|-3| = 3$ **−3 is 3 units from 0.**

b) $|2.5| - 2.5$ **2.5 is 2.5 units from 0.**

c) $|0| = 0$ **0 is 0 units from itself.**

Note that absolute value is never negative.

Study Tip

Organize Your Work

When doing homework, consider using a spiral notebook or a three-ring binder. Because your course will probably include graphing, consider purchasing a notebook filled with graph paper. Try to write legibly in it, labeling each section and each exercise and showing all steps. Legible, well-organized work will make it easier for those who read your work to give you constructive feedback.

Order

We use inequality symbols to indicate how two real numbers compare with each other. For any two numbers on the number line, the one to the left is said to be less than, or smaller than, the one to the right. In the figure below, we have $-6 < -1$ (since -6 is to the left of -1) and $|-6| > |-1|$ (since 6 is to the right of 1).

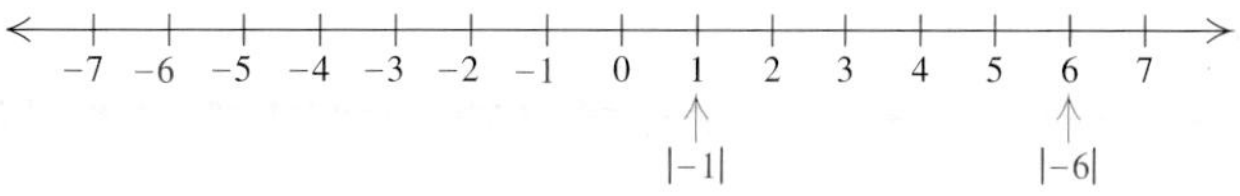

EXAMPLE 2 Determine whether each inequality is a true statement: **(a)** $-7 < -2$; **(b)** $1 > -4$; **(c)** $-3 \geq -2$.

SOLUTION

a) $-7 < -2$ is *true* because -7 is to the left of -2 on the number line.

b) $1 > -4$ is *true* because 1 is to the right of -4.

c) $-3 \geq -2$ is *false* because -3 is to the left of -2.

Addition, Subtraction, and Opposites

We are now ready to review the addition of real numbers.

Addition of Two Real Numbers

1. *Positive numbers*: Add the numbers. The result is positive.
2. *Negative numbers*: Add absolute values. Make the answer negative.
3. *A negative and a positive number*: If the numbers have the same absolute value, the answer is 0. Otherwise, subtract the smaller absolute value from the larger one:
   a) If the positive number has the greater absolute value, make the answer positive.
   b) If the negative number has the greater absolute value, make the answer negative.
4. *One number is zero*: The sum is the other number.

EXAMPLE 3 Add: **(a)** $-9 + (-5)$; **(b)** $-3.24 + 8.7$; **(c)** $-\frac{3}{4} + \frac{1}{3}$.

SOLUTION

a) $-9 + (-5)$ — We add the absolute values, getting 14. The answer is *negative*, -14.

b) $-3.24 + 8.7$ — The absolute values are 3.24 and 8.7. Subtract 3.24 from 8.7 to get 5.46. The positive number is further from 0, so the answer is *positive*, 5.46.

c) $-\frac{3}{4} + \frac{1}{3} = -\frac{9}{12} + \frac{4}{12}$ — The absolute values are $\frac{9}{12}$ and $\frac{4}{12}$. Subtract to get $\frac{5}{12}$. The negative number is further from 0, so the answer is *negative*, $-\frac{5}{12}$.

When numbers like 7 and -7 are added, the result is 0. Such numbers are called **opposites,** or **additive inverses,** of one another. The sum of two additive inverses is the **additive identity,** 0.

The Law of Opposites For any two numbers a and $-a$,

$$a + (-a) = 0.$$

(The sum of two opposites is 0.)

EXAMPLE 4 Find the opposite: **(a)** -17.5; **(b)** $\frac{4}{5}$; **(c)** 0.

SOLUTION

a) The opposite of -17.5 is 17.5 because $-17.5 + 17.5 = 0$.

b) The opposite of $\frac{4}{5}$ is $-\frac{4}{5}$ because $\frac{4}{5} + \left(-\frac{4}{5}\right) = 0$.

c) The opposite of 0 is 0 because $0 + 0 = 0$.

To name the opposite, we use the symbol "$-$" and read the symbolism $-a$ as "the opposite of a."

CAUTION! $-a$ does not necessarily denote a negative number. In particular, when a represents a *negative* number, $-a$ is *positive*.

EXAMPLE 5 Find $-x$ for the following: **(a)** $x = -2$; **(b)** $x = \frac{3}{4}$.

SOLUTION

a) If $x = -2$, then $-x = -(-2) = 2$. **The opposite of -2 is 2.**

b) If $x = \frac{3}{4}$, then $-x = -\frac{3}{4}$. **The opposite of $\frac{3}{4}$ is $-\frac{3}{4}$.**

Using the notation of opposites, we can formally define absolute value.

Absolute Value

$$|x| = \begin{cases} x, & \text{if } x \geq 0, \\ -x, & \text{if } x < 0 \end{cases}$$

(When x is nonnegative, the absolute value of x is x. When x is negative, the absolute value of x is the opposite of x. Thus, $|x|$ is never negative.)

A negative number is said to have a negative "sign" and a positive number a positive "sign." To subtract, we can add an opposite. Thus we sometimes say that we "change the sign of the number being subtracted and then add."

EXAMPLE 6 Subtract: **(a)** $5 - 9$; **(b)** $-1.2 - (-3.7)$; **(c)** $-\frac{4}{5} - \frac{2}{3}$.

SOLUTION

a) $5 - 9 = 5 + (-9)$ **Change the sign and add.**
$= -4$

b) $-1.2 - (-3.7) = -1.2 + 3.7$ **Instead of *subtracting* −3.7, we *add* 3.7.**
$= 2.5$

c) $-\frac{4}{5} - \frac{2}{3} = -\frac{4}{5} + \left(-\frac{2}{3}\right)$
$= -\frac{12}{15} + \left(-\frac{10}{15}\right)$ **Finding a common denominator**
$= -\frac{22}{15}$

Multiplication, Division, and Reciprocals

In this text, we direct graphing-calculator exploration of mathematical concepts with Interactive Discovery features like the one that follows. Such explorations are a part of the development of the material presented and should be performed as you read the text.

When real numbers are multiplied, the sign of the product depends on the signs of the factors.

Interactive Discovery

Consider the following patterns of products. If the patterns are to continue, what should the missing products be? Use a calculator to check your answers. How can we determine the sign of a product from the signs of the factors?

1.
$3 \cdot 2 = 6$
$3 \cdot 1 = 3$
$3 \cdot 0 = 0$
$3 \cdot (-1) = ?$
$3 \cdot (-2) = ?$

2.
$(-5) \cdot 2 = -10$
$(-5) \cdot 1 = -5$
$(-5) \cdot 0 = 0$
$(-5) \cdot (-1) = ?$
$(-5) \cdot (-2) = ?$

You may have noticed that when one factor is positive and one is negative, the product is negative. When both factors are positive or both are negative, the product is positive.

Division is defined in terms of multiplication. For example, $10 \div (-2) = -5$ because $(-5)(-2) = 10$. Thus the rules for division are just like those for multiplication.

Multiplication or Division of Two Real Numbers

1. To multiply or divide two numbers with *unlike signs*, multiply or divide their absolute values. The answer is *negative*.
2. To multiply or divide two numbers with the *same sign*, multiply or divide their absolute values. The answer is *positive*.

EXAMPLE 7 Multiply or divide: **(a)** $\left(-\frac{2}{3}\right)\left(-\frac{3}{8}\right)$; **(b)** $20 \div (-4)$; **(c)** $\frac{-45}{-15}$.

SOLUTION

a) $\left(-\frac{2}{3}\right)\left(-\frac{3}{8}\right) = \frac{6}{24} = \frac{1}{4}$ — **Multiply absolute values. The answer is positive.**

b) $20 \div (-4) = -5$ — **Divide absolute values. The answer is negative.**

c) $\frac{-45}{-15} = 3$ — **Divide absolute values. The answer is positive.**

Note that since

$$\frac{-8}{2} = \frac{8}{-2} = -\frac{8}{2} = -4,$$

we have the following generalization.

The Sign of a Fraction

For any number a and any nonzero number b,

$$\frac{-a}{b} = \frac{a}{-b} = -\frac{a}{b}.$$

Recall that

$$\frac{a}{b} = \frac{a}{1} \cdot \frac{1}{b} = a \cdot \frac{1}{b}.$$

That is, rather than divide by b, we can multiply by $\frac{1}{b}$. Provided that b is not 0, the numbers b and $\frac{1}{b}$ are called **reciprocals,** or **multiplicative inverses,** of each other. The product of two multiplicative inverses is the **multiplicative identity,** 1.

The Law of Reciprocals For any two numbers a and $\frac{1}{a}$ $(a \neq 0)$,

$$a \cdot \frac{1}{a} = 1.$$

(The product of reciprocals is 1.)

Every number except 0 has exactly one reciprocal. The number 0 has no reciprocal.

EXAMPLE 8 Find the reciprocal: **(a)** $\frac{7}{8}$; **(b)** $-\frac{3}{4}$; **(c)** -8.

SOLUTION

a) The reciprocal of $\frac{7}{8}$ is $\frac{8}{7}$ because $\frac{7}{8} \cdot \frac{8}{7} = 1$.

b) The reciprocal of $-\frac{3}{4}$ is $-\frac{4}{3}$.

c) The reciprocal of -8 is $\frac{1}{-8}$, or $-\frac{1}{8}$.

To divide, we can multiply by a reciprocal. We sometimes say that we "invert and multiply."

EXAMPLE 9 Divide: **(a)** $-\frac{1}{4} \div \frac{3}{5}$; **(b)** $-\frac{6}{7} \div (-10)$.

SOLUTION

a) $-\frac{1}{4} \div \frac{3}{5} = -\frac{1}{4} \cdot \frac{5}{3}$ **"Inverting" $\frac{3}{5}$ and changing division to multiplication**

$= -\frac{5}{12}$

b) $-\frac{6}{7} \div (-10) = -\frac{6}{7} \cdot \left(-\frac{1}{10}\right) = \frac{6}{70}$, or $\frac{3}{35}$

Thus far, we have not divided by 0 or, equivalently, had a denominator of 0. There is a reason for this. Suppose 5 were divided by 0. The answer would have to be a number that, when multiplied by 0, gave 5. But any number times 0 is 0. Thus we cannot divide 5 or any other nonzero number by 0.

What if we divide 0 by 0? In this case, our solution would need to be some number that, when multiplied by 0, gave 0. But then *any* number would work as a solution to $0 \div 0$. This could lead to contradictions, so we agree to exclude division of 0 by 0 also.

Division by Zero We never divide by 0. If asked to divide a nonzero number by 0, we say that the answer is *undefined*. If asked to divide 0 by 0, we say that the answer is *indeterminate*.

The rules for order of operations discussed in Section 1.1 apply to *all* real numbers, regardless of their signs.

EXAMPLE 10 Simplify: **(a)** $(-5)^2$; **(b)** -5^2.

SOLUTION An exponent is written immediately after the base. Thus, in $(-5)^2$, the base is (-5), and in -5^2, the base is 5.

a) $(-5)^2 = 25$ **Squaring −5**

b) $-5^2 = -25$ **Squaring 5 and then taking the opposite**

Note that $(-5)^2 \neq -5^2$.

EXAMPLE 11 Simplify: $7 - 5^2 + 6 \div 2(-5)^2$.

SOLUTION

$7 - 5^2 + 6 \div 2(-5)^2$	$= 7 - 25 + 6 \div 2 \cdot 25$	**Simplifying 5^2 and $(-5)^2$**
	$= 7 - 25 + 3 \cdot 25$	**Dividing**
	$= 7 - 25 + 75$	**Multiplying**
	$= -18 + 75$	**Subtracting**
	$= 57$	**Adding**

Besides parentheses, brackets, and braces, groupings may be indicated by a fraction bar, absolute-value symbol, or radical sign $(\sqrt{\ })$.

EXAMPLE 12 Calculate: $\dfrac{12|7 - 9| + 4 \cdot 5}{(-3)^4 + 2^3}$.

SOLUTION We simplify the numerator and the denominator before we divide the results:

$$\begin{aligned}\frac{12|7 - 9| + 4 \cdot 5}{(-3)^4 + 2^3} &= \frac{12|-2| + 20}{81 + 8} \\ &= \frac{12(2) + 20}{89} \\ &= \frac{44}{89}. \qquad \textbf{Multiplying and adding. This shows } 44 \div 89.\end{aligned}$$

Entering Expressions

Graphing calculators have different keys for writing negatives and subtracting. The key labeled (−) is used to create a negative sign, whereas the one labeled − is used for subtraction. Using the wrong key may result in an ERR: SYNTAX message.

Some grouping symbols, such as a fraction bar and an absolute-value symbol, may require parentheses to indicate the grouping.

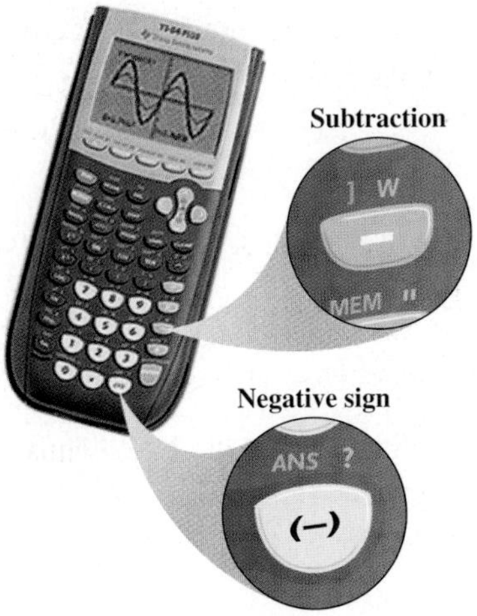

Menus The absolute-value option is accessed using a **menu,** a list of options that appears when a key is pressed. For example, pressing MATH results in a screen like the one on the left below. Four menu titles, or **submenus,** are listed across the top of the screen. In the screen on the left, the MATH submenu is highlighted and the options in that menu are listed. We refer to this submenu as MATH MATH, which means we first press MATH and then choose the MATH submenu.

```
MATH NUM CPX PRB
1:▸Frac
2:▸Dec
3:³
4:³√(
5:ˣ√
6:fMin(
7↓fMax(
```

```
MATH NUM CPX PRB
1:abs(
2:round(
3:iPart(
4:fPart(
5:int(
6:min(
7↓max(
```

Use the left and right arrow keys to highlight the desired menu. In the screen shown on the right above, the NUM menu is highlighted. The options in the highlighted menu appear on the screen. Note in both figures that the menus contain more options than can fit on a screen, as indicated by the arrows in entry 7. The remaining options will appear as the down arrow is pressed.

(*continued*)

To copy an item to the home screen, highlight its number using the up or down arrow keys and press ENTER, or simply press the number of the item.

Ans The most recent result is stored in the calculator's memory as Ans (short for Answer). When this appears on the screen, it has the value of the last result calculated on the home screen.

Converting from Decimal to Fraction Notation Many calculators can convert decimal notation to fraction notation. In the MATH MATH submenu shown above, option 1, ▶Frac, will return the simplified form of the equivalent fraction. Typically, this will work only for fractions with denominators less than 1000.

EXAMPLE 13 Calculate: $\dfrac{14 - 3|-16 + 38|}{4|-2^4 - 3^2|}$.

SOLUTION There is no fraction bar on the calculator, so the expression must be rewritten using parentheses:

$$(14 - 3|-16 + 38|) \div (4|-2^4 - 3^2|).$$

For most graphing calculators, $|x|$ is written abs(x) and is accessed through a menu. We press MATH ▷ 1 to copy abs on the screen. The expressions within the absolute-value symbols must be enclosed in parentheses. Some calculators will supply the left parenthesis; for these, you must close the expression with a right parenthesis.

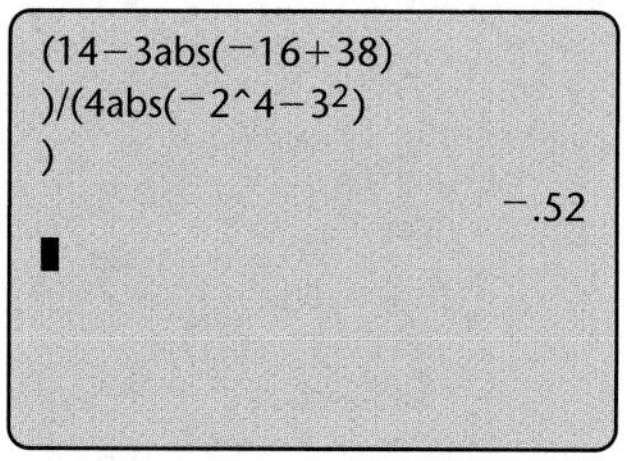

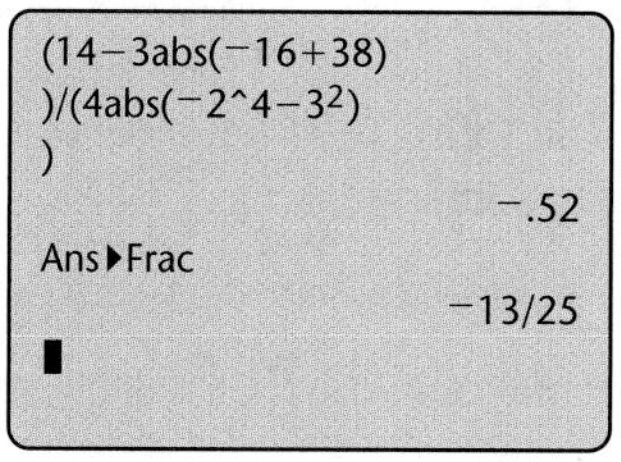

When we press MATH 1 to copy ▶Frac to the screen, the answer, -0.52, is converted to $-\frac{13}{25}$.

Example 13 demonstrates that being able to use a calculator does not make it any less important to understand the mathematics being studied. A solid understanding of the rules for order of operations is necessary in order to locate parentheses properly in an expression.

1.2 EXERCISE SET

FOR EXTRA HELP

MathXL | MyMathLab | InterAct Math | AW Math Tutor Center | Video Lectures on CD: Disc 1 | Student's Solutions Manual

Concept Reinforcement *Classify each of the following as either true or false.*

1. The sum of two negative numbers is always negative.
2. The product of two negative numbers is always negative.
3. The product of a negative number and a positive number is always negative.
4. The sum of a negative number and a positive number is always negative.
5. The sum of a negative number and a positive number is always positive.
6. If a and b are negative, with $a < b$, then $|a| > |b|$.
7. If a and b are positive, with $a < b$, then $|a| > |b|$.
8. The terms *additive inverses* and *opposites* have the same meaning.
9. Reciprocals have the same sign.
10. Every real number has an opposite.

Find each absolute value.

11. $|-9|$ **12.** $|-7|$ **13.** $|6|$

14. $|47|$ **15.** $|-6.2|$ **16.** $|-7.9|$

17. $|0|$ **18.** $|3\frac{3}{4}|$ **19.** $|1\frac{7}{8}|$

20. $|7.24|$ **21.** $|-4.21|$ **22.** $|-5.309|$

Determine whether each inequality is a true statement.

23. $-6 \leq -2$ **24.** $-1 \leq -5$

25. $-9 > 1$ **26.** $7 \geq -2$

27. $3 \geq -5$ **28.** $9 \leq 9$

29. $-8 < -3$ **30.** $7 > -8$

31. $-4 \geq -4$ **32.** $2 < 2$

33. $-5 < -5$ **34.** $-2 > -12$

Add.

35. $3 + 9$ **36.** $8 + 3$

37. $-3 + (-9)$ -12 **38.** $-8 + (-3)$

39. $-3.9 + 2.7$ **40.** $-1.9 + 7.3$

41. $\frac{2}{7} + \left(-\frac{3}{5}\right)$ **42.** $\frac{3}{8} + \left(-\frac{2}{5}\right)$

43. $-3.26 + (-5.8)$ **44.** $-2.1 + (-7.5)$

45. $-\frac{1}{9} + \frac{2}{3}$ **46.** $-\frac{1}{2} + \frac{4}{5}$

47. $0 + (-4.5)$ **48.** $-3.19 + 0$

49. $-7.24 + 7.24$ **50.** $-9.46 + 9.46$

51. $15.9 + (-22.3)$ **52.** $21.7 + (-28.3)$

Find the opposite, or additive inverse.

53. 3.14 **54.** 5.43 **55.** $-4\frac{1}{3}$

56. $2\frac{3}{5}$ **57.** 0 **58.** $-2\frac{3}{4}$

Find $-x$ for each of the following.

59. $x = 9$ **60.** $x = 3$

61. $x = -2.7$ **62.** $x = -1.9$

63. $x = 1.79$ **64.** $x = 3.14$

65. $x = 0$ **66.** $x = -7$

Subtract.

67. $8 - 5$ **68.** $10 - 3$

69. $5 - 8$ **70.** $3 - 10$

71. $-5 - (-12)$ **72.** $-3 - (-9)$

73. $-5 - 14$ **74.** $-9 - 8$

75. $2.7 - 5.8$ **76.** $3.7 - 4.2$

77. $-\frac{3}{5} - \frac{1}{2}$ **78.** $-\frac{2}{3} - \frac{1}{5}$

Aha! **79.** $-3.9 - (-3.9)$ **80.** $-5.4 - (-4.3)$

81. $0 - (-7.9)$ **82.** $0 - 5.3$

Multiply.

83. $(-5)6$

84. $(-4)7$

85. $(-4)(-9)$

86. $(-7)(-8)$

87. $(4.2)(-5)$

88. $(3.5)(-8)$

89. $\frac{3}{7}(-1)$

90. $-1 \cdot \frac{2}{5}$

91. $(-17.45) \cdot 0$

92. 15.2×0

93. $(-3.2) \times (-1.7)$

94. $(1.9) \cdot (4.3)$

Divide.

95. $\frac{-10}{-2}$

96. $\frac{-15}{-3}$

97. $\frac{-100}{20}$

98. $\frac{-50}{5}$

99. $\frac{73}{-1}$

100. $\frac{-62}{1}$

101. $\frac{0}{-7}$

102. $\frac{0}{-11}$

Find the reciprocal, or multiplicative inverse, if it exists.

103. 4

104. 3

105. -9

106. -5

107. $\frac{2}{3}$

108. $\frac{4}{7}$

109. $-\frac{3}{11}$

110. 0

Divide.

111. $\frac{2}{3} \div \frac{4}{5}$

112. $\frac{2}{7} \div \frac{6}{5}$

113. $-\frac{3}{5} \div \frac{1}{2}$

114. $\left(-\frac{4}{7}\right) \div \frac{1}{3}$

115. $\left(-\frac{2}{9}\right) \div (-8)$

116. $\left(-\frac{2}{11}\right) \div (-6)$

Aha! **117.** $-\frac{12}{7} \div \left(-\frac{12}{7}\right)$

118. $\left(-\frac{2}{7}\right) \div (-1)$

Calculate using the rules for order of operations. If an expression is undefined, state this. Check using a calculator.

119. $9 - (8 - 3 \cdot 2^3)$

120. $19 - (4 + 2 \cdot 3^2)$

121. $\dfrac{5 \cdot 2 - 4^2}{27 - 2^4}$

122. $\dfrac{7 \cdot 3 - 5^2}{9 + 4 \cdot 2}$

123. $\dfrac{3^4 - (5 - 3)^4}{8 - 2^3}$

124. $\dfrac{4^3 - (7 - 4)^2}{3^2 - 7}$

125. $\dfrac{(2 - 3)^3 - 5|2 - 4|}{7 - 2 \cdot 5^2}$

126. $\dfrac{8 \div 4 \cdot 6|4^2 - 5^2|}{9 - 4 + 11 - 4^2}$

127. $|2^2 - 7|^3 + 4$

128. $|-2 - 3| \cdot 4^2 - 3$

129. $32 - (-5)^2 + 15 \div (-3) \cdot 2$

130. $43 - (-9 + 2)^2 + 18 \div 6 \cdot (-2)$

131. $17 - \sqrt{11 - (3 + 4)} \div [-5 - (-6)]^2$

132. $15 - 1 + \sqrt{5^2 - (3 + 1)^2(-1)}$

Match each algebraic expression with the correct series of keystrokes. Check your answer by calculating the expression both by hand and by using the calculator.

133. $\dfrac{5(3 - 7) + 4^3}{(-2 - 3)^2}$

134. $(5(3 - 7) + 4)^3 \div (-2) - 3^2$

135. $5(3 - 7) + 4^3 \div (-2 - 3)^2$

136. $\dfrac{5(3 - 7) + 4^3}{-2 - 3^2}$

a) (5 (3 − 7) + 4 ^ 3) ÷ (((−) 2 − 3) x²) ENTER

b) (5 (3 − 7) + 4 ^ 3) ÷ ((−) 2 − 3 x²) ENTER

c) (5 (3 − 7) + 4) ^ 3 ÷ (−) 2 − 3 x² ENTER

d) 5 (3 − 7) + 4 ^ 3 ÷ ((−) 2 − 3) x² ENTER

TW **137.** Describe in your own words a method for determining the sign of the sum of a positive number and a negative number.

TW **138.** Explain in your own words the difference between the opposite of a number and the reciprocal of a number.

Skill Maintenance

To the student and the instructor: Exercises included for Skill Maintenance review skills previously studied in the text. Generally these exercises provide preparation for the next section of the text. The section(s) in which these types of exercises first appeared is shown in brackets. Answers to all Skill Maintenance exercises appear at the back of the book.

Evaluate. [1.1]

139. $2(x + 5)$ and $2x + 10$, for $x = 3$

140. $2a - 3$ and $a - 3 + a$, for $a = 7$

Synthesis

TW **141.** Explain in your own words why 7/0 is undefined.

TW **142.** Explain in your own words why 0 has an opposite but not a reciprocal.

Insert one pair of parentheses to convert each false statement into a true statement.

143. $8 - 5^3 + 9 = 36$

144. $2 \cdot 7 + 3^2 \cdot 5 = 104$

145. $5 \cdot 2^3 \div 3 - 4^4 = 40$

146. $2 - 7 \cdot 2^2 + 9 = -11$

147. Find the greatest value of a for which $|a| \geq 6.2$ and $a < 0$.

1.3 Equivalent Algebraic Expressions

The Commutative, Associative, and Distributive Laws ■ Combining Like Terms ■ Checking by Evaluating

In algebra, we are often interested in manipulating expressions without changing their value. In this section, we look at some properties of real numbers that allow us to write *equivalent expressions*.

Equivalent Expressions Two expressions that have the same value for all possible replacements are called *equivalent expressions.*

Study Tip

Do the Exercises

- Usually an instructor assigns some odd-numbered exercises. When you complete these, you can check your answers at the back of the book. If you miss any, closely examine your work and, if necessary, seek help (see the Study Tip on p.11).
- Whether or not your instructor assigns the even-numbered exercises, try to do some on your own. Check your answers later with a friend or your instructor.

The Commutative, Associative, and Distributive Laws

When a pair of real numbers are added or multiplied, the order in which the numbers are written does not affect the result.

The Commutative Laws For any real numbers a and b,

$a + b = b + a$; (for Addition)

$a \cdot b = b \cdot a$. (for Multiplication)

The commutative laws provide one way of writing equivalent expressions.

EXAMPLE 1 Use a commutative law to write an expression equivalent to $7x + 9$.

SOLUTION Using the commutative law of addition, we have

$$7x + 9 = 9 + 7x.$$

We can also use the commutative law of multiplication to write

$$7 \cdot x + 9 = x \cdot 7 + 9.$$

The expressions $7x + 9$, $9 + 7x$, and $x7 + 9$ are all equivalent. They name the same number for any replacement of x.

The *associative laws* also enable us to form equivalent expressions. We use the associative laws to change the *grouping* of numbers being added or multiplied.

The Associative Laws For any real numbers a, b, and c,

$$a + (b + c) = (a + b) + c; \qquad a \cdot (b \cdot c) = (a \cdot b) \cdot c.$$

(for Addition) (for Multiplication)

EXAMPLE 2 Write an expression equivalent to $(3x + 7y) + 9z$, using the associative law of addition.

SOLUTION We have

$$(3x + 7y) + 9z = 3x + (7y + 9z).$$

The expressions $(3x + 7y) + 9z$ and $3x + (7y + 9z)$ are equivalent. They name the same number for any replacements of x, y, and z.

EXAMPLE 3 Use the commutative and associative laws to write an expression equivalent to

$$\frac{5}{x} \cdot (yz).$$

SOLUTION Answers may vary. Here we use the associative law first and then the commutative law.

$$\frac{5}{x} \cdot (yz) = \left(\frac{5}{x} \cdot y\right) \cdot z \qquad \text{Using the associative law of multiplication}$$

$$= \left(y \cdot \frac{5}{x}\right) \cdot z \qquad \text{Using the commutative law of multiplication inside the parentheses}$$

The *distributive law* that follows provides still another way of forming equivalent expressions. In essence, the distributive law allows us to rewrite the *product* of a and $b + c$ as the *sum* of ab and ac.

The Distributive Law For any numbers a, b, and c,

$$a(b + c) = ab + ac.$$

Since $b - c = b + (-c)$, it is also true by the distributive law that $a(b - c) = ab - ac$.

Student Notes

The commutative, associative, and distributive laws are used so often in this course that it is worth the effort to memorize them.

EXAMPLE 4 Obtain an expression equivalent to $5x(y + 4)$ by multiplying.

SOLUTION We use the distributive law to get

$$\begin{aligned} 5x(y + 4) &= 5xy + 5x \cdot 4 && \textbf{Using the distributive law} \\ &= 5xy + 5 \cdot 4 \cdot x && \textbf{Using the commutative law of multiplication} \\ &= 5xy + 20x. && \textbf{Simplifying} \end{aligned}$$

The expressions $5x(y + 4)$ and $5xy + 20x$ are equivalent. They name the same number for any replacements of x and y.

When we reverse what we did in Example 4, we say that we are **factoring** an expression. This allows us to rewrite a sum or a difference as a product.

EXAMPLE 5 Obtain an expression equivalent to $3x - 6$ by factoring.

SOLUTION We use the distributive law to get

$$3x - 6 = 3 \cdot x - 3 \cdot 2 = 3(x - 2).$$

In Example 5, since the product of 3 and $x - 2$ is $3x - 6$, we say that 3 and $x - 2$ are **factors** of $3x - 6$. Thus "factor" can act as a noun or as a verb.

Combining Like Terms

In an expression like $8a^5 + 17 + \frac{4}{b} + (-6a^3b)$, the parts that are separated by addition signs are called *terms*. A **term** is a number, a variable, a product of numbers and/or variables, or a quotient of numbers and/or variables. Thus, $8a^5$, 17, $\frac{4}{b}$, and $-6a^3b$ are terms in $8a^5 + 17 + \frac{4}{b} + (-6a^3b)$. When terms have variable factors that are exactly the same, we refer to those terms as **like,** or **similar, terms.** Thus, $3x^2y$ and $-7x^2y$ are similar terms, but $3x^2y$ and $4xy^2$ are not. We can often simplify expressions by **combining,** or **collecting, like terms.**

EXAMPLE 6 Combine like terms: $3a + 5a^2 + 4a + a^2$.

SOLUTION

$$\begin{aligned} 3a + 5a^2 + 4a + a^2 &= 3a + 4a + 5a^2 + a^2 && \textbf{Using the commutative law} \\ &= (3 + 4)a + (5 + 1)a^2 && \textbf{Using the distributive law. Note that } a^2 = 1a^2. \\ &= 7a + 6a^2 \end{aligned}$$

Sometimes we must use the distributive law to remove grouping symbols before combining like terms. Remember to remove the innermost grouping symbols first.

EXAMPLE 7 Simplify: $3x + 2[4 + 5(x + 2y)]$.

SOLUTION

$$3x + 2[4 + 5(x + 2y)] = 3x + 2[4 + 5x + 10y] \quad \text{Using the distributive law}$$
$$= 3x + 8 + 10x + 20y \quad \text{Using the distributive law again}$$
$$= 13x + 8 + 20y \quad \text{Combining like terms}$$

The product of a number and -1 is its opposite, or additive inverse. For example,

$$-1 \cdot 8 = -8 \qquad \text{(the opposite of 8).}$$

Thus we have $-8 = -1 \cdot 8$, and in general, $-x = -1 \cdot x$. We can use this fact along with the distributive law when parentheses are preceded by a negative sign or subtraction.

EXAMPLE 8 Simplify $-(a - b)$, using multiplication by -1.

SOLUTION We have

$$-(a - b) = -1 \cdot (a - b) \quad \text{Replacing } - \text{ with multiplication by } -1$$
$$= -1 \cdot a - (-1) \cdot b \quad \text{Using the distributive law}$$
$$= -a - (-b) \quad \text{Replacing } -1 \cdot a \text{ with } -a \text{ and } (-1) \cdot b \text{ with } -b$$
$$= -a + b, \text{ or } b - a. \quad \text{Try to go directly to this step.}$$

The expressions $-(a - b)$ and $b - a$ are equivalent. They represent the same number for all replacements of a and b.

Example 8 illustrates a useful shortcut worth remembering:

The opposite of $a - b$ is $-a + b$, or $b - a$.

EXAMPLE 9 Simplify: $9x - 5y - (5x + y - 7)$.

SOLUTION

$$9x - 5y - (5x + y - 7) = 9x - 5y - 5x - y + 7 \quad \text{Using the distributive law}$$
$$= 4x - 6y + 7 \quad \text{Combining like terms}$$

EXAMPLE 10 Simplify by combining like terms:

$$3x^2 - 2\{3[x - 5x^2 - 4(x + 3) + 7] - x\}.$$

SOLUTION

$3x^2 - 2\{3[x - 5x^2 - 4(x + 3) + 7] - x\}$

$= 3x^2 - 2\{3[x - 5x^2 - 4x - 12 + 7] - x\}$	**Multiplying to remove the innermost parentheses using the distributive law**
$= 3x^2 - 2\{3[-5x^2 - 3x - 5] - x\}$	**Combining like terms inside the brackets**
$= 3x^2 - 2\{-15x^2 - 9x - 15 - x\}$	**Using the distributive law to remove the brackets**
$= 3x^2 - 2\{-15x^2 - 10x - 15\}$	**Combining like terms**
$= 3x^2 + 30x^2 + 20x + 30$	**Using the distributive law**
$= 33x^2 + 20x + 30$	**Combining like terms**

Editing Expressions

It is possible to correct an error or change a value in an expression by using the *arrow*, *insert*, and *delete* keys.

As the expression is being typed, use the arrow keys to move the cursor to the character you want to change. Pressing INS (the 2nd option associated with the DEL key) changes the calculator between the INSERT and OVERWRITE modes. The OVERWRITE mode is often indicated by a rectangular cursor and the INSERT mode by an underscore cursor. The following table shows how to make changes.

Insert a character in front of the cursor.	In the INSERT mode, press the character you wish to insert.
Replace the character under the cursor.	In the OVERWRITE mode, press the character you want as the replacement.
Delete the character under the cursor.	Press DEL.

After you have evaluated an expression by pressing ENTER, the expression can be recalled to the screen by pressing ENTRY. (ENTRY is the 2nd option associated with the ENTER key.) Then it can be edited as described above. Pressing ENTER will then evaluate the edited expression.

EXAMPLE 11 Evaluate $13 - 5[x + 6(4 - x) - 18]$ for $x = 1.3$ and for $x = -5$.

SOLUTION To evaluate the expression for $x = 1.3$, we type $13 - 5(1.3 + 6(4 - 1.3) - 18)$ and press ENTER. Then, to evaluate the expression for $x = -5$, we press ENTRY and replace each occurrence of 1.3 with -5. Since 1.3 consists of three characters and -5 consists of two characters, we delete one character of 1.3 and overwrite the other two with -5. Remember to use the (–) key to enter the negative sign. Then we press ENTER.

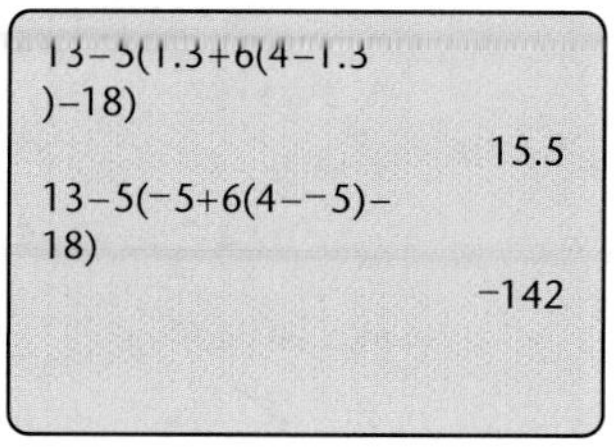

We see that $13 - 5[x + 6(4 - x) - 18]$ is 15.5 when $x = 1.3$ and -142 when $x = -5$.

Storing a Value to a Variable

The STO› key is used to store a value to a variable name. Then the value can be substituted for the variable in an algebraic expression. To let $X = -5$, press (–) 5 STO› X,T,Θ,n ENTER. Any letter can be used as a variable by pressing ALPHA and the key associated with that letter.

To see the value of X, press X,T,Θ,n ENTER. To evaluate the expression $2X + 1$ for $X = -5$, enter the expression as written and press ENTER. The value of the expression is -9.

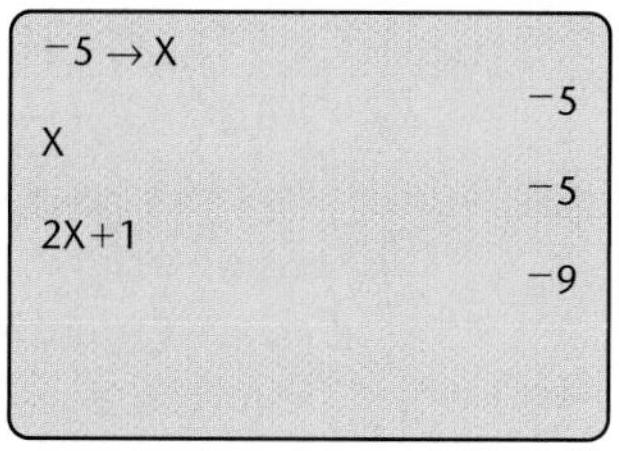

EXAMPLE 12 Use STO› to evaluate $13 - 5[x + 6(4 - x) - 18]$ for $x = 1.3$ and for $x = -5$.

SOLUTION We store the 1.3 to X and then enter the expression as written and press ENTER. Then we store -5 to X, press ENTRY twice to recall the expression, and press ENTER. We see that the value of the expression is 15.5 when $x = 1.3$ and -142 when $x = -5$.

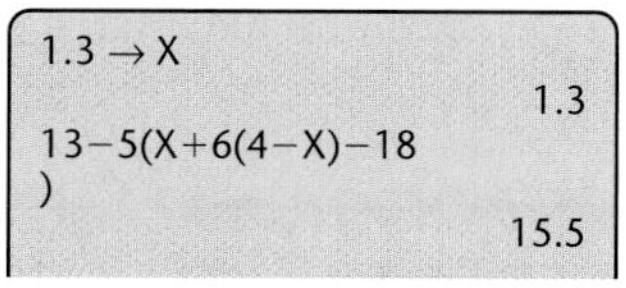

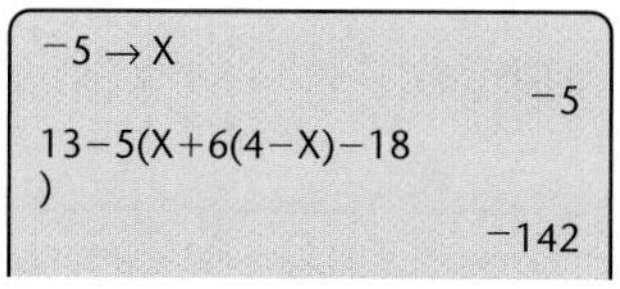

Checking by Evaluating

Recall that equivalent expressions have the same value for all possible replacements. Thus one way to check whether two expressions are equivalent is to evaluate both expressions using the same replacements for the variables.

Interactive Discovery

1. Evaluate both expressions in the following table for the given values for x, and determine whether the expressions appear to be equivalent.

x	$3x + 5x^2 + 4x + x^2$	$7x + 5x^2$	Equivalent?
0			
5			
−3			

2. Next, evaluate both expressions in the following table for the given values for x, and determine whether these expressions appear to be equivalent.

x	$2x - 3[4 - 6(x - 1)]$	$20x - 30$	Equivalent?
0			
5			
−3			

Can two expressions that are *not* equivalent have the same value for *some* possible replacements?

Checking by Evaluating

To check whether two expressions are equivalent, evaluate both expressions using the same replacements for the variables.

1. If the values are different, the expressions *are not* equivalent.
2. If the values are the same, the expressions *may be* equivalent.
3. Evaluating expressions for several different replacements makes the check more certain.

CAUTION! Checking by evaluating can be very useful when a quick, partial check is needed. However, showing that two expressions have the same value for one replacement *does not prove* that they are equivalent.

EXAMPLE 13 Check Example 9 by evaluating the expressions.

SOLUTION We evaluate the given expression, $9x - 5y - (5x + y - 7)$, and the simplified expression, $4x - 6y + 7$. We choose the replacement values of -1 for x and 2 for y.

$$9x - 5y - (5x + y - 7) = 9(-1) - 5(2) - (5(-1) + 2 - 7)$$
Replacing x with -1 and y with 2

$$= -9 - 10 - (-5 + 2 - 7)$$ **Multiplying**

$$= -9 - 10 - (-10)$$ **Simplifying within the parentheses**

$$= -9 - 10 + 10$$ **Removing parentheses**

$$= -9$$ **Adding and subtracting from left to right**

$$4x - 6y + 7 = 4(-1) - 6(2) + 7$$ **Replacing x with -1 and y with 2**

$$= -4 - 12 + 7$$ **Multiplying**

$$= -9$$ **Adding and subtracting from left to right**

```
-1→X
                 -1
2→Y
                  2
```

```
9X-5Y-(5X+Y-7)
                 -9
4X-6Y+7
                 -9
```

The computations can also be performed using a graphing calculator, as shown in the figures at left. Since $-9 = -9$, we have a partial check.

1.3 EXERCISE SET

FOR EXTRA HELP

MathXL | MyMathLab | InterAct Math | AW Math Tutor Center | Video Lectures on CD: Disc 1 | Student's Solutions Manual

Concept Reinforcement *Classify each statement as an illustration of a commutative law, an associative law, or the distributive law.*

1. $2x + 7 + 3x = 2x + 3x + 7$

2. $2x + 4 = 2(x + 2)$

3. $3 \cdot y \cdot 2 = 3 \cdot 2 \cdot y$

4. $115 + (-15 + 63) = (115 + (-15)) + 63$

5. $14(2x - 1) = 28x - 14$

6. $4(\frac{1}{2} \cdot 3) = (4 \cdot \frac{1}{2}) \cdot 3$

Write an equivalent expression using a commutative law. Answers may vary.

7. $4a + 7b$

8. $6 + xy$

9. $(7x)y$

10. $-9(ab)$

Write an equivalent expression using an associative law.

11. $(3x)y$

12. $-7(ab)$

13. $x + (2y + 5)$

14. $(3y + 4) + 10$

Write an equivalent expression using the distributive law.

15. $7(t + 2)$

16. $8(x + 1)$

17. $4(x - y)$

18. $9(a - b)$

19. $-5(2a + 3b)$

20. $-2(3c + 5d)$

21. $9a(b - c + d)$

22. $5x(y - z + w)$

Find an equivalent expression by factoring.

23. $5x + 50$

24. $7a + 7b$

25. $3p - 9$

26. $15x - 3$

27. $7x - 21y + 14z$

28. $6y - 9x - 3w$

Aha! 29. $255 - 34b$

30. $132a + 33$

31. $xy + x$

32. $ab + b$

List the terms of each expression.

33. $4x - 5y + 3$

34. $2a + 7b - 13$

35. $x^2 - 6x - 7$

36. $-3y^2 + 6y + 10$

Simplify to form an equivalent expression by combining like terms. Use the distributive law as needed.

37. $3x + 7x$

38. $9x + 3x$

39. $7rt - 9rt$

40. $3ab + 7ab$

41. $9t^2 + t^2$

42. $7a^2 + a^2$

43. $12a - a$

44. $15x - x$

45. $n - 8n$

46. $x - 6x$

47. $5x - 3x + 8x$

48. $3x - 11x + 2x$

49. $4x - 2x^2 + 3x$

50. $9a - 5a^2 + 4a$

51. $6a + 7a^2 - a + 4a^2$

52. $9x + 2x^3 + 5x - 6x^2$

53. $4x - 7 + 18x + 25$

54. $13p + 5 - 4p + 7$

55. $-7t^2 + 3t + 5t^3 - t^3 + 2t^2 - t$

56. $-9n + 8n^2 + n^3 - 2n^2 - 3n + 4n^3$

57. $7a - (2a + 5)$

58. $x - (5x + 9)$

59. $m - (m - 1)$

60. $5a - (4a - 3)$

61. $3d - 7 - (5 - 2d)$

62. $8x - 9 - (7 + 8x)$

63. $-2(x + 3) - 5(x - 4)$

64. $-9(y + 7) - 6(y - 3)$

65. $4x - 7(2x - 3)$

66. $9y - 4(5y - 6)$

67. $9a - [7 - 5(7a - 3)]$

68. $12b - [9 - 7(5b - 6)]$

69. $5\{-2a + 3[4 - 2(3a + 5)]\}$

70. $7\{-7x + 8[5 - 3(4x + 6)]\}$

71. $2y + \{7[3(2y - 5) - (8y + 7)] + 9\}$

72. $7b - \{6[4(3b - 7) - (9b + 10)] + 11\}$

Evaluate both expressions in each pair for $x = 1.3$, $x = -3$, and $x = 0$. Then determine whether the expressions appear to be equivalent.

73. $2x - 3(x + 5)$,
$-x + 15$

74. $2(x + 5) - 5(x + 2)$,
$7x$

75. $4(x + 3)$,
$4x + 12$

76. $7x + 1$,
$7(x + 1)$

TW 77. What is the difference between the associative law of multiplication and the distributive law?

TW 78. Write a sentence in which the word "factor" appears once as a verb and once as a noun.

Skill Maintenance

Find each absolute value. [1.2]

79. $|-3|$

80. $|3.59|$

81. Find the opposite of -35. [1.2]

82. Find the reciprocal of -35. [1.2]

Synthesis

TW 83. Explain how the distributive and commutative laws can be used to rewrite $3x + 6y + 4x + 2y$ as $7x + 8y$.

TW 84. Lee simplifies the expression $a(b + c)$ by omitting the parentheses and writing $ab + c$. Are these expressions equivalent? Why or why not?

Simplify.

85. $11(a - 3) + 12a - \{6[4(3b - 7) - (9b + 10)] + 11\}$

86. $-3[9(x - 4) + 5x] - 8\{3[5(3y + 4)] - 12\}$

87. $z - \{2z + [3z - (4z + 5z) - 6z] + 7z\} - 8z$

88. $x + [f - (f + x)] + [x - f] + 3x$

89. $x - \{x + 1 - [x + 2 - (x - 3 - \{x + 4 - [x - 5 + (x - 6)]\})]\}$

90. Use the commutative, associative, and distributive laws to show that $5(a + bc)$ is equivalent to $c(b5) + a5$. Use only one law in each step of your work.

TW 91. Are subtraction and division commutative? Why or why not?

TW 92. Are subtraction and multiplication associative? Why or why not?

1.4 Exponential and Scientific Notation

The Product and Quotient Rules ■ The Zero Exponent ■ Negative Integers as Exponents ■ Simplifying $(a^m)^n$ ■ Raising a Product or a Quotient to a Power ■ Scientific Notation ■ Significant Digits and Rounding

Study Tip

Seeking Help Off Campus

Are you aware of all the supplements that exist for this textbook? See the preface for a description of each supplement: the *Student's Solutions Manual,* a complete set of lessons in video or CD form, tutorial software, and www.MyMathLab.com. Through this Web site, you can find additional practice resources, helpful ideas, and links to other learning resources.

In Section 1.1, we introduced exponential notation. We now develop rules for manipulating exponents and determine what zero and negative integers will mean as exponents.

The Product and Quotient Rules

Note that the expression $x^3 \cdot x^4$ can be rewritten as follows:

$$x^3 \cdot x^4 = \underbrace{x \cdot x \cdot x}_{\text{3 factors}} \cdot \underbrace{x \cdot x \cdot x \cdot x}_{\text{4 factors}}$$

$$= \underbrace{x \cdot x \cdot x \cdot x \cdot x \cdot x \cdot x}_{\text{7 factors}}$$

$$= x^7.$$

This result is generalized in the *product rule.*

Multiplying with Like Bases: The Product Rule For any number a and any positive integers m and n,

$$a^m \cdot a^n = a^{m+n}.$$

(When multiplying, if the bases are the same, keep the base and add the exponents.)

Student Notes

Be careful to distinguish between how coefficients and exponents are handled. For example, in Example 1(b), the product of the coefficients 5 and 3 is 15, whereas the product of b^3 and b^5 is b^8.

EXAMPLE 1 Multiply and simplify: **(a)** $m^5 \cdot m^7$; **(b)** $(5a^2b^3)(3a^4b^5)$.

SOLUTION

a) $m^5 \cdot m^7 = m^{5+7} = m^{12}$ **Multiplying by adding exponents**

b) $(5a^2b^3)(3a^4b^5) = 5 \cdot 3 \cdot a^2 \cdot a^4 \cdot b^3 \cdot b^5$ **Using the associative and commutative laws**

$= 15a^{2+4}b^{3+5}$ **Multiplying; using the product rule**

$= 15a^6b^8$

CAUTION! $5^8 \cdot 5^6 = 5^{14}$; $5^8 \cdot 5^6 \neq 25^{14}$.

Next, we simplify a quotient:

$$\frac{x^8}{x^3} = \frac{x \cdot x \cdot x \cdot x \cdot x \cdot x \cdot x \cdot x}{x \cdot x \cdot x} \quad \begin{array}{l} \longleftarrow \text{8 factors} \\ \longleftarrow \text{3 factors} \end{array}$$

$$= \frac{x \cdot x \cdot x}{x \cdot x \cdot x} \cdot x \cdot x \cdot x \cdot x \cdot x \qquad \textbf{Note that } x^3/x^3 \textbf{ is 1.}$$

$$= x \cdot x \cdot x \cdot x \cdot x \qquad \longleftarrow \text{5 factors}$$

$$= x^5.$$

The generalization of this result is the *quotient rule.*

Dividing with Like Bases: The Quotient Rule For any nonzero number a and any positive integers m and n, $m > n$,

$$\frac{a^m}{a^n} = a^{m-n}.$$

(When dividing, if the bases are the same, keep the base and subtract the exponent of the denominator from the exponent of the numerator.)

EXAMPLE 2 Divide and simplify: **(a)** $\frac{r^9}{r^3}$; **(b)** $\frac{10x^{11}y^5}{2x^4y^3}$.

SOLUTION

a) $\frac{r^9}{r^3} = r^{9-3} = r^6$ **Using the quotient rule**

b) $\frac{10x^{11}y^5}{2x^4y^3} = 5 \cdot x^{11-4} \cdot y^{5-3}$ **Dividing; using the quotient rule**

$= 5x^7y^2$

CAUTION! $\frac{7^8}{7^2} = 7^6$; $\frac{7^8}{7^2} \neq 7^4$.

The Zero Exponent

Suppose now that the bases in the numerator and the denominator are identical and are both raised to the same power. On the one hand, any (nonzero) expression divided by itself is equal to 1. For example,

$$\frac{t^5}{t^5} = 1 \quad \text{and} \quad \frac{6^4}{6^4} = 1.$$

On the other hand, the quotient rule tells us to subtract exponents when dividing powers with the same base. If we now allow $m = n$, we have

$$\frac{t^5}{t^5} = t^{5-5} = t^0 \quad \text{and} \quad \frac{6^4}{6^4} = 6^{4-4} = 6^0.$$

This suggests that t^5/t^5 equals both 1 *and* t^0. It also suggests that $6^4/6^4$ equals both 1 *and* 6^0. This leads to the following definition.

The Zero Exponent For any nonzero real number a,

$$a^0 = 1.$$

(Any nonzero number raised to the zero power is 1. 0^0 is undefined.)

EXAMPLE 3 Evaluate each of the following for $x = 2.9$: **(a)** x^0; **(b)** $-x^0$; **(c)** $(-x)^0$.

SOLUTION

a) $x^0 = 2.9^0 = 1$ — **Using the definition of 0 as an exponent**

b) $-x^0 = -2.9^0 = -1$ — **The exponent 0 pertains only to the 2.9.**

c) $(-x)^0 = (-2.9)^0 = 1$ — **Because of the parentheses, the base here is −2.9.**

Parts (b) and (c) of Example 3 illustrate an important result:

Since $-a^n$ means $-1 \cdot a^n$, in general, $-a^n \neq (-a)^n$.*

Negative Integers as Exponents

Later in this text we will explain what numbers like $\frac{2}{9}$ or $\sqrt{2}$ mean as exponents. Until then, integer exponents will suffice.

To develop a definition for negative integer exponents, we simplify $5^3/5^7$ two ways. First we proceed as in arithmetic:

$$\frac{5^3}{5^7} = \frac{5 \cdot 5 \cdot 5}{5 \cdot 5 \cdot 5 \cdot 5 \cdot 5 \cdot 5 \cdot 5} = \frac{5 \cdot 5 \cdot 5 \cdot 1}{5 \cdot 5 \cdot 5 \cdot 5 \cdot 5 \cdot 5 \cdot 5}$$

$$= \frac{5 \cdot 5 \cdot 5}{5 \cdot 5 \cdot 5} \cdot \frac{1}{5 \cdot 5 \cdot 5 \cdot 5}$$

$$= \frac{1}{5^4}.$$

*When n is odd, it *does* follow that $-a^n = (-a)^n$. However, when n is even, we always have $-a^n \neq (-a)^n$, since $-a^n$ is always negative and $(-a)^n$ is always positive. We assume $a \neq 0$.

Were we to apply the quotient rule for any integer exponents, we would have

$$\frac{5^3}{5^7} = 5^{3-7} = 5^{-4}.$$

These two expressions for $5^3/5^7$ suggest that

$$5^{-4} = \frac{1}{5^4}.$$

This leads to the definition of integer exponents, which includes negative exponents.

Integer Exponents For any real number a that is nonzero and any integer n,

$$a^{-n} = \frac{1}{a^n}.$$

(The numbers a^{-n} and a^n are reciprocals of each other.)

The definitions above preserve the following pattern:

$4^3 = 4 \cdot 4 \cdot 4,$	
$4^2 = 4 \cdot 4,$	**Dividing both sides by 4**
$4^1 = 4,$	**Dividing both sides by 4**
$4^0 = 1,$	**Dividing both sides by 4**
$4^{-1} = \frac{1}{4},$	**Dividing both sides by 4**
$4^{-2} = \frac{1}{4 \cdot 4} = \frac{1}{4^2}.$	**Dividing both sides by 4**

CAUTION! A negative exponent does not, in itself, indicate that an expression is negative. As shown above,

$$4^{-2} \neq 4(-2).$$

EXAMPLE 4 Express using positive exponents and, if possible, simplify.

a) 3^{-2} **b)** $5x^{-4}y^3$ **c)** $\dfrac{1}{7^{-2}}$

SOLUTION

a) $3^{-2} = \dfrac{1}{3^2} = \dfrac{1}{9}$

b) $5x^{-4}y^3 = 5\left(\dfrac{1}{x^4}\right)y^3 = \dfrac{5y^3}{x^4}$

c) Since $\dfrac{1}{a^n} = a^{-n}$, we have

$$\frac{1}{7^{-2}} = 7^{-(-2)} = 7^2, \text{ or } 49.$$ **Remember: n can be a negative integer.**

The result from part (c) above can be generalized.

Factors and Negative Exponents

For any nonzero real numbers a and b and any integers m and n,

$$\frac{a^{-n}}{b^{-m}} = \frac{b^m}{a^n}.$$

(A factor can be moved to the other side of the fraction bar if the sign of the exponent is changed.)

EXAMPLE 5 Write an equivalent expression without negative exponents:

$$\frac{-5^2x^{-2}y^{-5}}{z^{-4}w^{-3}}.$$

SOLUTION We can move a factor to the other side of the fraction bar if we change the sign of its exponent:

$$\frac{-5^2x^{-2}y^{-5}}{z^{-4}w^{-3}} = \frac{-25z^4w^3}{x^2y^5}.$$ **Note that -5^2 does not have a negative exponent, so it is not moved.**

The product and quotient rules apply for all integer exponents.

EXAMPLE 6 Simplify: **(a)** $7^{-3} \cdot 7^8$; **(b)** $\dfrac{b^{-5}}{b^{-4}}$.

SOLUTION

a) $7^{-3} \cdot 7^8 = 7^{-3+8}$ **Using the product rule**

$= 7^5$

b) $\dfrac{b^{-5}}{b^{-4}} = b^{-5-(-4)} = b^{-1}$ **Using the quotient rule**

$= \dfrac{1}{b}$ **Writing the answer without a negative exponent**

Example 6(b) can also be simplified as follows:

$$\frac{b^{-5}}{b^{-4}} = \frac{b^4}{b^5} = b^{4-5} = b^{-1} = \frac{1}{b}.$$

Simplifying $(a^m)^n$

Next, consider an expression like $(3^4)^2$:

$$\begin{aligned}(3^4)^2 &= (3^4)(3^4) && \textbf{We are raising } 3^4 \textbf{ to the second power.}\\ &= (3 \cdot 3 \cdot 3 \cdot 3)(3 \cdot 3 \cdot 3 \cdot 3) \\ &= 3 \cdot 3 \cdot 3 \cdot 3 \cdot 3 \cdot 3 \cdot 3 \cdot 3 && \textbf{Using the associative law}\\ &= 3^8.\end{aligned}$$

Note that in this case, we could have multiplied the exponents:

$$(3^4)^2 = 3^{4 \cdot 2} = 3^8.$$

Likewise, $(y^8)^3 = (y^8)(y^8)(y^8) = y^{24}$. Once again, we get the same result if we multiply the exponents:

$$(y^8)^3 = y^{8 \cdot 3} = y^{24}.$$

The Power Rule For any real number a and any integers m and n,

$$(a^m)^n = a^{mn}.$$

(To raise a power to a power, multiply the exponents.)

EXAMPLE 7 Simplify: **(a)** $(3^5)^4$; **(b)** $(y^{-5})^7$; **(c)** $(a^{-3})^{-7}$.

SOLUTION

a) $(3^5)^4 = 3^{5 \cdot 4} = 3^{20}$

b) $(y^{-5})^7 = y^{-5 \cdot 7} = y^{-35} = \dfrac{1}{y^{35}}$

c) $(a^{-3})^{-7} = a^{(-3)(-7)} = a^{21}$

Raising a Product or a Quotient to a Power

When an expression inside parentheses is raised to a power, the inside expression is the base. Let's compare $2a^3$ and $(2a)^3$.

$$2a^3 = 2 \cdot a \cdot a \cdot a; \qquad \begin{aligned}(2a)^3 &= (2a)(2a)(2a)\\ &= 2 \cdot 2 \cdot 2 \cdot a \cdot a \cdot a\\ &= 2^3a^3 = 8a^3\end{aligned}$$

We see that $2a^3$ and $(2a)^3$ are *not* equivalent. Note also that to simplify $(2a)^3$ we can raise each factor to the power 3. This leads to the following rule.

Raising a Product to a Power For any integer n, and any real numbers a and b for which $(ab)^n$ exists,

$$(ab)^n = a^n b^n.$$

(To raise a product to a power, raise each factor to that power.)

EXAMPLE 8 Simplify: **(a)** $(-2x)^3$; **(b)** $(-3x^5y^{-1})^{-4}$.

SOLUTION

a) $(-2x)^3 = (-2)^3 \cdot x^3$ **Raising each factor to the third power**

$= -8x^3$ $\quad$ $(-2)^3 = (-2)(-2)(-2) = -8$

b) $(-3x^5y^{-1})^{-4} = (-3)^{-4}(x^5)^{-4}(y^{-1})^{-4}$ **Raising each factor to the negative fourth power**

$= \dfrac{1}{(-3)^4} \cdot x^{-20}y^4$ **Multiplying powers; writing** $(-3)^{-4}$ **as** $\dfrac{1}{(-3)^4}$

$= \dfrac{1}{81} \cdot \dfrac{1}{x^{20}} \cdot y^4$

$= \dfrac{y^4}{81x^{20}}$

There is a similar rule for raising a quotient to a power.

Raising a Quotient to a Power For any integer n, and any real numbers a and b for which $\frac{a}{b}$, a^n, and b^n exist,

$$\left(\frac{a}{b}\right)^n = \frac{a^n}{b^n}.$$

(To raise a quotient to a power, raise both the numerator and the denominator to that power.)

EXAMPLE 9 Simplify: **(a)** $\left(\dfrac{x^2}{2}\right)^4$; **(b)** $\left(\dfrac{y^2z^3}{5}\right)^{-3}$.

SOLUTION

a) $\left(\dfrac{x^2}{2}\right)^4 = \dfrac{(x^2)^4}{2^4} = \dfrac{x^8}{16}$ $\quad$ $\leftarrow 2 \cdot 4 = 8$; $\leftarrow 2^4 = 16$

b) $\left(\dfrac{y^2z^3}{5}\right)^{-3} = \dfrac{(y^2z^3)^{-3}}{5^{-3}}$

$= \dfrac{5^3}{(y^2z^3)^3}$ **Moving factors to the other side of the fraction bar and reversing the sign of those powers**

$= \dfrac{125}{y^6z^9}$

The rule for raising a quotient to a power allows us to derive a useful result for manipulating negative exponents:

$$\left(\frac{\boldsymbol{a}}{\boldsymbol{b}}\right)^{-n} = \frac{a^{-n}}{b^{-n}} = \frac{b^n}{a^n} = \left(\frac{\boldsymbol{b}}{\boldsymbol{a}}\right)^n.$$

Using this result, we can simplify Example 9(b) as follows:

$$\left(\frac{y^2z^3}{5}\right)^{-3} = \left(\frac{5}{y^2z^3}\right)^3 \quad \textbf{Taking the reciprocal of the base and changing the exponent's sign}$$

$$= \frac{5^3}{(y^2z^3)^3} = \frac{125}{y^6z^9}.$$

Definitions and Properties of Exponents

The following summary assumes that no denominators are 0 and that 0^0 is not considered. For any integers m and n,

1 as an exponent:	$a^1 = a$
0 as an exponent:	$a^0 = 1$
Negative exponents:	$a^{-n} = \frac{1}{a^n}$
	$\frac{a^{-n}}{b^{-m}} = \frac{b^m}{a^n}$
	$\left(\frac{a}{b}\right)^{-n} = \left(\frac{b}{a}\right)^n$
The Product Rule:	$a^m \cdot a^n = a^{m+n}$
The Quotient Rule:	$\frac{a^m}{a^n} = a^{m-n}$
The Power Rule:	$(a^m)^n = a^{mn}$
Raising a product to a power:	$(ab)^n = a^nb^n$
Raising a quotient to a power:	$\left(\frac{a}{b}\right)^n = \frac{a^n}{b^n}$

Scientific Notation

Very large and very small numbers that occur in science and other fields are often written in **scientific notation,** using exponents.

Scientific Notation *Scientific notation* for a number is an expression of the form $N \times 10^m$, where N is in decimal notation, $1 \leq N < 10$, and m is an integer.

Note that $10^b/10^b = 10^b \cdot 10^{-b} = 1$. To convert a number to scientific notation, we can multiply by 1, writing 1 in the form $10^b/10^b$ or $10^b \cdot 10^{-b}$.

EXAMPLE 10 Population Projections. It has been estimated that in 2020, the world population will be 7,516,000,000 (*Source*: U.S. Bureau of the Census, International Data Base). Write scientific notation for this number.

SOLUTION To write 7,516,000,000 as 7.516×10^m for some integer m, we must move the decimal point in 7,516,000,000 to the left 9 places. This can be accomplished by dividing—and then multiplying—by 10^9:

$$7{,}516{,}000{,}000 = \frac{7{,}516{,}000{,}000}{10^9} \cdot 10^9 \qquad \textbf{Multiplying by 1: } \frac{10^9}{10^9} = 1$$

$$= 7.516 \times 10^9. \qquad \textbf{This is scientific notation.}$$

EXAMPLE 11 Write scientific notation for the mass of a grain of sand:

0.0648 gram (g).

SOLUTION To write 0.0648 as 6.48×10^m for some integer m, we must move the decimal point 2 places to the right. To do this, we multiply—and then divide—by 10^2:

$$0.0648 = \frac{0.0648 \times 10^2}{10^2} \qquad \textbf{Multiplying by 1: } \frac{10^2}{10^2} = 1$$

$$= \frac{6.48}{10^2}$$

$$= 6.48 \times 10^{-2}\ \text{g}. \qquad \textbf{Writing scientific notation}$$

Try to make conversions to scientific notation mentally if possible. In doing so, remember that negative powers of 10 are used to represent small numbers and positive powers of 10 are used to represent large numbers.

EXAMPLE 12 Convert mentally to decimal notation: **(a)** 4.371×10^7; **(b)** 1.73×10^{-5}.

SOLUTION

a) $4.371 \times 10^7 = 43{,}710{,}000$ **Moving the decimal point 7 places to the right**

b) $1.73 \times 10^{-5} = 0.0000173$ **Moving the decimal point 5 places to the left**

EXAMPLE 13 Convert mentally to scientific notation: **(a)** 82,500,000; **(b)** 0.0000091.

SOLUTION

a) $82{,}500{,}000 = 8.25 \times 10^7$ ***Check*: Multiplying 8.25 by 10^7 moves the decimal point 7 places to the right.**

b) $0.0000091 = 9.1 \times 10^{-6}$ ***Check*: Multiplying 9.1 by 10^{-6} moves the decimal point 6 places to the left.**

Significant Digits and Rounding

In the world of science, it is important to know just how accurate a measurement is. For example, the measurement 5.12×10^3 km is more precise than the measurement 5.1×10^3 km. We say that 5.12×10^3 has three **significant digits** whereas 5.1×10^3 has only two significant digits. If 5.1×10^3, or 5100, includes no rounding in the tens column, we would indicate that by writing 5.10×10^3.

> When two or more measurements written in scientific notation are multiplied or divided, the result should be rounded so that it has the same number of significant digits as the measurement with the fewest significant digits. Rounding should be performed at the *end* of the calculation.

Thus,

$$(\underbrace{3.1}_{\text{2 digits}} \times 10^{-3}\text{ mm})(\underbrace{2.45}_{\text{3 digits}} \times 10^{-4}\text{ mm}) = 7.595 \times 10^{-7}\text{ mm}^2$$

should be rounded to

$$\overbrace{7.6}^{\text{2 digits}} \times 10^{-7}\text{ mm}^2.$$

> When two or more measurements written in scientific notation are added or subtracted, the result should be rounded so that it has as many decimal places as the measurement with the fewest decimal places.

For example,

$$1.\underbrace{6354}_{\text{4 decimal places}} \times 10^4\text{ km} + 2.\underbrace{078}_{\text{3 decimal places}} \times 10^4\text{ km} = 3.7134 \times 10^4\text{ km}$$

should be rounded to

$$3.\overbrace{713}^{\text{3 decimal places}} \times 10^4\text{ km}.$$

EXAMPLE 14 Multiply and write scientific notation for the answer:

$(7.2 \times 10^5)(4.3 \times 10^9)$.

SOLUTION We have

$$(7.2 \times 10^5)(4.3 \times 10^9) = (7.2 \times 4.3)(10^5 \times 10^9)$$ **Using the commutative and associative laws**

$$= 30.96 \times 10^{14},$$ **Adding exponents**

To find scientific notation for this result, we convert 30.96 to scientific notation and simplify:

$$30.96 \times 10^{14} = (3.096 \times 10^1) \times 10^{14}$$
$$= 3.096 \times 10^{15}$$
$$\approx 3.1 \times 10^{15}.$$ **Rounding to 2 significant digits**

EXAMPLE 15 Divide and write scientific notation for the answer:

$$\frac{3.48 \times 10^{-7}}{4.64 \times 10^6}.$$

SOLUTION

$$\frac{3.48 \times 10^{-7}}{4.64 \times 10^6} = \frac{3.48}{4.64} \times \frac{10^{-7}}{10^6}$$ **Separating factors. Our answer must have 3 significant digits.**

$$= 0.75 \times 10^{-13}$$ **Subtracting exponents; simplifying**

$$= (7.5 \times 10^{-1}) \times 10^{-13}$$ **Converting 0.75 to scientific notation**

$$= 7.50 \times 10^{-14}$$ **Adding exponents. We write 7.50 to indicate 3 significant digits.**

Exponents and Scientific Notation

To simplify an exponential expression like 3^5, we can use a calculator's exponentiation key, usually labeled ^. If the exponent is 2, we can also use the x^2 key. If it is -1, we can use the x^{-1} key. If the exponent is a single number, not an expression, we do not need to enclose it in parentheses.

Graphing calculators will accept entries using scientific notation, and will normally write very large or very small numbers using scientific notation. The EE key, the 2nd option associated with the , key, is used to enter scientific notation. On the calculator screen, a notation like E22 represents $\times 10^{22}$.

EXAMPLE 16 Calculate 3^5, $(-4.7)^2$, and $(-8)^{-1}$.

SOLUTION To calculate 3^5, we press [3] [^] [5] [ENTER].

To calculate $(-4.7)^2$, we press [(] [(−)] [4] [.] [7] [)] [^] [2] [ENTER] or [(] [(−)] [4] [.] [7] [)] [x²] [ENTER].

To calculate $(-8)^{-1}$, we press [(] [(−)] [8] [)] [x⁻¹] [ENTER] or [(] [(−)] [8] [)] [^] [(−)] [1] [ENTER].

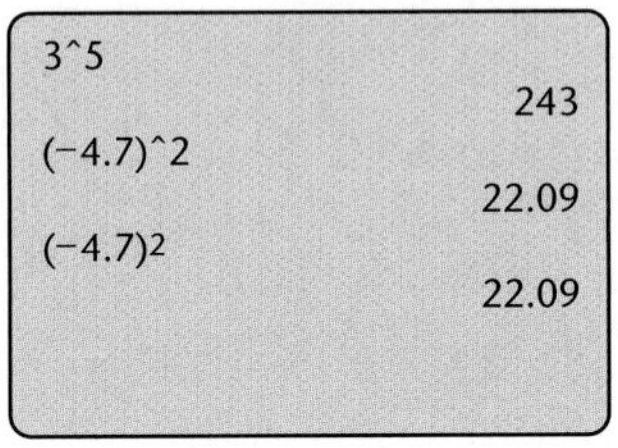

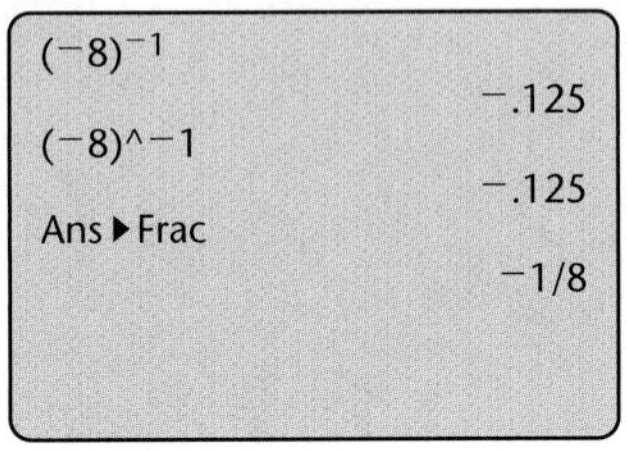

We see that $3^5 = 243$, $(-4.7)^2 = 22.09$, and $(-8)^{-1} = -0.125$, or $-\frac{1}{8}$.

EXAMPLE 17 Calculate $(7.5 \times 10^8)(1.2 \times 10^{-14})$.

SOLUTION We press [7] [.] [5] [2ND] [EE] [8] [×] [1] [.] [2] [2ND] [EE] [(−)] [1] [4] [ENTER].

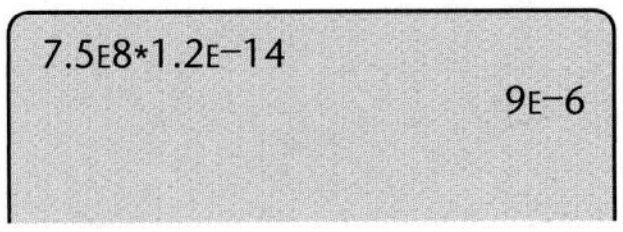

The result shown is then read as 9×10^{-6}. Using two significant digits, we write the answer as 9.0×10^{-6}.

1.4 EXERCISE SET

FOR EXTRA HELP

MathXL | MyMathLab | InterAct Math | AW Math Tutor Center | Video Lectures on CD: Disc 1 | Student's Solutions Manual

Concept Reinforcement *In each of Exercises 1–10, state whether the equation is an example of the product rule, the quotient rule, the power rule, raising a product to a power, or raising a quotient to a power.*

1. $(a^6)^4 = a^{24}$

2. $\left(\frac{5}{7}\right)^4 = \frac{5^4}{7^4}$

3. $(5x)^7 = 5^7x^7$

4. $\frac{m^9}{m^3} = m^6$

5. $m^6 \cdot m^4 = m^{10}$

6. $(5^2)^7 = 5^{14}$

7. $\left(\frac{a}{4}\right)^7 = \frac{a^7}{4^7}$

8. $(ab)^{10} = a^{10}b^{10}$

9. $\frac{x^{10}}{x^2} = x^8$

10. $r^5 \cdot r^7 = r^{12}$

Concept Reinforcement *State whether scientific notation for each of the following numbers would include a positive or a negative power of 10.*

11. The length of an Olympic marathon, in centimeters

12. The thickness of a cat's whisker, in meters

13. The mass of a hydrogen atom, in grams

14. The mass of a pickup truck, in grams

15. The time between leap years, in seconds

16. The time between a bird's heartbeats, in hours

Multiply and simplify. Leave the answer in exponential notation.

17. $5^6 \cdot 5^4$

18. $6^3 \cdot 6^5$

19. $m^9 \cdot m^0$

20. $x^0 \cdot x^5$

21. $6x^5 \cdot 3x^2$

22. $4a^3 \cdot 2a^7$

23. $(-2m^4)(-8m^9)$

24. $(-2a^5)(7a^4)$

25. $(x^3y^4)(x^7y^6z^0)$

26. $(m^6n^5)(m^4n^7p^0)$

Divide and simplify.

27. $\frac{a^9}{a^3}$

28. $\frac{x^{12}}{x^3}$

29. $\frac{12t^7}{4t^2}$

30. $\frac{20a^{20}}{5a^4}$

31. $\frac{32x^8y^5}{8x^2y}$

32. $\frac{35x^7y^8}{7xy^2}$

33. $\frac{28x^{10}y^9z^8}{-7x^2y^3z^2}$

34. $\frac{18x^8y^6z^7}{-3x^2y^3z}$

Evaluate each of the following for $x = -2$.

35. $-x^0$

36. $(-x)^0$

37. $(4x)^0$

38. $4x^0$

Simplify.

39. $(-2)^4$

40. $(-3)^4$

41. -2^4

42. -3^4

43. $(-4)^{-2}$

44. $(-5)^{-2}$

45. -4^{-2}

46. -5^{-2}

47. -1^{-8}

Write an equivalent expression without negative exponents and, if possible, simplify.

48. a^{-3}

49. n^{-6}

50. $\frac{1}{5^{-3}}$

51. $\frac{1}{2^{-6}}$

52. $8x^{-3}$

53. $7x^{-3}$

54. $3a^8b^{-6}$

55. $5a^{-7}b^4$

56. $\frac{z^{-4}}{3x^5}$

57. $\frac{y^{-5}}{x^{-3}}$

58. $\frac{x^{-2}y^7}{z^{-4}}$

59. $\frac{y^4z^{-3}}{x^{-2}}$

Write an equivalent expression with negative exponents.

60. $\frac{1}{8^4}$

61. $\frac{1}{(-5)^6}$

62. x^5

63. $4x^2$

64. $-4y^5$

65. $\dfrac{1}{(5y)^3}$

66. $\dfrac{1}{3y^4}$

Simplify. If negative exponents appear in the answer, write a second answer using only positive exponents.

67. $8^{-2} \cdot 8^{-4}$

68. $9^{-1} \cdot 9^{-6}$

69. $b^2 \cdot b^{-5}$

70. $a^4 \cdot a^{-3}$

71. $a^{-3} \cdot a^4 \cdot a^2$

72. $x^{-8} \cdot x^5 \cdot x^3$

73. $(5a^{-2}b^{-3})(2a^{-4}b)$

74. $(3a^{-5}b^{-7})(2ab^{-2})$

75. $\dfrac{10^{-3}}{10^6}$

76. $\dfrac{12^{-4}}{12^8}$

77. $\dfrac{2^{-7}}{2^{-5}}$

78. $\dfrac{9^{-4}}{9^{-6}}$

79. $\dfrac{y^4}{y^{-5}}$

80. $\dfrac{a^3}{a^{-2}}$

81. $\dfrac{24a^5b^3}{-8a^4b}$

82. $\dfrac{9a^2}{3ab^3}$

83. $\dfrac{-6x^{-2}y^4z^8}{24x^{-5}y^6z^{-3}}$

84. $\dfrac{8a^6b^{-4}c^8}{32a^{-4}b^5c^9}$

85. $(x^4)^3$

86. $(a^3)^2$

87. $(9^3)^{-4}$

88. $(8^4)^{-3}$

89. $(t^{-8})^{-5}$

90. $(x^{-4})^{-3}$

91. $(5x^4y)^2$

92. $(5ab^2)^3$

93. $\dfrac{(x^5)^2(x^{-3})^4}{(x^2)^3}$

94. $\dfrac{(a^{-2})^3(a^4)^2}{(a^3)^{-3}}$

95. $\dfrac{(2a^3)^3 4a^{-3}}{(a^2)^5}$

96. $\dfrac{(3x^2)^3 2x^{-4}}{(x^4)^2}$

Aha! **97.** $(8x^{-3}y^2)^{-4}(8x^{-3}y^2)^4$

98. $(2a^{-1}b^3)^{-2}(2a^{-1}b^3)^{-2}$

99. $\dfrac{(3x^3y^4)^3}{6xy^3}$

100. $\dfrac{(5a^3b)^2}{10a^2b}$

101. $\left(\dfrac{-4x^4y^{-2}}{5x^{-1}y^4}\right)^{-4}$

102. $\left(\dfrac{2x^3y^{-2}}{3y^{-3}}\right)^3$

Aha! **103.** $\left(\dfrac{4a^3b^{-9}}{6a^{-2}b^5}\right)^0$

104. $\left(\dfrac{5x^0y^{-7}}{2x^{-2}y^4}\right)^{-2}$

To the student and the instructor: The symbol indicates an exercise designed to be solved with a calculator.

Evaluate using a calculator.

105. -8^4

106. $(-8)^4$

107. $(-2)^{-4}$

108. -2^{-4}

109. $3^4 5^{-3}$

110. $\left(\dfrac{2}{3}\right)^{-5}$

Convert to decimal notation.

111. 4×10^{-4}

112. 5×10^{-5}

113. 6.73×10^8

114. 9.24×10^7

115. 8.923×10^{-10}

116. 7.034×10^{-2}

117. 9.03×10^{10}

118. 9.001×10^{10}

Convert to scientific notation.

119. 47,000,000,000

120. 2,600,000,000,000

121. 0.000000016

122. 0.000000263

123. 407,000,000,000

124. 3,090,000,000,000

125. 0.000000603

126. 0.00000000802

Write scientific notation for the number represented on each calculator screen.

127. 5.02E18

128. 1.067E−6

129. −3.05E−10

130. −5.968E27

Simplify and write scientific notation for the answer. Use the correct number of significant digits.

131. $(2.3 \times 10^6)(4.2 \times 10^{-11})$

132. $(6.5 \times 10^3)(5.2 \times 10^{-8})$

133. $(2.34 \times 10^{-8})(5.7 \times 10^{-4})$

134. $(4.26 \times 10^{-6})(8.2 \times 10^{-6})$

Aha! **135.** $(2.0 \times 10^{6})(3.02 \times 10^{-6})$

136. $(7.04 \times 10^{-9})(9.01 \times 10^{-7})$

137. $\dfrac{5.1 \times 10^{6}}{3.4 \times 10^{3}}$

138. $\dfrac{8.5 \times 10^{8}}{3.4 \times 10^{5}}$

139. $\dfrac{7.5 \times 10^{-9}}{2.5 \times 10^{-4}}$

140. $\dfrac{12.6 \times 10^{8}}{4.2 \times 10^{-3}}$

141. $\dfrac{1.23 \times 10^{8}}{6.87 \times 10^{-13}}$

142. $\dfrac{4.95 \times 10^{-3}}{1.64 \times 10^{10}}$

143. $5.9 \times 10^{23} + 6.3 \times 10^{23}$

144. $7.8 \times 10^{-34} + 5.4 \times 10^{-34}$

TW **145.** Explain why $(-1)^n = 1$ for any even number n.

TW **146.** List two advantages of using scientific notation. Answers may vary.

Skill Maintenance

147. Subtract: $-\frac{5}{6} - \left(-\frac{3}{4}\right)$. [1.2]

148. Multiply: $(-7.2)(-4.3)$. [1.2]

149. Multiply: $-2(4x - 6y)$. [1.3]

150. Factor: $8x - 10$. [1.3]

Synthesis

TW **151.** Explain why $(-17)^{-8}$ is positive.

TW **152.** Is the following true or false, and why?

$$5^{-6} > 4^{-9}$$

TW **153.** Some numbers exceed the limits of the calculator. Enter 1.3×10^{-1000} and 1.3×10^{1000} and explain the results.

Simplify. Assume that all variables represent nonzero integers.

154. $\dfrac{12a^{x-2}}{3a^{2x+2}}$

155. $\dfrac{-12x^{a+1}}{4x^{2-a}}$

156. $(3^{a+2})^{a}$

157. $(12^{3-a})^{2b}$

158. $\dfrac{4x^{2a+3}y^{2b-1}}{2x^{a+1}y^{b+1}}$

159. $\dfrac{25x^{a+b}y^{b-a}}{-5x^{a-b}y^{b+a}}$

160. Compare $8 \cdot 10^{-90}$ and $9 \cdot 10^{-91}$. Which is the larger value? How much larger? Write scientific notation for the difference.

161. Write the reciprocal of 8.00×10^{-23} in scientific notation.

162. Evaluate: $(4096)^{0.05}(4096)^{0.2}$.

163. What is the ones digit in 513^{128}?

164. Write $\frac{4}{32}$ in decimal notation, simplified fraction notation, and scientific notation.

165. A grain of sand is placed on the first square of a chessboard, two grains on the second square, four grains on the third, eight on the fourth, and so on. Without a calculator, use scientific notation to approximate the number of grains of sand required for the 64th square. (*Hint*: Use the fact that $2^{10} \approx 10^{3}$.)

Collaborative Corner

Paired Problem Solving

Focus: Problem solving, scientific notation, and unit conversion

Time: 15–25 minutes

Group Size: 3

ACTIVITY

Given that the earth's average distance from the sun is 1.5×10^{11} meters, determine the earth's orbital speed around the sun in miles per hour. Assume a circular orbit and use the following guidelines.

1. Each group should spend about 10 minutes attempting to solve this problem. Do not worry if the solution is not found.
2. Group members should each describe the interactions that took place, answering these three questions:
   a) What successful strategies were used?
   b) What unsuccessful strategies were used?
   c) What recommendations can you make for students working together to solve a problem?
3. Each group should then share their observations with each other and report to the class as a whole what they feel are their most significant observations.

1.5 Graphs

Points and Ordered Pairs ■ Quadrants and Scale ■ Graphs of Equations ■ Nonlinear Equations

It has often been said that a picture is worth a thousand words. As we turn our attention to the study of graphs, we discover that in mathematics this is quite literally the case. Graphs are a compact means of displaying information and provide a visual approach to problem solving.

Points and Ordered Pairs

Whereas on a number line, each point corresponds to a number, on a plane, each point corresponds to a pair of numbers. The idea of using two perpendicular number lines, called **axes** (pronounced ak-sēz; singular, **axis**) to identify points in a plane is commonly attributed to the great French mathematician and philosopher René Descartes (1596–1650). In honor of Descartes, this representation is also called the **Cartesian coordinate system.** The variable x is normally represented on the horizontal axis and the variable y on the vertical axis, so we will also refer to the ***x*, *y*-coordinate system.**

To label a point that appears on the x, y-coordinate system, we write a pair of numbers in the form (x, y), where x is the number located on the x-axis directly above or below the point and y is the number located on the y-axis to the left or right of the point (see the dashed lines in the figure below). Thus, $(2, 3)$ and $(3, 2)$ are different points. Because the order in which the numbers are listed is important, these are called **ordered pairs**. The numbers in an ordered pair are called **coordinates**. In $(-4, 3)$, the *first coordinate* is -4 and the *second coordinate** is 3. The point with coordinates $(0, 0)$ is called the **origin.**

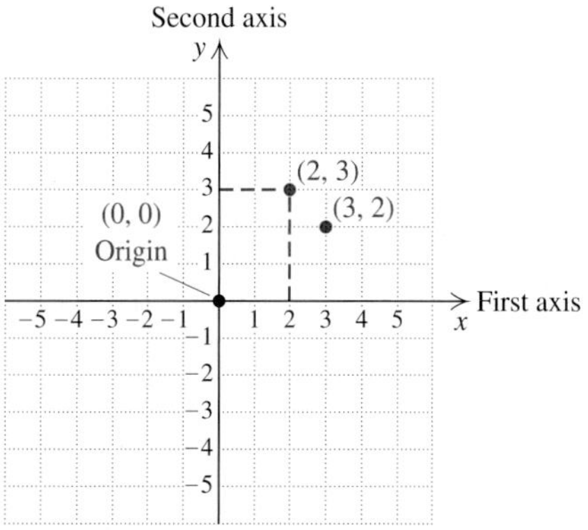

EXAMPLE 1 Plot the points $(-4, 3)$, $(-5, -3)$, $(0, 4)$, and $(2.5, 0)$.

SOLUTION To plot $(-4, 3)$, note that the first coordinate, -4, tells us the distance in the first, or horizontal, direction. We go 4 units *left* of the origin. From that location, we go 3 units *up*. The point $(-4, 3)$ is then marked, or "plotted."

The points $(-5, -3)$, $(0, 4)$, and $(2.5, 0)$ are also plotted below.

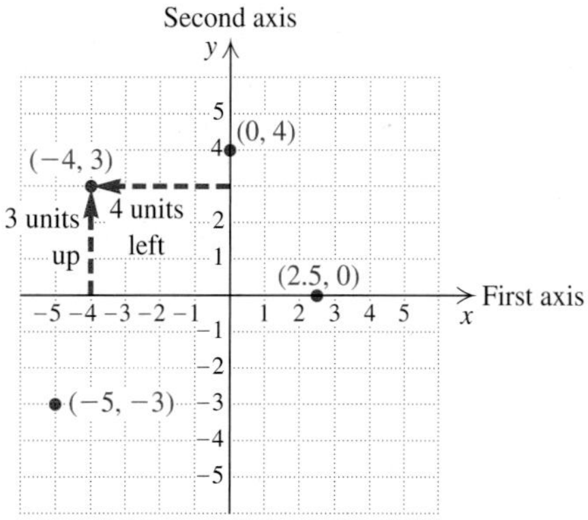

*The first coordinate is sometimes called the **abscissa** and the second coordinate the **ordinate.**

Quadrants and Scale

The axes divide the plane into four regions called **quadrants,** as shown here.

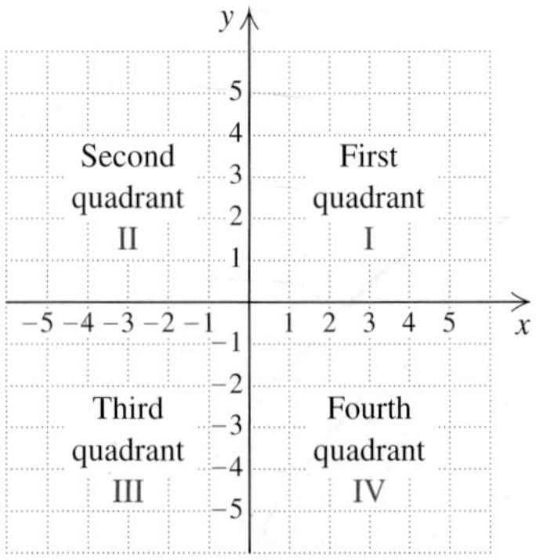

In region I (the *first* quadrant), both coordinates of a point are positive. In region II (the *second* quadrant), the first coordinate is negative and the second coordinate is positive. In the third quadrant, both coordinates are negative, and in the fourth quadrant, the first coordinate is positive and the second coordinate is negative.

Points with one or more 0's as coordinates, such as $(0, -6)$, $(4, 0)$, and $(0, 0)$, are on axes and *not* in quadrants.

The coordinate plane extends without end in all directions. We draw only part of it when we plot points. Although it is standard to show portions of all four quadrants, as in the graphs above, it may be more practical to show a different portion of the plane in order to display information more clearly. Sometimes a different *scale* is selected for each axis.

EXAMPLE 2 Plot the points $(10, 44)$, $(95, 120)$, $(55, 130)$, and $(70, 15)$.

SOLUTION The smallest first coordinate is 10 and the largest is 95. The smallest second coordinate is 15 and the largest is 130. Thus we need show only the first quadrant. We must show at least 95 units of the first axis and at least 130 units of the second axis. Because it would be impractical to label the axes with all the natural numbers, we will label only every tenth unit. We say that we use a scale of 10.

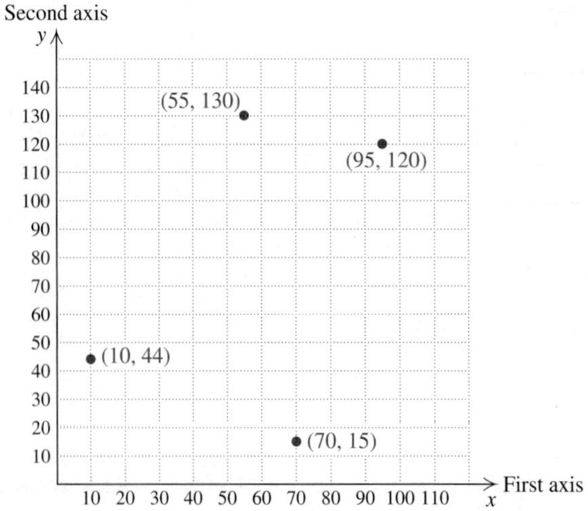

Windows

On a graphing calculator, the rectangular portion of the screen in which a graph appears is called the **viewing window.** Windows are described by four numbers of the form [L, R, B, T], representing the **L**eft and **R**ight endpoints of the x-axis and the **B**ottom and **T**op endpoints of the y-axis. The **standard viewing window** is the window determined by the settings $[-10, 10, -10, 10]$.

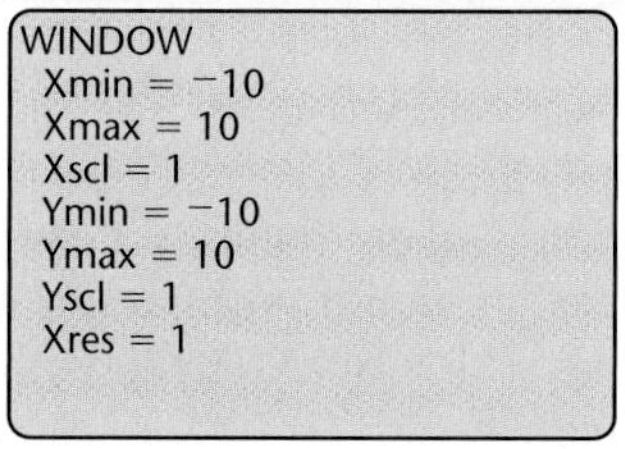

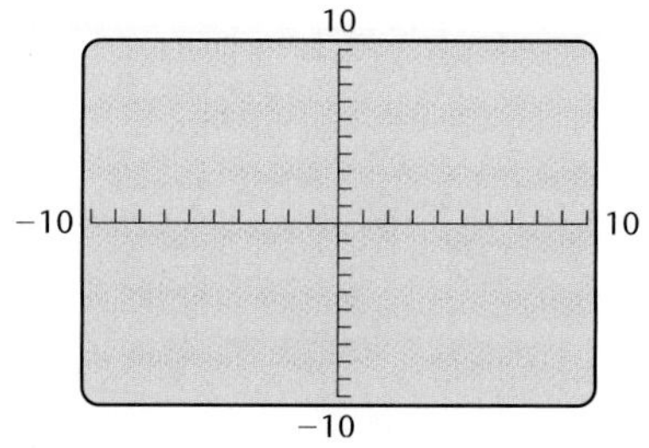

We press WINDOW to set the window dimensions and GRAPH to display the graph. Xmin is the smallest x-value that will be displayed on the screen, and Xmax is the largest. Similarly, the values of Ymin and Ymax determine the bottom and top endpoints of the vertical axis shown. The scales for the axes are set using Xscl and Yscl. Xres indicates the pixel resolution, which we generally set as Xres $= 1$. In this text, the window dimensions are written outside the graphs.

Inappropriate window dimensions may result in an error. For example, the message ERR: WINDOW RANGE occurs when Xmin is not less than Xmax.

Graphs of Equations

If an equation has two variables, its solutions are pairs of numbers. When such a solution is written as an ordered pair, the first number listed in the pair generally replaces the variable that occurs first alphabetically.

EXAMPLE 3 Determine whether the pairs $(4, 2)$, $(-1, -4)$, and $(2, 5)$ are solutions of the equation $y = 3x - 1$.

SOLUTION To determine whether each pair is a solution, we replace x with the first coordinate and y with the second coordinate. When the replacements make the equation true, we say that the ordered pair is a solution.

$y = 3x - 1$		$y = 3x - 1$		$y = 3x - 1$	
2	$3(4) - 1$	-4	$3(-1) - 1$	5	$3(2) - 1$
	$12 - 1$		$-3 - 1$		$6 - 1$
$2 \stackrel{?}{=} 11$		$-4 \stackrel{?}{=} -4$		$5 \stackrel{?}{=} 5$	

Since $2 = 11$ is *false*, the pair $(4, 2)$ *is not* a solution.

Since $-4 = -4$ is *true*, the pair $(-1, -4)$ *is* a solution.

Since $5 = 5$ is *true*, the pair $(2, 5)$ *is* a solution.

In fact, there is an infinite number of solutions of $y = 3x - 1$. We can use a graph as a convenient way of representing these solutions. Thus to **graph** an equation means to make a drawing that represents all of its solutions.

EXAMPLE 4 Graph the equation $y = x$.

SOLUTION We label the horizontal axis as the x-axis and the vertical axis as the y-axis.

Next, we find some ordered pairs that are solutions of the equation. In this case, no calculations are necessary. Here are a few pairs that satisfy the equation $y = x$:

$$(0,0), \quad (1,1), \quad (5,5), \quad (-1,-1), \quad (-6,-6).$$

Plotting these points, we can see that if we were to plot a hundred solutions, the dots would appear to form a line. Observing the pattern, we can draw the line with a ruler. The line is the graph of the equation $y = x$. We label the line $y = x$.

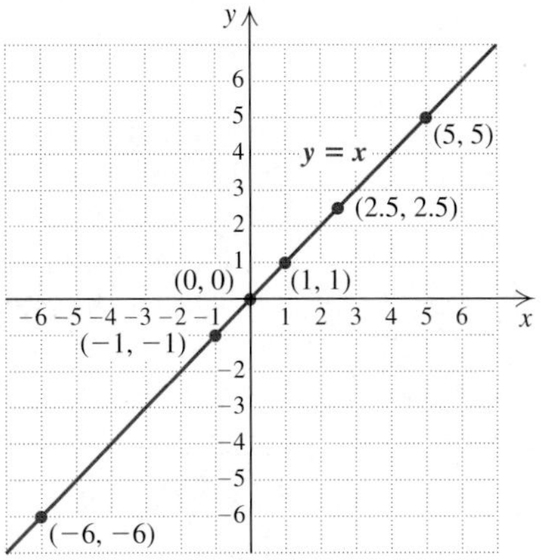

Note that the coordinates of *any* point on the line—for example, $(2.5, 2.5)$—satisfy the equation $y = x$. The line continues indefinitely in both directions, so we draw it to the edge of the grid.

Entering and Graphing Equations

Equations are entered using the equation-editor screen, often accessed by pressing Y=. The first part of each equation, "Y=," is supplied by the calculator, including a subscript that identifies the equation. An equation can be cleared by positioning the cursor on the equation and pressing CLEAR. On many calculators, a symbol before the Y indicates the graph style, and a highlighted = indicates that the equation selected is to be graphed. An equation can be selected or deselected by positioning the cursor on the = and pressing ENTER. Selected equations are then graphed by pressing GRAPH. *Note*: If the window is not set appropriately, the graph may not appear on the screen at all.

Study Tip

There's More Than One Way

A number of examples throughout this text are worked using two approaches. The steps are shown side by side, as in Example 6 on this page. You should read both methods and compare the steps and the results.

EXAMPLE 5 Graph $y = 2x$ using a graphing calculator.

SOLUTION After pressing (Y=) and clearing any other equations present, we enter $y = 2x$. The standard $[-10, 10, -10, 10]$ window is a good choice for this graph. We check the window dimensions and then graph the equation.

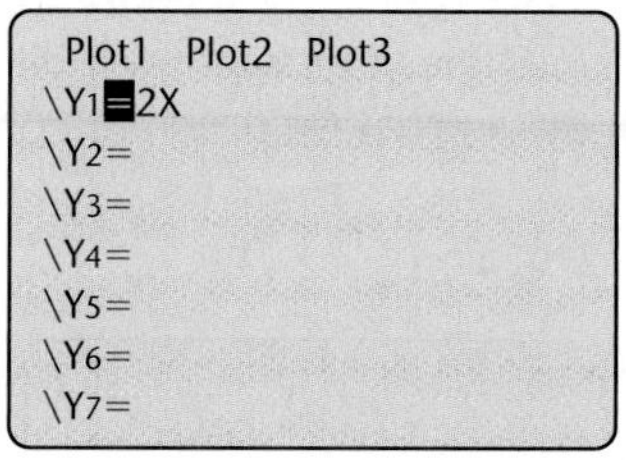

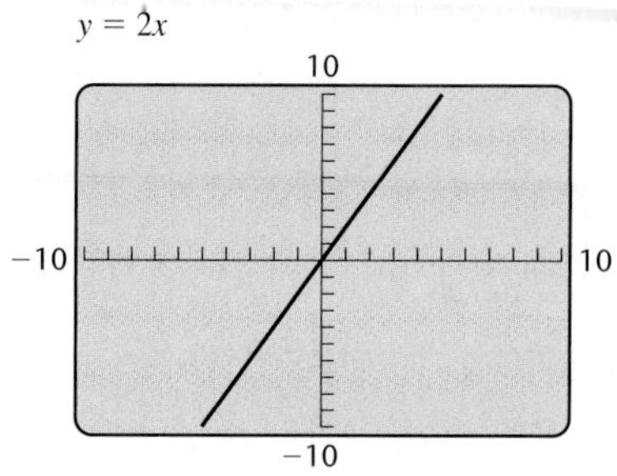

The equation in the next example is graphed both by hand and by using a graphing calculator. Let's compare the processes and the results.

EXAMPLE 6 Graph the equation $y = -\frac{1}{2}x$.

SOLUTION

BY HAND

We find some ordered pairs that are solutions. This time we list the pairs in a table. To find an ordered pair, we can choose *any* number for x and then determine y. By choosing even integers for x, we can avoid fractions when calculating y. For example, if we choose 4 for x, we get $y = \left(-\frac{1}{2}\right)(4)$, or -2. If x is -6, we get $y = \left(-\frac{1}{2}\right)(-6)$, or 3. We find several ordered pairs, plot them, and draw the line.

x	y	(x, y)
4	−2	(4, −2)
−6	3	(−6, 3)
0	0	(0, 0)
2	−1	(2, −1)

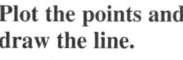

Choose any x.

Compute y.

Form the pair.

Plot the points and draw the line.

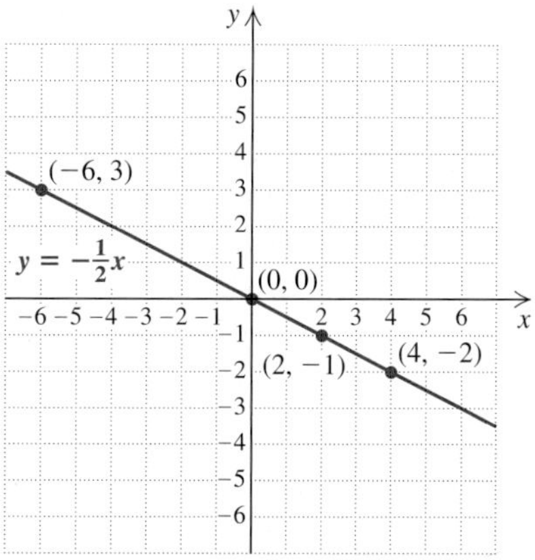

WITH A GRAPHING CALCULATOR

To enter the equation, we access the equation editor and clear any equations present. We then enter the equation as $y = -(1/2)x$. Remember to use (−) for the negative sign. Also note that some calculators require parentheses around a fraction coefficient. The standard viewing window is a good choice for this graph.

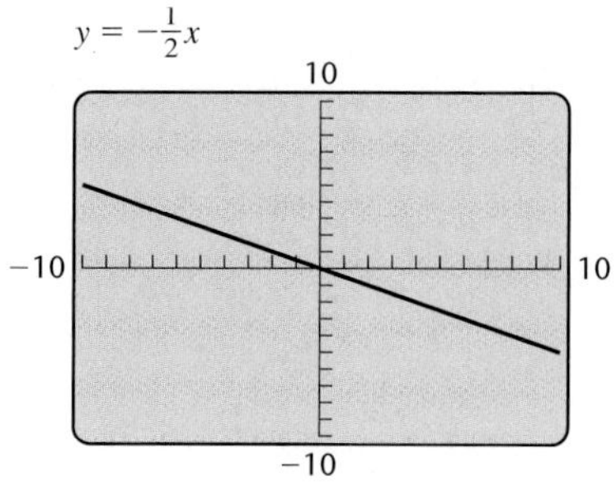

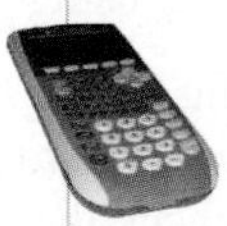

Choosing a Viewing Window

A standard viewing window is not always the best window to use. Choosing an appropriate viewing window for a graph can be challenging. There is generally no one "correct" window; the choice can vary according to personal preference and can also be dictated by the portion of the graph that you need to see.

The following screens show the equation $y = -x + 20$ graphed using various viewing windows. Note that the standard viewing window, shown on the left below, does not contain any portion of the line.

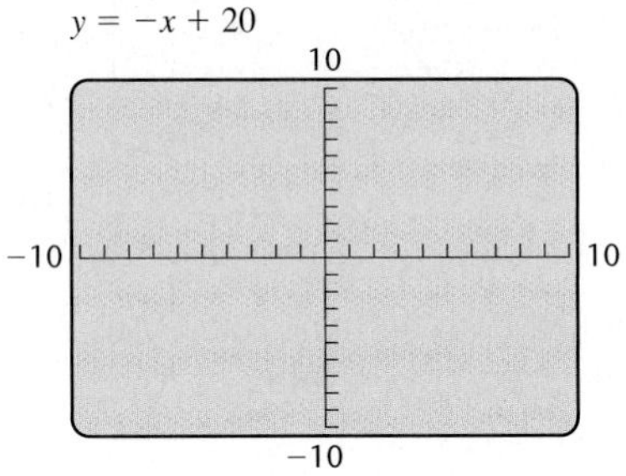

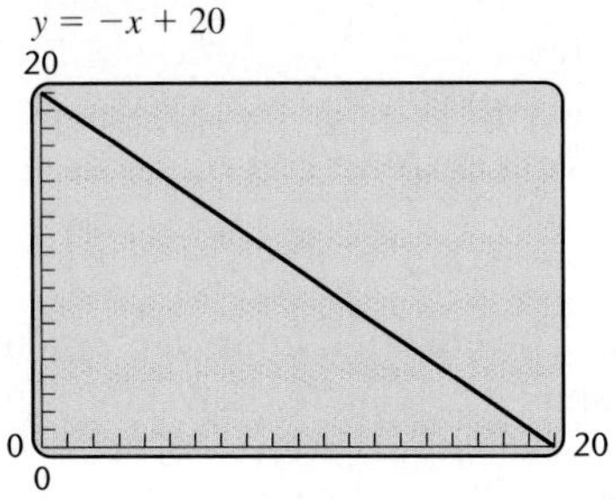

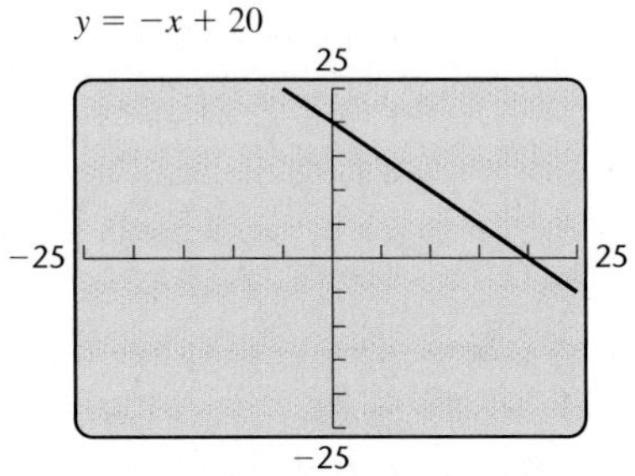

Xscl = 5, Yscl = 5

The windows in the middle and on the right above are both appropriate choices for a viewing window.

Pressing ZOOM can help in setting window dimensions; for example, choosing ZStandard from the menu will graph the selected equations using the standard viewing window.

ZOOM MEMORY
1: ZBox
2: Zoom In
3: Zoom Out
4: ZDecimal
5: ZSquare
6: ZStandard
7↓ZTrig

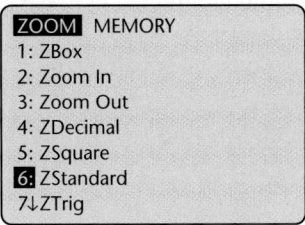

We need not create a table of values in order to graph an equation using a graphing calculator. However, such tables are useful in many situations, including the choice of a viewing window. To list ordered pairs that are solutions, we can use the TABLE feature of a graphing calculator.

Tables

A TABLE feature lists ordered pairs that are solutions of an equation. Since the value of y *depends* on the choice of the value for x, we say that y is the **dependent variable** and x is the **independent variable.**

After entering an equation on the equation-editor screen, we can view a table of solutions. A table is set up by pressing (TBLSET). (TBLSET is the 2nd feature associated with the (WINDOW) key.)

If we want to choose the values for the independent variable, we set Indpnt to Ask, as shown on the left below. Then we can create a table of values as we did when graphing by hand in Example 6. We first enter the equation $y_1 = -(1/2)x$, and then press (TABLE). (TABLE is the 2nd feature associated with the (GRAPH) key.) Entering the x-values 4, −6, 0, and 2 gives us the table shown on the right below.

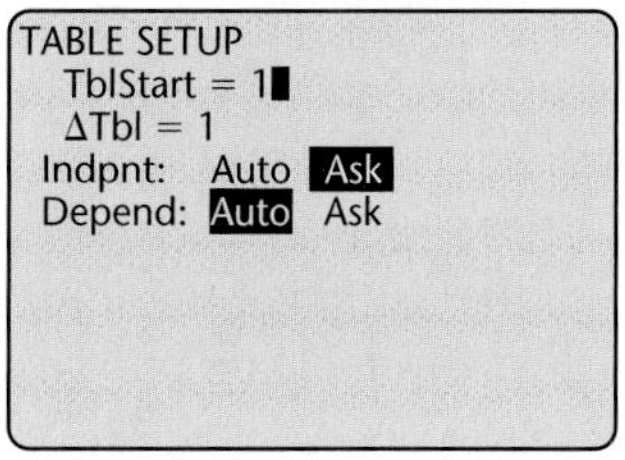

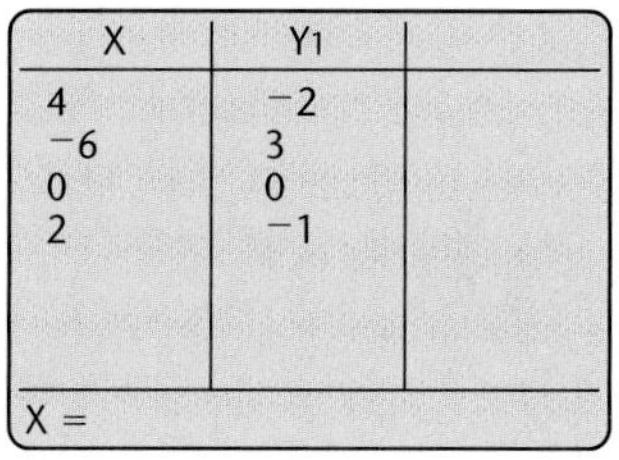

Student Notes

The Δ in ΔTbl is the upper-case Greek letter delta. Both Δ and δ, the lower-case delta, are often used in mathematics to represent a change or difference in the value of a variable.

If Indpnt is set to Auto, the calculator will provide values for x, beginning with the value specified as TblStart and continuing by adding the value of ΔTbl to the preceding value for x.

To create a table of ordered pairs that are solutions of the equation $y = -\frac{1}{2}x$, first enter the equation $y = -(1/2)x$ and then set up the table, as shown on the left below. Then press (TABLE) to view the table. We see that some solutions of the equation are $(-3, 1.5)$, $(-2, 1)$, $(-1, 0.5)$, and so on. Pressing the up and down arrow keys allows us to scroll through the table.

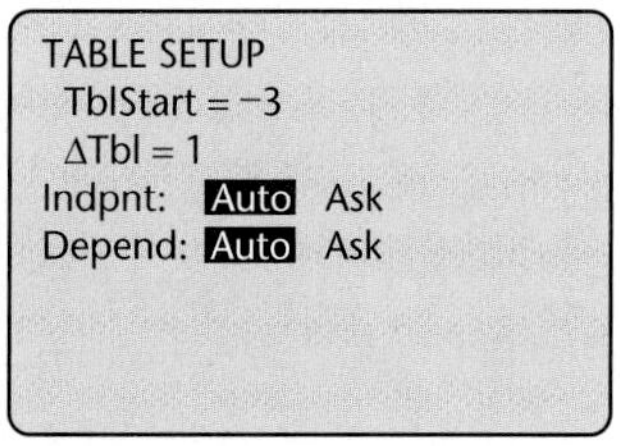

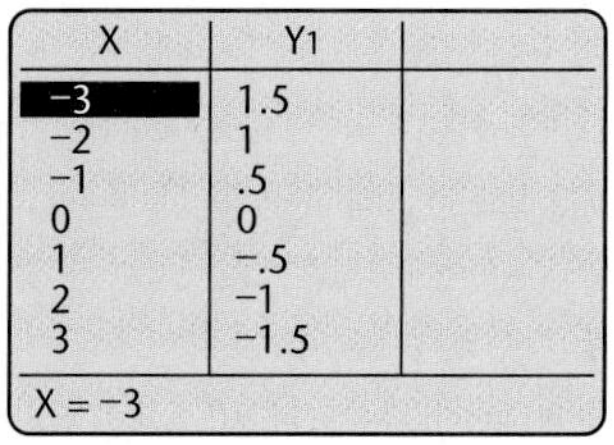

A table with Indpnt set to Auto is a valuable tool for determining an appropriate viewing window.

Nonlinear Equations

As you can see, the graphs in Examples 4–6 are straight lines. We refer to any equation whose graph is a straight line as a **linear equation**. Linear equations are discussed in more detail in Chapter 2. Many equations, however, are not linear. When ordered pairs that are solutions of such an equation are plotted, the pattern formed is not a straight line. Let's look at some of these **nonlinear equations.**

EXAMPLE 7 Graph: $y = x^2 - 5$.

SOLUTION We select numbers for x and find the corresponding values for y. For example, if we choose -2 for x, we get $y = (-2)^2 - 5 = 4 - 5 = -1$. The table lists several ordered pairs.

x	y
0	−5
−1	−4
1	−4
−2	−1
2	−1
−3	4
3	4

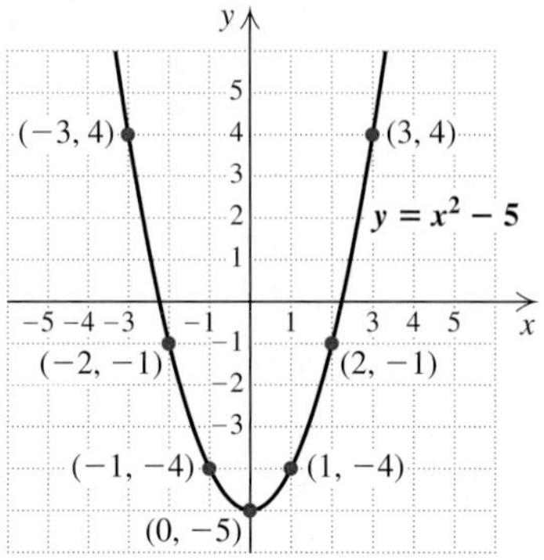

Next, we plot the points. The more points plotted, the clearer the shape of the graph becomes. Since the value of $x^2 - 5$ grows rapidly as x moves away from the origin, the graph rises steeply on either side of the y-axis.

Curves similar to the one in Example 7 are studied in detail in Chapter 8.

EXAMPLE 8 Use a graphing calculator to create a table of solutions of $y = |x|$ for integer values of x beginning at -3. Then graph the equation. Compare the values in the table with the graph.

SOLUTION We first enter the equation on the equation-editor screen as $y = \text{abs}(x)$. On many calculators, abs(is in the MATH NUM menu. We then create a table of values by setting Indpnt to Auto, letting TblStart $= -3$ and ΔTbl $= 1$. We use a standard viewing window for the graph.

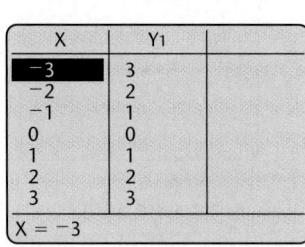

X	Y1
−3	3
−2	2
−1	1
0	0
1	1
2	2
3	3

X = −3

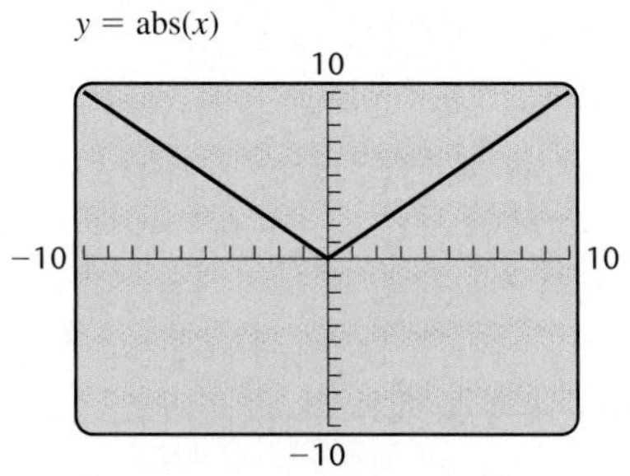

From the table, we see that the absolute value of a positive number is the same as the absolute value of its opposite. Thus, for example, the x-values 3 and -3 both are paired with the y-value 3. Note that the graph is V-shaped and centered at the origin.

Connecting the Concepts

INTERPRETING GRAPHS: SOLUTIONS

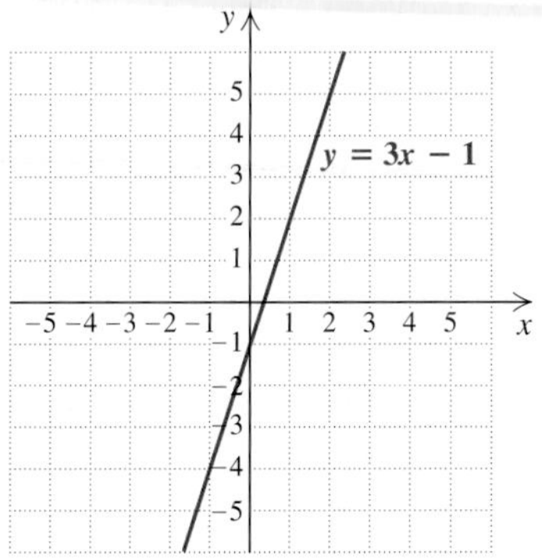

A solution of an equation like $5 = 3x - 1$ is a number such as 2. The graph of the solution of this equation is a point on the number line.

A solution of an equation like $y = 3x - 1$ is an ordered pair such as $(2, 5)$. The graph of the solutions of this equation is a line in a coordinate plane, as shown in the figure at left.

1.5 EXERCISE SET

FOR EXTRA HELP

MathXL | MyMathLab | InterAct Math | AW Math Tutor Center | Video Lectures on CD: Disc 1 | Student's Solutions Manual

Concept Reinforcement *Complete each of the following statements.*

1. In the fourth quadrant, a point's first coordinate is always positive and its second coordinate is always ______________.

2. In the third quadrant, a point's first coordinate is always ______________ and its second coordinate is always negative.

3. The two perpendicular number lines that are used for graphing are called ______________.

4. Because the order in which the numbers are listed is important, numbers listed in the form (x, y) are called ______________ pairs.

5. To graph an equation means to make a drawing that represents all ______________ of the equation.

6. An equation whose graph is a straight line is said to be a(n) ______________ equation.

Give the coordinates of each point.

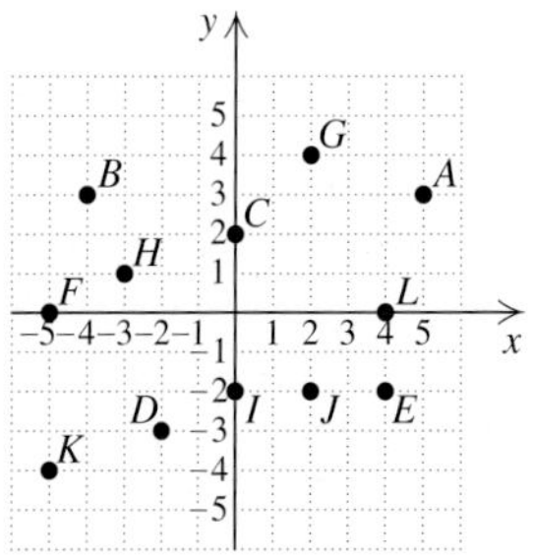

7. *A, B, C, D, E,* and *F*

8. *G, H, I, J, K,* and *L*

Plot the points. Label each point with the indicated letter.

9. $A(3, 0)$, $B(4, 2)$, $C(5, 4)$, $D(6, 6)$, $E(3, -4)$, $F(3, -3)$, $G(3, -2)$, $H(3, -1)$

10. $A(1, 1)$, $B(2, 3)$, $C(3, 5)$, $D(4, 7)$, $E(-2, 1)$, $F(-2, 2)$, $G(-2, 3)$, $H(-2, 4)$, $J(-2, 5)$, $K(-2, 6)$

11. Plot the points $M(2, 3)$, $N(5, -3)$, and $P(-2, -3)$. Draw $\overline{MN}$, $\overline{NP}$, and $\overline{MP}$. ($\overline{MN}$ means the line segment from M to N.) What kind of geometric figure is formed? What is its area?

12. Plot the points $Q(-4, 3)$, $R(5, 3)$, $S(2, -1)$, and $T(-7, -1)$. Draw $\overline{QR}$, $\overline{RS}$, $\overline{ST}$, and $\overline{TQ}$. What kind of figure is formed? What is its area?

Name the quadrant in which each point is located.

13. $(-4, 1)$

14. $(2, 17)$

15. $(-6, -7)$

16. $(4, -8)$

17. $\left(3, \frac{1}{2}\right)$

18. $(-1, -7)$

19. $(6.9, -2)$

20. $(-4, 31)$

Determine whether each ordered pair is a solution of the given equation. Remember to use alphabetical order for substitution.

21. $(1, -1)$; $y = 3x - 4$

22. $(2, 5)$; $y = 4x - 3$

23. $(2, 4)$; $5s - t = 8$

24. $(1, 3)$; $4p - q = 1$

25. $(3, 5)$; $4x - y = 7$

26. $(2, 7)$; $5x - y = 3$

27. $\left(0, \frac{3}{5}\right)$; $6a + 5b = 3$

28. $\left(0, \frac{3}{2}\right)$; $3f + 4g = 6$

29. $(2, -1)$; $4r - 2s = 10$

30. $(2, -4)$; $5w + 2z = 2$

31. $(5, 3)$; $x - 3y = -4$

32. $(1, 2)$; $2x - 5y = -6$

33. $(3, -1)$; $y = 3x^2$

34. $(2, 4)$; $2r^2 - s = 5$

35. $(2, 3)$; $5s^2 - t = 7$

36. $(2, 3)$; $y = x^3 - 5$

Graph by hand.

37. $y = x + 4$

38. $y = x + 3$

39. $y = -x$

40. $y = 3x$

41. $y = 3x - 1$

42. $y = -4x + 1$

43. $y = -2x + 3$

44. $y = -3x + 1$

Aha! **45.** $y + 2x = 3$

46. $y + 3x = 1$

Using a graphing calculator, create a table of solutions for integer values of x beginning at −3. Then graph.

47. $y = -\frac{3}{2}x + 1$

48. $y = -\frac{2}{3}x - 2$

49. $y = \frac{3}{4}x + 1$

50. $y = \frac{3}{4}x + 2$

51. $y = -x^2$

52. $y = x^2$

53. $y = x^2 - 3$

54. $y = x^2 + 2$

55. $y = |x| + 2$

56. $y = -|x|$

57. $y = 3 - x^2$

58. $y = 4 - x^2$

59. $y = x^3$

60. $y = x^3 - 2$

Graph each equation using both viewing windows indicated. Determine which window best shows not only the shape of the graph but also where it crosses the x- and y-axes.

61. $y = x - 15$

a) $[-10, 10, -10, 10]$, Xscl $= 1$, Yscl $= 1$
b) $[-20, 20, -20, 20]$, Xscl $= 5$, Yscl $= 5$

62. $y = -3x + 30$

a) $[-10, 10, -10, 10]$, Xscl $= 1$, Yscl $= 1$
b) $[-20, 20, -20, 40]$, Xscl $= 5$, Yscl $= 5$

63. $y = 5x^2 - 8$

a) $[-10, 10, -10, 10]$, Xscl $= 1$, Yscl $= 1$
b) $[-3, 3, -3, 3]$, Xscl $= 1$, Yscl $= 1$

64. $y = \frac{1}{10}x^2 + \frac{1}{3}$

a) $[-10, 10, -10, 10]$, Xscl $= 1$, Yscl $= 1$
b) $[-0.5, 0.5, -0.5, 0.5]$, Xscl $= 0.1$, Yscl $= 0.1$

65. $y = 4x^3 - 12$

a) $[-10, 10, -10, 10]$, Xscl $= 1$, Yscl $= 1$
b) $[-5, 5, -20, 10]$, Xscl $= 1$, Yscl $= 5$

66. $y = |4x^3 - 12|$

a) $[-10, 10, -10, 10]$, Xscl $= 1$, Yscl $= 1$
b) $[-3, 3, 0, 20]$, Xscl $= 1$, Yscl $= 5$

67. Determine which of the equations in the odd-numbered exercises 37–65 are linear.

68. Determine which of the equations in the even-numbered exercises 38–66 are linear.

TW **69.** What can be said about the location of two points that have the same first coordinates and second coordinates that are opposites of each other?

TW **70.** Examine Example 7 and explain why it is unwise to draw a graph after plotting just two points.

Skill Maintenance

Tell whether the number is a solution of the given equation or inequality. [1.1]

71. $3x - 5 = 10$; 5

72. $4y \geq 18$; 6

73. $3n - 7 < 5$; 4

74. $2t + 3 = 5$; 2

Evaluate.

75. $5s - 3t$, for $s = 2$ and $t = 4$ [1.2]

76. $(2m + n)^2$, for $m = 3$ and $n = 1$ [1.1]

Aha! **77.** $(5 - x)^4(x + 2)^3$, for $x = -2$ [1.2]

78. $2x^2 + 4x - 9$, for $x = -3$ [1.2]

Synthesis

TW **79.** Without making a drawing, how can you tell that the graph of $y = x - 30$ passes through three quadrants?

TW **80.** At what point will the line passing through $(a, -1)$ and $(a, 5)$ intersect the line that passes through $(-3, b)$ and $(2, b)$? Why?

TW **81.** Graph $y = 6x$, $y = 3x$, $y = \frac{1}{2}x$, $y = -6x$, $y = -3x$, and $y = -\frac{1}{2}x$ using the same set of axes or viewing window, and compare the slants of the lines. Describe the pattern that relates the slant of the line to the multiplier of x.

TW **82.** Using the same set of axes or viewing window, graph $y = 2x$, $y = 2x - 3$, and $y = 2x + 3$. Describe the pattern relating each line to the number that is added to $2x$.

83. Which of the following equations have $\left(-\frac{1}{3}, \frac{1}{4}\right)$ as a solution?

a) $-\frac{3}{2}x - 3y = -\frac{1}{4}$
b) $8y - 15x = \frac{7}{2}$
c) $0.16y = -0.09x + 0.1$
d) $2(-y + 2) - \frac{1}{4}(3x - 1) = 4$

84. If $(2, -3)$ and $(-5, 4)$ are the endpoints of a diagonal of a square, what are the coordinates of the other two vertices? What is the area of the square?

85. If $(-10, -2)$, $(-3, 4)$, and $(6, 4)$ are the coordinates of three consecutive vertices of a parallelogram, what are the coordinates of the fourth vertex?

86. One value of y for the equation $y = 3.2x - 5$ is -11.4. Use the TABLE feature of a graphing calculator to determine the x-value that is paired with -11.4.

87. Graph each of the following equations and determine which appear to be linear. Try to find a way to tell if an equation is linear without graphing it.

a) $y = 3x + 2$
b) $y = \frac{1}{2}x^2 - 5$
c) $y = 8$
d) $y = 4 - \frac{1}{5}x$
e) $y = |3 - x|$
f) $y = 4x^3$

88. The graph of $y = 0.5x^2 - 15x + 64$ crosses the x-axis twice. Determine a viewing window that shows both intersections of $y = 0.5x^2 - 15x + 64$ and the x-axis.

1.6 Solving Equations and Formulas

Solving equations is an essential part of problem solving in algebra. In this section, we review and practice solving basic equations.

Study Tip

To Err Is Human

It is no coincidence that the students who experience the greatest success in this course work in pencil. We all make mistakes and by using pencil and eraser we are more willing to admit to ourselves that something needs to be rewritten. Please work with a pencil and eraser if you aren't doing so already.

Equivalent Equations

In Section 1.1, we saw that the solution of $1.7 + x = 11.0$ is 9.3. That is, when x is replaced with 9.3, the equation $1.7 + x = 11.0$ is a true statement. It is important to know how to find such a solution using the principles of algebra. These principles are used to produce *equivalent equations* from which solutions are easily found.

Equivalent Equations Two equations are *equivalent* if they have the same solution(s).

EXAMPLE 1 Determine whether $4x = 12$ and $10x = 30$ are equivalent equations.

SOLUTION The equation $4x = 12$ is true only when x is 3. Similarly, $10x = 30$ is true only when x is 3. Since both equations have the same solution, they are equivalent.

EXAMPLE 2 Determine whether $x + 4 = 7$ and $x = 3$ are equivalent equations.

SOLUTION Each equation has only one solution, the number 3. Thus the equations are equivalent.

EXAMPLE 3 Determine whether $3x = 4x$ and $3/x = 4/x$ are equivalent equations.

SOLUTION Note that 0 is a solution of $3x = 4x$. Since neither $3/x$ nor $4/x$ is defined for $x = 0$, the equations $3x = 4x$ and $3/x = 4/x$ are *not* equivalent.

The Addition and Multiplication Principles

Suppose that a and b represent the same number and that some number c is added to a. If c is also added to b, we will get two equal sums, since a and b are the same number. The same is true if we multiply both a and b by c. In this manner, we can produce equivalent equations.

The Addition and Multiplication Principles for Equations For any real numbers a, b, and c:

a) $a = b$ is equivalent to $a + c = b + c$;
b) $a = b$ is equivalent to $a \cdot c = b \cdot c$, provided $c \neq 0$.

As shown in Examples 4 and 5, either a or b usually represents a variable expression.

Student Notes

The addition and multiplication principles can be used even when 0 appears on one side of an equation. Thus to solve $y - 4.7 = 0$, we would add 4.7 to both sides.

EXAMPLE 4 Solve: $y - 4.7 = 13.9$.

SOLUTION We have

$$y - 4.7 = 13.9$$
$$y - 4.7 + 4.7 = 13.9 + 4.7 \quad \textbf{Using the addition principle; adding 4.7}$$
$$y + 0 = 13.9 + 4.7 \quad \textbf{Using the law of opposites}$$
$$y = 18.6$$

Check:

$y - 4.7 = 13.9$		
$18.6 - 4.7$	13.9	**Substituting 18.6 for y**
$13.9 \stackrel{?}{=} 13.9$		TRUE

The solution is 18.6.

In Example 4, why did we add 4.7 to both sides? Because 4.7 is the opposite of -4.7 and we wanted y alone on one side of the equation. Adding 4.7 gave us $y + 0$, or just y, on the left side. This led to the equivalent equation, $y = 18.6$, from which the solution, 18.6, is immediately apparent.

EXAMPLE 5 Solve: $\frac{2}{5}x = \frac{9}{10}$.

SOLUTION We have

$$\frac{2}{5}x = \frac{9}{10}$$
$$\frac{5}{2} \cdot \frac{2}{5}x = \frac{5}{2} \cdot \frac{9}{10} \quad \textbf{Using the multiplication principle, we multiply by } \tfrac{5}{2}\textbf{, the reciprocal of } \tfrac{2}{5}.$$
$$1x = \frac{45}{20} \quad \textbf{Using the law of reciprocals}$$
$$x = \frac{9}{4}. \quad \textbf{Simplifying}$$

The check is left to the student. The solution is $\frac{9}{4}$.

In Example 5, why did we multiply by $\frac{5}{2}$? Because $\frac{5}{2}$ is the reciprocal of $\frac{2}{5}$ and we wanted x alone on one side of the equation. When we multiplied by $\frac{5}{2}$, we got $1x$, or just x, on the left side. This led to the equivalent equation $x = \frac{9}{4}$, from which the solution, $\frac{9}{4}$, is clear.

There is no need for a subtraction or division principle because subtraction can be regarded as adding opposites and division can be regarded as multiplying by reciprocals.

Connecting the Concepts

It is important to distinguish between *equivalent expressions* and *equivalent equations.*

In Examples 4 and 5, we used the addition and multiplication principles to write a sequence of equivalent equations that led to an equation for which the solution was clear. Because the equations were equivalent, the solution of the last equation was also a solution of the original equation.

In Section 1.3, we used the commutative, associative, and distributive laws to write a sequence of equivalent, and increasingly simpler, expressions that take on the same value when the variables are replaced with numbers. This is how we "simplify" an expression.

Equivalent Equations

$$\begin{aligned} y - 4.7 &= 13.9 \\ y - 4.7 + 4.7 &= 13.9 + 4.7 \\ y + 0 &= 13.9 + 4.7 \\ y &= 18.6 \end{aligned}$$

Each line here is a complete equation. Because they are equivalent, all four equations share the same solution.

Equivalent Expressions

$$\begin{aligned} &3x + 2[4 + 5(x + 2y)] \\ &= 3x + 2[4 + 5x + 10y] \\ &= 3x + 8 + 10x + 20y \\ &= 13x + 8 + 20y \end{aligned}$$

Each line here is an expression that is equivalent to those written above or below. There is no equation to "solve."

Often, as in the next example, we merge these ideas by forming an equivalent equation by replacing part of an equation with an equivalent expression.

EXAMPLE 6 Solve: $5x - 2(x - 5) = 7x - 2$.

SOLUTION We have

$5x - 2(x - 5) = 7x - 2$	
$5x - 2x + 10 = 7x - 2$	**Using the distributive law**
$3x + 10 = 7x - 2$	**Combining like terms**
$3x + 10 - 3x = 7x - 2 - 3x$	**Using the addition principle; adding $-3x$, the opposite of $3x$, to both sides**
$10 = 4x - 2$	**Combining like terms**
$10 + 2 = 4x - 2 + 2$	**Using the addition principle**
$12 = 4x$	**Simplifying**
$\frac{1}{4} \cdot 12 = \frac{1}{4} \cdot 4x$	**Using the multiplication principle; multiplying both sides by $\frac{1}{4}$, the reciprocal of 4**
$3 = x.$	**Using the law of reciprocals; simplifying**

CHECK BY HAND

$5x - 2(x - 5)$	$= 7x - 2$
$5 \cdot 3 - 2(3 - 5)$	$7 \cdot 3 - 2$
$15 - 2(-2)$	$21 - 2$
$15 + 4$	19
$19 \stackrel{?}{=} 19$	TRUE

The solution is 3.

CHECK USING A GRAPHING CALCULATOR

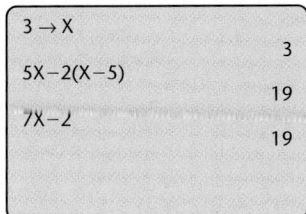

```
3→X
                3
5X−2(X−5)
               19
7X−2
               19
```

Since both $5x - 2(x - 5)$ and $7x - 2$ have a value of 19 when $x = 3$, the answer checks. The solution is 3.

Don't rush to solve equations in your head. Work neatly, keeping in mind that the number of steps in a solution is less important than producing a simpler, yet equivalent, equation in each step.

Types of Equations

In Examples 4, 5, and 6, we solved *linear equations*. A linear equation in one variable—say, x—is an equation equivalent to one of the form $ax = b$ with a and b constants and $a \neq 0$. The variable in a linear equation is always raised to the first power.

Every equation falls into one of three categories. An **identity** is an equation, like $x + 5 = 3 + x + 2$, that is true for all replacements. A **contradiction** is an equation, like $n + 5 = n + 7$, that is *never* true. A **conditional equation,** like $2x + 5 = 17$, is sometimes true and sometimes false, depending on what the replacement of x is. Most of the equations examined in this text are conditional.

EXAMPLE 7 Solve each of the following equations and classify the equation as an identity, a contradiction, or a conditional equation.

a) $2x + 7 = 7(x + 1) - 5x$

b) $3x - 5 = 3(x - 2) + 4$

c) $3 - 8x = 5 - 7x$

SOLUTION

a) $2x + 7 = 7(x + 1) - 5x$

$2x + 7 = 7x + 7 - 5x$ **Using the distributive law**

$2x + 7 = 2x + 7$ **Combining like terms**

The equation $2x + 7 = 2x + 7$ is true regardless of what x is replaced with, so all real numbers are solutions. Note that $2x + 7 = 2x + 7$ is equivalent to $2x = 2x$, $7 = 7$, or $0 = 0$. All real numbers are solutions and the equation is an identity.

b)

$$\begin{aligned} 3x - 5 &= 3(x - 2) + 4 \\ 3x - 5 &= 3x - 6 + 4 && \textbf{Using the distributive law} \\ 3x - 5 &= 3x - 2 && \textbf{Combining like terms} \\ -3x + 3x - 5 &= -3x + 3x - 2 && \textbf{Using the addition principle} \\ -5 &= -2 \end{aligned}$$

Since the original equation is equivalent to $-5 = -2$, which is false for any choice of x, the original equation has no solution. There is no choice of x for which $3x - 5 = 3(x - 2) + 4$. The equation is a contradiction.

c)

$$\begin{aligned} 3 - 8x &= 5 - 7x \\ 3 - 8x + 7x &= 5 - 7x + 7x && \textbf{Using the addition principle} \\ 3 - x &= 5 && \textbf{Simplifying} \\ -3 + 3 - x &= -3 + 5 && \textbf{Using the addition principle} \\ -x &= 2 && \textbf{Simplifying} \\ x &= \frac{2}{-1}, \text{ or } -2 && \textbf{Dividing both sides by } -1 \textbf{ or multiplying both sides by } \frac{1}{-1}\textbf{, or } -1 \end{aligned}$$

There is one solution, -2. For other choices of x, the equation is false. This equation is conditional since it can be true or false, depending on the replacement for x.

We will sometimes refer to the set of solutions, or **solution set,** of a particular equation. Thus the solution set for Example 7(c) is $\{-2\}$. The solution set for Example 7(a) is simply $\mathbb{R}$, the set of all real numbers, and the solution set for Example 7(b) is the **empty set,** denoted $\varnothing$ or $\{\ \}$. As its name suggests, the empty set is the set containing no elements.

Solving Formulas

A *formula* is an equation that uses letters to represent a relationship between two or more quantities. Some important geometric formulas are $A = \pi r^2$ (for the area A of a circle of radius r), $C = \pi d$ (for the circumference C of a circle of diameter d), and $A = b \cdot h$ (for the area A of a parallelogram of height h and base length b).* A more complete list of geometric formulas appears on the inside back cover.

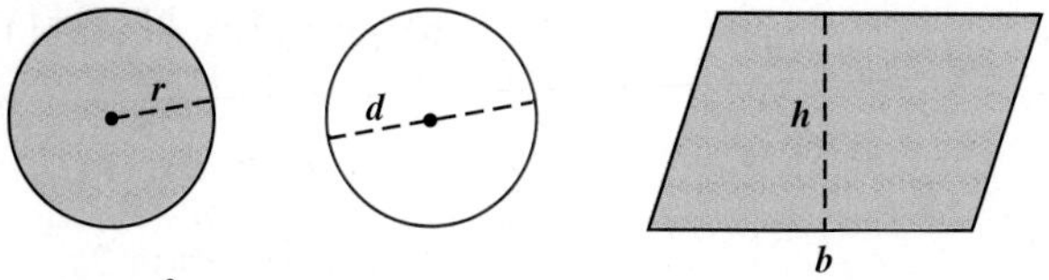

$A = \pi r^2$ $C = \pi d$ $A = b \cdot h$

*The Greek letter π, read "pi," is *approximately* 3.1415926535897932384626 4. Often 3.14 or 22/7 is used to approximate π when a calculator with a π key is unavailable.

Suppose we know the floor area and the width of a rectangular room and want to find the length. To do so, we could "solve" the formula $A = l \cdot w$ (Area = Length · Width) for l, using the same principles that we use for solving equations.

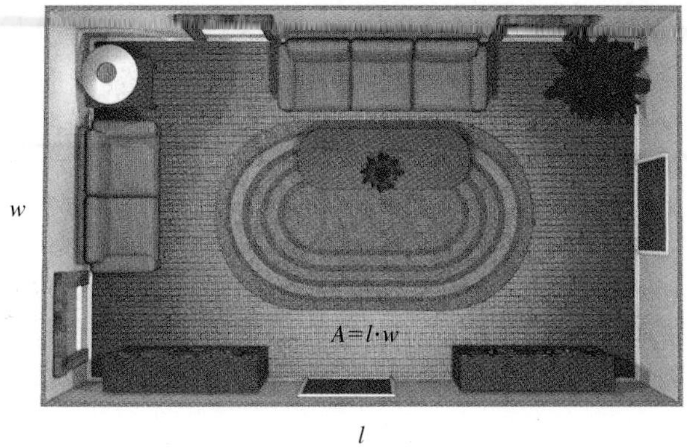

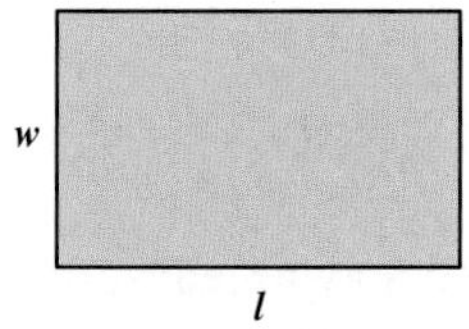

EXAMPLE 8 Area of a Rectangle. Solve the formula $A = l \cdot w$ for l.

SOLUTION

$$A = l \cdot w$$ **We want this letter alone.**

$$\frac{A}{w} = \frac{l \cdot w}{w}$$ **Dividing both sides by w, or multiplying both sides by $1/w$**

$$\frac{A}{w} = l \cdot \frac{w}{w}$$
$$\frac{A}{w} = l$$
Simplifying by removing a factor equal to 1: $\frac{w}{w} = 1$

Thus to find the length of a rectangular room, we can divide the area of the room by its width. Were we to do this calculation for a variety of rectangular rooms, the formula $l = A/w$ would be more convenient than repeatedly substituting into $A = l \cdot w$ and solving for l each time.

EXAMPLE 9 Simple Interest. The formula $I = Prt$ is used to determine the simple interest I earned when a principal of P dollars is invested for t years at an interest rate r. Solve this formula for t.

SOLUTION

$$I = Prt$$ **We want this letter alone.**

$$\frac{I}{Pr} = \frac{Prt}{Pr}$$ **Dividing both sides by Pr, or multiplying both sides by $\frac{1}{Pr}$**

$$\frac{I}{Pr} = t$$ **Simplifying by removing a factor equal to 1: $\frac{Pr}{Pr} = 1$**

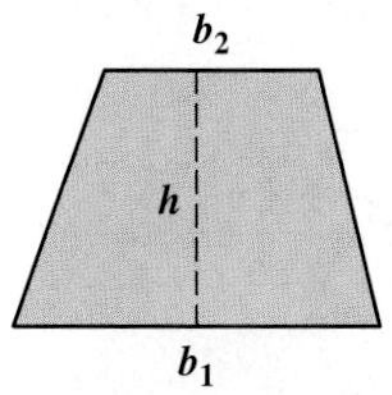

EXAMPLE 10 Area of a Trapezoid. A trapezoid is a geometric shape with four sides, exactly two of which, the bases, are parallel to each other. The formula for calculating the area A of a trapezoid with bases b_1 and b_2 (read "b sub one" and "b sub two") and height h is given by

$$A = \frac{h}{2}(b_1 + b_2),$$ **A derivation of this formula is outlined in Exercise 126 of this section.**

where the *subscripts* 1 and 2 distinguish one base from the other. Solve for b_1.

SOLUTION There are several ways to "remove" the parentheses. We could distribute $h/2$, but an easier approach is to multiply both sides by the reciprocal of $h/2$.

$$A = \frac{h}{2}(b_1 + b_2)$$

$$\frac{2}{h} \cdot A = \frac{2}{h} \cdot \frac{h}{2}(b_1 + b_2)$$ **Multiplying both sides by $\frac{2}{h}$ $\left(\text{or dividing by } \frac{h}{2}\right)$**

$$\frac{2A}{h} = b_1 + b_2$$ **Simplifying. The right side is "cleared" of fractions.**

$$\frac{2A}{h} - b_2 = b_1$$ **Adding $-b_2$ to both sides**

The similarities between solving formulas and solving equations can be seen below. In (a), we solve as we did before; in (b), we do not carry out all calculations; and in (c), we cannot carry out all calculations because the numbers are unknown. The same steps are used each time.

a)

$$9 = \frac{3}{2}(x + 5)$$
$$\frac{2}{3} \cdot 9 = \frac{2}{3} \cdot \frac{3}{2}(x + 5)$$
$$6 = x + 5$$
$$1 = x$$

b)

$$9 = \frac{3}{2}(x + 5)$$
$$\frac{2}{3} \cdot 9 = \frac{2}{3} \cdot \frac{3}{2}(x + 5)$$
$$\frac{2 \cdot 9}{3} = x + 5$$
$$\frac{2 \cdot 9}{3} - 5 = x$$

c)

$$A = \frac{h}{2}(b_1 + b_2)$$
$$\frac{2}{h} \cdot A = \frac{2}{h} \cdot \frac{h}{2}(b_1 + b_2)$$
$$\frac{2A}{h} = b_1 + b_2$$
$$\frac{2A}{h} - b_2 = b_1$$

EXAMPLE 11 Accumulated Simple Interest. The formula $A = P + Prt$ gives the amount A that a principal of P dollars will be worth in t years when invested at simple interest rate r. Solve the formula for P.

SOLUTION We have

$A = P + Prt$ — **We want this letter alone.**

$A = P(1 + rt)$ — **Factoring (using the distributive law) to combine like terms**

$\dfrac{A}{1 + rt} = \dfrac{P(1 + rt)}{1 + rt}$ — **Dividing both sides by $1 + rt$, or multiplying both sides by $\dfrac{1}{1 + rt}$**

$\dfrac{A}{1 + rt} = P.$ — **Simplifying**

This last equation can be used to determine how much should be invested at interest rate r in order to have A dollars t years later.

Note in Example 11 that the factoring enabled us to write P once rather than twice. This is comparable to combining like terms when solving an equation like $16 = x + 7x$.

Most graphing calculators require us to solve for the dependent variable. An equation written in another form must be solved for y before it can be entered into a graphing calculator.

EXAMPLE 12 Graph: $3x - 4y = 2y + 7$.

SOLUTION In order to graph $3x - 4y = 2y + 7$ using a graphing calculator, we must first solve for y:

$3x - 4y = 2y + 7$ — **We want this letter alone.**

$3x - 4y - 2y = 7$ — **Adding $-2y$ to both sides**

$3x - 6y = 7$ — **Combining like terms**

$-6y = -3x + 7$ — **Adding $-3x$ to both sides**

$y = \dfrac{-3x + 7}{-6}.$ — **Dividing both sides by -6**

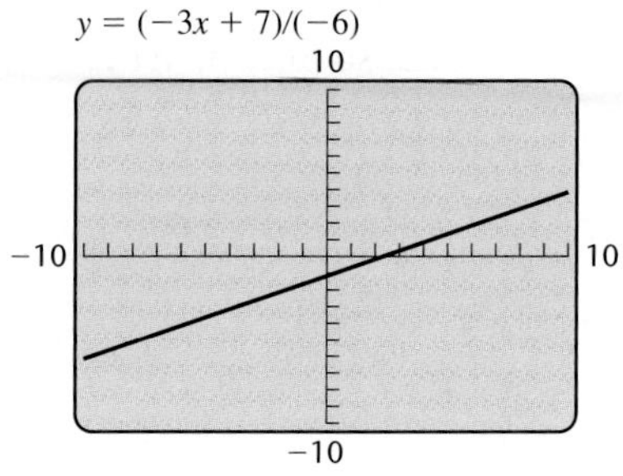

Now we can enter the equation and graph it, as shown at left. Since $y = (-3x + 7)/(-6)$ is equivalent to $3x - 4y = 2y + 7$, their graphs are the same.

You may find the following summary useful.

To Solve a Formula for a Specified Letter

1. Get all terms with the letter being solved for on one side of the equation and all other terms on the other side, using the addition principle. To do this may require removing parentheses.
   - To remove parentheses, either divide both sides by the multiplier in front of the parentheses or use the distributive law.
2. When all terms with the specified letter are on the same side, factor (if necessary) so that the variable is written only once.
3. Solve for the letter in question by dividing both sides by the multiplier of that letter.

1.6 EXERCISE SET

FOR EXTRA HELP

MathXL | MyMathLab | InterAct Math | AW Math Tutor Center | Video Lectures on CD: Disc 1 | Student's Solutions Manual

Concept Reinforcement *Complete each of the following statements.*

1. An equation in x of the form $ax = b$ is a(n) ______________ equation.
2. A(n) ______________ is an equation that is never true.
3. A formula is a(n) ______________ that uses letters to represent a relationship between two or more quantities.
4. The formula $A = \pi r^2$ is used to calculate the ______________ of a circle.
5. The formula $C = \pi d$ is used to calculate the ______________ of a circle.
6. The formula ______________ is used to calculate the perimeter of a rectangle of length l and width w.
7. The formula ______________ is used to calculate the area of a parallelogram of height h and base length b.
8. The formula $l = A/w$ can be used to determine the ______________ of a rectangle, given its area and width.
9. In the formula for the area of a trapezoid, $A = \frac{h}{2}(b_1 + b_2)$, the numbers 1 and 2 are referred to as ______________.
10. When two or more terms on the same side of a formula contain the letter for which we are solving, we can ______________ so that the letter is written only once.

Determine whether the two equations in each pair are equivalent.

11. $t + 5 = 11$ and $3t = 18$
12. $t - 3 = 7$ and $3t = 24$
13. $12 - x = 3$ and $2x = 20$
14. $3x - 4 = 8$ and $3x = 12$
15. $5x = 2x$ and $\frac{4}{x} = 3$
16. $6 = 2x$ and $5 = \frac{2}{3 - x}$

Solve. Be sure to check.

17. $x - 2.9 = 13.4$

18. $y + 4.3 = 11.2$

19. $8t = 72$

20. $9t = 63$

21. $4x - 12 = 60$

22. $4x - 6 = 70$

23. $\frac{3}{5}n + 2 = 17$

24. $\frac{2}{7}n + 1 = 9$

25. $2y - 11 = 37$

26. $3x - 13 = 29$

27. $6x + 3x = 54$

28. $3x + 7x = 150$

29. $\frac{2}{3}y - \frac{1}{4}y = 5$

30. $\frac{3}{5}t - \frac{1}{2}t = 3$

31. $5t - 13t = -32$

32. $-9y - 5y = 28$

33. $3(x + 4) = 7x$

34. $3(y + 5) = 8y$

35. $70 = 10(3t - 2)$

36. $27 = 9(5y - 2)$

37. $1.8(n - 2) = 9$

38. $2.1(x - 3) = 8.4$

39. $5y - (2y - 10) = 25$

40. $8x - (3x - 5) = 40$

41. $7y - 1 = 23 - 5y$

42. $14t + 20 = 8t - 22$

43. $\frac{1}{5} + \frac{3}{10}x = \frac{4}{5}$

44. $-\frac{5}{2}x + \frac{1}{2} = -18$

45. $\frac{9}{10}y - \frac{7}{10} = \frac{21}{5}$

46. $\frac{4}{5}t - \frac{3}{10} = \frac{2}{5}$

47. $7r - 2 + 5r = 6r + 6 - 4r$

48. $9m - 15 - 2m = 6m - 1 - m$

49. $\frac{2}{3}(x - 2) - 1 = \frac{1}{4}(x - 3)$

50. $\frac{1}{4}(6t + 48) - 20 = -\frac{1}{3}(4t - 72)$

51. $5 + 2(x - 3) = 2[5 - 4(x + 2)]$

52. $3[2 - 4(x - 1)] = 3 - 4(x + 2)$

Find each solution set. Then classify each equation as a conditional equation, an identity, or a contradiction.

53. $7x - 2 - 3x = 4x$

54. $3t + 5 + t = 5 + 4t$

55. $2 + 9x = 3(4x + 1) - 1$

56. $4 + 7x = 7(x + 1)$

Aha! **57.** $-9t + 2 = -9t - 7(6 \div 2(49) + 8)$

58. $-9t + 2 = 2 - 9t - 5(8 \div 4(1 + 3^4))$

59. $2\{9 - 3[-2x - 4]\} = 12x + 42$

60. $3\{7 - 2[7x - 4]\} = -40x + 45$

Solve.

61. $d = rt$, for r (a distance formula)

62. $d = rt$, for t

63. $F = ma$, for a (a physics formula)

64. $A = lw$, for w (an area formula)

65. $W = EI$, for I (an electricity formula)

66. $W = EI$, for E

67. $V = lwh$, for h (a volume formula)

68. $I = Prt$, for r (a formula for interest)

69. $L = \dfrac{k}{d^2}$, for k
(a formula for intensity of sound or light)

70. $F = \dfrac{mv^2}{r}$, for m (a physics formula)

71. $G = w + 150n$, for n
(a formula for the gross weight of a bus)

72. $P = b + 0.5t$, for t (a formula for parking prices)

73. $2w + 2h + l = p$, for l
(a formula used when shipping boxes)

74. $2w + 2h + l = p$, for w

75. $Ax + By = C$, for y (a formula for graphing lines)

76. $P = 2l + 2w$, for l (a perimeter formula)

77. $C = \frac{5}{9}(F - 32)$, for F (a temperature formula)

78. $T = \frac{3}{10}(I - 12{,}000)$, for I (a tax formula)

79. $A = \dfrac{h}{2}(b_1 + b_2)$, for b_2 (an area formula)

80. $A = \dfrac{h}{2}(b_1 + b_2)$, for h (an area formula)

81. $v = \dfrac{d_2 - d_1}{t}$, for t (a physics formula)
(*Hint*: Multiply by t to "clear" fractions.)

82. $v = \dfrac{s_2 - s_1}{m}$, for m

83. $v = \dfrac{d_2 - d_1}{t}$, for d_1

84. $v = \dfrac{s_2 - s_1}{m}$, for s_1

85. $r = m + mnp$, for m

86. $p = x - xyz$, for x

87. $y = ab - ac^2$, for a

88. $d = mn - mp^3$, for m

Solve for y.

89. $3x + 6y = 9$

90. $4x - 7y = 6$

91. $x = y - 7$

92. $2 = 4x - y$

93. $y - 3(x + 2) = 4 + 2y$

94. $3x - 2(x + y) = y - x$

95. $4y + x^2 = x + 1$

96. $2x^2 - y + 3x = 0$

97. *Investing.* Janos has $2600 to invest for 6 months. If he needs the money to earn $156 in that time, at what rate of simple interest must Janos invest?

98. *Banking.* Yvonne plans to buy a one-year certificate of deposit (CD) that earns 7% simple interest. If she needs the CD to earn $110, how much should Yvonne invest?

99. *Geometry.* The area of a parallelogram is 78 cm². The base of the figure is 13 cm. What is the height?

100. *Geometry.* The area of a parallelogram is 72 cm². The height of the figure is 6 cm. How long is the base?

Projected Birth Weight. Ultrasonic images of 29-week-old fetuses can be used to predict weight. One model, developed by Thurnau, is* $P = 9.337da - 299$; *a second model, developed by Weiner,† is* $P = 94.593c + 34.227a - 2134.616$. *For both formulas, P represents the estimated fetal weight in grams, d the diameter of the fetal head in centimeters, c the circumference of the fetal head in centimeters, and a the circumference of the fetal abdomen in centimeters.*

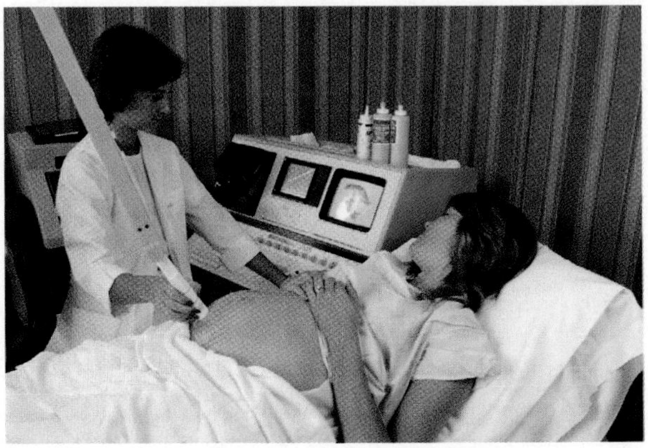

101. Solve Thurnau's model for d and use that equation to estimate the diameter of a fetus' head at 29 weeks when the estimated weight is 1614 g and the circumference of the fetal abdomen is 24.1 cm.

102. Solve Weiner's model for c and use that equation to estimate the circumference of a fetus' head at 29 weeks when the estimated weight is 1277 g and the circumference of the fetal abdomen is 23.4 cm.

TW **103.** As the first step in solving

$$2x + 5 = -3,$$

Pat multiplies both sides by $\frac{1}{2}$. Is this incorrect? Why or why not?

*Thurnau, G. R., R. K. Tamura, R. E. Sabbagha, et al. *Am. J. Obstet Gynecol* 1983; **145:** 557.

†Weiner, C. P., R. E. Sabbagha, N. Vaisrub, et al. *Obstet Gynecol* 1985; **65:** 812.

TW **104.** Explain how an identity can be easily altered so that it becomes a contradiction.

Skill Maintenance

Perform the indicated operations. [1.2]

105. $-2 + 3(-5) - 7$

106. $(-2 + 3)(-5 - 7)$

107. $\dfrac{10 \div 2(-4)}{2 + 3}$

108. $10 \div 2(-4) \div 2 + 3$

Synthesis

TW **109.** Explain why $a = b$ is *not* equivalent to $ac = bc$ when $c = 0$.

TW **110.** Explain the difference between equivalent expressions and equivalent equations.

Solve and check.

111. $4.23x - 17.898 = -1.65x - 42.454$

112. $-0.00458y + 1.7787 = 13.002y - 1.005$

113. $8x - \{3x - [2x - (5x - (7x - 1))]\} = 8x + 7$

114. $6x - \{5x - [7x - (4x - (3x + 1))]\} = 6x + 5$

115. $17 - 3\{5 + 2[x - 2]\} + 4\{x - 3(x + 7)\} = 9\{x + 3[2 + 3(4 - x)]\}$

116. $23 - 2\{4 + 3[x - 1]\} + 5\{x - 2(x + 3)\} = 7\{x - 2[5 - (2x + 3)]\}$

TW **117.** Create an equation for which it is preferable to use the multiplication principle *before* using the addition principle. Explain why it is best to solve the equation in this manner.

Solve.

118. $s = v_i t + \frac{1}{2}at^2$, for a

119. $A = 4lw + w^2$, for l

120. $\dfrac{P_1V_1}{T_1} = \dfrac{P_2V_2}{T_2}$, for T_2

121. $\dfrac{P_1V_1}{T_1} = \dfrac{P_2V_2}{T_2}$, for T_1

122. $\dfrac{b}{a - b} = c$, for b

123. $m = \dfrac{(d/e)}{(e/f)}$, for d

124. $\dfrac{a}{a + b} = c$, for a

Aha! **125.** $s + \dfrac{s + t}{s - t} = \dfrac{1}{t} + \dfrac{s + t}{s - t}$, for t

TW **126.** To derive the formula for the area of a trapezoid, consider the area of two congruent trapezoids, one of which is upside down.

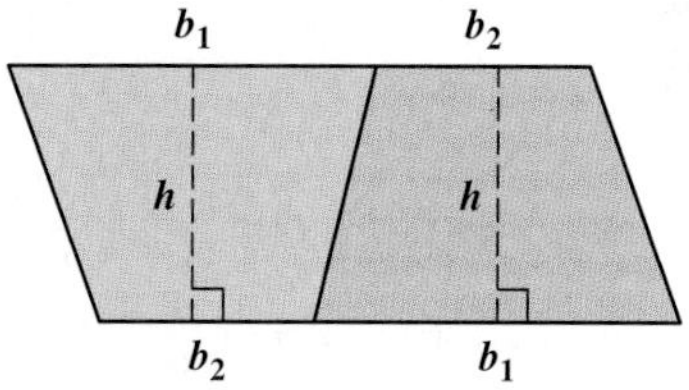

Explain why the total area of the two trapezoids is given by $h(b_1 + b_2)$. Then explain why the area of a trapezoid is given by $\dfrac{h}{2}(b_1 + b_2)$.

Graph.

127. $y - 2(x^2 + y) = 0$

128. $2x^3 - y + 7 = 4y$

129. $|x + 2| - 3y = 4y$

130. $2x^3 - 3y = 7(x - y)$

1.7 Introduction to Problem Solving and Models

The Five-Step Strategy ■ Translating to Algebraic Expressions ■ Problem Solving ■ Mathematical Models

Study Tip

Seeking Help on Campus

Your college or university probably has resources to support your learning.

1. There may be a learning lab or tutoring center for drop-in tutoring.
2. There may be study skills workshops or group tutoring sessions.
3. Often, there is a bulletin board or network for locating experienced private tutors.
4. You might be able to find classmates interested in forming a study group.
5. Consider an often overlooked resource: your instructor. Find out your instructor's office hours and visit if you need additional help. Many instructors welcome student e-mail, which can be a very effective way of receiving help.

We now begin to study and practice the "art" of problem solving. Although we are interested mainly in using algebra to solve problems, much of what we say here applies to solving all kinds of problems.

What do we mean by a *problem*? Perhaps you've already used algebra to solve some "real-world" problems. What procedure did you use? Is there an approach that can be used to solve problems of a more general nature? These are questions that we will answer in this section.

In this text, we do not restrict the use of the word "problem" to computations involving arithmetic or algebra, such as $589 + 437 = a$ or $3x + 5x = 9$. Here, a problem is simply a question to which we wish to find an answer. Perhaps this can best be illustrated with some sample problems:

1. Can I afford to rent a bigger apartment?
2. If I exercise twice a week and eat 3000 calories a day, will I lose weight?
3. Do I have enough time to take 4 courses while working 20 hours a week?
4. My fishing boat travels 12 km/h in still water. How long will it take me to cruise 25 km upstream if the river's current is 3 km/h?

Although these problems differ, there is a strategy that can be applied to all of them.

The Five-Step Strategy

Since you have already studied some algebra, you have some experience with problem solving. The following steps constitute a strategy that you may already have used and comprise a sound strategy for problem solving in general.

Five Steps for Problem Solving with Algebra

1. *Familiarize* yourself with the problem.
2. *Translate* to mathematical language.
3. *Carry out* some mathematical manipulation.
4. *Check* your possible answer in the original problem.
5. *State* the answer clearly.

Of the five steps, perhaps the most important is the first: becoming familiar with the problem situation. Here are some ways in which this can be done.

The First Step in Problem Solving with Algebra

To familiarize yourself with the problem:

1. If the problem is written, read it carefully. Then read it again, perhaps aloud. Verbalize the problem to yourself.
2. List the information given and restate the question being asked.
3. Select a variable or variables to represent any unknown(s) and clearly state what each variable represents. Be descriptive! For example, let $t =$ the flight time, in hours; let $p =$ Paul's weight, in pounds; and so on.
4. Find additional information. Look up formulas or definitions with which you are not familiar. Geometric formulas appear on the inside back cover of this text; important words appear in the index. Consult an expert in the field or a reference librarian.
5. Create a table, using variables, in which both known and unknown information is listed. Look for possible patterns.
6. Make and label a drawing.
7. Estimate an answer and check to see whether it is correct.

EXAMPLE 1 How might you familiarize yourself with the situation of Problem 1: "Can I afford to rent a bigger apartment?"

SOLUTION Clearly more information is needed to solve this problem. You might:

a) Estimate the rent of some apartments in which you are interested.

b) Examine what your savings are and how your income is budgeted.

c) Determine how much rent you can afford and whether you would consider having a roommate.

When enough information is known, it might be wise to make a chart or table to help you reach an answer.

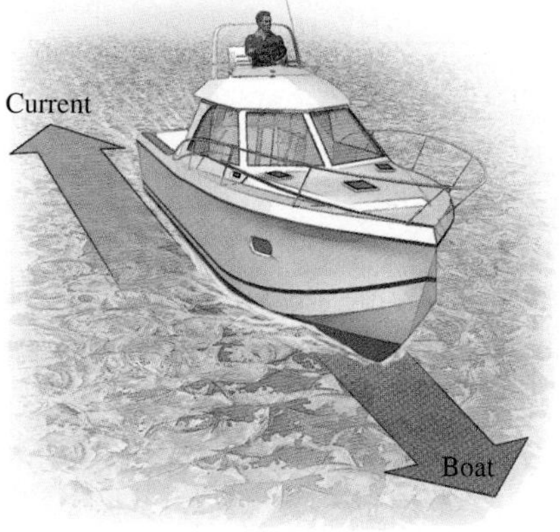

EXAMPLE 2 How might you familiarize yourself with Problem 4: "How long will it take the boat to cruise 25 km upstream?"

SOLUTION First read the question *very* carefully. This may even involve speaking aloud. You may need to reread the problem several times to fully understand what information is given and what information is required. A sketch is often helpful.

To gain more familiarity with the problem, we should determine, possibly with the aid of outside references, what relationships exist among the various quantities in the problem. With some effort, it can be learned that the current's speed should be subtracted from the boat's speed in still water to determine the boat's speed going upstream. We also need to find or recall an extremely important formula:

Distance = Speed × Time. **It is important to remember this equation.**

We organize the information in a table, letting t = the number of hours required for the boat to cruise 25 km upstream.

Distance to be Traveled	25 km
Speed of Boat Upstream	$12 - 3 = 9$ km/h
Time Required	t

Student Notes

It is extremely helpful to write down exactly what each variable represents before attempting to form an equation.

At this point we might try a guess. Suppose the boat traveled upstream for 2 hr. The boat would have then traveled

$$\underset{\text{Speed}}{\underset{\uparrow}{9\frac{\text{km}}{\text{hr}}}} \times \underset{\text{Time}}{\underset{\uparrow}{2\text{ hr}}} = \underset{\text{Distance}}{\underset{\uparrow}{18\text{ km.}}}$$

Note that $\frac{\textbf{km}}{\textbf{hr}} \cdot \textbf{hr} = \textbf{km}$**.**

Since $18 \neq 25$, our guess is wrong. Still, examining how we checked our guess sheds light on how to translate the problem to an equation. A better guess, when multiplied by 9, would yield a number closer to 25.

The second step in problem solving is to translate the situation to mathematical language. In algebra, this often means forming an equation.

The Second Step in Problem Solving with Algebra

Translate the problem to mathematical language. This is sometimes done by writing an algebraic expression, but most often in this text it is done by translating to an equation.

In the third step of our process, we work with the results of the first two steps. Often this requires us to use the algebra that we have studied.

The Third Step in Problem Solving with Algebra

Carry out some mathematical manipulation. If you have translated to an equation, this means to solve the equation.

To complete the problem-solving process, we should always **check** our solution and then **state** the solution in a clear and precise manner. To check, we make sure that our answer is reasonable and that all the conditions of the original problem are satisfied. If our answer checks, we write a complete English sentence stating the solution. The five steps are listed again below. Try to apply them regularly in your work.

Five Steps for Problem Solving with Algebra

1. *Familiarize* yourself with the problem.
2. *Translate* to mathematical language.
3. *Carry out* some mathematical manipulation.
4. *Check* your possible answer in the original problem.
5. *State* the answer clearly.

Translating to Algebraic Expressions

In order to translate problems to mathematical language, we must first be able to translate phrases to algebraic expressions. To do this, we need to know which words correspond to which symbols.

Key Words

Addition	Subtraction	Multiplication	Division
add	subtract	multiply	divide
sum of	difference of	product of	divided by
plus	minus	times	quotient of
increased by	decreased by	twice	ratio
more than	less than	of	per

When the value of a number is not given, we represent that number with a variable.

Phrase	Algebraic Expression
Five *more than* some number	$n + 5$
Half *of* a number	$\frac{1}{2}t$, or $\frac{t}{2}$
Five *more than* three *times* some number	$3p + 5$
The *difference of* x and y	$x - y$
Six *less than* the *product of* two numbers	$rs - 6$
Seventy-six percent *of* some number	$0.76z$, or $\frac{76}{100}z$

EXAMPLE 3 Translate to an algebraic expression:

Five less than forty-three percent of the quotient of two numbers.

SOLUTION We let r and s represent the two numbers.

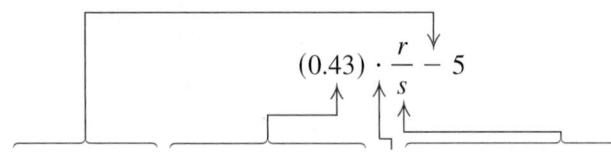

$$(0.43) \cdot \frac{r}{s} - 5$$

Five less than forty-three percent of the quotient of two numbers

Problem Solving

At this point, our study of algebra has just begun. Thus we have few algebraic tools with which to work problems. As the number of tools in our algebraic "toolbox" increases, so will the difficulty of the problems we can solve. For now our problems may seem simple; however, to gain practice with the problem-solving process, you should try to use all five steps. Later some steps may be shortened or combined.

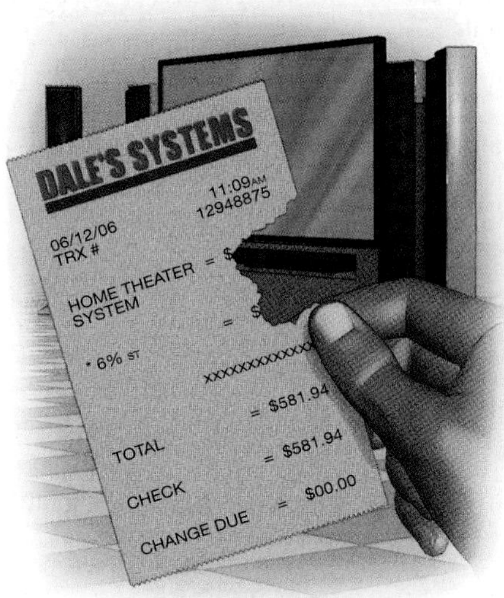

EXAMPLE 4 Purchasing. Maya pays $581.94 for a home theater system. If the price paid includes a 6% sales tax, what is the price of the system itself?

SOLUTION

1. **Familiarize.** First, we familiarize ourselves with the problem. Note that tax is calculated from, and then added to, the item's price. Let's guess that the home theater system's price is $500. To check the guess, we calculate the amount of tax, (0.06)($500) = $30, and add it to $500:

$$\$500 + (0.06)(\$500) = \$500 + \$30$$
$$= \$530. \qquad \mathbf{\$530 \neq \$581.94}$$

Our guess was too low, but the manner in which we checked the guess will guide us in the next step. We let

h = the home theater system's price, in dollars.

2. **Translate.** Our guess leads us to the following translation:

Rewording:	The system's price	plus	6% sales tax	is	the price with sales tax.
	↓	↓	↓	↓	↓
Translating:	h	$+$	$(0.06)h$	$=$	$\$581.94$

3. **Carry out.** Next, we carry out some mathematical manipulation:

$$h + (0.06)h = 581.94$$
$$1.06h = 581.94 \qquad \textbf{Combining like terms}$$
$$\frac{1}{1.06} \cdot 1.06h = \frac{1}{1.06} \cdot 581.94 \qquad \textbf{Using the multiplication principle}$$
$$h = 549.$$

4. **Check.** To check the answer in the original problem, note that the tax on a home theater system costing \$549 would be (0.06)(\$549) = \$32.94. When this is added to \$549, we have

$$\$549 + \$32.94, \quad \text{or} \quad \$581.94.$$

Thus, \$549 checks in the original problem.

5. **State.** We clearly state the answer: The home theater system itself costs \$549.

EXAMPLE 5 Home Maintenance. In an effort to make their home more energy-efficient, Alma and Drew purchased 200 in. of 3M Press-In-Place™ window glazing. This will be just enough to outline their two square skylights. If the length of the sides of the larger skylight is $1\frac{1}{2}$ times the length of the sides of the smaller one, how should the glazing be cut?

SOLUTION

1. **Familiarize.** Note that the *perimeter* of (distance around) each square is four times the length of a side. Furthermore, if s represents the length of a side of the smaller square, then $\left(1\frac{1}{2}\right)s$ represents the length of a side of the larger square. We make a drawing and note that the two perimeters must add up to 200 in.

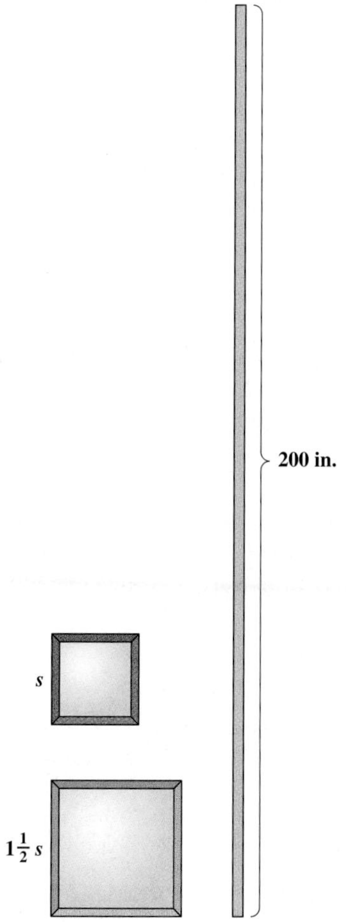

Perimeter of a square = 4 · *length of a side*

2. **Translate.** Rewording the problem can help us translate:

Rewording:	The perimeter of one square	plus	the perimeter of the other	is	200 in.
	↓	↓	↓	↓	↓
Translating:	$4s$	$+$	$4\left(1\frac{1}{2}s\right)$	$=$	200

3. **Carry out.** We solve the equation:

$$4s + 4\left(1\tfrac{1}{2}s\right) = 200$$
$$4s + 6s = 200 \qquad \textbf{Simplifying}$$
$$10s = 200 \qquad \textbf{Combining like terms}$$
$$s = \frac{1}{10} \cdot 200 \qquad \textbf{Multiplying both sides by } \tfrac{1}{10}$$
$$s = 20. \qquad \textbf{Simplifying}$$

4. **Check.** If 20 is the length of the smaller side, then $\left(1\frac{1}{2}\right)(20) = 30$ is the length of the larger side. The two perimeters would then be

$$4 \cdot 20 \text{ in.} = 80 \text{ in.} \quad \text{and} \quad 4 \cdot 30 \text{ in.} = 120 \text{ in.}$$

Since 80 in. + 120 in. = 200 in., our answer checks.

5. **State.** The glazing should be cut into two pieces, one 80 in. long and the other 120 in. long.

We cannot stress too much the importance of labeling the variables in your problem. In Example 5, solving for s is not enough: We need to find $4s$ and $4\left(1\frac{1}{2}s\right)$ to determine the numbers we are after.

Scientific notation can be useful in problem solving.

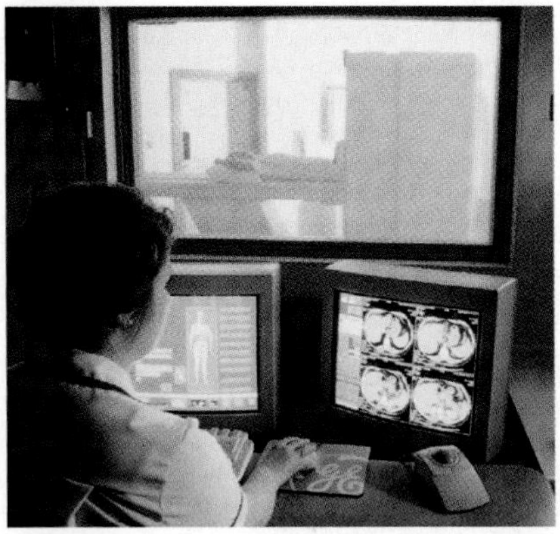

EXAMPLE 6 Information Technology. In 2003, the University of California, Berkeley, estimated that in the previous year, approximately 5 exabytes of new information were generated by the worldwide population of 6.3 billion people. If 1 exabyte is 10^{12} megabytes, find the average number of megabytes of information generated per person in 2002.

SOLUTION

1. **Familiarize.** If necessary, we can consult an outside reference to confirm that one billion is 1,000,000,000, or 10^9. Thus, 6.3 billion is 6.3×10^9. Note also that to find an average we need to divide. We let a = the average number of megabytes of information generated, per person, in 2002.

2. **Translate.** To find the average amount of information generated per person, we divide the total amount of information generated by the number representing the world population:

$$a = \frac{5.0 \times 10^{12} \text{ megabytes}}{6.3 \times 10^9 \text{ people}}.$$

3. **Carry out.** We calculate and write scientific notation for the result:

$$\begin{aligned} a &= \frac{5.0 \times 10^{12} \text{ megabytes}}{6.3 \times 10^9 \text{ people}} \\ &= \frac{5.0}{6.3} \times \frac{10^{12} \text{ megabytes}}{10^9 \text{ people}} \\ &\approx 0.79 \times 10^3 \text{ megabytes/person} && \textbf{Rounding to 2 significant digits} \\ &\approx 7.9 \times 10^{-1} \times 10^3 \text{ megabytes/person} && \textbf{Writing scientific} \\ &\approx 7.9 \times 10^2 \text{ megabytes/person.} && \textbf{notation} \end{aligned}$$

4. **Check.** To check, we multiply our answer, the average number of megabytes per person, by the worldwide population:

$$\underbrace{(7.9 \times 10^2 \text{ megabytes per person})}_{\text{Average amount of information generated}}\underbrace{(6.3 \times 10^9 \text{ people})}_{\text{Worldwide population}}$$

$$\begin{aligned} &= (7.9 \times 6.3)(10^2 \times 10^9)\frac{\text{megabytes}}{\text{person}} \cdot \text{people} \\ &= 49.77 \times 10^{11} \text{ megabytes} \\ &\approx 5.0 \times 10^1 \times 10^{11} \text{ megabytes} && \textbf{Rounding to 2 significant digits} \\ &\approx 5.0 \times 10^{12} \text{ megabytes.} \end{aligned}$$

Our answer checks.

5. **State.** An average of 7.9×10^2 megabytes of information was generated by each person in the world in 2002.

Mathematical Models

When we translate a problem into mathematical language, we say that we *model* the problem. A **mathematical model** is a representation, using mathematics, of a real-world situation. In problem solving, a mathematical model is formed in the *Translate* step.

A model can be an equation, as illustrated in the earlier examples of this section. A formula can also be a model of a situation. Equations and formulas are examples of *algebraic* models.

A graph is another kind of mathematical model. Graphs can represent information in a concise way, and can help to visualize situations and show relationships between two quantities. We can graph algebraic equations and read information from the graph. We can also graph data and use the graph to answer questions without actually forming an algebraic model.

Coordinates and Points

Coordinates of ordered pairs are entered as *data* in lists, using the STAT menu. To enter or change data, press STAT and then choose the EDIT option. The lists of numbers will appear as three columns on the screen. If there are already numbers in the lists, clear them by moving the cursor to the title of the list (L1, L2, and so on) and pressing CLEAR ENTER.

To enter a number in a list, move the cursor to the correct position, type in the number, and press ENTER. Enter the first coordinates of the ordered pairs as one list and the second coordinates as another list. The coordinates of each point should be at the same position on both lists. Note that a DIM MISMATCH error will occur if there are not the same number of items in each list.

To plot the points, first turn on the STAT PLOT feature. Press STAT PLOT. (STAT PLOT is the 2nd feature associated with the Y= key.) The calculator allows several different sets of points to be displayed at once. See the screen on the left below. Choose the Plot you wish to define; if you are simply plotting one set of points, choose Plot1 by pressing 1. Then turn Plot1 on by positioning the cursor over On and pressing ENTER, as shown on the right below.

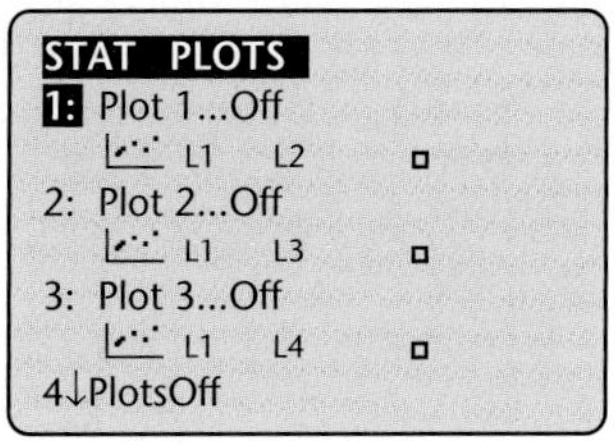

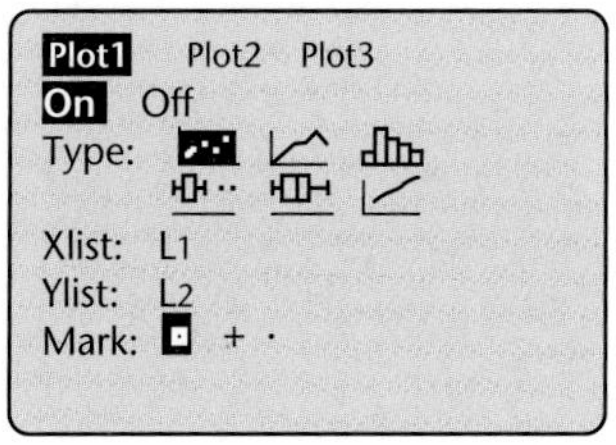

The remaining items on the screen define the plot. Use the down arrow key to move to the next item.

(continued)

The points entered can be used to make a line graph or a bar graph as well as to graph points. The screen on the right on the preceding page shows six available types of graphs. To plot points, choose the first type of graph shown, a scatter diagram or scatterplot. The second option in the list is a line graph, in which the points are connected. The third type is a bar graph. The last three types will not be discussed in this course.

The next item on the screen, Xlist, should be the list in which the first coordinates were entered, probably L1, and Ylist the list in which the second coordinates were entered, probably L2. List names can be selected by pressing (LIST). (LIST is the 2nd option associated with the (STAT) key.)

The last choice on the screen is the type of mark used to plot the points. Different marks can be used to distinguish among several sets of data.

When the STAT PLOT feature has been set correctly, choose window dimensions that will allow all the points to be seen and press (GRAPH). The ZoomStat option of the ZOOM menu will choose an appropriate window automatically. After graphing, pressing (TRACE) displays the coordinates of the point indicated by the cursor. The left and right arrow keys move the cursor along a graph or from point to point.

When you no longer wish to plot a set of data, turn off STAT PLOT by pressing (STAT PLOT), choosing the appropriate Plot, and highlighting Off. To turn off all the plots, press (4) to choose the PlotsOff option and press (ENTER).

CAUTION! The graphing calculator will attempt to graph any equations selected in the Y= screen and any points described by an active Plot. Be sure to clear or deselect any unwanted equations and turn off any unwanted Plots before graphing.

EXAMPLE 7 Airline Consumer Complaints. The following table lists the number of complaints filed by consumers against airlines between 1995 and 2004 (*Source*: U.S. Department of Transportation, Aviation Consumer Protection Division). Use the data to draw a line graph.

Year	Number of Complaints
1995	4,629
1996	5,782
1997	6,394
1998	7,980
1999	17,345
2000	20,564
2001	14,076
2002	7,697
2003	4,601
2004	5,863

SOLUTION We enter the years in list L1 and the number of complaints in L2, as shown in the figure on the left below. Pressing the down arrow will show the rest of the list. We turn the Plot feature On and choose the second type of graph, the line graph. The Xlist should be L1 and the Ylist should be L2. After clearing or deselecting any equations present in the equation-editor screen, we choose the ZoomStat option from the ZOOM menu to give the graph shown on the right below. Note that there are no axes shown on the screen, since the window dimensions chosen do not include 0. Only a portion of the first quadrant is shown.

L1	L2	L3 1
1995	4629	------
1996	5782	
1997	6394	
1998	7980	
1999	17345	
2000	20564	
2001	14076	

L1(1) = 1995

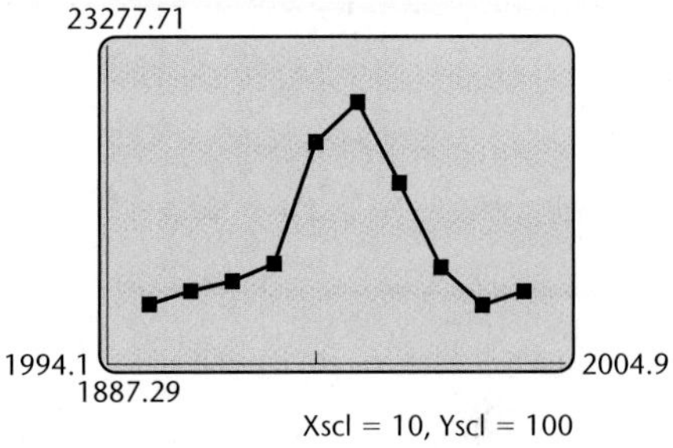

With no axes shown, it is difficult to read a value visually from the graph. We could press TRACE and the left and right arrow keys to show the coordinates of the points. We could also enter and graph the data differently. Instead of using the actual year, we will let x represent the number of years after 1995. This type of representation is often done to make the numbers with which we are working smaller. This change is shown in the lists on the left below. Instead of using ZoomStat, we will choose window dimensions of $[-1, 10, 0, 25000]$, with Yscl $= 5000$. The graph is shown on the right below. Note that the axes are shown in this graph.

L1	L2	L3 1
0	4629	------
1	5782	
2	6394	
3	7980	
4	17345	
5	20564	
6	14076	

L1(1) = 0

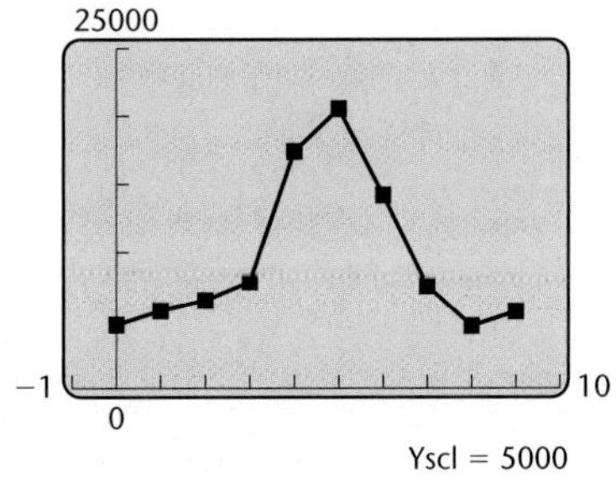

There is often more than one way to approach a problem. Some problems can be approached both *algebraically* and *graphically*. There are other valid approaches as well. *Verbalizing* a problem can be surprisingly helpful in solving it. Many problems can also be approached *numerically*. We will often show more than one valid approach when solving problems in this text.

EXAMPLE 8 Three numbers are such that the second is 6 less than three times the first and the third is 2 more than two-thirds the first. The sum of the three numbers is 150. Find the largest of the three numbers.

SOLUTION We proceed according to the five-step process.

1. **Familiarize.** A *verbal* approach to the problem might include reading the problem aloud, rephrasing each statement, and discussing the problem with another student. From such an approach, we can discover that the second and third numbers are described in terms of the first.

   We can then approach the problem *numerically* by forming a table of some possible numbers and checking their sums.

First Number	Second Number	Third Number	Sum
30	$3 \cdot 30 - 6 = 84$	$\frac{2}{3} \cdot 30 + 2 = 22$	136
45	$3 \cdot 45 - 6 = 129$	$\frac{2}{3} \cdot 45 + 2 = 32$	206

   From this table, we see that the number we are looking for will be between 30 and 45. We also note a pattern that will enable us to describe the problem *algebraically*. If we let x represent the first number, then $3x - 6$ represents the second number, and $\frac{2}{3}x + 2$ represents the third number. The sum of these numbers must be 150.

2. **Translate.** We reword the problem and translate to an equation.

   First number plus second number plus third number is 150.

   $$x \quad + \quad (3x - 6) \quad + \quad \left(\tfrac{2}{3}x + 2\right) \quad = \quad 150$$

3. **Carry out.** We can solve the equation, or we can graph the equation and estimate an answer from the graph.

ALGEBRAIC APPROACH

We solve the equation:

$$x + 3x - 6 + \tfrac{2}{3}x + 2 = 150 \qquad \text{Leaving off unnecessary parentheses}$$
$$\left(4 + \tfrac{2}{3}\right)x - 4 = 150 \qquad \text{Combining like terms}$$
$$\tfrac{14}{3}x - 4 = 150$$
$$\tfrac{14}{3}x = 154 \qquad \text{Adding 4 to both sides}$$
$$x = \tfrac{3}{14} \cdot 154 \qquad \text{Multiplying both sides by } \tfrac{3}{14}$$
$$x = 33.$$

Since x represents the first number, we have:

First: 33;

Second: $3x - 6 = 3 \cdot 33 - 6 = 93$;

Third: $\frac{2}{3}x + 2 = \frac{2}{3} \cdot 33 + 2 = 24$.

GRAPHICAL APPROACH

We graph the equation $y = x + 3x - 6 + \frac{2}{3}x + 2$. We want to find the x-value for which $y = 150$, so we choose a viewing window with Ymax > 150. We know from the *Familiarize* step that x is between 30 and 45. An appropriate window is thus [0, 50, 0, 200], with Xscl = 5 and Yscl = 50. We locate 150 on the vertical axis, move across to the graph, and move down to the x-axis to estimate a value of 33.

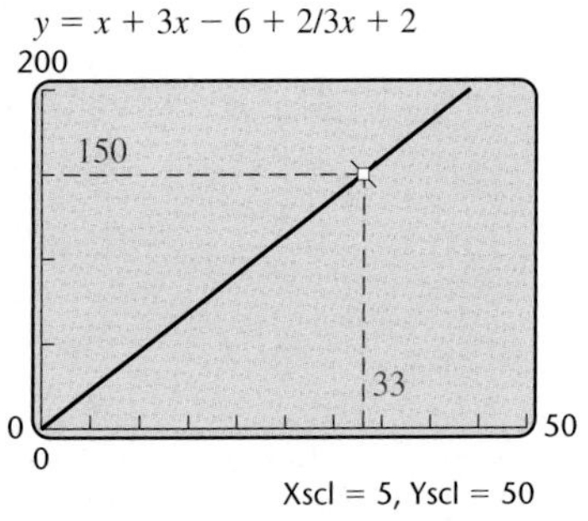

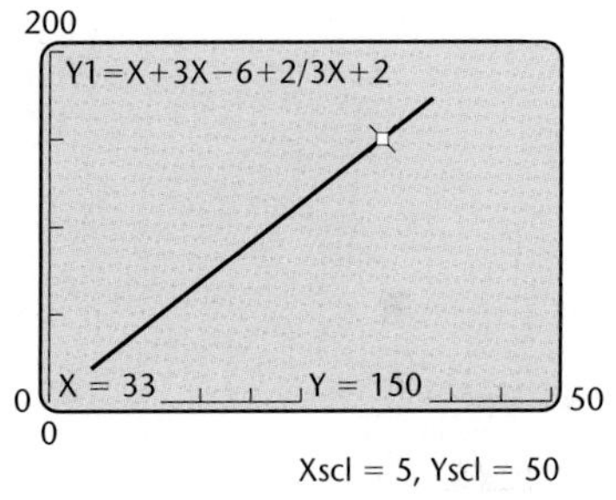

By using the Value option of the CALC menu and letting X = 33, we can verify that Y = 150 when X = 33.

Since X represents the first number, we have

First: 33;

Second: $3x - 6 = 3 \cdot 33 - 6 = 93$;

Third: $\frac{2}{3}x + 2 = \frac{2}{3} \cdot 33 + 2 = 24$.

4. **Check.** We return to the original problem. There are three numbers: 33, 93, and 24. Is the second number 6 less than three times the first?

$$3 \times 33 - 6 = 99 - 6 = 93$$

The answer is *yes*.

Is the third number 2 more than two-thirds the first?

$$\frac{2}{3} \times 33 + 2 = 22 + 2 = 24$$

The answer is *yes*.

Is the sum of the three numbers 150?

$$33 + 93 + 24 = 150$$

The answer is *yes*. The numbers do check.

5. **State.** The problem asks us to find the largest number, so the answer is: "The largest of the three numbers is 93."

Student Notes

Always read the problem again before writing your final answer. In Example 8, the answer is only one number, not three.

CAUTION! In Example 8, although the equation $x = 33$ enables us to find the largest number, 93, the number 33 is *not* the solution of the problem. By clearly labeling our variable in the first step, we can avoid thinking that the variable always represents the solution of the problem.

Connecting the Concepts

INTERPRETING GRAPHS: READING GRAPHS

When a graph is used to model a problem, information can be read directly from the graph. Listed here are a few kinds of information that a graph contains.

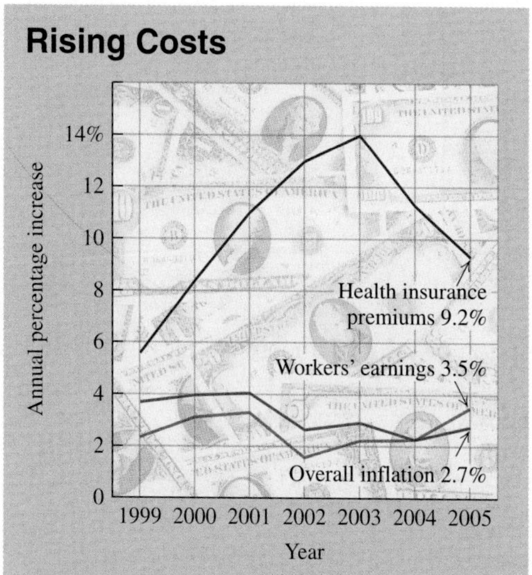

Source: Kaiser Family Foundation/Health Research and Educational Trust

1. *Labels and units.* When reading a graph, look first at any labels. The title of the graph and the labels on the axes tell what information the graph represents. The scales on each axis indicate the numbers represented by a point. If the graph represents an equation, that equation will appear as a label.

   The title of the figure shown here indicates that it represents rising costs. There are three graphs shown in the figure. One represents health insurance premiums, one workers' earnings, and one overall inflation. The horizontal axis is labeled "Year," and the vertical axis is labeled "Annual percentage increase." Thus a second coordinate of 14 represents a 14% increase in cost.

2. *Relationships.* The coordinates of a point on a graph indicate a relationship between the quantities represented on the horizontal and vertical axes. For the graph labeled "Health insurance premiums," the point (2003, 14) indicates that health insurance premiums rose 14% in 2003. Since all three graphs are drawn on the same set of axes, the figure also shows relationships between increases in health insurance premiums, workers' earnings, and overall inflation. We can see that while workers' earnings have approximately kept pace with inflation, health insurance premiums have risen much more quickly.

3. *Trends.* As the graph continues from left to right, the quantity represented on the horizontal axis increases. The quantity represented on the vertical axis may increase, decrease, or remain constant. If the graph rises from left to right, the quantity on the vertical axis is increasing; if the graph falls from left to right, the quantity is decreasing; if the graph is level, the quantity is not changing, so it is constant.

   The health insurance premiums graph indicates that premiums increased at a higher rate each year between 1999 and 2003. For these years, the percentage increase itself increased. Between 2003 and 2005, the percentage increase decreased. Note that this does not mean that the premiums themselves decreased; it means instead that they increased at a slower rate.

4. *Values.* If we know one coordinate of a point on a graph, we can find the other coordinate. To find the percentage increase in overall inflation in 2001, we locate 2001 on the horizontal axis and move up to the graph labeled "Overall inflation." We then move horizontally to the corresponding value on the vertical axis. We see that in 2001, the overall inflation rate was 4%.

5. *Maximums and minimums.* A "high" point or a "low" point on a graph indicates a maximum or minimum value for the quantity on the vertical axis. The highest point on the health insurance premiums graph is approximately (2003, 14). Thus the maximum percentage increase for health insurance premiums was 14% in 2003.

Visualizing the Graph

Match each phrase with the graph that it describes.

1. Increasing

2. Decreasing

3. Constant

4. Maximum y-value of 4

5. Minimum y-value of -4

6. $\{(-2,-2), (-1,-1), (0,0), (1,1), (2,2), (3,1)\}$

7. $\{(2,-3), (1,-4), (0,0), (-2,-2), (-4,-5)\}$

8. First increases, then is constant, then decreases

9. First increases, then is constant, then increases

10. First decreases, then increases, then is constant

Answers on page A-3

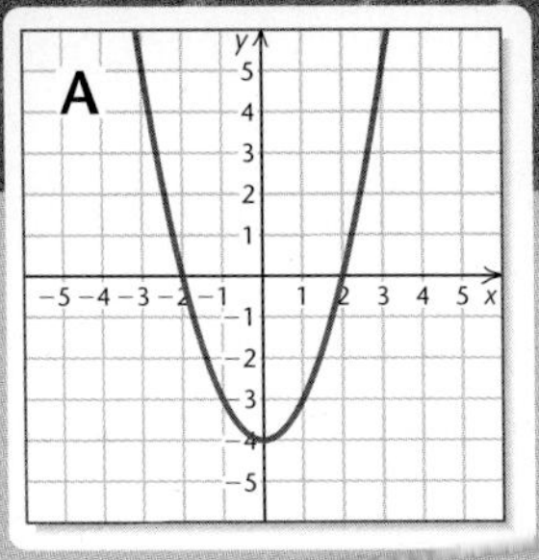

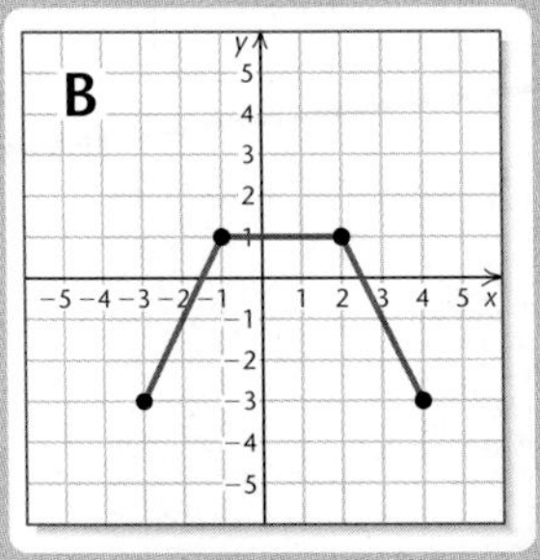

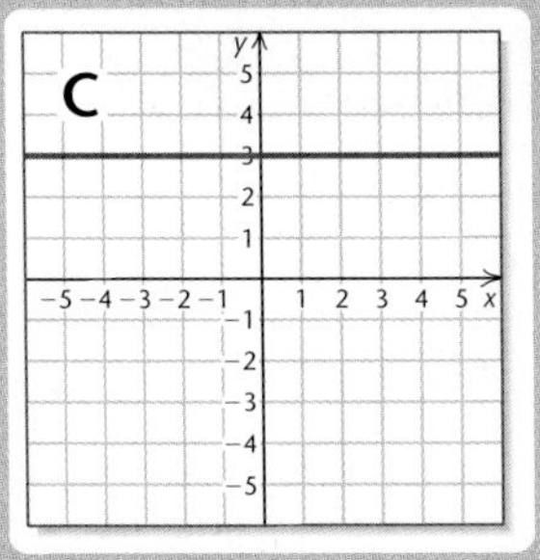

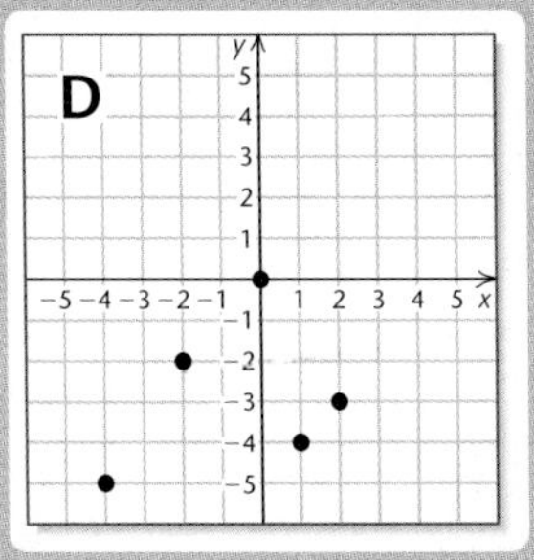

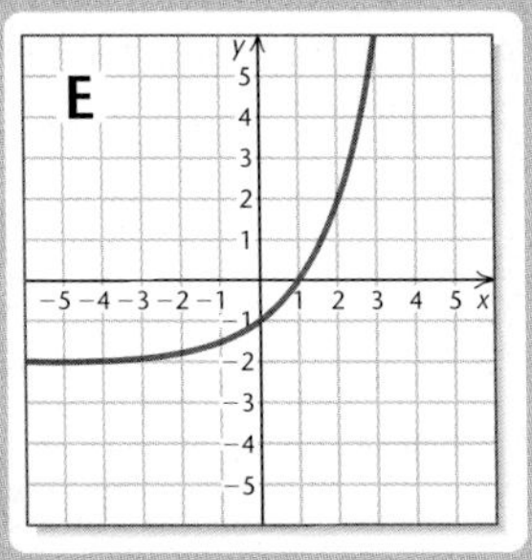

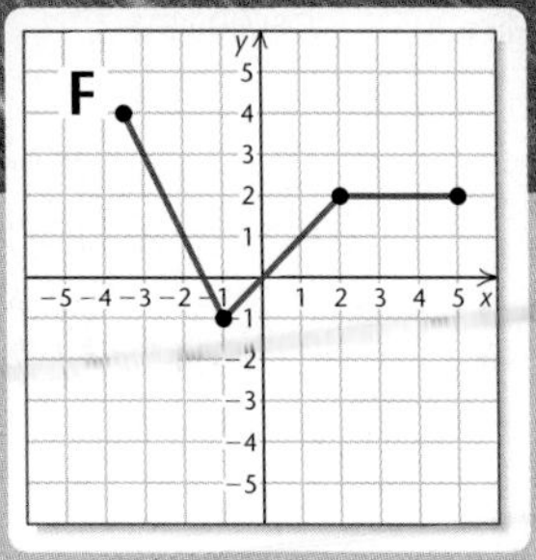

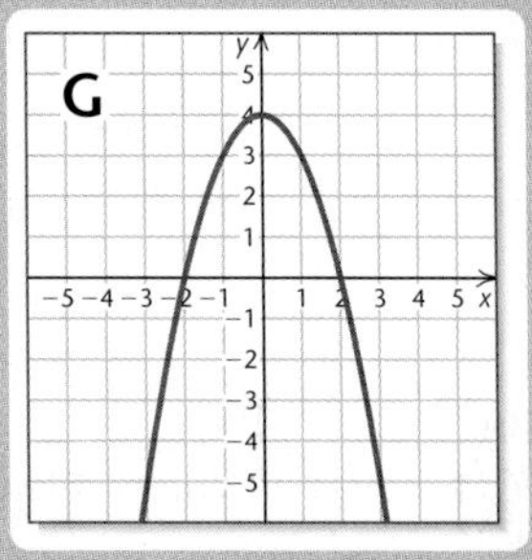

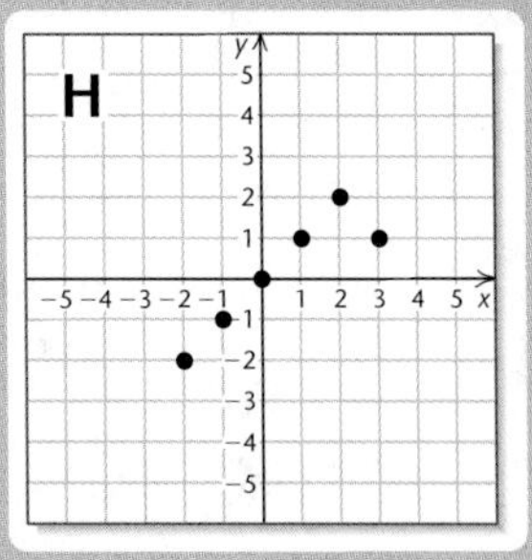

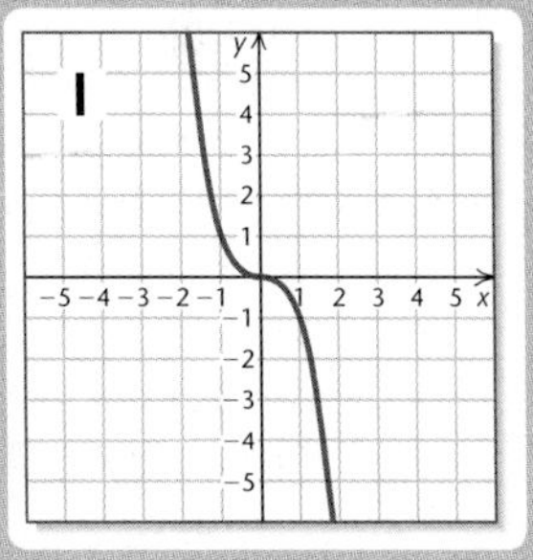

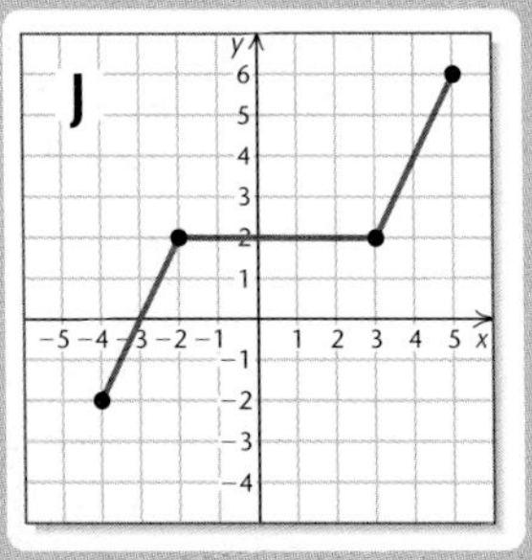

1.7 EXERCISE SET

FOR EXTRA HELP

MathXL | MyMathLab | InterAct Math | AW Math Tutor Center | Video Lectures on CD: Disc 1 | Student's Solutions Manual

Use mathematical symbols to translate each phrase.

1. Six less than some number

2. Four more than some number

3. Twelve times a number

4. Twice a number

5. Sixty-five percent of some number

6. Thirty-nine percent of some number

7. Nine more than twice a number

8. Six less than half of a number

9. Eight more than ten percent of some number

10. Five less than six percent of some number

11. One less than the difference of two numbers

12. One more than the product of two numbers

13. Ninety miles per every four gallons of gas

14. One hundred words per every sixty seconds

For each problem, familiarize yourself with the situation. Then translate to mathematical language. You need not actually solve the problem; just carry out the first two steps of the five-step strategy. You will be asked to complete some of the solutions as Exercises 45–52.

15. The sum of two numbers is 65. One of the numbers is 7 more than the other. What are the numbers?

16. The sum of two numbers is 83. One of the numbers is 11 more than the other. What are the numbers?

17. *Swimming.* An Olympic swimmer can swim at a sustained rate of 5 km/h in still water. The Lazy River flows at a rate of 2.3 km/h. How long would it take an Olympic swimmer to swim 1.8 km upstream?

18. *Boating.* The *Delta Queen* paddleboat tours the Mississippi River near New Orleans, Louisiana. It is not uncommon for the *Delta Queen* to run 7 mph in still water and for the Mississippi to flow at a rate of 3 mph (*Source*: *Delta Queen* information). At these rates, how long will it take the boat to cruise 2 mi upstream?

19. *Moving Sidewalks.* The moving sidewalk in O'Hare Airport is 300 ft long and moves at a rate of 5 ft/sec. If Alida walks at a rate of 4 ft/sec, how long will it take her to walk the length of the moving sidewalk?

20. *Aviation.* A Cessna airplane traveling 390 km/h in still air encounters a 65-km/h headwind. How long will it take the plane to travel 725 km into the wind?

21. *Angles in a Triangle.* The degree measures of the angles in a triangle are three consecutive integers. Find the measures of the angles.

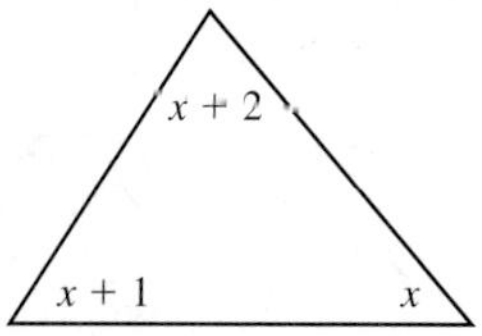

22. *Pricing.* Becker Lumber gives contractors a 10% discount on all orders. After the discount, a contractor's order cost \$279. What was the original cost of the order?

23. *Pricing.* The Sound Connection prices packages of CDRs by raising the wholesale price 50% and adding \$1.50. What must a package's wholesale price be if it is being sold for \$22.50?

24. *Pricing.* Miller Oil offers a 5% discount to customers who pay promptly for an oil delivery. The Blancos promptly paid \$142.50 for their December oil bill. What would the cost have been had they not promptly paid?

25. *Cruising Altitude.* A Boeing 747 has been instructed to climb from its present altitude of 8000 ft to a cruising altitude of 29,000 ft. If the plane ascends at a rate of 3500 ft/min, how long will it take to reach the cruising altitude?

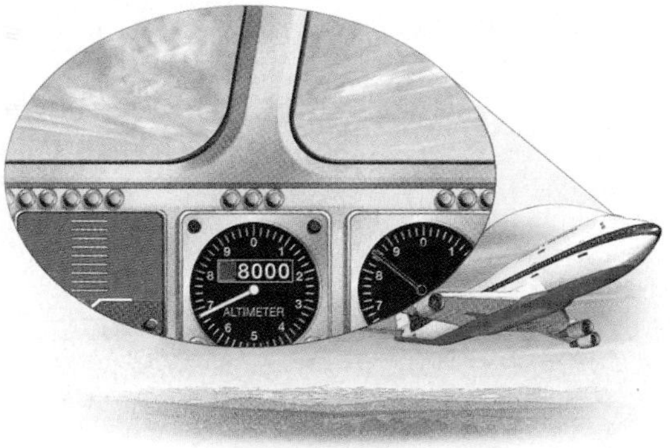

26. A piece of wire 10 m long is to be cut into two pieces, one of them $\frac{2}{3}$ as long as the other. How should the wire be cut?

27. *Angles in a Triangle.* One angle of a triangle is three times as great as a second angle. The third angle measures 12° less than twice the second angle. Find the measures of the angles.

28. *Angles in a Triangle.* One angle of a triangle is four times as great as a second angle. The third angle measures 5° more than twice the second angle. Find the measures of the angles.

29. Find two consecutive even integers such that two times the first plus three times the second is 76.

30. Find three consecutive odd integers such that the sum of the first, twice the second, and three times the third is 70.

31. A steel rod 90 cm long is to be cut into two pieces, each to be bent to make an equilateral triangle. The length of a side of one triangle is to be twice the length of a side of the other. How should the rod be cut?

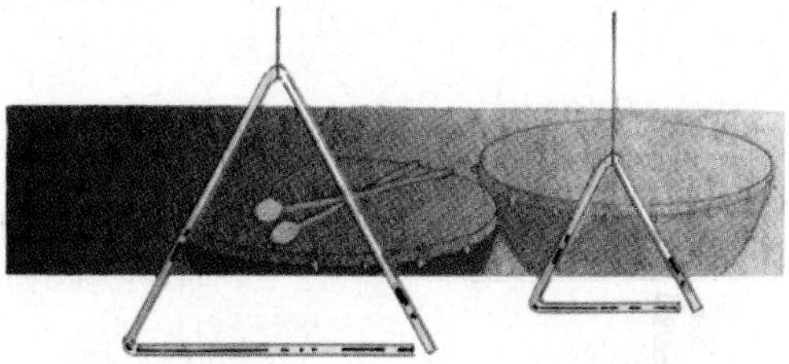

32. A piece of wire 100 cm long is to be cut into two pieces, and those pieces are each to be bent to make a square. The area of one square is to be 144 cm^2 greater than that of the other. How should the wire be cut? (*Remember*: Do not solve.)

33. *Rescue Calls.* Rescue crews working for Stockton Rescue average 3 calls per shift. After his first four shifts, Brian had received 5, 2, 1, and 3 calls. How many calls will Brian need on his next shift if he is to average 3 calls per shift?

34. *Test Scores.* Deirdre's scores on five tests are 93, 89, 72, 80, and 96. What must the score be on her next test so that the average will be 88?

Solve each problem. Use all five problem-solving steps.

35. *Pricing.* The price that Ruth paid for her graphing calculator, \$84, is less than what Tony paid by \$13. How much did Tony pay for his graphing calculator?

36. *Class Size.* The number of students in James' class, 35, is greater than the number in Rose's class by 12. How many students are in Rose's class?

37. *Public Health.* In 2050, the number of diagnosed cases of diabetes in the United States is predicted to reach 29 million. This is approximately $\frac{13}{5}$ of the number of cases in 2000. (*Source*: U.S. Centers for Disease Control) How many cases had been diagnosed in 2000?

38. *Home Improvement.* In 2003, painting the interior of a house increased the average sale price of the home by about \$2500. This was about $\frac{5}{3}$ of what the average paint job cost. (*Source*: Based on information from HomeGain.com) What was the cost of the average paint job?

39. Officer Reid wrote up 9 more tickets than Officer Schultz did. Together they wrote up 35 tickets. How many tickets did Officer Reid write?

40. On an evening shift, Lily waited on 7 fewer tables than Keith did. Together they waited on 25 tables. How many tables did Lily wait on?

41. The length of a rectangular mirror is three times its width, and its perimeter is 120 cm. Find the length and the width of the mirror.

42. The length of a rectangular tile is twice its width, and its perimeter is 21 cm. Find the length and the width of the tile.

43. The width of a rectangular greenhouse is one-fourth its length, and its perimeter is 130 m. Find the length and the width of the greenhouse.

44. The width of a rectangular garden is one-third its length, and its perimeter is 32 m. Find the dimensions of the garden.

45. Solve the problem of Exercise 17.

46. Solve the problem of Exercise 18.

47. Solve the problem of Exercise 28.

48. Solve the problem of Exercise 27.

49. Solve the problem of Exercise 24.

50. Solve the problem of Exercise 22.

51. Solve the problem of Exercise 23.

52. Solve the problem of Exercise 29.

Solve.

53. *Radioactivity.* The lightest known particle in the universe, a neutrino has a maximum mass of 1.8×10^{-36} kg. What is the smallest number of neutrinos that could have the same mass as an alpha particle of mass 3.62×10^{-27} kg that results from the decay of radon? (*Source*: *Guinness Book of World Records* 2004)

54. *Printing and Engraving.* A ton of five-dollar bills is worth \$4,540,000. How many pounds does a five-dollar bill weigh?

55. *Telecommunications.* A wire will be used for 375 km of transmission line. The wire has a diameter of 1.2 cm. What is the volume of wire needed for the line?

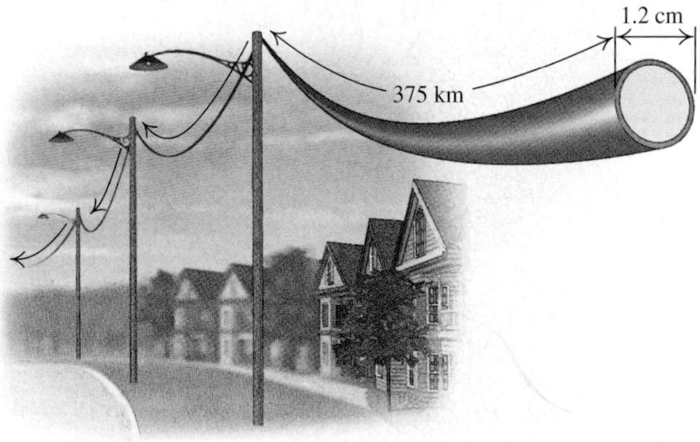

56. *High-Tech Fibers.* A carbon nanotube is a thin cylinder of carbon atoms that, pound for pound, is stronger than steel and may one day be used in clothing (*Source*: *The Indianapolis Star*, 6/15/03). With a diameter of about 4.0×10^{-10} in., a fiber can be made 100 yd long. Find the volume of such a fiber.

57. *Biology.* An average of 4.55×10^{11} bacteria live in each pound of U.S. mud. There are 60.0 drops in one teaspoon and 6.0 teaspoons in an ounce. (*Source*: *Harper's Magazine*, April 1996, p. 13) How many bacteria live in a drop of U.S. mud?

58. *Astronomy.* If a star 5.9×10^{14} mi from the earth were to explode today, its light would not reach us for 100 yr. How far does light travel in 13 weeks?

59. *Home Maintenance.* The thickness of a sheet of plastic is measured in *mils,* where 1 mil $= \frac{1}{1000}$ in. To help conserve heat, the foundation of a 24-ft by 32-ft rectangular home is covered with a 4-ft high sheet of 8-mil plastic. Find the volume of plastic used.

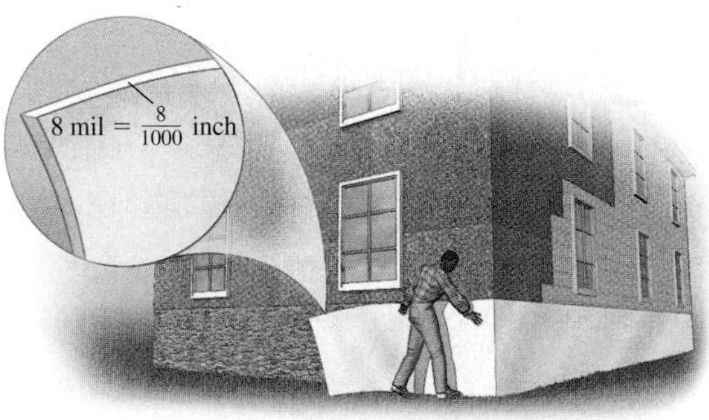

60. *Office Supplies.* A ream of copier paper weighs 2.25 kg. How much does a sheet of copier paper weigh?

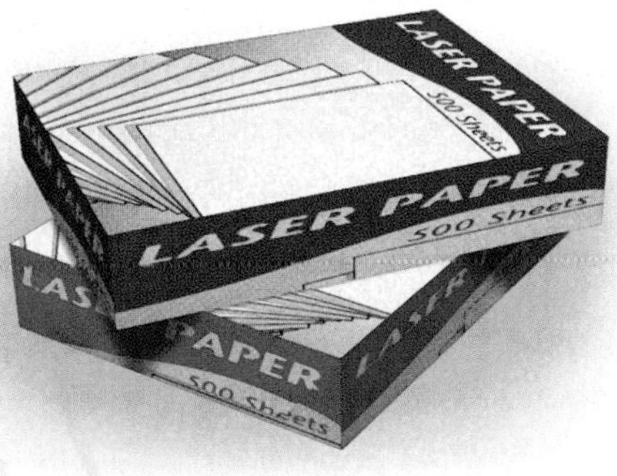

*Heart Attacks and Cholesterol. For Exercises 61 and 62, use the following graph, which shows the annual heart attack rate per 10,000 men on the basis of blood cholesterol level.**

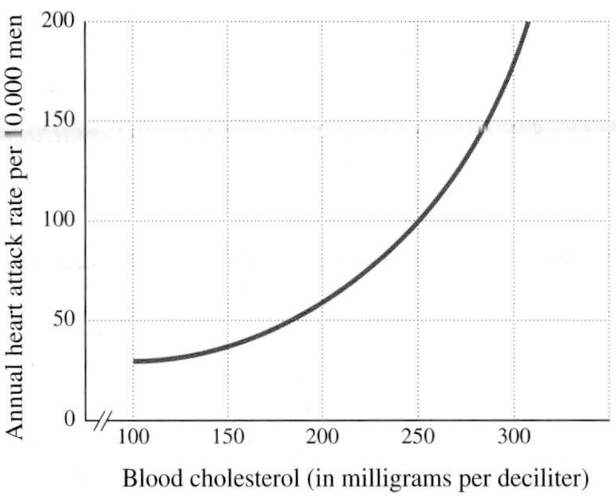

61. Approximate the annual heart attack rate for those men whose blood cholesterol level is 225 mg/dl.

62. Approximate the annual heart attack rate for those men whose blood cholesterol level is 275 mg/dl.

Gasoline Prices. For Exercises 63–66, use the following graph, which shows the average U.S. gasoline prices for regular unleaded gasoline.

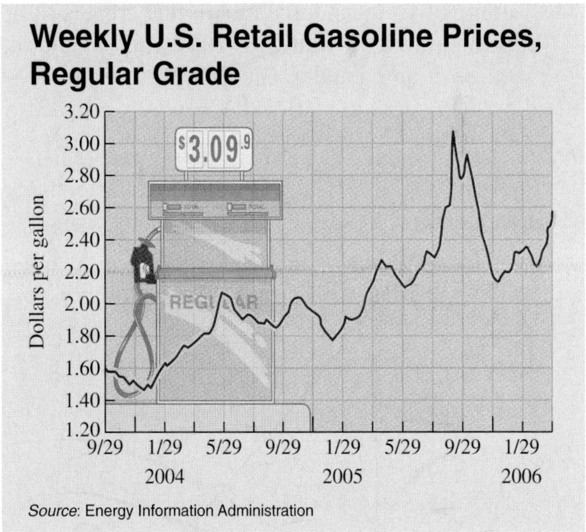

63. Approximate the price of regular unleaded gasoline on March 29, 2005.

64. Approximate the price of regular unleaded gasoline on March 29, 2006.

65. When was regular unleaded gasoline at its maximum price?

66. When was regular unleaded gasoline at its minimum price?

*Copyright 1989, CSPI. Adapted from *Nutrition Action Healthletter* (1875 Connecticut Avenue, N.W., Suite 300, Washington, DC 20009-5728. $24 for 10 issues).

Energy-Saving Lightbulbs. Compact fluorescent (CFL) lightbulbs can be used in most places that conventional incandescent bulbs are, but they use a fraction of the electricity. The following figure lists the CFL wattage and the incandescent wattage required to create the same amount of light. (*Source*: Westinghouse Lighting Corporation)

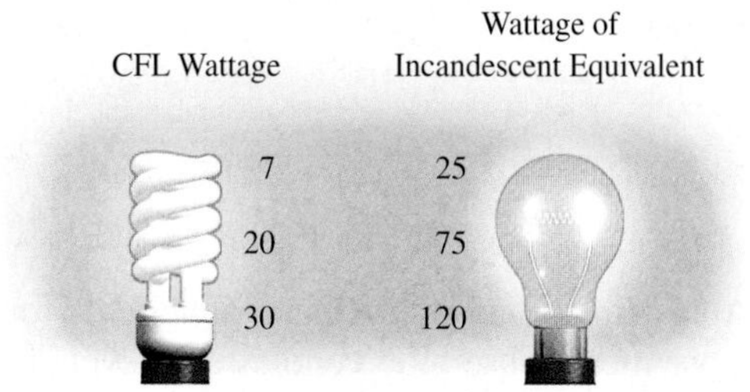

67. Use the data in the figure above to draw a graph and to estimate the wattage of an incandescent bulb that creates light equivalent to a 15-watt CFL bulb. Then predict the wattage of an incandescent bulb that creates light equivalent to a 35-watt CFL bulb.

68. Use the graph from Exercise 67 to estimate the wattage of an incandescent bulb that creates light equivalent to a 26-watt CFL bulb. Then predict the wattage of an incandescent bulb that creates light equivalent to a 40-watt CFL bulb.

69. *Endangered Species.* The following table lists the numbers of U.S. species of mammals considered endangered species during various years (*Source*: U.S. Fish and Wildlife Service). Use the data to draw a line graph with a graphing calculator.

Year	Number of Species of Endangered Mammals
1980	32
1985	43
1990	53
1993	56
1997	57
1999	61
2001	64
2003	65
2005	68

70. *Child Expenditure.* The following table lists annual expenditures on a child in 2004 by families with an income of less than $41,700 (*Source*: Department of Agriculture, Center for Nutrition Policy and Promotion, *Expenditures on Children by Families*, 2004 *Annual Report*). Use the data to draw a line graph with a graphing calculator.

Age of Child	Annual Expenditure
1	$7040
4	7210
7	7250
10	7220
13	8070
16	8000

U.S. Farms. The number of U.S. farms f, in millions, can be approximated by the equation

$$f = -\frac{1}{100}t + 2.3,$$

where t is the number of years after 1990. For example, $t = 0$ corresponds to 1990, $t = 1$ corresponds to 1991, and so on. (*Source*: U.S. Department of Agriculture, National Agricultural Statistics Service)

71. Graph the equation and use the graph to estimate how many farms there were in the United States in 2005.

72. Use the graph from Exercise 71 to approximate in what year there will be 2.0 million farms in the United States.

Watching TV. The average number of hours that an adult watches broadcast TV per year can be approximated by the equation

$$H = -0.7x^3 + 12.1x^2 - 60.5x + 870,$$

where x is the number of years after 2000, and x is between 0 *and* 8 (*Source*: Veronis Suhler Stevenson, New York, NY, *Communications Industry Forecast & Report*).

73. Graph the equation and use the graph to estimate in what year or years adults watched TV for 800 hr per year.

74. Use the graph from Exercise 73 to approximate the average number of hours per year that an adult watched TV in 2005.

Researchers at Yale University have suggested that the following graphs may represent three different aspects of love.*

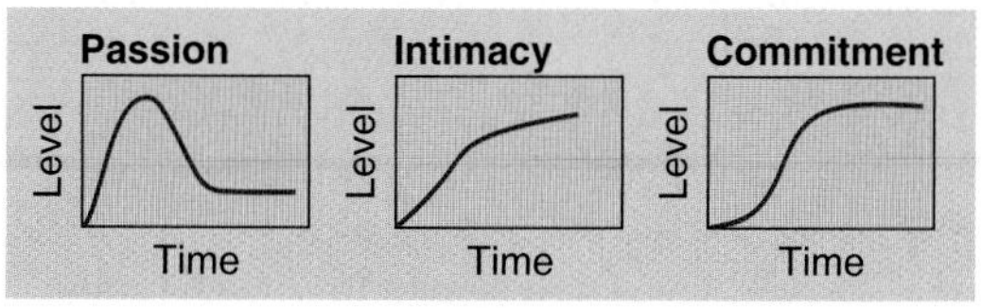

TW **75.** In what unit would you measure time if the horizontal length of each graph were ten units? Why?

TW **76.** Do you agree with the researchers that these graphs should be shaped as they are? Why or why not?

Skill Maintenance

Simplify. [1.4]

77. $(-2)^4$

78. -2^4

79. $z^{-2} \cdot z^6$

80. $\dfrac{z^{-2}}{z^6}$

81. $2(x^2y^3)^4$

82. $(2x^2y^3)^4$

Synthesis

TW **83.** How can a guess or estimate help prepare you for the *Translate* step in problem solving?

TW **84.** Describe at least two benefits of using a graph to model a situation.

Translate to an algebraic expression.

85. The quotient of the sum of two numbers and their difference

86. Three times the sum of the cubes of two numbers

87. Half of the difference of the squares of two numbers

88. The product of the difference of two numbers and their sum

89. *Test Scores.* Tico's scores on four tests are 83, 91, 78, and 81. How many points above his current average must Tico score on the next test in order to raise his average 2 points?

*From "A Triangular Theory of Love," by R. J. Sternberg, 1986, *Psychological Review*, **93**(2), 119–135. Copyright 1986 by the American Psychological Association, Inc. Reprinted by permission.

90. *Geometry.* The height and sides of a triangle are four consecutive integers. The height is the first integer, and the base is the third integer. The perimeter of the triangle is 42 in. Find the area of the triangle.

91. *Home Prices.* Panduski's real estate prices increased 6% from 2001 to 2002 and 2% from 2002 to 2003. From 2003 to 2004, prices dropped 1%. If a house sold for \$117,743 in 2004, what was its worth in 2001? (Round to the nearest dollar.)

92. *Adjusted Wages.* Blanche's salary is reduced $n\%$ during a period of financial difficulty. By what number should her salary be multiplied in order to bring it back to where it was before the reduction?

Cigarette Smoking. Use the following graph for Exercises 93–98. Note that there are two vertical scales. Use the left scale for the graph showing per-capita cigarette consumption and the right scale for the lung-cancer death rates.

Cigarette Use and U.S. Lung Cancer Deaths

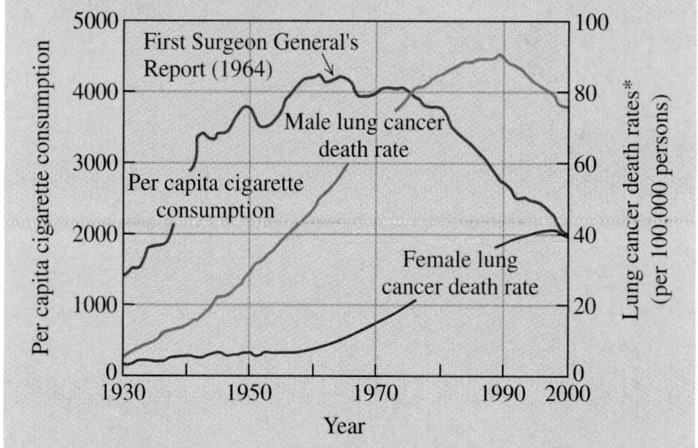

*Age-adjusted to the year 2000 U.S. standard population.

Source: Adapted from *NCI Cancer Bulletin*, Aug. 16, 2005 (www.cancer.gov).

93. What was the per-capita cigarette consumption in 2000?

94. What was the female lung-cancer death rate in 2000?

95. What appears to be the result in per-capita cigarette consumption of the first Surgeon General's Report in 1964?

96. In what year was the maximum lung-cancer death rate for males?

97. If the per-capita consumption rate continues to fall as it has from 1970 to 2000, when will no one be smoking?

TW **98.** Why do you think the lung-cancer death rates continued to climb after the per-capita consumption rates began to decline?

99. Match each sentence with the most appropriate of the four graphs shown below.

a) Carpooling to work, Terry spent 10 min on local streets, then 20 min cruising on the freeway, and then 5 min on local streets to his office.

b) For her commute to work, Sharon drove 10 min to the train station, rode the express for 20 min, and then walked for 5 min to her office.

c) For his commute to school, Roger walked 10 min to the bus stop, rode the express for 20 min, and then walked for 5 min to his class.

d) Coming home from school, Kristy waited 10 min for the school bus, rode the bus for 20 min, and then walked 5 min to her house.

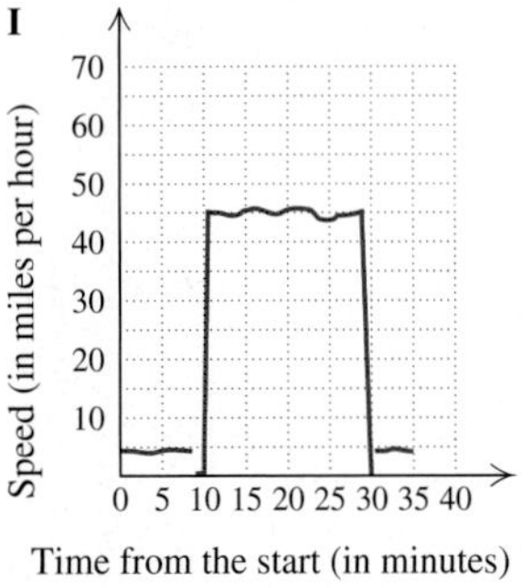

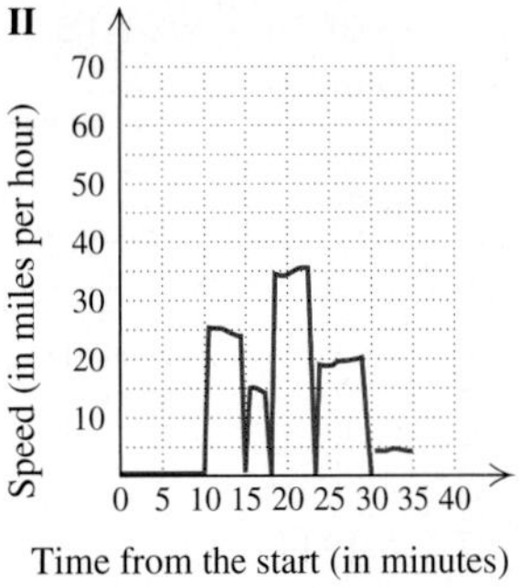

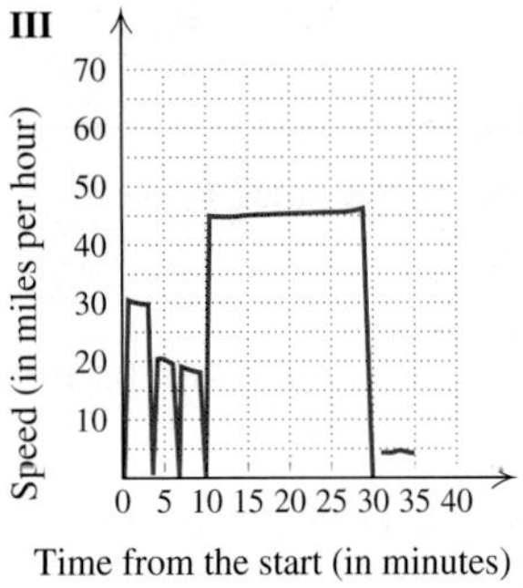

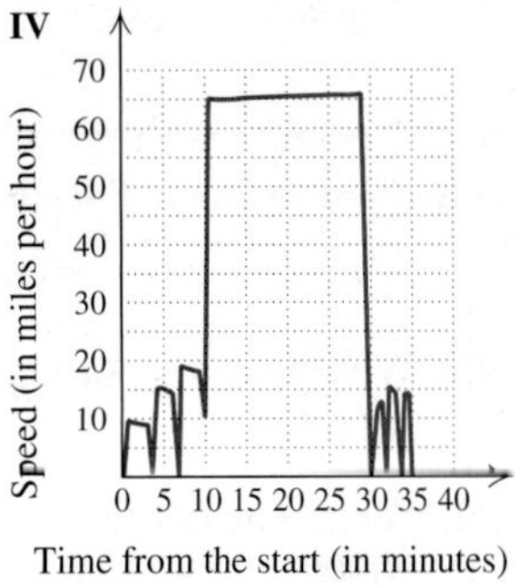

100. Match each sentence with the most appropriate of the four graphs shown.

a) Roberta worked part time until September, full time until December, and overtime until Christmas.

b) Clyde worked full time until September, half time until December, and full time until Christmas.

c) Clarissa worked overtime until September, full time until December, and overtime until Christmas.

d) Doug worked part time until September, half time until December, and full time until Christmas.

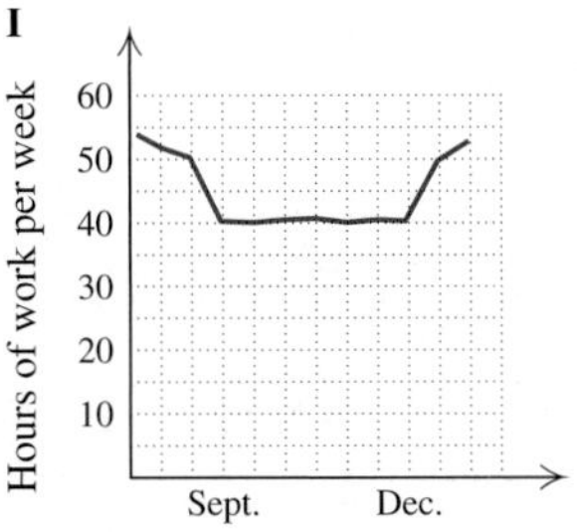

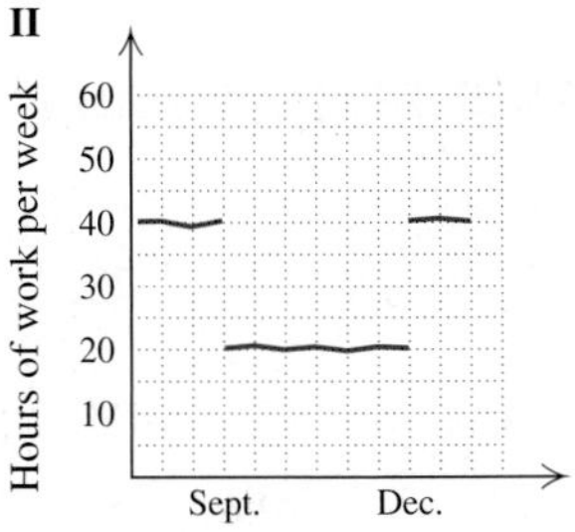

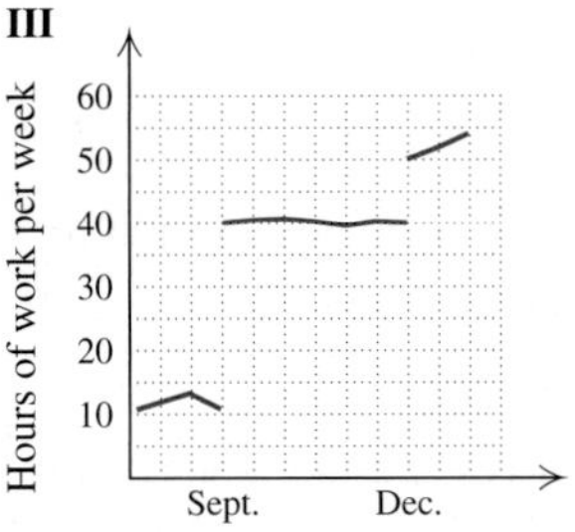

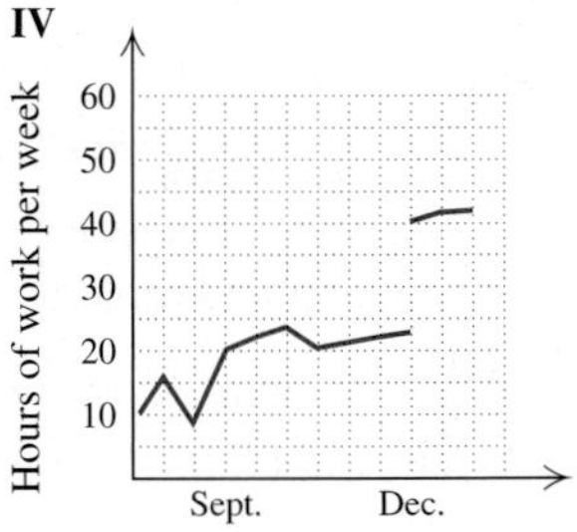

1 Chapter Summary and Review

KEY TERMS AND DEFINITIONS

EXPRESSIONS

Algebraic expression, p. 2 A collection of **variables** and **constants** on which operations are performed.

Factor, pp. 3, 26 As a noun, multipliers in a product. As a verb, to write as a product.

Term, p. 26 A number, a variable, or a product or quotient of numbers and variables. Terms are separated in an expression by + signs.

Like terms, p. 26 Terms with exactly the same variable factors. These can be **combined** or **collected** in an expression.

Equivalent expressions, p. 24 Expressions that have the same value for all allowable replacements.

EQUATIONS AND INEQUALITIES

Equation, p. 2 A sentence formed by placing an = sign between two expressions. Every equation is either an **identity,** a **contradiction,** or a **conditional equation.**

Identity: $x + 1 = x + 1$ True for all replacements

Contradiction: $x + 1 = x + 2$ Never true

Conditional equation: $x + 1 = 2$ True for $x = 1$

Inequality, p. 5 A sentence formed by placing an inequality sign between two expressions.

Equivalent equations, p. 60 Equations with the same solutions.

Solution, pp. 5, 64 A replacement that makes an equation or an inequality true. The set of all solutions of an equation is its **solution set.**

Linear equation, p. 55 An equation whose graph is a straight line. If an equation's graph is not a straight line, it is a **nonlinear equation.**

SETS OF NUMBERS

Natural numbers, p. 6 $\{1, 2, 3, \ldots\}$

Whole numbers, p. 6 $\{0, 1, 2, 3, \ldots\}$

Integers, p. 6 $\{\ldots, -3, -2, -1, 0, 1, 2, 3, \ldots\}$

Rational numbers, p. 7 Numbers that can be expressed as an integer divided by a nonzero integer.

Irrational numbers, p. 8 Real numbers that cannot be expressed as a ratio of integers.

Real numbers, p. 8 The set of all rational and irrational numbers.

INVERSES

Opposites, additive inverses, p. 16 Two numbers whose sum is 0; 2 and -2 are opposites, or additive inverses.

Reciprocals, multiplicative inverses, p. 18 Two numbers whose product is 1; 2 and $\frac{1}{2}$ are reciprocals, or multiplicative inverses.

SET NOTATION

Set, p. 7 A collection of objects. The **empty set,** $\varnothing$ or $\{ \}$, contains no elements.

FIVE-STEP PROBLEM-SOLVING STRATEGY

1. *Familiarize* yourself with the problem.
2. *Translate* to mathematical language.
3. *Carry out* some mathematical manipulation.
4. *Check* your possible answer in the original problem.
5. *State* the answer clearly.

GEOMETRY FORMULAS

Area of a rectangle:	$A = lw$
Area of a square:	$A = s^2$
Area of a parallelogram:	$A = bh$
Area of a trapezoid:	$A = \frac{h}{2}(b_1 + b_2)$
Area of a triangle:	$A = \frac{1}{2}bh$
Area of a circle:	$A = \pi r^2$
Circumference of a circle:	$C = \pi d$
Volume of a cube:	$V = s^3$
Volume of a right circular cylinder:	$V = \pi r^2 h$
Perimeter of a square:	$P = 4s$
Distance traveled:	$d = rt$
Simple interest:	$I = Prt$

IMPORTANT CONCEPTS

[Section references appear in brackets.]

Concept	Example
An algebraic expression is **evaluated** by **substituting** values for the variables and carrying out the operations.	Evaluate $2 + 3x \div 5y$ for $x = -10$ and $y = 2$. $2 + 3x \div 5y = 2 + 3(-10) \div 5(2)$ $= 2 + (-30) \div 5(2)$ $= 2 + (-6)(2)$ $= 2 + (-12)$ $= -10$ [1.1]
A set can be written using **roster notation** or **set-builder notation.**	Both of the following describe the same set. *Roster notation*: $\{0, 1, 2, 3, 4\}$ *Set-builder notation*: $\{x \mid x \text{ is a whole number less than } 5\}$ [1.1]
The **absolute value** of a number is defined by $\lvert x\rvert = \begin{cases} x, & \text{if } x \geq 0, \\ -x, & \text{if } x < 0. \end{cases}$	Since $3 \geq 0$, $\lvert 3\rvert = 3$. Since $-3 < 0$, $\lvert -3\rvert = -(-3) = 3$. [1.2]
To **add** two real numbers, use the rules on p. 15.	$-8 + (-3) = -11$; $-8 + 3 = -5$; $8 + (-3) = 5$; $-8 + 8 = 0$ [1.2]
To **subtract** two real numbers, add the opposite of the number being subtracted.	$-10 - 12 = -10 + (-12) = -22$; $-10 - (-12) = -10 + 12 = 2$ [1.2]

(continued)

To **multiply** or **divide** two real numbers, use the rules on p. 17. Division by 0 is undefined.	$(-5)(-2) = 10$; $30 \div (-6) = -5$; $-3 \div 0$ is undefined [1.2]
For any number a and any nonzero number b, $\frac{-a}{b} = \frac{a}{-b} = -\frac{a}{b}$.	$\frac{-3}{4} = \frac{3}{-4} = -\frac{3}{4}$ [1.2]
To **evaluate exponential expressions,** use the rules on p. 40.	$3^1 = 3$; $3^{-5} \cdot 3^9 = 3^{-5+9} = 3^4$; $3^0 = 1$; $(3^{-4})^2 = 3^{(-4)(2)} = 3^{-8} = \frac{1}{3^8}$; $3^{-2} = \frac{1}{3^2} = \frac{1}{9}$; $\frac{3^4}{3^{-1}} = 3^{4-(-1)} = 3^5$; $\frac{3^{-7}}{x^{-5}} = \frac{x^5}{3^7}$; $(3x^5)^4 = 3^4(x^5)^4 = 81x^{20}$; $\left(\frac{3}{x}\right)^{-2} = \left(\frac{x}{3}\right)^2$; $\left(\frac{3}{x}\right)^6 = \frac{3^6}{x^6}$ [1.4]
Scientific notation: $N \times 10^m$, $1 \le N < 10$	$4100 = 4.1 \times 10^3$; $0.005 = 5 \times 10^{-3}$ [1.4]
To perform multiple operations, use the **rules for order of operations** on p. 4.	$-3 + (3-5)^3 \div 4(-1) = -3 + (-2)^3 \div 4(-1)$ $= -3 + (-8) \div 4(-1)$ $= -3 + (-2)(-1)$ $= -3 + 2$ $= -1$ [1.2]
The law of **opposites:** $a + (-a) = 0$ The law of **reciprocals:** $a \cdot \frac{1}{a} = 1$, $a \neq 0$	4 is the opposite of -4 since $4 + (-4) = 0$. $-\frac{1}{4}$ is the reciprocal of -4 since $\left(-\frac{1}{4}\right)(-4) = 1$. [1.2]
Commutative laws: $a + b = b + a$; $ab = ba$	$3 + (-5) = -5 + 3$ $8(10) = 10(8)$ [1.3]
Associative laws: $a + (b + c) = (a + b) + c$; $a(bc) = (ab)c$	$-5 + (5 + 6) = (-5 + 5) + 6$ $2 \cdot (5 \cdot 9) = (2 \cdot 5) \cdot 9$ [1.3]
Distributive law: $a(b + c) = ab + ac$	$4(x + 2) = 4 \cdot x + 4 \cdot 2 = 4x + 8$ [1.3]
The **addition principle for equations:** $a = b$ is equivalent to $a + c = b + c$.	$x + 5 = -2$ is equivalent to $x + 5 + (-5) = -2 + (-5)$ [1.6]

(continued)

The **multiplication principle for equations:** $a = b$ is equivalent to $ac = bc$, for $c \neq 0$.	$-\frac{1}{3}x = 7$ is equivalent to $(-3)\left(-\frac{1}{3}x\right) = (-3)(7)$ [1.6]
To **graph** an equation means to make a drawing that represents all its solutions. Solutions of equations are represented with **ordered pairs** and graphed on a **coordinate system.**	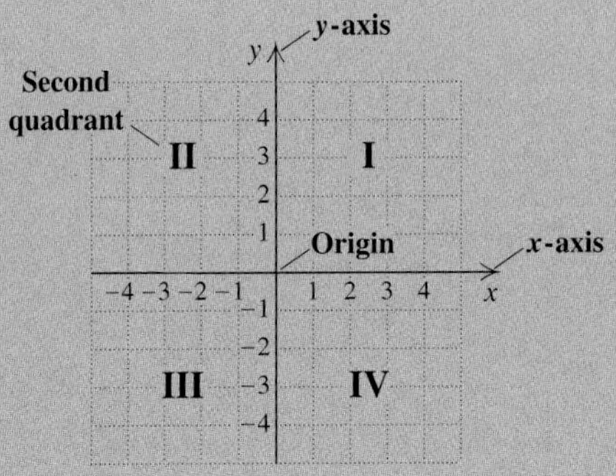 [1.5]

Review Exercises

The following review exercises are for practice. Answers are at the back of the book. If you need to, restudy the section indicated in red next to the exercise or the direction line that precedes it.

Concept Reinforcement *In each of Exercises 1–10, match the expression or equation with an equivalent expression or equation from the column on the right.*

1. ____ $2x - 1 = 9$ [1.6]
2. ____ $2x - 1$ [1.3]
3. ____ $\frac{3}{4}x = 5$ [1.6]
4. ____ $\frac{3}{4}x - 5$ [1.3]
5. ____ $2(x + 7)$ [1.3]
6. ____ $2(x + 7) = 6$ [1.6]
7. ____ $4x - 3 + 2x = 5$ [1.6]
8. ____ $4x - 3 + 2x$ [1.3]
9. ____ $6 + 2x$ [1.3]
10. ____ $6 = 2x$ [1.6]

a) $2 + \frac{3}{4}x - 7$
b) $2x + 14 = 6$
c) $6x - 3$
d) $2(3 + x)$
e) $2x = 10$
f) $6x - 3 = 5$
g) $5x - 1 - 3x$
h) $3 = x$
i) $2x + 14$
j) $\frac{4}{3} \cdot \frac{3}{4}x = \frac{4}{3} \cdot 5$

11. Evaluate
$$7x^2 - 5y \div zx$$
for $x = -2$, $y = 3$, and $z = -5$. [1.2]

12. Name the set consisting of the first five odd natural numbers using both roster notation and set-builder notation. [1.1]

13. Find the area of a triangular flag that has a base of 50 cm and a height of 70 cm. [1.1]

Tell whether each number is a solution of the given equation or inequality. [1.1]

14. $10 - 3x = 1$; **(a)** 7; **(b)** 3

15. $5a + 2 \leq 7$; **(a)** 0; **(b)** 1

Find the absolute value. [1.2]

16. $|-9.3|$ **17.** $|4.09|$ **18.** $|0|$

Perform the indicated operation. [1.2]

19. $-6.5 + (-3.7)$ **20.** $\left(-\frac{4}{5}\right) + \left(\frac{1}{7}\right)$

21. $\left(-\frac{1}{3}\right) + \frac{4}{5}$ **22.** $-7.9 - 3.6$

23. $-\frac{2}{3} - \left(-\frac{1}{2}\right)$ **24.** $12.5 - 17.9$

25. $(-4.2)(-3)$ **26.** $\left(-\frac{2}{3}\right)\left(\frac{5}{8}\right)$

27. $\dfrac{72.8}{-8}$

28. $-7 \div \dfrac{4}{3}$

29. Find $-a$ if $a = -4.01$. [1.2]

Use a commutative law to write an equivalent expression. [1.3]

30. $9 + a$

31. $7y$

32. $5x + y$

Use an associative law to write an equivalent expression. [1.3]

33. $(4 + a) + b$

34. $(xy)3$

35. Obtain an expression that is equivalent to $7mn + 14m$ by factoring. [1.3]

36. Combine like terms: $3x^3 - 6x^2 + x^3 + 5$. [1.3]

37. Simplify: $7x - 4[2x + 3(5 - 4x)]$. [1.3]

38. Multiply and simplify: $(5a^2b^7)(-2a^3b)$. [1.4]

39. Divide and simplify: $\dfrac{12x^3y^8}{3x^2y^2}$. [1.4]

40. Evaluate a^0, a^2, and $-a^2$ for $a = -5.3$. [1.4]

Simplify. Do not use negative exponents in the answer. [1.4]

41. $3^{-4} \cdot 3^7$

42. $(5a^2)^3$

43. $(-2a^{-3}b^2)^{-3}$

44. $\left(\dfrac{x^2y^3}{z^4}\right)^{-2}$

45. $\left(\dfrac{2a^{-2}b}{4a^3b^{-3}}\right)^4$

Simplify. [1.2]

46. $\dfrac{7(5 - 2 \cdot 3) - 3^2}{4^2 - 3^2}$

47. $1 - (2 - 5)^2 + 5 \div 10 \cdot 4^2$

48. Convert 0.000000103 to scientific notation. [1.4]

49. One *parsec* (a unit that is used in astronomy) is 30,860,000,000,000 km. Write scientific notation for this number. [1.4]

Simplify and write scientific notation for each answer. [1.4]

50. $(8.7 \times 10^{-9}) \times (4.3 \times 10^{15})$

51. $\dfrac{1.2 \times 10^{-12}}{1.5 \times 10^{-7}}$

Determine whether the ordered pair is a solution. [1.5]

52. $(3, 7)$; $4p - q = 5$

53. $(-2, 4)$; $x - 2y = 12$

54. $\left(0, -\frac{1}{2}\right)$, $3a - 4b = 2$

55. $(8, -2)$; $3c + 2d = 28$

Graph. [1.5]

56. $y = -3x + 2$

57. $y = -x^2 + 1$

58. $y = 3 - |x|$

59. $y = 6$

60. Using a graphing calculator, complete the following table for the equation

$$y = 2|x| - 3.$$

Then graph. [1.5]

X	Y1	
−3		
−2		
−1		
0		
1		
2		
3		

X = −3

Solve. If the solution set is ∅ or ℝ, classify the equation as a contradiction or as an identity. [1.6]

61. $x - 8.9 = 2.7$

62. $\frac{2}{3}a = 9$

63. $-9x + 4(2x - 3) = 5(2x - 3) + 7$

64. $3(x - 4) + 2 = x + 2(x - 5)$

65. $5t - (7 - t) = 4t + 2(9 + t)$

66. Solve for m: $P = m/S$. [1.6]

67. Solve for x: $c = mx - rx$. [1.6]

68. Translate to an equation but do not solve: 15 more than twice a number is 21. [1.7]

69. A number is 19 less than another number. The sum of the numbers is 115. Find the smaller number. [1.7]

70. One angle of a triangle measures three times the second angle. The third angle measures twice the second angle. Find the measures of the angles. [1.7]

71. The volume of a film canister is 28.26 cm^3. If the radius of the canister is 1.5 cm, determine the height. Use 3.14 for π. [1.7]

72. A sheet of plastic shrink wrap has a thickness of 0.00015 mm. The sheet is 1.2 m by 79 m. Use scientific notation to find the volume of the sheet. [1.7]

73. *Tropical Storms.* The following graph shows the number of North Atlantic tropical storms and hurricanes for several years (*Source*: National Hurricane Center).

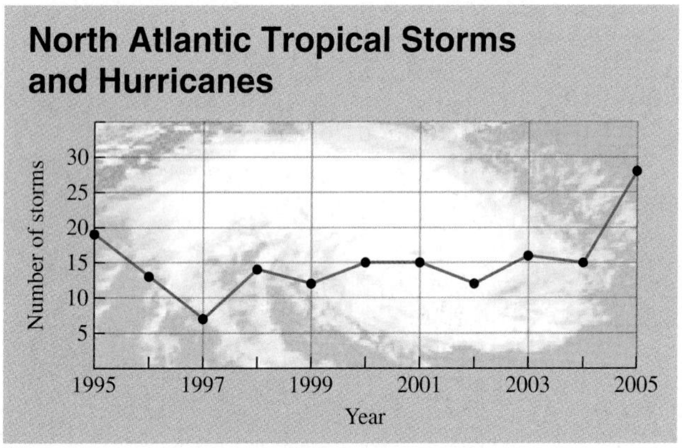

a) How many tropical storms or hurricanes were there in 1998? [1.7]
b) In what year were there the fewest tropical storms or hurricanes? [1.7]
c) Tropical storms and hurricanes are given names. If there are more than 25 storms, the National Hurricane Center titles the remaining storms using the Greek alphabet. In what year or years were there storms called by Greek alphabet letters? [1.7]
d) What was the difference between the highest number of storms in a year and the lowest number? [1.7]

74. The following table lists the U.S. crude oil output, in millions of barrels a day, for several years. Use a graphing calculator to draw a line graph. [1.7]

Year	Crude Oil Output (in millions of barrels)
1960	7.0
1970	9.6
1980	8.5
1990	7.3
2005	5.1

Source: *The Indianapolis Star*, April 28, 2006

Synthesis

TW **75.** Describe a method that could be used to write equations that have no solution. [1.6]

TW **76.** Explain how the distributive law can be used when combining like terms. [1.3]

77. If the smell of gasoline is detectable at 3 parts per billion, what percent of the air is occupied by the gasoline? [1.7]

78. Evaluate $a + b(c - a^2)^0 + (abc)^{-1}$ for $a = 3, b = -2$, and $c = -4$. [1.1], [1.4]

79. What's a better deal: a 13-in. diameter pizza for \$8 or a 17-in. diameter pizza for \$11? Explain. [1.7]

80. The surface area of a cube is 486 cm^2. Find the volume of the cube. [1.7]

81. Solve for z: $m = \dfrac{x}{y - z}$. [1.6]

82. Simplify: $\dfrac{(3^{-2})^a \cdot (3^b)^{-2a}}{(3^{-2})^b \cdot (9^{-b})^{-3a}}$. [1.4]

83. Each of Ray's test scores counts three times as much as a quiz score. If after 4 quizzes Ray's average is 82.5, what score does he need on the first test in order to raise his average to 85? [1.7]

84. Fill in the following blank so as to ensure that the equation is an identity. [1.6]

$$5x - 7(x + 3) - 4 = 2(7 - x) + \underline{\qquad}$$

85. Replace the blank with one term to ensure that the equation is a contradiction. [1.6]

$$20 - 7[3(2x + 4) - 10] = 9 - 2(x - 5) + \underline{\qquad}$$

86. Use the commutative law for addition once and the distributive law twice to show that

$$a2 + cb + cd + ad = a(d + 2) + c(b + d).$$

[1.3]

87. Find an irrational number between $\frac{1}{2}$ and $\frac{3}{4}$. [1.1]

Chapter Test 1

1. Evaluate $a^3 - 5b + b \div ac$ for $a = -2$, $b = 6$, and $c = 3$.

2. The base of a triangular stamp measures 3 cm and its height 2.5 cm. Find the area of the stamp.

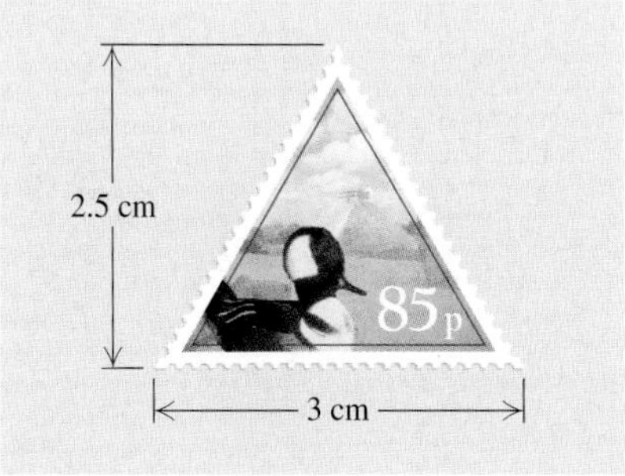

3. Tell whether each number is a solution of the equation $14 - 5x = 4$.
   a) 1
   b) 2
   c) 0

Perform the indicated operation.

4. $-25 + (-16)$

5. $-10.5 + 6.8$

6. $6.21 + (-8.32)$

7. $29.5 - 43.7$

8. $-17.8 - 25.4$

9. $-6.4(5.3)$

10. $-\frac{7}{3} - \left(-\frac{3}{4}\right)$

11. $-\frac{2}{7}\left(-\frac{5}{14}\right)$

12. $\frac{-42.6}{-7.1}$

13. $\frac{2}{5} \div \left(-\frac{3}{10}\right)$

14. Simplify: $5 + (1 - 3)^2 - 7 \div 2^2 \cdot 6$.

15. Use a commutative law to write an expression equivalent to $7x + y$.

16. Combine like terms: $4y - 10 - 7y - 19$.

17. Simplify: $9x - 3(2x - 5) - 7$.

Simplify. Do not use negative exponents in the answer.

18. $(12x^{-4}y^{-7})(-6x^{-6}y)$

19. -3^{-2}

20. $(-6x^2y^{-4})^{-2}$

21. $\left(\frac{2x^3y^{-6}}{-4y^{-2}}\right)^2$

22. $(5x^3y)^0$

Simplify and write scientific notation for the answer.

23. $(9.05 \times 10^{-3})(2.22 \times 10^{-5})$

24. $\frac{5.6 \times 10^7}{2.8 \times 10^{-3}}$

25. $\frac{1.0614 \times 10^{-5}}{3.48 \times 10^{-10}}$

Determine whether the ordered pair is a solution.

26. $(0, -5)$; $x + 4y = -20$

27. $(1, -4)$; $-2p + 5q = 18$

Graph.

28. $y = -5x + 4$

29. $y = -2x^2 + 3$

30. Create a table of solutions of the equation
$$y = 10 - x^2$$
for integer values of x from -3 to 3. Then graph.

Solve. If the solution set is $\mathbb{R}$ or $\varnothing$, classify the equation as an identity or a contradiction.

31. $10x - 7 = 38x + 49$

32. $13t - (5 - 2t) = 5(3t - 1)$

33. Solve for P_2: $\frac{P_1V_1}{T_1} = \frac{P_2V_2}{T_2}$.

34. Translate to an algebraic expression:

    Three more than the product of two numbers.

35. Linda's scores on five tests are 84, 80, 76, 96, and 80. What must Linda score on the sixth test so that her average will be 85?

36. Find three consecutive odd integers such that the sum of four times the first, three times the second, and two times the third is 167.

37. Binary stars orbit around each other and can be as close to each other as the earth is to the sun (*Source*: Abell, George O., *Exploration of the Universe*, 3rd ed. New York: Holt, Rinehart and Winston, 1975). If a pair of binary stars are 1.5×10^8 km apart, what is the length of their orbit? (Assume the orbit is circular.)

Gas Mileage. The following graph shows the gas mileage of a truck traveling at different speeds.

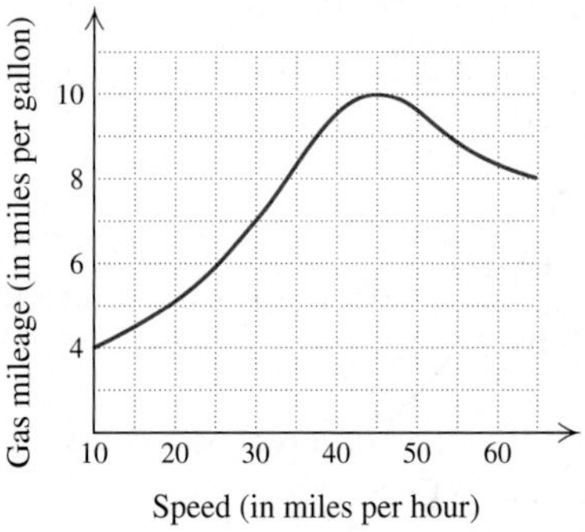

38. At what speed is the gas mileage highest?

39. What is the gas mileage when the truck is traveling at 30 mph?

Synthesis

Simplify.

40. $(4x^{3a}y^{b+1})^{2c}$

41. $\dfrac{-27a^{x+1}}{3a^{x-2}}$

42. $\dfrac{(-16x^{x-1}y^{y-2})(2x^{x+1}y^{y+1})}{(-7x^{x+2}y^{y+2})(8x^{x-2}y^{y-1})}$

2 Functions, Linear Equations, and Models

2.1 Functions
2.2 Linear Functions: Slope, Graphs, and Models
2.3 Another Look at Linear Graphs
2.4 Introduction to Curve Fitting: Point–Slope Form
2.5 Domains and the Algebra of Functions

A *function* is a certain kind of relationship between sets. Functions are very important in mathematics in general, and in problem solving in particular. In this chapter, you will learn what we mean by a function and then begin to use functions to solve problems.

A function that can be described by a linear equation is called a *linear function*. We will study graphs of linear equations in detail and will use linear equations and functions to model and solve applications.

APPLICATION *Registered Nurses.*

Data from the National Sample Survey of Registered Nurses, conducted every four years, indicate that the number of registered nurses in the United States who are working full time has increased steadily. Use linear regression to fit a linear function to the data, and use the function to predict the number of registered nurses working full time in 2008.

Year	Number of Registered Nurses Working Full Time
1980	850,000
1984	975,000
1988	1,100,000
1992	1,225,000
1996	1,500,000
2000	1,600,000
2004	1,700,000

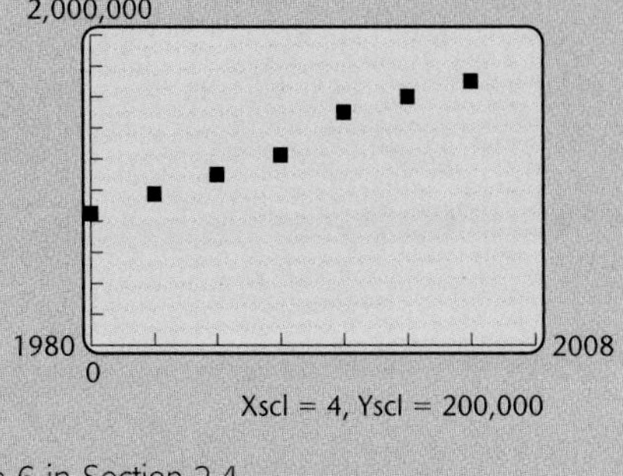

This problem appears as Example 6 in Section 2.4.

2.1 Functions

Functions and Graphs ■ Function Notation and Equations

We now develop the idea of a *function*—one of the most important concepts in mathematics. A function is a special kind of correspondence between two sets. For example,

To each person in a class	there corresponds	a date of birth.
To each bar code in a store	there corresponds	a price.
To each real number	there corresponds	the cube of that number.

In each example, the first set is called the **domain.** The second set is called the **range.** For any member of the domain, there is *exactly one* member of the range to which it corresponds. This kind of correspondence is called a **function.**

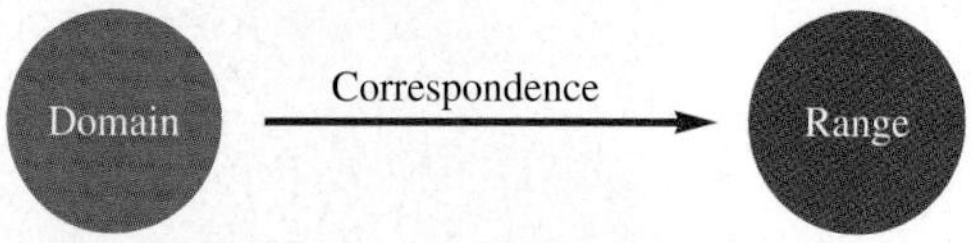

Note in the list above that although two members of a class may have the same date of birth, and two bar codes may correspond to the same price, each correspondence is still a function since every member of the domain is paired with exactly one member of the range.

EXAMPLE 1 Determine whether each correspondence is a function.

a)

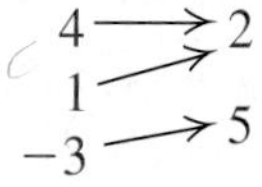

b)

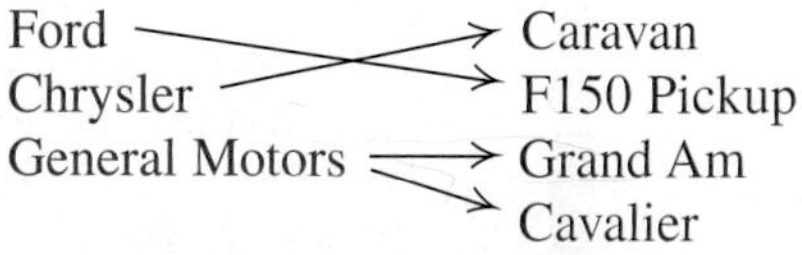

SOLUTION

a) The correspondence *is* a function because each member of the domain corresponds to *exactly one* member of the range.

b) The correspondence *is not* a function because a member of the domain (General Motors) corresponds to more than one member of the range.

Function

A *function* is a correspondence between a first set, called the *domain*, and a second set, called the *range*, such that each member of the domain corresponds to *exactly one* member of the range.

EXAMPLE 2 Determine whether each correspondence is a function.

	Domain	**Correspondence**	Range
a)	An elevator full of people	Each person's weight	A set of positive numbers
b)	$\{-2, 0, 1, 2\}$	Each number's square	$\{0, 1, 4\}$
c)	Authors of best-selling books	The titles of books written by each author	A set of book titles

SOLUTION

a) The correspondence *is* a function, because each person has *only one* weight.

b) The correspondence *is* a function, because each number has *only one* square.

c) The correspondence *is not* a function, because some authors have written *more than one* book.

Study Tip

Step by Step

The *Student's Solutions Manual* is an excellent resource if you need additional help with an exercise in the exercise sets. It contains step-by-step solutions to the odd-numbered exercises in each exercise set.

Functions and Graphs

The functions in Examples 1(a) and 2(b) can be expressed as sets of ordered pairs. Example 1(a) can be written $\{(-3, 5), (1, 2), (4, 2)\}$ and Example 2(b) can be written $\{(-2, 4), (0, 0), (1, 1), (2, 4)\}$. We can graph these functions as follows.

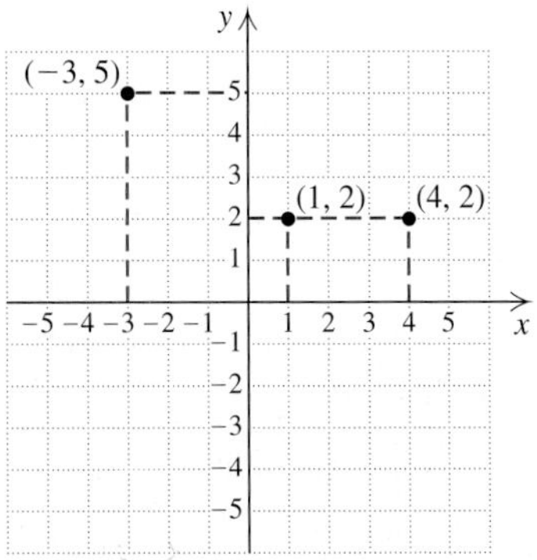

The function $\{(-3, 5), (1, 2), (4, 2)\}$
Domain is $\{-3, 1, 4\}$
Range is $\{5, 2\}$

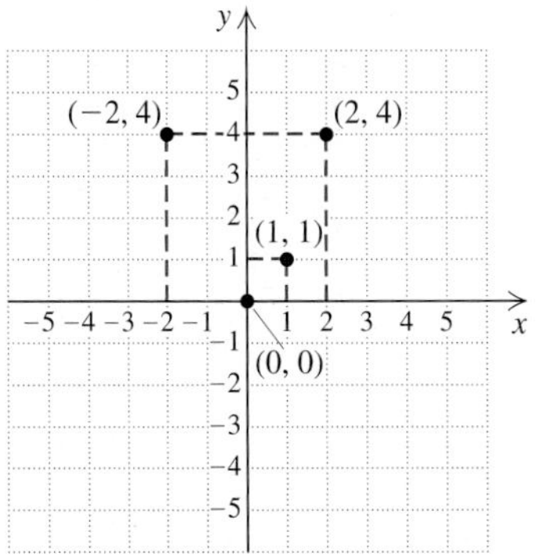

The function $\{(-2, 4), (0, 0), (1, 1), (2, 4)\}$
Domain is $\{-2, 0, 1, 2\}$
Range is $\{4, 0, 1\}$

When a function is given as a set of ordered pairs, the domain is the set of all first coordinates and the range is the set of all second coordinates. Functions are generally represented by lower- or upper-case letters.

EXAMPLE 3 Find the domain and the range of the function f shown here.

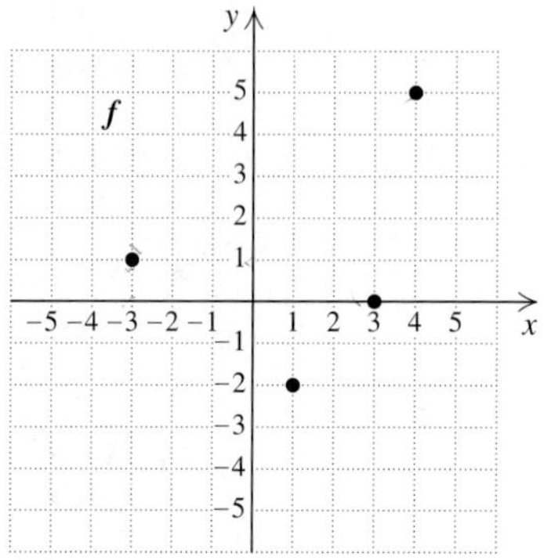

SOLUTION Here f can be written $\{(-3,1),(1,-2),(3,0),(4,5)\}$. The domain is the set of all first coordinates, $\{-3,1,3,4\}$, and the range is the set of all second coordinates, $\{1,-2,0,5\}$. We can also find the domain and the range directly from the graph, without first listing all pairs.

EXAMPLE 4 For the function f shown here, determine each of the following.

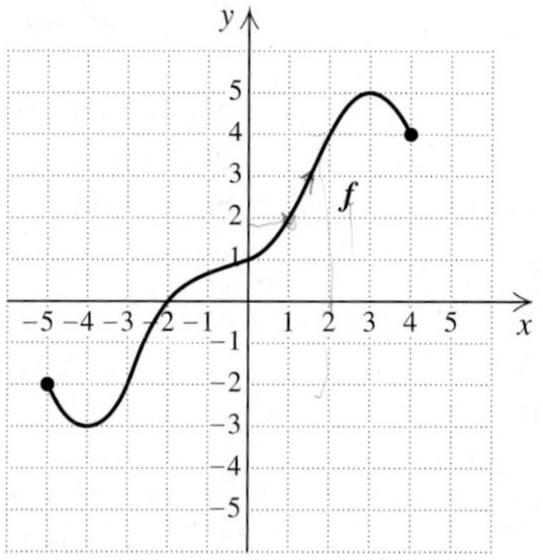

a) What member of the range is paired with 2

b) The domain of f

c) What member of the domain is paired with -3

d) The range of f

SOLUTION

a) To determine what member of the range is paired with 2, we locate 2 on the horizontal axis (this is where the domain is located). Next, we find the point directly above 2 on the graph of f. From that point, we can look to the vertical axis to find the corresponding y-coordinate, 4. The "input" 2 has the "output" 4.

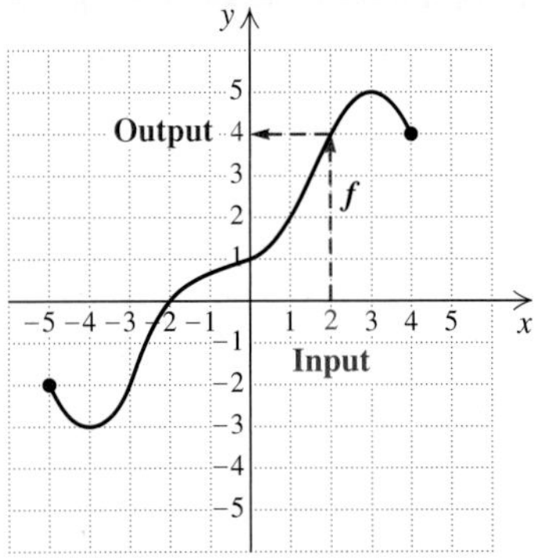

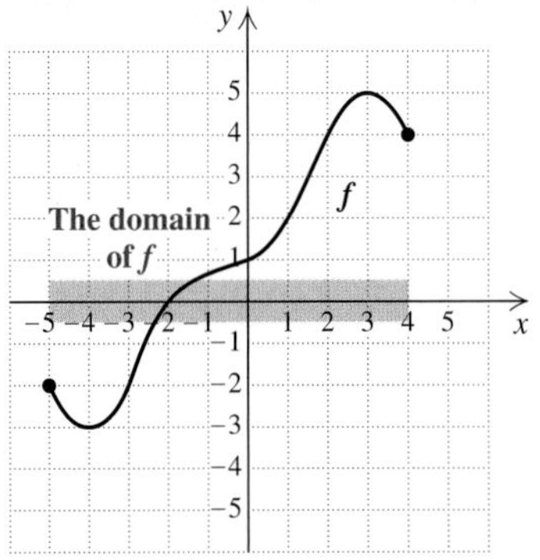

b) The domain of the function is the set of all x-values that are used in the points of the curve (see the figure on the right above). Because there are no breaks in the graph of f, these extend continuously from -5 to 4 and can be viewed as the curve's shadow, or *projection*, on the x-axis. Thus the domain is $\{x \mid -5 \le x \le 4\}$.

c) To determine what member of the domain is paired with -3, we locate -3 on the vertical axis. (This is where the range is located; see below.) From there we look left and right to the graph of f to find any points for which -3 is the second coordinate. One such point exists, $(-4, -3)$. We note that -4 is the only element of the domain paired with -3.

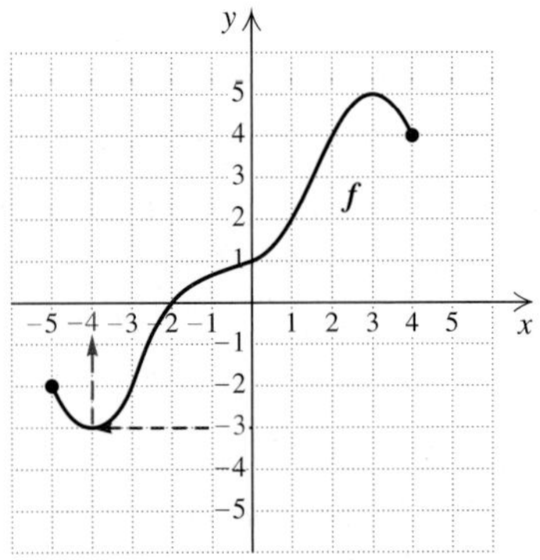

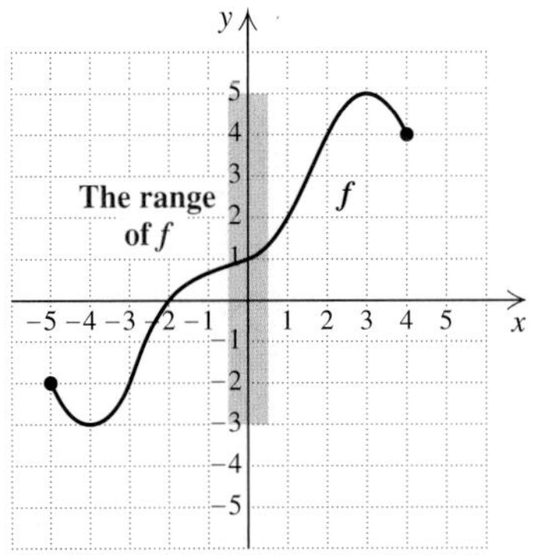

d) The range of the function is the set of all y-values that are in the graph. (See the graph on the right above.) These extend continuously from -3 to 5, and can be viewed as the curve's projection on the y-axis. Thus the range is $\{y \mid -3 \le y \le 5\}$.

Note that if a graph contains two or more points with the same first coordinate, that graph cannot represent a function (otherwise one member of the domain would correspond to more than one member of the range). This observation is the basis of the *vertical-line test.*

> **The Vertical-Line Test** If it is possible for a vertical line to cross a graph more than once, then the graph is not the graph of a function.

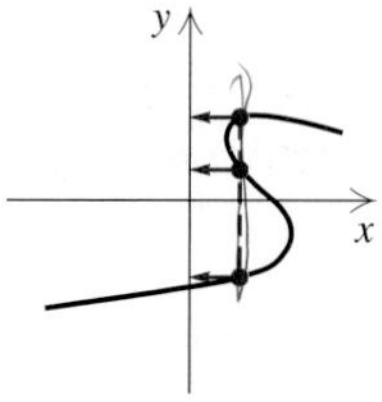

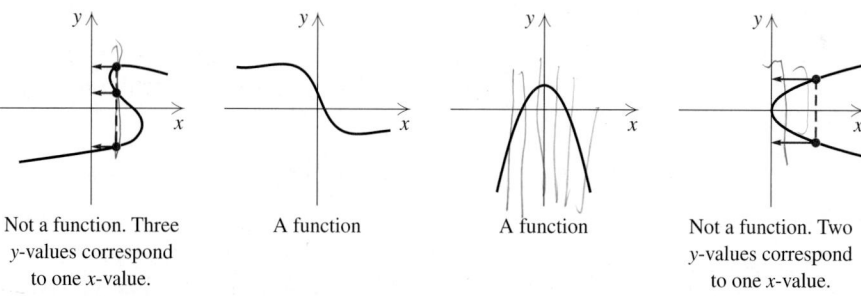

Not a function. Three y-values correspond to one x-value.

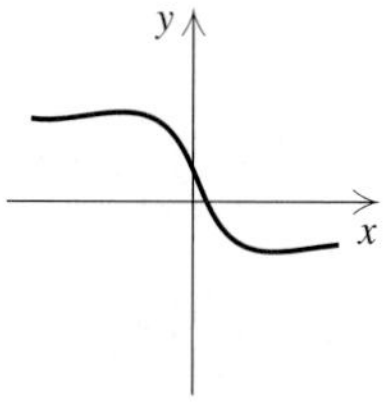

A function

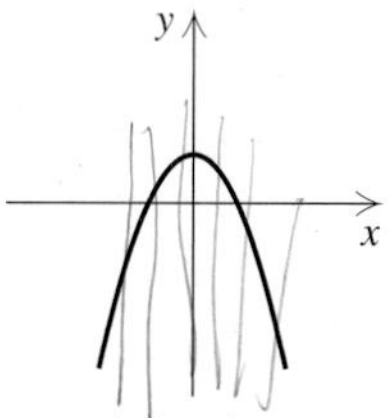

A function

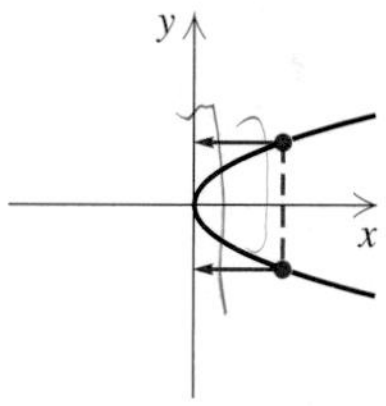

Not a function. Two y-values correspond to one x-value.

Graphs that do not represent functions still do represent **relations.**

> **Relation**
>
> A *relation* is a correspondence between a first set, called the *domain*, and a second set, called the *range*, such that each member of the domain corresponds to *at least one* member of the range.

Thus, although the correspondences and graphs above are not all functions, they *are* all relations.

Function Notation and Equations

We often think of an element of the domain of a function as an **input** and its corresponding element of the range as an **output.** In Example 3, the function f is written

$$\{(-3,1),(1,-2),(3,0),(4,5)\}.$$

Thus, for an input of -3, the corresponding output is 1, and for an input of 3, the corresponding output is 0.

We use *function notation* to indicate what output corresponds to a given input. For the function f defined above, we write

$$f(-3)=1, \quad f(1)=-2, \quad f(3)=0, \quad \text{and} \quad f(4)=5.$$

The notation $f(x)$ is read "f of x," "f at x," or "the value of f at x." If x is an input, then $f(x)$ is the corresponding output.

CAUTION! ***f(x) does not mean f times x.***

Most functions are described by equations. For example, $f(x) = 2x + 3$ describes the function that takes an input x, multiplies it by 2, and then adds 3.

Input

$$f(x) = \underbrace{2x}_{\text{Double}} \quad \underbrace{+\,3}_{\text{Add 3}}$$

To calculate the output $f(4)$, we take the input 4, double it, and add 3 to get 11. That is, we substitute 4 into the formula for $f(x)$:

$$f(4) = 2 \cdot 4 + 3$$
$$= 11. \longleftarrow \text{Output}$$

To understand function notation, it can help to imagine a "function machine." Think of putting an input into the machine. For the function $f(x) = 2x + 3$, the machine will double the input and then add 3. The result will be the output.

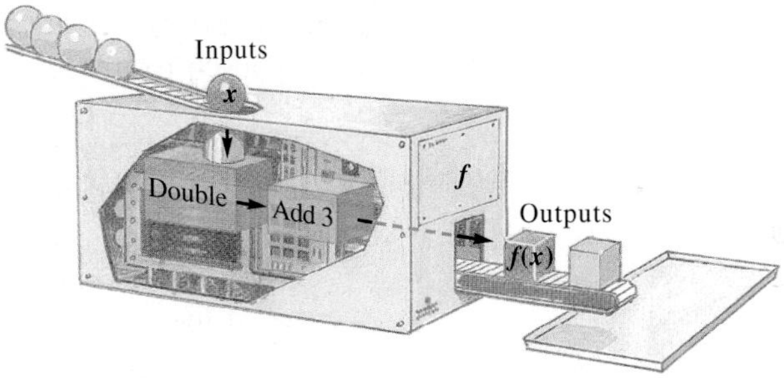

Sometimes, in place of $f(x) = 2x + 3$, we write $y = 2x + 3$, where it is understood that the value of y, the *dependent variable*, depends on our choice of x, the *independent variable*. To understand why $f(x)$ notation is so useful, consider two equivalent statements:

a) If $f(x) = 2x + 3$, then $f(4) = 11$.

b) If $\quad y = 2x + 3$, then the value of y is 11 when x is 4.

The notation used in part (a) is far more concise and emphasizes that x is the independent variable.

EXAMPLE 5 Find each indicated function value.

a) $f(5)$, for $f(x) = 3x + 2$

b) $g(-2)$, for $g(r) = 5r^2 + 3r$

c) $h(4)$, for $h(x) = 7$

d) $F(a + 1)$, for $F(x) = 3x + 2$

e) $F(a) + 1$, for $F(x) = 3x + 2$

SOLUTION Finding function values is much like evaluating an algebraic expression.

a) $f(x) = 3x + 2$
$f(5) = 3 \cdot 5 + 2 = 17$

b) $g(r) = 5r^2 + 3r$
$g(-2) = 5(-2)^2 + 3(-2)$
$= 5 \cdot 4 - 6 = 14$

c) For the function given by $h(x) = 7$, all inputs share the same output, 7. Therefore, $h(4) = 7$. The function h is an example of a *constant function.*

d) $F(x) = 3x + 2$
$F(a + 1) = 3(a + 1) + 2$ **The input is $a + 1$.**
$= 3a + 3 + 2 = 3a + 5$

e) $F(x) = 3x + 2$
$F(a) + 1 = [3(a) + 2] + 1$ **The input is a.**
$= [3a + 2] + 1 = 3a + 3$

Note that whether we write $f(x) = 3x + 2$, or $f(t) = 3t + 2$, or $f(■) = 3■ + 2$, we still have $f(5) = 17$. The variable in the parentheses (the independent variable) is the same as the variable in the algebraic expression. The letter chosen for the independent variable is not as important as the algebraic manipulations to which it is subjected.

Function notation is often used in formulas. For example, to emphasize that the area A of a circle is a function of its radius r, instead of

$$A = \pi r^2,$$

we can write

$$A(r) = \pi r^2.$$

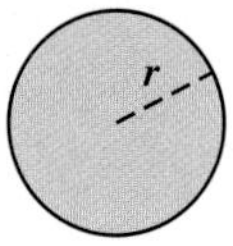

Function Notation

The values of a function that is described by an equation can be found directly using a graphing calculator. For example, if

$$f(x) = x^2 - 6x + 7$$

is entered as

$$y_1 = x^2 - 6x + 7,$$

then $f(2)$ can be calculated by evaluating Y1(2). After entering the function, move to the home screen by pressing QUIT. The notation "Y1" can be found by pressing VARS, choosing the Y-VARS submenu, selecting the FUNCTION option, and then selecting Y1. After Y1 appears on the home screen, press (2) ENTER to evaluate Y1(2).

Function values can also be found using a table with Indpnt set to Ask. They can also be read from the graph of a function using the VALUE option of the CALC menu.

EXAMPLE 6 For $f(a) = 2a^2 - 3a + 1$, find $f(3)$ and $f(-5.1)$.

SOLUTION We first enter the function into the graphing calculator, replacing a with x and the notation $f(a)$ with Y_1. The equation $Y_1 = 2x^2 - 3x + 1$ represents the same function, with the understanding that the Y_1-values are the outputs and the x-values are the inputs; $Y_1(x)$ is equivalent to $f(a)$.

To find $f(3)$, we enter $Y_1(3)$ and find that $f(3) = 10$.

```
Plot1  Plot2  Plot3
\Y1=2X²−3X+1
\Y2=
\Y3=
```

```
Y1(3)
                10
Y1(−5.1)
             68.32
```

X	Y1
3	10
−5.1	68.32

To find $f(-5.1)$, we can press ENTRY and edit the previous entry. Or, using a table, we can find both $f(3)$ and $f(-5.1)$, as shown on the right above. We see that $f(-5.1) = 68.32$.

We can also find function values from the graph of the function.

EXAMPLE 7 Find $g(2)$ for $g(x) = 2x - 5$.

SOLUTION We enter and graph the function, using the standard viewing window. Most graphing calculators have a VALUE option in the CALC menu that will calculate a function's value given a value of x. For most calculators, that x-value must be shown in the viewing window; that is, Xmin $\le x \le$ Xmax. In this case, when $x = 2$, $y = -1$, so $g(2) = -1$.

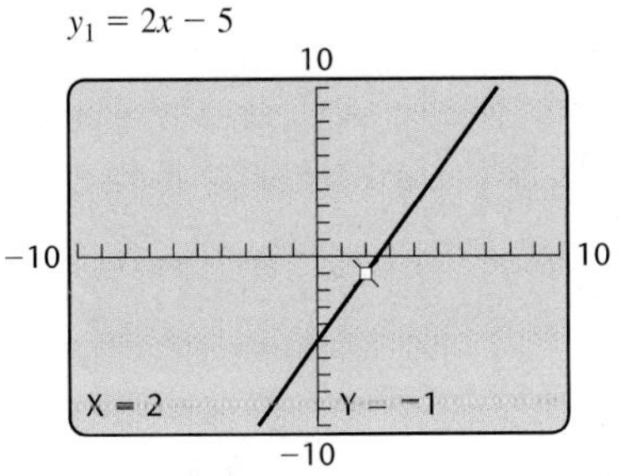

When we find a function value, we are determining an output that corresponds to a given input. To do this, we evaluate the expression for the given value of the input.

Sometimes we want to determine an input that corresponds to a given output. To do this, we solve an equation.

EXAMPLE 8 Let $f(x) = 3x - 7$.

a) What output corresponds to an input of 5?

b) What input corresponds to an output of 5?

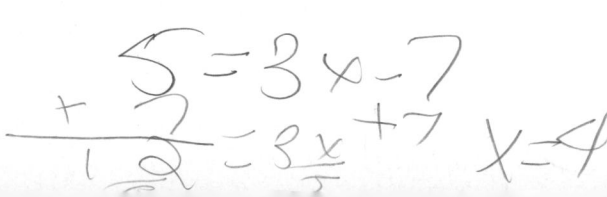

SOLUTION

a) We find $f(5)$:

$$\begin{aligned} f(x) &= 3x - 7 \\ f(5) &= 3(5) - 7 && \textbf{The input is 5. We substitute 5 for } x. \\ &= 15 - 7 \\ &= 8. && \textbf{Carrying out the calculations} \end{aligned}$$

The output 8 corresponds to the input 5; that is, $f(5) = 8$.

b) We find the value of x for which $f(x) = 5$:

$$\begin{aligned} f(x) &= 3x - 7 \\ 5 &= 3x - 7 && \textbf{The output is 5. We substitute 5 for } f(x). \\ 12 &= 3x \\ 4 &= x. && \textbf{Solving for } x \end{aligned}$$

The input 4 corresponds to the output 5; that is, $f(4) = 5$.

We are often interested in finding what inputs correspond to an output of 0. Such inputs are called the **zeros** of the function.

> **Zero of a Function** A zero of a function is an input whose corresponding output is 0. If a is a zero of the function f, then $f(a) = 0$.

The y-coordinate of a point is 0 when the point is on the x-axis. Thus a zero of a function is the x-coordinate of any points at which its graph crosses the x-axis.

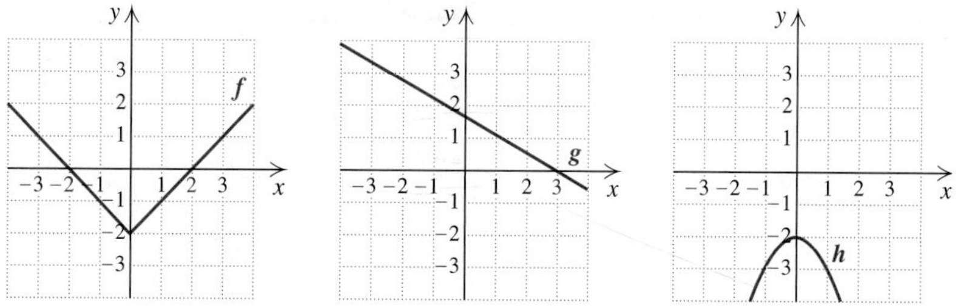

Two zeros: -2 and 2
$f(-2) = 0$; $f(2) = 0$

One zero: 3
$g(3) = 0$

No zeros

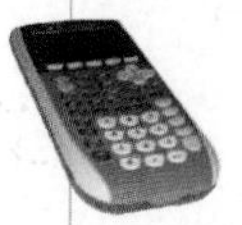

Zeros of a Function

We can determine any zeros of a function using the ZERO option in the CALC menu of a graphing calculator. After graphing the function and choosing the ZERO option, we will be prompted for a Left Bound, a Right Bound, and a Guess. Since there may be more than one zero of a function, the left and right bounds indicate which zero we are currently finding. We examine the graph to find any places where it appears that the graph touches or crosses the x-axis, and then find those x-values one at a time. By using the arrow keys or entering a value on the keyboard, we choose an x-value less than the zero for the left bound, an x-value more than the zero for the right bound, and a value close to the zero for the Guess.

EXAMPLE 9 Find the zeros of the function given by $f(x) = \frac{2}{3}x + 5$.

ALGEBRAIC APPROACH

We want to find any x-values for which $f(x) = 0$, so we substitute 0 for $f(x)$ and solve:

$$f(x) = \tfrac{2}{3}x + 5$$

$$0 = \tfrac{2}{3}x + 5 \qquad \textbf{Substituting 0 for } f(x)$$

$$-\tfrac{2}{3}x = 5 \qquad \textbf{Subtracting } -\tfrac{2}{3}x \textbf{ from both sides}$$

$$\left(-\tfrac{3}{2}\right)\left(-\tfrac{2}{3}x\right) = \left(-\tfrac{3}{2}\right)(5) \qquad \textbf{Multiplying by the reciprocal of } -\tfrac{2}{3}$$

$$x = -\tfrac{15}{2}. \qquad \textbf{Simplifying}$$

The zero of the function is $-\frac{15}{2}$, or -7.5.

GRAPHICAL APPROACH

We graph the function and look for any points at which the graph crosses the x-axis. From the graph, we might estimate that there is one zero of the function, approximately -7.

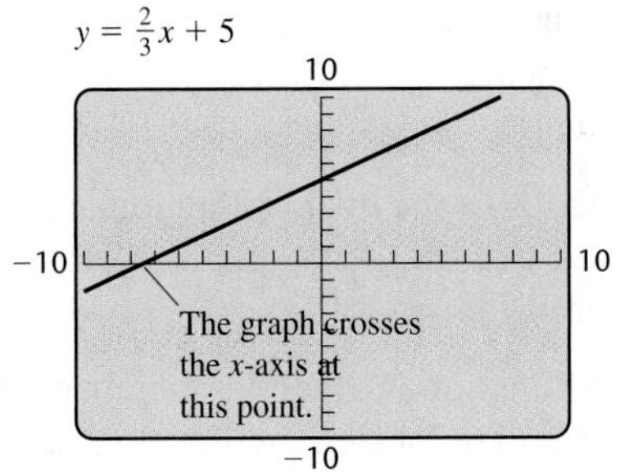

We press CALC 2 to choose the ZERO option of the CALC menu. When prompted, we use the number keys to choose -10 for a left bound, -5 for a right bound, and -7 for a guess. We can then read from the screen that -7.5 is the zero of the function.

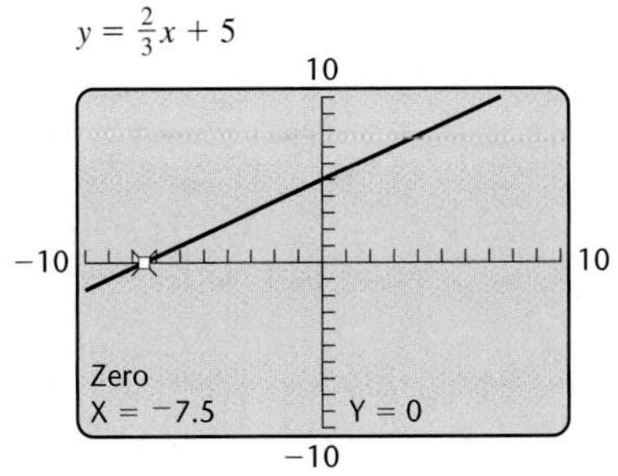

Note that the answer to Example 9 is a value for x. A zero of a function is an x-value, not a function value or an ordered pair.

2.1 EXERCISE SET

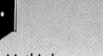

FOR EXTRA HELP

MathXL | MyMathLab | InterAct Math | AW Math Tutor Center | Video Lectures on CD: Disc 1 | Student's Solutions Manual

Concept Reinforcement *Complete each of the following sentences.*

1. For any function, the set of all inputs, or first values, is called the ______________.

2. For any function, the set of all outputs, or second values, is called the ______________.

3. In any function, each member of the domain is paired with ______________ one member of the range.

4. In a relation, a member of the domain may be paired with ______________ than one member of the range.

5. When a function is graphed, members of the domain are located on the ______________ axis.

6. When a function is graphed, members of the range are located on the ______________ axis.

7. The notation $f(3)$ is read ______________.

8. The vertical-line test can be used to determine whether or not a graph represents a(n) ______________.

Determine whether each correspondence is a function.

9.

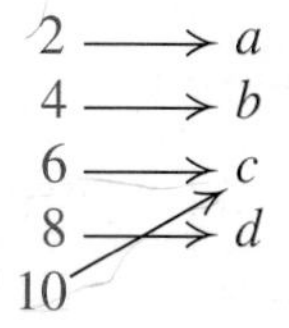

10.

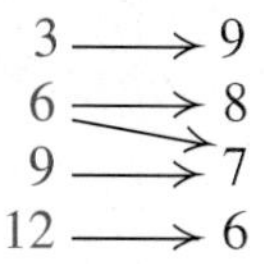

11. *Girl's age (in months)* — *Average daily weight gain (in grams)*

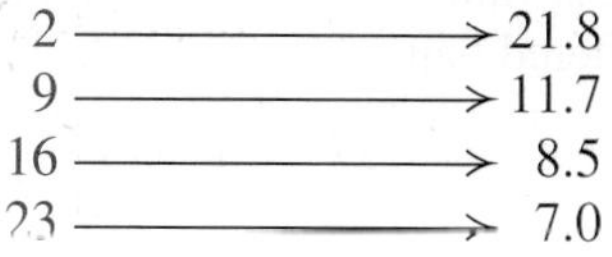

Source: *American Family Physician,* December 1993, p. 1435

12. *Boy's age (in months)* — *Average daily weight gain (in grams)*

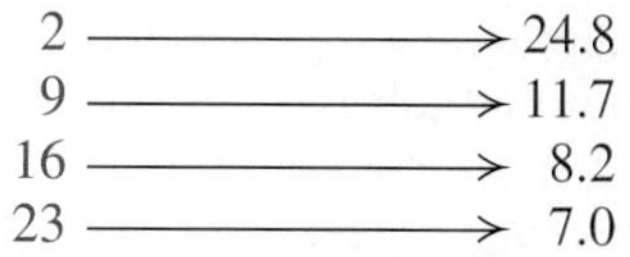

Source: *American Family Physician,* December 1993, p. 1435

13. *Celebrity* — *Birthday*

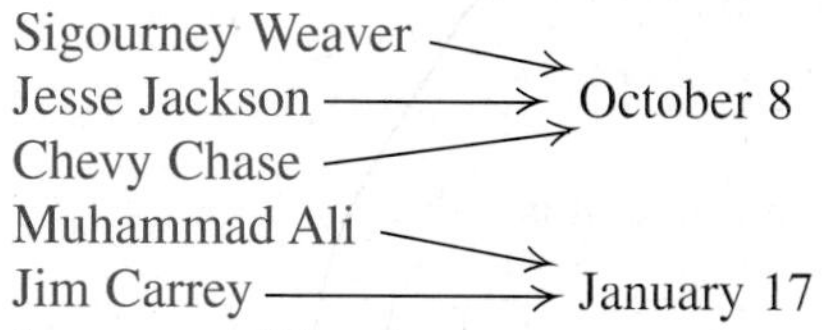

Source: www.kidsparties.com

14. *Predator* — *Prey*

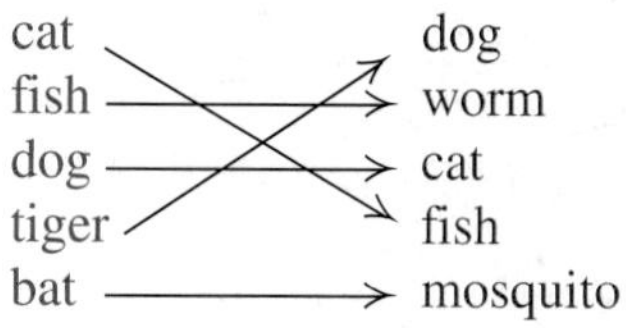

15. *Birthday* — *Celebrity*

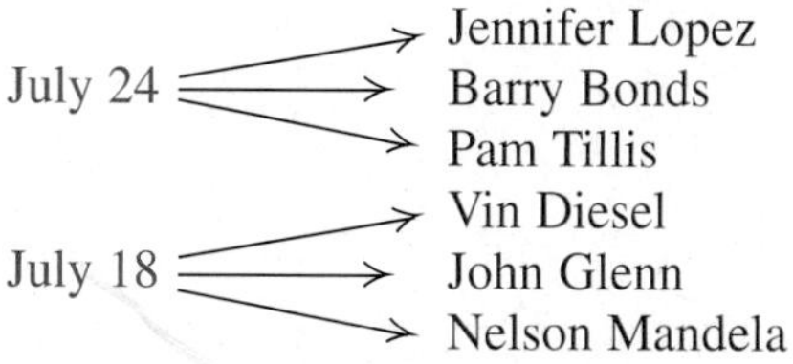

Source: www.leannesbirthdays.com and www.kidsparties.com

16. *State* — *Neighboring state*

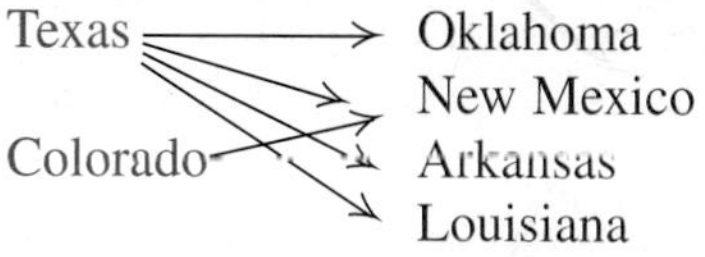

Determine whether each of the following is a function. Identify any relations that are not functions.

	Domain	*Correspondence*	*Range*
17.	A yard full of pumpkins	The price of each pumpkin	A set of prices
18.	The members of a rock band	An instrument the person can play	A set of instruments
19.	The players on a team	The uniform number of each player	A set of numbers
20.	A set of triangles	The area of each triangle	A set of numbers

For each graph of a function, determine **(a)** $f(1)$; **(b)** *the domain;* **(c)** *any x-values for which* $f(x) = 2$; *and* **(d)** *the range.*

21.

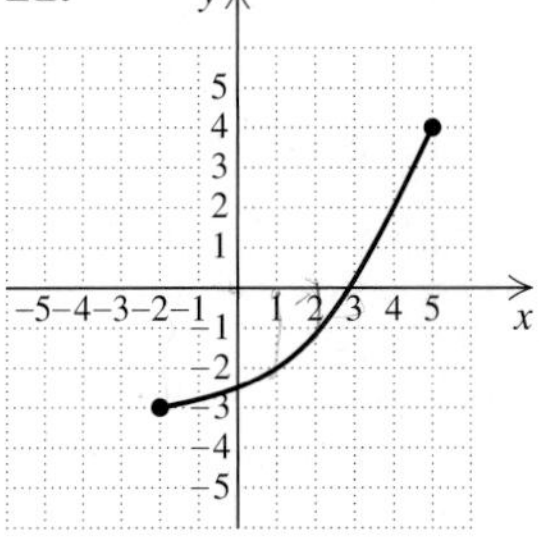

22.

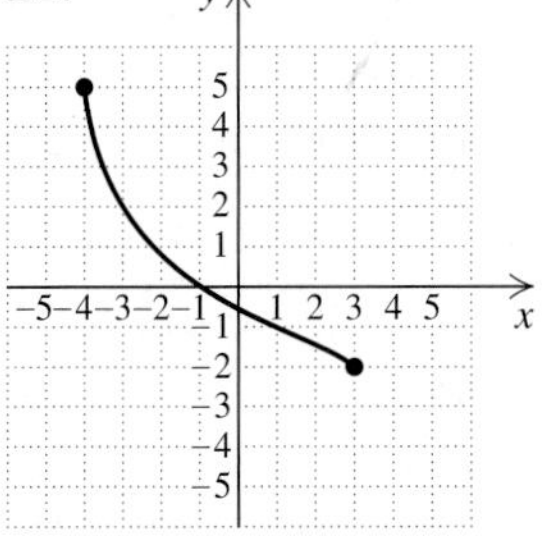

23.

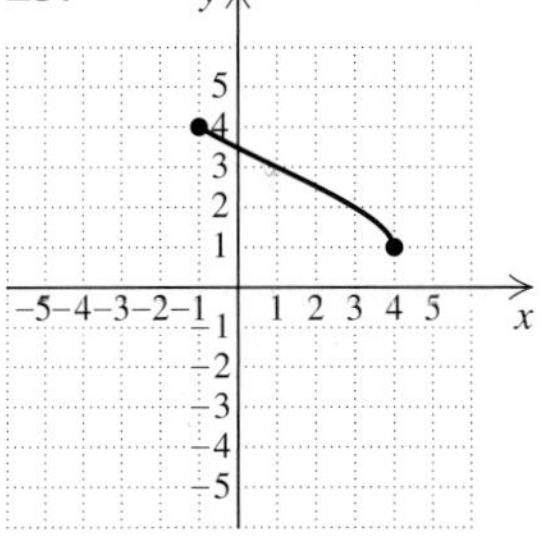

24.

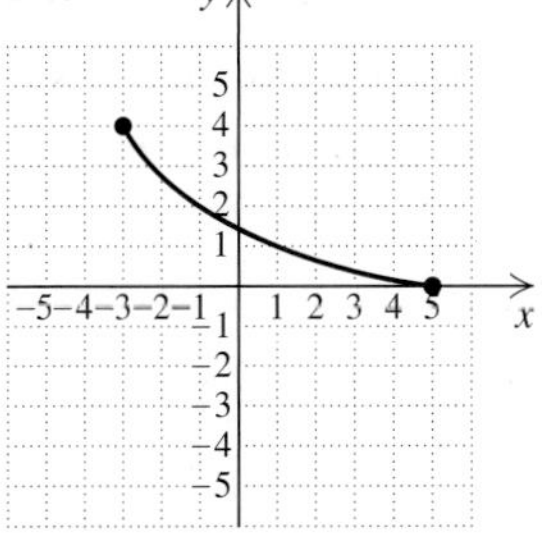

25.

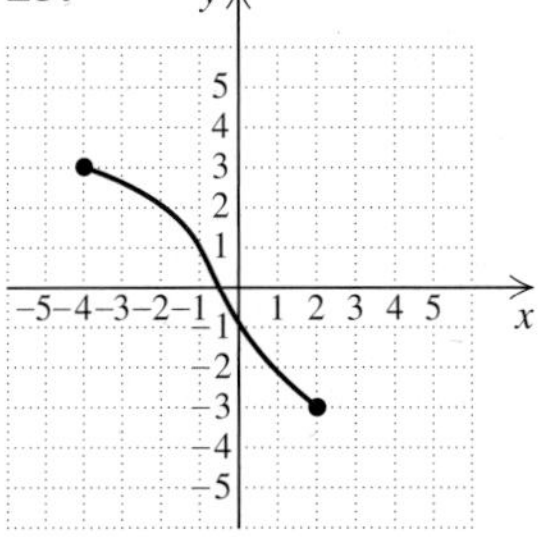

26.

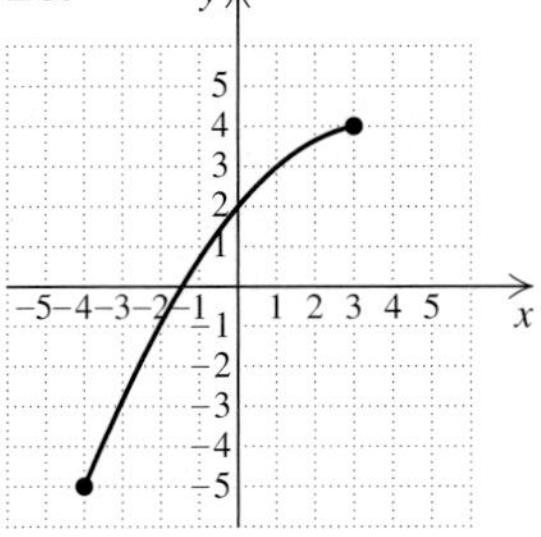

27.

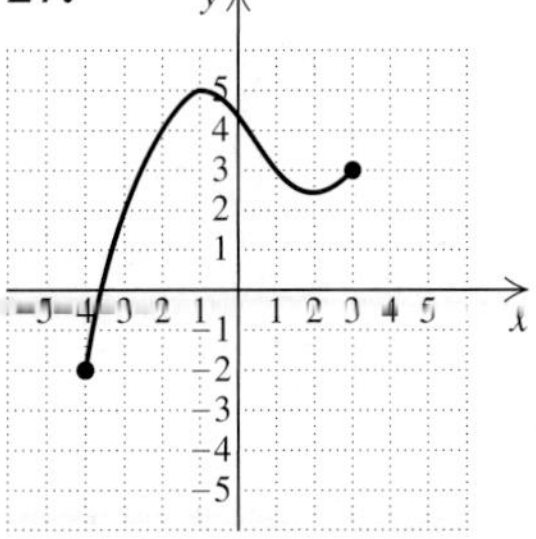

28.

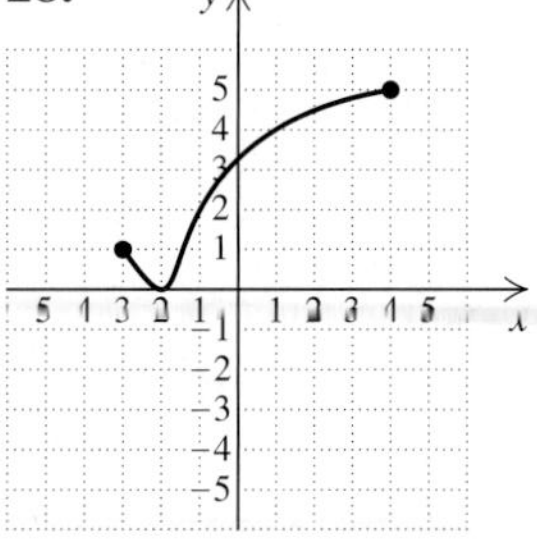

29.

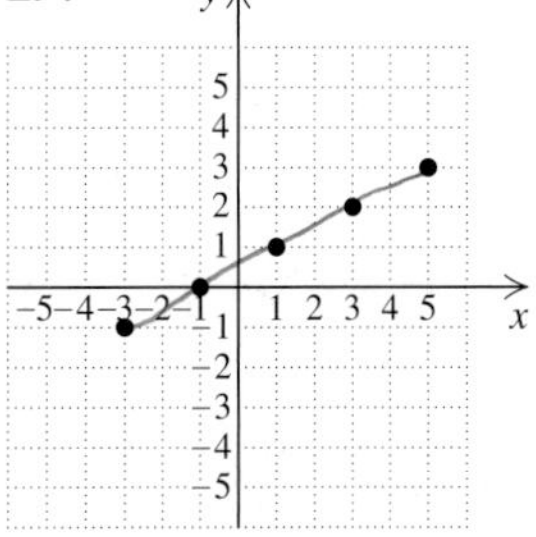

30.

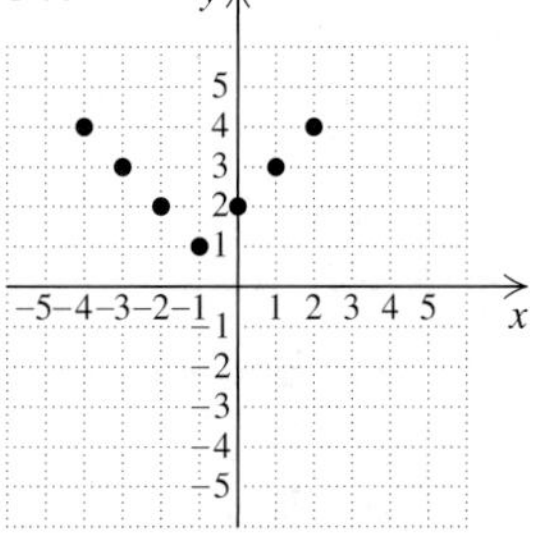

31.

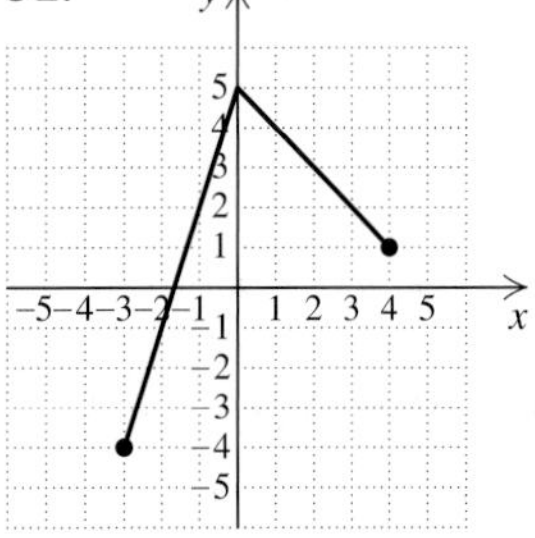

32.

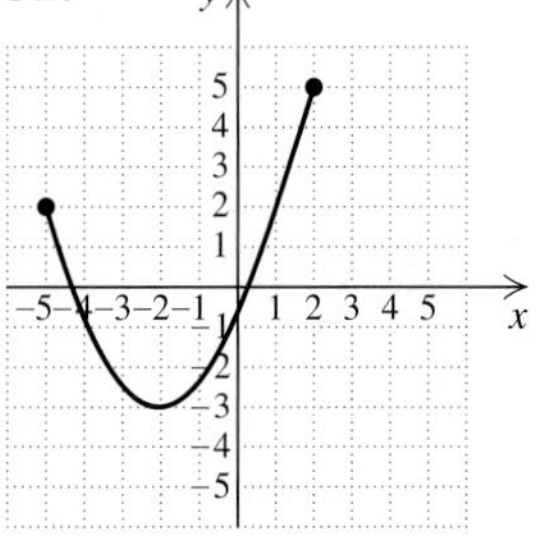

33.

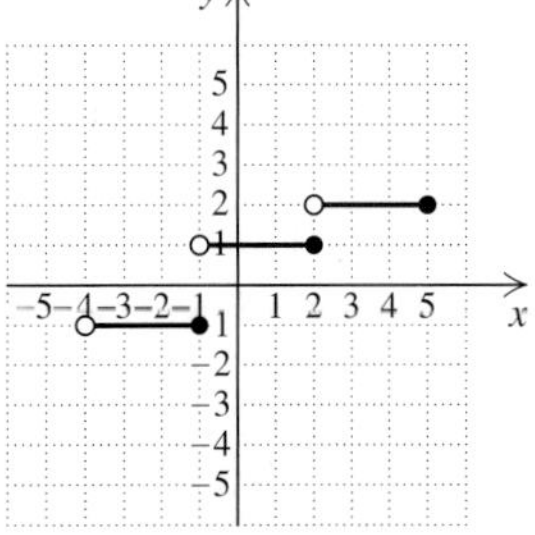

34.

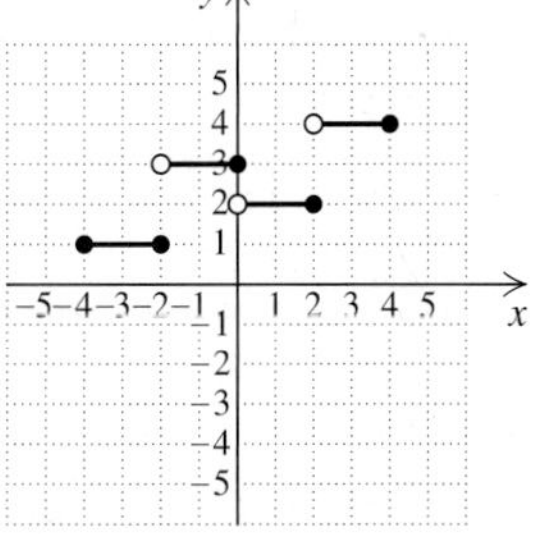

Determine whether each of the following is the graph of a function.

35.

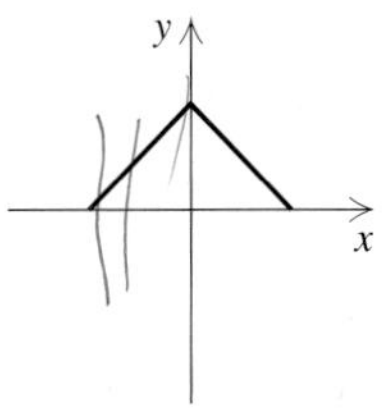

36.

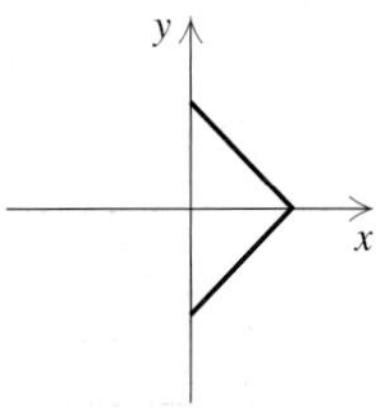

37.
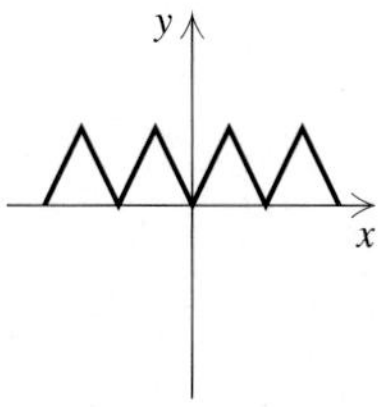

38.
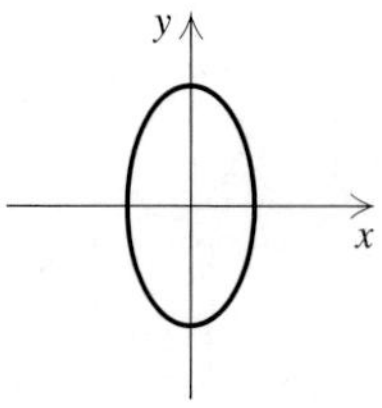

39.
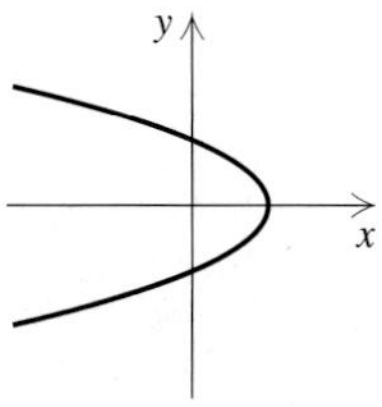

40.
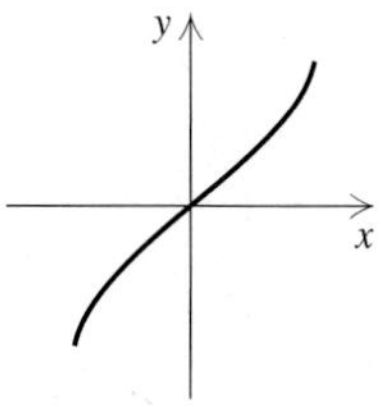

41.
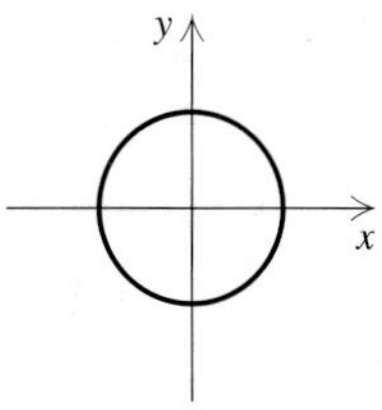

42.
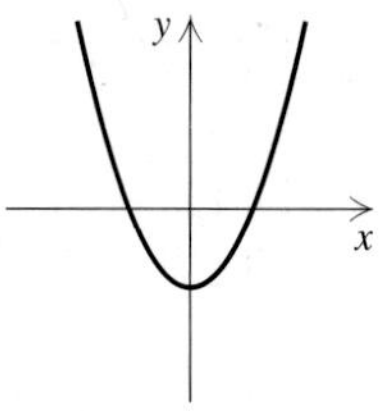

Find the function values.

43. $g(x) = 2x + 3$

a) $g(0)$ **b)** $g(-4)$ **c)** $g(-7)$
d) $g(8)$ **e)** $g(a + 2)$ **f)** $g(a) + 2$

44. $h(x) = 3x - 2$

a) $h(4)$ **b)** $h(8)$ **c)** $h(-3)$
d) $h(-4)$ **e)** $h(a - 1)$ **f)** $h(a) - 1$

45. $f(n) = 5n^2 + 4n$

a) $f(0)$ **b)** $f(-1)$ **c)** $f(3)$
Aha! **d)** $f(t)$ **e)** $f(2a)$ **f)** $2 \cdot f(a)$

46. $g(n) = 3n^2 - 2n$

a) $g(0)$ **b)** $g(-1)$ **c)** $g(3)$
d) $g(t)$ **e)** $g(2a)$ **f)** $2 \cdot g(a)$

47. $f(x) = \dfrac{x - 3}{2x - 5}$

a) $f(0)$ **b)** $f(4)$ **c)** $f(-1)$
d) $f(3)$ **e)** $f(x + 2)$

48. $s(x) = \dfrac{3x - 4}{2x + 5}$

a) $s(10)$ **b)** $s(2)$ **c)** $s(\frac{1}{2})$
d) $s(-1)$ **e)** $s(x + 3)$

The function A described by $A(s) = s^2\dfrac{\sqrt{3}}{4}$ *gives the area of an equilateral triangle with side s.*

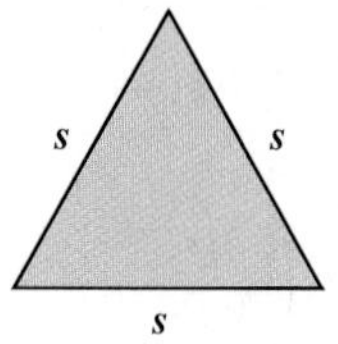

49. Find the area when a side measures 4 cm.

50. Find the area when a side measures 6 in.

The function V described by $V(r) = 4\pi r^2$ *gives the surface area of a sphere with radius r.*

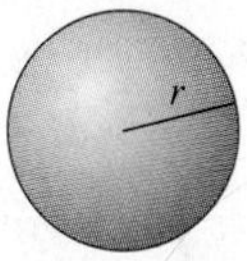

51. Find the surface area when the radius is 3 in.

52. Find the surface area when the radius is 5 cm.

53. *Pressure at Sea Depth.* The function $P(d) = 1 + (d/33)$ gives the pressure, in *atmospheres* (atm), at a depth of d feet in the sea. Note that $P(0) = 1$ atm, $P(33) = 2$ atm, and so on. Find the pressure at 20 ft, at 30 ft, and at 100 ft.

54. *Melting Snow.* The function $W(d) = 0.112d$ approximates the amount, in centimeters, of water that results from d centimeters of snow melting. Find the amount of water that results from snow melting from depths of 16 cm, 25 cm, and 100 cm.

Fill in the missing values in each table.

$f(x) = 2x - 5$

	x	$f(x)$
55.	8	
56.		13
57.		−5
58.	−4	

	$f(x) = \frac{1}{3}x + 4$	
	x	$f(x)$
59.		$\frac{1}{2}$
60.		$-\frac{1}{3}$
61.	$\frac{1}{2}$	
62.	$-\frac{1}{3}$	

63. If $f(x) = 4 - x$, for what input is the output 7?

64. If $f(x) = 5x + 1$, for what input is the output $\frac{1}{2}$?

65. If $f(x) = 0.1x - 0.5$, for what input is the output -3?

66. If $f(x) = 2.3 - 1.5x$, for what input is the output 10?

In Exercises 67–80, determine the zeros, if any, of each function.

67.

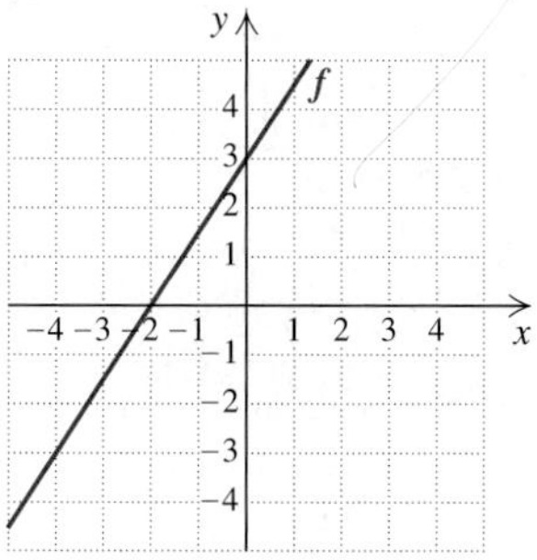

68.

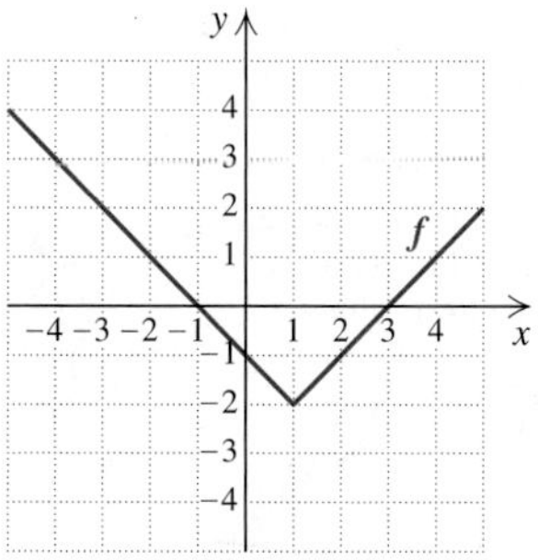

69.

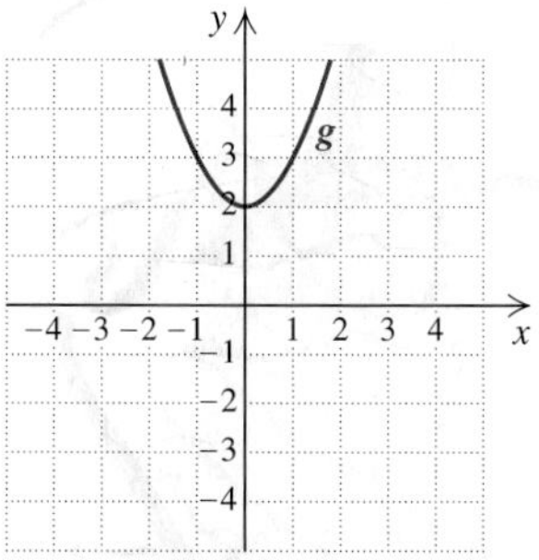

70.

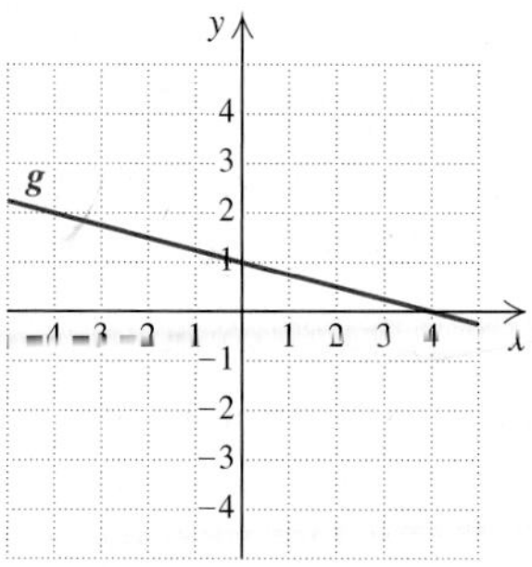

71.

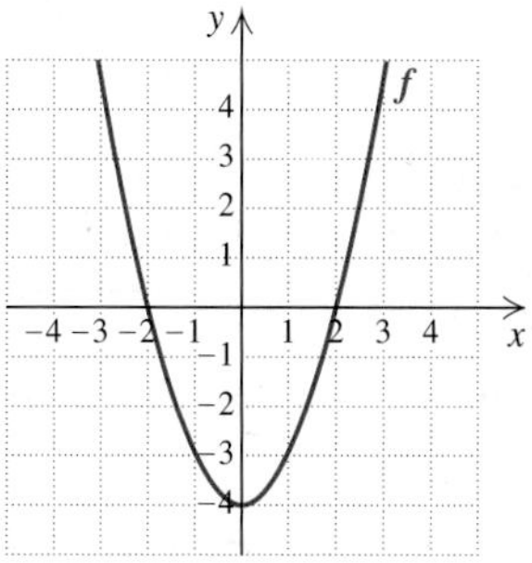

72.

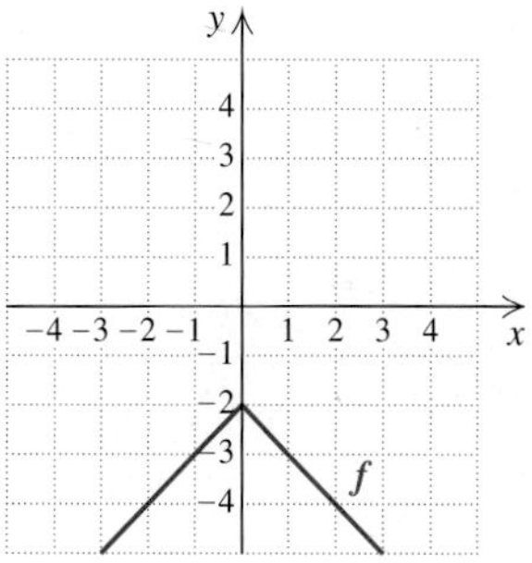

73. $f(x) = x - 5$

74. $f(x) = x + 3$

75. $f(x) = \frac{1}{2}x + 10$

76. $f(x) = \frac{2}{3}x - 6$

77. $f(x) = 2.7 - x$

78. $f(x) = 0.5 - x$

79. $f(x) = 3x + 7$

80. $f(x) = 5x - 8$

TW **81.** Explain the difference between finding $f(0)$ and finding the zeros of f.

TW **82.** Suppose a function f is defined by

$$f(x) = x^2 - 5x + 7.$$

Describe in your own words the operations performed on the input to produce the output.

Skill Maintenance

Graph by hand. [1.5]

83. $y = x + 3$

84. $y = x - 2$

85. $y = 2x$

86. $y = -2x$

Graph using a graphing calculator. [1.5]

87. $y = x^2$

88. $y = 0.1x - 5.7$

89. $y = |x|$

90. $y = \frac{1}{3}x - 15$

Synthesis

TW **91.** For the function given by $n(z) = ab + wz$, what is the independent variable? How can you tell?

TW **92.** Explain in your own words why every function is a relation, but not every relation is a function.

For Exercises 93 and 94, let $f(x) = 3x^2 - 1$ *and* $g(x) = 2x + 5$.

93. Find $f(g(-4))$ and $g(f(-4))$.

94. Find $f(g(-1))$ and $g(f(-1))$.

95. If f represents the function in Exercise 14, find $f(f(f(f(tiger))))$.

Pregnancy. For Exercises 96–99, use the following graph of a woman's "stress test." This graph shows the size of a pregnant woman's contractions as a function of time.

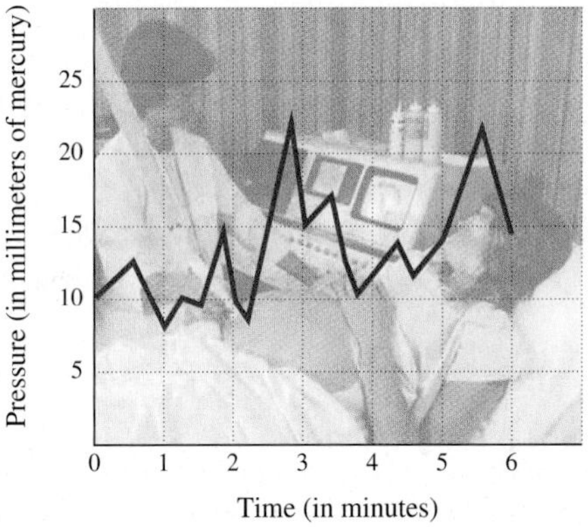

96. How large is the largest contraction that occurred during the test?

97. At what time during the test did the largest contraction occur?

TW **98.** On the basis of the information provided, how large a contraction would you expect 60 sec after the end of the test? Why?

99. What is the frequency of the largest contraction?

100. The *greatest integer function* $f(x) = [\![x]\!]$ is defined as follows: $[\![x]\!]$ is the greatest integer that is less than or equal to x. For example, if $x = 3.74$, then $[\![x]\!] = 3$; and if $x = -0.98$, then $[\![x]\!] = -1$. Graph the greatest integer function for $-5 \le x \le 5$. (The notation $f(x) = \text{int}(x)$, used in many graphing calculators, is often found in the MATH NUM submenu.)

101. Suppose that a function g is such that $g(-1) = -7$ and $g(3) = 8$. Find a formula for g if $g(x)$ is of the form $g(x) = mx + b$, where m and b are constants.

102. *Energy Expenditure.* On the basis of the information given below, determine what burns more energy: walking $4\frac{1}{2}$ mph for two hours or bicycling 14 mph for one hour.

Approximate energy expenditure by a 150-pound person during various activities

Activity	Calories per Hour
Walking, $2\frac{1}{2}$ mph	210
Bicycling, $5\frac{1}{2}$ mph	210
Walking, $3\frac{3}{4}$ mph	300
Bicycling, 13 mph	660

Source: Based on material prepared by Robert E. Johnson, M.D., Ph.D., and colleagues, University of Illinois.

2.2 Linear Functions: Slope, Graphs, and Models

Slope–Intercept Form of an Equation ■ Applications

The inside back cover illustrates the graphs of many different kinds of functions. In this section, we examine functions and real-life models with graphs that are straight lines. Such functions and their graphs are called *linear* and can be written in the form $f(x) = mx + b$, where m and b are constants.

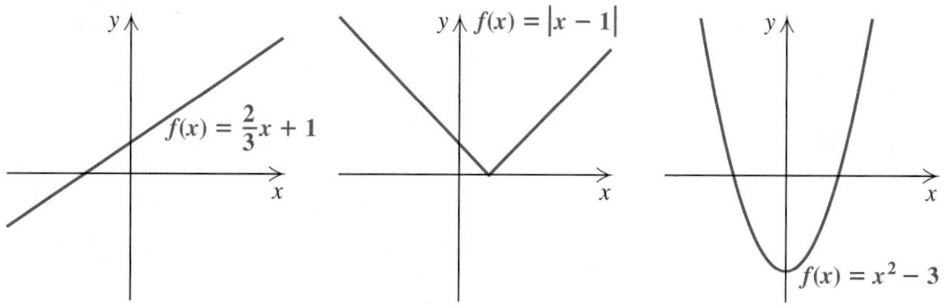

$f(x) = \frac{2}{3}x + 1$ is a *linear* function.

$f(x) = |x - 1|$ is a *nonlinear* function.

$f(x) = x^2 - 3$ is a *nonlinear* function.

Slope–Intercept Form of an Equation

We can learn much about functions by examining their graphs. Also, we can tell whether a function is linear by examining the equation that describes the function.

Interactive Discovery

Graph each of the following functions using a standard viewing window:

$$f(x) = 2x,$$
$$g(x) = -2x,$$
$$h(x) = 0.3x.$$

1. Do the graphs appear to be linear?

2. At what point do they cross the x-axis?

3. At what point do they cross the y-axis?

If your calculator has a Transfrm application, found in the APPS menu, run that application and enter $Y_1 = AX$ from the equation-editor screen. Then enter various values for A. When you are finished, select Transfrm again from the APPS menu and choose the Uninstall option.

4. Do all the graphs appear to be linear?

5. Do all the graphs go through the origin?

The pattern you may have observed is true in general.

A function given by an equation of the form $f(x) = mx$ is a *linear function*. Its graph is a straight line passing through the origin.

What happens to the graph of $f(x) = mx$ if we add a number b to get an equation of the form $f(x) = mx + b$?

Interactive Discovery

Graph the following equations on the same set of axes using a standard viewing window:

$$y_1 = 0.5x,$$
$$y_2 = 0.5x + 5,$$
$$y_3 = 0.5x - 7.$$

1. By examining the graphs and creating a table of values, explain how the values of y_2 and y_3 differ from those of y_1.
2. Predict what the graph of $y_4 = 0.5x + (-3.2)$ will look like. Test your description by graphing y_4 on the same set of axes as $y_1 = 0.5x$.

If your calculator has a Transfrm application, use it to enter $Y_1 = -3X + A$. Enter various values for A. Uninstall Transfrm when you are finished.

3. Describe what happens to the graph of $y_1 = -3x$ when a number b is added to $-3x$.

The pattern you may have observed is also true for other values of m.

EXAMPLE 1 Graph $y = 2x$ and $y = 2x + 3$ on the same set of axes.

SOLUTION We first make a table of solutions of both equations.

	y	y
x	$y = 2x$	$y = 2x + 3$
0	0	3
1	2	5
−1	−2	1
2	4	7
−2	−4	−1
3	6	9

Study Tip

Strike While the Iron's Hot

Make an effort to do your homework as soon as possible after each class. Make this part of your routine, choosing a time and a place where you can focus with a minimum of distractions.

We then plot these points. Drawing a blue line for $y = 2x + 3$ and a red line for $y = 2x$, we observe that the graph of $y = 2x + 3$ is simply the graph of $y = 2x$ shifted, or *translated*, 3 units up. The lines are parallel.

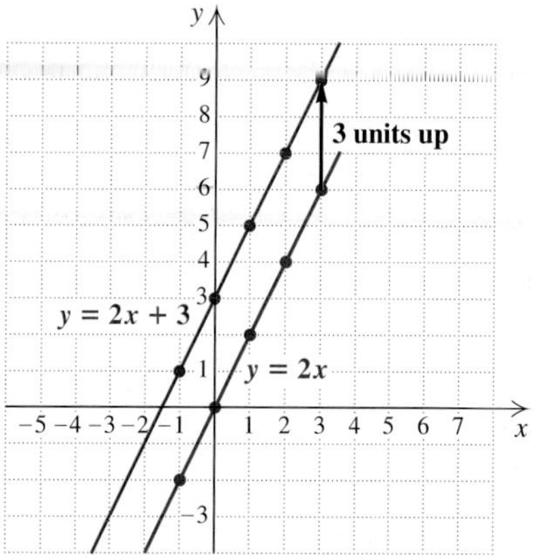

Note that the graph of $y = 2x + 3$ passes through the point $(0, 3)$. In general, we have the following.

The graph of $y = mx + b$, $b \neq 0$, is a line parallel to $y = mx$, passing through the point $(0, b)$.

The point $(0, b)$ is called the ***y*-intercept.** Often we refer to the number b as the y-intercept.

EXAMPLE 2 For each equation, find the y-intercept.

a) $y = -5x + 4$

b) $f(x) = 5.3x - 12$

SOLUTION

a) The y-intercept is $(0, 4)$, or simply 4.

b) The y-intercept is $(0, -12)$, or simply -12.

The number m, in the equation $y = mx + b$, is responsible for the slant of a line. Note that, for the graphs in Example 1, both equations have $m = 2$, and the slant of the red line seems to match the slant of the blue line. The following definition enables us to visualize this slant, or *slope*, of a line as a ratio of two lengths.

Slope The *slope* of the line passing through (x_1, y_1) and (x_2, y_2) is given by

$$m = \frac{\text{rise}}{\text{run}} = \frac{\text{vertical change}}{\text{horizontal change}}$$

$$= \frac{\text{the difference in } y}{\text{the difference in } x}$$

$$= \frac{y_2 - y_1}{x_2 - x_1} = \frac{y_1 - y_2}{x_1 - x_2}.$$

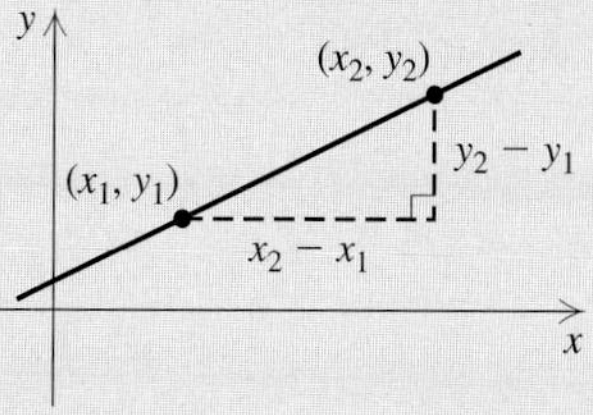

In the definition above, (x_1, y_1) and (x_2, y_2) — read "x sub-one, y sub-one and x sub-two, y sub-two" — represent two different points on a line. It does not matter which point is considered (x_1, y_1) and which is considered (x_2, y_2) so long as coordinates are subtracted in the same order in both the numerator and the denominator.

The letter m is traditionally used for slope. This usage has its roots in the French verb *monter*, to climb.

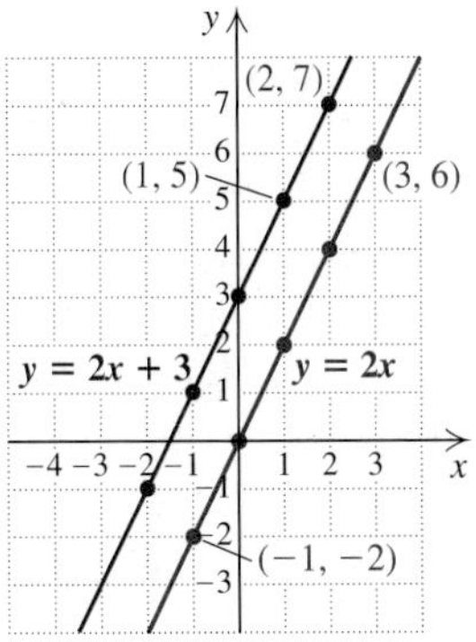

EXAMPLE 3 Find the slope of the lines drawn in Example 1.

SOLUTION To find the slope of a line, we can use the coordinates of any two points on that line. We use $(1, 5)$ and $(2, 7)$ to find the slope of the blue line in Example 1:

$$\text{Slope} = \frac{\text{rise}}{\text{run}} = \frac{\text{change in } y}{\text{change in } x} = \frac{y_2 - y_1}{x_2 - x_1} = \frac{7-5}{2-1} = 2.$$

To find the slope of the red line in Example 1, we use $(-1, -2)$ and $(3, 6)$:

$$\text{Slope} = \frac{\text{rise}}{\text{run}} = \frac{\text{change in } y}{\text{change in } x} = \frac{6-(-2)}{3-(-1)} = \frac{8}{4} = 2.$$

In Example 3, we found that the lines given by $y = 2x + 3$ and $y = 2x$ both have a slope of 2. This supports (but does not prove) the following:

The slope of any line written in the form $y = mx + b$ is m.

A proof of this result is outlined in Exercise 104 on page 132.

Note in the definition of slope that coordinates can be subtracted in either order, so long as the order is the same in the numerator and the denominator. Thus, in Example 3, the slope of the blue line can also be found by

$$\frac{y_1 - y_2}{x_1 - x_2} = \frac{5-7}{1-2} = \frac{-2}{-1} = 2.$$

Also note that the slope of a line is constant, so *any* two points on a line can be used to determine its slope.

EXAMPLE 4 Determine the slope of the line given by $y = \frac{2}{3}x + 4$, and graph the line.

SOLUTION Here $m = \frac{2}{3}$, so the slope is $\frac{2}{3}$. This means that from *any* point on the graph, we can locate a second point by simply going *up* 2 units (the *rise*) and *to the right* 3 units (the *run*). Where do we start? Because the y-intercept, $(0, 4)$, is known to be on the graph, we calculate that

$(0 + 3, 4 + 2)$, or $(3, 6)$, is also on the graph. Knowing two points, we can draw the line.

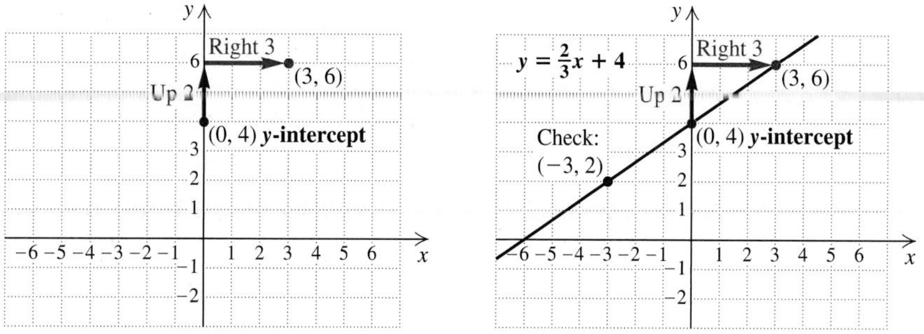

Important! To check the graph, we use some other value for x, say, -3, and determine y (in this case, 2). We plot that point and see that it *is* on the line. Were it not, we would know that some error had been made.

Slope–Intercept Form Any equation $y = mx + b$ has a graph that is a straight line. It goes through the *y-intercept* $(0, b)$ and has slope m.

Any equation of the form

$$y = mx + b$$

is said to be written in *slope–intercept form.*

EXAMPLE 5 Determine the slope and the y-intercept of the line given by $y = -\frac{1}{3}x + 2$.

SOLUTION The equation $y = -\frac{1}{3}x + 2$ is written in the form $y = mx + b$, with $m = -\frac{1}{3}$ and $b = 2$. Thus the slope is $-\frac{1}{3}$ and the y-intercept is $(0, 2)$.

Note that any graph of $y = mx + b$ will pass the vertical-line test and thus represents a function.

EXAMPLE 6 Find a linear function whose graph has slope $-\frac{2}{3}$ and y-intercept $(0, 4)$.

SOLUTION We use the slope–intercept form, $f(x) = mx + b$:

$f(x) = -\frac{2}{3}x + 4.$ **Substituting $-\frac{2}{3}$ for m and 4 for b; using function notation**

Interactive Discovery

To see the effect of the sign of m on the graph of $y = mx + b$, graph the following equations on the same set of axes using the standard viewing window:

$$y_1 = x + 1, \qquad y_3 = 2x + 1,$$
$$y_2 = -x + 1, \qquad y_4 = -2x + 1.$$

1. Which graphs slant up from left to right?
2. Which graphs slant down from left to right?
3. Compare the graphs of y_1 and y_3. Which graph is steeper?

If your calculator has a Transfrm application, use it to enter $Y_1 = AX + B$. Choose a value for B and enter various values for A. Choose some negative values and some positive values. Uninstall Transfrm when you are finished.

4. Describe how the sign of m affects the graph of $y = mx + b$.
5. Describe how different positive values of m affect the graph of $y = mx + b$.

When the slope of a line is positive, the line slants up from left to right. When the slope of a line is negative, the line slants down from left to right. The larger the absolute value of the slope, the steeper the slant.

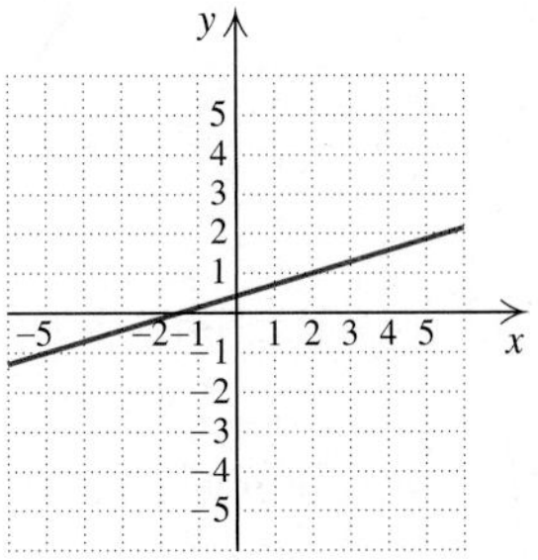

$m = \frac{2}{7}$

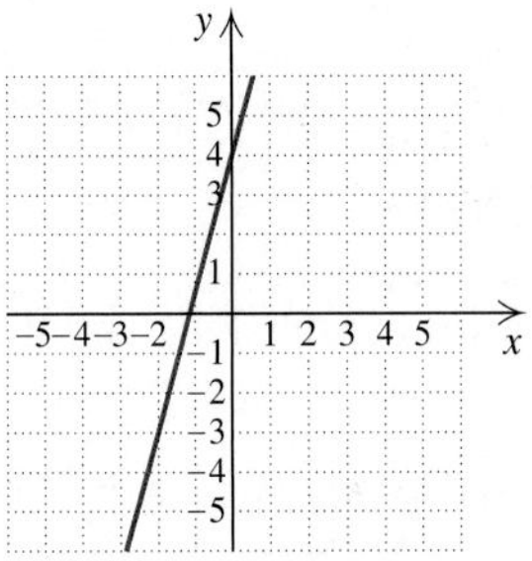

$m = \frac{7}{2}$

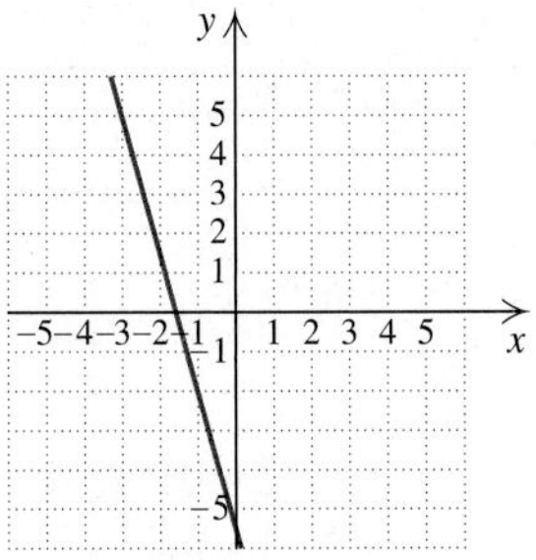

$m = -\frac{7}{2}$

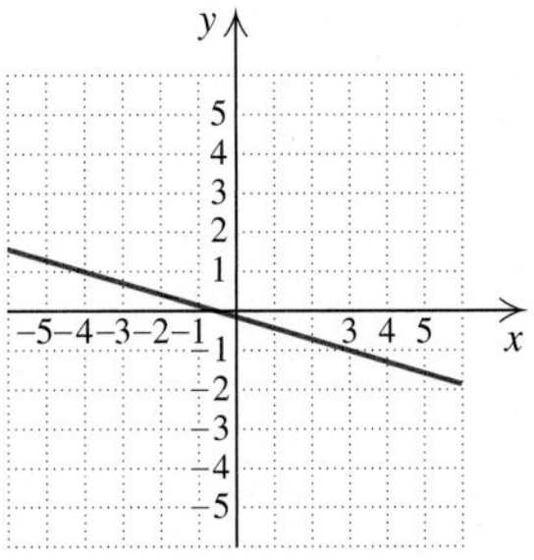

$m = -\frac{2}{7}$

Student Notes

Slope–intercept form is not only important—it is convenient to use. The sooner you feel comfortable writing, reading, and graphing slope–intercept form, the easier your work will become.

EXAMPLE 7 Graph: $f(x) = -\frac{1}{2}x + 5$.

SOLUTION The y-intercept is $(0, 5)$. The slope is $-\frac{1}{2}$, or $\frac{-1}{2}$. From the y-intercept, we go *down* 1 unit and *to the right* 2 units. That gives us the point $(2, 4)$. We can now draw the graph.

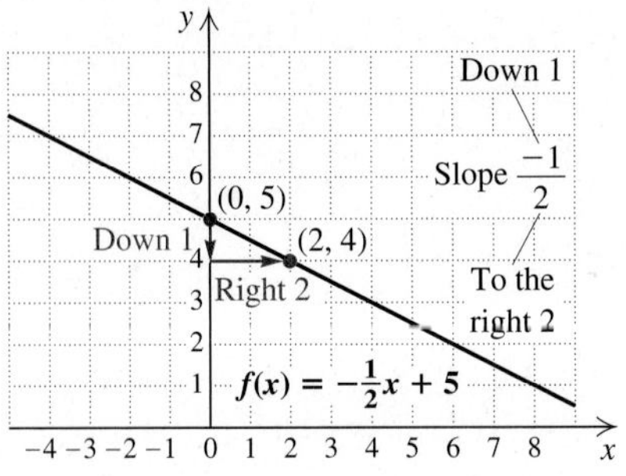

As a new type of check, we rewrite the slope in another form and find another point:

$$-\frac{1}{2} = \frac{1}{-2}.$$

Thus we can go *up* 1 unit and then *to the left* 2 units. This gives the point $(-2, 6)$. Since $(-2, 6)$ is on the line, we have a check.

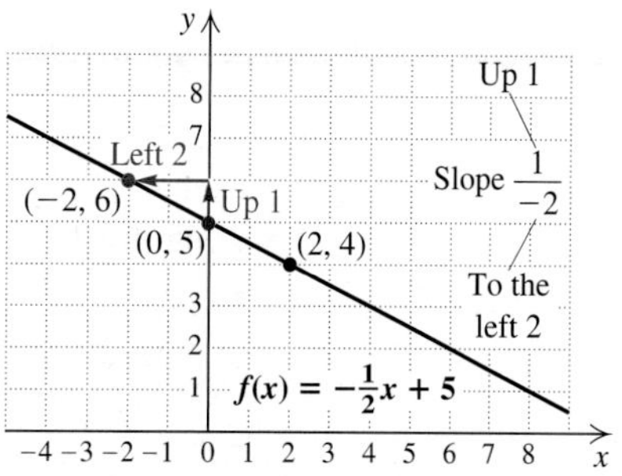

Note that $m = -\frac{1}{2}$ is negative. Thus the graph slants downward from left to right.

Often the easiest way to graph an equation is to rewrite it in slope–intercept form and then proceed as in Examples 4 and 7 above.

EXAMPLE 8 Determine the slope and the y-intercept for the equation $5x - 4y = 8$. Then graph.

SOLUTION We convert to slope–intercept form:

$$5x - 4y = 8$$

$$-4y = -5x + 8 \qquad \textbf{Adding } -5x \textbf{ to both sides}$$

$$y = -\tfrac{1}{4}(-5x + 8) \qquad \textbf{Multiplying both sides by } -\tfrac{1}{4}$$

$$y = \tfrac{5}{4}x - 2. \qquad \textbf{Using the distributive law}$$

Because we have an equation of the form $y = mx + b$, we know that the slope is $\frac{5}{4}$ and the y-intercept is $(0, -2)$. We plot $(0, -2)$, and from there we go *up* 5 units and *to the right* 4 units, and plot a second point at $(4, 3)$. We then draw the graph.

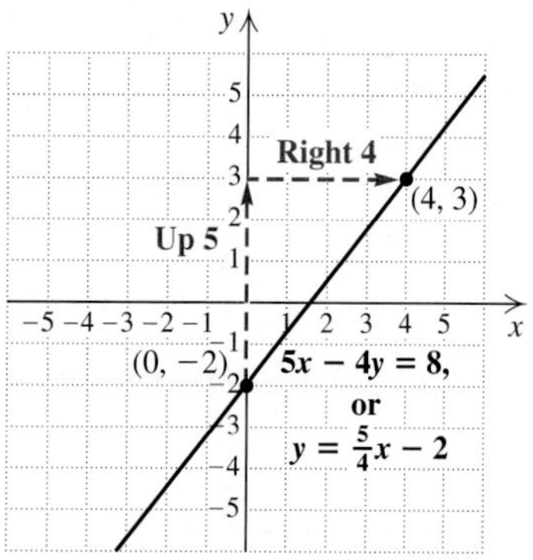

Note that $m = \frac{5}{4}$ is positive. Thus the graph slants upward from left to right.

To check that the line is drawn correctly, we calculate the coordinates of another point on the line. For $x = 2$, we have

$$\begin{aligned} 5 \cdot 2 - 4y &= 8 \\ 10 - 4y &= 8 \\ -4y &= -2 \\ y &= \frac{1}{2}. \end{aligned}$$

Thus, $\left(2, \frac{1}{2}\right)$ should appear on the graph. Since it *does* appear to be on the line, we have a check.

When graphing an equation, check first to see if it is linear. If it can be written in slope–intercept form, $y = mx + b$, it is linear. Then either plot at least two points from a table, or plot the y-intercept and find more points using the slope. Then draw the line. If the equation is not linear, more work must be done to determine the shape of the graph.

Applications

Because slope is a ratio that indicates how a change in the vertical direction corresponds to a change in the horizontal direction, it has many real-world applications. Foremost is the use of slope to represent a *rate of change*.

EXAMPLE 9 College Faculty. The number of full-time faculty in institutions of higher education is increasing linearly. Use the following graph to find the rate at which this number is growing.

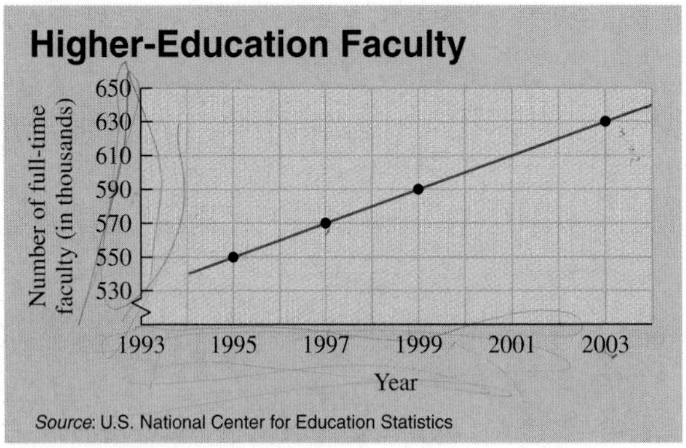

SOLUTION Note that the amounts on the horizontal axis are given in years and those on the vertical axis are given in thousands of faculty mem bers. Thus the rate of change will be given in thousands of faculty members per year. Since the graph is linear, we can use any pair of points to determine the rate of change. We choose the points (2003, 630) and (1997, 570), which gives us

$$\text{Rate of change} = \frac{630 \text{ thousand faculty} - 570 \text{ thousand faculty}}{2003 - 1997}$$
$$= \frac{60 \text{ thousand faculty}}{6 \text{ years}}$$
$$= 10 \text{ thousand faculty members per year.}$$

The number of full-time faculty in U.S. institutions of higher education is growing at a rate of 10,000 members per year.

EXAMPLE 10 Real Estate. The number of houses sold often depends on the mortgage interest rate, which in turn depends on the Federal Funds rate. During one month in 2005, the Federal Funds rate was 3.78% and 6,180,000 existing homes were sold. During another month, the Federal Funds rate was 3.62% and 6,290,000 existing homes were sold. (*Source*: *Mortgage Banking*, April 2006) At what rate did the sales of existing homes change?

SOLUTION The rate at which the sales of existing homes changed is given by

$$\text{Rate of change} = \frac{\text{Change in sales of homes}}{\text{Change in interest rate}}$$
$$= \frac{6{,}180{,}000 \text{ homes} - 6{,}290{,}000 \text{ homes}}{3.78\% - 3.62\%}$$
$$= \frac{-110{,}000 \text{ homes}}{0.16\%}$$
$$= -687{,}500 \text{ homes per percent change in interest rate.}$$

The number of existing homes sold each month decreased by 687,500 for every 1% increase in the Federal Funds rate.

EXAMPLE 11 Running Speed. Stephanie runs 10 km during each workout. For the first 7 km, her pace is twice as fast as it is for the last 3 km. Which of the following graphs best describes Stephanie's workout?

A.

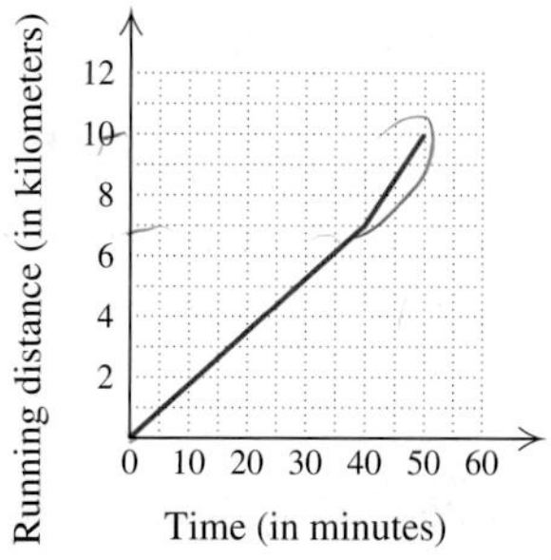

B.

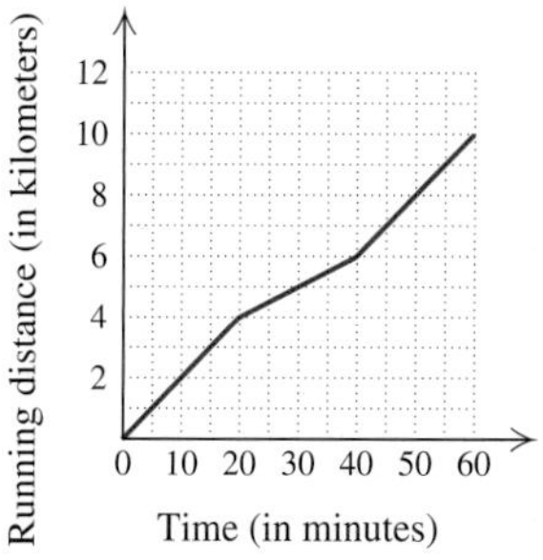

C.

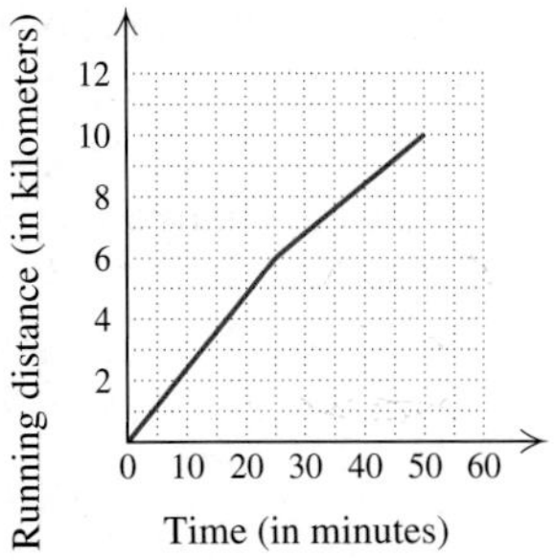

D.

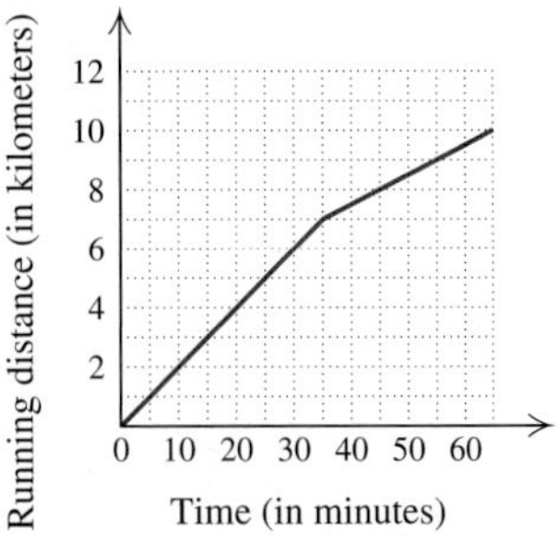

SOLUTION The slopes in graph A increase as we move to the right. This would indicate that Stephanie ran faster for the *last* part of her workout. Thus graph A is not the correct one.

The slopes in graph B indicate that Stephanie slowed down in the middle of her run and then resumed her original speed. Thus graph B does not correctly model the situation either.

According to graph C, Stephanie slowed down not at the 7-km mark, but at the 6-km mark. Thus graph C is also incorrect.

Graph D indicates that Stephanie ran the first 7 km in 35 min, a rate of 0.2 km/min. It also indicates that she ran the final 3 km in 30 min, a rate of 0.1 km/min. This means that Stephanie's rate was twice as fast for the first 7 km, so graph D provides a correct description of her workout.

Linear functions arise continually in today's world. As with the rate problems appearing in Examples 9–11, it is critical to use proper units in all answers.

EXAMPLE 12 Salvage Value. Tyline Electric uses the function

$$S(t) = -700t + 3500$$

to determine the *salvage value* $S(t)$, in dollars, of a color photocopier t years after its purchase.

a) What do the numbers -700 and 3500 signify?

b) How long will it take the copier to *depreciate* completely?

c) What is the domain of S?

SOLUTION Drawing, or at least visualizing, a graph can be useful here.

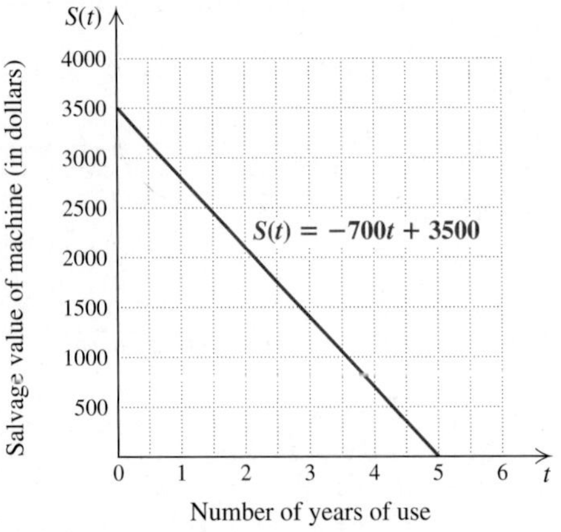

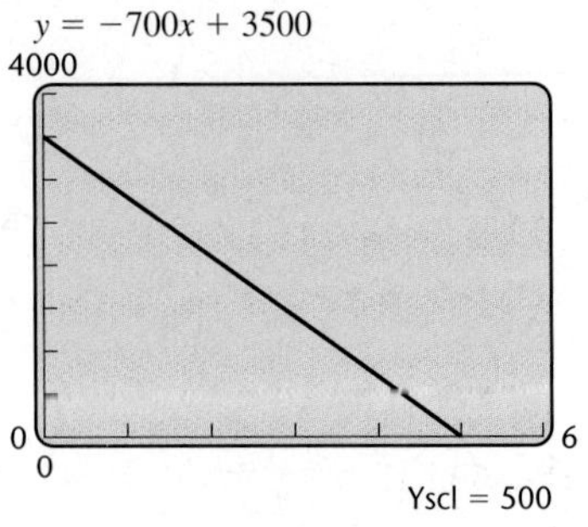

a) At time $t = 0$, we have $S(0) = -700 \cdot 0 + 3500 = 3500$. Thus the number 3500 signifies the original cost of the copier, in dollars.

This function is written in slope–intercept form. Since the output is measured in dollars and the input in years, the number -700 signifies that the value of the copier is declining at a rate of \$700 per year.

b) The copier will have depreciated completely when its value drops to 0. To learn when this occurs, we determine when $S(t) = 0$:

$S(t) = 0$	**A graph is not always available.**
$-700t + 3500 = 0$	**Substituting $-700t + 3500$ for $S(t)$**
$-700t = -3500$	**Subtracting 3500 from both sides**
$t = 5.$	**Dividing both sides by -700**

The copier will have depreciated completely in 5 yr.

c) Neither the number of years of service nor the salvage value can be negative. In part (b) we found that after 5 yr the salvage value will have dropped to 0. Thus the domain of S is $\{t \mid 0 \le t \le 5\}$. Either graph above serves as a visual check of this result.

2.2 EXERCISE SET

FOR EXTRA HELP

MathXL | MyMathLab | InterAct Math | AW Math Tutor Center | Video Lectures on CD: Disc 1 | Student's Solutions Manual

Concept Reinforcement *In each of Exercises 1–6, match the word with the most appropriate choice from the column on the right.*

1. ___ Rise — **a)** $y = mx + b$

2. ___ Run — **b)** Shifted

3. ___ Slope — **c)** $\dfrac{\text{Difference in } y}{\text{Difference in } x}$

4. ___ y-intercept — **d)** Difference in x

5. ___ Slope–intercept form — **e)** Difference in y

6. ___ Translated — **f)** $(0, b)$

Graph.

7. $f(x) = 2x - 7$

8. $g(x) = 3x - 7$

9. $g(x) = -\frac{1}{3}x + 2$

10. $f(x) = -\frac{1}{2}x - 5$

11. $h(x) = \frac{2}{5}x - 4$

12. $h(x) = \frac{4}{5}x + 2$

Determine the y-intercept.

13. $y = 3x + 4$

14. $y = 4x - 9$

15. $g(x) = -4x - 3$

16. $g(x) = -5x + 7$

17. $y = -\frac{3}{8}x - 4.5$

18. $y = \frac{15}{7}x + 2.2$

19. $f(x) = 2.9x - 9$

20. $f(x) = -3.1x + 5$

21. $y = 37x + 204$

22. $y = -52x + 700$

For each pair of points, find the slope of the line containing them.

23. $(6, 9)$ and $(4, 5)$

24. $(8, 7)$ and $(2, -1)$

25. $(3, 8)$ and $(9, -4)$

26. $(17, -12)$ and $(-9, -15)$

27. $\left(-\frac{2}{3}, \frac{1}{2}\right)$ and $\left(-\frac{4}{5}, \frac{4}{5}\right)$

28. $\left(\frac{3}{4}, -\frac{2}{5}\right)$ and $\left(\frac{1}{3}, -\frac{1}{4}\right)$

29. $(-9.7, 43.6)$ and $(4.5, 43.6)$

30. $(-2.8, -3.1)$ and $(-1.8, -2.6)$

Aha! **31.** Use the slope and the y-intercept of each line to match each equation with the correct graph.

a) $y = 3x - 5$
b) $y = 0.7x + 1$
c) $y = -0.25x - 3$
d) $y = -4x + 2$

I

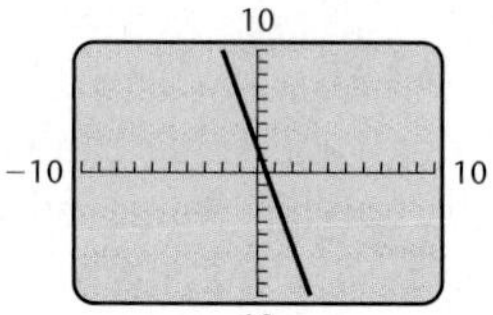

II

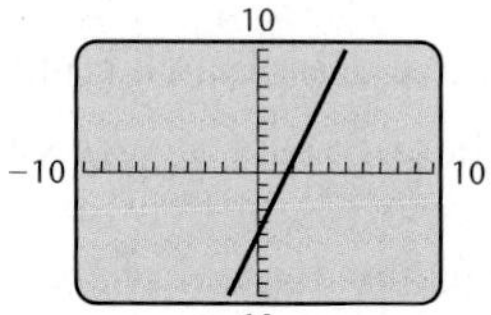

III

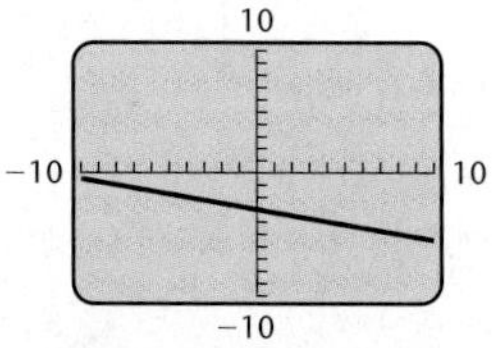

IV

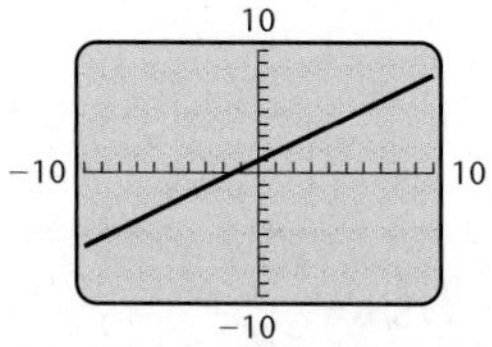

32. Use the slope and the y-intercept of each line to match each equation with the correct graph.

a) $y = \frac{1}{2}x - 5$
b) $y = 2x + 3$
c) $y = -3x + 1$
d) $y = -\frac{3}{4}x - 2$

I

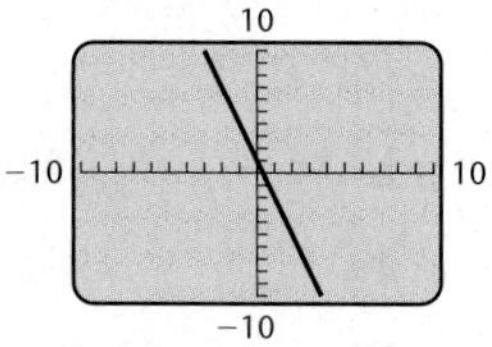

II

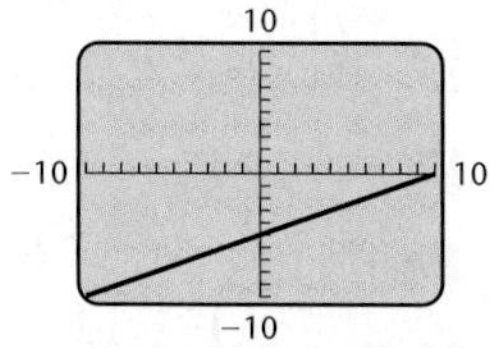

III

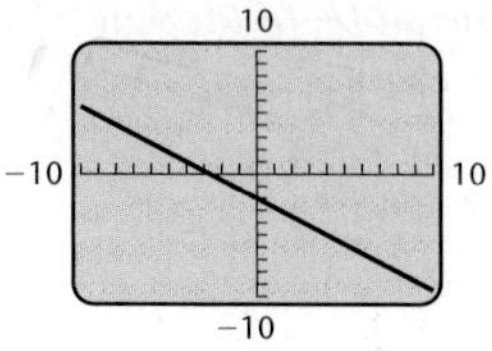

IV

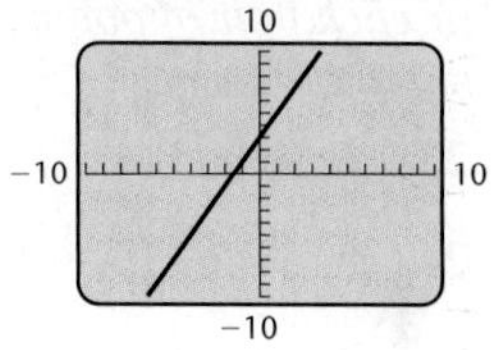

Determine the slope and the y-intercept. Then draw a graph. Be sure to check as in Example 4 or Example 7.

33. $y = \frac{5}{2}x + 3$

34. $y = \frac{2}{5}x + 4$

35. $f(x) = -\frac{5}{2}x + 2$

36. $f(x) = -\frac{2}{5}x + 3$

37. $2x - y = 5$

38. $2x + y = 4$

39. $F(x) = \frac{1}{3}x + 2$

40. $g(x) = -3x + 6$

41. $6y + 4x = 6$

42. $4y + 20 = x$

Aha! **43.** $g(x) = -0.25x$

44. $F(x) = 1.5x - 3$

45. $4x - 5y = 10$

46. $5x + 4y = 4$

47. $f(x) = \frac{5}{4}x - 2$

48. $f(x) = \frac{4}{3}x + 2$

49. $12 - 4f(x) = 3x$

50. $15 + 5f(x) = -2x$

Aha! **51.** $g(x) = 4.5$

52. $g(x) = \frac{3}{4}x$

Find a linear function whose graph has the given slope and y-intercept.

53. Slope $\frac{2}{3}$, y-intercept $(0, -9)$

54. Slope $-\frac{3}{4}$, y-intercept $(0, 12)$

55. Slope -6, y-intercept $(0, 2)$

56. Slope 2, y-intercept $(0, -1)$

57. Slope $-\frac{7}{9}$, y-intercept $(0, 5)$

58. Slope $-\frac{4}{11}$, y-intercept $(0, 9)$

59. Slope 5, y-intercept $\left(0, \frac{1}{2}\right)$

60. Slope 6, y-intercept $\left(0, \frac{2}{3}\right)$

For each graph, find the rate of change. Remember to use appropriate units. See Example 9.

61.

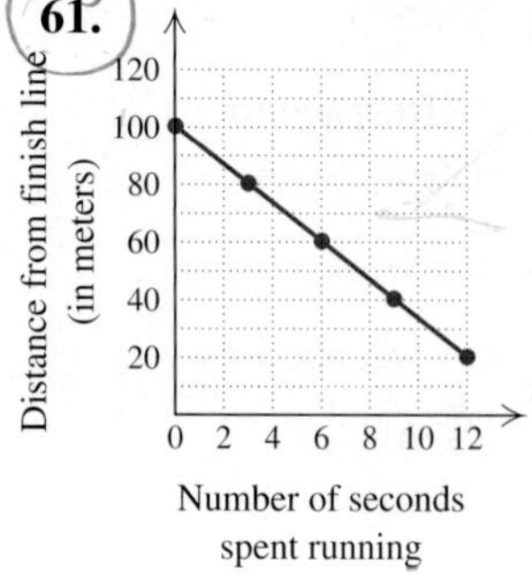

62.

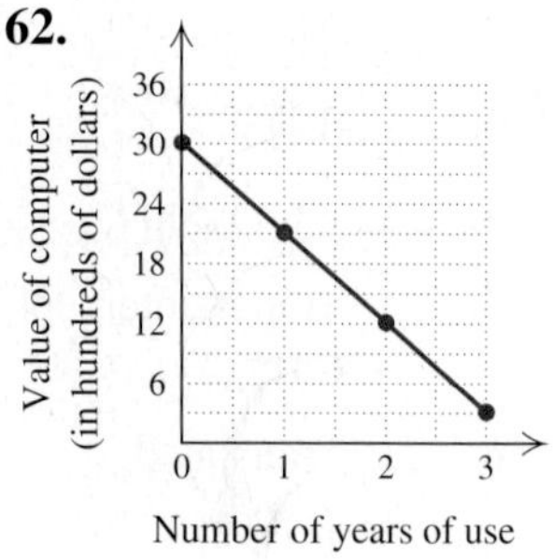

63.

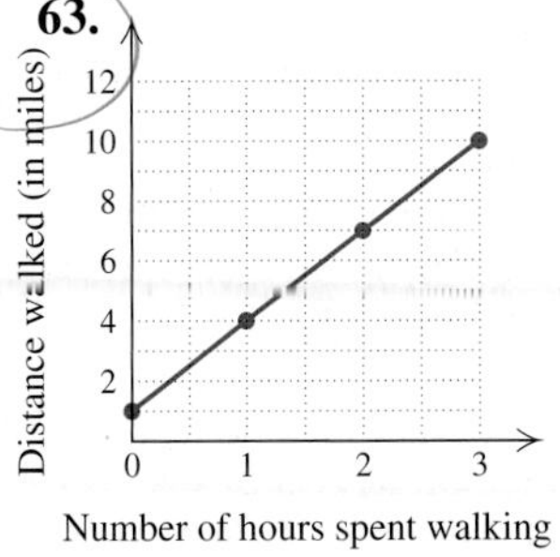

64.

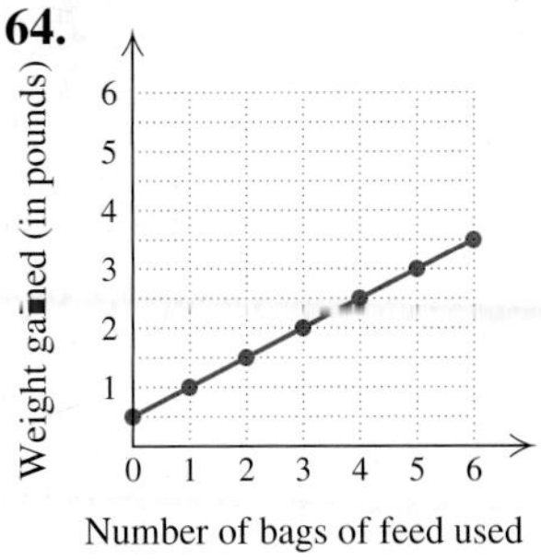

65.

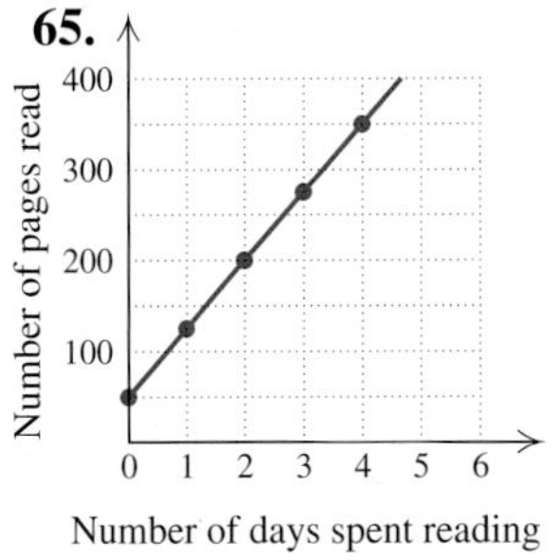

66.

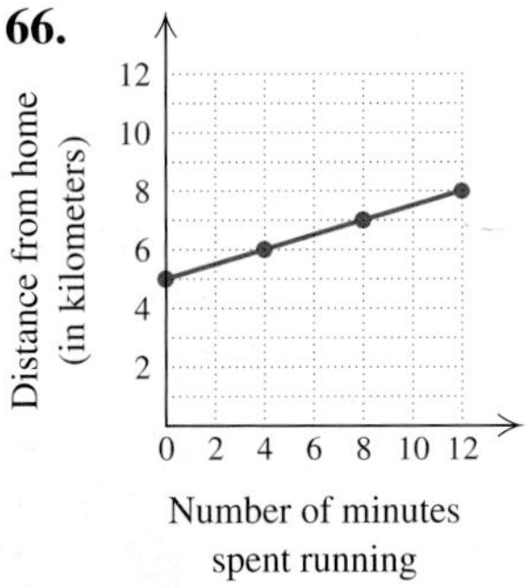

***67.**

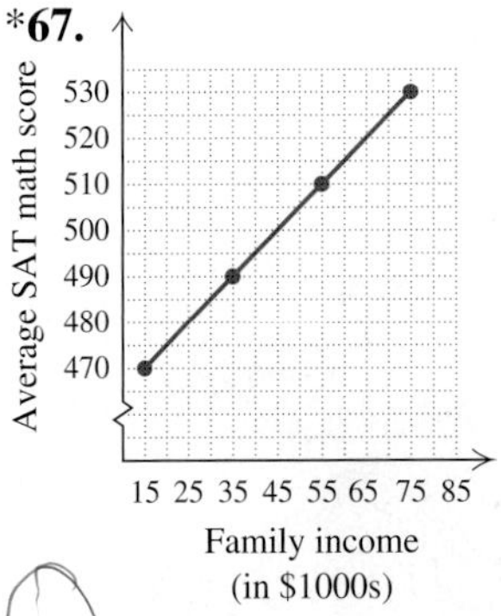

***68.**

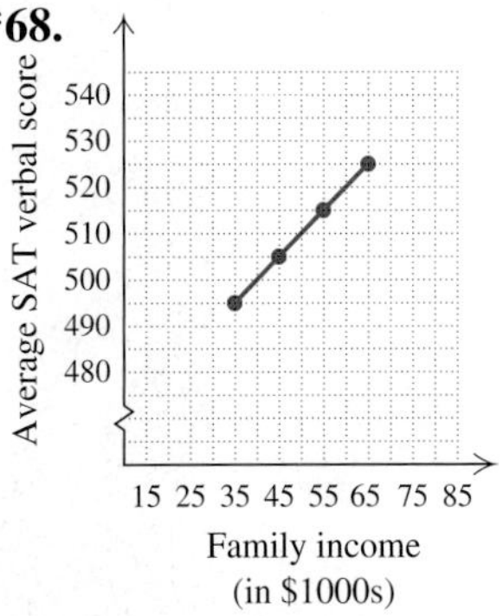

69. *Running Rate.* An ultra-marathoner passes the 15-mi point of a race after 2 hr and reaches the 22-mi point 56 min later. Assuming a constant rate, find the speed of the marathoner, in miles per hour.

70. *Skiing Rate.* A cross-country skier reaches the 3-km mark of a race in 15 min and the 12-km mark 45 min later. Assuming a constant rate, find the speed of the skier, in kilometers per hour.

71. *Rate of Descent.* A plane begins to descend to sea level from 12,000 ft after being airborne for $1\frac{1}{2}$ hr. The entire flight time is 2 hr 10 min. Determine the average rate of descent of the plane.

72. *Work Rate.* As a painter begins work, one-fourth of a house has already been painted. Eight hours later, the house is two-thirds done. Calculate the painter's work rate.

73. *Length of Cell-Phone Call.* In 1995, the average length of a cell-phone call was 2.15 min. By 2005, the average length was 3.00 min. (*Source*: Cellular Telecommunications Industry Association) Determine the rate at which the average length of a call increased.

74. *Wireline Phone Calls.* In 1998, 544 billion local calls were made using local exchange carriers. This number decreased to 426 billion calls in 2003. (*Source*: U.S. Federal Communications Commission) Determine the rate at which the number of local calls decreased.

75. *Nursing.* Match each sentence on the following page with the most appropriate of the four graphs shown.

I

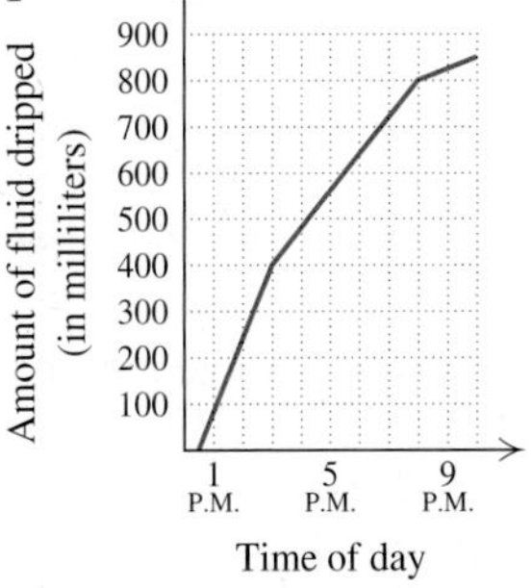

II

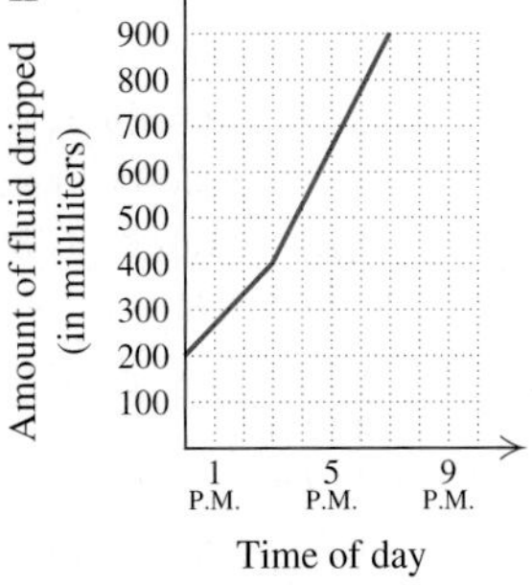

*Based on data from the College Board Online.

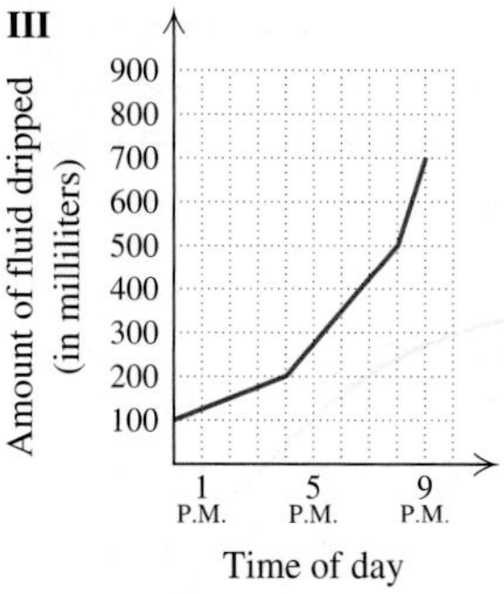

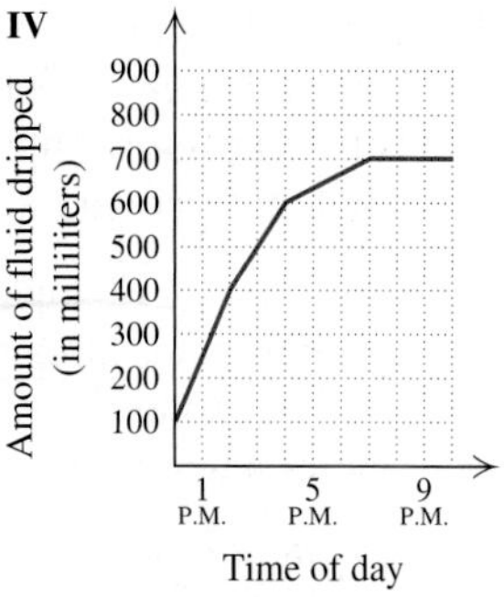

a) The rate at which fluids were given intravenously was doubled after 3 hr.
b) The rate at which fluids were given intravenously was gradually reduced to 0.
c) The rate at which fluids were given intravenously remained constant for 5 hr.
d) The rate at which fluids were given intravenously was gradually increased.

76. *Market Research.* Match each sentence with the most appropriate graph below.

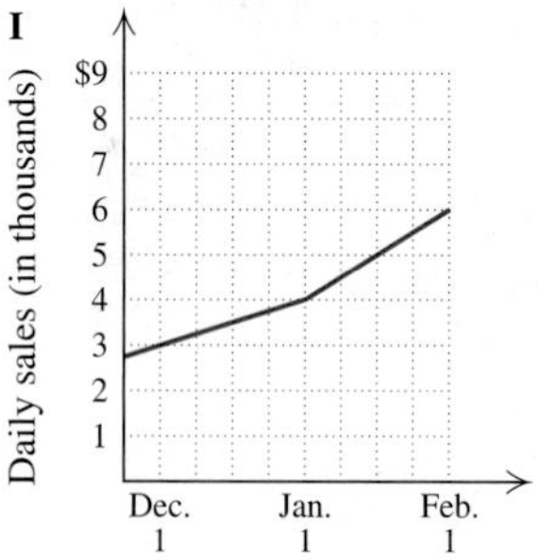

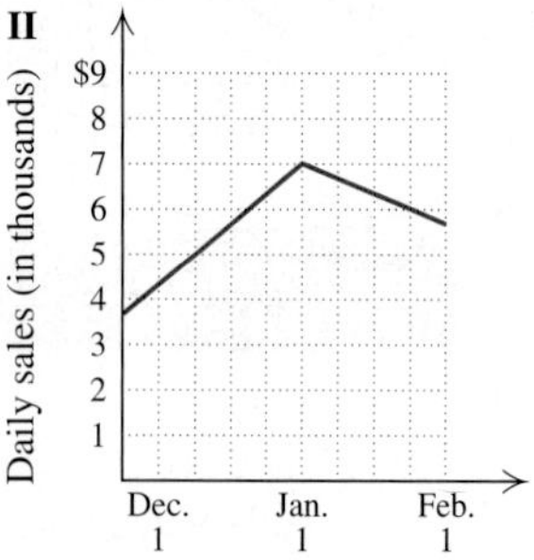

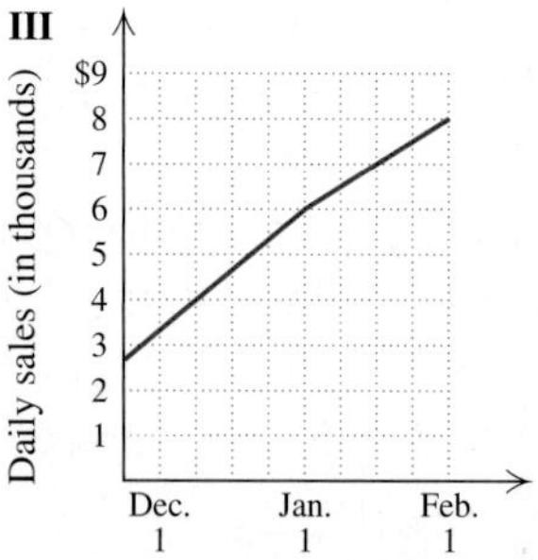

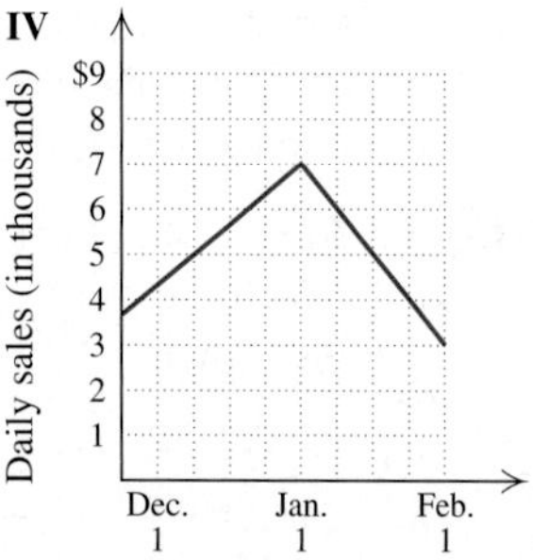

a) After January 1, daily sales continued to rise, but at a slower rate.
b) After January 1, sales decreased faster than they ever grew.
c) The rate of growth in daily sales doubled after January 1.
d) After January 1, daily sales decreased at half the rate that they grew in December.

In Exercises 77–86, each model is of the form $f(x) = mx + b$. In each case, determine what m and b signify.

77. *Catering.* When catering a party for x people, Jennette's Catering uses the formula $C(x) = 25x + 75$, where $C(x)$ is the cost of the party, in dollars.

78. *Weekly Pay.* Each salesperson at Knobby's Furniture is paid $P(x)$ dollars per week, where $P(x) = 0.05x + 200$ and x is the value of the salesperson's sales for the week.

79. *Hair Growth.* After Ty gets a "buzz cut," the length $L(t)$ of his hair, in inches, is given by $L(t) = \frac{1}{2}t + 1$, where t is the number of months after he gets the haircut.

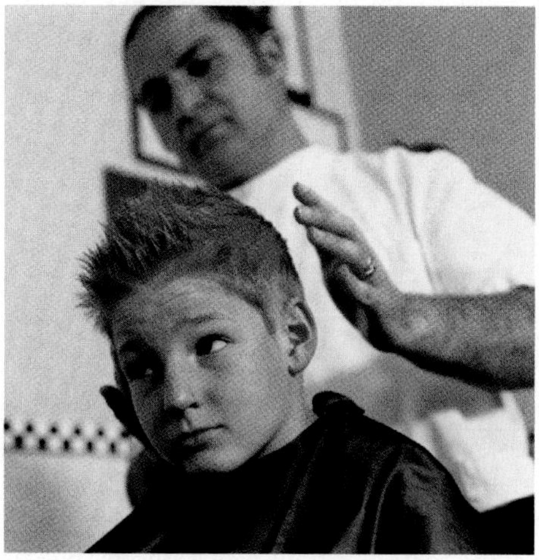

80. *Natural Gas Demand.* The demand, in quadrillions of Btu's, for natural gas is approximated by $D(t) = \frac{9}{25}t + 22$, where t is the number of years after 2000 (*Source*: Based on data from the U.S. Energy Information Administration).

81. *Life Expectancy of American Women.* The life expectancy of American women t years after 1970 is given by $A(t) = \frac{1}{7}t + 75.5$.

82. *Landscaping.* After being cut, the length $G(t)$ of the lawn, in inches, at Great Harrington Community College is given by $G(t) = \frac{1}{8}t + 2$, where t is the number of days since the lawn was cut.

83. *Cost of a Movie Ticket.* The average price $P(t)$, in dollars, of a movie ticket is given by $P(t) = 0.227t + 4.29$, where t is the number of years since 1995.

84. *Sales of Cotton Goods.* The function given by $f(t) = -0.5t + 7$ can be used to estimate the cash receipts, in billions of dollars, from farm marketing of cotton t years after 1995.

85. *Cost of a Taxi Ride.* The cost, in dollars, of a taxi ride during off-peak hours in New York City is given by $C(d) = 2d + 2.5$, where d is the number of miles traveled.

86. *Cost of Renting a Truck.* The cost, in dollars, of a one-day truck rental is given by $C(d) = 0.3d + 20$, where d is the number of miles driven.

87. *Salvage Value.* Green Glass Recycling uses the function given by $F(t) = -5000t + 90{,}000$ to determine the salvage value $F(t)$, in dollars, of a waste removal truck t years after it has been put into use.

a) What do the numbers -5000 and 90,000 signify?
b) How long will it take the truck to depreciate completely?
c) What is the domain of F?

88. *Salvage Value.* Consolidated Shirt Works uses the function given by $V(t) = -2000t + 15{,}000$ to determine the salvage value $V(t)$, in dollars, of a color separator t years after it has been put into use.

a) What do the numbers -2000 and 15,000 signify?
b) How long will it take the machine to depreciate completely?
c) What is the domain of V?

89. *Trade-In Value.* The trade-in value of a Homelite snowblower can be determined using the function given by $v(n) = -150n + 900$. Here $v(n)$ is the trade-in value, in dollars, after n winters of use.

a) What do the numbers -150 and 900 signify?
b) When will the trade-in value of the snowblower be \$300?
c) What is the domain of v?

90. *Trade-In Value.* The trade-in value of a John Deere riding lawnmower can be determined using the function given by $T(x) = -300x + 2400$. Here $T(x)$ is the trade-in value, in dollars, after x summers of use.

a) What do the numbers -300 and 2400 signify?
b) When will the value of the mower be \$1200?
c) What is the domain of T?

TW **91.** A student makes a mistake when using a graphing calculator to draw $4x + 5y = 12$ and the following screen appears. Use algebra to show that a mistake has been made. What do you think the mistake was?

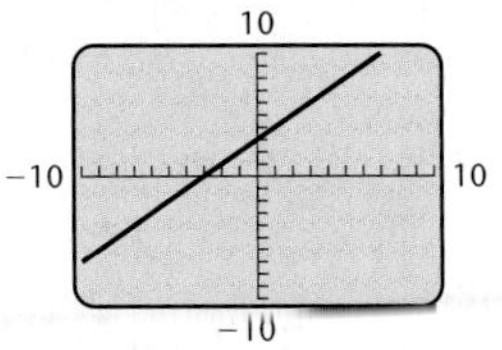

TW **92.** A student makes a mistake when using a graphing calculator to draw $5x - 2y = 3$ and the following screen appears. Use algebra to show that a mistake has been made. What do you think the mistake was?

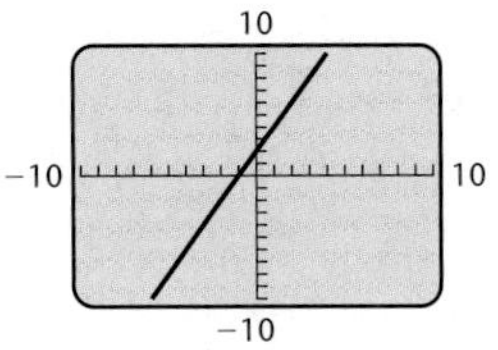

Skill Maintenance

Simplify.

93. $9\{2x - 3[5x + 2(-3x + y^0 - 2)]\}$ [1.3], [1.4]

94. $(-5a^2b^3)^3$ [1.4]

95. $(13m^2n^3)(-2m^5n)$ [1.4]

96. $2x - 3[y - 2(x + 3) + 10]$ [1.3]

Aha! **97.** $(4x^2y^3)(-3x^5y)^0$ [1.4]

98. $(2x^{-5}y^{-7})^{-2}$ [1.4]

Synthesis

TW **99.** *Economics.* In March of 2004, the federal debt could be modeled using $D(t) = mt + 7.1$, where $D(t)$ is in trillions of dollars t months after March and m is some constant (*Source*: www.publicdebt.treas.gov). If you were president of the United States, would you want m to be positive or negative? Why?

TW **100.** Hope claims that her firm's profits continue to go up, but the rate of increase is going down.

a) Sketch a graph that might represent her firm's profits as a function of time.
b) Explain why the graph can go up while the rate of increase goes down.

TW **101.** *Economics.* Examine the function given in Exercise 99. What units of measure must be used for m? Why?

In Exercises 102 and 103, assume that r, p, and s are constants and that x and y are variables. Determine the slope and the y-intercept.

102. $rx + py = s$

103. $rx + py = s - ry$

104. Let (x_1, y_1) and (x_2, y_2) be two distinct points on the graph of $y = mx + b$. Use the fact that both pairs are solutions of the equation to prove that m is the slope of the line given by $y = mx + b$. (*Hint*: Use the slope formula.)

Given that $f(x) = mx + b$, *classify each of the following as true or false.*

105. $f(c + d) = f(c) + f(d)$

106. $f(cd) = f(c)f(d)$

107. $f(kx) = kf(x)$

108. $f(c - d) = f(c) - f(d)$

109. Find k such that the line containing $(-3, k)$ and $(4, 8)$ is parallel to the line containing $(5, 3)$ and $(1, -6)$.

110. Match each sentence with the most appropriate graph below.

a) Annie drove 2 mi to a lake, swam 1 mi, and then drove 3 mi to a store.

b) During a preseason workout, Rico biked 2 mi, ran for 1 mi, and then walked 3 mi.

c) James bicycled 2 mi to a park, hiked 1 mi over the notch, and then took a 3-mi bus ride back to the park.

d) After hiking 2 mi, Marcy ran for 1 mi before catching a bus for the 3-mi ride into town.

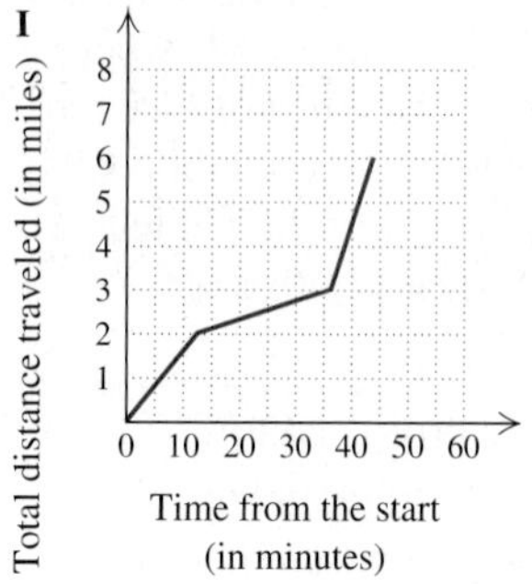

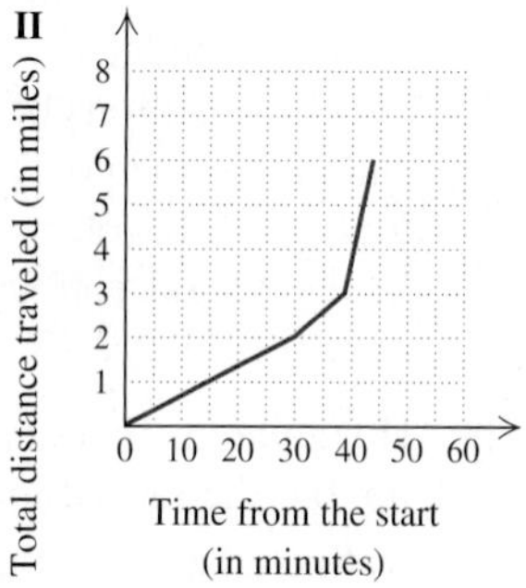

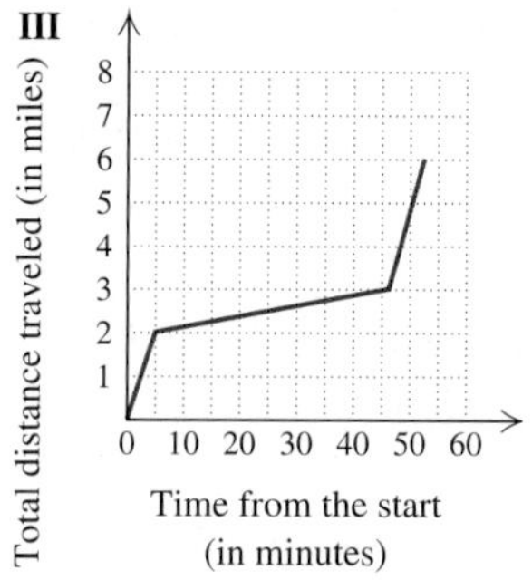

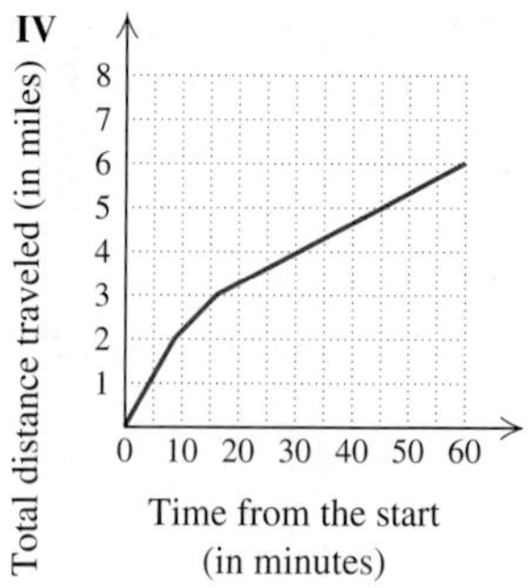

111. Find the slope of the line that contains the given pair of points.

a) $(5b, -6c)$, $(b, -c)$

b) (b, d), $(b, d + e)$

c) $(c + f, a + d)$, $(c - f, -a - d)$

112. *Cost of a Speeding Ticket.* The penalty schedule shown below is used to determine the cost of a speeding ticket in certain states. Use this schedule to graph the cost of a speeding ticket as a function of the number of miles per hour over the limit that a driver is going.

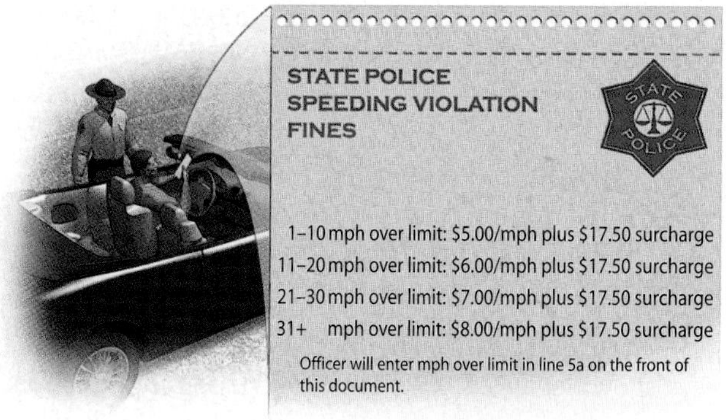
STATE POLICE
SPEEDING VIOLATION
FINES
STATE POLICE
1–10 mph over limit: $5.00/mph plus $17.50 surcharge
11–20 mph over limit: $6.00/mph plus $17.50 surcharge
21–30 mph over limit: $7.00/mph plus $17.50 surcharge
31+ mph over limit: $8.00/mph plus $17.50 surcharge
Officer will enter mph over limit in line 5a on the front of this document.

113. Graph the equations

$$y_1 = 1.4x + 2, \quad y_2 = 0.6x + 2,$$
$$y_3 = 1.4x + 5, \quad \text{and} \quad y_4 = 0.6x + 5$$

using a graphing calculator. If possible, use the SIMULTANEOUS mode so that you cannot tell which equation is being graphed first. Then decide which line corresponds to each equation.

2.3 Another Look at Linear Graphs

In Section 2.2, we graphed linear equations using slopes and y-intercepts. We now graph lines that have slope 0 or that have an undefined slope. We also graph lines using both x- and y-intercepts and learn how to recognize whether the graphs of two linear equations will be parallel or perpendicular.

Zero Slope and Lines with Undefined Slope

If two different points have the same second coordinate, what is the slope of the line joining them? In this case, we have $y_2 = y_1$, so

$$m = \frac{y_2 - y_1}{x_2 - x_1} = \frac{0}{x_2 - x_1} = 0.$$

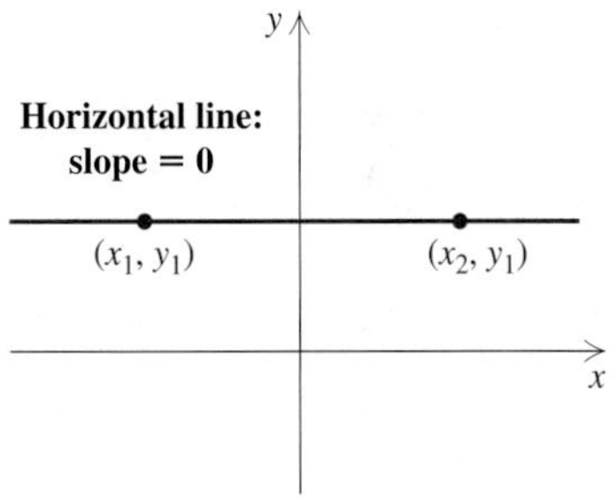

Slope of a Horizontal Line Every horizontal line has a slope of 0.

EXAMPLE 1 Use slope–intercept form to graph $f(x) = 3$.

SOLUTION Recall from Section 2.1 that a function of this type is called a *constant function.* Writing slope–intercept form,

$$f(x) = 0 \cdot x + 3,$$

we see that the y-intercept is $(0, 3)$ and the slope is 0. Thus we can graph f by plotting the point $(0, 3)$ and, from there, determining a slope of 0. Because $0 = 0/2$ (any nonzero number could be used in place of 2), we can draw the graph by going up 0 units and to the right 2 units. As a check, we also find some ordered pairs. Note that for any choice of x-value, $f(x)$ must be 3.

Study Tip

Talk It Over

If you are finding it difficult to master a particular topic or concept, talk about it with a classmate. Verbalizing your questions about the material might help clarify it for you. If your classmate is also having difficulty, it is possible that a majority of your classmates are confused and you can ask your instructor to explain the concept again.

x	$f(x)$
-1	3
0	3
2	3

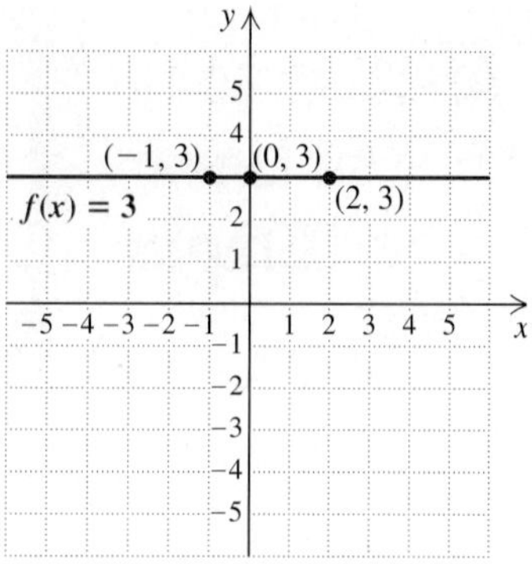

We see from Example 1 the following:

The graph of any constant function of the form $f(x) = b$ or $y = b$ is a horizontal line that crosses the y-axis at $(0, b)$.

Suppose that two different points are on a vertical line. They then have the same first coordinate. In this case, we have $x_2 = x_1$, so

$$m = \frac{y_2 - y_1}{x_2 - x_1} = \frac{y_2 - y_1}{0}.$$

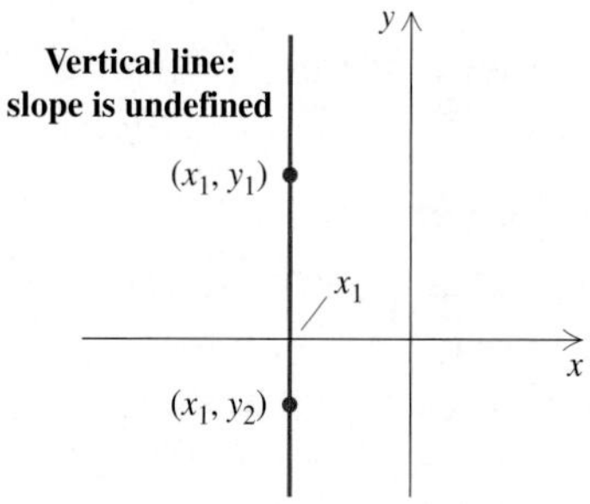

Since we cannot divide by 0, this is undefined. Note that when we say that $(y_2 - y_1)/0$ is undefined, it means that we have agreed to not attach any meaning to that expression.

Slope of a Vertical Line The slope of a vertical line is undefined.

EXAMPLE 2 Graph: $x = -2$.

SOLUTION With y missing, no matter which value of y is chosen, x must be -2. Thus the pairs $(-2, 3)$, $(-2, 0)$, and $(-2, -4)$ all satisfy the equation. The graph is a line parallel to the y-axis. Note that since y is missing, this equation cannot be written in slope–intercept form.

x	y
-2	3
-2	0
-2	-4

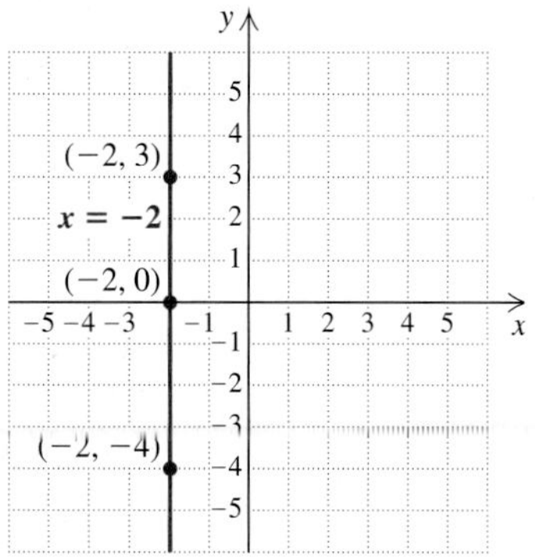

Example 2 shows the following:

The graph of any equation of the form $x = a$ is a vertical line that crosses the x-axis at $(a, 0)$.

EXAMPLE 3 Find the slope of each given line. If the slope is undefined, state this.

a) $3y + 2 = 14$

b) $2x = 10$

SOLUTION

a) We solve for y:

$$3y + 2 = 14$$

$$3y = 12 \quad \text{Subtracting 2 from both sides}$$

$$y = 4. \quad \text{Dividing both sides by 3}$$

The graph of $y = 4$ is a horizontal line. Since $3y + 2 = 14$ is equivalent to $y = 4$, the slope of the line $3y + 2 = 14$ is 0.

b) When y does not appear, we solve for x:

$$2x = 10$$

$$x = 5. \quad \text{Dividing both sides by 2}$$

The graph of $x = 5$ is a vertical line. Since $2x = 10$ is equivalent to $x = 5$, the slope of the line $2x = 10$ is undefined.

Graphing Using Intercepts

Any line that is not horizontal or vertical will cross both the x- and y-axes. We have already seen that the point at which a line crosses the y-axis is called the *y-intercept*. Similarly, the point at which a line crosses the x-axis is called the *x-intercept*. Any time the x- and y-intercepts are not both $(0, 0)$, they are two distinct points and can thus be used to draw the graph of the line. Recall that to find the y-intercept, we replace x with 0 and solve for y. To find the x-intercept, we replace y with 0 and solve for x.

To Determine Intercepts

The x-intercept is $(a, 0)$. To find a, let $y = 0$ and solve the original equation for x.

The y-intercept is $(0, b)$. To find b, let $x = 0$ and solve the original equation for y.

EXAMPLE 4 Graph the equation $3x + 2y = 12$ by using intercepts.

SOLUTION *To find the y-intercept, we let $x = 0$ and solve for y:*

$$3 \cdot 0 + 2y = 12$$
$$2y = 12$$
$$y = 6.$$

The y-intercept is $(0, 6)$.

To find the x-intercept, we let $y = 0$ and solve for x:

$$3x + 2 \cdot 0 = 12$$
$$3x = 12$$
$$x = 4.$$

The x-intercept is $(4, 0)$.

We plot the two intercepts and draw the line. A third point could be calculated and used as a check.

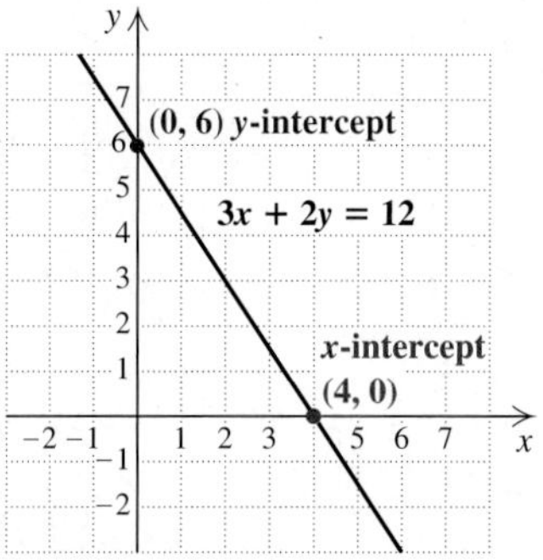

EXAMPLE 5 Graph $f(x) = 2x + 5$ by using intercepts.

SOLUTION Because the function is in slope–intercept form, we know that the y-intercept is $(0, 5)$. To find the x-intercept, we replace $f(x)$ with 0 and solve for x:

$$\begin{aligned} 0 &= 2x + 5 \\ -5 &= 2x \\ -\tfrac{5}{2} &= x. \end{aligned}$$

The x-intercept is $\left(-\frac{5}{2}, 0\right)$.

We plot the intercepts $(0, 5)$ and $\left(-\frac{5}{2}, 0\right)$ and draw the line. As a check, we can calculate the slope:

$$\begin{aligned} m &= \frac{5 - 0}{0 - \left(-\frac{5}{2}\right)} \\ &= \frac{5}{\frac{5}{2}} \\ &= 5 \cdot \frac{2}{5} \\ &= 2. \end{aligned}$$

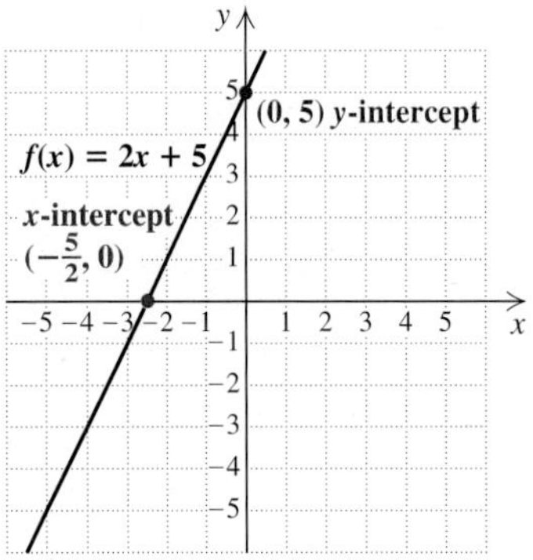

The slope is 2, as expected.

Intercepts

One way to find intercepts on a graphing calculator is to graph the function and then use the CALC menu.

Recall that to find the y-intercept, we let $x = 0$ and solve for y. The VALUE option of the CALC menu allows us to enter 0 for x, and the corresponding y-value, or $f(0)$, is given.

To find any x-intercepts, we let $y = 0$ and solve for x. The ZERO option of the CALC menu allows us to find values of x for which $f(x) = 0$.

EXAMPLE 6 Find the intercepts of the graph of the function

$$f(x) = -2x - 6.$$

ALGEBRAIC METHOD

The function is in slope–intercept form, so we know that the y-intercept is $(0, -6)$.

To find the x-intercept, we replace $f(x)$ with 0 and solve for x:

$$0 = -2x - 6$$
$$6 = -2x \qquad \textbf{Adding 6 to both sides}$$
$$-3 = x. \qquad \textbf{Dividing both sides by } -2$$

The x-intercept is $(-3, 0)$.

GRAPHICAL METHOD

We graph the function using a standard viewing window. To find the y-intercept, we choose the VALUE option from the CALC menu and enter 0 for x. We see that the y-intercept is $(0, -6)$.

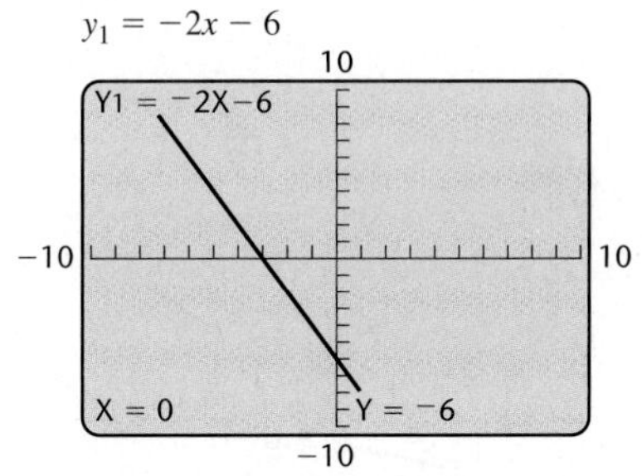

To find the x-intercept, we choose the ZERO option from the CALC menu. The graph appears to cross the x-axis somewhere between -4 and -2. We can use -4 for a left bound, -2 for a right bound, and -3 for a guess. The calculator returns a value of -3 for the zero of the function, so the x-intercept is $(-3, 0)$.

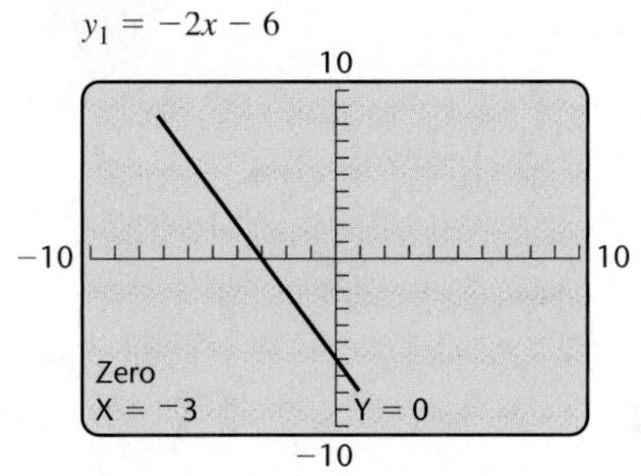

The intercepts of the graph of a function can help us determine an appropriate viewing window on a graphing calculator.

EXAMPLE 7 Determine a viewing window that shows the intercepts of the graph of the function $f(x) = 3x + 15$.

SOLUTION In this case the function is in slope–intercept form, so we know that the y-intercept is $(0, 15)$. To find the x-intercept, we replace $f(x)$ with 0 and solve for x:

$$0 = 3x + 15$$
$$-15 = 3x \quad \textbf{Subtracting 15 from both sides}$$
$$-5 = x.$$

The x-intercept is $(-5, 0)$.

A standard viewing window will not show the y-intercept, $(0, 15)$. Thus we adjust the Ymax value and choose a viewing window of $[-10, 10, -10, 20]$, with Yscl $= 5$. Other choices of window dimensions are also possible.

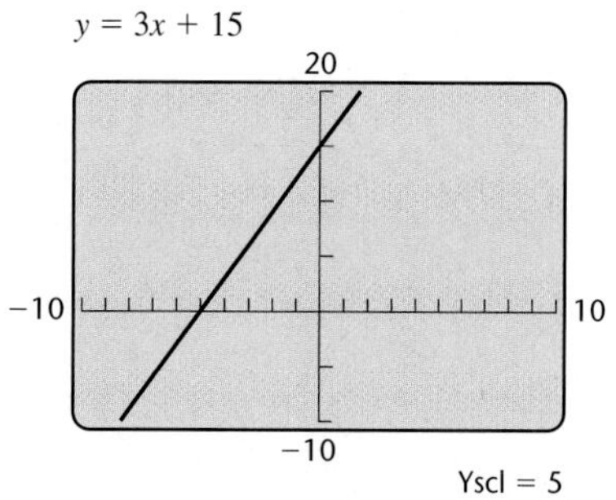

Parallel and Perpendicular Lines

If two lines are vertical, they are parallel. How can we tell whether nonvertical lines are parallel?

Interactive Discovery

1. Graph $y = 3x - 7$ and $y - 3x = 8$ on a graphing calculator using a standard viewing window and determine whether the lines appear to be parallel.
2. Again using a standard viewing window, graph $7x + 10y = 50$ and $4x + 5y = -25$. Do these lines appear to be parallel?
3. Now graph each pair of equations given above using the viewing window $[-100, 100, -100, 100]$, with Xscl $= 10$ and Yscl $= 10$. Do the lines still appear to be parallel?

Graphs of nonparallel lines may appear parallel in certain viewing windows. In order to determine whether two lines are parallel, we can look at their slopes.

Slope and Parallel Lines

Two lines are parallel if they have the same slope.

Note that the y-intercepts of parallel lines are different.

EXAMPLE 8 Determine whether the line passing through $(1, 7)$ and $(4, -2)$ is parallel to the line given by $f(x) = -3x + 4.2$.

SOLUTION The slope of the line passing through $(1, 7)$ and $(4, -2)$ is given by

$$m = \frac{7 - (-2)}{1 - 4} = \frac{9}{-3} = -3.$$

Since the graph of $f(x) = -3x + 4.2$ also has a slope of -3, the lines are parallel.

Two lines are perpendicular if they intersect at a right angle. If one line is vertical and another is horizontal, they are perpendicular. There are other instances in which two lines are perpendicular.

Consider a line $\overleftrightarrow{RS}$, as shown below, with slope a/b. Then think of rotating the figure 90° to get a line $\overleftrightarrow{R'S'}$ perpendicular to $\overleftrightarrow{RS}$.

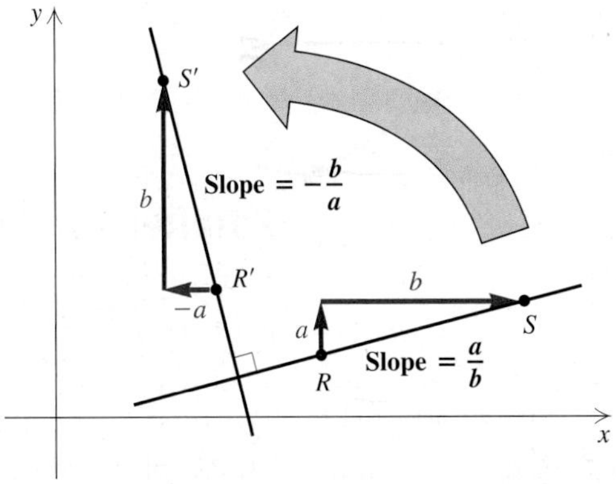

For the new line, the rise and the run are interchanged, but the run is now negative. Thus the slope of the new line is $-b/a$. Let's multiply the slopes:

$$\frac{a}{b}\left(-\frac{b}{a}\right) = -1.$$

This can help us determine which lines are perpendicular.

Slope and Perpendicular Lines

Two lines are perpendicular if the product of their slopes is -1 or if one line is vertical and the other is horizontal.

Thus, if one line has slope m ($m \neq 0$), the slope of a line perpendicular to it is $-1/m$. That is, we take the reciprocal of m ($m \neq 0$) and change the sign.

Squaring a Viewing Window

If the units on the x-axis are a different length than those on the y-axis, two lines that are perpendicular may not appear to be so when graphed. Since the screen on a graphing calculator is rectangular, the 20 units on the x-axis in a standard viewing window are longer than the 20 units on the y-axis in that window. Finding a viewing window with units the same length on both axes is called *squaring* the viewing window.

Windows can be squared by choosing the ZSQUARE option in the ZOOM menu. They can be squared manually by choosing the portions of the axes shown in the same proportion as the sides of the calculator screen.

EXAMPLE 9 Determine whether the lines given by the equations $3x - y = 7$ and $x + 3y = 1$ are perpendicular, and check by graphing.

SOLUTION To determine the slope of each line, we solve for y to find slope–intercept form:

$$3x - y = 7$$
$$-y = -3x + 7 \qquad \textbf{Adding } -3x \textbf{ to both sides}$$
$$y = 3x - 7; \qquad \textbf{Multiplying both sides by } -1$$
$$x + 3y = 1$$
$$3y = -x + 1 \qquad \textbf{Adding } -x \textbf{ to both sides}$$
$$y = -\tfrac{1}{3}x + \tfrac{1}{3}. \qquad \textbf{Multiplying both sides by } \tfrac{1}{3}$$

The slopes of the lines are 3 and $-\frac{1}{3}$. Since $3 \cdot \left(-\frac{1}{3}\right) = -1$, the lines are perpendicular.

To check, we graph $y_1 = 3x - 7$ and $y_2 = -\frac{1}{3}x + \frac{1}{3}$. The graphs are shown in a standard viewing window on the left below. Note that the lines do not appear to be perpendicular. If we press (ZOOM) (5), the lines are graphed in a squared viewing window as shown in the graph on the right below, and they do appear perpendicular. We can perform a better visual check by laying a corner of a piece of paper on the screen. If the lines are perpendicular, they should exactly fit the corner. Note that this visual check is only approximate; algebra is necessary to prove that lines are perpendicular.

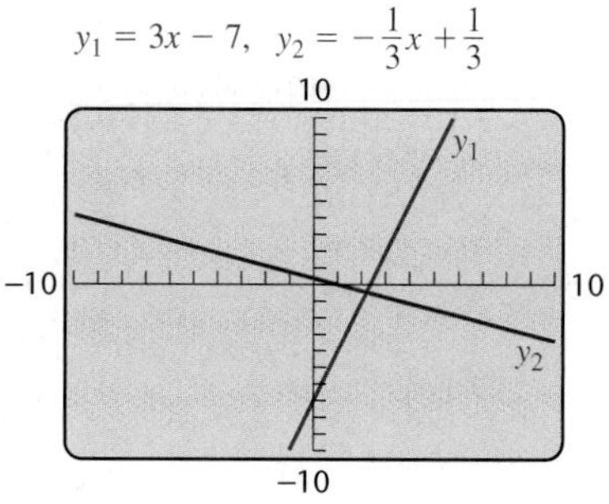

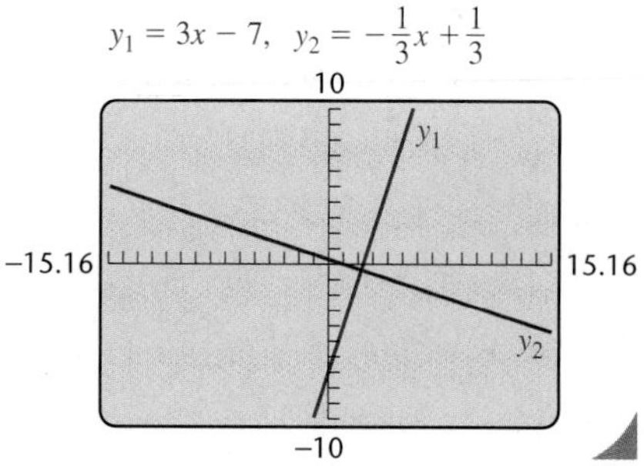

EXAMPLE 10 Consider the function given by $f(x) = \frac{2}{3}x - 8$.

a) Write an equation for a linear function g with a graph parallel to the graph of f and a y-intercept of $\left(0, \frac{3}{4}\right)$.

b) Write an equation for a linear function h with a graph perpendicular to the graph of f and a y-intercept of $\left(0, \frac{3}{4}\right)$.

SOLUTION

a) Since g is linear, it can be written in the form $g(x) = mx + b$. To find g, we must determine its slope and y-intercept. The slope of the line given by $f(x) = \frac{2}{3}x - 8$ is $\frac{2}{3}$. Therefore, the slope of a parallel line is $\frac{2}{3}$. Since we are given that the y-intercept of the graph of g is $\left(0, \frac{3}{4}\right)$, we have $g(x) = \frac{2}{3}x + \frac{3}{4}$.

b) Since the slope of the graph of f is $\frac{2}{3}$, the slope of a line perpendicular to the graph is $-\frac{3}{2}$. Thus, $m = -\frac{3}{2}$ and $b = \frac{3}{4}$, so $h(x) = -\frac{3}{2}x + \frac{3}{4}$.

Recognizing Linear Equations

Is every equation of the form $Ax + By = C$ linear? To find out, suppose that A and B are nonzero and solve for y:

$$Ax + By = C \qquad \textbf{A, B, and C are constants.}$$

$$By = -Ax + C \qquad \textbf{Adding } -Ax \textbf{ to both sides}$$

$$y = -\frac{A}{B}x + \frac{C}{B}. \qquad \textbf{Dividing both sides by } B$$

Since the last equation is a slope–intercept equation, we see that $Ax + By = C$ is a linear equation when $A \neq 0$ and $B \neq 0$.

But what if A or B (but not both) is 0? If A is 0, then $By = C$ and $y = C/B$. If B is 0, then $Ax = C$ and $x = C/A$. In the first case, the graph is a horizontal line; in the second case, the line is vertical. In either case, $Ax + By = C$ is a linear equation when A or B (but not both) is 0. We have now justified the following result.

The Standard Form of a Linear Equation Any equation $Ax + By = C$, where A, B, and C are real numbers and A and B are not both 0, has a graph that is a straight line.

Any equation of the form

$$Ax + By = C$$

is said to be written in *standard form.*

EXAMPLE 11 Determine whether the equation $y = x^2 - 5$ is linear.

SOLUTION We attempt to put the equation in standard form:

$$y = x^2 - 5$$
$$-x^2 + y = -5. \quad \textbf{Adding } -x^2 \textbf{ to both sides}$$

This last equation is not linear because it has an x^2-term.

We can see this as well from the graph of the equation below.

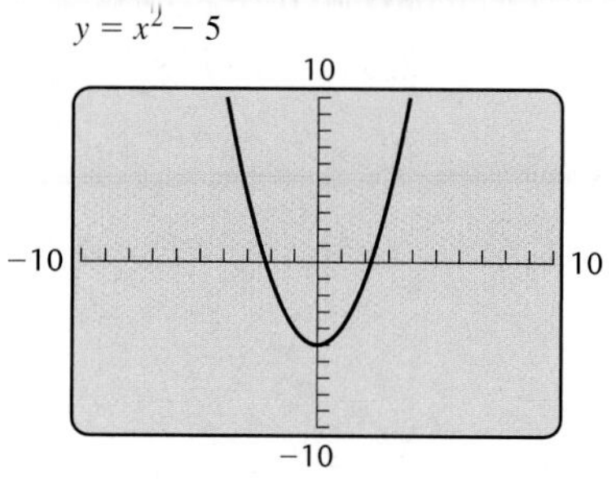

Only linear equations have graphs that are straight lines. Also, only linear graphs have a constant slope. Were you to try to calculate the slope between several pairs of points in Example 11, you would find that the slopes vary.

2.3 EXERCISE SET

FOR EXTRA HELP

MathXL | MyMathLab | InterAct Math | AW Math Tutor Center | Video Lectures on CD: Disc 1 | Student's Solutions Manual

Concept Reinforcement Complete each of the following statements.

1. Every ______________ line has a slope of 0.

2. The slope of a vertical line is ______________.

3. The graph of any equation of the form $x = a$ is a(n) ______________ line that crosses the x-axis at $(a, 0)$.

4. The graph of any function of the form $f(x) = b$ is a horizontal line that crosses the ______________ at $(0, b)$.

5. To find the x-intercept, we let $y =$ ______________ and solve the original equation for ______________.

6. To find the y-intercept, we let $x =$ ______________ and solve the original equation for ______________.

7. Two different lines with the same slope are ______________.

8. An equation like $4x + 3y = 8$ is said to be written in ______________ form.

9. Only ______________ equations have graphs that are straight lines.

10. The product of the slopes of two nonvertical perpendicular lines is ______________.

For each equation, find the slope. If the slope is undefined, state this.

11. $y - 9 = 3$
12. $x + 1 = 7$
13. $8x = 6$
14. $y - 3 = 5$
15. $3y = 28$
16. $19 = -6y$
17. $9 + x = 12$
18. $2x = 18$
19. $2x - 4 = 3$
20. $5y - 1 = 16$
21. $5y - 4 = 35$
22. $2x - 17 = 3$
23. $3y + x = 3y + 2$
24. $x - 4y = 12 - 4y$
25. $5x - 2 = 2x - 7$
26. $5y + 3 = y + 9$
Aha! **27.** $y = -\frac{2}{3}x + 5$
28. $y = -\frac{3}{2}x + 4$

Graph.

29. $y = 5$
30. $x = -1$
31. $x = 3$
32. $y = 2$
33. $4 \cdot f(x) = 20$
34. $6 \cdot g(x) = 12$
35. $3x = -15$
36. $2x = 10$
37. $4 \cdot g(x) + 3x = 12 + 3x$
38. $3 - f(x) = 2$

Find the intercepts. Then graph by using the intercepts, if possible, and a third point as a check.

39. $x + y = 4$
40. $x + y = 5$
41. $y = 2x + 6$
42. $y = 3x + 9$
43. $3x + 5y = -15$
44. $5x - 4y = 20$
45. $2x - 3y = 18$
46. $3x + 2y = -18$
47. $3y = 6x$
48. $5y = 15x$
49. $f(x) = 3x - 7$
50. $g(x) = 2x - 9$
51. $1.4y - 3.5x = -9.8$
52. $3.6x - 2.1y = 22.68$
53. $5x + 2g(x) = 7$
54. $3x - 4f(x) = 11$

For each function, determine which of the given viewing windows will show both intercepts.

55. $f(x) = 20 - 4x$
a) $[-10, 10, -10, 10]$
b) $[-5, 10, -5, 10]$
c) $[-10, 10, -10, 30]$
d) $[-10, 10, -30, 10]$

56. $g(x) = 3x + 7$
a) $[-10, 10, -10, 10]$
b) $[-1, 15, -1, 15]$
c) $[-15, 5, -15, 5]$
d) $[-10, 10, -30, 0]$

57. $p(x) = -35x + 7000$
a) $[-10, 10, -10, 10]$
b) $[-35, 0, 0, 7000]$
c) $[-1000, 1000, -1000, 1000]$
d) $[0, 500, 0, 10{,}000]$

58. $r(x) = 0.2 - 0.01x$
a) $[-10, 10, -10, 10]$
b) $[-5, 30, -1, 1]$
c) $[-1, 1, -5, 30]$
d) $[0, 0.01, 0, 0.2]$

Without graphing, tell whether the graphs of each pair of equations are parallel.

59. $x + 8 = y,$ $y - x = -5$
60. $2x - 3 = y,$ $y - 2x = 9$
61. $y + 9 = 3x,$ $3x - y = -2$
62. $y + 8 = -6x,$ $-2x + y = 5$
63. $f(x) = 3x + 9,$ $2y = -6x - 2$
64. $f(x) = -7x - 9,$ $-3y = 21x + 7$

Without graphing, tell whether the graphs of each pair of equations are perpendicular.

65. $f(x) = 4x - 3,$ $4y = 7 - x$
66. $2x - 5y = -3,$ $2x + 5y = 4$
67. $x + 2y = 7,$ $2x + 4y = 4$
68. $y = -x + 7,$ $f(x) = x + 3$

Write an equation for a linear function parallel to the given line with the given y-intercept.

69. $y = 3x - 2;\ (0, 9)$
70. $y = -5x + 7;\ (0, -2)$
71. $2x + y = 3;\ (0, -5)$
72. $3x = y + 10;\ (0, 1)$
73. $2x + 5y = 8;\ \left(0, -\frac{1}{3}\right)$
74. $3x - 6y = 4;\ \left(0, \frac{4}{5}\right)$
Aha! **75.** $3y = 12;\ (0, -5)$
76. $5 = 10y;\ (0, 12)$

Write an equation for a linear function perpendicular to the given line with the given y-intercept.

77. $y = x - 3;\ (0, 4)$
78. $y = 2x - 7;\ (0, -3)$
79. $2x + 3y = 6;\ (0, -4)$
80. $4x + 2y = 8;\ (0, 8)$
81. $5x - y = 13;\ \left(0, \frac{1}{5}\right)$
82. $2x - 5y = 7;\ \left(0, -\frac{1}{8}\right)$

Determine whether each equation is linear. Find the slope of any nonvertical lines.

83. $5x - 3y = 15$

84. $3x + 5y + 15 = 0$

85. $16 + 4y = 10$

86. $3x - 12 = 0$

87. $3g(x) = 6x^2$

88. $2x + 4f(x) = 8$

89. $3y = 7(2x - 4)$

90. $2(5 - 3x) = 5y$

91. $g(x) - \frac{1}{x} = 0$

92. $f(x) + \frac{1}{x} = 0$

93. $\frac{f(x)}{5} = x^2$

94. $\frac{g(x)}{2} = 3 + x$

TW **95.** *Engineering.* Wind friction, or *air resistance,* increases with speed. Following are some measurements made in a wind tunnel. Plot the data and explain why a linear function does or does not give an approximate fit.

Velocity (in kilometers per hour)	Force of Resistance (in newtons)
10	3
21	4.2
34	6.2
40	7.1
45	15.1
52	29.0

TW **96.** *Meteorology.* Wind chill is a measure of how cold the wind makes you feel. Below are some measurements of wind chill for a 15-mph breeze. How can you tell from the data that a linear function will give an approximate fit?

Temperature	15-mph Wind Chill
30°F	19°F
25°F	13°F
20°F	6°F
15°F	0°F
10°F	−7°F
5°F	−13°F
0°F	−19°F

Source: National Oceanic & Atmospheric Administration, as reported in USA TODAY.com, 2004

To the student and the instructor: Focused Review exercises such as the following replace Skill Maintenance exercises in selected exercise sets. These exercises focus on specific skills and often provide a mixed review connecting concepts of several sections or chapters.

Focused Review

Graph using the y-intercept and the slope. [2.2]

97. $y = \frac{1}{2}x + 3$

98. $f(x) = -x + 1$

Graph using intercepts. [2.3]

99. $3x - y = 3$

100. $2x + 3y = 6$

Graph each horizontal or vertical line. [2.3]

101. $f(x) = -2$

102. $x = 4$

Graph. [2.2], [2.3]

103. $5x + y = 5$

104. $y = \frac{1}{2}$

105. $f(x) = 2x - 3$

106. $x + y = 4$

Synthesis

TW **107.** Jim tries to avoid fractions as often as possible. Under what conditions will graphing using intercepts allow him to avoid fractions? Why?

TW **108.** Under what condition(s) will the x- and y-intercepts of a line coincide? What would the equation for such a line look like?

109. Give an equation, in standard form, for the line whose x-intercept is 5 and whose y-intercept is -4.

110. Find the x-intercept of $y = mx + b$, assuming that $m \neq 0$.

In Exercises 111–114, assume that r, p, and s are nonzero constants and that x and y are variables. Determine whether each equation is linear.

111. $rx + 3y = p^2 - s$

112. $py = sx - r^2y - 9$

113. $r^2x = py + 5$

114. $\frac{x}{r} - py = 17$

115. Suppose that two linear equations have the same y-intercept but that equation A has an x-intercept that is half the x-intercept of equation B. How do the slopes compare?

Consider the linear equation

$$ax + 3y = 5x - by + 8.$$

116. Find a and b if the graph is a horizontal line passing through (0, 4).

117. Find a and b if the graph is a vertical line passing through (4, 0).

118. Since a vertical line is not the graph of a function, many graphing calculators cannot graph equations of the form $x = a$. Some graphing calculators can draw vertical lines using the DRAW menu. Use the VERTICAL option of the DRAW menu to graph each of the following equations.

a) $x = 3.6$ **b)** $x = -1.52$
c) $3x - 5 = 7x + 2$ **d)** $2(x - 5) = x + 10$

119. A table of values can be used to determine whether two lines are parallel. If the difference between the y-values for two lines is the same for all x-values, the lines are parallel. Use the TABLE feature of a graphing calculator to determine whether each of the following pairs of lines is parallel.

a) $2.3x - 3.4y = 9.8$, $1.84x = 2.72y - 17.4$
b) $2.56y + 3.2x - 7.2 = 0$, $5.12y + 6.3x = 14.3$

2.4 Introduction to Curve Fitting: Point–Slope Form

Point–Slope Form ■ Interpolation and Extrapolation ■ Curve Fitting ■ Linear Regression

Specifying the slope of a line and one point through which the line passes enables us to draw the line. In this section, we study how this information can be used to produce an *equation* of the line.

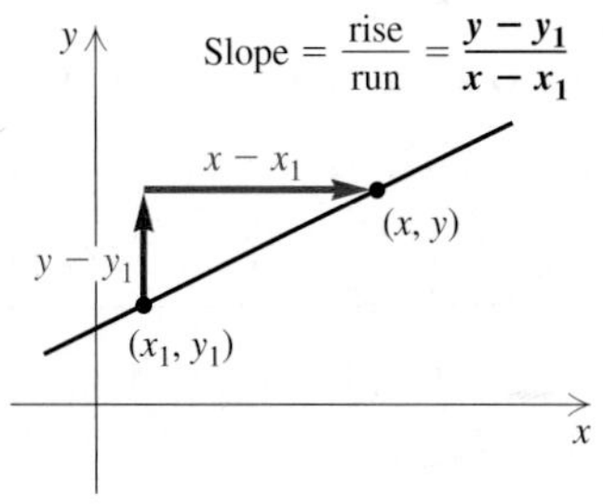

Point–Slope Form

Suppose that a line of slope m passes through the point (x_1, y_1). For any other point (x, y) to lie on this line, we must have

$$\frac{y - y_1}{x - x_1} = m.$$

It is tempting to use this last equation as an equation of the line of slope m that passes through (x_1, y_1). The problem with doing so is that when x and y are replaced with x_1 and y_1, we have $\frac{0}{0} = m$, a false equation. To avoid this difficulty, we multiply both sides by $x - x_1$ and simplify:

$$(x - x_1)\frac{y - y_1}{x - x_1} = m(x - x_1) \qquad \textbf{Multiplying both sides by } x - x_1$$

$$y - y_1 = m(x - x_1). \qquad \textbf{Removing a factor equal to 1: } \frac{x - x_1}{x - x_1} = 1$$

This is the *point–slope* form of a linear equation.

> **Point–Slope Form** Any equation $y - y_1 = m(x - x_1)$ has a graph that is a straight line. It passes through (x_1, y_1) and has slope m.
>
> Any equation of the form
>
> $$y - y_1 = m(x - x_1)$$
>
> is said to be written in *point–slope form.*

Student Notes

To help remember point–slope form, many students begin by writing the equation for slope and then clearing fractions. The equation in Example 1 would then be found as

$$\frac{y - 4}{x - 3} = -\frac{1}{2},$$

and

$$y - 4 = -\frac{1}{2}(x - 3).$$

EXAMPLE 1 Find and graph an equation of the line passing through $(3, 4)$ with slope $-\frac{1}{2}$.

SOLUTION We substitute in the point–slope equation:

$$y - y_1 = m(x - x_1)$$

$$y - 4 = -\tfrac{1}{2}(x - 3). \qquad \textbf{Substituting}$$

To graph this point–slope equation, we count off a slope of $-\frac{1}{2}$, starting at $(3, 4)$. Then we draw the line.

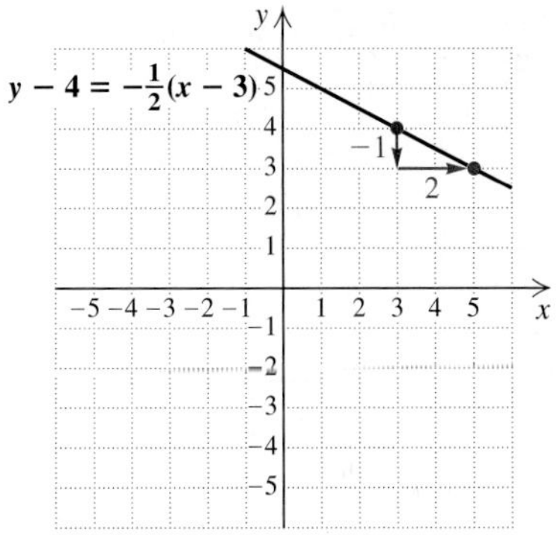

EXAMPLE 2 Find a linear function that has a graph passing through the points $(-1, -5)$ and $(3, -2)$.

SOLUTION We first determine the slope of the line and then write an equation in point–slope form. Note that

$$m = \frac{-5 - (-2)}{-1 - 3} = \frac{-3}{-4} = \frac{3}{4}.$$

Since the line passes through $(3, -2)$, we have

$$y - (-2) = \tfrac{3}{4}(x - 3) \qquad \textbf{Substituting into } y - y_1 = m(x - x_1)$$
$$y + 2 = \tfrac{3}{4}x - \tfrac{9}{4}. \qquad \textbf{Using the distributive law}$$

Before using function notation, we isolate y:

$$y = \tfrac{3}{4}x - \tfrac{9}{4} - 2 \qquad \textbf{Subtracting 2 from both sides}$$
$$y = \tfrac{3}{4}x - \tfrac{17}{4} \qquad -\tfrac{9}{4} - \tfrac{8}{4} = -\tfrac{17}{4}$$
$$f(x) = \tfrac{3}{4}x - \tfrac{17}{4}. \qquad \textbf{Using function notation}$$

You can check that using $(-1, -5)$ as (x_1, y_1) in $y - y_1 = \tfrac{3}{4}(x - x_1)$ will yield the same expression for $f(x)$.

Connecting the Concepts

We have now studied the slope–intercept, point–slope, and standard forms of a linear equation. These are the most common ways in which linear equations are written. Depending on what information we are given and what information we are seeking, one form may be more useful than the others. A referenced summary is given below.

Slope–intercept form, $y = mx + b$ **or** $f(x) = mx + b$	• Useful when an equation is needed and the slope and y-intercept are given. See Example 6 on p. 121. • Useful when a line's slope and y-intercepts are needed. See Example 5 on p. 121. • Useful when finding the zero of a function. See Example 9 on p. 111. • Commonly used for linear functions.
Standard form, $Ax + By = C$	• Allows for easy calculation of intercepts. See Example 4 on p. 136. • Will prove useful in future work. See Sections 3.1–3.3.
Point–slope form, $y - y_1 = m(x - x_1)$	• Useful when an equation is needed and the slope and a point on the line are given. See Example 1 on p. 147. • Useful when a linear function is needed and two points on its graph are given. See Example 2 on pp. 147–148. • Will prove useful in future work with curves and tangents in calculus.

Interpolation and Extrapolation

It is common to use known data to estimate other values. When the unknown point is *between* known points, this process is called **interpolation.** If the unknown point extends *beyond* the known points, the process is called **extrapolation.**

EXAMPLE 3 Elementary-School Math Proficiency. According to the National Assessment of Education Progress, the percentage of fourth-graders who are proficient in math has grown from 1992 to 2005, as shown in the following table. Estimate the percentage of fourth-graders who showed proficiency in 1998 and predict the percentage who will do so in 2007.

Year	Percentage of Fourth-Graders Performing at or above Proficiency
1992	18%
2000	24
2003	32
2005	36

Source: National Center for Education Statistics

SOLUTION

1. and **2. Familiarize** and **Translate.** The given information enables us to plot and connect four points. We let the horizontal axis represent the year and the vertical axis the percentage of fourth-graders demonstrating proficiency in math. We label the function itself P.

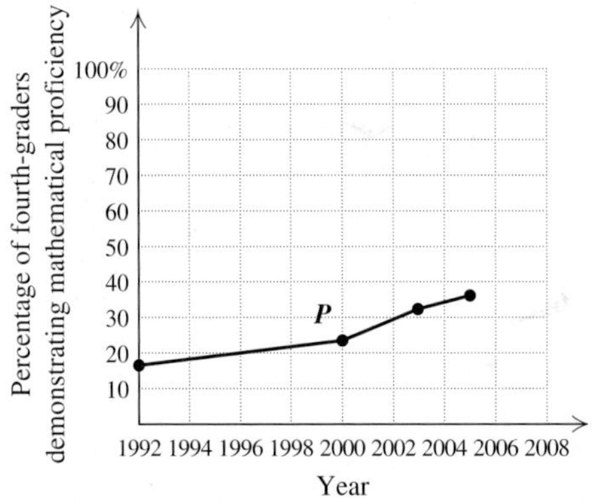

3. Carry out. To estimate the percentage of fourth-graders showing proficiency in math in 1998, we locate the point directly above the year 1998. We then estimate its second coordinate by moving horizontally from that point to the y-axis. Although our result is not exact, we see that $P(1998) \approx 22$. This process of estimation is *interpolation*. To predict the percentage of fourth-graders showing proficiency in 2007, we extend the graph and *extrapolate*. It appears that $P(2007) \approx 40$.

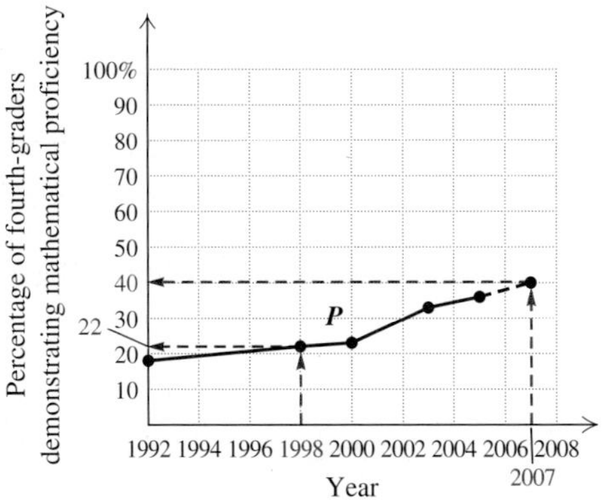

4. **Check.** A precise check requires consulting an outside information source. Since 22% is between 18% and 24% and 40% is greater than 36%, our estimates seem plausible.
5. **State.** In 1998, about 22% of all fourth-graders showed proficiency in math. By 2007, that figure is predicted to grow to 40%.

Connecting the Concepts

DATA INTERPRETATION: PREDICTIONS

In business and in many other fields, decisions are often made on the basis of a prediction of the future. Although extrapolation from past data is an important tool in making predictions, other factors should be considered as well.

Look, for example, at the following graph, which shows the number of pieces of priority mail handled by the U.S. Postal Service for 2000–2004.

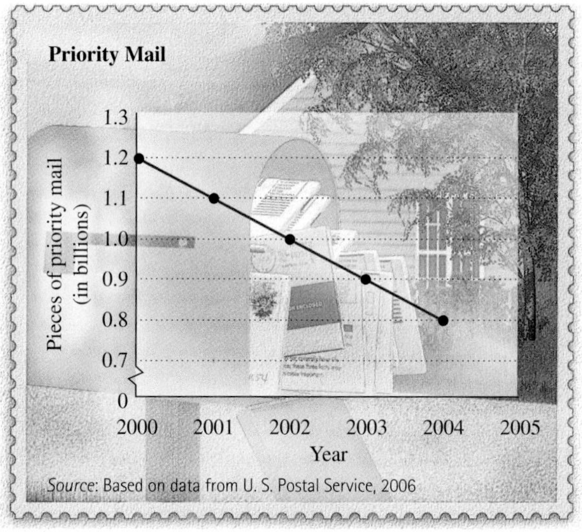

On the basis of past data, we might predict that the volume of priority mail will drop to 0.7 billion in 2005. However, it turns out that more pieces of priority mail were handled in 2005 than in 2004.

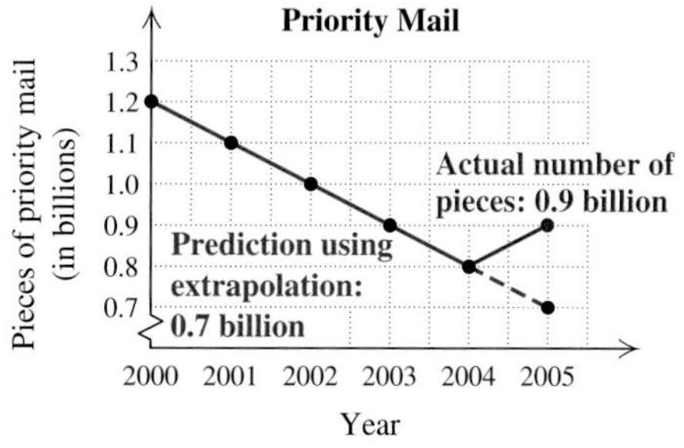

In this case, factors such as the economy, services offered, advertising, and competition contributed to the actual numbers. Although the past can be an excellent predictor of the future, any prediction is to some extent a guess.

Curve Fitting

Often, estimating values as we did in Example 3 is difficult or too imprecise. Another way to analyze data is to find an algebraic equation describing the data. This process is known as **curve fitting.** If the data can be plotted and appear to have a linear pattern, we can fit a linear equation to the data.

Study Tip

Form a Study Group

Consider forming a study group with some of your fellow students. Exchange telephone numbers, schedules, and any e-mail addresses so that you can coordinate study time for homework and tests.

EXAMPLE 4 Following are three graphs of sets of data. Determine whether each appears to be linear.

a) Fiber for Children

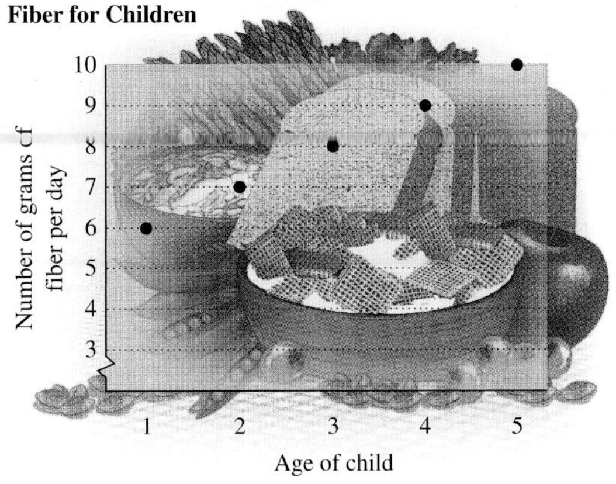

Source: Kellogg's

b) Winter Heating Costs

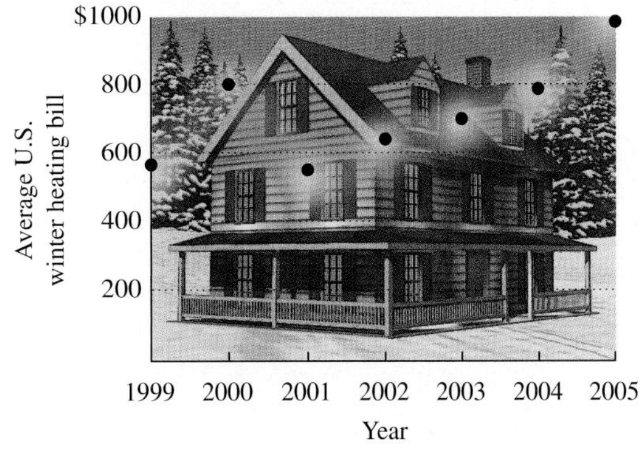

Source: Energy Information Administration, U.S. Department of Energy

c) Registered Nurses

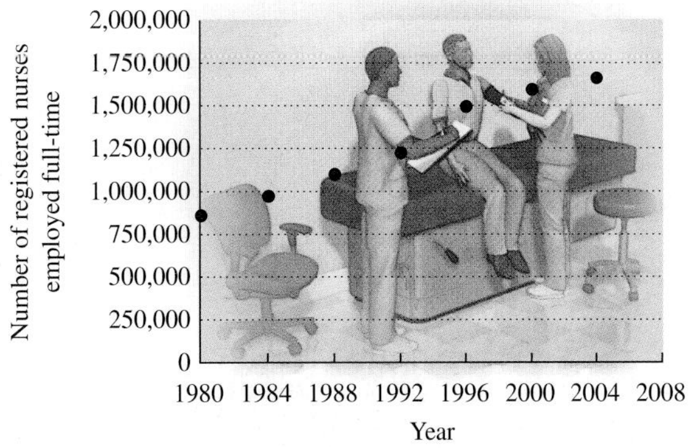

Source: National Sample Survey of Registered Nurses

SOLUTION In order for data to be linear, the points must lie, at least approximately, on a straight line. The rate of change is constant for a linear function, so the change in the quantity on the vertical axis should be about the same for each unit on the horizontal axis.

a) The points lie on a straight line, so the data are linear. The rate of change is constant—1 gram of fiber per year of age.

b) Note that the average heating cost increased, then decreased, then began to increase, but not at a constant rate. The points do not lie on a straight line. The data are not linear.

c) The points lie approximately on a straight line. The data appear to be linear.

If data are linear, we can fit a linear function to the data using the point–slope equation.

EXAMPLE 5 Television Viewing. In the second season of "American Idol," an average of 21.7 million people viewed each episode. By the fifth season, this number had grown to 30.3 million. (*Source*: *The Indianapolis Star*, 5/23/06) Assuming constant growth, what will be the average number of viewers per episode for the sixth season?

SOLUTION

1. **Familiarize.** Constant growth indicates a constant rate of change, so we can assume a linear relationship. We let v represent the average number of viewers per episode, in millions, and x the season. Since we will write the number of viewers as a function of the season, we write x on the horizontal axis and v on the vertical axis and form the pairs (2, 21.7) and (5, 30.3). After choosing suitable scales, we draw the graph.

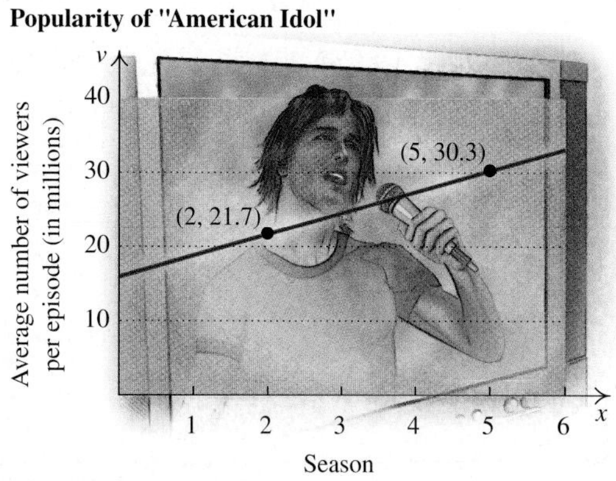

2. **Translate.** To find an equation relating v and x, we first find the slope of the line. This corresponds to the *growth rate*:

$$m = \frac{30.3 \text{ million viewers} - 21.7 \text{ million viewers}}{5 \text{ seasons} - 2 \text{ seasons}}$$

$$= \frac{8.6 \text{ million viewers}}{3 \text{ seasons}} \approx 2.9 \text{ million viewers per season.}$$

Next, we write point–slope form and solve for v:

$$v - 21.7 = 2.9(x - 2) \quad \textbf{Writing point–slope form}$$

$$v - 21.7 = 2.9x - 5.8 \quad \textbf{Using the distributive law}$$

$$v = 2.9x + 15.9. \quad \textbf{Adding 21.7 to both sides}$$

3. **Carry out.** Using function notation, we have

$$v(x) = 2.9x + 15.9.$$

To predict the average number of viewers per episode in the sixth season, we find

$$v(6) = 2.9 \cdot 6 + 15.9 = 33.3.$$

This represents 33.3 million viewers.

4. **Check.** To check, we can repeat our calculations. We could also extend the graph to see if the data (6, 33.3) appear to be on the line.
5. **State.** Assuming constant growth, there will be, on average, about 33.3 million viewers watching each episode of "American Idol" in its sixth season.

Linear Regression

Most real data that are considered linear do not lie exactly on a straight line. Were we to use the point–slope form to write an equation for the line, that equation would vary depending on our choice of points to use. We want to find the *best* line that fits the data.

The line that best describes a set of data may not actually go through any of the given points. There are different methods for finding an equation of a line that fits a set of data. These methods generally consider all the points, not just two, when fitting an equation to data. The most commonly used method is *linear regression.*

The development of the method of linear regression belongs to a later mathematics course, but most graphing calculators offer regression as a way of fitting a line or curve to a set of data.

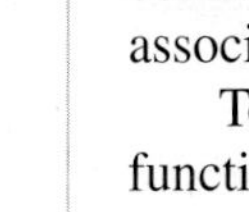

Linear Regression

Fitting a curve to a set of data is done using the STAT menu. Choose EDIT from the STAT EDIT menu, and enter the values of the independent variable in one list and the corresponding values of the dependent variable in another list. Then press **STAT** again, and choose the CALC menu. To fit a line to the data, choose the LINREG option. After copying LinReg(ax + b) to the home screen, enter the list names containing the data separated by commas, with the independent values first. (The list names are 2nd options associated with the number keys (1) through (6).)

To copy the equation found to the equation-editor screen, next enter the function name, using a comma after any list names. To execute the command, press **ENTER**.

(*continued*)

Lin Reg (ax+b) L1, L2, Y1

The command at left indicates that L1 contains the values for the independent variable, L2 contains the values for the dependent variable, and the equation is to be copied to Y1. If no list names are entered, the calculator assumes that the first list is L1 and the second is L2.

Using CATALOG you can turn DiagnosticOn or DiagnosticOff. If diagnostics are turned on, values of r^2 and r will appear on the screen along with the regression equation. These give an indication of how well the regression line fits the data. When r^2 is close to 1, the line is a good fit. We call r the *coefficient of correlation.*

EXAMPLE 6 Registered Nurses. Data from the National Sample Survey of Registered Nurses, conducted every four years, indicate that the number of registered nurses in the United States who are working full time has increased steadily. The information from the survey is shown in the following table and graph. Use linear regression to fit a linear function to the data. Graph the line with the data and use it to predict the number of registered nurses working full time in 2008.

Year	Number of Registered Nurses Working Full Time
1980	850,000
1984	975,000
1988	1,100,000
1992	1,225,000
1996	1,500,000
2000	1,600,000
2004	1,700,000

Registered Nurses

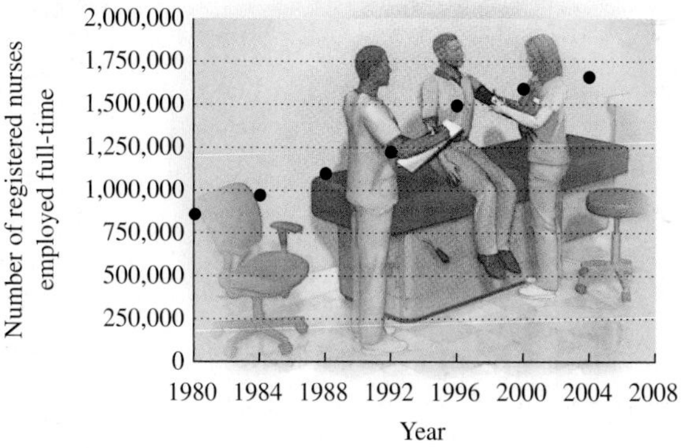

Source: National Sample Survey of Registered Nurses

SOLUTION To make the numbers easier to work with, we will redefine the year t as the number of years since 1980. Then 1980 corresponds to $t = 0$, 1984 corresponds to $t = 4$, and so on. We also let N represent the number of registered nurses working full time, in thousands. Thus the point $(1980, 850{,}000)$ becomes $(0, 850)$. We are looking for an equation of the form $N(t) = mt + b$.

We enter the data, with the number of years since 1980 as L1 and the number of full-time registered nurses, in thousands, as L2.

L1	L2	L3 1
0	850	------
4	975	
8	1100	
12	1225	
16	1500	
20	1600	
24	1700	

L1(1) = 0

Next, we make sure that Plot1 is turned on, and clear any equations listed in the Y= screen. Since years vary from 0 to 24 and the number of nurses (in thousands) varies from 850 to 1700, we set a viewing window of [0, 30, 0, 2000], with Xscl = 4 and Yscl = 200.

To calculate the equation, we choose the LinReg option in the STAT CALC menu. Since our lists are the default lists, L1 and L2, we need not give their names in the command. We select Y1 from the VARS Y-VARS function menu and press ENTER.

The screen on the left below indicates that the equation is

$$y = 37.5x + 828.5714286, \quad \text{or}$$

$$N(t) = 37.5t + 828.6. \qquad \textbf{Using function notation with } N \textbf{ and } t \textbf{ and rounding}$$

The screen in the middle shows the equation copied as Y1. Note that there are even more decimal places in the value for b than are shown in the screen on the left. Pressing GRAPH gives the screen on the right below, showing the points plotted and the line graphed.

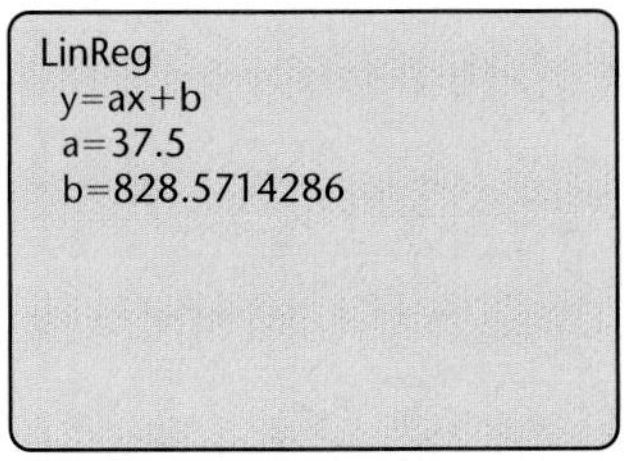

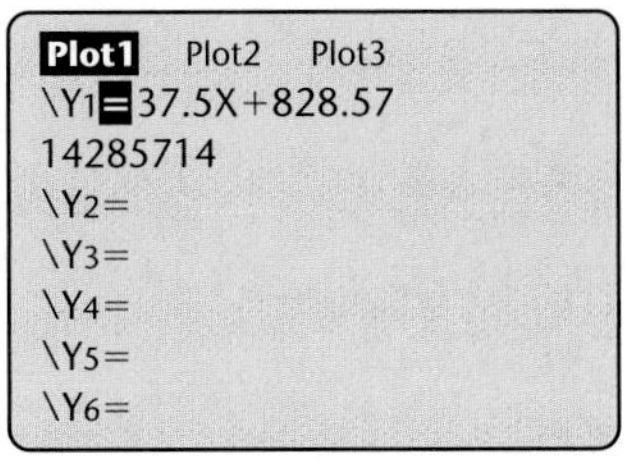

$y = 37.5x + 828.5714286$

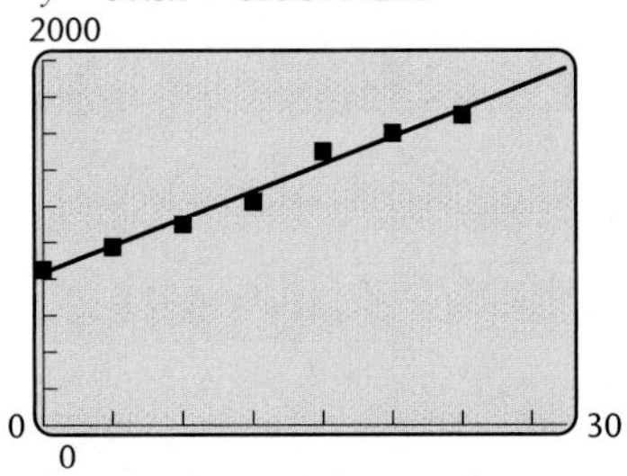

Xscl = 4, Yscl = 200

Since 2008 is 28 yr after 1980, we find $N(28)$ in order to predict the number of full-time registered nurses in 2008. We can do this using the VALUE option of the CALC menu for $x = 28$, a table for $x = 28$, or Y1(28) entered on the home screen. All three methods, shown below, give a value of approximately 1879. Thus we predict that the number of full-time registered nurses in 2008 will be 1,879,000.

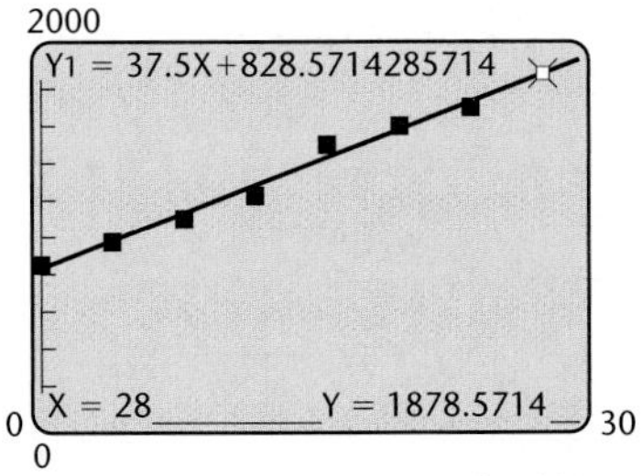

Xscl = 4, Yscl = 200

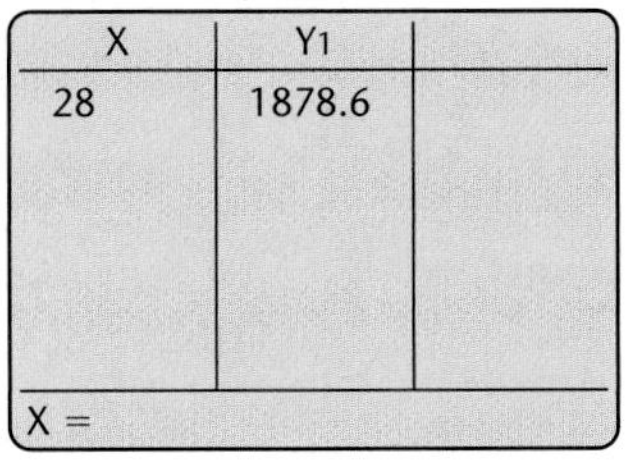

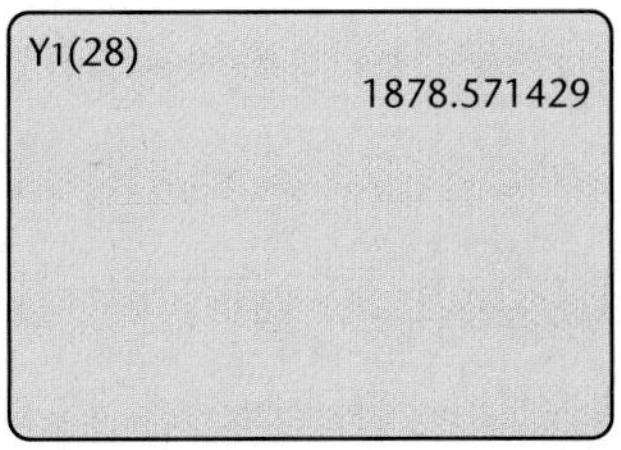

Visualizing the Graph

Match each equation or function with its graph.

1. $y = x + 4$

2. $y = 2x$

3. $y = 3$

4. $x = 3$

5. $f(x) = -\frac{1}{2}x$

6. $2x - 3y = 6$

7. $f(x) = -3x - 2$

8. $3x + 2y = 6$

9. $y - 3 = 2(x - 1)$

10. $y + 2 = \frac{1}{2}(x + 1)$

Answers on page A-9

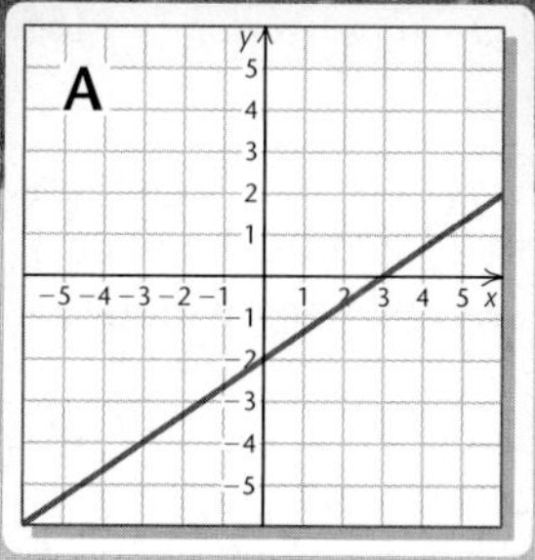

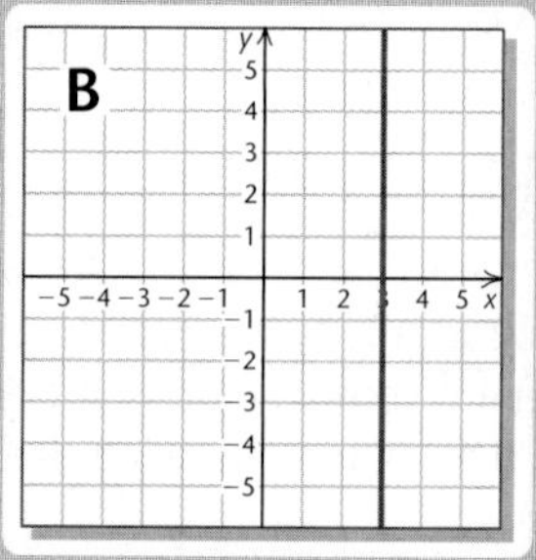

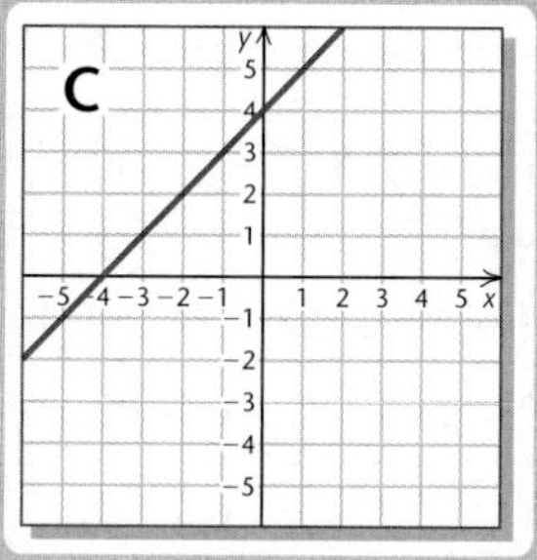

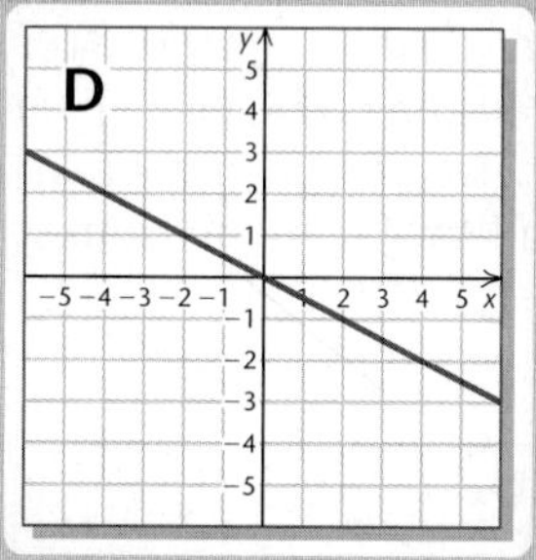

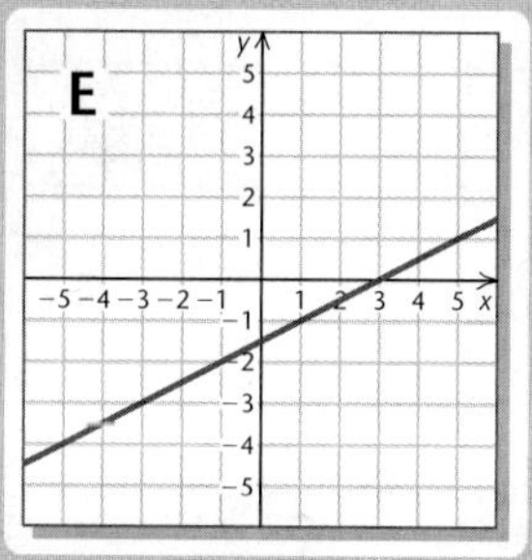

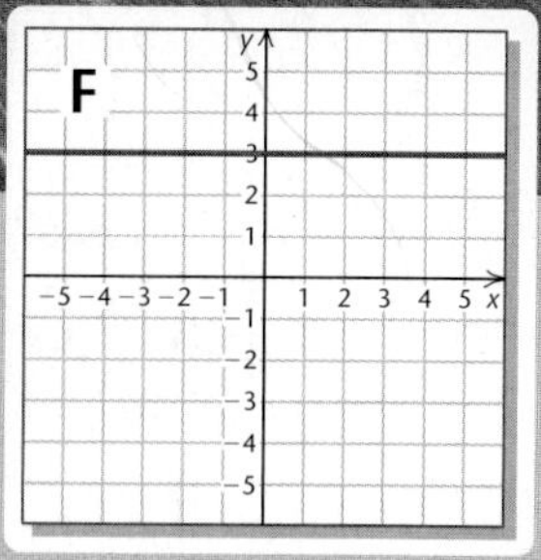

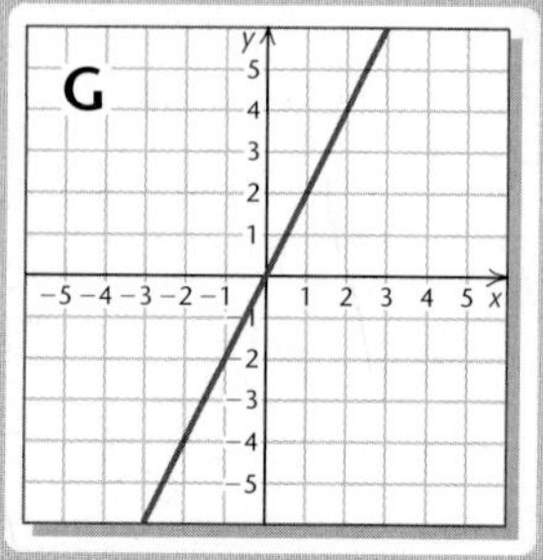

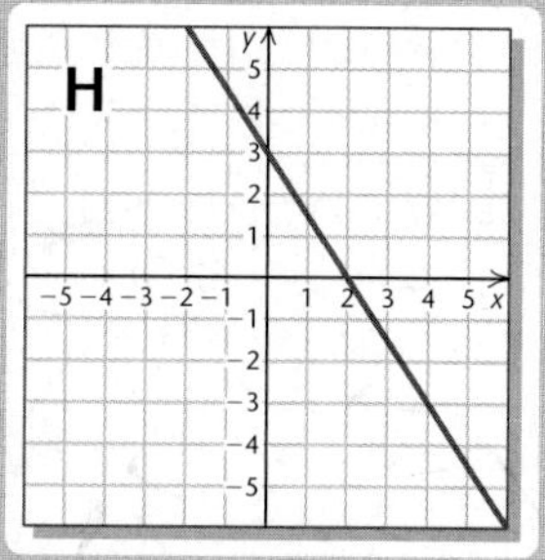

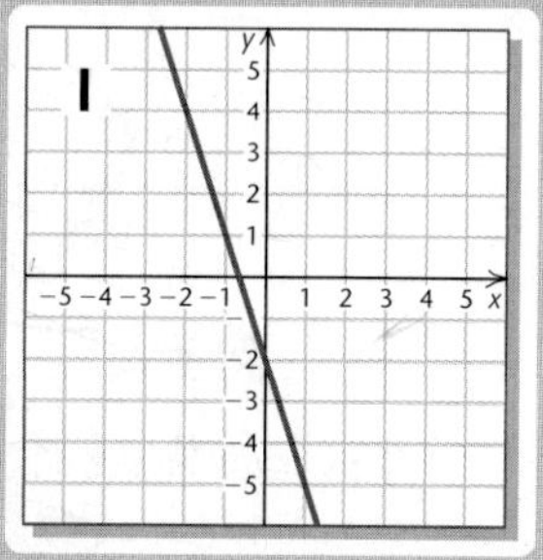

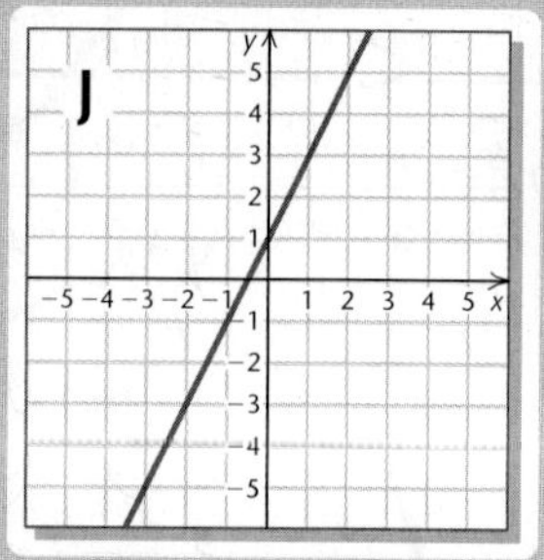

2.4 EXERCISE SET

FOR EXTRA HELP

MathXL | MyMathLab | InterAct Math | AW Math Tutor Center | Video Lectures on CD: Disc 1 | Student's Solutions Manual

Concept Reinforcement *Classify each statement as either true or false.*

1. The equation $y - 5 = -3(x - 7)$ is written in point–slope form.

2. The equation $y = -2x + 6$ is written in point–slope form.

3. Knowing the coordinates of just one point on a line is enough to write an equation of the line.

4. Knowing the coordinates of just two points on a line is enough to write an equation of the line.

5. The point–slope form gives enough information to graph the line.

6. The equations $y = 3x - 5$ and $3x - y = 5$ describe the same line.

7. Point–slope form can be used with either point that is used to calculate the slope of that line.

8. There are situations for which point–slope form is more convenient to use than slope–intercept form.

9. We use interpolation to predict a value that extends beyond known values.

10. Linear regression is used to fit an equation to data.

Find an equation in point–slope form of the line having the specified slope and containing the point indicated. Then graph the line.

11. $m = -2, (1, 4)$ **12.** $m = 5, (3, 1)$

13. $m = 3, (5, 2)$ **14.** $m = 2, (7, 3)$

15. $m = \frac{1}{2}, (-2, -4)$ **16.** $m = 1, (-5, -7)$

17. $m = -1, (8, 0)$ **18.** $m = -3, (-2, 0)$

For each point–slope equation listed, state the slope and a point on the graph.

19. $y - 9 = \frac{2}{7}(x - 8)$

20. $y - 3 = 9(x - 2)$

21. $y + 2 = -5(x - 7)$

22. $y - 4 = -\frac{2}{9}(x + 5)$

23. $y - 4 = -\frac{5}{3}(x + 2)$

24. $y + 7 = -4(x - 9)$

Aha! **25.** $y = \frac{4}{7}x$

26. $y = 3x$

Find an equation of the line having the specified slope and containing the indicated point. Write your final answer as a linear function in slope–intercept form. Then graph the line.

27. $m = 4, (2, -3)$ **28.** $m = -4, (-1, 5)$

29. $m = -\frac{3}{5}, (-4, 8)$ **30.** $m = -\frac{1}{5}, (-2, 1)$

31. $m = -0.6, (-3, -4)$ **32.** $m = 2.3, (4, -5)$

Aha! **33.** $m = \frac{2}{7}, (0, -6)$ **34.** $m = \frac{1}{4}, (0, 3)$

35. $m = \frac{3}{5}, (-4, 6)$ **36.** $m = -\frac{2}{7}, (6, -5)$

Find an equation of the line containing each pair of points. Write your final answer as a linear function in slope–intercept form.

37. $(1, 4)$ and $(5, 6)$

38. $(2, 6)$ and $(4, 1)$

39. $(2.5, -3)$ and $(6.5, 3)$

40. $(2, -1.3)$ and $(7, 1.7)$

Aha! **41.** $(1, 3)$ and $(0, -2)$

42. $(-3, 0)$ and $(0, -4)$

43. $(-2, -3)$ and $(-4, -6)$

44. $(-4, -7)$ and $(-2, -1)$

45. *Unemployment Rate.* The following table lists the unemployment rates in Indiana for several years. Use the data in the table to draw a graph. Then estimate the unemployment rate in Indiana in 2001 and in 2004.

Year	Unemployment Rate in Indiana (in percent)
2000	4.1
2002	6.9
2003	7.8

Source: American Community Survey

46. *Calories Burned.* The following table lists the number of calories burned when walking at 3.5 mph for walkers of various weights. Use the data in the table to draw a graph. Then estimate the number of calories burned by a walker weighing 150 lb and the number burned by a walker weighing 200 lb.

Weight (in pounds)	Calories Burned per Hour at 3.5 mph
120	71
140	83
160	93
180	107

Source: http://walking.about.com/cs/howtoloseweight/a/howcalburn.htm

47. *Calories Burned.* The following table lists the number of calories burned by a 140-lb person by walking at various speeds. Use the data in the table to draw a graph. Then estimate the number of calories burned by a 140-lb person walking at 4.0 mph and by a 140-lb person walking at 5.5 mph.

Speed (in miles per hour)	Calories Burned per Hour by a 140-lb Person
2.5	88
3.5	83
4.5	97
5.0	108

Source: http://walking.about.com/cs/howtoloseweight/a/howcalburn.htm

48. *Reading.* The following table lists the average annual amount spent per person on reading material for various years. Use the data in the table to draw a graph. Then estimate the amount spent per person on reading material in 2001 and in 2004.

Year	Amount Spent on Reading per Person
1996	$159
1998	161
2000	146
2002	139

Source: U.S. Bureau of Labor Statistics, *Consumer Expenditure Survey*

In Exercises 49–60, assume that a constant rate of change exists for each model formed.

49. *Dietary Trends.* In 1971, the average American woman consumed 1542 calories per day. By 2000, the figure had risen to 1877 calories per day. (*Source*: Centers for Disease Control and Prevention) Let $C(t)$ represent the average number of calories consumed per day by an American woman t years after 1971.

a) Find a linear function that fits the data.

b) Use the function from part (a) to predict the average number of calories consumed per day by an American woman in 2009.

c) When will the average number of calories consumed per day reach 2000?

50. *Dietary Trends.* In 1971, the average American man consumed 2450 calories per day. By 2000, the figure had risen to 2618 calories per day. (*Source*: Centers for Disease Control and Prevention) Let $C(t)$ represent the average number of calories consumed per day by an American man t years after 1971.

a) Find a linear function that fits the data.

b) Use the function from part (a) to predict the average number of calories consumed per day by an American man in 2008.

c) When will the average number of calories consumed per day reach 2750?

51. *Life Expectancy of Males in the United States.* In 1993, the life expectancy of males was 72.2 yr. In 2003, it was 74.8 yr. (*Source*: *Statistical Abstract of the United States,* 2006) Let $E(t)$ represent life expectancy and t the number of years since 1993.

a) Find a linear function that fits the data.

b) Use the function of part (a) to predict the life expectancy of males in 2009.

52. *Life Expectancy of Females in the United States.* In 1993, the life expectancy of females was 78.8 yr. In 2003, it was 80.1 yr. (*Source*: *Statistical Abstract of the United States,* 2006) Let $E(t)$ represent life expectancy and t the number of years since 1993.

a) Find a linear function that fits the data.

Aha! **b)** Use the function of part (a) to predict the life expectancy of females in 2010.

53. *PAC Contributions.* In 1994, Political Action Committees (PACs) contributed \$189.6 million to congressional candidates. In 2004, the figure rose to \$310.5 million. (*Source*: Congressional Research Service and Federal Election Commission) Let $A(t)$ represent the amount of PAC contributions, in millions, and t the number of years since 1994.

a) Find a linear function that fits the data.

b) Use the function of part (a) to predict the amount of PAC contributions in 2008.

54. *Consumer Demand.* Suppose that 6.5 million lb of coffee are sold when the price is \$8 per pound, and 4.0 million lb are sold when it is \$9 per pound.

a) Find a linear function that expresses the amount of coffee sold as a function of the price per pound.

b) Use the function of part (a) to predict how much consumers would be willing to buy at a price of \$6 per pound.

55. *Recycling.* In 1996, Americans recycled 57.3 million tons of solid waste. In 2003, the figure grew to 72.3 million tons. (*Source*: *Statistical Abstract of the United States*, 2006) Let $N(t)$ represent the number of tons recycled, in millions, and t the number of years since 1996.

a) Find a linear function that fits the data.

b) Use the function of part (a) to predict the amount recycled in 2010.

56. *Seller's Supply.* Suppose that suppliers are willing to sell 5.0 million lb of coffee at a price of \$8 per pound and 7.0 million lb at \$9 per pound.

a) Find a linear function that expresses the amount suppliers are willing to sell as a function of the price per pound.

b) Use the function of part (a) to predict how much suppliers would be willing to sell at a price of \$6 per pound.

57. *Online Travel Plans.* In 1999, about 48 million Americans used the Internet to find travel information. By 2006, that number had grown to about 79 million. (*Source*: Travel Industry Association of America) Let $N(t)$ represent the number of Americans using the Internet for travel information, in millions, t years after 1999.

a) Find a linear function that fits the data.

b) Use the function of part (a) to predict the number of Americans who will find travel information on the Internet in 2009.

c) In what year will 110 million Americans find travel information on the Internet?

58. *Records in the 100-meter Run.* In 1991, the record for the 100-m run was 9.86 sec. In 2005, it was 9.77 sec. (*Source*: International Athletic Federation) Let $R(t)$ represent the record in the 100-m run and t the number of years since 1991.

a) Find a linear function that fits the data.

b) Use the function of part (a) to predict the record in 2008 and in 2015.

c) When will the record be 9.6 sec?

59. *National Park Land.* In 2000, the National Park system consisted of about 78.2 million acres. By 2004, the figure had grown to 79.0 million acres. (*Source*: *Statistical Abstract of the United States,* 2006) Let $A(t)$ represent the amount of land in the National Park system, in millions of acres, t years after 2000.

a) Find a linear function that fits the data.

b) Use the function of part (a) to predict the amount of land in the National Park system in 2010.

60. *Pressure at Sea Depth.* The pressure 100 ft beneath the ocean's surface is approximately 4 atm (atmospheres), whereas at a depth of 200 ft, the pressure is about 7 atm.

a) Find a linear function that expresses pressure as a function of depth.

b) Use the function of part (a) to determine the pressure at a depth of 690 ft.

Determine whether the data in each graph appear to be linear.

61.

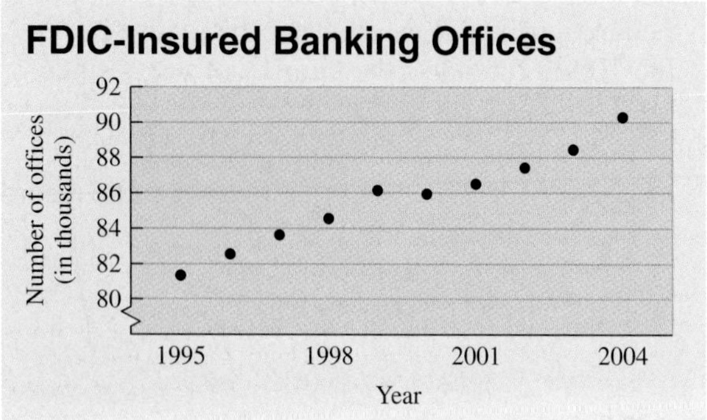

Source: U.S. Federal Deposit Insurance Corporation

62. **U.S. Shopping Centers**

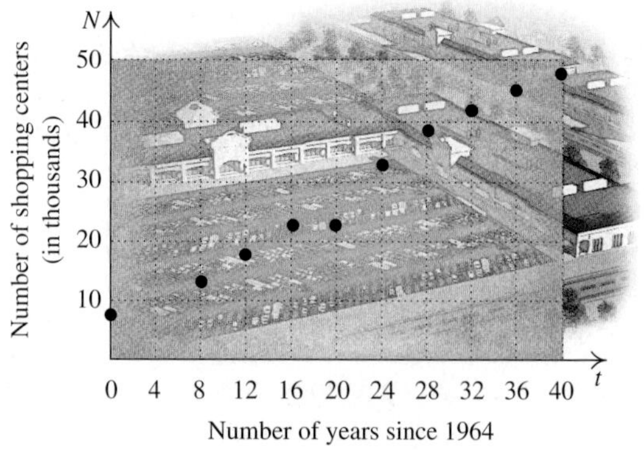

Source: International Council of Shopping Centers

63. **Amtrak Riders**

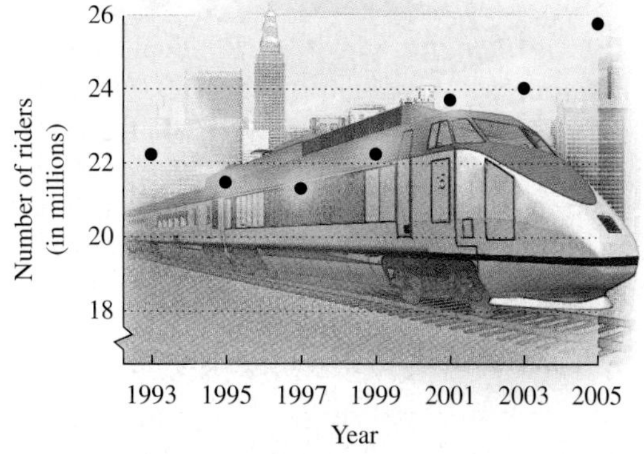

Source: National Association of Railroad Passengers

64.

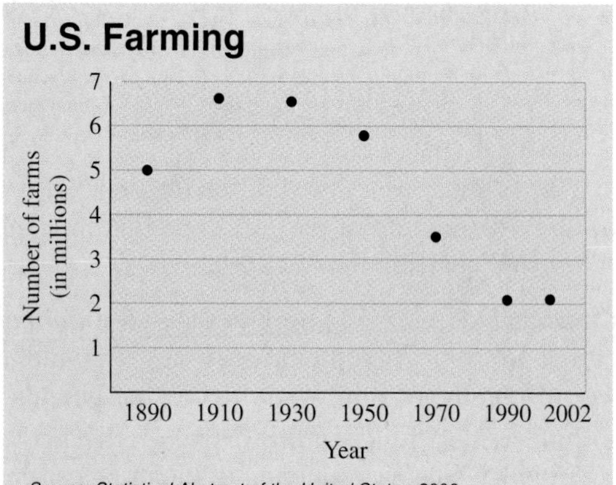

Source: *Statistical Abstract of the United States*, 2006

65.

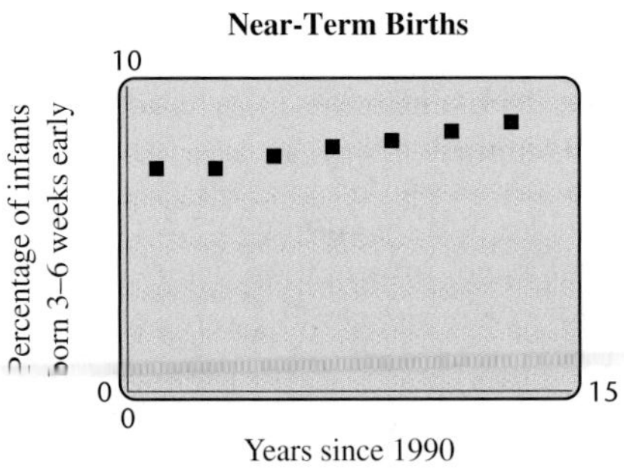

66.

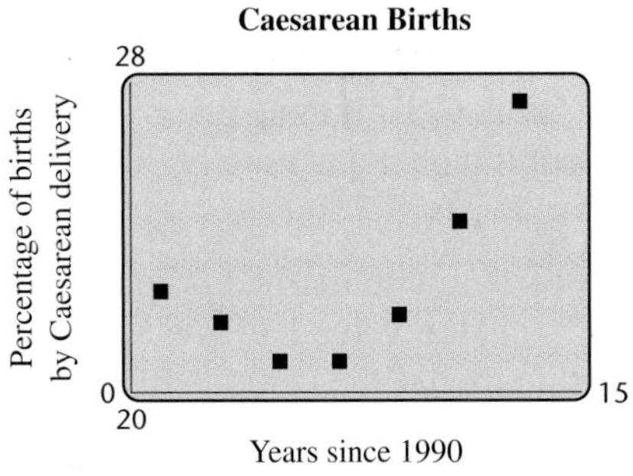

67. *Life Expectancy of Females in the United States.* The following table lists the life expectancy of women who were born in the United States in selected years.

Life expectancy of women

Year	Life Expectancy (in years)
1960	73.1
1970	74.4
1980	77.5
1990	78.8
2000	79.7
2003	80.1

Source: National Center for Health Statistics

a) Use linear regression to find a linear function that can be used to predict the life expectancy W of a woman as a function of the year in which she was born. (Let $x =$ the number of years since 1900.)

b) Predict the life expectancy of a woman in 2010 and compare your answer with the answer to Exercise 52.

68. *Life Expectancy of Males in the United States.* The following table lists the life expectancy of males born in the United States in selected years.

Life expectancy of men

Year	Life Expectancy (in years)
1960	66.6
1970	67.1
1980	70.0
1990	71.8
2000	74.4
2003	74.8

Source: National Center for Health Statistics

a) Use linear regression to find a linear function that can be used to predict the life expectancy M of a man as a function of the year in which he was born. (Let $x =$ the number of years since 1900.)

b) Predict the life expectancy of a man in 2009 and compare your answer with the answer to Exercise 51.

69. *Banking.* The following table lists the number of FDIC-insured institutions for several years.

Year	Number of FDIC-Insured Financial Institutions
1995	81,350
1996	82,578
1997	83,614
1998	84,587
1999	86,040
2000	85,952
2001	86,506
2002	87,429
2003	88,447
2004	90,267

Source: U.S. Federal Deposit Insurance Corporation

a) Use linear regression to find a linear function that can be used to predict the number of FDIC-insured financial institutions B as a function of the number of years x after 1995.

b) Estimate the number of FDIC-insured financial institutions in 2008.

70. *Near-Term Births.* The percent of infants who are born 3–6 weeks early has increased in recent years, as shown in the following table.

Year	Percent of Near-Term Births
1991	7.3
1993	7.3
1995	7.7
1997	8.0
1999	8.2
2001	8.5
2003	8.8

Source: National Center for Health Statistics

a) Use linear regression to find a linear function that can be used to predict the percent P of near-term births as a function of the number of years x after 1990.
b) Estimate the percent of near-term births in 2006.

TW **71.** Rosewood Graphics recently promised its employees 6% raises each year for the next 5 yr. Amy currently earns $30,000 a year. Can she use a linear function to predict her salary for the next 5 yr? Why or why not?

TW **72.** On the basis of your answers to Exercises 51 and 52, would you predict that at some point in the future the life expectancy of males will exceed that of females? Why or why not?

Focused Review

Find the slope of each line, if it exists.

73. $y = 3x - 5$ [2.2]

74. $y - 10 = -4(x + 6)$ [2.4]

75. $2x + 6y = 5$ [2.2]

76. $y = \frac{1}{2}$ [2.3]

Find the x- and y-intercepts of each line, if they exist.

77. $y = \frac{1}{2}x - 5$ [2.2], [2.3]

78. $y + 2 = -(x - 3)$ [2.4]

79. $4x + 5y = 20$ [2.3]

80. $y = -4$ [2.3]

Synthesis

TW **81.** In your answer to Exercise 59(a), what does each of the coefficients signify?

TW **82.** In 2004, about 66 million Americans used the Internet to find travel information. Does this new data point make your answer to Exercise 57(b) seem too low or too high? Why?

For Exercises 83–86, assume that a linear equation models each situation.

83. *Temperature Conversion.* Water freezes at 32° Fahrenheit and at 0° Celsius. Water boils at 212°F and at 100°C. What Celsius temperature corresponds to a room temperature of 70°F?

84. *Depreciation of a Computer.* After 6 mos of use, the value of Pearl's computer had dropped to $900. After 8 mos, the value had gone down to $750. How much did the computer cost originally?

85. *Cell-Phone Charges.* The total cost of Mel's cell phone was $230 after 5 mos of service and $390 after 9 mos. What costs had Mel already incurred when his service just began?

86. *Operating Expenses.* The total cost for operating Ming's Wings was $7500 after 4 mos and $9250 after 7 mos. Predict the total cost after 10 mos.

87. On the basis of the information given in Exercises 54 and 56, determine at what price the supply will equal the demand.

Write an equation of the line containing the specified point and parallel to the indicated line.

88. $(3, 7)$, $x + 2y = 6$

89. $(-1, 4)$, $3x - y = 7$

Write an equation of the line containing the specified point and perpendicular to the indicated line.

90. $(2, 5)$, $2x + y = -3$

91. $(4, 0)$, $x - 3y = 0$

92. Specify the domain of your answer to Exercise 54(a).

93. Specify the domain of your answer to Exercise 56(a).

94. For a linear function g, $g(3) = -5$ and $g(7) = -1$.
a) Find an equation for g.
b) Find $g(\ \ 2)$.
c) Find a such that $g(a) = 75$.

2.5 Domains and the Algebra of Functions

The Sum, Difference, Product, or Quotient of Two Functions ■ Domains and Graphs

We now examine four ways in which functions can be combined, followed by a reexamination of domains.

The Sum, Difference, Product, or Quotient of Two Functions

Suppose that a is in the domain of two functions, f and g. The input a is paired with $f(a)$ by f and with $g(a)$ by g. The outputs can then be added to get $f(a) + g(a)$.

Interactive Discovery

Enter $y_1 = x + 4$ and $y_2 = x^2 + 1$. Next, enter $y_3 = y_1 + y_2$.

1. Set up a table that shows the values of the three functions. If y_1 represents $f(x)$ and y_2 represents $g(x)$, what does y_3 represent?
2. Graph y_1, y_2, and y_3 using the same viewing window. How could you draw the graph of y_3 given the graphs of y_1 and y_2?
   Adding the expressions for y_1 and y_2 algebraically, we get
   $$(x + 4) + (x^2 + 1) = x^2 + x + 5.$$
3. Enter $y_4 = x^2 + x + 5$ and compare the values of y_3 and y_4 using a table or a graph. How are these functions related?

We see that if $f(x) = x + 4$ and $g(x) = x^2 + 1$, then $f(x) + g(x) = x^2 + x + 5$, which can be regarded as a "new" function, written $(f + g)(x)$.

The Algebra of Functions If f and g are functions and x is in the domain of both functions, then:

1. $(f + g)(x) = f(x) + g(x)$;
2. $(f - g)(x) = f(x) - g(x)$;
3. $(f \cdot g)(x) = f(x) \cdot g(x)$;
4. $(f/g)(x) = f(x)/g(x)$, provided $g(x) \neq 0$.

Study Tip

Test Preparation

The best way to prepare for taking tests is by working consistently throughout the course. That said, here are some extra suggestions.

- Make up your own practice test.
- Ask your instructor or former students for old exams to practice on.
- Review your notes and all homework that gave you difficulty.
- Make use of the Study Summary, Review Exercises, and Test at the end of each chapter.

EXAMPLE 1 For $f(x) = x^2 - x$ and $g(x) = x + 2$, find the following.

a) $(f + g)(3)$

b) $(f - g)(x)$ and $(f - g)(-1)$

c) $(f/g)(x)$ and $(f/g)(-4)$

d) $(f \cdot g)(3)$

SOLUTION

a) Since $f(3) = 3^2 - 3 = 6$ and $g(3) = 3 + 2 = 5$, we have

$$\begin{aligned}(f + g)(3) &= f(3) + g(3)\\ &= 6 + 5 \quad \textbf{Substituting}\\ &= 11.\end{aligned}$$

Alternatively, we could first find $(f + g)(x)$:

$$\begin{aligned}(f + g)(x) &= f(x) + g(x)\\ &= x^2 - x + x + 2\\ &= x^2 + 2. \quad \textbf{Combining like terms}\end{aligned}$$

Thus,

$$(f + g)(3) = 3^2 + 2 = 11. \quad \textbf{Our results match.}$$

b) We have

$$\begin{aligned}(f - g)(x) &= f(x) - g(x)\\ &= x^2 - x - (x + 2) \quad \textbf{Substituting}\\ &= x^2 - 2x - 2. \quad \textbf{Removing parentheses and combining like terms}\end{aligned}$$

Thus,

$$\begin{aligned}(f - g)(-1) &= (-1)^2 - 2(-1) - 2 \quad \textbf{Using } (f - g)(x) \textbf{ is faster than using } f(x) - g(x).\\ &= 1. \quad \textbf{Simplifying}\end{aligned}$$

c) We have

$$\begin{aligned}(f/g)(x) &= f(x)/g(x)\\ &= \frac{x^2 - x}{x + 2}. \quad \textbf{We assume that } x \neq -2.\end{aligned}$$

Thus,

$$\begin{aligned}(f/g)(-4) &= \frac{(-4)^2 - (-4)}{-4 + 2} \quad \textbf{Substituting}\\ &= \frac{20}{-2} = -10.\end{aligned}$$

d) Using our work in part (a), we have

$$(f \cdot g)(3) = f(3) \cdot g(3) = 6 \cdot 5 = 30.$$

It is also possible to compute $(f \cdot g)(3)$ by first multiplying $x^2 - x$ and $x + 2$ using methods we will discuss in Chapter 5.

Domains and Graphs

Although applications involving products and quotients of functions rarely appear in newspapers, situations involving sums or differences of functions often do appear in print. For example, the following graphs are similar to those published by the California Department of Education to promote breakfast programs in which students eat a balanced meal of fruit or juice, toast or cereal, and 2% or whole milk. The combination of carbohydrate, protein, and fat gives a sustained release of energy, delaying the onset of hunger for several hours.

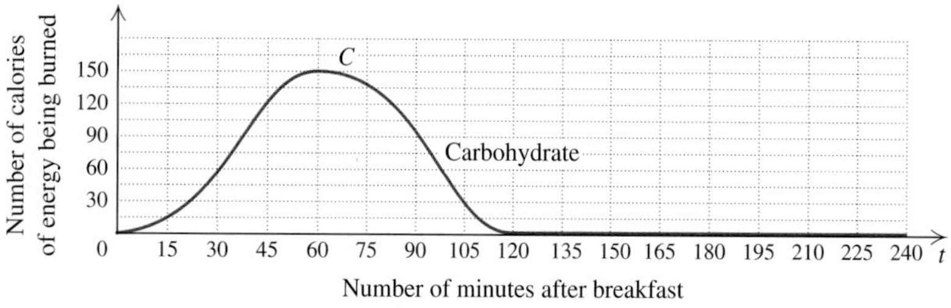

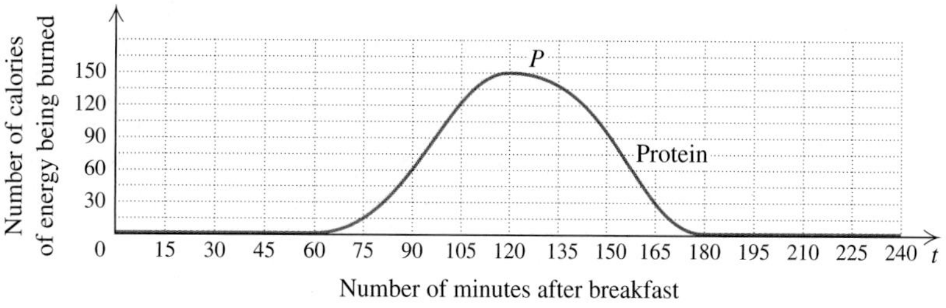

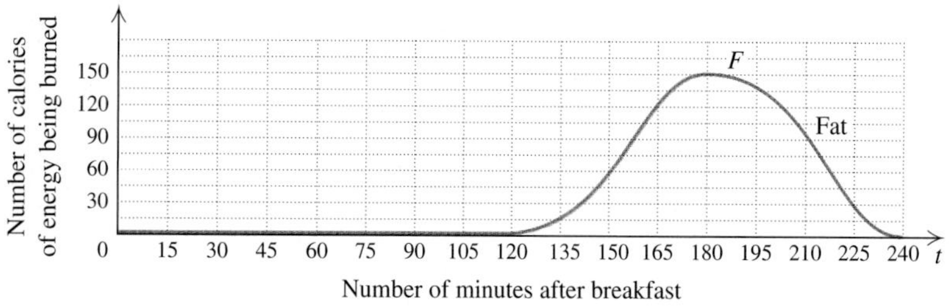

When the three graphs are superimposed, and the calorie expenditures added, it becomes clear that a balanced meal results in a steady, sustained supply of energy.

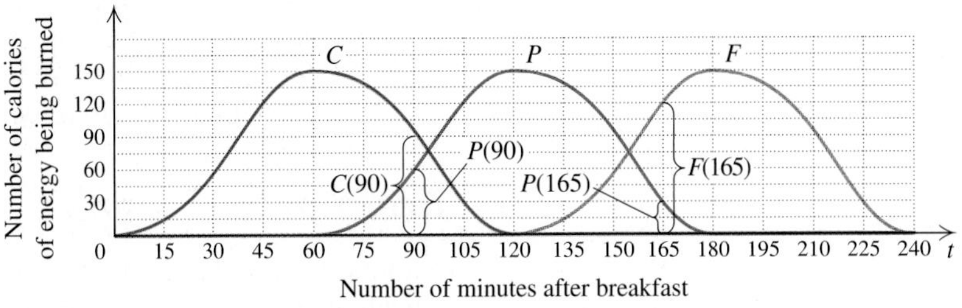

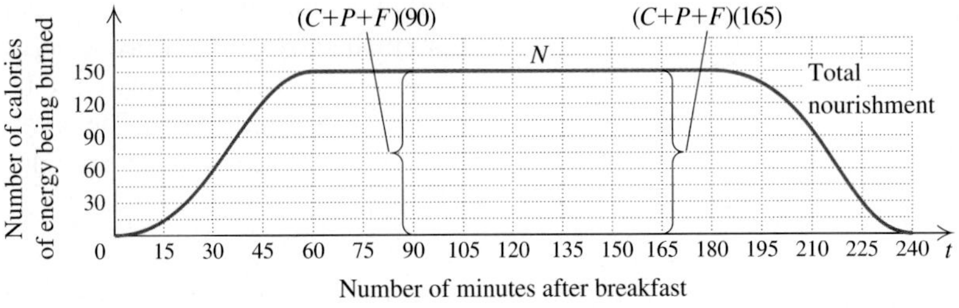

Note that for $t > 120$, we have $C(t) = 0$; for $t < 60$ or $t > 180$, we have $P(t) = 0$; and for $t < 120$, we have $F(t) = 0$. For any point on this last graph, we have

$$N(t) = (C + P + F)(t) = C(t) + P(t) + F(t).$$

To find $(f + g)(a)$, $(f - g)(a)$, $(f \cdot g)(a)$, or $(f/g)(a)$, we must first be able to find $f(a)$ and $g(a)$. Thus we need to ensure that a is in the domain of both f and g.

When a function is described by an equation, the domain is often unspecified. In such cases, the domain is the set of all numbers for which function values can be calculated.

EXAMPLE 2 For each equation, determine the domain of f.

a) $f(x) = |x|$

b) $f(x) = \dfrac{7}{2x - 6}$

SOLUTION

a) We ask ourselves, "Is there any number x for which we cannot compute $|x|$?" Since we can find the absolute value of *any* number, the answer is no. Thus the domain of f is $\mathbb{R}$, the set of all real numbers.

b) Is there any number x for which $\dfrac{7}{2x - 6}$ cannot be computed? Since $\dfrac{7}{2x - 6}$ cannot be computed when $2x - 6$ is 0, the answer is yes. To determine what x-value causes the denominator to be 0, we set up and solve an equation:

$2x - 6 = 0$	**Setting the denominator equal to 0**
$2x = 6$	**Adding 6 to both sides**
$x = 3.$	**Dividing both sides by 2**

Thus, 3 is *not* in the domain of f, whereas all other real numbers are. The domain of f is $\{x \mid x \text{ is a real number } \textit{and } x \neq 3\}$.

A graphing calculator can be used to determine whether a number is in a function's domain.

EXAMPLE 3 If $f(x) = \dfrac{7}{2x - 6}$, use a graphing calculator to determine whether 0, 5, and 3 are in the domain of the function.

SOLUTION To use a table, we enter $y = 7/(2x - 6)$ and set Indpnt to Ask in the Table Setup before viewing the table. To find the function values, we enter x-values of 0, 5, and 3.

From the table, we see that $f(0) \approx -1.167$ and $f(5) = 1.75$. The ERROR entry indicates that $f(3)$ is not defined. This tells us that 0 and 5 are in the domain of f and 3 is not.

X	Y1	
0	⁻1.167	
5	1.75	
3	ERROR	

Y1=ERROR

We can also use VALUE with the graph of a function and function notation on the home screen to determine whether a number is in the domain of a function.

If we use the VALUE option of the CALC menu for $x = 3$, no corresponding y-value will appear. If we attempt to find Y1(3) on the home screen, the calculator displays the following error. This also tells us that 3 is not in the domain of the function.

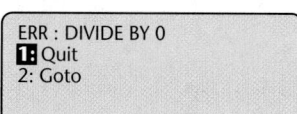

ERR : DIVIDE BY 0
1: Quit
2: Goto

CAUTION! A graphing calculator is useful for checking whether a particular value of x is in the domain of a function. However, algebraic methods are needed in order to determine the entire domain, since it is impossible to check every real number with a calculator.

Interactive Discovery

Let $f(x) = \dfrac{5}{x}$ and $g(x) = \dfrac{2x - 6}{x + 1}$. Enter $y_1 = f(x)$, $y_2 = g(x)$, and $y_3 = y_1 + y_2$. Create a table of values for the functions with TblStart $= -3$, ΔTbl $= 1$, and Indpnt set to Auto.

1. What number is not in the domain of f?
2. What number is not in the domain of g?
3. What numbers are not in the domain of $f + g$?

Now enter $y_4 = y_1 - y_2$, $y_5 = y_1 \cdot y_2$, and $y_6 = y_1/y_2$. Create a table of values for the functions.

4. Does the domain of $f - g$ appear to be the same as the domain of $f + g$?
5. Does the domain of $f \cdot g$ appear to be the same as the domain of $f + g$?
6. Does the domain of f/g appear to be the same as the domain of $f + g$?

In the Interactive Discovery above, because division by 0 is not defined, we have

$$\text{the domain of } f = \{x \mid x \text{ is a real number } \textit{and } x \neq 0\}$$

and

$$\text{the domain of } g = \{x \mid x \text{ is a real number } \textit{and } x \neq -1\}.$$

In order to find $f(a) + g(a)$, $f(a) - g(a)$, or $f(a) \cdot g(a)$, we must know that a is in *both* of the above domains. Thus,

$$\begin{aligned}\text{the domain of } f + g &= \text{the domain of } f - g = \text{the domain of } f \cdot g \\ &= \{x \mid x \text{ is a real number } \textit{and } x \neq 0 \textit{ and } x \neq -1\}.\end{aligned}$$

The domain of f/g also excludes the number 3, because $g(3) = 0$:

$$\begin{aligned}\text{the domain of } f/g = \{x \mid x \text{ is a real number } &\textit{and} \\ x \neq 0 \textit{ and } x \neq -1 \textit{ and } x \neq 3\}&.\end{aligned}$$

Determining the Domain

The domain of $f + g$, $f - g$, or $f \cdot g$ is the set of all values common to the domains of f and g.

The domain of f/g is the set of all values common to the domains of f and g, excluding any values for which $g(x)$ is 0.

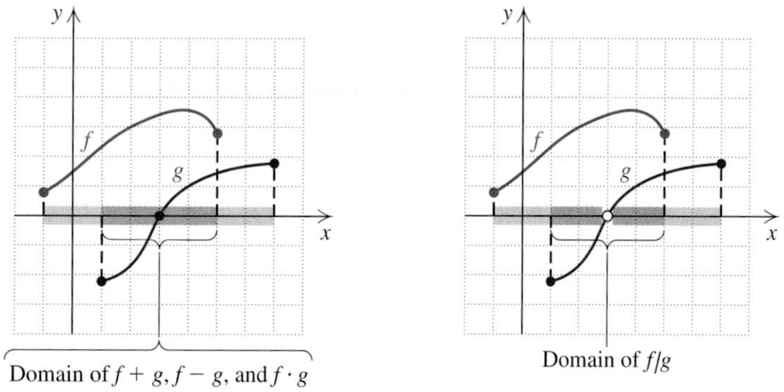

EXAMPLE 4 Given $f(x) = \dfrac{1}{x}$ and $g(x) = 2x - 7$, find the domains of $f + g$, $f - g$, $f \cdot g$, and f/g.

SOLUTION The domain of f is $\{x \mid x \text{ is a real number } \textit{and } x \neq 0\}$. The domain of g is $\mathbb{R}$. The domains of $f + g$, $f - g$, and $f \cdot g$ are the set of all elements common to both the domain of f and the domain of g. We have

$$\text{the domain of } f + g = \text{the domain of } f - g = \text{the domain of } f \cdot g$$
$$= \{x \mid x \text{ is a real number } \textit{and } x \neq 0\}.$$

The domain of f/g is $\{x \mid x \text{ is a real number } \textit{and } x \neq 0\}$, *with the additional restriction* that $g(x) \neq 0$. To determine what x-values would make $g(x) = 0$, we solve:

$$2x - 7 = 0 \qquad \textbf{Replacing } g(x) \textbf{ with } 2x - 7$$
$$2x = 7$$
$$x = \tfrac{7}{2}.$$

Since $g(x) = 0$ for $x = \frac{7}{2}$,

$$\text{the domain of } f/g = \left\{x \mid x \text{ is a real number } \textit{and } x \neq 0 \textit{ and } x \neq \tfrac{7}{2}\right\}.$$

Student Notes

The concern over a denominator being 0 arises throughout this course. Try to develop the habit of checking for any possible input-values that would create a denominator of 0 whenever you work with functions.

Division by 0 is not the only condition that can force restrictions on the domain of a function. In Chapter 7, we will examine functions similar to that given by $f(x) = \sqrt{x}$, for which the concern is taking the square root of a negative number.

2.5 EXERCISE SET

FOR EXTRA HELP

MathXL | MyMathLab | InterAct Math | AW Math Tutor Center | Video Lectures on CD: Disc 1 | Student's Solutions Manual

Concept Reinforcement Make each of the following sentences true by selecting the correct word for each blank.

1. If f and g are functions and x is in the __________ (range/domain) of both functions, then $(f + g)(x) = f(x) + g(x)$.

2. One way to compute $(f - g)(2)$ is to __________ (range/domain) $g(2)$ from $f(2)$.

3. One way to compute $(f - g)(2)$ is to simplify $f(x) - g(x)$ and then __________ (range/domain) the result for $x = 2$.

4. The domain of $f + g, f - g$, and $f \cdot g$ is the set of all values common to the __________ (range/domain) of f and g.

5. The domain of f/g is the set of all values common to the domains of f and g, __________ (range/domain) any values for which $g(x)$ is 0.

6. The height of $(f + g)(a)$ on a graph is the __________ (range/domain) of the heights of $f(a)$ and $g(a)$.

Let $f(x) = -3x + 1$ and $g(x) = x^2 + 2$. Find the following.

7. $f(2) + g(2)$

8. $f(-1) + g(-1)$

9. $f(5) - g(5)$

10. $f(4) - g(4)$

11. $f(-1) \cdot g(-1)$

12. $f(-2) \cdot g(-2)$

13. $f(-4)/g(-4)$

14. $f(3)/g(3)$

15. $g(1) - f(1)$

16. $g(2)/f(2)$

17. $(f + g)(x)$

18. $(g - f)(x)$

Let $F(x) = x^2 - 2$ and $G(x) = 5 - x$. Find the following.

19. $(F + G)(x)$

20. $(F + G)(a)$

21. $(F + G)(-4)$

22. $(F + G)(-5)$

23. $(F - G)(3)$

24. $(F - G)(2)$

25. $(F \cdot G)(-3)$

26. $(F \cdot G)(-4)$

27. $(F/G)(x)$

28. $(G - F)(x)$

29. $(F/G)(-2)$

30. $(F/G)(-1)$

In 2004, a study comparing high doses of the cholesterol-lowering drugs Lipitor and Pravachol indicated that patients taking Lipitor were significantly less likely to have heart attacks or require angioplasty or surgery.

In the graph below, $L(t)$ is the percentage of patients on Lipitor (80 mg) and $P(t)$ is the percentage of patients on Pravachol (40 mg) who suffered heart problems or death t years after beginning to take the medication (Source: New York Times, March 9, 2004).

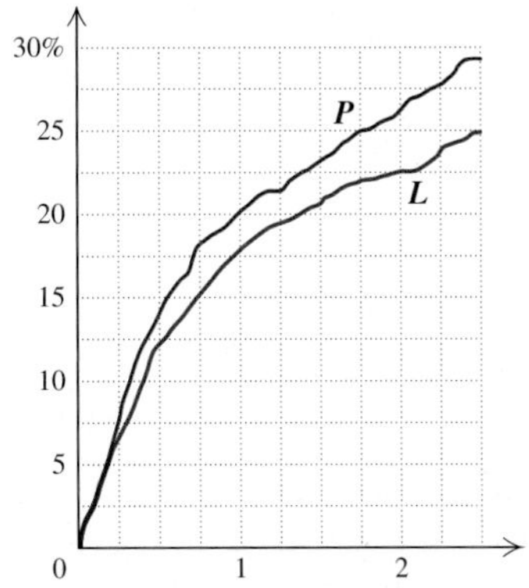

Years of follow-up of patients

Source: New England Journal of Medicine

31. Use estimates of $P(2)$ and $L(2)$ to estimate $(P - L)(2)$.

32. Use estimates of $P(1)$ and $L(1)$ to estimate $(P - L)(1)$.

The following graph shows the number of women, in millions, who had a child the previous year. Here $W(t)$ represents the number of women under 30 who gave birth in year t, $R(t)$ the number of women 30 and older

who gave birth in year t, and N(t) the total number of women who gave birth in year t.

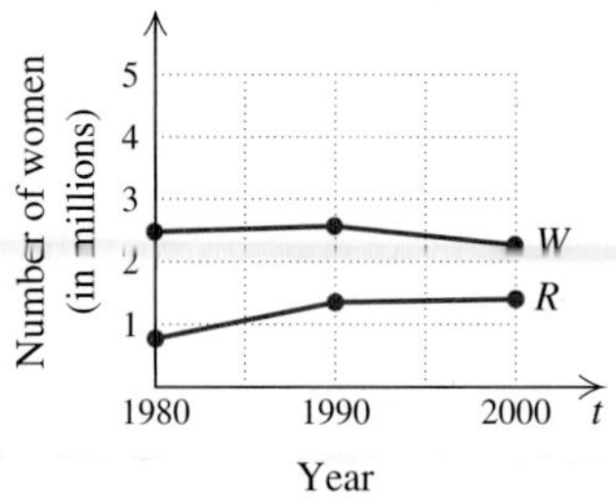

Source: U.S Bureau of the Census

33. Use estimates of $R(2000)$ and $W(2000)$ to estimate $N(2000)$.

34. Use estimates of $R(1990)$ and $W(1990)$ to estimate $N(1990)$.

Often function addition is represented by stacking the individual functions directly on top of each other. The graph below indicates how the three major airports servicing New York City have been utilized. The braces indicate the values of the individual functions.

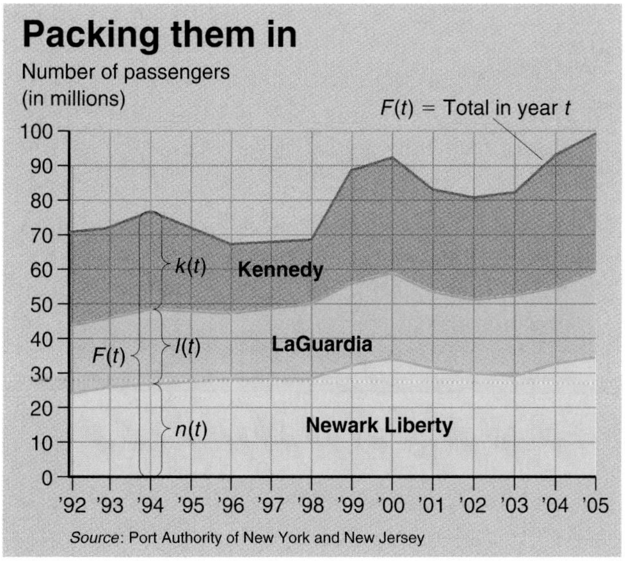

35. Estimate $(n + l)('98)$. What does it represent?

36. Estimate $(k + l)('98)$. What does it represent?

37. Estimate $F('02)$. What does it represent?

38. Estimate $F('01)$. What does it represent?

39. Estimate $(F - k)('02)$. What does it represent?

40. Estimate $(F - k)('01)$. What does it represent?

41. Find the domain of f.

a) $f(x) = \dfrac{5}{x - 3}$ **b)** $f(x) = \dfrac{7}{6 - x}$

c) $f(x) = 2x + 1$ **d)** $f(x) = x^2 + 3$

e) $f(x) = \dfrac{3}{2x - 5}$ **f)** $f(x) = |3x - 4|$

42. Find the domain of g.

a) $g(x) = \dfrac{3}{x - 1}$ **b)** $g(x) = |5 - x|$

c) $g(x) = \dfrac{9}{x + 3}$ **d)** $g(x) = \dfrac{4}{3x + 4}$

e) $g(x) = x^3 - 1$ **f)** $g(x) = 7x - 8$

For each pair of functions f and g, determine the domain of the sum, difference, and product of the two functions.

43. $f(x) = x^2$, $g(x) = 7x - 4$

44. $f(x) = 5x - 1$, $g(x) = 2x^2$

45. $f(x) = \dfrac{1}{x - 3}$, $g(x) = 4x^3$

46. $f(x) = 3x^2$, $g(x) = \dfrac{1}{x - 9}$

47. $f(x) = \dfrac{2}{x}$, $g(x) = x^2 - 4$

48. $f(x) = x^3 + 1$, $g(x) = \dfrac{5}{x}$

49. $f(x) = x + \dfrac{2}{x - 1}$, $g(x) = 3x^3$

50. $f(x) = 9 - x^2$, $g(x) = \dfrac{3}{x - 6} + 2x$

51. $f(x) = \dfrac{3}{x - 2}$, $g(x) = \dfrac{5}{4 - x}$

52. $f(x) = \dfrac{5}{x - 3}$, $g(x) = \dfrac{1}{x - 2}$

For each pair of functions f and g, determine the domain of f/g.

53. $f(x) = x^4$, $g(x) = x - 3$

54. $f(x) = 2x^3$, $g(x) = 5 - x$

55. $f(x) = 3x - 2$, $g(x) = 2x - 8$

56. $f(x) = 5 + x$, $g(x) = 6 - 2x$

57. $f(x) = \dfrac{3}{x - 4}$, $g(x) = 5 - x$

58. $f(x) = \dfrac{1}{2 - x}$, $g(x) = 7 - x$

For Exercises 59–66, consider the functions F and G as shown.

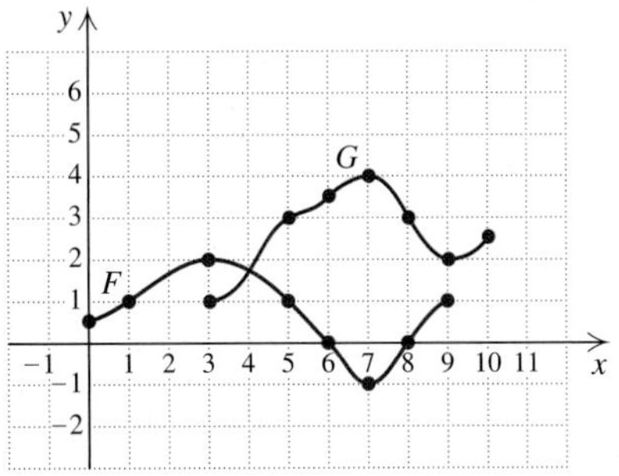

59. Determine $(F + G)(5)$ and $(F + G)(7)$.

60. Determine $(F \cdot G)(6)$ and $(F \cdot G)(9)$.

61. Determine $(G - F)(7)$ and $(G - F)(3)$.

62. Determine $(F/G)(3)$ and $(F/G)(7)$.

63. Find the domains of F, G, $F + G$, and F/G.

64. Find the domains of $F - G$, $F \cdot G$, and G/F.

65. Graph $F + G$.

66. Graph $G - F$.

*In the following graph, $W(t)$ represents the number of gallons of whole milk, $L(t)$ the number of gallons of lowfat milk, and $S(t)$ the number of gallons of skim milk consumed by the average American in year t.**

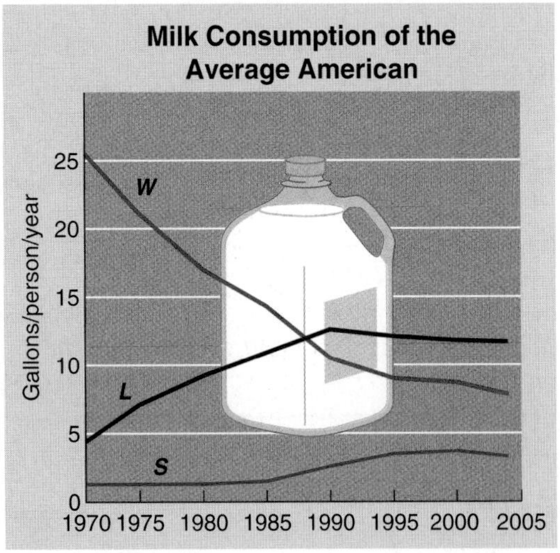

*Sources: Copyright 1990, CSPI. Adapted from *Nutrition Action Healthletter* (1875 Connecticut Avenue, N.W., Suite 300, Washington, DC 20009-5728. $24.00 for 10 issues); USDA Agricultural Fact Book 2000, USDA Economic Research Service; ers.usda.gov

TW **67.** From 1970 to 2004, did American milk consumption increase or decrease? Explain how you determined this.

TW **68.** Examine the graphs in Exercises 35–40. To what do you attribute the decline in $F(t)$ from 2000 to 2002?

Skill Maintenance

Solve. [1.6]

69. $4x - 7y = 8$, for x

70. $3x - 8y = 5$, for y

71. $5x + 2y = -3$, for y

72. $6x + 5y = -2$, for x

Translate each of the following. Do not solve. [1.7]

73. Five more than twice a number is 49.

74. Three less than half of some number is 57.

75. The sum of two consecutive integers is 145.

76. The difference between a number and its opposite is 20.

Synthesis

TW **77.** If $f(x) = c$, where c is some positive constant, describe how the graphs of $y = g(x)$ and $y = (f + g)(x)$ will differ.

TW **78.** Examine the graphs illustrating number of calories burned and explain how they might be modified to represent the absorption of 200 mg of Advil® taken four times a day.

79. Find the domain of f/g, if

$$f(x) = \frac{3x}{2x + 5} \quad \text{and} \quad g(x) = \frac{x^4 - 1}{3x + 9}.$$

80. Find the domain of F/G, if

$$F(x) = \frac{1}{x - 4} \quad \text{and} \quad G(x) = \frac{x^2 - 4}{x - 3}.$$

81. Sketch the graph of two functions f and g such that the domain of f/g is

$$\{x \mid -2 \le x \le 3 \text{ and } x \ne 1\}.$$

82. Find the domains of $f + g$, $f - g$, $f \cdot g$, and f/g, if

$$f = \{(-2, 1), (-1, 2), (0, 3), (1, 4), (2, 5)\}$$

and

$$g = \{(-4, 4), (-3, 3), (-2, 4), (-1, 0), (0, 5), (1, 6)\}.$$

83. Find the domain of m/n, if

$$m(x) = 3x \text{ for } -1 < x < 5$$

and

$$n(x) = 2x - 3.$$

84. For f and g as defined in Exercise 82, find $(f + g)(-2)$, $(f \cdot g)(0)$, and $(f/g)(1)$.

85. Write equations for two functions f and g such that the domain of $f + g$ is

$$\{x \mid x \text{ is a real number } and\ x \neq -2\ and\ x \neq 5\}.$$

86. Let $y_1 = 2.5x + 1.5$, $y_2 = x - 3$, and $y_3 = y_1/y_2$. For many calculators, depending on whether the CONNECTED or DOT mode is used, the graph of y_3 appears as follows.

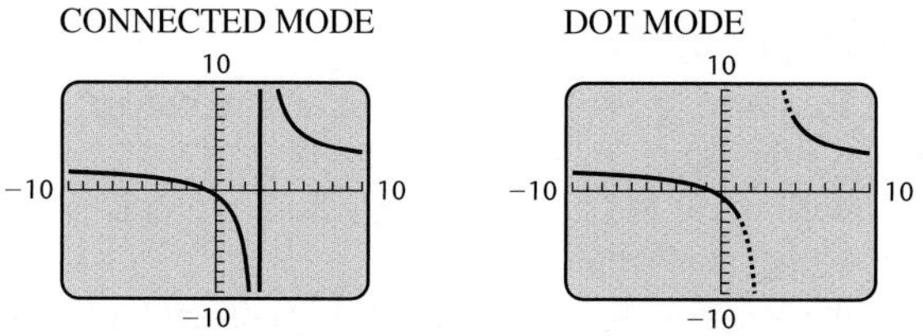

Use algebra to determine which graph more accurately represents y_3.

87. Use the TABLE feature on a graphing calculator to check your answers to Exercises 43, 49, 51, and 57.

88. Use the graphs of f and g, shown below, to match each of $(f + g)(x)$, $(f - g)(x)$, $(f \cdot g)(x)$, and $(f/g)(x)$ with its graph.

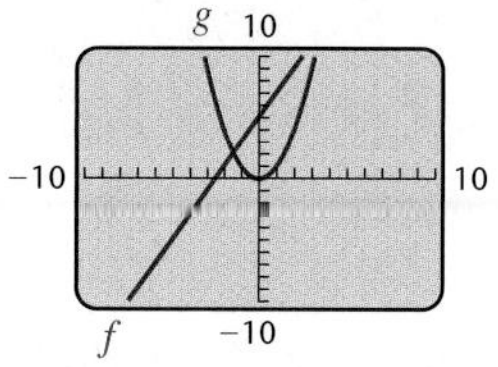

a) $(f + g)(x)$ **b)** $(f - g)(x)$
c) $(f \cdot g)(x)$ **d)** $(f/g)(x)$

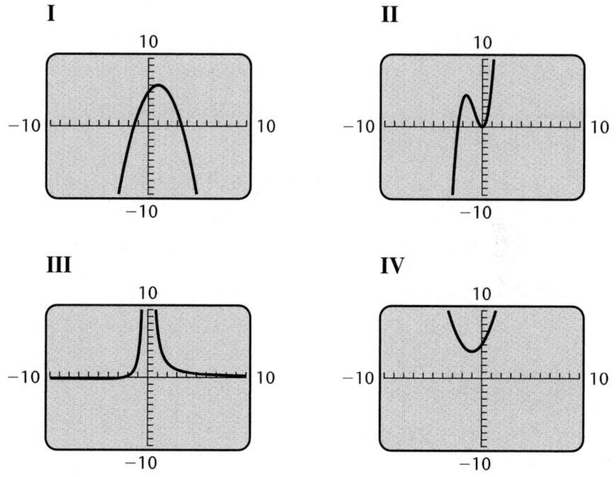

Collaborative Corner

Time On Your Hands

Focus: The algebra of functions
Time: 10–15 minutes
Group Size: 2–3

The graph and data at right chart the average retirement age $R(x)$ and life expectancy $E(x)$ of U.S. citizens in year x (*Source*: Bureau of Labor Statistics, courtesy of the Insurance Advisory Board).

ACTIVITY

1. Working as a team, perform the appropriate calculations and then graph $E - R$.
2. What does $(E - R)(x)$ represent? In what fields of study or business might the function $E - R$ prove useful?
3. Should E and R really be calculated separately for men and women? Why or why not?
4. What advice would you give to someone considering early retirement?

Year:	1955	1965	1975	1985	1995	2005e
Average Retirement Age:	67.3	64.9	63.2	62.8	62.7	61.5
Average Life Expectancy:	73.9	75.5	77.2	78.5	79.1	80.1

e = estimated

If You're Age 50, Consider This...

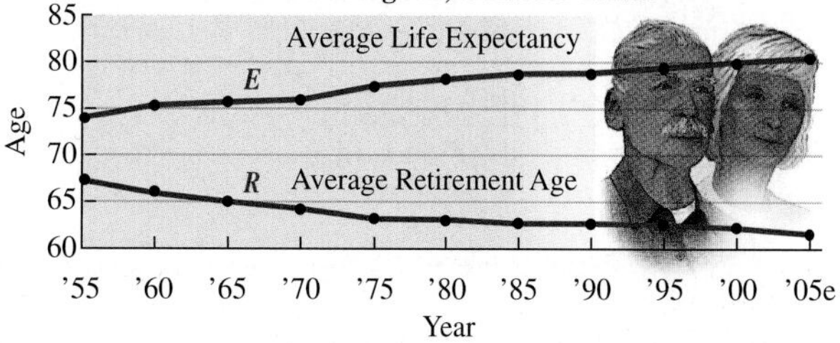

2 Chapter Summary and Review

KEY TERMS AND DEFINITIONS

FUNCTIONS

Function, p. 102 A correspondence between a set called the **domain** and a set called the **range,** such that each member of the domain corresponds to exactly one member of the range. A function is a special kind of **relation.**

Domain, p. 102 The set of all **inputs** of a function.

Range, p. 102 The set of all **outputs** of a function.

Zero, p. 110 An input value for a function that makes the output 0.

For $f(x) = x - 5$,

f is the **function;**

x is the **independent variable;**

5 is the **zero** since $f(5) = 0$;

values of x are the **inputs;**

values of $f(x)$ are the **outputs.**

The Algebra of Functions

1. $(f + g)(x) = f(x) + g(x)$
2. $(f - g)(x) = f(x) - g(x)$
3. $(f \cdot g)(x) = f(x) \cdot g(x)$
4. $(f/g)(x) = f(x)/g(x),\ g(x) \neq 0$

The vertical-line test, p. 106 A graph represents a function if it is not possible to draw a vertical line that intersects the graph more than once.

LINEAR FUNCTIONS

Constant function, p. 108 A function whose output is always the same; $f(x) = b$.

Linear function, p. 118 A function whose graph is a straight line; $f(x) = mx + b$.

***x*-intercept,** p. 135 A point at which a graph crosses the x-axis.

***y*-intercept,** p. 119 A point at which a graph crosses the y-axis.

Linear regression, p. 153 A method of finding a "best" line to fit a set of data.

Interpolation, p. 148 Estimating a value between two given values.

Extrapolation, p. 148 Predicting a future value on the basis of given data.

IMPORTANT CONCEPTS

[Section references appear in brackets.]

Concept	Example
Slope $= m = \dfrac{\text{rise}}{\text{run}} = \dfrac{\text{difference in } y}{\text{difference in } x} = \dfrac{y_2 - y_1}{x_2 - x_1}$	The slope of the line containing the points $(-1, -4)$ and $(2, -6)$ is $\dfrac{-4 - (-6)}{-1 - 2} = \dfrac{2}{-3} = -\dfrac{2}{3}$. [2.2]

(continued)

Horizontal line: Slope 0; equation $y = b$
Vertical line: Slope is undefined;
equation $x = a$

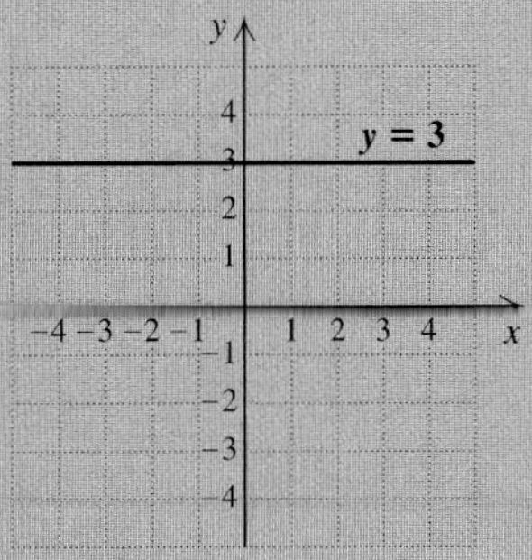

Horizontal line; slope is 0.

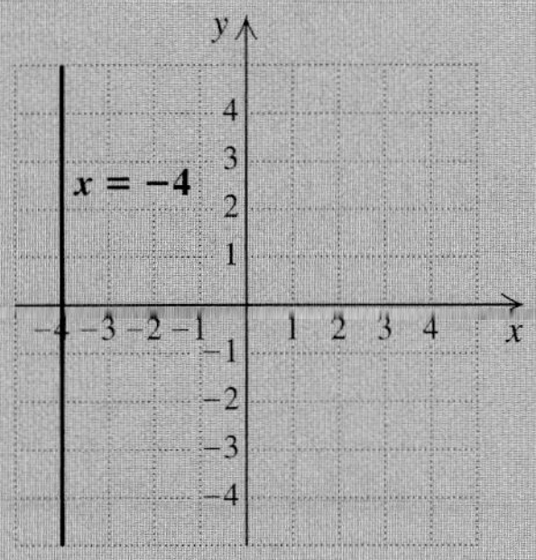

Vertical line; slope is undefined.

[2.3]

Parallel lines: The slopes are equal.
Perpendicular lines: The product of the slopes is -1.

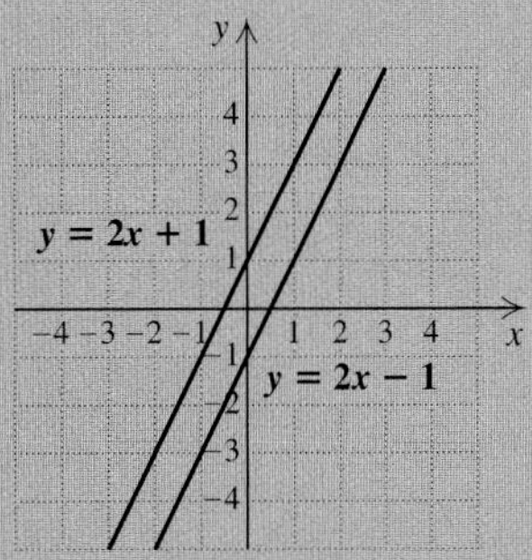

Both lines have slope 2.
The lines are parallel.

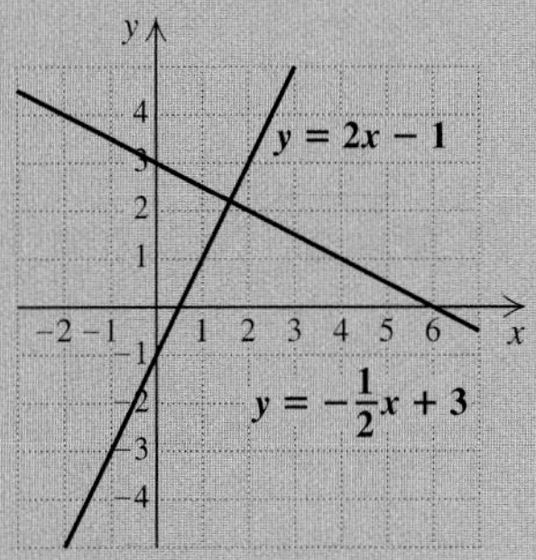

The product of the slopes is $2(-\frac{1}{2}) = -1$.
The lines are perpendicular.

[2.3]

The following are equations for the same line, in different forms:

Slope–intercept form: $y = mx + b$	Slope–intercept form: $y = -\frac{2}{3}x - 5$ [2.2] The slope is $-\frac{2}{3}$ and the y-intercept is $(0, -5)$.
Point–slope form: $y - y_1 = m(x - x_1)$	Point–slope form: $y + 7 = -\frac{2}{3}(x - 3)$ [2.4] The slope is $-\frac{2}{3}$, and $(3, -7)$ is a point on the line.
Standard form: $Ax + By = C$	Standard form: $2x + 3y = -15$ [2.3] The x-intercept is $\left(-\frac{15}{2}, 0\right)$ and the y-intercept is $(0, -5)$.

(continued)

Graphing lines using a point and a slope	Graph: $f(x) = \frac{2}{3}x - 1$. y-intercept is $(0, -1)$. Slope is $\frac{2}{3}$. Plot $(0, -1)$, and count off a slope: up 2 units and to the right 3 units. Draw the line. 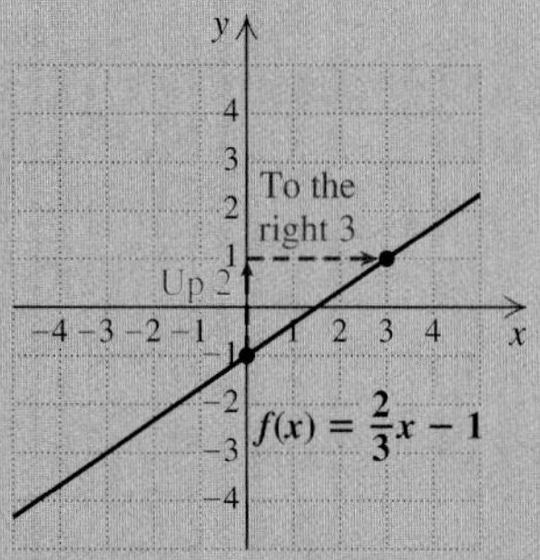 [2.2]
Graphing lines using two points	Graph: $3x + y = 3$. x-intercept is $(1, 0)$. y-intercept is $(0, 3)$. Plot the points and draw the line. 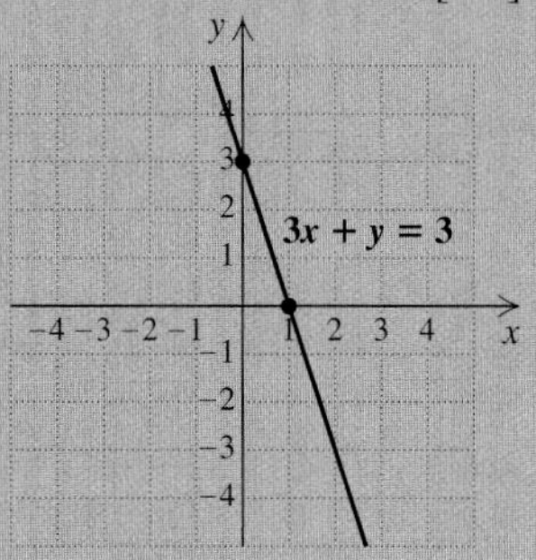 [2.3]

Review Exercises

Concept Reinforcement *Classify each statement as either true or false.*

1. A line's slope is a measure of how the line is slanted or tilted. [2.2]
2. Every line has a y-intercept. [2.3]
3. Every line has an x-intercept. [2.3]
4. No member of a function's range can be used in two different ordered pairs. [2.1]
5. The horizontal-line test is a quick way to determine whether a graph represents a function. [2.1]
6. The slope of a vertical line is undefined. [2.3]
7. The slope of the graph of a constant function is 0. [2.3]
8. Extrapolation is done to predict future values. [2.4]
9. $(f + g)(x) = f(x) + g(x)$ is not an example of the distributive law when f and g are functions. [2.5]
10. In order for $(f/g)(a)$ to exist, we must have $g(a) \neq 0$. [2.5]
11. For the following graph of f, determine **(a)** $f(2)$; **(b)** the domain of f; **(c)** any x-values for which $f(x) = 2$; and **(d)** the range of f. [2.1]

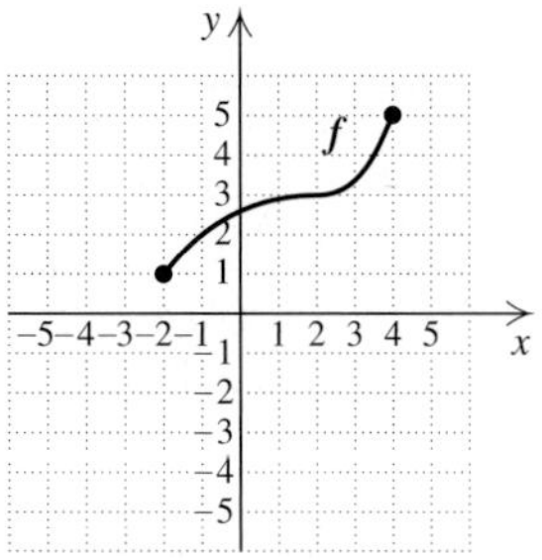

12. The function $A(t) = 0.233t + 5.87$ can be used to estimate the median age of cars in the United States t years after 1990 (*Source*: The Polk Co.). (In this context, a median age of 3 yr means that half the cars are more than 3 yr old and half are less.) Predict the median age of cars in 2010; that is, find $A(20)$. [2.1]

Determine whether each of the following is the graph of a function. [2.1]

13.

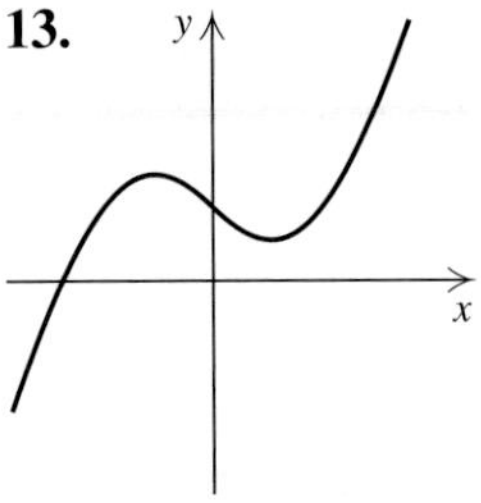

14.

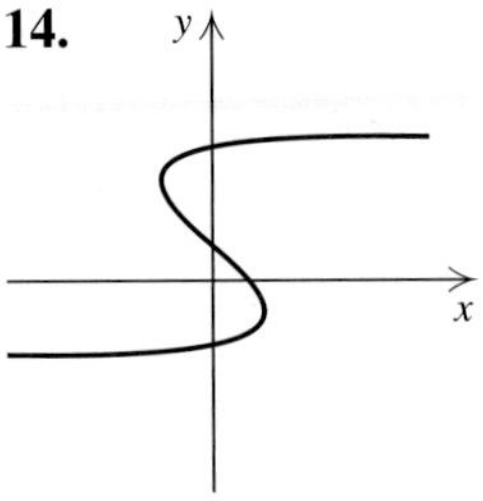

Let $g(x) = 3x - 1$. *Find the following.* [2.1]

15. $g(a + 1)$

16. $g(a) + 1$

17. $g(0)$

18. The input for which the output is 10

19. Any zeros of g

Find the slope and the y-intercept. Then draw the graph. [2.2]

20. $g(x) = -4x - 9$

21. $-6y + 2x = 14$

22. Find the rate of change for the graph shown. Be sure to use appropriate units. [2.2]

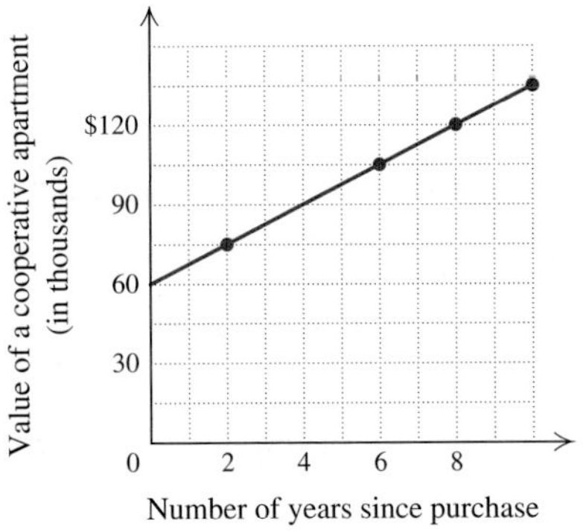

Find the slope of each line. If the slope is undefined, state this.

23. Containing the points $(4, 5)$ and $(-3, 1)$ [2.2]

24. Containing the points $(-16.4, 2.8)$ and $(-16.4, 3.5)$ [2.3]

25. *Recycling.* By August 1, 2005, there were 2938 Freecycle recycling groups. By June 1, 2006, the number had grown to 3579. (*Source*: www.freecycle.org) Calculate the rate at which Freecycle recycling groups were being established. [2.2]

26. The average cost of tuition at a state university t years after 1997 can be estimated by $C(t) = 645t + 9800$. What do the numbers 645 and 9800 signify? [2.2]

27. Find a linear function whose graph has slope $\frac{2}{7}$ and y-intercept $(0, -6)$. [2.2]

28. Graph using intercepts: $-2x + 4y = 8$. [2.3]

Graph.

29. $y = -3x + 2$ [2.2]

30. $y = 6$ [2.3]

31. $y + 1 = \frac{3}{4}(x - 5)$ [2.4]

32. $8x + 32 = 0$ [2.3]

33. $4x - y = 4$ [2.3]

34. Determine an appropriate viewing window for the graph of $f(x) = -\frac{1}{7}x + 14$. Answers may vary. [2.2]

Determine whether each pair of lines is parallel, perpendicular, or neither. [2.3]

35. $y + 5 = -x,$
$x - y = 2$

36. $3x - 5 = 7y,$
$7y - 3x = 7$

37. Write an equation for a linear function perpendicular to the line $y = x + 7$ and with a y-intercept of $(0, -3)$. [2.3]

Determine whether each of these is a linear equation. [2.3]

38. $2x - 7 = 0$

39. $3x - 8f(x) = 7$

40. $2a + 7b^2 = 3$

41. $2p - \frac{7}{q} = 1$

42. Find an equation in point–slope form of the line with slope -2 and containing $(-3, 4)$. [2.4]

43. Using function notation, write a slope–intercept equation for the line containing $(2, 5)$ and $(-4, -3)$. [2.4]

The following table lists the U.S. minimum hourly wage for various years.

Input, Year	Output, U.S. Minimum Hourly Wage
1955	$0.75
1980	3.10
2005	5.15

44. Use the data in the table to draw a graph and to estimate the U.S. minimum hourly wage in 1985. [2.4]

45. Use the graph from Exercise 44 to estimate the U.S. hourly minimum wage in 2009. [2.4]

46. *Records in the 200-Meter Run.* In 1983, the record for the 200-m run was 19.75 sec.* In 2003, it was 19.32 sec. (*Source*: runnersworld.com) Let $R(t)$ represent the record in the 200-m run and t the number of years since 1983. [2.4]

a) Find a linear function that fits the data.

b) Use the function of part (a) to predict the record in 2008 and in 2013.

Multiple Births. The following table lists the number of sets of twins born in the United States for various years.

Year	Number of Twin Births
1994	97,064
1996	100,750
1998	110,670
2000	118,916
2003	128,665

Source: Centers for Disease Control and Prevention

*Records are for elevations less than 1000 m.

47. Graph the data and determine whether the relationship appears to be linear. [2.4]

48. Use linear regression to find a linear function B that can be used to estimate the number of twin births as a function of t years after 1990. [2.4]

49. Use the function from Exercise 48 to estimate the number of twin births in 2009. [2.4]

Find the domain of f. [2.5]

50. $f(x) = \frac{5}{x + 3}$

51. $f(x) = \frac{x + 3}{5}$

Let $g(x) = 3x - 6$ and $h(x) = x^2 + 1$. Find the following. [2.5]

52. $(g \cdot h)(4)$

53. $(g - h)(-2)$

54. $(g/h)(-1)$

55. $(g + h)(x)$

56. The domains of $g + h$ and $g \cdot h$

57. The domain of h/g

Synthesis

TW **58.** Explain why every function is a relation, but not every relation is a function. [2.1]

TW **59.** Explain why the slope of a vertical line is undefined whereas the slope of a horizontal line is 0. [2.3]

60. Find the y-intercept of the function given by

$$f(x) + 3 = 0.17x^2 + (5 - 2x)^x - 7.$$

[1.4], [2.3]

61. Determine the value of a such that the lines

$$3x - 4y = 12 \quad \text{and} \quad ax + 6y = -9$$

are parallel. [2.3]

62. Homespun Jellies charges \$2.49 for each jar of preserves. Shipping charges are \$3.75 for handling, plus \$0.60 per jar. Find a linear function for determining the cost of x jars of preserves. [2.4]

63. Match each sentence with the most appropriate of the four graphs at right. [2.2]

a) Joni walks for 10 min to the train station, rides the train for 15 min, and then walks 5 min to the office.

b) During a workout, Phil bikes for 10 min, runs for 15 min, and then walks for 5 min.

c) Sam pilots his motorboat for 10 min to the middle of the lake, fishes for 15 min, and then motors for another 5 min to another spot.

d) Patti waits 10 min for her train, rides the train for 15 min, and then runs for 5 min to her job.

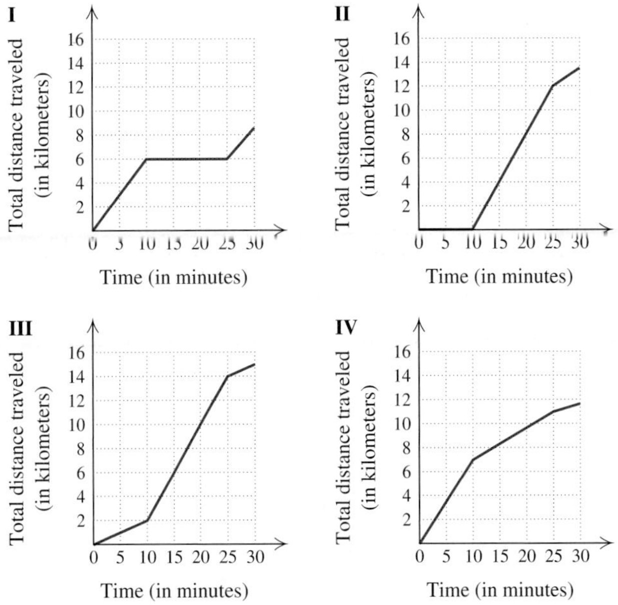

Chapter Test 2

1. For the following graph of f, determine **(a)** $f(-2)$; **(b)** the domain of f; **(c)** any x-value for which $f(x) = \frac{1}{2}$; and **(d)** the range of f.

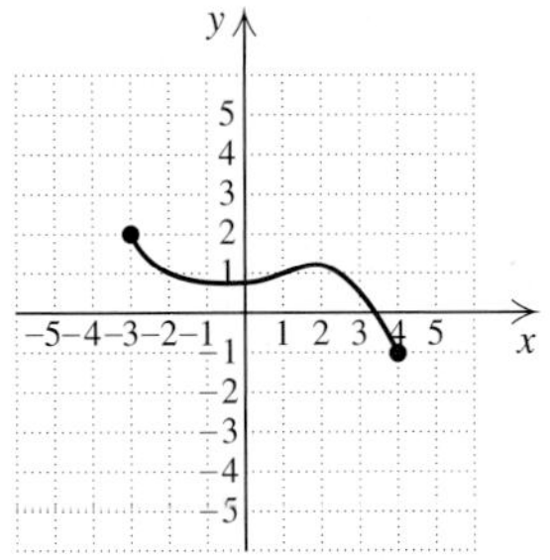

2. The function $S(t) = 1.24t + 34.5$ can be used to estimate the total U.S. sales of books, in billions of dollars, t years after 2000 (*Source*: Based on data from *Statistical Abstract of the United States,* 2006).

a) Predict the total U.S. sales of books in 2009.

b) What do the numbers 1.24 and 34.5 signify?

3. There were 48.5 million international visitors to the United States in 1999 and 46.1 million in 2004. Draw a graph and estimate the number of international visitors in 2002. (*Source*: Tourism Industries, International Trade Administration, U.S. Department of Commerce)

Find the slope and the y-intercept.

4. $f(x) = -\frac{3}{5}x + 12$

5. $-5y - 2x = 7$

Find the slope of the line containing the following points. If the slope is undefined, state this.

6. $(-2, -2)$ and $(6, 3)$

7. $(-3.1, 5.2)$ and $(-4.4, 5.2)$

8. Find the rate of change for the graph below. Use appropriate units.

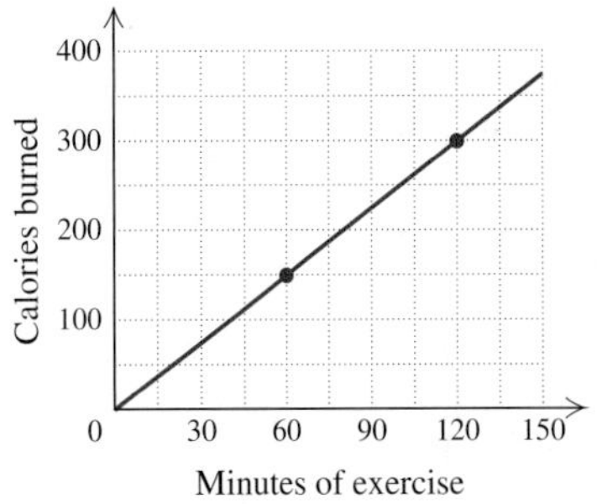

9. Find a linear function whose graph has slope -5 and y-intercept $(0, -1)$.

Graph by hand.

10. $y = -3x + 4$

11. $-2x + 5y = 12$

12. $y - 1 = -\frac{1}{2}(x + 4)$

13. $3 - x = 9$

14. Determine whether the standard viewing window shows the x- and y-intercepts of the graph of $f(x) = 2x + 9$.

Determine without graphing whether each pair of lines is parallel, perpendicular, or neither.

15. $4y + 2 = 3x$,
$-3x + 4y = -12$

16. $y = -2x + 5$,
$2y - x = 6$

17. Which of the following are linear equations?

a) $8x - 7 = 0$
b) $4y - 9x^2 = 2$
c) $2x - 5y = 3$

18. Find an equation in point–slope form of the line with slope 4 and containing $(-2, -4)$.

19. Use function notation to write an equation for the line containing $(3, -1)$ and $(4, -2)$.

20. If you rent a van for one day and drive it 250 mi, the cost is \$100. If you drive it 300 mi, the cost is \$115. Let $C(m) =$ the cost, in dollars, of driving m miles.

a) Find a linear function that fits the data.
b) Use the function to determine how much it will cost to rent the van for one day and drive it 500 mi.

21. *Accidental Deaths.* The number of accidental deaths in the United States per 100,000 population is dropping, as shown in the following table.

Year	Number of Accidental Deaths per 100,000
1910	84.4
1920	71.2
1930	80.5
1940	73.4
1950	60.3
1960	52.1
1970	56.2
1980	46.5
1990	36.9
2003	36.3

Source: U.S. Department of Health and Human Services, National Center for Health Statistics

a) Use linear regression to find a linear function A that can be used to predict the number of accidental deaths per 100,000 population as a function of the number of years x since 1900.
b) Predict the number of accidental deaths per 100,000 population in 2010.

22. Find the domain of $g(x) = \dfrac{x - 6}{2x + 1}$.

23. Find the following, given that $g(x) = -3x - 4$ and $h(x) = x^2 + 1$.

a) $h(-2)$
b) $g(0)$
c) $g(t) + 5$
d) $(g \cdot h)(3)$
e) Any zeros of $g(x)$
f) The domain of h/g

Synthesis

24. The function $f(t) = 5 + 15t$ can be used to determine a bicycle racer's location, in miles from the starting line, measured t hours after passing the 5-mi mark.

a) How far from the start will the racer be 1 hr and 40 min after passing the 5-mi mark?
b) Assuming a constant rate, how fast is the racer traveling?

Find an equation of the line.

25. Containing $(-3, 2)$ and parallel to the line $2x - 5y = 8$

26. Containing $(-3, 2)$ and perpendicular to the line $2x - 5y = 8$

27. The graph of the function $f(x) = mx + b$ contains the points $(r, 3)$ and $(7, s)$. Express s in terms of r if the graph is parallel to the line $3x - 2y = 7$.

28. Given that $f(x) = 5x^2 + 1$ and $g(x) = 4x - 3$, find an expression for $h(x)$ so that the domain of $f/g/h$ is

$$\left\{x \middle| x \text{ is a real number } and\ x \neq \tfrac{3}{4}\ and\ x \neq \tfrac{2}{7}\right\}.$$

Answers may vary.

3 Systems of Linear Equations and Problem Solving

The most difficult part of problem solving is nearly always translating the problem situation to mathematical language. Once a problem has been translated, the solution is generally straightforward. In this chapter, we study *systems of equations* and how to solve them using graphing, substitution, elimination, and matrices. Systems of equations often provide the easiest way to translate into mathematics real-world situations from fields such as psychology, sociology, business, education, engineering, and science.

APPLICATION *Bottled-Water Consumption.*

Americans are buying more bottled water and less regular (not diet) soft drink. The following table and graph show per-capita consumption of both beverages for several years. When will Americans consume as much bottled water per year as regular soft drink?

Year	Bottled Water (in gallons)	Soft Drink (in gallons)
2000	16.7	39.4
2001	18.2	39.0
2002	20.1	38.4
2003	21.6	37.5
2004	23.2	36.5

Source: USDA/Economic Research Service

$y_1 = 1.64x + 16.68,$
$y_2 = -0.73x + 39.62$

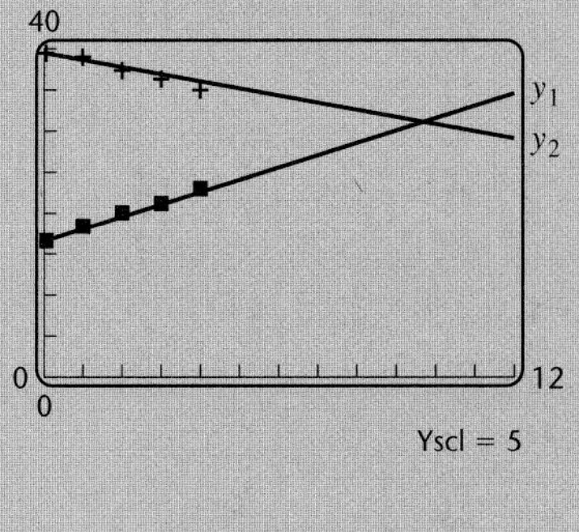

This problem appears as Example 7 in Section 3.1.

3.1 Systems of Equations in Two Variables

Translating ■ Identifying Solutions ■ Solving Systems Graphically ■ Models

Translating

Problems involving two unknown quantities are often solved most easily if we can first translate the situation to two equations in two unknowns.

EXAMPLE 1 Endangered Species. The number of species listed as threatened or endangered has more than tripled in the past 20 yr. In 2005, there were 749 species of plants considered threatened or endangered. The number considered threatened, or likely to become endangered in the foreseeable future, was 1 less than one-fourth of the number considered endangered, or in danger of becoming extinct. (*Source*: U.S. Fish and Wildlife Service) How many plant species were considered endangered and how many were considered threatened in 2005?

SOLUTION

1. **Familiarize.** Often statements of problems contain information that has no bearing on the question asked. In this case, the fact that the number of threatened or endangered species has tripled in the past 20 yr does not help us solve the problem. Instead, we need to focus on the number of endangered species and the number of threatened species in 2005. We let d represent the number of endangered plant species and t represent the number of threatened plant species in 2005.

Endangered Piper's Bellflowers. The endangered species Piper's bellflowers growing against a rock wall. Olympic National Park, Washington USA. © Darrell Gulin/CORBIS

Study Tip

Speak Up

Don't hesitate to ask questions in class at appropriate times. Most instructors welcome questions and encourage students to ask them. Other students in your class probably have the same questions you do.

2. **Translate.** There are two statements to translate. First, we look at the total number of endangered or threatened species of plants:

Rewording:	The number of endangered species	plus	the number of threatened species	was	749.
	↓	↓	↓	↓	↓
Translating:	d	$+$	t	$=$	749

The second statement compares the two amounts, d and t:

Rewording:	The number of threatened species	was	one less than one-fourth of the number of endangered species.
	↓	↓	↓
Translating:	t	$=$	$\frac{1}{4}d - 1$

We have now translated the problem to a pair, or **system, of equations:**

$$d + t = 749,$$

$$t = \frac{1}{4}d - 1.$$

We complete the solution of this problem in Section 3.3.

System of Equations A *system of equations* is a set of two or more equations, in two or more variables, for which a common solution is sought.

Problems like Example 1 *can* be solved using one variable; however, as problems become complicated, you will find that using more than one variable (and more than one equation) is often the preferable approach.

EXAMPLE 2 Purchasing. Recently the Woods County Art Center purchased 120 stamps for \$35.55. If the stamps were a combination of 24¢ postcard stamps and 39¢ first-class stamps, how many of each type were bought?

SOLUTION

1. **Familiarize.** To familiarize ourselves with this problem, let's guess that the art center bought 60 stamps at 24¢ each and 60 stamps at 39¢ each. The total cost would then be

$$60 \cdot \$0.24 + 60 \cdot \$0.39 = \$14.40 + \$23.40, \text{ or } \$37.80.$$

Since $\$37.80 \neq \$35.55$, our guess is incorrect. Rather than guess again, let's see how algebra can be used to translate the problem.

2. **Translate.** We let $p =$ the number of postcard stamps and $f =$ the number of first-class stamps. The information can be organized in a table, which will help with the translating.

Type of Stamp	Postcard	First-class	Total	
Number Sold	p	f	120	→ $p + f = 120$
Price	\$0.24	\$0.39		
Amount	\$0.24p	\$0.39f	\$35.55	→ $0.24p + 0.39f = 35.55$

The first row of the table and the first sentence of the problem indicate that a total of 120 stamps were bought:

$$p + f = 120.$$

Since each postcard stamp cost \$0.24 and p stamps were bought, $0.24p$ represents the amount paid, in dollars, for the postcard stamps. Similarly, $0.39f$ represents the amount paid, in dollars, for the first-class stamps. This leads to a second equation:

$$0.24p + 0.39f = 35.55.$$

Multiplying both sides by 100, we can clear the decimals. This gives the following system of equations as the translation:

$$p + f = 120,$$
$$24p + 39f = 3555.$$

We complete the solution of this problem in Section 3.3.

Identifying Solutions

A *solution* of a system of two equations in two variables is an ordered pair of numbers that makes *both* equations true.

EXAMPLE 3 Determine whether $(-4, 7)$ is a solution of the system

$$x + y = 3,$$
$$5x - y = -27.$$

SOLUTION Unless stated otherwise, we use alphabetical order of the variables. Thus we replace x with -4 and y with 7:

$$\begin{array}{c|c} x + y & 3 \\ \hline -4 + 7 & 3 \\ 3 & \stackrel{?}{=} 3 \quad \text{TRUE} \end{array} \qquad \begin{array}{c|c} 5x - y & -27 \\ \hline 5(-4) - 7 & -27 \\ -20 - 7 & \\ -27 & \stackrel{?}{=} -27 \quad \text{TRUE} \end{array}$$

The pair $(-4, 7)$ makes both equations true, so it is a solution of the system. We can also describe the solution by writing $x = -4$ and $y = 7$. Set notation can also be used to list the solution set $\{(-4, 7)\}$.

Solving Systems Graphically

Recall that the graph of an equation is a drawing that represents its solution set. If we graph the equations in Example 3, we find that $(-4, 7)$ is the only point common to both lines. Thus one way to solve a system of two equations is to graph both equations and identify any points of intersection. The coordinates of each point of intersection represent a solution of that system.

$$x + y = 3,$$
$$5x - y = -27$$

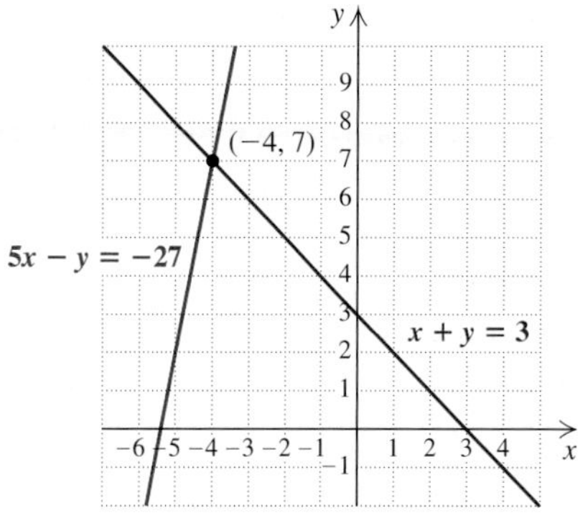

Point of Intersection

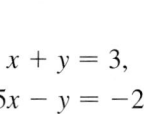

A graphing calculator can be used to determine the point of intersection of graphs. Often it provides a more accurate solution than graphs drawn by hand. We can trace along the graph of one of the functions to find the coordinates of the point of intersection, enlarging the graph using a ZOOM feature if necessary. Most graphing calculators can find the point of intersection directly using an INTERSECT feature, found in the CALC menu.

To find the point of intersection of two graphs, press **2ND** CALC and choose the INTERSECT option. Because more than two equations may be graphed, the questions FIRST CURVE? and SECOND CURVE? are used to identify the graphs with which we are concerned. Position the cursor on each graph, in turn, and press **ENTER**, using the up and down arrow keys if necessary.

The calculator then asks a third question, GUESS?. Since graphs may intersect at more than one point, we must visually identify the point of intersection in which we are interested and indicate the general location of that point. Enter a guess either by moving the cursor near the point of intersection and pressing **ENTER** or by typing a guess and pressing **ENTER**. The calculator then returns the coordinates of the point of intersection. At this point, the calculator variables X and Y contain the coordinates of the point of intersection. These values can be used to check an answer, and often they can be converted to fraction notation from the home screen using the Frac option of the MATH menu.

Most pairs of lines have exactly one point in common. We will soon see, however, that this is not always the case.

EXAMPLE 4 Solve graphically:

$$y - x = 1,$$
$$y + x = 3.$$

SOLUTION

BY HAND

We graph each equation using any method studied in Chapter 2. All ordered pairs from line L_1 in the graph below are solutions of the first equation. All ordered pairs from line L_2 are solutions of the second equation. The point of intersection has coordinates that make *both* equations true. Apparently, $(1, 2)$ is the solution. Graphing is not always accurate, so solving by graphing may yield approximate answers. Our check below shows that $(1, 2)$ is indeed the solution.

$$y - x = 1,$$
$$y + x = 3$$

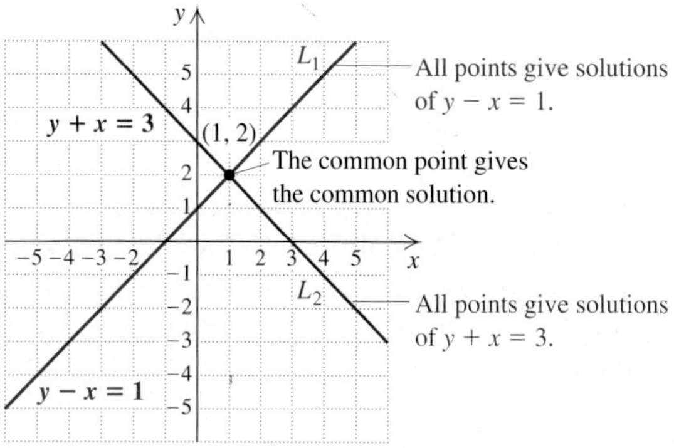

Check:

$y - x = 1$	
$2 - 1$	1
$1 \stackrel{?}{=} 1$	TRUE

$y + x = 3$	
$2 + 1$	3
$3 \stackrel{?}{=} 3$	TRUE

WITH A GRAPHING CALCULATOR

We first solve each equation for y:

$$y - x = 1 \qquad\qquad y + x = 3$$
$$y = x + 1; \qquad\qquad y = 3 - x.$$

Then we enter and graph $y_1 = x + 1$ and $y_2 = 3 - x$ using the same viewing window. After choosing the INTERSECT option, we indicate which two graphs, or *curves*, we are considering. Then we enter a guess. The coordinates of the point of intersection then appear at the bottom of the screen.

$y_1 = x + 1,\ y_2 = 3 - x$

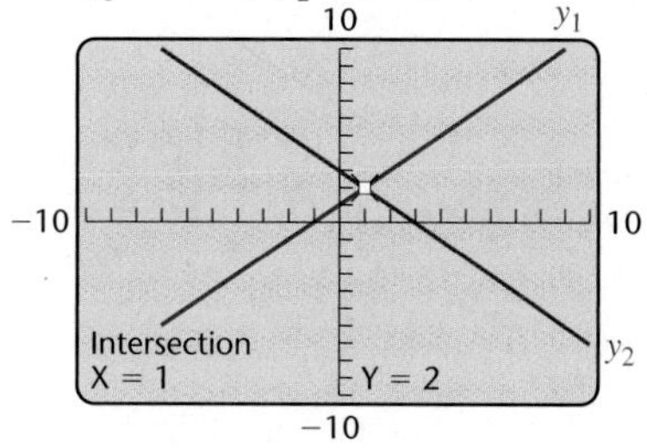

The point $(1, 2)$ must make both equations true in order to be a solution of the system. Since, in the calculator, X now has the value 1 and Y has the value 2, we can check from the home screen. The solution is $(1, 2)$.

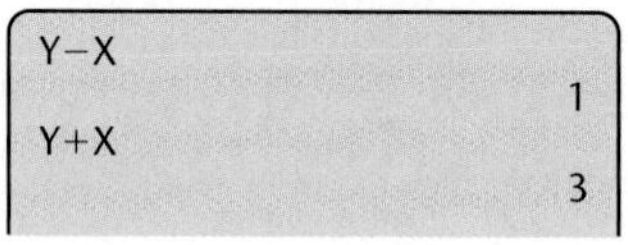

EXAMPLE 5 Solve each system graphically.

a) $y = -3x + 5,$
$y = -3x - 2$

b) $3y - 2x = 6,$
$-12y + 8x = -24$

SOLUTION

a) We graph the equations. The lines have the same slope, -3, and different y-intercepts, so they are parallel. There is no point at which they cross, so the system has no solution.

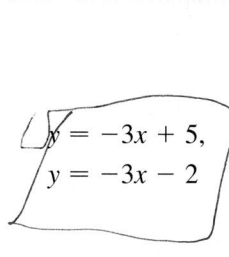

$y = -3x + 5,$
$y = -3x - 2$

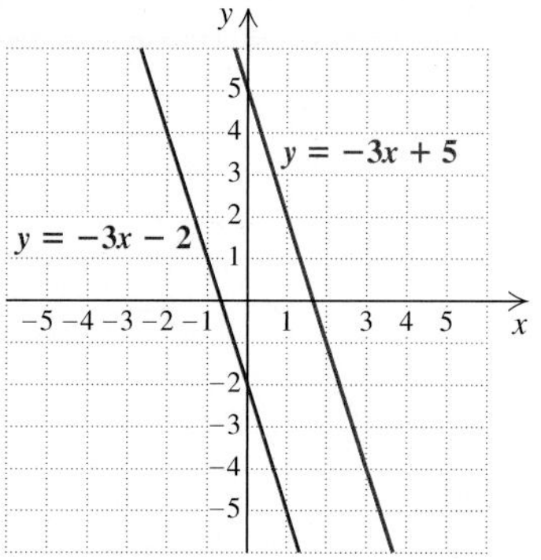

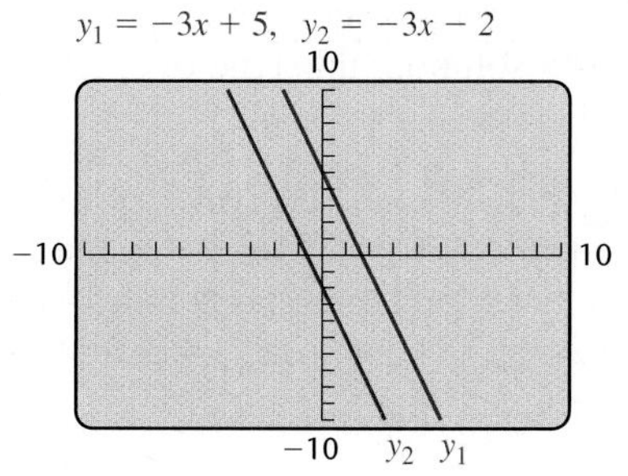

What happens when we try to solve this system using a graphing calculator? We enter and graph both equations as shown at left, noting that the graphs appear to be parallel. (This must be verified algebraically.) When we attempt to find the intersection using the INTERSECT feature, we get an error message.

b) We graph the equations and find that the same line is drawn twice. Thus any solution of one equation is a solution of the other. Each equation has an infinite number of solutions, so the system itself has an infinite number of solutions. We check one solution, (0, 2), which is the y-intercept of each equation.

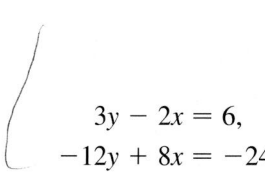

$3y - 2x = 6,$
$-12y + 8x = -24$

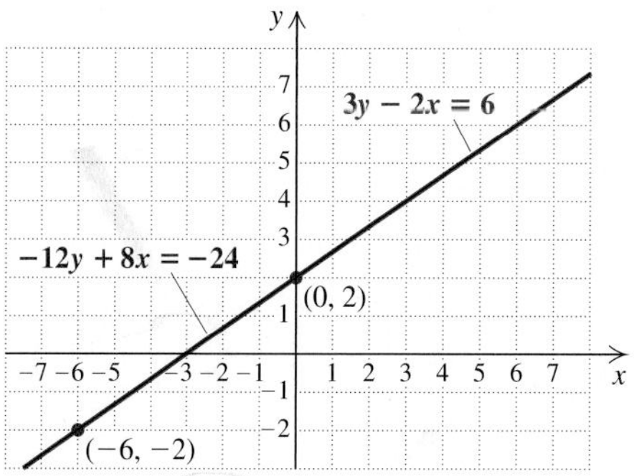

Check:

$3y - 2x = 6$	
$3(2) - 2(0)$	6
$6 - 0$	
$6 \stackrel{?}{=} 6$	TRUE

$-12y + 8x = -24$	
$-12(2) + 8(0)$	-24
$-24 + 0$	
$-24 \stackrel{?}{=} -24$	TRUE

Student Notes

Although the system in Example 5(b) is true for an infinite number of ordered pairs, those pairs must be of a certain form. Only pairs that are solutions of $3y - 2x = 6$ or $-12y + 8x = -24$ are solutions of the system. It is incorrect to think that *all* ordered pairs are solutions.

We can check that $(-6, -2)$ is another solution of both equations. In fact, any pair that is a solution of one equation is a solution of the other equation as well. Thus the solution set is

$$\{(x, y) \mid 3y - 2x = 6\}$$

or, in words, "the set of all pairs (x, y) for which $3y - 2x = 6$." Since the two equations are equivalent, we could have written instead $\{(x, y) \mid -12y + 8x = -24\}$.

If we attempt to find the intersection of the graphs of the equations in this system using INTERSECT, the graphing calculator will return as the intersection whatever point we choose as the guess. This is a point of intersection, as is any other point on the graph of the lines.

When we graph a system of two linear equations in two variables, one of the following three outcomes will occur.

1. The lines have one point in common, and that point is the only solution of the system (see Example 4). Any system that has at least one solution is said to be **consistent.**
2. The lines are parallel, with no point in common, and the system has no solution (see Example 5a). This type of system is called **inconsistent.**
3. The lines coincide, sharing the same graph. Because every solution of one equation is a solution of the other, the system has an infinite number of solutions (see Example 5b). Since it has a solution, this type of system is also consistent.

When one equation in a system can be obtained by multiplying both sides of another equation by a constant, the two equations are said to be **dependent.** Thus the equations in Example 5(b) are dependent, but those in Examples 4 and 5(a) are **independent.** For systems of three or more equations, the definitions of dependent and independent must be slightly modified.

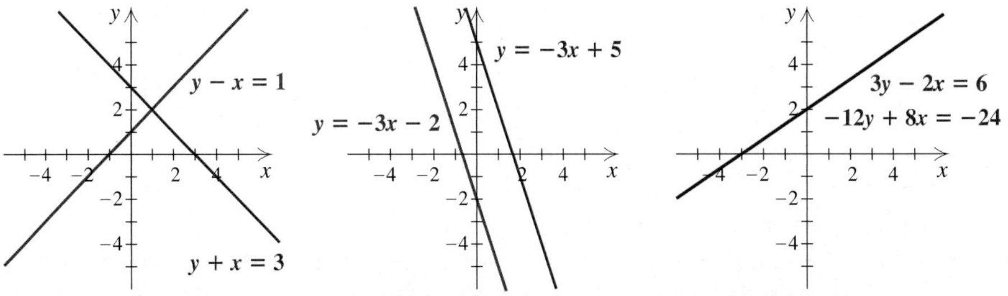

Graphs intersect at one point. The system is *consistent* and has one solution. Since neither equation is a multiple of the other, they are *independent.*

Graphs are parallel. The system is *inconsistent* because there is no solution. Since the equations are not equivalent, they are *independent.*

Equations have the same graph. The system is *consistent* and has an infinite number of solutions. The equations are *dependent* since they are equivalent.

Graphing calculators are especially useful when equations contain fractions or decimals or when the coordinates of the intersection are not integers.

EXAMPLE 6 Solve graphically:

$$3.45x + 4.21y = 8.39,$$
$$7.12x - 5.43y = 6.18.$$

SOLUTION First, we solve for y in each equation:

$$3.45x + 4.21y = 8.39$$
$$4.21y = 8.39 - 3.45x \quad \text{Subtracting } 3.45x \text{ from both sides}$$
$$y = (8.39 - 3.45x)/4.21; \quad \text{Dividing both sides by } 4.21$$

$$7.12x - 5.43y = 6.18$$
$$-5.43y = 6.18 - 7.12x \quad \text{Subtracting } 7.12x \text{ from both sides}$$
$$y = (6.18 - 7.12x)/(-5.43). \quad \text{Dividing both sides by } -5.43$$

It is not necessary to simplify further. We have the system

$$y = (8.39 - 3.45x)/4.21,$$
$$y = (6.18 - 7.12x)/(-5.43).$$

Next, we enter both equations and graph using the same viewing window. By using the INTERSECT feature in the CALC menu, we see that, to the nearest hundredth, the solution is (1.47, 0.79). Note that the coordinates found by the calculator are approximations.

$y_1 = (8.39 - 3.45x)/4.21,$
$y_2 = (6.18 - 7.12x)/(-5.43)$

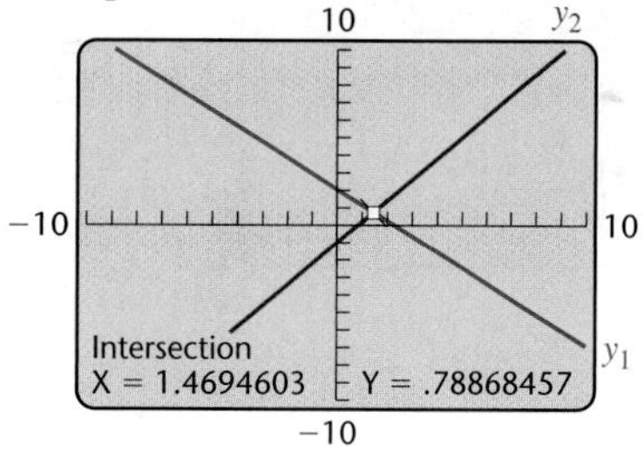

Models

Sometimes two or more sets of data can be modeled by a system of linear equations.

EXAMPLE 7 Bottled-Water Consumption. Americans are buying more bottled water and less regular (not diet) soft drink. The following table lists per-capita consumption of both beverages for several years. When will Americans consume as much bottled water per year as regular soft drink?

Year	Bottled Water (in gallons)	Soft Drink (in gallons)
2000	16.7	39.4
2001	18.2	39.0
2002	20.1	38.4
2003	21.6	37.5
2004	23.2	36.5

Source: USDA/Economic Research Service

SOLUTION We first enter and graph the data using a graphing calculator. To make the numbers smaller, we will let x represent the number of years after 2000. Thus we enter the number of years after 2000 as L_1, the number of gallons of bottled water as L_2, and the number of gallons of soft drink as L_3. A good viewing window, based on the data, is $[0, 8, 0, 40]$, with Yscl = 5. The left and middle screens below show the setup of the plots, with Plot1 graphing the bottled water data with boxes and Plot2 graphing the soft drink data with plus signs.

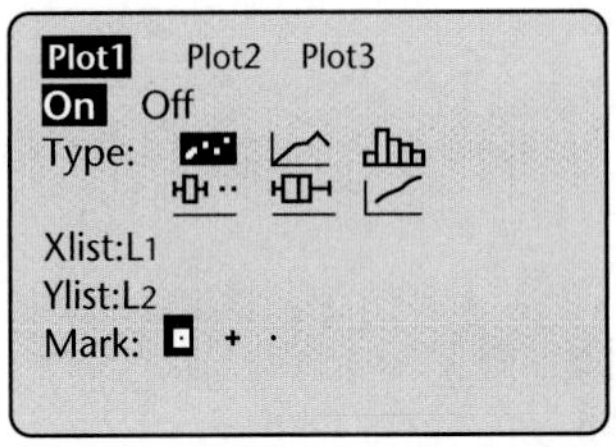

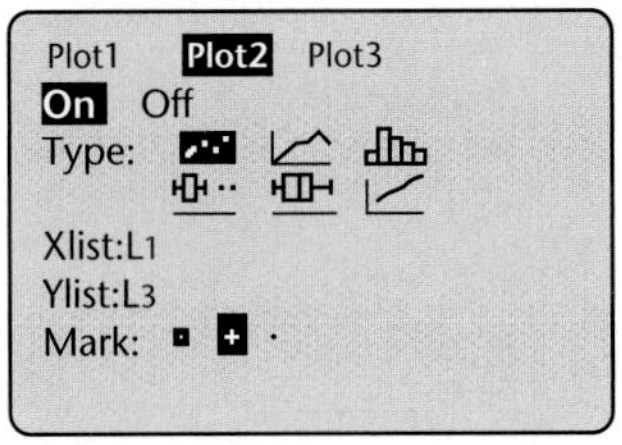

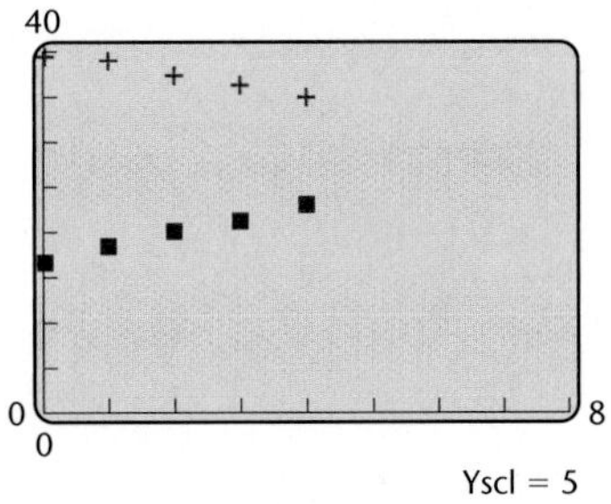

Both sets of data are approximately linear, and we can see that bottled water consumption is increasing and soft drink consumption is decreasing. It appears likely that the numbers will be equal in the future, if the trend continues. We use linear regression to fit a line to each set of data.

Since bottled water is entered as L_2, we first find the linear regression line when L_1 is the independent variable and L_2 is the dependent variable and store this equation as Y_1. Since L_1 and L_2 are the default lists, we did not need to specify them, although they are shown in the figure on the left below.

To find the second regression line, we must specify that L_1 is the independent variable and L_3 is the dependent variable. We store this equation to Y_2, as shown in the figure on the right below.

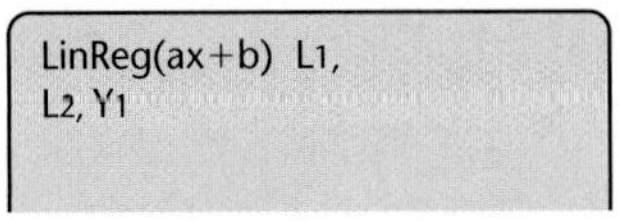

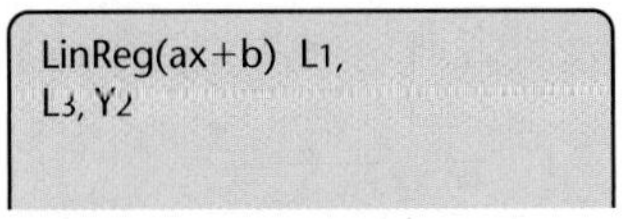

The screen on the left below gives us the two equations

$$y_1 = 1.64x + 16.68 \quad \text{and} \quad y_2 = -0.73x + 39.62.$$

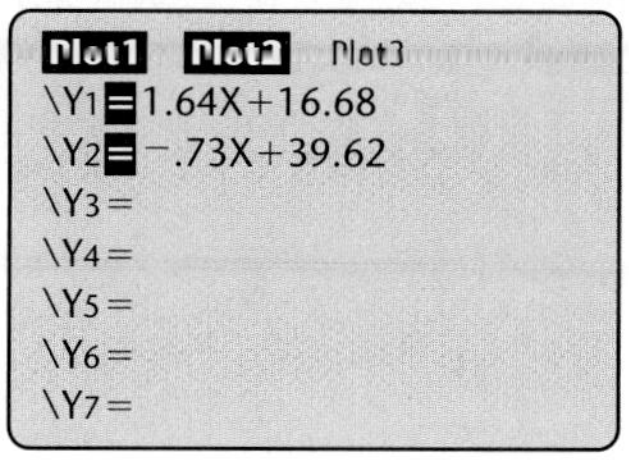

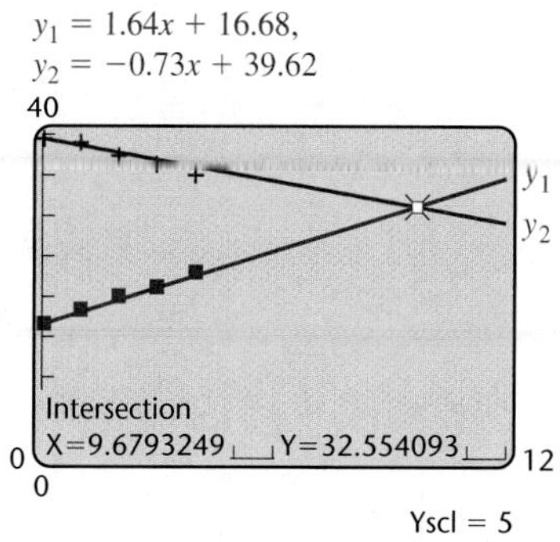

We graph the equations along with the data and determine the coordinates of the point of intersection. The x-coordinate of the point of intersection must be within the window dimensions, so we extend the window, as shown on the right above. The point of intersection is approximately (9.68, 32.55). Since x represents the number of years after 2000, this tells us that Americans will drink about 32.55 gal of bottled water and the same amount of soft drink about 9.7 yr after 2000. Rounding, we state that Americans will consume as much bottled water as regular soft drink in about 2010.

Graphing is helpful when solving systems because it allows us to "see" the solution. It can also be used on systems of nonlinear equations, and in many applications, it provides a satisfactory answer. However, graphing often lacks precision, especially when fraction or decimal solutions are involved. In Section 3.2, we will develop two algebraic methods of solving systems. Both methods produce exact answers.

Visualizing the Graph

Match each equation or system of equations with its graph.

1. $x + y = 2,$
   $x - y = 2$

2. $y = \frac{1}{3}x - 5$

3. $4x - 2y = -8$

4. $2x + y = 1,$
   $x + 2y = 1$

5. $8y + 32 = 0$

6. $f(x) = -x + 4$

7. $\frac{2}{3}x + y = 4$

8. $x = 4,$
   $y = 3$

9. $y = \frac{1}{2}x + 3,$
   $2y - x = 6$

10. $y = -x + 5,$
    $y = 3 - x$

Answers on page A-13

A
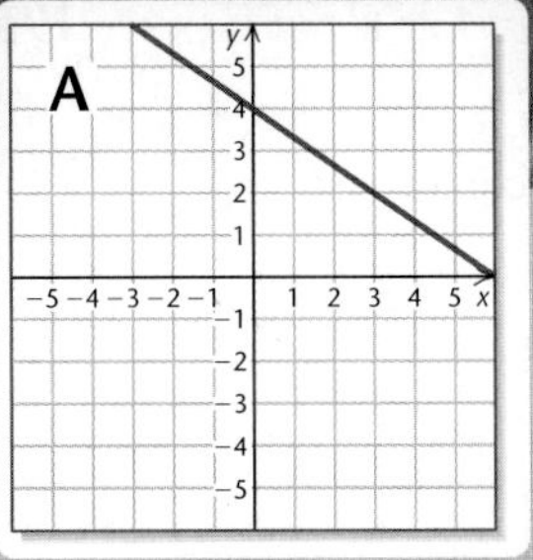

B
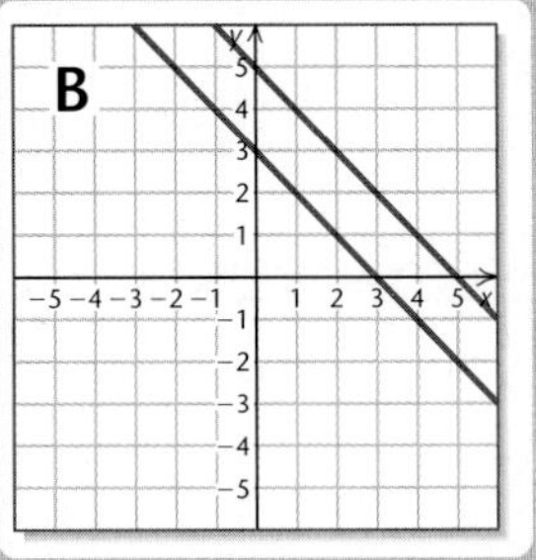

C
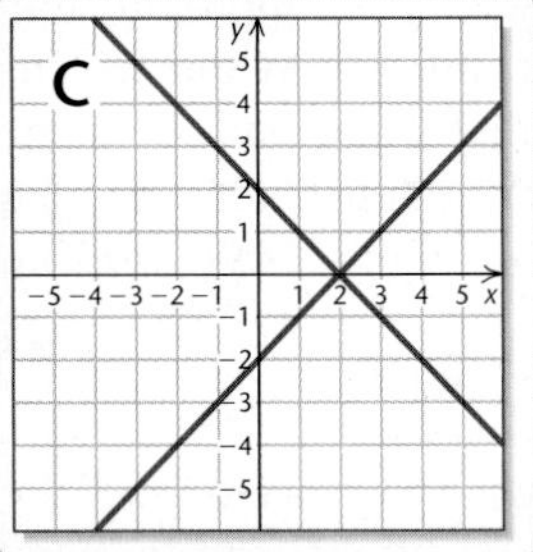

D
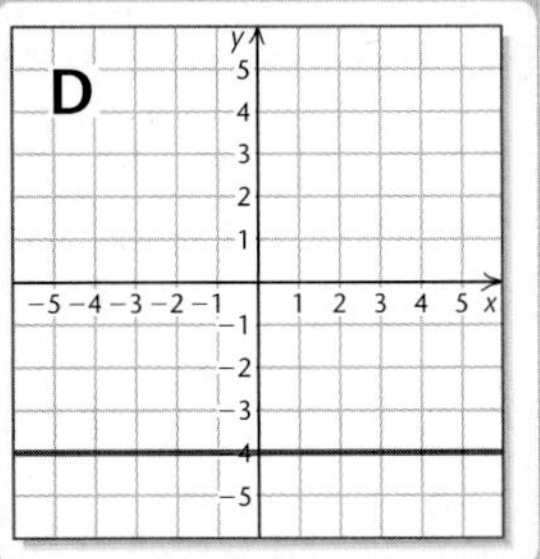

E
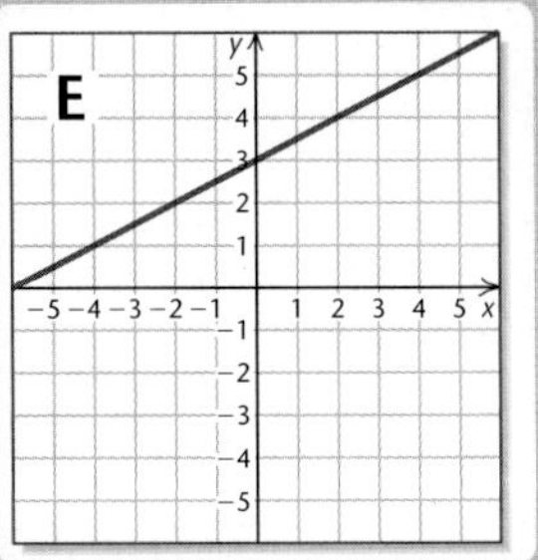

F
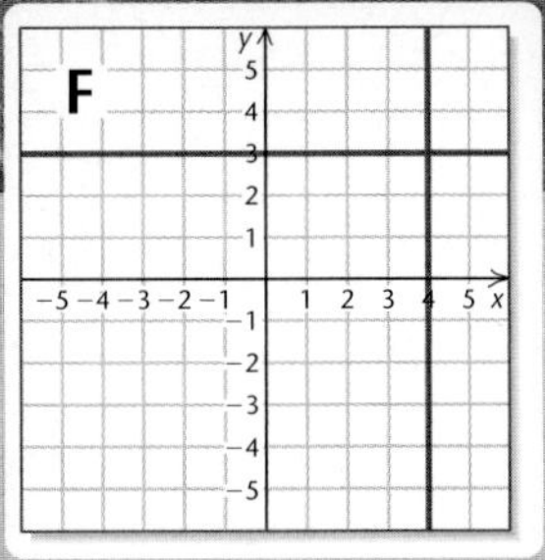

G
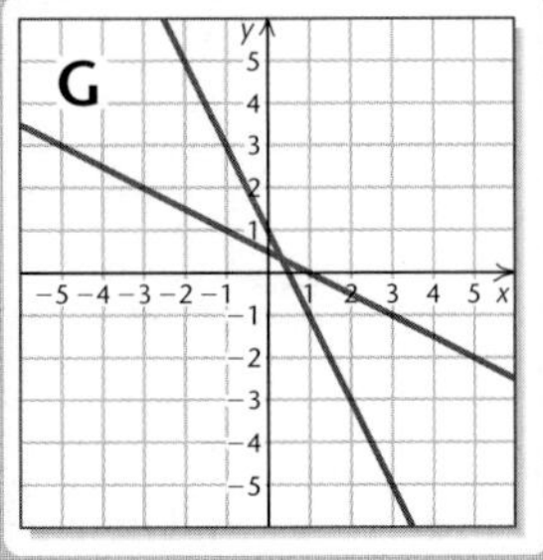

H
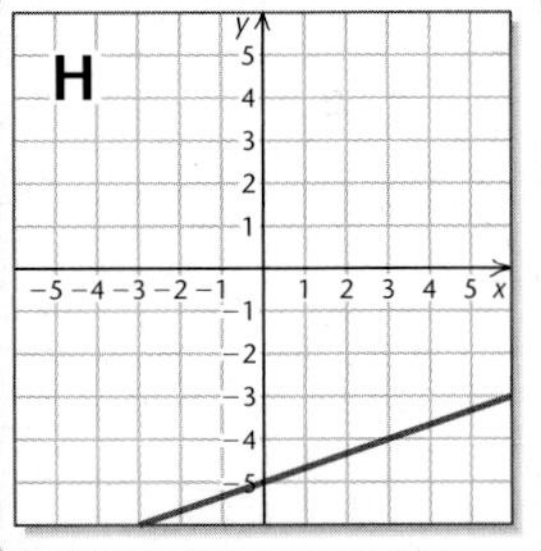

I
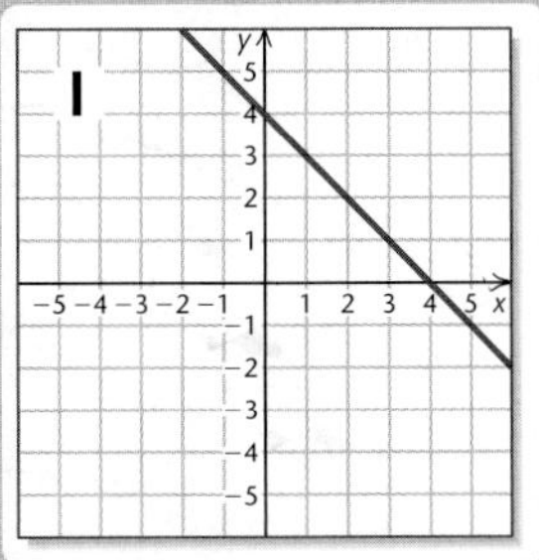

J
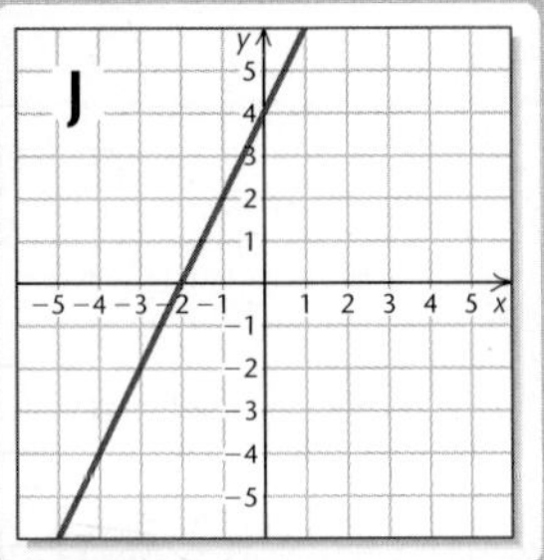

3.1 EXERCISE SET

FOR EXTRA HELP

MathXL | MyMathLab | InterAct Math | AW Math Tutor Center | Video Lectures on CD: Disc 2 | Student's Solutions Manual

Concept Reinforcement *Classify each statement as either true or false.*

1. Solutions of systems of equations in two variables are ordered pairs.

2. Every system of equations has at least one solution.

3. It is possible for a system of equations to have an infinite number of solutions.

4. The graphs of the equations in a system of two equations may coincide.

5. The graphs of the equations in a system of two equations could be parallel lines.

6. Any system of equations that has at most one solution is said to be consistent.

7. Any system of equations that has more than one solution is said to be inconsistent.

8. If one equation in a system can be obtained by multiplying both sides of another equation in that system by a constant, the two equations are said to be dependent.

Determine whether the ordered pair is a solution of the given system of equations. Remember to use alphabetical order of variables.

9. $(1, 2)$; $4x - y = 2,$ $10x - 3y = 4$

10. $(-1, -2)$; $2x + y = -4,$ $x - y = 1$

11. $(2, 5)$; $y = 3x - 1,$ $2x + y = 4$

12. $(-1, -2)$; $x + 3y = -7,$ $3x - 2y = 12$

13. $(1, 5)$; $x + y = 6,$ $y = 2x + 3$

14. $(5, 2)$; $a + b = 7,$ $2a - 8 = b$

Aha! **15.** $(3, 1)$; $3x + 4y = 13,$ $6x + 8y = 26$

16. $(4, -2)$; $-3x - 2y = -8,$ $8 = 3x + 2y$

Solve each system graphically. Be sure to check your solution. If a system has an infinite number of solutions, use set-builder notation to write the solution set. If a system has no solution, state this. Where appropriate, round to the nearest hundredth.

17. $x - y = 3,$ $x + y = 5$

18. $x + y = 4,$ $x - y = 2$

19. $3x + y = 5,$ $x - 2y = 4$

20. $2x - y = 4,$ $5x - y = 13$

21. $4y = x + 8,$ $3x - 2y = 6$

22. $4x - y = 9,$ $x - 3y = 16$

23. $x = y - 1,$ $2x = 3y$

24. $a = 1 + b,$ $b = 5 - 2a$

25. $x = -3,$ $y = 2$

26. $x = 4,$ $y = -5$

27. $t + 2s = -1,$ $s = t + 10$

28. $b + 2a = 2,$ $a = -3 - b$

29. $2b + a = 11,$ $a - b = 5$

30. $y = \frac{1}{3}x - 1,$ $4x - 3y = 18$

31. $y = -\frac{1}{4}x + 1,$ $2y = x - 4$

32. $6x - 2y = 2,$ $9x - 3y = 1$

33. $y - x = 5,$ $2x - 2y = 10$

34. $y = -x - 1,$ $4x - 3y = 24$

35. $y = 3 - x,$ $2x + 2y = 6$

36. $2x - 3y = 6,$ $3y - 2x = -6$

37. $y = -5.43x + 10.89,$ $y = 6.29x - 7.04$

38. $y = 123.52x + 89.32,$ $y = -89.22x + 33.76$

39. $2.6x - 1.1y = 4,$
$1.32y = 3.12x - 5.04$

40. $2.18x + 7.81y = 13.78,$
$5.79x - 3.45y = 8.94$

41. $0.2x - y = 17.5,$
$2y - 10.6x = 30$

42. $1.9x = 4.8y + 1.7,$
$12.92x + 23.8 = 32.64y$

43. For the systems in the odd-numbered exercises 17–41, which are consistent?

44. For the systems in the even-numbered exercises 18–42, which are consistent?

45. For the systems in the odd-numbered exercises 17–41, which contain dependent equations?

46. For the systems in the even-numbered exercises 18–42, which contain dependent equations?

Translate each problem situation in Exercises 47–60 to a system of equations. Do not attempt to solve, but save for later use.

47. The sum of two numbers is 50. The first number is 25% of the second number. What are the numbers?

48. The sum of two numbers is 40. The first number is 60% of the second number. What are the numbers?

49. *Nontoxic Furniture Polish.* A nontoxic wood furniture polish can be made by mixing mineral (or olive) oil with vinegar (*Sources*: Based on information from Chittenden Solid Waste District and *Clean House, Clean Planet* by Karen Logan). To make a 16-oz batch for a squirt bottle, Mabel uses an amount of mineral oil that is 4 oz more than twice the amount of vinegar. How much of each ingredient is required?

50. *Scholastic Aptitude Test.* Many high-school students take the Scholastic Aptitude Test. Each student receives two scores, a *verbal* score and a *math* score. In 2004–2005, the average total score of students was 1028, with the average math score exceeding the average verbal score by 12 points (*Source*: College Entrance Examination Board). What was the average verbal score and what was the average math score?

51. *Geometry.* Two angles are supplementary.* One angle is 3° less than twice the other. Find the measures of the angles.

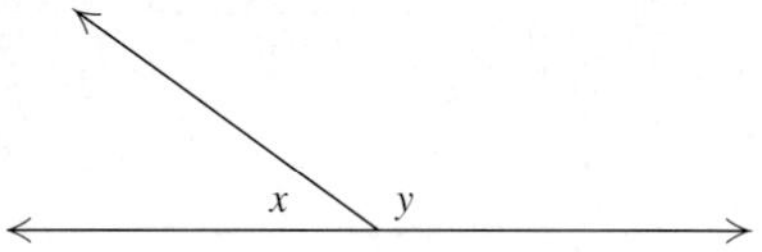

Supplementary angles

52. *Geometry.* Two angles are complementary.† The sum of the measures of the first angle and half the second angle is 64°. Find the measures of the angles.

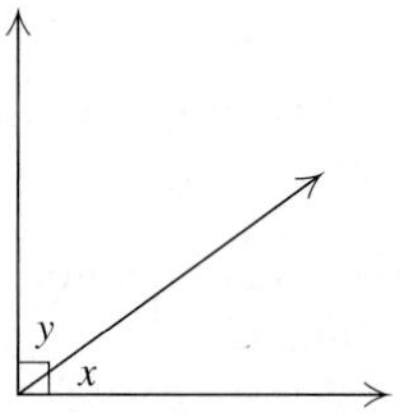

Complementary angles

53. *Basketball Scoring.* Wilt Chamberlain once scored 100 points, setting a record for points scored in an NBA game. Chamberlain took only two-point shots and (one-point) foul shots and made a total of 64 shots. How many shots of each type did he make?

54. *Basketball Scoring.* The Fenton College Cougars made 40 field goals in a recent basketball game, some 2-pointers and the rest 3-pointers. Altogether the 40 baskets counted for 89 points. How many of each type of field goal was made?

55. *Retail Sales.* Paint Town sold 45 paintbrushes, one kind at \$8.50 each and another at \$9.75 each. In all, \$398.75 was taken in for the brushes. How many of each kind were sold?

56. *Retail Sales.* Mountainside Fleece sold 40 neckwarmers. Polarfleece neckwarmers sold for \$9.90 each and wool ones sold for \$12.75 each. In all, \$421.65 was taken in for the neckwarmers. How many of each type were sold?

*The sum of the measures of two supplementary angles is 180°.
†The sum of the measures of two complementary angles is 90°.

57. *Sales of Pharmaceuticals.* In 2006, the Diabetic Express charged \$80.86 for a vial of Humalog insulin and \$83.70 for a vial of Novolog insulin. If a total of \$4125.36 was collected for 50 vials of insulin, how many vials of each type were sold?

58. *Fundraising.* The St. Mark's Community Barbecue served 250 dinners. A child's plate cost \$3.50 and an adult's plate cost \$7.00. A total of \$1347.50 was collected. How many of each type of plate was served?

59. *Court Dimensions.* The perimeter of a standard basketball court is 288 ft. The length is 44 ft longer than the width. Find the dimensions.

$P = 288$ ft

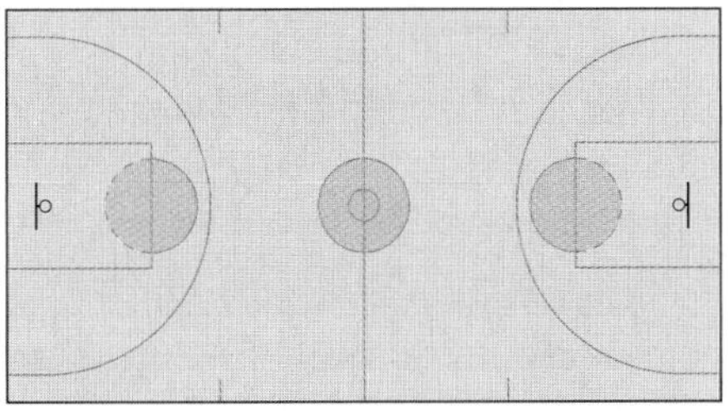

60. *Court Dimensions.* The perimeter of a standard tennis court used for doubles is 228 ft. The width is 42 ft less than the length. Find the dimensions.

61. *College Faculty.* The number of part-time faculty in institutions of higher learning is growing rapidly. The following table lists the number of full-time faculty and the number of part-time faculty for various years. Use linear regression to fit a line to each set of data, and use those equations to predict the year in which the number of part-time faculty and the number of full-time faculty will be the same.

Year	Number of Full-time Faculty (in thousands)	Number of Part-time Faculty (in thousands)
1980	450	236
1985	459	256
1991	536	291
1995	551	381
1999	591	437
2003	632	543

Source: U.S. National Center for Education Statistics

62. *Milk Cows.* The number of milk cows in Minnesota has decreased since 2000, while the number in New Mexico has increased, as shown in the following table. Use linear regression to fit a line to each set of data, and use those equations to predict the year in which the number of milk cows in the two states will be the same.

Year	Number of Milk Cows in Minnesota (in thousands)	Number of Milk Cows in New Mexico (in thousands)
2000	534	250
2001	510	268
2002	487	301
2003	473	317
2004	463	326

Source: U.S. Department of Agriculture

63. *Recycling.* In the United States, the amount of paper being recycled is slowly catching up to the amount of paper waste being generated, as shown in the following table. Use linear regression to fit a line to each set of data, and use those equations to predict the year in which the amount recycled will equal the amount generated.

Year	Amount of Paper Waste Generated (in millions of tons)	Amount of Paper Recycled (in millions of tons)
1990	72.7	27.8
1994	80.0	36.5
1998	84.1	41.6
2001	82.7	45.6
2003	83.1	48.1

Source: U.S. Bureau of Labor Statistics

64. *Grocery Stores.* Conventional grocery stores in the United States are being replaced by other types of food stores. One such type of store is the combination food and drug store, which contains a pharmacy and a greater variety of health and beauty aids than a conventional grocery store. Use linear regression to fit a line to each set of data in the following table, and use those equations to predict the year in which there will be the same number of combination food and drug stores as conventional grocery stores.

Year	Number of Conventional Grocery Stores (in thousands)	Number of Combination Food and Drug Stores (in thousands)
1990	13.2	1.6
1995	12.3	2.7
2000	9.9	3.7
2002	8.3	4.5
2003	8.9	5.0

Source: U.S. Department of Agriculture

TW **65.** Write a problem for a classmate to solve that requires writing a system of two equations. Devise the problem so that the solution is "The Lakers made 6 three-point baskets and 31 two-point baskets."

TW **66.** Write a problem for a classmate to solve that can be translated into a system of two equations. Devise the problem so that the solution is "Arnie took five 3-credit classes and two 4-credit classes."

Skill Maintenance

Solve. [1.6]

67. $2(4x - 3) - 7x = 9$

68. $6y - 3(5 - 2y) = 4$

69. $4x - 5x = 8x - 9 + 11x$

70. $8x - 2(5 - x) = 7x + 3$

Solve. [1.6]

71. $3x + 4y = 7$, for y

72. $2x - 5y = 9$, for y

Synthesis

Presidential Primaries. For Exercises 73 and 74, consider the following graph showing the results of a poll in which Iowans were asked which Democrat candidate for president they favored.

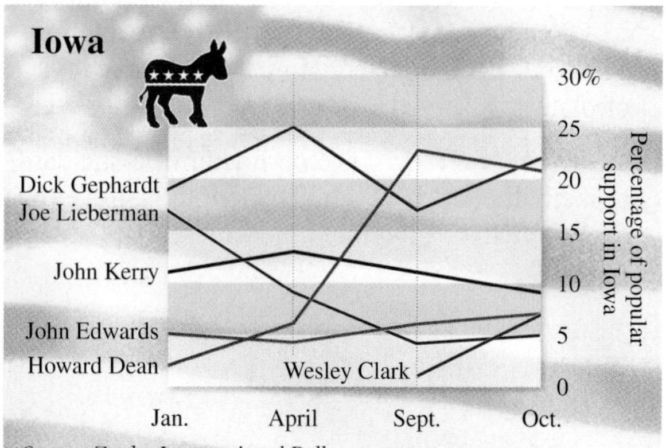

Source: Zogby International Polls

TW **73.** At what point in time could it have been said that no one was in fourth place? Explain.

TW **74.** At what point in time was there no clear leader? Explain how you reach this conclusion.

75. For each of the following conditions, write a system of equations.

a) (5, 1) is a solution.
b) There is no solution.
c) There is an infinite number of solutions.

76. A system of linear equations has $(1, -1)$ and $(-2, 3)$ as solutions. Determine:

a) a third point that is a solution, and
b) how many solutions there are.

77. The solution of the following system is $(4, -5)$. Find A and B.

$$Ax - 6y = 13,$$
$$x - By = -8.$$

Translate to a system of equations. Do not solve.

78. *Ages.* Burl is twice as old as his son. Ten years ago, Burl was three times as old as his son. How old are they now?

79. *Work Experience.* Lou and Juanita are mathematics professors at a state university. Together, they have 46 years of service. Two years ago, Lou had taught 2.5 times as many years as Juanita. How long has each taught at the university?

80. *Design.* A piece of posterboard has a perimeter of 156 in. If you cut 6 in. off the width, the length becomes four times the width. What are the dimensions of the original piece of posterboard?

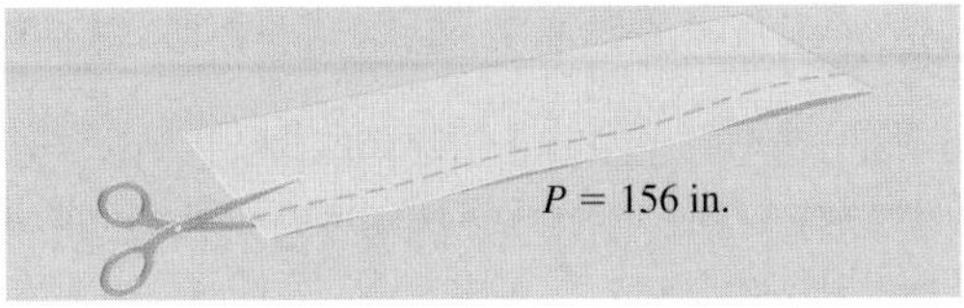

81. *Nontoxic Scouring Powder.* A nontoxic scouring powder is made up of 4 parts baking soda and 1 part vinegar. How much of each ingredient is needed for a 16-oz mixture?

Solve graphically.

82. $y = |x|,$
$x + 4y = 15$

83. $x - y = 0,$
$y = x^2$

In Exercises 84–87, match each system with the appropriate graph from the selections given.

a)

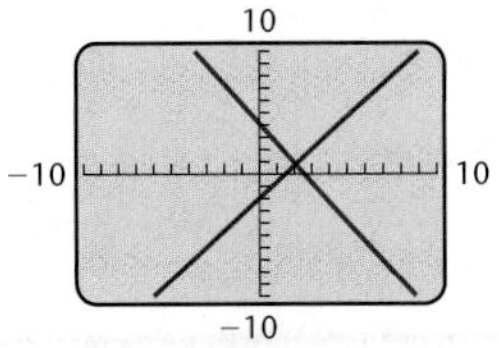

b)

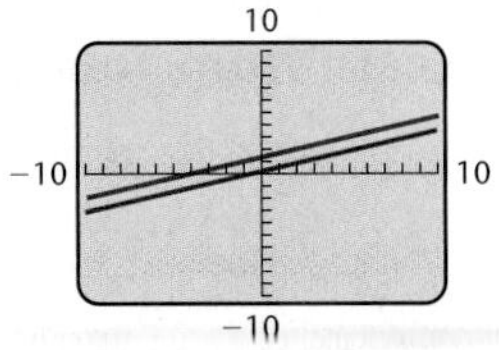

c)

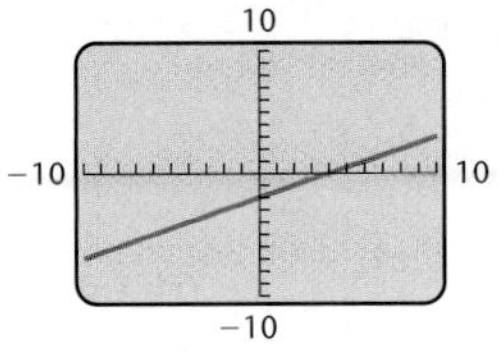

d)

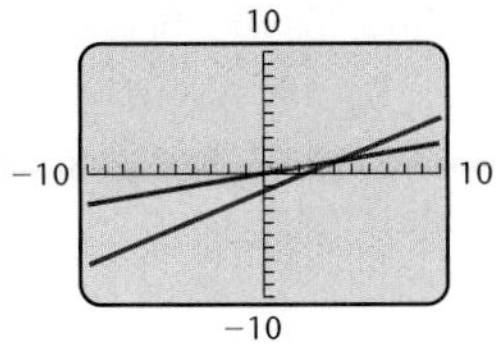

84. $x = 4y,$
$3x - 5y = 7$

85. $2x - 8 = 4y,$
$x - 2y = 4$

86. $8x + 5y = 20,$
$4x - 3y = 6$

87. $x = 3y - 4,$
$2x + 1 = 6y$

3.2 Solving by Substitution or Elimination

The Substitution Method ■ The Elimination Method ■ Comparing Methods

The Substitution Method

Algebraic (nongraphical) methods for solving systems are often superior to graphing, especially when fractions are involved. One algebraic method, the *substitution method,* relies on having a variable isolated.

EXAMPLE 1 Solve the system

$$x + y = 4, \quad (1)$$
$$x = y + 1. \quad (2)$$

For easy reference, we have numbered the equations.

Study Tip

Learn from Your Mistakes

Immediately after each quiz or test, write out a step-by-step solution to any questions you missed. Visit your professor during office hours or consult with a tutor for help with problems that are still giving you trouble. Misconceptions tend to resurface if they are not corrected as soon as possible.

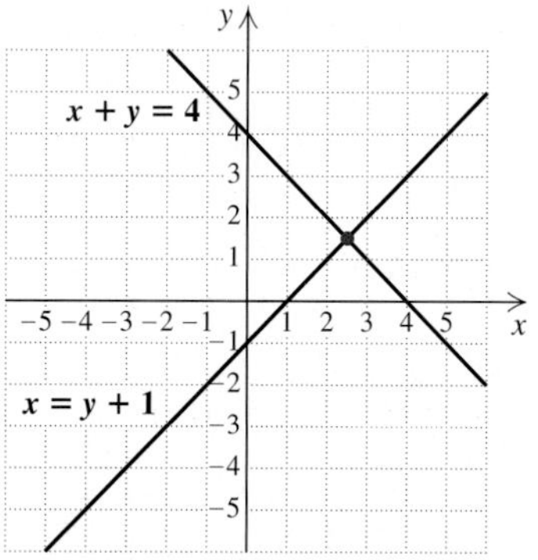

A visualization of Example 1. Note that the coordinates of the intersection are not obvious.

SOLUTION Equation (2) says that x and $y + 1$ name the same number. Thus we can substitute $y + 1$ for x in equation (1):

$$x + y = 4 \quad \textbf{Equation (1)}$$
$$(y + 1) + y = 4. \quad \textbf{Substituting } y + 1 \textbf{ for } x$$

We solve this last equation, using methods learned earlier:

$$(y + 1) + y = 4$$
$$2y + 1 = 4 \quad \textbf{Removing parentheses and combining like terms}$$
$$2y = 3 \quad \textbf{Subtracting 1 from both sides}$$
$$y = \tfrac{3}{2}. \quad \textbf{Dividing by 2}$$

We now return to the original pair of equations and substitute $\frac{3}{2}$ for y in either equation so that we can solve for x. For this problem, calculations are slightly easier if we use equation (2):

$$x = y + 1 \quad \textbf{Equation (2)}$$
$$= \tfrac{3}{2} + 1 \quad \textbf{Substituting } \tfrac{3}{2} \textbf{ for } y$$
$$= \tfrac{3}{2} + \tfrac{2}{2} = \tfrac{5}{2}.$$

We obtain the ordered pair $\left(\frac{5}{2}, \frac{3}{2}\right)$. A check ensures that it is a solution:

Check:

$$\begin{array}{c|c} x + y & 4 \\ \hline \frac{5}{2} + \frac{3}{2} & 4 \\ \frac{8}{2} & \\ 4 \stackrel{?}{=} 4 & \text{TRUE} \end{array} \qquad \begin{array}{c|c} x & y + 1 \\ \hline \frac{5}{2} & \frac{3}{2} + 1 \\ & \frac{3}{2} + \frac{2}{2} \\ \frac{5}{2} \stackrel{?}{=} \frac{5}{2} & \text{TRUE} \end{array}$$

Since $\left(\frac{5}{2}, \frac{3}{2}\right)$ checks, it is the solution.

The exact solution to Example 1 is difficult to find graphically because it involves fractions. Despite this, the graph shown does serve as a check and provides a visualization of the problem.

If neither equation in a system has a variable alone on one side, we first isolate a variable in one equation and then substitute.

EXAMPLE 2 Solve the system

$$2x + y = 6, \quad (1)$$
$$3x + 4y = 4. \quad (2)$$

SOLUTION First, we select an equation and solve for one variable. To isolate y, we can subtract $2x$ from both sides of equation (1):

$$2x + y = 6 \quad (1)$$
$$y = 6 - 2x. \quad (3) \quad \textbf{Subtracting } 2x \textbf{ from both sides}$$

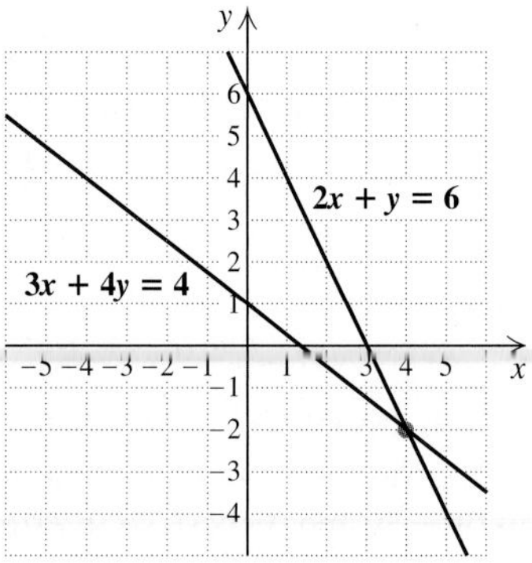

A visualization of Example 2

Next, we proceed as in Example 1, by substituting:

$$3x + 4(6 - 2x) = 4$$ **Substituting $6 - 2x$ for y in equation (2). Use parentheses!**

$$3x + 24 - 8x = 4$$ **Distributing to remove parentheses**

$$3x - 8x = 4 - 24$$ **Subtracting 24 from both sides**

$$-5x = -20$$

$$x = 4.$$ **Dividing both sides by -5**

Next, we substitute 4 for x in either equation (1), (2), or (3). It is easiest to use equation (3) because it has already been solved for y:

$$\begin{aligned} y &= 6 - 2x \\ &= 6 - 2(4) \\ &= 6 - 8 = -2. \end{aligned}$$

The pair $(4, -2)$ appears to be the solution. We check in equations (1) and (2).

Check:

$$\begin{array}{c|c} \multicolumn{2}{c}{2x + y = 6} \\ \hline 2(4) + (-2) & 6 \\ 8 - 2 & \\ \multicolumn{2}{r}{6 \stackrel{?}{=} 6 \quad \text{TRUE}} \end{array} \qquad \begin{array}{c|c} \multicolumn{2}{c}{3x + 4y = 4} \\ \hline 3(4) + 4(-2) & 4 \\ 12 - 8 & \\ \multicolumn{2}{r}{4 \stackrel{?}{=} 4 \quad \text{TRUE}} \end{array}$$

Since $(4, -2)$ checks, it is the solution.

Some systems have no solution, as we saw graphically in Section 3.1. How do we recognize such systems if we are solving by an algebraic method?

EXAMPLE 3 Solve the system

$$y = -3x + 5, \quad (1)$$
$$y = -3x - 2. \quad (2)$$

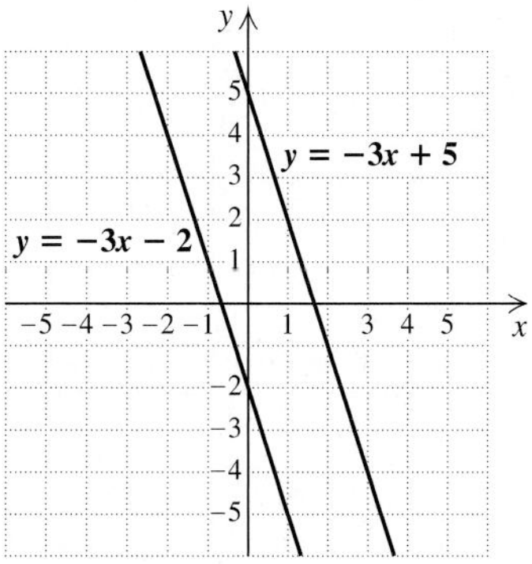

A visualization of Example 3

SOLUTION We solved this system graphically in Example 5(a) of Section 3.1, and found that the lines are parallel and the system has no solution. Let's now try to solve the system by substitution. Proceeding as in Example 1, we substitute $-3x - 2$ for y in the first equation:

$$-3x - 2 = -3x + 5$$ **Substituting $-3x - 2$ for y in equation (1)**

$$-2 = 5.$$ **Adding $3x$ to both sides; $-2 = 5$ is a contradiction.**

When we add $3x$ to get the x-terms on one side, the x-terms drop out and the result is a contradiction—that is, an equation that is always false. When solving algebraically yields a contradiction, we state that the system has no solution.

The Elimination Method

The *elimination method* for solving systems of equations makes use of the *addition principle*: If $a = b$, then $a + c = b + c$. Consider the following system:

$$\begin{aligned} 2x - 3y &= 0, \qquad (1) \\ -4x + 3y &= -1. \qquad (2) \end{aligned}$$

To see why the elimination method works well with this system, notice the $-3y$ in one equation and the $3y$ in the other. These terms are opposites. If we add all terms on the left side of the equations, the sum of $-3y$ and $3y$ is 0, so in effect, the variable y is "eliminated."

To use the addition principle with a system, note that according to equation (2), $-4x + 3y$ and -1 are the same number. Thus we can work vertically and add $-4x + 3y$ to the left side of equation (1) and -1 to the right side:

$$\begin{aligned} 2x - 3y &= 0 \qquad (1) \\ -4x + 3y &= -1 \qquad (2) \\ \hline -2x + 0y &= -1. \qquad \textbf{Adding} \end{aligned}$$

This eliminates the variable y, and leaves an equation with just one variable, x, for which we solve:

$$\begin{aligned} -2x &= -1 \\ x &= \tfrac{1}{2}. \end{aligned}$$

Next, we substitute $\frac{1}{2}$ for x in equation (1) and solve for y:

$$\begin{aligned} 2 \cdot \tfrac{1}{2} - 3y &= 0 \qquad \textbf{Substituting. We also could have used equation (2).} \\ 1 - 3y &= 0 \\ -3y &= -1, \text{ so } y = \tfrac{1}{3}. \end{aligned}$$

Check:

$$\begin{array}{c|c} \multicolumn{2}{c}{2x - 3y = 0} \\ \hline 2(\frac{1}{2}) - 3(\frac{1}{3}) & 0 \\ 1 - 1 & \\ \multicolumn{2}{c}{0 \stackrel{?}{=} 0 \quad \text{TRUE}} \end{array} \qquad \begin{array}{c|c} \multicolumn{2}{c}{-4x + 3y = -1} \\ \hline -4(\frac{1}{2}) + 3(\frac{1}{3}) & -1 \\ -2 + 1 & \\ \multicolumn{2}{c}{-1 \stackrel{?}{=} -1 \quad \text{TRUE}} \end{array}$$

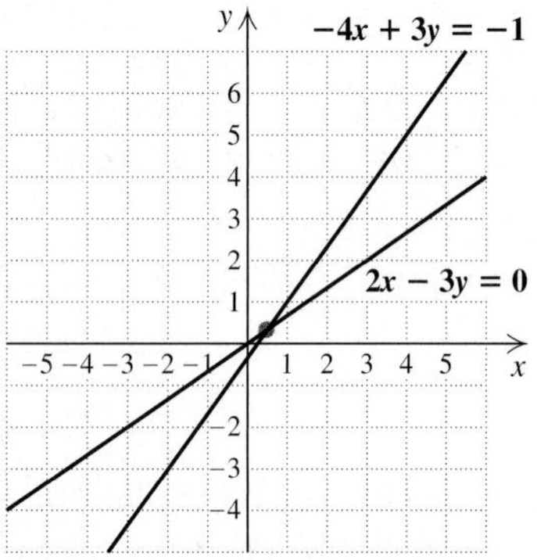

A visualization of the solution of
$2x - 3y = 0$
$-4x + 3y = -1$

Since $\left(\frac{1}{2}, \frac{1}{3}\right)$ checks, it is the solution. See also the graph at left.

To eliminate a variable, we must sometimes multiply before adding.

EXAMPLE 4 Solve the system

$$\begin{aligned} 5x + 4y &= 22, \qquad (1) \\ -3x + 8y &= 18. \qquad (2) \end{aligned}$$

Student Notes

It is wise to double-check each step of your work as you go along, rather than checking all steps after reaching the end of a problem. Finding and correcting an error as it occurs will save you time in the long run. One common error is to forget to multiply *both* sides of the equation when you use the multiplication principle.

$y_1 = (22 - 5x)/4, \ y_2 = (18 + 3x)/8$

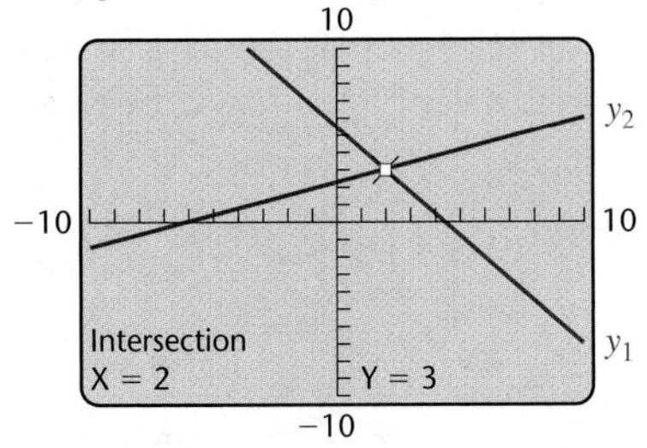

SOLUTION If we add the left sides of the two equations, we will not eliminate a variable. However, if the $4y$ in equation (1) were changed to $-8y$, we would. To accomplish this change, we multiply both sides of equation (1) by -2:

$$\begin{aligned} -10x - 8y &= -44 && \textbf{Multiplying both sides of equation (1) by } -2 \\ -3x + 8y &= \ \ 18 \\ \hline -13x + \ \ 0 &= -26 && \textbf{Adding} \\ x &= 2. && \textbf{Solving for } x \end{aligned}$$

Then

$$\begin{aligned} -3 \cdot 2 + 8y &= 18 && \textbf{Substituting 2 for } x \textbf{ in equation (2)} \\ -6 + 8y &= 18 \\ \left.\begin{aligned} 8y &= 24 \\ y &= 3. \end{aligned}\right\} && && \textbf{Solving for } y \end{aligned}$$

We obtain (2, 3), or $x = 2$, $y = 3$. We can check by substitution or by solving graphically, as shown in the figure at left. The solution is (2, 3).

Sometimes we must multiply twice in order to make two terms become opposites.

EXAMPLE 5 Solve the system

$$\begin{aligned} 2x + 3y &= 17, \quad &(1) \\ 5x + 7y &= 29. \quad &(2) \end{aligned}$$

SOLUTION We multiply so that the x-terms are eliminated.

$$\begin{aligned} 2x + 3y = 17, &\rightarrow \begin{matrix}\textbf{Multiplying both}\\ \textbf{sides by 5}\end{matrix} \rightarrow & 10x + 15y &= \ \ 85 \\ 5x + 7y = 29 &\rightarrow \begin{matrix}\textbf{Multiplying both}\\ \textbf{sides by } -2\end{matrix} \rightarrow & -10x - 14y &= -58 \\ & & \hline 0 + \ \ y &= \ \ 27 \quad \textbf{Adding} \\ & & y &= \ \ 27 \end{aligned}$$

Next, we substitute to find x:

$$\begin{aligned} 2x + 3 \cdot 27 &= 17 && \textbf{Substituting 27 for } y \textbf{ in equation (1)} \\ 2x + 81 &= 17 \\ \left.\begin{aligned} 2x &= -64 \\ x &= -32. \end{aligned}\right\} && && \textbf{Solving for } x \end{aligned}$$

Check:

$$\begin{array}{r|l} \multicolumn{2}{c}{2x + 3y = 17} \\ \hline 2(-32) + 3(27) & 17 \\ -64 + 81 & \\ 17 & \stackrel{?}{=} 17 \quad \text{TRUE} \end{array} \qquad \begin{array}{r|l} \multicolumn{2}{c}{5x + 7y = 29} \\ \hline 5(-32) + 7(27) & 29 \\ -160 + 189 & \\ 29 & \stackrel{?}{=} 29 \quad \text{TRUE} \end{array}$$

$y_1 = (17 - 2x)/3, \ y_2 = (29 - 5x)/7$

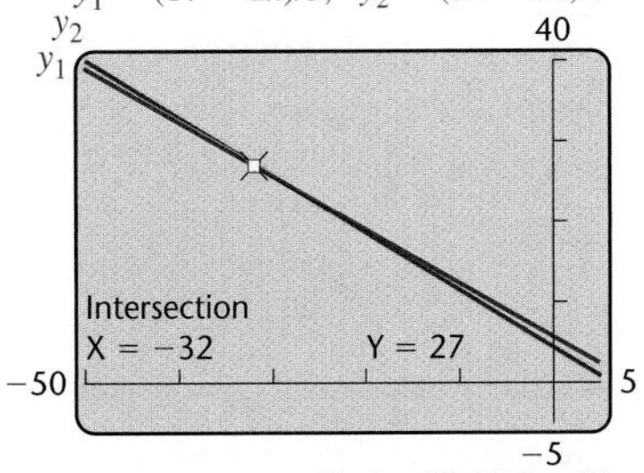

Xscl = 10, Yscl = 10

We can also check by using a graphing calculator, as shown in the figure at left. The solution is $(-32, 27)$, or $x = -32$, $y = 27$.

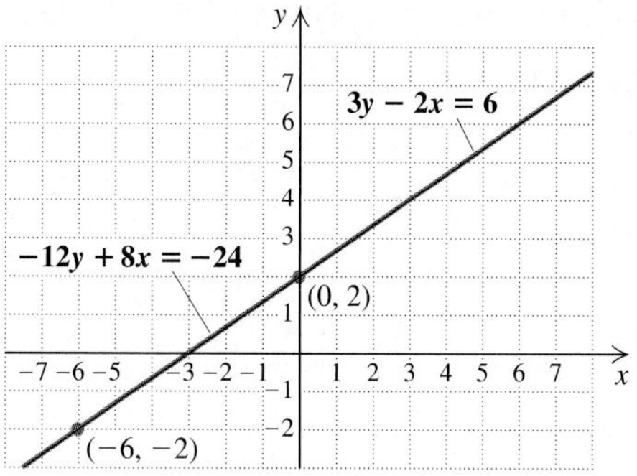

A visualization of Example 6

EXAMPLE 6 Solve the system

$$3y - 2x = 6, \qquad (1)$$
$$-12y + 8x = -24. \qquad (2)$$

SOLUTION We graphed this system in Example 5(b) of Section 3.1, and found that the lines coincide and the system has an infinite number of solutions. Suppose we were to solve this system using the elimination method:

$$\begin{aligned} 12y - 8x &= 24 &&\textbf{Multiplying both sides of equation (1) by 4}\\ -12y + 8x &= -24 \\ \hline 0 &= 0. &&\textbf{We obtain an identity; } \mathbf{0 = 0} \textbf{ is always true.}\end{aligned}$$

Note that both variables have been eliminated and what remains is an identity—that is, an equation that is always true. Any pair that is a solution of equation (1) is also a solution of equation (2). The equations are dependent and the solution set is infinite:

$$\{(x, y) \mid 3y - 2x = 6\}, \quad \text{or equivalently,} \quad \{(x, y) \mid -12y + 8x = -24\}.$$

Rules for Special Cases When solving a system of two linear equations in two variables:

1. If an identity is obtained, such as $0 = 0$, then the system has an infinite number of solutions. The equations are dependent and, since a solution exists, the system is consistent.*
2. If a contradiction is obtained, such as $0 = 7$, then the system has no solution. The system is inconsistent.

Should decimals or fractions appear, it often helps to *clear* before solving.

EXAMPLE 7 Solve the system

$$0.2x + 0.3y = 1.7,$$
$$\tfrac{1}{7}x + \tfrac{1}{5}y = \tfrac{29}{35}.$$

SOLUTION We have

$$0.2x + 0.3y = 1.7, \rightarrow \textbf{Multiplying both sides by 10} \rightarrow 2x + 3y = 17$$
$$\tfrac{1}{7}x + \tfrac{1}{5}y = \tfrac{29}{35} \rightarrow \textbf{Multiplying both sides by 35} \rightarrow 5x + 7y = 29.$$

We multiplied both sides of the first equation by 10 to clear the decimals. Multiplication by 35, the least common denominator, clears the fractions in the second equation. The problem now happens to be identical to Example 5. The solution is $(-32, 27)$, or $x = -32$, $y = 27$.

*Consistent systems and dependent equations are discussed in greater detail in Section 3.4.

Comparing Methods

Often it is helpful to use both an algebraic and a graphical method to solve a system of equations. In Examples 1–6, we solved each system algebraically and checked by solving graphically. A graph helps us visualize a solution that can be found exactly using algebra. On the other hand, algebra can help us determine an appropriate viewing window for a graph. Note that the graph of the system in Example 5 is shown in the viewing window $[-50, 5, -5, 40]$. If that system were graphed using a standard viewing window, the lines would appear to be parallel. When graphing a system of equations, we generally choose a viewing window that shows any points of intersection.

The following table is a summary that compares the graphical, substitution, and elimination methods for solving systems of equations.

Connecting the Concepts

We now have three different methods for solving systems of equations. Each method has certain strengths and weaknesses, as outlined below.

Method	Strengths	Weaknesses
Graphical	Solutions are displayed graphically. Can be used with any system that can be graphed. With a graphing calculator, can be used with any system with real-number coefficients.	May provide only approximate solutions. Solution may not appear on the part of the graph drawn. Systems with no solutions or an infinite number of solutions may be hard to detect.
Substitution	Yields exact solutions. Easy to use when a variable is alone on one side.	Introduces extensive computations with fractions when solving more complicated systems. Solutions are not displayed graphically.
Elimination	Yields exact solutions. Easy to use when fractions or decimals appear in the system. The preferred method for systems of 3 or more equations in 3 or more variables (see Section 3.4).	Solutions are not displayed graphically.

3.2 EXERCISE SET

FOR EXTRA HELP

MathXL | MyMathLab | InterAct Math | AW Math Tutor Center | Video Lectures on CD: Disc 2 | Student's Solutions Manual

Concept Reinforcement *In each of Exercises 1–6, match the system listed with the choice from the column on the right that would be a subsequent step in solving the system.*

1. ____ $-2x + 3y = 7,$
$2x + 5y = -8$

2. ____ $x = 4y - 5,$
$5x + 7y = 2$

3. ____ $3x + 4y = 13,$
$4x - 2y = 5$

4. ____ $8x + 6y = -15,$
$5x - 3y = 8$

5. ____ $y = 4x - 7,$
$6x + 3y = 19$

6. ____ $y = \frac{2}{3}x + 4,$
$y = -\frac{1}{5}x + 4$

a) $3x + 4y = 13,$
$8x - 4y = 10$

b) The lines intersect at (0, 4).

c) $6x + 3(4x - 7) = 19$

d) $8y = -1$

e) $5(4y - 5) + 7y = 2$

f) $8x + 6y = -15,$
$10x - 6y = 16$

For Exercises 7–54, if a system has an infinite number of solutions, use set-builder notation to write the solution set. If a system has no solution, state this.

Solve using the substitution method.

7. $y = 5 - 4x,$
$2x - 3y = 13$

8. $2y + x = 9,$
$x = 3y - 3$

9. $3x + 5y = 3,$
$x = 8 - 4y$

10. $9x - 2y = 3,$
$3x - 6 = y$

11. $3s - 4t = 14,$
$5s + t = 8$

12. $m - 2n = 16$
$4m + n = 1$

13. $4x - 2y = 6,$
$2x - 3 = y$

14. $t = 4 - 2s,$
$t + 2s = 6$

15. $-5s + t = 11,$
$4s + 12t = 4$

16. $5x + 6y = 14,$
$-3y + x = 7$

17. $2x + 2y = 2,$
$3x - y = 1$

18. $4p - 2q = 16,$
$5p + 7q = 1$

19. $x - 4y = 3,$
$5x + 3y = 4$

20. $2a + 2b = 5,$
$3a - b = 7$

21. $2x - 3 = y,$
$y - 2x = 1$

22. $a - 2b = 3,$
$3a = 6b + 9$

Solve using the elimination method.

23. $x + 3y = 7,$
$-x + 4y = 7$

24. $2x + y = 6,$
$x - y = 3$

25. $2x - y = -3,$
$x + y = 9$

26. $x - 2y = 6,$
$-x + 3y = -4$

27. $9x + 3y = -3,$
$2x - 3y = -8$

28. $6x - 3y = 18,$
$6x + 3y = -12$

29. $5x + 3y = 19,$
$2x - 5y = 11$

30. $3x + 2y = 3,$
$9x - 8y = -2$

31. $5r - 3s = 24,$
$3r + 5s = 28$

32. $5x - 7y = -16,$
$2x + 8y = 26$

33. $6s + 9t = 12,$
$4s + 6t = 5$

34. $10a + 6b = 8,$
$5a + 3b = 2$

35. $\frac{1}{2}x - \frac{1}{6}y = 3,$
$\frac{2}{5}x + \frac{1}{2}y = 2$

36. $\frac{1}{3}x + \frac{1}{5}y = 7,$
$\frac{1}{6}x - \frac{2}{5}y = -4$

37. $\frac{x}{2} + \frac{y}{3} = \frac{7}{6},$
$\frac{2x}{3} + \frac{3y}{4} = \frac{5}{4}$

38. $\frac{2x}{3} + \frac{3y}{4} = \frac{11}{12},$
$\frac{x}{3} + \frac{7y}{18} = \frac{1}{2}$

Aha! **39.** $12x - 6y = -15,$
$-4x + 2y = 5$

40. $8s + 12t = 16,$
$6s + 9t = 12$

41. $0.2a + 0.3b = 1,$
$0.3a - 0.2b = 4$

42. $-0.4x + 0.7y = 1.3,$
$0.7x - 0.3y = 0.5$

Solve using any appropriate method.

43. $a - 2b = 16,$
$b + 3 = 3a$

44. $5x - 9y = 7,$
$7y - 3x = -5$

45. $10x + y = 306,$
$10y + x = 90$

46. $3(a - b) = 15,$
$4a = b + 1$

47. $3y = x - 2,$
$x = 2 + 3y$

48. $x + 2y = 8,$
$x = 4 - 2y$

49. $3s - 7t = 5,$
$7t - 3s = 8$

50. $2s - 13t = 120,$
$-14s + 91t = -840$

51. $0.05x + 0.25y = 22,$
$0.15x + 0.05y = 24$

52. $1.3x - 0.2y = 12,$
$0.4x + 17y = 89$

53. $13a - 7b = 9,$
$2a - 8b = 6$

54. $3a - 12b = 9,$
$14a - 11b = 5$

In Exercises 55–58, determine which of the given viewing windows below shows the point of intersection of the graphs of the equations in the given system. Check by graphing.

a) [−5, 5, −5, 5]
b) [25, 50, 0, 10]
c) [0, 20, 0, 10]
d) [100, 200, 0, 100]

55. The system of Exercise 51

56. The system of Exercise 44

57. The system of Exercise 45

58. The system of Exercise 52

TW **59.** Describe a procedure that can be used to write an inconsistent system of equations.

TW **60.** Describe a procedure that can be used to write a system that has an infinite number of solutions.

Skill Maintenance

Solve. [1.7]

61. The fare for a taxi ride from Johnson Street to Elm Street is \$5.20. If the rate of the taxi is \$1.00 for the first $\frac{1}{2}$ mi and 30¢ for each additional $\frac{1}{4}$ mi, how far is it from Johnson Street to Elm Street?

62. A student's average after 4 tests is 78.5. What score is needed on the fifth test in order to raise the average to 80?

63. *Home Remodeling.* In a recent year, Americans spent \$35 billion to remodel bathrooms and kitchens. Twice as much was spent on kitchens as on bathrooms. (*Source*: *Indianapolis Star*) How much was spent on each?

64. A 480-m wire is cut into three pieces. The second piece is three times as long as the first. The third is four times as long as the second. How long is each piece?

65. *Car Rentals.* Badger Rent-A-Car rents a compact car at a daily rate of \$34.95 plus 10¢ per mile. A sales representative is allotted \$80 for car rental for one day. How many miles can she travel on the \$80 budget?

66. *Car Rentals.* Badger rents midsized cars at a rate of \$43.95 plus 10¢ per mile. A tourist has a car-rental budget of \$90 for one day. How many miles can he travel on the \$90 budget?

Synthesis

TW **67.** A student solving the system

$$17x + 19y = 102,$$
$$136x + 152y = 826$$

graphs both equations on a graphing calculator and gets the following screen. The student then (incorrectly) concludes that the equations are dependent and the solution set is infinite. How can algebra be used to convince the student that a mistake has been made?

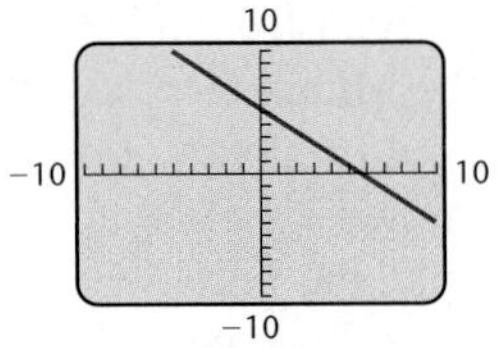

TW **68.** Some systems are more easily solved by substitution and some are more easily solved by elimination. What guidelines could be used to help someone determine which method to use?

69. If (1, 2) and (−3, 4) are two solutions of $f(x) = mx + b$, find m and b.

70. If $(0, -3)$ and $\left(-\frac{3}{2}, 6\right)$ are two solutions of $px - qy = -1$, find p and q.

71. Determine a and b for which $(-4, -3)$ is a solution of the system

$$ax + by = -26,$$
$$bx - ay = 7.$$

72. Solve for x and y in terms of a and b:

$$5x + 2y = a,$$
$$x - y = b.$$

Solve.

73. $$\frac{x + y}{2} - \frac{x - y}{5} = 1,$$
$$\frac{x - y}{2} + \frac{x + y}{6} = -2$$

74. $3.5x - 2.1y = 106.2,$
$4.1x + 16.7y = -106.28$

Each of the following is a system of nonlinear equations. However, each is reducible to linear, *since an appropriate substitution (say, u for 1/x and v for 1/y) yields a linear system. Make such a substitution, solve for the new variables, and then solve for the original variables.*

75. $\frac{2}{x} + \frac{1}{y} = 0,$
$\frac{5}{x} + \frac{2}{y} = -5$

76. $\frac{1}{x} - \frac{3}{y} = 2,$
$\frac{6}{x} + \frac{5}{y} = -34$

Collaborative Corner

How Many Two's? How Many Three's?

Focus: Systems of linear equations
Time: 20 minutes
Group Size: 3

The box score at right, from a 2006 NBA playoff game between the Miami Heat and the Detroit Pistons, contains information on how many field goals and free throws each player attempted and made. For example, the line "Wade 11-20 8-8 32" means that Miami's Dwayne Wade made 11 field goals out of 20 attempts and 8 free throws out of 8 attempts, for a total of 32 points. (Each free throw is worth 1 point and each field goal is worth either 2 or 3 points, depending on how far from the basket it was shot.)

ACTIVITY

1. Work as a group to develop a system of two equations in two unknowns that can be used to determine how many 2-pointers and how many 3-pointers were made by Detroit.
2. Each group member should solve the system from part (1) in a different way: one person algebraically, one person by making a table and methodically checking all combinations of 2- and 3-pointers, and one person by guesswork. Compare answers when this has been completed.
3. Determine, as a group, how many 2- and 3-pointers Miami made.

Miami (88)
Walker 3-12 4-4 11, Haslem 1-5 0-0 2, S. O'Neal 9-16 3-5 21, Wade 11-20 8-8 32, Williams 3-8 0-0 7, Posey 2-4 0-0 6, Payton 1-6 5-6 7, Mourning 1-2 0-0 2, S. Anderson 0-0 0-0 0
Totals 31-73 20-23 88
Detroit (92)
Prince 10-20 3-5 24, R. Wallace 6-12 2-2 16, B. Wallace 4-4 1-2 9, Hamilton 7-18 6-7 22, Billups, 5-13 7-7 18, Delfino 0-1 0-0 0, McDyess 1-4 1-4 3, D. Davis 0-0 0-0 0, Hunter 0-3 0-0 0, Delk 0-0 0-0 0
Totals 33-75 20-27 92

Miami	**12**	**25**	**19**	**32**	**88**
Detroit	**25**	**23**	**22**	**22**	**92**

3.3 Solving Applications: Systems of Two Equations

Total-Value and Mixture Problems ■ Motion Problems

Study Tip

Expect to be Challenged

Do not be surprised if your success rate drops some as you work on real-world problems. *This is normal.* Your success rate will increase as you gain experience with these types of problems and use some of the study tips already listed.

You are in a much better position to solve problems now that you know how systems of equations can be used. Using systems often makes the translating step easier.

EXAMPLE 1 Endangered Species. The number of species listed as threatened or endangered has more than tripled in the past 20 yr. In 2005, there were 749 species of plants considered threatened or endangered. The number considered threatened, or likely to become endangered in the foreseeable future, was 1 less than one-fourth of the number considered endangered, or in danger of becoming extinct. (*Source:* U.S. Fish and Wildlife Service) How many plant species were considered endangered and how many were considered threatened in 2005?

SOLUTION The *Familiarize* and *Translate* steps have been completed in Example 1 of Section 3.1. The resulting system of equations is

$$d + t = 749,$$
$$t = \tfrac{1}{4}d - 1,$$

where d is the number of endangered plant species and t is the number of threatened plant species in 2005.

3. Carry out. We solve the system of equations both algebraically and graphically.

ALGEBRAIC APPROACH

Since one equation already has a variable isolated, let's use the substitution method:

$$\begin{aligned} d + t &= 749 \\ d + \tfrac{1}{4}d - 1 &= 749 && \textbf{Substituting } \tfrac{1}{4}d - 1 \textbf{ for } t \\ \tfrac{5}{4}d - 1 &= 749 && \textbf{Combining like terms} \\ \tfrac{5}{4}d &= 750 && \textbf{Adding 1 to both sides} \\ d &= \tfrac{4}{5}\cdot 750 && \textbf{Multiplying both sides by } \tfrac{4}{5} \\ d &= 600. && \textbf{Simplifying} \end{aligned}$$

Next, using either of the original equations, we substitute and solve for t:

$$t = \tfrac{1}{4}\cdot 600 - 1 = 150 - 1 = 149.$$

We have $d = 600$, $t = 149$.

GRAPHICAL APPROACH

We let

$$d = x \quad \text{and} \quad t = y$$

and substitute. Then we graph $y_1 = 749 - x$ and $y_2 = \frac{1}{4}x - 1$ and find the point of intersection of the graphs. Since $x + y = 749$, and only positive values of x and y make sense in this problem, we choose a viewing window of [0, 750, 0, 750], with Xscl = 50 and Yscl = 50.

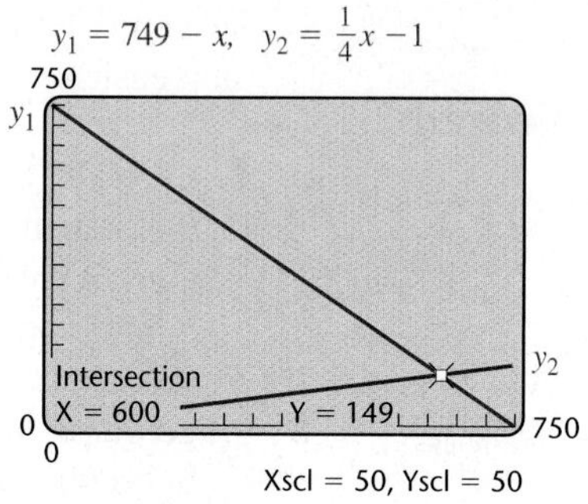

We have a solution of (600, 149).

4. **Check.** The sum of 600 and 149 is 749, so the total number of species is correct. Since 1 less than one-fourth of 600 is 150 − 1, or 149, the numbers check.

5. **State.** In 2005, there were 600 plant species considered endangered and 149 considered threatened.

Total-Value and Mixture Problems

EXAMPLE 2 Purchasing. Recently the Woods County Art Center purchased 120 stamps for \$35.55. If the stamps were a combination of 24¢ postcard stamps and 39¢ first-class stamps, how many of each type were bought?

SOLUTION The *Familiarize* and *Translate* steps were completed in Example 2 of Section 3.1.

3. **Carry out.** We are to solve the system of equations

$$p + f = 120, \qquad (1)$$

$$24p + 39f = 3555, \qquad (2) \quad \textbf{Working in cents rather than dollars}$$

where p is the number of postcard stamps bought and f is the number of first-class stamps bought.

ALGEBRAIC APPROACH

Because both equations are in the form $Ax + By = C$, let's use the elimination method to solve the system. We can eliminate p by multiplying both sides of equation (1) by -24 and adding them to the corresponding sides of equation (2):

$$-24p - 24f = -2880 \quad \textbf{Multiplying both sides of equation (1) by } -24$$
$$\underline{24p + 39f = 3555}$$
$$15f = 675 \quad \textbf{Adding}$$
$$f = 45. \quad \textbf{Solving for } f$$

To find p, we substitute 45 for f in equation (1) and then solve for p:

$$p + f = 120 \quad \textbf{Equation (1)}$$
$$p + 45 = 120 \quad \textbf{Substituting 45 for } f$$
$$p = 75. \quad \textbf{Solving for } p$$

We obtain (45, 75), or $f = 45$, $p = 75$.

GRAPHICAL APPROACH

We replace f with x and p with y and solve for y:

$$y + x = 120 \quad \textbf{Solving for } y \textbf{ in equation (1)}$$
$$y = 120 - x$$

and

$$24y + 39x = 3555 \quad \textbf{Solving for } y \textbf{ in equation (2)}$$
$$24y = 3555 - 39x$$
$$y = (3555 - 39x)/24.$$

Since the number of each kind of stamp is between 0 and 120, an appropriate viewing window is [0, 120, 0, 120].

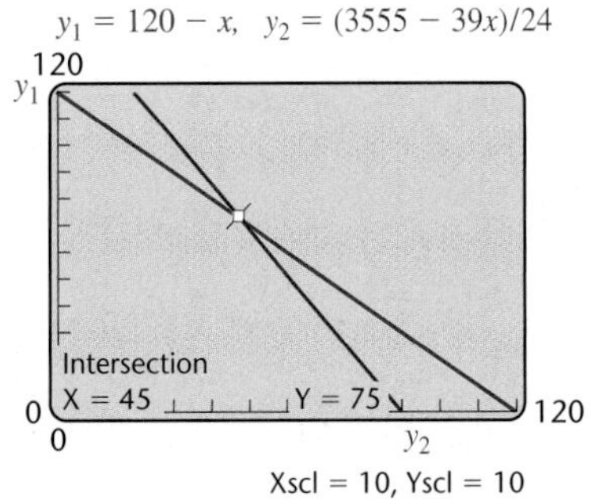

The point of intersection is (45, 75). Since f was replaced with x and p with y, we have $f = 45$, $p = 75$.

Student Notes

It is very important that you clearly label precisely what each variable represents. Not only will this assist you in writing equations, but it will help you to identify and state solutions.

4. **Check.** We check in the original problem. Recall that f is the number of first-class stamps and p the number of postcard stamps.

   Number of stamps: $f + p = 45 + 75 = 120$

   Cost of first-class stamps: $\$0.39f = 0.39 \times 45 = \$17.55$

   Cost of postcard stamps: $\$0.24p = 0.24 \times 75 = \underline{\$18.00}$

   Total $= \$35.55$

   The numbers check.

5. **State.** The art center bought 45 first-class stamps and 75 postcard stamps.

Example 2 involved two types of items (first-class stamps and postcard stamps), the quantity of each type bought, and the total value of the items. We refer to this type of problem as a *total-value problem*.

EXAMPLE 3 Blending Teas. Sonya's House of Tea sells loose Lapsang Souchong tea for 95¢ an ounce and Assam Gingia for \$1.43 an ounce. Sonya wants to make a 20-oz mixture of the two types, called Dragon Blend, that sells for \$1.10 an ounce. How much tea of each type should Sonya use?

SOLUTION

1. **Familiarize.** This problem is similar to Example 2. Rather than postcard stamps and first-class stamps, we have ounces of Assam Gingia and ounces of Lapsang Souchong. Instead of a different price for each type of stamp, we have a different price per ounce for each type of tea. Finally, rather than knowing the total cost of the stamps, we know the weight and the price per ounce of the mixture. Thus we can find the total value of the blend by multiplying 20 ounces times \$1.10, or 110¢ per ounce. Although we could make and check a guess, we proceed to let $l =$ the number of ounces of Lapsang Souchong and $a =$ the number of ounces of Assam Gingia.

2. **Translate.** Since a 20-oz batch is being made, we must have

$$l + a = 20.$$

To find a second equation, note that the total value of the 20-oz blend must match the combined value of the separate ingredients:

Rewording:	The value of the Lapsang Souchong	plus	the value of the Assam Gingia	is	the value of the Dragon Blend.
	↓	↓	↓	↓	↓
Translating:	$l \cdot 95$	$+$	$a \cdot 143$	$=$	$20 \cdot 110$

These equations can also be obtained from a table.

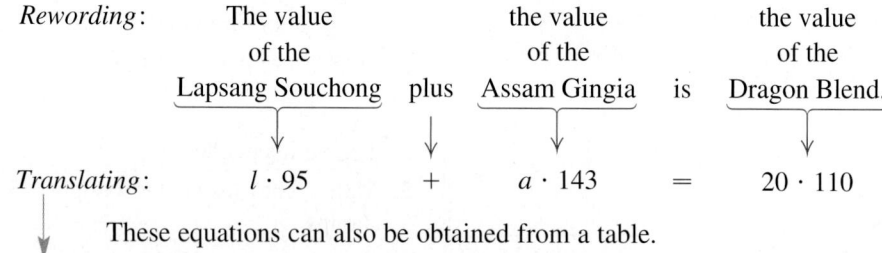

	Lapsang Souchong	Assam Gingia	Dragon Blend	
Number of Ounces	l	a	20	→ $l + a = 20$
Price per Ounce	95¢	143¢	110¢	
Value of Tea	$95l$	$143a$	$20 \cdot 110$, or 2200¢	→ $95l + 143a = 2200$

We have translated to a system of equations:

$$l + \quad a = 20, \qquad (1)$$
$$95l + 143a = 2200. \qquad (2)$$

3. **Carry out.** We solve the system both algebraically and graphically.

ALGEBRAIC APPROACH

We can solve using substitution. When equation (1) is solved for l, we have $l = 20 - a$. Substituting $20 - a$ for l in equation (2), we find a:

$95(20 - a) + 143a = 2200$	**Substituting**
$1900 - 95a + 143a = 2200$	**Using the distributive law**
$48a = 300$	**Combining like terms; subtracting 1900 from both sides**
$a = 6.25.$	**Dividing both sides by 48**

We have $a = 6.25$ and, from equation (1) above, $l + a = 20$. Thus, $l = 13.75$.

GRAPHICAL APPROACH

We replace a with x and l with y and solve for y, which gives us the system of equations

$$y = 20 - x,$$
$$y = (2200 - 143x)/95.$$

We know from the problem that the number of ounces of each kind of tea is between 0 and 20, so we choose the viewing window [0, 20, 0, 20], graph the equations, and find the point of intersection.

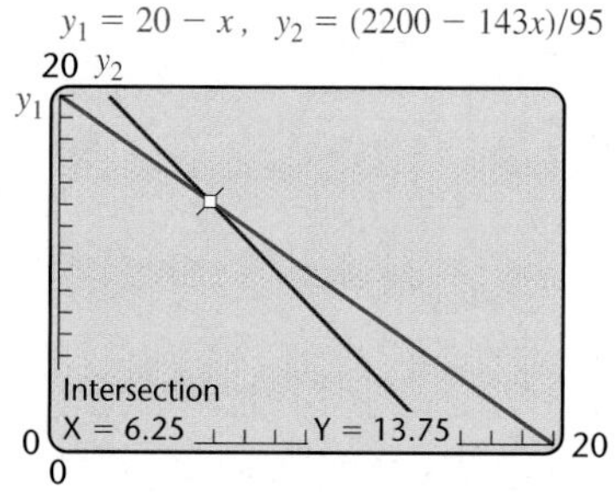

The point of intersection is (6.25, 13.75). Since a was replaced with x and l with y, we have $a = 6.25$, $l = 13.75$.

4. **Check.** If 13.75 oz of Lapsang Souchong and 6.25 oz of Assam Gingia are combined, a 20-oz blend will result. The value of 13.75 oz of Lapsang Souchong is 13.75(95¢) or 1306.25¢. The value of 6.25 oz of Assam Gingia is 6.25(143¢), or 893.75¢. Thus the combined value of the blend is 1306.25¢ + 893.75¢, or 2200¢, which is \$22. A 20-oz blend priced at \$1.10 an ounce would also be worth \$22, so our answer checks.

5. **State.** The Dragon Blend should be made by combining 13.75 oz of Lapsang Souchong with 6.25 oz of Assam Gingia.

EXAMPLE 4 Student Loans. Ranjay's student loans totaled \$9600. Part was a Perkins loan made at 5% interest and the rest was a Stafford loan made at 8% interest. After one year, Ranjay's loans accumulated \$633 in interest. What was the original amount of each loan?

SOLUTION

1. **Familiarize.** We begin with a guess. If \$7000 was borrowed at 5% and \$2600 was borrowed at 8%, the two loans would total \$9600. The interest would then be 0.05(\$7000), or \$350, and 0.08(\$2600), or \$208, for

a total of only \$558 in interest. Our guess was wrong, but checking the guess familiarized us with the problem. More than \$2600 was borrowed at the higher rate.

2. **Translate.** We let $p =$ the amount of the Perkins loan and $f =$ the amount of the Stafford loan. Next, we organize a table in which the entries in each column come from the formula for simple interest:

$$\textit{Principal} \cdot \textit{Rate} \cdot \textit{Time} = \textit{Interest}.$$

	Perkins Loan	Stafford Loan	Total	
Principal	p	f	\$9600	$\longrightarrow p + f = 9600$
Rate of Interest	5%	8%		
Time	1 yr	1 yr		
Interest	$0.05p$	$0.08f$	\$633	$\longrightarrow 0.05p + 0.08f = 633$

The total amount borrowed is found in the first row of the table:

$$p + f = 9600.$$

A second equation, representing the accumulated interest, can be found in the last row:

$$0.05p + 0.08f = 633, \quad \text{or} \quad 5p + 8f = 63{,}300. \qquad \textbf{Clearing decimals}$$

3. **Carry out.** The system can be solved by elimination:

$$\begin{aligned} p + f &= 9600, \\ 5p + 8f &= 63{,}300. \end{aligned} \quad \xrightarrow{\textbf{Multiplying both sides by } -5} \quad \begin{aligned} -5p - 5f &= -48{,}000 \\ 5p + 8f &= 63{,}300 \\ \hline 3f &= 15{,}300 \\ f &= 5100 \end{aligned}$$

$$\begin{aligned} p + f &= 9600 \longleftarrow f = 5100 \\ p + 5100 &= 9600 \\ p &= 4500. \end{aligned}$$

We find that $p = 4500$ and $f = 5100$.

4. **Check.** The total amount borrowed is \$4500 + \$5100, or \$9600. The interest on \$4500 at 5% for 1 yr is 0.05(\$4500), or \$225. The interest on \$5100 at 8% for 1 yr is 0.08(\$5100), or \$408. The total amount of interest is \$225 + \$408, or \$633, so the numbers check.

5. **State.** The Perkins loan was for \$4500 and the Stafford loan was for \$5100.

Problem-Solving Tip

When solving a problem, see if it is patterned or modeled after a problem that you have already solved.

Before proceeding to Example 5, briefly scan Examples 2–4 for similarities. Note that in each case, one of the equations in the system is a simple sum while the other equation represents a sum of products. Example 5 continues this pattern with what is commonly called a *mixture problem*.

EXAMPLE 5 Mixing Fertilizers. Sky Meadow Gardening, Inc., carries two brands of fertilizer containing nitrogen and water. "Gently Green" is 5% nitrogen and "Sun Saver" is 15% nitrogen. Sky Meadow Gardening needs to combine the two types of solutions in order to make 90 L of a solution that is 12% nitrogen. How much of each brand should be used?

SOLUTION

1. **Familiarize.** We make a drawing and then make a guess to gain familiarity with the problem.

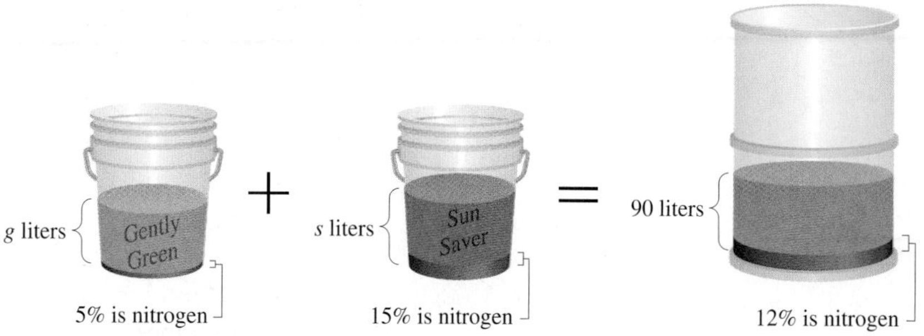

Suppose that 40 L of Gently Green and 50 L of Sun Saver are mixed. The resulting mixture will be the right size, 90 L, but will it be the right strength? To find out, note that 40 L of Gently Green would contribute $0.05(40) = 2$ L of nitrogen to the mixture while 50 L of Sun Saver would contribute $0.15(50) = 7.5$ L of nitrogen to the mixture. The total amount of nitrogen in the mixture would then be $2 + 7.5$, or 9.5 L. But we want 12% of 90, or 10.8 L, to be nitrogen. Thus our guess of 40 L and 50 L is incorrect. Still, checking our guess has familiarized us with the problem.

2. **Translate.** Let g = the number of liters of Gently Green and s = the number of liters of Sun Saver. The information can be organized in a table.

	Gently Green	Sun Saver	Mixture	
Number of Liters	g	s	90	→ $g + s = 90$
Percent of Nitrogen	5%	15%	12%	
Amount of Nitrogen	$0.05g$	$0.15s$	0.12×90, or 10.8 liters	→ $0.05g + 0.15s = 10.8$

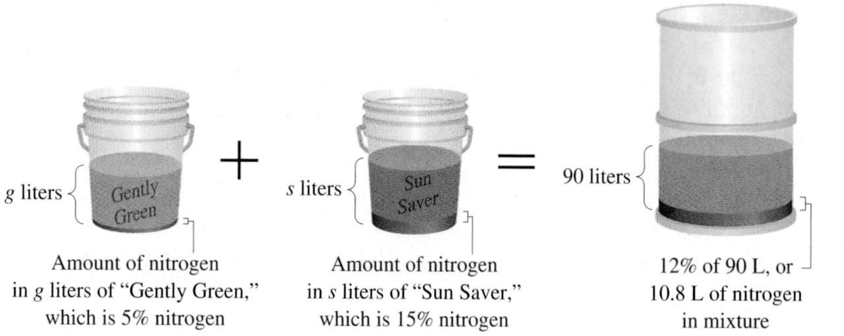

If we add g and s in the first row, we get one equation. It represents the total amount of mixture: $g + s = 90$.

If we add the amounts of nitrogen listed in the third row, we get a second equation. This equation represents the amount of nitrogen in the mixture: $0.05g + 0.15s = 10.8$.

After clearing decimals, we have translated the problem to the system

$$g + s = 90, \quad (1)$$
$$5g + 15s = 1080. \quad (2)$$

3. **Carry out.** We use the elimination method to solve the system:

$-5g - 5s = -450$	**Multiplying both sides of equation (1) by -5**
$5g + 15s = 1080$	
$10s = 630$	**Adding**
$s = 63;$	**Solving for s**
$g + 63 = 90$	**Substituting into equation (1)**
$g = 27.$	**Solving for g**

4. **Check.** Remember, g is the number of liters of Gently Green and s is the number of liters of Sun Saver.

Total amount of mixture: $g + s = 27 + 63 = 90$

Total amount of nitrogen: 5% of $27 + 15\%$ of $63 = 1.35 + 9.45 = 10.8$

Percentage of nitrogen in mixture: $\dfrac{\text{Total amount of nitrogen}}{\text{Total amount of mixture}} = \dfrac{10.8}{90} = 12\%$

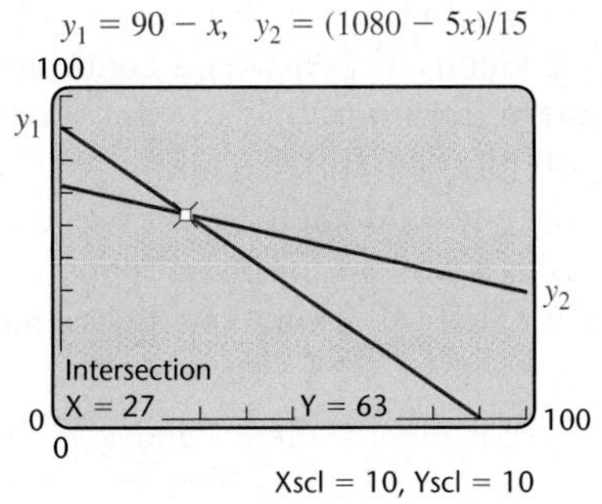

The numbers check in the original problem. We can also check graphically, as shown at left, where x replaces g and y replaces s.

5. **State.** Sky Meadow Gardening should mix 27 L of Gently Green with 63 L of Sun Saver.

Motion Problems

When a problem deals with distance, speed (rate), and time, recall the following.

Distance, Rate, and Time Equations If r represents rate, t represents time, and d represents distance, then

$$d = rt, \quad r = \frac{d}{t}, \quad \text{and} \quad t = \frac{d}{r}.$$

Be sure to remember at least one of these equations. The others can be obtained by multiplying or dividing on both sides as needed.

EXAMPLE 6 Train Travel. A Vermont Railways freight train, loaded with logs, leaves Boston, heading to Washington D.C., at a speed of 60 km/h. Two hours later, an Amtrak® Metroliner leaves Boston, bound for Washington D.C., on a parallel track at 90 km/h. At what point will the Metroliner catch up to the freight train?

SOLUTION

1. **Familiarize.** Let's make a guess—say, 180 km—and check to see if it is correct. The freight train, traveling 60 km/h, would travel 180 km in $\frac{180}{60} = 3$ hr. The Metroliner, traveling 90 km/h, would travel 180 km in $\frac{180}{90} = 2$ hr. Since 3 hr is *not* two hours more than 2 hr, our guess of 180 km is incorrect. Although our guess is wrong, we see that the time that the trains are running and the point at which they meet are both unknown. We let $t =$ the number of hours that the freight train is running before they meet and $d =$ the distance at which the trains meet. Since the freight train has a 2-hr head start, the Metroliner runs for $t - 2$ hours before catching up to the freight train, at which point both trains have traveled the same distance.

2. **Translate.** We can organize the information in a chart. The formula *Distance* = *Rate* · *Time* guides our choice of rows and columns.

	Distance	Rate	Time	
Freight Train	d	60	t	→ $d = 60t$
Metroliner	d	90	$t - 2$	→ $d = 90(t - 2)$

Using *Distance* = *Rate* · *Time* twice, we get two equations:

$$d = 60t, \quad (1)$$
$$d = 90(t - 2). \quad (2)$$

3. **Carry out.** We solve the system both algebraically and graphically.

ALGEBRAIC APPROACH

We solve the system using the substitution method:

$60t = 90(t - 2)$ **Substituting $60t$ for d in equation (2)**

$60t = 90t - 180$ **Using the distributive law**

$-30t = -180$

$t = 6.$

The time for the freight train is 6 hr, which means that the time for the Metroliner is $6 - 2$, or 4 hr. Remember that it is distance, not time, that the problem asked for. Thus for $t = 6$, we have $d = 60 \cdot 6 = 360$ km.

GRAPHICAL APPROACH

The variables used in the equations are d and t. Note that in both equations, d is given in terms of t. If we replace d with y and t with x, we can enter the equations directly. Since $y =$ distance and $x =$ time, we use a viewing window of $[0, 10, 0, 500]$. We graph $y_1 = 60x$ and $y_2 = 90(x - 2)$ and find the point of intersection.

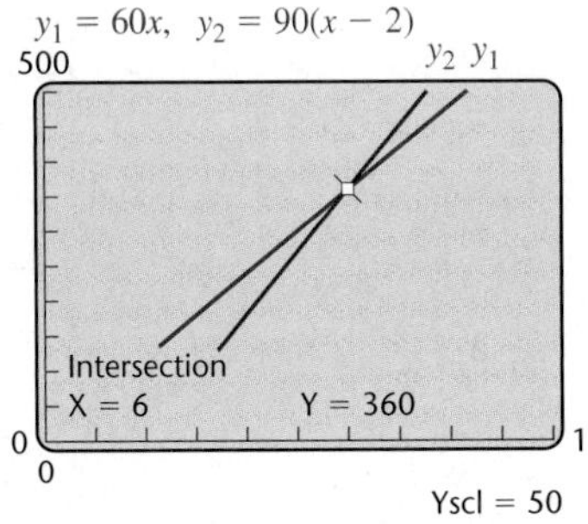

The point of intersection is $(6, 360)$. The problem asks for the distance that the trains travel. Recalling that $y =$ distance, we see that the distance that both trains travel is 360 km.

4. **Check.** At 60 km/h, the freight train will travel $60 \cdot 6$, or 360 km, in 6 hr. At 90 km/h, the Metroliner will travel $90 \cdot (6 - 2) = 360$ km in 4 hr. The numbers check.

5. **State.** The freight train will catch up to the Metroliner at a point 360 km from Boston.

EXAMPLE 7 Jet Travel. A Boeing 747-400 jet flies 4 hr west with a 60-mph tailwind. Returning *against* the wind takes 5 hr. Find the speed of the plane with no wind.

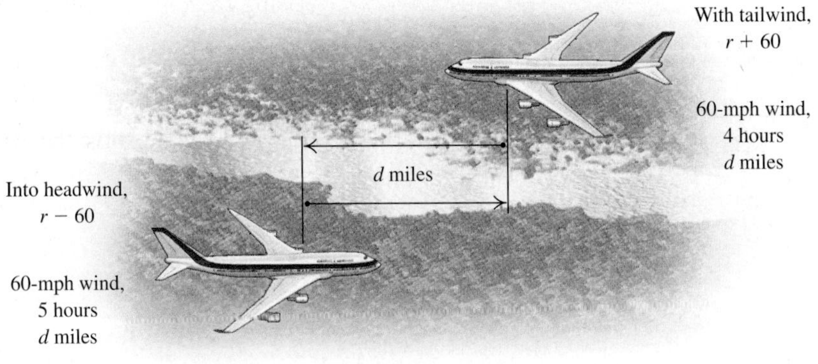

SOLUTION

1. **Familiarize.** We imagine the situation and make a drawing. Note that the wind *speeds up* the jet on the outbound flight, but *slows down* the jet on the return flight. Since the distances traveled each way must be the same, we can check a guess of the jet's speed with no wind. Suppose the speed of the jet with no wind is 400 mph. The jet would then fly $400 + 60 = 460$ mph with the wind and $400 - 60 = 340$ mph into the wind. In 4 hr, the jet would travel $460 \cdot 4 = 1840$ mi with the wind and $340 \cdot 5 = 1700$ mi against the wind. Since $1840 \neq 1700$, our guess of 400 mph is incorrect. Rather than guess again, let's have $r =$ the speed, in miles per hour, of the jet in still air. Then $r + 60 =$ the jet's speed with the wind and $r - 60 =$ the jet's speed against the wind. We also let $d =$ the distance traveled, in miles.

2. **Translate.** The information can be organized in a chart. The distances traveled are the same, so we use *Distance* = *Rate* (or *Speed*) · *Time*. Each row of the chart gives an equation.

	Distance	Rate	Time	
With Wind	d	$r + 60$	4	→ $d = (r + 60)4$
Against Wind	d	$r - 60$	5	→ $d = (r - 60)5$

The two equations constitute a system:

$$d = (r + 60)4, \quad (1)$$
$$d = (r - 60)5. \quad (2)$$

3. **Carry out.** We solve the system using substitution:

$(r - 60)5 = (r + 60)4$ **Substituting $(r - 60)5$ for d in equation (1)**

$5r - 300 = 4r + 240$ **Using the distributive law**

$r = 540.$ **Solving for r**

4. **Check.** When $r = 540$, the speed with the wind is $540 + 60 = 600$ mph, and the speed against the wind is $540 - 60 = 480$ mph. The distance with the wind, $600 \cdot 4 = 2400$ mi, matches the distance into the wind, $480 \cdot 5 = 2400$ mi, so we have a check.

5. **State.** The speed of the jet with no wind is 540 mph.

Tips for Solving Motion Problems

1. Draw a diagram using an arrow or arrows to represent distance and the direction of each object in motion.
2. Organize the information in a chart.
3. Look for times, distances, or rates that are the same. These often can lead to an equation.
4. Translating to a system of equations allows for the use of two variables.
5. Always make sure that you have answered the question asked.

3.3 EXERCISE SET

FOR EXTRA HELP

MathXL | MyMathLab | InterAct Math | AW Math Tutor Center | Video Lectures on CD: Disc 2 | Student's Solutions Manual

1.–14. For Exercises 1–14, solve Exercises 47–60 from pp. 194–195.

15. *Printing.* King Street Printing recently charged 1.9¢ per sheet of paper, but 2.4¢ per sheet for paper made of recycled fibers. Darren's bill for 150 sheets of paper was $3.41. How many sheets of each type were used?

16. *Photocopying.* Quick Copy recently charged 6¢ a page for copying pages that can be machine-fed and 18¢ a page for copying pages that must be hand-placed on the copier. If Lea's bill for 90 copies was $9.24, how many copies of each type were made?

17. *Lighting.* Booth Bros. Hardware charges $7.50 for a General Electric Biax Energy Saver light bulb and $5 for an SLi Lighting Cool White Energy Saver bulb. If Paul County Hospital purchased 200 such bulbs for $1150, how many of each type did they purchase?

18. *Office Supplies.* Barlow's Office Supply charges $16.75 for a box of Erase-A-Gel™ pens and $14.25 for a box of Icy™ automatic pencils. If Letsonville Community College purchased 120 such boxes for $1790, how many boxes of each type did they purchase?

19. *Sales.* Staples® recently sold a black Epson Stylus R340 ink cartridge for $17.35 and a black HP Officejet Pro K550 cartridge for $19.99. At the start of a recent fall semester, a combination of 50 of these cartridges was sold for a total of $907.10. How many of each type were purchased?

20. *Sales.* Staples® recently sold a wirebound graph-paper notebook for $2.50 and a college-ruled notebook made of recycled paper for $2.30. At the start of a recent spring semester, a combination of 50 of these notebooks was sold for a total of $118.60. How many of each type were sold?

21. *Blending Coffees.* The Bean Counter charges $9.00 per pound for Kenyan French Roast coffee and $8.00 per pound for Sumatran coffee. How much of each type should be used to make a 20-lb blend that sells for $8.40 per pound?

22. *Mixed Nuts.* Oh Nuts! sells cashews for $6.75 per pound and Brazil nuts for $5.00 per pound. How much of each type should be used to make a 50-lb mixture that sells for $5.70 per pound?

23. *Catering.* Casella's Catering is planning a wedding reception. The bride and groom would like to serve a nut mixture containing 25% peanuts. Casella has available mixtures that are either 40% or 10% peanuts. How much of each type should be mixed to get a 20-lb mixture that is 25% peanuts?

24. *Ink Remover.* Etch Clean Graphics uses one cleanser that is 25% acid and a second that is 50% acid. How many liters of each should be mixed to get 30 L of a solution that is 40% acid?

25. *Blending Granola.* Deep Thought Granola is 25% nuts and dried fruit. Oat Dream Granola is 10% nuts and dried fruit. How much of Deep Thought and how much of Oat Dream should be mixed to form a 20-lb batch of granola that is 19% nuts and dried fruit?

26. *Livestock Feed.* Soybean meal is 16% protein and corn meal is 9% protein. How many pounds of each should be mixed to get a 350-lb mixture that is 12% protein?

27. *Student Loans.* Lomasi's two student loans totaled $12,000. One of her loans was at 6% simple interest and the other at 9%. After one year, Lomasi owed $855 in interest. What was the amount of each loan?

28. *Investments.* An executive nearing retirement made two investments totaling $15,000. In one year, these investments yielded $1432 in simple interest. Part of the money was invested at 9% and the rest at 10%. How much was invested at each rate?

29. *Automotive Maintenance.* "Arctic Antifreeze" is 18% alcohol and "Frost No-More" is 10% alcohol. How many liters of each should be mixed to get 20 L of a mixture that is 15% alcohol?

30. *Chemistry.* E-Chem Testing has a solution that is 80% base and another that is 30% base. A technician needs 150 L of a solution that is 62% base. The 150 L will be prepared by mixing the two solutions on hand. How much of each should be used?

31. *Octane Ratings.* The octane rating of a gasoline is a measure of the amount of isooctane in the gas. The 2002 Dodge Neon RT requires 91-octane gasoline (*Sources*: Champlain Electric and Petroleum Equipment; Goss Dodge). How much 87-octane gas and 93-octane gas should Kasey mix in order to make 12 gal of 91-octane gas for her Neon RT?

32. *Octane Ratings.* The octane rating of a gasoline is a measure of the amount of isooctane in the gas. The 2005 Chrysler Crossfire requires 93-octane gasoline (*Sources*: Champlain Electric and Petroleum Equipment; Freedom Chrysler Plymouth). How much 87-octane gas and 95-octane gas should Ken mix in order to make 10 gal of 93-octane gas for his Crossfire?

33. *Food Science.* The following bar graph shows the milk fat percentages in three dairy products. How many pounds each of whole milk and cream should be mixed to form 200 lb of milk for cream cheese?

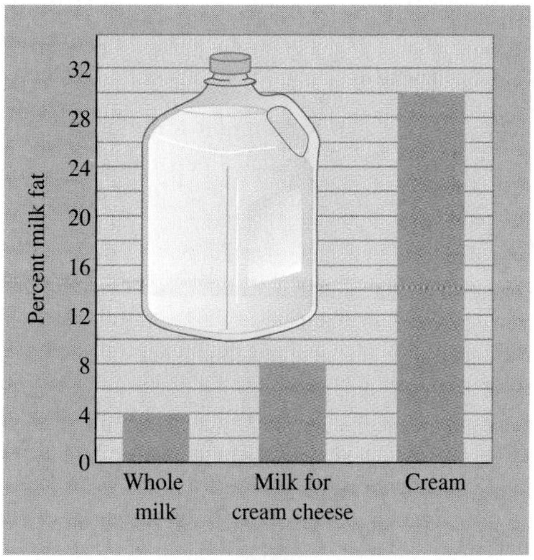

34. *Food Science.* How much lowfat milk (1% fat) and how much whole milk (4% fat) should be mixed to make 5 gal of reduced fat milk (2% fat)?

35. *Train Travel.* A train leaves Danville Junction and travels north at a speed of 75 km/h. Two hours later, an express train leaves on a parallel track and travels north at 125 km/h. How far from the station will they meet?

36. *Car Travel.* Two cars leave Salt Lake City, traveling in opposite directions. One car travels at a speed of 80 km/h and the other at 96 km/h. In how many hours will they be 528 km apart?

37. *Boating.* Mia's motorboat took 3 hr to make a trip downstream with a 6-mph current. The return trip against the same current took 5 hr. Find the speed of the boat in still water.

38. *Canoeing.* Alvin paddled for 4 hr with a 6-km/h current to reach a campsite. The return trip against the same current took 10 hr. Find the speed of Alvin's canoe in still water.

39. *Point of No Return.* A plane flying the 3458-mi trip from New York City to London has a 50-mph tailwind. The flight's *point of no return* is the point at which the flight time required to return to New York is the same as the time required to continue to London. If the speed of the plane in still air is 360 mph, how far is New York from the point of no return?

40. *Point of No Return.* A plane is flying the 2553-mi trip from Los Angeles to Honolulu into a 60-mph headwind. If the speed of the plane in still air is 310 mph, how far from Los Angeles is the plane's point of no return? (See Exercise 39.)

41. *Architecture.* The rectangular ground floor of the John Hancock building has a perimeter of 860 ft. The length is 100 ft more than the width. Find the length and the width.

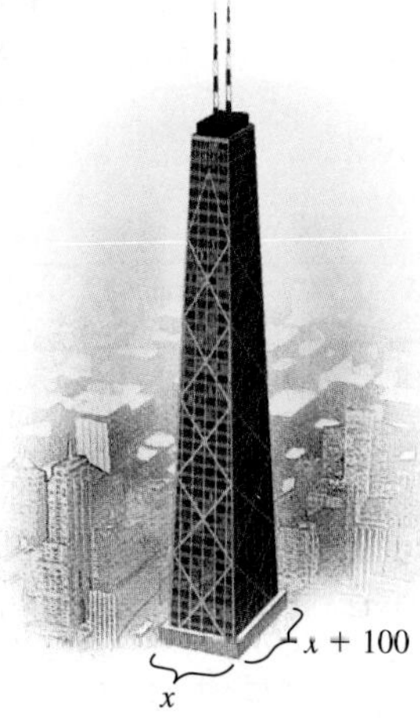
x + 100
x

42. *Real Estate.* The perimeter of a rectangular oceanfront lot is 190 m. The width is one-fourth of the length. Find the dimensions.

43. *Real Estate.* In 1996, the Simon Property Group and the DeBartolo Realty Corporation merged to form the largest real estate company in the United States, owning 183 shopping centers in 32 states. Prior to merging, Simon owned twice as many properties as DeBartolo. How many properties did each company own before the merger?

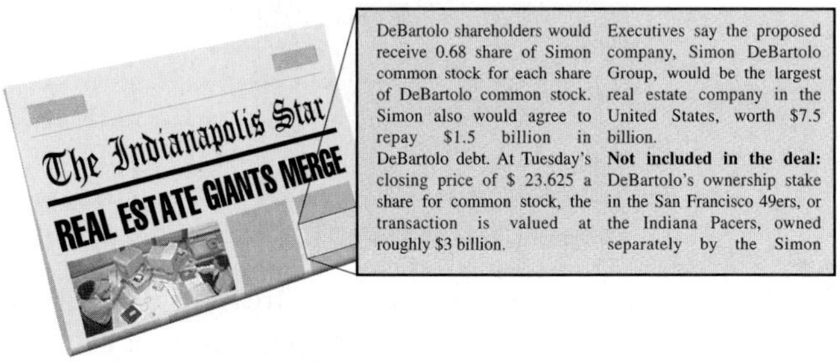
The Indianapolis Star
REAL ESTATE GIANTS MERGE
DeBartolo shareholders would receive 0.68 share of Simon common stock for each share of DeBartolo common stock. Simon also would agree to repay $1.5 billion in DeBartolo debt. At Tuesday's closing price of $ 23.625 a share for common stock, the transaction is valued at roughly $3 billion.
Executives say the proposed company, Simon DeBartolo Group, would be the largest real estate company in the United States, worth $7.5 billion.
Not included in the deal: DeBartolo's ownership stake in the San Francisco 49ers, or the Indiana Pacers, owned separately by the Simon

44. *Hockey Rankings.* Hockey teams receive 2 points for a win and 1 point for a tie. The Wildcats once won a championship with 60 points. They won 9 more games than they tied. How many wins and how many ties did the Wildcats have?

45. *Radio Airplay.* Roscoe must play 12 commercials during his 1-hr radio show. Each commercial is either 30 sec or 60 sec long. If the total commercial time during that hour is 10 min, how many commercials of each type does Roscoe play?

46. *DVD Rentals.* J. P.'s Video rents general-interest films for $3.00 each and children's films for $1.50 each. In one day, a total of $213 was taken in from the rental of 77 DVDs. How many of each type of DVD was rented?

47. *Making Change.* Cecilia makes a $9.25 purchase at the bookstore with a $20 bill. The store has no bills and gives her the change in quarters and fifty-cent pieces. There are 30 coins in all. How many of each kind are there?

48. *Teller Work.* Ashford goes to a bank and gets change for a $50 bill consisting of all $5 bills and $1 bills. There are 22 bills in all. How many of each kind are there?

TW **49.** In what ways are Examples 3 and 4 similar? In what sense are their systems of equations similar?

TW **50.** Write at least three study tips of your own for someone beginning this exercise set.

Skill Maintenance

Evaluate.

51. $2x - 3y + 12$, for $x = 5$ and $y = 2$ [1.1]

52. $7x - 4y + 9$, for $x = 2$ and $y = 3$ [1.1]

53. $5a - 7b + 3c$, for $a = -2$, $b = 3$, and $c = 1$ [1.1], [1.2]

54. $3a - 8b - 2c$, for $a = -4$, $b = -1$, and $c = 3$ [1.1], [1.2]

55. $4 - 2y + 3z$, for $y = \frac{1}{3}$ and $z = \frac{1}{4}$ [1.1]

56. $3 - 5y + 4z$, for $y = \frac{1}{2}$ and $z = \frac{1}{5}$ [1.1]

Synthesis

TW **57.** Suppose that in Example 3 you are asked only for the amount of Assam Gingia needed for the Dragon Blend. Would the method of solving the problem change? Why or why not?

TW **58.** Write a problem similar to Example 2 for a classmate to solve. Design the problem so that the solution is "The florist sold 14 hanging plants and 9 flats of petunias."

59. *Recycled Paper.* Unable to purchase 60 reams of paper that contains 20% post-consumer fiber, the Naylor School bought paper that was either 0% post-consumer fiber or 30% post-consumer fiber. How many reams of each should be purchased in order to use the same amount of post-consumer fiber as if the 20% post-consumer fiber paper were available?

60. *Retail.* Some of the world's best and most expensive coffee is Hawaii's Kona coffee. In order for coffee to be labeled "Kona Blend," it must contain at least 30% Kona beans. Bean Town Roasters has 40 lb of Mexican coffee. How much Kona coffee must they add if they wish to market it as Kona Blend?

61. *Automotive Maintenance.* The radiator in Michelle's car contains 6.3 L of antifreeze and water. This mixture is 30% antifreeze. How much of this mixture should she drain and replace with pure antifreeze so that there will be a mixture of 50% antifreeze?

62. *Exercise.* Natalie jogs and walks to school each day. She averages 4 km/h walking and 8 km/h jogging. From home to school is 6 km and Natalie makes the trip in 1 hr. How far does she jog in a trip?

63. *Book Sales.* A limited edition of a book published by a historical society was offered for sale to members. The cost was one book for $12 or two books for $20 (maximum of two per member). The society sold 880 books, for a total of $9840. How many members ordered two books?

64. The tens digit of a two-digit positive integer is 2 more than three times the units digit. If the digits are interchanged, the new number is 13 less than half the given number. Find the given integer. (*Hint*: Let $x =$ the tens-place digit and $y =$ the units-place digit; then $10x + y$ is the number.)

65. *Wood Stains.* Williams' Custom Flooring has 0.5 gal of stain that is 20% brown and 80% neutral. A customer orders 1.5 gal of a stain that is 60% brown and 40% neutral. How much pure brown stain and how much neutral stain should be added to the original 0.5 gal in order to make up the order?*

66. *Train Travel.* A train leaves Union Station for Central Station, 216 km away, at 9 A.M. One hour later, a train leaves Central Station for Union Station. They meet at noon. If the second train had started at 9 A.M. and the first train at 10:30 A.M., they would still have met at noon. Find the speed of each train.

67. *Fuel Economy.* Grady's station wagon gets 18 miles per gallon (mpg) in city driving and 24 mpg in highway driving. The car is driven 465 mi on 23 gal of gasoline. How many miles were driven in the city and how many were driven on the highway?

*This problem was suggested by Professor Chris Burditt of Yountville, California.

68. *Biochemistry.* Industrial biochemists routinely use a machine to mix a buffer of 10% acetone by adding 100% acetone to water. One day, instead of adding 5 L of acetone to create a vat of buffer, a machine added 10 L. How much additional water was needed to bring the concentration down to 10%?

69. See Exercise 65 above. Let $x =$ the amount of pure brown stain added to the original 0.5 gal. Find a function $P(x)$ that can be used to determine the percentage of brown stain in the 1.5-gal mixture. On a graphing calculator, draw the graph of P and use INTERSECT to confirm the answer to Exercise 65.

70. *Gender.* Phil and Phyllis are siblings. Phyllis has twice as many brothers as she has sisters. Phil has the same number of brothers as sisters. How many girls and how many boys are in the family?

3.4 Systems of Equations in Three Variables

Identifying Solutions ■ Solving Systems in Three Variables ■ Dependency, Inconsistency, and Geometric Considerations

Some problems translate directly to two equations. Others call for a translation to three or more equations. Here we learn how to solve systems of three linear equations. Later, we will use such systems in problem-solving situations.

Identifying Solutions

A **linear equation in three variables** is an equation equivalent to one in the form $Ax + By + Cz = D$, where A, B, C, and D are real numbers. We refer to the form $Ax + By + Cz = D$ as *standard form* for a linear equation in three variables.

A solution of a system of three equations in three variables is an ordered triple (x, y, z) that makes *all three* equations true.

EXAMPLE 1 Determine whether $\left(\frac{3}{2}, -4, 3\right)$ is a solution of the system

$$\begin{aligned} 4x - 2y - 3z &= 5, \\ -8x - y + z &= -5, \\ 2x + y + 2z &= 5. \end{aligned}$$

SOLUTION We substitute $\left(\frac{3}{2}, -4, 3\right)$ into the three equations, using alphabetical order.

BY HAND

We have the following:

$$\begin{array}{r|l}
\multicolumn{2}{c}{4x - 2y - 3z = 5} \\
\hline
4 \cdot \frac{3}{2} - 2(-4) - 3 \cdot 3 & 5 \\
6 + 8 - 9 & \\
\multicolumn{2}{r}{5 \stackrel{?}{=} 5 \quad \text{TRUE}}
\end{array}$$

$$\begin{array}{r|l}
\multicolumn{2}{c}{-8x - y + z = -5} \\
\hline
-8 \cdot \frac{3}{2} - (-4) + 3 & -5 \\
-12 + 4 + 3 & \\
\multicolumn{2}{r}{-5 \stackrel{?}{=} -5 \quad \text{TRUE}}
\end{array}$$

$$\begin{array}{r|l}
\multicolumn{2}{c}{2x + y + 2z = 5} \\
\hline
2 \cdot \frac{3}{2} + (-4) + 2 \cdot 3 & 5 \\
3 - 4 + 6 & \\
\multicolumn{2}{r}{5 \stackrel{?}{=} 5 \quad \text{TRUE}}
\end{array}$$

The triple makes all three equations true, so it is a solution.

USING A GRAPHING CALCULATOR

We store $\frac{3}{2}$ as X, -4 as Y, and 3 as Z, using the ALPHA key to enter Y and Z.

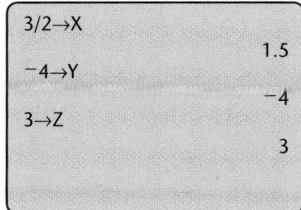

```
3/2→X
                 1.5
-4→Y
                  -4
3→Z
                   3
```

Now we enter the expression on the left side of each equation and press ENTER. If the value is the same as the right side of the equation, the ordered triple makes the equation true.

```
4X-2Y-3Z
                   5
-8X-Y+Z
                  -5
2X+Y+2Z
                   5
```

The triple makes all three equations true, so it is a solution.

Solving Systems in Three Variables

Graphical methods for solving linear equations in three variables are problematic, because a three-dimensional coordinate system is required and the graph of a linear equation in three variables is a plane. The substitution method *can* be used but becomes very cumbersome unless one or more of the equations has only two variables. Fortunately, the elimination method allows us to manipulate a system of three equations in three variables so that a simpler system of two equations in two variables is formed. Once that simpler system has been solved, we can substitute into one of the three original equations and solve for the third variable.

EXAMPLE 2 Solve the following system of equations:

$$\begin{aligned}
x + y + z &= 4, & (1)\\
x - 2y - z &= 1, & (2)\\
2x - y - 2z &= -1. & (3)
\end{aligned}$$

SOLUTION We select *any* two of the three equations and work to get one equation in two variables. Let's add equations (1) and (2):

$$\begin{array}{rl} x + y + z = 4 & (1) \\ x - 2y - z = 1 & (2) \\ \hline 2x - y \quad = 5. & (4) \quad \textbf{Adding to eliminate } z \end{array}$$

Next, we select a different pair of equations and eliminate the *same variable* we did above. Let's use equations (1) and (3) to again eliminate z. Be careful here! A common error is to eliminate a different variable in this step.

$$\begin{array}{l} x + y + z = 4, \\ 2x - y - 2z = -1 \end{array} \xrightarrow[\textbf{equation (1) by 2}]{\textbf{Multiplying both sides of}} \begin{array}{rl} 2x + 2y + 2z = 8 & \\ 2x - y - 2z = -1 & \\ \hline 4x + y \quad = 7 & (5) \end{array}$$

Now we solve the resulting system of equations (4) and (5). That solution will give us two of the numbers in the solution of the original system.

$$\begin{array}{rll} 2x - y = 5 & (4) & \\ 4x + y = 7 & (5) & \\ \hline 6x \quad = 12 & \textbf{Adding} & \\ x = 2 & & \end{array}$$

Note that we now have two equations in two variables. Had we not eliminated the same variable in both of the above steps, this would not be the case.

We can use either equation (4) or (5) to find y. We choose equation (5):

$$\begin{array}{rl} 4x + y = 7 & (5) \\ 4 \cdot 2 + y = 7 & \textbf{Substituting 2 for } x \textbf{ in equation (5)} \\ 8 + y = 7 & \\ y = -1. & \end{array}$$

We now have $x = 2$ and $y = -1$. To find the value for z, we use any of the original three equations and substitute to find the third number, z. Let's use equation (1) and substitute our two numbers in it:

$$\begin{array}{rl} x + y + z = 4 & (1) \\ 2 + (-1) + z = 4 & \textbf{Substituting 2 for } x \textbf{ and } -1 \textbf{ for } y \\ 1 + z = 4 & \\ z = 3. & \end{array}$$

We have obtained the triple $(2, -1, 3)$. It should check in *all three* equations:

$$\begin{array}{c|c} x + y + z & 4 \\ \hline 2 + (-1) + 3 & 4 \\ 4 \stackrel{?}{=} 4 & \text{TRUE} \end{array} \qquad \begin{array}{c|c} x - 2y - z & 1 \\ \hline 2 - 2(-1) - 3 & 1 \\ 1 \stackrel{?}{=} 1 & \text{TRUE} \end{array} \qquad \begin{array}{c|c} 2x - y - 2z & -1 \\ \hline 2 \cdot 2 - (-1) - 2 \cdot 3 & -1 \\ -1 \stackrel{?}{=} -1 & \text{TRUE} \end{array}$$

The solution is $(2, -1, 3)$.

Study Tip

Helping Others Will Help You Too

When you feel confident in your command of a topic, don't hesitate to help classmates experiencing trouble. Your understanding and retention of a concept will deepen when you explain it to someone else and your classmate will appreciate your help.

Solving Systems of Three Linear Equations

To use the elimination method to solve systems of three linear equations:

1. Write all equations in the standard form $Ax + By + Cz = D$.
2. Clear any decimals or fractions.
3. Choose a variable to eliminate. Then select two of the three equations and work to get one equation in which the selected variable is eliminated.
4. Next, use a different pair of equations and eliminate the same variable that you did in step (3).
5. Solve the system of equations that resulted from steps (3) and (4).
6. Substitute the solution from step (5) into one of the original three equations and solve for the third variable. Then check.

Student Notes

Because solving systems of three equations can be lengthy, it is important that you use plenty of paper, work in pencil, and double-check each step as you proceed.

EXAMPLE 3 Solve the system

$$\begin{aligned} 4x - 2y - 3z &= 5, \quad &(1)\\ -8x - y + z &= -5, \quad &(2)\\ 2x + y + 2z &= 5. \quad &(3) \end{aligned}$$

SOLUTION

1., 2. The equations are already in standard form with no fractions or decimals.

3. Next, select a variable to eliminate. We decide on y because the y-terms are opposites of each other in equations (2) and (3). We add:

$$\begin{aligned} -8x - y + z &= -5 \quad &(2)\\ 2x + y + 2z &= 5 \quad &(3)\\ \hline -6x \qquad + 3z &= 0. \quad &(4) \quad \textbf{Adding} \end{aligned}$$

4. We use another pair of equations to create a second equation in x and z. That is, we eliminate the same variable, y, as in step (3). We use equations (1) and (3):

$$\begin{aligned} 4x - 2y - 3z &= 5, \\ 2x + y + 2z &= 5 \end{aligned} \quad \xrightarrow[\textbf{of equation (3) by 2}]{\textbf{Multiplying both sides}} \quad \begin{aligned} 4x - 2y - 3z &= 5\\ 4x + 2y + 4z &= 10\\ \hline 8x \qquad + z &= 15. \quad (5) \end{aligned}$$

5. Now we solve the resulting system of equations (4) and (5). That allows us to find two parts of the ordered triple.

$$\begin{aligned} -6x + 3z &= 0, \\ 8x + z &= 15 \end{aligned} \quad \xrightarrow[\textbf{of equation (5) by } -3]{\textbf{Multiplying both sides}} \quad \begin{aligned} -6x + 3z &= 0\\ -24x - 3z &= -45\\ \hline -30x \qquad &= -45\\ x = \tfrac{-45}{-30} &= \tfrac{3}{2} \end{aligned}$$

We use equation (5) to find z:

$$\begin{aligned} 8x + z &= 15 && (5) \\ 8 \cdot \tfrac{3}{2} + z &= 15 && \textbf{Substituting } \tfrac{3}{2} \textbf{ for } x \\ 12 + z &= 15 \\ z &= 3. \end{aligned}$$

6. Finally, we use any of the original equations and substitute to find the third number, y. We choose equation (3):

$$\begin{aligned} 2x + y + 2z &= 5 && (3) \\ 2 \cdot \tfrac{3}{2} + y + 2 \cdot 3 &= 5 && \textbf{Substituting } \tfrac{3}{2} \textbf{ for } x \textbf{ and 3 for } z \\ 3 + y + 6 &= 5 \\ y + 9 &= 5 \\ y &= -4. \end{aligned}$$

The solution is $\left(\frac{3}{2}, -4, 3\right)$. The check was performed as Example 1.

Sometimes, certain variables are missing at the outset.

EXAMPLE 4 Solve the system

$$\begin{aligned} x + y + z &= 180, && (1) \\ x \quad\quad - z &= -70, && (2) \\ 2y - z &= 0. && (3) \end{aligned}$$

SOLUTION

1., 2. The equations appear in standard form with no fractions or decimals.

3., 4. Note that there is no x-term in equation (3) and no y-term in equation (2). It will save some steps if we eliminate one of these variables; we choose to eliminate y. Since, at the outset, we already have one equation with no y, we need only one more equation with y eliminated. We use equations (1) and (3):

$$\begin{aligned} x + y + z &= 180, \\ 2y - z &= 0 \end{aligned} \quad \xrightarrow[\textbf{of equation (1) by } -2]{\textbf{Multiplying both sides}} \quad \begin{aligned} -2x - 2y - 2z &= -360 \\ 2y - z &= 0 \\ \hline -2x \quad\quad - 3z &= -360. \quad (4) \end{aligned}$$

5., 6. Now we solve the resulting system of equations (2) and (4):

$$\begin{aligned} x - z &= -70, \\ -2x - 3z &= -360 \end{aligned} \quad \xrightarrow[\textbf{of equation (2) by 2}]{\textbf{Multiplying both sides}} \quad \begin{aligned} 2x - 2z &= -140 \\ -2x - 3z &= -360 \\ \hline -5z &= -500 \\ z &= 100. \end{aligned}$$

Continuing as in Examples 2 and 3, we get the solution $(30, 50, 100)$. The check is left to the student.

Dependency, Inconsistency, and Geometric Considerations

Each equation in Examples 2, 3, and 4 has a graph that is a plane in three dimensions. The solutions are points common to the planes of each system. Since three planes can have an infinite number of points in common or no points at all in common, we need to generalize the concept of *consistency*.

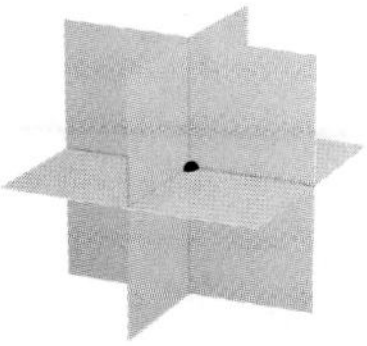

Planes intersect at one point. System is *consistent* and has one solution.

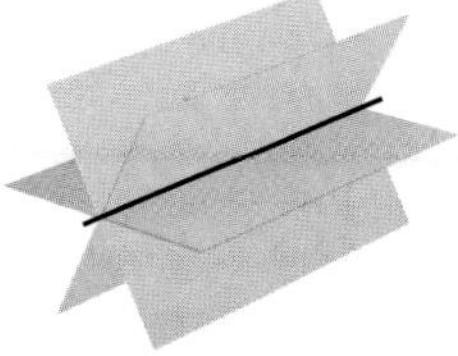

Planes intersect along a common line. System is *consistent* and has an infinite number of solutions.

Three parallel planes. System is *inconsistent;* it has no solution.

Planes intersect two at a time, with no point common to all three. System is *inconsistent;* it has no solution.

Consistency

A system of equations that has at least one solution is said to be **consistent.**

A system of equations that has no solution is said to be **inconsistent.**

EXAMPLE 5 Solve:

$$\begin{aligned} y + 3z &= 4, \quad &(1)\\ -x - y + 2z &= 0, \quad &(2)\\ x + 2y + z &= 1. \quad &(3) \end{aligned}$$

SOLUTION The variable x is missing in equation (1). By adding equations (2) and (3), we can find a second equation in which x is missing:

$$\begin{aligned} -x - y + 2z &= 0 \quad &(2)\\ x + 2y + z &= 1 \quad &(3)\\ \hline y + 3z &= 1. \quad &(4) \quad \textbf{Adding} \end{aligned}$$

Equations (1) and (4) form a system in y and z. We solve as before:

$$\begin{aligned} y + 3z &= 4, \xrightarrow[\textbf{of equation (1) by } -1]{\textbf{Multiplying both sides}} & -y - 3z &= -4\\ y + 3z &= 1 & y + 3z &= 1\\ & \textbf{This is a contraction.} \longrightarrow & 0 &= -3. \quad \textbf{Adding} \end{aligned}$$

Since we end up with a *false* equation, or contradiction, we know that the system has no solution. It is *inconsistent.*

The notion of *dependency* from Section 3.1 can also be extended.

EXAMPLE 6 Solve:

$$\begin{aligned} 2x + y + z &= 3, &\quad (1)\\ x - 2y - z &= 1, &\quad (2)\\ 3x + 4y + 3z &= 5. &\quad (3) \end{aligned}$$

SOLUTION Our plan is to first use equations (1) and (2) to eliminate z. Then we will select another pair of equations and again eliminate z:

$$\begin{array}{l} 2x + y + z = 3\\ \underline{x - 2y - z = 1}\\ 3x - y \qquad = 4. \qquad (4) \end{array}$$

Next, we use equations (2) and (3) to eliminate z again:

$$\begin{array}{l} x - 2y - z = 1,\\ 3x + 4y + 3z = 5 \end{array} \xrightarrow[\textbf{of equation (2) by 3}]{\textbf{Multiplying both sides}} \begin{array}{l} 3x - 6y - 3z = 3\\ \underline{3x + 4y + 3z = 5}\\ 6x - 2y \qquad = 8. \qquad (5) \end{array}$$

We now try to solve the resulting system of equations (4) and (5):

$$\begin{array}{l} 3x - y = 4,\\ 6x - 2y = 8 \end{array} \xrightarrow[\textbf{of equation (4) by } -2]{\textbf{Multiplying both sides}} \begin{array}{r} -6x + 2y = -8\\ \underline{6x - 2y = 8}\\ 0 = 0. \qquad (6) \end{array}$$

Equation (6), which is an identity, indicates that equations (1), (2), and (3) are *dependent*. This means that the original system of three equations is equivalent to a system of two equations. Removing equation (3) from the system does not affect the solution of the system.* In writing an answer to this problem, we simply state that "the equations are dependent."

Recall that when dependent equations appeared in Section 3.1, the solution sets were always infinite in size and were written in set-builder notation. There, all systems of dependent equations were *consistent*. This is not always the case for systems of three or more equations. The following figures illustrate some possibilities geometrically.

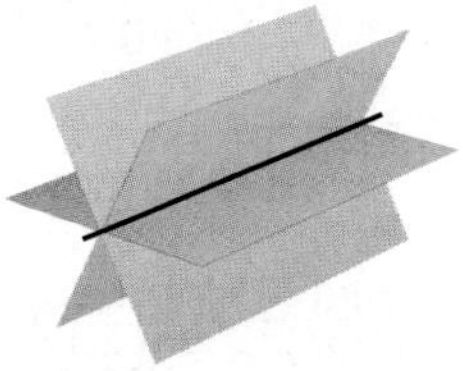

The planes intersect along a common line. The equations are dependent and the system is consistent. There is an infinite number of solutions.

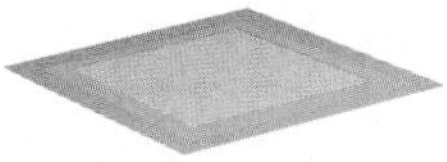

The planes coincide. The equations are dependent and the system is consistent. There is an infinite number of solutions.

Two planes coincide. The third plane is parallel. The equations are dependent and the system is inconsistent. There is no solution.

*A set of equations is dependent if at least one equation can be expressed as a sum of multiples of other equations in that set.

3.4 EXERCISE SET

FOR EXTRA HELP

MathXL | MyMathLab | InterAct Math | AW Math Tutor Center | Video Lectures on CD: Disc 2 | Student's Solutions Manual

Concept Reinforcement *Classify each statement as either true or false.*

1. $3x + 5y + 4z = 7$ is a linear equation in three variables.

2. It is not difficult to solve a system of three equations in three unknowns by graphing.

3. Every system of three equations in three unknowns has at least one solution.

4. If, when we are solving a system of three equations, a false equation results from adding a multiple of one equation to another, the system is inconsistent.

5. If, when we are solving a system of three equations, an identity results from adding a multiple of one equation to another, the equations are dependent.

6. Whenever a system of three equations contains dependent equations, there is an infinite number of solutions.

7. Determine whether $(2, -1, -2)$ is a solution of the system

$$\begin{aligned} x + y - 2z &= 5, \\ 2x - y - z &= 7, \\ -x - 2y + 3z &= 6. \end{aligned}$$

8. Determine whether $(1, -2, 3)$ is a solution of the system

$$\begin{aligned} x + y + z &= 2, \\ x - 2y - z &= 2, \\ 3x + 2y + z &= 2. \end{aligned}$$

Solve each system. If a system's equations are dependent or if there is no solution, state this.

9. $$\begin{aligned} 2x - y + z &= 10, \\ 4x + 2y - 3z &= 10, \\ x - 3y + 2z &= 8 \end{aligned}$$

10. $$\begin{aligned} x + y + z &= 6, \\ 2x - y + 3z &= 9, \\ -x + 2y + 2z &= 9 \end{aligned}$$

11. $$\begin{aligned} x - y + z &= 6, \\ 2x + 3y + 2z &= 2, \\ 3x + 5y + 4z &= 4 \end{aligned}$$

12. $$\begin{aligned} 2x - y - 3z &= -1, \\ 2x - y + z &= -9, \\ x + 2y - 4z &= 17 \end{aligned}$$

13. $$\begin{aligned} 6x - 4y + 5z &= 31, \\ 5x + 2y + 2z &= 13, \\ x + y + z &= 2 \end{aligned}$$

14. $$\begin{aligned} 2x - 3y + z &= 5, \\ x + 3y + 8z &= 22, \\ 3x - y + 2z &= 12 \end{aligned}$$

15. $$\begin{aligned} x + y + z &= 0, \\ 2x + 3y + 2z &= -3, \\ -x - 2y - z &= 1 \end{aligned}$$

16. $$\begin{aligned} 3a - 2b + 7c &= 13, \\ a + 8b - 6c &= -47, \\ 7a - 9b - 9c &= -3 \end{aligned}$$

17. $$\begin{aligned} 2x + y - 3z &= -4, \\ 4x - 2y + z &= 9, \\ 3x + 5y - 2z &= 5 \end{aligned}$$

18. $$\begin{aligned} 4x + y + z &= 17, \\ x - 3y + 2z &= -8, \\ 5x - 2y + 3z &= 5 \end{aligned}$$

19. $$\begin{aligned} 2x + y + 2z &= 11, \\ 3x + 2y + 2z &= 8, \\ x + 4y + 3z &= 0 \end{aligned}$$

20. $$\begin{aligned} 2x + y + z &= -2, \\ 2x - y + 3z &= 6, \\ 3x - 5y + 4z &= 7 \end{aligned}$$

21. $$\begin{aligned} -2x + 8y + 2z &= 4, \\ x + 6y + 3z &= 4, \\ 3x - 2y + z &= 0 \end{aligned}$$

22. $$\begin{aligned} x - y + z &= 4, \\ 5x + 2y - 3z &= 2, \\ 4x + 3y - 4z &= -2 \end{aligned}$$

23. $$\begin{aligned} 4x - y - z &= 4, \\ 2x + y + z &= -1, \\ 6x - 3y - 2z &= 3 \end{aligned}$$

24. $$\begin{aligned} a + 2b + c &= 1, \\ 7a + 3b - c &= -2, \\ a + 5b + 3c &= 2 \end{aligned}$$

25. $$\begin{aligned} r + \tfrac{3}{2}s + 6t &= 2, \\ 2r - 3s + 3t &= 0.5, \\ r + s + t &= 1 \end{aligned}$$

26. $$\begin{aligned} 5x + 3y + \tfrac{1}{2}z &= \tfrac{7}{2}, \\ 0.5x - 0.9y - 0.2z &= 0.3, \\ 3x - 2.4y + 0.4z &= -1 \end{aligned}$$

27. $$\begin{aligned} 4a + 9b &= 8, \\ 8a + 6c &= -1, \\ 6b + 6c &= -1 \end{aligned}$$

28. $$\begin{aligned} 3p + 2r &= 11, \\ q - 7r &= 4, \\ p - 6q &= 1 \end{aligned}$$

29. $$\begin{aligned} x + y + z &= 57, \\ -2x + y &= 3, \\ x - z &= 6 \end{aligned}$$

30. $$\begin{aligned} x + y + z &= 105, \\ 10y - z &= 11, \\ 2x - 3y &= 7 \end{aligned}$$

31. $$\begin{aligned} a - 3c &= 6, \\ b + 2c &= 2, \\ 7a - 3b - 5c &= 14 \end{aligned}$$

32. $$\begin{aligned} 2a - 3b &= 2, \\ 7a + 4c &= \tfrac{3}{4}, \\ 2c - 3b &= 1 \end{aligned}$$

Aha! **33.** $x + y + z = 83,$
$y = 2x + 3,$
$z = 40 + x$

34. $l + m = 7,$
$3m + 2n = 9,$
$4l + n = 5$

35. $x + z = 0,$
$x + y + 2z = 3,$
$y + z = 2$

36. $x + y = 0,$
$x + z = 1,$
$2x + y + z = 2$

37. $x + y + z = 1,$
$-x + 2y + z = 2,$
$2x - y = -1$

38. $y + z = 1,$
$x + y + z = 1,$
$x + 2y + 2z = 2$

TW **39.** Describe a method for writing an inconsistent system of three equations in three variables.

TW **40.** Abbie recommends that a frustrated classmate double- and triple-check each step of work when attempting to solve a system of three equations. Is this good advice? Why or why not?

Focused Review

41. Solve by graphing by hand. [3.1]

$x + y = 5,$
$2x + 3y = 6$

42. Solve using a graphing calculator. [3.1]

$2x - y = 5,$
$y = 3x + 7$

43. Solve using substitution. [3.2]

$y = \frac{1}{2}x + 3,$
$5x - 2y = 1$

44. Solve using elimination. [3.2]

$2x + y = 8,$
$3x - 2y = 1$

Solve.

45. $x + y - z = 4,$
$2x - y - z = 7,$
$x + 3y = 4z$ [3.4]

46. $1.5x - 3.2y = -10,$
$20x + 25y = 205$ [3.2]

Synthesis

TW **47.** Is it possible for a system of three linear equations to have exactly two ordered triples in its solution set? Why or why not?

TW **48.** Describe a procedure that could be used to solve a system of four equations in four variables.

Solve.

49. $\frac{x+2}{3} - \frac{y+4}{2} + \frac{z+1}{6} = 0,$
$\frac{x-4}{3} + \frac{y+1}{4} - \frac{z-2}{2} = -1,$
$\frac{x+1}{2} + \frac{y}{2} + \frac{z-1}{4} = \frac{3}{4}$

50. $w + x + y + z = 2,$
$w + 2x + 2y + 4z = 1,$
$w - x + y + z = 6,$
$w - 3x - y + z = 2$

51. $w + x - y + z = 0,$
$w - 2x - 2y - z = -5,$
$w - 3x - y + z = 4,$
$2w - x - y + 3z = 7$

For Exercises 52 and 53, let u represent $1/x$, v represent $1/y$, and w represent $1/z$. Solve for u, v, and w, and then solve for x, y, and z.

52. $\frac{2}{x} - \frac{1}{y} - \frac{3}{z} = -1,$
$\frac{2}{x} - \frac{1}{y} + \frac{1}{z} = -9,$
$\frac{1}{x} + \frac{2}{y} - \frac{4}{z} = 17$

53. $\frac{2}{x} + \frac{2}{y} - \frac{3}{z} = 3,$
$\frac{1}{x} - \frac{2}{y} - \frac{3}{z} = 9,$
$\frac{7}{x} - \frac{2}{y} + \frac{9}{z} = -39$

Determine k so that each system is dependent.

54. $x - 3y + 2z = 1,$
$2x + y - z = 3,$
$9x - 6y + 3z = k$

55. $5x - 6y + kz = -5,$
$x + 3y - 2z = 2,$
$2x - y + 4z = -1$

In each case, three solutions of an equation in x, y, and z are given. Find the equation.

56. $Ax + By + Cz = 12;$
$\left(1, \frac{3}{4}, 3\right), \left(\frac{4}{3}, 1, 2\right)$, and $(2, 1, 1)$

57. $z = b - mx - ny;$
$(1, 1, 2), (3, 2, -6)$, and $\left(\frac{3}{2}, 1, 1\right)$

58. Write an inconsistent system of equations that contains dependent equations.

Collaborative Corner

Finding the Preferred Approach

Focus: Systems of three linear equations
Time: 10–15 minutes
Group Size: 3

Consider the six steps outlined on p. 225 along with the following system:

$$2x + 4y = 3 - 5z,$$
$$0.3x = 0.2y + 0.7z + 1.4,$$
$$0.04x + 0.03y = 0.07 + 0.04z.$$

ACTIVITY

1. Working independently, each group member should solve the system above. One person should begin by eliminating x, one should first eliminate y, and one should first eliminate z. Write neatly so that others can follow your steps.
2. Once all group members have solved the system, compare your answers. If the answers do not check, exchange notebooks and check each other's work. If a mistake is detected, allow the person who made the mistake to make the repair.
3. Decide as a group which of the three approaches above (if any) ranks as easiest and which (if any) ranks as most difficult. Then compare your rankings with the other groups in the class.

3.5 Solving Applications: Systems of Three Equations

Applications of Three Equations in Three Unknowns

Study Tip

Use the Answer Section Carefully

When using the answers listed at the back of this book, try not to "work backward" from the answer. If you frequently require two or more attempts to answer an exercise correctly, you probably need to work more carefully and/or reread the section. Remember that on quizzes and tests you have only one attempt per problem and no answers section to check.

Solving systems of three or more equations is important in many applications. Such systems arise in the natural and social sciences, business, and engineering. In mathematics, purely numerical applications also arise.

EXAMPLE 1 The sum of three numbers is 4. The first number minus twice the second, minus the third is 1. Twice the first number minus the second, minus twice the third is -1. Find the numbers.

SOLUTION

1. **Familiarize.** There are three statements involving the same three numbers. Let's label these numbers x, y, and z.

2. **Translate.** We can translate directly as follows.

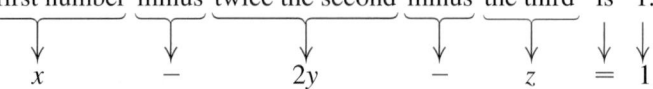

The sum of the three numbers is 4.

$$x + y + z = 4$$

The first number minus twice the second minus the third is 1.

$$x - 2y - z = 1$$

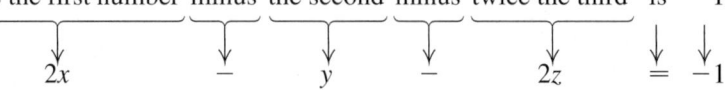

Twice the first number minus the second minus twice the third is −1.

$$2x - y - 2z = -1$$

We now have a system of three equations:

$$\begin{aligned} x + y + z &= 4, \\ x - 2y - z &= 1, \\ 2x - y - 2z &= -1. \end{aligned}$$

3. **Carry out.** We need to solve the system of equations. Note that we found the solution, $(2, -1, 3)$, in Example 2 of Section 3.4.
4. **Check.** The first statement of the problem says that the sum of the three numbers is 4. That checks, because $2 + (-1) + 3 = 4$. The second statement says that the first number minus twice the second, minus the third is 1: $2 - 2(-1) - 3 = 1$. That checks. The check of the third statement is left to the student.
5. **State.** The three numbers are 2, −1, and 3.

EXAMPLE 2 Architecture. In a triangular cross section of a roof, the largest angle is 70° greater than the smallest angle. The largest angle is twice as large as the remaining angle. Find the measure of each angle.

SOLUTION

1. **Familiarize.** The first thing we do is make a drawing, or a sketch.

Student Notes

It is quite likely that you are expected to remember that the sum of the measures of the angles in any triangle is 180°. You may want to ask your instructor which other formulas from geometry and elsewhere you are expected to know.

Since we don't know the size of any angle, we use x, y, and z to represent the three measures, from smallest to largest. Recall that the measures of the angles in any triangle add up to 180°.

2. **Translate.** This geometric fact about triangles gives us one equation:

$$x + y + z = 180.$$

Two of the statements can be translated almost directly.

$$\underbrace{\text{The largest angle}}_{z} \;\underbrace{\text{is}}_{=}\; \underbrace{70^\circ \text{ greater than the smallest angle.}}_{x + 70}$$

$$\underbrace{\text{The largest angle}}_{z} \;\underbrace{\text{is}}_{=}\; \underbrace{\text{twice as large as the remaining angle.}}_{2y}$$

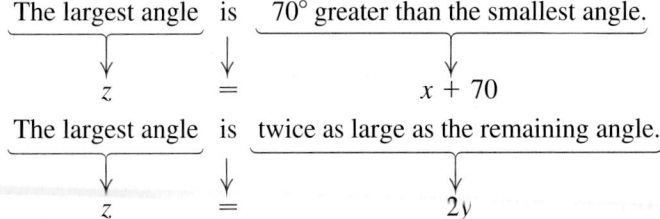

We now have a system of three equations:

$$\begin{aligned} x + y + z &= 180, \\ x + 70 &= z, \\ 2y &= z; \end{aligned} \qquad \text{or} \qquad \begin{aligned} x + y + z &= 180, \\ x \qquad - z &= -70, \\ 2y - z &= 0. \end{aligned}$$

Rewriting in standard form

3. **Carry out.** The system was solved in Example 4 of Section 3.4. The solution is (30, 50, 100).
4. **Check.** The sum of the numbers is 180, so that checks. The measure of the largest angle, 100°, is 70° greater than the measure of the smallest angle, 30°, so that checks. The measure of the largest angle is also twice the measure of the remaining angle, 50°. Thus we have a check.
5. **State.** The angles in the triangle measure 30°, 50°, and 100°.

EXAMPLE 3 Cholesterol Levels. Recent studies indicate that a child's intake of cholesterol should be no more than 300 mg per day. By eating 1 egg, 1 cupcake, and 1 slice of pizza, a child consumes 302 mg of cholesterol. A child who eats 2 cupcakes and 3 slices of pizza takes in 65 mg of cholesterol. By eating 2 eggs and 1 cupcake, a child consumes 567 mg of cholesterol. How much cholesterol is in each item?

SOLUTION

1. **Familiarize.** After reading the problem, it becomes clear that an egg contains considerably more cholesterol than the other foods. Let's guess that one egg contains 200 mg of cholesterol and one cupcake contains 50 mg. Because of the second sentence in the problem, it would follow that a slice of pizza contains 52 mg of cholesterol since $200 + 50 + 52 = 302$.

   To see if our guess satisfies the other statements in the problem, we find the amount of cholesterol that 2 cupcakes and 3 slices of pizza would contain: $2 \cdot 50 + 3 \cdot 52 = 256$. Since this does not match the 65 mg listed in the problem, our guess was incorrect. Rather than guess again, we examine how we checked our guess and let g, c, and s = the number of milligrams of cholesterol in an egg, a cupcake, and a slice of pizza, respectively.

2. **Translate.** Rewording some of the sentences, we can translate as follows:

The amount of cholesterol in 1 egg	plus	the amount of cholesterol in 1 cupcake	plus	the amount of cholesterol in 1 slice of pizza	is	302 mg.
g	$+$	c	$+$	s	$=$	302

The amount of cholesterol in 2 cupcakes	plus	the amount of cholesterol in 3 slices of pizza	is	65 mg.
$2c$	$+$	$3s$	$=$	65

The amount of cholesterol in 2 eggs	plus	the amount of cholesterol in 1 cupcake	is	567 mg.
$2g$	$+$	c	$=$	567

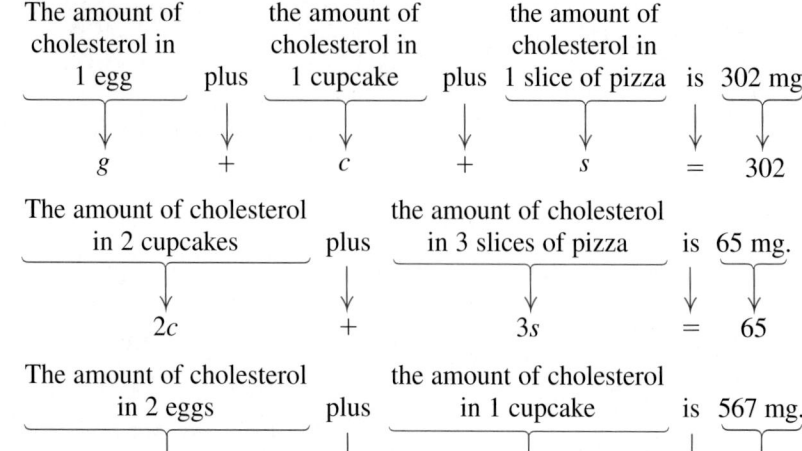

We now have a system of three equations:

$$\begin{aligned} g + c + s &= 302, \\ 2c + 3s &= 65, \\ 2g + c &= 567. \end{aligned}$$

3. **Carry out.** We solve and get $g = 274$, $c = 19$, and $s = 9$.

4. **Check.** The sum of 274, 19, and 9 is 302 so the total cholesterol in 1 egg, 1 cupcake, and 1 slice of pizza checks. Two cupcakes and three slices of pizza would contain $2 \cdot 19 + 3 \cdot 9 = 65$ mg, while two eggs and one cupcake would contain $2 \cdot 274 + 19 = 567$ mg of cholesterol. The answer checks.

5. **State.** An egg contains 274 mg of cholesterol, a cupcake contains 19 mg of cholesterol, and a slice of pizza contains 9 mg of cholesterol.

3.5 EXERCISE SET

FOR EXTRA HELP

MathXL | MyMathLab | InterAct Math | AW Math Tutor Center | Video Lectures on CD: Disc 2 | Student's Solutions Manual

Solve.

1. The sum of three numbers is 57. The second is 3 more than the first. The third is 6 more than the first. Find the numbers.

2. The sum of three numbers is 5. The first number minus the second plus the third is 1. The first minus the third is 3 more than the second. Find the numbers.

3. The sum of three numbers is 26. Twice the first minus the second is 2 less than the third. The third is the second minus three times the first. Find the numbers.

4. The sum of three numbers is 105. The third is 11 less than ten times the second. Twice the first is 7 more than three times the second. Find the numbers.

5. *Geometry.* In triangle ABC, the measure of angle B is three times that of angle A. The measure of angle C is 20° more than that of angle A. Find the angle measures.

6. *Geometry.* In triangle ABC, the measure of angle B is twice the measure of angle A. The measure of angle C is 80° more than that of angle A. Find the angle measures.

7. *Health Insurance.* In 2004, UNICARE® health insurance for adults under the age of 30 cost \$121/month for a couple, \$107/month for an adult with one child, and \$164/month for a couple with one child (*Source*: UNICARE Life and Health Insurance Company® advertisement). On the basis of the information given, find the monthly rate for an individual adult, a spouse, and a child.

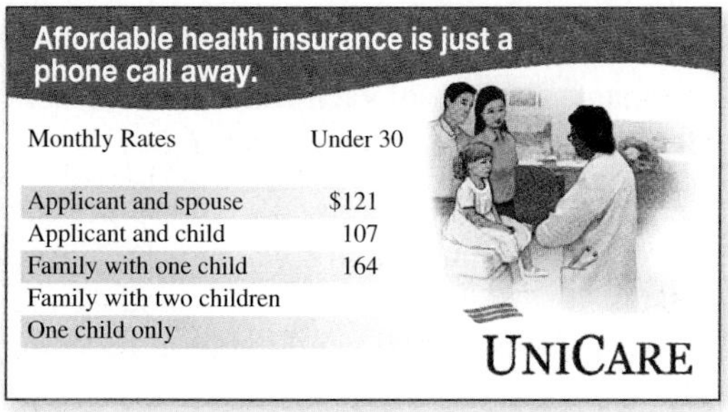

Monthly Rates	Under 30
Applicant and spouse	\$121
Applicant and child	107
Family with one child	164
Family with two children	
One child only	

8. *Health Insurance.* In 2004, UNICARE® health insurance cost \$160/month for a 35–39-year-old adult and spouse, \$145/month for a 35–39-year-old adult with one child, and \$245/month for a 35–39-year-old adult with a spouse and child (*Source*: UNICARE Life and Health Insurance Company® advertisement). On the basis of the information given, find the monthly rates for a 39-year-old adult, a 39-year-old's spouse, and a 39-year-old's child.

9. *Nutrition.* Most nutritionists now agree that a healthy adult diet should include 25–35 g of fiber each day (*Sources*: usda.gov; InteliHealth.com). A breakfast of 2 bran muffins, 1 banana, and a 1-cup serving of Wheaties® contains 9 g of fiber; a breakfast of 1 bran muffin, 2 bananas, and a 1-cup serving of Wheaties® contains 10.5 g of fiber; and a breakfast of 2 bran muffins and a 1-cup serving of Wheaties® contains 6 g of fiber. How much fiber is in each of these foods?

10. *Nutrition.* Refer to Exercise 9. A breakfast consisting of 2 pancakes and a 1-cup serving of strawberries contains 4.5 g of fiber, whereas a breakfast of 2 pancakes and a 1-cup serving of Cheerios® contains 4 g of fiber. When a meal consists of 1 pancake, a 1-cup serving of Cheerios®, and a 1-cup serving of strawberries, it contains 7 g of fiber. (*Source*: InteliHealth.com) How much fiber is in each of these foods?

Aha! **11.** *Automobile Pricing.* The basic model of a 2006 Jeep Grand Cherokee Laredo (2WD) with a power sunroof cost $28,215. When equipped with four-wheel drive (4WD) and a sunroof, the vehicle's price rose to $30,185. The cost of the basic model with 4WD was $29,385. Find the basic price, the cost of 4WD, and the cost of a sunroof.

12. *Lens Production.* When Sight-Rite's three polishing machines, A, B, and C, are all working, 5700 lenses can be polished in one week. When only A and B are working, 3400 lenses can be polished in one week. When only B and C are working, 4200 lenses can be polished in one week. How many lenses can be polished in a week by each machine?

13. *Welding Rates.* Elrod, Dot, and Wendy can weld 74 linear feet per hour when working together. Elrod and Dot together can weld 44 linear feet per hour, while Elrod and Wendy can weld 50 linear feet per hour. How many linear feet per hour can each weld alone?

14. *Telemarketing.* Sven, Tillie, and Isaiah can process 740 telephone orders per day. Sven and Tillie together can process 470 orders, while Tillie and Isaiah together can process 520 orders per day. How many orders can each person process alone?

15. *Coffee Prices.* Roz worked at a Starbucks® coffee shop where a 12-oz cup of coffee cost $1.40, a 16-oz cup cost $1.60, and a 20-oz cup cost $1.70. During one busy period, Roz served 55 cups of coffee, emptying six 144-oz "brewers" while collecting a total of $85.90. How many cups of each size did Roz fill?

16. *Advertising.* In a recent year, U.S. companies spent a total of $106.5 billion on newspaper, television, and radio ads. The total amount spent on television and radio ads was $18.7 billion more than the amount spent on newspaper ads alone. The amount spent on newspaper ads was $28.8 billion more than what was spent on radio ads. (*Sources*: NAA (newspapers); McCann–Erickson Inc. (television and radio)). How much was spent on each form of advertising?

17. *Restaurant Management.* McDonald's® recently sold small soft drinks for $1, medium soft drinks for $1.15, and large soft drinks for $1.30. During a lunch-time rush, Chris sold 40 soft drinks for a total of $45.25. The number of small and large drinks, combined, was 10 fewer than the number of medium drinks. How many drinks of each size were sold?

18. *Investments.* A business class divided an imaginary investment of $80,000 among three mutual funds. The first fund grew by 10%, the second by 6%, and the third by 15%. Total earnings were $8850. The earnings from the first fund were $750 more than the earnings from the third. How much was invested in each fund?

19. *Nutrition.* A dietician in a hospital prepares meals under the guidance of a physician. Suppose that for a particular patient a physician prescribes a meal to have 800 calories, 55 g of protein, and 220 mg of vitamin C. The dietician prepares a meal of roast beef, baked potatoes, and broccoli according to the data in the following table.

Serving Size	Calories	Protein (in grams)	Vitamin C (in milligrams)
Roast Beef, 3 oz	300	20	0
Baked Potato, 1	100	5	20
Broccoli, 156 g	50	5	100

How many servings of each food are needed in order to satisfy the doctor's orders?

20. *Nutrition.* Repeat Exercise 19 but replace the broccoli with asparagus, for which a 180-g serving contains 50 calories, 5 g of protein, and 44 mg of vitamin C. Which meal would you prefer eating?

21. *World Population Growth.* The world population is projected to be 9.1 billion in 2050. At that time, there are expected to be approximately 3 billion more people in Asia than in Africa. The population for the rest of the world will be approximately 0.1 billion more than half the population of Asia. (*Sources*: U.S. Bureau of the Census; *Burlington Free Press* 3/23/04) Find the projected populations of Asia, Africa, and the rest of the world in 2050.

22. *Crying Rate.* The sum of the average number of times a man, a woman, and a one-year-old child cry each month is 56.7. A woman cries 3.9 more times than a man. The average number of times a one-year-old cries per month is 43.3 more than the average number of times combined that a man and a woman cry. What is the average number of times per month that each cries?

23. *Basketball Scoring.* The New York Knicks recently scored a total of 92 points on a combination of 2-point field goals, 3-point field goals, and 1-point foul shots. Altogether, the Knicks made 50 baskets and 19 more 2-pointers than foul shots. How many shots of each kind were made?

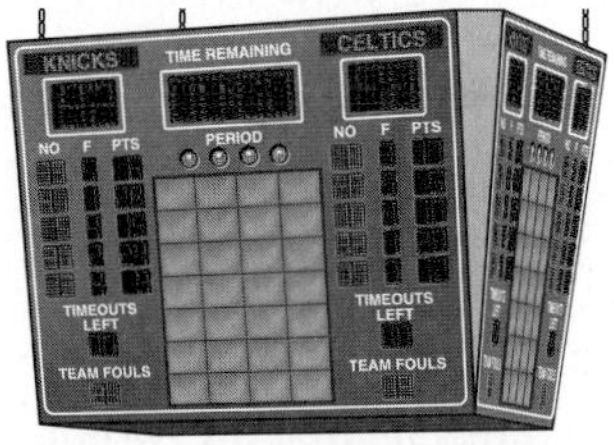

24. *History.* Find the year in which the first U.S. transcontinental railroad was completed. The following are some facts about the number. The sum of the digits in the year is 24. The ones digit is 1 more than the hundreds digit. Both the tens and the ones digits are multiples of 3.

TW **25.** Problems like Exercises 15 and 17 could be classified as total-value problems. How do these problems differ from the total-value problems of Section 3.3?

TW **26.** Write a problem for a classmate to solve. Design the problem so that it translates to a system of three equations in three variables.

Skill Maintenance

Simplify. [1.2], [1.3]

27. $5(-3) + 7$

28. $-4(-6) + 9$

29. $-6(8) + (-7)$

30. $7(-9) + (-8)$

31. $-7(2x - 3y + 5z)$

32. $-6(4a + 7b - 9c)$

33. $-4(2a + 5b) + 3a + 20b$

34. $3(2x - 7y) + 5x + 21y$

Synthesis

TW **35.** Consider Exercise 23. Suppose there were no foul shots made. Would there still be a solution? Why or why not?

TW **36.** Consider Exercise 15. Suppose Roz collected \$46. Could the problem still be solved? Why or why not?

37. *Health Insurance.* In 2004, UNICARE® health insurance for a 35–39-year-old and his or her spouse cost \$160/month. That rate increased to \$203/month if a child were included and \$243/month if two children were included. The rate dropped to \$145/month for just the applicant and one child. (*Source*: UNICARE Life and Health Insurance Company® advertisement) Find the separate costs for insuring the applicant, the spouse, the first child, and the second child.

38. Find a three-digit positive integer such that the sum of all three digits is 14, the tens digit is 2 more than the ones digit, and if the digits are reversed, the number is unchanged.

39. *Ages.* Tammy's age is the sum of the ages of Carmen and Dennis. Carmen's age is 2 more than the sum of the ages of Dennis and Mark. Dennis's age is four times Mark's age. The sum of all four ages is 42. How old is Tammy?

40. *Ticket Revenue.* A magic show's audience of 100 people consists of adults, students, and children. The ticket prices are \$10 for adults, \$3 for students, and 50¢ for children. The total amount of money taken in is \$100. How many adults, students, and children are in attendance? Does there seem to be some information missing? Do some more careful reasoning.

41. *Sharing Raffle Tickets.* Hal gives Tom as many raffle tickets as Tom first had and Gary as many as Gary first had. In like manner, Tom then gives Hal and Gary as many tickets as each then has. Similarly, Gary gives Hal and Tom as many tickets as each then has. If each finally has 40 tickets, with how many tickets does Tom begin?

42. Find the sum of the angle measures at the tips of the star in this figure.

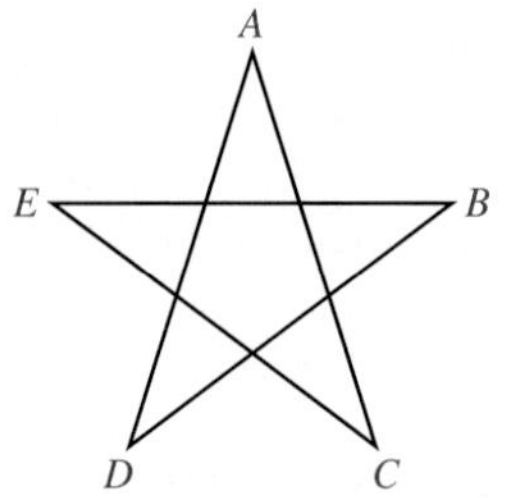

3.6 Elimination Using Matrices

Matrices and Systems ■ Row-Equivalent Operations

In solving systems of equations, we perform computations with the constants. The variables play no important role until the end. Thus we can simplify writing a system by omitting the variables. For example, the system

$$\begin{aligned} 3x + 4y &= 5, \\ x - 2y &= 1 \end{aligned} \quad \text{simplifies to} \quad \begin{matrix} 3 & 4 & 5 \\ 1 & -2 & 1 \end{matrix}$$

if we do not write the variables, the operation of addition, and the equals signs. Note that the coefficients of the x-terms are written first, the coefficients of the y-terms are written next, and the constant terms are written last. Equations must be written in standard form $Ax + By = C$ before they are written without variables.

Study Tip

Double-check the Numbers

Solving problems is challenging enough, without miscopying information. Always double-check that you have accurately transferred numbers from the correct exercise in the exercise set.

Matrices and Systems

In the example above, we have written a rectangular array of numbers. Such an array is called a **matrix** (plural, **matrices**). We ordinarily write brackets around matrices. The following are matrices:

$$\begin{bmatrix} -3 & 1 \\ 0 & 5 \end{bmatrix}, \quad \begin{bmatrix} 2 & 0 & -1 & 3 \\ -5 & 2 & 7 & -1 \\ 4 & 5 & 3 & 0 \end{bmatrix}, \quad \begin{bmatrix} 2 & 3 \\ 7 & 15 \\ -2 & 23 \\ 4 & 1 \end{bmatrix}.$$

The individual numbers are called *elements* or *entries.*

The **rows** of a matrix are horizontal, and the **columns** are vertical.

$$A = \begin{bmatrix} 5 & -2 & -2 \\ 1 & 0 & 1 \\ 4 & -3 & 2 \end{bmatrix} \begin{matrix} \leftarrow \text{row } 1 \\ \leftarrow \text{row } 2 \\ \leftarrow \text{row } 3 \end{matrix}$$

$\uparrow$ column 1 $\quad\uparrow$ column 2 $\quad\uparrow$ column 3

Let's see how matrices can be used to solve a system.

EXAMPLE 1 Solve the system

$$5x - 4y = -1,$$
$$-2x + 3y = 2.$$

As an aid for understanding, we list the corresponding system in the margin.

$$5x - 4y = -1,$$
$$-2x + 3y = 2$$

SOLUTION We write a matrix using only coefficients and constants, listing x-coefficients in the first column and y-coefficients in the second. Note that in each matrix a dashed line separates the coefficients from the constants:

$$\left[\begin{array}{rr:r} 5 & -4 & -1 \\ -2 & 3 & 2 \end{array}\right].$$

Our goal is to transform

$$\left[\begin{array}{rr:r} 5 & -4 & -1 \\ -2 & 3 & 2 \end{array}\right] \quad \text{into the form} \quad \left[\begin{array}{cc:c} a & b & c \\ 0 & d & e \end{array}\right].$$

The variables x and y can then be reinserted to form equations from which we can complete the solution.

We do calculations that are similar to those that we would do if we wrote the entire equations. The first step is to multiply and/or interchange the rows so that each number in the first column below the first number is a multiple of that number. Here that means multiplying Row 2 by 5. This corresponds to multiplying both sides of the second equation by 5.

$$5x - 4y = -1,$$
$$-10x + 15y = 10$$

$$\left[\begin{array}{rr:r} 5 & -4 & -1 \\ -10 & 15 & 10 \end{array}\right] \qquad \textbf{New Row 2 = 5(Row 2 from above)}$$

Next, we multiply the first row by 2, add this to Row 2, and write that result as the "new" Row 2. This corresponds to multiplying the first equation by 2 and adding the result to the second equation in order to eliminate a variable. Write out these computations as necessary—we perform them mentally.

$$5x - 4y = -1,$$
$$7y = 8$$

$$\left[\begin{array}{rr:r} 5 & -4 & -1 \\ 0 & 7 & 8 \end{array}\right]$$

2(5 −4 ⋮ −1) = (10 −8 ⋮ −2) and
(10 −8 ⋮ −2) + (−10 15 ⋮ 10) = (0 7 ⋮ 8)
New Row 2 = 2(Row 1) + (Row 2)

If we now reinsert the variables, we have

$$5x - 4y = -1, \qquad (1)$$
$$7y = 8. \qquad (2)$$

We can now proceed as before, solving equation (2) for y:

$$7y = 8 \qquad (2)$$
$$y = \tfrac{8}{7}.$$

Next, we substitute $\frac{8}{7}$ for y in equation (1):

$$5x - 4y = -1 \qquad (1)$$
$$5x - 4 \cdot \tfrac{8}{7} = -1 \qquad \textbf{Substituting } \tfrac{8}{7} \textbf{ for } y \textbf{ in equation (1)}$$
$$x = \tfrac{5}{7}. \qquad \textbf{Solving for } x$$

The solution is $\left(\frac{5}{7}, \frac{8}{7}\right)$. The check is left to the student.

All the systems of equations shown in the margin by Example 1 are **equivalent systems of equations;** that is, they all have the same solution.

EXAMPLE 2 Solve the system

$$\begin{aligned} 2x - y + 4z &= -3, \\ x \qquad - 4z &= 5, \\ 6x - y + 2z &= 10. \end{aligned}$$

SOLUTION We first write a matrix, using only the constants. Where there are missing terms, we must write 0's:

$$\left[\begin{array}{ccc|c} 2 & -1 & 4 & -3 \\ 1 & 0 & -4 & 5 \\ 6 & -1 & 2 & 10 \end{array}\right].$$

Note that the x-coefficients are in column 1, the y-coefficients in column 2, the z-coefficients in column 3, and the constant terms in the last column.

$$\begin{aligned} 2x - y + 4z &= -3, \\ x \qquad - 4z &= 5, \\ 6x - y + 2z &= 10 \end{aligned}$$

Our goal is to transform the matrix to one of the form

$$\left[\begin{array}{ccc|c} a & b & c & d \\ 0 & e & f & g \\ 0 & 0 & h & i \end{array}\right].$$

$$\begin{aligned} ax + by + cz &= d, \\ ey + fz &= g, \\ hz &= i \end{aligned}$$

A matrix of this form can be rewritten as a system of equations that is equivalent to the original system, and from which a solution can be easily found.

The first step is to multiply and/or interchange the rows so that each number in the first column is a multiple of the first number in the first row. In this case, we do so by interchanging Rows 1 and 2:

$$\left[\begin{array}{ccc|c} 1 & 0 & -4 & 5 \\ 2 & -1 & 4 & -3 \\ 6 & -1 & 2 & 10 \end{array}\right]$$

This corresponds to interchanging the first two equations.

$$\begin{aligned} x \qquad - 4z &= 5, \\ 2x - y + 4z &= -3, \\ 6x - y + 2z &= 10 \end{aligned}$$

Next, we multiply the first row by -2, add it to the second row, and replace Row 2 with the result:

$$\left[\begin{array}{ccc|c} 1 & 0 & -4 & 5 \\ 0 & -1 & 12 & -13 \\ 6 & -1 & 2 & 10 \end{array}\right].$$

$-2(1 \quad 0 \quad -4 \mid 5) = (-2 \quad 0 \quad 8 \mid -10)$ and $(-2 \quad 0 \quad 8 \mid -10) + (2 \quad -1 \quad 4 \mid -3) = (0 \quad -1 \quad 12 \mid -13)$

$$\begin{aligned} x \qquad - 4z &= 5, \\ -y + 12z &= -13, \\ 6x - y + 2z &= 10 \end{aligned}$$

Now we multiply the first row by -6, add it to the third row, and replace Row 3 with the result:

$$\left[\begin{array}{ccc|c} 1 & 0 & -4 & 5 \\ 0 & -1 & 12 & -13 \\ 0 & -1 & 26 & -20 \end{array}\right].$$

$-6(1 \quad 0 \quad -4 \mid 5) = (-6 \quad 0 \quad 24 \mid -30)$ and $(-6 \quad 0 \quad 24 \mid -30) + (6 \quad -1 \quad 2 \mid 10) = (0 \quad -1 \quad 26 \mid -20)$

$$\begin{aligned} x \qquad - 4z &= 5, \\ -y + 12z &= -13, \\ -y + 26z &= -20 \end{aligned}$$

$$\begin{aligned} x \qquad - 4z &= 5, \\ -y + 12z &= -13, \\ 14z &= -7 \end{aligned}$$

Next, we multiply Row 2 by -1, add it to the third row, and replace Row 3 with the result:

$$\left[\begin{array}{ccc|c} 1 & 0 & -4 & 5 \\ 0 & -1 & 12 & -13 \\ 0 & 0 & 14 & -7 \end{array}\right].$$

$-1(0 \quad -1 \quad 12 \mid -13) = (0 \quad 1 \quad -12 \mid 13)$ and $(0 \quad 1 \quad -12 \mid 13) + (0 \quad -1 \quad 26 \mid -20) = (0 \quad 0 \quad 14 \mid -7)$

Reinserting the variables gives us

$$\begin{aligned} x \qquad - 4z &= 5, \\ -y + 12z &= -13, \\ 14z &= -7. \end{aligned}$$

We now solve this last equation for z and get $z = -\frac{1}{2}$. Next, we substitute $-\frac{1}{2}$ for z in the preceding equation and solve for y: $-y + 12\left(-\frac{1}{2}\right) = -13$, so $y = 7$. Since there is no y-term in the first equation of this last system, we need only substitute $-\frac{1}{2}$ for z to solve for x: $x - 4\left(-\frac{1}{2}\right) = 5$, so $x = 3$. The solution is $\left(3, 7, -\frac{1}{2}\right)$. The check is left to the student.

The operations used in the preceding example correspond to those used to produce equivalent systems of equations. We call the matrices **row-equivalent** and the operations that produce them **row-equivalent operations.**

Row-Equivalent Operations

Student Notes

Note that row-equivalent matrices are not *equal*. It is the solutions of the corresponding equations that are the same.

Row-Equivalent Operations

Each of the following row-equivalent operations produces a row-equivalent matrix:

a) Interchanging any two rows.
b) Multiplying all elements of a row by a nonzero constant.
c) Replacing a row with the sum of that row and a multiple of another row.

The best overall method for solving systems of equations is by row-equivalent matrices; even computers are programmed to use them. Matrices are part of a branch of mathematics known as linear algebra. They are also studied in many courses in finite mathematics.

When we solved the systems in Examples 1 and 2, we used row-equivalent operations to write an equivalent system of equations that we could solve without using elimination. We can continue to use row-equivalent operations to write a row-equivalent matrix in **reduced row-echelon form,** from which the solution of the system can often be read directly.

Reduced Row-Echelon Form

A matrix is in reduced row-echelon form if:

1. All rows consisting entirely of zeros are at the bottom of the matrix.
2. The first nonzero number in any nonzero row is 1, called a leading 1.
3. The leading 1 in any row is farther to the left than the leading 1 in any lower row.
4. Each column that contains a leading 1 has zeros everywhere else.

EXAMPLE 3 Solve the system

$$\begin{aligned} 5x - 4y &= -1, \\ -2x + 3y &= 2 \end{aligned}$$

by writing a reduced row-echelon matrix.

SOLUTION We began this solution in Example 1, and used row-equivalent operations to write the row-equivalent matrix:

$$\begin{aligned} 5x - 4y &= -1, \\ 7y &= 8 \end{aligned}$$

$$\left[\begin{array}{cc|c} 5 & -4 & -1 \\ 0 & 7 & 8 \end{array}\right].$$

Before we can write reduced row-echelon form, the first nonzero number in any nonzero row must be 1. We multiply Row 1 by $\frac{1}{5}$ and Row 2 by $\frac{1}{7}$:

$$\begin{aligned} x - \tfrac{4}{5}y &= -\tfrac{1}{5}, \\ y &= \tfrac{8}{7} \end{aligned}$$

$$\left[\begin{array}{cc|c} 1 & -\frac{4}{5} & -\frac{1}{5} \\ 0 & 1 & \frac{8}{7} \end{array}\right].$$ **New Row 1 = $\frac{1}{5}$(Row 1 from above)**
New Row 2 = $\frac{1}{7}$(Row 2 from above)

There are no zero rows, there is a leading 1 in each row, and the leading 1 in Row 1 is farther to the left than the leading 1 in Row 2. Thus conditions (1)–(3) for reduced row-echelon form have been met.

To satisfy condition (4), we must obtain a 0 in Row 1, Column 2, above the leading 1 in Row 2. We multiply Row 2 by $\frac{4}{5}$ and add it to Row 1:

$$\begin{aligned} x &= \tfrac{5}{7}, \\ y &= \tfrac{8}{7} \end{aligned}$$

$$\left[\begin{array}{cc|c} 1 & 0 & \frac{5}{7} \\ 0 & 1 & \frac{8}{7} \end{array}\right].$$

$\frac{4}{5}\left(0 \quad 1 \mid \frac{8}{7}\right) = \left(0 \quad \frac{4}{5} \mid \frac{32}{35}\right)$ and
$\left(0 \quad \frac{4}{5} \mid \frac{32}{35}\right) + \left(1 \quad -\frac{4}{5} \mid -\frac{1}{5}\right) = \left(1 \quad 0 \mid \frac{5}{7}\right)$
New Row 1 = $\frac{4}{5}$(Row 2) + Row 1

This matrix is in reduced row-echelon form. If we reinsert the variables, we have

$$\begin{aligned} x &= \tfrac{5}{7}, \\ y &= \tfrac{8}{7}. \end{aligned}$$

The solution, $\left(\frac{5}{7}, \frac{8}{7}\right)$, can be read directly from the last column of the matrix.

Finding reduced row-echelon form often involves extensive calculations with fractions or decimals. Thus it is common to use a computer or a graphing calculator to store and manipulate matrices.

Matrix Operations

On many graphing calculators, matrix operations are accessed by pressing (MATRIX). (On some calculators, MATRIX is the 2nd feature associated with the (x^{-1}) key.) The MATRIX menu has three submenus: NAMES, MATH, and EDIT.

The EDIT menu allows matrices to be entered. For example, to enter

$$A = \begin{bmatrix} 1 & 2 & 5 \\ -3 & 0 & 4 \end{bmatrix},$$

choose option [A] from the MATRIX EDIT menu. Enter the **dimensions,** or number of rows and columns, of the matrix first, listing the number of rows before the number of columns. Thus the dimensions of A are 2×3, read "2 by 3." Then enter each element of A by pressing the number and (ENTER). The notation $2, 3 = 4$ indicates that the 3rd entry of the 2nd row is 4. Matrices are generally entered row by row rather than column by column.

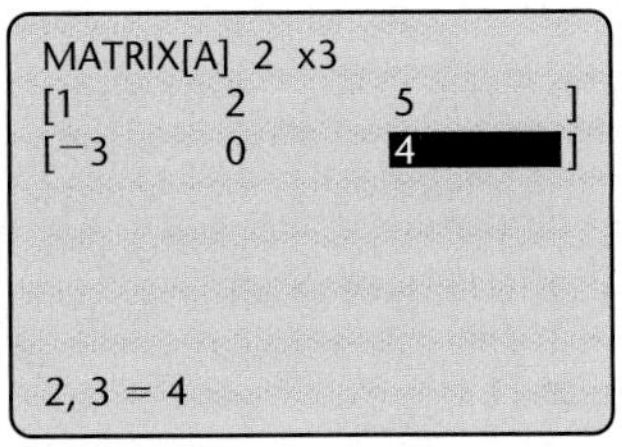

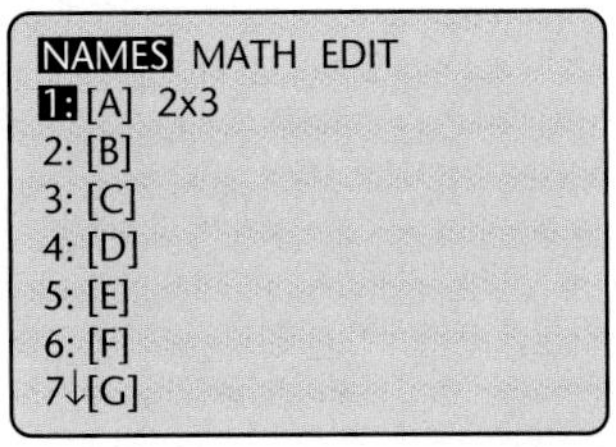

After entering the matrix, exit the matrix editor by pressing (2ND) (QUIT). To access a matrix once it has been entered, use the MATRIX NAMES submenu as shown above on the right. The brackets around A indicate that it is a matrix.

The submenu MATRIX MATH lists operations that can be performed on matrices.

EXAMPLE 4 Solve the following system using a graphing calculator:

$$2x + 5y - 8z = 7,$$
$$3x + 4y - 3z = 8,$$
$$5y - 2x = 9.$$

SOLUTION Before writing a matrix to represent this system, we rewrite the third equation in the form $ax + by + cz = d$:

$$2x + 5y - 8z = 7,$$
$$3x + 4y - 3z = 8,$$
$$-2x + 5y + 0z = 9.$$

The matrix that represents this system is thus

$$\left[\begin{array}{rrr|r} 2 & 5 & -8 & 7 \\ 3 & 4 & -3 & 8 \\ -2 & 5 & 0 & 9 \end{array}\right].$$

We enter the matrix as A using the MATRIX EDIT menu, noting that its dimensions are 3×4. Once we have entered each element of the matrix, we return to the home screen to perform operations. The contents of matrix A can be displayed on the screen, using the MATRIX NAMES menu.

```
[A]
          [[2 5 -8 7]
           [3 4 -3 8]
           [-2 5 0 9]]
```

Each of the row-equivalent operations can be performed using the MATRIX MATH menu. On many calculators, it is also possible to go directly to the reduced row-echelon form. To find this form, from the home screen, we choose the RREF option from the MATRIX MATH menu and then choose [A] from the MATRIX NAMES menu. Finally, press MATH 1 ENTER to write the entries in the reduced row-echelon form using fraction notation.

```
rref ( [A] )▸Frac
   [[1 0 0 1/2]
    [0 1 0 2]
    [0 0 1 1/2]]
```

The reduced row-echelon form shown on the screen above is equivalent to the system of equations

$$x = \tfrac{1}{2},$$
$$y = 2,$$
$$z = \tfrac{1}{2}.$$

The solution of the system is thus $\left(\frac{1}{2}, 2, \frac{1}{2}\right)$, which can be read directly from the last column of the reduced row-echelon matrix.

Recall that some systems of equations are inconsistent, and some sets of equations are dependent. When these cases occur, the reduced row-echelon form of the corresponding matrix will have a row or a column of zeros. For example, the matrix

$$\left[\begin{array}{cc|c} 1 & 0 & 4 \\ 0 & 0 & 0 \end{array}\right]$$

is in reduced row-echelon form. The second row translates to the equation $0 = 0$, and indicates that the equations in the system are dependent. The second row of the matrix

$$\left[\begin{array}{cc|c} 1 & 0 & 4 \\ 0 & 0 & 6 \end{array}\right]$$

translates to the equation $0 = 6$. This system is inconsistent.

3.6 EXERCISE SET

FOR EXTRA HELP

MathXL | MyMathLab | InterAct Math | AW Math Tutor Center | Video Lectures on CD: Disc 2 | Student's Solutions Manual

Concept Reinforcement *Complete each of the following statements.*

1. The rows of a matrix are ______________ and the ______________ are vertical.

2. Multiplying the numbers in a row of a matrix by a constant corresponds to multiplying both sides of a(n) ______________ by a constant.

3. Each number in a matrix is called a(n) ______________ or element.

4. The plural of the word matrix is ______________.

5. To solve a system using matrices, we can replace any row by the sum of that row and a(n) ______________ of another row.

6. In the final step of solving a system of equations, the leftmost column has zeros in all rows except the ______________ one.

Solve using matrices.

7. $9x - 2y = 5,$
$3x - 3y = 11$

8. $4x + y = 7,$
$5x - 3y = 13$

9. $x + 4y = 8,$
$3x + 5y = 3$

10. $x + 4y = 5,$
$-3x + 2y = 13$

11. $6x - 2y = 4,$
$7x + y = 13$

12. $3x + 4y = 7,$
$-5x + 2y = 10$

13. $3x + 2y + 2z = 3,$
$x + 2y - z = 5,$
$2x - 4y + z = 0$

14. $4x - y - 3z = 19,$
$8x + y - z = 11,$
$2x + y + 2z = -7$

15. $p - 2q - 3r = 3,$
$2p - q - 2r = 4,$
$4p + 5q + 6r = 4$

16. $x + 2y - 3z = 9,$
$2x - y + 2z = -8,$
$3x - y - 4z = 3$

17. $3p + 2r = 11,$
$q - 7r = 4,$
$p - 6q = 1$

18. $4a + 9b = 8,$
$8a + 6c = -1,$
$6b + 6c = -1$

19. $2x + 2y - 2z - 2w = -10,$
$w + y + z + x = -5,$
$x - y + 4z + 3w = -2,$
$w - 2y + 2z + 3x = -6$

20. $-w - 3y + z + 2x = -8,$
$x + y - z - w = -4,$
$w + y + z + x = 22,$
$x - y - z - w = -14$

Solve using matrices.

21. *Coin Value.* A collection of 42 coins consists of dimes and nickels. The total value is \$3.00. How many dimes and how many nickels are there?

22. *Coin Value.* A collection of 43 coins consists of dimes and quarters. The total value is \$7.60. How many dimes and how many quarters are there?

23. *Mixed Granola.* Grace sells two kinds of granola. One is worth \$4.05 per pound and the other is worth \$2.70 per pound. She wants to blend the two granolas to get a 15-lb mixture worth \$3.15 per pound. How much of each kind of granola should be used?

24. *Trail Mix.* Phil mixes nuts worth \$1.60 per pound with oats worth \$1.40 per pound to get 20 lb of trail mix worth \$1.54 per pound. How many pounds of nuts and how many pounds of oats should be used?

25. *Investments.* Elena receives \$212 per year in simple interest from three investments totaling \$2500. Part is invested at 7%, part at 8%, and part at 9%. There is \$1100 more invested at 9% than at 8%. Find the amount invested at each rate.

26. *Investments.* Miguel receives \$306 per year in simple interest from three investments totaling \$3200. Part is invested at 8%, part at 9%, and part at 10%. There is \$1900 more invested at 10% than at 9%. Find the amount invested at each rate.

TW **27.** Explain how you can recognize dependent equations when solving with matrices.

TW **28.** Explain how you can recognize an inconsistent system when solving with matrices.

Skill Maintenance

Simplify. [1.2]

29. $5(-3) - (-7)4$

30. $8(-5) - (-2)9$

31. $-2(5 \cdot 3 - 4 \cdot 6) - 3(2 \cdot 7 - 15) + 4(3 \cdot 8 - 5 \cdot 4)$

32. $6(2 \cdot 7 - 3(-4)) - 4(3(-8) - 10) + 5(4 \cdot 3 - (-2)7)$

Synthesis

TW **33.** If the matrices

$$\left[\begin{array}{cc|c} a_1 & b_1 & c_1 \\ d_1 & e_1 & f_1 \end{array}\right] \quad \text{and} \quad \left[\begin{array}{cc|c} a_2 & b_2 & c_2 \\ d_2 & e_2 & f_2 \end{array}\right]$$

share the same solution, does it follow that the corresponding entries are all equal to each other ($a_1 = a_2$, $b_1 = b_2$, etc.)? Why or why not?

TW **34.** Explain how the row-equivalent operations make use of the addition, multiplication, and distributive properties.

35. The sum of the digits in a four-digit number is 10. Twice the sum of the thousands digit and the tens digit is 1 less than the sum of the other two digits. The tens digit is twice the thousands digit. The ones digit equals the sum of the thousands digit and the hundreds digit. Find the four-digit number.

36. Solve for x and y:

$$ax + by = c,$$
$$dx + ey = f.$$

3.7 Determinants and Cramer's Rule

Determinants of 2×2 Matrices ■ Cramer's Rule: 2×2 Systems ■ Cramer's Rule: 3×3 Systems

Study Tip

Find the Highlights

If you do not already own one, consider purchasing a highlighter to use as you read this text and work on the exercises. Often the best time to highlight an important sentence or step in an example is after you have read through the section the first time.

Determinants of 2×2 Matrices

When a matrix has m rows and n columns, it is called an "m by n" matrix. Thus its *dimensions* are denoted by $m \times n$. If a matrix has the same number of rows and columns, it is called a **square matrix**. Associated with every square matrix is a number called its **determinant,** defined as follows for 2×2 matrices.

2 × 2 Determinants The determinant of a two-by-two matrix $\begin{bmatrix} a & c \\ b & d \end{bmatrix}$ is denoted $\begin{vmatrix} a & c \\ b & d \end{vmatrix}$ and is defined as follows:

$$\begin{vmatrix} a & c \\ b & d \end{vmatrix} = ad - bc.$$

EXAMPLE 1 Evaluate: $\begin{vmatrix} 2 & -5 \\ 6 & 7 \end{vmatrix}$.

SOLUTION We multiply and subtract as follows:

$$\begin{vmatrix} 2 & -5 \\ 6 & 7 \end{vmatrix} = 2 \cdot 7 - 6 \cdot (-5) = 14 + 30 = 44.$$

Cramer's Rule: 2×2 Systems

One of the many uses for determinants is in solving systems of linear equations in which the number of variables is the same as the number of equations and the constants are not all 0. Let's consider a system of two equations:

$$a_1x + b_1y = c_1,$$
$$a_2x + b_2y = c_2.$$

If we use the elimination method, a series of steps can show that

$$x = \frac{c_1b_2 - c_2b_1}{a_1b_2 - a_2b_1} \quad \text{and} \quad y = \frac{a_1c_2 - a_2c_1}{a_1b_2 - a_2b_1}.$$

These fractions can be rewritten using determinants.

Cramer's Rule: 2 × 2 Systems The solution of the system

$$a_1x + b_1y = c_1,$$
$$a_2x + b_2y = c_2,$$

if it is unique, is given by

$$x = \frac{\begin{vmatrix} c_1 & b_1 \\ c_2 & b_2 \end{vmatrix}}{\begin{vmatrix} a_1 & b_1 \\ a_2 & b_2 \end{vmatrix}}, \qquad y = \frac{\begin{vmatrix} a_1 & c_1 \\ a_2 & c_2 \end{vmatrix}}{\begin{vmatrix} a_1 & b_1 \\ a_2 & b_2 \end{vmatrix}}.$$

These formulas apply only if the denominator is not 0. If the denominator *is* 0, then one of two things happens:

1. If the denominator is 0 and the numerators are also 0, then the equations in the system are dependent.
2. If the denominator is 0 and at least one numerator is not 0, then the system is inconsistent.

To use Cramer's rule, we find the determinants and compute x and y as shown above. Note that the denominators are identical and the coefficients of x and y appear in the same position as in the original equations. In the numerator of x, the constants c_1 and c_2 replace a_1 and a_2. In the numerator of y, the constants c_1 and c_2 replace b_1 and b_2.

EXAMPLE 2 Solve using Cramer's rule:

$$2x + 5y = 7,$$
$$5x - 2y = -3.$$

SOLUTION We have

$$x = \frac{\begin{vmatrix} 7 & 5 \\ -3 & -2 \end{vmatrix}}{\begin{vmatrix} 2 & 5 \\ 5 & -2 \end{vmatrix}} \qquad \text{Using Cramer's rule}$$

$$= \frac{7(-2) - (-3)5}{2(-2) - 5 \cdot 5} = -\frac{1}{29}$$

and

$$y = \frac{\begin{vmatrix} 2 & 7 \\ 5 & -3 \end{vmatrix}}{\begin{vmatrix} 2 & 5 \\ 5 & -2 \end{vmatrix}} \qquad \text{Using Cramer's rule}$$

$$= \frac{2(-3) - 5 \cdot 7}{-29} = \frac{41}{29}. \qquad \text{The denominator is the same as in the expression for } x.$$

The solution is $\left(-\frac{1}{29}, \frac{41}{29}\right)$. The check is left to the student.

Cramer's Rule: 3 × 3 Systems

Cramer's rule can be extended for systems of three linear equations. However, before doing so, we must define what a 3 × 3 determinant is.

3 × 3 Determinants The determinant of a three-by-three matrix is defined as follows:

Subtract. Add.

$$\begin{vmatrix} a_1 & b_1 & c_1 \\ a_2 & b_2 & c_2 \\ a_3 & b_3 & c_3 \end{vmatrix} = a_1\begin{vmatrix} b_2 & c_2 \\ b_3 & c_3 \end{vmatrix} - a_2\begin{vmatrix} b_1 & c_1 \\ b_3 & c_3 \end{vmatrix} + a_3\begin{vmatrix} b_1 & c_1 \\ b_2 & c_2 \end{vmatrix}.$$

Student Notes

Cramer's rule and the evaluation of determinants rely on patterns. The specific formulas are less important than the patterns that they represent.

Note that the a's come from the first column. Note too that the 2 × 2 determinants above can be obtained by crossing out the row and the column in which the a occurs.

For a_1:

$$\begin{vmatrix} \cancel{a_1} & \cancel{b_1} & \cancel{c_1} \\ \cancel{a_2} & b_2 & c_2 \\ \cancel{a_3} & b_3 & c_3 \end{vmatrix}$$

For a_2:

$$\begin{vmatrix} \cancel{a_1} & b_1 & c_1 \\ \cancel{a_2} & \cancel{b_2} & \cancel{c_2} \\ \cancel{a_3} & b_3 & c_3 \end{vmatrix}$$

For a_3:

$$\begin{vmatrix} \cancel{a_1} & b_1 & c_1 \\ \cancel{a_2} & b_2 & c_2 \\ \cancel{a_3} & \cancel{b_3} & \cancel{c_3} \end{vmatrix}$$

EXAMPLE 3 Evaluate:

$$\begin{vmatrix} -1 & 0 & 1 \\ -5 & 1 & -1 \\ 4 & 8 & 1 \end{vmatrix}.$$

SOLUTION We have

Subtract. Add.

$$\begin{vmatrix} -1 & 0 & 1 \\ -5 & 1 & -1 \\ 4 & 8 & 1 \end{vmatrix} = -1\begin{vmatrix} 1 & -1 \\ 8 & 1 \end{vmatrix} - (-5)\begin{vmatrix} 0 & 1 \\ 8 & 1 \end{vmatrix} + 4\begin{vmatrix} 0 & 1 \\ 1 & -1 \end{vmatrix}$$

$$= -1(1 + 8) + 5(0 - 8) + 4(0 - 1)$$ **Evaluating the three determinants**

$$= -9 - 40 - 4 = -53.$$

Cramer's Rule: 3 × 3 Systems The solution of the system

$$a_1x + b_1y + c_1z = d_1,$$
$$a_2x + b_2y + c_2z = d_2,$$
$$a_3x + b_3y + c_3z = d_3$$

can be found using the following determinants:

$$D = \begin{vmatrix} a_1 & b_1 & c_1 \\ a_2 & b_2 & c_2 \\ a_3 & b_3 & c_3 \end{vmatrix}, \quad D_x = \begin{vmatrix} d_1 & b_1 & c_1 \\ d_2 & b_2 & c_2 \\ d_3 & b_3 & c_3 \end{vmatrix},$$

D contains only coefficients. In D_x, the d's replace the a's.

$$D_y = \begin{vmatrix} a_1 & d_1 & c_1 \\ a_2 & d_2 & c_2 \\ a_3 & d_3 & c_3 \end{vmatrix}, \quad D_z = \begin{vmatrix} a_1 & b_1 & d_1 \\ a_2 & b_2 & d_2 \\ a_3 & b_3 & d_3 \end{vmatrix}.$$

In D_y, the d's replace the b's. In D_z, the d's replace the c's.

If a unique solution exists, it is given by

$$x = \frac{D_x}{D}, \quad y = \frac{D_y}{D}, \quad z = \frac{D_z}{D}.$$

EXAMPLE 4 Solve using Cramer's rule:

$$x - 3y + 7z = 13,$$
$$x + y + z = 1,$$
$$x - 2y + 3z = 4.$$

SOLUTION We compute D, D_x, D_y, and D_z:

$$D = \begin{vmatrix} 1 & -3 & 7 \\ 1 & 1 & 1 \\ 1 & -2 & 3 \end{vmatrix} = -10; \quad D_x = \begin{vmatrix} 13 & -3 & 7 \\ 1 & 1 & 1 \\ 4 & -2 & 3 \end{vmatrix} = 20;$$

$$D_y = \begin{vmatrix} 1 & 13 & 7 \\ 1 & 1 & 1 \\ 1 & 4 & 3 \end{vmatrix} = -6; \quad D_z = \begin{vmatrix} 1 & -3 & 13 \\ 1 & 1 & 1 \\ 1 & -2 & 4 \end{vmatrix} = -24.$$

Then

$$x = \frac{D_x}{D} = \frac{20}{-10} = -2;$$
$$y = \frac{D_y}{D} = \frac{-6}{-10} = \frac{3}{5};$$
$$z = \frac{D_z}{D} = \frac{-24}{-10} = \frac{12}{5}.$$

The solution is $\left(-2, \frac{3}{5}, \frac{12}{5}\right)$. The check is left to the student.

In Example 4, we need not have evaluated D_z. Once x and y were found, we could have substituted them into one of the equations to find z.

To use Cramer's rule, we divide by D, provided $D \neq 0$. If $D = 0$ and at least one of the other determinants is not 0, then the system is inconsistent. If *all* the determinants are 0, then the equations in the system are dependent.

Determinants

Determinants can be evaluated using the DET(option of the MATRIX MATH menu. After entering the matrix, we go to the home screen and select the determinant operation. Then we enter the name of the matrix using the MATRIX NAMES menu. The graphing calculator will return the value of the determinant of the matrix. For example, if

$$\mathbf{A} = \begin{bmatrix} 1 & 6 & -1 \\ -3 & -5 & 3 \\ 0 & 4 & 2 \end{bmatrix},$$

we have

```
det([A])
              26
```
.

3.7 EXERCISE SET

FOR EXTRA HELP

MathXL | MyMathLab | InterAct Math | AW Math Tutor Center | Video Lectures on CD: Disc 2 | Student's Solutions Manual

Concept Reinforcement *Classify each of the following as either true or false.*

1. A square matrix has the same number of rows and columns.

2. A 3×4 matrix has 3 rows and 4 columns.

3. Cramer's rule exists only for 2×2 systems.

4. Whenever Cramer's rule yields a denominator that is 0, the system has no solution.

5. Whenever Cramer's rule yields a numerator that is 0, the equations are dependent.

6. Cramer's rule allows us to solve some systems that could not be solved any other way.

Evaluate.

7. $\begin{vmatrix} 5 & 1 \\ 2 & 4 \end{vmatrix}$

8. $\begin{vmatrix} 3 & 2 \\ 2 & -3 \end{vmatrix}$

9. $\begin{vmatrix} 6 & -9 \\ 2 & 3 \end{vmatrix}$

10. $\begin{vmatrix} 3 & 2 \\ -7 & 5 \end{vmatrix}$

11. $\begin{vmatrix} 1 & 4 & 0 \\ 0 & -1 & 2 \\ 3 & -2 & 1 \end{vmatrix}$

12. $\begin{vmatrix} 3 & 0 & -2 \\ 5 & 1 & 2 \\ 2 & 0 & -1 \end{vmatrix}$

13. $\begin{vmatrix} -1 & -2 & -3 \\ 3 & 4 & 2 \\ 0 & 1 & 2 \end{vmatrix}$

14. $\begin{vmatrix} 1 & 2 & 2 \\ 2 & 1 & 0 \\ 3 & 3 & 1 \end{vmatrix}$

15. $\begin{vmatrix} -4 & -2 & 3 \\ -3 & 1 & 2 \\ 3 & 4 & -2 \end{vmatrix}$

16. $\begin{vmatrix} 2 & -1 & 1 \\ 1 & 2 & -1 \\ 3 & 4 & -3 \end{vmatrix}$

Solve using Cramer's rule.

17. $5x + 8y = 1,$
$3x + 7y = 5$

18. $3x - 4y = 6,$
$5x + 9y = 10$

19. $5x - 4y = -3,$
$7x + 2y = 6$

20. $-2x + 4y = 3,$
$3x - 7y = 1$

21. $3x - y + 2z = 1,$
$x - y + 2z = 3,$
$-2x + 3y + z = 1$

22. $3x + 2y - z = 4,$
$3x - 2y + z = 5,$
$4x - 5y - z = -1$

23. $2x - 3y + 5z = 27,$
$x + 2y - z = -4,$
$5x - y + 4z = 27$

24. $x - y + 2z = -3,$
$x + 2y + 3z = 4,$
$2x + y + z = -3$

25. $r - 2s + 3t = 6,$
$2r - s - t = -3,$
$r + s + t = 6$

26. $a - 3c = 6,$
$b + 2c = 2,$
$7a - 3b - 5c = 14$

TW 27. What is it about Cramer's rule that makes it useful?

TW 28. Which version of Cramer's rule do you find more useful: the version for 2 × 2 systems or the version for 3 × 3 systems? Why?

Focused Review

29. Solve by graphing by hand. [3.1]

$x + y = -2,$
$y = 2 - 5x$

30. Solve using a graphing calculator. [3.1]

$8y = 14 - 3x,$
$5y = 8 + x$

31. Solve using substitution. [3.2]

$x + y = 3,$
$y = 2x - 1$

32. Solve using elimination. [3.2]

$2x + 3y = 10,$
$5x - 3y = 4$

33. Solve using matrices. [3.6]

$x - 3y = 7,$
$2x + y = 8$

34. Solve using Cramer's rule. [3.7]

$2x + y = 1,$
$3x - 2y = 10$

Synthesis

TW 35. Cramer's rule states that if $a_1x + b_1y = c_1$ and $a_2x + b_2y = c_2$ are dependent, then

$$\begin{vmatrix} a_1 & b_1 \\ a_2 & b_2 \end{vmatrix} = 0.$$

Explain why this will always happen.

TW 36. Under what conditions can a 3 × 3 system of linear equations be consistent but unable to be solved using Cramer's rule?

Solve.

37. $\begin{vmatrix} y & -2 \\ 4 & 3 \end{vmatrix} = 44$

38. $\begin{vmatrix} 2 & x & -1 \\ -1 & 3 & 2 \\ -2 & 1 & 1 \end{vmatrix} = -12$

39. $\begin{vmatrix} m + 1 & -2 \\ m - 2 & 1 \end{vmatrix} = 27$

40. Show that an equation of the line through (x_1, y_1) and (x_2, y_2) can be written

$$\begin{vmatrix} x & y & 1 \\ x_1 & y_1 & 1 \\ x_2 & y_2 & 1 \end{vmatrix} = 0.$$

3.8 Business and Economics Applications

Break-Even Analysis ■ Supply and Demand

Study Tip

Try to Look Ahead

If you are able to at least skim through an upcoming section before your instructor covers that lesson, you will be better able to focus on what is being emphasized in class. Similarly, if you can begin studying for a quiz or test a day or two before you really *need* to, you will reap great rewards for doing so.

CAUTION! Do not confuse "cost" with "price." When we discuss the *cost* of an item, we are referring to what it costs to produce the item. The *price* of an item is what a consumer pays to purchase the item and is used when calculating revenue.

Break-Even Analysis

When a company manufactures x units of a product, it spends money. This is **total cost** and can be thought of as a function C, where $C(x)$ is the total cost of producing x units. When the company sells x units of the product, it takes in money. This is **total revenue** and can be thought of as a function R, where $R(x)$ is the total revenue from the sale of x units. **Total profit** is the money taken in less the money spent, or total revenue minus total cost. Total profit from the production and sale of x units is a function P given by

$$\textbf{Profit} = \textbf{Revenue} - \textbf{Cost}, \quad \text{or} \quad P(x) = R(x) - C(x).$$

If $R(x)$ is greater than $C(x)$, there is a gain and $P(x)$ is positive. If $C(x)$ is greater than $R(x)$, there is a loss and $P(x)$ is negative. When $R(x) = C(x)$, the company breaks even.

There are two kinds of costs. First, there are costs like rent, insurance, machinery, and so on. These costs, which must be paid whether a product is produced or not, are called *fixed costs*. When a product is being produced, there are costs for labor, materials, marketing, and so on. These are called *variable costs*, because they vary according to the amount being produced. The sum of the fixed cost and the variable cost gives the *total cost* of producing a product.

EXAMPLE 1 Manufacturing Lamps. Ergs, Inc., is planning to make a new lamp. Fixed costs will be \$90,000, and it will cost \$15 to produce each lamp (variable costs). Each lamp sells for \$26.

a) Find the total cost $C(x)$ of producing x lamps.

b) Find the total revenue $R(x)$ from the sale of x lamps.

c) Find the total profit $P(x)$ from the production and sale of x lamps.

d) What profit will the company realize from the production and sale of 3000 lamps? of 14,000 lamps?

e) Graph the total-cost, total-revenue, and total-profit functions using the same set of axes. Determine the break-even point.

SOLUTION

a) Total cost is given by

$$C(x) = (\text{Fixed costs}) \text{ plus } (\text{Variable costs}),$$
$$\text{or} \quad C(x) = 90{,}000 + 15x,$$

where x is the number of lamps produced.

b) Total revenue is given by

$R(x) = 26x.$ **\$26 times the number of lamps sold. We assume that every lamp produced is sold.**

c) Total profit is given by

$P(x) = R(x) - C(x)$ **Profit is revenue minus cost.**

$$= 26x - (90{,}000 + 15x)$$
$$= 11x - 90{,}000.$$

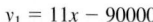

d) Profits will be

$$P(3000) = 11 \cdot 3000 - 90{,}000 = -\$57{,}000$$

when 3000 lamps are produced and sold, and

$$P(14{,}000) = 11 \cdot 14{,}000 - 90{,}000 = \$64{,}000$$

when 14,000 lamps are produced and sold. These values can also be found using a graphing calculator, as shown in the figure at left. Thus the company loses money if only 3000 lamps are sold, but makes money if 14,000 are sold.

$y_1 = 11x - 90000$

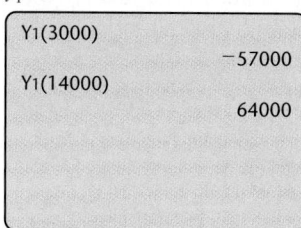

```
Y1(3000)
                -57000
Y1(14000)
                 64000
```

e) The graphs of each of the three functions are shown below:

$R(x) = 26x,$ **This represents the revenue function.**

$C(x) = 90{,}000 + 15x,$ **This represents the cost function.**

$P(x) = 11x - 90{,}000.$ **This represents the profit function.**

$R(x)$, $C(x)$, and $P(x)$ are all in dollars.

The revenue function has a graph that goes through the origin and has a slope of 26. The cost function has an intercept on the \$-axis of 90,000 and has a slope of 15. The profit function has an intercept on the \$-axis of −90,000 and has a slope of 11. It is shown by the dashed line. The red dashed line shows a "negative" profit, which is a loss. (That is what is known as "being in the red.") The black dashed line shows a "positive" profit, or gain. (That is what is known as "being in the black.")

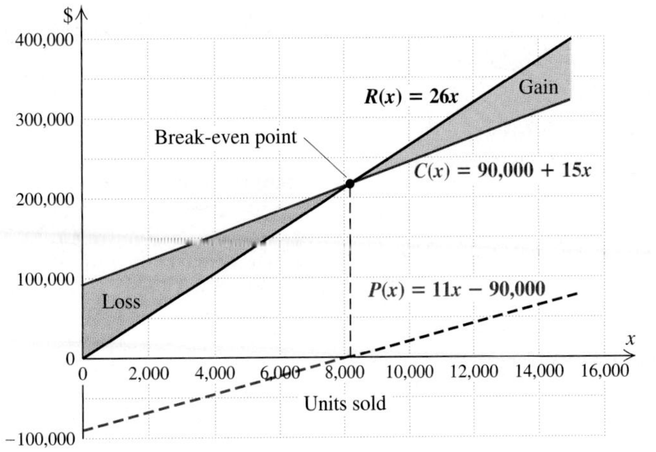

Gains occur where the revenue is greater than the cost. Losses occur where the revenue is less than the cost. The **break-even point** occurs where the graphs of R and C cross. Thus to find the break-even point, we solve a system:

$$R(x) = 26x,$$
$$C(x) = 90{,}000 + 15x.$$

Since both revenue and cost are in *dollars* and they are equal at the break-even point, the system can be rewritten as

$$d = 26x, \qquad (1)$$
$$d = 90{,}000 + 15x \qquad (2)$$

and solved using substitution:

$$26x = 90{,}000 + 15x \qquad \textbf{Substituting } 26x \textbf{ for } d \textbf{ in equation (2)}$$
$$11x = 90{,}000$$
$$x \approx 8181.8.$$

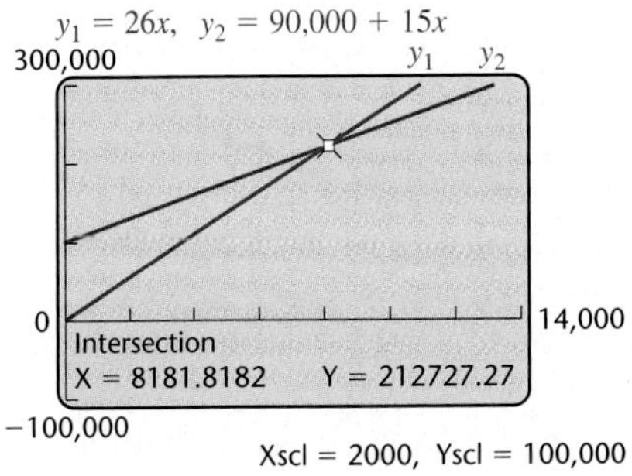

The system can also be solved using a graphing calculator, as shown in the figure at left.

The firm will break even if it produces and sells about 8182 lamps (8181 will yield a tiny loss and 8182 a tiny gain), and takes in a total of $R(8182) = 26 \cdot 8182 = \$212{,}732$ in revenue. Note that the x-coordinate of the break-even point can also be found by solving $P(x) = 0$. The break-even point is (8182 lamps, \$212,732).

Student Notes

If you plan to study business or economics, you may want to consult the material in this section when these topics arise in your other courses.

Supply and Demand

As the price of coffee varies, the amount sold varies. The table and graph below show that *consumers will demand less as the price goes up.*

Demand function, *D*

Price, p, per Kilogram	Quantity, $D(p)$ (in millions of kilograms)
\$ 8.00	25
9.00	20
10.00	15
11.00	10
12.00	5

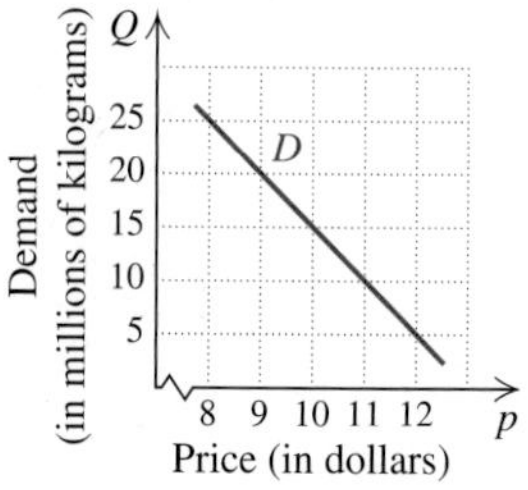

As the price of coffee varies, the amount available varies. The table and graph below show that *sellers will supply more as the price goes up.*

Supply function, *S*

Price, p, per Kilogram	Quantity, $S(p)$ (in millions of kilograms)
\$ 9.00	5
9.50	10
10.00	15
10.50	20
11.00	25

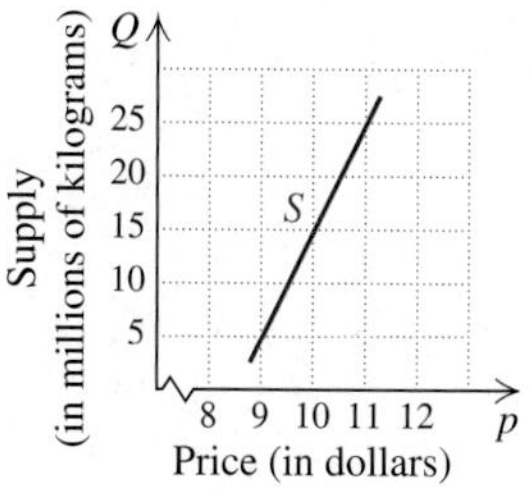

Let's look at the above graphs together. We see that as price increases, demand decreases. As price increases, supply increases. The point of intersection is called the **equilibrium point.** At that price, the amount that the seller will supply is the same amount that the consumer will buy. The situation is analogous to a buyer and a seller negotiating the price of an item. The equilibrium point is the price and quantity that they finally agree on.

Any ordered pair of coordinates from the graph is (price, quantity), because the horizontal axis is the price axis and the vertical axis is the quantity axis. If D is a demand function and S is a supply function, then the equilibrium point is where demand equals supply:

$$D(p) = S(p).$$

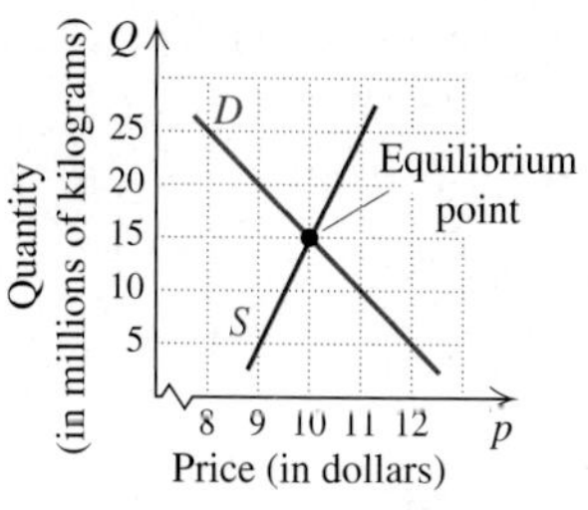

EXAMPLE 2 Find the equilibrium point for the demand and supply functions given:

$$D(p) = 1000 - 60p, \qquad (1)$$
$$S(p) = 200 + 4p. \qquad (2)$$

SOLUTION Since both demand and supply are *quantities* and they are equal at the equilibrium point, we rewrite the system as

$$q = 1000 - 60p, \qquad (1)$$
$$q = 200 + 4p. \qquad (2)$$

We substitute $200 + 4p$ for q in equation (1) and solve:

$$200 + 4p = 1000 - 60p \qquad \textbf{Substituting } 200 + 4p \textbf{ for } q \textbf{ in equation (1)}$$
$$200 + 64p = 1000 \qquad \textbf{Adding } 60p \textbf{ to both sides}$$
$$64p = 800 \qquad \textbf{Adding } -200 \textbf{ to both sides}$$
$$p = \tfrac{800}{64} = 12.5.$$

Thus the equilibrium price is \$12.50 per unit.

To find the equilibrium quantity, we substitute \$12.50 into either $D(p)$ or $S(p)$. We use $S(p)$:

$$S(12.5) = 200 + 4(12.5) = 200 + 50 = 250.$$

Thus the equilibrium quantity is 250 units, and the equilibrium point is (\$12.50, 250). The graph at left confirms the solution.

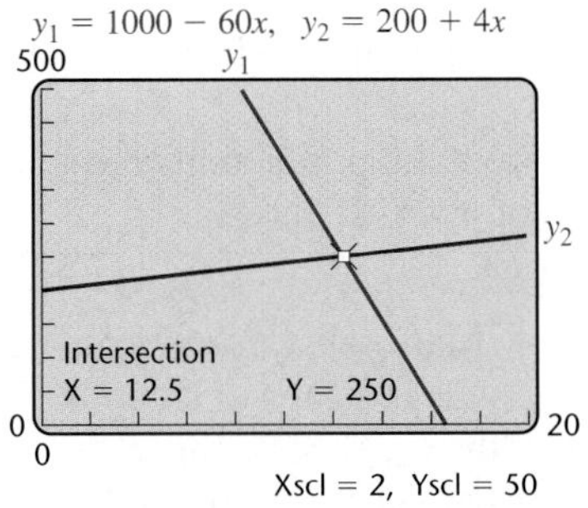

3.8 EXERCISE SET

FOR EXTRA HELP

MathXL · MyMathLab · InterAct Math · AW Math Tutor Center · Video Lectures on CD: Disc 2 · Student's Solutions Manual

Concept Reinforcement *In each of Exercises 1–8, match the word or phrase with the most appropriate choice from the column on the right.*

1. ____ Total cost

2. ____ Total revenue

3. ____ Total profit

4. ____ Fixed costs

5. ____ Variable costs

6. ____ Break-even point

7. ____ Equilibrium point

8. ____ Price

a) The amount of money that a company takes in

b) The sum of fixed costs and variable costs

c) The point at which total revenue equals total cost

d) What consumers pay per item

e) The difference between total revenue and total cost

f) What companies spend whether or not a product is produced

g) The point at which supply equals demand

h) The costs that vary according to the number of items produced

For each of the following pairs of total-cost and total-revenue functions, find **(a)** *the total-profit function and* **(b)** *the break-even point.*

9. $C(x) = 45x + 300{,}000;$
$R(x) = 65x$

10. $C(x) = 25x + 270{,}000;$
$R(x) = 70x$

11. $C(x) = 10x + 120{,}000;$
$R(x) = 60x$

12. $C(x) = 30x + 49{,}500;$
$R(x) = 85x$

13. $C(x) = 40x + 22{,}500;$
$R(x) = 85x$

14. $C(x) = 20x + 10{,}000;$
$R(x) = 100x$

15. $C(x) = 22x + 16{,}000;$
$R(x) = 40x$

16. $C(x) = 15x + 75{,}000;$
$R(x) = 55x$

Aha! **17.** $C(x) = 75x + 100{,}000;$
$R(x) = 125x$

18. $C(x) = 20x + 120{,}000;$
$R(x) = 50x$

Find the equilibrium point for each of the following pairs of demand and supply functions.

19. $D(p) = 1000 - 10p,$
$S(p) = 230 + p$

20. $D(p) = 2000 - 60p,$
$S(p) = 460 + 94p$

21. $D(p) = 760 - 13p,$
$S(p) = 430 + 2p$

22. $D(p) = 800 - 43p,$
$S(p) = 210 + 16p$

23. $D(p) = 7500 - 25p,$
$S(p) = 6000 + 5p$

24. $D(p) = 8800 - 30p,$
$S(p) = 7000 + 15p$

25. $D(p) = 1600 - 53p,$
$S(p) = 320 + 75p$

26. $D(p) = 5500 - 40p,$
$S(p) = 1000 + 85p$

Solve.

27. *Computer Manufacturing.* Biz.com Electronics is planning to introduce a new line of computers. The fixed costs for production are \$125,300. The variable costs for producing each computer are \$450. The revenue from each computer is \$800. Find the following.

a) The total cost $C(x)$ of producing x computers
b) The total revenue $R(x)$ from the sale of x computers
c) The total profit $P(x)$ from the production and sale of x computers
d) The profit or loss from the production and sale of 100 computers; of 400 computers
e) The break-even point

28. *Manufacturing CD Players.* SoundGen, Inc., is planning to manufacture a new type of CD player. The fixed costs for production are \$22,500. The variable costs for producing each CD player are estimated to be \$40. The revenue from each CD player is to be \$85. Find the following.

a) The total cost $C(x)$ of producing x CD players
b) The total revenue $R(x)$ from the sale of x CD players
c) The total profit $P(x)$ from the production and sale of x CD players
d) The profit or loss from the production and sale of 3000 CD players; of 400 CD players
e) The break-even point

29. *Manufacturing Caps.* Martina's Custom Printing is planning on adding painter's caps to its product line. For the first year, the fixed costs for setting up production are \$16,404. The variable costs for producing a dozen caps are \$6.00. The revenue on each dozen caps will be \$18.00. Find the following.

a) The total cost $C(x)$ of producing x dozen caps
b) The total revenue $R(x)$ from the sale of x dozen caps
c) The total profit $P(x)$ from the production and sale of x dozen caps
d) The profit or loss from the production and sale of 3000 dozen caps; of 1000 dozen caps
e) The break-even point

30. *Sport Coat Production.* Sarducci's is planning a new line of sport coats. For the first year, the fixed costs for setting up production are \$10,000. The variable costs for producing each coat are \$30. The revenue from each coat is to be \$80. Find the following.

a) The total cost $C(x)$ of producing x coats
b) The total revenue $R(x)$ from the sale of x coats
c) The total profit $P(x)$ from the production and sale of x coats
d) The profit or loss from the production and sale of 2000 coats; of 50 coats
e) The break-even point

31. *Dog Food Production.* Great Foods will soon begin producing a new line of puppy food. The marketing department predicts that the demand function will be $D(p) = -14.97p + 987.35$ and the supply function will be $S(p) = 98.55p - 5.13$.

a) To the nearest cent, what price per unit should be charged in order to have equilibrium between supply and demand?

b) The production of the puppy food involves \$5265 in fixed costs and \$2.10 per unit in variable costs. If the price per unit is the value you found in part (a), how many units must be sold in order to break even?

32. *Computer Production.* Number Solutions Computers is planning a new line of computers, each of which will sell for \$970. For the first year, the fixed costs in setting up production are \$1,235,580 and the variable costs for each computer are \$697.

a) What is the break-even point? (Round to the nearest whole number.)

b) The marketing department at Number Solutions is not sure that \$970 is the best price. Their demand function for the new computers is given by $D(p) = -304.5p + 374{,}580$ and their supply function is given by $S(p) = 788.7p - 576{,}504$. What price p would result in equilibrium between supply and demand?

TW **33.** In Example 1, the slope of the line representing Revenue is the sum of the slopes of the other two lines. This is not a coincidence. Explain why.

TW **34.** Variable costs and fixed costs are often compared to the slope and the y-intercept, respectively, of an equation for a line. Explain why you feel this analogy is or is not valid.

Skill Maintenance

Solve. [1.6]

35. $3x - 9 = 27$

36. $4x - 7 = 53$

37. $4x - 5 = 7x - 13$

38. $2x + 9 = 8x - 15$

39. $7 - 2(x - 8) = 14$

40. $6 - 4(3x - 2) = 10$

Synthesis

TW **41.** Ian claims that since his fixed costs are \$1000, he need sell only 20 birdbaths at \$50 each in order to break even. Does this sound plausible? Why or why not?

TW **42.** In this section, we examined supply and demand functions for coffee. Does it seem realistic to you for the graph of D to have a constant slope? Why or why not?

43. *Yo-yo Production.* Bing Boing Hobbies is willing to produce 100 yo-yo's at \$2.00 each and 500 yo-yo's at \$8.00 each. Research indicates that the public will buy 500 yo-yo's at \$1.00 each and 100 yo-yo's at \$9.00 each. Find the equilibrium point.

44. *Loudspeaker Production.* Fidelity Speakers, Inc., has fixed costs of \$15,400 and variable costs of \$100 for each pair of speakers produced. If the speakers sell for \$250 a pair, how many pairs of speakers must be produced (and sold) in order to have enough profit to cover the fixed costs of two additional facilities? Assume that all fixed costs are identical.

45. *Peanut Butter.* The following table lists the data for supply and demand of an 18-oz jar of peanut butter at various prices.

Price	Supply (in millions)	Demand (in millions)
\$1.59	23.4	22.5
1.29	19.2	24.8
1.69	26.8	22.2
1.19	18.4	29.7
1.99	30.7	19.3

a) Use linear regression to find the supply function $S(p)$ for suppliers of peanut butter at price p.

b) Use linear regression to find the demand function $D(p)$ for consumers of peanut butter at price p.

c) Find the equilibrium point.

46. *Funnel Cakes.* Each year, the Harvey County Fair sets prices for concession vendors. The following table lists the data for supply and demand of a funnel cake at different prices.

Price	Supply (in thousands)	Demand (in thousands)
$1.50	3.6	5.4
1.75	4.8	5.2
2.00	6.2	5.0
2.50	7.2	4.2
2.25	7.1	4.0

a) Use linear regression to find the supply function $S(p)$ for suppliers of funnel cakes at price p.
b) Use linear regression to find the demand function $D(p)$ for consumers of funnel cakes at price p.
c) Find the equilibrium point.

Collaborative Corner

Patient Profits

Focus: Cost, revenue, and profit models

Time: 20 minutes

Group Size: 2–4

Dr. Bill Marks has been charting the operating cost and the revenue of his dental practice. The following table lists his data for the first six months of a year.

Month	Number of Patients	Costs	Revenue
January	252	$13,948	$12,493
February	174	12,742	8,750
March	310	14,678	15,125
April	298	14,620	14,137
May	369	15,683	17,930
June	342	15,365	17,138

ACTIVITY

1. Divide the group into two smaller groups, with one examining the cost data and the other the revenue data. Each group should do the following.
   a) Enter and graph the data, treating either cost or revenue as a function of the number of patients. Confirm that the relationship appears to be linear.
   b) Use linear regression to find either the monthly cost $C(x)$ of or the monthly revenue $R(x)$ from treating x patients.
2. Working together, use the results of step (1) to find the monthly profit $P(x)$ from treating x patients.
3. Using the profit function, estimate the profit or loss from treating 310 patients and compare it with the profit for March using the data in the table.
4. Find the break-even point.
5. Use the cost function to estimate the fixed costs and the variable costs.
6. In order to accept more patients, Dr. Marks must hire another part-time hygienist at a monthly salary of $2000.
   a) Find the new monthly cost $C_1(x)$ and profit $P_1(x)$.
   b) Determine the new break-even point.
   c) At least how many patients must the hygienist see monthly in order to cover the extra costs?

3 Chapter Summary and Review

KEY TERMS AND DEFINITIONS

SYSTEMS

System of equations, p. 183 Two or more equations that are to be solved simultaneously. A system is **consistent** if it has at least one solution. Otherwise it is **inconsistent.** The equations in a system are **dependent** if one of them can be removed without changing the solution set. Otherwise, they are **independent.**

Matrix (matrices), p. 238 A rectangular array of numbers. The **elements,** or **entries,** in a matrix are arranged in **rows** and **columns.**

APPLICATIONS

Distance *d*, rate *r*, and time *t* (p. 214) are related by

$$d = rt, \quad r = \frac{d}{t}, \quad \text{and} \quad t = \frac{d}{r}.$$

Total profit, p. 253 **Total revenue** (the amount taken in) minus **total cost** (the amount spent).

Break-even point, p. 255 The point at which total revenue equals total cost, or the point at which total profit is 0.

Equilibrium point, p. 256 The point at which **demand** equals **supply.**

IMPORTANT CONCEPTS

[Section references appear in brackets.]

<table>
<tr><th>Concept</th><th>Example</th></tr>
<tr><td>Solve systems of equations graphically.</td><td>
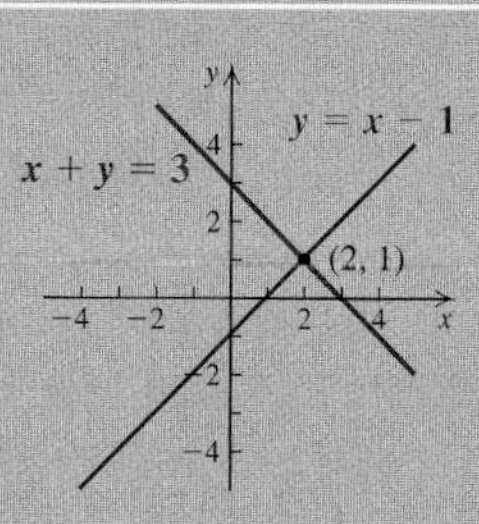

$x + y = 3,$
$y = x - 1$
The graphs intersect at (2, 1).
The solution is (2, 1).
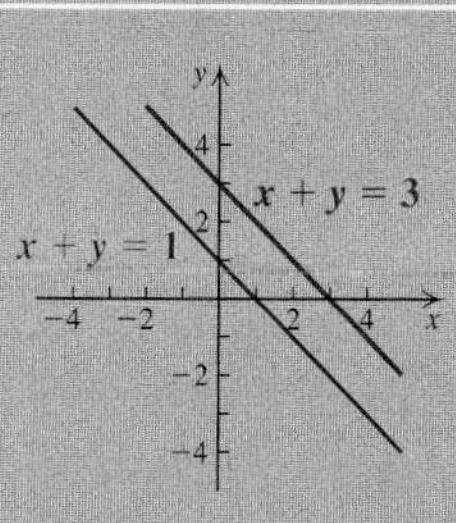

$x + y = 3,$
$x + y = 1$
The graphs do not intersect.
There is no solution.
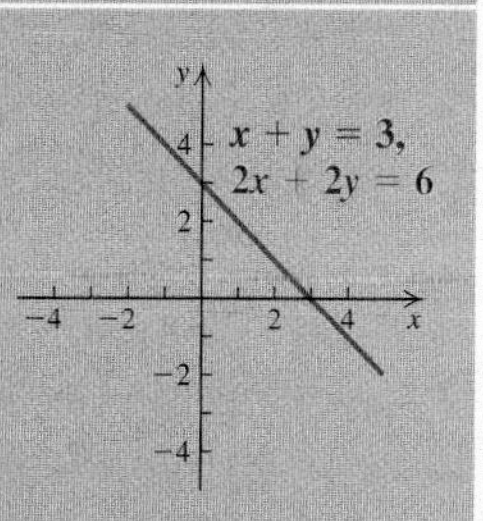

$x + y = 3,$
$2x + 2y = 6$
The graphs are the same.
The solution set is
$\{(x, y) \mid x + y = 3\}$.

[3.1]</td></tr>
</table>

(*continued*)

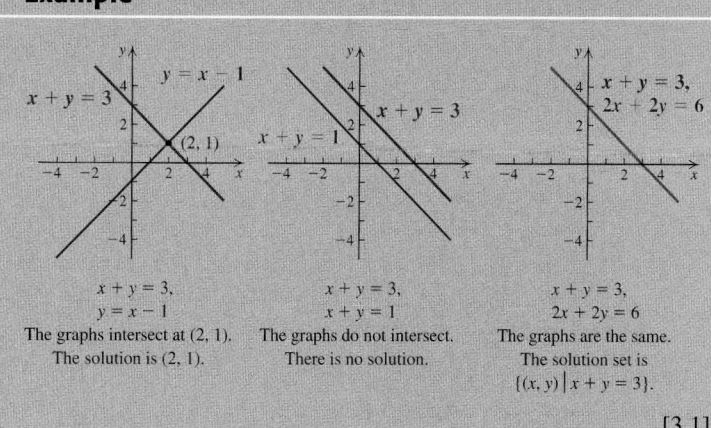

Solve systems of equations using substitution.	Solve: $2x + 3y = 8,$ $x = y + 1.$ *Substitute and solve for y:* $2(y + 1) + 3y = 8$ $2y + 2 + 3y = 8$ $y = \frac{6}{5}.$ *Substitute and solve for x:* $x = y + 1$ $x = \frac{6}{5} + 1$ $x = \frac{11}{5}.$ The solution is $\left(\frac{11}{5}, \frac{6}{5}\right)$. [3.2]
Solve systems of equations using elimination.	Solve: $4x - 2y = 6,$ $3x + y = 7.$ *Eliminate y and solve for x:* $4x - 2y = 6$ $\underline{6x + 2y = 14}$ $10x = 20$ $x = 2.$ *Substitute and solve for y:* $3x + y = 7$ $3 \cdot 2 + y = 7$ $y = 1.$ The solution is $(2, 1)$. [3.2]
Solve systems of equations using matrices.	Solve: $x + 4y = 1,$ $2x - y = 3.$ *Write as a matrix in row-echelon form.* $\left[\begin{array}{cc\|c} 1 & 4 & 1 \\ 2 & -1 & 3 \end{array}\right]$ $\left[\begin{array}{cc\|c} 1 & 4 & 1 \\ 0 & -9 & 1 \end{array}\right]$ *Rewrite as equations and solve:* $-9y = 1$ $y = -\frac{1}{9}$ $x + 4\left(-\frac{1}{9}\right) = 1$ $x = \frac{13}{9}.$ The solution is $\left(\frac{13}{9}, -\frac{1}{9}\right)$. [3.6]
Solve systems of three equations.	Solve: $x + y - z = 3,$ $-x + y + 2z = -5,$ $2x - y - 3z = 9$ *Eliminate x using two equations:* $x + y - z = 3$ $\underline{-x + y + 2z = -5}$ $2y + z = -2.$ *Eliminate x again using two different equations:* $-2x - 2y + 2z = -6$ $\underline{2x - y - 3z = 9}$ $-3y - z = 3.$ *Solve the system of two equations for y and z:* $2y + z = -2$ $\underline{-3y - z = 3}$ $-y = 1$ $y = -1$ $2(-1) + z = -2$ $z = 0.$ *Substitute and solve for x:* $x + y - z = 3$ $\underline{x + (-1) - 0 = 3}$ $x = 4.$ The solution is $(4, -1, 0)$. [3.4]

(continued)

Find the determinant of a matrix.	$\begin{bmatrix} 2 & 3 \\ -1 & 5 \end{bmatrix} = 2 \cdot 5 - (-1)(3) = 13$	[3.7]
Solve systems of equations using Cramer's rule.	Solve: $x - 3y = 7,$ $2x + 5y = 4.$ $x = \dfrac{\begin{vmatrix} 7 & -3 \\ 4 & 5 \end{vmatrix}}{\begin{vmatrix} 1 & -3 \\ 2 & 5 \end{vmatrix}}$ $y = \dfrac{\begin{vmatrix} 1 & 7 \\ 2 & 4 \end{vmatrix}}{\begin{vmatrix} 1 & -3 \\ 2 & 5 \end{vmatrix}}$ $x = \frac{47}{11}$ $y = \frac{-10}{11}$ The solution is $\left(\frac{47}{11}, -\frac{10}{11}\right)$.	[3.7]

Review Exercises

Concept Reinforement *Complete each of the following sentences.*

1. The system

$$5x + 3y = 7,$$
$$y = 2x + 1$$

is most easily solved using the ______________ method. [3.2]

2. The system

$$-2x + 3y = 8,$$
$$2x + 2y = 7$$

is most easily solved using the ______________ method. [3.2]

3. A weakness in using graphs to solve a system is that when solutions involve fractions or decimals, the graph may yield only a(n) ______________ solution. [3.2]

4. When one equation in a system is a multiple of another equation in that system, the equations are said to be ______________. [3.1]

5. A system for which there is no solution is said to be ______________. [3.1]

6. When using elimination to solve a system of two equations, if an identity is obtained, we know that there is a(n) ______________ number of solutions. [3.2]

7. When we are graphing to solve a system of two equations, if there is no solution, the lines will be ______________. [3.1]

8. When a matrix has the same number of rows and columns, it is said to be ______________. [3.7]

9. Cramer's rule is a formula in which the numerator and the denominator of each fraction is a(n) ______________. [3.7]

10. At the break-even point, the value of the profit function is ______________. [3.8]

For Exercises 11–19, if a system has an infinite number of solutions, use set-builder notation to write the solution set. If a system has no solution, state this. Solve graphically. [3.1]

11. $3x + 2y = -4,$
$y = 3x + 7$

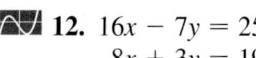

12. $16x - 7y = 25,$
$8x + 3y = 19$

Solve using the substitution method. [3.2]

13. $9x - 6y = 2,$
$x = 4y + 5$

14. $y = x + 2,$
$y - x = 8$

15. $x - 3y = -2,$
$7y - 4x = 6$

Solve using the elimination method. [3.2]

16. $8x - 2y = 10,$
$-4y - 3x = -17$

17. $4x - 7y = 18,$
$9x + 14y = 40$

18. $3x - 5y = -4,$
$5x - 3y = 4$

19. $1.5x - 3 = -2y,$
$3x + 4y = 6$

Solve. [3.3]

20. Luther bought two equally priced DVDs and one videocassette for \$48. If he had purchased one DVD and two videocassettes, he would have spent \$3 less. What is the price of a DVD? What is the price of a videocassette?

21. A freight train leaves Houston at midnight traveling north at a speed of 44 mph. One hour later, a passenger train, going 55 mph, travels north from Houston on a parallel track. How many hours will the passenger train travel before it overtakes the freight train?

22. Yolanda wants 14 L of fruit punch that is 10% juice. At the store, she finds punch that is 15% juice and punch that is 8% juice. How much of each should she purchase?

Solve. If a system's equations are dependent or if there is no solution, state this.

23. $x + 4y + 3z = 2,$
$2x + y + z = 10,$
$-x + y + 2z = 8$
[3.4]

24. $4x + 2y - 6z = 34,$
$2x + y + 3z = 3,$
$6x + 3y - 3z = 37$
[3.4]

25. $2x - 5y - 2z = -4,$
$7x + 2y - 5z = -6,$
$-2x + 3y + 2z = 4$
[3.4]

26. $-5x + 5y = -6,$
$2x - 2y = 4$
[3.2]

27. $3x + y = 2,$
$x + 3y + z = 0,$
$x + z = 2$ [3.4]

Solve.

28. In triangle ABC, the measure of angle A is four times the measure of angle C, and the measure of angle B is 45° more than the measure of angle C. What are the measures of the angles of the triangle? [3.5]

29. *Nontoxic Floor Wax.* A nontoxic floor wax can be made from lemon juice and food-grade linseed oil. The amount of oil should be twice the amount of lemon juice. How much of each ingredient is needed to make 32 oz of floor wax? (The mix should be spread with a rag and buffed when dry.) [3.3]

30. *Lumber Production.* Denison Lumber can convert logs into either lumber or plywood. In a given day, the mill turns out 42 pallets of plywood and lumber. It makes a profit of \$75 on a pallet of lumber and \$120 on a pallet of plywood. How many pallets of each type must be produced and sold in order to make a profit of \$3735? [3.3]

Solve using matrices. Show your work. [3.6]

31. $3x + 4y = -13,$
$5x + 6y = 8$

32. $3x - y + z = -1,$
$2x + 3y + z = 4,$
$5x + 4y + 2z = 5$

Evaluate. [3.7]

33. $\begin{vmatrix} -2 & 4 \\ -3 & 5 \end{vmatrix}$

34. $\begin{vmatrix} 2 & 3 & 0 \\ 1 & 4 & -2 \\ 2 & -1 & 5 \end{vmatrix}$

Solve using Cramer's rule. Show your work. [ ]

35. $2x + 3y = 6,$
$x - 4y = 14$

36. $2x + y + z = -2,$
$2x - y + 3z = 6,$
$3x - 5y + 4z = 7$

37. Find the equilibrium point for the demand and supply functions

$$S(p) = 60 + 7p$$

and

$$D(p) = 120 - 13p.\ [3.8]$$

38. Auriel is beginning to produce organic honey. For the first year, the fixed costs for setting up production are \$9000. The variable costs for producing each pint of honey are \$0.75. The revenue from each pint of honey is \$5.25. Find the following. [3.8]

a) The total cost $C(x)$ of producing x pints of honey

b) The total revenue $R(x)$ from the sale of x pints of honey

c) The total profit $P(x)$ from the production and sale of x pints of honey

d) The profit or loss from the production and sale of 1500 pints of honey; of 5000 pints of honey

e) The break-even point

Synthesis

TW **39.** How would you go about solving a problem that involves four variables? [3.5]

TW **40.** Explain how a system of equations can be both dependent and inconsistent. [3.4]

41. Auriel is quitting a job that pays $27,000 a year to make honey (see Exercise 38). How many pints of honey must she produce and sell in order to make the same amount that she made in the job she left? [3.8]

42. Solve graphically:

$$y = x + 2,$$
$$y = x^2 + 2. \text{ [3.1]}$$

43. The graph of $f(x) = ax^2 + bx + c$ contains the points $(-2, 3)$, $(1, 1)$, and $(0, 3)$. Find a, b, and c and give a formula for the function. [3.5]

Chapter Test 3

1. Solve graphically:

$$2x + y = 8,$$
$$y - x = 2.$$

Solve, if possible, using the substitution method.

2. $x + 3y = -8,$
$4x - 3y = 23$

3. $2x + 4y = -6,$
$y = 3x - 9$

Solve, if possible, using the elimination method.

4. $4x - 6y = 3,$
$6x - 4y = -3$

5. $4y + 2x = 18,$
$3x + 6y = 26$

6. The perimeter of a rectangle is 96. The length of the rectangle is 6 less than twice the width. Find the dimensions of the rectangle.

7. Pepperidge Farm® Goldfish is a snack food for which 40% of its calories come from fat. Rold Gold® Pretzels receive 9% of their calories from fat. How many grams of each would be needed to make 620 g of a snack mix for which 15% of the calories are from fat?

Solve. If a system's equations are dependent or if there is no solution, state this.

8. $-3x + y - 2z = 8,$
$-x + 2y - z = 5,$
$2x + y + z = -3$

9. $6x + 2y - 4z = 15,$
$-3x - 4y + 2z = -6,$
$4x - 6y + 3z = 8$

10. $2x + 2y = 0,$
$4x + 4z = 4,$
$2x + y + z = 2$

11. $3x + 3z = 0,$
$2x + 2y = 2,$
$3y + 3z = 3$

Solve using matrices.

12. $7x - 8y = 10,$
$9x + 5y = -2$

13. $x + 3y - 3z = 12,$
$3x - y + 4z = 0,$
$-x + 2y - z = 1$

Evaluate.

14. $\begin{vmatrix} 4 & -2 \\ 3 & 7 \end{vmatrix}$

15. $\begin{vmatrix} 3 & 4 & 2 \\ 2 & -5 & 4 \\ 4 & 5 & -3 \end{vmatrix}$

16. Solve using Cramer's rule:

$$8x - 3y = 5,$$
$$2x + 6y = 3.$$

17. An electrician, a carpenter, and a plumber are hired to work on a house. The electrician earns $21 per hour, the carpenter $19.50 per hour, and the plumber $24 per hour. The first day on the job, they worked a total of 21.5 hr and earned a total of $469.50. If the plumber worked 2 more hours than the carpenter did, how many hours did each work?

18. Find the equilibrium point for the demand and supply functions

$$D(p) = 79 - 8p \quad \text{and} \quad S(p) = 37 + 6p.$$

19. Kick Back, Inc., is producing a new hammock. For the first year, the fixed costs for setting up production are $40,000. The variable costs for producing each hammock are $25. The revenue from each hammock is $70. Find the following.

a) The total cost $C(x)$ of producing x hammocks

b) The total revenue $R(x)$ from the sale of x hammocks

c) The total profit $P(x)$ from the production and sale of x hammocks

d) The profit or loss from the production and sale of 300 hammocks; of 900 hammocks

e) The break-even point

Synthesis

20. The graph of the function $f(x) = mx + b$ contains the points $(-1, 3)$ and $(-2, -4)$. Find m and b.

21. At a county fair, an adult's ticket sold for $5.50, a senior citizen's ticket for $4.00, and a child's ticket for $1.50. On opening day, the number of adults' and senior citizens' tickets sold was 30 more than the number of children's tickets sold. The number of adults' tickets sold was 6 more than four times the number of senior citizens' tickets sold. Total receipts from the ticket sales were $11,219.50. How many of each type of ticket were sold?

1–3 Cumulative Review

Solve.

1. U.S. bicycle sales rose from 15 million in 1995 to 20 million in 2005 (*Sources*: National Bicycle Dealers Association; U.S. Department of Transportation). Find the rate of change of bicycle sales. [2.2]

2. A 150-lb person will burn 240 calories an hour when riding a bicycle at 6 mph. The same person will burn 410 calories an hour when cycling at 12 mph. (*Source*: American Heart Association) [2.4]

a) Find an equation for a linear function that can be used to estimate the number of calories burned as a function of cycling speed.

b) Use the function of part (a) to estimate the number of calories burned by a 150-lb person cycling at 10 mph.

3. The following table lists the number of calories burned per hour while riding a bicycle at 12 mph for persons of various weights. [2.4]

Weight (in pounds)	Calories Burned Per Hour at 12 mph
100	270
150	410
200	534

Source: American Heart Association

a) Use linear regression to find a function that can be used to estimate the number of calories burned as a function of weight.

b) Use the function of part (a) to estimate the number of calories burned per hour by a 135-lb person cycling at 12 mph.

4. Americans spent an estimated $238 billion on home remodeling in 2006. This was $\frac{17}{15}$ of the amount spent on remodeling in 2005. (*Source*: National Association of Home Builders' Remodelers Council) How much was spent on remodeling in 2005? [1.7]

5. Ethan's scores on four tests are 83, 92, 100, and 85. What must the score be on the fifth test so that the average will be 90? [1.7]

6. The perimeter of a rectangle is 32 cm. If five times the width equals three times the length, what are the dimensions of the rectangle? [3.3]

7. There are 4 more nickels than dimes in a bank. The total amount of money in the bank is $2.45. How many of each type of coin are in the bank? [3.3]

8. One month Lori and Jon spent $680 for electricity, rent, and telephone. The electric bill was $\frac{1}{4}$ of the rent and the rent was $400 more than the phone bill. How much was the electric bill? [3.5]

9. "Soakem" is 34% salt and the rest water. "Rinsem" is 61% salt and the rest water. How many ounces of each would be needed to obtain 120 oz of a mixture that is 50% salt? [3.3]

10. A hockey team played 64 games one season. It won 15 more games than it tied and lost 10 more games than it won. How many games did it win? lose? tie? [3.5]

11. Reggie, Jenna, and Achmed are counting calories. For lunch one day, Reggie ate two cookies and a banana, for a total of 260 calories. Jenna had a cup of yogurt and a banana, for a total of 245 calories. Achmed ate a cookie, a cup of yogurt, and two bananas, for a total of 415 calories. How many calories are in each item? [3.5]

Simplify. Do not leave negative exponents in your answers. [1.4]

12. $x^4 \cdot x^{-6} \cdot x^{13}$

13. $(4x^{-3}y^2)(-10x^4y^{-7})$

14. $(6x^2y^3)^2(-2x^0y^4)^3$

15. $\frac{y^4}{y^{-6}}$

16. $\frac{-10a^7b^{-11}}{25a^{-4}b^{22}}$

17. $\left(\frac{3x^4y^{-2}}{4x^{-5}}\right)^4$

18. $(1.95 \times 10^{-3})(5.73 \times 10^8)$

19. $\frac{2.42 \times 10^5}{6.05 \times 10^{-2}}$

20. Solve $A = \frac{1}{2}h(b + t)$ for b. [1.6]

21. Determine whether $(-3, 4)$ is a solution of $5a - 2b = -23$. [1.5]

Graph.

22. $f(x) = -2x + 8$ [2.2]

23. $y = x^2 - 1$ [1.5]

24. $4x + 16 = 0$ [2.3]

25. $-3x + 2y = 6$ [2.3]

26. Find the slope and the y-intercept of the line with equation $-4y + 9x = 12$. [2.2]

27. Find the slope, if it exists, of the line containing the points $(2, 7)$ and $(-1, 3)$. [2.2]

28. Find an equation of the line with slope -3 and containing the point $(2, -11)$. [2.4]

29. Find an equation of the line containing the points $(-6, 3)$ and $(4, 2)$. [2.4]

30. Determine whether the lines are parallel, perpendicular, or neither:

$$2x = 4y + 7,$$
$$x - 2y = 5. \text{ [2.3]}$$

31. Find an equation of the line with y-intercept $(0, 5)$ and perpendicular to the line $x - 2y = 5$. [2.3]

32. For the graph of f shown, determine the domain, the range, $f(-3)$, and any value of x for which $f(x) = 5$. [2.1]

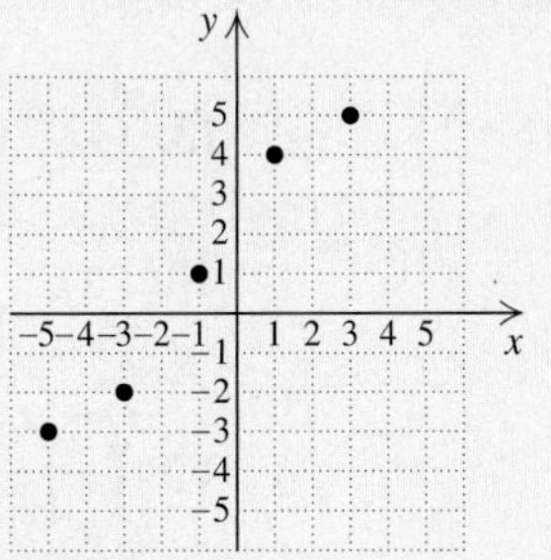

33. Determine the domain of the function given by

$$f(x) = \frac{7}{2x - 1}. \text{ [2.5]}$$

Given $g(x) = 4x - 3$ *and* $h(x) = -2x^2 + 1$, *find the following function values.*

34. $h(4)$ [2.1]

35. $-g(0)$ [2.1]

36. $(g \cdot h)(-1)$ [2.5]

37. $g(a) - h(2a)$ [2.5]

Solve.

38. $x + 9.4 = -12.6$ [1.6]

39. $-2.4x = -48$ [1.6]

40. $\frac{3}{8}x + 7 = -14$ [1.6]

41. $-3 + 5x = 2x + 15$ [1.6]

42. $3n - (4n - 2) = 7$ [1.6]

43. $6y - 5(3y - 4) = 10$ [1.6]

44. $14 + 2c = -3(c + 4) - 6$ [1.6]

45. $5x - [4 - 2(6x - 1)] = 12$ [1.6]

46. $3x + y = 4,$
$6x - y = 5$ [3.2]

47. $4x + 4y = 4,$
$5x - 3y = -19$ [3.2]

48. $6x - 10y = -22,$
$-11x - 15y = 27$ [3.2]

49. $x + y + z = -5,$
$2x + 3y - 2z = 8,$
$x - y + 4z = -21$ [3.4]

50. $2x + 5y - 3z = -11,$
$-5x + 3y - 2z = -7,$
$3x - 2y + 5z = 12$ [3.4]

Evaluate. [3.7]

51. $\begin{vmatrix} 2 & -3 \\ 4 & 1 \end{vmatrix}$

52. $\begin{vmatrix} 1 & 0 & 1 \\ -1 & 2 & 1 \\ 2 & 1 & 3 \end{vmatrix}$

Synthesis

53. Simplify: $(6x^{a+2}y^{b+2})(-2x^{a-2}y^{y+1})$. [1.4]

54. Chaney Chevrolet discovers that when \$1000 is spent on radio advertising, weekly sales increase by \$101,000. When \$1250 is spent on radio advertising, weekly sales increase by \$126,000. Assuming that sales increase according to a linear equation, by what amount would sales increase when \$1500 is spent on radio advertising? [2.4]

55. Given that $f(x) = mx + b$ and that $f(5) = -3$ when $f(-4) = 2$, find m and b. [2.4], [3.3]

4 Inequalities and Problem Solving

Inequalities are mathematical sentences containing symbols such as < (is less than). Principles similar to those used for solving equations enable us to solve inequalities and the problems that translate to inequalities. In this chapter, we develop procedures for solving a variety of inequalities and systems of inequalities.

APPLICATION *Earnings Ratio.*

The ratio of median earnings of women to median earnings of men is called the female-to-male earnings ratio. A ratio of 0.77, or 77:100, means that women earn $77 for every $100 that men earn. This ratio is shown for various years in the following table. Use linear regression to find a linear function that can be used to estimate the female-to-male earnings ratio x years after 1980. Then use the function to predict the years in which the female-to-male earnings ratio will be 0.9 or higher.

Year	Female-to-Male Earnings Ratio
1980	0.60
1985	0.64
1990	0.71
1995	0.72
2000	0.75
2004	0.77

Source: U.S. Bureau of the Census

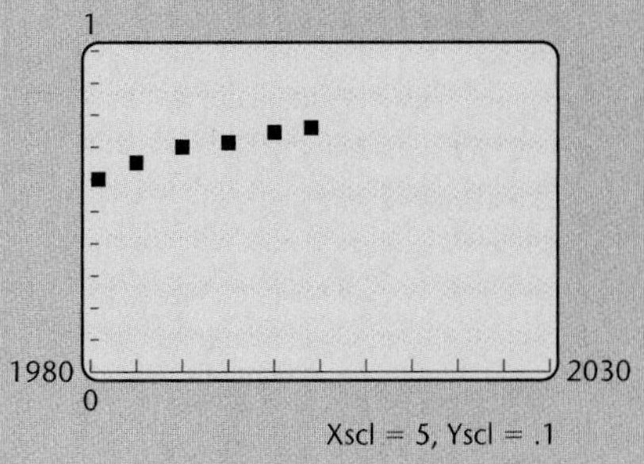

This problem appears as Example 8 in Section 4.2.

4.1 Inequalities and Applications

Solutions of Inequalities ■ Interval Notation and Graphs ■ The Addition Principle for Inequalities ■ The Multiplication Principle for Inequalities ■ Using the Principles Together ■ Problem Solving

Study Tip

Create Your Own Glossary

Understanding mathematical terminology is essential for success in any math course. Consider writing your own glossary of important words toward the back of your notebook. Often, just the act of writing out a word's definition can help you remember what the word means.

Solutions of Inequalities

We now extend our equation-solving skills to the solving of inequalities. An **inequality** is any sentence containing $<$, $>$, $\leq$, $\geq$, or $\neq$ (see Section 1.1)—for example,

$$-2 < a, \quad x > 4, \quad x + 3 \leq 6, \quad 7y \geq 10y - 4, \quad \text{and} \quad 5x \neq 10.$$

Any replacement for the variable that makes an inequality true is called a **solution.** The set of all solutions is called the **solution set.** When all solutions of an inequality are found, we say that we have **solved** the inequality.

EXAMPLE 1 Determine whether the given number is a solution of the inequality.

a) $x + 3 < 6$; 5 **b)** $-3 > -5 - 2x$; 1

SOLUTION

a) We substitute to get $5 + 3 < 6$, or $8 < 6$, a false sentence. Thus, 5 *is not* a solution.

b) We substitute to get $-3 > -5 - 2 \cdot 1$, or $-3 > -7$, a true sentence. Thus 1 *is* a solution.

The solutions of the inequality $x < 4$ are all real numbers less than 4. We can write this set using *set-builder* notation (see Section 1.1):

$$\{x \mid x < 4\}.$$

This is read

"The set of all x such that x is less than 4."

Note that

$x < 4$ is an inequality, and

$\{x \mid x < 4\}$ is the solution set of the inequality.

Interval Notation and Graphs

Another way to write solutions of an inequality in one variable is to use **interval notation.** Interval notation uses parentheses, ( ), and brackets, [ ].

If a and b are real numbers such that $a < b$, we define the **open interval** (a, b) as the set of all numbers x for which $a < x < b$. Thus,

$$(a, b) = \{x \mid a < x < b\}. \quad \textbf{Parentheses are used to exclude endpoints.}$$

We graph an open interval by marking the endpoints a and b with parentheses to indicate that the endpoints are not included. Then we shade the portion of the number line between a and b.

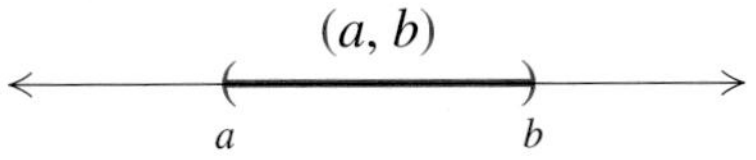

CAUTION! Do not confuse the *interval* (a, b) with the *ordered pair* (a, b). The context in which the notation appears usually makes the meaning clear.

The **closed interval [*a*, *b*]** is defined as the set of all numbers x for which $a \leq x \leq b$. Thus,

$$[a, b] = \{x \mid a \leq x \leq b\}.$$ **Brackets are used to include endpoints.**

Its graph includes the endpoints, as indicated by the brackets:*

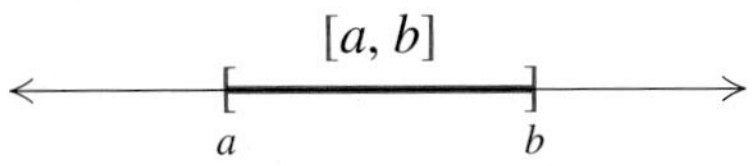

There are two kinds of **half-open intervals,** defined as follows:

1. $(a, b] = \{x \mid a < x \leq b\}$. This is open on the left. Its graph is as follows:

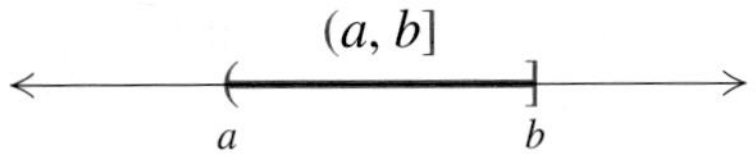

2. $[a, b) = \{x \mid a \leq x < b\}$. This is open on the right. Its graph is as follows:

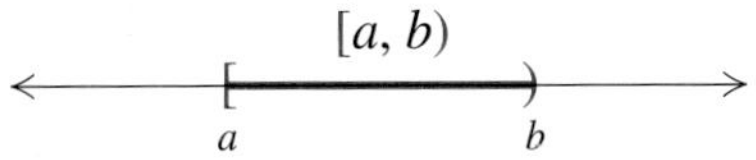

Student Notes

You may have noticed which inequality signs in set-builder notation correspond to brackets and which correspond to parentheses. The relationship could be written informally as

$$\leq \quad \geq \qquad [\;]$$
$$< \quad > \qquad (\;).$$

We use the symbols ∞ and $-\infty$ to represent positive and negative infinity, respectively. The symbol ∞ is used when there is no upper limit to the set of numbers, and $-\infty$ is used when there is no lower limit. Thus the notation (a, ∞)

---

*Some books use the representations 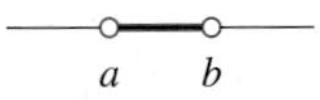 and 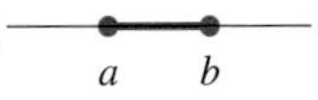 instead of, respectively, 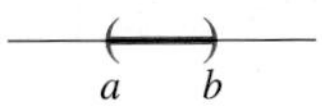 and 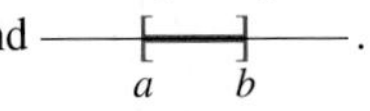.

represents the set of all real numbers greater than a, and $(-\infty, a)$ represents the set of all real numbers less than a.

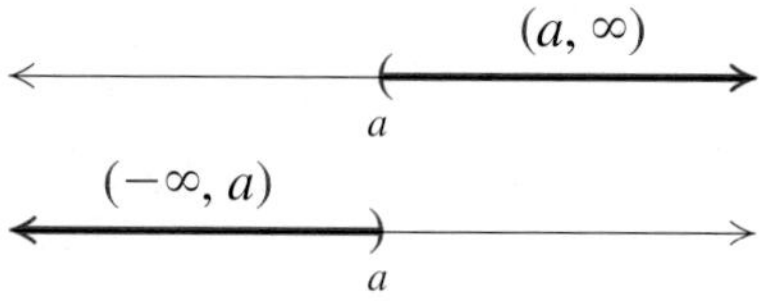

The notations $[a, \infty)$ and $(-\infty, a]$ are used when we want to include the endpoint a.

EXAMPLE 2 Graph $y \geq -2$ on the number line and write the solution set using both set-builder and interval notations.

SOLUTION Using set-builder notation, we write the solution set as $\{y \mid y \geq -2\}$.

Using interval notation, we write the solution set as $[-2, \infty)$.

To graph the solution, we shade all numbers to the right of -2 and use a bracket to indicate that -2 is also a solution.

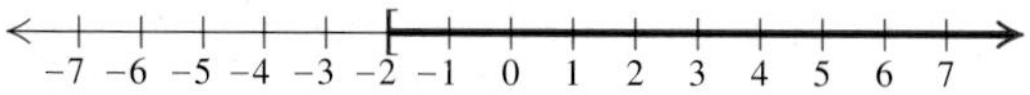

The Addition Principle for Inequalities

Two inequalities are *equivalent* if they have the same solution set. For example, the inequalities $x > 4$ and $4 < x$ are equivalent. Just as the addition principle for equations produces equivalent equations, the addition principle for inequalities produces equivalent inequalities.

The Addition Principle for Inequalities For any real numbers a, b, and c:

$a < b$ is equivalent to $a + c < b + c$;

$a > b$ is equivalent to $a + c > b + c$.

Similar statements hold for $\leq$ and $\geq$.

As with equations, we try to get the variable alone on one side in order to determine solutions easily.

EXAMPLE 3 Solve and graph: **(a)** $x + 5 > 1$; **(b)** $4x - 1 \geq 5x - 2$.

SOLUTION

a) $x + 5 > 1$

$x + 5 + (-5) > 1 + (-5)$

$x > -4$ **Using the addition principle to add −5 to both sides**

When an inequality—like this last one—has an infinite number of solutions, we cannot possibly check them all. Instead, we can perform a partial check by substituting one member of the solution set (here we use -1) into the original inequality:

Check:	$x + 5 > 1$	
	$-1 + 5$	1
	$4 \overset{?}{>} 1$	TRUE

Since $4 > 1$ is true, we have our check. The solution set is $\{x \mid x > -4\}$, or $(-4, \infty)$. The graph is as follows:

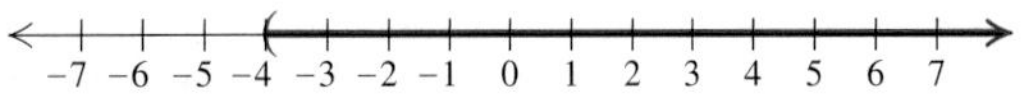

b)
$$4x - 1 \ge 5x - 2$$
$$4x - 1 + 2 \ge 5x - 2 + 2 \qquad \textbf{Adding 2 to both sides}$$
$$4x + 1 \ge 5x \qquad \textbf{Simplifying}$$
$$4x + 1 - 4x \ge 5x - 4x \qquad \textbf{Adding } -4x \textbf{ to both sides}$$
$$1 \ge x \qquad \textbf{Simplifying}$$

We know that $1 \ge x$ has the same meaning as $x \le 1$. You can check that any number less than or equal to 1 is a solution. The solution set is $\{x \mid 1 \ge x\}$ or, more commonly, $\{x \mid x \le 1\}$. Using interval notation, we write the solution set as $(-\infty, 1]$. The graph is as follows:

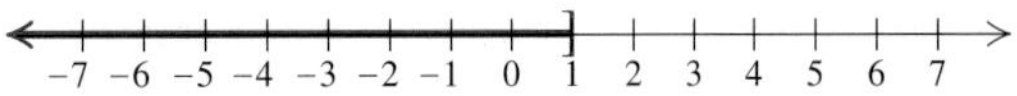

The Multiplication Principle for Inequalities

The multiplication principle for inequalities differs from the multiplication principle for equations.

Consider this true inequality: $4 < 9$. If we multiply both sides of $4 < 9$ by 2, we get another true inequality:

$$4 \cdot 2 < 9 \cdot 2, \quad \text{or} \quad 8 < 18.$$

If we multiply both sides of $4 < 9$ by -2, we get a false inequality:

$$\text{FALSE} \longrightarrow 4(-2) < 9(-2), \quad \text{or} \quad -8 < -18. \longleftarrow \text{FALSE}$$

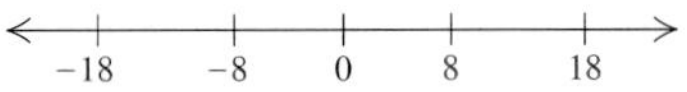

This is because multiplication (or division) by a negative number changes the sign of the number being multiplied (or divided). When the signs of both numbers in an inequality are changed, the position of the numbers on the number line with respect to each other is reversed.

$$-8 > -18. \longleftarrow \text{TRUE}$$

The < symbol has been reversed!

The Multiplication Principle for Inequalities For any real numbers a and b, and for any *positive* number c,

$a < b$ is equivalent to $ac < bc$;

$a > b$ is equivalent to $ac > bc$.

For any real numbers a and b, and for any *negative* number c,

$a < b$ is equivalent to $ac > bc$;

$a > b$ is equivalent to $ac < bc$.

Similar statements hold for $\le$ and $\ge$.

Since division by c is the same as multiplication by $1/c$, there is no need for a separate division principle.

CAUTION! Remember that whenever we multiply or divide both sides of an inequality by a negative number, we must reverse the inequality symbol.

EXAMPLE 4 Solve and graph: **(a)** $3a < \frac{3}{4}$; **(b)** $-5x \ge -80$.

SOLUTION

a) $3a < \frac{3}{4}$

The symbol stays the same because $\frac{1}{3}$ is positive.

$\frac{1}{3} \cdot 3a < \frac{1}{3} \cdot \frac{3}{4}$ **Multiplying both sides by $\frac{1}{3}$ or dividing both sides by 3**

$a < \frac{1}{4}$

Any number less than $\frac{1}{4}$ is a solution.

The solution set is $\{a \mid a < \frac{1}{4}\}$, or $(-\infty, \frac{1}{4})$. The graph is as follows:

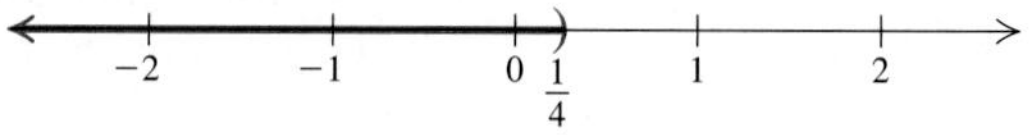

Student Notes

Remember to reverse the inequality symbol as soon as both sides are multiplied or divided by a negative number. Don't wait until after the multiplication or division has been carried out to reverse the symbol.

b) $-5x \ge -80$

The symbol must be reversed because −5 is negative.

$\frac{-5x}{-5} \le \frac{-80}{-5}$ **Dividing both sides by −5 or multiplying both sides by $-\frac{1}{5}$**

$x \le 16$

The solution set is $\{x \mid x \le 16\}$, or $(-\infty, 16]$. The graph is as follows:

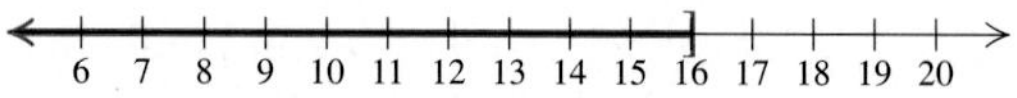

Using the Principles Together

We use the addition and multiplication principles together in solving inequalities in much the same way as in solving equations.

EXAMPLE 5 Solve.

a) $16 - 7y \geq 10y - 4$

b) $-3(x + 8) - 5x > 4x - 9$

SOLUTION

a)

$$16 - 7y \geq 10y - 4$$

$$-16 + 16 - 7y \geq -16 + 10y - 4 \qquad \textbf{Adding } -16 \textbf{ to both sides}$$

$$-7y \geq 10y - 20$$

$$-10y + (-7y) \geq -10y + 10y - 20 \qquad \textbf{Adding } -10y \textbf{ to both sides}$$

$$-17y \geq -20$$

The symbol must be reversed.

$$-\tfrac{1}{17} \cdot (-17y) \leq -\tfrac{1}{17} \cdot (-20) \qquad \textbf{Multiplying both sides by } -\tfrac{1}{17} \textbf{ or dividing both sides by } -17$$

$$y \leq \tfrac{20}{17}$$

The solution set is $\{y \mid y \leq \frac{20}{17}\}$, or $\left(-\infty, \frac{20}{17}\right]$.

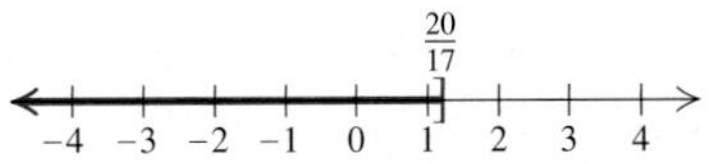

b)

$$-3(x + 8) - 5x > 4x - 9$$

$$-3x - 24 - 5x > 4x - 9 \qquad \textbf{Using the distributive law}$$

$$-24 - 8x > 4x - 9$$

$$-24 - 8x + 8x > 4x - 9 + 8x \qquad \textbf{Adding } 8x \textbf{ to both sides}$$

$$-24 > 12x - 9$$

$$-24 + 9 > 12x - 9 + 9 \qquad \textbf{Adding 9 to both sides}$$

$$-15 > 12x$$

The symbol stays the same.

$$-\tfrac{5}{4} > x \qquad \textbf{Dividing by 12 and simplifying}$$

The solution set is $\{x \mid -\frac{5}{4} > x\}$, or $\{x \mid x < -\frac{5}{4}\}$, or $\left(-\infty, -\frac{5}{4}\right)$.

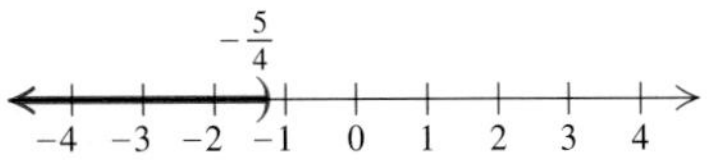

Problem Solving

Many problem-solving situations translate to inequalities. In addition to "is less than" and "is more than," other phrases are commonly used.

Important Words	Sample Sentence	Translation
is at least	Max is at least 5 years old	$m \geq 5$
are at most	There are at most 6 people in the car.	$n \leq 6$
cannot exceed	Total weight in the elevator cannot exceed 2000 pounds.	$w \leq 2000$
must exceed	The speed must exceed 15 mph.	$s > 15$
is between	Heather's income is between \$23,000 and \$35,000.	$23{,}000 < h < 35{,}000$
no more than	Bing weighs no more than 90 pounds	$w \leq 90$
no less than	Saul would accept no less than \$5000 for his used car.	$t \geq 5000$

EXAMPLE 6 Domestic Oil Production. The yearly U.S. production of crude oil $C(t)$, in millions of barrels, t years after 1990, can be approximated by the equation

$$C(t) = -53.5t + 2683$$

(*Source*: Based on data from the *Statistical Abstract of the United States* 2003). Determine (using an inequality) those years for which domestic production will be less than 1750 million barrels.

SOLUTION

1. **Familiarize.** We already have a formula. To become more familiar with it, we might make a substitution for t. Suppose we want to predict production after 20 years, in 2010. We substitute 20 for t:

   $$C(20) = -53.5 \cdot 20 + 2683 = 1613.$$

   We see that by 2010, production will be less than 1750 million barrels. To predict the exact years in which fewer than 1750 million barrels will be produced, we could check other substitutions. Instead, we proceed to the next step.

2. **Translate.** We are asked to find the years for which U.S. oil production $C(t)$ will be *less than* 1750 million barrels. Thus we have

   $$C(t) < 1750.$$

   We replace $C(t)$ with $-53.5t + 2683$ to find the times t that solve the inequality:

   $$-53.5t + 2683 < 1750. \quad \textbf{Substituting}$$

3. **Carry out.** We solve the inequality:

$$-53.5t + 2683 < 1750$$

$$-53.5t < -933 \quad \textbf{Adding } -2683 \textbf{ to both sides}$$

$$t > 17.44. \quad \textbf{Dividing both sides by } -53.5\textbf{, reversing the symbol, and rounding}$$

4. **Check.** A partial check is to substitute a value for t greater than 17.44. We did that in the *Familiarize* step.

5. **State.** U.S. oil production will fall below 1750 million barrels about 17.4 years after 1990, or in 2007, and will remain below 1750 million for all years after that.

EXAMPLE 7 Job Offers. After graduation, Rose had two job offers in sales:

Uptown Fashions: A salary of \$600 per month, plus a commission of 4% of sales;

Ergo Designs: A salary of \$800 per month, plus a commission of 6% of sales in excess of \$10,000.

If sales always exceed \$10,000, for what amount of sales would Uptown Fashions provide higher pay?

SOLUTION

1. **Familiarize.** Listing the given information in a table will be helpful.

Uptown Fashions Monthly Income	Ergo Designs Monthly Income
\$600 salary 4% of sales *Total*: \$600 + 4% of sales	\$800 salary 6% of sales over \$10,000 *Total*: \$800 + 6% of sales over \$10,000

Next, suppose that Rose sold a certain amount—say, \$12,000—in one month. Which plan would be better? Working for Uptown, she would earn \$600 plus 4% of \$12,000, or

$$600 + 0.04(12{,}000) = \$1080.$$

Since with Ergo Designs commissions are paid only on sales in excess of \$10,000, Rose would earn \$800 plus 6% of (\$12,000 − \$10,000), or

$$800 + 0.06(2000) = \$920.$$

This shows that for monthly sales of \$12,000, Uptown pays better. Similar calculations will show that for sales of \$30,000 a month, Ergo pays better. To determine *all* values for which Uptown pays more money, we must solve an inequality that is based on the calculations above.

2. **Translate.** We let S = the amount of monthly sales, in dollars, and will assume $S > 10{,}000$ so that both plans will pay a commission. Examining the calculations in the *Familiarize* step, we see that monthly income from

Uptown is $600 + 0.04S$ and from Ergo is $800 + 0.06(S - 10{,}000)$. We want to find all values of S for which

Income from Uptown	is greater than	income from Ergo
↓	↓	↓
$600 + 0.04S$	$>$	$800 + 0.06(S - 10{,}000)$.

3. **Carry out.** We solve the inequality:

$$600 + 0.04S > 800 + 0.06(S - 10{,}000)$$
$$600 + 0.04S > 800 + 0.06S - 600 \qquad \textbf{Using the distributive law}$$
$$600 + 0.04S > 200 + 0.06S \qquad \textbf{Combining like terms}$$
$$400 > 0.02S \qquad \textbf{Subtracting 200 and 0.04S from both sides}$$
$$20{,}000 > S, \text{ or } S < 20{,}000. \qquad \textbf{Dividing both sides by 0.02}$$

4. **Check.** The above steps indicate that income from Uptown Fashions is higher than income from Ergo Designs for sales less than \$20,000. In the *Familiarize* step, we saw that for sales of \$12,000, Uptown pays more. Since $12{,}000 < 20{,}000$, this is a partial check.

5. **State.** When monthly sales are less than \$20,000, Uptown Fashions provides the higher pay.

4.1 EXERCISE SET

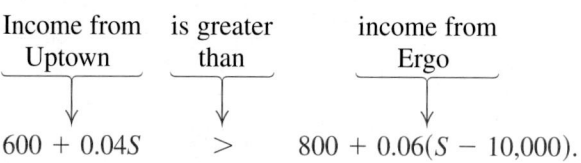

FOR EXTRA HELP

MathXL | MyMathLab | InterAct Math | AW Math Tutor Center | Video Lectures on CD: Disc 2 | Student's Solutions Manual

Concept Reinforcement Classify each of the following as equivalent inequalities, equivalent equations, equivalent expressions, or not equivalent.

1. $x - 7 > -2,\ x > 5$

2. $t + 3 < 1,\ t < 2$

3. $5x + 7 = 6 - 3x,\ 8x + 7 = 6$

4. $2(4x + 1),\ 8x + 2$

5. $-4t \le 12,\ t \le -3$

6. $\frac{3}{5}a + \frac{1}{5} = 2,\ 3a + 1 = 10$

7. $6a + 9,\ 3(2a + 3)$

8. $-4x \ge -8,\ x \ge 2$

9. $-\frac{1}{2}x < 7,\ x > 14$

10. $-\frac{1}{3}t \le -5,\ t \ge 15$

Determine whether the given numbers are solutions of the inequality.

11. $x - 3 \ge 5;\ -4, 0, 8, 13$

12. $3x + 5 \le -10;\ -5, -10, 0, 27$

13. $t - 6 > 2t - 1;\ 0, -8, -9, -3$

14. $5y - 9 < 3 - y;\ 2, -3, 0, 3$

Graph each inequality, and write the solution set using both interval notation and set-builder notation.

15. $y < 6$

16. $x > 4$

17. $x \geq -4$

18. $t \leq 6$

19. $t > -3$

20. $y < -3$

21. $x \leq -7$

22. $x \geq -6$

Solve. Then graph.

23. $x + 8 > 2$

24. $x + 5 > 2$

25. $a + 7 \leq -13$

26. $a + 9 \leq -12$

27. $x - 8 \leq 9$

28. $t + 14 \geq 9$

29. $y - 9 > -18$

30. $y - 8 > -14$

31. $y - 20 \leq -6$

32. $x - 11 \leq -2$

33. $9t < -81$

34. $8x \geq 24$

35. $0.3x < -18$

36. $0.5x < 25$

37. $-9x \geq -8.1$

38. $-8y \leq 3.2$

39. $-\frac{3}{4}x \geq -\frac{5}{8}$

40. $-\frac{5}{6}y \leq -\frac{3}{4}$

41. $\dfrac{2x + 7}{5} < -9$

42. $\dfrac{5y + 13}{4} > -2$

43. $\dfrac{3t - 7}{-4} \leq 5$

44. $\dfrac{2t - 9}{-3} \geq 7$

45. Let $f(x) = 2x + 1$ and $g(x) = x + 7$. Find all values of x for which $f(x) \geq g(x)$.

46. Let $f(x) = 5 - x$ and $g(x) = 4x - 5$. Find all values of x for which $f(x) \geq g(x)$.

47. Let $f(x) = 7 - 3x$ and $g(x) = 2x - 3$. Find all values of x for which $f(x) \leq g(x)$.

48. Let $f(x) = 8x - 9$ and $g(x) = 3x - 11$. Find all values of x for which $f(x) \leq g(x)$.

49. Let $f(x) = 2x - 7$ and $g(x) = 5x - 9$. Find all values of x for which $f(x) < g(x)$.

50. Let $f(x) = 0.4x + 5$ and $g(x) = 1.2x - 4$. Find all values of x for which $g(x) \geq f(x)$.

51. Let $f(x) = \frac{3}{8} + 2x$ and $g(x) = 3x - \frac{1}{8}$. Find all values of x for which $g(x) \geq f(x)$.

52. Let $f(x) = 2x + 1$ and $g(x) = -\frac{1}{2}x + 6$. Find all values of x for which $f(x) < g(x)$.

Solve.

53. $4(3y - 2) \geq 9(2y + 5)$

54. $4m + 7 \geq 14(m - 3)$

55. $5(t - 3) + 4t < 2(7 + 2t)$

56. $2(4 + 2x) > 2x + 3(2 - 5x)$

57. $5[3m - (m + 4)] > -2(m - 4)$

58. $8x - 3(3x + 2) - 5 \geq 3(x + 4) - 2x$

59. $19 - (2x + 3) \leq 2(x + 3) + x$

60. $13 - (2c + 2) \geq 2(c + 2) + 3c$

61. $\frac{1}{4}(8y + 4) - 17 < -\frac{1}{2}(4y - 8)$

62. $\frac{1}{3}(6x + 24) - 20 > -\frac{1}{4}(12x - 72)$

63. $2[8 - 4(3 - x)] - 2 \geq 8[2(4x - 3) + 7] - 50$

64. $5[3(7 - t) - 4(8 + 2t)] - 20 \leq -6[2(6 + 3t) - 4]$

Solve.

65. *Truck Rentals.* Campus Entertainment rents a truck for \$45 plus 20¢ per mile. A budget of \$75 has been set for the rental. For what mileages will they not exceed the budget?

66. *Truck Rentals.* Metro Concerts can rent a truck for either \$55 with unlimited mileage or \$29 plus 40¢ per mile. For what mileages would the unlimited mileage plan save money?

67. *Insurance Claims.* After a serious automobile accident, most insurance companies will replace the damaged car with a new one if repair costs exceed 80% of the NADA, or "blue-book," value of the car. Miguel's car recently sustained \$9200 worth of damage but was not replaced. What was the blue-book value of his car?

68. *Phone Rates.* A long-distance telephone call using Down East Calling costs 10 cents for the first minute and 8 cents for each additional minute. The same call, placed on Long Call Systems, costs 15 cents for the first minute and 6 cents for each additional minute. For what length phone calls is Down East Calling less expensive?

69. *Checking-Account Rates.* The Hudson Bank offers two checking-account plans. Their Anywhere plan charges 20¢ per check whereas their Acu-checking plan costs \$2 per month plus 12¢ per check. For what numbers of checks per month will the Acu-checking plan cost less?

70. *Moving Costs.* Musclebound Movers charges \$85 plus \$40 an hour to move households across town. Champion Moving charges \$60 an hour for cross-town moves. For what lengths of time is Champion more expensive?

71. *Wages.* Toni can be paid in one of two ways:

Plan A: A salary of \$400 per month, plus a commission of 8% of gross sales;

Plan B: A salary of \$610 per month, plus a commission of 5% of gross sales.

For what amount of gross sales should Toni select plan A?

72. *Wages.* Branford can be paid for his masonry work in one of two ways:

Plan A: \$300 plus \$9.00 per hour;

Plan B: Straight \$12.50 per hour.

Suppose that the job takes n hours. For what values of n is plan B better for Branford?

73. *Wedding Costs.* The Arnold Inn offers two plans for wedding parties. Under plan A, the inn charges \$30 for each person in attendance. Under plan B, the inn charges \$1300 plus \$20 for each person in excess of the first 25 who attend. For what size parties will plan B cost less? (Assume that more than 25 guests will attend.)

74. *Insurance Benefits.* Bayside Insurance offers two plans. Under plan A, Giselle would pay the first \$50 of her medical bills and 20% of all bills after that. Under plan B, Giselle would pay the first \$250 of bills, but only 10% of the rest. For what amount of medical bills will plan B save Giselle money? (Assume that her bills will exceed \$250.)

75. Abriana rented a compact car for a business trip. At the time of rental, she was given the option of prepaying for an entire tank of gasoline at \$3.099 per gallon, or waiting until her return and paying \$6.34 per gallon for enough gasoline to fill the tank. If the tank holds 14 gal, how many gallons can she use and still save money by choosing the second option?

76. Refer to Exercise 75. If Abriana's rental car gets 30 mpg, how many miles must she drive in order to make the first option more economical?

TW **77.** Explain in your own words why the inequality symbol must be reversed when both sides of an inequality are multiplied by a negative number.

TW **78.** Why isn't roster notation used to write solutions of inequalities?

Skill Maintenance

Find the domain of f. [2.5]

79. $f(x) = \dfrac{3}{x - 2}$

80. $f(x) = \dfrac{x - 5}{4x + 12}$

81. $f(x) = \dfrac{5x}{7 - 2x}$

82. $f(x) = \dfrac{x + 3}{9 - 4x}$

Simplify. [1.3]

83. $9x - 2(x - 5)$

84. $8x + 7(2x - 1)$

Synthesis

TW **85.** A Presto photocopier costs \$510 and an Exact Image photocopier costs \$590. Write a problem that involves the cost of the copiers, the cost per page of photocopies, and the number of copies for which the Presto machine is the more expensive machine to own.

TW **86.** Explain how the addition principle can be used to avoid ever needing to multiply or divide both sides of an inequality by a negative number.

Solve for x and y. Assume that a, b, c, d, and m are positive constants.

87. $3ax + 2x \geq 5ax - 4$; assume $a > 1$

88. $6by - 4y \leq 7by + 10$

89. $a(by - 2) \geq b(2y + 5)$; assume $a > 2$

90. $c(6x - 4) < d(3 + 2x)$; assume $3c > d$

91. $c(2 - 5x) + dx > m(4 + 2x)$; assume $5c + 2m < d$

92. $a(3 - 4x) + cx < d(5x + 2)$;
assume $c > 4a + 5d$

Determine whether each statement is true or false. If false, give an example that shows this.

93. For any real numbers a, b, c, and d, if $a < b$ and $c < d$, then $a - c < b - d$.

94. For all real numbers x and y, if $x < y$, then $x^2 < y^2$.

TW **95.** Are the inequalities

$$x < 3 \quad \text{and} \quad x + \frac{1}{x} < 3 + \frac{1}{x}$$

equivalent? Why or why not?

TW **96.** Are the inequalities

$$x < 3 \quad \text{and} \quad 0 \cdot x < 0 \cdot 3$$

equivalent? Why or why not?

Solve. Then graph.

97. $x + 5 \leq 5 + x$

98. $x + 8 < 3 + x$

99. $x^2 > 0$

Collaborative Corner

Reduce, Reuse, and Recycle

Focus: Inequalities and problem solving
Time: 15–20 minutes
Group Size: 2

In the United States, the amount of solid waste (rubbish) being recycled is slowly catching up to the amount being generated. In 1993, each person generated, on average, 4.3 lb of solid waste every day, of which 0.9 lb was recycled. In 2003, each person generated, on average, 4.4 lb of solid waste, of which 1.4 lb was recycled. (*Sources*: U.S. Bureau of the Census; *Statistical Abstract of the United States* 2000 and 2006)

ACTIVITY

Assume that the amount of solid waste being generated and the amount recycled are both increasing linearly. One group member should find a linear function w for which $w(t)$ represents the number of pounds of waste generated per person per day t years after 1993. The other group member should find a linear function r for which $r(t)$ represents the number of pounds recycled per person per day t years after 1993. Finally, working together, the group should determine those years for which the amount recycled will meet or exceed the amount generated.

4.2 Solving Equations and Inequalities by Graphing

Solving Equations Graphically: The Intersect Method ■ Solving Equations Graphically: The Zero Method ■ Solving Inequalities Graphically ■ Applications

Study Tip

If You Must Miss a Class

Occasionally you may be forced to miss an upcoming class. It is usually best to alert your instructor to this situation as soon as possible. He or she may permit you to make up a missed quiz or test *if you provide enough advance notice.* Make an effort to find out what assignment will be given in your absence and try your best to learn the material on your own beforehand so that you can ask questions in advance of your absence. Sometimes it is possible to attend another section of the course or view a video for the lesson that you must miss.

Recall that to *solve* an equation or inequality means to find all the replacements for the variable that make the equation or inequality true. We have seen how to solve algebraically; we now use a graphical method to solve.

Solving Equations Graphically: The Intersect Method

To see how solutions of equations are related to graphs, consider the graphs of the functions given by $f(x) = 2x + 5$ and $g(x) = -3$.

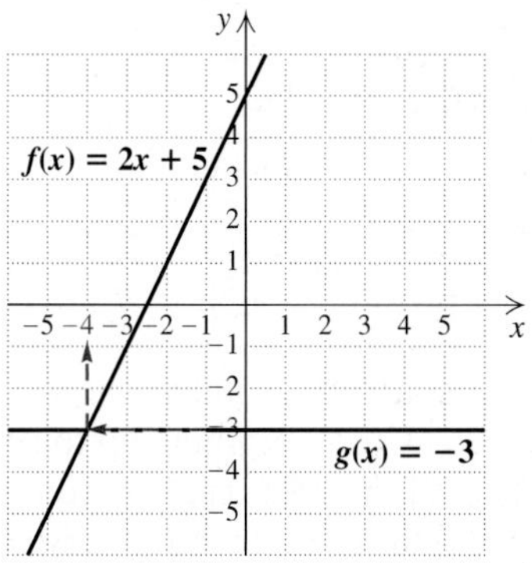

At the point at which the graphs intersect, $f(x) = g(x)$. Thus, for that particular x-value, we have $2x + 5 = -3$. In other words, the solution of $2x + 5 = -3$ is the x-coordinate of the point of intersection of the graphs of $f(x) = 2x + 5$ and $g(x) = -3$. Careful inspection suggests that -4 is that x-value. To check, note that $f(-4) = 2(-4) + 5 = -3$.

EXAMPLE 1 Solve graphically: $\frac{1}{2}x + 3 = 2$.

SOLUTION To find the x-value for which $\frac{1}{2}x + 3$ will equal 2, we graph $f(x) = \frac{1}{2}x + 3$ and $g(x) = 2$ on the same set of axes. Since the intersection appears to be $(-2, 2)$, the solution is apparently -2.

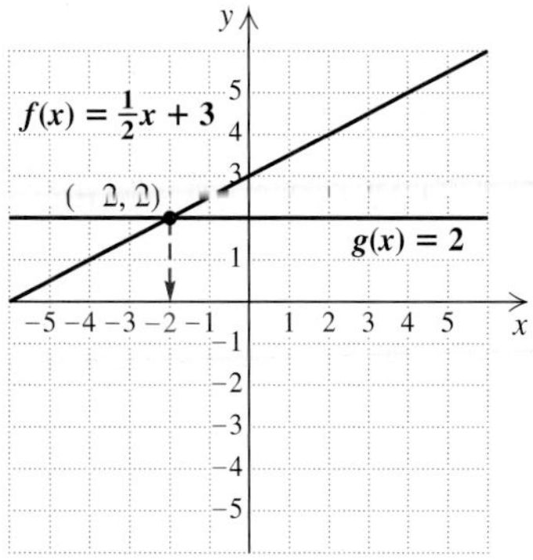

Check:

$$\begin{array}{c|c} \frac{1}{2}x + 3 & 2 \\ \hline \frac{1}{2}(-2) + 3 & 2 \\ -1 + 3 & \\ 2 & \stackrel{?}{=} 2 \quad \text{TRUE} \end{array}$$

The solution is -2.

Connecting the Concepts

INTERPRETING GRAPHS: POINT OF INTERSECTION

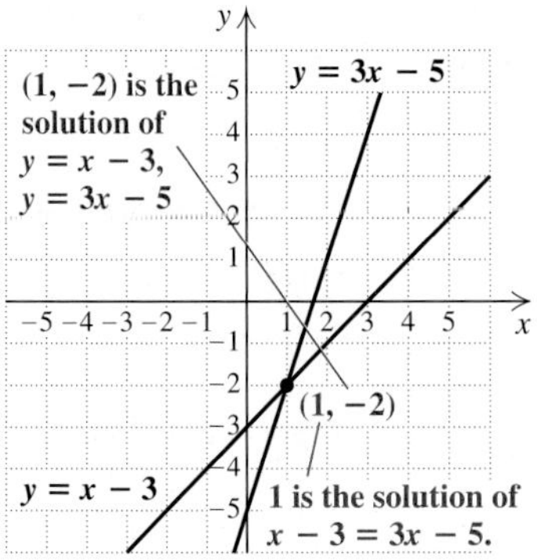

In Section 3.1, we solved systems of linear equations by graphing two equations and determining their point of intersection. The solution of the system was given by the coordinates of the point of intersection. For example, to solve the system

$$y = x - 3,$$
$$y = 3x - 5,$$

we graph $y = x - 3$ and $y = 3x - 5$. The solution of the system is the ordered pair $(1, -2)$. Since there are two variables in the system, the solution will contain a value for each variable.

In this section, we are solving linear equations in one variable by graphing two functions and looking for points of intersection. For example, to solve

$$x - 3 = 3x - 5,$$

we graph $f(x) = x - 3$ and $g(x) = 3x - 5$. The x-coordinate of the point of intersection, 1, gives the solution of the equation. Since there is one variable in the equation, the solution is a single number.

EXAMPLE 2 Solve: $-\frac{3}{4}x + 6 = 2x - 1$.

ALGEBRAIC APPROACH

We have

$$-\tfrac{3}{4}x + 6 = 2x - 1$$

$$-\tfrac{3}{4}x + 6 - 6 = 2x - 1 - 6 \qquad \textbf{Subtracting 6 from both sides}$$

$$-\tfrac{3}{4}x = 2x - 7 \qquad \textbf{Simplifying}$$

$$-\tfrac{3}{4}x - 2x = 2x - 7 - 2x \qquad \textbf{Subtracting } 2x \textbf{ from both sides}$$

$$-\tfrac{3}{4}x - \tfrac{8}{4}x = -7 \qquad \textbf{Simplifying}$$

$$-\tfrac{11}{4}x = -7 \qquad \textbf{Combining like terms}$$

$$-\tfrac{4}{11}\left(-\tfrac{11}{4}x\right) = -\tfrac{4}{11}(-7) \qquad \textbf{Multiplying both sides by } -\tfrac{4}{11}$$

$$x = \tfrac{28}{11}. \qquad \textbf{Simplifying}$$

Check: $-\frac{3}{4}x + 6 = 2x - 1$

$-\frac{3}{4}\left(\frac{28}{11}\right) + 6$	$2\left(\frac{28}{11}\right) - 1$
$-\frac{21}{11} + \frac{66}{11}$	$\frac{56}{11} - \frac{11}{11}$

$\frac{45}{11} \stackrel{?}{=} \frac{45}{11}$ TRUE

The solution is $\frac{28}{11}$.

GRAPHICAL APPROACH

We graph $f(x) = -\frac{3}{4}x + 6$ and $g(x) = 2x - 1$ on the same set of axes. It appears that the lines intersect at $(2.5, 4)$.

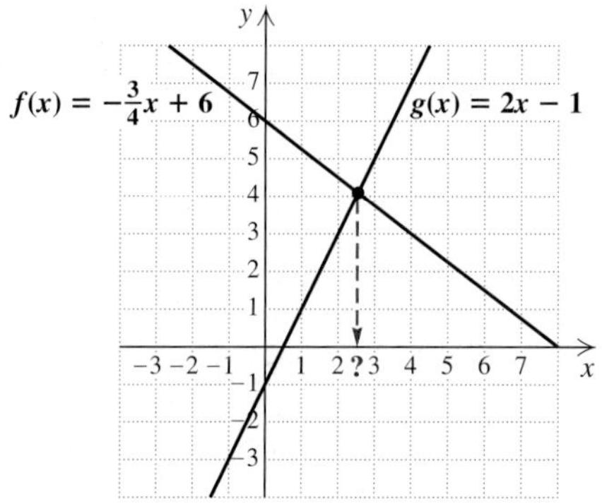

Check: $-\frac{3}{4}x + 6 = 2x - 1$

$-\frac{3}{4}(2.5) + 6$	$2(2.5) - 1$
$-1.875 + 6$	$5 - 1$

$4.125 \stackrel{?}{=} 4$ FALSE

Our check shows that 2.5 is *not* the solution, although it may not be off by much. To find the exact solution graphically, we need a method that will determine coordinates more precisely.

> **CAUTION!** When using a hand-drawn graph to solve an equation, it is important to use graph paper and to work as neatly as possible. Use a straightedge when drawing lines and be sure to erase any mistakes completely.

We can use the INTERSECT option of the CALC menu on a graphing calculator to find the point of intersection.

EXAMPLE 3 Solve using a graphing calculator: $-\frac{3}{4}x + 6 = 2x - 1$.

SOLUTION In Example 2, we saw that the graphs of $f(x) = -\frac{3}{4}x + 6$ and $g(x) = 2x - 1$ intersect near the point $(2.5, 4)$. To determine more precisely the point of intersection, we graph $y_1 = -\frac{3}{4}x + 6$ and $y_2 = 2x - 1$ using the same viewing window. After choosing the INTERSECT feature, we indicate which two graphs, or *curves*, we are considering. Then we enter a guess. The coordinates of the point of intersection then appear at the bottom of the screen.

$y_1 = -\frac{3}{4}x + 6,\ y_2 = 2x - 1$

It appears from the screen above that the solution is 2.5454545. To check, we evaluate both sides of the equation $-\frac{3}{4}x + 6 = 2x - 1$ for this value of x. The first coordinate of the point of intersection is stored as X, so we evaluate Y1(X) and Y2(X) from the home screen, as shown on the left below.

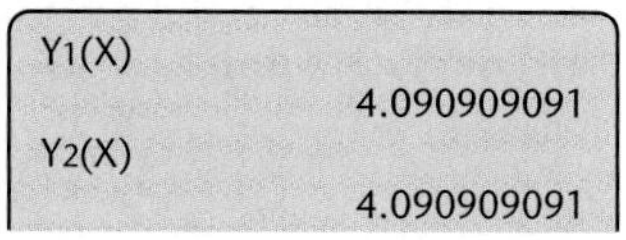

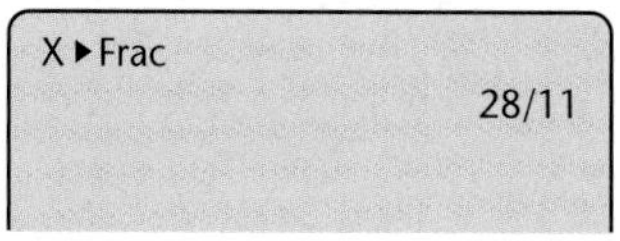

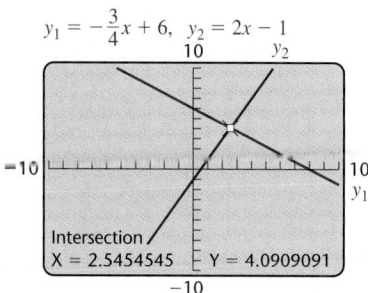

Although the check shows that 2.5454545 is the solution, it is actually an approximation of the solution. In some cases, the calculator can give us the exact answer. Since the x-coordinate of the point of intersection is stored as X, converting X to fraction notation will give an exact solution. From the screen on the right above, we see that $\frac{28}{11}$ is the solution of the equation.

Solving Equations Graphically: The Zero Method

When we are solving an equation graphically, it can be challenging to determine a portion of the x, y-coordinate plane that contains the point of intersection. The Zero method makes that determination easier because we are interested only in the point at which a graph crosses the x-axis.

To solve an equation using the Zero method, we use the addition principle to get zero on one side of the equation. Then we look for the intersection of the line and the x-axis, or the *x-intercept* of the graph. If using function notation, we look for the *zero* of the function.

EXAMPLE 4 Solve graphically, using the Zero method: $2x - 5 = 4x - 11$.

SOLUTION We first get 0 on one side of the equation:

$$2x - 5 = 4x - 11$$
$$-2x - 5 = -11 \quad \textbf{Subtracting } 4x \textbf{ from both sides}$$
$$-2x + 6 = 0. \quad \textbf{Adding 11 to both sides}$$

We then graph $f(x) = -2x + 6$, and find the x-intercept of the graph.

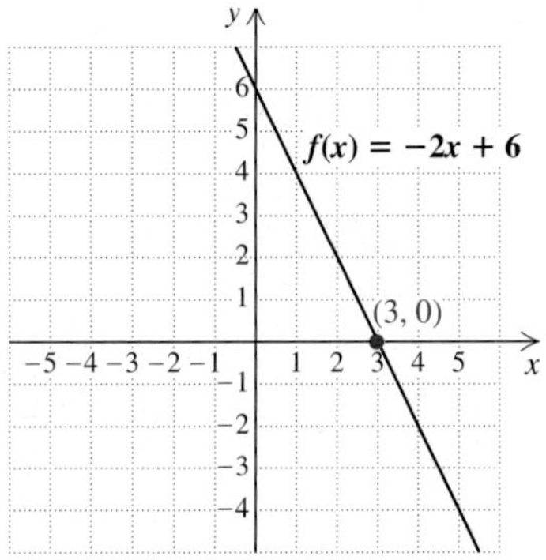

The x-intercept of the graph appears to be $(3, 0)$. We check 3 in the original equation.

Check:

$2x - 5$	$= 4x - 11$
$2 \cdot 3 - 5$	$4 \cdot 3 - 11$
$6 - 5$	$12 - 11$
1	$\stackrel{?}{=} 1$ TRUE

The solution is 3.

From knowing the point of intersection of the graph in Example 4 with the x-axis, we can make all three of the following statements:

The x-intercept of the graph of $f(x) = -2x + 6$ is $(3, 0)$.

The zero of $f(x) = -2x + 6$ is 3, since $f(3) = 0$.

The solution of the equation $2x - 5 = 4x - 11$ is 3.

We can use the ZERO option of the CALC menu on a graphing calculator to solve an equation.

EXAMPLE 5 Solve $3 - 8x = 5 - 7x$ using both the Intersect and Zero methods.

GRAPHICAL SOLUTION: INTERSECT METHOD

We graph $y_1 = 3 - 8x$ and $y_2 = 5 - 7x$ and determine the coordinates of any point of intersection.

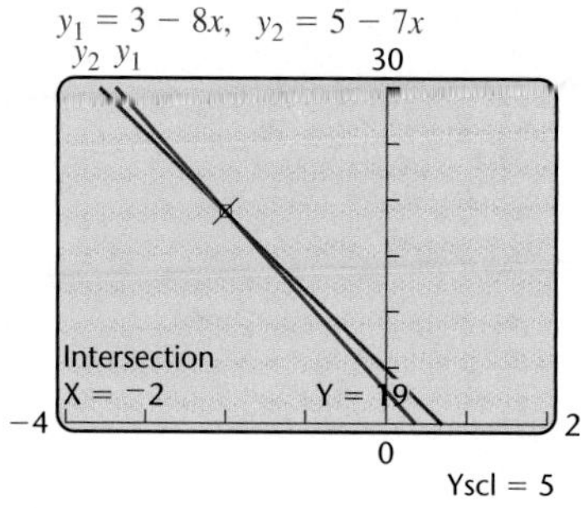

The point of intersection is $(-2, 19)$.
The solution is -2.

GRAPHICAL SOLUTION: ZERO METHOD

We first get zero on one side of the equation:

$$3 - 8x = 5 - 7x$$

$$-2 - 8x = -7x \qquad \textbf{Subtracting 5 from both sides}$$

$$-2 - x = 0. \qquad \textbf{Adding } 7x \textbf{ to both sides}$$

Then we graph $y = -2 - x$ and determine the point for which $y = 0$, or the x-intercept of the graph. We do this using the ZERO option in the CALC menu.

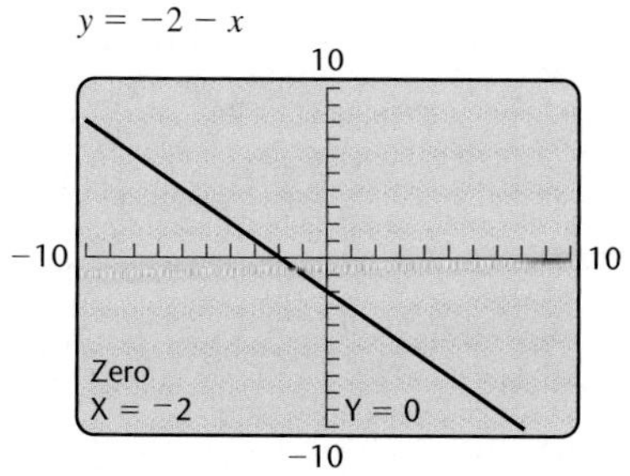

The solution is the first coordinate of the x-intercept, or -2.

Solving Inequalities Graphically

Consider the graphs of the functions $f(x) = 2x + 4$ and $g(x) = -x + 1$. The function values are *equal* at the point of intersection, $(-1, 2)$.

Graphs of $f(x)$ and $g(x)$

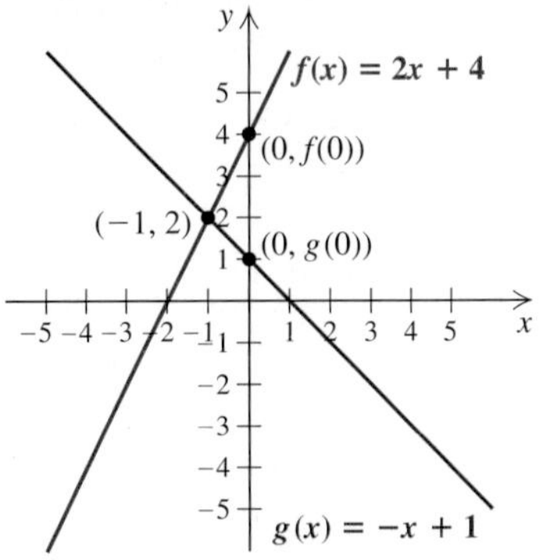

Graph of solution set of $f(x) = g(x)$

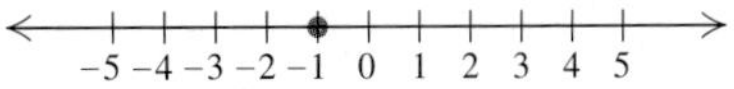

Graph of solution set of $f(x) < g(x)$

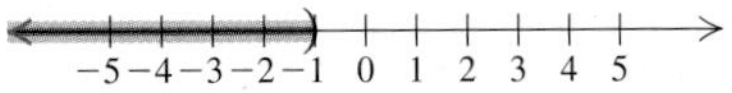

Graph of solution set of $f(x) > g(x)$

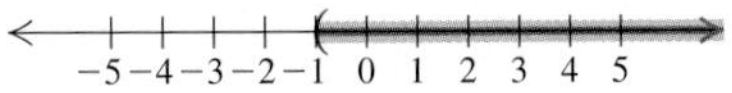

At all x-values except -1, either

$$f(x) > g(x) \quad \text{or} \quad f(x) < g(x).$$

Note from the graphs that $f(x) > g(x)$ when the graph of f lies above the graph of g. Also, $f(x) < g(x)$ when the graph of f lies below the graph of g.

Compare $f(0)$ and $g(0)$. Note from the graphs that $f(0)$ lies above $g(0)$. This can be seen also by finding $f(0)$ and $g(0)$:

$$f(0) = 2 \cdot 0 + 4 = 4,$$
$$g(0) = -0 + 1 = 1.$$

Since $4 > 1, f(0) > g(0)$.

In fact, for all values of x greater than $-1, f(x) > g(x)$. For all values of x less than $-1, f(x) < g(x)$. In this way, the point of intersection of the graphs marks the endpoint of the solution set of an inequality.

EXAMPLE 6 Solve graphically: $16 - 7x \geq 10x - 4$.

SOLUTION We let $y_1 = 16 - 7x$ and $y_2 = 10x - 4$, and graph y_1 and y_2 in the window $[-5, 5, -5, 15]$.

From the graph, we see that $y_1 = y_2$ at the point of intersection. To the left of the point of intersection, $y_1 > y_2$. You can check this by calculating the values of both y_1 and y_2 for an x-value to the left of the point of intersection—say, for $x = 0.5$. Since $12.5 > 1$, we know that $y_1 > y_2$ when $x = 0.5$. Similarly, we see that $y_1 < y_2$ to the right of the point of intersection.

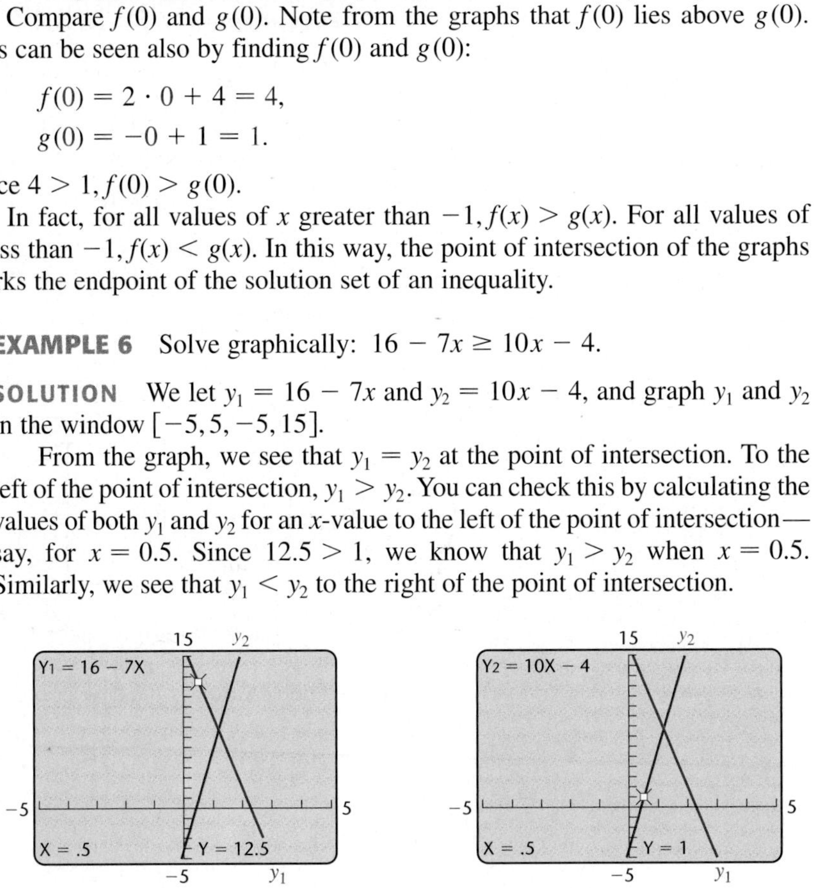

The solution set will be all x-values to the left of the point of intersection, as well as the x-coordinate of the point of intersection. Using INTERSECT, we find that the x-coordinate of the point of intersection is approximately 1.1764706. Thus the solution set is approximately $(-\infty, 1.1764706]$, or converting to fraction notation, $\left(-\infty, \frac{20}{17}\right]$. Compare this solution with the algebraic solution in Example 5(a) of Section 4.1.

On many graphing calculators, the interval that is the solution set can be indicated by using the VARS and TEST keys. To find the solution of Example 6, we also enter and graph $y_3 = y_1 \geq y_2$ (the "$\geq$" symbol can be found in the TEST menu). Where this is true, the value of y_3 will be 1, and where it is false, the value will be 0. The solution set is thus displayed as an interval, shown by a horizontal line 1 unit above the x-axis. The endpoint of the interval corresponds to the intersection of the graphs of the equations. For some calculators, using DOT mode for y_3 will result in a more accurate graph.

$y_1 = 16 - 7x,\ y_2 = 10x - 4,$
$y_3 = y_1 \geqslant y_2$

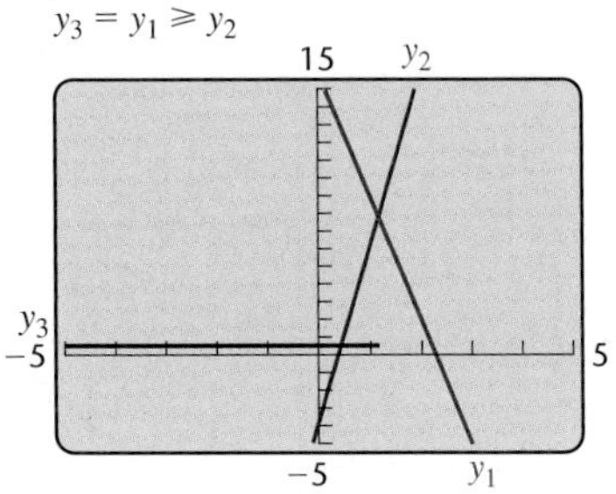

Applications

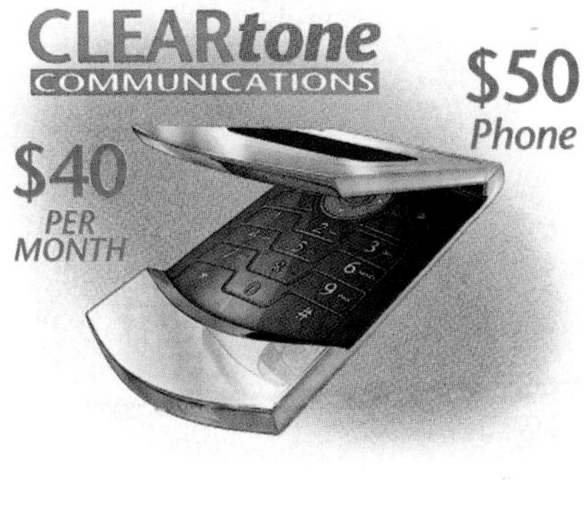

EXAMPLE 7 Cost Projections. Cleartone Communications charges \$50 for a cell phone and \$40 per month for calls made under its Call Anywhere plan. Formulate and graph a mathematical model for the cost. Then use the model to estimate the time required for the total cost to reach \$250.

SOLUTION

1. **Familiarize.** The problem describes a situation in which a monthly fee is charged after an initial purchase has been made. After 1 month of service, the total cost will be $\$50 + \$40 = \$90$. After 2 months, the total cost will be $\$50 + \$40 \cdot 2 = \$130$. This can be generalized in a model if we let $C(t)$ represent the total cost, in dollars, for t months of service.

2. Translate. We reword and translate as follows:

Rewording:	The total cost	is	the cost of the phone	plus	\$40 per month.
	↓	↓	↓	↓	↓
Translating:	$C(t)$	$=$	50	$+$	$40 \cdot t$

where $t \geq 0$ (since there cannot be a negative number of months).

3. Carry out. To estimate the time required for the total cost to reach \$250, we are estimating the solution of

$$40t + 50 = 250. \quad \textbf{Replacing } C(t) \textbf{ with 250}$$

We do this by graphing $C(t) = 40t + 50$ and $y = 250$ and looking for the point of intersection. On a graphing calculator, we let $y_1 = 40x + 50$ and $y_2 = 250$, and adjust the window dimensions to include the point of intersection. Using either a hand-drawn graph or one generated by a graphing calculator, we find a point of intersection at $(5, 250)$.

Thus we estimate that it takes 5 months for the total cost to reach \$250.

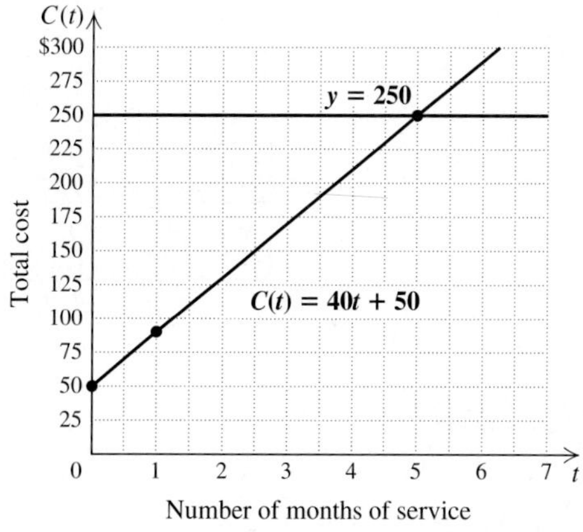

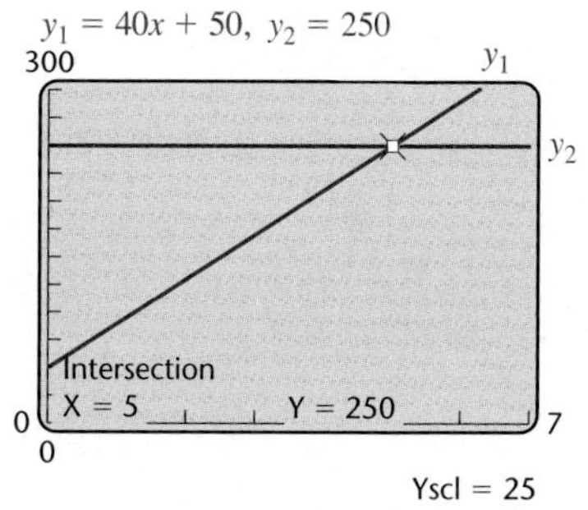

4. Check. We evaluate:

$$\begin{aligned} C(5) &= 40 \cdot 5 + 50 \\ &= 200 + 50 \\ &= 250. \end{aligned}$$

Our estimate turns out to be precise.

5. State. It takes 5 months for the total cost to reach \$250.

EXAMPLE 8 Earnings Ratio. The ratio of median earnings of women to median earnings of men is called the female-to-male earnings ratio. A ratio of 0.77, or 77:100, means that women earn \$77 for every \$100 that men earn. This ratio is shown for various years in the following table. Use linear regression to find a linear function that can be used to estimate the female-to-male

earnings ratio x years after 1980. Then use the function to predict the years in which the female-to-male earnings ratio will be 0.9 or higher.

Year	Female-to-Male Earnings Ratio
1980	0.60
1985	0.64
1990	0.71
1995	0.72
2000	0.75
2004	0.77

Source: U.S. Bureau of the Census

SOLUTION We let x represent the number of years after 1980 and y the female-to-male earnings ratio, and graph the data to determine whether the relationship appears to be linear.

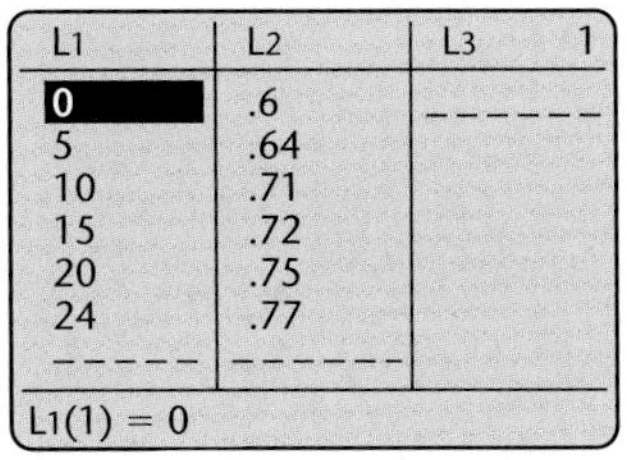

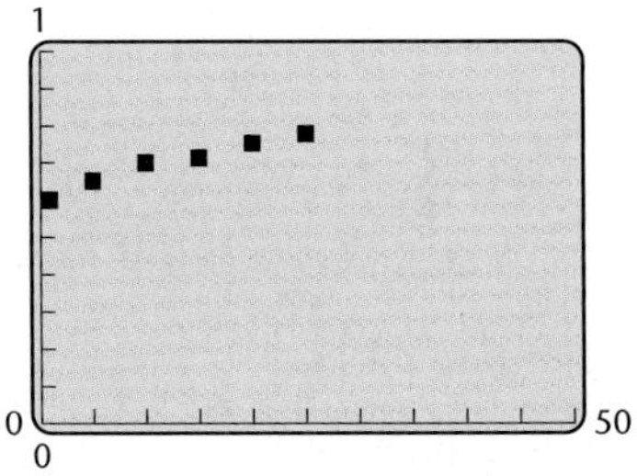

Xscl = 5, Yscl = .1

The data are approximately linear. We find a linear equation that fits the data, as shown in the figure on the left below. Since we want to find the years in which the ratio is 0.9 or higher, we graph $y_1 = 0.0070241935x + 0.6117016129$ and $y_2 = 0.9$. The point of intersection, shown in the figure on the right below, tells us that the ratio will be 0.9 approximately 41 yr after 1980, or in about 2021. Since $y_1 > y_2$ to the right of the point of intersection, we predict that the female-to-male earnings ratio will be 0.9 or higher in 2021 and after, or for years after 2020.

$y_1 = 0.0070241935x + 0.6117016129$,
$y_2 = 0.9$

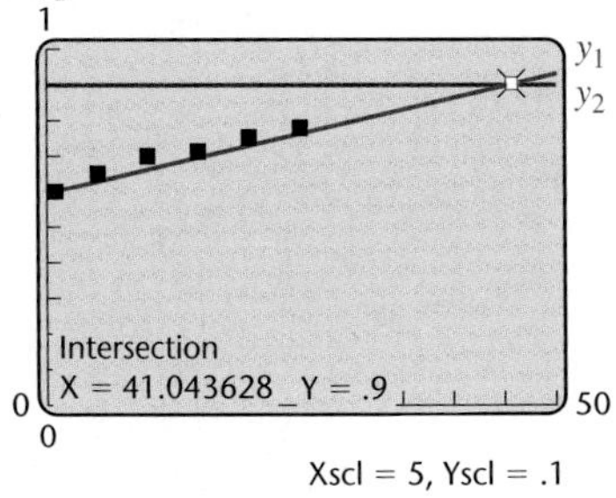

Xscl = 5, Yscl = .1

4.2 EXERCISE SET

FOR EXTRA HELP

MathXL | MyMathLab | InterAct Math | AW Math Tutor Center | Video Lectures on CD: Disc 2 | Student's Solutions Manual

Concept Reinforcement *Exercises 1–8 refer to the following graph. Match each description with the appropriate answer from the column on the right.*

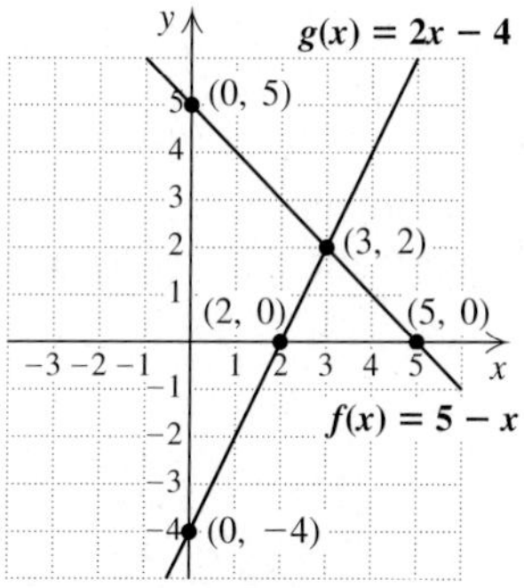

1. The solution of $5 - x = 2x - 4$
2. The zero of g
3. The solution of $5 - x = 0$
4. The x-intercept of the graph of f
5. The y-intercept of the graph of g
6. The solution of $y = 5 - x,$ $y = 2x - 4$
7. The solution of $5 - x > 2x - 4$
8. The solution of $5 - x \le 2x - 4$

a) $(0, -4)$
b) $(5, 0)$
c) $(3, 2)$
d) 2
e) 3
f) 5
g) $(-\infty, 3)$
h) $[3, \infty)$

Estimate the solution of each equation from the associated graph.

9. $2x - 1 = -5$

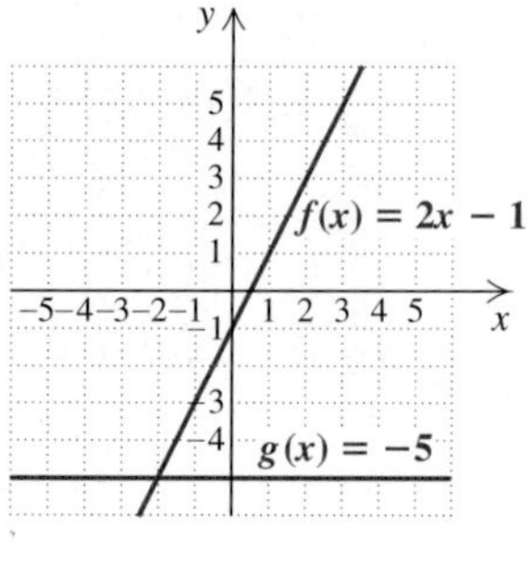

10. $\frac{1}{2}x + 1 = 3$

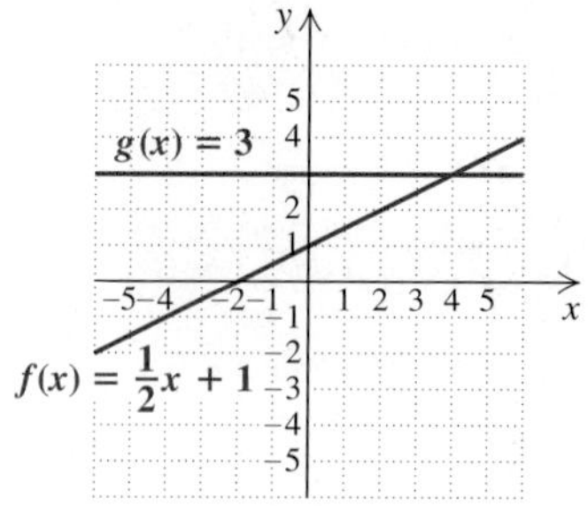

11. $2x + 3 = x - 1$

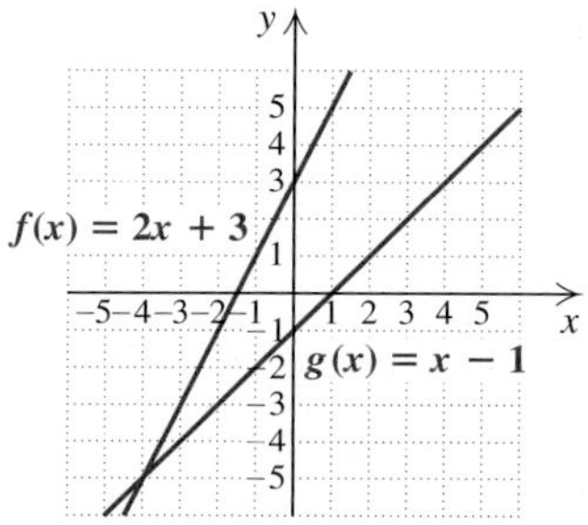

12. $2 - x = 3x - 2$

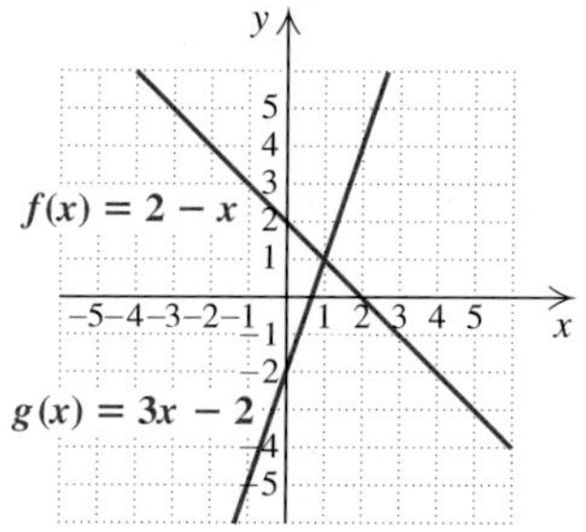

13. $\frac{1}{2}x + 3 = x - 1$

$y_1 = (1/2)x + 3, \;\; y_2 = x - 1$

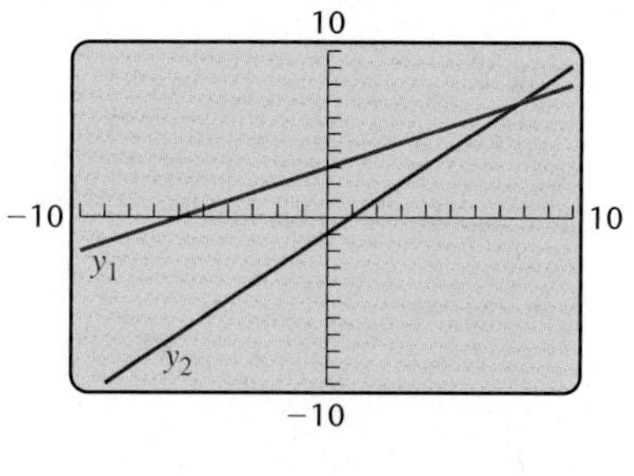

14. $2x - \frac{1}{2} = x + \frac{5}{2}$

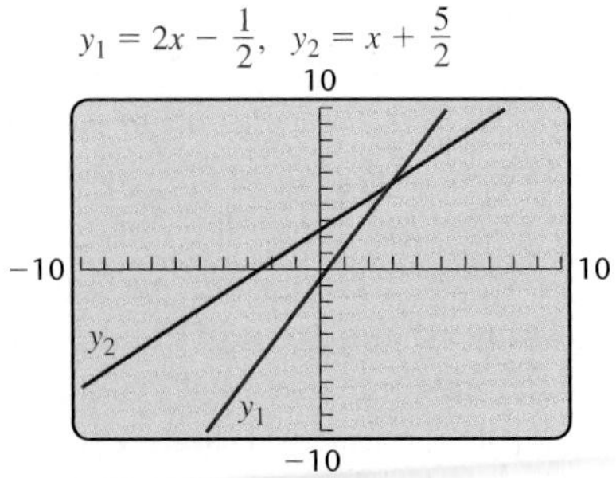

15. $f(x) = g(x)$

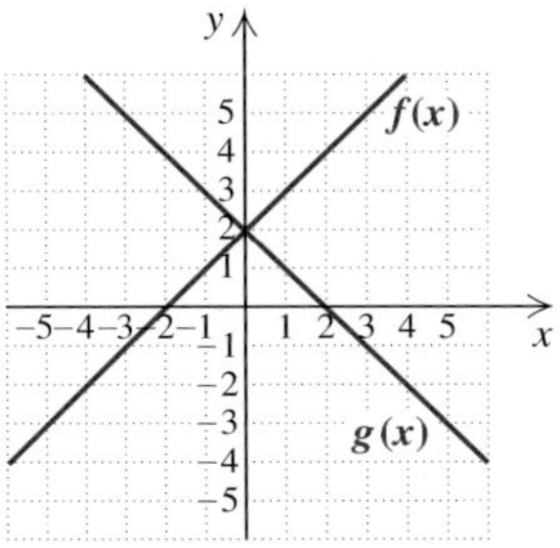

16. $f(x) = g(x)$

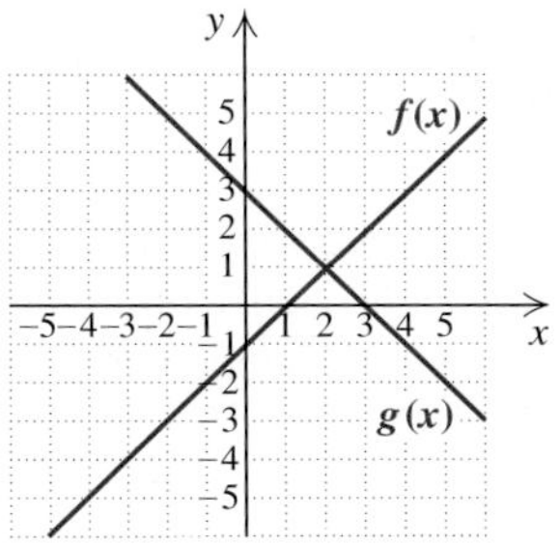

17. Estimate the value of x for which $y_1 = y_2$.

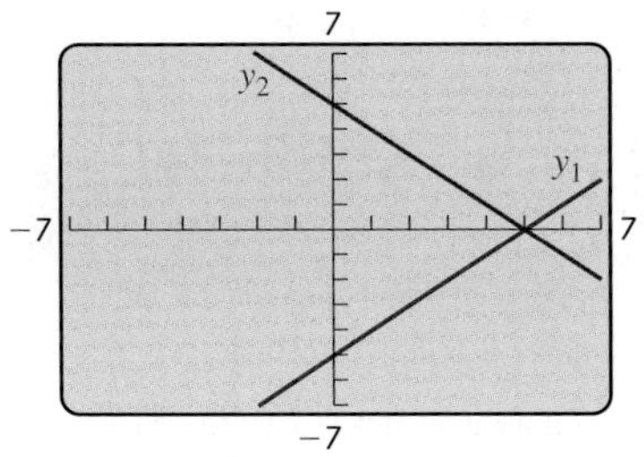

18. Estimate the value of x for which $y_1 = y_2$.

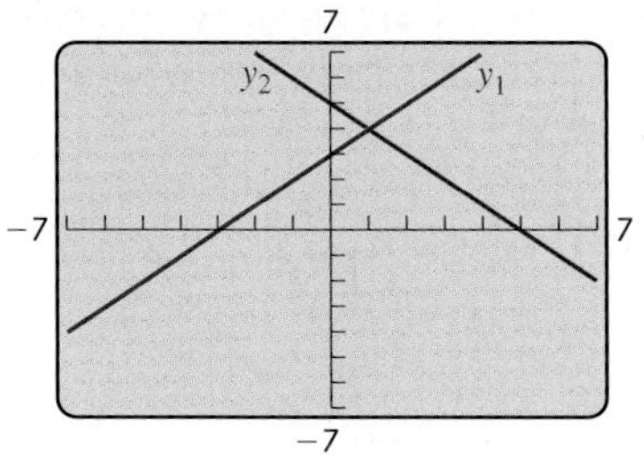

Solve graphically.

19. $x - 3 = 4$

20. $x + 4 = 6$

21. $2x + 1 = 7$

22. $3x - 5 = 1$

23. $\frac{1}{3}x - 2 = 1$

24. $\frac{1}{2}x + 3 = -1$

25. $x + 3 = 5 - x$

26. $x - 7 = 3x - 3$

27. $5 - \frac{1}{2}x = x - 4$

28. $3 - x = \frac{1}{2}x - 3$

29. $2x - 1 = -x + 3$

30. $-3x + 4 = 3x - 4$

Solve each inequality using the given graph.

31. $f(x) \geq g(x)$

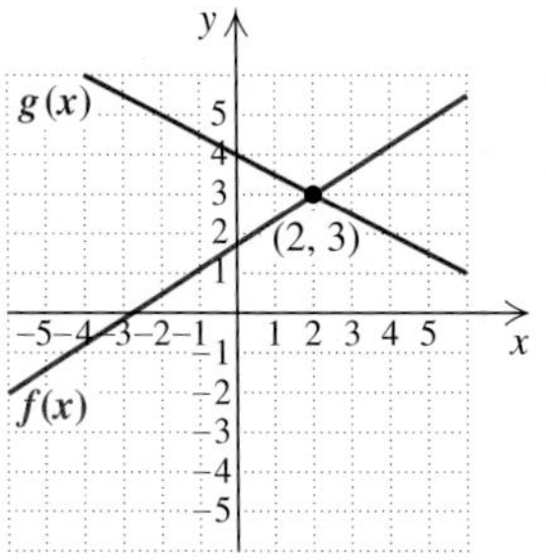

32. $f(x) < g(x)$

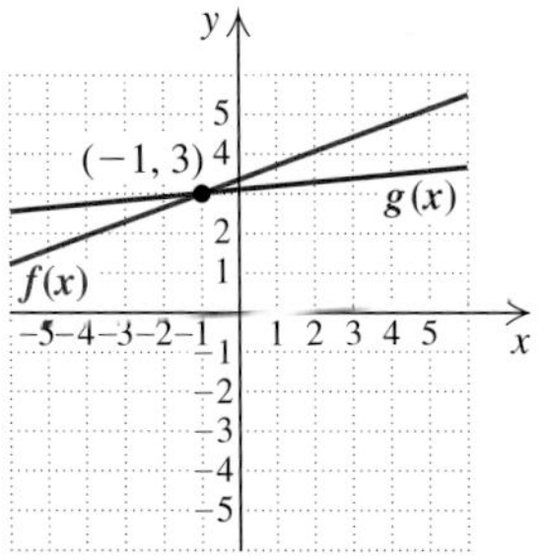

33. $y_1 < y_2$

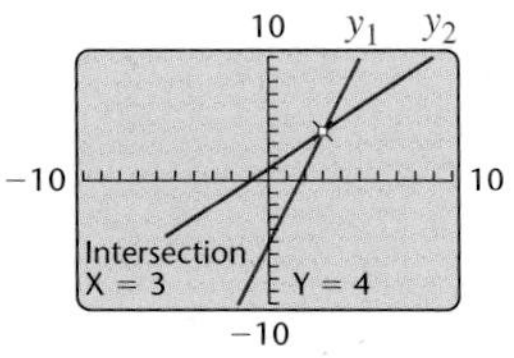

34. $y_1 \geq y_2$

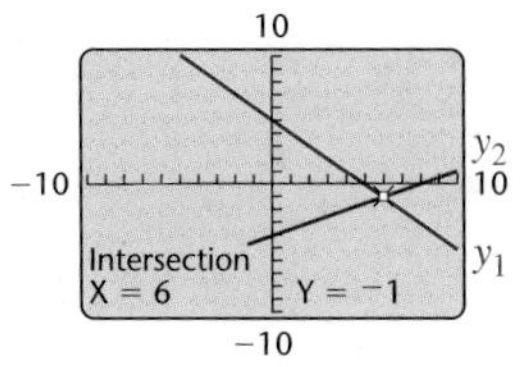

35. The graphs of $f(x) = 2x + 1$, $g(x) = -\frac{1}{2}x + 3$, and $h(x) = x - 1$ are as shown below. Solve each inequality, referring only to the figure.

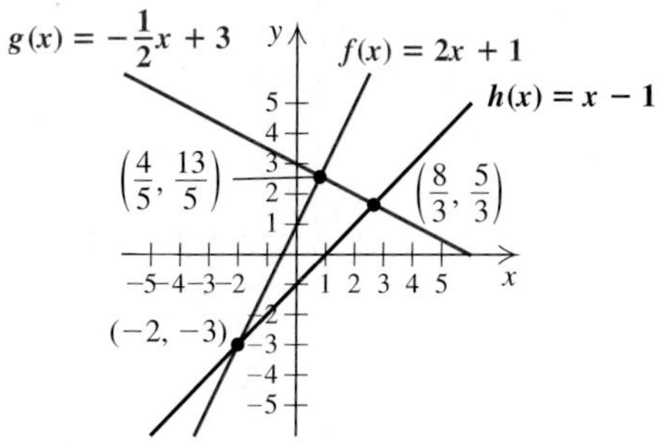

a) $2x + 1 \leq x - 1$
b) $x - 1 > -\frac{1}{2}x + 3$
c) $-\frac{1}{2}x + 3 < 2x + 1$

36. The graphs of $y_1 = -\frac{1}{2}x + 5$, $y_2 = x - 1$, and $y_3 = 2x - 3$ are as shown below. Solve each inequality, referring only to the figure.

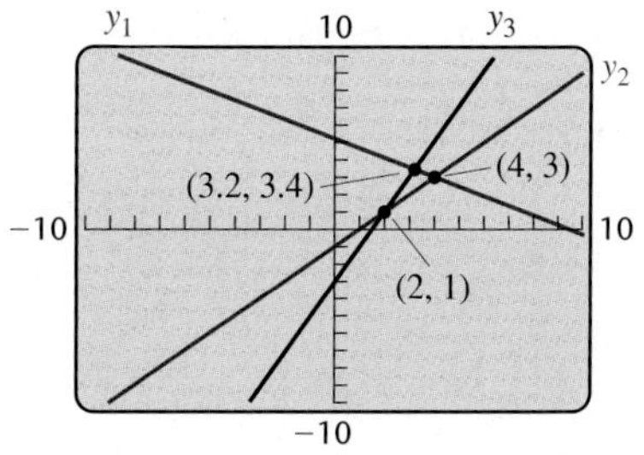

a) $-\frac{1}{2}x + 5 > x - 1$
b) $x - 1 \leq 2x - 3$
c) $2x - 3 \geq -\frac{1}{2}x + 5$

Solve graphically.

37. $x - 3 < 4$

38. $x + 4 \geq 6$

39. $2x - 3 \geq 1$

40. $3x + 1 < 1$

41. $x + 3 > 2x - 5$

42. $3x - 5 \leq 3 - x$

43. $\frac{1}{2}x - 2 \leq 1 - x$

44. $x + 5 > \frac{1}{3}x - 1$

45. $4x + 7 \leq 3 - 5x$

46. $5x + 6 < 8x - 11$

Use a graph to estimate the solution in each of the following. Be sure to use graph paper and a straightedge if graphing by hand.

47. *Healthcare Costs.* Under one particular CIGNA health-insurance plan, an individual pays the first \$100 of each hospital stay plus $\frac{1}{5}$ of all charges in excess of \$100 (*Source*: VT State Chamber of Commerce membership information from VACE, 2004). By approximately how much did Gerry's hospital bill exceed \$100, if the hospital stay cost him a total of \$2200?

48. *Indiana Toll Road.* To drive an automobile on the Indiana Toll Road costs approximately 21¢ plus 3¢ per mile (*Source*: www.in.gov/dot/motoristinfo). Estimate how far Gina has driven on the Toll Road if her toll is \$2.50.

49. *Telephone Charges.* Skytone Calling charges \$20 for a telephone and \$25 per month under its economy plan. Estimate the time required for the total cost to reach \$145.

50. *Cell-Phone Charges.* The Cellular Connection charges \$30 for a cell phone and \$40 per month under its economy plan. Estimate the time required for the total cost to reach \$190.

51. *Parking Fees.* Karla's Parking charges \$3.00 to park plus 50¢ for each 15-min unit of time. Estimate how long someone can park for \$7.50.*

*A more precise, nonlinear model of Exercise 51 appears in Exercise 83.

52. *Cost of a Road Call.* Dave's Foreign Auto Village charges \$50 for a road call plus \$15 for each 15-min unit of time. Estimate the time required for a road call that cost \$140.*

53. *Cost of a FedEx Delivery.* In 2006, for Standard delivery to the closest zone of packages weighing from 100 to 499 lb, FedEx charged \$114 plus \$1.14 for each pound over 100 (*Source*: www.fedex.com). Estimate the weight of a package that cost \$171 to ship.

54. *Copying Costs.* For each copy of a town report, a FedEx Kinko's Office and Print Center charged \$3.45 for a coil binding and 15¢ for each page that was a double-sided copy (*Source*: FedEx Kinko's price list brochure). Estimate the number of pages in a coil-bound town report that cost \$6.90 per copy. Assume that every page in the report was a double-sided copy.

55. *Show Business.* A band receives \$750 plus 15% of receipts over \$750 for playing a club date. If a club charges a \$6 cover charge, how many people must attend in order for the band to receive at least \$1200?

56. *Temperature Conversion.* The function

$$C(F) = \tfrac{5}{9}(F - 32)$$

can be used to find the Celsius temperature $C(F)$ that corresponds to $F°$ Fahrenheit.

a) Gold is solid at Celsius temperatures less than 1063°C. Find the Fahrenheit temperatures for which gold is solid.

b) Silver is solid at Celsius temperatures less than 960.8°C. Find the Fahrenheit temperatures for which silver is solid.

57. *Manufacturing.* Ergs, Inc., is planning to make a new kind of radio. Fixed costs will be \$90,000, and variable costs will be \$15 for the production of each radio. The total-cost function for x radios is

$$C(x) = 90{,}000 + 15x.$$

The company makes \$26 in revenue for each radio sold. The total-revenue function for x radios is

$$R(x) = 26x.$$

(See Section 3.8.)

a) When $R(x) < C(x)$, the company loses money. Find the values of x for which the company loses money.

b) When $R(x) > C(x)$, the company makes a profit. Find the values of x for which the company makes a profit.

58. *Publishing.* The demand and supply functions for a locally produced poetry book are approximated by

$$D(p) = 2000 - 60p \quad \text{and}$$
$$S(p) = 460 + 94p,$$

where p is the price in dollars (see Section 3.8).

a) Find those values of p for which demand exceeds supply.

b) Find those values of p for which demand is less than supply.

59. *Winter Olympic Games.* The number of nations participating in the Winter Olympic Games has been increasing over the years, as shown in the following table. Use linear regression to find a linear function that can be used to predict the number of nations participating x years after 1988. Then predict those years in which more than 100 nations will participate in the games.

Year	Number of Nations Participating in Winter Olympic Games
1988	57
1992	64
1994	67
1998	72
2002	77
2006	80

Source: Olympics.org

*A more precise, nonlinear model of Exercise 52 appears in Exercise 84.

60. *Trampoline Injuries.* The following table lists the number of trampoline injuries in the United States for various years. Use linear regression to find a linear function that can be used to estimate the number t of trampoline injuries x years after 1990. Then predict those years in which the number of trampoline injuries will exceed 150,000.

Year	Number of Injuries
1990	29,600
1992	39,000
1995	58,400
1997	82,722
2002	89,393

Source: National Safety Council, Itasca, IL, *Injury Facts*

61. *Media Usage.* The following table lists the number of hours spent per person per year reading a newspaper and playing video games. Use linear regression to find two linear functions that can be used to estimate the number of hours n spent reading a newspaper and the number of hours v spent playing video games x years after 2000. Then predict those years in which a person will spend more time playing video games than reading a newspaper.

Year	Daily Newspapers (in hours per person per year)	Video Games (in hours per person per year)
2000	180	59
2001	177	60
2002	175	64
2003	171	69
2004*	169	71
2005*	168	75
2006*	165	81
2007*	165	86
2008*	164	98

*Projected

Source: Veronis Suhler Stevenson, New York, NY

62. *Basketball.* The following table lists the attendance for NCAA and professional basketball for various years. Use linear regression to find two linear functions that can be used to estimate the attendance n for the NCAA and the attendance p for professional basketball x years after 1985. Then predict those years in which professional basketball attendance will exceed NCAA basketball attendance.

Year	NCAA Basketball Attendance (in thousands)	Professional Basketball Attendance (in thousands)
1985	26,584	11,534
1990	28,741	18,586
1995	28,548	19,883
2000	29,025	21,503
2002	29,395	21,571
2004	30,761	22,953

Sources: NCAA; NBA

63. *High Jump.* The following table lists the world record for the men's high jump for various years. Use linear regression to find a linear function that can be used to predict the high-jump world record x years after 1900. Then predict those years in which the world record will exceed 2.5 m.

Year	High Jump (in meters)	Athlete
1912	2.00	George L. Horine (USA)
1924	2.04	Harold M. Osborn (USA)
1934	2.06	Walter Marty (USA)
1941	2.11	Lester Steers (USA)
1957	2.162	Yuriy Styepanov (USSR)
1970	2.29	Ni Zhinquin (China)
1980	2.36	Gerd Wessig (Germany)
1993	2.45	Javier Sotomayor (Cuba)

Source: www.iaaf.org

64. *Mile Run.* The following table lists the world record for the mile run for various years. Use linear regression to find a linear function that can be used to predict the world record for the mile run x years after 1954. Then predict those years in which the world record will be less than 3.5 min.

Year	Time for Mile (in minutes)	Athlete
1954	3.99	Sir Roger Bannister (Great Britain)
1962	3.9017	Peter Snell (New Zealand)
1975	3.8233	John Walker (New Zealand)
1981	3.7888	Sebastian Coe (Great Britain)
1993	3.7398	Noureddine Morceli (Algeria)
1999	3.7188	Hican El Guerrouj (Morocco)

Source: www.iaaf.org

TW **65.** Examine the data in Exercise 61. How can you tell without graphing that at some point people will spend more time playing video games than reading the newspaper?

TW **66.** Darnell used a graphical method to solve $2x - 3 = x + 4$. He stated that the solution was $(7, 11)$. Can this answer be correct? What mistake do you think Darnell is making?

Skill Maintenance

Solve. [1.6]

67. $0.5x - 2.34 + 2.4x = 7.8x - 9$

68. $5x + 7x = -144$

69. A piece of wire 32.8 ft long is to be cut into two pieces, and those pieces are each to be bent to make a square. The length of a side of one square is to be 2.2 ft greater than the length of a side of the other. How should the wire be cut? [1.7]

70. *Inventory.* The Freeport College store paid \$1728 for an order of 45 calculators. The store paid \$9 for each scientific calculator. The others, all graphing calculators, cost the store \$58 each. How many of each type of calculator was ordered? [3.3]

71. *Insulation.* The Mazzas' attic required three and a half times as much insulation as did the Kranepools'. Together, the two attics required 36 rolls of insulation. How much insulation did each attic require? [3.3]

72. *Sales of Food.* High Flyin' Wings charges \$12 for a bucket of chicken wings and \$7 for a chicken dinner. After filling 28 orders for buckets and dinners, High Flyin' Wings had collected \$281. How many buckets and how many dinners did they sell? [3.3]

Synthesis

TW **73.** Explain why, when solving an equation graphically, the x-coordinate of the point of intersection gives the solution of the equation.

TW **74.** Explain the difference between "solving by graphing" and "graphing the solution set."

Estimate the solution(s) from the associated graph.

75. $f(x) = g(x)$

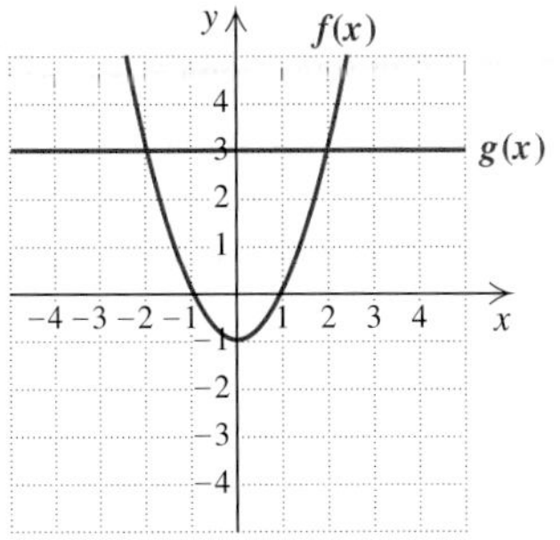

76. $f(x) = g(x)$

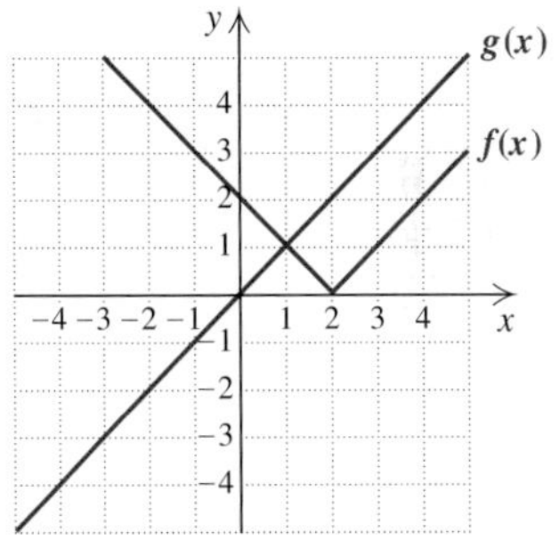

Solve graphically. Be sure to check.

77. $2x = |x + 1|$

78. $x - 1 = |4 - 2x|$

79. $\frac{1}{2}x = 3 - |x|$

80. $2 - |x| = 1 - 3x$

81. $x^2 = x + 2$

82. $x^2 = x$

83. (Refer to Exercise 51.) It costs as much to park at Karla's for 16 min as it does for 29 min. Thus the linear graph drawn in the solution of Exercise 51 is not a precise representation of the situation. Draw a graph with a series of "steps" that more accurately reflects the situation.

84. (Refer to Exercise 52.) A 32-min road call with Dave's costs the same as a 44-min road call. Thus the linear graph drawn in the solution of Exercise 52 is not a precise representation of the situation. Draw a graph with a series of "steps" that more accurately reflects the situation.

4.3 Intersections, Unions, and Compound Inequalities

Intersections of Sets and Conjunctions of Sentences ■ Unions of Sets and Disjunctions of Sentences ■ Interval Notation and Domains

Two inequalities joined by the word "and" or the word "or" are called **compound inequalities.** Thus, "$2x - 7 < 3$ *or* $x - 1 > 4$" and "$7x < 9$ *and* $x - 1 > -5$" are two examples of compound inequalities. In order to discuss how to solve compound inequalities, we must first study ways in which sets can be combined.

Intersections of Sets and Conjunctions of Sentences

The **intersection** of two sets A and B is the set of all elements that are common to both A and B. We denote the intersection of sets A and B as

$$A \cap B.$$

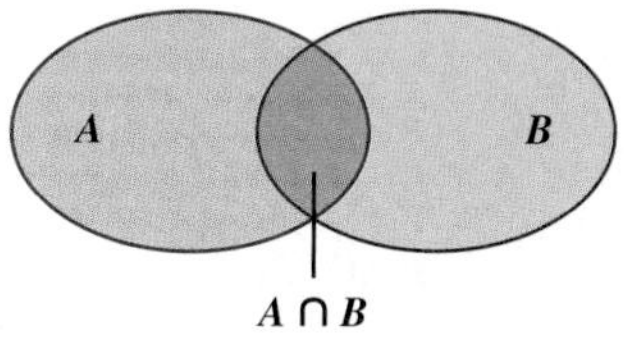

The intersection of two sets is represented by the purple region shown in the figure at left. For example, if $A = \{$all students who are more than 5′4″ tall$\}$ and $B = \{$all students who weigh more than 120 lb$\}$, then $A \cap B = \{$all students who are more than 5′4″ tall and weigh more than 120 lb$\}$.

EXAMPLE 1 Find the intersection: $\{1, 2, 3, 4, 5\} \cap \{-2, -1, 0, 1, 2, 3\}$.

SOLUTION The numbers 1, 2, and 3 are common to both sets, so the intersection is $\{1, 2, 3\}$.

When two or more sentences are joined by the word *and* to make a compound sentence, the new sentence is called a **conjunction** of the sentences. The following is a conjunction of inequalities:

$$-2 < x \quad and \quad x < 1.$$

The word *and* means that *both* sentences must be true if the conjunction is to be true.

A number is a solution of a conjunction if it is a solution of *both* of the separate parts. For example, -1 is a solution because it is a solution of $-2 < x$ as well as $x < 1$.

Below we show the graph of $-2 < x$, followed by the graph of $x < 1$, and finally the graph of the conjunction $-2 < x$ *and* $x < 1$. *Note that the solution set of a conjunction is the intersection of the solution sets of the individual sentences.*

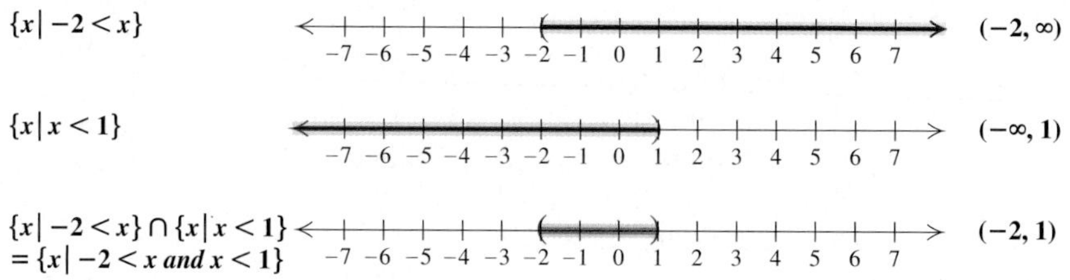

Study Tip

Two Books Are Better Than One

Many students find it helpful to use a second book as a reference when studying. Perhaps you or a friend own a text from a previous math course that can serve as a resource. Often professors have older texts that they will happily give away. Library book sales and thrift shops can also be excellent sources for extra books. Saving your text when you finish a math course can provide you with an excellent aid for your next course.

Because there are numbers that are both greater than -2 and less than 1, the conjunction $-2 < x$ *and* $x < 1$ can be abbreviated by $-2 < x < 1$. Thus the interval $(-2, 1)$ can be represented as $\{x \mid -2 < x < 1\}$, the set of all numbers that are *simultaneously* greater than -2 *and* less than 1. Note that for $a < b$,

$$a < x \quad \textit{and} \quad x < b \quad \textbf{can be abbreviated} \quad a < x < b;$$

and, equivalently,

$$b > x \quad \textit{and} \quad x > a \quad \textbf{can be abbreviated} \quad b > x > a.$$

EXAMPLE 2 Solve and graph; $-1 \leq 2x + 5 < 13$.

SOLUTION This inequality is an abbreviation for the conjunction

$$-1 \leq 2x + 5 \quad \textit{and} \quad 2x + 5 < 13.$$

We solve both algebraically and graphically.

ALGEBRAIC APPROACH

The word *and* corresponds to set *intersection*. To solve the conjunction, we solve each of the two inequalities separately and then find the intersection of the solution sets:

$-1 \leq 2x + 5$	*and*	$2x + 5 < 13$	
$-6 \leq 2x$	*and*	$2x < 8$	**Subtracting 5 from both sides of each inequality**
$-3 \leq x$	*and*	$x < 4.$	**Dividing both sides of each inequality by 2**

These steps are sometimes combined as follows:

$$-1 \leq 2x + 5 < 13$$

$$-1 - 5 \leq 2x + 5 - 5 < 13 - 5 \quad \textbf{Subtracting 5 from all three regions}$$

$$-6 \leq 2x < 8$$

$$-3 \leq x < 4. \quad \textbf{Dividing by 2 in all three regions}$$

The solution set is $\{x \mid -3 \leq x < 4\}$, or, in interval notation, $[-3, 4)$. The graph is the intersection of the two separate solution sets.

$\{x \mid -3 \leq x\}$, or $[-3, \infty)$

$\{x \mid x < 4\}$, or $(-\infty, 4)$

$\{x \mid -3 \leq x\} \cap \{x \mid x < 4\}$ $= \{x \mid -3 \leq x < 4\}$, or $[-3, 4)$

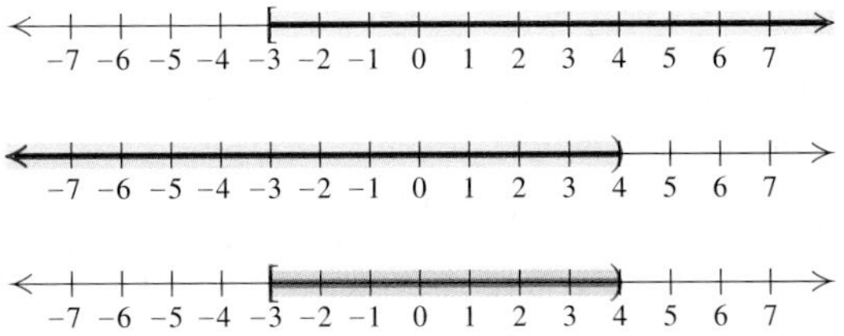

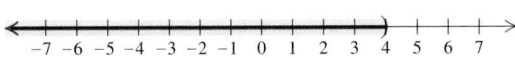

GRAPHICAL APPROACH

We graph the equations $y_1 = -1$, $y_2 = 2x + 5$, and $y_3 = 13$, and determine those x-values for which $y_1 \leq y_2$ *and* $y_2 < y_3$.

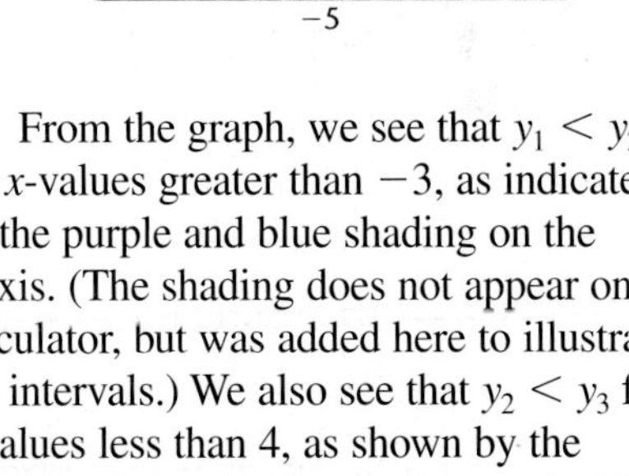

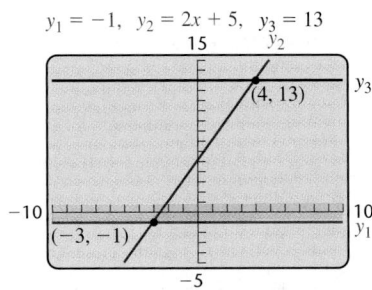

From the graph, we see that $y_1 < y_2$ for x-values greater than -3, as indicated by the purple and blue shading on the x-axis. (The shading does not appear on the calculator, but was added here to illustrate the intervals.) We also see that $y_2 < y_3$ for x-values less than 4, as shown by the purple and red shading on the x-axis. The solution set, indicated by the purple shading, is the intersection of these sets, as well as the number -3. It includes all x-values for which the line $y_2 = 2x + 5$ is both on or above the line $y_1 = -1$ *and* below the line $y_3 = 13$. This can be written $\{x \mid -3 \leq x < 4\}$, or $[-3, 4)$.

CAUTION! The abbreviated form of a conjunction, like $-3 \le x < 4$, can be written only if both inequality symbols point in the same direction.

Connecting the Concepts

Graphs of $y = 2x + 1$ and $y = 4 - x$

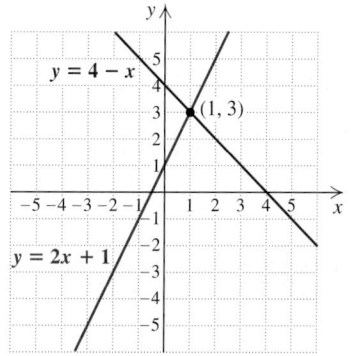

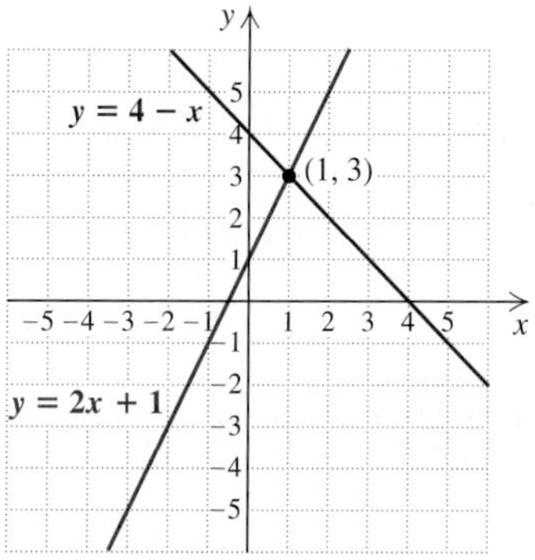

Graph of solution set of $2x + 1 = 4 - x$

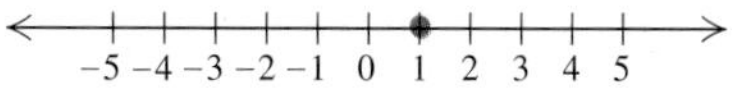

{1}

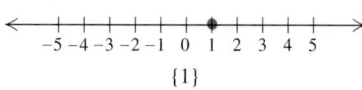

Graph of solution set of
$y = 2x + 1$,
$y = 4 - x$

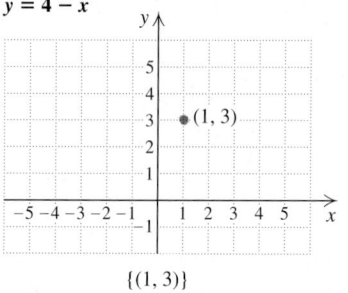

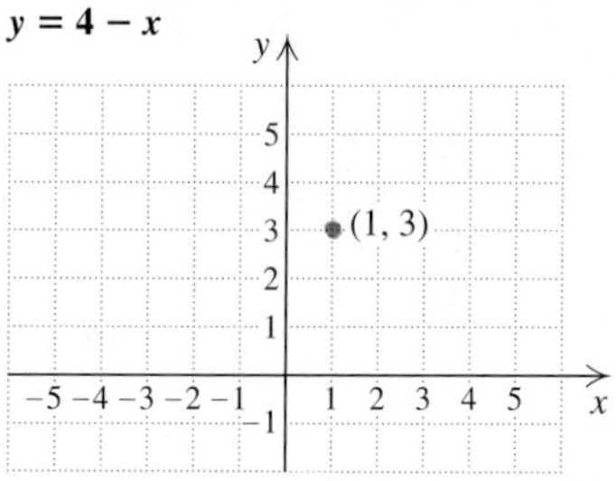

{(1, 3)}

Graph of solution set of $2x + 1 < 4 - x$

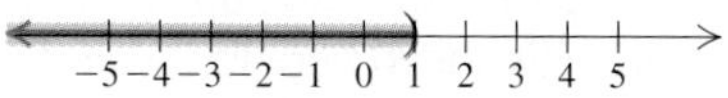

$\{x \mid x < 1\}$

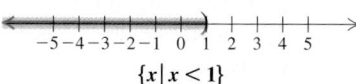

INTERPRETING GRAPHS: GRAPHS OF SOLUTION SETS

When solving an equation or an inequality graphically, it is important to distinguish the graphs used in the solution from the graph of the solution set. For example, we can use the graphs of the equations

$$y = 2x + 1 \quad \text{and} \quad y = 4 - x,$$

shown at left, to solve several types of equations and inequalities.

In this chapter, we solve linear equations by finding the x-coordinate of a point of intersection. To solve the equation

$$2x + 1 = 4 - x,$$

we graph $y = 2x + 1$ and $y = 4 - x$ as above. The point of intersection is $(1, 3)$, and the solution set of the equation is $\{1\}$. The graphs of the equations were used to find the solution; the graph of the solution set is simply a point on the number line.

In Chapter 3, we solved systems of linear equations by finding the coordinates of a point of intersection. To solve the system

$$y = 2x + 1,$$
$$y = 4 - x,$$

we graph the equations as above. The solution set of the system is $\{(1, 3)\}$. The graphs of the equations were used to find the solution; the graph of the solution set is a point on the x, y-plane.

In this chapter, we use the graphs of equations to solve inequalities. To solve the inequality

$$2x + 1 < 4 - x,$$

we graph $y = 2x + 1$ and $y = 4 - x$ as above. Since the graph of $y = 2x + 1$ lies below the graph of $y = 4 - x$ to the left of the point of intersection, the solution set is $\{x \mid x < 1\}$. The graphs of the equations were used to find the solution; the graph of the solution set is an interval on the number line.

EXAMPLE 3 Solve and graph: $2x - 5 \ge -3$ *and* $5x + 2 \ge 17$.

SOLUTION We first solve each inequality separately, retaining the word *and*:

$$2x - 5 \ge -3 \quad \textit{and} \quad 5x + 2 \ge 17$$
$$2x \ge 2 \quad \textit{and} \quad 5x \ge 15$$
$$x \ge 1 \quad \textit{and} \quad x \ge 3.$$

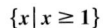

$y_1 = 2x - 5,\ y_2 = 5x + 2,$
$y_3 = -3,\ y_4 = 17$

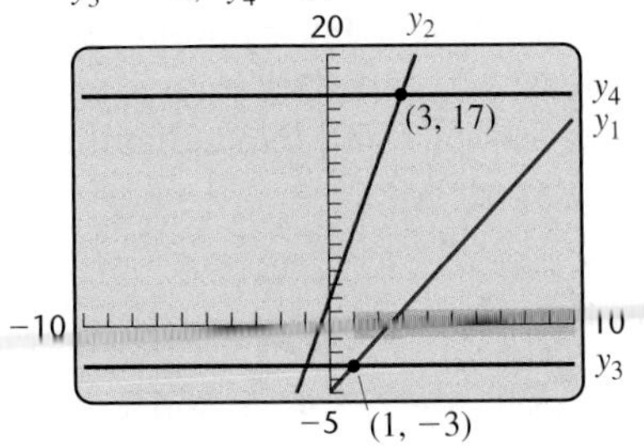

Next, we find the intersection of the two separate solution sets.

$\{x \mid x \geq 1\}$

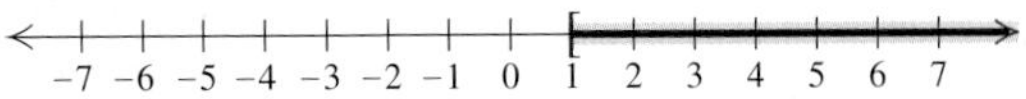

$\{x \mid x \geq 3\}$

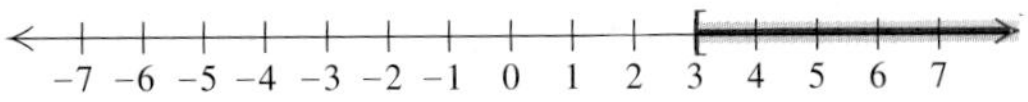

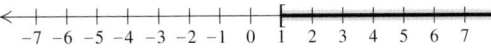

$\{x \mid x \geq 1\} \cap \{x \mid x \geq 3\}$
$= \{x \mid x \geq 3\}$

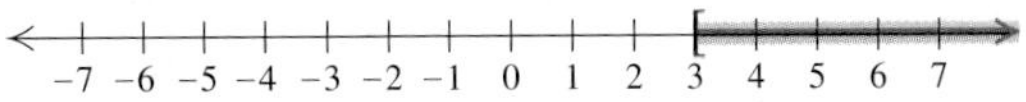

The numbers common to both sets are those greater than or equal to 3. Thus the solution set is $\{x \mid x \geq 3\}$, or, in interval notation, $[3, \infty)$. You should check that any number in $[3, \infty)$ satisfies the conjunction whereas numbers outside $[3, \infty)$ do not. The graph at left serves as another check.

Mathematical Use of the Word "and" The word "and" corresponds to "intersection" and to the symbol "$\cap$". Any solution of a conjunction must make each part of the conjunction true.

Sometimes there is no way to solve both parts of a conjunction at once.

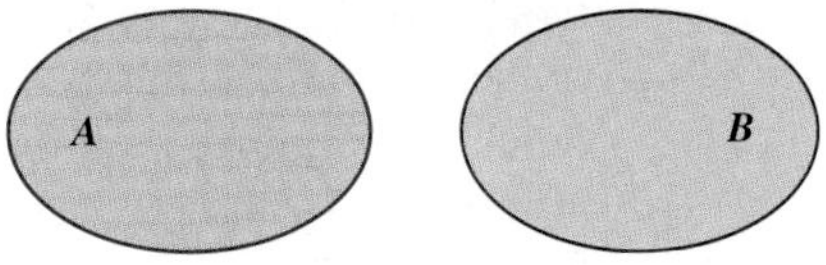

$A \cap B = \varnothing$

When $A \cap B = \varnothing$, A and B are said to be *disjoint*.

EXAMPLE 4 Solve and graph: $2x - 3 > 1$ *and* $3x - 1 < 2$.

SOLUTION We solve each inequality separately:

$$\begin{aligned} 2x - 3 > 1 &\quad \textit{and} \quad 3x - 1 < 2 \\ 2x > 4 &\quad \textit{and} \quad 3x < 3 \\ x > 2 &\quad \textit{and} \quad x < 1. \end{aligned}$$

The solution set is the intersection of the individual inequalities.

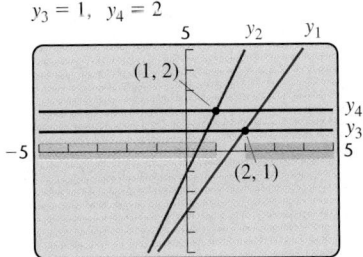

$y_1 = 2x - 3,\ y_2 = 3x - 1,$
$y_3 = 1,\ y_4 = 2$

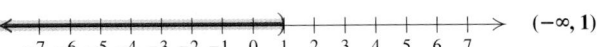

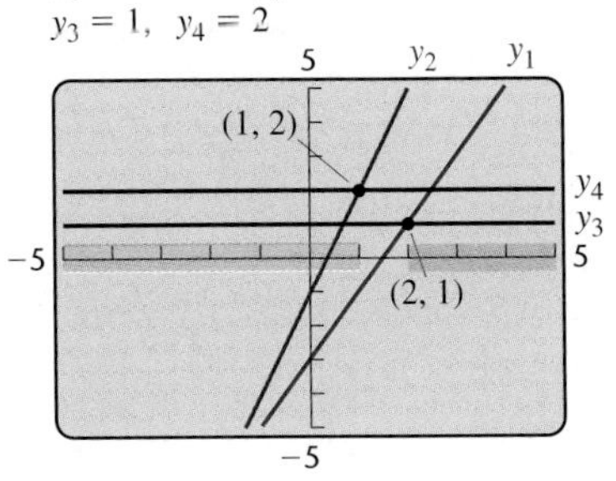

$\{x \mid x > 2\}$

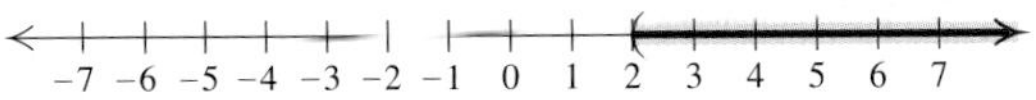

(2, ∞)

$\{x \mid x < 1\}$

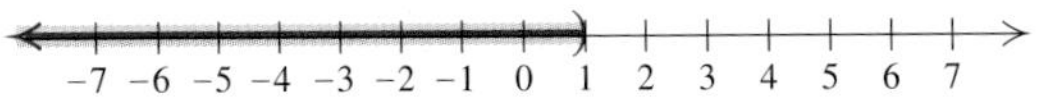

(−∞, 1)

$\{x \mid x > 2\} \cap \{x \mid x < 1\}$
$= \{x \mid x > 2 \textit{ and } x < 1\} = \varnothing$

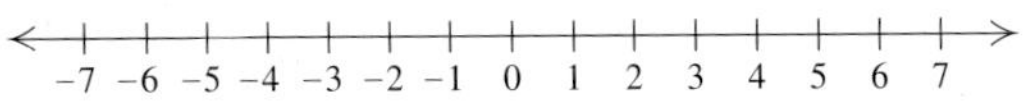

$\varnothing$

Since no number is both greater than 2 and less than 1, the solution set is the empty set, $\varnothing$. The graph at left confirms that the solution set is empty.

Unions of Sets and Disjunctions of Sentences

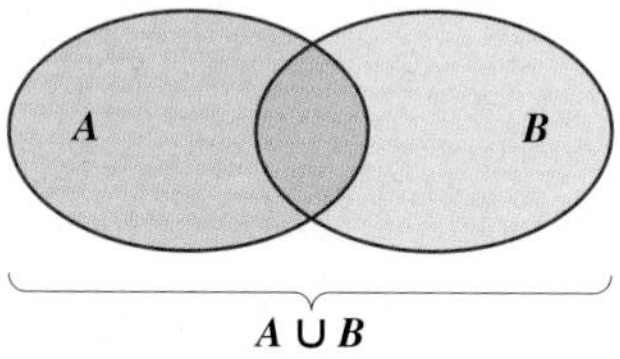

The **union** of two sets A and B is the collection of elements belonging to A and/or B. We denote the union of A and B by

$$A \cup B.$$

The union of two sets is often pictured as shown at left. For example, if $A = \{\text{all parents}\}$ and $B = \{\text{all people who are at least 30 years old}\}$, then $A \cup B = \{\text{all people who are parents } or \text{ who are at least 30 years old}\}$. Note that this set includes people who are parents *and* at least 30 years old.

EXAMPLE 5 Find the union: $\{2, 3, 4\} \cup \{3, 5, 7\}$.

SOLUTION The numbers in either or both sets are 2, 3, 4, 5, and 7, so the union is $\{2, 3, 4, 5, 7\}$.

Student Notes

Remember that the union or intersection of two sets is itself a set and should be written with set braces.

When two or more sentences are joined by the word *or* to make a compound sentence, the new sentence is called a **disjunction** of the sentences. Here is an example:

$$x < -3 \quad or \quad x > 3.$$

The word *or* is generally used to mean "one or the other but not both." You may have to decide whether to take an algebra class at 10:00 *or* at 2:00; you would not take it at both times. Mathematicians, however, use *or* to mean "one or the other or possibly both." Sometimes we do use *or* this way. Your major may require you to take a philosophy course *or* a psychology course. You must take at least one of these courses, but you would still fulfill the requirements if you took both.

A number is a solution of a disjunction if it is a solution of at least one of the separate parts. For example, -5 is a solution of the disjunction above since -5 is a solution of $x < -3$. Below we show the graph of $x < -3$, followed by the graph of $x > 3$, and finally the graph of the disjunction $x < -3$ *or* $x > 3$. *Note that the solution set of a disjunction is the union of the solution sets of the individual sentences.*

$\{x \mid x < -3\}$ — $(-\infty, -3)$

$\{x \mid x > 3\}$ — $(3, \infty)$

$\{x \mid x < -3\} \cup \{x \mid x > 3\} = \{x \mid x < -3 \text{ or } x > 3\}$ — $(-\infty, -3) \cup (3, \infty)$

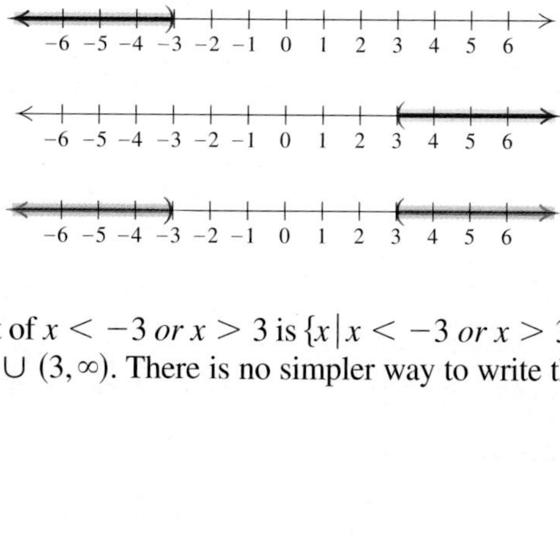

The solution set of $x < -3$ *or* $x > 3$ is $\{x \mid x < -3 \text{ or } x > 3\}$, or, in interval notation, $(-\infty, -3) \cup (3, \infty)$. There is no simpler way to write the solution.

> **Mathematical Use of the Word "or"** The word "or" corresponds to "union" and to the symbol "∪". For a number to be a solution of a disjunction, it must be in *at least one* of the solution sets of the individual sentences.

EXAMPLE 6 Solve and graph: $7 + 2x < -1$ *or* $13 - 5x \leq 3$.

SOLUTION We solve each inequality separately, retaining the word *or*:

$$7 + 2x < -1 \quad or \quad 13 - 5x \leq 3$$
$$2x < -8 \quad or \quad -5x \leq -10$$

Dividing by a negative and reversing the symbol

$$x < -4 \quad or \quad x \geq 2.$$

To find the solution set of the disjunction, we consider the individual graphs. We graph $x < -4$ and then $x \geq 2$. Then we take the union of the graphs.

$y_1 = 7 + 2x,\ y_2 = 13 - 5x,$
$y_3 = -1,\ y_4 = 3$

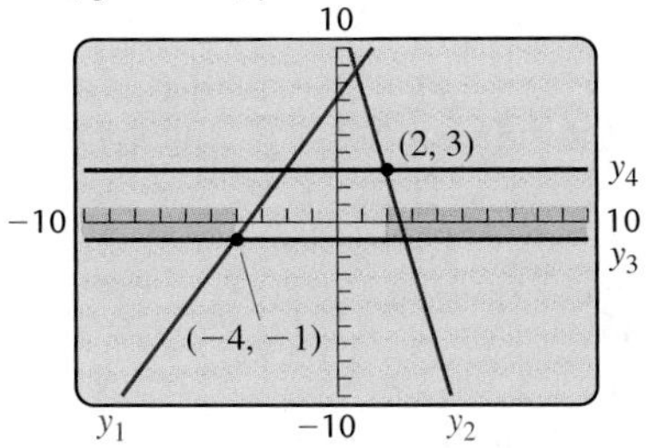

$\{x \mid x < -4\}$ **$(-\infty, -4)$**

$\{x \mid x \geq 2\}$ **$[2, \infty)$**

$\{x \mid x < -4\} \cup \{x \mid x \geq 2\}$
$= \{x \mid x < -4$ *or* $x \geq 2\}$ **$(-\infty, -4) \cup [2, \infty)$**

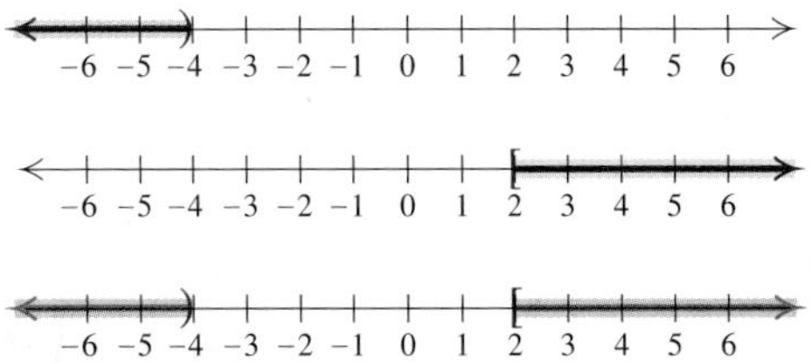

The solution set is $\{x \mid x < -4$ *or* $x \geq 2\}$, or $(-\infty, -4) \cup [2, \infty)$. This is confirmed by the graph at left.

> **CAUTION!** A compound inequality like
>
> $$x < -4 \quad or \quad x \geq 2,$$
>
> as in Example 6, *cannot* be expressed as $2 \leq x < -4$ because to do so would be to say that x is *simultaneously* less than -4 and greater than or equal to 2. No number is both less than -4 *and* greater than 2, but many are less than -4 *or* greater than 2.

EXAMPLE 7 Solve: $-2x - 5 < -2$ *or* $x - 3 < -10$.

SOLUTION We solve the individual inequalities separately, retaining the word *or*:

$$-2x - 5 < -2 \quad or \quad x - 3 < -10$$
$$-2x < 3 \quad or \quad x < -7$$

Dividing by a negative and reversing the symbol **Keep the word "or."**

$$x > -\tfrac{3}{2} \quad or \quad x < -7.$$

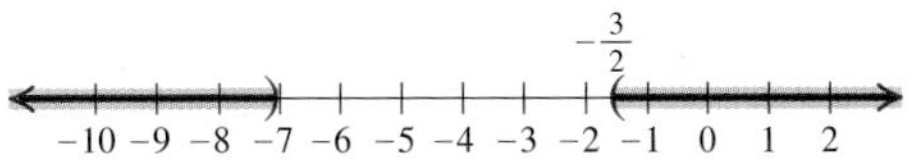

The solution set is $\left\{x \mid x < -7 \text{ or } x > -\frac{3}{2}\right\}$, or $(-\infty, -7) \cup \left(-\frac{3}{2}, \infty\right)$.

EXAMPLE 8 Solve: $3x - 11 < 4$ *or* $4x + 9 \geq 1$.

SOLUTION We solve the individual inequalities separately, retaining the word *or:*

$$\begin{aligned} 3x - 11 &< 4 \quad \text{or} \quad 4x + 9 \geq 1 \\ 3x &< 15 \quad \text{or} \quad \quad 4x \geq -8 \\ x &< 5 \quad \text{or} \quad \quad x \geq -2. \end{aligned}$$

Keep the word "or."

To find the solution set, we first look at the individual graphs.

$\{x \mid x < 5\}$

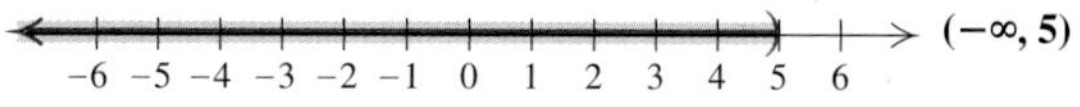

$\{x \mid x \geq -2\}$

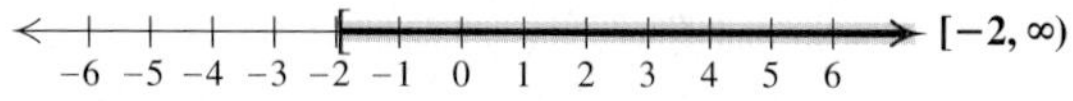

$\{x \mid x < 5\} \cup \{x \mid x \geq -2\}$
$= \{x \mid x < 5 \text{ or } x \geq -2\}$

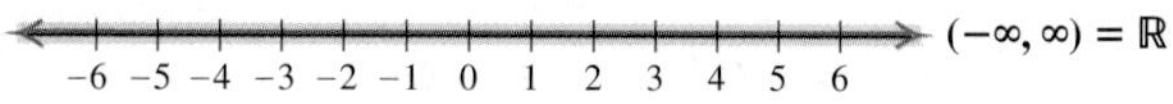

Since *all* numbers are less than 5 or greater than or equal to -2, the two sets fill the entire number line. Thus the solution set is $\mathbb{R}$, the set of all real numbers.

Interval Notation and Domains

In Section 2.5, we saw that if $g(x) = \dfrac{5x - 2}{3x - 7}$, then the domain of $g = \{x \mid x$ is a real number *and* $x \neq \frac{7}{3}\}$. We can now represent such a set using interval notation:

$$\{x \mid x \text{ is a real number } and\ x \neq \tfrac{7}{3}\} = \left(-\infty, \tfrac{7}{3}\right) \cup \left(\tfrac{7}{3}, \infty\right).$$

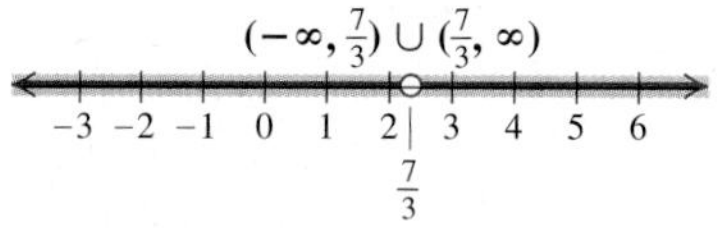

EXAMPLE 9 Use interval notation to write the domain of f if $f(x) = \sqrt{x + 2}$.

SOLUTION The expression $\sqrt{x + 2}$ is not a real number when $x + 2$ is negative. Thus the domain of f is the set of all x-values for which $x + 2 \geq 0$:

$x + 2 \geq 0$ **$x + 2$ cannot be negative.**

$x \geq -2.$ **Adding -2 to both sides**

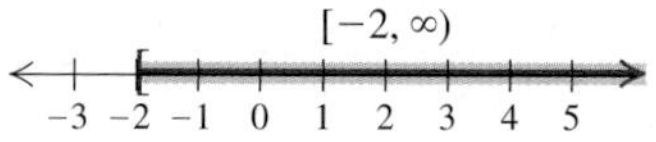

We have the domain of $f = \{x \mid x \geq -2\} = [-2, \infty)$.

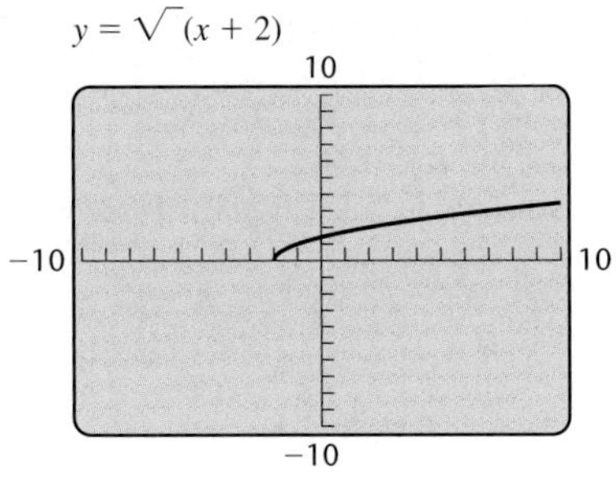

We can partially check that $[-2, \infty)$ is the domain of f either by using a table of values or by viewing the graph of $f(x)$. By tracing the curve, we can confirm that no y-value is given for x-values less than -2.

The check in Example 9 is a partial check, because we cannot check every number in the interval. Also, if a function is not defined at a point, that "hole" in the graph will not be seen on a graphing calculator. Keeping those limitations in mind, know that a graph is still an efficient and visual check of a domain.

Interactive Discovery

Consider the functions $f(x) = \sqrt{3 - x}$ and $g(x) = \sqrt{x + 1}$. Enter these in a graphing calculator as $y_1 = \sqrt{\ }(3 - x)$ and $y_2 = \sqrt{\ }(x + 1)$.

1. Determine algebraically the domain of f and the domain of g. Then graph y_1 and y_2, and trace each curve to verify the domains.
2. Graph $y_1 + y_2$, $y_1 - y_2$, and $y_1 \cdot y_2$. Use the graphs to determine the domains of $f + g$, $f - g$, and $f \cdot g$.
3. How can the domains of the sum, the difference, and the product of f and g be found algebraically?

Domain of the Sum, Difference, or Product of Functions
The domain of the sum, the difference, or the product of the functions f and g is the intersection of the domains of f and g.

EXAMPLE 10 Find the domain of $f + g$ if $f(x) = \sqrt{2x - 5}$ and $g(x) = \sqrt{x + 1}$.

SOLUTION We first find the domain of f and the domain of g. The domain of f is the set of all x-values for which $2x - 5 \geq 0$, or $\left\{x \mid x \geq \frac{5}{2}\right\}$, or $\left[\frac{5}{2}, \infty\right)$. Similarly, the domain of g is $\{x \mid x \geq -1\}$, or $[-1, \infty)$. The intersection of the domains is $\left\{x \mid x \geq \frac{5}{2}\right\}$, or $\left[\frac{5}{2}, \infty\right)$. We can confirm this at least approximately by tracing the graph of $f + g$.

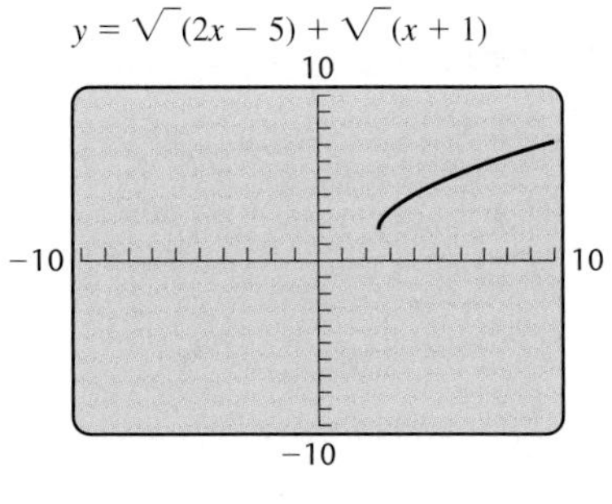

Thus the

$$\text{Domain of } f + g = \left\{x \mid x \geq \tfrac{5}{2}\right\}, \text{ or } \left[\tfrac{5}{2}, \infty\right).$$

4.3 EXERCISE SET

Concept Reinforcement *In each of Exercises 1–10, match the set with the most appropriate choice from the column on the right.*

1. ____ $\{x \mid x < -2 \text{ or } x > 2\}$

2. ____ $\{x \mid x < -2 \text{ and } x > 2\}$

3. ____ $\{x \mid x > -2\} \cap \{x \mid x < 2\}$

4. ____ $\{x \mid x \leq -2\} \cup \{x \mid x \geq 2\}$

5. ____ $\{x \mid x \leq -2\} \cup \{x \mid x \leq 2\}$

6. ____ $\{x \mid x \leq -2\} \cap \{x \mid x \leq 2\}$

7. ____ $\{x \mid x \geq -2\} \cap \{x \mid x \geq 2\}$

8. ____ $\{x \mid x \geq -2\} \cup \{x \mid x \geq 2\}$

9. ____ $\{x \mid x \leq 2\}$ *and* $\{x \mid x \geq -2\}$

10. ____ $\{x \mid x \leq 2\}$ *or* $\{x \mid x \geq -2\}$

a) 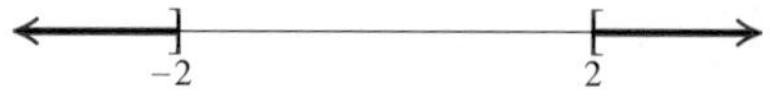

b) 

c) 

d) 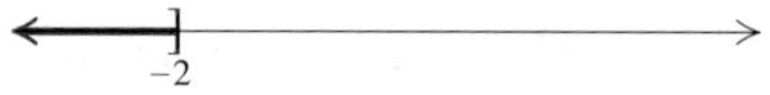

e) 

f) 

g)  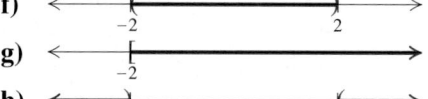

h) 

i) $\mathbb{R}$

j) $\varnothing$

Find each indicated intersection or union.

11. $\{5, 9, 11\} \cap \{9, 11, 18\}$

12. $\{2, 4, 8\} \cup \{8, 9, 10\}$

13. $\{0, 5, 10, 15\} \cup \{5, 15, 20\}$

14. $\{2, 5, 9, 13\} \cap \{5, 8, 10\}$

15. $\{a, b, c, d, e, f\} \cap \{b, d, f\}$

16. $\{a, b, c\} \cup \{a, c\}$

17. $\{r, s, t\} \cup \{r, u, t, s, v\}$

18. $\{m, n, o, p\} \cap \{m, o, p\}$

19. $\{3, 6, 9, 12\} \cap \{5, 10, 15\}$

20. $\{1, 5, 9\} \cup \{4, 6, 8\}$

21. $\{3, 5, 7\} \cup \varnothing$

22. $\{3, 5, 7\} \cap \varnothing$

Graph and write interval notation for each compound inequality.

23. $3 < x < 7$

24. $0 \leq y \leq 4$

25. $-6 \leq y \leq -2$

26. $-9 \leq x < -5$

27. $x < -1$ *or* $x > 4$

28. $x < -5$ *or* $x > 1$

29. $x \leq -2$ *or* $x > 1$

30. $x \leq -5$ *or* $x > 2$

31. $-4 \leq -x < 2$

32. $x > -7$ *and* $x < -2$

33. $x > -2$ *and* $x < 4$

34. $3 > -x \geq -1$

35. $5 > a$ *or* $a > 7$

36. $t \geq 2$ *or* $-3 > t$

37. $x \geq 5$ *or* $-x \geq 4$

38. $-x < 3$ *or* $x < -6$

39. $7 > y$ *and* $y \geq -3$

40. $6 > -x \geq 0$

41. $x < 7$ *and* $x \geq 3$

42. $x \geq -3$ *and* $x < 3$

Aha! **43.** $t < 2$ *or* $t < 5$

44. $t > 4$ *or* $t > -1$

Solve and graph each solution set.

45. $-2 < t + 1 < 8$

46. $-3 < t + 1 \leq 5$

47. $2 < x + 3$ *and* $x + 1 \leq 5$

48. $-1 < x + 2$ *and* $x - 4 < 3$

49. $-7 \leq 2a - 3$ *and* $3a + 1 < 7$

50. $-4 \leq 3n + 5$ *and* $2n - 3 \leq 7$

Aha! **51.** $x + 7 \leq -2$ *or* $x + 7 \geq -3$

52. $x + 5 < -3$ *or* $x + 5 \geq 4$

53. $5 > \dfrac{x - 3}{4} > 1$

54. $3 \geq \dfrac{x - 1}{2} \geq -4$

55. $-7 \leq 4x + 5 \leq 13$

56. $-4 \leq 2x + 3 \leq 15$

57. $2 \leq f(x) \leq 8$, where $f(x) = 3x - 1$

58. $7 \geq g(x) \geq -2$, where $g(x) = 3x - 5$

59. $-21 \leq f(x) < 0$, where $f(x) = -2x - 7$

60. $4 > g(t) \geq 2$, where $g(t) = -3t - 8$

61. $f(x) \leq 2$ *or* $f(x) \geq 8$, where $f(x) = 3x - 1$

62. $g(x) \leq -2$ *or* $g(x) \geq 10$, where $g(x) = 3x - 5$

63. $f(x) < -3$ *or* $f(x) > 5$, where $f(x) = 2x - 7$

64. $g(x) < -7$ *or* $g(x) > 7$, where $g(x) = 3x + 5$

65. $6 > 2a - 1$ *or* $-4 \leq -3a + 2$

66. $3a - 7 > -10$ *or* $5a + 2 \leq 22$

67. $a + 3 < -2$ *and* $3a - 4 < 8$

68. $1 - a < -2$ *and* $2a + 1 > 9$

69. $3x + 2 < 2$ *or* $4 - 2x < 14$

70. $2x - 1 > 5$ *or* $3 - 2x \geq 7$

71. $2t - 7 \leq 5$ *or* $5 - 2t > 3$

72. $5 - 3a \leq 8$ *or* $2a + 1 > 7$

73. Use the accompanying graph of $f(x) = 2x - 5$ to solve $-7 < 2x - 5 < 7$.

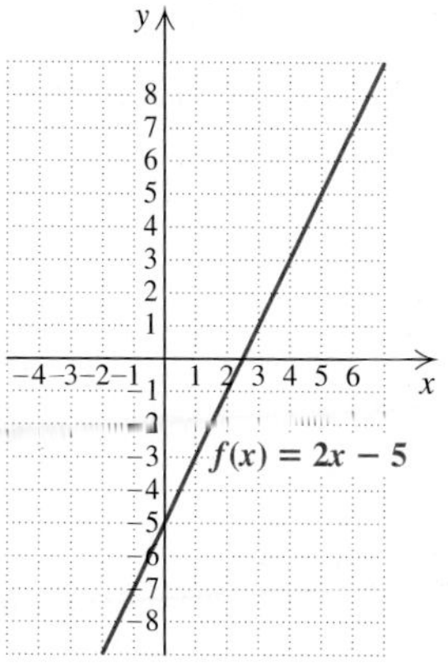

74. Use the accompanying graph of $g(x) = 4 - x$ to solve $4 - x < -2$ *or* $4 - x > 7$.

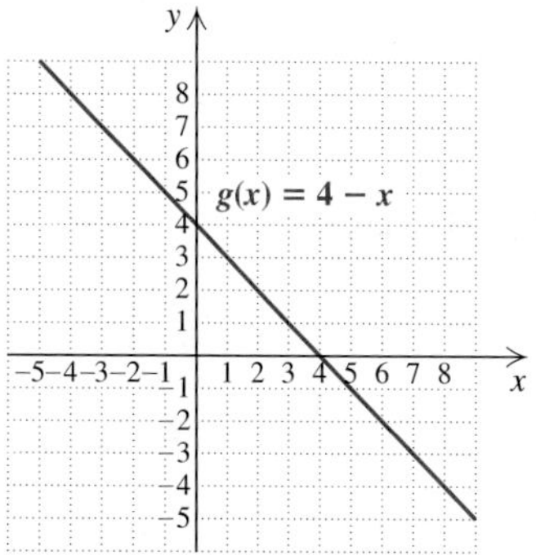

For $f(x)$ *as given, use interval notation to write the domain of* f.

75. $f(x) = \dfrac{9}{x + 8}$

76. $f(x) = \dfrac{2}{x + 3}$

77. $f(x) = \sqrt{x - 6}$

78. $f(x) = \sqrt{x - 2}$

79. $f(x) = \dfrac{x + 3}{2x - 8}$

80. $f(x) = \dfrac{x - 1}{3x + 6}$

81. $f(x) = \sqrt{2x + 7}$

82. $f(x) = \sqrt{8 - 5x}$

83. $f(x) = \sqrt{8 - 2x}$

84. $f(x) = \sqrt{10 - 2x}$

Use interval notation to write each domain.

85. The domain of $f + g$, if $f(x) = \sqrt{x - 5}$ and $g(x) = \sqrt{\frac{1}{2}x + 1}$

86. The domain of $f - g$, if $f(x) = \sqrt{x + 3}$ and $g(x) = \sqrt{2x - 1}$

87. The domain of $f \cdot g$, if $f(x) = \sqrt{3 - x}$ and $g(x) = \sqrt{3x - 2}$

88. The domain of $f + g$, if $f(x) = \sqrt{3 - 4x}$ and $g(x) = \sqrt{x + 2}$

TW **89.** Why can the conjunction $2 < x$ *and* $x < 5$ be rewritten as $2 < x < 5$, but the disjunction $2 < x$ *or* $x < 5$ cannot be rewritten as $2 < x < 5$?

TW **90.** Can the solution set of a disjunction be empty? Why or why not?

Skill Maintenance

Graph.

91. $y = 5$ [2.3]

92. $y = -2$ [2.3]

93. $f(x) = |x|$ [2.1]

94. $g(x) = x - 1$ [2.2]

Solve each system graphically. [3.1]

95. $y = x - 3,$
$y = 5$

96. $y = x + 2,$
$y = -3$

Synthesis

TW **97.** What can you conclude about a, b, c, and d, if $[a, b] \cup [c, d] = [a, d]$? Why?

TW **98.** What can you conclude about a, b, c, and d, if $[a, b] \cap [c, d] = [a, b]$? Why?

99. *Childless Women.* On the basis of trends from the late 1900s, the function given by

$$P(t) = 0.44t + 10.2$$

can be used to estimate the percentage of U.S. women age 40–44, $P(t)$, t years after 1980, who have not given birth (*Sources*: Based on data from the U.S. Bureau of the Census; *Deseret Morning News*, 11/03). For what years will the percentage of childless 40–44-year-old women be between 19 and 30 percent?

100. *Pressure at Sea Depth.* The function given by

$$P(d) = 1 + \frac{d}{33}$$

gives the pressure, in atmospheres (atm), at a depth of d feet in the sea. For what depths d is the pressure at least 1 atm and at most 7 atm?

101. *Converting Dress Sizes.* The function given by

$$f(x) = 2(x + 10)$$

can be used to convert dress sizes x in the United States to dress sizes $f(x)$ in Italy. For what dress sizes in the United States will dress sizes in Italy be between 32 and 46?

102. *Solid-Waste Generation.* The function given by

$$w(t) = 0.01t + 4.3$$

can be used to estimate the number of pounds of solid waste, $w(t)$, produced daily, on average, by each person in the United States, t years after 1991. For what years will waste production range from 4.5 to 4.75 lb per person per day?

103. *Records in the Women's 100-m Dash.* Florence Griffith Joyner set a world record of 10.49 sec in the women's 100-m dash in 1988. The function given by

$$R(t) = -0.0433t + 10.49$$

can be used to predict the world record in the women's 100-m dash t years after 1988. (*Sources*: *Guinness Book of World Records* 2004; www.Runnersworld.com) Predict (using an inequality) those years for which the world record was between 11.5 and 10.8 sec. (Measure from the middle of 1988.)

104. *Temperatures of Liquids.* The formula

$$C = \frac{5}{9}(F - 32)$$

can be used to convert Fahrenheit temperatures F to Celsius temperatures C.

a) Gold is liquid for Celsius temperatures C such that $1063° \le C < 2660°$. Find a comparable inequality for Fahrenheit temperatures.

b) Silver is liquid for Celsius temperatures C such that $960.8° \le C < 2180°$. Find a comparable inequality for Fahrenheit temperatures.

105. *Minimizing Tolls.* A \$3.00 toll is charged to cross the bridge from Sanibel Island to mainland Florida. A six-month pass, costing \$15.00, reduces the toll to \$0.50. A one-year pass, costing \$150, allows for free crossings. (*Source*: www.leewayinfo.com) How many crossings per year does it take, on average, for two consecutive six-month passes to be the most economical choice? Assume a constant number of trips per month.

Solve and graph.

106. $4a - 2 \le a + 1 \le 3a + 4$

107. $4m - 8 > 6m + 5$ *or* $5m - 8 < -2$

108. $x - 10 < 5x + 6 \le x + 10$

109. $3x < 4 - 5x < 5 + 3x$

Determine whether each sentence is true or false for all real numbers a, b, and c.

110. If $-b < -a$, then $a < b$.

111. If $a \le c$ and $c \le b$, then $b > a$.

112. If $a < c$ and $b < c$, then $a < b$.

113. If $-a < c$ and $-c > b$, then $a > b$.

For f(x) as given, use interval notation to write the domain of f.

114. $f(x) = \dfrac{\sqrt{5 + 2x}}{x - 1}$

115. $f(x) = \dfrac{\sqrt{3 - 4x}}{x + 7}$

116. On many graphing calculators, the TEST key provides access to inequality symbols, while the LOGIC option of that same key accesses the conjunction *and* and the disjunction *or*. Thus, if $y_1 = x > -2$ and $y_2 = x < 4$, Exercise 33 can be checked by forming the expression $y_3 = y_1$ *and* y_2. The interval(s) in the solution set appears as a horizontal line 1 unit above the x-axis. (Be careful to "deselect" y_1 and y_2 so that only y_3 is drawn.)

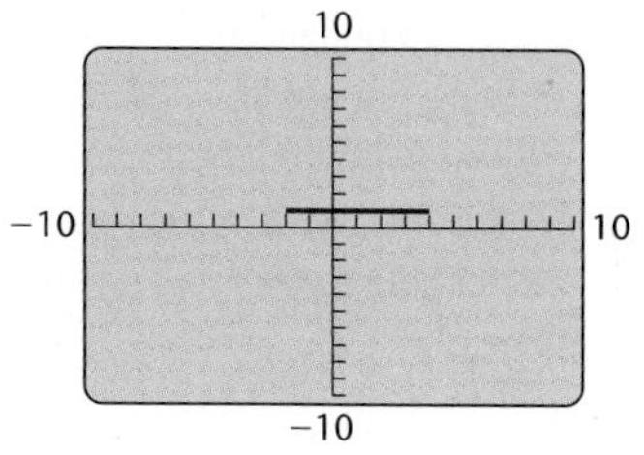

4.4 Absolute-Value Equations and Inequalities

Equations with Absolute Value ■ Inequalities with Absolute Value

Equations with Absolute Value

Recall from Section 1.2 the definition of absolute value.

Absolute Value The absolute value of x, denoted $|x|$, is defined as

$$|x| = \begin{cases} x, & \text{if } x \geq 0, \\ -x, & \text{if } x < 0. \end{cases}$$

(When x is nonnegative, the absolute value of x is x. When x is negative, the absolute value of x is the opposite of x.)

To better understand this definition, suppose x is -5. Then $|x| = |-5| = 5$, and 5 is the opposite of -5. This shows that when x represents a negative number, the absolute value of x is positive.

Since distance is always nonnegative, we can think of a number's absolute value as its distance from zero on the number line.

Study Tip

Understand Your Mistakes

When you receive a graded quiz, test, or assignment back from your instructor, it is important to review and understand what your mistakes were. Try not to simply file away old papers without first understanding how to correct your mistakes. If you are not able to do this on your own, seek help.

EXAMPLE 1 Find the solution set: **(a)** $|x| = 4$; **(b)** $|x| = 0$; **(c)** $|x| = -7$.

SOLUTION

a) We interpret $|x| = 4$ to mean that the number x is 4 units from zero on the number line. There are two such numbers, 4 and -4. Thus the solution set is $\{-4, 4\}$.

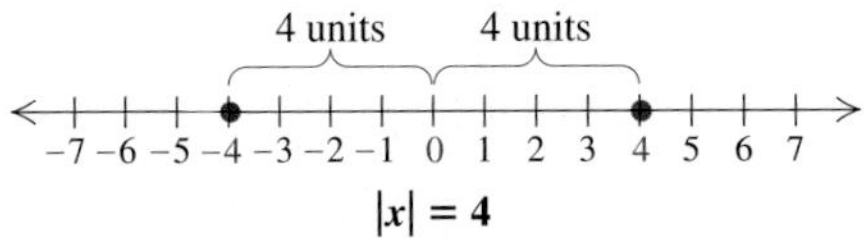

A second way to visualize this problem is to graph $f(x) = |x|$ (see Section 2.1). We also graph $g(x) = 4$. The x-values of the points of intersection are the solutions of $|x| = 4$.

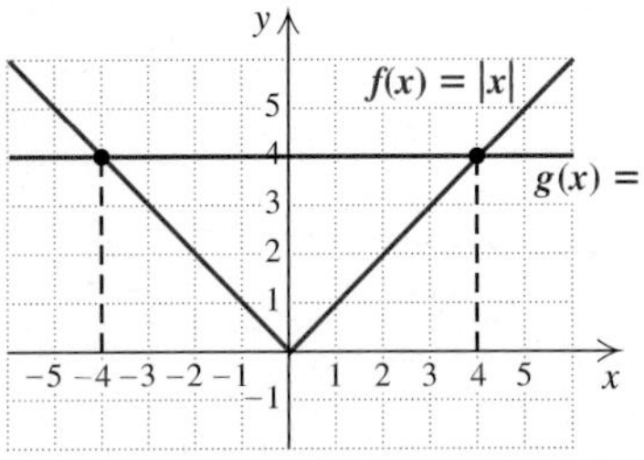

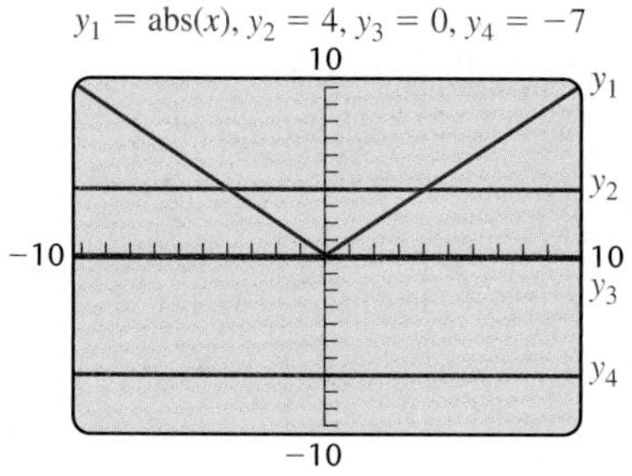

b) We interpret $|x| = 0$ to mean that x is 0 units from zero on the number line. The only number that satisfies this is 0 itself. Thus the solution set is $\{0\}$.

c) Since distance is always nonnegative, it doesn't make sense to talk about a number that is -7 units from zero. Remember: The absolute value of a number is never negative. Thus, $|x| = -7$ has no solution; the solution set is $\varnothing$.

The graph at left illustrates that, in Example 1, there are two solutions of $|x| = 4$, one solution of $|x| = 0$, and no solutions of $|x| = -7$.

Other equations involving absolute value can be solved graphically.

EXAMPLE 2 Solve: $|x - 3| = 5$.

SOLUTION We let $y_1 = \text{abs}(x - 3)$ and $y_2 = 5$. The solution set of the equation consists of the x-coordinates of any points of intersection of the graphs of y_1 and y_2. We see that the graphs intersect at two points. We use INTERSECT twice to find the coordinates of the points of intersection. Each time, we enter a GUESS that is close to the point we are looking for.

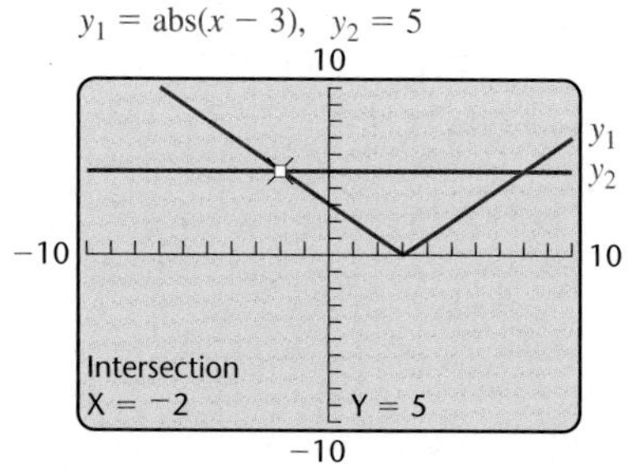

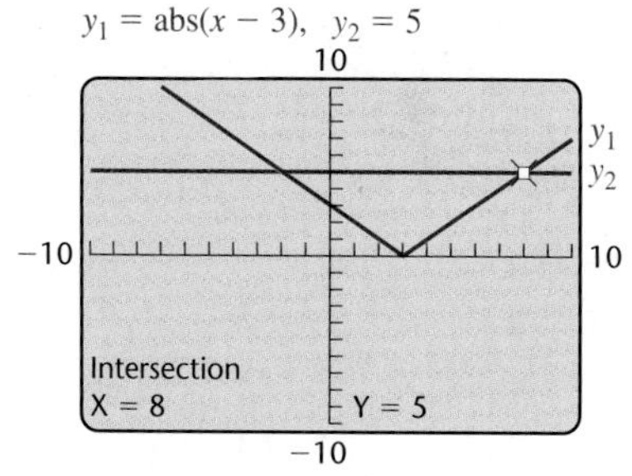

The graphs intersect at $(-2, 5)$ and $(8, 5)$. Thus the solutions are -2 and 8. To check, we substitute each in the original equation.

Check:

For -2:

$\lvert x - 3\rvert = 5$	
$\lvert -2 - 3\rvert$	5
$\lvert -5\rvert$	
$5 \overset{?}{=} 5$	TRUE

For 8:

$\lvert x - 3\rvert = 5$	
$\lvert 8 - 3\rvert$	5
$\lvert 5\rvert$	
$5 \overset{?}{=} 5$	TRUE

The solution set is $\{-2, 8\}$.

We can solve equations involving absolute value algebraically using the following principle.

Student Notes

The absolute-value principle will be used with a variety of replacements for X. Make sure that the principle, as stated here, makes sense to you before going further.

The Absolute-Value Principle for Equations

For any positive number p and any algebraic expression X:

a) The solutions of $|X| = p$ are those numbers that satisfy

$$X = -p \quad or \quad X = p.$$

b) The equation $|X| = 0$ is equivalent to the equation $X = 0$.

c) The equation $|X| = -p$ has no solution.

EXAMPLE 3 Find the solution set: **(a)** $|2x + 5| = 13$; **(b)** $|4 - 7x| = -8$.

a)

ALGEBRAIC APPROACH

We use the absolute-value principle:

$$|X| = p$$

$$|2x + 5| = 13 \quad \textbf{Substituting}$$

$$2x + 5 = -13 \quad or \quad 2x + 5 = 13$$

$$2x = -18 \quad or \quad 2x = 8$$

$$x = -9 \quad or \quad x = 4.$$

The solutions are -9 and 4.

Check:

For -9:

$\lvert 2x + 5\rvert = 13$	
$\lvert 2(-9) + 5\rvert$	13
$\lvert -18 + 5\rvert$	
$\lvert -13\rvert$	
$13 \overset{?}{=} 13$ TRUE	

For 4:

$\lvert 2x + 5\rvert = 13$	
$\lvert 2 \cdot 4 + 5\rvert$	13
$\lvert 8 + 5\rvert$	
$\lvert 13\rvert$	
$13 \overset{?}{=} 13$ TRUE	

The number $2x + 5$ is 13 units from zero if x is replaced with -9 or 4. The solution set is $\{-9, 4\}$.

GRAPHICAL APPROACH

We graph $y_1 = \text{abs}(2x + 5)$ and $y_2 = 13$ and use INTERSECT to find any points of intersection. The graphs intersect at $(-9, 13)$ and $(4, 13)$. The x-coordinates, -9 and 4, are the solutions.

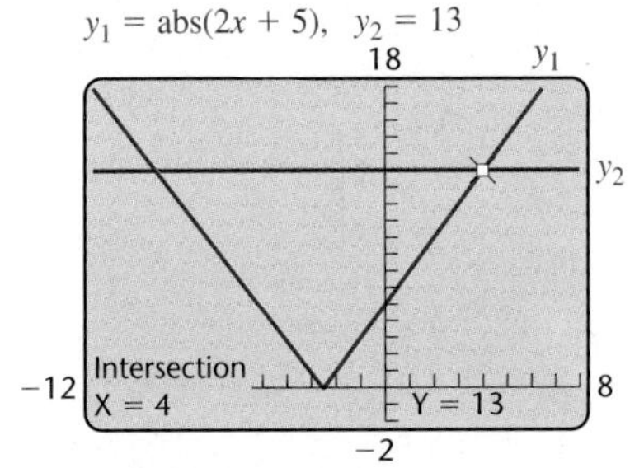

The algebraic solution serves as a check. The solution set is $\{-9, 4\}$.

b)

ALGEBRAIC APPROACH

The absolute-value principle reminds us that absolute value is always nonnegative. Thus the equation $|4 - 7x| = -8$ has no solution. The solution set is $\varnothing$.

GRAPHICAL APPROACH

We graph $y_1 = \text{abs}(4 - 7x)$ and $y_2 = -8$. There are no points of intersection, so the solution set is $\varnothing$.

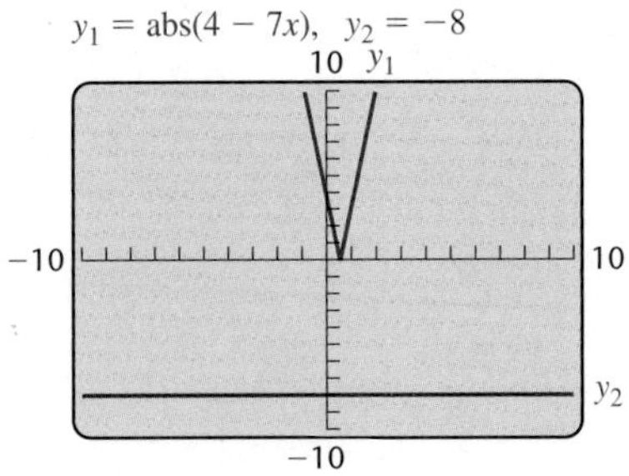

To use the absolute-value principle, we must be sure that the absolute-value expression is alone on one side of the equation.

EXAMPLE 4 Given that $f(x) = 2|x + 3| + 1$, find all x for which $f(x) = 15$.

ALGEBRAIC APPROACH

Since we are looking for $f(x) = 15$, we substitute:

$$f(x) = 15$$

$$2|x + 3| + 1 = 15 \quad \textbf{Replacing } f(x) \textbf{ with } 2|x + 3| + 1$$

$$2|x + 3| = 14 \quad \textbf{Subtracting 1 from both sides}$$

$$|x + 3| = 7 \quad \textbf{Dividing both sides by 2}$$

$$x + 3 = -7 \quad or \quad x + 3 = 7 \quad \textbf{Replacing } X \textbf{ with } x + 3 \textbf{ and } p \textbf{ with 7 in the absolute-value principle}$$

$$x = -10 \quad or \quad x = 4.$$

Check: $f(-10) = 2|-10 + 3| + 1 = 2|-7| + 1$
$= 2 \cdot 7 + 1 = 15;$

$f(4) = 2|4 + 3| + 1 = 2|7| + 1$
$= 2 \cdot 7 + 1 = 15$

The solution set is $\{-10, 4\}$.

GRAPHICAL APPROACH

We graph $y_1 = 2\text{ abs}(x + 3) + 1$ and $y_2 = 15$ and determine the coordinates of any points of intersection. Choosing a viewing window that includes all points of intersection may involve some trial and error.

$y_1 = 2\text{ abs}(x + 3) + 1, \; y_2 = 15$

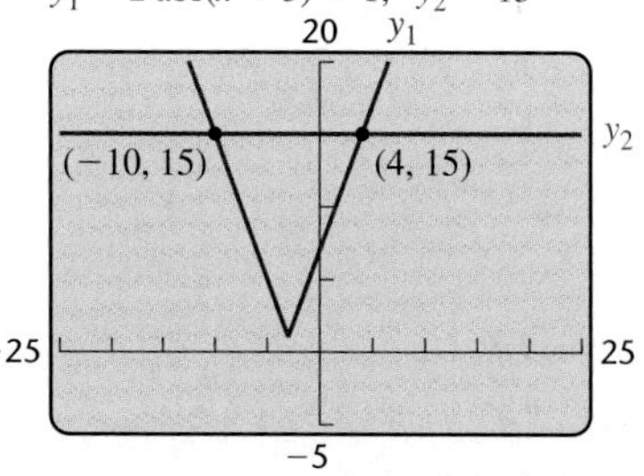

Xscl = 5, Yscl = 5

The solutions are the x-coordinates of the points of intersection, -10 and 4. The solution set is $\{-10, 4\}$.

EXAMPLE 5 Solve: $|x - 2| = 3$.

SOLUTION Because this equation is of the form $|a - b| = c$, it can be solved two ways.

Method 1. We interpret $|x - 2| = 3$ as stating that the number $x - 2$ is 3 units from zero. Using the absolute-value principle, we replace X with $x - 2$ and p with 3:

$$|X| = p$$

$$|x - 2| = 3$$

$$x - 2 = -3 \quad or \quad x - 2 = 3 \quad \textbf{Using the absolute-value principle}$$

$$x = -1 \quad or \quad x = 5.$$

Method 2. This approach is helpful in calculus. The expressions $|a - b|$ and $|b - a|$ can be used to represent the *distance between a and b on the number line*. For example, the distance between 7 and 8 is given by $|8 - 7|$ or $|7 - 8|$. From this viewpoint, the equation $|x - 2| = 3$ states that the distance between x and 2 is 3 units. We draw a number line and locate all numbers that are 3 units from 2.

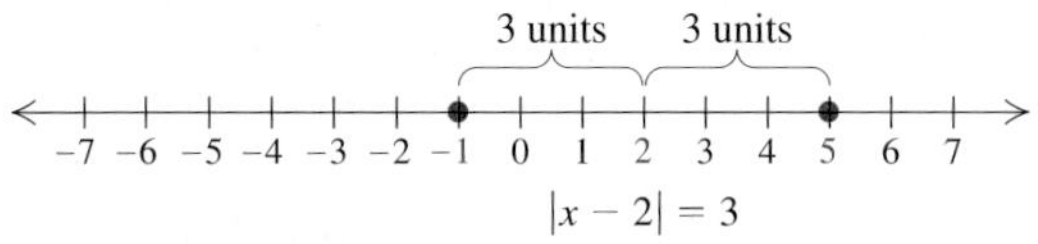

The solutions of $|x-2|=3$ are -1 and 5.

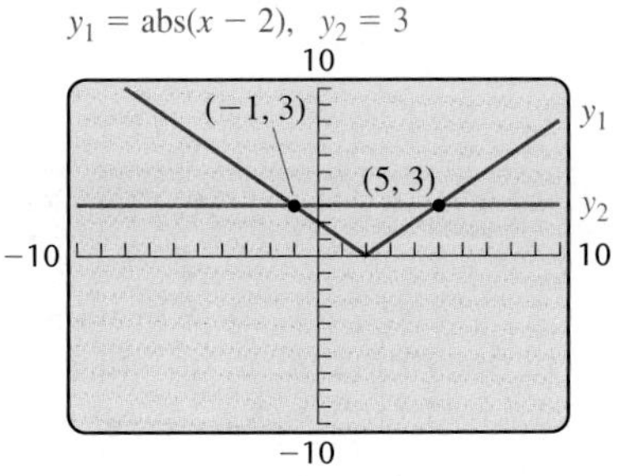

Check: The check consists of observing that both methods give the same solutions. The graphical solution shown at left serves as another check. The solution set is $\{-1, 5\}$.

Sometimes an equation has two absolute-value expressions. Consider $|a|=|b|$. This means that a and b are the same distance from zero.

If a and b are the same distance from zero, then either they are the same number or they are opposites.

For any algebraic expressions X and Y:

If $|X| = |Y|$, then $X = Y$ or $X = -Y$.

EXAMPLE 6 Solve: $|2x-3|=|x+5|$.

ALGEBRAIC APPROACH

The given equation tells us that $2x-3$ and $x+5$ are the same distance from zero. This means that they are either the same number or opposites.

This assumes these numbers are the same. | **This assumes these numbers are opposites.**

$$2x-3 = x+5 \quad or \quad 2x-3 = -(x+5)$$
$$x-3=5 \quad or \quad 2x-3=-x-5$$
$$x=8 \quad or \quad 3x-3=-5$$
$$3x=-2$$
$$x=-\tfrac{2}{3}$$

The solutions are 8 and $-\frac{2}{3}$. The solution set is $\left\{-\frac{2}{3}, 8\right\}$.

GRAPHICAL APPROACH

We graph $y_1 = \text{abs}(2x-3)$ and $y_2 = \text{abs}(x+5)$ and look for any points of intersection. The x-coordinates of the points of intersection are -0.6666667 and 8.

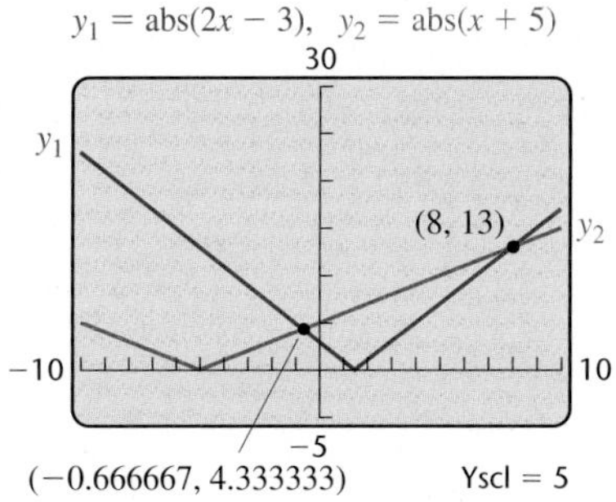

The solution set is $\{-0.6666667, 8\}$. Converted to fraction notation, the solution set is $\left\{-\frac{2}{3}, 8\right\}$.

Inequalities with Absolute Value

Our methods for solving equations with absolute value can be adapted for solving inequalities. Inequalities of this sort arise regularly in more advanced courses.

EXAMPLE 7 Solve $|x| < 4$. Then graph.

SOLUTION The solutions of $|x| < 4$ are all numbers whose *distance from zero is less than* 4. By substituting or by looking at the number line, we can see that numbers like $-3, -2, -1, -\frac{1}{2}, -\frac{1}{4}, 0, \frac{1}{4}, \frac{1}{2}, 1, 2,$ and 3 are all solutions. In fact, the solutions are all the numbers between -4 and 4. The solution set is $\{x \mid -4 < x < 4\}$. In interval notation, the solution set is $(-4, 4)$. The graph is as follows:

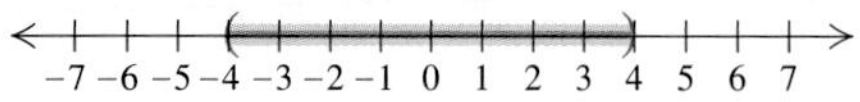

$|x| < 4$

We can also visualize Example 7 by graphing $f(x) = |x|$ and $g(x) = 4$, as in Example 1. The solution set consists of all x-values for which $(x, f(x))$ is below the horizontal line $g(x) = 4$. These x-values comprise the interval $(-4, 4)$.

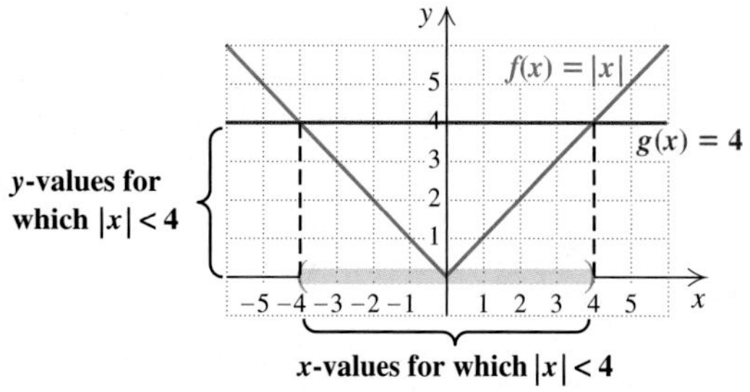

EXAMPLE 8 Solve $|x| \geq 4$. Then graph.

SOLUTION The solutions of $|x| \geq 4$ are all numbers that are at least 4 units from zero—in other words, those numbers x for which $x \leq -4$ *or* $4 \leq x$. The solution set is $\{x \mid x \leq -4 \text{ or } x \geq 4\}$. In interval notation, the solution set is $(-\infty, -4] \cup [4, \infty)$. We can check mentally with numbers like $-4.1, -5, 4.1$, and 5. The graph is as follows:

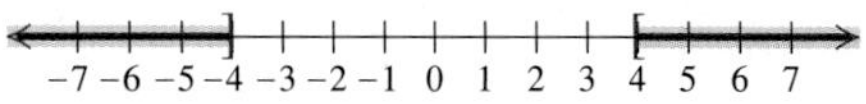

$|x| \geq 4$

As with Examples 1 and 7, Example 8 can be visualized by graphing $f(x) = |x|$ and $g(x) = 4$. The solution set of $|x| \geq 4$ consists of all x-values for which $(x, f(x))$ is on or above the horizontal line $g(x) = 4$. These x-values comprise $(-\infty, -4] \cup [4, \infty)$.

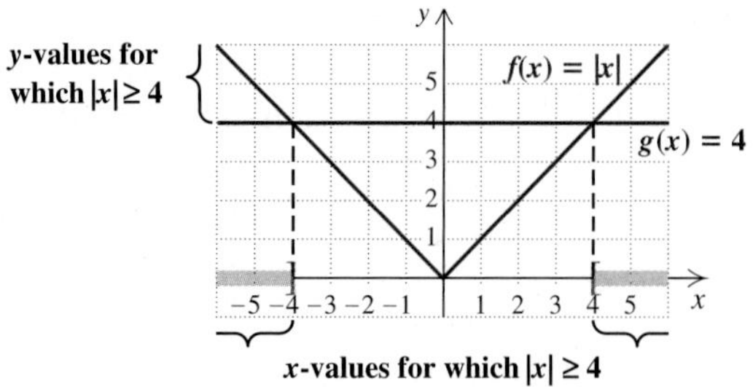

Examples 1, 7, and 8 illustrate three types of problems in which absolute-value symbols appear. The following is a general principle for solving such problems.

Principles for Solving Absolute-Value Problems For any positive number p and any expression X:

a) The solutions of $|X| = p$ are those numbers that satisfy $X = -p \text{ or } X = p$.

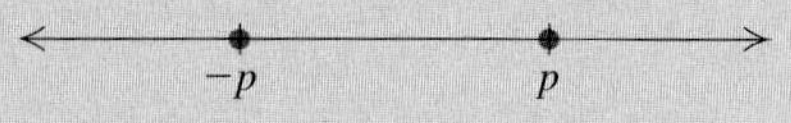

b) The solutions of $|X| < p$ are those numbers that satisfy $-p < X < p$.

c) The solutions of $|X| > p$ are those numbers that satisfy $X < -p \text{ or } p < X$.

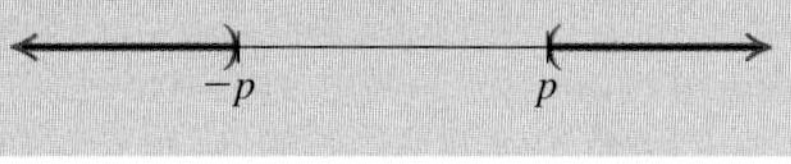

Student Notes

Resist the temptation to simply ignore the absolute-value symbol and solve the resulting equation or inequality. Doing so will lead to only *part* of the solution.

Of course, if p is negative, any value of X will satisfy the inequality $|X| > p$ because absolute value is never negative. By the same reasoning, $|X| < p$ has no solution when p is not positive. Thus, $|2x - 7| > -3$ is true for any real number x, and $|2x - 7| < -3$ has no solution.

Note that an inequality of the form $|X| < p$ corresponds to a *con*junction, whereas an inequality of the form $|X| > p$ corresponds to a *dis*junction.

EXAMPLE 9 Solve: $|3x - 2| < 4$.

ALGEBRAIC APPROACH

The number $3x - 2$ must be less than 4 units from zero. This is of the form $|X| < p$, so part (b) of the priniciples listed above applies.

$|X| < p$

$|3x - 2| < 4$ **Replacing X with $3x - 2$ and p with 4**

$-4 < 3x - 2 < 4$ **The number $3x - 2$ must be within 4 units of zero.**

$-2 < 3x < 6$ **Adding 2**

$-\frac{2}{3} < x < 2$ **Multiplying by $\frac{1}{3}$**

The solution set is $\{x \mid -\frac{2}{3} < x < 2\}$, or, in interval notation, $(-\frac{2}{3}, 2)$. The graph is as follows:

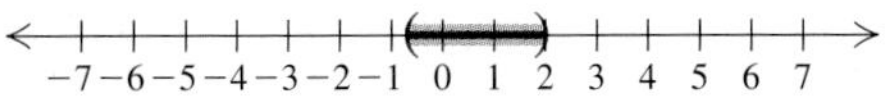

$|3x - 2| < 4$

GRAPHICAL APPROACH

We graph $y_1 = \text{abs}(3x - 2)$ and $y_2 = 4$.

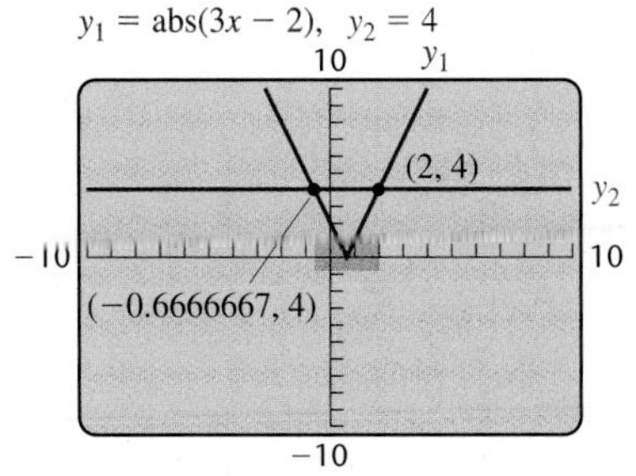

The solution set consists of the x-values for which $y_1 < y_2$, and y_1 is less than y_2 between the points $(-0.6666667, 4)$ and $(2, 4)$. The solution set of the inequality is thus the interval $(-0.6666667, 2)$, as indicated by the shading on the x-axis. Converted to fraction notation, the solution set is $(-\frac{2}{3}, 2)$.

EXAMPLE 10 Given that $f(x) = |4x + 2|$, find all x for which $f(x) \geq 6$.

ALGEBRAIC APPROACH

We have

$f(x) \geq 6$,

or $|4x + 2| \geq 6$. **Substituting**

To solve, we use part (c) of the principles listed above. In this case, X is $4x + 2$ and p is 6.

$|X| \geq p$

$|4x + 2| \geq 6$ **Replacing X with $4x + 2$ and p with 6**

$4x + 2 \leq -6$ *or* $6 \leq 4x + 2$

The number $4x + 2$ must be at least 6 units from zero.

$4x \leq -8$ *or* $4 \leq 4x$ **Adding -2**

$x \leq -2$ *or* $1 \leq x$ **Multiplying by $\frac{1}{4}$**

The solution set is $\{x \mid x \leq -2 \text{ or } x \geq 1\}$, or, in interval notation, $(-\infty, -2] \cup [1, \infty)$. The graph is as follows:

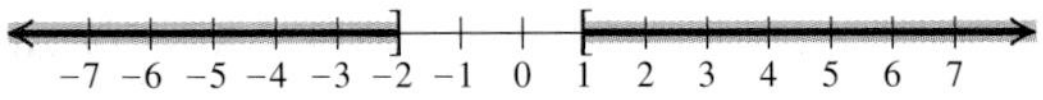

$|4x + 2| \geq 6$

GRAPHICAL APPROACH

To find all values of x for which $|4x + 2| \geq 6$, we graph $y_1 = \text{abs}(4x + 2)$ and $y_2 = 6$ and determine the points of intersection of the graphs.

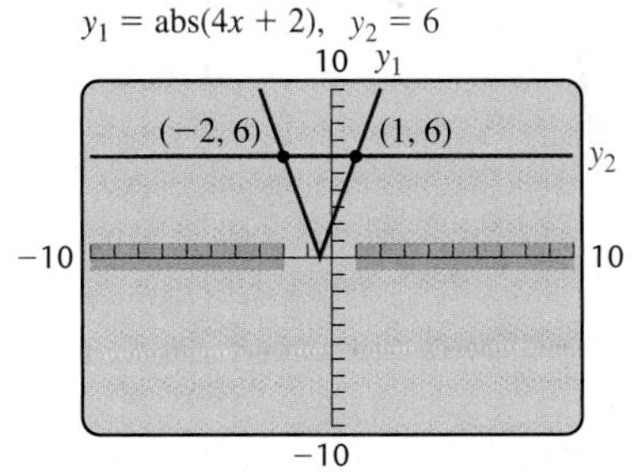

The solution set consists of all x-values for which the graph of $y_1 = |4x + 2|$ is *on or above* the graph of $y_2 = 6$. The graph of y_1 lies *on* the graph of y_2 when $x = -2$ or $x = 1$. The graph of y_1 is *above* the graph of y_2 when $x < -2$ or $x > 1$. Thus the solution set is $(-\infty, -2] \cup [1, \infty)$, as indicated by the shading on the x-axis.

4.4 EXERCISE SET

FOR EXTRA HELP

MathXL | MyMathLab | InterAct Math | AW Math Tutor Center | Video Lectures on CD: Disc 2 | Student's Solutions Manual

Concept Reinforcement *Classify each of the following as either true or false.*

1. If x is negative, then $|x| = -x$.
2. $|x|$ is never negative.
3. $|x|$ is always positive.
4. The distance between a and b can be expressed as $|a - b|$.
5. The number a is $|a|$ units from 0.
6. There are two solutions of $|3x - 8| = 17$.
7. There is no solution of $|4x + 9| > -5$.
8. All real numbers are solutions of $|2x - 7| < -3$.

Use the following graph to solve Exercises 9–14.

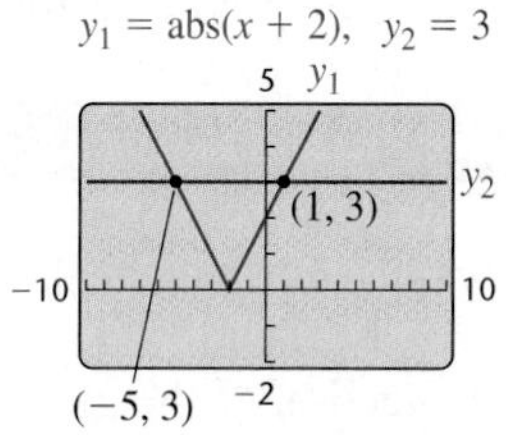

9. $|x + 2| = 3$
10. $|x + 2| \leq 3$
11. $|x + 2| < 3$
12. $|x + 2| > 3$
13. $|x + 2| \geq 3$
14. $|x + 2| = -1$

Solve.

15. $|x| = 7$
16. $|x| = 9$
17. Aha! $|x| = -6$
18. $|x| = -3$
19. $|p| = 0$
20. $|y| = 7.3$
21. $|t| = 5.5$
22. $|m| = 0$
23. $|2x - 3| = 4$
24. $|5x + 2| = 7$
25. $|3x - 5| = -8$
26. $|7x - 2| = -9$
27. $|x - 2| = 6$
28. $|x - 3| = 8$
29. $|x - 5| = 3$
30. $|x - 6| = 1$
31. $|x - 7| = 9$
32. $|x - 4| = 5$
33. $|5x| - 3 = 37$
34. $|2y| - 5 = 13$
35. $7|q| - 2 = 9$
36. $7|z| + 2 = 16$
37. $\left|\frac{2x - 1}{3}\right| = 5$
38. $\left|\frac{4 - 5x}{6}\right| = 3$
39. $|m + 5| + 9 = 16$
40. $|t - 7| + 1 = 4$
41. $5 - 2|3x - 4| = -5$
42. $3|2x - 5| - 7 = -1$
43. Let $f(x) = |2x + 6|$. Find all x for which $f(x) = 8$.
44. Let $f(x) = |2x + 4|$. Find all x for which $f(x) = 10$.
45. Let $f(x) = |x| - 3$. Find all x for which $f(x) = 5.7$.
46. Let $f(x) = |x| + 7$. Find all x for which $f(x) = 18$.
47. Let $f(x) = \left|\frac{3x - 2}{5}\right|$. Find all x for which $f(x) = 2$.
48. Let $f(x) = \left|\frac{1 - 2x}{3}\right|$. Find all x for which $f(x) = 1$.

Solve.

49. $|x + 4| = |2x - 7|$
50. $|3x + 5| = |x - 6|$
51. $|x + 4| = |x - 3|$
52. $|x - 9| = |x + 6|$
53. $|3a - 1| = |2a + 4|$
54. $|5t + 7| = |4t + 3|$
55. Aha! $|n - 3| = |3 - n|$
56. $|y - 2| = |2 - y|$
57. $|7 - a| = |a + 5|$
58. $|6 - t| = |t + 7|$
59. $\left|\frac{1}{2}x - 5\right| = \left|\frac{1}{4}x + 3\right|$
60. $\left|2 - \frac{2}{3}x\right| = \left|4 + \frac{7}{8}x\right|$

Solve and graph.

61. $|a| \leq 9$

62. $|x| < 2$

63. $|x| > 8$

64. $|a| \geq 3$

65. $|t| > 0$

66. $|t| \geq 1.7$

67. $|x - 1| < 4$

68. $|x - 1| < 3$

69. $|x + 2| \leq 6$

70. $|x + 4| \leq 1$

71. $|x - 3| + 2 > 7$

72. $|x - 4| + 5 > 2$

Aha! **73.** $|2y - 9| > -5$

74. $|3y - 4| > 8$

75. $|3a - 4| + 2 \geq 8$

76. $|2a - 5| + 1 \geq 9$

77. $|y - 3| < 12$

78. $|p - 2| < 3$

79. $9 - |x + 4| \leq 5$

80. $12 - |x - 5| \leq 9$

81. $|4 - 3y| > 8$

82. $|7 - 2y| < -6$

Aha! **83.** $|5 - 4x| < -6$

84. $7 + |4a - 5| \leq 26$

85. $\left|\frac{2 - 5x}{4}\right| \geq \frac{2}{3}$

86. $\left|\frac{1 + 3x}{5}\right| > \frac{7}{8}$

87. $|m + 3| + 8 \leq 14$

88. $|t - 7| + 3 \geq 4$

89. $25 - 2|a + 3| > 19$

90. $30 - 4|a + 2| > 12$

91. Let $f(x) = |2x - 3|$. Find all x for which $f(x) \leq 4$.

92. Let $f(x) = |5x + 2|$. Find all x for which $f(x) \leq 3$.

93. Let $f(x) = 5 + |3x - 4|$. Find all x for which $f(x) \geq 16$.

94. Let $f(x) = |2 - 9x|$. Find all x for which $f(x) \geq 25$.

95. Let $f(x) = 7 + |2x - 1|$. Find all x for which $f(x) < 16$.

96. Let $f(x) = 5 + |3x + 2|$. Find all x for which $f(x) < 19$.

TW **97.** Explain in your own words why -7 is not a solution of $|x| < 5$.

TW **98.** Explain in your own words why $[6, \infty)$ is only part of the solution of $|x| \geq 6$.

Focused Review

Solve.

99. $3(x + 1) = 5(x - 2) + 1$ [1.6]

100. $3(x + 1) \leq 5(x - 2) + 1$ [4.1]

101. $y = 3(x + 1)$,
$y = 5(x - 2) + 1$ [3.2]

102. $3(x + 1) \leq 2$ *and* $5(x - 2) + 1 \leq -4$ [4.3]

103. $|x + 5| < 6$ [4.4]

104. $|x + 5| = 6$ [4.4]

Synthesis

TW **105.** Is it possible for an equation in x of the form $|ax + b| = c$ to have exactly one solution? Why or why not?

TW **106.** Isabel is using the following graph to solve $|x - 3| < 4$. How can you tell that a mistake has been made?

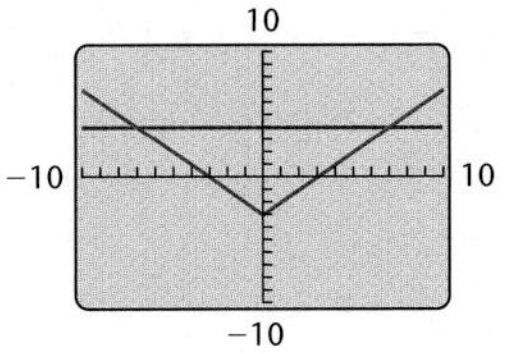

107. From the definition of absolute value, $|x| = x$ only when $x \geq 0$. Solve $|3t - 5| = 3t - 5$ using this same reasoning.

Solve.

108. $|3x - 5| = x$

109. $|x + 2| > x$

110. $2 \leq |x - 1| \leq 5$

111. $|5t - 3| = 2t + 4$

112. $t - 2 \leq |t - 3|$

Find an equivalent inequality with absolute value.

113. $-3 < x < 3$

114. $-5 \leq y \leq 5$

115. $x \leq -6$ *or* $6 \leq x$

116. $x < -4$ *or* $4 < x$

117. $x < -8$ *or* $2 < x$

118. $-5 < x < 1$

119. x is less than 2 units from 7.

120. x is less than 1 unit from 5.

Write an absolute-value inequality for which the interval shown is the solution.

121. 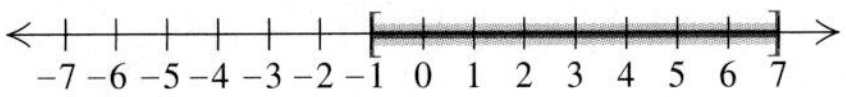

122. 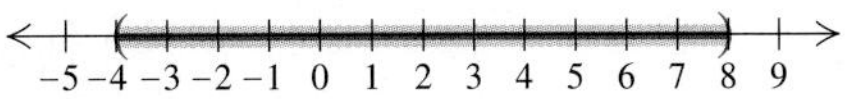

123. 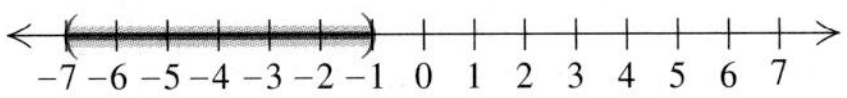

124. 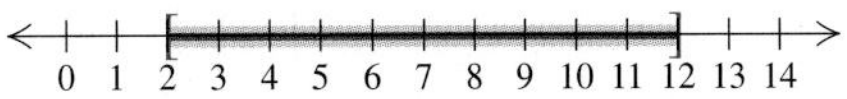

125. *Bungee Jumping.* A bungee jumper is bouncing up and down so that her distance d above a river satisfies the inequality $|d\text{-}60 \text{ ft}| \leq 10$ ft (see the figure below). If the bridge from which she jumped is 150 ft above the river, how far is the bungee jumper from the bridge at any given time?

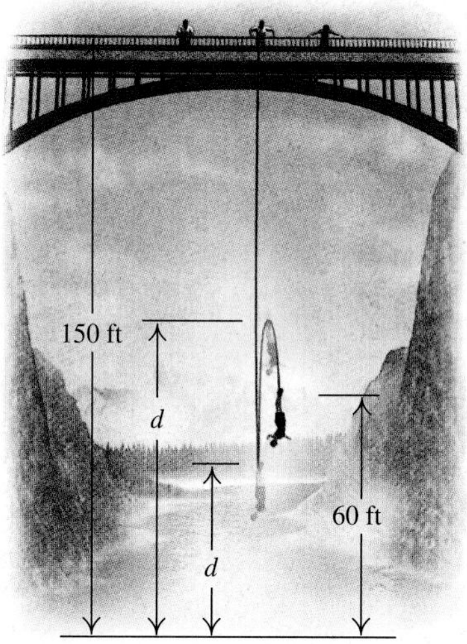

126. *Water Level.* Depending on how dry or wet the weather has been, water in a well will rise and fall. The distance d that a well's water level is below the ground satisfies the inequality $|d - 15| \leq 2.5$ (see the figure below).

a) Solve for d.

b) How tall a column of water is in the well at any given time?

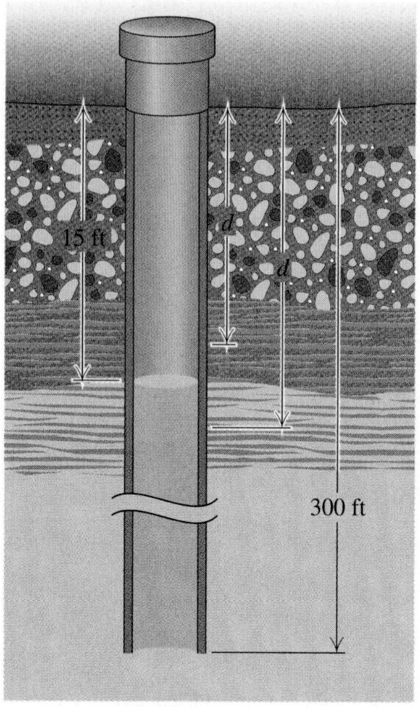

4.5 Inequalities in Two Variables

Graphs of Linear Inequalities ■ Systems of Linear Inequalities

In Section 4.1, we graphed inequalities in one variable on the number line. Now we graph inequalities in two variables on a plane.

Graphs of Linear Inequalities

When the equals sign in a linear equation is replaced with an inequality sign, a **linear inequality** is formed. Solutions of linear inequalities are ordered pairs.

Study Tip

Add Your Voice to the Author's

If you own your text, consider using it as a notebook. Since many instructors closely parallel the book, you may find it helpful to make notes on the appropriate page as he or she is lecturing.

EXAMPLE 1 Determine whether $(-3, 2)$ and $(6, -7)$ are solutions of the inequality $5x - 4y > 13$.

SOLUTION Below, on the left, we replace x with -3 and y with 2. On the right, we replace x with 6 and y with -7.

$5x - 4y > 13$	
$5(-3) - 4 \cdot 2$	13
$-15 - 8$	
$-23 \overset{?}{>} 13$	FALSE

Since $-23 > 13$ is false, $(-3, 2)$ is not a solution.

$5x - 4y > 13$	
$5(6) - 4(-7)$	13
$30 + 28$	
$58 \overset{?}{>} 13$	TRUE

Since $58 > 13$ is true, $(6, -7)$ is a solution.

The graph of a linear equation is a straight line. The graph of a linear inequality is a *half-plane*, bordered by the graph of the *related equation.* To find an inequality's related equation, we simply replace the inequality sign with an equals sign.

EXAMPLE 2 Graph: $y \le x$.

SOLUTION We first graph the related equation $y = x$. Every solution of $y = x$ is an ordered pair, like $(3, 3)$, in which both coordinates are the same. The graph of $y = x$ is shown on the left below. Since the inequality symbol is $\le$, the line is drawn solid and is part of the graph of $y \le x$.

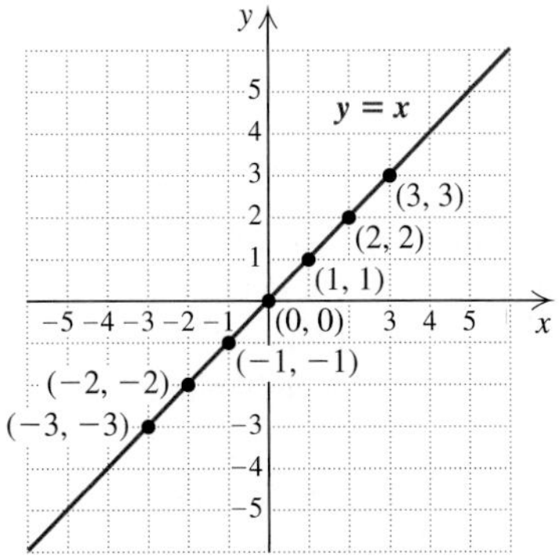

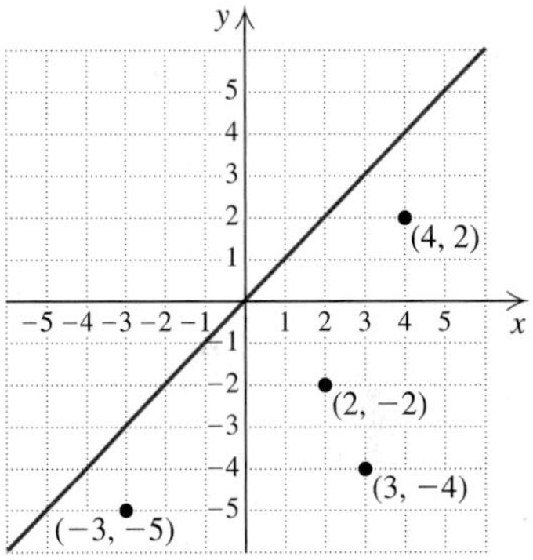

Note that in the graph on the right above each ordered pair on the half-plane below $y = x$ contains a y-coordinate that is less than the x-coordinate. All these pairs represent solutions of $y \le x$. We check one pair, $(4, 2)$, as follows:

$y \le x$		
2	4	TRUE

It turns out that *any* point on the same side of $y = x$ as $(4, 2)$ is also a solution. Thus, if one point in a half-plane is a solution, then *all* points in that half-plane are solutions. We finish drawing the solution set by shading the half-plane below $y = x$. The complete solution set consists of the shaded half-plane as well as the boundary line itself.

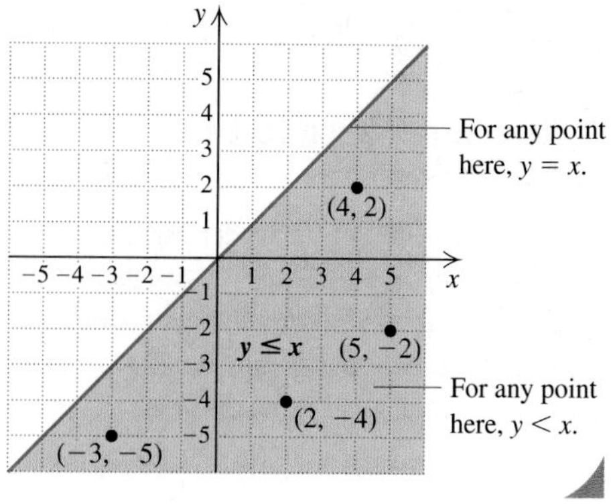

From Example 2, we see that for any inequality of the form $y \le f(x)$ or $y < f(x)$, we shade *below* the graph of $y = f(x)$.

EXAMPLE 3 Graph: $8x + 3y > 24$.

SOLUTION First, we sketch the graph of $8x + 3y = 24$. Since the inequality sign is $>$, points on this line do not represent solutions of the inequality, so the line is drawn dashed. Points representing solutions of $8x + 3y > 24$ are in either the half-plane above the line or the half-plane below the line. To determine which, we select a point that is not on the line and determine whether it is a solution of $8x + 3y > 24$. Let's use $(-3, 4)$ as this *test point*:

$$\begin{array}{c|c} \multicolumn{2}{c}{8x + 3y > 24} \\ \hline 8(-3) + 3 \cdot 4 & 24 \\ -24 + 12 & \\ \end{array}$$

$$-12 \overset{?}{>} 24 \quad \text{FALSE}$$

Since $-12 > 24$ is *false*, $(-3, 4)$ is not a solution. Thus no point in the half-plane containing $(-3, 4)$ is a solution. The points in the other half-plane *are* solutions, so we shade that half-plane and obtain the graph shown at right.

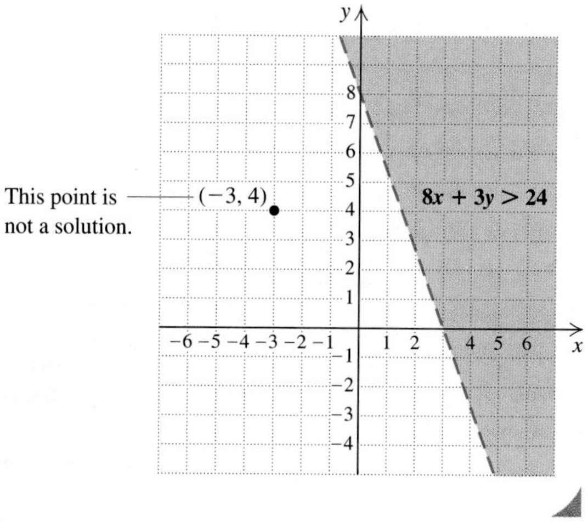

Steps for Graphing Linear Inequalities

1. Replace the inequality sign with an equals sign and graph this line as the boundary. If the inequality symbol is $<$ or $>$, draw the line dashed. If the inequality symbol is $\leq$ or $\geq$, draw the line solid.
2. The graph consists of a half-plane on one side of the line and, if the line is solid, the line as well.
   a) If the inequality is of the form $y < mx + b$ or $y \leq mx + b$, shade *below* the line. If the inequality is of the form $y > mx + b$ or $y \geq mx + b$, shade *above* the line.
   b) If y is not isolated, either solve for y and proceed as in part (a) or select a test point not on the line. If the test point represents a solution of the inequality, shade the half-plane containing the point. If it does not, shade the other half-plane.

Linear Inequalities

On most graphing calculators, an inequality like $y < \frac{6}{5}x + 3.49$ can be drawn by entering $(6/5)x + 3.49$ as y_1, moving the cursor to the GraphStyle icon just to the left of y_1, pressing ENTER until ◣ appears, and then pressing GRAPH.

Many newer calculators have an INEQUALZ program that is accessed using the APPS key. Running this program allows us to write inequalities at the Y= screen by pressing ALPHA and then one of the five keys just below the screen, as shown on the left below.

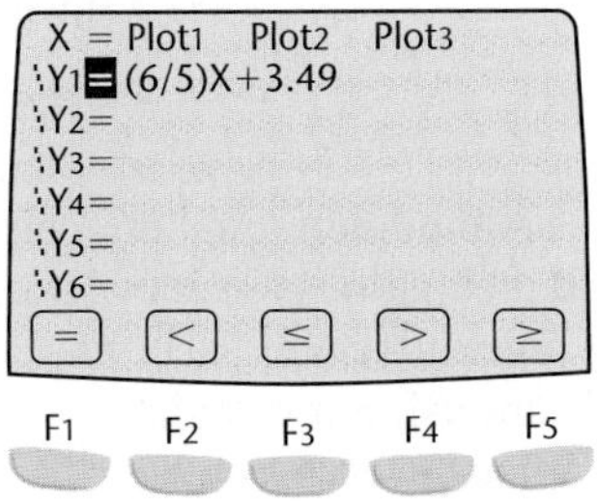

$y_1 < (6/5)x + 3.49$, or
◣$y_1 = (6/5)x + 3.49$

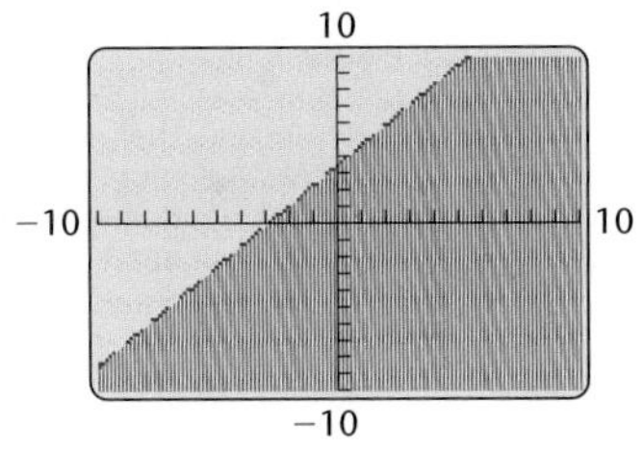

Although the graphs should be identical regardless of the method used, on the newer calculators the boundary line appears dashed when $<$ or $>$ is selected. The graph of the inequality is shown on the right above. Most calculators will shade this area with black vertical lines. Inequalities containing $\leq$ or $\geq$ are handled in a similar manner.

When you are finished with the INEQUALZ application, select it again from the APPS menu and quit the program.

EXAMPLE 4 Graph: $6x - 2y < 12$.

SOLUTION We graph both by hand and using a graphing calculator.

BY HAND

We could graph $6x - 2y = 12$ and use a test point, as in Example 3. Instead, let's solve $6x - 2y < 12$ for y:

$6x - 2y < 12$

$-2y < -6x + 12$ **Adding $-6x$ to both sides**

$y > 3x - 6.$ **Dividing both sides by -2 and reversing the < symbol**

The graph consists of the half plane above the dashed boundary line $y = 3x - 6$ (see the graph below).

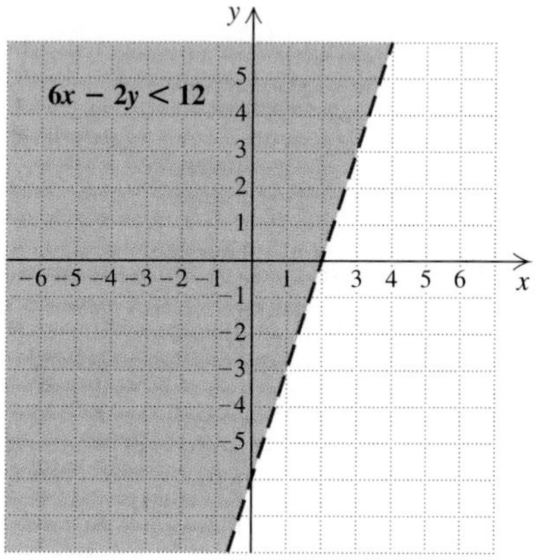

USING A GRAPHING CALCULATOR

We solve the inequality for y:

$6x - 2y < 12$

$-2y < -6x + 12$

$y > 3x - 6.$ **Reversing the < symbol**

We enter the boundary equation, $y = 3x - 6$, and use the INEQUALZ application to graph.

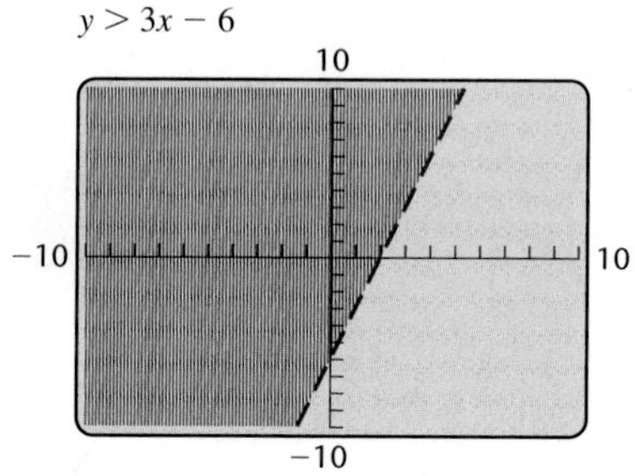

Many calculators do not draw a dashed line, so the graph of $y > 3x - 6$ appears to be the same as the graph of $y \geq 3x - 6$.

Quit the INEQUALZ application when you are finished.

Although a graphing calculator can be a very useful tool, some equations and inequalities are actually quicker to graph by hand. It is important to be able to quickly sketch basic linear equations and inequalities by hand. Also, when we are using a calculator to graph equations, knowing the basic shape of the graph can serve as a check that the equation was entered correctly.

EXAMPLE 5 Graph $x > -3$ on a plane.

SOLUTION There is a missing variable in this inequality. If we graph the inequality on a line, its graph is as follows:

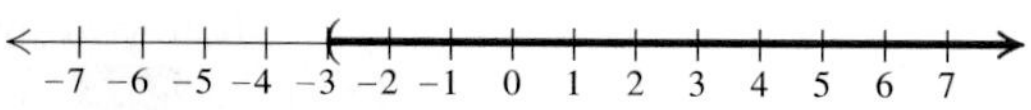

However, we can also write this inequality as $x + 0y > -3$ and graph it on a plane. We can use the same technique as in the examples above. First, we graph the related equation $x = -3$ on a plane. Then we test some point, say, $(2, 5)$:

$$\begin{array}{c|c} x + 0y & > -3 \\ \hline 2 + 0 \cdot 5 & -3 \\ 2 & \overset{?}{>} -3 \quad \text{TRUE} \end{array}$$

Since $(2, 5)$ is a solution, all points in the half-plane containing $(2, 5)$ are solutions. We shade that half-plane. Another approach is to simply note that the solutions of $x > -3$ are all pairs with first coordinates greater than -3.

Although many graphing calculators can graph this inequality from the DRAW menu, it is simpler to graph it by hand.

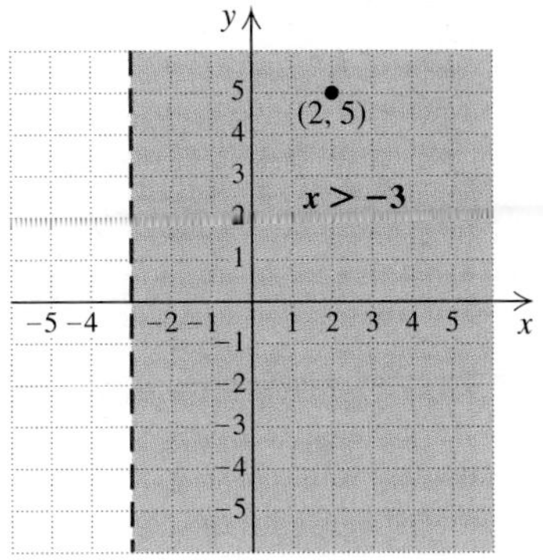

EXAMPLE 6 Graph $y \leq 4$ on a plane.

SOLUTION The inequality is of the form $y \leq mx + b$ (with $m = 0$), so we shade below the solid horizontal line representing $y = 4$.

This inequality can also be graphed by drawing $y = 4$ and testing a point above or below the line. The half-plane below $y = 4$ should be shaded.

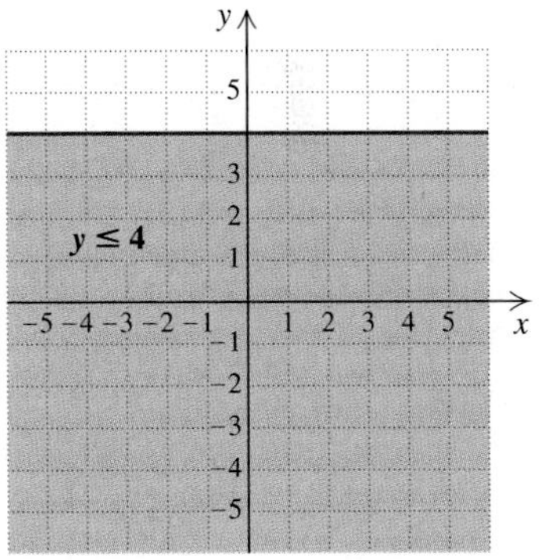

Systems of Linear Inequalities

To graph a system of equations, we graph the individual equations and then find the intersection of the individual graphs. We do the same thing for a system of inequalities—that is, we graph each inequality and find the intersection of the individual graphs.

EXAMPLE 7 Graph the system

$$x + y \leq 4,$$
$$x - y < 4.$$

SOLUTION To graph $x + y \leq 4$, we graph $x + y = 4$ using a solid line. Since the test point $(0, 0)$ *is* a solution and $(0, 0)$ is below the line, we shade the half-plane below the graph red. The arrows near the ends of the line are another way of indicating the half-plane containing solutions.

Next, we graph $x - y < 4$. We graph $x - y = 4$ using a dashed line and consider $(0, 0)$ as a test point. Again, $(0, 0)$ is a solution, so we shade that side of the line blue. The solution set of the system is the region that is shaded purple (both red and blue) and part of the line $x + y = 4$.

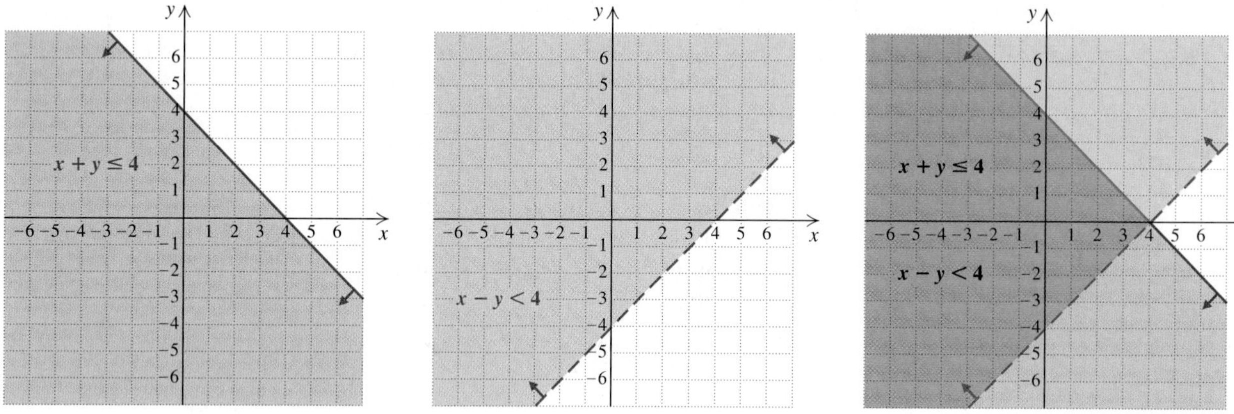

Student Notes

If you don't use differently colored pencils or pens to shade different regions, consider using a pencil to make slashes that tilt in different directions in each region, similar to the way shading is done on a graphing calculator. You may also find it useful to attach arrows to the lines, as in the examples shown.

EXAMPLE 8 Graph: $-2 < x \leq 3$.

SOLUTION This is a system of inequalities:

$$-2 < x,$$
$$x \leq 3.$$

We graph the equation $-2 = x$, and see that the graph of the first inequality is the half-plane to the right of the boundary $-2 = x$. It is shaded red.

We graph the second inequality, starting with the line $x = 3$, and find that its graph is the line and also the half-plane to its left. It is shaded blue.

The solution set of the system is the region that is the intersection of the individual graphs. Since it is shaded both blue and red, it appears to be purple. All points in this region have x-coordinates that are greater than -2 but do not exceed 3.

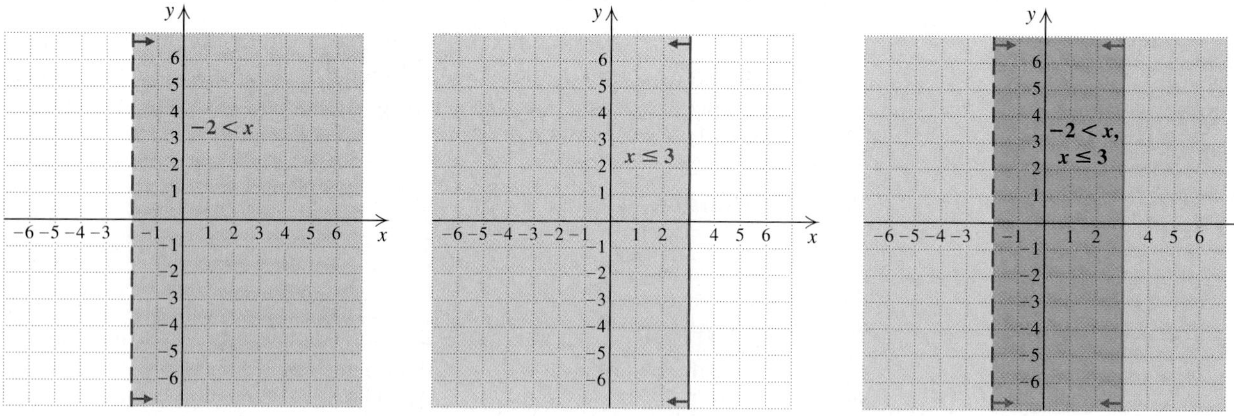

A system of inequalities may have a graph that consists of a polygon and its interior. The corners of such a graph are called *vertices* (singular, *vertex*).

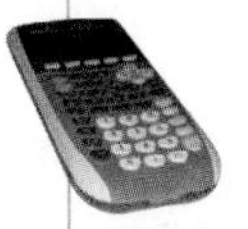

Systems of Linear Inequalities

Systems of inequalities can be graphed by solving for y and then graphing each inequality. The graph of the system will be the intersection of the shaded regions. To graph systems directly using the INEQUALZ application, enter the correct inequalities, press GRAPH, and then press ALPHA and Shades (F1 or F2). At the SHADES menu, select Ineq Intersection to see the final graph. To find the vertices, or points of intersection, select Pol-Trace from the graph menu.

EXAMPLE 9 Graph the system of inequalities. Find the coordinates of any vertices formed.

$$6x - 2y \le 12, \quad (1)$$
$$y - 3 \le 0, \quad (2)$$
$$x + y \ge 0. \quad (3)$$

SOLUTION We graph both by hand and using a graphing calculator.

BY HAND

We graph the boundaries

$$6x - 2y = 12,$$
$$y - 3 = 0,$$

and $$x + y = 0$$

using solid lines. The regions for each inequality are indicated by the arrows near the ends of the lines. We note where the regions overlap and shade the region of solutions purple.

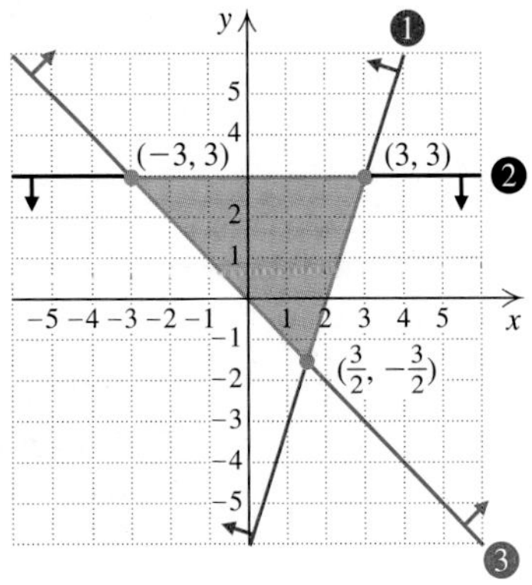

To find the vertices, we solve three different systems of two equations. The system of boundary equations from inequalities (1) and (2) is

$$6x - 2y = 12,$$
$$y - 3 = 0.$$

Solving, we obtain the vertex $(3, 3)$.

The system of boundary equations from inequalities (1) and (3) is

$$6x - 2y = 12,$$
$$x + y = 0.$$

Solving, we obtain the vertex $\left(\frac{3}{2}, -\frac{3}{2}\right)$.

The system of boundary equations from inequalities (2) and (3) is

$$y - 3 = 0,$$
$$x + y = 0.$$

Solving, we obtain the vertex $(-3, 3)$.

USING A GRAPHING CALCULATOR

First, we solve each inequality for y and obtain the equivalent system of inequalities

$$y_1 \geq (12 - 6x)/(-2),$$
$$y_2 \leq 3,$$
$$y_3 \geq -x.$$

We run the INEQUALZ application, enter each inequality, and press GRAPH. We then press ALPHA and F1 or F2 to choose Shades. Selecting the Ineq Intersection option results in the graph shown on the left below.

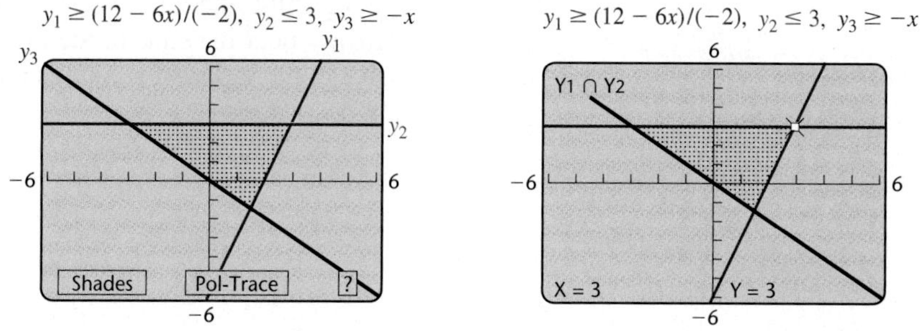

The vertices are the points of intersection of the graphs of the boundary equations y_1 and y_2, y_1 and y_3, and y_2 and y_3. We can find them by choosing Pol-Trace. Pressing the arrow keys moves the cursor to different vertices. The vertices are $(3, 3)$, $(-3, 3)$, and $(1.5, -1.5)$.

Graphs of systems of inequalities and the coordinates of vertices are used to solve a variety of problems using a branch of mathematics called *linear programming.*

Connecting the Concepts

We have now solved a variety of equations, inequalities, systems of equations, and systems of inequalities. In each case, there are different ways to represent the solution. Below is a list of the different types of problems we have solved, along with illustrations of each type.

Type	Example	Solution	Graph
Linear equations in one variable	$2x - 8 = 3(x + 5)$	A number	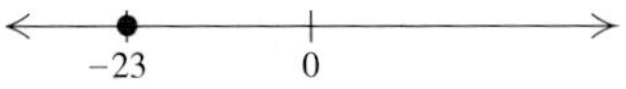
Linear inequalities in one variable	$-3x + 5 > 2$	A set of numbers; an interval	
Linear equations in two variables	$2x + y = 7$	A set of ordered pairs; a line	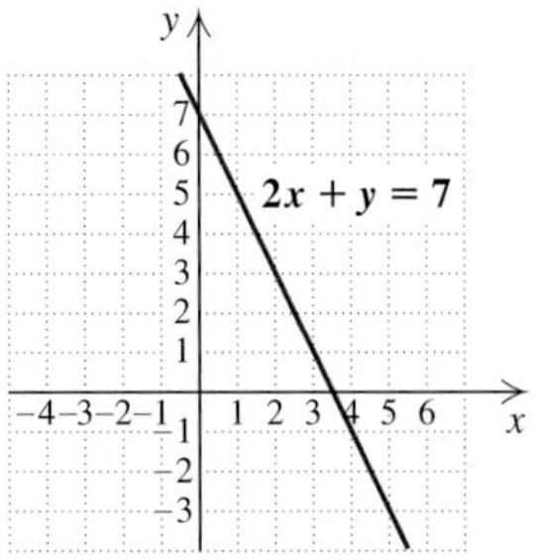
Linear inequalities in two variables	$x + y \geq 4$	A set of ordered pairs; a half-plane	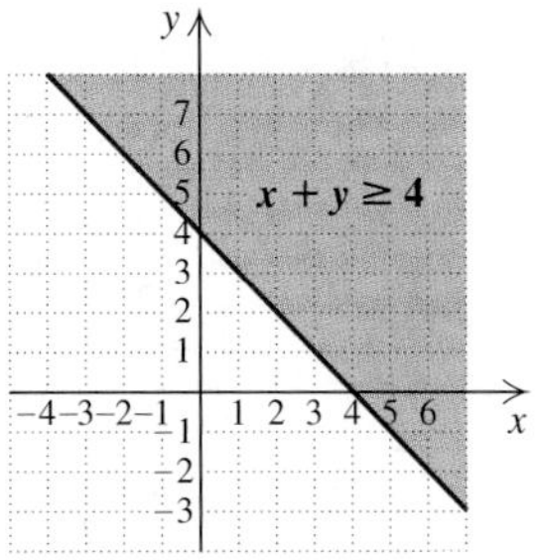
System of equations in two variables	$x + y = 3,$ $5x - y = -27$	An ordered pair or a (possibly empty) set of ordered pairs	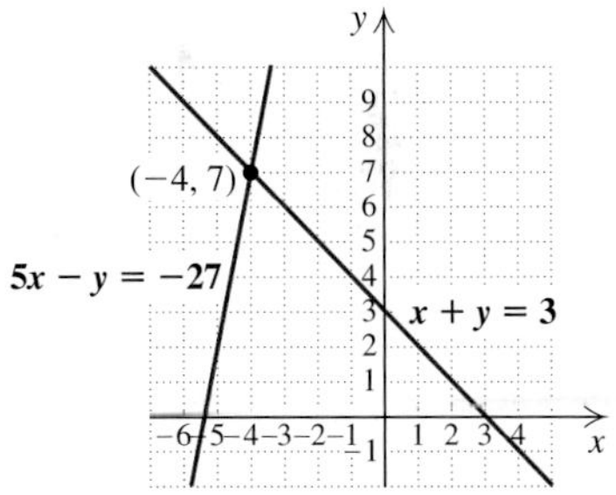
System of inequalities in two variables	$6x - 2y \leq 12,$ $y - 3 \leq 0,$ $x + y \geq 0$	A set of ordered pairs; a region of a plane	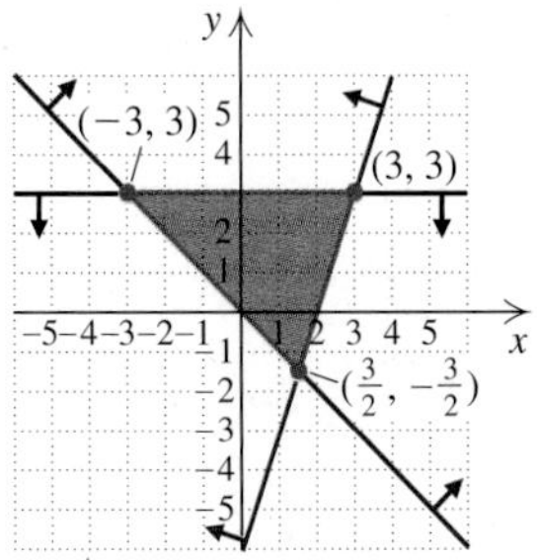

Keeping in mind how these solutions vary and what their graphs look like will help you as you progress further in this book and in mathematics in general.

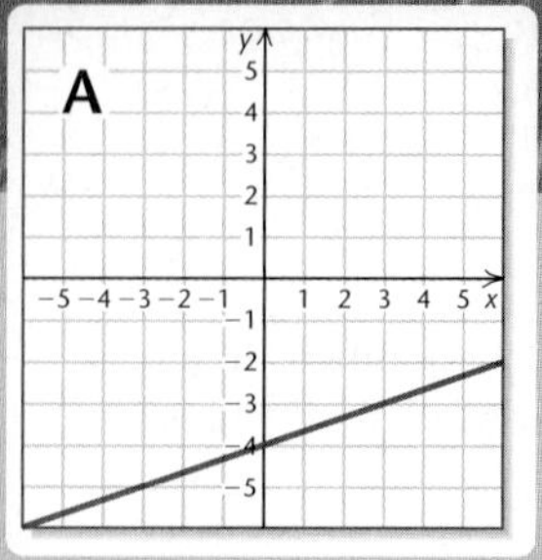

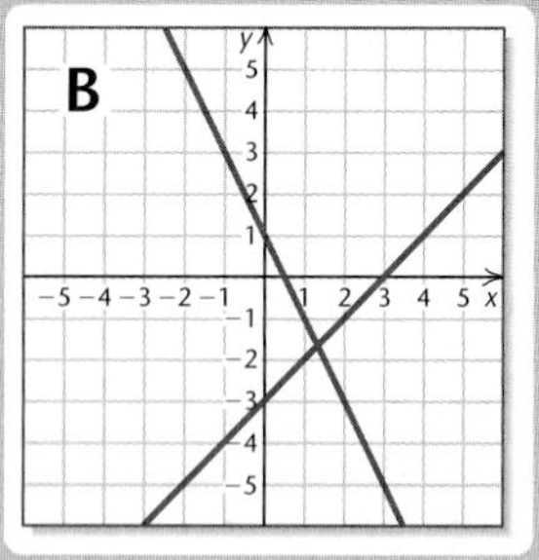

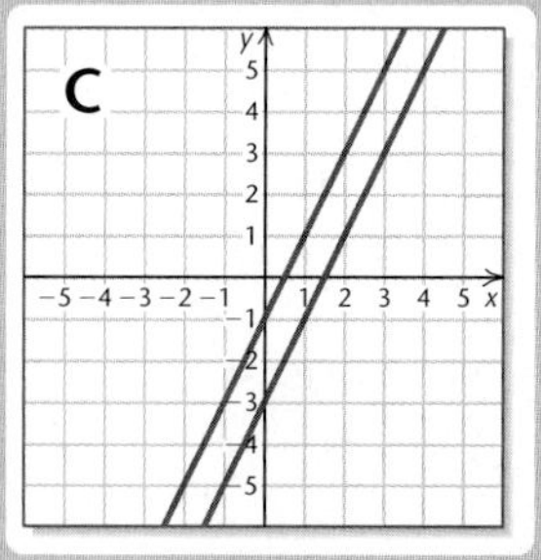

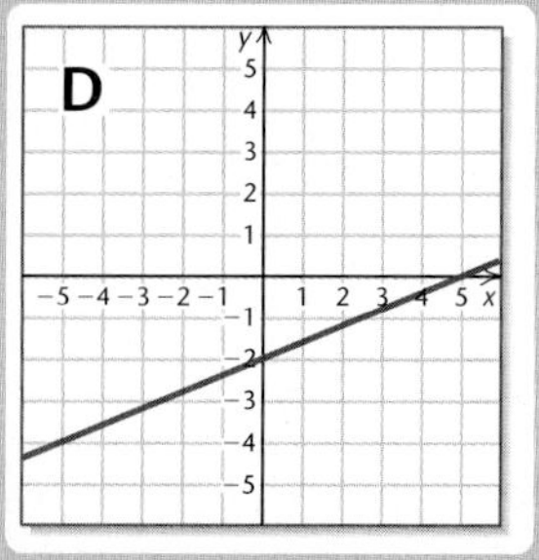

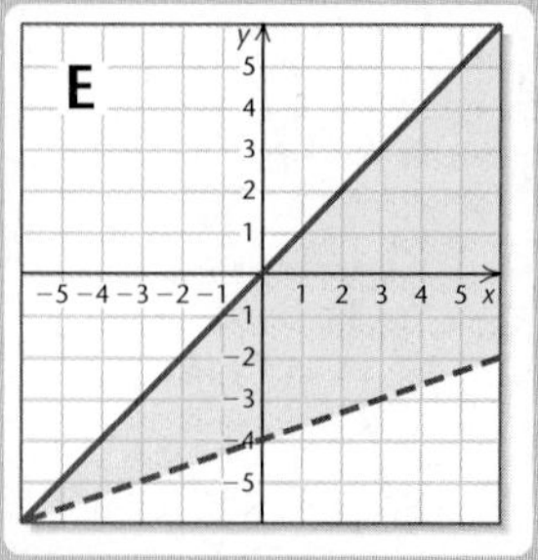

Visualizing the Graph

Match each equation, inequality, or system of equations or inequalities to its graph.

1. $x - y = 3,$
   $2x + y = 1$

2. $3x - y \leq 5$

3. $x > -3$

4. $y = \frac{1}{3}x - 4$

5. $y > \frac{1}{3}x - 4,$
   $y \leq x$

6. $x = y$

7. $y = 2x - 1,$
   $y = 2x - 3$

8. $2x - 5y = 10$

9. $x + y \leq 3,$
   $2y \leq x + 1$

10. $y = \frac{3}{2}$

Answers on page A-18

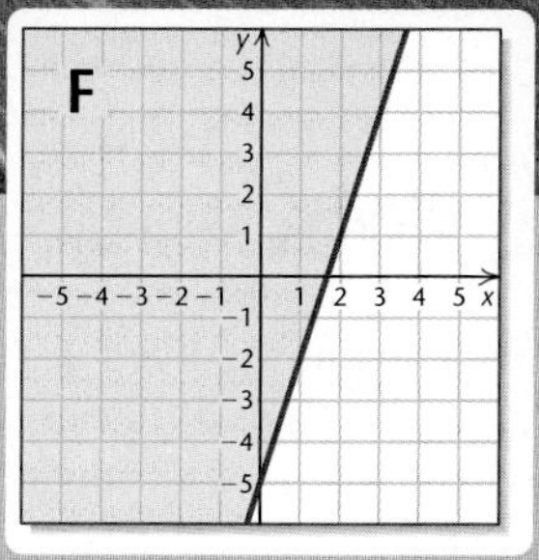

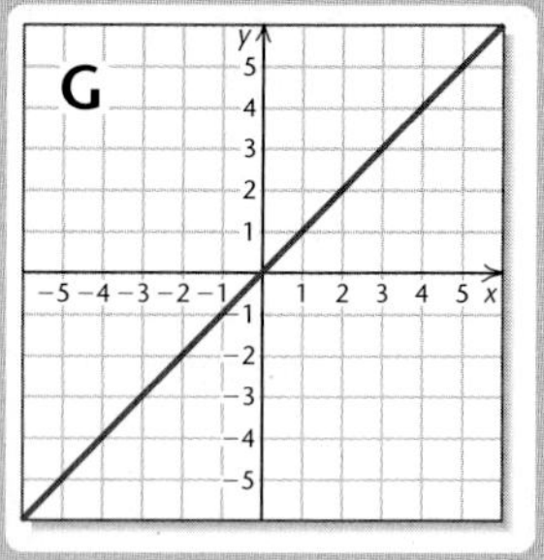

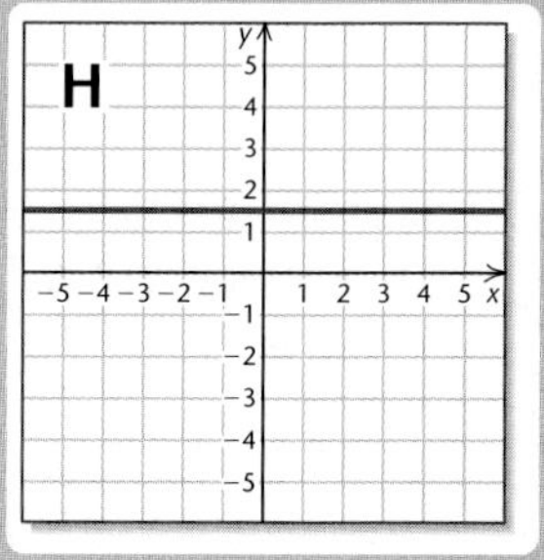

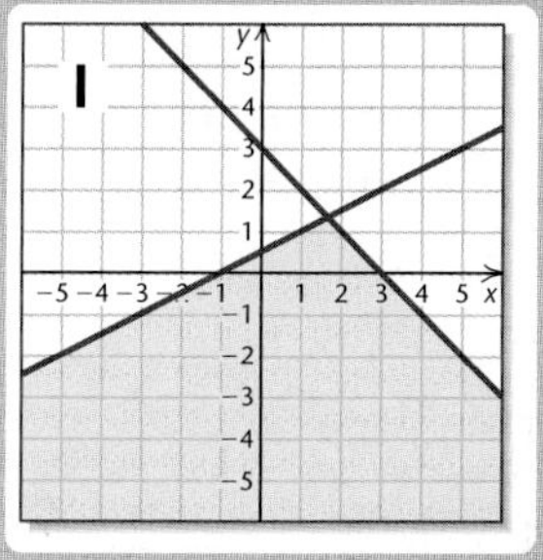

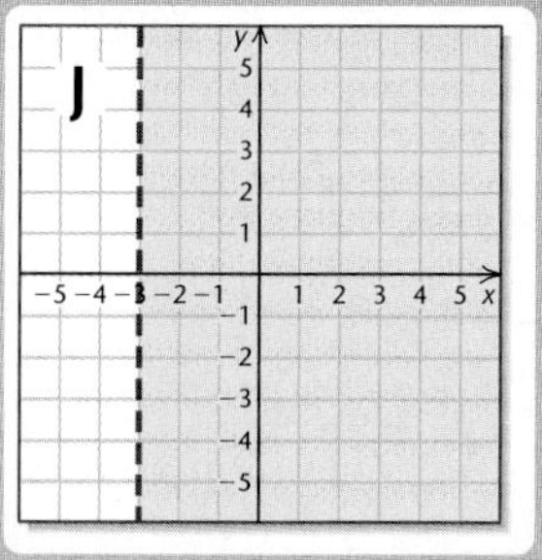

4.5 EXERCISE SET

FOR EXTRA HELP

MathXL | MyMathLab | InterAct Math | AW Math Tutor Center | Video Lectures on CD: Disc 2 | Student's Solutions Manual

Concept Reinforcement *In each of Exercises 1–6, match the phrase with the most appropriate choice from the column on the right.*

1. ____ A solution of a linear inequality

2. ____ The graph of a linear inequality

3. ____ The graph of a system of linear inequalities

4. ____ Often a convenient test point

5. ____ The name for the corners of a graph of a system of linear inequalities

6. ____ A dashed line

a) $(0, 0)$

b) Vertices

c) A half-plane

d) The intersection of two or more half-planes

e) An ordered pair that satisfies the inequality

f) Indicates the line is not part of the solution

Determine whether each ordered pair is a solution of the given inequality.

7. $(-4, 2)$; $2x + 3y < -1$

8. $(3, -6)$; $4x + 2y \leq -2$

9. $(8, 14)$; $2y - 3x \geq 9$

10. $(7, 20)$; $3x - y > -1$

Graph on a plane.

11. $y > \frac{1}{2}x$

12. $y > 2x$

13. $y \geq x - 3$

14. $y < x + 3$

15. $y \leq x + 5$

16. $y > x - 2$

17. $x - y \leq 4$

18. $x + y < 4$

19. $2x + 3y < 6$

20. $3x + 4y \leq 12$

21. $2y - x \leq 4$

22. $2y - 3x > 6$

23. $2x - 2y \geq 8 + 2y$

24. $3x - 2 \leq 5x + y$

25. $y \geq 3$

26. $x < -5$

27. $x \leq 6$

28. $y > -3$

29. $-2 < y < 7$

30. $-4 < y < -1$

31. $-4 \leq x \leq 2$

32. $-3 \leq y \leq 4$

33. $0 \leq y \leq 3$

34. $0 \leq x \leq 6$

Graph using a graphing calculator.

35. $y > x + 3.5$

36. $7y \leq 2x + 5$

37. $8x - 2y < 11$

38. $11x + 13y + 4 \geq 0$

Graph each system.

39. $y > -x,$ $y < x + 2$

40. $y < x,$ $y > -x + 1$

41. $y \geq x,$ $y \geq 2x - 4$

42. $y \geq x,$ $y \leq -x + 4$

43. $y \leq -3,$ $x \geq -1$

44. $y \geq -3,$ $x \geq 1$

45. $x > -4,$ $y < -2x + 3$

46. $x < 3,$ $y > -3x + 2$

47. $y \leq 5,$ $y \geq -x + 4$

48. $y \geq -2,$ $y \geq x + 3$

49. $x + y \leq 6,$ $x - y \leq 4$

50. $x + y < 1,$ $x - y < 2$

51. $y + 3x > 0,$ $y + 3x < 2$

52. $y - 2x \geq 1,$ $y - 2x \leq 3$

Graph each system of inequalities. Find the coordinates of any vertices formed.

53. $y \leq 2x - 3,$ $y \geq -2x + 1,$ $x \leq 5$

54. $2y - x \leq 2,$ $y - 3x \geq -4,$ $y \geq -1$

55. $x + 2y \leq 12,$ $2x + y \leq 12,$ $x \geq 0,$ $y \geq 0$

56. $x - y \leq 2,$ $x + 2y \geq 8,$ $y \leq 4$

57. $8x + 5y \leq 40,$ $x + 2y \leq 8,$ $x \geq 0,$ $y \geq 0$

58. $4y - 3x \geq -12,$ $4y + 3x \geq -36,$ $y \leq 0,$ $x \leq 0$

59. $y - x \geq 2,$ $y - x \leq 4,$ $2 \leq x \leq 5$

60. $3x + 4y \geq 12,$ $5x + 6y \leq 30,$ $1 \leq x \leq 3$

TW **61.** In Example 7, is the point (4, 0) part of the solution set? Why or why not?

TW **62.** When graphing linear inequalities, Ron makes a habit of always shading above the line when the symbol $\geq$ is used. Is this wise? Why or why not?

Skill Maintenance

Solve.

63. *Catering.* Sandy's Catering needs to provide 10 lb of mixed nuts for a wedding reception. Peanuts cost \$2.50 per pound and fancy nuts cost \$7 per pound. If \$40 has been allocated for nuts, how many pounds of each type should be mixed? [3.3]

64. *Household Waste.* The Hendersons generate two and a half times as much trash as their neighbors, the Savickis. Together, the two households produce 14 bags of trash each month. How much trash does each household produce? [3.3]

65. *Paid Admissions.* There were 203 tickets sold for a volleyball game. For activity-card holders the price was \$2.50, and for noncard holders the price was \$4. The total amount of money collected was \$620. How many of each type of ticket were sold? [3.3]

66. *Paid Admissions.* There were 200 tickets sold for a women's basketball game. Tickets for students were \$4 each and for adults were \$6 each. The total amount collected was \$1060. How many of each type of ticket were sold? [3.3]

67. *Landscaping.* Grass seed is being spread on a triangular traffic island. If the grass seed can cover an area of 200 ft^2 and the island's base is 16 ft long, how tall a triangle can the seed fill? [1.6]

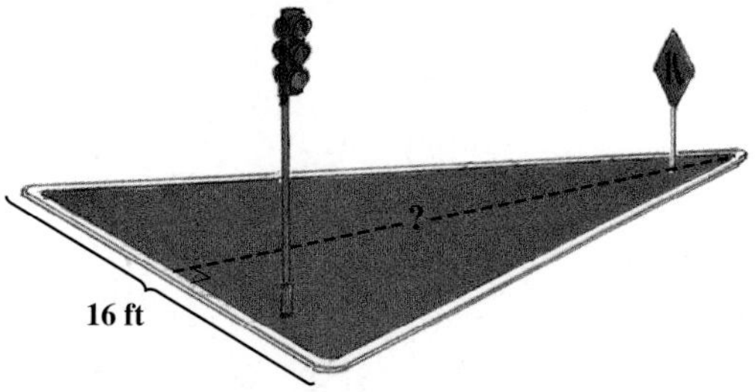

68. *Interest Rate.* What rate of interest is required for a principal of \$1280 to earn \$17.60 in half a year? [1.6]

Synthesis

TW **69.** Explain how a system of linear inequalities could have a solution set containing exactly one pair.

TW **70.** Do all systems of linear inequalities have solutions? Why or why not?

Graph.

71. $x + y > 8,$
$x + y \leq -2$

72. $x + y \geq 1,$
$-x + y \geq 2,$
$x \geq -2,$
$y \geq 2,$
$y \leq 4,$
$x \leq 2$

73. $x - 2y \leq 0,$
$-2x + y \leq 2,$
$x \leq 2,$
$y \leq 2,$
$x + y \leq 4$

74. Write four systems of four inequalities that describe a 2-unit by 2-unit square that has (0, 0) as one of the vertices.

75. *Luggage Size.* Unless an additional fee is paid, most major airlines will not check any luggage for which the sum of the item's length, width, and height exceeds 62 in. The U.S. Postal Service will ship a package only if the sum of the package's length and girth (distance around its midsection) does not exceed 130 in. (*Sources*: U.S. Postal Service; www.case2go.com) Video Promotions is ordering several 30-in. long cases that will be both mailed and checked as luggage. Using w and h for width and height (in inches), respectively, write and graph an inequality that represents all acceptable combinations of width and height.

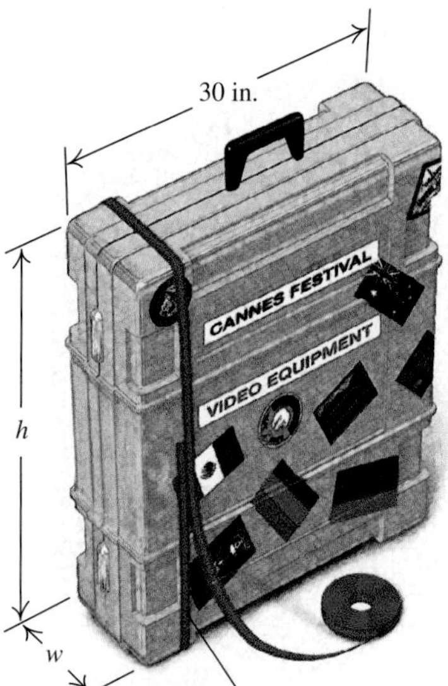

76. *Hockey Wins and Losses.* The Skating Stars figure that they need at least 60 points for the season in order to make the playoffs. A win is worth 2 points and a tie is worth 1 point. Graph a system of inequalities that describes the situation. (*Hint*: Let w = the number of wins and t = the number of ties.)

77. *Elevators.* Many elevators have a capacity of 1 metric ton (1000 kg). Suppose that c children, each weighing 35 kg, and a adults, each 75 kg, are on an elevator. Graph a system of inequalities that indicates when the elevator is overloaded.

78. *Widths of a Basketball Floor.* Sizes of basketball floors vary due to building sizes and other constraints such as cost. The length L is to be at most 94 ft and the width W is to be at most 50 ft. Graph a system of inequalities that describes the possible dimensions of a basketball floor.

Write a system of inequalities for each region shown.

79.

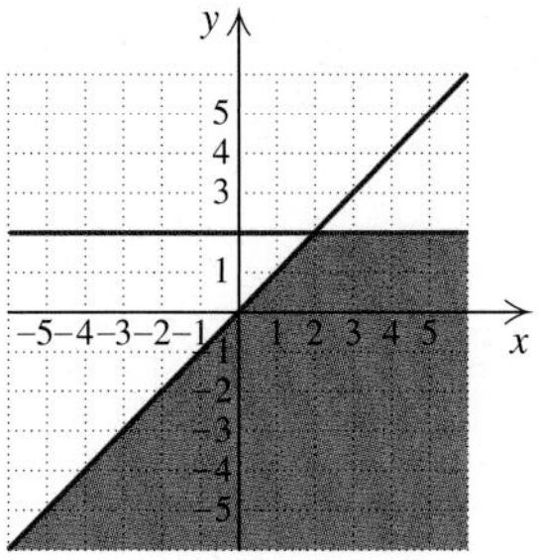

80.

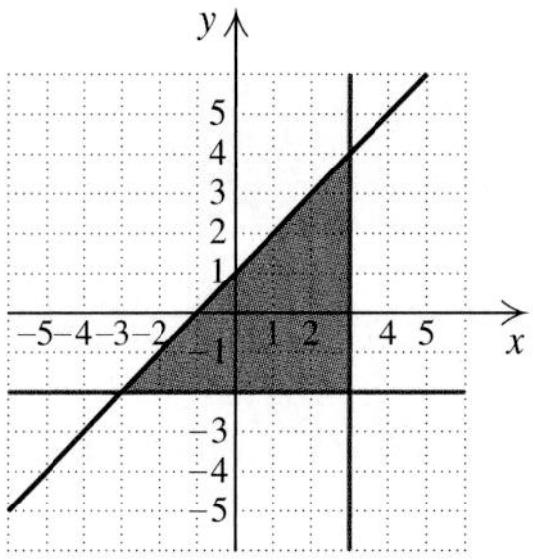

81.

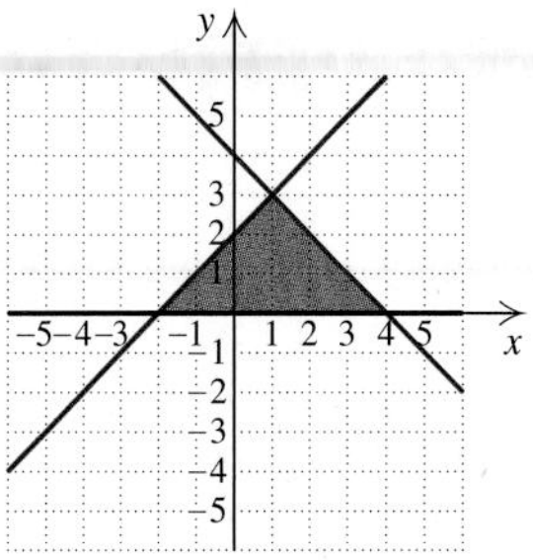

82.

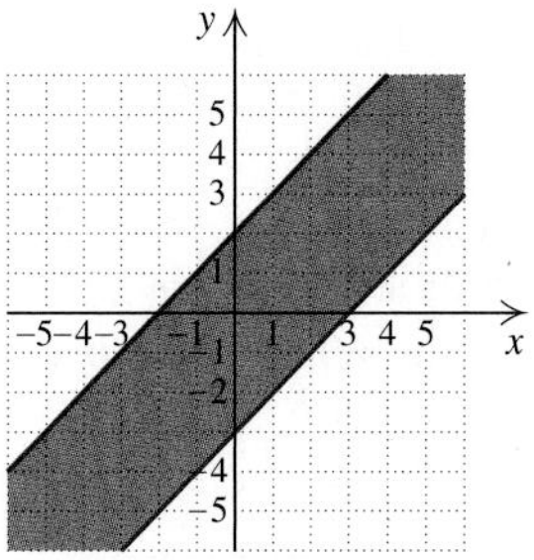

4

Chapter Summary and Review

KEY TERMS AND DEFINITIONS

INEQUALITIES

Inequality, p. 270 Any sentence containing $<$, $>$, $\leq$, $\geq$, or $\neq$.

Compound inequality, p. 298 Two or more inequalities that are combined using the word *and* or the word *or*.

Conjunction, p. 298 Two statements joined by the word *and*. The solution set of a conjunction is the **intersection,** indicated by $\cap$, of the solution sets of the individual statements.

Disjunction. p. 298 Two statements joined by the word *or*. The solution set of a disjunction is the **union,** indicated by $\cup$, of the solution sets of the individual statements.

GRAPHS OF INEQUALITIES

Linear inequality, p. 320 An inequality whose related equation is a linear equation.

Half-plane, p. 321 The region to one side of a boundary line.

Vertices (singular, vertex), p. 327 The points of intersection of boundary lines in a system of inequalities.

IMPORTANT CONCEPTS

[Section references appear in brackets.]

Concept	Example
The Addition Principle for Inequalities For any real numbers a, b, and c: $a < b$ is equivalent to $a + c < b + c$; $a > b$ is equivalent to $a + c > b + c$. Similar statements hold for $\leq$ and $\geq$.	$x + 3 \leq 5$ is equivalent to $x + 3 - 3 \leq 5 - 3$ $x \leq 2$. [4.1]

(*continued*)

The Multiplication Principle for Inequalities

For any real numbers a and b, and for any *positive* number c,

$a < b$ is equivalent to $ac < bc$;

$a > b$ is equivalent to $ac > bc$.

For any real numbers a and b, and for any *negative* number c,

$a < b$ is equivalent to $ac > bc$;

$a > b$ is equivalent to $ac < bc$.

Similar statements hold for $\le$ and $\ge$.

$3x > 9$ is equivalent to

$\frac{1}{3} \cdot 3x > \frac{1}{3} \cdot 9$ **The inequality symbol does not change, because $\frac{1}{3}$ is positive.**

$x > 3$

$-3x > 9$ is equivalent to

$-\frac{1}{3} \cdot (-3x) < -\frac{1}{3} \cdot 9$ **The inequality symbol is reversed, because $-\frac{1}{3}$ is negative.**

$x < -3$. [4.1]

Solution sets of inequalities can be **graphed** and written in **interval notation.**

Interval Notation	Set-builder Notation
(a, b)	$\{x \mid a < x < b\}$
$[a, b]$	$\{x \mid a \le x \le b\}$
$[a, b)$	$\{x \mid a \le x < b\}$
$(a, b]$	$\{x \mid a < x \le b\}$
(a, ∞)	$\{x \mid a < x\}$
$(-\infty, a)$	$\{x \mid x < a\}$

Graph

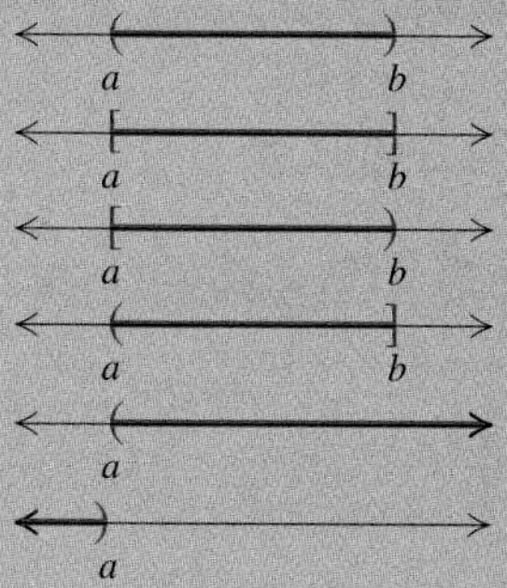

[4.1]

The Absolute-Value Principles for Equations and Inequalities

For any positive number p and any algebraic expression X:

a) The solutions of $|X| = p$ are those numbers that satisfy $X = -p$ *or* $X = p$.
b) The solutions of $|X| < p$ are those numbers that satisfy $-p < X < p$.
c) The solutions of $|X| > p$ are those numbers that satisfy $X < -p$ *or* $p < X$.

If $|X| = 0$, then $X = 0$. If p is negative, then $|X| = p$ and $|X| < p$ have no solution, and any value of X will satisfy $|X| > p$.

Using part (a):

$|x + 3| = 4$

$x + 3 = 4$ *or* $x + 3 = -4$

Using part (b):

$|x + 3| < 4$

$-4 < x + 3 < 4$

Using part (c):

$|x + 3| \ge 4$

$x + 3 \le -4$ *or* $4 \le x + 3$ [4.4]

(*continued*)

Equations and inequalities can be solved graphically using either the **Intersect** or **Zero** method.

Solve graphically: $x + 1 = 2x - 3$.

Intersect method

Graph $y = x + 1$ and $y = 2x - 3$.

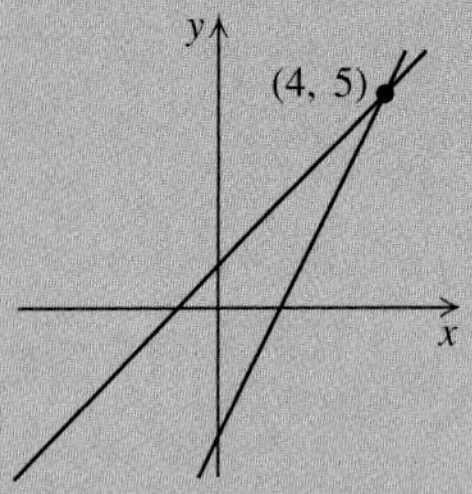

The solution is 4.

Zero method

$x + 1 = 2x - 3$
$0 = x - 4$
Graph $y = x - 4$.

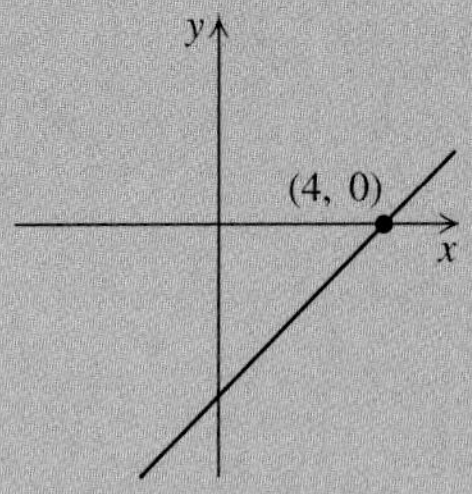

The solution is 4. [4.2]

Review Exercises

Concept Reinforcement Classify each of the following as either true or false.

1. The addition and multiplication principles for inequalities are used to write equivalent inequalities. [4.1]
2. It is always true that if $a > b$, then $ac > bc$. [4.1]
3. To solve an equation graphically using the Zero method, we find the x-intercept of a graph. [4.2]
4. The solution of $|3x - 5| \leq 8$ is a closed interval. [4.4]
5. The inequality $2 < 5x + 1 < 9$ is equivalent to $2 < 5x + 1$ *or* $5x + 1 < 9$. [4.3]
6. The solution set of a disjunction is the union of two solution sets. [4.3]
7. The equation $|x| = -p$ has no solution when p is positive. [4.4]
8. $|f(x)| > 3$ is equivalent to $f(x) < -3$ *or* $f(x) > 3$. [4.4]
9. A test point is used to determine whether the line in a linear inequality is drawn solid or dashed. [4.5]
10. The graph of a system of linear inequalities is always a half-plane. [4.5]

Graph each inequality and write the solution set using both set-builder notation and interval notation. [4.1]

11. $x \leq -2$
12. $a + 7 \leq -14$
13. $y - 5 \geq -12$
14. $4y > -15$
15. $-0.3y < 9$
16. $-6x - 5 < 4$
17. $-\frac{1}{2}x - \frac{1}{4} > \frac{1}{2} - \frac{1}{4}x$
18. $0.3y - 7 < 2.6y + 15$
19. $-2(x - 5) \geq 6(x + 7) - 12$
20. Let $f(x) = 3x - 5$ and $g(x) = 11 - x$. Find all values of x for which $f(x) \leq g(x)$. [4.1]

Solve. [4.1]

21. Rose can choose between two summer jobs. She can work as a checker in a discount store for \$8.40 an hour, or she can mow lawns for \$12.00 an hour. In order to mow lawns, she must buy a \$450 lawnmower. For how many hours must Rose work in order for the mowing to be more profitable than checking?

22. Clay is going to invest \$9000, part at 3% and the rest at 3.5%. What is the most he can invest at 3% and still be guaranteed \$300 in interest each year?

Solve graphically. [4.2]

23. $x - 3 = 3x + 5$

24. $x + 1 \geq \frac{1}{2}x - 2$

25. Find the intersection:

$\{1, 2, 5, 6, 9\} \cap \{1, 3, 5, 9\}$. [4.3]

26. Find the union:

$\{1, 2, 5, 6, 9\} \cup \{1, 3, 5, 9\}$. [4.3]

Graph and write interval notation. [4.3]

27. $x \leq 3$ *and* $x > -5$

28. $x \leq 3$ *or* $x > -5$

Solve and graph each solution set. [4.3]

29. $-4 < x + 8 \leq 5$

30. $-15 < -4x - 5 < 0$

31. $3x < -9$ *or* $-5x < -5$

32. $2x + 5 < -17$ *or* $-4x + 10 \leq 34$

33. $2x + 7 \leq -5$ *or* $x + 7 \geq 15$

34. $f(x) < -5$ *or* $f(x) > 5$, where $f(x) = 3 - 5x$

For f(x) as given, use interval notation to write the domain of f. [4.3]

35. $f(x) = \dfrac{2x}{x - 8}$

36. $f(x) = \sqrt{x + 5}$

37. $f(x) = \sqrt{8 - 3x}$

Solve. [4.4]

38. $|x| = 5$

39. $|t| \geq 3.5$

40. $|x - 3| = 7$

41. $|2x + 5| < 12$

42. $|3x - 4| \geq 15$

43. $|2x + 5| = |x - 9|$

44. $|5n + 6| = -8$

45. $\left|\dfrac{x + 4}{6}\right| \leq 2$

46. $2|x - 5| - 7 > 3$

47. Let $f(x) = |3x - 5|$. Find all x for which $f(x) < 0$. [4.4]

48. Graph $x - 2y \geq 6$ on a plane. [4.5]

Graph each system of inequalities. Find the coordinates of any vertices formed. [4.5]

49. $x + 3y > -1,$
$x + 3y < 4$

50. $x - 3y \leq 3,$
$x + 3y \geq 9,$
$y \leq 6$

Synthesis

TW **51.** Explain in your own words why $|X| = p$ has two solutions when p is positive and no solution when p is negative. [4.4]

TW **52.** Explain why the graph of the solution of a system of linear inequalities is the intersection, not the union, of the individual graphs. [4.5]

53. Solve: $|2x + 5| \leq |x + 3|$. [4.4]

54. Classify as true or false: If $x < 3$, then $x^2 < 9$. If false, give an example showing why. [4.1]

55. Just-For-Fun manufactures marbles with a 1.1-cm diameter and a ±0.03-cm manufacturing tolerance, or allowable variation in diameter. Write the tolerance as an inequality with absolute value. [4.4]

56. The Twinrocker Paper Company makes paper by hand. Each sheet is between 18 thousandths and 25 thousandths of an inch thick. Write the thickness t of a sheet of handmade paper as an inequality with absolute value. [4.4]

Chapter Test 4

Graph each inequality and write the solution set using both set-builder notation and interval notation.

1. $x - 2 < 10$

2. $-0.6y < 30$

3. $-4y - 3 \geq 5$

4. $3a - 5 \leq -2a + 6$

5. $3(7 - x) < 2x + 5$

6. $-8(2x + 3) + 6(4 - 5x) \geq 2(1 - 7x) - 4(4 + 6x)$

7. Let $f(x) = -5x - 1$ and $g(x) = -9x + 3$. Find all values of x for which $f(x) > g(x)$.

8. Lia can rent a van for either \$40 with unlimited mileage or \$30 with 100 free miles and an extra charge of 15¢ for each mile over 100. For what numbers of miles traveled would the unlimited mileage plan save Lia money?

9. A refrigeration repair company charges \$80 for the first half-hour of work and \$60 for each additional hour. Blue Mountain Camp has budgeted \$200 to repair its walk-in cooler. For what lengths of a service call will the budget not be exceeded?

Solve graphically.

10. $2x + 1 = 3x - 2$

11. $x + 2 > \frac{1}{3}x$

12. Find the intersection:

$\{1, 3, 5, 7, 9\} \cap \{3, 5, 11, 13\}$.

13. Find the union:

$\{1, 3, 5, 7, 9\} \cup \{3, 5, 11, 13\}$.

14. Write the domain of f using interval notation if $f(x) = \sqrt{8 - 2x}$.

Solve and graph each solution set.

15. $-2 < x - 3 < 5$

16. $-11 \leq -5t - 2 < 0$

17. $3x - 2 < 7$ *or* $x - 2 > 4$

18. $-3x > 12$ *or* $4x > -10$

19. $-\frac{1}{3} \leq \frac{1}{6}x - 1 < \frac{1}{4}$

20. $|x| = 13$

21. $|a| > 7$

22. $|3x - 1| < 7$

23. $|-5t - 3| \geq 10$

24. $|2 - 5x| = -12$

25. $g(x) < -3$ *or* $g(x) > 3$, where $g(x) = 4 - 2x$

26. Let $f(x) = |x + 10|$ and $g(x) = |x - 12|$. Find all values of x for which $f(x) = g(x)$.

Graph each system of inequalities. Find the coordinates of any vertices formed.

27. $x + y \geq 3,$
$x - y \geq 5$

28. $2y - x \geq -7,$
$2y + 3x \leq 15,$
$y \leq 0,$
$x \leq 0$

Synthesis

Solve. Write the solution set using interval notation.

29. $|2x - 5| \leq 7$ *and* $|x - 2| \geq 2$

30. $7x < 8 - 3x < 6 + 7x$

31. Write an absolute-value inequality for which the interval shown is the solution.

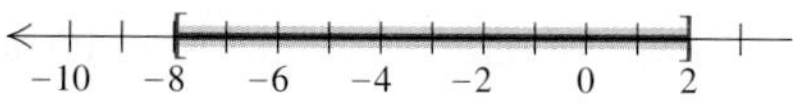

5

Polynomials and Polynomial Functions

We have already used polynomials, like $2x + 5$ or $x^2 - 1$, in this text. In many ways, polynomials are central to the study of algebra. In this chapter, we clearly define polynomials, learn how to manipulate them, and examine polynomial functions both algebraically and graphically.

APPLICATION *College Tuition.*

The average in-state tuition for a public four-year college for various years is shown in the following table and graph. Fit a quartic polynomial function to the data, and use the function to predict public-college tuition in 2008.

YEAR	IN-STATE PUBLIC COLLEGE TUITION
1985	\$1386
1988	1726
1991	2159
1994	2820
1997	3323
2000	3768
2003	4686
2005*	5948

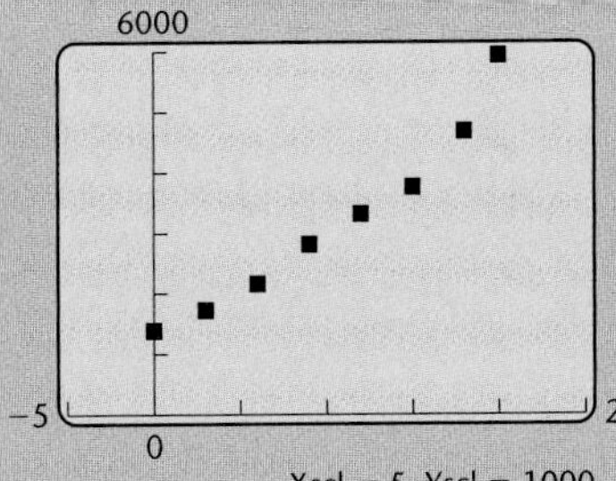

*Preliminary data

Source: National Center for Education Statistics, *Digest of Education Statistics*

This problem appears as Example 4 in Section 5.8.

5.1 Introduction to Polynomials and Polynomial Functions

Algebraic Expressions and Polynomials ■ Polynomial Functions ■ Adding Polynomials ■ Opposites and Subtraction

We now begin the study of *polynomials* and *polynomial functions.* Earlier in this text, we examined several other types of functions.

Connecting the Concepts

FAMILIES OF GRAPHS OF FUNCTIONS

Often, the shape of a graph of a function can be predicted by examining the equation describing the function.

We examined *linear functions* and their graphs in Chapter 2. Functions f that can be expressed in the form $f(x) = mx + b$ have graphs that are straight lines. Their slope, or rate of change, is constant. We can think of linear functions as a "family" of functions, with the function $f(x) = x$ being the simplest such function.

We have also studied some nonlinear functions. In Chapter 4, we solved equations and inequalities containing absolute-value symbols by examining graphs of absolute-value functions. An *absolute-value function* of the form $f(x) = |ax + b|$, where $a \neq 0$, has a graph similar to that of $f(x) = |x|$.

Another nonlinear function that we considered in Chapter 4 when examining domains is a square-root function. A *square-root function* of the form $f(x) = \sqrt{ax + b}$, where $a \neq 0$, has a graph similar to the graph of $f(x) = \sqrt{x}$.

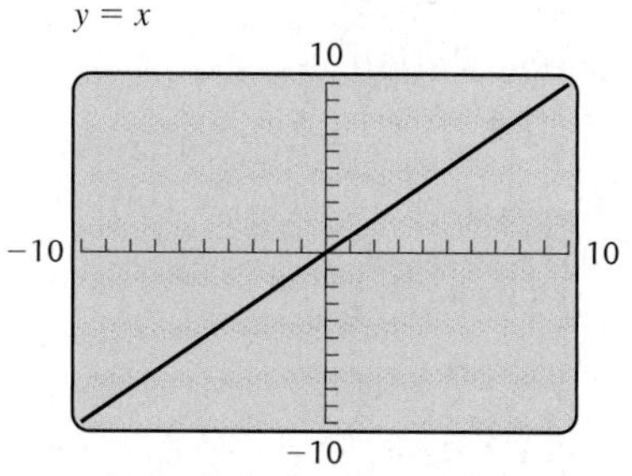

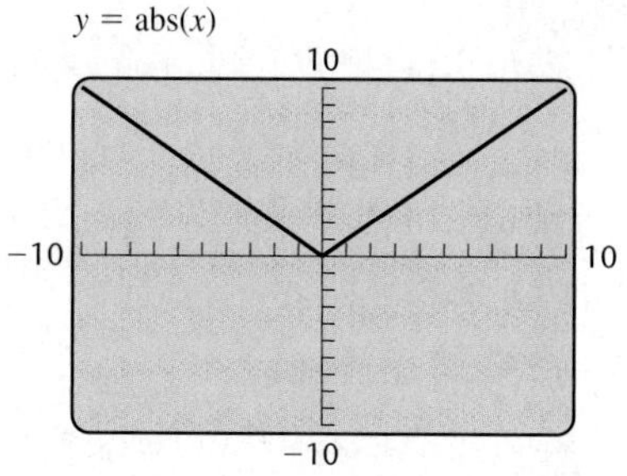

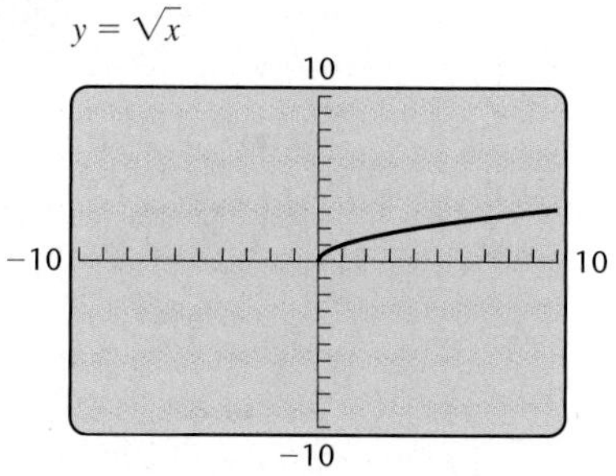

These basic functions, along with others that we will develop throughout the text, form a **library of functions.** Knowing the general shape of the graphs of different types of functions will help in analyzing both data and graphs.

In this chapter, we look at polynomial functions. The graphs of polynomial functions have certain common characteristics.

Interactive Discovery

The following table lists some polynomial functions and some nonpolynomial functions. Graph each function and compare the graphs.

Polynomial Functions	Nonpolynomial Functions
$f(x) = x^2 + 3x + 5$	$f(x) = \|x - 4\|$
$f(x) = 4$	$f(x) = 1 + \sqrt{2x - 5}$
$f(x) = -0.5x^4 + 5x - 2.3$	$f(x) = \frac{x - 7}{2x}$

What are some characteristics of graphs of polynomial functions?

Study Tip

Keep Your Focus

When studying with someone else, it can be tempting to prolong conversation that has little to do with mathematics. If you see that this may happen, explain to your partner(s) that you enjoy the nonmathematical conversation, but would enjoy it more later—after the math work has been completed.

You may have noticed the following:

The graph of a polynomial function is "smooth," that is, there are no sharp corners.

The graph of a polynomial function is continuous, that is, there are no holes or breaks.

The domain of a polynomial function, unless otherwise specified, is all real numbers.

There are other characteristics of polynomial functions, but before discussing them, we need to establish some terminology.

Algebraic Expressions and Polynomials

In Chapter 1, we introduced algebraic expressions like

$$5x^2, \quad \frac{3}{x^2 + 5}, \quad 9a^3b^4, \quad 3x^{-2}, \quad 6x^2 + 3x + 1, \quad -9, \quad \text{and} \quad 5 - 2x.$$

Of the expressions listed, $5x^2$, $9a^3b^4$, $3x^{-2}$, and -9 are examples of *terms.* A **term** is simply a number or a variable raised to a power or a product of numbers and variables raised to powers. The power used may be 1 and thus not written.

When all variables in a term are raised to whole-number powers, the term is a **monomial.** Of the terms listed above, $5x^2$, $9a^3b^4$, and -9 are monomials. Since -2 is not a whole number, $3x^{-2}$ is *not* a monomial. The **degree** of a monomial is the sum of the exponents of the variables. Thus, $5x^2$ has degree 2 and $9a^3b^4$ has degree 7. Nonzero constant terms, like -9, can be written $-9x^0$ and therefore have degree 0. The term 0 itself is said to have no degree.

The number 5 is said to be the **coefficient** of $5x^2$. Thus the coefficient of $9a^3b^4$ is 9 and the coefficient of $-2x$ is -2. The coefficient of a constant term is just that constant.

A **polynomial** is a monomial or a sum of monomials. Of the expressions listed, $5x^2$, $9a^3b^4$, $6x^2 + 3x + 1$, -9, and $5 - 2x$ are polynomials. In fact, with the exception of $9a^3b^4$, these are all polynomials *in one variable.* The expression $9a^3b^4$ is a *polynomial in two variables.* Note that $5 - 2x$ is the sum of 5 and $-2x$. Thus, 5 and $-2x$ are the terms in the polynomial $5 - 2x$.

The **leading term** of a polynomial is the term of highest degree. Its coefficient is called the **leading coefficient.** The **degree of a polynomial** is the same as the degree of its leading term.

EXAMPLE 1 For each polynomial given, find the degree of each term, the degree of the polynomial, the leading term, and the leading coefficient.

a) $2x^3 + 8x^2 - 17x - 3$

b) $6x^2 + 8x^2y^3 - 17xy - 24xy^2z^4 + 2y + 3$

SOLUTION

a) $2x^3 + 8x^2 - 17x - 3$ **b)** $6x^2 + 8x^2y^3 - 17xy - 24xy^2z^4 + 2y + 3$

Term	$2x^3$	$8x^2$	$-17x$	-3	$6x^2$	$8x^2y^3$	$-17xy$	$-24xy^2z^4$	$2y$	3
Degree	3	2	1	0	2	5	2	7	1	0
Leading Term	$2x^3$				$-24xy^2z^4$					
Leading Coefficient	2				-24					
Degree of Polynomial	3				7					

A polynomial of degree 0 or 1 is called **linear.** A polynomial in one variable is said to be **quadratic** if it is of degree 2, **cubic** if it is of degree 3, and **quartic** if it is of degree 4.

The following are some names for certain kinds of polynomials.

Type	**Definition**	**Examples**					
Monomial	A polynomial of one term	4	$-3p$	$5x^2$	$-7a^2b^3$	0	xyz
Binomial	A polynomial of two terms	$2x + 7$	$a - 3b$	$5x^2 + 7y^3$			
Trinomial	A polynomial of three terms	$x^2 - 7x + 12$	$4a^2 + 2ab + b^2$				

We generally arrange polynomials in one variable so that the exponents *decrease* from left to right. This is called **descending order.** Some polynomials may be written with exponents *increasing* from left to right, which is **ascending order.** Generally, if an exercise is written in one kind of order, the answer is written in that same order.

EXAMPLE 2 Arrange in ascending order: $12 + 2x^3 - 7x + x^2$.

SOLUTION

$$12 + 2x^3 - 7x + x^2 = 12 - 7x + x^2 + 2x^3$$

Polynomials in several variables can be arranged with respect to the powers of one of the variables.

EXAMPLE 3 Arrange in descending powers of x:

$$y^4 + 2 - 5x^2 + 3x^3y + 7xy^2.$$

SOLUTION

$$y^4 + 2 - 5x^2 + 3x^3y + 7xy^2 = 3x^3y - 5x^2 + 7xy^2 + y^4 + 2$$

Polynomial Functions

A *polynomial function* is a function in which ordered pairs are determined by evaluating a polynomial. For example, the function P given by

$$P(x) = 5x^7 + 3x^5 - 4x^2 - 5$$

is an example of a polynomial function. To evaluate a polynomial function, we substitute a number for the variable just as in Chapter 2. In this text, we limit ourselves to polynomial functions in one variable.

EXAMPLE 4 Find $P(-5)$ for the polynomial function given by $P(x) = -x^2 + 4x - 1$.

SOLUTION We use several methods to evaluate the function.

Using algebraic substitution. We substitute -5 for x and carry out the operations using the rules for order of operations:

$$\begin{aligned} P(-5) &= -(-5)^2 + 4(-5) - 1 \\ &= -25 - 20 - 1 \\ &= -46. \end{aligned}$$

CAUTION! Note that $-(-5)^2 = -25$. We square the input first and then take its opposite.

Using function notation on a graphing calculator. We let $y_1 = -x^2 + 4x - 1$. We enter the function into the graphing calculator and evaluate $Y_1(-5)$ using the Y-VARS menu.

```
Y1(-5)
                -46
```

Using a table. If $y_1 = -x^2 + 4x - 1$, we can find the value of y_1 for $x = -5$ by setting Indpnt to Ask in the TABLE SETUP. If Depend is set to Auto, the y-value will appear when the x-value is entered. If Depend is set to Ask, we position the cursor in the y-column and press ENTER to see the corresponding y-value.

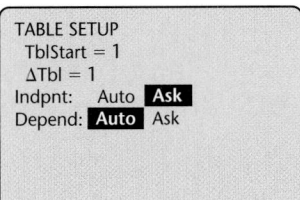

```
TABLE SETUP
 TblStart = 1
 ΔTbl = 1
Indpnt:  Auto [Ask]
Depend: [Auto] Ask
```

X	Y1	
−5	−46	

X =

Using the graph of the function. We graph $y_1 = -x^2 + 4x - 1$ and find the value of y_1 for $x = -5$ using VALUE in the CALC menu.

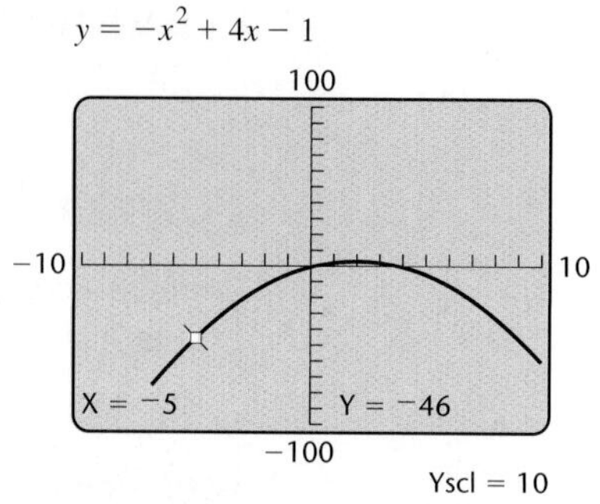

No matter which method we use, we have $P(-5) = -46$.

EXAMPLE 5 Medicine. Ibuprofen is a medication used to relieve pain. The polynomial function

$$M(t) = 0.5t^4 + 3.45t^3 - 96.65t^2 + 347.7t, \quad 0 \le t \le 6$$

can be used to estimate the number of milligrams of ibuprofen in the bloodstream t hours after 400 mg of the medication has been swallowed (*Source*: Based on data from Dr. P. Carey, Burlington, VT).

a) How many milligrams of ibuprofen are in the bloodstream 2 hr after 400 mg has been swallowed?

b) Use the graph below to estimate $M(4)$.

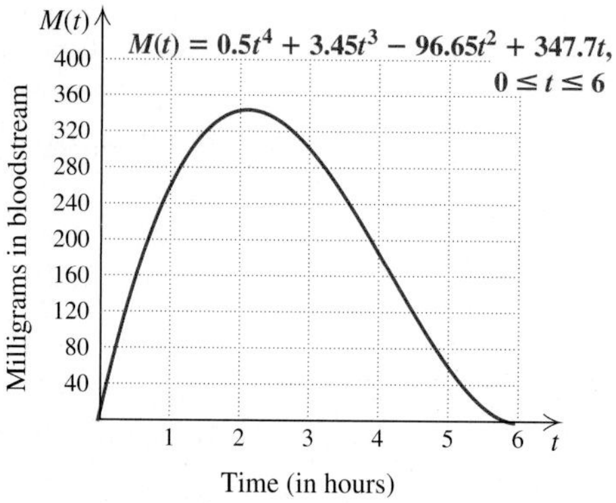

SOLUTION

a) We evaluate the function for $t = 2$:

$$\begin{aligned} M(2) &= 0.5(2)^4 + 3.45(2)^3 - 96.65(2)^2 + 347.7(2) \\ &= 0.5(16) + 3.45(8) - 96.65(4) + 347.7(2) \\ &= 8 + 27.6 - 386.6 + 695.4 \\ &= 344.4. \end{aligned}$$

We carry out the calculation using the rules for order of operations.

Approximately 344 mg of ibuprofen is in the bloodstream 2 hr after 400 mg has been swallowed.

b) To estimate $M(4)$, the amount in the bloodstream after 4 hr, we locate 4 on the horizontal axis. From there we move vertically to the graph of the function and then horizontally to the $M(t)$-axis, as shown below. This locates a value of about $M(4) \approx 190$.

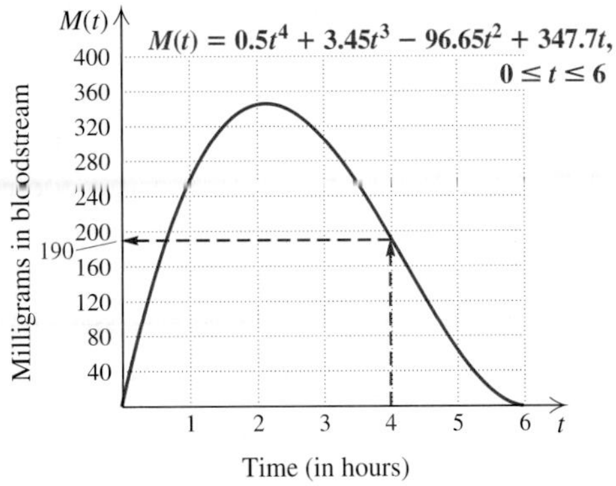

After 4 hr, approximately 190 mg of ibuprofen is still in a person's bloodstream, assuming an original dosage of 400 mg.

Recall from Section 2.5 that the domain of a function, when not specified, is the set of all real numbers for which the function is defined. If no restrictions are given, a polynomial function is defined for all real numbers. In other words,

The domain of a polynomial function is $(-\infty, \infty)$.

Note in Example 5 that the domain of the function M is restricted by the context of the problem. Since ibuprofen cannot have a negative concentration, values of t that make M negative are not in the domain of the function for this problem. Thus the domain of M, as estimated from its graph, is approximately $[0, 6]$.

For a polynomial function with an unrestricted domain, the range may be $(-\infty, \infty)$, or there may be a maximum or a minimum value of the function. We can estimate the range of a function from its graph. In Example 5, the values of M vary from a minimum of 0 to a maximum of approximately 345. Thus we estimate that the range of M is about $[0, 345]$.

EXAMPLE 6 Estimate the range of each of the following functions from its graph.

a)

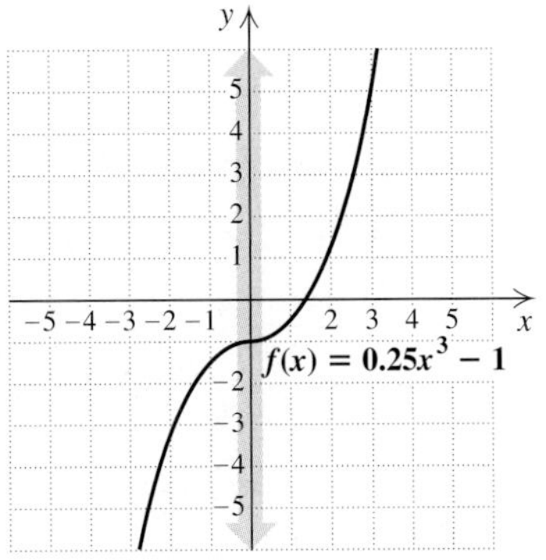

b)

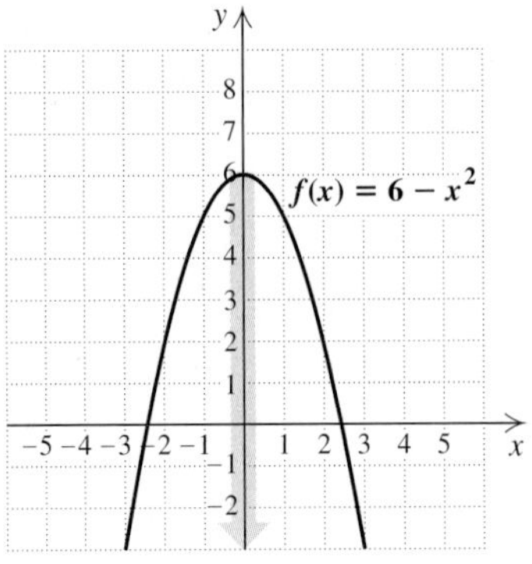

c)

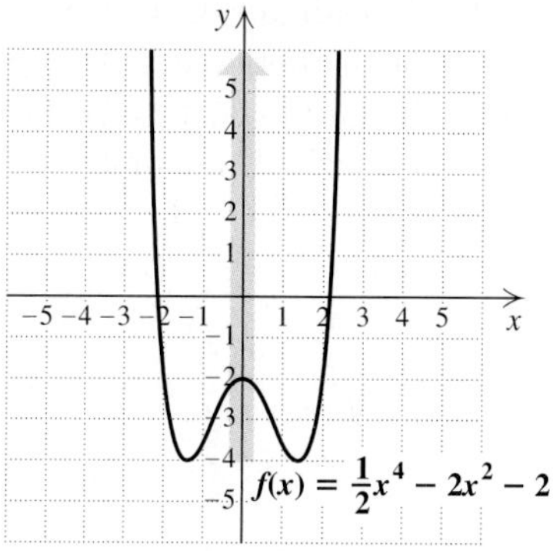

SOLUTION The domain of each function represented in the graphs is $(-\infty, \infty)$.

a) Since there is no maximum or minimum value indicated on the graph of the first function, we estimate its range to be $(-\infty, \infty)$. The range is indicated by the shading on the y-axis.

b) The second function has a maximum value of about 6, and no minimum is indicated, so its range is about $(-\infty, 6]$.

c) The third function has a minimum value of about -4 and no maximum value, so its range is $[-4, \infty)$.

When we are estimating the range of a polynomial function from its graph, care must be taken to show enough of the graph to see its behavior. As you learn more about graphs of polynomial functions, you will be able to more readily choose appropriate scales or viewing windows. In this chapter, we will supply the graph or an appropriate viewing window for the graph of a polynomial function.

EXAMPLE 7 Graph each of the following functions in the standard viewing window and estimate the range of the function.

a) $f(x) = x^3 - 4x^2 + 5$ **b)** $g(x) = x^4 - 4x^2 + 5$

SOLUTION

a) The graph of $y = x^3 - 4x^2 + 5$ is shown on the left below. It appears that the graph extends downward and upward indefinitely, so the range of the function is $(-\infty, \infty)$.

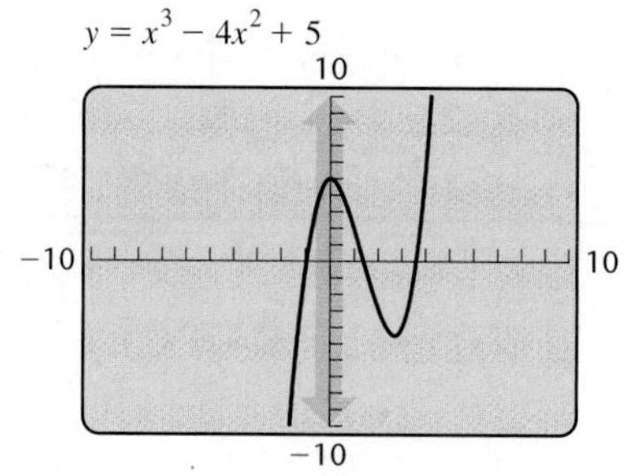

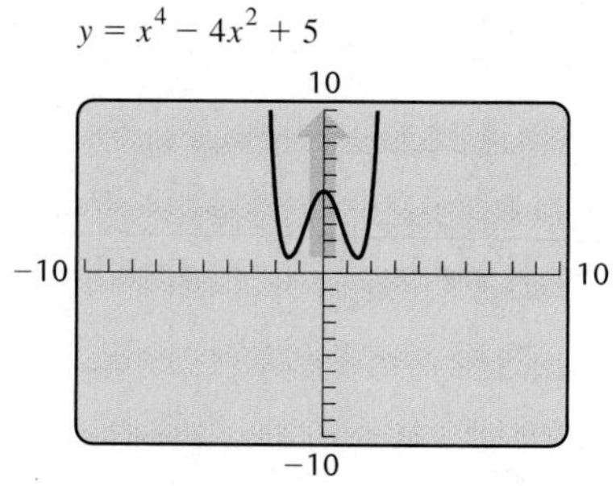

b) The graph of $y = x^4 - 4x^2 + 5$ is shown on the right at the bottom of the preceding page. The graph extends upward indefinitely but does not go below a y-value of 1. (This can be confirmed by tracing along the graph.) Since there is a minimum function value of 1 and no maximum function value, the range of the function is $[1, \infty)$.

Adding Polynomials

Recall from Section 1.3 that when two terms have the same variable(s) raised to the same power(s), they are **similar,** or **like, terms** and can be "combined" or "collected."

EXAMPLE 8 Combine like terms.

a) $3x^2 - 4y + 2x^2$

b) $9x^3 + 5x - 4x^2 - 2x^3 + 5x^2$

c) $3x^2y + 5xy^2 - 3x^2y - xy^2$

SOLUTION

a) $3x^2 - 4y + 2x^2 = 3x^2 + 2x^2 - 4y$ **Rearranging terms using the commutative law for addition**

$= (3 + 2)x^2 - 4y$ **Using the distributive law**

$= 5x^2 - 4y$

b) $9x^3 + 5x - 4x^2 - 2x^3 + 5x^2 = 7x^3 + x^2 + 5x$ **We usually perform the middle steps mentally and write just the answer.**

c) $3x^2y + 5xy^2 - 3x^2y - xy^2 = 4xy^2$

The sum of two polynomials can be found by writing a plus sign between them and then combining like terms.

EXAMPLE 9 Add: $(-3x^3 + 2x - 4) + (4x^3 + 3x^2 + 2)$.

SOLUTION

$$(-3x^3 + 2x - 4) + (4x^3 + 3x^2 + 2) = x^3 + 3x^2 + 2x - 2$$

Checking

A graphing calculator can be used to check whether two algebraic expressions given by y_1 and y_2 are equivalent. Some of the methods listed below have been described before; here they are listed together as a summary.

1. *Comparing graphs.* Graph both y_1 and y_2 on the same set of axes using the SEQUENTIAL mode. If the expressions are equivalent, the graphs will be identical. Using the PATH graphstyle for y_2 allows us to see if the graph of y_2 is being traced over the graph of y_1. Two different graphs may appear identical in certain windows, so this provides only a partial check.

(continued)

2. *Subtracting expressions.* Let $y_3 = y_1 - y_2$. If the expressions are equivalent, the graph of y_3 will be $y = 0$, or the x-axis. Using the PATH graphstyle or TRACE allows us to see if the graph of y_3 is the x-axis.
3. *Comparing values.* Viewing a table allows us to compare values of y_1 and y_2. If the expressions are equivalent, the values will be the same for any given x-value. Again, this is only a partial check, but it becomes more certain as we use more x-values.
4. *Comparing both graphs and values.* Many graphing calculators will split the viewing screen and show both a graph and a table. This choice is usually made using MODE. In the bottom line of the screen on the left below, the FULL mode indicates a full-screen table or graph. In the HORIZ mode, the screen is split in half horizontally, and in the G-T mode, the screen is split vertically. In the screen in the middle below, a graph and a table of values are shown for $y = x^2 - 1$ using the HORIZ mode. The screen on the right shows the same graph and table in the G-T mode.

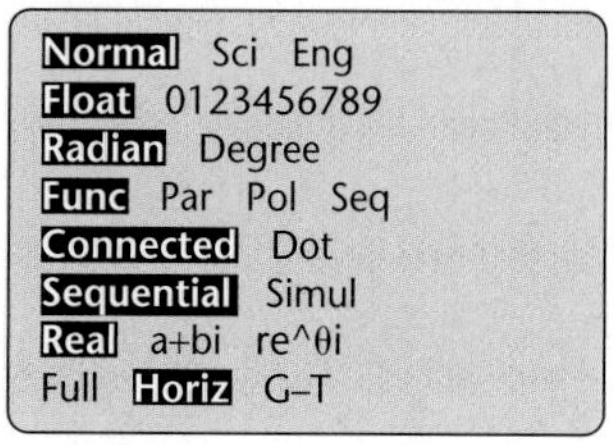

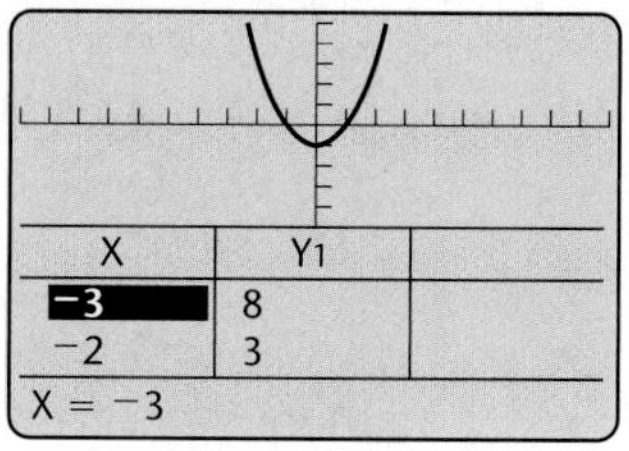

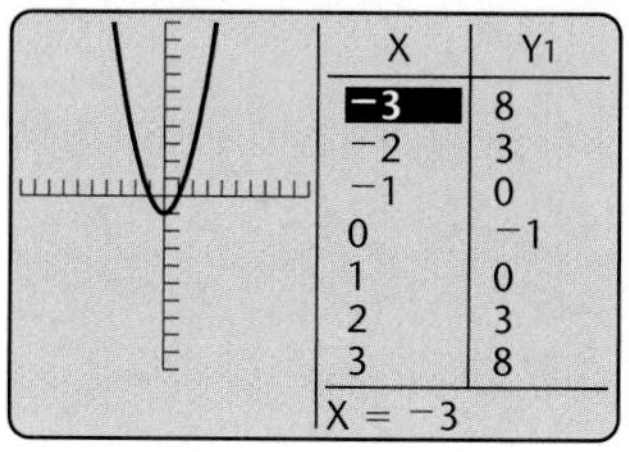

Interactive Discovery

In Example 9, we added polynomials to obtain the identity

$$(-3x^3 + 2x - 4) + (4x^3 + 3x^2 + 2) = x^3 + 3x^2 + 2x - 2.$$

Let $y_1 = (-3x^3 + 2x - 4) + (4x^3 + 3x^2 + 2)$

and $y_2 = x^3 + 3x^2 + 2x - 2.$

1. If the addition is correct, how should the graphs of y_1 and y_2 compare? Check the addition by graphing y_1 and y_2.
2. Now let $y_3 = y_2 - y_1$. If our addition is correct, what will be the graph of y_3? Check Example 9 by graphing y_3.
3. As a third check, compare the values of y_1 and y_2 using a table.

Now check the addition

$$(-2x^4 + 3x^2 + 5) + (6x^4 - 2x - 8) = 4x^4 + 3x^2 - 2x.$$

Let $y_1 = (-2x^4 + 3x^2 + 5) + (6x^4 - 2x - 8)$

and $y_2 = 4x^4 + 3x^2 - 2x.$

4. Graph y_1 and y_2 using a viewing window of $[-10, 10, -100, 100]$. Do the graphs appear to coincide?

5. Use a screen split horizontally to compare the graphs with a table of values. Are the values of y_1 and y_2 the same for any given x-value?

6. Graph $y_3 = y_1 - y_2$ using a standard viewing window. Is the graph of y_3 the x-axis? Use a screen split vertically to compare a table of values with the graph. What advantages does this check have over graphing y_1 and y_2?

7. Is the addition $(-2x^4 + 3x^2 + 5) + (6x^4 - 2x - 8) = 4x^4 + 3x^2 - 2x$ correct?

Graphs and tables of values provide good checks of the results of algebraic manipulation. When checking by graphing, remember that two different graphs might not appear different in some viewing windows. It is also possible for two different functions to have the same value for several entries in a table. However, when enough values of two polynomial functions are the same, a table does provide a foolproof check.

If the values of two nth-degree polynomial functions are the same for $n + 1$ or more values, then the polynomials are equivalent.

Using columns for addition is sometimes helpful. To do so, we write the polynomials one under the other, listing like terms under one another and leaving spaces for missing terms.

EXAMPLE 10 Add: $4x^3 + 4x - 5$ and $-x^3 + 7x^2 - 2$. Check using a table of values.

SOLUTION We have

$$\begin{array}{r} 4x^3 + 4x - 5 \\ -x^3 + 7x^2 - 2 \\ \hline 3x^3 + 7x^2 + 4x - 7 \end{array}$$

X	Y1	Y2
0	−7	−7
1	7	7
2	53	53
3	149	149
4	313	313
5	563	563
6	917	917

X = 0

To check, we let $y_1 = (4x^3 + 4x - 5) + (-x^3 + 7x^2 - 2)$ and $y_2 = 3x^3 + 7x^2 + 4x - 7$. Because the polynomials are of degree 3, we need to show that $y_1 = y_2$ for four different x-values. This can be accomplished quickly by creating a table of values, as shown at left. The answer checks.

Note that if a graphing calculator were not available, we would need to show, manually, that $y_1 = y_2$ for four different x-values.

EXAMPLE 11 Add: $13x^3y + 3x^2y - 5y$ and $x^3y + 4x^2y - 3xy$.

SOLUTION

$$(13x^3y + 3x^2y - 5y) + (x^3y + 4x^2y - 3xy) = 14x^3y + 7x^2y - 3xy - 5y$$

Opposites and Subtraction

If the sum of two polynomials is 0, the polynomials are *opposites*, or *additive inverses*, of each other. For example,

$$(3x^2 - 5x + 2) + (-3x^2 + 5x - 2) = 0,$$

so the opposite of $(3x^2 - 5x + 2)$ must be $(-3x^2 + 5x - 2)$. We can say the same thing using algebraic symbolism, as follows:

$$\underbrace{\text{The opposite of}}_{-}\quad \underbrace{(3x^2 - 5x + 2)}_{(3x^2 - 5x + 2)}\quad \underbrace{\text{is}}_{=}\quad \underbrace{(-3x^2 + 5x - 2).}_{-3x^2 + 5x - 2}$$

To form the opposite of a polynomial, we can think of distributing the "−" sign, or multiplying each term of the polynomial by -1, and removing the parentheses. The effect is to change the sign of each term in the polynomial.

The Opposite of a Polynomial The *opposite* of a polynomial P can be written as $-P$ or, equivalently, by replacing each term with its opposite.

EXAMPLE 12 Write two equivalent expressions for the opposite of

$$7xy^2 - 6xy - 4y + 3.$$

SOLUTION

a) The opposite of $7xy^2 - 6xy - 4y + 3$ can be written as

$$-(7xy^2 - 6xy - 4y + 3). \qquad \textbf{Writing the opposite of } P \textbf{ as } -P$$

b) The opposite of $7xy^2 - 6xy - 4y + 3$ can also be written as

$$-7xy^2 + 6xy + 4y - 3. \qquad \textbf{Multiplying each term by } -1$$

To subtract a polynomial, we add its opposite.

EXAMPLE 13 Subtract: $(-5x^2 + 4) - (2x^2 + 3x - 1)$.

SOLUTION We have

$$\begin{aligned}
&(-5x^2 + 4) - (2x^2 + 3x - 1) \\
&\quad = (-5x^2 + 4) + (-2x^2 - 3x + 1) && \textbf{Adding the opposite} \\
&\quad = -5x^2 + 4 - 2x^2 - 3x + 1 && \textbf{Try to go directly to this step.} \\
&\quad = -7x^2 - 3x + 5. && \textbf{Combining like terms}
\end{aligned}$$

We can check the result using graphs or a table of values.

With practice, you will find that you can skip more steps, by mentally taking the opposite of each term and then combining like terms.

To use columns for subtraction, we mentally change the signs of the terms being subtracted.

EXAMPLE 14 Subtract:

$$(4x^2y - 6x^3y^2 + x^2y^2) - (4x^2y + x^3y^2 + 3x^2y^3 - 8x^2y^2).$$

SOLUTION

Write: (Subtract)

$$\begin{array}{l} 4x^2y - 6x^3y^2 \qquad\qquad + x^2y^2 \\ \underline{-(4x^2y + x^3y^2 + 3x^2y^3 - 8x^2y^2)} \end{array}$$

Think: (Add)

$$\begin{array}{l} 4x^2y - 6x^3y^2 \qquad\qquad\quad + x^2y^2 \\ \underline{-4x^2y - x^3y^2 - 3x^2y^3 + 8x^2y^2} \\ \qquad\quad -7x^3y^2 - 3x^2y^3 + 9x^2y^2 \end{array}$$

Take the opposite of each term mentally and add.

5.1 EXERCISE SET

FOR EXTRA HELP

MathXL | MyMathLab | InterAct Math | AW Math Tutor Center | Video Lectures on CD: Disc 3 | Student's Solutions Manual

Concept Reinforcement *In each of Exercises 1–10, match the item with the best example of that item from the column on the right.*

1. ____ A binomial
2. ____ A trinomial
3. ____ A monomial
4. ____ A third-degree polynomial
5. ____ A polynomial written in ascending powers of t
6. ____ A term that is not a monomial
7. ____ A polynomial with a leading term of degree 5
8. ____ A polynomial with a leading coefficient of 5
9. ____ A polynomial in three variables
10. ____ A polynomial containing similar terms

a) $9a^7$

b) $6s^2 - 2t + 4st^2 - st^3$

c) $4t^{-2}$

d) $t^4 - st + s^3$

e) $7t^3 - 13 + 5t^4 - 2t$

f) $4x + 6xy - 3z + 2y^2z^2$

g) $5 + a$

h) $4t^3 + 7t - 8t^2 + 5$

i) $8st^3 - 6s^2t + 4st^3 - 2$

j) $7s^3t^2 - 4s^2t + 3st^2 + 1$

Determine the degree of each term and the degree of the polynomial.

11. $-6x^5 - 8x^3 + x^2 + 3x - 4$

12. $t^3 - 5t^2 + 2t + 7$

13. $y^3 + 2y^7 + x^2y^4 - 8$

14. $-2u^2 + 3v^5 - u^3v^4 - 7$

15. $a^5 + 4a^2b^4 + 6ab + 4a - 3$

16. $8p^6 + 2p^4t^4 - 7p^3t + 5p^2 - 14$

Arrange in descending order. Then find the leading term and the leading coefficient.

17. $3 - 5y + 6y^2 + 11y^3 - 18y^4$

18. $9 - 3x - 10x^4 + 7x^2$

19. $a + 5a^3 - a^7 - 19a^2 + 8a^5$

20. $a^3 - 7 + 11a^4 + a^9 - 5a^2$

Arrange in ascending powers of x.

21. $6x - 9 + 3x^4 - 5x^2$

22. $-3x^4 + 4x - x^3 + 9$

23. $5x^2y^2 - 9xy + 8x^3y^2 - 5x^4$

24. $4ax - 7ab + 4x^6 - 7ax^2$

Evaluate each polynomial for $x = 4$.

25. $-7x + 5$

26. $4x - 13$

27. $x^3 - 5x^2 + x$

28. $7 - x + 3x^2$

Evaluate each polynomial function for $x = -1$.

29. $f(x) = -5x^3 + 3x^2 - 4x - 3$

30. $g(x) = -4x^3 + 2x^2 + 5x - 7$

Find the specified function values.

31. Find $F(2)$ and $F(5)$: $F(x) = 2x^2 - 6x - 9$.

32. Find $P(4)$ and $P(0)$: $P(x) = 3x^2 - 2x + 7$.

33. Find $Q(-3)$ and $Q(0)$:

$$Q(y) = -8y^3 + 7y^2 - 4y - 9.$$

34. Find $G(3)$ and $G(-1)$:

$$G(x) = -6x^2 - 5x + x^3 - 1.$$

35. *Freestyle Skiing.* During the 2004 Winter X-Games, Simon Dumont established a world record by launching himself 22 ft above the edge of a halfpipe. His speed $v(t)$, in miles per hour, upon re-entering the pipe can be approximated by $v(t) = 10.9t$, where t is the number of seconds for which he was airborne. Dumont was airborne for approximately 2.34 sec. (*Source*: Based on data from EXPN.com) How fast was he going when he re-entered the halfpipe?

36. *Cliff Diving.* The speed $v(t)$, in miles per hour, at which a swimmer enters the water can be approximated by $v(t) = 21.82t$, where t is the number of seconds the diver is falling. Cliff divers in Acapulco, Mexico, often are in the air for a total of 5 sec. (*Source*: Based on information from www.guinnessworldrecords.com) Estimate the speed of such a cliff diver when he or she enters the water.

37. *Skateboarding.* The distance $s(t)$, in feet, traveled by a body falling freely from rest in t seconds is approximated by

$$s(t) = 16t^2.$$

In 2002, Adil Dyani set a record for the highest skateboard drop into a quarterpipe. Dyani was airborne for about 1.36 sec. (*Source*: Based on data from *Guinness Book of World Records* 2004) How far did he drop?

38. A paintbrush falls from a scaffold and takes 3 sec to hit the ground (see Exercise 37). How high is the scaffold?

Electing Officers. For a club consisting of p people, the number of ways N in which a president, vice president, and treasurer can be elected can be determined using the function given by

$$N(p) = p^3 - 3p^2 + 2p.$$

39. The Southside Rugby Club has 20 members. In how many ways can they elect a president, vice president, and treasurer?

40. The Stage Right drama club has 12 members. In how many ways can a president, vice president, and treasurer be elected?

NASCAR Attendance. Attendance at NASCAR auto races has grown rapidly over a 10-yr span. Attendance A, in millions, can be approximated by the polynomial function given by

$$A(x) = 0.0024x^3 - 0.005x^2 + 0.31x + 3,$$

where x is the number of years since 1989. Use the following graph for Exercises 41–44.

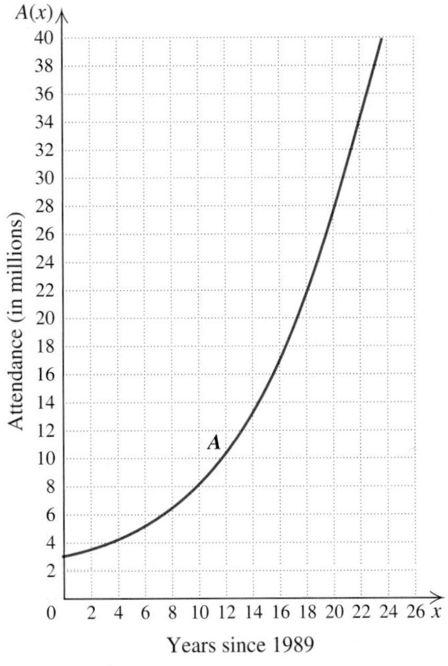

Source: NASCAR

41. Estimate the attendance at NASCAR races in 2009.

42. Estimate the attendance at NASCAR races in 2006.

43. Approximate $A(8)$.

44. Approximate $A(12)$.

45. *Stacking Spheres.* In 2004, the journal *Annals of Mathematics* accepted a proof of the so-called Kepler Conjecture: that the most efficient way to pack spheres is in the shape of a square pyramid (*Source*: *The New York Times* 4/6/04).

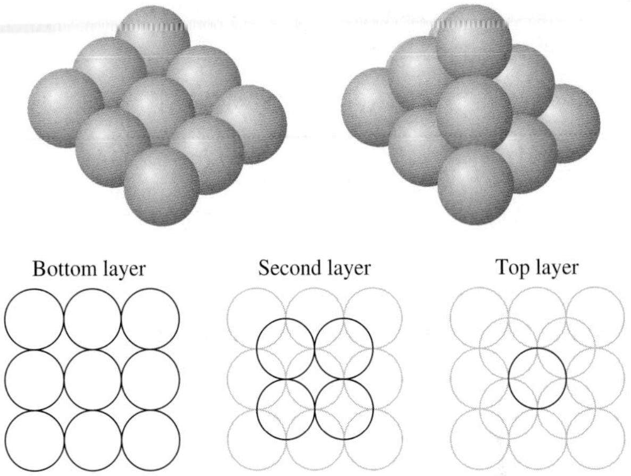

The number N of balls in the stack is given by the polynomial function

$$N(x) = \tfrac{1}{3}x^3 + \tfrac{1}{2}x^2 + \tfrac{1}{6}x,$$

where x is the number of layers. Use both the function and the figure to find $N(3)$. Then calculate the number of oranges in a pyramid with 5 layers.

46. *Stacking Cannonballs.* The function in Exercise 45 was discovered by Thomas Harriot, assistant to Sir Walter Raleigh, when preparing for an expedition at sea (*Source*: *The New York Times* 4/7/04). How many cannonballs did they pack if there were 10 layers to their pyramid?

Veterinary Science. Gentamicin is an antibiotic frequently used by veterinarians. The concentration, in micrograms per milliliter (mcg/mL), of Gentamicin in a horse's bloodstream t hours after injection can be approximated by the polynomial function

$$C(t) = -0.005t^4 + 0.003t^3 + 0.35t^2 + 0.5t$$

(*Source*: Michele Tulis, DVM, telephone interview). *Use the following graph for Exercises 47–50.*

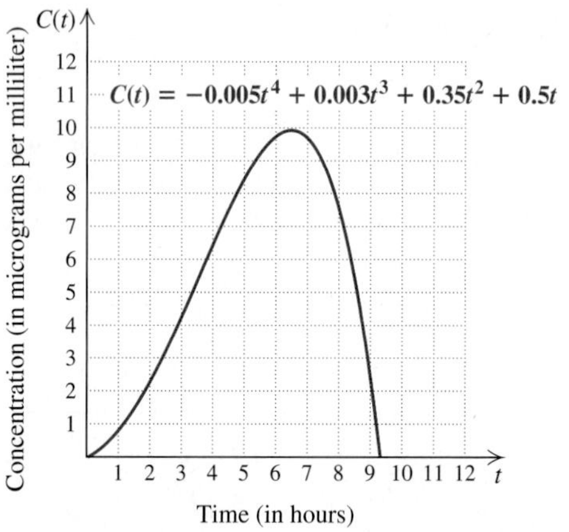

47. Estimate the concentration, in mcg/mL, of Gentamicin in the bloodstream 2 hr after injection.

48. Estimate the concentration, in mcg/mL, of Gentamicin in the bloodstream 4 hr after injection.

49. Approximate the range of C.

50. Approximate the domain of C.

Surface Area of a Right Circular Cylinder. *The surface area of a right circular cylinder is given by the polynomial*

$$2\pi rh + 2\pi r^2,$$

where h is the height, r is the radius of the base, and h and r are given in the same units.

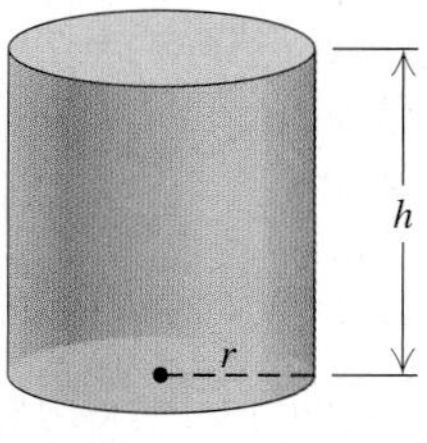

51. A 16-oz beverage can has height 6.3 in. and radius 1.2 in. Find the surface area of the can. (Use a calculator with a (π) key or use 3.141592654 for π.)

52. A 12-oz beverage can has height 4.7 in. and radius 1.2 in. Find the surface area of the can. (Use a calculator with a (π) key or use 3.141592654 for π.)

Total Revenue. *An electronics firm is marketing a new kind of DVD player. The firm determines that when it sells x DVD players, its total revenue is*

$$R(x) = 280x - 0.4x^2 \text{ dollars.}$$

53. What is the total revenue from the sale of 75 DVD players?

54. What is the total revenue from the sale of 100 DVD players?

Total Cost. *The electronics firm determines that the total cost, in dollars, of producing x DVD players is given by*

$$C(x) = 5000 + 0.6x^2.$$

55. What is the total cost of producing 75 DVD players?

56. What is the total cost of producing 100 DVD players?

Estimate the range of each function from its graph.

57.

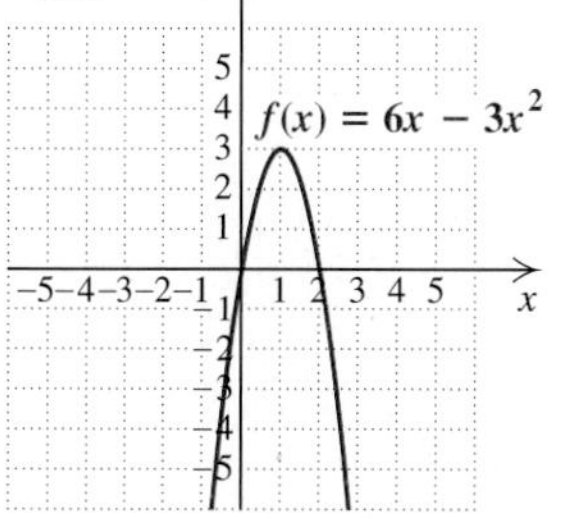

58.

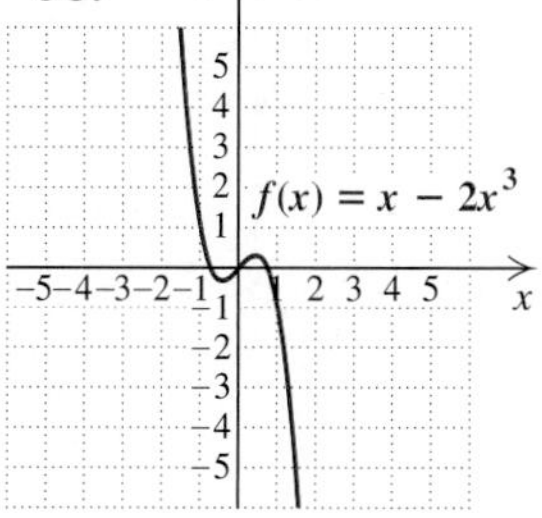

59.

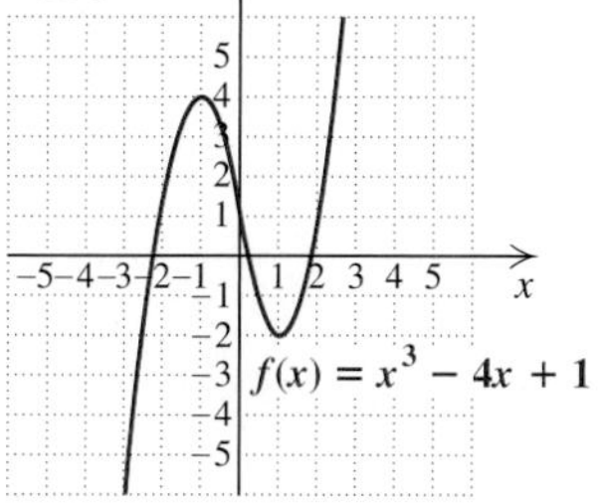

60.

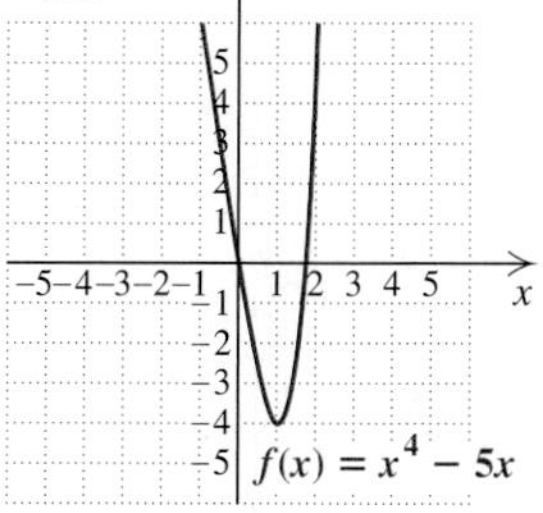

61.

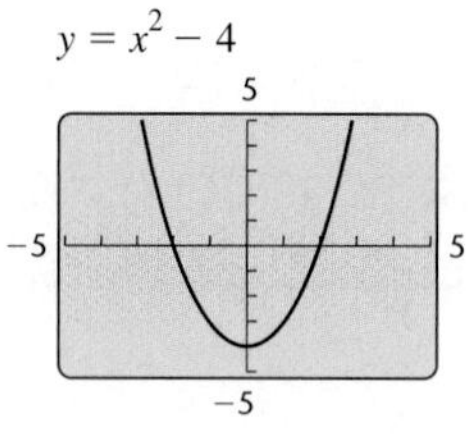

62.

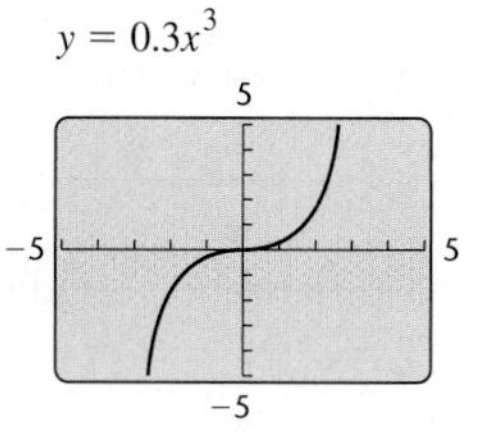

63. $y = x^4 - 5x^3 + x - 2$

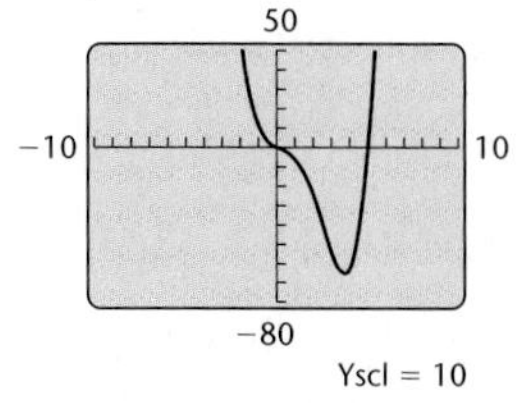

64. $y = 0.15x^3 - 5x^2 + 7$

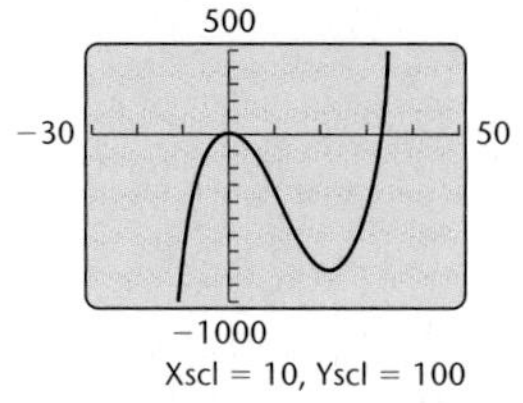

Use a graphing calculator to graph each polynomial function in the indicated viewing window, and estimate its range.

65. $f(x) = x^2 + 2x + 1, [-10, 10, -10, 10]$

66. $p(x) = x^2 + x - 6, [-10, 10, -10, 10]$

67. $q(x) = -2x^2 + 5, [-10, 10, -10, 10]$

68. $g(x) = 1 - x^2, [-10, 10, -10, 10]$

69. $p(x) = -2x^3 + x + 5, [-10, 10, -10, 10]$

70. $f(x) = -x^4 + 2x^3 - 10, [-5, 5, -30, 10]$, Yscl $= 5$

71. $g(x) = x^4 + 2x^3 - 5, [-5, 5, -10, 10]$

72. $q(x) = x^5 - 2x^4, [-5, 5, -5, 5]$

Combine like terms to write an equivalent expression.

73. $5a + 6 - 4 + 2a^3 - 6a + 2$

74. $6x + 13 - 8 - 7x + 5x^2 + 10$

75. $3a^2b + 4b^2 - 9a^2b - 7b^2$

76. $5x^2y^2 + 4x^3 - 8x^2y^2 - 12x^3$

77. $9x^2 - 3xy + 12y^2 + x^2 - y^2 + 5xy + 4y^2$

78. $a^2 - 2ab + b^2 + 9a^2 + 5ab - 4b^2 + a^2$

Add.

79. $(7x - 5y + 3z) + (9x + 12y - 8z)$

80. $(a^2 - 3b^2 + 4c^2) + (-5a^2 + 2b^2 - c^2)$

81. $(x^2 + 2x - 3xy - 7) + (-3x^2 - x + 2xy + 6)$

82. $(3a^2 - 2b + ab + 6) + (-a^2 + 5b - 5ab - 2)$

83. $(8x^2y - 3xy^2 + 4xy) + (-2x^2y - xy^2 + xy)$

84. $(9ab - 3ac + 5bc) + (13ab - 15ac - 8bc)$

85. $(2r^2 + 12r - 11) + (6r^2 - 2r + 4) + (r^2 - r - 2)$

86. $(5x^2 + 19x - 23) + (7x^2 - 2x + 1) + (-x^2 - 9x + 8)$

87. $\left(\frac{1}{8}xy - \frac{3}{5}x^3y^2 + 4.3y^3\right) + \left(-\frac{1}{3}xy - \frac{3}{4}x^3y^2 - 2.9y^3\right)$

88. $\left(\frac{2}{3}xy + \frac{5}{6}xy^2 + 5.1x^2y\right) + \left(-\frac{4}{5}xy + \frac{3}{4}xy^2 - 3.4x^2y\right)$

Write two expressions, one with parentheses and one without, for the opposite of each polynomial.

89. $5x^3 - 7x^2 + 3x - 9$

90. $-8y^4 - 18y^3 + 4y - 7$

91. $-12y^5 + 4ay^4 - 7by^2$

92. $7ax^3y^2 - 8by^4 - 7abx - 12ay$

Subtract.

93. $(7x - 5) - (-3x + 4)$

94. $(8y + 2) - (-6y - 5)$

95. $(-3x^2 + 2x + 9) - (x^2 + 5x - 4)$

96. $(-7y^2 + 5y + 6) - (4y^2 + 3y - 2)$

97. $(8a - 3b + c) - (2a + 3b - 4c)$

98. $(9r - 5s - t) - (7r - 5s + 3t)$

99. $(6a^2 + 5ab - 4b^2) - (8a^2 - 7ab + 3b^2)$

100. $(4y^2 - 13yz - 9z^2) - (9y^2 - 6yz + 3z^2)$

101. $(6ab - 4a^2b + 6ab^2) - (3ab^2 - 10ab - 12a^2b)$

102. $(10xy - 4x^2y^2 - 3y^3) - (-9x^2y^2 + 4y^3 - 7xy)$

103. $\left(\frac{5}{8}x^4 - \frac{1}{4}x^2 - \frac{1}{2}\right) - \left(-\frac{3}{8}x^4 + \frac{3}{4}x^2 + \frac{1}{2}\right)$

104. $\left(\frac{5}{6}y^4 - \frac{1}{2}y^2 - 7.8y\right) - \left(-\frac{3}{8}y^4 + \frac{3}{4}y^2 + 3.4y\right)$

Perform the indicated operations.

105. $(5x^2 + 7) - (2x^2 + 1) + (x^2 + 2x)$

106. $(3t^2 + 5) - (t^2 + 2) + (4t^2 + 3t)$

107. $(8r^2 - 6r) - (2r - 6) + (5r^2 - 7)$

108. $(7s^2 - 5s) - (4s - 1) + (3s^2 - 5)$

Aha! **109.** $(x^2 - 4x + 7) + (3x^2 - 9) - (x^2 - 4x + 7)$

110. $(t^2 - 5t + 6) + (5t - 8) - (t^2 + 3t - 4)$

Total Profit. Total profit is defined as total revenue minus total cost. In Exercises 111 and 112, R(x) and C(x) are the revenue and the cost, respectively, from the sale of x futons.

111. If $R(x) = 280x - 0.4x^2$ and $C(x) = 5000 + 0.6x^2$, find the profit from the sale of 70 futons.

112. If $R(x) = 280x - 0.7x^2$ and $C(x) = 8000 + 0.5x^2$, find the profit from the sale of 100 futons.

In Exercises 113–116, tell which of the following statements are true for the given polynomials.

a) *The graphs of y_1 and y_2 coincide.*
b) *The graphs of y_1 and y_2 differ.*
c) $y_1 - y_2 = 0$
d) $y_1 - y_2 \neq 0$

113. $y_1 = (5x^3 + 2x + 3) + (3x^3 - 1)$,
$y_2 = 8x^3 + 2x + 2$

114. $y_1 = (7x^2 - 9x + 3) - (x^3 - 9x + 3)$,
$y_2 = -x^3 + 7x^2$

115. $y_1 = (x^2 + 8x + 1) - (-x^2 + 3x + 5)$,
$y_2 = 5x - 4$

116. $y_1 = (x^4 + x^2 + 4) + (2x^3 - 3x^2 - 7)$,
$y_2 = 3x^4 - 2x^2 - 3$

TW **117.** Is the sum of two binomials always a binomial? Why or why not?

TW **118.** Ani claims that she can add any two polynomials but finds subtraction difficult. What advice would you offer her?

Skill Maintenance

Simplify. [1.4]

119. $(x^2)^5$

120. $(t^3)^4$

121. $(5x^3)^2$

122. $(7t^4)^2$

123. $x^5 \cdot x^4$

124. $a^2 \cdot a^6$

Synthesis

TW **125.** Write a problem in which revenue and cost functions are given and a profit function, $P(x)$, is required. Devise the problem so that $P(0) < 0$ and $P(100) > 0$.

TW **126.** A student who is trying to graph

$$p(x) = 0.05x^4 - x^2 + 5$$

gets the following screen. How can the student tell at a glance that a mistake has been made?

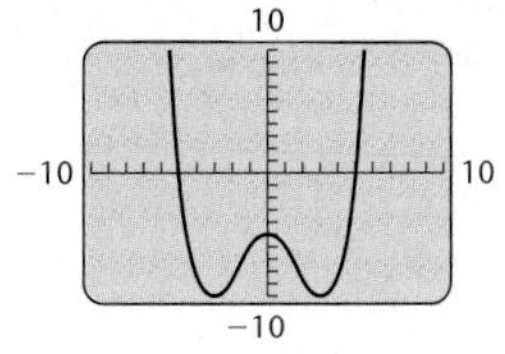

For $P(x)$ and $Q(x)$ as given, find the following.

$P(x) = 13x^5 - 22x^4 - 36x^3 + 40x^2 - 16x + 75$,
$Q(x) = 42x^5 - 37x^4 + 50x^3 - 28x^2 + 34x + 100$

127. $2[P(x)] + Q(x)$

128. $3[P(x)] - Q(x)$

129. $2[Q(x)] - 3[P(x)]$

130. $4[P(x)] + 3[Q(x)]$

131. *Volume of a Display.* The number of spheres in a triangular pyramid with x layers is given by the function

$$N(x) = \tfrac{1}{6}x^3 + \tfrac{1}{2}x^2 + \tfrac{1}{3}x.$$

The volume of a sphere of radius r is given by the function

$$V(r) = \tfrac{4}{3}\pi r^3,$$

where π can be approximated as 3.14.

Chocolate Heaven has a window display of truffles piled in a triangular pyramid formation 5 layers deep. If the diameter of each truffle is 3 cm, find the volume of chocolate in the display.

132. If one large truffle were to have the same volume as the display of truffles in Exercise 131, what would be its diameter?

133. Find a polynomial function that gives the outside surface area of a box like this one, with an open top and dimensions as shown.

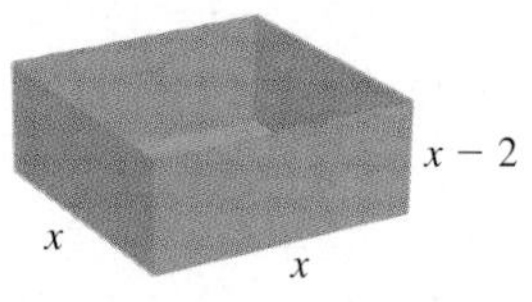

134. Develop a formula for the surface area of a right circular cylinder in which h is the height, in *centimeters*, and r is the radius, in *meters*. (See Exercises 51 and 52.)

Perform the indicated operation. Assume that the exponents are natural numbers.

135. $(2x^{2a} + 4x^{a} + 3) + (6x^{2a} + 3x^{a} + 4)$

136. $(3x^{6a} - 5x^{5a} + 4x^{3a} + 8) - (2x^{6a} + 4x^{4a} + 3x^{3a} + 2x^{2a})$

137. $(2x^{5b} + 4x^{4b} + 3x^{3b} + 8) - (x^{5b} + 2x^{3b} + 6x^{2b} + 9x^{b} + 8)$

Collaborative Corner

How Many Handshakes?

Focus: Polynomial functions

Time: 20 minutes

Group Size: 5

ACTIVITY

1. All group members should shake hands with each other. Without "double counting," determine how many handshakes occurred.
2. Complete the table in the next column.
3. Join another group to determine the number of handshakes for a group of size 10.
4. Try to find a function of the form $H(n) = an^2 + bn$, for which $H(n)$ is the number of different handshakes that are possible in a group of n people. Make sure that $H(n)$ produces all of the values in the table above. (*Hint*: Use the table to twice select n and $H(n)$. Then solve the resulting system of equations for a and b.)

Group Size	Number of Handshakes
1	
2	
3	
4	
5	

5.2 Multiplication of Polynomials

Multiplying Monomials ▪ Multiplying Monomials and Binomials ▪ Multiplying Any Two Polynomials ▪ The Product of Two Binomials: FOIL ▪ Squares of Binomials ▪ Products of Sums and Differences ▪ Function Notation

Study Tip

You've Got Mail

Many students overlook an excellent opportunity to get questions cleared up—e-mail. If your instructor makes his or her e-mail address available, consider using it to get help. Often, just the act of writing out your question brings some clarity. If you do use e-mail, allow some time for your instructor to reply.

Just like numbers, polynomials can be multiplied. The product of two polynomials $P(x)$ and $Q(x)$ is a polynomial $R(x)$ that gives the same value as $P(x) \cdot Q(x)$ for any replacement of x.

Multiplying Monomials

To multiply monomials, we first multiply their coefficients. Then we multiply the variables using the rules for exponents and the commutative and associative laws. With practice, we can work mentally, writing only the answer.

EXAMPLE 1 Multiply and simplify.

a) $(-8x^4y^7)(5x^3y^2)$

b) $(-2x^2yz^5)(-6x^5y^{10}z^2)$

SOLUTION

a) $(-8x^4y^7)(5x^3y^2) = -8 \cdot 5 \cdot x^4 \cdot x^3 \cdot y^7 \cdot y^2$ **Using the associative and commutative laws**

$= -40x^{4+3}y^{7+2}$ **Multiplying coefficients; adding exponents**

$= -40x^7y^9$

b) $(-2x^2yz^5)(-6x^5y^{10}z^2) = (-2)(-6) \cdot x^2 \cdot x^5 \cdot y \cdot y^{10} \cdot z^5 \cdot z^2$

$= 12x^7y^{11}z^7$ **Multiplying coefficients; adding exponents**

Multiplying Monomials and Binomials

The distributive law is the basis for multiplying polynomials other than monomials. We first multiply a monomial and a binomial.

EXAMPLE 2 Multiply: **(a)** $2x(3x - 5)$; **(b)** $3a^2b(a^2 - b^2)$.

SOLUTION

a) $2x(3x - 5) = 2x \cdot 3x - 2x \cdot 5$ **Using the distributive law**

$= 6x^2 - 10x$ **Multiplying monomials**

b) $3a^2b(a^2 - b^2) = 3a^2b \cdot a^2 - 3a^2b \cdot b^2$ **Using the distributive law**

$= 3a^4b - 3a^2b^3$

X	Y1	Y2
0	0	0
1	−4	−4
2	4	4
3	24	24
4	56	56
5	100	100
6	156	156
X = 0		

We can use graphs or tables to check the multiplication of polynomials in one variable. For example, we can check the product found in Example 2(a) by letting $y_1 = 2x(3x - 5)$ and $y_2 = 6x^2 - 10x$. The table of values for y_1 and y_2 at left serves as a check.

The distributive law is also used when multiplying two binomials. In this case, however, we begin by distributing a *binomial* rather than a monomial. With practice, some of the following steps can be combined.

EXAMPLE 3 Multiply: $(y^3 - 5)(2y^3 + 4)$.

SOLUTION

$(y^3 - 5)(2y^3 + 4) = (y^3 - 5)2y^3 + (y^3 - 5)4$ **"Distributing" the $y^3 - 5$**

$= 2y^3(y^3 - 5) + 4(y^3 - 5)$ **Using the commutative law for multiplication. Try to do this step mentally.**

$= 2y^3 \cdot y^3 - 2y^3 \cdot 5 + 4 \cdot y^3 - 4 \cdot 5$ **Using the distributive law (twice)**

$= 2y^6 - 10y^3 + 4y^3 - 20$ **Multiplying the monomials**

$= 2y^6 - 6y^3 - 20$ **Combining like terms**

Multiplying Any Two Polynomials

Repeated use of the distributive law enables us to multiply *any* two polynomials, regardless of how many terms are in each.

EXAMPLE 4 Multiply: $(p + 2)(p^4 - 2p^3 + 3)$.

SOLUTION We can use the distributive law from right to left if we wish:

$$\begin{aligned}
&(p + 2)(p^4 - 2p^3 + 3)\\
&\quad = p(p^4 - 2p^3 + 3) + 2(p^4 - 2p^3 + 3)\\
&\quad = p \cdot p^4 - p \cdot 2p^3 + p \cdot 3 + 2 \cdot p^4 - 2 \cdot 2p^3 + 2 \cdot 3\\
&\quad = p^5 - 2p^4 + 3p + 2p^4 - 4p^3 + 6\\
&\quad = p^5 - 4p^3 + 3p + 6. \qquad \textbf{Combining like terms}
\end{aligned}$$

$y_1 = (x + 2)(x^4 - 2x^3 + 3)$,
$y_2 = x^5 - 4x^3 + 3x + 6$

X	Y1	Y2
0	6	6
1	6	6
2	12	12
3	150	150
4	786	786
5	2646	2646
6	6936	6936

X = 0

In order for these polynomials to be equivalent, they must have the same value for at least 6 x-values, because the degree is 5. The table at left shows that this is true.

The Product of Two Polynomials

The *product* of two polynomials $P(x)$ and $Q(x)$ is found by multiplying each term of $P(x)$ by every term of $Q(x)$ and then combining like terms.

It is also possible to stack the polynomials, multiplying each term at the top by every term below, keeping like terms in columns, and leaving spaces for missing terms. Then we add just as we do in long multiplication with numbers.

EXAMPLE 5 Multiply: $(5x^3 + x - 4)(-2x^2 + 3x + 6)$.

SOLUTION

$$\begin{array}{rrrrrrl}
 & & 5x^3 & & +\ x & -\ 4 & \\
 & & & -2x^2 & +\ 3x & +\ 6 & \\
\hline
 & & 30x^3 & & +\ 6x & -\ 24 & \textbf{Multiplying by 6}\\
 & 15x^4 & & +\ 3x^2 & -\ 12x & & \textbf{Multiplying by } \mathbf{3x}\\
-10x^5 & & -\ 2x^3 & +\ 8x^2 & & & \textbf{Multiplying by } \mathbf{-2x^2}\\
\hline
-10x^5 & +\ 15x^4 & +\ 28x^3 & +\ 11x^2 & -\ 6x & -\ 24 & \textbf{Adding}
\end{array}$$

$y_1 = (5x^3 + x - 4)(-2x^2 + 3x + 6)$,
$y_2 = -10x^5 + 15x^4 + 28x^3 + 11x^2 - 6x - 24$,
$y_3 = y_1 - y_2$

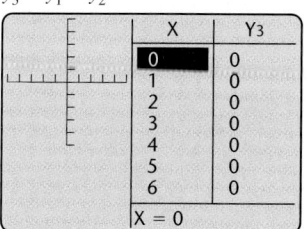

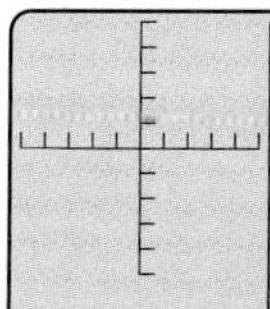

X	Y3
0	0
1	0
2	0
3	0
4	0
5	0
6	0

X = 0

We check by showing that the difference of the original expression and the resulting product is 0, as shown in the graph and table at left.

The Product of Two Binomials: FOIL

We now consider what are called *special products*. These products of polynomials occur often and can be simplified using shortcuts that we now develop.

To find a faster special-product rule for the product of two binomials, consider $(x + 7)(x + 4)$. We multiply each term of $(x + 7)$ by each term of $(x + 4)$:

$$(x + 7)(x + 4) = x \cdot x + x \cdot 4 + 7 \cdot x + 7 \cdot 4.$$

A visualization of $(x + 7)(x + 4)$ using areas

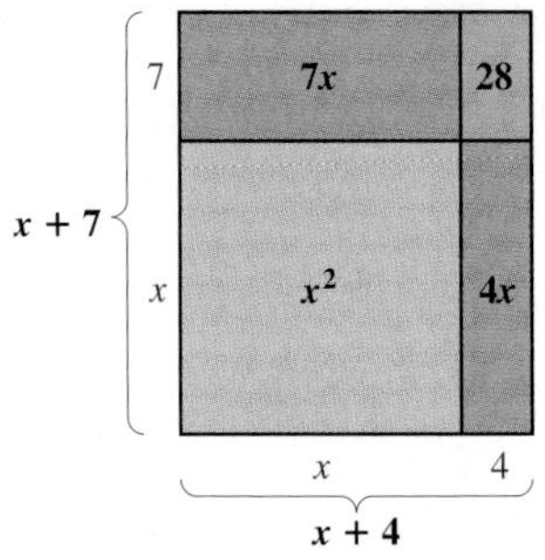

This multiplication illustrates a pattern that occurs whenever two binomials are multiplied:

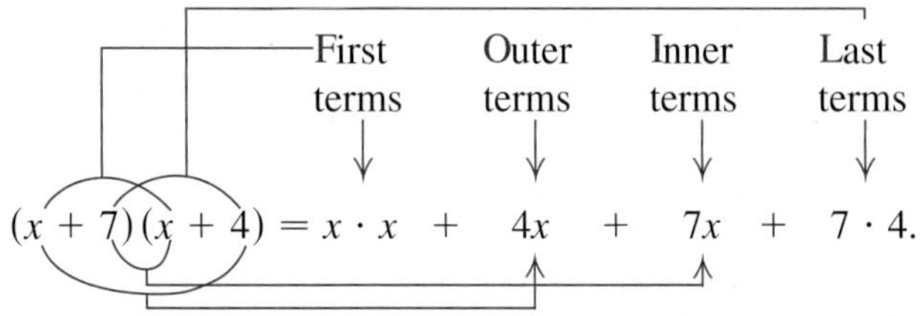

We use the mnemonic device FOIL to remember this method for multiplying.

The FOIL Method To multiply two binomials $A + B$ and $C + D$, multiply the First terms AC, the Outer terms AD, the Inner terms BC, and then the Last terms BD. Then combine like terms, if possible.

$$(A + B)(C + D) = AC + AD + BC + BD$$

1. Multiply First terms: AC.
2. Multiply Outer terms: AD.
3. Multiply Inner terms: BC.
4. Multiply Last terms: BD.

↓
FOIL

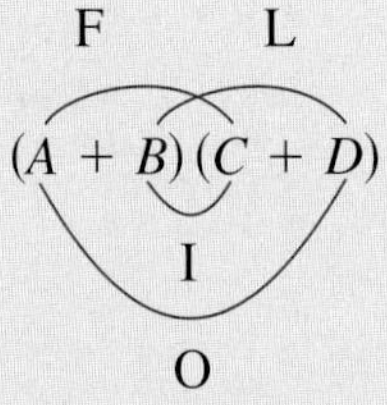

EXAMPLE 6 Multiply.

a) $(x + 5)(x - 8)$ **b)** $(2x + 3y)(x - 4y)$

c) $(t + 2)(t - 4)(t + 5)$

SOLUTION

F O I L

a) $(x + 5)(x - 8) = x^2 - 8x + 5x - 40$

$= x^2 - 3x - 40$ **Combining like terms**

b) $(2x + 3y)(x - 4y) = 2x^2 - 8xy + 3xy - 12y^2$ **Using FOIL**

$= 2x^2 - 5xy - 12y^2$ **Combining like terms**

c) $(t + 2)(t - 4)(t + 5) = (t^2 - 4t + 2t - 8)(t + 5)$ **Using FOIL**

$= (t^2 - 2t - 8)(t + 5)$

$= (t^2 - 2t - 8) \cdot t + (t^2 - 2t - 8) \cdot 5$

Using the distributive law

$= t^3 - 2t^2 - 8t + 5t^2 - 10t - 40$

$= t^3 + 3t^2 - 18t - 40$ **Combining like terms**

Squares of Binomials

Interactive Discovery

Determine whether each of the following is an identity.

1. $(x + 3)^2 = x^2 + 9$
2. $(x + 3)^2 = x^2 + 6x + 9$
3. $(x - 3)^2 = x^2 - 6x - 9$
4. $(x - 3)^2 = x^2 - 9$
5. $(x - 3)^2 = x^2 - 6x + 9$

A visualization of $(A + B)^2$ using areas

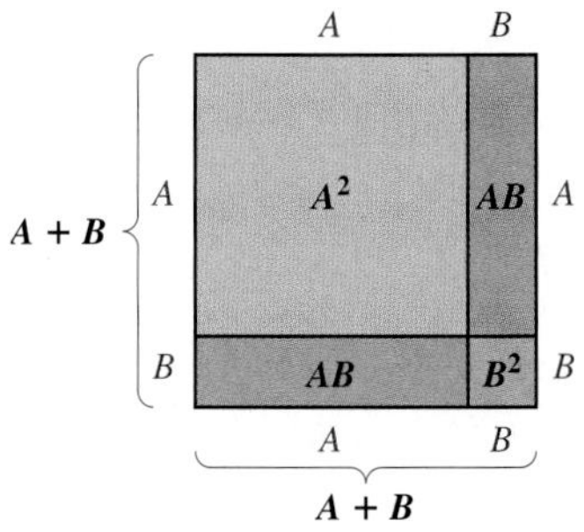

A fast method for squaring binomials can be developed using FOIL:

$$\begin{aligned}(A + B)^2 &= (A + B)(A + B)\\ &= A^2 + AB + AB + B^2 &&\textbf{Note that } AB \textbf{ occurs twice.}\\ &= A^2 + 2AB + B^2;\end{aligned}$$

$$\begin{aligned}(A - B)^2 &= (A - B)(A - B)\\ &= A^2 - AB - AB + B^2 &&\textbf{Note that } -AB \textbf{ occurs twice.}\\ &= A^2 - 2AB + B^2.\end{aligned}$$

Squaring a Binomial

$$(A + B)^2 = A^2 + 2AB + B^2;$$
$$(A - B)^2 = A^2 - 2AB + B^2$$

The square of a binomial is the square of the first term, plus twice the product of the two terms, plus the square of the last term.

Trinomials that can be written in the form $A^2 + 2AB + B^2$ or $A^2 - 2AB + B^2$ are called *perfect-square trinomials.*

It can help to remember the words of the rules and say them while multiplying.

EXAMPLE 7 Multiply: **(a)** $(y - 5)^2$; **(b)** $(2x + 3y)^2$; **(c)** $\left(\frac{1}{2}x - 3y^4\right)^2$.

SOLUTION

$$(A - B)^2 = A^2 - 2 \cdot A \cdot B + B^2$$
$$\downarrow \quad \downarrow \quad\quad \downarrow \quad\quad \downarrow \quad \downarrow \quad \downarrow$$

a) $(y - 5)^2 = y^2 - 2 \cdot y \cdot 5 + 5^2$ **Note that $-2 \cdot y \cdot 5$ is twice the product of y and -5.**

$$= y^2 - 10y + 25$$

b) $(2x + 3y)^2 = (2x)^2 + 2 \cdot 2x \cdot 3y + (3y)^2$

$$= 4x^2 + 12xy + 9y^2 \qquad \textbf{Raising a product to a power}$$

c) $\left(\frac{1}{2}x - 3y^4\right)^2 = \left(\frac{1}{2}x\right)^2 - 2 \cdot \frac{1}{2}x \cdot 3y^4 + (3y^4)^2$ **$2 \cdot \frac{1}{2}x \cdot (-3y^4) = -2 \cdot \frac{1}{2}x \cdot 3y^4$**

$= \frac{1}{4}x^2 - 3xy^4 + 9y^8$ **Raising a product to a power; multiplying exponents**

CAUTION! Note that $(y - 5)^2 \neq y^2 - 5^2$. (To see this, replace y with 6 and note that $(6 - 5)^2 = 1^2 = 1$ and $6^2 - 5^2 = 36 - 25 = 11$.) More generally,

$$(A + B)^2 \neq A^2 + B^2 \quad \text{and} \quad (A - B)^2 \neq A^2 - B^2.$$

Products of Sums and Differences

Another pattern emerges when we are multiplying a sum and a difference of the same two terms.

Interactive Discovery

Determine whether each of the following is an identity.

1. $(x + 3)(x - 3) = x^2 - 9$
2. $(x + 3)(x - 3) = x^2 + 9$
3. $(x + 3)(x - 3) = x^2 - 6x + 9$
4. $(x + 3)(x - 3) = x^2 + 6x + 9$

Note the following:

$$\begin{aligned}(A + B)(A - B) &= \overset{\text{F}}{A^2} - \overset{\text{O}}{AB} + \overset{\text{I}}{AB} - \overset{\text{L}}{B^2}\\ &= A^2 - B^2. \qquad \mathbf{-AB + AB = 0}\end{aligned}$$

Student Notes

Being able to quickly multiply products of the form $(A + B)(A - B)$ or $(A - B)(A + B)$ is a skill heavily relied on in Section 5.6.

The Product of a Sum and a Difference

$$(A + B)(A - B) = A^2 - B^2$$ This is called a *difference of two squares.*

The product of the sum and the difference of the same two terms is the square of the first term minus the square of the second term.

EXAMPLE 8 Multiply.

a) $(y + 5)(y - 5)$

b) $(2xy^2 + 3x)(2xy^2 - 3x)$

c) $(0.2t - 1.4m)(0.2t + 1.4m)$

d) $\left(\frac{2}{3}n - m^3\right)\left(\frac{2}{3}n + m^3\right)$

SOLUTION

$$(A + B)(A - B) = A^2 - B^2$$
$$\downarrow \quad \downarrow \quad \downarrow \quad \downarrow \quad \downarrow \quad \downarrow$$

a) $(y + 5)(y - 5) = y^2 - 5^2$ **Replacing A with y and B with 5**

$= y^2 - 25$ **Try to do problems like this mentally.**

b) $(2xy^2 + 3x)(2xy^2 - 3x) = (2xy^2)^2 - (3x)^2$

$= 4x^2y^4 - 9x^2$ **Raising a product to a power**

c) $(0.2t - 1.4m)(0.2t + 1.4m) = (0.2t)^2 - (1.4m)^2$

$= 0.04t^2 - 1.96m^2$

d) $\left(\frac{2}{3}n - m^3\right)\left(\frac{2}{3}n + m^3\right) = \left(\frac{2}{3}n\right)^2 - (m^3)^2$

$= \frac{4}{9}n^2 - m^6$

EXAMPLE 9 Multiply.

a) $(5y + 4 + 3x)(5y + 4 - 3x)$ **b)** $(3xy^2 + 4y)(-3xy^2 + 4y)$

c) $(2t + 3)^2 - (t - 1)(t + 1)$

SOLUTION

a) By far the easiest way to multiply $(5y + 4 + 3x)(5y + 4 - 3x)$ is to note that it is in the form $(A + B)(A - B)$:

$(5y + 4 + 3x)(5y + 4 - 3x) = (5y + 4)^2 - (3x)^2$ **Try to be alert for situations like this.**

$= 25y^2 + 40y + 16 - 9x^2.$

We can also multiply $(5y + 4 + 3x)(5y + 4 - 3x)$ using columns, but not as quickly.

b) $(3xy^2 + 4y)(-3xy^2 + 4y) = (4y + 3xy^2)(4y - 3xy^2)$ **Rewriting**

$= (4y)^2 - (3xy^2)^2$

$= 16y^2 - 9x^2y^4$

c) $(2t + 3)^2 - (t - 1)(t + 1) = 4t^2 + 12t + 9 - (t^2 - 1)$ **Squaring a binomial; simplifying $(t - 1)(t + 1)$**

$= 4t^2 + 12t + 9 - t^2 + 1$ **Subtracting**

$= 3t^2 + 12t + 10$ **Combining like terms**

Function Notation

Let's stop for a moment and look back at what we have done in this section. We have shown, for example, that

$$(x - 2)(x + 2) = x^2 - 4;$$

that is, $x^2 - 4$ and $(x - 2)(x + 2)$ are equivalent expressions.

From the viewpoint of functions, if

$$f(x) = x^2 - 4 \quad \text{and} \quad g(x) = (x - 2)(x + 2),$$

then for any given input x, the outputs $f(x)$ and $g(x)$ are identical. Thus the graphs of these functions are identical and we say that f and g represent the same function. Functions like these are graphed in detail in Chapter 8.

x	$f(x)$	$g(x)$
3	5	5
2	0	0
1	−3	−3
0	−4	−4
−1	−3	−3
−2	0	0
−3	5	5

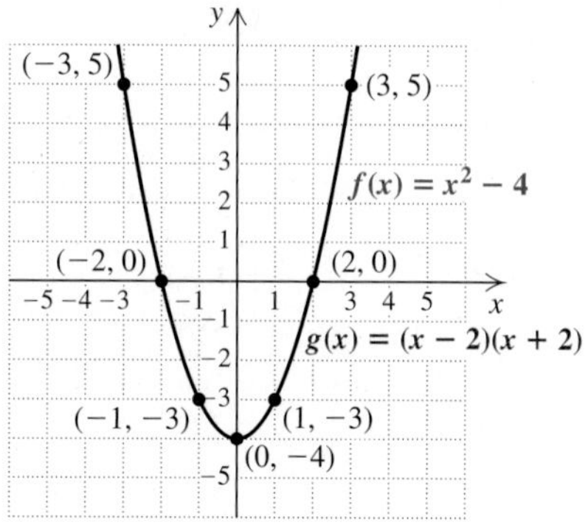

Our work with multiplying can be used when evaluating functions.

EXAMPLE 10 Given $f(x) = x^2 - 4x + 5$, find and simplify each of the following.

a) $f(a) + 3$ **b)** $f(a + 3)$ **c)** $f(a + h) - f(a)$

SOLUTION

a) To find $f(a) + 3$, we replace x with a to find $f(a)$. Then we add 3 to the result:

$$\begin{aligned} f(a) + 3 &= a^2 - 4a + 5 + 3 \quad \textbf{Evaluating } f(a) \\ &= a^2 - 4a + 8. \end{aligned}$$

b) To find $f(a + 3)$, we replace x with $a + 3$. Then we simplify:

$$\begin{aligned} f(a + 3) &= (a + 3)^2 - 4(a + 3) + 5 \\ &= a^2 + 6a + 9 - 4a - 12 + 5 \\ &= a^2 + 2a + 2. \end{aligned}$$

CAUTION! Note from parts (a) and (b) that, in general,

$$f(a + 3) \neq f(a) + 3.$$

c) To find $f(a + h)$ and $f(a)$, we replace x with $a + h$ and a, respectively:

$$\begin{aligned} f(a + h) - f(a) &= [(a + h)^2 - 4(a + h) + 5] - [a^2 - 4a + 5] \\ &= [a^2 + 2ah + h^2 - 4a - 4h + 5] - [a^2 - 4a + 5] \\ &= a^2 + 2ah + h^2 - 4a - 4h + 5 - a^2 + 4a - 5 \\ &= 2ah + h^2 - 4h. \end{aligned}$$

5.2 EXERCISE SET

FOR EXTRA HELP

MathXL | MyMathLab | InterAct Math | AW Math Tutor Center | Video Lectures on CD: Disc 3 | Student's Solutions Manual

Concept Reinforcement *Classify each of the following as either true or false.*

1. The product of two monomials is a monomial.
2. The product of $5a^2$ and $3a^4$ is $15a^8$.
3. The product of a monomial and a binomial is found using the distributive law.
4. To simplify the product of two binomials, we often need to combine like terms.
5. FOIL can be used whenever a monomial and a binomial are multiplied.
6. The square of a binomial is a difference of two squares.
7. The product of two polynomials of the form $(A - B)(A + B)$ is a difference of two squares.
8. In general, $f(a + 5) \neq f(a) + 5$.

Multiply.

9. $8a^2 \cdot 5a$
10. $-5x^3 \cdot 2x$
11. $5x(-4x^2y)$
12. $-3ab^2(2a^2b^2)$
13. $(4x^3y^2)(-5x^2y^4)$
14. $(7a^2bc^4)(-8ab^3c^2)$
15. $7x(3 - x)$
16. $3a(a^2 - 4a)$
17. $5cd(4c^2d - 5cd^2)$
18. $a^2(2a^2 - 5a^3)$
19. $(x + 3)(x + 5)$
20. $(t + 1)(t + 4)$
21. $(t - 2)(t + 7)$
22. $(x - 5)(x + 6)$
23. $(2a + 3)(4a + 1)$
24. $(3r + 4)(2r + 1)$
25. $(5x - 4y)(2x + y)$
26. $(4t + 3)(2t - 7)$
27. $(x + 2)(x^2 - 3x + 1)$
28. $(a + 3)(a^2 - 4a + 2)$
29. $(t - 5)(t^2 + 2t - 3)$
30. $(x - 4)(x^2 + x - 7)$
31. $(a^2 + a - 1)(a^2 + 4a - 5)$
32. $(x^2 - 2x + 1)(x^2 + x + 2)$
33. $(x + 3)(x^2 - 3x + 9)$
34. $(y + 4)(y^2 - 4y + 16)$
35. $(a - b)(a^2 + ab + b^2)$
36. $(x - y)(x^2 + xy + y^2)$
37. $\left(t - \frac{1}{3}\right)\left(t - \frac{1}{4}\right)$
38. $\left(x - \frac{1}{2}\right)\left(x - \frac{1}{5}\right)$
39. $(r + 3)(r + 2)(r - 1)$
40. $(t + 4)(t + 1)(t - 2)$
41. $(x + 5)^2$
42. $(t + 6)^2$
43. $(1.2t + 3s)(2.5t - 5s)$
44. $(30a - 0.24b)(0.2a + 10b)$
45. $(2y - 7)^2$
46. $(3x - 4)^2$
47. $(5a - 3b)^2$
48. $(7x - 4y)^2$
49. $(2a^3 - 3b^2)^2$
50. $(3s^2 + 4t^3)^2$
51. $(x^3y^4 + 5)^2$
52. $(a^4b^2 + 3)^2$

53. Let $P(x) = 3x^2 - 5$ and $Q(x) = 4x^2 - 7x + 1$. Find $P(x) \cdot Q(x)$.

54. Let $P(x) = x^2 - x + 1$ and $Q(x) = x^3 + x^2 + 5$. Find $P(x) \cdot Q(x)$.

55. Let $P(x) = 5x - 2$. Find $P(x) \cdot P(x)$.

56. Let $Q(x) = 3x^2 + 1$. Find $Q(x) \cdot Q(x)$.

57. Let $F(x) = 2x - \frac{1}{3}$. Find $[F(x)]^2$.

58. Let $G(x) = 5x - \frac{1}{2}$. Find $[G(x)]^2$.

Multiply.

59. $(c + 7)(c - 7)$

60. $(x - 3)(x + 3)$

61. $(4x + 1)(4x - 1)$

62. $(3 - 2x)(3 + 2x)$

63. $(3m - 2n)(3m + 2n)$

64. $(3x + 5y)(3x - 5y)$

65. $(x^3 + yz)(x^3 - yz)$

66. $(4a^3 + 5ab)(4a^3 - 5ab)$

67. $(-mn + m^2)(mn + m^2)$

68. $(-3b + a^2)(3b + a^2)$

69. $(x + 7)^2 - (x + 3)(x - 3)$

70. $(t + 5)^2 - (t - 4)(t + 4)$

71. $(2m - n)(2m + n) - (m - 2n)^2$

72. $(3x + y)(3x - y) - (2x + y)^2$

Aha! **73.** $(a + b + 1)(a + b - 1)$

74. $(m + n + 2)(m + n - 2)$

75. $(2x + 3y + 4)(2x + 3y - 4)$

76. $(3a - 2b + c)(3a - 2b - c)$

77. *Compounding Interest.* Suppose that P dollars is invested in a savings account at interest rate i, compounded annually, for 2 yr. The amount A in the account after 2 yr is given by

$$A = P(1 + i)^2.$$

Find an equivalent expression for A.

78. *Compounding Interest.* Suppose that P dollars is invested in a savings account at interest rate i, compounded semiannually, for 1 yr. The amount A in the account after 1 yr is given by

$$A = P\left(1 + \frac{i}{2}\right)^2.$$

Find an equivalent expression for A.

79. Given $f(x) = x^2 + 5$, find and simplify.
a) $f(t - 1)$
b) $f(a + h) - f(a)$
c) $f(a) - f(a - h)$

80. Given $f(x) = x^2 + 7$, find and simplify.
a) $f(p + 1)$
b) $f(a + h) - f(a)$
c) $f(a) - f(a - h)$

TW **81.** Find two binomials whose product is $x^2 - 25$ and explain how you decided on those two binomials.

TW **82.** Find two binomials whose product is $x^2 - 6x + 9$ and explain how you decided on those two binomials.

Skill Maintenance

Solve. [1.6]

83. $ab + ac = d$, for a

84. $xy + yz = w$, for y

85. $mn + m = p$, for m

86. $rs + s = t$, for s

87. *Value of Coins.* There are 50 dimes in a roll of dimes, 40 nickels in a roll of nickels, and 40 quarters in a roll of quarters. Kacie has 13 rolls of coins, which have a total value of \$89. There are three more rolls of dimes than nickels. How many of each type of roll does she have? [3.5]

88. *Wages.* Takako worked a total of 17 days last month at her father's restaurant. She earned \$50 a day during the week and \$60 a day during the weekend. Last month Takako earned \$940. How many weekdays did she work? [3.3]

Synthesis

TW **89.** We have seen that $(a - b)(a + b) = a^2 - b^2$. Explain how this result can be used to develop a fast way of multiplying $95 \cdot 105$.

TW **90.** A student incorrectly claims that since $2x^2 \cdot 2x^2 = 4x^4$, it follows that $5x^5 \cdot 5x^5 = 25x^{25}$. What mistake is the student making?

Multiply. Assume that variables in exponents represent natural numbers.

91. $(x^2 + y^n)(x^2 - y^n)$

92. $(a^n + b^n)^2$

93. $x^2y^3(5x^n + 4y^n)$

94. $a^nb^m(7a^2 - 3b^3)$

95. $(t^n + 3)(t^{2n} - 2t^n + 1)$

96. $(x^n - 4)(x^{2n} + 3x^n - 2)$

Aha! **97.** $(a - b + c - d)(a + b + c + d)$

98. $[(a + b)(a - b)][5 - (a + b)][5 + (a + b)]$

99. $\left(\frac{2}{3}x + \frac{1}{3}y + 1\right)\left(\frac{2}{3}x - \frac{1}{3}y - 1\right)$

100. $(x^2 - 3x + 5)(x^2 + 3x + 5)$

101. $(x^a + y^b)(x^a - y^b)(x^{2a} + y^{2b})$

102. $(x - 1)(x^2 + x + 1)(x^3 + 1)$

103. $(x^{a-b})^{a+b}$

104. $(M^{x+y})^{x+y}$

Aha! **105.** $(x - a)(x - b)(x - c) \cdots (x - z)$

106. Given $f(x) = x^2 + 7$, find and simplify

$$\frac{f(a + h) - f(a)}{h}.$$

107. Given $g(x) = x^2 - 9$, find and simplify

$$\frac{g(a + h) - g(a)}{h}.$$

108. Draw rectangles similar to those on p. 360 to show that $(x + 2)(x + 5) = x^2 + 7x + 10$.

109. Use a graphing calculator to determine whether each of the following is an identity.

a) $(x - 1)^2 = x^2 - 1$
b) $(x - 2)(x + 3) = x^2 + x - 6$
c) $(x - 1)^3 = x^3 - 3x^2 + 3x - 1$
d) $(x + 1)^4 = x^4 + 1$
e) $(x + 1)^4 = x^4 + 4x^3 + 8x^2 + 4x + 1$

Collaborative Corner

Algebra and Number Tricks

Focus: Polynomial multiplication
Time: 15–20 minutes
Group Size: 2

Consider the following dialogue:

Jinny: Cal, let me do a number trick with you. Think of a number between 1 and 7. I'll have you perform some manipulations to this number, you'll tell me the result, and I'll tell you your number.

Cal: Okay. I've thought of a number.

Jinny: Good. Write it down so I can't see it, double it, and then subtract x from the result.

Cal: Hey, this is algebra!

Jinny: I know. Now square your binomial and subtract x^2.

Cal: How did you know I had an x^2? I *thought* this was rigged!

Jinny: It is. Now, divide by 4 and tell me either your constant term or your x-term. I'll tell you the other term and the number you chose.

Cal: Okay. The constant term is 16.

Jinny: Then the other term is $-4x$ and the number you chose is 4.

Cal: You're right! How did you do it?

ACTIVITY

1. Each group member should follow Jinny's instructions. Then determine how Jinny determined Cal's number and the other term.
2. Suppose that, at the end, Cal told Jinny the x-term. How would Jinny have determined Cal's number and the other term?
3. Would Jinny's "trick" work with *any* real number? Why do you think she specified numbers between 1 and 7?
4. Each group member should create a new number "trick" and perform it on the other group member. Be sure to include a variable so that both members can gain practice with polynomials.

5.3 Polynomial Equations and Factoring

Whenever two polynomials are set equal to each other, the result is a **polynomial equation.** In this section, we learn how to solve such equations both graphically and algebraically by *factoring*.

Graphical Solutions

EXAMPLE 1 Solve: $x^2 = 6x$.

SOLUTION We can find real-number solutions of a polynomial equation by finding the points of intersection of two graphs.

Alternatively, we can rewrite the equation so that one side is 0 and then find the x-intercepts of one graph, or the zeros of a function. Both methods were discussed in Section 4.2.

INTERSECT METHOD

To solve $x^2 = 6x$ using the first method, we graph $y_1 = x^2$ and $y_2 = 6x$ and find the coordinates of any points of intersection.

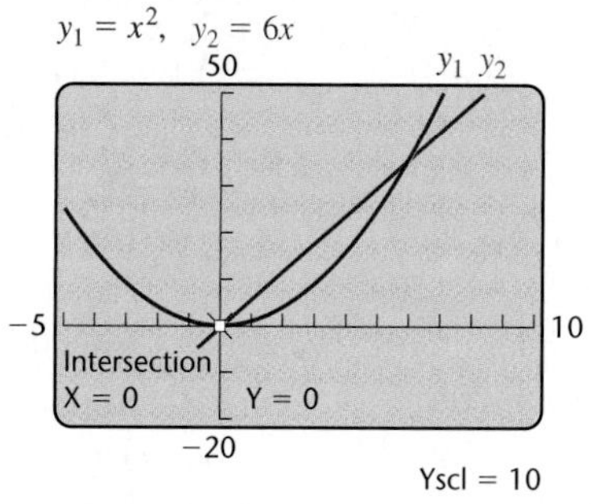

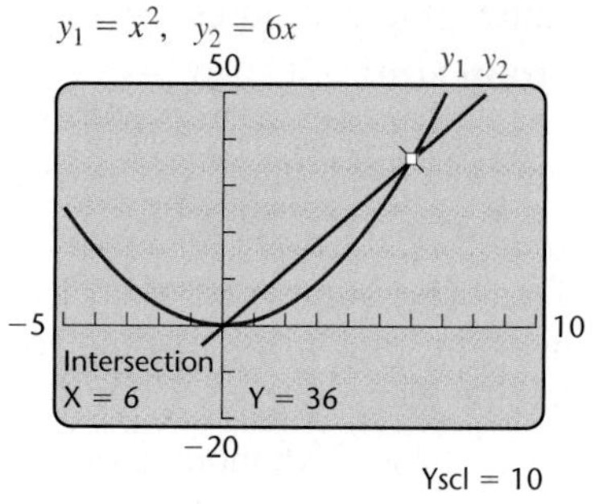

There are two points of intersection of the graphs, so there are two solutions of the equation. The x-coordinates of the points of intersection are 0 and 6, so the solutions of the equation are 0 and 6.

ZERO METHOD

Using the second method, we rewrite the equation so that one side is 0:

$$x^2 = 6x$$
$$x^2 - 6x = 0. \quad \textbf{Adding } -6x \textbf{ to both sides}$$

To solve $x^2 - 6x = 0$, we graph the function $f(x) = x^2 - 6x$ and look for values of x for which $f(x) = 0$, or the zeros of f. These correspond to the x-intercepts of the graph.

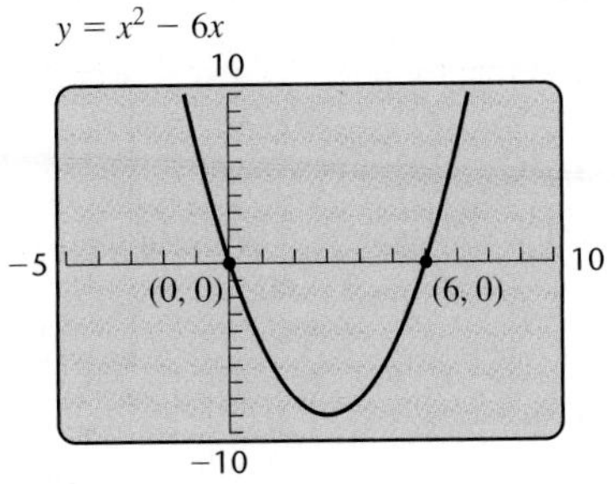

The solutions of the equation are the x-coordinates of the x-intercepts of the graph, 0 and 6. Both 0 and 6 check in the original equation.

Study Tip

Math Web Sites

If for some reason you feel uneasy about asking an instructor or tutor for help, and have already considered the Addison-Wesley Math Tutor Center (see the Preface), you may benefit from visiting the Web sites www.hotmath.com, www.mathforum.org/dr.math, www.sosmath.com, and www.webmath.com. These sites are not specifically for adult learners, but can be very helpful nonetheless. (*Source*: *The Burlington Free Press*, 4/30/02, p. 8 B)

In Example 1, the values 0 and 6 are called **zeros** of the function $f(x) = x^2 - 6x$. They are also referred to as **roots** of the equation $f(x) = 0$.

Zeros and Roots The x-values for which a function $f(x)$ is 0 are called the *zeros* of the function.

The x-values for which an equation such as $f(x) = 0$ is true are called the *roots* of the equation.

In this chapter, we consider only the real-number zeros of functions. We can solve, or find the roots of, the equation $f(x) = 0$ by finding the zeros of the function f.

EXAMPLE 2 Find the zeros of the function given by

$$f(x) = x^3 - 3x^2 - 4x + 12.$$

SOLUTION First, we graph the equation $y = x^3 - 3x^2 - 4x + 12$, choosing a viewing window that shows the x-intercepts of the graph. It may require trial and error to choose an appropriate viewing window. We might try the standard viewing window first, as shown on the left below. It appears that there are three zeros, but we cannot see the shape of the graph for x-values between -2 and 1; Ymax should be increased. We might next try a viewing window of $[-10, 10, -100, 100]$, with Yscl $= 10$, as shown in the middle below. Here we get a better idea of the overall shape of the graph, but we cannot see all three x-intercepts clearly.

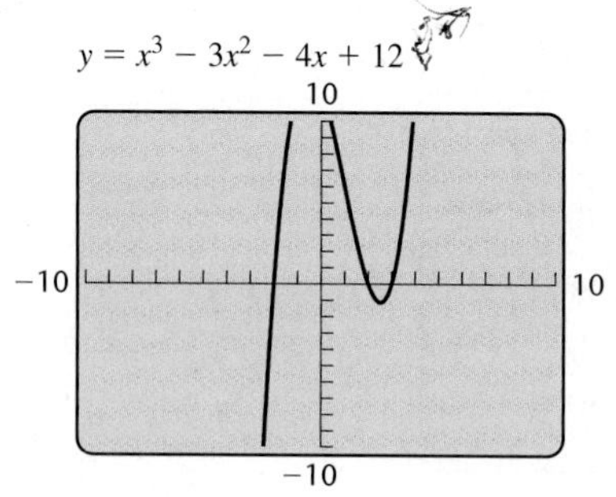

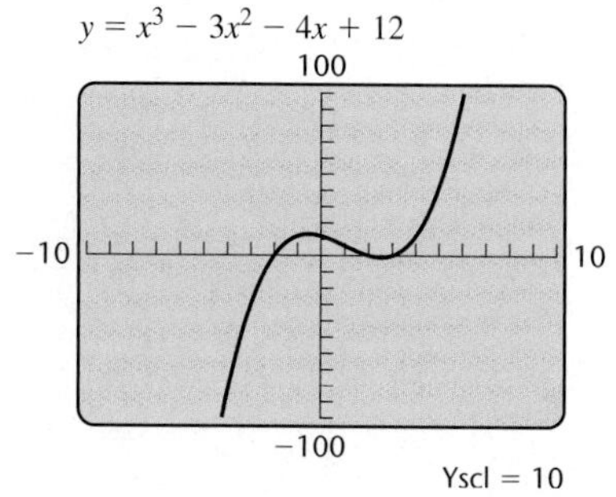

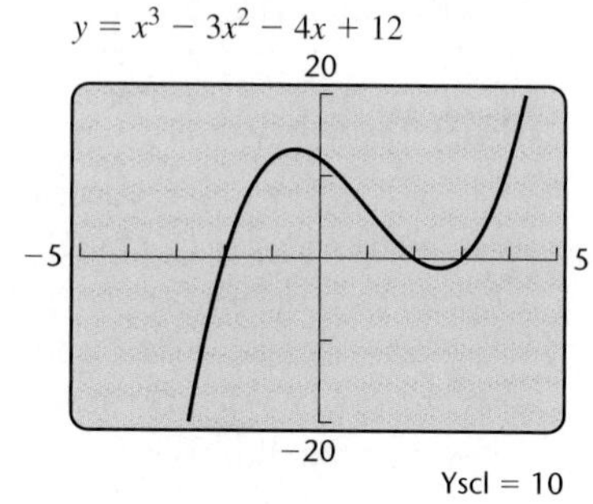

We magnify the portion of the graph close to the origin by making the window dimensions smaller. The window shown on the right above, $[-5, 5, -20, 20]$, is a good choice for viewing the zeros of this function. There are other good choices as well. The zeros of the function seem to be about -2, 2, and 3.

To find the zero that appears to be about -2, we first choose the ZERO option from the CALC menu. We then choose a Left Bound to the left of -2 on the x-axis. Next, we choose a Right Bound to the right of -2 on the x-axis. For a Guess, we choose an x-value close to -2. We see that -2 is indeed a zero of the function f.

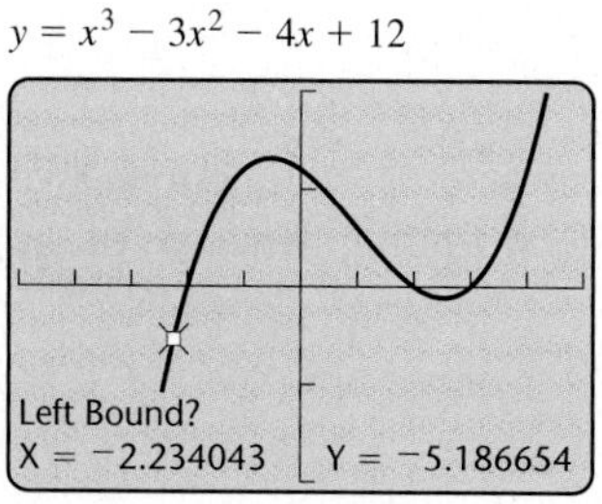

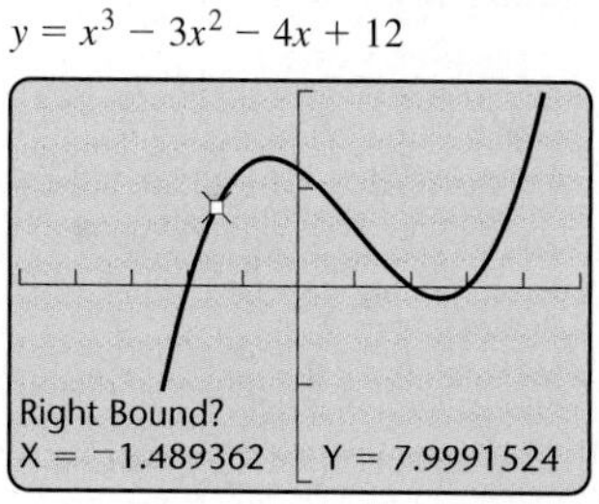

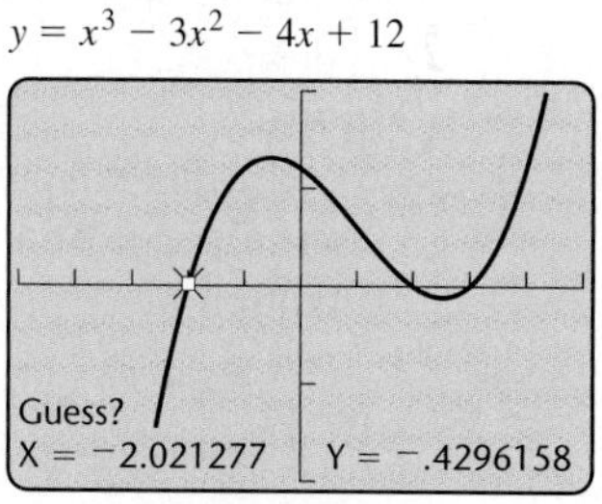

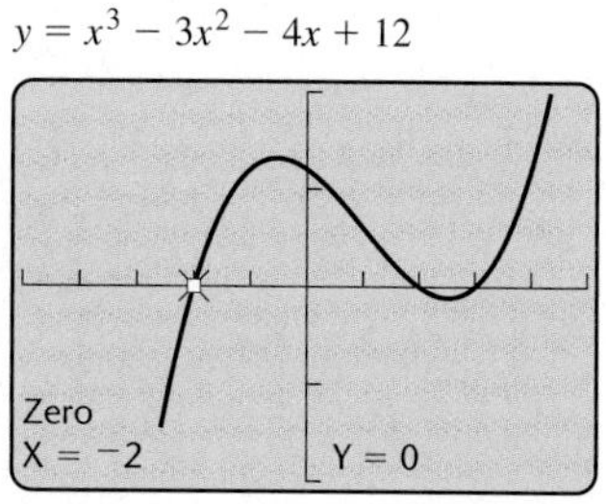

Using the same procedure for each of the other two zeros, we find that the zeros of the function $f(x) = x^3 - 3x^2 - 4x + 12$ are -2, 2, and 3.

We saw in the discussion after Example 1 that $f(x) = x^2 - 6x$ has 2 zeros. We also saw in Example 2 that $f(x) = x^3 - 3x^2 - 4x + 12$ has 3 zeros. We can tell something about the number of zeros of a polynomial function from its degree.

Interactive Discovery

Each of the following is a second-degree polynomial (quadratic) function. Use a graph to determine the number of real-number zeros of each function.

1. $f(x) = x^2 - 2x + 1$
2. $g(x) = x^2 + 9$
3. $h(x) = 2x^2 + 7x - 4$

Each of the following is a third-degree polynomial (cubic) function. Use a graph to determine the *number* of real-number zeros of each function.

4. $f(x) = 2x^3 - 5x^2 - 3x$
5. $g(x) = x^3 - 7x^2 + 16x \quad 12$
6. $h(x) = x^3 - 4x^2 + 5x - 20$
7. Compare the degree of each function with the number of real-number zeros of that function. What conclusion can you draw?

A second-degree polynomial function will have 0, 1, or 2 real-number zeros. A third-degree polynomial function will have 1, 2, or 3 real-number zeros. This result can be generalized, although we will not prove it here.

An nth-degree polynomial function will have at most n zeros.

"Seeing" all the zeros of a polynomial function when solving an equation graphically depends on a good choice of viewing window. Many polynomial equations can be solved algebraically. One principle used in solving polynomial equations is the *principle of zero products.*

The Principle of Zero Products

When we multiply two or more numbers, the product is 0 if any one of those numbers (factors) is 0. Conversely, if a product is 0, then at least one of the factors must be 0. This property of 0 gives us a new principle for solving equations.

The Principle of Zero Products For any real numbers a and b: If $ab = 0$, then $a = 0$ or $b = 0$. If $a = 0$ or $b = 0$, then $ab = 0$.

When a polynomial is written as a product, we say that it is *factored.* The product is called a *factorization* of the polynomial. For example,

$$3x(x + 4) \quad \text{is a factorization of} \quad 3x^2 + 12x$$

since $3x(x + 4)$ is a product and

$$3x(x + 4) = 3x^2 + 12x.$$

The polynomials $3x$ and $x + 4$ are *factors* of $3x^2 + 12x$. The zeros of a polynomial function are related to the factorization of the polynomial.

Consider the function f given by $f(x) = 3x^2 + 12x$. If $g(x) = 3x$ and $h(x) = x + 4$, then $f(x) = g(x) \cdot h(x)$. The graph on the left below shows that the zero of g is 0. The graph in the middle shows that the zero of h is -4. On the right, we see that the zeros of f are 0 and -4.

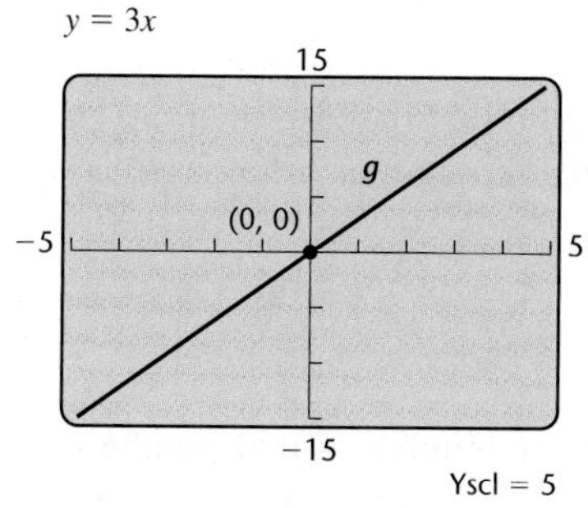

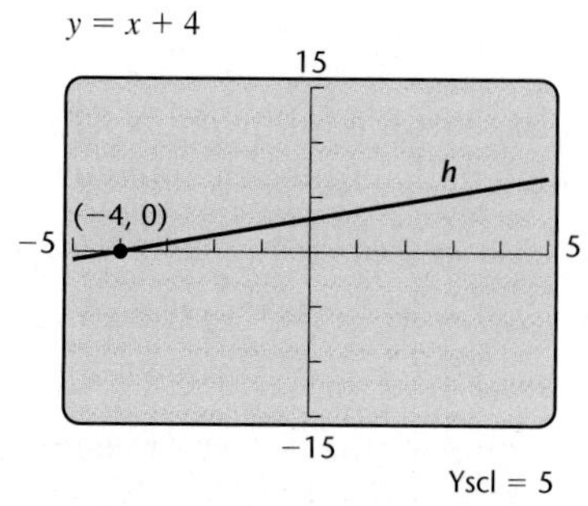

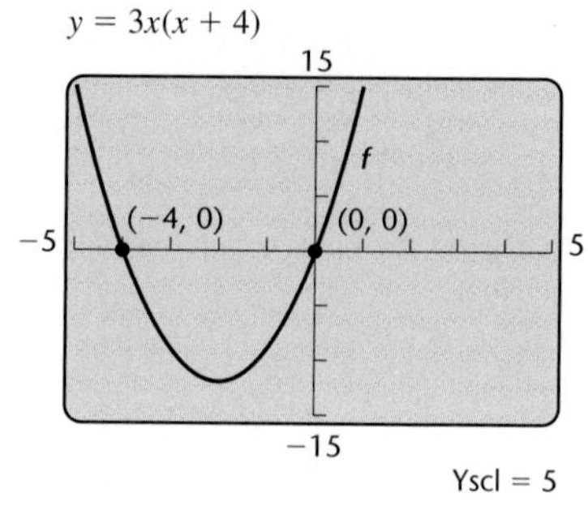

The zeros of a polynomial function are the zeros of the functions described by the factors of the polynomial.

Thus we can use the principle of zero products to solve a polynomial equation. In order to do so, we must have 0 on one side of the equation and the other side must be factored.

EXAMPLE 3 Solve: $(x - 3)(x + 2) = 0$.

SOLUTION The principle of zero products says that in order for $(x - 3)(x + 2)$ to be 0, at least one factor must be 0. Thus,

$x - 3 = 0$ *or* $x + 2 = 0$. **Using the principle of zero products**

Each of these linear equations is then solved separately:

$x = 3$ *or* $x = -2$.

We check both algebraically and graphically, as follows.

ALGEBRAIC CHECK

For 3:

$$\begin{array}{c|c} (x - 3)(x + 2) & 0 \\ \hline (3 - 3)(3 + 2) & 0 \\ 0(5) & \\ 0 & \stackrel{?}{=} 0 \end{array}$$

TRUE

For -2:

$$\begin{array}{c|c} (x - 3)(x + 2) & 0 \\ \hline (-2 - 3)(-2 + 2) & 0 \\ (-5)(0) & \\ 0 & \stackrel{?}{=} 0 \end{array}$$

TRUE

GRAPHICAL CHECK

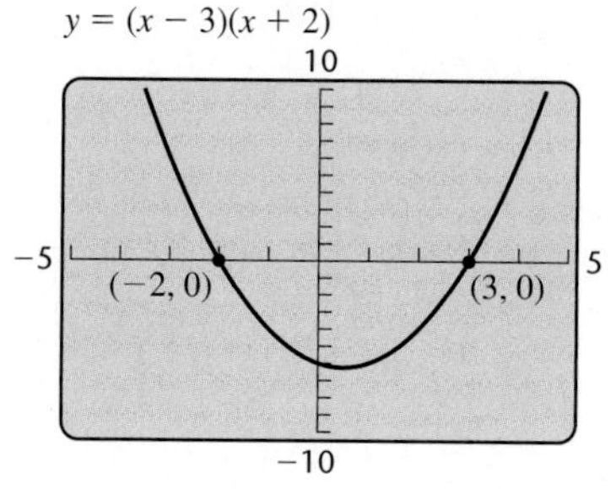

The solutions are -2 and 3.

EXAMPLE 4 Given that $f(x) = x(3x + 2)$, find all the values of a for which $f(a) = 0$.

SOLUTION We are looking for all numbers a for which $f(a) = 0$. Since $f(a) = a(3a + 2)$, we must have

$$a(3a + 2) = 0 \qquad \textbf{Setting } f(a) \textbf{ equal to 0}$$

$$a = 0 \quad or \quad 3a + 2 = 0 \qquad \textbf{Using the principle of zero products}$$

$$a = 0 \quad or \quad a = -\tfrac{2}{3}.$$

We check by evaluating $f(0)$ and $f\left(-\frac{2}{3}\right)$. One way to check with a graphing calculator is to enter $y_1 = x(3x + 2)$ and calculate Y1(0) and Y1(−2/3).

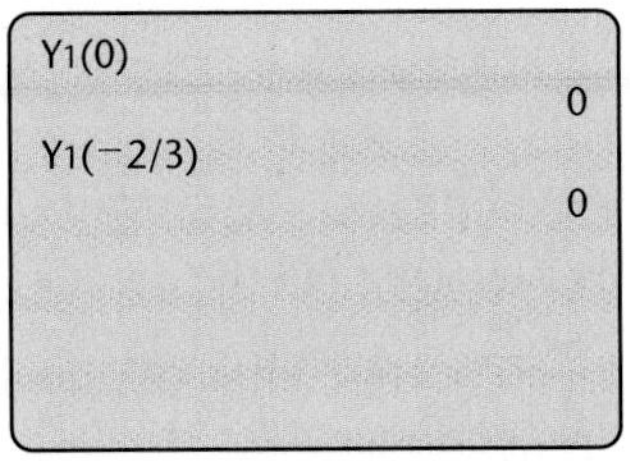

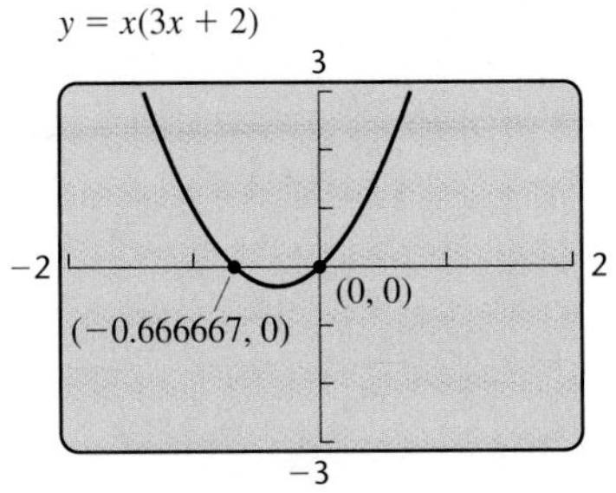

We can also graph $f(x) = x(3x + 2)$ and observe that its zeros are $-\frac{2}{3}$ and 0. Thus, to have $f(a) = 0$, we must have $a = 0$ or $a = -\frac{2}{3}$.

If the polynomial in an equation of the form $p(x) = 0$ is not in factored form, we must factor it before we can use the principle of zero products to solve the equation. To **factor** an expression means to write an equivalent expression that is a product.

Terms with Common Factors

When factoring a polynomial, we look for factors common to every term and then use the distributive law.

EXAMPLE 5 Factor out a common factor: $6y^2 - 18$.

SOLUTION We have

$$6y^2 - 18 = 6 \cdot y^2 - 6 \cdot 3 \qquad \textbf{Noting that 6 is a common factor}$$

$$= 6(y^2 - 3). \qquad \textbf{Using the distributive law}$$

Check: $6(y^2 - 3) = 6y^2 - 18$.

Suppose in Example 5 that the common factor 2 were used:

$$6y^2 - 18 = 2 \cdot 3y^2 - 2 \cdot 9 \qquad \textbf{2 is a common factor.}$$

$$= 2(3y^2 - 9). \qquad \textbf{Using the distributive law}$$

Note that $3y^2 - 9$ itself has a common factor, 3. It is standard practice to factor out the *largest*, or *greatest*, *common factor*, so that the polynomial factor cannot be factored any further. Thus, by now factoring out 3, we can complete the factorization:

$$6y^2 - 18 = 2(3y^2 - 9)$$
$$= 2 \cdot 3(y^2 - 3) = 6(y^2 - 3).$$

Remember to multiply the two common factors: $2 \cdot 3 = 6$.

To find the greatest common factor of a polynomial, we multiply the greatest common factor of the coefficients by the greatest common factor of the variable factors appearing in every term. Thus, to find the greatest common factor of $30x^4 + 20x^5$, we multiply the greatest common factor of 30 and 20, which is 10, by the greatest common factor of x^4 and x^5, which is x^4:

$$30x^4 + 20x^5 = 10 \cdot 3 \cdot x^4 + 10 \cdot 2 \cdot x^4 \cdot x$$
$$= 10x^4(3 + 2x).$$

The greatest common factor is $10x^4$.

EXAMPLE 6 Write an expression equivalent to $8p^6q^2 - 4p^5q^3 + 10p^4q^4$ by factoring out the greatest common factor.

SOLUTION First, we look for the greatest positive common factor in the coefficients:

$$8, -4, 10 \longrightarrow \text{Greatest common factor} = 2.$$

Second, we look for the greatest common factor in the powers of p:

$$p^6, p^5, p^4 \longrightarrow \text{Greatest common factor} = p^4.$$

Third, we look for the greatest common factor in the powers of q:

$$q^2, q^3, q^4 \longrightarrow \text{Greatest common factor} = q^2.$$

Thus, $2p^4q^2$ is the greatest common factor of the given polynomial. Then

$$8p^6q^2 - 4p^5q^3 + 10p^4q^4 = 2p^4q^2 \cdot 4p^2 - 2p^4q^2 \cdot 2pq + 2p^4q^2 \cdot 5q^2$$
$$= 2p^4q^2(4p^2 - 2pq + 5q^2).$$

As a final check, note that

$$2p^4q^2(4p^2 - 2pq + 5q^2) = 2p^4q^2 \cdot 4p^2 - 2p^4q^2 \cdot 2pq + 2p^4q^2 \cdot 5q^2$$
$$= 8p^6q^2 - 4p^5q^3 + 10p^4q^4,$$

which is the original polynomial. Since $4p^2 - 2pq + 5q^2$ has no common factor, we know that $2p^4q^2$ is the greatest common factor.

The polynomials in Examples 5 and 6 have been **factored completely.** They cannot be factored further. The factors in the resulting factorizations are said to be **prime polynomials.**

When the leading coefficient is a negative number, we generally factor out a common factor with a negative coefficient.

EXAMPLE 7 Write an equivalent expression by factoring out a common factor with a negative coefficient.

a) $-4x - 24$

b) $-2x^3 + 6x^2 - 2x$

SOLUTION

a) $-4x - 24 = -4(x + 6)$

b) $-2x^3 + 6x^2 - 2x = -2x(x^2 - 3x + 1)$ **The 1 is essential.**

EXAMPLE 8 Height of a Thrown Object. Suppose that a baseball is thrown upward with an initial velocity of 64 ft/sec and an initial height of 0 ft. Its height in feet, $h(t)$, after t seconds is given by

$$h(t) = -16t^2 + 64t.$$

Find an equivalent expression for $h(t)$ by factoring out a common factor with a negative coefficient.

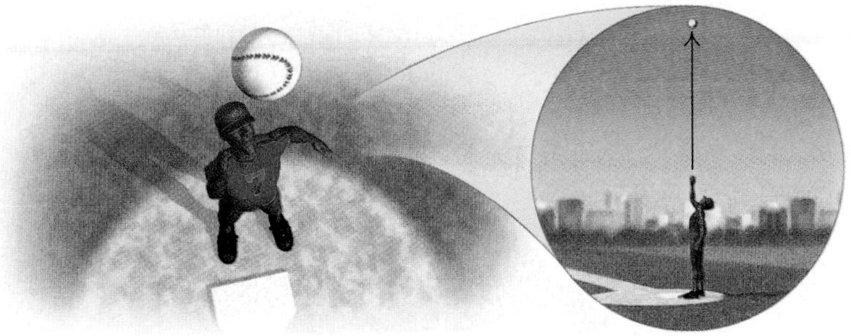

SOLUTION We factor out $-16t$ as follows:

$$h(t) = -16t^2 + 64t = -16t(t - 4).$$ ***Check:*** $-16t \cdot t = -16t^2$ **and** $-16t(-4) = 64t$.

Note that we can obtain function values using either expression for $h(t)$, since factoring forms equivalent expressions. For example,

$$h(1) = -16 \cdot 1^2 + 64 \cdot 1 = 48$$

and $$h(1) = -16 \cdot 1(1 - 4) = 48.$$ **Using the factorization**

$y_1 = -16x^2 + 64x$,
$y_2 = (-16x)(x - 4)$

X	Y1	Y2
0	0	0
1	48	48
2	64	64
3	48	48
4	0	0
5	−80	−80
6	−192	−192
X = 0		

We can evaluate the expressions $-16t^2 + 64t$ and $-16t(t - 4)$ in Example 8 using any value for t. The results should always match. Thus a quick partial check of any factorization is to evaluate the factorization and the original polynomial for one or two convenient replacements. The check in Example 8 becomes foolproof if three replacements are used. Recall that, in general, an nth-degree factorization is correct if it checks for $n + 1$ different replacements. The table shown at left confirms that the factorization is correct.

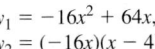

Factoring by Grouping

The largest common factor is sometimes a binomial.

EXAMPLE 9 Factor: $(a - b)(x + 5) + (a - b)(x - y^2)$.

SOLUTION Here the largest common factor is the binomial $a - b$:

$$(a - b)(x + 5) + (a - b)(x - y^2) = (a - b)[(x + 5) + (x - y^2)]$$
$$= (a - b)[2x + 5 - y^2].$$

Often, in order to identify a common binomial factor in a polynomial with four terms, we must regroup into two groups of two terms each.

EXAMPLE 10 Write an equivalent expression by factoring.

a) $y^3 + 3y^2 + 4y + 12$ **b)** $4x^3 - 15 + 20x^2 - 3x$

SOLUTION

a)

$y^3 + 3y^2 + 4y + 12 = (y^3 + 3y^2) + (4y + 12)$	**Each grouping has a common factor.**
$= y^2(y + 3) + 4(y + 3)$	**Factoring out a common factor from each binomial**
$= (y + 3)(y^2 + 4)$	**Factoring out $y + 3$**

b) When we try grouping $4x^3 - 15 + 20x^2 - 3x$ as

$$(4x^3 - 15) + (20x^2 - 3x),$$

we are unable to factor $4x^3 - 15$. When this happens, we can rearrange the polynomial and try a different grouping:

$4x^3 - 15 + 20x^2 - 3x = 4x^3 + 20x^2 - 3x - 15$	**Using the commutative law to rearrange the terms**
$= 4x^2(x + 5) - 3(x + 5)$	**By factoring out -3, we see that $x + 5$ is a common factor.**
$= (x + 5)(4x^2 - 3).$	

In Example 8 of Section 1.3 (see p. 27), we saw that

$$b - a, \quad -(a - b), \quad \text{and} \quad -1(a - b)$$

are equivalent. Remembering this can help anytime we wish to reverse subtraction (see the third step below).

EXAMPLE 11 Factor: $ax - bx + by - ay$.

SOLUTION We have

$ax - bx + by - ay = (ax - bx) + (by - ay)$	**Grouping**
$= x(a - b) + y(b - a)$	**Factoring each binomial**
$= x(a - b) + y(-1)(a - b)$	**Factoring out -1 to reverse $b - a$**
$= x(a - b) - y(a - b)$	**Simplifying**
$= (a - b)(x - y).$	**Factoring out $a - b$**

Check: To check, note that $a - b$ and $x - y$ are both prime and that

$$(a - b)(x - y) = ax - ay - bx + by = ax - bx + by - ay.$$

Student Notes

In Example 11, make certain that you understand why -1 or $-y$ must be factored from $by - ay$.

Some polynomials with four terms, like $x^3 + x^2 + 3x - 3$, are prime. Not only is there no common monomial factor, but no matter how we group terms, there is no common binomial factor:

$$x^3 + x^2 + 3x - 3 = x^2(x + 1) + 3(x - 1); \quad \textbf{No common factor}$$
$$x^3 + 3x + x^2 - 3 = x(x^2 + 3) + (x^2 - 3); \quad \textbf{No common factor}$$
$$x^3 - 3 + x^2 + 3x = (x^3 - 3) + x(x + 3). \quad \textbf{No common factor}$$

Factoring and Equations

Factoring can help us solve polynomial equations.

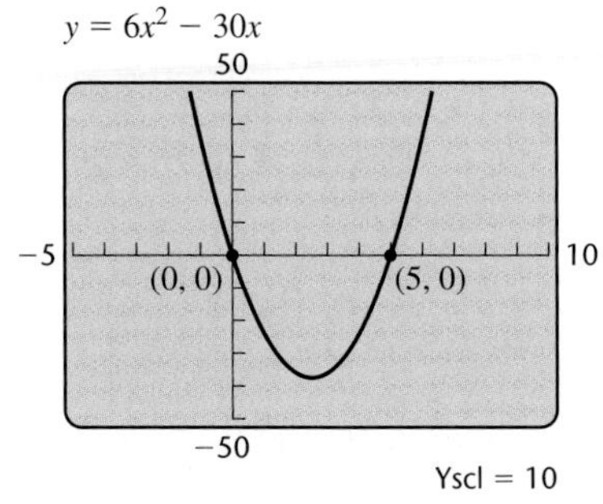

EXAMPLE 12 Solve: $6x^2 = 30x$.

SOLUTION We can use the principle of zero products if there is a 0 on one side of the equation and the other side is in factored form:

$$6x^2 = 30x$$
$$6x^2 - 30x = 0 \quad \textbf{Subtracting } 30x\textbf{. One side is now 0.}$$
$$6x(x - 5) = 0 \quad \textbf{Factoring}$$
$$6x = 0 \quad or \quad x - 5 = 0 \quad \textbf{Using the principle of zero products}$$
$$x = 0 \quad or \quad x = 5.$$

We check by substitution or graphically, as shown in the figure at left. The solutions are 0 and 5.

Connecting the Concepts

When we factor an expression such as $6x^2 - 30x$, we are finding an equivalent expression:

$$6x^2 - 30x = 6x(x - 5).$$

We are *not* "solving" the expression. Remember that we can *factor expressions* and *solve equations*. In the process of solving an equation, we may factor an expression within the equation:

$$6x^2 - 30x = 0 \quad \textbf{This is an equation that we can attempt to solve.}$$
$$6x(x - 5) = 0. \quad \textbf{We factor the expression } 6x^2 - 30x\textbf{.}$$

Do not make the mistake of attempting to solve an expression.

To Use the Principle of Zero Products

1. Write an equivalent equation with 0 on one side, using the addition principle.
2. Factor the nonzero side of the equation.
3. Set each factor that is not a constant equal to 0.
4. Solve the resulting equations.

5.3 EXERCISE SET

FOR EXTRA HELP

MathXL | MyMathLab | InterAct Math | AW Math Tutor Center | Video Lectures on CD: Disc 3 | Student's Solutions Manual

Concept Reinforcement *Classify each statement as either true or false.*

1. The largest common factor of $10x^4 + 15x^2$ is $5x$.

2. The largest common factor of a polynomial always has the same degree as the polynomial itself.

3. It is possible for a polynomial to contain several different common factors.

4. When the leading coefficient of a polynomial is negative, we generally factor out a common factor with a negative coefficient.

5. A polynomial is not prime if it contains a common factor other than 1 or -1.

6. It is possible for a polynomial with four terms to factor into a product of two binomials.

7. The expressions $b - a$, $-(a - b)$, and $-1(a - b)$ are all equivalent.

8. The complete factorization of $12x^3 - 20x^2$ is $4x(3x^2 - 5x)$.

In Exercises 9 and 10, use the graph to solve $f(x) = 0$.

9.

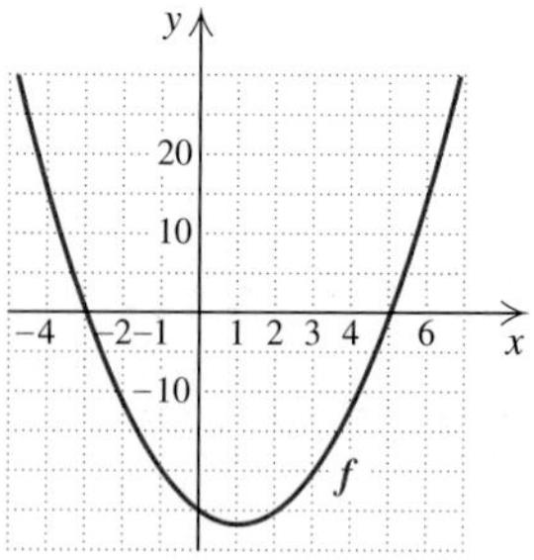

10.

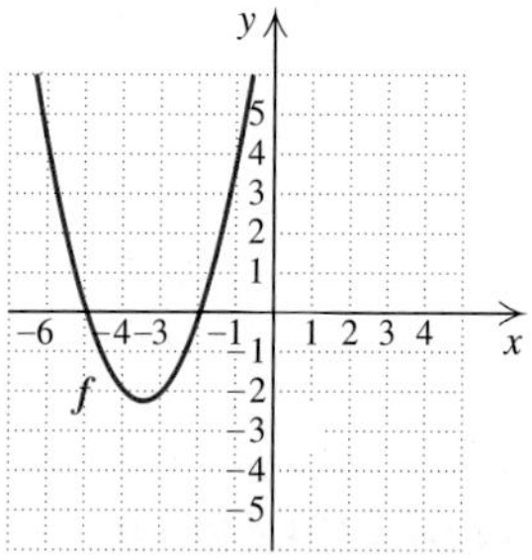

In Exercises 11 and 12, use the graph to find the zeros of the function f.

11.

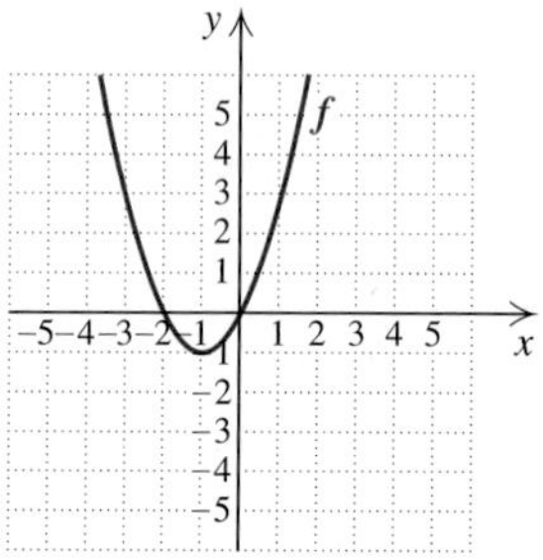

12.

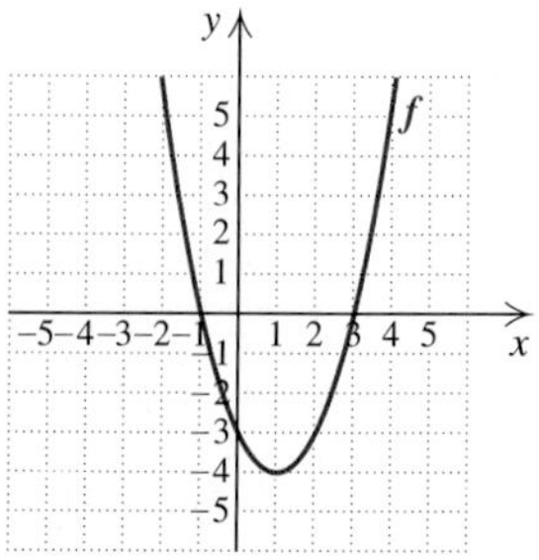

13. Use the following graph to solve $x^2 + 2x = 3$.

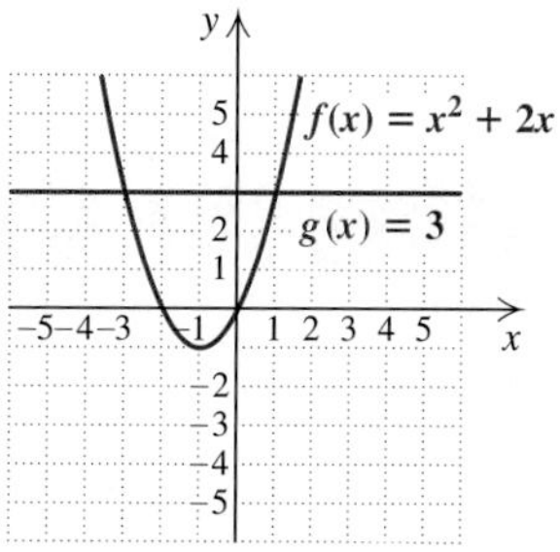

14. Use the following graph to solve $x^2 = 4$.

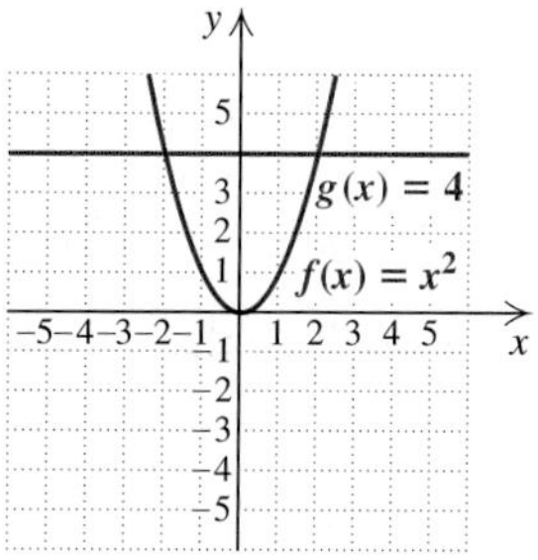

15. Use the following graph to solve $x^2 + 2x - 8 = 0$.

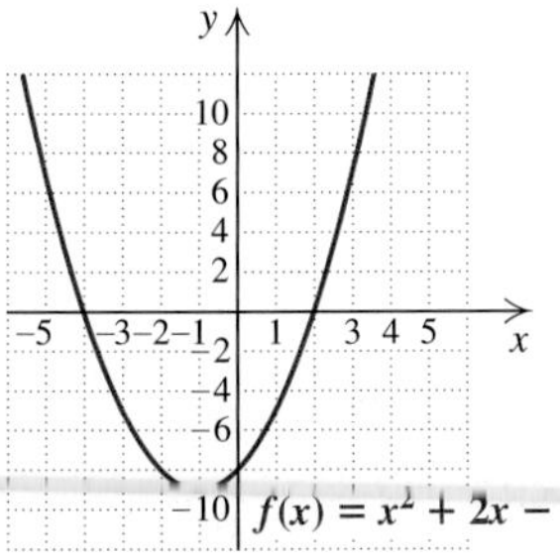

16. Use the following graph to find the zeros of the function given by $f(x) = x^2 - 2x + 1$.

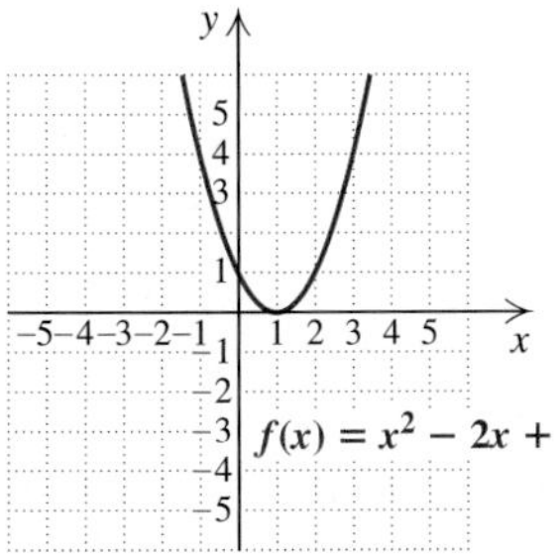

Solve using a graphing calculator.

17. $x^2 = 5x$

18. $2x^2 = 20x$

19. $4x = x^2 + 3$

20. $x^2 = 1$

21. $x^2 + 150 = 25x$

22. $2x^2 + 25 = 51x$

23. $x^3 - 3x^2 + 2x = 0$

24. $x^3 + 2x^2 = x + 2$

25. $x^3 - 3x^2 - 198x + 1080 = 0$

26. $2x^3 + 25x^2 - 282x + 360 = 0$

27. $21x^2 + 2x - 3 = 0$

28. $66x^2 - 49x - 5 = 0$

Find the zeros of each function.

29. $f(x) = x^2 - 4x - 45$

30. $g(x) = x^2 + x - 20$

31. $p(x) = 2x^2 - 13x - 7$

32. $f(x) = 6x^2 + 17x + 6$

33. $f(x) = x^3 - 2x^2 - 3x$

34. $r(x) = 3x^3 - 12x$

Aha! *Match each graph to the corresponding function in Exercises 35–38.*

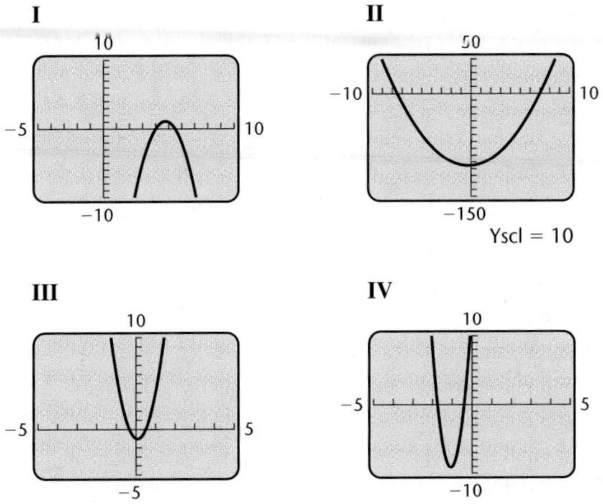

35. $f(x) = (2x - 1)(3x + 1)$

36. $f(x) = (2x + 15)(x - 7)$

37. $f(x) = (4 - x)(2x - 11)$

38. $f(x) = (5x + 2)(4x + 7)$

Tell whether each of the following is an expression or an equation.

39. $x^2 + 6x + 9$

40. $x^3 = x^2 - x + 3$

41. $3x^2 = 3x$

42. $x^4 + 3x^3 + x^2$

43. $2x^3 + x^2 = 0$

44. $5x^4 + 5x$

Write an equivalent expression by factoring out the greatest common factor.

45. $2t^2 + 8t$

46. $3y^2 + 6y$

47. $y^3 + 9y^2$

48. $x^3 + 8x^2$

49. $15x^2 - 5x^4$

50. $8y^2 + 4y^4$

51. $4x^2y - 12xy^2$

52. $5x^2y^3 + 15x^3y^2$

53. $3y^2 - 3y - 9$

54. $5x^2 - 5x + 15$

55. $6ab - 4ad + 12ac$

56. $8xy + 10xz - 14xw$

57. $9x^3y^6z^2 - 12x^4y^4z^4 + 15x^2y^5z^3$

58. $14a^4b^3c^5 + 21a^3b^5c^4 - 35a^4b^4c^3$

Write an equivalent expression by factoring out a factor with a negative coefficient.

59. $-5x + 35$

60. $-5x - 40$

61. $-6y - 72$

62. $-8t + 72$

63. $-2x^2 + 4x - 12$

64. $-2x^2 + 12x + 40$

65. $3y - 24x$

66. $7x - 56y$

67. $7s - 14t$

68. $5r - 10s$

69. $-x^2 + 5x - 9$

70. $-p^3 - 4p^2 + 11$

71. $-a^4 + 2a^3 - 13a$

72. $-m^3 - m^2 + m - 2$

Write an equivalent expression by factoring.

73. $a(b - 5) + c(b - 5)$

74. $r(t - 3) - s(t - 3)$

75. $(x + 7)(x - 1) + (x + 7)(x - 2)$

76. $(a + 5)(a - 2) + (a + 5)(a + 1)$

77. $a^2(x - y) + 5(y - x)$

78. $5x^2(x - 6) + 2(6 - x)$

79. $ac + ad + bc + bd$

80. $xy + xz + wy + wz$

81. $b^3 - b^2 + 2b - 2$

82. $y^3 - y^2 + 3y - 3$

83. $a^3 - 3a^2 + 6 - 2a$

84. $t^3 + 6t^2 - 2t - 12$

85. $72x^3 - 36x^2 + 24x$

86. $12a^4 - 21a^3 - 9a^2$

87. $x^6 - x^5 - x^3 + x^4$

88. $y^4 - y^3 - y + y^2$

89. $2y^4 + 6y^2 + 5y^2 + 15$

90. $2xy - x^2y - 6 + 3x$

91. *Height of a Baseball.* A baseball is popped up with an upward velocity of 72 ft/sec. Its height in feet, $h(t)$, after t seconds is given by

$$h(t) = -16t^2 + 72t.$$

a) Find an equivalent expression for $h(t)$ by factoring out a common factor with a negative coefficient.

b) Perform a partial check of part (a) by evaluating both expressions for $h(t)$ at $t = 1$.

92. *Height of a Rocket.* A model rocket is launched upward with an initial velocity of 96 ft/sec. Its height in feet, $h(t)$, after t seconds is given by

$$h(t) = -16t^2 + 96t.$$

a) Find an equivalent expression for $h(t)$ by factoring out a common factor with a negative coefficient.

b) Check your factoring by evaluating both expressions for $h(t)$ at $t = 1$.

93. *Airline Routes.* When an airline links n cities so that from any one city it is possible to fly directly to each of the other cities, the total number of direct routes is given by

$$R(n) = n^2 - n.$$

Find an equivalent expression for $R(n)$ by factoring out a common factor.

94. *Surface Area of a Silo.* A silo is a structure that is shaped like a right circular cylinder with a half sphere on top. The surface area of a silo of height h and radius r (including the area of the base) is given by the polynomial $2\pi rh + \pi r^2$. Find an equivalent expression by factoring out a common factor.

95. *Total Profit.* When x hundred CD players are sold, Rolics Electronics collects a profit of $P(x)$, where

$$P(x) = x^2 - 3x,$$

and $P(x)$ is in thousands of dollars. Find an equivalent expression by factoring out a common factor.

96. *Total Profit.* After t weeks of production, Claw Foot, Inc., is making a profit of $P(t) = t^2 - 5t$ from sales of their surfboards. Find an equivalent expression by factoring out a common factor.

97. *Total Revenue.* Urban Sounds is marketing a new MP3 player. The firm determines that when it sells x units, the total revenue R is given by the polynomial function

$$R(x) = 280x - 0.4x^2 \text{ dollars.}$$

Find an equivalent expression for $R(x)$ by factoring out $0.4x$.

98. *Total Cost.* Urban Sounds determines that the total cost C of producing x MP3 players is given by the polynomial function

$$C(x) = 0.18x + 0.6x^2.$$

Find an equivalent expression for $C(x)$ by factoring out $0.6x$.

99. *Counting Spheres in a Pile.* The number N of spheres in a triangular pile like the one shown here is a polynomial function given by

$$N(x) = \tfrac{1}{6}x^3 + \tfrac{1}{2}x^2 + \tfrac{1}{3}x,$$

where x is the number of layers and $N(x)$ is the number of spheres. Find an equivalent expression for $N(x)$ by factoring out $\frac{1}{6}$.

100. *Number of Games in a League.* If there are n teams in a league and each team plays every other team once, we can find the total number of games played by using the polynomial function $f(n) = \frac{1}{2}n^2 - \frac{1}{2}n$. Find an equivalent expression by factoring out $\frac{1}{2}$.

101. *High-fives.* When a team of n players all give each other high-fives, a total of $H(n)$ hand slaps occurs, where

$$H(n) = \tfrac{1}{2}n^2 - \tfrac{1}{2}n.$$

Find an equivalent expression by factoring out $\frac{1}{2}n$.

102. *Number of Diagonals.* The number of diagonals of a polygon having n sides is given by the polynomial function

$$P(n) = \tfrac{1}{2}n^2 - \tfrac{3}{2}n.$$

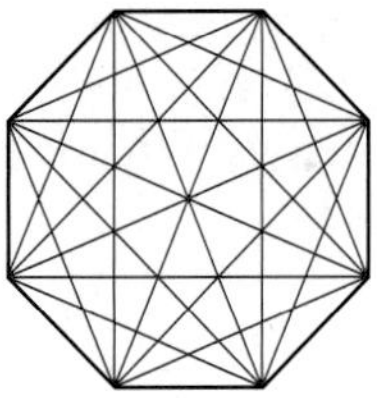

Find an equivalent expression for $P(n)$ by factoring out $\frac{1}{2}$.

Solve using the principle of zero products.

103. $x(x + 1) = 0$

104. $5x(x - 2) = 0$

105. $x^2 - 3x = 0$

106. $2x^2 + 8x = 0$

107. $-5x^2 = 15x$

108. $2x - 4x^2 = 0$

109. $12x^4 + 4x^3 = 0$

110. $21x^3 = 7x^2$

TW **111.** Under what conditions would it be easier to evaluate a polynomial *after* it has been factored?

TW **112.** Gabrielle claims that the zeros of the function given by $f(x) = x^4 - 3x^2 + 7x + 20$ are -1, 1, 2, 4, and 5. How can you tell, without performing any calculations, that she cannot be correct?

Skill Maintenance

Simplify. [1.2]

113. $2(-3) + 4(-5)$

114. $-7(-2) + 5(-3)$

115. $4(-6) - 3(2)$

116. $5(-3) + 2(-2)$

117. *Geometry.* The perimeter of a triangle is 174. The lengths of the three sides are consecutive even numbers. What are the lengths of the sides of the triangle? [1.7]

118. *Manufacturing.* In a factory, there are three machines A, B, and C. When all three are running, they produce 222 suitcases per day. If A and B work but C does not, they produce 159 suitcases per day. If B and C work but A does not, they produce 147 suitcases. What is the daily production of each machine? [3.5]

Synthesis

TW **119.** Is it true that if the coefficients and exponents of a polynomial are all prime numbers, then the polynomial itself is prime? Why or why not?

TW **120.** Following Example 8, we stated that checking the factorization of a second-degree polynomial by making a single replacement is only a *partial* check. Write an *incorrect* factorization and explain how evaluating both the polynomial and the factorization might not catch the mistake.

121. Use the results of Exercise 15 to factor $x^2 + 2x - 8$.

122. Use the results of Exercise 16 to factor $x^2 - 2x + 1$.

Complete each of the following.

123. $x^5y^4 + \_\_\_\_ = x^3y(\_\_\_\_ + xy^5)$

124. $a^3b^7 - \_\_\_\_ = \_\_\_\_(ab^4 - c^2)$

Write an equivalent expression by factoring.

125. $rx^2 - rx + 5r + sx^2 - sx + 5s$

126. $3a^2 + 6a + 30 + 7a^2b + 14ab + 70b$

127. $a^4x^4 + a^4x^2 + 5a^4 + a^2x^4 + a^2x^2 + 5a^2 + 5x^4 + 5x^2 + 25$
(*Hint*: Use three groups of three.)

Write an equivalent expression by factoring out the smallest power of x in each of the following.

128. $x^{-8} + x^{-4} + x^{-6}$

129. $x^{-6} + x^{-9} + x^{-3}$

130. $x^{3/4} + x^{1/2} - x^{1/4}$

131. $x^{1/3} - 5x^{1/2} + 3x^{3/4}$

132. $x^{-3/2} + x^{-1/2}$

133. $x^{-5/2} + x^{-3/2}$

134. $x^{-3/4} - x^{-5/4} + x^{-1/2}$

135. $x^{-4/5} - x^{-7/5} + x^{-1/3}$

Write an equivalent expression by factoring. Assume that all exponents are natural numbers.

136. $2x^{3a} + 8x^a + 4x^{2a}$

137. $3a^{n+1} + 6a^n - 15a^{n+2}$

138. $4x^{a+b} + 7x^{a-b}$

139. $7y^{2a+b} - 5y^{a+b} + 3y^{a+2b}$

5.4 Equations Containing Trinomials of the Type $x^2 + bx + c$

Factoring Trinomials of the Type $x^2 + bx + c$ ■ Equations Containing Trinomials ■ Zeros and Factoring

In this section, we expand our list of the types of polynomials that we can factor so that we can solve a wider variety of polynomial equations. We begin by factoring trinomials of the type $x^2 + bx + c$ and then solve equations containing such trinomials. We then see how to use zeros of a function to factor a polynomial.

Factoring Trinomials of the Type $x^2 + bx + c$

When trying to factor trinomials of the type $x^2 + bx + c$, we can use a trial-and-error procedure.

Study Tip

Pace Yourself

Most instructors agree that it is better for a student to study for one hour four days in a week, than to study once a week for four hours. Of course, the total weekly study time will vary from student to student. It is common to expect an average of two hours of homework for each hour of class time.

Constant Term Positive

Recall the FOIL method of multiplying two binomials:

$$\begin{aligned} & \quad\;\; \text{F} \quad\;\; \text{O} \quad\;\; \text{I} \quad\;\; \text{L} \\ (x + 3)(x + 5) &= x^2 + \underbrace{5x + 3x}_{\downarrow} + 15 \\ &= x^2 + \quad 8x \quad + 15. \end{aligned}$$

Because the leading coefficient in each binomial is 1, the leading coefficient in the product is also 1. To factor $x^2 + 8x + 15$, we think of FOIL: The first term, x^2, is the product of the First terms of two binomial factors, so the first term in each binomial must be x. The challenge is to find two numbers p and q such that

$$\begin{aligned} x^2 + 8x + 15 &= (x + p)(x + q) \\ &= x^2 + qx + px + pq. \end{aligned}$$

Note that the Outer and Inner products, qx and px, can be written as $(p + q)x$. The Last product, pq, will be a constant. Thus the numbers p and q must be selected so that their product is 15 and their sum is 8. In this case, we know from above that these numbers are 3 and 5. The factorization is

$$(x + 3)(x + 5), \quad \text{or} \quad (x + 5)(x + 3). \qquad \textbf{Using a commutative law}$$

In general, to factor $x^2 + (p + q)x + pq$, we use FOIL in reverse:

$$x^2 + (p + q)x + pq = (x + p)(x + q).$$

EXAMPLE 1 Write an equivalent expression by factoring: $x^2 + 9x + 8$.

SOLUTION We think of FOIL in reverse. The first term of each factor is x. We are looking for numbers p and q such that:

$$x^2 + 9x + 8 = (x + p)(x + q) = x^2 + (p + q)x + pq.$$

Thus we search for factors of 8 whose sum is 9.

Pair of Factors	Sum of Factors
2, 4	6
1, 8	9 ← The numbers we need are 1 and 8.

The factorization is thus $(x + 1)(x + 8)$. The student should check by multiplying to confirm that the product is the original trinomial.

When factoring trinomials with a leading coefficient of 1, it suffices to consider all pairs of factors along with their sums, as we did above. At times, however, you may be tempted to form factors without calculating any sums. It is essential that you check any attempt made in this manner! For example, if we attempt the factorization

$$x^2 + 9x + 8 \stackrel{?}{=} (x + 2)(x + 4),$$

a check reveals that

$$(x + 2)(x + 4) = x^2 + 6x + 8 \neq x^2 + 9x + 8.$$

This type of trial-and-error procedure becomes easier to use with time. As you gain experience, you will find that many trials can be performed mentally.

When the constant term of a trinomial is positive, the constant terms in the binomial factors must either both be positive or both be negative. This ensures a positive product. The sign used is that of the trinomial's middle term.

EXAMPLE 2 Factor: $t^2 - 9t + 20$.

SOLUTION Since the constant term is positive and the coefficient of the middle term is negative, we look for a factorization of 20 in which both factors are negative. Their sum must be -9.

Pair of Factors	Sum of Factors
$-1, -20$	-21
$-2, -10$	-12
$-4, -5$	-9 ← The numbers we need are -4 and -5.

The factorization is $(t - 4)(t - 5)$. We check by comparing values using a table, as shown at left.

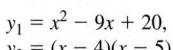

$y_1 = x^2 - 9x + 20,$
$y_2 = (x - 4)(x - 5)$

X	Y1	Y2
0	20	20
1	12	12
2	6	6
3	2	2
4	0	0
5	0	0
6	2	2

X = 0

Constant Term Negative

When the constant term of a trinomial is negative, we look for one negative factor and one positive factor. The sum of the factors must still be the coefficient of the middle term.

EXAMPLE 3 Factor: $x^3 - x^2 - 30x$.

SOLUTION *Always* look first for a common factor! This time there is one, x. We factor it out:

$$x^3 - x^2 - 30x = x(x^2 - x - 30).$$

Now we consider $x^2 - x - 30$. We need a factorization of -30 in which one factor is positive, the other factor is negative, and the sum of the factors is -1. Since the sum is to be negative, the negative factor must have the greater absolute value. Thus we need consider only the following pairs of factors.

Pair of Factors	Sum of Factors
1, −30	−29
3, −10	−7
5, −6	−1 ← **The numbers we need are 5 and −6.**

The factorization of $x^2 - x - 30$ is $(x + 5)(x - 6)$. *Don't forget to include the factor that was factored out earlier*! In this case, the factorization of the original trinomial is $x(x + 5)(x - 6)$. We check by graphing $y_1 = x^3 - x^2 - 30x$ and $y_2 = x(x + 5)(x - 6)$, as shown at left.

$y_1 = x^3 - x^2 - 30x,\ y_2 = x(x + 5)(x - 6)$

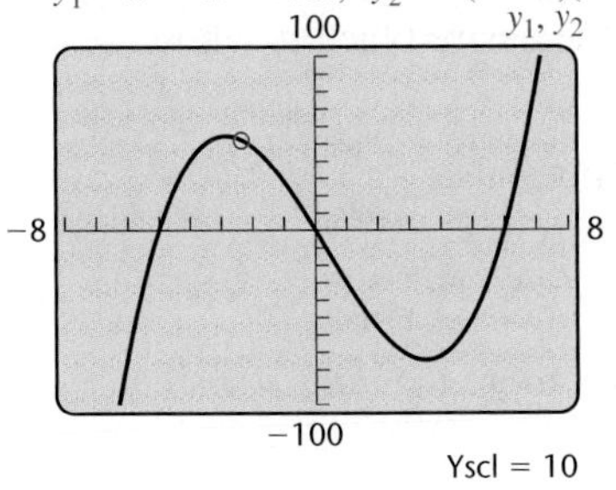

EXAMPLE 4 Factor: $2x^2 + 34x - 220$.

SOLUTION *Always* look first for a common factor! This time we can factor out 2:

$$2x^2 + 34x - 220 = 2(x^2 + 17x - 110).$$

We next look for a factorization of -110 in which one factor is positive, the other factor is negative, and the sum of the factors is 17. Since the sum is to be positive, we examine only pairs of factors in which the positive term has the larger absolute value.

Pair of Factors	Sum of Factors
−1, 110	109
−2, 55	53
−5, 22	17 ← **The numbers we need are −5 and 22.**

The factorization of $x^2 + 17x - 110$ is $(x - 5)(x + 22)$. The factorization of the original trinomial, $2x^2 + 34x - 220$, is $2(x - 5)(x + 22)$.

Check: $2(x - 5)(x + 22) = 2[x^2 + 17x - 110]$
$= 2x^2 + 34x - 220.$

Some polynomials are not factorable using integers.

EXAMPLE 5 Factor: $x^2 - x - 7$.

SOLUTION There are no factors of -7 whose sum is -1. This trinomial is *not* factorable into binomials with integer coefficients. Although $x^2 - x - 7$ can be factored using more advanced techniques, for our purposes the polynomial is **prime.**

Tips for Factoring $x^2 + bx + c$

1. If necessary, rewrite the trinomial in descending order.
2. Find a pair of factors that have c as their product and b as their sum. Remember the following:
   - If c is positive, its factors will have the same sign as b.
   - If c is negative, one factor will be positive and the other will be negative. Select the factors such that the factor with the larger absolute value is the factor with the same sign as b.
   - If the sum of the two factors is the opposite of b, changing the signs of both factors will give the desired factors whose sum is b.
3. Check the result by multiplying the binomials.

These tips also apply when a trinomial has more than one variable.

EXAMPLE 6 Factor: $x^2 - 2xy - 48y^2$.

SOLUTION We look for numbers p and q such that

$$x^2 - 2xy - 48y^2 = (x + py)(x + qy).$$

The x's and y's can be written in the binomials in advance.

Our thinking is much the same as if we were factoring $x^2 - 2x - 48$. We look for factors of -48 whose sum is -2. Those factors are 6 and -8. Thus,

$$x^2 - 2xy - 48y^2 = (x + 6y)(x - 8y).$$

The check is left to the student.

Equations Containing Trinomials

We can now use our new factoring skills to solve some polynomial equations.

CAUTION! In Example 7, we are solving an equation. Do not try to solve an expression!

EXAMPLE 7 Solve: $x^2 + 9x + 8 = 0$.

ALGEBRAIC APPROACH

We use the principle of zero products:

$$x^2 + 9x + 8 = 0$$

$(x + 1)(x + 8) = 0$ **Using the factorization from Example 1**

$x + 1 = 0$ *or* $x + 8 = 0$ **Using the principle of zero products**

$x = -1$ *or* $x = -8.$ **Solving each equation for x**

The solutions are -1 and -8.

GRAPHICAL APPROACH

The real-number solutions of $x^2 + 9x + 8 = 0$ are the first coordinates of the x-intercepts of the graph of $f(x) = x^2 + 9x + 8$.

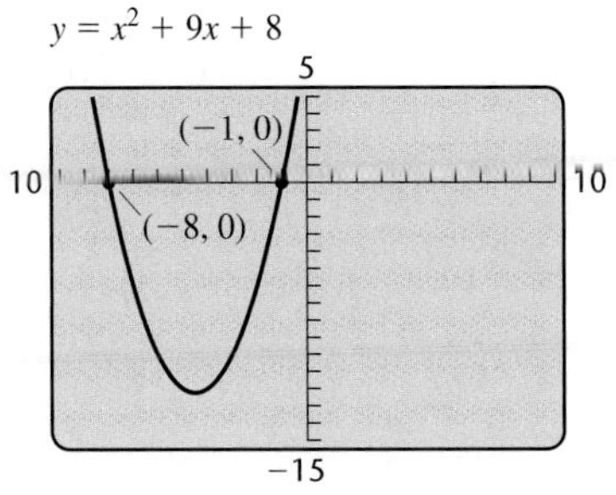

We find that -1 and -8 are solutions.

Check:

For -1:

$$\begin{array}{l|l} x^2 + 9x + 8 & 0 \\ \hline (-1)^2 + 9(-1) + 8 & 0 \\ 1 - 9 + 8 & \\ 0 \stackrel{?}{=} 0 & \text{TRUE} \end{array}$$

For -8:

$$\begin{array}{l|l} x^2 + 9x + 8 & 0 \\ \hline (-8)^2 + 9(-8) + 8 & 0 \\ 64 - 72 + 8 & \\ 0 \stackrel{?}{=} 0 & \text{TRUE} \end{array}$$

The solutions are -1 and -8.

EXAMPLE 8 Solve: $(t - 10)(t + 1) = -24$.

ALGEBRAIC APPROACH

Note that the left side of the equation is factored. It may be tempting to set both factors equal to -24 and solve, but this is *not* correct. The principle of zero products requires 0 on one side of the equation. We begin by multiplying the left side.

$(t - 10)(t + 1) = -24$

$t^2 - 9t - 10 = -24$ **Multiplying**

$t^2 - 9t + 14 = 0$ **Adding 24 to both sides to get 0 on one side**

$(t - 2)(t - 7) = 0$ **Factoring**

$t - 2 = 0$ *or* $t - 7 = 0$ **Using the principle of zero products**

$t = 2$ *or* $t = 7$

The solutions are 2 and 7.

GRAPHICAL APPROACH

There is no need to multiply. We let $y_1 = (x - 10)(x + 1)$ and $y_2 = -24$ and look for any points of intersection of the graphs. The x-coordinates of these points are the solutions of the equation.

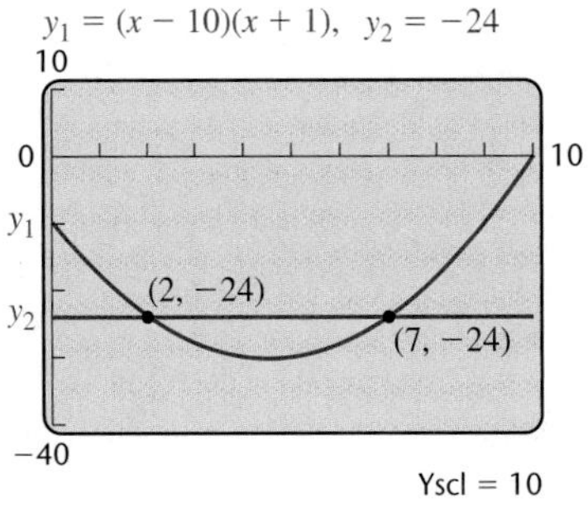

We find that 2 and 7 are solutions.

EXAMPLE 9 Solve: $x^3 = x^2 + 30x$.

ALGEBRAIC APPROACH

In order to solve $x^3 = x^2 + 30x$ using the principle of zero products, we must rewrite the equation with 0 on one side.

We have

$$x^3 = x^2 + 30x$$

$$x^3 - x^2 - 30x = 0 \quad \textbf{Getting 0 on one side}$$

$$x(x-6)(x+5) = 0 \quad \textbf{Using the factorization from Example 3}$$

$$x = 0 \quad or \quad x - 6 = 0 \quad or \quad x + 5 = 0 \quad \textbf{Using the principle of zero products}$$

$$x = 0 \quad or \quad x = 6 \quad or \quad x = -5.$$

The solutions are 0, 6, and -5.

GRAPHICAL APPROACH

We let $y_1 = x^3$ and $y_2 = x^2 + 30x$ and look for points of intersection of the graphs.

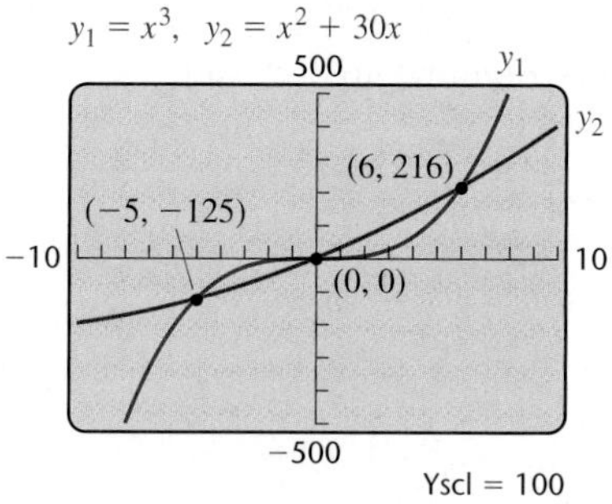

The viewing window $[-10, 10, -500, 500]$ shows three points of intersection. Since the polynomial equation is of degree 3, we know there will be no more than 3 solutions. The x-coordinates of the points of intersection are -5, 0, and 6.

We check by substituting $-5, 0$, and 6 into the original equation. The solutions are $-5, 0$, and 6.

Zeros and Factoring

Note in Example 7 the relationship between the factors of $x^2 + 9x + 8$ and the solutions of the equation $x^2 + 9x + 8 = 0$. We can use the principle of zero products "in reverse" to factor a polynomial and to write a function with given zeros.

EXAMPLE 10 Factor: $x^2 - 13x - 608$.

SOLUTION The factorization of the trinomial $x^2 - 13x - 608$ will be in the form

$$x^2 - 13x - 608 = (x + p)(x + q).$$

We could use trial and error to find p and q, but there are many factors of 608. Thus we factor the trinomial using the principle of zero products.

Note that the roots of the equation

$$x^2 - 13x - 608 = 0$$

are the zeros of the function $f(x) = x^2 - 13x - 608$. We graph the function and find the zeros.

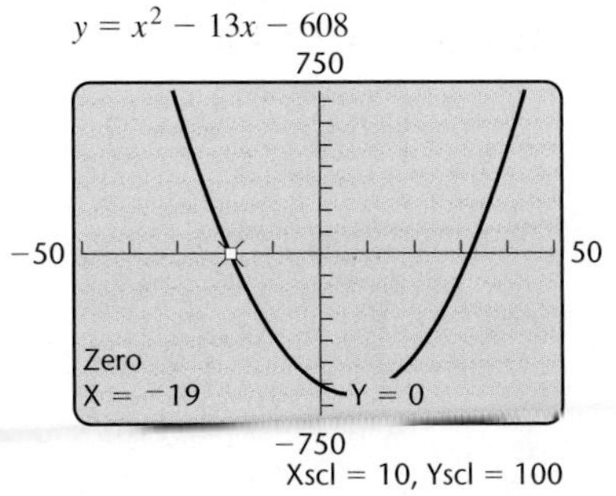

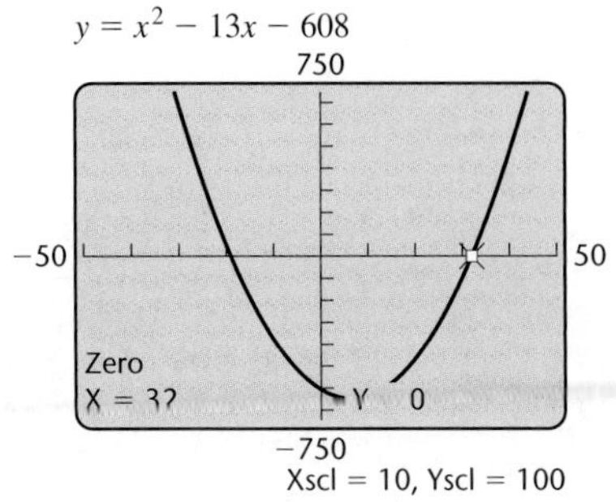

The zeros are -19 and 32. Each zero is a root of $x^2 - 13x - 608 = 0$. We also know that -19 is a zero of the linear function $g(x) = x + 19$ and that 32 is a zero of the linear function $h(x) = x - 32$. This suggests that $x + 19$ and $x - 32$ are factors of $x^2 - 13x - 608$. Using the principle of zero products in reverse, we now have

$$x^2 - 13x - 608 = (x + 19)(x - 32).$$

Multiplication indicates that the factorization is correct.

Not every trinomial is factorable. The roots of $x^2 - 2x - 2 = 0$ are irrational; thus $x^2 - 2x - 2$ cannot be factored using integers. The equation $x^2 - x + 1 = 0$ has no real roots, and is also prime. We will study equations like these in Chapter 8.

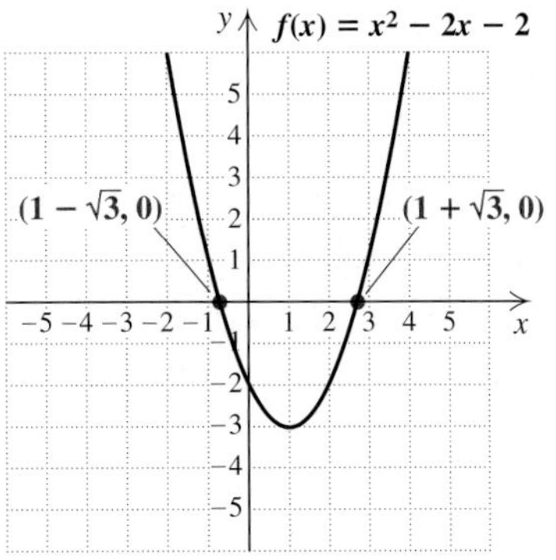

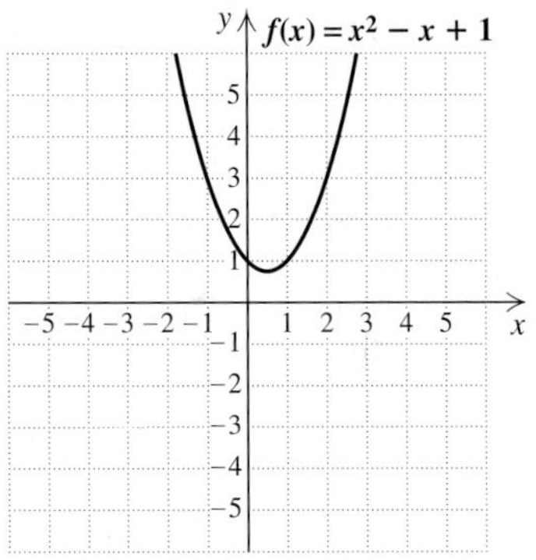

We can also use the principle of zero products to write a function whose zeros are given.

EXAMPLE 11 Write a polynomial function $f(x)$ whose zeros are -1, 0, and 3.

SOLUTION Each zero of the polynomial function f is a zero of a linear factor of the polynomial. We write a linear function for each given zero:

-1 is a zero of $g(x) = x + 1$;

0 is a zero of $h(x) = x$;

and 3 is a zero of $k(x) = x - 3$.

Thus a polynomial function with zeros -1, 0, and 3 is

$$f(x) = (x + 1) \cdot x \cdot (x - 3);$$

multiplying gives us

$$f(x) = x^3 - 2x^2 - 3x.$$

5.4 EXERCISE SET

FOR EXTRA HELP

MathXL | MyMathLab | InterAct Math | AW Math Tutor Center | Video Lectures on CD: Disc 3 | Student's Solutions Manual

Concept Reinforcement *Classify each of the following as either true or false.*

1. When factoring any polynomial, it is always best to look first for a common factor.
2. Whenever the sum of a negative number and a positive number is negative, the negative number has the greater absolute value.
3. Whenever the product of a pair of factors is negative, the factors have the same sign.
4. If $p + q = -17$, then $-p + (-q) = 17$.
5. If b and c are positive and $x^2 + bx + c$ can be factored as $(x + p)(x + q)$, then it follows that p and q are both negative.
6. If b is negative and c is positive and $x^2 + bx + c$ can be factored as $(x + p)(x + q)$, then it follows that p and q are both positive.
7. If 1 is a zero of a polynomial function, then $x - 1$ is a factor of the polynomial.
8. If $x - 2$ is a factor of $p(x)$, then $p(2) = 0$.

Factor. If a polynomial is prime, state this.

9. $x^2 + 8x + 12$
10. $x^2 + 6x + 5$
11. $t^2 + 8t + 15$
12. $y^2 + 12y + 27$
13. $x^2 - 27 - 6x$
14. $t^2 - 15 - 2t$
15. $2n^2 - 20n + 50$
16. $2a^2 - 16a + 32$
17. $a^3 - a^2 - 72a$
18. $x^3 + 3x^2 - 54x$
19. $14x + x^2 + 45$
20. $12y + y^2 + 32$
21. $3x + x^2 - 10$
22. $x + x^2 - 6$
23. $3x^2 + 15x + 18$
24. $5y^2 + 40y + 35$
25. $56 + x - x^2$
26. $32 + 4y - y^2$
27. $32y + 4y^2 - y^3$
28. $56x + x^2 - x^3$
29. $x^4 + 11x^3 - 80x^2$
30. $y^4 + 5y^3 - 84y^2$
31. $x^2 + 12x + 13$
32. $x^2 - 3x + 7$
33. $p^2 - 5pq - 24q^2$
34. $x^2 + 12xy + 27y^2$
35. $y^2 + 8yz + 16z^2$
36. $x^2 - 14xy + 49y^2$
37. $p^4 + 80p^3 + 79p^2$
38. $x^4 + 50x^3 + 49x^2$

39. Use the results of Exercise 9 to solve $x^2 + 8x + 12 = 0$.

40. Use the results of Exercise 10 to solve $x^2 + 6x + 5 = 0$.

41. Use the results of Exercise 15 to solve $2n^2 + 50 = 20n$.

42. Use the results of Exercise 26 to solve $32 + 4y = y^2$.

43. Use the results of Exercise 17 to solve $a^3 - a^2 = 72a$.

44. Use the results of Exercise 28 to solve $56x + x^2 = x^3$.

In Exercises 45–48, use the graph to solve the given equation. Check by substituting into the equation.

Aha! **45.** $x^2 + 4x - 5 = 0$

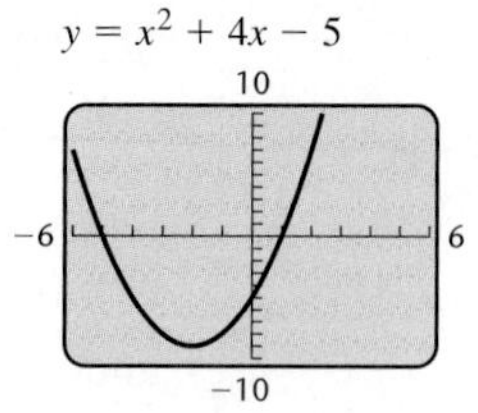

46. $x^2 - x - 6 = 0$

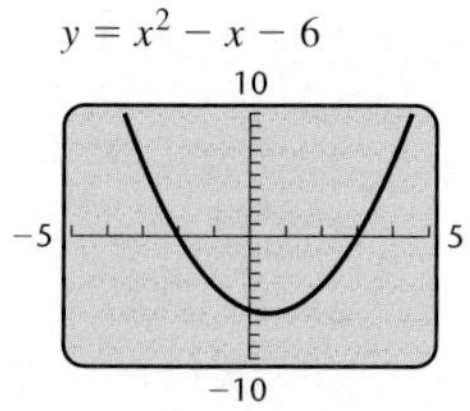

47. $x^2 + x - 6 = 0$

$y = x^2 + x - 6$

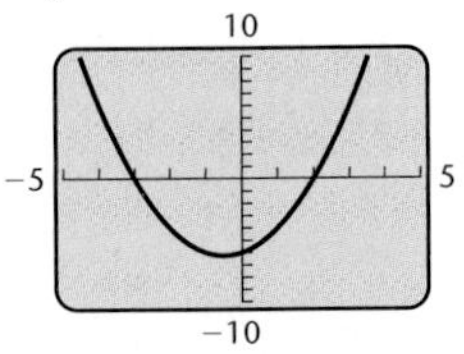

48. $x^2 + 8x + 15 = 0$

$y = x^2 + 8x + 15$

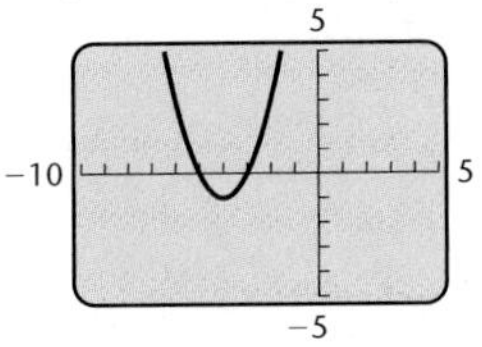

Find the zeros of each function.

49. $f(x) = x^2 - 4x - 45$

50. $f(x) = x^2 + x - 20$

51. $r(x) = x^3 - 2x^2 - 3x$

52. $g(x) = 3x^2 + 21x + 30$

Solve.

53. $x^2 + 4x = 45$

54. $t^2 - 3t = 28$

55. $x^2 - 9x = 0$

56. $a^2 + 18a = 0$

57. $a^3 - 3a^2 = 40a$

58. $x^3 - 2x^2 = 63x$

59. $(x - 3)(x + 2) = 14$

60. $(z + 4)(z - 2) = -5$

61. $35 - x^2 = 2x$

62. $40 - x^2 + 3x = 0$

In Exercises 63 and 64, use the graph to factor the given polynomial.

63. $x^2 + 10x - 264$

64. $x^2 + 16x - 336$

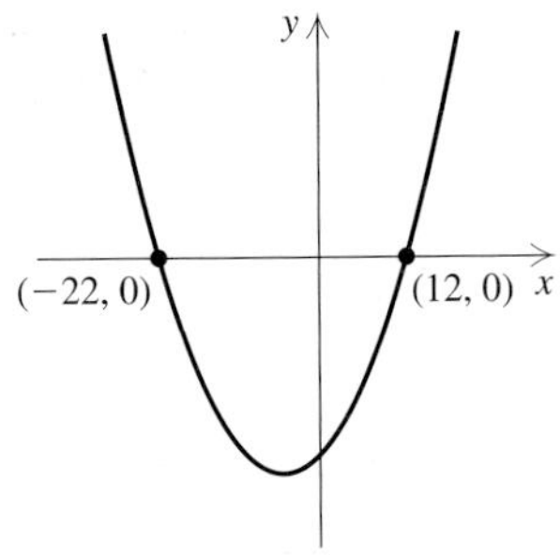

$f(x) = x^2 + 10x - 264$

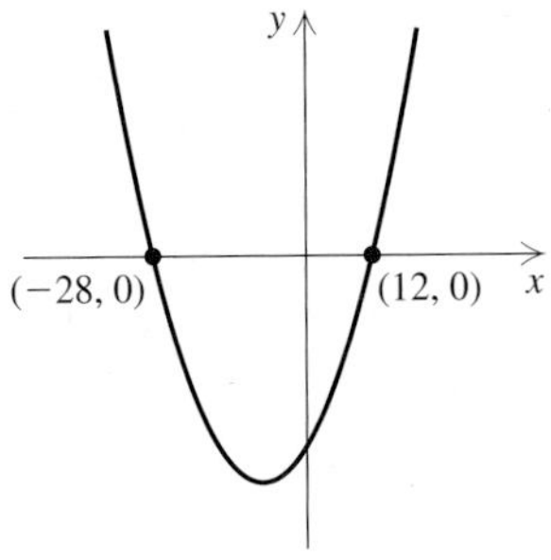

$f(x) = x^2 + 16x - 336$

In Exercises 65–68, use a graph to help factor each polynomial.

65. $x^2 + 40x + 384$

66. $x^2 - 13x - 300$

67. $x^2 + 26x - 2432$

68. $x^2 - 46x + 504$

Write a polynomial function that has the given zeros. Answers may vary.

69. $-1, 2$

70. $2, 5$

71. $-7, -10$

72. $8, -3$

73. $0, 1, 2$

74. $-3, 0, 5$

TW 75. Sandra says that she will never miss a point of intersection when solving graphically because she always uses a $[-100, 100, -100, 100]$ window. Is she correct? Why or why not?

TW 76. How can one conclude that $x^2 - 59x + 6$ is a prime polynomial without performing any trials?

Focused Review

Factor completely.

77. $9x^5 - 24x^3 + 3x^2$ [5.3]

78. $x^4 + 3x^3 + 2x^2$ [5.4]

79. $xy + 3x + 2y + 6$ [5.3]

80. $(x + 2)(x - 7) + (x - 5)(x - 7)$ [5.3]

81. $45x^3y^4 + 30x^2y^5 + 60x^4y^3$ [5.3]

82. $-2t^3 - 4t + 8$ [5.3]

83. $xw - xz - wy + yz$ [5.3]

84. $x^2 + 15x + 50$ [5.4]

Synthesis

TW 85. Explain how the following graph of

$$y = x^2 + 3x - 2 - (x - 2)(x + 1)$$

can be used to show that

$$x^2 + 3x - 2 \neq (x - 2)(x + 1).$$

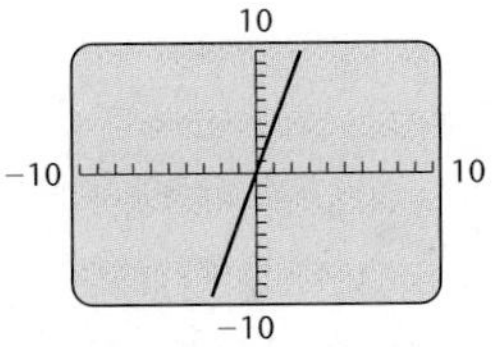

TW 86. Explain why Example 11 reads "Write *a* polynomial function $f(x)$" and not "Write *the* polynomial function $f(x)$."

87. Use the following graph of $f(x) = x^2 - 2x - 3$ to solve $x^2 - 2x - 3 = 0$ and to solve $x^2 - 2x - 3 < 5$.

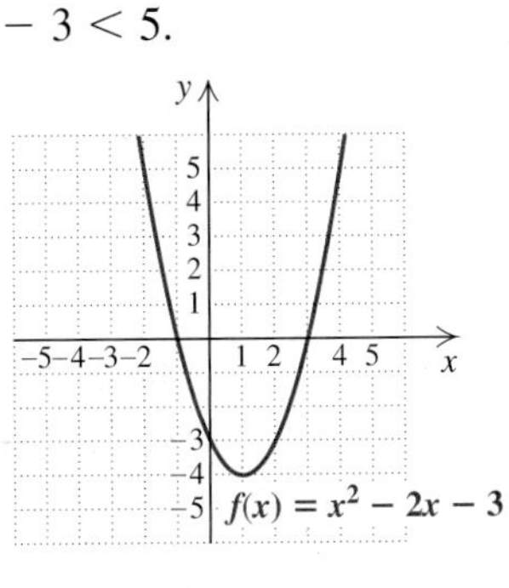

88. Use the following graph of $g(x) = -x^2 - 2x + 3$ to solve $-x^2 - 2x + 3 = 0$ and to solve $-x^2 - 2x + 3 \geq -5$.

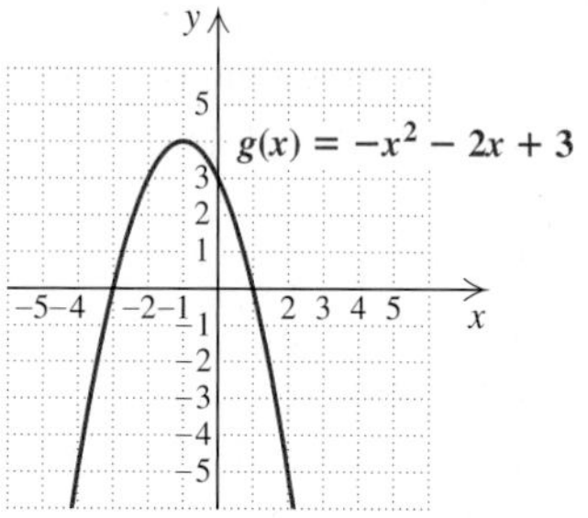

89. Find a polynomial function f for which $f(2) = 0$, $f(-1) = 0, f(3) = 0$, and $f(0) = 30$.

90. Find a polynomial function g for which $g(-3) = 0, g(1) = 0, g(5) = 0$, and $g(0) = 45$.

In Exercises 91–94, use a graphing calculator to find any solutions that exist accurate to two decimal places.

91. $-x^2 + 13.80x = 47.61$

92. $-x^2 + 3.63x + 34.34 = x^2$

93. $x^3 - 3.48x^2 + x = 3.48$

94. $x^2 + 4.68 = 1.2x$

Factor. Assume that variables in exponents represent positive integers.

95. $2a^4b^6 - 3a^2b^3 - 20ab^2$

96. $5x^8y^6 + 35x^4y^3 + 60$

97. $x^2 - \frac{4}{25} + \frac{3}{5}x$

98. $y^2 - \frac{8}{49} + \frac{2}{7}y$

99. $y^2 + 0.4y - 0.05$

100. $x^{2a} + 5x^a - 24$

101. $x^2 + ax + bx + ab$

102. $bdx^2 + adx + bcx + ac$

103. $a^2p^{2a} + a^2p^a - 2a^2$

Aha! **104.** $(x + 3)^2 - 2(x + 3) - 35$

105. Find all integers m for which $x^2 + mx + 75$ can be factored.

106. Find all integers q for which $x^2 + qx - 32$ can be factored.

107. One factor of $x^2 - 345x - 7300$ is $x + 20$. Find the other factor.

5.5 Equations Containing Trinomials of the Type $ax^2 + bx + c$

Factoring Trinomials of the Type $ax^2 + bx + c$ ■ Equations and Functions

Factoring Trinomials of the Type $ax^2 + bx + c$

Now we look at trinomials in which the leading coefficient is not 1. We consider two factoring methods. Use what works best for you or what your instructor chooses for you.

Method 1: Reversing FOIL

We first consider the **FOIL method** for factoring trinomials of the type

$$ax^2 + bx + c, \text{ where } a \neq 1.$$

Consider the following multiplication.

$$\begin{aligned}(3x + 2)(4x + 5) &= \overset{F}{12x^2} + \underbrace{\overset{O}{15x} + \overset{I}{8x}} + \overset{L}{10}\\ &= 12x^2 + \quad 23x \quad + 10\end{aligned}$$

To factor $12x^2 + 23x + 10$, we must reverse what we just did. We look for two binomials whose product is this trinomial. The product of the First terms must be $12x^2$. The product of the Outer terms plus the product of the Inner terms must be $23x$. The product of the Last terms must be 10. We know from the preceding discussion that the factorization is

$$(3x + 2)(4x + 5).$$

In general, however, finding such an answer involves trial and error. We use the following method.

Study Tip

Leave a Trail

Students sometimes make the mistake of viewing their supporting work as "scrap" work. Most instructors regard your reasoning as more important than your final answer. Try to organize your supporting work so that your instructor (and you as well) can follow your steps. Instead of erasing work you are not pleased with, consider simply crossing it out so that you (and others) can study and learn from these attempts.

To Factor $ax^2 + bx + c$ by Reversing FOIL

1. Factor out the largest common factor, if one exists. Here we assume none does.
2. Find two **F**irst terms whose product is ax^2:

   $$(\square x + \quad)(\square x + \quad) = ax^2 + bx + c.$$
   FOIL

3. Find two **L**ast terms whose product is c:

   $$(\quad x + \square)(\quad x + \square) = ax^2 + bx + c.$$
   FOIL

4. Repeat steps (2) and (3) until a combination is found for which the sum of the **O**uter and **I**nner products is bx:

   $$(\square x + \square)(\square x + \square) = ax^2 + bx + c.$$
   I, O — FOIL

EXAMPLE 1 Factor: $3x^2 + 10x - 8$.

SOLUTION

1. First, note that there is no common factor (other than 1 or -1).
2. Next, factor the first term, $3x^2$. The only possibility for factors is $3x \cdot x$. Thus, if a factorization exists, it must be of the form

   $$(3x + \square)(x + \square).$$

   We need to find the right numbers for the blanks.
3. Note that the constant term, -8, can be factored as $(-8)(1)$, $8(-1)$, $(-2)4$, and $2(-4)$, as well as $(1)(-8)$, $(-1)8$, $4(-2)$, and $(-4)2$.

4. Find a pair of factors for which the sum of the Outer and Inner products is the middle term, $10x$. Each possibility should be checked by multiplying:

$$(3x - 8)(x + 1) = 3x^2 - 5x - 8. \qquad \mathbf{O + I = 3x + (-8x) = -5x}$$

This gives a middle term with a negative coefficient. Since a positive coefficient is needed, a second possibility must be tried:

$$(3x + 8)(x - 1) = 3x^2 + 5x - 8. \qquad \mathbf{O + I = -3x + 8x = 5x}$$

Note that changing the signs of the two constant terms changes only the sign of the middle term. We try again:

$$(3x - 2)(x + 4) = 3x^2 + 10x - 8. \qquad \textbf{This is what we wanted.}$$

Thus the desired factorization is $(3x - 2)(x + 4)$.

Student Notes

Keep your work organized so that you can see what you have already considered. When factoring $6x^2 - 19x + 10$, we can list all possibilities and cross out those in which a common factor appears:

$(3x - 10)(2x - 1)$,
~~$(3x - 1)(2x - 10)$~~,
~~$(3x - 5)(2x - 2)$~~,
$(3x - 2)(2x - 5)$,
~~$(6x - 10)(x - 1)$~~,
$(6x - 1)(x - 10)$,
$(6x - 5)(x - 2)$,
~~$(6x - 2)(x - 5)$~~.

By being organized and not erasing, we can see that there are only four possible factorizations.

$y_1 = 6x^6 - 19x^5 + 10x^4$,
$y_2 = x^4(3x - 2)(2x - 5)$

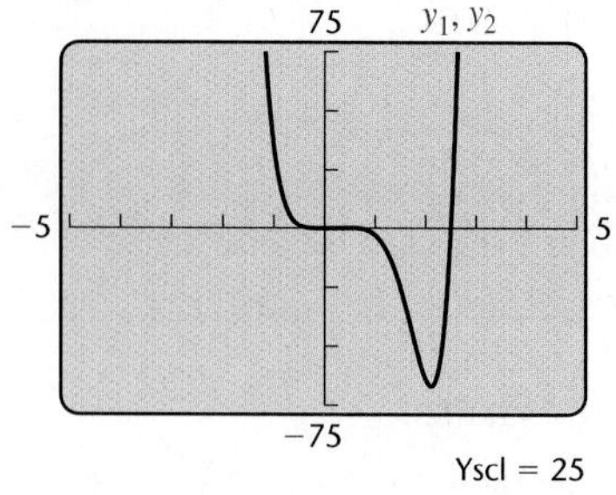

EXAMPLE 2 Factor: $6x^6 - 19x^5 + 10x^4$.

SOLUTION

1. First, factor out the common factor x^4:

$$x^4(6x^2 - 19x + 10).$$

2. Note that $6x^2 = 6x \cdot x$ and $6x^2 = 3x \cdot 2x$. Thus, $6x^2 - 19x + 10$ may factor into

$$(3x + \blacksquare)(2x + \blacksquare) \quad \text{or} \quad (6x + \blacksquare)(x + \blacksquare).$$

3. We factor the last term, 10. The possibilities are $10 \cdot 1$, $(-10)(-1)$, $5 \cdot 2$, and $(-5)(-2)$, as well as $1 \cdot 10$, $(-1)(-10)$, $2 \cdot 5$, and $(-2)(-5)$.

4. There are 8 possibilities for *each* factorization in step (2). We need factors for which the sum of the Outer and Inner products is the middle term, $-19x$. Since the x-coefficient is negative, we consider pairs of negative factors. Each possible factorization must be checked by multiplying:

$$(3x - 10)(2x - 1) = 6x^2 - 23x + 10.$$

We try again:

$$(3x - 5)(2x - 2) = 6x^2 - 16x + 10.$$

Actually this last attempt could have been rejected by simply noting that $2x - 2$ has a common factor, 2. Since the *largest* common factor was removed in step (1), no other common factors can exist. We try again, reversing the -5 and -2:

$$(3x - 2)(2x - 5) = 6x^2 - 19x + 10. \qquad \textbf{This is what we wanted.}$$

The factorization of $6x^2 - 19x + 10$ is $(3x - 2)(2x - 5)$. *But do not forget the common factor*! We must include it to get the complete factorization of the original trinomial:

$$6x^6 - 19x^5 + 10x^4 = x^4(3x - 2)(2x - 5).$$

The graphs of the original polynomial and the factorization coincide, as shown in the figure at left.

Tips for Factoring $ax^2 + bx + c$ with FOIL

1. If the largest common factor has been factored out of the original trinomial, then no binomial factor can have a common factor (other than 1 or -1).
2. If a and c are both positive, then the signs in the factors will be the same as the sign of b.
3. When a possible factoring produces the opposite of the desired middle term, reverse the signs of the constants in the factors.
4. Be systematic about your trials. Keep track of those possibilities that you have tried and those that you have not.

Keep in mind that this method of factoring involves trial and error. With practice, you will find yourself making fewer and better guesses.

Method 2: The *ac*-Method

The second method for factoring trinomials of the type $ax^2 + bx + c$, $a \neq 1$, is known as the ***ac*-method.** It involves not only trial and error and FOIL but also factoring by grouping. We know that

$$\begin{aligned} x^2 + 7x + 10 &= x^2 + 2x + 5x + 10 \\ &= x(x + 2) + 5(x + 2) \\ &= (x + 2)(x + 5), \end{aligned}$$

but what if the leading coefficient is not 1? Consider $6x^2 + 23x + 20$. The method is similar to what we just did with $x^2 + 7x + 10$, but we need two more steps.* First, multiply the leading coefficient, 6, and the constant, 20, to get 120. Then find a factorization of 120 in which the sum of the factors is the coefficient of the middle term: 23. The middle term is then split into a sum or difference using these factors.

$6x^2 + 23x + 20$

(1) Multiply 6 and 20: $6 \cdot 20 = 120$.
(2) Factor 120: $120 = 8 \cdot 15$, and $8 + 15 = 23$.
(3) Split the middle term: $23x = 8x + 15x$.
(4) Factor by grouping.

We factor by grouping as follows:

$$\begin{aligned} 6x^2 + 23x + 20 &= 6x^2 + 8x + 15x + 20 \\ &= 2x(3x + 4) + 5(3x + 4) \\ &= (3x + 4)(2x + 5). \end{aligned}$$

Factoring by grouping

*The rationale behind these steps is outlined in Exercise 115.

To Factor $ax^2 + bx + c$ Using the *ac*-Method

1. Make sure that any common factors have been factored out.
2. Multiply the leading coefficient a and the constant c.
3. Find a pair of factors, p and q, such that $pq = ac$ and $p + q = b$.
4. Rewrite the middle term of the trinomial, bx, as $px + qx$.
5. Factor by grouping.

EXAMPLE 3 Factor: $3x^2 + 10x - 8$.

SOLUTION

1. First, look for a common factor. There is none (other than 1 or -1).
2. Multiply the leading coefficient and the constant, 3 and -8.

   $3(-8) = -24.$

3. Try to factor -24 so that the sum of the factors is 10:

   $-24 = 12(-2)$ and $12 + (-2) = 10.$

4. Split $10x$ using the results of step (3):

   $10x = 12x - 2x.$

5. Finally, factor by grouping:

$$3x^2 + 10x - 8 = 3x^2 + 12x - 2x - 8 \qquad \textbf{Substituting } 12x - 2x \textbf{ for } 10x$$
$$= 3x(x + 4) - 2(x + 4)$$
$$= (x + 4)(3x - 2). \qquad \textbf{Factoring by grouping}$$

We check the solution by multiplying and using a table.

ALGEBRAIC CHECK

$$(x + 4)(3x - 2) = 3x^2 - 2x + 12x - 8$$
$$= 3x^2 + 10x - 8$$

CHECK BY EVALUATING

$y_1 = 3x^2 + 10x - 8,$
$y_2 = (x + 4)(3x - 2)$

X	Y1	Y2
0	−8	−8
1	5	5
2	24	24
3	49	49
4	80	80
5	117	117
6	160	160

X = 0

Pairs of Factors	Sum of Factors
1, −120	−119
2, −60	−58
3, −40	−37
4, −30	−26
5, −24	−19
6, −20	−14
8, −15	−7
10, −12	−2

EXAMPLE 4 Factor: $6x^4 - 116x^3 - 80x^2$.

SOLUTION

1. First, factor out the greatest common factor, if any. The expression $2x^2$ is common to all three terms: $2x^2(3x^2 - 58x - 40)$.
2. To factor $3x^2 - 58x - 40$, we first multiply the leading coefficient, 3, and the constant, -40: $3(-40) = -120$.
3. Next, try to factor -120 so that the sum of the factors is -58. Since -58 is negative, the negative factor of -120 must have the larger absolute value. We see from the table at left that the factors we need are 2 and -60.
4. Split the middle term, $-58x$, using the results of step (3): $-58x = 2x - 60x$.
5. Factor by grouping:

$$3x^2 - 58x - 40 = 3x^2 + 2x - 60x - 40 \quad \textbf{Substituting } 2x - 60x \textbf{ for } -58x$$
$$= x(3x + 2) - 20(3x + 2)$$
$$= (3x + 2)(x - 20). \quad \textbf{Factoring by grouping}$$

The factorization of $3x^2 - 58x - 40$ is $(3x + 2)(x - 20)$. *But don't forget the common factor*! We must include it to factor the original trinomial:

$$6x^4 - 116x^3 - 80x^2 = 2x^2(3x + 2)(x - 20).$$

Check: $2x^2(3x + 2)(x - 20) = 2x^2(3x^2 - 58x - 40)$
$= 6x^4 - 116x^3 - 80x^2$.

The complete factorization is $2x^2(3x + 2)(x - 20)$.

Equations and Functions

We now use our new factoring skill to solve a polynomial equation. We factor a polynomial *expression* and use the principle of zero products to solve the *equation*.

EXAMPLE 5 Solve: $6x^6 - 19x^5 + 10x^4 = 0$.

SOLUTION We note at the outset that the polynomial is of degree 6, so there will be at most 6 solutions of the equation.

ALGEBRAIC APPROACH

We solve as follows:

$$6x^6 - 19x^5 + 10x^4 = 0$$
$$x^4(3x - 2)(2x - 5) = 0 \quad \textbf{Factoring as in Example 2}$$
$$x^4 = 0 \quad or \quad 3x - 2 = 0 \quad or \quad 2x - 5 = 0 \quad \textbf{Using the principle of zero products}$$
$$x = 0 \quad or \quad x = \tfrac{2}{3} \quad or \quad x = \tfrac{5}{2}. \quad \textbf{Solving for } x; \; x^4 = x \cdot x \cdot x \cdot x = 0 \textbf{ when } x = 0$$

The solutions are 0, $\frac{2}{3}$, and $\frac{5}{2}$.

GRAPHICAL APPROACH

We find the x-intercepts of the function

$$f(x) = 6x^6 - 19x^5 + 10x^4$$

using the ZERO option of the CALC menu.

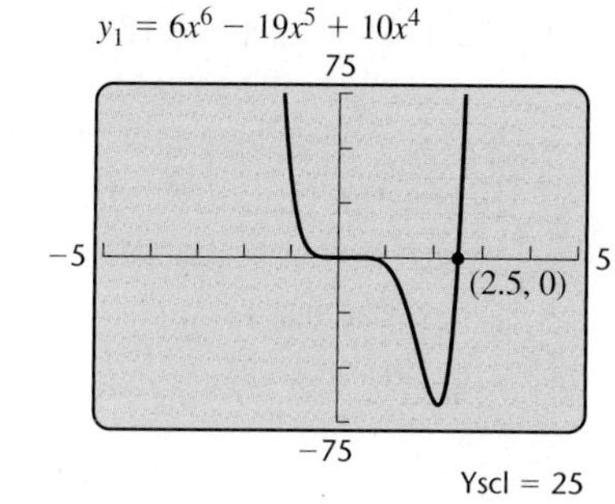

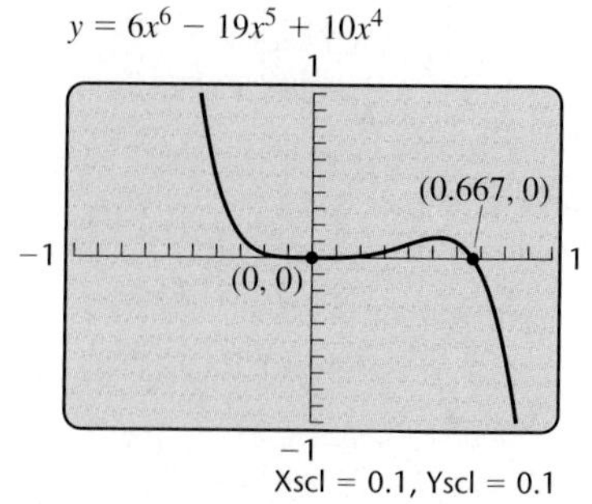

Using the viewing window $[-5, 5, -75, 75]$ shown on the left above, we see that 2.5 is a zero of the function. From this window, we cannot tell how many times the graph of the function intersects the x-axis between -1 and 1. We magnify that portion of the x-axis, as shown on the right above. The x-coordinates of the x-intercepts are 0, 0.667, and 2.5. Since $\frac{5}{2} = 2.5$ and $\frac{2}{3} \approx 0.667$, the solutions are 0, $\frac{2}{3}$, and $\frac{5}{2}$.

Example 5 illustrates two disadvantages of solving equations graphically: (1) It can be easy to "miss" solutions in some viewing windows and (2) solutions may be given as approximations.

Our work with factoring can help us when we are working with functions.

EXAMPLE 6 Given that $f(x) = 3x^2 - 4x$, find all values of a for which $f(a) = 4$.

SOLUTION We want all numbers a for which $f(a) = 4$. Since $f(a) = 3a^2 - 4a$, we must have

$3a^2 - 4a = 4$		**Setting $f(a)$ equal to 4**
$3a^2 - 4a - 4 = 0$		**Getting 0 on one side**
$(3a + 2)(a - 2) = 0$		**Factoring**
$3a + 2 = 0$	*or*	$a - 2 = 0$
$a = -\frac{2}{3}$	*or*	$a = 2$.

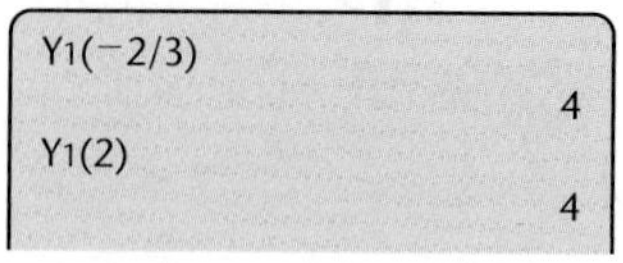

To check using a graphing calculator, we enter $y_1 = 3x^2 - 4x$ and calculate Y1$(-2/3)$ and Y1(2). To have $f(a) = 4$, we must have $a = -\frac{2}{3}$ or $a = 2$.

EXAMPLE 7 Find the domain of F if $F(x) = \dfrac{x - 2}{x^2 + 2x - 15}$.

SOLUTION The domain of F is the set of all values for which

$$\frac{x - 2}{x^2 + 2x - 15}$$

is a real number. Since division by 0 is undefined, $F(x)$ cannot be calculated for any x-value for which the denominator, $x^2 + 2x - 15$, is 0. To make sure these values are *excluded*, we solve:

$$x^2 + 2x - 15 = 0 \quad \textbf{Setting the denominator equal to 0}$$
$$(x - 3)(x + 5) = 0 \quad \textbf{Factoring}$$
$$x - 3 = 0 \quad or \quad x + 5 = 0$$
$$x = 3 \quad or \quad x = -5. \quad \textbf{These are the values to } \textbf{\textit{exclude.}}$$

X	Y1	
3	ERROR	
−5	ERROR	
0	.13333	
.5	.10909	
8	.09231	
127	.00764	
−3	.41667	

X = 3

We can check to see whether either 3 or −5 is in the domain of F using a graphing calculator, as shown at left. We enter $y = (x - 2)/(x^2 + 2x - 15)$ and set up a table with Indpnt set to Ask. When the x-values of 3 and −5 are supplied, the graphing calculator should return an error message, indicating that those numbers are not in the domain of the function. There will be a function value given for any other value of x.

The domain of F is $\{x \mid x$ is a real number *and* $x \neq -5$ *and* $x \neq 3\}$, or, in interval notation, $(-\infty, -5) \cup (-5, 3) \cup (3, \infty)$.

5.5 EXERCISE SET

FOR EXTRA HELP

MathXL | MyMathLab | InterAct Math | AW Math Tutor Center | Video Lectures on CD: Disc 3 | Student's Solutions Manual

Concept Reinforcement *In each of Exercises 1–8, match the polynomial with one of its factors from the column on the right.*

1. $2x^2 - 7x - 15$ **a)** $5x + 2$

2. $3x^2 + 4x - 7$ **b)** $5x + 3$

3. $6x^2 + 7x + 2$ **c)** $3x + 7$

4. $10x^2 - x - 2$ **d)** $2x + 7$

5. $3x^2 + 4x - 15$ **e)** $3x + 2$

6. $2x^2 + 9x + 7$ **f)** $2x + 3$

7. $10x^2 + 9x - 7$ **g)** $3x - 5$

8. $15x^2 + 14x + 3$ **h)** $5x + 7$

Factor.

9. $6x^2 - 5x - 25$

10. $3x^2 - 16x - 12$

11. $10y^3 - 12y - 7y^2$

12. $6x^3 - 15x - x^2$

13. $24a^2 - 14a + 2$

14. $3a^2 - 10a + 8$

15. $35y^2 + 34y + 8$

16. $9a^2 + 18a + 8$

17. $4t + 10t^2 - 6$

18. $8x + 30x^2 - 6$

19. $8x^2 - 16 - 28x$

20. $18x^2 - 24 - 6x$

21. $14x^4 - 19x^3 - 3x^2$

22. $70x^4 - 68x^3 + 16x^2$

23. $12a^2 - 4a - 16$

24. $12a^2 - 14a - 20$

25. $9x^2 + 15x + 4$

26. $6y^2 - y - 2$

Aha! **27.** $4x^2 + 15x + 9$

28. $2y^2 + 7y + 6$

29. $-8t^2 - 8t + 30$

30. $-36a^2 + 21a - 3$

31. $8 - 6z - 9z^2$

32. $3 + 35a - 12a^2$

33. $18xy^3 + 3xy^2 - 10xy$

34. $3x^3y^2 - 5x^2y^2 - 2xy^2$

35. $24x^2 - 2 - 47x$

36. $15z^2 - 10 - 47z$

37. $63x^3 + 111x^2 + 36x$

38. $50t^3 + 115t^2 + 60t$

39. $48x^4 + 4x^3 - 30x^2$

40. $40y^4 + 4y^2 - 12$

41. $12a^2 - 17ab + 6b^2$

42. $20p^2 - 23pq + 6q^2$

43. $2x^2 + xy - 6y^2$

44. $8m^2 - 6mn - 9n^2$

45. $6x^2 - 29xy + 28y^2$

46. $10p^2 + 7pq - 12q^2$

47. $9x^2 - 30xy + 25y^2$

48. $4p^2 + 12pq + 9q^2$

49. $9x^2y^2 + 5xy - 4$

50. $7a^2b^2 + 13ab + 6$

51. Use the results of Exercise 9 to solve $6x^2 - 5x - 25 = 0$.

52. Use the results of Exercise 10 to solve $3x^2 - 16x - 12 = 0$.

53. Use the results of Exercise 31 to solve $9z^2 + 6z = 8$.

54. Use the results of Exercise 32 to solve $3 + 35a = 12a^2$.

55. Use the results of Exercise 37 to solve $63x^3 + 111x^2 + 36x = 0$.

56. Use the results of Exercise 38 to solve $50t^3 + 115t^2 + 60t = 0$.

Solve.

57. $3x^2 - 8x + 4 = 0$

58. $9x^2 - 15x + 4 = 0$

59. $4t^3 + 11t^2 + 6t = 0$

60. $8n^3 + 10n^2 + 3n = 0$

61. $6x^2 = 13x + 5$

62. $40x^2 + 43x = 6$

63. $x(5 + 12x) = 28$

64. $a(1 + 21a) = 10$

65. Find the zeros of the function given by

$$f(x) = 2x^2 - 13x - 7.$$

66. Find the zeros of the function given by

$$g(x) = 6x^2 + 13x + 6.$$

67. Let $f(x) = x^2 + 12x + 40$. Find a such that $f(a) = 8$.

68. Let $f(x) = x^2 + 14x + 50$. Find a such that $f(a) = 5$.

69. Let $g(x) = 2x^2 + 5x$. Find a such that $g(a) = 12$.

70. Let $g(x) = 2x^2 - 15x$. Find a such that $g(a) = -7$.

71. Let $h(x) = 12x + x^2$. Find a such that $h(a) = -27$.

72. Let $h(x) = 4x - x^2$. Find a such that $h(a) = -32$.

Find the domain of the function f given by each of the following.

73. $f(x) = \dfrac{3}{x^2 - 4x - 5}$

74. $f(x) = \dfrac{2}{x^2 - 7x + 6}$

75. $f(x) = \dfrac{x - 5}{9x - 18x^2}$

76. $f(x) = \dfrac{1 + x}{3x - 15x^2}$

77. $f(x) = \dfrac{7}{5x^3 - 35x^2 + 50x}$

78. $f(x) = \dfrac{3}{2x^3 - 2x^2 - 12x}$

TW **79.** Emily has factored a polynomial as $(a - b)(x - y)$, while Jorge has factored the same polynomial as $(b - a)(y - x)$. Who is correct, and why?

TW **80.** Austin says that the domain of the function

$$F(x) = \frac{x + 3}{3x^2 - x - 2}$$

is $\left\{-\frac{2}{3}, 1\right\}$. Is he correct? Why or why not?

Focused Review

Factor.

81. $x^2 + 4x + 3$ [5.4]

82. $2x^2 - 3x - 5$ [5.5]

83. $x^2y + 2xy + xy^2 + 2y^2$ [5.3]

84. $15x^3 + 4x^2 - 3x$ [5.5]

85. $12x^2 - 5x - 3$ [5.5]

86. $2x^2 - 10x - 28$ [5.4]

87. $ac - ad - bc + bd$ [5.3]

88. $12x^3y^4 - 24x^2y^5 + 4x^2y^3$ [5.3]

89. $(a - 2)(a + b) + (a - 2)(2a - b)$ [5.3]

90. $-8x^2 + 8x + 8$ [5.3]

Synthesis

TW **91.** Suppose that $(rx + p)(sx - q) = ax^2 - bx + c$ is true. Explain how this can be used to factor $ax^2 + bx + c$.

TW **92.** Describe in your own words an approach that can be used to factor any "nonprime" trinomial of the form $ax^2 + bx + c$.

Use a graph to help factor each polynomial.

93. $4x^2 + 120x + 675$

94. $4x^2 + 164x + 1197$

95. $3x^3 + 150x^2 - 3672x$

96. $5x^4 + 20x^3 - 1600x^2$

Solve.

97. $(8x + 11)(12x^2 - 5x - 2) = 0$

98. $(x + 1)^3 = (x - 1)^3 + 26$

99. $(x - 2)^3 = x^3 - 2$

Factor. Assume that variables in exponents represent positive integers. If a polynomial is prime, state this.

100. $9x^2y^2 - 12xy - 2$

101. $18a^2b^2 - 3ab - 10$

102. $16x^2y^3 + 20xy^2 + 4y$

103. $16a^2b^3 + 25ab^2 + 9$

104. $9t^8 + 12t^4 + 4$

105. $25t^{10} - 10t^5 + 1$

106. $-15x^{2m} + 26x^m - 8$

107. $20x^{2n} + 16x^n + 3$

108. $3(a + 1)^{n+1}(a + 3)^2 - 5(a + 1)^n(a + 3)^3$

109. $7(t - 3)^{2n} + 5(t - 3)^n - 2$

110. $6(x - 7)^2 + 13(x - 7) - 5$

111. $2a^4b^6 - 3a^2b^3 - 20$

112. $5x^8y^6 + 35x^4y^3 + 60$

113. $4x^{2a} - 4x^a - 3$

114. $2ar^2 + 4asr + as^2 - asr$

115. To better understand factoring $ax^2 + bx + c$ by the ac-method, suppose that

$$ax^2 + bx + c = (mx + r)(nx + s).$$

Show that if $P = ms$ and $Q = rn$, then $P + Q = b$ and $PQ = ac$.

5.6 Equations Containing Perfect-Square Trinomials and Differences of Squares

Perfect-Square Trinomials ■ Differences of Squares ■ More Factoring by Grouping ■ Solving Equations

We now introduce a faster way to factor trinomials that are squares of binomials. A method for factoring differences of squares is also developed. These factoring methods enable us to solve more types of polynomial equations using the principle of zero products.

Perfect-Square Trinomials

Consider the trinomial

$$x^2 + 6x + 9.$$

To factor it, we can proceed as in Section 5.4 and look for factors of 9 that add to 6. These factors are 3 and 3 and the factorization is

$$x^2 + 6x + 9 = (x + 3)(x + 3) = (x + 3)^2.$$

Note that the result is the square of a binomial. Because of this, we call $x^2 + 6x + 9$ a **perfect-square trinomial.** Although trial and error can be used to factor a perfect-square trinomial, once recognized, a perfect-square trinomial can be quickly factored.

Student Notes

If you're not already quick to recognize that $1^2 = 1$, $2^2 = 4$, $3^2 = 9$, $4^2 = 16$, $5^2 = 25$, $6^2 = 36$, $7^2 = 49$, $8^2 = 64$, $9^2 = 81$, $10^2 = 100$, $11^2 = 121$, and $12^2 = 144$, this is a good time to familiarize yourself with these numbers.

To Recognize a Perfect-Square Trinomial

- Two terms must be squares, such as A^2 and B^2.
- There must be no minus sign before A^2 or B^2.
- The remaining term must be $2AB$ or its opposite, $-2AB$.

EXAMPLE 1 Determine whether each polynomial is a perfect-square trinomial.

a) $x^2 + 10x + 25$ **b)** $4x + 16 + 3x^2$ **c)** $100y^2 + 81 - 180y$

SOLUTION

a)
- Two of the terms in $x^2 + 10x + 25$ are squares: x^2 and 25.
- There is no minus sign before either x^2 or 25.
- The remaining term, $10x$, is twice the product of the square roots, x and 5.

Thus, $x^2 + 10x + 25$ *is* a perfect square.

b) In $4x + 16 + 3x^2$, only one term, 16, is a square ($3x^2$ is not a square because 3 is not a perfect square; $4x$ is not a square because x is not a square).

Thus, $4x + 16 + 3x^2$ *is not* a perfect square.

c) It can help to first write the polynomial in descending order:

$$100y^2 - 180y + 81.$$

- Two of the terms, $100y^2$ and 81, are squares.
- There is no minus sign before either $100y^2$ or 81.
- If the product of the square roots, $10y$ and 9, is doubled, we get the opposite of the remaining term: $2(10y)(9) = 180y$ (the opposite of $-180y$).

Thus, $100y^2 + 81 - 180y$ *is* a perfect-square trinomial.

To factor a perfect-square trinomial, we reuse the patterns that we learned in Section 5.2.

Study Tip

Fill In Your Blanks

Don't hesitate to write out any missing steps that you'd like to see included. For instance, in Example 1(c), we state that $100y^2$ is a square. To solidify your understanding, you may want to write in $10y \cdot 10y = 100y^2$.

Factoring a Perfect-Square Trinomial

$$A^2 + 2AB + B^2 = (A + B)^2;$$
$$A^2 - 2AB + B^2 = (A - B)^2$$

CAUTION! In Example 2, we are factoring expressions. These are *not* equations and thus we do not try to solve them.

EXAMPLE 2 Factor.

a) $x^2 - 10x + 25$ **b)** $16y^2 + 49 + 56y$ **c)** $-20xy + 4y^2 + 25x^2$

SOLUTION

a) $x^2 - 10x + 25 = (x - 5)^2$ **We find the square terms and write the square roots with a minus sign between them.**

Note the sign!

As always, any factorization can be checked by multiplying:

$$(x - 5)^2 = (x - 5)(x - 5) = x^2 - 5x - 5x + 25 = x^2 - 10x + 25.$$

b) $16y^2 + 49 + 56y = 16y^2 + 56y + 49$ **Using a commutative law**

$= (4y + 7)^2$ **We find the square terms and write the square roots with a plus sign between them.**

The check is left to the student.

c) $-20xy + 4y^2 + 25x^2 = 4y^2 - 20xy + 25x^2$ **Writing descending order with respect to *y***

$= (2y - 5x)^2$

This square can also be expressed as

$$25x^2 - 20xy + 4y^2 = (5x - 2y)^2.$$

The student should confirm that both factorizations check.

When factoring, always look first for a factor common to all the terms.

EXAMPLE 3 Factor: **(a)** $2x^2 - 12xy + 18y^2$; **(b)** $-4y^2 - 144y^8 + 48y^5$.

SOLUTION

a) We first look for a common factor. This time, there is a common factor, 2.

$2x^2 - 12xy + 18y^2 = 2(x^2 - 6xy + 9y^2)$ **Factoring out the 2**

$= 2(x - 3y)^2$ **Factoring the perfect-square trinomial**

The check is left to the student.

b) $-4y^2 - 144y^8 + 48y^5 = -4y^2(1 + 36y^6 - 12y^3)$ **Factoring out the common factor**

$= -4y^2(36y^6 - 12y^3 + 1)$ **Changing order. Note that $(y^3)^2 = y^6$.**

$= -4y^2(6y^3 - 1)^2$ **Factoring the perfect-square trinomial**

Check: $-4y^2(6y^3 - 1)^2 = -4y^2(6y^3 - 1)(6y^3 - 1)$
$$= -4y^2(36y^6 - 12y^3 + 1)$$
$$= -144y^8 + 48y^5 - 4y^2$$
$$= -4y^2 - 144y^8 + 48y^5$$

The factorization $-4y^2(6y^3 - 1)^2$ checks.

Differences of Squares

When an expression like $x^2 - 9$ is recognized as a difference of two squares, we can reverse another pattern first seen in Section 5.2.

Factoring a Difference of Two Squares

$$A^2 - B^2 = (A + B)(A - B)$$

To factor a difference of two squares, write the product of the sum and the difference of the quantities being squared.

EXAMPLE 4 Factor: **(a)** $x^2 - 9$; **(b)** $25y^6 - 49x^2$.

SOLUTION

a) $x^2 - 9 = x^2 - 3^2 = (x + 3)(x - 3)$

$$A^2 - B^2 = (A + B)(A - B)$$
$$\downarrow \quad \downarrow \quad \downarrow \quad \downarrow \quad \downarrow \quad \downarrow$$

b) $25y^6 - 49x^2 = (5y^3)^2 - (7x)^2 = (5y^3 + 7x)(5y^3 - 7x)$

As always, the first step in factoring is to look for common factors.

EXAMPLE 5 Factor: **(a)** $5 - 5x^2y^6$; **(b)** $16x^4y - 81y$.

SOLUTION

a) $5 - 5x^2y^6 = 5(1 - x^2y^6)$ **Factoring out the common factor**

$= 5[1^2 - (xy^3)^2]$ **Rewriting x^2y^6 as a quantity squared**

$= 5(1 + xy^3)(1 - xy^3)$ **Factoring the difference of squares**

Check: $5(1 + xy^3)(1 - xy^3) = 5(1 - xy^3 + xy^3 - x^2y^6)$
$$= 5(1 - x^2y^6) = 5 - 5x^2y^6$$

The factorization $5(1 + xy^3)(1 - xy^3)$ checks.

b) $16x^4y - 81y = y(16x^4 - 81)$ **Factoring out the common factor**

$= y[(4x^2)^2 - 9^2]$

$= y(4x^2 + 9)(4x^2 - 9)$ **Factoring the difference of squares**

$= y(4x^2 + 9)(2x + 3)(2x - 3)$ **Factoring $4x^2 - 9$, which is itself a difference of squares**

The check is left to the student.

Note in Example 5(b) that $4x^2 - 9$ *could* be factored further. Whenever a factor itself can be factored, do so. We say that we have factored completely when none of the factors can be factored further.

More Factoring by Grouping

Sometimes, when factoring a polynomial with four terms, we may be able to factor further.

EXAMPLE 6 Factor: $x^3 + 3x^2 - 4x - 12$.

SOLUTION

$$x^3 + 3x^2 - 4x - 12 = x^2(x + 3) - 4(x + 3) \quad \textbf{Factoring by grouping}$$
$$= (x + 3)(x^2 - 4) \quad \textbf{Factoring out } x + 3$$
$$= (x + 3)(x + 2)(x - 2) \quad \textbf{Factoring } x^2 - 4$$

A difference of squares can have four or more terms. For example, one of the squares may be a trinomial. In this case, a type of grouping can be used.

EXAMPLE 7 Factor: **(a)** $x^2 + 6x + 9 - y^2$; **(b)** $a^2 - b^2 + 8b - 16$.

SOLUTION

a) $x^2 + 6x + 9 - y^2 = (x^2 + 6x + 9) - y^2$ **Grouping as a perfect-square trinomial minus y^2 to show a difference of squares**

$$= (x + 3)^2 - y^2$$
$$= (x + 3 + y)(x + 3 - y)$$

b) Grouping $a^2 - b^2 + 8b - 16$ into two groups of two terms does not yield a common binomial factor, so we look for a perfect-square trinomial. In this case, the perfect-square trinomial is being subtracted from a^2:

$$a^2 - b^2 + 8b - 16 = a^2 - (b^2 - 8b + 16) \quad \textbf{Factoring out } -1 \textbf{ and rewriting as subtraction}$$
$$= a^2 - (b - 4)^2 \quad \textbf{Factoring the perfect-square trinomial}$$
$$= (a + (b - 4))(a - (b - 4)) \quad \textbf{Factoring a difference of squares}$$
$$= (a + b - 4)(a - b + 4) \quad \textbf{Removing parentheses}$$

Solving Equations

We can now solve polynomial equations involving differences of squares and perfect-square trinomials.

EXAMPLE 8 Solve: $x^3 + 3x^2 = 4x + 12$.

SOLUTION We first note that the equation is a third-degree polynomial equation. Thus it will have 3 or fewer solutions.

ALGEBRAIC APPROACH

We have

$$x^3 + 3x^2 = 4x + 12$$

$$x^3 + 3x^2 - 4x - 12 = 0 \quad \textbf{Getting 0 on one side}$$

$$(x + 3)(x + 2)(x - 2) = 0 \quad \textbf{Factoring; using the results of Example 6}$$

$$x + 3 = 0 \quad or \quad x + 2 = 0 \quad or \quad x - 2 = 0$$

Using the principle of zero products

$$x = -3 \quad or \quad x = -2 \quad or \quad x = 2.$$

The solutions are -3, -2, and 2.

GRAPHICAL APPROACH

We let $f(x) = x^3 + 3x^2$ and $g(x) = 4x + 12$ and look for any points of intersection of the graphs of f and g.

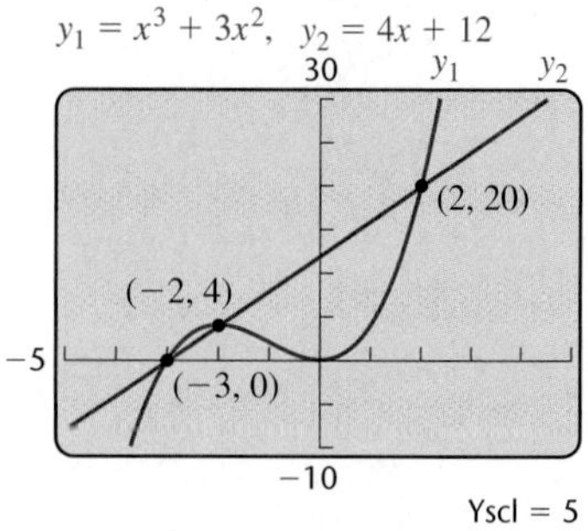

The x-coordinates of the points of intersection are -3, -2, and 2. These are the solutions of the equation.

As a partial check, note that both methods yield the same solutions. Substituting -3, -2, and 2 into the original equation, we see that all three numbers check. The solutions are -3, -2, and 2.

EXAMPLE 9 Find the zeros of the function given by $f(x) = x^3 + x^2 - x - 1$.

SOLUTION This function is a third-degree polynomial function, so there will be at most 3 zeros. To find the zeros of the function, we find the roots of the equation $x^3 + x^2 - x - 1 = 0$.

ALGEBRAIC APPROACH

We first factor the polynomial:

$$x^3 + x^2 - x - 1 = 0$$

$$x^2(x + 1) - 1(x + 1) = 0 \quad \textbf{Factoring by grouping}$$

$$(x + 1)(x^2 - 1) = 0$$

$$(x + 1)(x + 1)(x - 1) = 0. \quad \textbf{Factoring a difference of squares}$$

Then we use the principle of zero products:

$$x + 1 = 0 \quad or \quad x + 1 = 0 \quad or \quad x - 1 = 0$$

$$x = -1 \quad or \quad x = -1 \quad or \quad x = 1.$$

The solutions are -1 and 1. The numbers -1 and 1 check. The zeros are -1 and 1.

GRAPHICAL APPROACH

We graph $f(x) = x^3 + x^2 - x - 1$ and find the zeros of the function.

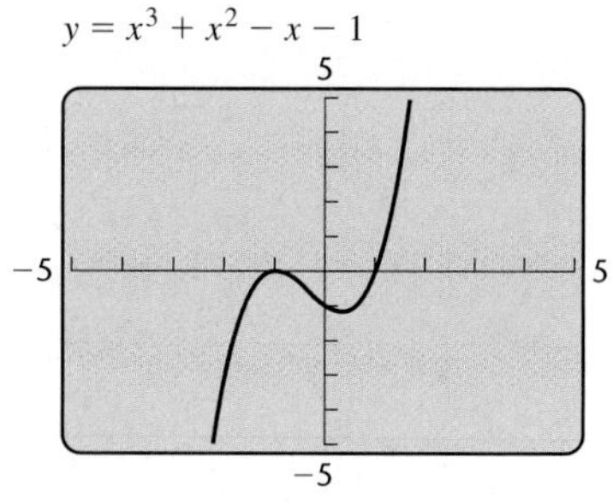

The zeros are -1 and 1.

In Example 9, the factor $(x + 1)$ appeared twice in the factorization. When a factor appears two or more times, we say that we have a **repeated root.** If the factor occurs twice, we say that we have a **double root,** or a **root of multiplicity two.**

In this chapter, we have emphasized the relationship between factoring and equation solving. There are many polynomial equations, however, that cannot be solved by factoring. Nonetheless, we can find approximations of any real-number solutions using a graphing calculator.

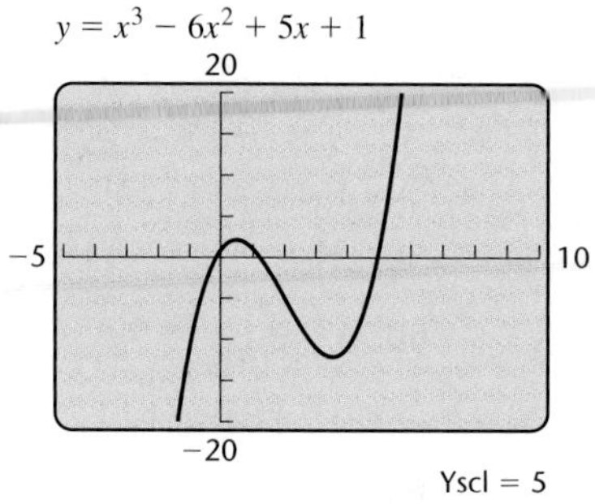

X	Y1
−.166	9E−5
1.217	9.5E−4
4.949	.00328

X = −.166

EXAMPLE 10 Solve: $x^3 - 6x^2 + 5x + 1 = 0$.

SOLUTION The polynomial $x^3 - 6x^2 + 5x + 1$ is prime; it cannot be factored using rational coefficients. We solve the equation by graphing $f(x) = x^3 - 6x^2 + 5x + 1$ and finding the zeros of the function. Since the polynomial is of degree 3, we need not look for more than 3 zeros.

There are three x-intercepts of the graph. We use the ZERO feature to find each root, or zero. Since the solutions are irrational, the calculator will give approximations. Rounding each to the nearest thousandth, we have the solutions -0.166, 1.217, and 4.949. We check the solutions using a table (see the figure at left). Note that the function values are close to 0 for each approximate solution.

We have considered both algebraic and graphical methods of solving polynomial equations. It is important to understand and be able to use both methods. Some of the advantages and disadvantages of each method are given in the following table.

	Advantages	**Disadvantages**
Algebraic Method	• Can find exact answers • Works well when the polynomial is in factored form or can be readily factored • Can be used to find solutions that are not real numbers (see Chapter 8)	• Cannot be used if the polynomial is not factorable • Can be difficult to use if factorization is not readily apparent
Graphical Method	• Does not require the polynomial to be factored • Can visualize solutions	• Easy to miss solutions if an appropriate viewing window is not chosen • Gives approximations of solutions • For most graphing calculators, only real-number solutions can be found.

5.6 EXERCISE SET

FOR EXTRA HELP

MathXL | MyMathLab | InterAct Math | AW Math Tutor Center | Video Lectures on CD: Disc 3 | Student's Solutions Manual

Concept Reinforcement *Classify each of the following as a perfect-square trinomial, a difference of two squares, a polynomial having a common factor, or none of these.*

1. $9t^2 - 49$

2. $25x^2 - 20x + 4$

3. $36x^2 - 12x + 1$

4. $36a^2 - 25$

5. $4r^2 + 8r + 9$

6. $9x^2 - 12$

7. $4x^2 + 8x + 10$

8. $t^2 - 6t + 8$

9. $4t^2 + 9s^2 + 12st$

10. $9rt^2 - 5rt + 6r$

Factor completely.

11. $t^2 + 6t + 9$

12. $x^2 - 8x + 16$

13. $a^2 - 14a + 49$

14. $a^2 + 16a + 64$

15. $4a^2 - 16a + 16$

16. $2a^2 + 8a + 8$

17. $y^2 + 36 + 12y$

18. $y^2 + 36 - 12y$

19. $-18y^2 + y^3 + 81y$

20. $24a^2 + a^3 + 144a$

21. $2x^2 - 40x + 200$

22. $32x^2 + 48x + 18$

23. $1 - 8d + 16d^2$

24. $64 + 25y^2 - 80y$

25. $y^3 + 8y^2 + 16y$

26. $a^3 - 10a^2 + 25a$

27. $0.25x^2 + 0.30x + 0.09$

28. $0.04x^2 - 0.28x + 0.49$

29. $p^2 - 2pq + q^2$

30. $m^2 + 2mn + n^2$

31. $25a^2 + 30ab + 9b^2$

32. $49p^2 - 84pq + 36q^2$

33. $5a^2 - 10ab + 5b^2$

34. $4t^2 - 8tr + 4r^2$

35. $y^2 - 100$

36. $x^2 - 16$

37. $m^2 - 64$

38. $p^2 - 49$

39. $p^2q^2 - 25$

40. $a^2b^2 - 81$

41. $8x^2 - 8y^2$

42. $6x^2 - 6y^2$

43. $7xy^4 - 7xz^4$

44. $25ab^4 - 25az^4$

45. $4a^3 - 49a$

46. $9x^4 - 25x^2$

47. $3x^8 - 3y^8$

48. $9a^4 - a^2b^2$

49. $9a^4 - 25a^2b^4$

50. $16x^6 - 121x^2y^4$

51. $\frac{1}{49} - x^2$

52. $\frac{1}{16} - y^2$

53. $(a + b)^2 - 9$

54. $(p + q)^2 - 25$

55. $x^2 - 6x + 9 - y^2$

56. $a^2 - 8a + 16 - b^2$

57. $t^3 + 8t^2 - t - 8$

58. $x^3 - 7x^2 - 4x + 28$

59. $r^3 - 3r^2 - 9r + 27$

60. $t^3 + 2t^2 - 4t - 8$

61. $m^2 - 2mn + n^2 - 25$

62. $x^2 + 2xy + y^2 - 9$

63. $36 - (x + y)^2$

64. $49 - (a + b)^2$

65. $r^2 - 2r + 1 - 4s^2$

66. $c^2 + 4cd + 4d^2 - 9p^2$

Aha! **67.** $16 - a^2 - 2ab - b^2$

68. $9 - x^2 - 2xy - y^2$

69. $x^3 + 5x^2 - 4x - 20$

70. $t^3 + 6t^2 - 9t - 54$

71. $a^3 - ab^2 - 2a^2 + 2b^2$

72. $p^2q - 25q + 3p^2 - 75$

Solve. Round any irrational solutions to the nearest thousandth.

73. $a^2 + 1 = 2a$

74. $r^2 + 16 = 8r$

75. $2x^2 - 24x + 72 = 0$

76. $-t^2 - 16t - 64 = 0$

77. $x^2 - 9 = 0$

78. $r^2 - 64 = 0$

79. $a^2 = \frac{1}{25}$

80. $x^2 = \frac{1}{100}$

81. $8x^3 + 1 = 4x^2 + 2x$

82. $27x^3 + 18x^2 = 12x + 8$

83. $x^3 + 3 = 3x^2 + x$

84. $x^3 + x^2 = 16x + 16$

85. $x^2 - 3x - 7 = 0$

86. $x^2 - 5x + 1 = 0$

87. $2x^2 + 8x + 1 = 0$

88. $3x^2 + x - 1 = 0$

89. $x^3 + 3x^2 + x - 1 = 0$

90. $x^3 + x^2 + x - 1 = 0$

91. Let $f(x) = x^2 - 12x$. Find a such that $f(a) = -36$.

92. Let $g(x) = x^2$. Find a such that $g(a) = 144$.

Find the zeros of each function.

93. $f(x) = x^2 - 16$

94. $f(x) = x^2 - 8x + 16$

95. $f(x) = 2x^2 + 4x + 2$

96. $f(x) = 3x^2 - 27$

97. $f(x) = x^3 - 2x^2 - x + 2$

98. $f(x) = x^3 + x^2 - 4x - 4$

TW **99.** Are the product and power rules for exponents (see Section 1.4) important when factoring differences of squares? Why or why not?

TW **100.** Describe a procedure that could be used to find a polynomial with four terms that can be factored as a difference of two squares.

Focused Review

Factor completely.

101. $6x^2 - 17x + 5$ [5.5]

102. $16x^2 - 1$ [5.6]

103. $16x^2 + 8x + 1$ [5.6]

104. $3x^3 - 3x^2 - 6x$ [5.4]

105. $x^3 + 3x^2 - 2x - 6$ [5.3]

106. $(2x + 1)^2 - 16$ [5.6]

107. $\frac{1}{64} - x^2$ [5.6]

108. $100 - t^2 - 2st - s^2$ [5.6]

109. $12x^3y - 24x^2y^2 + 12xy^3$ [5.6]

110. $-3x^4 + 18x^3 + 3x$ [5.3]

Synthesis

TW **111.** Without finding the entire factorization, determine the number of factors of $x^{256} - 1$. Explain how you arrived at your answer.

TW **112.** Under what conditions can a sum of two squares be factored?

Factor completely. Assume that variables in exponents represent positive integers.

113. $-\frac{8}{27}r^2 - \frac{10}{9}rs - \frac{1}{6}s^2 + \frac{2}{3}rs$

114. $\frac{1}{36}x^8 + \frac{2}{9}x^4 + \frac{4}{9}$

115. $0.09x^8 + 0.48x^4 + 0.64$

116. $a^2 + 2ab + b^2 - c^2 + 6c - 9$

117. $r^2 - 8r - 25 - s^2 - 10s + 16$

118. $x^{2a} - y^2$

119. $x^{4a} - y^{2b}$

Aha! **120.** $4y^{4a} + 20y^{2a} + 20y^{2a} + 100$

121. $25y^{2a} - (x^{2b} - 2x^b + 1)$

122. $8(a - 3)^2 - 64(a - 3) + 128$

123. $3(x + 1)^2 + 12(x + 1) + 12$

124. $5c^{100} - 80d^{100}$

125. $9x^{2n} - 6x^n + 1$

126. $c^{2w+1} + 2c^{w+1} + c$

127. $(t + 1)^2 - 4(t + 1) + 3$

128. $(a - 3)^2 - 8(a - 3) + 16$

129. $m^2 + 4mn + 4n^2 + 5m + 10n$

130. $s^2 - 4st + 4t^2 + 4s - 8t + 4$

131. If $P(x) = x^2$, use factoring to simplify $P(a + h) - P(a)$.

132. If $P(x) = x^4$, use factoring to simplify $P(a + h) - P(a)$.

133. *Volume of Carpeting.* The volume of a carpet that is rolled up can be estimated by the polynomial $\pi R^2h - \pi r^2h$.

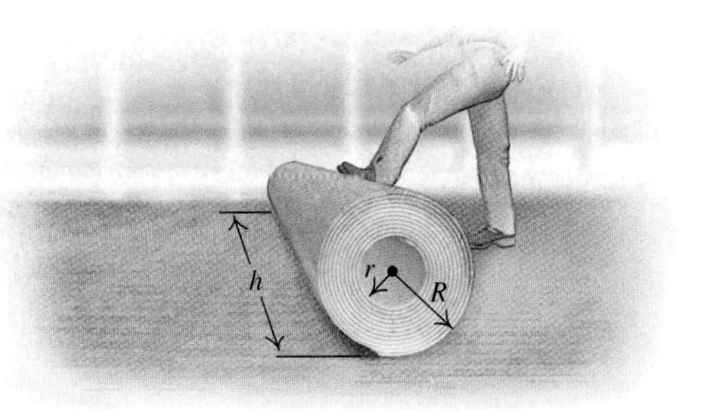

a) Factor the polynomial.

b) Use both the original and the factored forms to find the volume of a roll for which $R = 50$ cm, $r = 10$ cm, and $h = 4$ m. Use 3.14 for π.

5.7 Equations Containing Sums or Differences of Cubes

Factoring Sums or Differences of Cubes ■ Solving Equations

Factoring Sums or Differences of Cubes

We have seen that a difference of two squares can be factored but (unless a common factor exists) a *sum* of two squares is usually prime. The situation is different with cubes: The difference *or sum* of two cubes can always be factored. To see this, consider the following products:

$$\begin{aligned}(A + B)(A^2 - AB + B^2) &= A(A^2 - AB + B^2) + B(A^2 - AB + B^2)\\ &= A^3 - A^2B + AB^2 + A^2B - AB^2 + B^3\\ &= A^3 + B^3 \qquad \textbf{Combining like terms}\end{aligned}$$

and

$$\begin{aligned}(A - B)(A^2 + AB + B^2) &= A(A^2 + AB + B^2) - B(A^2 + AB + B^2)\\ &= A^3 + A^2B + AB^2 - A^2B - AB^2 - B^3\\ &= A^3 - B^3. \qquad \textbf{Combining like terms}\end{aligned}$$

These products allow us to factor a sum or a difference of two cubes. Note how the location of the $+$ and $-$ signs changes.

Factoring a Sum or a Difference of Two Cubes

$$A^3 + B^3 = (A + B)(A^2 - AB + B^2);$$
$$A^3 - B^3 = (A - B)(A^2 + AB + B^2)$$

When factoring a sum or a difference of cubes, it can be helpful to remember that $1^3 = 1, 2^3 = 8, 3^3 = 27, 4^3 = 64, 5^3 = 125, 6^3 = 216$, and so on. We say that 2 is the *cube root* of 8, that 3 is the cube root of 27, and so on.

Study Tip

A Good Night's Sleep

Don't try studying when you are tired. A good night's sleep or a 10-minute "power nap" can often make a problem suddenly seem much easier to solve.

EXAMPLE 1 Write an equivalent expression by factoring: $x^3 - 27$.

SOLUTION We first observe that

$$x^3 - 27 = x^3 - 3^3.$$

Next, in one set of parentheses, we write the first cube root, x, minus the second cube root, 3:

$$(x - 3)(\qquad).$$

To get the other factor, we think of $x - 3$ and do the following:

Square the first term: x^2.
Multiply the terms and then change the sign: $3x$.
Square the second term: $(-3)^2$, or 9.

$$(x - 3)(x^2 + 3x + 9).$$

Check: $(x - 3)(x^2 + 3x + 9) = x^3 + 3x^2 + 9x - 3x^2 - 9x - 27$
$= x^3 - 27$ **Combining like terms**

Thus, $x^3 - 27 = (x - 3)(x^2 + 3x + 9)$.

In Example 1, note that $x^2 + 3x + 9$ cannot be factored further. As a general rule, unless A and B share a common factor, the trinomials $A^2 + AB + B^2$ and $A^2 - AB + B^2$ are prime.

EXAMPLE 2 Factor.

a) $125x^3 + y^3$ **b)** $m^6 + 64$
c) $128y^7 - 250x^6y$ **d)** $r^6 - s^6$

SOLUTION

a) We have

$$125x^3 + y^3 = (5x)^3 + y^3.$$

In one set of parentheses, we write the cube root of the first term, $5x$, plus the cube root of the second term, y:

$$(5x + y)(\qquad).$$

To get the other factor, we think of $5x + y$ and do the following:

Square the first term: $(5x)^2$, or $25x^2$.
Multiply the terms and then change the sign: $-5xy$.
Square the second term: y^2.

$$(5x + y)(25x^2 - 5xy + y^2).$$

Student Notes

If you think of $A^3 - B^3$ as $A^3 + (-B)^3$, it is then sufficient to remember only the pattern for factoring a sum of two cubes. Be sure to simplify your result if you do this.

Check: $(5x + y)(25x^2 - 5xy + y^2) = 125x^3 - 25x^2y + 5xy^2 + 25x^2y - 5xy^2 + y^3$

$= 125x^3 + y^3$ **Combining like terms**

Thus, $125x^3 + y^3 = (5x + y)(25x^2 - 5xy + y^2)$.

b) We have

$$m^6 + 64 = (m^2)^3 + 4^3. \quad \textbf{Rewriting as quantities cubed}$$

Next, we reuse the pattern used in part (a) above:

$$A^3 + B^3 = (A + B)(A^2 - A \cdot B + B^2)$$
$$\downarrow \quad \downarrow$$
$$(m^2)^3 + 4^3 = (m^2 + 4)((m^2)^2 - m^2 \cdot 4 + 4^2)$$
$$= (m^2 + 4)(m^4 - 4m^2 + 16).$$

The check is left to the student.

c) We have

$$128y^7 - 250x^6y = 2y(64y^6 - 125x^6) \quad \textbf{Remember: \textit{Always} look for a common factor.}$$
$$= 2y[(4y^2)^3 - (5x^2)^3]. \quad \textbf{Rewriting as quantities cubed}$$

To factor $(4y^2)^3 - (5x^2)^3$, we again use the pattern in Example 1:

$$A^3 - B^3 = (A - B)(A^2 + A \cdot B + B^2)$$
$$\downarrow \quad \downarrow \quad \downarrow \quad \downarrow \quad \downarrow \quad \downarrow \quad \downarrow \quad \downarrow$$
$$(4y^2)^3 - (5x^2)^3 = (4y^2 - 5x^2)((4y^2)^2 + 4y^2 \cdot 5x^2 + (5x^2)^2)$$
$$= (4y^2 - 5x^2)(16y^4 + 20x^2y^2 + 25x^4).$$

The check is left to the student. We have

$$128y^7 - 250x^6y = 2y(4y^2 - 5x^2)(16y^4 + 20x^2y^2 + 25x^4).$$

d) We have

$$r^6 - s^6 = (r^3)^2 - (s^3)^2$$
$$= (r^3 + s^3)(r^3 - s^3) \quad \textbf{Factoring a difference of two \textit{squares}}$$
$$= (r + s)(r^2 - rs + s^2)(r - s)(r^2 + rs + s^2).$$
Factoring the sum and difference of two cubes

To check, read the steps in reverse order and inspect the multiplication.

In Example 2(d), suppose we first factored $r^6 - s^6$ as a difference of two cubes:

$$(r^2)^3 - (s^2)^3 = (r^2 - s^2)(r^4 + r^2s^2 + s^4)$$
$$= (r + s)(r - s)(r^4 + r^2s^2 + s^4).$$

In this case, we might have missed some factors; $r^4 + r^2s^2 + s^4$ can be factored as $(r^2 - rs + s^2)(r^2 + rs + s^2)$, but we probably would never have suspected that such a factorization exists. Given a choice, it is generally better to factor as a difference of squares before factoring as a sum or a difference of cubes.

Try to remember the following.

Useful Factoring Facts

Sum of cubes: $A^3 + B^3 = (A + B)(A^2 - AB + B^2)$

Difference of cubes: $A^3 - B^3 = (A - B)(A^2 + AB + B^2)$

Difference of squares: $A^2 - B^2 = (A + B)(A - B)$

There is no formula for factoring a sum of squares.

Solving Equations

We can now solve equations involving sums or differences of cubes.

EXAMPLE 3 Solve: $x^3 = 27$.

SOLUTION We rewrite the equation to get 0 on one side and factor:

$$x^3 = 27$$

$$x^3 - 27 = 0 \quad \text{Subtracting 27 from both sides}$$

$$(x - 3)(x^2 + 3x + 9) = 0. \quad \text{Using the factorization from Example 1}$$

The second-degree factor, $x^2 + 3x + 9$, does not factor using real coefficients. When we apply the principle of zero products, we have

$$x - 3 = 0 \quad or \quad x^2 + 3x + 9 = 0. \quad \text{Using the principle of zero products}$$

$$x = 3 \quad \text{We cannot factor } x^2 + 3x + 9.$$

We have one solution, 3. In Chapter 8, we will learn how to solve equations like $x^2 + 3x + 9 = 0$. For now, we can solve the equation $x^3 = 27$ graphically, by looking for any points of intersection of the graphs of $y_1 = x^3$ and $y_2 = 27$.

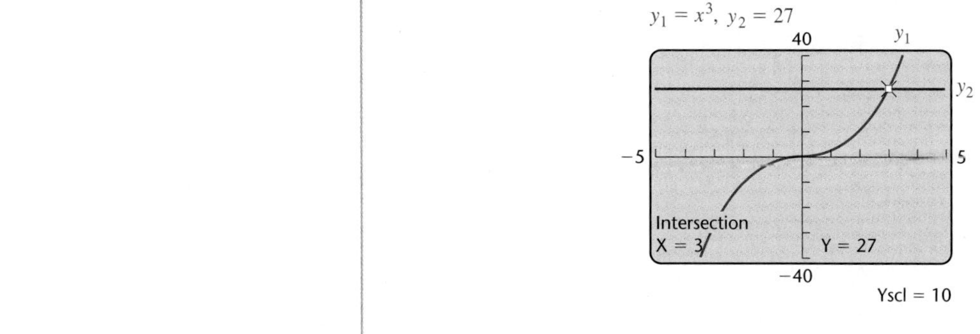

It appears that the only real-number solution is the one found algebraically. The solution of $x^3 = 27$ is 3.

5.7 EXERCISE SET

FOR EXTRA HELP

MathXL | MyMathLab | InterAct Math | AW Math Tutor Center | Video Lectures on CD: Disc 3 | Student's Solutions Manual

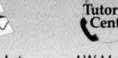

Concept Reinforcement *Classify each binomial as either a sum of two cubes, a difference of two cubes, a difference of two squares, both a difference of two cubes and a difference of two squares, a prime polynomial, or none of these.*

1. $8x^3 - 27$

2. $64t^3 + 27$

3. $9x^4 - 25$

4. $9x^2 + 25$

5. $1000t^3 + 1$

6. $m^3 - 8n^3$

7. $25x^2 + 8x$

8. $64t^6 - 1$

9. $s^{12} - t^6$

10. $14x^3 - 2x$

Factor completely.

11. $x^3 + 64$

12. $t^3 + 27$

13. $z^3 - 1$

14. $x^3 - 8$

15. $x^3 - 27$

16. $m^3 - 64$

17. $27x^3 + 1$

18. $8a^3 + 1$

19. $64 - 125x^3$

20. $27 - 8t^3$

21. $27y^3 + 64$

22. $8x^3 + 27$

23. $x^3 - y^3$

24. $y^3 - z^3$

25. $a^3 + \frac{1}{8}$

26. $x^3 + \frac{1}{27}$

27. $8t^3 - 8$

28. $2y^3 - 128$

29. $54x^3 + 2$

30. $8a^3 + 1000$

31. $ab^3 + 125a$

32. $rs^3 + 64r$

33. $5x^3 - 40z^3$

34. $2y^3 - 54z^3$

35. $x^3 + 0.001$

36. $y^3 + 0.125$

37. $64x^6 - 8t^6$

38. $125c^6 - 8d^6$

39. $2y^4 - 128y$

40. $3z^5 - 3z^2$

41. $z^6 - 1$

42. $t^6 + 1$

43. $t^6 + 64y^6$

44. $p^6 - q^6$

45. $x^{12} - y^3z^{12}$

46. $a^9 + b^{12}c^{15}$

Solve.

47. $x^3 + 1 = 0$

48. $t^3 - 8 = 0$

49. $8x^3 = 27$

50. $64x^3 + 27 = 0$

51. $2t^3 - 2000 = 0$

52. $375 = 24x^3$

TW **53.** How could you use factoring to convince someone that $x^3 + y^3 \neq (x + y)^3$?

TW **54.** Is the following statement true or false and why? If A^3 and B^3 have a common factor, then A and B have a common factor.

Focused Review

Factor completely.

55. $5x^3 - 5z^3$ [5.7]

56. $x^2 + 4x - 21$ [5.4]

57. $-8x^3z + 24x^2z^2 - 16xz^3 - 8xz$ [5.3]

58. $1 + 1000x^3$ [5.7]

59. $3x^4 - 3z^4$ [5.6]

60. $4x^2 - 20x + 25$ [5.6]

61. $x^6 - 1$ [5.7]

62. $x^3 + x^2 + x + 1$ [5.3]

63. $3x^3 + 60x^2 + 300x$ [5.6]

64. $x^4y^3 + 8x^3y^3 + 12x^2y^3$ [5.4]

65. $12x^2 + 25x + 12$ [5.5]

66. $x^2 + 2x + 1 - y^2$ [5.6]

Synthesis

TW **67.** Explain how the geometric model below can be used to verify the formula for factoring $a^3 - b^3$.

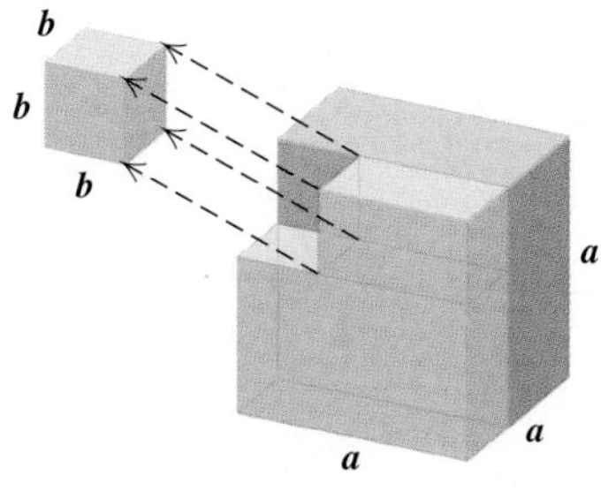

TW **68.** Explain how someone could construct a binomial that is both a difference of two cubes and a difference of two squares.

Factor.

69. $x^{6a} - y^{3b}$

70. $2x^{3a} + 16y^{3b}$

Aha! **71.** $(x + 5)^3 + (x - 5)^3$

72. $\frac{1}{16}x^{3a} + \frac{1}{2}y^{6a}z^{9b}$

73. $5x^3y^6 - \frac{5}{8}$

74. $x^3 - (x + y)^3$

75. $x^{6a} - (x^{2a} + 1)^3$

76. $(x^{2a} - 1)^3 - x^{6a}$

77. $t^4 - 8t^3 - t + 8$

78. If $P(x) = x^3$, use factoring to simplify
$P(a + h) - P(a)$.

79. If $Q(x) = x^6$, use factoring to simplify
$Q(a + h) - Q(a)$.

80. Using one viewing window, graph the following.
a) $f(x) = x^3$
b) $g(x) = x^3 - 8$
c) $h(x) = (x - 2)^3$

5.8 Applications of Polynomial Equations

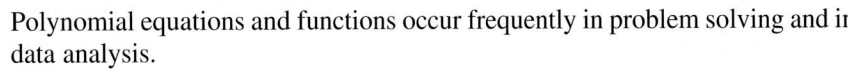

Problem Solving ■ Fitting Polynomial Functions to Data

Polynomial equations and functions occur frequently in problem solving and in data analysis.

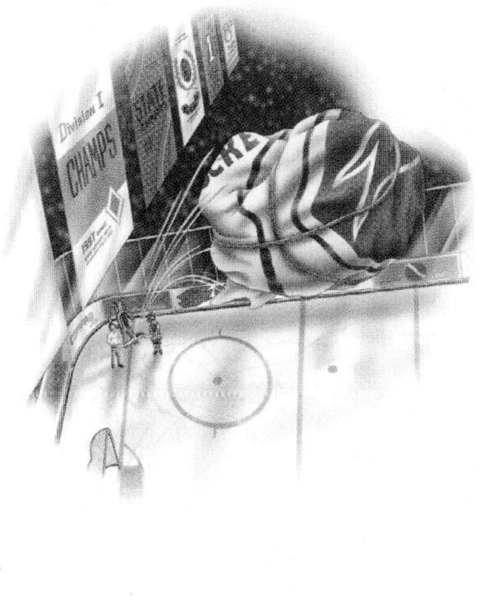

Problem Solving

Some problems can be translated to polynomial equations that we can now solve. The problem-solving process is the same as that used for other kinds of problems.

EXAMPLE 1 Prize Tee Shirts. During intermission at sporting events, it has become common for team mascots to launch tightly rolled tee shirts into the stands. The height $h(t)$, in feet, of an airborne tee shirt t seconds after being launched can be approximated by

$$h(t) = -15t^2 + 75t + 10.$$

After peaking, a rolled-up tee shirt is caught by a fan 70 ft above ground level. How long was the tee shirt in the air?

SOLUTION

1. **Familiarize.** We make a drawing and label it, using the information provided (see the figure on the following page). If we wanted to, we could

evaluate $h(t)$ for a few values of t. Note that t cannot be negative, since it represents time from launch.

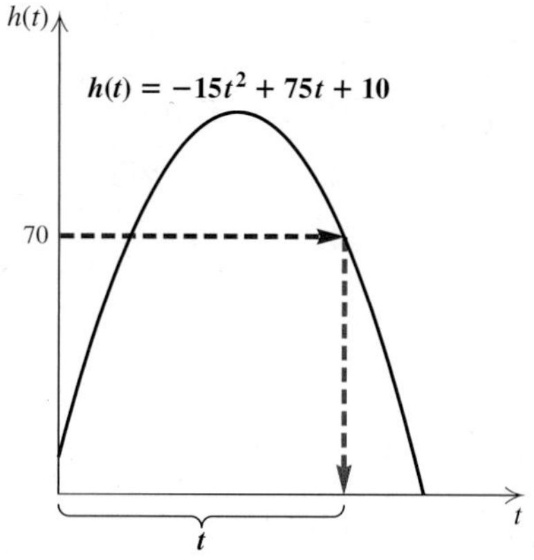

2. **Translate.** The relevant function has been provided. Since we are asked to determine how long it will take for the shirt to reach someone 70 ft above ground level, we are interested in the value of t for which $h(t) = 70$:

$$-15t^2 + 75t + 10 = 70.$$

3. **Carry out.** We solve both algebraically and graphically, as follows.

ALGEBRAIC APPROACH

We solve by factoring:

$$-15t^2 + 75t + 10 = 70$$

$$-15t^2 + 75t - 60 = 0 \quad \textbf{Subtracting 70 from both sides}$$

$$\left.\begin{aligned}-15(t^2 - 5t + 4) &= 0\\ -15(t-4)(t-1) &= 0\end{aligned}\right\} \quad \textbf{Factoring}$$

$$t - 4 = 0 \quad or \quad t - 1 = 0$$

$$t = 4 \quad or \quad t = 1.$$

The solutions appear to be 4 and 1.

GRAPHICAL APPROACH

We graph $y_1 = -15x^2 + 75x + 10$ and $y_2 = 70$ and look for any points of intersection of the graphs.

$y_1 = -15x^2 + 75x + 10, \; y_2 = 70$

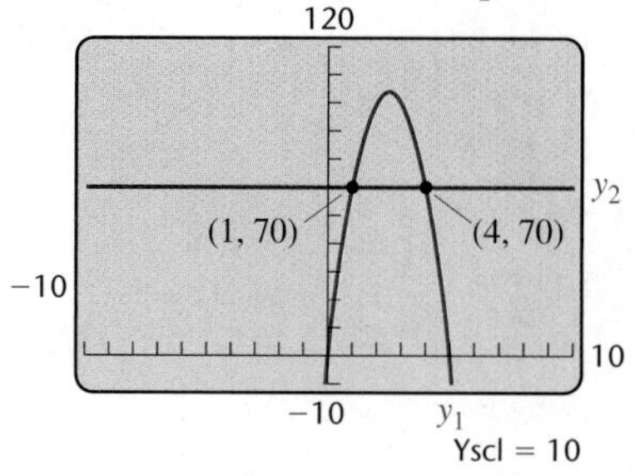

Using the INTERSECT option of the CALC menu, we see that the graphs intersect at $(1, 70)$ and $(4, 70)$. Thus the solutions of the equation are 1 and 4.

4. **Check.** We have

$$h(4) = -15 \cdot 4^2 + 75 \cdot 4 + 10 = -240 + 300 + 10 = 70 \text{ ft};$$
$$h(1) = -15 \cdot 1^2 + 75 \cdot 1 + 10 = -15 + 75 + 10 = 70 \text{ ft}.$$

Both 4 and 1 check. However, the problem states that the tee shirt is caught after peaking. Thus we reject 1 since that would indicate when the height of the tee shirt was 70 ft on the way *up*.

5. **State.** The tee shirt was in the air for 4 sec before being caught 70 ft above ground level.

Study Tip

Finishing a Chapter

Try to be aware of when you near the end of a chapter. Sometimes the end of a chapter signals the arrival of a quiz or test. Almost always, the end of a chapter indicates the end of a particular area of study. Because future work will likely rely on your mastery of the chapter completed, make use of the chapter review and test to solidify your understanding before moving onward.

The following problem involves the **Pythagorean theorem,** which relates the lengths of the sides of a right triangle. A **right triangle** has a 90°, or right, angle, which is denoted in the triangle by the symbol ⌜ or ⌝. The longest side, opposite the 90° angle, is called the **hypotenuse.** The other sides, called **legs,** form the two sides of the right angle.

The Pythagorean Theorem

In any right triangle, if a and b are the lengths of the legs and c is the length of the hypotenuse, then

$$a^2 + b^2 = c^2.$$

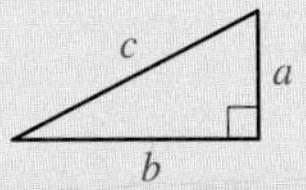

The symbol ⌜ denotes a 90° angle.

EXAMPLE 2 Carpentry. In order to build a deck at a right angle to their house, Lucinda and Felipe decide to place a stake in the ground a precise distance from the back wall of their house. This stake will combine with two marks on the house to form a right triangle. From a course in geometry, Lucinda remembers that there are three consecutive integers that can work as sides of a right triangle. Find the measurements of that triangle.

SOLUTION

1. **Familiarize.** Recall that x, $x + 1$, and $x + 2$ can be used to represent three unknown consecutive integers. Since $x + 2$ is the largest number, it must represent the hypotenuse. The legs serve as the sides of the right angle, so one leg must be formed by the marks on the house. We make a drawing in which

x = the distance between the marks on the house,

$x + 1$ = the length of the other leg,

and

$x + 2$ = the length of the hypotenuse.

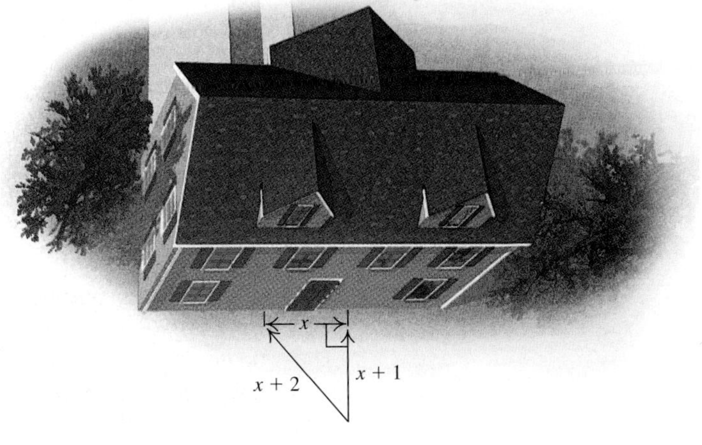

2. **Translate.** Applying the Pythagorean theorem, we translate as follows:

$$a^2 + b^2 = c^2$$
$$x^2 + (x + 1)^2 = (x + 2)^2.$$

3. **Carry out.** We solve both algebraically and graphically.

ALGEBRAIC APPROACH

We solve the equation as follows:

$$x^2 + (x + 1)^2 = (x + 2)^2$$
$$x^2 + (x^2 + 2x + 1) = x^2 + 4x + 4 \quad \text{Squaring the binomials}$$
$$2x^2 + 2x + 1 = x^2 + 4x + 4 \quad \text{Combining like terms}$$
$$x^2 - 2x - 3 = 0 \quad \text{Subtracting } x^2 + 4x + 4 \text{ from both sides}$$
$$(x - 3)(x + 1) = 0 \quad \text{Factoring}$$
$$x - 3 = 0 \quad or \quad x + 1 = 0 \quad \text{Using the principle of zero products}$$
$$x = 3 \quad or \quad x = -1.$$

GRAPHICAL APPROACH

We graph $y_1 = x^2 + (x + 1)^2$ and $y_2 = (x + 2)^2$ and look for the points of intersection of the graphs. A standard $[-10, 10, -10, 10]$ viewing window shows one point of intersection.

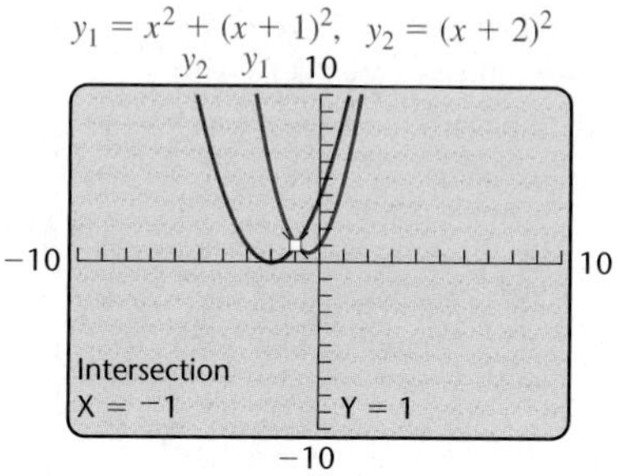

To see if the graphs intersect at another point, we use a larger viewing window. The window $[-10, 10, -10, 50]$ shows that there is indeed another point of intersection.

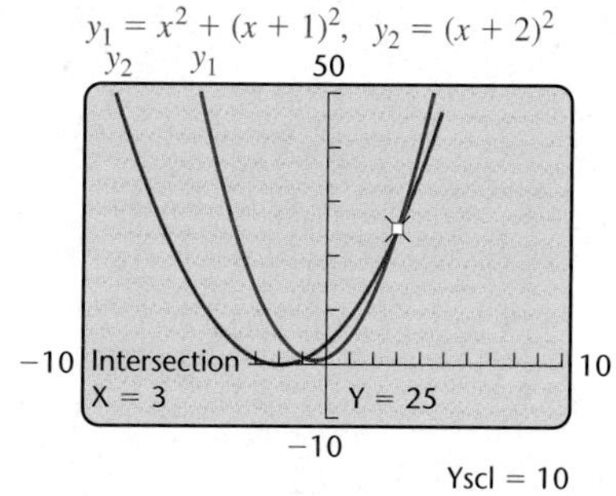

The points of intersection are $(-1, 1)$ and $(3, 25)$, so the solutions are -1 and 3.

4. **Check.** The integer -1 cannot be a length of a side because it is negative. For $x = 3$, we have $x + 1 = 4$, and $x + 2 = 5$. Since $3^2 + 4^2 = 5^2$, the lengths 3, 4, and 5 determine a right triangle. Thus, 3, 4, and 5 check.

5. **State.** Lucinda and Felipe should use a triangle with sides having a ratio of 3:4:5. Thus, if the marks on the house are 3 yd apart, they should locate the stake at the point in the yard that is precisely 4 yd from one mark and 5 yd from the other mark.

EXAMPLE 3 Display of a Sports Card. A valuable sports card is 4 cm wide and 5 cm long. The card is to be sandwiched by two pieces of Lucite, each of which is $5\frac{1}{2}$ times the area of the card. Determine the dimensions of the Lucite that will ensure a uniform border around the card.

SOLUTION

1. **Familiarize.** We make a drawing and label it, using x to represent the width of the border, in centimeters. Since the border extends uniformly around the entire card, the length of the Lucite must be $5 + 2x$ and the width must be $4 + 2x$.

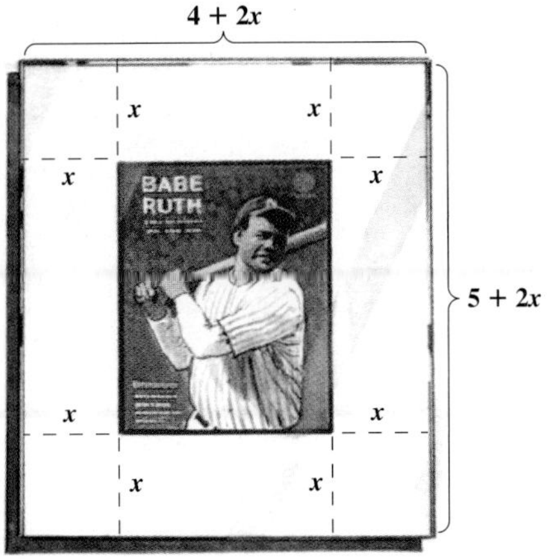

2. **Translate.** We rephrase the information given and translate as follows:

$$\underbrace{\text{Area of Lucite}}_{(5 + 2x)(4 + 2x)} \quad \underbrace{\text{is}}_{=} \quad \underbrace{5\tfrac{1}{2}\text{ times}}_{5\frac{1}{2}\,\cdot} \quad \underbrace{\text{area of card.}}_{5 \cdot 4}$$

3. **Carry out.** We solve both algebraically and graphically, as follows.

ALGEBRAIC APPROACH

We solve the equation:

$$(5 + 2x)(4 + 2x) = 5\tfrac{1}{2} \cdot 5 \cdot 4$$

$$20 + 10x + 8x + 4x^2 = 110 \qquad \textbf{Multiplying}$$

$$4x^2 + 18x - 90 = 0 \qquad \textbf{Finding standard form}$$

$$2x^2 + 9x - 45 = 0 \qquad \textbf{Multiplying by } \tfrac{1}{2} \textbf{ on both sides}$$

$$(2x + 15)(x - 3) = 0 \qquad \textbf{Factoring}$$

$$2x + 15 = 0 \quad \textit{or} \quad x - 3 = 0 \qquad \textbf{Principle of zero products}$$

$$x = -7\tfrac{1}{2} \quad \textit{or} \quad x = 3.$$

GRAPHICAL APPROACH

We graph $y_1 = (5 + 2x)(4 + 2x)$ and $y_2 = 5.5(5)(4)$ and look for the points of intersection of the graphs. Since we have $y_2 = 110$, we use a viewing window with Ymax > 110.

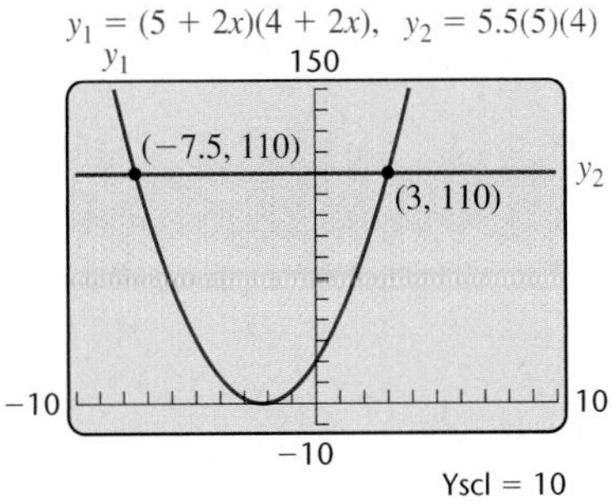

Using INTERSECT, we find the points of intersection $(-7.5, 110)$ and $(3, 110)$. The solutions are -7.5 and 3.

4. **Check.** We check 3 in the original problem. (Note that $-7\frac{1}{2}$ is not a solution because measurements cannot be negative.) If the border is 3 cm wide, the Lucite will have a length of $5 + 2 \cdot 3$, or 11 cm, and a width of $4 + 2 \cdot 3$, or 10 cm. The area of the Lucite is thus $11 \cdot 10$, or 110 cm^2. Since the area of the card is 20 cm^2 and 110 cm^2 is $5\frac{1}{2}$ times 20 cm^2, the number 3 checks.
5. **State.** Each piece of Lucite should be 11 cm long and 10 cm wide.

Fitting Polynomial Functions to Data

In Chapters 2 and 3, we modeled data using linear functions, or polynomial functions of degree 1. Some data that are not linear can be modeled using polynomial functions with higher degrees.

When data are not linear, it can be difficult to choose what type of function would best model the data. Graphs of quadratic functions are *parabolas*. We will study these graphs in detail in Chapter 8. Data that, when graphed, follow one of these patterns may best be modeled with a quadratic function.

Graphs of quadratic equations

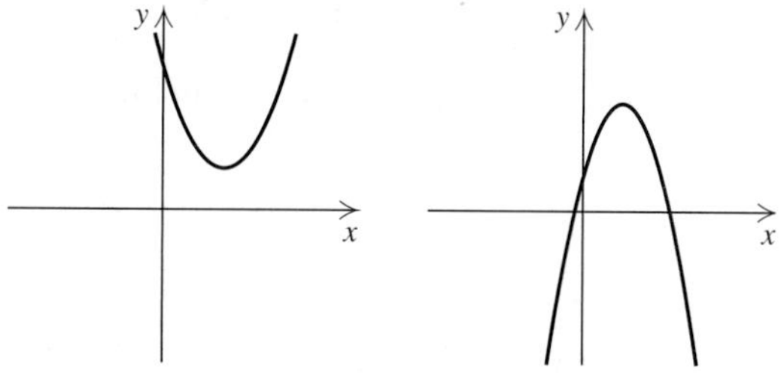

Graphs of cubic functions can follow any of the general shapes below. Polynomials of higher degree have even more possibilities for graphs. Choosing a model may involve forming the model and noting how well it matches the data. Another consideration is whether estimates and predictions made using the model are reasonable. In this section, we will indicate what type of model to use for each set of data.

Graphs of cubic equations

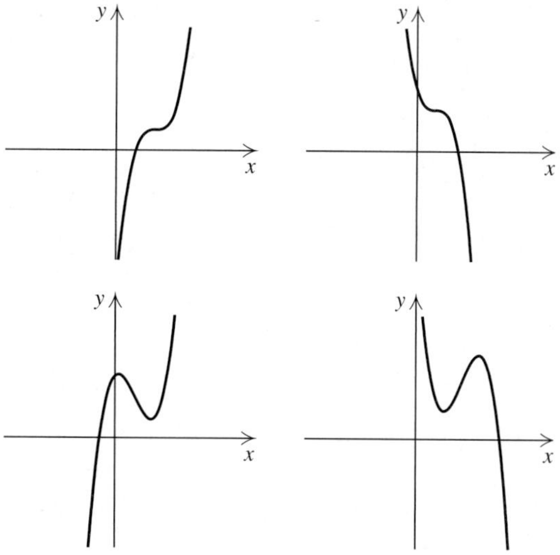

EXAMPLE 4 College Tuition. The average in-state tuition for a public four-year college for various years is shown in the following table and graph.

a) Fit a quartic (degree 4) polynomial function to the data.

b) Use the function found in part (a) to estimate public-college tuition in 2004 and to predict public-college tuition in 2008.

c) In what year will public-college tuition be $12,000?

Year	In-State Public-College Tuition
1985	$1386
1988	1726
1991	2159
1994	2820
1997	3323
2000	3768
2003	4686
2005*	5948

*Preliminary data
Source: National Center for Education Statistics, *Digest of Education Statistics*

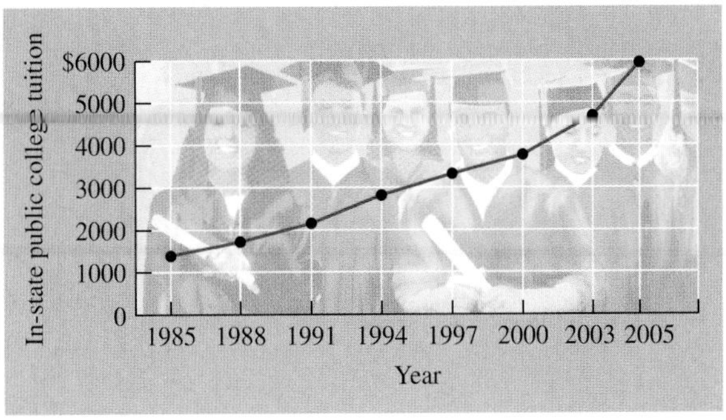

SOLUTION

a) We enter the data in a graphing calculator. To avoid large numbers, we let x = the number of years since 1985. The years are entered as L1 and the tuition as L2.

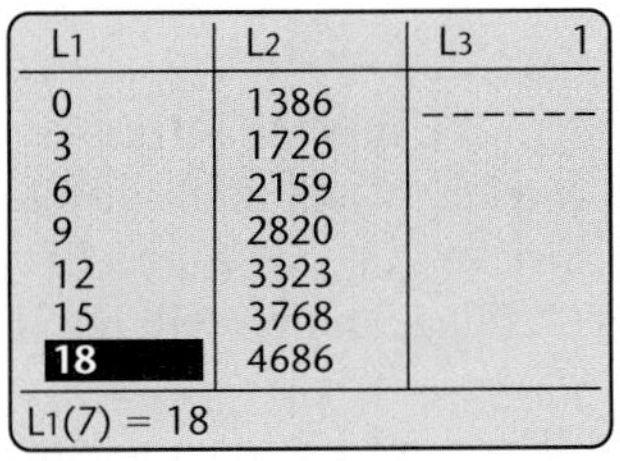

L1 L2 L3 1
0 1386
3 1726
6 2159
9 2820
12 3323
15 3768
18 4686
L1(7) = 18

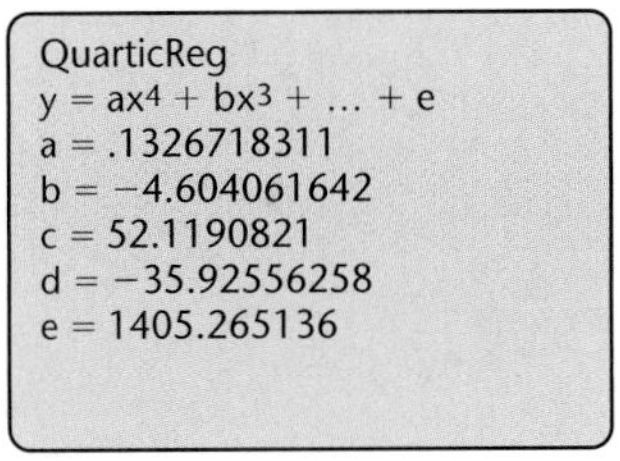

QuarticReg
y = ax4 + bx3 + ... + e
a = .1326718311
b = −4.604061642
c = 52.1190821
d = −35.92556258
e = 1405.265136

From the graph, we see that the data are not linear. We fit a quartic function to the data, choosing the QuartReg option from the STAT CALC menu. Pressing STAT ▷ 7 VARS ▷ 1 1 ENTER will calculate the regression equation and copy it to Y1. Rounding the coefficients, shown in the figure on the right above, to the nearest thousandth, we find that the quartic function that fits the data is

$$f(x) = 0.133x^4 - 4.604x^3 + 52.119x^2 - 35.926x + 1405.265.$$

b) To estimate and predict tuition for years not given, we will use a table with Indpnt set to Ask. Since 2004 is 19 yr after 1985, and 2008 is 23 yr after 1985, we enter 19 and 23 for X in the table. On the basis of this model, we estimate that public-college tuition was \$5248 in 2004 and will be \$9259 in 2008.

X	Y1
19	5248.3
23	9259.4

c) To determine the year in which public-college tuition will be \$12,000, we solve the equation $f(x) = 12{,}000$. Since $f(x)$ has been copied as Y_1, we let $Y_2 = 12{,}000$ and graph both equations. To see the scatterplot of the data as well, we turn on Plot1 and check that Xlist is L_1 and Ylist is L_2. We choose window settings that will show the given data as well as the intersection of the graphs of Y_1 and Y_2. For this situation, we use a setting of $[0, 30, 0, 15000]$, with Xscl = 5 and Yscl = 1000.

$$y_1 = .1326718311x^4 - 4.604061642x^3 + 52.1190821x^2 - 35.92556258x + 1405.265136,$$
$$y_2 = 12{,}000$$

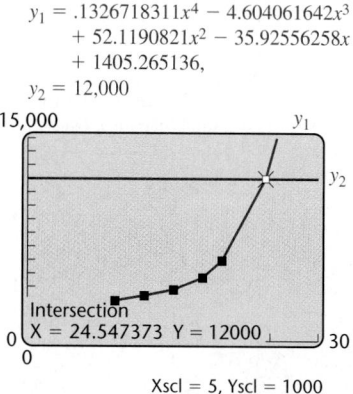

Xscl = 5, Yscl = 1000

Using the INTERSECT option of the CALC menu, we find that the coordinates of the point of intersection are $(24.547373, 12{,}000)$. Rounding, we have $x = 25$. Since 25 yr after 1985 is 2010, we estimate that public-college tuition will be \$12,000 in 2010.

When we use data to predict values not given, we assume that the trends shown by the data are consistent for values not shown. In general, estimating values between given data points (interpolation) is more reliable than predicting values beyond the last data point (extrapolation). However, the results of both interpolation and extrapolation depend on the assumptions we made.

In Example 4, we predicted that public-college tuition will be \$12,000 in 2010. This assumes that tuition continues to rise as quickly as it has in the years before 2005. Before 2010, we cannot know whether that assumption is valid.

Many personal, business, and government decisions are made on the basis of predictions of the future. Mathematics and curve fitting can be helpful in making those predictions, but we must keep in mind the underlying assumptions and the knowledge that a prediction is not a fact.

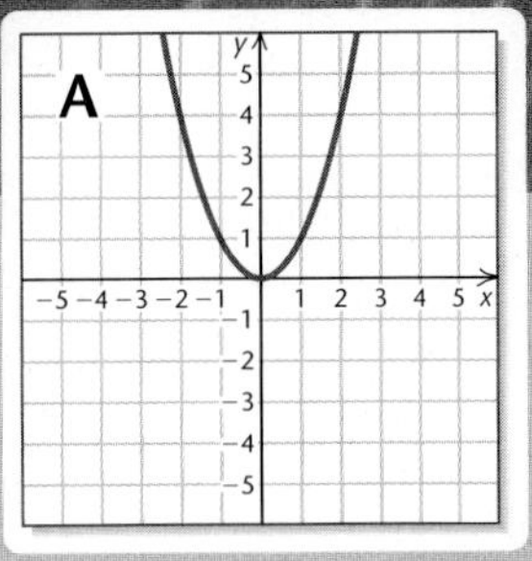

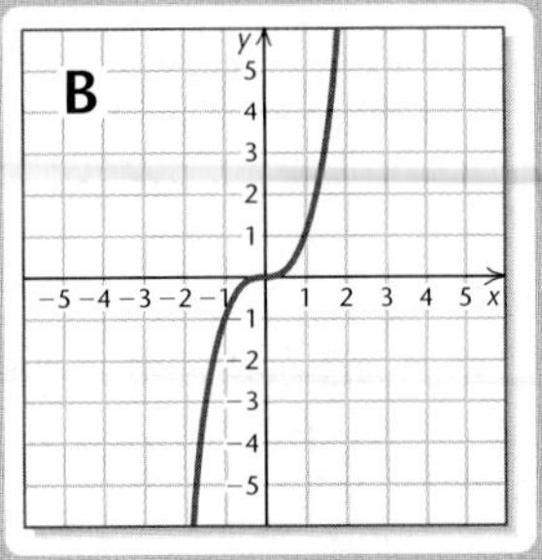

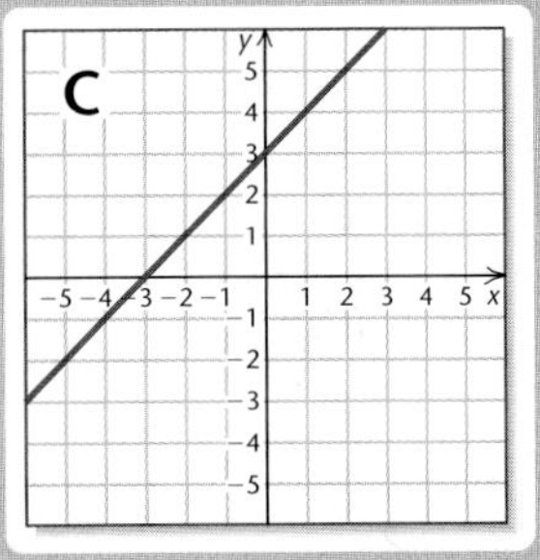

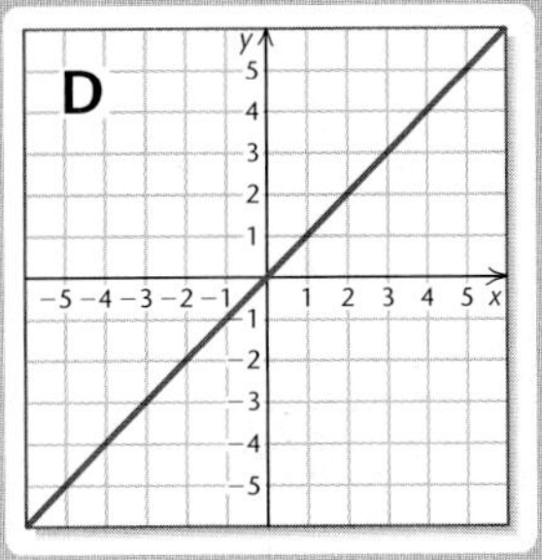

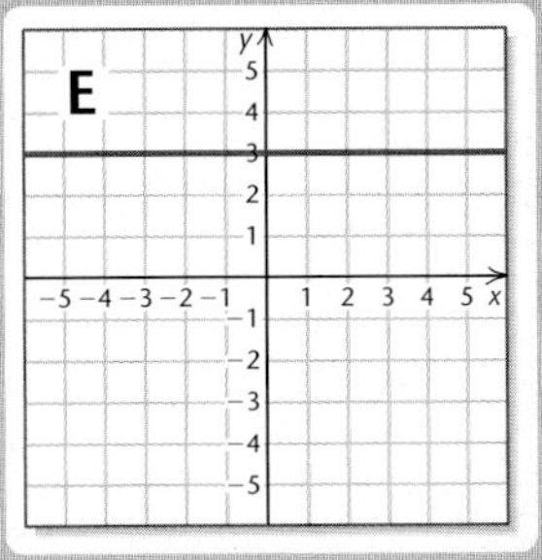

Visualizing the Graph

Match each equation with its graph.

1. $y = x$

2. $y = |x|$

3. $y = x^2$

4. $y = x^3$

5. $y = 3$

6. $y = x + 3$

7. $y = x - 3$

8. $y = 2x$

9. $y = -2x$

10. $y = 2x + 3$

Answers on page A-24

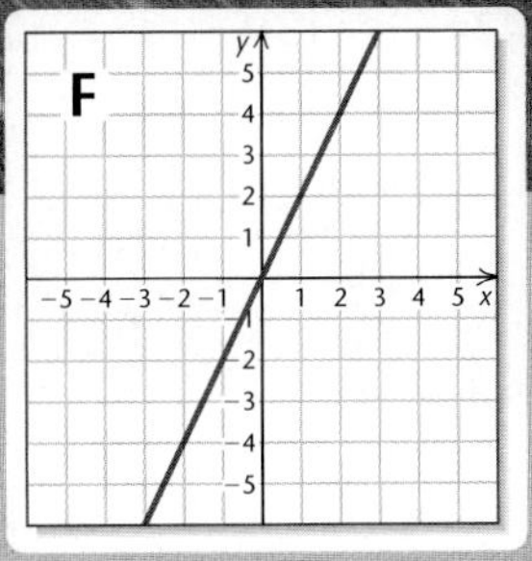

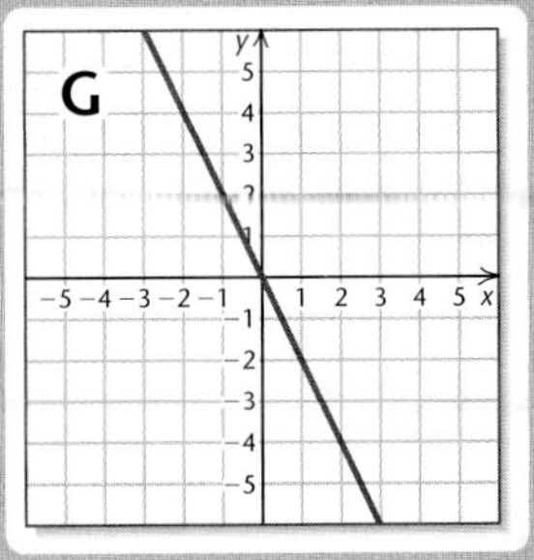

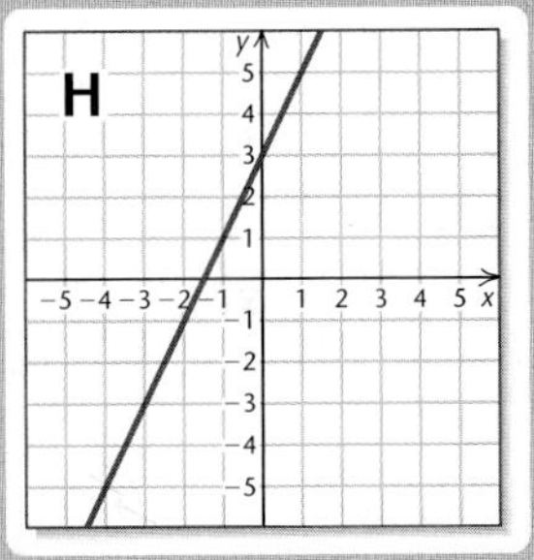

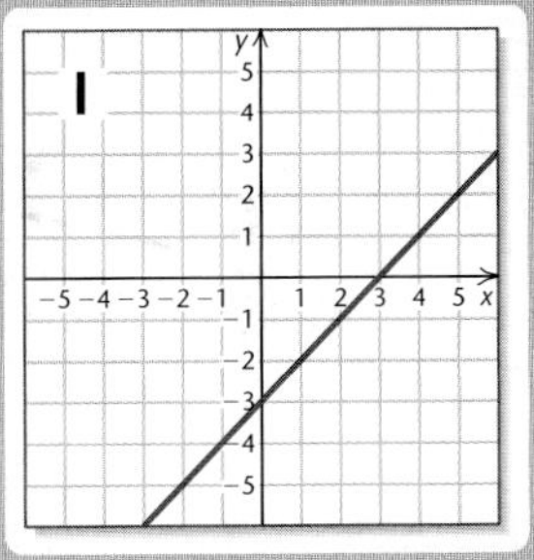

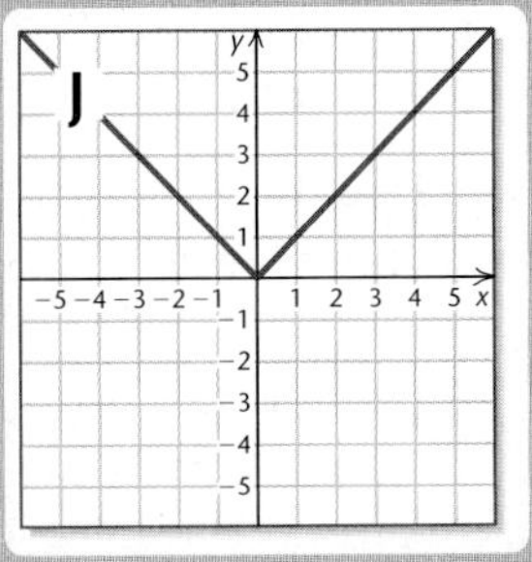

5.8 EXERCISE SET

FOR EXTRA HELP

MathXL | MyMathLab | InterAct Math | AW Math Tutor Center | Video Lectures on CD: Disc 3 | Student's Solutions Manual

Solve.

1. The square of a number plus the number is 132. What is the number?

2. The square of a number plus the number is 156. What is the number?

3. A photo is 5 cm longer than it is wide. Find the length and the width if the area is 84 cm^2.

4. An envelope is 4 cm longer than it is wide. The area is 96 cm^2. Find the length and the width.

5. *Geometry.* If each of the sides of a square is lengthened by 4 m, the area becomes 49 m^2. Find the length of a side of the original square.

6. *Geometry.* If each of the sides of a square is lengthened by 6 cm, the area becomes 144 cm^2. Find the length of a side of the original square.

7. *Framing a Picture.* A picture frame measures 12 cm by 20 cm, and 84 cm^2 of picture shows. Find the width of the frame.

8. *Framing a Picture.* A picture frame measures 14 cm by 20 cm, and 160 cm^2 of picture shows. Find the width of the frame.

9. *Landscaping.* A rectangular lawn measures 60 ft by 80 ft. Part of the lawn is torn up to install a sidewalk of uniform width around it. The area of the new lawn is 2400 ft^2. How wide is the sidewalk?

10. *Landscaping.* A rectangular garden is 30 ft by 40 ft. Part of the garden is removed in order to install a walkway of uniform width around it. The area of the new garden is one-half the area of the old garden. How wide is the walkway?

11. Three consecutive even integers are such that the square of the third is 76 more than the square of the second. Find the three integers.

12. Three consecutive even integers are such that the square of the first plus the square of the third is 136. Find the three integers.

13. *Tent Design.* The triangular entrance to a tent is 2 ft taller than it is wide. The area of the entrance is 12 ft^2. Find the height and the base.

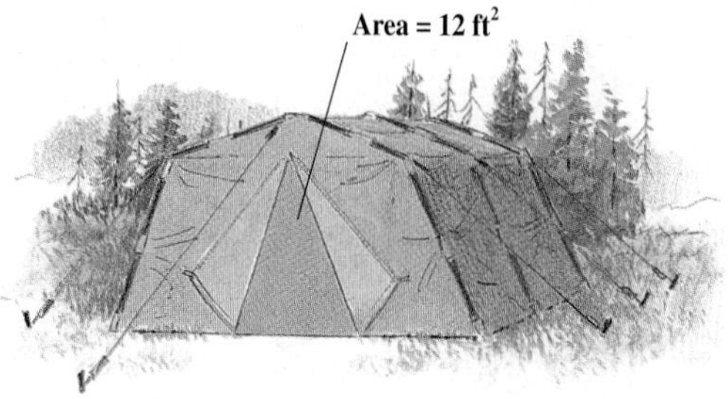

14. *Antenna Wires.* A wire is stretched from the ground to the top of an antenna tower, as shown. The wire is 20 ft long. The height of the tower is 4 ft greater than the distance d from the tower's base to the bottom of the wire. Find the distance d and the height of the tower.

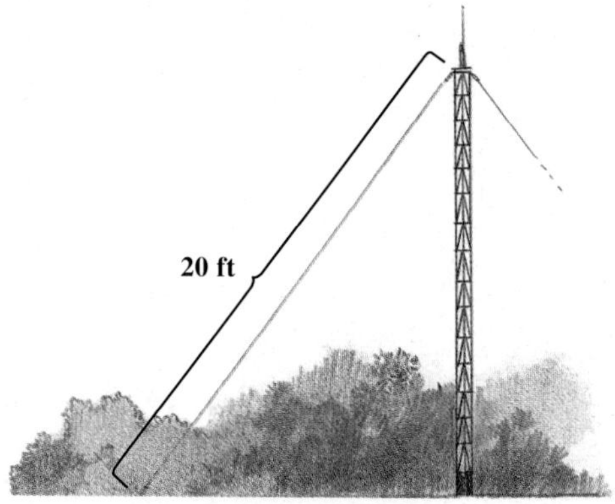

15. *Sailing.* A triangular sail is 9 m taller than it is wide. The area is 56 m^2. Find the height and the base of the sail.

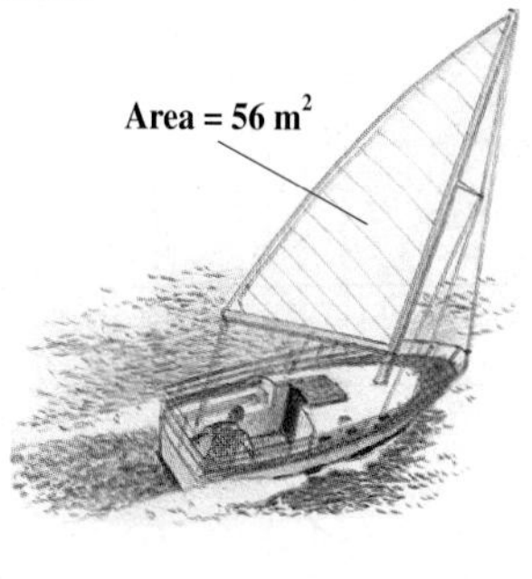

16. *Parking Lot Design.* A rectangular parking lot is 50 ft longer than it is wide. Determine the dimensions of the parking lot if it measures 250 ft diagonally.

17. *Ladder Location.* The foot of an extension ladder is 9 ft from a wall. The height that the ladder reaches on the wall and the length of the ladder are consecutive integers. How long is the ladder?

18. *Ladder Location.* The foot of an extension ladder is 10 ft from a wall. The ladder is 2 ft longer than the height that it reaches on the wall. How far up the wall does the ladder reach?

19. *Garden Design.* Ignacio is planning a garden that is 25 m longer than it is wide. The garden will have an area of 7500 m^2. What will its dimensions be?

20. *Garden Design.* A flower bed is to be 3 m longer than it is wide. The flower bed will have an area of 108 m^2. What will its dimensions be?

21. *Cabinet Making.* Dovetail Woodworking determines that the revenue R, in thousands of dollars, from the sale of x sets of cabinets is given by $R(x) = 2x^2 + x$. If the cost C, in thousands of dollars, of producing x sets of cabinets is given by $C(x) = x^2 - 2x + 10$, how many sets must be produced and sold in order for the company to break even?

22. *Camcorder Production.* Suppose that the cost of making x video cameras is $C(x) = \frac{1}{9}x^2 + 2x + 1$, where $C(x)$ is in thousands of dollars. If the revenue from the sale of x video cameras is given by $R(x) = \frac{5}{36}x^2 + 2x$, where $R(x)$ is in thousands of dollars, how many cameras must be sold in order for the firm to break even?

23. *Prize Tee Shirts.* Using the model in Example 1, determine how long a tee shirt has been airborne if it is caught on the way *up* by a fan 100 ft above ground level.

24. *Prize Tee Shirts.* Using the model in Example 1, determine how long a tee shirt has been airborne if it is caught on the way *down* by a fan 10 ft above ground level.

25. *Fireworks Displays.* Fireworks are typically launched from a mortar with an upward velocity (initial speed) of about 64 ft/sec. The height $h(t)$, in feet, of a "weeping willow" display, t seconds after having been launched from an 80-ft-high rooftop, is given by

$$h(t) = -16t^2 + 64t + 80.$$

After how long will the cardboard shell from the fireworks reach the ground?

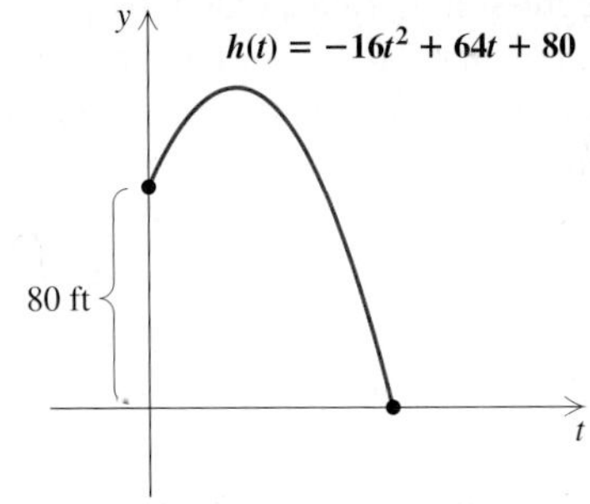

26. *Safety Flares.* Suppose that a flare is launched upward with an initial velocity of 80 ft/sec from a height of 224 ft. Its height in feet, $h(t)$, after t seconds is given by

$$h(t) = -16t^2 + 80t + 224.$$

After how long will the flare reach the ground?

27. *Home-healthcare.* The number of home-healthcare patients in the United States per 10,000 people, t years after 1992, can be approximated by $N(t) = -2t^2 + 15t + 51$ (*Source*: Based on data from Centers for Disease Control, 2003). How many years after 1992 were there about 69 home-healthcare patients per 10,000 people?

28. *Publishing.* The number of books B, in thousands, published in the United States t years after 2000 can be approximated by $B(t) = -8t^2 + 62t + 60$ (*Source*: Based on data from R. R. Bowker). How many years after 2000 were there about 170,000 books published in the United States?

29. *Baseball.* The following table lists the average value, in millions of dollars, of multiyear contracts for free-agent pitchers.

Year	Average Value of Multiyear Contract of Free-agent MLB Pitchers (in millions)
2001	$21.4
2002	17.8
2003	10.3
2004	12.1
2005	17.2
2006	22.6

Source: *USA TODAY* 12/7/05

a) Use regression to find a quadratic polynomial function C that can be used to estimate the average value of a multiyear contract x years after 2000.
b) Estimate the average value of a multiyear contract in 2008.
c) In what year after 2000 will the average value of a multiyear contract for a free-agent pitcher be $30 million?

30. *Olympics.* The following table lists the number of female athletes participating in the winter Olympic games for various years.

Year	Number of Female Athletes in the Winter Olympic Games
1924	11
1936	80
1952	109
1964	199
1976	231
1984	274
1994	522
2006	960

Source: Olympics.org

a) Use regression to find a cubic polynomial function F that can be used to estimate the number of female athletes competing in the winter Olympic games x years after 1900.
b) Estimate the number of female athletes competing in 2010.
c) In what year will 1200 female athletes compete in the winter Olympic games?

31. *Life Insurance Premiums.* The following table lists monthly life insurance premiums for a $500,000 policy for females of various ages.

Age	Monthly Premium for $500,000 Life Insurance Policy for Females
35	$14
40	18
45	24
50	33
55	48
60	70
65	110
70	188

Source: Insure.com

a) Use regression to find a cubic polynomial function L that can be used to estimate the monthly life insurance premium for a female of age x.
b) Estimate the monthly premium for a female of age 58.
c) For what age is the monthly premium $100?

32. *Health Insurance Premiums.* The following table lists the percent of increase from the previous year in health insurance premiums in the United States.

Year	Percent of Increase in Health Insurance Premiums
1988	12.0%
1989	18.0
1990	14.0
1993	8.5
1996	0.8
1999	5.3
2000	8.2
2001	10.9
2003	13.9
2004	11.2

Source: Kaiser Family Foundation

a) Use regression to find a quartic polynomial function H that can be used to estimate the percent of increase in health insurance premiums x years after 1988.

b) Estimate the percent of increase in premiums in 2002.

c) In what year or years after 1988 is the percent of increase 5%?

33. *Biological Sciences.* The following table lists the number of bachelor's degrees earned in the biological and biomedical sciences for various years.

Year	Number of Bachelor's Degrees Earned in Biological and Biomedical Sciences
1990	37,204
1992	42,781
1994	51,157
1996	60,750
1998	65,583
2000	63,005
2002	59,415
2004	61,509

Source: U.S. National Council for Education Statistics

a) Use regression to find a quartic polynomial function B that can be used to estimate the number of bachelor's degrees earned in the biological and biomedical sciences x years after 1990.

b) Estimate the number of degrees earned in 2006.

c) In what year or years were approximately 62,000 degrees earned?

34. *Engineering.* The following table lists the number of bachelor's degrees earned in engineering and engineering technologies for various years.

Year	Number of Bachelor's Degrees Earned in Engineering and Engineering Technologies
1986	97,122
1990	82,480
1994	78,662
1998	74,649
2002	74,679
2004	78,227

Source: U.S. National Council for Education Statistics

a) Use regression to find a quadratic polynomial function E that can be used to estimate the number of bachelor's degrees earned in engineering and engineering technologies x years after 1980.

b) Estimate the number of degrees earned in 2010.

c) In what year after 1986 will aproximately 100,000 degrees be earned?

TW **35.** Tyler disregards any negative solutions that he finds when solving applied problems. Is this approach correct? Why or why not?

TW **36.** Write a chart of the population of two imaginary cities. Devise the numbers in such a way that one city has linear growth and the other has nonlinear growth.

Focused Review

Solve.

37. $3(x - 7) + 5x = 4x - (7 - x)$ [1.6]

38. $2x - 14 + 9x > -8x + 16 + 10x$ [4.1]

39. $x - y + z = 6,$
$2x + y - z = 0,$
$x + 2y + z = 3$ [3.4]

40. $|5 - 7x| \geq 9$ [4.4]

41. $|5 - 7x| \leq 9$ [4.4]

42. $5 - 7x > -9 + 12x$ [4.1]

43. $|x + 3| = 29$ [4.4]

44. $2x - 3y = 8,$
$x + 3y = 10$ [3.2]

45. $x^2 = \frac{1}{9}$ [5.6]

46. $x^2 - 6 = 5x$ [5.4]

Synthesis

TW **47.** Explain how you might decide what kind of function would fit a particular set of data.

TW **48.** Compare the data in Exercise 33 with that in Exercise 34. Describe the trends occurring between 1990 and 2004. How might this affect decisions made by college administrations?

49. *Box Construction.* A rectangular piece of tin is twice as long as it is wide. Squares 2 cm on a side are cut out of each corner, and the ends are turned up to make a box whose volume is 480 cm³. What are the dimensions of the piece of tin?

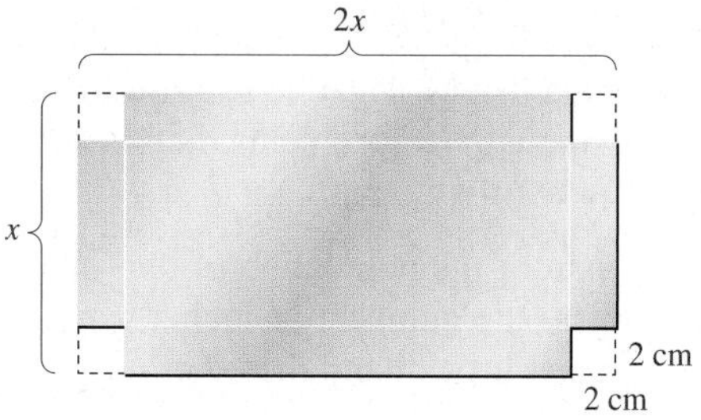

50. *Navigation.* A tugboat and a freighter leave the same port at the same time at right angles. The freighter travels 7 km/h slower than the tugboat. After 4 hr, they are 68 km apart. Find the speed of each boat.

51. *Skydiving.* During the first 13 sec of a jump, a skydiver falls approximately $11.12t^2$ feet in t seconds. A small heavy object (with less wind resistance) falls about $15.4t^2$ feet in t seconds. Suppose that a skydiver jumps from 30,000 ft, and 1 sec later a camera falls out of the airplane. How long will it take the camera to catch up to the skydiver?

52. *Olympics.* The following table lists the number of male athletes participating in the winter Olympic games for various years.

Year	Number of Male Athletes in the Winter Olympic Games
1924	247
1936	566
1952	585
1964	892
1976	892
1984	998
1994	1215
2006	1548

Source: Olympics.org

a) Use regression to find a linear function M that can be used to estimate the number of male athletes competing in the winter Olympic games x years after 1900.

b) Using the model found in part (a) and that found in Exercise 30, predict the first year in which there will be more female athletes than male athletes in the winter Olympic games.

Collaborative Corner

Diet and Health

Focus: Polynomial functions and data analysis
Time: 20 minutes
Group Size: 3

A country's average life expectancy and infant mortality rates are related to the diet of its citizens. The following table lists daily caloric intake, life expectancy, and infant mortality rates for several countries.

Country	Daily Caloric Intake, c	Life Expectancy, L	Infant Mortality Rate, M (in number of deaths per 1000 births)
Argentina	2880	72.7	15.7
Bangladesh	2021	61.7	64.3
Brazil	2751	71.4	30.1
Canada	3482	80.0	4.8
Ethiopia	1667	48.7	96.9
Germany	3443	78.5	4.2
Mexico	3181	74.9	21.7
United States	3671	77.4	6.6

Sources: Statistical Package for Social Sciences; U.S. Bureau of the Census

ACTIVITY

1. Consider life expectancy as a function of daily caloric intake.

a) Use regression to fit a linear function, a quadratic function, and a cubic function to the data, with each group member modeling one type of function.
b) Use each of the three functions to estimate the life expectancy for a country with a daily caloric intake of 1500 calories; of 4500 calories.
c) Decide as a group which of the three functions best models the situation.
d) Is there an "optimal" daily caloric intake; that is, one for which the life expectancy is greatest?

2. Consider infant mortality rate as a function of daily caloric intake.

a) Use regression to fit a linear function, a quadratic function, and a cubic function to the data, with each group member modeling one type of function.
b) Use each of the three functions to estimate the infant mortality rate for a country with a daily caloric intake of 1500 calories; of 4500 calories.
c) Decide as a group which of the three functions best models the situation.
d) Is there an "optimal" daily caloric intake; that is, one for which the infant mortality rate is lowest?

5

Chapter Summary and Review

KEY TERMS AND DEFINITIONS

POLYNOMIALS

Polynomial, p. 341 A monomial or a sum of monomials. All exponents of variables in a polynomial are whole numbers.

Term, p. 341 A number, a variable, or a product or a quotient of numbers and/or variables. The number in a term is its **coefficient.** The **degree** of a term is the number of variable factors.

Term	Coefficient	Degree
$-8t^2$	-8	2
$\frac{1}{3}xy^4$	$\frac{1}{3}$	5
7	7	0

Leading term, p. 342 The term in a polynomial with the highest degree. The coefficient of the leading term is the **leading coefficient,** and the degree of the leading term is the **degree of the polynomial.**

Descending order, p. 342 An arrangement of a polynomial with terms of higher degree written first. If the terms with lower degree are written first, the polynomial is in **ascending order.**

FACTORING

Factor completely, p. 374 To write a polynomial as a product in which no factor can itself be factored.

Prime polynomial, p. 374 A polynomial that cannot be factored.

SOLVING

Root, p. 369 A solution of an equation that is set equal to 0.

Zero, p. 369 An input for which the function output is 0.

Repeated root, p. 407 A root of an equation that appears in more than one factor. If the root is repeated twice, it is a **root of multiplicity two.**

A STRATEGY FOR FACTORING

A. Always factor out the greatest common factor, if possible.

B. Once the greatest common factor has been factored out, *count the number of terms* in the other factor:

Two terms: Try factoring as a difference of squares first. Next, try factoring as a sum or a difference of cubes.

Three terms: Try factoring as a perfect-square trinomial. Next, try trial and error, either reversing the FOIL method or using the *ac*-method.

Four or more terms: Try factoring by grouping and factoring out a common binomial factor. Next, try grouping into a difference of squares, one of which is a trinomial.

C. Always *factor completely*. If a factor with more than one term can itself be factored further, do so.

D. *Check* the factorization by multiplying.

IMPORTANT CONCEPTS

[Section references appear in brackets.]

<table>
<tr><th>Concept</th><th>Example</th></tr>
<tr><td>Polynomials are classified by the number of terms.</td><td>Monomial (one term): $4x^3$
Binomial (two terms): $x^2 - 5$
Trinomial (three terms): $3t^3 + 2t - 10$ [5.1]</td></tr>
<tr><td>Polynomials are classified by degree.</td><td>Constant (degree 0): -9
Linear (degree 1): $\frac{1}{3}x + 8$
Quadratic (degree 2): $x^2 + 2x + 3$
Cubic (degree 3): $-x^3 - 5x$
Quartic (degree 4): $15x^4 + x^2 + 1$ [5.1]</td></tr>
<tr><td>Add polynomials by combining like terms.

Subtract polynomials by adding the opposite of the polynomial being subtracted.

Multiply polynomials by multiplying each term of one polynomial by each term of the other.</td><td>$(2x^2 - 3x + 7) + (5x^3 + 3x - 9)$
$= 2x^2 + (-3x) + 7 + 5x^3 + 3x + (-9)$
$= 5x^3 + 2x^2 - 2$
$(2x^2 - 3x + 7) - (5x^3 + 3x - 9)$
$= 2x^2 - 3x + 7 - 5x^3 - 3x + 9$
$= -5x^3 + 2x^2 - 6x + 16$
$(x + 2)(x^2 - x - 1) = x \cdot x^2 - x \cdot x - x \cdot 1$
$+ 2 \cdot x^2 - 2 \cdot x - 2 \cdot 1$
$= x^3 - x^2 - x + 2x^2 - 2x - 2$
$= x^3 + x^2 - 3x - 2$ [5.1], [5.2]</td></tr>
<tr><td>Some polynomials can be factored by grouping.</td><td>$x^3 - x^2 - 2x + 2 = x^2(x - 1) - 2(x - 1)$
$= (x - 1)(x^2 - 2)$ [5.3]</td></tr>
<tr><td>Factoring by Reversing FOIL
Form products of two binomials.
The product of the First terms is ax^2.
The product of the Last terms is c.
Binomials with common factors can be eliminated.
Check remaining products.</td><td>$6x^2 + x - 2$
$= (3x + 2)(2x - 1)$
~~$(6x + 2)(x - 1)$~~ Common factor
~~$(6x - 2)(x + 1)$~~ Common factor
$(6x + 1)(x - 2)$ $6x^2 - 11x - 2$
$(6x - 1)(x + 2)$ $6x^2 + 11x - 2$
Correct product ⟶ $(3x + 2)(2x - 1)$ $6x^2 + x - 2$
$(3x - 2)(2x + 1)$ $6x^2 - x - 2$
~~$(3x + 1)(2x - 2)$~~ Common factor
~~$(3x - 1)(2x + 2)$~~ Common factor [5.5]</td></tr>
<tr><td>Factoring Using the ac-Method
Multiply a and c.
Factor ac so that the sum of the factors is b.
Split the middle term.
Factor by grouping.</td><td>$15x^2 - 2x - 8$ $15 \cdot (-8) = -120$
$= 15x^2 - 12x + 10x - 8$ $-120 = 10 \cdot (-12)$
$= 3x(5x - 4) + 2(5x - 4)$ $10 + (-12) = -2$
$= (5x - 4)(3x + 2)$ [5.5]</td></tr>
</table>

(continued)

Factoring Formulas

Trinomial Square:

$A^2 + 2AB + B^2 = (A + B)^2$

$A^2 - 2AB + B^2 = (A - B)^2$

$x^2 + 20x + 100 = (x + 10)^2$

$4y^2z^2 - 4yz + 1 = (2yz - 1)^2$

Difference of Two Squares:

$A^2 - B^2 = (A + B)(A - B)$

$x^2 - 81 = (x + 9)(x - 9)$

Sum of Two Cubes:

$A^3 + B^3 = (A + B)(A^2 - AB + B^2)$

$y^3 + 27 = (y + 3)(y^2 - 3y + 9)$

Difference of Two Cubes:

$A^3 - B^3 = (A - B)(A^2 + AB + B^2)$

$1000 - t^3 = (10 - t)(100 + 10t + t^2)$

[5.6], [5.7]

The Principle of Zero Products

For any real numbers a and b:

If $ab = 0$, then $a = 0$ or $b = 0$.
If $a = 0$ or $b = 0$, then $ab = 0$.

$6x^2 + x = 2$

$6x^2 + x - 2 = 0$ **Writing with 0 on one side**

$(3x + 2)(2x - 1) = 0$ **Factoring**

$3x + 2 = 0$ *or* $2x - 1 = 0$ **Principle of zero products**

$3x = -2$ *or* $2x = 1$

$x = -\frac{2}{3}$ *or* $x = \frac{1}{2}$

The solutions are $-\frac{2}{3}$ and $\frac{1}{2}$.

[5.3]

Pythagorean Theorem

In any right triangle, if a and b are the lengths of the legs and c is the length of the hypotenuse, then $a^2 + b^2 = c^2$.

For the right triangle shown,

$x^2 + (x + 1)^2 = 5^2$.

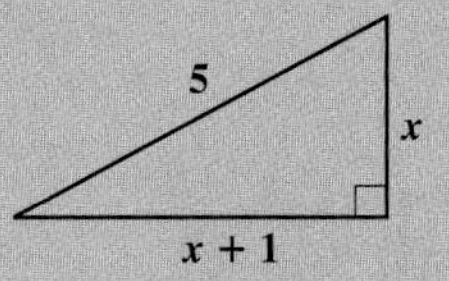

[5.8]

The **roots** of an equation, the **zeros** of a function, and the ***x*-intercepts** of the graph of the function are related.

The roots of $x^2 - 4 = 0$ are -2 and 2.
The zeros of $f(x) = x^2 - 4$ are -2 and 2.
The x-intercepts of the graph of $f(x) = x^2 - 4$ are $(-2, 0)$ and $(2, 0)$.

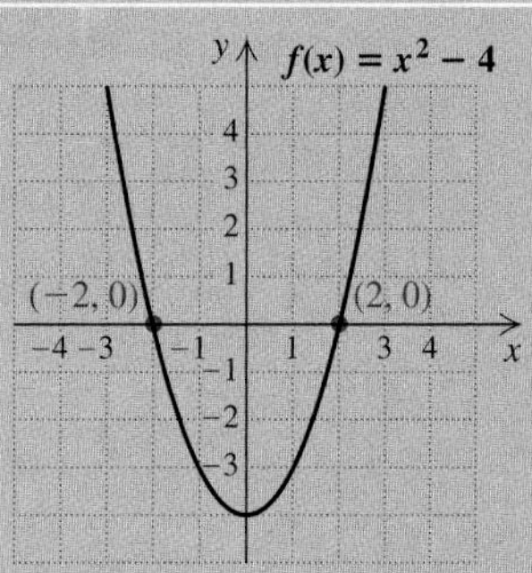

[5.3]–[5.7]

Review Exercises

Concept Reinforcement *In each of Exercises 1–10, match the item with the most appropriate choice from the column on the right.*

1. ___ A polynomial with four terms [5.1]
2. ___ A difference of two squares [5.6]
3. ___ A perfect-square trinomial [5.6]
4. ___ A difference of two cubes [5.7]
5. ___ The principle of zero products [5.3]
6. ___ A quadratic equation [5.8]
7. ___ A term that is not a monomial [5.1]
8. ___ The longest side in any right triangle [5.8]
9. ___ A polynomial written in ascending order [5.1]
10. ___ A polynomial that cannot be factored [5.3]

a) $5x + 2x^2 - 4x^3$
b) $3x^{-1}$
c) If $a \cdot b = 0$, then $a = 0$ or $b = 0$.
d) Prime
e) $9 - t^2$
f) Hypotenuse
g) $8x^3 - 4x^2 + 12x + 14$
h) $27 - t^3$
i) $2x^2 - 4x = 7$
j) $4a^2 - 12a + 9$

11. Given the polynomial
$$2xy^6 - 7x^8y^3 + 2x^3 + 9,$$
determine the degree of each term and the degree of the polynomial. [5.1]

12. Given the polynomial
$$3x - 5x^3 + 2x^2 + 9,$$
arrange in descending order and determine the leading term and the leading coefficient. [5.1]

13. Arrange in ascending powers of x:
$$8x^6y - 7x^8y^3 + 2x^3 - 3x^2. \text{ [5.1]}$$

14. Find $P(0)$ and $P(-1)$:
$$P(x) = x^3 - x^2 + 4x. \text{ [5.1]}$$

15. Evaluate the polynomial function for $x = -2$:
$$P(x) = 4 - 2x - x^2. \text{ [5.1]}$$

Combine like terms. [5.1]

16. $6 - 4a + a^2 - 2a^3 - 10 + a$

17. $4x^2y - 3xy^2 - 5x^2y + xy^2$

Add. [5.1]

18. $(-7x^3 - 4x^2 + 3x + 2) + (5x^3 + 2x + 6x^2 + 1)$

19. $(3x^4 + 3x^3 - 8x + 9) + (-6x^4 + 4x + 7 + 3x)$

20. $(-9xy^2 - xy + 6x^2y) + (-5x^2y - xy + 4xy^2)$

Subtract. [5.1]

21. $(8x - 5) - (-6x + 2)$

22. $(4a - b + 3c) - (6a - 7b - 4c)$

23. $(8x^2 - 4xy + y^2) - (2x^2 + 3xy - 2y^2)$

Simplify as indicated. [5.2]

24. $(3x^2y)(-6xy^3)$

25. $(x^4 - 2x^2 + 3)(x^4 + x^2 - 1)$

26. $(4ab + 3c)(2ab - c)$

27. $(2x + 5y)(2x - 5y)$

28. $(3x - 4y)^2$

29. $(x + 3)(2x - 1)$

30. $(x^2 + 4y^3)^2$

31. $(3t - 5)^2 - (2t + 3)^2$

32. $\left(x - \frac{1}{3}\right)\left(x - \frac{1}{6}\right)$

Factor.

33. $7x^2 + 6x$ [5.3]

34. $9y^4 - 3y^2$ [5.3]

35. $15x^4 - 18x^3 + 21x^2 - 9x$ [5.3]

36. $a^2 - 12a + 27$ [5.4]

37. $3m^2 + 14m + 8$ [5.5]

38. $25x^2 + 20x + 4$ [5.6]

39. $4y^2 - 16$ [5.6]

40. $5x^2 + x^3 - 14x$ [5.4]

41. $ax + 2bx - ay - 2by$ [5.3]

42. $3y^3 + 6y^2 - 5y - 10$ [5.3]

43. $a^4 - 81$ [5.6]

44. $4x^4 + 4x^2 + 20$ [5.3]

45. $27x^3 - 8$ [5.7]

46. $0.064b^3 - 0.125c^3$ [5.7]

47. $y^5 + y$ [5.3]

48. $2z^8 - 16z^6$ [5.3]

49. $54x^6y - 2y$ [5.7]

50. $36x^2 - 120x + 100$ [5.6]

51. $6t^2 + 17pt + 5p^2$ [5.5]

52. $x^3 + 2x^2 - 9x - 18$ [5.6]

53. $a^2 - 2ab + b^2 - 4t^2$ [5.6]

54. The following graph is that of a polynomial function $p(x)$. Use the graph to find the zeros of p. [5.3]

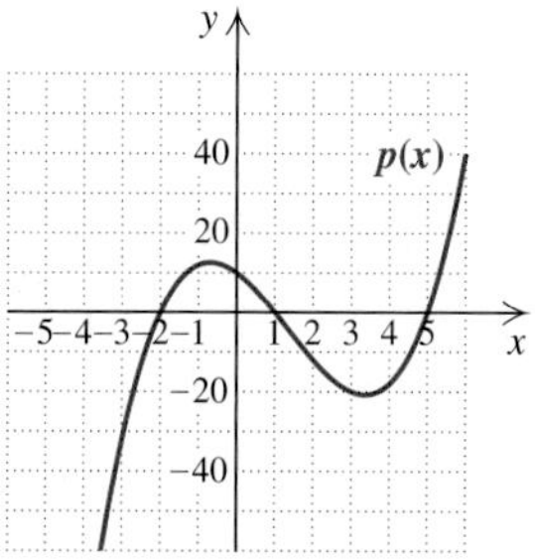

55. Find the zeros of the function given by
$f(x) = x^2 - 11x + 28$. [5.4]

Solve. Where appropriate, round answers to the nearest thousandth.

56. $x^2 - 20x = -100$ [5.6]

57. $6b^2 - 13b + 6 = 0$ [5.5]

58. $8t^2 = 14t$ [5.3]

59. $r^2 = 16$ [5.6]

60. $a^3 = 4a^2 + 21a$ [5.4]

61. $(x - 1)(x - 4) = 10$ [5.4]

62. $x^3 - 5x^2 - 16x + 80 = 0$ [5.6]

63. $x^2 + 180 = 27x$ [5.4]

64. $x^2 - 2x = 6$ [5.4]

65. Let $f(x) = x^2 - 7x - 40$. Find a such that $f(a) = 4$. [5.5]

66. Find the domain of the function f given by

$$f(x) = \frac{x - 3}{3x^2 + 19x - 14}. \quad [5.5]$$

67. The area of a square is 5 more than four times the length of a side. What is the length of a side of the square? [5.8]

68. The sum of the squares of three consecutive odd numbers is 83. Find the numbers. [5.8]

69. A photograph is 3 in. longer than it is wide. When a 2-in. border is placed around the photograph, the total area of the photograph and the border is 108 in^2. Find the dimensions of the photograph. [5.8]

70. Tim is designing a rectangular garden with a width of 8 ft. The path that leads diagonally across the garden is 2 ft longer than the length of the garden. How long is the path? [5.8]

71. *Labor Force.* The following table lists the percent of the U.S. population age 65 and older participating in the labor force. [5.8]

Year	Labor Force Participation Rate, 65 and Older
1970	25.2
1980	16.8
1990	15.7
2000	17.3
2004	20.3

Source: U.S. Bureau of Labor Statistics

a) Use regression to find a quadratic polynomial function P that can be used to estimate the labor participation rate for those 65 and older x years after 1970.

b) Estimate the participation rate in 2010.

c) In what year or years is the participation rate 24%?

Synthesis

TW **72.** Explain how to find the roots of a polynomial function from its graph. [5.3]

TW **73.** Explain in your own words why there must be a 0 on one side of an equation before you can use the principle of zero products. [5.3]

Factor. [5.7]

74. $128x^6 - 2y^6$

75. $(x - 1)^3 - (x + 1)^3$

Multiply.

76. $[a - (b - 1)][(b - 1)^2 + a(b - 1) + a^2]$ [5.2], [5.7]

77. $(z^{n^2})^{n^3}(z^{4n^3})^{n^2}$ [5.2]

78. Solve: $(x + 1)^3 = x^2(x + 1)$. [5.6]

Chapter Test 5

Given the polynomial $3xy^3 - 4x^2y + 5x^5y^4 - 2x^4y$.

1. Determine the degree of the polynomial.

2. Arrange in descending powers of x.

3. Determine the leading term of the polynomial $8a - 2 + a^2 - 4a^3$.

4. Given $P(x) = 2x^3 + 3x^2 - x + 4$, find $P(0)$ and $P(-2)$.

5. Given $P(x) = x^2 - 5x$, find and simplify

$$P(a + h) - P(a).$$

6. Combine like terms:

$$5xy - 2xy^2 - 2xy + 5xy^2.$$

Add.

7. $(-6x^3 + 3x^2 - 4y) + (3x^3 - 2y - 7y^2)$

8. $(5m^3 - 4m^2n - 6mn^2 - 3n^3) + (9mn^2 - 4n^3 + 2m^3 + 6m^2n)$

Subtract.

9. $(9a - 4b) - (3a + 4b)$

10. $(6y^2 - 2y - 5y^3) - (4y^2 - 7y - 6y^3)$

Multiply.

11. $(-4x^2y)(-16xy^2)$

12. $(6a - 5b)(2a + b)$

13. $(x - y)(x^2 - xy - y^2)$

14. $(2x^3 + 5)^2$

15. $(4y - 9)^2$

16. $(x - 2y)(x + 2y)$

Factor.

17. $15x^2 - 5x^4$

18. $y^3 + 5y^2 - 4y - 20$

19. $p^2 - 12p - 28$

20. $12m^2 + 20m + 3$

21. $9y^2 - 25$

22. $3r^3 - 3$

23. $9x^2 + 25 - 30x$

24. $x^8 - y^8$

25. $y^2 + 8y + 16 - 100t^2$

26. $20a^2 - 5b^2$

27. $24x^2 - 46x + 10$

28. $16a^7b + 54ab^7$

29. $4y^4x + 36yx^2 + 8y^2x^3 - 16xy$

30. The graph shown below is that of the polynomial function $p(x)$. Use the graph to determine the zeros of p.

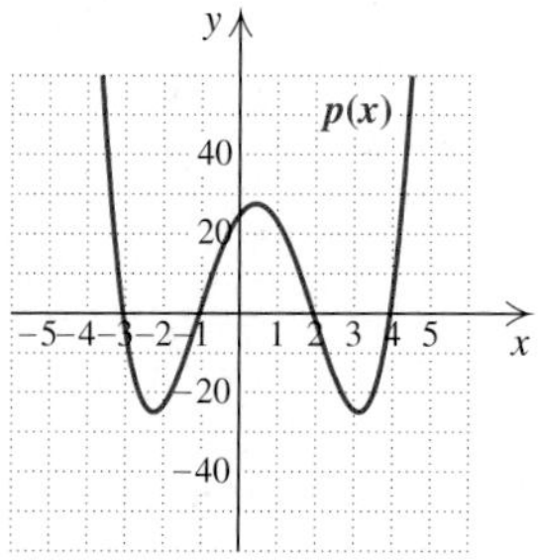

31. Find the zeros of the function given by $f(x) = 2x^2 - 11x - 40$.

Solve. Where appropriate, round solutions to the nearest thousandth.

32. $x^2 - 18 = 3x$

33. $5t^2 = 125$

34. $2x^2 + 21 = -17x$

35. $9x^2 + 3x = 0$

36. $x^2 + 81 = 18x$

37. $x^2(x + 1) = 8x$

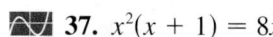

38. Let $f(x) = 3x^2 - 15x + 11$. Find a such that $f(a) = 11$.

39. Find the domain of the function f given by

$$f(x) = \frac{3 - x}{x^2 + 2x + 1}.$$

40. A photograph is 3 cm longer than it is wide. Its area is 40 cm^2. Find its length and its width.

41. To celebrate a town's centennial, fireworks are launched over a lake off a dam 36 ft above the water. The height of a display, t seconds after it has been launched, is given by

$$h(t) = -16t^2 + 64t + 36.$$

After how long will the shell from the fireworks reach the water?

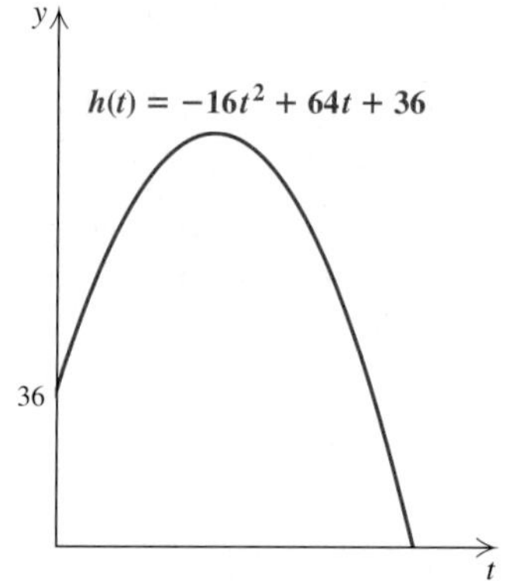

42. *IRS Revenue.* The following table lists the amount of IRS enforcement revenue, in billions of dollars, for various years.

Year	IRS Enforcement Revenue (in billions)
2001	\$33.8
2002	34.1
2003	37.6
2004	43.1
2005	47.3

Source: *INC. Magazine*, March 2006

a) Use regression to find a quadratic polynomial function E that can be used to estimate the IRS enforcement revenue x years after 2000.
b) Estimate the enforcement revenue in 2009.
c) In what year will the enforcement revenue be \$100 billion?

Synthesis

43. a) Multiply: $(x^2 + x + 1)(x^3 - x^2 + 1)$.
b) Factor: $x^5 + x + 1$.

44. Factor: $6x^{2n} - 7x^n - 20$.

6 Rational Expressions, Equations, and Functions

Like fractions in arithmetic, a rational expression is an expression that indicates division. In this chapter, we add, subtract, multiply, and divide rational expressions and use them in equations and functions. We then use rational expressions to solve problems that we could not have solved before.

APPLICATION *Commuter Travel.*

One factor influencing urban planning is VMT, or vehicle miles traveled. Areas with a high population density have fewer VMT per household than those with a lower density. This occurs in part because employment and shopping are available closer to higher density areas. The following table lists annual VMT per household for various densities for a typical urban area. Find an equation of variation that describes the data.

Population Density	Annual VMT
25	12,000
50	6,000
100	3,000
200	1,500

Source: Based on information from http://www.sflcv.org/density

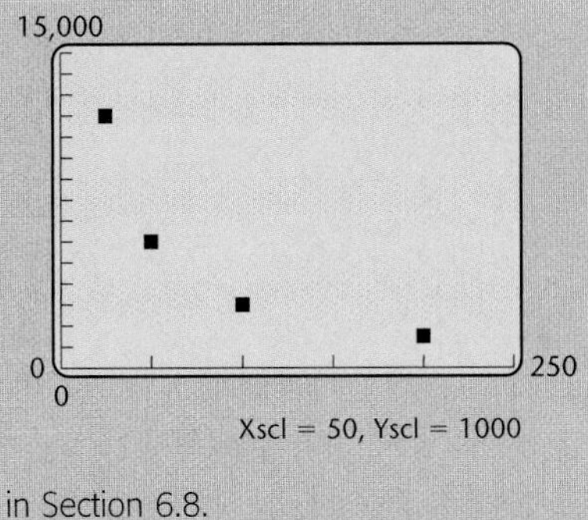

This problem appears as Example 10 in Section 6.8.

6.1 Rational Expressions and Functions: Multiplying and Dividing

Rational Functions ■ Multiplying ■ Simplifying Rational Expressions and Functions ■ Dividing and Simplifying ■ Vertical Asymptotes

An expression that consists of a polynomial divided by a nonzero polynomial is called a **rational expression.** The following are examples of rational expressions:

$$\frac{3}{4}, \quad \frac{x}{y}, \quad \frac{9}{a+b}, \quad \frac{x^2+7xy-4}{x^3-y^3}, \quad \frac{1+z^3}{1-z^6}.$$

Study Tip

Try an Exercise Break

Often the best way to regain your energy or focus is to take a break from your studies in order to exercise. Jogging, biking, or walking briskly are just a few of the ways in which you can improve your concentration for when you return to your studies.

Rational Functions

Like polynomials, certain rational expressions are used to describe functions. Such functions are called **rational functions.**

Following is the graph of the rational function

$$f(x) = \frac{1}{x}.$$

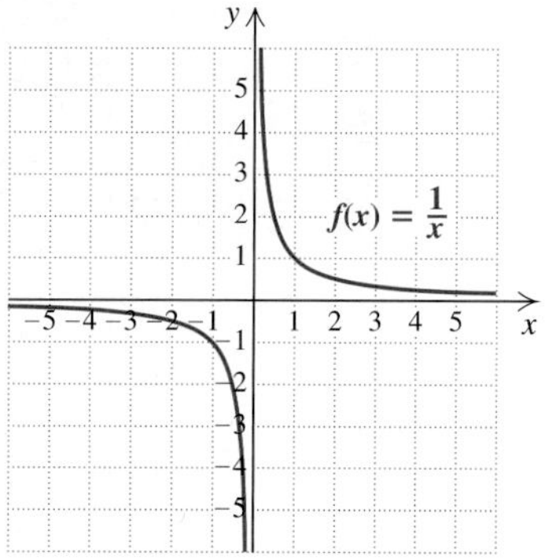

Graphs of rational functions vary widely in shape, but some general statements can be drawn from the graph shown above.

1. Graphs of rational functions may not be continuous—that is, there may be a break in the graph. The graph of $f(x) = 1/x$ consists of two unconnected parts, with a break at $x = 0$.
2. The domain of a rational function may not include all real numbers. The domain of $f(x) = 1/x$ does not include 0.

Although we will not consider graphs of rational functions in detail in this course, you should add the graph of $f(x) = 1/x$ to your library of functions.

EXAMPLE 1 The function given by

$$H(t) = \frac{t^2 + 5t}{2t + 5}$$

gives the time, in hours, for two machines, working together, to complete a job that the first machine could do alone in t hours and the other machine could do in $t + 5$ hours. How long will the two machines, working together, require for the job if the first machine alone would take **(a)** 1 hour? **(b)** 7 hours?

SOLUTION

a) $H(1) = \frac{1^2 + 5 \cdot 1}{2 \cdot 1 + 5} = \frac{1 + 5}{2 + 5} = \frac{6}{7}$ hr

b) $H(7) = \frac{7^2 + 5 \cdot 7}{2 \cdot 7 + 5} = \frac{49 + 35}{14 + 5} = \frac{84}{19}$, or $4\frac{8}{19}$ hr

$y_1 = (x^2 + 5x)/(2x + 5)$

X	Y1	
−2.5	ERR:	

In Section 5.5, we found that the domain of a rational function must exclude any numbers for which the denominator is 0. For a function like H above, the denominator is 0 when t is $-\frac{5}{2}$, so the domain of H is $\left(-\infty, -\frac{5}{2}\right) \cup \left(-\frac{5}{2}, \infty\right)$. Note from the table at left that H is not defined when $t = -\frac{5}{2}$.

Multiplying

The calculations that are performed with rational expressions resemble those performed in arithmetic.

Products of Rational Expressions To multiply two rational expressions, multiply numerators and multiply denominators:

$$\frac{A}{B} \cdot \frac{C}{D} = \frac{AC}{BD}, \quad \text{where } B \neq 0,\ D \neq 0.$$

EXAMPLE 2 Multiply: $\frac{x + 1}{y - 3} \cdot \frac{x^2}{y + 1}$.

SOLUTION

$$\frac{x + 1}{y - 3} \cdot \frac{x^2}{y + 1} = \frac{(x + 1)x^2}{(y - 3)(y + 1)}$$ **Multiplying the numerators and multiplying the denominators**

Recall from arithmetic that multiplication by 1 can be used to find equivalent expressions:

$$\frac{3}{5} = \frac{3}{5} \cdot \frac{2}{2}$$ **Multiplying by $\frac{2}{2}$, which is 1**

$$= \frac{6}{10}.$$ **$\frac{3}{5}$ and $\frac{6}{10}$ represent the same number.**

Similarly, multiplication by 1 can be used to find equivalent rational expressions:

$$\left.\begin{aligned}\frac{x-5}{x+2} &= \frac{x-5}{x+2}\cdot\frac{x+3}{x+3}\\ &= \frac{(x-5)(x+3)}{(x+2)(x+3)}.\end{aligned}\right\}$$

Multiplying by $\frac{x+3}{x+3}$, which, provided $x \neq -3$, is 1

The expressions

$$\frac{x-5}{x+2} \quad\text{and}\quad \frac{(x-5)(x+3)}{(x+2)(x+3)}$$

are equivalent: So long as x is replaced with a number other than -2 or -3, both expressions represent the same number. For example, if $x = 4$, then

$$\frac{x-5}{x+2} = \frac{4-5}{4+2} = \frac{-1}{6}$$

and

$$\frac{(x-5)(x+3)}{(x+2)(x+3)} = \frac{(4-5)(4+3)}{(4+2)(4+3)} = \frac{-1\cdot 7}{6\cdot 7} = \frac{-7}{42} = \frac{-1}{6}.$$

Simplifying Rational Expressions and Functions

As in arithmetic, rational expressions are *simplified* by "removing" a factor equal to 1. This reverses the process shown above:

$$\frac{6}{10} = \frac{3\cdot 2}{5\cdot 2} = \frac{3}{5}\cdot\frac{2}{2} = \frac{3}{5}.$$

We "removed" the factor that equals 1: $\frac{2}{2} = 1$.

Similarly,

$$\frac{(x-5)(x+3)}{(x+2)(x+3)} = \frac{x-5}{x+2}\cdot\frac{x+3}{x+3} = \frac{x-5}{x+2}.$$

We "removed" the factor that equals 1: $\frac{x+3}{x+3} = 1$.

Because rational expressions often appear when we are writing functions, it is important that the function's domain not be changed as a result of simplifying. For example, the domain of the function given by

$$F(x) = \frac{(x-5)(x+3)}{(x+2)(x+3)}$$

is assumed to be all real numbers for which the denominator is nonzero. Thus,

Domain of $F = \{x \mid x \neq -2, x \neq -3\}$, or $(-\infty, -3) \cup (-3, -2) \cup (-2, \infty)$.

Above, we wrote $\frac{(x-5)(x+3)}{(x+2)(x+3)}$ in simplified form as $\frac{x-5}{x+2}$. There is a serious problem with stating that these are equivalent. The difficulty arises from the fact that, unless we specify otherwise, the domain of the function given by

$G(x) = \frac{x-5}{x+2}$ is assumed to be $\{x \mid x \neq -2\}$, or $(-\infty, -2) \cup (-2, \infty)$.

Thus, as presently written, the domain of G includes -3, but the domain of F does not. This problem is easily addressed by specifying

$$\frac{(x-5)(x+3)}{(x+2)(x+3)} = \frac{x-5}{x+2} \qquad \textit{with } x \neq -3.$$

EXAMPLE 3 Write the function given by

$$f(t) = \frac{7t^2 + 21t}{14t}$$

in simplified form.

SOLUTION We first factor the numerator and the denominator, looking for the largest factor common to both. Once the greatest common factor is found, we use it to write 1 and simplify:

$$f(t) = \frac{7t^2 + 21t}{14t}$$ **Note that the domain of $f = \{t \mid t \neq 0\}$.**

$$= \frac{7t(t+3)}{7 \cdot 2 \cdot t}$$ **Factoring. The greatest common factor is $7t$.**

$$= \frac{7t}{7t} \cdot \frac{t+3}{2}$$ **Rewriting as a product of two rational expressions**

$$= 1 \cdot \frac{t+3}{2}$$ **For $t \neq 0$, we have $\frac{7t}{7t} = 1$. Try to do this step mentally.**

$$= \frac{t+3}{2}, \quad t \neq 0.$$ **Removing the factor 1. To keep the same domain, we specify that $t \neq 0$.**

X	Y1	Y2
−3	0	0
−2	.5	.5
−1	1	1
0	ERROR	1.5
1	2	2
2	2.5	2.5
3	3	3

X = −3

Thus simplified form is $f(t) = \dfrac{t+3}{2}$, *with* $t \neq 0$.

As a partial check, we let $y_1 = (7x^2 + 21x)/(14x)$ and $y_2 = (x+3)/2$ and compare values in a table. We let TblStart $= -3$ and $\triangle$Tbl $= 1$, and set Indpnt to Auto. If numbers restricted from the domain are not integers, we would set up the table differently. Note that the y-values are the same for any x-value except 0.

A rational expression is said to be **simplified** when no factors equal to 1 can be removed. Often, this requires two or more steps. For example, suppose we remove $7/7$ instead of $(7t)/(7t)$ in Example 3. We would then have

$$\frac{7t^2 + 21t}{14t} = \frac{7(t^2 + 3t)}{7 \cdot 2t} = \frac{t^2 + 3t}{2t}.$$

Removing a factor equal to 1: $\frac{7}{7} = 1$. Note that $t \neq 0$.

Here, since another common factor remains, we need to simplify further:

$$\frac{t^2 + 3t}{2t} = \frac{t(t+3)}{t \cdot 2} = \frac{t+3}{2}, \quad t \neq 0.$$

Removing another factor equal to 1: $t/t = 1$. The rational expression is now simplified. We must still specify $t \neq 0$.

EXAMPLE 4 Write the function given by

$$g(x) = \frac{x^2 + 3x - 10}{2x^2 - 3x - 2}$$

in simplified form.

SOLUTION We have

$$g(x) = \frac{x^2 + 3x - 10}{2x^2 - 3x - 2}$$

$$= \frac{(x - 2)(x + 5)}{(2x + 1)(x - 2)}$$ **Factoring the numerator and the denominator. Note that $x \neq -\frac{1}{2}$ and $x \neq 2$.**

$$= \frac{x - 2}{x - 2} \cdot \frac{x + 5}{2x + 1}$$ **Rewriting as a product of two rational expressions**

$$= \frac{x + 5}{2x + 1}, \; x \neq -\frac{1}{2}, 2.$$ **Removing a factor equal to 1: $\frac{x - 2}{x - 2} = 1$**

Thus, $g(x) = \frac{x + 5}{2x + 1}, \; x \neq -\frac{1}{2}, 2.$

The concern about domain applies only to work with functions.

EXAMPLE 5 Simplify: **(a)** $\frac{9x^2 + 6xy - 3y^2}{12x^2 - 12y^2}$; **(b)** $\frac{4 - t}{3t - 12}$.

SOLUTION

a) $$\frac{9x^2 + 6xy - 3y^2}{12x^2 - 12y^2} = \frac{3(x + y)(3x - y)}{12(x + y)(x - y)}$$ **Factoring the numerator and the denominator**

$$= \frac{3(x + y)}{3(x + y)} \cdot \frac{3x - y}{4(x - y)}$$ **Rewriting as a product of two rational expressions**

$$= \frac{3x - y}{4(x - y)}$$ **Removing a factor equal to 1: $\frac{3(x + y)}{3(x + y)} = 1$**

For purposes of later work, we usually do not multiply out the numerator and the denominator of the simplified expression.

b) $$\frac{4 - t}{3t - 12} = \frac{4 - t}{3(t - 4)}$$ **Factoring**

Since $4 - t$ is the opposite of $t - 4$, we factor out -1 to reverse the subtraction:

$$\frac{4 - t}{3t - 12} = \frac{-1(t - 4)}{3(t - 4)}$$ **$4 - t = -1(-4 + t) = -1(t - 4)$**

$$= \frac{-1}{3} \cdot \frac{t - 4}{t - 4}$$

$$= -\frac{1}{3}.$$ **Removing a factor equal to 1: $\frac{t - 4}{t - 4} = 1$**

Canceling

"Canceling" is a shortcut often used for removing a factor equal to 1 when working with fractions. With caution, we mention it as a possible way to speed up your work. Canceling removes factors equal to 1 in products. It *cannot* be done in sums or when adding expressions together. If your instructor permits canceling (not all do), it must be done with care and understanding. Example 5(a) might have been done faster as follows:

$$\frac{9x^2 + 6xy - 3y^2}{12x^2 - 12y^2} = \frac{\cancel{3}\cancel{(x+y)}(3x - y)}{\cancel{3} \cdot 4\cancel{(x+y)}(x - y)}$$

When a factor that equals 1 is found, it is "canceled" as shown.

$$= \frac{3x - y}{4(x - y)}.$$

Removing a factor equal to 1: $\frac{3(x + y)}{3(x + y)} = 1$

CAUTION! Canceling is often performed incorrectly:

$$\frac{\cancel{x} + 3}{\cancel{x}} = 3, \qquad \frac{\cancel{4}x + 3}{\cancel{2}} = 2x + 3, \qquad \frac{\cancel{5}}{\cancel{5} + x} = \frac{1}{x}$$

Incorrect! Incorrect! Incorrect!

To check that these are not equivalent, substitute a number for x or compare graphs.

In each situation, the expressions canceled are *not* both factors. For example, 4 is a factor of $4(x + 3)$, but it is *not* a factor of $4x + 3$. If you can't factor, you can't cancel! When in doubt, don't cancel!

If, after multiplying two rational expressions, it is possible to simplify, it is common practice to do so.

EXAMPLE 6 Multiply. Then simplify by removing a factor equal to 1.

a) $\frac{x + 2}{x - 3} \cdot \frac{x^2 - 4}{x^2 + x - 2}$ **b)** $\frac{1 - a^3}{a^2} \cdot \frac{a^5}{a^2 - 1}$

SOLUTION

a) $\frac{x + 2}{x - 3} \cdot \frac{x^2 - 4}{x^2 + x - 2} = \frac{(x + 2)(x^2 - 4)}{(x - 3)(x^2 + x - 2)}$ **Multiplying the numerators and also the denominators**

$= \frac{(x + 2)(x - 2)(x + 2)}{(x - 3)(x + 2)(x - 1)}$ **Factoring the numerator and the denominator and finding common factors**

$= \frac{(x + 2)\cancel{(x + 2)}(x - 2)}{(x - 3)\cancel{(x + 2)}(x - 1)}$ **Removing a factor equal to 1:** $\frac{x + 2}{x + 2} = 1$

$= \frac{(x + 2)(x - 2)}{(x - 3)(x - 1)}$ **Simplifying**

b) $\dfrac{1-a^3}{a^2} \cdot \dfrac{a^5}{a^2-1} = \dfrac{(1-a^3)a^5}{a^2(a^2-1)}$

$= \dfrac{(1-a)(1+a+a^2)a^5}{a^2(a-1)(a+1)}$ **Factoring a difference of cubes and a difference of squares**

$= \dfrac{-1(a-1)(1+a+a^2)a^5}{a^2(a-1)(a+1)}$ **Factoring out -1 reverses the subtraction.**

$= \dfrac{(a-1)a^2 \cdot a^3(-1)(1+a+a^2)}{(a-1)a^2(a+1)}$ **Rewriting a^5 as $a^2 \cdot a^3$; removing a factor equal to 1: $\dfrac{(a-1)a^2}{(a-1)a^2} = 1$**

$= \dfrac{-a^3(1+a+a^2)}{a+1}$ **Simplifying**

As in Example 5, there is no need for us to multiply out the numerator or the denominator of the final result.

Dividing and Simplifying

Two expressions are reciprocals of each other if their product is 1. As in arithmetic, to find the reciprocal of a rational expression, we interchange numerator and denominator.

The reciprocal of $\dfrac{x}{x^2+3}$ is $\dfrac{x^2+3}{x}$.

The reciprocal of $y-8$ is $\dfrac{1}{y-8}$.

Student Notes

The procedures covered in this chapter are by their nature rather long. As is often the case in mathematics, it usually helps to write out each step as you do the problems. Use plenty of paper and, if you are using lined paper, consider using two spaces at a time, writing the fraction bar on a line of the paper.

Quotients of Rational Expressions For any rational expressions A/B and C/D, with $B, C, D \neq 0$,

$$\frac{A}{B} \div \frac{C}{D} = \frac{A}{B} \cdot \frac{D}{C}.$$

(To divide two rational expressions, multiply by the reciprocal of the divisor. We often say that we "*invert* and multiply.")

EXAMPLE 7 Divide. Simplify by removing a factor equal to 1 if possible.

$$\frac{x-2}{x+1} \div \frac{x+5}{x-3}$$

SOLUTION

$\dfrac{x-2}{x+1} \div \dfrac{x+5}{x-3} = \dfrac{x-2}{x+1} \cdot \dfrac{x-3}{x+5}$ **Multiplying by the reciprocal of the divisor**

$= \dfrac{(x-2)(x-3)}{(x+1)(x+5)}$ **Multiplying the numerators and the denominators**

EXAMPLE 8 Write the function given by

$$g(a) = \frac{a^2 - 2a + 1}{a + 1} \div \frac{a^3 - a}{a - 2}$$

in simplified form, and list all restrictions on the domain.

SOLUTION The denominators of the rational expressions are $a + 1$ and $a - 2$, so the domain of g cannot contain -1 or 2. Also, division by $(a^3 - a)/(a - 2)$ is defined only when $(a^3 - a)/(a - 2)$ is nonzero. This expression will be zero when the numerator is zero, so we solve $a^3 - a = 0$:

$$\begin{aligned}
a^3 - a &= 0 \\
a(a^2 - 1) &= 0 \\
a(a + 1)(a - 1) &= 0 \\
a = 0 \quad \textit{or} \quad a + 1 &= 0 \quad \textit{or} \quad a - 1 = 0 \\
a = 0 \quad \textit{or} \quad a &= -1 \quad \textit{or} \quad a = 1.
\end{aligned}$$

Thus the domain of g cannot contain 0, -1, or 1.

$$\begin{aligned}
g(a) &= \frac{a^2 - 2a + 1}{a + 1} \div \frac{a^3 - a}{a - 2} \\
&= \frac{a^2 - 2a + 1}{a + 1} \cdot \frac{a - 2}{a^3 - a} && \textbf{Multiplying by the reciprocal of the divisor. Note that } a \neq -1, 0, 1, 2. \\
&= \frac{(a^2 - 2a + 1)(a - 2)}{(a + 1)(a^3 - a)} && \textbf{Multiplying the numerators and the denominators} \\
&= \frac{(a - 1)(a - 1)(a - 2)}{(a + 1)a(a + 1)(a - 1)} && \textbf{Factoring the numerator and the denominator} \\
&= \frac{(a - 1)\cancel{(a - 1)}(a - 2)}{(a + 1)a(a + 1)\cancel{(a - 1)}} && \textbf{Removing a factor equal to 1: } \frac{a - 1}{a - 1} = 1 \\
&= \frac{(a - 1)(a - 2)}{a(a + 1)(a + 1)}, \quad a \neq -1, 0, 1, 2 && \textbf{Simplifying}
\end{aligned}$$

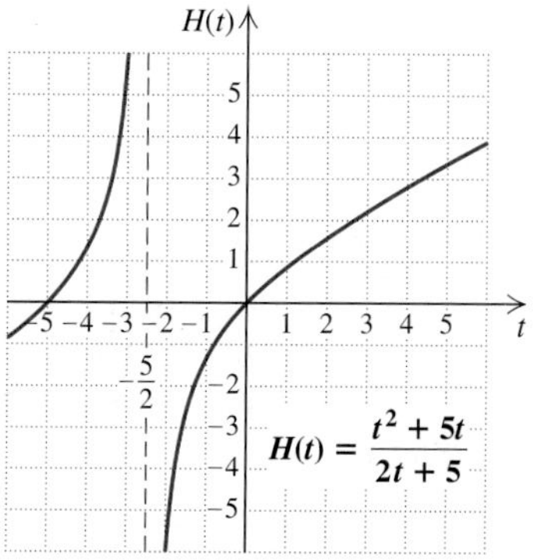

Vertical Asymptotes

Consider the graph of the function in Example 1,

$$H(t) = \frac{t^2 + 5t}{2t + 5}.$$

We consider all values for t, although negative values would not be considered for time. Note that the graph consists of two unconnected "branches." Since $-\frac{5}{2}$ is not in the domain of H, a vertical line drawn at $-\frac{5}{2}$ does not touch the graph of H. The line $t = -\frac{5}{2}$ is called a **vertical asymptote.** As t gets closer to, or *approaches*, $-\frac{5}{2}$, $H(t)$ approaches the asymptote.

Now consider the graph of

$$H(t) = \frac{t^2 + 5t}{2t + 5}$$

as drawn using some graphing calculators. The vertical line that appears on the screen on the left below is not part of the graph, nor should it be considered a vertical asymptote. Since a graphing calculator graphs an equation by plotting points and connecting them, the vertical line is actually connecting a point just to the left of the line $x = -\frac{5}{2}$ with a point just to the right of the line $x = -\frac{5}{2}$. Such a vertical line may or may not appear for values not in the domain of a function.

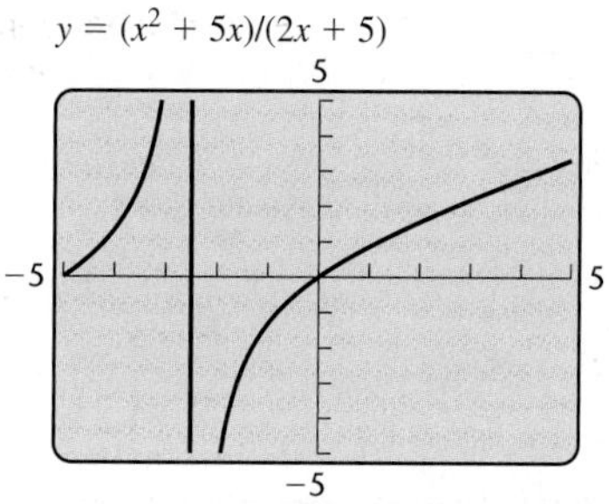

CONNECTED mode

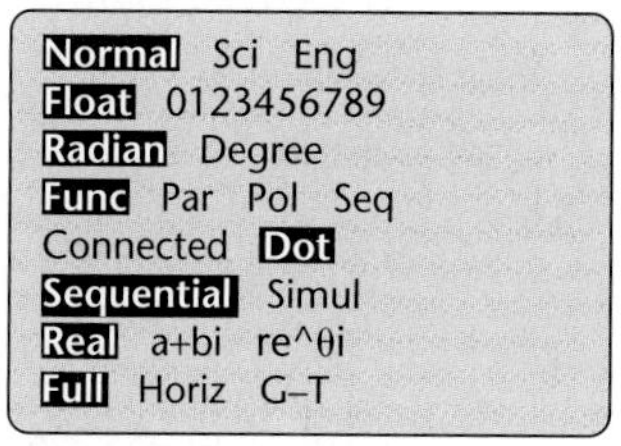

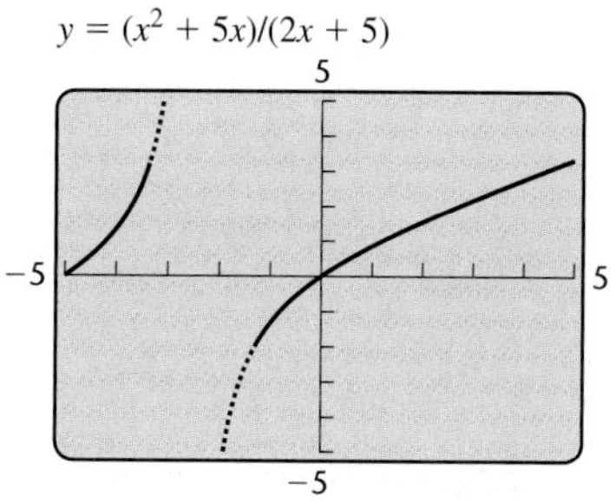

DOT mode

If we tell the calculator not to connect the points that it plots, the vertical line will not appear. Pressing MODE and changing from CONNECTED to DOT gives us the graph on the right above.

Not every number excluded from the domain of a function corresponds to a vertical asymptote.

Interactive Discovery

Let

$$f(x) = \frac{(x + 2)(x - 2)}{(2x + 1)(x - 2)} \quad \text{and} \quad g(x) = \frac{x + 2}{2x + 1}.$$

1. Determine the domain of f and the domain of g.
2. Graph $f(x)$. How many vertical asymptotes does the graph appear to have? What is the vertical asymptote?
3. Graph $g(x)$ and determine the vertical asymptote.

If a function $f(x)$ is described by a *simplified* rational expression, and a is a number that makes the denominator 0, then $x = a$ is a vertical asymptote of the graph of $f(x)$.

EXAMPLE 9 Determine the vertical asymptotes of the graph of

$$f(x) = \frac{9x^2 + 6x - 3}{12x^2 - 12}.$$

SOLUTION We first simplify the rational expression describing the function:

$$\frac{9x^2 + 6x - 3}{12x^2 - 12} = \frac{3(x + 1)(3x - 1)}{3 \cdot 4(x + 1)(x - 1)} \qquad \textbf{Factoring the numerator and the denominator}$$

$$= \frac{3(x + 1)}{3(x + 1)} \cdot \frac{3x - 1}{4(x - 1)} \qquad \textbf{Factoring the rational expression}$$

$$= \frac{3x - 1}{4(x - 1)}. \qquad \textbf{Removing a factor equal to 1}$$

The denominator of the simplified expression, $4(x - 1)$, is 0 when $x = 1$. Thus, $x = 1$ is a vertical asymptote of the graph. Although the domain of the function also excludes -1, there is no asymptote at $x = -1$, only a "hole." This hole should be indicated on a hand-drawn graph but it may not be obvious on a graphing-calculator screen. A table will show that -1 is not in the domain of this function.

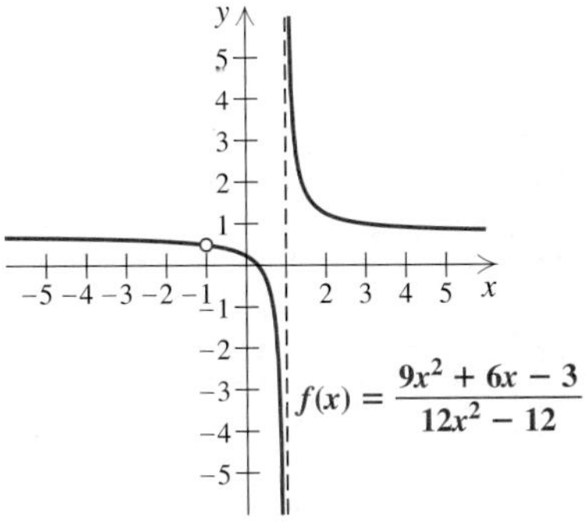

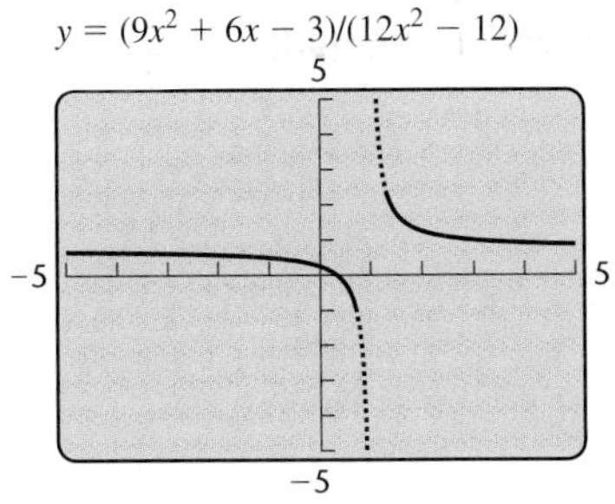

X	Y1
−3	.625
−2	.58333
−1	ERROR
0	.25
1	ERROR
2	1.25
3	1
X = −3	

Visualizing the Graph

Match each equation, inequality, or set of equations or inequalities with its graph.

1. $y = -2$

2. $y = x^3$

3. $y = x^2$

4. $y = \dfrac{1}{x}$

5. $y = x + 1$

6. $y = |x + 1|$

7. $y \leq x + 1$

8. $y > 2x - 3$

9. $y = x + 1,$
   $y = 2x - 3$

10. $y \leq x + 1,$
    $y > 2x - 3$

Answers on page A-25

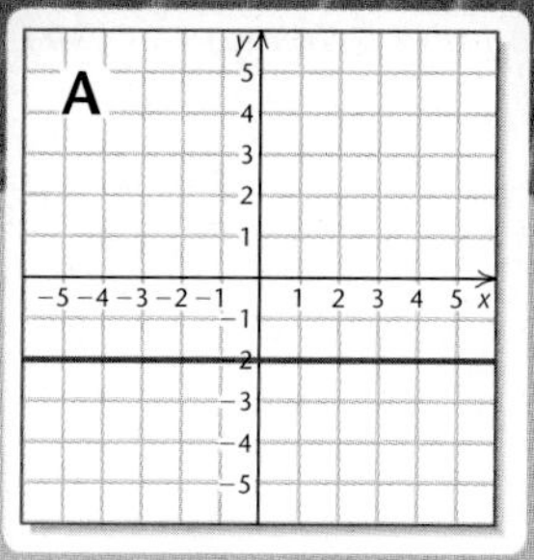

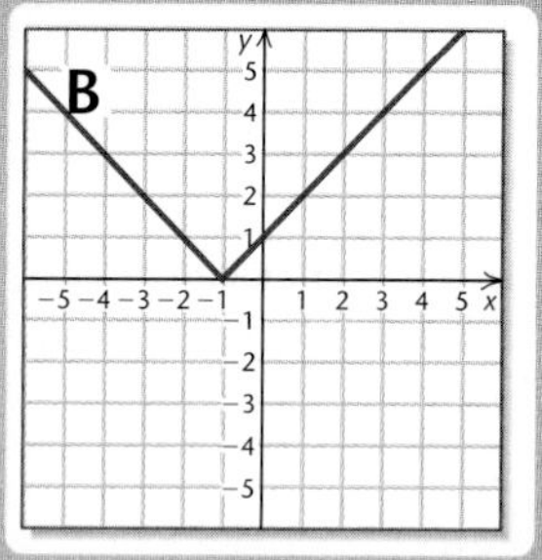

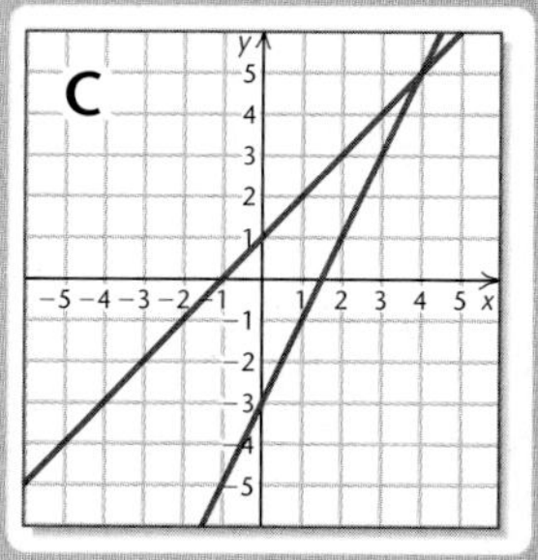

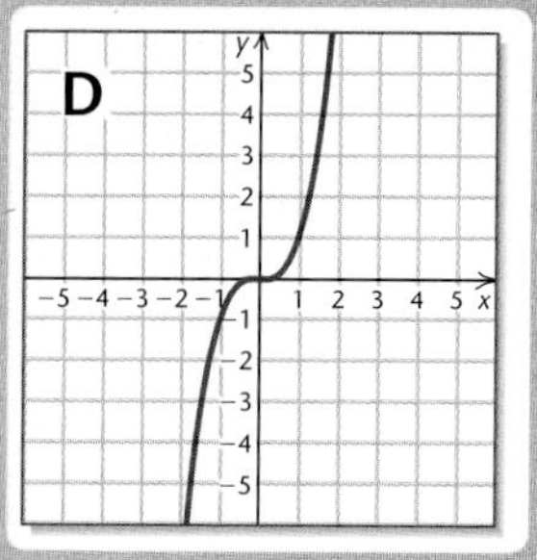

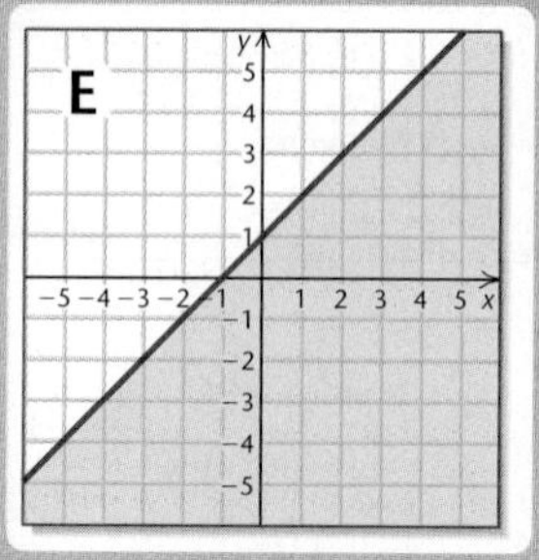

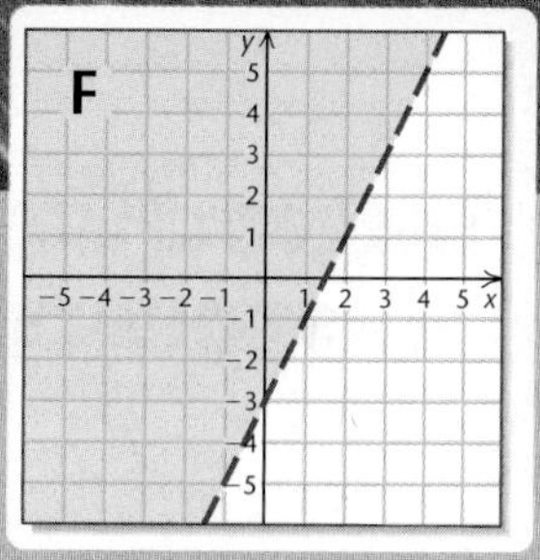

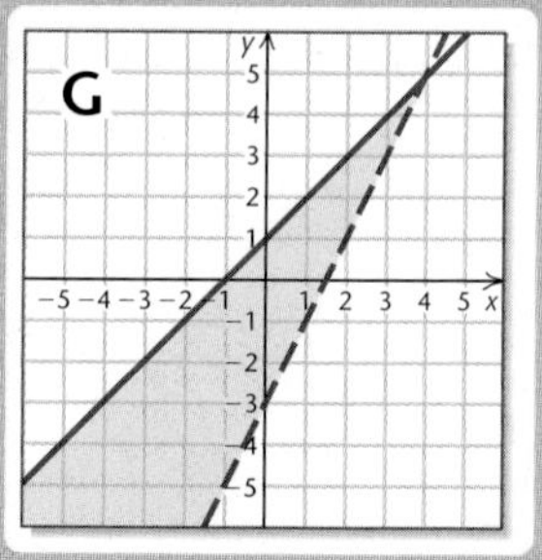

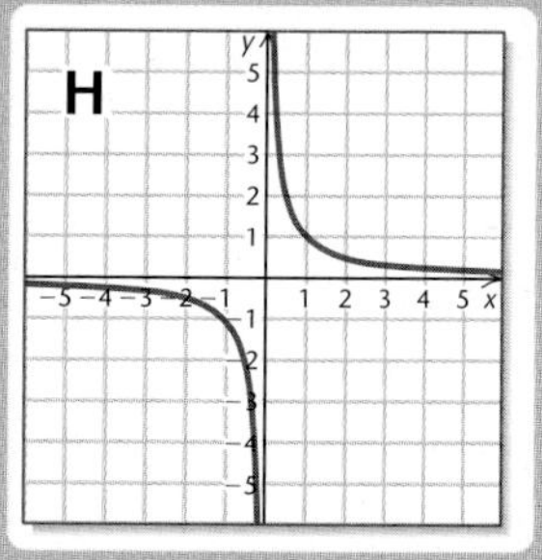

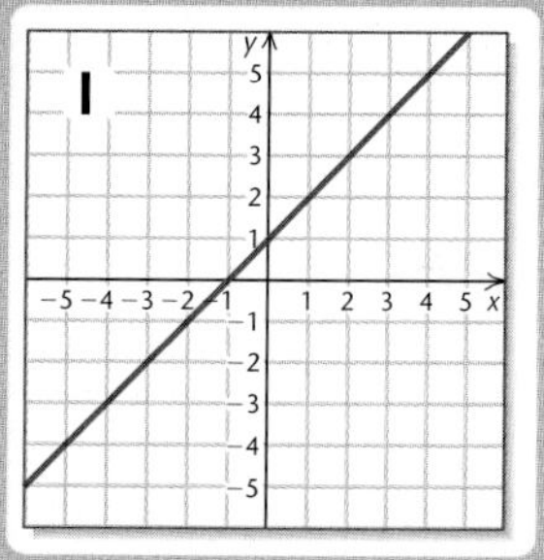

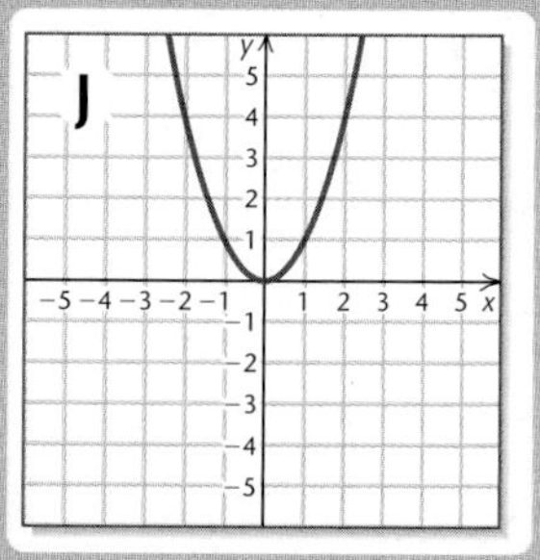

6.1 EXERCISE SET

FOR EXTRA HELP

MathXL | MyMathLab | InterAct Math | AW Math Tutor Center | Video Lectures on CD: Disc 3 | Student's Solutions Manual

Concept Reinforcement *In each of Exercises 1–10, match the function described with the appropriate domain from the column on the right.*

1. ___ $f(x) = \dfrac{2 - x}{x - 5}$

2. ___ $g(x) = \dfrac{x + 2}{x + 5}$

3. ___ $h(x) = \dfrac{x - 5}{x - 2}$

4. ___ $f(x) = \dfrac{x + 5}{x + 2}$

5. ___ $g(x) = \dfrac{x - 3}{(x - 2)(x - 5)}$

6. ___ $h(x) = \dfrac{x - 3}{(x + 2)(x + 5)}$

7. ___ $f(x) = \dfrac{x + 3}{(x - 2)(x + 5)}$

8. ___ $g(x) = \dfrac{x + 3}{(x + 2)(x - 5)}$

9. ___ $h(x) = \dfrac{(x - 2)(x - 3)}{x + 3}$

10. ___ $f(x) = \dfrac{(x + 2)(x + 3)}{x - 3}$

a) $\{x \mid x \neq -5, x \neq 2\}$

b) $\{x \mid x \neq 3\}$

c) $\{x \mid x \neq -2\}$

d) $\{x \mid x \neq -3\}$

e) $\{x \mid x \neq 5\}$

f) $\{x \mid x \neq -2, x \neq 5\}$

g) $\{x \mid x \neq 2\}$

h) $\{x \mid x \neq -2, x \neq -5\}$

i) $\{x \mid x \neq 2, x \neq 5\}$

j) $\{x \mid x \neq -5\}$

Photo Developing. *Rik usually takes 3 hr more than Pearl does to process a day's orders at Liberty Place Photo. If Pearl takes t hr to process a day's orders, the function given by*

$$H(t) = \frac{t^2 + 3t}{2t + 3}$$

can be used to determine how long it would take if they worked together.

11. How long will it take them, working together, to complete a day's orders if Pearl can process the orders alone in 5 hr?

12. How long will it take them, working together, to complete a day's orders if Pearl can process the orders alone in 7 hr?

For each rational function, find the function values indicated, provided the value exists.

13. $v(t) = \dfrac{4t^2 - 5t + 2}{t + 3}$; $v(0), v(-2), v(7)$

14. $f(x) = \dfrac{5x^2 + 4x - 12}{6 - x}$; $f(0), f(-1), f(3)$

15. $g(x) = \dfrac{2x^3 - 9}{x^2 - 4x + 4}$; $g(0), g(2), g(-1)$

16. $r(t) = \dfrac{t^2 - 5t + 4}{t^2 - 9}$; $r(1), r(2), r(-3)$

Multiply to obtain equivalent expressions. Do not simplify. Assume that all denominators are nonzero.

17. $\dfrac{4x}{4x} \cdot \dfrac{x - 3}{x + 2}$

18. $\dfrac{3 - a^2}{a - 7} \cdot \dfrac{-1}{-1}$

19. $\dfrac{t - 2}{t + 3} \cdot \dfrac{-1}{-1}$

20. $\dfrac{x - 4}{x + 5} \cdot \dfrac{x - 5}{x - 5}$

Simplify by removing a factor equal to 1.

21. $\dfrac{15x}{5x^2}$

22. $\dfrac{7a^3}{21a}$

23. $\dfrac{18t^3s^2}{27t^7s^9}$

24. $\dfrac{8y^5z^3}{4y^9z^4}$

25. $\dfrac{2a - 10}{2}$

26. $\dfrac{3a + 12}{3}$

27. $\dfrac{15}{25a - 30}$

28. $\dfrac{21}{6x - 9}$

29. $\dfrac{3x - 12}{3x + 15}$

30. $\dfrac{4y - 20}{4y + 12}$

Write simplified form for each of the following. Be sure to list all restrictions on the domain, as in Example 4.

31. $f(x) = \dfrac{3x + 21}{x^2 + 7x}$

32. $f(x) = \dfrac{5x + 20}{x^2 + 4x}$

33. $g(x) = \dfrac{x^2 - 9}{5x + 15}$

34. $g(x) = \dfrac{8x - 16}{x^2 - 4}$

35. $h(x) = \dfrac{4 - x}{5x - 20}$

36. $h(x) = \dfrac{7 - x}{3x - 21}$

37. $f(t) = \dfrac{t^2 - 16}{t^2 - 8t + 16}$

38. $f(t) = \dfrac{t^2 - 25}{t^2 + 10t + 25}$

39. $g(t) = \dfrac{21 - 7t}{3t - 9}$

40. $g(t) = \dfrac{12 - 6t}{5t - 10}$

41. $h(t) = \dfrac{t^2 + 5t + 4}{t^2 - 8t - 9}$

42. $h(t) = \dfrac{t^2 - 3t - 4}{t^2 + 9t + 8}$

43. $f(x) = \dfrac{9x^2 - 4}{3x - 2}$

44. $f(x) = \dfrac{4x^2 - 1}{2x - 1}$

45. $g(t) = \dfrac{16 - t^2}{t^2 - 8t + 16}$

46. $g(p) = \dfrac{25 - p^2}{p^2 + 10p + 25}$

Multiply and, if possible, simplify.

47. $\dfrac{5a^3}{3b} \cdot \dfrac{7b^3}{10a^7}$

48. $\dfrac{25a}{9b^8} \cdot \dfrac{3b^5}{5a^2}$

49. $\dfrac{8x - 16}{5x} \cdot \dfrac{x^3}{5x - 10}$

50. $\dfrac{5t^3}{4t - 8} \cdot \dfrac{6t - 12}{10t}$

51. $\dfrac{x^2 - 16}{x^2} \cdot \dfrac{x^2 - 4x}{x^2 - x - 12}$

52. $\dfrac{y^2 + 10y + 25}{y^2 - 9} \cdot \dfrac{y^2 + 3y}{y + 5}$

53. $\dfrac{7a - 14}{4 - a^2} \cdot \dfrac{5a^2 + 6a + 1}{35a + 7}$

54. $\dfrac{a^2 - 1}{2 - 5a} \cdot \dfrac{15a - 6}{a^2 + 5a - 6}$

Aha! **55.** $\dfrac{t^3 - 4t}{t - t^4} \cdot \dfrac{t^4 - t}{4t - t^3}$

56. $\dfrac{x^2 - 6x + 9}{12 - 4x} \cdot \dfrac{x^6 - 9x^4}{x^3 - 3x^2}$

57. $\dfrac{c^3 + 8}{c^5 - 4c^3} \cdot \dfrac{c^6 - 4c^5 + 4c^4}{c^2 - 2c + 4}$

58. $\dfrac{x^3 - 27}{x^4 - 9x^2} \cdot \dfrac{x^5 - 6x^4 + 9x^3}{x^2 + 3x + 9}$

59. $\dfrac{a^3 - b^3}{3a^2 + 9ab + 6b^2} \cdot \dfrac{a^2 + 2ab + b^2}{a^2 - b^2}$

60. $\dfrac{x^3 + y^3}{x^2 + 2xy - 3y^2} \cdot \dfrac{x^2 - y^2}{3x^2 + 6xy + 3y^2}$

Divide and, if possible, simplify.

61. $\dfrac{9x^5}{8y^2} \div \dfrac{3x}{16y^9}$

62. $\dfrac{16a^7}{3b^5} \div \dfrac{8a^3}{6b}$

63. $\dfrac{5x + 10}{x^8} \div \dfrac{x + 2}{x^3}$

64. $\dfrac{3y + 15}{y^7} \div \dfrac{y + 5}{y^2}$

65. $\dfrac{25x^2 - 4}{x^2 - 9} \div \dfrac{2 - 5x}{x + 3}$

66. $\dfrac{4a^2 - 1}{a^2 - 4} \div \dfrac{2a - 1}{2 - a}$

67. $\dfrac{5y - 5x}{15y^3} \div \dfrac{x^2 - y^2}{3x + 3y}$

68. $\dfrac{x^2 - y^2}{4x + 4y} \div \dfrac{3y - 3x}{12x^2}$

69. $\dfrac{x^2 - 16}{x^2 - 10x + 25} \div \dfrac{3x - 12}{x^2 - 3x - 10}$

70. $\dfrac{y^2 - 36}{y^2 - 8y + 16} \div \dfrac{3y - 18}{y^2 - y - 12}$

71. $\dfrac{x^3 - 64}{x^3 + 64} \div \dfrac{x^2 - 16}{x^2 - 4x + 16}$

72. $\dfrac{8y^3 - 27}{64y^3 - 1} \div \dfrac{4y^2 - 9}{16y^2 + 4y + 1}$

Write simplified form for each of the following. Be sure to list all restrictions on the domain.

73. $f(t) = \dfrac{t^2 - 16}{4t + 12} \cdot \dfrac{t + 3}{t - 4}$

74. $g(m) = \dfrac{m^2 - 9}{3m + 3} \cdot \dfrac{m + 3}{m - 3}$

75. $g(x) = \dfrac{x^2 - 2x - 35}{2x^3 - 3x^2} \cdot \dfrac{4x^3 - 9x}{7x - 49}$

76. $h(t) = \dfrac{t^2 - 10t + 9}{t^2 - 1} \cdot \dfrac{1 - t^2}{t^2 - 5t - 36}$

77. $f(x) = \dfrac{x^2 - 4}{x^3} \div \dfrac{x^5 - 2x^4}{x + 4}$

78. $g(x) = \dfrac{x^2 - 9}{x^2} \div \dfrac{x^5 + 3x^4}{x + 2}$

79. $h(n) = \dfrac{n^3 + 3n}{n^2 - 9} \div \dfrac{n^2 + 5n - 14}{n^2 + 4n - 21}$

80. $f(x) = \dfrac{x^3 + 4x}{x^2 - 16} \div \dfrac{x^2 + 8x + 15}{x^2 + x - 20}$

Perform the indicated operations and, if possible, simplify.

81. $\dfrac{4x^2 - 9y^2}{8x^3 - 27y^3} \div \dfrac{4x + 6y}{3x - 9y} \cdot \dfrac{4x^2 + 6xy + 9y^2}{4x^2 - 8xy + 3y^2}$

82. $\dfrac{5x^2 - 5y^2}{27x^3 + 8y^3} \div \dfrac{x^2 - 2xy + y^2}{9x^2 - 6xy + 4y^2} \cdot \dfrac{6x + 4y}{10x - 15y}$

83. $\dfrac{a^3 - ab^2}{2a^2 + 3ab + b^2} \cdot \dfrac{4a^2 - b^2}{a^2 - 2ab + b^2} \div \dfrac{a^2 + a}{a - 1}$

84. $\dfrac{2x + 4y}{2x^2 + 5xy + 2y^2} \cdot \dfrac{4x^2 - y^2}{8x^2 - 8} \div \dfrac{x^2 + 4xy + 4y^2}{x^2 - 6xy + 9y^2}$

Determine the vertical asymptotes of the graph of each function.

85. $f(x) = \dfrac{3x - 12}{3x + 15}$

86. $f(x) = \dfrac{4x - 20}{4x + 12}$

87. $g(x) = \dfrac{12 - 6x}{5x - 10}$

88. $r(x) = \dfrac{21 - 7x}{3x - 9}$

89. $t(x) = \dfrac{x^3 + 3x^2}{x^2 + 6x + 9}$

90. $g(x) = \dfrac{x^2 - 4}{2x^2 - 5x + 2}$

91. $f(x) = \dfrac{x^2 - x - 6}{x^2 - 6x + 8}$

92. $f(x) = \dfrac{x^2 + 2x + 1}{x^2 - 2x + 1}$

In Exercises 93–98, match each function with one of the following graphs.

a)

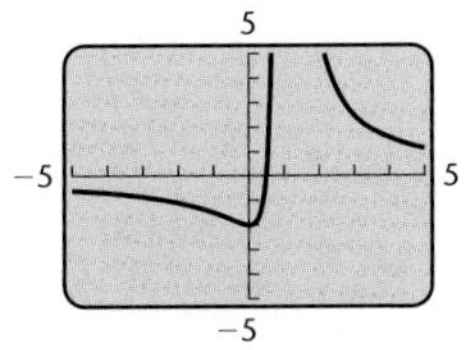

b)

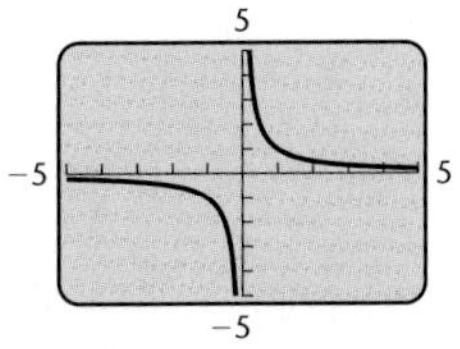

c)

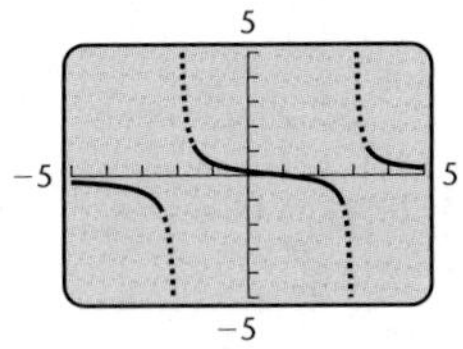

d)

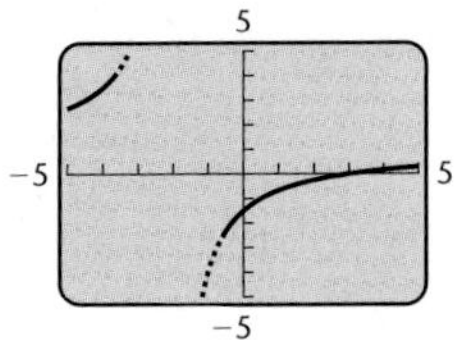

e)

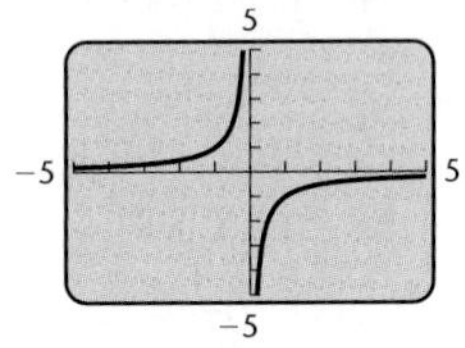

f)

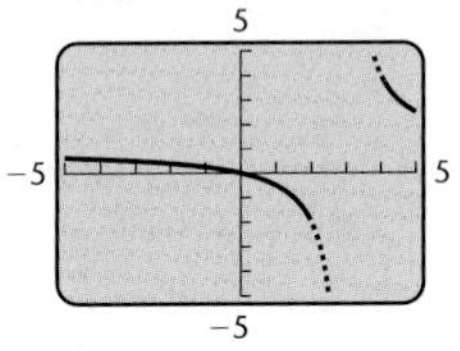

93. $h(x) = \dfrac{1}{x}$

94. $q(x) = -\dfrac{1}{x}$

95. $f(x) = \dfrac{x}{x-3}$

96. $g(x) = \dfrac{x-3}{x+2}$

97. $r(x) = \dfrac{4x-2}{x^2-2x+1}$

98. $t(x) = \dfrac{x-1}{x^2-x-6}$

TW **99.** Explain why the graphs of $f(x) = 5x$ and $g(x) = \dfrac{5x^2}{x}$ differ.

TW **100.** Nancy *incorrectly* simplifies $\dfrac{x+2}{x}$ as

$$\frac{x+2}{x} = \frac{\cancel{x}+2}{\cancel{x}} = 1 + 2 = 3.$$

She insists this is correct because it checks when x is replaced with 1. Explain her misconception.

Skill Maintenance

Simplify.

101. $\dfrac{3}{10} - \dfrac{8}{15}$ [1.2]

102. $\dfrac{3}{8} - \dfrac{7}{10}$ [1.2]

103. $\dfrac{2}{3} \cdot \dfrac{5}{7} - \dfrac{5}{7} \cdot \dfrac{1}{6}$ [1.2]

104. $\dfrac{4}{7} \cdot \dfrac{1}{5} - \dfrac{3}{10} \cdot \dfrac{2}{7}$ [1.2]

105. $(8x^3 - 5x^2 + 6x + 2) - (4x^3 + 2x^2 - 3x + 7)$ [5.1]

106. $(6t^4 + 9t^3 - t^2 + 4t) - (8t^4 - 2t^3 - 6t + 3)$ [5.1]

Synthesis

TW **107.** To check Example 3, Kara lets

$$y_1 = \frac{7x^2 + 21x}{14x} \quad \text{and} \quad y_2 = \frac{x+3}{2}.$$

Since the graphs of y_1 and y_2 appear to be identical, Kara believes that the domains of the functions described by y_1 and y_2 are the same, $\mathbb{R}$. How could you convince Kara otherwise?

TW **108.** Tony *incorrectly* argues that since

$$\frac{a^2-4}{a-2} = \frac{a^2}{a} + \frac{-4}{-2} = a + 2$$

is correct, it follows that

$$\frac{x^2+9}{x+1} = \frac{x^2}{x} + \frac{9}{1} = x + 9.$$

Explain his misconception.

109. Calculate the slope of the line passing through $(a, f(a))$ and $(a + h, f(a + h))$ for the function f given by $f(x) = x^2 + 5$. Be sure your answer is simplified.

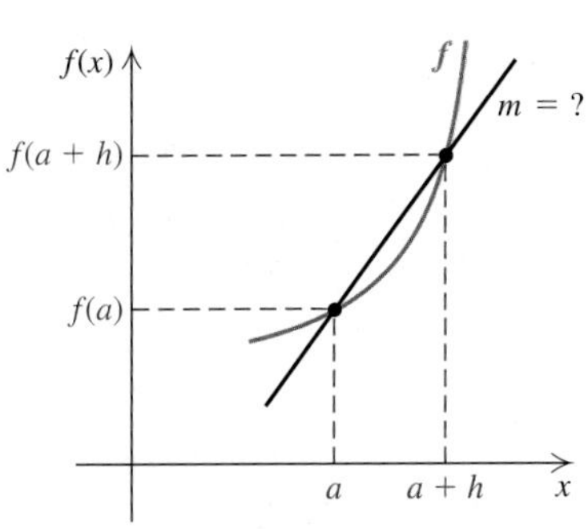

110. Calculate the slope of the line passing through the points $(a, f(a))$ and $(a + h, f(a + h))$ for the function f given by $f(x) = 3x^2$. Be sure your answer is simplified.

111. Let

$$g(x) = \frac{2x + 3}{4x - 1}.$$

Determine each of the following.

a) $g(x + h)$

b) $g(2x - 2) \cdot g(x)$

c) $g\left(\frac{1}{2}x + 1\right) \cdot g(x)$

112. Graph the function given by

$$f(x) = \frac{x^2 - 9}{x - 3}.$$

(*Hint*: Determine the domain of f and simplify.)

Perform the indicated operations and simplify.

113. $\dfrac{r^2 - 4s^2}{r + 2s} \div (r + 2s)^2\left(\dfrac{2s}{r - 2s}\right)^2$

114. $\dfrac{d^2 - d}{d^2 - 6d + 8} \cdot \dfrac{d - 2}{d^2 + 5d} \div \left(\dfrac{5d^2}{d^2 - 9d + 20}\right)^2$

Aha! **115.** $\dfrac{6t^2 - 26t + 30}{8t^2 - 15t - 21} \cdot \dfrac{5t^2 - 9t - 15}{6t^2 - 14t - 20} \div \dfrac{5t^2 - 9t - 15}{6t^2 - 14t - 20}$

Simplify.

116. $\dfrac{m^2 - t^2}{m^2 + t^2 + m + t + 2mt}$

117. $\dfrac{a^3 - 2a^2 + 2a - 4}{a^3 - 2a^2 - 3a + 6}$

118. $\dfrac{x^3 + x^2 - y^3 - y^2}{x^2 - 2xy + y^2}$

119. $\dfrac{u^6 + v^6 + 2u^3v^3}{u^3 - v^3 + u^2v - uv^2}$

120. $\dfrac{x^5 - x^3 + x^2 - 1 - (x^3 - 1)(x + 1)^2}{(x^2 - 1)^2}$

121. Let

$$f(x) = \frac{4}{x^2 - 1} \quad \text{and} \quad g(x) = \frac{4x^2 + 8x + 4}{x^3 - 1}.$$

Find each of the following.

a) $(f \cdot g)(x)$

b) $(f/g)(x)$

c) $(g/f)(x)$

Determine the domain and the range of each function from its graph.

122.

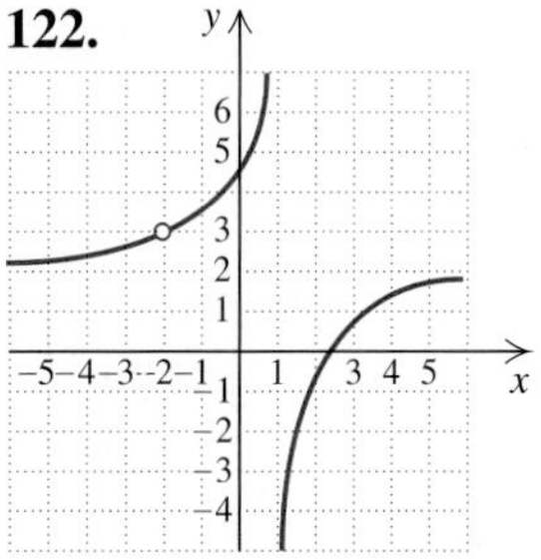

123.

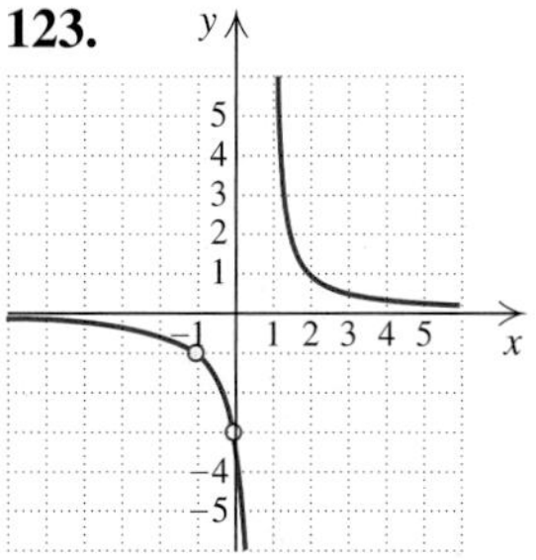

124.

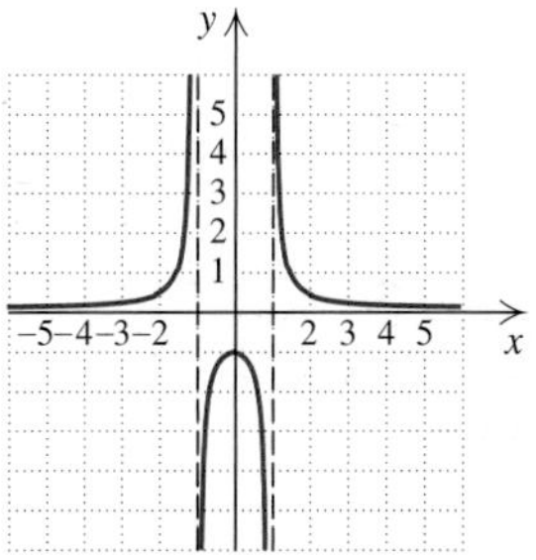

6.2 Rational Expressions and Functions: Adding and Subtracting

When Denominators Are the Same ■ When Denominators Are Different

Rational expressions are added in much the same way as the fractions of arithmetic.

When Denominators Are the Same

Addition and Subtraction with Like Denominators To add or subtract when denominators are the same, add or subtract the numerators and keep the same denominator.

$$\frac{A}{C} + \frac{B}{C} = \frac{A+B}{C} \quad \text{and} \quad \frac{A}{C} - \frac{B}{C} = \frac{A-B}{C}, \quad \text{where } C \neq 0$$

EXAMPLE 1 Add: $\dfrac{3+x}{x} + \dfrac{4}{x}$.

SOLUTION We have

$$\frac{3+x}{x} + \frac{4}{x} = \frac{3+x+4}{x} = \frac{x+7}{x}.$$

Because x is not a factor of both the numerator and the denominator, the result cannot be simplified.

X	Y1	Y2
−2	−2.5	−2.5
−1	−6	−6
0	ERROR	ERROR
1	8	8
2	4.5	4.5
3	3.3333	3.3333
4	2.75	2.75
X = −2		

To check, we let $y_1 = (3 + x)/x + 4/x$ and $y_2 = (x + 7)/x$. The table at left shows that $y_1 = y_2$ for all x not equal to 0.

CAUTION! Using a table can indicate that two expressions are equivalent, but it does not verify that one of them is completely simplified. A rational expression is simplified when its numerator and its denominator do not contain any common factors.

EXAMPLE 2 Add: $\dfrac{4x^2 - 5xy}{x^2 - y^2} + \dfrac{2xy - y^2}{x^2 - y^2}$.

Study Tip

A Text is Not Light Reading

Do not expect a math text to read like a magazine or novel. On one hand, most assigned readings in a math text consist of only a few pages. On the other hand, every sentence and word is important and should make sense. If they don't, ask for help as soon as possible.

SOLUTION

$$\frac{4x^2 - 5xy}{x^2 - y^2} + \frac{2xy - y^2}{x^2 - y^2} = \frac{4x^2 - 3xy - y^2}{x^2 - y^2}$$ **Adding the numerators and combining like terms. The denominator is unchanged.**

$$= \frac{(x - y)(4x + y)}{(x - y)(x + y)}$$ **Factoring the numerator and the denominator and looking for common factors**

$$= \frac{\cancel{(x - y)}(4x + y)}{\cancel{(x - y)}(x + y)}$$ **Removing a factor equal to 1: $\frac{x-y}{x-y} = 1$**

$$= \frac{4x + y}{x + y}$$ **Simplifying**

Recall that a fraction bar is a grouping symbol. The next example shows that when a numerator is subtracted, care must be taken to subtract, or change the sign of, *each* term in that polynomial.

EXAMPLE 3 If

$$f(x) = \frac{4x + 5}{x + 3} - \frac{x - 2}{x + 3},$$

find a simplified form of $f(x)$ and list all restrictions on the domain.

SOLUTION

$$f(x) = \frac{4x + 5}{x + 3} - \frac{x - 2}{x + 3}$$ **Note that $x \neq -3$.**

$$= \frac{4x + 5 - (x - 2)}{x + 3}$$ **The parentheses remind us to subtract *both* terms.**

$$= \frac{4x + 5 - x + 2}{x + 3}$$

$$= \frac{3x + 7}{x + 3}, \quad x \neq -3$$

When Denominators Are Different

In order to add rational expressions such as

$$\frac{7}{12xy^2} + \frac{8}{15x^3y} \quad \text{or} \quad \frac{x}{x^2 - y^2} + \frac{y}{x^2 - 4xy + 3y^2},$$

we must first find common denominators. As in arithmetic, our work is easier when we use the *least common multiple* (LCM) of the denominators.

Least Common Multiple To find the least common multiple (LCM) of two or more expressions, find the prime factorization of each expression and form a product that contains each factor the greatest number of times that it occurs in any one prime factorization.

EXAMPLE 4 Find the least common multiple of each pair of polynomials.

a) $21x$ and $3x^2$ **b)** $x^2 + x - 12$ and $x^2 - 16$

SOLUTION

a) We write the prime factorizations of $21x$ and $3x^2$:

$$21x = 3 \cdot 7 \cdot x \quad \text{and} \quad 3x^2 = 3 \cdot x \cdot x.$$

The factors 3, 7, and x must appear in the LCM if $21x$ is to be a factor of the LCM. The other polynomial, $3x^2$, is not a factor of $3 \cdot 7 \cdot x$ because the prime factors of $3x^2$—namely, 3, x, and x—do not all appear in $3 \cdot 7 \cdot x$. However, if $3 \cdot 7 \cdot x$ is multiplied by another factor of x, a product is formed that contains both $21x$ and $3x^2$ as factors:

$21x$ is a factor.

$$\text{LCM} = 3 \cdot 7 \cdot x \cdot x = 21x^2.$$

$3x^2$ is a factor.

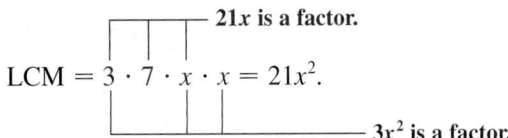

Note that each factor (3, 7, and x) is used the greatest number of times that it occurs as a factor of either $21x$ or $3x^2$. The LCM is $3 \cdot 7 \cdot x \cdot x$, or $21x^2$.

b) We factor both expressions:

$$x^2 + x - 12 = (x - 3)(x + 4),$$
$$x^2 - 16 = (x + 4)(x - 4).$$

The LCM must contain each polynomial as a factor. By multiplying the factors of $x^2 + x - 12$ by $x - 4$, we form a product that contains both $x^2 + x - 12$ and $x^2 - 16$ as factors:

$x^2 + x - 12$ is a factor.

$$\text{LCM} = (x - 3)(x + 4)(x - 4).$$ **There is no need to multiply this out.**

$x^2 - 16$ is a factor.

To Add or Subtract Rational Expressions

1. Determine the *least common denominator* (LCD) by finding the least common multiple of the denominators.
2. Rewrite each of the original rational expressions, as needed, in an equivalent form that has the LCD.
3. Add or subtract the resulting rational expressions, as indicated.
4. Simplify the result, if possible, and list any restrictions, on the domain of functions.

EXAMPLE 5 Add: $\dfrac{2}{21x} + \dfrac{5}{3x^2}$.

SOLUTION In Example 4(a), we found that the LCD is $3 \cdot 7 \cdot x \cdot x$, or $21x^2$. We now multiply each rational expression by 1, using expressions for 1 that give us the LCD in each expression. To determine what to use, ask "$21x$ times what is $21x^2$?" and "$3x^2$ times what is $21x^2$?" The answers are x and 7, respectively, so we multiply by x/x and $7/7$:

$$\frac{2}{21x} \cdot \frac{x}{x} + \frac{5}{3x^2} \cdot \frac{7}{7} = \frac{2x}{21x^2} + \frac{35}{21x^2}$$ **We now have a common denominator.**

$$= \frac{2x + 35}{21x^2}.$$ **This expression cannot be simplified.**

EXAMPLE 6 Add: $\dfrac{x^2}{x^2 + 2xy + y^2} + \dfrac{2x - 2y}{x^2 - y^2}$.

SOLUTION To find the LCD, we first factor the denominators. We also factor numerators to see if we can simplify either expression. In this case, the rightmost rational expression can be simplified:

$$\frac{x^2}{x^2 + 2xy + y^2} + \frac{2x - 2y}{x^2 - y^2} = \frac{x^2}{(x + y)(x + y)} + \frac{2(x - y)}{(x + y)(x - y)}$$ **Factoring**

$$= \frac{x^2}{(x + y)(x + y)} + \frac{2}{x + y}.$$ **Removing a factor equal to 1:** $\dfrac{x - y}{x - y} = 1$

Note that the LCM of $(x + y)(x + y)$ and $(x + y)$ is $(x + y)(x + y)$. The factor missing in the denominator of the second expression is $x + y$. We multiply that expression by 1, using $(x + y)/(x + y)$. Then we add and, if possible, simplify.

$$\frac{x^2}{(x + y)(x + y)} + \frac{2}{x + y} = \frac{x^2}{(x + y)(x + y)} + \frac{2}{x + y} \cdot \frac{x + y}{x + y}$$

$$= \frac{x^2}{(x + y)(x + y)} + \frac{2x + 2y}{(x + y)(x + y)}$$ **We now have the LCD.**

$$= \frac{x^2 + 2x + 2y}{(x + y)(x + y)}$$ **Since the numerator cannot be factored, we cannot simplify further.**

EXAMPLE 7 Subtract: $\dfrac{2y + 1}{y^2 - 7y + 6} - \dfrac{y + 3}{y^2 - 5y - 6}$.

SOLUTION

$$\frac{2y + 1}{y^2 - 7y + 6} - \frac{y + 3}{y^2 - 5y - 6}$$

$$= \frac{2y + 1}{(y - 6)(y - 1)} - \frac{y + 3}{(y - 6)(y + 1)}$$ **The LCD is $(y - 6)(y - 1)(y + 1)$.**

$$= \frac{2y + 1}{(y - 6)(y - 1)} \cdot \frac{y + 1}{y + 1} - \frac{y + 3}{(y - 6)(y + 1)} \cdot \frac{y - 1}{y - 1}$$

Multiplying by 1 to get the LCD in each expression

$$= \frac{(2y + 1)(y + 1) - (y + 3)(y - 1)}{(y - 6)(y - 1)(y + 1)}$$

$$= \frac{2y^2 + 3y + 1 - (y^2 + 2y - 3)}{(y-6)(y-1)(y+1)}$$ **The parentheses are important.**

$$= \frac{2y^2 + 3y + 1 - y^2 - 2y + 3}{(y-6)(y-1)(y+1)}$$

$$= \frac{y^2 + y + 4}{(y-6)(y-1)(y+1)}$$ **The numerator cannot be factored. We leave the denominator in factored form.**

EXAMPLE 8 Add: $\frac{3}{8a} + \frac{1}{-8a}$.

SOLUTION

$$\frac{3}{8a} + \frac{1}{-8a} = \frac{3}{8a} + \frac{-1}{-1} \cdot \frac{1}{-8a}$$

When denominators are opposites, we multiply one rational expression by $-1/-1$ to get the LCD.

$$= \frac{3}{8a} + \frac{-1}{8a} = \frac{2}{8a}$$

$$= \frac{\not{2} \cdot 1}{\not{2} \cdot 4a} = \frac{1}{4a}$$ **Simplifying by removing a factor equal to 1: $\frac{2}{2} = 1$**

EXAMPLE 9 Subtract: $\frac{5x}{x-2y} - \frac{3y-7}{2y-x}$.

SOLUTION

$$\frac{5x}{x-2y} - \frac{3y-7}{2y-x} = \frac{5x}{x-2y} - \frac{-1}{-1} \cdot \frac{3y-7}{2y-x}$$ **Note that $x - 2y$ and $2y - x$ are opposites.**

$$= \frac{5x}{x-2y} - \frac{7-3y}{x-2y}$$ **Performing the multiplication.** ***Note***: **$-1(2y - x) = -2y + x = x - 2y$.**

$$= \frac{5x - (7 - 3y)}{x - 2y}$$

$$= \frac{5x - 7 + 3y}{x - 2y}$$

Subtracting. The parentheses are important.

In Example 9, you may have noticed that when $3y - 7$ is multiplied by -1 and subtracted, the result is $-7 + 3y$, which is equivalent to the original $3y - 7$. Thus, instead of multiplying the numerator by -1 and then subtracting, we could have simply *added* $3y - 7$ to $5x$, as in the following:

$$\frac{5x}{x-2y} - \frac{3y-7}{2y-x} = \frac{5x}{x-2y} + (-1) \cdot \frac{3y-7}{2y-x}$$ **Rewriting subtraction as addition**

$$= \frac{5x}{x-2y} + \frac{1}{-1} \cdot \frac{3y-7}{2y-x}$$ **Writing -1 as $\frac{1}{-1}$**

$$= \frac{5x}{x-2y} + \frac{3y-7}{x-2y}$$ **The opposite of $2y - x$ is $x - 2y$.**

$$= \frac{5x + 3y - 7}{x - 2y}.$$ **This checks with the answer to Example 9.**

EXAMPLE 10 Find simplified form for the function given by

$$f(x) = \frac{2x}{x^2 - 4} + \frac{5}{2 - x} - \frac{1}{2 + x}$$

and list all restrictions on the domain.

SOLUTION We have

$$\frac{2x}{x^2 - 4} + \frac{5}{2 - x} - \frac{1}{2 + x}$$

$$= \frac{2x}{(x - 2)(x + 2)} + \frac{5}{2 - x} - \frac{1}{2 + x}$$ **Factoring. Note that $x \neq -2, 2$.**

$$= \frac{2x}{(x - 2)(x + 2)} + \frac{-1}{-1} \cdot \frac{5}{(2 - x)} - \frac{1}{x + 2}$$ **Multiplying by $-1/-1$ since $2 - x$ is the opposite of $x - 2$**

$$= \frac{2x}{(x - 2)(x + 2)} + \frac{-5}{x - 2} - \frac{1}{x + 2}$$ **The LCD is $(x - 2)(x + 2)$.**

$$= \frac{2x}{(x - 2)(x + 2)} + \frac{-5}{x - 2} \cdot \frac{x + 2}{x + 2} - \frac{1}{x + 2} \cdot \frac{x - 2}{x - 2}$$ **Multiplying by 1 to get the LCD**

$$= \frac{2x - 5(x + 2) - (x - 2)}{(x - 2)(x + 2)}$$

$$= \frac{2x - 5x - 10 - x + 2}{(x - 2)(x + 2)}$$

$$= \frac{-4x - 8}{(x - 2)(x + 2)}$$

$$= \frac{-4(x + 2)}{(x - 2)(x + 2)}$$

$$= \frac{-4\cancel{(x + 2)}}{(x - 2)\cancel{(x + 2)}}$$ **Removing a factor equal to 1: $\frac{x + 2}{x + 2} = 1, x \neq -2$**

$$= \frac{-4}{x - 2}, \text{ or } -\frac{4}{x - 2}, x \neq \pm 2.$$

An equivalent answer is $\frac{4}{2 - x}$. It is found by writing $-\frac{4}{x - 2}$ as $\frac{4}{-(x - 2)}$ and then using the distributive law to remove parentheses.

Student Notes

Adding or subtracting rational expressions can involve many steps. Therefore, it is important to double-check each step of your work as you proceed. Waiting until the end of the problem to check your work is generally a less efficient use of your time.

Our work in Example 10 indicates that if

$$f(x) = \frac{2x}{x^2 - 4} + \frac{5}{2 - x} - \frac{1}{2 + x} \quad \text{and} \quad g(x) = \frac{-4}{x - 2},$$

then, for $x \neq -2$ and $x \neq 2$, we have $f = g$. Note that whereas the domain of f includes all real numbers except -2 or 2, the domain of g excludes only 2. This is illustrated in the following graphs. Methods for drawing such graphs by hand are discussed in more advanced courses. The graphs are for visualization only.

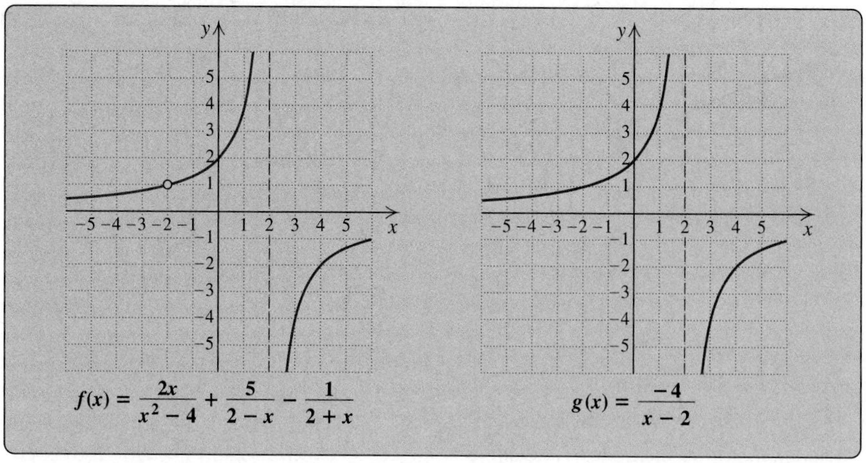

A computer-generated visualization of Example 10

A quick, partial check of any simplification is to evaluate both the original and the simplified expressions for a convenient choice of x. For instance, to check Example 10, if $x = 1$, we have

$$f(1) = \frac{2 \cdot 1}{1^2 - 4} + \frac{5}{2 - 1} - \frac{1}{2 + 1} = \frac{2}{-3} + \frac{5}{1} - \frac{1}{3} = 5 - \frac{3}{3} = 4$$

and

$$g(1) = \frac{-4}{1 - 2} = \frac{-4}{-1} = 4.$$

Since both functions include the pair $(1, 4)$, our algebra was *probably* correct. Although this is only a partial check (on rare occasions, an incorrect answer might "check"), because it is so easy to perform, it is nonetheless very useful. Further evaluation provides a more definitive check.

6.2 EXERCISE SET

FOR EXTRA HELP

MathXL | MyMathLab | InterAct Math | AW Math Tutor Center | Video Lectures on CD: Disc 3 | Student's Solutions Manual

Concept Reinforcement *Classify each of the following statements as either true or false.*

1. To add or subtract rational expressions, a common denominator is required.

2. To find the least common denominator, we find the least common multiple of the denominators.

3. It is rarely necessary to factor in order to find a common denominator.

4. The sum of two rational expressions is the sum of the numerators over the sum of the denominators.

5. The least common multiple of two expressions always contains each of those expressions as a factor.

6. The least common multiple of two expressions is always the product of those two expressions.

7. To add two rational expressions, it is often necessary to multiply at least one of those expressions by a form of 1.

8. After two rational expressions are added, it is unnecessary to simplify the result.

Perform the indicated operations. Simplify when possible.

9. $\frac{3}{2y} + \frac{5}{2y}$

10. $\frac{5}{3a} + \frac{7}{3a}$

11. $\frac{5}{3m^2n^2} - \frac{4}{3m^2n^2}$

12. $\frac{1}{4a^2b} - \frac{5}{4a^2b}$

13. $\frac{x - 3y}{x + y} + \frac{x + 5y}{x + y}$

14. $\frac{a - 5b}{a + b} + \frac{a + 7b}{a + b}$

15. $\frac{3t + 2}{t - 4} - \frac{t - 2}{t - 4}$

16. $\frac{4y + 2}{y - 2} - \frac{y - 3}{y - 2}$

Find simplified form for f(x) and list all restrictions on the domain.

17. $f(x) = \frac{9x}{x + 3} + \frac{4}{5x}$

18. $f(x) = \frac{7x}{x + 2} + \frac{3}{2x}$

19. $f(x) = \frac{5}{x^2 - 9} - \frac{2}{x^2 - 9}$

20. $f(x) = \frac{8}{x^2 - 4} - \frac{1}{x^2 - 4}$

21. $f(x) = \frac{7}{3x^2} + \frac{2}{x^2 - 1}$

22. $f(x) = \frac{9}{4x^3} + \frac{3}{x^2 - 4}$

23. $f(x) = \frac{2}{x - 1} + \frac{3}{1 - x}$

24. $f(x) = \frac{4}{x - 7} + \frac{2}{7 - x}$

Perform the indicated operations. Simplify when possible.

25. $\frac{5 - 4x}{x^2 - 6x - 7} + \frac{5x - 4}{x^2 - 6x - 7}$

26. $\frac{3 - 2x}{x^2 - 5x + 4} + \frac{3x - 4}{x^2 - 5x + 4}$

27. $\frac{2a - 5}{a^2 - 9} - \frac{3a - 8}{a^2 - 9}$

28. $\frac{3a - 2}{a^2 - 25} - \frac{4a - 7}{a^2 - 25}$

29. $\frac{s^2}{r - s} + \frac{r^2}{s - r}$

30. $\frac{a^2}{a - b} + \frac{b^2}{b - a}$

31. $\frac{2}{a} - \frac{5}{-a}$

32. $\frac{7}{x} - \frac{8}{-x}$

33. $\frac{y - 4}{y^2 - 25} - \frac{9 - 2y}{25 - y^2}$

34. $\frac{x - 7}{x^2 - 16} - \frac{x - 1}{16 - x^2}$

35. $\frac{y^2 - 5}{y^4 - 81} + \frac{4}{81 - y^4}$

36. $\frac{t^2 + 3}{t^4 - 16} + \frac{7}{16 - t^4}$

37. $\frac{r - 6s}{r^3 - s^3} - \frac{5s}{s^3 - r^3}$

38. $\frac{m - 3n}{m^3 - n^3} - \frac{2n}{n^3 - m^3}$

39. $\frac{a + 2}{a - 4} + \frac{a - 2}{a + 3}$

40. $\frac{a + 3}{a - 5} + \frac{a - 2}{a + 4}$

41. $4 + \frac{x - 3}{x + 1}$

42. $3 + \frac{y + 2}{y - 5}$

43. $\frac{4xy}{x^2 - y^2} + \frac{x - y}{x + y}$

44. $\frac{5ab}{a^2 - b^2} + \frac{a + b}{a - b}$

45. $\frac{8}{2x^2 - 7x + 5} + \frac{3x + 2}{2x^2 - x - 10}$

46. $\frac{7}{3y^2 + y - 4} + \frac{9y + 2}{3y^2 - 2y - 8}$

47. $\frac{4}{x + 1} + \frac{x + 2}{x^2 - 1} + \frac{3}{x - 1}$

48. $\frac{-2}{y + 2} + \frac{5}{y - 2} + \frac{y + 3}{y^2 - 4}$

49. $\frac{x + 6}{5x + 10} - \frac{x - 2}{4x + 8}$

50. $\dfrac{a+3}{5a+25} - \dfrac{a-1}{3a+15}$

51. $\dfrac{5ab}{a^2-b^2} - \dfrac{a-b}{a+b}$

52. $\dfrac{6xy}{x^2-y^2} - \dfrac{x+y}{x-y}$

53. $\dfrac{x}{x^2+9x+20} - \dfrac{4}{x^2+7x+12}$

54. $\dfrac{x}{x^2+11x+30} - \dfrac{5}{x^2+9x+20}$

55. $\dfrac{3y}{y^2-7y+10} - \dfrac{2y}{y^2-8y+15}$

56. $\dfrac{5x}{x^2-6x+8} - \dfrac{3x}{x^2-x-12}$

57. $\dfrac{2x+1}{x-y} + \dfrac{5x^2-5xy}{x^2-2xy+y^2}$

58. $\dfrac{2-3a}{a-b} + \dfrac{3a^2+3ab}{a^2-b^2}$

59. $\dfrac{3y+2}{y^2+5y-24} + \dfrac{7}{y^2+4y-32}$

60. $\dfrac{3x+2}{x^2-7x+10} + \dfrac{2x}{x^2-8x+15}$

61. $\dfrac{2y-6}{y^2-9} - \dfrac{y}{y-1} + \dfrac{y^2+2}{y^2+2y-3}$

62. $\dfrac{x-1}{x^2-1} - \dfrac{x}{x-2} + \dfrac{x^2+2}{x^2-x-2}$

Aha! 63. $\dfrac{5y}{1-4y^2} - \dfrac{2y}{2y+1} + \dfrac{5y}{4y^2-1}$

64. $\dfrac{4x}{x^2-1} + \dfrac{3x}{1-x} - \dfrac{4}{x-1}$

Find simplified form for f(x) and list all restrictions on the domain.

65. $f(x) = 2 + \dfrac{x}{x-3} - \dfrac{18}{x^2-9}$

66. $f(x) = 5 + \dfrac{x}{x+2} - \dfrac{8}{x^2-4}$

67. $f(x) = \dfrac{3x-1}{x^2+2x-3} - \dfrac{x+4}{x^2-16}$

68. $f(x) = \dfrac{3x-2}{x^2+2x-24} - \dfrac{x-3}{x^2-9}$

69. $f(x) = \dfrac{1}{x^2+5x+6} - \dfrac{2}{x^2+3x+2} - \dfrac{1}{x^2+5x+6}$

70. $f(x) = \dfrac{2}{x^2-5x+6} - \dfrac{4}{x^2-2x-3} + \dfrac{2}{x^2+4x+3}$

TW 71. Janine found that the sum of two rational expressions was $(3-x)/(x-5)$. The answer given at the back of the book is $(x-3)/(5-x)$. Is Janine's answer incorrect? Why or why not?

TW 72. When two rational expressions are added or subtracted, should the numerator of the result be factored? Why or why not?

Skill Maintenance

Simplify. Use only positive exponents in your answer. [1.4]

73. $\dfrac{15x^{-7}y^{12}z^4}{35x^{-2}y^6z^{-3}}$

74. $\dfrac{21a^{-4}b^6c^8}{27a^{-2}b^{-5}c}$

Simplify. [1.2]

75. $\dfrac{\frac{2}{3}}{\frac{6}{7}}$

76. $\dfrac{-\frac{5}{8}}{\frac{3}{4}}$

77. *Value of Coins.* There are 50 dimes in a roll of dimes, 40 nickels in a roll of nickels, and 40 quarters in a roll of quarters. Robert has a total of 12 rolls of coins with a total value of \$70.00. If he has 3 more rolls of nickels than dimes, how many of each roll of coins does he have? [3.5]

78. *Audiotapes.* Anna wants to buy tapes for her work at the campus radio station. She needs some 30-min tapes and some 60-min tapes. If she buys 12 tapes with a total recording time of 10 hr, how many tapes of each length did she buy? [3.3]

Synthesis

TW 79. Many students make the mistake of always multiplying denominators when looking for a common denominator. Use Example 7 to explain why this approach can yield results that are more difficult to simplify.

TW **80.** Is the sum of two rational expressions always a rational expression? Why or why not?

81. *Prescription Drugs.* After visiting her doctor, Corinna went to the pharmacy for a two-week supply of Zyrtec®, a 20-day supply of Albuterol, and a 30-day supply of Pepcid®. Corinna refills each prescription as soon as her supply runs out. How long will it be until she can refill all three prescriptions on the same day?

82. *Astronomy.* The earth, Jupiter, Saturn, and Uranus all revolve around the sun. The earth takes 1 yr, Jupiter 12 yr, Saturn 30 yr, and Uranus 84 yr. How frequently do these four planets line up with each other?

83. *Music.* To duplicate a common African polyrhythm, a drummer needs to play sextuplets (6 beats per measure) on a tom-tom while simultaneously playing quarter notes (4 beats per measure) on a bass drum. Into how many equally sized parts must a measure be divided, in order to precisely execute this rhythm?

84. *Home Appliances.* Refrigerators last an average of 20 yr, clothes washers about 14 yr, and dishwashers about 10 yr (*Source*. U.S. Department of Energy). In 1980, Westgate College bought new refrigerators for its dormitories. In 1990, the college bought new clothes washers and dishwashers. Predict the year in which the college will need to replace all three types of appliances at once.

Find the LCM.

85. $x^8 - x^4,\ x^5 - x^2,\ x^5 - x^3,\ x^5 + x^2$

86. $2a^3 + 2a^2b + 2ab^2,\ a^6 - b^6,$
$2b^2 + ab - 3a^2,\ 2a^2b + 4ab^2 + 2b^3$

87. The LCM of two expressions is $8a^4b^7$. One of the expressions is $2a^3b^7$. List all the possibilities for the other expression.

88. Determine the domain and the range of the function graphed below.

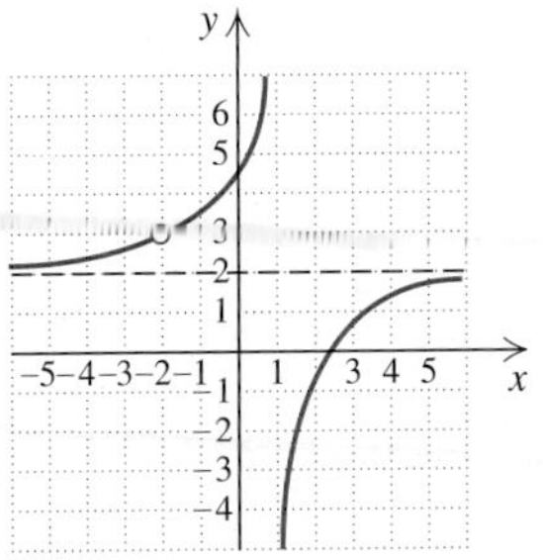

If

$$f(x) = \frac{x^3}{x^2 - 4} \quad and \quad g(x) = \frac{x^2}{x^2 + 3x - 10},$$

find each of the following.

89. $(f + g)(x)$ **90.** $(f - g)(x)$

91. $(f \cdot g)(x)$ **92.** $(f/g)(x)$

93. The domain of $f + g$ **94.** The domain of f/g

Perform the indicated operations and simplify.

95. $5(x - 3)^{-1} + 4(x + 3)^{-1} - 2(x + 3)^{-2}$

96. $4(y - 1)(2y - 5)^{-1} + 5(2y + 3)(5 - 2y)^{-1}$
$+ (y - 4)(2y - 5)^{-1}$

97. $\dfrac{x + 4}{6x^2 - 20x} \cdot \left(\dfrac{x}{x^2 - x - 20} + \dfrac{2}{x + 4}\right)$

98. $\dfrac{x^2 - 7x + 12}{x^2 - x - 29/3} \cdot \left(\dfrac{3x + 2}{x^2 + 5x - 24} + \dfrac{7}{x^2 + 4x - 32}\right)$

99. $\dfrac{8t^5}{2t^2 - 10t + 12} \div \left(\dfrac{2t}{t^2 - 8t + 15} - \dfrac{3t}{t^2 - 7t + 10}\right)$

100. $\dfrac{9t^3}{3t^3 - 12t^2 + 9t} \div \left(\dfrac{t + 4}{t^2 - 9} - \dfrac{3t - 1}{t^2 + 2t - 3}\right)$

Use algebra and a graphing calculator to determine the domain and estimate the range of each function.

101. $f(x) = 2 + \dfrac{x - 3}{x + 1}$

102. $g(x) = \dfrac{2}{(x + 1)^2} + 5$

103. $r(x) = \dfrac{1}{x^2} + \dfrac{1}{(x - 1)^2}$

6.3 Complex Rational Expressions

Multiplying by 1 ■ Dividing Two Rational Expressions

Study Tip

If You Miss a Class

If you miss a class, find out what the new assignment is and try to determine what problems were reviewed, what new concepts were discussed, and what announcements (regarding quizzes, hand-in assignments, schedule of class meetings, etc.) you may have missed. If possible, borrow notes from a classmate who attended.

A **complex rational expression** is a rational expression that contains rational expressions within its numerator and/or its denominator. Here are some examples:

$$\frac{x + \frac{5}{x}}{4x}, \quad \frac{\frac{x-y}{x+y}}{\frac{2x-y}{3x+y}}, \quad \frac{\frac{7x}{3} - \frac{4}{x}}{\frac{5x}{6} + \frac{8}{3}}, \quad \frac{\frac{r}{6} + \frac{r^2}{144}}{\frac{r}{12}}.$$

The rational expressions within each complex rational expression are red.

Complex rational expressions arise in a variety of real-world applications. For example, the last complex rational expression in the list above is used when calculating the size of certain loan payments.

Two methods are used to simplify complex rational expressions. Determining restrictions on variables may now require the solution of equations not yet studied. *Thus for this section we will not state restrictions on variables.*

Method 1: Multiplying by 1

One method of simplifying a complex rational expression is to multiply the entire expression by 1. To write 1, we use the LCD of the rational expressions within the complex rational expression.

Student Notes

In Example 1, we found the LCD for four rational expressions. To find this LCD, you may prefer to (1) find one LCD for the two expressions in the numerator, (2) find a second LCD for the expressions in the denominator, and (3) find the LCM of those two LCDs.

EXAMPLE 1 Simplify:

$$\frac{\frac{1}{a^3b} + \frac{1}{b}}{\frac{1}{a^2b^2} - \frac{1}{b^2}}.$$

SOLUTION The denominators within the complex rational expression are a^3b, b, a^2b^2, and b^2. Thus the LCD is a^3b^2. We multiply by 1, using $(a^3b^2)/(a^3b^2)$:

$$\frac{\frac{1}{a^3b} + \frac{1}{b}}{\frac{1}{a^2b^2} - \frac{1}{b^2}} = \frac{\frac{1}{a^3b} + \frac{1}{b}}{\frac{1}{a^2b^2} - \frac{1}{b^2}} \cdot \frac{a^3b^2}{a^3b^2}$$

Multiplying by 1, using the LCD

$$= \frac{\left(\frac{1}{a^3b} + \frac{1}{b}\right)a^3b^2}{\left(\frac{1}{a^2b^2} - \frac{1}{b^2}\right)a^3b^2}$$

Multiplying the numerator and the denominator. Remember to use parentheses.

$$= \frac{\frac{1}{a^3b} \cdot a^3b^2 + \frac{1}{b} \cdot a^3b^2}{\frac{1}{a^2b^2} \cdot a^3b^2 - \frac{1}{b^2} \cdot a^3b^2}$$ **Using the distributive law to carry out the multiplications**

$$= \frac{\frac{\cancel{a^3b}}{\cancel{a^3b}} \cdot b + \frac{\cancel{b}}{\cancel{b}} \cdot a^3b}{\frac{\cancel{a^2b^2}}{\cancel{a^2b^2}} \cdot a - \frac{\cancel{b^2}}{\cancel{b^2}} \cdot a^3}$$ **Removing factors that equal 1. Study this carefully.**

$$= \frac{b + a^3b}{a - a^3}$$ **Simplifying**

$$= \frac{b(1 + a^3)}{a(1 - a^2)}$$ **Factoring**

$$= \frac{b\cancel{(1 + a)}(1 - a + a^2)}{a\cancel{(1 + a)}(1 - a)}$$ **Factoring further and identifying a factor that equals 1**

$$= \frac{b(1 - a + a^2)}{a(1 - a)}.$$ **Simplifying**

Using Multiplication by 1 to Simplify a Complex Rational Expression

1. Find the LCD of all rational expressions *within* the complex rational expression.
2. Multiply the complex rational expression by 1, writing 1 as the LCD divided by itself.
3. Distribute and simplify so that the numerator and the denominator of the complex rational expression are polynomials.
4. Factor and, if possible, simplify.

Note that use of the LCD when multiplying by a form of 1 clears the numerator and the denominator of the complex rational expression of all rational expressions.

EXAMPLE 2 Simplify:

$$\frac{\frac{3}{2x - 2} - \frac{1}{x + 1}}{\frac{1}{x - 1} + \frac{x}{x^2 - 1}}.$$

SOLUTION In this case, to find the LCD, we have to factor first:

$$\frac{\frac{3}{2x - 2} - \frac{1}{x + 1}}{\frac{1}{x - 1} + \frac{x}{x^2 - 1}} = \frac{\frac{3}{2(x - 1)} - \frac{1}{x + 1}}{\frac{1}{x - 1} + \frac{x}{(x - 1)(x + 1)}}$$ **The LCD is $2(x - 1)(x + 1)$.**

$$= \frac{\dfrac{3}{2(x-1)} - \dfrac{1}{x+1}}{\dfrac{1}{x-1} + \dfrac{x}{(x-1)(x+1)}} \cdot \frac{2(x-1)(x+1)}{2(x-1)(x+1)}$$ **Multiplying by 1, using the LCD**

$$= \frac{\dfrac{3}{2(x-1)} \cdot 2(x-1)(x+1) - \dfrac{1}{x+1} \cdot 2(x-1)(x+1)}{\dfrac{1}{x-1} \cdot 2(x-1)(x+1) + \dfrac{x}{(x-1)(x+1)} \cdot 2(x-1)(x+1)}$$ **Using the distributive law**

$$= \frac{\dfrac{\cancel{2(x-1)}}{\cancel{2(x-1)}} \cdot 3(x+1) - \dfrac{\cancel{x+1}}{\cancel{x+1}} \cdot 2(x-1)}{\dfrac{\cancel{x-1}}{\cancel{x-1}} \cdot 2(x+1) + \dfrac{\cancel{(x-1)(x+1)}}{\cancel{(x-1)(x+1)}} \cdot 2x}$$ **Removing factors that equal 1**

$$= \frac{3(x+1) - 2(x-1)}{2(x+1) + 2x}$$ **Simplifying**

$$= \frac{3x + 3 - 2x + 2}{2x + 2 + 2x}$$ **Using the distributive law**

$$= \frac{x+5}{4x+2}.$$ **Combining like terms**

Entering Rational Expressions

We can check the simplification of rational expressions in one variable using a graphing calculator. To enter a complex rational expression, we use parentheses around the entire numerator and parentheses around the entire denominator. If necessary, we also use parentheses around numerators and denominators of rational expressions within the complex rational expression. Following are some examples of complex rational expressions rewritten with parentheses.

Complex Rational Expression	Expression Rewritten with Parentheses
$\dfrac{x + \dfrac{5}{x}}{4x}$	$(x + 5/x)/(4x)$
$\dfrac{\dfrac{3}{2x-2} - \dfrac{1}{x+1}}{\dfrac{1}{x-1} + \dfrac{x}{x^2-1}}$	$(3/(2x-2) - 1/(x+1))/(1/(x-1) + x/(x^2-1))$

The following may help in placing parentheses properly.

1. Close each set of parentheses: For every left parenthesis, there should be a right parenthesis.
2. Remember the rules for order of operations when deciding how to place parentheses.
3. When in doubt, place parentheses around every numerator and around every denominator. These extra parentheses will make the expression look more complicated, but will not change its value.

Method 2: Dividing Two Rational Expressions

Another method for simplifying complex rational expressions involves first adding or subtracting, as necessary, to obtain one rational expression in the numerator and one rational expression in the denominator. The problem is then a division of two rational expressions.

EXAMPLE 3 Simplify:

$$\frac{\dfrac{3}{x} - \dfrac{2}{x^2}}{\dfrac{3}{x-2} + \dfrac{1}{x^2}}.$$

SOLUTION

$$\frac{\dfrac{3}{x} - \dfrac{2}{x^2}}{\dfrac{3}{x-2} + \dfrac{1}{x^2}} = \frac{\dfrac{3}{x}\cdot\dfrac{x}{x} - \dfrac{2}{x^2}}{\dfrac{3}{x-2}\cdot\dfrac{x^2}{x^2} + \dfrac{1}{x^2}\cdot\dfrac{x-2}{x-2}}$$

Multiplying $3/x$ by 1 to obtain x^2 as a common denominator

Multiplying by 1, twice, to obtain $x^2(x-2)$ as a common denominator

$$= \frac{\dfrac{3x}{x^2} - \dfrac{2}{x^2}}{\dfrac{3x^2}{(x-2)x^2} + \dfrac{x-2}{x^2(x-2)}}$$

There is now a common denominator in the numerator and a common denominator in the denominator of the complex rational expression.

$$= \frac{\dfrac{3x-2}{x^2}}{\dfrac{3x^2+x-2}{(x \quad 2)x^2}}$$

Subtracting in the numerator and adding in the denominator. We now have one rational expression divided by another rational expression.

$$= \frac{3x-2}{x^2} \div \frac{3x^2+x-2}{(x-2)x^2}$$

Rewriting with a division symbol

$$= \frac{3x-2}{x^2} \cdot \frac{(x-2)x^2}{3x^2+x-2}$$

To divide, multiply by the reciprocal of the divisor.

$$= \frac{\cancel{(3x-2)}(x-2)\cancel{x^2}}{\cancel{x^2}\cancel{(3x-2)}(x+1)}$$

Factoring and removing a factor equal to 1: $\dfrac{x^2(3x-2)}{x^2(3x-2)} = 1$

$$= \frac{x-2}{x+1}$$

$y_1 = (3/x - 2/x^2)/(3/(x-2) + 1/x^2)$,
$y_2 = (x-2)/(x+1)$

X	Y1	Y2
−3	2.5	2.5
−2	4	4
−1	ERROR	ERROR
0	ERROR	−2
1	−.5	−.5
2	ERROR	0
3	.25	.25

X = −3

As a partial check, we let $y_1 = (3/x - 2/x^2)/(3/(x-2) + 1/x^2)$ and $y_2 = (x-2)/(x+1)$. A table of values, like that shown at left, indicates that the values of y_1 and y_2 are the same for all x-values for which both expressions are defined.

Using Division to Simplify a Complex Rational Expression

1. Add or subtract, as necessary, to get one rational expression in the numerator.
2. Add or subtract, as necessary, to get one rational expression in the denominator.
3. Perform the indicated division (invert the divisor and multiply).
4. Simplify, if possible, by removing any factors that equal 1.

EXAMPLE 4 Simplify:

$$\frac{1 + \frac{2}{x}}{1 - \frac{4}{x^2}}.$$

SOLUTION We have

$$\frac{1 + \frac{2}{x}}{1 - \frac{4}{x^2}} = \frac{\frac{x}{x} + \frac{2}{x}}{\frac{x^2}{x^2} - \frac{4}{x^2}}$$ **Finding a common denominator** (numerator); **Finding a common denominator** (denominator)

$$= \frac{\frac{x+2}{x}}{\frac{x^2-4}{x^2}}$$ **Adding in the numerator**; **Subtracting in the denominator**

$$= \frac{x+2}{x} \cdot \frac{x^2}{x^2-4}$$ **Multiplying by the reciprocal of the divisor**

$$= \frac{(x+2) \cdot x^2}{x(x+2)(x-2)}$$ **Factoring. Remember to simplify when possible.**

$$= \frac{\cancel{(x+2)}\cancel{x} \cdot x}{\cancel{x}\cancel{(x+2)}(x-2)}$$ **Removing a factor equal to 1:** $\frac{(x+2)x}{(x+2)x} = 1$

$$= \frac{x}{x-2}.$$ **Simplifying**

As a quick, partial check, we select a convenient value for x—say, 1:

$$\frac{1 + \frac{2}{1}}{1 - \frac{4}{1^2}} = \frac{1+2}{1-4} = \frac{3}{-3} = -1$$ **We evaluated the original expression for $x = 1$.**

and

$$\frac{1}{1-2} = \frac{1}{-1} = -1.$$ **We evaluated the simplified expression for $x = 1$.**

Since both expressions yield the same result, our simplification is probably correct. More evaluation would provide a more definitive check and can be done using the TABLE feature of a graphing calculator.

If negative exponents occur, we first find an equivalent expression using positive exponents and then proceed as in the preceding examples.

EXAMPLE 5 Simplify:

$$\frac{a^{-1}+b^{-1}}{a^{-3}+b^{-3}}.$$

SOLUTION

$$\frac{a^{-1}+b^{-1}}{a^{-3}+b^{-3}} = \frac{\frac{1}{a}+\frac{1}{b}}{\frac{1}{a^3}+\frac{1}{b^3}}$$ **Rewriting with positive exponents. We continue, using method 2.**

$$= \frac{\frac{1}{a}\cdot\frac{b}{b}+\frac{1}{b}\cdot\frac{a}{a}}{\frac{1}{a^3}\cdot\frac{b^3}{b^3}+\frac{1}{b^3}\cdot\frac{a^3}{a^3}}$$ **Finding a common denominator** / **Finding a common denominator**

$$= \frac{\frac{b}{ab}+\frac{a}{ab}}{\frac{b^3}{a^3b^3}+\frac{a^3}{a^3b^3}}$$

$$= \frac{\frac{b+a}{ab}}{\frac{b^3+a^3}{a^3b^3}}$$ **Adding in the numerator** / **Adding in the denominator**

$$= \frac{b+a}{ab}\cdot\frac{a^3b^3}{b^3+a^3}$$ **Multiplying by the reciprocal of the divisor**

$$= \frac{(b+a)\cdot ab\cdot a^2b^2}{ab(b+a)(b^2-ab+a^2)}$$ **Factoring and looking for common factors**

$$= \frac{\cancel{(b+a)}\cdot \cancel{ab}\cdot a^2b^2}{\cancel{ab}\cancel{(b+a)}(b^2-ab+a^2)}$$ **Removing a factor equal to 1: $\frac{(b+a)ab}{(b+a)ab} = 1$**

$$= \frac{a^2b^2}{b^2-ab+a^2}$$

There is no one method that is best to use. When it is little or no work to write an expression as a quotient of two rational expressions, the second method is probably easier to use. On the other hand, some expressions require fewer steps if we use the first method. Either method can be used with any complex rational expression.

6.3 EXERCISE SET

FOR EXTRA HELP

MathXL | MyMathLab | InterAct Math | AW Math Tutor Center | Video Lectures on CD: Disc 3 | Student's Solutions Manual

Concept Reinforcement *In each of Exercises 1–6, match the complex rational expression with an equivalent expression from the column on the right.*

1. ___ $\dfrac{\frac{5}{x}+\frac{3}{y}}{\frac{3}{x}-\frac{2}{y}}$

2. ___ $\dfrac{\frac{5}{x+y}}{\frac{3}{x-y}}$

3. ___ $\dfrac{\frac{3}{x}+\frac{2}{y}}{\frac{5}{x}-\frac{3}{y}}$

4. ___ $\dfrac{\frac{3}{x}-\frac{2}{y}}{\frac{5}{x}+\frac{3}{y}}$

5. ___ $\dfrac{\frac{5}{x}-\frac{3}{y}}{\frac{3}{x}+\frac{2}{y}}$

6. ___ $\dfrac{\frac{3}{x-y}}{\frac{5}{x+y}}$

a) $\dfrac{3}{x-y} \div \dfrac{5}{x+y}$

b) $\dfrac{\frac{5}{x}\cdot\frac{y}{y}-\frac{3}{y}\cdot\frac{x}{x}}{\frac{3}{x}\cdot\frac{y}{y}+\frac{2}{y}\cdot\frac{x}{x}}$

c) $\dfrac{\frac{3}{x}-\frac{2}{y}}{\frac{5}{x}+\frac{3}{y}}\cdot\dfrac{xy}{xy}$

d) $\dfrac{5}{x+y} \div \dfrac{3}{x-y}$

e) $\dfrac{\frac{5}{x}\cdot\frac{y}{y}+\frac{3}{y}\cdot\frac{x}{x}}{\frac{3}{x}\cdot\frac{y}{y}-\frac{2}{y}\cdot\frac{x}{x}}$

f) $\dfrac{\frac{3}{x}+\frac{2}{y}}{\frac{5}{x}-\frac{3}{y}}\cdot\dfrac{xy}{xy}$

Simplify. If possible, use a second method, evaluation, or a graphing calculator as a check.

7. $\dfrac{\frac{x+2}{x-1}}{\frac{x+4}{x-3}}$

8. $\dfrac{\frac{x-1}{x+3}}{\frac{x-6}{x+2}}$

9. $\dfrac{\frac{5}{a}-\frac{4}{b}}{\frac{2}{a}+\frac{3}{b}}$

10. $\dfrac{\frac{2}{r}-\frac{3}{t}}{\frac{4}{r}+\frac{5}{t}}$

11. $\dfrac{\frac{3}{z}+\frac{2}{y}}{\frac{4}{z}-\frac{1}{y}}$

12. $\dfrac{\frac{6}{x}+\frac{7}{y}}{\frac{7}{x}-\frac{6}{y}}$

13. $\dfrac{\frac{a^2-b^2}{ab}}{\frac{a-b}{b}}$

14. $\dfrac{\frac{x^2-y^2}{xy}}{\frac{x-y}{y}}$

15. $\dfrac{1-\frac{2}{3x}}{x-\frac{4}{9x}}$

16. $\dfrac{\frac{3x}{y}-x}{2y-\frac{y}{x}}$

17. $\dfrac{x^{-1}+y^{-1}}{\frac{x^2-y^2}{xy}}$

18. $\dfrac{a^{-1}+b^{-1}}{\frac{a^2-b^2}{ab}}$

19. $\dfrac{\dfrac{1}{a-h}-\dfrac{1}{a}}{h}$

20. $\dfrac{\dfrac{1}{x+h}-\dfrac{1}{x}}{h}$

21. $\dfrac{\dfrac{a^2-4}{a^2+3a+2}}{\dfrac{a^2-5a-6}{a^2-6a-7}}$

22. $\dfrac{\dfrac{x^2-x-12}{x^2-2x-15}}{\dfrac{x^2+8x+12}{x^2-5x-14}}$

23. $\dfrac{\dfrac{x}{x^2+3x-4}-\dfrac{1}{x^2+3x-4}}{\dfrac{x}{x^2+6x+8}+\dfrac{3}{x^2+6x+8}}$

24. $\dfrac{\dfrac{x}{x^2+5x-6}+\dfrac{6}{x^2+5x-6}}{\dfrac{x}{x^2-5x+4}-\dfrac{2}{x^2-5x+4}}$

25. $\dfrac{\dfrac{1}{y}+2}{\dfrac{1}{y}-3}$

26. $\dfrac{7+\dfrac{1}{a}}{\dfrac{1}{a}-3}$

27. $\dfrac{y+y^{-1}}{y-y^{-1}}$

28. $\dfrac{x-x^{-1}}{x+x^{-1}}$

29. $\dfrac{\dfrac{1}{x-2}+\dfrac{3}{x-1}}{\dfrac{2}{x-1}+\dfrac{5}{x-2}}$

30. $\dfrac{\dfrac{2}{y-3}+\dfrac{1}{y+1}}{\dfrac{3}{y+1}+\dfrac{4}{y-3}}$

31. $\dfrac{a(a+3)^{-1}-2(a-1)^{-1}}{a(a+3)^{-1}-(a-1)^{-1}}$

32. $\dfrac{a(a+2)^{-1}-3(a-3)^{-1}}{a(a+2)^{-1}-(a-3)^{-1}}$

33. $\dfrac{\dfrac{2}{a^2-1}+\dfrac{1}{a+1}}{\dfrac{3}{a^2-1}+\dfrac{2}{a-1}}$

34. $\dfrac{\dfrac{3}{a^2-9}+\dfrac{2}{a+3}}{\dfrac{4}{a^2-9}+\dfrac{1}{a+3}}$

35. $\dfrac{\dfrac{5}{x^2-4}-\dfrac{3}{x-2}}{\dfrac{4}{x^2-4}-\dfrac{2}{x+2}}$

36. $\dfrac{\dfrac{4}{x^2-1}-\dfrac{3}{x+1}}{\dfrac{5}{x^2-1}-\dfrac{2}{x-1}}$

37. $\dfrac{\dfrac{y}{y^2-4}+\dfrac{5}{4-y^2}}{\dfrac{y^2}{y^2-4}+\dfrac{25}{4-y^2}}$

38. $\dfrac{\dfrac{y}{y^2-1}+\dfrac{3}{1-y^2}}{\dfrac{y^2}{y^2-1}+\dfrac{9}{1-y^2}}$

39. $\dfrac{\dfrac{y^2}{y^2-25}-\dfrac{y}{y-5}}{\dfrac{y}{y^2-25}-\dfrac{1}{y+5}}$

40. $\dfrac{\dfrac{y^2}{y^2-9}-\dfrac{y}{y+3}}{\dfrac{y}{y^2-9}-\dfrac{1}{y-3}}$

41. $\dfrac{\dfrac{a}{a+2}+\dfrac{5}{a}}{\dfrac{a}{2a+4}+\dfrac{1}{3a}}$

42. $\dfrac{\dfrac{a}{a+3}+\dfrac{4}{5a}}{\dfrac{a}{2a+6}+\dfrac{3}{a}}$

43. $\dfrac{\dfrac{1}{x^2-3x+2}+\dfrac{1}{x^2-4}}{\dfrac{1}{x^2+4x+4}+\dfrac{1}{x^2-4}}$

44. $\dfrac{\dfrac{1}{x^2+3x+2}+\dfrac{1}{x^2-1}}{\dfrac{1}{x^2-1}+\dfrac{1}{x^2-4x+3}}$

45. $\dfrac{\dfrac{3}{a^2-4a+3}+\dfrac{3}{a^2-5a+6}}{\dfrac{3}{a^2-3a+2}+\dfrac{3}{a^2+3a-10}}$

46. $\dfrac{\dfrac{1}{a^2+7a+10}-\dfrac{2}{a^2-7a+12}}{\dfrac{2}{a^2-a-6}-\dfrac{1}{a^2+a-20}}$

Aha! **47.** $\dfrac{\dfrac{y}{y^2-4}-\dfrac{2y}{y^2+y-6}}{\dfrac{2y}{y^2+y-6}-\dfrac{y}{y^2-4}}$

48. $\dfrac{\dfrac{y}{y^2-1}-\dfrac{3y}{y^2+5y+4}}{\dfrac{3y}{y^2-1}-\dfrac{y}{y^2-4y+3}}$

49. $\dfrac{\dfrac{3}{x^2+2x-3}-\dfrac{1}{x^2-3x-10}}{\dfrac{3}{x^2-6x+5}-\dfrac{1}{x^2+5x+6}}$

50. $\dfrac{\dfrac{1}{a^2+7a+12}+\dfrac{1}{a^2+a-6}}{\dfrac{1}{a^2+2a-8}+\dfrac{1}{a^2+5a+4}}$

TW **51.** Michael *incorrectly* simplifies

$$\frac{a+b^{-1}}{a+c^{-1}} \text{ as } \frac{a+c}{a+b}.$$

What mistake is he making and how could you convince him that this is incorrect?

TW **52.** To simplify a complex rational expression in which the sum of two fractions is divided by the difference of the same two fractions, which method is easier? Why?

Focused Review

Simplify.

53. $\dfrac{-6+3x}{5} \div \dfrac{4x-8}{25}$ [6.1]

54. $\dfrac{t}{t-3} - \dfrac{3}{4t-12}$ [6.2]

55. $\dfrac{7}{a^2+a-2} + \dfrac{a}{a^2-4a+3}$ [6.2]

56. $\dfrac{t^2+2t-3}{t^2+4t-5} \cdot \dfrac{t^2-3t-10}{t^2+5t+6}$ [6.1]

57. $\dfrac{\dfrac{2}{x^2y}+\dfrac{3}{xy^2}}{\dfrac{3}{xy^2}+\dfrac{2}{x^2y}}$ [6.3]

58. $\dfrac{x-2+\dfrac{1}{x}}{x-5+\dfrac{4}{x}}$ [6.3]

Synthesis

TW **59.** In arithmetic, we are taught that

$$\frac{a}{b} \div \frac{c}{d} = \frac{a}{b} \cdot \frac{d}{c}$$

(to divide by a fraction, we invert and multiply). Use method 1 to explain *why* we do this.

TW **60.** Use algebra to determine the domain of the function given by

$$f(x) = \frac{\dfrac{1}{x-2}}{\dfrac{x}{x-2}-\dfrac{5}{x-2}}.$$

Then explain how a graphing calculator could be used to check your answer.

Simplify.

61. $\dfrac{5x^{-2}+10x^{-1}y^{-1}+5y^{-2}}{3x^{-2}-3y^{-2}}$

62. $(a^2-ab+b^2)^{-1}(a^2b^{-1}+b^2a^{-1}) \times (a^{-2}-b^{-2})(a^{-2}+2a^{-1}b^{-1}+b^{-2})^{-1}$

63. *Astronomy.* When two galaxies are moving in opposite directions at velocities v_1 and v_2, an observer in one of the galaxies would see the other galaxy receding at speed

$$\frac{v_1+v_2}{1+\dfrac{v_1v_2}{c^2}},$$

where c is the speed of light. Determine the observed speed if v_1 and v_2 are both one-fourth the speed of light.

Find and simplify

$$\frac{f(x+h)-f(x)}{h}$$

for each rational function f in Exercises 64–67.

64. $f(x) = \dfrac{2}{x^2}$

65. $f(x) = \dfrac{3}{x}$

66. $f(x) = \dfrac{x}{1 - x}$

67. $f(x) = \dfrac{2x}{1 + x}$

68. If

$$F(x) = \frac{3 + \dfrac{1}{x}}{2 - \dfrac{8}{x^2}},$$

find the domain of F.

69. If

$$G(x) = \frac{x - \dfrac{1}{x^2 - 1}}{\dfrac{1}{9} - \dfrac{1}{x^2 - 16}},$$

find the domain of G.

70. Find the reciprocal of y if

$$y = x^2 + x + 1 + \frac{1}{x} + \frac{1}{x^2}.$$

71. For $f(x) = \dfrac{2}{2 + x}$, find $f(f(a))$.

72. For $g(x) = \dfrac{x + 3}{x - 1}$, find $g(g(a))$.

73. Let

$$f(x) = \left[\frac{\dfrac{x + 3}{x - 3} + 1}{\dfrac{x + 3}{x - 3} - 1}\right]^4.$$

Find a simplified form of $f(x)$ and specify the domain of f.

74. *Financial Planning.* Alexis wishes to invest a portion of each month's pay in an account that pays 7.5% interest. If he wants to have \$30,000 in the account after 10 yr, the amount invested each month is given by

$$\frac{30{,}000 \cdot \dfrac{0.075}{12}}{\left(1 + \dfrac{0.075}{12}\right)^{120} - 1}.$$

Find the amount of Alexis' monthly investment.

Collaborative Corner

Which Method Is Better?

Focus: Complex rational expressions
Time: 10–15 minutes
Group Size: 3–4

ACTIVITY

Consider the steps in Examples 2 and 3 for simplifying a complex rational expression by each of the two methods. Then, work as a group to simplify

$$\frac{\dfrac{5}{x + 1} - \dfrac{1}{x}}{\dfrac{2}{x^2} + \dfrac{4}{x}}$$

subject to the following conditions.

1. The group should predict which method will more easily simplify this expression.
2. Using the method selected in part (1), one group member should perform the first step in the simplification and then pass the problem on to another member of the group. That person then checks the work, performs the next step, and passes the problem on to another group member. If a mistake is found, the problem should be passed to the person who made the mistake for repair. This process continues until, eventually, the simplification is complete.
3. At the same time that part (2) is being performed, another group member should perform the first step of the solution using the method not selected in part (1). He or she should then pass the problem to another group member and so on, just as in part (2).
4. What method *was* easier? Why? Compare your responses with those of other groups.

6.4 Rational Equations

Solving Rational Equations

Connecting the Concepts

It is not unusual to learn a skill that enables us to write equivalent expressions and then to put that skill to work solving a new type of equation. That is precisely what we are now doing: Sections 6.1–6.3 have been devoted to writing equivalent expressions in which rational expressions appeared. There we used least common denominators to add or subtract. Here in Section 6.4, we return to the task of solving equations, but this time the equations contain rational expressions and the LCD is used as part of the multiplication principle.

Study Tip

Does More Than One Solution Exist?

Keep in mind that many problems—in math and elsewhere—have more than one solution. When asked to solve an equation, we are expected to find any and all solutions of the equation.

Solving Rational Equations

In Sections 6.1–6.3, we learned how to *simplify expressions*. We now learn to *solve* a new type of *equation*. A **rational equation** is an equation that contains one or more rational expressions. Here are some examples:

$$\frac{2}{3} - \frac{5}{6} = \frac{1}{t}, \qquad \frac{a-1}{a-5} = \frac{4}{a^2-25}, \qquad x^3 + \frac{6}{x} = 5.$$

As you will see in Section 6.5, equations of this type occur frequently in applications. To solve rational equations, recall that one way to *clear fractions* from an equation is to multiply both sides of the equation by the LCD.

To Solve a Rational Equation

Multiply both sides of the equation by the LCD. This is called *clearing fractions* and produces an equation similar to those we have already solved.

Recall that division by 0 is undefined. Note, too, that variables usually appear in at least one denominator of a rational equation. Thus certain numbers can often be ruled out as possible solutions before we even attempt to solve a given rational equation.

EXAMPLE 1 Solve: $\dfrac{x+4}{3x} + \dfrac{x+8}{5x} = 2.$

ALGEBRAIC APPROACH

Because the left side of this equation is undefined when x is 0, we state at the outset that $x \neq 0$. Next, we multiply both sides of the equation by the LCD, $3 \cdot 5 \cdot x$, or $15x$:

$$15x\left(\frac{x+4}{3x} + \frac{x+8}{5x}\right) = 15x \cdot 2$$ **Multiplying by the LCD to clear fractions**

$$15x \cdot \frac{x+4}{3x} + 15x \cdot \frac{x+8}{5x} = 15x \cdot 2$$ **Using the distributive law**

$$\frac{5 \cdot 3x \cdot (x+4)}{3x} + \frac{3 \cdot 5x \cdot (x+8)}{5x} = 30x$$ **Locating factors equal to 1**

$$5(x+4) + 3(x+8) = 30x$$ **Removing factors equal to 1:** $\dfrac{3x}{3x} = 1$; $\dfrac{5x}{5x} = 1$

We have solved equations like this before.

$$5x + 20 + 3x + 24 = 30x$$ **Using the distributive law**

$$8x + 44 = 30x$$

$$44 = 22x$$

$$2 = x.$$ **This should check since $x \neq 0$.**

GRAPHICAL APPROACH

We let $y_1 = (x+4)/(3x) + (x+8)/(5x)$ and $y_2 = 2$. The solutions of the equation will be the x-coordinates of any points of intersection.

$y_1 = (x+4)/(3x) + (x+8)/(5x)$, $y_2 = 2$

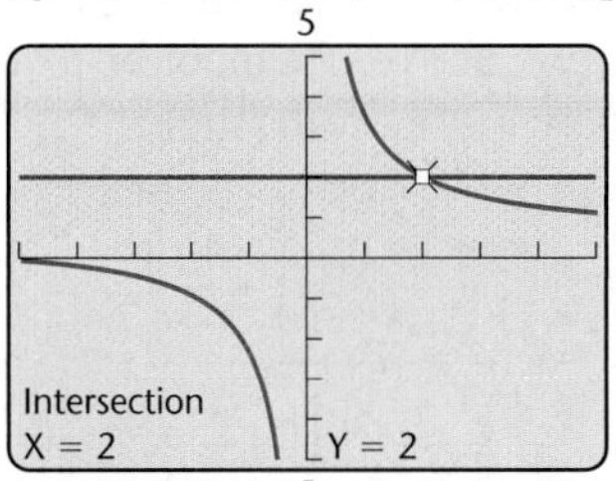

It appears from the graph that there is one point of intersection, (2, 2). The solution is 2.

Student Notes

When we are solving rational equations, as in Examples 1–5, the LCD is never actually used as a denominator. Rather, it is used as a multiplier to eliminate the denominators in the equation. This differs from how the LCD is used in Section 6.2.

Check:

$$\frac{x+4}{3x} + \frac{x+8}{5x} = 2$$

$$\frac{2+4}{3 \cdot 2} + \frac{2+8}{5 \cdot 2} \quad \Big| \quad 2$$

$$\frac{6}{6} + \frac{10}{10}$$

$$2 \stackrel{?}{=} 2 \quad \text{TRUE}$$

The number 2 is the solution.

It is important that when we clear fractions, all denominators are eliminated. This produces an equation without rational expressions, which is easier to solve.

EXAMPLE 2 Solve: $\dfrac{x-1}{x-5} = \dfrac{4}{x-5}.$

ALGEBRAIC APPROACH

To ensure that neither denominator is 0, we state at the outset the restriction that $x \neq 5$. Then we proceed as before, multiplying both sides by the LCD, $x - 5$:

$$(x-5)\cdot\frac{x-1}{x-5} = (x-5)\cdot\frac{4}{x-5}$$
$$x - 1 = 4$$
$$x = 5. \quad \textbf{But recall that } x \neq 5.$$

In this case, it is important to remember that, because of the restriction above, 5 cannot be a solution. A check confirms the necessity of that restriction.

Check:

$$\frac{x-1}{x-5} = \frac{4}{x-5}$$

$\frac{5-1}{5-5}$	$\frac{4}{5-5}$

$$\frac{4}{0} \stackrel{?}{=} \frac{4}{0} \quad \textbf{Division by 0 is undefined.}$$

This equation has no solution.

GRAPHICAL APPROACH

We graph $y_1 = (x-1)/(x-5)$ and $y_2 = 4/(x-5)$ and look for any points of intersection of the graphs.

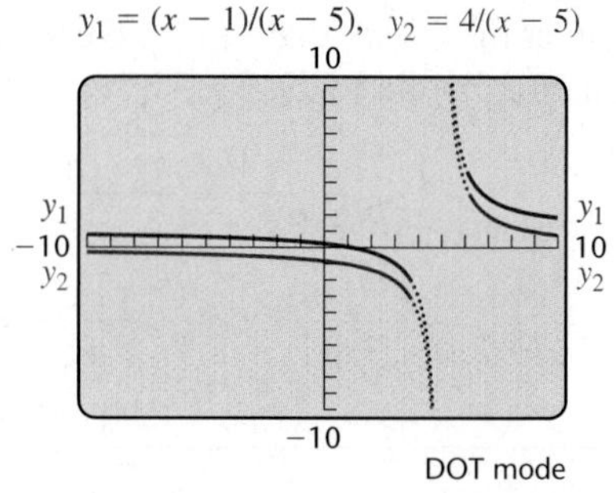

It appears that the graphs may intersect around $x = 4$ or $x = 6$. If we use INTERSECT, the calculator returns an error message. This in itself does not tell us that there is no point of intersection, because we do not see all the graph in the viewing window. In order to state positively that there is no solution, we must solve the equation with algebra.

To see why 5 is not a solution of Example 2, note that the multiplication principle for equations requires that we multiply both sides by a *nonzero* number. When both sides of an equation are multiplied by an expression containing variables, it is possible that certain replacements will make that expression equal to 0. Thus it is safe to say that *if* a solution of

$$\frac{x-1}{x-5} = \frac{4}{x-5}$$

exists, then it is also a solution of $x - 1 = 4$. We *cannot* conclude that every solution of $x - 1 = 4$ is a solution of the original equation.

CAUTION! When solving rational equations, do not forget to list any restrictions as part of the first step. Refer to the restriction(s) as you proceed.

Example 2 illustrates a disadvantage of graphical methods of solving equations. It can be difficult to tell whether two graphs actually intersect. Also, it is easy to "miss" seeing a solution. This is especially true for rational equations, where there are usually two or more branches of the graph. Solving by graphing does make a good check, however, of an algebraic solution.

EXAMPLE 3 Solve: $\dfrac{x^2}{x-3} = \dfrac{9}{x-3}$.

SOLUTION Note that $x \neq 3$. We multiply both sides by the LCD, $x - 3$:

$y_1 = x^2/(x-3),\ y_2 = 9/(x-3)$

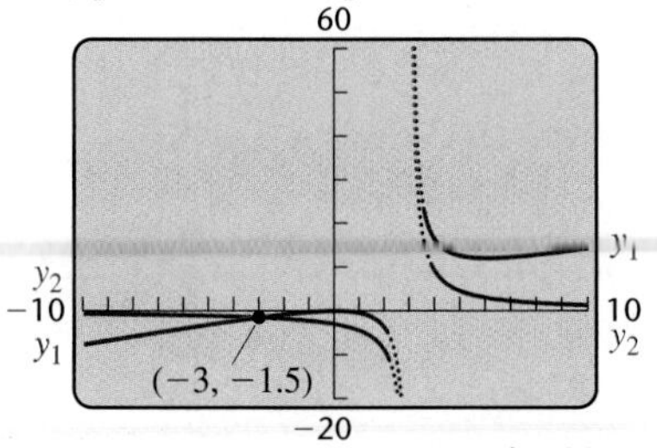

Yscl = 10
DOT mode

$$(x-3)\cdot\frac{x^2}{x-3} = (x-3)\cdot\frac{9}{x-3}$$

$$x^2 = 9 \qquad \textbf{Simplifying}$$

$$x^2 - 9 = 0 \qquad \textbf{Getting 0 on one side}$$

$$(x-3)(x+3) = 0 \qquad \textbf{Factoring}$$

$$x = 3 \quad or \quad x = -3. \qquad \textbf{Using the principle of zero products}$$

Although 3 is a solution of $x^2 = 9$, it must be rejected as a solution of the rational equation because of the restriction stated in red above. You should check that -3 *is* a solution despite the fact that 3 is not. This check can also be done graphically, as shown at left.

EXAMPLE 4 Solve: $\dfrac{2}{x+5} + \dfrac{1}{x-5} = \dfrac{16}{x^2-25}$.

SOLUTION To find all restrictions and to assist in finding the LCD, we factor:

$$\frac{2}{x+5} + \frac{1}{x-5} = \frac{16}{(x+5)(x-5)}. \qquad \textbf{Factoring } x^2 - 25$$

Note that $x \neq -5$ and $x \neq 5$. We multiply by the LCD, $(x+5)(x-5)$, and then use the distributive law:

$$(x+5)(x-5)\left(\frac{2}{x+5} + \frac{1}{x-5}\right) = (x+5)(x-5)\cdot\frac{16}{(x+5)(x-5)}$$

$$(x+5)(x-5)\frac{2}{x+5} + (x+5)(x-5)\frac{1}{x-5} = \frac{(x+5)(x-5)16}{(x+5)(x-5)}$$

$$2(x-5) + (x+5) = 16$$

$$2x - 10 + x + 5 = 16$$

$$3x - 5 = 16$$

$$3x = 21$$

$$x = 7.$$

A check will confirm that the solution is 7.

EXAMPLE 5 Let $f(x) = x + \dfrac{6}{x}$. Find all values of a for which $f(a) = 5$.

SOLUTION Since $f(a) = a + \dfrac{6}{a}$, the problem asks that we find all values of a for which

$$a + \frac{6}{a} = 5.$$

First note that $a \neq 0$. To solve for a, we multiply both sides of the equation by the LCD, a:

$$a\left(a + \frac{6}{a}\right) = 5 \cdot a \quad \text{Multiplying both sides by } a\text{. Parentheses are important.}$$

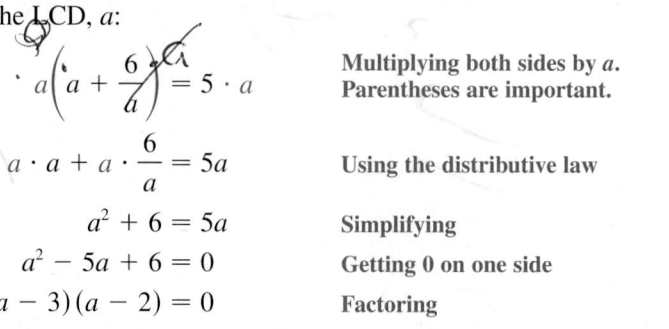

$$a \cdot a + a \cdot \frac{6}{a} = 5a \quad \text{Using the distributive law}$$

$$a^2 + 6 = 5a \quad \text{Simplifying}$$

$$a^2 - 5a + 6 = 0 \quad \text{Getting 0 on one side}$$

$$(a - 3)(a - 2) = 0 \quad \text{Factoring}$$

$$a = 3 \quad or \quad a = 2. \quad \text{Using the principle of zero products}$$

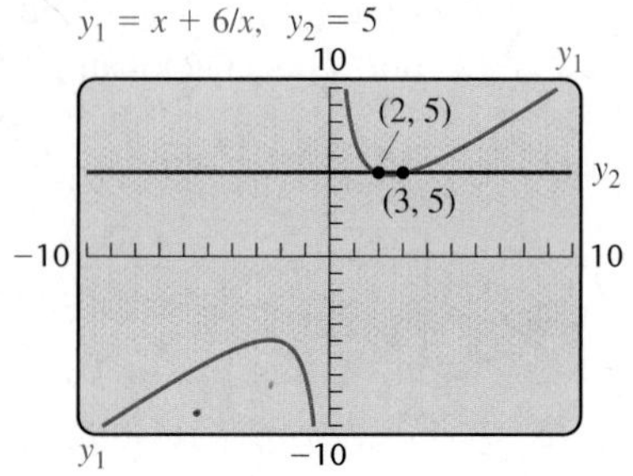

Check:

$$f(3) = 3 + \frac{6}{3} = 3 + 2 = 5;$$

$$f(2) = 2 + \frac{6}{2} = 2 + 3 = 5.$$

The solutions are 2 and 3.

The graph at left shows the points of intersection of the graph of $f(x)$ and the line $y = 5$. It also shows that we have two solutions of $f(x) = 5$, 2 and 3. For $a = 2$ or $a = 3$, we have $f(a) = 5$.

6.4 EXERCISE SET

FOR EXTRA HELP

MathXL | MyMathLab | InterAct Math | AW Math Tutor Center | Video Lectures on CD: Disc 3 | Student's Solutions Manual

Concept Reinforcement *Classify each of the following as either an expression or an equation.*

1. $\frac{5x}{x + 2} - \frac{3}{x} = 7$

2. $\frac{3}{x - 4} + \frac{2}{x + 4}$

3. $\frac{4}{t^2 - 1} + \frac{3}{t + 1}$

4. $\frac{2}{t^2 - 1} + \frac{3}{t + 1} = 5$

5. $\frac{2}{x + 7} + \frac{6}{5x} = 4$

6. $\frac{5}{2x} - \frac{3}{x^2} = 7$

7. $\frac{t + 3}{t - 4} = \frac{t - 5}{t - 7}$

8. $\frac{7t}{2t - 3} \div \frac{3t}{2t + 3}$

9. $\frac{5x}{x^2 - 4} \cdot \frac{7}{x^2 - 5x + 4}$

10. $\frac{7x}{2 - x} = \frac{3}{4 - x}$

Solve. If no solution exists, state this.

11. $\frac{t}{2} + \frac{t}{3} = 7$

12. $\frac{y}{4} + \frac{y}{3} = 8$

13. $\frac{5}{8} - \frac{1}{x} = \frac{3}{4}$

14. $\frac{1}{3} - \frac{1}{t} = \frac{5}{6}$

15. $\frac{2x + 1}{3} + \frac{x + 2}{5} = \frac{4x}{15}$

16. $\frac{3x + 2}{2} + \frac{x + 4}{8} = \frac{2x}{5}$

17. $\frac{2}{3} - \frac{1}{t} = \frac{7}{3t}$

18. $\frac{1}{2} - \frac{2}{t} = \frac{3}{2t}$

19. $\frac{4}{x - 1} + \frac{5}{6} = \frac{2}{3x - 3}$

20. $\frac{3}{2x + 10} + \frac{5}{4} = \frac{7}{x + 5}$

Aha! **21.** $\frac{2}{6} + \frac{1}{2x} = \frac{1}{3}$

22. $\frac{12}{15} - \frac{1}{3x} = \frac{4}{5}$

23. $y + \frac{4}{y} = -5$

24. $t + \frac{6}{t} = -5$

25. $x - \frac{12}{x} = 4$

26. $y - \frac{14}{y} = 5$

27. $\frac{t - 1}{t - 3} = \frac{2}{t - 3}$

28. $\frac{x - 2}{x - 4} = \frac{2}{x - 4}$

29. $\frac{x}{x - 5} = \frac{25}{x^2 - 5x}$

30. $\frac{t}{t - 6} = \frac{36}{t^2 - 6t}$

31. $\frac{5}{4t} = \frac{7}{5t - 2}$

32. $\frac{3}{x - 2} = \frac{5}{x + 4}$

Aha! **33.** $\frac{x^2 + 4}{x - 1} = \frac{5}{x - 1}$

34. $\frac{x^2 - 1}{x + 2} = \frac{3}{x + 2}$

35. $\frac{6}{a + 1} = \frac{a}{a - 1}$

36. $\frac{4}{a - 7} = \frac{-2a}{a + 3}$

37. $\frac{60}{t - 5} - \frac{18}{t} = \frac{40}{t}$

38. $\frac{50}{t - 2} - \frac{16}{t} = \frac{30}{t}$

39. $\frac{3}{x - 3} + \frac{5}{x + 2} = \frac{5x}{x^2 - x - 6}$

40. $\frac{2}{x - 2} + \frac{1}{x + 4} = \frac{x}{x^2 + 2x - 8}$

41. $\frac{3}{x} + \frac{x}{x + 2} = \frac{4}{x^2 + 2x}$

42. $\frac{x}{x + 1} + \frac{5}{x} = \frac{1}{x^2 + x}$

43. $\frac{5}{x + 2} - \frac{3}{x - 2} = \frac{2x}{4 - x^2}$

44. $\frac{y + 3}{y + 2} - \frac{y}{y^2 - 4} = \frac{y}{y - 2}$

45. $\frac{3}{x^2 - 6x + 9} + \frac{x - 2}{3x - 9} = \frac{x}{2x - 6}$

46. $\frac{3 - 2y}{y + 1} - \frac{10}{y^2 - 1} = \frac{2y + 3}{1 - y}$

In Exercises 47–52, a rational function f is given. Find all values of a for which f(a) is the indicated value.

47. $f(x) = 2x - \frac{15}{x}$; $f(a) = 7$

48. $f(x) = 2x - \frac{6}{x}$; $f(a) = 1$

49. $f(x) = \frac{x - 5}{x + 1}$; $f(a) = \frac{3}{5}$

50. $f(x) = \frac{x - 3}{x + 2}$; $f(a) = \frac{1}{5}$

51. $f(x) = \frac{12}{x} - \frac{12}{2x}$; $f(a) = 8$

52. $f(x) = \frac{6}{x} - \frac{6}{2x}$; $f(a) = 5$

For each pair of functions f and g, find all values of a for which f(a) = g(a).

53. $f(x) = \frac{3x - 1}{x^2 - 7x + 10}$,
$g(x) = \frac{x - 1}{x^2 - 4} + \frac{2x + 1}{x^2 - 3x - 10}$

54. $f(x) = \frac{2x + 5}{x^2 + 4x + 3}$,
$g(x) = \frac{x + 2}{x^2 - 9} + \frac{x - 1}{x^2 - 2x - 3}$

55. $f(x) = \frac{2}{x^2 - 8x + 7}$,
$g(x) = \frac{3}{x^2 - 2x - 3} - \frac{1}{x^2 - 1}$

56. $f(x) = \frac{4}{x^2 + 3x - 10}$,
$g(x) = \frac{3}{x^2 - x - 12} + \frac{1}{x^2 + x - 6}$

TW **57.** Below are unlabeled graphs of $f(x) = x + 2$ and $g(x) = (x^2 - 4)/(x - 2)$. How could you determine which graph represents f and which graph represents g?

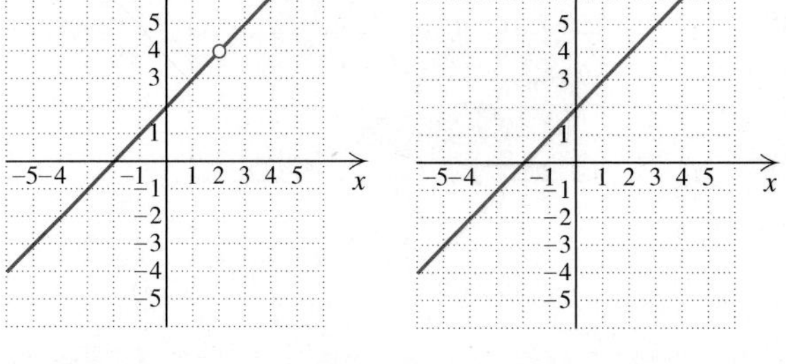

TW **58.** Explain how one can easily produce rational equations for which no solution exists. (*Hint*: Examine Example 2.)

Focused Review

Solve.

59. $3x - 2(5 - x) = 4(x + 7)$ [1.6]

60. $5 - 3x < 7x - 2$ [4.1]

61. $|2x - 7| \leq 8$ [4.4]

62. $3x + 5 < 10$ *or* $x - 3 > 7$ [4.3]

63. $2x - y = 8,$
$y = 3x - 5$ [3.2]

64. $3x - 2y + 5z = 18,$
$2x + y - z = -5,$
$-x + 3y + z = 12$ [3.4]

65. $|4x - 3| = 10$ [4.4]

66. $\frac{6}{x} + 1 = x$ [6.4]

67. $(x + 3)(x - 4) = 8$ [5.4]

68. $2x^3 + 9x^2 = 5x$ [5.5]

Synthesis

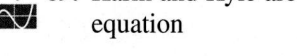

TW **69.** Karin and Kyle are working together to solve the equation

$$\frac{x}{x^2 - x - 6} = 2x - 1.$$

Karin obtains the first graph below using the DOT mode of a graphing calculator, while Kyle obtains the second graph at the top of the next column using the CONNECTED mode. Which approach is the better method of showing the solutions? Why?

$y_1 = x/(x^2 - x - 6), \; y_2 = 2x - 1$

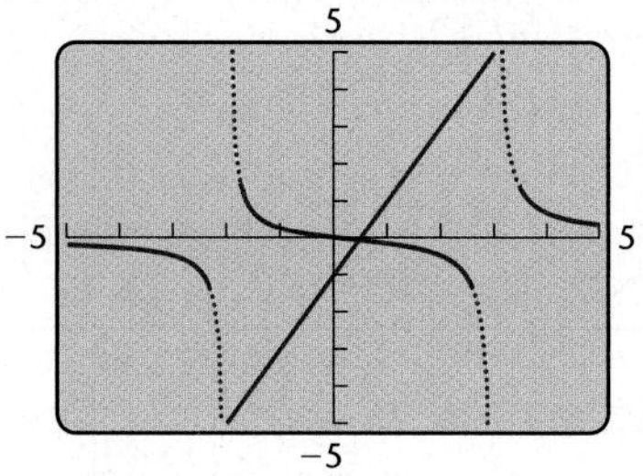

$y_1 = x/(x^2 - x - 6), \; y_2 = 2x - 1$

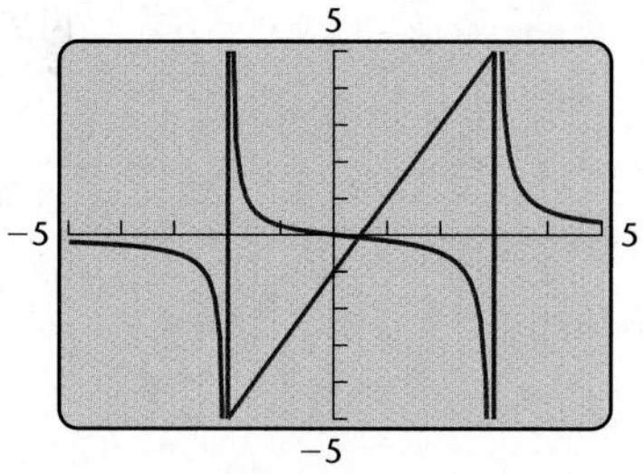

TW **70.** Is the following statement true or false: "For any real numbers a, b, and c, if $ac = bc$, then $a = b$"? Explain why you answered as you did.

For each pair of functions f and g, find all values of a for which $f(a) = g(a)$.

71. $f(x) = \dfrac{x - \frac{2}{3}}{x + \frac{1}{2}}, \; g(x) = \dfrac{x + \frac{2}{3}}{x - \frac{3}{2}}$

72. $f(x) = \dfrac{2 - \frac{x}{4}}{2}, \; g(x) = \dfrac{\frac{x}{4} - 2}{\frac{x}{2} + 2}$

73. $f(x) = \dfrac{x + 3}{x + 2} - \dfrac{x + 4}{x + 3}, \; g(x) = \dfrac{x + 5}{x + 4} - \dfrac{x + 6}{x + 5}$

74. $f(x) = \dfrac{1}{1 + x} + \dfrac{x}{1 - x}, \; g(x) = \dfrac{1}{1 - x} - \dfrac{x}{1 + x}$

75. $f(x) = \dfrac{0.793}{x} + 18.15, \; g(x) = \dfrac{6.034}{x} - 43.17$

76. $f(x) = \dfrac{2.315}{x} - \dfrac{12.6}{17.4}, \; g(x) = \dfrac{6.71}{x} + 0.763$

Recall that identities are true for any possible replacement of the variable(s). Determine whether each of the following equations is an identity.

77. $\dfrac{x^2 + 6x - 16}{x - 2} = x + 8, \; x \neq 2$

78. $\dfrac{x^3 + 8}{x^2 - 4} = \dfrac{x^2 - 2x + 4}{x - 2}, \; x \neq -2, x \neq 2$

6.5 Solving Applications Using Rational Equations

Problems Involving Work ■ Problems Involving Motion

Now that we are able to solve rational equations, it is possible to solve problems that we could not have handled before. The five problem-solving steps remain the same.

Problems Involving Work

EXAMPLE 1 The roof of Grady and Cleo's townhouse needs to be reshingled. Grady can do the job alone in 8 hr and Cleo can do the job alone in 10 hr. How long will it take the two of them, working together, to reshingle the roof?

SOLUTION

1. **Familiarize.** We familiarize ourselves with the problem by considering two *incorrect* ways of translating the problem to mathematical language.

   a) One *incorrect* way to translate the problem is to add the two times:

   8 hr + 10 hr = 18 hr.

   Think about this. Grady can do the job *alone* in 8 hr. If Grady and Cleo work together, whatever time it takes them must be *less* than 8 hr.

   b) Another *incorrect* approach is to assume that each person reshingles half of the roof. Were this the case,

   Grady would reshingle $\frac{1}{2}$ the roof in $\frac{1}{2}$(8 hr), or 4 hr,

   and

   Cleo would reshingle $\frac{1}{2}$ the roof in $\frac{1}{2}$(10 hr), or 5 hr.

   But time would be wasted since Grady would finish 1 hr before Cleo. Were Grady to help Cleo after completing his half, the entire job would take between 4 and 5 hr. This information provides a partial check on any answer we get—the answer should be between 4 and 5 hr.

   Let's consider how much of the job each person completes in 1 hr, 2 hr, 3 hr, and so on. Since Grady takes 8 hr to reshingle the entire roof, in 1 hr he reshingles $\frac{1}{8}$ of the roof. Since Cleo takes 10 hr to reshingle the entire roof, in 1 hr she reshingles $\frac{1}{10}$ of the roof. Thus Grady works at a rate of $\frac{1}{8}$ roof per hour, and Cleo works at a rate of $\frac{1}{10}$ roof per hour.

   Working together, Grady and Cleo reshingle $\frac{1}{8} + \frac{1}{10}$ roof in 1 hr, so their rate—as a team—is $\frac{1}{8} + \frac{1}{10} = \frac{5}{40} + \frac{4}{40} = \frac{9}{40}$ roof per hour.

Study Tip

Take Advantage of Free Checking

It is always wise to check an answer, if it is possible to do so. When an applied problem is being solved, it is usually possible to check an answer in the equation from which it came. While such a check can be helpful, it is even more important to check the answer with the words of the original problem.

In 2 hr, Grady reshingles $\frac{1}{8} \cdot 2$ of the roof and Cleo reshingles $\frac{1}{10} \cdot 2$ of the roof. Working together, they reshingle

$\frac{1}{8} \cdot 2 + \frac{1}{10} \cdot 2$, or $\frac{9}{20}$ of the roof in 2 hr. **Note that $\frac{9}{40} \cdot 2 = \frac{9}{20}$.**

We are looking for the time required to reshingle 1 entire roof, not just a part of it. To find this time, we form a table and extend the pattern developed above.

Time	Fraction of the Roof Reshingled		
	By Grady	**By Cleo**	**Together**
1 hr	$\frac{1}{8}$	$\frac{1}{10}$	$\frac{1}{8} + \frac{1}{10}$, or $\frac{9}{40}$
2 hr	$\frac{1}{8} \cdot 2$	$\frac{1}{10} \cdot 2$	$\left(\frac{1}{8} + \frac{1}{10}\right)2$, or $\frac{9}{40} \cdot 2$, or $\frac{9}{20}$
3 hr	$\frac{1}{8} \cdot 3$	$\frac{1}{10} \cdot 3$	$\left(\frac{1}{8} + \frac{1}{10}\right)3$, or $\frac{9}{40} \cdot 3$, or $\frac{27}{40}$
t hr	$\frac{1}{8} \cdot t$	$\frac{1}{10} \cdot t$	$\left(\frac{1}{8} + \frac{1}{10}\right)t$, or $\frac{9}{40} \cdot t$

2. **Translate.** From the table, we see that t must be some number for which

Fraction of roof done by Grady in t hr; Fraction of roof done by Cleo in t hr

$$\frac{1}{8} \cdot t + \frac{1}{10} \cdot t = 1,$$

or

$$\frac{t}{8} + \frac{t}{10} = 1.$$

3. **Carry out.** We solve the equation both algebraically and graphically.

ALGEBRAIC APPROACH

We solve the equation:

$$\frac{t}{8} + \frac{t}{10} = 1$$

$$40\left(\frac{t}{8} + \frac{t}{10}\right) = 40 \cdot 1 \qquad \textbf{Multiplying by the LCD}$$

$$\frac{40t}{8} + \frac{40t}{10} = 40 \qquad \textbf{Distributing the 40}$$

$$5t + 4t = 40 \qquad \textbf{Simplifying}$$

$$9t = 40$$

$$t = \frac{40}{9}, \text{ or } 4\frac{4}{9}.$$

GRAPHICAL APPROACH

We graph $y_1 = x/8 + x/10$ and $y_2 = 1$.

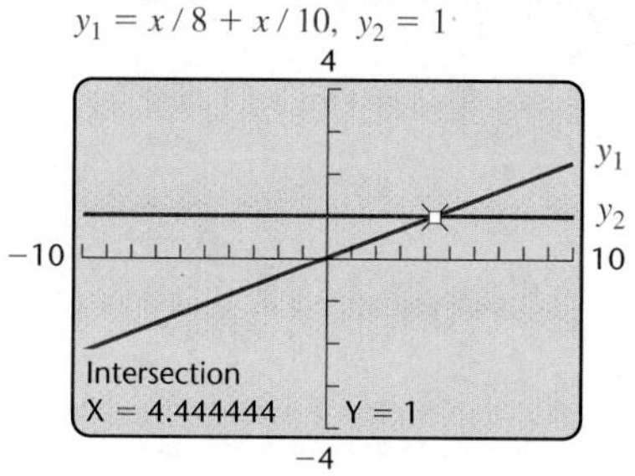

We can convert X to fraction notation and write the point of intersection as $\left(\frac{40}{9}, 1\right)$. The solution is $\frac{40}{9}$, or about 4.4.

4. **Check.** In $\frac{40}{9}$ hr, Grady reshingles $\frac{1}{8} \cdot \frac{40}{9}$, or $\frac{5}{9}$, of the roof and Cleo reshingles $\frac{1}{10} \cdot \frac{40}{9}$, or $\frac{4}{9}$, of the roof. Together, they reshingle $\frac{5}{9} + \frac{4}{9}$, or 1 roof. The fact that our solution is between 4 and 5 hr (see step 1 above) is also a check.
5. **State.** It will take $4\frac{4}{9}$ hr for Grady and Cleo, working together, to reshingle the roof.

EXAMPLE 2 It takes Pepe 9 hr longer than Wendy to rebuild an engine. Working together, they can do the job in 20 hr. How long would it take each, working alone, to rebuild an engine?

SOLUTION

1. **Familiarize.** Unlike Example 1, this problem does not provide us with the times required by the individuals to do the job alone. Let's have $w =$ the number of hours it would take Wendy working alone and $w + 9 =$ the number of hours it would take Pepe working alone.
2. **Translate.** Using the same reasoning as in Example 1, we see that Wendy completes $\frac{1}{w}$ of the job in 1 hr and Pepe completes $\frac{1}{w+9}$ of the job in 1 hr. In 2 hr, Wendy completes $\frac{1}{w} \cdot 2$ of the job and Pepe completes $\frac{1}{w+9} \cdot 2$ of the job. We are told that, working together, Wendy and Pepe can complete the entire job in 20 hr. This gives the following:

Fraction of job done by Wendy in 20 hr; Fraction of job done by Pepe in 20 hr

$$\frac{1}{w} \cdot 20 + \frac{1}{w+9} \cdot 20 = 1,$$

or

$$\frac{20}{w} + \frac{20}{w+9} = 1.$$

3. **Carry out.** We solve the equation both algebraically and graphically.

ALGEBRAIC APPROACH

We have

$$\frac{20}{w} + \frac{20}{w+9} = 1$$

$$w(w+9)\left(\frac{20}{w} + \frac{20}{w+9}\right) = w(w+9)1 \qquad \textbf{Multiplying by the LCD}$$

$$(w+9)20 + w \cdot 20 = w(w+9) \qquad \textbf{Distributing and simplifying}$$

$$40w + 180 = w^2 + 9w$$

$$0 = w^2 - 31w - 180 \qquad \textbf{Getting 0 on one side}$$

$$0 = (w-36)(w+5) \qquad \textbf{Factoring}$$

$$w - 36 = 0 \quad or \quad w + 5 = 0 \qquad \textbf{Principle of zero products}$$

$$w = 36 \quad or \quad w = -5.$$

GRAPHICAL APPROACH

We graph $y_1 = 20/x + 20/(x + 9)$ and $y_2 = 1$ and find the points of intersection. If we use the window $[-8, 8, -10, 10]$, we see that there is a point of intersection near the x-value -5. It also appears that the graphs may intersect at a point off the screen to the right. Thus we adjust the window to $[-8, 50, 0, 5]$ and see that there is indeed another point of intersection.

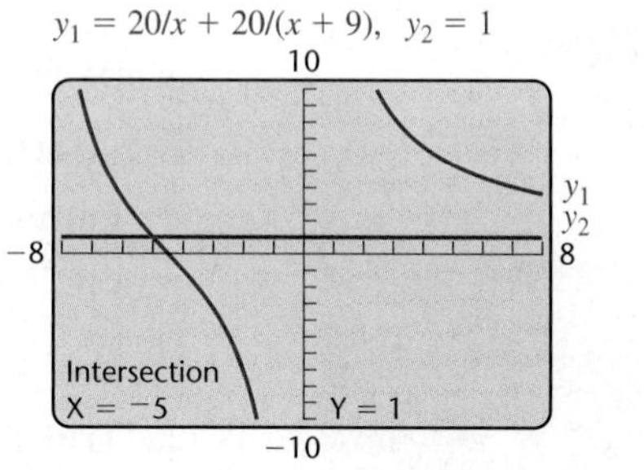

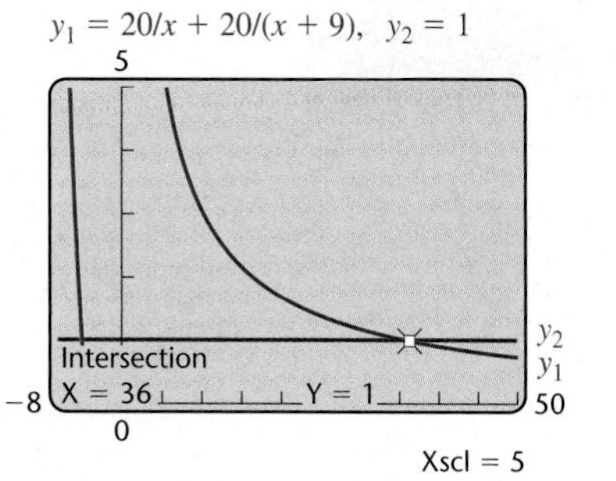

We have $x = 36$ or $x = -5$.

4. **Check.** Since negative time has no meaning in the problem, -5 is not a solution to the original problem. The number 36 checks since, if Wendy takes 36 hr alone and Pepe takes $36 + 9 = 45$ hr alone, in 20 hr they would have rebuilt

$$\frac{20}{36} + \frac{20}{45} = \frac{5}{9} + \frac{4}{9} = 1 \text{ complete engine.}$$

5. **State.** It would take Wendy 36 hr to rebuild an engine alone, and Pepe 45 hr.

The equations used in Examples 1 and 2 can be generalized as follows.

Modeling Work Problems

If

$a =$ the time needed for A to complete the work alone,

$b =$ the time needed for B to complete the work alone, and

$t =$ the time needed for A and B to complete the work together,

then

$$\frac{t}{a} + \frac{t}{b} = 1.$$

The following are equivalent equations that can also be used:

$$\frac{1}{a} \cdot t + \frac{1}{b} \cdot t = 1 \quad \text{and} \quad \frac{1}{a} + \frac{1}{b} = \frac{1}{t}.$$

Problems Involving Motion

Problems dealing with distance, rate (or speed), and time are called **motion problems.** To translate them, we use either the basic motion formula, $d = rt$, or the formulas $r = d/t$ or $t = d/r$, which can be derived from $d = rt$.

EXAMPLE 3 On her road bike, Ignacia bikes 15 km/h faster than Franklin does on his mountain bike. In the time it takes Ignacia to travel 80 km, Franklin travels 50 km. Find the speed of each bicyclist.

SOLUTION

1. **Familiarize.** Let's guess that Franklin is going 10 km/h. Ignacia would then be traveling $10 + 15$, or 25 km/h. At 25 km/h, she would travel 80 km in $\frac{80}{25} = 3.2$ hr. Going 10 km/h, Franklin would cover 50 km in $\frac{50}{10} = 5$ hr. Since $3.2 \neq 5$, our guess was wrong, but we can see that if $r =$ the rate, in kilometers per hour, of Franklin's bike, then the rate of Ignacia's bike $= r + 15$.

   Making a drawing and constructing a table can be helpful.

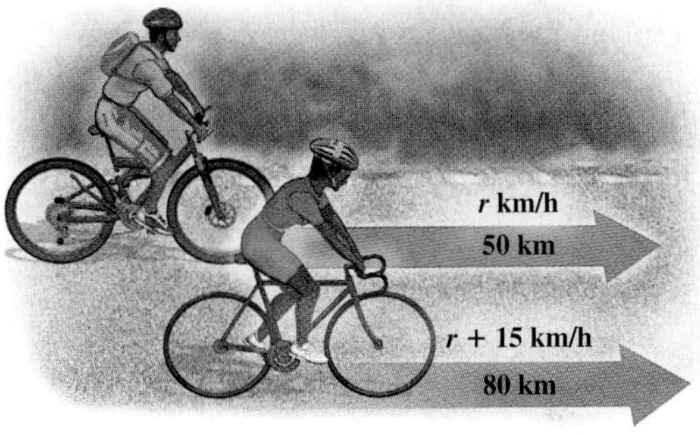

	Distance	Speed	Time
Franklin's Mountain Bike	50	r	t
Ignacia's Road Bike	80	$r + 15$	t

2. **Translate.** By looking at how we checked our guess, we see that in the **Time** column of the table, the t's can be replaced, using the formula $Time = Distance/Rate$, as follows.

	Distance	Speed	Time
Franklin's Mountain Bike	50	r	$50/r$
Ignacia's Road Bike	80	$r + 15$	$80/(r + 15)$

Since we are told that the times must be the same, we can write an equation:

$$\frac{50}{r} = \frac{80}{r + 15}.$$

3. **Carry out.** We solve the equation both algebraically and graphically.

ALGEBRAIC APPROACH

We have

$$\frac{50}{r} = \frac{80}{r + 15}$$

$$r(r + 15)\frac{50}{r} = r(r + 15)\frac{80}{r + 15} \quad \textbf{Multiplying by the LCD}$$

$$50r + 750 = 80r \quad \textbf{Simplifying}$$

$$750 = 30r$$

$$25 = r.$$

GRAPHICAL APPROACH

We graph the equations $y_1 = 50/x$ and $y_2 = 80/(x + 15)$, and look for a point of intersection. By adjusting the window dimensions and using the INTERSECT feature, we see that the graphs intersect at the point $(25, 2)$.

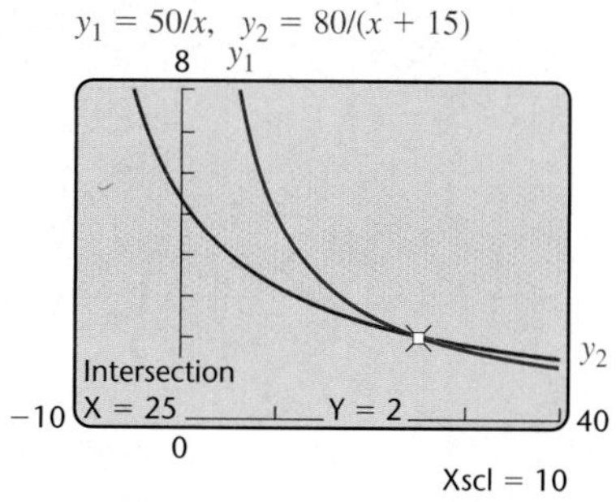

The solution is 25.

4. **Check.** If our answer checks, Franklin's mountain bike is going 25 km/h and Ignacia's road bike is going $25 + 15 = 40$ km/h.

   Traveling 80 km at 40 km/h, Ignacia is riding for $\frac{80}{40} = 2$ hr. Traveling 50 km at 25 km/h, Franklin is riding for $\frac{50}{25} = 2$ hr. Our answer checks since the two times are the same.

5. **State.** Ignacia's speed is 40 km/h, and Franklin's speed is 25 km/h.

In the following example, although the distance is the same in both directions, the key to the translation lies in an additional piece of given information.

EXAMPLE 4 A Hudson River tugboat goes 10 mph in still water. It travels 24 mi upstream and 24 mi back in a total time of 5 hr. What is the speed of the current? (*Sources*: Based on information from the Department of the Interior, U.S. Geological Survey, and *The Tugboat Captain*, Montgomery County Community College)

SOLUTION

1. **Familiarize.** Let's guess that the speed of the current is 4 mph. The tugboat would then be moving $10 - 4 = 6$ mph upstream and $10 + 4 = 14$ mph downstream. The tugboat would require $\frac{24}{6} = 4$ hr to travel 24 mi upstream and $\frac{24}{14} = 1\frac{5}{7}$ hr to travel 24 mi downstream. Since the total time, $4 + 1\frac{5}{7} = 5\frac{5}{7}$ hr, is not the 5 hr mentioned in the problem, we know that our guess is wrong.

   Suppose that the current's speed $= c$ mph. The tugboat's speed would then be $10 - c$ mph going upstream and $10 + c$ mph going downstream.

   A drawing and table can help display the information.

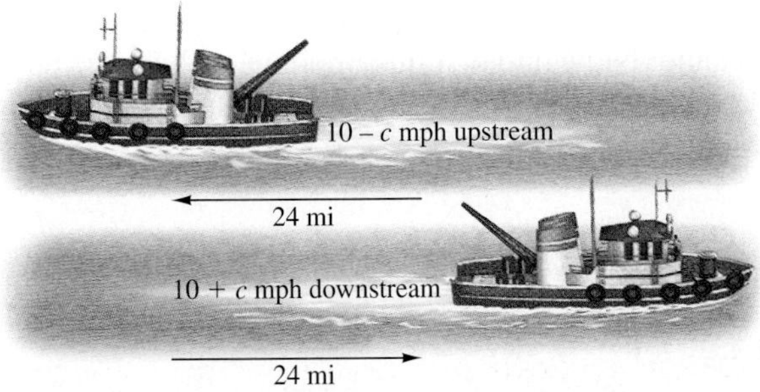

	Distance	Speed	Time
Upstream	24	$10 - c$	t_1
Downstream	24	$10 + c$	t_2

2. **Translate.** From examining our guess, we see that the time traveled can be represented using the formula *Time* = *Distance*/*Rate*:

	Distance	Speed	Time
Upstream	24	$10 - c$	$24/(10 - c)$
Downstream	24	$10 + c$	$24/(10 + c)$

Since the total time upstream and back is 5 hr, we use the last column of the table to form an equation:

$$\frac{24}{10 - c} + \frac{24}{10 + c} = 5.$$

3. **Carry out.** We solve the equation both algebraically and graphically.

ALGEBRAIC APPROACH

We have

$$\frac{24}{10-c}+\frac{24}{10+c}=5$$

$$(10-c)(10+c)\left[\frac{24}{10-c}+\frac{24}{10+c}\right]=(10-c)(10+c)5$$

Multiplying by the LCD

$$24(10+c)+24(10-c)=(100-c^2)5$$

$$480=500-5c^2 \quad \textbf{Simplifying}$$

$$5c^2-20=0$$

$$5(c^2-4)=0$$

$$5(c-2)(c+2)=0$$

$$c=2 \quad or \quad c=-2.$$

GRAPHICAL APPROACH

We graph

$y_1 = 24/(10 - x) + 24/(10 + x)$ and

$y_2 = 5$

and see that there are two points of intersection.

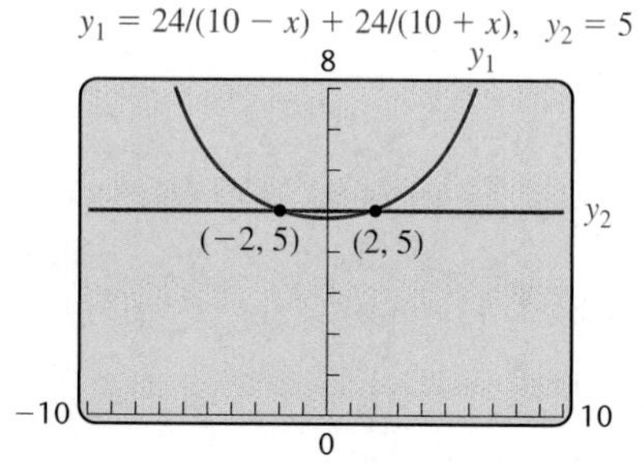

The solutions are -2 and 2.

4. **Check.** Since speed cannot be negative in this problem, -2 cannot be a solution. You should confirm that 2 checks in the original problem.
5. **State.** The speed of the current is 2 mph.

6.5 EXERCISE SET

FOR EXTRA HELP

MathXL MyMathLab InterAct Math AW Math Tutor Center Video Lectures on CD: Disc 3 Student's Solutions Manual

Solve.

1. The reciprocal of 3, plus the reciprocal of 6, is the reciprocal of what number?
2. The reciprocal of 5, plus the reciprocal of 7, is the reciprocal of what number?
3. The sum of a number and 6 times its reciprocal is -5. Find the number.
4. The sum of a number and 21 times its reciprocal is -10. Find the number.
5. The reciprocal of the product of two consecutive integers is $\frac{1}{42}$. Find the two integers.
6. The reciprocal of the product of two consecutive integers is $\frac{1}{72}$. Find the two integers.
7. *Photo Processing.* Nadine can develop a day's accumulation of film orders in 5 hr. Willy, a new employee, needs 9 hr to complete the same job. Working together, how long will it take them to do the job?
8. *Home Restoration.* Cedric can refinish the floor of an apartment in 8 hr. Carolyn can refinish the floor in 6 hr. How long will it take them, working together, to refinish the floor?

9. *Filling a Pool.* The San Paulo community swimming pool can be filled in 12 hr if water enters through a pipe alone or in 30 hr if water enters through a hose alone. If water is entering through both the pipe and the hose, how long will it take to fill the pool?

10. *Filling a Tank.* A community water tank can be filled in 18 hr by the town office well alone and in 22 hr by the high school well alone. How long will it take to fill the tank if both wells are working?

11. *Pumping Water.* A $\frac{1}{3}$ HP Craftsman Thermoplastic sump pump can remove water from Helen's flooded basement in 44 min. The $\frac{1}{2}$ HP Simer Thermoplastic sump pump can complete the same job in 36 min. (*Sources*: Based on data from Sears, Northern Tool & Electric Co., and Home Depot) How long would it take the two pumps together to pump out the basement?

12. *Hotel Management.* The Honeywell HQ17 air cleaner can clean the air in a 12-ft by 14-ft conference room in 10 min. The HQ174 can clean the air in a room of the same size in 6 min. How long would it take the two machines together to clean the air in such a room?

13. *Photocopiers.* The HP Officejet 6110 takes twice the time required by the Canon Imageclass D680 to photocopy brochures for the New Bretton Arts Council year-end concert (*Source*: Manufacturers' marketing brochures). If working together the two machines can complete the job in 24 min, how long would it take each machine, working alone, to copy the brochures?

14. *Computer Printers.* The HP Laser Jet 9000 works twice as fast as the Laser Jet 2300 (*Source*: www.hewlettpackard.com). If the machines work together, a university can produce all its staff manuals in 15 hr. Find the time it would take each machine, working alone, to complete the same job.

15. *Hotel Management.* The Blueair 402 can purify the air in a conference hall in 10 fewer minutes than it takes the Panasonic F-P20HU1 to do the same job (*Source*: Based on information from manufacturers' websites). Together the two machines can purify the air in the conference hall in $\frac{120}{7}$, or $17\frac{1}{7}$ min. How long would it take each machine, working alone, to purify the air in the room?

16. *Cutting Firewood.* Jake can cut and split a cord of firewood in 6 fewer hr than Skyler can. When they work together, it takes them 4 hr. How long would it take each of them to do the job alone?

17. *Forest Fires.* The Erickson Air-Crane helicopter can scoop water and douse a certain forest fire four times as fast as an S-58T helicopter (*Sources*: Based on information from www.emergency.com and www.arishelicopters.com). Working together, the two helicopters can douse the fire in 8 hr. How long would it take each helicopter, working alone, to douse the fire?

18. *Waxing a Car.* Rosita can wax her car in 2 hr. When she works together with Helga, they can wax the car in 45 min. How long would it take Helga, working by herself, to wax the car?

19. *Newspaper Delivery.* Zsuzanna can deliver papers three times as fast as Stan can. If they work together, it takes them 1 hr. How long would it take each to deliver the papers alone?

20. *Sorting Recyclables.* Together, it takes John and Deb 2 hr 55 min to sort recyclables. Alone, John would require 2 more hr than Deb. How long would it take Deb to do the job alone? (*Hint*: Convert minutes to hours or hours to minutes.)

21. *Paving.* Together, Larry and Mo require 4 hr 48 min to pave a driveway. Alone, Larry would require 4 hr more than Mo. How long would it take Mo to do the job alone? (*Hint*: Convert minutes to hours.)

22. *Painting.* Sara takes 3 hr longer to paint a floor than it takes Kate. When they work together, it takes them 2 hr. How long would each take to do the job alone?

23. *Kayaking.* The speed of the current in Catamount Creek is 3 mph. Zeno can kayak 4 mi upstream in the same time it takes him to kayak 10 mi downstream. What is the speed of Zeno's kayak in still water?

24. *Boating.* The current in the Lazy River moves at a rate of 4 mph. Monica's dinghy motors 6 mi upstream in the same time it takes to motor 12 mi downstream. What is the speed of the dinghy in still water?

25. *Moving Sidewalks.* Newark Airport's moving sidewalk moves at a speed of 1.7 ft/sec. Walking on the moving sidewalk, Benny can travel 120 ft forward in the same time it takes to travel 52 ft in the opposite direction. How fast would Benny be walking on a nonmoving sidewalk?

26. *Moving Sidewalks.* The moving sidewalk at O'Hare Airport in Chicago moves 1.8 ft/sec. Walking on the moving sidewalk, Camille travels 105 ft forward in the time it takes to travel 51 ft in the opposite direction. How fast would Camille be walking on a nonmoving sidewalk?

27. *Train Speed.* The speed of the A&M freight train is 14 mph less than the speed of the A&M passenger train. The passenger train travels 400 mi in the same time that the freight train travels 330 mi. Find the speed of each train.

28. *Walking.* Rosanna walks 2 mph slower than Simone. In the time it takes Simone to walk 8 mi, Rosanna walks 5 mi. Find the speed of each person.

Aha! **29.** *Bus Travel.* A local bus travels 7 mph slower than the express. The express travels 45 mi in the time it takes the local to travel 38 mi. Find the speed of each bus.

30. *Train Speed.* The A train goes 12 mph slower than the E train. The A train travels 230 mi in the same time that the E train travels 290 mi. Find the speed of each train.

31. *Boating.* Laverne's Mercruiser travels 15 km/h in still water. She motors 140 km downstream in the same time it takes to travel 35 km upstream. What is the speed of the river?

32. *Boating.* Audrey's paddleboat travels 2 km/h in still water. The boat is paddled 4 km downstream in the same time it takes to go 1 km upstream. What is the speed of the river?

33. *Shipping.* A barge moves 7 km/h in still water. It travels 45 km upriver and 45 km downriver in a total time of 14 hr. What is the speed of the current?

34. *Moped Speed.* Jaime's moped travels 8 km/h faster than Mara's. Jaime travels 69 km in the same time that Mara travels 45 km. Find the speed of each person's moped.

35. *Aviation.* A Citation II Jet travels 350 mph in still air and flies 487.5 mi into the wind and 487.5 mi with the wind in a total of 2.8 hr (*Source*: Eastern Air Charter). Find the wind speed.

36. *Canoeing.* Al paddles 55 m per minute in still water. He paddles 150 m upstream and 150 m downstream in a total time of 5.5 min. What is the speed of the current?

37. *Train Travel.* A freight train covered 120 mi at a certain speed. Had the train been able to travel 10 mph faster, the trip would have been 2 hr shorter. How fast did the train go?

38. *Boating.* Julia's Boston Whaler cruised 45 mi upstream and 45 mi back in a total of 8 hr. The speed of the river is 3 mph. Find the speed of the boat in still water.

TW **39.** Two steamrollers are paving a parking lot. Working together, will the two steamrollers take less than half as long as the slower steamroller would working alone? Why or why not?

TW **40.** Two fuel lines are filling a freighter with oil. Will the faster fuel line take more or less than twice as long to fill the freighter by itself? Why?

Skill Maintenance

Simplify.

41. $\dfrac{35a^6b^8}{7a^2b^2}$ [1.4]

42. $\dfrac{20x^9y^6}{4x^3y^2}$ [1.4]

43. $\dfrac{36s^{15}t^{10}}{9s^5t^2}$ [1.4]

44. $6x^4 - 3x^2 + 9x - (8x^4 + 4x^2 - 2x)$ [5.1]

45. $2(x^3 + 4x^2 - 5x + 7) - 5(2x^3 - 4x^2 + 3x - 1)$ [5.1]

46. $9x^4 + 7x^3 + x^2 - 8 - (-2x^4 + 3x^2 + 4x + 2)$ [5.1]

Synthesis

TW **47.** Write a work problem for a classmate to solve. Devise the problem so that the solution is "Liane and Michele will take 4 hr to complete the job, working together."

TW **48.** Write a work problem for a classmate to solve. Devise the problem so that the solution is "Jen takes 5 hr and Pablo takes 6 hr to complete the job alone."

49. *Filling a Bog.* The Norwich cranberry bog can be filled in 9 hr and drained in 11 hr. How long will it take to fill the bog if the drainage gate is left open?

50. *Filling a Tub.* Justine's hot tub can be filled in 10 min and drained in 8 min. How long will it take to empty a full tub if the water is left on?

51. Refer to Exercise 24. How long will it take Monica to motor 3 mi downstream?

52. Refer to Exercise 23. How long will it take Zeno to kayak 5 mi downstream?

53. *Escalators.* Together, a 100-cm wide escalator and a 60-cm wide escalator can empty a 1575-person auditorium in 14 min. The wider escalator moves twice as many people as the narrower one. (*Source*: *McGraw-Hill Encyclopedia of Science and Technology*) How many people per hour does the 60-cm wide escalator move?

54. *Aviation.* A Coast Guard plane has enough fuel to fly for 6 hr, and its speed in still air is 240 mph. The plane departs with a 40-mph tailwind and returns to the same airport flying into the same wind. How far from the airport can the plane travel under these conditions? (Assume that the plane can use all its fuel.)

55. *Boating.* Shoreline Travel operates a 3-hr paddle-boat cruise on the Missouri River. If the speed of the boat in still water is 12 mph, how far upriver can the pilot travel against a 5-mph current before it is time to turn around?

56. *Travel by Car.* Melissa drives to work at 50 mph and arrives 1 min late. She drives to work at 60 mph and arrives 5 min early. How far does Melissa live from work?

57. *Photocopying.* The printer in an admissions office can print a 500-page document in 50 min, while the printer in the business office can print the same document in 40 min. If the two printers work together to print the document, with the faster machine starting on page 1 and the slower machine working backwards from page 500, at what page will the two machines meet to complete the job?

58. At what time after 4:00 will the minute hand and the hour hand of a clock first be in the same position?

59. At what time after 10:30 will the hands of a clock first be perpendicular?

Average speed is defined as total distance divided by total time.

60. Lenore drove 200 km. For the first 100 km of the trip, she drove at a speed of 40 km/h. For the second half of the trip, she traveled at a speed of 60 km/h. What was the average speed of the entire trip? (It was *not* 50 km/h.)

61. For the first 50 mi of a 100-mi trip, Chip drove 40 mph. What speed would he have to travel for the last half of the trip so that the average speed for the entire trip would be 45 mph?

6.6 Division of Polynomials

Dividing by a Monomial ■ Dividing by a Polynomial

Study Tip

Get a Terrific Seat

Your classtime is very precious. To make the most of that time, try to seat yourself at or near the front of the classroom. This makes it easier to participate and helps minimize distractions.

A rational expression indicates division. Division of polynomials, like division of real numbers, relies on our multiplication and subtraction skills.

Dividing by a Monomial

To divide a monomial by a monomial, we divide coefficients and, if the bases are the same, subtract exponents (see Section 1.4).

$$\frac{45x^{10}}{3x^4} = 15x^{10-4} = 15x^6, \qquad \frac{8a^2b^5}{-2ab^2} = -4a^{2-1}b^{5-2} = -4ab^3.$$

(Dividend: $45x^{10}$; Divisor: $3x^4$; Quotient: $15x^6$ and $-4ab^3$)

To divide a polynomial by a monomial, we regard the division as a sum of quotients of monomials. This uses the fact that since

$$\frac{A}{C} + \frac{B}{C} = \frac{A+B}{C}, \quad \text{we know that} \quad \frac{A+B}{C} = \frac{A}{C} + \frac{B}{C}.$$

EXAMPLE 1 Divide $12x^3 + 8x^2 + x + 4$ by $4x$.

SOLUTION

$$(12x^3 + 8x^2 + x + 4) \div (4x) = \frac{12x^3 + 8x^2 + x + 4}{4x} \quad \text{Writing a rational expression}$$

$$= \frac{12x^3}{4x} + \frac{8x^2}{4x} + \frac{x}{4x} + \frac{4}{4x} \quad \text{Writing as a sum of quotients}$$

$$= 3x^2 + 2x + \frac{1}{4} + \frac{1}{x} \quad \text{Performing the four indicated divisions}$$

EXAMPLE 2 Divide: $(8x^4y^5 - 3x^3y^4 + 5x^2y^3) \div (-x^2y^3)$.

SOLUTION

$$\frac{8x^4y^5 - 3x^3y^4 + 5x^2y^3}{-x^2y^3} = \frac{8x^4y^5}{-x^2y^3} - \frac{3x^3y^4}{-x^2y^3} + \frac{5x^2y^3}{-x^2y^3} \quad \text{Try to perform this step mentally.}$$

$$= -8x^2y^2 + 3xy - 5$$

Division by a Monomial

To divide a polynomial by a monomial, divide each term of the polynomial by the monomial.

Dividing by a Polynomial

When the divisor has more than one term, we use a procedure very similar to long division in arithmetic.

EXAMPLE 3 Divide $2x^2 - 7x - 15$ by $x - 5$.

SOLUTION We have

$$\begin{array}{r} 2x \\ x-5\overline{)2x^2 - 7x - 15} \\ -(2x^2 - 10x) \\ \hline 3x - 15 \end{array}$$

Divide $2x^2$ by x: $2x^2/x = 2x$.

Multiply $x - 5$ by $2x$.

Subtract by mentally changing signs and adding: $-7x + 10x = 3x$.

We next divide the leading term of this remainder, $3x$, by the leading term of the divisor, x.

$$\begin{array}{r} 2x + 3 \\ x-5\overline{)2x^2 - 7x - 15} \\ 2x^2 - 10x \\ \hline 3x - 15 \\ -(3x - 15) \\ \hline 0 \end{array}$$

Divide $3x$ by x: $3x/x = 3$.

Multiply $x - 5$ by 3.

Subtract. Our remainder is now 0.

Check: $(x - 5)(2x + 3) = 2x^2 - 7x - 15$. The answer checks.

The quotient is $2x + 3$.

To understand why we perform long division as we do, note that Example 3 amounts to "filling in" an unknown polynomial:

$$(x - 5)(\ ?\ ) = 2x^2 - 7x - 15.$$

We see that $2x$ must be in the unknown polynomial if we are to get the first term, $2x^2$, from the multiplication. To see what else is needed, note that

$$(x - 5)(2x \quad) = 2x^2 - 10x \neq 2x^2 - 7x - 15.$$

The $2x$ can be regarded as a (poor) approximation of the quotient that we are seeking. To see how far off the approximation is, we subtract:

$$\begin{array}{r} 2x^2 - 7x - 15 \\ -(2x^2 - 10x) \\ \hline 3x - 15 \end{array}$$

Note where this appeared in the long division above.

This is the first remainder.

To get the needed terms, $3x - 15$, we need another term in the unknown polynomial. We use 3 because $(x - 5) \cdot 3$ is $3x - 15$:

$$(x - 5)(2x + 3) = 2x^2 - 10x + 3x - 15 = 2x^2 - 7x - 15.$$

Now when we subtract the product $(x - 5)(2x + 3)$ from $2x^2 - 7x - 15$, the remainder is 0.

If a nonzero remainder occurs, when do we stop dividing? We continue until the degree of the remainder is less than the degree of the divisor.

EXAMPLE 4 Divide $x^2 + 5x + 8$ by $x + 3$.

SOLUTION We have

$$\begin{array}{r l}
x & \\
x + 3\overline{)x^2 + 5x + 8} & \text{Divide the first term of the dividend by the first term of the divisor: } x^2/x = x. \\
\underline{x^2 + 3x} \leftarrow & \text{Multiply } x \text{ above by } x + 3. \\
2x + 8 & \text{Subtract: } 5x - 3x = 2x.
\end{array}$$

Note that

$$x^2 + 5x + 8 - (x^2 + 3x) = x^2 + 5x + 8 - x^2 - 3x.$$

Remember: To subtract, add the opposite (change the sign of every term, then add).

We now focus on the current remainder, $2x + 8$, and repeat the process:

$$\begin{array}{r l}
x + 2 \leftarrow & \text{Divide the first term by the first term: } 2x/x = 2. \\
x + 3\overline{)x^2 + 5x + 8} & \\
\underline{x^2 + 3x} & \\
2x + 8 \leftarrow & 2x + 8 \text{ is the first remainder.} \\
\underline{2x + 6} \leftarrow & \text{Multiply 2 by } x + 3. \\
2 \leftarrow & \text{Subtract: } (2x + 8) - (2x + 6).
\end{array}$$

The quotient is $x + 2$, with remainder 2. Note that the degree of the remainder is 0 and the degree of the divisor, $x + 3$, is 1. Since $0 < 1$, the process stops.

Check: $(x + 3)(x + 2) + 2 = x^2 + 5x + 6 + 2$ **Add the remainder to the product.**

$$= x^2 + 5x + 8$$

We write our answer as $x + 2$, R 2, or as

$$\text{Quotient} + \frac{\text{Remainder}}{\text{Divisor}}$$

$$x + 2 + \frac{2}{x + 3}.$$

This is how answers are listed at the back of the book.

Student Notes

See if, after studying Examples 3 and 4, you can explain why long division of numbers is performed as it is.

The last answer in Example 4 can also be checked by multiplying:

$$(x + 3)\left[(x + 2) + \frac{2}{x + 3}\right] = (x + 3)(x + 2) + (x + 3)\frac{2}{x + 3}$$

Using the distributive law

$$= x^2 + 5x + 6 + 2$$

$$= x^2 + 5x + 8.$$ **This was the dividend in Example 4.**

You may have noticed that it is helpful to have all polynomials written in descending order.

Tips for Dividing Polynomials

1. Arrange polynomials in descending order.
2. If there are missing terms in the dividend, either write them with 0 coefficients or leave space for them.
3. Continue the long division process until the degree of the remainder is less than the degree of the divisor.

EXAMPLE 5 Divide: $(9a^2 + a^3 - 5) \div (a^2 - 1)$.

SOLUTION We rewrite the problem in descending order:

$$(a^3 + 9a^2 - 5) \div (a^2 - 1).$$

Thus,

$$\begin{array}{r@{}l}
 & a \quad + 9 \\
\cline{2-2}
a^2 - 1 & \big) a^3 + 9a^2 + 0a - 5 \\
 & a^3 \qquad\quad - a \\
\cline{2-2}
 & \qquad 9a^2 + a - 5 \\
 & \qquad 9a^2 \qquad\;\; - 9 \\
\cline{2-2}
 & \qquad\qquad\quad a + 4
\end{array}$$

When there is a missing term in the dividend, we can write it in, as shown here, or leave space, as in Example 6 below.

Subtracting: $0a - (-a) = a$.

The degree of the remainder is less than the degree of the divisor, so we are finished.

The answer is $a + 9 + \dfrac{a + 4}{a^2 - 1}$.

EXAMPLE 6 Let $f(x) = 125x^3 - 8$ and $g(x) = 5x - 2$. If $F(x) = (f/g)(x)$, find a simplified expression for $F(x)$ and list all restrictions on the domain.

SOLUTION Recall that $(f/g)(x) = f(x)/g(x)$. Thus,

$$F(x) = \frac{125x^3 - 8}{5x - 2}$$

and

$$\begin{array}{r@{}l}
 & 25x^2 + 10x + 4 \\
\cline{2-2}
5x - 2 & \big) 125x^3 \qquad\qquad\qquad - 8 \\
 & 125x^3 - 50x^2 \\
\cline{2-2}
 & \qquad\quad 50x^2 \qquad\qquad - 8 \\
 & \qquad\quad 50x^2 - 20x \\
\cline{2-2}
 & \qquad\qquad\qquad 20x - 8 \\
 & \qquad\qquad\qquad 20x - 8 \\
\cline{2-2}
 & \qquad\qquad\qquad\qquad\; 0
\end{array}$$

Leaving space for the missing terms

Subtracting: $125x^3 - (125x^3 - 50x^2) = 50x^2$

Subtracting

Note that, because $F(x) = f(x)/g(x)$, $g(x)$ cannot be 0. Since $g(x)$ is 0 for $x = \frac{2}{5}$ (check this), we have

$$F(x) = 25x^2 + 10x + 4, \quad \text{provided } x \neq \tfrac{2}{5}.$$

6.6 EXERCISE SET

FOR EXTRA HELP

MathXL | MyMathLab | InterAct Math | AW Math Tutor Center | Video Lectures on CD: Disc 3 | Student's Solutions Manual

Concept Reinforcement *Classify each statement as either true or false.*

1. To divide a polynomial by a monomial, we divide each term of the polynomial by the monomial.
2. To divide a monomial by a monomial, we can subtract coefficients and divide exponents.
3. When the divisor has more than one term, polynomial division is very similar to long division in arithmetic.
4. In long division, we subtract to find how far off our approximation of the quotient is.
5. When we are dividing by $2x + 3$, the remainder in our answer may include a variable of x.
6. When we are dividing by $x^2 - 5$, the remainder in our answer may include a variable.

Divide and check.

7. $\dfrac{32x^6 + 18x^5 - 27x^2}{6x^2}$
8. $\dfrac{30y^8 - 15y^6 + 40y^4}{5y^4}$
9. $\dfrac{21a^3 + 7a^2 - 3a - 14}{7a}$
10. $\dfrac{-25x^3 + 20x^2 - 3x + 7}{5x}$
11. $\dfrac{18t^4 - 15t + 21}{-3t}$
12. $\dfrac{14a^5 - 21a + 7}{-7a}$
13. $\dfrac{16y^4z^2 - 8y^6z^4 + 12y^8z^3}{-4y^4z}$
14. $\dfrac{6p^2q^2 - 9p^2q + 12pq^2}{-3pq}$
15. $(16y^3 - 9y^2 - 8y) \div (2y^2)$
16. $(6a^4 + 9a^2 - 8) \div (2a)$
17. $(15x^7 - 21x^4 - 3x^2) \div (-3x^2)$
18. $(36y^6 - 18y^4 - 12y^2) \div (-6y)$
19. $(a^2b - a^3b^3 - a^5b^5) \div (a^2b)$
20. $(x^3y^2 - x^3y^3 - x^4y^2) \div (x^2y^2)$
21. Aha! $(x^2 + 10x + 21) \div (x + 7)$
22. $(y^2 - 8y + 16) \div (y - 4)$
23. $(a^2 - 8a - 16) \div (a + 4)$
24. $(y^2 - 10y - 25) \div (y - 5)$
25. $(x^2 - 9x + 21) \div (x - 4)$
26. $(x^2 - 11x + 23) \div (x - 6)$
27. $(y^2 - 25) \div (y + 5)$
28. $(a^2 - 81) \div (a - 9)$
29. $(y^3 - 4y^2 + 3y - 6) \div (y - 2)$
30. $(x^3 - 5x^2 + 4x - 7) \div (x - 3)$
31. $(2x^3 + 3x^2 - x - 3) \div (x + 2)$
32. $(3x^3 - 5x^2 - 3x - 2) \div (x - 2)$
33. $(a^3 - a + 10) \div (a - 4)$
34. $(x^3 - x + 6) \div (x + 2)$
35. $(10y^3 + 6y^2 - 9y + 10) \div (5y - 2)$
36. $(6x^3 - 11x^2 + 11x - 2) \div (2x - 3)$
37. $(2x^4 - x^3 - 5x^2 + x - 6) \div (x^2 + 2)$
38. $(3x^4 + 2x^3 - 11x^2 - 2x + 5) \div (x^2 - 2)$

For Exercises 39–46, $f(x)$ and $g(x)$ are as given. Find a simplified expression for $F(x)$ if $F(x) = (f/g)(x)$. (See Example 6.) Be sure to list all restrictions on the domain of $F(x)$.

39. $f(x) = 8x^3 - 27,\ g(x) = 2x - 3$

40. $f(x) = 64x^3 + 8,\ g(x) = 4x + 2$

41. $f(x) = 6x^2 - 11x - 10,\ g(x) = 3x + 2$

42. $f(x) = 8x^2 - 22x - 21,\ g(x) = 2x - 7$

43. $f(x) = x^4 - 24x^2 - 25,\ g(x) = x^2 - 25$

44. $f(x) = x^4 - 3x^2 - 54,\ g(x) = x^2 - 9$

45. $f(x) = 8x^2 - 3x^4 - 2x^3 + 2x^5 - 5,\ g(x) = x^2 - 1$

46. $f(x) = 4x - x^3 - 10x^2 + 3x^4 - 8,\ g(x) = x^2 - 4$

TW 47. Explain how factoring could be used to solve Example 6.

TW 48. Explain how to construct a polynomial of degree 4 that has a remainder of 3 when divided by $x + 1$.

Focused Review

Perform the indicated operations and simplify.

49. $(5x^3 + x^2 - x - 3) + (2x^2 + 5x - 8)$ [5.1]

50. $(5x^3 + x^2 - x - 3) - (2x^2 + 5x - 8)$ [5.1]

51. $(2x - 7)(3x + 5)$ [5.2]

52. $\left(\frac{1}{3}x + 10\right)\left(\frac{1}{3}x - 10\right)$ [5.2]

53. $(4x - 8)^2$ [5.2]

54. $(24x^4 - 8x^3 + 4x^2) \div (-4x^2)$ [6.6]

55. $(x^2 + 3x + 1) \div (x + 1)$ [6.6]

56. $(x + 1)(x - 2) - (x - 3)(x + 5)$ [5.1], [5.2]

Synthesis

TW 57. Explain how to construct a polynomial of degree 4 that has a remainder of 2 when divided by $x + c$.

TW 58. Do addition, subtraction, and multiplication of polynomials always result in a polynomial? Does division? Why or why not?

Divide.

59. $(4a^3b + 5a^2b^2 + a^4 + 2ab^3) \div (a^2 + 2b^2 + 3ab)$

60. $(x^4 - x^3y + x^2y^2 + 2x^2y - 2xy^2 + 2y^3) \div (x^2 - xy + y^2)$

61. $(a^7 + b^7) \div (a + b)$

62. Find k such that when $x^3 - kx^2 + 3x + 7k$ is divided by $x + 2$, the remainder is 0.

63. When $x^2 - 3x + 2k$ is divided by $x + 2$, the remainder is 7. Find k.

64. Let

$$f(x) = \frac{3x + 7}{x + 2}.$$

a) Use division to find an expression equivalent to $f(x)$. Then graph f.

b) On the same set of axes, sketch both $g(x) = 1/(x + 2)$ and $h(x) = 1/x$.

c) How do the graphs of f, g, and h compare?

65. Use a graphing calculator to check Example 3 by setting $y_1 = (2x^2 - 7x - 15)/(x - 5)$ and $y_2 = 2x + 3$. Then press either TRACE (after selecting the ZOOM ZDECIMAL option) or TABLE (with TblMin = 0 and ΔTbl = 1) to show that $y_1 \neq y_2$ for $x = 5$.

TW 66. Use a graphing calculator to check Exercise 23. Let $y_1 = (x^2 - 8x - 16)/(x + 4)$ and $y_2 = x - 12 + 32/(x + 4)$ and compare graphs or tables. Then graph $y_3 = x - 12$ using the same window. Note that $x - 12$ is the linear part of the quotient. How is the graph of y_3 related to that of y_1?

6.7 Synthetic Division

Streamlining Long Division ■ The Remainder Theorem

Streamlining Long Division

To divide a polynomial by a binomial of the type $x - a$, we can streamline the usual procedure to develop a process called *synthetic division*.

Compare the following. In each stage, we attempt to write a bit less than in the previous stage, while retaining enough essentials to solve the problem. At the end, we will return to the usual polynomial notation.

Stage 1

When a polynomial is written in descending order, the coefficients provide the essential information:

$$
\begin{array}{r|l}
 & 4x^2 + 5x + 11 \\ \hline
x - 2 & 4x^3 - 3x^2 + x + 7 \\
 & \underline{4x^3 - 8x^2} \\
 & \quad 5x^2 + x \\
 & \quad \underline{5x^2 - 10x} \\
 & \qquad 11x + 7 \\
 & \qquad \underline{11x - 22} \\
 & \qquad\qquad 29
\end{array}
\qquad\qquad
\begin{array}{r|l}
 & 4 + 5 + 11 \\ \hline
1 - 2 & 4 - 3 + 1 + 7 \\
 & \underline{4 - 8} \\
 & \quad 5 + 1 \\
 & \quad \underline{5 - 10} \\
 & \qquad 11 + 7 \\
 & \qquad \underline{11 - 22} \\
 & \qquad\qquad 29
\end{array}
$$

Because the leading coefficient in the divisor is 1, each time we multiply the divisor by a term in the answer, the leading coefficient of that product duplicates a coefficient in the answer. In the next stage, we don't bother to duplicate these numbers. We also show where -2 is used and drop the 1 from the divisor.

Stage 2

$$
\begin{array}{r|l}
 & 4x^2 + 5x + 11 \\ \hline
x - 2 & 4x^3 - 3x^2 + x + 7 \\
 & \underline{4x^3 - 8x^2} \\
 & \quad 5x^2 + x \\
 & \quad \underline{5x^2 - 10x} \\
 & \qquad 11x + 7 \\
 & \qquad \underline{11x - 22} \\
 & \qquad\qquad 29
\end{array}
\qquad\qquad
\begin{array}{r|ll}
 & 4 + 5 + 11 & \\ \hline
-2 & 4 - 3 + 1 + 7 & \\
 & \underline{\;\;- 8} & \leftarrow \textbf{Multiply: } -2 \cdot 4 = -8. \\
 & \quad 5 + 1 & \leftarrow \textbf{Subtract: } -3 - (-8) = 5. \\
 & \quad \underline{\;\;- 10} & \leftarrow \textbf{Multiply: } -2 \cdot 5 = -10. \\
 & \qquad 11 + 7 & \leftarrow \textbf{Subtract: } 1 - (-10) = 11. \\
 & \qquad \underline{\;\;- 22} & \leftarrow \textbf{Multiply: } -2 \cdot 11 = -22. \\
 & \qquad\qquad 29 & \leftarrow \textbf{Subtract: } 7 - (-22) = 29.
\end{array}
$$

To simplify further, we now reverse the sign of the -2 in the divisor and, in exchange, *add* at each step in the long division.

Study Tip

If a Question Stumps You

Unless you know the material "cold," don't be surprised if a quiz or test includes a question that you feel unsure about. Should this happen, do not get rattled—simply skip the question and continue with the quiz or test, returning to the trouble spot after the other questions have been answered.

Stage 3

$$\begin{array}{r}
4x^2 + 5x + 11 \\
x - 2\overline{)4x^3 - 3x^2 + x + 7} \\
\underline{4x^3 - 8x^2} \quad\quad\quad\quad \\
5x^2 + x \quad\quad \\
\underline{5x^2 - 10x} \quad\quad \\
11x + 7 \\
\underline{11x - 22} \\
29
\end{array}$$

$$\begin{array}{l}
\downarrow 4 + 5 + 11 \\
2\overline{)4 - 3 + 1 + 7}
\end{array}$$

$2\overline{)4 - 3 + 1 + 7}$	**Replace the −2 with 2.**
$\underline{8}$	**Multiply: 2 · 4 = 8.**
$5 + 1$	**Add: −3 + 8 = 5.**
$\underline{10}$	**Multiply: 2 · 5 = 10.**
$11 + 7$	**Add: 1 + 10 = 11.**
$\underline{22}$	**Multiply: 2 · 11 = 22.**
29	**Add: 7 + 22 = 29.**

The blue numbers can be eliminated if we look at the red numbers instead.

Stage 4

$$\begin{array}{r}
4x^2 + 5x + 11 \\
x - 2\overline{)4x^3 - 3x^2 + x + 7} \\
\underline{4x^3 - 8x^2} \quad\quad\quad\quad \\
5x^2 + x \quad\quad \\
\underline{5x^2 - 10x} \quad\quad \\
11x + 7 \\
\underline{11x - 22} \\
29
\end{array}
\qquad
\begin{array}{rrrr}
 & 4 & 5 & 11 \\
2\overline{)4} & -3 & 1 & 7 \\
 & \underline{8} & \underline{10} & \underline{22} \\
 & 5 & 11 & 29
\end{array}$$

Don't lose sight of how the products 8, 10, and 22 are found. Also, note that the 5 and 11 preceding the remainder 29 coincide with the 5 and 11 following the 4 on the top line. By writing a 4 to the left of 5 on the bottom line, we can eliminate the top line in stage 4 and read our answer from the bottom line. This final stage is commonly called **synthetic division.**

Stage 5

$$\begin{array}{rrrr}
 & 4 & 5 & 11 \\
2\overline{)4} & -3 & 1 & 7 \\
 & \underline{8} & \underline{10} & \underline{22} \\
 & 5 & 11 & 29
\end{array}
\qquad
\begin{array}{r|rrrr}
2 & 4 & -3 & 1 & 7 \\
 & & 8 & 10 & 22 \\
\hline
 & 4 & 5 & 11 & 29
\end{array}$$

- 29 ← **This is the remainder.**
- 11 ← **This is the zero-degree coefficient.**
- 5 ← **This is the first-degree coefficient.**
- 4 ← **This is the second-degree coefficient.**

The quotient is $4x^2 + 5x + 11$. The remainder is 29.

Student Notes

You will not need to write out all five stages when performing synthetic division on your own. We show the steps to help you understand the reasoning behind the method.

> Remember that in order for this method to work, the divisor must be of the form $x - a$, that is, a variable minus a constant. Both the coefficient and the exponent of the variable must be 1.

EXAMPLE 1 Use synthetic division to divide:

$$(x^3 + 6x^2 - x - 30) \div (x - 2).$$

SOLUTION

$$\begin{array}{r|rrrr} 2 & 1 & 6 & -1 & -30 \\ & & & & \\ \hline & 1 & & & \end{array}$$

Write the 2 of $x - 2$ and the coefficients of the dividend. Bring down the first coefficient.

$$\begin{array}{r|rrrr} 2 & 1 & 6 & -1 & -30 \\ & & 2 & & \\ \hline & 1 & 8 & & \end{array}$$

Multiply 1 by 2 to get 2.
Add 6 and 2.

$$\begin{array}{r|rrrr} 2 & 1 & 6 & -1 & -30 \\ & & 2 & 16 & \\ \hline & 1 & 8 & 15 & \end{array}$$

Multiply 8 by 2.
Add −1 and 16.

$$\begin{array}{r|rrrr} 2 & 1 & 6 & -1 & -30 \\ & & 2 & 16 & 30 \\ \hline & 1 & 8 & 15 & 0 \end{array}$$

Multiply 15 by 2 and add.

Since the remainder is 0, we have

$$(x^3 + 6x^2 - x - 30) \div (x - 2) = x^2 + 8x + 15.$$

We can check this with a table, letting

$$y_1 = (x^3 + 6x^2 - x - 30) \div (x - 2) \quad \text{and} \quad y_2 = x^2 + 8x + 15.$$

The values of both expressions are the same except for $x = 2$, when y_1 is not defined.

X	Y1	Y2
−2	3	3
−1	8	8
0	15	15
1	24	24
2	ERROR	35
3	48	48
4	63	63

X = −2

The answer is $x^2 + 8x + 15$ with R 0, or just $x^2 + 8x + 15$.

EXAMPLE 2 Use synthetic division to divide.

a) $(2x^3 + 7x^2 - 5) \div (x + 3)$

b) $(10x^2 - 13x + 3x^3 - 20) \div (4 + x)$

SOLUTION

a) $(2x^3 + 7x^2 - 5) \div (x + 3)$

The dividend has no x-term, so we need to write 0 for its coefficient of x. Note that $x + 3 = x - (-3)$, so we write -3 inside the $\underline{\quad}|$.

$$\begin{array}{r|rrrr} -3 & 2 & 7 & 0 & -5 \\ & & -6 & -3 & 9 \\ \hline & 2 & 1 & -3 & 4 \end{array}$$

The answer is $2x^2 + x - 3$, with R 4, or $2x^2 + x - 3 + \dfrac{4}{x + 3}$.

b) We first rewrite $(10x^2 - 13x + 3x^3 - 20) \div (4 + x)$ in descending order:

$$(3x^3 + 10x^2 - 13x - 20) \div (x + 4).$$

Next, we use synthetic division. Note that $x + 4 = x - (-4)$.

$$\begin{array}{r|rrrr} -4 & 3 & 10 & -13 & -20 \\ & & -12 & 8 & 20 \\ \hline & 3 & -2 & -5 & 0 \end{array}$$

The answer is $3x^2 - 2x - 5$.

The Remainder Theorem

When a polynomial function $f(x)$ is divided by $x - a$, the remainder is related to the function value $f(a)$. Compare the following from Examples 1 and 2.

Polynomial Function	Divisor	Remainder	Function Value
Example 1: $f(x) = x^3 + 6x^2 - x - 30$	$x - 2$	0	$f(2) = 0$
Example 2(a): $g(x) = 2x^3 + 7x^2 - 5$	$x + 3$, or $x - (-3)$	4	$g(-3) = 4$
Example 2(b): $p(x) = 10x^2 - 13x + 3x^3 - 20$	$4 + x$, or $x - (-4)$	0	$p(-4) = 0$

When the remainder is 0, the divisor $x - a$ is a factor of the polynomial. Thus we can write $f(x) = x^3 + 6x^2 - x - 30$ in Example 1 as $f(x) = (x - 2)(x^2 + 8x + 15)$. Then

$$\begin{aligned} f(2) &= (2 - 2)(2^2 + 8 \cdot 2 + 15) \\ &= 0(2^2 + 8 \cdot 2 + 15) = 0. \end{aligned}$$

Thus, when the remainder is 0 after division by $x - a$, the function value $f(a)$ is also 0. Remarkably, this pattern extends to nonzero remainders as well. The fact that the remainder and the function value coincide is predicted by the remainder theorem.

The Remainder Theorem The remainder obtained by dividing $P(x)$ by $x - r$ is $P(r)$.

A proof of this result is outlined in Exercise 39.

EXAMPLE 3 Let $f(x) = 8x^5 - 6x^3 + x - 8$. Use synthetic division to find $f(2)$.

SOLUTION The remainder theorem tells us that $f(2)$ is the remainder when $f(x)$ is divided by $x - 2$. We use synthetic division to find that remainder:

$$\begin{array}{r|rrrrrr} 2 & 8 & 0 & -6 & 0 & 1 & -8 \\ & & 16 & 32 & 52 & 104 & 210 \\ \hline & 8 & 16 & 26 & 52 & 105 & \big| \; 202 \end{array}$$

Note that to find $f(2)$, we use 2 in the synthetic division.

Although the bottom line can be used to find the quotient for the division $(8x^5 - 6x^3 + x - 8) \div (x - 2)$, what we are really interested in is the remainder. It tells us that $f(2) = 202$.

In practice, the remainder theorem is often used to check divison. Thus Example 2(a) can be checked by evaluating $g(-3)$, with $g(x) = 2x^3 + 7x^2 - 5$. Since $g(-3) = 4$ (check this) and the remainder in Example 2(a) is also 4, our divison was probably correct.

6.7 EXERCISE SET

FOR EXTRA HELP

MathXL | MyMathLab | InterAct Math | AW Math Tutor Center | Video Lectures on CD: Disc 3 | Student's Solutions Manual

Concept Reinforcement *Classify each statement as either true or false.*

1. If $x - 2$ is a factor of some polynomial $P(x)$, then $P(2) = 0$.

2. If $p(3) = 0$ for some polynomial $p(x)$, then $x - 3$ is a factor of $p(x)$.

3. If $P(-5) = 39$ and $P(x) = x^3 + 7x^2 + 3x + 4$, then

$$\begin{array}{r|rrrr} -5 & 1 & 7 & 3 & 4 \\ & & -5 & -10 & 35 \\ \hline & 1 & 2 & -7 & \big| \; 39. \end{array}$$

4. In order for $f(x)/g(x)$ to exist, $g(x)$ must be 0.

5. In order to use synthetic division, we must be sure that the divisor is of the form $x - a$.

6. Synthetic division can be used in problems in which long division could not be used.

7. Both $3(-4)^3 + 10(-4)^2 - 13(-4) - 20 = 0$ and

$$\begin{array}{r|rrrr} -4 & 3 & 10 & -13 & -20 \\ & & -12 & 8 & 20 \\ \hline & 3 & -2 & -5 & \big| \; 0 \end{array}$$

indicate that for $p(x) = 3x^3 + 10x^2 - 13x - 20$, $p(-4) = 0$.

8. Synthetic division can be used to show that for any polynomial function P, $P(a)$ and $P(-a)$ are opposites.

Use synthetic division to divide.

9. $(x^3 - 2x^2 + 2x - 7) \div (x - 1)$

10. $(x^3 - 2x^2 + 2x - 7) \div (x + 1)$

11. $(a^2 + 8a + 11) \div (a + 3)$

12. $(a^2 + 8a + 11) \div (a + 5)$

13. $(x^3 - 7x^2 - 13x + 3) \div (x + 2)$

14. $(x^3 - 7x^2 - 13x + 3) \div (x - 2)$

15. $(3x^3 + 7x^2 - 4x + 3) \div (x + 3)$

16. $(3x^3 + 7x^2 - 4x + 3) \div (x - 3)$

17. $(y^3 - 3y + 10) \div (y - 2)$

18. $(x^3 - 2x^2 + 8) \div (x + 2)$

19. $(x^5 - 32) \div (x - 2)$

20. $(y^5 - 1) \div (y - 1)$

21. $(3x^3 + 1 - x + 7x^2) \div \left(x + \frac{1}{3}\right)$

22. $(8x^3 - 1 + 7x - 6x^2) \div \left(x - \frac{1}{2}\right)$

Use synthetic division to find the indicated function value.

23. $f(x) = 5x^4 + 12x^3 + 28x + 9;\ f(-3)$

24. $g(x) = 3x^4 - 25x^2 - 18;\ g(3)$

25. $P(x) = 6x^4 - x^3 - 7x^2 + x + 2;\ P(-1)$

26. $F(x) = 3x^4 + 8x^3 + 2x^2 - 7x - 4;\ F(-2)$

27. $f(x) = x^4 - x^3 - 19x^2 + 49x - 30;\ f(4)$

28. $p(x) = x^4 + 7x^3 + 11x^2 - 7x - 12;\ p(2)$

TW **29.** Why is it that we *add* when performing synthetic division, but *subtract* when performing long division?

TW **30.** Explain how synthetic division could be useful when attempting to factor a polynomial.

Skill Maintenance

Solve. [1.6]

31. $9 + cb = a - b$, for b

32. $8 + ac = bd + ab$, for a

Find the domain of f.

33. $f(x) = \dfrac{5}{3x^2 - 75}$ [5.6]

34. $f(x) = \dfrac{7}{2x^2 + 7x - 9}$ [5.5]

Graph.

35. $y - 2 = \frac{3}{4}(x + 1)$ [2.4]

36. $y = -\frac{4}{3}x + 2$ [2.2]

Synthesis

TW **37.** Let $Q(x)$ be a polynomial function with $p(x)$ a factor of $Q(x)$. If $p(3) = 0$, does it follow that $Q(3) = 0$? Why or why not? If $Q(3) = 0$, does it follow that $p(3) = 0$? Why or why not?

TW **38.** What adjustments must be made if synthetic division is to be used to divide a polynomial by a binomial of the form $ax + b$, with $a > 1$?

39. To prove the remainder theorem, note that any polynomial $P(x)$ can be rewritten as $(x - r) \cdot Q(x) + R$, where $Q(x)$ is the quotient polynomial that arises when $P(x)$ is divided by $x - r$, and R is some constant (the remainder).

a) How do we know that R must be a constant?

b) Show that $P(r) = R$ (this says that $P(r)$ is the remainder when $P(x)$ is divided by $x - r$).

40. Let $f(x) = 6x^3 - 13x^2 - 79x + 140$. Find $f(4)$ and then solve the equation $f(x) = 0$.

41. Let $f(x) = 4x^3 + 16x^2 - 3x - 45$. Find $f(-3)$ and then solve the equation $f(x) = 0$.

42. Use the TRACE feature on a graphing calculator to check your answer to Exercise 40.

43. Use the TRACE feature on a graphing calculator to check your answer to Exercise 41.

Nested Evaluation. One way to evaluate a polynomial function like $P(x) = 3x^4 - 5x^3 + 4x^2 - 1$ *is to successively factor out x as shown:*

$$P(x) = x(x(x(3x - 5) + 4) + 0) - 1.$$

Computations are then performed using this "nested" form of P(x).

44. Use nested evaluation to find $f(4)$ in Exercise 40. Note the similarities to the calculations performed with synthetic division.

45. Use nested evaluation to find $f(-3)$ in Exercise 41. Note the similarities to the calculations performed with synthetic division.

6.8 Formulas, Applications, and Variation

Formulas ■ Direct Variation ■ Inverse Variation ■ Joint and Combined Variation ■ Models

Formulas

Formulas occur frequently as mathematical models. Many formulas contain rational expressions, and to solve such formulas for a specified letter, we proceed as when solving rational equations.

EXAMPLE 1 Electronics. The formula

$$\frac{1}{R} = \frac{1}{r_1} + \frac{1}{r_2}$$

is used by electricians to determine the resistance R of two resistors r_1 and r_2 connected in parallel.* Solve for r_1.

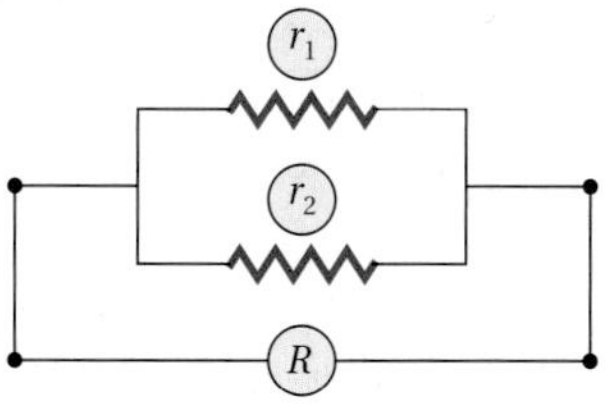

SOLUTION We use the same approach as in Section 6.4:

$$Rr_1r_2 \cdot \frac{1}{R} = Rr_1r_2 \cdot \left(\frac{1}{r_1} + \frac{1}{r_2}\right)$$ **Multiplying both sides by the LCD**

$$Rr_1r_2 \cdot \frac{1}{R} = Rr_1r_2 \cdot \frac{1}{r_1} + Rr_1r_2 \cdot \frac{1}{r_2}$$ **Multiplying to remc parentheses**

$$r_1r_2 = Rr_2 + Rr_1.$$ **Simplifying by removing factors equal to 1: $\frac{R}{R} = 1; \frac{r_1}{r_1} = 1; \frac{r_2}{r_2} = 1$**

At this point it is tempting to multiply by $1/r_2$ to get r_1 alone on the left, *but* note that there is an r_1 on the right. We must get all the terms involving r_1 on the *same side* of the equation.

$$r_1r_2 - Rr_1 = Rr_2$$ **Subtracting Rr_1 from both sides**

$$r_1(r_2 - R) = Rr_2$$ **Factoring out r_1 in order to combine like terms**

$$r_1 = \frac{Rr_2}{r_2 - R}$$ **Dividing both sides by $r_2 - R$ to get r_1 alone**

This formula can be used to calculate r_1 whenever R and r_2 are known.

*Recall that the subscripts 1 and 2 merely indicate that r_1 and r_2 are different variables representing similar quantities.

Study Tip

Visualize the Steps

If you have completed all assignments and are studying for a quiz or test, don't feel that you need to redo every assigned problem. A more productive use of your time would be simply to reread the assigned problems, making certain that you can visualize the steps that lead to a solution. When you are unsure of how to solve a problem, redo that problem in its entirety, seeking outside help as needed.

EXAMPLE 2 Astronomy. The formula

$$\frac{V^2}{R^2} = \frac{2g}{R + h}$$

is used to find a satellite's *escape velocity* V, where R is a planet's radius, h is the satellite's height above the planet, and g is the planet's gravitational constant. Solve for h.

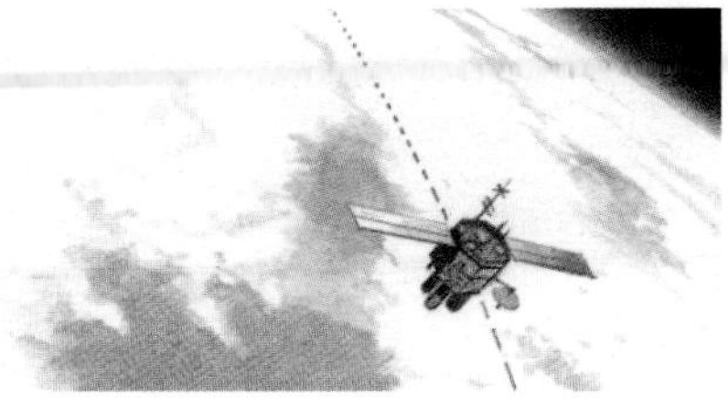

SOLUTION We first clear fractions by multiplying by the LCD, which is $R^2(R + h)$:

$$\frac{V^2}{R^2} = \frac{2g}{R + h}$$

$$R^2(R + h)\frac{V^2}{R^2} = R^2(R + h)\frac{2g}{R + h}$$

$$\frac{R^2(R + h)V^2}{R^2} = \frac{R^2(R + h)2g}{R + h}$$

$$(R + h)V^2 = R^2 \cdot 2g. \quad \text{Removing factors equal to 1: } \frac{R^2}{R^2} = 1 \text{ and } \frac{R + h}{R + h} = 1$$

Remember: We are solving for h. Although we *could* distribute V^2, since h appears only within the factor $R + h$, it is easier to divide both sides by V^2:

$$\frac{(R + h)V^2}{V^2} = \frac{2R^2g}{V^2} \quad \text{Dividing both sides by } V^2$$

$$R + h = \frac{2R^2g}{V^2} \quad \text{Removing a factor equal to 1: } \frac{V^2}{V^2} = 1$$

$$h = \frac{2R^2g}{V^2} - R. \quad \text{Subtracting } R \text{ from both sides}$$

The last equation can be used to determine the height of a satellite above a planet when the planet's radius and gravitational constant, along with the satellite's escape velocity, are known.

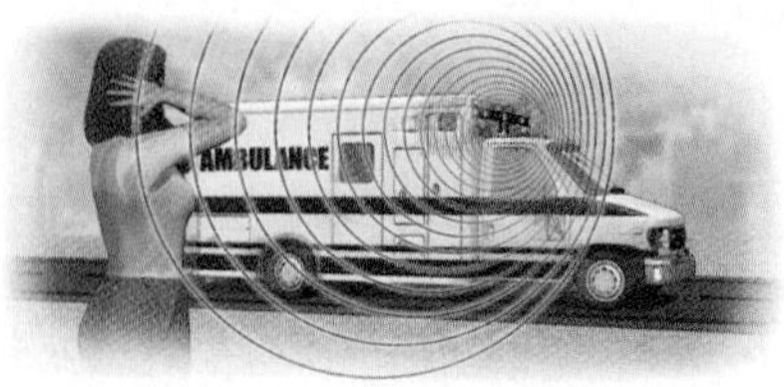

EXAMPLE 3 Acoustics (the Doppler Effect). The formula

$$f = \frac{sg}{s + v}$$

is used to determine the frequency f of a sound that is moving at velocity v toward a listener who hears the sound as frequency g. Here s is the speed of sound in a particular medium. Solve for s.

SOLUTION We first clear fractions by multiplying by the LCD, $s + v$:

$$f \cdot (s + v) = \frac{sg}{s + v}(s + v)$$

$$fs + fv = sg.$$ **The variable for which we are solving, s, appears on both sides, forcing us to distribute on the left side.**

Next, we must get all terms containing s on one side:

$$fv = sg - fs$$ **Subtracting fs from both sides**

$$fv = s(g - f)$$ **Factoring out s**

$$\frac{fv}{g - f} = s.$$ **Dividing both sides by $g - f$**

Since s is isolated on one side, we have solved for s. This last equation can be used to determine the speed of sound whenever f, v, and g are known.

Student Notes

The steps used to solve equations are precisely the same steps used to solve formulas. If you feel "rusty" in this regard, study the earlier section in which this type of equation first appeared. Then make sure that you can consistently solve those equations before returning to the work with formulas.

To Solve a Rational Equation for a Specified Variable

1. If necessary, multiply both sides by the LCD to clear fractions.
2. Multiply, as needed, to remove parentheses.
3. Get all terms with the specified variable alone on one side.
4. Factor out the specified variable if it is in more than one term.
5. Multiply or divide on both sides to isolate the specified variable.

Variation

To extend our study of formulas and functions, we now examine three real-world situations: direct variation, inverse variation, and combined variation.

Direct Variation

A registered nurse earns \$22 per hour. In 1 hr, \$22 is earned. In 2 hr, \$44 is earned. In 3 hr, \$66 is earned, and so on. This gives rise to a set of ordered pairs:

$$(1, 22), (2, 44), (3, 66), (4, 88), \text{ and so on.}$$

Note that the ratio of earnings E to time t is $\frac{22}{1}$ in every case.

If a situation gives rise to pairs of numbers in which the ratio is constant, we say that there is **direct variation.** Here earnings *vary directly* as the time:

We have $\dfrac{E}{t} = 22$, so $E = 22t$ or, using function notation, $E(t) = 22t$.

Direct Variation When a situation gives rise to a linear function of the form $f(x) = kx$, or $y = kx$, where k is a nonzero constant, we say that there is *direct variation*, that *y varies directly* as x, or that *y is proportional to x*. The number k is called the *variation constant*, or *constant of proportionality*.

Note that for $k > 0$, any equation of the form $y = kx$ indicates that as x increases, y increases as well.

EXAMPLE 4 Find the variation constant and an equation of variation if y varies directly as x, and $y = 32$ when $x = 2$.

SOLUTION We know that $(2, 32)$ is a solution of $y = kx$. Therefore,

$$32 = k \cdot 2 \qquad \textbf{Substituting}$$

$$\frac{32}{2} = k, \quad \text{or} \quad k = 16. \qquad \textbf{Solving for } k$$

The variation constant is 16. The equation of variation is $y = 16x$. The notation $y(x) = 16x$ or $f(x) = 16x$ is also used.

EXAMPLE 5 Water from Melting Snow. The number of centimeters W of water produced from melting snow varies directly as the number of centimeters S of snow. Meteorologists know that under certain conditions, 150 cm of snow will melt to 16.8 cm of water. The average annual snowfall in Alta, Utah, is 500 in. Assuming the above conditions, how much water will replace the 500 in. of snow?

SOLUTION

1. **Familiarize.** Because of the phrase "W ... varies directly as ... S," we express the amount of water as a function of the amount of snow. Thus, $W(S) = kS$, where k is the variation constant. Knowing that 150 cm of snow becomes 16.8 cm of water, we have $W(150) = 16.8$. Because we are using ratios, it does not matter whether we work in inches or centimeters, provided the same units are used for W and S.

Alta, Utah

2. **Translate.** We find the variation constant using the data and then use it to write the equation of variation:

$$W(S) = kS$$
$$W(150) = k \cdot 150 \quad \textbf{Replacing } S \textbf{ with 150}$$
$$16.8 = k \cdot 150 \quad \textbf{Replacing } W(150) \textbf{ with 16.8}$$
$$\frac{16.8}{150} = k \quad \textbf{Solving for } k$$
$$0.112 = k. \quad \textbf{This is the variation constant.}$$

The equation of variation is $W(S) = 0.112S$. This is the translation.

3. **Carry out.** To find how much water 500 in. of snow will become, we compute $W(500)$:

$$W(S) = 0.112S$$
$$W(500) = 0.112(500) \quad \textbf{Substituting 500 for } S$$
$$W = 56.$$

4. **Check.** To check, we could reexamine all our calculations. Note that our answer seems reasonable since $500/56$ and $150/16.8$ are equal.

5. **State.** Alta's 500 in. of snow will become 56 in. of water.

Inverse Variation

To see what we mean by inverse variation, suppose a bus is traveling 20 mi. At 20 mph, the trip will take 1 hr. At 40 mph, it will take $\frac{1}{2}$ hr. At 60 mph, it will take $\frac{1}{3}$ hr, and so on. This gives rise to pairs of numbers, all having the same product:

$$(20, 1), \left(40, \tfrac{1}{2}\right), \left(60, \tfrac{1}{3}\right), \left(80, \tfrac{1}{4}\right), \quad \text{and so on.}$$

Note that the product of each pair of numbers is 20. Whenever a situation gives rise to pairs of numbers for which the product is constant, we say that there is **inverse variation.** Since $r \cdot t = 20$, the time t, in hours, required for the bus to travel 20 mi at r mph is given by

$$t = \frac{20}{r} \quad \text{or, using function notation,} \quad t(r) = \frac{20}{r}.$$

Inverse Variation When a situation gives rise to a rational function of the form $f(x) = k/x$, or $y = k/x$, where k is a nonzero constant, we say that there is *inverse variation*, that *y varies inversely as x*, or that *y is inversely proportional to x*. The number k is called the *variation constant*, or *constant of proportionality*.

Note that for $k > 0$, any equation of the form $y = k/x$ indicates that as x increases, y decreases.

EXAMPLE 6 Find the variation constant and an equation of variation if y varies inversely as x, and $y = 32$ when $x = 0.2$.

SOLUTION We know that $(0.2, 32)$ is a solution of

$$y = \frac{k}{x}.$$

Therefore,

$$32 = \frac{k}{0.2} \quad \textbf{Substituting}$$
$$(0.2)32 = k$$
$$6.4 = k. \quad \textbf{Solving for } k$$

The variation constant is 6.4. The equation of variation is

$$y = \frac{6.4}{x}.$$

Many real-life problems translate to an equation of inverse variation.

EXAMPLE 7 Ultraviolet Index. The ultraviolet, or UV, index is a measure that indicates the strength of the sun's rays in a particular locale. For those people whose skin is quite sensitive, a UV rating of 6 will cause sunburn after 10 min (*Source*: *The Electronic Textbook of Dermatology* found at www.telemedicine.org, January 2004). Given that the number of minutes it takes to burn, t, varies inversely as the UV rating u, how long will it take a highly sensitive person to burn on a day with a UV rating of 4?

SOLUTION

1. **Familiarize.** Because of the phrase "… varies inversely as the UV rating," we express the amount of time needed to burn as a function of the UV rating: $t(u) = k/u$.
2. **Translate.** We use the given information to solve for k. Then we use that result to write the equation of variation.

$$t(u) = \frac{k}{u} \quad \textbf{Using function notation}$$
$$t(6) = \frac{k}{6} \quad \textbf{Replacing } u \textbf{ with 6}$$
$$10 = \frac{k}{6} \quad \textbf{Replacing } t(6) \textbf{ with 10}$$
$$60 = k \quad \textbf{Solving for } k\textbf{, the variation constant}$$

The equation of variation is $t(u) = 60/u$. This is the translation.

3. **Carry out.** To find how long it would take a highly sensitive person to burn on a day with a UV index of 4, we calculate $t(4)$:

$$t(4) = \frac{60}{4} = 15. \quad t = 15 \textbf{ when } u = 4$$

4. **Check.** We could now recheck each step. Note that, as expected, as the UV rating goes *down*, the time it takes to burn goes *up*.
5. **State.** On a day with a UV rating of 4, a highly sensitive person will begin to burn after 15 min of exposure.

Joint and Combined Variation

When a variable varies directly with more than one other variable, we say that there is *joint variation*. For example, in the formula for the volume of a right circular cylinder, $V = \pi r^2 h$, we say that V varies *jointly* as h and the square of r.

Joint Variation y varies *jointly* as x and z if, for some nonzero constant k, $y = kxz$.

EXAMPLE 8 Find an equation of variation if y varies jointly as x and z, and $y = 30$ when $x = 2$ and $z = 3$.

SOLUTION We have

$$y = kxz,$$

so

$$30 = k \cdot 2 \cdot 3$$
$$k = 5. \quad \textbf{The variation constant is 5.}$$

The equation of variation is $y = 5xz$.

Joint variation is one form of *combined variation*. In general, when a variable varies directly and/or inversely, at the same time, with more than one other variable, there is **combined variation.** Examples 8 and 9 are both examples of combined variation.

EXAMPLE 9 Find an equation of variation if y varies jointly as x and z and inversely as the square of w, and $y = 105$ when $x = 3$, $z = 20$, and $w = 2$.

SOLUTION The equation of variation is of the form

$$y = k \cdot \frac{xz}{w^2},$$

so, substituting, we have

$$105 = k \cdot \frac{3 \cdot 20}{2^2}$$
$$105 = k \cdot 15$$
$$k = 7.$$

Thus,

$$y = 7 \cdot \frac{xz}{w^2}.$$

Models

We may be able to recognize from a set of data whether two quantities vary directly or inversely. A graph like the one on the left below indicates that the quantities represented vary directly. The graph on the right below represents quantities that vary inversely. If we know the type of variation involved, we can choose one data point and calculate an equation of variation.

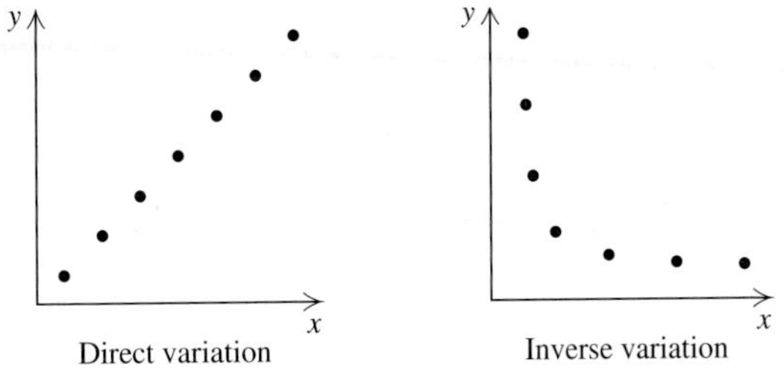

Direct variation Inverse variation

EXAMPLE 10 Commuter Travel. One factor influencing urban planning is VMT, or vehicle miles traveled. Areas with a high population density have fewer VMT per household than those with a lower density. This occurs in part because employment and shopping are available closer to higher density areas. The following table lists annual VMT per household for various densities for a typical urban area.

Population Density (in number of households per residential acre)	Annual VMT per Household
25	12,000
50	6,000
100	3,000
200	1,500

Source: Based on information from http://www.sflcv.org/density

a) Determine whether the data indicate direct variation or inverse variation.

b) Find an equation of variation that describes the data.

c) Use the equation to estimate the annual VMT per household for areas with 10 households per residential acre.

SOLUTION

a) We graph the data, letting x represent the population density, in number of households per residential acre, and y the annual VMT per household.

The points approximate the graph of a function of the type $f(x) = k/x$. Annual VMT per household varies inversely as the population density.

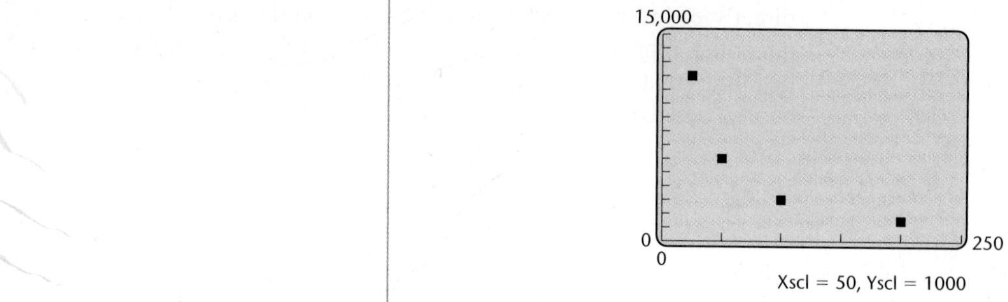

b) To find an equation of variation, we choose one point. We will use the point $(50, 6000)$.

$$y = \frac{k}{x} \quad \textbf{An equation of inverse variation}$$

$$6000 = \frac{k}{50} \quad \textbf{Substituting 50 for } \boldsymbol{x} \textbf{ and 6000 for } \boldsymbol{y}$$

$$300{,}000 = k \quad \textbf{Multiplying both sides by 50}$$

We now have an equation of inverse variation, $y = 300{,}000/x$. This equation can also be written $y = 300{,}000x^{-1}$. The PwrReg option of the STAT CALC menu will return equations of this type, as shown in the figure on the right below. If we graph the equation along with the data, we can see that it does match the data.

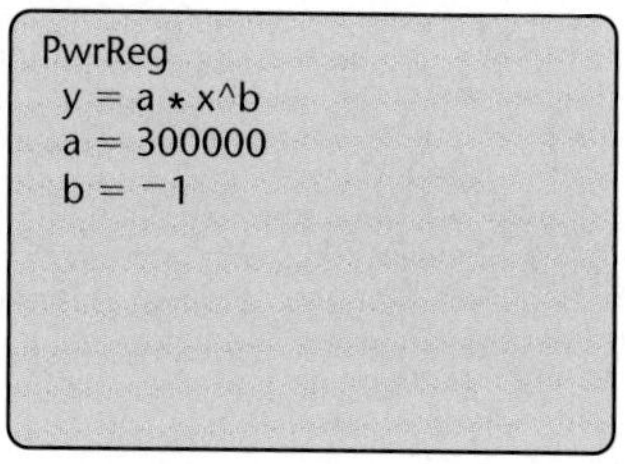

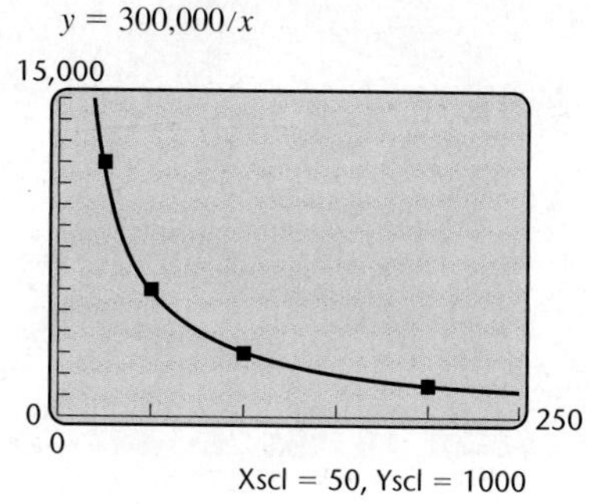

c) To find the annual VMT per household when the population density is 10 households per residential acre, we substitute 10 for x in the equation of variation and calculate y:

$$y = \frac{300{,}000}{x} \quad \textbf{The equation of variation}$$

$$y = \frac{300{,}000}{10} \quad \textbf{Substituting}$$

$$y = 30{,}000.$$

When the population density is 10 households per residential acre, each household will drive 30,000 miles per year.

6.8 EXERCISE SET

FOR EXTRA HELP

MathXL | MyMathLab | InterAct Math | AW Math Tutor Center | Video Lectures on CD: Disc 3 | Student's Solutions Manual

Concept Reinforcement *Complete each of the following statements.*

1. To solve a formula for a variable appearing in a denominator, we multiply both sides of the equation by the ______________.

2. All terms containing the variable being solved for must ultimately be on the ______________ side of the equation.

3. If the variable being solved for appears in more than one term on the same side of the equation, it is usually necessary to ______________.

4. If y varies directly as x, then $y = kx$ and k is called the ______________ of ______________.

Concept Reinforcement *Determine whether each situation represents direct or inverse variation.*

5. Two painters can scrape a house in 9 hr, whereas three painters can scrape the house in 6 hr.

6. Fanny planted 5 bulbs in 20 min and 7 bulbs in 28 min.

7. Glen swam 2 laps in 7 min and 6 laps in 21 min.

8. It took 2 band members 80 min to set up for a show; with 4 members working, it took 40 min.

9. It took 3 hr for 4 volunteers to wrap the campus' collection of Toys for Tots, but only 1.5 hr with 8 volunteers working.

10. Fatuma's air conditioner cooled off 1000 ft^3 in 10 min and 3000 ft^3 in 30 min.

Solve each formula for the specified variable.

11. $f = \dfrac{L}{d}$; d

12. $\dfrac{W_1}{W_2} = \dfrac{d_1}{d_2}$; W_1

13. $s = \dfrac{(v_1 + v_2)t}{2}$; v_1

14. $s = \dfrac{(v_1 + v_2)t}{2}$; t

15. $\dfrac{t}{a} + \dfrac{t}{b} = 1$; b

16. $\dfrac{1}{R} = \dfrac{1}{r_1} + \dfrac{1}{r_2}$; R

17. $I = \dfrac{2V}{R + 2r}$; R

18. $I = \dfrac{2V}{R + 2r}$; r

19. $R = \dfrac{gs}{g + s}$; g

20. $K = \dfrac{rt}{r - t}$; t

21. $I = \dfrac{nE}{R + nr}$; n

22. $I = \dfrac{nE}{R + nr}$; r

23. $\dfrac{1}{p} + \dfrac{1}{q} = \dfrac{1}{f}$; q

24. $\dfrac{1}{p} + \dfrac{1}{q} = \dfrac{1}{f}$; p

25. $S = \dfrac{H}{m(t_1 - t_2)}$; t_1

26. $S = \dfrac{H}{m(t_1 - t_2)}$; H

27. $\dfrac{E}{e} = \dfrac{R + r}{r}$; r

28. $\dfrac{E}{e} = \dfrac{R + r}{R}$; R

29. $S = \dfrac{a}{1 - r}$; r

30. $S = \dfrac{a - ar^n}{1 - r}$; a

31. $c = \dfrac{f}{(a + b)c}$; $a + b$

32. $d = \dfrac{g}{d(c + f)}$; $c + f$

33. *Interest.* The formula

$$P = \frac{A}{1 + r}$$

is used to determine what principal P should be invested for one year at $(100 \cdot r)\%$ simple interest in order to have A dollars after a year. Solve for r.

34. *Taxable Interest.* The formula

$$I_t = \frac{I_f}{1 - T}$$

gives the *taxable interest rate* I_t equivalent to the *tax-free interest rate* I_f for a person in the $(100 \cdot T)\%$ tax bracket. Solve for T.

35. *Average Speed.* The formula

$$v = \frac{d_2 - d_1}{t_2 - t_1}$$

gives an object's average speed v when that object has traveled d_1 miles in t_1 hours and d_2 miles in t_2 hours. Solve for t_2.

36. *Average Acceleration.* The formula

$$a = \frac{v_2 - v_1}{t_2 - t_1}$$

gives a vehicle's *average acceleration* when its velocity changes from v_1 at time t_1 to v_2 at time t_2. Solve for t_1.

37. *Planetary Orbits.* The formula

$$\frac{x^2}{a^2} + \frac{y^2}{b^2} = 1$$

can be used to plot a planet's elliptical orbit of width $2a$ and length $2b$. Solve for b^2.

38. *Work Rate.* The formula

$$\frac{1}{t} = \frac{1}{a} + \frac{1}{b}$$

gives the total time t required for two workers to complete a job, if the workers' individual times are a and b. Solve for t.

39. *Semester Average.* The formula

$$A = \frac{2Tt + Qq}{2T + Q}$$

gives a student's average A after T tests and Q quizzes, where each test counts as 2 quizzes, t is the test average, and q is the quiz average. Solve for Q.

40. *Astronomy.* The formula

$$L = \frac{dR}{D - d},$$

where D is the diameter of the sun, d is the diameter of the earth, R is the earth's distance from the sun, and L is some fixed distance, is used in calculating when lunar eclipses occur. Solve for D.

Find the variation constant and an equation of variation if y varies directly as x and the following conditions apply.

41. $y = 28$ when $x = 4$

42. $y = 5$ when $x = 12$

43. $y = 3.4$ when $x = 2$

44. $y = 2$ when $x = 5$

45. $y = 2$ when $x = \frac{1}{3}$

46. $y = 0.9$ when $x = 0.5$

Find the variation constant and an equation of variation in which y varies inversely as x, and the following conditions exist.

47. $y = 3$ when $x = 20$

48. $y = 16$ when $x = 4$

49. $y = 28$ when $x = 4$

50. $y = 9$ when $x = 5$

51. $y = 27$ when $x = \frac{1}{3}$

52. $y = 81$ when $x = \frac{1}{9}$

53. *Use of Aluminum Cans.* The number N of aluminum cans used each year varies directly as the number of people using the cans. If 250 people use 60,000 cans in one year, how many cans are used each year in Dallas, which has a population of 1,008,000?

54. *Hooke's Law.* Hooke's law states that the distance d that a spring is stretched by a hanging object varies directly as the mass m of the object. If the distance is 20 cm when the mass is 3 kg, what is the distance when the mass is 5 kg?

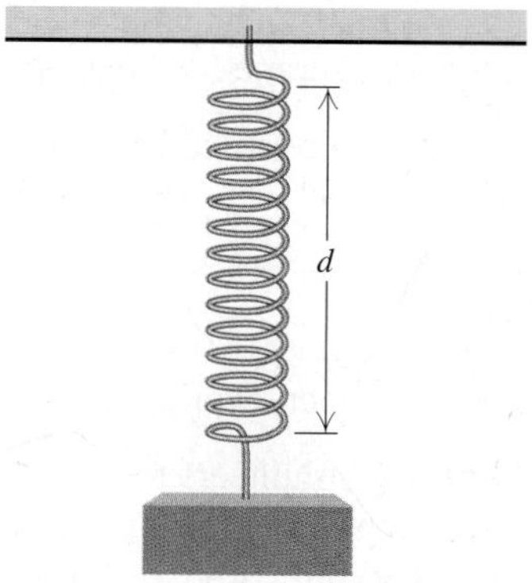

55. *Ohm's Law.* The electric current I, in amperes, in a circuit varies directly as the voltage V. When 15 volts are applied, the current is 5 amperes. What is the current when 18 volts are applied?

56. *Pumping Rate.* The time t required to empty a tank varies inversely as the rate r of pumping. If a Briggs and Stratton pump can empty a tank in 45 min at the rate of 600 kL/min, how long will it take the pump to empty the tank at 1000 kL/min?

57. *Work Rate.* The time T required to do a job varies inversely as the number of people P working. It takes 5 hr for 7 volunteers to pick up rubbish from 1 mi of roadway. How long would it take 10 volunteers to complete the job?

58. *Weekly Allowance.* According to Fidelity Investments *Investment Vision Magazine*, the average weekly allowance A of children varies directly as their grade level, G. In a recent year, the average allowance of a 9th-grade student was \$13.50 per week (*Source*: Based on data from www.kidsmoney.org). What was the average weekly allowance of a 4th-grade student?

Aha! **59.** *Mass of Water in a Human.* The number of kilograms W of water in a human body varies directly as the mass of the body. A 96-kg person contains 64 kg of water. How many kilograms of water are in a 48-kg person?

60. *Weight on Mars.* The weight M of an object on Mars varies directly as its weight E on Earth. A person who weighs 95 lb on Earth weighs 38 lb on Mars. How much would a 100-lb person weigh on Mars?

61. *Bicycling.* The number of calories burned by a bicyclist is directly proportional to the time spent bicycling. At 10 mph, it takes 30 min to burn 150 calories (*Source*: Physical Activity and Health: A Report of the Surgeon General). How long would it take to burn 250 calories when biking 10 mph?

62. *Wavelength and Frequency.* The wavelength W of a radio wave varies inversely as its frequency F. A wave with a frequency of 1200 kilohertz has a length of 300 meters. What is the length of a wave with a frequency of 800 kilohertz?

63. *Ultraviolet Index.* At an ultraviolet, or UV, rating of 4, those people who are moderately sensitive to the sun will burn in 70 min (*Source*: *The Electronic Textbook of Dermatology* at www.telemedicine.org). Given that the number of minutes it takes to burn, t, varies inversely with the UV rating, u, how long will it take moderately sensitive people to burn when the UV rating is 14?

64. *Current and Resistance.* The current I in an electrical conductor varies inversely as the resistance R of the conductor. If the current is $\frac{1}{2}$ ampere when the resistance is 240 ohms, what is the current when the resistance is 540 ohms?

65. *Air Pollution.* The average U.S. household of 2.6 people released 1.1 tons of carbon monoxide into the environment in a recent year (*Sources*: Based on data from the U.S. Environmental Protection Agency and the U.S. Bureau of the Census). How many tons were released nationally? Use 289,000,000 as the U.S. population.

66. *Relative Aperture.* The relative aperture, or f-stop, of a 23.5-mm lens is directly proportional to the focal length F of the lens. If a lens with a 150-mm focal length has an f-stop of 6.3, find the f-stop of a 23.5-mm lens with a focal length of 80 mm.

Find an equation of variation in which:

67. y varies directly as the square of x, and $y = 6$ when $x = 3$.

68. y varies directly as the square of x, and $y = 0.15$ when $x = 0.1$.

69. y varies inversely as the square of x, and $y = 6$ when $x = 3$.

70. y varies inversely as the square of x, and $y = 0.15$ when $x = 0.1$.

71. y varies jointly as x and the square of z, and $y = 105$ when $x = 14$ and $z = 5$.

72. y varies jointly as x and z and inversely as w, and $y = \frac{3}{2}$ when $x = 2$, $z = 3$, and $w = 4$.

73. y varies jointly as w and the square of x and inversely as z, and $y = 49$ when $w = 3$, $x = 7$, and $z = 12$.

74. y varies directly as x and inversely as w and the square of z, and $y = 4.5$ when $x = 15$, $w = 5$, and $z = 2$.

75. *Electrical Safety.* The amount of time t needed for an electrical shock to stop a 150-lb person's heart from beating varies inversely as the square of the current flowing through the body. It is known that a 0.089-amp current is deadly to a 150-lb person after 3.4 sec. (*Source*: Safety Consulting Services) How long would it take a 0.096-amp current to be deadly?

76. *Stopping Distance of a Car.* The stopping distance d of a car after the brakes have been applied varies directly as the square of the speed r (*Source*: Based on data from Edmunds.com). Once the brakes are applied, a car traveling 60 mph can stop in 138 ft. What stopping distance corresponds to a speed of 40 mph?

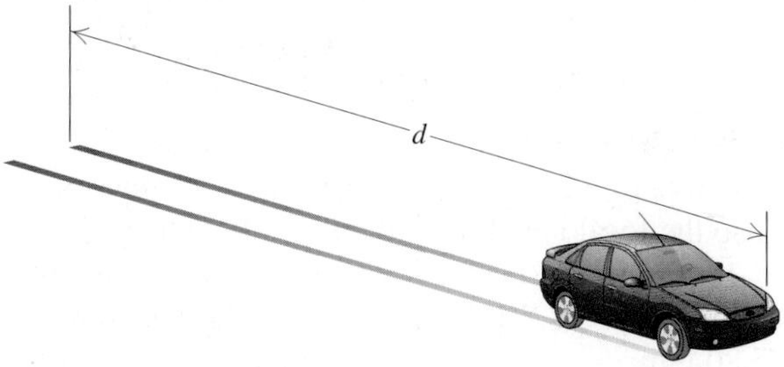

77. *Volume of a Gas.* The volume V of a given mass of a gas varies directly as the temperature T and inversely as the pressure P. If $V = 231\text{ cm}^3$ when $T = 300°\text{K}$ (Kelvin) and $P = 20\text{ lb/cm}^2$, what is the volume when $T = 320°\text{K}$ and $P = 16\text{ lb/cm}^2$?

78. *Intensity of a Signal.* The intensity I of a television signal varies inversely as the square of the distance d from the transmitter. If the intensity is 25 W/m^2 at a distance of 2 km, what is the intensity 6.25 km from the transmitter?

79. *Atmospheric Drag.* Wind resistance, or atmospheric drag, tends to slow down moving objects. Atmospheric drag W varies jointly as an object's surface area A and velocity v. If a car traveling at a speed of 40 mph with a surface area of 37.8 ft^2 experiences a drag of 222 N (Newtons), how fast must a car with 51 ft^2 of surface area travel in order to experience a drag force of 430 N?

80. *Drag Force.* The drag force F on a boat varies jointly as the wetted surface area A and the square of the velocity of the boat. If a boat traveling 6.5 mph experiences a drag force of 86 N when the wetted surface area is 41.2 ft^2, find the wetted surface area of a boat traveling 8.2 mph with a drag force of 94 N.

81. *Ultraviolet Index.* The following table shows the safe exposure time for people with less sensitive skin (see Example 7).

UV Index	Safe Exposure Time (in minutes)
2	120
4	75
6	50
8	35
10	25

a) Determine whether the data indicate direct variation or inverse variation.
b) Find an equation of variation that approximates the data. Use the data point (6, 50).
c) Use the equation to predict the safe exposure time for people with less sensitive skin when the UV rating is 3.

82. *Perceived Height.* The following table shows the perceived height y of a pole when the observer is x feet from the pole.

Distance from Pole (in feet)	Perceived Height of Pole (in feet)
5	4
10	2
15	$1\frac{1}{3}$
20	1
40	0.5

a) Determine whether the data indicate direct variation or inverse variation.
b) Find an equation of variation that approximates the data. Use the data point $(10, 2)$.
c) Use the equation to predict the perceived height of the pole when the observer is 2 ft from the pole.

83. *Mail Order.* The following table lists the number of persons y ordering merchandise by telephone and the number of persons x ordering merchandise by mail for various age groups.

Age	Number Ordering by Mail (in millions)	Number Ordering by Telephone (in millions)
18–24	6.146	5.255
25–34	11.813	12.775
35–44	14.294	16.650
45–54	11.294	13.703
55–64	6.948	8.275
65+	10.541	9.358

Source: Simmons Market Research Bureau, *Study of Media and Markets*, New York, NY

a) Determine whether the number of people ordering by mail varies directly or inversely as the number ordering by telephone.

b) Find an equation of variation that approximates the data. Use the data point (11.813, 12.775).

c) Use the equation to predict the number of persons ordering by telephone if 8 million order by mail.

84. *Motor Vehicle Registrations.* The following table lists the number of automobile registrations y and the number of drivers licensed x for various states.

State	Number of Driver's Licenses (in millions)	Number of Automobiles Registered (in millions)
Alabama	2.063	3.434
Colorado	1.843	2.946
Georgia	4.033	5.316
Idaho	0.502	0.863
Maryland	2.622	3.178
Texas	7.456	13.323
Virginia	3.774	4.787

a) Determine whether the number of automobiles registered varies directly or inversely as the number of driver's licenses.

b) Find an equation of variation that approximates the data. Use the data point (2.063, 3.434).

c) Use the equation to estimate the number of automobiles registered in Hawaii, where there are 0.45 million licensed drivers.

TW **85.** If two quantities vary directly, as in Exercise 83, does this mean that one is "caused" by the other? Why or why not?

TW **86.** If y varies directly as x, does doubling x cause y to be doubled as well? Why or why not?

Skill Maintenance

Find the domain of f. [2.5]

87. $f(x) = \dfrac{2x - 1}{x^2 + 1}$

88. $f(x) = |2x - 1|$

89. Graph on a plane: $6x - y < 6$. [4.5]

90. If $f(x) = x^3 - x$, find $f(2a)$. [2.1]

91. Factor: $t^3 + 8b^3$. [5.7]

92. Solve: $6x^2 = 11x + 35$. [5.5]

Synthesis

TW **93.** Suppose that the number of customer complaints is inversely proportional to the number of employees hired. Will a firm reduce the number of complaints more by expanding from 5 to 10 employees, or from 20 to 25? Explain. Consider using a graph to help justify your answer.

TW **94.** Why do you think subscripts are used in Exercises 13 and 25 but not in Exercises 27 and 28?

95. *Escape Velocity.* A satellite's escape velocity is 6.5 mi/sec, the radius of the earth is 3960 mi, and the earth's gravitational constant is 32.2 ft/sec^2. How far is the satellite from the surface of the earth? (See Example 2.)

96. The *harmonic mean* of two numbers a and b is a number M such that the reciprocal of M is the average of the reciprocals of a and b. Find a formula for the harmonic mean.

97. *Health-care.* Young's rule for determining the size of a particular child's medicine dosage c is

$$c = \frac{a}{a + 12} \cdot d,$$

where a is the child's age and d is the typical adult dosage (*Source*: Olsen, June Looby, Leon J. Ablon, and Anthony Patrick Giangrasso, *Medical Dosage Calculations*, 6th ed.). If a child's age is doubled, the dosage increases. Find the ratio of the larger dosage to the smaller dosage. By what percent does the dosage increase?

98. Solve for x:

$$x^2\left(1 - \frac{2pq}{x}\right) = \frac{2p^2q^3 - pq^2x}{-q}.$$

99. *Average Acceleration.* The formula

$$a = \frac{\dfrac{d_4 - d_3}{t_4 - t_3} - \dfrac{d_2 - d_1}{t_2 - t_1}}{t_4 - t_2}$$

can be used to approximate average acceleration, where the d's are distances and the t's are the corresponding times. Solve for t_1.

100. If y varies inversely as the cube of x and x is multiplied by 0.5, what is the effect on y?

101. *Intensity of Light.* The intensity I of light from a bulb varies directly as the wattage of the bulb and inversely as the square of the distance d from the bulb. If the wattage of a light source and its distance from reading matter are both doubled, how does the intensity change?

102. Describe in words the variation represented by $W = \dfrac{km_1M_1}{d^2}$. Assume k is a constant.

103. *Tension of a Musical String.* The tension T on a string in a musical instrument varies jointly as the string's mass per unit length m, the square of its length l, and the square of its fundamental frequency f. A 2-m long string of mass 5 gm/m with a fundamental frequency of 80 has a tension of 100 N. How long should the same string be if its tension is going to be changed to 72 N?

104. *Volume and Cost.* A peanut butter jar in the shape of a right circular cylinder is 4 in. high and 3 in. in diameter and sells for \$1.20. If we assume that cost is proportional to volume, how much should a jar 6 in. high and 6 in. in diameter cost?

105. *Golf Distance Finder.* A device used in golf to estimate the distance d to a hole measures the size s that the 7-ft pin *appears* to be in a viewfinder. The viewfinder uses the principle, diagrammed here, that s gets bigger when d gets smaller. If $s = 0.56$ in. when $d = 50$ yd, find an equation of variation that expresses d as a function of s. What is d when $s = 0.40$ in.?

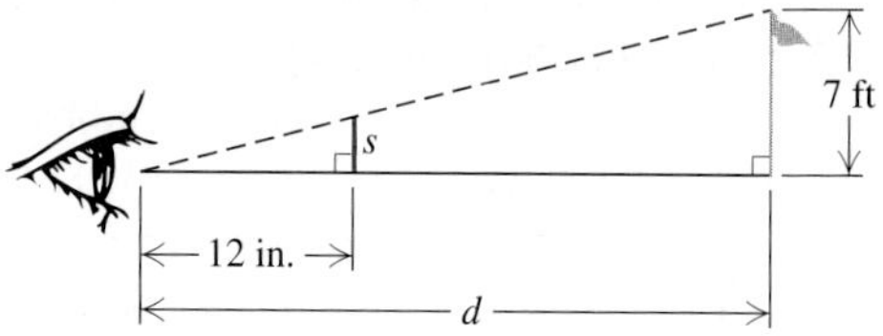

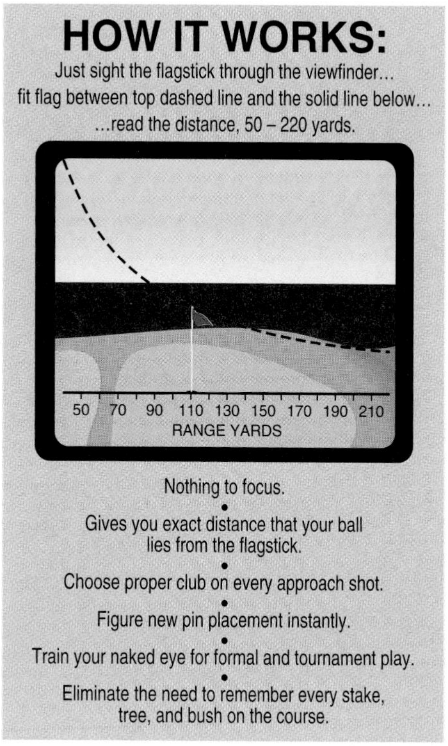
HOW IT WORKS:
Just sight the flagstick through the viewfinder...
fit flag between top dashed line and the solid line below...
...read the distance, 50 – 220 yards.
50 70 90 110 130 150 170 190 210
RANGE YARDS
Nothing to focus.
Gives you exact distance that your ball lies from the flagstick.
Choose proper club on every approach shot.
Figure new pin placement instantly.
Train your naked eye for formal and tournament play.
Eliminate the need to remember every stake, tree, and bush on the course.

6 Chapter Summary and Review

KEY TERMS AND DEFINITIONS

RATIONAL EXPRESSIONS AND EQUATIONS

Rational expression, p. 438 A quotient of two polynomials. The domain of a **rational function** does not include values that make a denominator zero.

Least common denominator, LCD, p. 456 The least common multiple of the denominators.

APPLICATIONS

Variation

y varies directly as x if there is some nonzero constant k such that $y = kx$.

y varies inversely as x if there is some nonzero constant k such that $y = k/x$.

y varies jointly as x and z if there is some nonzero constant k such that $y = kxz$.

Modeling Work Problems

If

$a =$ the time needed for A to complete the work alone,

$b =$ the time needed for B to complete the work alone, and

$t =$ the time needed for A and B to complete the work together,

then

$$\frac{t}{a} + \frac{t}{b} = 1 \quad \text{and} \quad \frac{1}{a} \cdot t + \frac{1}{b} \cdot t = 1$$

$$\text{and} \quad \frac{1}{a} + \frac{1}{b} = \frac{1}{t}.$$

Motion Formula

$$d = rt \quad \text{or} \quad r = d/t \quad \text{or} \quad t = d/r$$

IMPORTANT CONCEPTS

[Section references appear in brackets.]

Concept	Example
The graph of a rational function may approach one or more **vertical asymptotes.**	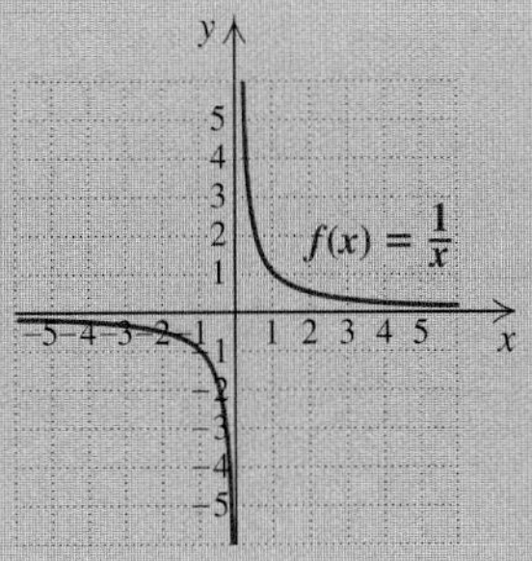 Domain: $(-\infty, 0) \cup (0, \infty)$ Range: $(-\infty, 0) \cup (0, \infty)$ Vertical asymptote: $x = 0$ [6.1]

(*continued*)

Addition: $\dfrac{A}{C} + \dfrac{B}{C} = \dfrac{A + B}{C}$

Subtraction: $\dfrac{A}{C} - \dfrac{B}{C} = \dfrac{A - B}{C}$

You must have a common denominator to add or subtract.

Multiplication: $\dfrac{A}{B} \cdot \dfrac{C}{D} = \dfrac{AC}{BD}$

Division: $\dfrac{A}{B} \div \dfrac{C}{D} = \dfrac{A}{B} \cdot \dfrac{D}{C}$

$$\frac{3}{x} + \frac{2}{x + 1} = \frac{3x + 3}{x(x + 1)} + \frac{2x}{x(x + 1)} = \frac{5x + 3}{x(x + 1)}$$ [6.2]

$$\frac{2x + 5}{x - 3} - \frac{x - 7}{x - 3} = \frac{(2x + 5) - (x - 7)}{x - 3}$$

$$= \frac{2x + 5 - x + 7}{x - 3} = \frac{x + 12}{x - 3}$$ [6.2]

$$\frac{2x}{x + 3} \cdot \frac{x^2 + 3x}{4x^2} = \frac{\cancel{2} \cdot \cancel{x} \cdot \cancel{x} \cdot \cancel{(x + 3)}}{\cancel{(x + 3)} \cdot \cancel{2} \cdot 2 \cdot \cancel{x} \cdot \cancel{x}} = \frac{1}{2}$$ [6.1]

$$\frac{x}{x + 1} \div \frac{3}{x + 1} = \frac{x}{x + 1} \cdot \frac{x + 1}{3} = \frac{x\cancel{(x + 1)}}{\cancel{(x + 1)}3} = \frac{x}{3}$$ [6.1]

Complex rational expressions contain one or more rational expressions within the numerator and/or denominator. They can be simplified either by multiplying by a form of 1 to clear the fractions or by dividing two rational expressions (pp. 465 and 468).

Multiplying by 1:

$$\frac{\frac{4}{x} + \frac{4}{y}}{\frac{2}{x} + \frac{6}{y}} = \frac{\frac{4}{x} + \frac{4}{y}}{\frac{2}{x} + \frac{6}{y}} \cdot \frac{xy}{xy} = \frac{\frac{4\cancel{x}y}{\cancel{x}} + \frac{4x\cancel{y}}{\cancel{y}}}{\frac{2\cancel{x}y}{\cancel{x}} + \frac{6x\cancel{y}}{\cancel{y}}}$$

$$= \frac{4y + 4x}{2y + 6x} = \frac{\cancel{2} \cdot 2 \cdot (y + x)}{\cancel{2}(y + 3x)} = \frac{2(y + x)}{y + 3x}$$

Dividing two rational expressions:

$$\frac{\frac{4}{x} + \frac{4}{y}}{\frac{2}{x} + \frac{6}{y}} = \frac{\frac{4y}{xy} + \frac{4x}{xy}}{\frac{2y}{xy} + \frac{6x}{xy}} = \frac{\frac{4y + 4x}{xy}}{\frac{2y + 6x}{xy}}$$

$$= \frac{4y + 4x}{xy} \cdot \frac{xy}{2y + 6x}$$

$$= \frac{\cancel{2} \cdot 2(y + x) \cdot \cancel{x} \cdot \cancel{y}}{\cancel{x} \cdot \cancel{y} \cdot \cancel{2}(y + 3x)} = \frac{2(y + x)}{y + 3x}$$ [6.3]

Multiplying a **rational equation** by the LCD will clear the equation of fractions. All potential solutions must be checked in the original equation.

$$\frac{1}{x} = \frac{x}{4}$$

$$4x\left(\frac{1}{x}\right) = 4x\left(\frac{x}{4}\right)$$

$$4 = x^2$$

$x = 2$ *or* $x = -2$ **Both 2 and −2 check. The solutions are 2 and −2.** [6.4]

The Remainder Theorem

The remainder obtained by dividing $P(x)$ by $x - r$ is $P(r)$.

The remainder can be found using **synthetic division.**

```
2| 3  -4   0    6
       6   4    8
   ----------------
   3   2   4 | 14
```

$$(3x^3 - 4x^2 + 6) \div (x - 2)$$

$$= 3x^2 + 2x + 4 + \frac{14}{x - 2}$$

If $P(x) = 3x^3 - 4x^2 + 6$, then $P(2) = 14$. [6.7]

Review Exercises

Concept Reinforcement *Classify each of the following statements as either true or false.*

1. If $f(x) = \dfrac{x-3}{x^2-4}$, the domain of f is assumed to be $\{x \mid x \neq -2, x \neq 2\}$. [6.1]

2. We can simplify a rational expression whenever a term that appears in the numerator also appears in the denominator. [6.1]

3. To add rational expressions in which the denominators are opposites, we use the product of the denominators as the least common denominator. [6.2]

4. A complex rational expression can always be simplified by multiplying its numerator and its denominator by the LCD of those expressions. [6.3]

5. Checking the solution of a rational equation is no more important than checking the solution of any other equation. [6.4]

6. If Hannah can do a job alone in t_1 hr and Jacob can do the same job in t_2 hr, then working together it will take them $(t_1 + t_2)/2$ hr. [6.5]

7. If Carlie swims 5 km/h in still water and heads into a current of 2 km/h, her speed will change to 3 km/h. [6.5]

8. To divide a polynomial by a monomial, we divide the monomial by each term in the polynomial. [6.6]

9. To divide two polynomials using synthetic division, we must make sure that the divisor is of the form $x - a$. [6.7]

10. If x varies inversely as y, then there exists some constant k for which $x = k/y$. [6.8]

11. If

$$f(t) = \frac{t^2 - 3t + 2}{t^2 - 9},$$

find the following function values. [6.1]

a) $f(0)$
b) $f(-1)$
c) $f(2)$

Find the LCD. [6.2]

12. $\dfrac{5}{6x^3}, \dfrac{15}{16x^2}$

13. $\dfrac{x-2}{x^2+x-20}, \dfrac{x-3}{x^2+3x-10}$

Perform the indicated operations and, if possible, simplify.

14. $\dfrac{x^2}{x-3} - \dfrac{9}{x-3}$ [6.2]

15. $\dfrac{3a^2b^3}{5c^3d^2} \cdot \dfrac{15c^9d^4}{9a^7b}$ [6.1]

16. $\dfrac{5}{6m^2n^3p} + \dfrac{7}{9mn^4p^2}$ [6.2]

17. $\dfrac{x^3-8}{x^2-25} \cdot \dfrac{x^2+10x+25}{x^2+2x+4}$ [6.1]

18. $\dfrac{x^2-3x-18}{x^2+5x+6} \div \dfrac{x^2-16}{x^3-64}$ [6.1]

19. $\dfrac{x}{x^2+5x+6} - \dfrac{2}{x^2+3x+2}$ [6.2]

20. $\dfrac{-4xy}{x^2-y^2} + \dfrac{x+y}{x-y}$ [6.2]

21. $\dfrac{2x^2}{x-y} + \dfrac{2y^2}{y-x}$ [6.2]

22. $\dfrac{3}{y+4} - \dfrac{y}{y-1} + \dfrac{y^2+3}{y^2+3y-4}$ [6.2]

Find simplified form for $f(x)$ and list all restrictions on the domain.

23. $f(x) = \dfrac{4x-2}{x^2-5x+4} - \dfrac{3x+2}{x^2-5x+4}$ [6.2]

24. $f(x) = \dfrac{x+8}{x+5} \cdot \dfrac{2x+10}{x^2-64}$ [6.1]

25. $f(x) = \dfrac{9x^2-1}{x^2-9} \div \dfrac{3x+1}{x+3}$ [6.1]

Simplify. [6.3]

26. $\dfrac{\dfrac{4}{x} - 4}{\dfrac{9}{x} - 9}$

27. $\dfrac{\dfrac{3}{a} + \dfrac{3}{b}}{\dfrac{6}{a^3} + \dfrac{6}{b^3}}$

28. $\dfrac{\dfrac{y^2 + 4y - 77}{y^2 - 10y + 25}}{\dfrac{y^2 - 5y - 14}{y^2 - 25}}$

29. $\dfrac{\dfrac{5}{x^2 - 9} - \dfrac{3}{x + 3}}{\dfrac{4}{x^2 + 6x + 9} + \dfrac{2}{x - 3}}$

Solve. [6.4]

30. $\dfrac{3}{x} + \dfrac{7}{x} = 5$

31. $\dfrac{5}{3x + 2} = \dfrac{3}{2x}$

32. $\dfrac{4x}{x + 1} + \dfrac{4}{x} + 9 = \dfrac{4}{x^2 + x}$

33. $\dfrac{x + 6}{x^2 + x - 6} + \dfrac{x}{x^2 + 4x + 3} = \dfrac{x + 2}{x^2 - x - 2}$

34. If

$$f(x) = \frac{2}{x - 1} + \frac{2}{x + 2},$$

find all a for which $f(a) = 1$. [6.4]

Solve. [6.5]

35. Kim can arrange the books for a book sale in 9 hr. Kelly can set up for the same book sale in 12 hr. How long would it take them, working together, to set up for the book sale?

36. A research company uses personal computers to process data while the owner is not using the computer. A Pentium 4 3.0-gigahertz processor can process a megabyte of data in 15 sec less time than a Celeron 2.53-gigahertz processor. Working together, the computers can process a megabyte of data in 18 sec. How long does it take each computer to process one megabyte of data?

37. The Black River's current is 6 mph. A boat travels 50 mi downstream in the same time that it takes to travel 30 mi upstream. What is the speed of the boat in still water?

38. A car and a motorcycle leave a rest area at the same time, with the car traveling 8 mph faster than the motorcycle. The car then travels 105 mi in the time it takes the motorcycle to travel 93 mi. Find the speed of each vehicle.

Divide. [6.6]

39. $(30r^2s^3 + 25r^2s^2 - 20r^3s^3) \div (5r^2s)$

40. $(y^3 + 125) \div (y + 5)$

41. $(4x^3 + 3x^2 - 5x - 2) \div (x^2 + 1)$

42. Divide using synthetic division: [6.7]

$$(x^3 + 3x^2 + 2x - 6) \div (x - 3).$$

43. If $f(x) = 4x^3 - 6x^2 - 9$, use synthetic division to find $f(5)$. [6.7]

Solve. [6.8]

44. $R = \dfrac{gs}{g + s}$, for s

45. $S = \dfrac{H}{m(t_1 - t_2)}$, for m

46. $\dfrac{1}{ac} = \dfrac{2}{ab} - \dfrac{3}{bc}$, for c

47. $T = \dfrac{A}{v(t_2 - t_1)}$, for t_1

48. The amount of waste generated by a family varies directly as the number of people in the family. The average U.S. family has 3.14 people and generates 13.8 lb of waste daily (*Sources*: Based on data from the U.S. Bureau of the Census and the *U.S. Statistical Abstract* 2003). How many pounds of waste would be generated daily by a family of 5?

49. A warning dye is used by people in lifeboats to aid search planes. The volume V of the dye used varies directly as the square of the diameter d of the circular patch of water formed by the dye. If 4 L of dye is required for a 10-m wide circle, how much dye is needed for a 40-m wide circle?

50. Find an equation of variation in which y varies inversely as x, and $y = 3$ when $x = \frac{1}{4}$.

51. The following table lists the size y of one serving of breakfast cereal for the corresponding number of servings x obtained from a given box. [6.8]

Number of Servings	Size of Serving (in ounces)
1	24
4	6
6	4
12	2
16	1.5

a) Determine whether the data indicate direct variation or inverse variation.
b) Find an equation of variation that fits the data. Use the data point (12, 2).
c) Use the equation to estimate the size of each serving when 8 servings are obtained from the box.

Synthesis

TW 52. Discuss at least three different uses of the LCD studied in this chapter. [6.2], [6.3], [6.4]

TW 53. Explain the difference between a rational expression and a rational equation. [6.1], [6.4]

Solve.

54. $\frac{5}{x-13} - \frac{5}{x} = \frac{65}{x^2 - 13x}$ [6.4]

55. $\dfrac{\frac{x}{x^2-25} + \frac{2}{x-5}}{\frac{3}{x-5} - \frac{4}{x^2-10x+25}} = 1$ [6.3], [6.4]

56. A Xeon 3.6-gigahertz processor can process a megabyte of data in 20 sec. How long would it take a Xeon working together with the Pentium 4 and Celeron processors (see Exercise 36) to process a megabyte of data? [6.5]

Chapter Test 6

Simplify.

1. $\frac{t+1}{t+3} \cdot \frac{3t+9}{4t^2-4}$

2. $\frac{x^3+27}{x^2-16} \div \frac{x^2+8x+15}{x^2+x-20}$

3. Find the LCD:

$$\frac{3x}{x^2+5x-24}, \quad \frac{x+1}{x^2-12x+27}.$$

Perform the indicated operation and simplify when possible.

4. $\frac{25x}{x+5} + \frac{x^3}{x+5}$

5. $\frac{3a^2}{a-b} - \frac{3b^2-6ab}{b-a}$

6. $\frac{4ab}{a^2-b^2} + \frac{a^2+b^2}{a+b}$

7. $\frac{6}{x^3-64} - \frac{4}{x^2-16}$

Find simplified form for $f(x)$ and list all restrictions on the domain.

8. $f(x) = \frac{4}{x+3} - \frac{x}{x-2} + \frac{x^2+4}{x^2+x-6}$

9. $f(x) = \frac{x^2-1}{x+2} \div \frac{x^2-2x}{x^2+x-2}$

Simplify.

10. $\dfrac{\frac{2}{a} + \frac{3}{b}}{\frac{5}{ab} + \frac{1}{a^2}}$

11. $\dfrac{\frac{x^2-5x-36}{x^2-36}}{\frac{x^2+x-12}{x^2-12x+36}}$

12. $\dfrac{\frac{2}{x+3} - \frac{1}{x^2-3x+2}}{\frac{3}{x-2} + \frac{4}{x^2+2x-3}}$

Solve.

13. $\dfrac{4}{2x-5} = \dfrac{6}{5x+3}$

14. $\dfrac{t+11}{t^2-t-12} + \dfrac{1}{t-4} = \dfrac{4}{t+3}$

For Exercises 15 and 16, let $f(x) = \dfrac{x+3}{x-1}$.

15. Find $f(2)$ and $f(-3)$.

16. Find all a for which $f(a) = 7$.

17. Kyla can install a vinyl kitchen floor in 3.5 hr. Brock can perform the same job in 4.5 hr. How long will it take them, working together, to install the vinyl?

Divide.

18. $(16ab^3c - 10ab^2c^2 + 12a^2b^2c) \div (4a^2b)$

19. $(y^2 - 20y + 64) \div (y - 6)$

20. $(6x^4 + 3x^2 + 5x + 4) \div (x^2 + 2)$

21. Divide using synthetic division:
$(x^3 + 5x^2 + 4x - 7) \div (x - 4)$.

22. If $f(x) = 3x^4 - 5x^3 + 2x - 7$, use synthetic division to find $f(4)$.

23. Solve $A = \dfrac{h(b_1 + b_2)}{2}$ for b_1.

24. The product of the reciprocals of two consecutive integers is $\frac{1}{30}$. Find the integers.

25. Emma bicycles 12 mph with no wind. Against the wind, she bikes 8 mi in the same time that it takes to bike 14 mi with the wind. What is the speed of the wind?

26. The number of workers n needed to clean a stadium after a game varies inversely as the amount of time t allowed for the cleanup. If it takes 25 workers to clean the stadium when there are 6 hr allowed for the job, how many workers are needed if the stadium must be cleaned in 5 hr?

27. The surface area of a balloon varies directly as the square of its radius. The area is 325 in^2 when the radius is 5 in. What is the area when the radius is 7 in.?

Synthesis

28. Let
$$f(x) = \frac{1}{x+3} + \frac{5}{x-2}.$$
Find all a for which $f(a) = f(a+5)$.

29. Solve: $\dfrac{6}{x-15} - \dfrac{6}{x} = \dfrac{90}{x^2 - 15x}$.

30. Find the x- and y-intercepts for the function given by
$$f(x) = \frac{\dfrac{5}{x+4} - \dfrac{3}{x-2}}{\dfrac{2}{x-3} + \dfrac{1}{x+4}}.$$

31. One summer, Hans mowed 4 lawns for every 3 lawns mowed by his brother Franz. Together, they mowed 98 lawns. How many lawns did each mow?

1–6 Cumulative Review

1. *Meal Preparation.* In 1993, on average, adults spent 49 min per weeknight preparing dinner. This time decreased to 31 min per weeknight by 2003. (*Source*: *Parade Magazine*, 11/16/03) What was the rate of decrease? [2.2]

2. *Frequent Fliers.* The number of unredeemed frequent flier miles has increased from 5.5 trillion in 1999 to 14.2 trillion in 2005 (*Source*: WebFlyer.com). [2.4]
   a) Find a linear function $F(t)$ that fits the data.
   b) Use the function of part (a) to predict the number of unredeemed frequent flier miles in 2007.

3. *Automobiles.* The following table lists the number of automobiles in the United States for various years. [2.4]

Year	Registered Automobiles (in millions)
1960	67.9
1970	98.1
1980	139.8
1990	179.3
2005	220.0

Source: *The Indianapolis Star,* 4/28/06

a) Use linear regression to find a function $a(t)$ that can be used to estimate the number of automobiles in the United States t years after 1960.

b) Use the function of part (a) to estimate the number of automobiles in the United States in 2015.

4. *Fuel Economy.* The following table lists the fuel consumption and expenditures by the fuel economy of the vehicle. [6.8]

Fuel Economy (in miles per gallon)	Fuel Consumption (in billions)	Fuel Expenditures (in billions)
10.9 or less	2.3 gal	\$3.0
11 to 12.9	2.0	2.5
13 to 15.9	16.3	21.7
16 to 18.9	26.6	35.5
19 to 21.9	26.1	34.9
22 to 24.9	22.1	29.4
25 to 29.9	13.6	18.1
30 or more	4.1	5.2

Source: Energy Information Administration/Household Vehicles Energy Use

a) Use a graph to show that fuel expenditure varies directly as fuel consumption.

b) Find an equation of variation that describes the data. Use the point (26.6, 35.5).

c) What does the variation constant signify?

5. *College Football.* The greatest payout to a college team for a bowl game in 2005–2006 was \$18.3 million. The smallest payout was \$750,000. Fourteen teams received either the greatest or smallest payout, for a total of \$63.15 million. How many teams received \$18.3 million and how many received \$750,000? [3.3]

In Exercises 6–9, match each equation with one of the following graphs.

a)

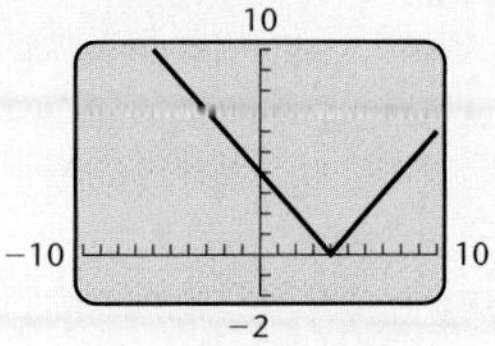

b)

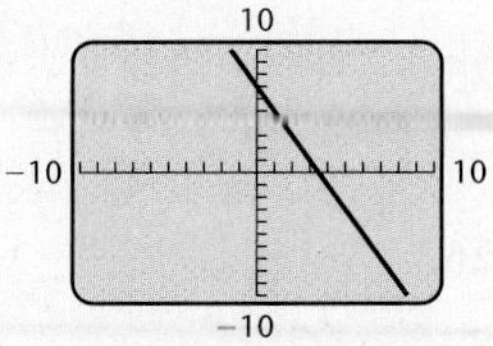

c)

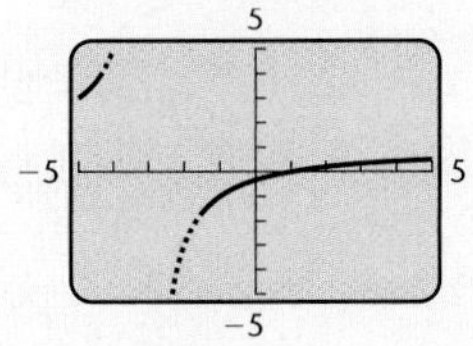

d)

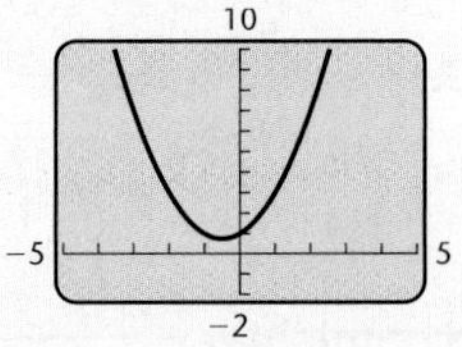

6. $y = -2x + 7$ [1.5]

7. $y = |x - 4|$ [4.4]

8. $y = x^2 + x + 1$ [5.1]

9. $y = \dfrac{x - 1}{x + 3}$ [6.1]

Graph on a plane.

10. $4x \geq 5y + 20$ [4.5]

11. $y = \frac{1}{3}x - 2$ [2.2]

12. Evaluate

$$\frac{2x - y^2}{x + y}$$

for $x = 3$ and $y = -4$. [1.1], [1.2]

13. Convert to scientific notation: 5,760,000,000. [1.4]

14. Determine the slope and the y-intercept for the line given by $7x - 4y = 12$. [2.2]

15. Find an equation for the line that passes through the points $(-1, 7)$ and $(2, -3)$. [2.4]

Perform the indicated operations and simplify.

16. $(2x^2 - 3x + 1) + (6x - 3x^3 + 7x^2 - 4)$ [5.1]

17. $(5x^3y^2)(-3xy^2)$ [5.2]

18. $(3a + b - 2c) - (-4b + 3c - 2a)$ [5.1]

19. $(5x^2 - 2x + 1)(3x^2 + x - 2)$ [5.2]

20. $(2x^2 - y)^2$ [5.2]

21. $(2x^2 - y)(2x^2 + y)$ [5.2]

22. $(-5m^3n^2 - 3mn^3) + (-4m^2n^2 + 4m^3n^2) - (2mn^3 - 3m^2n^2)$ [5.1]

23. $\dfrac{y^2 - 36}{2y + 8} \cdot \dfrac{y + 4}{y + 6}$ [6.1]

24. $\dfrac{x^4 - 1}{x^2 - x - 2} \div \dfrac{x^2 + 1}{x - 2}$ [6.1]

25. $\dfrac{5ab}{a^2 - b^2} + \dfrac{a + b}{a - b}$ [6.2]

26. $\dfrac{2}{m + 1} + \dfrac{3}{m - 5} - \dfrac{m^2 - 1}{m^2 - 4m - 5}$ [6.2]

27. $y - \dfrac{2}{3y}$ [6.2]

28. Simplify: $\dfrac{\frac{1}{x} - \frac{1}{y}}{x + y}$. [6.3]

29. Divide: $(9x^3 + 5x^2 + 2) \div (x + 2)$. [6.6]

Factor.

30. $4x^3 + 18x^2$ [5.3]

31. $x^2 + 8x - 84$ [5.4]

32. $16y^2 - 81$ [5.6]

33. $64x^3 + 8$ [5.7]

34. $t^2 - 16t + 64$ [5.6]

35. $x^6 - x^2$ [5.6]

36. $0.027b^3 - 0.008c^3$ [5.7]

37. $20x^2 + 7x - 3$ [5.5]

38. $3x^2 - 17x - 28$ [5.5]

39. $x^5 - x^3y + x^2y - y^2$ [5.3]

40. If

$$f(x) = \frac{x - 2}{x - 5},$$

find the following.

a) $f(3)$ [6.1] **b)** The domain of f [2.5]

41. Write the domain of f using interval notation if $f(x) = \sqrt{x - 7}$. [4.3]

42. If $f(x) = x^2 - 4$ and $g(x) = x^2 - 7x + 10$, find the domain of f/g. [2.5], [5.4]

Solve.

43. $8x = 1 + 16x^2$ [5.6]

44. $625 = 49y^2$ [5.6]

45. $20 > 2 - 6x$ [4.1]

46. $\frac{1}{3}x - \frac{1}{5} \geq \frac{1}{5}x - \frac{1}{3}$ [4.1]

47. $-8 < x + 2 < 15$ [4.3]

48. $3x - 2 < -6$ *or* $x + 3 > 9$ [4.3]

49. $|x| > 6.4$ [4.4]

50. $|4x - 1| \leq 14$ [4.4]

51. $\dfrac{2}{n} - \dfrac{7}{n} = 3$ [6.4]

52. $\dfrac{6}{x - 5} = \dfrac{2}{2x}$ [6.4]

53. $\dfrac{3x}{x - 2} - \dfrac{6}{x + 2} = \dfrac{24}{x^2 - 4}$ [6.4]

54. $\dfrac{3x^2}{x + 2} + \dfrac{5x - 22}{x - 2} = \dfrac{-48}{x^2 - 4}$ [6.4]

55. $5x - 2y = -23,$
$3x + 4y = 7$ [3.2]

56. $-3x + 4y + z = -5,$
$x - 3y - z = 6,$
$2x + 3y + 5z = -8$ [3.4]

57. Let $f(x) = |3x - 5|$. Find all values of x for which $f(x) = 2$. [4.4]

Solve.

58. $5m - 3n = 4m + 12$, for n [1.6]

59. $P = \dfrac{3a}{a + b}$, for a [6.8]

Find the solutions of each equation, inequality, or system from the given graph.

60. $10 - 5x = 3x + 2$ [4.2] **61.** $f(x) \leq g(x)$ [4.2]

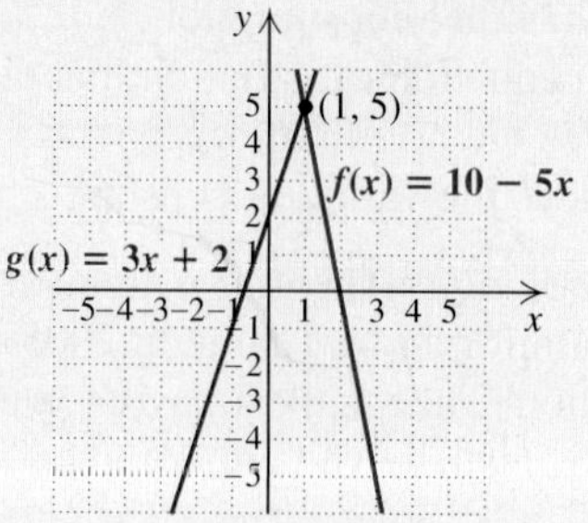

62. $y = x - 5,$
$y = 1 - 2x$ [3.1]

63. $x - 5 = 1 - 2x$ [4.2]

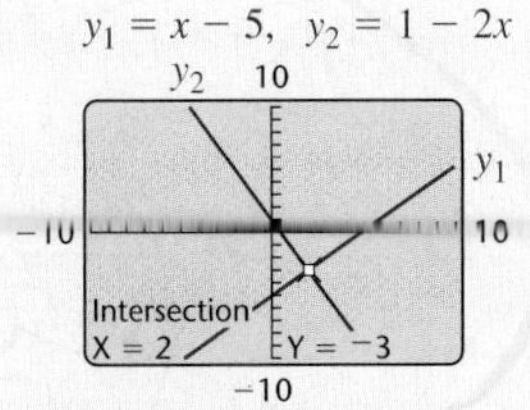

64. $x^2 - 2x + 1 = 0$ [5.3]

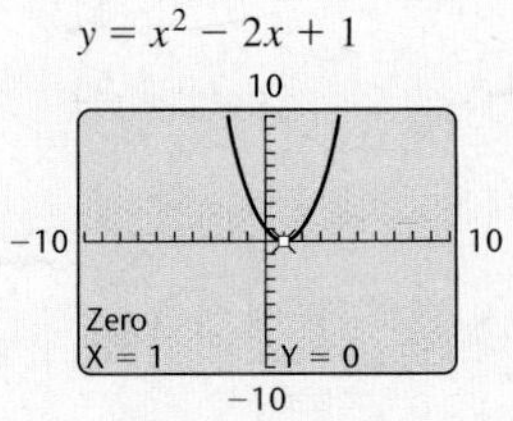

65. $|x - 3| \leq 2$ [4.4]

66. $|x - 3| = 2$ [4.4]

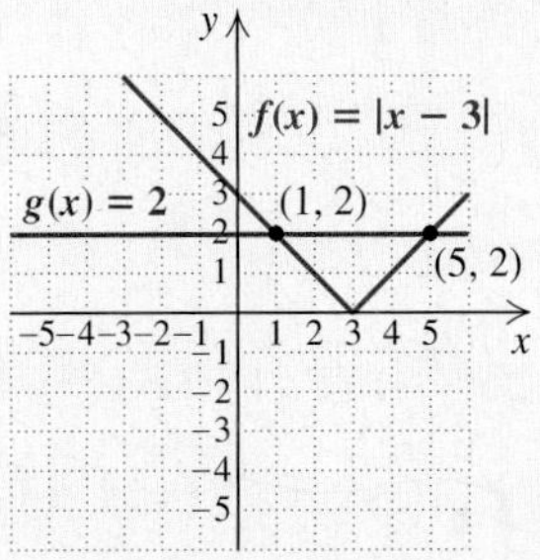

67. The sum of three numbers is 20. The first number is 3 less than twice the third number. The second number minus the third number is −7. What are the numbers? [3.5]

68. A digital data circuit can transmit a particular set of data in 4 sec. An analog phone circuit can transmit the same data in 20 sec. How long would it take, working together, for both circuits to transmit the data? [6.5]

69. The floor area of a rental trailer is rectangular. The length is 3 ft more than the width. A rug of area 54 ft^2 exactly fills the floor of the trailer. Find the perimeter of the trailer. [5.8]

70. The sum of the squares of three consecutive even integers is equal to 8 more than three times the square of the second number. Find the integers. [5.8]

71. The volume of wood V in a tree trunk varies jointly as the height h and the square of the girth g (girth is distance around). If the volume is 35 ft^3 when the height is 20 ft and the girth is 5 ft, what is the height when the volume is 85.75 ft^3 and the girth is 7 ft? [6.8]

Synthesis

72. Multiply: $(x - 4)^3$. [5.2]

73. Find all roots for $f(x) = x^4 - 34x^2 + 225$. [5.6]

Solve.

74. $4 \leq |3 - x| \leq 6$ [4.3], [4.4]

75. $\dfrac{18}{x - 9} + \dfrac{10}{x + 5} = \dfrac{28x}{x^2 - 4x - 45}$ [6.4]

76. $16x^3 = x$ [5.6]

7 Exponents and Radical Functions

In this chapter, we learn about square roots, cube roots, fourth roots, and so on. These roots are studied in connection with the manipulation of radical expressions and the solution of real-world applications. Exponents that are fractions are also studied and are used to ease some of our work with radicals. The chapter closes with an examination of the complex-number system.

APPLICATION *Firefighting.*

The number of gallons per minute discharged from a fire hose depends on the diameter of the hose and the nozzle pressure. The following table and graph show the amount of water flow for a 2-in. diameter solid bore nozzle at various nozzle pressures. Determine whether the graph indicates that a radical function can be used to model water flow.

Nozzle Pressure (in pounds per square inch, psi)	Water Flow (in gallons per) minute, gpm)
40	752
60	921
80	1063
100	1188
120	1302
150	1455
200	1681

Source: www.firetactics.com

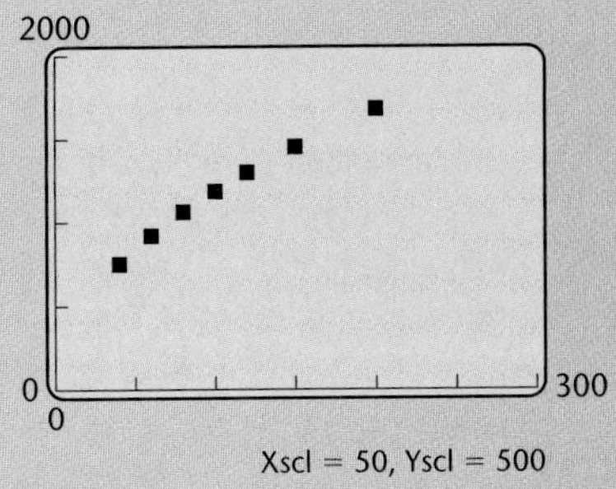

This problem appears as Example 14 in Section 7.1.

7.1 Radical Expressions, Functions, and Models

Square Roots and Square Root Functions ■ Expressions of the Form $\sqrt{a^2}$ ■ Cube Roots ■ Odd and Even *n*th Roots ■ Radical Functions and Models

In this section, we consider roots, such as square roots and cube roots. We look at the symbolism that is used and ways in which expressions can be manipulated to get equivalent expressions. All of this will be important in problem solving.

Square Roots and Square Root Functions

When a number is multiplied by itself, we say that the number is squared. Often we need to know what number was squared in order to produce some value a. If such a number can be found, we call that number a *square root* of a.

Square Root The number c is a *square root* of a if $c^2 = a$.

For example,

9 has -3 and 3 as square roots because $(-3)^2 = 9$ and $3^2 = 9$.

25 has -5 and 5 as square roots because $(-5)^2 = 25$ and $5^2 = 25$.

-4 does not have a real-number square root because there is no real number c such that $c^2 = -4$.

Note that every positive number has two square roots, whereas 0 has only itself as a square root. Negative numbers do not have real-number square roots, although later in this chapter we introduce the *complex-number* system in which such square roots do exist.

EXAMPLE 1 Find the two square roots of 64.

SOLUTION The square roots of 64 are 8 and -8, because $8^2 = 64$ and $(-8)^2 = 64$.

Whenever we refer to *the* square root of a number, we mean the nonnegative square root of that number. This is often referred to as the *principal square root* of the number.

Principal Square Root The *principal square root* of a nonnegative number is its nonnegative square root. The symbol $\sqrt{}$ is called a *radical sign* and is used to indicate the principal square root of the number over which it appears.

Study Tip

Advance Planning Pays Off

The best way to prepare for a final exam is to do so over a period of at least two weeks. First review each chapter, studying the terms, formulas, problems, properties, and procedures in the chapter Summary and Review. Then retake your quizzes and tests. If you miss any questions, spend extra time reviewing the corresponding topics. Watch the DVDs that accompany the text or use the accompanying tutorial software. Also consider participating in a study group or attending a tutoring or review session.

Student Notes

It is important to remember the difference between *the* square root of 9 and *a* square root of 9. *A* square root of 9 means either 3 or -3, but *the* square root of 9, or $\sqrt{9}$, means the principal square root of 9, or 3.

EXAMPLE 2 Simplify each of the following.

a) $\sqrt{25}$ **b)** $\sqrt{\frac{25}{64}}$

c) $-\sqrt{64}$ **d)** $\sqrt{0.0049}$

SOLUTION

a) $\sqrt{25} = 5$ **$\sqrt{\ }$ indicates the principal square root. Note that $\sqrt{25} \neq -5$.**

b) $\sqrt{\frac{25}{64}} = \frac{5}{8}$ **Since $\left(\frac{5}{8}\right)^2 = \frac{25}{64}$**

c) $-\sqrt{64} = -8$ **Since $\sqrt{64} = 8$, $-\sqrt{64} = -8$.**

d) $\sqrt{0.0049} = 0.07$ **$(0.07)(0.07) = 0.0049$**

In addition to being read as "the principal square root of a," $\sqrt{a}$ is also read as "the square root of a," or simply "root a" or "radical a." Any expression in which a radical sign appears is called a *radical expression*. The following are radical expressions:

$$\sqrt{5}, \quad \sqrt{a}, \quad -\sqrt{3x}, \quad \sqrt{\frac{y^2 + 7}{y}}, \quad \sqrt{x} + 8.$$

The expression under the radical sign is called the **radicand.** In the expressions above, the radicands are 5, a, $3x$, $(y^2 + 7)/y$, and x.

All but the most basic calculators give values for square roots. For example, to calculate $\sqrt{5}$ on most graphing calculators, we press (√) (5) ( )) (ENTER). On some calculators, (√) is pressed after 5 has been entered. A calculator will display an approximation like

2.23606798

for $\sqrt{5}$. The exact value of $\sqrt{5}$ is not given by any repeating or terminating decimal. The same is true for the square root of any whole number that is not a perfect square. We discussed such *irrational numbers* in Chapter 1.

The square-root function, given by

$$f(x) = \sqrt{x},$$

has the interval $[0, \infty)$ as its domain. We can draw its graph by selecting convenient values for x and calculating the corresponding outputs. Once these ordered pairs have been graphed, a smooth curve can be drawn.

As x increases, the output $\sqrt{x}$ increases. The range of $f(x)$ is $[0, \infty)$.

$f(x) = \sqrt{x}$

x	$\sqrt{x}$	$(x, f(x))$
0	0	(0, 0)
1	1	(1, 1)
4	2	(4, 2)
9	3	(9, 3)

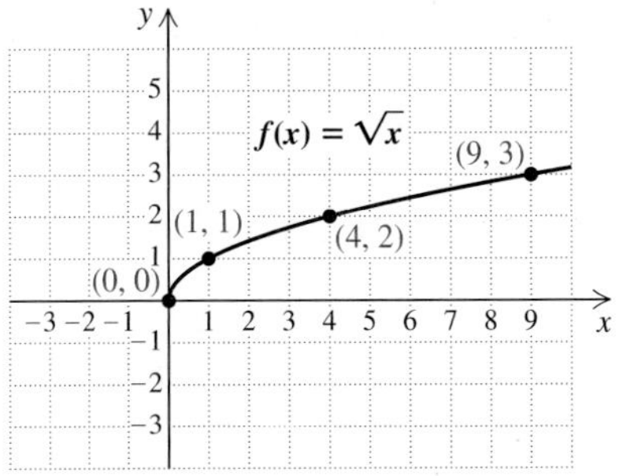

EXAMPLE 3 For each function, find the indicated function value.

a) $f(x) = \sqrt{3x - 2};\ f(1)$

b) $g(z) = -\sqrt{6z + 4};\ g(3)$

SOLUTION

a) $f(1) = \sqrt{3 \cdot 1 - 2}$ **Substituting**

$= \sqrt{1} = 1$ **Simplifying**

b) $g(3) = -\sqrt{6 \cdot 3 + 4}$ **Substituting**

$= -\sqrt{22}$ **Simplifying. This answer is exact.**

≈ -4.69041576 **Using a calculator to approximate $\sqrt{22}$**

Expressions of the Form $\sqrt{a^2}$

It is tempting to write $\sqrt{a^2} = a$, but the next example shows that, as a rule, this is untrue.

EXAMPLE 4 Evaluate $\sqrt{x^2}$ for the following values: **(a)** 5; **(b)** 0; **(c)** -5.

SOLUTION

a) $\sqrt{5^2} = \sqrt{25} = 5$ — **Same**

b) $\sqrt{0^2} = \sqrt{0} = 0$ — **Same**

c) $\sqrt{(-5)^2} = \sqrt{25} = 5$ — **Opposites** **Note that $\sqrt{(-5)^2} \neq -5$.**

In Example 4, we saw that $\sqrt{5^2} = 5$ and $\sqrt{(-5)^2} = 5$. Recall that $|5| = 5$ and $|-5| = 5$. Thus evaluating $\sqrt{a^2}$ is just like evaluating $|a|$.

Interactive Discovery

Use graphs or tables to determine which of the following are identities. Be sure to enclose the entire radicand in parentheses.

1. $\sqrt{x^2} = x$
2. $\sqrt{x^2} = -x$
3. $\sqrt{x^2} = |x|$
4. $\sqrt{(x + 3)^2} = x + 3$
5. $\sqrt{(x + 3)^2} = |x + 3|$
6. $\sqrt{x^8} = x^4$

In general, we cannot say that $\sqrt{x^2} = x$. However, we can simplify $\sqrt{x^2}$ using absolute value.

Simplifying $\sqrt{a^2}$ For any real number a,

$$\sqrt{a^2} = |a|.$$

(The principal square root of a^2 is the absolute value of a.)

When a radicand is the square of a variable expression, like $(x + 5)^2$ or $36t^2$, absolute-value signs are needed when simplifying. We use absolute-value signs unless we know that the expression being squared is nonnegative. This assures that our result is never negative.

EXAMPLE 5 Simplify each expression. Assume that the variable can represent any real number.

a) $\sqrt{(x + 1)^2}$ **b)** $\sqrt{x^2 - 8x + 16}$

c) $\sqrt{a^8}$ **d)** $\sqrt{t^6}$

SOLUTION

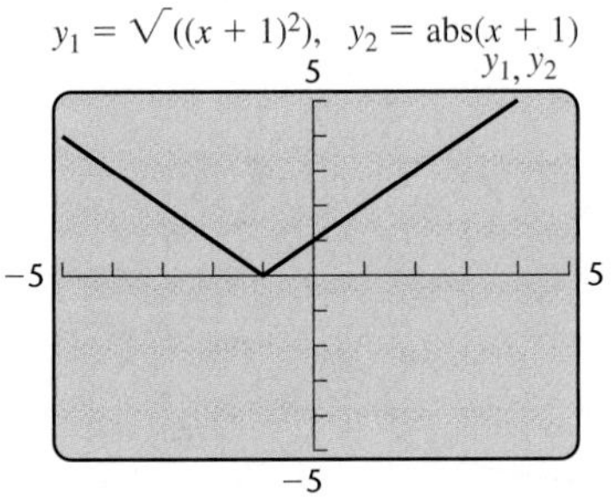

a) $\sqrt{(x + 1)^2} = |x + 1|$ **Since $x + 1$ might be negative (for example, if $x = -3$), absolute-value notation is necessary.**

The graph at left confirms the identity.

b) $\sqrt{x^2 - 8x + 16} = \sqrt{(x - 4)^2} = |x - 4|$ **Since $x - 4$ might be negative, absolute-value notation is necessary.**

c) Note that $(a^4)^2 = a^8$ and that a^4 is never negative. Thus,

$$\sqrt{a^8} = |a^4| = a^4. \quad \textbf{Absolute-value notation is unnecessary here.}$$

d) Note that $(t^3)^2 = t^6$. Thus,

$$\sqrt{t^6} = |t^3|. \quad \textbf{Since } t^3 \textbf{ might be negative, absolute-value notation is necessary.}$$

If we know that no radicands have been formed by raising negative quantities to even powers, we do not need absolute-value signs.

EXAMPLE 6 Simplify each expression. Assume that no radicands were formed by raising negative quantities to even powers.

a) $\sqrt{y^2}$ **b)** $\sqrt{a^{10}}$ **c)** $\sqrt{9x^2 - 6x + 1}$

SOLUTION

a) $\sqrt{y^2} = y$ **We are assuming that y is nonnegative, so no absolute-value notation is necessary. When y is negative, $\sqrt{y^2} \neq y$.**

b) $\sqrt{a^{10}} = a^5$ **Assuming that a^5 is nonnegative. Note that $(a^5)^2 = a^{10}$.**

c) $\sqrt{9x^2 - 6x + 1} = \sqrt{(3x - 1)^2} = 3x - 1$ **Assuming that $3x - 1$ is nonnegative**

Cube Roots

We often need to know what number was cubed in order to produce a certain value. When such a number is found, we say that we have found a *cube root*. For example,

$$2 \text{ is the cube root of } 8 \text{ because } 2^3 = 2 \cdot 2 \cdot 2 = 8;$$

$$-4 \text{ is the cube root of } -64 \text{ because } (-4)^3 = (-4)(-4)(-4) = -64.$$

Cube Root The number c is the *cube root* of a if $c^3 = a$. In symbols, we write $\sqrt[3]{a}$ to denote the cube root of a.

The cube-root function, given by

$$f(x) = \sqrt[3]{x}$$

has $\mathbb{R}$ as its domain. We can draw its graph by selecting convenient values for x and calculating the corresponding outputs. Once these ordered pairs have been graphed, a smooth curve can be drawn. Note that the range of $f(x)$ is also $\mathbb{R}$.

$f(x) = \sqrt[3]{x}$

x	$\sqrt[3]{x}$	$(x, f(x))$
0	0	(0, 0)
1	1	(1, 1)
8	2	(8, 2)
−1	−1	(−1, −1)
−8	−2	(−8, −2)

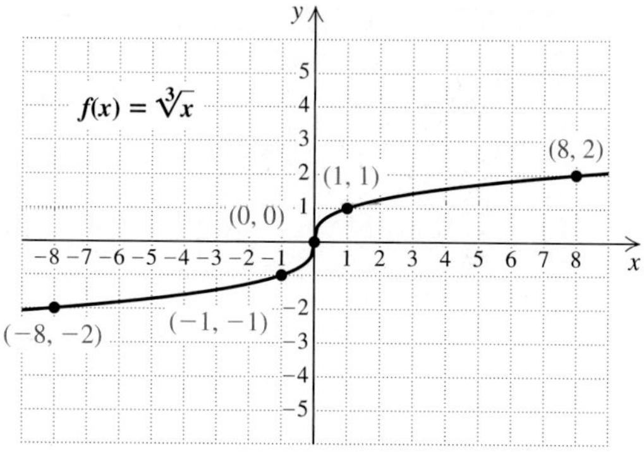

In the real-number system, every number has exactly one cube root. The cube root of a positive number is positive, the cube root of a negative number is negative, and the cube root of 0 is 0.

The following graphs illustrate that

$$\sqrt{x^2} = |x| \quad \text{and} \quad \sqrt[3]{x^3} = x.$$

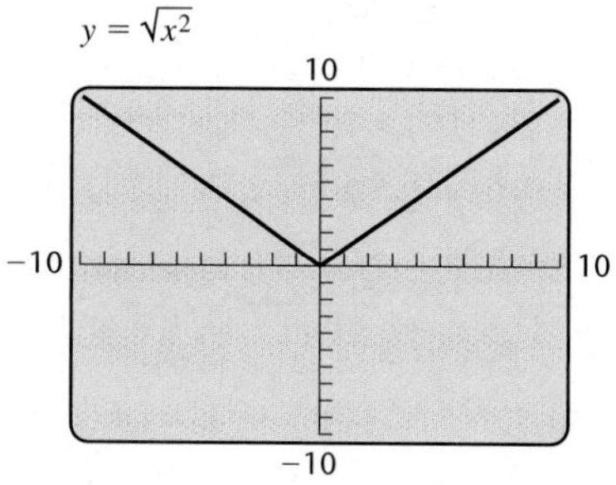

This is the graph of $y = |x|$. Note that $\sqrt{x^2}$ is never negative.

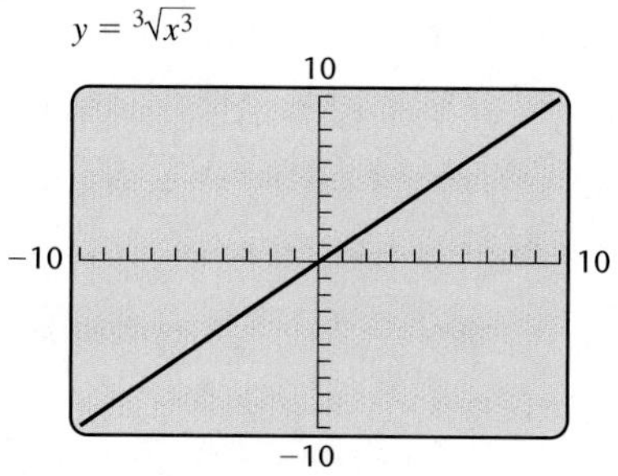

This is the graph of $y = x$. Note that $y = \sqrt[3]{x^3}$ can be negative.

EXAMPLE 7 For each function, find the indicated function value.

a) $f(y) = \sqrt[3]{y}$; $f(125)$

b) $g(x) = \sqrt[3]{x - 3}$; $g(-24)$

SOLUTION

a) $f(125) = \sqrt[3]{125} = 5$ **Since $5 \cdot 5 \cdot 5 = 125$**

b) $g(-24) = \sqrt[3]{-24 - 3}$
$= \sqrt[3]{-27}$
$= -3$ **Since $(-3)(-3)(-3) = -27$**

EXAMPLE 8 Simplify: $\sqrt[3]{-8y^3}$.

SOLUTION

$\sqrt[3]{-8y^3} = -2y$ **Since $(-2y)(-2y)(-2y) = -8y^3$**

Odd and Even *n*th Roots

The fourth root of a number a is the number c for which $c^4 = a$. There are also 5th roots, 6th roots, and so on. We write $\sqrt[n]{a}$ for the nth root. The number n is called the **index** (plural, **indices**). When the index is 2, we generally do not write it.

EXAMPLE 9 Find each of the following.

a) $\sqrt[5]{32}$ **b)** $\sqrt[5]{-32}$
c) $-\sqrt[5]{32}$ **d)** $-\sqrt[5]{-32}$

SOLUTION

a) $\sqrt[5]{32} = 2$ **Since $2^5 = 32$**
b) $\sqrt[5]{-32} = -2$ **Since $(-2)^5 = -32$**
c) $-\sqrt[5]{32} = -2$ **Taking the opposite of $\sqrt[5]{32}$**
d) $-\sqrt[5]{-32} = -(-2) = 2$ **Taking the opposite of $\sqrt[5]{-32}$**

Every number has just one real root when n is odd. Odd roots of positive numbers are positive and odd roots of negative numbers are negative. Absolute-value signs are not used when finding odd roots.

EXAMPLE 10 Find each of the following.

a) $\sqrt[7]{x^7}$ **b)** $\sqrt[9]{(x - 1)^9}$

SOLUTION

a) $\sqrt[7]{x^7} = x$ **b)** $\sqrt[9]{(x - 1)^9} = x - 1$

When the index n is even, we say that we are taking an *even root*. Every positive real number has two real nth roots when n is even—one positive and one negative. Negative numbers do not have real nth roots when n is even.

When n is even, the notation $\sqrt[n]{a}$ indicates the nonnegative nth root. Thus, to write even nth roots, absolute-value signs are often necessary.

EXAMPLE 11 Simplify each expression, if possible. Assume that variables can represent any real number.

a) $\sqrt[4]{16}$

b) $-\sqrt[4]{16}$

c) $\sqrt[4]{-16}$

d) $\sqrt[4]{81x^4}$

e) $\sqrt[6]{(y+7)^6}$

SOLUTION

a) $\sqrt[4]{16} = 2$ — **Since $2^4 = 16$**

b) $-\sqrt[4]{16} = -2$ — **Taking the opposite of $\sqrt[4]{16}$**

c) $\sqrt[4]{-16}$ cannot be simplified. — **$\sqrt[4]{-16}$ is not a real number.**

d) $\sqrt[4]{81x^4} = 3|x|$ — **Use absolute-value notation since x could represent a negative number.**

e) $\sqrt[6]{(y+7)^6} = |y+7|$ — **Use absolute-value notation since $y + 7$ could be negative.**

We summarize as follows.

Simplifying *n*th roots

n	a	$\sqrt[n]{a}$	$\sqrt[n]{a^n}$
Even	Positive	Positive	$\lvert a \rvert$ (or a)
	Negative	Not a real number	$\lvert a \rvert$ (or $-a$)
Odd	Positive	Positive	a
	Negative	Negative	a

Radical Functions and Models

A **radical function** is a function that can be described by a radical expression.

Radical Expressions and Functions

When entering radical expressions on a graphing calculator, care must be taken to place parentheses properly. For example, we enter

$$f(x) = \frac{3.5 + \sqrt{4.5 - 6x}}{5}$$

on the Y= screen as (3 . 5 + 2ND √ 4 . 5 − 6 X,T,Θ,n)) ÷ 5. The outer parentheses enclose the numerator of the expression. The first right parenthesis) indicates the end of the radicand; the left parenthesis of the radicand is supplied by the calculator when √ is pressed.

(continued)

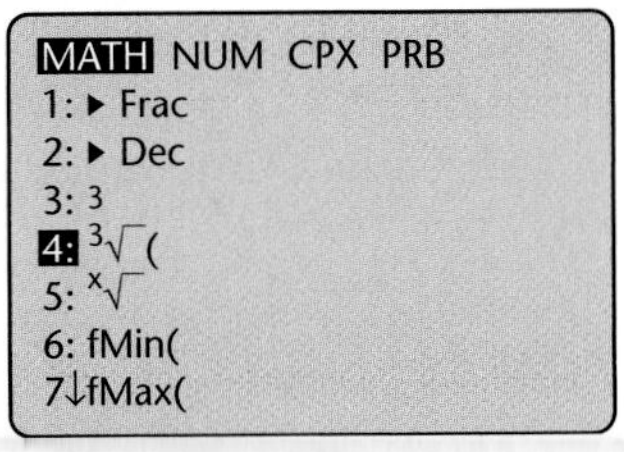

Cube roots can be entered using the $\sqrt[3]{}$(option in the MATH MATH menu, as shown at left. For an index other than 2 or 3, first enter the index, then choose the $\sqrt[x]{}$ option from the MATH MATH menu, and then enter the radicand. The $\sqrt[x]{}$ option may not supply the left parenthesis.

If we can determine the domain of the function algebraically, we can use that information to choose an appropriate viewing window. For example, compare the following graphs of the function $f(x) = 2\sqrt{15 - x}$. Note that the domain of f is $\{x | x \le 15\}$, or $(-\infty, 15]$.

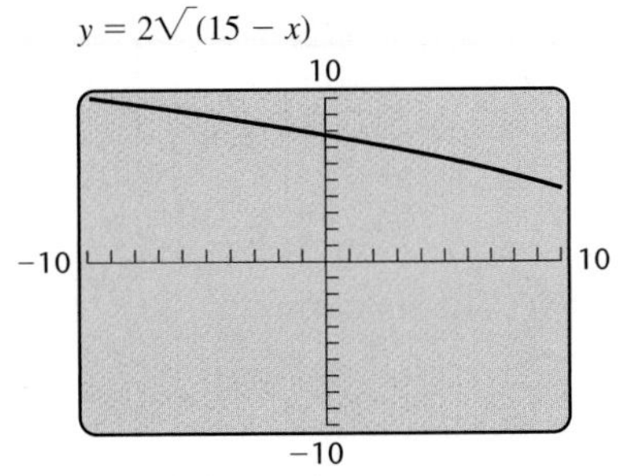

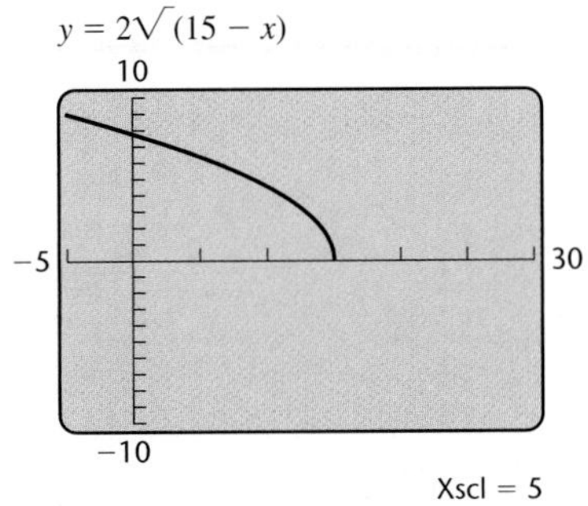

The graph on the left above uses the standard viewing window. The domain of the function is not clear from the graph, and the shape looks almost like a straight line. The viewing window for the graph on the right above is $[-5, 30, -10, 10]$. The part of the x-axis shown there includes the endpoint of the interval that gives the domain of the function. Knowing the domain of the function helps us choose Xmin and Xmax. For now, choosing appropriate values of Ymin and Ymax may require some trial and error.

If a function is given by a radical expression with an odd index, the domain is the set of all real numbers. If a function is given by a radical expression with an even index, the domain is the set of replacements for which the radicand is nonnegative.

EXAMPLE 12 Find the domain of the function given by each of the following equations. Check by graphing the function. Then, from the graph, estimate the range of the function.

a) $q(x) = \sqrt{-x}$

b) $t(x) = \sqrt{2x - 5} - 3$

c) $f(x) = \sqrt{x^2 + 1}$

SOLUTION

a) We find all values of x for which the radicand is nonnegative:

$$-x \ge 0$$

$$x \le 0. \quad \textbf{Multiplying by } -1\textbf{; reversing the direction of the inequality}$$

The domain is $\{x | x \leq 0\}$, or $(-\infty, 0]$, as indicated on the graph by the shading on the x-axis. The range appears to be $[0, \infty)$, as indicated by the shading on the y-axis. This can also be seen by examining a table of values.

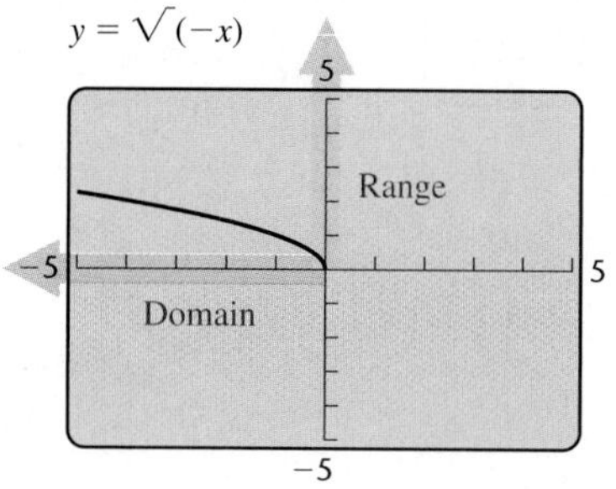

b) The function $t(x) = \sqrt{2x - 5} - 3$ is defined when $2x - 5 \geq 0$:

$$2x - 5 \geq 0$$
$$2x \geq 5 \quad \textbf{Adding 5}$$
$$x \geq \tfrac{5}{2}. \quad \textbf{Dividing by 2}$$

The domain is $\{x | x \geq \frac{5}{2}\}$, or $[\frac{5}{2}, \infty)$, as indicated on the graph by the shading on the x-axis. The range appears to be $[-3, \infty)$, as indicated by the shading on the y-axis.

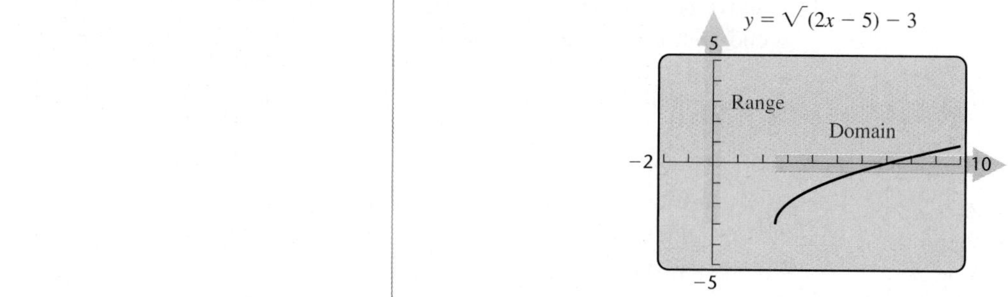

c) The radicand in $f(x) = \sqrt{x^2 + 1}$ is $x^2 + 1$. We must have

$$x^2 + 1 \geq 0$$
$$x^2 \geq -1.$$

Since x^2 is nonnegative for all real numbers x, the inequality is true for all real numbers. The domain is $(-\infty, \infty)$. The range appears to be $[1, \infty)$.

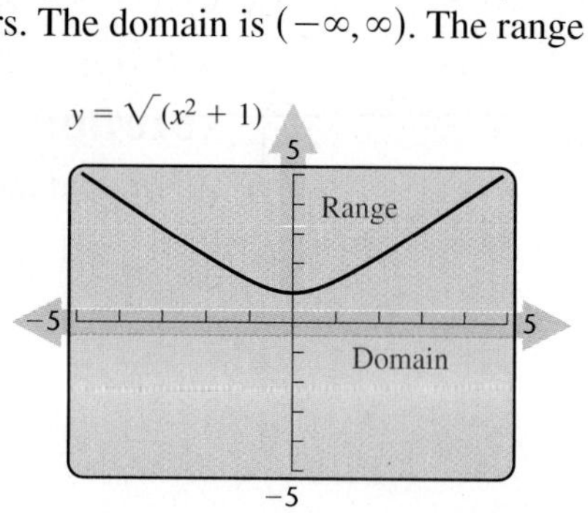

EXAMPLE 13 Determine the domain of the function given by

$$g(x) = \sqrt[6]{7 - 3x}.$$

SOLUTION Since the index is even, the radicand, $7 - 3x$, must be nonnegative. We solve the inequality:

$$7 - 3x \geq 0 \qquad \text{We cannot find the 6th root of a negative number.}$$
$$-3x \geq -7$$
$$x \leq \tfrac{7}{3}. \qquad \text{Multiplying both sides by } -\tfrac{1}{3} \text{ and reversing the inequality}$$

Thus,

the domain of g is $\left\{x \mid x \leq \frac{7}{3}\right\}$, or $\left(-\infty, \frac{7}{3}\right]$.

Some situations can be modeled using radical functions. The graphs of radical functions can have many different shapes. However, radical functions given by equations of the form

$$r(x) = \sqrt{ax + b}$$

will have the general shape of the graph of $f(x) = \sqrt{x}$.

We can determine whether a radical function might fit a set of data by plotting the points.

EXAMPLE 14 Firefighting. The number of gallons per minute discharged from a fire hose depends on the diameter of the hose and the nozzle pressure. The following table lists the amount of water flow for a 2-in. diameter solid bore nozzle at various nozzle pressures. Graph the data and determine whether a radical function can be used to model water flow.

Nozzle Pressure (in pounds per square inch, psi)	Water Flow (in gallons per minute, GPM)
40	752
60	921
80	1063
100	1188
120	1302
150	1455
200	1681

Source: www.firetactics.com

SOLUTION We graph the data, entering the nozzle pressure as L1 and the water flow as L2. The data appear to follow the pattern of the graph of a radical function. We determine that a radical function could be used to model water flow.

L1	L2	L3 1
40	752	------
60	921	
80	1063	
100	1188	
120	1302	
150	1455	
200	1681	

L1(1) = 40

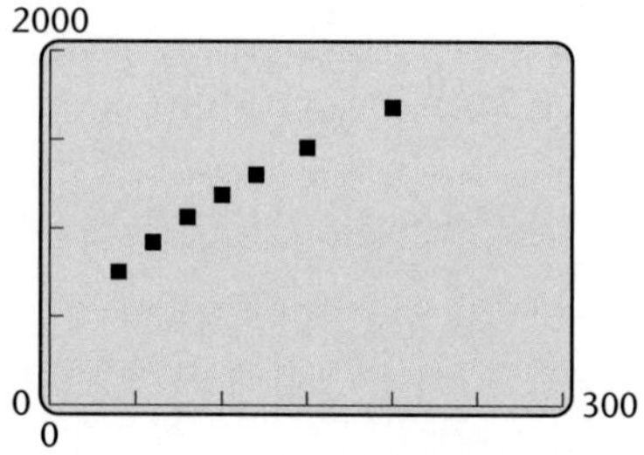

Xscl = 50, Yscl = 500

EXAMPLE 15 Firefighting. Water flow, as described in Example 14, depends on the diameter of the hose and the nozzle pressure. For a 2-in. diameter solid bore nozzle, the water flow is given by

$$f(x) = 118.8\sqrt{x},$$

where $f(x)$ is the water flow in gallons per minute (GPM) when the nozzle pressure is x pounds per square inch (psi) (*Source*: Houston Fire Department Continuing Education, www.houstontx.gov). Find the water flow when the nozzle pressure is 50 psi and when it is 175 psi.

SOLUTION We enter the equation and graph it along with the data in Example 14 to show that the model fits the data. Using a table with Indpnt set to Ask, we determine that when the nozzle pressure is 50 psi, the water flow is approximately 840 GPM, and at 175 psi, the water flow is approximately 1572 GPM.

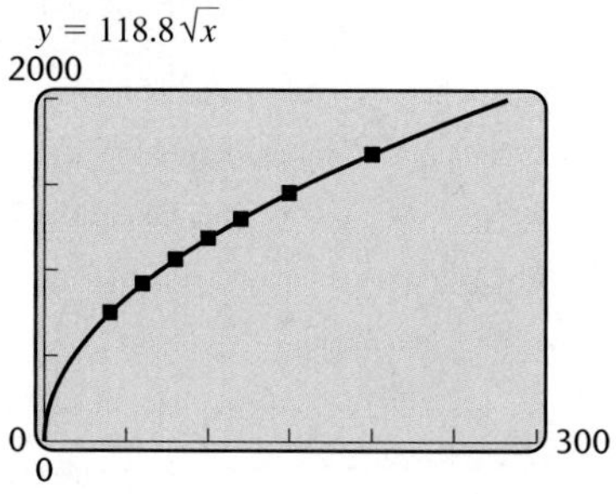

Xscl = 50, Yscl = 500

X	Y1	
50	840.04	
175	1571.6	

X =

Visualizing the Graph

Match each function with its graph.

1. $f(x) = 2x - 5$

2. $f(x) = x^2 - 1$

3. $f(x) = \sqrt{x - 2}$

4. $f(x) = x - 2$

5. $f(x) = -\frac{1}{3}x$

6. $f(x) = x^3$

7. $f(x) = 4 - x$

8. $f(x) = |2x - 5|$

9. $f(x) = -2$

10. $f(x) = -\frac{1}{3}x + 4$

Answers on page A-29

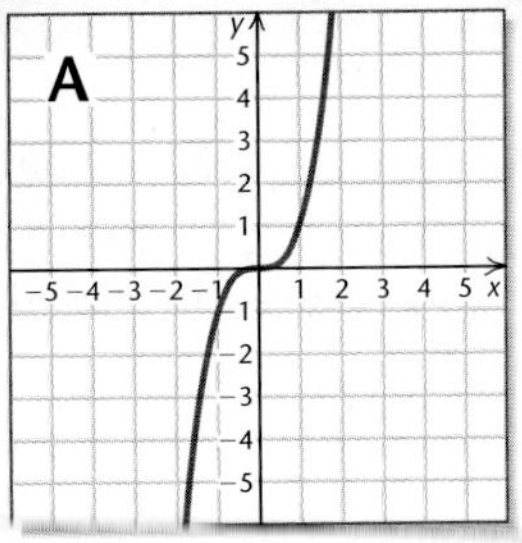

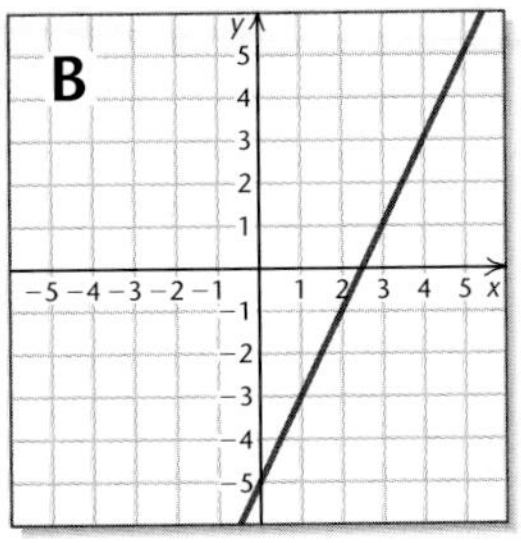

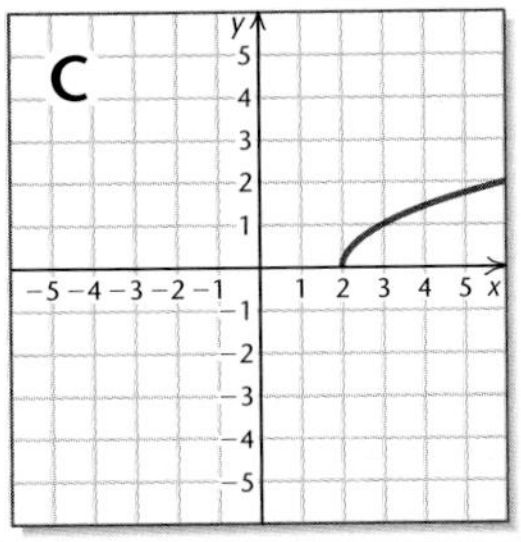

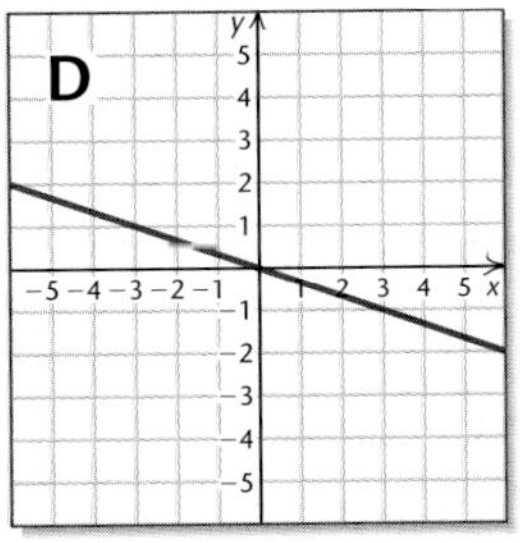

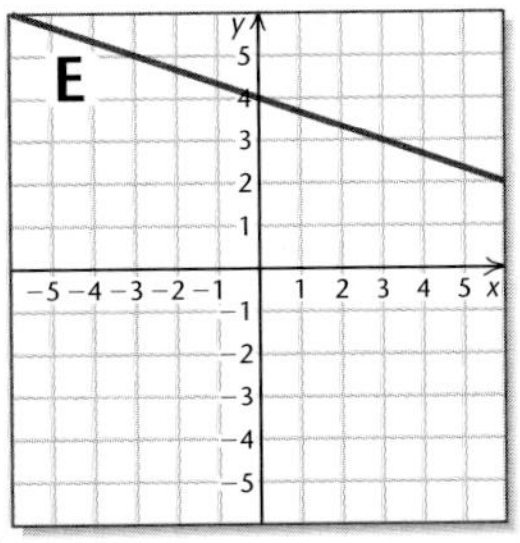

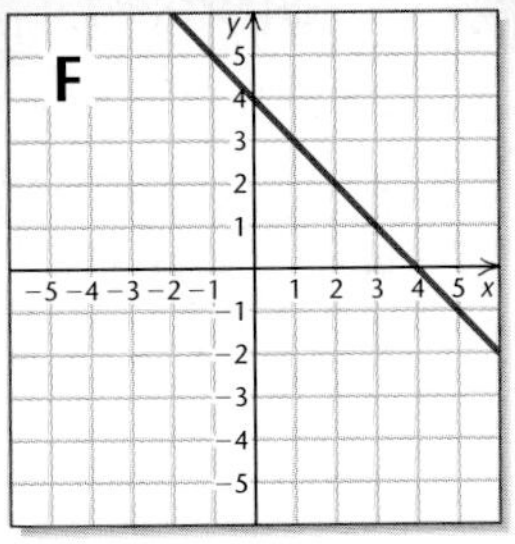

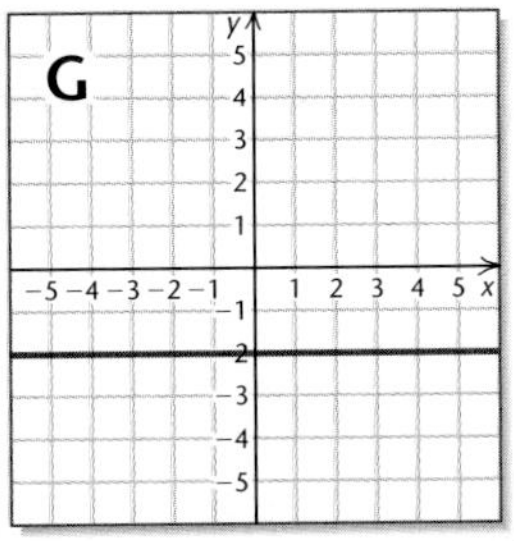

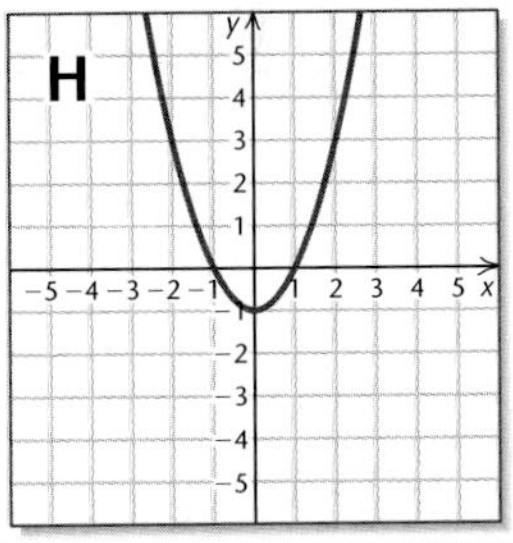

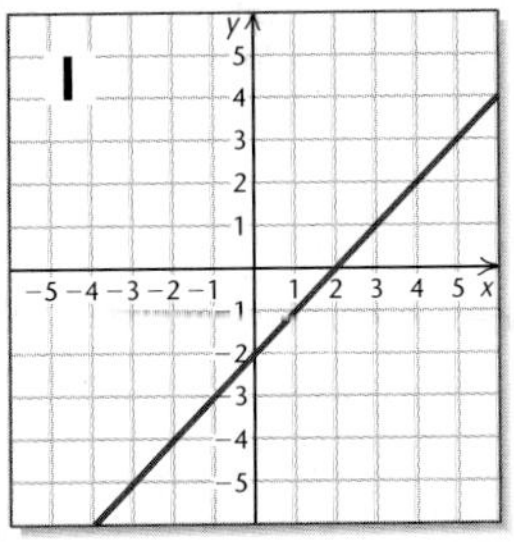

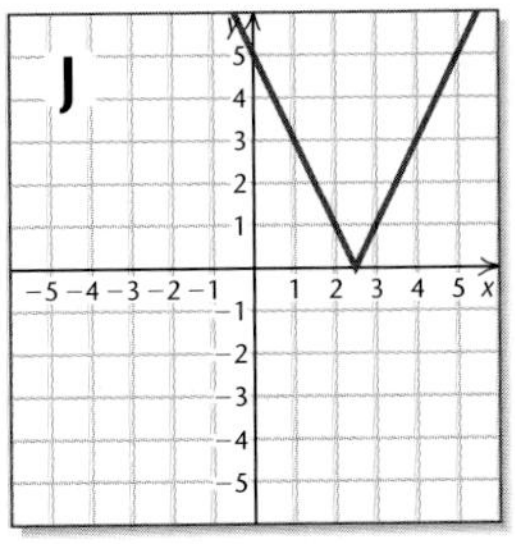

7.1 EXERCISE SET

FOR EXTRA HELP

MathXL | MyMathLab | InterAct Math | AW Math Tutor Center | Video Lectures on CD: Disc 4 | Student's Solutions Manual

Concept Reinforcement *Select the appropriate word to complete each of the following.*

1. Every positive number has ________ square root(s). (one/two)

2. The principal square root is never ________. (negative/positive)

3. For any ________ number a, we have $\sqrt{a^2} = a$. (negative/positive)

4. For any ________ number a, we have $\sqrt{a^2} = -a$. (negative/positive)

5. If a is a whole number that is not a perfect square, then $\sqrt{a}$ is a(n) ________ number. (irrational/rational)

6. The domain of the function f given by $f(x) = \sqrt[3]{x}$ is all ________ numbers. (whole/real/positive)

7. If $\sqrt[4]{x}$ is a real number, then x must be ________. (negative/positive/nonnegative)

8. If $\sqrt[3]{x}$ is negative, then x must be ________. (negative/positive)

For each number, find all of its square roots.

9. 49 **10.** 81

11. 144 **12.** 9

13. 400 **14.** 2500

15. 900 **16.** 225

Simplify.

17. $-\sqrt{\dfrac{36}{49}}$ **18.** $-\sqrt{\dfrac{361}{9}}$

19. $\sqrt{441}$ **20.** $\sqrt{196}$

21. $-\sqrt{\dfrac{16}{81}}$ **22.** $-\sqrt{\dfrac{81}{144}}$

23. $\sqrt{0.04}$ **24.** $\sqrt{0.36}$

25. $-\sqrt{0.0025}$ **26.** $\sqrt{0.0144}$

Identify the radicand and the index for each expression.

27. $5\sqrt{p^2 + 4}$ **28.** $-7\sqrt{y^2 - 8}$

29. $x^2y^3\sqrt[3]{\dfrac{x}{y+4}}$ **30.** $a^2b^3\sqrt[3]{\dfrac{a}{a^2 - b}}$

For each function, find the specified function value, if it exists. If it does not exist, state this.

31. $f(t) = \sqrt{5t - 10}$; $f(6), f(2), f(1), f(-1)$

32. $g(x) = \sqrt{x^2 - 25}$; $g(-6), g(3), g(6), g(13)$

33. $t(x) = -\sqrt{2x + 1}$; $t(4), t(0), t(-1), t\left(-\frac{1}{2}\right)$

34. $p(z) = \sqrt{2z^2 - 20}$; $p(4), p(3), p(-5), p(0)$

35. $f(t) = \sqrt{t^2 + 1}$; $f(0), f(-1), f(-10)$

36. $g(x) = -\sqrt{(x + 1)^2}$; $g(-3), g(4), g(-5)$

37. $g(x) = \sqrt{x^3 + 9}$; $g(-2), g(-3), g(3)$

38. $f(t) = \sqrt{t^3 - 10}$; $f(2), f(3), f(4)$

Simplify. Remember to use absolute-value notation when necessary. If a root cannot be simplified, state this.

39. $\sqrt{36x^2}$ **40.** $\sqrt{25t^2}$

41. $\sqrt{(-6b)^2}$ **42.** $\sqrt{(-7c)^2}$

43. $\sqrt{(8 - t)^2}$ **44.** $\sqrt{(a + 3)^2}$

45. $\sqrt{y^2 + 16y + 64}$ **46.** $\sqrt{x^2 - 4x + 4}$

47. $\sqrt{4x^2 + 28x + 49}$ **48.** $\sqrt{9x^2 - 30x + 25}$

49. $\sqrt[4]{256}$ **50.** $-\sqrt[4]{625}$

51. $\sqrt[5]{-1}$ **52.** $-\sqrt[5]{3^5}$

53. $\sqrt[5]{-\dfrac{32}{243}}$ **54.** $\sqrt[5]{-\dfrac{1}{32}}$

55. $\sqrt[6]{x^6}$ **56.** $\sqrt[8]{y^8}$

57. $\sqrt[4]{(6a)^4}$ **58.** $\sqrt[4]{(7b)^4}$

59. $\sqrt[10]{(-6)^{10}}$

60. $\sqrt[12]{(-10)^{12}}$

61. $\sqrt[414]{(a + b)^{414}}$

62. $\sqrt[1976]{(2a + b)^{1976}}$

63. $\sqrt{a^{22}}$

64. $\sqrt{x^{10}}$

65. $\sqrt{-25}$

66. $\sqrt{-16}$

Simplify. Assume that no radicands were formed by raising negative quantities to even powers.

67. $\sqrt{16x^2}$

68. $\sqrt{25t^2}$

69. $\sqrt{(3t)^2}$

70. $\sqrt{(7c)^2}$

71. $\sqrt{(a + 1)^2}$

72. $\sqrt{(5 + b)^2}$

73. $\sqrt{4x^2 + 8x + 4}$

74. $\sqrt{9x^2 + 36x + 36}$

75. $\sqrt{9t^2 - 12t + 4}$

76. $\sqrt{25t^2 - 20t + 4}$

77. $\sqrt[3]{27}$

78. $-\sqrt[3]{64}$

79. $\sqrt[4]{16x^4}$

80. $\sqrt[4]{81x^4}$

81. $\sqrt[3]{-216}$

82. $-\sqrt[5]{-100{,}000}$

83. $-\sqrt[3]{-125y^3}$

84. $-\sqrt[3]{-64x^3}$

85. $\sqrt{t^{18}}$

86. $\sqrt{a^{14}}$

87. $\sqrt{(x - 2)^8}$

88. $\sqrt{(x + 3)^{10}}$

For each function, find the specified function value, if it exists. If it does not exist, state this.

89. $f(x) = \sqrt[3]{x + 1}$; $f(7), f(26), f(-9), f(-65)$

90. $g(x) = -\sqrt[3]{2x - 1}$; $g(0), g(-62), g(-13), g(63)$

91. $g(t) = \sqrt[4]{t - 3}$; $g(19), g(-13), g(1), g(84)$

92. $f(t) = \sqrt[4]{t + 1}$; $f(0), f(15), f(-82), f(80)$

Determine the domain of each function described.

93. $f(x) = \sqrt{x - 6}$

94. $g(x) = \sqrt{x + 8}$

95. $g(t) = \sqrt[4]{t + 8}$

96. $f(x) = \sqrt[4]{x - 9}$

97. $g(x) = \sqrt[4]{2x - 10}$

98. $g(t) = \sqrt[3]{2t - 6}$

99. $f(t) = \sqrt[5]{8 - 3t}$

100. $f(t) = \sqrt[6]{4 - 3t}$

101. $h(z) = -\sqrt[6]{5z + 2}$

102. $d(x) = -\sqrt[4]{7x - 5}$

Aha! 103. $f(t) = 7 + \sqrt[8]{t^8}$

104. $g(t) = 9 + \sqrt[6]{t^6}$

Determine algebraically the domain of each function described. Then use a graphing calculator to confirm your answer and to estimate the range.

105. $f(x) = \sqrt{5 - x}$

106. $g(x) = \sqrt{2x + 1}$

107. $f(x) = 1 - \sqrt{x + 1}$

108. $g(x) = 2 + \sqrt{3x - 5}$

109. $g(x) = 3 + \sqrt{x^2 + 4}$

110. $f(x) = 5 - \sqrt{3x^2 + 1}$

In Exercises 111–114, match each function with one of the following graphs without using a calculator.

a)

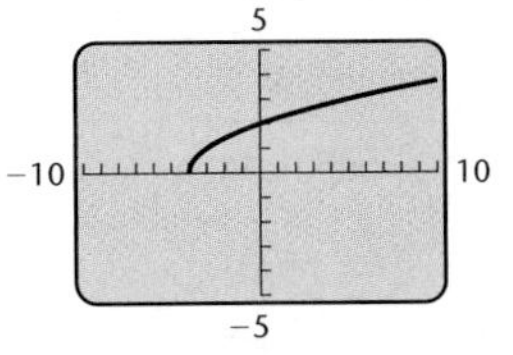

b)

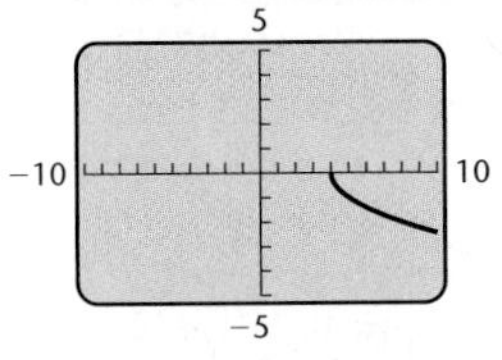

c)

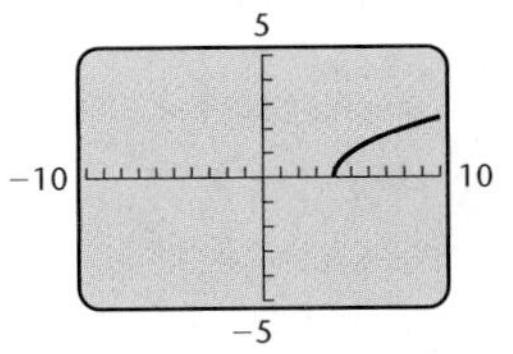

d)

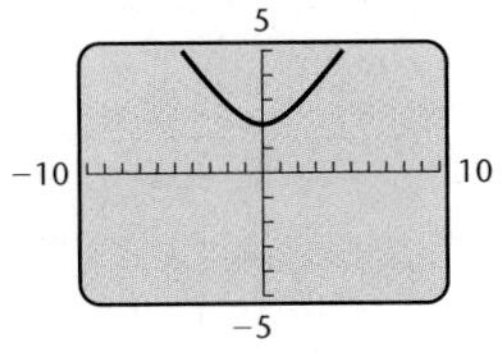

111. $f(x) = \sqrt{x - 4}$

112. $g(x) = \sqrt{x + 4}$

113. $h(x) = \sqrt{x^2 + 4}$

114. $f(x) = -\sqrt{x - 4}$

In Exercises 115–120, determine whether a radical function would be a good model of the given situation.

115. *Wind Chill.* When the wind is blowing, the air temperature feels lower than the actual temperature. This is referred to as the wind-chill temperature. The following table lists wind-chill temperatures for various wind speeds at a thermometer reading of 15°F.

Wind Speed (in miles per hour)	Wind-Chill Temperature (in degrees Fahrenheit)
5	7
10	3
15	0
20	−2
25	−4
30	−5
35	−7
40	−8

Source: National Weather Service

116. *Farm Size.* The following table lists the average size of United States' farms for various years from 1940 to 2002.

Year	Average Size of Farm (in acres)
1940	175
1960	303
1980	426
1997	431
2002	441

Source: U.S. Department of Agriculture

117. *Koi Growth.* Koi, a popular fish for backyard pools, grow from $\frac{1}{40}$ cm when newly hatched to an average length of 80 cm. The following table lists the length of a koi at various ages.

Age (in months)	Length (in centimeters)
1	2.9
2	5.0
4	9.1
13	24.9
16	29.3
30	45.8
36	51.1
48	59.3
60	65.2
72	69.4

Source: www.coloradokoi.com

118. *Cancer Research.* The following table lists the amount of federal funds allotted to the National Cancer Institute for cancer research in the United States from 2003 to 2007.

Year	Funds (in billions)
2003	$4.59
2004	4.74
2005	4.83
2006	4.79
2007*	4.75

*Requested
Source: National Cancer Institute

119. *Postage.* The following table lists the cost to send a Media Mail Package through the U.S. Postal Service in 2006 for packages of various weights.

Weight (in pounds)	Cost
1	$1.59
2	2.07
3	2.55
4	3.03
5	3.51
6	3.99

Source: U.S. Postal Service

120. *Cable Television.* The following table lists the number of households served by cable television from 1980 to 2002.

Year	Number of Households Served by Cable Television (in millions)
1980	17.7
1985	39.9
1990	54.9
1993	58.8
1994	60.5
1996	64.6
1998	67.6
2002	73.0

Sources: Nielsen Media Research; energycommerce.house.gov

121. *Koi Growth.* The length, in centimeters, of a koi of age x months can be estimated using the function

$$f(x) = 0.27 + \sqrt{71.94x - 164.41}$$

(see Exercise 117). Use the function to estimate the length of a koi at 8 months and at 20 months.

122. *Cable Television.* The number of households h, in millions, served by cable television x years after 1970 can be modeled by the function

$$h(x) = 11.67 + \sqrt{166.67x - 1659.72}$$

(see Exercise 120). Use the function to estimate the number of households served by cable television in 1992 and in 2005.

TW **123.** Explain how to write the negative square root of a number using radical notation.

TW **124.** Does the square root of a number's absolute value always exist? Why or why not?

Skill Maintenance

Simplify. Do not use negative exponents in your answer. [1.4]

125. $(a^3b^2c^5)^3$

126. $(5a^7b^8)(2a^3b)$

127. $(2a^{-2}b^3c^{-4})^{-3}$

128. $(5x^{-3}y^{-1}z^2)^{-2}$

129. $\dfrac{8x^{-2}y^5}{4x^{-6}z^{-2}}$

130. $\dfrac{10a^{-6}b^{-7}}{2a^{-2}c^{-3}}$

Synthesis

TW **131.** If the domain of $f = [1, \infty)$ and the range of $f = [2, \infty)$, find a possible expression for $f(x)$ and explain how such an expression is formulated.

TW **132.** Kelly obtains the following graph of

$$f(x) = \sqrt{x^2 - 4x - 12}$$

and concludes that the domain of f is $(-\infty, -2]$. Is she correct? If not, what mistake is she making?

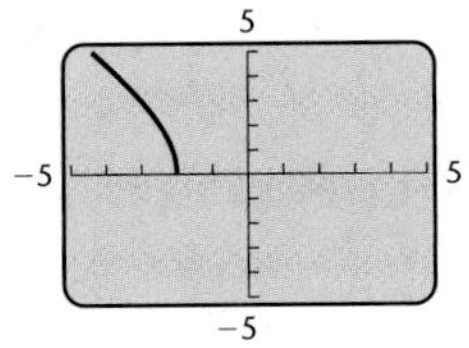

TW **133.** Could the following situation possibly be modeled using a radical function? Why or why not? "For each year, the yield increases. The amount of increase is smaller each year."

TW **134.** Could the following situation possibly be modeled using a radical function? Why or why not? "For each year, the costs increase. The amount of increase is the same each year."

135. *Spaces in a Parking Lot.* A parking lot has attendants to park the cars. The number N of stalls needed for waiting cars before attendants can get to them is given by the formula $N = 2.5\sqrt{A}$, where A is the number of arrivals in peak hours. Find the number of spaces needed for the given number of arrivals in peak hours: **(a)** 25; **(b)** 36; **(c)** 49; **(d)** 64.

Determine the domain of each function described. Then draw the graph of each function.

136. $g(x) = \sqrt{x} + 5$

137. $f(x) = \sqrt{x + 5}$

138. $f(x) = \sqrt{x - 2}$

139. $g(x) = \sqrt{x} - 2$

140. Find the domain of f if

$$f(x) = \frac{\sqrt{x+3}}{\sqrt[4]{2-x}}.$$

141. Find the domain of g if

$$g(x) = \frac{\sqrt[4]{5-x}}{\sqrt[6]{x+4}}.$$

142. Find the domain of F if $F(x) = \dfrac{x}{\sqrt{x^2 - 5x - 6}}$.

143. Examine the graph of the data in Exercise 118. What type of function could be used to model the data?

144. Examine the graph of the data in Exercise 119. What type of function could be used to model the data?

7.2 Rational Numbers as Exponents

Rational Exponents ■ Negative Rational Exponents ■ Laws of Exponents ■ Simplifying Radical Expressions

Study Tip

It Looks Familiar

Sometimes a topic that seems familiar—perhaps the material in this section—reappears and students relax, thinking the material is review. Try to avoid this tendency. Often a new wrinkle is included that will catch the unsuspecting student by surprise.

In Section 1.1, we considered the natural numbers as exponents. Our discussion of exponents was expanded to include all integers in Section 1.4. We now expand the study still further—to include all rational numbers. This will give meaning to expressions like $a^{1/3}$, $7^{-1/2}$, and $(3x)^{4/5}$. Such notation will help us simplify certain radical expressions.

Rational Exponents

Consider $a^{1/2} \cdot a^{1/2}$. To add exponents when multiplying, it must follow that $a^{1/2} \cdot a^{1/2} = a^{1/2+1/2}$, or a^1. This suggests that $a^{1/2}$ is a square root of a. Similarly, $a^{1/3} \cdot a^{1/3} \cdot a^{1/3} = a^{1/3+1/3+1/3}$, or a^1, so $a^{1/3}$ should mean $\sqrt[3]{a}$.

> $\mathbf{a^{1/n} = \sqrt[n]{a}}$ $\quad a^{1/n}$ means $\sqrt[n]{a}$. When a is nonnegative, n can be any natural number greater than 1. When a is negative, n must be odd.

Note that the denominator of the exponent becomes the index and the base becomes the radicand.

EXAMPLE 1 Write an equivalent expression using radical notation.

a) $x^{1/2}$ **b)** $(-8)^{1/3}$

c) $(abc)^{1/5}$ **d)** $(25x^{16})^{1/2}$

SOLUTION

a) $x^{1/2} = \sqrt{x}$

b) $(-8)^{1/3} = \sqrt[3]{-8} = -2$

c) $(abc)^{1/5} = \sqrt[5]{abc}$

The denominator of the exponent becomes the index. The base becomes the radicand. Recall that for square roots, the index 2 is understood without being written.

d) $(25x^{16})^{1/2} = \sqrt{25x^{16}} = 5x^8$

EXAMPLE 2 Write an equivalent expression using exponential notation.

a) $\sqrt[5]{9xy}$ **b)** $\sqrt[7]{\dfrac{x^3y}{4}}$ **c)** $\sqrt{5x}$

SOLUTION Parentheses are required to indicate the base.

a) $\sqrt[5]{9xy} = (9xy)^{1/5}$

b) $\sqrt[7]{\dfrac{x^3y}{4}} = \left(\dfrac{x^3y}{4}\right)^{1/7}$

The index becomes the denominator of the exponent. The radicand becomes the base.

c) $\sqrt{5x} = (5x)^{1/2}$ The index 2 is understood without being written. We assume $x \geq 0$.

> **CAUTION!** When converting from radical notation to exponential notation, parentheses are necessary to indicate the base.
>
> $\sqrt{5x} = (5x)^{1/2}$

Rational Exponents

We can enter a radical expression in radical notation or by using rational exponents.

To use rational exponents, enter the radicand enclosed in parentheses, then press [^], and then enter the rational exponent, also enclosed in parentheses. If the rational exponent is written as a single decimal number, the parentheses around the exponent are unnecessary. If decimal notation for a rational exponent must be rounded, fraction notation should be used.

EXAMPLE 3 Graph: $f(x) = \sqrt[4]{2x - 7}$.

SOLUTION We can enter the equation in radical notation using $\sqrt[x]{\ }$. Alternatively, we can rewrite the radical expression using a rational exponent:

$$f(x) = (2x - 7)^{1/4}, \quad \text{or} \quad f(x) = (2x - 7)^{0.25}.$$

Then we let $y = (2x - 7)$ ^ $(1/4)$ or $y = (2x - 7)$ ^ $.25$. The screen on the left below shows all three forms of the equation; they are equivalent.

Knowing the domain of the function can help us determine an appropriate viewing window. Since the index is even, the domain is the set of all x for which the radicand is nonnegative, or $\left[\frac{7}{2}, \infty\right)$. We choose a viewing window of $[-1, 10, -1, 5]$. This will show the axes and the first quadrant. The graph is shown on the right below.

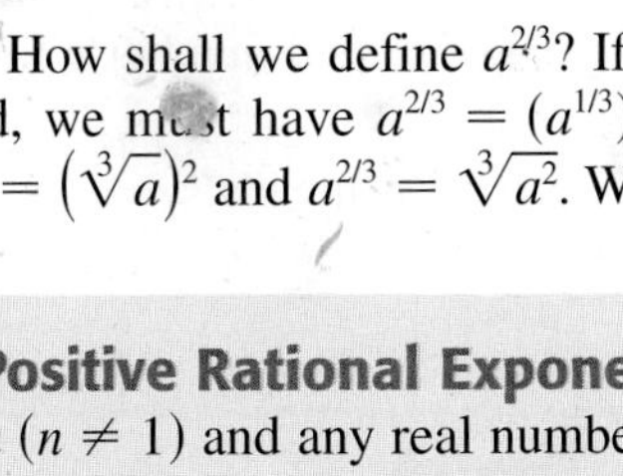

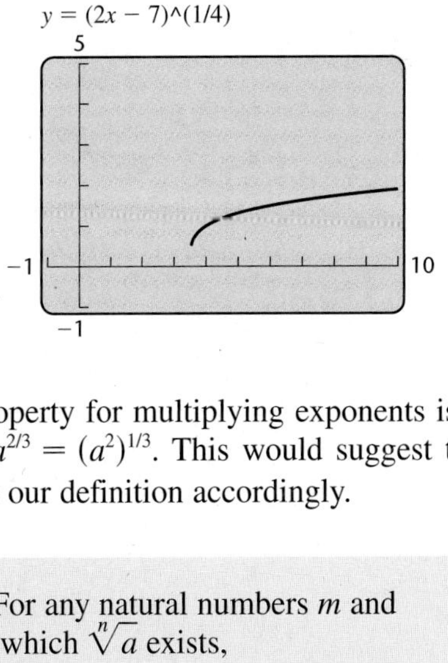

How shall we define $a^{2/3}$? If the property for multiplying exponents is to hold, we must have $a^{2/3} = (a^{1/3})^2$ and $a^{2/3} = (a^2)^{1/3}$. This would suggest that $a^{2/3} = (\sqrt[3]{a})^2$ and $a^{2/3} = \sqrt[3]{a^2}$. We make our definition accordingly.

> **Positive Rational Exponents** For any natural numbers m and n $(n \neq 1)$ and any real number a for which $\sqrt[n]{a}$ exists,
>
> $$a^{m/n} \text{ means } (\sqrt[n]{a})^m, \text{ or } \sqrt[n]{a^m}.$$

Student Notes

It is important to remember both meanings of $a^{m/n}$. When the root of the base a is known, $(\sqrt[n]{a})^m$ is generally easier to work with. When it is not known, $\sqrt[n]{a^m}$ is often more convenient.

EXAMPLE 4 Write an equivalent expression using radical notation and simplify.

a) $27^{2/3}$ **b)** $25^{3/2}$

SOLUTION

a) $27^{2/3}$ means $(\sqrt[3]{27})^2$ or, equivalently, $\sqrt[3]{27^2}$. Let's see which is easier to simplify:

$$(\sqrt[3]{27})^2 = 3^2 = 9; \qquad \sqrt[3]{27^2} = \sqrt[3]{729} = 9.$$

The simplification on the left is probably easier for most people.

b) $25^{3/2}$ means $(\sqrt[2]{25})^3$ or, equivalently, $\sqrt[2]{25^3}$ (the index 2 is normally omitted). Since $\sqrt{25}$ is more commonly known than $\sqrt{25^3}$, we use that form:

$$25^{3/2} = (\sqrt{25})^3 = 5^3 = 125.$$

EXAMPLE 5 Write an equivalent expression using exponential notation.

a) $\sqrt[3]{9^4}$ **b)** $(\sqrt[4]{7xy})^5$

SOLUTION

a) $\sqrt[3]{9^4} = 9^{4/3}$

b) $(\sqrt[4]{7xy})^5 = (7xy)^{5/4}$

The index becomes the denominator of the fraction that is the exponent.

Rational roots of numbers can be approximated on a calculator.

EXAMPLE 6 Approximate $\sqrt[5]{(-23)^3}$. Round to the nearest thousandth.

SOLUTION We first rewrite the expression using a rational exponent:

$$\sqrt[5]{(-23)^3} = (-23)^{3/5}.$$

Using a calculator, we have

$$(-23)\wedge(3/5) \approx -6.562.$$

Negative Rational Exponents

Recall that $x^{-2} = \frac{1}{x^2}$. Negative rational exponents behave similarly.

Negative Rational Exponents For any rational number m/n and any nonzero real number a for which $a^{m/n}$ exists,

$$a^{-m/n} \text{ means } \frac{1}{a^{m/n}}.$$

CAUTION! A negative exponent does not indicate that the expression in which it appears is negative.

EXAMPLE 7 Write an equivalent expression with positive exponents and, if possible, simplify.

a) $9^{-1/2}$ **b)** $(5xy)^{-4/5}$ **c)** $64^{-2/3}$

d) $4x^{-2/3}y^{1/5}$ **e)** $\left(\dfrac{3r}{7s}\right)^{-5/2}$

SOLUTION

a) $9^{-1/2} = \dfrac{1}{9^{1/2}}$ **$9^{-1/2}$ is the reciprocal of $9^{1/2}$.**

Since $9^{1/2} = \sqrt{9} = 3$, the answer simplifies to $\dfrac{1}{3}$.

b) $(5xy)^{-4/5} = \dfrac{1}{(5xy)^{4/5}}$ **$(5xy)^{-4/5}$ is the reciprocal of $(5xy)^{4/5}$.**

c) $64^{-2/3} = \dfrac{1}{64^{2/3}}$ **$64^{-2/3}$ is the reciprocal of $64^{2/3}$.**

Since $64^{2/3} = (\sqrt[3]{64})^2 = 4^2 = 16$, the answer simplifies to $\dfrac{1}{16}$.

d) $4x^{-2/3}y^{1/5} = 4 \cdot \dfrac{1}{x^{2/3}} \cdot y^{1/5} = \dfrac{4y^{1/5}}{x^{2/3}}$

e) In Section 1.4, we found that $(a/b)^{-n} = (b/a)^n$. This property holds for *any* negative exponent:

$$\left(\frac{3r}{7s}\right)^{-5/2} = \left(\frac{7s}{3r}\right)^{5/2}.$$ **Writing the reciprocal of the base and changing the sign of the exponent**

Laws of Exponents

The same laws hold for rational exponents as for integer exponents.

Laws of Exponents For any real numbers a and b and any rational exponents m and n for which a^m, a^n, and b^m are defined:

1. $a^m \cdot a^n = a^{m+n}$ In multiplying, add exponents if the bases are the same.
2. $\dfrac{a^m}{a^n} = a^{m-n}$ In dividing, subtract exponents if the bases are the same. (Assume $a \neq 0$.)
3. $(a^m)^n = a^{m \cdot n}$ To raise a power to a power, multiply the exponents.
4. $(ab)^m = a^m b^m$ To raise a product to a power, raise each factor to the power and multiply.

EXAMPLE 8 Use the laws of exponents to simplify.

a) $3^{1/5} \cdot 3^{3/5}$ **b)** $a^{1/4}/a^{1/2}$

c) $(7.2^{2/3})^{3/4}$ **d)** $(a^{-1/3}b^{2/5})^{1/2}$

SOLUTION

a) $3^{1/5} \cdot 3^{3/5} = 3^{1/5+3/5} = 3^{4/5}$ **Adding exponents**

b) $\dfrac{a^{1/4}}{a^{1/2}} = a^{1/4-1/2} = a^{1/4-2/4}$ **Subtracting exponents after finding a common denominator**

$= a^{-1/4}$, or $\dfrac{1}{a^{1/4}}$ **$a^{-1/4}$ is the reciprocal of $a^{1/4}$.**

c) $(7.2^{2/3})^{3/4} = 7.2^{(2/3)(3/4)} = 7.2^{6/12}$ **Multiplying exponents**

$= 7.2^{1/2}$ **Using arithmetic to simplify the exponent**

d) $(a^{-1/3}b^{2/5})^{1/2} = a^{(-1/3)(1/2)} \cdot b^{(2/5)(1/2)}$ **Raising a product to a power and multiplying exponents**

$= a^{-1/6}b^{1/5}$, or $\dfrac{b^{1/5}}{a^{1/6}}$

Simplifying Radical Expressions

Many radical expressions can be simplified using rational exponents.

To Simplify Radical Expressions

1. Convert radical expressions to exponential expressions.
2. Use arithmetic and the laws of exponents to simplify.
3. Convert back to radical notation as needed.

EXAMPLE 9 Use rational exponents to simplify. Do not use exponents that are fractions in the final answer.

a) $\sqrt[6]{(5x)^3}$

b) $\sqrt[5]{t^{20}}$

c) $\left(\sqrt[3]{ab^2c}\right)^{12}$

d) $\sqrt{\sqrt[3]{x}}$

SOLUTION

a) $\sqrt[6]{(5x)^3} = (5x)^{3/6}$ **Converting to exponential notation**

$= (5x)^{1/2}$ **Simplifying the exponent**

$= \sqrt{5x}$ **Returning to radical notation**

To check on a graphing calculator, we let $y_1 = \sqrt[6]{(5x)^3}$ and $y_2 = \sqrt{5x}$ and compare values in a table. If we scroll through the table (see the figure at left), we see that $y_1 = y_2$, so our simplification is probably correct.

$y_1 = 6\sqrt[x]{\ }((5x)\wedge 3),\ y_2 = \sqrt{\ }(5x)$

X	Y1	Y2
0	0	0
1	2.2361	2.2361
2	3.1623	3.1623
3	3.873	3.873
4	4.4721	4.4721
5	5	5
6	5.4772	5.4772

X = 0

b) $\sqrt[5]{t^{20}} = t^{20/5}$ **Converting to exponential notation**

$= t^4$ **Simplifying the exponent**

c) $\left(\sqrt[3]{ab^2c}\right)^{12} = (ab^2c)^{12/3}$ **Converting to exponential notation**

$= (ab^2c)^4$ **Simplifying the exponent**

$= a^4b^8c^4$ **Using the laws of exponents**

$y_1 = \sqrt{(\sqrt[3]{(x)})}$, $y_2 = x^\wedge(1/6)$

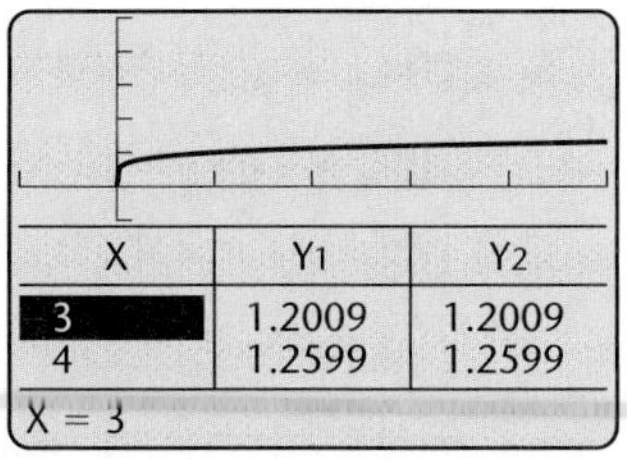

d) $\sqrt{\sqrt[3]{x}} = \sqrt{x^{1/3}}$ **Converting the radicand to exponential notation**

$= (x^{1/3})^{1/2}$ **Try to go directly to this step.**

$= x^{1/6}$ **Using the laws of exponents**

$= \sqrt[6]{x}$ **Returning to radical notation**

We can check by graphing $y_1 = \sqrt{\sqrt[3]{x}}$ and $y_2 = \sqrt[6]{x}$. The graphs coincide, as we also see by scrolling through the table of values shown at left.

7.2 EXERCISE SET

FOR EXTRA HELP

MathXL | MyMathLab | InterAct Math | AW Math Tutor Center | Video Lectures on CD: Disc 4 | Student's Solutions Manual

Concept Reinforcement *In each of Exercises 1–8, match the expression with the equivalent expression from the column on the right.*

1. ___ $x^{2/5}$

2. ___ $x^{5/2}$

3. ___ $x^{-5/2}$

4. ___ $x^{-2/5}$

5. ___ $x^{1/5} \cdot x^{2/5}$

6. ___ $(x^{1/5})^{5/2}$

7. ___ $\sqrt[5]{x^4}$

8. ___ $(\sqrt[4]{x})^5$

a) $x^{3/5}$

b) $(\sqrt[5]{x})^4$

c) $\sqrt{x^5}$

d) $x^{1/2}$

e) $\dfrac{1}{(\sqrt{x})^5}$

f) $\sqrt[4]{x^5}$

g) $\sqrt[5]{x^2}$

h) $\dfrac{1}{(\sqrt[5]{x})^2}$

Note: Assume for all exercises that even roots are of nonnegative quantities and that all denominators are nonzero.

Write an equivalent expression using radical notation and, if possible, simplify.

9. $x^{1/6}$

10. $y^{1/5}$

11. $16^{1/2}$

12. $8^{1/3}$

13. $81^{1/4}$

14. $64^{1/6}$

15. $9^{1/2}$

16. $25^{1/2}$

17. $(xyz)^{1/3}$

18. $(ab)^{1/4}$

19. $(a^2b^2)^{1/5}$

20. $(x^3y^3)^{1/4}$

21. $t^{2/5}$

22. $b^{3/2}$

23. $16^{3/4}$

24. $4^{7/2}$

25. $27^{4/3}$

26. $9^{5/2}$

27. $(81x)^{3/4}$

28. $(125a)^{2/3}$

29. $(25x^4)^{3/2}$

30. $(9y^6)^{3/2}$

Write an equivalent expression using exponential notation.

31. $\sqrt[3]{20}$

32. $\sqrt[3]{19}$

33. $\sqrt{17}$

34. $\sqrt{6}$

35. $\sqrt{x^3}$

36. $\sqrt{a^5}$

37. $\sqrt[5]{m^2}$

38. $\sqrt[5]{n^4}$

39. $\sqrt[4]{cd}$

40. $\sqrt[5]{xy}$

41. $\sqrt[5]{xy^2z}$

42. $\sqrt[7]{x^3y^2z^2}$

43. $(\sqrt{3mn})^3$

44. $(\sqrt[3]{7xy})^4$

45. $(\sqrt[7]{8x^2y})^5$

46. $(\sqrt[6]{2a^5b})^7$

47. $\dfrac{2x}{\sqrt[3]{z^2}}$

48. $\dfrac{3a}{\sqrt[5]{c^2}}$

Write an equivalent expression with positive exponents and, if possible, simplify.

49. $x^{-1/3}$

50. $y^{-1/4}$

51. $(2rs)^{-3/4}$

52. $(5xy)^{-5/6}$

53. $\left(\dfrac{1}{16}\right)^{-3/4}$

54. $\left(\dfrac{1}{8}\right)^{-2/3}$

55. $\dfrac{2c}{a^{-3/5}}$

56. $\dfrac{3b}{a^{-5/7}}$

57. $5x^{-2/3}y^{4/5}z$

58. $2a^{3/4}b^{-1/2}c^{2/3}$

59. $3^{-5/2}a^3b^{-7/3}$

60. $2^{-1/3}x^4y^{-2/7}$

61. $\left(\dfrac{2ab}{3c}\right)^{-5/6}$

62. $\left(\dfrac{7x}{8yz}\right)^{-3/5}$

63. $\dfrac{6a}{\sqrt[4]{b}}$

64. $\dfrac{7x}{\sqrt[3]{z}}$

Graph using a graphing calculator.

65. $f(x) = \sqrt[4]{x+7}$

66. $g(x) = \sqrt[5]{4-x}$

67. $r(x) = \sqrt[7]{3x-2}$

68. $q(x) = \sqrt[6]{2x+3}$

69. $f(x) = \sqrt[6]{x^3}$

70. $g(x) = \sqrt[8]{x^2}$

Approximate. Round to the nearest thousandth.

71. $\sqrt[5]{9}$

72. $\sqrt[6]{13}$

73. $\sqrt[4]{10}$

74. $\sqrt[7]{-127}$

75. $\sqrt[3]{(-3)^5}$

76. $\sqrt[10]{(1.5)^6}$

Use the laws of exponents to simplify. Do not use negative exponents in any answers.

77. $7^{3/4} \cdot 7^{1/8}$

78. $11^{2/3} \cdot 11^{1/2}$

79. $\dfrac{3^{5/8}}{3^{-1/8}}$

80. $\dfrac{8^{7/11}}{8^{-2/11}}$

81. $\dfrac{5.2^{-1/6}}{5.2^{-2/3}}$

82. $\dfrac{2.3^{-3/10}}{2.3^{-1/5}}$

83. $(10^{3/5})^{2/5}$

84. $(5^{5/4})^{3/7}$

85. $a^{2/3} \cdot a^{5/4}$

86. $x^{3/4} \cdot x^{1/3}$

Aha! **87.** $(64^{3/4})^{4/3}$

88. $(27^{-2/3})^{3/2}$

89. $(m^{2/3}n^{-1/4})^{1/2}$

90. $(x^{-1/3}y^{2/5})^{1/4}$

Use rational exponents to simplify. Do not use fraction exponents in the final answer.

91. $\sqrt[6]{x^4}$

92. $\sqrt[6]{a^2}$

93. $\sqrt[4]{a^{12}}$

94. $\sqrt[3]{x^{15}}$

95. $\sqrt[5]{a^{10}}$

96. $\sqrt[6]{x^{18}}$

97. $(\sqrt[7]{xy})^{14}$

98. $(\sqrt[3]{ab})^{15}$

99. $\sqrt[4]{(7a)^2}$

100. $\sqrt[8]{(3x)^2}$

101. $(\sqrt[8]{2x})^6$

102. $(\sqrt[10]{3a})^5$

103. $\sqrt[3]{\sqrt[6]{a}}$

104. $\sqrt[4]{\sqrt{x}}$

105. $\sqrt[4]{(xy)^{12}}$

106. $\sqrt{(ab)^6}$

107. $(\sqrt[5]{a^2b^4})^{15}$

108. $(\sqrt[3]{x^2y^5})^{12}$

109. $\sqrt[3]{\sqrt[4]{xy}}$

110. $\sqrt[5]{\sqrt{2a}}$

TW **111.** If $f(x) = (x+5)^{1/2}(x+7)^{-1/2}$, find the domain of f. Explain how you found your answer.

TW **112.** Explain why $\sqrt[3]{x^6} = x^2$ for any value of x, whereas $\sqrt[2]{x^6} = x^3$ only when $x \geq 0$.

Skill Maintenance

Simplify. [5.2]

113. $3x(x^3 - 2x^2) + 4x^2(2x^2 + 5x)$

114. $5t^3(2t^2 - 4t) - 3t^4(t^2 - 6t)$

115. $(3a - 4b)(5a + 3b)$

116. $(7x - y)^2$

117. *Real Estate Taxes.* For homes under \$100,000, the property transfer tax in Vermont is 0.5% of the selling price. Find the selling price of a home that had a transfer tax of \$467.50. [1.7]

118. What numbers are their own squares? [5

Synthesis

TW **119.** Let $f(x) = 5x^{-1/3}$. Under what condition will we have $f(x) > 0$? Why?

TW **120.** If $g(x) = x^{3/n}$, in what way does the domain of g depend on whether n is odd or even?

Use rational exponents to simplify.

121. $\sqrt{x\sqrt[3]{x^2}}$

122. $\sqrt[4]{\sqrt[3]{8x^3y^6}}$

123. $\sqrt[12]{p^2 + 2pq + q^2}$

*Music. The function given by $f(x) = k2^{x/12}$ can be used to determine the frequency, in cycles per second, of a musical note that is x half-steps above a note with frequency k.**

124. The frequency of concert A for a trumpet is 440 cycles per second. Find the frequency of the A that is two octaves (24 half-steps) above concert A (few trumpeters can reach this note.)

*This application was inspired by information provided by Dr. Homer B. Tilton of Pima Community College East.

125. Show that the G that is 7 half-steps (a "perfect fifth") above middle C (262 cycles per second) has a frequency that is about 1.5 times that of middle C.

126. Show that the C sharp that is 4 half-steps (a "major third") above concert A (see Exercise 124) has a frequency that is about 25% greater than that of concert A.

127. *Baseball.* The statistician Bill James has found that a baseball team's winning percentage P can be approximated by

$$P = \frac{r^{1.83}}{r^{1.83} + \sigma^{1.83}},$$

where r is the total number of runs scored by that team and σ is the total number of runs scored by their opponents (*Source*: M. Bittinger, *One Man's Journey Through Mathematics*. Boston: Addison-Wesley, 2004). During a recent season, the San Francisco Giants scored 799 runs and their opponents scored 749 runs. Use James's formula to predict the Giants' winning percentage (the team actually won 55.6% of their games).

128. *Road Pavement Messages.* In a psychological study, it was determined that the proper length L of the letters of a word printed on pavement is given by

$$L = \frac{0.000169d^{2.27}}{h},$$

where d is the distance of a car from the lettering and h is the height of the eye above the surface of the road. All units are in meters. This formula says that if a person is h meters above the surface of the road and is to be able to recognize a message d meters away, that message will be the most recognizable if the length of the letters is L. Find L to the nearest tenth of a meter, given d and h.

a) $h = 1$ m, $d = 60$ m
b) $h = 0.9906$ m, $d = 75$ m
c) $h = 2.4$ m, $d = 80$ m
d) $h = 1.1$ m, $d = 100$ m

129. *Physics.* The equation $m = m_0(1 - v^2c^{-2})^{-1/2}$, developed by Albert Einstein, is used to determine the mass m of an object that is moving v meters per second and has mass m_0 before the motion begins. The constant c is the speed of light, approximately 3×10^8 m/sec. Suppose that a particle with mass 8 mg is accelerated to a speed of $\frac{9}{5} \times 10^8$ m/sec. Without using a calculator, find the new mass of the particle.

130. A person's body surface area (BSA) can be approximated by the DuBois formula

$$\text{BSA} = 0.007184w^{0.425}h^{0.725},$$

where w is mass, in kilograms, h is height, in centimeters, and BSA is in square meters (*Source*: www.halls.md). What is the BSA of a child who is 122 cm tall and has a mass of 29.5 kg?

131. Using a graphing calculator, select the MODE SIMUL and the FORMAT EXPROFF. Then graph

$$y_1 = x^{1/2}, \quad y_2 = 3x^{2/5},$$
$$y_3 = x^{4/7}, \quad \text{and} \quad y_4 = \tfrac{1}{5}x^{3/4}.$$

Looking only at coordinates, match each graph with its equation.

Collaborative Corner

Are Equivalent Fractions Equivalent Exponents?

Focus: Functions and rational exponents
Time: 10–20 minutes
Group Size: 3
Materials: Graph paper

In arithmetic, we have seen that $\frac{1}{3}$, $\frac{1}{6} \cdot 2$, and $2 \cdot \frac{1}{6}$ all represent the same number. Interestingly,

$$f(x) = x^{1/3},$$
$$g(x) = (x^{1/6})^2, \quad \text{and}$$
$$h(x) = (x^2)^{1/6}$$

represent three *different* functions.

ACTIVITY

1. Selecting a variety of values for x and using the definition of positive rational exponents, one group member should graph f, a second group member should graph g, and a third group member should graph h. Be sure to check whether negative x-values are in the domain of the function.
2. Compare the three graphs and check each other's work. How and why do the graphs differ?
3. Decide as a group which graph, if any, would best represent the graph of $k(x) = x^{2/6}$. Then be prepared to explain your reasoning to the entire class. (*Hint*: Study the definition of $a^{m/n}$ on p. 547 carefully.)

7.3 Multiplying Radical Expressions

Multiplying Radical Expressions ■ Simplifying by Factoring ■ Multiplying and Simplifying

Study Tip

Interested in Something Extra?

Students interested in extra-credit projects are often pleasantly surprised to find that their instructor will not only provide a possible project for them to pursue, but might even tailor that project to the student's interests. Remember that if you do an extra-credit project, you must still attend to the regular course work.

Multiplying Radical Expressions

Note that $\sqrt{4}\sqrt{25} = 2 \cdot 5 = 10$. Also $\sqrt{4 \cdot 25} = \sqrt{100} = 10$. Likewise,

$$\sqrt[3]{27}\,\sqrt[3]{8} = 3 \cdot 2 = 6 \quad \text{and} \quad \sqrt[3]{27 \cdot 8} = \sqrt[3]{216} = 6.$$

These examples suggest the following.

The Product Rule for Radicals For any real numbers $\sqrt[n]{a}$ and $\sqrt[n]{b}$,

$$\sqrt[n]{a} \cdot \sqrt[n]{b} = \sqrt[n]{a \cdot b}.$$

(The product of two nth roots is the nth root of the product of the two radicands.)

Rational exponents can be used to derive this rule:

$$\sqrt[n]{a} \cdot \sqrt[n]{b} = a^{1/n} \cdot b^{1/n} = (a \cdot b)^{1/n} = \sqrt[n]{a \cdot b}.$$

EXAMPLE 1 Multiply.

a) $\sqrt{3} \cdot \sqrt{5}$

b) $\sqrt{x+3}\,\sqrt{x-3}$

c) $\sqrt[3]{4} \cdot \sqrt[3]{5}$

d) $\sqrt[4]{\frac{y}{5}} \cdot \sqrt[4]{\frac{7}{x}}$

SOLUTION

a) When no index is written, roots are understood to be square roots with an unwritten index of two. We apply the product rule:

$$\sqrt{3} \cdot \sqrt{5} = \sqrt{3 \cdot 5}$$
$$= \sqrt{15}.$$

b) $\sqrt{x+3}\,\sqrt{x-3} = \sqrt{(x+3)(x-3)}$ **The product of two square roots is the square root of the product.**

$$= \sqrt{x^2 - 9}$$

CAUTION!

$\sqrt{x^2 - 9} \neq \sqrt{x^2} - \sqrt{9}.$

c) Both $\sqrt[3]{4}$ and $\sqrt[3]{5}$ have indices of three, so to multiply we can use the product rule:

$$\sqrt[3]{4} \cdot \sqrt[3]{5} = \sqrt[3]{4 \cdot 5} = \sqrt[3]{20}.$$

d) $\sqrt[4]{\frac{y}{5}} \cdot \sqrt[4]{\frac{7}{x}} = \sqrt[4]{\frac{y}{5} \cdot \frac{7}{x}} = \sqrt[4]{\frac{7y}{5x}}$ **In Section 7.4, we discuss other ways to write answers like this.**

CAUTION! The product rule for radicals applies only when radicals have the same index:

$$\sqrt[n]{a} \cdot \sqrt[m]{b} \neq \sqrt[nm]{a \cdot b}.$$

Connecting the Concepts

INTERPRETING GRAPHS: DOMAINS OF RADICAL FUNCTIONS

Although, as we saw in Example 1(b), $\sqrt{x+3}\,\sqrt{x-3} = \sqrt{x^2-9}$, it is not true that $f(x) = \sqrt{x+3}\,\sqrt{x-3}$ and $g(x) = \sqrt{x^2-9}$ represent the same function. The domains of f and g are not the same.

Since f is written as the product of two rational expressions, both $\sqrt{x+3}$ and $\sqrt{x-3}$ must be defined for all inputs for f. Thus we must have

$$x + 3 \ge 0 \quad \textit{and} \quad x - 3 \ge 0$$
$$x \ge -3 \quad \textit{and} \quad x \ge 3.$$

The domain of f is thus $\{x \mid x \ge -3 \textit{ and } x \ge 3\}$, or $[3, \infty)$.

The function g is defined when $x^2 - 9 \ge 0$. Although we do not solve inequalities of this type until Chapter 8, note that when $x^2 - 9 \ge 0$, $x^2 \ge 9$. This is true when $|x| \ge 3$, or for $\{x \mid x \le -3 \textit{ or } x \ge 3\}$. Thus the domain of g is $(-\infty, -3] \cup [3, \infty)$.

These domains are confirmed by the following graphs. Note that the graphs do coincide where the domains coincide. We say that the expressions are equivalent because they represent the same number for all possible replacements for *both* expressions.

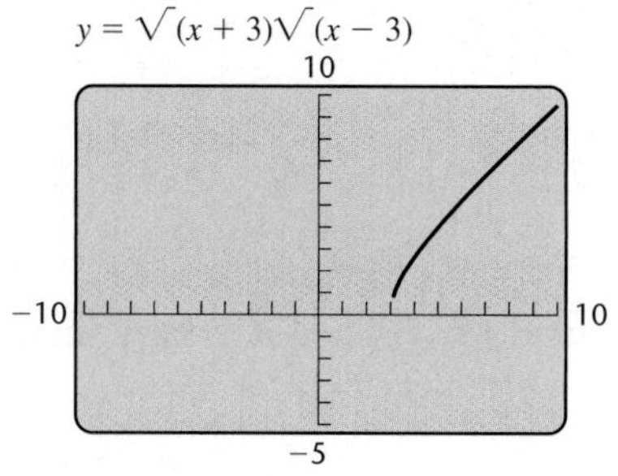

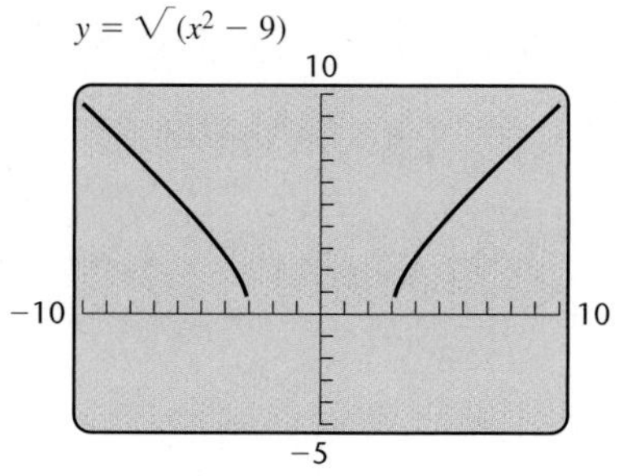

Simplifying by Factoring

An integer p is a *perfect square* if there exists a rational number q for which $q^2 = p$. We say that p is a *perfect cube* if $q^3 = p$ for some rational number q. In general, p is a *perfect nth power* if $q^n = p$ for some rational number q. The product rule allows us to simplify $\sqrt[n]{ab}$ when a or b is a perfect nth power.

Using the Product Rule to Simplify

$\sqrt[n]{ab} = \sqrt[n]{a} \cdot \sqrt[n]{b}$.

($\sqrt[n]{a}$ and $\sqrt[n]{b}$ must both be real numbers.)

To illustrate, suppose we wish to simplify $\sqrt{20}$. Since this is a *square* root, we check to see if there is a factor of 20 that is a perfect square. There is one, 4, so we express 20 as $4 \cdot 5$ and use the product rule.

$$\begin{aligned}\sqrt{20} &= \sqrt{4 \cdot 5} && \textbf{Factoring the radicand (4 is a perfect square)}\\ &= \sqrt{4} \cdot \sqrt{5} && \textbf{Factoring into two radicals}\\ &= 2\sqrt{5} && \textbf{Taking the square root of 4}\end{aligned}$$

To Simplify a Radical Expression with Index *n* by Factoring

1. Express the radicand as a product in which one factor is the largest perfect *n*th power possible.
2. Take the *n*th root of each factor.
3. Simplification is complete when no radicand has a factor that is a perfect *n*th power.

It is often safe to assume that a radicand does not represent a negative number raised to an even power. We will henceforth make this assumption—unless functions are involved—and discontinue use of absolute-value notation when taking even roots.

EXAMPLE 2 Simplify by factoring: **(a)** $\sqrt{200}$; **(b)** $\sqrt{18x^2y}$; **(c)** $\sqrt[3]{72}$; **(d)** $\sqrt[4]{162x^6}$.

SOLUTION

a) $\sqrt{200} = \sqrt{100 \cdot 2}$ **100 is the largest perfect-square factor of 200.**

$= \sqrt{100} \cdot \sqrt{2} = 10\sqrt{2}$

b) $\sqrt{18x^2y} = \sqrt{9 \cdot 2 \cdot x^2 \cdot y}$ **$9x^2$ is the largest perfect-square factor of $18x^2y$.**

$= \sqrt{9x^2} \cdot \sqrt{2y}$ **Factoring into two radicals**

$= 3x\sqrt{2y}$ **Taking the square root of $9x^2$**

c) $\sqrt[3]{72} = \sqrt[3]{8 \cdot 9}$ **8 is the largest perfect-cube (third-power) factor of 72.**

$= \sqrt[3]{8} \cdot \sqrt[3]{9} = 2\sqrt[3]{9}$

Let's look at this example another way. We write a complete factorization and look for triples of factors. Each triple of factors makes a cube:

$\sqrt[3]{72} = \sqrt[3]{\underline{2 \cdot 2 \cdot 2} \cdot 3 \cdot 3}$ **Each triple of factors is a cube.**

$= 2\sqrt[3]{3 \cdot 3} = 2\sqrt[3]{9}.$

d) $\sqrt[4]{162x^6} = \sqrt[4]{81 \cdot 2 \cdot x^4 \cdot x^2}$ **$81 \cdot x^4$ is the largest perfect fourth-power factor of $162x^6$.**

$= \sqrt[4]{81x^4} \cdot \sqrt[4]{2x^2}$ **Factoring into two radicals**

$= 3x\sqrt[4]{2x^2}$ **Taking fourth roots**

Let's look at this example another way. We write a complete factorization and look for quadruples of factors. Each quadruple makes a perfect fourth power:

$\sqrt[4]{162x^6} = \sqrt[4]{\underline{3 \cdot 3 \cdot 3 \cdot 3} \cdot 2 \cdot \underline{x \cdot x \cdot x \cdot x} \cdot x \cdot x}$ **Each quadruple of factors is a power of 4.**

$= 3 \cdot x \cdot \sqrt[4]{2 \cdot x \cdot x} = 3x\sqrt[4]{2x^2}.$

EXAMPLE 3 If $f(x) = \sqrt{3x^2 - 6x + 3}$, find a simplified form for $f(x)$.

SOLUTION

$$\begin{aligned} f(x) &= \sqrt{3x^2 - 6x + 3} \\ &= \sqrt{3(x^2 - 2x + 1)} \\ &= \sqrt{(x-1)^2 \cdot 3} \end{aligned}$$ **Factoring the radicand; $x^2 - 2x + 1$ is a perfect square.**

$$= \sqrt{(x-1)^2} \cdot \sqrt{3}$$ **Factoring into two radicals**

$$= |x - 1|\sqrt{3}$$ **Taking the square root of $(x - 1)^2$**

We can check Example 3 by graphing $y_1 = \sqrt{3x^2 - 6x + 3}$ and $y_2 = |x - 1|\sqrt{3}$, as shown in the graph on the left below.

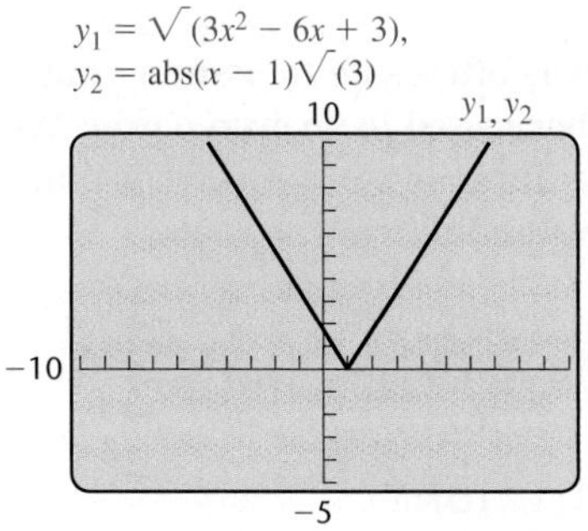

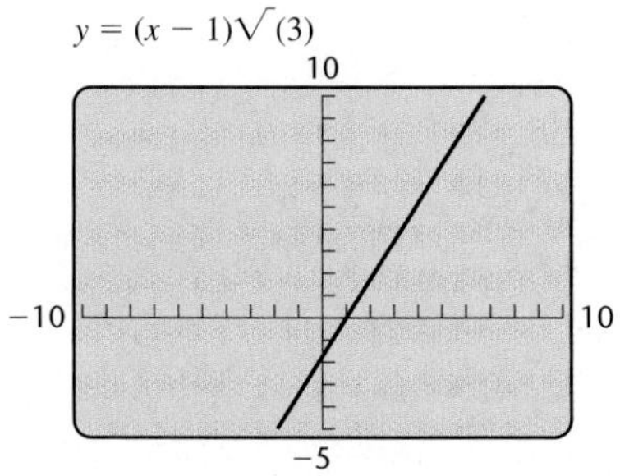

It appears that the graphs coincide, and scrolling through a table of values would show us that this is indeed the case. Note from the graph on the right above that the absolute-value sign is important since the graph of $y = (x - 1)\sqrt{3}$ is different from the graph of $y = \sqrt{3x^2 - 6x + 3}$.

EXAMPLE 4 Simplify: **(a)** $\sqrt{x^7y^{11}z^9}$; **(b)** $\sqrt[3]{-16a^7b^{14}}$.

SOLUTION

a) There are many ways to factor $x^7y^{11}z^9$. Because of the square root (index of 2), we identify the largest exponents that are multiples of 2:

$$\sqrt{x^7y^{11}z^9} = \sqrt{x^6 \cdot x \cdot y^{10} \cdot y \cdot z^8 \cdot z}$$ **Using the largest even powers of x, y, and z**

$$= \sqrt{x^6}\sqrt{y^{10}}\sqrt{z^8}\sqrt{xyz}$$ **Factoring into several radicals**

$$= x^{6/2}\,y^{10/2}\,z^{8/2}\sqrt{xyz}$$ **Converting to fraction exponents**

$$= x^3y^5z^4\sqrt{xyz}.$$

Check: $(x^3y^5z^4\sqrt{xyz})^2 = (x^3)^2(y^5)^2(z^4)^2(\sqrt{xyz})^2$
$= x^6 \cdot y^{10} \cdot z^8 \cdot xyz = x^7y^{11}z^9$

Our check shows that $x^3y^5z^4\sqrt{xyz}$ is the square root of $x^7y^{11}z^9$.

b) Since the cube root of a negative number is negative, we can write

$$\sqrt[3]{-16a^7b^{14}} = -\sqrt[3]{16a^7b^{14}}.$$

There are many ways to factor $16a^7b^{14}$. Because of the cube root (index of 3), we identify factors with the largest exponents that are multiples of 3:

$$-\sqrt[3]{16a^7b^{14}} = -\sqrt[3]{8 \cdot 2 \cdot a^6 \cdot a \cdot b^{12} \cdot b^2}$$ **Using the largest perfect-cube factors**

$$= -\sqrt[3]{8}\,\sqrt[3]{a^6}\,\sqrt[3]{b^{12}}\,\sqrt[3]{2ab^2}$$ **Factoring into several radicals**

$$= -2 \cdot a^{6/3} \cdot b^{12/3}\sqrt[3]{2ab^2}$$ **Converting to fraction exponents**

$$= -2a^2b^4\sqrt[3]{2ab^2}$$

As a check, let's redo the problem using a complex factorization of the radicand:

$$\sqrt[3]{-16a^7b^{14}} = -\sqrt[3]{\underline{2 \cdot 2 \cdot 2} \cdot 2 \cdot \underline{a \cdot a \cdot a} \cdot \underline{a \cdot a \cdot a} \cdot a \cdot \underline{b \cdot b \cdot b} \cdot \underline{b \cdot b \cdot b} \cdot \underline{b \cdot b \cdot b} \cdot \underline{b \cdot b \cdot b} \cdot b \cdot b}$$

Each triple of factors makes a cube.

$$= -2 \cdot a \cdot a \cdot b \cdot b \cdot b \cdot b \cdot \sqrt[3]{2 \cdot a \cdot b \cdot b}$$

$$= -2a^2b^4\sqrt[3]{2ab^2}.$$ **Our answer checks.**

> To simplify an nth root, identify factors in the radicand with exponents that are multiples of n.

Multiplying and Simplifying

We have used the product rule for radicals to find products and also to simplify radical expressions. For some radical expressions, it is possible to do both: First find a product and then simplify.

EXAMPLE 5 Multiply and simplify.

a) $\sqrt{15}\,\sqrt{6}$ **b)** $3\sqrt[3]{25} \cdot 2\sqrt[3]{5}$ **c)** $\sqrt[4]{8x^3y^5}\,\sqrt[4]{4x^2y^3}$

SOLUTION

a) $\sqrt{15}\,\sqrt{6} = \sqrt{15 \cdot 6}$ **Multiplying radicands**

$= \sqrt{90} = \sqrt{9 \cdot 10}$ **9 is a perfect square.**

$= 3\sqrt{10}$

b) $3\sqrt[3]{25} \cdot 2\sqrt[3]{5} = 3 \cdot 2 \cdot \sqrt[3]{25 \cdot 5}$ **Using a commutative law; multiplying radicands**

$= 6 \cdot \sqrt[3]{125}$ **125 is a perfect cube.**

$= 6 \cdot 5$, or 30

c) $\sqrt[4]{8x^3y^5}\,\sqrt[4]{4x^2y^3} = \sqrt[4]{32x^5y^8}$ **Multiplying radicands**

$= \sqrt[4]{16x^4y^8 \cdot 2x}$ **Identifying perfect fourth-power factors**

$= \sqrt[4]{16}\,\sqrt[4]{x^4}\,\sqrt[4]{y^8}\,\sqrt[4]{2x}$ **Factoring into radicals**

$= 2xy^2\sqrt[4]{2x}$ **Finding the fourth roots; assume $x \geq 0$.**

The checks are left to the student.

Student Notes

To multiply $\sqrt{x} \cdot \sqrt{x}$, remember what $\sqrt{x}$ represents and go directly to the product, x. For $x \geq 0$,

$\sqrt{x} \cdot \sqrt{x} = x$,

$(\sqrt{x})^2 = x$, and

$\sqrt{x^2} = x$.

7.3 EXERCISE SET

FOR EXTRA HELP

MathXL | MyMathLab | InterAct Math | AW Math Tutor Center | Video Lectures on CD: Disc 4 | Student's Solutions Manual

Concept Reinforcement *Classify each of the following statements as either true or false.*

1. For any real numbers $\sqrt[n]{a}$ and $\sqrt[n]{b}$, $\sqrt[n]{a} \cdot \sqrt[n]{b} = \sqrt[n]{ab}$.
2. For any real numbers $\sqrt[n]{a}$ and $\sqrt[n]{b}$, $\sqrt[n]{a} + \sqrt[n]{b} = \sqrt[n]{a+b}$.
3. For any real numbers $\sqrt[n]{a}$ and $\sqrt[m]{b}$, $\sqrt[n]{a} \cdot \sqrt[m]{b} = \sqrt[nm]{ab}$.
4. For $x > 0$, $\sqrt{x^2 - 9} = x - 3$.
5. The expression $\sqrt[3]{X}$ is not simplified if X contains a factor that is a perfect cube.
6. It is often possible to simplify $\sqrt{A \cdot B}$ even though $\sqrt{A}$ and $\sqrt{B}$ cannot be simplified.

Multiply.

7. $\sqrt{5}\sqrt{7}$
8. $\sqrt{10}\sqrt{7}$
9. $\sqrt[3]{7}\sqrt[3]{2}$
10. $\sqrt[3]{2}\sqrt[3]{5}$
11. $\sqrt[4]{6}\sqrt[4]{3}$
12. $\sqrt[4]{8}\sqrt[4]{9}$
13. $\sqrt{2x}\sqrt{13y}$
14. $\sqrt{5a}\sqrt{6b}$
15. $\sqrt[5]{8y^3}\sqrt[5]{10y}$
16. $\sqrt[5]{9t^2}\sqrt[5]{2t}$
17. $\sqrt{y-b}\sqrt{y+b}$
18. $\sqrt{x-a}\sqrt{x+a}$
19. $\sqrt[3]{0.7y}\sqrt[3]{0.3y}$
20. $\sqrt[3]{0.5x}\sqrt[3]{0.2x}$
21. $\sqrt[5]{x-2}\sqrt[5]{(x-2)^2}$
22. $\sqrt[4]{x-1}\sqrt[4]{x^2+x+1}$
23. $\sqrt{\dfrac{7}{t}}\sqrt{\dfrac{s}{11}}$
24. $\sqrt{\dfrac{x}{6}}\sqrt{\dfrac{7}{y}}$
25. $\sqrt[7]{\dfrac{x-3}{4}}\sqrt[7]{\dfrac{5}{x+2}}$
26. $\sqrt[6]{\dfrac{a}{b-2}}\sqrt[6]{\dfrac{3}{b+2}}$

Simplify by factoring.

27. $\sqrt{18}$
28. $\sqrt{50}$
29. $\sqrt{27}$
30. $\sqrt{45}$
31. $\sqrt{8}$
32. $\sqrt{75}$
33. $\sqrt{198}$
34. $\sqrt{325}$
35. $\sqrt{36a^4b}$
36. $\sqrt{175y^8}$
37. $\sqrt[3]{8x^3y^2}$
38. $\sqrt[3]{27ab^6}$
39. $\sqrt[3]{-16x^6}$
40. $\sqrt[3]{-32a^6}$

Find a simplified form of $f(x)$. Assume that x can be any real number.

41. $f(x) = \sqrt[3]{125x^5}$
42. $f(x) = \sqrt[3]{16x^6}$
43. $f(x) = \sqrt{49(x-3)^2}$
44. $f(x) = \sqrt{81(x-1)^2}$
45. $f(x) = \sqrt{5x^2 - 10x + 5}$
46. $f(x) = \sqrt{2x^2 + 8x + 8}$

Simplify. Assume that no radicands were formed by raising negative numbers to even powers.

47. $\sqrt{a^6b^7}$
48. $\sqrt{x^6y^9}$
49. $\sqrt[3]{x^5y^6z^{10}}$
50. $\sqrt[3]{a^6b^7c^{13}}$
51. $\sqrt[5]{-32a^7b^{11}}$
52. $\sqrt[4]{16x^5y^{11}}$
53. $\sqrt[5]{x^{13}y^8z^{17}}$
54. $\sqrt[5]{a^6b^8c^9}$
55. $\sqrt[3]{-80a^{14}}$
56. $\sqrt[4]{810x^9}$

Multiply and simplify.

57. $\sqrt{6}\sqrt{3}$
58. $\sqrt{15}\sqrt{5}$
59. $\sqrt{15}\sqrt{21}$
60. $\sqrt{10}\sqrt{14}$
61. $\sqrt[3]{9}\sqrt[3]{3}$
62. $\sqrt[3]{2}\sqrt[3]{4}$
63. **Aha!** $\sqrt{18a^3}\sqrt{18a^3}$
64. $\sqrt{75x^7}\sqrt{75x^7}$
65. $\sqrt[3]{5a^2}\sqrt[3]{2a}$
66. $\sqrt[3]{7x}\sqrt[3]{3x^2}$
67. $\sqrt{2x^5}\sqrt{10x^2}$
68. $\sqrt{5a^7}\sqrt{15a^3}$
69. $\sqrt[3]{s^2t^4}\sqrt[3]{s^4t^6}$
70. $\sqrt[3]{x^2y^4}\sqrt[3]{x^2y^6}$
71. $\sqrt[3]{(x+5)^2}\sqrt[3]{(x+5)^4}$
72. $\sqrt[3]{(a-b)^5}\sqrt[3]{(a-b)^7}$

73. $\sqrt[4]{20a^3b^7}\sqrt[4]{4a^2b^5}$

74. $\sqrt[4]{9x^7y^2}\sqrt[4]{9x^2y^9}$

75. $\sqrt[5]{x^3(y+z)^6}\sqrt[5]{x^3(y+z)^4}$

76. $\sqrt[5]{a^3(b-c)^4}\sqrt[5]{a^7(b-c)^4}$

TW 77. Why do we need to know how to multiply radical expressions before learning how to simplify radical expressions?

TW 78. Why is it incorrect to say that, in general, $\sqrt{x^2} = x$?

Skill Maintenance

Perform the indicated operation and, if possible, simplify. [6.2]

79. $\dfrac{3x}{16y} + \dfrac{5y}{64x}$

80. $\dfrac{2}{a^3b^4} + \dfrac{6}{a^4b}$

81. $\dfrac{4}{x^2-9} - \dfrac{7}{2x-6}$

82. $\dfrac{8}{x^2-25} - \dfrac{3}{2x-10}$

Simplify. [1.4]

83. $\dfrac{9a^4b^7}{3a^2b^5}$

84. $\dfrac{12a^2b^7}{4ab^2}$

Synthesis

TW 85. Explain why it is true that $\sqrt[n]{ab} = \sqrt[n]{a}\cdot\sqrt[n]{b}$.

TW 86. Is the equation $\sqrt{(2x+3)^8} = (2x+3)^4$ always, sometimes, or never true? Why?

87. *Radar Range.* The function given by

$$R(x) = \frac{1}{2}\sqrt[4]{\frac{x\cdot 3.0\times 10^6}{\pi^2}}$$

can be used to determine the maximum range $R(x)$, in miles, of an ARSR-3 surveillance radar with a peak power of x watts (*Source*: Introduction to RADAR Techniques, Federal Aviation Administration, 1988). Determine the maximum radar range when the peak power is 5×10^4 watts.

88. *Speed of a Skidding Car.* Police can estimate the speed at which a car was traveling by measuring its skid marks. The function given by

$$r(L) = 2\sqrt{5L}$$

can be used, where L is the length of a skid mark, in feet, and $r(L)$ is the speed, in miles per hour. Find the exact speed and an estimate (to the nearest tenth mile per hour) for the speed of a car that left skid marks **(a)** 20 ft long; **(b)** 70 ft long; **(c)** 90 ft long. See also Exercise 102.

89. *Wind Chill Temperature.* When the temperature is T degrees Celsius and the wind speed is v meters per second, the *wind chill temperature*, T_w, is the temperature (with no wind) that it feels like. Here is a formula for finding wind chill temperature:

$$T_w = 33 - \frac{(10.45 + 10\sqrt{v} - v)(33 - T)}{22}.$$

Estimate the wind chill temperature (to the nearest tenth of a degree) for the given actual temperatures and wind speeds.

a) $T = 7°C$, $v = 8$ m/sec
b) $T = 0°C$, $v = 12$ m/sec
c) $T = -5°C$, $v = 14$ m/sec
d) $T = -23°C$, $v = 15$ m/sec

Simplify. Assume that all variables are nonnegative.

90. $(\sqrt{r^3t})^7$

91. $(\sqrt[3]{25x^4})^4$

92. $(\sqrt[3]{a^2b^4})^5$

93. $(\sqrt{a^3b^5})^7$

Draw and compare the graphs of each group of equations.

94. $f(x) = \sqrt{x^2+2x+1}$,
$g(x) = x+1$,
$h(x) = |x+1|$

95. $f(x) = \sqrt{x^2-2x+1}$,
$g(x) = x-1$,
$h(x) = |x-1|$

96. If $f(t) = \sqrt{t^2-3t-4}$, what is the domain of f?

97. What is the domain of g, if $g(x) = \sqrt{x^2-6x+8}$?

Solve.

98. $\sqrt[3]{5x^{k+1}}\sqrt[3]{25x^k} = 5x^7$, for k

99. $\sqrt[5]{4a^{3k+2}}\sqrt[5]{8a^{6-k}} = 2a^4$, for k

100. Use a graphing calculator to check your answers to Exercises 21 and 41.

TW **101.** Rony is puzzled. When he uses a graphing calculator to graph $y = \sqrt{x} \cdot \sqrt{x}$, he gets the following screen. Explain why Rony did not get the complete line $y = x$.

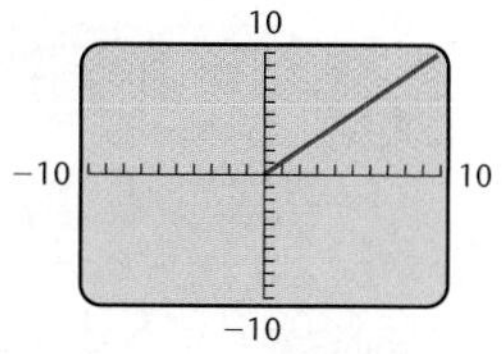

TW **102.** Does a car traveling twice as fast as another car leave a skid mark that is twice as long? (See Exercise 88.) Why or why not?

7.4 Dividing Radical Expressions

Dividing and Simplifying ■ Rationalizing Denominators and Numerators (Part 1)

Dividing and Simplifying

Just as the root of a product can be expressed as the product of two roots, the root of a quotient can be expressed as the quotient of two roots. For example,

$$\sqrt[3]{\frac{27}{8}} = \frac{3}{2} \quad \text{and} \quad \frac{\sqrt[3]{27}}{\sqrt[3]{8}} = \frac{3}{2}.$$

This example suggests the following.

The Quotient Rule for Radicals For any real numbers $\sqrt[n]{a}$ and $\sqrt[n]{b}$, $b \neq 0$,

$$\sqrt[n]{\frac{a}{b}} = \frac{\sqrt[n]{a}}{\sqrt[n]{b}}.$$

Study Tip

Professors Are Human

Even the best professors sometimes make mistakes. If, as you review your notes, you find that something doesn't make sense, it may be due to your instructor having made a mistake. If, after double-checking, you still perceive a mistake, politely ask him or her about it. In doing so, you may be helping *everyone* in your class.

Remember that an nth root is simplified when its radicand has no factors that are perfect nth powers. Recall too that we assume that no radicands represent negative quantities raised to an even power.

EXAMPLE 1 Simplify by taking the roots of the numerator and the denominator.

a) $\sqrt[3]{\frac{27}{125}}$

b) $\sqrt{\frac{25}{y^2}}$

SOLUTION

a) $\sqrt[3]{\dfrac{27}{125}} = \dfrac{\sqrt[3]{27}}{\sqrt[3]{125}} = \dfrac{3}{5}$ — **Taking the cube roots of the numerator and the denominator**

b) $\sqrt{\dfrac{25}{y^2}} = \dfrac{\sqrt{25}}{\sqrt{y^2}} = \dfrac{5}{y}$ — **Taking the square roots of the numerator and the denominator. Assume $y > 0$.**

As in Section 7.3, any radical expressions appearing in the answers should be simplified as much as possible.

EXAMPLE 2 Simplify: **(a)** $\sqrt{\dfrac{16x^3}{y^8}}$; **(b)** $\sqrt[3]{\dfrac{27y^{14}}{8x^3}}$.

SOLUTION

a) $\sqrt{\dfrac{16x^3}{y^8}} = \dfrac{\sqrt{16x^3}}{\sqrt{y^8}}$

$= \dfrac{\sqrt{16x^2 \cdot x}}{\sqrt{y^8}} = \dfrac{4x\sqrt{x}}{y^4}$ — **Simplifying the numerator and the denominator**

b) $\sqrt[3]{\dfrac{27y^{14}}{8x^3}} = \dfrac{\sqrt[3]{27^{14}}}{\sqrt[3]{8x^3}}$

$= \dfrac{\sqrt[3]{27y^{12}y^2}}{\sqrt[3]{8x^3}} = \dfrac{\sqrt[3]{27y^{12}}\,\sqrt[3]{y^2}}{\sqrt[3]{8x^3}} = \dfrac{3y^4\sqrt[3]{y^2}}{2x}$ — **Simplifying the numerator and the denominator**

If we read from right to left, the quotient rule tells us that to divide two radical expressions that have the same index, we can divide the radicands.

EXAMPLE 3 Divide and, if possible, simplify.

a) $\dfrac{\sqrt{80}}{\sqrt{5}}$ b) $\dfrac{5\sqrt[3]{32}}{\sqrt[3]{2}}$ c) $\dfrac{\sqrt{72xy}}{2\sqrt{2}}$ d) $\dfrac{\sqrt[4]{18a^9b^5}}{\sqrt[4]{3b}}$

SOLUTION

a) $\dfrac{\sqrt{80}}{\sqrt{5}} = \sqrt{\dfrac{80}{5}} = \sqrt{16} = 4$

Because the indices match, we can divide the radicands.

b) $\dfrac{5\sqrt[3]{32}}{\sqrt[3]{2}} = 5\sqrt[3]{\dfrac{32}{2}} = 5\sqrt[3]{16}$

$= 5\sqrt[3]{8 \cdot 2}$ — **8 is the largest perfect-cube factor of 16.**

$= 5\sqrt[3]{8}\,\sqrt[3]{2} = 5 \cdot 2\sqrt[3]{2}$

$= 10\sqrt[3]{2}$

c) $\dfrac{\sqrt{72xy}}{2\sqrt{2}} = \dfrac{1}{2}\sqrt{\dfrac{72xy}{2}}$

$= \dfrac{1}{2}\sqrt{36xy} = \dfrac{1}{2} \cdot 6\sqrt{xy} = 3\sqrt{xy}$

Because the indices match, we can divide the radicands.

d) $\dfrac{\sqrt[4]{18a^9b^5}}{\sqrt[4]{3b}} = \sqrt[4]{\dfrac{18a^9b^5}{3b}}$

$= \sqrt[4]{6a^9b^4} = \sqrt[4]{a^8b^4}\,\sqrt[4]{6a}$ **Note that 8 is the largest power less than 9 that is a multiple of the index 4.**

$= a^2b\sqrt[4]{6a}$ ***Partial check:*** $(a^2b)^4 = a^8b^4$

Rationalizing Denominators and Numerators (Part 1)*

When a radical expression appears in a denominator, it can be useful to find an equivalent expression in which the denominator no longer contains a radical.† The procedure for finding such an expression is called **rationalizing the denominator.** We carry this out by multiplying by 1 in either of two ways.

One way is to multiply by 1 *under* the radical to make the denominator of the radicand a perfect power.

EXAMPLE 4 Rationalize each denominator.

a) $\sqrt{\dfrac{7}{3}}$ **b)** $\sqrt[3]{\dfrac{5}{16}}$

SOLUTION

a) We multiply by 1 under the radical, using $\frac{3}{3}$. We do this so that the denominator of the radicand will be a perfect square:

$\sqrt{\dfrac{7}{3}} = \sqrt{\dfrac{7}{3} \cdot \dfrac{3}{3}}$ **Multiplying by 1 under the radical**

$= \sqrt{\dfrac{21}{9}}$ **The denominator, 9, is now a perfect square.**

$= \dfrac{\sqrt{21}}{\sqrt{9}} = \dfrac{\sqrt{21}}{3}.$

b) Note that $16 = 4^2$. Thus, to make the denominator a perfect cube, we multiply under the radical by $\frac{4}{4}$:

$\sqrt[3]{\dfrac{5}{16}} = \sqrt[3]{\dfrac{5}{4 \cdot 4} \cdot \dfrac{4}{4}}$ **Since the index is 3, we need 3 identical factors in the denominator.**

$= \sqrt[3]{\dfrac{20}{4^3}}$ **The denominator is now a perfect cube.**

$= \dfrac{\sqrt[3]{20}}{\sqrt[3]{4^3}} = \dfrac{\sqrt[3]{20}}{4}.$

---

*Denominators and numerators with two terms are rationalized in Section 7.5.

†See Exercise 73 on p. 568.

Another way to rationalize a denominator is to multiply by 1 *outside* the radical.

EXAMPLE 5 Rationalize each denominator.

a) $\sqrt{\dfrac{4}{5b}}$ **b)** $\dfrac{\sqrt[3]{a}}{\sqrt[3]{9x}}$

SOLUTION

a) We rewrite the expression as a quotient of two radicals. Then we simplify and multiply by 1:

$$\sqrt{\frac{4}{5b}} = \frac{\sqrt{4}}{\sqrt{5b}} = \frac{2}{\sqrt{5b}} \qquad \textbf{We assume } b > 0.$$

$$= \frac{2}{\sqrt{5b}} \cdot \frac{\sqrt{5b}}{\sqrt{5b}} \qquad \textbf{Multiplying by 1}$$

$$= \frac{2\sqrt{5b}}{\left(\sqrt{5b}\right)^2} \qquad \textbf{Try to do this step mentally.}$$

$$= \frac{2\sqrt{5b}}{5b}.$$

b) To rationalize the denominator $\sqrt[3]{9x}$, note that $9x$ is $3 \cdot 3 \cdot x$. In order for this radicand to be a cube, we need another factor of 3 and two more factors of x. Thus we multiply by 1, using $\sqrt[3]{3x^2}/\sqrt[3]{3x^2}$:

$$\frac{\sqrt[3]{a}}{\sqrt[3]{9x}} = \frac{\sqrt[3]{a}}{\sqrt[3]{9x}} \cdot \frac{\sqrt[3]{3x^2}}{\sqrt[3]{3x^2}} \qquad \textbf{Multiplying by 1}$$

$$= \frac{\sqrt[3]{3ax^2}}{\sqrt[3]{27x^3}} \longleftarrow \textbf{This radicand is now a perfect cube.}$$

$$= \frac{\sqrt[3]{3ax^2}}{3x}.$$

Sometimes in calculus it is necessary to rationalize a numerator. To do so, we multiply by 1 to make the radicand in the *numerator* a perfect power.

EXAMPLE 6 Rationalize each numerator.

a) $\sqrt{\dfrac{7}{5}}$ **b)** $\dfrac{\sqrt[3]{4a^2}}{\sqrt[3]{5b}}$

SOLUTION

a) $$\sqrt{\frac{7}{5}} = \sqrt{\frac{7}{5} \cdot \frac{7}{7}} \qquad \textbf{Multiplying by 1 under the radical. We also could have multiplied by } \sqrt{7}/\sqrt{7} \textbf{ outside the radical.}$$

$$= \sqrt{\frac{49}{35}} \qquad \textbf{The numerator is now a perfect square.}$$

$$= \frac{\sqrt{49}}{\sqrt{35}} = \frac{7}{\sqrt{35}}$$

b) $\dfrac{\sqrt[3]{4a^2}}{\sqrt[3]{5b}} = \dfrac{\sqrt[3]{4a^2}}{\sqrt[3]{5b}} \cdot \dfrac{\sqrt[3]{2a}}{\sqrt[3]{2a}}$ **Multiplying by 1**

$= \dfrac{\sqrt[3]{8a^3}}{\sqrt[3]{10ba}}$ ← **This radicand is now a perfect cube.**

$= \dfrac{2a}{\sqrt[3]{10ab}}$

7.4 EXERCISE SET

FOR EXTRA HELP

MathXL | MyMathLab | InterAct Math | AW Math Tutor Center | Video Lectures on CD: Disc 4 | Student's Solutions Manual

Concept Reinforcement *In each of Exercises 1–8, match the expression with an equivalent expression from the column on the right. Assume a, b > 0.*

1. ___ $\sqrt[3]{\dfrac{a^2}{b^6}}$ a) $\dfrac{\sqrt[5]{a^2}\sqrt[5]{b^2}}{\sqrt[5]{b^5}}$

2. ___ $\dfrac{\sqrt[3]{a^6}}{\sqrt[3]{b^9}}$ b) $\dfrac{a^2}{b^3}$

3. ___ $\sqrt[5]{\dfrac{a^6}{b^4}}$ c) $\sqrt{\dfrac{a \cdot b}{b^3 \cdot b}}$

4. ___ $\sqrt{\dfrac{a}{b^3}}$ d) $\sqrt{a}$

5. ___ $\dfrac{\sqrt[5]{a^2}}{\sqrt[5]{b^2}}$ e) $\dfrac{\sqrt[3]{a^2}}{b^2}$

6. ___ $\dfrac{\sqrt{5a^4}}{\sqrt{5a^3}}$ f) $\sqrt[5]{\dfrac{a^6 b}{b^4 \cdot b}}$

7. ___ $\dfrac{\sqrt[5]{a^2}}{\sqrt[5]{b^3}}$ g) $2a$

8. ___ $\sqrt[4]{\dfrac{16a^6}{a^2}}$ h) $\dfrac{\sqrt[5]{a^2 b^3}}{\sqrt[5]{b^5}}$

Simplify by taking the roots of the numerator and the denominator. Assume all variables represent positive numbers.

9. $\sqrt{\dfrac{36}{25}}$

10. $\sqrt{\dfrac{100}{81}}$

11. $\sqrt[3]{\dfrac{64}{27}}$

12. $\sqrt[3]{\dfrac{343}{1000}}$

13. $\sqrt{\dfrac{49}{y^2}}$

14. $\sqrt{\dfrac{121}{x^2}}$

15. $\sqrt{\dfrac{36y^3}{x^4}}$

16. $\sqrt{\dfrac{25a^5}{b^6}}$

17. $\sqrt[3]{\dfrac{27a^4}{8b^3}}$

18. $\sqrt[3]{\dfrac{64x^7}{216y^6}}$

19. $\sqrt[4]{\dfrac{16a^4}{b^4c^8}}$

20. $\sqrt[4]{\dfrac{81x^4}{y^8z^4}}$

21. $\sqrt[4]{\dfrac{a^5b^8}{c^{10}}}$

22. $\sqrt[4]{\dfrac{x^9y^{12}}{z^6}}$

23. $\sqrt[5]{\dfrac{32x^6}{y^{11}}}$

24. $\sqrt[5]{\dfrac{243a^9}{b^{13}}}$

25. $\sqrt[6]{\dfrac{x^6y^8}{z^{15}}}$

26. $\sqrt[6]{\dfrac{a^9b^{12}}{c^{13}}}$

Divide and, if possible, simplify. Assume all variables represent positive numbers.

27. $\dfrac{\sqrt{35x}}{\sqrt{7x}}$

28. $\dfrac{\sqrt{28y}}{\sqrt{4y}}$

29. $\dfrac{\sqrt[3]{270}}{\sqrt[3]{10}}$

30. $\dfrac{\sqrt[3]{40}}{\sqrt[3]{5}}$

31. $\dfrac{\sqrt{40xy^3}}{\sqrt{8x}}$

32. $\dfrac{\sqrt{56ab^3}}{\sqrt{7a}}$

33. $\dfrac{\sqrt[3]{96a^4b^2}}{\sqrt[3]{12a^2b}}$

34. $\dfrac{\sqrt[3]{189x^5y^7}}{\sqrt[3]{7x^2y^2}}$

35. $\dfrac{\sqrt{100ab}}{5\sqrt{2}}$

36. $\dfrac{\sqrt{75ab}}{3\sqrt{3}}$

37. $\dfrac{\sqrt[4]{48x^9y^{13}}}{\sqrt[4]{3xy^{-2}}}$

38. $\dfrac{\sqrt[5]{64a^{11}b^{28}}}{\sqrt[5]{2ab^{-2}}}$

39. $\dfrac{\sqrt[3]{x^3 - y^3}}{\sqrt[3]{x - y}}$

40. $\dfrac{\sqrt[3]{r^3 + s^3}}{\sqrt[3]{r + s}}$

***Hint*: Factor and then simplify.**

Rationalize each denominator. Assume all variables represent positive numbers.

41. $\sqrt{\dfrac{3}{2}}$

42. $\sqrt{\dfrac{6}{7}}$

43. $\dfrac{2\sqrt{5}}{7\sqrt{3}}$

44. $\dfrac{3\sqrt{5}}{2\sqrt{2}}$

45. $\sqrt[3]{\dfrac{16}{9}}$

46. $\sqrt[3]{\dfrac{2}{9}}$

47. $\dfrac{\sqrt[3]{3a}}{\sqrt[3]{5c}}$

48. $\dfrac{\sqrt[3]{7x}}{\sqrt[3]{3y}}$

49. $\dfrac{\sqrt[3]{5y^4}}{\sqrt[3]{6x^4}}$

50. $\dfrac{\sqrt[3]{3a^4}}{\sqrt[3]{7b^2}}$

51. $\sqrt[3]{\dfrac{2}{x^2y}}$

52. $\sqrt[3]{\dfrac{5}{ab^2}}$

53. $\sqrt{\dfrac{7a}{18}}$

54. $\sqrt{\dfrac{3x}{10}}$

55. $\sqrt{\dfrac{9}{20x^2y}}$

56. $\sqrt{\dfrac{7}{32a^2b}}$

Aha! **57.** $\sqrt{\dfrac{10ab^2}{72a^3b}}$

58. $\sqrt{\dfrac{21x^2y}{75xy^5}}$

Rationalize each numerator. Assume all variables represent positive numbers.

59. $\dfrac{\sqrt{5}}{\sqrt{7x}}$

60. $\dfrac{\sqrt{10}}{\sqrt{3x}}$

61. $\sqrt{\dfrac{14}{21}}$

62. $\sqrt{\dfrac{12}{15}}$

63. $\dfrac{4\sqrt{13}}{3\sqrt{7}}$

64. $\dfrac{5\sqrt{21}}{2\sqrt{5}}$

65. $\dfrac{\sqrt[3]{7}}{\sqrt[3]{2}}$

66. $\dfrac{\sqrt[3]{5}}{\sqrt[3]{4}}$

67. $\sqrt{\dfrac{7x}{3y}}$

68. $\sqrt{\dfrac{7a}{6b}}$

69. $\sqrt[3]{\dfrac{2a^5}{5b}}$

70. $\sqrt[3]{\dfrac{2a^4}{7b}}$

71. $\sqrt{\dfrac{x^3y}{2}}$

72. $\sqrt{\dfrac{ab^5}{3}}$

TW **73.** Explain why it is easier to approximate

$$\frac{\sqrt{2}}{2} \quad \text{than} \quad \frac{1}{\sqrt{2}}$$

if no calculator is available and $\sqrt{2} \approx 1.414213562$.

TW **74.** A student *incorrectly* claims that

$$\frac{5 + \sqrt{2}}{\sqrt{18}} = \frac{5 + \sqrt{1}}{\sqrt{9}} = \frac{5 + 1}{3}.$$

How could you convince the student that a mistake has been made? How would you explain the correct way of rationalizing the denominator?

Skill Maintenance

Multiply. [6.1]

75. $\dfrac{3}{x - 5} \cdot \dfrac{x - 1}{x + 5}$

76. $\dfrac{7}{x+4} \cdot \dfrac{x-2}{x-4}$

Simplify.

77. $\dfrac{a^2 - 8a + 7}{a^2 - 49}$ [6.1]

78. $\dfrac{t^2 + 9t - 22}{t^2 - 4}$ [6.1]

79. $(5a^3b^4)^3$ [1.4]

80. $(3x^4)^2(5xy^3)^2$ [1.4]

Synthesis

TW **81.** Is the quotient of two irrational numbers always an irrational number? Why or why not?

TW **82.** Is it possible to understand how to rationalize a denominator without knowing how to multiply rational expressions? Why or why not?

83. *Pendulums.* The *period* of a pendulum is the time it takes to complete one cycle, swinging to and fro. For a pendulum that is L centimeters long, the period T is given by the formula

$$T = 2\pi\sqrt{\frac{L}{980}},$$

where T is in seconds. Find, to the nearest hundredth of a second, the period of a pendulum of length **(a)** 65 cm; **(b)** 98 cm; **(c)** 120 cm. Use a calculator's π key if possible.

Perform the indicated operations.

84. $\dfrac{7\sqrt{a^2b}\,\sqrt{25xy}}{5\sqrt{a^{-4}b^{-1}}\,\sqrt{49x^{-1}y^{-3}}}$

85. $\dfrac{(\sqrt[3]{81mn^2})^2}{(\sqrt[3]{mn})^2}$

86. $\dfrac{\sqrt{44x^2y^9z}\,\sqrt{22y^9z^6}}{(\sqrt{11xy^8z^2})^2}$

87. $\sqrt{a^2 - 3} - \dfrac{a^2}{\sqrt{a^2 - 3}}$

88. $5\sqrt{\dfrac{x}{y}} + 4\sqrt{\dfrac{y}{x}} - \dfrac{3}{\sqrt{xy}}$

89. Provide a reason for each step in the following derivation of the quotient rule:

$$\sqrt[n]{\frac{a}{b}} = \left(\frac{a}{b}\right)^{1/n} \quad \_\_\_\_\_\_\_\_\_\_$$
$$= \frac{a^{1/n}}{b^{1/n}}$$
$$= \frac{\sqrt[n]{a}}{\sqrt[n]{b}}$$

90. Show that $\dfrac{\sqrt[n]{a}}{\sqrt[n]{b}}$ is the nth root of $\dfrac{a}{b}$ by raising it to the nth power and simplifying.

91. Let $f(x) = \sqrt{18x^3}$ and $g(x) = \sqrt{2x}$. Find $(f/g)(x)$ and specify the domain of f/g.

92. Let $f(t) = \sqrt{2t}$ and $g(t) = \sqrt{50t^3}$. Find $(f/g)(t)$ and specify the domain of f/g.

93. Let $f(x) = \sqrt{x^2 - 9}$ and $g(x) = \sqrt{x - 3}$. Find $(f/g)(x)$ and specify the domain of f/g.

7.5 Expressions Containing Several Radical Terms

Adding and Subtracting Radical Expressions ■ Products and Quotients of Two or More Radical Terms ■ Rationalizing Denominators and Numerators (Part 2) ■ Terms with Differing Indices

Radical expressions like $6\sqrt{7} + 4\sqrt{7}$ or $(\sqrt{a} + \sqrt{b})(\sqrt{a} - \sqrt{b})$ contain more than one *radical term* and can sometimes be simplified.

Adding and Subtracting Radical Expressions

When two radical expressions have the same indices and radicands, they are said to be **like radicals.** Like radicals can be combined (added or subtracted) in much the same way that we combined like terms earlier in this text.

Study Tip

Review Material on Your Own

Never hesitate to review earlier material in which you feel a lack of confidence. For example, if you feel unsure about how to multiply with fraction notation, be sure to review that material before studying any new material that involves multiplication of fractions. Doing (and checking) some practice problems from that section is also a good way to sharpen any skills that may not have been used for a while.

EXAMPLE 1 Simplify by combining like radical terms.

a) $6\sqrt{7} + 4\sqrt{7}$

b) $\sqrt[3]{2} - 7x\sqrt[3]{2} + 5\sqrt[3]{2}$

c) $6\sqrt[5]{4x} + 3\sqrt[5]{4x} - \sqrt[3]{4x}$

SOLUTION

a) $6\sqrt{7} + 4\sqrt{7} = (6 + 4)\sqrt{7}$ **Using the distributive law (factoring out $\sqrt{7}$)**

$= 10\sqrt{7}$ **You can think: 6 square roots of 7 plus 4 square roots of 7 results in 10 square roots of 7.**

b) $\sqrt[3]{2} - 7x\sqrt[3]{2} + 5\sqrt[3]{2} = (1 - 7x + 5)\sqrt[3]{2}$ **Factoring out $\sqrt[3]{2}$**

$= (6 - 7x)\sqrt[3]{2}$ **These parentheses are important!**

c) $6\sqrt[5]{4x} + 3\sqrt[5]{4x} - \sqrt[3]{4x} = (6 + 3)\sqrt[5]{4x} - \sqrt[3]{4x}$ **Try to do this step mentally.**

$= 9\sqrt[5]{4x} - \sqrt[3]{4x}$ **Because the indices differ, we are done.**

Our ability to simplify radical expressions can help us to find like radicals even when, at first, it may appear that none exists.

EXAMPLE 2 Simplify by combining like radical terms, if possible.

a) $3\sqrt{8} - 5\sqrt{2}$

b) $9\sqrt{5} - 4\sqrt{3}$

c) $\sqrt[3]{2x^6y^4} + 7\sqrt[3]{2y}$

SOLUTION

a)
$$\left.\begin{aligned} 3\sqrt{8} - 5\sqrt{2} &= 3\sqrt{4 \cdot 2} - 5\sqrt{2} \\ &= 3\sqrt{4} \cdot \sqrt{2} - 5\sqrt{2} \\ &= 3 \cdot 2 \cdot \sqrt{2} - 5\sqrt{2} \end{aligned}\right\} \text{ Simplifying } \sqrt{8}$$

$= 6\sqrt{2} - 5\sqrt{2}$

$= \sqrt{2}$ **Combining like radicals**

b) $9\sqrt{5} - 4\sqrt{3}$ cannot be simplified.

c) $\sqrt[3]{2x^6y^4} + 7\sqrt[3]{2y} = \sqrt[3]{x^6y^3 \cdot 2y} + 7\sqrt[3]{2y}$

$= \sqrt[3]{x^6y^3} \cdot \sqrt[3]{2y} + 7\sqrt[3]{2y}$ **Simplifying $\sqrt[3]{2x^6y^4}$**

$= x^2y \cdot \sqrt[3]{2y} + 7\sqrt[3]{2y}$

$= (x^2y + 7)\sqrt[3]{2y}$ **Factoring to combine like radical terms**

Products and Quotients of Two or More Radical Terms

Radical expressions often contain factors that have more than one term. The procedure for multiplying out such expressions is similar to finding products of polynomials. Some products will yield like radical terms, which we can now combine.

EXAMPLE 3 Multiply.

a) $\sqrt{3}(x - \sqrt{5})$ **b)** $\sqrt[3]{y}(\sqrt[3]{y^2} + \sqrt[3]{2})$

c) $(4\sqrt{3} + \sqrt{2})(\sqrt{3} - 5\sqrt{2})$ **d)** $(\sqrt{a} + \sqrt{b})(\sqrt{a} - \sqrt{b})$

SOLUTION

a) $\sqrt{3}(x - \sqrt{5}) = \sqrt{3} \cdot x - \sqrt{3} \cdot \sqrt{5}$ **Using the distributive law**

$= x\sqrt{3} - \sqrt{15}$ **Multiplying radicals**

b) $\sqrt[3]{y}(\sqrt[3]{y^2} + \sqrt[3]{2}) = \sqrt[3]{y} \cdot \sqrt[3]{y^2} + \sqrt[3]{y} \cdot \sqrt[3]{2}$ **Using the distributive law**

$= \sqrt[3]{y^3} + \sqrt[3]{2y}$ **Multiplying radicals**

$= y + \sqrt[3]{2y}$ **Simplifying $\sqrt[3]{y^3}$**

c) $(4\sqrt{3} + \sqrt{2})(\sqrt{3} - 5\sqrt{2}) = \overset{F}{4(\sqrt{3})^2} - \overset{O}{20\sqrt{3} \cdot \sqrt{2}} + \overset{I}{\sqrt{2} \cdot \sqrt{3}} - \overset{L}{5(\sqrt{2})^2}$

$= 4 \cdot 3 - 20\sqrt{6} + \sqrt{6} - 5 \cdot 2$ **Multiplying radicals**

$= 12 - 20\sqrt{6} + \sqrt{6} - 10$

$= 2 - 19\sqrt{6}$ **Combining like terms**

d) $(\sqrt{a} + \sqrt{b})(\sqrt{a} - \sqrt{b}) = (\sqrt{a})^2 - \sqrt{a}\sqrt{b} + \sqrt{a}\sqrt{b} - (\sqrt{b})^2$ **Using FOIL**

$= a - b$ **Combining like terms**

In Example 3(d) above, you may have noticed that since the outer and inner products in the multiplication are opposites, the result, $a - b$, is not itself a radical expression. Pairs of radical terms, like $\sqrt{a} + \sqrt{b}$ and $\sqrt{a} - \sqrt{b}$, are called **conjugates.**

Rationalizing Denominators and Numerators (Part 2)

The use of conjugates allows us to rationalize denominators or numerators with two terms.

EXAMPLE 4 Rationalize each denominator: **(a)** $\dfrac{4}{\sqrt{3}+x}$; **(b)** $\dfrac{4+\sqrt{2}}{\sqrt{5}-\sqrt{2}}$.

SOLUTION

a)
$$\frac{4}{\sqrt{3}+x} = \frac{4}{\sqrt{3}+x}\cdot\frac{\sqrt{3}-x}{\sqrt{3}-x}$$
Multiplying by 1, using the conjugate of $\sqrt{3}+x$, which is $\sqrt{3}-x$

$$= \frac{4(\sqrt{3}-x)}{(\sqrt{3}+x)(\sqrt{3}-x)}$$
Multiplying numerators and denominators

$$= \frac{4(\sqrt{3}-x)}{(\sqrt{3})^2-x^2}$$
Using FOIL in the denominator

$$= \frac{4\sqrt{3}-4x}{3-x^2}$$
Simplifying. No radicals remain in the denominator.

b)
$$\frac{4+\sqrt{2}}{\sqrt{5}-\sqrt{2}} = \frac{4+\sqrt{2}}{\sqrt{5}-\sqrt{2}}\cdot\frac{\sqrt{5}+\sqrt{2}}{\sqrt{5}+\sqrt{2}}$$
Multiplying by 1, using the conjugate of $\sqrt{5}-\sqrt{2}$, which is $\sqrt{5}+\sqrt{2}$

$$= \frac{(4+\sqrt{2})(\sqrt{5}+\sqrt{2})}{(\sqrt{5}-\sqrt{2})(\sqrt{5}+\sqrt{2})}$$
Multiplying numerators and denominators

$$= \frac{4\sqrt{5}+4\sqrt{2}+\sqrt{2}\sqrt{5}+(\sqrt{2})^2}{(\sqrt{5})^2-(\sqrt{2})^2}$$
Using FOIL

$$= \frac{4\sqrt{5}+4\sqrt{2}+\sqrt{10}+2}{5-2}$$
Squaring in the denominator and the numerator

$$= \frac{4\sqrt{5}+4\sqrt{2}+\sqrt{10}+2}{3}$$
No radicals remain in the denominator.

```
(4+√(2))/(√(5)-√
(2))
                    6.587801273
(4√(5)+4√(2)+√(1
0)+2)/3
                    6.587801273
```

We can check by approximating the value of the original expression and the value of the rationalized expression, as shown at left. Care must be taken to place the parentheses properly. The approximate values are the same, so we have a check.

To rationalize a numerator with two terms, we use the conjugate of the numerator.

EXAMPLE 5 Rationalize the numerator: $\dfrac{4+\sqrt{2}}{\sqrt{5}-\sqrt{2}}$.

SOLUTION We have

$$\frac{4+\sqrt{2}}{\sqrt{5}-\sqrt{2}} = \frac{4+\sqrt{2}}{\sqrt{5}-\sqrt{2}}\cdot\frac{4-\sqrt{2}}{4-\sqrt{2}}$$
Multiplying by 1, using the conjugate of $4+\sqrt{2}$, which is $4-\sqrt{2}$

$$= \frac{16-(\sqrt{2})^2}{4\sqrt{5}-\sqrt{5}\sqrt{2}-4\sqrt{2}+(\sqrt{2})^2}$$

$$= \frac{14}{4\sqrt{5}-\sqrt{10}-4\sqrt{2}+2}.$$

```
(4+√(2))/(√(5)-√
(2))
                    6.587801273
14/(4√(5)-√(10)-
4√(2)+2)
                    6.587801273
```

We check by comparing the values of the original expression and the expression with a rationalized numerator, as shown at left.

Terms with Differing Indices

To multiply or divide radical terms with different indices, we can convert to exponential notation, use the rules for exponents, and then convert back to radical notation.

Student Notes

Expressions similar to the one in Example 6 are most easily simplified by rewriting the expression using exponents in place of radicals. After simplifying, remember to write your final result in radical notation. In general, if a problem is presented in one form, it is expected that the final result be presented in the same form.

EXAMPLE 6 Divide and, if possible, simplify: $\dfrac{\sqrt[4]{(x+y)^3}}{\sqrt{x+y}}$.

SOLUTION

$$\frac{\sqrt[4]{(x+y)^3}}{\sqrt{x+y}} = \frac{(x+y)^{3/4}}{(x+y)^{1/2}}$$ **Converting to exponential notation**

$$= (x+y)^{3/4-1/2}$$ **Since the bases are identical, we can subtract exponents:** $\frac{3}{4} - \frac{1}{2} = \frac{3}{4} - \frac{2}{4} = \frac{1}{4}$.

$$= (x+y)^{1/4}$$
$$= \sqrt[4]{x+y}$$ **Converting back to radical notation**

To Simplify Products or Quotients with Differing Indices

1. Convert all radical expressions to exponential notation.
2. When the bases are identical, subtract exponents to divide and add exponents to multiply. This may require finding a common denominator.
3. Convert back to radical notation and, if possible, simplify.

EXAMPLE 7 Multiply and simplify: $\sqrt{x^3}\,\sqrt[3]{x}$.

SOLUTION

$$\sqrt{x^3}\,\sqrt[3]{x} = x^{3/2} \cdot x^{1/3}$$ **Converting to exponential notation**

$$= x^{11/6}$$ **Adding exponents:** $\frac{3}{2} + \frac{1}{3} = \frac{9}{6} + \frac{2}{6}$

$$= \sqrt[6]{x^{11}}$$ **Converting back to radical notation**

$$= \sqrt[6]{x^6}\,\sqrt[6]{x^5}$$
$$= x\sqrt[6]{x^5}$$ **Simplifying**

EXAMPLE 8 If $f(x) = \sqrt[3]{x^2}$ and $g(x) = \sqrt{x} + \sqrt[4]{x}$, find $(f \cdot g)(x)$.

SOLUTION Recall from Section 2.5 that $(f \cdot g)(x) = f(x) \cdot g(x)$. Thus,

$$(f \cdot g)(x) = \sqrt[3]{x^2}\left(\sqrt{x} + \sqrt[4]{x}\right)$$ ***x* is assumed to be nonnegative.**

$$= x^{2/3}(x^{1/2} + x^{1/4})$$ **Converting to exponential notation**

$$= x^{2/3} \cdot x^{1/2} + x^{2/3} \cdot x^{1/4}$$ **Using the distributive law**

$$= x^{2/3+1/2} + x^{2/3+1/4}$$ **Adding exponents**

$$= x^{7/6} + x^{11/12}$$ $\frac{2}{3} + \frac{1}{2} = \frac{4}{6} + \frac{3}{6}; \frac{2}{3} + \frac{1}{4} = \frac{8}{12} + \frac{3}{12}$

$$= \sqrt[6]{x^7} + \sqrt[12]{x^{11}}$$ **Converting back to radical notation**

$$= \sqrt[6]{x^6}\,\sqrt[6]{x} + \sqrt[12]{x^{11}}$$
$$= x\sqrt[6]{x} + \sqrt[12]{x^{11}}$$ **Simplifying**

If factors are raised to powers that share a common denominator, we can write the final result as a single radical expression.

EXAMPLE 9 Divide and, if possible, simplify: $\dfrac{\sqrt[3]{a^2b^4}}{\sqrt{ab}}$.

SOLUTION

$$\begin{aligned}
\frac{\sqrt[3]{a^2b^4}}{\sqrt{ab}} &= \frac{(a^2b^4)^{1/3}}{(ab)^{1/2}} && \textbf{Converting to exponential notation}\\
&= \frac{a^{2/3}b^{4/3}}{a^{1/2}b^{1/2}} && \textbf{Using the product and power rules}\\
&= a^{2/3-1/2}b^{4/3-1/2} && \textbf{Subtracting exponents}\\
&= a^{1/6}b^{5/6}\\
&= \sqrt[6]{a}\,\sqrt[6]{b^5} && \textbf{Converting to radical notation}\\
&= \sqrt[6]{ab^5} && \textbf{Using the product rule for radicals}
\end{aligned}$$

7.5 EXERCISE SET

FOR EXTRA HELP

MathXL | MyMathLab | InterAct Math | AW Math Tutor Center | Video Lectures on CD: Disc 4 | Student's Solutions Manual

Concept Reinforcement For each of Exercises 1–6, fill in the blanks by selecting from the following words (which may be used more than once).

radicand(s)

indices

conjugate(s)

base(s)

denominator(s)

numerator(s)

1. To add radical expressions, the ______________ and the ______________ must be the same.

2. To multiply radical expressions, the ______________ must be the same.

3. To find a product by adding exponents, the ______________ must be the same.

4. To add rational expressions, the ______________ must be the same.

5. To rationalize the ______________ of $\dfrac{\sqrt{a+0.1}-\sqrt{a}}{0.1}$, we multiply by a form of 1, using the ______________ of $\sqrt{a+0.1}-\sqrt{a}$, or $\sqrt{a+0.1}+\sqrt{a}$, to write 1.

6. To find a quotient by subtracting exponents, the ______________ must be the same.

Add or subtract. Simplify by combining like radical terms, if possible. Assume that all variables and radicands represent positive real numbers.

7. $2\sqrt{5} + 7\sqrt{5}$

8. $4\sqrt{7} + 2\sqrt{7}$

9. $7\sqrt[3]{4} - 5\sqrt[3]{4}$

10. $14\sqrt[5]{2} - 6\sqrt[5]{2}$

11. $\sqrt[3]{y} + 9\sqrt[3]{y}$

12. $9\sqrt[4]{t} - 3\sqrt[4]{t}$

13. $8\sqrt{2} - 6\sqrt{2} + 5\sqrt{2}$

14. $\sqrt{6} + 8\sqrt{6} - 3\sqrt{6}$

15. $9\sqrt[3]{7} - \sqrt{3} + 4\sqrt[3]{7} + 2\sqrt{3}$

16. $5\sqrt{7} - 8\sqrt[4]{11} + \sqrt{7} + 9\sqrt[4]{11}$

17. $4\sqrt{27} - 3\sqrt{3}$

18. $9\sqrt{50} - 4\sqrt{2}$

19. $3\sqrt{45} + 7\sqrt{20}$

20. $5\sqrt{12} + 16\sqrt{27}$

21. $3\sqrt[3]{16} + \sqrt[3]{54}$

22. $\sqrt[3]{27} - 5\sqrt[3]{8}$

23. $\sqrt{5a} + 2\sqrt{45a^3}$

24. $4\sqrt{3x^3} - \sqrt{12x}$

25. $\sqrt[3]{6x^4} + \sqrt[3]{48x}$

26. $\sqrt[3]{54x} - \sqrt[3]{2x^4}$

27. $\sqrt{4a - 4} + \sqrt{a - 1}$

28. $\sqrt{9y + 27} + \sqrt{y + 3}$

29. $\sqrt{x^3 - x^2} + \sqrt{9x - 9}$

30. $\sqrt{4x - 4} - \sqrt{x^3 - x^2}$

Multiply. Assume all variables represent nonnegative real numbers.

31. $\sqrt{3}(4 + \sqrt{3})$

32. $\sqrt{7}(3 - \sqrt{7})$

33. $3\sqrt{5}(\sqrt{5} - \sqrt{2})$

34. $4\sqrt{2}(\sqrt{3} - \sqrt{5})$

35. $\sqrt{2}(3\sqrt{10} - 2\sqrt{2})$

36. $\sqrt{3}(2\sqrt{5} - 3\sqrt{4})$

37. $\sqrt[3]{3}(\sqrt[3]{9} - 4\sqrt[3]{21})$

38. $\sqrt[3]{2}(\sqrt[3]{4} - 2\sqrt[3]{32})$

39. $\sqrt[3]{a}(\sqrt[3]{a^2} + \sqrt[3]{24a^2})$

40. $\sqrt[3]{x}(\sqrt[3]{3x^2} - \sqrt[3]{81x^2})$

41. $(2 + \sqrt{6})(5 - \sqrt{6})$

42. $(4 - \sqrt{5})(2 + \sqrt{5})$

43. $(\sqrt{2} + \sqrt{7})(\sqrt{3} - \sqrt{7})$

44. $(\sqrt{7} - \sqrt{2})(\sqrt{5} + \sqrt{2})$

45. $(3 - \sqrt{5})(3 + \sqrt{5})$

46. $(6 - \sqrt{7})(6 + \sqrt{7})$

47. $(\sqrt{6} + \sqrt{8})(\sqrt{6} - \sqrt{8})$

48. $(\sqrt{5} + \sqrt{3})(\sqrt{5} - \sqrt{3})$

49. $(3\sqrt{7} + 2\sqrt{5})(2\sqrt{7} - 4\sqrt{5})$

50. $(4\sqrt{5} - 3\sqrt{2})(2\sqrt{5} + 4\sqrt{2})$

51. $(2 + \sqrt{3})^2$

52. $(3 + \sqrt{7})^2$

53. $(\sqrt{3} - \sqrt{2})^2$

54. $(\sqrt{5} - \sqrt{3})^2$

55. $(\sqrt{2t} + \sqrt{5})^2$

56. $(\sqrt{3x} - \sqrt{2})^2$

57. $(3 - \sqrt{x + 5})^2$

58. $(4 + \sqrt{x - 3})^2$

59. $(2\sqrt[4]{7} - \sqrt[4]{6})(3\sqrt[4]{9} + 2\sqrt[4]{5})$

60. $(4\sqrt[3]{3} + \sqrt[3]{10})(2\sqrt[3]{7} + 5\sqrt[3]{6})$

Rationalize each denominator.

61. $\dfrac{5}{4 - \sqrt{3}}$

62. $\dfrac{3}{4 - \sqrt{7}}$

63. $\dfrac{2 + \sqrt{5}}{6 + \sqrt{3}}$

64. $\dfrac{1 + \sqrt{2}}{3 + \sqrt{5}}$

65. $\dfrac{\sqrt{a}}{\sqrt{a} + \sqrt{b}}$

66. $\dfrac{\sqrt{z}}{\sqrt{x} - \sqrt{z}}$

Aha! **67.** $\dfrac{\sqrt{7} - \sqrt{3}}{\sqrt{3} - \sqrt{7}}$

68. $\dfrac{\sqrt{7} + \sqrt{5}}{\sqrt{5} + \sqrt{2}}$

69. $\dfrac{3\sqrt{2} - \sqrt{7}}{4\sqrt{2} + 2\sqrt{5}}$

70. $\dfrac{5\sqrt{3} - \sqrt{11}}{2\sqrt{3} - 5\sqrt{2}}$

Rationalize each numerator. If possible, simplify your result.

71. $\dfrac{\sqrt{7} + 2}{5}$

72. $\dfrac{\sqrt{3} + 1}{4}$

73. $\dfrac{\sqrt{6} - 2}{\sqrt{3} + 7}$

74. $\dfrac{\sqrt{10} + 4}{\sqrt{2} - 3}$

75. $\dfrac{\sqrt{x} - \sqrt{y}}{\sqrt{x} + \sqrt{y}}$

76. $\dfrac{\sqrt{a} + \sqrt{b}}{\sqrt{a} - \sqrt{b}}$

77. $\dfrac{\sqrt{a + h} - \sqrt{a}}{h}$

78. $\dfrac{\sqrt{x - h} - \sqrt{x}}{h}$

Perform the indicated operation and simplify. Assume all variables represent positive real numbers.

79. $\sqrt{a}\,\sqrt[4]{a^3}$

80. $\sqrt[3]{x^2}\,\sqrt[6]{x^5}$

81. $\sqrt[5]{b^2}\,\sqrt{b^3}$

82. $\sqrt[4]{a^3}\,\sqrt[3]{a^2}$

83. $\sqrt{xy^3}\,\sqrt[3]{x^2y}$

84. $\sqrt[5]{a^3b}\,\sqrt{ab}$

85. $\sqrt[4]{9ab^3}\,\sqrt{3a^4b}$

86. $\sqrt{2x^3y^3}\,\sqrt[3]{4xy^2}$

87. $\sqrt{a^4b^3c^4}\,\sqrt[3]{ab^2c}$

88. $\sqrt[3]{xy^2z}\,\sqrt{x^3yz^2}$

89. $\dfrac{\sqrt[3]{a^2}}{\sqrt[4]{a}}$

90. $\dfrac{\sqrt[3]{x^2}}{\sqrt[5]{x}}$

91. $\dfrac{\sqrt[4]{x^2y^3}}{\sqrt[3]{xy}}$

92. $\dfrac{\sqrt[5]{a^4b}}{\sqrt[3]{ab}}$

93. $\dfrac{\sqrt{ab^3}}{\sqrt[5]{a^2b^3}}$

94. $\dfrac{\sqrt[5]{x^3y^4}}{\sqrt{xy}}$

95. $\dfrac{\sqrt[4]{(3x-1)^3}}{\sqrt[5]{(3x-1)^3}}$

96. $\dfrac{\sqrt[3]{(2+5x)^2}}{\sqrt[4]{2+5x}}$

97. $\dfrac{\sqrt[3]{(2x+1)^2}}{\sqrt[5]{(2x+1)^2}}$

98. $\dfrac{\sqrt[4]{(5+3x)^3}}{\sqrt[3]{(5+3x)^2}}$

99. $\sqrt[3]{x^2y}\left(\sqrt{xy} - \sqrt[5]{xy^3}\right)$

100. $\sqrt[4]{a^2b}\left(\sqrt[3]{a^2b} - \sqrt[5]{a^2b^2}\right)$

101. $\left(m + \sqrt[3]{n^2}\right)\left(2m + \sqrt[4]{n}\right)$

102. $\left(r - \sqrt[4]{s^3}\right)\left(3r - \sqrt[5]{s}\right)$

In Exercises 103–106, $f(x)$ and $g(x)$ are as given. Find $(f \cdot g)(x)$. Assume all variables represent nonnegative real numbers.

103. $f(x) = \sqrt[4]{x},\ g(x) = \sqrt[4]{2x} - \sqrt[4]{x^{11}}$

104. $f(x) = \sqrt[4]{x^7} + \sqrt[4]{3x^2},\ g(x) = \sqrt[4]{x}$

105. $f(x) = x + \sqrt{7},\ g(x) = x - \sqrt{7}$

106. $f(x) = x - \sqrt{2},\ g(x) = x + \sqrt{6}$

Let $f(x) = x^2$. Find each of the following.

107. $f\left(5 + \sqrt{2}\right)$

108. $f\left(7 + \sqrt{3}\right)$

109. $f\left(\sqrt{3} - \sqrt{5}\right)$

110. $f\left(\sqrt{6} - \sqrt{3}\right)$

TW **111.** In what way(s) is combining like radical terms similar to combining like terms that are monomials?

TW **112.** Why do we need to know how to multiply radical expressions before learning how to add them?

Focused Review

Perform the indicated operation and simplify. Assume all variables represent positive real numbers.

113. $5\sqrt{3} + 2\sqrt{3}$ [7.5]

114. $5\sqrt{3} \cdot 2\sqrt{3}$ [7.3]

115. $\dfrac{\sqrt{200xy}}{2\sqrt{2y}}$ [7.4]

116. $\sqrt[3]{x^2y^4z} \cdot \sqrt[3]{x^4y^2z}$ [7.3]

117. $\left(5\sqrt{3} + \sqrt{x}\right)\left(5\sqrt{3} - \sqrt{x}\right)$ [7.5]

118. $\sqrt{2x}\,\sqrt[4]{8x^3}$ [7.5]

Synthesis

TW **119.** Ramon *incorrectly* writes

$$\sqrt[5]{x^2} \cdot \sqrt{x^3} = x^{2/5} \cdot x^{3/2} = \sqrt[5]{x^3}.$$

What mistake do you suspect he is making?

TW **120.** After examining the expression $\sqrt[4]{25xy^3}\,\sqrt{5x^4y}$, Dyan (correctly) concludes that x and y are both nonnegative. Explain how she could reach this conclusion.

Find a simplified form for $f(x)$. Assume $x \geq 0$.

121. $f(x) = \sqrt{20x^2 + 4x^3} - 3x\sqrt{45 + 9x} + \sqrt{5x^2 + x^3}$

122. $f(x) = \sqrt{x^3 - x^2} + \sqrt{9x^3 - 9x^2} - \sqrt{4x^3 - 4x^2}$

123. $f(x) = \sqrt[4]{x^5 - x^4} + 3\sqrt[4]{x^9 - x^8}$

124. $f(x) = \sqrt[4]{16x^4 + 16x^5} - 2\sqrt[4]{x^8 + x^9}$

Simplify.

125. $\frac{1}{2}\sqrt{36a^5bc^4} - \frac{1}{2}\sqrt[3]{64a^4bc^6} + \frac{1}{6}\sqrt{144a^3bc^6}$

126. $7x\sqrt{(x+y)^3} - 5xy\sqrt{x+y} - 2y\sqrt{(x+y)^3}$

127. $\sqrt{27a^5(b+1)}\,\sqrt[3]{81a(b+1)^4}$

128. $\sqrt{8x(y+z)^5}\,\sqrt[3]{4x^2(y+z)^2}$

129. $\dfrac{\dfrac{1}{\sqrt{w}} - \sqrt{w}}{\dfrac{\sqrt{w}+1}{\sqrt{w}}}$

130. $\dfrac{1}{4+\sqrt{3}} + \dfrac{1}{\sqrt{3}} + \dfrac{1}{\sqrt{3}-4}$

Express each of the following as the product of two radical expressions.

131. $x - 5$

132. $y - 7$

133. $x - a$

Multiply.

134. $\sqrt{9 + 3\sqrt{5}}\,\sqrt{9 - 3\sqrt{5}}$

135. $\left(\sqrt{x+2} - \sqrt{x-2}\right)^2$

7.6 Solving Radical Equations

The Principle of Powers ■ Equations with Two Radical Terms

Connecting the Concepts

In Sections 7.1–7.5, we learned how to manipulate radical expressions as well as expressions containing rational exponents. We performed this work to find *equivalent expressions.*

Now that we know how to work with radicals and rational exponents, we can learn how to solve a new type of equation. As in our earlier work with equations, finding *equivalent equations* will be part of our strategy. What is different, however, is that now we will use a step that does not always produce equivalent equations. Checking solutions will therefore be more important than ever.

The Principle of Powers

A **radical equation** is an equation in which the variable appears in a radicand. Examples are

$$\sqrt[3]{2x} + 1 = 5, \quad \sqrt{a} + \sqrt{a-2} = 7, \quad \text{and} \quad 4 - \sqrt{3x+1} = \sqrt{6-x}.$$

To solve such equations, we need a new principle. Suppose $a = b$ is true. If we square both sides, we get another true equation: $a^2 = b^2$. This can be generalized.

The Principle of Powers If $a = b$, then $a^n = b^n$ for any exponent n.

Note that the principle of powers is an "if–then" statement. The statement obtained by interchanging the two parts of the sentence—"if $a^n = b^n$ for some exponent n, then $a = b$"—*is not always true*. For example, "if $x = 3$, then $x^2 = 9$" is true, but the statement "if $x^2 = 9$, then $x = 3$" is *not* true when x is replaced with -3.

Interactive Discovery

Solve each pair of equations graphically, and compare the solution sets.

1. $x = 3;\ x^2 = 9$
2. $x = -2;\ x^2 = 4$
3. $\sqrt{x} = 5;\ x = 25$
4. $\sqrt{x} = -3;\ x = 9$

We see that every solution of $x = a$ is a solution of $x^2 = a^2$, but not every solution of $x^2 = a^2$ is necessarily a solution of $x = a$. A similar statement can be made for any even exponent n; for example, the solution sets of $x^4 = 3^4$ and $x = 3$ are not the same. For this reason, when we raise both sides of an equation to an even power, it will be essential for us to check the answer in the original equation. We *will* find all the solutions; we *may* also find additional numbers that are not solutions of the original equation.

EXAMPLE 1 Solve: $\sqrt{x} - 3 = 4$.

ALGEBRAIC APPROACH

Before using the principle of powers, we need to isolate the radical term:

$$\sqrt{x} - 3 = 4$$

$$\sqrt{x} = 7 \qquad \text{Isolating the radical by adding 3 to both sides}$$

$$(\sqrt{x})^2 = 7^2 \qquad \text{Using the principle of powers}$$

$$x = 49.$$

Check:

$$\begin{array}{c|c} \sqrt{x} - 3 & 4 \\ \hline \sqrt{49} - 3 & 4 \\ 7 - 3 & \\ 4 & \stackrel{?}{=} 4 \quad \text{TRUE} \end{array}$$

The solution is 49.

GRAPHICAL APPROACH

We let $f(x) = \sqrt{x} - 3$and $g(x) = 4$. Since the domain of f is $[0, \infty)$, we try a viewing window of $[-1, 10, -10, 10]$.

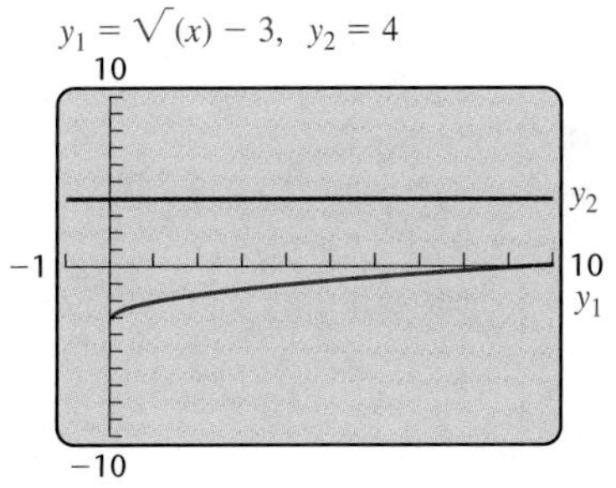

It appears as though the graphs may intersect at an x-value greater than 10. We adjust the viewing window to $[-10, 100, -5, 5]$ and find the point of intersection.

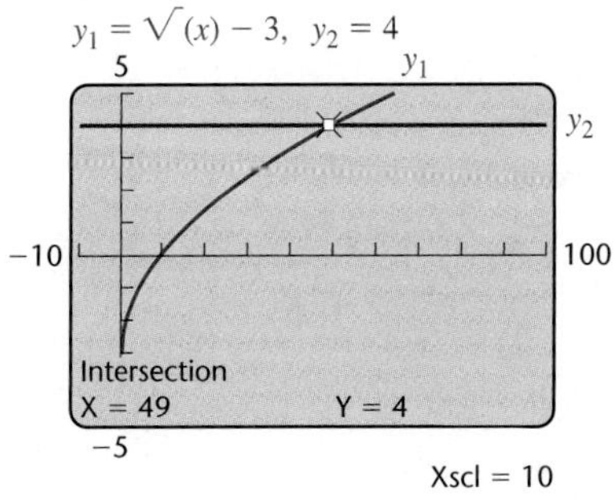

We have a solution of 49, which we can check by substituting into the original equation, as shown in the algebraic approach.

The solution is 49.

EXAMPLE 2 Solve: $\sqrt{x} - 5 = -7$.

ALGEBRAIC APPROACH

We have

$$\sqrt{x} - 5 = -7$$

$$\sqrt{x} = -2 \quad \textbf{Isolating the radical by adding 5 to both sides}$$

The equation $\sqrt{x} = -2$ has no solution because the principal square root of a number is never negative. We continue as in Example 1 for comparison.

$$(\sqrt{x})^2 = (-2)^2 \quad \textbf{Using the principle of powers (squaring)}$$

$$x = 4.$$

Check:

$\sqrt{x} - 5$	$= -7$
$\sqrt{4} - 5$	-7
$2 - 5$	
-3	$\stackrel{?}{=} -7$ FALSE

The number 4 does not check. Thus, $\sqrt{x} - 5 = -7$ has no solution.

GRAPHICAL APPROACH

We let $f(x) = \sqrt{x} - 5$ and $g(x) = -7$ and graph.

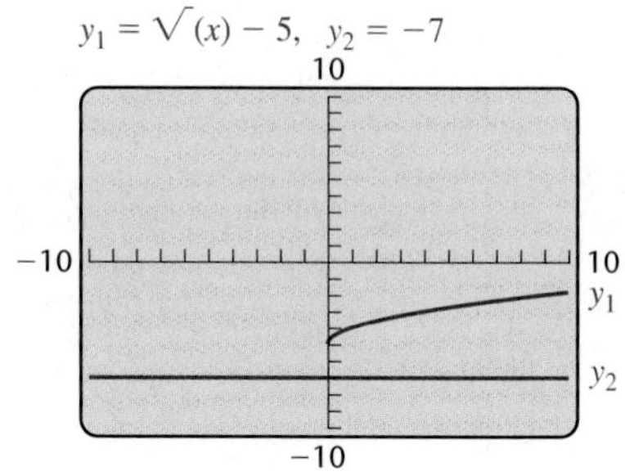

The graphs do not intersect. There is no real-number solution.

CAUTION! Raising both sides of an equation to an even power may not produce an equivalent equation. In this case, a check is essential.

Note in Example 2 that $x = 4$ has solution 4 but $\sqrt{x} - 5 = -7$ has *no* solution. Thus the equations $x = 4$ and $\sqrt{x} - 5 = -7$ are *not* equivalent.

To Solve an Equation with a Radical Term

1. Isolate the radical term on one side of the equation.
2. Use the principle of powers and solve the resulting equation.
3. Check any possible solution in the original equation.

EXAMPLE 3 Solve: $x = \sqrt{x + 7} + 5$.

ALGEBRAIC APPROACH

$$x = \sqrt{x + 7} + 5$$

$$x - 5 = \sqrt{x + 7}$$ **Isolating the radical by subtracting 5 from both sides**

$$(x - 5)^2 = (\sqrt{x + 7})^2$$

$$x^2 - 10x + 25 = x + 7$$ **Using the principle of powers; squaring both sides**

$$x^2 - 11x + 18 = 0$$ **Adding $-x - 7$ to both sides to write the quadratic equation in standard form**

$$(x - 9)(x - 2) = 0$$ **Factoring**

$$x = 9 \quad or \quad x = 2$$ **Using the principle of zero products**

The possible solutions are 9 and 2. Let's check.

Check:

For 9:

$$x = \sqrt{x + 7} + 5$$

$$9 \;\Big|\; \sqrt{9 + 7} + 5$$

$$9 \stackrel{?}{=} 9 \qquad \text{TRUE}$$

For 2:

$$x = \sqrt{x + 7} + 5$$

$$2 \;\Big|\; \sqrt{2 + 7} + 5$$

$$2 \stackrel{?}{=} 8 \qquad \text{FALSE}$$

Since 9 checks but 2 does not, the solution is 9.

GRAPHICAL APPROACH

We graph $y_1 = x$ and $y_2 = \sqrt{\ }(x + 7) + 5$ and determine any points of intersection.

$y_1 = x, \; y_2 = \sqrt{\ }(x + 7) + 5$

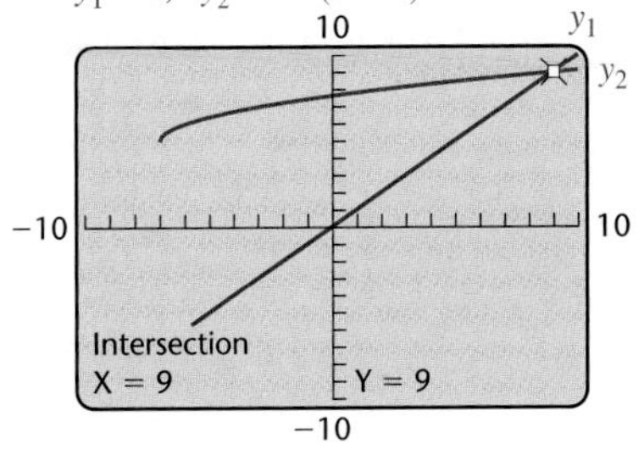

We obtain a solution of 9.

It is important to isolate a radical term before using the principle of powers. Suppose in Example 3 that both sides of the equation were squared *before* isolating the radical. We then would have had the expression $(\sqrt{x + 7} + 5)^2$ or $x + 7 + 10\sqrt{x + 7} + 25$ in the third step of the algebraic approach, and the radical would have remained in the problem.

EXAMPLE 4 Solve: $(2x + 1)^{1/3} + 5 = 0$.

ALGEBRAIC APPROACH

We need not use radical notation to solve:

$$(2x + 1)^{1/3} + 5 = 0$$

$$(2x + 1)^{1/3} = -5$$ **Subtracting 5 from both sides**

$$[(2x + 1)^{1/3}]^3 = (-5)^3$$ **Cubing both sides**

$$(2x + 1)^1 = (-5)^3$$ **Multiplying exponents. Try to do this mentally.**

$$2x + 1 = -125$$

$$2x = -126$$ **Subtracting 1 from both sides**

$$x = -63.$$

Because both sides were raised to an *odd* power, we need check only the accuracy of our work.

Study Tip

Plan Your Future

As you register for next semester's courses, be careful to consider your work and family commitments. Before signing up, speak to faculty and other students to estimate how demanding each course is. If in doubt, it is usually better to take one fewer course than one too many.

Check:

$$\begin{array}{r|l}
\multicolumn{2}{c}{(2x+1)^{1/3}+5=0} \\
\hline
[2(-63)+1]^{1/3}+5 & 0 \\
[-126+1]^{1/3}+5 & \\
(-125)^{1/3}+5 & \\
\sqrt[3]{-125}+5 & \\
-5+5 & \\
0 & \overset{?}{=} 0 \quad \text{TRUE}
\end{array}$$

GRAPHICAL APPROACH

We let $f(x) = (2x + 1)^{1/3} + 5$ and determine any values of x for which $f(x) = 0$. We might start by graphing f using a standard viewing window, as shown on the left below.

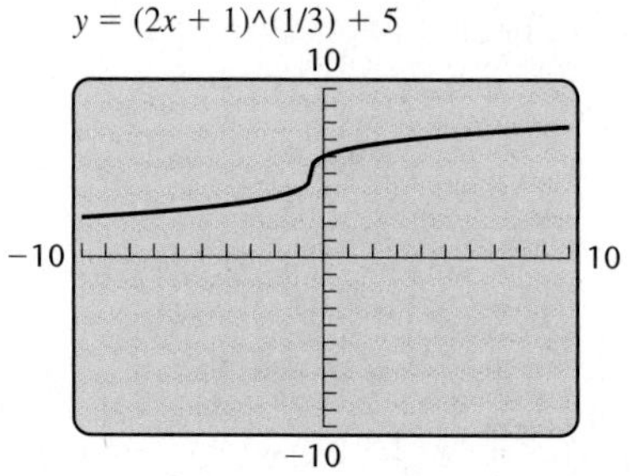

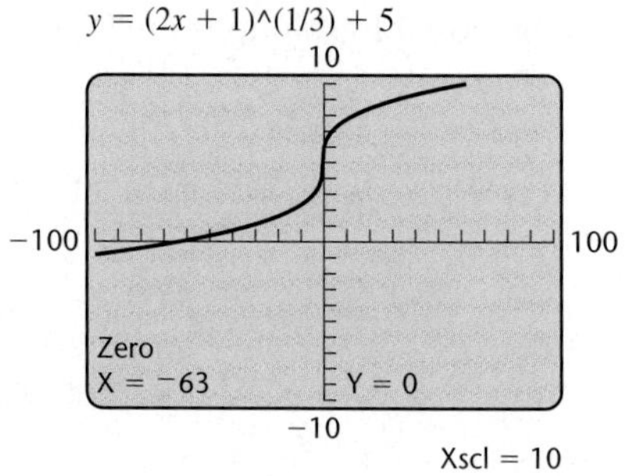

It appears that if the graph of f does have an x-intercept, it will be at a value of x less than -10. Let's try a viewing window of $[-100, 100, -10, 10]$, as shown on the right above. We find that $f(x) = 0$ when $x = -63$.

Equations with Two Radical Terms

A strategy for solving equations with two or more radical terms is as follows.

To Solve an Equation with Two or More Radical Terms

1. Isolate one of the radical terms.
2. Use the principle of powers.
3. If a radical remains, perform steps (1) and (2) again.
4. Solve the resulting equation.
5. Check possible solutions in the original equation.

EXAMPLE 5 Solve: $\sqrt{2x-5} = 1 + \sqrt{x-3}$.

ALGEBRAIC APPROACH

We have

$$\sqrt{2x-5} = 1 + \sqrt{x-3}$$

$$\left(\sqrt{2x-5}\right)^2 = \left(1 + \sqrt{x-3}\right)^2$$ **One radical is already isolated. We square both sides.**

This is like squaring a binomial. We square 1, then find twice the product of 1 and $\sqrt{x-3}$, and then the square of $\sqrt{x-3}$.

$$2x - 5 = 1 + 2\sqrt{x-3} + \left(\sqrt{x-3}\right)^2$$

$$2x - 5 = 1 + 2\sqrt{x-3} + (x-3)$$

$$x - 3 = 2\sqrt{x-3}$$ **Isolating the remaining radical term**

$$(x-3)^2 = \left(2\sqrt{x-3}\right)^2$$ **Squaring both sides**

$$x^2 - 6x + 9 = 4(x-3)$$ **Remember to square both the 2 and the $\sqrt{x-3}$ on the right side.**

$$x^2 - 6x + 9 = 4x - 12$$

$$x^2 - 10x + 21 = 0$$

$$(x-7)(x-3) = 0$$ **Factoring**

$$x = 7 \quad or \quad x = 3.$$ **Using the principle of zero products**

The possible solutions are 7 and 3. Both 7 and 3 check in the original equation and are the solutions.

GRAPHICAL APPROACH

We let $f(x) = \sqrt{2x-5}$ and $g(x) = 1 + \sqrt{x-3}$. The domain of f is $\left[\frac{5}{2}, \infty\right)$, and the domain of g is $[3, \infty)$. Since the intersection of their domains is $[3, \infty)$, any point of intersection will occur in that interval. We choose a viewing window of $[-1, 10, -1, 5]$.

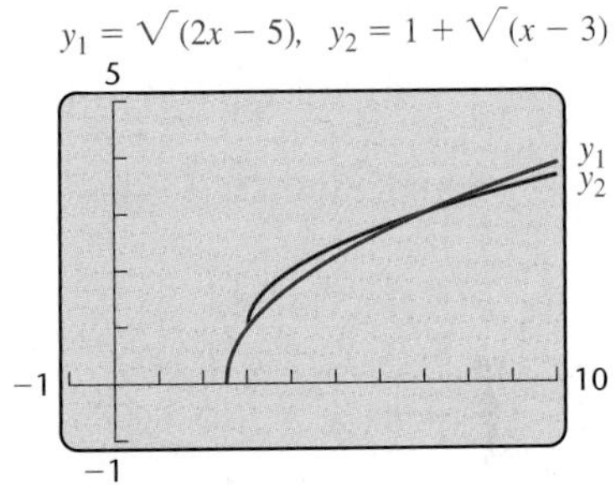

The graphs appear to intersect when x is approximately 3 and again when x is approximately 7. The points of intersection are $(3, 1)$ and $(7, 3)$. To confirm that there are no other points of intersection, we could examine the graphs using a larger viewing window.

CAUTION! A common error in solving equations like

$$\sqrt{2x-5} = 1 + \sqrt{x-3}$$

is to obtain $1 + (x-3)$ as the square of the right side. This is wrong because $(A+B)^2 \neq A^2 + B^2$. Placing parentheses around each side when squaring serves as a reminder to square the entire expression:

$$\left(1 + \sqrt{x-3}\right)^2 = \left(1 + \sqrt{x-3}\right)\left(1 + \sqrt{x-3}\right)$$
$$= 1 + 2\sqrt{x-3} + \left(\sqrt{x-3}\right)^2.$$

EXAMPLE 6 Let $f(x) = \sqrt{x+5} - \sqrt{x-3}$. Find all x-values for which $f(x) = 2$.

ALGEBRAIC APPROACH

We must have $f(x) = 2$, or

$\sqrt{x+5} - \sqrt{x-3} = 2.$ **Substituting for $f(x)$**

To solve, we isolate one radical term and square both sides:

$$\sqrt{x+5} = 2 + \sqrt{x-3}$$ **Adding $\sqrt{x-3}$. This isolates one of the radical terms.**

$$\left(\sqrt{x+5}\right)^2 = \left(2 + \sqrt{x-3}\right)^2$$ **Using the principle of powers (squaring both sides)**

$$x + 5 = 4 + 4\sqrt{x-3} + (x - 3)$$ **Using $(A + B)^2 = A^2 + 2AB + B^2$**

$$5 = 1 + 4\sqrt{x-3}$$ **Adding $-x$ and combining like terms**

$$4 = 4\sqrt{x-3}$$
$$1 = \sqrt{x-3}$$ **Isolating the remaining radical term**

$$1^2 = \left(\sqrt{x-3}\right)^2$$ **Squaring both sides**

$$1 = x - 3$$
$$4 = x.$$

GRAPHICAL APPROACH

We graph $y_1 = \sqrt{\ }(x + 5) - \sqrt{\ }(x - 3)$ and $y_2 = 2$ and determine the coordinates of any points of intersection.

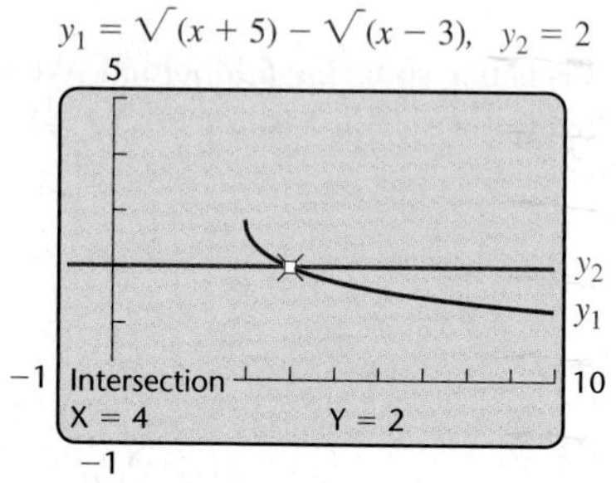

The solution is the x-coordinate of the point of intersection, which is 4.

Check: $f(4) = \sqrt{4+5} - \sqrt{4-3} = \sqrt{9} - \sqrt{1} = 3 - 1 = 2.$

We have $f(x) = 2$ when $x = 4$.

7.6 EXERCISE SET

FOR EXTRA HELP

MathXL | MyMathLab | InterAct Math | AW Math Tutor Center | Video Lectures on CD: Disc 4 | Student's Solutions Manual

Concept Reinforcement *Classify each of the following as either true or false.*

1. If $x^2 = 25$, then $x = 5$.
2. If $t = 7$, then $t^2 = 49$.
3. If $\sqrt{x} = 3$, then $\left(\sqrt{x}\right)^2 = 3^2$.
4. If $x^2 = 36$, then $x = 6$.
5. $\sqrt{x} - 8 = 7$ is equivalent to $\sqrt{x} = 15$.
6. $\sqrt{t} + 5 = 8$ is equivalent to $\sqrt{t} = 3$.

Solve.

7. $\sqrt{5x - 2} = 7$
8. $\sqrt{3x - 2} = 6$
9. $\sqrt{3x} + 1 = 6$
10. $\sqrt{2x} - 1 = 2$
11. $\sqrt{y + 1} - 5 = 8$
12. $\sqrt{x - 2} - 7 = -4$
13. $\sqrt{x - 7} + 3 = 10$
14. $\sqrt{y + 4} + 6 = 7$
15. $\sqrt[3]{x + 5} = 2$
16. $\sqrt[3]{x - 2} = 3$
17. $\sqrt[4]{y - 1} = 3$
18. $\sqrt[4]{x + 3} = 2$
19. $3\sqrt{x} = x$
20. $8\sqrt{y} = y$

21. $2y^{1/2} - 7 = 9$

22. $3x^{1/2} + 12 = 9$

23. $\sqrt[3]{x} = -3$

24. $\sqrt[3]{y} = -4$

25. $t^{1/3} - 2 = 3$

26. $x^{1/4} - 2 = 1$

Aha! 27. $(y - 3)^{1/2} = -2$

28. $(x + 2)^{1/2} = -4$

29. $\sqrt[4]{3x + 1} - 4 = -1$

30. $\sqrt[4]{2x + 3} - 5 = -2$

31. $(x + 7)^{1/3} = 4$

32. $(y - 7)^{1/4} = 3$

33. $\sqrt[3]{3y + 6} + 7 = 8$

34. $\sqrt[3]{6x + 9} + 5 = 2$

35. $\sqrt{3t + 4} = \sqrt{4t + 3}$

36. $\sqrt{2t - 7} = \sqrt{3t - 12}$

37. $3(4 - t)^{1/4} = 6^{1/4}$

38. $2(1 - x)^{1/3} = 4^{1/3}$

39. $3 + \sqrt{5 - x} = x$

40. $x = \sqrt{x - 1} + 3$

41. $\sqrt{4x - 3} = 2 + \sqrt{2x - 5}$

42. $3 + \sqrt{z - 6} = \sqrt{z + 9}$

43. $\sqrt{20 - x} + 8 = \sqrt{9 - x} + 11$

44. $4 + \sqrt{10 - x} = 6 + \sqrt{4 - x}$

45. $\sqrt{x + 2} + \sqrt{3x + 4} = 2$

46. $\sqrt{6x + 7} - \sqrt{3x + 3} = 1$

47. If $f(x) = \sqrt{x} + \sqrt{x - 9}$, find any x for which $f(x) = 1$.

48. If $g(x) = \sqrt{x} + \sqrt{x - 5}$, find any x for which $g(x) = 5$.

49. If $f(t) = \sqrt{t - 2} - \sqrt{4t + 1}$, find any t for which $f(t) = -3$.

50. If $g(t) = \sqrt{2t + 7} - \sqrt{t + 15}$, find any t for which $g(t) = -1$.

51. If $f(x) = \sqrt{2x - 3}$ and $g(x) = \sqrt{x + 7} - 2$, find any x for which $f(x) = g(x)$.

52. If $f(x) = 2\sqrt{3x + 6}$ and $g(x) = 5 + \sqrt{4x + 9}$, find any x for which $f(x) = g(x)$.

53. If $f(t) = 4 - \sqrt{t - 3}$ and $g(t) = (t + 5)^{1/2}$, find any t for which $f(t) = g(t)$.

54. If $f(t) = 7 + \sqrt{2t - 5}$ and $g(t) = 3(t + 1)^{1/2}$, find any t for which $f(t) = g(t)$.

TW 55. Explain in your own words why it is important to check your answers when using the principle of powers.

TW 56. The principle of powers is an "if–then" statement that becomes false when the sentence parts are interchanged. Give an example of another such if–then statement from everyday life (answers will vary).

Focused Review

Solve.

57. $x^2 = x + 12$ [5.4]

58. $\frac{1}{3}x - \frac{1}{4} = 5$ [1.6]

59. $|2x - 3| > 5$ [4.4]

60. $\sqrt{x - 3} + 2 = 7$ [7.6]

61. $4 - x \leq 2x + 3$ [4.1]

62. $x^2 = 36$ [5.6]

63. $2(x - 1)^{1/3} = -4$ [7.6]

64. $\dfrac{1}{x} + \dfrac{1}{x - 1} = \dfrac{2}{x - 3}$ [6.4]

65. $2x - 3y = 8,$
$x + 2y = 5$ [3.2]

66. $2x - 3y + z = 0,$
$4x + 5y + z = 1,$
$2x + y - 4z = 5$ [3.4]

Synthesis

TW 67. Describe a procedure that could be used to create radical equations that have no solution.

TW 68. Is checking essential when the principle of powers is used with an odd power n? Why or why not?

Steel Manufacturing. In the production of steel and other metals, the temperature of the molten metal is so great that conventional thermometers melt. Instead, sound is transmitted across the surface of the metal to a receiver on the far side and the speed of the sound is measured. The formula

$$S(t) = 1087.7\sqrt{\frac{9t + 2617}{2457}}$$

gives the speed of sound $S(t)$, in feet per second, at a temperature of t degrees Celsius.

69. Find the temperature of a blast furnace where sound travels 1880 ft/sec.

70. Find the temperature of a blast furnace where sound travels 1502.3 ft/sec.

71. Solve the above equation for t.

Automotive Repair. For an engine with a displacement of 2.8 L, *the function given by*

$$d(n) = 0.75\sqrt{2.8n}$$

can be used to determine the diameter size of the carburetor's opening, in millimeters. Here n is the number of rpm's at which the engine achieves peak performance. (*Source*: macdizzy.com)

72. If a carburetor's opening is 81 mm, for what number of rpm's will the engine produce peak power?

73. If a carburetor's opening is 84 mm, for what number of rpm's will the engine produce peak power?

Escape Velocity. A formula for the escape velocity v of a satellite is

$$v = \sqrt{2gr}\sqrt{\frac{h}{r + h}},$$

where g is the force of gravity, r is the planet or star's radius, and h is the height of the satellite above the planet or star's surface.

74. Solve for h.

75. Solve for r.

Sighting to the Horizon. The function $D(h) = 1.2\sqrt{h}$ can be used to approximate the distance D, in miles, that a person can see to the horizon from a height h, in feet.

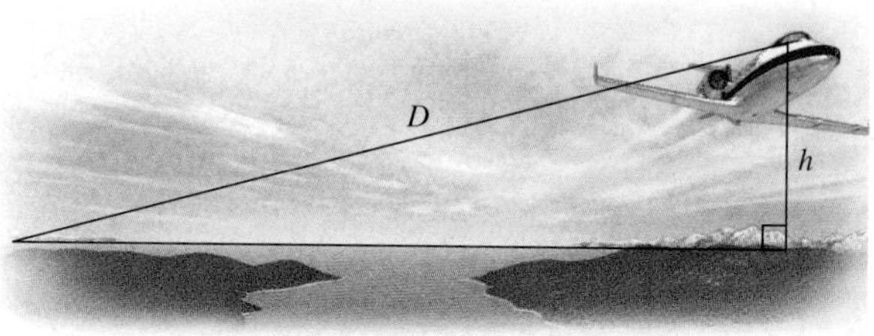

76. How far above sea level must a pilot fly in order to see a horizon that is 180 mi away?

77. How high above sea level must a sailor climb in order to see 10.2 mi out to sea?

Solve.

78. $\left(\frac{z}{4} - 5\right)^{2/3} = \frac{1}{25}$

79. $\frac{x + \sqrt{x + 1}}{x - \sqrt{x + 1}} = \frac{5}{11}$

80. $\sqrt{\sqrt{y} + 49} = 7$

81. $(z^2 + 17)^{3/4} = 27$

82. $x^2 - 5x - \sqrt{x^2 - 5x - 2} = 4$
(*Hint*: Let $u = x^2 - 5x - 2$.)

83. $\sqrt{8 - b} = b\sqrt{8 - b}$

Without graphing, determine the x-intercepts of the graphs given by each of the following.

84. $f(x) = \sqrt{x - 2} - \sqrt{x + 2} + 2$

85. $g(x) = 6x^{1/2} + 6x^{-1/2} - 37$

86. $f(x) = (x^2 + 30x)^{1/2} - x - (5x)^{1/2}$

TW **87.** Saul is trying to solve Exercise 77 using a graphing calculator. Without resorting to trial and error, how can he determine a suitable viewing window for finding the solution?

Collaborative Corner

Tailgater Alert

Focus: Radical equations and problem solving
Time: 15–25 minutes
Group size: 2–3
Materials: Calculators

The faster a car is traveling, the more distance it needs to stop. Thus it is important for drivers to allow sufficient space between their vehicle and the vehicle in front of them. Police recommend that for each 10 mph of speed, a driver allow 1 car length. Thus a driver going 30 mph should have at least 3 car lengths between his or her vehicle and the one in front.

In Exercise Set 7.3, the function $r(L) = 2\sqrt{5L}$ was used to find the speed, in miles per hour, that a car was traveling when it left skid marks L feet long.

ACTIVITY

1. Each group member should estimate the length of a car in which he or she frequently travels. (Each should use a different length, if possible.)
2. Using a calculator as needed, each group member should complete the table at right. Column 1 gives a car's speed s, column 2 lists the minimum amount of space between cars traveling s miles per hour, as recommended by police. Column 3 is the speed that a vehicle *could* travel were it forced to stop in the distance listed in column 2, using the above function.

Column 1 s (in miles per hour)	Column 2 $L(s)$ (in feet)	Column 3 $r(L)$ (in miles per hour)
20		
30		
40		
50		
60		
70		

3. Determine whether there are any speeds at which the "1 car length per 10 mph" guideline might not suffice. On what reasoning do you base your answer? Compare tables to determine how car length affects the results. What recommendations would your group make to a new driver?

7.7 Geometric Applications

Using the Pythagorean Theorem ■ Two Special Triangles

Using the Pythagorean Theorem

There are many kinds of problems that involve powers and roots. Many also involve right triangles and the Pythagorean theorem, which we studied in Section 5.8 and restate here.

Study Tip

Making Sketches

One need not be an artist to make highly useful mathematical sketches. That said, it is important to make sure that your sketches are drawn accurately enough to represent the relative sizes within each shape. For example, if one side of a triangle is clearly the longest, make sure your drawing reflects this.

The Pythagorean Theorem* In any right triangle, if a and b are the lengths of the legs and c is the length of the hypotenuse, then

$$a^2 + b^2 = c^2.$$

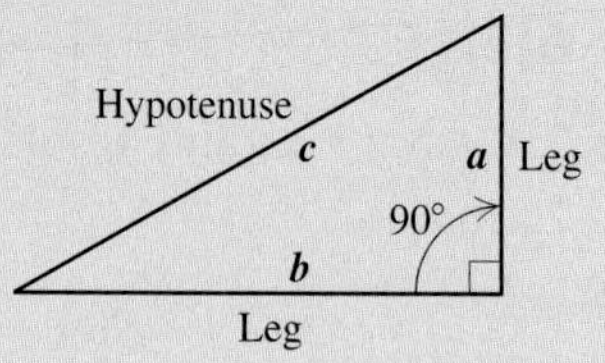

In using the Pythagorean theorem, we often make use of the following principle.

The Principle of Square Roots For any nonnegative real number n,

If $x^2 = n$, then $x = \sqrt{n}$ or $x = -\sqrt{n}$.

For most real-world applications involving length or distance, $-\sqrt{n}$ is not needed.

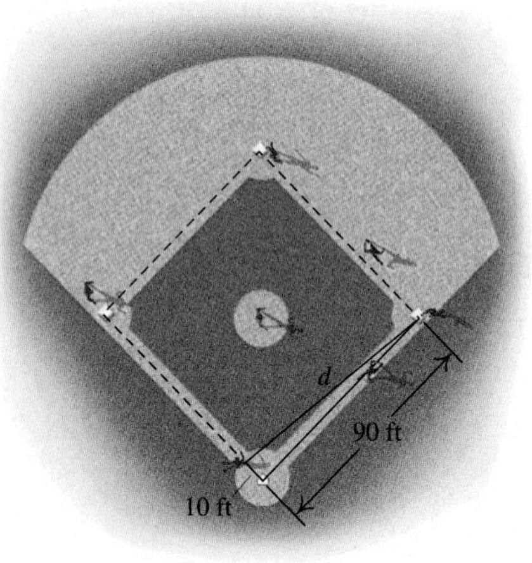

EXAMPLE 1 Baseball. A baseball diamond is actually a square 90 ft on a side. Suppose a catcher fields a ball while standing on the third-base line 10 ft from home plate. How far is the catcher's throw to first base? Give an exact answer and an approximation to three decimal places.

SOLUTION We make a drawing and let $d =$ the distance, in feet, to first base. Note that a right triangle is formed in which the leg from home plate to first base measures 90 ft and the leg from home plate to where the catcher fields the ball measures 10 ft.

We substitute these values into the Pythagorean theorem to find d:

$$d^2 = 90^2 + 10^2$$
$$d^2 = 8100 + 100$$
$$d^2 = 8200.$$

We now use the principle of square roots: If $d^2 = 8200$, then $d = \sqrt{8200}$ or $d = -\sqrt{8200}$. Since d represents a length, it follows that d is the positive square root of 8200:

$d = \sqrt{8200}$ ft **This is an exact answer.**

$d \approx 90.554$ ft. **Using a calculator for an approximation**

*The converse of the Pythagorean theorem also holds. That is, if a, b, and c are the lengths of the sides of a triangle and $a^2 + b^2 = c^2$, then the triangle is a right triangle.

EXAMPLE 2 Guy Wires. The base of a 40-ft-long guy wire is located 15 ft from the telephone pole that it is anchoring. How high up the pole does the guy wire reach? Give an exact answer and an approximation to three decimal places.

SOLUTION We make a drawing and let h = the height, in feet, to which the guy wire reaches. A right triangle is formed in which one leg measures 15 ft and the hypotenuse measures 40 ft. Using the Pythagorean theorem, we have

$$\begin{aligned} h^2 + 15^2 &= 40^2 \\ h^2 + 225 &= 1600 \\ h^2 &= 1375 \\ h &= \sqrt{1375}. \end{aligned}$$

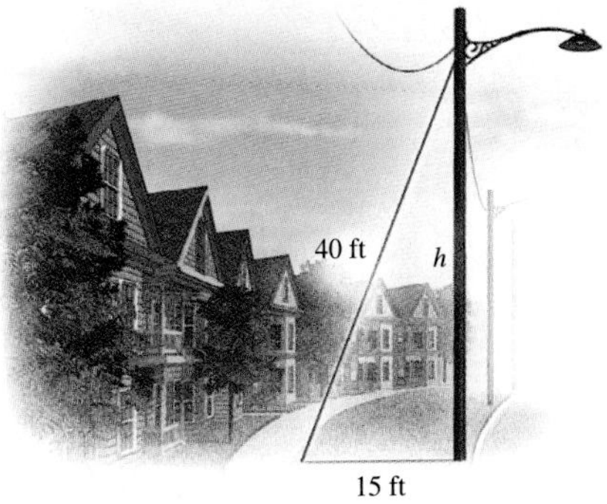

Exact answer:

$h = \sqrt{1375}$ ft

Approximation:

$h \approx 37.081$ ft **Using a calculator**

Two Special Triangles

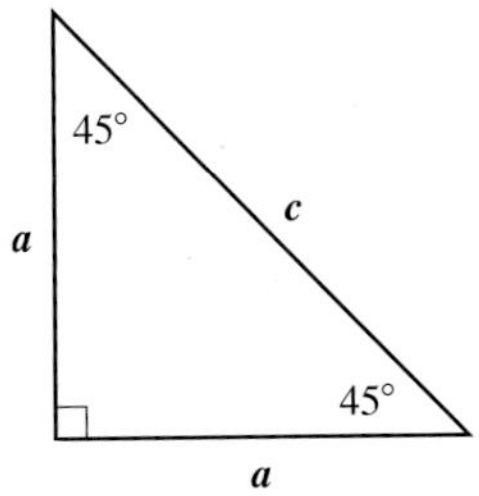

When both legs of a right triangle are the same size, we call the triangle an *isosceles right triangle*, as shown at left. If one leg of an isosceles right triangle has length a, we can find a formula for the length of the hypotenuse as follows:

$c^2 = a^2 + b^2$

$c^2 = a^2 + a^2$ **Because the triangle is isosceles, both legs are the same size: $a = b$.**

$c^2 = 2a^2.$ **Combining like terms**

Next, we use the principle of square roots. Because a, b, and c are lengths, there is no need to consider negative square roots or absolute values. Thus,

$c = \sqrt{2a^2}$ **Using the principle of square roots**

$= \sqrt{a^2 \cdot 2} = a\sqrt{2}.$

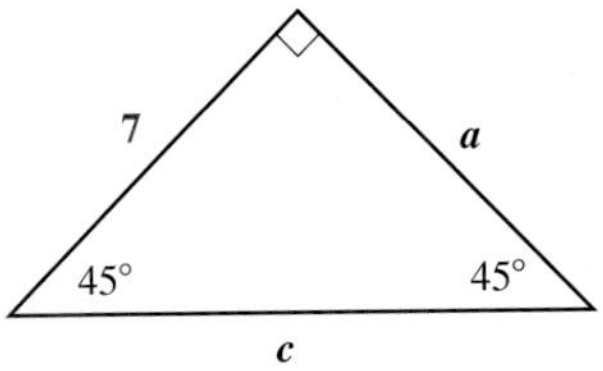

EXAMPLE 3 One leg of an isosceles right triangle measures 7 cm. Find the length of the hypotenuse. Give an exact answer and an approximation to three decimal places.

SOLUTION We substitute:

$c = a\sqrt{2}$ **This equation is worth memorizing.**

$= 7\sqrt{2}.$

Exact answer: $c = 7\sqrt{2}$ cm

Approximation: $c \approx 9.899$ cm **Using a calculator**

When the hypotenuse of an isosceles right triangle is known, the lengths of the legs can be found.

EXAMPLE 4 The hypotenuse of an isosceles right triangle is 5 ft long. Find the length of a leg. Give an exact answer and an approximation to three decimal places.

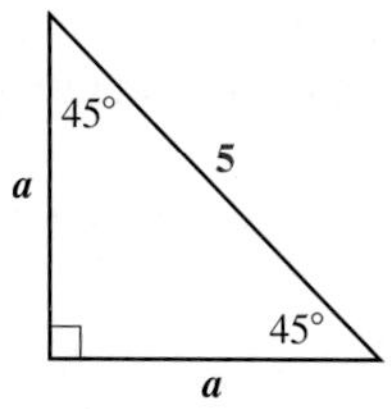

SOLUTION We replace c with 5 and solve for a:

$$5 = a\sqrt{2} \quad \text{Substituting 5 for } c \text{ in } c = a\sqrt{2}$$

$$\frac{5}{\sqrt{2}} = a \quad \text{Dividing both sides by } \sqrt{2}$$

$$\frac{5\sqrt{2}}{2} = a. \quad \text{Rationalize the denominator if desired.}$$

Exact answer: $a = \frac{5}{\sqrt{2}}$ ft, or $\frac{5\sqrt{2}}{2}$ ft

Approximation: $a \approx 3.536$ ft **Using a calculator**

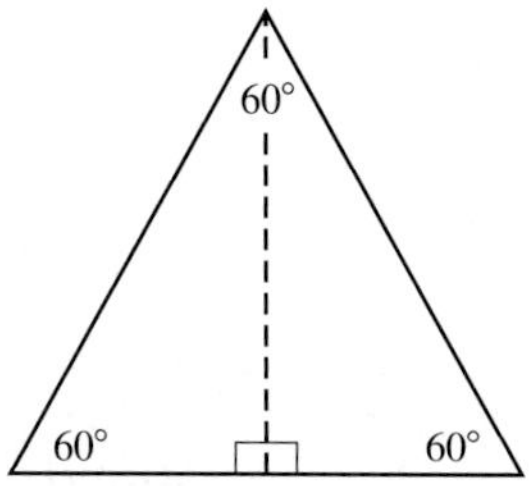

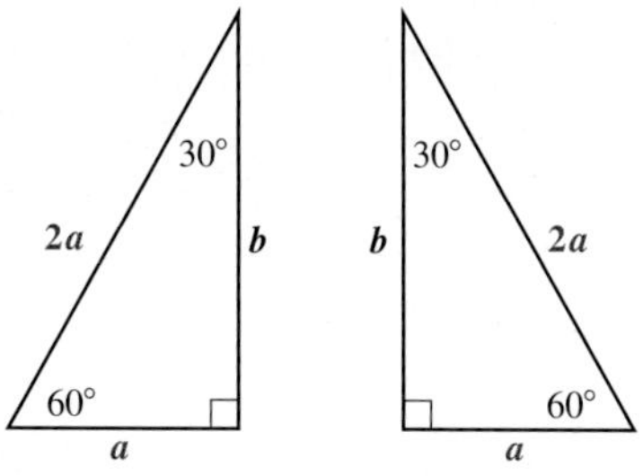

A second special triangle is known as a 30°–60°–90° right triangle, so named because of the measures of its angles. Note that in an equilateral triangle, all sides have the same length and all angles are 60°. An altitude, drawn dashed in the figure, bisects, or splits in half, one angle and one side. Two 30°–60°–90° right triangles are thus formed.

Because of the way in which the altitude is drawn, if a represents the length of the shorter leg in a 30°–60°–90° right triangle, then $2a$ represents the length of the hypotenuse. We have

$$a^2 + b^2 = (2a)^2 \quad \text{Using the Pythagorean theorem}$$

$$a^2 + b^2 = 4a^2$$

$$b^2 = 3a^2 \quad \text{Subtracting } a^2 \text{ from both sides}$$

$$b = \sqrt{3a^2}$$

$$b = \sqrt{a^2 \cdot 3}$$

$$b = a\sqrt{3}.$$

EXAMPLE 5 The shorter leg of a 30°–60°–90° right triangle measures 8 in. Find the lengths of the other sides. Give exact answers and, where appropriate, an approximation to three decimal places.

SOLUTION The hypotenuse is twice as long as the shorter leg, so we have

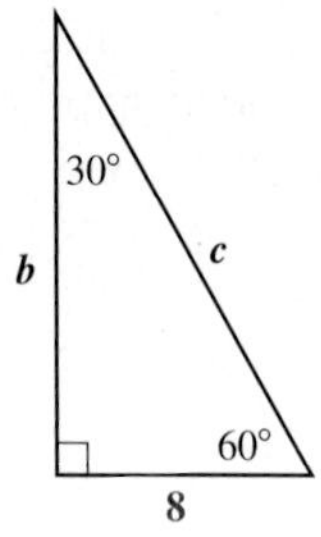

$$c = 2a \quad \text{This relationship is worth memorizing.}$$

$$= 2 \cdot 8 = 16 \text{ in.} \quad \text{This is the length of the hypotenuse.}$$

The length of the longer leg is the length of the shorter leg times $\sqrt{3}$. This gives us

$$b = a\sqrt{3} \quad \text{This is also worth memorizing.}$$

$$= 8\sqrt{3} \text{ in.} \quad \text{This is the length of the longer leg.}$$

Exact answer: $c = 16$ in., $b = 8\sqrt{3}$ in.

Approximation: $b \approx 13.856$ in.

EXAMPLE 6 The length of the longer leg of a 30°–60°–90° right triangle is 14 cm. Find the length of the hypotenuse. Give an exact answer and an approximation to three decimal places.

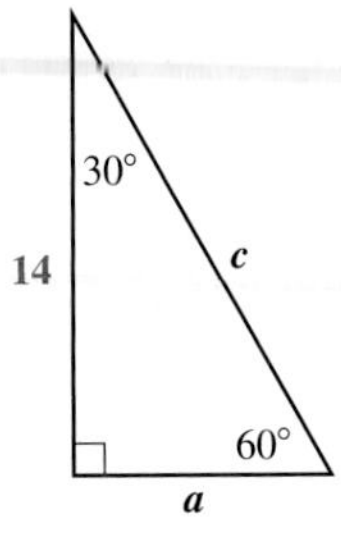

SOLUTION The length of the hypotenuse is twice the length of the shorter leg. We first find a, the length of the shorter leg, by using the length of the longer leg:

$$14 = a\sqrt{3} \qquad \textbf{Substituting 14 for } b \textbf{ in } b = a\sqrt{3}$$

$$\frac{14}{\sqrt{3}} = a. \qquad \textbf{Dividing by } \sqrt{3}$$

Since the hypotenuse is twice as long as the shorter leg, we have

$$\begin{aligned} c &= 2a \\ &= 2 \cdot \frac{14}{\sqrt{3}} \qquad \textbf{Substituting} \\ &= \frac{28}{\sqrt{3}} \text{ cm.} \end{aligned}$$

Exact answer: $c = \frac{28}{\sqrt{3}}$ cm, or $\frac{28\sqrt{3}}{3}$ cm if the denominator is rationalized.

Approximation: $c \approx 16.166$ cm

Student Notes

Perhaps the easiest way to remember the important results listed in the adjacent box is to write out, on your own, the derivations shown on pp. 587 and 588.

Lengths Within Isosceles and 30°–60°–90° Right Triangles

The length of the hypotenuse in an isosceles right triangle is the length of a leg times $\sqrt{2}$.

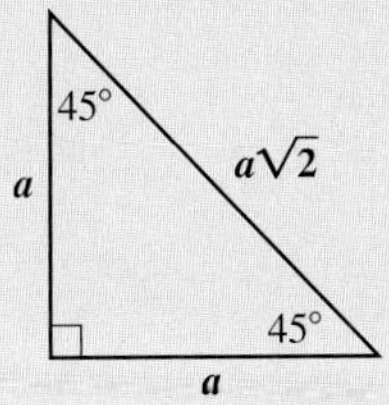

The length of the longer leg in a 30°–60°–90° right triangle is the length of the shorter leg times $\sqrt{3}$. The hypotenuse is twice as long as the shorter leg.

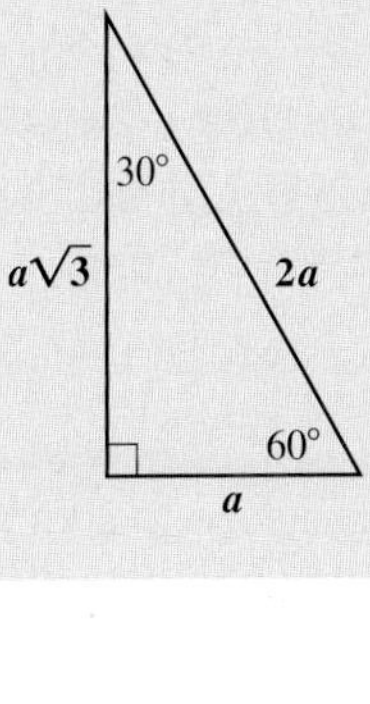

7.7 EXERCISE SET

FOR EXTRA HELP

MathXL | MyMathLab | InterAct Math | AW Math Tutor Center | Video Lectures on CD: Disc 4 | Student's Solutions Manual

Concept Reinforcement *Complete each of the following sentences.*

1. In any ____________ triangle, the square of the length of the ____________ is the sum of the squares of the lengths of the legs.

2. The shortest side of a right triangle is always one of the two ____________.

3. The principle of ____________ ____________ states that if $x^2 = n$, then $x = \sqrt{n}$ or $x = -\sqrt{n}$.

4. In a(n) ____________ right triangle, both legs have the same length.

5. In a(n) ___-___-___ right triangle, the hypotenuse is twice as long as the shorter ____________.

6. If both legs have measure a, then the hypotenuse measures ____________.

In a right triangle, find the length of the side not given. Give an exact answer and, where appropriate, an approximation to three decimal places.

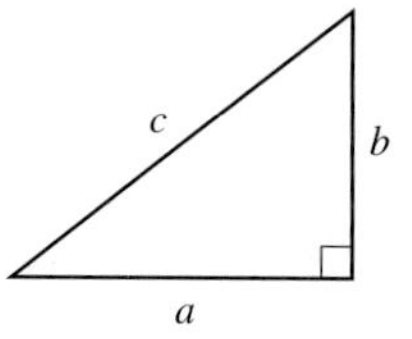

7. $a = 5, b = 3$

8. $a = 8, b = 10$

Aha! 9. $a = 9, b = 9$

10. $a = 10, b = 10$

11. $b = 12, c = 13$

12. $a = 5, c = 12$

In Exercises 13–28, give an exact answer and, where appropriate, an approximation to three decimal places.

13. A right triangle's hypotenuse is 8 m and one leg is $4\sqrt{3}$ m. Find the length of the other leg.

14. A right triangle's hypotenuse is 6 cm and one leg is $\sqrt{5}$ cm. Find the length of the other leg.

15. The hypotenuse of a right triangle is $\sqrt{20}$ in. and one leg measures 1 in. Find the length of the other leg.

16. The hypotenuse of a right triangle is $\sqrt{15}$ ft and one leg measures 2 ft. Find the length of the other leg.

Aha! 17. One leg of a right triangle is 1 m and the hypotenuse measures $\sqrt{2}$ m. Find the length of the other leg.

18. One leg of a right triangle is 1 yd and the hypotenuse measures 2 yd. Find the length of the other leg.

19. *Bicycling.* Clare routinely bicycles across a rectangular parking lot on her way to class. If the lot is 200 ft long and 150 ft wide, how far does Clare travel when she rides across the lot diagonally?

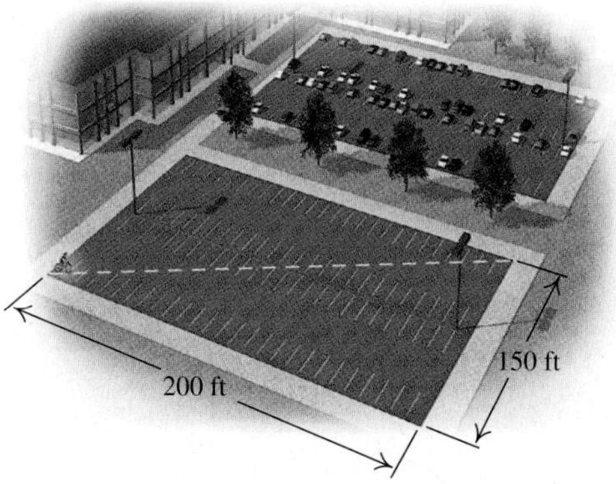

20. *Guy Wire.* How long is a guy wire if it reaches from the top of a 15-ft pole to a point on the ground 10 ft from the pole?

21. *Softball.* A slow-pitch softball diamond is actually a square 65 ft on a side. How far is it from home plate to second base?

22. *Baseball.* Suppose the catcher in Example 1 makes a throw to second base from the same location. How far is that throw?

23. *Television Sets.* What does it mean to refer to a 20-in. TV set or a 25-in. TV set? Such units refer to

the diagonal of the screen. A 20-in. TV set has a width of 16 in. What is its height?

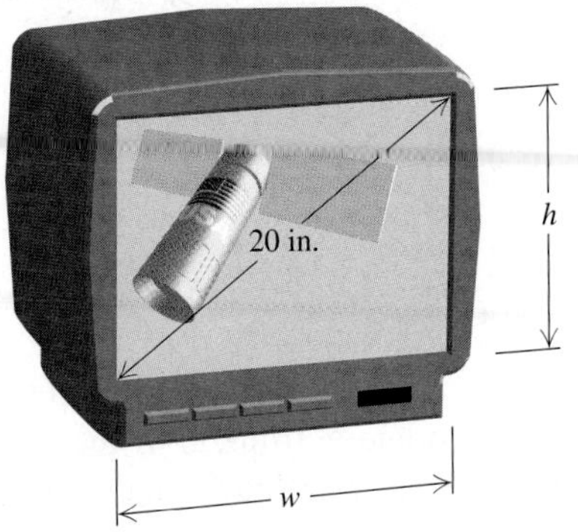

24. *Television Sets.* A 25-in. TV set has a screen with a height of 15 in. What is its width? (See Exercise 23.)

25. *Speaker Placement.* A stereo receiver is in a corner of a 12-ft by 14-ft room. Speaker wire will run under a rug, diagonally, to a speaker in the far corner. If 4 ft of slack is required on each end, how long a piece of wire should be purchased?

26. *Distance over Water.* To determine the width of a pond, a surveyor locates two stakes at either end of the pond and uses instrumentation to place a third stake so that the distance across the pond is the length of a hypotenuse. If the third stake is 90 m from one stake and 70 m from the other, what is the distance across the pond?

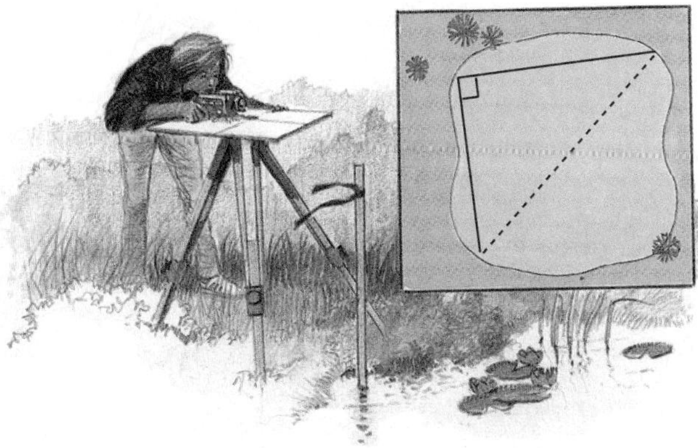

27. *Walking.* Students at Compkin Community College have worn a path that cuts diagonally across the campus "quad." If the quad is actually a rectangle that Marissa measured to be 70 paces long and 40 paces wide, how many paces will Marissa save by using the diagonal path?

28. *Crosswalks.* The diagonal crosswalk at the intersection of State St. and Main St. is the hypotenuse of a triangle in which the crosswalks across State St. and Main St. are the legs. If State St. is 28 ft wide and Main St. is 40 ft wide, how much shorter is the distance traveled by pedestrians using the diagonal crosswalk?

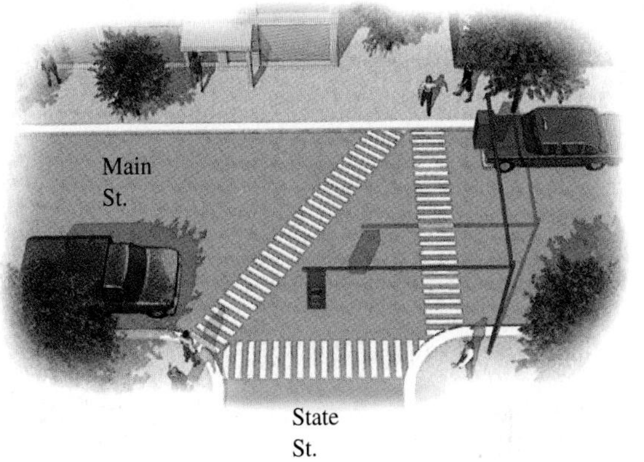

For each triangle, find the missing length(s). Give an exact answer and, where appropriate, an approximation to three decimal places.

29.

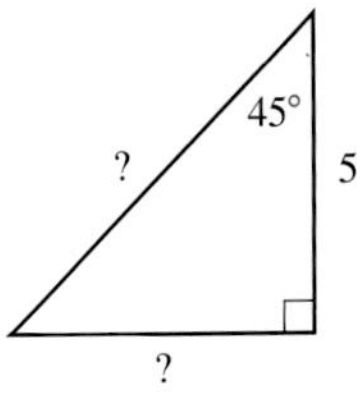

30.

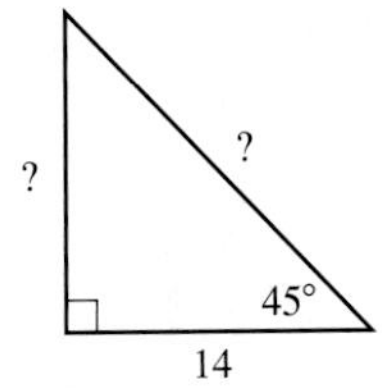

31.

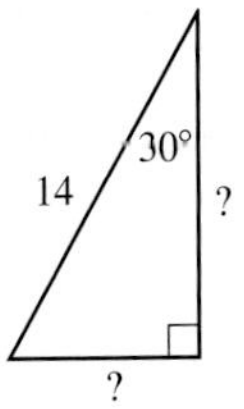

32.

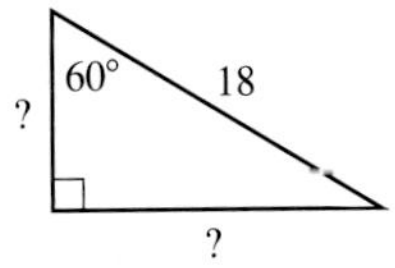

33.

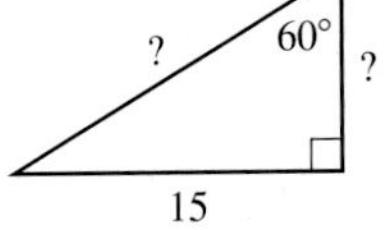

34.

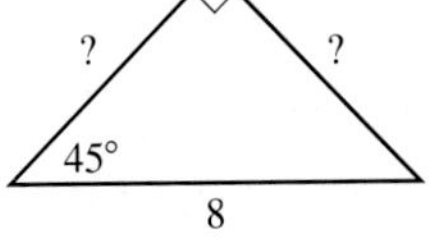

35.

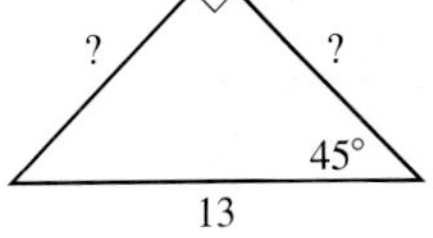

36.

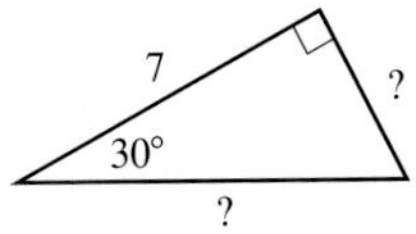

37. 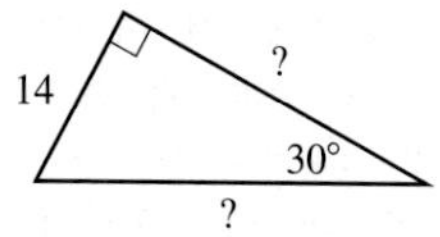

38. 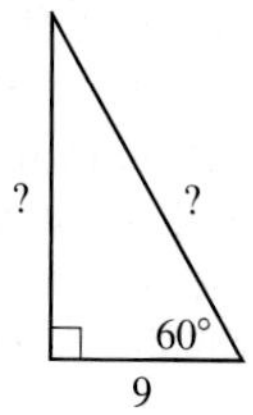

39. 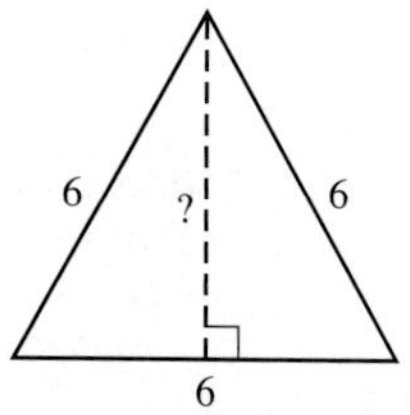

40. 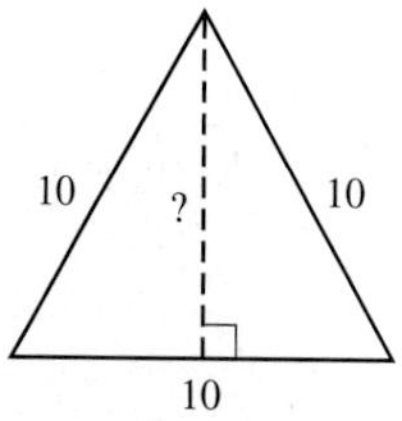

41. 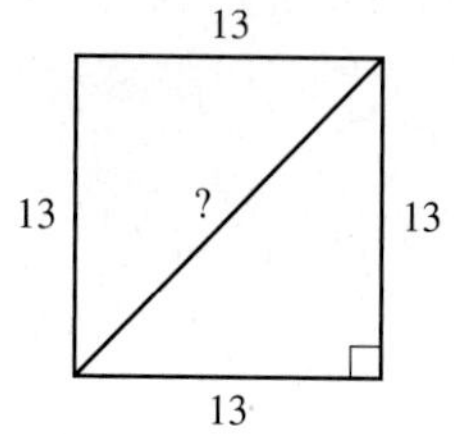

42. 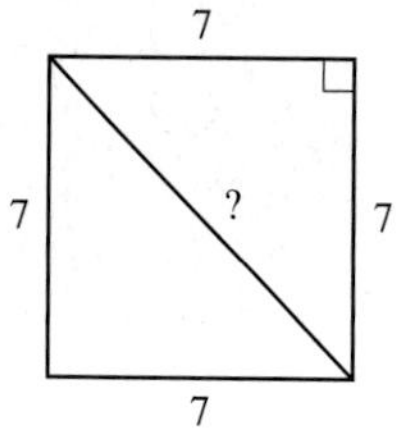

43. 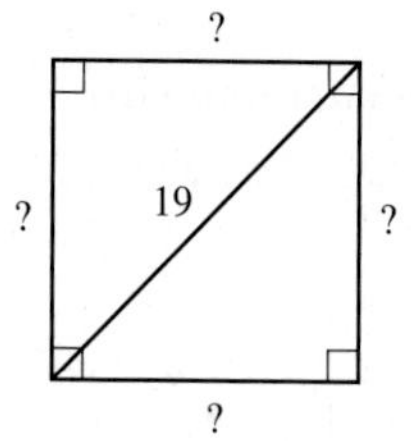

44. 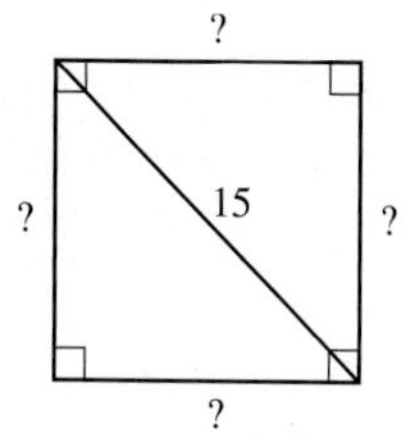

In Exercises 45–50, give an exact answer and, where appropriate, an approximation to three decimal places.

45. *Bridge Expansion.* During the summer heat, a 2-mi bridge expands 2 ft in length. If we assume that the bulge occurs straight up the middle, how high is the bulge? (The answer may surprise you. Most bridges have expansion spaces to avoid such buckling.)

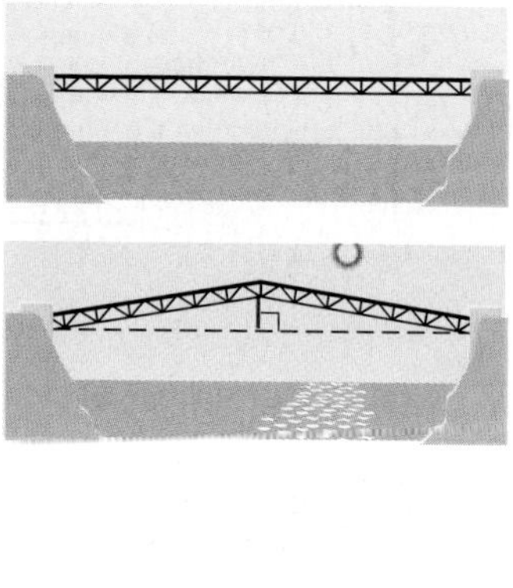

46. Triangle ABC has sides of lengths 25 ft, 25 ft, and 30 ft. Triangle PQR has sides of lengths 25 ft, 25 ft, and 40 ft. Which triangle, if either, has the greater area and by how much?

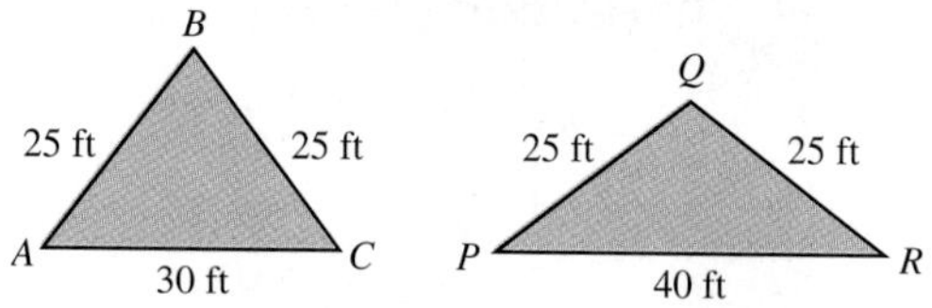

47. *Camping Tent.* The entrance to a pup tent is the shape of an equilateral triangle. If the base of the tent is 4 ft wide, how tall is the tent?

48. The length and the width of a rectangle are given by consecutive integers. The area of the rectangle is 90 cm^2. Find the length of a diagonal of the rectangle.

49. Find all points on the y-axis of a Cartesian coordinate system that are 5 units from the point (3, 0).

50. Find all points on the x-axis of a Cartesian coordinate system that are 5 units from the point (0, 4).

TW **51.** Write a problem for a classmate to solve in which the solution is: "The height of the tepee is $5\sqrt{3}$ yd."

TW **52.** Write a problem for a classmate to solve in which the solution is: "The height of the window is $15\sqrt{3}$ ft."

Skill Maintenance

Graph.

53. $f(x) = \frac{2}{3}x - 5$ [2.2]

54. $g(x) = 3x + 4$ [2.2]

55. $F(x) < -2x + 4$ [4.5]

56. $G(x) > -3x + 2$ [4.5]

57. $f(x) = -2$ [2.3]

58. $g(x) = |x| + 1$ [2.1]

Synthesis

TW **59.** Are there any right triangles, other than those with sides measuring 3, 4, and 5, that have consecutive numbers for the lengths of the sides? Why or why not?

TW **60.** If a 30°–60°–90° triangle and an isosceles right triangle have the same perimeter, which will have the greater area? Why?

61. The perimeter of a regular hexagon is 72 cm. Determine the area of the shaded region shown.

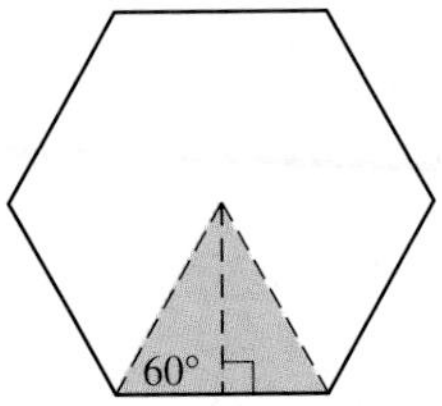

62. If the perimeter of a regular hexagon is 120 ft, what is its area? (*Hint*: See Exercise 61.)

63. Each side of a regular octagon has length s. Find a formula for the distance d between the parallel sides of the octagon.

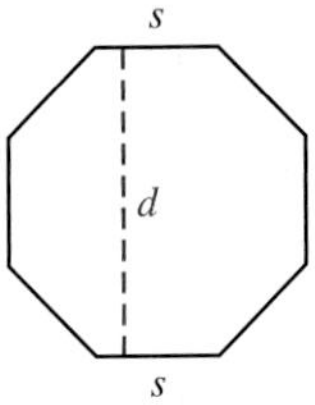

64. *Roofing.* Kit's home, which is 24 ft wide and 32 ft long, needs a new roof. By counting clapboards that are 4 in. apart, Kit determines that the peak of the roof is 6 ft higher than the sides. If one packet of shingles covers 100 ft^2, how many packets will the job require?

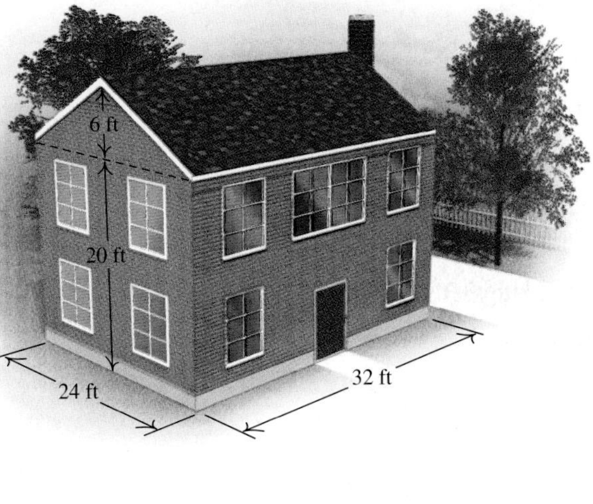

65. *Painting.* (Refer to Exercise 64.) A gallon of Benjamin Moore® exterior acrylic paint covers 450–500 ft^2. If Kit's house has dimensions as shown above, how many gallons of paint should be bought to paint the house? What assumption(s) is made in your answer?

66. *Contracting.* Oxford Builders has an extension cord on their generator that permits them to work, with electricity, anywhere in a circular area of 3850 ft^2. Find the dimensions of the largest square room they could work on without having to relocate the generator to reach each corner of the floor plan.

67. *Contracting.* Cleary Construction has a hose attached to their insulation blower that permits them to work, with electricity, anywhere in a circular area of 6160 ft^2. Find the dimensions of the largest square room with 12-ft ceilings in which they could reach all corners with the hose while leaving the blower centrally located. Assume that the blower sits on the floor.

68. A cube measures 5 cm on each side. How long is the diagonal that connects two opposite corners of the cube? Give an exact answer.

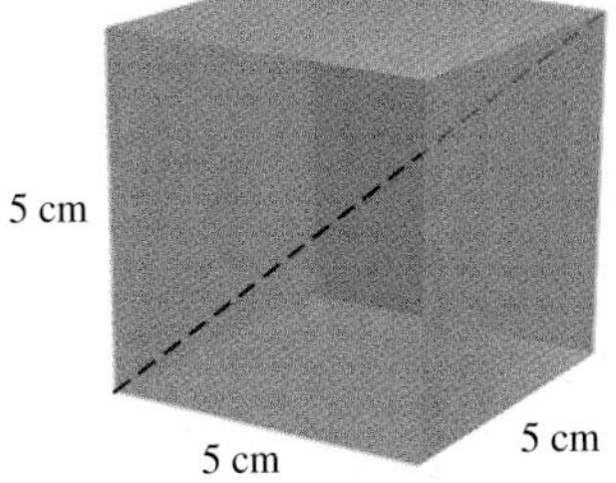

7.8 The Complex Numbers

Imaginary and Complex Numbers ■ Addition and Subtraction ■ Multiplication ■ Conjugates and Division ■ Powers of *i*

Imaginary and Complex Numbers

Negative numbers do not have square roots in the real-number system. However, a larger number system that contains the real-number system is designed so that negative numbers *do* have square roots. That system is called the **complex-number system,** and it will allow us to solve equations like $x^2 + 1 = 0$. The complex-number system makes use of i, a number that is, by definition, a square root of -1.

The Number *i* i is the unique number for which $i = \sqrt{-1}$ and $i^2 = -1$.

We can now define the square root of a negative number as follows:

$$\sqrt{-p} = \sqrt{-1}\,\sqrt{p} = i\sqrt{p} \text{ or } \sqrt{p}i, \quad \text{for any positive number } p.$$

EXAMPLE 1 Express in terms of i: **(a)** $\sqrt{-7}$; **(b)** $\sqrt{-16}$; **(c)** $-\sqrt{-13}$; **(d)** $-\sqrt{-50}$.

SOLUTION

a) $\sqrt{-7} = \sqrt{-1 \cdot 7} = \sqrt{-1} \cdot \sqrt{7} = i\sqrt{7}$, or $\sqrt{7}i$

b) $\sqrt{-16} = \sqrt{-1 \cdot 16} = \sqrt{-1} \cdot \sqrt{16} = i \cdot 4 = 4i$

c) $-\sqrt{-13} = -\sqrt{-1 \cdot 13} = -\sqrt{-1} \cdot \sqrt{13} = -i\sqrt{13}$, or $-\sqrt{13}i$

d) $-\sqrt{-50} = -\sqrt{-1} \cdot \sqrt{25} \cdot \sqrt{2} = -i \cdot 5 \cdot \sqrt{2}$
$= -5i\sqrt{2}$, or $-5\sqrt{2}i$

***i* is *not* under the radical.**

Imaginary Numbers An *imaginary number* is a number that can be written in the form $a + bi$, where a and b are real numbers and $b \neq 0$.

Don't let the name "imaginary" fool you. Imaginary numbers appear in fields such as engineering and the physical sciences. The following are examples of imaginary numbers:

$5 + 4i$, **Here $a = 5, b = 4$.**

$\sqrt{3} - \pi i$, **Here $a = \sqrt{3}, b = -\pi$.**

$\sqrt{7}i$ **Here $a = 0, b = \sqrt{7}$.**

Study Tip

Studying Together by Phone

Working with a classmate over the telephone can be a very effective way to receive or give help. The fact that you cannot point to figures on paper forces you to verbalize the mathematics and the act of speaking mathematics will improve your understanding of the material. In some cases it may be more effective to study with a classmate over the phone than in person.

The union of the set of all imaginary numbers and the set of all real numbers is the set of all **complex numbers.**

> **Complex Numbers** A *complex number* is any number that can be written in the form $a + bi$, where a and b are real numbers. (Note that a and b both can be 0.)

The following are examples of complex numbers:

$7 + 3i$ (here $a \neq 0, b \neq 0$); $\quad$ $4i$ (here $a = 0, b \neq 0$);

8 (here $a \neq 0, b = 0$); $\quad$ 0 (here $a = 0, b = 0$).

Complex numbers like $17i$ or $4i$, in which $a = 0$ and $b \neq 0$, are imaginary numbers with no real part. Such numbers are called *pure imaginary numbers*.

Note that when $b = 0$, we have $a + 0i = a$, so every real number is a complex number. The relationships among various real and complex numbers are shown below.

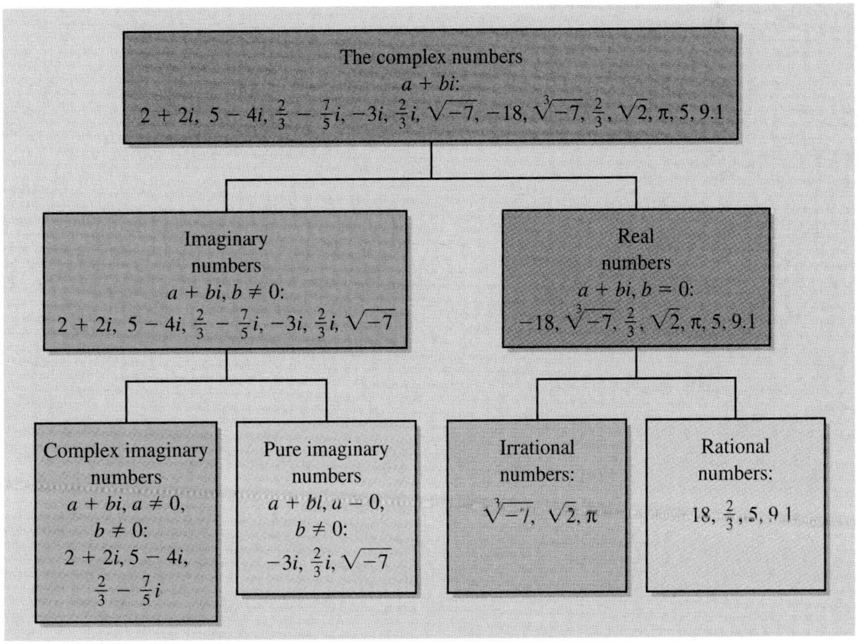

Note that although $\sqrt{-7}$ and $\sqrt[3]{-7}$ are both complex numbers, $\sqrt{-7}$ is imaginary whereas $\sqrt[3]{-7}$ is real.

Addition and Subtraction

The commutative, associative, and distributive laws hold for complex numbers. Thus we can add and subtract them as we do binomials.

EXAMPLE 2 Add or subtract and simplify.

a) $(8 + 6i) + (3 + 2i)$ **b)** $(4 + 5i) - (6 - 3i)$

SOLUTION

a) $(8 + 6i) + (3 + 2i) = (8 + 3) + (6i + 2i)$ **Combining the real parts and the imaginary parts**

$= 11 + (6 + 2)i = 11 + 8i$

b) $(4 + 5i) - (6 - 3i) = (4 - 6) + [5i - (-3i)]$ **Note that the 6 and the $-3i$ are *both* being subtracted.**

$= -2 + 8i$

Multiplication

To multiply square roots of negative real numbers, we first express them in terms of i. For example,

$$\begin{aligned}\sqrt{-2} \cdot \sqrt{-5} &= \sqrt{-1} \cdot \sqrt{2} \cdot \sqrt{-1} \cdot \sqrt{5} \\ &= i \cdot \sqrt{2} \cdot i \cdot \sqrt{5} \\ &= i^2 \cdot \sqrt{10} \\ &= -1\sqrt{10} = -\sqrt{10} \text{ is correct!}\end{aligned}$$

Student Notes

The rule developed in Section 7.3, $\sqrt[n]{a} \cdot \sqrt[n]{b} = \sqrt[n]{a \cdot b}$, does *not* apply when n is 2 and either a or b is negative. Indeed this condition is stated on p. 554 when it is specified that $\sqrt[n]{a}$ and $\sqrt[n]{b}$ are both *real* numbers.

CAUTION! With complex numbers, simply multiplying radicands is *incorrect* when both radicands are negative: $\sqrt{-2} \cdot \sqrt{-5} \neq \sqrt{10}$.

With this in mind, we can now multiply complex numbers.

EXAMPLE 3 Multiply and simplify. When possible, write answers in the form $a + bi$.

a) $\sqrt{-16} \cdot \sqrt{-25}$ **b)** $\sqrt{-5} \cdot \sqrt{-7}$ **c)** $-3i \cdot 8i$

d) $-4i(3 - 5i)$ **e)** $(1 + 2i)(1 + 3i)$

SOLUTION

a)
$$\begin{aligned}\sqrt{-16} \cdot \sqrt{-25} &= \sqrt{-1} \cdot \sqrt{16} \cdot \sqrt{-1} \cdot \sqrt{25} \\ &= i \cdot 4 \cdot i \cdot 5 \\ &= i^2 \cdot 20 \\ &= -1 \cdot 20 \qquad i^2 = -1 \\ &= -20\end{aligned}$$

b)
$$\begin{aligned}\sqrt{-5} \cdot \sqrt{-7} &= \sqrt{-1} \cdot \sqrt{5} \cdot \sqrt{-1} \cdot \sqrt{7} \qquad \textbf{Try to do this step mentally.} \\ &= i \cdot \sqrt{5} \cdot i \cdot \sqrt{7} \\ &= i^2 \cdot \sqrt{35} \\ &= -1 \cdot \sqrt{35} \qquad i^2 = -1 \\ &= -\sqrt{35}\end{aligned}$$

c) $-3i \cdot 8i = -24 \cdot i^2$
$= -24 \cdot (-1)$ $\quad i^2 = -1$
$= 24$

d) $-4i(3 - 5i) = -4i \cdot 3 + (-4i)(-5i)$ **Using the distributive law**
$= -12i + 20i^2$
$= -12i - 20$ $\quad i^2 = -1$
$= -20 - 12i$ **Writing in the form $a + bi$**

e) $(1 + 2i)(1 + 3i) = 1 + 3i + 2i + 6i^2$ **Multiplying each term of $1 + 3i$ by every term of $1 + 2i$ (FOIL)**
$= 1 + 3i + 2i - 6$ $\quad i^2 = -1$
$= -5 + 5i$ **Combining like terms**

Conjugates and Division

Recall that the conjugate of $4 + \sqrt{2}$ is $4 - \sqrt{2}$.

Conjugates of complex numbers are defined in a similar manner.

Conjugate of a Complex Number The *conjugate* of a complex number $a + bi$ is $a - bi$, and the *conjugate* of $a - bi$ is $a + bi$.

EXAMPLE 4 Find the conjugate of each number.

a) $-3 + 7i$ **b)** $14 - 5i$ **c)** $4i$

SOLUTION

a) $-3 + 7i$ The conjugate is $-3 - 7i$.
b) $14 - 5i$ The conjugate is $14 + 5i$.
c) $4i$ The conjugate is $-4i$. Note that $4i = 0 + 4i$.

The product of a complex number and its conjugate is a real number.

EXAMPLE 5 Multiply: $(5 + 7i)(5 - 7i)$.

SOLUTION

$(5 + 7i)(5 - 7i) = 5^2 - (7i)^2$ **Using $(A + B)(A - B) = A^2 - B^2$**
$= 25 - 49i^2$
$= 25 - 49(-1)$ $\quad i^2 = -1$
$= 25 + 49 = 74$

Conjugates are used when dividing complex numbers. The procedure is much like that used to rationalize denominators in Section 7.5.

EXAMPLE 6 Divide and simplify to the form $a + bi$.

a) $\dfrac{-5 + 9i}{1 - 2i}$ **b)** $\dfrac{7 + 3i}{5i}$

SOLUTION

a) To divide and simplify $(-5 + 9i)/(1 - 2i)$, we multiply by 1, using the conjugate of the denominator to form 1:

$$\frac{-5 + 9i}{1 - 2i} = \frac{-5 + 9i}{1 - 2i} \cdot \frac{1 + 2i}{1 + 2i}$$ **Multiplying by 1 using the conjugate of the denominator in the symbol for 1**

$$= \frac{-5 - 10i + 9i + 18i^2}{1^2 - 4i^2}$$ **Multiplying using FOIL**

$$= \frac{-5 - i - 18}{1 - 4(-1)}$$ $i^2 = -1$

$$= \frac{-23 - i}{5}$$

$$= -\frac{23}{5} - \frac{1}{5}i$$

Writing in the form $a + bi$; note that $\dfrac{X + Y}{Z} = \dfrac{X}{Z} + \dfrac{Y}{Z}$

b) The conjugate of $5i$ is $-5i$, so we *could* multiply by $-5i/(-5i)$. However, when the denominator is a pure imaginary number, it is easiest if we multiply by i/i:

$$\frac{7 + 3i}{5i} = \frac{7 + 3i}{5i} \cdot \frac{i}{i}$$ **Multiplying by 1 using i/i. We can also use the conjugate of $5i$ to write $-5i/(-5i)$.**

$$= \frac{7i + 3i^2}{5i^2}$$ **Multiplying**

$$= \frac{7i + 3(-1)}{5(-1)}$$ $i^2 = -1$

$$= \frac{7i - 3}{-5} = \frac{-3}{-5} + \frac{7}{-5}i, \text{ or } \frac{3}{5} - \frac{7}{5}i.$$ **Writing in the form $a + bi$**

Powers of i

Answers to problems involving complex numbers are generally written in the form $a + bi$. In the following discussion, we show why there is no need to use powers of i (other than 1) when writing answers.

Recall that -1 raised to an *even* power is 1, and -1 raised to an *odd* power is -1. Simplifying powers of i can then be done by using the fact that $i^2 = -1$ and expressing the given power of i in terms of i^2. Consider the following:

$$\begin{aligned}
i^2 &= -1, \\
i^3 &= i^2 \cdot i = (-1)i = -i, \\
i^4 &= (i^2)^2 = (-1)^2 = 1, \\
i^5 &= i^4 \cdot i = (i^2)^2 \cdot i = (-1)^2 \cdot i = i, \\
i^6 &= (i^2)^3 = (-1)^3 = -1.
\end{aligned}$$

The pattern is now repeating.

The powers of i cycle themselves through the values i, -1, $-i$, and 1. Even powers of i are -1 or 1 whereas odd powers of i are i or $-i$.

EXAMPLE 7 Simplify: **(a)** i^{18}; **(b)** i^{24}.

SOLUTION

a) $i^{18} = (i^2)^9$ **Using the power rule**

$= (-1)^9 = -1$ **Raising −1 to a power**

b) $i^{24} = (i^2)^{12} = (-1)^{12} = 1$

To simplify i^n when n is odd, we rewrite i^n as $i^{n-1} \cdot i$.

EXAMPLE 8 Simplify: **(a)** i^{29}; **(b)** i^{75}.

SOLUTION

a) $i^{29} = i^{28}i^1$ **Using the product rule. This is a key step when i is raised to an odd power.**

$= (i^2)^{14}i$ **Using the power rule**

$= (-1)^{14}i = 1 \cdot i = i$

b) $i^{75} = i^{74}i^1$ **Using the product rule**

$= (i^2)^{37}i$ **Using the power rule**

$= (-1)^{37}i = -1 \cdot i = -i$

Complex Numbers

Many graphing calculators can perform operations with complex numbers. To do so, first choose the a + bi setting in the MODE screen.

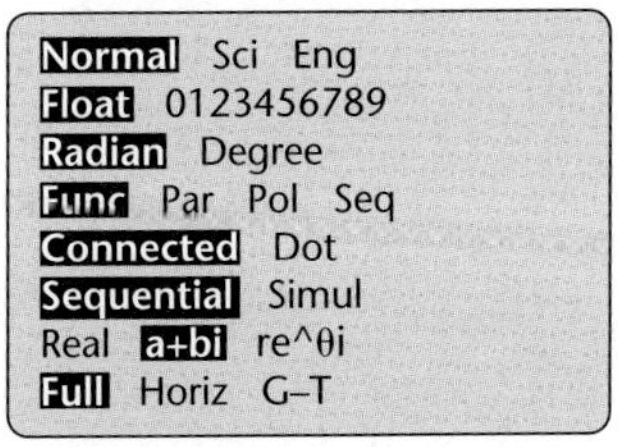

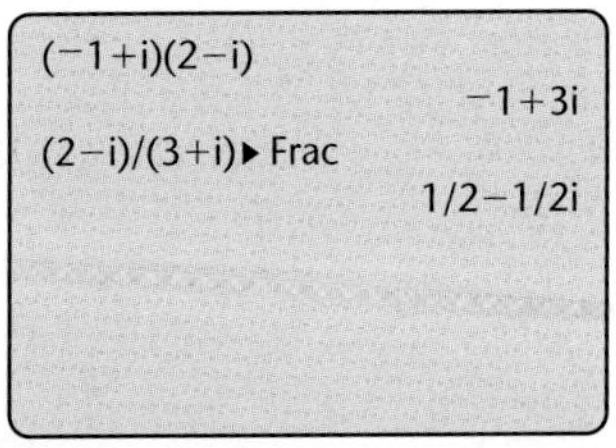

Now complex numbers can be entered as they are written. Press (i) for the complex number i. ((i) is often the 2nd option associated with the (.) key.) For example, to calculate $(-1 + i)(2 - i)$, press ((−) 1 + i) (2 − i) ENTER. As seen in the screen on the right above, the result is $-1 + 3i$. To calculate $\frac{2 - i}{3 + i}$ and write the result using fraction notation, press (2 − i) ÷ (3 + i) MATH 1 ENTER. Note the need for parentheses around the numerator and the denominator of the expression. The result appears in the form $a + bi$; i is not in the denominator of the fraction. The result is $\frac{1}{2} - \frac{1}{2}i$.

7.8 EXERCISE SET

FOR EXTRA HELP

MathXL | MyMathLab | InterAct Math | AW Math Tutor Center | Video Lectures on CD: Disc 4 | Student's Solutions Manual

Concept Reinforcement *Classify each statement as either true or false.*

1. Imaginary numbers are so named because they have no real-world applications.
2. Every real number is imaginary, but not every imaginary number is real.
3. Every imaginary number is a complex number, but not every complex number is imaginary.
4. Every real number is a complex number, but not every complex number is real.
5. Addition and subtraction of complex numbers have much in common with addition and subtraction of polynomials.
6. The product of a complex number and its conjugate is always a real number.
7. The square of a complex number is always a real number.
8. The quotient of two complex numbers is always a complex number.

Express in terms of i.

9. $\sqrt{-36}$
10. $\sqrt{-25}$
11. $\sqrt{-13}$
12. $\sqrt{-19}$
13. $\sqrt{-18}$
14. $\sqrt{-98}$
15. $\sqrt{-3}$
16. $\sqrt{-4}$
17. $\sqrt{-81}$
18. $\sqrt{-27}$
19. $-\sqrt{-300}$
20. $-\sqrt{-75}$
21. $6 - \sqrt{-84}$
22. $4 - \sqrt{-60}$
23. $-\sqrt{-76} + \sqrt{-125}$
24. $\sqrt{-4} + \sqrt{-12}$
25. $\sqrt{-18} - \sqrt{-100}$
26. $\sqrt{-72} - \sqrt{-25}$

Perform the indicated operation and simplify. Write each answer in the form $a + bi$.

27. $(6 + 7i) + (5 + 3i)$
28. $(4 - 5i) + (3 + 9i)$
29. $(9 + 8i) - (5 + 3i)$
30. $(9 + 7i) - (2 + 4i)$
31. $(7 - 4i) - (5 - 3i)$
32. $(5 - 3i) - (9 + 2i)$
33. $(-5 - i) - (7 + 4i)$
34. $(-2 + 6i) - (-7 + i)$
35. $7i \cdot 6i$
36. $6i \cdot 9i$
37. $(-4i)(-6i)$
38. $7i \cdot (-8i)$
39. $\sqrt{-36}\sqrt{-9}$
40. $\sqrt{-49}\sqrt{-25}$
41. $\sqrt{-5}\sqrt{-2}$
42. $\sqrt{-6}\sqrt{-7}$
43. $\sqrt{-6}\sqrt{-21}$
44. $\sqrt{-15}\sqrt{-10}$
45. $5i(2 + 6i)$
46. $2i(7 + 3i)$
47. $-7i(3 - 4i)$
48. $-4i(6 - 5i)$
49. $(1 + i)(3 + 2i)$
50. $(1 + 5i)(4 + 3i)$
51. $(6 - 5i)(3 + 4i)$
52. $(5 - 6i)(2 + 5i)$
53. $(7 - 2i)(2 - 6i)$
54. $(-4 + 5i)(3 - 4i)$
55. $(-2 + 3i)(-2 + 5i)$
56. $(-3 + 6i)(-3 + 4i)$
57. $(-5 - 4i)(3 + 7i)$

58. $(2 + 9i)(-3 - 5i)$

59. $(4 - 2i)^2$

60. $(1 - 2i)^2$

61. $(2 + 3i)^2$

62. $(3 + 2i)^2$

63. $(-2 + 3i)^2$

64. $(-5 - 2i)^2$

65. $\dfrac{7}{4 + i}$

66. $\dfrac{3}{8 + i}$

67. $\dfrac{2}{3 - 2i}$

68. $\dfrac{4}{2 - 3i}$

69. $\dfrac{2i}{5 + 3i}$

70. $\dfrac{3i}{4 + 2i}$

71. $\dfrac{5}{6i}$

72. $\dfrac{4}{7i}$

73. $\dfrac{5 - 3i}{4i}$

74. $\dfrac{2 + 7i}{5i}$

Aha! **75.** $\dfrac{7i + 14}{7i}$

76. $\dfrac{6i + 3}{3i}$

77. $\dfrac{4 + 5i}{3 - 7i}$

78. $\dfrac{5 + 3i}{7 - 4i}$

79. $\dfrac{2 + 3i}{2 + 5i}$

80. $\dfrac{3 + 2i}{4 + 3i}$

81. $\dfrac{3 - 2i}{4 + 3i}$

82. $\dfrac{5 - 2i}{3 + 6i}$

Simplify.

83. i^7

84. i^{11}

85. i^{24}

86. i^{35}

87. i^{42}

88. i^{64}

89. i^9

90. $(-i)^{71}$

91. $(-i)^6$

92. $(-i)^4$

93. $(5i)^3$

94. $(-3i)^5$

95. $i^2 + i^4$

96. $5i^5 + 4i^3$

TW **97.** Is the product of two imaginary numbers always an imaginary number? Why or why not?

TW **98.** In what way(s) are conjugates of complex numbers similar to the conjugates used in Section 7.5?

Skill Maintenance

Factor.

99. $12x^2 - 23x + 10$ [5.5]

100. $x^3 - 3x^2 - 10x$ [5.4]

101. $20x^2y^3 + 15x^3y^2 - 35x^4y^3$ [5.3]

102. $x^2y - 3y - x^3 + 3x$ [5.3]

103. $14x^4y - 14y^5$ [5.6]

104. $27x^3 - 64y^3$ [5.7]

Synthesis

TW **105.** Is the set of real numbers a subset of the set of complex numbers? Why or why not?

TW **106.** Is the union of the set of imaginary numbers and the set of real numbers the set of complex numbers? Why or why not?

A function g is given by

$$g(z) = \frac{z^4 - z^2}{z - 1}.$$

107. Find $g(3i)$.

108. Find $g(1 + i)$.

109. Find $g(5i - 1)$.

110. Find $g(2 - 3i)$.

111. Evaluate

$$\frac{1}{w - w^2} \quad \text{for} \quad w = \frac{1 - i}{10}.$$

Simplify.

112. $\dfrac{i^5 + i^6 + i^7 + i^8}{(1 - i)^4}$

113. $(1 - i)^3(1 + i)^3$

114. $\dfrac{5 - \sqrt{5}i}{\sqrt{5}i}$

115. $\dfrac{6}{1 + \dfrac{3}{i}}$

116. $\left(\dfrac{1}{2} - \dfrac{1}{3}i\right)^2 - \left(\dfrac{1}{2} + \dfrac{1}{3}i\right)^2$

117. $\dfrac{i - i^{38}}{1 + i}$

7 Chapter Summary and Review

KEY TERMS AND DEFINITIONS

RADICALS

Square root, p. 530 c is a square root of a if $c^2 = a$. $\sqrt{a}$ means the **principal,** or nonnegative, square root of a.

Cube root, p. 534 c is a cube root of a if $c^3 = a$. $\sqrt[3]{a}$ means the cube root of a.

***n*th root,** p. 535 c is an nth root of a if $c^n = a$. The nth root of a is written

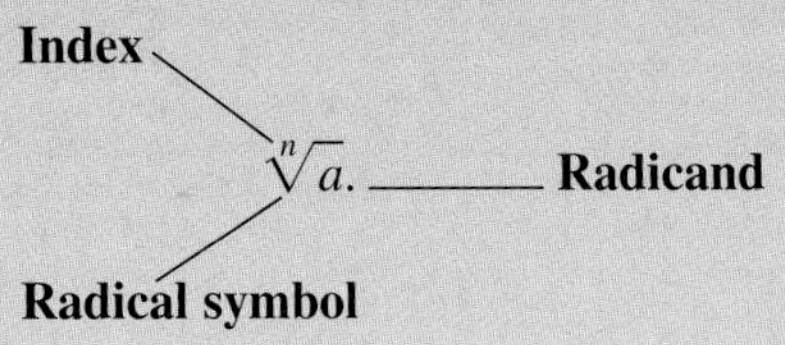

$\sqrt[n]{a^n} = |a|$ when n is even.

$\sqrt[n]{a^n} = a$ when n is odd.

COMPLEX NUMBERS

Complex number, p. 595 Any number that can be written in the form $a + bi$, where a and b are real numbers and $i = \sqrt{-1}$. A complex number is real when $b = 0$, is **imaginary** when $b \neq 0$, and is **pure imaginary** when $a = 0$.

APPLICATIONS

Pythagorean Theorem, p. 586 For any right triangle with legs of lengths a and b and hypotenuse of length c,

$$a^2 + b^2 = c^2.$$

Special triangles, p. 589 The length of the hypotenuse in an isosceles right triangle is the length of a leg times $\sqrt{2}$.

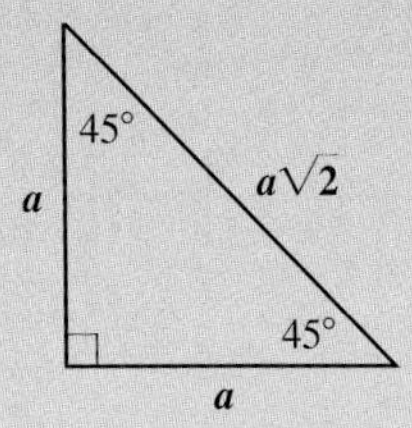

The length of the longer leg in a 30°–60°–90° right triangle is the length of the shorter leg times $\sqrt{3}$. The hypotenuse is twice as long as the shorter leg.

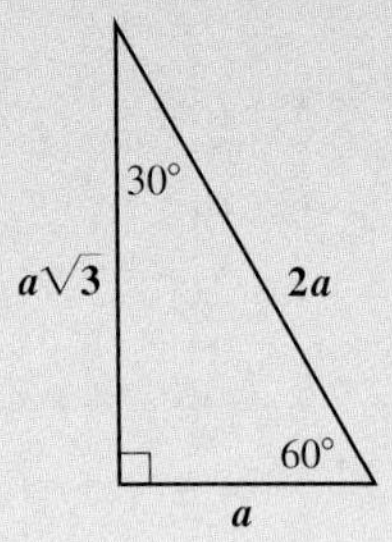

IMPORTANT CONCEPTS

[Section references appear in brackets.]

Concept	Example
The domain of a **radical function** with an even index is the set of all inputs for which the radicand is nonnegative.	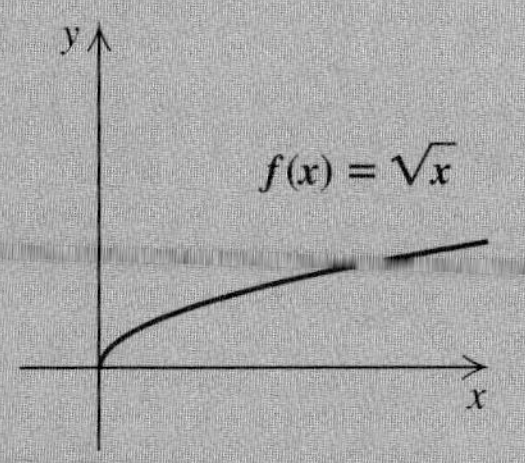 Domain of f: $[0, \infty)$ Range of f: $[0, \infty)$ [7.1]
Radical notation can be written using **rational exponents**. $a^{1/n}$ means $\sqrt[n]{a}$. $a^{m/n}$ means $(\sqrt[n]{a})^m$ or $\sqrt[n]{a^m}$. When a is negative, n must be odd.	$7^{1/10} = \sqrt[10]{7}$ $\quad x^{1/2} = \sqrt{x}$ $3^{4/5} = \sqrt[5]{3^4} = (\sqrt[5]{3})^4$ [7.2]
Radical expressions can often be simplified. **1.** Simplify by factoring. **2.** Use rational exponents to simplify. **3.** Combine **like radical terms.**	$\sqrt[3]{a^6 b} = \sqrt[3]{a^6}\sqrt[3]{b} = a^2\sqrt[3]{b}$ [7.3] $\sqrt[3]{p} \cdot \sqrt[4]{q^3} = p^{1/3} \cdot q^{3/4}$ $= p^{4/12} \cdot q^{9/12} = \sqrt[12]{p^4 q^9}$ [7.5] $\sqrt{8} + 3\sqrt{2} = \sqrt{4} \cdot \sqrt{2} + 3\sqrt{2}$ $= 2\sqrt{2} + 3\sqrt{2} = 5\sqrt{2}$ [7.5]
Radicands of radical expressions can be multiplied and divided if the indices are the same. **The Product Rule for Radicals** For any real numbers $\sqrt[n]{a}$ and $\sqrt[n]{b}$, $\sqrt[n]{a}\,\sqrt[n]{b} = \sqrt[n]{a \cdot b}$. **The Quotient Rule for Radicals** For any real numbers $\sqrt[n]{a}$ and $\sqrt[n]{b}$, $b \neq 0$, $\sqrt[n]{\frac{a}{b}} = \frac{\sqrt[n]{a}}{\sqrt[n]{b}}$.	$\sqrt{15}\sqrt{35} = \sqrt{15 \cdot 35}$ $= \sqrt{3 \cdot 5 \cdot 5 \cdot 7} = 5\sqrt{21}$ [7.3] $\sqrt[3]{\frac{8x^3}{27}} = \frac{\sqrt[3]{8x^3}}{\sqrt[3]{27}} = \frac{2x}{3}$; $\frac{\sqrt{20}}{\sqrt{2}} = \sqrt{\frac{20}{2}} = \sqrt{10}$ [7.4]

(continued)

A denominator is **rationalized** when it does not contain any irrational numbers. A denominator with two terms is rationalized using the **conjugate** of the denominator.	$\frac{1}{\sqrt{2}} = \frac{1}{\sqrt{2}} \cdot \frac{\sqrt{2}}{\sqrt{2}} = \frac{\sqrt{2}}{2}$ [7.4] $\frac{3}{2+\sqrt{5}} = \frac{3}{2+\sqrt{5}} \cdot \frac{2-\sqrt{5}}{2-\sqrt{5}}$ $= \frac{6-3\sqrt{5}}{4-5} = \frac{6-3\sqrt{5}}{-1} = -6+3\sqrt{5}$ [7.5]
To solve a radical equation, use the principle of powers and the steps on pp. 578 and 581. Solutions must be checked in the original equation. **The Principle of Powers** If $a = b$, then $a^n = b^n$ for any exponent n.	$x - 7 = \sqrt{x-5}$ $(x-7)^2 = (\sqrt{x-5})^2$ $x^2 - 14x + 49 = x - 5$ $x^2 - 15x + 54 = 0$ $(x-6)(x-9) = 0$ $x = 6$ *or* $x = 9$ Only 9 checks and is the solution. $2 + \sqrt{x} = \sqrt{x+8}$ $(2+\sqrt{x})^2 = (\sqrt{x+8})^2$ $4 + 4\sqrt{x} + x = x + 8$ $4\sqrt{x} = 4$ $\sqrt{x} = 1$ $(\sqrt{x})^2 = (1)^2$ $x = 1$ 1 checks and is the solution. [7.6]
Complex numbers can be added, subtracted, multiplied, and divided.	$(3+2i) + (4-7i) = 7 - 5i;$ $(8+6i) - (5+2i) = 3 + 4i;$ $(2+3i)(4-i) = 8 - 2i + 12i - 3i^2$ $= 8 + 10i - 3(-1) = 11 + 10i;$ $\frac{1-4i}{3-2i} = \frac{1-4i}{3-2i} \cdot \frac{3+2i}{3+2i}$ $= \frac{3+2i-12i-8i^2}{9+6i-6i-4i^2}$ $= \frac{3-10i-8(-1)}{9-4(-1)}$ $= \frac{11-10i}{13} = \frac{11}{13} - \frac{10}{13}i.$ [7.8]

Review Exercises

Concept Reinforcement *Classify each of the following as either true or false.*

1. $\sqrt{ab} = \sqrt{a} \cdot \sqrt{b}$ for any real numbers $\sqrt{a}$ and $\sqrt{b}$. [7.3]
2. $\sqrt{a^2} = a$, for any real number a. [7.1]
3. $\sqrt[3]{a^3} = a$, for any real number a. [7.1]
4. $x^{2/5}$ means $\sqrt[5]{x^2}$ and $(\sqrt[5]{x})^2$. [7.2]
5. A hypotenuse is never shorter than either leg. [7.7]
6. $i^{13} = (i^2)^6 i = i$. [7.8]
7. Some radical equations have no solution. [7.6]
8. If $f(x) = \sqrt{x - 5}$, then the domain of f is the set of all nonnegative real numbers. [7.1]

Simplify. [7.1]

9. $\sqrt{\dfrac{49}{9}}$

10. $-\sqrt{0.25}$

Let $f(x) = \sqrt{2x - 7}$. *Find the following.* [7.1]

11. $f(16)$
12. The domain of f
13. *Dairy Farming.* As a calf grows, it needs more milk for nourishment. The number of pounds of milk, M, required by a calf weighing x pounds can be estimated using the formula
$$M = -5 + \sqrt{6.7x - 444}.$$
(*Source*: www.ext.vt.edu) Estimate the number of pounds of milk required by a calf of the given weight: **(a)** 300 lb; **(b)** 100 lb; **(c)** 200 lb; **(d)** 400 lb. [7.1]

Simplify. Assume that each variable can represent any real number.

14. $\sqrt{25t^2}$ [7.1]
15. $\sqrt{(c + 8)^2}$ [7.1]
16. $\sqrt{x^2 - 6x + 9}$ [7.1]
17. $\sqrt{4x^2 + 4x + 1}$ [7.1]
18. $\sqrt[5]{-32}$ [7.1]
19. $\sqrt[3]{-\dfrac{64x^6}{27}}$ [7.4]
20. $\sqrt[4]{x^{12}y^8}$ [7.3]
21. $\sqrt[6]{64x^{12}}$ [7.3]
22. Write an equivalent expression using exponential notation: $(\sqrt[3]{5ab})^4$. [7.2]
23. Write an equivalent expression using radical notation: $(16a^6)^{3/4}$. [7.2]

Use rational exponents to simplify. Assume $x, y \geq 0$. [7.2]

24. $\sqrt{x^6y^{10}}$

25. $(\sqrt[6]{x^2y})^2$

Simplify. Do not use negative exponents in the answers. [7.2]

26. $(x^{-2/3})^{3/5}$

27. $\dfrac{7^{-1/3}}{7^{-1/2}}$

28. If $f(x) = \sqrt{25(x - 6)^2}$, find a simplified form for $f(x)$. [7.3]

Perform the indicated operation and, if possible, simplify. Write all answers using radical notation.

29. $\sqrt{2x}\sqrt{3y}$ [7.3]
30. $\sqrt[3]{a^5b}\sqrt[3]{27b}$ [7.3]
31. $\sqrt[3]{-24x^{10}y^8}\sqrt[3]{18x^7y^4}$ [7.3]
32. $\dfrac{\sqrt[3]{60xy^3}}{\sqrt[3]{10x}}$ [7.4]
33. $\dfrac{\sqrt{75x}}{2\sqrt{3}}$ [7.4]
34. $\sqrt[4]{\dfrac{48a^{11}}{c^8}}$ [7.4]
35. $5\sqrt[3]{x} + 2\sqrt[3]{x}$ [7.5]
36. $2\sqrt{75} - 9\sqrt{3}$ [7.5]
37. $\sqrt[3]{8x^4} + \sqrt[3]{xy^6}$ [7.5]
38. $\sqrt{50} + 2\sqrt{18} + \sqrt{32}$ [7.5]
39. $(\sqrt{3} - 3\sqrt{8})(\sqrt{5} + 2\sqrt{8})$ [7.5]

40. $\sqrt[4]{x}\sqrt{x}$ [7.5]

41. $\dfrac{\sqrt[3]{x^2}}{\sqrt[4]{x}}$ [7.5]

42. If $f(x) = x^2$, find $f(a - \sqrt{2})$. [7.5]

43. Rationalize the denominator:

$$\frac{4\sqrt{5}}{\sqrt{2} + \sqrt{3}}.\ [7.5]$$

44. Rationalize the numerator of the expression in Exercise 43. [7.5]

Solve. [7.6]

45. $\sqrt{y+6} - 2 = 3$

46. $(x+1)^{1/3} = -5$

47. $1 + \sqrt{x} = \sqrt{3x - 3}$

48. If $f(x) = \sqrt[4]{x+2}$, find a such that $f(a) = 2$. [7.6]

Solve. Give an exact answer and, where appropriate, an approximation to three decimal places. [7.7]

49. The diagonal of a square has length 10 cm. Find the length of a side of the square.

50. A skate-park jump has a ramp that is 6 ft long and is 2 ft high. How long is its base?

51. Find the missing lengths. Give exact answers and, where appropriate, an approximation to three decimal places.

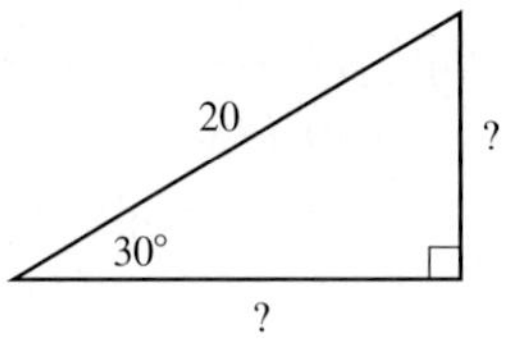

52. Express in terms of i and simplify: $-\sqrt{-8}$. [7.8]

53. Add: $(-4 + 3i) + (2 - 12i)$. [7.8]

54. Subtract: $(9 - 7i) - (3 - 8i)$. [7.8]

Simplify. [7.8]

55. $(2 + 5i)(2 - 5i)$

56. i^{18}

57. Simplify: $(6 - 3i)(2 - i)$. [7.8]

58. Divide and simplify to the form $a + bi$:

$$\frac{7 - 2i}{3 + 4i}.\ [7.8]$$

Synthesis

TW **59.** What makes some complex numbers real and others imaginary? [7.8]

TW **60.** Explain why $\sqrt[n]{x^n} = |x|$ when n is even, but $\sqrt[n]{x^n} = x$ when n is odd. [7.1]

61. Solve:

$$\sqrt{11x + \sqrt{6 + x}} = 6.\ [7.6]$$

62. Simplify:

$$\frac{2}{1 - 3i} - \frac{3}{4 + 2i}.\ [7.8]$$

63. Write a quotient of two imaginary numbers that is a real number (answers may vary). [7.8]

Chapter Test 7

Simplify. Assume that variables can represent any real number.

1. $\sqrt{50}$

2. $\sqrt[3]{-\frac{8}{x^6}}$

3. $\sqrt{81a^2}$

4. $\sqrt{x^2 - 8x + 16}$

5. $\sqrt[5]{x^{12}y^8}$

6. $\sqrt{\frac{25x^2}{36y^4}}$

7. $\sqrt[3]{3z}\sqrt[3]{5y^2}$

8. $\frac{\sqrt[5]{x^3y^4}}{\sqrt[5]{xy^2}}$

9. $\sqrt[4]{x^3y^2}\sqrt{xy}$

10. $\frac{\sqrt[5]{a^2}}{\sqrt[4]{a}}$

11. $8\sqrt{2} - 2\sqrt{2}$

12. $\sqrt{x^4y} + \sqrt{9y^3}$

13. $(7 + \sqrt{x})(2 - 3\sqrt{x})$

14. Write an equivalent expression using exponential notation: $\sqrt{7xy}$.

15. Write an equivalent expression using radical notation: $(4a^3b)^{5/6}$.

16. If $f(x) = \sqrt{2x - 10}$, determine the domain of f.

17. If $f(x) = x^2$, find $f(5 + \sqrt{2})$.

18. Rationalize the denominator:

$$\frac{\sqrt{3}}{5 + \sqrt{2}}.$$

Solve.

19. $x = \sqrt{2x - 5} + 4$

20. $\sqrt{x} = \sqrt{x + 1} - 5$

21. *Falling Object.* The number of seconds t that it takes for an object to fall x meters when thrown down at a velocity of 9.5 meters per second is given by the function.

$$t(x) = -0.9694 + \sqrt{0.9397 + 0.2041x}.$$

A rock is thrown from the top of Turkey Bluff, 90 m above the Gasconade River in Missouri, at a velocity of 9.5 meters per second. After how many seconds will the rock hit the water?

Solve. Give exact answers and approximations to three decimal places.

22. One leg of a 30°–60°–90° right triangle is 7 cm long. Find the possible lengths of the other leg.

23. A referee jogs diagonally from one corner of a 50-ft by 90-ft basketball court to the far corner. How far does she jog?

24. Express in terms of i and simplify: $\sqrt{-50}$.

25. Subtract: $(9 + 8i) - (-3 + 6i)$.

26. Multiply: $\sqrt{-16}\sqrt{-36}$.

27. Multiply. Write the answer in the form $a + bi$.

$$(4 - i)^2$$

28. Divide and simplify to the form $a + bi$:

$$\frac{-2 + i}{3 - 5i}.$$

29. Simplify: i^{37}.

Synthesis

30. Solve:

$$\sqrt{2x - 2} + \sqrt{7x + 4} = \sqrt{13x + 10}.$$

31. Simplify:

$$\frac{1 - 4i}{4i(1 + 4i)^{-1}}.$$

32. Drake's Discount Shoe Center has two locations. The sign at the original location is shaped like an isosceles right triangle. The sign at the newer location is shaped like a 30°–60°–90° triangle. The hypotenuse of each sign measures 6 ft. Which sign has the greater area and by how much? (Round to three decimal places.)

8

Quadratic Functions and Equations

In translating problem situations to mathematics, we often obtain a function or equation containing a second-degree polynomial in one variable. Such functions or equations are said to be *quadratic*. In this chapter, we examine a variety of equations, inequalities, and applications for which we will need to solve quadratic equations or graph quadratic functions.

APPLICATION *Sonoma Sunshine.*

The percent of days each month during which the sun shines in Sonoma, California, can be modeled by a quadratic function, as indicated by the table and graph below. Find a quadratic function that fits the data.

Month	Percent of Days with Sunshine
1 (Jan.)	46
3 (March)	75
5 (May)	90
7 (July)	98
9 (Sep.)	92
11 (Nov.)	65

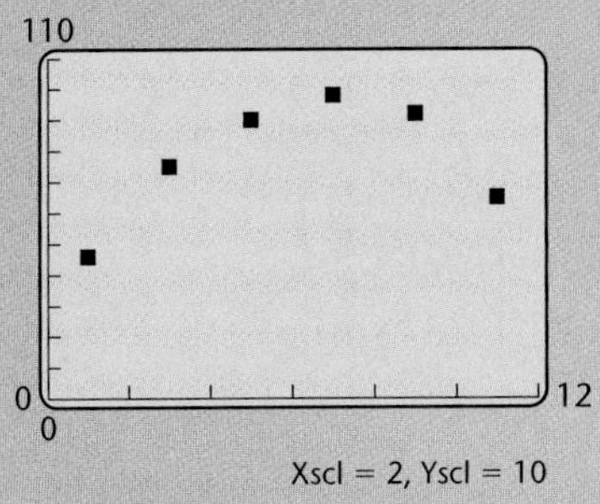

This problem appears as Example 4 in Section 8.8.

8.1 Quadratic Equations

The Principle of Square Roots ■ Completing the Square ■ Problem Solving

Study Tip

Choosing Tunes

While many people prefer working in silence, it is not always possible to do so. Whether listening for pleasure or to block out nearby noise, you may find that music without lyrics is most conducive to focusing your concentration on your studies. At least one study has shown that classical background music can improve one's ability to concentrate.

In Chapter 5, we solved *quadratic equations* like $x^2 = 10 + 3x$ by graphing and by factoring. One way to solve by graphing is to rewrite the equation above, for example, as $x^2 - 3x - 10 = 0$ and then graph the *quadratic function* given by $f(x) = x^2 - 3x - 10$. The solutions of the equation, -2 and 5, are the first coordinates of the x-intercepts of the graph of f.

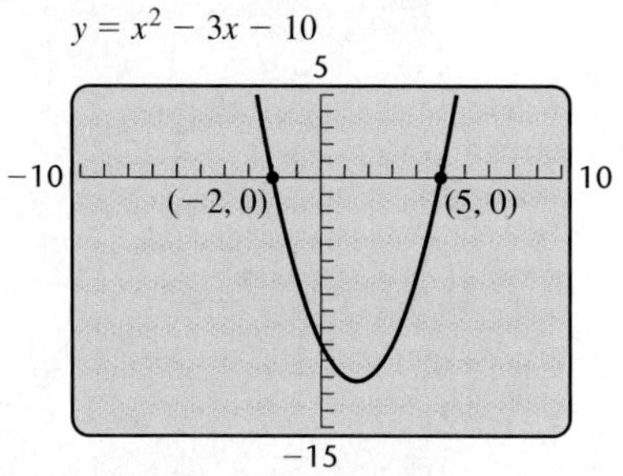

A quadratic equation will have no, one, or two real-number solutions. We can see this by examining the graphs of quadratic functions.

Interactive Discovery

Graph each function and determine the number of x-intercepts.

1. $f(x) = x^2$
2. $g(x) = -x^2$
3. $h(x) = (x - 2)^2$
4. $p(x) = 2x^2 + 1$
5. $f(x) = -1.5x^2 + x - 3$
6. $g(x) = 4x^2 - 2x - 7$
7. Describe the shape of the graph of a quadratic function.

Although we will study graphs of quadratic functions in detail in Section 8.6, we can make some general observations here. The graph of a quadratic function is a cup-shaped curve called a *parabola*. It can open upward, like the graph of $f(x) = x^2$, or downward, like the graph of $g(x) = -x^2$.

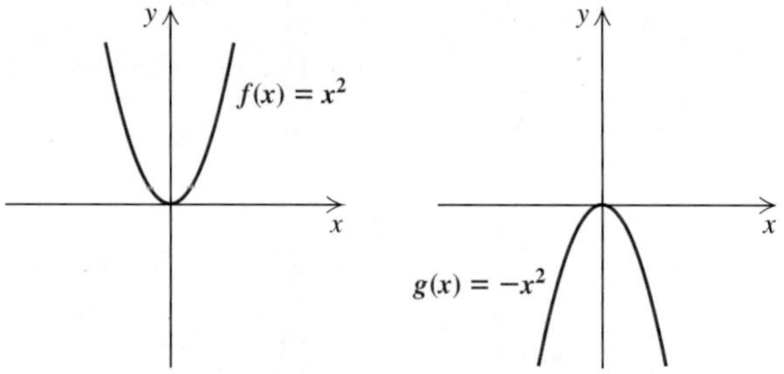

The graph of a quadratic function can have no, one, or two x-intercepts, as illustrated below. Thus a quadratic equation can have no, one, or two real-number roots. Quadratic equations can have imaginary-number roots. These must be found using algebraic methods.

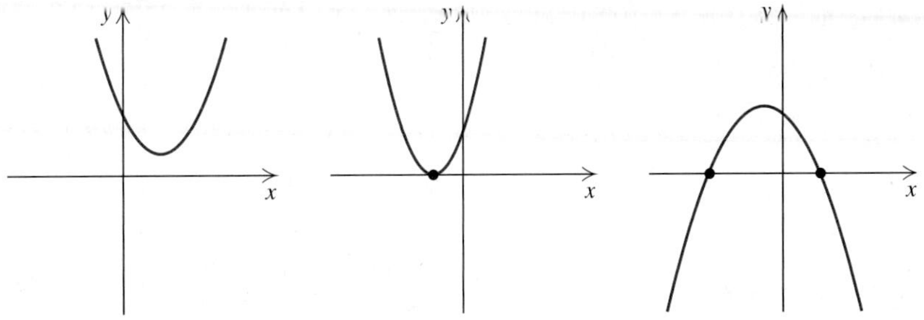

No x-intercepts
No real-number roots

One x-intercept
One real-number root

Two x-intercepts
Two real-number roots

To solve a quadratic equation by factoring, we write the equation in the *standard form* $ax^2 + bx + c = 0$, factor, and use the principle of zero products.

EXAMPLE 1 Solve: $3x^2 = 2 - x$.

SOLUTION We have

$$3x^2 = 2 - x$$

$$3x^2 + x - 2 = 0 \quad \textbf{Adding } -2 + x \textbf{ to both sides to obtain standard form}$$

$$(3x - 2)(x + 1) = 0 \quad \textbf{Factoring}$$

$$3x - 2 = 0 \quad or \quad x + 1 = 0 \quad \textbf{Using the principle of zero products}$$

$$3x = 2 \quad or \quad x = -1$$

$$x = \tfrac{2}{3} \quad or \quad x = -1.$$

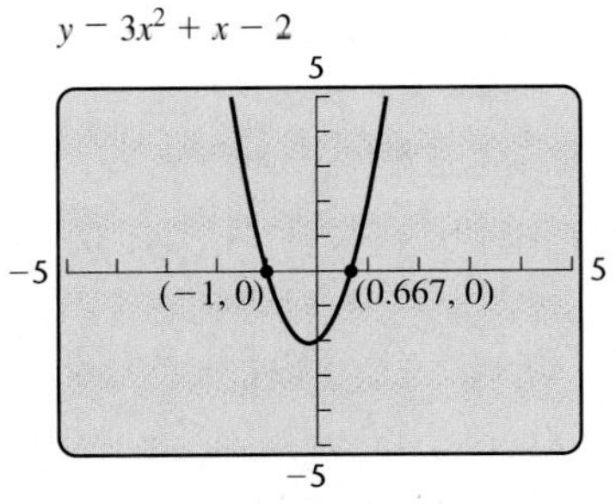

We check by substituting -1 and $\frac{2}{3}$ into the original equation. A graphical solution (see the figure at left) provides another check.

Check: For -1:

$$\begin{array}{c|c} \multicolumn{2}{c}{3x^2 = 2 - x} \\ \hline 3(-1)^2 & 2 - (-1) \\ 3 \cdot 1 & 2 + 1 \\ \multicolumn{2}{c}{3 \stackrel{?}{=} 3} \end{array} \quad \text{TRUE}$$

For $\frac{2}{3}$:

$$\begin{array}{c|c} \multicolumn{2}{c}{3x^2 = 2 - x} \\ \hline 3(\frac{2}{3})^2 & 2 - \frac{2}{3} \\ 3 \cdot \frac{4}{9} & \frac{6}{3} - \frac{2}{3} \\ \multicolumn{2}{c}{\frac{4}{3} \stackrel{?}{=} \frac{4}{3}} \end{array} \quad \text{TRUE}$$

The solutions are -1 and $\frac{2}{3}$.

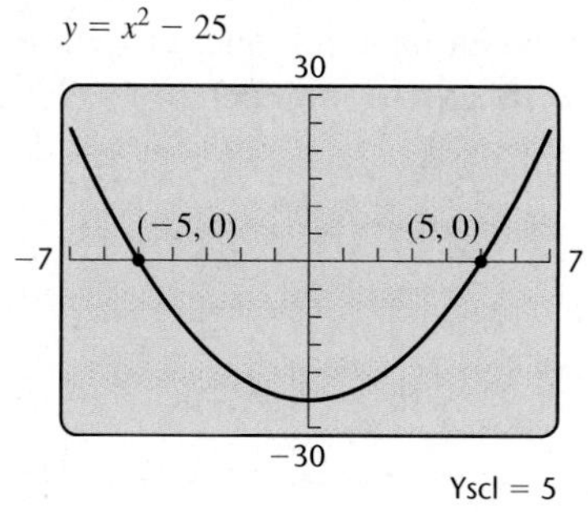

EXAMPLE 2 Solve: $x^2 = 25$.

SOLUTION We have

$$\begin{aligned} x^2 &= 25 & \\ x^2 - 25 &= 0 & \quad \textbf{Writing in standard form} \\ (x-5)(x+5) &= 0 & \quad \textbf{Factoring} \\ x - 5 = 0 \quad &or \quad x + 5 = 0 & \quad \textbf{Using the principle of zero products} \\ x = 5 \quad &or \quad x = -5. & \end{aligned}$$

The solutions are 5 and −5. The graph at left confirms the solutions.

The Principle of Square Roots

Consider the equation $x^2 = 25$ again. We know from Chapter 7 that the number 25 has two real-number square roots, 5 and −5, the solutions of the equation in Example 2. Thus square roots can provide a quick method for solving equations of the type $x^2 = k$.

> **The Principle of Square Roots** For any real number k, if $x^2 = k$, then $x = \sqrt{k}$ or $x = -\sqrt{k}$.

EXAMPLE 3 Solve: $3x^2 = 6$. Give exact answers and approximations to three decimal places.

SOLUTION We have

$$\begin{aligned} 3x^2 &= 6 & \\ x^2 &= 2 & \quad \textbf{Isolating } x^2 \\ x = \sqrt{2} \quad &or \quad x = -\sqrt{2}. & \quad \textbf{Using the principle of square roots} \end{aligned}$$

We often use the symbol $\pm\sqrt{2}$ to represent the two numbers $\sqrt{2}$ and $-\sqrt{2}$.

Check: For $\sqrt{2}$:

$$\begin{array}{c|c} \multicolumn{2}{c}{3x^2 = 6} \\ \hline 3(\sqrt{2})^2 & 6 \\ 3 \cdot 2 & \\ 6 \stackrel{?}{=} 6 & \text{TRUE} \end{array}$$

For $-\sqrt{2}$:

$$\begin{array}{c|c} \multicolumn{2}{c}{3x^2 = 6} \\ \hline 3(-\sqrt{2})^2 & 6 \\ 3 \cdot 2 & \\ 6 \stackrel{?}{=} 6 & \text{TRUE} \end{array}$$

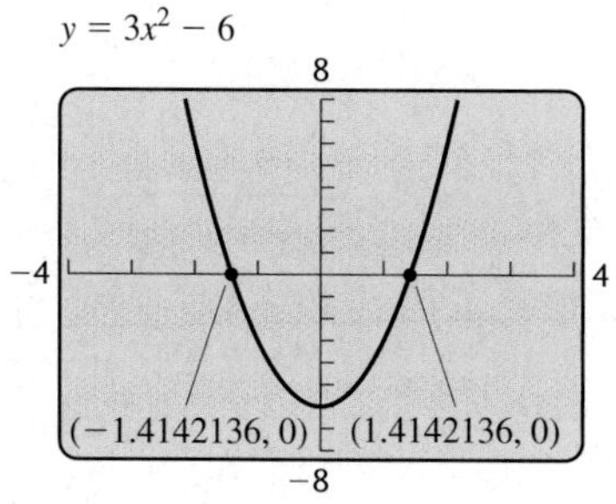

The solutions are $\sqrt{2}$ and $-\sqrt{2}$, or $\pm\sqrt{2}$, which round to 1.414 and −1.414. A graphical solution, shown at left, yields approximate solutions, since $\sqrt{2}$ is irrational. The solutions found graphically correspond to those found algebraically.

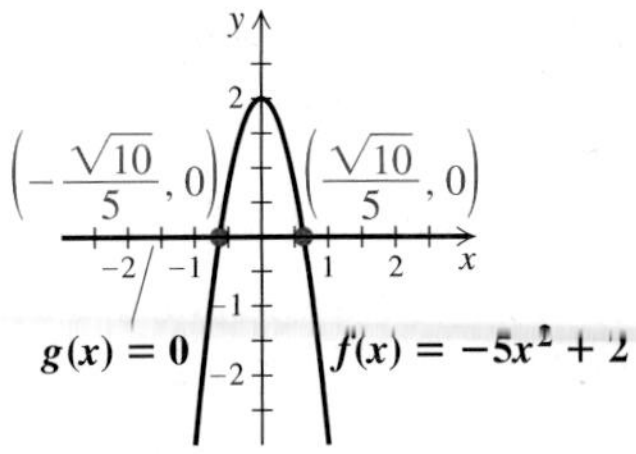

A visualization of Example 4

EXAMPLE 4 Solve: $-5x^2 + 2 = 0$.

SOLUTION We have

$$-5x^2 + 2 = 0$$

$$x^2 = \frac{2}{5} \qquad \text{Isolating } x^2$$

$$x = \sqrt{\frac{2}{5}} \quad or \quad x = -\sqrt{\frac{2}{5}}. \qquad \text{Using the principle of square roots}$$

The solutions are $\sqrt{\frac{2}{5}}$ and $-\sqrt{\frac{2}{5}}$. This can also be written as $\pm\sqrt{\frac{2}{5}}$, or, if we rationalize the denominator, $\pm\frac{\sqrt{10}}{5}$. The checks are left to the student.

Sometimes we get solutions that are imaginary numbers.

EXAMPLE 5 Solve: $4x^2 + 9 = 0$.

SOLUTION We have

$$4x^2 + 9 = 0$$

$$x^2 = -\tfrac{9}{4} \qquad \text{Isolating } x^2$$

$$x = \sqrt{-\tfrac{9}{4}} \quad or \quad x = -\sqrt{-\tfrac{9}{4}} \qquad \text{Using the principle of square roots}$$

$$x = \sqrt{\tfrac{9}{4}}\sqrt{-1} \quad or \quad x = -\sqrt{\tfrac{9}{4}}\sqrt{-1}$$

$$x = \tfrac{3}{2}i \quad or \quad x = -\tfrac{3}{2}i. \qquad \text{Recall that } \sqrt{-1} = i.$$

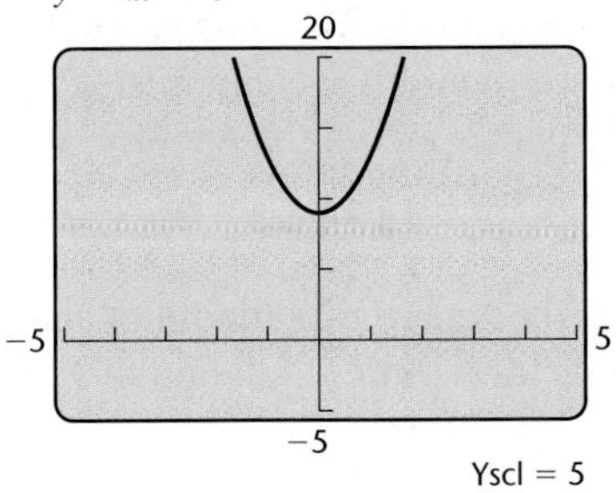

Check:

For $\frac{3}{2}i$:

$4x^2 + 9 = 0$	
$4\left(\frac{3}{2}i\right)^2 + 9$	0
$4 \cdot \frac{9}{4} \cdot i^2 + 9$	
$9(-1) + 9$	

$0 \stackrel{?}{=} 0$ TRUE

For $-\frac{3}{2}i$:

$4x^2 + 9 = 0$	
$4\left(-\frac{3}{2}i\right)^2 + 9$	0
$4 \cdot \frac{9}{4} \cdot i^2 + 9$	
$9(-1) + 9$	

$0 \stackrel{?}{=} 0$ TRUE

The solutions are $\frac{3}{2}i$ and $-\frac{3}{2}i$, or $\pm\frac{3}{2}i$. The graph at left confirms that there are no real-number solutions, because there are no x-intercepts.

The principle of square roots can be restated in a more general form that pertains to algebraic expressions other than just x.

The Principle of Square Roots (Generalized Form) For any real number k and any algebraic expression X,

If $X^2 = k$, then $X = \sqrt{k}$ or $X = -\sqrt{k}$.

EXAMPLE 6 Let $f(x) = (x - 2)^2$. Find all x-values for which $f(x) = 7$.

SOLUTION We are asked to find all x-values for which

$$f(x) = 7,$$

or

$$(x - 2)^2 = 7. \quad \textbf{Substituting } (x-2)^2 \textbf{ for } f(x)$$

We solve both algebraically and graphically.

ALGEBRAIC APPROACH

The generalized principle of square roots gives us

$$x - 2 = \sqrt{7} \quad or \quad x - 2 = -\sqrt{7}$$
$$x = 2 + \sqrt{7} \quad or \quad x = 2 - \sqrt{7}.$$

Check: $f(2 + \sqrt{7}) = (2 + \sqrt{7} - 2)^2 = (\sqrt{7})^2 = 7.$

Similarly,

$$f(2 - \sqrt{7}) = (2 - \sqrt{7} - 2)^2 = (-\sqrt{7})^2 = 7.$$

Rounded to the nearest thousandth,

$$2 + \sqrt{7} \approx 4.646 \quad \text{and} \quad 2 - \sqrt{7} \approx -0.646,$$

so the graphical solutions match the algebraic solutions.

The solutions are $2 + \sqrt{7}$ and $2 - \sqrt{7}$, or simply $2 \pm \sqrt{7}$.

GRAPHICAL APPROACH

We graph the equations $y_1 = (x - 2)^2$ and $y_2 = 7$. If there are any real-number x-values for which $f(x) = 7$, they will be the x-coordinates of the points of intersection of the graphs.

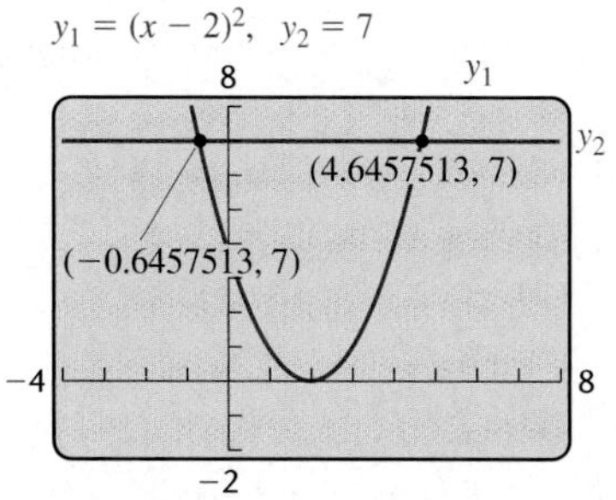

Rounded to the nearest thousandth, the solutions are approximately -0.646 and 4.646.

We can check the algebraic solutions using the VALUE option of the CALC menu. If we enter $2 + \sqrt{7}$ or $2 - \sqrt{7}$ for x, the corresponding value of y_1 will be 7, so the solutions check.

In Example 6, one side of the equation is the square of a binomial and the other side is a constant. Sometimes we must factor in order to obtain this form.

EXAMPLE 7 Solve: $x^2 + 6x + 9 = 2$.

SOLUTION We have

$$x^2 + 6x + 9 = 2 \quad \textbf{The left side is the square of a binomial.}$$
$$(x + 3)^2 = 2 \quad \textbf{Factoring}$$
$$x + 3 = \sqrt{2} \quad or \quad x + 3 = -\sqrt{2} \quad \textbf{Using the principle of square roots}$$
$$x = -3 + \sqrt{2} \quad or \quad x = -3 - \sqrt{2}. \quad \textbf{Adding } -3 \textbf{ to both sides}$$

```
Y1(-3 + √(2))
                    2
Y1(-3 - √(2))
                    2
```

One way to check the solutions on a graphing calculato[r] ... $y_1 = x^2 + 6x + 9$ and then evaluate $y_1(-3 + \sqrt{2})$ and $y_1(-3 - \sqrt{2})$, ... shown at left. The second calculation can be entered quickly by copying the first entry and editing it.

The solutions are $-3 + \sqrt{2}$ and $-3 - \sqrt{2}$, or $-3 \pm \sqrt{2}$.

Completing the Square

By using a method called *completing the square*, we can use the principle of square roots to solve *any* quadratic equation.

EXAMPLE 8 Solve: $x^2 + 6x + 4 = 0$.

SOLUTION We have

$$x^2 + 6x + 4 = 0$$

$$x^2 + 6x = -4 \qquad \textbf{Subtracting 4 from both sides}$$

$$x^2 + 6x + 9 = -4 + 9 \qquad \textbf{Adding 9 to both sides. We explain this shortly.}$$

$$(x + 3)^2 = 5 \qquad \textbf{Factoring the perfect-square trinomial}$$

$$x + 3 = \pm\sqrt{5} \qquad \textbf{Using the principle of square roots. Remember that } \pm\sqrt{5} \textbf{ represents two numbers.}$$

$$x = -3 \pm \sqrt{5}. \qquad \textbf{Adding } -3 \textbf{ to both sides}$$

```
-3+√(5)→X
              -.7639320225
X2+6X+4
                         0
```

We check by storing $-3 + \sqrt{5}$ as x and evaluating $x^2 + 6x + 4$, as shown at left. We can repeat the process for $-3 - \sqrt{5}$ to check the second solution. The solutions are $-3 + \sqrt{5}$ and $-3 - \sqrt{5}$.

In Example 8, we chose to add 9 to both sides because it creates a perfect-square trinomial on the left side. The 9 was determined by taking half of the coefficient of x and squaring it.

To help see why this procedure works, examine the following drawings.

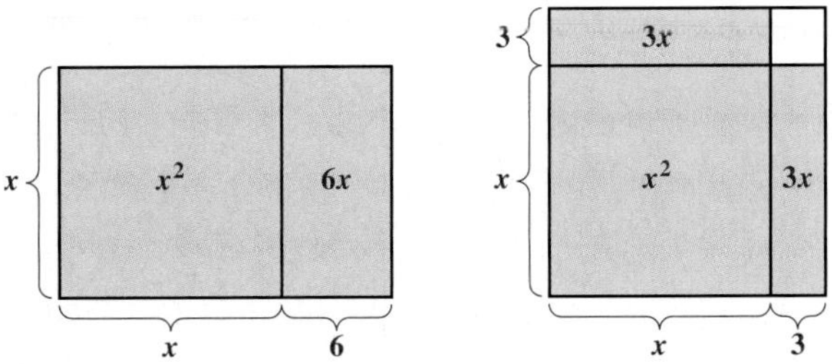

Note that the shaded areas in both figures represent the same area, $x^2 + 6x$. However, only the figure on the right, in which the $6x$ is halved, can be converted into a square with the addition of a constant term. The constant 9 is the "missing" piece that *completes* the square.

To complete the square for $x^2 + bx$, we add $\left(\frac{b}{2}\right)^2$.

Example 9, which follows, provides practice in finding numbers that complete the square. We will then use this skill to solve equations.

Student Notes

In problems like Examples 9(b) and (c), it is best to avoid decimal notation. Most students have an easier time recognizing $\frac{9}{64}$ as $\left(\frac{3}{8}\right)^2$ than regarding 0.140625 as 0.375^2.

EXAMPLE 9 Replace the blanks in each equation with constants to form a true equation.

a) $x^2 + 14x + \_\_\_\_ = (x + \_\_\_\_)^2$

b) $x^2 - 5x + \_\_\_\_ = (x - \_\_\_\_)^2$

c) $x^2 + \frac{3}{4}x + \_\_\_\_ = (x + \_\_\_\_)^2$

SOLUTION We take half of the coefficient of x and square it.

a) Half of 14 is 7, and $7^2 = 49$. Thus, $x^2 + 14x + 49$ is a perfect-square trinomial and is equivalent to $(x + 7)^2$. We have

$$x^2 + 14x + 49 = (x + 7)^2.$$

b) Half of -5 is $-\frac{5}{2}$, and $\left(-\frac{5}{2}\right)^2 = \frac{25}{4}$. Thus, $x^2 - 5x + \frac{25}{4}$ is a perfect-square trinomial and is equivalent to $\left(x - \frac{5}{2}\right)^2$. We have

$$x^2 - 5x + \frac{25}{4} = \left(x - \frac{5}{2}\right)^2.$$

c) Half of $\frac{3}{4}$ is $\frac{3}{8}$, and $\left(\frac{3}{8}\right)^2 = \frac{9}{64}$. Thus, $x^2 + \frac{3}{4}x + \frac{9}{64}$ is a perfect-square trinomial and is equivalent to $\left(x + \frac{3}{8}\right)^2$. We have

$$x^2 + \frac{3}{4}x + \frac{9}{64} = \left(x + \frac{3}{8}\right)^2.$$

We can now use the method of completing the square to solve equations similar to Example 8.

EXAMPLE 10 Solve: **(a)** $x^2 - 8x - 7 = 0$; **(b)** $x^2 + 5x - 3 = 0$.

SOLUTION

a)

$x^2 - 8x - 7 = 0$	
$x^2 - 8x = 7$	**Adding 7 to both sides. We can now complete the square on the left side.**
$x^2 - 8x + 16 = 7 + 16$	**Adding 16 to both sides to complete the square: $\frac{1}{2}(-8) = -4$, and $(-4)^2 = 16$**
$(x - 4)^2 = 23$	**Factoring and simplifying**
$x - 4 = \pm\sqrt{23}$	**Using the principle of square roots**
$x = 4 \pm \sqrt{23}$	**Adding 4 to both sides**

The solutions are $4 - \sqrt{23}$ and $4 + \sqrt{23}$, or $4 \pm \sqrt{23}$. The checks are left to the student.

CAUTION! In Example 10(a), be sure to add 16 to *both sides* of the equation.

b)
$$
\begin{aligned}
x^2 + 5x - 3 &= 0 \\
x^2 + 5x \quad &= 3 && \textbf{Adding 3 to both sides} \\
x^2 + 5x + \frac{25}{4} &= 3 + \frac{25}{4} && \textbf{Completing the square: } \tfrac{1}{2}\cdot 5 = \tfrac{5}{2}\textbf{, and } \left(\tfrac{5}{2}\right)^2 = \tfrac{25}{4} \\
\left(x + \frac{5}{2}\right)^2 &= \frac{37}{4} && \textbf{Factoring and simplifying} \\
x + \frac{5}{2} &= \pm\frac{\sqrt{37}}{2} && \textbf{Using the principle of square roots and the quotient rule for radicals} \\
x = -\frac{5}{2} \pm \frac{\sqrt{37}}{2} \quad &or \quad x = \frac{-5 \pm \sqrt{37}}{2} && \textbf{Adding } -\tfrac{5}{2} \textbf{ to both sides}
\end{aligned}
$$

The checks are left to the student. The solutions are $-\frac{5}{2} \pm \frac{\sqrt{37}}{2}$, or $\frac{-5 \pm \sqrt{37}}{2}$.

Before we complete the square, the x^2-coefficient must be 1. When it is not 1, we divide both sides of the equation by whatever that coefficient may be.

EXAMPLE 11 Find the x-intercepts of the function given by

$$f(x) = 3x^2 + 7x - 2.$$

SOLUTION The value of $f(x)$ must be 0 at any x-intercepts. Thus,

$$
\begin{aligned}
f(x) &= 0 && \textbf{We set } f(x) \textbf{ equal to 0.} \\
3x^2 + 7x - 2 &= 0 && \textbf{Substituting} \\
3x^2 + 7x \quad &= 2 && \textbf{Adding 2 to both sides} \\
x^2 + \frac{7}{3}x \quad &= \frac{2}{3} && \textbf{Dividing both sides by 3} \\
x^2 + \frac{7}{3}x + \frac{49}{36} &= \frac{2}{3} + \frac{49}{36} && \textbf{Completing the square: } \left(\tfrac{1}{2}\cdot\tfrac{7}{3}\right)^2 = \tfrac{49}{36} \\
\left(x + \frac{7}{6}\right)^2 &= \frac{73}{36} && \textbf{Factoring and simplifying} \\
x + \frac{7}{6} &= \pm\frac{\sqrt{73}}{6} && \textbf{Using the principle of square roots and the quotient rule for radicals} \\
x = -\frac{7}{6} \pm \frac{\sqrt{73}}{6}, \quad &\text{or} \quad \frac{-7 \pm \sqrt{73}}{6}. && \textbf{Adding } -\tfrac{7}{6} \textbf{ to both sides}
\end{aligned}
$$

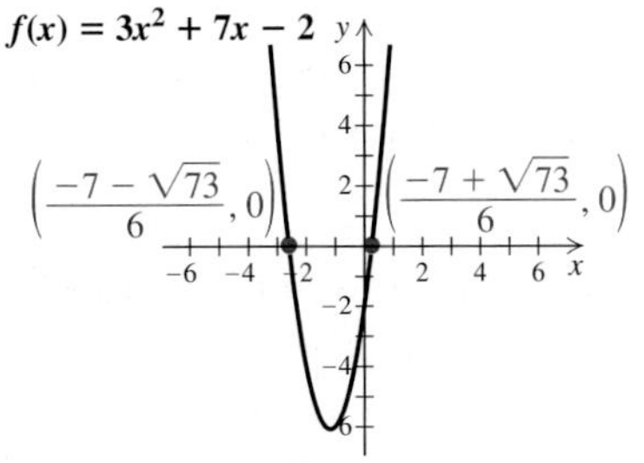

A visualization of Example 11

The x-intercepts are

$$\left(\frac{-7 - \sqrt{73}}{6}, 0\right) \text{ and } \left(\frac{-7 + \sqrt{73}}{6}, 0\right).$$

The checks are left to the student.

The procedure used in Example 11 is important because it can be used to solve *any* quadratic equation.

To Solve a Quadratic Equation in x by Completing the Square

1. Isolate the terms with variables on one side of the equation, and arrange them in descending order.
2. Divide both sides by the coefficient of x^2 if that coefficient is not 1.
3. Complete the square by taking half of the coefficient of x and adding its square to both sides.
4. Express the trinomial as the square of a binomial (factor the trinomial) and simplify the other side.
5. Use the principle of square roots (find the square roots of both sides).
6. Solve for x by adding or subtracting on both sides.

Problem Solving

After one year, an amount of money P, invested at 4% per year, is worth 104% of P, or $P(1.04)$. If that amount continues to earn 4% interest per year, after the second year the investment will be worth 104% of $P(1.04)$, or $P(1.04)^2$. This is called **compounding interest** since after the first time period, interest is earned on both the initial investment *and* the interest from the first time period. Continuing the above pattern, we see that after the third year, the investment will be worth 104% of $P(1.04)^2$. Generalizing, we have the following.

The Compound-Interest Formula If an amount of money P is invested at interest rate r, compounded annually, then in t years, it will grow to the amount A given by

$$A = P(1 + r)^t. \quad (r \text{ is written in decimal notation.})$$

We can use quadratic equations to solve certain interest problems.

EXAMPLE 12 Investment Growth. Rosa invested \$4000 at interest rate r, compounded annually. In 2 yr, it grew to \$4410. What was the interest rate?

SOLUTION

1. **Familiarize.** We are already familiar with the compound-interest formula. If we were not, we would need to consult an outside source.
2. **Translate.** The translation consists of substituting into the formula:

$$A = P(1 + r)^t$$

$$4410 = 4000(1 + r)^2. \quad \textbf{Substituting}$$

3. **Carry out.** We solve for r both algebraically and graphically.

ALGEBRAIC APPROACH

We have

$$4410 = 4000(1 + r)^2$$

$$\frac{4410}{4000} = (1 + r)^2 \quad \textbf{Dividing both sides by 4000}$$

$$\frac{441}{400} = (1 + r)^2 \quad \textbf{Simplifying}$$

$$\pm\sqrt{\frac{441}{400}} = 1 + r \quad \textbf{Using the principle of square roots}$$

$$\pm\frac{21}{20} = 1 + r \quad \textbf{Simplifying}$$

$$-\frac{20}{20} \pm \frac{21}{20} = r \quad \textbf{Adding } -1\textbf{, or } -\tfrac{20}{20}\textbf{, to both sides}$$

$$\frac{1}{20} = r \quad \text{or} \quad -\frac{41}{20} = r.$$

Since r represents an interest rate, we convert to decimal notation. We now have

$$r = 0.05 \quad \text{or} \quad r = -2.05.$$

GRAPHICAL APPROACH

We let $y_1 = 4410$ and $y_2 = 4000(1 + x)^2$. Using the window $[-3, 1, 0, 5000]$, we see that the graphs intersect at two points. The first coordinates of the points of intersection are -2.05 and 0.05.

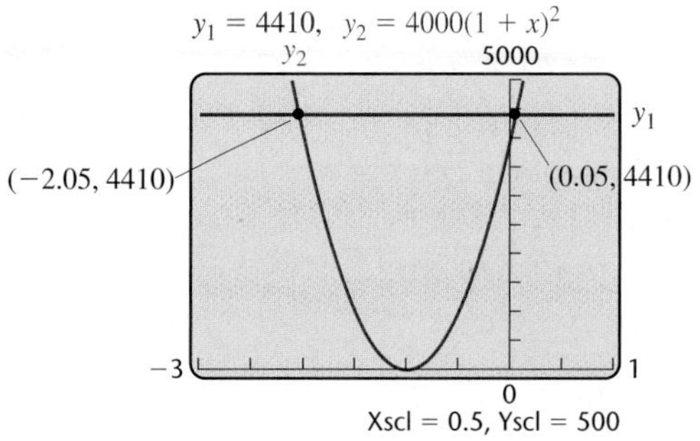

4. **Check.** Since the rate cannot be negative, we need only check 0.05, or 5%. If \$4000 were invested at 5% interest, compounded annually, then in 2 yr it would grow to $4000(1.05)^2$, or \$4410. The number 0.05 checks.
5. **State.** The interest rate was 5%.

EXAMPLE 13 Free-Falling Objects. The formula $s = 16t^2$ is used to approximate the distance s, in feet, that an object falls freely from rest in t seconds. Ireland's Cliffs of Moher are 702 ft tall (*Source*: Based on data from 4windstravel.com). How long will it take a stone to fall from the top? Round to the nearest tenth of a second.

SOLUTION

1. **Familiarize.** We make a drawing to help visualize the problem and agree to disregard air resistance.
2. **Translate.** We substitute into the formula:

$$s = 16t^2$$
$$702 = 16t^2.$$

3. **Carry out.** We solve for t:

$$702 = 16t^2$$
$$\frac{702}{16} = t^2$$
$$43.875 = t^2$$
$$\sqrt{43.875} = t \quad \textbf{Using the principle of square roots; rejecting the negative square root since } t \textbf{ cannot be negative in this problem}$$
$$6.6 \approx t. \quad \textbf{Using a calculator and rounding to the nearest tenth}$$

4. **Check.** Since $16(6.6)^2 \approx 697 \approx 702$, our answer checks.
5. **State.** It takes about 6.6 sec for a stone to fall freely from the Cliffs of Moher.

8.1 EXERCISE SET

FOR EXTRA HELP

MathXL | MyMathLab | InterAct Math | AW Math Tutor Center | Video Lectures on CD: Disc 4 | Student's Solutions Manual

Concept Reinforcement *Complete each of the following to form true statements.*

1. The principle of square roots states that if $x^2 = k$, then $x =$ ____ or $x =$ ____.
2. If $(x + 5)^2 = 49$, then $x + 5 =$ ____ or $x + 5 =$ ____.
3. If $t^2 + 6t + 9 = 17$, then $(\_\_\_\_)^2 = 17$ and ____ $= \pm\sqrt{17}$.
4. The equations $x^2 + 8x +$ ____ $= 23$ and $x^2 + 8x = 7$ are equivalent.
5. The expressions $t^2 + 10t +$ ____ and $(t +$ ____$)^2$ are equivalent.
6. The expressions $x^2 - 6x +$ ____ and $(x -$ ____$)^2$ are equivalent.

Determine the number of real-number solutions of each equation from the given graph.

7. $x^2 + x - 12 = 0$

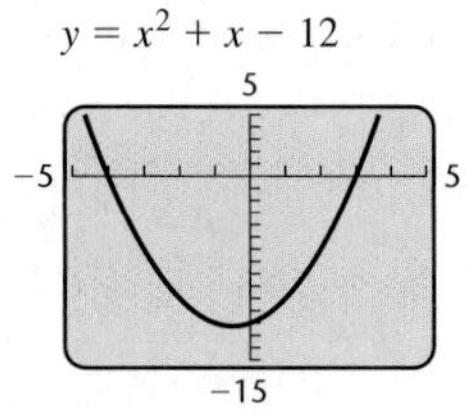

8. $-3x^2 - x - 7 = 0$

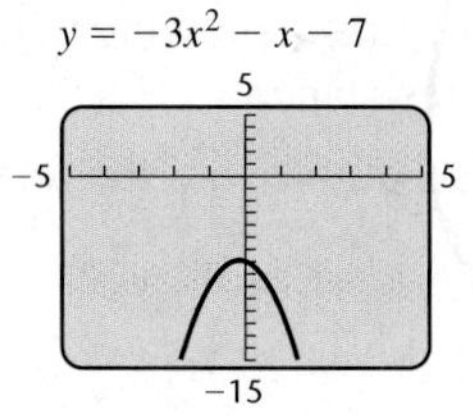

9. $4x^2 + 9 = 12x$

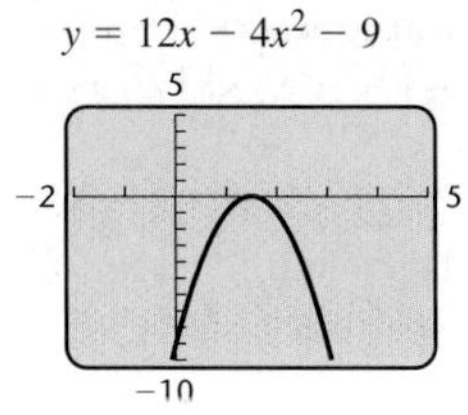

10. $2x^2 + 3 = 6x$

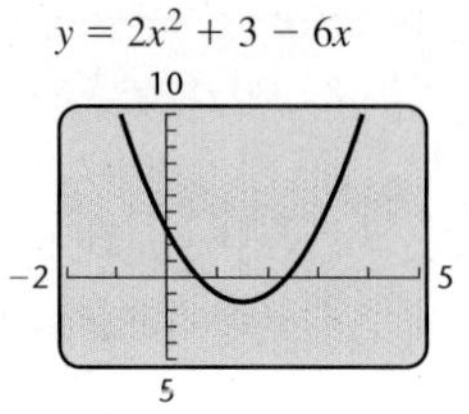

11. $f(x) = 0$

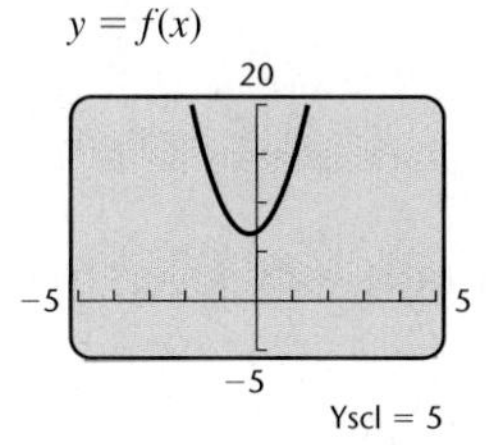

12. $f(x) = 0$

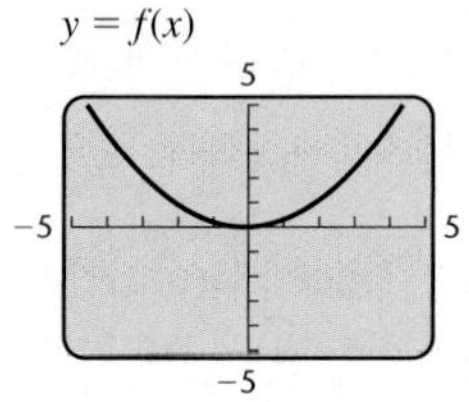

Solve.

13. $4x^2 = 20$
14. $7x^2 = 21$
15. $9x^2 + 16 = 0$
16. $25x^2 + 4 = 0$
17. $5t^2 - 7 = 0$
18. $3t^2 - 2 = 0$
19. $(x - 1)^2 = 49$
20. $(x + 2)^2 = 25$
21. $(a - 13)^2 = 18$
22. $(a + 5)^2 = 8$
23. $(x + 1)^2 = -9$
24. $(x - 1)^2 = -49$
25. $\left(y + \frac{3}{4}\right)^2 = \frac{17}{16}$
26. $\left(t + \frac{3}{2}\right)^2 = \frac{7}{2}$
27. $x^2 - 10x + 25 = 64$
28. $x^2 - 6x + 9 = 100$
29. Let $f(x) = (x - 5)^2$. Find x such that $f(x) = 16$.
30. Let $g(x) = (x - 2)^2$. Find x such that $g(x) = 25$.
31. Let $F(t) = (t + 4)^2$. Find t such that $F(t) = 13$.
32. Let $f(t) = (t + 6)^2$. Find t such that $f(t) = 15$.
33. **Aha!** Let $g(x) = x^2 + 14x + 49$. Find x such that $g(x) = 49$.
34. Let $F(x) = x^2 + 8x + 16$. Find x such that $F(x) = 9$.

Replace the blanks in each equation with constants to complete the square and form a true equation.

35. $x^2 + 16x +$ ____ $= (x +$ ____$)^2$
36. $x^2 + 8x +$ ____ $= (x +$ ____$)^2$

37. $t^2 - 10t + \_\_ = (t - \_\_)^2$

38. $t^2 - 6t + \_\_ = (t - \_\_)^2$

39. $x^2 + 3x + \_\_ = (x + \_\_)^2$

40. $x^2 + 7x + \_\_ = (x + \_\_)^2$

41. $t^2 - 9t + \_\_ = (t - \_\_)^2$

42. $t^2 - 3t + \_\_ = (t - \_\_)^2$

43. $x^2 + \frac{2}{5}x + \_\_ = (x + \_\_)^2$

44. $x^2 + \frac{2}{3}x + \_\_ = (x + \_\_)^2$

45. $t^2 - \frac{5}{6}t + \_\_ = (t - \_\_)^2$

46. $t^2 - \frac{5}{3}t + \_\_ = (t - \_\_)^2$

Solve by completing the square. Show your work.

47. $x^2 + 6x = 7$

48. $x^2 + 8x = 9$

49. $t^2 - 10t = -24$

50. $t^2 - 10t = -21$

51. $x^2 + 10x + 9 = 0$

52. $x^2 + 8x + 7 = 0$

53. $t^2 + 8t - 3 = 0$

54. $t^2 + 6t - 5 = 0$

Complete the square to find the x-intercepts of each function given by the equation listed.

55. $f(x) = x^2 + 6x + 7$

56. $f(x) = x^2 + 5x + 3$

57. $g(x) = x^2 + 12x + 25$

58. $g(x) = x^2 + 4x + 2$

59. $f(x) = x^2 - 10x - 22$

60. $f(x) = x^2 - 8x - 10$

Solve by completing the square. Remember to first divide, as in Example 11, to make sure that the coefficient of x^2 is 1.

61. $9x^2 + 18x = -8$

62. $4x^2 + 8x = -3$

63. $3x^2 - 5x - 2 = 0$

64. $2x^2 - 5x - 3 = 0$

65. $5x^2 + 4x - 3 = 0$

66. $4x^2 + 3x - 5 = 0$

67. Find the x-intercepts of the function given by $f(x) = 4x^2 + 2x - 3$.

68. Find the x-intercepts of the function given by $f(x) = 3x^2 + x - 5$.

69. Find the x-intercepts of the function given by $g(x) = 2x^2 - 3x - 1$.

70. Find the x-intercepts of the function given by $g(x) = 3x^2 - 5x - 1$.

Interest. Use $A = P(1 + r)^t$ to find the interest rate in Exercises 71–76. Refer to Example 12.

71. \$2000 grows to \$2420 in 2 yr

72. \$2560 grows to \$2890 in 2 yr

73. \$1280 grows to \$1805 in 2 yr

74. \$1000 grows to \$1440 in 2 yr

75. \$6250 grows to \$6760 in 2 yr

76. \$6250 grows to \$7290 in 2 yr

Free-Falling Objects. Use $s = 16t^2$ for Exercises 77–80. Refer to Example 13 and neglect air resistance.

77. Suspended 1053 ft above the water, the bridge over Colorado's Royal Gorge is the world's highest suspension bridge (*Source*: *The Guinness Book of Records*). How long would it take an object to fall freely from the bridge?

78. The CN Tower in Toronto, at 1815 ft, is the world's tallest self-supporting tower on land (no guy wires) (*Source*: *The Guinness Book of Records*). How long would it take an object to fall freely from the top?

79. The Sears Tower in Chicago is 1454 ft tall. How long would it take an object to fall freely from the top?

80. The Gateway Arch in St. Louis is 630 ft high (*Source*: www.icivilengineer.com). How long would it take an object to fall freely from the top?

TW **81.** Explain in your own words a sequence of steps that can be used to solve any quadratic equation in the quickest way.

TW **82.** Describe how to write a quadratic equation that can be solved algebraically but not graphically.

Skill Maintenance

Evaluate. [1.2]

83. $at^2 - bt$, for $a = 3$, $b = 5$, and $t = 4$

84. $mn^2 - mp$, for $m = -2$, $n = 7$, and $p = 3$

Simplify. [7.3]

85. $\sqrt[3]{270}$

86. $\sqrt{80}$

Let $f(x) = \sqrt{3x - 5}$. *Find each of the following.* [7.1]

87. $f(10)$

88. $f(18)$

Synthesis

TW **89.** What would be better: to receive 3% interest every 6 months or to receive 6% interest every 12 months? Why?

TW **90.** Example 12 was solved with a graphing calculator by graphing each side of

$$4410 = 4000(1 + r)^2.$$

How could you determine, from a reading of the problem, a suitable viewing window?

Find b such that each trinomial is a square.

91. $x^2 + bx + 81$

92. $x^2 + bx + 49$

93. If $f(x) = 2x^5 - 9x^4 - 66x^3 + 45x^2 + 280x$ and $x^2 - 5$ is a factor of $f(x)$, find all a for which $f(a) = 0$.

94. If

$$f(x) = \left(x - \tfrac{1}{3}\right)(x^2 + 6)$$

and

$$g(x) = \left(x - \tfrac{1}{3}\right)\left(x^2 - \tfrac{2}{3}\right),$$

find all a for which $(f + g)(a) = 0$.

95. *Boating.* A barge and a fishing boat leave a dock at the same time, traveling at a right angle to each other. The barge travels 7 km/h slower than the fishing boat. After 4 hr, the boats are 68 km apart. Find the speed of each boat.

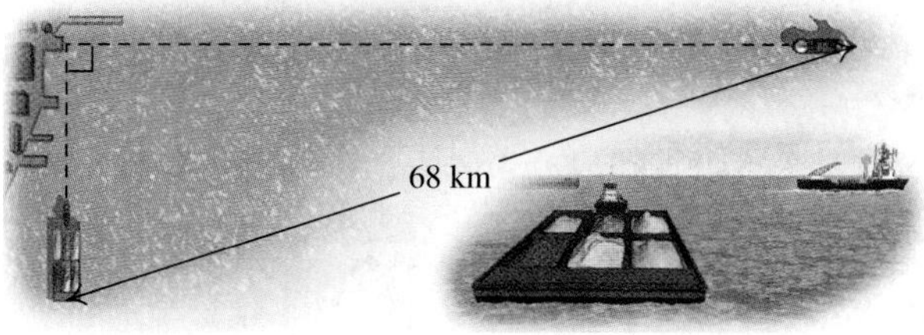

96. Find three consecutive integers such that the square of the first plus the product of the other two is 67.

8.2 The Quadratic Formula

Solving Using the Quadratic Formula ■ Approximating Solutions

There are at least two reasons for learning to complete the square. One is to help graph certain equations that appear later in this chapter. Another is to develop a general formula for solving quadratic equations.

Solving Using the Quadratic Formula

Each time we solve by completing the square, the procedure is the same. In mathematics, when a procedure is repeated many times, a formula can often be developed to speed up our work.

We begin with a quadratic equation in standard form,

$$ax^2 + bx + c = 0,$$

Study Tip

Know It "By Heart"

When memorizing something like the quadratic formula, try to first understand and write out the derivation. Doing this will help you remember the formula.

with $a > 0$. For $a < 0$, a slightly different derivation is needed (see Exercise 58), but the result is the same. Let's solve by completing the square. As the steps are performed, compare them with Example 11 on p. 617.

$$ax^2 + bx = -c \qquad \textbf{Adding } -c \textbf{ to both sides}$$

$$x^2 + \frac{b}{a}x = -\frac{c}{a} \qquad \textbf{Dividing both sides by } a$$

Half of $\frac{b}{a}$ is $\frac{b}{2a}$ and $\left(\frac{b}{2a}\right)^2$ is $\frac{b^2}{4a^2}$. We add $\frac{b^2}{4a^2}$ to both sides:

$$x^2 + \frac{b}{a}x + \frac{b^2}{4a^2} = -\frac{c}{a} + \frac{b^2}{4a^2} \qquad \textbf{Adding } \frac{b^2}{4a^2} \textbf{ to complete the square}$$

$$\left(x + \frac{b}{2a}\right)^2 = -\frac{4ac}{4a^2} + \frac{b^2}{4a^2} \qquad \textbf{Factoring on the left side; finding a common denominator on the right side}$$

$$\left(x + \frac{b}{2a}\right)^2 = \frac{b^2 - 4ac}{4a^2}$$

$$x + \frac{b}{2a} = \pm\frac{\sqrt{b^2 - 4ac}}{2a} \qquad \textbf{Using the principle of square roots and the quotient rule for radicals; since } a > 0, \sqrt{4a^2} = 2a$$

$$x = \frac{-b \pm \sqrt{b^2 - 4ac}}{2a}. \qquad \textbf{Adding } -\frac{b}{2a} \textbf{ to both sides}$$

It is important that you remember the quadratic formula and know how to use it.

The Quadratic Formula The solutions of $ax^2 + bx + c = 0$, $a \neq 0$, are given by

$$x = \frac{-b \pm \sqrt{b^2 - 4ac}}{2a}.$$

EXAMPLE 1 Solve $5x^2 + 8x = -3$ using the quadratic formula.

SOLUTION We first find standard form and determine a, b, and c:

$$5x^2 + 8x + 3 = 0; \qquad \textbf{Adding 3 to both sides to get 0 on one side}$$

$$a = 5, \quad b = 8, \quad c = 3.$$

Next, we use the quadratic formula:

$$x = \frac{-b \pm \sqrt{b^2 - 4ac}}{2a}$$

$$x = \frac{-8 \pm \sqrt{8^2 - 4 \cdot 5 \cdot 3}}{2 \cdot 5} \qquad \textbf{Substituting}$$

Be sure to write the fraction bar all the way across.

$$x = \frac{-8 \pm \sqrt{64 - 60}}{10}$$

$$x = \frac{-8 \pm \sqrt{4}}{10} = \frac{-8 \pm 2}{10}$$

$$x = \frac{-8 + 2}{10} \quad or \quad x = \frac{-8 - 2}{10}$$

$$x = \frac{-6}{10} \quad or \quad x = \frac{-10}{10}$$

$$x = -\frac{3}{5} \quad or \quad x = -1.$$

The solutions are $-\frac{3}{5}$ and -1. The checks are left to the student.

Because $5x^2 + 8x + 3$ can be factored as $(5x + 3)(x + 1)$, the quadratic formula may not have been the fastest way of solving Example 1. However, because the quadratic formula works for *any* quadratic equation, we need not spend too much time struggling to solve a quadratic equation by factoring.

To Solve a Quadratic Equation

1. If the equation can be easily written in the form $ax^2 = p$ or $(x + k)^2 = d$, use the principle of square roots as in Section 8.1.
2. If step (1) does not apply, write the equation in the form $ax^2 + bx + c = 0$.
3. Try factoring and using the principle of zero products.
4. If factoring seems to be difficult or impossible, use the quadratic formula. Completing the square can also be used, but is usually slower.

The solutions of a quadratic equation can always be found using the quadratic formula. They cannot always be found by factoring.

Recall that a second-degree polynomial in one variable is said to be quadratic. Similarly, a second-degree polynomial function in one variable is said to be a **quadratic function.**

EXAMPLE 2 For the quadratic function given by $f(x) = 3x^2 - 6x - 4$, find all x for which $f(x) = 0$.

SOLUTION We substitute and solve for x:

$$f(x) = 0$$

$$3x^2 - 6x - 4 = 0 \qquad \text{Substituting. You can try to solve this by factoring.}$$

$$a = 3, \quad b = -6, \quad c = -4.$$

We then substitute into the quadratic formula:

$$x = \frac{-(-6) \pm \sqrt{(-6)^2 - 4 \cdot 3 \cdot (-4)}}{2 \cdot 3}$$

$$= \frac{6 \pm \sqrt{36 + 48}}{6}$$

$$= \frac{6 \pm \sqrt{84}}{6}$$

$$= \frac{6}{6} \pm \frac{\sqrt{84}}{6}$$ **Note that 4 is a perfect-square factor of 84.**

$$= 1 \pm \frac{\sqrt{4}\sqrt{21}}{6}$$ **$84 = 4 \cdot 21$**

$$= 1 \pm \frac{2\sqrt{21}}{6}$$

$$= 1 \pm \frac{\sqrt{21}}{3}.$$ **Simplifying**

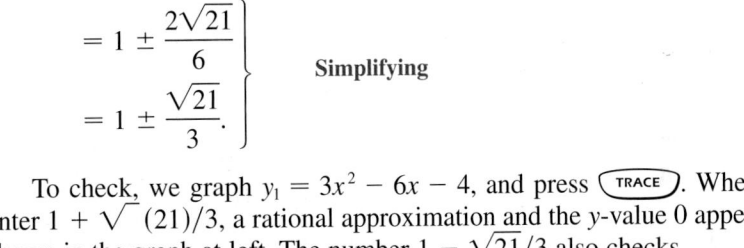

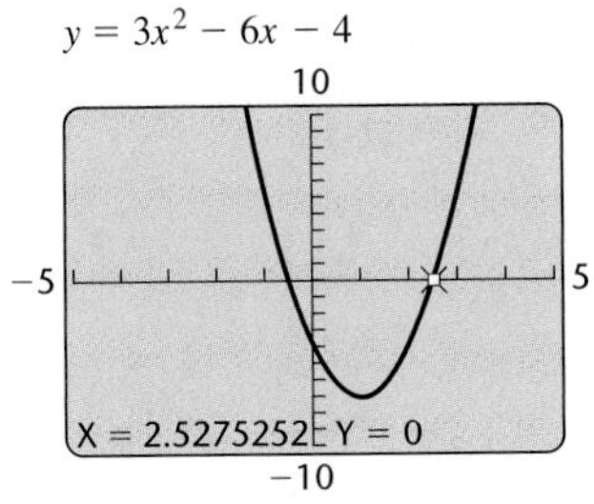

To check, we graph $y_1 = 3x^2 - 6x - 4$, and press TRACE. When we enter $1 + \sqrt{\ }(21)/3$, a rational approximation and the y-value 0 appear, as shown in the graph at left. The number $1 - \sqrt{21}/3$ also checks.

The solutions are $1 - \frac{\sqrt{21}}{3}$ and $1 + \frac{\sqrt{21}}{3}$.

Some quadratic equations have solutions that are imaginary numbers.

EXAMPLE 3 Solve: $x^2 + 2 = -x$.

SOLUTION We first find standard form:

$$x^2 + x + 2 = 0.$$ **Adding x to both sides**

Since we cannot factor $x^2 + x + 2$, we use the quadratic formula with $a = 1$, $b = 1$, and $c = 2$:

$$x = \frac{-1 \pm \sqrt{1^2 - 4 \cdot 1 \cdot 2}}{2 \cdot 1}$$ **Substituting**

$$= \frac{-1 \pm \sqrt{1 - 8}}{2} = \frac{-1 \pm \sqrt{-7}}{2}$$

$$= \frac{-1 \pm i\sqrt{7}}{2}, \text{ or } -\frac{1}{2} \pm \frac{\sqrt{7}}{2}i$$

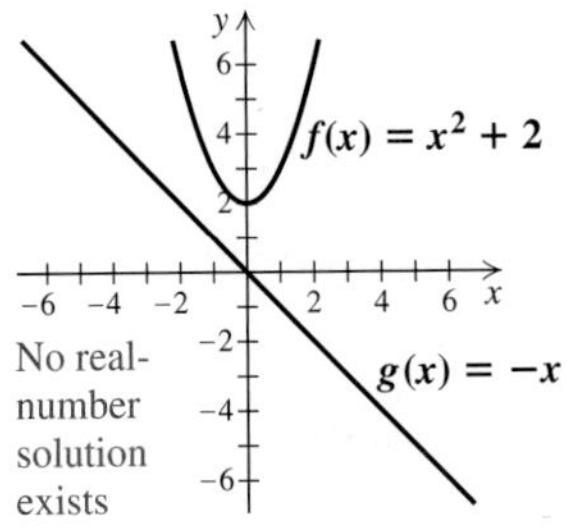

A visualization of Example 3

The solutions are $-\frac{1}{2} - \frac{\sqrt{7}}{2}i$ and $-\frac{1}{2} + \frac{\sqrt{7}}{2}i$. The checks are left to the student.

Note from the graph at left that the graphs of $f(x) = x^2 + 2$ and $g(x) = -x$ do not intersect. Thus there are no real-number solutions of the equation.

The quadratic formula can be used to solve certain rational equations.

EXAMPLE 4 If $f(x) = 2 + \frac{7}{x}$ and $g(x) = \frac{4}{x^2}$, find all x for which $f(x) = g(x)$.

SOLUTION We set $f(x)$ equal to $g(x)$ and solve:

$$f(x) = g(x)$$

$$2 + \frac{7}{x} = \frac{4}{x^2}. \quad \textbf{Substituting. Note that } x \neq 0.$$

This is a rational equation similar to those in Section 6.4. To solve, we multiply both sides by the LCD, x^2:

$$x^2\left(2 + \frac{7}{x}\right) = x^2 \cdot \frac{4}{x^2}$$

$$2x^2 + 7x = 4 \quad \textbf{Simplifying}$$

$$2x^2 + 7x - 4 = 0. \quad \textbf{Subtracting 4 from both sides}$$

We have

$$a = 2, \quad b = 7, \quad \text{and} \quad c = -4.$$

Substituting then gives us

$$x = \frac{-7 \pm \sqrt{7^2 - 4 \cdot 2 \cdot (-4)}}{2 \cdot 2}$$

$$= \frac{-7 \pm \sqrt{49 + 32}}{4} = \frac{-7 \pm \sqrt{81}}{4} = \frac{-7 \pm 9}{4}$$

$$x = \frac{-7 + 9}{4} = \frac{1}{2} \quad or \quad x = \frac{-7 - 9}{4} = -4. \quad \textbf{Both answers should check since } x \neq 0.$$

You can confirm that $f\left(\frac{1}{2}\right) = g\left(\frac{1}{2}\right)$ and $f(-4) = g(-4)$. The solutions are $\frac{1}{2}$ and -4.

Approximating Solutions

When the solution of an equation is irrational, a rational-number approximation is often useful. This is often the case in real-world applications similar to those found in Section 8.3.

Student Notes

It is important that you understand both the rules for order of operations *and* the manner in which your calculator applies those rules.

EXAMPLE 5 Use a calculator to approximate the solutions of Example 2.

SOLUTION On most graphing calculators, the following sequence of keystrokes can be used to approximate $1 + \sqrt{21}/3$:

1 + √ 2 1) ÷ 3 ENTER.

Similar keystrokes can be used to approximate $1 - \sqrt{21}/3$.

The solutions are approximately 2.527525232 and −0.5275252317.

8.2 EXERCISE SET

FOR EXTRA HELP

MathXL | MyMathLab | InterAct Math | AW Math Tutor Center | Video Lectures on CD: Disc 4 | Student's Solutions Manual

Concept Reinforcement *Classify each of the following as either true or false.*

1. The quadratic formula can be used to solve *any* quadratic equation.
2. The quadratic formula does not work if solutions are imaginary numbers.
3. Solving by factoring is always slower than using the quadratic formula.
4. The steps used to derive the quadratic formula are the same as those used when solving by completing the square.
5. A quadratic equation can have as many as four solutions.
6. It is possible for a quadratic equation to have no real-number solutions.

Solve.

7. $x^2 + 7x - 3 = 0$
8. $x^2 - 7x + 4 = 0$
9. $3p^2 = 18p - 6$
10. $3u^2 = 8u - 5$
11. $x^2 + x + 1 = 0$
12. $x^2 + x + 2 = 0$
13. $x^2 + 13 = 4x$
14. $x^2 + 13 = 6x$
15. $h^2 + 4 = 6h$
16. $r^2 + 3r = 8$
17. $\frac{1}{x^2} - 3 = \frac{8}{x}$
18. $\frac{9}{x} - 2 = \frac{5}{x^2}$
19. $3x + x(x - 2) = 4$
20. $4x + x(x - 3) = 5$
21. $12t^2 + 9t = 1$
22. $15t^2 + 7t = 2$
23. $25x^2 - 20x + 4 = 0$
24. $36x^2 + 84x + 49 = 0$
25. $7x(x + 2) + 5 = 3x(x + 1)$
26. $5x(x - 1) - 7 = 4x(x - 2)$
27. $14(x - 4) - (x + 2) = (x + 2)(x - 4)$
28. $11(x - 2) + (x - 5) = (x + 2)(x - 6)$
29. $5x^2 = 13x + 17$
30. $25x = 3x^2 + 28$
31. $x^2 + 9 = 4x$
32. $x^2 + 7 = 3x$
33. $x^3 - 8 = 0$ (*Hint*: Factor the difference of cubes. Then use the quadratic formula.)
34. $x^3 + 1 = 0$
35. Let $f(x) = 3x^2 - 5x + 2$. Find x such that $f(x) = 0$.
36. Let $g(x) = 4x^2 - 2x - 3$. Find x such that $g(x) = 0$.
37. Let
$$f(x) = \frac{7}{x} + \frac{7}{x + 4}.$$
Find all x for which $f(x) = 1$.
38. Let
$$g(x) = \frac{2}{x} + \frac{2}{x + 3}.$$
Find all x for which $g(x) = 1$.
39. Let
$$F(x) = \frac{x + 3}{x} \quad \text{and} \quad G(x) = \frac{x - 4}{3}.$$
Find all x for which $F(x) = G(x)$.
40. Let
$$f(x) = \frac{3 - x}{4} \quad \text{and} \quad g(x) = \frac{1}{4x}.$$
Find all x for which $f(x) = g(x)$.
41. Let
$$f(x) = \frac{15 - 2x}{6} \quad \text{and} \quad g(x) = \frac{3}{x}.$$
Find all x for which $f(x) = g(x)$.

42. Let

$$f(x) = x + 5 \quad \text{and} \quad g(x) = \frac{3}{x - 5}.$$

Find all x for which $f(x) = g(x)$.

Solve. Use a calculator to approximate, as precisely as possible, the solutions as rational numbers.

43. $x^2 + 4x - 7 = 0$

44. $x^2 + 6x + 4 = 0$

45. $x^2 - 6x + 4 = 0$

46. $x^2 - 4x + 1 = 0$

47. $2x^2 - 3x - 7 = 0$

48. $3x^2 - 3x - 2 = 0$

TW **49.** Are there any equations that can be solved by the quadratic formula but not by completing the square? Why or why not?

TW **50.** If you had to choose between remembering the method of completing the square and remembering the quadratic formula, which would you choose? Why?

Skill Maintenance

Simplify.

51. $\dfrac{x^2 + xy}{2x}$ [6.1]

52. $\dfrac{a^3 - ab^2}{ab}$ [6.1]

53. $\sqrt{27a^2b^5} \cdot \sqrt{6a^3b}$ [7.3]

54. $\sqrt{8a^3b} \cdot \sqrt{12ab^5}$ [7.3]

55. $\dfrac{\dfrac{3}{x-1}}{\dfrac{1}{x+1} + \dfrac{2}{x-1}}$ [6.3]

56. $\dfrac{\dfrac{4}{a^2b}}{\dfrac{3}{a} - \dfrac{4}{b^2}}$ [6.3]

Synthesis

TW **57.** Suppose you had a large number of quadratic equations to solve and none of the equations had a constant term. Would you use factoring or the quadratic formula to solve these equations? Why?

TW **58.** If $a < 0$ and $ax^2 + bx + c = 0$, then $-a$ is positive and the equivalent equation, $-ax^2 - bx - c = 0$, can be solved using the quadratic formula.

a) Find this solution, replacing a, b, and c in the formula with $-a$, $-b$, and $-c$ from the equation.

b) How does the result of part (a) indicate that the quadratic formula "works" regardless of the sign of a?

For Exercises 59–61, let

$$f(x) = \frac{x^2}{x - 2} + 1 \quad \textit{and} \quad g(x) = \frac{4x - 2}{x - 2} + \frac{x + 4}{2}.$$

59. Find the x-intercepts of the graph of f.

60. Find the x-intercepts of the graph of g.

61. Find all x for which $f(x) = g(x)$.

Solve.

62. $x^2 - 0.75x - 0.5 = 0$

63. $z^2 + 0.84z - 0.4 = 0$

64. $(1 + \sqrt{3})x^2 - (3 + 2\sqrt{3})x + 3 = 0$

65. $\sqrt{2}x^2 + 5x + \sqrt{2} = 0$

66. $ix^2 - 2x + 1 = 0$

67. One solution of $kx^2 + 3x - k = 0$ is -2. Find the other.

TW **68.** Can a graph be used to solve *any* quadratic equation? Why or why not?

TW **69.** Solve Example 2 graphically and compare with the algebraic solution. Which method is faster? Which method is more precise?

TW **70.** Solve Example 4 graphically and compare with the algebraic solution. Which method is faster? Which method is more precise?

8.3 Applications Involving Quadratic Equations

Solving Problems ■ Solving Formulas

Solving Problems

As we found in Section 6.5, some problems translate to rational equations. The solution of such rational equations can involve quadratic equations.

EXAMPLE 1 Motorcycle Travel. Madison rode her motorcycle 300 mi at a certain average speed. Had she averaged 10 mph more, the trip would have taken 1 hr less. Find the average speed of the motorcycle.

SOLUTION

1. **Familiarize.** We make a drawing, labeling it with the information provided. As in Section 6.5, we can create a table. We let r represent the rate, in miles per hour, and t the time, in hours, for Madison's trip.

Distance	Speed	Time
300	r	t
300	$r+10$	$t-1$

$\longrightarrow r = \dfrac{300}{t}$

$\longrightarrow r + 10 = \dfrac{300}{t-1}$

Recall that the definition of speed, $r = d/t$, relates the three quantities.

2. **Translate.** From the first two lines of the table, we obtain

$$r = \frac{300}{t} \quad \text{and} \quad r + 10 = \frac{300}{t-1}.$$

Study Tip

Keep Your Book Handy

Although most of your studying should be done in large blocks of time, you can supplement this by using smaller blocks of time for reviewing material. By keeping your text and notebook nearby (in your car or book bag), you can review course material in a doctor's waiting room, while sitting through a youngster's soccer practice, while riding a bus, or whenever you find yourself with some free time.

3. **Carry out.** A system of equations has been formed. We substitute for r from the first equation into the second and solve the resulting equation:

$$\frac{300}{t} + 10 = \frac{300}{t-1}$$ **Substituting $300/t$ for r**

$$t(t-1)\cdot\left[\frac{300}{t} + 10\right] = t(t-1)\cdot\frac{300}{t-1}$$ **Multiplying by the LCD**

$$\cancel{t}(t-1)\cdot\frac{300}{\cancel{t}} + t(t-1)\cdot 10 = t\cancel{(t-1)}\cdot\frac{300}{\cancel{t-1}}$$ **Using the distributive law and removing factors that equal 1: $\frac{t}{t} = 1; \frac{t-1}{t-1} = 1$**

$$\left.\begin{aligned} 300(t-1) + 10(t^2 - t) &= 300t \\ 300t - 300 + 10t^2 - 10t &= 300t \\ 10t^2 - 10t - 300 &= 0 \end{aligned}\right\}$$ **Rewriting in standard form**

$$t^2 - t - 30 = 0$$ **Multiplying by $\frac{1}{10}$ or dividing by 10**

$$(t-6)(t+5) = 0$$ **Factoring**

$$t = 6 \quad or \quad t = -5.$$ **Principle of zero products**

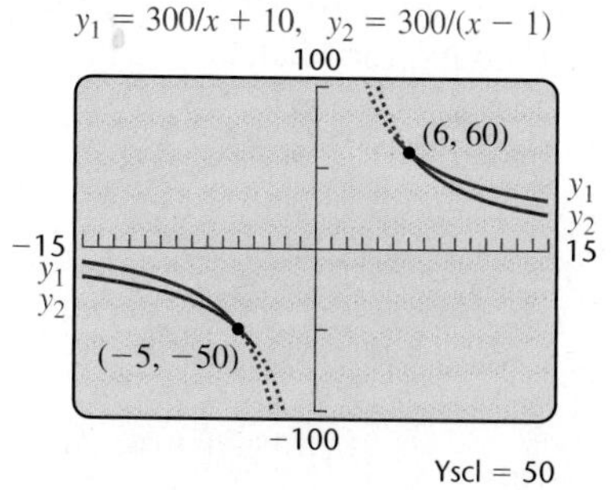

4. **Check.** As a partial check, we graph $y_1 = 300/x + 10$ and $y_2 = 300/(x-1)$. The graph at left shows that the solutions are −5 and 6.

Note that we have solved for t, not r as required. Since negative time has no meaning here, we disregard the −5 and use 6 hr to find r:

$$r = \frac{300 \text{ mi}}{6 \text{ hr}} = 50 \text{ mph}.$$

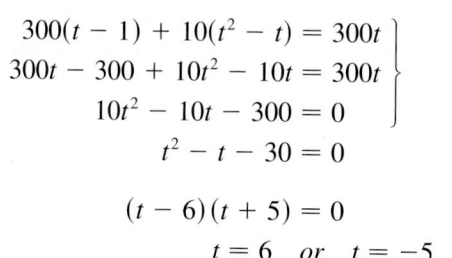

> **CAUTION!** Always make sure that you find the quantity asked for in the problem.

To see if 50 mph checks, we increase the speed 10 mph to 60 mph and see how long the trip would have taken at that speed:

$$t = \frac{d}{r} = \frac{300 \text{ mi}}{60 \text{ mph}} = 5 \text{ hr}.$$ **Note that mi/mph = mi ÷ $\frac{\text{mi}}{\text{hr}}$ = $\cancel{\text{mi}}\cdot\frac{\text{hr}}{\cancel{\text{mi}}}$ = hr.**

This is 1 hr less than the trip actually took, so the answer checks.

5. **State.** Madison's motorcycle traveled at an average speed of 50 mph.

Solving Formulas

Recall that to solve a formula for a certain letter, we use the principles for solving equations to get that letter alone on one side.

EXAMPLE 2 Period of a Pendulum. The time T required for a pendulum of length l to swing back and forth (complete one period) is given by the formula $T = 2\pi\sqrt{l/g}$, where g is the earth's gravitational constant. Solve for l.

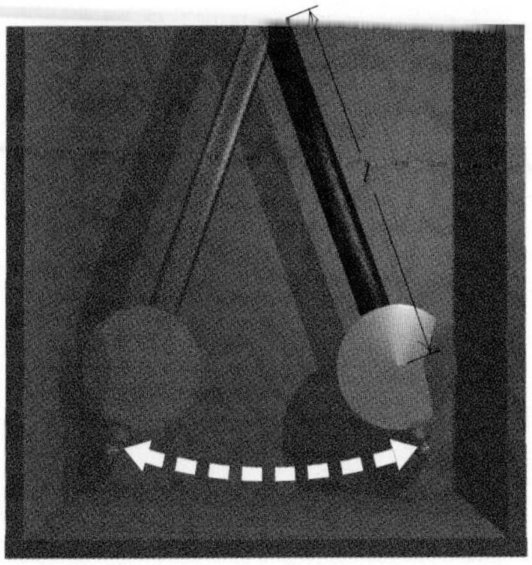

SOLUTION We have

$$T = 2\pi\sqrt{\frac{l}{g}} \qquad \textbf{This is a radical equation (see Section 7.6).}$$

$$T^2 = \left(2\pi\sqrt{\frac{l}{g}}\right)^2 \qquad \textbf{Principle of powers (squaring both sides)}$$

$$T^2 = 2^2\pi^2\frac{l}{g}$$

$$gT^2 = 4\pi^2 l \qquad \textbf{Multiplying both sides by } g \textbf{ to clear fractions}$$

$$\frac{gT^2}{4\pi^2} = l. \qquad \textbf{Dividing both sides by } 4\pi^2$$

We now have l alone on one side and l does not appear on the other side, so the formula is solved for l.

In formulas for which variables represent nonnegative numbers, there is no need for absolute-value signs when taking square roots.

EXAMPLE 3 Hang Time.* An athlete's *hang time* is the amount of time that the athlete can remain airborne when jumping. A formula relating an athlete's vertical leap V, in inches, to hang time T, in seconds, is $V = 48T^2$. Solve for T.

---

*This formula is taken from an article by Peter Brancazio, "The Mechanics of a Slam Dunk," *Popular Mechanics*, November 1991. Courtesy of Professor Peter Brancazio, Brooklyn College.

SOLUTION

$$48T^2 = V$$

$$T^2 = \frac{V}{48} \quad \textbf{Dividing by 48 to get } T^2 \textbf{ alone}$$

$$T = \frac{\sqrt{V}}{\sqrt{48}} \quad \textbf{Using the principle of square roots and the quotient rule for radicals. We assume } V, T \geq 0.$$

$$= \frac{\sqrt{V}}{\sqrt{16}\sqrt{3}} = \frac{\sqrt{V}}{4\sqrt{3}}$$

$$= \frac{\sqrt{V}}{4\sqrt{3}} \cdot \frac{\sqrt{3}}{\sqrt{3}} = \frac{\sqrt{3V}}{12}.$$

Rationalizing the denominator

EXAMPLE 4 Falling Distance. An object tossed downward with an initial speed (velocity) of v_0 will travel a distance of s meters, where $s = 4.9t^2 + v_0t$ and t is measured in seconds. Solve for t.

SOLUTION Since t is squared in one term and raised to the first power in the other term, the equation is quadratic in t.

$$4.9t^2 + v_0t = s$$

$$4.9t^2 + v_0t - s = 0 \quad \textbf{Writing standard form}$$

$$a = 4.9, \quad b = v_0, \quad c = -s$$

$$t = \frac{-v_0 \pm \sqrt{v_0^2 - 4(4.9)(-s)}}{2(4.9)} \quad \textbf{Using the quadratic formula}$$

Since the negative square root would yield a negative value for t, we use only the positive root:

$$t = \frac{-v_0 + \sqrt{v_0^2 + 19.6s}}{9.8}.$$

The following list of steps should help you when solving formulas for a given letter. Try to remember that when solving a formula, you use the same approach that you would to solve an equation.

Student Notes

After identifying which numbers to use as a, b, and c, be careful to replace only the *letters* in the quadratic formula.

To Solve a Formula for a Letter—Say, b

1. Clear fractions and use the principle of powers, as needed. Perform these steps until radicals containing b are gone and b is not in any denominator.
2. Combine all like terms.
3. If the only power of b is b^1, the equation can be solved as in Sections 1.6 and 6.8.
4. If b^2 appears but b does not, solve for b^2 and use the principle of square roots to solve for b.
5. If there are terms containing both b and b^2, put the equation in standard form and use the quadratic formula.

8.3 EXERCISE SET

FOR EXTRA HELP

MathXL | MyMathLab | InterAct Math | AW Math Tutor Center | Video Lectures on CD: Disc 4 | Student's Solutions Manual

Solve.

1. *Car Trips.* During the first part of a trip, Raena's Honda traveled 120 mi at a certain speed. Raena then drove another 100 mi at a speed that was 10 mph slower. If the total time of Raena's trip was 4 hr, what was her speed on each part of the trip?

2. *Canoeing.* During the first part of a canoe trip, Alex covered 60 km at a certain speed. He then traveled 24 km at a speed that was 4 km/h slower. If the total time for the trip was 8 hr, what was the speed on each part of the trip?

3. *Car Trips.* Petra's Plymouth travels 200 mi averaging a certain speed. If the car had gone 10 mph faster, the trip would have taken 1 hr less. Find Petra's average speed.

4. *Car Trips.* Sandi's Subaru travels 280 mi averaging a certain speed. If the car had gone 5 mph faster, the trip would have taken 1 hr less. Find Sandi's average speed.

5. *Air Travel.* A Cessna flies 600 mi at a certain speed. A Beechcraft flies 1000 mi at a speed that is 50 mph faster, but takes 1 hr longer. Find the speed of each plane.

6. *Air Travel.* A turbo-jet flies 50 mph faster than a super-prop plane. If a turbo-jet goes 2000 mi in 3 hr less time than it takes the super-prop to go 2800 mi, find the speed of each plane.

7. *Bicycling.* Naoki bikes the 40 mi to Hillsboro averaging a certain speed. The return trip is made at a speed that is 6 mph slower. Total time for the round trip is 14 hr. Find Naoki's average speed on each part of the trip.

8. *Car Speed.* On a sales trip, Gail drives the 600 mi to Richmond averaging a certain speed. The return trip is made at an average speed that is 10 mph slower. Total time for the round trip is 22 hr. Find Gail's average speed on each part of the trip.

9. *Navigation.* The Hudson River flows at a rate of 3 mph. A patrol boat travels 60 mi upriver and returns in a total time of 9 hr. What is the speed of the boat in still water?

10. *Navigation.* The current in a typical Mississippi River shipping route flows at a rate of 4 mph. In order for a barge to travel 24 mi upriver and then return in a total of 5 hr, approximately how fast must the barge be able to travel in still water?

11. *Filling a Pool.* A well and a spring are filling a swimming pool. Together, they can fill the pool in 4 hr. The well, working alone, can fill the pool in 6 hr less time than the spring. How long would the spring take, working alone, to fill the pool?

12. *Filling a Tank.* Two pipes are connected to the same tank. Working together, they can fill the tank in 2 hr. The larger pipe, working alone, can fill the tank in 3 hr less time than the smaller one. How long would the smaller one take, working alone, to fill the tank?

13. *Paddleboats.* Ellen paddles 1 mi upstream and 1 mi back in a total time of 1 hr. The speed of the river is 2 mph. Find the speed of Ellen's paddleboat in still water.

14. *Rowing.* Dan rows 10 km upstream and 10 km back in a total time of 3 hr. The speed of the river is 5 km/h. Find Dan's speed in still water.

Solve each formula for the indicated letter. Assume that all variables represent nonnegative numbers.

15. $A = 4\pi r^2$, for r
(Surface area of a sphere of radius r)

16. $A = 6s^2$, for s
(Surface area of a cube with sides of length s)

17. $A = 2\pi r^2 + 2\pi rh$, for r
(Surface area of a right cylindrical solid with radius r and height h)

18. $F = \dfrac{Gm_1m_2}{r^2}$, for r
(Law of gravity)

19. $N = \dfrac{kQ_1Q_2}{s^2}$, for s
(Number of phone calls between two cities)

20. $A = \pi r^2$, for r
(Area of a circle)

21. $T = 2\pi\sqrt{\dfrac{l}{g}}$, for g
(A pendulum formula)

22. $a^2 + b^2 = c^2$, for b
(Pythagorean formula in two dimensions)

23. $a^2 + b^2 + c^2 = d^2$, for c
(Pythagorean formula in three dimensions)

24. $N = \dfrac{k^2 - 3k}{2}$, for k
(Number of diagonals of a polygon with k sides)

25. $s = v_0t + \dfrac{gt^2}{2}$, for t
(A motion formula)

26. $A = \pi r^2 + \pi rs$, for r
(Surface area of a cone)

27. $N = \frac{1}{2}(n^2 - n)$, for n
(Number of games if n teams play each other once)

28. $A = A_0(1 - r)^2$, for r
(A business formula)

29. $V = 3.5\sqrt{h}$, for h
(Distance to horizon from a height)

30. $W = \sqrt{\dfrac{1}{LC}}$, for L
(An electricity formula)

Aha! **31.** $at^2 + bt + c = 0$, for t
(An algebraic formula)

32. $A = P_1(1 + r)^2 + P_2(1 + r)$, for r
(Amount in an account when P_1 is invested for 2 yr and P_2 for 1 yr at interest rate r)

Solve.

33. *Falling Distance.* (Use $4.9t^2 + v_0t = s$.)

a) A bolt falls off an airplane at an altitude of 500 m. Approximately how long does it take the bolt to reach the ground?

b) A ball is thrown downward at a speed of 30 m/sec from an altitude of 500 m. Approximately how long does it take the ball to reach the ground?

c) Approximately how far will an object fall in 5 sec, when thrown downward at an initial velocity of 30 m/sec from a plane?

34. *Falling Distance.* (Use $4.9t^2 + v_0t = s$.)

a) A ring is dropped from a helicopter at an altitude of 75 m. Approximately how long does it take the ring to reach the ground?

b) A coin is tossed downward with an initial velocity of 30 m/sec from an altitude of 75 m. Approximately how long does it take the coin to reach the ground?

c) Approximately how far will an object fall in 2 sec, if thrown downward at an initial velocity of 20 m/sec from a helicopter?

35. *Bungee Jumping.* Jesse is tied to one end of a 40-m elasticized (bungee) cord. The other end of the cord is tied to the middle of a bridge. If Jesse jumps off the bridge, for how long will he fall before the cord begins to stretch? (Use $4.9t^2 = s$.)

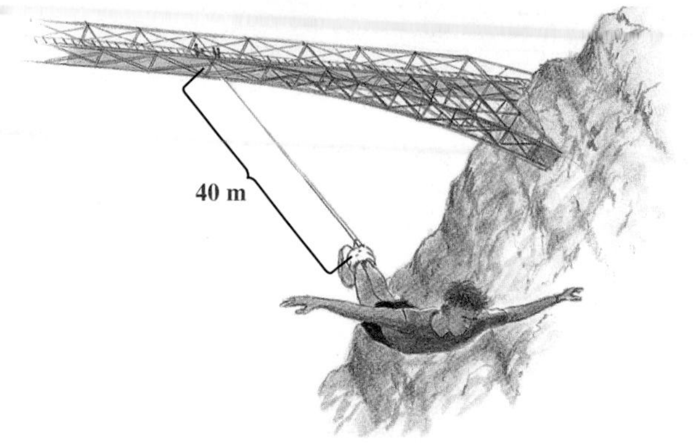

36. *Bungee Jumping.* Sheila is tied to a bungee cord (see Exercise 35) and falls for 2.5 sec before her cord begins to stretch. How long is the bungee cord?

37. *Hang Time.* The NBA's Steve Francis reportedly has a vertical leap of 45 in. (*Source*: www.maximonline.com). What is his hang time? (Use $V = 48T^2$.)

38. *League Schedules.* In a bowling league, each team plays each of the other teams once. If a total of 66 games is played, how many teams are in the league? (See Exercise 27.)

For Exercises 39 and 40, use $4.9t^2 + v_0t = s$.

39. *Downward Speed.* An object thrown downward from a 100-m cliff travels 51.6 m in 3 sec. What was the initial velocity of the object?

40. *Downward Speed.* An object thrown downward from a 200-m cliff travels 91.2 m in 4 sec. What was the initial velocity of the object?

For Exercises 41 and 42, use
$A = P_1(1 + r)^2 + P_2(1 + r)$. *(See Exercise 32.)*

41. *Compound Interest.* A firm invests \$3000 in a savings account for 2 yr. At the beginning of the second year, an additional \$1700 is invested. If a total of \$5253.70 is in the account at the end of the second year, what is the annual interest rate?

42. *Compound Interest.* A business invests \$10,000 in a savings account for 2 yr. At the beginning of the second year, an additional \$3500 is invested. If a total of \$15,569.75 is in the account at the end of the second year, what is the annual interest rate?

TW 43. Marti is tied to a bungee cord that is twice as long as the cord tied to Pedro. Will Marti's fall take twice as long as Pedro's before their cords begin to stretch? Why or why not? (See Exercises 35 and 36.)

TW 44. Under what circumstances would a negative value for t, time, have meaning?

Focused Review

Solve.

45. *Credits.* Tivon transferred to Oak College with 32 credits, and is taking 16 credits a semester. Alexis transferred with 38 credits, and is taking 14 credits a semester. Determine, in terms of an inequality, when Tivon will have more credits than Alexis. [4.1]

46. *Donuts.* South Street Bakers charges \$1.10 for a cream-filled donut and 85¢ for a glazed donut. On a recent Sunday, a total of 90 glazed and cream-filled donuts were sold for \$88.00. How many of each type were sold? [3.3]

47. *Picture Frames.* A rectangular picture frame measures 10 in. by 13 in., and 88 in^2 of picture shows. Find the width of the frame. [5.8]

48. *Fast Food.* Jaime can process customer orders twice as fast as Cheri, a newly hired employee. Working together, they process one customer's order in 2 min. How long would it take each of them, working alone, to process the order? [6.5]

49. *File Download.* Ethan has downloaded 254 kilobytes of a file. He has twice as much left to download. How big is the file he is downloading? [1.6]

50. *Commuting.* Jennifer commutes 90 mi to her work in Washington D.C., averaging a certain speed. If she could average 15 mph faster, the trip would take $\frac{1}{2}$ hr less. Find Jennifer's average speed. [8.3]

Synthesis

TW **51.** Write a problem for a classmate to solve. Devise the problem so that **(a)** the solution is found after solving a rational equation and **(b)** the solution is "The express train travels 90 mph."

TW **52.** In what ways do the motion problems of this section (like Example 1) differ from the motion problems in Chapter 6 (see p. 485)?

53. *Biochemistry.* The equation

$$A = 6.5 - \frac{20.4t}{t^2 + 36}$$

is used to calculate the acid level A in a person's blood t minutes after sugar is consumed. Solve for t.

54. *Special Relativity.* Einstein found that an object with initial mass m_0 and traveling velocity v has mass

$$m = \frac{m_0}{\sqrt{1 - \frac{v^2}{c^2}}},$$

where c is the speed of light. Solve the formula for c.

55. Find a number for which the reciprocal of 1 less than the number is the same as 1 more than the number.

56. *Purchasing.* A discount store bought a quantity of beach towels for \$250 and sold all but 15 at a profit of \$3.50 per towel. With the total amount received, the manager could buy 4 more than twice as many as were bought before. Find the cost per towel.

57. *Art and Aesthetics.* For over 2000 yr, artists, sculptors, and architects have regarded the proportions of a "golden" rectangle as visually appealing. A rectangle of width w and length l is considered "golden" if

$$\frac{w}{l} = \frac{l}{w + l}.$$

Solve for l.

58. *Diagonal of a Cube.* Find a formula that expresses the length of the three-dimensional diagonal of a cube as a function of the cube's surface area.

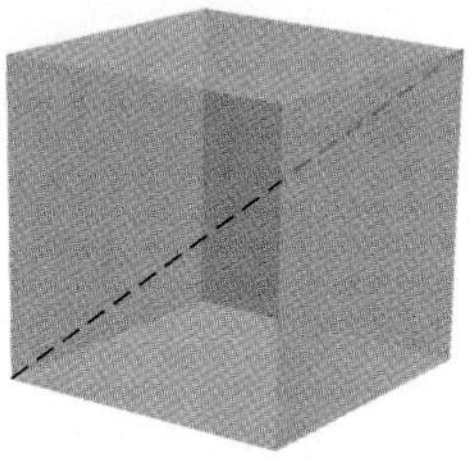

59. Solve for n:

$$mn^4 - r^2pm^3 - r^2n^2 + p = 0.$$

60. *Surface Area.* Find a formula that expresses the diameter of a right cylindrical solid as a function of its surface area and its height. (See Exercise 17.)

61. A sphere is inscribed in a cube as shown in the figure below. Express the surface area of the sphere as a function of the surface area S of the cube. (See Exercise 15.)

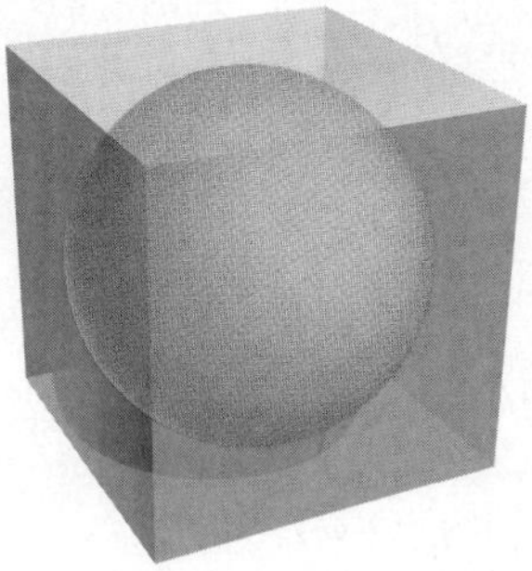

8.4 Studying Solutions of Quadratic Equations

The Discriminant ■ Writing Equations from Solutions

The Discriminant

It is sometimes enough to know what *type* of number a solution will be, without actually solving the equation. Suppose we want to know if $4x^2 + 7x - 15 = 0$ has rational solutions (and thus can be solved by factoring). Using the quadratic formula, we would have

$$x = \frac{-b \pm \sqrt{b^2 - 4ac}}{2a}$$
$$= \frac{-7 \pm \sqrt{7^2 - 4 \cdot 4 \cdot (-15)}}{2 \cdot 4}.$$

Note that the radicand, $7^2 - 4 \cdot 4 \cdot (-15)$, determines what type of number the solutions will be. Since $7^2 - 4 \cdot 4 \cdot (-15) = 49 - 16(-15) = 289$, and since 289 is a perfect square $\left(\sqrt{289} = 17\right)$, we know that the solutions of the equation will be two rational numbers. This means that $4x^2 + 7x - 15 = 0$ *can* be solved by factoring.

It is $b^2 - 4ac$, known as the **discriminant,** that determines what type of number the solutions of a quadratic equation are. If a, b, and c are rational, then:

- When $b^2 - 4ac$ simplifies to 0, it doesn't matter if we use $+\sqrt{b^2 - 4ac}$ or $-\sqrt{b^2 - 4ac}$; we get the same solution twice. Thus, when the discriminant is 0, there is one *repeated* solution and it is rational.

  Example: $9x^2 + 6x + 1 = 0 \rightarrow b^2 - 4ac = 6^2 - 4 \cdot 9 \cdot 1 = 0.$

- When $b^2 - 4ac$ is positive, there are two different real-number solutions: If $b^2 - 4ac$ is a perfect square, these solutions are rational numbers.

  Example: $6x^2 + 5x + 1 = 0 \rightarrow b^2 - 4ac = 5^2 - 4 \cdot 6 \cdot 1 = 1.$

- When $b^2 - 4ac$ is positive, but not a perfect square, there are two irrational solutions and they are conjugates of each other (see p. 570).

  Example: $2x^2 + 4x + 1 = 0 \rightarrow b^2 - 4ac = 4^2 - 4 \cdot 2 \cdot 1 = 8.$

- When $b^2 - 4ac$ is negative, there are two imaginary-number solutions and they are complex conjugates of each other.

  Example: $3x^2 + 2x + 1 = 0 \rightarrow b^2 - 4ac = 2^2 - 4 \cdot 3 \cdot 1 = -8.$

Note that any equation for which $b^2 - 4ac$ is a perfect square can be solved by factoring.

Study Tip

When Something Seems Easy

Every so often, you may encounter a lesson that you remember from a previous math course. When this occurs, make sure that *all* of that lesson is understood and review any tricky spots.

Discriminant $b^2 - 4ac$	Nature of Solutions
0	One solution; a rational number
Positive Perfect square Not a perfect square	Two different real-number solutions Solutions are rational. Solutions are irrational conjugates.
Negative	Two different imaginary-number solutions (complex conjugates)

Note that all quadratic equations have one or two solutions. They can always be found algebraically; only real-number solutions can be found graphically.

EXAMPLE 1 For each equation, determine what type of number the solutions are and how many solutions exist.

a) $9x^2 - 12x + 4 = 0$ **b)** $x^2 + 5x + 8 = 0$

c) $2x^2 + 7x - 3 = 0$

SOLUTION

a) For $9x^2 - 12x + 4 = 0$, we have

$$a = 9, \quad b = -12, \quad c = 4.$$

We substitute and compute the discriminant:

$$b^2 - 4ac = (-12)^2 - 4 \cdot 9 \cdot 4$$
$$= 144 - 144 = 0.$$

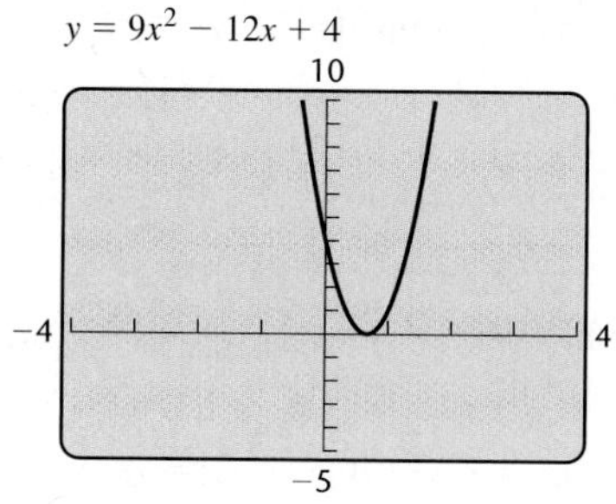

There is exactly one solution, and it is rational. This indicates that $9x^2 - 12x + 4 = 0$ can be solved by factoring. The graph at left confirms that $9x^2 - 12x + 4 = 0$ has just one solution.

b) For $x^2 + 5x + 8 = 0$, we have

$$a = 1, \quad b = 5, \quad c = 8.$$

We substitute and compute the discriminant:

$$b^2 - 4ac = 5^2 - 4 \cdot 1 \cdot 8$$
$$= 25 - 32 = -7.$$

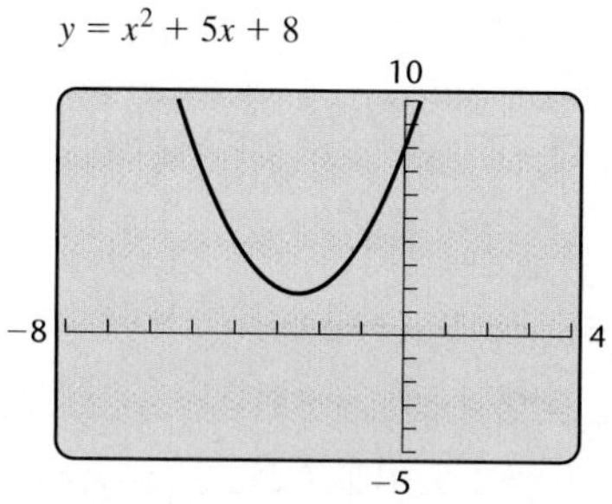

Since the discriminant is negative, there are two imaginary-number solutions that are complex conjugates of each other. As the graph at left shows, there are no x-intercepts of the graph of $y = x^2 + 5x + 8$, and thus no real solutions of the equation $x^2 + 5x + 8 = 0$.

c) For $2x^2 + 7x - 3 = 0$, we have

$$a = 2, \quad b = 7, \quad c = -3;$$
$$b^2 - 4ac = 7^2 - 4 \cdot 2(-3)$$
$$= 49 - (-24) = 73.$$

$y = 2x^2 + 7x - 3$

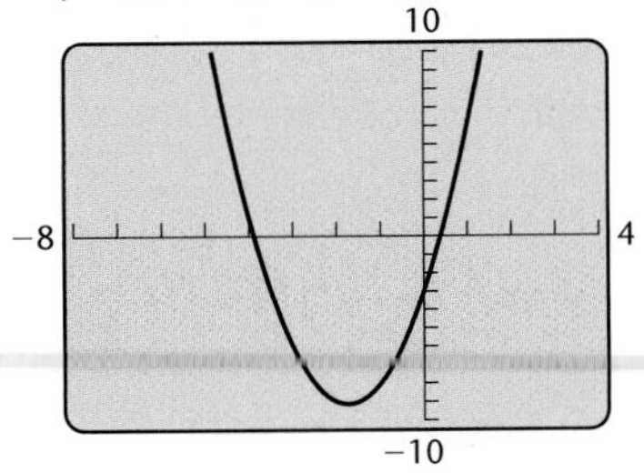

The discriminant is a positive number that is not a perfect square. Thus there are two irrational solutions that are conjugates of each other. From the graph at left, we see that there are two solutions of the equation. We cannot tell from simply observing the graph whether the solutions are rational or irrational.

As we saw in Example 1, discriminants can be used to determine the number of real-number solutions of $ax^2 + bx + c = 0$. This can be used as an aid in graphing.

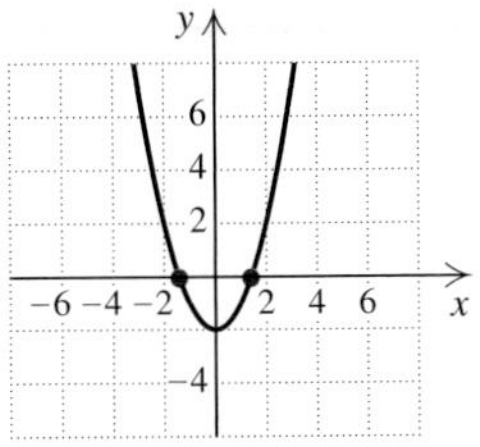

$y = ax^2 + bx + c$
$b^2 - 4ac > 0$
Two real solutions
of $ax^2 + bx + c = 0$
Two x-intercepts

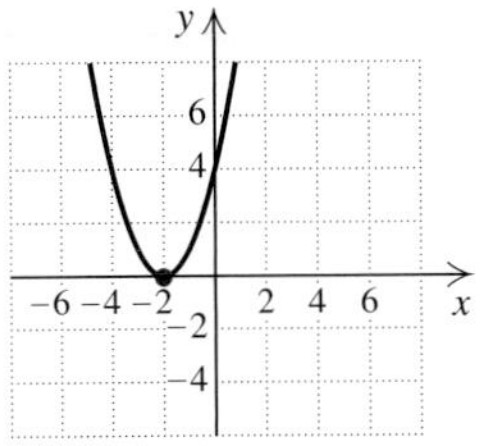

$y = ax^2 + bx + c$
$b^2 - 4ac = 0$
One real solution
of $ax^2 + bx + c = 0$
One x-intercept

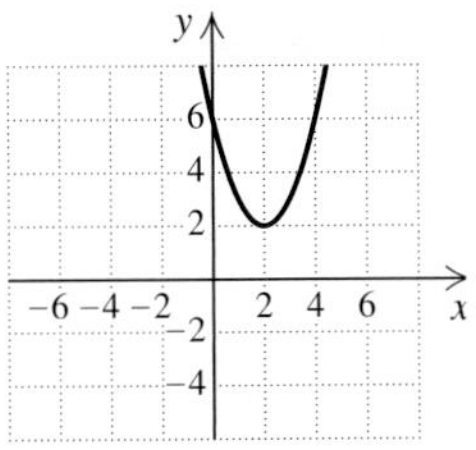

$y = ax^2 + bx + c$
$b^2 - 4ac < 0$
No real solutions
of $ax^2 + bx + c = 0$
No x-intercept

Writing Equations from Solutions

We know by the principle of zero products that $(x - 2)(x + 3) = 0$ has solutions 2 and -3. If we know the solutions of an equation, we can write an equation, using the principle in reverse.

EXAMPLE 2 Find an equation for which the given numbers are solutions.

a) 3 and $-\frac{2}{5}$
b) $2i$ and $-2i$
c) $5\sqrt{7}$ and $-5\sqrt{7}$
d) -4, 0, and 1

SOLUTION

a)

$$x = 3 \quad or \quad x = -\tfrac{2}{5}$$

$$x - 3 = 0 \quad or \quad x + \tfrac{2}{5} = 0 \qquad \textbf{Getting 0's on one side}$$

$$(x - 3)\left(x + \tfrac{2}{5}\right) = 0 \qquad \textbf{Using the principle of zero products (multiplying)}$$

$$x^2 + \tfrac{2}{5}x - 3x - 3 \cdot \tfrac{2}{5} = 0 \qquad \textbf{Multiplying}$$

$$x^2 - \tfrac{13}{5}x - \tfrac{6}{5} = 0 \qquad \textbf{Combining like terms}$$

$$5x^2 - 13x - 6 = 0 \qquad \textbf{Multiplying both sides by 5 to clear fractions}$$

Note that multiplying both sides by the LCD, 5, clears the equations of fractions. Had we preferred, we could have multiplied $x + \frac{2}{5} = 0$ by 5, thus clearing fractions *before* using the principle of zero products.

b)

$$x = 2i \quad or \quad x = -2i$$
$$x - 2i = 0 \quad or \quad x + 2i = 0 \qquad \textbf{Getting 0's on one side}$$
$$(x - 2i)(x + 2i) = 0 \qquad \textbf{Using the principle of zero products (multiplying)}$$
$$x^2 - (2i)^2 = 0 \qquad \textbf{Finding the product of a sum and difference}$$
$$x^2 - 4i^2 = 0$$
$$x^2 + 4 = 0 \qquad i^2 = -1$$

c)

$$x = 5\sqrt{7} \quad or \quad x = -5\sqrt{7}$$
$$x - 5\sqrt{7} = 0 \quad or \quad x + 5\sqrt{7} = 0 \qquad \textbf{Getting 0's on one side}$$
$$\left(x - 5\sqrt{7}\right)\left(x + 5\sqrt{7}\right) = 0 \qquad \textbf{Using the principle of zero products}$$
$$x^2 - \left(5\sqrt{7}\right)^2 = 0 \qquad \textbf{Finding the product of a sum and difference}$$
$$x^2 - 25 \cdot 7 = 0$$
$$x^2 - 175 = 0$$

d)

$$x = -4 \quad or \quad x = 0 \quad or \quad x = 1$$
$$x + 4 = 0 \quad or \quad x = 0 \quad or \quad x - 1 = 0 \qquad \textbf{Getting 0's on one side}$$
$$(x + 4)x(x - 1) = 0 \qquad \textbf{Using the principle of zero products}$$
$$x(x^2 + 3x - 4) = 0 \qquad \textbf{Multiplying}$$
$$x^3 + 3x^2 - 4x = 0$$

8.4 EXERCISE SET

FOR EXTRA HELP

MathXL | MyMathLab | InterAct Math | AW Math Tutor Center | Video Lectures on CD: Disc 4 | Student's Solutions Manual

Concept Reinforcement *Complete each of the following.*

1. In the quadratic formula, the expression $b^2 - 4ac$ is called the ______________.

2. When $b^2 - 4ac$ is 0, there is/are ______________ solution(s).

3. When $b^2 - 4ac$ is positive, there is/are ______________ solution(s).

4. When $b^2 - 4ac$ is negative, there is/are ______________ solution(s).

5. When $b^2 - 4ac$ is a perfect square, the answers are ______________ numbers.

6. When $b^2 - 4ac$ is negative, the answer(s) is/are ______________ numbers.

For each equation, determine what type of number the solutions are and how many solutions exist.

7. $x^2 - 7x + 5 = 0$

8. $x^2 - 5x + 3 = 0$

9. $x^2 + 3 = 0$

10. $x^2 + 5 = 0$

11. $x^2 - 5 = 0$

12. $x^2 - 3 = 0$

13. $4x^2 + 8x - 5 = 0$

14. $4x^2 - 12x + 9 = 0$

15. $x^2 + 4x + 6 = 0$

16. $x^2 - 2x + 4 = 0$

17. $9t^2 - 48t + 64 = 0$

18. $6t^2 - 19t - 20 = 0$

19. $10x^2 - x - 2 = 0$

20. $6x^2 + 5x - 4 = 0$

Aha! **21.** $9t^2 - 3t = 0$

22. $4m^2 + 7m = 0$

23. $x^2 + 4x = 8$

24. $x^2 + 5x = 9$

25. $2a^2 - 3a = -5$

26. $3a^2 + 5 = 7a$

27. $y^2 + \frac{9}{4} = 4y$

28. $x^2 = \frac{1}{2}x - \frac{3}{5}$

Write a quadratic equation having the given numbers as solutions.

29. $-7, 3$

30. $-6, 4$

31. 3, only solution (*Hint*: It must be a repeated solution.)

32. -5, only solution

33. $-1, -3$

34. $-2, -5$

35. $5, \frac{3}{4}$

36. $4, \frac{2}{3}$

37. $-\frac{1}{4}, -\frac{1}{2}$

38. $\frac{1}{2}, \frac{1}{3}$

39. $2.4, -0.4$

40. $-0.6, 1.4$

41. $-\sqrt{3}, \sqrt{3}$

42. $-\sqrt{7}, \sqrt{7}$

43. $2\sqrt{5}, -2\sqrt{5}$

44. $3\sqrt{2}, -3\sqrt{2}$

45. $4i, -4i$

46. $3i, -3i$

47. $2 - 7i, 2 + 7i$

48. $5 - 2i, 5 + 2i$

49. $3 - \sqrt{14}, 3 + \sqrt{14}$

50. $2 - \sqrt{10}, 2 + \sqrt{10}$

51. $1 - \frac{\sqrt{21}}{3}, 1 + \frac{\sqrt{21}}{3}$

52. $\frac{5}{4} - \frac{\sqrt{33}}{4}, \frac{5}{4} + \frac{\sqrt{33}}{4}$

Write a third-degree equation having the given numbers as solutions.

53. $-2, 1, 5$

54. $-5, 0, 2$

55. $-1, 0, 3$

56. $-2, 2, 3$

TW **57.** Under what condition(s) is the discriminant *not* the fastest way to determine how many and what type of solutions exist?

TW **58.** While solving a quadratic equation of the form $ax^2 + bx + c = 0$ with a graphing calculator, Shawn-Marie gets the following screen. How could the sign of the discriminant help her check the graph?

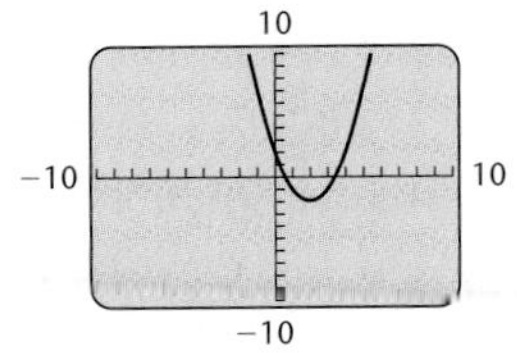

Skill Maintenance

Simplify. [1.4]

59. $(3a^2)^4$

60. $(4x^3)^2$

Find the x-intercepts of the graph of f. [5.4]

61. $f(x) = x^2 - 7x - 8$

62. $f(x) = x^2 - 6x + 8$

63. During a one-hour television show, there were 12 commercials. Some of the commercials were 30 sec long and the others were 60 sec long. The amount of time for 30-sec commercials was 6 min less than the total number of minutes of commercial time during the show. How many 30-sec commercials were used? [3.3]

64. Graph: $y = -\frac{3}{7}x + 4$. [2.2]

Synthesis

TW **65.** If we assume that a quadratic equation has integers for coefficients, will the product of the solutions always be a real number? Why or why not?

TW **66.** Can a fourth-degree equation have exactly three irrational solutions? Why or why not?

67. The graph of an equation of the form

$$y = ax^2 + bx + c$$

is a curve similar to the one shown below. Determine a, b, and c from the information given.

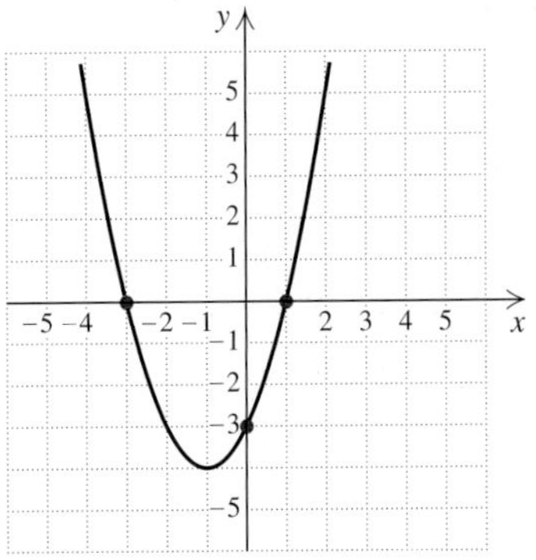

68. Show that the product of the solutions of $ax^2 + bx + c = 0$ is c/a.

For each equation under the given condition, **(a)** *find k and* **(b)** *find the other solution.*

69. $kx^2 - 2x + k = 0$; one solution is -3

70. $x^2 - kx + 2 = 0$; one solution is $1 + i$

71. $x^2 - (6 + 3i)x + k = 0$; one solution is 3

72. Show that the sum of the solutions of $ax^2 + bx + c = 0$ is $-b/a$.

73. Show that whenever there is just one solution of $ax^2 + bx + c = 0$, that solution is of the form $-b/(2a)$.

74. Find h and k, where $3x^2 - hx + 4k = 0$, the sum of the solutions is -12, and the product of the solutions is 20. (*Hint*: See Exercises 68 and 72.)

75. Suppose that $f(x) = ax^2 + bx + c$, with $f(-3) = 0$, $f(\frac{1}{2}) = 0$, and $f(0) = -12$. Find a, b, and c.

76. Find an equation for which $2 - \sqrt{3}$, $2 + \sqrt{3}$, $5 - 2i$, and $5 + 2i$ are solutions.

Aha! **77.** Find an equation with integer coefficients for which $1 - \sqrt{5}$ and $3 + 2i$ are two of the solutions.

TW **78.** A discriminant that is a perfect square indicates that factoring can be used to solve the quadratic equation. Why?

8.5 Equations Reducible to Quadratic

Recognizing Equations in Quadratic Form ■ Radical and Rational Equations

Study Tip

Use Your Words

Although you may at first feel a bit hesitant to do so, when a new word or phrase (such as *reducible to quadratic*) arises, try to use it in conversation with classmates, your instructor, or a study partner. Doing so will help you in your reading, deepen your level of understanding, and increase your confidence.

Recognizing Equations in Quadratic Form

Certain equations that are not really quadratic can be thought of in such a way that they can be solved as quadratic. For example, because the square of x^2 is x^4, the equation $x^4 - 9x^2 + 8 = 0$ is said to be "quadratic in x^2":

$$x^4 - 9x^2 + 8 = 0$$

$$\downarrow$$

$$(x^2)^2 - 9(x^2) + 8 = 0 \qquad \textbf{Thinking of } x^4 \textbf{ as } (x^2)^2$$

$$\downarrow$$

$$u^2 - 9u + 8 = 0. \qquad \textbf{To make this clearer, write } u \textbf{ instead of } x^2.$$

The equation $u^2 - 9u + 8 = 0$ can be solved by factoring or by the quadratic formula. Then, remembering that $u = x^2$, we can solve for x. Equations that can be solved like this are *reducible to quadratic*, or *in quadratic form*.

EXAMPLE 1 Solve: $x^4 - 9x^2 + 8 = 0$.

SOLUTION We solve both algebraically and graphically.

ALGEBRAIC APPROACH

Let $u = x^2$. Then we solve by substituting u for x^2 and u^2 for x^4:

$$u^2 - 9u + 8 = 0$$
$$(u - 8)(u - 1) = 0 \qquad \textbf{Factoring}$$
$$u - 8 = 0 \quad or \quad u - 1 = 0 \qquad \textbf{Principle of zero products}$$
$$u = 8 \quad or \quad u = 1.$$

We replace u with x^2 and solve these equations:

$$x^2 = 8 \quad or \quad x^2 = 1$$
$$x = \pm\sqrt{8} \quad or \quad x = \pm 1$$
$$x = \pm 2\sqrt{2} \quad or \quad x = \pm 1.$$

To check, note that for both $x = 2\sqrt{2}$ and $-2\sqrt{2}$, we have $x^2 = 8$ and $x^4 = 64$. Similarly, for both $x = 1$ and -1, we have $x^2 = 1$ and $x^4 = 1$. Thus instead of making four checks, we need make only two.

Check:

For $\pm 2\sqrt{2}$:

$$\begin{array}{c|c} x^4 - 9x^2 + 8 & 0 \\ \hline (\pm 2\sqrt{2})^4 - 9(\pm 2\sqrt{2})^2 + 8 & 0 \\ 64 - 9 \cdot 8 + 8 & \end{array}$$
$$0 \stackrel{?}{=} 0 \quad \text{TRUE}$$

For ± 1:

$$\begin{array}{c|c} x^4 - 9x^2 + 8 & 0 \\ \hline (\pm 1)^4 - 9(\pm 1)^2 + 8 & 0 \\ 1 - 9 + 8 & \end{array}$$
$$0 \stackrel{?}{=} 0 \quad \text{TRUE}$$

The solutions are $1, -1, 2\sqrt{2}$, and $-2\sqrt{2}$.

GRAPHICAL APPROACH

We let $f(x) = x^4 - 9x^2 + 8$ and determine the zeros of the function.

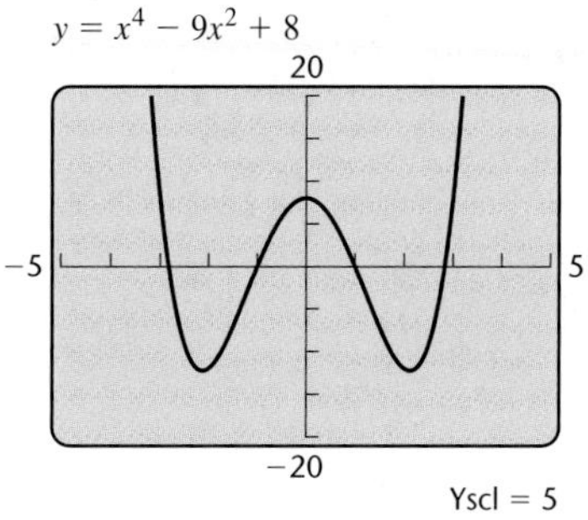

It appears as though the graph has 4 x-intercepts. Recall from Chapter 5 that a fourth-degree polynomial function has at most 4 zeros, so we know that we have not missed any zeros. Using ZERO four times, we obtain the solutions -2.8284271, -1, 1, and 2.8284271. The solutions are ± 1 and, approximately, ± 2.8284271. Note that since $2\sqrt{2} \approx 2.8284271$, the solutions found graphically are the same as those found algebraically.

Student Notes

To recognize an equation in quadratic form, note that there will be two terms with variables. The exponent of the first term is twice the exponent of the second term. It is the variable expression in the *second* term that is written as u.

Example 1 can be solved directly by factoring:

$$x^4 - 9x^2 + 8 = 0$$
$$(x^2 - 1)(x^2 - 8) = 0$$
$$x^2 - 1 = 0 \quad or \quad x^2 - 8 = 0$$
$$x^2 = 1 \quad or \quad x^2 = 8$$
$$x = \pm 1 \quad or \quad x = \pm 2\sqrt{2}.$$

There is nothing wrong with this approach. However, in the examples that follow, you will note that it becomes increasingly difficult to solve the equation without first making a substitution.

Radical and Rational Equations

Sometimes rational equations, radical equations, or equations containing exponents that are fractions are reducible to quadratic. It is especially important that answers to these equations be checked in the original equation.

EXAMPLE 2 Solve: $x - 3\sqrt{x} - 4 = 0$.

SOLUTION This radical equation could be solved using the method discussed in Section 7.6. However, if we note that the square of $\sqrt{x}$ is x, we can regard the equation as "quadratic in $\sqrt{x}$."

We let $u = \sqrt{x}$ and consequently $u^2 = x$:

$$x - 3\sqrt{x} - 4 = 0$$

$$u^2 - 3u - 4 = 0 \qquad \textbf{Substituting}$$

$$(u - 4)(u + 1) = 0$$

$$u = 4 \quad or \quad u = -1. \qquad \textbf{Using the principle of zero products}$$

CAUTION! A common error is to solve for u but forget to solve for x. Remember to solve for the *original* variable!

Next, we replace u with $\sqrt{x}$ and solve these equations:

$$\sqrt{x} = 4 \quad or \quad \sqrt{x} = -1.$$

Squaring gives us $x = 16$ or $x = 1$ and also makes checking essential.

Check: For 16:

$x - 3\sqrt{x} - 4 = 0$	
$16 - 3\sqrt{16} - 4$	0
$16 - 3 \cdot 4 - 4$	

$0 \stackrel{?}{=} 0$ TRUE

For 1:

$x - 3\sqrt{x} - 4 = 0$	
$1 - 3\sqrt{1} - 4$	0
$1 - 3 \cdot 1 - 4$	

$-6 \stackrel{?}{=} 0$ FALSE

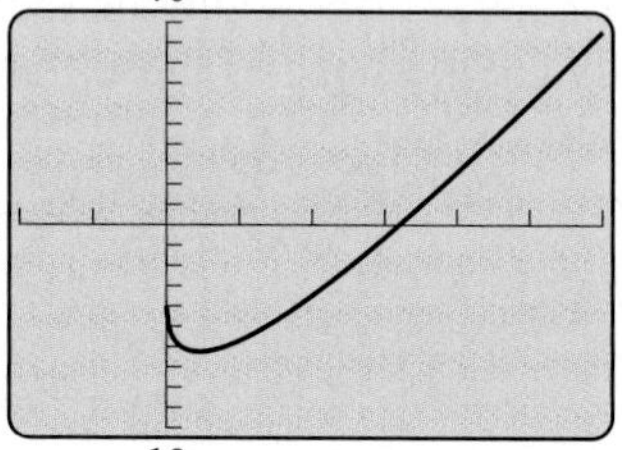

The number 16 checks, but 1 does not. Had we noticed that $\sqrt{x} = -1$ has no solution (since principal roots are never negative), we could have solved only the equation $\sqrt{x} = 4$. The graph at left provides further evidence that although 16 is a solution, -1 is not. The solution is 16.

EXAMPLE 3 Find the x-intercepts of the graph of

$$f(x) = (x^2 - 1)^2 - (x^2 - 1) - 2.$$

SOLUTION The x-intercepts occur where $f(x) = 0$ so we must have

$$(x^2 - 1)^2 - (x^2 - 1) - 2 = 0. \qquad \textbf{Setting } f(x) \textbf{ equal to 0}$$

The equation is quadratic in $x^2 - 1$, so we let $u = x^2 - 1$ and $u^2 = (x^2 - 1)^2$:

$$u^2 - u - 2 = 0 \qquad \textbf{Substituting in } (x^2 - 1)^2 - (x^2 - 1) - 2 = 0$$

$$(u - 2)(u + 1) = 0$$

$$u = 2 \quad or \quad u = -1. \qquad \textbf{Using the principle of zero products}$$

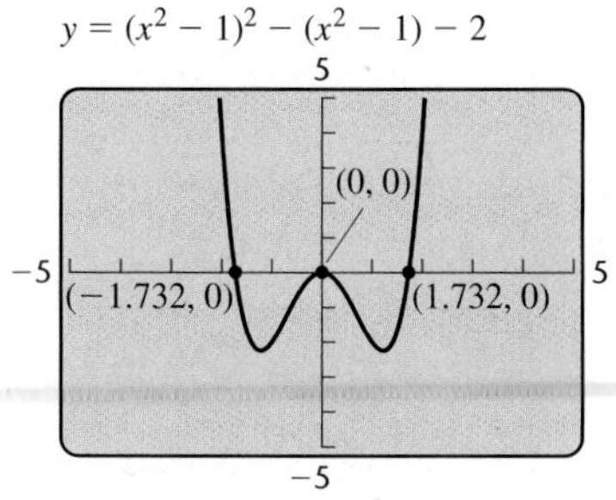

Next, we replace u with $x^2 - 1$ and solve these equations:

$$x^2 - 1 = 2 \quad or \quad x^2 - 1 = -1$$
$$x^2 = 3 \quad or \quad x^2 = 0 \qquad \textbf{Adding 1 to both sides}$$
$$x = \pm\sqrt{3} \quad or \quad x = 0. \qquad \textbf{Using the principle of square roots}$$

The x intercepts occur at $(-\sqrt{3}, 0)$, $(0, 0)$, and $(\sqrt{3}, 0)$. These solutions are confirmed by the graph at left.

Sometimes great care must be taken in deciding what substitution to make.

Student Notes

When it is difficult to decide what substitution to make, try letting u represent one variable expression in the equation. If u^2 is the other variable expression, then you have found an appropriate substitution.

In Example 4, you might try two substitutions:

$$u = m^{-2} \Rightarrow u^2 = m^{-4};$$
$$u = m^{-1} \Rightarrow u^2 = m^{-2}.$$

Only for $u = m^{-1}$ are both u and u^2 in the equation.

EXAMPLE 4 Solve: $m^{-2} - 6m^{-1} + 4 = 0$.

SOLUTION Note that the square of m^{-1} is $(m^{-1})^2$, or m^{-2}. This allows us to regard the equation as quadratic in m^{-1}.

We let $u = m^{-1}$ and $u^2 = m^{-2}$:

$$u^2 - 6u + 4 = 0 \qquad \textbf{Substituting}$$
$$u = \frac{-(-6) \pm \sqrt{(-6)^2 - 4 \cdot 1 \cdot 4}}{2 \cdot 1} \qquad \textbf{Using the quadratic formula}$$
$$\left.\begin{aligned} u &= \frac{6 \pm \sqrt{20}}{2} = \frac{2 \cdot 3 \pm 2\sqrt{5}}{2} \\ u &= 3 \pm \sqrt{5}. \end{aligned}\right\} \qquad \textbf{Simplifying}$$

Next, we replace u with m^{-1} and solve:

$$m^{-1} = 3 \pm \sqrt{5}$$
$$\frac{1}{m} = 3 \pm \sqrt{5} \qquad \textbf{Recall that } m^{-1} = \frac{1}{m}.$$
$$1 = m(3 \pm \sqrt{5}) \qquad \textbf{Multiplying both sides by } m$$
$$\frac{1}{3 \pm \sqrt{5}} = m. \qquad \textbf{Dividing both sides by } 3 \pm \sqrt{5}$$

We can check both solutions as follows.

Check:

For $1/(3 - \sqrt{5})$:

$$m^{-2} - 6m^{-1} + 4 = 0$$
$$\left(\frac{1}{3 - \sqrt{5}}\right)^{-2} - 6\left(\frac{1}{3 - \sqrt{5}}\right)^{-1} + 4 \;\Big|\; 0$$
$$(3 - \sqrt{5})^2 - 6(3 - \sqrt{5}) + 4$$
$$9 - 6\sqrt{5} + 5 - 18 + 6\sqrt{5} + 4$$
$$0 \stackrel{?}{=} 0 \quad \text{TRUE}$$

For $1/(3 + \sqrt{5})$:

$$m^{-2} - 6m^{-1} + 4 = 0$$
$$\left(\frac{1}{3 + \sqrt{5}}\right)^{-2} - 6\left(\frac{1}{3 + \sqrt{5}}\right)^{-1} + 4 \;\Big|\; 0$$
$$(3 + \sqrt{5})^2 - 6(3 + \sqrt{5}) + 4$$
$$9 + 6\sqrt{5} + 5 - 18 - 6\sqrt{5} + 4$$
$$0 \stackrel{?}{=} 0 \quad \text{TRUE}$$

We could also check the solutions using a calculator by entering $y_1 = x^{-2} - 6x^{-1} + 4$ and evaluating y_1 for both numbers, as shown on the left below.

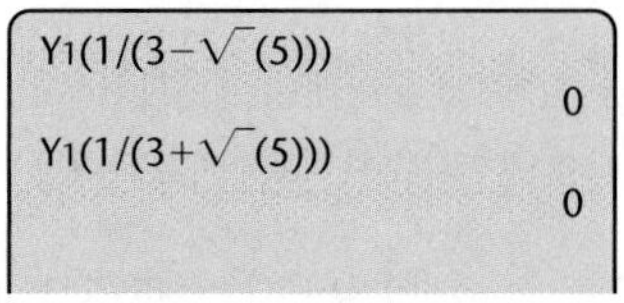

$y = x^{-2} - 6x^{-1} + 4$

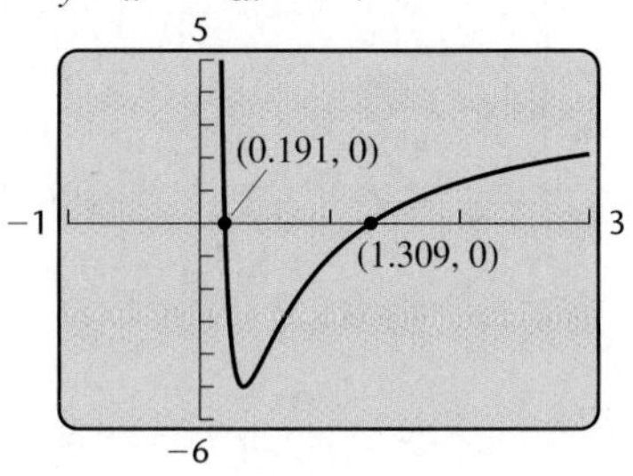

As another check, we solve the equation graphically. The first coordinates of the x-intercepts of the graph on the right above are approximate solutions.

Both numbers check. The solutions are $1/(3 - \sqrt{5})$ and $1/(3 + \sqrt{5})$, or approximately 1.309 and 0.191.

EXAMPLE 5 Solve: $t^{2/5} - t^{1/5} - 2 = 0$.

SOLUTION Note that the square of $t^{1/5}$ is $(t^{1/5})^2$, or $t^{2/5}$. The equation is therefore quadratic in $t^{1/5}$, so we let $u = t^{1/5}$ and $u^2 = t^{2/5}$:

$$u^2 - u - 2 = 0 \quad \textbf{Substituting}$$
$$(u - 2)(u + 1) = 0$$
$$u = 2 \quad or \quad u = -1. \quad \textbf{Using the principle of zero products}$$

Now we replace u with $t^{1/5}$ and solve:

$$t^{1/5} = 2 \quad or \quad t^{1/5} = -1$$
$$t = 32 \quad or \quad t = -1. \quad \textbf{Principle of powers; raising to the 5th power}$$

Check:

For 32:

$t^{2/5} - t^{1/5} - 2 = 0$	
$32^{2/5} - 32^{1/5} - 2$	0
$(32^{1/5})^2 - 32^{1/5} - 2$	
$2^2 - 2 - 2$	
$0 \stackrel{?}{=} 0$ TRUE	

For −1:

$t^{2/5} - t^{1/5} - 2 = 0$	
$(-1)^{2/5} - (-1)^{1/5} - 2$	0
$[(-1)^{1/5}]^2 - (-1)^{1/5} - 2$	
$(-1)^2 - (-1) - 2$	
$0 \stackrel{?}{=} 0$ TRUE	

$y = x^{2/5} - x^{1/5} - 2$

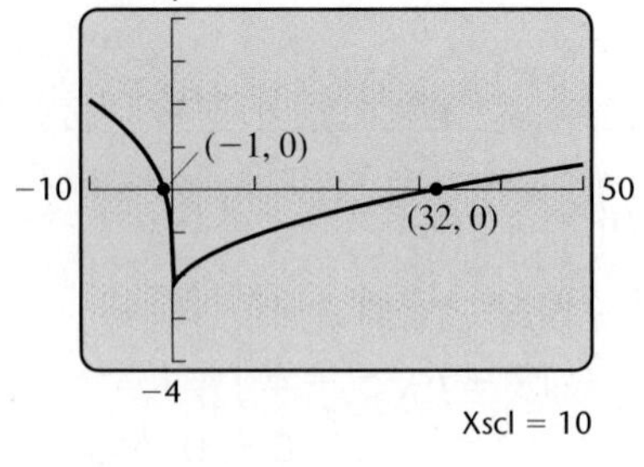

Both numbers check and are confirmed by the graph at left. The solutions are 32 and −1.

The following tips may prove useful.

To Solve an Equation That Is Reducible to Quadratic

1. The equation is quadratic in form if the variable factor in one term is the square of the variable factor in the other variable term.
2. Write down any substitutions that you are making.
3. Whenever you make a substitution, be sure to solve for the variable that is used in the original equation.
4. Check possible answers in the original equation.

8.5 EXERCISE SET

FOR EXTRA HELP

MathXL | MyMathLab | InterAct Math | AW Math Tutor Center | Video Lectures on CD: Disc 4 | Student's Solutions Manual

Concept Reinforcement *In each of Exercises 1–8, match the equation with an appropriate substitution from the column on the right that could be used to reduce the equation to quadratic form.*

1. ___ $4x^6 - 2x^3 + 1 = 0$ **a)** $u = x^{-1/3}$
2. ___ $3x^4 + 4x^2 - 7 = 0$ **b)** $u = x^{1/3}$
3. ___ $5x^8 + 2x^4 - 3 = 0$ **c)** $u = x^{-2}$
4. ___ $2x^{2/3} - 5x^{1/3} + 4 = 0$ **d)** $u = x^2$
5. ___ $3x^{4/3} + 4x^{2/3} - 7 = 0$ **e)** $u = x^{-2/3}$
6. ___ $2x^{-2/3} + x^{-1/3} + 6 = 0$ **f)** $u = x^3$
7. ___ $4x^{-4/3} - 2x^{-2/3} + 3 = 0$ **g)** $u = x^{2/3}$
8. ___ $3x^{-4} + 4x^{-2} - 2 = 0$ **h)** $u = x^4$

Solve.

9. $x^4 - 5x^2 + 4 = 0$
10. $x^4 - 10x^2 + 9 = 0$
11. $x^4 - 9x^2 + 20 = 0$
12. $x^4 - 12x^2 + 27 = 0$
13. $4t^4 - 19t^2 + 12 = 0$
14. $9t^4 - 14t^2 + 5 = 0$
15. $r - 2\sqrt{r} - 6 = 0$
16. $s - 4\sqrt{s} - 1 = 0$
17. $(x^2 - 7)^2 - 3(x^2 - 7) + 2 = 0$
18. $(x^2 - 1)^2 - 5(x^2 - 1) + 6 = 0$
19. $(1 + \sqrt{x})^2 + 5(1 + \sqrt{x}) + 6 = 0$
20. $(3 + \sqrt{x})^2 + 3(3 + \sqrt{x}) - 10 = 0$
21. $x^{-2} - x^{-1} - 6 = 0$
22. $2x^{-2} - x^{-1} - 1 = 0$
23. $4x^{-2} + x^{-1} - 5 = 0$
24. $m^{-2} + 9m^{-1} - 10 = 0$
25. $t^{2/3} + t^{1/3} - 6 = 0$
26. $w^{2/3} - 2w^{1/3} - 8 = 0$
27. $y^{1/3} - y^{1/6} - 6 = 0$
28. $t^{1/2} + 3t^{1/4} + 2 = 0$
29. $t^{1/3} + 2t^{1/6} = 3$
30. $m^{1/2} + 6 = 5m^{1/4}$
31. $(3 - \sqrt{x})^2 - 10(3 - \sqrt{x}) + 23 = 0$
32. $(5 + \sqrt{x})^2 - 12(5 + \sqrt{x}) + 33 = 0$

33. $16\left(\frac{x-1}{x-8}\right)^2 + 8\left(\frac{x-1}{x-8}\right) + 1 = 0$

34. $9\left(\frac{x+2}{x+3}\right)^2 - 6\left(\frac{x+2}{x+3}\right) + 1 = 0$

Find all x-intercepts of the given function f. If none exists, state this.

35. $f(x) = 5x + 13\sqrt{x} - 6$

36. $f(x) = 3x + 10\sqrt{x} - 8$

37. $f(x) = (x^2 - 3x)^2 - 10(x^2 - 3x) + 24$

38. $f(x) = (x^2 - 6x)^2 - 2(x^2 - 6x) - 35$

39. $f(x) = x^{2/5} + x^{1/5} - 6$

40. $f(x) = x^{1/2} - x^{1/4} - 6$

Aha! **41.** $f(x) = \left(\frac{x^2+2}{x}\right)^4 + 7\left(\frac{x^2+2}{x}\right)^2 + 5$

42. $f(x) = \left(\frac{x^2+1}{x}\right)^4 + 4\left(\frac{x^2+1}{x}\right)^2 + 12$

TW **43.** To solve $25x^6 - 10x^3 + 1 = 0$, Don lets $u = 5x^3$ and Robin lets $u = x^3$. Can they both be correct? Why or why not?

TW **44.** While trying to solve $0.05x^4 - 0.8 = 0$ with a graphing calculator, Murray gets the following screen. Can Murray solve this equation with a graphing calculator? Why or why not?

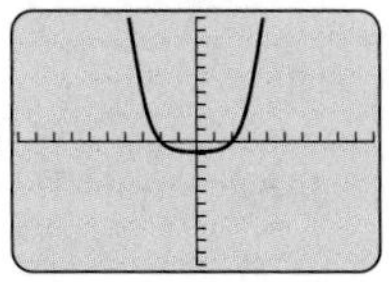

Focused Review

Solve.

45. $x^2 + x = 6$ [5.4]

46. $2x^2 - 3x + 4 = 0$ [8.2]

47. $(x - 7)^2 = 5$ [8.1]

48. $4x^2 + x = 6$ [8.2]

49. $x^4 - 5x^2 + 6$ [8.5]

50. $2(3 + \sqrt{x})^2 + 5(3 + \sqrt{x}) + 2 = 0$ [8.5]

Synthesis

TW **51.** Describe a procedure that could be used to solve any equation of the form $ax^4 + bx^2 + c = 0$.

TW **52.** Describe a procedure that could be used to write an equation that is quadratic in $3x^2 - 1$. Then explain how the procedure could be adjusted to write equations that are quadratic in $3x^2 - 1$ and have no real-number solution.

Solve.

53. $5x^4 - 7x^2 + 1 = 0$

54. $3x^4 + 5x^2 - 1 = 0$

55. $(x^2 - 4x - 2)^2 - 13(x^2 - 4x - 2) + 30 = 0$

56. $(x^2 - 5x - 1)^2 - 18(x^2 - 5x - 1) + 65 = 0$

57. $\frac{x}{x-1} - 6\sqrt{\frac{x}{x-1}} - 40 = 0$

58. $\left(\sqrt{\frac{x}{x-3}}\right)^2 - 24 = 10\sqrt{\frac{x}{x-3}}$

59. $a^5(a^2 - 25) + 13a^3(25 - a^2) + 36a(a^2 - 25) = 0$

60. $a^3 - 26a^{3/2} - 27 = 0$

61. $x^6 - 28x^3 + 27 = 0$

62. $x^6 + 7x^3 - 8 = 0$

8.6 Quadratic Functions and Their Graphs

The Graph of $f(x) = ax^2$ ■ The Graph of $f(x) = a(x - h)^2$ ■ The Graph of $f(x) = a(x - h)^2 + k$

We have already used quadratic functions when we solved equations earlier in this chapter. In this section and the next, we learn to graph such functions.

The Graph of $f(x) = ax^2$

The most basic quadratic function is $f(x) = x^2$.

EXAMPLE 1 Graph: $f(x) = x^2$.

SOLUTION We choose some values for x and compute $f(x)$ for each. Then we plot the ordered pairs and connect them with a smooth curve.

x	$f(x) = x^2$	$(x, f(x))$
−3	9	(−3, 9)
−2	4	(−2, 4)
−1	1	(−1, 1)
0	0	(0, 0)
1	1	(1, 1)
2	4	(2, 4)
3	9	(3, 9)

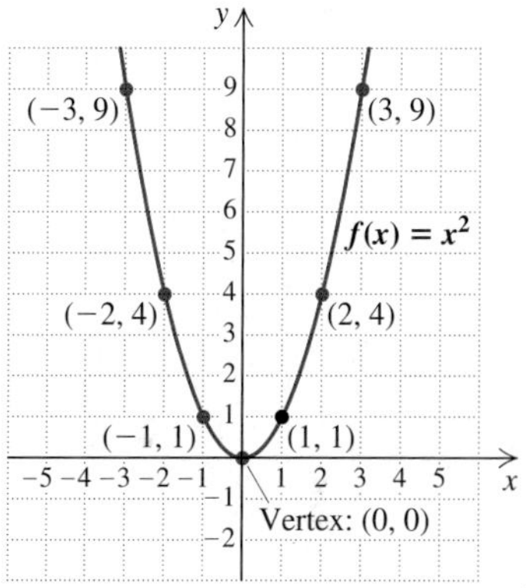

All quadratic functions have graphs similar to the one in Example 1. Such curves are called *parabolas*. They are U-shaped and symmetric with respect to a vertical line known as the parabola's *axis of symmetry*. For the graph of $f(x) = x^2$, the y-axis (or the line $x = 0$) is the axis of symmetry. Were the paper folded on this line, the two halves of the curve would match. The point $(0, 0)$ is known as the *vertex* of this parabola.

Since x^2 is a polynomial, we know that the domain of $f(x) = x^2$ is the set of all real numbers, or $(-\infty, \infty)$. We see from the graph and the table in Example 1 that the range of f is $\{y \mid y \geq 0\}$, or $[0, \infty)$. The function has a *minimum value* of 0. This minimum value occurs when $x = 0$.

Now let's consider a quadratic function of the form $g(x) = ax^2$. How does the constant a affect the graph of the function?

Study Tip

Take a Peek Ahead

Try to at least glance at the next section of material that will be covered in class. This will make it easier to concentrate on your instructor's lecture instead of trying to write everything down.

Student Notes

By paying attention to the symmetry of each parabola and the location of the vertex, you save yourself considerable work. Note too that if the x^2-coefficient is a, the x-values 1 unit to the right or left of the vertex are paired with the y-value a units above the vertex. Thus the graph of $y = \frac{3}{2}x^2$ includes the points $\left(-1, \frac{3}{2}\right)$ and $\left(1, \frac{3}{2}\right)$.

Interactive Discovery

Graph each of the following equations along with $y_1 = x^2$ in a $[-5, 5, -10, 10]$ window. If your calculator has a Transfrm application, run that application and graph $y_2 = Ax^2$. Then enter the values for A indicated in Exercises 1–6. For each equation, answer the following questions.

a) What is the vertex of the graph of y_2?

b) What is the axis of symmetry of the graph of y_2?

c) Does the graph of y_2 open upward or downward?

d) Is the parabola narrower or wider than the parabola given by $y_1 = x^2$?

1. $y_2 = 3x^2$ **2.** $y_2 = \frac{1}{3}x^2$ **3.** $y_2 = 0.2x^2$

4. $y_2 = -x^2$ **5.** $y_2 = -4x^2$ **6.** $y_2 = -\frac{2}{3}x^2$

7. Describe the effect of multiplying x^2 by a when $a > 1$ and when $0 < a < 1$.

8. Describe the effect of multiplying x^2 by a when $a < -1$ and when $-1 < a < 0$.

You may have noticed that for $f(x) = ax^2$, the constant a, when $|a| \neq 1$, made the graph wider or narrower than the graph of $g(x) = x^2$. When a was negative, the graph opened downward.

Note that if a parabola opens upward ($a > 0$), the function value, or y-value, at the vertex is a least, or *minimum,* value. That is, it is less than the y-value at any other point on the graph. If the parabola opens downward ($a < 0$), the function value at the vertex is a greatest, or *maximum,* value.

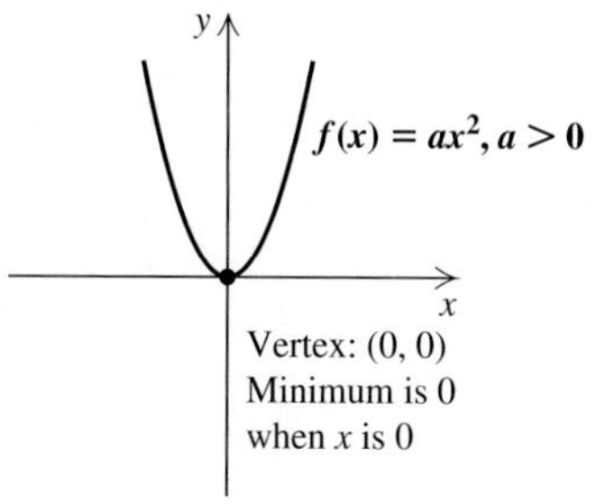

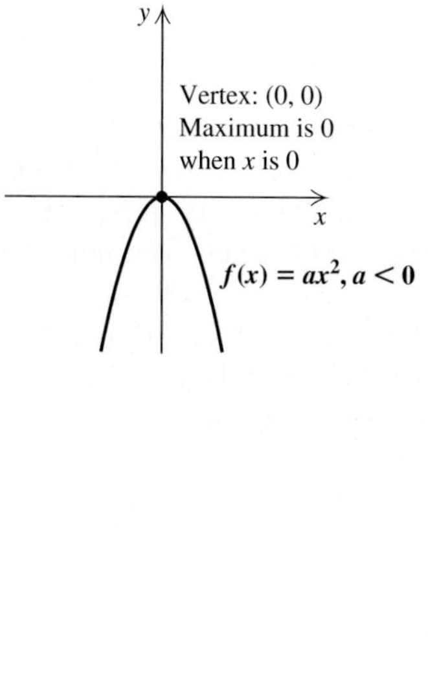

Graphing $f(x) = ax^2$

The graph of $f(x) = ax^2$ is a parabola with $x = 0$ as its axis of symmetry. Its vertex is the origin. The domain of f is $(-\infty, \infty)$.

For $a > 0$, the parabola opens upward. The range of the function is $[0, \infty)$. A minimum function value of 0 occurs when $x = 0$.

For $a < 0$, the parabola opens downward. The range of the function is $(-\infty, 0]$. A maximum function value of 0 occurs when $x = 0$.

If $|a|$ is greater than 1, the parabola is narrower than $y = x^2$.

If $|a|$ is between 0 and 1, the parabola is wider than $y = x^2$.

EXAMPLE 2 Graph: $g(x) = \frac{1}{2}x^2$.

SOLUTION The function is in the form $g(x) = ax^2$, where $a = \frac{1}{2}$. The vertex is $(0, 0)$ and the axis of symmetry is $x = 0$. Since $\frac{1}{2} > 0$, the parabola opens upward. The parabola is wider than $y = x^2$ since $|\frac{1}{2}|$ is between 0 and 1.

To graph the function, we calculate and plot points. Because we know the general shape of the graph, we can sketch the graph using only a few points.

x	$g(x) = \frac{1}{2}x^2$
-3	$\frac{9}{2}$
-2	2
-1	$\frac{1}{2}$
0	0
1	$\frac{1}{2}$
2	2
3	$\frac{9}{2}$

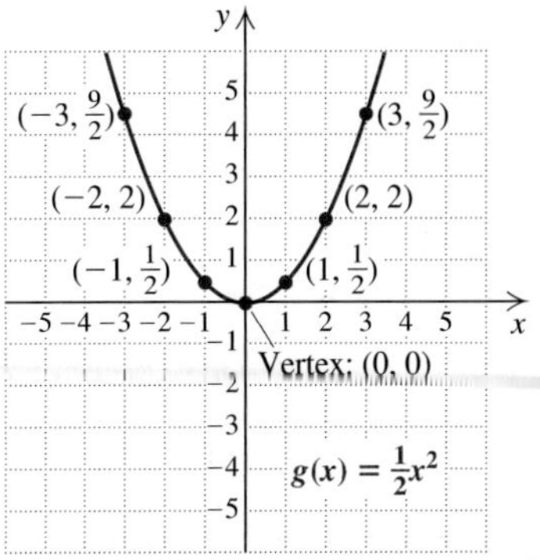

Note in Example 2 that since a parabola is symmetric, we can calculate function values for inputs to one side of the vertex and then plot the mirror images of those points on the other "half" of the graph.

The Graph of $f(x) = a(x - h)^2$

We could now consider graphs of

$$f(x) = ax^2 + bx + c,$$

where b and c are not both 0. In effect, we will do so, but in a different form. It turns out to be convenient to first graph $f(x) = a(x - h)^2$, where h is some constant. This allows us to observe similarities to the graphs drawn above.

Interactive Discovery

Graph each of the following pairs of equations in a $[-5, 5, -10, 10]$ window. For each pair, answer the following questions.

a) What is the vertex of the graph of y_2?

b) What is the axis of symmetry of the graph of y_2?

c) Is the graph of y_2 narrower, wider, or the same shape as the graph of y_1?

1. $y_1 = x^2;\ y_2 = (x - 3)^2$

2. $y_1 = 2x^2;\ y_2 = 2(x + 1)^2$

3. $y_1 = -\frac{1}{2}x^2;\ y_2 = -\frac{1}{2}\left(x - \frac{3}{2}\right)^2$

4. $y_1 = -3x^2;\ y_2 = -3(x + 2)^2$

5. Describe the effect of h on the graph of $g(x) = a(x - h)^2$.

The graph of $g(x) = a(x - h)^2$ looks just like the graph of $f(x) = ax^2$, except that it is moved, or *translated*, $|h|$ units horizontally.

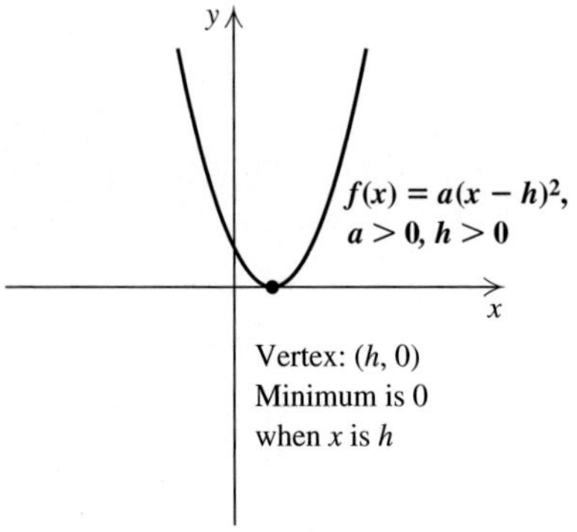

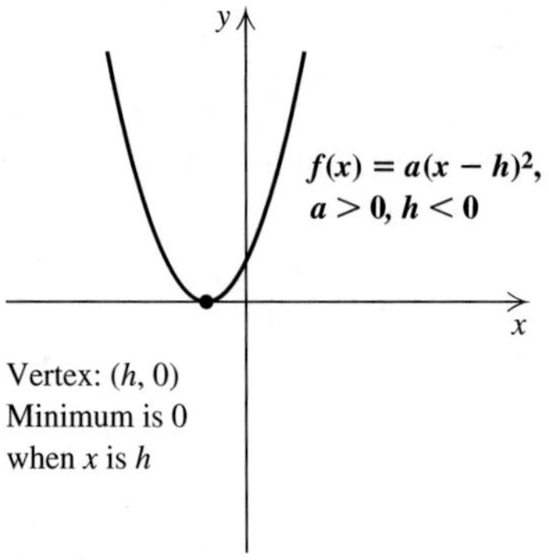

Graphing $f(x) = a(x - h)^2$

The graph of $f(x) = a(x - h)^2$ has the same shape as the graph of $y = ax^2$. The domain of f is $(-\infty, \infty)$.

If h is positive, the graph of $y = ax^2$ is shifted h units to the right.

If h is negative, the graph of $y = ax^2$ is shifted $|h|$ units to the left.

The vertex is $(h, 0)$, and the axis of symmetry is $x = h$.

For $a > 0$, the range of f is $[0, \infty)$. A minimum function value of 0 occurs when $x = h$.

For $a < 0$, the range of f is $(-\infty, 0]$. A maximum function value of 0 occurs when $x = h$.

EXAMPLE 3 Graph $g(x) = -2(x + 3)^2$, and determine the maximum value of $g(x)$ and where it occurs.

SOLUTION We rewrite the equation as $g(x) = -2[x - (-3)]^2$. In this case, $a = -2$ and $h = -3$, so the graph looks like that of $y = 2x^2$ translated 3 units to the left and, since $-2 < 0$, it opens downward. The vertex is $(-3, 0)$, and the axis of symmetry is $x = -3$. Plotting points as needed, we obtain the graph shown below. We say that g is a *reflection* of the graph of $f(x) = 2(x + 3)^2$ across the x-axis.

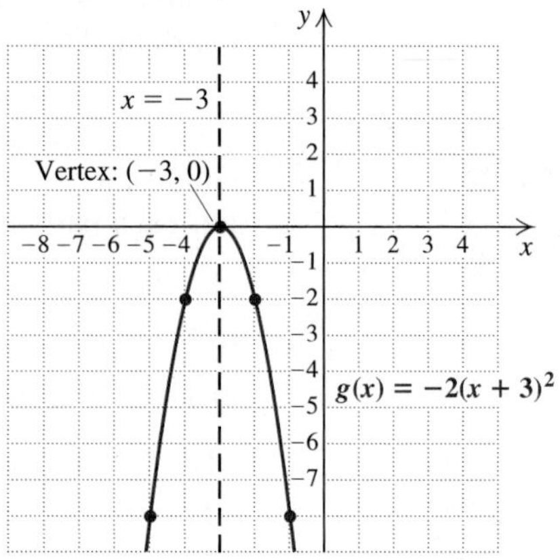

The maximum value of $g(x)$ is 0 when $x = -3$.

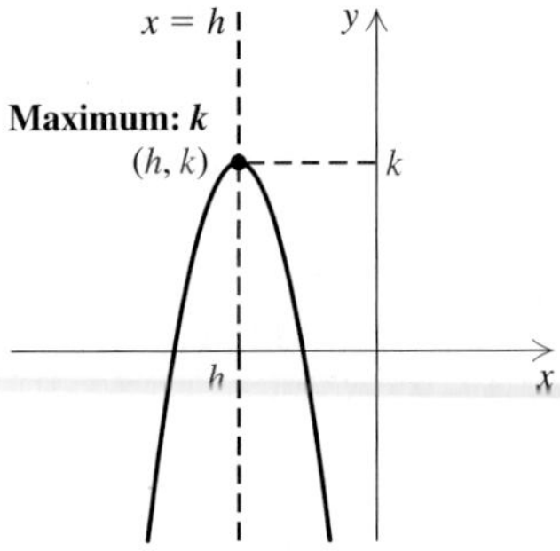

$f(x) = a(x - h)^2 + k,$
$a < 0$

The Graph of $f(x) = a(x - h)^2 + k$

If we add a positive constant k to the graph of $f(x) = a(x - h)^2$, the graph is moved up. If we add a negative constant k, the graph is moved down. The axis of symmetry for the parabola remains at $x = h$, but the vertex will be at (h, k).

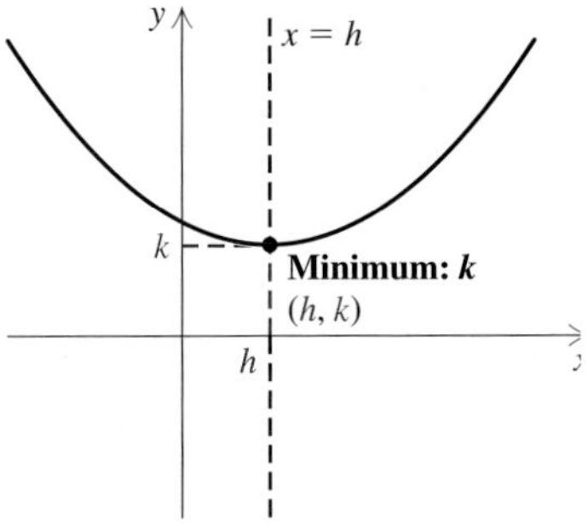

$f(x) = a(x - h)^2 + k,$
$a < 0$

Graphing $f(x) = a(x - h)^2 + k$

The graph of $f(x) = a(x - h)^2 + k$ has the same shape as the graph of $y = a(x - h)^2$.

If k is positive, the graph of $y = a(x - h)^2$ is shifted k units up.

If k is negative, the graph of $y = a(x - h)^2$ is shifted $|k|$ units down.

The vertex is (h, k), and the axis of symmetry is $x = h$.

The domain of f is $(-\infty, \infty)$.

For $a > 0$, the range of f is $[k, \infty)$. The minimum function value is k, which occurs when $x = h$.

For $a < 0$, the range of f is $(-\infty, k]$. The maximum function value is k, which occurs when $x = h$.

EXAMPLE 4 Graph $g(x) = (x - 3)^2 - 5$, and find the minimum function value and the range of g.

SOLUTION The graph will look like that of $f(x) = (x - 3)^2$ but shifted 5 units down. You can confirm this by plotting some points.

The vertex is now $(3, -5)$, and the minimum function value is -5. The range of g is $[-5, \infty)$.

x	$g(x) = (x - 3)^2 - 5$	
0	4	
1	-1	
2	-4	
3	-5	← Vertex
4	-4	
5	-1	
6	4	

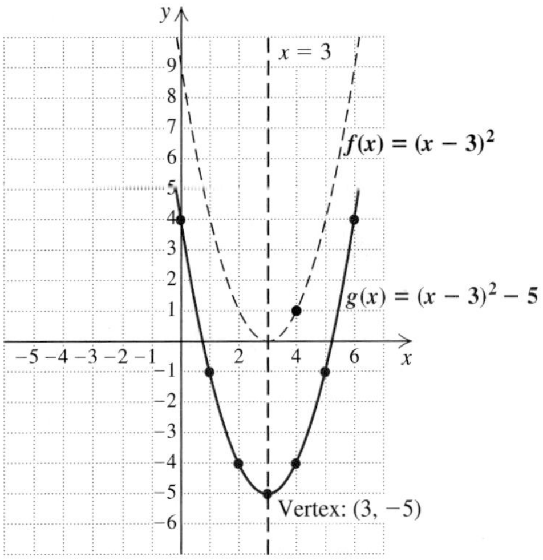

EXAMPLE 5 Graph $h(x) = \frac{1}{2}(x - 3)^2 + 5$, and find the minimum function value and the range of h.

SOLUTION The graph looks just like that of $f(x) = \frac{1}{2}x^2$ but moved 3 units to the right and 5 units up. The vertex is $(3, 5)$, and the axis of symmetry is $x = 3$. We draw $f(x) = \frac{1}{2}x^2$ and then shift the curve over and up.

By plotting some points, we have a check. The minimum function value is 5, and the range is $[5, \infty)$.

x	$h(x) = \frac{1}{2}(x-3)^2 + 5$	
0	$9\frac{1}{2}$	
1	7	
3	5	← Vertex
5	7	
6	$9\frac{1}{2}$	

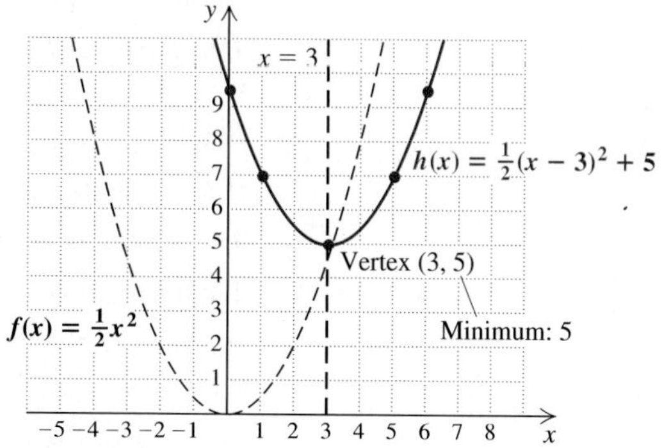

EXAMPLE 6 Graph $f(x) = -2(x + 3)^2 + 5$. Find the vertex, the axis of symmetry, the maximum or minimum function value, and the range of f.

SOLUTION We first express the equation in the equivalent form

$$f(x) = -2[x - (-3)]^2 + 5.$$

The graph looks like that of $y = -2x^2$ translated 3 units to the left and 5 units up. The vertex is $(-3, 5)$, and the axis of symmetry is $x = -3$. Since -2 is negative, we know that 5, the second coordinate of the vertex, is the maximum function value. The range of f is $(-\infty, 5]$.

We compute a few points as needed, selecting convenient x-values on either side of the vertex. The graph is shown here.

x	$f(x) = -2(x+3)^2 + 5$	
−4	3	
−3	5	← Vertex
−2	3	

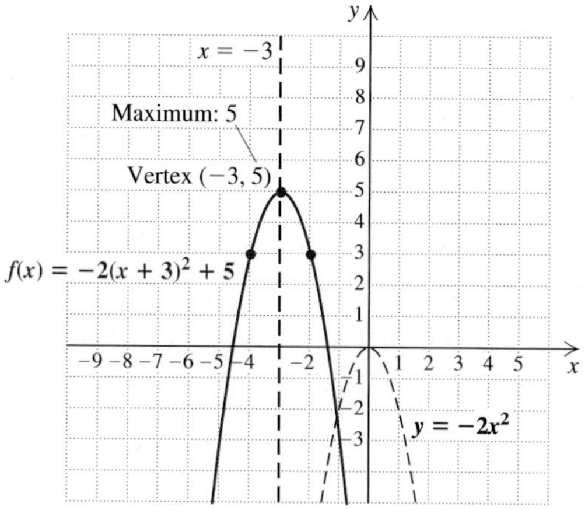

Because we can read the vertex and determine the orientation directly from the equation, it is often faster to graph $y = a(x - h)^2 + k$ by hand than by using a graphing calculator.

8.6 EXERCISE SET

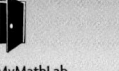

FOR EXTRA HELP

MathXL | MyMathLab | InterAct Math | AW Math Tutor Center | Video Lectures on CD: Disc 4 | Student's Solutions Manual

Concept Reinforcement *In each of Exercises 1–8, match the equation with the corresponding graph from those shown.*

1. ____ $f(x) = 2(x - 1)^2 + 3$

2. ____ $f(x) = -2(x - 1)^2 + 3$

3. ____ $f(x) = 2(x + 1)^2 + 3$

4. ____ $f(x) = 2(x - 1)^2 - 3$

5. ____ $f(x) = -2(x + 1)^2 + 3$

6. ____ $f(x) = -2(x + 1)^2 - 3$

7. ____ $f(x) = 2(x + 1)^2 - 3$

8. ____ $f(x) = -2(x - 1)^2 - 3$

a)
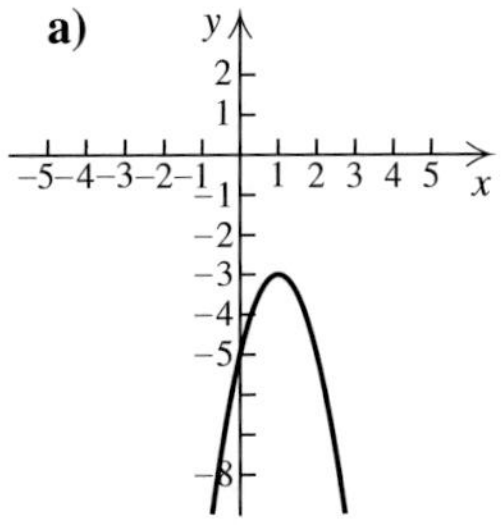

b)
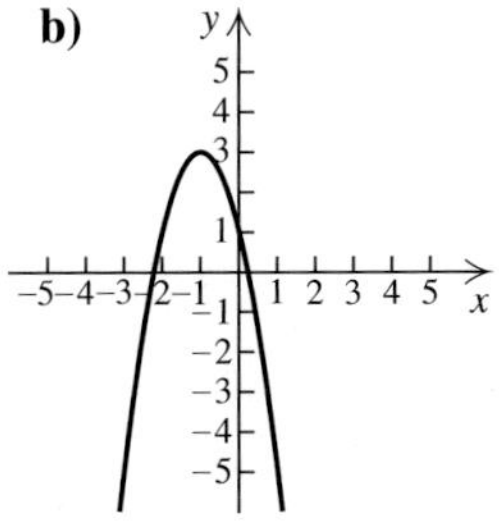

c)
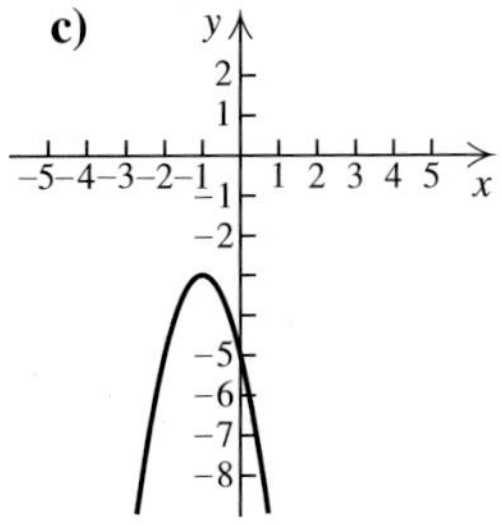

d)
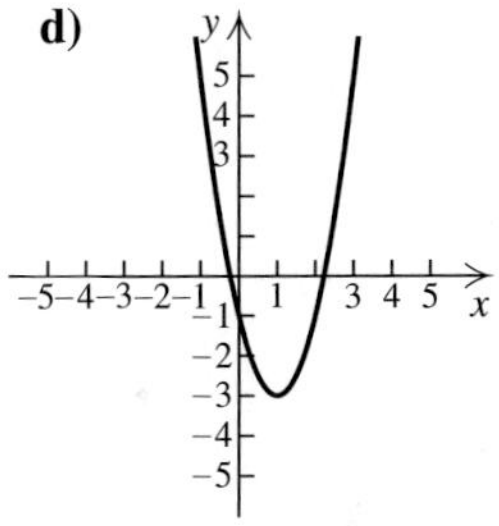

e)
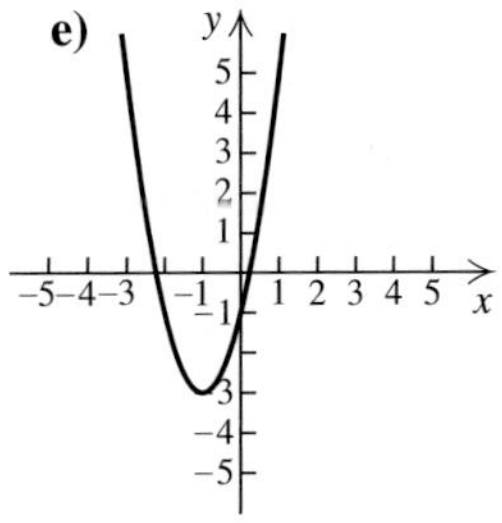

f)
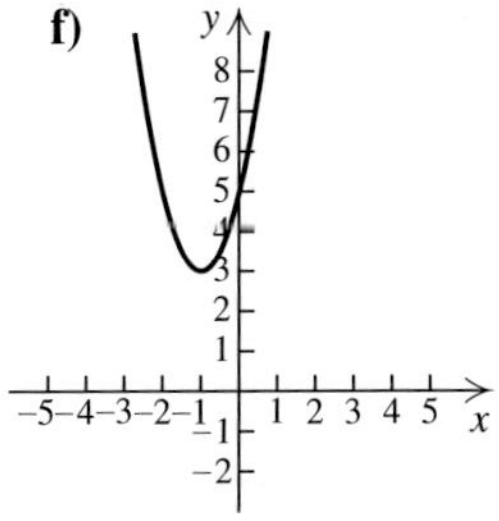

g)
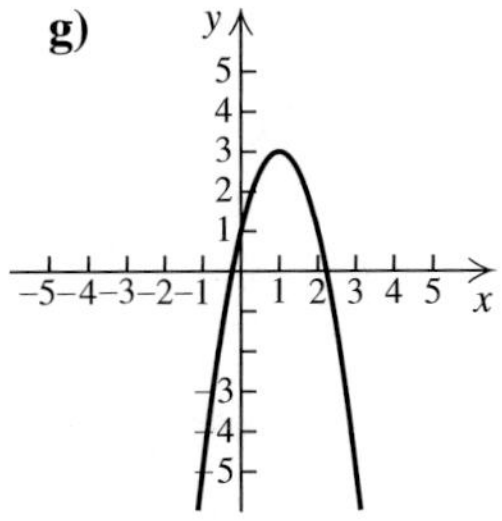

h)
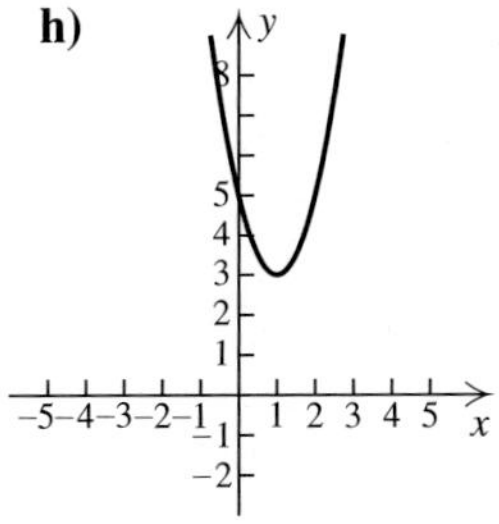

For each graph of a quadratic function

$$f(x) = a(x - h)^2 + k$$

in Exercises 9–14:

a) *Tell whether a is positive or negative.*
b) *Determine the vertex.*
c) *Determine the axis of symmetry.*
d) *Determine the range.*

9.

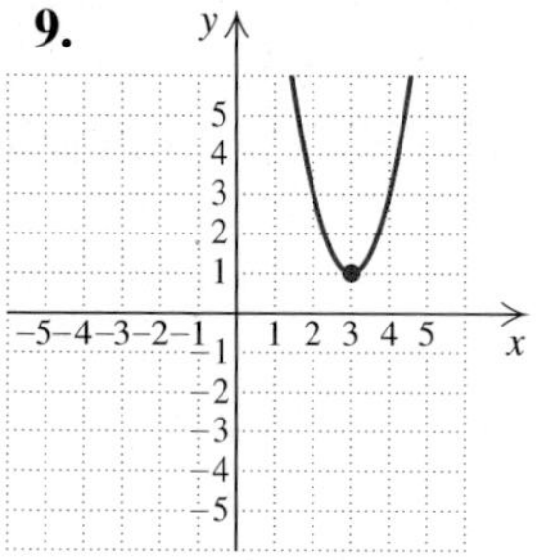

10.

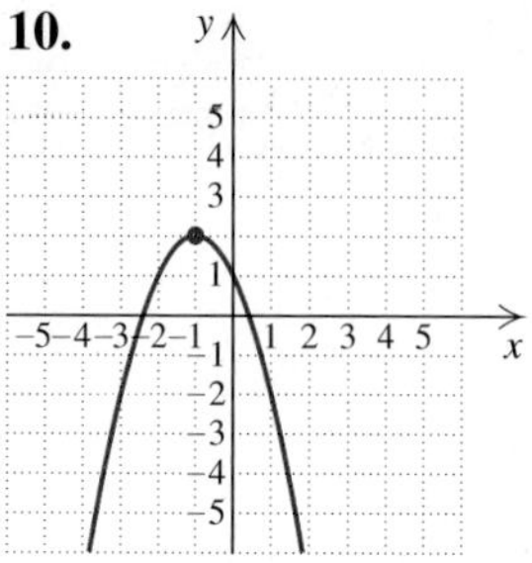

11.

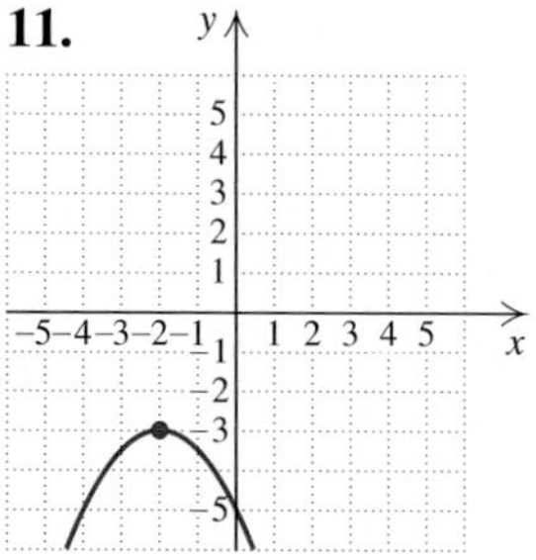

12.

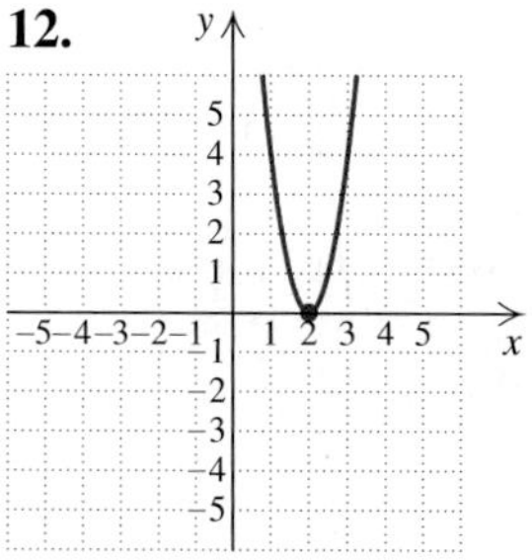

13.

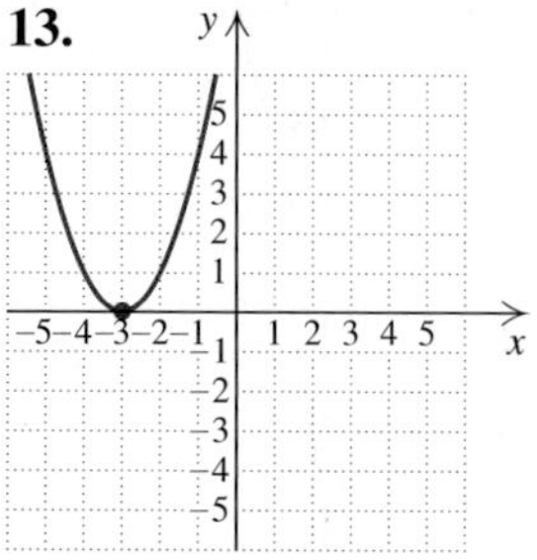

14.

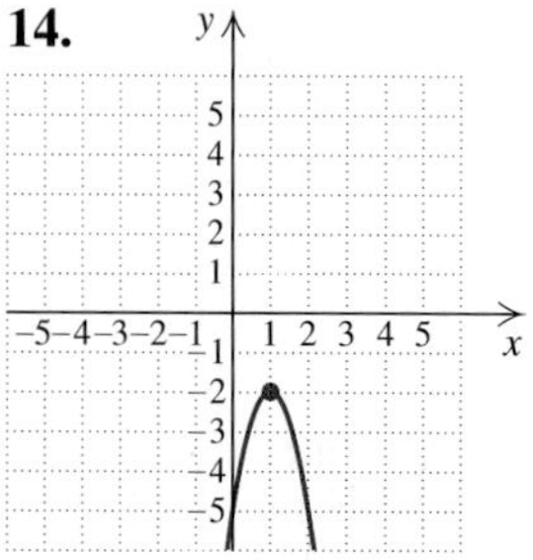

Graph.

15. $f(x) = x^2$

16. $f(x) = -x^2$

17. $f(x) = -2x^2$

18. $f(x) = -3x^2$

19. $g(x) = \frac{1}{3}x^2$

20. $g(x) = \frac{1}{4}x^2$

Aha! **21.** $h(x) = -\frac{1}{3}x^2$

22. $h(x) = -\frac{1}{4}x^2$

23. $f(x) = \frac{5}{2}x^2$

24. $f(x) = \frac{3}{2}x^2$

For each of the following, graph the function, label the vertex, and draw the axis of symmetry.

25. $g(x) = (x + 1)^2$

26. $g(x) = (x + 4)^2$

27. $f(x) = (x - 2)^2$

28. $f(x) = (x - 1)^2$

29. $h(x) = (x - 3)^2$

30. $h(x) = (x - 4)^2$

31. $f(x) = -(x + 1)^2$

32. $f(x) = -(x - 1)^2$

33. $g(x) = -(x - 2)^2$

34. $g(x) = -(x + 4)^2$

35. $f(x) = 2(x + 1)^2$

36. $f(x) = 2(x + 4)^2$

37. $h(x) = -\frac{1}{2}(x - 4)^2$

38. $h(x) = -\frac{3}{2}(x - 2)^2$

39. $f(x) = \frac{1}{2}(x - 1)^2$

40. $f(x) = \frac{1}{3}(x + 2)^2$

41. $f(x) = -2(x + 5)^2$

42. $f(x) = 2(x + 7)^2$

43. $h(x) = -3\left(x - \frac{1}{2}\right)^2$

44. $h(x) = -2\left(x + \frac{1}{2}\right)^2$

For each of the following, graph the function and find the vertex, the axis of symmetry, and the maximum value or the minimum value.

45. $f(x) = (x - 5)^2 + 2$

46. $f(x) = (x + 3)^2 - 2$

47. $f(x) = (x + 1)^2 - 3$

48. $f(x) = (x - 1)^2 + 2$

49. $g(x) = (x + 4)^2 + 1$

50. $g(x) = -(x - 2)^2 - 4$

51. $h(x) = -2(x - 1)^2 - 3$

52. $h(x) = -2(x + 1)^2 + 4$

53. $f(x) = 2(x + 4)^2 + 1$

54. $f(x) = 2(x - 5)^2 - 3$

55. $g(x) = -\frac{3}{2}(x - 1)^2 + 4$

56. $g(x) = \frac{3}{2}(x + 2)^2 - 3$

Without graphing, find the vertex, the axis of symmetry, and the maximum value or the minimum value.

57. $f(x) = 6(x - 8)^2 + 7$

58. $f(x) = 4(x + 5)^2 - 6$

59. $h(x) = -\frac{2}{7}(x + 6)^2 + 11$

60. $h(x) = -\frac{3}{11}(x - 7)^2 - 9$

61. $f(x) = 7\left(x + \frac{1}{4}\right)^2 - 13$

62. $f(x) = 6\left(x - \frac{1}{4}\right)^2 + 15$

63. $f(x) = \sqrt{2}(x + 4.58)^2 + 65\pi$

64. $f(x) = 4\pi(x - 38.2)^2 - \sqrt{34}$

TW **65.** While trying to graph $y = -\frac{1}{2}x^2 + 3x + 1$, Omar gets the following screen. How can Omar tell at a glance that a mistake has been made?

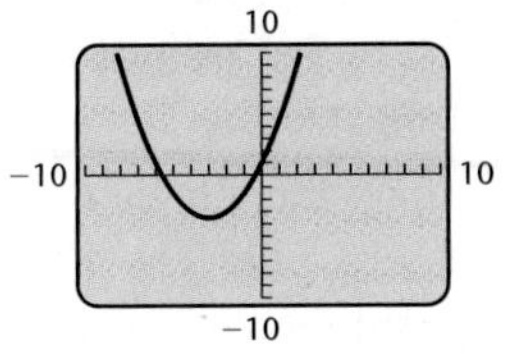

TW **66.** Explain, without plotting points, why the graph of $y = (x + 2)^2$ looks like the graph of $y = x^2$ translated 2 units to the left.

Skill Maintenance

Graph using intercepts. [2.3]

67. $2x - 7y = 28$

68. $6x - 3y = 36$

Solve each system. [3.2]

69. $3x + 4y = -19,$
$7x - 6y = -29$

70. $5x + 7y = 9,$
$3x - 4y = -11$

Replace the blanks with constants to form a true equation. [8.1]

71. $x^2 + 5x + \_\_\_ = (x + \_\_\_)^2$

72. $x^2 - 9x + \_\_\_ = (x - \_\_\_)^2$

Synthesis

TW **73.** Before graphing a quadratic function, Sophie always plots five points. First, she calculates and plots the coordinates of the vertex. Then she plots *four* more points after calculating *two* more ordered pairs. How is this possible?

TW **74.** If the graphs of $f(x) = a_1(x - h_1)^2 + k_1$ and $g(x) = a_2(x - h_2)^2 + k_2$ have the same shape, what, if anything, can you conclude about the a's, the h's, and the k's? Why?

Write an equation for a function having a graph with the same shape as the graph of $f(x) = \frac{3}{5}x^2$, but with the given point as the vertex.

75. (4, 1)

76. (2, 6)

77. (3, −1)

78. (5, −6)

79. (−2, −5)

80. (−4, −2)

For each of the following, write the equation of the parabola that has the shape of $f(x) = 2x^2$ or $g(x) = -2x^2$ and has a maximum or minimum value at the specified point.

81. Minimum: (2, 0)

82. Minimum: (−4, 0)

83. Maximum: (0, 3)

84. Maximum: (3, 8)

Find an equation for a quadratic function F that satisfies the following conditions.

85. The graph of F is the same shape as the graph of f, where $f(x) = 3(x + 2)^2 + 7$, and $F(x)$ is a minimum at the same point that $g(x) = -2(x - 5)^2 + 1$ is a maximum.

86. The graph of F is the same shape as the graph of f, where $f(x) = -\frac{1}{3}(x - 2)^2 + 7$, and $F(x)$ is a maximum at the same point that $g(x) = 2(x + 4)^2 - 6$ is a minimum.

Functions other than parabolas can be translated. When calculating $f(x)$, if we replace x with $x - h$, where h is a constant, the graph will be moved horizontally. If we replace $f(x)$ with $f(x) + k$, the graph will be moved vertically. Use the graph below for Exercises 87–92.

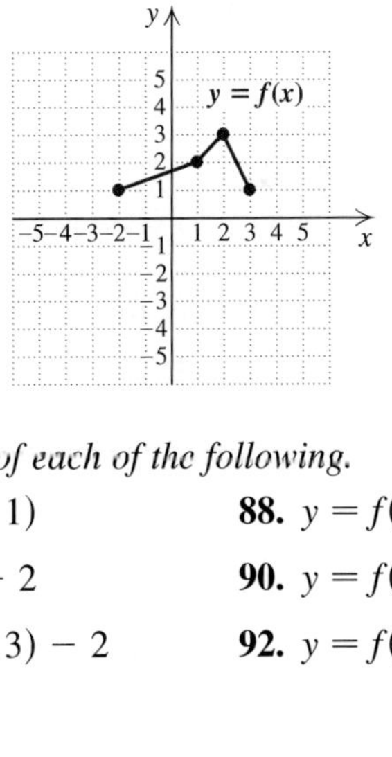

Draw a graph of each of the following.

87. $y = f(x - 1)$

88. $y = f(x + 2)$

89. $y = f(x) + 2$

90. $y = f(x) - 3$

91. $y = f(x + 3) - 2$

92. $y = f(x - 3) + 1$

Collaborative Corner

Match the Graph

Focus: Graphing quadratic functions
Time: 15–20 minutes
Group Size: 6
Materials: Index cards

ACTIVITY

1. On each of six index cards, write one of the following equations:

$$y = \tfrac{1}{2}(x - 3)^2 + 1; \quad y = \tfrac{1}{2}(x - 1)^2 + 3;$$
$$y = \tfrac{1}{2}(x + 1)^2 - 3; \quad y = \tfrac{1}{2}(x + 3)^2 + 1;$$
$$y = \tfrac{1}{2}(x + 3)^2 - 1; \quad y = \tfrac{1}{2}(x + 1)^2 + 3.$$

2. Fold each index card and mix up the six cards in a hat or bag. Then, one by one, each group member should select one of the equations. Do not let anyone see your equation.
3. Each group member should carefully graph the equation selected. Make the graph large enough so that when it is finished, it can be easily viewed by the rest of the group. Be sure to scale the axes and label the vertex, but **do not label the graph with the equation used.**
4. When all group members have drawn a graph, place the graphs in a pile. The group should then match and agree on the correct equation for each graph *with no help from the person who drew the graph.* If a mistake has been made and a graph has no match, determine what its equation *should* be.
5. Compare your group's labeled graphs with those of other groups to reach consensus within the class on the correct label for each graph.

8.7 More About Graphing Quadratic Functions

Finding the Vertex ■ Finding Intercepts

Finding the Vertex

By *completing the square* (see Section 8.1), we can rewrite any polynomial $ax^2 + bx + c$ in the form $a(x - h)^2 + k$. Once that has been done, the procedures discussed in Section 8.6 will enable us to graph any quadratic function.

EXAMPLE 1 Graph: $g(x) = x^2 - 6x + 4$.

SOLUTION We have

$$g(x) = x^2 - 6x + 4$$
$$= (x^2 - 6x) + 4.$$

Study Tip

Use What You Know

An excellent and common strategy for solving any new type of problem is to rewrite the problem in an equivalent form that we already know how to solve. Although this is not always feasible, when it is—as in most of the problems in this section—it can make a new topic much easier to learn.

To complete the square inside the parentheses, we take half the x-coefficient, $\frac{1}{2} \cdot (-6) = -3$, and square it to get $(-3)^2 = 9$. Then we add $9 - 9$ inside the parentheses:

$$g(x) = (x^2 - 6x + 9 - 9) + 4 \qquad \textbf{The effect is of adding 0.}$$
$$= (x^2 - 6x + 9) + (-9 + 4) \qquad \textbf{Using the associative law of addition to regroup}$$
$$= (x - 3)^2 - 5. \qquad \textbf{Factoring and simplifying}$$

This equation appeared as Example 4 of Section 8.6. The graph is that of $f(x) = x^2$ translated 3 units to the right and 5 units down. The vertex is $(3, -5)$.

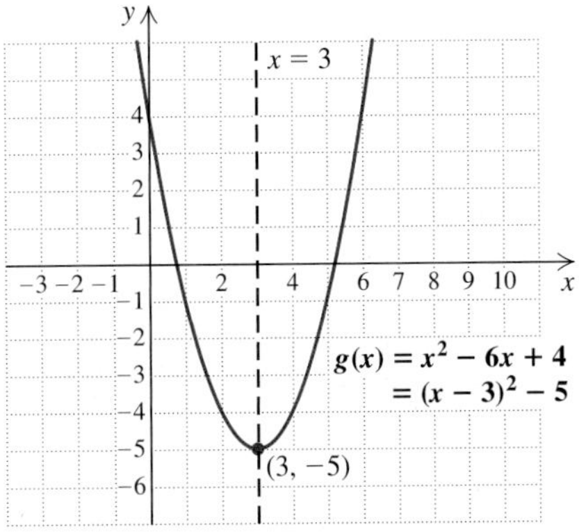

When the leading coefficient is not 1, we factor out that number from the first two terms. Then we complete the square and use the distributive law.

EXAMPLE 2 Graph: $f(x) = 3x^2 + 12x + 13$.

SOLUTION Since the coefficient of x^2 is not 1, we need to factor out that number—in this case, 3—from the first two terms. Remember that we want the form $f(x) = a(x - h)^2 + k$:

$$f(x) = 3x^2 + 12x + 13 = 3(x^2 + 4x) + 13.$$

Now we complete the square as before. We take half of the x-coefficient, $\frac{1}{2} \cdot 4 = 2$, and square it: $2^2 = 4$. Then we add $4 - 4$ inside the parentheses:

$$f(x) = 3(x^2 + 4x + 4 - 4) + 13. \qquad \textbf{Adding 4 − 4, or 0, inside the parentheses}$$

The distributive law allows us to separate the -4 from the perfect-square trinomial so long as it is multiplied by 3:

$$f(x) = 3(x^2 + 4x + 4) + 3(-4) + 13 \qquad \textbf{This leaves a perfect-square trinomial inside the parentheses.}$$

The −4 was added inside the parentheses. To separate it, we *must* multiply it by 3.

$$= 3(x + 2)^2 + 1. \qquad \textbf{Factoring and simplifying}$$

The vertex is $(-2, 1)$, and the axis of symmetry is $x = -2$. The coefficient of x^2 is 3, so the graph is narrow and opens upward. We choose a few x-values on either side of the vertex, compute y-values, and then graph the parabola.

x	$f(x) = 3(x + 2)^2 + 1$
-2	1 ← Vertex
-3	4
-1	4

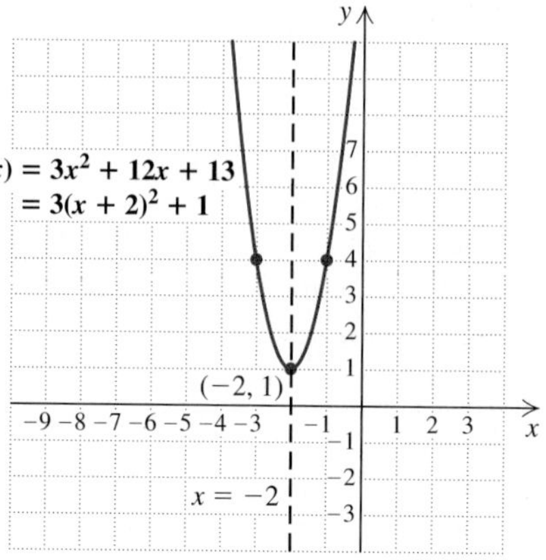

EXAMPLE 3 Graph: $f(x) = -2x^2 + 10x - 7$.

SOLUTION We first find the vertex by completing the square. To do so, we factor out -2 from the first two terms of the expression. This makes the coefficient of x^2 inside the parentheses 1:

$$f(x) = -2x^2 + 10x - 7$$
$$= -2(x^2 - 5x) - 7.$$

Now we complete the square as before. We take half of the x-coefficient and square it to get $\frac{25}{4}$. Then we add $\frac{25}{4} - \frac{25}{4}$ inside the parentheses:

$$f(x) = -2\left(x^2 - 5x + \tfrac{25}{4} - \tfrac{25}{4}\right) - 7$$
$$= -2\left(x^2 - 5x + \tfrac{25}{4}\right) + (-2)\left(-\tfrac{25}{4}\right) - 7 \quad \textbf{Multiplying by } -2\textbf{, using the distributive law, and regrouping}$$
$$= -2\left(x - \tfrac{5}{2}\right)^2 + \tfrac{11}{2}. \quad \textbf{Factoring and simplifying}$$

The vertex is $\left(\frac{5}{2}, \frac{11}{2}\right)$, and the axis of symmetry is $x = \frac{5}{2}$. The coefficient of x^2, -2, is negative, so the graph opens downward. We plot a few points on either side of the vertex, including the y-intercept, $f(0)$, and graph the parabola.

x	$f(x)$
$\frac{5}{2}$	$\frac{11}{2}$ ← Vertex
0	-7 ← y-intercept
1	1
4	1

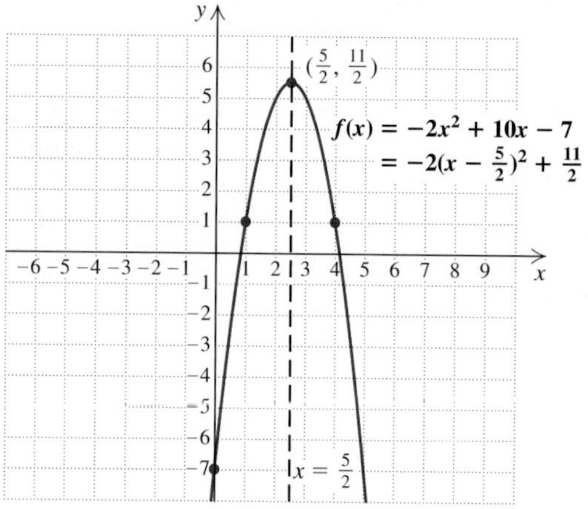

The method used in Examples 1–3 can be generalized to find a formula for locating the vertex. We complete the square as follows:

$$f(x) = ax^2 + bx + c$$

$$= a\left(x^2 + \frac{b}{a}x\right) + c. \quad \textbf{Factoring } a \textbf{ out of the first two terms. Check by multiplying.}$$

Half of the x-coefficient, $\frac{b}{a}$, is $\frac{b}{2a}$. We square it to get $\frac{b^2}{4a^2}$ and add $\frac{b^2}{4a^2} - \frac{b^2}{4a^2}$ inside the parentheses. Then we distribute the a and regroup terms:

$$f(x) = a\left(x^2 + \frac{b}{a}x + \frac{b^2}{4a^2} - \frac{b^2}{4a^2}\right) + c$$

$$= a\left(x^2 + \frac{b}{a}x + \frac{b^2}{4a^2}\right) + a\left(-\frac{b^2}{4a^2}\right) + c \quad \textbf{Using the distributive law}$$

$$= a\left(x + \frac{b}{2a}\right)^2 + \frac{-b^2}{4a} + \frac{4ac}{4a} \quad \textbf{Factoring and finding a common denominator}$$

$$= a\left[x - \left(-\frac{b}{2a}\right)\right]^2 + \frac{4ac - b^2}{4a}.$$

Thus we have the following.

The Vertex of a Parabola The vertex of the parabola given by $f(x) = ax^2 + bx + c$ is

$$\left(-\frac{b}{2a}, f\left(-\frac{b}{2a}\right)\right), \quad \text{or} \quad \left(-\frac{b}{2a}, \frac{4ac - b^2}{4a}\right).$$

The x-coordinate of the vertex is $-b/(2a)$. The axis of symmetry is $x = -b/(2a)$. The second coordinate of the vertex is most commonly found by computing $f\left(-\frac{b}{2a}\right)$.

Student Notes

An excellent way to remember a formula is to understand its derivation. Check with your instructor to determine what, if any, formulas you will be expected to remember.

Let's reexamine Example 3 to see how we could have found the vertex directly. From the formula above,

$$\text{the } x\text{-coordinate of the vertex is } -\frac{b}{2a} = -\frac{10}{2(-2)} = \frac{5}{2}.$$

Substituting $\frac{5}{2}$ into $f(x) = -2x^2 + 10x - 7$, we find the second coordinate of the vertex:

$$f\left(\tfrac{5}{2}\right) = -2\left(\tfrac{5}{2}\right)^2 + 10\left(\tfrac{5}{2}\right) - 7$$

$$= -2\left(\tfrac{25}{4}\right) + 25 - 7$$

$$= -\tfrac{25}{2} + 18$$

$$= -\tfrac{25}{2} + \tfrac{36}{2} = \tfrac{11}{2}.$$

The vertex is $\left(\frac{5}{2}, \frac{11}{2}\right)$. The axis of symmetry is $x = \frac{5}{2}$.

Note that a quadratic function has a maximum or a minimum value at the vertex of its graph. Thus determining the maximum or minimum value of the function allows us to find the vertex of the graph.

Maximums and Minimums

On most graphing calculators, we can find a maximum or minimum function value for any given interval. This MAXIMUM or MINIMUM feature is often found in the CALC menu. (CALC is the 2nd option associated with the TRACE key.)

To find a maximum or a minimum, enter and graph the function, choosing a viewing window that will show the vertex. Next, press CALC and choose either the MAXIMUM or MINIMUM option in the menu. The graphing calculator will find the maximum or minimum function value over a specified interval, so the left and right endpoints, or bounds, of the interval must be entered as well as a guess near where the maximum or minimum occurs. The calculator will return the coordinates of the point for which the function value is a maximum or minimum within the interval.

EXAMPLE 4 Use a graphing calculator to determine the vertex of the graph of the function given by $f(x) = -2x^2 + 10x - 7$.

Student Notes

It is easy to press a wrong key when using a calculator. Always check to see if your answer is reasonable. For example, if the MINIMUM option were chosen in Example 4 instead of MAXIMUM, one endpoint of the interval would have been returned instead of the maximum. In this case, noting that the highest point on the graph is marked is a quick check that we did indeed find the maximum.

SOLUTION The coefficient of x^2 is negative, so the parabola opens downward and the function has a maximum value. We enter and graph the function, choosing a window that will show the vertex.

We choose the MAXIMUM option from the CALC menu. The calculator will prompt us first for a left bound. After visually locating the vertex, we either move the cursor to a point left of the vertex using the left and right arrow keys or type a value of x that is less than the x-value at the vertex. (See the graph on the left below.) After pressing ENTER, we choose a right bound in a similar manner (as shown in the graph on the right), and press ENTER again. Finally, we enter a guess that is close to the vertex (as in the graph on the left at the top of the next page), and press ENTER. (Many calculators will use the right bound as a guess if we simply press ENTER.) The calculator indicates that the coordinates of the vertex are (2.5, 5.5). (See the graph on the right at the top of the next page.)

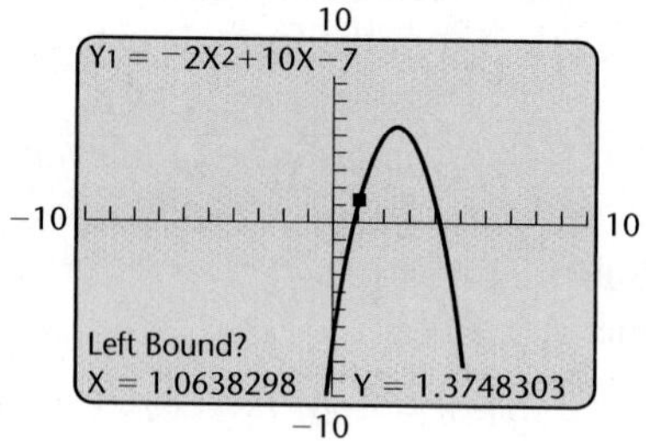

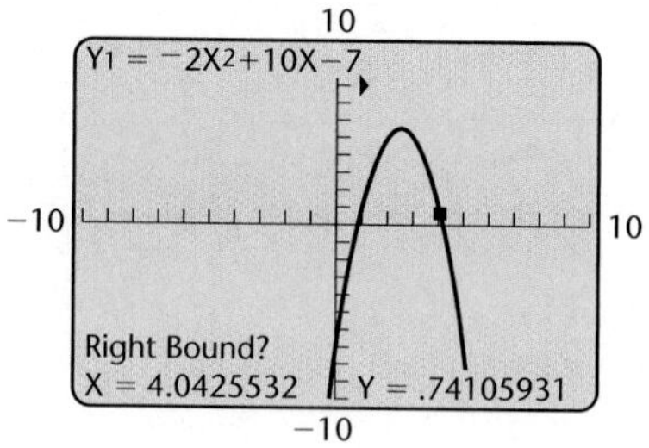

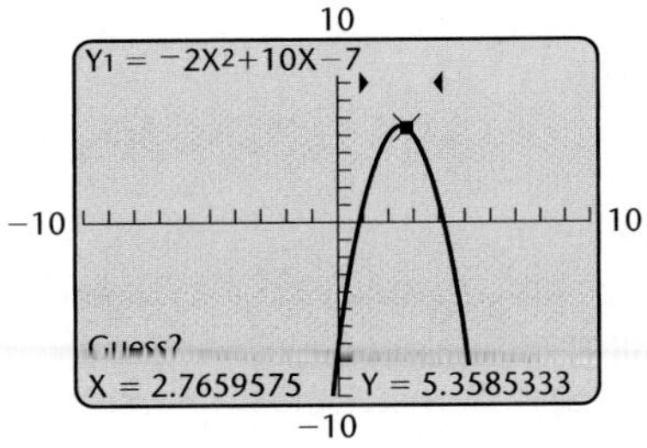

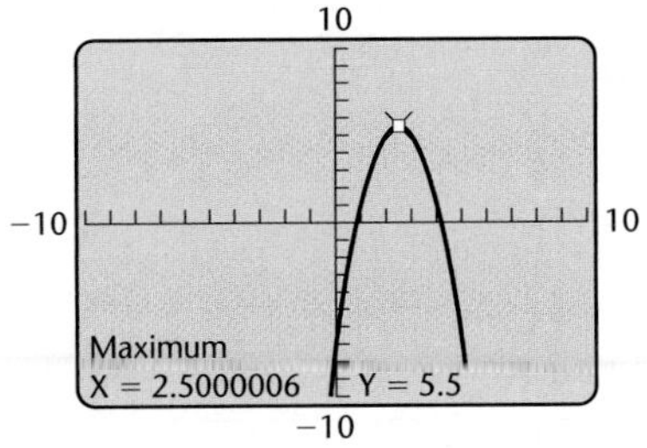

In Example 3, we found the coordinates of the vertex of the graph of $f(x) = -2x^2 + 10x - 7$, $\left(\frac{5}{2}, \frac{11}{2}\right)$, by completing the square. On p. 661, we also found these coordinates using the formula for the vertex of a parabola. The coordinates found using a calculator in Example 4 were in decimal notation. Because of the calculator's method used to find the maximum function value, the coordinates may not be exact and can indeed vary slightly for choices of windows. Since $2.5000006 \approx \frac{5}{2}$, and since $5.5 = \frac{11}{2}$, the coordinates check.

We have actually developed three methods for finding the vertex. One is by completing the square, the second is by using a formula, and the third is by using a graphing calculator. You should check with your instructor about which method to use.

Finding Intercepts

For any function, the y-intercept occurs at $f(0)$. Therefore, for $f(x) = ax^2 + bx + c$, the y-intercept is simply $(0, c)$. To find x-intercepts, we look for points where $y = 0$ or $f(x) = 0$. Thus, for $f(x) = ax^2 + bx + c$, the x-intercepts occur at those x-values for which $ax^2 + bx + c = 0$.

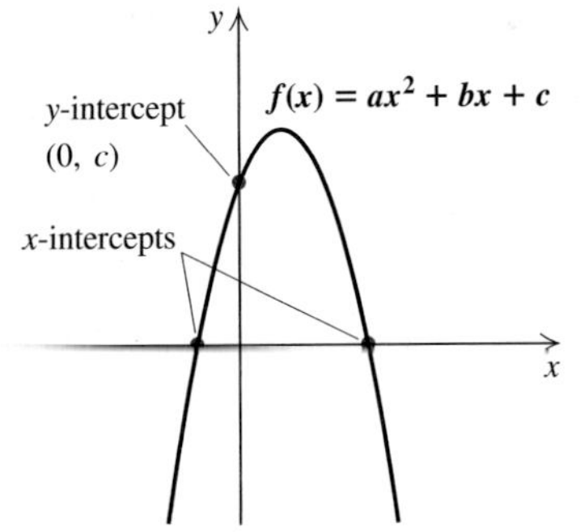

EXAMPLE 5 Find the x- and y-intercepts of the graph of

$$f(x) = x^2 - 2x - 2.$$

SOLUTION The y-intercept is simply $(0, f(0))$, or $(0, -2)$. To find the x-intercepts, we solve the equation

$$0 = x^2 - 2x - 2.$$

We are unable to factor $x^2 - 2x - 2$, so we use the quadratic formula and get $x = 1 \pm \sqrt{3}$. Thus the x-intercepts are $\left(1 - \sqrt{3}, 0\right)$ and $\left(1 + \sqrt{3}, 0\right)$.

If graphing, we would approximate, to get $(-0.7, 0)$ and $(2.7, 0)$.

Connecting the Concepts

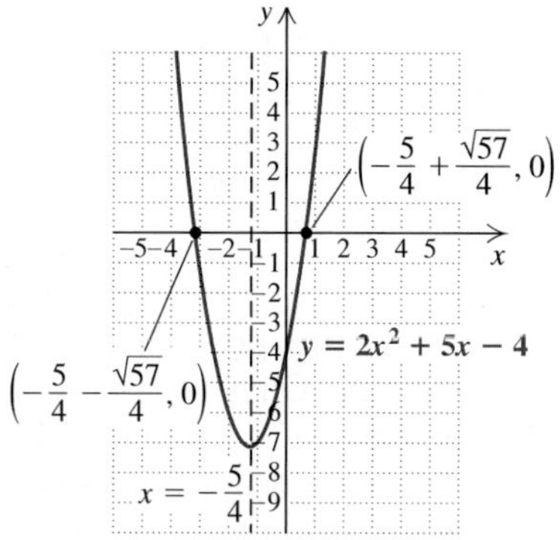

Because the graph of a quadratic equation is symmetric, the x-intercepts of the graph, if they exist, will be symmetric with respect to the axis of symmetry. This symmetry can be seen directly if the x-intercepts are found using the quadratic formula.

For example, the x-intercepts of the graph of $y = 2x^2 + 5x - 4$ are

$$\left(-\frac{5}{4}+\frac{\sqrt{57}}{4}, 0\right) \text{ and } \left(-\frac{5}{4}-\frac{\sqrt{57}}{4}, 0\right).$$

For this equation, the axis of symmetry is $x = -\frac{5}{4}$ and the x-intercepts are $\frac{\sqrt{57}}{4}$ units to the left and right of $-\frac{5}{4}$ on the x-axis.

The Graph of a Quadratic Function Given by $f(x) = ax^2 + bx + c$ or $f(x) = a(x - h)^2 + k$

The graph is a parabola.

The vertex is (h, k) or $\left(-\frac{b}{2a}, f\left(-\frac{b}{2a}\right)\right)$.

The axis of symmetry is $x = h$.

The y-intercept of the graph is $(0, c)$.

The x-intercepts can be found by solving $ax^2 + bx + c = 0$.

If $b^2 - 4ac > 0$, there are two x-intercepts.
If $b^2 - 4ac = 0$, there is one x-intercept.
If $b^2 - 4ac < 0$, there are no x-intercepts.

The domain of the function is $(-\infty, \infty)$.

If a is positive: The graph opens upward.
The function has a minimum value, given by k.
This occurs when $x = h$.
The range of the function is $[k, \infty)$.

If a is negative: The graph opens downward.
The function has a maximum value, given by k.
This occurs when $x = h$.
The range of the function is $(-\infty, k]$.

Visualizing the Graph

Match each function with its graph.

1. $f(x) = 3x^2$

2. $f(x) = x^2 - 4$

3. $f(x) = (x - 4)^2$

4. $f(x) = x - 4$

5. $f(x) = -2x^2$

6. $f(x) = x + 3$

7. $f(x) = |x + 3|$

8. $f(x) = (x + 3)^2$

9. $f(x) = \sqrt{x + 3}$

10. $f(x) = (x + 3)^2 - 4$

Answers on page A-36

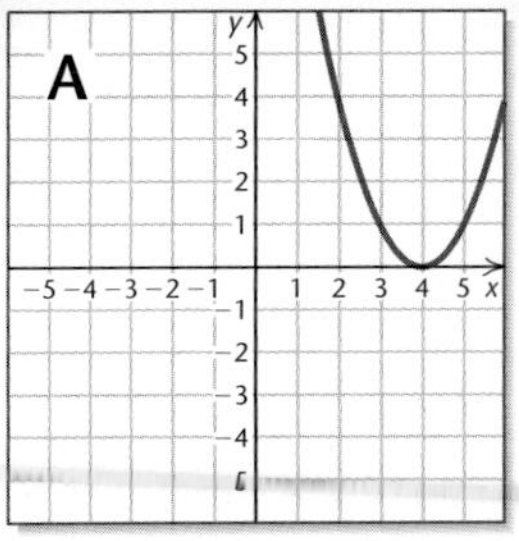

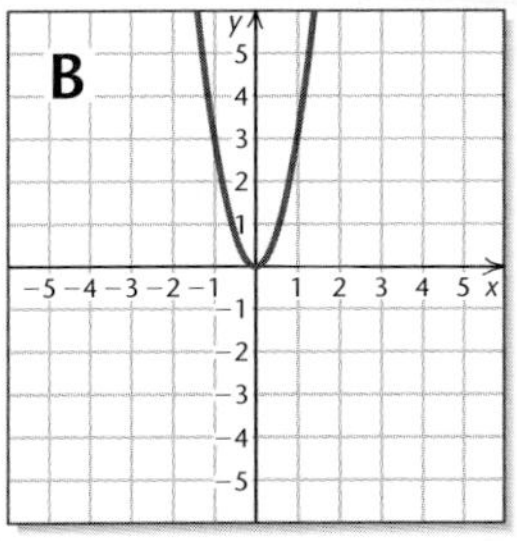

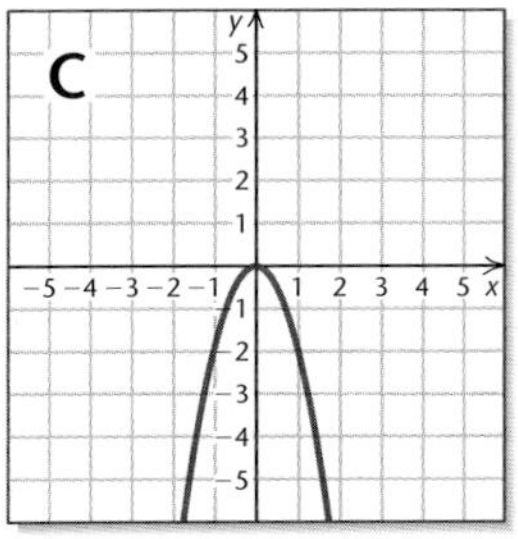

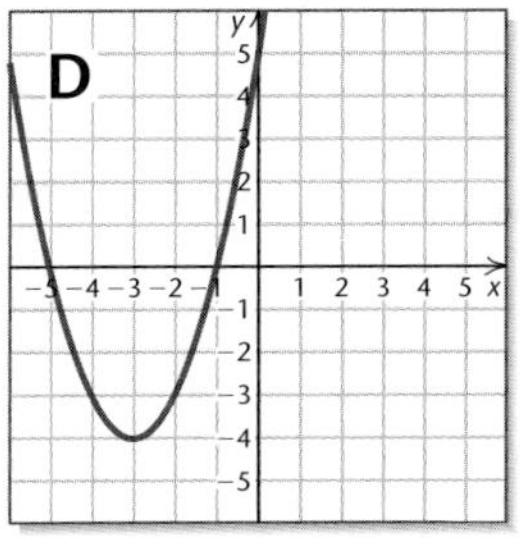

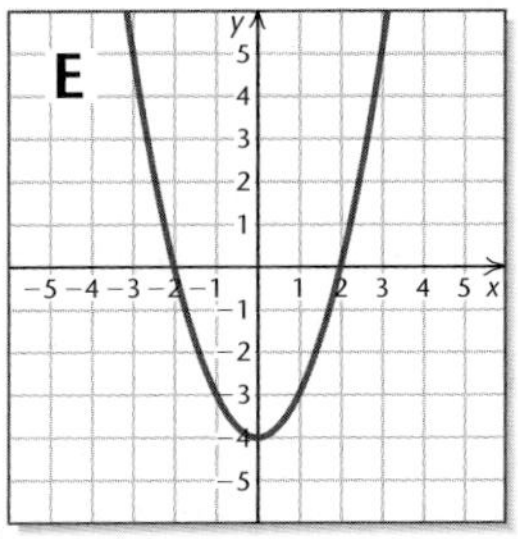

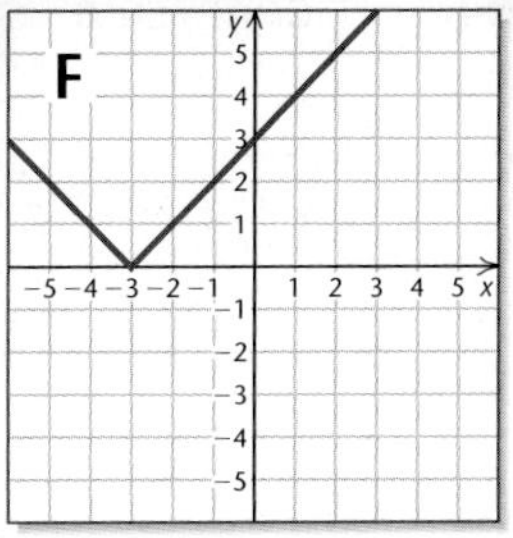

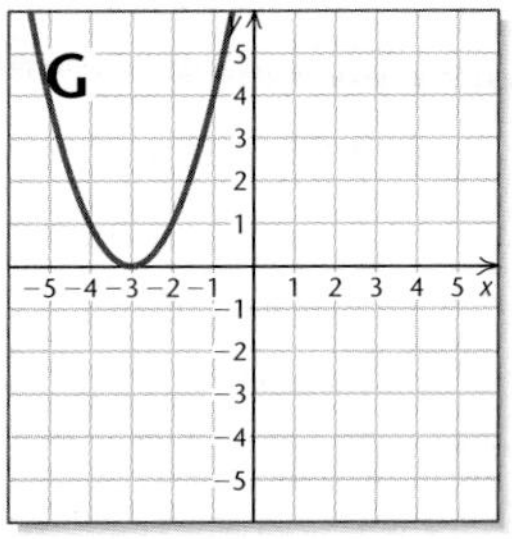

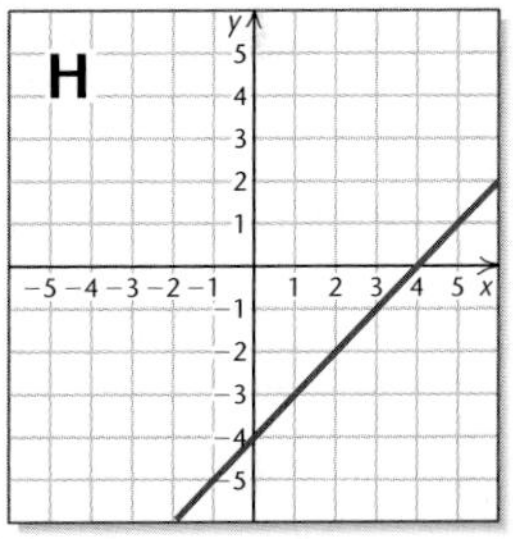

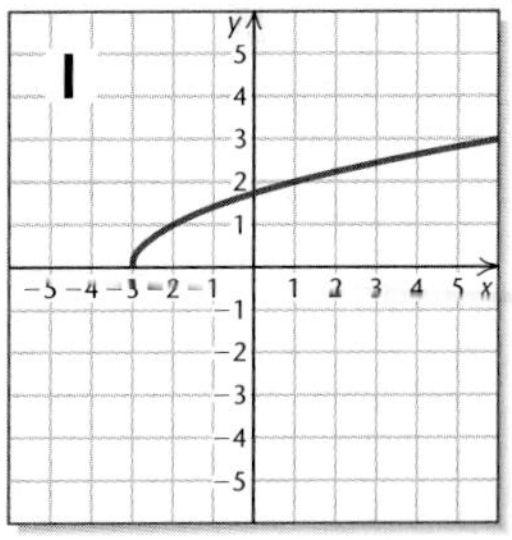

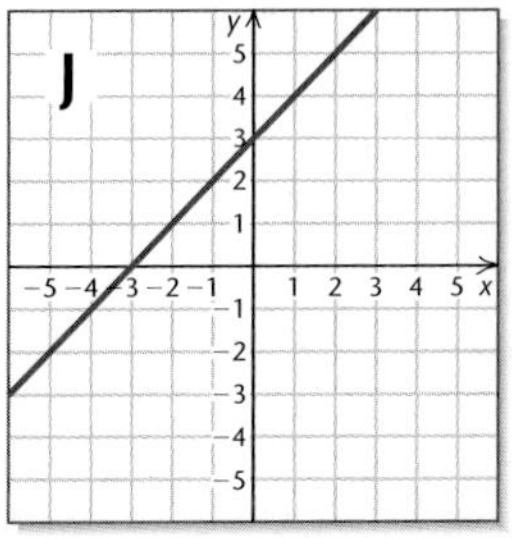

8.7 EXERCISE SET

FOR EXTRA HELP

MathXL | MyMathLab | InterAct Math | AW Math Tutor Center | Video Lectures on CD: Disc 4 | Student's Solutions Manual

Concept Reinforcement *Complete each of the following.*

1. The expressions $x^2 + 6x + 5$ and $(x^2 + 6x + 9) - \_\_\_ + 5$ are equivalent.

2. The expressions $x^2 + 8x + 3$ and $(x^2 + 8x + \_\_\_) - 16 + 3$ are equivalent.

3. The functions given by $f(x) = 2x^2 + 12x - 7$ and $f(x) = 2(x^2 + 6x + \_\_\_) - 18 - 7$ are equivalent.

4. The functions given by $g(x) = 3x^2 - 6x - 5$ and $g(x) = 3(x^2 - 2x + \_\_\_) - 3 - 5$ are equivalent.

5. The graph of $f(x) = -2(x - \_\_\_)^2 + 7$ has its vertex at (3, 7).

6. The graph of $g(x) = -3(x - 4)^2 + \_\_\_$ has its vertex at (4, 1).

7. The graph of $f(x) = 2\left(x - \_\_\_\right)^2 + \_\_\_$ has its axis of symmetry at $x = \frac{5}{2}$ and has its vertex at $\left(\frac{5}{2}, -4\right)$.

8. The graph of $g(x) = \frac{1}{2}(x - \_\_\_)^2 + \frac{7}{2}$ has $x = -2$ as its axis of symmetry and ___ as its vertex.

For each quadratic function, **(a)** *write the function in the form* $f(x) = a(x - h)^2 + k$ *and* **(b)** *find the vertex and the axis of symmetry.*

9. $f(x) = x^2 - 4x + 3$

10. $f(x) = x^2 + 6x + 13$

11. $f(x) = -x^2 + 3x - 10$

12. $f(x) = x^2 + 5x + 6$

13. $f(x) = 2x^2 - 7x + 1$

14. $f(x) = -2x^2 + 5x - 1$

For each quadratic function, **(a)** *find the vertex and the axis of symmetry and* **(b)** *graph the function.*

15. $f(x) = x^2 + 4x + 5$

16. $f(x) = x^2 + 2x - 5$

17. $g(x) = x^2 - 6x + 13$

18. $g(x) = x^2 - 4x + 5$

19. $f(x) = x^2 + 8x + 20$

20. $f(x) = x^2 - 10x + 21$

21. $h(x) = 2x^2 - 16x + 25$

22. $h(x) = 2x^2 + 16x + 23$

23. $f(x) = -x^2 + 2x + 5$

24. $f(x) = -x^2 - 2x + 7$

25. $g(x) = x^2 + 3x - 10$

26. $g(x) = x^2 + 5x + 4$

27. $f(x) = 3x^2 - 24x + 50$

28. $f(x) = 4x^2 + 8x - 3$

29. $h(x) = x^2 + 7x$

30. $h(x) = x^2 - 5x$

31. $f(x) = -2x^2 - 4x - 6$

32. $f(x) = -3x^2 + 6x + 2$

33. $g(x) = 2x^2 - 8x + 3$

34. $g(x) = 2x^2 + 5x - 1$

35. $f(x) = -3x^2 + 5x - 2$

36. $f(x) = -3x^2 - 7x + 2$

37. $h(x) = \frac{1}{2}x^2 + 4x + \frac{19}{3}$

38. $h(x) = \frac{1}{2}x^2 - 3x + 2$

Use a graphing calculator to find the vertex of the graph of each function.

39. $f(x) = x^2 + x - 6$

40. $f(x) = x^2 + 2x - 5$

41. $f(x) = 5x^2 - x + 1$

42. $f(x) = -4x^2 - 3x + 7$

43. $f(x) = -0.2x^2 + 1.4x - 6.7$

44. $f(x) = 0.5x^2 + 2.4x + 3.2$

Find the x- and y-intercepts. If no x-intercepts exist, state this.

45. $f(x) = x^2 - 6x + 3$

46. $f(x) = x^2 + 5x + 2$

47. $g(x) = -x^2 + 2x + 3$

48. $g(x) = x^2 - 6x + 9$

Aha! 49. $f(x) = x^2 - 9x$

50. $f(x) = x^2 - 7x$

51. $h(x) = -x^2 + 4x - 4$

52. $h(x) = 4x^2 - 12x + 3$

53. $f(x) = 2x^2 - 4x + 6$

54. $f(x) = x^2 - x + 2$

For each quadratic function, find **(a)** *the maximum or minimum value and* **(b)** *the x- and y-intercepts. Round to the nearest hundredth.*

55. $f(x) = 2.31x^2 - 3.135x - 5.89$

56. $f(x) = -18.8x^2 + 7.92x + 6.18$

57. $g(x) = -1.25x^2 + 3.42x - 2.79$

58. $g(x) = 0.45x^2 - 1.72x + 12.92$

TW 59. Does the graph of every quadratic function have a y-intercept? Why or why not?

TW 60. Is it possible for the graph of a quadratic function to have only one x-intercept if the vertex is off the x-axis? Why or why not?

Focused Review

Graph.

61. $f(x) = \frac{3}{2}x$ [2.2]

62. $f(x) = -\frac{2}{3}x$ [2.2]

63. $f(x) = \dfrac{2}{x}$ [2.1]

64. $f(x) = \dfrac{3}{x}$ [2.1]

65. $f(x) = 3x - 1$ [2.2]

66. $f(x) = (x - 2)^2 - 1$ [8.6]

67. $f(x) = x^2 - x - 6$ [8.7]

68. $f(x) = 2x^2 + x + 1$ [8.7]

Synthesis

TW 69. If the graphs of two quadratic functions have the same x-intercepts, will they also have the same vertex? Why or why not?

TW 70. Suppose that the graph of $f(x) = ax^2 + bx + c$ has $(x_1, 0)$ and $(x_2, 0)$ as x-intercepts. Explain why the graph of $g(x) = -ax^2 - bx - c$ will also have $(x_1, 0)$ and $(x_2, 0)$ as x-intercepts.

71. Graph the function
$$f(x) = x^2 - x - 6.$$
Then use the graph to approximate solutions to each of the following equations.
a) $x^2 - x - 6 = 2$
b) $x^2 - x - 6 = -3$

72. Graph the function
$$f(x) = \frac{x^2}{2} + x - \frac{3}{2}.$$
Then use the graph to approximate solutions to each of the following equations.
a) $\dfrac{x^2}{2} + x - \dfrac{3}{2} = 0$
b) $\dfrac{x^2}{2} + x - \dfrac{3}{2} = 1$
c) $\dfrac{x^2}{2} + x - \dfrac{3}{2} = 2$

Find an equivalent equation of the type
$$f(x) = a(x - h)^2 + k.$$

73. $f(x) = mx^2 - nx + p$

74. $f(x) = 3x^2 + mx + m^2$

75. A quadratic function has $(-1, 0)$ as one of its intercepts and $(3, -5)$ as its vertex. Find an equation for the function.

76. A quadratic function has $(4, 0)$ as one of its intercepts and $(-1, 7)$ as its vertex. Find an equation for the function.

Graph.

77. $f(x) = |x^2 - 1|$

78. $f(x) = |x^2 - 3x - 4|$

79. $f(x) = |2(x - 3)^2 - 5|$

8.8 Problem Solving and Quadratic Functions

Maximum and Minimum Problems ■ Fitting Quadratic Functions to Data

Let's look now at some of the many situations in which quadratic functions are used for problem solving.

Maximum and Minimum Problems

We have seen that for any quadratic function f, the value of $f(x)$ at the vertex is either a maximum or a minimum. Thus problems in which a quantity must be maximized or minimized can often be solved by finding the coordinates of a vertex, assuming the problem can be modeled with a quadratic function.

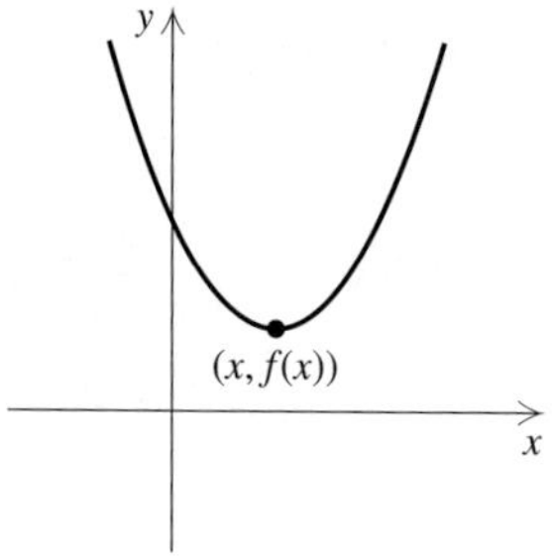

$f(x)$ at the vertex a minimum

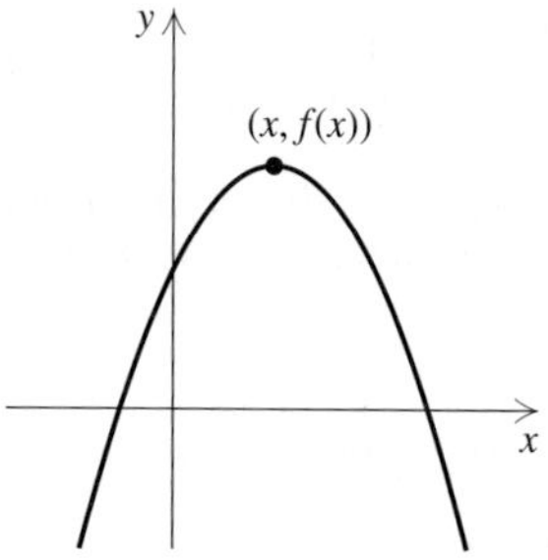

$f(x)$ at the vertex a maximum

Study Tip

Don't Fight Sleep

Study time is too precious to waste. If you find yourself with heavy eyelids, don't continue studying even if you're nearly finished. A problem that may require 20 minutes to complete when you are sleepy often takes 5 minutes when you are fresh and rested.

EXAMPLE 1 Newborn Calves. The number of pounds of milk per day recommended for a calf that is x weeks old can be approximated by $p(x)$, where $p(x) = -0.2x^2 + 1.3x + 6.2$ (*Source*: C. Chaloux, University of Vermont, 1998). When is a calf's milk consumption greatest and how much milk does it consume at that time?

SOLUTION

$y = -0.2x^2 + 1.3x + 6.2$

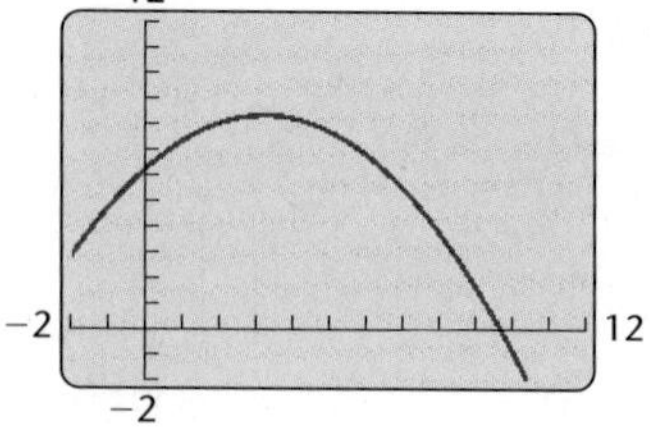

1., 2. **Familiarize** and **Translate.** We are given the function for milk consumption by a calf. Note that it is a quadratic function of x, the calf's age in weeks. Since the coefficient of x^2 is negative, the calf's consumption increases and then decreases. The graph at left confirms this.

3. **Carry out.** We can either complete the square,

$$\begin{aligned} p(x) &= -0.2x^2 + 1.3x + 6.2 \\ &= -0.2(x^2 - 6.5x) + 6.2 \\ &= -0.2(x^2 - 6.5x + 3.25^2 - 3.25^2) + 6.2 && \textbf{Completing the square; } \frac{6.5}{2} = 3.25 \\ &= -0.2(x^2 - 6.5x + 3.25^2) + (-0.2)(-3.25^2) + 6.2 \\ &= -0.2(x - 3.25)^2 + 8.3125, && \textbf{Factoring and simplifying} \end{aligned}$$

or we can use $-b/(2a) = -1.3/(-0.4) = 3.25$. Using a calculator, we find that

$$p(3.25) = -0.2(3.25)^2 + 1.3(3.25) + 6.2 = 8.3125.$$

4. **Check.** Both of the approaches in step (3) indicate that a maximum occurs when $x = 3.25$, or $3\frac{1}{4}$. We could also use the MAXIMUM feature on a graphing calculator as a check.
5. **State.** A calf's milk consumption is greatest when the calf is $3\frac{1}{4}$ weeks old. At that time, it drinks about 8.3 lb of milk per day.

EXAMPLE 2 Swimming Area. A lifeguard has 100 m of roped-together flotation devices with which to cordon off a rectangular swimming area at Lakeside Beach. If the shoreline forms one side of the rectangle, what dimensions will maximize the size of the area for swimming?

SOLUTION

1. **Familiarize.** We make a drawing and label it, letting w = the width of the rectangle, in meters, and l = the length of the rectangle, in meters.

Recall that Area $= l \cdot w$ and Perimeter $= 2w + 2l$. Since the beach forms one length of the rectangle, the flotation devices comprise three sides. Thus,

$$2w + l = 100.$$

To get a better feel for the problem, we can look at some possible dimensions for a rectangular area that can be enclosed with 100 m of flotation devices. All possibilities are chosen so that $2w + l = 100$.

We can make a table by hand or use a graphing calculator. To use a calculator, we let x represent the width of the swimming area. Solving $2w + l = 100$ for l, we have $l = 100 - 2w$. Thus, if $y_1 = 100 - 2x$, then the area is $y_2 = x \cdot y_1$.

$y_1 = 100 - 2x$, $y_2 = x \cdot y_1$

X	Y1	Y2
20	60	1200
22	56	1232
24	52	1248
26	48	1248
28	44	1232
30	40	1200
32	36	1152

X = 20

What choice of X will maximize Y2?

l	w	Rope Length	Area
40 m	30 m	100 m	1200 m^2
30 m	35 m	100 m	1050 m^2
20 m	40 m	100 m	800 m^2
.	.	.	.
.	.	.	.
.	.	.	.

What choice of l and w will maximize A?

The table helps us understand the problem, but the maximum area may or may not be one of the values shown. To find the maximum area, we use algebra.

2. **Translate.** We have two equations: One guarantees that all 100 m of flotation devices are used; the other expresses area in terms of length and width.

$$2w + l = 100,$$
$$A = l \cdot w$$

3. **Carry out.** We solve the system of equations both algebraically and graphically.

ALGEBRAIC APPROACH

We need to express A as a function of l or w but not both. To do so, we solve for l in the first equation to obtain $l = 100 - 2w$. Substituting for l in the second equation, we get a quadratic function:

$$A = (100 - 2w)w \quad \text{Substituting for } l$$
$$= 100w - 2w^2. \quad \text{This represents a parabola opening downward, so a maximum exists.}$$

Factoring and completing the square, we get

$$A = -2(w^2 - 50w + 625 - 625) \quad \text{We could also use the vertex formula.}$$
$$= -2(w - 25)^2 + 1250. \quad \text{This suggests a maximum of 1250 m}^2 \text{ when } w = 25 \text{ m.}$$

The maximum area, 1250 m^2, occurs when $w = 25$ m and $l = 100 - 2(25)$, or 50 m.

GRAPHICAL APPROACH

As in the *Familiarize* step, we let x represent the width of the area. Then the length is given by $y_1 = 100 - 2x$ and the area is $y_2 = x \cdot y_1$, or $y = 100x - 2x^2$. The maximum area is the second coordinate of the vertex of the graph of $y = 100x - 2x^2$.

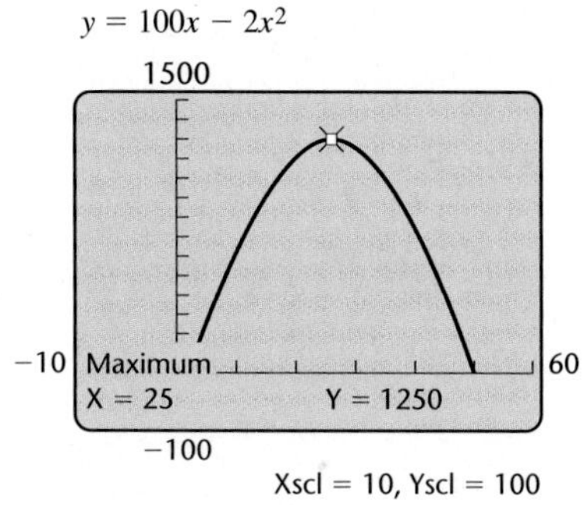

The maximum is 1250 m^2 when the width, x, is 25 m and the length, $100 - 2x$, is 50 m.

4. **Check.** Note that 1250 m^2 is greater than any of the values for A found in the *Familiarize* step. To be more certain, we could check values other than those used in that step. For example, if $w = 26$ m, then $l = 100 - 2 \cdot 26 = 48$ m, and $A = 26 \cdot 48 = 1248$ m^2. Since 1250 m^2 is greater than 1248 m^2, it appears that we have a maximum.

5. **State.** The largest rectangular area for swimming that can be enclosed is 25 m by 50 m.

Fitting Quadratic Functions to Data

We can now model some real-world situations using quadratic functions. As always, before attempting to fit an equation to data, we should graph the data and compare the result to the shape of the graphs of different types of functions and ask what type of function seems appropriate.

Connecting the Concepts

The general shapes of the graphs of many functions that we have studied are shown below.

Linear function:
$f(x) = mx + b$

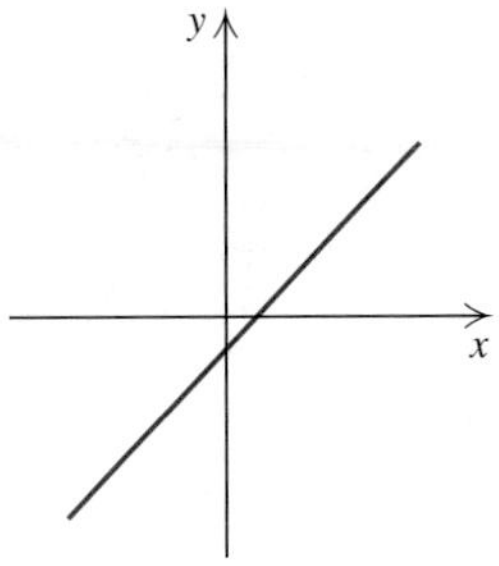

Absolute-value function:
$f(x) = |x|$

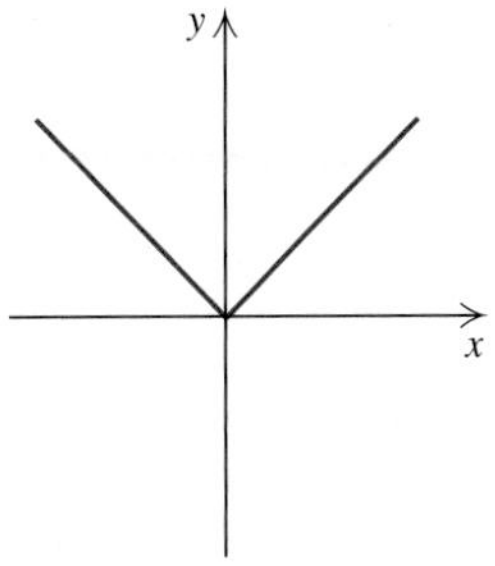

Rational function:
$f(x) = \frac{1}{x}$

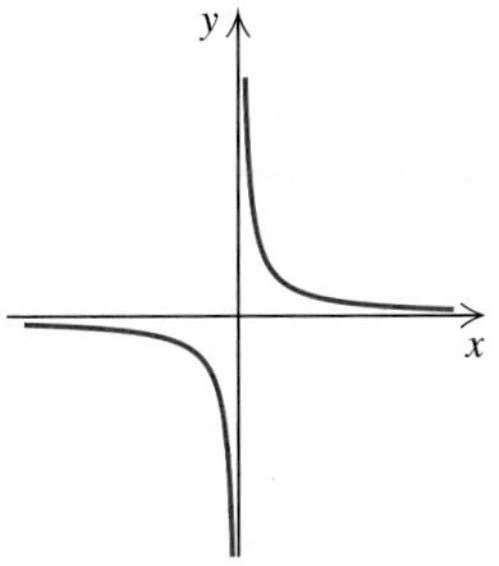

Radical function:
$f(x) = \sqrt{x}$

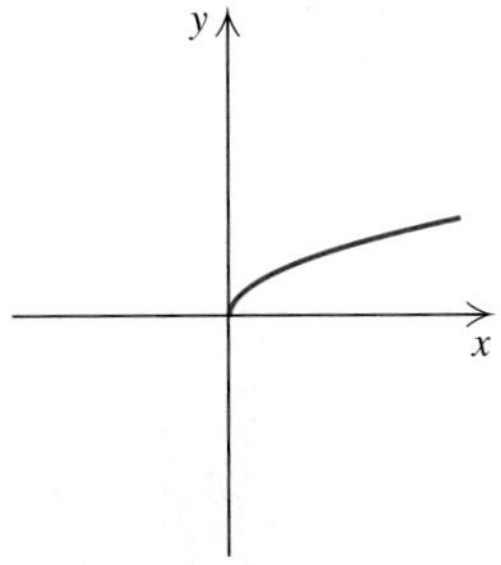

Quadratic function:
$f(x) = ax^2 + bx + c, a > 0$

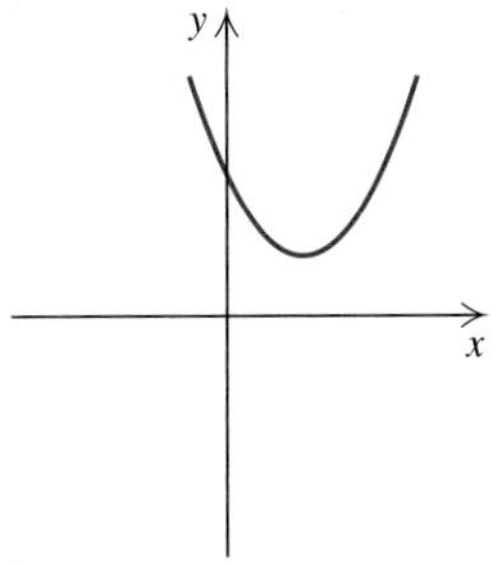

Quadratic function:
$f(x) = ax^2 + bx + c, a < 0$

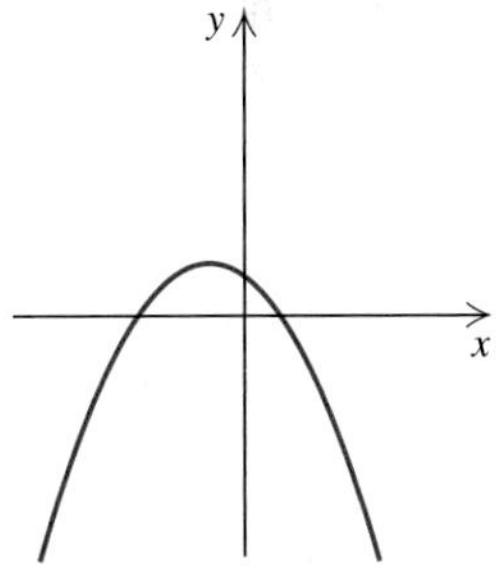

We see that in order for a quadratic function to fit a set of data, the graph of the data must approximate the shape of a parabola. Data that resemble one half of a parabola might also be modeled using a quadratic function with a restricted domain, as illustrated in the figure below.

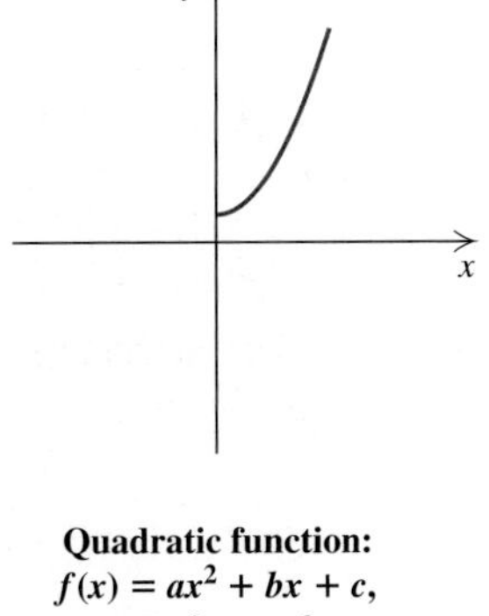

Quadratic function:
$f(x) = ax^2 + bx + c,$
$a > 0, x \geq 0$

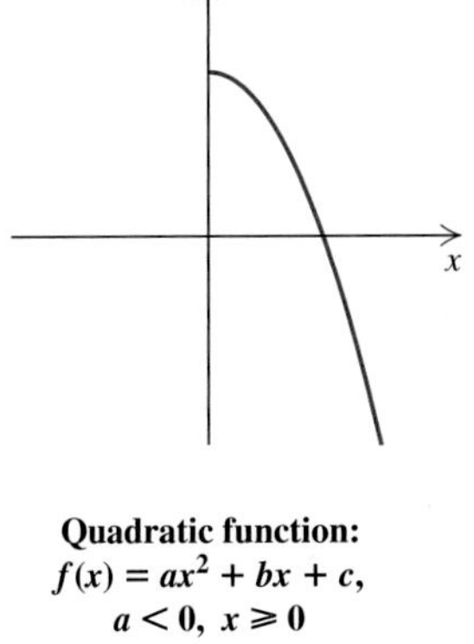

Quadratic function:
$f(x) = ax^2 + bx + c,$
$a < 0, x \geq 0$

EXAMPLE 3 Determine whether a quadratic function can be used to model each of the following situations.

a) Sonoma Sunshine. The percent of days in each month that the sun shines in Sonoma, California.

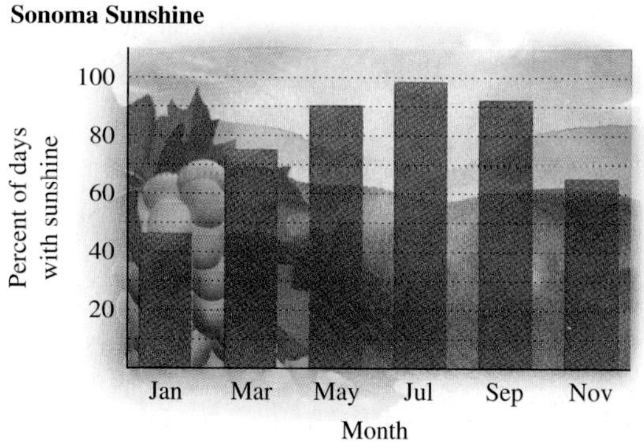

Source: www.city-data.com

b) Sonoma Precipitation. The amount of rainfall each month in Sonoma, California.

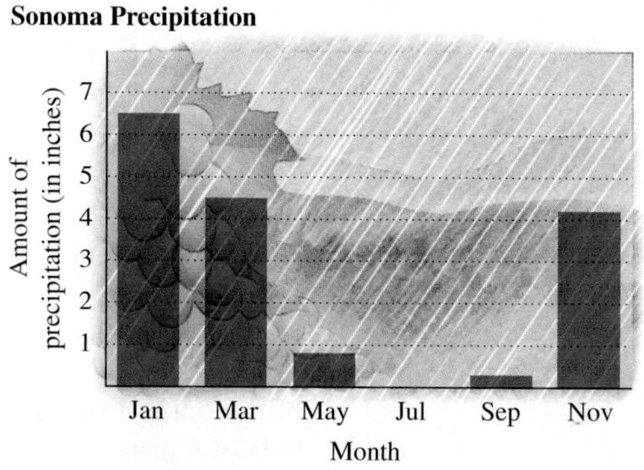

Source: www.city-data.com

c) Vehicles. The number of vehicles registered in the United States.

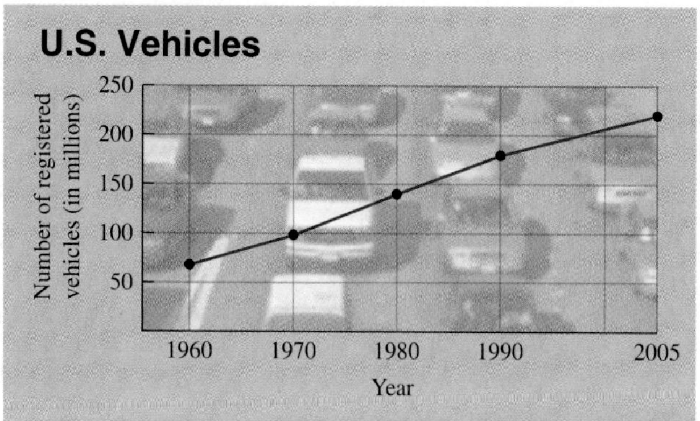

Source: The Indianapolis Star, 4/28/06

d) Alternative-Fueled Vehicles. The number of hybrid-electric passenger vehicles sold in the United States.

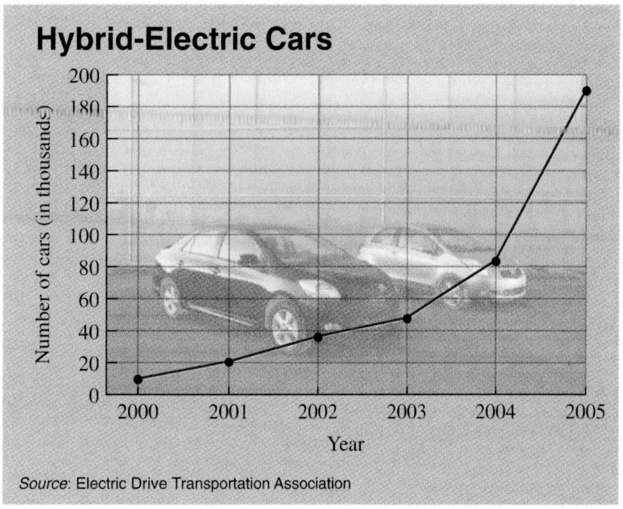

e) Olympic Volunteers. The number of volunteers at the Winter Olympics.

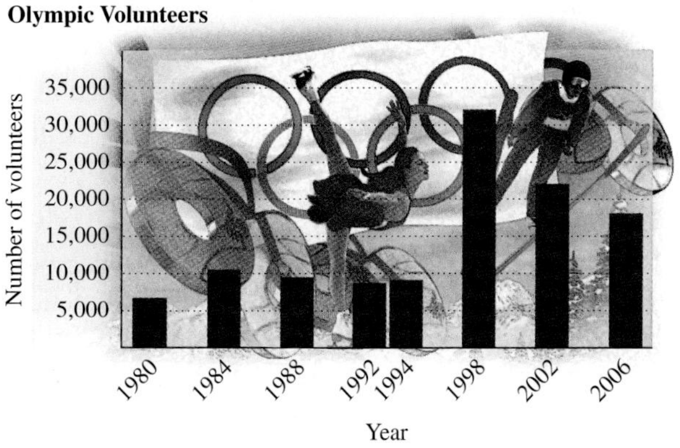

SOLUTION

a) The data rise and then fall, resembling a parabola that opens downward. The situation could be modeled by a quadratic function $f(x) = ax^2 + bx + c, a < 0$.

b) The data fall and then rise, resembling a parabola that opens upward. A quadratic function $f(x) = ax^2 + bx + c,\ a > 0$, might be used as a model for this situation.

c) The data appear nearly linear. A linear function is a better model for this situation than a quadratic function.

d) The data resemble the right half of a parabola that opens upward. We could use a quadratic function $f(x) = ax^2 + bx + c,\ a > 0,\ x \geq 0$, to model the situation.

e) The data do not resemble a parabola. A quadratic function would not be a good model for this situation.

EXAMPLE 4 Sonoma Sunshine. The percent of days each month during which the sun shines in Sonoma, California, can be modeled by a quadratic function, as seen in Example 3(a). To do this, we let $x =$ the number of the month, where 1 represents January, 2 represents February, and so on.

Month	Percent of Days with Sunshine
1 (Jan.)	46
3 (March)	75
5 (May)	90
7 (July)	98
9 (Sep.)	92
11 (Nov.)	65

a) Use the data points (1, 46), (5, 90), and (11, 65) to fit a quadratic function $f(x)$ to the data.

b) Use the function from part (a) to estimate the percent of days in April that the sun shines in Sonoma.

c) Use the REGRESSION feature of a graphing calculator to fit a quadratic function $g(x)$ to the data.

d) Use the function from part (c) to estimate the percent of days in April that the sun shines in Sonoma, and compare the estimate with the estimate from part (b).

SOLUTION

a) Only one parabola will go through any three noncollinear points. The quadratic function found will depend on the choice of those points. We are looking for a function of the form $f(x) = ax^2 + bx + c$, where $f(x)$ is the percent of days with sunshine in month x. To solve for a, b, and c, we can use any of the methods discussed in Chapter 3 for solving a system of three equations. Here we use elimination, and substitute the given values of x and $f(x)$ from the three data points listed:

$46 = a(1)^2 + b(1) + c,$ **Using the data point (1, 46)**

$90 = a(5)^2 + b(5) + c,$ **Using the data point (5, 90)**

$65 = a(11)^2 + b(11) + c.$ **Using the data point (11, 65)**

After simplifying, we have the system of three equations

$$46 = a + b + c, \quad (1)$$
$$90 = 25a + 5b + c, \quad (2)$$
$$65 = 121a + 11b + c. \quad (3)$$

We multiply equation (1) by -25 and add that to equation (2) to eliminate a. We also multiply equation (1) by -121 and add that to equation (3) to eliminate a again.

$$\begin{aligned} -1150 &= -25a - 25b - 25c \\ 90 &= 25a + 5b + c \\ \hline -1060 &= \quad\quad - 20b - 24c \end{aligned} \qquad \begin{aligned} -5566 &= -121a - 121b - 121c \\ 65 &= 121a + 11b + c \\ \hline -5501 &= \quad\quad - 110b - 120c \end{aligned}$$

We obtain the system of equations

$$-1060 = -20b - 24c,$$
$$-5501 = -110b - 120c.$$

We can divide both sides of the first equation by -4, and both sides of the second equation by -1, to obtain

$$265 = 5b + 6c, \qquad (4)$$
$$5501 = 110b + 120c. \qquad (5)$$

To solve, we multiply equation (4) by -22 and add:

$$\begin{aligned} -5830 &= -110b - 132c \\ 5501 &= 110b + 120c \\ \hline -329 &= \quad\quad - 12c \\ \tfrac{329}{12} &= c. \end{aligned}$$

Next, we solve for b, using equation (4) above:

$265 = 5b + 6\left(\frac{329}{12}\right)$ **Substituting**

$265 = 5b + \frac{329}{2}$ **Multiplying**

$530 = 10b + 329$ **Clearing fractions**

$201 = 10b$

$\frac{201}{10} = b.$

Finally, we use equation (1) to solve for a:

$46 = a + \frac{201}{10} + \frac{329}{12}$ **Substituting**

$\frac{2760}{60} = a + \frac{1206}{60} + \frac{1645}{60}$ **The LCD is 60.**

$-\frac{91}{60} = a.$ **Solving for *a***

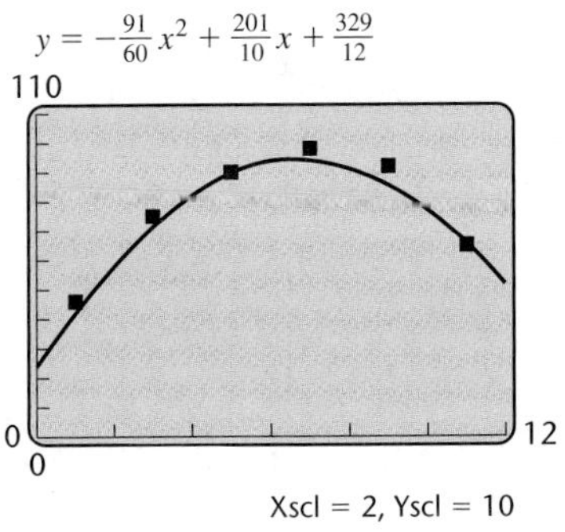

Thus the function

$$f(x) = -\frac{91}{60}x^2 + \frac{201}{10}x + \frac{329}{12}$$

fits the three points given. As a partial check, we graph the function along with the data, as shown at left. The function goes through the three given points.

b) Since April is month 4, $f(4)$ will give an estimate of the percent of days of sunshine in April:

$$f(x) = -\frac{91}{60}(4)^2 + \frac{201}{10}(4) + \frac{329}{12} \approx 84.$$

According to this model, the sun shines on about 84% of April days in Sonoma.

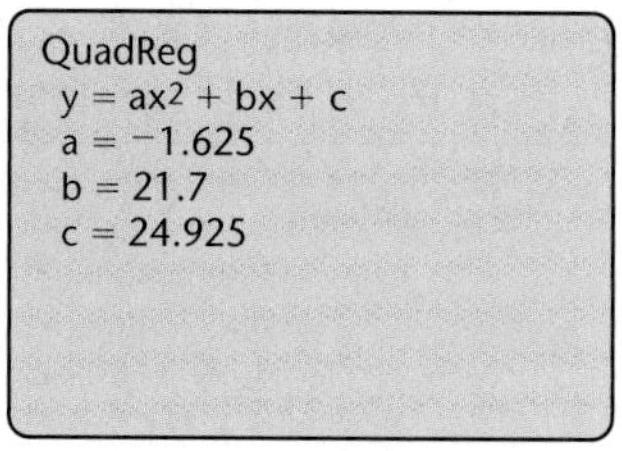

$y = -1.625x^2 + 21.7x + 24.925$

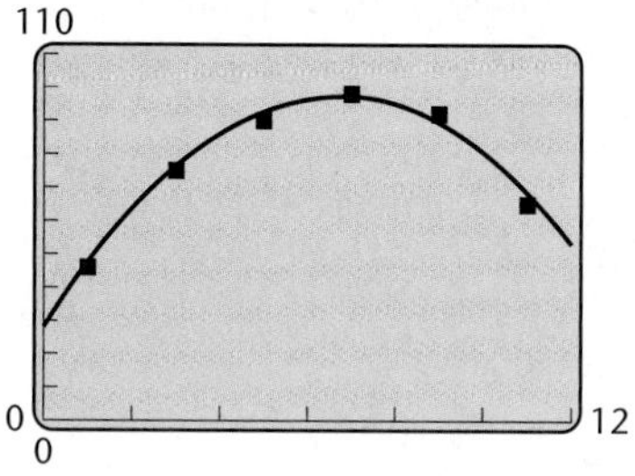

Xscl = 2, Yscl = 10

c) Using regression allows us to consider all the data given in the problem, not just three points. After entering the data, we choose the QUADREG option in the STAT CALC menu.

From the first screen at left, we obtain the quadratic function

$$g(x) = -1.625x^2 + 21.7x + 24.925.$$

On the second screen at left, we graph $g(x)$, along with the data points.

d) To estimate the percent of days of sunshine in April, we evaluate $g(4)$ and obtain $g(4) \approx 86$. According to this model, the sun will shine about 86% of April days in Sonoma. This estimate is slightly higher than the 84% found in part (b). The second model, which took into consideration all the data, is a better fit for the data. Without further information, it is difficult to tell which model more accurately describes the sunshine pattern in Sonoma.

8.8 EXERCISE SET

FOR EXTRA HELP

MathXL | MyMathLab | InterAct Math | AW Math Tutor Center | Video Lectures on CD: Disc 4 | Student's Solutions Manual

Concept Reinforcement In each of Exercises 1–6, match the description with the graph that displays that characteristic.

1. ____ A minimum value of $f(x)$ exists.

2. ____ A maximum value of $f(x)$ exists.

3. ____ No maximum or minimum value of $f(x)$ exists.

4. ____ The data points appear to suggest a linear model.

5. ____ The data points appear to suggest a quadratic model with a maximum.

6. ____ The data points appear to suggest a quadratic model with a minimum.

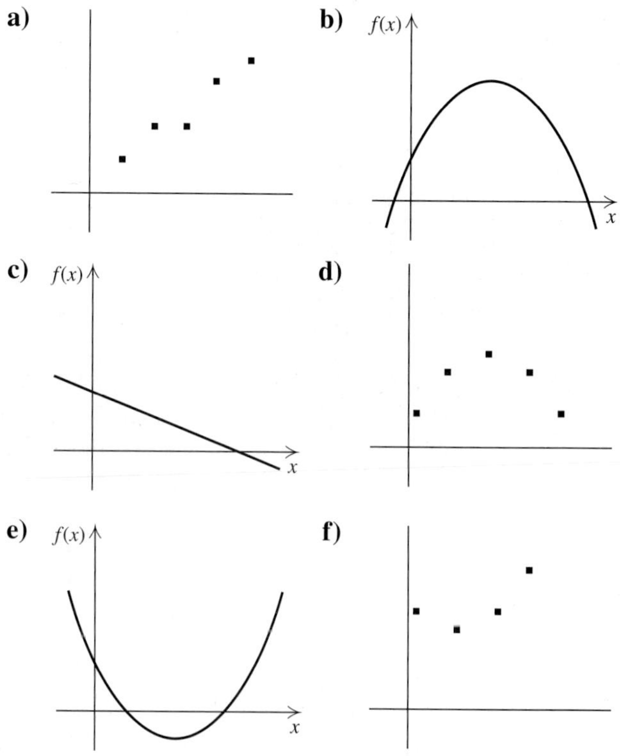

Solve.

7. *Ticket Sales.* The number of tickets sold each day for an upcoming Los Lobos show can be approximated by

$$N(x) = -0.4x^2 + 9x + 11,$$

where x is the number of days since the concert was first announced. When will daily ticket sales peak and how many tickets will be sold that day?

8. *Stock Prices.* The value of a share of I. J. Solar can be represented by $V(x) = x^2 - 6x + 13$, where x is the number of months after January 2004. What is the lowest value $V(x)$ will reach, and when did that occur?

9. *Minimizing Cost.* Sweet Harmony Crafts has determined that when x hundred Dobros are built, the average cost per Dobro can be estimated by

$$C(x) = 0.1x^2 - 0.7x + 2.425,$$

where $C(x)$ is in hundreds of dollars. What is the minimum average cost per Dobro and how many Dobros should be built to achieve that minimum?

10. *Maximizing Profit.* Recall that total profit P is the difference between total revenue R and total cost C. Given $R(x) = 1000x - x^2$ and $C(x) = 3000 + 20x$, find the total profit, the maximum value of the total profit, and the value of x at which it occurs.

11. *Furniture Design.* A furniture builder is designing a rectangular end table with a perimeter of 128 in. What dimensions will yield the maximum area?

12. *Architecture.* An architect is designing an atrium for a hotel. The atrium is to be rectangular with a perimeter of 720 ft of brass piping. What dimensions will maximize the area of the atrium?

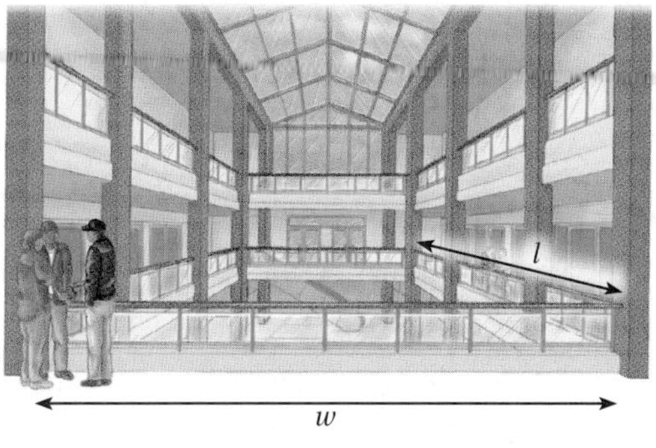

13. *Patio Design.* A stone mason has enough stones to enclose a rectangular patio with 60 ft of perimeter, assuming that the attached house forms one side of the rectangle. What is the maximum area that the mason can enclose? What should the dimensions of the patio be in order to yield this area?

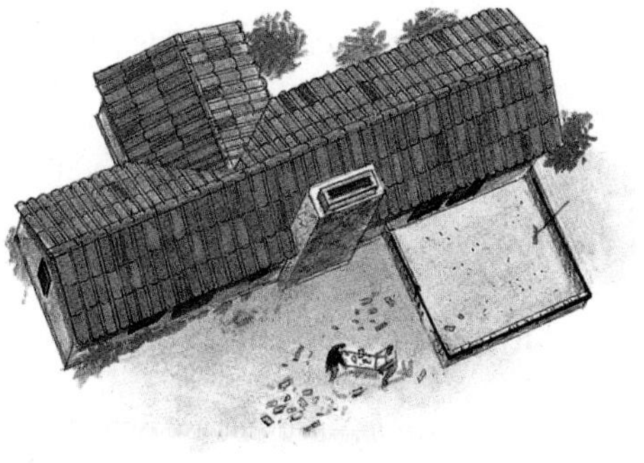

14. *Garden Design.* Ginger is fencing in a rectangular garden, using the side of her house as one side of the rectangle. What is the maximum area that she can enclose with 40 ft of fence? What should the dimensions of the garden be in order to yield this area?

15. *Molding Plastics.* Economite Plastics plans to produce a one-compartment vertical file by bending the long side of an 8-in. by 14-in. sheet of plastic along two lines to form a U shape. How tall should the file be in order to maximize the volume that the file can hold?

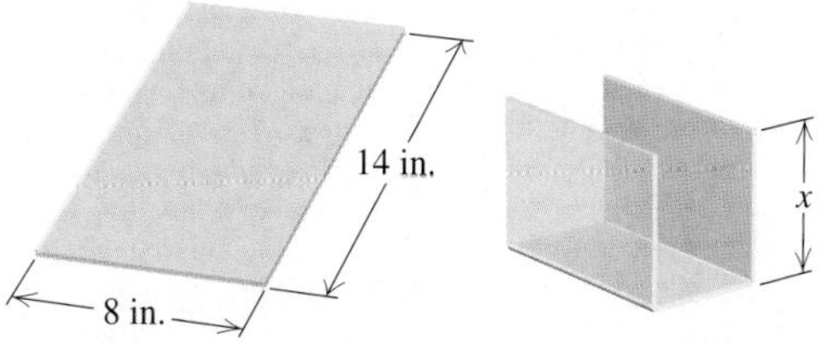

16. *Composting.* A rectangular compost container is to be formed in a corner of a fenced yard, with 8 ft of chicken wire completing the other two sides of the rectangle. If the chicken wire is 3 ft high, what dimensions of the base will maximize the container's volume?

17. What is the maximum product of two numbers that add to 18? What numbers yield this product?

18. What is the maximum product of two numbers that add to 26? What numbers yield this product?

19. What is the minimum product of two numbers that differ by 8? What are the numbers?

20. What is the minimum product of two numbers that differ by 7? What are the numbers?

Aha! **21.** What is the maximum product of two numbers that add to -10? What numbers yield this product?

22. What is the maximum product of two numbers that add to -12? What numbers yield this product?

Choosing Models. For the scatterplots and graphs in Exercises 23–32, determine which, if any, of the following functions might be used as a model for the data: Linear, with $f(x) = mx + b$; quadratic, with $f(x) = ax^2 + bx + c, a > 0$; quadratic, with $f(x) = ax^2 + bx + c, a < 0$; neither quadratic nor linear.

23.

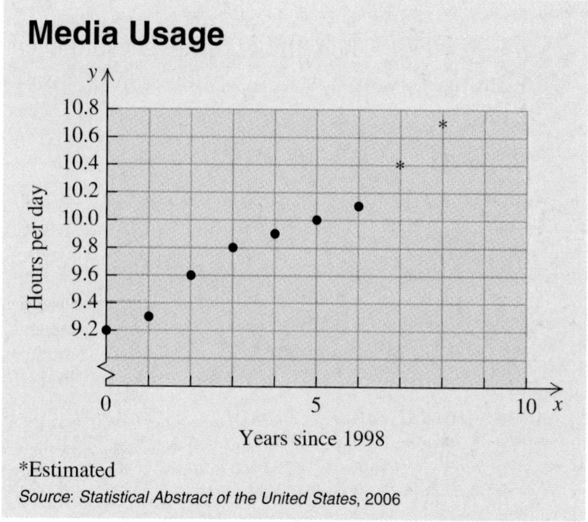

24.

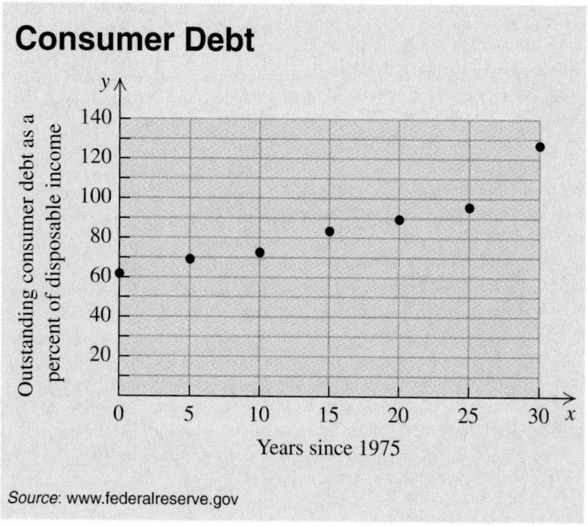

25.

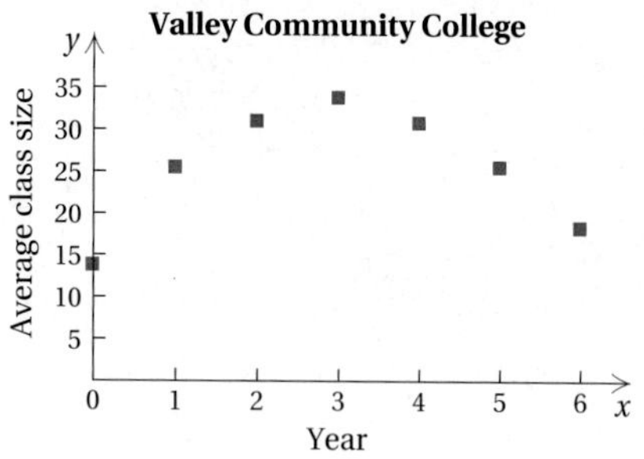

26.

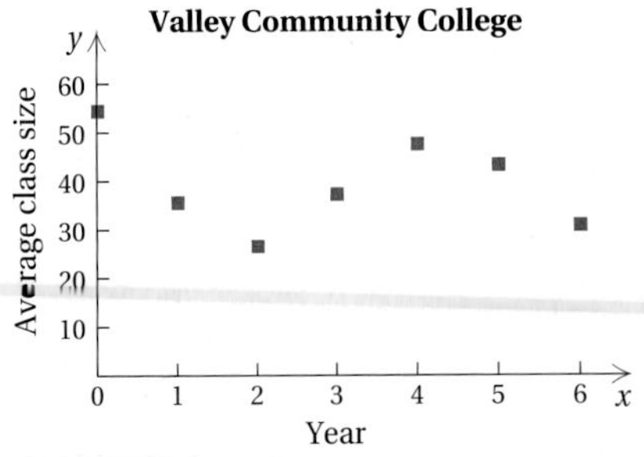

27.

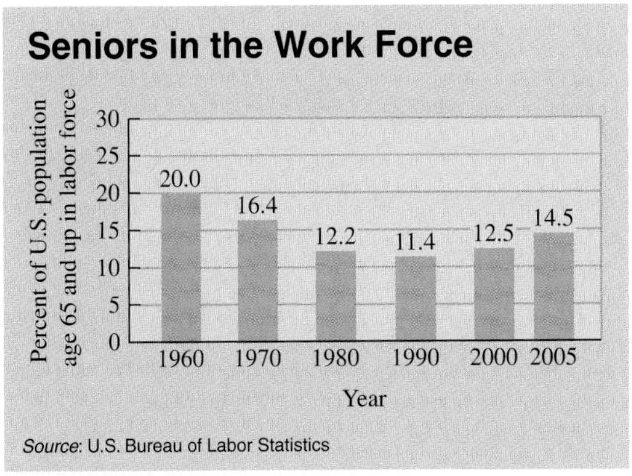

28.

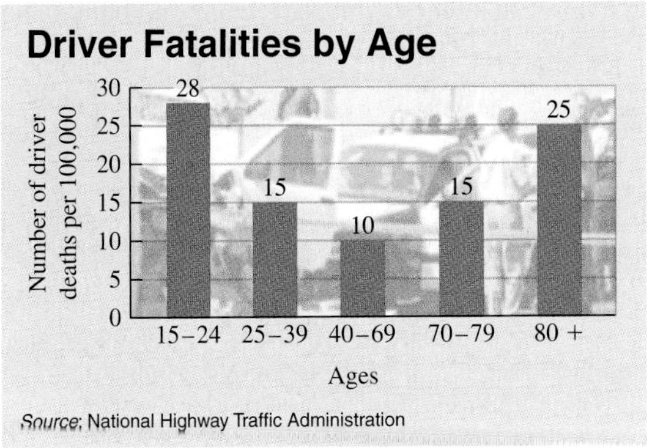

29.

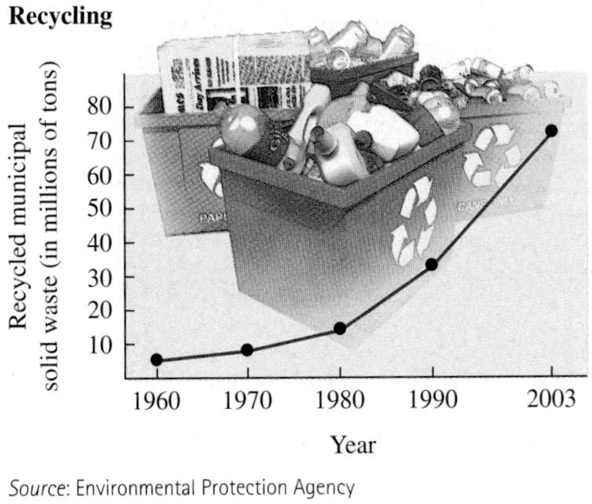

30.

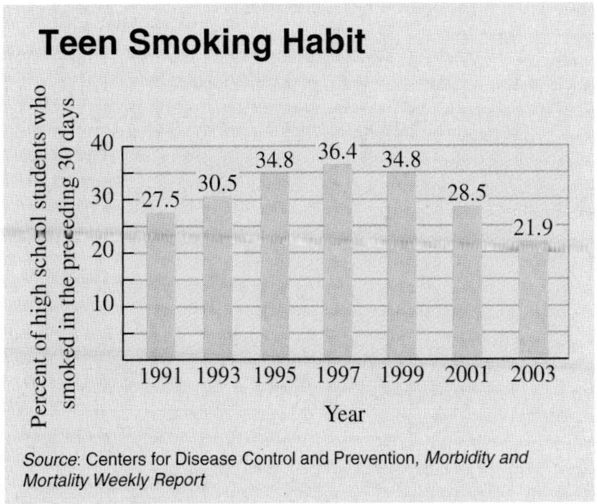

31.

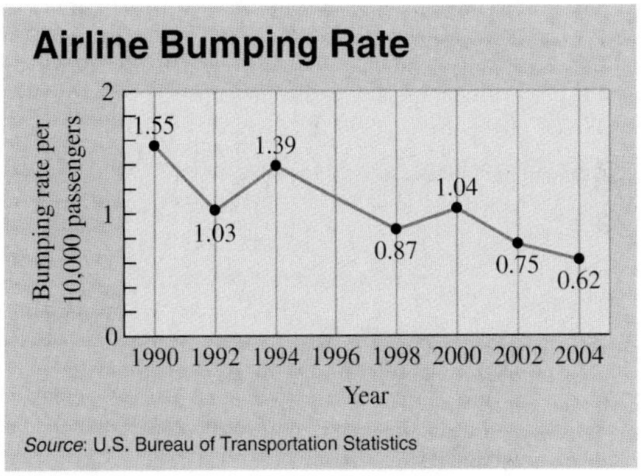

32.

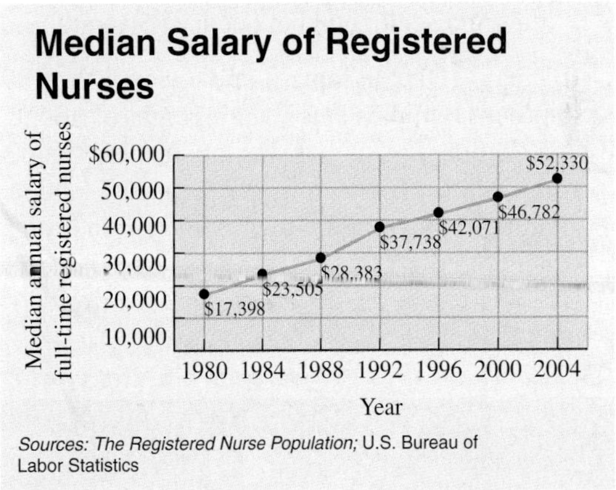

Find a quadratic function that fits the set of data points.

33. $(1, 4), (-1, -2), (2, 13)$

34. $(1, 4), (-1, 6), (-2, 16)$

35. $(2, 0), (4, 3), (12, -5)$

36. $(-3, -30), (3, 0), (6, 6)$

37. a) Find a quadratic function that fits the following data.

Travel Speed (in kilometers per hour)	Number of Nighttime Accidents (for every 200 million kilometers driven)
60	400
80	250
100	250

b) Use the function to estimate the number of nighttime accidents that occur at 50 km/h.

38. a) Find a quadratic function that fits the following data.

Travel Speed (in kilometers per hour)	Number of Daytime Accidents (for every 200 million kilometers driven)
60	100
80	130
100	200

b) Use the function to estimate the number of daytime accidents that occur at 50 km/h.

39. *Archery.* The Olympic flame tower at the 1992 Summer Olympics was lit at a height of about 27 m by a flaming arrow that was launched about 63 m from the base of the tower. If the arrow landed about 63 m beyond the tower, find a quadratic function that expresses the height h of the arrow as a function of the distance d that it traveled horizontally.

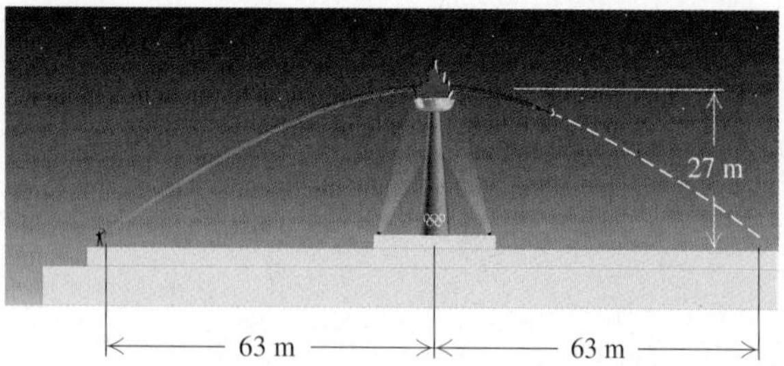

40. *Pizza Prices.* Pizza Unlimited has the following prices for pizzas.

Diameter	Price
8 in.	\$ 6.00
12 in.	\$ 8.50
16 in.	\$11.50

Is price a quadratic function of diameter? It probably should be, because the price should be proportional to the area, and the area is a quadratic function of the diameter. (The area of a circular region is given by $A = \pi r^2$ or $(\pi/4) \cdot d^2$.)

a) Express price as a quadratic function of diameter using the data points (8, 6), (12, 8.50), and (16, 11.50).

b) Use the function to find the price of a 14-in. pizza.

41. *Hydrology.* The drawing below shows the cross section of a river. Typically rivers are deepest in the middle, with the depth decreasing to 0 at the edges. A hydrologist measures the depths D, in feet, of a river at distances x, in feet, from one bank. The results are listed in the table below.

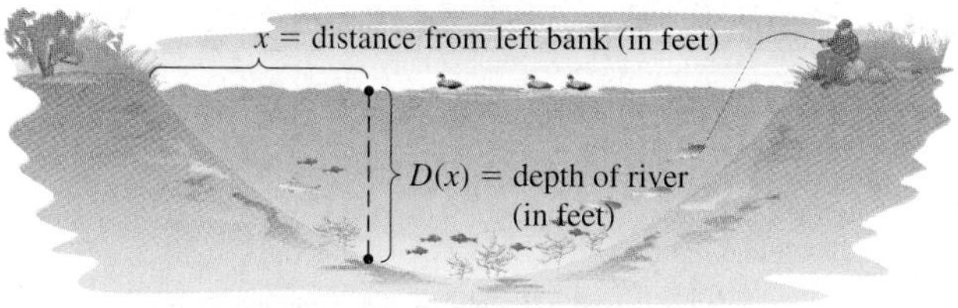

Distance x, from the Left Bank (in feet)	Depth, D, of the River (in feet)
0	0
15	10.2
25	17
50	20
90	7.2
100	0

a) Use regression to find a quadratic function that fits the data.

b) Use the function to estimate the depth of the river 70 ft from the left bank.

42. *Work Force.* The graph in Exercise 27 indicates that the percent of the U.S. population age 65 and older that is in the work force is increasing after several decades of decrease.

a) Use regression to find a quadratic function that can be used to estimate the percent p of the population 65 and older that is in the work force x years after 1960.

b) Use the function found in part (a) to predict the percent of those 65 and older who will be in the work force in 2010.

43. *Alternative-Fueled Vehicles.* The number of hybrid-electric passenger vehicles sold in the United States during several years is shown in the table below. As we saw in Example 3, a quadratic function can be used to model this data.

Year	Number of Hybrid-Electric Passenger Vehicles
2000	9,367
2001	20,287
2002	35,961
2003	47,525
2004	83,153
2005	189,916

Source: Electric Drive Transportation Association

a) Use regression to find a quadratic function that can be used to estimate the number of hybrid-electric vehicles v sold x years after 2000.

b) Use the function found in part (a) to predict the number of hybrid-electric vehicles sold in 2008.

44. *Teen Smoking.* The graph in Exercise 30 indicates that the percent of high school students who smoke is decreasing after several years of increase.

a) Use regression to find a quadratic function that can be used to estimate the percent t of high school students who smoked in the preceding 30 days x years after 1990.

b) Use the function found in part (a) to estimate the percent of high school students who smoked in the preceding 30 days in 2005.

TW 45. Does every nonlinear function have a minimum or a maximum value? Why or why not?

TW 46. Explain how the leading coefficient of a quadratic function can be used to determine if a maximum or a minimum function value exists.

Skill Maintenance

Simplify.

47. $\dfrac{x}{x^2 + 17x + 72} - \dfrac{8}{x^2 + 15x + 56}$ [6.2]

48. $\dfrac{x^2 - 9}{x^2 - 8x + 7} \div \dfrac{x^2 + 6x + 9}{x^2 - 1}$ [6.1]

Solve. [4.1]

49. $5x - 9 < 31$

50. $3x - 8 \geq 22$

51. *Registered Nurses' Salary.* The average annual salary of full-time registered nurses is shown in Exercise 32. Use the points (4, 23,505) and (20, 46,782) to find a linear function that can be used to estimate the average annual salary r of registered nurses x years after 1980. [2.4]

52. Use the data in Exercise 32 and regression to find a linear function that can be used to predict the average annual salary r of registered nurses x years after 1980. [2.4]

Synthesis

TW 53. Write a problem for a classmate to solve. Design the problem so that its solution requires finding a minimum or maximum function value.

TW 54. Explain what restrictions should be placed on the quadratic functions developed in Exercises 37 and 42 and why such restrictions are needed.

55. *Bridge Design.* The cables supporting a straight-line suspension bridge are nearly parabolic in shape. Suppose that a suspension bridge is being designed with concrete supports 160 ft apart and with vertical cables 30 ft above road level at the midpoint of the bridge and 80 ft above road level at a point 50 ft from the midpoint of the bridge. How long are the longest vertical cables?

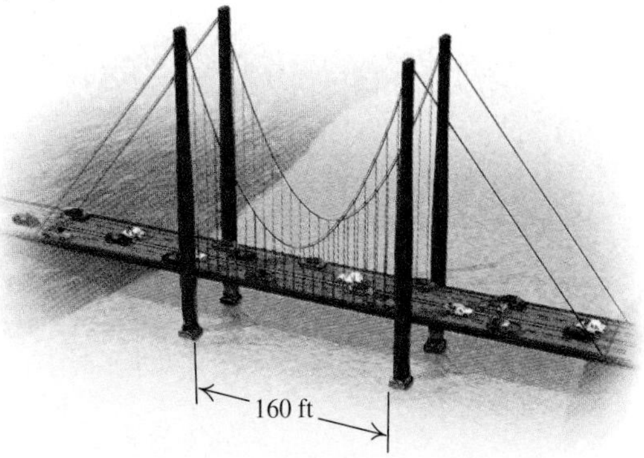

56. *Trajectory of a Launched Object.* The height above the ground of a launched object is a quadratic function of the time that it is in the air. Suppose that a flare is launched from a cliff 64 ft above sea level. If 3 sec after being launched the flare is again level with the cliff, and if 2 sec after that it lands in the sea, what is the maximum height that the flare will reach?

57. *Norman Window.* A *Norman window* is a rectangle with a semicircle on top. Big Sky Windows is designing a Norman window that will require 24 ft of trim. What dimensions will allow the maximum amount of light to enter a house?

58. *Crop Yield.* An orange grower finds that she gets an average yield of 40 bushels (bu) per tree when she plants 20 trees on an acre of ground. Each time she adds a tree to an acre, the yield per tree decreases by 1 bu, due to congestion. How many trees per acre should she plant for maximum yield?

59. *Cover Charges.* When the owner of Sweet Sounds charges a $10 cover charge, an average of 80 people will attend a show. For each 25¢ increase in admission price, the average number attending decreases by 1. What should the owner charge in order to make the most money?

60. *Minimizing Area.* A 36-in. piece of string is cut into two pieces. One piece is used to form a circle while the other is used to form a square. How should the string be cut so that the sum of the areas is a minimum?

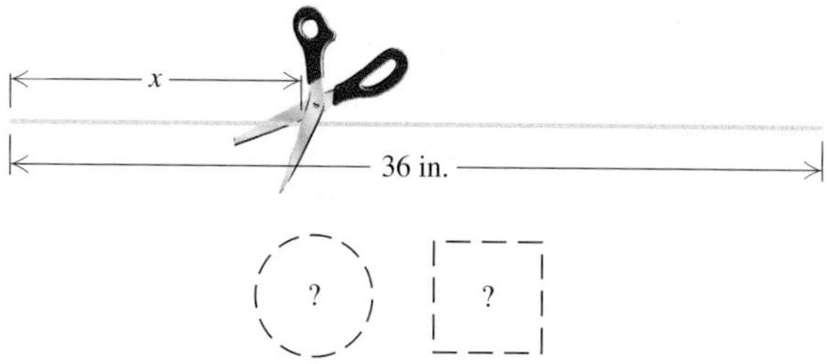

8.9 Polynomial and Rational Inequalities

Quadratic and Other Polynomial Inequalities ■ Rational Inequalities

Quadratic and Other Polynomial Inequalities

Inequalities like the following are called *polynomial inequalities*:

$$x^3 - 5x > x^2 + 7, \qquad 4x - 3 < 9, \qquad 5x^2 - 3x + 2 \geq 0.$$

Second-degree polynomial inequalities in one variable are called *quadratic inequalities*. To solve polynomial inequalities, we often focus attention on where the outputs of a polynomial function are positive and where they are negative.

Study Tip

Map Out Your Day

As the semester winds down and term papers are due, it becomes more critical than ever that you manage your time wisely. If you aren't already doing so, consider writing out an hour-by-hour schedule for each day and then abide by it as much as possible.

EXAMPLE 1 Solve: $x^2 + 3x - 10 > 0$.

SOLUTION Consider the "related" function $f(x) = x^2 + 3x - 10$ and its graph. Its graph opens upward since the leading coefficient is positive.

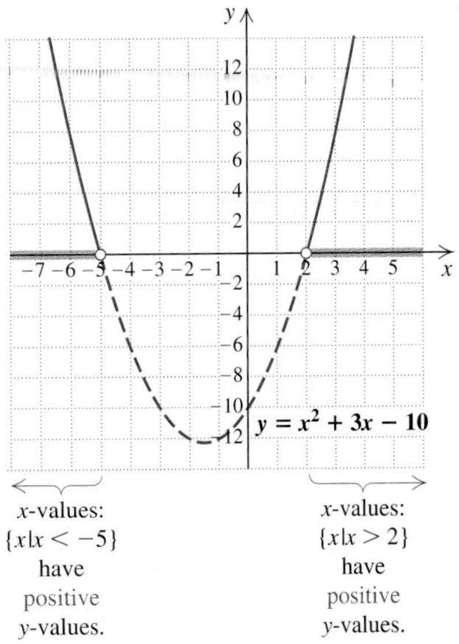

When $x^2 + 3x - 10 > 0$, $f(x)$, or y, will be positive.

Values of y will be positive to the left and right of the x-intercepts, as shown. To find the intercepts, we set the polynomial equal to 0 and solve:

$$x^2 + 3x - 10 = 0$$
$$(x + 5)(x - 2) = 0$$
$$x + 5 = 0 \quad or \quad x - 2 = 0$$
$$x = -5 \quad or \quad x = 2.$$

Thus the solution set of the inequality is

$$\{x \mid x < -5 \text{ or } x > 2\}, \quad \text{or} \quad (-\infty, -5) \cup (2, \infty).$$

Any inequality with 0 on one side can be solved by considering a graph of the related function and finding intercepts as in Example 1. Sometimes the quadratic formula is needed to find the intercepts.

EXAMPLE 2 Solve: $x^2 - 2x \le 2$.

SOLUTION We first find standard form with 0 on one side:

$$x^2 - 2x - 2 \le 0. \quad \textbf{This is equivalent to the original inequality.}$$

If $f(x) = x^2 - 2x - 2$, then $x^2 - 2x - 2 \le 0$ when $f(x)$ is 0 or negative. The graph of $f(x) = x^2 - 2x - 2$ is a parabola opening upward. Values of $f(x)$ are negative for x-values between the x-intercepts. We find the x-intercepts by solving $f(x) = 0$:

$$x = \frac{-b \pm \sqrt{b^2 - 4ac}}{2a}$$

$$x = \frac{-(-2) \pm \sqrt{(-2)^2 - 4 \cdot 1(-2)}}{2 \cdot 1} \qquad \textbf{Substituting}$$

$$x = \frac{2 \pm \sqrt{12}}{2} = \frac{2}{2} \pm \frac{2\sqrt{3}}{2} = 1 \pm \sqrt{3}.$$

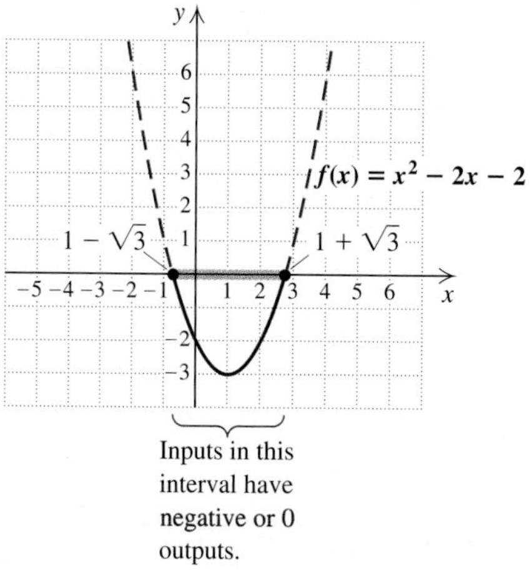

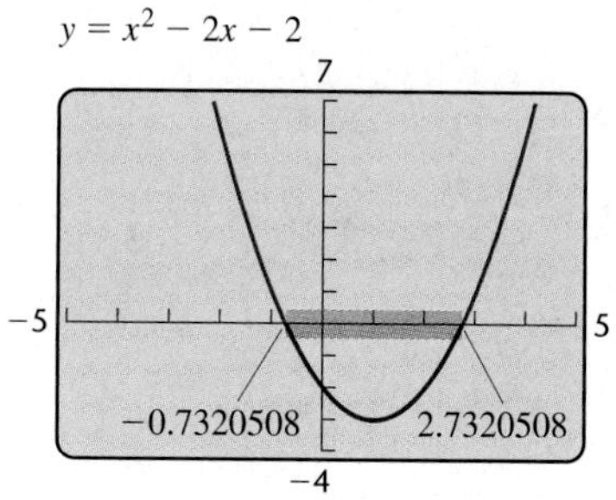

We can also find the x-intercepts using a graphing calculator. Because the numbers are irrational, this procedure will result in an approximation.

At the x-intercepts, $1 - \sqrt{3}$ and $1 + \sqrt{3}$, the value of $f(x)$ is 0. Thus the solution set of the inequality is

$$\left[1 - \sqrt{3}, 1 + \sqrt{3}\right], \quad \text{or} \quad \left\{x \middle| 1 - \sqrt{3} \le x \le 1 + \sqrt{3}\right\}.$$

If the solutions are found graphically, the solution set is approximately $[-0.7320508, 2.7320508]$.

In Example 2, it was not essential to draw the graph. The important information came from finding the x-intercepts and the sign of $f(x)$ on each side of those intercepts. We now solve a third-degree polynomial inequality, without graphing, by locating the x-intercepts, or *zeros*, of f and then using *test points* to determine the sign of $f(x)$ over each interval of the x-axis.

EXAMPLE 3 For $f(x) = 5x^3 + 10x^2 - 15x$, find all x-values for which $f(x) > 0$.

SOLUTION We first solve the related equation:

$$5x^3 + 10x^2 - 15x = 0$$
$$5x(x^2 + 2x - 3) = 0$$
$$5x(x + 3)(x - 1) = 0$$
$$5x = 0 \quad or \quad x + 3 = 0 \quad or \quad x - 1 = 0$$
$$x = 0 \quad or \quad x = -3 \quad or \quad x = 1.$$

The zeros of f are -3, 0, and 1. These zeros divide the number line, or x-axis, into four intervals: A, B, C, and D.

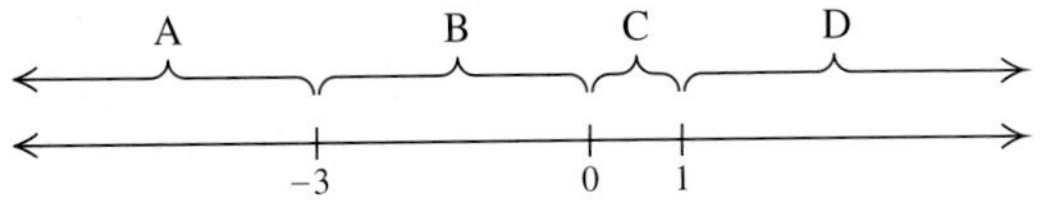

Next, selecting one convenient test value from each interval, we determine the sign of $f(x)$ for that interval. Within each interval, the sign of $f(x)$ cannot change. If it did, there would need to be another zero in that interval.

We choose -4 for a test value from interval A, -1 from interval B, 0.5 from interval C, and 2 from interval D. We enter $y_1 = 5x^3 + 10x^2 - 15x$ and use a table to evaluate the polynomial for each test value.

X	Y1
−4	−100
−1	20
.5	−4.375
2	50

X = −4

We are interested only in the signs of each function value. From the table, we see that $f(-4)$ and $f(0.5)$ are negative and that $f(-1)$ and $f(2)$ are positive. We indicate on the number line the sign of $f(x)$ in each interval.

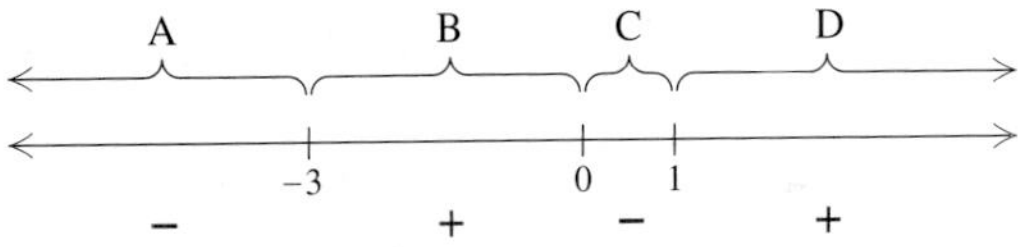

Recall that we are looking for all x for which $5x^3 + 10x^2 - 15x > 0$. The calculations above indicate that $f(x)$ is positive for any number in intervals B and D. The solution set of the original inequality is

$$(-3, 0) \cup (1, \infty), \quad \text{or} \quad \{x \mid -3 < x < 0 \text{ or } x > 1\}.$$

$y = 5x^3 + 10x^2 - 15x$

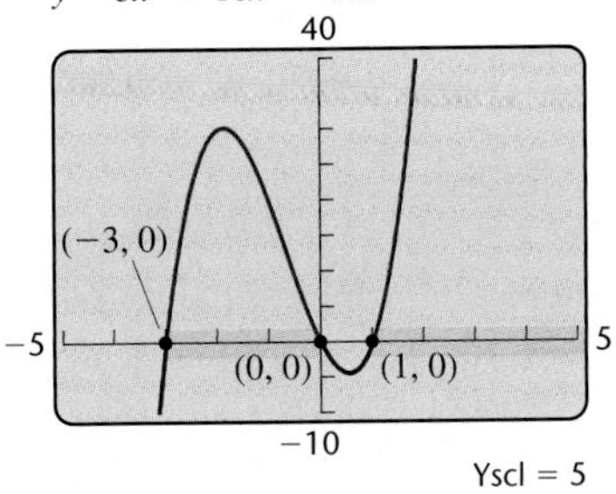

The method of Example 3 works because polynomial function values can change signs only when the graph of the function crosses the x-axis. The graph of $f(x) = 5x^3 + 10x^2 - 15x$ illustrates the solution of Example 3. We can see from the graph at left that the function values are positive in the intervals $(-3, 0)$ and $(1, \infty)$.

EXAMPLE 4 Solve: $x^4 + x^3 - 2x^2 < 0$.

SOLUTION We first solve the related equation:

$$x^4 + x^3 - 2x^2 = 0$$
$$x^2(x^2 + x - 2) = 0$$
$$x^2(x + 2)(x - 1) = 0$$

$$x^2 = 0 \quad or \quad x + 2 = 0 \quad or \quad x - 1 = 0$$
$$x = 0 \quad or \quad x = -2 \quad or \quad x = 1.$$

The function $p(x) = x^4 + x^3 - 2x^2$ has zeros at -2, 0, and 1. We graph the function and determine the sign of $p(x)$ over each interval of the number line. The solution will include those intervals where $p(x)$ is negative.

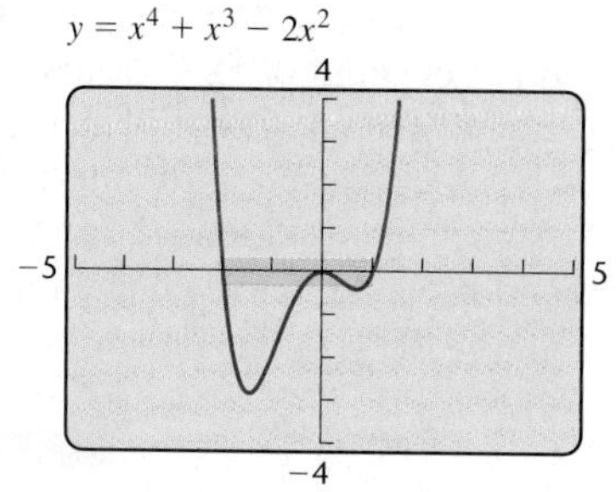

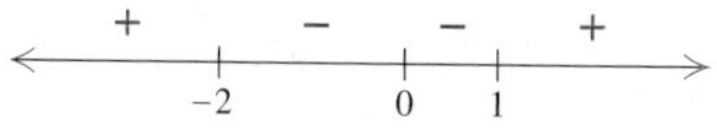

We see from the graph that $p(x)$ is negative in the intervals $(-2, 0)$ and $(0, 1)$. The solution set is $(-2, 0) \cup (0, 1)$.

In Example 4, if the inequality symbol were $\leq$, the endpoints of the intervals would be included in the solution set. The solution of $x^4 + x^3 - 2x^2 \leq 0$ is $[-2, 0] \cup [0, 1]$, or simply $[-2, 1]$.

To Solve a Polynomial Inequality

1. Add or subtract to get 0 on one side and solve the related polynomial equation $p(x) = 0$.
2. Use the numbers found in step (1) to divide the number line into intervals.
3. Using a test value from each interval or the graph of the related function, determine the sign of $p(x)$ over each interval.
4. Select the interval(s) for which the inequality is satisfied and write set-builder notation or interval notation for the solution set. Include the endpoints of the intervals when $\leq$ or $\geq$ is used.

Note that if the polynomial cannot be factored, a graphing calculator can be used to at least approximate any x-intercepts of the graph of the function, as in Example 2.

Rational Inequalities

Inequalities involving rational expressions are called **rational inequalities.** Like polynomial inequalities, rational inequalities can be solved using test values. Unlike polynomials, however, rational expressions often have values for which the expression is undefined.

EXAMPLE 5 Solve: $\dfrac{x-3}{x+4} \geq 2.$

SOLUTION We write the related equation by changing the $\geq$ symbol to $=$:

$$\frac{x-3}{x+4} = 2. \qquad \textbf{Note that } x \neq -4.$$

We show both algebraic and graphical approaches.

ALGEBRAIC APPROACH

We first solve the related equation:

$$(x+4) \cdot \frac{x-3}{x+4} = (x+4) \cdot 2 \qquad \text{Multiplying both sides by the LCD, } x+4$$

$$x - 3 = 2x + 8$$

$$-11 = x. \qquad \text{Solving for } x$$

In the case of rational inequalities, we must always find any values that make the denominator 0. As noted at the beginning of the example, $x \neq -4$.

Now we use -11 and -4 to divide the number line into intervals:

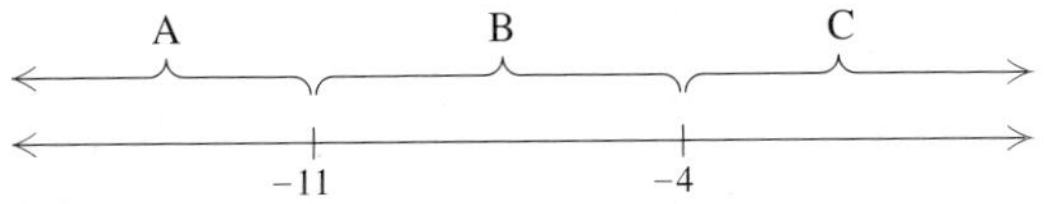

We test a number in each interval to see where the original inequality is satisfied:

$$\frac{x-3}{x+4} \geq 2.$$

If we enter $y_1 = (x-3)/(x+4)$ and $y_2 = 2$, the inequality is satisfied when $y_1 \geq y_2$.

X	Y1	Y2
-15	1.6364	2
-8	2.75	2
1	-.4	2
X =		

$y_1 < y_2$; **−15 *is not* a solution.**
$y_1 > y_2$; **−8 *is* a solution.**
$y_1 < y_2$; **1 *is not* a solution.**

The solution set includes the interval B. The endpoint -11 is included because the inequality symbol is $\geq$ and -11 is a solution of the related equation. The number -4 is *not* included because $(x-3)/(x+4)$ is undefined for $x = -4$. Thus the solution set of the original inequality is

$$[-11, -4), \quad \text{or} \quad \{x \mid -11 \leq x < -4\}.$$

GRAPHICAL APPROACH

We graph $y_1 = (x-3)/(x+4)$ and $y_2 = 2$ using DOT mode. The inequality is true for those values of x for which the graph of y_1 is above or intersects the graph of y_2, or where $y_1 \geq y_2$.

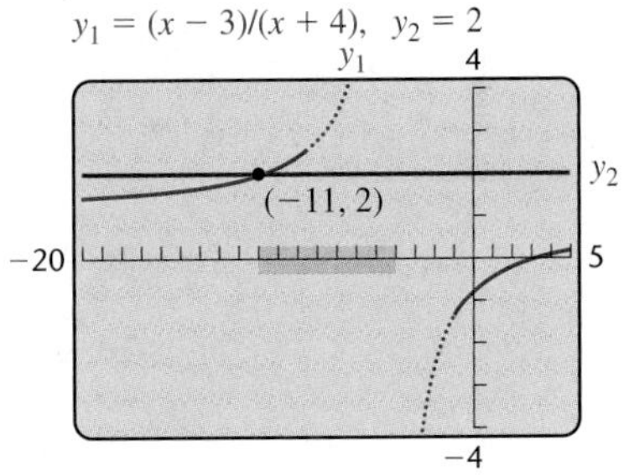

DOT mode

We see that

$$y_1 = y_2 \quad \text{when } x = -11$$

and

$$y_1 > y_2 \quad \text{when } -11 < x < -4.$$

Thus the solution set is $[-11, -4)$.

To Solve a Rational Inequality

1. Change the inequality symbol to an equals sign and solve the related equation.
2. Find any replacements for which the rational expression is undefined.
3. Use the numbers found in steps (1) and (2) to divide the number line into intervals.
4. Substitute a test value from each interval into the inequality. If the number is a solution, then the interval to which it belongs is part of the solution set.
5. Select the interval(s) and any endpoints for which the inequality is satisfied and write set-builder or interval notation for the solution set. If the inequality symbol is $\leq$ or $\geq$, then the solutions from step (1) are also included in the solution set. Those numbers found in step (2) should be excluded from the solution set, even if they are solutions from step (1).

To Solve a Rational Inequality: Graphical Approach

1. Let $y_1 =$ one side of the inequality and $y_2 =$ the other side of the inequality. Examine the graph to determine the intervals that satisfy the inequality.
2. Select the intervals for which the inequality is satisfied. If the inequality symbol is $\leq$ or $\geq$, include any endpoints that also satisfy the inequality. Write set-builder notation or interval notation for the solution set.

8.9 EXERCISE SET

FOR EXTRA HELP

MathXL | MyMathLab | InterAct Math | AW Math Tutor Center | Video Lectures on CD: Disc 4 | Student's Solutions Manual

Concept Reinforcement *Classify each of the following as either true or false.*

1. The solution of $(x - 3)(x + 2) \leq 0$ is $[-2, 3]$.

2. The solution of $(x + 5)(x - 4) \geq 0$ is $[-5, 4]$.

3. The solution of $(x - 1)(x - 6) > 0$ is $\{x \mid x < 1 \text{ or } x > 6\}$.

4. The solution of $(x + 4)(x + 2) < 0$ is $(-4, -2)$.

5. To solve $\frac{x - 5}{x + 4} \geq 0$ using intervals, we divide the number line into the intervals $(-\infty, -4)$, $(-4, 5)$, and $(5, \infty)$.

6. To solve $\frac{x + 2}{x \quad 3} < 0$ using intervals, we divide the number line into the intervals $(-\infty, -2)$, $(-2, 3)$, and $(3, \infty)$.

7. The solution of $\dfrac{3}{x-5} \leq 0$ is $[5, \infty)$.

8. The solution of $\dfrac{-2}{x+3} \geq 0$ is $(-\infty, -3]$.

Determine the solution set of each inequality from the given graph.

9. $p(x) \leq 0$

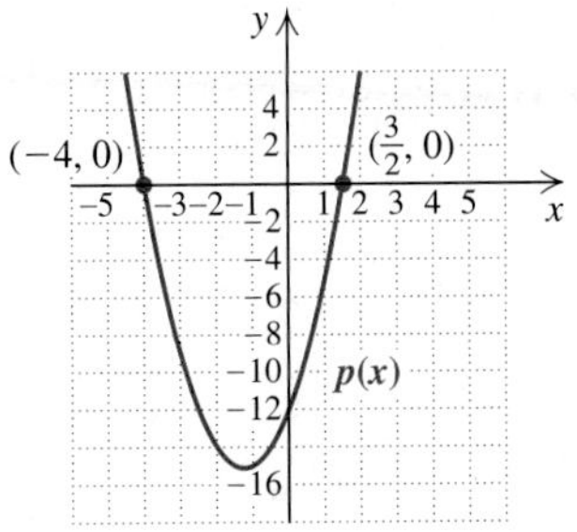

10. $p(x) < 0$

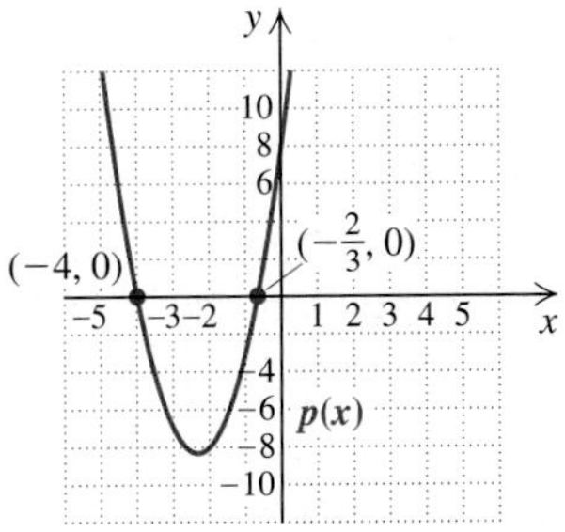

11. $x^4 + 12x > 3x^3 + 4x^2$

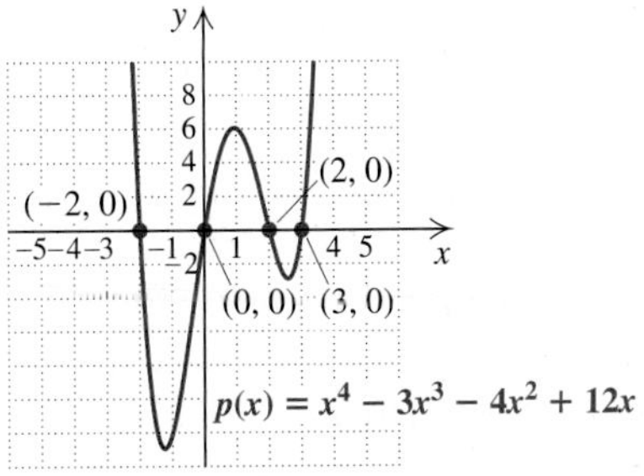

12. $x^4 + x^3 \geq 6x^2$

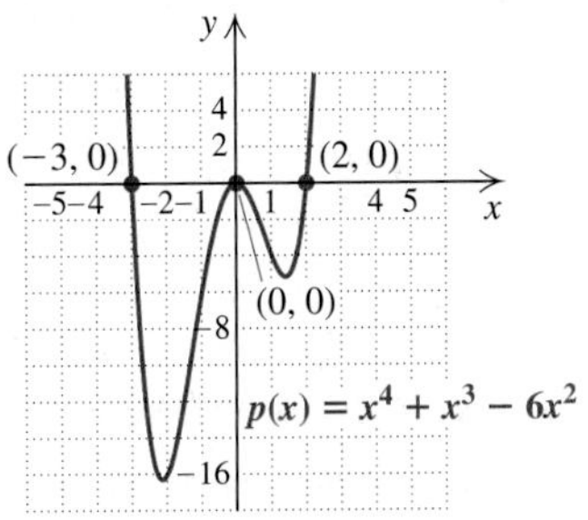

13. $\dfrac{x-1}{x+2} < 3$

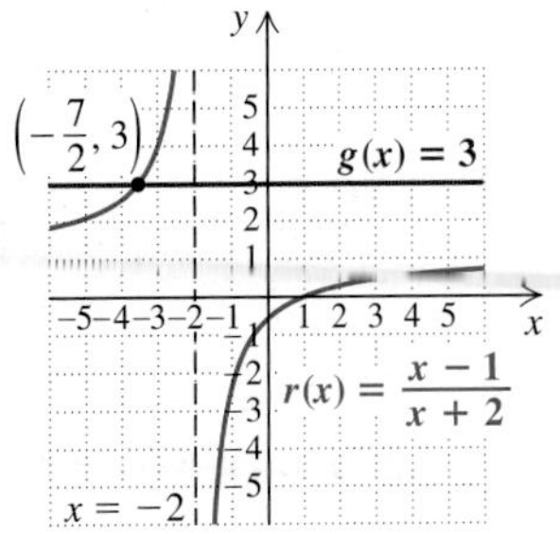

14. $\dfrac{2x-1}{x-5} \geq 1$

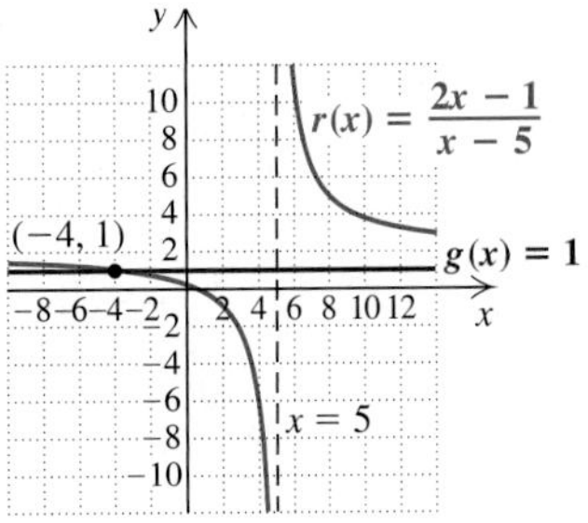

Solve.

15. $(x+4)(x-3) < 0$

16. $(x-5)(x+2) > 0$

17. $(x+7)(x-2) \geq 0$

18. $(x-1)(x+4) \leq 0$

19. $x^2 - x - 2 > 0$

20. $x^2 + x - 2 < 0$

Aha! **21.** $x^2 + 4x + 4 < 0$

22. $x^2 + 6x + 9 < 0$

23. $x^2 - 4x < 12$

24. $x^2 + 6x > -8$

25. $3x(x+2)(x-2) < 0$

26. $5x(x+1)(x-1) > 0$

27. $(x-1)(x+2)(x-4) \geq 0$

28. $(x+3)(x+2)(x-1) < 0$

29. $4.32x^2 - 3.54x - 5.34 \leq 0$

30. $7.34x^2 - 16.55x - 3.89 \geq 0$

31. $x^3 - 2x^2 - 5x + 6 < 0$

32. $\frac{1}{3}x^3 - x + \frac{2}{3} > 0$

33. For $f(x) = x^2 - 1$, find all x-values for which $f(x) \leq 3$.

34. For $f(x) = x^2 - 20$, find all x-values for which $f(x) > 5$.

35. For $g(x) = (x - 2)(x - 3)(x + 1)$, find all x-values for which $g(x) > 0$.

36. For $g(x) = (x + 3)(x - 2)(x + 1)$, find all x-values for which $g(x) < 0$.

37. For $F(x) = x^3 - 7x^2 + 10x$, find all x-values for which $F(x) \leq 0$.

38. For $G(x) = x^3 - 8x^2 + 12x$, find all x-values for which $G(x) \geq 0$.

Solve.

39. $\dfrac{1}{x + 5} < 0$

40. $\dfrac{1}{x + 4} > 0$

41. $\dfrac{x + 1}{x - 3} \geq 0$

42. $\dfrac{x - 2}{x + 4} \leq 0$

43. $\dfrac{x + 1}{x + 6} \geq 1$

44. $\dfrac{x - 1}{x - 2} \leq 1$

45. $\dfrac{(x - 2)(x + 1)}{x - 5} \leq 0$

46. $\dfrac{(x + 4)(x - 1)}{x + 3} \geq 0$

47. $\dfrac{x}{x + 3} \geq 0$

48. $\dfrac{x - 2}{x} \leq 0$

49. $\dfrac{x - 5}{x} < 1$

50. $\dfrac{x}{x - 1} > 2$

51. $\dfrac{x - 1}{(x - 3)(x + 4)} \leq 0$

52. $\dfrac{x + 2}{(x - 2)(x + 7)} \geq 0$

53. For $f(x) = \dfrac{5 - 2x}{4x + 3}$, find all x-values for which $f(x) \geq 0$.

54. For $g(x) = \dfrac{2 + 3x}{2x - 4}$, find all x-values for which $g(x) \geq 0$.

55. For $G(x) = \dfrac{1}{x - 2}$, find all x-values for which $G(x) \leq 1$.

56. For $F(x) = \dfrac{1}{x - 3}$, find all x-values for which $F(x) \leq 2$.

TW 57. Explain how any quadratic inequality can be solved by examining a parabola.

TW 58. Describe a method for creating a quadratic inequality for which there is no solution.

Focused Review

Solve.

59. $2(x - 3) \geq 4 - (x + 1)$ [4.1]

60. $|2x + 3| > 8$ [4.4]

61. $|2 - 5x| \leq 10$ [4.4]

62. $|1.8x - 5.6| < -3$ [4.4]

63. $x^2 - 6x < 7$ [8.9]

64. $\dfrac{1}{x} \leq 1$ [8.9]

Synthesis

TW 65. Step (5) on p. 688 states that even when the inequality symbol is $\leq$ or $\geq$, the solutions from step (1) are not always part of the solution set. Why?

TW 66. Describe a method that could be used to create quadratic inequalities that have $(-\infty, a] \cup [b, \infty)$ as the solution set.

Find each solution set.

67. $x^2 + 2x < 5$

68. $x^4 + 2x^2 \geq 0$

69. $x^4 + 3x^2 \leq 0$

70. $\left|\dfrac{x + 2}{x - 1}\right| \leq 3$

71. *Total Profit.* Derex, Inc., determines that its total-profit function is given by

$$P(x) = -3x^2 + 630x - 6000.$$

a) Find all values of x for which Derex makes a profit.

b) Find all values of x for which Derex loses money.

72. *Height of a Thrown Object.* The function

$$S(t) = -16t^2 + 32t + 1920$$

gives the height S, in feet, of an object thrown from a cliff that is 1920 ft high. Here t is the time, in seconds, that the object is in the air.

a) For what times does the height exceed 1920 ft?

b) For what times is the height less than 640 ft?

73. *Number of Handshakes.* There are n people in a room. The number N of possible handshakes by the people is given by the function

$$N(n) = \frac{n(n-1)}{2}.$$

For what number of people n is $66 \leq N \leq 300$?

74. *Number of Diagonals.* A polygon with n sides has D diagonals, where D is given by the function

$$D(n) = \frac{n(n-3)}{2}.$$

Find the number of sides n if

$$27 \leq D \leq 230.$$

Use a graphing calculator to graph each function and find solutions of $f(x) = 0$. Then solve the inequalities $f(x) < 0$ and $f(x) > 0$.

75. $f(x) = x + \frac{1}{x}$

76. $f(x) = x - \sqrt{x}, x \geq 0$

77. $f(x) = \frac{x^3 - x^2 - 2x}{x^2 + x - 6}$

78. $f(x) = x^4 - 4x^3 - x^2 + 16x - 12$

79. *HMOs.* The percent of the population of the United States enrolled in an HMO (Health Maintenance Organization) increased in the 1990s but began to decrease in 2000. The following table lists the percent enrolled for several years.

Year	Percent Enrolled in an HMO
1994	17.3%
1996	22.3
1998	28.6
2000	30.0
2002	26.4
2004	23.4

Source: Centers for Disease Control and Prevention

a) Use regression to find a quadratic function that can be used to estimate the percent h of the population enrolled in an HMO x years after 1994.

b) Use the function to predict the years after 1994 for which less than 20% of the population will be enrolled in an HMO.

8 Chapter Summary and Review

KEY TERMS AND DEFINITIONS

QUADRATIC EQUATIONS

Quadratic equation, p. 610 Any equation that can be written in the form $ax^2 + bx + c = 0$, with a, b, and c constant.

Discriminant, p. 637 The radicand, $b^2 - 4ac$, in the quadratic formula.

$b^2 - 4ac = 0 \rightarrow$ One solution, a rational number

$b^2 - 4ac > 0 \rightarrow$ Two real solutions, both rational if $b^2 - 4ac$ is a perfect square

$b^2 - 4ac < 0 \rightarrow$ Two imaginary-number solutions

Vertex of a parabola, $f(x) = ax^2 + bx + c$, p. 661

$$\left(-\frac{b}{2a}, f\left(-\frac{b}{2a}\right)\right), \quad \text{or} \quad \left(-\frac{b}{2a}, \frac{4ac - b^2}{4a}\right)$$

APPLICATIONS

Compound interest

$$A = P(1 + r)^t$$

Free-fall distance (in feet)

$$s = 16t^2$$

IMPORTANT CONCEPTS

[Section references appear in brackets.]

Concept	Example
Graphs of quadratic equations are **parabolas.**	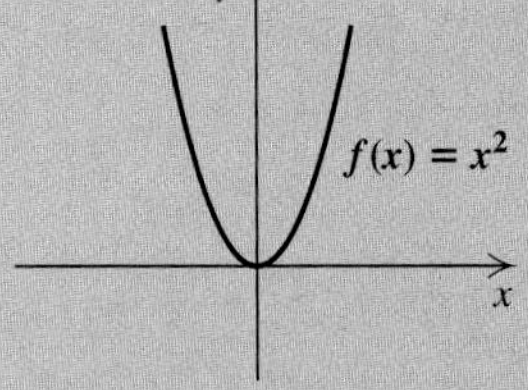 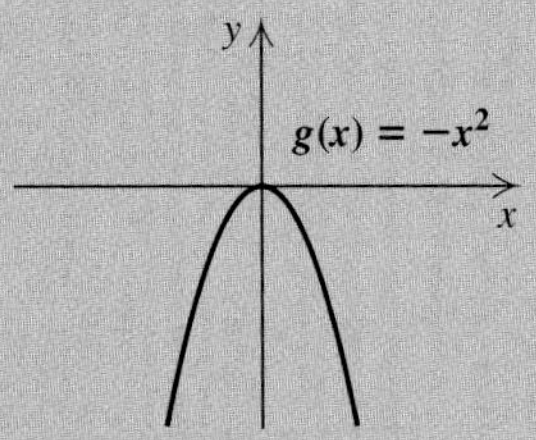 [8.1]
The **principle of square roots** can be used to solve a quadratic equation if a perfect-square trinomial is on one side and a constant is on the other side. For any real number k, if $x^2 = k$, then $x = \sqrt{k}$ or $x = -\sqrt{k}$.	$x^2 - 8x + 16 = 25$ $(x - 4)^2 = 25$ $x - 4 = -5$ *or* $x - 4 = 5$ $x = -1$ *or* $x = 9$ [8.1]

(continued)

Factoring and using the **principle of zero products** is the easiest method to use to solve a quadratic equation if you can factor the polynomial.	$x^2 - 3x - 10 = 0$ $(x + 2)(x - 5) = 0$ $x + 2 = 0 \quad or \quad x - 5 = 0$ $x = -2 \quad or \quad x = 5$ [8.1]
Completing the square can be used to solve any quadratic equation. Calculations can be lengthy.	$x^2 + 6x = 1$ $x^2 + 6x + (\frac{6}{2})^2 = 1 + (\frac{6}{2})^2$ $x^2 + 6x + 9 = 1 + 9$ $(x + 3)^2 = 10$ $x + 3 = \pm\sqrt{10}$ $x = -3 \pm \sqrt{10}$ [8.1]
The **quadratic formula** can be used to solve any quadratic equation. If $ax^2 + bx + c = 0$, then $x = \dfrac{-b \pm \sqrt{b^2 - 4ac}}{2a}$.	$x^2 - 2x - 5 = 0$ $x = \dfrac{-(-2) \pm \sqrt{(-2)^2 - 4(1)(-5)}}{2 \cdot 1}$ $x = \dfrac{2 \pm \sqrt{4 + 20}}{2}$ $x = \dfrac{2}{2} \pm \dfrac{2\sqrt{6}}{2}$ $x = 1 \pm \sqrt{6}$ [8.2]
Equations that are **reducible to quadratic** or **quadratic in form** can be solved by making an appropriate substitution, solving a quadratic equation, and then solving for the original variable.	$x^4 - 10x^2 + 9 = 0$ Let $u = x^2$. Then $u^2 = x^4$. $u^2 - 10u + 9 = 0$ $(u - 9)(u - 1) = 0$ $u - 9 = 0 \quad or \quad u - 1 = 0$ $u = 9 \quad or \quad u = 1$ $x^2 = 9 \quad or \quad x^2 = 1$ $x = \pm 3 \quad or \quad x = \pm 1.$ [8.5]
Polynomial and **rational inequalities** are solved by dividing the x-axis into intervals and using test values from each interval to identify the solution set.	$(x + 2)(x - 3) < 0$ Let $f(x) = (x + 2)(x - 3)$. Solve: $(x + 2)(x - 3) = 0$. $x = -2 \quad or \quad x = 3.$ Test $-3, 0, 4$: A: $f(-3)$ is positive. B: $f(0)$ is negative. C: $f(4)$ is positive. 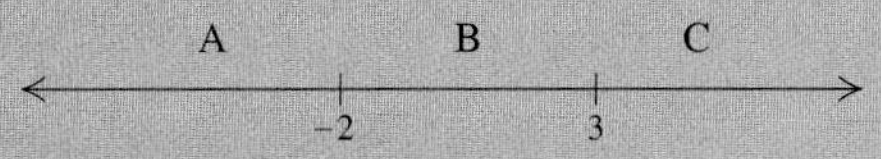 Any number in interval B is a solution. The endpoints are not solutions. The solution set is $(-2, 3)$, or $\{x \mid -2 < x < 3\}$. [8.9]

(*continued*)

The graph of $f(x) = ax^2 + bx + c = a(x - h)^2 + k$ is a parabola.

The vertex is (h, k).

The axis of symmetry is $x = h$.

The domain of f is $(-\infty, \infty)$.

If $a > 0$:

The parabola opens up.

The minimum function value is k.

The range of f is $[k, \infty)$.

If $a < 0$:

The parabola opens down.

The maximum function value is k.

The range of f is $(-\infty, k]$.

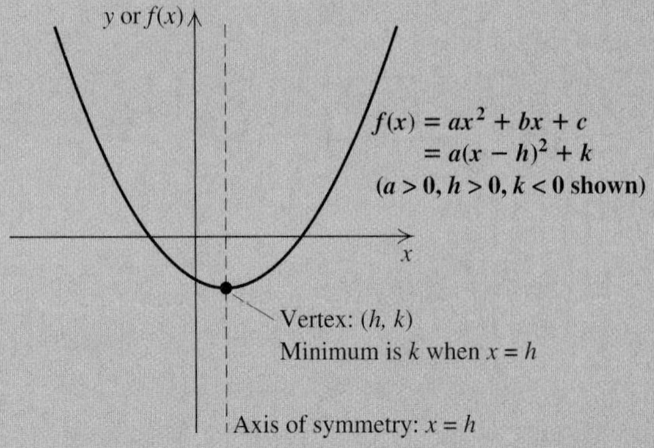

[8.6], [8.7]

Review Exercises

Classify each statement as either true or false.

1. Every quadratic equation has two different solutions. [8.4]

2. Every quadratic equation has at least one solution. [8.4]

3. If an equation cannot be solved by completing the square, it cannot be solved by the quadratic formula. [8.2]

4. A negative discriminant indicates two imaginary-number solutions of a quadratic equation. [8.4]

5. Certain radical or rational equations can be written in quadratic form. [8.5]

6. The graph of $f(x) = 2(x + 3)^2 - 4$ has its vertex at $(3, -4)$. [8.6]

7. The graph of $g(x) = 5x^2$ has $x = 0$ as its axis of symmetry. [8.6]

8. The graph of $f(x) = -2x^2 + 1$ has no minimum value. [8.6]

9. The zeros of $g(x) = x^2 - 9$ are -3 and 3. [8.6]

10. To solve a polynomial inequality, we often must solve a polynomial equation. [8.9]

11. Given the following graph of $f(x) = ax^2 + bx + c$.
    a) State the number of real-number solutions of $ax^2 + bx + c = 0$. [8.1]
    b) State whether a is positive or negative. [8.6]
    c) Determine the minimum value of f. [8.6]

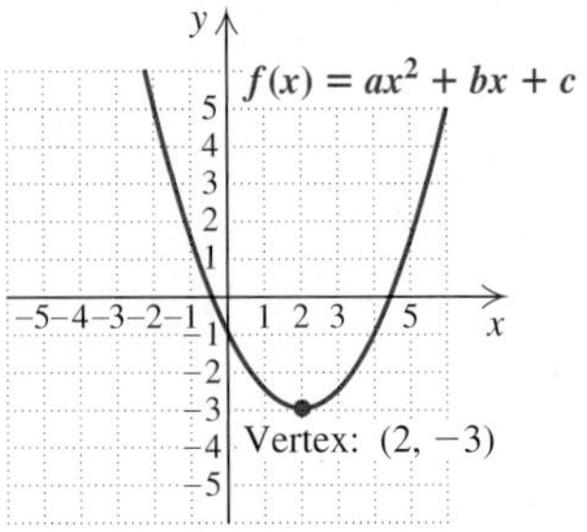

Solve.

12. $4x^2 - 9 = 0$ [8.1]

13. $8x^2 + 6x = 0$ [8.1]

14. $x^2 - 12x + 36 = 9$ [8.1]

15. $x^2 - 4x + 8 = 0$ [8.2]

16. $x(3x + 4) = 4x(x - 1) + 15$ [8.2]

17. $x^2 + 9x = 1$ [8.2]

18. $x^2 - 5x - 2 = 0$. Use a calculator to approximate the solutions with rational numbers. [8.2]

19. Let $f(x) = 4x^2 - 3x - 1$. Find x such that $f(x) = 0$. [8.2]

Replace the blanks with constants to form a true equation. [8.1]

20. $x^2 - 12x + \_\_\_ = (x - \_\_\_)^2$

21. $x^2 + \frac{3}{5}x + \_\_\_ = (x + \_\_\_)^2$

22. Solve by completing the square. Show your work.

$$x^2 - 6x + 1 = 0 \text{ [8.1]}$$

23. \$2500 grows to \$3025 in 2 yr. Use the formula $A = P(1 + r)^t$ to find the interest rate. [8.1]

24. The Peachtree Center Plaza in Atlanta, Georgia, is 723 ft tall. Use $s = 16t^2$ to approximate how long it would take an object to fall from the top. [8.1]

Solve. [8.3]

25. A corporate pilot must fly from company headquarters to a manufacturing plant and back in 4 hr. The distance between headquarters and the plant is 300 mi. If there is a 20-mph headwind going and a 20-mph tailwind returning, how fast must the plane be able to travel in still air?

26. Working together, Erica and Shawna can answer a day's worth of customer-service e-mails in 4 hr. Working alone, Erica takes 6 hr longer than Shawna. How long would it take Shawna to answer the e-mails alone?

For each equation, determine whether the solutions are real or imaginary. If they are real, specify whether they are rational or irrational. [8.4]

27. $x^2 + 3x - 6 = 0$

28. $x^2 + 2x + 5 = 0$

29. Write a quadratic equation having the solutions $\sqrt{5}$ and $-\sqrt{5}$. [8.4]

30. Write a quadratic equation having -4 as its only solution. [8.4]

31. Find all x-intercepts of the graph of $f(x) = x^4 - 13x^2 + 36$. [8.5]

Solve. [8.5]

32. $15x^{-2} - 2x^{-1} - 1 = 0$

33. $(x^2 - 4)^2 - (x^2 - 4) - 6 = 0$

34. **a)** Graph: $f(x) = -3(x + 2)^2 + 4$. [8.6]
b) Label the vertex.
c) Draw the axis of symmetry.
d) Find the maximum or the minimum value.

35. For the function given by $f(x) = 2x^2 - 12x + 23$: [8.7]
a) find the vertex and the axis of symmetry;
b) graph the function.

36. Find the x- and y-intercepts of

$$f(x) = x^2 - 9x + 14. \text{ [8.7]}$$

37. Solve $N = 3\pi\sqrt{1/p}$ for p. [8.3]

38. Solve $2A + T = 3T^2$ for T. [8.3]

Determine which, if any, of the following functions might be used as a model for the data: Linear, with $f(x) = mx + b$; quadratic, with $f(x) = ax^2 + bx + c$, $a > 0$; quadratic, with $f(x) = ax^2 + bx + c$, $a < 0$; neither quadratic nor linear. [8.8]

39.

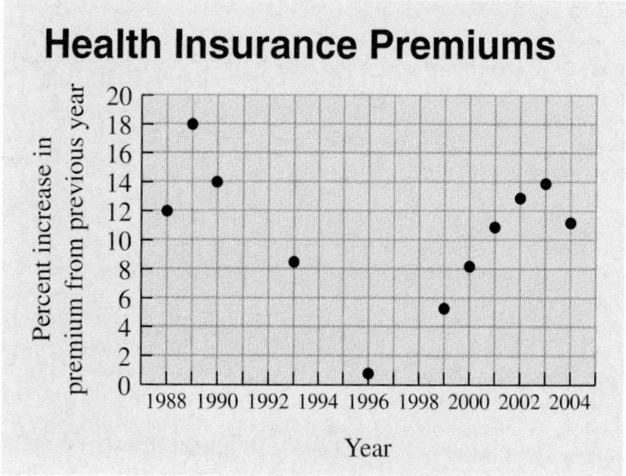

40.

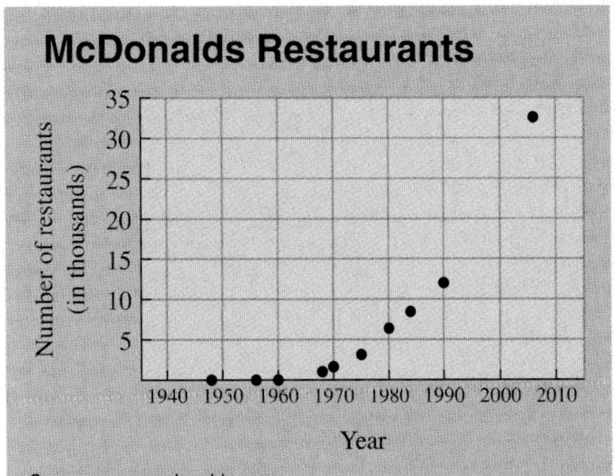

Source: www.mcdonalds.com

41.

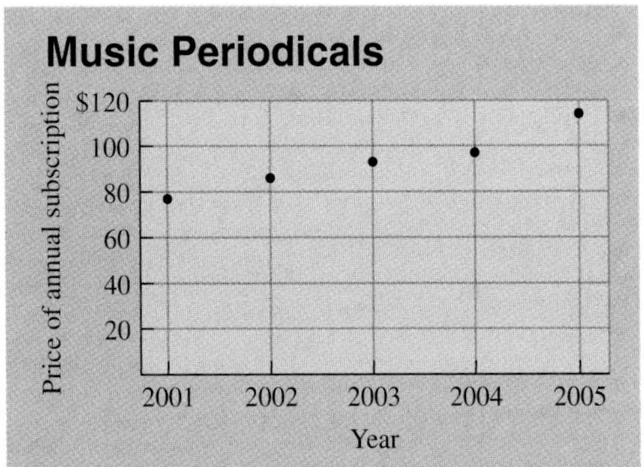

Source: *Statistical Abstract of the United States*, 2006

42. Eastgate Consignments wants to build a rectangular area in a corner for children to play in while their parents shop. They have 30 ft of low fencing. What is the maximum area they can enclose? What dimensions will yield this area? [8.8]

43. *McDonalds.* The following table lists the number of McDonalds restaurants for various years (see Exercise 40). [8.8]

Year	Number of McDonalds Restaurants
1948	1
1956	14
1960	228
1968	1,000
1970	1,600
1975	3,076
1980	6,263
1984	8,300
1990	11,800
2006	32,000

a) Use the data points (0, 1), (20, 1000), and (36, 8300) to find a quadratic function that can be used to estimate the number M of McDonalds restaurants x years after 1948.

b) Use the function to estimate the number of McDonalds restaurants in 2010.

44. a) Use the REGRESSION feature of a graphing calculator and all the data in Exercise 43 to find a quadratic function that can be used to estimate the number M of McDonalds restaurants x years after 1948. [8.8]

b) Use the function to estimate the number of McDonalds restaurants in 2010. [8.8]

Solve. [8.9]

45. $x^3 - 3x > 2x^2$

46. $\dfrac{x-5}{x+3} \le 0$

Synthesis

TW **47.** Discuss two ways in which completing the square was used in this chapter. [8.1], [8.2], [8.7]

TW **48.** Compare the results of Exercises 43(b) and 44(b). Which model do you think gives a better prediction? [8.8]

49. A quadratic function has x-intercepts at -3 and 5. If the y-intercept is at -7, find an equation for the function. [8.7]

50. Find h and k if, for $3x^2 - hx + 4k = 0$, the sum of the solutions is 20 and the product of the solutions is 80. [8.4]

51. The average of two positive integers is 171. One of the numbers is the square root of the other. Find the integers. [8.5]

52. *Starbucks.* The following table lists the number of Starbucks stores for various years. [8.8]

Year	Number of Starbucks Stores
2001	3475
2002	4242
2003	5201
2004	6132
2005	7950

Source: www.starbucks.com

a) Use the REGRESSION feature of a graphing calculator to find a quadratic function that can be used to estimate the number S of Starbucks stores x years after 1948. Use 1948 in order to compare the number of Starbucks stores and McDonalds restaurants (Exercise 44).

b) Use the function from part (a) and the function from Exercise 44(a) to predict the years in which there will be more Starbucks stores than McDonalds restaurants.

Chapter Test 8

1. Given the following graph of $f(x) = ax^2 + bx + c$.

a) State the number of real-number solutions of $ax^2 + bx + c = 0$.

b) State whether a is positive or negative.

c) Determine the maximum function value of f.

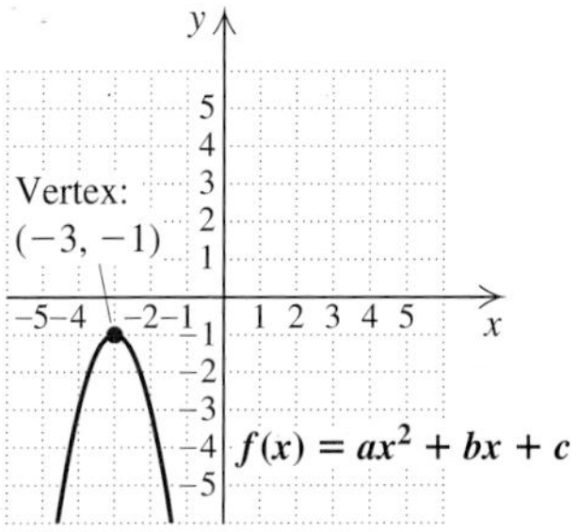

Solve.

2. $3x^2 - 16 = 0$

3. $4x(x - 2) - 3x(x + 1) = -18$

4. $x^2 + x + 1 = 0$

5. $2x + 5 = x^2$

6. $x^{-2} - x^{-1} = \frac{3}{4}$

7. $x^2 + 3x = 5$. Use a calculator to approximate the solutions with rational numbers.

8. Let $f(x) = 12x^2 - 19x - 21$. Find x such that $f(x) = 0$.

Replace the blanks with constants to form a true equation.

9. $x^2 - 16x + \underline{\quad} = (x - \underline{\quad})^2$

10. $x^2 + \frac{2}{7}x + \underline{\quad} = \left(x + \underline{\quad}\right)^2$

11. Solve by completing the square. Show your work.

$$x^2 + 10x + 15 = 0$$

Solve.

12. The Connecticut River flows at a rate of 4 km/h for the length of a popular scenic route. In order for a cruiser to travel 60 km upriver and then return in a total of 8 hr, how fast must the boat be able to travel in still water?

13. Brock and Ian can assemble a swing set in $1\frac{1}{2}$ hr. Working alone, it takes Ian 4 hr longer than Brock to assemble the swing set. How long would it take Brock, working alone, to assemble the swing set?

14. Determine the type of number that the solutions of $x^2 + 5x + 17 = 0$ will be.

15. Write a quadratic equation having solutions -2 and $\frac{1}{3}$.

16. Find all x-intercepts of the graph of

$$f(x) = (x^2 + 4x)^2 + 2(x^2 + 4x) - 3.$$

17. **a)** Graph: $f(x) = 4(x - 3)^2 + 5$.
b) Label the vertex.
c) Draw the axis of symmetry.
d) Find the maximum or the minimum function value.

18. For the function $f(x) = 2x^2 + 4x - 6$:
a) find the vertex and the axis of symmetry;
b) graph the function.

19. Find the x- and y-intercepts of

$$f(x) = x^2 - x - 6.$$

20. Solve $V = \frac{1}{3}\pi(R^2 + r^2)$ for r. Assume all variables are positive.

21. State whether the graph appears to represent a linear function, a quadratic function, or neither.

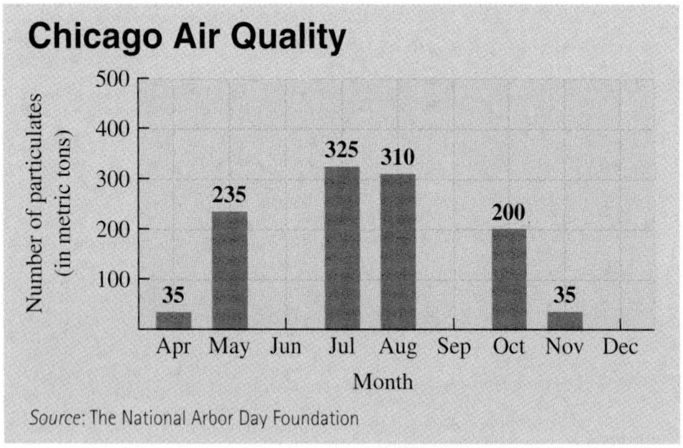

22. Jay's Custom Pickups has determined that when x hundred truck caps are built, the average cost per cap is given by

$$C(x) = 0.2x^2 - 1.3x + 3.4025,$$

where $C(x)$ is in hundreds of dollars. What is the minimum cost per truck cap and how many caps should be built to achieve that minimum?

23. Trees improve air quality in part by retaining airborne particles, called particulates, from the air. The following table shows the number of metric tons of particulates retained by trees in Chicago during the spring, summer, and autumn months. (See Exercise 21.)

Month		Amount of Particulates Retained (in metric tons)
April	(0)	35
May	(1)	235
July	(3)	325
August	(4)	310
October	(6)	200
November	(7)	35

Source: The National Arbor Day Foundation

Use the data points $(0, 35)$, $(4, 310)$, and $(6, 200)$ to find a quadratic function that can be used to estimate the number of metric tons p of particulates retained x months after April.

24. Use regression to find a quadratic function that fits the data in Exercise 23.

Solve.

25. $x^2 + 5x \leq 6$

26. $x - \dfrac{1}{x} > 0$

Synthesis

27. One solution of $kx^2 + 3x - k = 0$ is -2. Find the other solution.

28. Find a fourth-degree polynomial equation, with integer coefficients, for which $2 - \sqrt{3}$ and $5 - i$ are solutions.

29. Find a polynomial equation, with integer coefficients, for which 5 is a repeated root and $\sqrt{2}$ and $\sqrt{3}$ are solutions.

9 Exponential and Logarithmic Functions

The functions that we consider in this chapter are interesting not only from a purely intellectual point of view, but also for their rich applications to many fields. We will look at applications such as compound interest and population growth, to name just two.

The theory centers on functions with variable exponents (*exponential functions*). We study these functions, their inverses, and their properties.

APPLICATION *Cruise-Ship Passengers.*

In 1970, cruise lines carried approximately 500,000 passengers. This number was projected to increase to 15 million in 2006. (*Sources*: Cruise Lines International; *Annual Cruise Review* 2004) Use the data in the following table to estimate the number of cruise-ship passengers in 2012.

Year	Number of Passengers (in millions)
1970	0.5
1980	1.4
1990	3.6
2000	6.5
2004	13.4
2006	15 (proj.)

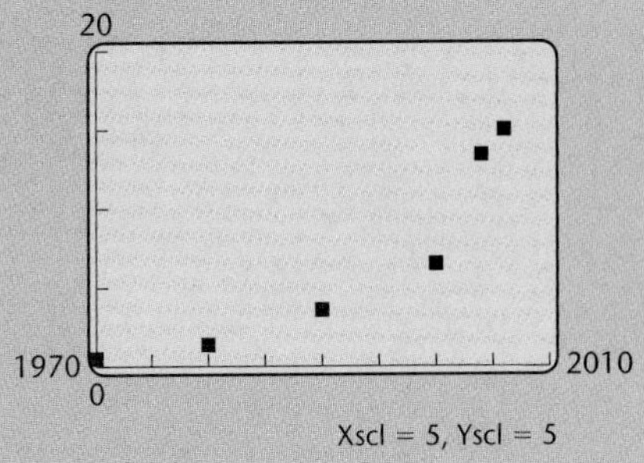

This problem appears as Example 9 in Section 9.7.

9.1 Composite and Inverse Functions

Composite Functions ▪ Inverses and One-to-One Functions ▪ Finding Formulas for Inverses ▪ Graphing Functions and Their Inverses ▪ Inverse Functions and Composition

Study Tip

Divide and Conquer

In longer sections of reading, there are almost always subsections. Rather than feel obliged to read the entire assignment at once, make use of the subsections as natural resting points. Reading two or three subsections and then taking a break can increase your comprehension and is thus an efficient use of your time.

Composite Functions

In the real world, functions frequently occur in which some quantity depends on a variable that, in turn, depends on another variable. For instance, a firm's profits may depend on the number of items the firm produces, which may in turn depend on the number of employees hired. Functions like this are called **composite functions.**

For example, the function g that gives a correspondence between women's shoe sizes in the United States and those in Italy is given by $g(x) = 2x + 24$, where x is the U.S. size and $g(x)$ is the Italian size. Thus a U.S. size 4 corresponds to a shoe size of $g(4) = 2 \cdot 4 + 24$, or 32, in Italy.

A different function gives a correspondence between women's shoe sizes in Italy and those in Britain. This particular function is given by $f(x) = \frac{1}{2}x - 14$, where x is the Italian size and $f(x)$ is the corresponding British size. Thus an Italian size 32 corresponds to a British size $f(32) = \frac{1}{2} \cdot 32 - 14$, or 2.

It seems reasonable to conclude that a shoe size of 4 in the United States corresponds to a size of 2 in Britain and that some function h describes this correspondence. Can we find a formula for h? If we look at the following tables, we might guess that such a formula is $h(x) = x - 2$, and that is indeed correct. But, for more complicated formulas, we would need to use algebra.

g: $g(x) = 2x + 24$ (U.S. → Italy)

f: $f(x) = \frac{1}{2}x - 14$ (Italy → Britain)

U.S.	Italy	Britain
4	32	2
5	34	3
6	36	4
7	38	5

h: $h(x) = ?$ (U.S. → Britain)

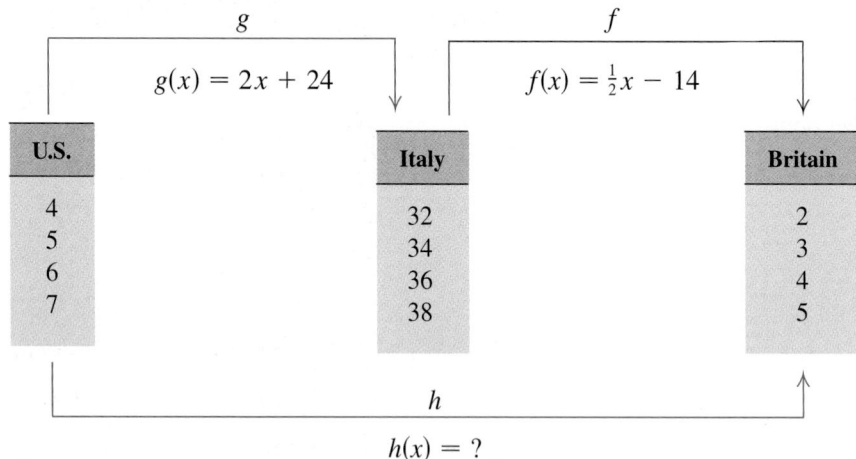

Size x shoes in the United States correspond to size $g(x)$ shoes in Italy, where

$$g(x) = 2x + 24.$$

Size n shoes in Italy correspond to size $f(n)$ shoes in Britain. Thus size $g(x)$ shoes in Italy correspond to size $f(g(x))$ shoes in Britain. Since the x in the expression $f(g(x))$ represents a U.S. shoe size, we can find the British shoe size that corresponds to a U.S. size x as follows:

$$f(g(x)) = f(2x + 24) = \frac{1}{2} \cdot (2x + 24) - 14 \quad \text{Using } g(x) \text{ as an input}$$
$$= x + 12 - 14 = x - 2.$$

This gives a formula for h: $h(x) = x - 2$. Thus U.S. size 4 corresponds to British size $h(4) = 4 - 2$, or 2. The function h is the *composition* of f and g and is denoted $f \circ g$ (read "the composition of f and g," "f composed with g," or "f circle g").

Composition of Functions The *composite function* $f \circ g$, the *composition* of f and g, is defined as

$$(f \circ g)(x) = f(g(x)).$$

We can visualize the composition of functions as follows.

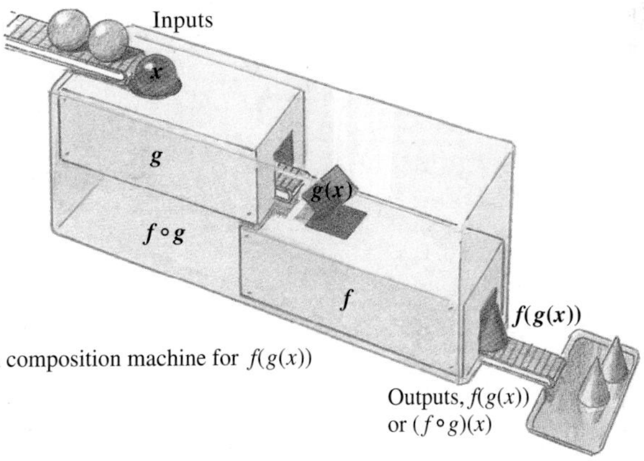

A composition machine for $f(g(x))$

EXAMPLE 1 Given $f(x) = 3x$ and $g(x) = 1 + x^2$:

a) Find $(f \circ g)(5)$ and $(g \circ f)(5)$.

b) Find $(f \circ g)(x)$ and $(g \circ f)(x)$.

SOLUTION Consider each function separately:

$f(x) = 3x$ **This function multiplies each input by 3.**

and $g(x) = 1 + x^2$. **This function adds 1 to the square of each input.**

a) To find $(f \circ g)(5)$, we first find $g(5)$ and then use that as an input for f:

$$(f \circ g)(5) = f(g(5)) = f(1 + 5^2) \quad \text{Using } g(x) = 1 + x^2$$
$$= f(26) = 3 \cdot 26 = 78. \quad \text{Using } f(x) = 3x$$

To find $(g \circ f)(5)$, we first find $f(5)$ and then use that as an input for g:

$$(g \circ f)(5) = g(f(5)) = g(3 \cdot 5) \quad \text{Note that } f(5) = 3 \cdot 5 = 15.$$
$$= g(15) = 1 + 15^2 = 1 + 225 = 226.$$

b) We find $(f \circ g)(x)$ by substituting $g(x)$ for x in the equation for $f(x)$:

$$(f \circ g)(x) = f(g(x)) = f(1 + x^2) \qquad \textbf{Using } g(x) = 1 + x^2$$
$$= 3 \cdot (1 + x^2) = 3 + 3x^2. \qquad \textbf{Using } f(x) = 3x. \textit{ These } \textbf{parentheses indicate multiplication.}$$

To find $(g \circ f)(x)$, we substitute $f(x)$ for x in the equation for $g(x)$:

$$(g \circ f)(x) = g(f(x)) = g(3x) \qquad \textbf{Substituting } 3x \textbf{ for } f(x)$$
$$= 1 + (3x)^2 = 1 + 9x^2.$$

As a check, note that $(g \circ f)(5) = 1 + 9 \cdot 5^2 = 1 + 9 \cdot 25 = 226$, as expected from part (a) above.

Example 1 shows that, in general, $(f \circ g)(5) \neq (g \circ f)(5)$ and $(f \circ g)(x) \neq (g \circ f)(x)$.

EXAMPLE 2 Given $f(x) = \sqrt{x}$ and $g(x) = x - 1$, find $(f \circ g)(x)$ and $(g \circ f)(x)$.

SOLUTION

$$(f \circ g)(x) = f(g(x)) = f(x - 1) = \sqrt{x - 1} \qquad \textbf{Using } g(x) = x - 1$$
$$(g \circ f)(x) = g(f(x)) = g(\sqrt{x}) = \sqrt{x} - 1 \qquad \textbf{Using } f(x) = \sqrt{x}$$

To check using a graphing calculator, let

$y_1 = f(x) = \sqrt{}(x)$,

$y_2 = g(x) = x - 1$,

$y_3 = \sqrt{}(x - 1)$,

and $y_4 = y_1(y_2(x))$.

If our work is correct, y_4 will be equivalent to y_3. We form a table of values, selecting only y_3 and y_4. The table on the left below indicates that the functions are probably equivalent.

X	Y3	Y4
1	0	0
1.5	.70711	.70711
2	1	1
2.5	1.2247	1.2247
3	1.4142	1.4142
3.5	1.5811	1.5811
4	1.7321	1.7321
X = 1		

X	Y5	Y6
1	0	0
1.5	.22474	.22474
2	.41421	.41421
2.5	.58114	.58114
3	.73205	.73205
3.5	.87083	.87083
4	1	1
X = 1		

Next, we let $y_5 = \sqrt{}(x) - 1$ and $y_6 = y_2(y_1(x))$ and select only y_5 and y_6. The table on the right above indicates that the functions are probably equivalent. Thus we conclude that

$$(f \circ g)(x) = \sqrt{x - 1} \quad \text{and} \quad (g \circ f)(x) = \sqrt{x} - 1.$$

Functions that are not defined by an equation can still be composed when the output of the first function is in the domain of the second function.

EXAMPLE 3 Use the following table to find each value, if possible.

a) $(y_2 \circ y_1)(3)$ **b)** $(y_1 \circ y_2)(3)$

X	Y1	Y2
0	−2	3
1	2	5
2	−1	7
3	1	9
4	0	11
5	0	13
6	−3	15
X = 0		

SOLUTION

a) Since $(y_2 \circ y_1)(3) = y_2(y_1(3))$, we first find $y_1(3)$. We locate 3 in the x-column and then move across to the y_1-column to find $y_1(3) = 1$.

X	Y1	Y2
0	−2	3
1	2	5
2	−1	7
3	1	9
4	0	11
5	0	13
6	−3	15
Y2 = 5		

The output 1 now becomes an input for the function y_2:

$$y_2(y_1(3)) = y_2(1).$$

To find $y_2(1)$, we locate 1 in the x-column and then move across to the y_2-column to find $y_2(1) = 5$. Thus, $(y_2 \circ y_1)(3) = 5$.

b) Since $(y_1 \circ y_2)(3) = y_1(y_2(3))$, we find $y_2(3)$ by locating 3 in the x-column and moving across to the y_2-column. We have $y_2(3) = 9$, so

$$(y_1 \circ y_2)(3) = y_1(y_2(3)) = y_1(9).$$

However, y_1 is not defined for $x = 9$, so $(y_1 \circ y_2)(3)$ is undefined.

In fields ranging from chemistry to geology and economics, one needs to recognize how a function can be regarded as the composition of two "simpler" functions. This is sometimes called *de*composition.

EXAMPLE 4 If $h(x) = (7x + 3)^2$, find f and g such that $h(x) = (f \circ g)(x)$.

SOLUTION To find $h(x)$, we can think of first forming $7x + 3$ and then squaring. This suggests that $g(x) = 7x + 3$ and $f(x) = x^2$. We check by forming the composition:

$$(f \circ g)(x) = f(g(x)) = f(7x + 3) = (7x + 3)^2 = h(x), \text{ as desired.}$$

Student Notes

Throughout this chapter, keep in mind that functions are defined using dummy variables that represent an arbitrary member of the domain. Thus, $f(x) = x^2 - 3x$ and $f(t) = t^2 - 3t$ describe the same function f.

There are other less "obvious" answers. For example, if

$$f(x) = (x - 1)^2 \quad \text{and} \quad g(x) = 7x + 4,$$

then

$$\begin{aligned}(f \circ g)(x) &= f(g(x)) = f(7x + 4)\\ &= (7x + 4 - 1)^2 = (7x + 3)^2 = h(x).\end{aligned}$$

Inverses and One-to-One Functions

Let's consider the following two functions. We think of them as relations, or correspondences.

Countries and their capitals

Domain (Set of Inputs)	Range (Set of Outputs)
Australia →	Canberra
China →	Beijing
Cyprus →	Nicosia
Egypt →	Cairo
Haiti →	Port-au-Prince
Peru →	Lima

U.S. Senators and their states

Domain (Set of Inputs)	Range (Set of Outputs)
Bunning →	Kentucky
McConnell →	(Kentucky)
Kerry →	Massachusetts
Kennedy →	(Massachusetts)
Smith →	Oregon
Wyden →	(Oregon)

Suppose we reverse the arrows. We obtain what is called the **inverse relation.** Are these inverse relations functions?

Countries and their capitals

Range (Set of Outputs)	Domain (Set of Inputs)
Australia	← Canberra
China	← Beijing
Cyprus	← Nicosia
Egypt	← Cairo
Haiti	← Port-au-Prince
Peru	← Lima

U.S. Senators and their states

Range (Set of Outputs)	Domain (Set of Inputs)
Bunning	← Kentucky
McConnell	← (Kentucky)
Kerry	← Massachusetts
Kennedy	← (Massachusetts)
Smith	← Oregon
Wyden	← (Oregon)

Recall that for each input, a function provides exactly one output. However, it is possible for different inputs to correspond to the same output. Only when this possibility is *excluded* will the inverse be a function. For the functions listed above, this means the inverse of the "Capital" correspondence is a function, but the inverse of the "U.S. Senator" correspondence is not.

In the Capital function, different inputs have different outputs, so it is a **one-to-one function.** In the U.S. Senator function, *Kerry* and *Kennedy* are both paired with *Massachusetts*. Thus the U.S. Senator function is not one-to-one.

One-To-One Function A function f is *one-to-one* if different inputs have different outputs. That is, if for a and b in the domain of f with $a \neq b$, we have $f(a) \neq f(b)$, then the function f is one-to-one. If a function is one-to-one, then its inverse correspondence is also a function.

How can we tell graphically whether a function is one-to-one?

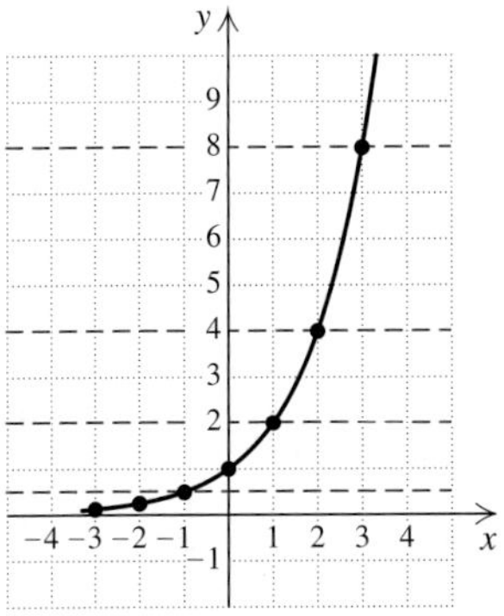

EXAMPLE 5 Shown at left is the graph of a function similar to those we will study in Section 9.2. Determine whether the function is one-to-one and thus has an inverse that is a function.

SOLUTION A function is one-to-one if different inputs have different outputs—that is, if no two x-values have the same y-value. For this function, we cannot find two x-values that have the same y-value. Note that this means that no horizontal line can be drawn so that it crosses the graph more than once. The function is one-to-one so its inverse is a function.

The graph of every function must pass the vertical-line test. In order for a function to have an inverse that is a function, it must pass the *horizontal-line test* as well.

The Horizontal-Line Test If it is impossible to draw a horizontal line that intersects a function's graph more than once, then the function is one-to-one. For every one-to-one function, an inverse function exists.

EXAMPLE 6 Determine whether the function $f(x) = x^2$ is one-to-one and thus has an inverse that is a function.

SOLUTION The graph of $f(x) = x^2$ is shown here. Many horizontal lines cross the graph more than once. For example, the line $y = 4$ crosses where the first coordinates are -2 and 2. Although these are different inputs, they have the same output. That is, $-2 \neq 2$, but $f(-2) = f(2) = 4$.

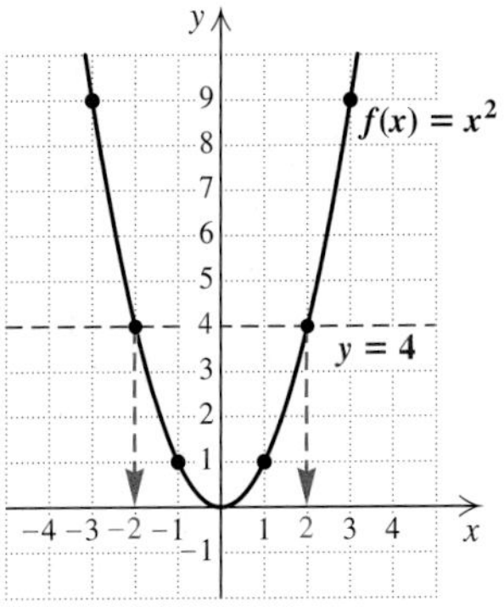

Thus the function is not one-to-one and no inverse function exists.

Finding Formulas for Inverses

When the inverse of f is also a function, it is denoted f^{-1} (read "f-inverse").

CAUTION! The -1 in f^{-1} is *not* an exponent!

Suppose a function is described by a formula. If its inverse is a function, how do we find a formula for that inverse? For any equation in two variables, if we interchange the variables, we form an equation of the inverse correspondence. If it is a function, we proceed as follows to find a formula for f^{-1}.

To Find a Formula for f^{-1}

First make sure that f is one-to-one. Then:

1. Replace $f(x)$ with y.
2. Interchange x and y. (This gives the inverse function.)
3. Solve for y.
4. Replace y with $f^{-1}(x)$. (This is inverse function notation.)

EXAMPLE 7 Determine whether each function is one-to-one and if it is, find a formula for $f^{-1}(x)$.

a) $f(x) = x + 2$ **b)** $f(x) = 2x - 3$

SOLUTION

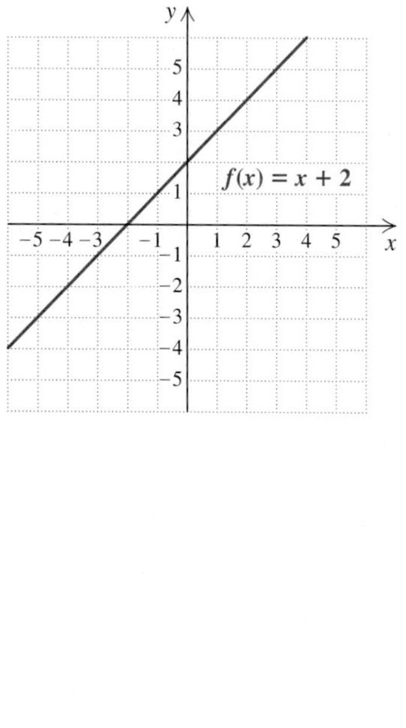

a) The graph of $f(x) = x + 2$ is shown at left. It passes the horizontal-line test, so it is one-to-one. Thus its inverse is a function.

1. Replace $f(x)$ with y: $y = x + 2$.

2. Interchange x and y: $x = y + 2$. **This gives the inverse function.**

3. Solve for y: $x - 2 = y$.

4. Replace y with $f^{-1}(x)$: $f^{-1}(x) = x - 2$. **We also "reversed" the equation.**

In this case, the function f adds 2 to all inputs. Thus, to "undo" f, the function f^{-1} must subtract 2 from its inputs.

b) The function $f(x) = 2x - 3$ is also linear. Any linear function that is not constant will pass the horizontal-line test. Thus, f is one-to-one.

1. Replace $f(x)$ with y: $y = 2x - 3$.

2. Interchange x and y: $x = 2y - 3$.

3. Solve for y: $x + 3 = 2y$

$$\frac{x+3}{2} = y.$$

4. Replace y with $f^{-1}(x)$: $f^{-1}(x) = \dfrac{x+3}{2}$.

In this case, the function f doubles all inputs and then subtracts 3. Thus, to "undo" f, the function f^{-1} adds 3 to each input and then divides by 2.

Graphing Functions and Their Inverses

How do the graphs of a function and its inverse compare?

EXAMPLE 8 Graph $f(x) = 2x - 3$ and $f^{-1}(x) = (x + 3)/2$ on the same set of axes. Then compare.

SOLUTION The graph of each function follows. Note that the graph of f^{-1} can be drawn by reflecting the graph of f across the line $y = x$. That is, if we graph $f(x) = 2x - 3$ in wet ink and fold the paper along the line $y = x$, the graph of $f^{-1}(x) = (x + 3)/2$ will appear as the impression made by f.

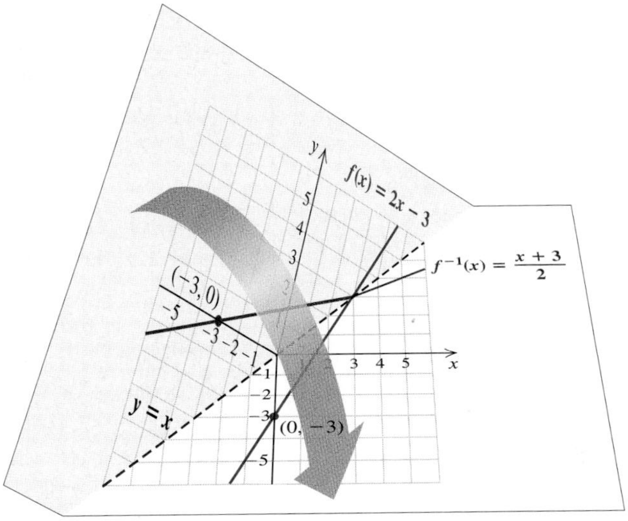

When x and y are interchanged to find a formula for the inverse, we are, in effect, reflecting or flipping the graph of $f(x) = 2x - 3$ across the line $y = x$. For example, when the coordinates of the y-intercept of the graph of f, $(0, -3)$, are reversed, we get $(-3, 0)$, the x-intercept of the graph of f^{-1}.

Visualizing Inverses

The graph of f^{-1} is a reflection of the graph of f across the line $y = x$.

EXAMPLE 9 Consider $g(x) = x^3 + 2$.

a) Determine whether the function is one-to-one.

b) If it is one-to-one, find a formula for its inverse.

c) Graph the inverse, if it exists.

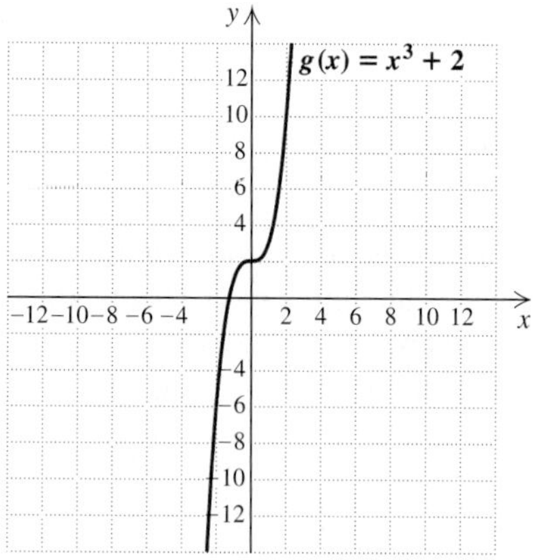

SOLUTION

a) The graph of $g(x) = x^3 + 2$ is shown at left. It passes the horizontal-line test and thus has an inverse.

b) 1. Replace $g(x)$ with y: $y = x^3 + 2$. **Using $g(x) = x^3 + 2$**

2. Interchange x and y: $x = y^3 + 2$.

3. Solve for y: $x - 2 = y^3$

$\sqrt[3]{x - 2} = y$. **Each number has only one cube root, so we can solve for y.**

4. Replace y with $g^{-1}(x)$: $g^{-1}(x) = \sqrt[3]{x - 2}$.

c) To find the graph, we reflect the graph of $g(x) = x^3 + 2$ across the line $y = x$, as we did in Example 8. We can also substitute into $g^{-1}(x) = \sqrt[3]{x - 2}$ and plot points. Note that $(2, 10)$ is on the graph of g, whereas $(10, 2)$ is on the graph of g^{-1}. The graphs of g and g^{-1} are shown together below.

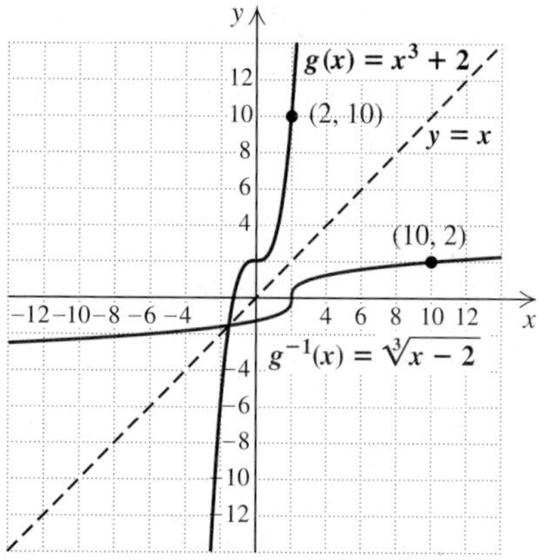

Student Notes

Since we interchange x and y to find an equation for an inverse, the domain of the function is the range of the inverse, and the range of the function is the domain of the inverse.

Inverse Functions and Composition

Let's consider inverses of functions in terms of function machines. Suppose that a one-to-one function f is programmed into a machine. If the machine has a reverse switch, when the switch is thrown, the machine performs the inverse function f^{-1}. Inputs then enter at the opposite end, and the entire process is reversed.

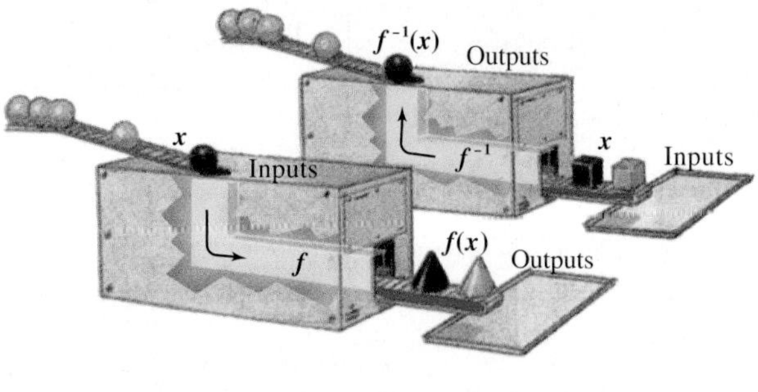

Consider $g(x) = x^3 + 2$ and $g^{-1}(x) = \sqrt[3]{x - 2}$ from Example 9. For the input 3,

$$g(3) = 3^3 + 2 = 27 + 2 = 29.$$

The output is 29. Now we use 29 for the input in the inverse:

$$g^{-1}(29) = \sqrt[3]{29 - 2} = \sqrt[3]{27} = 3.$$

The function g takes 3 to 29. The inverse function g^{-1} takes the number 29 back to 3.

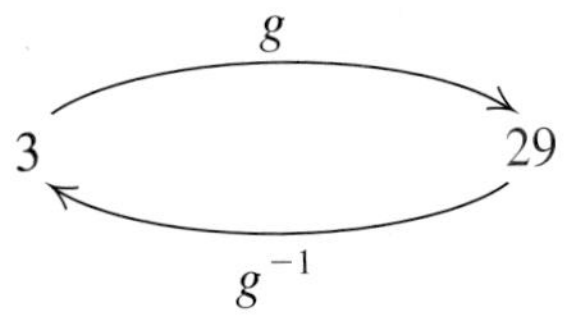

In general, for any output $f(x)$, the function f^{-1} takes that output back to x. Similarly, for any output $f^{-1}(x)$, the function f takes that output back to x.

Composition and Inverses If a function f is one-to-one, then f^{-1} is the unique function for which

$$(f^{-1} \circ f)(x) = f^{-1}(f(x)) = x \quad \text{and}$$
$$(f \circ f^{-1})(x) = f(f^{-1}(x)) = x.$$

EXAMPLE 10 Let $f(x) = 2x + 1$. Show that

$$f^{-1}(x) = \frac{x - 1}{2}.$$

SOLUTION We find $(f^{-1} \circ f)(x)$ and $(f \circ f^{-1})(x)$ and check to see that each is x.

$$(f^{-1} \circ f)(x) = f^{-1}(f(x)) = f^{-1}(2x + 1)$$
$$= \frac{(2x + 1) - 1}{2}$$
$$= \frac{2x}{2} = x$$

$$(f \circ f^{-1})(x) = f(f^{-1}(x)) = f\left(\frac{x - 1}{2}\right)$$
$$= 2 \cdot \frac{x - 1}{2} + 1$$
$$= x - 1 + 1 = x$$

Inverse Functions: Graphing and Composition

In Example 9, we found that the inverse of $y_1 = x^3 + 2$ is $y_2 = \sqrt[3]{x - 2}$. There are several ways to check this result using a graphing calculator.

1. We can check visually by graphing both functions, along with the line $y = x$, on a "squared" set of axes. The graph of y_2 *appears* to be the reflection of the graph of y_1 across the line $y = x$. This is only a partial check since we are comparing graphs visually.

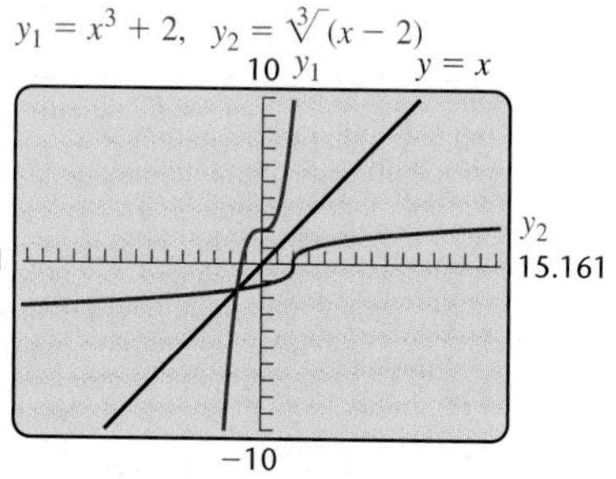

2. A second, more precise, visual check begins with graphing $y_1 = x^3 + 2$ using a squared viewing window. We then press DRAW 8 VARS ▷ 1 1 ENTER to select the DrawInv option of the DRAW menu and graph the inverse. The resulting graph should coincide with the graph of y_2.
3. For a third check, note that if y_2 is the inverse of y_1, then $(y_2 \circ y_1)(x) = x$ and $(y_1 \circ y_2)(x) = x$. We enter $y_3 = y_2(y_1(x))$ and $y_4 = y_1(y_2(x))$, and form a table to compare x, y_3, and y_4. Note that for the values shown, $y_3 = y_4 = x$.

TblStart = −3, ΔTbl = 1

X	Y3	Y4
−3	−3	−3
−2	−2	−2
−1	−1	−1
0	0	0
1	1	1
2	2	2
3	3	3

X = −3

9.1 EXERCISE SET

FOR EXTRA HELP

MathXL | MyMathLab | InterAct Math | AW Math Tutor Center | Video Lectures on CD: Disc 5 | Student's Solutions Manual

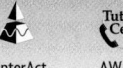

Concept Reinforcement *Classify each statement as either true or false.*

1. The composition of two functions f and g is written $f \circ g$.
2. The notation $(f \circ g)(x)$ means $f(g(x))$.
3. If $f(x) = x^2$ and $g(x) = x + 3$, then $(g \circ f)(x) = (x + 3)^2$.
4. If $f(2) = 15$ and $g(15) = 25$, then $(g \circ f)(2) = 25$.
5. The function f is one-to-one if $f(1) = 1$.
6. For f^{-1} to be a function, f cannot be one-to-one.
7. The function f is the inverse of f^{-1}.
8. If g and h are inverses of each other, then $(g \circ h)(x) = x$.

Find $(f \circ g)(1)$, $(g \circ f)(1)$, $(f \circ g)(x)$, and $(g \circ f)(x)$.

9. $f(x) = x^2 + 1;\ g(x) = 2x - 3$
10. $f(x) = 2x + 1;\ g(x) = x^2 - 5$
11. $f(x) = x - 3;\ g(x) = 2x^2 - 7$
12. $f(x) = 3x^2 + 4;\ g(x) = 4x - 1$
13. $f(x) = x + 7;\ g(x) = 1/x^2$
14. $f(x) = 1/x^2;\ g(x) = x + 2$
15. $f(x) = \sqrt{x};\ g(x) = x + 3$
16. $f(x) = 10 - x;\ g(x) = \sqrt{x}$
17. $f(x) = \sqrt{4x};\ g(x) = 1/x$
18. $f(x) = \sqrt{x + 3};\ g(x) = 13/x$
19. $f(x) = x^2 + 4;\ g(x) = \sqrt{x - 1}$
20. $f(x) = x^2 + 8;\ g(x) = \sqrt{x + 17}$

Use the following table to find each value, if possible.

X	Y1	Y2
−3	−4	1
−2	−1	−2
−1	2	−3
0	5	−2
1	8	1
2	11	6
3	14	11
X =		

21. $(y_1 \circ y_2)(-3)$
22. $(y_2 \circ y_1)(-3)$
23. $(y_1 \circ y_2)(-1)$
24. $(y_2 \circ y_1)(-1)$
25. $(y_2 \circ y_1)(1)$
26. $(y_1 \circ y_2)(1)$

Use the following table to find each value, if possible.

x	$f(x)$	$g(x)$
1	0	1
2	3	5
3	2	8
4	6	5
5	4	1

27. $(f \circ g)(2)$
28. $(g \circ f)(4)$
29. $f(g(3))$
30. $g(f(5))$

Find $f(x)$ and $g(x)$ such that $h(x) = (f \circ g)(x)$. Answers may vary.

31. $h(x) = (7 + 5x)^2$
32. $h(x) = (3x - 1)^2$
33. $h(x) = \sqrt{2x + 7}$
34. $h(x) = \sqrt{5x + 2}$
35. $h(x) = \dfrac{2}{x - 3}$
36. $h(x) = \dfrac{3}{x} + 4$
37. $h(x) = \dfrac{1}{\sqrt{7x + 2}}$
38. $h(x) = \sqrt{x - 7} - 3$
39. $h(x) = \dfrac{1}{\sqrt{3x}} + \sqrt{3x}$
40. $h(x) = \dfrac{1}{\sqrt{2x}} - \sqrt{2x}$

Determine whether each function is one-to-one.

Aha! **41.** $f(x) = x - 5$

42. $f(x) = 5 - 2x$

43. $f(x) = x^2 + 1$

44. $f(x) = 1 - x^2$

45.

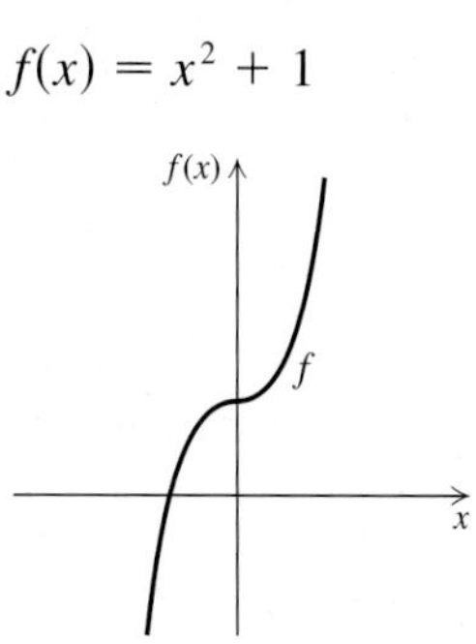

46.

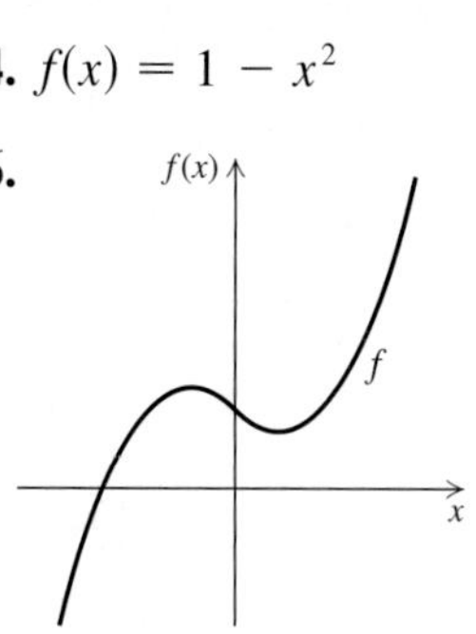

47.

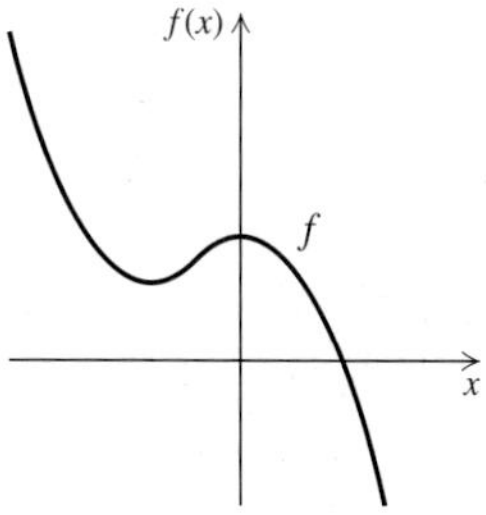

48.

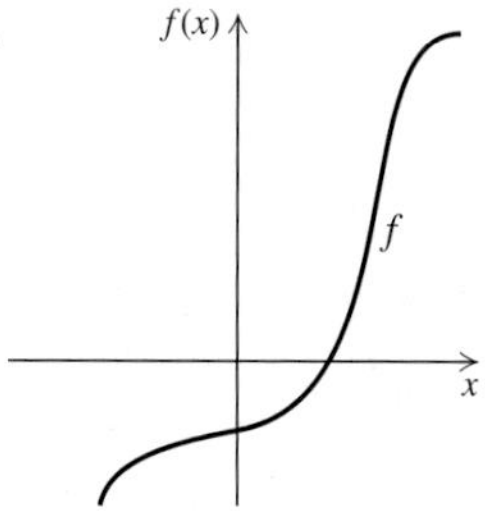

For each function, **(a)** *determine whether it is one-to-one;* **(b)** *if it is one-to-one, find a formula for the inverse.*

49. $f(x) = x + 4$

50. $f(x) = x + 2$

51. $f(x) = 2x$

52. $f(x) = 3x$

53. $g(x) = 3x - 1$

54. $g(x) = 2x - 5$

55. $f(x) = \frac{1}{2}x + 1$

56. $f(x) = \frac{1}{3}x + 2$

57. $g(x) = x^2 + 5$

58. $g(x) = x^2 - 4$

59. $h(x) = -2x + 4$

60. $h(x) = -3x + 1$

Aha! **61.** $f(x) = \dfrac{1}{x}$

62. $f(x) = \dfrac{3}{x}$

63. $G(x) = 4$

64. $H(x) = 2$

65. $f(x) = \dfrac{2x + 1}{3}$

66. $f(x) = \dfrac{3x + 2}{5}$

67. $f(x) = x^3 - 5$

68. $f(x) = x^3 + 7$

69. $g(x) = (x - 2)^3$

70. $g(x) = (x + 7)^3$

71. $f(x) = \sqrt{x}$

72. $f(x) = \sqrt{x - 1}$

Graph each function and its inverse using the same set of axes.

73. $f(x) = \frac{2}{3}x + 4$

74. $g(x) = \frac{1}{4}x + 2$

75. $f(x) = x^3 + 1$

76. $f(x) = x^3 - 1$

77. $g(x) = \frac{1}{2}x^3$

78. $g(x) = \frac{1}{3}x^3$

79. $F(x) = -\sqrt{x}$

80. $f(x) = \sqrt{x}$

81. $f(x) = -x^2, x \geq 0$

82. $f(x) = x^2 - 1, x \leq 0$

83. Let $f(x) = \sqrt[3]{x - 4}$. Use composition to show that $f^{-1}(x) = x^3 + 4$.

84. Let $f(x) = 3/(x + 2)$. Use composition to show that

$$f^{-1}(x) = \frac{3}{x} - 2.$$

85. Let $f(x) = (1 - x)/x$. Use composition to show that

$$f^{-1}(x) = \frac{1}{x + 1}.$$

86. Let $f(x) = x^3 - 5$. Use composition to show that $f^{-1}(x) = \sqrt[3]{x + 5}$.

Use a graphing calculator to help determine whether or not the given pairs of functions are inverses of each other.

87. $f(x) = 0.75x^2 + 2;\ g(x) = \sqrt{\dfrac{4(x - 2)}{3}}$

88. $f(x) = 1.4x^3 + 3.2;\ g(x) = \sqrt[3]{\dfrac{x - 3.2}{1.4}}$

89. $f(x) = \sqrt{2.5x + 9.25}$;
$g(x) = 0.4x^2 - 3.7, x \geq 0$

90. $f(x) = 0.8x^{1/2} + 5.23$;
$g(x) = 1.25(x^2 - 5.23), x \geq 0$

In Exercises 91 and 92 here and on the following page, match the graph of each function in Column A with the graph of its inverse in Column B.

91. Column A

(1)

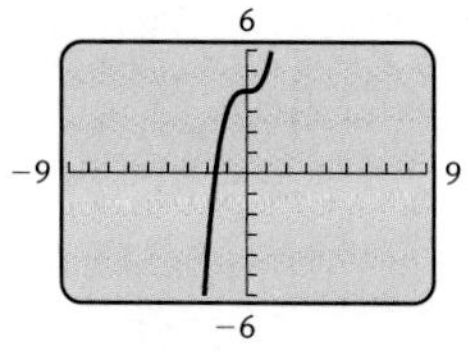

Column B

A.

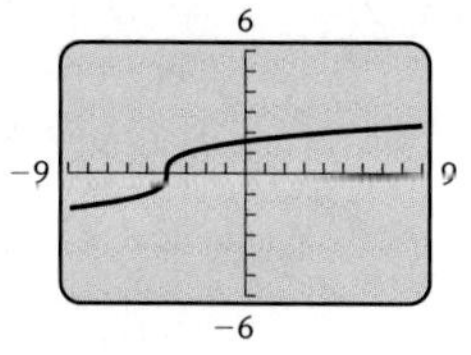

(2)

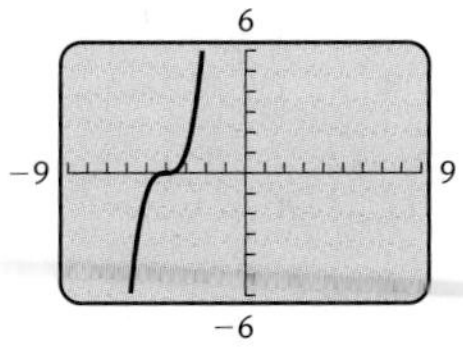

B.

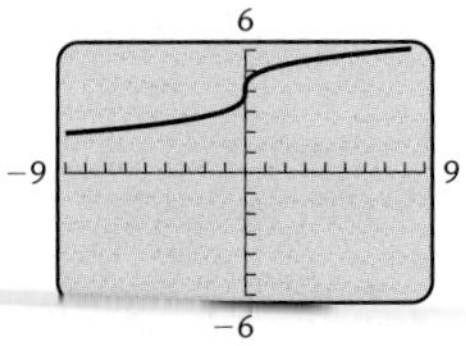

(3)

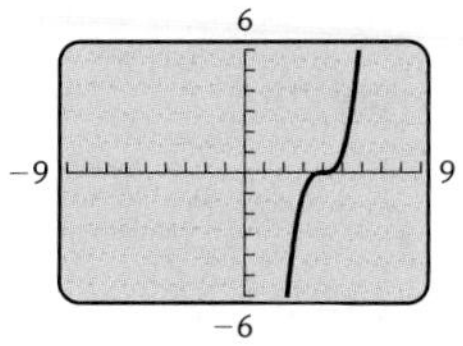

C.

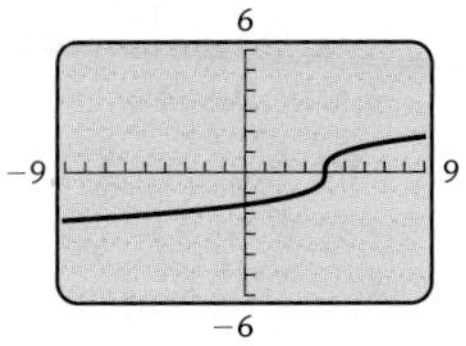

(4)

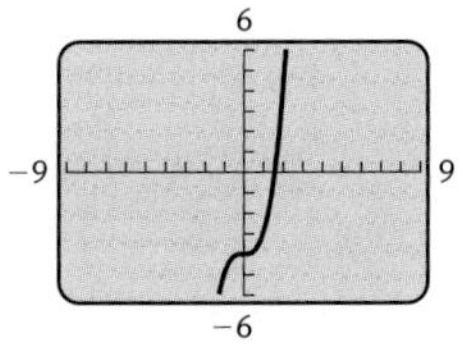

D.

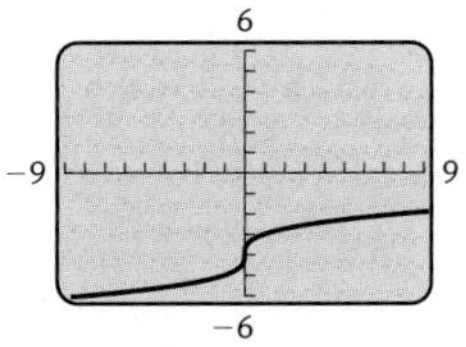

92. Column A

Column B

(1)

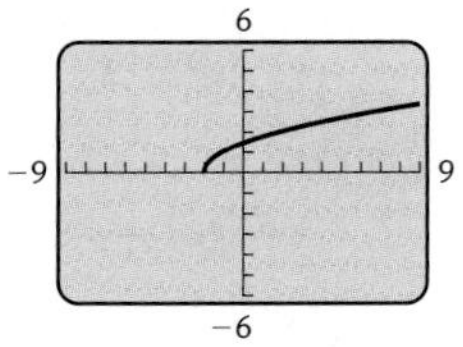

A.

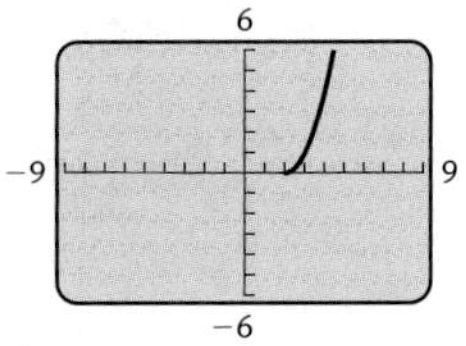

(2)

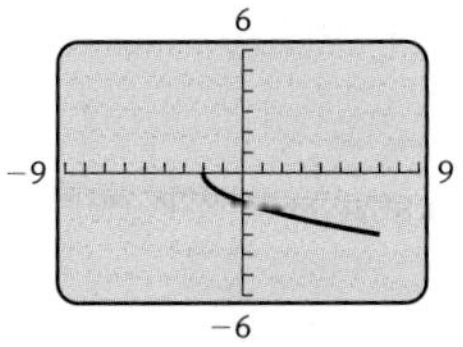

B.

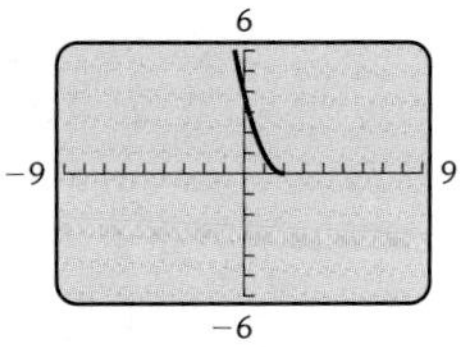

(3)

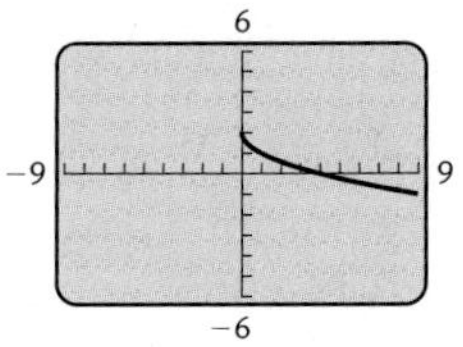

C.

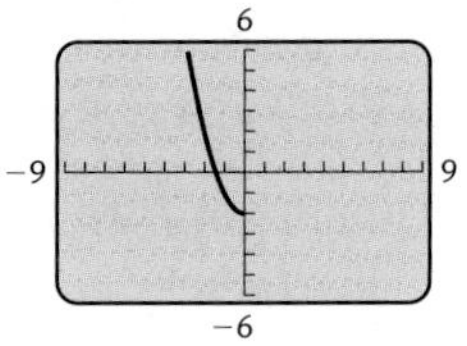

(4)

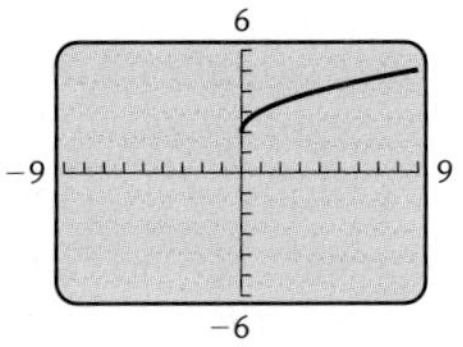

D.

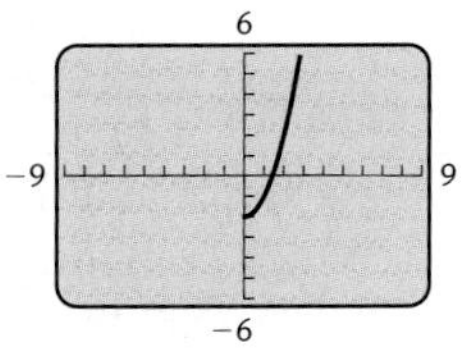

93. *Dress Sizes in the United States and France.* A size-6 dress in the United States is size 38 in France. A function that converts dress sizes in the United States to those in France is

$$f(x) = x + 32.$$

a) Find the dress sizes in France that correspond to sizes 8, 10, 14, and 18 in the United States.

b) Determine whether this function has an inverse that is a function. If so, find a formula for the inverse.

c) Use the inverse function to find dress sizes in the United States that correspond to sizes 40, 42, 46, and 50 in France.

94. *Dress Sizes in the United States and Italy.* A size-6 dress in the United States is size 36 in Italy. A function that converts dress sizes in the United States to those in Italy is

$$f(x) = 2(x + 12).$$

a) Find the dress sizes in Italy that correspond to sizes 8, 10, 14, and 18 in the United States.

b) Determine whether this function has an inverse that is a function. If so, find a formula for the inverse.

c) Use the inverse function to find dress sizes in the United States that correspond to sizes 40, 44, 52, and 60 in Italy.

TW **95.** Is there a one-to-one relationship between the numbers and letters on the keypad of a telephone? Why or why not?

TW **96.** Mathematicians usually try to select "logical" words when forming definitions. Does the term "one-to-one" seem logical? Why or why not?

Focused Review

Let $f(x) = 2x + \sqrt{x}$ *and* $g(x) = x^2$. *Find each of the following.*

97. $f(3)$ [2.1]

98. $f(-3)$ [2.1]

99. $(f + g)(x)$ [2.5]

100. $(f/g)(x)$ [2.5]

101. The domain of f [7.1]

102. The domain of g [8.1]

103. The domain of $f + g$ [2.5]

104. The domain of f/g [2.5]

105. $(f \circ g)(x)$ [9.1]

106. $(g \circ f)(x)$ [9.1]

Synthesis

TW **107.** The function $V(t) = 750(1.2)^t$ is used to predict the value, $V(t)$, of a certain rare stamp t years from 2005. Do not calculate $V^{-1}(t)$, but explain how V^{-1} could be used.

TW **108.** An organization determines that the cost per person of chartering a bus is given by the function

$$C(x) = \frac{100 + 5x}{x},$$

where x is the number of people in the group and $C(x)$ is in dollars. Determine $C^{-1}(x)$ and explain how this inverse function could be used.

For Exercises 109 and 110, graph the inverse of f.

109.

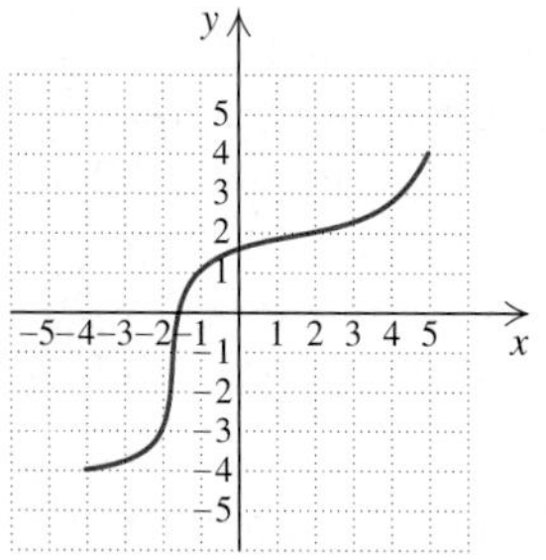

110.

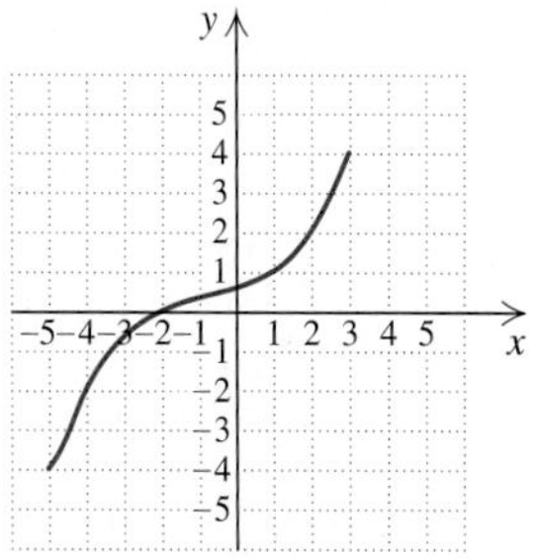

111. *Dress Sizes in France and Italy.* Use the information in Exercises 93 and 94 to find a function for the French dress size that corresponds to a size x dress in Italy.

112. *Dress Sizes in Italy and France.* Use the information in Exercises 93 and 94 to find a function for the Italian dress size that corresponds to a size x dress in France.

TW **113.** What relationship exists between the answers to Exercises 111 and 112? Explain how you determined this.

114. Show that function composition is associative by showing that $((f \circ g) \circ h)(x) = (f \circ (g \circ h))(x)$.

115. Show that if $h(x) = (f \circ g)(x)$, then $h^{-1}(x) = (g^{-1} \circ f^{-1})(x)$. (*Hint*: Use Exercise 114.)

116. Match each function in Column A with its inverse from Column B.

Column A	Column B
(1) $y = 5x^3 + 10$	**A.** $y = \dfrac{\sqrt[3]{x} - 10}{5}$
(2) $y = (5x + 10)^3$	**B.** $y = \sqrt[3]{\dfrac{x}{5}} - 10$
(3) $y = 5(x + 10)^3$	**C.** $y = \sqrt[3]{\dfrac{x - 10}{5}}$
(4) $y = (5x)^3 + 10$	**D.** $y = \dfrac{\sqrt[3]{x - 10}}{5}$

TW **117.** Examine the following table. Is it possible that f and g could be inverses of each other? Why or why not?

x	$f(x)$	$g(x)$
6	6	6
7	6.5	8
8	7	10
9	7.5	12
10	8	14
11	8.5	16
12	9	18

118. Assume in Exercise 117 that f and g are both linear functions. Find equations for $f(x)$ and $g(x)$. Are f and g inverses of each other?

119. Let $c(w)$ represent the cost of mailing a package that weighs w pounds. Let $f(n)$ represent the weight, in pounds, of n copies of a certain book. Explain what $(c \circ f)(n)$ represents.

120. Let $g(a)$ represent the number of gallons of sealant needed to seal a bamboo floor with area a. Let $c(s)$ represent the cost of s gallons of sealant. Which composition makes sense: $(c \circ g)(a)$ or $(g \circ c)(s)$? What does it represent?

*The following graphs show the rate of flow R, in liters per minute, of blood from the heart in a man who bicycles for 20 min, and the pressure P, in millimeters of mercury, in the artery leading to the lungs for a rate of blood flow R from the heart.**

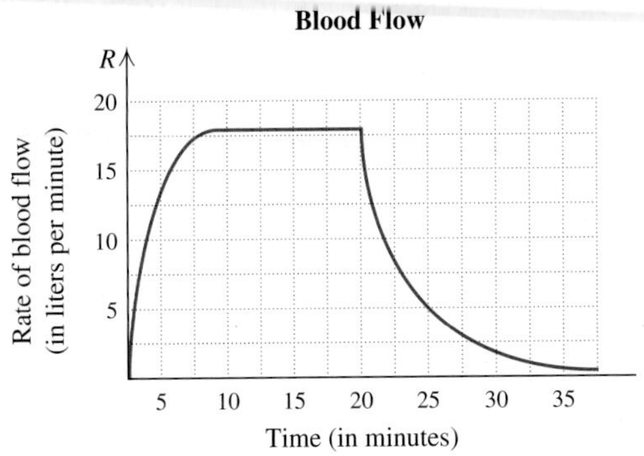

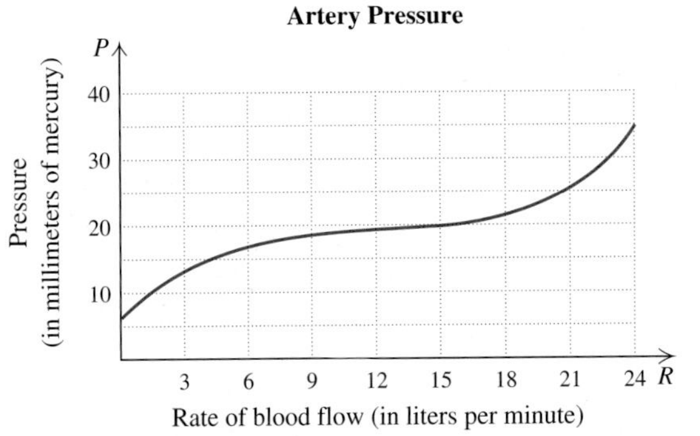

121. Estimate $P(R(10))$.

122. Explain what $P(R(10))$ represents.

123. Estimate $P^{-1}(20)$.

124. Explain what $P^{-1}(20)$ represents.

125. *Real Estate.* The following table lists the number of existing homes sold and the Federal Funds rate for several months in 2005 and in 2006 (*Source*: Mortgage Banking, April 2006)

Month	Existing Homes Sold (in thousands of units)	Federal Funds Rate
Sept. 2005	6290	3.62%
Oct. 2005	6180	3.78
Nov. 2005	6150	4.00
Dec. 2005	5860	4.16
Jan. 2006	5770	4.29

a) Use linear regression to find a linear function that can be used to predict the number of existing homes h sold as a function of the number of months t after September 2005. Let $y_1 = h(t)$.

b) Use linear regression to find a linear function that can be used to predict the Federal Funds rate r as a function of the number of months t after September 2005. Let $y_2 = r(t)$.

c) Use linear regression to find a linear function that can be used to predict the number of existing homes H sold as a function of the Federal Funds rate r. Let $y_3 = H(r)$.

d) Let $y_4 = y_3(y_2(x))$. Use a table of values to compare the functions y_1 and y_4. Use composition of functions to write the relationship between the functions h, r, and H.

---

*This problem was suggested by Kandace Kling, Portland Community College, Sylvania, Oregon.

9.2 Exponential Functions

Graphing Exponential Functions ■ Equations with x and y Interchanged ■ Applications of Exponential Functions

Composite and inverse functions, as shown in Section 9.1, are included in this chapter because they are needed in order to understand the logarithmic functions that appear in Section 9.3. Here in Section 9.2, we introduce a new type of function, the *exponential function*, so that we can study both it and its inverse in Sections 9.3–9.7.

Consider the graph below. The rapidly rising curve approximates the graph of an *exponential function*. We now consider such functions and some of their applications.

Study Tip

Know Your Machine

Whether you use a scientific or a graphing calculator, it is a wise investment of time to study the user's manual. If you cannot find a paper manual to consult, an electronic version can usually be found, online, at the manufacturer's website. Experimenting by pressing various combinations of keystrokes is also useful.

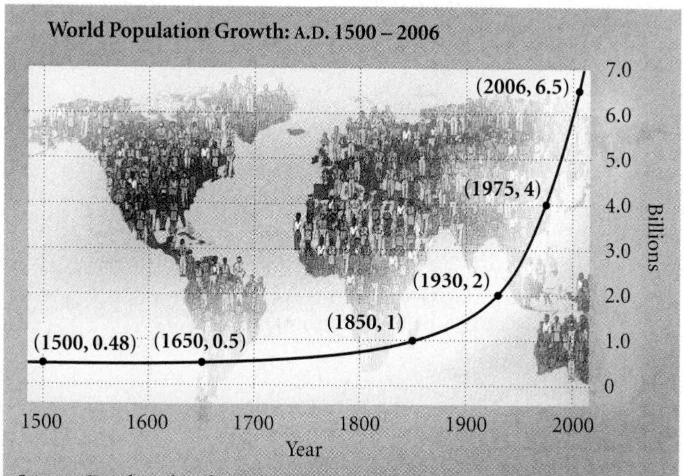

Sources: Based on data from *The World Almanac* 2000 and U.S. Bureau of the Census, Population Division

Graphing Exponential Functions

In Chapter 7, we studied exponential expressions with rational-number exponents, such as

$$5^{1/4}, \quad 3^{-3/4}, \quad 7^{2.34}, \quad 5^{1.73}.$$

For example, $5^{1.73}$, or $5^{173/100}$, represents the 100th root of 5 raised to the 173rd power. What about expressions with irrational exponents, such as $5^{\sqrt{3}}$ or $7^{-\pi}$? To attach meaning to $5^{\sqrt{3}}$, consider a rational approximation, r, of $\sqrt{3}$. As r gets closer to $\sqrt{3}$, the value of 5^r gets closer to some real number p.

r closes in on $\sqrt{3}$.	5^r closes in on some real number p.
$1.7 < r < 1.8$	$15.426 \approx 5^{1.7} < p < 5^{1.8} \approx 18.119$
$1.73 < r < 1.74$	$16.189 \approx 5^{1.73} < p < 5^{1.74} \approx 16.452$
$1.732 < r < 1.733$	$16.241 \approx 5^{1.732} < p < 5^{1.733} \approx 16.267$

We define $5^{\sqrt{3}}$ to be the number p. To eight decimal places,

$$5^{\sqrt{3}} \approx 16.24245082.$$

Any positive irrational exponent can be interpreted in a similar way. Negative irrational exponents are then defined using reciprocals. Thus, so long as a is positive, a^x has meaning for *any* real number x. All of the laws of exponents still hold, but we will not prove that here. We now define an *exponential function.*

Exponential Function The function $f(x) = a^x$, where a is a positive constant, $a \neq 1$, is called the *exponential function*, base a.

We require the base a to be positive to avoid imaginary numbers that would result from taking even roots of negative numbers. The restriction $a \neq 1$ is made to exclude the constant function $f(x) = 1^x$, or $f(x) = 1$.

The following are examples of exponential functions:

$$f(x) = 2^x, \quad f(x) = \left(\tfrac{1}{3}\right)^x, \quad f(x) = 5^{-3x}.$$ **Note that $5^{-3x} = (5^{-3})^x$.**

Like polynomial functions, the domain of an exponential function is the set of all real numbers. Unlike polynomial functions, exponential functions have a variable exponent. Because of this, graphs of exponential functions either rise or fall dramatically.

EXAMPLE 1 Graph the exponential function given by $y = f(x) = 2^x$.

SOLUTION We compute some function values, thinking of y as $f(x)$, and list the results in a table. It is a good idea to start by letting $x = 0$.

$$f(0) = 2^0 = 1; \qquad f(-1) = 2^{-1} = \frac{1}{2^1} = \frac{1}{2};$$
$$f(1) = 2^1 = 2;$$
$$f(2) = 2^2 = 4; \qquad f(-2) = 2^{-2} = \frac{1}{2^2} = \frac{1}{4};$$
$$f(3) = 2^3 = 8; \qquad f(-3) = 2^{-3} = \frac{1}{2^3} = \frac{1}{8}$$

x	y, or $f(x)$
0	1
1	2
2	4
3	8
−1	$\frac{1}{2}$
−2	$\frac{1}{4}$
−3	$\frac{1}{8}$

Next, we plot these points and connect them with a smooth curve.

The curve comes very close to the x-axis, but does not touch or cross it.

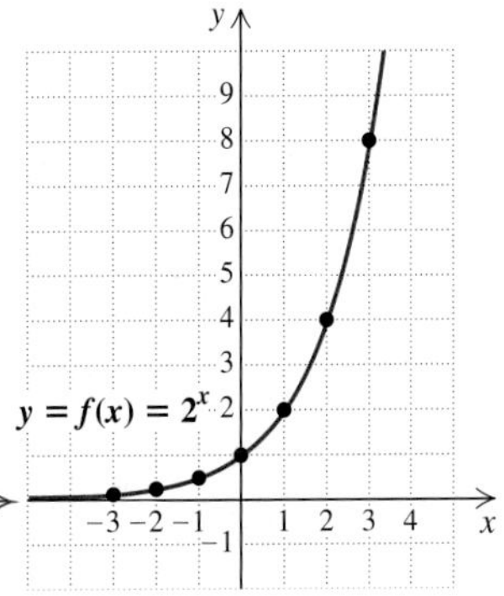

Be sure to plot enough points to determine how steeply the curve rises.

Note that as x increases, the function values increase without bound. As x decreases, the function values decrease, getting very close to 0. The x-axis, or the line $y = 0$, is a horizontal *asymptote*, meaning that the curve gets closer and closer to this line the further we move to the left.

EXAMPLE 2 Graph: $y = f(x) = \left(\frac{1}{2}\right)^x$.

SOLUTION We compute some function values, thinking of y as $f(x)$, and list the results in a table. Before we do this, note that

$$y = f(x) = \left(\tfrac{1}{2}\right)^x = (2^{-1})^x = 2^{-x}.$$

Then we have

$$f(0) = 2^{-0} = 1;$$

$$f(1) = 2^{-1} = \frac{1}{2^1} = \frac{1}{2};$$

$$f(2) = 2^{-2} = \frac{1}{2^2} = \frac{1}{4};$$

$$f(3) = 2^{-3} = \frac{1}{2^3} = \frac{1}{8};$$

$$f(-1) = 2^{-(-1)} = 2^1 = 2;$$

$$f(-2) = 2^{-(-2)} = 2^2 = 4;$$

$$f(-3) = 2^{-(-3)} = 2^3 = 8.$$

x	y, or $f(x)$
0	1
1	$\frac{1}{2}$
2	$\frac{1}{4}$
3	$\frac{1}{8}$
-1	2
-2	4
-3	8

Next, we plot these points and connect them with a smooth curve. This curve is a mirror image, or *reflection*, of the graph of $y = 2^x$ (see Example 1) across the y-axis. The line $y = 0$ is again an asymptote.

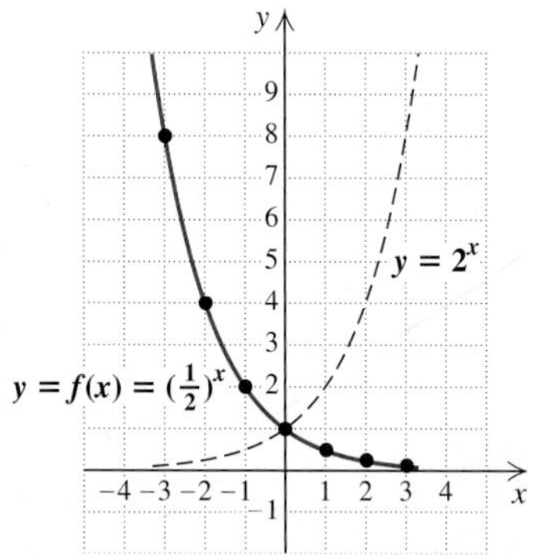

Interactive Discovery

If the values of $f(x)$ increase as x increases, we say that f is an *increasing function.* The function in Example 1, $f(x) = 2^x$, is increasing. If the values of $f(x)$ decrease as x increases, we say that f is a *decreasing function.* The function in Example 2, $f(x) = \left(\frac{1}{2}\right)^x$, is decreasing.

1. Graph each of the following functions, and determine whether the graph increases or decreases from left to right.

a) $f(x) = 3^x$ **b)** $g(x) = 4^x$

c) $h(x) = 1.5^x$ **d)** $r(x) = \left(\frac{1}{3}\right)^x$

e) $t(x) = 0.75^x$

2. How can you tell from the number a in $f(x) = a^x$ whether the graph of f increases or decreases? $a > 1$, $0 < a < 1$,

3. Compare the graphs of f, g, and h. Which curve is steepest?

4. Compare the graphs of r and t. Which curve is steeper?

We can make the following observations.

A. For $a > 1$, the graph of $f(x) = a^x$ increases from left to right. The greater the value of a, the steeper the curve. (See the figure on the left below.)

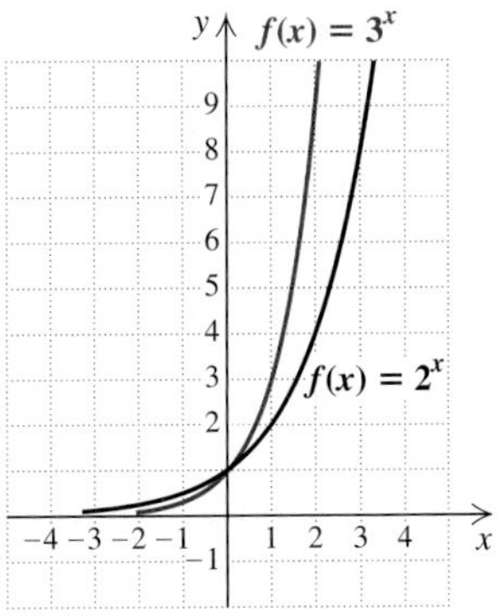

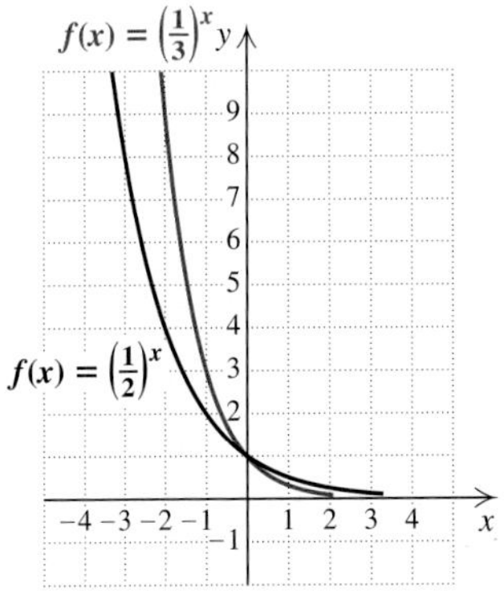

B. For $0 < a < 1$, the graph of $f(x) = a^x$ decreases from left to right. For smaller values of a, the curve becomes steeper. (See the figure on the right above.)

C. All graphs of $f(x) = a^x$ go through the y-intercept $(0, 1)$.

D. All graphs of $f(x) = a^x$ have the x-axis as the asymptote.

E. If $f(x) = a^x$, with $a > 0$, $a \neq 1$, the domain of f is all real numbers, and the range of f is all positive real numbers.

F. For $a > 0$, $a \neq 1$, the function given by $f(x) = a^x$ is one-to-one. Its graph passes the horizontal-line test.

Student Notes

When using translations, make sure that you are shifting in the correct direction. When in doubt, substitute a value for x and make some calculations.

EXAMPLE 3 Graph: $y = f(x) = 2^{x-2}$.

SOLUTION We construct a table of values. Then we plot the points and connect them with a smooth curve. Here $x - 2$ is the *exponent*.

$$f(0) = 2^{0-2} = 2^{-2} = \frac{1}{4}; \qquad f(-1) = 2^{-1-2} = 2^{-3} = \frac{1}{8};$$

$$f(1) = 2^{1-2} = 2^{-1} = \frac{1}{2}; \qquad f(-2) = 2^{-2-2} = 2^{-4} = \frac{1}{16}$$

$$f(2) = 2^{2-2} = 2^0 = 1;$$

$$f(3) = 2^{3-2} = 2^1 = 2;$$

$$f(4) = 2^{4-2} = 2^2 = 4;$$

x	y, or $f(x)$
0	$\frac{1}{4}$
1	$\frac{1}{2}$
2	1
3	2
4	4
−1	$\frac{1}{8}$
−2	$\frac{1}{16}$

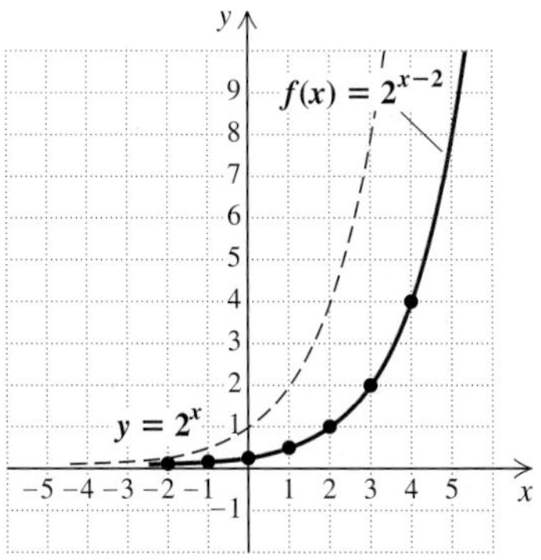

The graph looks just like the graph of $y = 2^x$, but it is translated 2 units to the right. The y-intercept of $y = 2^x$ is $(0, 1)$. The y-intercept of $y = 2^{x-2}$ is $\left(0, \frac{1}{4}\right)$. The line $y = 0$ is again the asymptote.

As we saw in Example 3, the graph of $f(x) = 2^{x-2}$ looks like the graph of $y = 2^x$ translated 2 units to the right. In general, the graph of $f(x) = 2^{x-h}$ will look like the graph of $y = 2^x$ translated right or left. Similarly, the graph of $f(x) = 2^x + k$ will look like the graph of 2^x translated up or down.

This observation is true in general.

The graph of $f(x) = a^{x-h} + k$ looks like the graph of $y = a^x$ translated $|h|$ units left or right and $|k|$ units up or down.

- If $h > 0$, $y = a^x$ is translated h units right.
- If $h < 0$, $y = a^x$ is translated $|h|$ units left.
- If $k > 0$, $y = a^x$ is translated k units up.
- If $k < 0$, $y = a^x$ is translated $|k|$ units down.

Equations with *x* and *y* Interchanged

It will be helpful in later work to be able to graph an equation in which the x and the y in $y = a^x$ are interchanged.

EXAMPLE 4 Graph: $x = 2^y$.

SOLUTION Note that x is alone on one side of the equation. To find ordered pairs that are solutions, we choose values for y and then compute values for x:

For $y = 0$, $x = 2^0 = 1$.
For $y = 1$, $x = 2^1 = 2$.
For $y = 2$, $x = 2^2 = 4$.
For $y = 3$, $x = 2^3 = 8$.
For $y = -1$, $x = 2^{-1} = \frac{1}{2}$.
For $y = -2$, $x = 2^{-2} = \frac{1}{4}$.
For $y = -3$, $x = 2^{-3} = \frac{1}{8}$.

x	y
1	0
2	1
4	2
8	3
$\frac{1}{2}$	-1
$\frac{1}{4}$	-2
$\frac{1}{8}$	-3

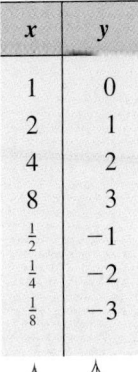

(1) Choose values for y.
(2) Compute values for x.

We plot the points and connect them with a smooth curve.

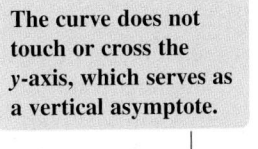

The curve does not touch or cross the y-axis, which serves as a vertical asymptote.

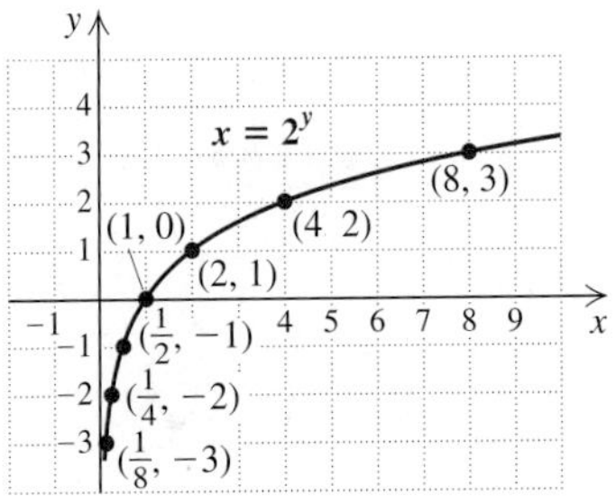

Note too that this curve looks just like the graph of $y = 2^x$, except that it is reflected across the line $y = x$, as shown here.

We have graphed $y = 2^x$ and its inverse, $x = 2^y$.

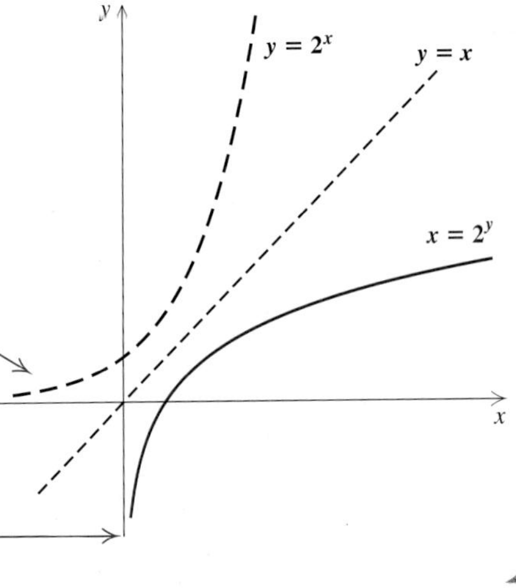

Applications of Exponential Functions

EXAMPLE 5 Interest Compounded Annually. The amount of money A that a principal P will be worth after t years at interest rate i, compounded annually, is given by the formula

$$A = P(1 + i)^t.$$ **You might review Example 12 in Section 8.1.**

Suppose that \$100,000 is invested at 8% interest, compounded annually.

a) Find a function for the amount in the account after t years.

b) Find the amount of money in the account at $t = 0$, $t = 4$, $t = 8$, and $t = 10$.

c) Graph the function.

SOLUTION

a) If $P = \$100{,}000$ and $i = 8\% = 0.08$, we can substitute these values and form the following function:

$$A(t) = \$100{,}000(1 + 0.08)^t \quad \text{Using } A = P(1 + i)^t$$
$$= \$100{,}000(1.08)^t.$$

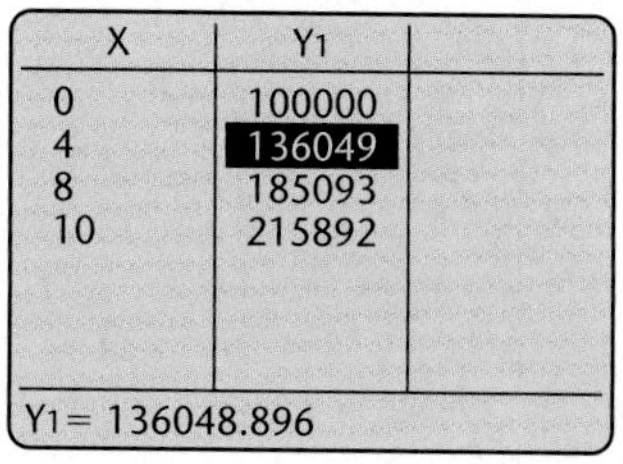

b) An efficient way to find the function values using a graphing calculator is to let $y_1 = 100{,}000(1.08)^\wedge x$ and use the TABLE feature with Indpnt set to Ask, as shown at left. We highlight a table entry if we want to view the function value to more decimal places than is shown in the table. Note that large values in a table will be written in scientific notation.

Alternatively, we can use $y_1(\ )$ notation to evaluate the function for each value. Using either procedure gives us

$$A(0) = \$100{,}000,$$
$$A(4) \approx \$136{,}048.90,$$
$$A(8) \approx \$185{,}093.02,$$
and $$A(10) \approx \$215{,}892.50.$$

c) We can use the function values computed in part (b), and others if we wish, to draw the graph by hand. Whether we are graphing by hand or using a graphing calculator, the axes will be scaled differently because of the large function values.

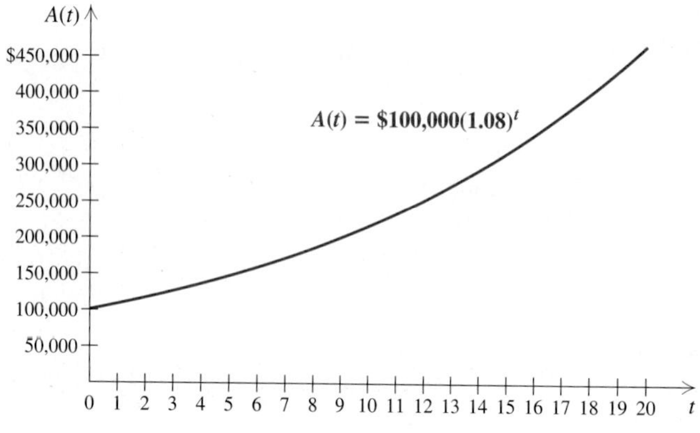

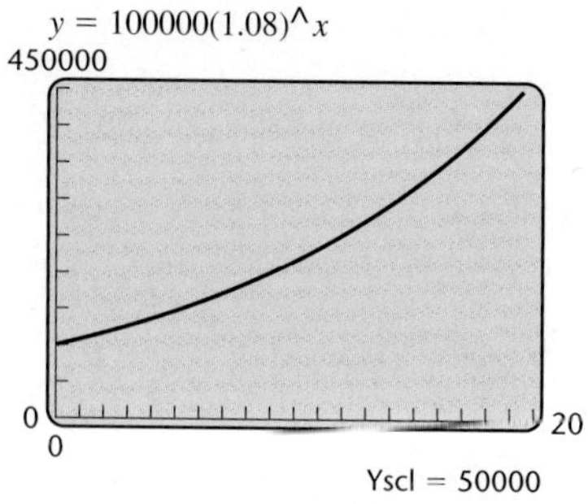

Connecting the Concepts

INTERPRETING GRAPHS: LINEAR AND EXPONENTIAL GROWTH

When the graph of a quantity over time approximates the graph of an exponential function, as in the World Population graph on p. 716, we say that there is *exponential* growth. This is different from *linear* growth, which we examined in Chapter 2.

Linear functions increase (or decrease) by a constant amount. The rate of change (slope) of a linear graph is constant. The rate of change of an exponential graph is not constant.

To illustrate the difference between linear and exponential functions, compare the salaries for two hypothetical jobs, shown in the following table and graph. Note that the starting salaries are the same, but the salary for job A increases by a constant amount (linearly) and the salary for job B increases by a constant percent (exponentially).

Note that the salary for job A is larger for the first few years only. The rate of change of the salary for job B increases each year, since the increase is based on a larger salary each year.

	Job A	Job B
Starting salary	\$50,000	\$50,000
Guaranteed raise	\$5000 per year	8% per year
Salary function	$f(x) = 50{,}000 + 5000x$	$g(x) = 50{,}000(1.08)^x$

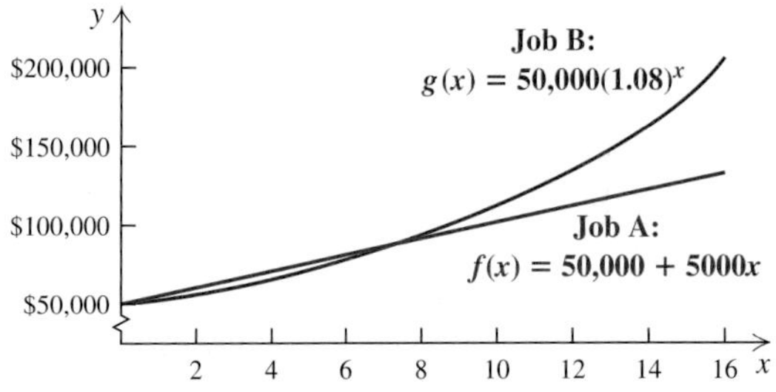

9.2 EXERCISE SET

FOR EXTRA HELP

MathXL | MyMathLab | InterAct Math | AW Math Tutor Center | Video Lectures on CD: Disc 5 | Student's Solutions Manual

Concept Reinforcement *Classify each of the following as either true or false.*

1. The graph of $f(x) = a^x$ always passes through the point (0, 1).

2. The graph of $g(x) = \left(\frac{1}{2}\right)^x$ gets closer and closer to the x-axis as x gets larger and larger.

3. The graph of $f(x) = 2^{x-3}$ looks just like the graph of $y = 2^x$, but it is translated 3 units to the right.

4. The graph of $g(x) = 2^x - 3$ looks just like the graph of $y = 2^x$, but it is translated 3 units up.

5. The graph of $y = 3^x$ gets close to, but never touches, the y-axis.

6. The graph of $x = 3^y$ gets close to, but never touches, the y-axis.

In each of Exercises 7–10 is the graph of a function $f(x) = a^x$. Determine from the graph whether $a > 1$ or $0 < a < 1$.

7.
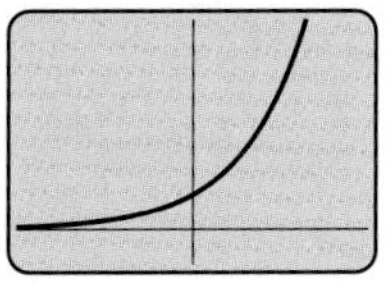

8.
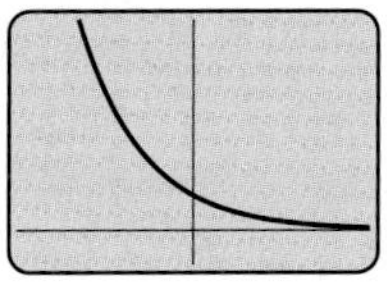

9.

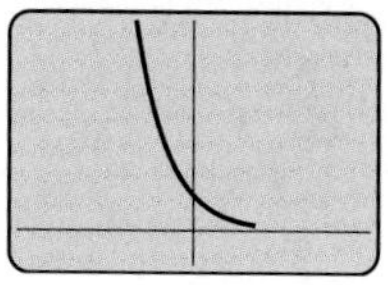

10.

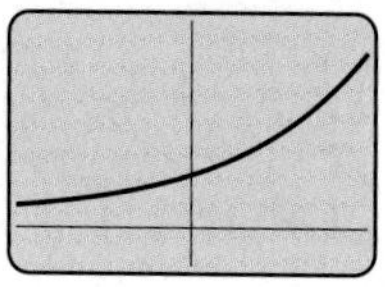

Graph.

11. $y = f(x) = 2^x$

12. $y = f(x) = 3^x$

13. $y = 5^x$

14. $y = 6^x$

15. $y = 2^x + 3$

16. $y = 2^x + 1$

17. $y = 3^x - 1$

18. $y = 3^x - 2$

19. $y = 2^{x-1}$

20. $y = 2^{x-2}$

21. $y = 2^{x+3}$

22. $y = 2^{x+1}$

23. $y = \left(\frac{1}{5}\right)^x$

24. $y = \left(\frac{1}{4}\right)^x$

25. $y = \left(\frac{1}{2}\right)^x$

26. $y = \left(\frac{1}{3}\right)^x$

27. $y = 2^{x-3} - 1$

28. $y = 2^{x+1} - 3$

29. $y = 1.7^x$

30. $y = 4.8^x$

31. $y = 0.15^x$

32. $y = 0.98^x$

33. $x = 3^y$

34. $x = 6^y$

35. $x = 2^{-y}$

36. $x = 3^{-y}$

37. $x = 5^y$

38. $x = 4^y$

39. $x = \left(\frac{3}{2}\right)^y$

40. $x = \left(\frac{4}{3}\right)^y$

Graph each pair of equations using the same set of axes.

41. $y = 3^x,\ x = 3^y$

42. $y = 2^x,\ x = 2^y$

43. $y = \left(\frac{1}{2}\right)^x,\ x = \left(\frac{1}{2}\right)^y$

44. $y = \left(\frac{1}{4}\right)^x,\ x = \left(\frac{1}{4}\right)^y$

Aha! *In Exercises 45–50, match each equation with one of the following graphs.*

a)

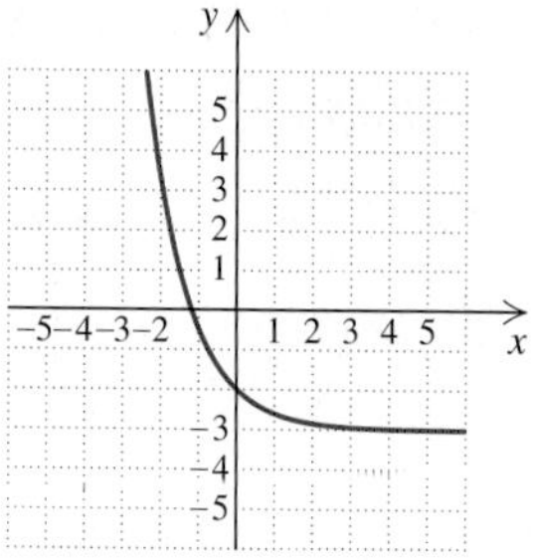

b)

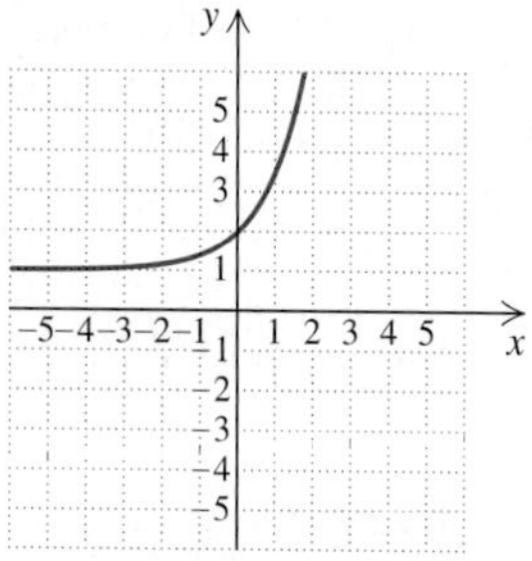

c)

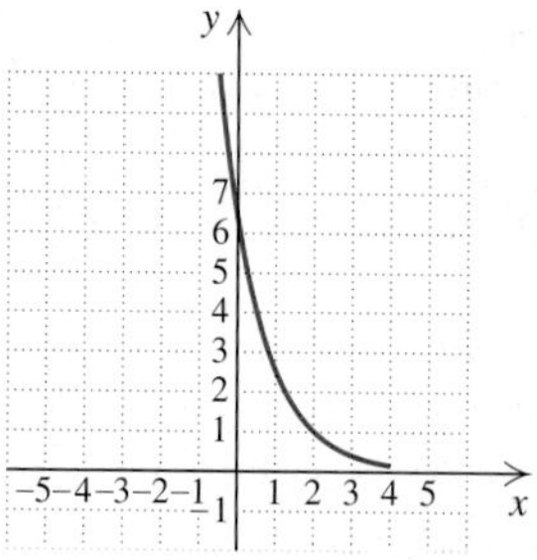

d)

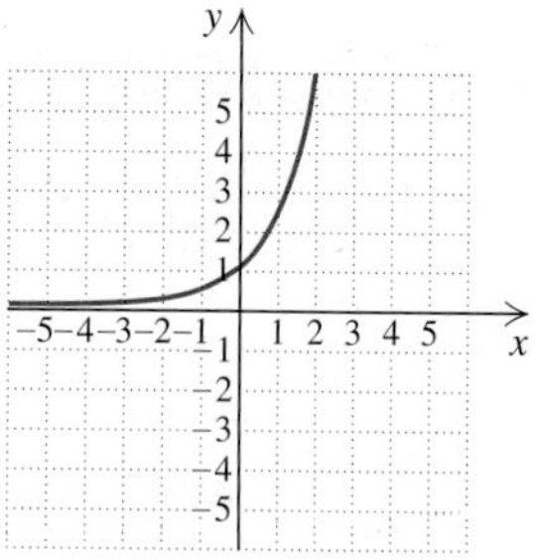

e)

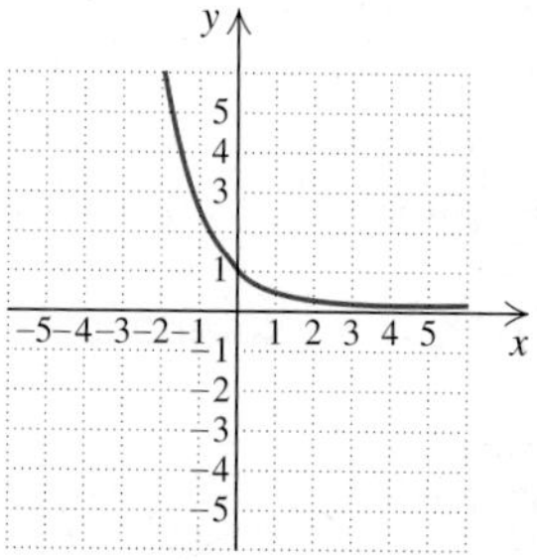

f)

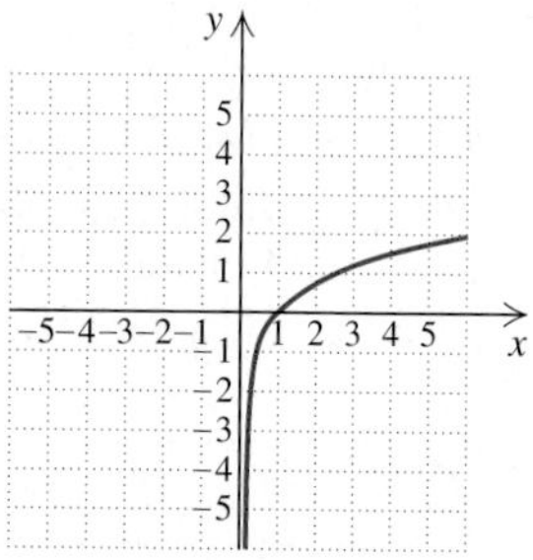

45. $y = \left(\frac{5}{2}\right)^x$

46. $y = \left(\frac{2}{5}\right)^x$

47. $x = \left(\frac{5}{2}\right)^y$

48. $y = \left(\frac{2}{5}\right)^x - 3$

49. $y = \left(\frac{2}{5}\right)^{x-2}$

50. $y = \left(\frac{5}{2}\right)^x + 1$

Solve.

51. *Population Growth.* The world population $P(t)$, in billions, t years after 1980, can be approximated by

$$P(t) = 4.495(1.015)^t$$

(*Source*: Based on data from U.S. Bureau of the Census, International Data Base).

a) Predict the world population in 2008, in 2012, and in 2016.

b) Graph the function.

52. *Growth of Bacteria.* The bacteria *Escherichia coli* are commonly found in the human bladder. Suppose that 3000 of the bacteria are present at time $t = 0$. Then t minutes later, the number of bacteria present can be approximated by

$$N(t) = 3000(2)^{t/20}.$$

a) How many bacteria will be present after 10 min? 20 min? 30 min? 40 min? 60 min?

b) Graph the function.

53. *Smoking Cessation.* The percentage of smokers P who, with telephone counseling to quit smoking, are still successful t months later can be approximated by

$$P(t) = 21.4(0.914)^t$$

(*Sources*: *New England Journal of Medicine*; data from California's Smokers' Hotline).

a) Estimate the percentage of smokers receiving telephone counseling who are successful in quitting for 1 month, 3 months, and 1 year.

b) Graph the function.

54. *Smoking Cessation.* The percentage of smokers P who, without telephone counseling, have successfully quit smoking for t months (see Exercise 53) can be approximated by

$$P(t) = 9.02(0.93)^t$$

(*Sources*: *New England Journal of Medicine*; data from California's Smokers' Hotline).

a) Estimate the percentage of smokers not receiving telephone counseling who are successful in quitting for 1 month, 3 months, and 1 year.

b) Graph the function.

55. *Marine Biology.* Due to excessive whaling prior to the mid 1970s, the humpback whale is considered an endangered species. The worldwide population of humpbacks, $P(t)$, in thousands, t years after 1900 ($t < 70$) can be approximated by*

$$P(t) = 150(0.960)^t.$$

a) How many humpback whales were alive in 1930? in 1960?

b) Graph the function.

56. *Salvage Value.* A photocopier is purchased for \$5200. Its value each year is about 80% of the value of the preceding year. Its value, in dollars, after t years is given by the exponential function

$$V(t) = 5200(0.8)^t.$$

a) Find the value of the machine after 0 yr, 1 yr, 2 yr, 5 yr, and 10 yr.

b) Graph the function.

57. *Marine Biology.* As a result of preservation efforts in most countries in which whaling was common, the humpback whale population has grown since the 1970s. The worldwide population of humpbacks, $P(t)$, in thousands, t years after 1982 can be approximated by*

$$P(t) = 5.5(1.08)^t.$$

a) How many humpback whales were alive in 1992? in 2006?

b) Graph the function.

*Based on information from the American Cetacean Society, 2006, and the ASK Archive, 1998.

58. *Recycling Aluminum Cans.* About $\frac{1}{2}$ of all aluminum cans will be recycled. A beverage company distributes 250,000 cans. The number in use after t years is given by the exponential function

$$N(t) = 250{,}000\left(\tfrac{1}{2}\right)^t$$

(*Source*; The Aluminum Association, Inc., May 2004).

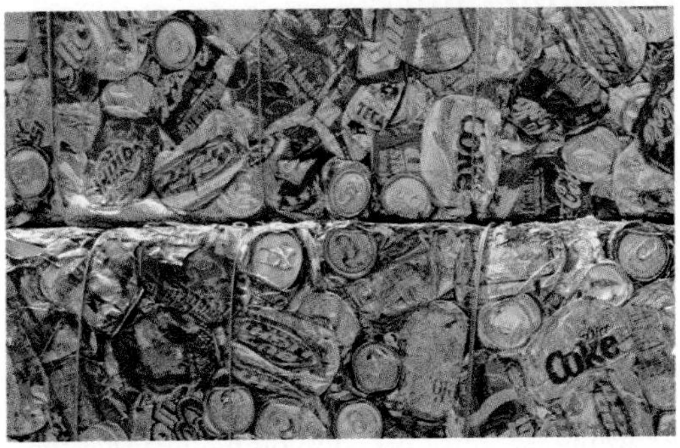

a) How many cans are still in use after 0 yr? 1 yr? 4 yr? 10 yr?

b) Graph the function.

59. *Spread of Zebra Mussels.* Beginning in 1988, infestations of zebra mussels began to threaten water treatment facilities, power plants, and entire ecosystems in North America.† The function

$$A(t) = 10 \cdot 34^t$$

can be used to estimate the number of square centimeters of lake bottom that will be covered with mussels t years after an infestation covering 10 cm^2 first occurs.

a) How many square centimeters of lake bottom will be covered with mussels 5 yr after an infestation covering 10 cm^2 first appears? 7 yr after the infestation first appears?

b) Graph the function.

60. *Cell Phones.* The number N of cell phones in use in the United States, in millions, can be estimated by

$$N(t) = 7.12(1.3)^t,$$

where t is the number of years after 1990 (*Source*: Cellular Telecommunications and Internet Association).

a) Estimate the number of cell phones in use in 1995, in 2005, and in 2010.

b) Graph the function.

†Many thanks to Dr. Gerald Mackie of the Department of Zoology at the University of Guelph in Ontario for the background information for this exercise.

TW **61.** Without using a calculator, explain why 2^{π} must be greater than 8 but less than 16.

TW **62.** Suppose that \$1000 is invested for 5 yr at 7% interest, compounded annually. In what year will the most interest be earned? Why?

Skill Maintenance

Simplify.

63. 5^{-2} [1.4]

64. 2^{-5} [1.4]

65. $1000^{2/3}$ [7.2]

66. $25^{-3/2}$ [7.2]

67. $\dfrac{10a^8b^7}{2a^2b^4}$ [1.4]

68. $\dfrac{24x^6y^4}{4x^2y^3}$ [1.4]

Synthesis

TW **69.** Examine Exercise 60. Do you believe that the equation for the number of cell phones in use in the United States will be accurate 20 yr from now? Why or why not?

TW **70.** Why was it necessary to discuss irrational exponents before graphing exponential functions?

Determine which of the two given numbers is larger. Do not use a calculator.

71. $\pi^{1.3}$ or $\pi^{2.4}$

72. $\sqrt{8^3}$ or $8^{\sqrt{3}}$

Graph.

73. $y = 2^x + 2^{-x}$

74. $y = \left|\left(\frac{1}{2}\right)^x - 1\right|$

75. $y = |2^x - 2|$

76. $y = 2^{-(x-1)^2}$

77. $y = |2^{x^2} - 1|$

78. $y = 3^x + 3^{-x}$

Graph both equations using the same set of axes.

79. $y = 3^{-(x-1)}$, $x = 3^{-(y-1)}$

80. $y = 1^x$, $x = 1^y$

81. *Sales of MP3 Players.* Sales of MP3 players have grown from \$80 million in 2000 to \$205 million in 2002 to \$1204 million in 2004 (*Source*: Consumer Electronics Association). Use the ExpReg option in the STAT CALC menu to find an exponential function that can be used to estimate sales of MP3 players x years after 2000. Then use that function to predict sales in 2008.

Collaborative Corner

The True Cost of a New Car

Focus: Car loans and exponential functions

Time: 30 minutes

Group Size: 2

Materials: Calculators with exponentiation keys

The formula

$$M = \frac{Pr}{1 - (1 + r)^{-n}}$$

is used to determine the payment size, M, when a loan of P dollars is to be repaid in n equally sized monthly payments. Here r represents the monthly interest rate. Loans repaid in this fashion are said to be *amortized* (spread out equally) over a period of n months.

ACTIVITY

1. Suppose one group member is selling the other a car for \$2600, financed at 1% interest per month for 24 months. What should be the size of each monthly payment?
2. Suppose both group members are shopping for the same model new car. Each group member visits a different dealer. One dealer offers the car for \$13,000 at 10.5% interest (0.00875 monthly interest) for 60 months (no down payment). The other dealer offers the same car for \$12,000, but at 12% interest (0.01 monthly interest) for 48 months (no down payment).
   a) Determine the monthly payment size for each offer. Then determine the total amount paid for the car under each offer. How much of each total is interest?
   b) Work together to find the annual interest rate for which the total cost of 60 monthly payments for the \$13,000 car would equal the total amount paid for the \$12,000 car (as found in part a above).

9.3 Logarithmic Functions

Graphs of Logarithmic Functions ■ Common Logarithms ■ Equivalent Equations ■ Solving Certain Logarithmic Equations

We are now ready to study inverses of exponential functions. These functions have many applications and are referred to as *logarithm*, or *logarithmic, functions*.

Graphs of Logarithmic Functions

Consider the exponential function $f(x) = 2^x$. Like all exponential functions, f is one-to-one. Can a formula for f^{-1} be found?

To answer this, we use the method of Section 9.1:

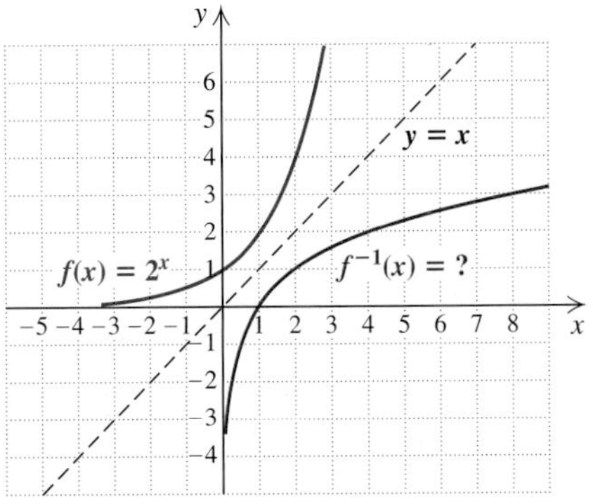

1. Replace $f(x)$ with y: $y = 2^x$.
2. Interchange x and y: $x = 2^y$.
3. Solve for y: $y =$ the power to which we raise 2 to get x.
4. Replace y with $f^{-1}(x)$: $f^{-1}(x) =$ the power to which we raise 2 to get x.

We now define a new symbol to replace the words "the power to which we raise 2 to get x":

$\log_2 x$, read "the logarithm, base 2, of x," or "log, base 2, of x," means "the exponent to which we raise 2 to get x."

Thus if $f(x) = 2^x$, then $f^{-1}(x) = \log_2 x$. Note that $f^{-1}(8) = \log_2 8 = 3$, because 3 is *the exponent to which we raise* 2 *to get* 8.

EXAMPLE 1 Simplify: **(a)** $\log_2 32$; **(b)** $\log_2 1$; **(c)** $\log_2 \frac{1}{8}$.

SOLUTION

a) Think of $\log_2 32$ as the exponent to which we raise 2 to get 32. That exponent is 5. Therefore, $\log_2 32 = 5$.

b) We ask ourselves: "To what exponent do we raise 2 in order to get 1?" That exponent is 0 (recall that $2^0 = 1$). Thus, $\log_2 1 = 0$.

c) To what exponent do we raise 2 in order to get $\frac{1}{8}$? Since $2^{-3} = \frac{1}{8}$, we have $\log_2 \frac{1}{8} = -3$.

Study Tip

When a Turn Is Trouble

Occasionally a page turn can interrupt one's thoughts and slow one's progress through challenging material. As a student, one of your authors found it helpful to rewrite (in pencil) the last equation or sentence appearing on one page at the very top of the next page.

Although numbers like $\log_2 13$ can be only approximated, we must remember that $\log_2 13$ represents *the exponent to which we raise* 2 *to get* 13. That is, $2^{\log_2 13} = 13$. A calculator can be used to show that $\log_2 13 \approx 3.7$ and $2^{3.7} \approx 13$. Later in this chapter, we will use a calculator to find such approximations.

For any exponential function $f(x) = a^x$, the inverse is called a **logarithmic function, base *a*.** The graph of the inverse can be drawn by reflecting the graph of $f(x) = a^x$ across the line $y = x$. It will be helpful to remember that the inverse of $f(x) = a^x$ is given by $f^{-1}(x) = \log_a x$.

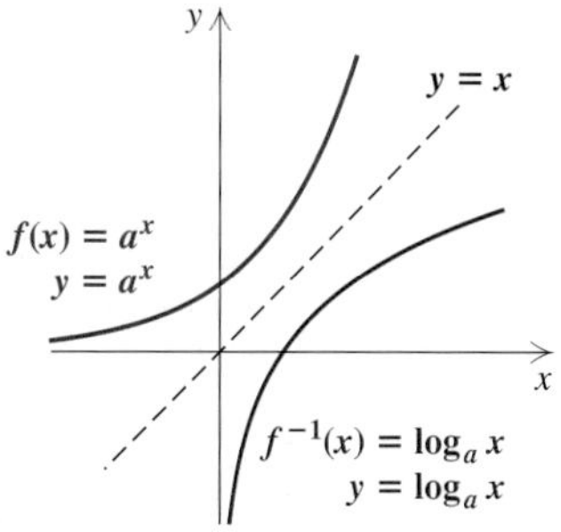

Student Notes

As an aid in remembering what $\log_a x$ means, note that a is called the *base*, just as it is the base in $a^y = x$.

The Meaning of $\log_a x$

For $x > 0$ and a a positive constant other than 1, $\log_a x$ is the exponent to which a must be raised in order to get x. Thus,

$$\log_a x = m \text{ means } a^m = x$$

or equivalently,

$\log_a x$ is that unique exponent for which $a^{\log_a x} = x$.

It is important to remember that *a logarithm is an exponent.* It might help to repeat several times: "The logarithm, base a, of a number x is the exponent to which a must be raised in order to get x."

EXAMPLE 2 Simplify: $7^{\log_7 85}$.

SOLUTION Remember that $\log_7 85$ is the exponent to which 7 is raised to get 85. Raising 7 to that exponent, we have

$$7^{\log_7 85} = 85.$$

Because logarithmic and exponential functions are inverses of each other, the result in Example 2 should come as no surprise: If $f(x) = \log_7 x$, then

for $f(x) = \log_7 x$, we have $f^{-1}(x) = 7^x$
and $f^{-1}(f(x)) = f^{-1}(\log_7 x) = 7^{\log_7 x} = x$.

Thus, $f^{-1}(f(85)) = 7^{\log_7 85} = 85$.

The following is a comparison of exponential and logarithmic functions.

Exponential Function	Logarithmic Function
$y = a^x$	$x = a^y$
$f(x) = a^x$	$g(x) = \log_a x$
$a > 0, a \neq 1$	$a > 0, a \neq 1$
The domain is $\mathbb{R}$.	The range is $\mathbb{R}$.
The range is $(0, \infty)$.	The domain is $(0, \infty)$.
$f^{-1}(x) = \log_a x$	$g^{-1}(x) = a^x$
The graph of $f(x)$ contains the points $(0, 1)$ and $(1, a)$.	The graph of $g(x)$ contains the points $(1, 0)$ and $(a, 1)$.

Student Notes

Since $g(x) = \log_a x$ is the inverse of $f(x) = a^x$, the domain of f is the range of g, and the range of f is the domain of g.

EXAMPLE 3 Graph: $y = f(x) = \log_5 x$.

SOLUTION If $y = \log_5 x$, then $5^y = x$. We can find ordered pairs that are solutions by choosing values for y and computing the x-values.

For $y = 0$, $x = 5^0 = 1$.
For $y = 1$, $x = 5^1 = 5$.
For $y = 2$, $x = 5^2 = 25$.
For $y = -1$, $x = 5^{-1} = \frac{1}{5}$.
For $y = -2$, $x = 5^{-2} = \frac{1}{25}$.

(1) Select y.
(2) Compute x.

x, or 5^y	y
1	0
5	1
25	2
$\frac{1}{5}$	-1
$\frac{1}{25}$	-2

This table shows the following:

$$\left.\begin{array}{l}\log_5 1 = 0; \\ \log_5 5 = 1; \\ \log_5 25 = 2; \\ \log_5 \frac{1}{5} = -1; \\ \log_5 \frac{1}{25} = -2.\end{array}\right\}$$

These can all be checked using the equations above.

We plot the set of ordered pairs and connect the points with a smooth curve. The graphs of $y = 5^x$ and $y = x$ are shown only for reference.

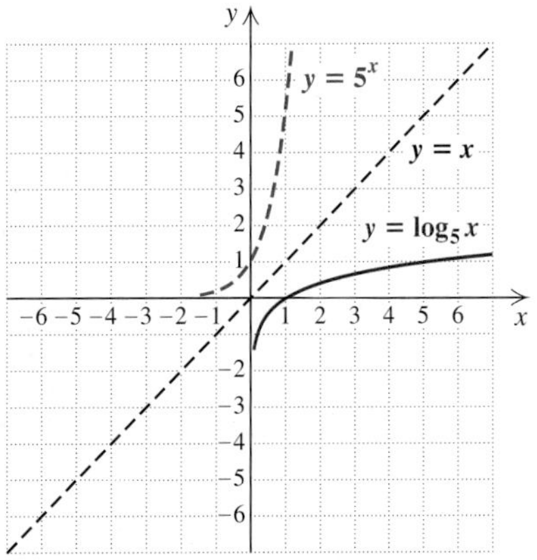

Common Logarithms

Any positive number other than 1 can serve as the base of a logarithmic function. However, some numbers are easier to use than others.

Base-10 logarithms, called **common logarithms,** are useful because they have the same base as our "commonly" used decimal system. Before calculators became so widely available, common logarithms helped with tedious calculations. In fact, that is why logarithms were developed. Before the advent of calculators, tables were developed to list common logarithms. Today we find common logarithms using calculators.

The abbreviation **log,** with no base written, is generally understood to mean logarithm base 10, or a common logarithm. Thus,

$\log 17$ means $\log_{10} 17$. **It is important to remember this abbreviation.**

Common Logarithms

To find the common logarithm of a number on most graphing calculators, press [LOG], the number, and then [ENTER]. Some calculators automatically supply the left parenthesis when [LOG] is pressed. Of these, some require that a right parenthesis be entered and some assume one is present at the end of the expression. Even if a calculator does not require an ending parenthesis, it is a good idea to include one.

For example, if we are using a calculator that supplies the left parenthesis, log 5 can be found by pressing [LOG] [5] [)] [ENTER]. Note from the screen on the left below that $\log 5 \approx 0.6990$.

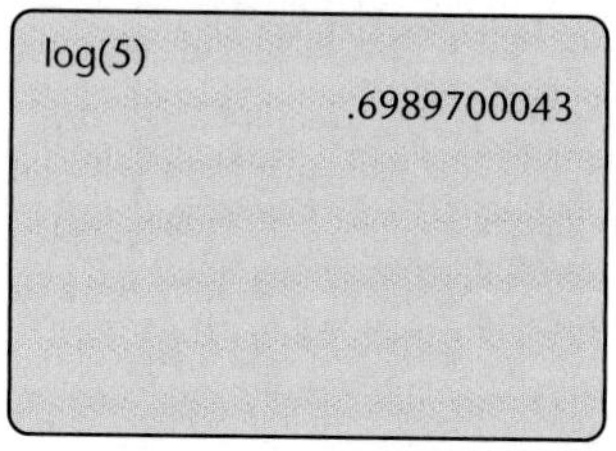

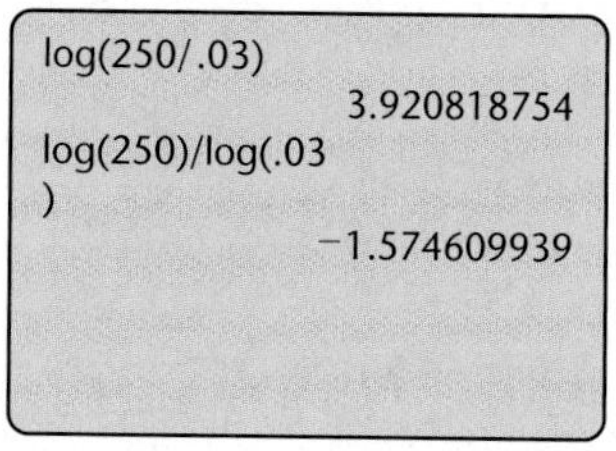

Parentheses must be placed carefully when evaluating expressions containing logarithms. For example, to evaluate

$$\log \frac{250}{0.03},$$

press [LOG] [2] [5] [0] [÷] [.] [0] [3] [)] [ENTER]. To evaluate

$$\frac{\log 250}{\log 0.03},$$

press [LOG] [2] [5] [0] [)] [÷] [LOG] [.] [0] [3] [)] [ENTER]. Note from the screen on the right above that $\log \frac{250}{0.03} \approx 3.9208$ and that $\frac{\log 250}{\log 0.03} \approx -1.5746$.

(continued)

The inverse of a logarithmic function is an exponential function. Because of this, on many calculators the LOG key serves as the 10^x key after the 2ND key is pressed. As with logarithms, some calculators automatically supply a left parenthesis before the exponent. Using such a calculator, we find $10^{3.417}$ by pressing 10^x 3 . 4 1 7) ENTER. The result is approximately 2612.1614. If we then press LOG ANS) ENTER, the result is 3.417. This illustrates that $f(x) = \log x$ and $g(x) = 10^x$ are inverse functions.

```
10^(3.417)
                 2612.161354
log(Ans)
                       3.417
```

A graphing calculator can quickly draw graphs of logarithmic functions, base 10. It is important to remember to place parentheses properly when entering logarithmic expressions.

EXAMPLE 4 Graph: $f(x) = \log \dfrac{x}{5} + 1$.

SOLUTION We enter $y = \log (x/5) + 1$. Since logarithms of negative numbers are not defined, we set the window accordingly. The graph is shown below. Note that although on the calculator the graph appears to touch the y-axis, a table of values or TRACE would show that the y-axis is indeed an asymptote.

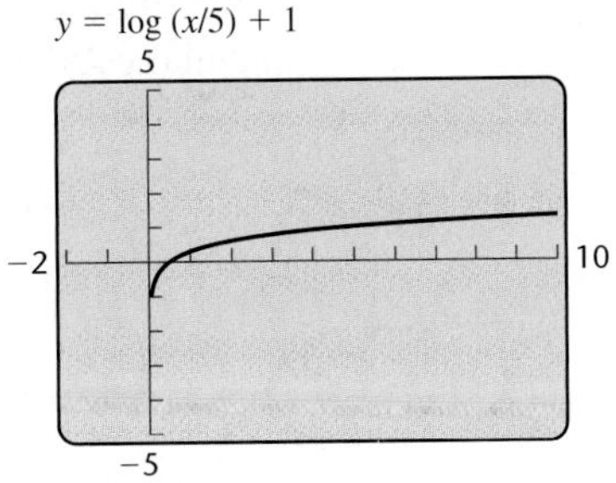

Equivalent Equations

We use the definition of logarithm to rewrite a *logarithmic equation* as an equivalent *exponential equation* or the other way around:

$$m = \log_a x \quad \text{is equivalent to} \quad a^m = x.$$

CAUTION! Do not forget this relationship! It is probably the most important definition in the chapter. Many times this definition will be used to justify a property we are considering.

EXAMPLE 5 Rewrite each as an equivalent exponential equation: **(a)** $y = \log_3 5$; **(b)** $-2 = \log_a 7$; **(c)** $a = \log_b d$.

SOLUTION

a) $y = \log_3 5$ is equivalent to $3^y = 5$ — **The logarithm is the exponent.** **The base remains the base.**

b) $-2 = \log_a 7$ is equivalent to $a^{-2} = 7$

c) $a = \log_b d$ is equivalent to $b^a = d$

EXAMPLE 6 Rewrite each as an equivalent logarithmic equation: **(a)** $8 = 2^x$; **(b)** $y^{-1} = 4$; **(c)** $a^b = c$.

SOLUTION

a) $8 = 2^x$ is equivalent to $x = \log_2 8$ — **The exponent is the logarithm.** **The base remains the base.**

b) $y^{-1} = 4$ is equivalent to $-1 = \log_y 4$

c) $a^b = c$ is equivalent to $b = \log_a c$

Solving Certain Logarithmic Equations

Logarithmic equations are often solved by rewriting them as equivalent exponential equations.

EXAMPLE 7 Solve: **(a)** $\log_2 x = -3$; **(b)** $\log_x 16 = 2$.

SOLUTION

a) $\log_2 x = -3$

$2^{-3} = x$ **Rewriting as an exponential equation**

$\frac{1}{8} = x$ **Computing 2^{-3}**

Check: $\log_2 \frac{1}{8}$ is the exponent to which 2 is raised to get $\frac{1}{8}$. Since that exponent is -3, we have a check. The solution is $\frac{1}{8}$.

b) $\log_x 16 = 2$

$x^2 = 16$ **Rewriting as an exponential equation**

$x = 4$ *or* $x = -4$ **Principle of square roots**

Check: $\log_4 16 = 2$ because $4^2 = 16$. Thus, 4 is a solution of $\log_x 16 = 2$. Because all logarithmic bases must be positive, -4 cannot be a solution. Logarithmic bases must be positive because logarithms are defined using exponential functions that require positive bases. The solution is 4.

One method for solving certain logarithmic and exponential equations relies on the following property, which results from the fact that exponential functions are one-to-one.

The Principle of Exponential Equality For any real number b, where $b \neq -1, 0,$ or 1,

$$b^x = b^y \quad \text{is equivalent to} \quad x = y.$$

(Powers of the same base are equal if and only if the exponents are equal.)

EXAMPLE 8 Solve: **(a)** $\log_{10} 1000 = x$; **(b)** $\log_4 1 = t$.

SOLUTION

a) We rewrite $\log_{10} 1000 = x$ in exponential form and solve:

$10^x = 1000$ **Rewriting as an exponential equation**

$10^x = 10^3$ **Writing 1000 as a power of 10**

$x = 3.$ **Equating exponents**

Check: This equation can also be solved directly by determining the exponent to which we raise 10 in order to get 1000. In both cases, we find that $\log_{10} 1000 = 3$, so we have a check. The solution is 3.

b) We rewrite $\log_4 1 = t$ in exponential form and solve:

$4^t = 1$ **Rewriting as an exponential equation**

$4^t = 4^0$ **Writing 1 as a power of 4. This can be done mentally.**

$t = 0.$ **Equating exponents**

Check: As in part (a), this equation can be solved directly by determining the exponent to which we raise 4 in order to get 1. In both cases, we find that $\log_4 1 = 0$, so we have a check. The solution is 0.

Example 8 illustrates an important property of logarithms.

$\log_a 1$

The logarithm, base a, of 1 is always 0: $\log_a 1 = 0$.

This follows from the fact that $a^0 = 1$ is equivalent to the logarithmic equation $\log_a 1 = 0$. Thus, $\log_{10} 1 = 0$, $\log_7 1 = 0$, and so on.

Another property results from the fact that $a^1 = a$. This is equivalent to the equation $\log_a a = 1$.

> **$\log_a a$**
>
> The logarithm, base a, of a is always 1: $\log_a a = 1$.

Thus, $\log_{10} 10 = 1$, $\log_8 8 = 1$, and so on.

9.3 EXERCISE SET

FOR EXTRA HELP

MathXL | MyMathLab | InterAct Math | AW Math Tutor Center | Video Lectures on CD: Disc 5 | Student's Solutions Manual

Concept Reinforcement In each of Exercises 1–8, match the expression or equation with an equivalent expression or equation from the column on the right.

1. ____ $\log_5 25$	**a)** 1
2. ____ $2^5 = x$	**b)** x
3. ____ $\log_5 5$	**c)** $x^5 = 27$
4. ____ $\log_2 1$	**d)** $\log_2 x = 5$
5. ____ $\log_5 5^x$	**e)** $\log_2 8 = x$
6. ____ $\log_x 27 = 5$	**f)** $\log_x 5 = -2$
7. ____ $8 = 2^x$	**g)** 2
8. ____ $x^{-2} = 5$	**h)** 0

Simplify.

9. $\log_{10} 1000$

10. $\log_{10} 100$

11. $\log_2 16$

12. $\log_2 8$

13. $\log_3 81$

14. $\log_3 27$

15. $\log_4 \frac{1}{16}$

16. $\log_4 \frac{1}{4}$

17. $\log_7 \frac{1}{7}$

18. $\log_7 \frac{1}{49}$

19. $\log_5 625$

20. $\log_5 125$

21. $\log_8 8$

22. $\log_7 1$

23. $\log_8 1$

24. $\log_8 8$

Aha! **25.** $\log_9 9^5$

26. $\log_9 9^{10}$

27. $\log_{10} 0.01$

28. $\log_{10} 0.1$

29. $\log_9 3$

30. $\log_{16} 4$

31. $\log_9 27$

32. $\log_{16} 64$

33. $\log_{1000} 100$

34. $\log_{27} 9$

35. $5^{\log_5 7}$

36. $6^{\log_6 13}$

Graph by hand.

37. $y = \log_{10} x$

38. $y = \log_2 x$

39. $y = \log_3 x$

40. $y = \log_7 x$

41. $f(x) = \log_6 x$

42. $f(x) = \log_4 x$

43. $f(x) = \log_{2.5} x$

44. $f(x) = \log_{1/2} x$

Graph both functions using the same set of axes.

45. $f(x) = 3^x$, $f^{-1}(x) = \log_3 x$

46. $f(x) = 4^x$, $f^{-1}(x) = \log_4 x$

Use a calculator to find each of the following rounded to four decimal places.

47. $\log 4$

48. $\log 5$

49. $\log 13{,}400$

50. $\log 93{,}100$

51. $\log 0.527$

52. $\log 0.493$

Use a calculator to find each of the following rounded to four decimal places.

53. $10^{2.3}$

54. $10^{0.173}$

55. $10^{-2.9523}$

56. $10^{4.8982}$

57. $10^{0.0012}$

58. $10^{-3.89}$

Graph using a graphing calculator.

59. $\log (x + 2)$

60. $\log (x - 5)$

61. $\log (1 - 2x)$

62. $\log (3x + 2.7)$

63. $\log (x^2)$

64. $\log (x^2 + 1)$

Rewrite each of the following as an equivalent exponential equation. Do not solve.

65. $t = \log_5 9$

66. $h = \log_7 10$

67. $\log_5 25 = 2$

68. $\log_6 6 = 1$

69. $\log_{10} 0.1 = -1$

70. $\log_{10} 0.01 = -2$

71. $\log_{10} 7 = 0.845$

72. $\log_{10} 3 = 0.4771$

73. $\log_c m = 8$

74. $\log_b n = 23$

75. $\log_t Q = r$

76. $\log_m P = a$

77. $\log_e 0.25 = -1.3863$

78. $\log_e 0.989 = -0.0111$

79. $\log_r T = -x$

80. $\log_c M = -w$

Rewrite each of the following as an equivalent logarithmic equation. Do not solve.

81. $10^2 = 100$

82. $10^4 = 10{,}000$

83. $4^{-5} = \frac{1}{1024}$

84. $5^{-3} = \frac{1}{125}$

85. $16^{3/4} = 8$

86. $8^{1/3} = 2$

87. $10^{0.4771} = 3$

88. $10^{0.3010} = 2$

89. $z^m = 6$

90. $m^n = r$

91. $p^m = V$

92. $Q^t = x$

93. $e^3 = 20.0855$

94. $e^2 = 7.3891$

95. $e^{-4} = 0.0183$

96. $e^{-2} = 0.1353$

Solve.

97. $\log_3 x = 2$

98. $\log_4 x = 3$

99. $\log_5 125 = x$

100. $\log_4 64 = x$

101. $\log_2 16 = x$

102. $\log_3 27 = x$

103. $\log_x 7 = 1$

104. $\log_x 8 = 1$

105. $\log_3 x = -2$

106. $\log_2 x = -1$

107. $\log_{32} x = \frac{2}{5}$

108. $\log_8 x = \frac{2}{3}$

TW **109.** Express in words what number is represented by $\log_b c$.

TW **110.** Is it true that $2 = b^{\log_b 2}$? Why or why not?

Skill Maintenance

Simplify.

111. $\dfrac{x^{12}}{x^4}$ [1.4]

112. $\dfrac{u^{15}}{a^3}$ [1.4]

113. $(a^4b^6)(a^3b^2)$ [1.4]

114. $(x^3y^5)(x^2y^7)$ [1.4]

115. $\dfrac{\dfrac{3}{x} - \dfrac{2}{xy}}{\dfrac{2}{x^2} + \dfrac{1}{xy}}$ [6.3]

116. $\dfrac{\dfrac{4 + x}{x^2 + 2x + 1}}{\dfrac{3}{x + 1} - \dfrac{2}{x + 2}}$ [6.3]

Synthesis

TW **117.** Would a manufacturer be pleased or unhappy if sales of a product grew logarithmically? Why?

TW **118.** Explain why the number $\log_2 13$ must be between 3 and 4.

119. Graph both equations using the same set of axes:

$$y = \left(\tfrac{3}{2}\right)^x, \qquad y = \log_{3/2} x.$$

Graph by hand.

120. $y = \log_2(x - 1)$

121. $y = \log_3 |x + 1|$

Solve.

122. $|\log_3 x| = 2$

123. $\log_4 (3x - 2) = 2$

124. $\log_8 (2x + 1) = -1$

125. $\log_{10} (x^2 + 21x) = 2$

Simplify.

126. $\log_{1/4} \frac{1}{64}$

127. $\log_{1/5} 25$

128. $\log_{81} 3 \cdot \log_3 81$

129. $\log_{10} (\log_4 (\log_3 81))$

130. $\log_2 (\log_2 (\log_4 256))$

131. Show that $b^x = b^y$ is *not* equivalent to $x = y$ for $b = 0$ or $b = 1$.

TW **132.** If $\log_b a = x$, does it follow that $\log_a b = 1/x$? Why or why not?

9.4 Properties of Logarithmic Functions

Logarithms of Products ■ Logarithms of Powers ■ Logarithms of Quotients ■ Using the Properties Together

Logarithmic functions are important in many applications and in more advanced mathematics. We now establish some basic properties that are useful in manipulating expressions involving logarithms.

Interactive Discovery

For each of the following, use a calculator to determine which is the equivalent logarithmic expression.

1. $\log(20 \cdot 5)$
- **a)** $\log(20) \cdot \log(5)$
- **b)** $5 \cdot \log(20)$
- **c)** $\log(20) + \log(5)$

2. $\log(20^5)$
- **a)** $\log(20) \cdot \log(5)$
- **b)** $5 \cdot \log(20)$
- **c)** $\log(20) - \log(5)$

3. $\log\left(\frac{20}{5}\right)$
- **a)** $\log(20) \cdot \log(5)$
- **b)** $\log(20)/\log(5)$
- **c)** $\log(20) - \log(5)$

4. $\log\left(\frac{1}{20}\right)$
- **a)** $-\log(20)$
- **b)** $\log(1)/\log(20)$
- **c)** $\log(1) \cdot \log(20)$

In this section, we will state and prove the results you may have observed. As their proofs reveal, the properties of logarithms are related to the properties of exponents.

Logarithms of Products

The first property we discuss is related to the product rule for exponents: $a^m \cdot a^n = a^{m+n}$. Its proof appears immediately after Example 2.

The Product Rule for Logarithms For any positive numbers M, N, and a $(a \neq 1)$,

$$\log_a(MN) = \log_a M + \log_a N.$$

(The logarithm of a product is the sum of the logarithms of the factors.)

EXAMPLE 1 Express as an equivalent expression that is a sum of logarithms: $\log_2 (4 \cdot 16)$.

SOLUTION We have

$$\log_2 (4 \cdot 16) = \log_2 4 + \log_2 16. \quad \textbf{Using the product rule for logarithms}$$

As a check, note that

$$\log_2 (4 \cdot 16) = \log_2 64 = 6 \qquad 2^6 = 64$$

and that

$$\log_2 4 + \log_2 16 = 2 + 4 = 6. \qquad 2^2 = 4 \text{ and } 2^4 = 16$$

EXAMPLE 2 Express as an equivalent expression that is a single logarithm: $\log_b 7 + \log_b 5$.

SOLUTION We have

$$\log_b 7 + \log_b 5 = \log_b (7 \cdot 5) \quad \textbf{Using the product rule for logarithms}$$
$$= \log_b 35.$$

A PROOF OF THE PRODUCT RULE. Let $\log_a M = x$ and $\log_a N = y$. Converting to exponential equations, we have $a^x = M$ and $a^y = N$.

Now we multiply the left sides and the right sides of the equations to obtain

$$MN = a^x \cdot a^y, \quad \text{or} \quad MN = a^{x+y}.$$

Converting back to a logarithmic equation, we get

$$\log_a (MN) = x + y.$$

Recalling what x and y represent, we get

$$\log_a (MN) = \log_a M + \log_a N.$$

■

Study Tip

Go as Slow as Necessary

When new material seems challenging to you, do not pay attention to how quickly or slowly a classmate absorbs the concepts. We each move at our own speed when it comes to digesting new material.

Logarithms of Powers

The second basic property is related to the power rule for exponents: $(a^m)^n = a^{mn}$. Its proof follows Example 3.

The Power Rule for Logarithms For any positive numbers M and a ($a \neq 1$), and any real number p,

$$\log_a M^p = p \cdot \log_a M.$$

(The logarithm of a power of M is the exponent times the logarithm of M.)

To better understand the power rule, note that

$$\log_a M^3 = \log_a (M \cdot M \cdot M) = \log_a M + \log_a M + \log_a M = 3 \log_a M.$$

EXAMPLE 3 Use the power rule for logarithms to write an equivalent expression that is a product: **(a)** $\log_a 9^{-5}$; **(b)** $\log_7 \sqrt[3]{x}$.

SOLUTION

a) $\log_a 9^{-5} = -5 \log_a 9$ **Using the power rule for logarithms**

b) $\log_7 \sqrt[3]{x} = \log_7 x^{1/3}$ **Writing exponential notation**

$= \frac{1}{3} \log_7 x$ **Using the power rule for logarithms**

A PROOF OF THE POWER RULE. Let $x = \log_a M$. The equivalent exponential equation is $a^x = M$. Raising both sides to the pth power, we get

$$(a^x)^p = M^p, \quad \text{or} \quad a^{xp} = M^p. \qquad \textbf{Multiplying exponents}$$

Converting back to a logarithmic equation gives us

$$\log_a M^p = xp.$$

But $x = \log_a M$, so substituting, we have

$$\log_a M^p = (\log_a M)p = p \cdot \log_a M.$$

■

Logarithms of Quotients

The third property that we study is similar to the quotient rule for exponents: $\frac{a^m}{a^n} = a^{m-n}$. Its proof follows Example 5.

The Quotient Rule for Logarithms For any positive numbers M, N, and a $(a \neq 1)$,

$$\log_a \frac{M}{N} = \log_a M - \log_a N.$$

(The logarithm of a quotient is the logarithm of the dividend minus the logarithm of the divisor.)

To better understand the quotient rule, note that

$$\log_a\left(\frac{b^5}{b^3}\right) = \log_a b^2 = 2 \log_a b = 5 \log_a b - 3 \log_a b$$
$$= \log_a b^5 - \log_a b^3.$$

EXAMPLE 4 Express as an equivalent expression that is a difference of logarithms: $\log_t (6/U)$.

SOLUTION

$$\log_t \frac{6}{U} = \log_t 6 - \log_t U \qquad \textbf{Using the quotient rule for logarithms}$$

EXAMPLE 5 Express as an equivalent expression that is a single logarithm: $\log_b 17 - \log_b 27$.

SOLUTION

$$\log_b 17 - \log_b 27 = \log_b \frac{17}{27}$$ **Using the quotient rule for logarithms "in reverse"**

A PROOF OF THE QUOTIENT RULE. Our proof uses both the product and power rules:

$$\begin{aligned}
\log_a \frac{M}{N} &= \log_a MN^{-1} && \text{Rewriting } \frac{M}{N} \text{ as } MN^{-1} \\
&= \log_a M + \log_a N^{-1} && \text{Using the product rule for logarithms} \\
&= \log_a M + (-1)\log_a N && \text{Using the power rule for logarithms} \\
&= \log_a M - \log_a N.
\end{aligned}$$

■

Using the Properties Together

EXAMPLE 6 Express as an equivalent expression, using the individual logarithms of x, y, and z.

a) $\log_b \frac{x^3}{yz}$ **b)** $\log_a \sqrt[4]{\frac{xy}{z^3}}$

SOLUTION

a)
$$\begin{aligned}
\log_b \frac{x^3}{yz} &= \log_b x^3 - \log_b yz && \text{Using the quotient rule for logarithms} \\
&= 3 \log_b x - \log_b yz && \text{Using the power rule for logarithms} \\
&= 3 \log_b x - (\log_b y + \log_b z) && \text{Using the product rule for logarithms. Because of the subtraction, parentheses are essential.} \\
&= 3 \log_b x - \log_b y - \log_b z && \text{Using the distributive law}
\end{aligned}$$

b)
$$\begin{aligned}
\log_a \sqrt[4]{\frac{xy}{z^3}} &= \log_a \left(\frac{xy}{z^3}\right)^{1/4} && \text{Writing exponential notation} \\
&= \frac{1}{4} \cdot \log_a \frac{xy}{z^3} && \text{Using the power rule for logarithms} \\
&= \frac{1}{4}(\log_a xy - \log_a z^3) && \text{Using the quotient rule for logarithms. Parentheses are important.} \\
&= \frac{1}{4}(\log_a x + \log_a y - 3 \log_a z) && \text{Using the product and power rules for logarithms}
\end{aligned}$$

CAUTION! Because the product and quotient rules replace one term with two, parentheses are often necessary, as in Example 6.

EXAMPLE 7 Express as an equivalent expression that is a single logarithm.

a) $\frac{1}{2} \log_a x - 7 \log_a y + \log_a z$ **b)** $\log_a \frac{b}{\sqrt{x}} + \log_a \sqrt{bx}$

SOLUTION

a) $\frac{1}{2}\log_a x - 7\log_a y + \log_a z$

$= \log_a x^{1/2} - \log_a y^7 + \log_a z$ **Using the power rule for logarithms**

$= \left(\log_a \sqrt{x} - \log_a y^7\right) + \log_a z$ **Using parentheses to emphasize the order of operations; $x^{1/2} = \sqrt{x}$**

$= \log_a \frac{\sqrt{x}}{y^7} + \log_a z$ **Using the quotient rule for logarithms**

$= \log_a \frac{z\sqrt{x}}{y^7}$ **Using the product rule for logarithms**

b) $\log_a \frac{b}{\sqrt{x}} + \log_a \sqrt{bx} = \log_a \frac{b \cdot \sqrt{bx}}{\sqrt{x}}$ **Using the product rule for logarithms**

$= \log_a b\sqrt{b}$ **Removing a factor equal to 1: $\frac{\sqrt{x}}{\sqrt{x}} = 1$**

$= \log_a b^{3/2}$, or $\frac{3}{2}\log_a b$ **Since $b\sqrt{b} = b^1 \cdot b^{1/2}$**

If we know the logarithms of two different numbers (to the same base), the properties allow us to calculate other logarithms.

EXAMPLE 8 Given $\log_a 2 = 0.431$ and $\log_a 3 = 0.683$, calculate a numerical value for each of the following, if possible.

a) $\log_a 6$ **b)** $\log_a \frac{2}{3}$ **c)** $\log_a 81$

d) $\log_a \frac{1}{3}$ **e)** $\log_a 2a$ **f)** $\log_a 5$

SOLUTION

a) $\log_a 6 = \log_a (2 \cdot 3) = \log_a 2 + \log_a 3$ **Using the product rule for logarithms**

$= 0.431 + 0.683 = 1.114$

Check: $a^{1.114} = a^{0.431} \cdot a^{0.683} = 2 \cdot 3 = 6$

b) $\log_a \frac{2}{3} = \log_a 2 - \log_a 3$ **Using the quotient rule for logarithms**

$= 0.431 - 0.683 = -0.252$

c) $\log_a 81 = \log_a 3^4 = 4\log_a 3$ **Using the power rule for logarithms**

$= 4(0.683) = 2.732$

d) $\log_a \frac{1}{3} = \log_a 1 - \log_a 3$ **Using the quotient rule for logarithms**

$= 0 - 0.683 = -0.683$

e) $\log_a 2a = \log_a 2 + \log_a a$ **Using the product rule for logarithms**

$= 0.431 + 1 = 1.431$

f) $\log_a 5$ *cannot be found using these properties.* ($\log_a 5 \neq \log_a 2 + \log_a 3$)

A final property follows from the product rule: Since $\log_a a^k = k \log_a a$, and $\log_a a = 1$, we have $\log_a a^k = k$.

The Logarithm of the Base to an Exponent For any base a,

$$\log_a a^k = k.$$

(The logarithm, base a, of a to an exponent is the exponent.)

This property also follows from the definition of logarithm: k is the exponent to which you raise a in order to get a^k.

EXAMPLE 9 Simplify: **(a)** $\log_3 3^7$; **(b)** $\log_{10} 10^{-5.2}$.

SOLUTION

a) $\log_3 3^7 = 7$ — **7 is the exponent to which you raise 3 in order to get 3^7.**

b) $\log_{10} 10^{-5.2} = -5.2$

We summarize the properties covered in this section as follows.

For any positive numbers M, N, and a $(a \neq 1)$:

$$\log_a (MN) = \log_a M + \log_a N; \qquad \log_a M^p = p \cdot \log_a M;$$

$$\log_a \frac{M}{N} = \log_a M - \log_a N; \qquad \log_a a^k = k.$$

9.4 EXERCISE SET

FOR EXTRA HELP

MathXL | MyMathLab | InterAct Math | AW Math Tutor Center | Video Lectures on CD: Disc 5 | Student's Solutions Manual

Concept Reinforcement *In each of Exercises 1–6, match the expression with an equivalent expression from the column on the right.*

1. ___ $\log_7 20$	**a)** $\log_7 5 - \log_7 4$
2. ___ $\log_7 5^4$	**b)** 1
3. ___ $\log_7 \frac{5}{4}$	**c)** 0
4. ___ $\log_7 7$	**d)** $\log_7 30$
5. ___ $\log_7 1$	**e)** $\log_7 5 + \log_7 4$
6. ___ $\log_7 5 + \log_7 6$	**f)** $4 \log_7 5$

Express as an equivalent expression that is a sum of logarithms.

7. $\log_3 (81 \cdot 27)$

8. $\log_2 (16 \cdot 32)$

9. $\log_4 (64 \cdot 16)$

10. $\log_5 (25 \cdot 125)$

11. $\log_c (rst)$

12. $\log_t (3ab)$

Express as an equivalent expression that is a single logarithm.

13. $\log_a 5 + \log_a 14$

14. $\log_b 65 + \log_b 2$

15. $\log_c t + \log_c y$

16. $\log_t H + \log_t M$

Express as an equivalent expression that is a product.

17. $\log_a r^8$

18. $\log_b t^5$

19. $\log_c y^6$

20. $\log_{10} y^7$

21. $\log_b C^{-3}$

22. $\log_c M^{-5}$

Express as an equivalent expression that is a difference of two logarithms.

23. $\log_2 \frac{25}{13}$

24. $\log_3 \frac{23}{9}$

25. $\log_b \dfrac{m}{n}$

26. $\log_a \dfrac{y}{x}$

Express as an equivalent expression that is a single logarithm.

27. $\log_a 17 - \log_a 6$

28. $\log_b 32 - \log_b 7$

29. $\log_b 36 - \log_b 4$

30. $\log_a 26 - \log_a 2$

31. $\log_a 7 - \log_a 18$

32. $\log_b 5 - \log_b 13$

Express as an equivalent expression, using the individual logarithms of w, x, y, and z.

33. $\log_a (xyz)$

34. $\log_a (wxy)$

35. $\log_a (x^3z^4)$

36. $\log_a (x^2y^5)$

37. $\log_a (x^2y^{-2}z)$

38. $\log_a (xy^2z^{-3})$

39. $\log_a \dfrac{x^4}{y^3z}$

40. $\log_a \dfrac{x^4}{yz^2}$

41. $\log_b \dfrac{xy^2}{wz^3}$

42. $\log_b \dfrac{w^2x}{y^3z}$

43. $\log_a \sqrt{\dfrac{x^7}{y^5z^8}}$

44. $\log_c \sqrt[3]{\dfrac{x^4}{y^3z^2}}$

45. $\log_a \sqrt[3]{\dfrac{x^6y^3}{a^2z^7}}$

46. $\log_a \sqrt[4]{\dfrac{x^8y^{12}}{a^3z^5}}$

Express as an equivalent expression that is a single logarithm and, if possible, simplify.

47. $8 \log_a x + 3 \log_a z$

48. $2 \log_b m + \frac{1}{2} \log_b n$

49. $\log_a x^2 - 2 \log_a \sqrt{x}$

50. $\log_a \dfrac{a}{\sqrt{x}} - \log_a \sqrt{ax}$

51. $\frac{1}{2} \log_a x + 5 \log_a y - 2 \log_a x$

52. $\log_a 2x + 3(\log_a x - \log_a y)$

53. $\log_a (x^2 - 4) - \log_a (x + 2)$

54. $\log_a (2x + 10) - \log_a (x^2 - 25)$

Given $\log_b 3 = 0.792$ and $\log_b 5 = 1.161$. If possible, calculate numerical values for each of the following.

55. $\log_b 15$

56. $\log_b \frac{5}{3}$

57. $\log_b \frac{3}{5}$

58. $\log_b \frac{1}{3}$

59. $\log_b \frac{1}{5}$

60. $\log_b \sqrt{b}$

61. $\log_b \sqrt{b^3}$

62. $\log_b 3b$

63. $\log_b 8$

64. $\log_b 45$

Simplify.

Aha! **65.** $\log_t t^7$

66. $\log_p p^4$

67. $\log_e e^m$

68. $\log_Q Q^{-2}$

Find each of the following.

Aha! **69.** $\log_5 (125 \cdot 625)$

70. $\log_3 (9 \cdot 81)$

71. $\log_2 \left(\dfrac{128}{16}\right)$

72. $\log_3 \left(\dfrac{243}{27}\right)$

TW **73.** A student *incorrectly* reasons that

$$\log_b \frac{1}{x} = \log_b \frac{x}{xx}$$
$$= \log_b x - \log_b x + \log_b x = \log_b x.$$

What mistake has the student made?

TW **74.** How could you convince someone that

$$\log_a c \neq \log_c a?$$

Skill Maintenance

Graph. [7.1]

75. $f(x) = \sqrt{x} - 3$

76. $g(x) = \sqrt{x} + 2$

77. $g(x) = \sqrt[3]{x} + 1$

78. $f(x) = \sqrt[3]{x} - 1$

Simplify. [1.4]

79. $(a^3b^2)^5(a^2b^7)$

80. $(x^5y^3z^2)(x^2yz^2)^3$

Synthesis

TW **81.** Is it possible to express $\log_b \dfrac{x}{5}$ as an equivalent expression that is a difference of two logarithms without using the quotient rule? Why or why not?

TW **82.** Is it true that $\log_a x + \log_b x = \log_{ab} x$? Why or why not?

Express as an equivalent expression that is a single logarithm and, if possible, simplify.

83. $\log_a (x^8 - y^8) - \log_a (x^2 + y^2)$

84. $\log_a (x + y) + \log_a (x^2 - xy + y^2)$

Express as an equivalent expression that is a sum or difference of logarithms and, if possible, simplify.

85. $\log_a \sqrt{1 - s^2}$

86. $\log_a \dfrac{c - d}{\sqrt{c^2 - d^2}}$

87. If $\log_a x = 2$, $\log_a y = 3$, and $\log_a z = 4$, what is

$$\log_a \frac{\sqrt[3]{x^2 z}}{\sqrt[3]{y^2 z^{-2}}}?$$

88. If $\log_a x = 2$, what is $\log_a (1/x)$?

89. If $\log_a x = 2$, what is $\log_{1/a} x$?

Classify each of the following as true or false. Assume a, x, P, and $Q > 0$, $a \neq 1$.

90. $\log_a \left(\dfrac{P}{Q}\right)^x = x \log_a P - \log_a Q$

91. $\log_a (Q + Q^2) - \log_a Q = \log_a (Q + 1)$

92. Use graphs to show that

$$\log x^2 \neq \log x \cdot \log x.$$

(*Note*: log means $\log_{10}$.)

9.5 Natural Logarithms and Changing Bases

The Base e and Natural Logarithms ■ Changing Logarithmic Bases ■ Graphs of Exponential and Logarithmic Functions, Base e

There are logarithm bases that fit into certain applications more naturally than others. We have already looked at logarithms with one such base, base-10 logarithms, or common logarithms. Another logarithm base widely used today is an irrational number named e.

The Base e and Natural Logarithms

When interest is computed n times a year, the compound interest formula is

$$A = P\left(1 + \frac{r}{n}\right)^{nt},$$

where A is the amount that an initial investment P will be worth after t years at interest rate r. Suppose that \$1 is invested at 100% interest for 1 year (no bank would pay this). The preceding formula becomes a function A defined in terms of the number of compounding periods n:

$$A(n) = \left(1 + \frac{1}{n}\right)^n.$$

Interactive Discovery

1. What happens to the function values of $A(n) = \left(1 + \frac{1}{n}\right)^n$ as n gets larger? To find out, use the TABLE feature with Indpnt set to Ask to fill in the following table. Round each entry to six decimal places.

n	$A(n) = \left(1 + \frac{1}{n}\right)^n$
1 (compounded annually)	$\left(1 + \frac{1}{1}\right)^1$, or \$2.00
2 (compounded semiannually)	$\left(1 + \frac{1}{2}\right)^2$, or \$?
3	
4 (compounded quarterly)	
12 (compounded monthly)	
52 (compounded weekly)	
365 (compounded daily)	
8760 (compounded hourly)	

2. Which of the following statements appears to be true?

a) $A(n)$ gets very large as n gets very large.

b) $A(n)$ gets very small as n gets very large.

c) $A(n)$ approaches a certain number as n gets very large.

The numbers in the table approach a very important number in mathematics, called e. Because e is irrational, its decimal representation does not terminate or repeat.

The Number e $e \approx 2.7182818284\ldots$

Logarithms base e are called **natural logarithms,** or **Napierian logarithms,** in honor of John Napier (1550–1617), who first "discovered" logarithms.

The abbreviation "ln" is generally used with natural logarithms. Thus,

$$\ln 53 \quad \text{means} \quad \log_e 53.$$

Remember: $\log x$ means $\log_{10} x$, and $\ln x$ means $\log_e x$.

Study Tip

Is Your Answer Reasonable?

It is always a good idea—especially when using a calculator—to check that your answer is reasonable. It is easy for an incorrect calculation or keystroke to result in an answer that is clearly too big or too small.

Natural Logarithms

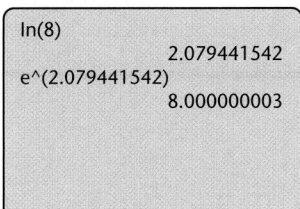

Natural logarithms are entered on a graphing calculator much like common logarithms are (see pp. 730–731). For example, if we are using a calculator that supplies the left parenthesis, ln 8 can be found by pressing [LN] [8] [)] [ENTER]. Note from the screen below that $\ln 8 \approx 2.0794$.

```
ln(8)
                2.079441542
e^(2.079441542)
                8.000000003
```

On many calculators, the [LN] key serves as the [e^x] key after the [2ND] key is pressed. Some calculators automatically supply a left parenthesis before the exponent. If such a calculator is used,

$$e^{2.079441542}$$

is found by pressing [e^x] [2] [.] [0] [7] [9] [4] [4] [1] [5] [4] [2] [)] [ENTER]. As shown on the screen above, $e^{2.079441542} \approx 8$, as we would expect, since $f(x) = e^x$ is the inverse function of $g(x) = \ln x$.

Changing Logarithmic Bases

Most calculators can find both common logarithms and natural logarithms. To find a logarithm with some other base, a conversion formula is usually needed.

The Change-of-Base Formula For any logarithmic bases a and b, and any positive number M,

$$\log_b M = \frac{\log_a M}{\log_a b}.$$

(To find the log, base b, we typically compute $\log M/\log b$ or $\ln M/\ln b$.)

PROOF. Let $x = \log_b M$. Then,

$$b^x = M \qquad \textbf{$\log_b M = x$ is equivalent to $b^x = M$.}$$

$$\log_a b^x = \log_a M \qquad \textbf{Taking the logarithm, base a, on both sides}$$

$$x \log_a b = \log_a M \qquad \textbf{Using the power rule for logarithms}$$

$$x = \frac{\log_a M}{\log_a b}. \qquad \textbf{Dividing both sides by $\log_a b$}$$

But at the outset we stated that $x = \log_b M$. Thus, by substitution, we have

$$\log_b M = \frac{\log_a M}{\log_a b}. \qquad \textbf{This is the change-of-base formula.}$$

■

EXAMPLE 1 Find $\log_5 8$ using the change-of-base formula.

SOLUTION We use the change-of-base formula with $a = 10$, $b = 5$, and $M = 8$:

$$\log_5 8 = \frac{\log_{10} 8}{\log_{10} 5} \qquad \textbf{Substituting into } \log_b M = \frac{\log_a M}{\log_a b}$$

$$\approx \frac{0.903089987}{0.6989700043} \qquad \textbf{Using LOG twice}$$

$$\approx 1.2920. \qquad$$ **When using a calculator, it is best not to round before dividing.**

The figure below shows the computation using a graphing calculator.

```
log(8)/log(5)
                1.292029674
```

To check, note that $\ln 8/\ln 5 \approx 1.2920$. We can also use a calculator to verify that $5^{1.2920} \approx 8$.

EXAMPLE 2 Find $\log_4 31$.

SOLUTION As indicated in the check of Example 1, base e can also be used.

$$\log_4 31 = \frac{\log_e 31}{\log_e 4} \qquad \textbf{Substituting into } \log_b M = \frac{\log_a M}{\log_a b}$$

$$= \frac{\ln 31}{\ln 4} \qquad \textbf{Using LN twice}$$

$$\approx 2.4771. \qquad \textit{Check: } 4^{2.4771} \approx 31$$

```
ln(31)/ln(4)
                2.477098155
log(31)/log(4)
                2.477098155
4^Ans
                         31
```

The screen at left illustrates that we find the same solution using either natural or common logarithms.

Graphs of Exponential and Logarithmic Functions, Base *e*

EXAMPLE 3 Graph $f(x) = e^x$ and $g(x) = e^{-x}$ and state the domain and the range of f and g.

SOLUTION We use a calculator with an e^x key to find approximate values of e^x and e^{-x}. Using these values, we can graph the functions.

x	e^x	e^{-x}
0	1	1
1	2.7	0.4
2	7.4	0.1
−1	0.4	2.7
−2	0.1	7.4

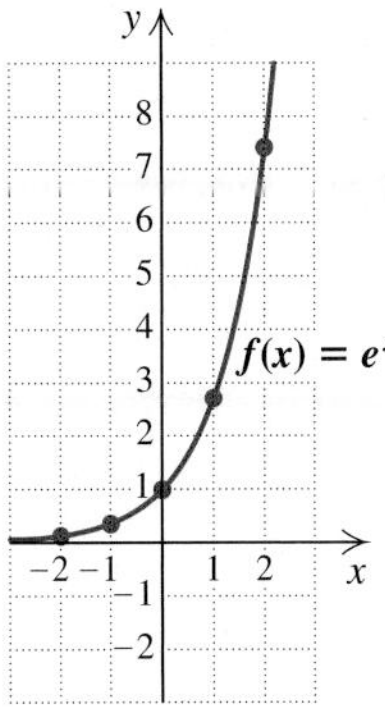

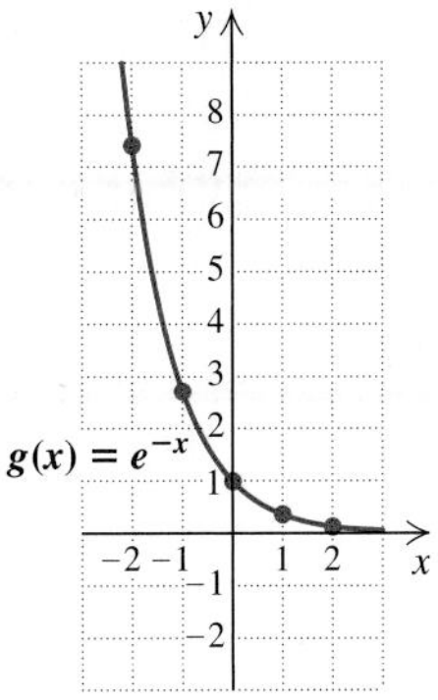

The domain of each function is $\mathbb{R}$ and the range of each function is $(0, \infty)$.

EXAMPLE 4 Graph $f(x) = e^{-x} + 2$ and state the domain and the range of f.

SOLUTION To graph by hand, we find some solutions with a calculator, plot them, and then draw the graph. For example, $f(2) = e^{-2} + 2 \approx 2.1$. The graph is exactly like the graph of $g(x) = e^{-x}$, but is translated 2 units up.

To graph using a graphing calculator, we enter $y = e \wedge (-x) + 2$ and choose a viewing window that shows more of the y-axis than the x-axis, since the function is exponential.

x	$e^{-x} + 2$
0	3
1	2.4
2	2.1
−1	4.7
−2	9.4

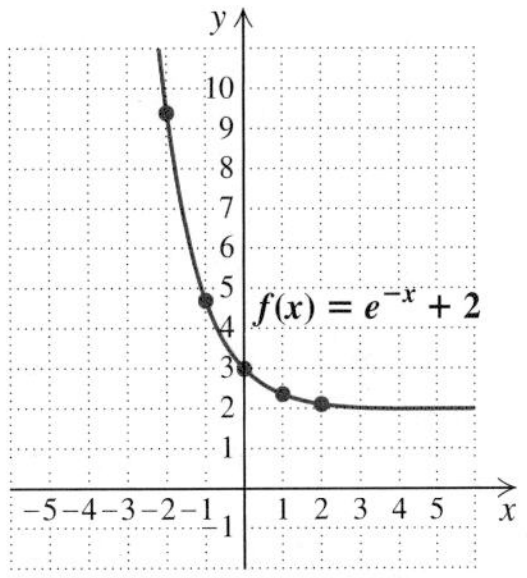

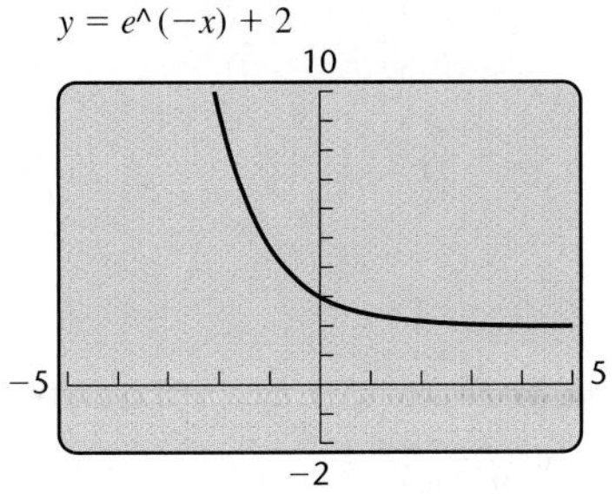

The domain of f is $\mathbb{R}$ and the range is $(2, \infty)$.

EXAMPLE 5 Graph and state the domain and the range of each function.

a) $g(x) = \ln x$ **b)** $f(x) = \ln(x + 3)$

SOLUTION

a) To graph by hand, we find some solutions with a calculator and then draw the graph. As expected, the graph is a reflection across the line $y = x$ of the graph of $y = e^x$. We can also graph $y = \ln(x)$ using a graphing calculator. Since a logarithmic function is the inverse of an exponential function, we choose a viewing window that shows more of the x-axis than the y-axis.

x	$\ln x$
1	0
4	1.4
7	1.9
0.5	−0.7

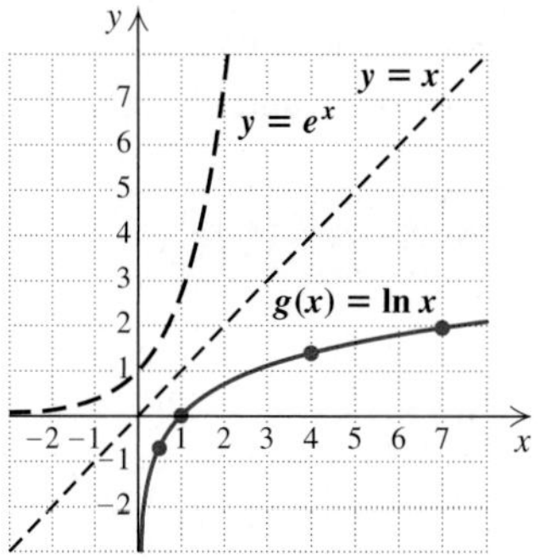

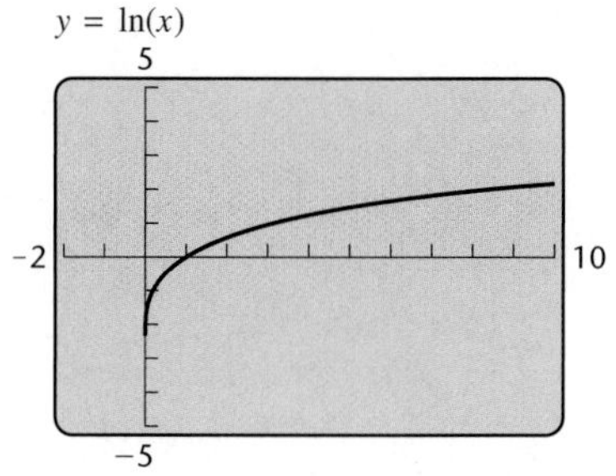

The domain of g is $(0, \infty)$ and the range is $\mathbb{R}$.

b) We find some solutions with a calculator, plot them, and draw the graph by hand. Note that the graph of $y = \ln(x + 3)$ is the graph of $y = \ln x$ translated 3 units to the left. To graph using a graphing calculator, we enter $y = \ln(x + 3)$.

x	$\ln(x + 3)$
0	1.1
1	1.4
2	1.6
3	1.8
4	1.9
−1	0.7
−2	0
−2.5	−0.7

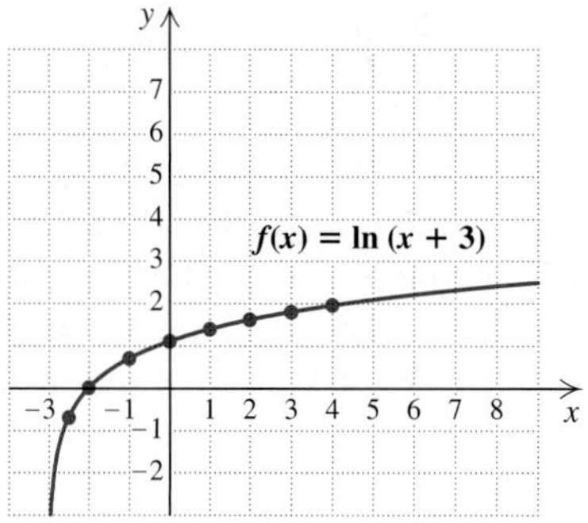

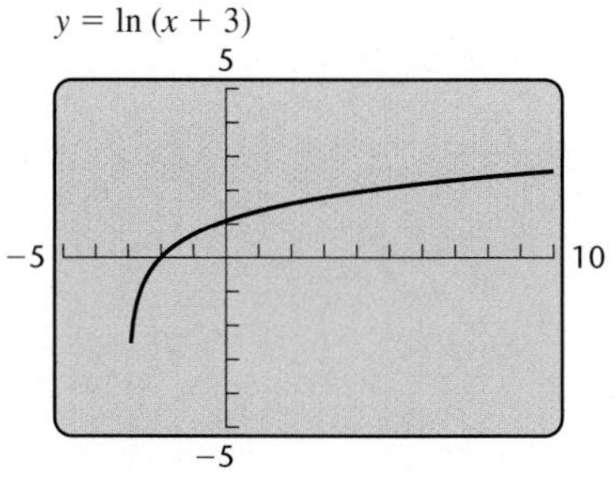

Since $x + 3$ must be positive, the domain of f is $(-3, \infty)$ and the range is $\mathbb{R}$.

Logarithmic functions with bases other than 10 or e can be graphed on a graphing calculator using the change-of-base formula.

EXAMPLE 6 Graph: $f(x) = \log_7 x + 2$.

SOLUTION We use the change-of-base formula with natural logarithms. (We would get the same graph if we used common logarithms.)

$$f(x) = \log_7 x + 2 \qquad \textbf{Note that this is not } \log_7 (x + 2).$$

$$= \frac{\ln x}{\ln 7} + 2 \qquad \textbf{Using the change-of-base formula for } \log_7 x$$

We graph $y = \ln(x)/\ln(7) + 2$.

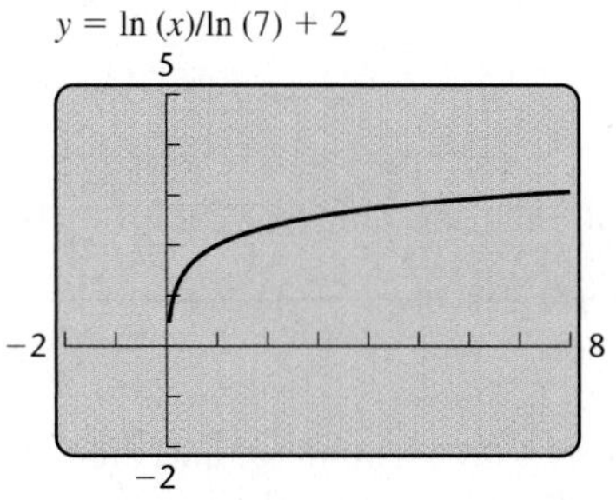

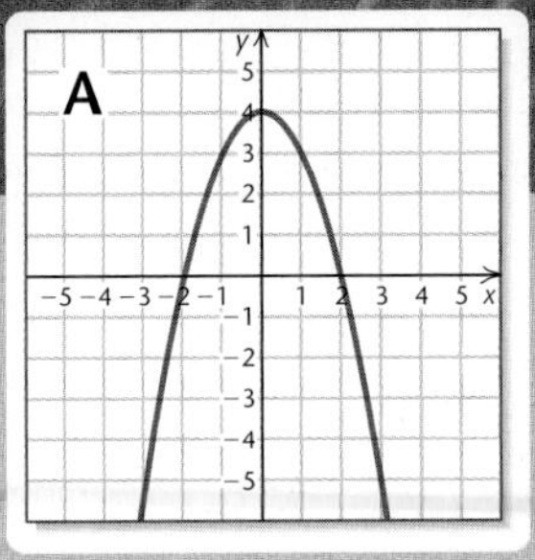

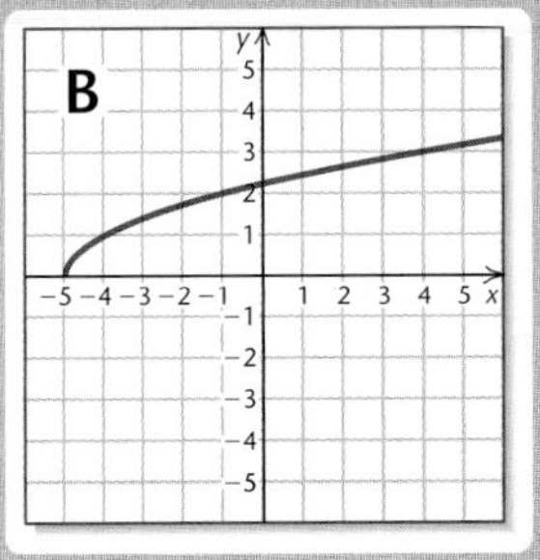

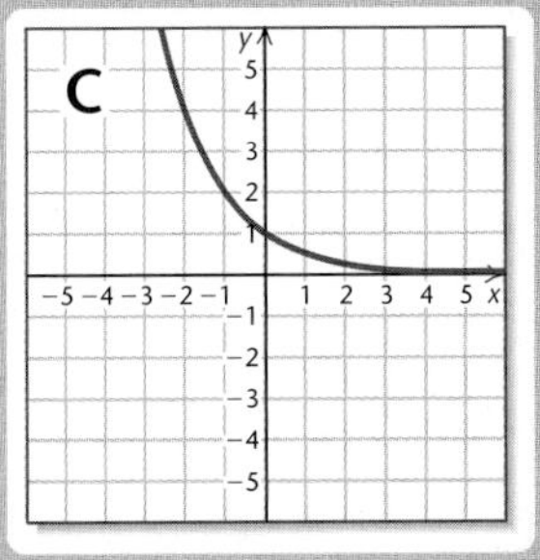

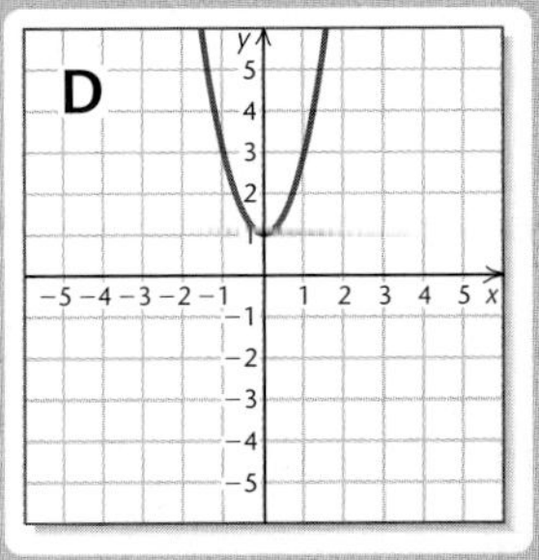

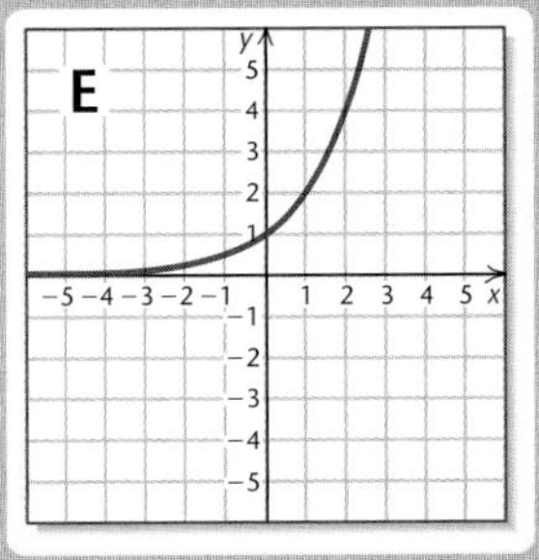

Visualizing the Graph

Match each function with its graph.

1. $f(x) = 2x - 3$

2. $f(x) = 2x^2 + 1$

3. $f(x) = \sqrt{x + 5}$

4. $f(x) = |4 - x|$

5. $f(x) = \ln x$

6. $f(x) = 2^{-x}$

7. $f(x) = -4$

8. $f(x) = \log x + 3$

9. $f(x) = 2^x$

10. $f(x) = 4 - x^2$

Answers on page A-44

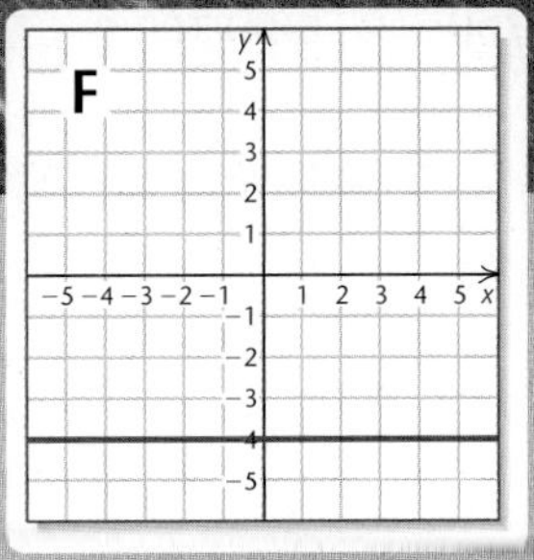

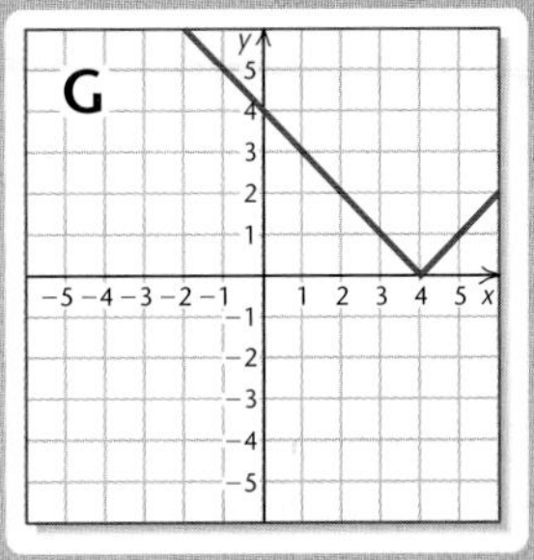

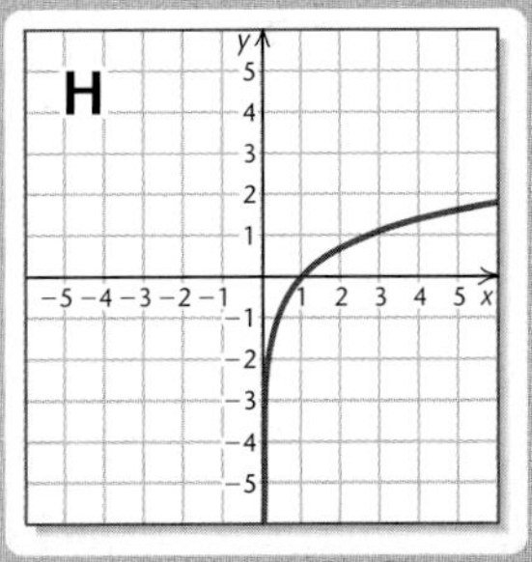

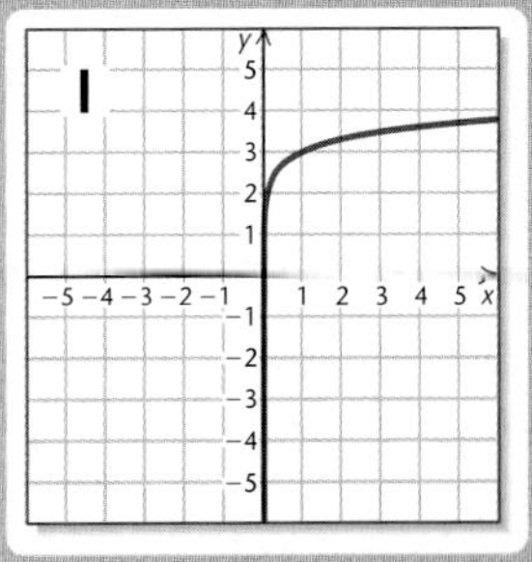

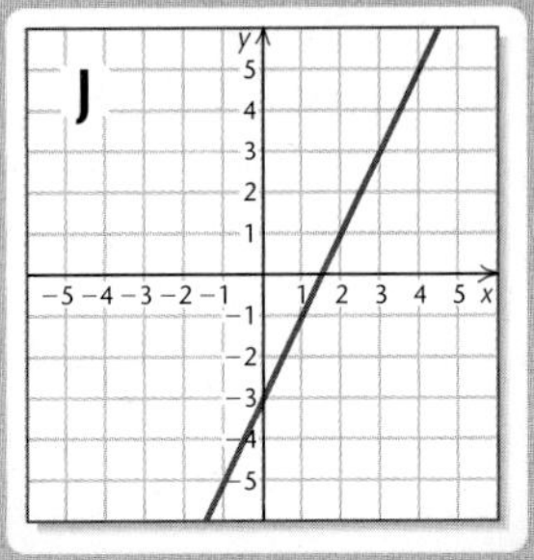

9.5 EXERCISE SET

FOR EXTRA HELP

MathXL | MyMathLab | InterAct Math | AW Math Tutor Center | Video Lectures on CD: Disc 5 | Student's Solutions Manual

Concept Reinforcement *Classify each of the following as either true or false.*

1. The expression log 23 means $\log_{10} 23$.
2. The expression ln 7 means $\log_e 7$.
3. The number e is approximately 2.7.
4. The expressions log 9 and log 18/log 2 are equivalent.
5. The expressions $\log_2 9$ and ln 9/ln 2 are equivalent.
6. The domain of the function given by $f(x) = \ln(x + 2)$ is $(-2, \infty)$.
7. The range of the function given by $g(x) = e^x$ is $(0, \infty)$.
8. The range of the function given by $f(x) = \ln x$ is $(-\infty, \infty)$.

Use a calculator to find each of the following to the nearest ten-thousandth.

9. log 6
10. log 5
11. log 72.8
12. log 73.9
13. (Aha!) log 1000
14. log 100
15. log 0.527
16. log 0.493
17. $\dfrac{\log 8200}{\log 150}$
18. $\dfrac{\log 5700}{\log 90}$
19. $10^{2.3}$
20. $10^{3.4}$
21. $10^{0.173}$
22. $10^{0.247}$
23. $10^{-2.9523}$
24. $10^{-3.2046}$
25. ln 5
26. ln 2
27. ln 57
28. ln 30
29. ln 0.0062
30. ln 0.00073
31. $\dfrac{\ln 2300}{0.08}$
32. $\dfrac{\ln 1900}{0.07}$
33. $e^{2.71}$
34. $e^{3.06}$
35. $e^{-3.49}$
36. $e^{-2.64}$
37. $e^{4.7}$
38. $e^{1.23}$

Find each of the following logarithms using the change-of-base formula. Round answers to the nearest ten-thousandth.

39. $\log_6 92$
40. $\log_3 78$
41. $\log_2 100$
42. $\log_7 100$
43. $\log_7 65$
44. $\log_5 42$
45. $\log_{0.5} 5$
46. $\log_{0.1} 3$
47. $\log_2 0.2$
48. $\log_2 0.08$
49. $\log_\pi 58$
50. $\log_\pi 200$

Graph by hand or using a graphing calculator and state the domain and the range of each function.

51. $f(x) = e^x$
52. $f(x) = e^{-x}$
53. $f(x) = e^x + 3$
54. $f(x) = e^x + 2$
55. $f(x) = e^x - 2$
56. $f(x) = e^x - 3$
57. $f(x) = 0.5e^x$
58. $f(x) = 2e^x$
59. $f(x) = 0.5e^{2x}$
60. $f(x) = 2e^{-0.5x}$
61. $f(x) = e^{x-3}$
62. $f(x) = e^{x-2}$
63. $f(x) = e^{x+2}$
64. $f(x) = e^{x+3}$
65. $f(x) = -e^x$
66. $f(x) = -e^{-x}$
67. $g(x) = \ln x + 1$
68. $g(x) = \ln x + 3$
69. $g(x) = \ln x - 2$
70. $g(x) = \ln x - 1$
71. $g(x) = 2 \ln x$
72. $g(x) = 3 \ln x$
73. $g(x) = -2 \ln x$
74. $g(x) = -\ln x$
75. $g(x) = \ln(x + 2)$
76. $g(x) = \ln(x + 1)$
77. $g(x) = \ln(x - 1)$
78. $g(x) = \ln(x - 3)$

Write an equivalent expression for the function that could be graphed using a graphing calculator. Then graph the function.

79. $f(x) = \log_5 x$

80. $f(x) = \log_3 x$

81. $f(x) = \log_2 (x - 5)$

82. $f(x) = \log_5 (2x + 1)$

83. $f(x) = \log_3 x + x$

84. $f(x) = \log_2 x - x + 1$

TW **85.** Using a calculator, Zeno *incorrectly* says that log 79 is between 4 and 5. How could you convince him, without using a calculator, that he is mistaken?

TW **86.** Examine Exercise 85. What mistake do you believe Zeno made?

Skill Maintenance

87. Find an equation of variation if y varies directly as x, and $y = 7.2$ when $x = 0.8$. [6.8]

88. Find an equation of variation if y varies inversely as x, and $y = 3.5$ when $x = 6.1$. [6.8]

Solve. [8.3]

89. $T = 2\pi\sqrt{L/32}$, for L

90. $E = mc^2$, for c
(Assume $E, m, c > 0$.)

91. Joni can key in a musical score in 2 hr. Miles takes 3 hr to key in the same score. How long would it take them, working together, to key in the score? [6.5]

TW **92.** The side exit at the Flynn Theater can empty a capacity crowd in 25 min. The main exit can empty a capacity crowd in 15 min. How long will it take to empty a capacity crowd when both exits are in use? [6.5]

Synthesis

TW **93.** In an attempt to solve $\ln x = 1.5$, Emma gets the following graph. How can Emma tell at a glance that she has made a mistake?

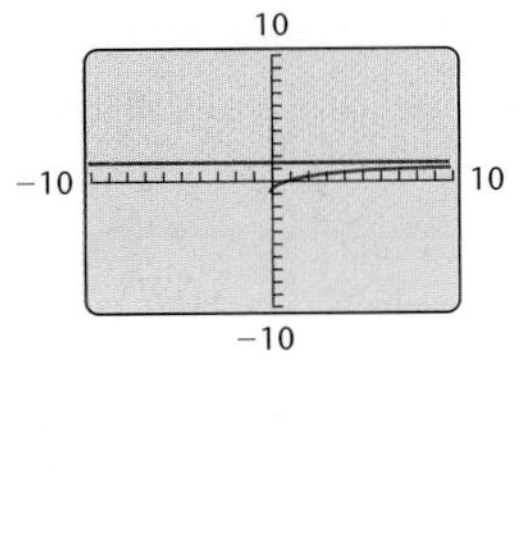

TW **94.** Explain how the graph of $f(x) = \ln x$ could be used to graph the function given by $g(x) = e^{x-1}$.

Knowing only that $\log 2 \approx 0.301$ *and* $\log 3 \approx 0.477$, *find each of the following.*

95. $\log_6 81$

96. $\log_9 16$

97. $\log_{12} 36$

98. Find a formula for converting common logarithms to natural logarithms.

99. Find a formula for converting natural logarithms to common logarithms.

Solve for x.

100. $\log (275x^2) = 38$

101. $\log (492x) = 5.728$

102. $\dfrac{3.01}{\ln x} = \dfrac{28}{4.31}$

103. $\log 692 + \log x = \log 3450$

For each function given below, **(a)** *determine the domain and the range,* **(b)** *set an appropriate window, and* **(c)** *draw the graph.*

104. $f(x) = 7.4e^x \ln x$

105. $f(x) = 3.4 \ln x - 0.25e^x$

106. $f(x) = x \ln (x - 2.1)$

107. $f(x) = 2x^3 \ln x$

9.6 Solving Exponential and Logarithmic Equations

Solving Exponential Equations ■ Solving Logarithmic Equations

Solving Exponential Equations

Equations with variables in exponents, such as $5^x = 12$ and $2^{7x} = 64$, are called **exponential equations.** In Section 9.3, we solved certain logarithmic equations by using the principle of exponential equality. We restate that principle below.

Study Tip

Abbrevs. Cn Help U Go Fst

If you take notes and have trouble keeping up with your instructor, use abbreviations to speed up your work. Consider standard abbreviations like "Ex" for "Example," "$\approx$" for "is approximately equal to," "$\therefore$" for "therefore," and "$\Rightarrow$" for "implies." Feel free to create your own abbreviations as well.

The Principle of Exponential Equality For any real number b, where $b \neq -1, 0,$ or 1,

$$b^x = b^y \quad \text{is equivalent to} \quad x = y.$$

(Powers of the same base are equal if and only if the exponents are equal.)

EXAMPLE 1 Solve: $4^{3x-5} = 16$.

SOLUTION Note that $16 = 4^2$. Thus we can write each side as a power of the same number:

$$4^{3x-5} = 4^2.$$

Since the base is the same, 4, the exponents must be the same. Thus,

$$3x - 5 = 2 \quad \textbf{Equating exponents}$$
$$3x = 7$$
$$x = \tfrac{7}{3}.$$

Check:

$$4^{3x-5} = 16$$

4^{3x-5}	16
$4^{3(7/3)-5}$	16
4^{7-5}	

$$4^2 \stackrel{?}{=} 16 \quad \text{TRUE}$$

The solution is $\frac{7}{3}$.

When it seems impossible to write both sides of an equation as powers of the same base, we use the following principle and write an equivalent logarithmic equation.

> **The Principle of Logarithmic Equality** For any logarithmic base a, and for $x, y > 0$,
>
> $$x = y \quad \text{is equivalent to} \quad \log_a x = \log_a y.$$
>
> (Two expressions are equal if and only if the logarithms of those expressions are equal.)

Because calculators can generally find only common or natural logarithms (without resorting to the change-of-base formula), we usually take the common or natural logarithm on both sides of the equation.

The principle of logarithmic equality, used together with the power rule for logarithms, allows us to solve equations with a variable in an exponent.

EXAMPLE 2 Solve: $7^{x-2} = 60$.

ALGEBRAIC APPROACH

We have

$$7^{x-2} = 60$$

$$\log 7^{x-2} = \log 60$$ **Using the principle of logarithmic equality to take the common logarithm on both sides. Natural logarithms also would work.**

$$(x-2)\log 7 = \log 60$$ **Using the power rule for logarithms**

$$x - 2 = \frac{\log 60}{\log 7}$$

> **CAUTION!** This is *not* $\log 60 - \log 7$.

$$x = \frac{\log 60}{\log 7} + 2$$ **Adding 2 to both sides**

$$x \approx 4.1041.$$ **Using a calculator and rounding to four decimal places**

Since $7^{4.1041-2} \approx 60.0027$, we have a check. The solution is $\dfrac{\log 60}{\log 7} + 2$, or approximately 4.1041.

GRAPHICAL APPROACH

We graph $y_1 = 7 \wedge (x-2)$ and $y_2 = 60$.

$y_1 = 7^{x-2}, \quad y_2 = 60$

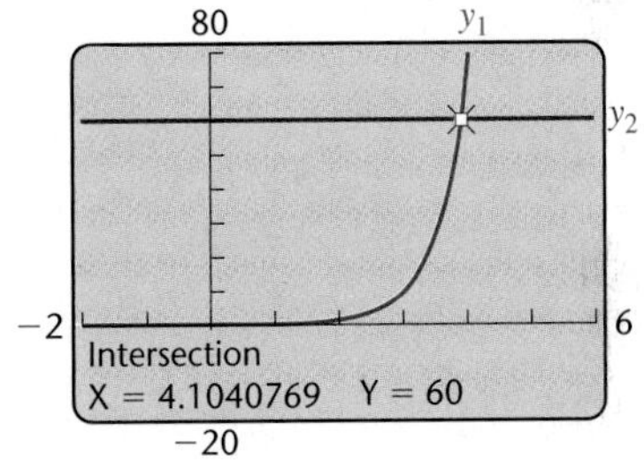

Yscl = 10

Rounded to four decimal places, the x-coordinate of the point of intersection is 4.1041. The solution is 4.1041.

EXAMPLE 3 Solve: $e^{0.06t} = 1500$.

ALGEBRAIC APPROACH

Since one side is a power of e, we take the *natural logarithm* on both sides:

$\ln e^{0.06t} = \ln 1500$	**Taking the natural logarithm on both sides**
$0.06t = \ln 1500$	**Finding the logarithm of the base to a power: $\log_a a^k = k$**
$t = \dfrac{\ln 1500}{0.06}$	**Dividing both sides by 0.06**
$\approx 121.887.$	**Using a calculator and rounding to three decimal places**

The solution is 121.887.

GRAPHICAL APPROACH

We graph $y_1 = e\wedge(0.06x)$ and $y_2 = 1500$. Since $y_2 = 1500$, we choose a value for Ymax that is greater than 1500. It may require trial and error to choose appropriate units for the x-axis.

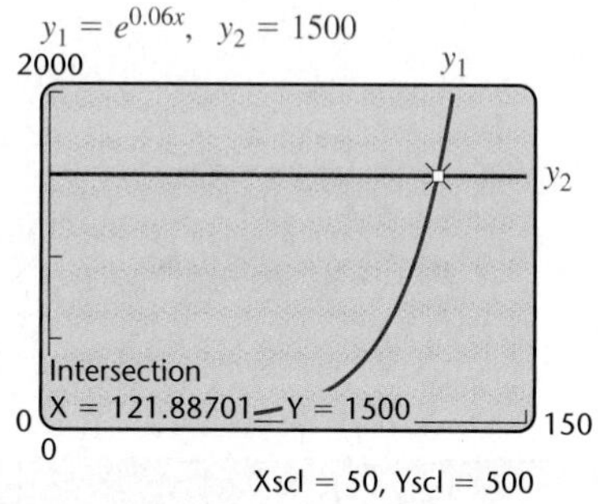

Rounded to three decimal places, the solution is 121.887.

To Solve an Equation of the Form $a^t = b$ for t

1. Take the logarithm (either natural or common) of both sides.
2. Use the power rule for exponents so that the variable is no longer written as an exponent.
3. Divide both sides by the coefficient of the variable to isolate the variable.
4. If appropriate, use a calculator to find an approximate solution in decimal form.

Some equations, like the one in Example 3, are readily solved algebraically. There are other exponential equations for which we do not have the tools to solve algebraically, but we can nevertheless solve graphically.

EXAMPLE 4 Solve: $xe^{3x-1} = 5$.

SOLUTION We graph $y_1 = xe\wedge(3x - 1)$ and $y_2 = 5$ and determine the coordinates of any points of intersection.

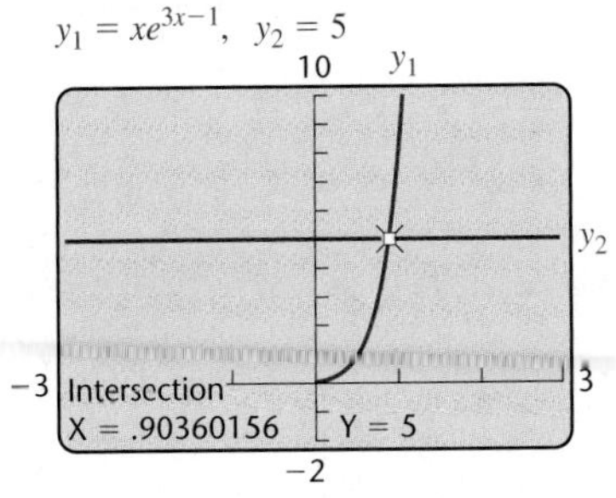

The x-coordinate of the point of intersection is approximately 0.90360156. Thus the solution is approximately 0.904.

Solving Logarithmic Equations

Equations containing logarithmic expressions are called **logarithmic equations.** We saw in Section 9.3 that certain logarithmic equations can be solved by writing an equivalent exponential equation.

EXAMPLE 5 Solve: **(a)** $\log_4 (8x - 6) = 3$; **(b)** $\ln 5x = 27$.

SOLUTION

a) $\log_4 (8x - 6) = 3$

$4^3 = 8x - 6$ **Writing the equivalent exponential equation**

$64 = 8x - 6$

$70 = 8x$ **Adding 6 to both sides**

$x = \frac{70}{8}$, or $\frac{35}{4}$.

Check:

$\log_4 (8x - 6) = 3$	
$\log_4 \left(8 \cdot \frac{35}{4} - 6\right)$	3
$\log_4 (2 \cdot 35 - 6)$	
$\log_4 64$	
$3 \overset{?}{=} 3$	TRUE

The solution is $\frac{35}{4}$.

b) $\ln 5x = 27$ **Remember: ln 5x means $\log_e 5x$.**

$e^{27} = 5x$ **Writing the equivalent exponential equation**

$\frac{e^{27}}{5} = x$ **This is a very large number.**

The solution is $\frac{e^{27}}{5}$. The check is left to the student.

Often the properties for logarithms are needed. The goal is to first write an equivalent equation in which the variable appears in just one logarithmic expression. We then isolate that expression and solve as in Example 5. Since logarithms of negative numbers are not defined, all possible solutions must be checked.

EXAMPLE 6 Solve: $\log x + \log (x - 3) = 1$.

ALGEBRAIC APPROACH

To increase understanding, we write in the base, 10.

$$\log_{10} x + \log_{10} (x - 3) = 1$$

$$\log_{10} [x(x - 3)] = 1 \quad \text{Using the product rule for logarithms to obtain a single logarithm}$$

$$x(x - 3) = 10^1 \quad \text{Writing an equivalent exponential equation}$$

$$x^2 - 3x = 10$$

$$x^2 - 3x - 10 = 0$$

$$(x + 2)(x - 5) = 0 \quad \text{Factoring}$$

$$x + 2 = 0 \quad \text{or} \quad x - 5 = 0 \quad \text{Using the principle of zero products}$$

$$x = -2 \quad \text{or} \quad x = 5$$

The possible solutions are -2 and 5.

GRAPHICAL APPROACH

We graph $y_1 = \log (x) + \log (x - 3)$ and $y_2 = 1$ and determine the coordinates of any points of intersection.

$y_1 = \log (x) + \log (x - 3), \quad y_2 = 1$

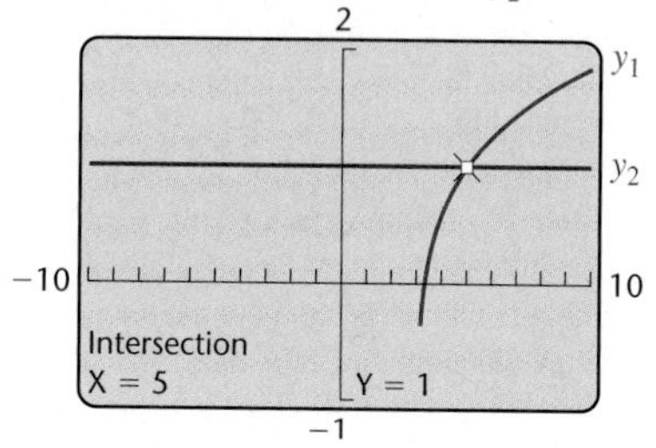

There is one point of intersection, (5, 1). The solution is the first coordinate of that point, or 5.

Note that the algebraic approach resulted in two possible solutions and the graphical approach in only one. To check, we substitute both -2 and 5 in the original equation.

Check:

For -2:

$$\frac{\log x + \log (x - 3) = 1}{\log (-2) + \log (-2 - 3) \ ? \ 1} \quad \text{FALSE}$$

For 5:

$$\log x + \log (x - 3) = 1$$

$$\log 5 + \log (5 - 3) \mid 1$$

$$\log 5 + \log 2$$

$$\log 10$$

$$1 \stackrel{?}{=} 1 \quad \text{TRUE}$$

The number -2 *does not check* because the logarithm of a negative number is undefined. The solution is 5.

Student Notes

It is essential that you remember the properties of logarithms from Section 9.4. Consider reviewing the properties before attempting to solve equations similar to those in Examples 6–8.

EXAMPLE 7 Solve: $\log_2 (x + 7) - \log_2 (x - 7) = 3$.

ALGEBRAIC APPROACH

We have

$$\log_2 (x + 7) - \log_2 (x - 7) = 3$$

$$\log_2 \frac{x + 7}{x - 7} = 3$$ **Using the quotient rule for logarithms to obtain a single logarithm**

$$\frac{x + 7}{x - 7} = 2^3$$ **Writing an equivalent exponential equation**

$$\frac{x + 7}{x - 7} = 8$$

$$x + 7 = 8(x - 7)$$ **Multiplying by the LCD, $x - 7$**

$$x + 7 = 8x - 56$$ **Using the distributive law**

$$63 = 7x$$

$$9 = x.$$ **Dividing by 7**

GRAPHICAL APPROACH

We first use the change-of-base formula to write the base-2 logarithms using common logarithms. Then we graph and determine the coordinates of any points of intersection.

$y_1 = \log (x + 7)/\log (2) - \log (x - 7)/\log (2)$,
$y_2 = 3$

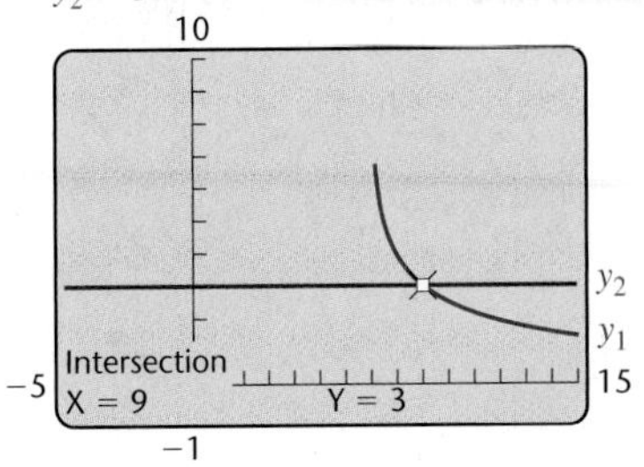

The graphs intersect at (9, 3). The solution is 9.

Check:

$\log_2 (x + 7) - \log_2 (x - 7)$	$= 3$
$\log_2 (9 + 7) - \log_2 (9 - 7)$	3
$\log_2 16 - \log_2 2$	
$4 - 1$	
$3 \stackrel{?}{=} 3$	TRUE

The solution is 9.

EXAMPLE 8 Solve: $\log_7 (x + 1) + \log_7 (x - 1) = \log_7 8$.

ALGEBRAIC APPROACH

We have

$$\log_7 (x + 1) + \log_7 (x - 1) = \log_7 8$$

$$\log_7 [(x + 1)(x - 1)] = \log_7 8$$ **Using the product rule for logarithms**

$$\log_7 (x^2 - 1) = \log_7 8$$ **Multiplying. Note that both sides are base-7 logarithms.**

$$x^2 - 1 = 8$$ **Using the principle of logarithmic equality. Study this step carefully.**

$$x^2 - 9 = 0$$

$$(x - 3)(x + 3) = 0$$ **Solving the quadratic equation**

$$x = 3 \quad or \quad x = -3.$$

The student should confirm that 3 checks but -3 does not. The solution is 3.

GRAPHICAL APPROACH

Using the change-of-base formula, we graph

$$y_1 = \log(x+1)/\log(7) + \log(x-1)/\log(7)$$

and

$$y_2 = \log(8)/\log(7).$$

$y_1 = \log(x+1)/\log(7) + \log(x-1)/\log(7)$,
$y_2 = \log(8)/\log(7)$

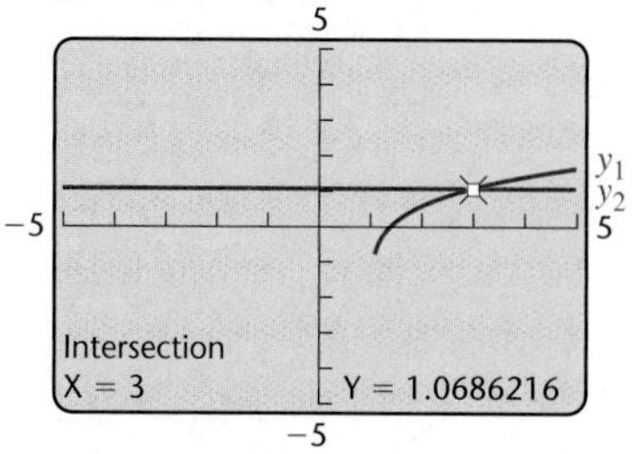

The graphs intersect at (3, 1.0686216). The solution is 3.

9.6 EXERCISE SET

FOR EXTRA HELP

MathXL | MyMathLab | InterAct Math | AW Math Tutor Center | Video Lectures on CD: Disc 5 | Student's Solutions Manual

Concept Reinforcement *In each of Exercises 1–8, match the equation with an equivalent equation from the column on the right that could be the next step in the solution process.*

1. ___ $5^x = 3$

2. ___ $e^{5x} = 3$

3. ___ $\ln x = 3$

4. ___ $\log_x 5 = 3$

5. ___ $\log_5 x + \log_5 (x-2) = 3$

6. ___ $\log_5 x - \log_5 (x-2) = 3$

7. ___ $\ln x - \ln (x-2) = 3$

8. ___ $\log x + \log (x-2) = 3$

a) $\ln e^{5x} = \ln 3$

b) $\log_5 (x^2 - 2x) = 3$

c) $\log (x^2 - 2x) = 3$

d) $\log_5 \dfrac{x}{x-2} = 3$

e) $\log 5^x = \log 3$

f) $e^3 = x$

g) $\ln \dfrac{x}{x-2} = 3$

h) $x^3 = 5$

Solve. Where appropriate, include approximations to the nearest thousandth.

9. $2^x = 19$

10. $2^x = 15$

11. $8^{x-1} = 17$

12. $4^{x+1} = 13$

13. $e^t = 1000$

14. $e^t = 100$

15. $e^{0.03t} + 2 = 7$

16. $e^{-0.07t} + 3 = 3.08$

17. $5 = 3^{x+1}$

18. $7 = 3^{x-1}$

Aha! **19.** $2^{x+3} = 16$

20. $4^{x+1} = 64$

21. $4.9^x - 87 = 0$

22. $7.2^x - 65 = 0$

23. $19 = 2e^{4x}$

24. $29 = 3e^{2x}$

25. $7 + 3e^{5x} = 13$

26. $4 + 5e^{4x} = 9$

Aha! **27.** $\log_3 x = 4$

28. $\log_2 x = 6$

29. $\log_2 x = -3$

30. $\log_5 x = 3$

31. $\ln x = 5$

32. $\ln x = 4$

33. $\log_8 x = \frac{1}{3}$

34. $\log_4 x = \frac{1}{2}$

35. $\ln 4x = 3$

36. $\ln 3x = 2$

37. $\log x = 2.5$

38. $\log x = 0.5$

39. $\ln (2x + 1) = 4$

40. $\ln (4x - 2) = 3$

Aha! **41.** $\ln x = 1$

42. $\log x = 1$

43. $5 \ln x = -15$

44. $3 \ln x = -3$

45. $\log_2 (8 - 6x) = 5$

46. $\log_5 (2x - 7) = 3$

47. $\log (x - 9) + \log x = 1$

48. $\log (x + 9) + \log x = 1$

49. $\log x - \log (x + 3) = 1$

50. $\log x - \log (x + 7) = -1$

51. $\log_4 (x + 3) = 2 + \log_4 (x - 5)$

52. $\log_2 (x + 3) = 4 + \log_2 (x - 3)$

53. $\log_7 (x + 1) + \log_7 (x + 2) = \log_7 6$

54. $\log_6 (x + 3) + \log_6 (x + 2) = \log_6 20$

55. $\log_5 (x + 4) + \log_5 (x - 4) = \log_5 20$

56. $\log_4 (x + 2) + \log_4 (x - 7) = \log_4 10$

57. $\ln (x + 5) + \ln (x + 1) = \ln 12$

58. $\ln (x - 6) + \ln (x + 3) = \ln 22$

59. $\log_2 (x - 3) + \log_2 (x + 3) = 4$

60. $\log_3 (x - 4) + \log_3 (x + 4) = 2$

61. $\log_{12} (x + 5) - \log_{12} (x - 4) = \log_{12} 3$

62. $\log_6 (x + 7) - \log_6 (x - 2) = \log_6 5$

63. $\log_2 (x - 2) + \log_2 x = 3$

64. $\log_4 (x + 6) - \log_4 x = 2$

65. $e^{0.5x} - 7 = 2x + 6$

66. $e^{-x} - 3 = x^2$

67. $\ln (3x) = 3x - 8$

68. $\ln (x^2) = -x^2$

69. Find the value of x for which the natural logarithm is the same as the common logarithm.

70. Find all values of x for which the common logarithm of the square of x is the same as the square of the common logarithm of x.

TW **71.** Could Example 2 have been solved by taking the natural logarithm on both sides? Why or why not?

TW **72.** Christina finds that the solution of $\log_3 (x + 4) = 1$ is -1, but rejects -1 as an answer because it is negative. What mistake is she making?

Focused Review

Solve.

73. $4x^2 - 25 = 0$ [5.6]

74. $5x^2 = 7x$ [5.3]

75. $|8 - 3x| = 4$ [4.4]

76. $9 - 13x = 3x + 9$ [1.6]

77. $x^{1/2} - 6x^{1/4} + 8 = 0$ [8.5]

78. $\sqrt{4x - 4} = \sqrt{x + 4} + 1$ [7.6]

79. $x^2 + 3x + 5 = 0$ [8.2]

80. $2^{x+5} = 8^{4x}$ [9.6]

81. $e^x = 1.5$ [9.6]

82. $\log x + \log (x - 3) = 1$ [9.6]

83. $\frac{4}{x} = 3 + x$ [6.4]

84. $\frac{x + 1}{x + 2} = \frac{x + 3}{x + 4}$ [6.4]

85. $5x - 3y = 16,$
$4x + 2y = 4$ [3.2]

86. $4a - 5b + c = 3,$
$3a - 4b + 2c = 3,$
$a + b - 7c = -2$ [3.4]

Synthesis

TW **87.** Can the principle of logarithmic equality be expanded to include all functions? That is, is the statement "$m = n$ is equivalent to $f(m) = f(n)$" true for any function f? Why or why not?

TW **88.** Explain how Exercises 37 and 38 could be solved using the graph of $f(x) = \log x$.

Solve. If no solution exists, state this.

89. $27^x = 81^{2x-3}$

90. $8^x = 16^{3x+9}$

91. $\log_x (\log_3 27) = 3$

92. $\log_6 (\log_2 x) = 0$

93. $x \log \frac{1}{8} = \log 8$

94. $\log_5 \sqrt{x^2 - 9} = 1$

95. $2^{x^2+4x} = \frac{1}{8}$

96. $\log (\log x) = 5$

97. $\log_5 |x| = 4$

98. $\log x^2 = (\log x)^2$

99. $\log \sqrt{2x} = \sqrt{\log 2x}$

100. $1000^{2x+1} = 100^{3x}$

101. $3^{x^2} \cdot 3^{4x} = \frac{1}{27}$

102. $3^{3x} \cdot 3^{x^2} = 81$

103. $\log x^{\log x} = 25$

104. $3^{2x} - 8 \cdot 3^x + 15 = 0$

105. $(81^{x-2})(27^{x+1}) = 9^{2x-3}$

106. $3^{2x} - 3^{2x-1} = 18$

107. Given that $2^y = 16^{x-3}$ and $3^{y+2} = 27^x$, find the value of $x + y$.

108. If $x = (\log_{125} 5)^{\log_5 125}$, what is the value of $\log_3 x$?

9.7 Applications of Exponential and Logarithmic Functions

Applications of Logarithmic Functions ■ Applications of Exponential Functions

We now consider applications of exponential and logarithmic functions.

Applications of Logarithmic Functions

EXAMPLE 1 Sound Levels. To measure the volume, or "loudness," of a sound, the *decibel* scale is used. The loudness L, in decibels (dB), of a sound is given by

$$L = 10 \cdot \log \frac{I}{I_0},$$

where I is the intensity of the sound, in watts per square meter (W/m^2), and $I_0 = 10^{-12}$ W/m^2. (I_0 is approximately the intensity of the softest sound that can be heard by the human ear.)

a) The intensity of sound inside a racecar reaches 10^1 W/m^2. How loud, in decibels, is the sound level?

b) The Occupational Safety and Health Administration (OSHA) considers sound levels of 85 dB and above unsafe. What is the intensity of such sounds?

Danica Patrick, the first woman to lead an Indianapolis 500

Study Tip

Sorting by Type

When a section contains a variety of problems, try to sort them out by type. For instance, interest compounded continuously, population growth, and the spread of a virus can all be regarded as one type of problem: exponential growth. Once you know how to solve this type of problem, you can focus on determining which problems fall into this category. The solution should then follow in a rather straightforward manner.

SOLUTION

a) To find the loudness, in decibels, we use the above formula:

$$\begin{aligned} L &= 10 \cdot \log \frac{I}{I_0} \\ &= 10 \cdot \log \frac{10^1}{10^{-12}} && \textbf{Substituting} \\ &= 10 \cdot \log 10^{13} && \textbf{Subtracting exponents} \\ &= 10 \cdot 13 && \log 10^a = a \\ &= 130. \end{aligned}$$

The sound inside a racecar reaches 130 dB.

b) We substitute and solve for I:

$$\begin{aligned} L &= 10 \cdot \log \frac{I}{I_0} \\ 85 &= 10 \cdot \log \frac{I}{10^{-12}} && \textbf{Substituting} \\ 8.5 &= \log \frac{I}{10^{-12}} && \textbf{Dividing both sides by 10} \\ 8.5 &= \log I - \log 10^{-12} && \textbf{Using the quotient rule for logarithms} \\ 8.5 &= \log I - (-12) && \log 10^a = a \\ -3.5 &= \log I && \textbf{Adding } -12 \textbf{ to both sides} \\ 10^{-3.5} &= I. && \textbf{Converting to an exponential equation} \end{aligned}$$

Earplugs would be recommended for sounds with intensities exceeding $10^{-3.5}$ W/m^2.

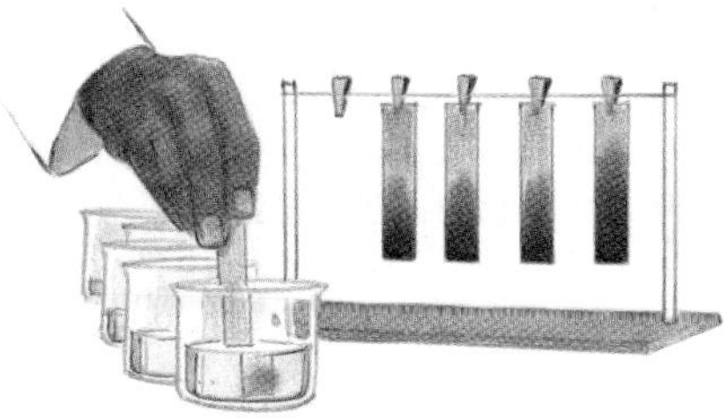

EXAMPLE 2 Chemistry: pH of Liquids. In chemistry, the pH of a liquid is a measure of its acidity. We calculate pH as follows:

$$\text{pH} = -\log\,[\text{H}^+],$$

where $[\text{H}^+]$ is the hydrogen ion concentration in moles per liter.

a) The hydrogen ion concentration of human blood is normally about 3.98×10^{-8} moles per liter. Find the pH.

b) The pH of seawater is about 8.3. Find the hydrogen ion concentration.

SOLUTION

a) To find the pH of blood, we use the above formula:

$$\begin{aligned} \text{pH} &= -\log\,[\text{H}^+] \\ &= -\log\,[3.98 \times 10^{-8}] \\ &\approx -(-7.400117) && \textbf{Using a calculator} \\ &\approx 7.4. \end{aligned}$$

The pH of human blood is normally about 7.4.

b) We substitute and solve for $[H^+]$:

$$\begin{aligned} 8.3 &= -\log [H^+] && \textbf{Using pH} = -\log [H^+] \\ -8.3 &= \log [H^+] && \textbf{Dividing both sides by } -1 \\ 10^{-8.3} &= [H^+] && \textbf{Converting to an exponential equation} \\ 5.01 \times 10^{-9} &\approx [H^+]. && \textbf{Using a calculator; writing scientific notation} \end{aligned}$$

The hydrogen ion concentration of seawater is about 5.01×10^{-9} moles per liter.

Applications of Exponential Functions

EXAMPLE 3 Interest Compounded Annually. Suppose that \$25,000 is invested at 4% interest, compounded annually. (See Example 5 in Section 9.2.) In t years, it will grow to the amount A given by the function

$$A(t) = 25{,}000(1.04)^t.$$

a) How long will it take to accumulate \$80,000 in the account?

b) Find the amount of time it takes for the \$25,000 to double itself.

SOLUTION

a) We set $A(t) = 80{,}000$ and solve for t:

$y_1 = 25{,}000(1.04)^x, \; y_2 = 80{,}000$

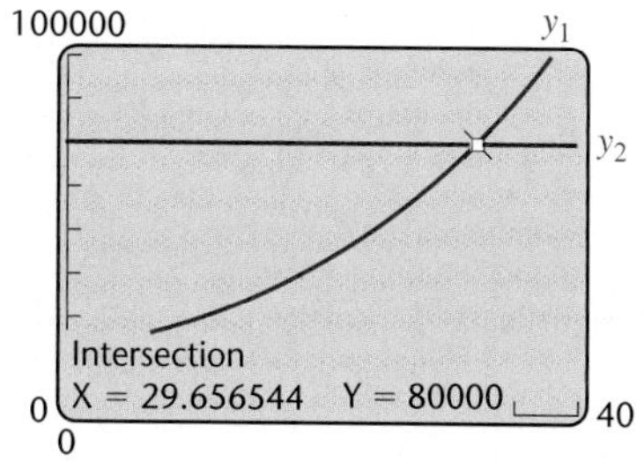

Xscl = 5, Yscl = 10,000

$$\begin{aligned} 80{,}000 &= 25{,}000(1.04)^t \\ \frac{80{,}000}{25{,}000} &= 1.04^t && \textbf{Dividing both sides by 25,000} \\ 3.2 &= 1.04^t \\ \log 3.2 &= \log 1.04^t && \textbf{Taking the common logarithm on both sides} \\ \log 3.2 &= t \log 1.04 && \textbf{Using the power rule for logarithms} \\ \frac{\log 3.2}{\log 1.04} &= t && \textbf{Dividing both sides by log 1.04} \\ 29.7 &\approx t. && \textbf{Using a calculator} \end{aligned}$$

Remember that when doing a calculation like this on a calculator, it is best to wait until the end to round off. We check by solving graphically, as shown at left. At an interest rate of 4% per year, it will take about 29.7 yr for \$25,000 to grow to \$80,000.

b) To find the *doubling time*, we replace $A(t)$ with 50,000 and solve for t:

$$\begin{aligned} 50{,}000 &= 25{,}000(1.04)^t \\ 2 &= (1.04)^t && \textbf{Dividing both sides by 25,000} \\ \log 2 &= \log (1.04)^t && \textbf{Taking the common logarithm on both sides} \\ \log 2 &= t \log 1.04 && \textbf{Using the power rule for logarithms} \\ t &= \frac{\log 2}{\log 1.04} \approx 17.7. && \textbf{Dividing both sides by log 1.04 and using a calculator} \end{aligned}$$

At an interest rate of 4% per year, the doubling time is about 17.7 yr.

Student Notes

Study the different steps in the solution of Example 3(b). Note that if 50,000 and 25,000 are replaced with 8000 and 4000, the doubling time is unchanged.

Like investments, populations often grow exponentially.

Exponential Growth

An **exponential growth model** is a function of the form

$$P(t) = P_0e^{kt}, \quad k > 0,$$

where P_0 is the population at time 0, $P(t)$ is the population at time t, and k is the **exponential growth rate** for the situation. The **doubling time** is the amount of time necessary for the population to double in size.

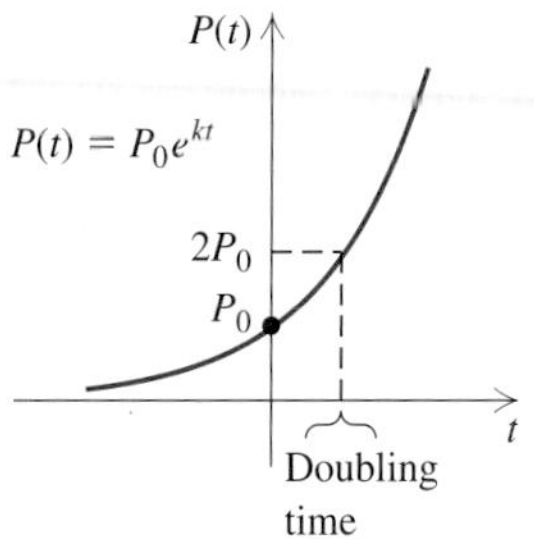

The exponential growth rate is the rate of growth of a population at any *instant* in time. Since the population is continually growing, the percent of total growth after one year will exceed the exponential growth rate.

EXAMPLE 4 Growth of Zebra Mussel Populations. Zebra mussels, inadvertently imported from Europe, began fouling North American waters in 1988. These mussels are so prolific that lake and river bottoms, as well as water intake pipes, can become blanketed with them, altering an entire ecosystem. In 2000, a portion of the Hudson River contained an average of 10 zebra mussels per square mile. The exponential growth rate was 340% per year.

a) Find the exponential growth function that models the data.

b) Predict the number of mussels per square mile in 2007.

SOLUTION

a) In 2000, at $t = 0$, the population was $10/\text{mi}^2$. We substitute 10 for P_0 and 340%, or 3.4, for k. This gives the exponential growth function

$$P(t) = 10e^{3.4t}.$$

b) In 2007, we have $t = 7$ (since 7 yr have passed since 2000). To find the population in 2007, we compute $P(7)$:

$$P(7) = 10e^{3.4(7)} \quad \text{Using } P(t) = 10e^{3.4t} \text{ from part (a)}$$
$$= 10e^{23.8}$$
$$\approx 217{,}000{,}000{,}000. \quad \text{Using a calculator}$$

The population of zebra mussels in the specified portion of the Hudson River will reach approximately 217,000,000,000 per square mile in 2007.

EXAMPLE 5 Spread of a Computer Virus. The number of computers infected by a virus t hours after it first appears usually increases exponentially. In 2004, the "MyDoom" worm spread from 100 computers to about 100,000 computers in 24 hr (*Source*: Based on data from IDG News Service).

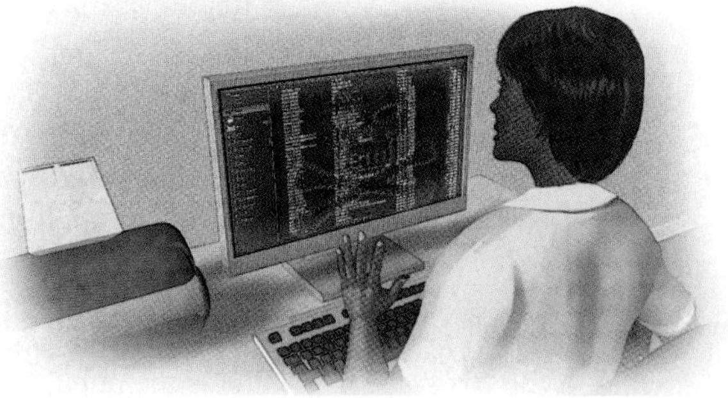

a) Find the exponential growth rate and the exponential growth function.

b) Assuming exponential growth, estimate how long it took the MyDoom worm to infect 9000 computers.

SOLUTION

a) We use $N(t) = N_0 e^{kt}$, where t is the number of hours since the first 100 computers were infected. Substituting 100 for N_0 gives

$$N(t) = 100e^{kt}.$$

To find the exponential growth rate, k, note that after 24 hr, 100,000 computers had been infected:

$$\left.\begin{aligned} N(24) &= 100e^{k \cdot 24} \\ 100{,}000 &= 100e^{24k} \end{aligned}\right\} \quad \textbf{Substituting}$$

$$1000 = e^{24k} \qquad \textbf{Dividing both sides by 100}$$

$$\ln 1000 = \ln e^{24k} \qquad \textbf{Taking the natural logarithm on both sides}$$

$$\ln 1000 = 24k \qquad \ln e^a = a$$

$$\frac{\ln 1000}{24} = k \qquad \textbf{Dividing both sides by 24}$$

$$0.288 \approx k. \qquad \textbf{Using a calculator and rounding}$$

The exponential growth rate is 28.8% and the exponential growth function is given by $N(t) = 100e^{0.288t}$.

b) To estimate how long it took for 9000 computers to be infected, we replace $N(t)$ with 9000 and solve for t:

$$9000 = 100e^{0.288t}$$

$$90 = e^{0.288t} \qquad \textbf{Dividing both sides by 100}$$

$$\ln 90 = \ln e^{0.288t} \qquad \textbf{Taking the natural logarithm on both sides}$$

$$\ln 90 = 0.288t \qquad \ln e^a = a$$

$$\frac{\ln 90}{0.288} = t \qquad \textbf{Dividing both sides by 0.288}$$

$$15.6 \approx t. \qquad \textbf{Using a calculator}$$

Rounding up to 16, we see that, according to this model, it took about 16 hr for 9000 computers to be infected.

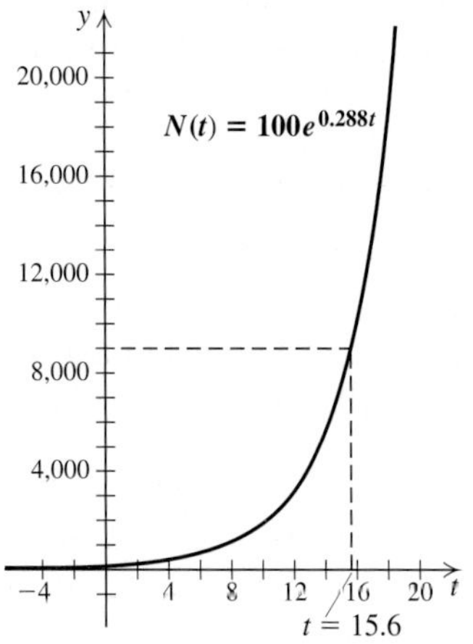

A graphical solution of Example 5

EXAMPLE 6 Interest Compounded Continuously. When an amount of money P_0 is invested at interest rate k, compounded *continuously*, interest is computed every "instant" and added to the original amount. The balance $P(t)$, after t years, is given by the exponential growth model

$$P(t) = P_0 e^{kt}.$$

a) Suppose that \$30,000 is invested and grows to \$44,754.75 in 5 yr. Find the exponential growth function.

b) What is the doubling time?

SOLUTION

a) We have $P(0) = 30{,}000$. Thus the exponential growth function is

$$P(t) = 30{,}000e^{kt}, \quad \text{where } k \text{ must still be determined.}$$

Knowing that for $t = 5$ we have $P(5) = 44{,}754.75$, it is possible to solve for k:

$$44{,}754.75 = 30{,}000e^{k(5)} = 30{,}000e^{5k}$$

$$\frac{44{,}754.75}{30{,}000} = e^{5k} \qquad \textbf{Dividing both sides by 30,000}$$

$$1.491825 = e^{5k}$$

$$\ln 1.491825 = \ln e^{5k} \qquad \textbf{Taking the natural logarithm on both sides}$$

$$\ln 1.491825 = 5k \qquad \ln e^a = a$$

$$\frac{\ln 1.491825}{5} = k \qquad \textbf{Dividing both sides by 5}$$

$$0.08 \approx k. \qquad \textbf{Using a calculator and rounding}$$

The interest rate is about 0.08, or 8%, compounded continuously. Because interest is being compounded continuously, the yearly interest rate is a bit more than 8%. The exponential growth function is

$$P(t) = 30{,}000e^{0.08t}.$$

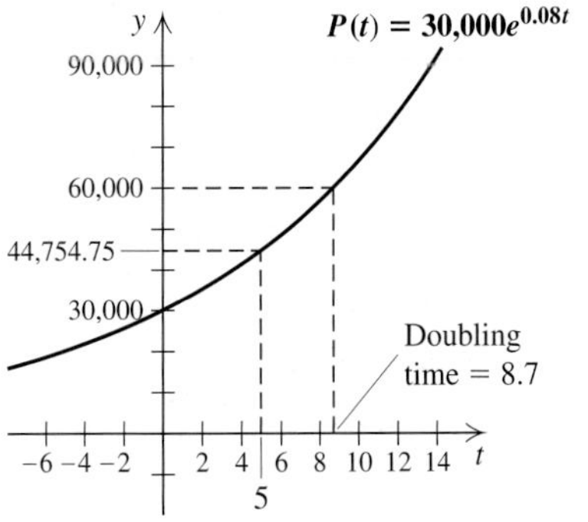

A graphical solution of Example 6

b) To find the doubling time T, we replace $P(T)$ with 60,000 and solve for T:

$$60{,}000 = 30{,}000e^{0.08T}$$

$$2 = e^{0.08T} \qquad \textbf{Dividing both sides by 30,000}$$

$$\ln 2 = \ln e^{0.08T} \qquad \textbf{Taking the natural logarithm on both sides}$$

$$\ln 2 = 0.08T \qquad \ln e^a = a$$

$$\frac{\ln 2}{0.08} = T \qquad \textbf{Dividing both sides by 0.08}$$

$$8.7 \approx T. \qquad \textbf{Using a calculator and rounding}$$

Thus the original investment of \$30,000 will double in about 8.7 yr.

For any specified interest rate, continuous compounding gives the highest yield and the shortest doubling time.

In some real-life situations, a quantity or population is *decreasing* or *decaying* exponentially.

Exponential Decay

An **exponential decay model** is a function of the form

$$P(t) = P_0 e^{-kt}, \quad k > 0,$$

where P_0 is the quantity present at time 0, $P(t)$ is the amount present at time t, and k is the **decay rate.** The **half-life** is the amount of time necessary for half of the quantity to decay.

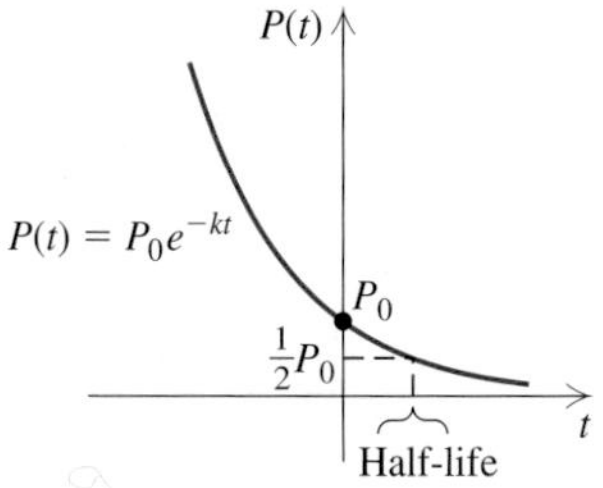

Chaco Canyon, New Mexico

EXAMPLE 7 Carbon Dating. The radioactive element carbon-14 has a half-life of 5750 yr. The percentage of carbon-14 present in the remains of organic matter can be used to determine the age of that organic matter. Recently, while digging in Chaco Canyon, New Mexico, archaeologists found corn pollen that had lost 38.1% of its carbon-14. The age of this corn pollen was evidence that Indians had been cultivating crops in the Southwest centuries earlier than scientists had thought. What was the age of the pollen? (*Source*: *American Anthropologist*)

SOLUTION We first find k. To do so, we use the concept of half-life. When $t = 5750$ (the half-life), $P(t)$ will be half of P_0. Then

$$0.5P_0 = P_0 e^{-k(5750)} \qquad \text{Substituting in } P(t) = P_0 e^{-kt}$$
$$0.5 = e^{-5750k} \qquad \text{Dividing both sides by } P_0$$
$$\ln 0.5 = \ln e^{-5750k} \qquad \text{Taking the natural logarithm on both sides}$$
$$\ln 0.5 = -5750k \qquad \ln e^a = a$$
$$\frac{\ln 0.5}{-5750} = k \qquad \text{Dividing}$$
$$0.00012 \approx k. \qquad \text{Using a calculator and rounding}$$

Now we have a function for the decay of carbon-14:

$$P(t) = P_0 e^{-0.00012t}. \longleftarrow \text{This completes the first part of our solution.}$$

If the corn pollen has lost 38.1% of its carbon-14 from an initial amount P_0, then 100% − 38.1%, or 61.9%, of P_0 is still present. To find the age t of the pollen, we solve this equation for t:

$$0.619P_0 = P_0 e^{-0.00012t} \qquad \text{We want to find } t \text{ for which } P(t) = 0.619P_0.$$
$$0.619 = e^{-0.00012t} \qquad \text{Dividing both sides by } P_0$$
$$\ln 0.619 = \ln e^{-0.00012t} \qquad \text{Taking the natural logarithm on both sides}$$
$$\ln 0.619 = -0.00012t \qquad \ln e^a = a$$
$$\frac{\ln 0.619}{-0.00012} = t \qquad \text{Dividing}$$
$$4000 \approx t. \qquad \text{Using a calculator}$$

The pollen is about 4000 yr old.

The equation

$$P(t) = P_0 e^{-0.00012t}$$

can be used for any subsequent carbon-dating problem.

By looking at the graph of a set of data, we can tell whether a population or other quantity is growing or decaying exponentially.

EXAMPLE 8 For each of the following graphs, determine whether an exponential function might fit the data.

a)

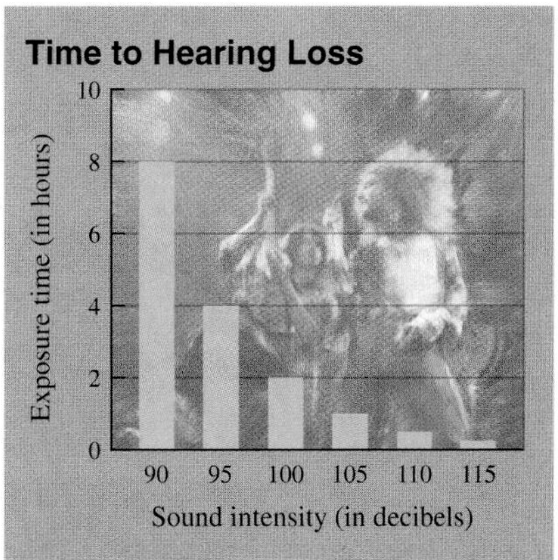

Source: American Hearing Research Foundation

b)

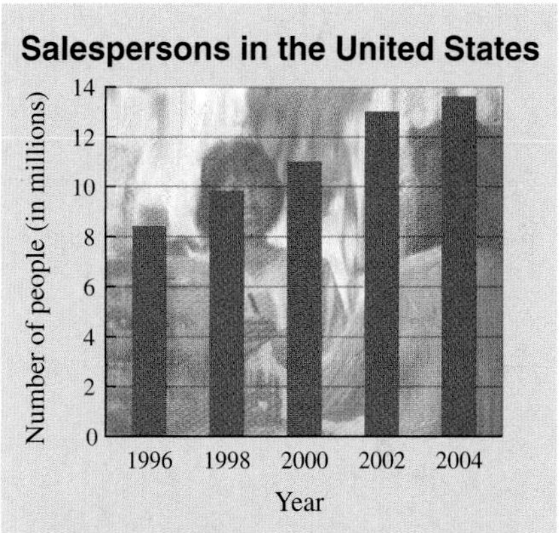

Source: Direct Selling Association

c)

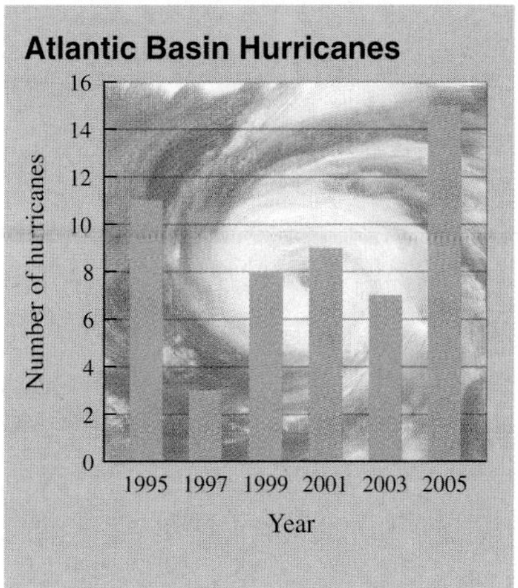

Source: NOAA

d)

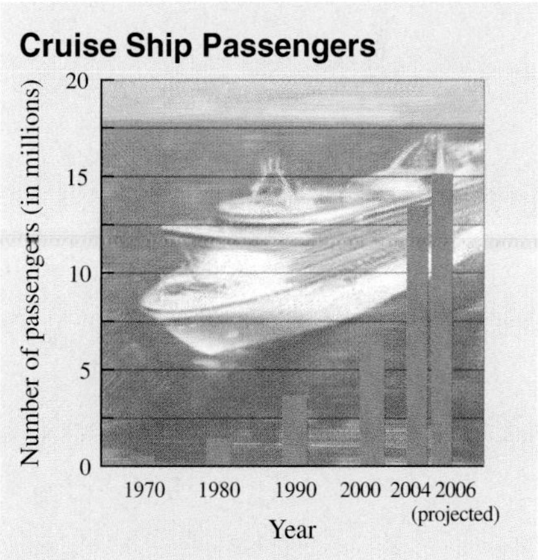

Sources: Cruise Lines International; *Annual Cruise Line Review*, 2004

Student Notes

Like linear functions, two points determine an exponential function. It is important to determine whether data follow a linear or an exponential model (if either) before fitting a function to the data.

SOLUTION

a) As the sound intensity increases, the safe exposure time decreases. The amount of decrease gets smaller as the intensity increases. It appears that an exponential decay function might fit the data.

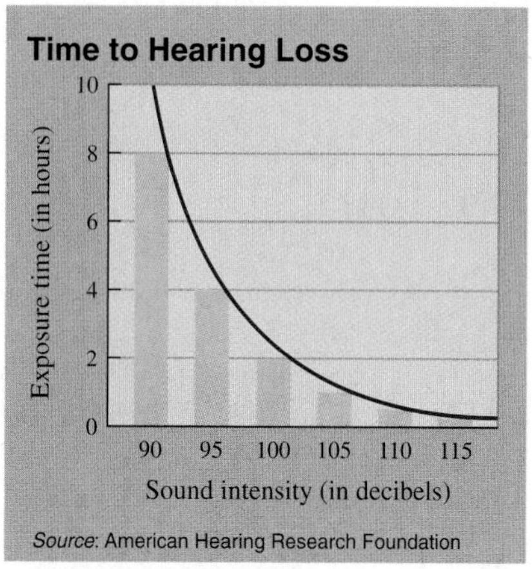

b) The number of salespersons increased between 1996 and 2004 at approximately the same rate each year. It does not appear that an exponential function models the data; instead, a linear function might be appropriate.

c) The number of hurricanes first fell, then rose, then fell again, then rose again. This does not fit an exponential model.

d) The number of cruise-ship passengers increased from 1970 to 2006, and the amount of yearly growth also increased during that time. This suggests that an exponential growth function could be used to model this situation.

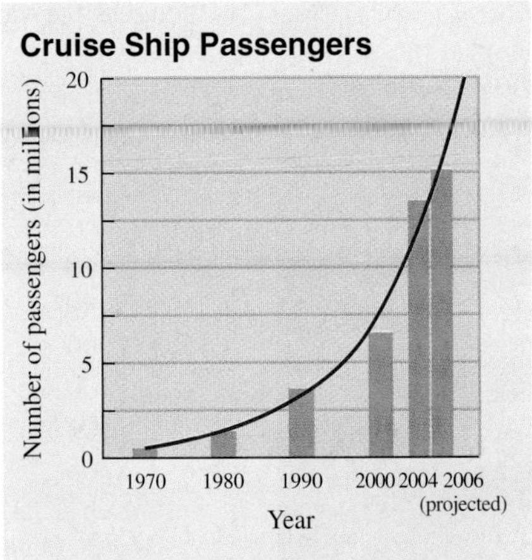

Sources: Cruise Lines International; *Annual Cruise Line Review*, 2004

Exponential Regression

To fit an exponential function to a set of data, we use the ExpReg option in the STAT CALC menu. After entering the data, with the independent variable as L1 and the dependent variable as L2, choose the ExpReg option. For the data shown on the left below, the calculator will return a screen like that shown on the right below.

L1	L2	L3 1
1	12	------
2	60	
3	300	
4	1500	
5	7500	
6	37500	
------	-------	

L1(1)=1

ExpReg
y=a*b^x
a=2.4
b=5

The exponential function found is of the form $f(x) = ab^x$.

We can use the function in this form or convert it to an exponential function of the form $f(x) = ae^{kx}$. The number a remains unchanged. To find k, we solve the equation $b^x = e^{kx}$ for k:

$$b^x = e^{kx}$$

$$b^x = (e^k)^x \quad \textbf{Writing both sides as a power of } x$$

$$b = e^k \quad \textbf{The bases must be equal.}$$

$$\ln b = \ln e^k \quad \textbf{Taking the natural logarithm on both sides}$$

$$\ln b = k. \quad \ln e^k = k$$

Thus to write $f(x) = 2.4(5)^x$ in the form $f(x) = ae^{kx}$, we can write $f(x) = 2.4e^{(\ln 5)x}$, or

$$f(x) \approx 2.4e^{1.609x}.$$

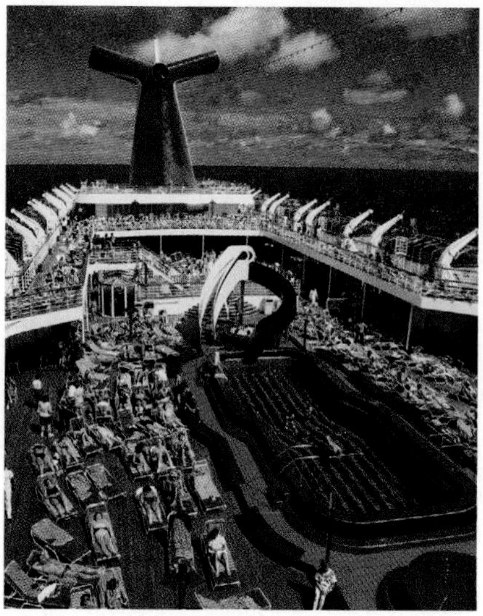

EXAMPLE 9 Cruise-Ship Passengers. In 1970, cruise lines carried approximately 500,000 passengers. This number was projected to increase to 15 million in 2006. The following table and graph show the number of passengers for various years.

Year	Number of Passengers (in millions)
1970	0.5
1980	1.4
1990	3.6
2000	6.5
2004	13.4
2006	15 (proj.)

Sources: Cruise Lines International; *Annual Cruise Review* 2004

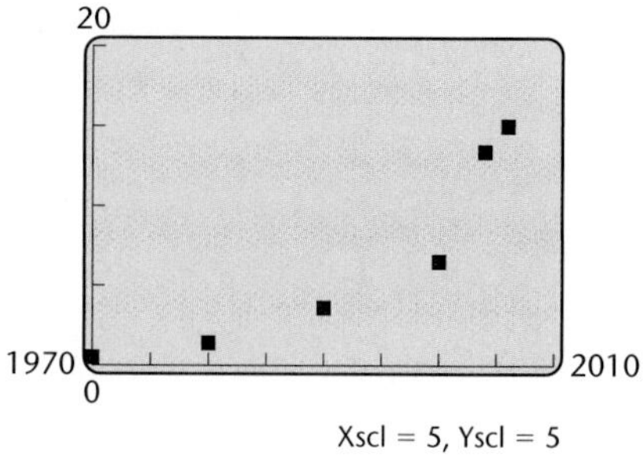

a) Use regression to fit an exponential function to the data and graph the function.

b) Determine the exponential growth rate.

c) The cruise industry projects that in 2012 there will be 20 million cruise-ship passengers. Use the exponential function from part (a) to estimate the number of passengers in 2012. How does this estimate correspond to the cruise-industry projection?

SOLUTION

a) We let x represent the number of years since 1970, and enter the data, as shown on the left below. Then we use regression to find the function and copy it as y_1:

$$f(x) = 0.5240654694(1.096501437)^x.$$

The graph of the function, along with the data points, is shown on the right below.

L1	L2	L3 1
0	.5	------
10	1.4	
20	3.6	
30	6.5	
34	13.4	
36	15	
------	------	

L1(1) = 0

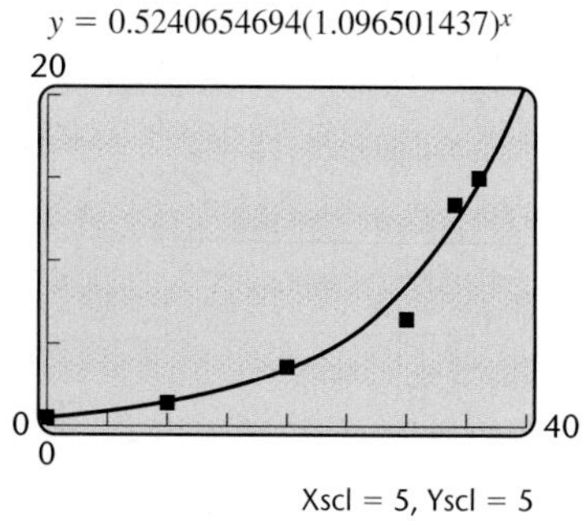

b) The exponential growth rate is the number k in the equation $f(x) = ae^{kx}$. To write $f(x) = 0.5240654694(1.096501437)^x$ in the form $f(x) = ae^{kx}$, we find k:

$$k = \ln b \approx \ln 1.096501437 \approx 0.0921245994.$$

Thus the exponential growth rate is approximately 0.092, or 9.2%.

c) Since the year 2012 is 42 yr after 1970, we find $f(42)$ using a table of values or Y1(42):

$$f(42) \approx 25.$$

Using the function, we estimate that there will be 25 million cruise-ship passengers in 2012. This is 5 million more than the cruise-industry projection for that year.

Connecting the Concepts

We can now add the exponential functions $f(t) = P_0e^{kt}$ and $f(t) = P_0e^{-kt}$, $k > 0$, and the logarithmic function $f(x) = \log_b x$, $b > 1$, to our library of functions.

Linear function:
$f(x) = mx + b$

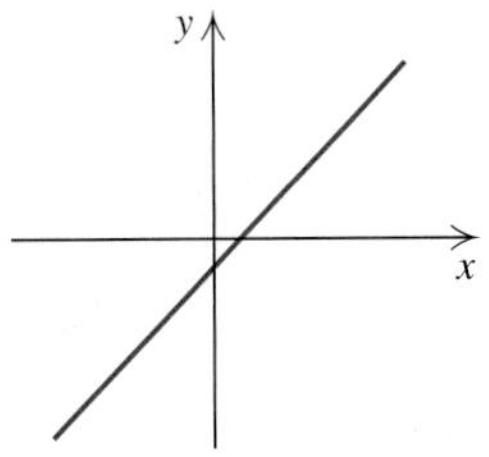

Absolute-value function:
$f(x) = |x|$

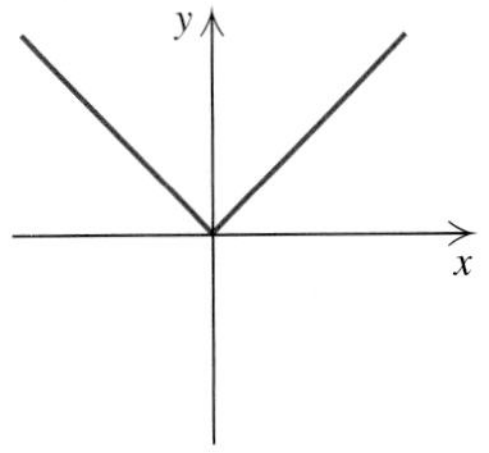

Rational function:
$f(x) = \frac{1}{x}$

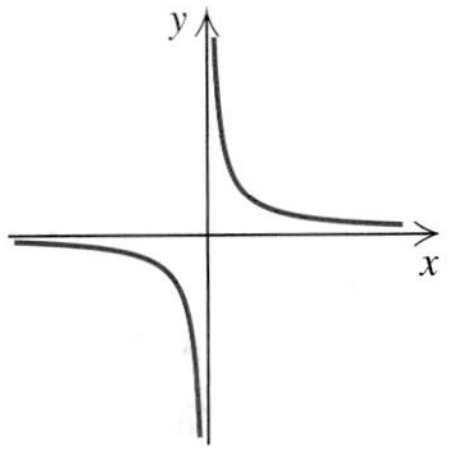

Radical function:
$f(x) = \sqrt{x}$

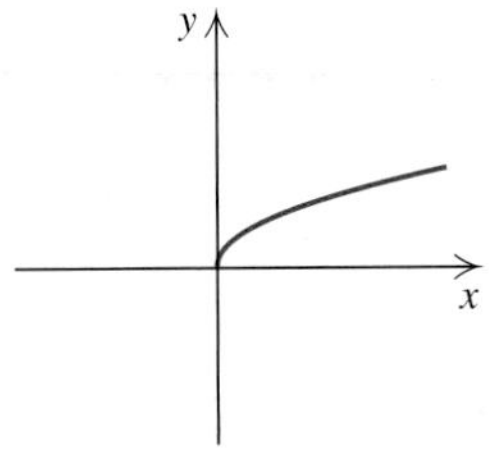

Quadratic function:
$f(x) = ax^2 + bx + c, a > 0$

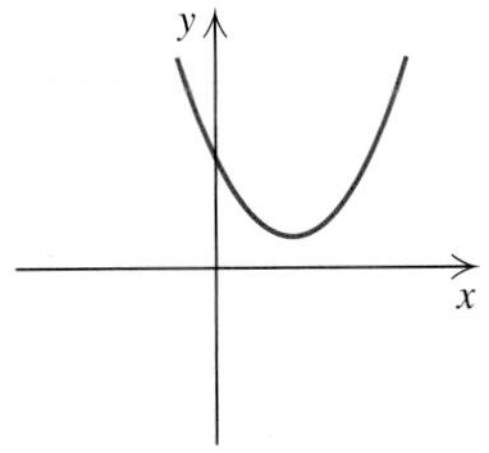

Quadratic function:
$f(x) = ax^2 + bx + c, a < 0$

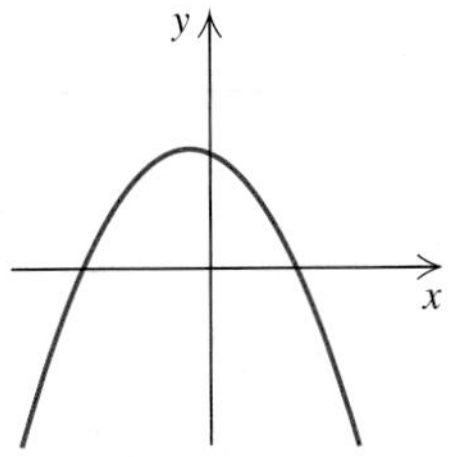

Exponential growth function:
$f(t) = P_0e^{kt}, k > 0$

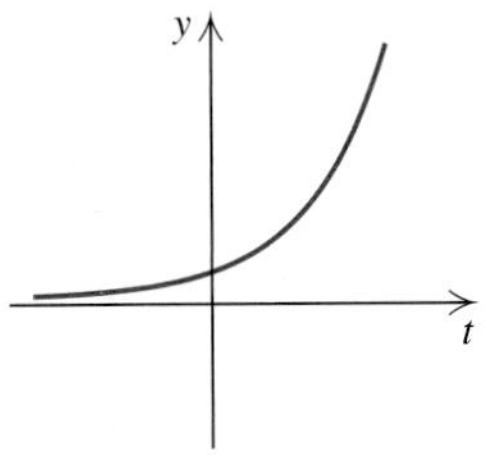

Exponential decay function:
$f(t) = P_0e^{-kt}, k > 0$

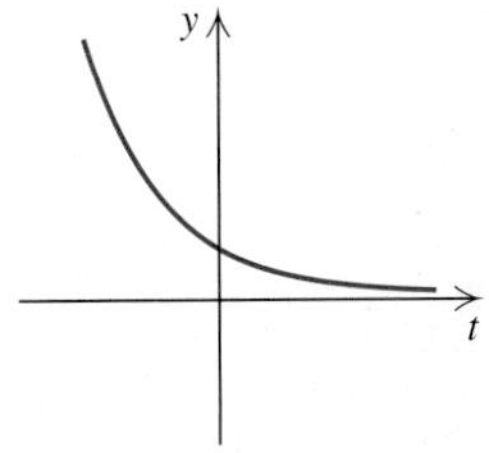

Logarithmic function:
$f(x) = \log_b x, b > 1$

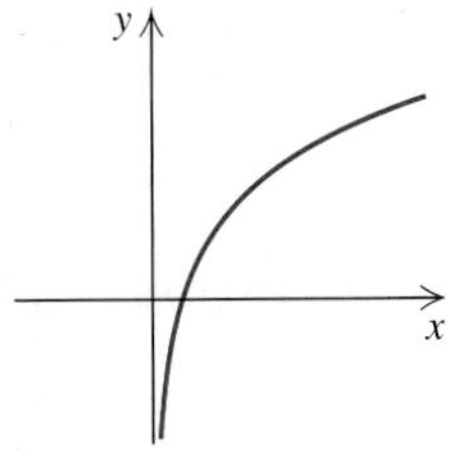

9.7 EXERCISE SET

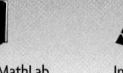

FOR EXTRA HELP

MathXL | MyMathLab | InterAct Math | AW Math Tutor Center | Video Lectures on CD: Disc 5 | Student's Solutions Manual

Solve.

1. *Digital Marketing.* The amount spent by U.S. companies on digital marketing, in billions of dollars, t years after 2000 can be estimated by

$$d(t) = 1.47(1.56)^t$$

(*Source*: Based on data from Jupiter Media Matrix).

a) Determine the year in which companies first spent \$10 billion on digital marketing.

b) What is the doubling time for digital marketing spending?

2. *Cell Phones.* The number of cell phones in use in the United States, in millions, t years after 1995 can be estimated by

$$N(t) = 36.9(1.2)^t$$

(*Source*: Cellular Telecommunications and Internet Association).

a) In what year did the number of cell phones in use first reach 200 million?

b) What is the doubling time for the number of cell phones in use?

3. *Skateboarding.* The number of skateboarders of age x, in thousands, can be approximated by

$$N(x) = 1089(0.9)^x, \quad 7 \le x \le 75$$

(*Sources*: Based on figures from the National Sporting Goods Association and *Statistical Abstract of the United States*, 2006).

a) Estimate the number of 21-year-old skateboarders.

b) At what age are there only 6300 skateboarders?

4. *Recycling Aluminum Cans.* Approximately one-half of all aluminum cans distributed will be recycled each year. A beverage company distributes 250,000 cans. The number still in use after t years is given by the function

$$N(t) = 250{,}000\left(\tfrac{1}{2}\right)^t$$

(*Source*: The Aluminum Association, Inc., May 2004).

a) After how many years will 60,000 cans still be in use?

b) After what amount of time will only 1000 cans still be in use?

5. *Student Loan Repayment.* A college loan of \$29,000 is made at 3% interest, compounded annually. After t years, the amount due, A, is given by the function

$$A(t) = 29{,}000(1.03)^t.$$

a) After what amount of time will the amount due reach \$35,000?

b) Find the doubling time.

6. *Spread of a Rumor.* The number of people who have heard a rumor increases exponentially. If all who hear a rumor repeat it to two people a day, and if 20 people start the rumor, the number of people N who have heard the rumor after t days is given by

$$N(t) = 20(3)^t.$$

a) After what amount of time will 1000 people have heard the rumor?

b) What is the doubling time for the number of people who have heard the rumor?

7. *Telephone Lines.* As more Americans make cell phones their *only* phones, the percent of phone lines that are land lines has been shrinking. The percent of U.S. phone lines that are land lines $P(t)$, in use t years after 2000, can be estimated by

$$P(t) - 63.03(0.95)^t$$

(*Sources*: Based on data from Federal Communications Commission; Cellular Telecommunications and Internet Association).

a) In what year did the percentage of phones that are land lines drop below 50%?
b) In what year will the percentage of phones that are land lines drop below 25%?

8. *Smoking.* The percentage of smokers who received telephone counseling and had successfully quit smoking for t months is given by

$$P(t) = 21.4(0.914)^t$$

(*Sources*: *New England Journal of Medicine*; data from California's Smoker's Hotline).

a) In what month will 15% of those who quit and used telephone counseling still be smoke-free?
b) In what month will 5% of those who quit and used phone counseling still be smoke-free?

9. *Marine Biology.* As a result of preservation efforts in countries in which whaling was once common, the humpback whale population has grown since the 1970s. The worldwide population $P(t)$, in thousands, t years after 1982 can be estimated by

$$P(t) = 5.5(1.08)^t.$$

a) In what year will the humpback whale population reach 40,000?
b) Find the doubling time.

10. *World Population.* The world population $P(t)$, in billions, t years after 1980 can be approximated by

$$P(t) = 4.495(1.015)^t$$

(*Sources*: Based on data from U.S. Bureau of the Census; International Data Base).

a) In what year will the world population reach 7 billion?
b) Find the doubling time.

Use $\text{pH} = -\log[H^+]$ *for Exercises 11–14.*

11. *Chemistry.* The hydrogen ion concentration of fresh-brewed coffee is about 1.3×10^{-5} moles per liter. Find the pH.

12. *Chemistry.* The hydrogen ion concentration of milk is about 1.6×10^{-7} moles per liter. Find the pH.

13. *Medicine.* When the pH of a patient's blood drops below 7.4, a condition called *acidosis* sets in. Acidosis can be deadly when the patient's pH reaches 7.0. What would the hydrogen ion concentration of the patient's blood be at that point?

14. *Medicine.* When the pH of a patient's blood rises above 7.4, a condition called *alkalosis* sets in. Alkalosis can be deadly when the patient's pH reaches 7.8. What would the hydrogen ion concentration of the patient's blood be at that point?

Use $L = 10 \cdot \log \frac{I}{I_0}$ *for Exercises 15–18, where* $I_0 = 10^{-12}\ \text{W/m}^2$.

15. *Audiology.* The intensity of sound in normal conversation is about $3.2 \times 10^{-6}\ \text{W/m}^2$. How loud in decibels is this sound level?

16. *Audiology.* The intensity of a riveter at work is about $3.2 \times 10^{-3}\ \text{W/m}^2$. How loud in decibels is this sound level?

17. *Music.* The band U2 recently performed and sound measurements of 105 dB were recorded. What is the intensity of such sounds?

18. *Music.* The band Strange Folk performed in Burlington, VT, and reached sound levels of 111 dB (*Source*: Melissa Garrido, *Burlington Free Press*). What is the intensity of such sounds?

19. *Stellar Magnitude.* The apparent stellar magnitude m of a star with received intensity I is given by

$$m(I) = -(19 + 2.5 \cdot \log I),$$

where I is in W/m^2 (*Source*: The Columbus Optical SETI Observatory). The smaller the apparent stellar magnitude, the brighter the star appears.

a) The intensity of light received from the sun is 1390 W/m^2. What is the apparent stellar magnitude of the sun?

b) The 5-m diameter Hale telescope on Mt. Palomar can detect a star with magnitude +23. What is the received intensity of light from such a star?

20. *Richter Scale.* The Richter scale, developed in 1935, has been used for years to measure earthquake magnitude. The Richter magnitude of an earthquake m is given by the formula

$$m(A) = \log\frac{A}{A_0},$$

where A is the maximum amplitude of the earthquake and A_0 is a constant. What is the magnitude on the Richter scale of an earthquake with an amplitude that is a million times A_0?

Use $P(t) = P_0e^{kt}$ *for Exercises 21 and 22.*

21. *Interest Compounded Continuously.* Suppose that P_0 is invested in a savings account where interest is compounded continuously at 2.5% per year.

a) Express $P(t)$ in terms of P_0 and 0.025.

b) Suppose that \$5000 is invested. What is the balance after 1 yr? after 2 yr?

c) When will an investment of \$5000 double itself?

22. *Interest Compounded Continuously.* Suppose that P_0 is invested in a savings account where interest is compounded continuously at 3.1% per year.

a) Express $P(t)$ in terms of P_0 and 0.031.

b) Suppose that \$1000 is invested. What is the balance after 1 yr? after 2 yr?

c) When will an investment of \$1000 double itself?

23. *Population Growth.* In 2006, the population of the United States was 300 million and the exponential growth rate was 0.9% per year (*Source*: U.S. Bureau of the Census).

a) Find the exponential growth function.

b) Predict the U.S. population in 2010.

c) When will the U.S. population reach 325 million?

24. *World Population Growth.* In 2006, the world population was 6.5 billion and the exponential growth rate was 1.1% per year (*Source*: U.S. Bureau of the Census).

a) Find the exponential growth function.

b) Predict the world population in 2010.

c) When will the world population be 8.0 billion?

25. *iPod Sales.* The number of iPods sold since January 1, 2003, has grown at an exponential growth rate of 11.2% per month (*Source*: Based on data from iLounge.com and the register.co.uk). What is the doubling time for iPod sales?

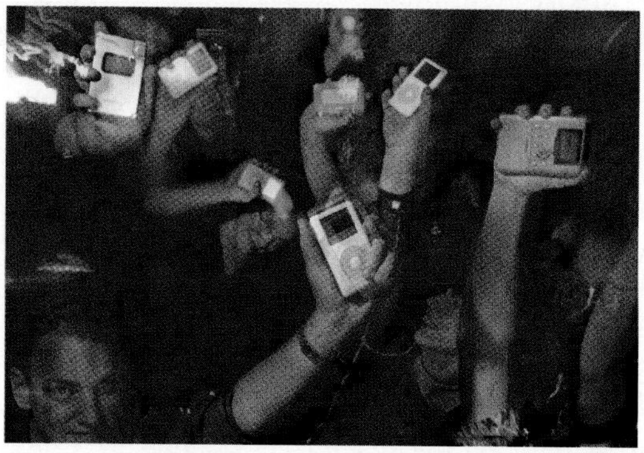

26. *Population Growth.* The exponential growth rate of the population of Somalia is 3.4% per year (one of the highest in the world) (*Source*: *Time Almanac* 2006). What is the doubling time?

27. *World Population.* The function

$$Y(x) = 67.17 \ln\frac{x}{4.5}$$

can be used to estimate the number of years $Y(x)$ after 1980 required for the world population to reach x billion people (*Sources*: Based on data from U.S. Bureau of the Census; International Data Base).

a) In what year will the world population reach 7 billion?

b) In what year will the world population reach 8 billion?

c) Graph the function.

28. *Marine Biology.* The function

$$Y(x) = 13 \ln \frac{x}{5.5}$$

can be used to estimate the number of years $Y(x)$ after 1982 required for the world's humpback whale population to reach x thousand whales.

a) In what year will the whale population reach 40,000?

b) In what year will the whale population reach 50,000?

c) Graph the function.

29. *Forgetting.* Students in an English class took a final exam. They took equivalent forms of the exam at monthly intervals thereafter. The average score $S(t)$, in percent, after t months was found to be given by

$$S(t) = 68 - 20 \log (t + 1), \quad t \geq 0.$$

a) What was the average score when they initially took the test, $t = 0$?

b) What was the average score after 4 months? after 24 months?

c) Graph the function.

d) After what time t was the average score 50%?

30. *Advertising.* A model for advertising response is given by

$$N(a) = 2000 + 500 \log a, \quad a \geq 1,$$

where $N(a)$ is the number of units sold and a is the amount spent on advertising, in thousands of dollars.

a) How many units were sold after spending \$1000 ($a = 1$) on advertising?

b) How many units were sold after spending \$8000?

c) Graph the function.

d) How much would have to be spent in order to sell 5000 units?

31. *MP3 Sales.* Sales of MP3 players have grown exponentially since 2002, when 1.7 million units were sold. This increased to 11.0 million units in 2005. (*Source*: NPD Group)

a) Find an exponential growth function that fits the data.

b) Predict the number of units sold in 2007.

32. *iPod Sales.* Sales of iPods have grown exponentially since January 1, 2003, at which time 656,000 units had been sold. Approximately 34 months later, in November 2005, the 30-millionth unit was sold. (*Sources*: Based on data from iLounge.com and the register.co.uk)

a) Find an exponential growth function that fits the data.

b) Predict the number of units sold as of June 2006.

33. *Decline in Farmland.* The number of acres of farmland in the United States has decreased from 987 million acres in 1990 to 938 million acres in 2002 (*Source*: *Statistical Abstract of the United States*, 2006). Assume the number of acres of farmland is decreasing exponentially.

a) Find the value k, and write an equation for an exponential function that can be used to predict the number of acres of U.S. farmland t years after 1990.

b) Predict the number of acres of farmland in 2008.

c) In what year will there be only 800 million acres of U.S. farmland remaining?

34. *Decline in Cases of Mumps.* The number of cases of mumps has dropped exponentially from 900 in 1995 to 300 in 2001 (*Source*: U.S. Centers for Disease Control and Prevention).

a) Find the value k, and write an exponential function that can be used to estimate the number of cases t years after 1995.

b) Estimate the number of cases of mumps in 2008.

c) In what year (theoretically) will there be only 1 case of mumps?

35. *Archaeology.* When archaeologists found the Dead Sea Scrolls, they determined that the linen wrapping had lost 22.3% of its carbon-14. How old is the linen wrapping? (See Example 7.)

36. *Archaeology.* In 1996, researchers found an ivory tusk that had lost 18% of its carbon-14. How old was the tusk? (See Example 7.)

37. *Chemistry.* The exponential decay rate of iodine-131 is 9.6% per day. What is its half-life?

38. *Chemistry.* The decay rate of krypton-85 is 6.3% per year. What is its half-life?

39. *Home Construction.* The chemical urea formaldehyde was found in some insulation used in houses built during the mid to late 1960s. Unknown at the time was the fact that urea formaldehyde emitted toxic fumes as it decayed. The half-life of urea formaldehyde is 1 yr. What is its decay rate?

40. *Plumbing.* Lead pipes and solder are often found in older buildings. Unfortunately, as lead decays, toxic chemicals can get in the water resting in the pipes. The half-life of lead is 22 yr. What is its decay rate?

41. *Value of a Sports Card.* Legend has it that because he objected to smoking, and because his first baseball card was issued in cigarette packs, the great shortstop Honus Wagner halted production of his card before many were produced. One of these cards was purchased in 1991 by hockey great Wayne Gretzky (and a partner) for $451,000. The same card was sold in 2000 for $1.1 million. For the following questions, assume that the card's value increases exponentially, as it has for many years.

a) Find the exponential growth rate k, and determine an exponential function V that can be used to estimate the dollar value, $V(t)$, of the card t years after 1991.
b) Estimate the value of the card in 2006.
c) What is the doubling time for the value of the card?
d) In what year will the value of the card first exceed $3,000,000?

42. *Art Masterpieces.* In August 2004, a collector paid $104,168,000, for Pablo Picasso's "Garçon à la Pipe." The same painting sold for $30,000 in 1950. (*Source*: BBC News, 5/6/04)

a) Find the exponential growth rate k, and determine the exponential growth function V, for which $V(t)$ is the painting's value, in millions of dollars, t years after 1950.
b) Estimate the value of the painting in 2009.
c) What is the doubling time for the value of the painting?
d) How long after 1950 will the value of the painting be $1 billion?

In Exercises 43–46, determine whether an exponential function might fit the data.

43.

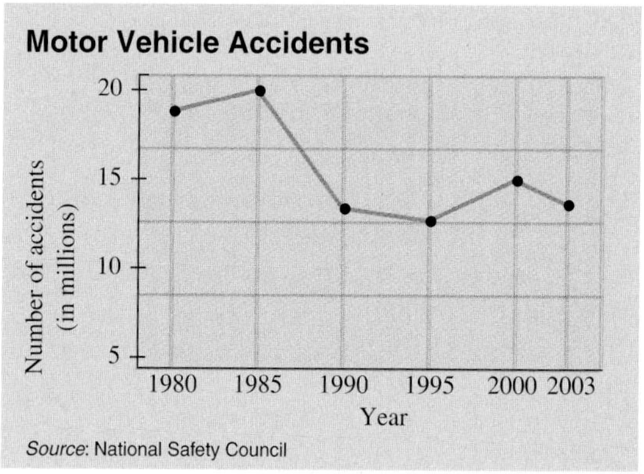

44.

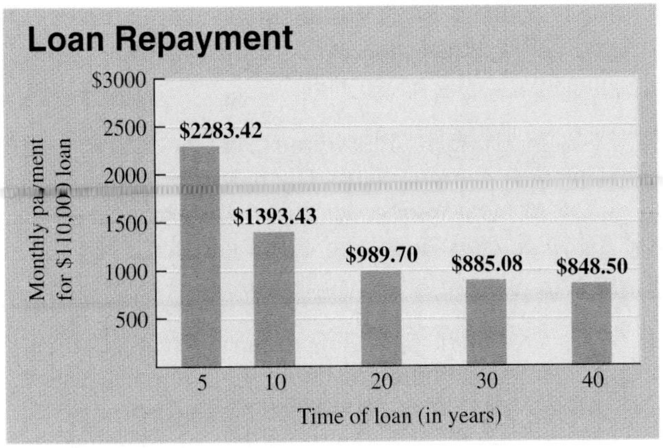

45. Boston Red Sox infield roof box seats

Year	Season Price
1980	$ 8
1991	16
2000	45
2003	65
2006	90

Sources: harvard magazine.com, boston.com, and redsox.mlb.com

46. World automobile production

Year	Number of Automobiles Produced (in millions)
1950	8
1960	13
1970	23
1980	29
1990	36
2002	41

Sources: ibike.org and mindfully.org

47. *Baseball.* The price of an infield roof box seat for Boston Red Sox fans has been growing exponentially. The table in Exercise 45 shows the price per game for season ticket holders for various years.

a) Use regression to find an exponential function that can be used to estimate the price p of an infield roof box seat x years after 1980.

b) Determine the exponential growth rate.

c) Predict the price of an infield roof box seat in 2010.

48. *World Population.* The population of the world is growing exponentially, as shown by the graph on p. 716.

a) Use the four most recent data points and regression to fit an exponential function of the form $P(t) = P_0e^{kt}$ to the data, where t is the year and P is in billions.

b) Predict the world population in 2050.

49. *Hearing Loss.* Prolonged exposure to high noise levels can lead to hearing loss. The following table lists the safe exposure time, in hours, for different sound intensities.

Sound Intensity (in decibels)	Safe Exposure Time (in hours)
90	8
100	2
105	1
110	0.5
115	0.25

Source: American Hearing Research Foundation

a) Use regression to fit an exponential function of the form $f(x) = ab^x$ to the data.

b) Estimate the safe exposure time for a sound intensity of 95 dB.

50. *Loan Repayment.* The size of a monthly loan repayment decreases exponentially as the time of the loan increases. The table in Exercise 44 shows the size of monthly loan payments for a $110,000 loan at a fixed interest rate for various lengths of loans.

a) Use regression to fit an exponential function of the form $f(x) = ab^x$ to the data.

b) Estimate the monthly loan payment for a 15-yr $110,000 loan.

TW **51.** Will the model used to predict the number of cell phones in Exercise 2 still be realistic in 2020? Why or why not?

TW **52.** Examine the restriction on t in Exercise 29.

a) What upper limit might be placed on t?

b) In practice, would this upper limit ever be enforced? Why or why not?

Skill Maintenance

Graph. [8.7]

53. $y = x^2 - 8x$

54. $y = x^2 - 5x - 6$

55. $f(x) = 3x^2 - 5x - 1$

56. $g(x) = 2x^2 - 6x + 3$

Solve by completing the square. [8.1]

57. $x^2 - 8x = 7$

58. $x^2 + 10x = 6$

Synthesis

TW **59.** *Atmospheric Pressure.* Atmospheric pressure P at altitude a is given by

$$P(a) = P_0 e^{-0.00005a},$$

where P_0 is the pressure at sea level $\approx 14.7 \text{ lb/in}^2$ (pounds per square inch). Explain how a barometer, or some other device for measuring atmospheric pressure, can be used to find the height of a skyscraper.

TW **60.** Write a problem for a classmate to solve in which information is provided and the classmate is asked to find an exponential growth function. Make the problem as realistic as possible.

61. *Sports Salaries.* As part of Alex Rodriguez's 10-yr \$252-million contract with the New York Yankees, he will receive \$24 million in 2010 (part from the Yankees and part from his former team, the Texas Rangers) (*Source*: *The San Francisco Chronicle*). How much money would need to be invested in 2004, at 4% interest compounded continuously, in order to have \$24 million for Rodriguez in 2010? (This is much like finding what \$24 million in 2010 is worth in 2004 dollars.)

62. *Supply and Demand.* The supply and demand for the sale of stereos by Sound Ideas are given by

$$S(x) = e^x \quad \text{and} \quad D(x) = 162{,}755e^{-x},$$

where $S(x)$ is the price at which the company is willing to supply x stereos and $D(x)$ is the demand price for a quantity of x stereos. Find the equilibrium point. (For reference, see Section 3.8.)

63. Use Exercise 7 to form a model for the percentage of U.S. phone lines that are cellular t years after 2000.

TW **64.** Use the model developed in Exercise 63 to predict the percentage of U.S. phone lines that will be cellular in 2020. Does your prediction seem plausible? Why or why not?

65. *Nuclear Energy.* Plutonium-239 (Pu-239) is used in nuclear energy plants. The half-life of Pu-239 is 24,360 yr. (*Source*: *Microsoft Encarta 97 Encyclopedia*) How long will it take for a fuel rod of Pu-239 to lose 90% of its radioactivity?

66. *Growth of Bacteria.* The bacteria *Escherichia coli* (*E. coli*) are commonly found in the human bladder. Suppose that 3000 of the bacteria are present at time $t = 0$. Then t minutes later, the number of bacteria present is

$$N(t) = 3000(2)^{t/20}.$$

If 100,000,000 bacteria accumulate, a bladder infection can occur. If, at 11:00 A.M., a patient's bladder contains 25,000 *E. coli* bacteria, at what time can infection occur?

67. Show that for exponential growth at rate k, the doubling time T is given by $T = \dfrac{\ln 2}{k}$.

68. Show that for exponential decay at rate k, the half-life T is given by $T = \dfrac{\ln 2}{k}$.

69. *Sports Medicine.* The temperature of the ankle surface begins to increase logarithmically a few minutes after ice is applied (*Source*: *Journal of Athletic Training*, Vol 41, Number 2, June 2006). After 10 min, the temperature is 15.5°C. After 20 min, the temperature is 21.1°C. If the ice is then removed, in 10 more min the temperature will be 23.6°C, and in 10 more min, the temperature will be 24.9°C.

a) Let $t = 0$ represent the time at which the ice is applied. Enter the data into a table and graph the data.

b) Use the LnReg option of the STAT CALC menu to find a logarithmic function that could be used to estimate the ankle surface temperature t min after the ice is applied.

c) The surface temperature before ice is applied was 27.6°C. How many minutes after the ice was applied will the surface temperature be back to 27.6°C?

Logistic Curves. *Realistically, most quantities that are growing exponentially eventually level off. The quantity may continue to increase, but at a decreasing rate. This pattern of growth can be modeled by a logistic function*

$$f(x) = \frac{c}{1 + ae^{-bx}}.$$

The general shape of this family of functions is shown by the following graph.

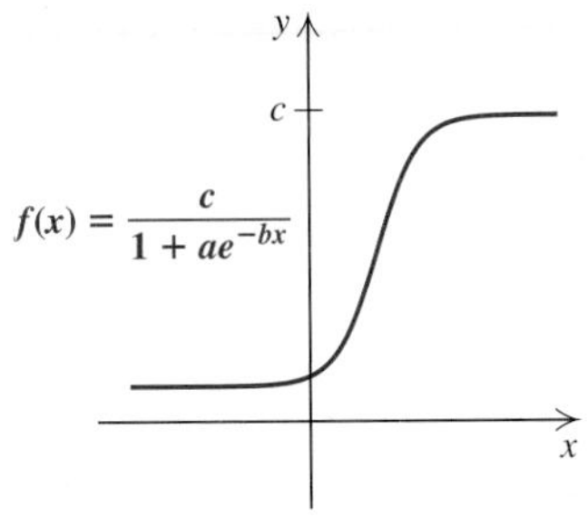

Many graphing calculators can fit a logistic function to a set of data.

70. *Internet Access.* The percentage of U.S. households with Internet access for various years is listed in the following table.

Year	Percentage of Households with Internet Access
1997	18.6
1998	26.2
2000	41.5
2001	50.3
2003	54.6

Source: www.ntia.doc.gov

a) Find a logistic function

$$f(x) = \frac{c}{1 + ae^{-bx}}$$

that could be used to estimate the percentage of U.S. households with Internet access x years after 1997.

b) Use the function found in part (a) to predict the percentage of U.S. households with Internet access in 2010.

TW **71.** *Heart Transplants.* In 1967, Dr. Christiaan Barnard of South Africa stunned the world by performing the first heart transplant. Since that time, the operation's popularity has both grown and declined, as shown in the table below.

Year	Number of Heart Transplants Worldwide
1982	189
1983	318
1984	669
1985	1189
1986	2167
1987	2720
1988	3157
1989	3378
1990	4016
1991	4186
1992	4199
1993	4346
1994	4402
1995	4314
1996	4128
1997	4039
1998	3744
1999	3419
2000	3246
2001	3122
2002	3265
2003	3020

Source: International Society for Heart & Lung Transplantation

a) Using 1982 as $t = 0$, graph the data.

b) Does it appear that an exponential function might have ever served as an appropriate model for these data? If so, for what years would this have been the case?

c) Considering *all* the data points on your graph, which would be the most appropriate model: a linear, a quadratic, or an exponential function? Why?

9 Chapter Summary and Review

KEY TERMS AND DEFINITIONS

FUNCTIONS

One-to-one function, p. 704 A function for which different inputs have different outputs; a function whose graph passes the **horizontal-line test.**

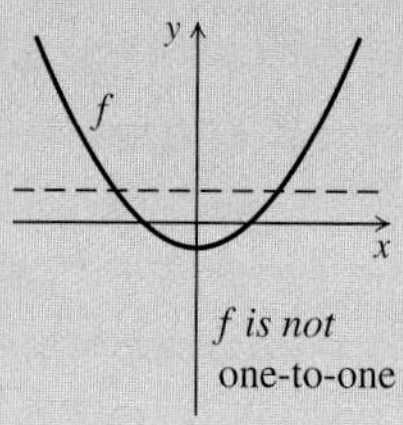

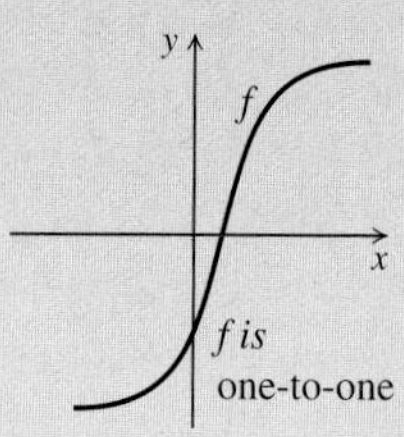

Inverse function, p. 704 A function formed by interchanging the members of the domain and the members of the range of a function. A one-to-one function has an inverse.

EXPONENTIAL AND LOGARITHMIC FUNCTIONS

Logarithm, p. 728 $\log_a x$ is the exponent to which a must be raised in order to get x. $\log_a x = m$ means $a^m = x$. *A logarithm is an exponent.*

Exponential function, p. 716

$f(x) = a^x$

Domain: $\mathbb{R}$

Range: $(0, \infty)$

Logarithmic function, p. 728

$g(x) = \log_a x$

Domain: $(0, \infty)$

Range: $\mathbb{R}$

Common logarithm, p. 730 A logarithm, base 10.
Natural logarithm, p. 744 A logarithm, base e.
***e*,** p. 744 An irrational number; $e \approx 2.718281828$.

APPLICATIONS

Interest compounded annually

$$A = P(1 + i)^t$$

Interest compounded continuously

$$P(t) = P_0e^{kt}$$

Loudness of sound

$$L = 10 \cdot \log \frac{I}{I_0}$$

pH

$$\text{pH} = -\log[H^+]$$

Carbon dating

$$P(t) = P_0e^{-0.00012t}$$

IMPORTANT CONCEPTS

[Section references appear in brackets.]

Concept	Example
The **composition** of f and g is defined as $(f \circ g)(x) = f(g(x))$.	If $f(x) = x^2 + 1$ and $g(x) = x - 5$, then $(f \circ g)(x) = f(g(x))$ $= f(x - 5)$ $= (x - 5)^2 + 1$ $= x^2 - 10x + 25 + 1$ $= x^2 - 10x + 26$. [9.1]
If f is one-to-one, it is possible to find its inverse: **1.** Replace $f(x)$ with y. **2.** Interchange x and y. **3.** Solve for y. **4.** Replace y with $f^{-1}(x)$.	If $f(x) = 2x - 3$, find $f^{-1}(x)$. **1.** $y = 2x - 3$ **2.** $x = 2y - 3$ **3.** $x + 3 = 2y$ $\frac{x+3}{2} = y$ **4.** $\frac{x+3}{2} = f^{-1}(x)$ [9.1]
The graph of $f(x) = a^x$ contains the points $(0, 1)$ and $(1, a)$. It increases if $a > 1$ and decreases if $0 < a < 1$. The graph of $g(x) = \log_a x$ contains the points $(1, 0)$ and $(a, 1)$.	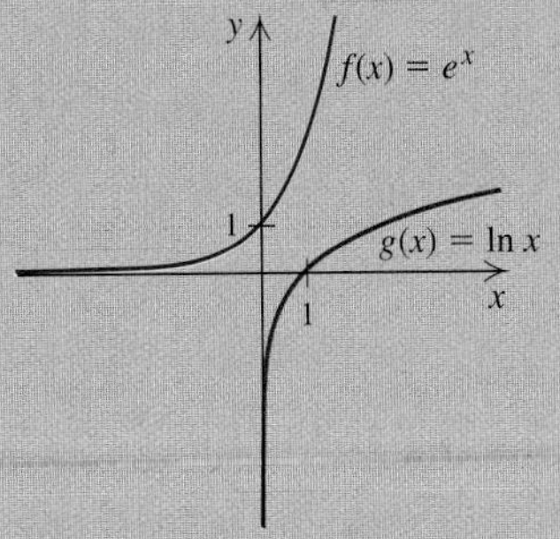 [9.2], [9.3], [9.5]

(continued)

Properties of Logarithms $\log_a (MN) = \log_a M + \log_a N$ $\log_a \frac{M}{N} = \log_a M - \log_a N$ $\log_a M^p = p \cdot \log_a M$ $\log_a 1 = 0$ $\log_a a = 1$ $\log_a a^k = k$ $\log M = \log_{10} M$ $\ln M = \log_e M$ $\log_b M = \frac{\log_a M}{\log_a b}$	$\log_7 10 = \log_7 5 + \log_7 2$ $\log_5 \frac{14}{3} = \log_5 14 - \log_5 3$ $\log_8 5^{12} = 12 \log_8 5$ $\log_9 1 = 0$ $\log_4 4 = 1$ $\log_3 3^8 = 8$ $\log 43 = \log_{10} 43$ $\ln 37 = \log_e 37$ $\log_6 31 = \frac{\log 31}{\log 6} = \frac{\ln 31}{\ln 6}$	[9.3], [9.4], [9.5]
The Principle of Exponential Equality For any real number b, $b \neq -1, 0$, or 1: $b^x = b^y$ is equivalent to $x = y$.	$25 = 5^x$ $5^2 = 5^x$ $2 = x$	[9.6]
The Principle of Logarithmic Equality For any logarithm base a, and for $x, y > 0$: $x = y$ is equivalent to $\log_a x = \log_a y$.	$83 = 7^x$ $\log 83 = \log 7^x$ $\log 83 = x \log 7$ $\frac{\log 83}{\log 7} = x$	[9.6]
Exponential Growth Model $P(t) = P_0 e^{kt}, k > 0$ P_0 is the population at time 0. $P(t)$ is the population at time t. k is the **exponential growth rate.** The **doubling time** is the amount of time necessary for the population to double in size.	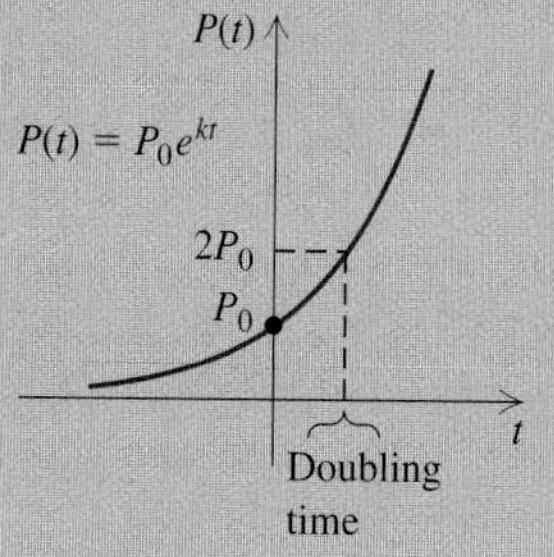	[9.7]
Exponential Decay Model $P(t) = P_0 e^{-kt}, k > 0$ P_0 is the quantity present at time 0. $P(t)$ is the amount present at time t. k is the **exponential decay rate.** The **half-life** is the amount of time necessary for half of the quantity to decay.	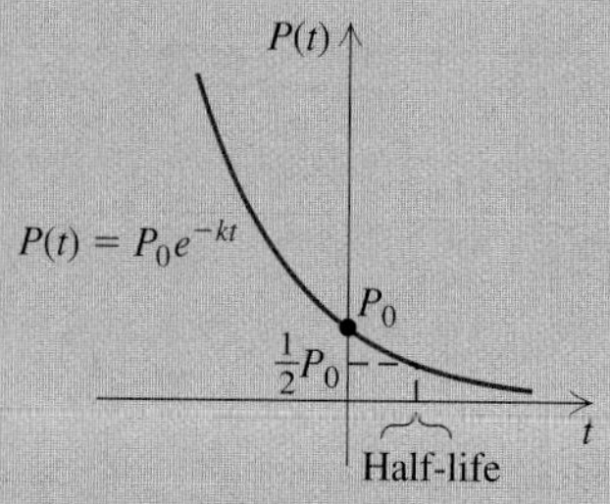	[9.7]

Review Exercises

Concept Reinforcement *In each of Exercises 1–10, classify the statement as either true or false.*

1. The functions given by $f(x) = e^x$ and $g(x) = \ln x$ are inverses of each other. [9.5]

2. A function's doubling time is the amount of time t for which $f(t) = 2f(0)$. [9.7]

3. A radioactive isotope's half-life is the amount of time t for which $f(t) = \frac{1}{2} f(0)$. [9.7]

4. $\ln(ab) = \ln a - \ln b$ [9.4]

5. $\log x^a = x \ln a$ [9.4]

6. $\log_a \dfrac{m}{n} = \log_a m - \log_a n$ [9.4]

7. For $f(x) = 3^x$, the domain of f is $[0, \infty)$. [9.2]

8. For $g(x) = \log_2 x$, the domain of g is $[0, \infty)$. [9.3]

9. The function F is not one-to-one if $F(-2) = F(5)$. [9.1]

10. The function g is one-to-one if it passes the vertical-line test. [9.1]

11. Find $(f \circ g)(x)$ and $(g \circ f)(x)$ if $f(x) = x^2 + 1$ and $g(x) = 2x - 3$. [9.1]

12. If $h(x) = \sqrt{3 - x}$, find $f(x)$ and $g(x)$ such that $h(x) = (f \circ g)(x)$. Answers may vary. [9.1]

13. Determine whether $f(x) = 4 - x^2$ is one-to-one. [9.1]

Find a formula for the inverse of each function. [9.1]

14. $f(x) = x - 8$

15. $g(x) = \dfrac{3x + 1}{2}$

16. $f(x) = 27x^3$

Graph by hand.

17. $f(x) = 3^x + 1$ [9.2]

18. $x = \left(\frac{1}{4}\right)^y$ [9.2]

19. $y = \log_5 x$ [9.3]

Simplify. [9.3]

20. $\log_3 9$

21. $\log_{10} \frac{1}{100}$

22. $\log_5 5^7$

23. $\log_9 3$

Rewrite as an equivalent logarithmic equation. [9.3]

24. $10^{-2} = \frac{1}{100}$

25. $25^{1/2} = 5$

Rewrite as an equivalent exponential equation. [9.3]

26. $\log_4 16 = x$

27. $\log_8 1 = 0$

Express as an equivalent expression using the individual logarithms of x, y, and z. [9.4]

28. $\log_a x^4 y^2 z^3$

29. $\log_a \dfrac{x^5}{yz^2}$

30. $\log \sqrt[4]{\dfrac{z^2}{x^3 y}}$

Express as an equivalent expression that is a single logarithm and, if possible, simplify. [9.4]

31. $\log_a 7 + \log_a 8$

32. $\log_a 72 - \log_a 12$

33. $\frac{1}{2} \log a - \log b - 2 \log c$

34. $\frac{1}{3}[\log_a x - 2 \log_a y]$

Simplify. [9.4]

35. $\log_m m$

36. $\log_m 1$

37. $\log_m m^{17}$

Given $\log_a 2 = 1.8301$ *and* $\log_a 7 = 5.0999$, *find each of the following.* [9.4]

38. $\log_a 14$

39. $\log_a \frac{2}{7}$

40. $\log_a 28$

41. $\log_a 3.5$

42. $\log_a \sqrt{7}$

43. $\log_a \frac{1}{4}$

Use a calculator to find each of the following to the nearest ten-thousandth. [9.3], [9.5]

44. $\log 75$

45. $10^{1.789}$

46. $\ln 0.05$

47. $e^{-0.98}$

Find each of the following logarithms using the change-of-base formula. Round answers to the nearest ten-thousandth. [9.5]

48. $\log_5 2$

49. $\log_{12} 70$

Graph and state the domain and the range of each function. [9.5]

50. $f(x) = e^x - 1$

51. $g(x) = 0.6 \ln x$

Solve. Where appropriate, include approximations to the nearest ten-thousandth. [9.6]

52. $2^x = 32$

53. $3^x = \frac{1}{9}$

54. $\log_3 x = -4$

55. $\log_x 16 = 4$

56. $\log x = -3$

57. $3 \ln x = -6$

58. $4^{2x-5} = 19$

59. $2^{x^2} \cdot 2^{4x} = 32$

60. $4^x = 8.3$

61. $e^{-0.1t} = 0.03$

62. $3e^{2x} = 6$

63. $\log_3 (2x - 5) = 1$

64. $\log_4 x + \log_4 (x - 6) = 2$

65. $\log x + \log (x - 15) = 2$

66. $\log_3 (x - 4) = 3 - \log_3 (x + 4)$

67. In a business class, students were tested at the end of the course with a final exam. They were then tested again 6 months later. The forgetting formula was determined to be

$$S(t) = 82 - 18 \log (t + 1),$$

where t is the time, in months, after taking the final exam. [9.7]

a) Determine the average score when they first took the exam (when $t = 0$).

b) What was the average score after 6 months?

c) After what time was the average score 54?

68. A color photocopier is purchased for \$5200. Its value each year is about 80% of its value in the preceding year. Its value in dollars after t years is given by the exponential function

$$V(t) = 5200(0.8)^t. \text{ [9.7]}$$

a) After what amount of time will the copier's value be \$1200?

b) After what amount of time will the copier's value be half the original value?

69. *Cell Phones.* In 1997, there were 107.8 million cell phones sold, worldwide. This number grew exponentially to 816.6 million in 2005. (*Source*: Gartner Dataquest) [9.7]

a) Find the value k, and write an exponential function that describes the number of cell phones sold t years after 1997.

b) Predict the number of cell phones that will be sold in 2009.

c) In what year will 1.5 billion cell phones be sold?

70. *MLB Salaries.* The average salary of a Major League baseball player has grown exponentially. Average salaries for various years are listed in the following table. [9.7]

Year	Average Salary
1970	\$ 29,803
1980	143,756
1990	578,930
2000	1,998,034
2005	2,632,655

Source: Major League Baseball

a) Use regression to find an exponential function of the form $f(x) = ab^x$ that can be used to estimate the average salary of a Major League baseball player x years after 1970.

b) Use the function of part (a) to estimate the average salary of a Major League baseball player in 2006.

c) The average salary of a Major League baseball player in 2006 was \$2,866,544. By what percent was the prediction of part (b) off?

71. The value of Jose's stock market portfolio doubled in 3 yr. What was the exponential growth rate? [9.7]

72. How long will it take \$7600 to double itself if it is invested at 4.2%, compounded continuously? [9.7]

73. How old is a skull that has lost 34% of its carbon-14? (Use $P(t) = P_0e^{-0.00012t}$.) [9.7]

74. What is the pH of a substance if its hydrogen ion concentration is 2.3×10^{-7} moles per liter? (Use pH $= -\log [H^+]$.) [9.7]

75. The intensity of the sound of water at the foot of the Niagara Falls is about 10^{-3} W/m^2.* How loud in decibels is this sound level? [9.7]

$$\left(\text{Use } L = 10 \cdot \log \frac{I}{10^{-12}}.\right)$$

Synthesis

TW 76. Explain why negative numbers do not have logarithms. [9.3]

TW 77. Explain why taking the natural or common logarithm on each side of an equation produces an equivalent equation. [9.6]

Solve. [9.6]

78. $\ln (\ln x) = 3$

79. $2^{x^2+4x} = \frac{1}{8}$

80. Solve the system:

$$5^{x+y} = 25,$$
$$2^{2x-y} = 64. \text{ [9.6]}$$

81. *Blogs.* The doubling time for the number of blogs is 6 months (*Source*: Technorati, *State of the Blogosphere,* April 2006). At the beginning of 2003, there were about 0.6 million blogs. Find an exponential function that can be used to predict the number of blogs t months after January 2003. [9.7]

***Sound and Hearing,* Life Science Library. (New York: Time Incorporated, 1965), p. 173.

Chapter Test 9

1. Find $(f \circ g)(x)$ and $(g \circ f)(x)$ if $f(x) = x + x^2$ and $g(x) = 2x + 1$.

2. If

$$h(x) = \frac{1}{2x^2 + 1},$$

find $f(x)$ and $g(x)$ such that $h(x) = (f \circ g)(x)$. Answers may vary.

3. Determine whether $f(x) = |x - 3|$ is one-to-one.

Find a formula for the inverse of each function.

4. $f(x) = 3x + 4$

5. $g(x) = (x + 1)^3$

Graph by hand.

6. $f(x) = 2^x - 3$

7. $g(x) = \log_7 x$

Simplify.

8. $\log_5 125$

9. $\log_{100} 10$

10. $3^{\log_3 18}$

Rewrite as an equivalent logarithmic equation.

11. $4^{-3} = \frac{1}{64}$

12. $256^{1/2} = 16$

Rewrite as an equivalent exponential equation.

13. $m = \log_7 49$

14. $\log_3 81 = 4$

15. Express as an equivalent expression using the individual logarithms of a, b, and c:

$$\log \frac{a^3 b^{1/2}}{c^2}.$$

16. Express as an equivalent expression that is a single logarithm:

$$\tfrac{1}{3} \log_a x + 2 \log_a z.$$

Simplify.

17. $\log_p p$

18. $\log_t t^{23}$

19. $\log_c 1$

Given $\log_a 2 = 0.301$, $\log_a 6 = 0.778$, *and* $\log_a 7 = 0.845$, *find each of the following.*

20. $\log_a 14$

21. $\log_a 3$

22. $\log_a 16$

Use a calculator to find each of the following to the nearest ten-thousandth.

23. $\log 12.3$

24. $10^{-0.8}$

25. $\ln 0.035$

26. $e^{4.8}$

27. Find $\log_3 14$ using the change-of-base formula. Round to the nearest ten-thousandth.

Graph and state the domain and the range of each function.

28. $f(x) = e^x + 3$

29. $g(x) = \ln(x - 4)$

Solve. Where appropriate, include approximations to the nearest ten-thousandth.

30. $2^x = \frac{1}{32}$

31. $\log_x 25 = 2$

32. $\log_4 x = \frac{1}{2}$

33. $\log x = 4$

34. $5^{4-3x} = 87$

35. $7^x = 1.2$

36. $\ln x = \frac{1}{4}$

37. $\log(x - 3) + \log(x + 1) = \log 5$

38. The average walking speed R of people living in a city of population P, in thousands, is given by $R = 0.37 \ln P + 0.05$, where R is in feet per second.

a) The population of Tucson, Arizona, is 512,000. Find the average walking speed.

b) Philadelphia, Pennsylvania, has an average walking speed of about 2.76 ft/sec. Find the population.

39. The population of Nigeria was about 128.8 million in 2005, and the exponential growth rate was 2.4% per year.

a) Write an exponential function describing the population of Nigeria.

b) What will the population be in 2007? in 2012?

c) When will the population be 175 million?

d) What is the doubling time?

40. The average cost of a year at a private four-year college grew exponentially from \$18,039 in 1997 to \$23,503 in 2003 (*Source*: National Center for Education Statistics, Digest of Education Statistics, 2003).

a) Find the value k, and write an exponential function that approximates the cost of a year of college t years after 1997.

b) Predict the cost of a year of college in 2010.

c) In what year will the average cost of college be \$50,000?

41. *Online Advertising.* The amount of money spent in online advertising grew exponentially from 2000 through 2006, as listed in the following table.

Year	Online Advertising Spending (in billions)
2000	\$ 5.4
2001	5.7
2002	6.8
2003	8.6
2004	10.6
2005	12.9
2006	15.4

Source: Jupiter Media Metrix

a) Use regression to find an exponential function of the form $f(x) = ab^x$ that can be used to predict online advertising spending t years after 2000.

b) Use the function to estimate online advertising spending in 2008.

42. An investment with interest compounded continuously doubled itself in 15 yr. What is the interest rate?

43. How old is an animal bone that has lost 43% of its carbon-14? (Use $P(t) = P_0 e^{-0.00012t}$.)

44. The sound of traffic at a busy intersection averages 75 dB. What is the intensity of such a sound? $\left(\text{Use } L = 10 \cdot \log \frac{I}{I_0}.\right)$

45. The hydrogen ion concentration of water is 1.0×10^{-7} moles per liter. What is the pH? (Use $\text{pH} = -\log[H^+]$.)

Synthesis

46. Solve: $\log_5 |2x - 7| = 4$.

47. If $\log_a x = 2$, $\log_a y = 3$, and $\log_a z = 4$, find

$$\log_a \frac{\sqrt[3]{x^2 z}}{\sqrt[3]{y^2 z^{-1}}}.$$

1–9 Cumulative Review

Travel. *Prices for a massage at Passport Travel Spa in an airport waiting area are shown in the following graph.*

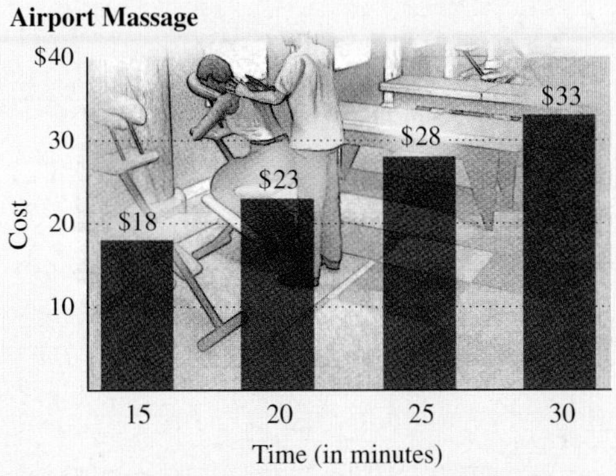

Source: Passport Travel Spa, Indianapolis International Airport

1. Determine whether the data can be modeled by a linear, quadratic, or exponential function. [2.4]

2. Find the rate of change, in dollars per minute. [2.2]

3. Find a linear function $f(x) = mx + b$ that fits the data, where f is the cost of the massage and x is the length, in minutes. [2.4]

4. Use the function in Exercise 3 to estimate the cost of a 10-min massage. [2.4]

5. For the function in Exercise 3, what do the values of m and b signify? [2.3]

Life Insurance. *The monthly premium for a term life insurance policy increases as the age of the insured person increases. The following graph shows some premiums for a $250,000 policy for a male.*

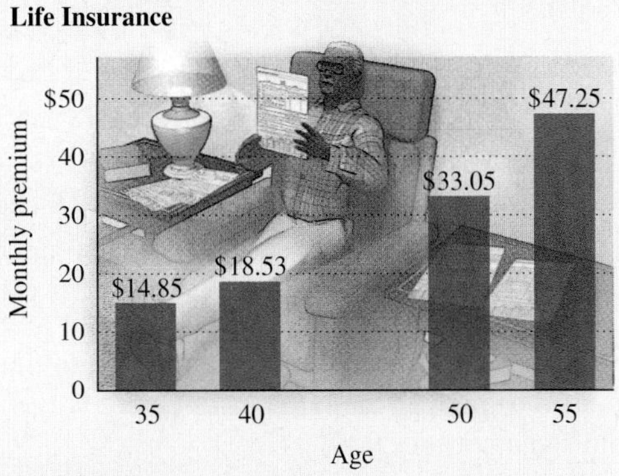

Source: Insurance company advertisement

6. Determine whether the data can be modeled by a linear, quadratic, or exponential function. [8.8], [9.7]

7. Use regression to find an exponential function of the form $f(x) = ab^x$ that can be used to estimate the monthly insurance premium m for a male who is x years old. [9.7]

8. Use the function in Exercise 7 to estimate the monthly premium for a 45-year-old male. [9.7]

College Texts. The number of college textbooks sold each year from 2000–2005 is listed in the following graph.

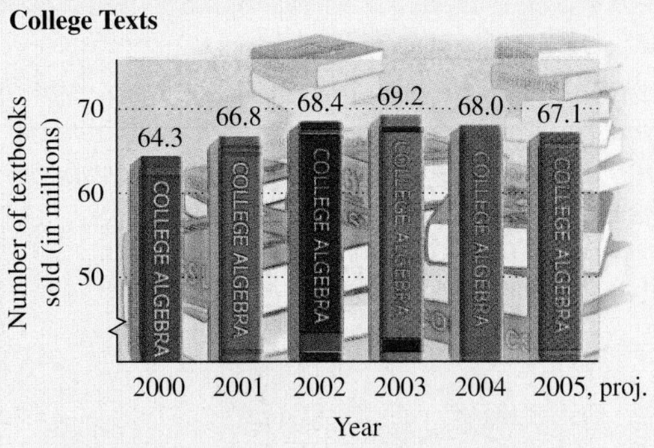

Source: *Book Industry Trends*, 2005

9. Determine whether the data can be modeled by a linear, quadratic, or exponential function. [8.8]

10. Use regression to find a quadratic function that can be used to predict the number of college texts c sold t years after 2000. [8.8]

11. Use the function in Exercise 10 to estimate the number of college texts sold in 2006. [8.8]

12. The perimeter of a rectangular garden is 112 m. The length is 16 m more than the width. Find the length and the width. [1.7]

13. In triangle ABC, the measure of angle B is three times the measure of angle A. The measure of angle C is 105° greater than the measure of angle A. Find the angle measures. [1.7]

14. Good's Candies of Indiana makes all their chocolates by hand. It takes Anne 10 min to coat a tray of candies in chocolate. It takes Clay 12 min to coat a tray of candies. How long would it take Anne and Clay, working together, to coat the candies? [6.5]

15. Joe's Thick and Tasty salad dressing gets 45% of its calories from fat. The Light and Lean dressing gets 20% of its calories from fat. How many ounces of each should be mixed in order to get 15 oz of dressing that gets 30% of its calories from fat? [3.3]

16. A cruise ship can move at a speed of 21 mph in still water. The ship travels 113 mi down the Amazon River in the same time that it takes to travel 55 mi upriver. What is the speed of the Amazon? [6.5]

17. What is the minimum product of two numbers whose difference is 14? What are the numbers that yield this product? [8.8]

In Exercises 18–21, match each function with one of the following graphs.

a)

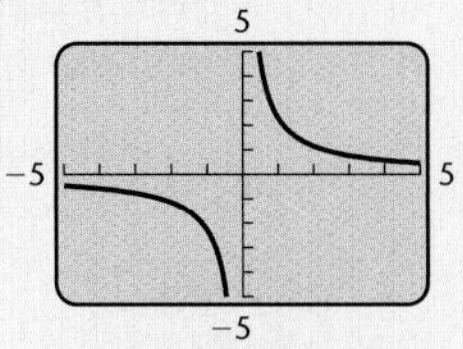

b)

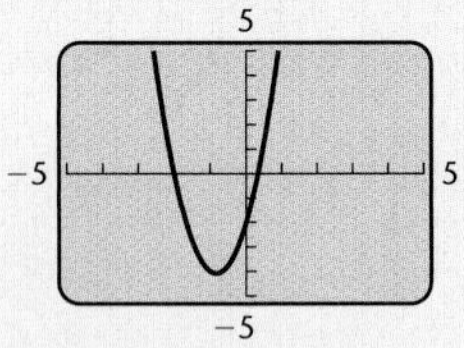

c)

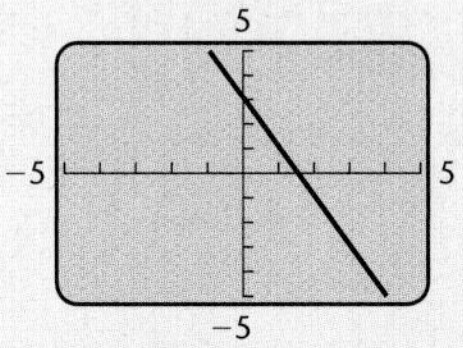

d)

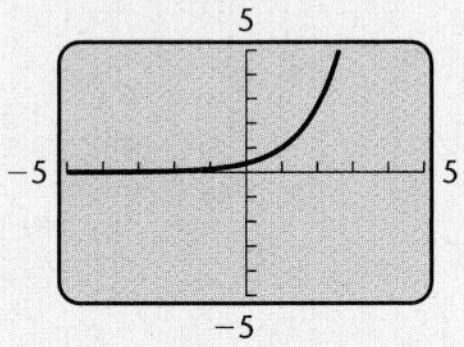

18. $f(x) = -2x + 3.1$ [2.3]

19. $p(x) = 3x^2 + 5x - 2$ [8.7]

20. $h(x) = \dfrac{2.3}{x}$ [6.1]

21. $g(x) = e^{x-1}$ [9.5]

22. Find the domain of the function f given by

$$f(x) = \frac{-4}{3x^2 - 5x - 2}. \text{ [5.5]}$$

23. For the function described by

$$h(x) = -3x^2 + 4x + 8,$$

find $h(-2)$. [2.1]

24. Find the inverse of f if $f(x) = 9 - 2x$. [9.1]

25. Find a linear function with a graph that contains the points $(0, -8)$ and $(-1, 2)$. [2.4]

26. Find an equation of the line whose graph has a y-intercept of $(0, 7)$ and is perpendicular to the line given by $2x + y = 6$. [2.3]

Graph by hand.

27. $5x = 15 + 3y$ [2.4]

28. $y = 2x^2 - 4x - 1$ [8.7]

29. $y = \log_3 x$ [9.3]

30. $y = 3^x$ [9.2]

31. $-2x - 3y \le 12$ [4.5]

32. Graph: $f(x) = 2(x + 3)^2 + 1$. [8.6]
   a) Label the vertex.
   b) Draw the axis of symmetry.
   c) Find the maximum or minimum value.

33. Graph $f(x) = 2e^x$ and determine the domain and the range. [9.5]

34. Evaluate $\dfrac{x^0 + y}{-z}$ for $x = 6$, $y = 9$, and $z = -5$. [1.1], [1.4]

Simplify.

35. $\left|-\frac{5}{2} + \left(-\frac{7}{2}\right)\right|$ [1.2]

36. $(-2x^2y^{-3})^{-4}$ [1.4]

37. $(-5x^4y^{-3}z^2)(-4x^2y^2)$ [1.4]

38. $\dfrac{3x^4y^6z^{-2}}{-9x^4y^2z^3}$ [1.4]

39. $4x - 3 - 2[5 - 3(2 - x)]$ [1.3]

40. $3^3 + 2^2 - (32 \div 4 - 16 \div 8)$ [1.1]

Perform the indicated operations and simplify.

41. $(5p^2q^3 + 6pq - p^2 + p) + (2p^2q^3 + p^2 - 5pq - 9)$ [5.1]

42. $(11x^2 - 6x - 3) - (3x^2 + 5x - 2)$ [5.1]

43. $(3x^2 - 2y)^2$ [5.2]

44. $(5a + 3b)(2a - 3b)$ [5.2]

45. $\dfrac{x^2 + 8x + 16}{2x + 6} \div \dfrac{x^2 + 3x - 4}{x^2 - 9}$ [6.1]

46. $\dfrac{1 + \dfrac{3}{x}}{x - 1 - \dfrac{12}{x}}$ [6.3]

47. $\dfrac{a^2 - a - 6}{a^3 - 27} \cdot \dfrac{a^2 + 3a + 9}{6}$ [6.1]

48. $\dfrac{3}{x + 6} - \dfrac{2}{x^2 - 36} + \dfrac{4}{x - 6}$ [6.2]

Factor.

49. $xy + 2xz - xw$ [5.3]

50. $8 - 125x^3$ [5.7]

51. $6x^2 + 8xy - 8y^2$ [5.5]

52. $x^4 - 4x^3 + 7x - 28$ [5.3]

53. $2m^2 + 12mn + 18n^2$ [5.6]

54. $x^4 - 16y^4$ [5.6]

55. Divide: $(x^4 - 5x^3 + 2x^2 - 6) \div (x - 3)$. [6.6]

56. Multiply $(5.2 \times 10^4)(3.5 \times 10^{-6})$. Write scientific notation for the answer. [1.4]

For the radical expressions that follow, assume that all variables represent positive numbers.

57. Divide and simplify: $\dfrac{\sqrt[3]{40xy^8}}{\sqrt[3]{5xy}}$. [7.4]

58. Multiply and simplify: $\sqrt{7xy^3} \cdot \sqrt{28x^2y}$. [7.3]

59. Write as an equivalent expression without rational exponents: $(27a^6b)^{4/3}$. [7.4]

60. Rationalize the denominator: $\dfrac{3 - \sqrt{y}}{2 - \sqrt{y}}$. [7.5]

61. Divide and simplify: $\dfrac{\sqrt{x + 5}}{\sqrt[5]{x + 5}}$. [7.5]

62. Multiply these complex numbers: $\left(2 - i\sqrt{3}\right)\left(6 + 2i\sqrt{3}\right)$. [7.8]

63. Add: $(8 + 2i) + (5 - 3i)$. [7.8]

64. Express in terms of logarithms of a, b, and c: $\log\left(\dfrac{a^2c^3}{b}\right)$. [9.4]

65. Express as a single logarithm: $3 \log x - \frac{1}{2} \log y - 2 \log z$. [9.4]

66. Convert to an exponential equation: $\log_a 5 = x$. [9.3]

67. Convert to a logarithmic equation: $x^3 = t$. [9.3]

Find each of the following using a calculator. Round to the nearest ten-thousandth. [9.3], [9.5]

68. $\log 0.05566$

69. $10^{2.89}$

70. $\ln 12.78$

71. $e^{-1.4}$

Solve.

72. $5(2x - 3) = 9 - 5(2 - x)$ [1.6]

73. $4x - 3y = 15$,
$3x + 5y = 4$ [3.2]

74. $\begin{aligned} x + y - 3z &= -1, \\ 2x - y + z &= 4, \\ -x - y + z &= 1 \end{aligned}$ [3.4]

75. $x(x - 3) = 10$ [5.3]

76. $\frac{7}{x^2 - 5x} - \frac{2}{x - 5} = \frac{4}{x}$ [6.4]

77. $\frac{8}{x + 1} + \frac{11}{x^2 - x + 1} = \frac{24}{x^3 + 1}$ [6.4], [8.2]

78. $\sqrt{4 - 5x} = 2x - 1$ [7.6]

79. $\sqrt[3]{2x} = 1$ [7.6]

80. $3x^2 + 75 = 0$ [8.1]

81. $x - 8\sqrt{x} + 15 = 0$ [8.5]

82. $x^4 - 13x^2 + 36 = 0$ [8.5]

83. $\log_7 x = 1$ [9.3]

84. $\log_x 36 = 2$ [9.3]

85. $9^x = 27$ [9.6]

86. $3^{5x} = 7$ [9.6]

87. $\ln x - \ln (x - 8) = 1$ [9.6]

88. $x^2 + 4x > 5$ [8.9]

89. If $f(x) = x^2 + 6x$, find a such that $f(a) = 11$. [8.2]

90. If $f(x) = |2x - 3|$, find all x for which $f(x) \geq 7$. [4.4]

Solve.

91. $D = \frac{ab}{b + a}$, for a [6.8]

92. $\frac{1}{p} + \frac{1}{q} = \frac{1}{f}$, for q [6.8]

93. $M = \frac{2}{3}(A + B)$, for B [1.6]

Students in a biology class just took a final exam. A formula for determining what the average exam grade on a similar test will be t months later is

$$S(t) = 78 - 15 \log (t + 1).$$

94. The average score when the students first took the exam occurs when $t = 0$. Find the students' average score on the final exam. [9.7]

95. What would the average score be on a retest after 4 months? [9.7]

The population of Kenya was 33.8 million in 2005, and the exponential growth rate was 2.6% per year.

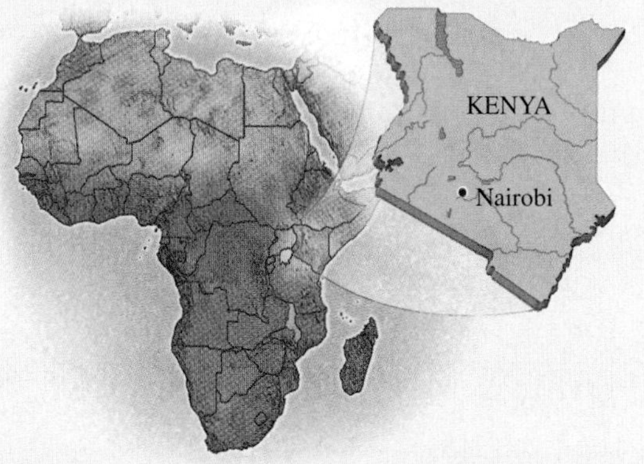

96. Write an exponential function describing the growth of the population of Kenya. [9.7]

97. Predict what the population will be in 2008 and in 2014. [9.7]

98. What is the doubling time of the population? [9.7]

99. y varies directly as the square of x and inversely as z, and $y = 2$ when $x = 5$ and $z = 100$. What is y when $x = 3$ and $z = 4$? [6.8]

Synthesis

Solve.

100. $\frac{5}{3x - 3} + \frac{10}{3x + 6} = \frac{5x}{x^2 + x - 2}$ [6.4]

101. $\log \sqrt{3x} = \sqrt{\log 3x}$ [9.6]

102. A train travels 280 mi at a certain speed. If the speed had been increased by 5 mph, the trip could have been made in 1 hr less time. Find the actual speed. [8.3]

10 Conic Sections

The ellipse described on this page is one example of a *conic section,* meaning that it can be regarded as a cross section of a cone. This chapter presents a variety of applications and equations with graphs that are conic sections. We have already worked with two conic sections, *lines* and *parabolas,* in Chapters 2 and 8.

APPLICATION *Astronomy*

To model the earth's orbit around the sun, we can draw axes showing the sun at the origin. The following table and graph indicate that the earth's orbit is an ellipse with the sun at one focus. If the sun is 2,500,000 km from the center of the ellipse, what is the maximum distance of the earth from the sun?

LOCATION OF EARTH (UNITS ARE IN MILLIONS OF KILOMETERS)

(147.1, 0)	(−152.1, 0)
(100, 113.4)	(−125, −78.2)
(25, 147.7)	(−50, −140.1)
(0, 149.56)	(0, −149.56)
(−50, 140.1)	(25, −147.7)
(−125, 78.2)	(100, −113.4)

Based on information from gsfc.nasa.gov

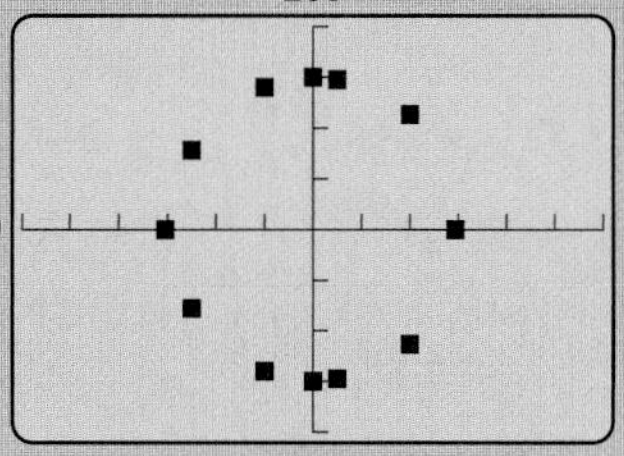

Xscl = 50, Yscl = 50

This problem appears as Exercise 57 in Exercise Set 10.2.

10.1 Conic Sections: Parabolas and Circles

Parabolas ■ The Distance and Midpoint Formulas ■ Circles

This section and the next two examine curves formed by cross sections of cones. These curves are all graphs of $Ax^2 + By^2 + Cxy + Dx + Ey + F = 0$. The constants A, B, C, D, E, and F determine which of the following shapes will serve as the graph.

Study Tip

Don't Give Up Now!

Often students are tempted to ease up once the end of a course is in sight. Don't let this happen to you! You have invested a great deal of time and energy thus far. Don't tarnish that hard effort by doing anything less than your best work as the course winds down.

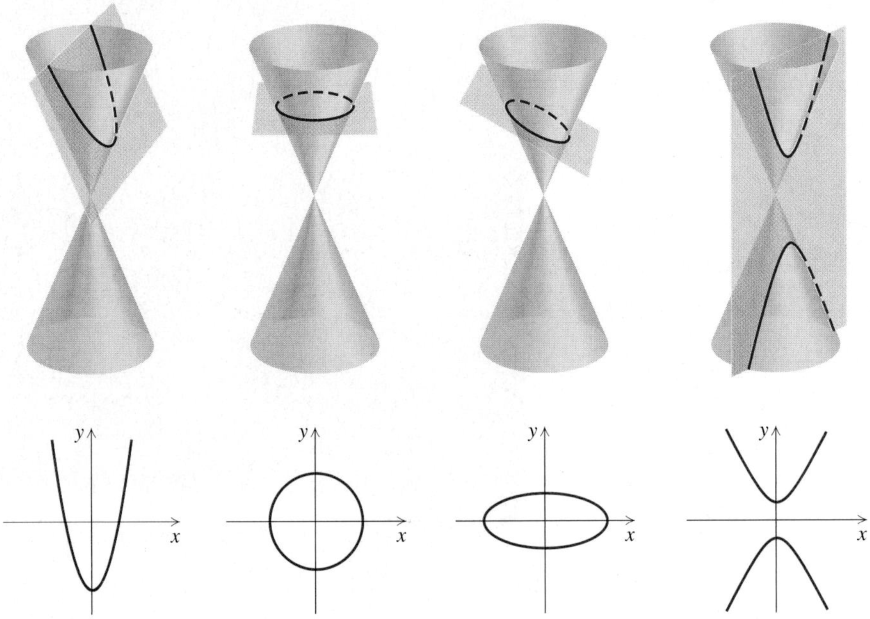

Parabola   Circle   Ellipse   Hyperbola

Parabolas

When a cone is cut as shown in the first figure above, the conic section formed is a **parabola.** Parabolas have many applications in electricity, mechanics, and optics. A cross section of a contact lens or satellite dish is a parabola, and arches that support certain bridges are parabolas.

Equation of a Parabola A parabola with a vertical axis of symmetry opens upward or downward and has an equation that can be written in the form

$$y = ax^2 + bx + c.$$

A parabola with a horizontal axis of symmetry opens to the right or left and has an equation that can be written in the form

$$x = ay^2 + by + c.$$

Parabolas with equations of the form $f(x) = ax^2 + bx + c$ were graphed in Chapter 8.

EXAMPLE 1 Graph: $y = x^2 - 4x + 9$.

SOLUTION To locate the vertex, we can use either of two approaches. One way is to complete the square:

$$y = (x^2 - 4x) + 9$$ **Note that half of −4 is −2, and $(-2)^2 = 4$.**

$$= (x^2 - 4x + 4 - 4) + 9$$ **Adding and subtracting 4**

$$= (x^2 - 4x + 4) + (-4 + 9)$$ **Regrouping**

$$= (x - 2)^2 + 5.$$ **Factoring and simplifying**

The vertex is (2, 5).

A second way to find the vertex is to recall that the x-coordinate of the vertex of the parabola given by $y = ax^2 + bx + c$ is $-b/(2a)$:

$$x = -\frac{b}{2a} = -\frac{-4}{2(1)} = 2.$$

To find the y-coordinate of the vertex, we substitute 2 for x:

$$y = x^2 - 4x + 9 = 2^2 - 4(2) + 9 = 5.$$

Either way, the vertex is (2, 5). Next, we calculate and plot some points on each side of the vertex. As expected for a positive coefficient of x^2, the graph opens upward.

x	y	
2	5	← **Vertex**
0	9	← ***y*-intercept**
1	6	
3	6	
4	9	

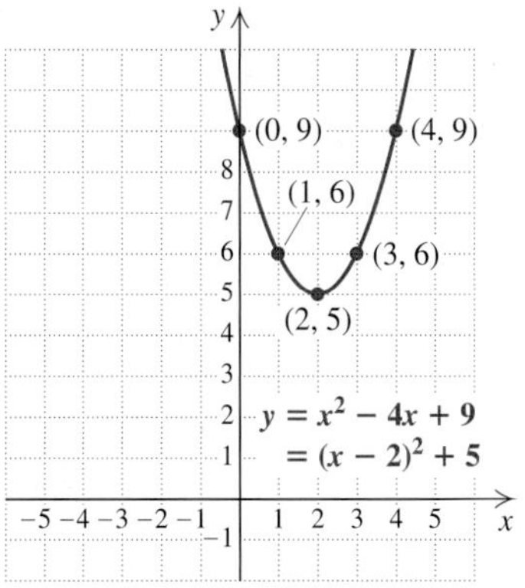

To Graph an Equation of the Form $y = ax^2 + bx + c$

1. Find the vertex (h, k) either by completing the square to find an equivalent equation

   $$y = a(x - h)^2 + k,$$

   or by using $-b/(2a)$ to find the x-coordinate and substituting to find the y-coordinate.
2. Choose other values for x on each side of the vertex, and compute the corresponding y-values.
3. The graph opens upward for $a > 0$ and downward for $a < 0$.

Equations of the form $x = ay^2 + by + c$ represent horizontal parabolas. These parabolas open to the right for $a > 0$, open to the left for $a < 0$, and have axes of symmetry parallel to the x-axis.

EXAMPLE 2 Graph: $x = y^2 - 4y + 9$.

SOLUTION This equation is like that in Example 1 but with x and y interchanged. The vertex is $(5, 2)$ instead of $(2, 5)$. To find ordered pairs, we choose values for y on each side of the vertex. Then we compute values for x. Note that the x- and y-values of the table in Example 1 are now switched. You should confirm that, by completing the square, we get $x = (y - 2)^2 + 5$.

x	y	
5	2	← **Vertex**
9	0	← ***x*-intercept**
6	1	
6	3	
9	4	

(1) Choose these values for *y*.
(2) Compute these values for *x*.

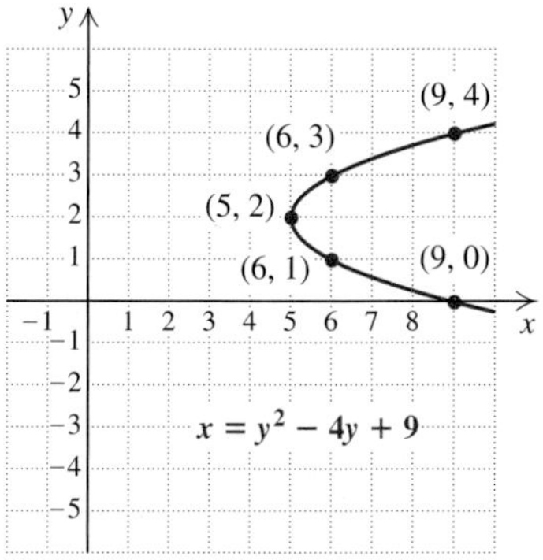

To Graph an Equation of the Form $x = ay^2 + by + c$

1. Find the vertex (h, k) either by completing the square to find an equivalent equation

   $$x = a(y - k)^2 + h,$$

   or by using $-b/(2a)$ to find the y-coordinate and substituting to find the x-coordinate.
2. Choose other values for y that are above and below the vertex, and compute the corresponding x-values.
3. The graph opens to the right if $a > 0$ and to the left if $a < 0$.

EXAMPLE 3 Graph: $x = -2y^2 + 10y - 7$.

SOLUTION We find the vertex by completing the square:

$$\begin{aligned} x &= -2y^2 + 10y - 7 \\ &= -2(y^2 - 5y \qquad) - 7 \\ &= -2\left(y^2 - 5y + \tfrac{25}{4}\right) - 7 - (-2)\tfrac{25}{4} && \tfrac{1}{2}(-5) = \tfrac{-5}{2};\ \left(\tfrac{-5}{2}\right)^2 = \tfrac{25}{4};\ \text{we add and subtract } (-2)\tfrac{25}{4}. \\ &= -2\left(y - \tfrac{5}{2}\right)^2 + \tfrac{11}{2}. && \text{Factoring and simplifying} \end{aligned}$$

The vertex is $\left(\frac{11}{2}, \frac{5}{2}\right)$.

For practice, we also find the vertex by first computing its y-coordinate, $-b/(2a)$, and then substituting to find the x-coordinate:

$$y = -\frac{b}{2a} = -\frac{10}{2(-2)} = \frac{5}{2}$$

$$x = -2y^2 + 10y - 7 = -2\left(\tfrac{5}{2}\right)^2 + 10\left(\tfrac{5}{2}\right) - 7 = \tfrac{11}{2}.$$

To find ordered pairs, we choose values for y on each side of the vertex and then compute values for x. A table is shown below, together with the graph. The graph opens to the left because the y^2-coefficient, -2, is negative.

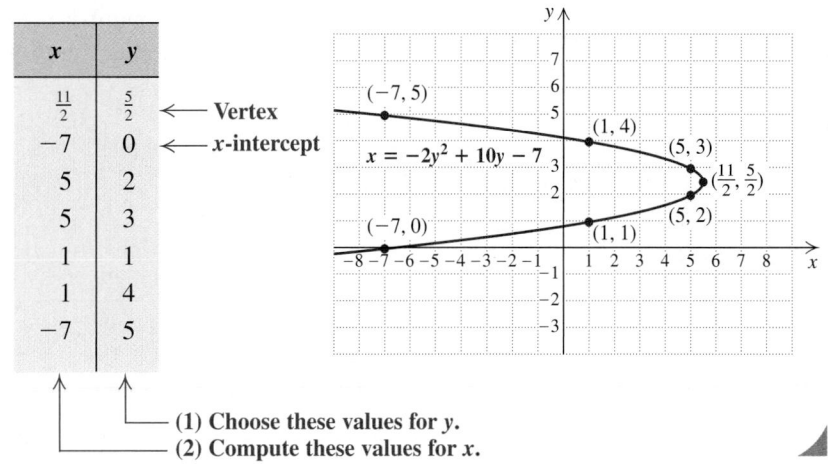

x	y	
$\frac{11}{2}$	$\frac{5}{2}$	← **Vertex**
-7	0	← ***x*-intercept**
5	2	
5	3	
1	1	
1	4	
-7	5	

(1) Choose these values for *y*.
(2) Compute these values for *x*.

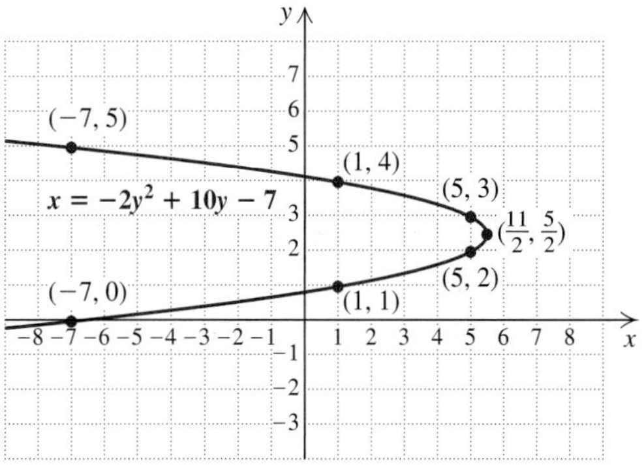

The Distance and Midpoint Formulas

If two points are on a horizontal line, they have the same second coordinate. We can find the distance between them by subtracting their first coordinates. This difference may be negative, depending on the order in which we subtract. So, to make sure we get a positive number, we take the absolute value of this difference. The distance between the points (x_1, y_1) and (x_2, y_1) on a horizontal line is thus $|x_2 - x_1|$. Similarly, the distance between the points (x_2, y_1) and (x_2, y_2) on a vertical line is $|y_2 - y_1|$.

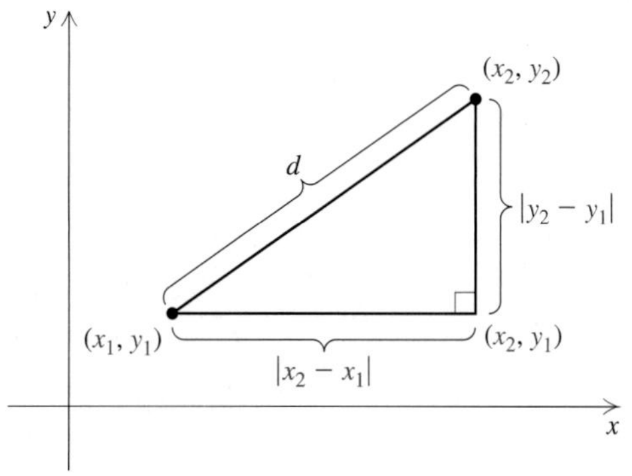

Now consider *any* two points (x_1, y_1) and (x_2, y_2). If $x_1 \neq x_2$ and $y_1 \neq y_2$, these points, along with the point (x_2, y_1), describe a right triangle. The lengths of the legs are $|x_2 - x_1|$ and $|y_2 - y_1|$. We find d, the length of the hypotenuse, by using the Pythagorean theorem:

$$d^2 = |x_2 - x_1|^2 + |y_2 - y_1|^2.$$

Since the square of a number is the same as the square of its opposite, we can replace the absolute-value signs with parentheses:

$$d^2 = (x_2 - x_1)^2 + (y_2 - y_1)^2.$$

Taking the principal square root, we have a formula for distance.

The Distance Formula The distance d between any two points (x_1, y_1) and (x_2, y_2) is given by

$$d = \sqrt{(x_2 - x_1)^2 + (y_2 - y_1)^2}.$$

EXAMPLE 4 Find the distance between $(5, -1)$ and $(-4, 6)$. Find an exact answer and an approximation to three decimal places.

SOLUTION We substitute into the distance formula:

$$d = \sqrt{(-4 - 5)^2 + [6 - (-1)]^2} \quad \textbf{Substituting}$$
$$= \sqrt{(-9)^2 + 7^2}$$
$$= \sqrt{130} \quad \textbf{This is exact.}$$
$$\approx 11.402. \quad \textbf{Using a calculator for an approximation}$$

The distance formula is needed to develop the formula for a circle, which follows, and to verify certain properties of conic sections. It is also needed to verify a formula for the coordinates of the *midpoint* of a segment connecting two points. We state the midpoint formula and leave its proof to the exercises.

Student Notes

To help remember the formulas correctly, note that the distance formula (a variation on the Pythagorean theorem) involves both subtraction and addition, whereas the midpoint formula does not include any subtraction.

The Midpoint Formula If the endpoints of a segment are (x_1, y_1) and (x_2, y_2), then the coordinates of the midpoint are

$$\left(\frac{x_1 + x_2}{2}, \frac{y_1 + y_2}{2}\right).$$

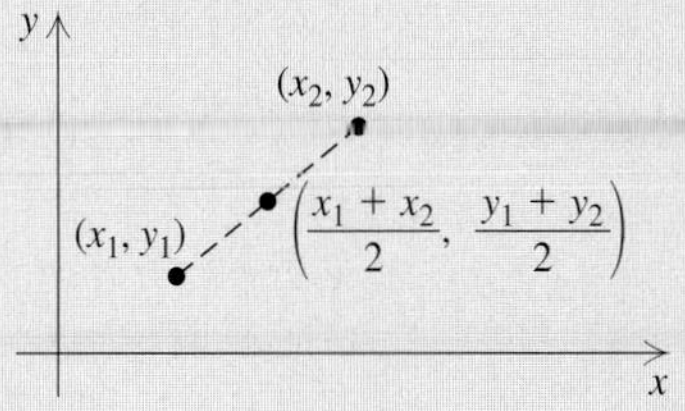

(To locate the midpoint, average the x-coordinates and average the y-coordinates.)

EXAMPLE 5 Find the midpoint of the segment with endpoints $(-2, 3)$ and $(4, -6)$.

SOLUTION Using the midpoint formula, we obtain

$$\left(\frac{-2 + 4}{2}, \frac{3 + (-6)}{2}\right), \quad \text{or} \quad \left(\frac{2}{2}, \frac{-3}{2}\right), \quad \text{or} \quad \left(1, -\frac{3}{2}\right).$$

Circles

One conic section, the **circle,** is a set of points in a plane that are a fixed distance r, called the **radius** (plural, **radii**), from a fixed point (h, k), called the **center.** Note that the word radius can mean either any segment connecting a point on a circle to the center or the length of such a segment. If (x, y) is on the circle, then by the definition of a circle and the distance formula, it follows that

$$r = \sqrt{(x - h)^2 + (y - k)^2}.$$

Squaring both sides gives the equation of a circle in standard form.

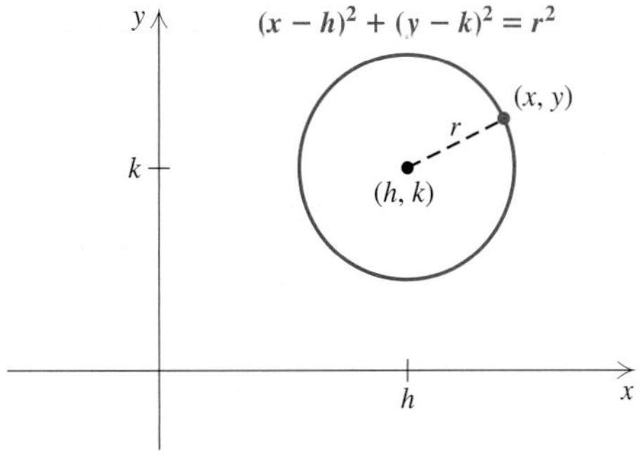

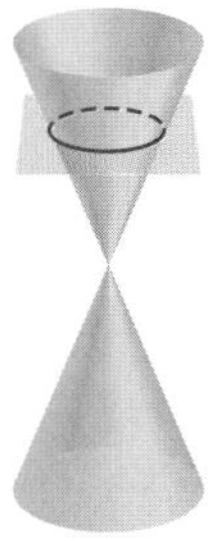

Equation of a Circle (Standard Form)

The equation of a circle, centered at (h, k), with radius r, is given by

$$(x - h)^2 + (y - k)^2 = r^2.$$

Note that for $h = 0$ and $k = 0$, the circle is centered at the origin. Otherwise, the circle is translated $|h|$ units horizontally and $|k|$ units vertically.

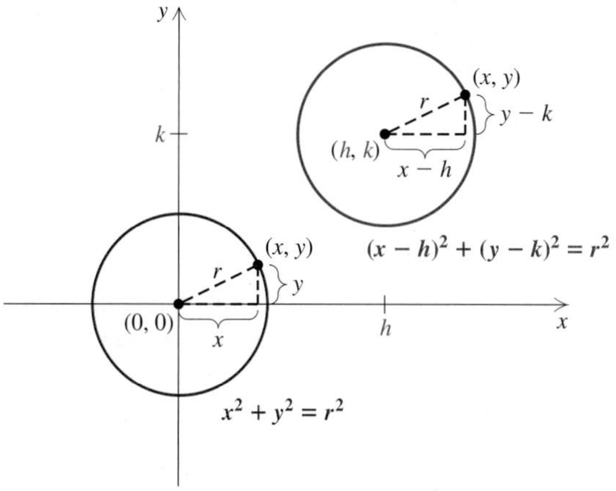

EXAMPLE 6 Find an equation of the circle having center (4, 5) and radius 6.

SOLUTION Using the standard form, we obtain

$$(x - 4)^2 + (y - 5)^2 = 6^2, \qquad \text{Using } (x - h)^2 + (y - k)^2 = r^2$$

or

$$(x - 4)^2 + (y - 5)^2 = 36.$$

EXAMPLE 7 Find the center and the radius and then graph each circle.

a) $(x - 2)^2 + (y + 3)^2 = 4^2$

b) $x^2 + y^2 + 8x - 2y + 15 = 0$

SOLUTION

a) We write standard form:

$$(x - 2)^2 + [y - (-3)]^2 = 4^2.$$

The center is $(2, -3)$ and the radius is 4. To graph, we plot the points $(2, 1)$, $(2, -7)$, $(-2, -3)$, and $(6, -3)$, which are, respectively, 4 units above, below, left, and right of $(2, -3)$. We then either sketch a circle by hand or use a compass.

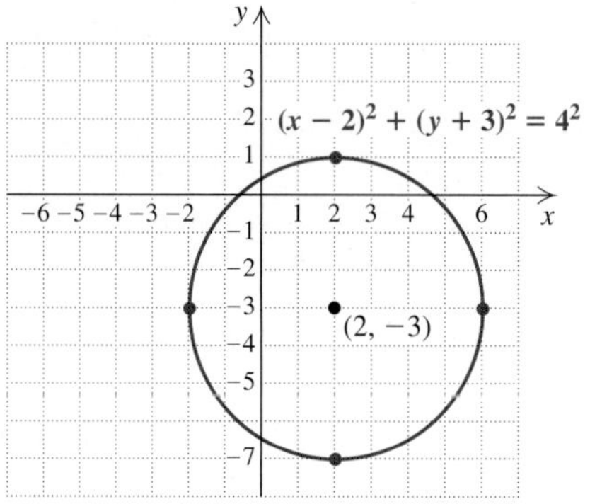

b) To write the equation $x^2 + y^2 + 8x - 2y + 15 = 0$ in standard form, we complete the square twice, once with $x^2 + 8x$ and once with $y^2 - 2y$:

$$x^2 + y^2 + 8x - 2y + 15 = 0$$

$$x^2 + 8x \quad + y^2 - 2y \quad = -15$$ **Grouping the x-terms and the y-terms, adding -15 to both sides**

$$x^2 + 8x + 16 + y^2 - 2y + 1 = -15 + 16 + 1$$ **Adding $\left(\frac{8}{2}\right)^2$, or 16, and $\left(-\frac{2}{2}\right)^2$ or 1, to both sides to get standard form**

$$(x + 4)^2 + (y - 1)^2 = 2$$ **Factoring**

$$[x - (-4)]^2 + (y - 1)^2 = \left(\sqrt{2}\right)^2.$$ **Writing standard form**

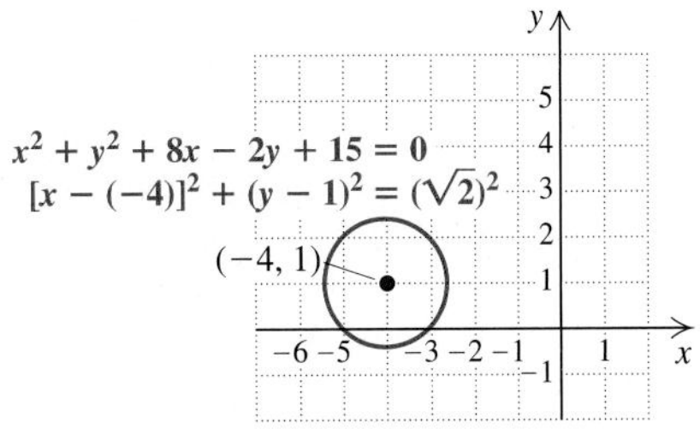

The center is $(-4, 1)$ and the radius is $\sqrt{2}$.

Graphing Circles

Because most graphing calculators can graph only functions, graphing the equation of a circle usually requires two steps:

1. Solve the equation for y. The result will include a $\pm$ sign in front of a radical.
2. Graph two functions, one for the $+$ sign and the other for the $-$ sign, on the same set of axes.

For example, to graph $(x - 3)^2 + (y + 1)^2 = 16$, solve for $y + 1$ and then y:

$$(y + 1)^2 = 16 - (x - 3)^2$$

$$y + 1 = \pm\sqrt{16 - (x - 3)^2}$$

$$y = -1 \pm \sqrt{16 - (x - 3)^2}.$$

Then

$$y_1 = -1 + \sqrt{16 - (x - 3)^2}$$

and

$$y_2 = -1 - \sqrt{16 - (x - 3)^2}.$$

(continued)

When both functions are graphed (in a "squared" window to eliminate distortion), the result is as follows.

$$y_1 = -1 + \sqrt{16 - (x - 3)^2}$$
$$y_2 = -1 - \sqrt{16 - (x - 3)^2}$$

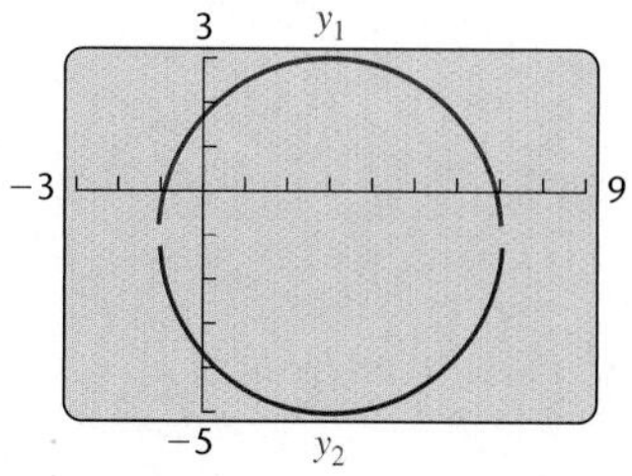

Circles can also be drawn using the CIRCLE option of the DRAW menu, but such graphs cannot be traced. Many calculators have a Conics application, accessed by pressing APPS. If we use this program, equations of circles in standard form can be graphed directly, and then Traced, by entering only the values of h, k, and r. Equations in the form $Ax^2 + Ay^2 + Bx + Cy + D = 0$ are also circles and can be graphed directly by entering the values of A, B, C, and D.

10.1 EXERCISE SET

FOR EXTRA HELP

MathXL | MyMathLab | InterAct Math | AW Math Tutor Center | Video Lectures on CD: Disc 5 | Student's Solutions Manual

Concept Reinforcement *In each of Exercises 1–8, match the equation with the center or vertex of its graph, listed in the column on the right.*

1. ____ $(x-2)^2 + (y+5)^2 = 9$	**a)** Vertex: $(-2, 5)$
2. ____ $(x+2)^2 + (y-5)^2 = 9$	**b)** Vertex: $(5, -2)$
3. ____ $(x-5)^2 + (y+2)^2 = 9$	**c)** Vertex: $(2, -5)$
4. ____ $(x+5)^2 + (y-2)^2 = 9$	**d)** Vertex: $(-5, 2)$
5. ____ $y = (x-2)^2 - 5$	**e)** Center: $(-2, 5)$
6. ____ $y = (x-5)^2 - 2$	**f)** Center: $(2, -5)$
7. ____ $x = (y-2)^2 - 5$	**g)** Center: $(5, -2)$
8. ____ $x = (y-5)^2 - 2$	**h)** Center: $(-5, 2)$

Graph. Be sure to label each vertex.

9. $y = -x^2$

10. $y = 2x^2$

11. $y = -x^2 + 4x - 5$

12. $x = 4 - 3y - y^2$

13. $x = y^2 - 4y + 2$

14. $y = x^2 + 2x + 3$

15. $x = y^2 + 3$

16. $x = 2y^2$

17. $x = -\frac{1}{2}y^2$

18. $x = y^2 - 1$

19. $x = -y^2 - 4y$

20. $x = y^2 + y - 6$

21. $x = 4 - y - y^2$

22. $y = x^2 + 2x + 1$

23. $y = x^2 - 2x + 1$

24. $y = -\frac{1}{2}x^2$

25. $x = -y^2 + 2y - 1$

26. $x = -y^2 - 2y + 3$

27. $x = -2y^2 - 4y + 1$

28. $x = 2y^2 + 4y - 1$

Find the distance between each pair of points. Where appropriate, find an approximation to three decimal places.

29. $(1, 6)$ and $(5, 9)$

30. $(1, 10)$ and $(7, 2)$

31. $(0, -7)$ and $(3, -4)$

32. $(6, 2)$ and $(6, -8)$

33. $(-4, 4)$ and $(6, -6)$

34. $(5, 21)$ and $(-3, 1)$

Aha! **35.** $(8.6, -3.4)$ and $(-9.2, -3.4)$

36. $(5.9, 2)$ and $(3.7, -7.7)$

37. $\left(\frac{5}{7}, \frac{1}{14}\right)$ and $\left(\frac{1}{7}, \frac{11}{14}\right)$

38. $(0, \sqrt{7})$ and $(\sqrt{6}, 0)$

39. $(-\sqrt{6}, \sqrt{2})$ and $(0, 0)$

40. $(\sqrt{5}, -\sqrt{3})$ and $(0, 0)$

41. $(-4, -2)$ and $(-7, -11)$

42. $(-3, -7)$ and $(-1, -5)$

Find the midpoint of each segment with the given endpoints.

43. $(-7, 6)$ and $(9, 2)$

44. $(6, 7)$ and $(7, -9)$

45. $(2, -1)$ and $(5, 8)$

46. $(-1, 2)$ and $(1, -3)$

47. $(-8, -5)$ and $(6, -1)$

48. $(8, -2)$ and $(-3, 4)$

49. $(-3.4, 8.1)$ and $(2.9, -8.7)$

50. $(4.1, 6.9)$ and $(5.2, -6.9)$

51. $\left(\frac{1}{6}, -\frac{3}{4}\right)$ and $\left(-\frac{1}{3}, \frac{5}{6}\right)$

52. $\left(-\frac{4}{5}, -\frac{2}{3}\right)$ and $\left(\frac{1}{8}, \frac{3}{4}\right)$

53. $(\sqrt{2}, -1)$ and $(\sqrt{3}, 4)$

54. $(9, 2\sqrt{3})$ and $(-4, 5\sqrt{3})$

Find an equation of the circle satisfying the given conditions.

55. Center $(0, 0)$, radius 6

56. Center $(0, 0)$, radius 5

57. Center $(7, 3)$, radius $\sqrt{5}$

58. Center $(5, 6)$, radius $\sqrt{2}$

59. Center $(-4, 3)$, radius $4\sqrt{3}$

60. Center $(-2, 7)$, radius $2\sqrt{5}$

61. Center $(-7, -2)$, radius $5\sqrt{2}$

62. Center $(-5, -8)$, radius $3\sqrt{2}$

63. Center $(0, 0)$, passing through $(-3, 4)$

64. Center $(3, -2)$, passing through $(11, -2)$

65. Center $(-4, 1)$, passing through $(-2, 5)$

66. Center $(-1, -3)$, passing through $(-4, 2)$

Find the center and the radius of each circle. Then graph the circle.

67. $x^2 + y^2 = 64$

68. $x^2 + y^2 = 36$

69. $(x + 1)^2 + (y + 3)^2 = 36$

70. $(x - 2)^2 + (y + 3)^2 = 4$

71. $(x - 4)^2 + (y + 3)^2 = 10$

72. $(x + 5)^2 + (y - 1)^2 = 15$

73. $x^2 + y^2 = 10$

74. $x^2 + y^2 = 7$

75. $(x - 5)^2 + y^2 = \frac{1}{4}$

76. $x^2 + (y - 1)^2 = \frac{1}{25}$

77. $x^2 + y^2 + 8x - 6y - 15 = 0$

78. $x^2 + y^2 + 6x - 4y - 15 = 0$

79. $x^2 + y^2 - 8x + 2y + 13 = 0$

80. $x^2 + y^2 + 6x + 4y + 12 = 0$

81. $x^2 + y^2 + 10y - 75 = 0$

82. $x^2 + y^2 - 8x - 84 = 0$

83. $x^2 + y^2 + 7x - 3y - 10 = 0$

84. $x^2 + y^2 - 21x - 33y + 17 = 0$

85. $36x^2 + 36y^2 = 1$

86. $4x^2 + 4y^2 = 1$

Graph using a graphing calculator.

87. $x^2 + y^2 - 16 = 0$

88. $4x^2 + 4y^2 = 100$

89. $x^2 + y^2 + 14x - 16y + 54 = 0$

90. $x^2 + y^2 - 10x - 11 = 0$

TW **91.** Describe a procedure that would use the distance formula to determine whether three points, (x_1, y_1), (x_2, y_2), and (x_3, y_3), are vertices of a right triangle.

TW **92.** Does the graph of an equation of a circle include the point that is the center? Why or why not?

Skill Maintenance

Solve. [6.4]

93. $\frac{x}{4} + \frac{5}{6} = \frac{2}{3}$

94. $\frac{t}{6} - \frac{1}{9} = \frac{7}{12}$

95. A rectangle 10 in. long and 6 in. wide is bordered by a strip of uniform width. If the perimeter of the larger rectangle is twice that of the smaller rectangle, what is the width of the border? [1.7]

96. One airplane flies 60 mph faster than another. To fly a certain distance, the faster plane takes 4 hr and the slower plane takes 4 hr and 24 min. What is the distance? [3.3]

Solve each system. [3.2]

97. $3x - 8y = 5,$
$2x + 6y = 5$

98. $4x - 5y = 9,$
$12x - 10y = 18$

Synthesis

TW **99.** Outline a procedure that would use the distance formula to determine whether three points, (x_1, y_1), (x_2, y_2), and (x_3, y_3), are collinear (lie on the same line).

TW **100.** Why does the discussion of the distance formula precede the discussion of circles?

Find an equation of a circle satisfying the given conditions.

101. Center $(3, -5)$ and tangent to (touching at one point) the y-axis

102. Center $(-7, -4)$ and tangent to the x-axis

103. The endpoints of a diameter are $(7, 3)$ and $(-1, -3)$.

104. Center $(-3, 5)$ with a circumference of 8π units

105. Find the point on the y-axis that is equidistant from $(2, 10)$ and $(6, 2)$.

106. Find the point on the x-axis that is equidistant from $(-1, 3)$ and $(-8, -4)$.

107. *Wrestling.* The equation $x^2 + y^2 = \frac{81}{4}$, where x and y represent the number of meters from the center, can be used to draw the outer circle on a wrestling mat used in International, Olympic, and World Championship wrestling. The equation $x^2 + y^2 = 16$ can be used to draw the inner edge of the red zone. (*Source*: Based on data from the Government of Western Australia) Find the area of the red zone.

108. *Snowboarding.* Each side edge of the 156-cm Salomon Forecast ERA snowboard is an arc of a circle with a "running length" of 1220 mm and a "sidecut depth" of 22.7 mm (see the figure below).

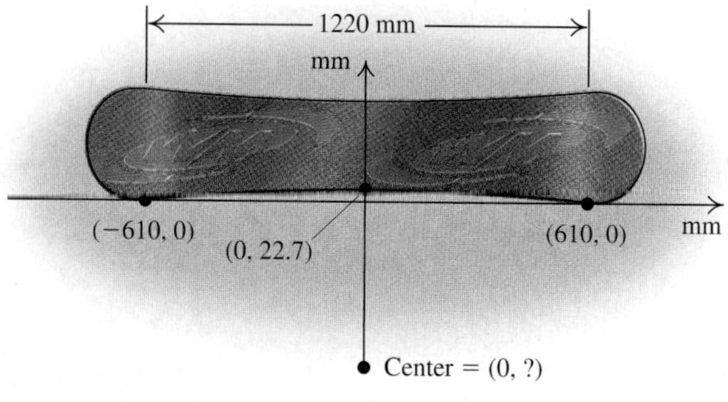

a) Using the coordinates shown, locate the center of the circle. (*Hint*: Equate distances.)

b) What radius is used for the edge of the board?

109. *Snowboarding.* The Academy Merit 157 snowboard has a running length of 1220 mm and a sidecut depth of 24.2 mm (see Exercise 108). What radius is used for the edge of this snowboard?

110. *Skiing.* The Völkl Supersport S5 Ti ski, when lying flat and viewed from above, has edges that are arcs of a circle. (Actually, each edge is made of two arcs of slightly different radii. The arc for the rear half of the ski edge has a slightly larger radius.)

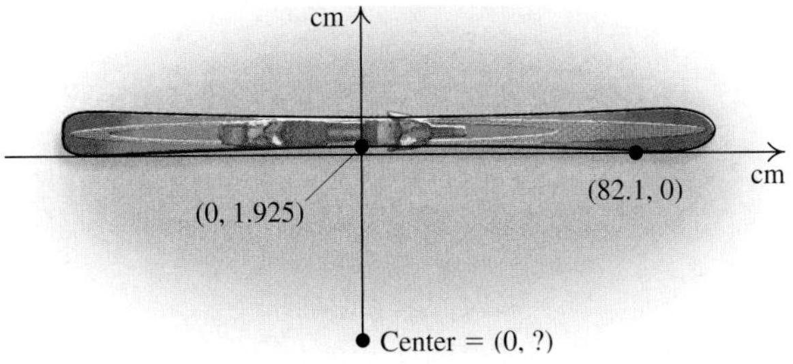

a) Using the coordinates shown, locate the center of the circle. (*Hint*: Equate distances.)

b) What radius is used for the arc passing through (0, 1.925) and (82.1, 0)?

111. *Doorway Construction.* Ace Carpentry needs to cut an arch for the top of an entranceway. The arch needs to be 8 ft wide and 2 ft high. To draw the arch, the carpenters will use a stretched string with chalk attached at an end as a compass.

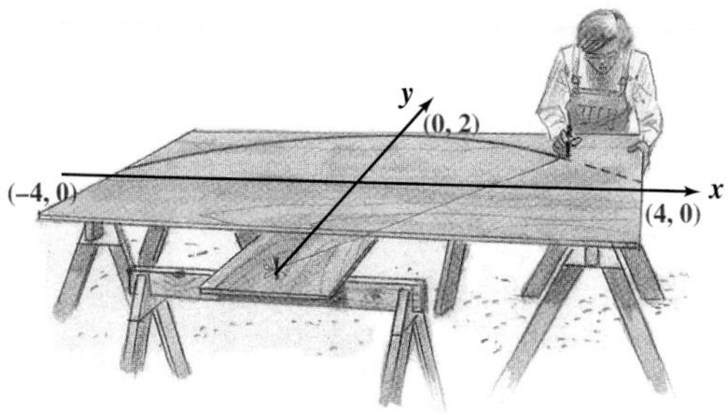

a) Using a coordinate system, locate the center of the circle.

b) What radius should the carpenters use to draw the arch?

112. *Archaeology.* During an archaeological dig, Martina finds the bowl fragment shown below. What was the original diameter of the bowl?

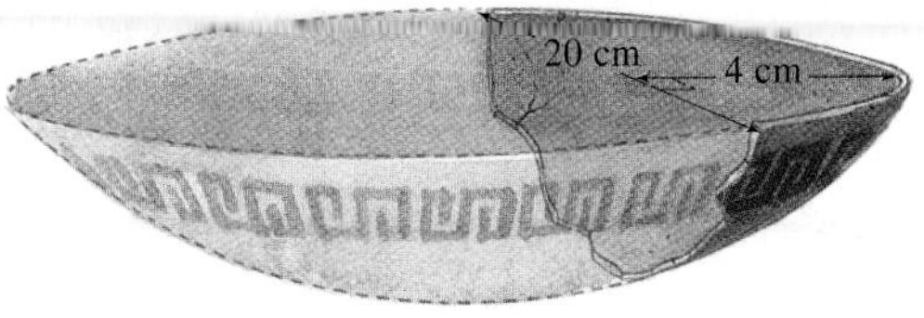

113. *Ferris Wheel Design.* A ferris wheel has a radius of 24.3 ft. Assuming that the center is 30.6 ft off the ground and that the origin is below the center, as in the following figure, find an equation of the circle.

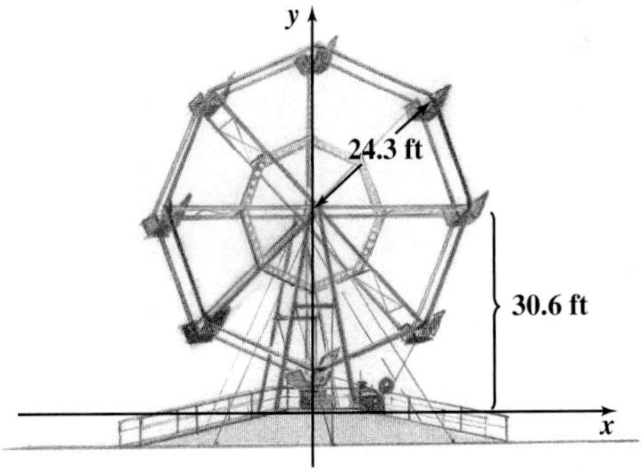

114. Use a graph of the equation $x = y^2 - y - 6$ to approximate to the nearest tenth the solutions of each of the following equations.

a) $y^2 - y - 6 = 2$ (*Hint*: Graph $x = 2$ on the same set of axes as the graph of $x = y^2 - y - 6$.)

b) $y^2 - y - 6 = -3$

115. *Power of a Motor.* The horsepower of a certain kind of engine is given by the formula

$$H = \frac{D^2N}{2.5},$$

where N is the number of cylinders and D is the diameter, in inches, of each piston. Graph this equation, assuming that $N = 6$ (a six-cylinder engine). Let D run from 2.5 to 8.

116. Prove the midpoint formula by showing that

i) the distance from (x_1, y_1) to

$$\left(\frac{x_1 + x_2}{2}, \frac{y_1 + y_2}{2}\right)$$

equals the distance from (x_2, y_2) to

$$\left(\frac{x_1 + x_2}{2}, \frac{y_1 + y_2}{2}\right);$$

and

ii) the points

$$(x_1, y_1), \left(\frac{x_1 + x_2}{2}, \frac{y_1 + y_2}{2}\right),$$

and

(x_2, y_2)

lie on the same line (see Exercise 99).

117. If the equation $x^2 + y^2 - 6x + 2y - 6 = 0$ is written as $y^2 + 2y + (x^2 - 6x - 6) = 0$, it can be regarded as quadratic in y.

a) Use the quadratic formula to solve for y.

b) Show that the graph of your answer to part (a) coincides with the graph on p. 800.

TW **118.** How could a graphing calculator best be used to help you sketch the graph of an equation of the form $x = ay^2 + by + c$?

TW **119.** Why should a graphing calculator's window be "squared" before graphing a circle?

10.2 Conic Sections: Ellipses

Ellipses Centered at (0, 0) ■ Ellipses Centered at (h, k)

When a cone is cut at an angle, as shown below, the conic section formed is an *ellipse*. To draw an ellipse, stick two tacks in a piece of cardboard. Then tie a loose string to the tacks, place a pencil as shown, and draw an oval by moving the pencil while stretching the string tight.

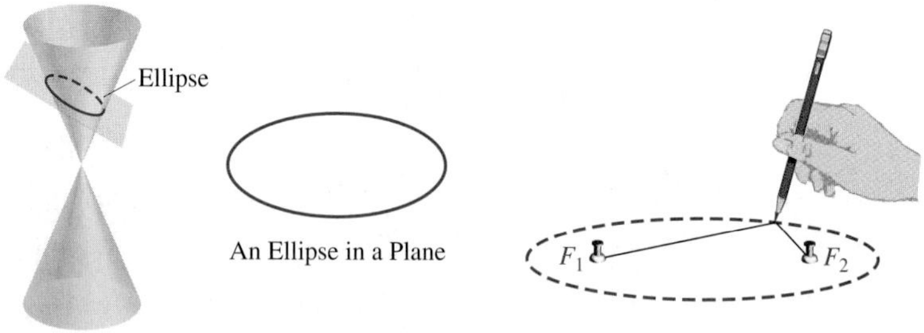

Ellipses Centered at (0, 0)

An **ellipse** is defined as the set of all points in a plane for which the sum of the distances from two fixed points F_1 and F_2 is constant. The points F_1 and F_2 are called **foci** (pronounced fō-sī), the plural of focus. In the figure above, the tacks are at the foci and the length of the string is the constant sum of the distances. The midpoint of the segment F_1F_2 is the **center.** The equation of an ellipse is as follows. Its derivation is left to the exercises.

Study Tip

Preparing for the Final Exam

It is never too early to begin studying for a final exam. If you have at least three days, consider the following:

- Reviewing the highlighted or boxed information in each chapter;
- Studying the Chapter Tests, Review Exercises, Cumulative Reviews, and Study Summaries;
- Re-taking all quizzes and tests that have been returned to you;
- Attending any review sessions being offered;
- Organizing or joining a study group;
- Using the video or software supplements, or asking a tutor or professor about any trouble spots;
- Asking for previous final exams (and answers) for practice.

Equation of an Ellipse Centered at the Origin The equation of an ellipse centered at the origin and symmetric with respect to both axes is

$$\frac{x^2}{a^2} + \frac{y^2}{b^2} = 1, \quad a, b > 0. \qquad \text{(Standard form)}$$

To graph an ellipse centered at the origin, it helps to first find the intercepts. If we replace x with 0, we can find the y-intercepts:

$$\frac{0^2}{a^2} + \frac{y^2}{b^2} = 1$$

$$\frac{y^2}{b^2} = 1$$

$$y^2 = b^2 \quad \text{or} \quad y = \pm b.$$

Thus the y-intercepts are $(0, b)$ and $(0, -b)$. Similarly, the x-intercepts are $(a, 0)$ and $(-a, 0)$. If $a > b$, the ellipse is said to be horizontal and $(-a, 0)$ and $(a, 0)$ are referred to as the **vertices** (singular, **vertex**). If $b > a$, the ellipse is said to be vertical and $(0, -b)$ and $(0, b)$ are then the vertices.

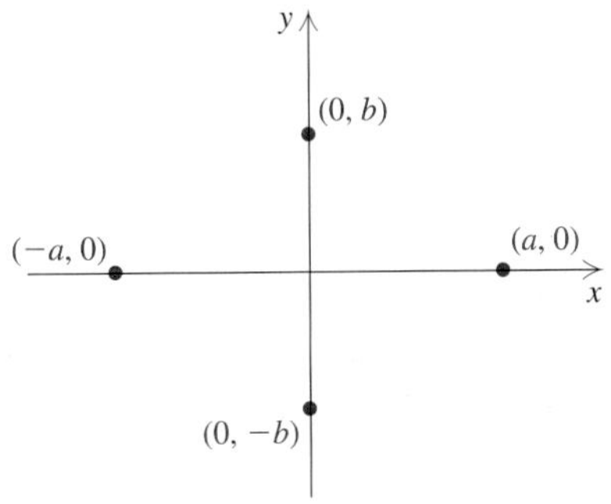

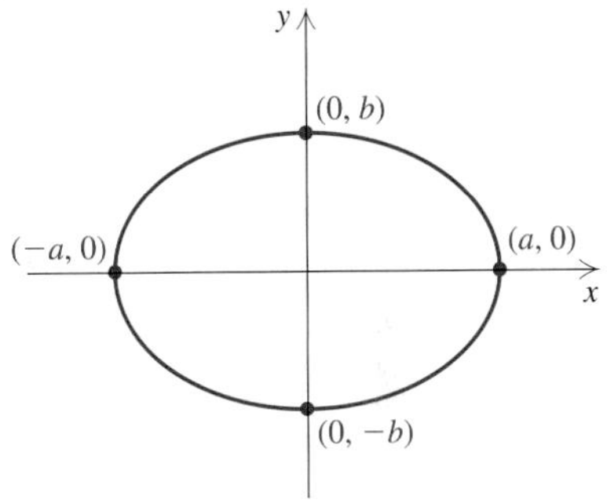

Plotting these four points and drawing an oval-shaped curve, we graph the ellipse. If a more precise graph is desired, we can plot more points.

Using a and b to Graph an Ellipse For the ellipse

$$\frac{x^2}{a^2} + \frac{y^2}{b^2} = 1,$$

the x-intercepts are $(-a, 0)$ and $(a, 0)$. The y-intercepts are $(0, -b)$ and $(0, b)$. For $a^2 > b^2$, the ellipse is horizontal. For $b^2 > a^2$, the ellipse is vertical.

EXAMPLE 1 Graph the ellipse

$$\frac{x^2}{4} + \frac{y^2}{9} = 1.$$

SOLUTION Note that

$$\frac{x^2}{4} + \frac{y^2}{9} = \frac{x^2}{2^2} + \frac{y^2}{3^2}.$$ **Identifying a and b. Since $b > a$, the ellipse is vertical.**

Thus the x-intercepts are $(-2, 0)$ and $(2, 0)$, and the y-intercepts are $(0, -3)$ and $(0, 3)$. We plot these points and connect them with an oval-shaped curve. To plot two other points, we let $x = 1$ and solve for y:

$$\frac{1^2}{4} + \frac{y^2}{9} = 1$$

$$36\left(\frac{1}{4} + \frac{y^2}{9}\right) = 36 \cdot 1$$

$$36 \cdot \frac{1}{4} + 36 \cdot \frac{y^2}{9} = 36$$

$$9 + 4y^2 = 36$$

$$4y^2 = 27$$

$$y^2 = \frac{27}{4}$$

$$y = \pm\sqrt{\frac{27}{4}}$$

$$y \approx \pm 2.6.$$

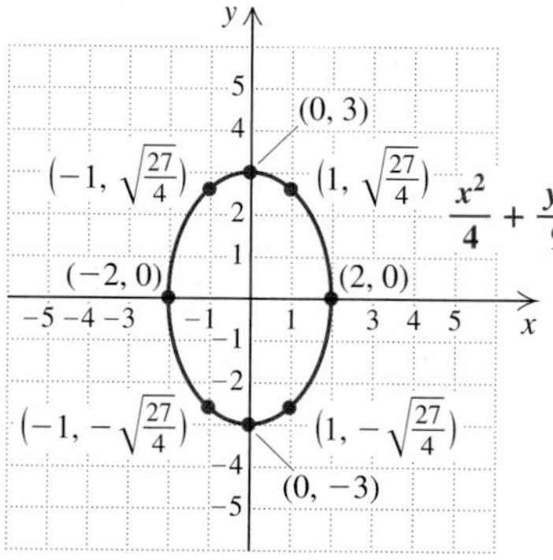

Thus, $(1, 2.6)$ and $(1, -2.6)$ can also be used to draw the graph. Similarly, the points $(-1, 2.6)$ and $(-1, -2.6)$ should appear on the graph.

Student Notes

Note that any equation of the form $Ax^2 + By^2 = C$ can be rewritten as an equivalent equation in standard form. The graph is an ellipse.

EXAMPLE 2 Graph: $4x^2 + 25y^2 = 100$.

SOLUTION To write the equation in standard form, we divide both sides by 100 to get 1 on the right side:

$$\frac{4x^2 + 25y^2}{100} = \frac{100}{100}$$ **Dividing by 100 to get 1 on the right side**

$$\left.\begin{aligned}\frac{4x^2}{100} + \frac{25y^2}{100} &= 1\\ \frac{x^2}{25} + \frac{y^2}{4} &= 1\end{aligned}\right\}$$ **Simplifying**

$$\frac{x^2}{5^2} + \frac{y^2}{2^2} = 1.$$ $a = 5, b = 2$

The x-intercepts are $(-5, 0)$ and $(5, 0)$, and the y-intercepts are $(0, -2)$ and $(0, 2)$. We plot the intercepts and connect them with an oval-shaped curve. Other points can also be computed and plotted as shown at left.

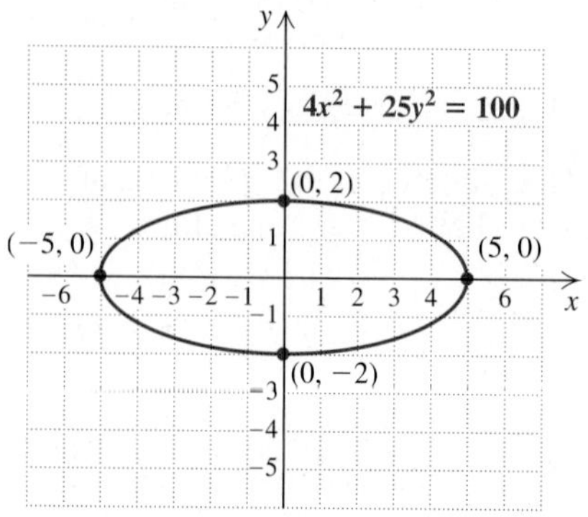

Ellipses Centered at (h, k)

Horizontal and vertical translations, similar to those used in Chapter 8, can be used to graph ellipses that are not centered at the origin.

Equation of an Ellipse Centered at (h, k) The standard form of a horizontal or vertical ellipse centered at (h, k) is

$$\frac{(x-h)^2}{a^2} + \frac{(y-k)^2}{b^2} = 1.$$

The vertices are $(h + a, k)$ and $(h - a, k)$ if horizontal; $(h, k + b)$ and $(h, k - b)$ if vertical.

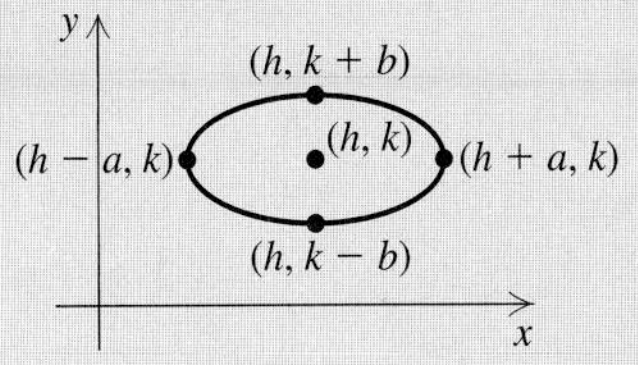

EXAMPLE 3 Graph the ellipse

$$\frac{(x-1)^2}{4} + \frac{(y+5)^2}{9} = 1.$$

SOLUTION Note that

$$\frac{(x-1)^2}{4} + \frac{(y+5)^2}{9} = \frac{(x-1)^2}{2^2} + \frac{(y+5)^2}{3^2}.$$

Thus, $a = 2$ and $b = 3$. To determine the center of the ellipse, (h, k), note that

$$\frac{(x-1)^2}{2^2} + \frac{(y+5)^2}{3^2} = \frac{(x-1)^2}{2^2} + \frac{(y-(-5))^2}{3^2}.$$

Thus the center is $(1, -5)$. We plot the points 2 units to the left and right of center, as well as the points 3 units above and below center. These are the points $(3, -5)$, $(-1, -5)$, $(1, -2)$, and $(1, -8)$. The graph of the ellipse is shown at left.

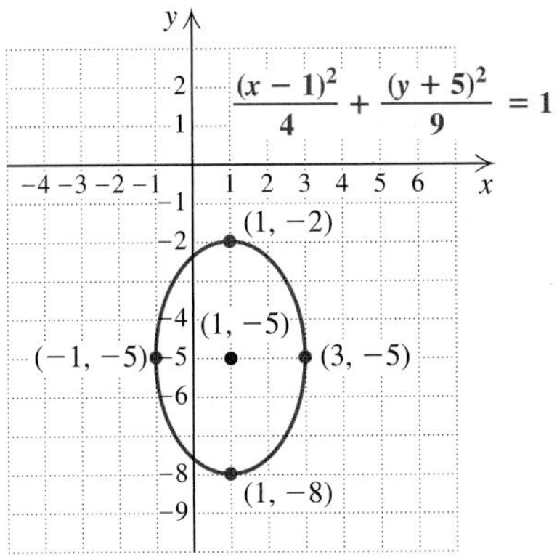

Note that this ellipse is the same as the ellipse in Example 1 but translated 1 unit to the right and 5 units down.

Ellipses have many applications. Communications satellites move in elliptical orbits with the earth as a focus while the earth itself follows an elliptical path around the sun. A medical instrument, the lithotripter, uses shock waves originating at one focus to crush a kidney stone located at the other focus.

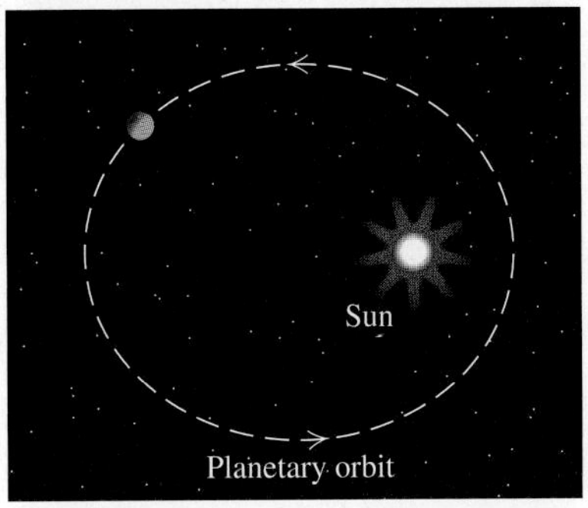

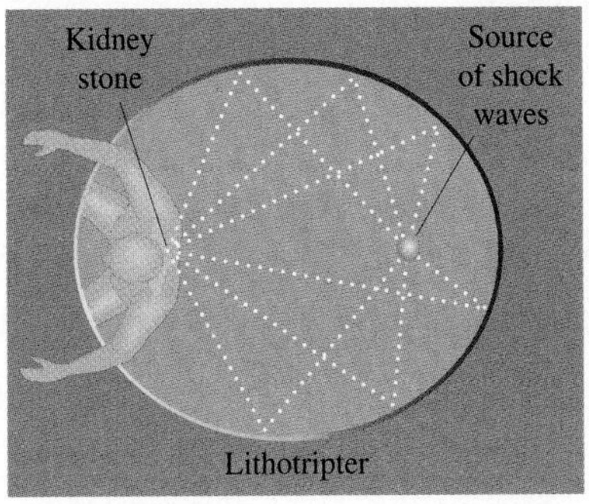

In some buildings, an ellipsoidal ceiling creates a "whispering gallery" in which a person at one focus can whisper and still be heard clearly at the other focus. This happens because sound waves coming from one focus are all reflected to the other focus. Similarly, light waves bouncing off an ellipsoidal mirror are used in a dentist's or surgeon's reflector light. The light source is located at one focus while the patient's mouth or surgical field is at the other.

Graphing Ellipses

Graphing an ellipse on a graphing calculator is much like graphing a circle: We graph it in two pieces after solving for y.

To graph the ellipse given by the equation $4x^2 + 25y^2 = 100$, we first solve for y:

$$\begin{aligned} 4x^2 + 25y^2 &= 100 \\ 25y^2 &= 100 - 4x^2 \\ y^2 &= 4 - \tfrac{4}{25}x^2 \\ y &= \pm\sqrt{4 - \tfrac{4}{25}x^2}. \end{aligned}$$

Then, using a squared window, we graph.

$y_1 = -\sqrt{4 - \tfrac{4}{25}x^2}, \quad y_2 = \sqrt{4 - \tfrac{4}{25}x^2}$

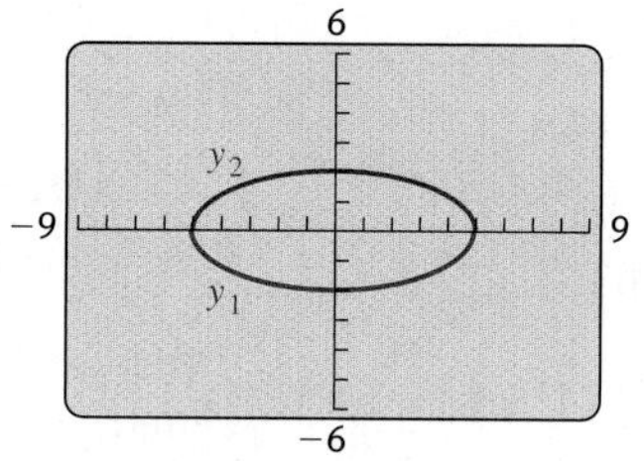

On many calculators, pressing APPS and selecting Conics and then Ellipse accesses a program in which equations in standard form of ellipses centered at (h, k) can be graphed directly.

10.2 EXERCISE SET

FOR EXTRA HELP

MathXL | MyMathLab | InterAct Math | AW Math Tutor Center | Video Lectures on CD: Disc 5 | Student's Solutions Manual

Concept Reinforcement *Classify each of the following as either true or false.*

1. The graph of $\dfrac{x^2}{9} + \dfrac{y^2}{25} = 1$ includes the points $(-3, 0)$ and $(3, 0)$.

2. The graph of $\dfrac{x^2}{36} + \dfrac{y^2}{25} = 1$ includes the points $(0, -5)$ and $(0, 5)$.

3. The graph of $\dfrac{x^2}{28} + \dfrac{y^2}{48} = 1$ is a vertical ellipse.

4. The graph of $\frac{x^2}{30} + \frac{y^2}{20} = 1$ is a vertical ellipse.

5. The graph of $\frac{x^2}{25} - \frac{y^2}{9} = 1$ is a horizontal ellipse.

6. The graph of $\frac{-x^2}{20} + \frac{y^2}{16} = 1$ is a horizontal ellipse.

7. The graph of $\frac{(x+3)^2}{25} + \frac{(y-2)^2}{36} = 1$ is an ellipse centered at $(-3, 2)$.

8. The graph of $\frac{(x-2)^2}{49} + \frac{(y+5)^2}{9} = 1$ is an ellipse centered at $(2, -5)$.

Graph each of the following equations.

9. $\frac{x^2}{1} + \frac{y^2}{9} = 1$

10. $\frac{x^2}{9} + \frac{y^2}{1} = 1$

11. $\frac{x^2}{25} + \frac{y^2}{9} = 1$

12. $\frac{x^2}{16} + \frac{y^2}{25} = 1$

13. $4x^2 + 9y^2 = 36$

14. $9x^2 + 4y^2 = 36$

15. $16x^2 + 9y^2 = 144$

16. $9x^2 + 16y^2 = 144$

17. $2x^2 + 3y^2 = 6$

18. $5x^2 + 7y^2 = 35$

Aha! **19.** $5x^2 + 5y^2 = 125$

20. $8x^2 + 5y^2 = 80$

21. $3x^2 + 7y^2 - 63 = 0$

22. $3x^2 + 8y^2 - 72 = 0$

23. $8x^2 = 96 - 3y^2$

24. $6y^2 = 24 - 8x^2$

25. $16x^2 + 25y^2 = 1$

26. $9x^2 + 4y^2 = 1$

27. $\frac{(x-3)^2}{9} + \frac{(y-2)^2}{25} = 1$

28. $\frac{(x-2)^2}{25} + \frac{(y-4)^2}{9} = 1$

29. $\frac{(x+4)^2}{16} + \frac{(y-3)^2}{49} = 1$

30. $\frac{(x+5)^2}{4} + \frac{(y-2)^2}{36} = 1$

31. $12(x-1)^2 + 3(y+4)^2 = 48$
(*Hint*: Divide both sides by 48.)

32. $4(x-6)^2 + 9(y+2)^2 = 36$

Aha! **33.** $4(x+3)^2 + 4(y+1)^2 - 10 = 90$

34. $9(x+6)^2 + (y+2)^2 - 20 = 61$

TW **35.** Is the center of an ellipse part of the ellipse itself? Why or why not?

TW **36.** Can an ellipse ever be the graph of a function? Why or why not?

Skill Maintenance

Solve.

37. $\frac{3}{x-2} - \frac{5}{x-2} = 9$ [6.4]

38. $\frac{7}{x+3} - \frac{2}{x+3} = 8$ [6.4]

39. $\frac{x}{x-4} - \frac{3}{x-5} = \frac{2}{x-4}$ [8.2]

40. $\frac{7}{x-3} - \frac{x}{x-2} = \frac{4}{x-2}$ [8.2]

41. $9 - \sqrt{2x+1} = 7$ [7.6]

42. $5 - \sqrt{x+3} = 9$ [7.6]

Synthesis

TW **43.** An eccentric person builds a pool table in the shape of an ellipse with a hole at one focus and a tiny dot at the other. Guests are amazed at how many bank shots the owner of the pool table makes. Explain why this occurs.

TW **44.** Can a circle be considered a special type of ellipse? Why or why not?

Find an equation of an ellipse that contains the following points.

45. $(-9, 0)$, $(9, 0)$, $(0, -11)$, and $(0, 11)$

46. $(-7, 0)$, $(7, 0)$, $(0, -5)$, and $(0, 5)$

47. $(-2, -1)$, $(6, -1)$, $(2, -4)$, and $(2, 2)$

48. $(4, 3)$, $(-6, 3)$, $(-1, -1)$ and $(-1, 7)$

49. *Theatrical Lighting.* The spotlight on a violin soloist casts an ellipse of light on the floor below her that is 6 ft wide and 10 ft long. Find an equation of that ellipse if the performer is in its center, x is the distance from the performer to the side of the ellipse, and y is the distance from the performer to the top of the ellipse.

50. *Astronomy.* The maximum distance of the planet Mars from the sun is 2.48×10^8 mi. The minimum distance is 3.46×10^7 mi. The sun is at one focus of the elliptical orbit. Find the distance from the sun to the other focus.

51. Let $(-c, 0)$ and $(c, 0)$ be the foci of an ellipse. Any point $P(x, y)$ is on the ellipse if the sum of the distances from the foci to P is some constant. Use $2a$ to represent this constant.

a) Show that an equation for the ellipse is given by

$$\frac{x^2}{a^2} + \frac{y^2}{a^2 - c^2} = 1.$$

b) Substitute b^2 for $a^2 - c^2$ to get standard form.

52. *President's Office.* The Oval Office of the President of the United States is an ellipse 31 ft wide and 38 ft long. Show in a sketch precisely where the President and an adviser could sit to best hear each other using the room's acoustics. (*Hint*: See Exercise 51(b) and the discussion following Example 3.)

53. *Dentistry.* The light source in a dental lamp shines against a reflector that is shaped like a portion of an ellipse in which the light source is one focus of the ellipse. Reflected light enters a patient's mouth at the other focus of the ellipse. If the ellipse from which the reflector was formed is 2 ft wide and 6 ft long, how far should the patient's mouth be from the light source? (*Hint*: See Exercise 51(b).)

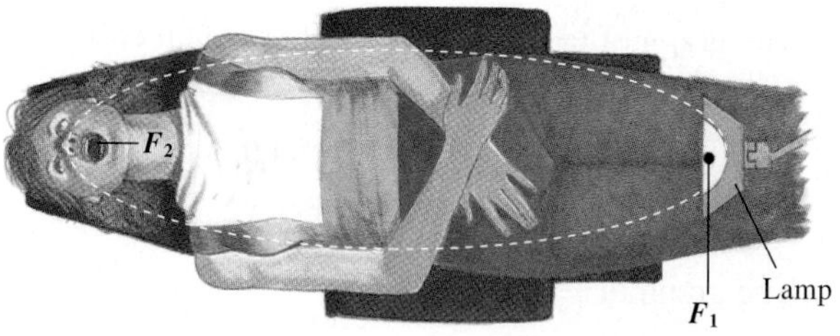

54. *Firefighting.* The size and shape of certain forest fires can be approximated as the union of two "half-ellipses." For the blaze modeled below, the equation of the smaller ellipse—the part of the fire moving *into* the wind—is

$$\frac{x^2}{40{,}000} + \frac{y^2}{10{,}000} = 1.$$

The equation of the other ellipse—the part moving *with* the wind—is

$$\frac{x^2}{250{,}000} + \frac{y^2}{10{,}000} = 1.$$

Determine the width and the length of the fire. (*Source for figure*: "Predicting Wind-Driven Wild Land Fire Size and Shape," Hal E. Anderson, Research Paper INT-305, U.S. Department of Agriculture, Forest Service, February 1983)

For each of the following equations, complete the square as needed and find an equivalent equation in standard form. Then graph the ellipse.

55. $x^2 - 4x + 4y^2 + 8y - 8 = 0$

56. $4x^2 + 24x + y^2 - 2y - 63 = 0$

Astronomy. To model the earth's orbit around the sun, we begin by drawing axes showing the sun at the origin. The following table and graph show the location of the earth at various times during the year.

Location of Earth (units are in millions of kilometers)	
(147.1, 0)	(−152.1, 0)
(100, 113.4)	(−125, −78.2)
(25, 147.7)	(−50, −140.1)
(0, 149.56)	(0, −149.56)
(−50, 140.1)	(25, −147.7)
(−125, 78.2)	(100, −113.4)

Based on information from gsfc.nasa.gov

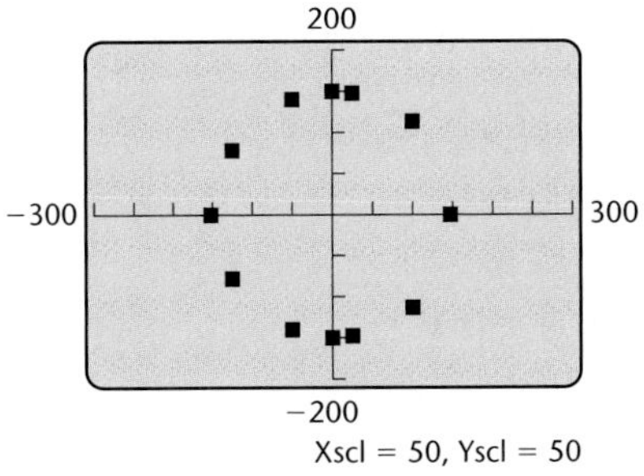

Although the points appear to form a circle, the shape of the orbit is actually an ellipse, with the sun at one focus.

57. If the sun is 2,500,000 km from the center of the ellipse, what is the maximum distance of the earth from the sun?

58. An *astronomical unit*, which has been defined using the mean distance of the earth from the sun, is about 149,600,000 km. What is the maximum distance of the earth from the sun in astronomical units?

Collaborative Corner

A Cosmic Path

Focus: Ellipses

Time: 20–30 minutes

Group Size: 2

Materials: Scientific calculators

In March 1996, the comet Hyakutake came within 21 million mi of the sun, and closer to Earth than any comet in over 500 yr (*Source*: Associated Press newspaper story, 3/20/96). Hyakutake is traveling in an elliptical orbit with the sun at one focus. The comet's average speed is about 100,000 mph (it actually goes much faster near its foci and slower as it gets further from the foci) and one orbit takes about 15,000 yr. (Astronomers estimate the time at 10,000–20,000 yr.)

ACTIVITY

1. The elliptical orbit of Hyakutake is so elongated that the distance traveled in one orbit can be estimated by $4a$ (see the following figure). Use the information above to estimate the distance, in millions of miles, traveled in one orbit. Then determine a.

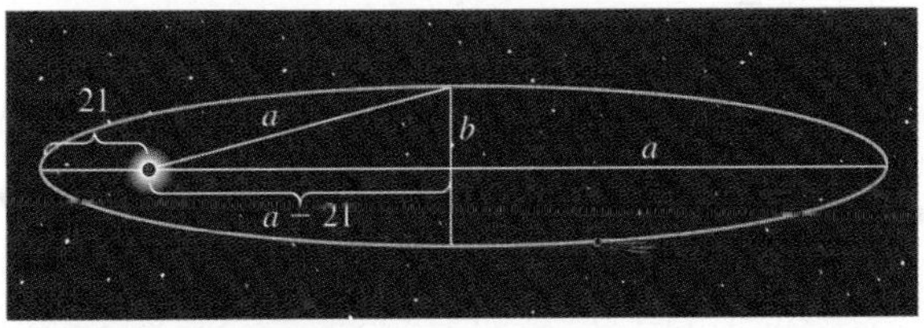

Units are millions of miles.

2. Using the figure above, express b^2 as a function of a. Then solve for b using the value found for a in part (1).
3. Approximately how far will Hyakutake be from the sun at the most distant part of its orbit?
4. Repeat parts (1)–(3), with one group member using the lower estimate of orbit time (10,000 yr) and the other using the upper estimate of orbit time (20,000 yr). By how much do the three answers to part (3) vary?

10.3 Conic Sections: Hyperbolas

Hyperbolas ■ Hyperbolas (Nonstandard Form) ■ Classifying Graphs of Equations

Study Tip

Listen Up!

Many professors make a point of telling their classes what topics or chapters will (or will not) be covered on the final exam. Take special note of this information and focus your studying around it.

Hyperbolas

A **hyperbola** looks like a pair of parabolas, but the shapes are actually different. A hyperbola has two **vertices** and the line through the vertices is known as the **axis.** The point halfway between the vertices is called the **center.** The two curves that comprise a hyperbola are called **branches.**

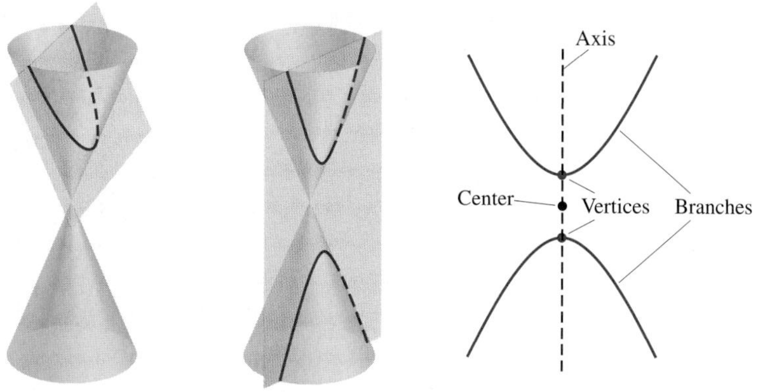

Parabola   Hyperbola in three dimensions   Hyperbola in a plane

Equation of a Hyperbola Centered at the Origin A hyperbola with its center at the origin* has its equation as follows:

$$\frac{x^2}{a^2} - \frac{y^2}{b^2} = 1 \quad \text{(Horizontal axis);}$$

$$\frac{y^2}{b^2} - \frac{x^2}{a^2} = 1 \quad \text{(Vertical axis).}$$

Note that both equations have a 1 on the right-hand side and a subtraction symbol between the terms. For the discussion that follows, we assume $a, b > 0$.

*Hyperbolas with horizontal or vertical axes and centers *not* at the origin are discussed in Exercises 59–64.

To graph a hyperbola, it helps to begin by graphing two lines called **asymptotes.** Although the asymptotes themselves are not part of the graph, they serve as guidelines for an accurate sketch.

As a hyperbola gets farther away from the origin, it gets closer and closer to its asymptotes. The larger $|x|$ gets, the closer the graph gets to an asymptote. The asymptotes act to "constrain" the graph of a hyperbola. Parabolas are *not* constrained by any asymptotes.

Asymptotes of a Hyperbola

For hyperbolas with equations as shown below, the asymptotes are the lines

$$y = \frac{b}{a}x \quad \text{and} \quad y = -\frac{b}{a}x.$$

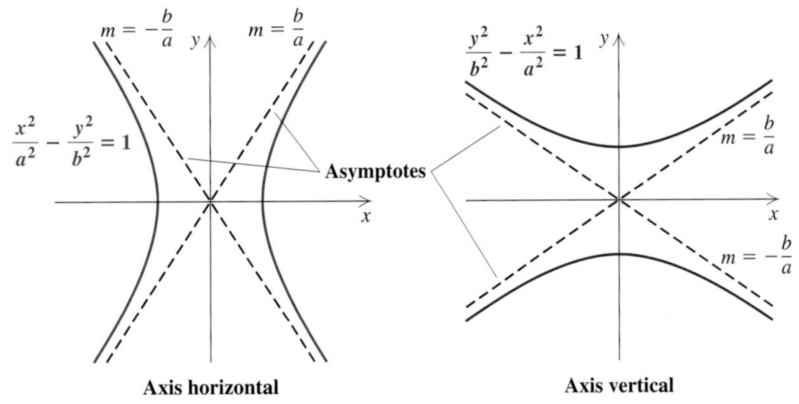

In Section 10.2, we used a and b to determine the width and the length of an ellipse. For hyperbolas, a and b are used to determine the base and the height of a rectangle that can be used as an aid in sketching asymptotes and locating vertices. This is illustrated in the following example.

EXAMPLE 1 Graph: $\frac{x^2}{4} - \frac{y^2}{9} = 1$.

SOLUTION Note that

$$\frac{x^2}{4} - \frac{y^2}{9} = \frac{x^2}{2^2} - \frac{y^2}{3^2}, \qquad \textbf{Identifying } \boldsymbol{a} \textbf{ and } \boldsymbol{b}$$

so $a = 2$ and $b = 3$. The asymptotes are thus

$$y = \frac{3}{2}x \quad \text{and} \quad y = -\frac{3}{2}x.$$

To help us sketch asymptotes and locate vertices, we use a and b—in this case, 2 and 3—to form the pairs $(-2, 3)$, $(2, 3)$, $(2, -3)$, and $(-2, -3)$. We plot these pairs and lightly sketch a rectangle. The asymptotes pass through the corners and, since this is a horizontal hyperbola, the vertices are where the rectangle intersects the x-axis. Finally, we draw the hyperbola, as shown below.

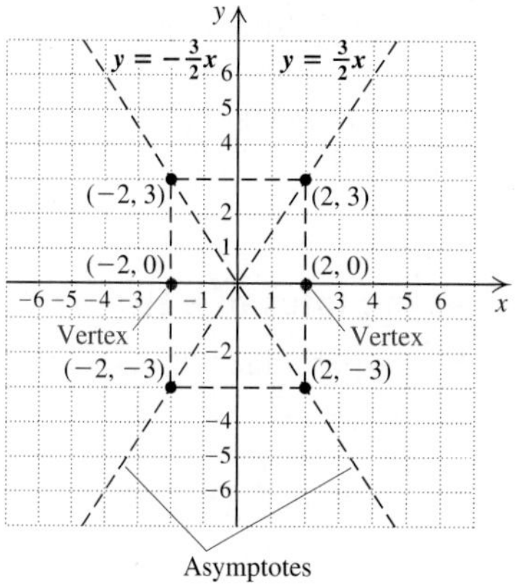

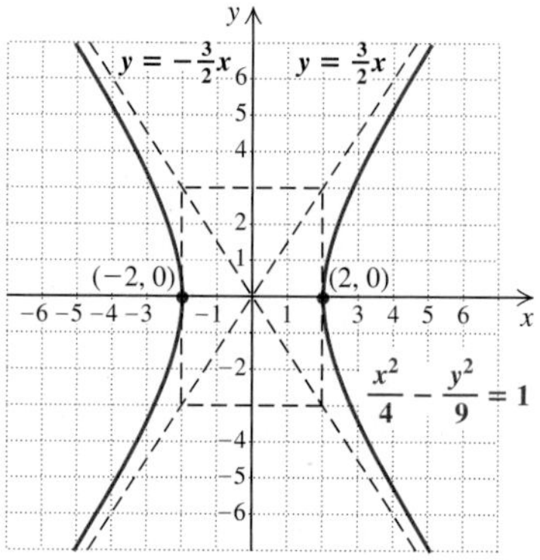

EXAMPLE 2 Graph: $\frac{y^2}{36} - \frac{x^2}{4} = 1$.

SOLUTION Note that

$$\frac{y^2}{36} - \frac{x^2}{4} = \frac{y^2}{6^2} - \frac{x^2}{2^2} = 1.$$

Whether the hyperbola is horizontal or vertical is determined by the nonnegative term. Here there is a *y* in this term, so the hyperbola is vertical.

Student Notes

Regarding the orientation of a hyperbola, you may find it helpful to think as follows: "The axis is parallel to the x-axis if $\frac{x^2}{a^2}$ is the positive term. The axis is parallel to the y-axis if $\frac{y^2}{b^2}$ is the positive term."

Using ± 2 as x-coordinates and ± 6 as y-coordinates, we plot $(2, 6)$, $(2, -6)$, $(-2, 6)$, and $(-2, -6)$, and lightly sketch a rectangle through them. The asymptotes pass through the corners (see the figure on the left below). Since the hyperbola is vertical, its vertices are $(0, 6)$ and $(0, -6)$. Finally, we draw curves through the vertices toward the asymptotes, as shown below.

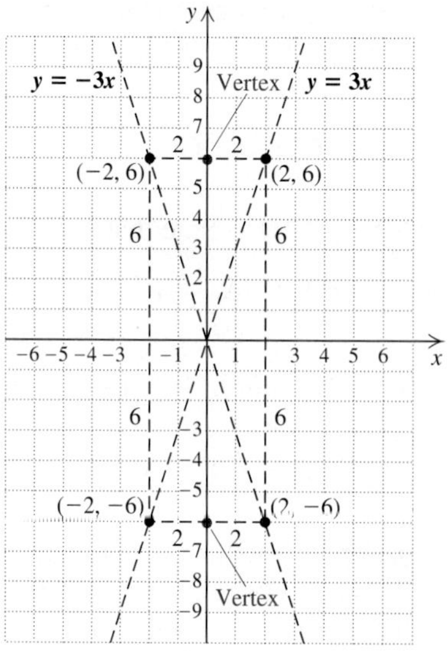

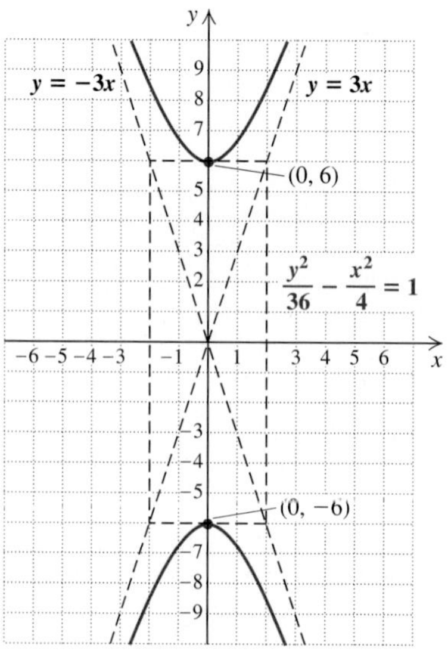

Hyperbolas (Nonstandard Form)

The equations for hyperbolas just examined are the standard ones, but there are other hyperbolas. We consider some of them.

Equation of a Hyperbola in Nonstandard Form Hyperbolas having the x- and y-axes as asymptotes have equations as follows:

$xy = c$, where c is a nonzero constant.

EXAMPLE 3 Graph: $xy = -8$.

SOLUTION We first solve for y:

$$y = -\frac{8}{x}. \qquad \textbf{Dividing both sides by } x\textbf{. Note that } x \neq 0.$$

Next, we find some solutions, keeping the results in a table. Note that x cannot be 0 and that for large values of $|x|$, y will be close to 0. Thus the x- and y-axes serve as asymptotes. We plot the points and draw two curves.

x	y
2	−4
−2	4
4	−2
−4	2
1	−8
−1	8
8	−1
−8	1

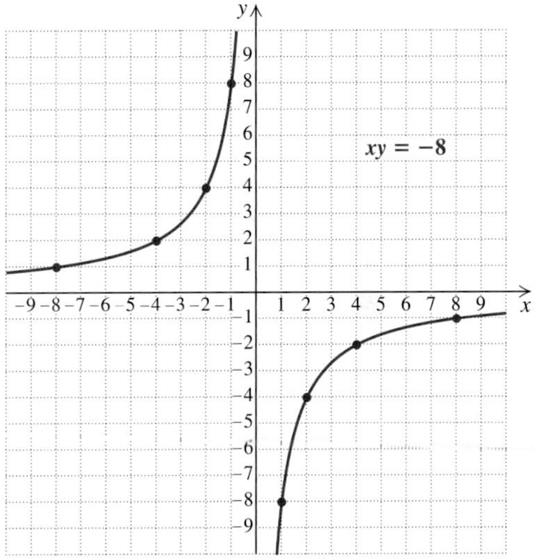

Hyperbolas have many applications. A jet breaking the sound barrier creates a sonic boom with a wave front the shape of a cone. The intersection of the cone with the ground is one branch of a hyperbola. Some comets travel in hyperbolic orbits, and a cross section of many lenses is hyperbolic in shape.

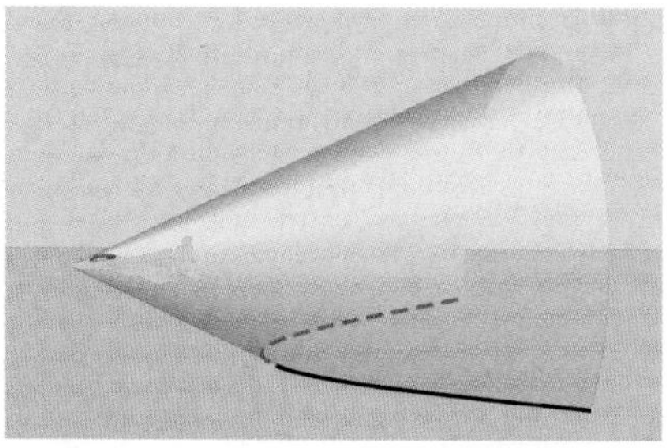

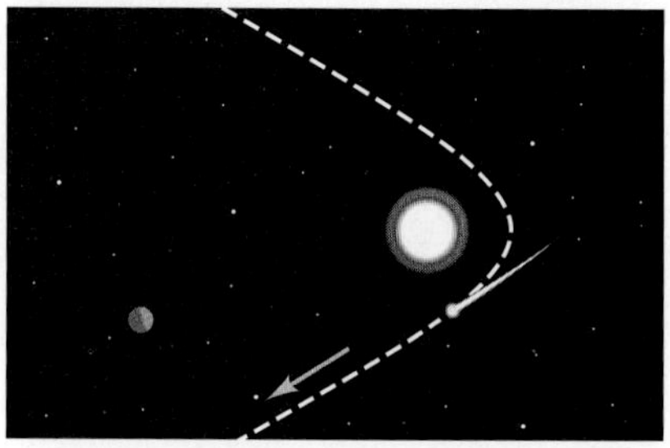

Graphing Hyperbolas

Graphing a hyperbola on a graphing calculator is much like graphing a circle or an ellipse.

To graph the hyperbola given by the equation

$$\frac{x^2}{25} - \frac{y^2}{49} = 1.$$

we solve for y, which gives the equations

$$y_1 = \frac{\sqrt{49x^2 - 1225}}{5} = \frac{7}{5}\sqrt{x^2 - 25}$$

and $$y_2 = \frac{-\sqrt{49x^2 - 1225}}{5} = -\frac{7}{5}\sqrt{x^2 - 25},$$

or $$y_2 = -y_1.$$

When the two pieces are drawn on the same squared window, the result is as shown. The gaps occur where the graph is nearly vertical.

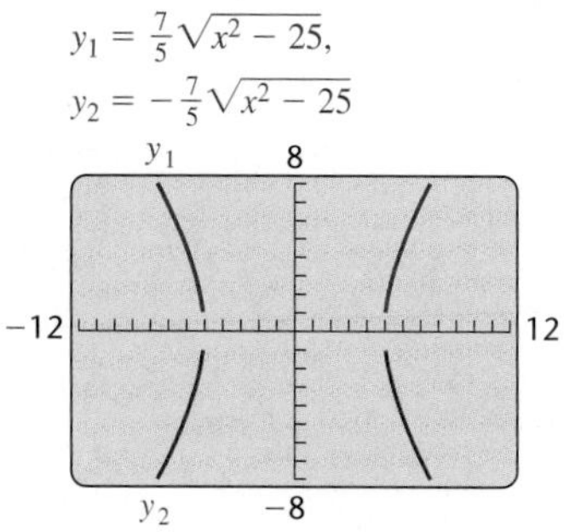

On many calculators, pressing APPS and selecting Conics and then Hyperbola accesses a program in which equations in standard form of hyperbolas centered at (h, k) can be graphed directly. (See Exercises 59–64.)

Classifying Graphs of Equations

Connecting the Concepts

We summarize the equations and the graphs of the conic sections studied. The examples resume on p. 819.

PARABOLA

$$y = ax^2 + bx + c,\ a > 0$$
$$= a(x - h)^2 + k$$

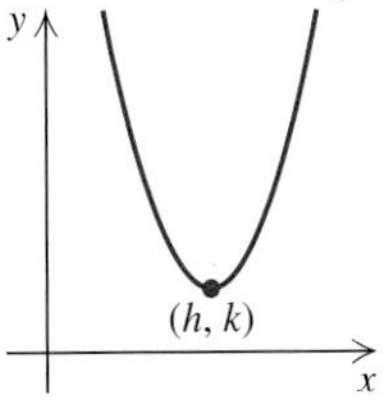

$$y = ax^2 + bx + c,\ a < 0$$
$$= a(x - h)^2 + k$$

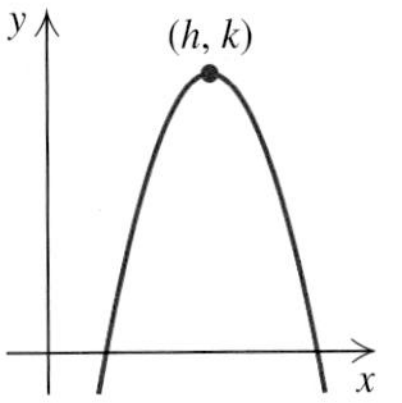

$$x = ay^2 + by + c,\ a > 0$$
$$= a(y - k)^2 + h$$

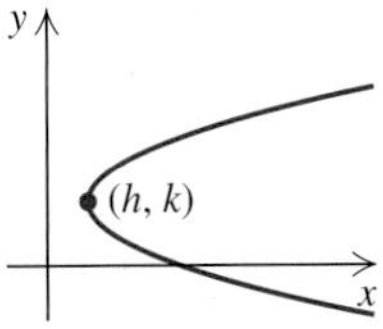

$$x = ay^2 + by + c,\ a < 0$$
$$= a(y - k)^2 + h$$

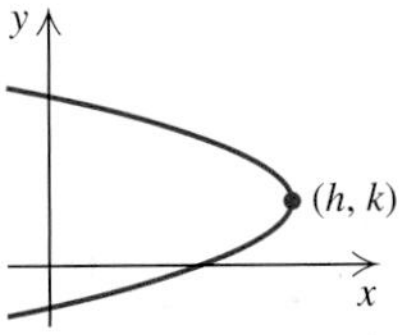

CIRCLE

Center at the origin:

$$x^2 + y^2 = r^2$$

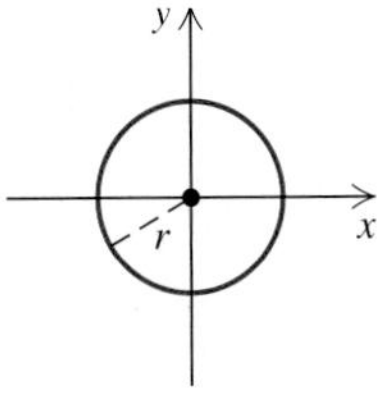

Center at (h, k):

$$(x - h)^2 + (y - k)^2 = r^2$$

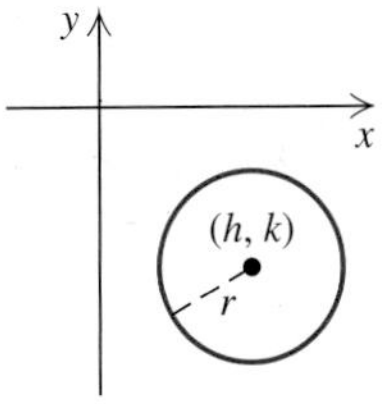

(continued)

HYPERBOLA

Center at the origin:

$$\frac{x^2}{a^2} - \frac{y^2}{b^2} = 1$$

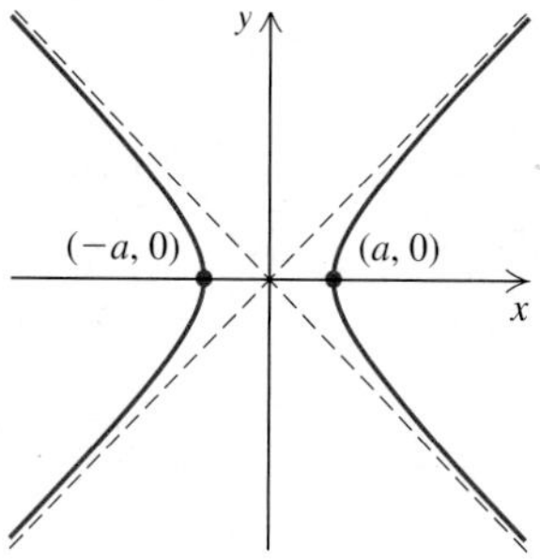

$$\frac{y^2}{b^2} - \frac{x^2}{a^2} = 1$$

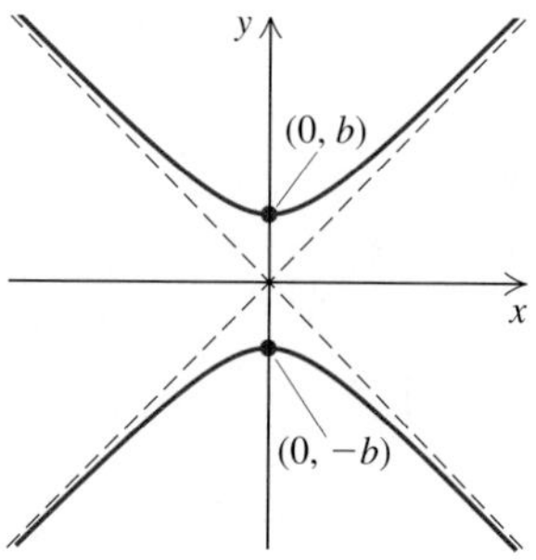

$xy = c,\ c > 0$

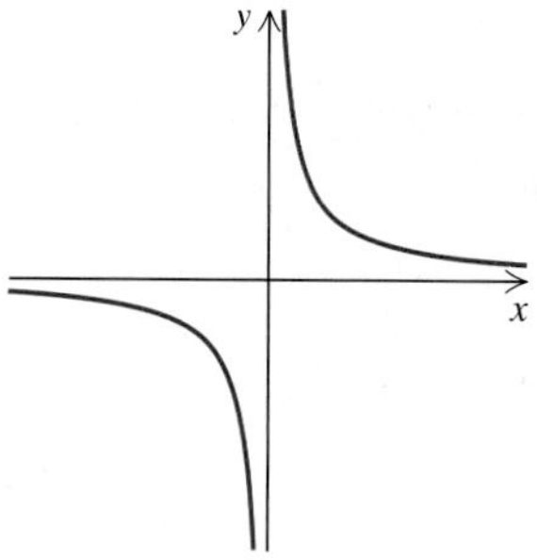

$xy = c,\ c < 0$

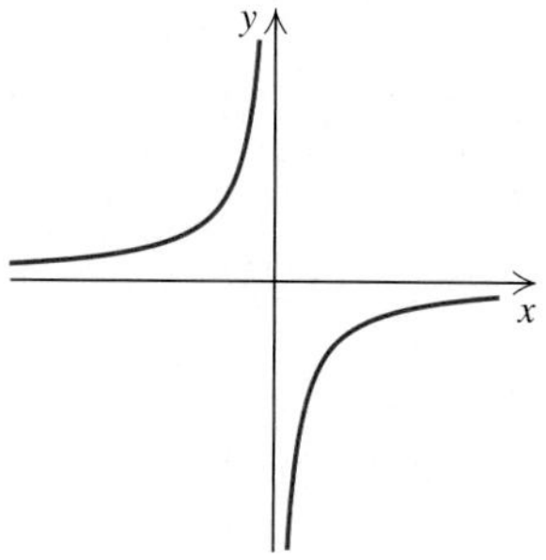

Center at (h, k)*:

$$\frac{(x-h)^2}{a^2} - \frac{(y-k)^2}{b^2} = 1$$

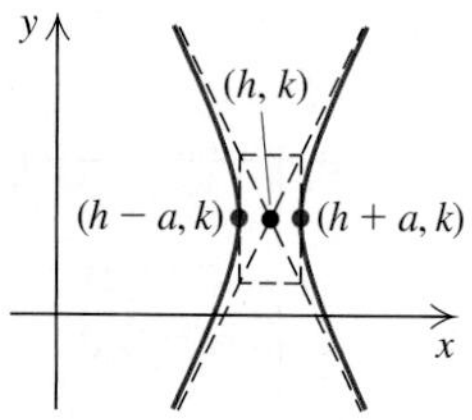

$$\frac{(y-k)^2}{b^2} - \frac{(x-h)^2}{a^2} = 1$$

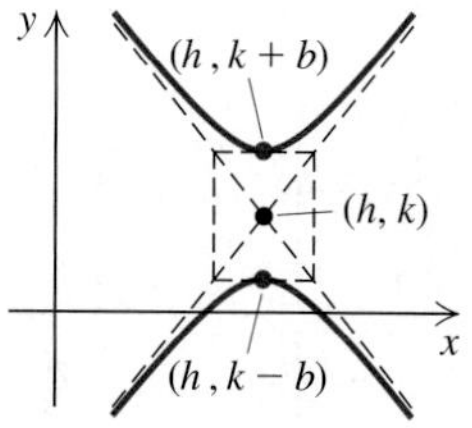

(*continued*)

*See Exercises 59–64.

ELLIPSE

Center at the origin:

$$\frac{x^2}{a^2} + \frac{y^2}{b^2} = 1$$

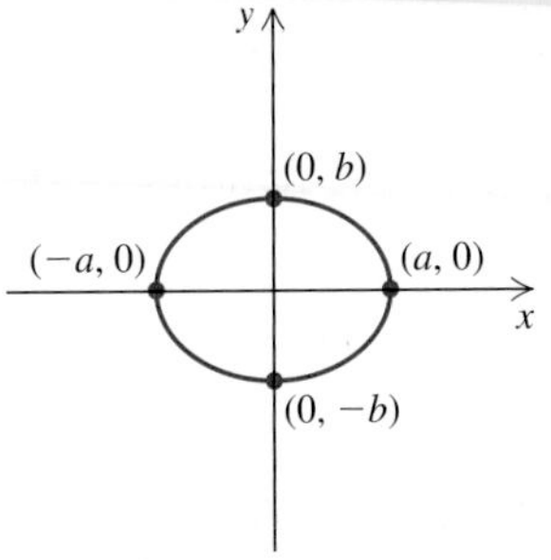

Center at (h, k):

$$\frac{(x - h)^2}{a^2} + \frac{(y - k)^2}{b^2} = 1$$

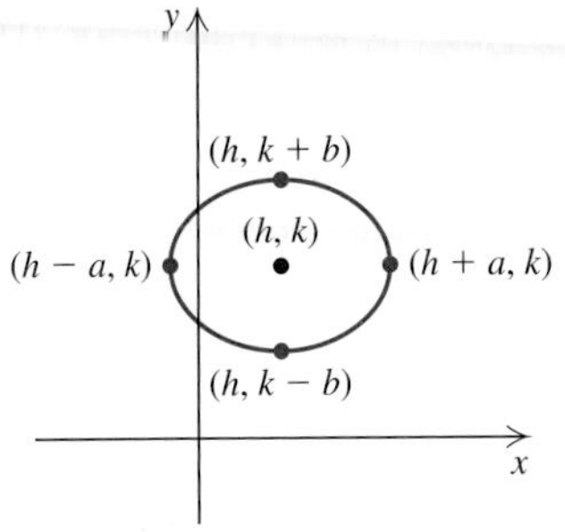

At the beginning of this chapter, we stated that the conic sections represent graphs of $Ax^2 + By^2 + Cxy + Dx + Ey + F = 0$ (we have assumed $C = 0$):

If A or B (but not both) is 0, the equation can be written in the form $y = ax^2 + bx + c$ or $x = ay^2 + by + c$, and represents a parabola.

If $A = B$, the equation can be written in the form $x^2 + y^2 = c$ or $(x - h)^2 + (y - k)^2 = r^2$, and represents a circle.

If $A \neq B$, but both A and B have the same sign, the equation can be written in the form $b^2x^2 + a^2y^2 = c$ or $b^2(x - h)^2 + a^2(y - k)^2 = c$, and represents an ellipse.

If A and B have opposite signs, the equation can be written in the form $b^2x^2 - a^2y^2 = c$ or $b^2(x - h)^2 - a^2(y - k)^2 = c$, and represents a hyperbola.

Algebraic manipulations may be needed to express an equation in one of the preceding forms.

EXAMPLE 4 Classify the graph of each equation as a circle, an ellipse, a parabola, or a hyperbola.

a) $5x^2 = 20 - 5y^2$ **b)** $x + 3 + 8y = y^2$

c) $x^2 = y^2 + 4$ **d)** $x^2 = 16 - 4y^2$

SOLUTION

a) We get the terms with variables on one side by adding $5y^2$:

$$5x^2 + 5y^2 = 20.$$

Since x and y are *both* squared, we do not have a parabola. The fact that the squared terms are *added* tells us that we do not have a hyperbola. Do we have a circle? To find out, we need to get $x^2 + y^2$ by itself. We can do that by factoring the 5 out of both terms on the left and then dividing by 5:

$$5(x^2 + y^2) = 20 \quad \textbf{Factoring out 5}$$
$$x^2 + y^2 = 4 \quad \textbf{Dividing both sides by 5}$$
$$x^2 + y^2 = 2^2. \quad \textbf{This is an equation for a circle.}$$

We can see that the graph is a circle with center at the origin and radius 2.

b) The equation $x + 3 + 8y = y^2$ has only one variable squared, so we solve for the other variable:

$$x = y^2 - 8y - 3. \quad \textbf{This is an equation for a parabola.}$$

The graph is a horizontal parabola that opens to the right.

c) In $x^2 = y^2 + 4$, both variables are squared, so the graph is not a parabola. We subtract y^2 on both sides and divide by 4 to obtain

$$\frac{x^2}{2^2} - \frac{y^2}{2^2} = 1. \quad \textbf{This is an equation for a hyperbola.}$$

The minus sign here indicates that the graph of this equation is a hyperbola. Because it is the x^2-term that is nonnegative, the hyperbola is horizontal.

d) In $x^2 = 16 - 4y^2$, both variables are squared, so the graph cannot be a parabola. We obtain the following equivalent equation:

$$x^2 + 4y^2 = 16.$$

If the coefficients of the terms were the same, we would have the graph of a circle, as in part (a), but they are not. Dividing both sides by 16 yields

$$\frac{x^2}{16} + \frac{y^2}{4} = 1. \quad \textbf{This is an equation for an ellipse.}$$

The graph of this equation is a horizontal ellipse.

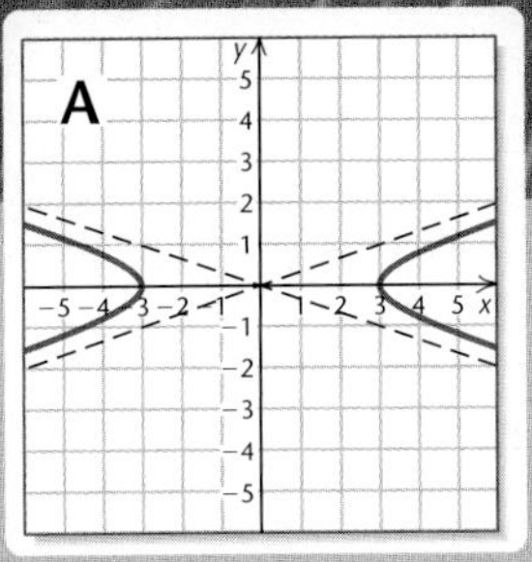

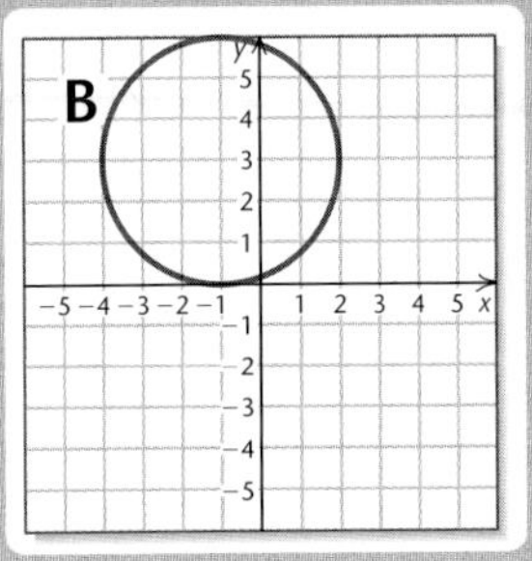

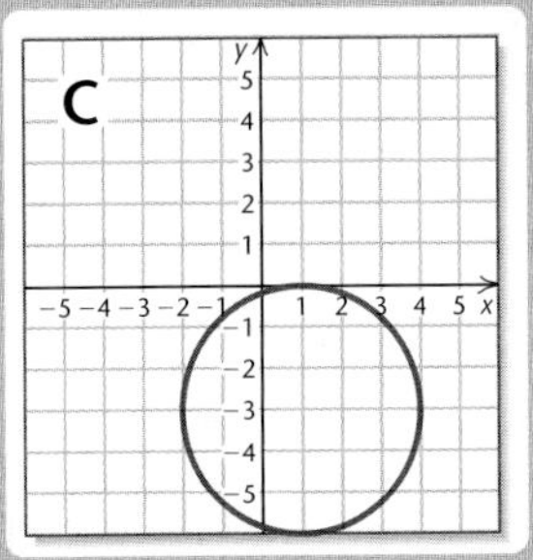

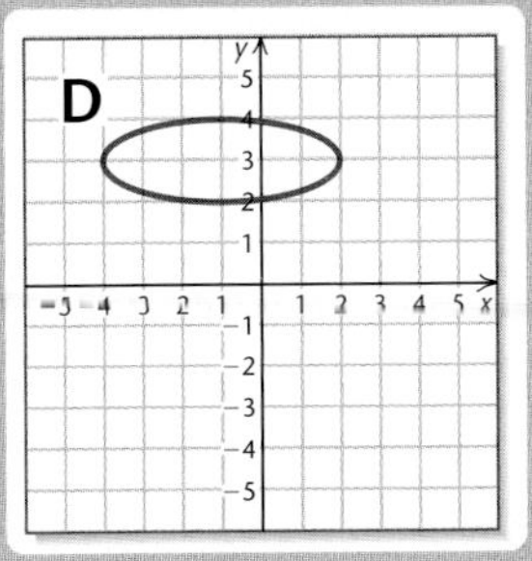

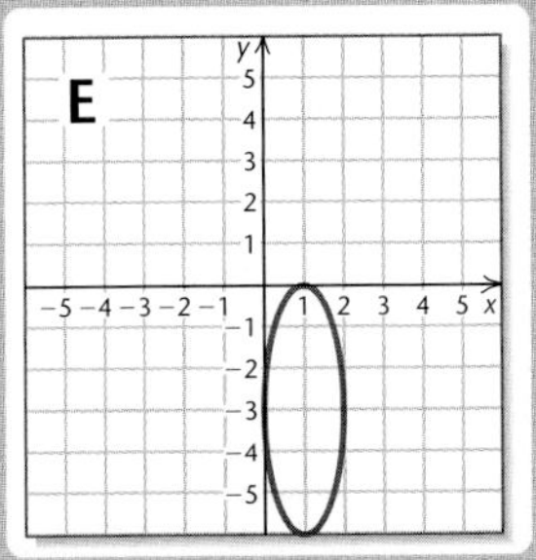

Visualizing the Graph

Match each equation with its graph.

1. $(x - 1)^2 + (y + 3)^2 = 9$

2. $\dfrac{x^2}{9} - \dfrac{y^2}{1} = 1$

3. $y = (x - 1)^2 - 3$

4. $(x + 1)^2 + (y - 3)^2 = 9$

5. $x = (y - 1)^2 - 3$

6. $\dfrac{(x + 1)^2}{9} + \dfrac{(y - 3)^2}{1} = 1$

7. $xy = 3$

8. $y = -(x + 1)^2 + 3$

9. $\dfrac{y^2}{9} - \dfrac{x^2}{1} = 1$

10. $\dfrac{(x - 1)^2}{1} + \dfrac{(y + 3)^2}{9} = 1$

Answers on page A-52

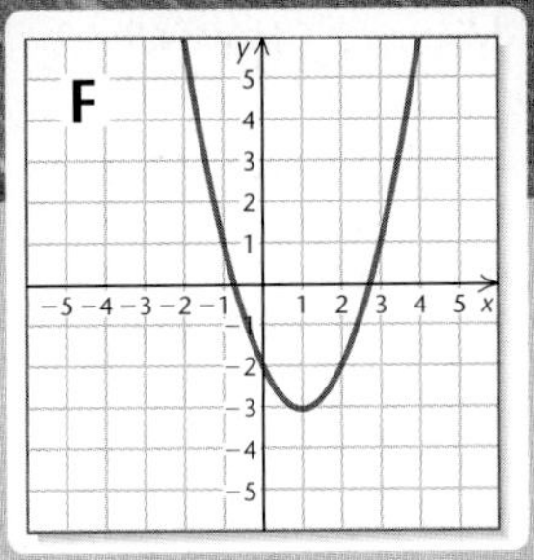

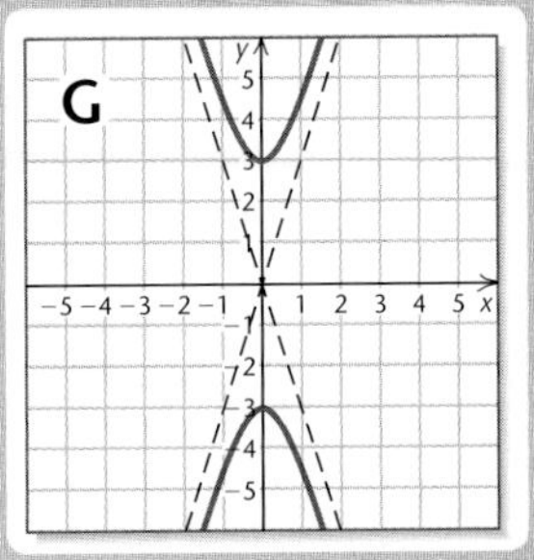

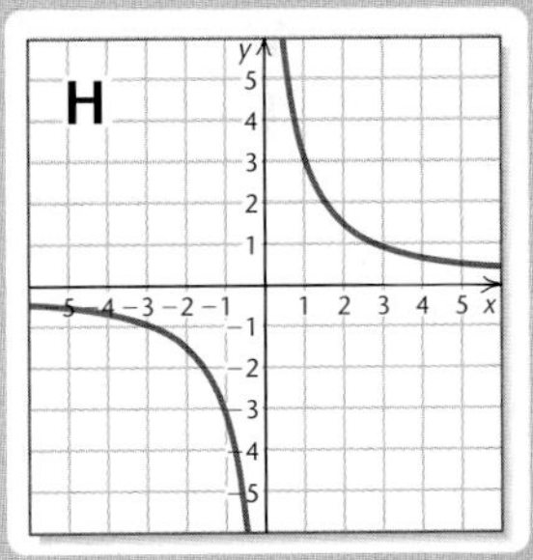

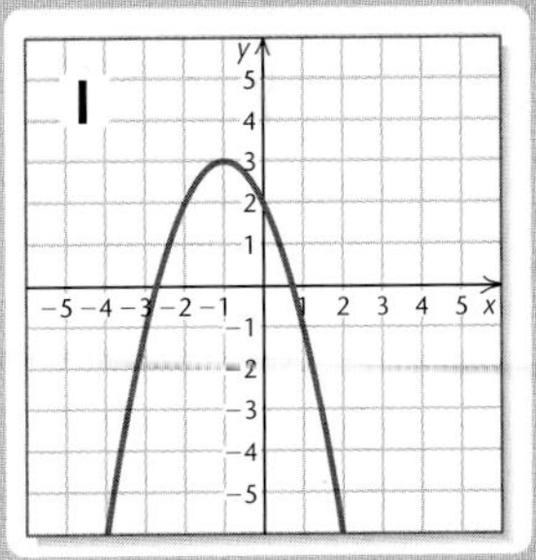

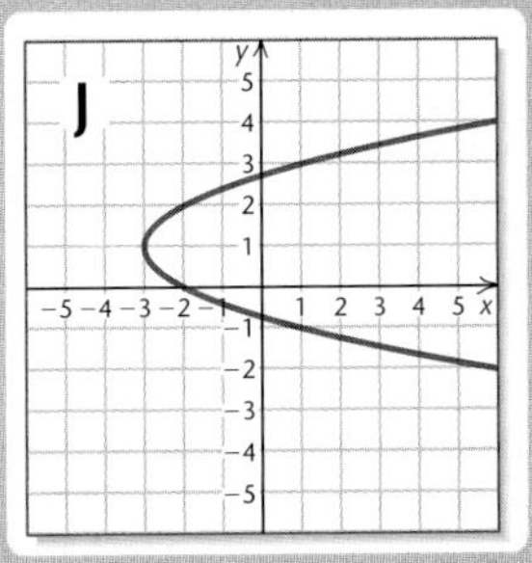

10.3 EXERCISE SET

FOR EXTRA HELP

MathXL | MyMathLab | InterAct Math | AW Math Tutor Center | Video Lectures on CD: Disc 5 | Student's Solutions Manual

Concept Reinforcement *In each of Exercises 1–8, match the conic section with the equation in the column on the right that represents that type of conic section.*

1. ____ A hyperbola with a horizontal axis
2. ____ A hyperbola with a vertical axis
3. ____ An ellipse with its center not at the origin
4. ____ An ellipse with its center at the origin
5. ____ A circle with its center at the origin
6. ____ A circle with its center not at the origin
7. ____ A parabola opening upward or downward
8. ____ A parabola opening to the right or left

a) $\frac{x^2}{10} + \frac{y^2}{12} = 1$

b) $(x + 1)^2 + (y - 3)^2 = 30$

c) $y - x^2 = 5$

d) $\frac{x^2}{9} - \frac{y^2}{10} = 1$

e) $x - 2y^2 = 3$

f) $\frac{y^2}{20} - \frac{x^2}{35} = 1$

g) $3x^2 + 3y^2 = 75$

h) $\frac{(x-1)^2}{10} + \frac{(y-4)^2}{8} = 1$

Graph each hyperbola. Label all vertices and sketch all asymptotes.

9. $\frac{y^2}{16} - \frac{x^2}{16} = 1$

10. $\frac{x^2}{9} - \frac{y^2}{9} = 1$

11. $\frac{x^2}{4} - \frac{y^2}{25} = 1$

12. $\frac{y^2}{16} - \frac{x^2}{9} = 1$

13. $\frac{y^2}{36} - \frac{x^2}{9} = 1$

14. $\frac{x^2}{25} - \frac{y^2}{36} = 1$

15. $y^2 - x^2 = 25$

16. $x^2 - y^2 = 4$

17. $25x^2 - 16y^2 = 400$

18. $4y^2 - 9x^2 = 36$

Graph.

19. $xy = -5$

20. $xy = 5$

21. $xy = 4$

22. $xy = -9$

23. $xy = -2$

24. $xy = -1$

25. $xy = 1$

26. $xy = 2$

Classify each of the following as the equation of a circle, an ellipse, a parabola, or a hyperbola.

27. $x^2 + y^2 - 6x + 4y - 30 = 0$

28. $y + 9 = 3x^2$

29. $9x^2 + 4y^2 - 36 = 0$

30. $x + 3y = 2y^2 - 1$

31. $4x^2 - 9y^2 - 72 = 0$

32. $y^2 + x^2 = 8$

33. $x^2 + y^2 = 2x + 4y + 4$

34. $2y + 13 + x^2 = 8x - y^2$

35. $4x^2 = 64 - y^2$

36. $y = \frac{7}{x}$

37. $x - \frac{8}{y} = 0$

38. $x - 4 = y^2 - 3y$

39. $y + 6x = x^2 + 5$

40. $x^2 = 16 + y^2$

41. $9y^2 = 36 + 4x^2$

42. $3x^2 + 5y^2 + x^2 - y^2 + 49$

43. $3x^2 + y^2 - x = 2x^2 - 9x + 10y + 40$

44. $4y^2 + 20x^2 + 1 = 8y - 5x^2$

45. $16x^2 + 5y^2 - 12x^2 + 8y^2 - 3x + 4y = 568$

46. $56x^2 - 17y^2 = 234 - 13x^2 - 38y^2$

TW 47. What does graphing hyperbolas have in common with graphing ellipses?

TW 48. Is it possible for a hyperbola to represent the graph of a function? Why or why not?

Focused Review

Graph.

49. $(x - 1)^2 + (y + 2)^2 = 9$ [10.1]

50. $xy = -4$ [10.3]

51. $4x^2 + y^2 = 16$ [10.2]

52. $4x^2 - y^2 = 16$ [10.3]

53. $x = y^2 + 2y + 3$ [10.1]

54. $y = x^2 + 2x + 3$ [10.1]

Synthesis

TW 55. What is it in the equation of a hyperbola that controls how wide open the branches are? Explain your reasoning.

TW 56. If, in

$$\frac{x^2}{a^2} - \frac{y^2}{b^2} = 1,$$

$a = b$, what are the asymptotes of the graph? Why?

Find an equation of a hyperbola satisfying the given conditions.

57. Having intercepts $(0, 6)$ and $(0, -6)$ and asymptotes $y = 3x$ and $y = -3x$

58. Having intercepts $(8, 0)$ and $(-8, 0)$ and asymptotes $y = 4x$ and $y = -4x$

The standard equations for horizontal or vertical hyperbolas centered at (h, k) are as follows:

$$\frac{(x - h)^2}{a^2} - \frac{(y - k)^2}{b^2} = 1$$

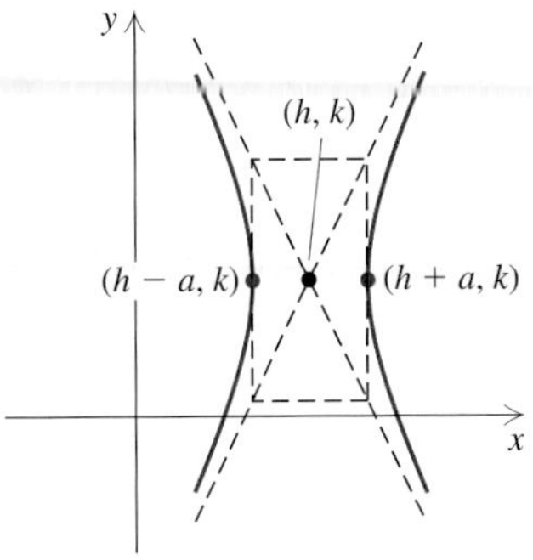

$$\frac{(y - k)^2}{b^2} - \frac{(x - h)^2}{a^2} = 1$$

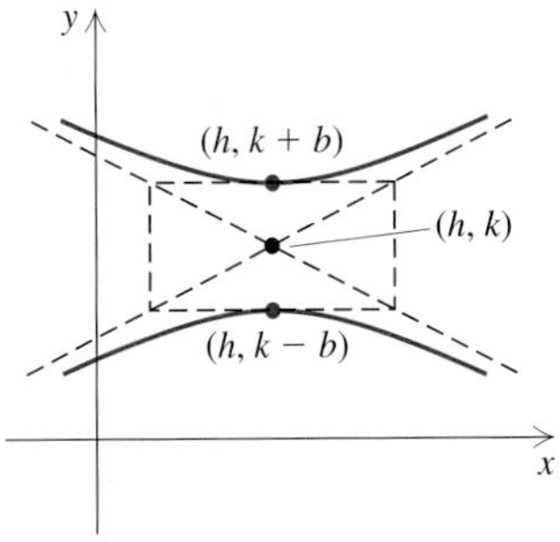

The vertices are as labeled and the asymptotes are

$$y - k = \frac{b}{a}(x - h) \quad \textit{and} \quad y - k = -\frac{b}{a}(x - h).$$

For each of the following equations of hyperbolas, complete the square, if necessary, and write in standard form. Find the center, the vertices, and the asymptotes. Then graph the hyperbola.

59. $\dfrac{(x - 5)^2}{36} - \dfrac{(y - 2)^2}{25} = 1$

60. $\dfrac{(x - 2)^2}{9} - \dfrac{(y - 1)^2}{4} = 1$

61. $8(y + 3)^2 - 2(x - 4)^2 = 32$

62. $25(x - 4)^2 - 4(y + 5)^2 = 100$

63. $4x^2 - y^2 + 24x + 4y + 28 = 0$

64. $4y^2 - 25x^2 - 8y - 100x - 196 = 0$

10.4 Nonlinear Systems of Equations

Systems Involving One Nonlinear Equation ▪ Systems of Two Nonlinear Equations ▪ Problem Solving

The equations appearing in systems of two equations have thus far in our discussion always been linear. We now consider systems of two equations in which at least one equation is nonlinear.

Study Tip

Summing It All Up

In preparation for a final exam, many students find it helpful to write up a few pages of notes that represent the most important concepts of the course. After doing so, it is a good idea to try to condense those notes down to just one page. This exercise will help you focus on the most important material.

Systems Involving One Nonlinear Equation

Suppose that a system consists of an equation of a circle and an equation of a line. In what ways can the circle and the line intersect? The figures below represent three ways in which the situation can occur. We see that such a system will have 0, 1, or 2 real solutions.

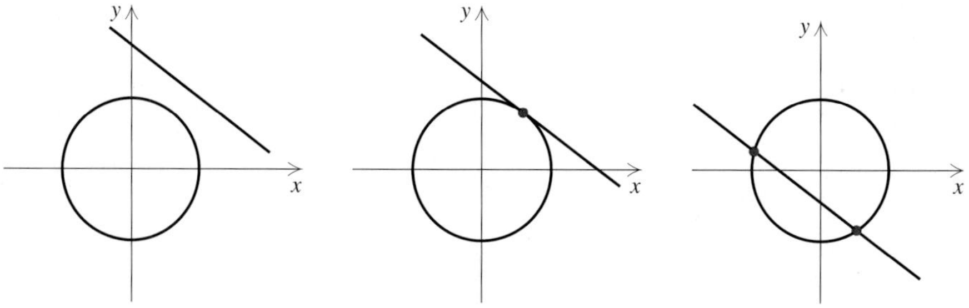

0 real solutions   1 real solution   2 real solutions

Recall that graphing, *elimination*, and *substitution* were all used to solve systems of linear equations. To solve systems in which one equation is of first degree and one is of second degree, it is preferable to use the *substitution* method.

EXAMPLE 1 Solve the system

$x^2 + y^2 = 25,$ (1) (The graph is a circle.)

$3x - 4y = 0.$ (2) (The graph is a line.)

SOLUTION First, we solve the linear equation, (2), for x:

$x = \frac{4}{3}y.$ (3) **We could have solved for y instead.**

Then we substitute $\frac{4}{3}y$ for x in equation (1) and solve for y:

$$\left(\tfrac{4}{3}y\right)^2 + y^2 = 25$$
$$\tfrac{16}{9}y^2 + y^2 = 25$$
$$\tfrac{25}{9}y^2 = 25$$
$$y^2 = 9 \quad \textbf{Multiplying both sides by } \tfrac{9}{25}$$
$$y = \pm 3. \quad \textbf{Using the principle of square roots}$$

Student Notes

Be sure to either list each solution of a system as an ordered pair or separately state the value of each variable. Remember that many of these systems will have more than one solution.

Now we substitute these numbers for y in equation (3) and solve for x:

$$\text{for } y = 3, \quad x = \tfrac{4}{3}(3) = 4;$$
$$\text{for } y = -3, \quad x = \tfrac{4}{3}(-3) = -4.$$

Check: For (4, 3):

$x^2 + y^2 = 25$	
$4^2 + 3^2$	25
$16 + 9$	
$25 \stackrel{?}{=} 25$	TRUE

$3x - 4y = 0$	
$3(4) - 4(3)$	0
$12 - 12$	
$0 \stackrel{?}{=} 0$	TRUE

It is left to the student to confirm that $(-4, -3)$ also checks in both equations.

The pairs (4, 3) and $(-4, -3)$ check, so they are solutions. We can see the solutions in the graph. Intersections occur at (4, 3) and $(-4, -3)$.

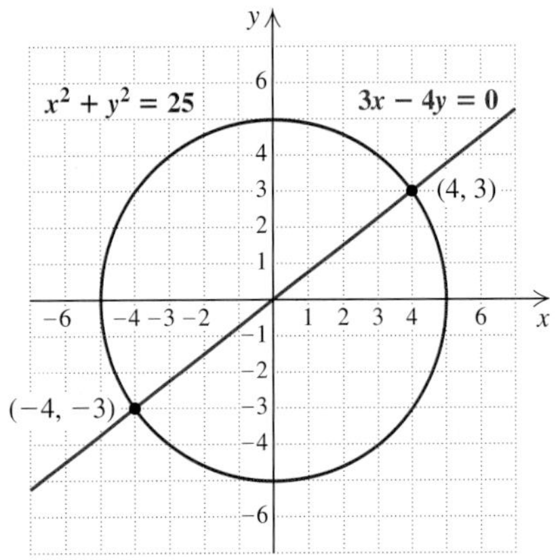

Even if we do not know what the graph of each equation in a system looks like, the algebraic approach of Example 1 can still be used.

EXAMPLE 2 Solve the system

$$y + 3 = 2x, \qquad (1)$$
$$x^2 + 2xy = -1 \qquad (2)$$

SOLUTION First, we solve the linear equation (1) for y:

$$y = 2x - 3. \qquad (3)$$

Then we substitute $2x - 3$ for y in equation (2) and solve for x:

$$x^2 + 2x(2x - 3) = -1$$
$$x^2 + 4x^2 - 6x = -1$$
$$5x^2 - 6x + 1 = 0$$
$$(5x - 1)(x - 1) = 0 \qquad \textbf{Factoring}$$
$$5x - 1 = 0 \quad or \quad x - 1 = 0 \qquad \textbf{Using the principle of zero products}$$
$$x = \tfrac{1}{5} \quad or \quad x = 1.$$

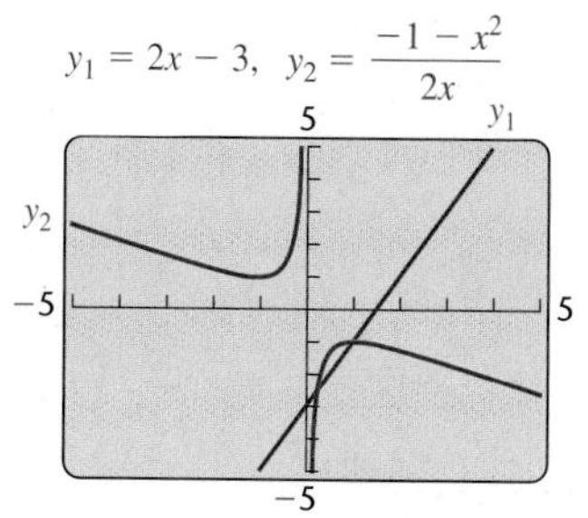

Now we substitute these numbers for x in equation (3) and solve for y:

$$\text{for } x = \tfrac{1}{5}, \quad y = 2\left(\tfrac{1}{5}\right) - 3 = -\tfrac{13}{5};$$
$$\text{for } x = 1, \quad y = 2(1) - 3 = -1.$$

Using a graphing calculator, as shown at left, we find that the solutions are $(0.2, -2.6)$ and $(1, -1)$. You can confirm that $\left(\frac{1}{5}, -\frac{13}{5}\right)$ and $(1, -1)$ check, so they are both solutions.

EXAMPLE 3 Solve the system

$$x + y = 5, \quad (1) \quad \text{(The graph is a line.)}$$
$$y = 3 - x^2. \quad (2) \quad \text{(The graph is a parabola.)}$$

SOLUTION We substitute $3 - x^2$ for y in the first equation:

$$x + 3 - x^2 = 5$$
$$-x^2 + x - 2 = 0 \quad \textbf{Adding } -5 \textbf{ to both sides and rearranging}$$
$$x^2 - x + 2 = 0. \quad \textbf{Multiplying both sides by } -1$$

Since $x^2 - x + 2$ does not factor, we need the quadratic formula:

$$x = \frac{-b \pm \sqrt{b^2 - 4ac}}{2a}$$
$$= \frac{-(-1) \pm \sqrt{(-1)^2 - 4 \cdot 1 \cdot 2}}{2(1)} \quad \textbf{Substituting}$$
$$= \frac{1 \pm \sqrt{1 - 8}}{2} = \frac{1 \pm \sqrt{-7}}{2} = \frac{1}{2} \pm \frac{\sqrt{7}}{2}i.$$

Solving equation (1) for y gives us $y = 5 - x$. Substituting values for x gives

$$y = 5 - \left(\frac{1}{2} + \frac{\sqrt{7}}{2}i\right) = \frac{9}{2} - \frac{\sqrt{7}}{2}i \quad \text{and}$$
$$y = 5 - \left(\frac{1}{2} - \frac{\sqrt{7}}{2}i\right) = \frac{9}{2} + \frac{\sqrt{7}}{2}i.$$

The solutions are

$$\left(\frac{1}{2} + \frac{\sqrt{7}}{2}i, \frac{9}{2} - \frac{\sqrt{7}}{2}i\right) \quad \text{and} \quad \left(\frac{1}{2} - \frac{\sqrt{7}}{2}i, \frac{9}{2} + \frac{\sqrt{7}}{2}i\right).$$

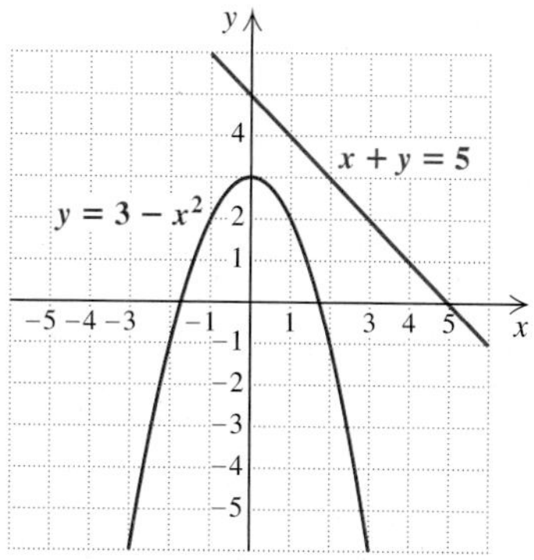

There are no real-number solutions. Note in the figure at left that the graphs do not intersect. Getting only nonreal solutions tells us that the graphs do not intersect.

Systems of Two Nonlinear Equations

We now consider systems of two second-degree equations. Graphs of such systems can involve any two conic sections. The following figure shows some ways in which a circle and a hyperbola can intersect.

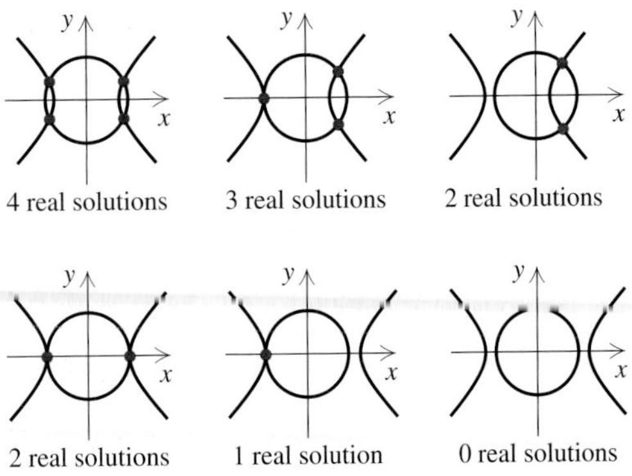

To solve systems of two second-degree equations, we either substitute or eliminate. The elimination method is generally better when both equations are of the form $Ax^2 + By^2 = C$. Then we can eliminate an x^2- or y^2-term in a manner similar to the procedure used in Chapter 3.

EXAMPLE 4 Solve the system

$$2x^2 + 5y^2 = 22, \quad (1)$$
$$3x^2 - y^2 = -1. \quad (2)$$

SOLUTION Here we multiply equation (2) by 5 and then add:

$$\begin{aligned} 2x^2 + 5y^2 &= 22 \\ \underline{15x^2 - 5y^2} &= -5 && \textbf{Multiplying both sides of equation (2) by 5} \\ 17x^2 \qquad &= 17 && \textbf{Adding} \\ x^2 &= 1 \\ x &= \pm 1. \end{aligned}$$

Whether x is -1 or 1, we have $x^2 = 1$. Thus we can simultaneously substitute 1 and -1 for x in equation (2):

$$\left.\begin{aligned} 3 \cdot (\pm 1)^2 - y^2 &= -1 \\ 3 - y^2 &= -1 \\ -y^2 &= -4 \end{aligned}\right\} \quad \textbf{Since } (-1)^2 = 1^2 \textbf{, we can evaluate for } x = -1 \textbf{ and } x = 1 \textbf{ simultaneously.}$$

$$y^2 = 4 \quad \text{or} \quad y = \pm 2.$$

Thus, if $x = 1$, then $y = 2$ or $y = -2$; and if $x = -1$, then $y = 2$ or $y = -2$. The four possible solutions are $(1, 2)$, $(1, -2)$, $(-1, 2)$, and $(-1, -2)$.

Check: Since $(2)^2 = (-2)^2$ and $(1)^2 = (-1)^2$, we can check all four pairs at once.

$2x^2 + 5y^2 = 22$	
$2(\pm 1)^2 + 5(\pm 2)^2$	22
$2 + 20$	
$22 \stackrel{?}{=} 22$	TRUE

$3x^2 - y^2 = -1$	
$3(\pm 1)^2 - (\pm 2)^2$	-1
$3 - 4$	
$-1 \stackrel{?}{=} -1$	TRUE

The graph below serves as a visual check. The solutions are $(1, 2)$, $(1, -2)$, $(-1, 2)$, and $(-1, -2)$.

$y_1 = \sqrt{(22 - 2x^2)/5},\ y_2 = -y_1,$

$y_3 = \sqrt{3x^2 + 1},\ y_4 = -y_3$

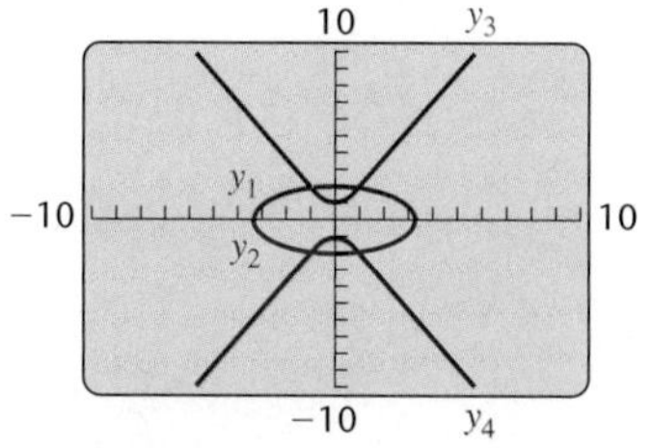

When a product of variables is in one equation and the other equation is of the form $Ax^2 + By^2 = C$, we often solve for a variable in the equation with the product and then use substitution.

EXAMPLE 5 Solve the system

$$x^2 + 4y^2 = 20, \quad (1)$$
$$xy = 4. \quad (2)$$

SOLUTION First, we solve equation (2) for y:

$$y = \frac{4}{x}. \qquad \textbf{Dividing both sides by } x\textbf{. Note that } x \neq 0.$$

Then we substitute $4/x$ for y in equation (1) and solve for x:

$$x^2 + 4\left(\frac{4}{x}\right)^2 = 20$$
$$x^2 + \frac{64}{x^2} = 20$$
$$x^4 + 64 = 20x^2 \qquad \textbf{Multiplying by } x^2$$
$$x^4 - 20x^2 + 64 = 0 \qquad \textbf{Obtaining standard form. This equation is reducible to quadratic.}$$
$$(x^2 - 4)(x^2 - 16) = 0 \qquad \textbf{Factoring. If you prefer, let } u = x^2 \textbf{ and substitute.}$$
$$(x - 2)(x + 2)(x - 4)(x + 4) = 0 \qquad \textbf{Factoring again}$$
$$x = 2 \quad or \quad x = -2 \quad or \quad x = 4 \quad or \quad x = -4. \qquad \textbf{Using the principle of zero products}$$

Since $y = 4/x$, for $x = 2$, we have $y = 4/2$, or 2. Thus, $(2, 2)$ is a solution. Similarly, $(-2, -2)$, $(4, 1)$, and $(-4, -1)$ are solutions. You can show that all four pairs check.

Problem Solving

We now consider applications that can be modeled by a system of equations in which at least one equation is not linear.

EXAMPLE 6 Architecture. For a college gymnasium, an architect wants to lay out a rectangular piece of land that has a perimeter of 204 m and an area of 2565 m^2. Find the dimensions of the piece of land.

SOLUTION

1. **Familiarize.** We make a drawing and label it, letting l = the length and w = the width, both in meters.

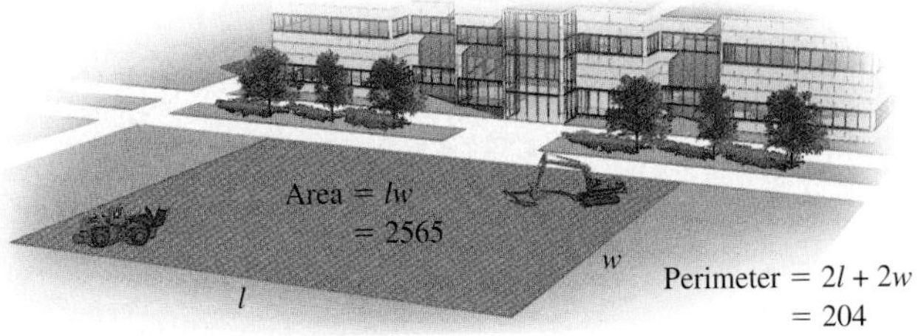

2. **Translate.** We then have the following translation:

$$\text{Perimeter: } 2w + 2l = 204;$$
$$\text{Area: } lw = 2565.$$

3. **Carry out.** We solve the system

$$2w + 2l = 204,$$
$$lw = 2565.$$

Solving the second equation for l gives us $l = 2565/w$. Then we substitute $2565/w$ for l in the first equation and solve for w:

$$2w + 2\left(\frac{2565}{w}\right) = 204$$
$$2w^2 + 2(2565) = 204w \quad \textbf{Multiplying both sides by } w$$
$$2w^2 - 204w + 2(2565) = 0 \quad \textbf{Standard form}$$
$$w^2 - 102w + 2565 = 0 \quad \textbf{Multiplying by } \tfrac{1}{2}$$
$$w = \frac{-(-102) \pm \sqrt{(-102)^2 - 4 \cdot 1 \cdot 2565}}{2 \cdot 1}$$
$$w = \frac{102 \pm \sqrt{144}}{2} = \frac{102 \pm 12}{2}$$
$$w = 57 \quad or \quad w = 45.$$

Factoring could be used instead of the quadratic formula, but the numbers are quite large.

If $w = 57$, then $l = 2565/w = 2565/57 = 45$. If $w = 45$, then $l = 2565/w = 2565/45 = 57$. Since length is usually considered to be longer than width, we have the solution $l = 57$ and $w = 45$, or $(57, 45)$.

4. **Check.** If $l = 57$ and $w = 45$, the perimeter is $2 \cdot 57 + 2 \cdot 45$, or 204. The area is $57 \cdot 45$, or 2565. The numbers check.

5. **State.** The length is 57 m and the width is 45 m.

EXAMPLE 7 HDTV Dimensions. High-definition television (HDTV) offers greater clarity than conventional television. The Kaplans' new HDTV screen has an area of 1296 in^2 and has a $\sqrt{3033}$-in. (about 55-in.) diagonal screen. Find the width and the length of the screen.

SOLUTION

1. **Familiarize.** We make a drawing and label it. Note the right triangle in the figure. We let l = the length and w = the width, both in inches.

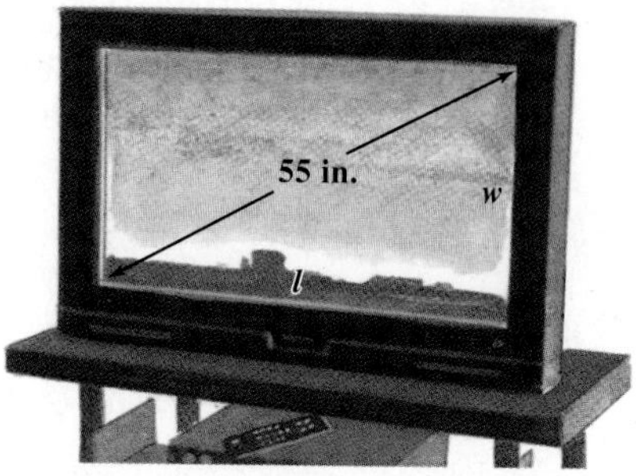

2. **Translate.** We translate to a system of equations:

$$l^2 + w^2 = \sqrt{3033}^2, \quad \textbf{Using the Pythagorean theorem}$$
$$lw = 1296. \quad \textbf{Using the formula for the area of a rectangle}$$

3. **Carry out.** We solve the system

$$\left.\begin{aligned} l^2 + w^2 &= 3033, \\ lw &= 1296 \end{aligned}\right\} \quad \textbf{You should complete the solution of this system.}$$

to get (48, 27), (27, 48), $(-48, -27)$, and $(-27, -48)$.

4. **Check.** Measurements must be positive and length is usually greater than width, so we check only (48, 27). In the right triangle, $48^2 + 27^2 = 2304 + 729 = 3033$ or $\sqrt{3033}^2$. The area is $48 \cdot 27 = 1296$, so our answer checks.

5. **State.** The length is 48 in. and the width is 27 in.

10.4 EXERCISE SET

FOR EXTRA HELP

MathXL | MyMathLab | InterAct Math | AW Math Tutor Center | Video Lectures on CD: Disc 5 | Student's Solutions Manual

Concept Reinforcement *Classify each statement as either true or false.*

1. A system of equations that represent a line and an ellipse can have 0, 1, or 2 solutions.

2. A system of equations that represent a parabola and a circle can have up to 4 solutions.

3. A system of equations that represent a hyperbola and a circle can have no fewer than 2 solutions.

4. A system of equations that represent an ellipse and a line has either 0 or 2 solutions.

5. Systems containing one first-degree equation and one second-degree equation are most easily solved using the substitution method.

6. Systems containing two second-degree equations of the form $Ax^2 + By^2 = C$ are most easily solved using the elimination method.

Solve. Remember that graphs can be used to confirm all real solutions.

7. $x^2 + y^2 = 25,$
$y - x = 1$

8. $x^2 + y^2 = 100,$
$y - x = 2$

9. $4x^2 + 9y^2 = 36,$
$3y + 2x = 6$

10. $9x^2 + 4y^2 = 36,$
$3x + 2y = 6$

11. $y^2 = x + 3,$
$2y = x + 4$

12. $y = x^2,$
$3x = y + 2$

13. $x^2 - xy + 3y^2 = 27,$
$x - y = 2$

14. $2y^2 + xy + x^2 = 7,$
$x - 2y = 5$

15. $x^2 + 4y^2 = 25,$
$x + 2y = 7$

16. $x^2 - y^2 = 16,$
$x - 2y = 1$

17. $x^2 - xy + 3y^2 = 5,$
$x - y = 2$

18. $m^2 + 3n^2 = 10,$
$m - n = 2$

19. $3x + y = 7,$
$4x^2 + 5y = 24$

20. $2y^2 + xy = 5,$
$4y + x = 7$

21. $a + b = 7,$
$ab = 4$

22. $p + q = -6,$
$pq = -7$

23. $2a + b = 1,$
$b = 4 - a^2$

24. $4x^2 + 9y^2 = 36,$
$x + 3y = 3$

25. $a^2 + b^2 = 89,$
$a - b = 3$

26. $xy = 4,$
$x + y = 5$

Aha! **27.** $y = x^2,$
$x = y^2$

28. $x^2 + y^2 = 25,$
$y^2 = x + 5$

29. $x^2 + y^2 = 9,$
$x^2 - y^2 = 9$

30. $y^2 - 4x^2 = 4,$
$4x^2 + y^2 = 4$

31. $x^2 + y^2 = 25,$
$xy = 12$

32. $x^2 - y^2 = 16,$
$x + y^2 = 4$

33. $x^2 + y^2 = 9,$
$25x^2 + 16y^2 = 400$

34. $x^2 + y^2 = 4,$
$9x^2 + 16y^2 = 144$

35. $x^2 + y^2 = 14,$
$x^2 - y^2 = 4$

36. $x^2 + y^2 = 16,$
$y^2 - 2x^2 = 10$

37. $x^2 + y^2 = 20,$
$xy = 8$

38. $x^2 + y^2 = 5,$
$xy = 2$

39. $x^2 + 4y^2 = 20,$
$xy = 4$

40. $x^2 + y^2 = 13,$
$xy = 6$

41. $2xy + 3y^2 = 7,$
$3xy - 2y^2 = 4$

42. $3xy + x^2 = 34,$
$2xy - 3x^2 = 8$

43. $4a^2 - 25b^2 = 0,$
$2a^2 - 10b^2 = 3b + 4$

44. $xy - y^2 = 2,$
$2xy - 3y^2 = 0$

45. $ab - b^2 = -4,$
$ab - 2b^2 = -6$

46. $x^2 - y = 5,$
$x^2 + y^2 = 25$

Solve.

47. *Computer Parts.* Dataport Electronics needs a rectangular memory board that has a perimeter of 28 cm and a diagonal of length 10 cm. What should the dimensions of the board be?

48. *Geometry.* A rectangle has an area of 2 yd^2 and a perimeter of 6 yd. Find its dimensions.

49. *Geometry.* A rectangle has an area of 20 in^2 and a perimeter of 18 in. Find its dimensions.

50. *Tile Design.* The New World tile company wants to make a new rectangular tile that has a perimeter of 6 in. and a diagonal of length $\sqrt{5}$ in. What should the dimensions of the tile be?

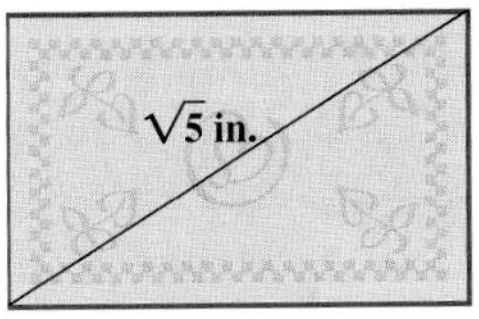

51. *Design of a Van.* The cargo area of a delivery van must be 60 ft^2, and the length of a diagonal must accommodate a 13-ft board. Find the dimensions of the cargo area.

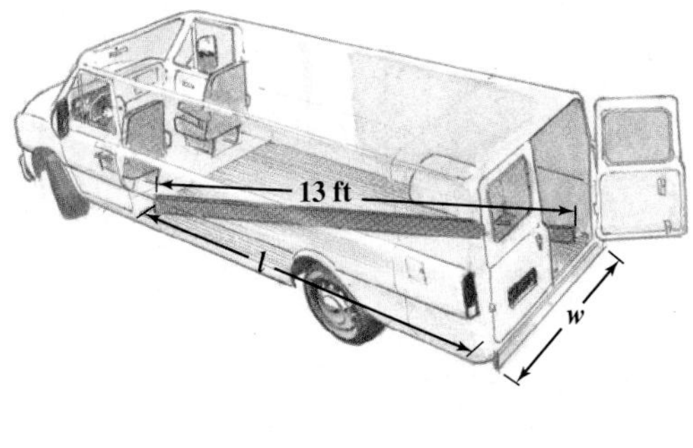

52. *Dimensions of a Rug.* The diagonal of a Persian rug is 25 ft. The area of the rug is 300 ft^2. Find the length and the width of the rug.

53. The product of two numbers is 60. The sum of their squares is 136. Find the numbers.

54. *Investments.* A certain amount of money saved for 1 yr at a certain interest rate yielded \$225 in interest. If \$750 more had been invested and the rate had been 1% less, the interest would have been the same. Find the principal and the rate.

55. *Garden Design.* A garden contains two square peanut beds. Find the length of each bed if the sum of their areas is 832 ft^2 and the difference of their areas is 320 ft^2.

56. The area of a rectangle is $\sqrt{3}$ m^2, and the length of a diagonal is 2 m. Find the dimensions.

57. The product of the lengths of the legs of a right triangle is 156. The hypotenuse has length $\sqrt{313}$. Find the lengths of the legs.

58. The area of a rectangle is $\sqrt{2}$ m^2, and the length of a diagonal is $\sqrt{3}$ m. Find the dimensions.

TW **59.** How can an understanding of conic sections be helpful when a system of nonlinear equations is being solved algebraically?

TW **60.** Suppose a system of equations is comprised of one linear equation and one nonlinear equation. Is it possible for such a system to have three solutions? Why or why not?

Skill Maintenance

Simplify. [1.4]

61. $(-1)^9(-2)^4$

62. $(-1)^{10}(-2)^5$

Evaluate each of the following. [1.2], [1.4]

63. $\dfrac{(-1)^k}{k-5}$, for $k = 6$

64. $\dfrac{(-1)^k}{k-5}$, for $k = 9$

65. $\dfrac{n}{2}(3 + n)$, for $n = 8$

66. $\dfrac{7(1 - r^2)}{1 - r}$, for $r = 3$

Synthesis

TW **67.** Write a problem that translates to a system of two equations. Design the problem so that at least one equation is nonlinear and so that no real solution exists.

TW **68.** Write a problem for a classmate to solve. Devise the problem so that a system of two nonlinear equations with exactly one real solution is solved.

69. Find the equation of a circle that passes through $(-2, 3)$ and $(-4, 1)$ and whose center is on the line $5x + 8y = -2$.

70. Find the equation of an ellipse centered at the origin that passes through the points $(2, -3)$ and $(1, \sqrt{13})$.

Solve.

71. $p^2 + q^2 = 13,$
$\dfrac{1}{pq} = -\dfrac{1}{6}$

72. $a + b = \dfrac{5}{6},$
$\dfrac{a}{b} + \dfrac{b}{a} = \dfrac{13}{6}$

73. *Fence Design.* A roll of chain-link fencing contains 100 ft of fence. The fencing is bent at a 90° angle to enclose a rectangular work area of 2475 ft^2, as shown. Determine the length and the width of the rectangle.

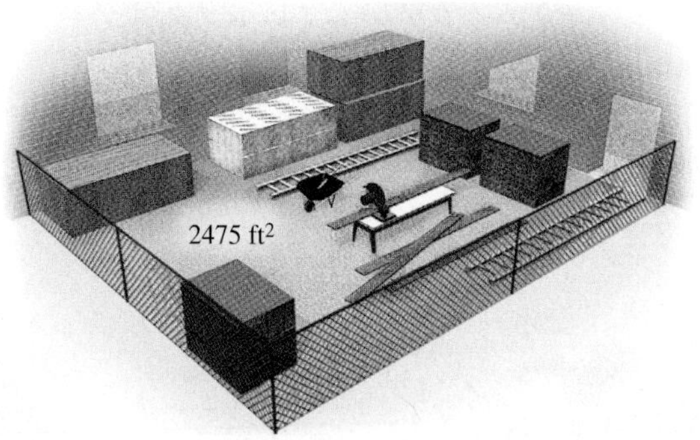

74. A piece of wire 100 cm long is to be cut into two pieces and those pieces are each to be bent to make a square. The area of one square is to be 144 cm^2 greater than that of the other. How should the wire be cut?

75. *Box Design.* Four squares with sides 5 in. long are cut from the corners of a rectangular metal sheet that has an area of 340 in^2. The edges are bent up to form an open box with a volume of 350 in^3. Find the dimensions of the box.

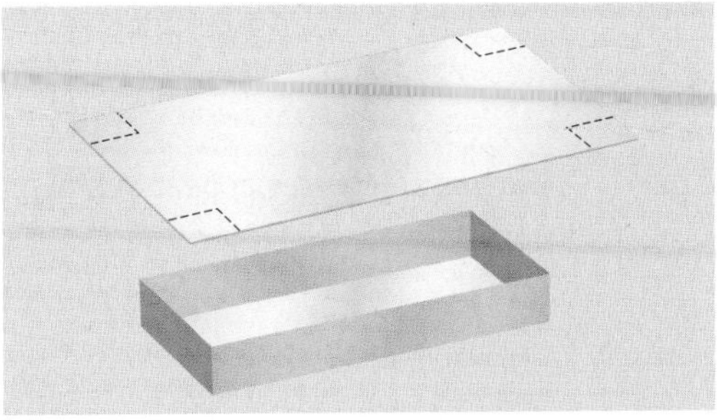

76. *Computer Screens.* The ratio of the length to the height of the screen on a computer monitor is 4 to 3. A Dell Inspiron notebook has a 15-in. diagonal screen. Find the dimensions of the screen.

77. *HDTV Screens.* The ratio of the length to the height of an HDTV screen (see Example 7) is 16 to 9. The Remton Lounge has an HDTV screen with a $\sqrt{4901}$-in. (about 70-in.) diagonal screen. Find the dimensions of the screen.

78. *Railing Sales.* Fireside Castings finds that the total revenue R from the sale of x units of railing is given by

$$R = 100x + x^2.$$

Fireside also finds that the total cost C of producing x units of the same product is given by

$$C = 80x + 1500.$$

A break-even point is a value of x for which total revenue is the same as total cost; that is, $R = C$. How many units must be sold to break even?

Solve using a graphing calculator. Round all values to two decimal places.

79. $4xy - 7 = 0,$
$x - 3y - 2 = 0$

80. $x^2 + y^2 = 14,$
$16x + 7y^2 = 0$

10 Chapter Summary and Review

IMPORTANT CONCEPTS

[Section references appear in brackets.]

Concept	Example
The Distance Formula The distance d between any two points (x_1, y_1) and (x_2, y_2) is given by $$d = \sqrt{(x_2 - x_1)^2 + (y_2 - y_1)^2}.$$	The distance between $(3, -4)$ and $(-1, -5)$ is $$\begin{aligned} d &= \sqrt{(-1 - 3)^2 + (-5 - (-4))^2} \\ &= \sqrt{(-4)^2 + (-1)^2} \\ &= \sqrt{16 + 1} \\ &= \sqrt{17}. \end{aligned}$$ [10.1]
The Midpoint Formula If the endpoints of a segment are (x_1, y_1) and (x_2, y_2), then the coordinates of the midpoint are $$\left(\frac{x_1 + x_2}{2}, \frac{y_1 + y_2}{2}\right).$$	The midpoint of the segment with endpoints $(3, -4)$ and $(-1, -5)$ is $$\left(\frac{3 + (-1)}{2}, \frac{-4 + (-5)}{2}\right), \quad \text{or} \quad \left(1, -\frac{9}{2}\right).$$ [10.1]
Parabola $y = ax^2 + bx + c$ Opens upward $(a > 0)$ or downward $(a < 0)$ $= a(x - h)^2 + k$ Vertex: (h, k) $x = ay^2 + by + c$ Opens right $(a > 0)$ or left $(a < 0)$ $= a(y - k)^2 + h$ Vertex: (h, k) 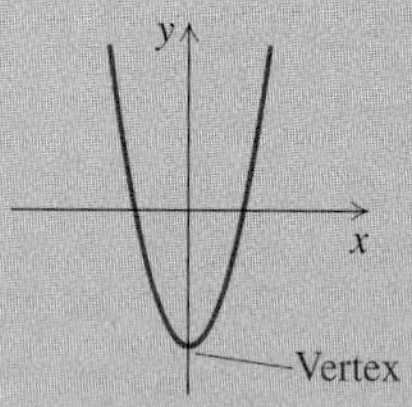 Parabola	$$\begin{aligned} x &= -y^2 + 4y - 1 \\ &= -(y^2 - 4y \quad\;\;) - 1 \\ &= -(y^2 - 4y + 4) - 1 - (-1)(4) \\ &= -(y - 2)^2 + 3 \qquad \boldsymbol{a = -1;} \textbf{ parabola opens left} \end{aligned}$$ 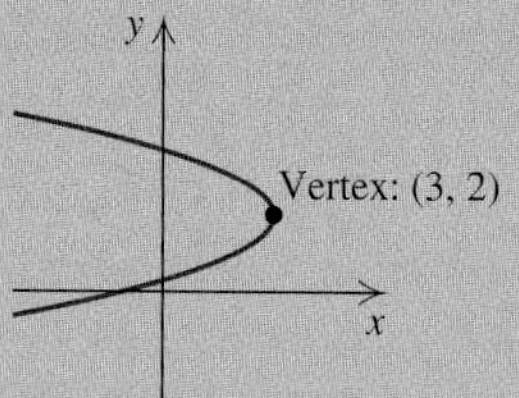 [10.1]

(*continued*)

Circle

$x^2 + y^2 = r^2$ Radius: r
Center: $(0, 0)$

$(x - h)^2 + (y - k)^2 = r^2$ Radius: r
Center: (h, k)

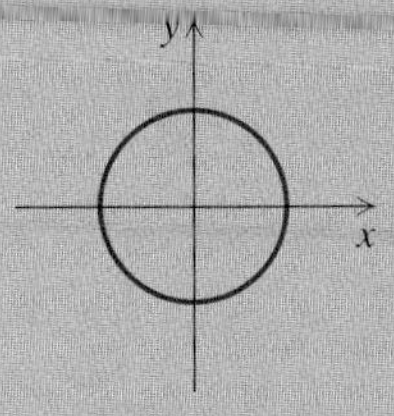

Circle

$$x^2 + y^2 + 2x - 6y + 6 = 0$$
$$x^2 + 2x \quad + y^2 - 6y \quad = -6$$
$$x^2 + 2x + 1 + y^2 - 6y + 9 = -6 + 1 + 9$$
$$(x + 1)^2 + (y - 3)^2 = 4$$
$$[x - (-1)]^2 + (y - 3)^2 = 2^2$$ Radius: 2
Center: $(-1, 3)$

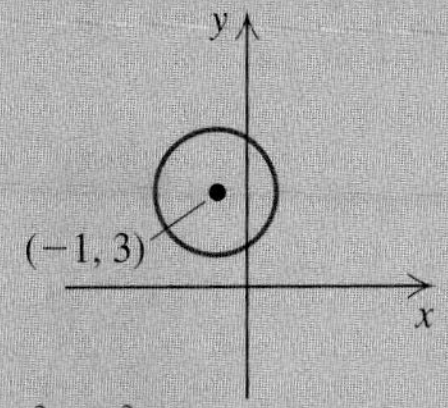

$x^2 + y^2 + 2x - 6y + 6 = 0$

[10.1]

Ellipse

$\dfrac{x^2}{a^2} + \dfrac{y^2}{b^2} = 1$ Center: $(0, 0)$

$\dfrac{(x - h)^2}{a^2} + \dfrac{(y - k)^2}{b^2} = 1$ Center: (h, k)

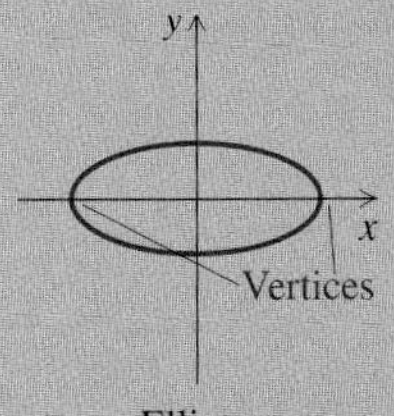

Ellipse

$$\frac{(x - 4)^2}{4} + \frac{(y + 1)^2}{9} = 1$$
$$\frac{(x - 4)^2}{2^2} + \frac{[y - (-1)]^2}{3^2} = 1$$ $3 > 2$;
ellipse is vertical

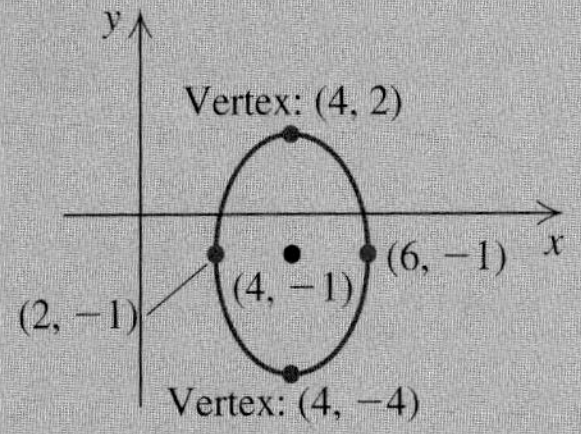

$\dfrac{(x - 4)^2}{4} + \dfrac{(y + 1)^2}{9} = 1$

[10.2]

Hyperbola

$\dfrac{x^2}{a^2} - \dfrac{y^2}{b^2} = 1$ Two branches opening right and left

$\dfrac{y^2}{b^2} - \dfrac{x^2}{a^2} = 1$ Two branches opening upward and downward

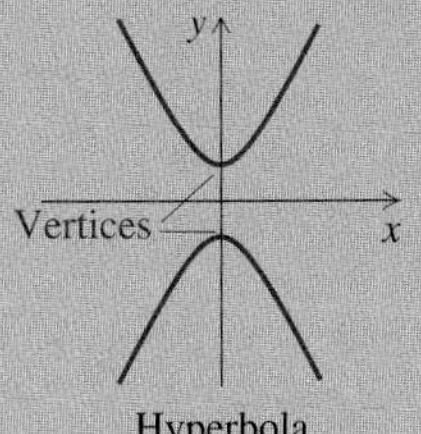

Hyperbola

$$\frac{x^2}{4} - \frac{y^2}{1} = 1$$
$$\frac{x^2}{2^2} - \frac{y^2}{1^2} = 1$$ Opens right and left

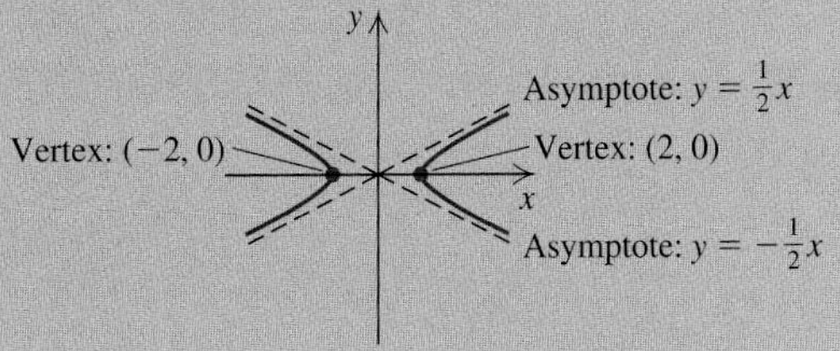

$\dfrac{x^2}{4} - \dfrac{y^2}{1} = 1$

[10.3]

Review Exercises

Concept Reinforcement Classify each statement as either true or false.

1. To use the distance formula, one must have an understanding of radical notation. [10.1]
2. The midpoint of the segment connecting (x_1, y_1) and (x_2, y_2) is $(x_1 + x_2, y_1 + y_2)$. [10.1]
3. The center of a circle is part of the circle itself. [10.1]
4. The foci of an ellipse always share the same second coordinate. [10.2]
5. Every parabola that opens upward or downward can represent the graph of a function. [10.1]
6. It is possible for a hyperbola to represent the graph of a function. [10.3]
7. Every system of nonlinear equations has at least one real solution. [10.4]
8. Both substitution and elimination can be used as methods for solving a system of nonlinear equations. [10.4]

Find the distance between each pair of points. Where appropriate, find an approximation to three decimal places. [10.1]

9. $(3, 6)$ and $(7, 6)$
10. $(-1, 1)$ and $(-5, 4)$
11. $(1.4, 3.6)$ and $(4.7, -5.3)$
12. $(2, 3a)$ and $(-1, a)$

Find the midpoint of the segment with the given endpoints. [10.1]

13. $(2, -1)$ and $(7, -1)$
14. $(-1, 10)$ and $(-5, 4)$
15. $(1, \sqrt{3})$ and $(\frac{1}{2}, -\sqrt{2})$
16. $(2, 3a)$ and $(-1, a)$

Find the center and the radius of each circle. [10.1]

17. $(x + 3)^2 + (y - 2)^2 = 7$
18. $(x - 5)^2 + y^2 = 49$
19. $x^2 + y^2 - 6x - 2y + 1 = 0$
20. $x^2 + y^2 + 8x - 6y = 10$
21. Find an equation of the circle with center $(-4, 3)$ and radius $4\sqrt{3}$. [10.1]
22. Find an equation of the circle with center $(7, -2)$ and radius $2\sqrt{5}$. [10.1]

Classify each equation as a circle, an ellipse, a parabola, or a hyperbola. Then graph. [10.1], [10.2], [10.3]

23. $5x^2 + 5y^2 = 80$
24. $9x^2 + 2y^2 = 18$
25. $y = -x^2 + 2x - 3$
26. $\frac{y^2}{9} - \frac{x^2}{4} = 1$
27. $xy = 9$
28. $x = y^2 + 2y - 2$
29. $\frac{(x + 1)^2}{3} + (y - 3)^2 = 1$
30. $x^2 + y^2 + 6x - 8y - 39 = 0$

Solve. [10.4]

31. $x^2 - y^2 = 33,$ $x + y = 11$
32. $x^2 - 2x + 2y^2 = 8,$ $2x + y = 6$
33. $x^2 - y = 3,$ $2x - y = 3$
34. $x^2 + y^2 = 25,$ $x^2 - y^2 = 7$
35. $x^2 - y^2 = 3,$ $y = x^2 - 3$
36. $x^2 + y^2 = 18,$ $2x + y = 3$
37. $x^2 + y^2 = 100,$ $2x^2 - 3y^2 = -120$
38. $x^2 + 2y^2 = 12,$ $xy = 4$

39. A rectangular bandstand has a perimeter of 38 m and an area of 84 m^2. What are the dimensions of the bandstand? [10.4]

40. One type of carton used by tableproducts.com exactly fits both a rectangular napkin of area 108 in^2 and a candle of length 15 in., laid diagonally on top of the napkin. Find the length and the width of the carton. [10.4]

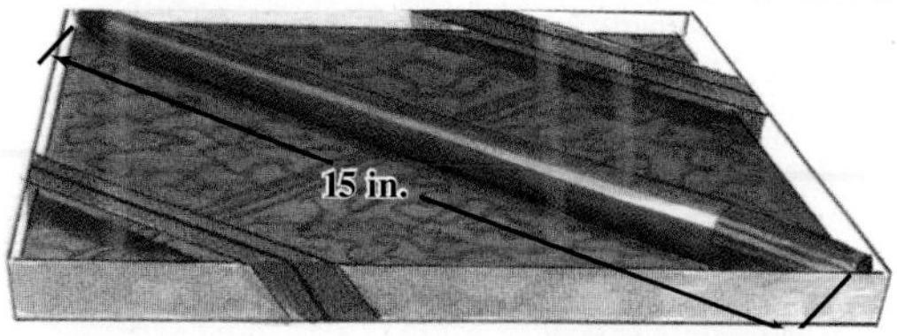

41. The perimeter of a square mirror is 12 cm more than the perimeter of another square mirror. Its area exceeds the area of the other by 39 cm^2. Find the perimeter of each mirror. [10.4]

42. The sum of the areas of two circles is 130π ft^2. The difference of the circumferences is 16π ft. Find the radius of each circle. [10.4]

Synthesis

TW 43. How does the graph of a hyperbola differ from the graph of a parabola? [10.1], [10.3]

TW 44. Explain why function notation rarely appears in this chapter, and list the graphs discussed for which function notation could be used. [10.1], [10.2], [10.3]

45. Solve: [10.4]

$$4x^2 - x - 3y^2 = 9,$$
$$-x^2 + x + y^2 = 2.$$

46. Find the points whose distance from $(8, 0)$ and from $(-8, 0)$ is 10. [10.1]

47. Find an equation of the circle that passes through $(-2, -4)$, $(5, -5)$, and $(6, 2)$. [10.1], [10.4]

48. Find an equation of the ellipse with the following intercepts: $(-9, 0)$, $(9, 0)$, $(0, -5)$, and $(0, 5)$. [10.2]

49. Find the point on the x-axis that is equidistant from $(-3, 4)$ and $(5, 6)$. [10.1]

Chapter Test 10

Find the distance between each pair of points. Where appropriate, find an approximation to three decimal places.

1. $(5, -1)$ and $(-4, 8)$

2. $(3, -a)$ and $(-3, a)$

Find the midpoint of the segment with the given endpoints.

3. $(4, -1)$ and $(-5, 8)$

4. $(3, -a)$ and $(-3, a)$

Find the center and the radius of each circle.

5. $(x + 5)^2 + (y - 1)^2 = 81$

6. $x^2 + y^2 + 4x - 6y + 4 = 0$

Classify the equation as a circle, an ellipse, a parabola, or a hyperbola. Then graph.

7. $y = x^2 - 4x - 1$

8. $x^2 + y^2 + 2x + 6y + 6 = 0$

9. $\dfrac{x^2}{16} - \dfrac{y^2}{9} = 1$

10. $16x^2 + 4y^2 = 64$

11. $xy = -5$

12. $x = -y^2 + 4y$

Solve.

13. $\dfrac{x^2}{4} + \dfrac{y^2}{9} = 1,$
$3x + 4y = 12$

14. $x^2 + y^2 = 16,$
$\dfrac{x^2}{16} - \dfrac{y^2}{9} = 1$

15. $x^2 - 2y^2 = 1,$
$xy = 6$

16. $x^2 + y^2 = 10,$
$x^2 = y^2 + 2$

17. A rectangular bookmark with diagonal of length $5\sqrt{5}$ has an area of 22. Find the dimensions of the bookmark.

18. Two squares are such that the sum of their areas is 8 m^2 and the difference of their areas is 2 m^2. Find the length of a side of each square.

19. A rectangular dance floor has a diagonal of length 40 ft and a perimeter of 112 ft. Find the dimensions of the dance floor.

20. Nikki invested a certain amount of money for 1 yr and earned \$72 in interest. Erin invested \$240 more than Nikki at an interest rate that was $\frac{5}{6}$ of the rate given to Nikki, but she earned the same amount of interest. Find the principal and the interest rate for Nikki's investment.

Synthesis

21. Find an equation of the ellipse passing through $(6, 0)$ and $(6, 6)$ with vertices at $(1, 3)$ and $(11, 3)$.

22. Find the point on the y-axis that is equidistant from $(-3, -5)$ and $(4, -7)$.

23. The sum of two numbers is 36, and the product is 4. Find the sum of the reciprocals of the numbers.

24. *Theatrical Production.* An E.T.C. spotlight for a college's production of *Hamlet* projects an ellipse of light on a stage that is 8 ft wide and 14 ft long. Find an equation of that ellipse if an actor is in its center and x represents the number of feet, horizontally, from the actor to the edge of the ellipse and y represents the number of feet, vertically, from the actor to the edge of the ellipse.

11 Sequences, Series, and the Binomial Theorem

The first three sections of this chapter are devoted to sequences and series. A sequence is simply an ordered list. For example, when a baseball coach writes a batting order, a sequence is being formed. When the members of a sequence are numbers, we can discuss their sum. Such a sum is called a series.

Section 11.4 presents the binomial theorem, which is used to expand expressions of the form $(a + b)^n$. Such an expression is itself a series.

APPLICATION *Aerobic Exercise.*

The following table and graph show the target heart rates during aerobic exercise for women of various ages. Determine whether or not the graph could be the graph of an arithmetic sequence.

Age	Target Heart Rate (in beats per minute)
20	150
40	135
60	120
80	105

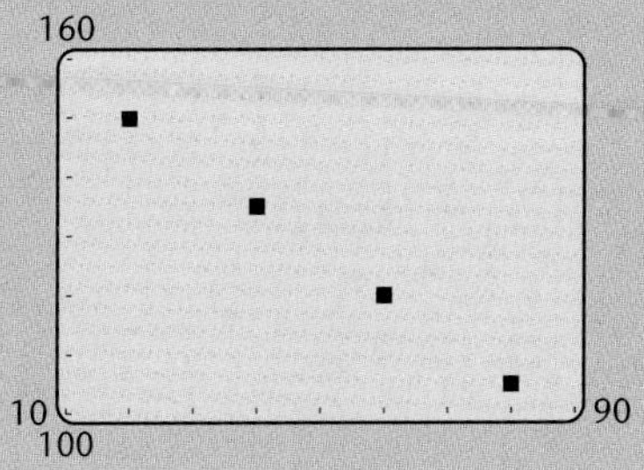

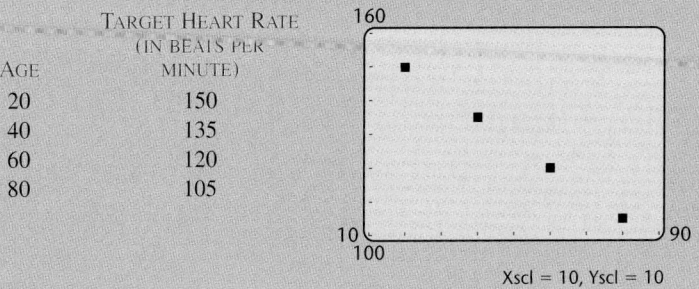

Xscl = 10, Yscl = 10

This problem appears as Exercise 73 in Exercise Set 11.2.

11.1 Sequences and Series

Sequences ■ Finding the General Term ■ Sums and Series ■ Sigma Notation ■ Graphs of Sequences

Study Tip

Crunch Time for the Final

It is always best to study for a final exam over several days or more. If you have only one or two days of study time, however, begin by studying the formulas, problems, properties, and procedures in each chapter's Summary and Review. Then do the exercises in the Cumulative Reviews. Make sure to attend a review session if one is offered.

Sequences

Suppose that \$10,000 is borrowed at 5%, compounded annually. The amount owed at the start of years 1, 2, 3, 4, and so on, is

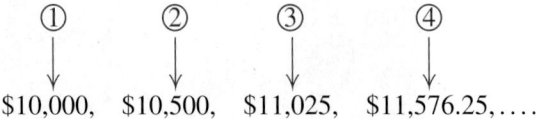

$$\overset{①}{\downarrow}\ \$10{,}000,\quad \overset{②}{\downarrow}\ \$10{,}500,\quad \overset{③}{\downarrow}\ \$11{,}025,\quad \overset{④}{\downarrow}\ \$11{,}576.25, \ldots.$$

We can regard this as a function that pairs 1 with \$10,000, 2 with \$10,500, 3 with \$11,025, and so on. A **sequence** (or **progression**) is thus a function, where the domain is a set of consecutive positive integers beginning with 1, and the range varies from sequence to sequence.

If we continue computing the amounts in the account forever, we obtain an **infinite sequence,** with function values

\$10,000, \$10,500, \$11,025, \$11,576.25, \$12,155.06, ...

The three dots at the end indicate that the sequence goes on without stopping. If we stop after a certain number of years, we obtain a **finite sequence:**

\$10,000, \$10,500, \$11,025, \$11,576.25.

Sequences An *infinite sequence* is a function having for its domain the set of natural numbers: $\{1, 2, 3, 4, 5, \ldots\}$.

A *finite sequence* is a function having for its domain a set of natural numbers: $\{1, 2, 3, 4, 5, \ldots, n\}$, for some natural number n.

As another example, consider the sequence given by

$$a(n) = 2^n, \quad \text{or} \quad a_n = 2^n.$$

The notation a_n means the same as $a(n)$ but is used more commonly with sequences. Some function values (also called *terms* of the sequence) follow:

$$a_1 = 2^1 = 2,$$
$$a_2 = 2^2 = 4,$$
$$a_3 = 2^3 = 8,$$
$$a_6 = 2^6 = 64.$$

The first term of the sequence is a_1, the fifth term is a_5, and the nth term, or **general term,** is a_n. This sequence can also be denoted in the following ways:

$$2, 4, 8, \ldots;$$

or $\quad 2, 4, 8, \ldots, 2^n, \ldots.$ **The 2^n emphasizes that the nth term of this sequence is found by raising 2 to the nth power.**

Sequences

When sequences are entered into a graphing calculator, the SEQUENCE MODE must be selected for some options. To do this, first press **MODE** and then move the cursor to the line that reads Func Par Pol Seq, the fourth line in the screen shown on the left below. Next, move the cursor to Seq and press **ENTER**. The functions in the equation-editor screen will be called $u(n)$ and $v(n)$ instead of y_1 and y_2, and the variable n is used instead of x; the variable is still entered using the **X,T,Θ,n** key.

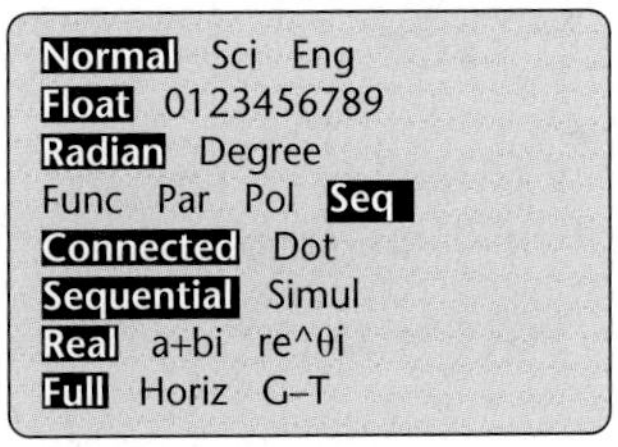

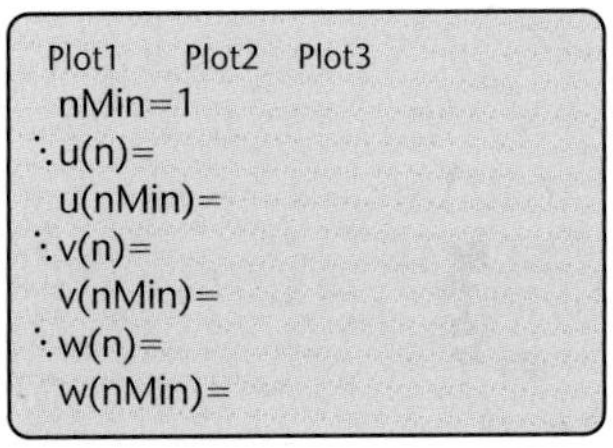

Most graphing calculators will write the terms of a sequence as a list without using the SEQUENCE mode. Usually found in the LIST OPS submenu, the seq option will list a finite sequence when the general term and the beginning and ending values of n are given. The cumSum option will list the cumulative sums of the elements of a sequence.

If the seq option is used, information must be supplied in the following order:

seq(general term, variable, value of n for the first term, value of n for the last term).

For example, to list the first 6 terms of the sequence given by

$$a_n = 2^n,$$

enter seq$(2^\wedge n, n, 1, 6)$. This is done by selecting seq under the LIST OPS menu, shown on the left at the top of the following page, and then pressing **2** **^** **X,T,Θ,n** **,** **X,T,Θ,n** **,** **1** **,** **6** **)** **ENTER**. Note that if this sequence of keystrokes is done in the FUNCTION mode, rather than the SEQUENCE mode, the variable used will be x, but the resulting sequence will be the same.

(continued)

The result is shown in the screen on the right below. The first 6 terms of the sequence are 2, 4, 8, 16, 32, 64.

```
NAMES OPS MATH
1: SortA(
2: SortD(
3: dim(
4: Fill(
5: seq(
6: cumSum(
7↓ΔList(
```

```
seq(2^n,n,1,6)
{2 4 8 16 32 64}
cumSum(seq(2^n,n
,1,6))
{2 6 14 30 62 1...
```

The cumSum (cumulative sums) option can be used in connection with the seq option. Instead of listing the sequence itself, cumSum lists the first term, then the sum of the first two terms, then the sum of the first three terms, and so on. The screen on the right above shows the cumulative sums of the sequence given by $a_n = 2^n$. The first term is 2, the sum of the first two terms is 6, the sum of the first three terms is 14, and so on. Pressing the right arrow key will show more cumulative sums of the sequence.

EXAMPLE 1 Find the first 4 terms and the 13th term of the sequence for which the general term is given by $a_n = (-1)^n n^2$.

SOLUTION We solve both algebraically and using a calculator.

BY HAND

We have $a_n = (-1)^n n^2$, so

$$\begin{aligned} a_1 &= (-1)^1 \cdot 1^2 = -1, \\ a_2 &= (-1)^2 \cdot 2^2 = 4, \\ a_3 &= (-1)^3 \cdot 3^2 = -9, \\ a_4 &= (-1)^4 \cdot 4^2 = 16, \\ a_{13} &= (-1)^{13} \cdot 13^2 = -169. \end{aligned}$$

USING A GRAPHING CALCULATOR

We let $u(n) = (-1)^\wedge n{*}n^2$.

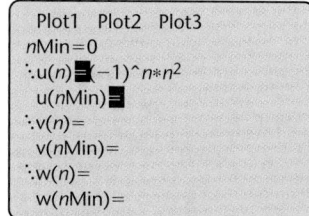

```
Plot1 Plot2 Plot3
nMin=0
⋱u(n)=(−1)^n*n²
 u(nMin)=
⋱v(n)=
 v(nMin)=
⋱w(n)=
 w(nMin)=
```

We set up a table with Indpnt set to Ask, and then supply 1, 2, 3, 4, and 13 as values for n.

n	u(n)	
1	−1	
2	4	
3	−9	
4	16	
13	−169	

n =

Note in Example 1 that the expression $(-1)^n$ causes the signs of the terms to alternate between positive and negative, depending on whether n is even or odd.

EXAMPLE 2 Use a graphing calculator to find the first 5 terms of the sequence for which the general term is given by $a_n = n/(n + 1)^2$.

SOLUTION We use the seq feature, supplying the formula for the general term, the variable, and the values of n for the first and last terms that we wish to calculate.

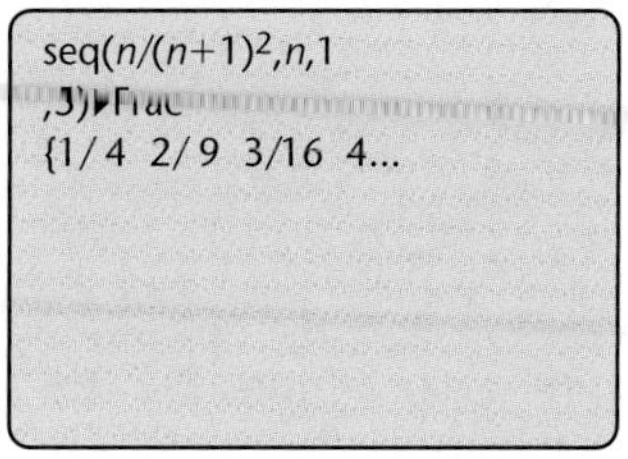
seq(n/(n+1)²,n,1
,5)▶Frac
{1/4 2/9 3/16 4...

Here we used ▶Frac to write each term in the sequence in fraction notation.

The first three terms are listed on the screen; the remaining can be found by pressing ▷. We have

$$a_1 = \tfrac{1}{4}, \quad a_2 = \tfrac{2}{9}, \quad a_3 = \tfrac{3}{16}, \quad a_4 = \tfrac{4}{25}, \quad \text{and} \quad a_5 = \tfrac{5}{36}.$$

Finding the General Term

When only the first few terms of a sequence are known, it is impossible to be certain what the general term is. Still, a prediction can be made by looking for a pattern.

EXAMPLE 3 For each sequence, predict the general term.

a) $1, 4, 9, 16, 25, \ldots$

b) $-1, 2, -4, 8, -16, \ldots$

c) $2, 4, 8, \ldots$

SOLUTION

a) $1, 4, 9, 16, 25, \ldots$

These are squares of consecutive positive integers, so the general term could be n^2.

b) $-1, 2, -4, 8, -16, \ldots$

These are powers of 2 with alternating signs, so the general term may be $(-1)^n[2^{n-1}]$. To check, note that 8 is the fourth term, and

$$\begin{aligned}(-1)^4[2^{4-1}] &= 1 \cdot 2^3 \\ &= 8.\end{aligned}$$

c) $2, 4, 8, \ldots$

We regard the pattern as powers of 2, in which case 16 would be the next term and 2^n the general term. The sequence could then be written with more terms as

$$2, 4, 8, 16, 32, 64, 128, \ldots .$$

In part (c) above, suppose that the second term is found by adding 2, the third term by adding 4, the next term by adding 6, and so on. In this case, 14 would be the next term and the sequence would be

$$2, 4, 8, 14, 22, 32, 44, 58, \ldots.$$

This illustrates that the fewer terms we are given, the greater the uncertainty about the nth term.

Sums and Series

Series Given the infinite sequence

$$a_1, a_2, a_3, a_4, \ldots, a_n, \ldots,$$

the sum of the terms

$$a_1 + a_2 + a_3 + \cdots + a_n + \cdots$$

is called an *infinite series* and is denoted S_∞. A *partial sum* is the sum of the first n terms:

$$a_1 + a_2 + a_3 + \cdots + a_n.$$

A partial sum is also called a *finite series* and is denoted S_n.

EXAMPLE 4 For the sequence −2, 4, −6, 8, −10, 12, −14, find: **(a)** S_2; **(b)** S_3; **(c)** S_7.

SOLUTION

a) $S_2 = -2 + 4 = 2$ **This is the sum of the first 2 terms.**

b) $S_3 = -2 + 4 + (-6) = -4$ **This is the sum of the first 3 terms.**

c) $S_7 = -2 + 4 + (-6) + 8 + (-10) + 12 + (-14) = -8$

This is the sum of the first 7 terms.

We can use a graphing calculator to find partial sums of a sequence for which the general term is given by a formula.

```
cumSum(seq((-1)^
n/(n+1),n,1,4))▶
Frac
{-1/2 -1/6 -5/1...
```

EXAMPLE 5 Use a graphing calculator to find S_1, S_2, S_3, and S_4 for the sequence in which the general term is given by $a_n = (-1)^n/(n + 1)$.

SOLUTION We use the cumSum and seq options in the LIST OPS menu and ▶Frac to write each sum in fraction notation. The first two sums, S_1 and S_2, are listed on the screen; S_3 and S_4 can be seen by pressing ▷. We have

$$S_1 = -\tfrac{1}{2}, \quad S_2 = -\tfrac{1}{6}, \quad S_3 = -\tfrac{5}{12}, \quad \text{and} \quad S_4 = -\tfrac{13}{60}.$$

Sigma Notation

When the general term of a sequence is known, the Greek letter Σ (capital sigma) can be used to write a series. For example, the sum of the first four terms of the sequence $3, 5, 7, 9, 11, \ldots, 2k + 1, \ldots$ can be named as follows, using *sigma notation*, or *summation notation*:

$$\sum_{k=1}^{4} (2k + 1).$$ **This represents** $(2 \cdot 1 + 1) + (2 \cdot 2 + 1) + (2 \cdot 3 + 1) + (2 \cdot 4 + 1)$**.**

This is read "the sum as k goes from 1 to 4 of $(2k + 1)$." The letter k is called the *index of summation*. The index of summation need not start at 1.

EXAMPLE 6 Write out and evaluate each sum.

a) $\sum_{k=1}^{5} k^2$ **b)** $\sum_{k=4}^{6} (-1)^k(2k)$ **c)** $\sum_{k=0}^{3} (2^k + 5)$

SOLUTION

a) $\sum_{k=1}^{5} k^2 = 1^2 + 2^2 + 3^2 + 4^2 + 5^2 = 1 + 4 + 9 + 16 + 25 = 55$

Evaluate k^2 for all integers from 1 through 5. Then add.

b) $\sum_{k=4}^{6} (-1)^k(2k) = (-1)^4(2 \cdot 4) + (-1)^5(2 \cdot 5) + (-1)^6(2 \cdot 6)$

$= 8 - 10 + 12 = 10$

c) $\sum_{k=0}^{3} (2^k + 5) = (2^0 + 5) + (2^1 + 5) + (2^2 + 5) + (2^3 + 5)$

$= 6 + 7 + 9 + 13 = 35$

Student Notes

A great deal of information is condensed into sigma notation. Be careful to pay attention to what values the index of summation will take on. Evaluate the expression following sigma, the general term, for each value and then add the results.

EXAMPLE 7 Write sigma notation for each sum.

a) $1 + 4 + 9 + 16 + 25$ **b)** $-1 + 3 - 5 + 7$

c) $3 + 9 + 27 + 81 + \cdots$

SOLUTION

a) $1 + 4 + 9 + 16 + 25$

Note that this is a sum of squares, $1^2 + 2^2 + 3^2 + 4^2 + 5^2$, so the general term is k^2. Sigma notation is

$$\sum_{k=1}^{5} k^2.$$ **The sum starts with 1^2 and ends with 5^2.**

Answers may vary here. For example, another—perhaps less obvious—way of writing $1 + 4 + 9 + 16 + 25$ is

$$\sum_{k=2}^{6} (k - 1)^2.$$

b) $-1 + 3 - 5 + 7$

Except for the alternating signs, this is the sum of the first four positive odd numbers. It is useful to know that $2k - 1$ is a formula for the kth positive odd number. It is also important to note that since $(-1)^k = 1$ when k is even and $(-1)^k = -1$ when k is odd, the factor $(-1)^k$ can be used to create the alternating signs. The general term is thus $(-1)^k(2k - 1)$, beginning with $k = 1$. Sigma notation is

$$\sum_{k=1}^{4} (-1)^k(2k - 1).$$

To check, we can evaluate $(-1)^k(2k - 1)$ using 1, 2, 3, and 4. Then we can write the sum of the four terms. We leave this to the student.

c) $3 + 9 + 27 + 81 + \cdots$

This is a sum of powers of 3, and it is also an infinite series. We use the symbol ∞ for infinity and write the series using sigma notation:

$$\sum_{k=1}^{\infty} 3^k.$$

Graphs of Sequences

Because the domain of a sequence is a set of integers, the graph of a sequence is a set of points that are not connected. We can use the DOT mode to graph a sequence when a formula for the general term is known.

EXAMPLE 8 Graph the sequence for which the general term is given by $a_n = (-1)^n/n$.

SOLUTION We let $u(n) = (-1)$^n/n. In SEQUENCE mode, the default graph mode is DOT. You may wish to examine a table of values to help set up the window dimensions for the graph. From the table on the left below, the terms appear to be between -1 and 1, so we let Ymin $= -1$ and Ymax $= 1$.

For the window settings, we must determine nMin and nMax, as well as Xmin, Xmax, Ymin, and Ymax. Here nMin is the smallest value of n for which we wish to evaluate the sequence, and nMax is the largest value. We let nMin $= 1$ and nMax $= 20$. If the graphing calculator also allows us to set PlotStart and PlotStep, we set each of those to 1.

n	u(n)
1	−1
2	.5
3	−.3333
4	.25
5	−.2
6	.16667
7	−.1429

$n = 1$

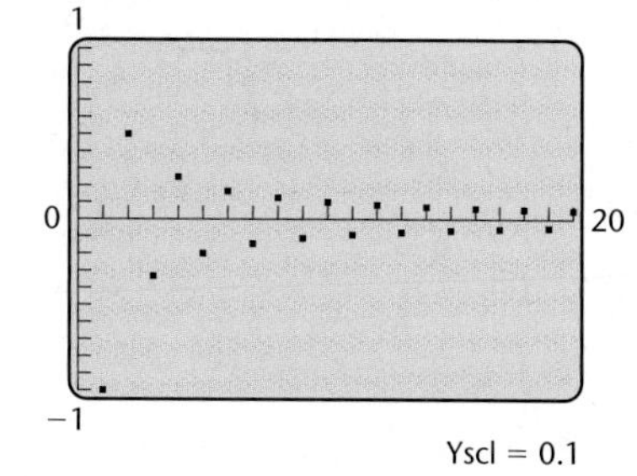

We see from the graph on the right above that the absolute value of the terms gets smaller as n gets larger. The graph also illustrates that the signs of the terms alternate.

11.1 EXERCISE SET

FOR EXTRA HELP

MathXL | MyMathLab | InterAct Math | AW Math Tutor Center | Video Lectures on CD: Disc 6 | Student's Solutions Manual

Concept Reinforcement *In each of Exercises 1–6, match the expression with the most appropriate expression from the column on the right.*

1. ___ $\sum_{k=1}^{4} k^2$
2. ___ $\sum_{k=3}^{6} (-1)^k$
3. ___ $5 + 10 + 15 + 20$
4. ___ $a_n = 5^n$
5. ___ $a_n = 3n + 2$
6. ___ $a_1 + a_2 + a_3$

a) $-1 + 1 + (-1) + 1$
b) $a_2 = 25$
c) $a_2 = 8$
d) $\sum_{k=1}^{4} 5k$
e) S_3
f) $1 + 4 + 9 + 16$

In each of the following, the nth term of a sequence is given. In each case, find the first 4 terms; the 10th term, a_{10}; and the 15th term, a_{15}.

7. $a_n = 2n + 3$
8. $a_n = 5n - 2$
9. $a_n = n^2 + 2$
10. $a_n = \dfrac{n}{n + 1}$
11. $a_n = \dfrac{n^2 - 1}{n^2 + 1}$
12. $a_n = n^2 - 2n$
13. $a_n = \left(-\dfrac{1}{2}\right)^{n-1}$
14. $a_n = n + \dfrac{1}{n}$
15. $a_n = (-1)^n(n + 3)$
16. $a_n = (-1)^n n^2$
17. $a_n = (-1)^n(n^3 - 1)$
18. $a_n = (-1)^{n+1}(3n - 5)$

Find the indicated term of each sequence.

19. $a_n = 2n - 3$; a_8
20. $a_n = 3n + 2$; a_8
21. $a_n = (3n + 1)(2n - 5)$; a_9
22. $a_n = (3n + 2)^2$; a_6
23. $a_n = (-1)^{n-1}(3.4n - 17.3)$; a_{12}
24. $a_n = (-2)^{n-2}(45.68 - 1.2n)$; a_{23}
25. $a_n = 3n^2(9n - 100)$; a_{11}
26. $a_n = 4n^2(2n - 39)$; a_{22}
27. $a_n = \left(1 + \dfrac{1}{n}\right)^2$; a_{20}
28. $a_n = \left(1 - \dfrac{1}{n}\right)^3$; a_{15}

Look for a pattern and then predict the general term, or nth term, a_n, of each sequence. Answers may vary.

29. $2, 4, 6, 8, 10, \ldots$
30. $1, 3, 5, 7, \ldots$
31. $1, -1, 1, -1, \ldots$
32. $-1, 1, -1, 1, \ldots$
33. $-1, 2, -3, 4, \ldots$
34. $1, -2, 3, -4, \ldots$
35. $3, 5, 7, 9, \ldots$
36. $4, 6, 8, 10, \ldots$
37. $-2, 6, -18, 54, \ldots$
38. $-2, 3, 8, 13, 18, \ldots$
39. $\frac{1}{2}, \frac{2}{3}, \frac{3}{4}, \frac{4}{5}, \frac{5}{6}, \ldots$
40. $1 \cdot 2, 2 \cdot 3, 3 \cdot 4, 4 \cdot 5, \ldots$
41. $5, 25, 125, 625, \ldots$
42. $4, 16, 64, 256, \ldots$
43. $-1, 4, -9, 16, \ldots$
44. $1, -4, 9, -16, \ldots$

Find the indicated partial sum for each sequence.

45. $1, -2, 3, -4, 5, -6, \ldots$; S_7
46. $1, -3, 5, -7, 9, -11, \ldots$; S_8
47. $2, 4, 6, 8, \ldots$; S_5
48. $1, \frac{1}{4}, \frac{1}{9}, \frac{1}{16}, \frac{1}{25}, \ldots$; S_5

Write out and evaluate each sum.

49. $\sum_{k=1}^{5} \frac{1}{2k}$

50. $\sum_{k=1}^{6} \frac{1}{2k - 1}$

51. $\sum_{k=0}^{4} 3^k$

52. $\sum_{k=4}^{7} \sqrt{2k + 1}$

53. $\sum_{k=2}^{8} \frac{k}{k - 1}$

54. $\sum_{k=2}^{5} \frac{k - 2}{k + 3}$

55. $\sum_{k=1}^{8} (-1)^{k+1}2^k$

56. $\sum_{k=1}^{7} (-1)^k 4^{k+1}$

57. $\sum_{k=0}^{5} (k^2 - 2k + 3)$

58. $\sum_{k=0}^{5} (k^2 - 3k + 4)$

59. $\sum_{k=3}^{5} \frac{(-1)^k}{k(k + 1)}$

60. $\sum_{k=3}^{7} \frac{k}{2^k}$

Rewrite each sum using sigma notation. Answers may vary.

61. $\frac{2}{3} + \frac{3}{4} + \frac{4}{5} + \frac{5}{6} + \frac{6}{7}$

62. $3 + 6 + 9 + 12 + 15$

63. $1 + 4 + 9 + 16 + 25 + 36$

64. $\frac{1}{1^2} + \frac{1}{2^2} + \frac{1}{3^2} + \frac{1}{4^2} + \frac{1}{5^2}$

65. $4 - 9 + 16 - 25 + \cdots + (-1)^n n^2$

66. $9 - 16 + 25 + \cdots + (-1)^{n+1} n^2$

67. $5 + 10 + 15 + 20 + 25 + \cdots$

68. $7 + 14 + 21 + 28 + 35 + \cdots$

69. $\frac{1}{1 \cdot 2} + \frac{1}{2 \cdot 3} + \frac{1}{3 \cdot 4} + \frac{1}{4 \cdot 5} + \cdots$

70. $\frac{1}{1 \cdot 2^2} + \frac{1}{2 \cdot 3^2} + \frac{1}{3 \cdot 4^2} + \frac{1}{4 \cdot 5^2} + \cdots$

TW **71.** The sequence $1, 4, 9, 16, \ldots$ can be written as $f(x) = x^2$ with the domain the set of all positive integers. Explain how the graph of f would compare with the graph of $y = x^2$.

TW **72.** Eric says he expects he will prefer sequences to functions because he dislikes fractions. Will his expectations prove correct? Why or why not?

Skill Maintenance

Evaluate. [1.1]

73. $\frac{7}{2}(a_1 + a_7)$, for $a_1 = 8$ and $a_7 = 14$

74. $a_1 + (n - 1)d$, for $a_1 = 3$, $n = 6$, and $d = 4$

Multiply. [5.2]

75. $(x + y)^3$

76. $(a - b)^3$

77. $(2a - b)^3$

78. $(2x + y)^3$

Synthesis

TW **79.** Explain why the equation

$$\sum_{k=1}^{n} (a_k + b_k) = \sum_{k=1}^{n} a_k + \sum_{k=1}^{n} b_k$$

is true for any positive integer n. What laws are used to justify this result?

TW **80.** Consider the sums

$$\sum_{k=1}^{5} 3k^2 \quad \text{and} \quad 3\sum_{k=1}^{5} k^2.$$

a) Which is easier to evaluate and why?

b) Is it true that

$$\sum_{k=1}^{n} ca_k = c\sum_{k=1}^{n} a_k?$$

Why or why not?

Some sequences are given by a recursive *definition. The value of the first term, a_1, is given, and then we are told how to find any subsequent term from the term preceding it. Find the first six terms of each of the following recursively defined sequences.*

81. $a_1 = 1$, $a_{n+1} = 5a_n - 2$

82. $a_1 = 0$, $a_{n+1} = a_n^2 + 3$

83. *Value of a Copier.* The value of a color photocopier is \$5200. Its scrap value each year is 75% of its value the year before. Give a sequence that lists the scrap value of the machine at the start of each year for a 10-yr period.

84. *Cell Biology.* A single cell of bacterium divides into two every 15 min. Suppose that the same rate of division is maintained for 4 hr. Give a sequence that lists the number of cells after successive 15-min periods.

85. Find S_{100} and S_{101} for the sequence in which $a_n = (-1)^n$.

Find the first five terms of each sequence; then find S_5.

86. $a_n = \frac{1}{2^n} \log 1000^n$

87. $a_n = i^n, i = \sqrt{-1}$

88. Find all values for x that solve the following:

$$\sum_{k=1}^{x} i^k = -1.$$

89. The nth term of a sequence is given by $a_n = n^5 - 14n^4 + 6n^3 + 416n^2 - 655n - 1050$. Use a graphing calculator with a TABLE feature to determine what term in the sequence is 6144.

90. To define a sequence recursively on a graphing calculator (see Exercises 81 and 82), the SEQ MODE is used. The general term u_n or v_n can be expressed in terms of u_{n-1} or v_{n-1} by pressing 2ND 7 or 2ND 8. The starting values of u_n, v_n, and n are set as one of the WINDOW variables.

Use recursion to determine how many handshakes will occur if a group of 50 people shake hands with one another. To develop the recursion formula, begin with a group of 2 and determine how many additional handshakes occur with the arrival of each new group member. (See the Collaborative Corner following Exercise Set 5.1 on p. 357.)

11.2 Arithmetic Sequences and Series

Arithmetic Sequences ▪ Sum of the First n Terms of an Arithmetic Sequence ▪ Problem Solving

Study Tip

Rest Before a Test

The final exam is probably the most important math test of the semester. Do yourself a favor and see to it that you get a good night's sleep the night before. Being well rested will help guarantee that you put forth your best work.

In this section, we concentrate on sequences and series that are said to be arithmetic (pronounced ar-ith-MET-ik).

Arithmetic Sequences

In an **arithmetic sequence** (or **progression**), any term (other than the first) can be found by adding the same number to its preceding term. For example, the sequence 2, 5, 8, 11, 14, 17, ... is arithmetic because adding 3 to any term produces the next term.

Arithmetic Sequence A sequence is *arithmetic* if there exists a number d, called the *common difference*, such that $a_{n+1} = a_n + d$ for any integer $n \geq 1$.

EXAMPLE 1 For each arithmetic sequence, identify the first term, a_1, and the common difference, d.

a) 4, 9, 14, 19, 24, ...

b) 27, 20, 13, 6, −1, −8, ...

Student Notes

When the terms are decreasing, the common difference is negative. Try determining the sign of the common difference before calculating it.

SOLUTION To find a_1, we simply use the first term listed. To find d, we choose any term beyond the first and subtract the preceding term from it.

Sequence	First Term, a_1	Common Difference, d
a) 4, 9, 14, 19, 24, ...	4	$5 \longleftarrow 9 - 4 = 5$
b) 27, 20, 13, 6, −1, −8, ...	27	$-7 \longleftarrow 20 - 27 = -7$

To find the common difference, we subtracted a_1 from a_2. Had we subtracted a_2 from a_3 or a_3 from a_4, we would have found the same values for d.

Check: As a check, note that when d is added to each term, the result is the next term in the sequence.

a) $4 + 5 = 9, \quad 9 + 5 = 14, \quad 14 + 5 = 19, \quad 19 + 5 = 24$

b) $27 + (-7) = 20, \quad 20 + (-7) = 13, \quad 13 + (-7) = 6,$
$6 + (-7) = -1, \quad -1 + (-7) = -8$

To develop a formula for the general, nth, term of any arithmetic sequence, we denote the common difference by d and write out the first few terms:

$$a_1,$$
$$a_2 = a_1 + d,$$
$$a_3 = a_2 + d = (a_1 + d) + d = a_1 + 2d, \quad \textbf{Substituting } a_1 + d \textbf{ for } a_2$$
$$a_4 = a_3 + d = (a_1 + 2d) + d = a_1 + 3d. \quad \textbf{Substituting } a_1 + 2d \textbf{ for } a_3$$

Note that the coefficient of d in each case is 1 less than the subscript.

Generalizing, we obtain the following formula.

> **To Find a_n for an Arithmetic Sequence**
>
> The nth term of an arithmetic sequence with common difference d is
>
> $$a_n = a_1 + (n - 1)d, \quad \text{for any integer } n \geq 1.$$

EXAMPLE 2 Find the 14th term of the arithmetic sequence 6, 9, 12, 15, ...

SOLUTION First we note that $a_1 = 6$, $d = 3$, and $n = 14$. Using the formula for the nth term of an arithmetic sequence, we have

$$a_n = a_1 + (n - 1)d$$
$$a_{14} = 6 + (14 - 1) \cdot 3 = 6 + 13 \cdot 3 = 6 + 39 = 45.$$

The 14th term is 45.

EXAMPLE 3 For the sequence in Example 2, which term is 300? That is, find n if $a_n = 300$.

SOLUTION We substitute into the formula for the nth term of an arithmetic sequence and solve for n:

$$\begin{aligned} a_n &= a_1 + (n - 1)d \\ 300 &= 6 + (n - 1) \cdot 3 \\ 300 &= 6 + 3n - 3 \\ 297 &= 3n \\ 99 &= n. \end{aligned}$$

The term 300 is the 99th term of the sequence.

Given two terms and their places in an arithmetic sequence, we can construct the sequence.

EXAMPLE 4 The 3rd term of an arithmetic sequence is 14, and the 16th term is 79. Find a_1 and d and construct the sequence.

SOLUTION We know that $a_3 = 14$ and $a_{16} = 79$. Thus we would have to add d 13 times to get from 14 to 79. That is,

$$14 + 13d = 79. \qquad \textbf{a_3 and a_{16} are 13 terms apart; } \mathbf{16 - 3 = 13}$$

Solving $14 + 13d = 79$, we obtain

$$\begin{aligned} 13d &= 65 && \textbf{Subtracting 14 from both sides} \\ d &= 5. && \textbf{Dividing both sides by 13} \end{aligned}$$

We subtract d twice from a_3 to get to a_1. Thus,

$$a_1 = 14 - 2 \cdot 5 = 4. \qquad \textbf{a_1 and a_3 are 2 terms apart; } \mathbf{3 - 1 = 2}$$

The sequence is 4, 9, 14, 19, ... Note that we could have subtracted d 15 times from a_{16} in order to find a_1.

In general, d should be subtracted $(n - 1)$ times from a_n in order to find a_1. What will the graph of an arithmetic sequence look like?

Interactive Discovery

Graph each of the following arithmetic sequences. What pattern do you observe?

1. $a_n = -5 + (n - 1)3$
2. $a_n = \frac{1}{2} + (n - 1)\frac{3}{2}$
3. $a_n = 60 + (n - 1)(-5)$
4. $a_n = -1.7 + (n - 1)(-0.2)$

The pattern you may have observed above is true in general:

The graph of an arithmetic sequence is a set of points that lie on a straight line.

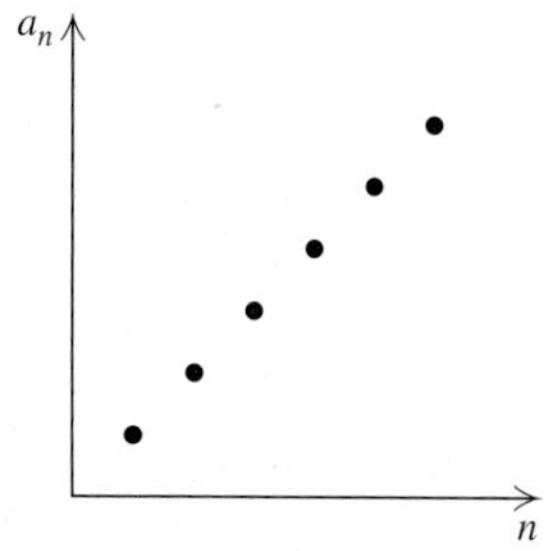

$a_n = a_1 + (n - 1)d, d > 0$

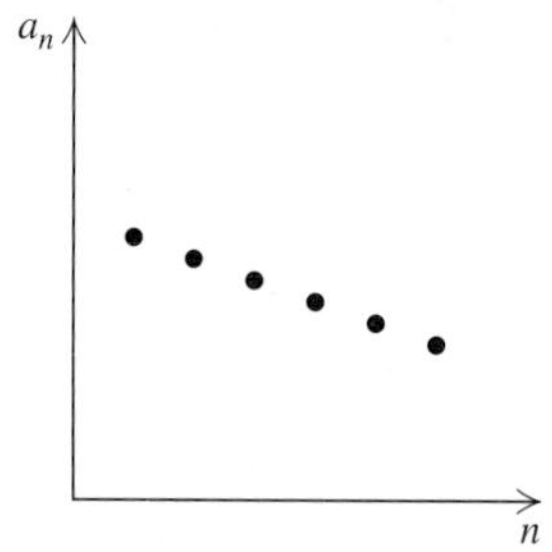

$a_n = a_1 + (n - 1)d, d < 0$

Sum of the First *n* Terms of an Arithmetic Sequence

When the terms of an arithmetic sequence are added, an **arithmetic series** is formed. To develop a formula for computing S_n when the series is arithmetic, we list the first n terms of the sequence as follows:

This is the next-to-last term. If you add d to this term, the result is a_n.

$$a_1, (a_1 + d), (a_1 + 2d), \ldots, (a_n - 2d), (a_n - d), a_n$$

This term is two terms back from the end. If you add d to this term, you get the next-to-last term, $a_n - d$.

Thus, S_n is given by

$$S_n = a_1 + (a_1 + d) + (a_1 + 2d) + \cdots + (a_n - 2d) + (a_n - d) + a_n.$$

Using a commutative law, we have a second equation:

$$S_n = a_n + (a_n - d) + (a_n - 2d) + \cdots + (a_1 + 2d) + (a_1 + d) + a_1.$$

Adding corresponding terms on each side of the above equations, we get

$$2S_n = [a_1 + a_n] + [(a_1 + d) + (a_n - d)] + [(a_1 + 2d) + (a_n - 2d)] + \cdots + [(a_n - 2d) + (a_1 + 2d)] + [(a_n - d) + (a_1 + d)] + [a_n + a_1].$$

This simplifies to

$$2S_n = [a_1 + a_n] + [a_1 + a_n] + [a_1 + a_n] + \cdots + [a_n + a_1] + [a_n + a_1] + [a_n + a_1].$$

There are n bracketed sums.

Since $[a_1 + a_n]$ is being added n times, it follows that

$$2S_n = n[a_1 + a_n].$$

Dividing both sides by 2 leads to the following formula.

Student Notes

The formula for the sum of an arithmetic sequence is very useful, but remember that it does not work for sequences that are not arithmetic.

To Find S_n for an Arithmetic Sequence

The sum of the first n terms of an arithmetic sequence is given by

$$S_n = \frac{n}{2}(a_1 + a_n).$$

EXAMPLE 5 Find the sum of the first 100 positive even numbers.

SOLUTION The sum is

$$2 + 4 + 6 + \cdots + 198 + 200.$$

This is the sum of the first 100 terms of the arithmetic sequence for which

$$a_1 = 2, \quad n = 100, \quad \text{and} \quad a_n = 200.$$

Substituting in the formula

$$S_n = \frac{n}{2}(a_1 + a_n),$$

we get

$$\begin{aligned} S_{100} &= \frac{100}{2}(2 + 200) \\ &= 50(202) = 10{,}100. \end{aligned}$$

The formula above is useful when we know the first and last terms, a_1 and a_n. To find S_n when a_n is unknown, but a_1, n, and d are known, we can use the formula $a_n = a_1 + (n - 1)d$ to calculate a_n and then proceed as in Example 5.

EXAMPLE 6 Find the sum of the first 15 terms of the arithmetic sequence 4, 7, 10, 13, ...

SOLUTION Note that

$$a_1 = 4, \quad n = 15, \quad \text{and} \quad d = 3.$$

Before using the formula for S_n, we find a_{15}:

$$\begin{aligned} a_{15} &= 4 + (15 - 1)3 && \textbf{Substituting into the formula for } a_n \\ &= 4 + 14 \cdot 3 = 46. \end{aligned}$$

Thus, knowing that $a_{15} = 46$, we have

$$\begin{aligned} S_{15} &= \tfrac{15}{2}(4 + 46) && \textbf{Using the formula for } S_n \\ &= \tfrac{15}{2}(50) = 375. \end{aligned}$$

Problem Solving

For some problem-solving situations, the translation may involve sequences or series. There is often a variety of ways in which a problem can be solved. You should use the one that is best or easiest for you. In this chapter, however, we will try to emphasize sequences and series and their related formulas.

EXAMPLE 7 Hourly Wages. Chris accepts a job managing a music store, starting with an hourly wage of \$14.60, and is promised a raise of 25¢ per hour every 2 months for 5 years. After 5 years of work, what will be Chris's hourly wage?

SOLUTION

1. **Familiarize.** It helps to write down the hourly wage for several two-month time periods.

Beginning:	14.60,
After two months:	14.85,
After four months:	15.10,

and so on.

What appears is a sequence of numbers: 14.60, 14.85, 15.10, ... Since the same amount is added each time, the sequence is arithmetic.

We list what we know about arithmetic sequences. The pertinent formulas are

$$a_n = a_1 + (n - 1)d$$

and

$$S_n = \frac{n}{2}(a_1 + a_n).$$

In this case, we are not looking for a sum, so we use the first formula. We want to determine the last term in a sequence. To do so, we need to know a_1, n, and d. From our list above, we see that

$$a_1 = 14.60 \quad \text{and} \quad d = 0.25.$$

What is n? That is, how many terms are in the sequence? After 1 year, there will have been 6 raises, since Chris gets a raise every 2 months. There are 5 years, so the total number of raises will be $5 \cdot 6$, or 30. Altogether, there will be 31 terms: the original wage and 30 increased rates.

2. **Translate.** We want to find a_n for the arithmetic sequence in which $a_1 = 14.60$, $n = 31$, and $d = 0.25$.

3. **Carry out.** Substituting in the formula for a_n gives us

$$a_{31} = 14.60 + (31 - 1) \cdot 0.25 = 22.10.$$

4. **Check.** We can check by redoing the calculations or we can calculate in a slightly different way for another check. For example, at the end of a year, there will be 6 raises, for a total raise of \$1.50. At the end of 5 years, the total raise will be $5 \times \$1.50$, or \$7.50. If we add that to the original wage of \$14.60, we obtain \$22.10. The answer checks.

5. **State.** After 5 years, Chris's hourly wage will be \$22.10.

EXAMPLE 8 Telephone Pole Storage. A stack of telephone poles has 30 poles in the bottom row. There are 29 poles in the second row, 28 in the next row, and so on. How many poles are in the stack if there are 5 poles in the top row?

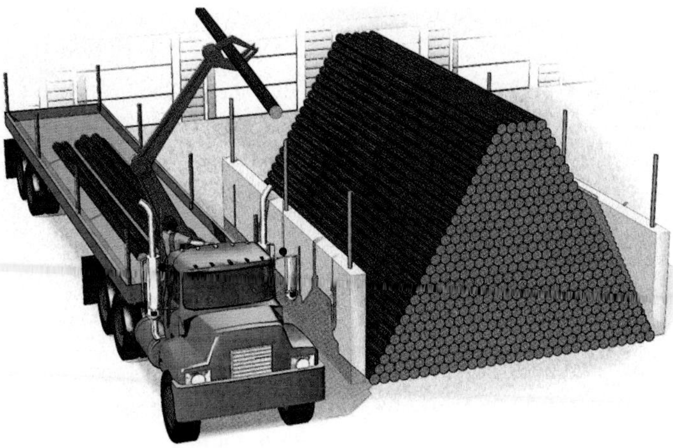

SOLUTION

1. **Familiarize.** The following figure shows the ends of the poles. There are 30 poles on the bottom and one fewer in each successive row. How many rows will there be?

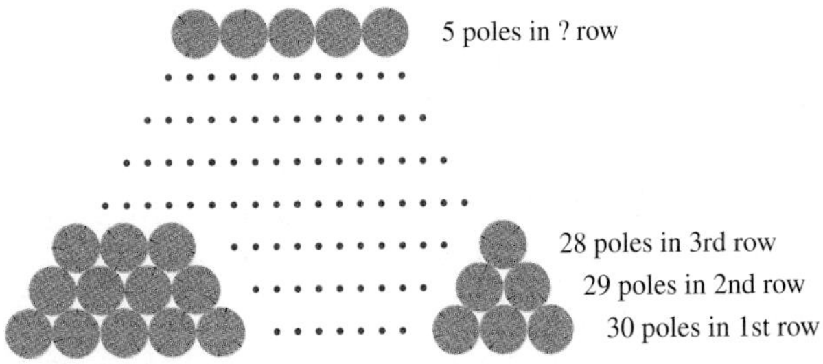

Note that there are $30 - 1 = 29$ poles in the 2nd row, $30 - 2 = 28$ poles in the 3rd row, $30 - 3 = 27$ poles in the 4th row, and so on. The pattern leads to $30 - 25 = 5$ poles in the 26th row.

The situation is represented by the equation

$30 + 29 + 28 + \cdots + 5.$ **There are 26 terms in this series.**

Thus we have an arithmetic series. We recall the formula

$$S_n = \frac{n}{2}(a_1 + a_n).$$

2. **Translate.** We want to find the sum of the first 26 terms of an arithmetic sequence in which $a_1 = 30$ and $a_{26} = 5$.
3. **Carry out.** Substituting into the above formula gives us

$$S_{26} = \tfrac{26}{2}(30 + 5) = 13 \cdot 35 = 455.$$

4. **Check.** In this case, we can check the calculations by doing them again. A longer, harder way would be to do the entire addition:

$$30 + 29 + 28 + \cdots + 5.$$

5. **State.** There are 455 poles in the stack.

11.2 EXERCISE SET

FOR EXTRA HELP

MathXL | MyMathLab | InterAct Math | AW Math Tutor Center | Video Lectures on CD: Disc 6 | Student's Solutions Manual

Concept Reinforcement *Classify each statement as either true or false.*

1. In an arithmetic sequence, the difference between any two consecutive terms is always the same.

2. In an arithmetic sequence, if $a_9 - a_8 = 4$, then $a_{13} - a_{12} = 4$ as well.

3. In an arithmetic sequence containing 17 terms, the common difference is $a_{17} - a_1$.

4. In an arithmetic sequence, if $a_1 = 8$ and $a_3 = 12$, then a_6 must be 16.

5. The sum of the first 20 terms of an arithmetic sequence can be found by knowing just a_1 and a_{20}.

6. The sum of the first 30 terms of an arithmetic sequence can be found by knowing just a_1 and d, the common difference.

7. The notation S_5 means $a_1 + a_5$.

8. For any arithmetic sequence, $S_9 = S_8 + d$, where d is the common difference.

Find the first term and the common difference.

9. $2, 6, 10, 14, \ldots$

10. $1.06, 1.12, 1.18, 1.24, \ldots$

11. $7, 3, -1, -5, \ldots$

12. $-8, -5, -2, 1, \ldots$

13. $\frac{3}{2}, \frac{9}{4}, 3, \frac{15}{4}, \ldots$

14. $\frac{3}{5}, \frac{1}{10}, -\frac{2}{5}, \ldots$

15. \$5.12, \$5.24, \$5.36, \$5.48, ...

16. \$214, \$211, \$208, \$205, ...

17. Find the 15th term of the arithmetic sequence $7, 10, 13, \ldots$.

18. Find the 17th term of the arithmetic sequence $6, 10, 14, \ldots$.

19. Find the 18th term of the arithmetic sequence $8, 2, -4, \ldots$.

20. Find the 14th term of the arithmetic sequence $3, \frac{7}{3}, \frac{5}{3}, \ldots$.

21. Find the 13th term of the arithmetic sequence \$1200, \$964.32, \$728.64, ...

22. Find the 10th term of the arithmetic sequence \$2345.78, \$2967.54, \$3589.30, ...

23. In the sequence of Exercise 17, what term is 82?

24. In the sequence of Exercise 18, what term is 126?

25. In the sequence of Exercise 19, what term is -328?

26. In the sequence of Exercise 20, what term is -27?

27. Find a_{17} when $a_1 = 2$ and $d = 5$.

28. Find a_{20} when $a_1 = 14$ and $d = -3$.

29. Find a_1 when $d = 4$ and $a_8 = 33$.

30. Find a_1 when $d = 8$ and $a_{11} = 26$.

31. Find n when $a_1 = 5$, $d = -3$, and $a_n = -76$.

32. Find n when $a_1 = 25$, $d = -14$, and $a_n = -507$.

33. For an arithmetic sequence in which $a_{17} = -40$ and $a_{28} = -73$, find a_1 and d. Write the first five terms of the sequence.

34. In an arithmetic sequence, $a_{17} = \frac{25}{3}$ and $a_{32} = \frac{95}{6}$. Find a_1 and d. Write the first five terms of the sequence.

Aha! **35.** Find a_1 and d if $a_{13} = 13$ and $a_{54} = 54$.

36. Find a_1 and d if $a_{12} = 24$ and $a_{25} = 50$.

37. Find the sum of the first 20 terms of the arithmetic series $1 + 5 + 9 + 13 + \cdots$.

38. Find the sum of the first 14 terms of the arithmetic series $11 + 7 + 3 + \cdots$.

39. Find the sum of the first 250 natural numbers.

40. Find the sum of the first 400 natural numbers.

41. Find the sum of the even numbers from 2 to 100, inclusive.

42. Find the sum of the odd numbers from 1 to 99, inclusive.

43. Find the sum of all multiples of 6 from 6 to 102, inclusive.

44. Find the sum of all multiples of 4 that are between 15 and 521.

45. An arithmetic series has $a_1 = 4$ and $d = 5$. Find S_{20}.

46. An arithmetic series has $a_1 = 9$ and $d = -3$. Find S_{32}.

Solve.

47. *Band Formations.* The South Brighton Drum and Bugle Corps has 7 marchers in the front row, 9 in the second row, 11 in the third row, and so on, for 15 rows. How many marchers are in the last row? How many marchers are there altogether?

48. *Gardening.* A gardener is planting bulbs near an entrance to a college. She has 39 plants in the front row, 35 in the second row, 31 in the third row, and so on. If the pattern is consistent, how many plants will be in the last row? How many plants will there be altogether?

49. *Archaeology.* Many ancient Mayan pyramids were constructed over a span of several generations. Each layer of the pyramid has a stone perimeter, enclosing a layer of dirt or debris on which a structure once stood. One drawing of such a pyramid indicates that the perimeter of the bottom layer contains 36 stones, the next level up contains 32 stones, and so on, up to the top row, which contains 4 stones. How many stones are in the pyramid?

50. *Telephone Pole Piles.* How many poles will be in a pile of telephone poles if there are 50 in the first layer, 49 in the second, and so on, until there are 6 in the top layer?

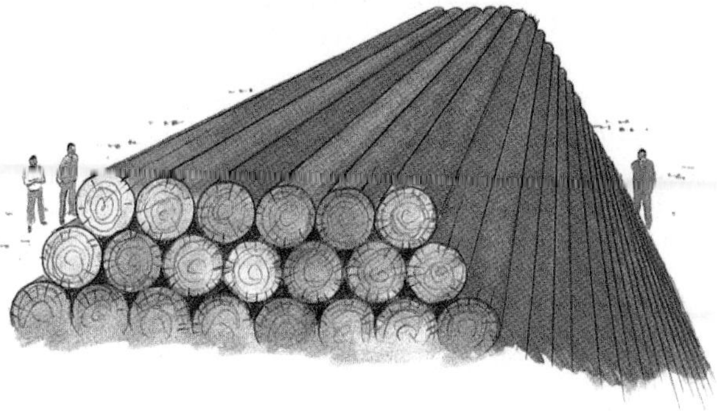

51. *Accumulated Savings.* If 10¢ is saved on October 1, another 20¢ on October 2, another 30¢ on October 3, and so on, how much is saved during October? (October has 31 days.)

52. *Accumulated Savings.* Renata saves money in an arithmetic sequence: $700 for the first year, another $850 the second, and so on, for 20 yr. How much does she save in all (disregarding interest)?

53. *Auditorium Design.* Theaters are often built with more seats per row as the rows move toward the back. The Sanders Amphitheater has 20 seats in the first row, 22 in the second, 24 in the third, and so on, for 19 rows. How many seats are in the amphitheater?

54. *Accumulated Savings.* Shirley sets up an investment such that it will return $5000 the first year, $6125 the second year, $7250 the third year, and so on, for 25 yr. How much in all is received from the investment?

TW **55.** It is said that as a young child, the mathematician Karl F. Gauss (1777–1855) was able to compute the sum $1 + 2 + 3 + \cdots + 100$ very quickly in his head. Explain how Gauss might have done this and present a formula for the sum of the first n natural numbers. (*Hint*: $1 + 99 = 100$.)

TW **56.** Is it true that if every number in a sequence is doubled and then added, the result is the same as if the numbers were first added and the sum then doubled? Why or why not?

Skill Maintenance

Simplify. [6.2]

57. $\dfrac{3}{10x} + \dfrac{2}{15x}$

58. $\dfrac{2}{9t} + \dfrac{5}{12t}$

Convert to an exponential equation. [9.3]

59. $\log_a P = k$

60. $\ln t = a$

Find an equation of the circle satisfying the given conditions. [10.1]

61. Center (0, 0), radius 9

62. Center $(-2, 5)$, radius $3\sqrt{2}$

Synthesis

TW **63.** Write a problem for a classmate to solve. Devise the problem so that its solution requires computing S_{17} for an arithmetic sequence.

TW **64.** The sum of the first n terms of an arithmetic sequence is also given by

$$S_n = \frac{n}{2}[2a_1 + (n - 1)d].$$

Use the earlier formulas for a_n and S_n to explain how this equation was developed.

65. A frog is at the bottom of a 100-ft well. With each jump, the frog climbs 4 ft, but then slips back 1 ft. How many jumps does it take for the frog to reach the top of the hole?

66. Find a formula for the sum of the first n consecutive odd numbers starting with 1:

$$1 + 3 + 5 + \cdots + (2n - 1).$$

67. In an arithmetic sequence, $a_1 = \$8760$ and $d = -\$798.23$. Find the first 10 terms of the sequence.

68. Find the sum of the first 10 terms of the sequence given in Exercise 67.

69. Prove that if p, m, and q are consecutive terms in an arithmetic sequence, then

$$m = \frac{p + q}{2}.$$

70. *Straight-Line Depreciation.* A company buys a color copier for \$5200 on January 1 of a given year. The machine is expected to last for 8 yr, at the end of which time its *trade-in*, or *salvage, value* will be \$1100. If the company figures the decline in value to be the same each year, then the trade-in values, after t years, $0 \le t \le 8$, form an arithmetic sequence given by

$$a_t = C - t\left(\frac{C - S}{N}\right),$$

where C is the original cost of the item, N the years of expected life, and S the salvage value.

a) Find the formula for a_t for the straight-line depreciation of the copier.

b) Find the salvage value after 0 yr, 1 yr, 2 yr, 3 yr, 4 yr, 7 yr, and 8 yr.

c) Find a formula that expresses a_t recursively.

71. Use your answer to Exercise 39 to find the sum of all integers from 501 through 750.

In Exercises 72–75, graph the data in each table, and state whether or not the graph could be the graph of an arithmetic sequence. If it can, find a formula for the general term of the sequence.

72. *Minimum Weight.*

Height (in inches)	Minimum Weight for Women (in pounds)
61	105
63	111
65	117
67	123
69	129

Source: www.netfit.co.uk

73. *Aerobic Exercise.*

Age	Target Heart Rate (in beats per minute)
20	150
40	135
60	120
80	105

74. *Video Games.*

Year	Hours Spent Playing Video Games (per person per year)
2000	59
2001	60
2002	64
2003	69
2004	71
2005	75

Source: Veronis Suhler Stevenson, *Communications Industry Forecast & Report*

75. *Rodeos.*

Year	Number of Professional Rodeos
2000	688
2001	668
2002	666
2003	657
2004	671

Source: Professional Rodeo Cowboys Association

11.3 Geometric Sequences and Series

Geometric Sequences ▪ Sum of the First *n* Terms of a Geometric Sequence ▪ Infinite Geometric Series ▪ Problem Solving

In an arithmetic sequence, a certain number is added to each term to get the next term. When each term in a sequence is *multiplied* by a certain fixed number to get the next term, the sequence is **geometric.** In this section, we examine *geometric sequences* (or *progressions*) and *geometric series*.

Geometric Sequences

Consider the sequence

$$2, 6, 18, 54, 162, \ldots.$$

If we multiply each term by 3, we obtain the next term. The multiplier is called the *common ratio* because it is found by dividing any term by the preceding term.

Geometric Sequence A sequence is *geometric* if there exists a number r, called the *common ratio*, for which

$$\frac{a_{n+1}}{a_n} = r, \quad \text{or} \quad a_{n+1} = a_n \cdot r, \quad \text{for any integer } n \geq 1.$$

EXAMPLE 1 For each geometric sequence, find the common ratio.

a) 4, 20, 100, 500, 2500, ...

b) 3, −6, 12, −24, 48, −96, ...

c) \$5200, \$3900, \$2925, \$2193.75, ...

Student Notes

Try determining the sign of the common ratio before calculating it.

SOLUTION

Sequence	Common Ratio	
a) 4, 20, 100, 500, 2500, ...	5	$\frac{20}{4} = 5, \frac{100}{20} = 5$, **and so on**
b) 3, −6, 12, −24, 48, −96, ...	−2	$\frac{-6}{3} = -2, \frac{12}{-6} = -2$, **and so on**
c) \$5200, \$3900, \$2925, \$2193.75, ...	0.75	$\frac{\$3900}{\$5200} = 0.75, \frac{\$2925}{\$3900} = 0.75$

To develop a formula for the general, or nth, term of a geometric sequence, let a_1 be the first term and let r be the common ratio. We write out the first few terms as follows:

$a_1,$

$a_2 = a_1 r,$

$a_3 = a_2 r = (a_1 r)r = a_1 r^2,$ **Substituting $a_1 r$ for a_2**

$a_4 = a_3 r = (a_1 r^2)r = a_1 r^3.$ **Substituting $a_1 r^2$ for a_3**

Note that the exponent is 1 less than the subscript.

Generalizing, we obtain the following.

To Find a_n for a Geometric Sequence

The nth term of a geometric sequence with common ratio r is given by

$$a_n = a_1 r^{n-1}, \quad \text{for any integer } n \geq 1.$$

EXAMPLE 2 Find the 7th term of the geometric sequence 4, 20, 100, ...

SOLUTION First we note that

$$a_1 = 4 \quad \text{and} \quad n = 7.$$

Study Tip

Ask to See Your Final

Once the course is over, many students neglect to find out how they fared on the final exam. Please don't overlook this valuable opportunity to extend your learning. It is important for you to find out what mistakes you may have made and to be certain no grading errors have occurred.

To find the common ratio, we can divide any term (other than the first) by the term preceding it. Since the second term is 20 and the first is 4,

$$r = \frac{20}{4}, \quad \text{or } 5.$$

The formula

$$a_n = a_1 r^{n-1}$$

gives us

$$a_7 = 4 \cdot 5^{7-1} = 4 \cdot 5^6 = 4 \cdot 15{,}625 = 62{,}500.$$

EXAMPLE 3 Find the 10th term of the geometric sequence

$$64, -32, 16, -8, \ldots.$$

SOLUTION First, we note that

$$a_1 = 64, \qquad n = 10, \quad \text{and} \quad r = \frac{-32}{64} = -\frac{1}{2}.$$

Then, using the formula for the nth term of a geometric sequence, we have

$$a_{10} = 64 \cdot \left(-\frac{1}{2}\right)^{10-1} = 64 \cdot \left(-\frac{1}{2}\right)^9 = 2^6 \cdot \left(-\frac{1}{2^9}\right) = -\frac{1}{2^3} = -\frac{1}{8}.$$

The 10th term is $-\frac{1}{8}$.

What does the graph of a geometric series look like?

Interactive Discovery

Graph each of the following geometric series. What patterns, if any, do you observe?

1. $a_n = 5 \cdot 2^{n-1}$
2. $a_n = \frac{1}{2} \cdot \left(\frac{5}{4}\right)^{n-1}$
3. $a_n = 2.3 \cdot (0.75)^{n-1}$
4. $a_n = 3\left(\frac{9}{10}\right)^{n-1}$

The pattern you may have observed is true in general for $r > 0$.

The graph of a geometric series is a set of points that lie on the graph of an exponential function.

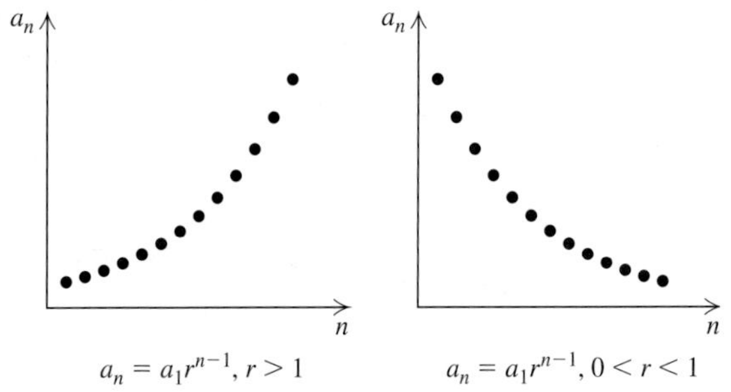

Student Notes

The three determining characteristics of a geometric sequence or series are the first term (a_1), the number of terms (n), and the common ratio (r). Be sure you understand how to use these characteristics to write out a sequence or a series.

Sum of the First *n* Terms of a Geometric Sequence

We next develop a formula for S_n when a sequence is geometric:

$$a_1, a_1r, a_1r^2, a_1r^3, \ldots, a_1r^{n-1}, \ldots.$$

The **geometric series** S_n is given by

$$S_n = a_1 + a_1r + a_1r^2 + \cdots + a_1r^{n-2} + a_1r^{n-1}. \qquad (1)$$

Multiplying both sides by r gives us

$$rS_n = a_1r + a_1r^2 + a_1r^3 + \cdots + a_1r^{n-1} + a_1r^n. \qquad (2)$$

When we subtract corresponding sides of equation (2) from equation (1), the color terms drop out, leaving

$$S_n - rS_n = a_1 - a_1r^n$$

$$S_n(1 - r) = a_1(1 - r^n), \qquad \textbf{Factoring}$$

or

$$S_n = \frac{a_1(1 - r^n)}{1 - r}. \qquad \textbf{Dividing both sides by } 1 - r$$

To Find S_n for a Geometric Sequence

The sum of the first n terms of a geometric sequence with common ratio r is given by

$$S_n = \frac{a_1(1 - r^n)}{1 - r}, \quad \text{for any } r \neq 1.$$

EXAMPLE 4 Find the sum of the first 7 terms of the geometric sequence 3, 15, 75, 375, ...

SOLUTION First, we note that

$$a_1 = 3, \quad n = 7, \quad \text{and} \quad r = \frac{15}{3} = 5.$$

Then, substituting in the formula $S_n = \dfrac{a_1(1 - r^n)}{1 - r}$, we have

$$S_7 = \frac{3(1 - 5^7)}{1 - 5} = \frac{3(1 - 78{,}125)}{-4}$$

$$= \frac{3(-78{,}124)}{-4}$$

$$= 58{,}593.$$

Infinite Geometric Series

Suppose we consider the sum of the terms of an infinite geometric sequence, such as 2, 4, 8, 16, 32, ... We get what is called an **infinite geometric series:**

$$2 + 4 + 8 + 16 + 32 + \cdots.$$

Here, as n increases, the sum of the first n terms, S_n, increases without bound. There are also infinite series that get closer and closer to some specific number. Here is an example:

$$\frac{1}{2} + \frac{1}{4} + \frac{1}{8} + \frac{1}{16} + \cdots + \frac{1}{2^n} + \cdots.$$

Let's consider S_n for the first four values of n:

$$S_1 = \tfrac{1}{2} = \tfrac{1}{2} = 0.5,$$
$$S_2 = \tfrac{1}{2} + \tfrac{1}{4} = \tfrac{3}{4} = 0.75,$$
$$S_3 = \tfrac{1}{2} + \tfrac{1}{4} + \tfrac{1}{8} = \tfrac{7}{8} = 0.875,$$
$$S_4 = \tfrac{1}{2} + \tfrac{1}{4} + \tfrac{1}{8} + \tfrac{1}{16} = \tfrac{15}{16} = 0.9375.$$

The denominator of the sum is 2^n, where n is the subscript of S. The numerator is $2^n - 1$.

Thus, for this particular series, we have

$$S_n = \frac{2^n - 1}{2^n} = \frac{2^n}{2^n} - \frac{1}{2^n} = 1 - \frac{1}{2^n}.$$

Note that the value of S_n is less than 1 for any value of n, but as n gets larger and larger, the value of $1/2^n$ gets closer to 0 and the value of S_n gets closer to 1. We can visualize S_n by considering the area of a square of area 1. For S_1, we shade half the square. For S_2, we shade half the square plus half the remaining part, or $\frac{1}{4}$. For S_3, we shade the parts shaded in S_2 plus half the remaining part. Again we see that the values of S_n will continue to get close to 1 (shading the complete square).

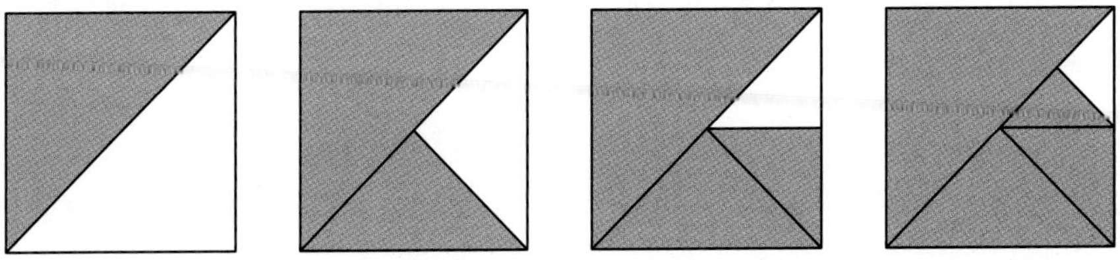

$S_1 = \frac{1}{2}$ $S_2 = \frac{3}{4}$ $S_3 = \frac{7}{8}$ $S_4 = \frac{15}{16}$

We say that 1 is the **limit** of S_n and that 1 is the sum of this infinite geometric series.

How can we tell if an infinite geometric series has a sum?

Interactive Discovery

For each of the following geometric sequences, **(a)** determine the common ratio, **(b)** list the first 10 partial sums, S_1 through S_{10}, and **(c)** estimate, if possible, S_∞.

1. $a_n = \left(\frac{1}{3}\right)^{n-1}$

2. $a_n = \left(-\frac{7}{2}\right)^{n-1}$

3. $a_n = 4(2)^{n-1}$

4. $a_n = 1.3(-0.4)^{n-1}$

5. What relationship do you observe between the common ratio and the existence of S_∞?

It can be shown (but we will not do it here) that the pattern you may have observed is true in general.

The sum of the terms of an infinite geometric sequence exists if and only if $|r| < 1$.

To develop a formula for the limit of an infinite geometric series, we first consider the sum of the first n terms:

$$S_n = \frac{a_1(1 - r^n)}{1 - r} = \frac{a_1 - a_1 r^n}{1 - r}. \qquad \textbf{Using the distributive law}$$

For $|r| < 1$, it follows that values of r^n get closer to 0 as n gets larger. (Check this by selecting a number between -1 and 1 and finding larger and larger powers on a calculator.) As r^n gets closer to 0, so does $a_1 r^n$. Thus, S_n gets closer to $a_1/(1 - r)$.

The Limit of an Infinite Geometric Series

When $|r| < 1$, the limit of an infinite geometric series is given by

$$S_\infty = \frac{a_1}{1 - r}. \qquad \text{(For } |r| \geq 1\text{, no limit exists.)}$$

EXAMPLE 5 Determine whether each series has a limit. If a limit exists, find it.

a) $1 + 3 + 9 + 27 + \cdots$

b) $-2 + 1 - \frac{1}{2} + \frac{1}{4} - \frac{1}{8} + \cdots$

SOLUTION

a) Here $r = 3$, so $|r| = |3| = 3$. Since $|r| \not< 1$, the series *does not* have a limit.

b) Here $r = -\frac{1}{2}$, so $|r| = \left|-\frac{1}{2}\right| = \frac{1}{2}$. Since $|r| < 1$, the series *does* have a limit. We find the limit by substituting into the formula for S_∞:

$$S_\infty = \frac{-2}{1 - \left(-\frac{1}{2}\right)} = \frac{-2}{\frac{3}{2}} = -2 \cdot \frac{2}{3} = -\frac{4}{3}.$$

EXAMPLE 6 Find fraction notation for 0.63636363....

SOLUTION We can express this as

$$0.63 + 0.0063 + 0.000063 + \cdots.$$

This is an infinite geometric series, where $a_1 = 0.63$ and $r = 0.01$. Since $|r| < 1$, this series has a limit:

$$S_\infty = \frac{a_1}{1 - r} = \frac{0.63}{1 - 0.01} = \frac{0.63}{0.99} = \frac{63}{99}.$$

Thus fraction notation for 0.63636363... is $\frac{63}{99}$, or $\frac{7}{11}$.

Problem Solving

For some problem-solving situations, the translation may involve geometric sequences or series.

EXAMPLE 7 Daily Wages. Suppose someone offered you a job for the month of September (30 days) under the following conditions. You will be paid \$0.01 for the first day, \$0.02 for the second, \$0.04 for the third, and so on, doubling your previous day's salary each day. How much would you earn? (Would you take the job? Make a guess before reading further.)

SOLUTION

1. **Familiarize.** You earn \$0.01 the first day, \$0.01(2) the second day, \$0.01(2)(2) the third day, and so on. Since each day's wages are a constant multiple of the previous day's wages, a geometric sequence is formed.
2. **Translate.** The amount earned is the geometric series

$$\$0.01 + \$0.01(2) + \$0.01(2^2) + \$0.01(2^3) + \cdots + \$0.01(2^{29}),$$

where

$$a_1 = \$0.01, \qquad n = 30, \quad \text{and} \quad r = 2.$$

3. **Carry out.** Using the formula

$$S_n = \frac{a_1(1 - r^n)}{1 - r},$$

we have

$$S_{30} = \frac{\$0.01(1 - 2^{30})}{1 - 2}$$

$$= \frac{\$0.01(-1{,}073{,}741{,}823)}{-1} \qquad \textbf{Using a calculator}$$

$$= \$10{,}737{,}418.23.$$

4. **Check.** The calculations can be repeated as a check.
5. **State.** The pay exceeds \$10.7 million for the month. Most people would probably take the job!

EXAMPLE 8 Loan Repayment. Francine's student loan is in the amount of \$6000. Interest is to be 9% compounded annually, and the entire amount is to be paid after 10 yr. How much is to be paid back?

SOLUTION

1. **Familiarize.** Suppose we let P represent any principal amount. At the end of one year, the amount owed will be $P + 0.09P$, or $1.09P$. That amount will be the principal for the second year. The amount owed at the end of the second year will be $1.09 \times \text{New principal} = 1.09(1.09P)$, or 1.09^2P. Thus the amount owed at the beginning of successive years is

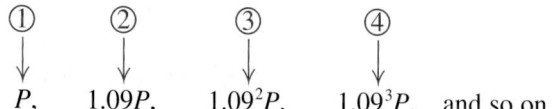

$$\underset{①}{P}, \quad \underset{②}{1.09P}, \quad \underset{③}{1.09^2P}, \quad \underset{④}{1.09^3P}, \quad \text{and so on.}$$

We have a geometric sequence. The amount owed at the beginning of the 11th year will be the amount owed at the end of the 10th year.

2. **Translate.** We have a geometric sequence with $a_1 = 6000$, $r = 1.09$, and $n = 11$. The appropriate formula is

$$a_n = a_1r^{n-1}.$$

3. **Carry out.** We substitute and calculate:

$$a_{11} = \$6000(1.09)^{11-1} = \$6000(1.09)^{10}$$
$$\approx \$14{,}204.18.$$ **Using a calculator and rounding to the nearest hundredth**

4. **Check.** A check, by repeating the calculations, is left to the student.
5. **State.** Francine will owe \$14,204.18 at the end of 10 yr.

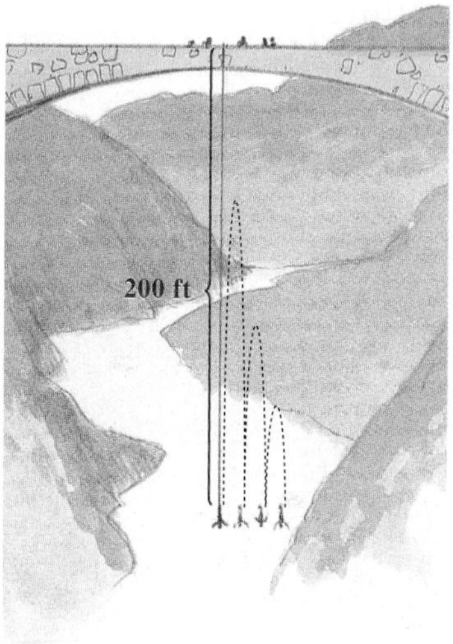

EXAMPLE 9 Bungee Jumping. A bungee jumper rebounds 60% of the height jumped. A bungee jump is made using a cord that stretches to 200 ft.

a) After jumping and then rebounding 9 times, how far has a bungee jumper traveled upward (the total rebound distance)?

b) Approximately how far will a jumper have traveled upward (bounced) before coming to rest?

SOLUTION

1. **Familiarize.** Let's do some calculations and look for a pattern.

First fall:	200 ft
First rebound:	0.6×200, or 120 ft
Second fall:	120 ft, or 0.6×200
Second rebound:	0.6×120, or $0.6(0.6 \times 200)$, which is 72 ft
Third fall:	72 ft, or $0.6(0.6 \times 200)$
Third rebound:	0.6×72, or $0.6(0.6(0.6 \times 200))$, which is 43.2 ft

The rebound distances form a geometric sequence:

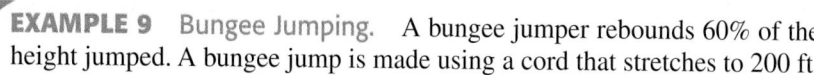

$$120, \quad 0.6 \times 120, \quad 0.6^2 \times 120, \quad 0.6^3 \times 120, \ldots.$$

2. Translate.

a) The total rebound distance after 9 bounces is the sum of a geometric sequence. The first term is 120 and the common ratio is 0.6. There will be 9 terms, so we can use the formula

$$S_n = \frac{a_1(1 - r^n)}{1 - r}.$$

b) Theoretically, the jumper will never stop bouncing. Realistically, the bouncing will eventually stop. To approximate the actual distance, we consider an infinite number of bounces and use the formula

$$S_\infty = \frac{a_1}{1 - r}.$$ **Since $r = 0.6$ and $|0.6| < 1$, we know that S_∞ exists.**

3. Carry out.

a) We substitute into the formula and calculate:

$$S_9 = \frac{120[1 - (0.6)^9]}{1 - 0.6} \approx 297.$$ **Using a calculator**

b) We substitute and calculate:

$$S_\infty = \frac{120}{1 - 0.6} = 300.$$

4. Check. We can do the calculations again.

5. State.

a) In 9 bounces, the bungee jumper will have traveled upward a total distance of about 297 ft.

b) The jumper will have traveled upward a total of about 300 ft before coming to rest.

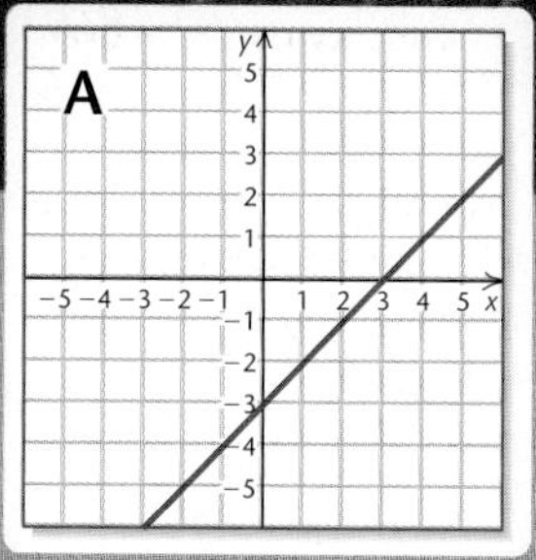

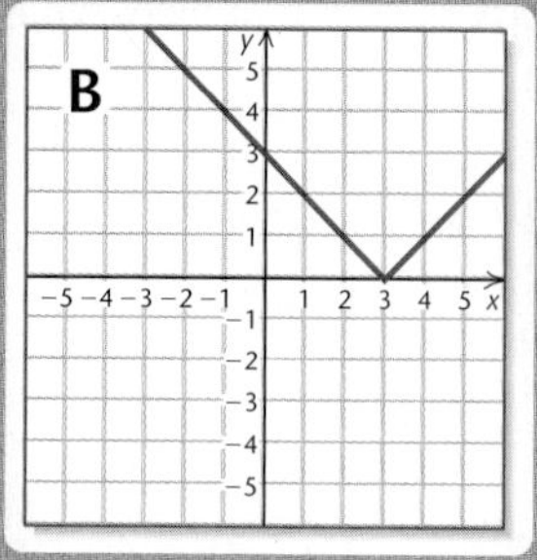

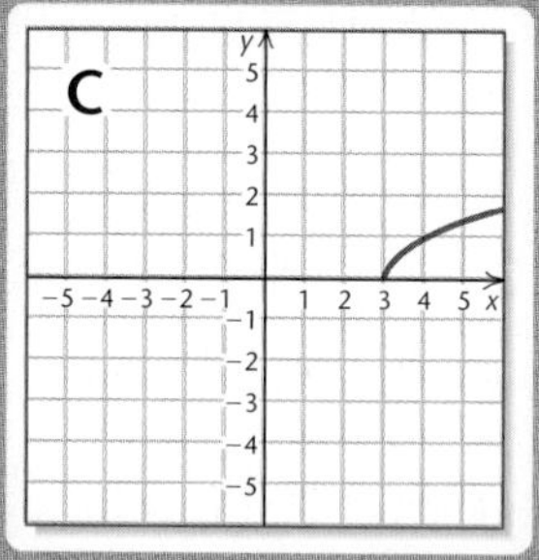

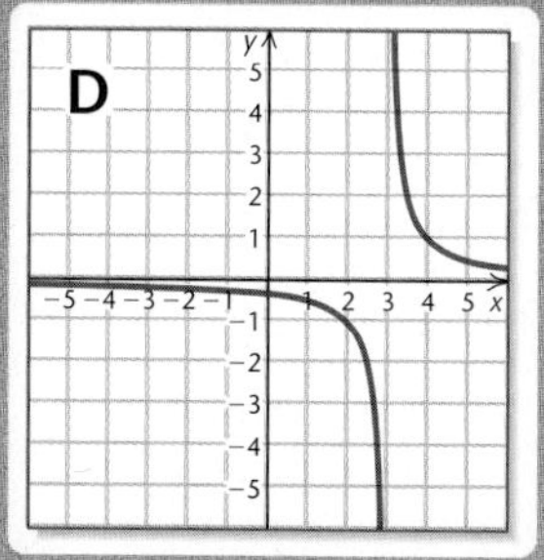

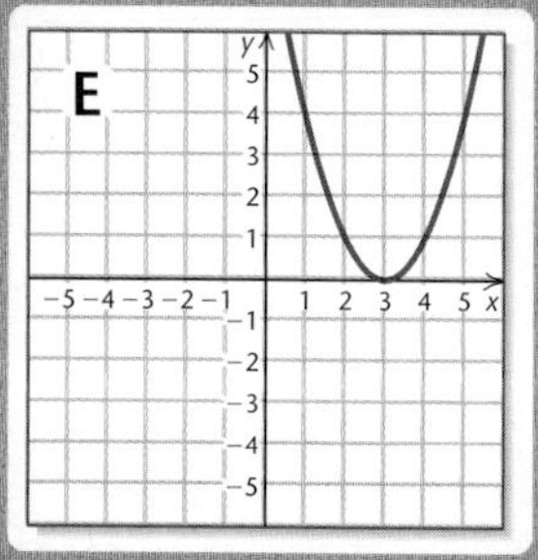

Visualizing the Graph

Match each equation with its graph.

1. $xy = 2$

2. $y = \log_2 x$

3. $y = x - 3$

4. $(x - 3)^2 + y^2 = 4$

5. $\dfrac{(x - 3)^2}{1} + \dfrac{y^2}{4} = 1$

6. $y = |x - 3|$

7. $y = (x - 3)^2$

8. $y = \dfrac{1}{x - 3}$

9. $y = 2^x$

10. $y = \sqrt{x - 3}$

Answers on page A-56

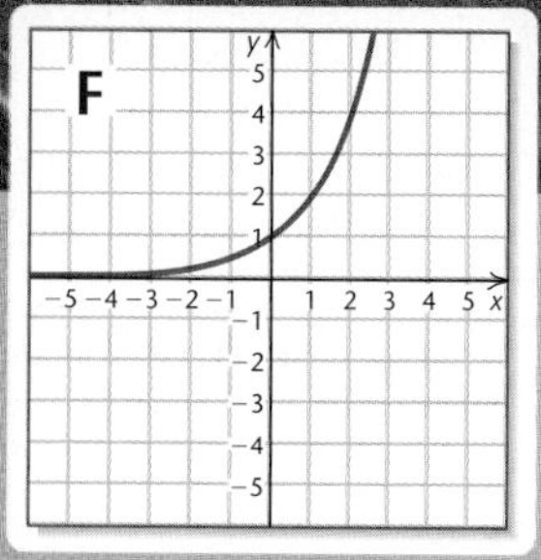

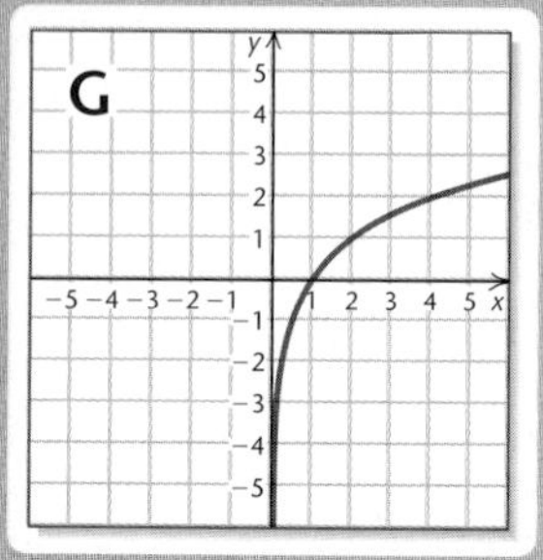

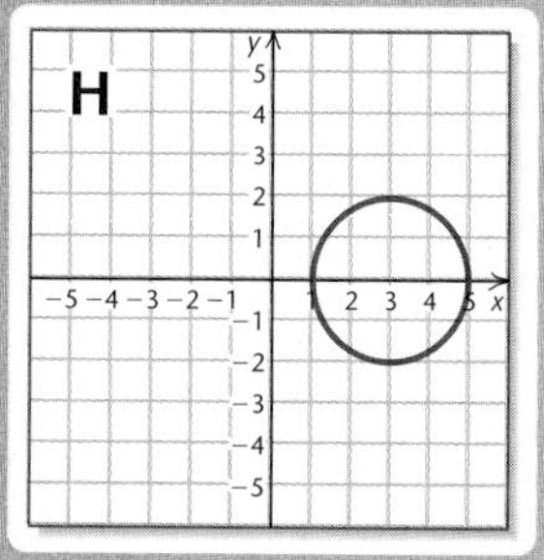

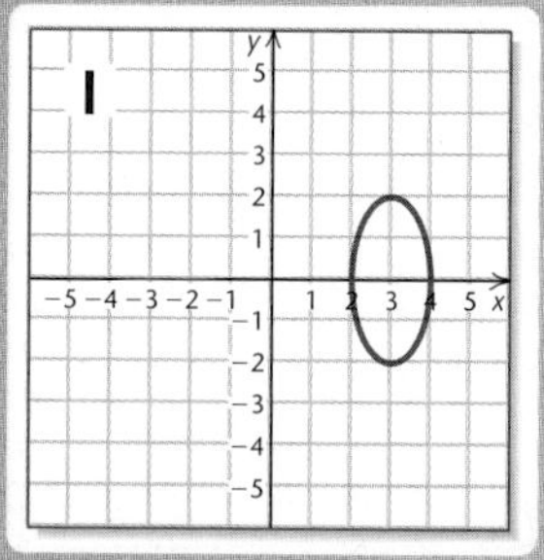

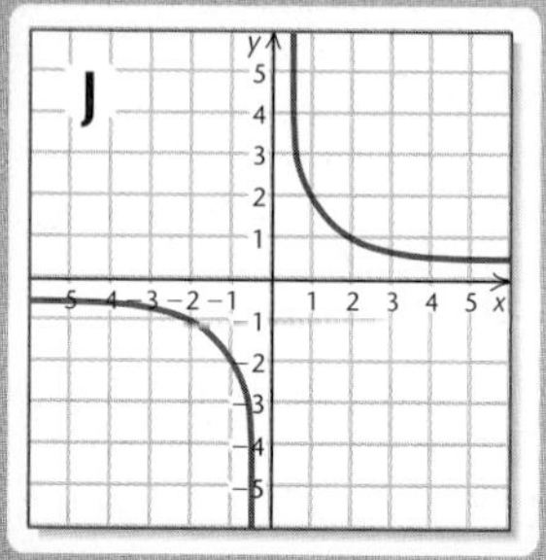

11.3 EXERCISE SET

FOR EXTRA HELP

MathXL | MyMathLab | InterAct Math | AW Math Tutor Center | Video Lectures on CD: Disc 6 | Student's Solutions Manual

Concept Reinforcement *Classify each of the following as an arithmetic sequence, a geometric sequence, an arithmetic series, a geometric series, or none of these.*

1. $2, 6, 18, 54, \dots$

2. $3, 5, 7, 9, \dots$

3. $1, 6, 11, 16, 21, \dots$

4. $5, 15, 45, 135, 405, \dots$

5. $4 + 20 + 100 + 500 + 2500 + 12{,}500$

6. $10 + 12 + 14 + 16 + 18 + 20$

7. $3 - \frac{3}{2} + \frac{3}{4} - \frac{3}{8} + \frac{3}{16} - \cdots$

8. $1 + \frac{1}{2} + \frac{1}{3} + \frac{1}{4} + \frac{1}{5} + \frac{1}{6} + \cdots$

Find the common ratio for each geometric sequence.

9. $7, 14, 28, 56, \dots$

10. $2, 6, 18, 54, \dots$

11. $6, -0.6, 0.06, -0.006, \dots$

12. $-5, -0.5, -0.05, -0.005, \dots$

13. $\frac{1}{2}, -\frac{1}{4}, \frac{1}{8}, -\frac{1}{16}, \dots$

14. $\frac{2}{3}, -\frac{4}{3}, \frac{8}{3}, -\frac{16}{3}, \dots$

15. $75, 15, 3, \frac{3}{5}, \dots$

16. $12, -4, \frac{4}{3}, -\frac{4}{9}, \dots$

17. $\frac{1}{m}, \frac{6}{m^2}, \frac{36}{m^3}, \frac{216}{m^4}, \dots$

18. $4, \frac{4m}{5}, \frac{4m^2}{25}, \frac{4m^3}{125}, \dots$

Find the indicated term for each geometric sequence.

19. $3, 6, 12, \dots$; the 7th term

20. $2, 8, 32, \dots$; the 9th term

21. $7, 7\sqrt{2}, 14, \dots$; the 10th term

22. $4, 4\sqrt{3}, 12, \dots$; the 8th term

23. $-\frac{8}{243}, \frac{8}{81}, -\frac{8}{27}, \dots$; the 14th term

24. $\frac{7}{625}, \frac{-7}{125}, \frac{7}{25}, \dots$; the 13th term

25. \$1000, \$1080, \$1166.40, . . . ; the 12th term

26. \$1000, \$1070, \$1144.90, . . . ; the 11th term

Find the nth, or general, term for each geometric sequence.

27. $1, 5, 25, 125, \dots$

28. $2, 4, 8, \dots$

29. $1, -1, 1, -1, \dots$

30. $\frac{1}{4}, \frac{1}{16}, \frac{1}{64}, \dots$

31. $\frac{1}{x}, \frac{1}{x^2}, \frac{1}{x^3}, \dots$

32. $5, \frac{5m}{2}, \frac{5m^2}{4}, \dots$

For Exercises 33–40, use the formula for S_n to find the indicated sum for each geometric series.

33. S_9 for $6 + 12 + 24 + \cdots$

34. S_6 for $16 - 8 + 4 - \cdots$

35. S_7 for $\frac{1}{18} - \frac{1}{6} + \frac{1}{2} - \cdots$

Aha! **36.** S_5 for $7 + 0.7 + 0.07 + \cdots$

37. S_8 for $1 + x + x^2 + x^3 + \cdots$

38. S_{10} for $1 + x^2 + x^4 + x^6 + \cdots$

39. S_{16} for \$200, \$200(1.06), \$200(1.06)^2, . . .

40. S_{23} for \$1000, \$1000(1.08), \$1000(1.08)^2, . . .

Determine whether each infinite geometric series has a limit. If a limit exists, find it.

41. $16 + 4 + 1 + \cdots$

42. $8 + 4 + 2 + \cdots$

43. $7 + 3 + \frac{9}{7} + \cdots$

44. $12 + 9 + \frac{27}{4} + \cdots$

45. $3 + 15 + 75 + \cdots$

46. $2 + 3 + \frac{9}{2} + \cdots$

47. $4 - 6 + 9 - \frac{27}{2} + \cdots$

48. $-6 + 3 - \frac{3}{2} + \frac{3}{4} - \cdots$

49. $0.43 + 0.0043 + 0.000043 + \cdots$

50. $0.37 + 0.0037 + 0.000037 + \cdots$

51. $\$500(1.02)^{-1} + \$500(1.02)^{-2} + \$500(1.02)^{-3} + \cdots$

52. $\$1000(1.08)^{-1} + \$1000(1.08)^{-2} + \$1000(1.08)^{-3} + \cdots$

Find fraction notation for each infinite sum. (Each can be regarded as an infinite geometric series.)

53. 0.7777...

54. 0.2222...

55. 8.3838...

56. 7.4747...

57. 0.15151515...

58. 0.12121212...

Solve. Use a calculator as needed for evaluating formulas.

59. *Rebound Distance.* A ping-pong ball is dropped from a height of 20 ft and always rebounds one-fourth of the distance fallen. How high does it rebound the 6th time?

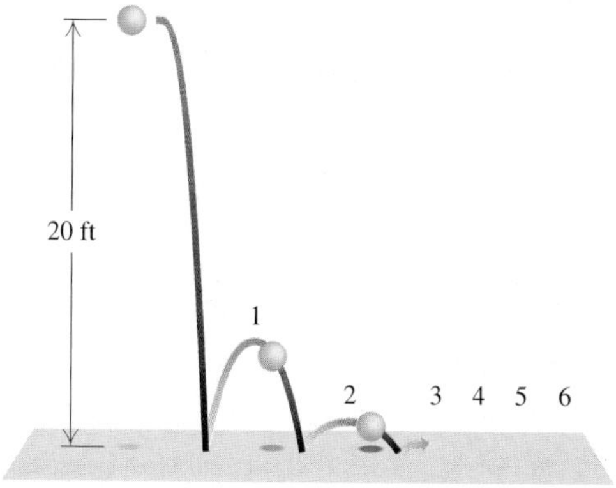

60. *Rebound Distance.* Approximate the total of the rebound heights of the ball in Exercise 59.

61. *Population Growth.* Yorktown has a current population of 100,000, and the population is increasing by 3% each year. What will the population be in 15 yr?

62. *Amount Owed.* Gilberto borrows $15,000. The loan is to be repaid in 13 yr at 8.5% interest, compounded annually. How much will be repaid at the end of 13 yr?

63. *Shrinking Population.* A population of 5000 fruit flies is dying off at a rate of 4% per minute. How many flies will be alive after 15 min?

64. *Shrinking Population.* For the population of fruit flies in Exercise 63, how long will it take for only 1800 fruit flies to remain alive? (*Hint*: Use logarithms.) Round to the nearest minute.

65. *Housing Units.* Approximately 534,000 new apartments and houses were built in the United States in 1991. Since then, the number has grown by about 5.35% per year. (*Sources*: Based on data from the U.S. Bureau of the Census and the U.S. Department of Housing and Urban Development) How many new apartments and houses were built in the United States from 1991 through 2004?

66. *Housing Units.* Approximately 144,000 new apartments and houses were built in the western United States in 1991. Since then, the number has grown by about 5.0% per year. (*Sources*: Based on data from the U.S. Bureau of the Census and the U.S. Department of Housing and Urban Development) How many new houses and apartments were built in the western United States from 1991 through 2004?

67. *Rebound Distance.* A superball dropped from the top of the Washington Monument (556 ft high) rebounds three-fourths of the distance fallen. How far (up and down) will the ball have traveled when it hits the ground for the 6th time?

68. *Rebound Distance.* Approximate the total distance that the ball of Exercise 67 will have traveled when it comes to rest.

69. *Stacking Paper.* Construction paper is about 0.02 in. thick. Beginning with just one piece, a stack is doubled again and again 10 times. Find the height of the final stack.

70. *Monthly Earnings.* Suppose you accepted a job for the month of February (28 days) under the following conditions. You will be paid \$0.01 the first day, \$0.02 the second, \$0.04 the third, and so on, doubling your previous day's salary each day. How much would you earn?

In Exercises 71–76, state whether the graph is that of an arithmetic sequence or a geometric sequence.

71.

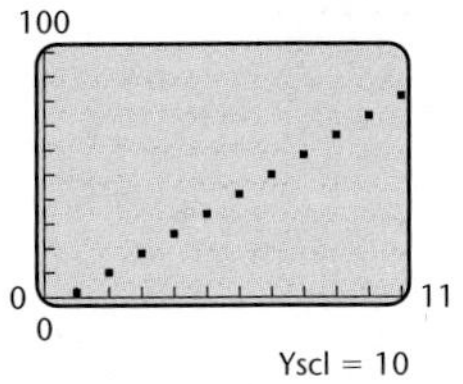

72.

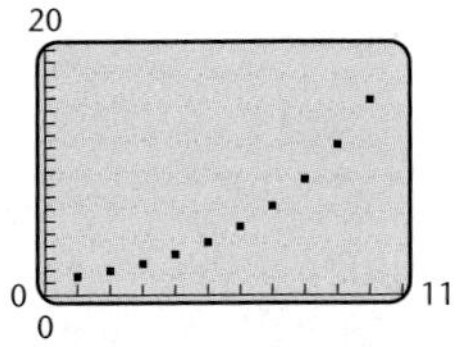

73.

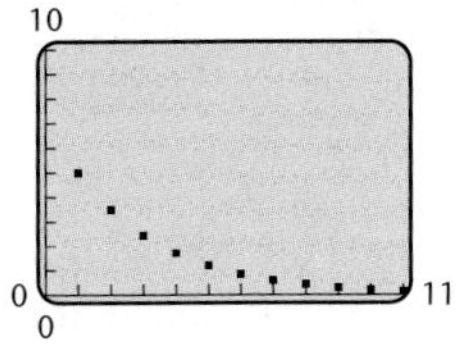

74.

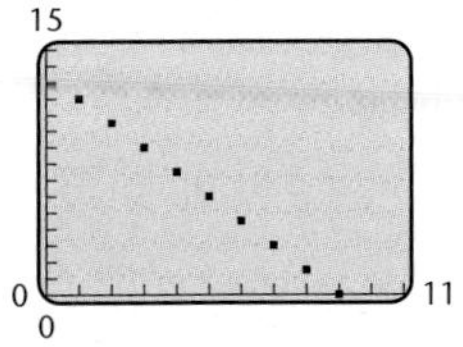

75.

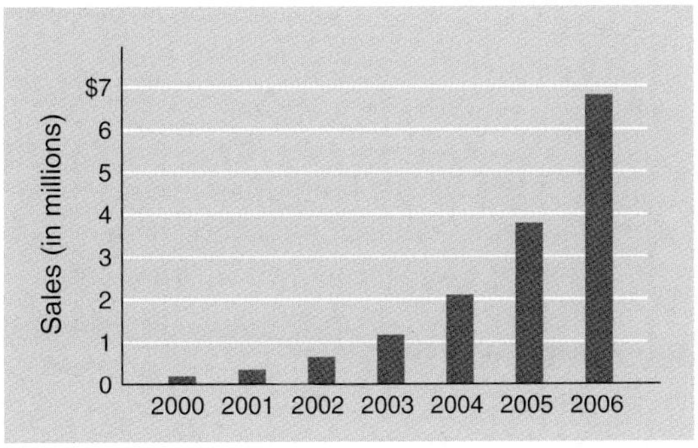

76.

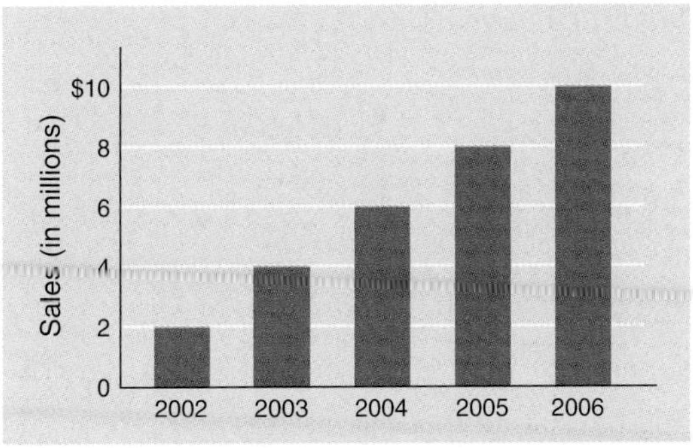

Aha! TW **77.** Under what circumstances is it possible for the 5th term of a geometric sequence to be greater than the 4th term but less than the 7th term?

TW **78.** When r is negative, a series is said to be *alternating*. Why do you suppose this terminology is used?

Focused Review

79. Find the first term and the common difference: 2, 4, 6, 8, ... [11.2]

80. Find the first term and the common ratio: 2, 4, 8, 16, ... [11.3]

81. Find the 10th term of the arithmetic sequence $8, 5, 2, -1, \ldots$. [11.2]

82. Find the 10th term of the geometric sequence $32, -16, 8, -4, \ldots$. [11.3]

83. Find S_{12} for the arithmetic series $1 + 4 + 7 + 10 + \cdots$. [11.2]

84. Find S_{12} for the geometric series $1 + 4 + 16 + 64 + \cdots$. [11.3]

Synthesis

TW **85.** Using Example 5 and Exercises 41–52, explain how the graph of a geometric sequence can be used to determine whether a geometric series has a limit.

TW **86.** The infinite series

$$S_\infty = 2 + \frac{1}{2} + \frac{1}{2 \cdot 3} + \frac{1}{2 \cdot 3 \cdot 4} + \frac{1}{2 \cdot 3 \cdot 4 \cdot 5} + \frac{1}{2 \cdot 3 \cdot 4 \cdot 5 \cdot 6} + \cdots$$

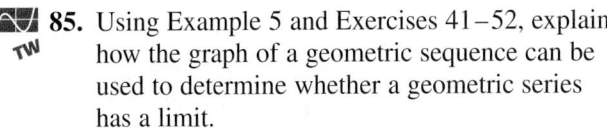

is not geometric, but it does have a sum. Using S_1, S_2, S_3, S_4, S_5, and S_6, make a conjecture about the value of S_∞ and explain your reasoning.

Calculate each of the following sums.

87. $\sum_{k=1}^{\infty} 6(0.9)^k$

88. $\sum_{k=1}^{\infty} 5(-0.7)^k$

89. Find the sum of the first n terms of

$$x^2 - x^3 + x^4 - x^5 + \cdots.$$

90. Find the sum of the first n terms of

$$1 + x + x^2 + x^3 + \cdots.$$

91. The sides of a square are each 16 cm long. A second square is inscribed by joining the midpoints of the sides, successively. In the second square we repeat the process, inscribing a third square. If this process is continued indefinitely, what is the sum of all of the areas of all the squares? (*Hint*: Use an infinite geometric series.) 512 cm^2

92. Show that 0.999... is 1.

Collaborative Corner

Bargaining for a Used Car

Focus: Geometric series
Time: 30 minutes
Group size: 2
Materials: Graphing calculators are optional.

ACTIVITY*

1. One group member ("the seller") has a car for sale and is asking \$3500. The second ("the buyer") offers \$1500. The seller splits the difference (\$2000 ÷ 2 = \$1000) and lowers the price to \$2500. The buyer then splits the difference again (\$1000 ÷ 2 = \$500) and counters with \$2000. Continue in this manner and stop when you are able to agree on the car's selling price to the nearest penny.
2. What should the buyer's initial offer be in order to achieve a purchase price of \$2000? (Check several guesses to find the appropriate initial offer.)

*This activity is based on the article, "Bargaining Theory, or Zeno's Used Cars," by James C. Kirby, *The College Mathematics Journal*, **27**(4), September 1996.

3. The seller's price in the bargaining above can be modeled recursively (see Exercises 81, 82, and 90 in Section 11.1) by the sequence

$$a_1 = 3500, \qquad a_n = a_{n-1} - \frac{d}{2^{2n-3}},$$

where d is the difference between the initial price and the first offer. Use this recursively defined sequence to solve parts (1) and (2) above either manually or by using the SEQ MODE and the TABLE feature of a graphing calculator.

4. The first four terms in the sequence in part (3) can be written as

$$a_1, \quad a_1 - \frac{d}{2}, \quad a_1 - \frac{d}{2} - \frac{d}{8},$$

$$a_1 - \frac{d}{2} - \frac{d}{8} - \frac{d}{32}.$$

Use the formula for the limit of an infinite geometric series to find a simple algebraic formula for the eventual sale price, P, when the bargaining process from above is followed. Verify the formula by using it to solve parts (1) and (2) above.

11.4 The Binomial Theorem

Binomial Expansion Using Pascal's Triangle ■ Binomial Expansion Using Factorial Notation

Connecting the Concepts

Sequences and series occur in many settings, some of which were mentioned in Sections 11.1–11.3. Although you may not have viewed it this way before, the expression $(x + y)^2$ can be regarded as a series: $x^2 + 2xy + y^2$.

In Chapter 5, we found that the expansion of $(x + y)^n$, for powers greater than 2, can be quite time-consuming. The reason for this extends all the way back to Chapter 1 and the rules for the order of operations and the properties of exponents: $(x + y)^n \neq x^n + y^n$. Since the terms in the expansion of $(x + y)^n$ have many uses, we devote this section to two methods that streamline the expansion of this important algebraic expression.

Study Tip

Make the Most of Each Minute

As your final exam nears, it is essential to make wise use of your time. If it is possible to reschedule haircuts, dentist appointments, and the like until after your exam(s), consider doing so. If you can, consider writing out your hourly schedule of daily activities leading up to the exam(s).

Binomial Expansion Using Pascal's Triangle

Consider the following expanded powers of $(a + b)^n$:

$$
\begin{aligned}
(a+b)^0 &= 1\\
(a+b)^1 &= a + b\\
(a+b)^2 &= a^2 + 2a^1b^1 + b^2\\
(a+b)^3 &= a^3 + 3a^2b^1 + 3a^1b^2 + b^3\\
(a+b)^4 &= a^4 + 4a^3b^1 + 6a^2b^2 + 4a^1b^3 + b^4\\
(a+b)^5 &= a^5 + 5a^4b^1 + 10a^3b^2 + 10a^2b^3 + 5a^1b^4 + b^5.
\end{aligned}
$$

Each expansion is a polynomial. There are some patterns to be noted:

1. There is one more term than the power of the binomial, n. That is, there are $n + 1$ terms in the expansion of $(a + b)^n$.
2. In each term, the sum of the exponents is the power to which the binomial is raised.
3. The exponents of a start with n, the power of the binomial, and decrease to 0 (since $a^0 = 1$, the last term has no factor of a). The first term has no factor of b, so powers of b start with 0 and increase to n.
4. The coefficients start at 1, increase through certain values, and then decrease through these same values back to 1. Let's study the coefficients further.

Suppose we wish to expand $(a + b)^8$. The patterns we noticed above indicate 9 terms in the expansion:

$$a^8 + c_1a^7b + c_2a^6b^2 + c_3a^5b^3 + c_4a^4b^4 + c_5a^3b^5 + c_6a^2b^6 + c_7ab^7 + b^8.$$

How can we determine the values for the c's? One method seems very simple, but it has some drawbacks. It involves writing down the coefficients in a triangular array as follows. We form what is known as **Pascal's triangle:**

$(a + b)^0$: 1
$(a + b)^1$: 1 1
$(a + b)^2$: 1 2 1
$(a + b)^3$: 1 3 3 1
$(a + b)^4$: 1 4 6 4 1
$(a + b)^5$: 1 5 10 10 5 1

There are many patterns in the triangle. Find as many as you can.

Perhaps you discovered a way to write the next row of numbers, given the numbers in the row above it. There are always 1's on the outside. Each remaining number is the sum of the two numbers above:

1
1 1
1 2 1
1 3 3 1
1 4 6 4 1
1 5 10 10 5 1
1 6 15 20 15 6 1

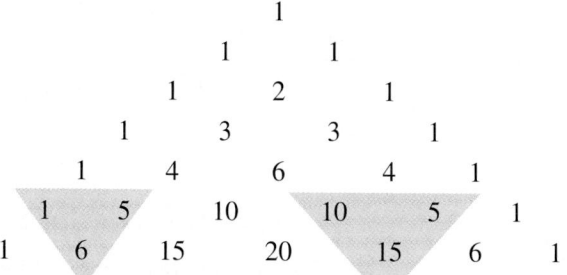

We see that in the bottom (seventh) row

the 1st and last numbers are 1;
the 2nd number is 1 + 5, or 6;
the 3rd number is 5 + 10, or 15;
the 4th number is 10 + 10, or 20;
the 5th number is 10 + 5, or 15; and
the 6th number is 5 + 1, or 6.

Thus the expansion of $(a + b)^6$ is

$$(a + b)^6 = 1a^6 + 6a^5b + 15a^4b^2 + 20a^3b^3 + 15a^2b^4 + 6ab^5 + 1b^6.$$

To expand $(a + b)^8$, we complete two more rows of Pascal's triangle:

1
1 1
1 2 1
1 3 3 1
1 4 6 4 1
1 5 10 10 5 1
1 6 15 20 15 6 1
1 7 21 35 35 21 7 1
1 8 28 56 70 56 28 8 1

Thus the expansion of $(a + b)^8$ has coefficients found in the 9th row above:

$$(a + b)^8 = 1a^8 + 8a^7b + 28a^6b^2 + 56a^5b^3 + 70a^4b^4 + 56a^3b^5 + 28a^2b^6 + 8ab^7 + 1b^8.$$

We can generalize our results as follows:

The Binomial Theorem (Form 1) For any binomial $a + b$ and any natural number n,

$$(a + b)^n = c_0a^nb^0 + c_1a^{n-1}b^1 + c_2a^{n-2}b^2 + \cdots + c_{n-1}a^1b^{n-1} + c_na^0b^n,$$

where the numbers $c_0, c_1, c_2, \ldots, c_n$ are from the $(n + 1)$st row of Pascal's triangle.

EXAMPLE 1 Expand: $(u - v)^5$.

SOLUTION Using the binomial theorem, we have $a = u$, $b = -v$, and $n = 5$. We use the 6th row of Pascal's triangle: 1 5 10 10 5 1. Thus,

$$\begin{aligned}(u - v)^5 &= [u + (-v)]^5 \qquad \textbf{Rewriting } u - v \textbf{ as a sum}\\ &= 1(u)^5 + 5(u)^4(-v)^1 + 10(u)^3(-v)^2 + 10(u)^2(-v)^3\\ &\quad + 5(u)^1(-v)^4 + 1(-v)^5\\ &= u^5 - 5u^4v + 10u^3v^2 - 10u^2v^3 + 5uv^4 - v^5.\end{aligned}$$

Note that the signs of the terms alternate between + and −. When $-v$ is raised to an odd power, the sign is −.

EXAMPLE 2 Expand: $\left(2t + \dfrac{3}{t}\right)^6$.

SOLUTION Note that $a = 2t$, $b = 3/t$, and $n = 6$. We use the 7th row of Pascal's triangle: 1 6 15 20 15 6 1. Thus,

$$\begin{aligned}\left(2t + \frac{3}{t}\right)^6 &= 1(2t)^6 + 6(2t)^5\left(\frac{3}{t}\right)^1 + 15(2t)^4\left(\frac{3}{t}\right)^2 + 20(2t)^3\left(\frac{3}{t}\right)^3\\ &\quad + 15(2t)^2\left(\frac{3}{t}\right)^4 + 6(2t)^1\left(\frac{3}{t}\right)^5 + 1\left(\frac{3}{t}\right)^6\\ &= 64t^6 + 6(32t^5)\left(\frac{3}{t}\right) + 15(16t^4)\left(\frac{9}{t^2}\right) + 20(8t^3)\left(\frac{27}{t^3}\right)\\ &\quad + 15(4t^2)\left(\frac{81}{t^4}\right) + 6(2t)\left(\frac{243}{t^5}\right) + \frac{729}{t^6}\\ &= 64t^6 + 576t^4 + 2160t^2 + 4320 + 4860t^{-2} + 2916t^{-4}\\ &\quad + 729t^{-6}.\end{aligned}$$

Binomial Expansion Using Factorial Notation

The drawback to using Pascal's triangle is that we must compute all the preceding rows in the table to obtain the row needed for the expansion in which we are interested. The following method avoids this difficulty. It will also enable us to find a specific term—say, the 8th term—without computing all the other terms in the expansion. This method is useful in such courses as finite mathematics, calculus, and statistics.

To develop the method, we need some new notation. Products of successive natural numbers, such as $6 \cdot 5 \cdot 4 \cdot 3 \cdot 2 \cdot 1$ and $8 \cdot 7 \cdot 6 \cdot 5 \cdot 4 \cdot 3 \cdot 2 \cdot 1$, have a special notation. For the product $6 \cdot 5 \cdot 4 \cdot 3 \cdot 2 \cdot 1$, we write $6!$, read "6 factorial."

Factorial Notation

For any natural number n,

$$n! = n(n-1)(n-2)\cdots(3)(2)(1).$$

Here are some examples:

$$\begin{aligned}
6! &= 6 \cdot 5 \cdot 4 \cdot 3 \cdot 2 \cdot 1 = 720,\\
5! &= 5 \cdot 4 \cdot 3 \cdot 2 \cdot 1 = 120,\\
4! &= 4 \cdot 3 \cdot 2 \cdot 1 = 24,\\
3! &= 3 \cdot 2 \cdot 1 = 6,\\
2! &= 2 \cdot 1 = 2,\\
1! &= 1 = 1.
\end{aligned}$$

We also define $0!$ to be 1 for reasons explained shortly.

To simplify expressions like

$$\frac{8!}{5!\,3!},$$

note that

$$8! = 8 \cdot 7 \cdot 6 \cdot 5 \cdot 4 \cdot 3 \cdot 2 \cdot 1 = 8 \cdot 7! = 8 \cdot 7 \cdot 6! = 8 \cdot 7 \cdot 6 \cdot 5!,$$

and so on.

CAUTION! $\dfrac{6!}{3!} \neq 2!$ To see this, note that

$$\frac{6!}{3!} = \frac{6 \cdot 5 \cdot 4 \cdot \cancel{3} \cdot \cancel{2} \cdot \cancel{1}}{\cancel{3} \cdot \cancel{2} \cdot \cancel{1}} = 6 \cdot 5 \cdot 4.$$

Student Notes

It is important to recognize factorial notation as representing a product with descending factors. Thus, 7!, $7 \cdot 6!$, and $7 \cdot 6 \cdot 5!$ all represent the same product.

EXAMPLE 3 Simplify: $\dfrac{8!}{5!\,3!}$.

SOLUTION

$$\frac{8!}{5!\,3!} = \frac{8 \cdot 7 \cdot 6 \cdot 5!}{5! \cdot 3 \cdot 2 \cdot 1} = 8 \cdot 7 \qquad \text{Removing a factor equal to 1: } \frac{6 \cdot 5!}{5! \cdot 3 \cdot 2} = 1$$

$$= 56$$

The following notation is used in our second formulation of the binomial theorem.

$\binom{n}{r}$ Notation

For n, r nonnegative integers with $n \geq r$,

$$\binom{n}{r}, \quad \text{read "}n\text{ choose }r\text{,"} \quad \text{means} \quad \frac{n!}{(n-r)!\,r!}.^{*}$$

EXAMPLE 4 Simplify: **(a)** $\binom{7}{2}$; **(b)** $\binom{9}{6}$; **(c)** $\binom{6}{6}$.

SOLUTION

a) $\binom{7}{2} = \dfrac{7!}{(7-2)!\,2!}$

$= \dfrac{7!}{5!\,2!} = \dfrac{7 \cdot 6 \cdot 5!}{5! \cdot 2 \cdot 1} = \dfrac{7 \cdot 6}{2}$ **Writing 7! as $7 \cdot 6 \cdot 5!$ to help with simplification**

$= 7 \cdot 3$

$= 21$

b) $\binom{9}{6} = \dfrac{9!}{3!\,6!}$

$= \dfrac{9 \cdot 8 \cdot 7 \cdot 6!}{3 \cdot 2 \cdot 1 \cdot 6!} = \dfrac{9 \cdot 8 \cdot 7}{3 \cdot 2}$ **Writing 9! as $9 \cdot 8 \cdot 7 \cdot 6!$**

$= 3 \cdot 4 \cdot 7$

$= 84$

c) $\binom{6}{6} = \dfrac{6!}{0!\,6!} = \dfrac{6!}{1 \cdot 6!}$ **Since $0! = 1$**

$= \dfrac{6!}{6!}$

$= 1$

*In many books and for many calculators, the notation ${}_nC_r$ is used instead of $\binom{n}{r}$.

Factorials and $\binom{n}{r}$

The PRB submenu of the MATH menu provides access to both factorial calculations and $\binom{n}{r}$, or ${}_nC_r$. In both cases, a number is entered on the home screen before the option is accessed.

To find 7!, press (7), select option 4: ! from the MATH PRB menu, and press ENTER.

To find $\binom{9}{2}$, press (9), select option 3: nCr from the MATH PRB menu, press (2), and then press ENTER.

The screen on the left below shows the options listed in the MATH PRB menu. From the screen on the right below, note that $7! = 5040$ and $\binom{9}{2} = {}_9C_2 = 36$.

```
MATH NUM CPX PRB
1: rand
2: nPr
3: nCr
4: !
5: randInt(
6: randNorm(
7: randBin(
```

```
7!
                5040
9 nCr 2
                  36
```

Now we can restate the binomial theorem using our new notation.

The Binomial Theorem (Form 2) For any binomial $a + b$ and any natural number n,

$$(a+b)^n = \binom{n}{0}a^n + \binom{n}{1}a^{n-1}b + \binom{n}{2}a^{n-2}b^2 + \cdots + \binom{n}{n}b^n.$$

EXAMPLE 5 Expand: $(3x + y)^4$.

SOLUTION We use the binomial theorem (Form 2) with $a = 3x$, $b = y$, and $n = 4$:

$$\begin{aligned}(3x+y)^4 &= \binom{4}{0}(3x)^4 + \binom{4}{1}(3x)^3y + \binom{4}{2}(3x)^2y^2 + \binom{4}{3}(3x)y^3 + \binom{4}{4}y^4 \\ &= \frac{4!}{4!\,0!}3^4x^4 + \frac{4!}{3!\,1!}3^3x^3y + \frac{4!}{2!\,2!}3^2x^2y^2 + \frac{4!}{1!\,3!}3xy^3 + \frac{4!}{0!\,4!}y^4 \\ &= 81x^4 + 108x^3y + 54x^2y^2 + 12xy^3 + y^4. \quad \textbf{Simplifying}\end{aligned}$$

EXAMPLE 6 Expand: $(x^2 - 2y)^5$.

SOLUTION In this case, $a = x^2$, $b = -2y$, and $n = 5$:

$$\begin{aligned}(x^2 - 2y)^5 &= \binom{5}{0}(x^2)^5 + \binom{5}{1}(x^2)^4(-2y) + \binom{5}{2}(x^2)^3(-2y)^2 \\ &\quad + \binom{5}{3}(x^2)^2(-2y)^3 + \binom{5}{4}(x^2)(-2y)^4 + \binom{5}{5}(-2y)^5 \\ &= \frac{5!}{5!0!}x^{10} + \frac{5!}{4!1!}x^8(-2y) + \frac{5!}{3!2!}x^6(-2y)^2 + \frac{5!}{2!3!}x^4(-2y)^3 \\ &\quad + \frac{5!}{1!4!}x^2(-2y)^4 + \frac{5!}{0!5!}(-2y)^5 \\ &= x^{10} - 10x^8y + 40x^6y^2 - 80x^4y^3 + 80x^2y^4 - 32y^5.\end{aligned}$$

Note that in the binomial theorem (Form 2), $\binom{n}{0}a^nb^0$ gives us the first term, $\binom{n}{1}a^{n-1}b^1$ gives us the second term, $\binom{n}{2}a^{n-2}b^2$ gives us the third term, and so on. This can be generalized to give a method for finding a specific term without writing the entire expansion.

Finding a Specific Term

When $(a + b)^n$ is expanded and written in descending powers of a, the $(r + 1)$st term is

$$\binom{n}{r}a^{n-r}b^r.$$

EXAMPLE 7 Find the 5th term in the expansion of $(2x - 3y)^7$.

SOLUTION First, we note that $5 = 4 + 1$. Thus, $r = 4$, $a = 2x$, $b = -3y$, and $n = 7$. Then the 5th term of the expansion is

$$\binom{7}{4}(2x)^{7-4}(-3y)^4, \quad \text{or} \quad \frac{7!}{3!4!}(2x)^3(-3y)^4, \quad \text{or} \quad 22{,}680x^3y^4.$$

It is because of the binomial theorem that $\binom{n}{r}$ is called a *binomial coefficient*. We can now explain why 0! is defined to be 1. In the binomial expansion, we want $\binom{n}{0}$ to equal 1 and we also want the definition

$$\binom{n}{r} = \frac{n!}{(n-r)!\,r!}$$

to hold for all whole numbers n and r. Thus we must have

$$\binom{n}{0} = \frac{n!}{(n-0)!\,0!} = \frac{n!}{n!\,0!} = 1.$$

This is satisfied only if 0! is defined to be 1.

11.4 EXERCISE SET

FOR EXTRA HELP

MathXL | MyMathLab | InterAct Math | AW Math Tutor Center | Video Lectures on CD: Disc 6 | Student's Solutions Manual

Concept Reinforcement *Complete each of the following.*

1. The last term in the expansion of $(x+2)^5$ is ________.

2. The expansion of $(x+y)^7$, when simplified, contains a total of ________ terms.

3. In the expansion of $(a+b)^9$, the exponents in each term add to ________.

4. The expression ________ represents $4 \cdot 3 \cdot 2 \cdot 1$.

5. The expression ________ represents $\dfrac{8!}{3!\,5!}$.

6. In the expansion of $(a+b)^{10}$, the coefficient of a^8b^2 is ________.

7. In the expansion of $(x+y)^9$, the coefficient of x^2y^7 is the same as the coefficient of ________.

8. The notation $\dbinom{10}{4}$ is read ________.

Simplify.

9. $9!$

10. $8!$

11. $11!$

12. $10!$

13. $\dfrac{8!}{6!}$

14. $\dfrac{7!}{4!}$

15. $\dfrac{9!}{5!}$

16. $\dfrac{10!}{7!}$

17. $\dbinom{7}{4}$

18. $\dbinom{8}{2}$

19. $\dbinom{9}{5}$

20. $\dbinom{10}{6}$

21. $\dbinom{30}{3}$

22. $\dbinom{20}{18}$

23. $\dbinom{40}{38}$

24. $\dbinom{35}{2}$

Expand. Use both of the methods shown in this section.

25. $(a-b)^4$

26. $(m+n)^5$

27. $(p+q)^7$

28. $(x-y)^6$

29. $(3c-d)^7$

30. $(x^2-3y)^5$

31. $(t^{-2}+2)^6$

32. $(3c-d)^6$

33. $(x-y)^5$

34. $(x-y)^3$

35. $\left(3s+\dfrac{1}{t}\right)^9$

36. $\left(x+\dfrac{2}{y}\right)^9$

37. $(x^3-2y)^5$

38. $(a^2-b^3)^5$

39. $(\sqrt{5}+t)^6$

40. $(\sqrt{3}-t)^4$

41. $\left(\dfrac{1}{\sqrt{x}}-\sqrt{x}\right)^6$

42. $(x^{-2}+x^2)^4$

Find the indicated term for each binomial expression.

43. 3rd, $(a+b)^6$

44. 6th, $(x+y)^7$

45. 12th, $(a-3)^{14}$

46. 11th, $(x-2)^{12}$

47. 5th, $(2x^3+\sqrt{y})^8$

48. 4th, $\left(\dfrac{1}{b^2}+c\right)^7$

49. Middle, $(2u+3v^2)^{10}$

50. Middle two, $(\sqrt{x}+\sqrt{3})^5$

Aha! **51.** 9th, $(x-y)^8$

52. 13th, $(a-\sqrt{b})^{12}$

TW **53.** Maya claims that she can calculate mentally the first two and the last two terms of the expansion of $(a+b)^n$ for any whole number n. How do you think she does this?

TW **54.** Without performing any calculations, explain why the expansions of $(x-y)^8$ and $(y-x)^8$ must be equal.

Skill Maintenance

Solve. [9.6]

55. $\log_2 x + \log_2(x - 2) = 3$

56. $\log_3(x + 2) - \log_3(x - 2) = 2$

57. $e^t = 280$

58. $\log_5 x^2 = 2$

Synthesis

TW **59.** Explain how someone can determine the x^2-term of the expansion of $\left(x - \frac{3}{x}\right)^{10}$ without calculating any other terms.

TW **60.** Devise two problems requiring the use of the binomial theorem. Design the problems so that one is solved more easily using Form 1 and the other is solved more easily using Form 2. Then explain what makes one form easier to use than the other in each case.

61. Show that there are exactly $\binom{5}{3}$ ways of choosing a subset of size 3 from $\{a, b, c, d, e\}$.

62. *Baseball.* At one point in July 2004, Barry Bonds of the San Francisco Giants had a batting average of .360 (*Source*: www.mlb.com). At that time, if someone were to randomly select 5 of his "at-bats," the probability of Bonds getting exactly 3 hits would be the 3rd term of the binomial expansion of $(0.360 + 0.640)^5$. Find that term and use a calculator to estimate the probability.

63. *Widows or Divorcees.* The probability that a woman will be either widowed or divorced is 85%. If 8 women are randomly selected, the probability that exactly 5 of them will be either widowed or divorced is the 6th term of the binomial expansion of $(0.15 + 0.85)^8$. Use a calculator to estimate that probability.

64. *Baseball.* In reference to Exercise 62, the probability that Bonds will get *at most* 3 hits is found by adding the last 4 terms of the binomial expansion of $(0.360 + 0.640)^5$. Find these terms and use a calculator to estimate the probability.

65. *Widows or Divorcees.* In reference to Exercise 63, the probability that *at least* 6 of the women will be widowed or divorced is found by adding the last three terms of the binomial expansion of $(0.15 + 0.85)^8$. Find these terms and use a calculator to estimate the probability.

66. Find the term of
$$\left(\frac{3x^2}{2} - \frac{1}{3x}\right)^{12}$$
that does not contain x.

67. Prove that
$$\binom{n}{r} = \binom{n}{n-r}$$
for any whole numbers n and r. Assume $r \leq n$.

68. Find the middle term of $(x^2 - 6y^{3/2})^6$.

69. Find the ratio of the 4th term of
$$\left(p^2 - \frac{1}{2}p\sqrt[3]{q}\right)^5$$
to the 3rd term.

70. Find the term containing $\frac{1}{x^{1/6}}$ of
$$\left(\sqrt[3]{x} - \frac{1}{\sqrt{x}}\right)^7.$$

Aha! **71.** Multiply: $(x^2 + 2xy + y^2)(x^2 + 2xy + y^2)^2(x + y)$.

72. What is the degree of $(x^3 + 2)^4$?

11 Chapter Summary and Review

KEY TERMS AND DEFINITIONS

SEQUENCES AND SERIES

Sequence, p. 840 An ordered list of numbers; the **general term,** is denoted a_n.

5, 7, 8, 11, 17 is a **finite sequence.**

6, 9, 12, 15, . . . is an **infinite sequence.**

Arithmetic sequence, p. 849 A sequence in which the difference between consecutive terms is constant.

Geometric sequence, p. 860 A sequence in which the ratio of consecutive terms is constant.

Series, p. 844 The sum of specified terms of a sequence. A **finite series,** or **partial sum,** S_n, is the sum of the first n terms. An **infinite series,** S_∞, is the sum of the terms of an infinite sequence.

BINOMIAL THEOREM

Binomial expansion, p. 873 The multiplied form of a binomial raised to a power.

Pascal's triangle, p. 873 An organized table of coefficients of $(a + b)^n$.

IMPORTANT CONCEPTS

[Section references appear in brackets.]

Concept	Example
Sigma or **Summation Notation** $\sum_{k=1}^{n} a_k$ — k is the **index of summation**	$\sum_{k=3}^{5}(-1)^k(k^2) = (-1)^3(3^2) + (-1)^4(4^2) + (-1)^5(5^2)$ $= -1 \cdot 9 + 1 \cdot 16 + (-1) \cdot 25$ $= -9 + 16 - 25 = -18$ [11.1]
Arithmetic Sequences and Series $a_{n+1} = a_n + d$ — d is the **common difference.** $a_n = a_1 + (n - 1)d$ — The nth term $S_n = \frac{n}{2}(a_1 + a_n)$ — The sum of the first n terms	For the arithmetic sequence 10, 7, 4, 1, . . .: $d = -3$; $a_7 = 10 + (7 - 1)(-3) = 10 - 18 = -8$; $S_7 = \frac{7}{2}(10 + (-8)) = \frac{7}{2}(2) = 7.$ [11.2]

(*continued*)

Geometric Sequences and Series	For the geometric sequence $25, -5, 1, -\frac{1}{5}, \ldots$:
$a_{n+1} = a_n \cdot r$ — r is the **common ratio.**	$r = -\frac{1}{5}$;
$a_n = a_1 r^{n-1}$ — The nth term	$a_7 = 25\left(-\frac{1}{5}\right)^{7-1} = 5^2 \cdot \frac{1}{5^6} = \frac{1}{625}$;
$S_n = \frac{a_1(1 - r^n)}{1 - r}$ — The sum of the first n terms	$S_7 = \frac{25\left(1 - \left(-\frac{1}{5}\right)^7\right)}{1 - \left(-\frac{1}{5}\right)} = \frac{5^2\left(\frac{78126}{5^7}\right)}{\frac{6}{5}} = \frac{13021}{625}$;
$S_\infty = \frac{a_1}{1 - r}, \; \lvert r \rvert < 1$ — Limit of an infinite geometric series	$S_\infty = \frac{25}{1-\left(-\frac{1}{5}\right)} = \frac{125}{6}$. [11.3]
Factorial Notation	
$n! = n(n-1)(n-2)\cdots 3 \cdot 2 \cdot 1$	$7! = 7 \cdot 6 \cdot 5 \cdot 4 \cdot 3 \cdot 2 \cdot 1 = 5040$ [11.4]
Binomial Coefficient	
$\binom{n}{r} = {}_nC_r = \frac{n!}{(n-r)!\,r!}$	$\binom{10}{3} = {}_{10}C_3 = \frac{10!}{7!\,3!} = \frac{10 \cdot 9 \cdot 8 \cdot 7!}{7! \cdot 3 \cdot 2 \cdot 1} = 120$ [11.4]
Binomial Theorem	$(1-2x)^3 = \binom{3}{0}1^3 + \binom{3}{1}1^2(-2x) + \binom{3}{2}1(-2x)^2 + \binom{3}{3}(-2x)^3$
$(a+b)^n = \binom{n}{0}a^n + \binom{n}{1}a^{n-1}b + \cdots + \binom{n}{n}b^n$	$= 1 \cdot 1 + 3 \cdot 1(-2x) + 3 \cdot 1 \cdot (4x^2) + 1 \cdot (-8x^3)$
$(r+1)$st term of $(a+b)^n$: $\binom{n}{r}a^{n-r}b^r$	$= 1 - 6x + 12x^2 - 8x^3$
	3rd term of $(1-2x)^3$:
	$\binom{3}{2}(1)^1(-2x)^2 = 12x^2 \quad (r = 2)$ [11.4]

Review Exercises

Concept Reinforcement *Classify each of the following as either true or false.*

1. The next term in the arithmetic sequence 10, 15, 20, ... is 35. [11.2]
2. The next term in the geometric sequence 2, 6, 18, 54, ... is 162. [11.3]
3. $\sum_{k=1}^{3} k^2$ means $1^2 + 2^2 + 3^2$. [11.1]
4. If $a_n = 3n - 1$, then $a_{17} = 19$. [11.1]
5. The nth term of an arithmetic sequence with common difference d is $a_n = a_1 + (n - 1)d$, for any integer $n \geq 1$. [11.2]
6. The nth term of a geometric sequence with common ratio r is given by $a_n = a_1 r^{n-1}$, for any integer $n \geq 1$. [11.3]
7. For any natural number n, $n! = n(n - 1)$. [11.4]
8. When simplified, the expansion of $(x + y)^{17}$ has 19 terms. [11.4]

Find the first four terms; the 8th term, a_8; and the 12th term, a_{12}. [11.1]

9. $a_n = 4n - 3$
10. $a_n = \dfrac{n - 1}{n^2 + 1}$

Predict the general term. Answers may vary. [11.1]

11. 7, 14, 21, 28, 35, ...
12. $-1, 3, -5, 7, -9, \ldots$

Write out and evaluate each sum. [11.1]

13. $\sum_{k=1}^{5} (-2)^k$
14. $\sum_{k=2}^{7} (1 - 2k)$

Rewrite using sigma notation. [11.1]

15. $4 + 8 + 12 + 16 + 20$
16. $\dfrac{-1}{2} + \dfrac{1}{4} + \dfrac{-1}{8} + \dfrac{1}{16} + \dfrac{-1}{32}$
17. Find the 14th term of the arithmetic sequence $-6, 1, 8, \ldots$. [11.2]
18. An arithmetic sequence has $a_1 = 11$ and $a_{13} = 43$. Find the common difference, d. [11.2]
19. An arithmetic sequence has $a_8 = 20$ and $a_{24} = 40$. Find the first term, a_1, and the common difference, d. [11.2]
20. Find the sum of the first 17 terms of the arithmetic series $-8 + (-11) + (-14) + \cdots$. [11.2]
21. Find the sum of all the multiples of 6 from 12 to 318, inclusive. [11.2]
22. Find the 20th term of the geometric sequence $2, 2\sqrt{2}, 4, \ldots$. [11.3]
23. Find the common ratio of the geometric sequence $2, \frac{4}{3}, \frac{8}{9}, \ldots$. [11.3]
24. Find the nth term of the geometric sequence $-2, 2, -2, \ldots$. [11.3]
25. Find the nth term of the geometric sequence $3, \frac{3}{4}x, \frac{3}{16}x^2, \ldots$. [11.3]
26. Find S_6 for the geometric series
$$3 + 12 + 48 + \cdots. \text{ [11.3]}$$
27. Find S_{12} for the geometric series
$$3x - 6x + 12x - \cdots. \text{ [11.3]}$$

Determine whether each infinite geometric series has a limit. If a limit exists, find it. [11.3]

28. $6 + 3 + 1.5 + 0.75 + \cdots$
29. $7 - 4 + \frac{16}{7} - \cdots$
30. $2 + (-2) + 2 + (-2) + \cdots$
31. $0.04 + 0.08 + 0.16 + 0.32 + \cdots$
32. $\$2000 + \$1900 + \$1805 + \$1714.75 + \cdots$
33. Find fraction notation for 0.555555.... [11.3]
34. Find fraction notation for 1.39393939.... [11.3]
35. Adam took a job working in a convenience store starting with an hourly wage of $11.50. He was promised a raise of 40¢ per hour every 3 mos for 8 yr. At the end of 8 yr, what will be his hourly wage? [11.2]

36. A stack of poles has 42 poles in the bottom row. There are 41 poles in the second row, 40 poles in the third row, and so on, ending with 1 pole in the top row. How many poles are in the stack? [11.2]

37. Stacey's student loan is in the amount of \$12,000. Interest is 4%, compounded annually, and the amount is to be paid off in 7 yr. How much is to be paid back? [11.3]

38. Find the total rebound distance of a ball, given that it is dropped from a height of 12 m and each rebound is one third of the preceding one. [11.3]

Simplify. [11.4]

39. $7!$

40. $\binom{8}{3}$

41. Find the 3rd term of $(a + b)^{20}$. [11.4]

42. Expand: $(x - 2y)^4$. [11.4]

Synthesis

TW **43.** What happens to a_n in a geometric sequence with $|r| < 1$, as n gets larger? Why? [11.3]

TW **44.** Compare the two forms of the binomial theorem given in the text. Under what circumstances would one be more useful than the other? [11.4]

45. Find the sum of the first n terms of the geometric series $1 - x + x^2 - x^3 + \cdots$. [11.3]

46. Expand: $(x^{-3} + x^3)^5$. [11.4]

Chapter Test 11

1. Find the first five terms and the 12th term of a sequence with general term $a_n = 6n - 5$.

2. Predict the general term of the sequence

$\frac{4}{3}, \frac{4}{9}, \frac{4}{27}, \ldots$.

3. Write out and evaluate:

$$\sum_{k=2}^{6} (3 - 2^k).$$

4. Rewrite using sigma notation:

$$1 + (-8) + 27 + (-64) + 125.$$

5. Find the 13th term, a_{13}, of the arithmetic sequence $9, 4, -1, \ldots$.

Assume arithmetic sequences for Questions 6 and 7.

6. Find the common difference d when $a_1 = 7$ and $a_7 = 9\frac{1}{4}$.

7. Find a_1 and d when $a_5 = 16$ and $a_{10} = -3$.

8. Find the sum of all the multiples of 12 from 24 to 240, inclusive.

9. Find the 6th term of the geometric sequence $72, 18, 4\frac{1}{2}, \ldots$.

10. Find the common ratio of the geometric sequence $22\frac{1}{2}, 15, 10, \ldots$.

11. Find the nth term of the geometric sequence $3, 9, 27, \ldots$.

12. Find the sum of the first nine terms of the geometric series

$$(1 + x) + (2 + 2x) + (4 + 4x) + \cdots.$$

Determine whether each infinite geometric series has a limit. If a limit exists, find it.

13. $0.5 + 0.25 + 0.125 + \cdots$

14. $0.5 + 1 + 2 + 4 + \cdots$

15. $\$1000 + \$80 + \$6.40 + \cdots$

16. Find fraction notation for $0.85858585\ldots$.

17. An auditorium has 31 seats in the first row, 33 seats in the second row, 35 seats in the third row, and so on, for 18 rows. How many seats are in the 17th row?

18. Lindsay's Uncle Ken gave her \$100 for her first birthday, \$200 for her second birthday, \$300 for her third birthday, and so on, until her eighteenth birthday. How much did he give her in all?

19. Each week the price of a \$15,000 boat will be reduced 5% of the previous week's price. If we assume that it is not sold, what will be the price after 10 weeks?

20. Find the total rebound distance of a ball that is dropped from a height of 18 m, with each rebound two thirds of the preceding one.

21. Simplify: $\binom{12}{9}$.

22. Expand: $(x^2 - 3y)^5$.

23. Find the 4th term in the expansion of $(a + x)^{12}$.

Synthesis

24. Find a formula for the sum of the first n even natural numbers:

$$2 + 4 + 6 + \cdots + 2n.$$

25. Find the sum of the first n terms of

$$1 + \frac{1}{x} + \frac{1}{x^2} + \frac{1}{x^3} + \cdots.$$

1–11 Cumulative Review

1. The diameter of plastic tubing needed for collecting sap from sugar maple trees depends on the slope of the land between the trees and the sugarhouse. The following table lists some recommended diameters for a 1000-tap line. Anna estimates that there is a fall of about 15 ft over the 500-ft run from the sugarbush to the sugarhouse. What diameter tubing should she use? [2.2]

Slope (in percent)	Diameter of Tubing (in inches)
25	3/4
20	3/4
10	1
5	1
3	1 1/4
1	1 1/2

Source: North American Maple Syrup Producers Manual Bulletin 856

2. Anna can tap the sugar maple trees in Southway Park in 21 hr. Emma can tap the trees in 14 hr. How long would it take them, working together, to tap the trees? [6.5]

3. A basketball team increases its score by 7 points in each of the two consecutive games after the home opener. If the team scored a total of 228 points in all three games, what was its score in the home opener? [1.7]

4. A music club offers two types of membership. Limited members pay a fee of \$10 a year and can buy CDs for \$10 each. Preferred members pay \$20 a year and can buy CDs for \$7.50 each. For what numbers of annual CD purchases would it be less expensive to be a preferred member? [4.1]

5. A pentagon with all five sides the same size has a perimeter equal to that of an octagon in which all eight sides are the same size. One side of the pentagon is 2 less than three times one side of the octagon. What is the perimeter of each figure? [3.3]

6. Roger's Organics mixes herbs that cost \$2.68 an ounce with herbs that cost \$4.60 an ounce to create a seasoning that costs \$3.80 an ounce. How many ounces of each herb should be mixed together to make 24 oz of the seasoning? [3.3]

7. An airplane can fly 190 mi with the wind in the same time it takes to fly 160 mi against the wind. The speed of the wind is 30 mph. How fast can the plane fly in still air? [6.5]

8. The centripetal force F of an object moving in a circle varies directly as the square of the velocity v and inversely as the radius r of the circle. If $F = 8$ when $v = 1$ and $r = 10$, what is F when $v = 2$ and $r = 16$? [6.8]

9. The Brighton recreation department plans to fence in a rectangular park next to a river. (Note that no fence will be needed along the river.) What is the area of the largest region that can be fenced in with 200 ft of fencing? [8.8]

10. The perimeter of a rectangular sign is 34 ft. The length of a diagonal is 13 ft. Find the dimensions of the sign. [10.4]

11. On Elyse's 9th birthday, her grandmother opened a savings account for her with \$500. The account pays 3% interest, compounded annually. If Elyse neither adds to nor withdraws any money from the bank, how much will be in the account on her 18th birthday? [11.3]

12. The number of households in Mexico has grown linearly from 21.3 million in 1998 to 23.7 million in 2003 (*Source*: Economist Intelligence Unit).

a) Write a linear function describing the number of households in Mexico t years after 1998. [2.4]
b) Predict the number of households in Mexico in 2010. [2.4]

13. The number of personal computers in Mexico has grown exponentially from 0.12 million in 1985 to 9.3 million in 2003 (*Source*: International Telecommunications Union).

a) Find the exponential growth rate, k, to three decimal places and write an exponential function describing the number of personal computers in Mexico t years after 1985. [9.7]
b) Predict the number of personal computers in Mexico in 2010. [9.7]

14. Find a linear equation whose graph has a y-intercept of $(0, -8)$ and is parallel to the line whose equation is $3x - y = 6$. [2.3]

Graph.

15. $3x - y = 7$ [2.3]

16. $y - 4 = -\frac{2}{3}(x - 1)$ [2.4]

17. $x^2 + y^2 = 100$ [10.1]

18. $\frac{x^2}{36} - \frac{y^2}{9} = 1$ [10.3]

19. $y = \log_2 x$ [9.3]

20. $f(x) = 2^x - 3$ [9.2]

21. $2x - 3y < -6$ [4.5]

22. Graph: $f(x) = -2(x - 3)^2 + 1$. [8.7]

a) Label the vertex.
b) Draw the axis of symmetry.
c) Find the maximum or minimum value.

23. Graph $f(x) = \sqrt{4 - x}$ and use the graph to estimate the domain and the range of f. [2.5]

24. For the function described by

$$f(x) = 3x^2 - 4x,$$

find $f(-2)$. [2.1]

Find the domain of each function.

25. $f(x) = \sqrt{5 - 3x}$ [7.1]

26. $g(x) = \frac{x - 4}{x^2 - 2x + 1}$ [5.5]

For each of the following graphs, **(a)** *determine whether the graph is that of a linear, quadratic, exponential, or logarithmic function and* **(b)** *use the graph to find the zeros of f.*

27.

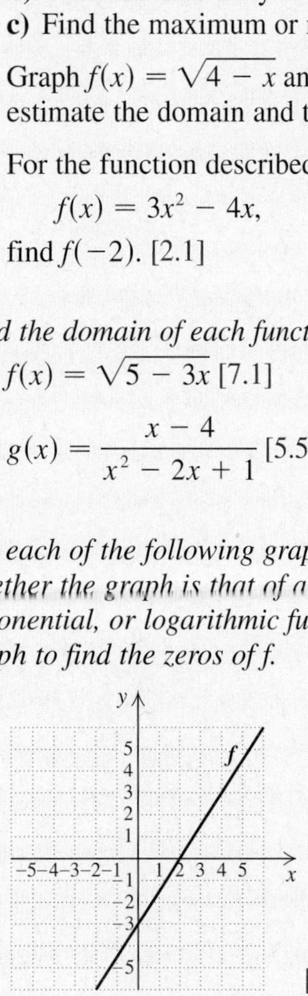

[2.4], [2.2]

28.

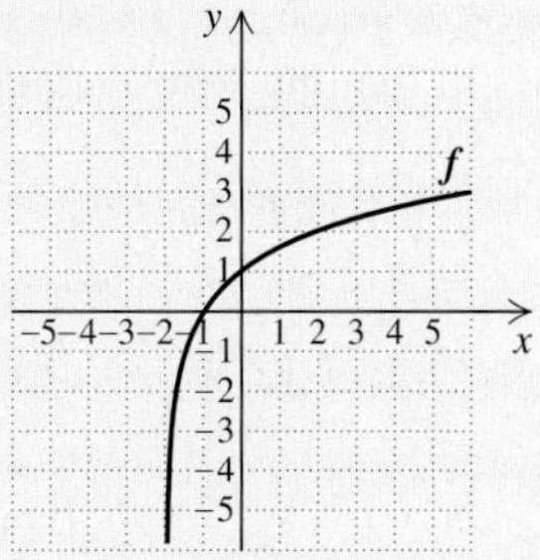

[9.3], [9.6]

29.

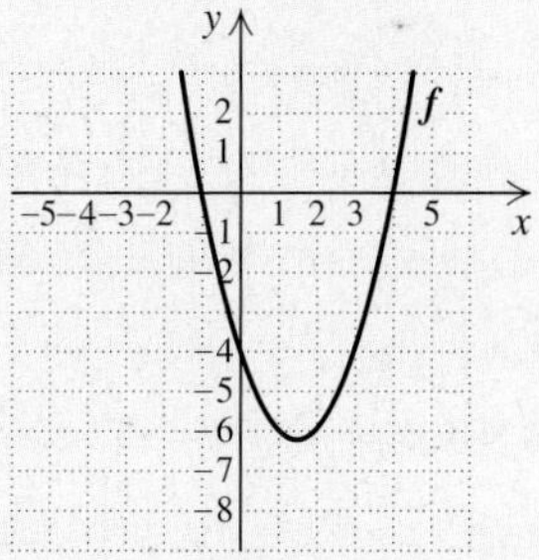

[8.7], [8.1]

30.

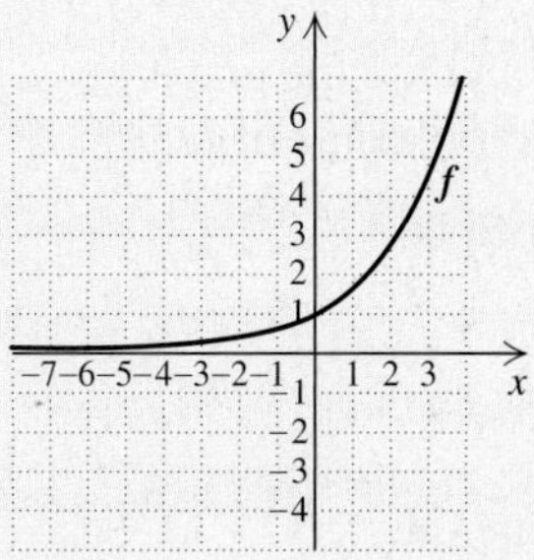

[9.2], [9.6]

Solve.

31. $8(x - 1) - 3(x - 2) = 1$ [1.6]

32. $\frac{6}{x} + \frac{6}{x + 2} = \frac{5}{2}$ [6.4]

33. $2x + 1 > 5$ *or* $x - 7 \leq 3$ [4.3]

34. $5x + 3y = -2,$
$3x + 5y = 2$ [3.2]

35. $x + y - z = 0,$
$3x + y + z = 6,$
$x - y + 2z = 5$ [3.4]

36. $3\sqrt{x - 1} = 5 - x$ [7.6]

37. $x^4 - 29x^2 + 100 = 0$ [8.5]

38. $x^2 + y^2 = 8,$
$x^2 - y^2 = 2$ [10.4]

39. $4^x = 7$ [9.6]

40. $\log (x^2 - 25) - \log (x + 5) = 3$ [9.6]

41. $\log_5 x = -2$ [9.6]

42. $7^{2x+3} = 49$ [9.6]

43. $|2x - 1| \leq 5$ [4.4]

44. $7x^2 + 14 = 0$ [8.1]

45. $x^2 + 4x = 3$ [8.2]

46. $y^2 + 3y > 10$ [8.9]

47. Let $f(x) = x^2 - 2x$. Find a such that $f(a) = 48$. [5.4]

48. If $f(x) = \sqrt{-x + 4} + 3$ and $g(x) = \sqrt{x - 2} + 3$, find a such that $f(a) = g(a)$. [7.6]

49. Solve $V = P - Prt$ for r. [1.6]

50. Solve $I = \frac{R}{R + r}$ for R. [6.8]

Simplify.

51. $(-7x^2y^3)(5x^4y^{-7})$ [1.4]

52. $|-3.5 + 9.8|$ [1.2]

53. $y - [3 - 4(5 - 2y) - 3y]$ [1.3]

54. $(10 \cdot 8 - 9 \cdot 7)^2 - 54 \div 9 - 3$ [1.1]

55. Evaluate

$$\frac{ab - ac}{bc}$$

for $a = -2$, $b = 3$, and $c = -4$. [1.1], [1.2]

Perform the indicated operations to create an equivalent expression. Be sure to simplify your result if possible.

56. $(5a^2 - 3ab - 7b^2) - (2a^2 + 5ab + 8b^2)$ [5.1]

57. $(-3x^2 + 4x^3 - 5x - 1) + (9x^3 - 4x^2 + 7 - x)$ [5.1]

58. $(2a - 1)(3a + 5)$ [5.2]

59. $(3a^2 - 5y)^2$ [5.2]

60. $\frac{1}{x - 2} - \frac{4}{x^2 - 4} + \frac{3}{x + 2}$ [6.2]

61. $\frac{x^2 - 6x + 8}{4x + 12} \cdot \frac{x + 3}{x^2 - 4}$ [6.1]

62. $\dfrac{3x + 3y}{5x - 5y} \div \dfrac{3x^2 + 3y^2}{5x^3 - 5y^3}$ [6.1]

63. $\dfrac{x - \dfrac{a^2}{x}}{1 + \dfrac{a}{x}}$ [6.3]

Factor, if possible, to form an equivalent expression.

64. $4x^2 - 12x + 9$ [5.6]

65. $27a^3 - 8$ [5.7]

66. $a^3 + 3a^2 - ab - 3b$ [5.3]

67. $15y^4 + 33y^2 - 36$ [5.5]

68. Divide:

$$(7x^4 - 5x^3 + x^2 - 4) \div (x - 2). \text{ [6.6]}$$

69. Multiply $(8.9 \times 10^{-17})(7.6 \times 10^4)$. Write scientific notation for the answer. [1.4]

70. Multiply and simplify: $\sqrt{8x}\,\sqrt{8x^3y}$. [7.3]

71. Simplify: $(25x^{4/3}y^{1/2})^{3/2}$. [7.2]

72. Divide and simplify:

$$\frac{\sqrt[3]{25x}}{\sqrt[3]{5y^2}}. \text{ [7.4]}$$

73. Write an equivalent expression by rationalizing the denominator:

$$\frac{1 - \sqrt{x}}{1 + \sqrt{x}}. \text{ [7.5]}$$

74. Multiply these complex numbers:

$$(3 + 2i)(4 - 7i). \text{ [7.8]}$$

75. Write a quadratic equation whose solutions are $5\sqrt{2}$ and $-5\sqrt{2}$. [8.4]

76. Find the center and the radius of the circle given by

$$x^2 + y^2 - 4x + 6y - 23 = 0. \text{ [10.1]}$$

77. Write an equivalent expression that is a single logarithm:

$$\tfrac{2}{3}\log_a x - \tfrac{1}{2}\log_a y + 5\log_a z. \text{ [9.4]}$$

78. Write an equivalent exponential equation: $\log_a c = 5$. [9.3]

Use a calculator to find each of the following. [9.3], [9.5]

79. $\log 5677.2$

80. $10^{-3.587}$

81. $\ln 5677.2$

82. $e^{-3.587}$

83. Find the distance between the points $(-1, -5)$ and $(2, -1)$. [10.1]

84. Find the 21st term of the arithmetic sequence $19, 12, 5, \ldots$. [11.2]

85. Find the sum of the first 25 terms of the arithmetic series $-1 + 2 + 5 + \cdots$. [11.2]

86. Find the general term of the geometric sequence $16, 4, 1, \ldots$. [11.3]

87. Find the 7th term of $(a - 2b)^{10}$. [11.4]

88. Find the sum of the first nine terms of the geometric series $x + 1.5x + 2.25x + \cdots$. [11.3]

The following graphs show the number of foreign students enrolled in U.S. colleges for years from 1985 to 2004. For each graph, **(a)** *determine whether the data would be best modeled by a linear, quadratic, or exponential function, and* **(b)** *use regression to find a function that can be used to predict the number of foreign students from that region or country t years after 1985.*

89. **Africa**

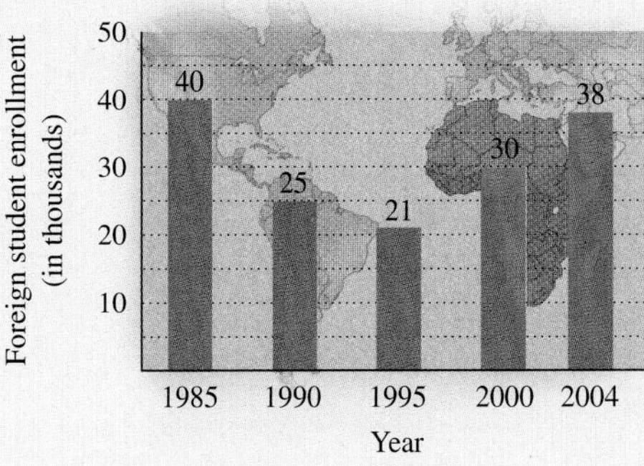

Source: Institute of International Education

[8.8]

90. **India**

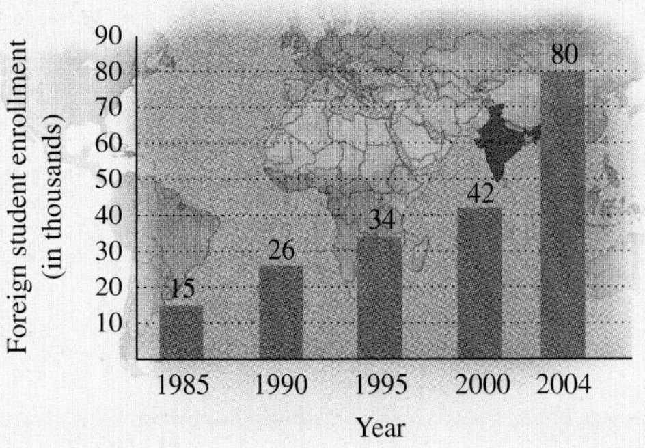

Source: Institute of International Education

[9.7]

91. **South Korea**

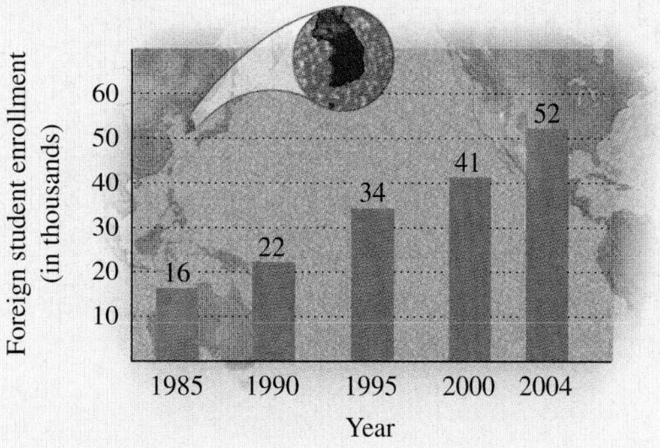

Source: Institute of International Education

[2.4]

92. If enrollment trends continue, for what years after 2004 will more students from Africa be enrolled in U.S. colleges than students from South Korea? [8.9]

Synthesis

Solve.

93. $\dfrac{9}{x} - \dfrac{9}{x+12} = \dfrac{108}{x^2 + 12x}$ [6.4]

94. $\log_2(\log_3 x) = 2$ [9.6]

95. y varies directly as the cube of x and x is multiplied by 0.5. What is the effect on y? [6.8]

96. Divide these complex numbers:

$$\frac{2\sqrt{6} + 4\sqrt{5}i}{2\sqrt{6} - 4\sqrt{5}i}.$$ [7.8]

97. Diaphantos, a famous mathematician, spent $\frac{1}{6}$ of his life as a child, $\frac{1}{12}$ as an adolescent, and $\frac{1}{7}$ as a bachelor. Five years after he was married, he had a son who died 4 years before his father at half his father's final age. How long did Diaphantos live? [3.5]

Appendix

A Working with Units

Calculating with Dimension Symbols ▪ Making Unit Changes ▪ Dimension Analysis

Quantities such as lengths, weights, and time are described using standard units such as centimeters, pounds, and seconds, respectively. It is important to know how to work with units when combining quantities, using formulas, and changing units. Analyzing the units used can even serve as a partial check of an answer to a problem.

Calculating with Dimension Symbols

Inside calculations, units can often be treated somewhat like variables. We can add, subtract, multiply, and divide quantities.

EXAMPLE 1 Rondel worked 8 hr on Friday and 9 hr on Saturday. How many hours did he work in all?

SOLUTION We can think of this addition as

$$8 \text{ hr} + 9 \text{ hr} = (8 + 9) \text{ hr} = 17 \text{ hr}.$$

Rondel worked 17 hr.

The addition in Example 1 is similar to the addition

$$8x + 9x = (8 + 9)x = 17x.$$

The 8 hr and 9 hr are, in this situation, "like terms." We could not have combined quantities with different units—say, 8 hr and 9 min—in this way.

When we substitute quantities into a formula, handling dimension symbols correctly should give us the correct unit for the answer.

EXAMPLE 2 Find the area of this square.

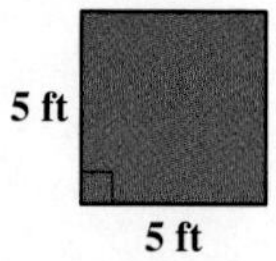

SOLUTION The formula for the area of a square is

$$A = s^2.$$

We treat the symbol "ft" as though it were a variable and obtain the correct unit:

$$\begin{aligned} A &= s^2 \\ &= (5\text{ ft})^2 \qquad \textbf{Substituting 5 ft for } s \\ &= 5\text{ ft} \cdot 5\text{ ft} \\ &= (5 \cdot 5)(\text{ft} \cdot \text{ft}) \\ &= 25\text{ ft}^2. \end{aligned}$$

EXAMPLE 3 Lorrie drove 150 km in 2 hr. What was her average speed?

SOLUTION Since speed is given by

$$r = \frac{d}{t},$$

the average speed r is

$$r = \frac{d}{t} = \frac{150\text{ km}}{2\text{ hr}} = \frac{150}{2} \cdot \frac{\text{km}}{\text{hr}} = 75\frac{\text{km}}{\text{hr}}. \qquad d = 150\text{ km};\ t = 2\text{ hr}$$

Lorrie's average speed was 75 km/h.

EXAMPLE 4 The amount of work done is given by the formula

work = *force* × *distance*.

How much work is done when a 100-lb rock is lifted 10 ft?

SOLUTION We substitute and calculate:

work = *force* × *distance* = 100 lb × 10 ft **Substituting**

= 100 × 10 × lb × ft

= 1000 foot-pounds. **A unit used to represent work is foot-pounds.**

EXAMPLE 5 The formula for the period T of a pendulum is $T = 2\pi\sqrt{\frac{l}{g}}$, where l is the length of the pendulum and g is the gravitational constant. Find the period of a pendulum of length 4 m.

SOLUTION Since l is given in meters, we use the value for g that is given in m/sec^2: $g \approx 9.8\ \text{m/sec}^2$. Substituting, we have

$$T = 2\pi\sqrt{\frac{l}{g}} \quad \textbf{Recall that } \pi \textbf{ is a constant with no unit.}$$

$$\approx 2\pi\sqrt{\frac{4\ \text{m}}{\frac{9.8\ \text{m}}{\text{sec}^2}}} \quad \textbf{Substituting}$$

$$= 2\pi\sqrt{4\ \text{m} \cdot \frac{\text{sec}^2}{9.8\ \text{m}}} \quad \textbf{Dividing by multiplying by the reciprocal}$$

$$= 2\pi\sqrt{\frac{4}{9.8}} \cdot \sqrt{\frac{\cancel{\text{m}} \cdot \text{sec}^2}{\cancel{\text{m}}}} \quad \sqrt{ab} = \sqrt{a}\sqrt{b}$$

$$= 2\pi\sqrt{\frac{4}{9.8}} \cdot \sqrt{\text{sec}^2} \quad \textbf{Removing a factor equal to 1: } \frac{\text{m}}{\text{m}} = 1$$

$$\approx 4.0\ \text{sec.} \quad \sqrt{\text{sec}^2} = \text{sec}$$

Making Unit Changes

It is often necessary to write a quantity using a different unit. Some unit equivalencies are given in the following tables.

American Units of Length	
1 ft	12 in.
1 yd	3 ft
1 mi	5280 ft

Metric Units of Length	
1 cm	10 mm
1 m	100 cm
1 km	1000 m

We can change units using substitution or multiplication by 1.

EXAMPLE 6 Convert 3 feet to inches.

SOLUTION

Substitution: We know that 1 ft = 12 in., so we substitute:

$$3\ \text{ft} = 3 \cdot 1\ \text{ft} \quad \textbf{Rewriting to make the substitution clear}$$

$$= 3 \cdot 12\ \text{in.} \quad \textbf{Substituting 12 in. for 1 ft}$$

$$= 36\ \text{in.}$$

Multiplication by 1: Two symbols for 1 can be written using the equivalency 1 ft = 12 in.:

$$\frac{1 \text{ ft}}{12 \text{ in.}} = 1 \quad \text{and} \quad \frac{12 \text{ in.}}{1 \text{ ft}} = 1.$$

We are converting from "ft" to "in." Since we want to eliminate the unit "ft" and end up with the unit "in.," we use the symbol for 1 with "ft" in the denominator and "in." in the numerator.

$$\begin{aligned} 3 \text{ ft} &= 3 \text{ ft} \cdot 1 \\ &= \frac{3 \text{ ft}}{1} \cdot \frac{12 \text{ in.}}{1 \text{ ft}} && \textbf{“ft” in the numerator and “ft” in the denominator will cancel.} \\ &= \frac{3 \cdot 12}{1 \cdot 1} \cdot \frac{\cancel{\text{ft}} \cdot \text{in.}}{\cancel{\text{ft}}} && \frac{\textbf{ft}}{\textbf{ft}} = \textbf{1} \\ &= 36 \text{ in.} \end{aligned}$$

We can multiply by 1 several times to make successive conversions.

EXAMPLE 7 The speed limit on route I-35 in Oklahoma City is $60\frac{\text{mi}}{\text{hr}}$. Convert this to $\frac{\text{ft}}{\text{sec}}$.

SOLUTION We write successive conversions, but do not carry out the operations until the last step.

$$\begin{aligned} 60\frac{\text{mi}}{\text{hr}} &= 60\frac{\text{mi}}{\text{hr}} \cdot \frac{5280 \text{ ft}}{1 \text{ mi}} && \textbf{Converting mi to ft. The unit is now ft/hr.} \\ &= 60\frac{\text{mi}}{\text{hr}} \cdot \frac{5280 \text{ ft}}{1 \text{ mi}} \cdot \frac{1 \text{ hr}}{60 \text{ min}} && \textbf{Converting hr to min. The unit is now ft/min.} \\ &= 60\frac{\text{mi}}{\text{hr}} \cdot \frac{5280 \text{ ft}}{1 \text{ mi}} \cdot \frac{1 \text{ hr}}{60 \text{ min}} \cdot \frac{1 \text{ min}}{60 \text{ sec}} && \textbf{Converting min to sec. The unit is now ft/sec.} \\ &= \frac{60 \cdot 5280 \cdot 1 \cdot 1}{1 \cdot 60 \cdot 60} \cdot \frac{\cancel{\text{mi}} \cdot \text{ft} \cdot \cancel{\text{hr}} \cdot \cancel{\text{min}}}{\cancel{\text{hr}} \cdot \cancel{\text{mi}} \cdot \cancel{\text{min}} \cdot \text{sec}} && \frac{\textbf{mi} \cdot \textbf{hr} \cdot \textbf{min}}{\textbf{mi} \cdot \textbf{hr} \cdot \textbf{min}} = \textbf{1} \\ &= 88\frac{\text{ft}}{\text{sec}} && \textbf{Carrying out the calculations} \end{aligned}$$

The speed limit is $88\frac{\text{ft}}{\text{sec}}$.

Dimension Analysis

By noting the units used in calculations, we can help assure the accuracy of our work. Some quantities may need to be written with different units in order to answer a problem correctly.

EXAMPLE 8 Alanna is driving at a speed of 65 mph. How far will she travel in 10 min?

SOLUTION We use the formula $d = rt$. Since the speed is given in miles per *hour*, we must write the time traveled, 10 min, in terms of hours.

$$d = rt = \frac{65 \text{ mi}}{\text{hr}} \cdot 10 \text{ min} \qquad \textbf{Substituting}$$

$$= \frac{65 \text{ mi}}{\text{hr}} \cdot \frac{10 \text{ min}}{1} \cdot \frac{1 \text{hr}}{60 \text{ min}} \qquad \textbf{Converting min to hr}$$

$$= \frac{650}{60} \text{ mi} = 10\frac{5}{6} \text{ mi} \qquad \frac{\textbf{min} \cdot \textbf{hr}}{\textbf{min} \cdot \textbf{hr}} = 1$$

Alanna will travel $10\frac{5}{6}$ mi.

Some formulas are derived using specific units. These units must be used in the calculations.

EXAMPLE 9 Body mass index (BMI), one measure of appropriate weight to height, is given by

$$\text{BMI} = \frac{w}{h^2} \times 703,$$

where w is weight, in pounds, and h is height, in inches. The index is rounded to the nearest whole number and does not have a unit. Jolene is 5 ft 3 in. tall and weighs 130 lb. What is her BMI?

SOLUTION We begin by examining the units used in the formula. Weights must be in pounds and height in inches. Jolene's weight is given in pounds, but we must convert her height to inches before substituting in the formula:

$$5 \text{ ft } 3 \text{ in.} = 5 \text{ ft} + 3 \text{ in.} \qquad \textbf{Writing as an addition}$$

$$= 5(12 \text{ in.}) + 3 \text{ in.} \qquad \textbf{Substituting 12 in. for 1 ft}$$

$$= 63 \text{ in.}$$

We substitute 130 for w and 63 for h and calculate:

$$\text{BMI} = \frac{w}{h^2} \times 703 = \frac{130}{63^2} \times 703 \approx 23.$$

Jolene's BMI is 23.

Sometimes we know that the result should be stated in certain units. For example, measurements of lengths are given in linear units, such as cm. Areas are measured in square units, such as cm^2, and volumes are measured in cubic units, such as cm^3. If the answer we get has the correct units, this provides a partial check of our calculations. Analysis of units used can even help us check the formulas we use.

EXAMPLE 10 Chad needs to calculate the area of a circular flower garden with a radius of 5 ft. He is certain that one of two formulas he remembers from geometry gives the area of a circle:

$$X = 2\pi r \quad \text{and} \quad X = \pi r^2.$$

Which could be the formula for the area of a circle?

SOLUTION Area is always expressed in square units. Chad can substitute 5 ft for r in both formulas and check the units.

Substituting in $X = 2\pi r$: $X = 2\pi(5 \text{ ft}) = 10\pi \text{ ft}$ **This is a measure of length.**

Substituting in $X = \pi r^2$: $X = \pi(5 \text{ ft})^2 = 25\pi \text{ ft}^2$ **This is a measure of area.**

Since the second formula gives the answer in square feet, $X = \pi r^2$ could be the formula for the area of a circle.

Working with Units

- Write all length quantities with the same units, all time quantities with the same units, and so on.
- Use units specified by any formulas.
- Check the answer units to see if they are appropriate.

A EXERCISE SET

FOR EXTRA HELP

MathXL MyMathLab InterAct Math AW Math Tutor Center

Perform the indicated operations.

1. $12 \text{ ft} + 8 \text{ ft}$

2. $350 \text{ mph} + 20 \text{ mph}$

3. $19 \text{ g} - 8 \text{ g}$

4. $14.2 \text{ lb} - 16 \text{ lb}$

5. $16 \text{ ft}^3 + 8 \text{ ft}^3$

6. $9 \text{ m}^2 - 3 \text{ m}^2$

7. $(4 \text{ cm})(3 \text{ cm})$

8. $(2 \text{ in.})(5 \text{ in.})$

9. $36 \text{ ft} \cdot \dfrac{1 \text{ yd}}{3 \text{ ft}}$

10. $55\dfrac{\text{mi}}{\text{hr}} \cdot 4 \text{ hr}$

11. $\dfrac{3 \text{ ft} \cdot 8 \text{ lb}}{6 \text{ sec}}$

12. $\dfrac{60 \text{ men} \cdot 8 \text{ hr}}{20 \text{ day}}$

Find the average speeds, given total distance and total time.

13. 160 km, 8 hr

14. 90 mi, 6 hr

15. 8.8 cm, 2 days

16. 1038 ft, 0.1 sec

Evaluate each formula using the given values.

17. $A = l \cdot w$; $l = 16 \text{ cm}$, $w = 3 \text{ cm}$
(Area of a rectangle)

18. $V = s^3$; $s = 10 \text{ ft}$
(Volume of a cube)

19. $C = 2\pi r;\ r = 4$ in.
(Circumference of a circle)

20. $A = \pi r^2;\ r = 4$ in.
(Area of a circle)

21. $s = \frac{1}{2}at^2;\ a = 5\text{ m/sec}^2, t = 10\text{ sec}$
(Distance traveled)

22. $\lambda = \dfrac{v}{f};\ v = 400\text{ m/sec}, f = 250/\text{sec}$
(Wavelength)
(*Note*: f is in cycles/sec, but cycles is not used in the formula.)

23. $S = \dfrac{I}{Mr};\ I = 324\text{ kg-cm}^2, M = 0.340\text{ kg}, r = 60\text{ cm}$
(Sweet spot of tennis racquet)

24. $\text{EBV} = w(\text{ABV});\ w = 50\text{ kg}, \text{ABV} = 6.5\text{ mL/kg}$
(Estimated blood volume)

25. $\text{KE} = \frac{1}{2}mv^2;\ m = 50\text{ kg}, v = 15\text{ m/sec}$
(Kinetic energy)

26. $v = \sqrt{\dfrac{2 \cdot \text{KE}}{m}};\ \text{KE} = 980\text{ kg-m}^2/\text{sec}^2, m = 50\text{ kg}$
(Velocity)

Make the unit changes.

27. 6.2 km to m

28. 82 cm to m

29. 35 mi/hr to ft/min

30. \$9 per hour to ¢/min

31. 1 billion sec to yr (Use 365 days = 1 yr.)

32. 2 days to sec

33. 216 in^2 to ft^2

34. 10 m^3 to cm^3

35. The speed of light is 186,000 mi/sec. What is the speed of light in mi/yr?

36. The American woodcock can fly as slowly as 5 mph. What is the speed in ft/sec?

Solve.

37. A piece of lumber is 12 ft long and 8 in. wide. How many square feet of wood is in the piece of lumber?

38. A *board foot* is an amount of wood equivalent to 1 ft square and 1 in. thick. How many board feet of wood is in an old barn timber that is 4 in. thick, 8 in. wide, and 14 ft long?

39. The number of calories needed daily for a woman of weight w, in kilograms, height h, in centimeters, and age a, in years, can be estimated by the formula

$$C = 655 + 9.6w + 1.8h - 4.7a.$$

If Tawny weighs 54.7 kg, is 1.6 m tall, and is 28 years old, what is her daily caloric requirement?

40. Dr. Thomas orders a Dopamine drip of 20 mcg/kg/min. Charmagne uses an 800-mg vial of Dopamine in 500 mL of solution for a patient weighing 70 kg. What should the drip rate of the solution be, in mL/hr? (*Hint*: 1 mg = 1000 mcg.)

For Exercises 41–44, determine which formula could be the correct one to use to find the given measure.

41. Volume of a cylinder
A. $X = \pi r^2 h$ **B.** $X = 2\pi rh + 2\pi r^2$

42. Surface area of a sphere
A. $X = \frac{4}{3}\pi r^3$ **B.** $X = 4\pi r^2$

43. Surface area of a cone
A. $X = \frac{1}{3}\pi r^2 h$ **B.** $X = \pi r^2 + \pi rs$

44. Slant height of a cone
A. $X = \pi r^2 + \pi rs$ **B.** $X = \sqrt{r^2 + h^2}$

Answers

Chapter 1

Exercise Set 1.1, pp. 12–14

1. Constant **2.** Variable **3.** Factors **4.** Evaluating **5.** Left **6.** Base; exponent **7.** Rational **8.** Irrational **9.** Terminating **10.** Repeating **11.** 25 **13.** 8 **15.** 27 **17.** 0 **19.** 5 **21.** 25 **23.** 225 **25.** 33 **27.** 7 **29.** 17.5 ft^2 **31.** 11.2 m^2 **33.** **(a)** Yes; **(b)** no; **(c)** no **35.** **(a)** No; **(b)** yes; **(c)** yes **37.** **(a)** Yes; **(b)** no; **(c)** no **39.** $\{a, e, i, o, u\}$, or $\{a, e, i, o, u, y\}$ **41.** $\{1, 3, 5, 7, ...\}$ **43.** $\{5, 10, 15, 20, ...\}$ **45.** $\{x \mid x$ is an odd number between 10 and 20$\}$ **47.** $\{x \mid x$ is a whole number less than 5$\}$ **49.** $\{n \mid n$ is a multiple of 5 between 7 and 79$\}$ **51.** **(a)** 0, 6; **(b)** -3, 0, 6; **(c)** $-8.7, -3, 0, \frac{2}{3}, 6$; **(d)** $\sqrt{7}$; **(e)** $-8.7, -3, 0, \frac{2}{3}, \sqrt{7}, 6$ **53.** **(a)** 0, 3; **(b)** -17, 0, 3; **(c)** $-17, -4.13, 0, \frac{5}{4}, 3$; **(d)** $\sqrt{77}$; **(e)** $-17, -4.13, 0, \frac{5}{4}, 3, \sqrt{77}$ **55.** False **57.** True **59.** False **61.** True **63.** True **65.** False **67.** 15 **69.** 41.4494 **71.** 25.125 **73.** 13,778 **75.** TW **77.** TW **79.** $\{0\}$ **81.** $\{5, 10, 15, 20, \ldots\}$ **83.** $\{1, 3, 5, 7, \ldots\}$ **85.**

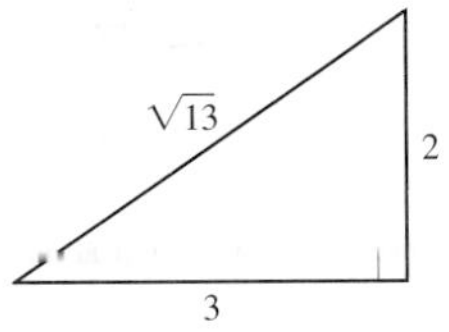

Interactive Discovery, p. 17

1. $-3, -6$ **2.** 5, 10

Exercise Set 1.2, pp. 22–24

1. True **2.** False **3.** True **4.** False **5.** False **6.** True **7.** False **8.** True **9.** True **10.** True **11.** 9 **13.** 6 **15.** 6.2 **17.** 0 **19.** $1\frac{7}{8}$ **21.** 4.21 **23.** True **25.** False **27.** True **29.** True **31.** True **33.** False **35.** 12 **37.** -12 **39.** -1.2 **41.** $-\frac{11}{35}$ **43.** -9.06 **45.** $\frac{5}{9}$ **47.** -4.5 **49.** 0 **51.** -6.4 **53.** -3.14 **55.** $4\frac{1}{3}$ **57.** 0 **59.** -9 **61.** 2.7 **63.** -1.79 **65.** 0 **67.** 3 **69.** -3 **71.** 7 **73.** -19 **75.** -3.1 **77.** $-\frac{11}{10}$ **79.** 0 **81.** 7.9 **83.** -30 **85.** 36 **87.** -21 **89.** $-\frac{3}{7}$ **91.** 0 **93.** 5.44 **95.** 5 **97.** -5 **99.** -73 **101.** 0 **103.** $\frac{1}{4}$ **105.** $-\frac{1}{9}$ **107.** $\frac{3}{2}$ **109.** $-\frac{11}{3}$ **111.** $\frac{5}{6}$ **113.** $-\frac{6}{5}$ **115.** $\frac{1}{36}$ **117.** 1 **119.** 25 **121.** $-\frac{6}{11}$ **123.** Undefined **125.** $\frac{11}{43}$ **127.** 31 **129.** -3 **131.** 15 **133.** (a) **135.** (d) **137.** TW **139.** 16; 16 **140.** 11; 11 **141.** TW **143.** $(8 - 5)^3 + 9 = 36$ **145.** $5 \cdot 2^3 \div (3 - 4)^4 = 40$ **147.** -6.2

Interactive Discovery, p. 30

1. 0, 0; 185, 160; 33, 24; not equivalent **2.** $-30, -30$; 70, 70; $-90, -90$; equivalent

Exercise Set 1.3, pp. 31–32

1. Commutative law **2.** Distributive law **3.** Commutative law **4.** Associative law **5.** Distributive law **6.** Associative law **7.** $7b + 4a$; $a4 + b7$; $4a + b7$ **9.** $y(7x)$; $(x7)y$ **11.** $3(xy)$ **13.** $(x + 2y) + 5$ **15.** $7t + 14$ **17.** $4x - 4y$ **19.** $-10a - 15b$ **21.** $9ab - 9ac + 9ad$ **23.** $5(x + 10)$ **25.** $3(p - 3)$ **27.** $7(x - 3y + 2z)$ **29.** $17(15 - 2b)$ **31.** $x(y + 1)$ **33.** $4x, -5y, 3$ **35.** $x^2, -6x, -7$ **37.** $10x$ **39.** $-2rt$ **41.** $10t^2$ **43.** $11a$ **45.** $-7n$ **47.** $10x$ **49.** $7x - 2x^2$ **51.** $5a + 11a^2$ **53.** $22x + 18$ **55.** $-5t^2 + 2t + 4t^3$ **57.** $5a - 5$ **59.** 1 **61.** $5d - 12$ **63.** $-7x + 14$ **65.** $-10x + 21$ **67.** $44a - 22$ **69.** $-100a - 90$ **71.** $-12y - 145$ **73.** First expression: $-16.3, -12, -15$; second expression: 13.7, 18, 15; not equivalent **75.** First expression: 17.2, 0, 12; second expression: 17.2, 0, 12; equivalent **77.** TW **79.** 3 **80.** 3.59 **81.** 35 **82.** $-\frac{1}{35}$

83. TW **85.** $23a - 18b + 184$ **87.** $-4z$ **89.** $-x + 19$ **91.** TW

Exercise Set 1.4, pp. 45–47

1. The power rule **2.** Raising a quotient to a power **3.** Raising a product to a power **4.** The quotient rule **5.** The product rule **6.** The power rule **7.** Raising a quotient to a power **8.** Raising a product to a power **9.** The quotient rule **10.** The product rule **11.** Positive power of 10 **12.** Negative power of 10 **13.** Negative power of 10 **14.** Positive power of 10 **15.** Positive power of 10 **16.** Negative power of 10 **17.** 5^{10} **19.** m^9 **21.** $18x^7$ **23.** $16m^{13}$ **25.** $x^{10}y^{10}$ **27.** a^6 **29.** $3t^5$ **31.** $4x^6y^4$ **33.** $-4x^8y^6z^6$ **35.** -1 **37.** 1 **39.** 16 **41.** -16 **43.** $\frac{1}{16}$ **45.** $-\frac{1}{16}$ **47.** -1 **49.** $\frac{1}{n^6}$ **51.** 64 **53.** $\frac{7}{x^3}$ **55.** $\frac{5b^4}{a^7}$ **57.** $\frac{x^3}{y^5}$ **59.** $\frac{x^2y^4}{z^3}$ **61.** $(-5)^{-6}$ **63.** $\frac{4}{x^{-2}}$ **65.** $(5y)^{-3}$ **67.** 8^{-6}, or $\frac{1}{8^6}$ **69.** b^{-3}, or $\frac{1}{b^3}$ **71.** a^3 **73.** $10a^{-6}b^{-2}$, or $\frac{10}{a^6b^2}$ **75.** 10^{-9}, or $\frac{1}{10^9}$ **77.** 2^{-2}, or $\frac{1}{2^2}$, or $\frac{1}{4}$ **79.** y^9 **81.** $-3ab^2$ **83.** $-\frac{1}{4}x^3y^{-2}z^{11}$, or $-\frac{x^3z^{11}}{4y^2}$ **85.** x^{12} **87.** 9^{-12}, or $\frac{1}{9^{12}}$ **89.** t^{40} **91.** $25x^8y^2$ **93.** x^{-8}, or $\frac{1}{x^8}$ **95.** $32a^{-4}$, or $\frac{32}{a^4}$ **97.** 1 **99.** $\frac{9x^8y^9}{2}$ **101.** $\frac{625}{256}x^{-20}y^{24}$, or $\frac{625y^{24}}{256x^{20}}$ **103.** 1 **105.** -4096 **107.** 0.0625 **109.** 0.648 **111.** 0.0004 **113.** 673,000,000 **115.** 0.0000000008923 **117.** 90,300,000,000 **119.** 4.7×10^{10} **121.** 1.6×10^{-8} **123.** 4.07×10^{11} **125.** 6.03×10^{-7} **127.** 5.02×10^{18} **129.** -3.05×10^{-10} **131.** 9.7×10^{-5} **133.** 1.3×10^{-11} **135.** 6.0 **137.** 1.5×10^3 **139.** 3.0×10^{-5} **141.** 1.79×10^{20} **143.** 12.2×10^{23}, or 1.22×10^{24} **145.** TW **147.** $-\frac{1}{12}$ **148.** 30.96 **149.** $-8x + 12y$ **150.** $2(4x - 5)$ **151.** TW **153.** TW **155.** $-3x^{2a-1}$ **157.** $12x^{6b-2ab}$ **159.** $-5x^{2b}y^{-2a}$ **161.** 1.25×10^{22} **163.** 1 **165.** 8×10^{18} grains

Exercise Set 1.5, pp. 57–59

1. Negative **2.** Negative **3.** Axes **4.** Ordered **5.** Solutions **6.** Linear **7.** (5, 3), (−4, 3), (0, 2), (−2, −3), (4, −2), and (−5, 0)

9.

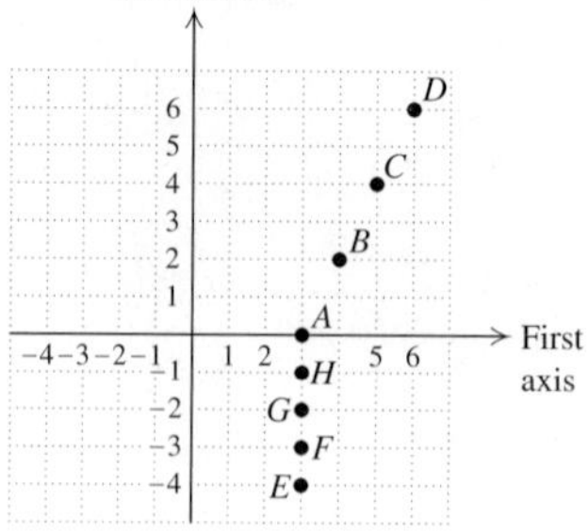

11. Triangle, 21 units^2

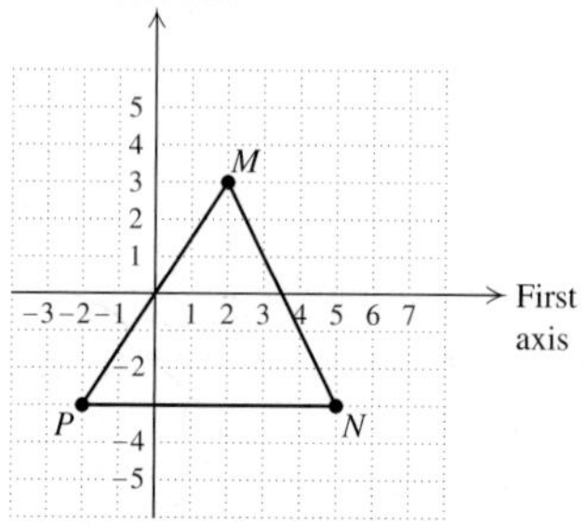

13. II **15.** III **17.** I **19.** IV **21.** Yes **23.** No **25.** Yes **27.** Yes **29.** Yes **31.** Yes **33.** No **35.** No

37.

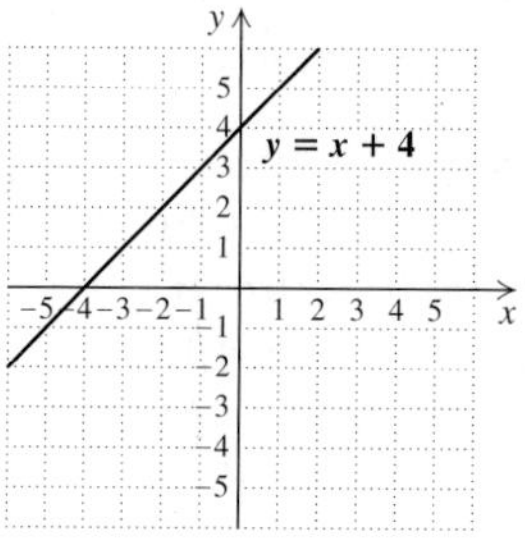

39.

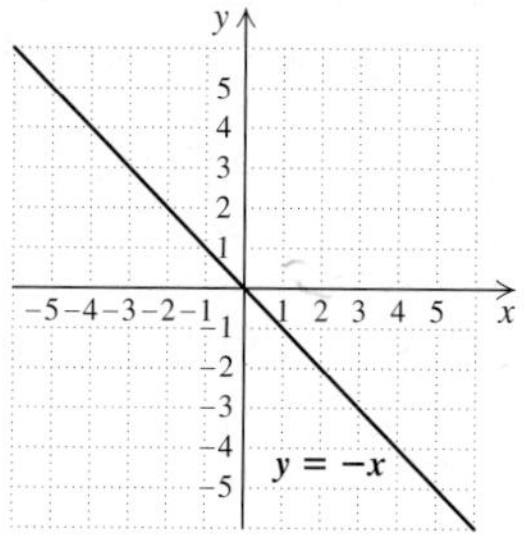

41.

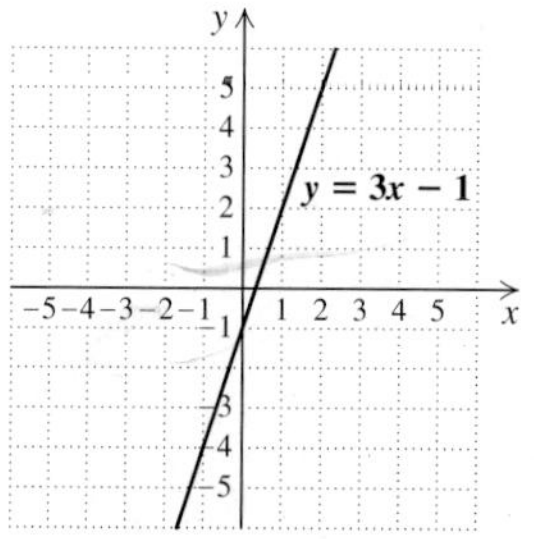

43.

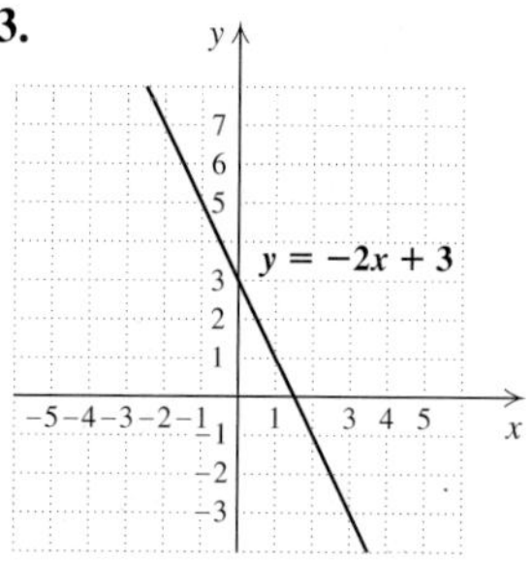

45.

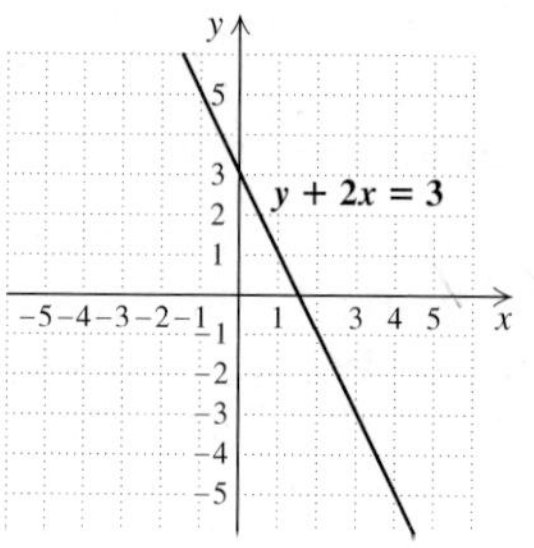

47.

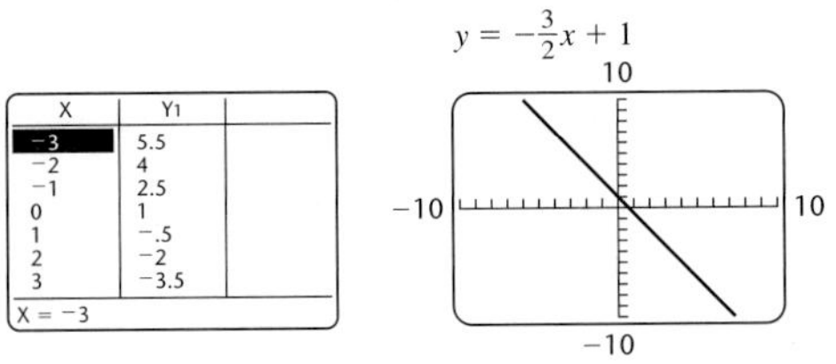

49.

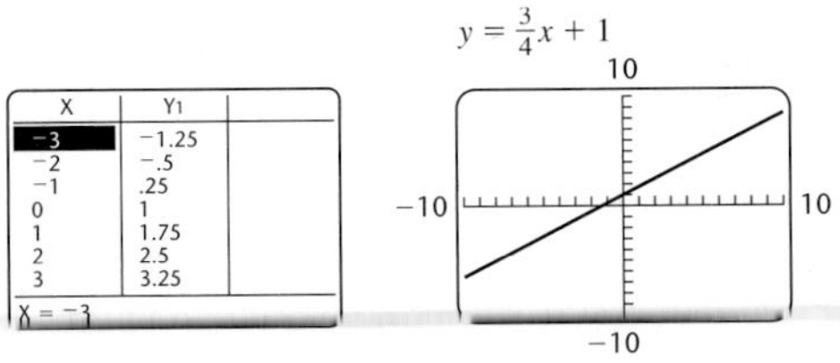

51.

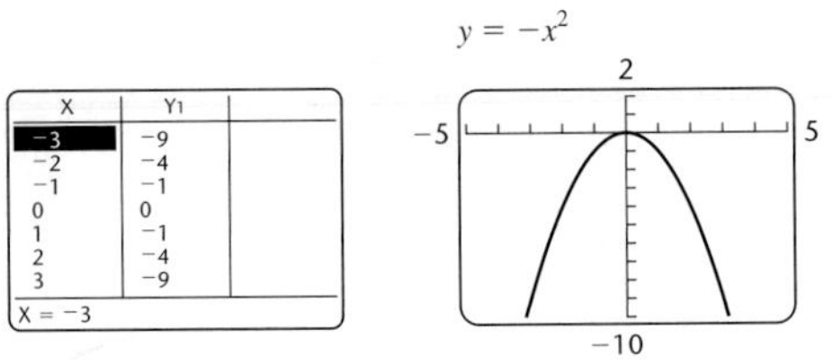

53.

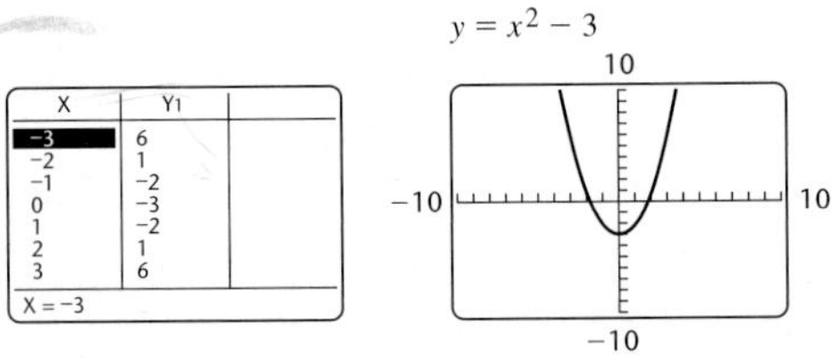

55.

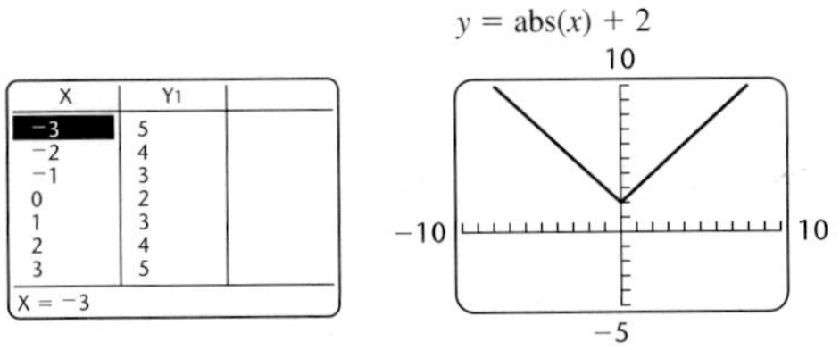

57.

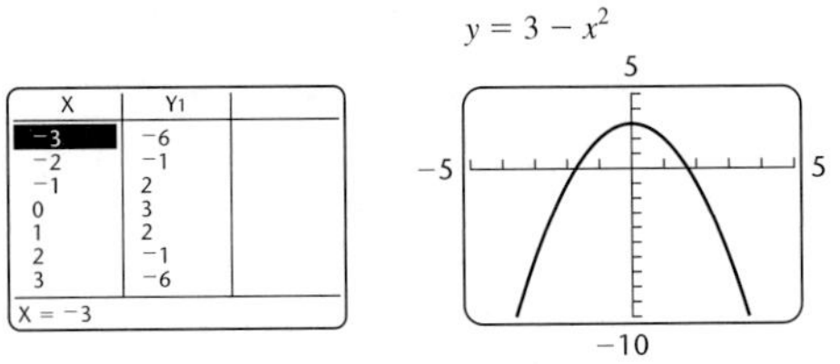

59.

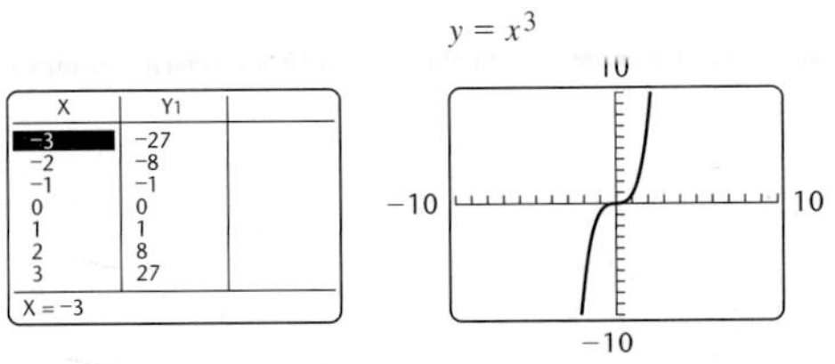

61. (b) **63.** (a) **65.** (b) **67.** Exercises 37–49 and 61 are linear. **69.** TW **71.** Yes **72.** Yes **73.** No **74.** No **75.** −2 **76.** 49 **77.** 0 **78.** −3 **79.** TW

81. TW **83.** (a), (d) **85.** (−1, −2), (−19, −2), or (13, 10) **87.** Equations (a), (c), and (d) appear to be linear.

Exercise Set 1.6, pp. 68–71

1. Linear **2.** Contradiction **3.** Equation **4.** Area **5.** Circumference **6.** $P = 2l + 2w$ **7.** $A = bh$ **8.** Length **9.** Subscripts **10.** Factor **11.** Equivalent **13.** Not equivalent **15.** Not equivalent **17.** 16.3 **19.** 9 **21.** 18 **23.** 25 **25.** 24 **27.** 6 **29.** 12 **31.** 4 **33.** 3 **35.** 3 **37.** 7 **39.** 5 **41.** 2 **43.** 2 **45.** $\frac{49}{9}$ **47.** $\frac{4}{5}$ **49.** $\frac{19}{5}$ **51.** $-\frac{1}{2}$ **53.** $\varnothing$; contradiction **55.** $\{0\}$; conditional **57.** $\varnothing$; contradiction **59.** $\mathbb{R}$; identity **61.** $r = \dfrac{d}{t}$ **63.** $a = \dfrac{F}{m}$ **65.** $I = \dfrac{W}{E}$ **67.** $h = \dfrac{V}{lw}$ **69.** $k = Ld^2$ **71.** $n = \dfrac{G - w}{150}$ **73.** $l = p - 2w - 2h$ **75.** $y = \dfrac{C - Ax}{B}$ **77.** $F = \frac{9}{5}C + 32$ **79.** $b_2 = \dfrac{2A}{h} - b_1$ **81.** $t = \dfrac{d_2 - d_1}{v}$ **83.** $d_1 = d_2 - vt$ **85.** $m = \dfrac{r}{1 + np}$ **87.** $a = \dfrac{y}{b - c^2}$ **89.** $y = \dfrac{3 - x}{2}$ **91.** $y = x + 7$ **93.** $y = -3x - 10$ **95.** $y = \dfrac{-x^2 + x + 1}{4}$ **97.** 12% **99.** 6 cm **101.** About 8.5 cm **103.** TW **105.** −24 **106.** −12 **107.** −4 **108.** −7 **109.** TW **111.** −4.176190476 **113.** 8 **115.** $\frac{224}{29}$ **117.** TW **119.** $l = \dfrac{A - w^2}{4w}$ **121.** $T_1 = \dfrac{P_1V_1T_2}{P_2V_2}$ **123.** $d = \dfrac{me^2}{f}$ **125.** $t = \dfrac{1}{s}$

127.

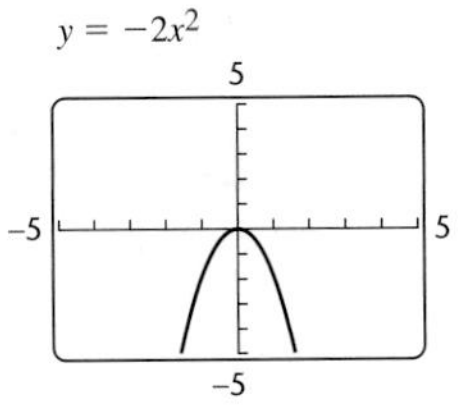

129.

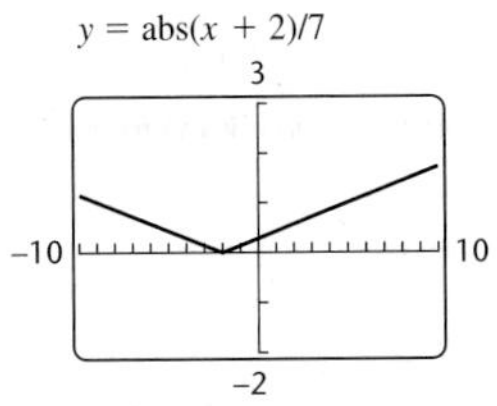

Visualizing the Graph, p. 85

1. E **2.** I **3.** C **4.** G **5.** A **6.** H **7.** D **8.** B **9.** J **10.** F

Exercise Set 1.7, pp. 86–92

1. Let n represent the number; $n - 6$
3. Let t represent the number; $12t$
5. Let x represent the number; $0.65x$, or $\frac{65}{100}x$
7. Let y represent the number; $2y + 9$
9. Let s represent the number; $0.1s + 8$, or $\frac{10}{100}s + 8$
11. Let m and n represent the numbers; $m - n - 1$
13. $90 \div 4$, or $\frac{90}{4}$ **15.** Let x and $x + 7$ represent the numbers; $x + (x + 7) = 65$ **17.** Let t represent the time, in hours, it will take the swimmer to swim 1.8 km upstream; $2.7t = 1.8$ **19.** Let t represent the time, in seconds, it takes Alida to walk the length of the sidewalk; $9t = 300$ **21.** Let x, $x + 1$, and $x + 2$ represent the angle measures; $x + (x + 1) + (x + 2) = 180$ **23.** Let w represent the wholesale price; $w + 0.5w + 1.50 = 22.50$ **25.** Let t represent the number of minutes spent climbing; $8000 + 3500t = 29{,}000$ **27.** Let x represent the measure of the second angle, in degrees; $3x + x + (2x - 12) = 180$
29. Let x represent the first even number; $2x + 3(x + 2) = 76$ **31.** Let s represent the length, in centimeters, of a side of the smaller triangle; $3s + 3 \cdot 2s = 90$
33. Let c represent Brian's calls on his next shift; $\dfrac{5 + 2 + 1 + 3 + c}{5} = 3$ **35.** \$97 **37.** Approximately 11.15 million cases **39.** 22 tickets **41.** Length: 45 cm; width: 15 cm **43.** Length: 52 m; width: 13 m **45.** $\frac{2}{3}$ hr
47. 100°, 25°, 55° **49.** \$150 **51.** \$14.00
53. 2.0×10^9 neutrinos **55.** $4.2 \times 10^1\ \text{m}^3$
57. 7.9×10^7 bacteria **59.** Approximately $5 \times 10^2\ \text{in}^3$, or $3 \times 10^{-1}\ \text{ft}^3$ **61.** 75 heart attacks per 10,000 men
63. About \$2.15 per gallon **65.** In August 2005
67. 60 watts; 140 watts

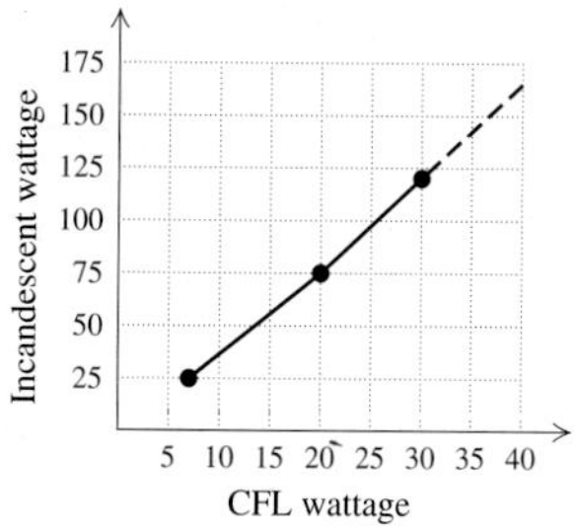

69.

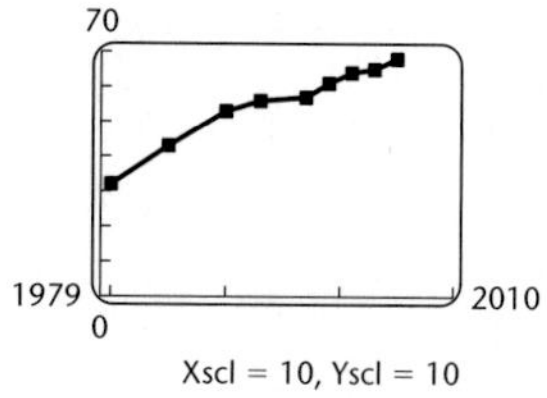

71.

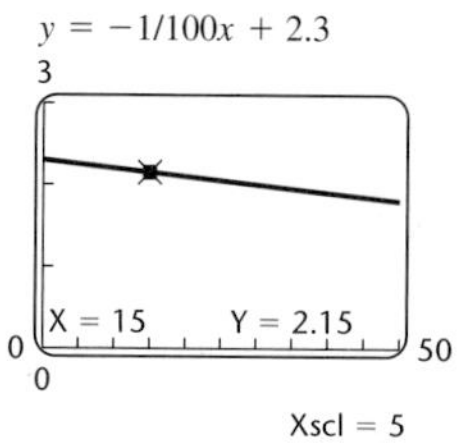

Approximately 2.15 million

73.

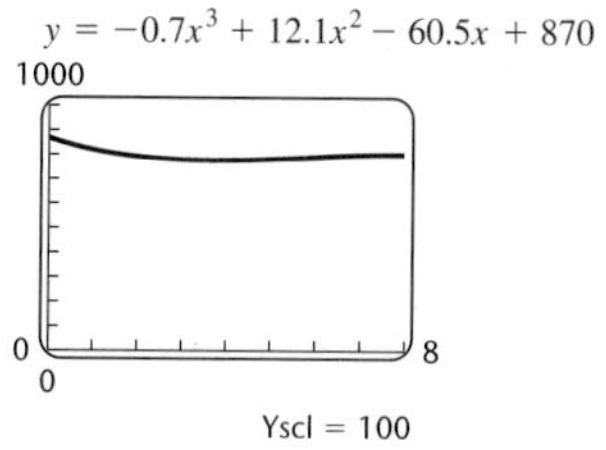

Approximately 1.6 yr and 7.1 yr after 2000, or in 2001 and in 2007

75. TW **77.** 16 **78.** −16 **79.** z^4 **80.** z^{-8}, or $\dfrac{1}{z^8}$
81. $2x^8y^{12}$ **82.** $16x^8y^{12}$ **83.** TW
85. Let a and b represent the numbers; $\dfrac{a + b}{a - b}$
87. Let r and s represent the numbers; $\dfrac{1}{2}(r^2 - s^2)$, or $\dfrac{r^2 - s^2}{2}$
89. 10 points **91.** \$110,000 **93.** About 2000 cigarettes per capita **95.** It began to decrease. **97.** By 2030
99. **(a)** IV; **(b)** III; **(c)** I; **(d)** II

Review Exercises: Chapter 1, pp. 96–98

1. (e) **2.** (g) **3.** (j) **4.** (a) **5.** (i) **6.** (b) **7.** (f)
8. (c) **9.** (d) **10.** (h) **11.** 22 **12.** $\{1, 3, 5, 7, 9\}$; $\{x \mid x \text{ is an odd natural number between 0 and 10}\}$
13. 1750 sq cm **14.** **(a)** No; **(b)** yes **15.** **(a)** Yes; **(b)** yes
16. 9.3 **17.** 4.09 **18.** 0 **19.** −10.2 **20.** $-\frac{23}{35}$
21. $\frac{7}{15}$ **22.** −11.5 **23.** $-\frac{1}{6}$ **24.** −5.4 **25.** 12.6
26. $-\frac{5}{12}$ **27.** −9.1 **28.** $-\frac{21}{4}$ **29.** 4.01 **30.** $a + 9$
31. $y \cdot 7$ **32.** $x \cdot 5 + y$, or $y + 5x$ **33.** $4 + (a + b)$
34. $x(y \cdot 3)$ **35.** $7m(n + 2)$ **36.** $4x^3 - 6x^2 + 5$
37. $47x - 60$ **38.** $-10a^5b^8$ **39.** $4xy^6$
40. 1; 28.09; −28.09 **41.** 3^3, or 27 **42.** $125a^6$
43. $-\dfrac{a^9}{8b^6}$ **44.** $\dfrac{z^8}{x^4y^6}$ **45.** $\dfrac{b^{16}}{16a^{20}}$ **46.** $-\frac{16}{7}$
47. 0 **48.** 1.03×10^{-7} **49.** 3.086×10^{13}
50. 3.7×10^7 **51.** 8.0×10^{-6} **52.** Yes **53.** No
54. Yes **55.** No

56.

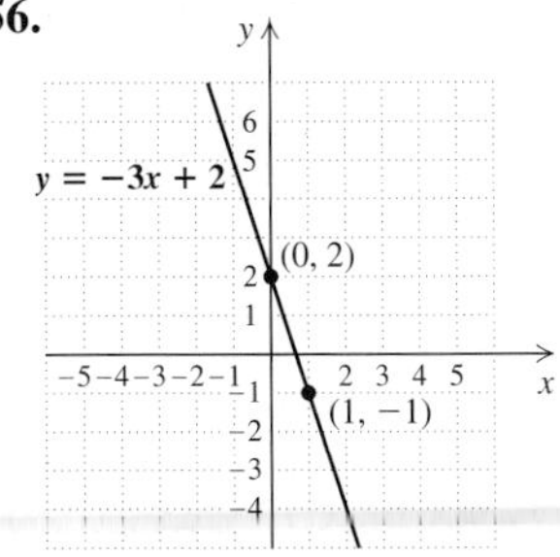

57.

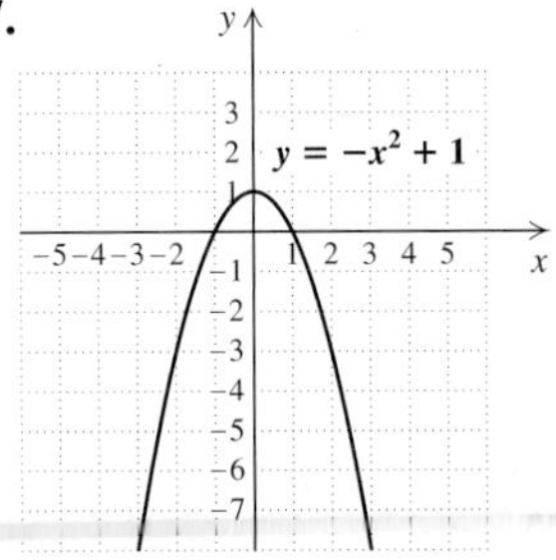

58.

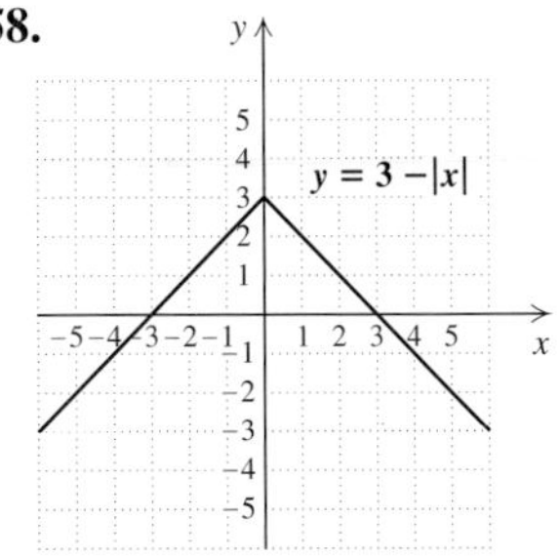

59.

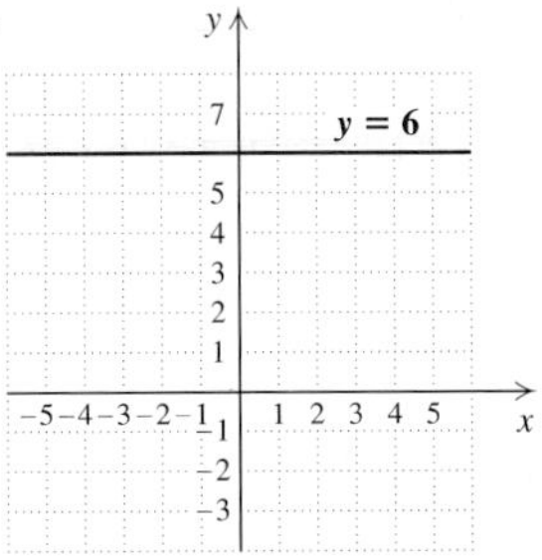

60. 3, 1, −1, −3, −1, 1, 3

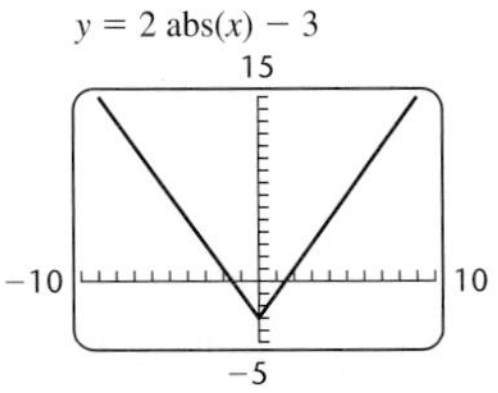

61. 11.6 **62.** $\frac{27}{2}$ **63.** $-\frac{4}{11}$ **64.** ℝ; identity

65. ∅; contradiction **66.** $m = PS$ **67.** $x = \dfrac{c}{m - r}$

68. Let x represent the number; $2x + 15 = 21$ **69.** 48 **70.** 90°, 30°, 60° **71.** 4 cm **72.** 1.4×10^4 mm³, or 1.4×10^{-5} m³ **73.** **(a)** About 14 storms; **(b)** 1997; **(c)** 2005; **(d)** There were about 21 more storms in 2005 than there were in 1997.

74.

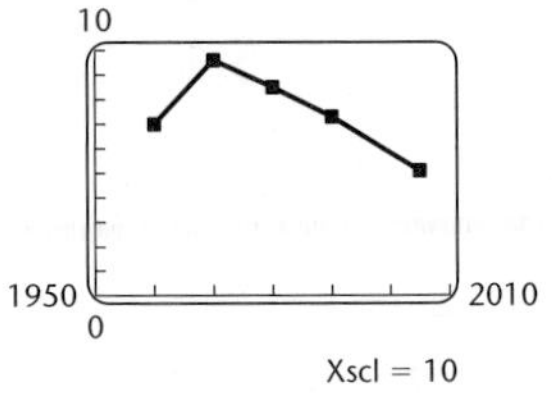

75. TW To write an equation that has no solution, begin with a simple equation that is false for any value of x, such as $x = x + 1$. Then add or multiply by the same quantities on both sides of the equation to construct a more complicated equation with no solution. **76.** TW Use the distributive law to rewrite a sum of like terms as a single term by first writing the sum as a product. For example, $2a + 5a = (2 + 5)a = 7a$. **77.** 0.0000003% **78.** $\frac{25}{24}$ **79.** The 17-in. pizza is a better deal. It costs about 5¢ per square inch; the 13-in. pizza costs about 6¢ per square inch. **80.** 729 cm³

81. $z = y - \dfrac{x}{m}$ **82.** $3^{-2a+2b-8ab}$ **83.** $88.\overline{3}$ **84.** −39

85. $-40x$ **86.** $a \cdot 2 + cb + cd + ad = ad + a \cdot 2 + cb + cd = a(d + 2) + c(b + d)$

87. $\sqrt{5}/4$; answers may vary

Test Chapter 1, pp. 99–100

1. [1.2] −17 **2.** [1.1] 3.75 cm² **3.** [1.1] **(a)** No, **(b)** yes, **(c)** no **4.** [1.2] −41 **5.** [1.2] −3.7 **6.** [1.2] −2.11 **7.** [1.2] −14.2 **8.** [1.2] −43.2 **9.** [1.2] −33.92 **10.** [1.2] $-\frac{19}{12}$ **11.** [1.2] $\frac{5}{49}$ **12.** [1.2] 6 **13.** [1.2] $-\frac{4}{3}$ **14.** [1.2] $-\frac{3}{2}$ **15.** [1.3] $y + 7x$, or $x \cdot 7 + y$, or $y + x \cdot 7$ **16.** [1.3] $-3y - 29$ **17.** [1.3] $3x + 8$

18. [1.4] $-\dfrac{72}{x^{10}y^6}$ **19.** [1.4] $-\frac{1}{9}$ **20.** [1.4] $\dfrac{y^8}{36x^4}$

21. [1.4] $\dfrac{x^6}{4y^8}$ **22.** [1.4] 1 **23.** [1.4] 2.01×10^{-7}

24. [1.4] 2.0×10^{10} **25.** [1.4] 3.05×10^4 **26.** [1.5] Yes **27.** [1.5] No

28. [1.5]

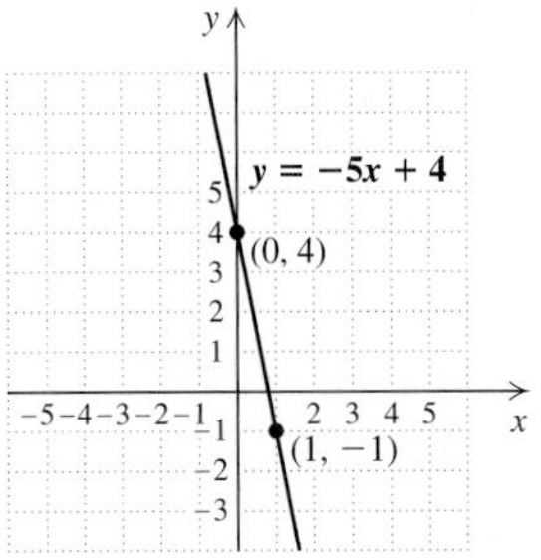

29. [1.5]

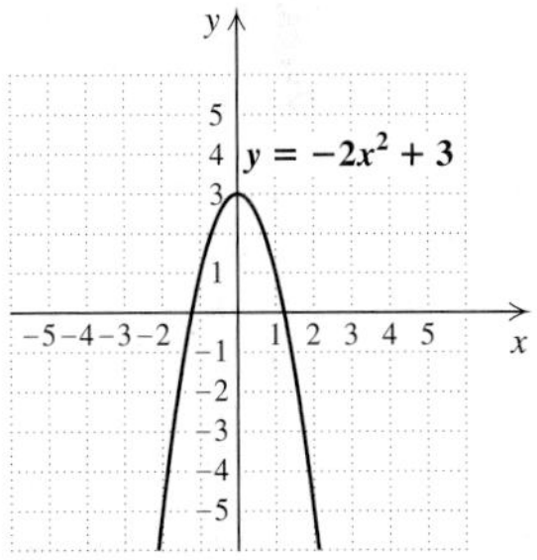

30. [1.5]

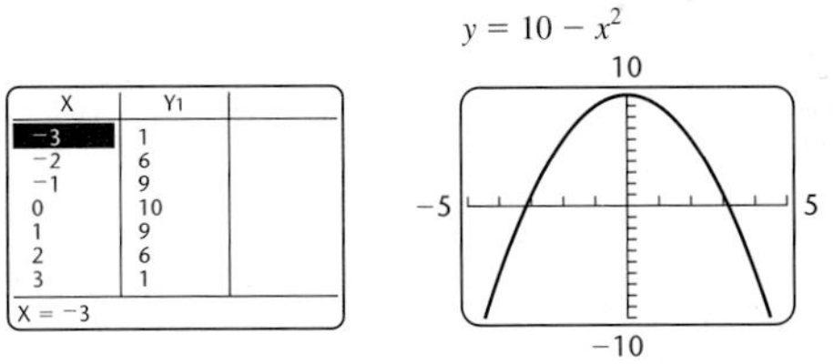

31. [1.6] −2 **32.** [1.6] ℝ; identity **33.** [1.6] $P_2 = \dfrac{P_1V_1T_2}{T_1V_2}$

34. [1.7] Let m and n represent the numbers; $mn + 3$, or $3 + mn$ **35.** [1.7] 94 **36.** [1.7] 17, 19, 21 **37.** [1.7] 9.4×10^8 km **38.** [1.7] 45 mph **39.** [1.7] 7 mpg

40. [1.4] $16^c x^{6ac} y^{2bc+2c}$ **41.** [1.4] $-9a^3$ **42.** [1.4] $\dfrac{4}{7y^2}$

Chapter 2

Exercise Set 2.1, pp. 112–116

1. Domain **2.** Range **3.** Exactly **4.** More **5.** Horizontal **6.** Vertical **7.** "f of 3," "f at 3," or "the value of f at 3" **8.** Function **9.** Yes **11.** Yes **13.** Yes **15.** No **17.** Function **19.** Function **21.** **(a)** -2; **(b)** $\{x \mid -2 \le x \le 5\}$; **(c)** 4; **(d)** $\{y \mid -3 \le y \le 4\}$ **23.** **(a)** 3; **(b)** $\{x \mid -1 \le x \le 4\}$; **(c)** 3; **(d)** $\{y \mid 1 \le y \le 4\}$ **25.** **(a)** -2; **(b)** $\{x \mid -4 \le x \le 2\}$; **(c)** -2; **(d)** $\{y \mid -3 \le y \le 3\}$ **27.** **(a)** 3; **(b)** $\{x \mid -4 \le x \le 3\}$; **(c)** -3; **(d)** $\{y \mid -2 \le y \le 5\}$ **29.** **(a)** 1; **(b)** $\{-3, -1, 1, 3, 5\}$; **(c)** 3; **(d)** $\{-1, 0, 1, 2, 3\}$ **31.** **(a)** 4; **(b)** $\{x \mid -3 \le x \le 4\}$; **(c)** -1, 3; **(d)** $\{y \mid -4 \le y \le 5\}$ **33.** **(a)** 1; **(b)** $\{x \mid -4 < x \le 5\}$; **(c)** $\{x \mid 2 < x \le 5\}$; **(d)** $\{-1, 1, 2\}$ **35.** Yes **37.** Yes **39.** No **41.** No **43.** **(a)** 3; **(b)** -5; **(c)** -11; **(d)** 19; **(e)** $2a + 7$; **(f)** $2a + 5$ **45.** **(a)** 0; **(b)** 1; **(c)** 57; **(d)** $5t^2 + 4t$; **(e)** $20a^2 + 8a$; **(f)** $10a^2 + 8a$ **47.** **(a)** $\frac{3}{5}$; **(b)** $\frac{1}{3}$; **(c)** $\frac{4}{7}$; **(d)** 0; **(e)** $\dfrac{x-1}{2x-1}$

49. $4\sqrt{3}\text{ cm}^2 \approx 6.93\text{ cm}^2$ **51.** $36\pi\text{ in}^2 \approx 113.10\text{ in}^2$

53. $1\frac{20}{33}$ atm; $1\frac{10}{11}$ atm; $4\frac{1}{33}$ atm **55.** 11 **57.** 0 **59.** $-\frac{21}{2}$ **61.** $\frac{25}{6}$ **63.** -3 **65.** -25 **67.** -2 **69.** None **71.** -2, 2 **73.** 5 **75.** -20 **77.** 2.7 **79.** $-\frac{7}{3}$ **81.** TW

83.

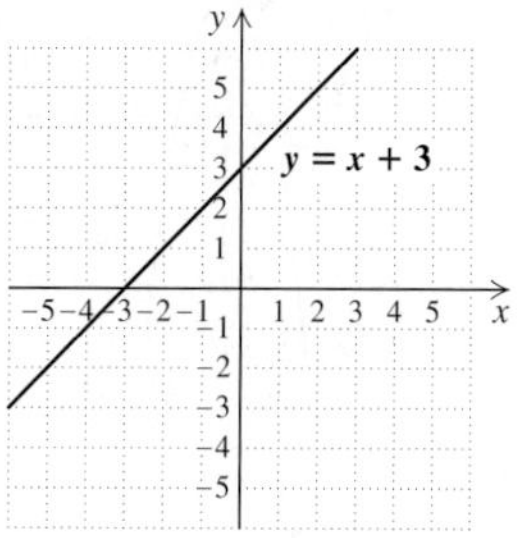

84.

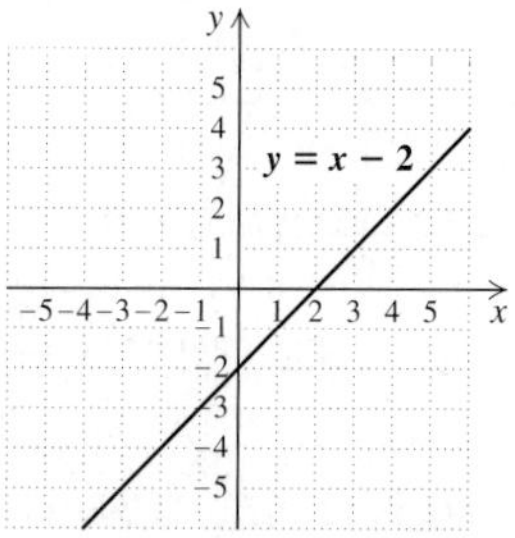

85.

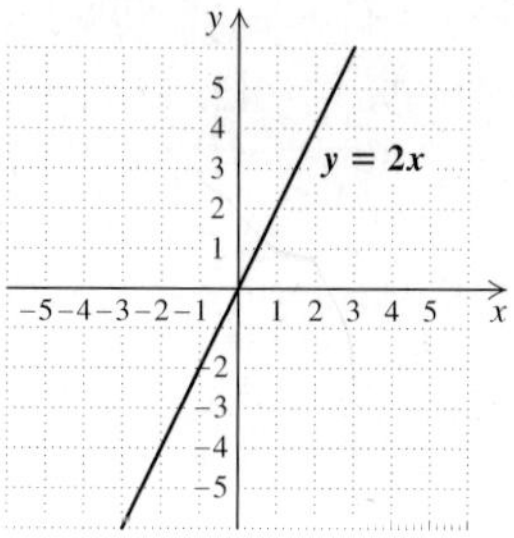

86.

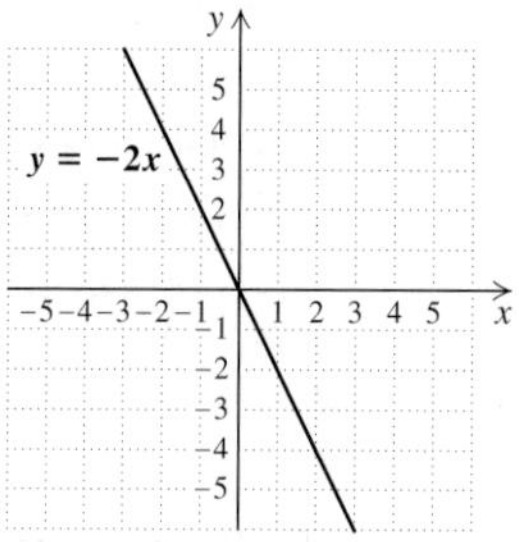

87.

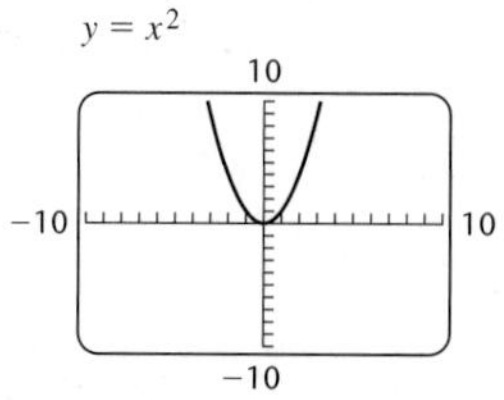

88.

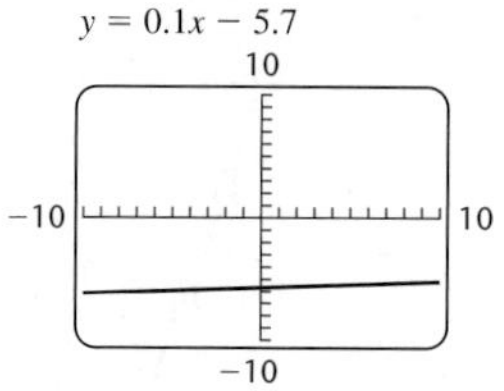

89.

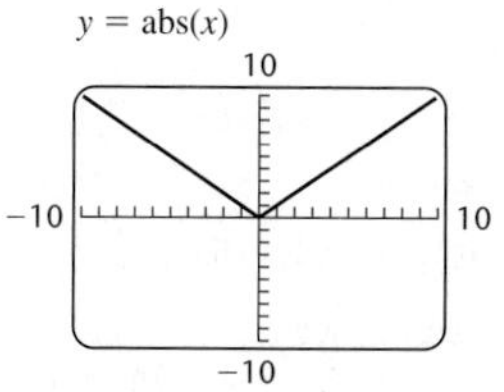

90.

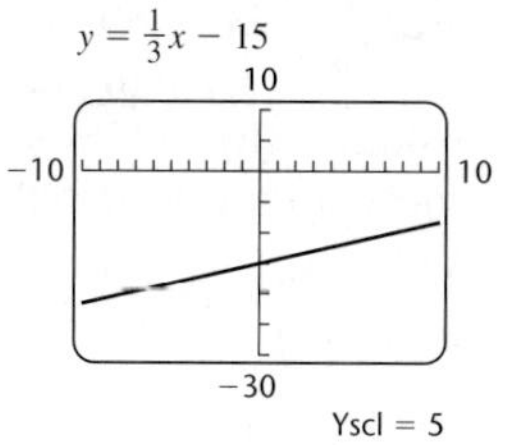

91. TW **93.** 26; 99 **95.** Worm **97.** About 2 min 50 sec **99.** 1 every 3 min **101.** $g(x) = \dfrac{15}{4}x - \dfrac{13}{4}$

Interactive Discovery, p. 117

1. Yes **2.** The origin, or $(0, 0)$ **3.** The origin, or $(0, 0)$ **4.** Yes **5.** Yes

Interactive Discovery, p. 118

1. The value of y_2 is 5 more than that of y_1 for the same value of x. The value of y_3 is 7 less than that of y_1. **2.** The graph of y_4 will look like the graph of y_1, shifted down 3.2 units. **3.** The graph is shifted up or down, depending on the sign of b.

Interactive Discovery, p. 122

1. y_1 and y_3 **2.** y_2 and y_4 **3.** y_3 **4.** The graph slants up if m is positive and down if m is negative. **5.** The larger m is, the steeper the graph.

Exercise Set 2.2, pp. 127–132

1. (e) **2.** (d) **3.** (c) **4.** (f) **5.** (a) **6.** (b)

7.

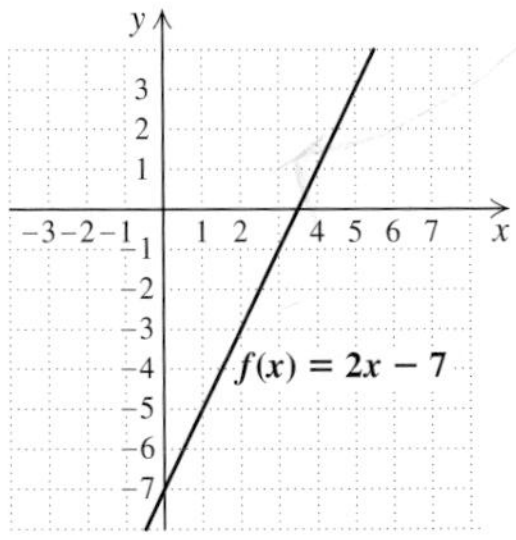

9.

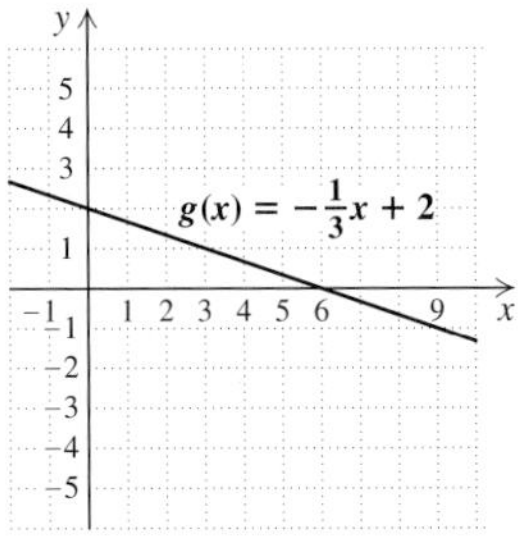

11.

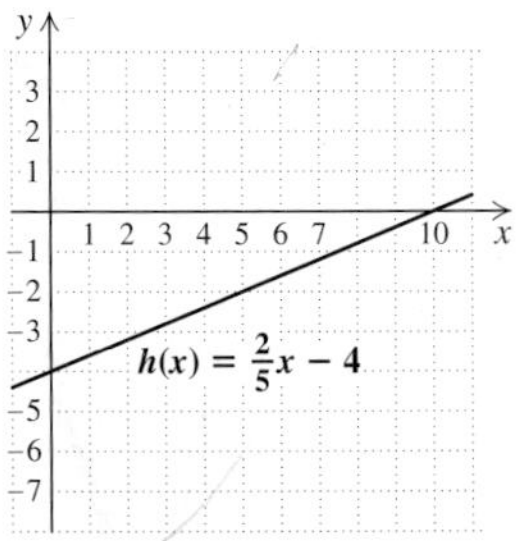

13. (0, 4) **15.** (0, −3) **17.** (0, −4.5) **19.** (0, −9) **21.** (0, 204) **23.** 2 **25.** −2 **27.** $-\frac{9}{4}$ **29.** 0

31. **(a)** II; **(b)** IV; **(c)** III; **(d)** I

33. Slope $\frac{5}{2}$; y-intercept: (0, 3)

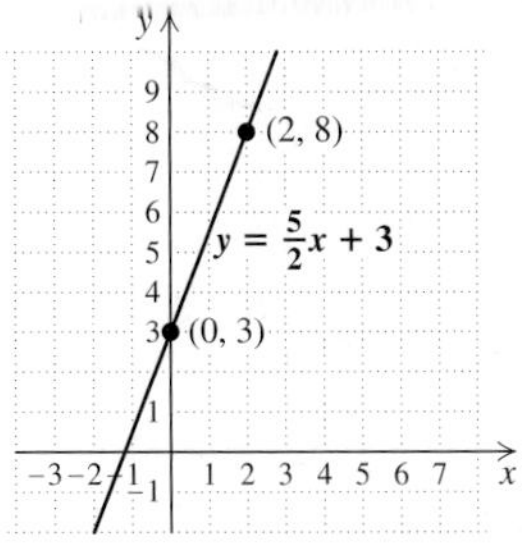

35. Slope $-\frac{5}{2}$; y-intercept: (0, 2)

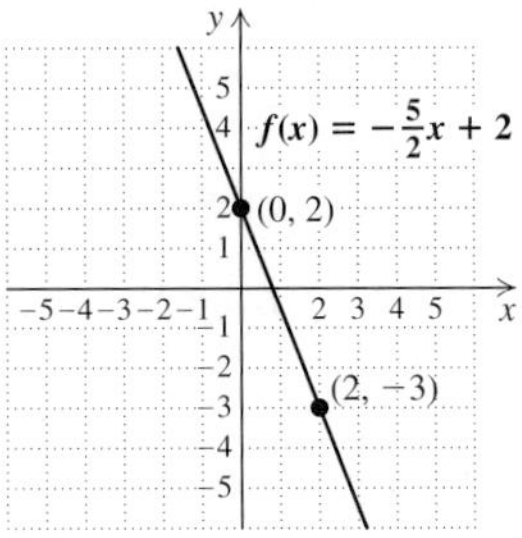

37. Slope 2; y-intercept: (0, −5)

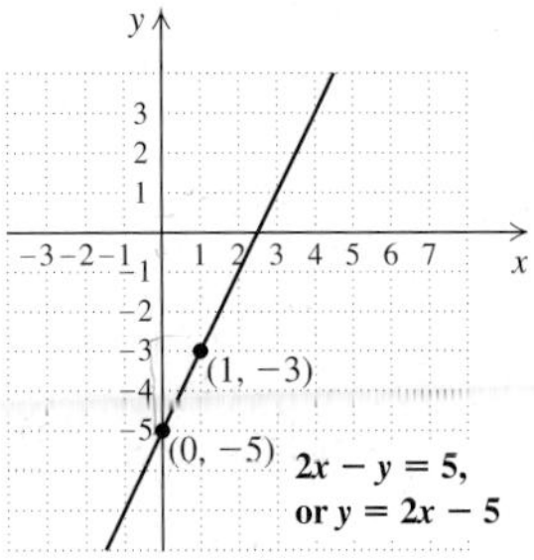

39. Slope $\frac{1}{3}$; y-intercept: (0, 2)

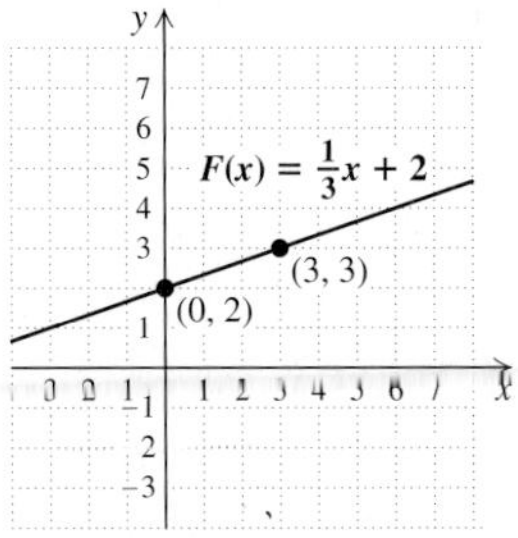

41. Slope $-\frac{2}{3}$; y-intercept: (0, 1)

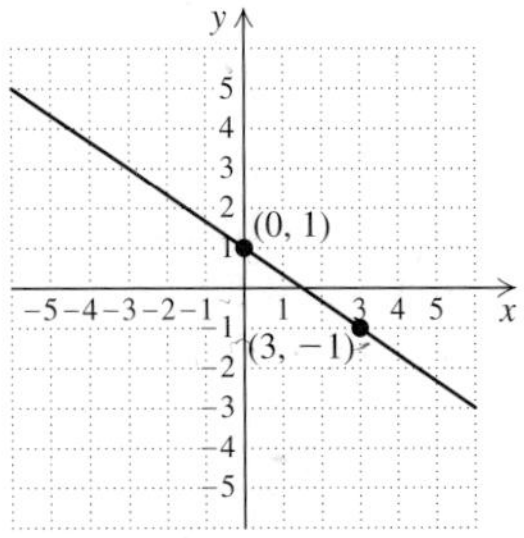

$6y + 4x = 6$, or $y = -\frac{2}{3}x + 1$

43. Slope −0.25; y-intercept: (0, 0)

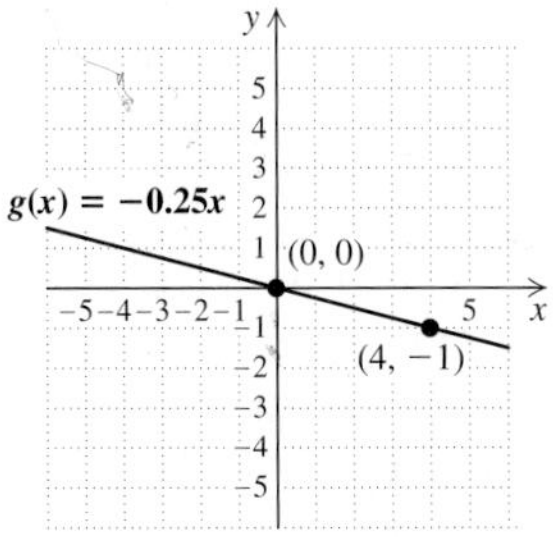

45. Slope $\frac{4}{5}$; y-intercept: (0, −2)

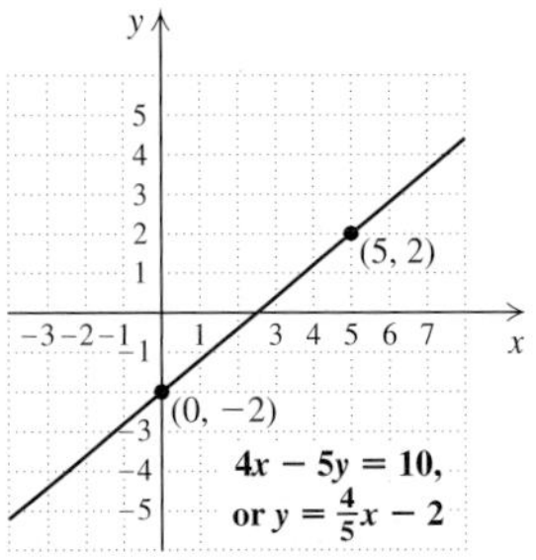

47. Slope $\frac{5}{4}$; y-intercept: (0, −2)

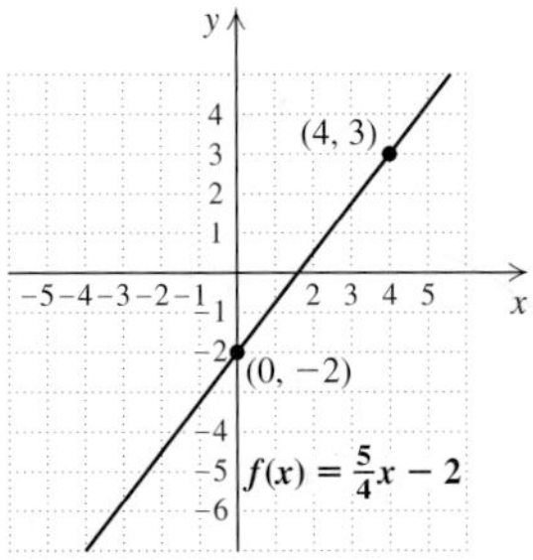

49. Slope $-\frac{3}{4}$; y-intercept: (0, 3)

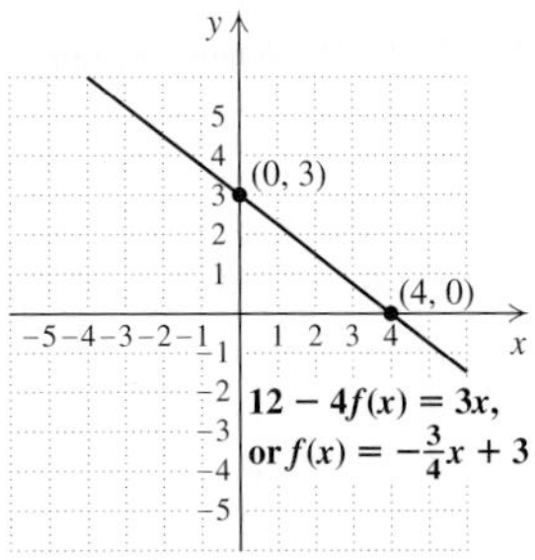

51. Slope 0; y-intercept: (0, 4.5)

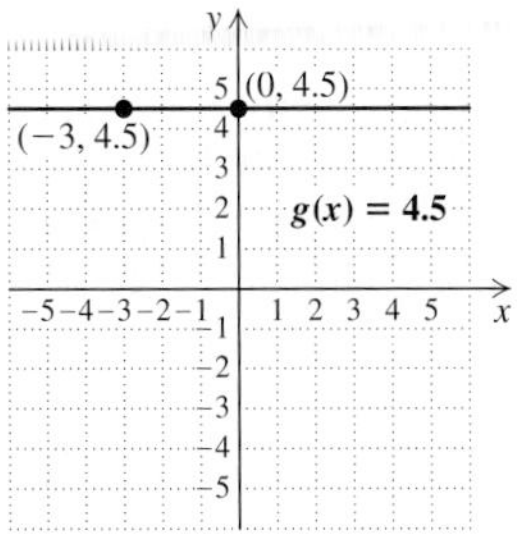

53. $f(x) = \frac{2}{3}x - 9$ **55.** $f(x) = -6x + 2$ **57.** $f(x) = -\frac{7}{9}x + 5$ **59.** $f(x) = 5x + \frac{1}{2}$ **61.** The distance from the finish line is decreasing at a rate of $6\frac{2}{3}$ m per second.

63. The distance is increasing at a rate of 3 mi per hour. **65.** The number of pages read is increasing at a rate of 5 pages per day. **67.** The average SAT math score is increasing at a rate of 1 point per thousand dollars of family income. **69.** 7.5 mph **71.** 300 ft/min
73. 0.085 min/yr **75.** **(a)** II; **(b)** IV; **(c)** I; **(d)** III
77. 25 signifies that the cost per person is \$25; 75 signifies that the setup cost for the party is \$75. **79.** $\frac{1}{2}$ signifies that Ty's hair grows $\frac{1}{2}$ in. per month; 1 signifies that his hair is 1 in. long when cut. **81.** $\frac{1}{7}$ signifies that the life expectancy of American women increases $\frac{1}{7}$ yr per year, for years after 1970; 75.5 signifies that the life expectancy in 1970 was 75.5 yr. **83.** 0.227 signifies that the price increases \$0.227 per year, for years since 1995; 4.29 signifies that the average cost of a movie ticket in 1995 was \$4.29. **85.** 2 signifies that the cost per mile of a taxi ride is \$2; 2.5 signifies that the minimum cost of a taxi ride is \$2.50. **87.** **(a)** -5000 signifies that the depreciation is \$5000 per year; 90,000 signifies that the original value of the truck was \$90,000; **(b)** 18 yr; **(c)** $\{t \mid 0 \le t \le 18\}$ **89.** **(a)** -150 signifies that the depreciation is \$150 per winter of use; 900 signifies that the original value of the snowblower was \$900; **(b)** after 4 winters of use; **(c)** $\{n \mid 0 \le n \le 6\}$ **91.** TW **93.** $45x + 54$
94. $-125a^6b^9$ **95.** $-26m^7n^4$ **96.** $8x - 3y - 12$
97. $4x^2y^3$ **98.** $\dfrac{x^{10}y^{14}}{4}$ **99.** TW **101.** TW

103. Slope: $-\dfrac{r}{r+p}$; y-intercept: $\left(0, \dfrac{s}{r+p}\right)$

105. False **107.** False **109.** $-\frac{31}{4}$ **111.** **(a)** $-\dfrac{5c}{4b}$; **(b)** undefined; **(c)** $\dfrac{a+d}{f}$ **113.**

Interactive Discovery, p. 139

1. The lines appear to be parallel. **2.** The lines appear to be parallel. **3.** The lines in Exercise 1 still appear parallel; the lines in Exercise 2 are not parallel.

Exercise Set 2.3, pp. 143–146

1. Horizontal **2.** Undefined **3.** Vertical **4.** y-axis
5. 0; x **6.** 0; y **7.** Parallel **8.** Standard **9.** Linear
10. -1 **11.** 0 **13.** Undefined **15.** 0 **17.** Undefined
19. Undefined **21.** 0 **23.** Undefined **25.** Undefined
27. $-\frac{2}{3}$

29.
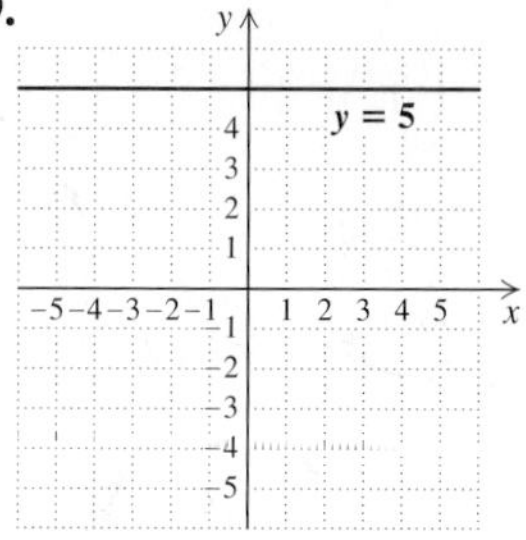

31.
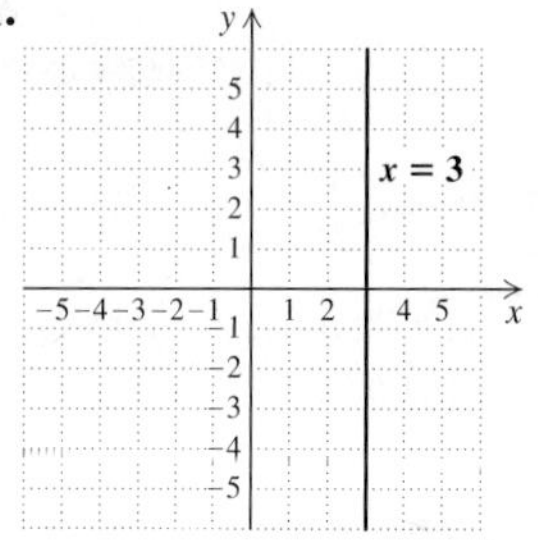

33.
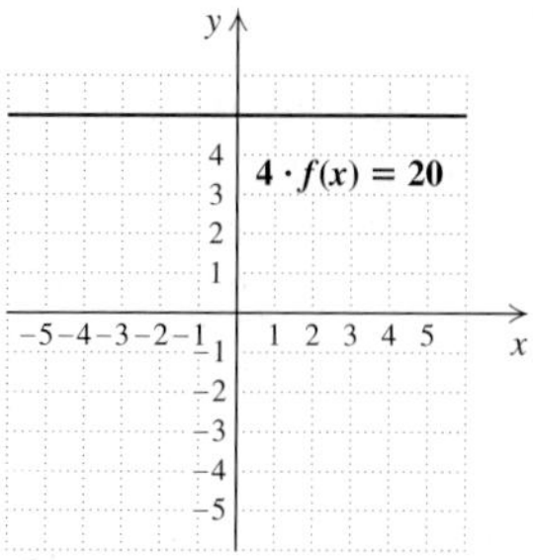

35.
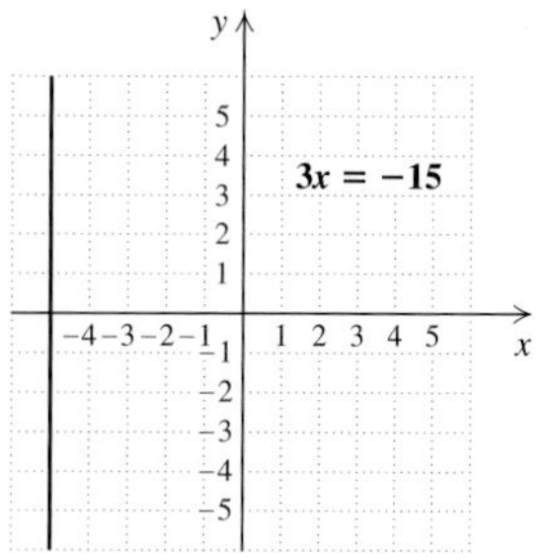

37.
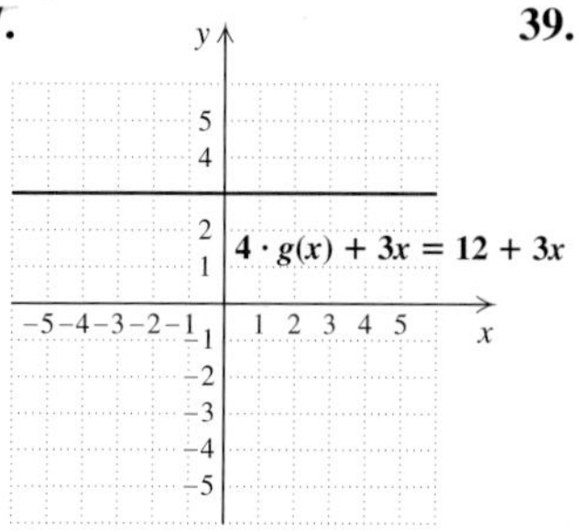

39.
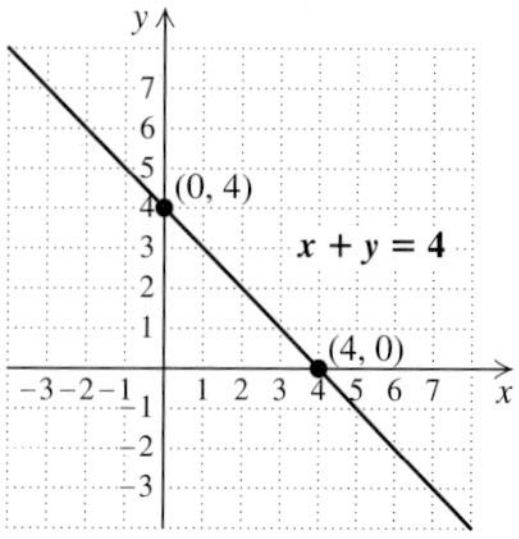

41.
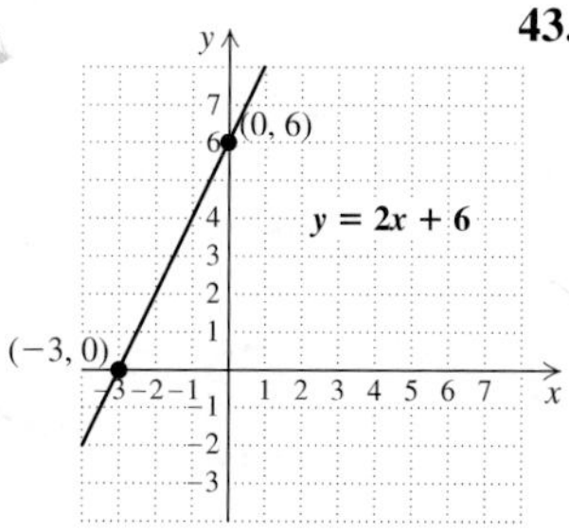

43.
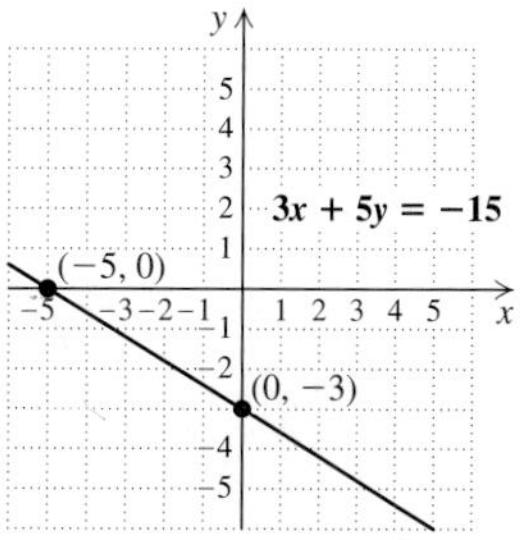

45.
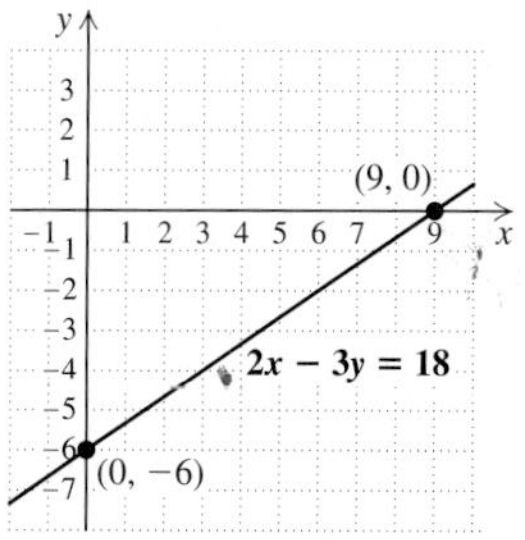

47.
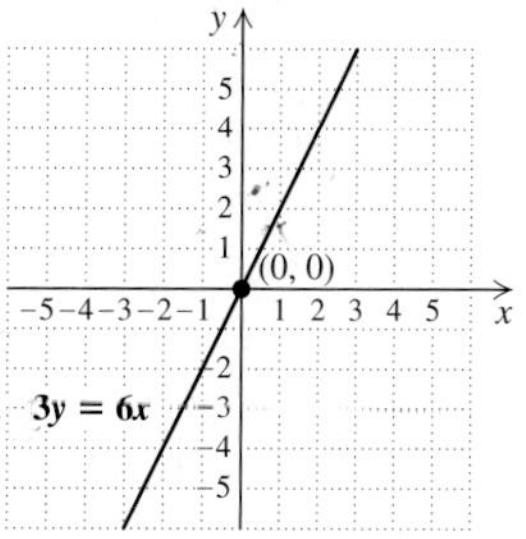

49.
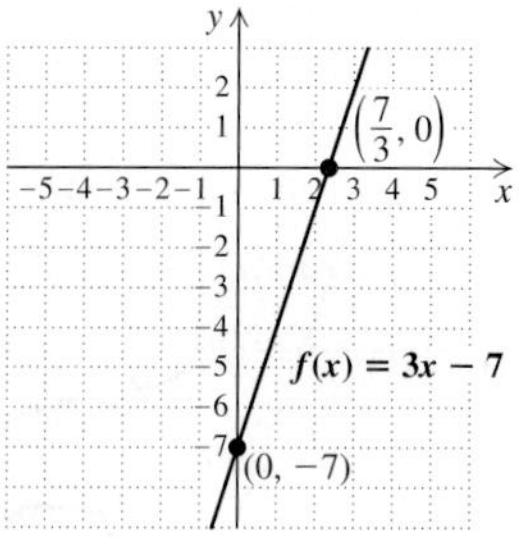

51.
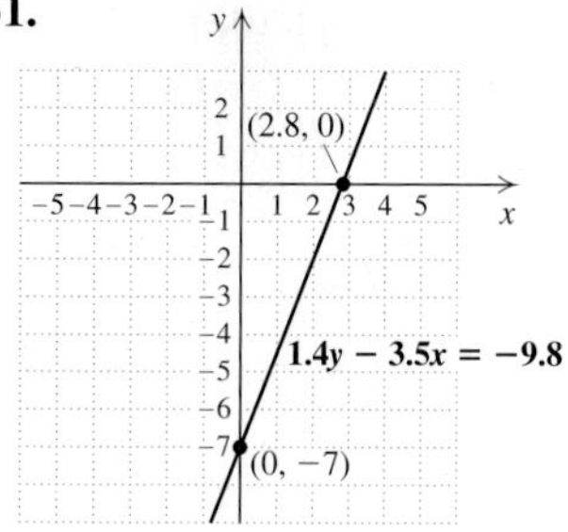

53.

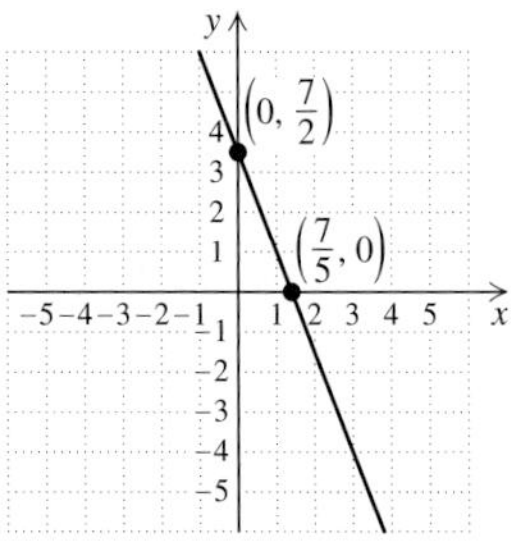

$5x + 2g(x) = 7$

55. (c) **57.** (d) **59.** Yes **61.** Yes **63.** No
65. Yes **67.** No **69.** $f(x) = 3x + 9$
71. $f(x) = -2x - 5$ **73.** $f(x) = -\frac{2}{5}x - \frac{1}{3}$
75. $f(x) = -5$ **77.** $f(x) = -x + 4$ **79.** $f(x) = \frac{3}{2}x - 4$
81. $f(x) = -\frac{1}{5}x + \frac{1}{5}$ **83.** Linear; $\frac{5}{3}$ **85.** Linear; 0
87. Not linear **89.** Linear; $\frac{14}{3}$ **91.** Not linear
93. Not linear **95.** TW

97.

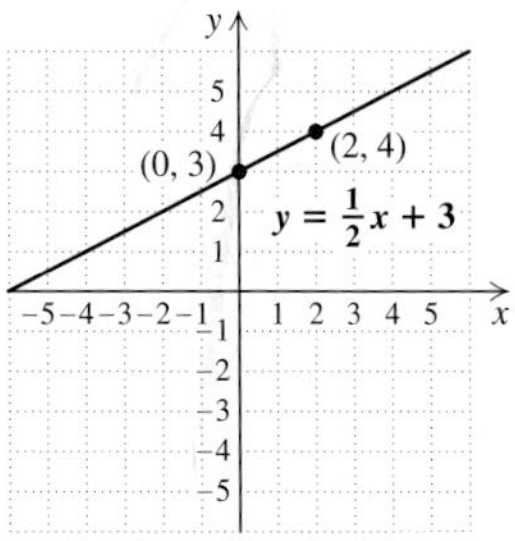

98.

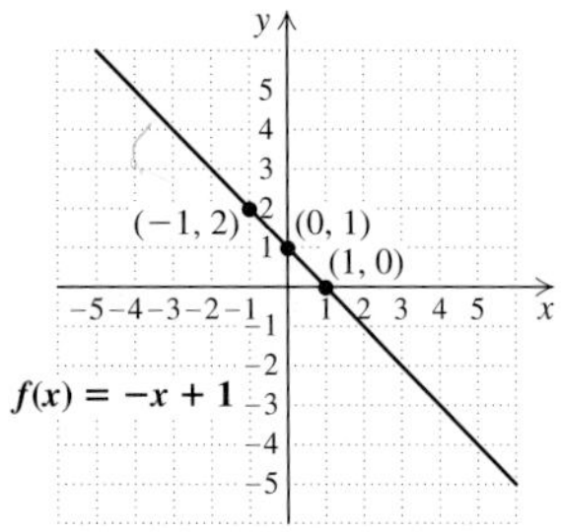

99.

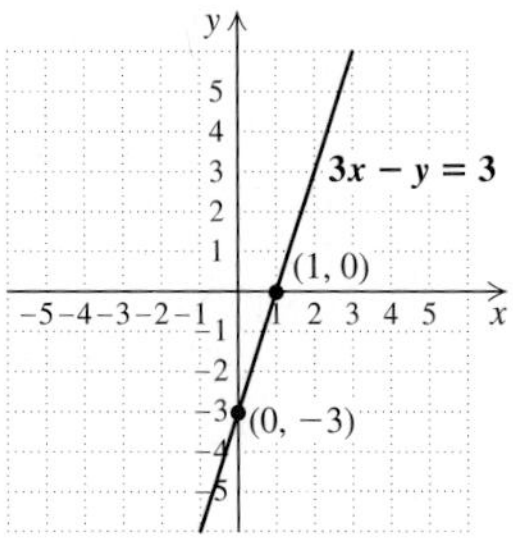

100.

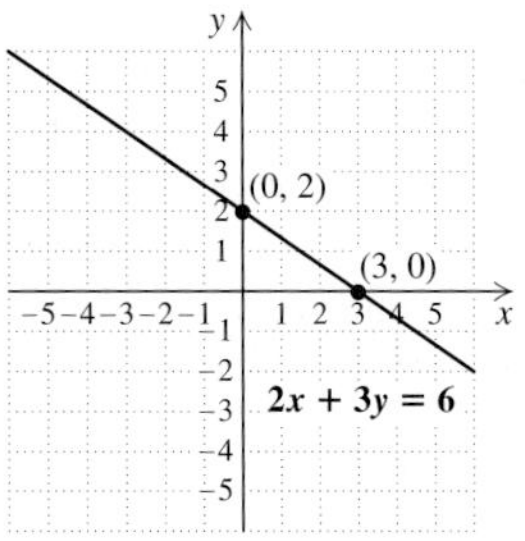

101.

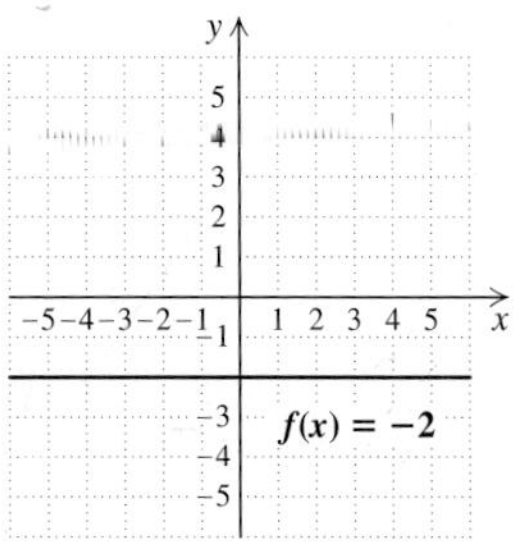

102.

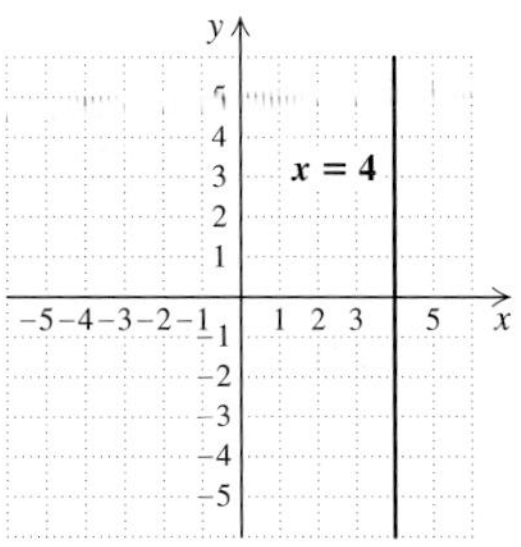

103.

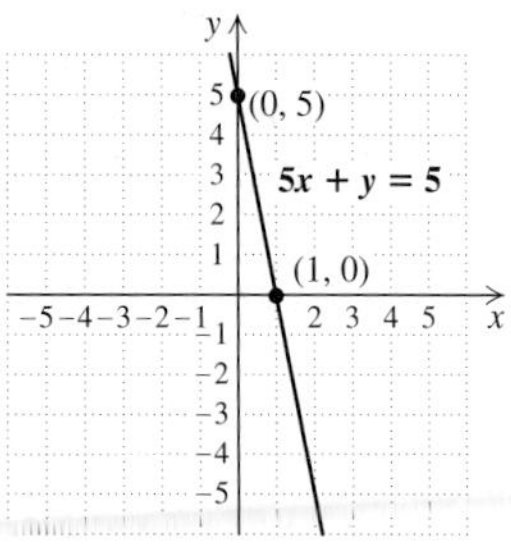

104.

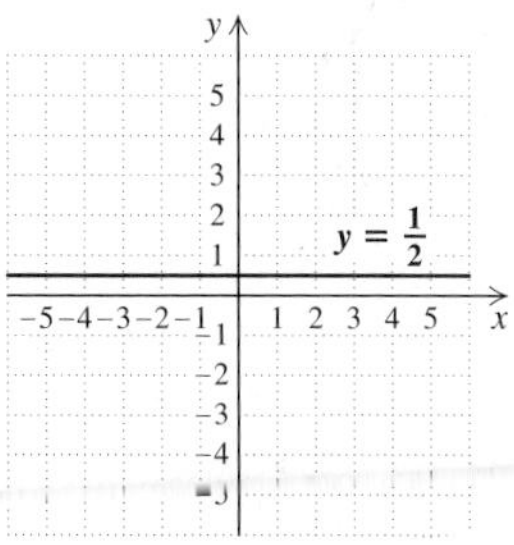

105.

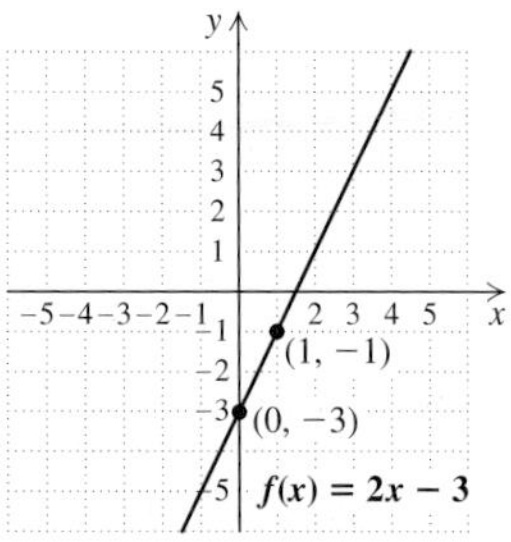

106.

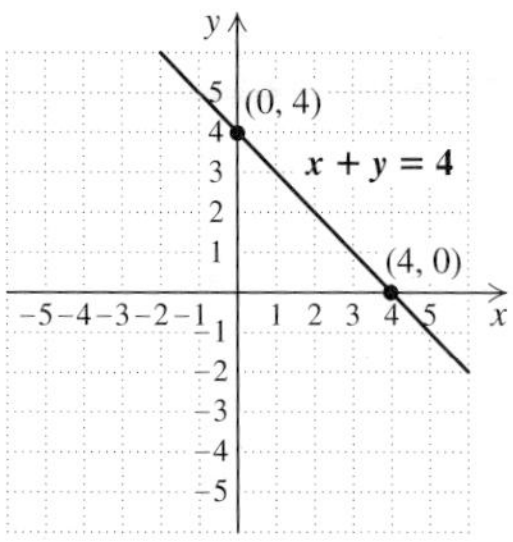

107. TW **109.** $4x - 5y = 20$ **111.** Linear
113. Linear **115.** The slope of equation B is $\frac{1}{2}$ the slope of equation A. **117.** $a = 7, b = -3$ **119.** **(a)** Yes; **(b)** no

Visualizing the Graph, p. 156

1. C **2.** G **3.** F **4.** B **5.** D **6.** A **7.** I **8.** H
9. J **10.** E

Exercise Set 2.4, pp. 157–162

1. True **2.** False **3.** False **4.** True **5.** True
6. True **7.** True **8.** True **9.** False **10.** True

11.

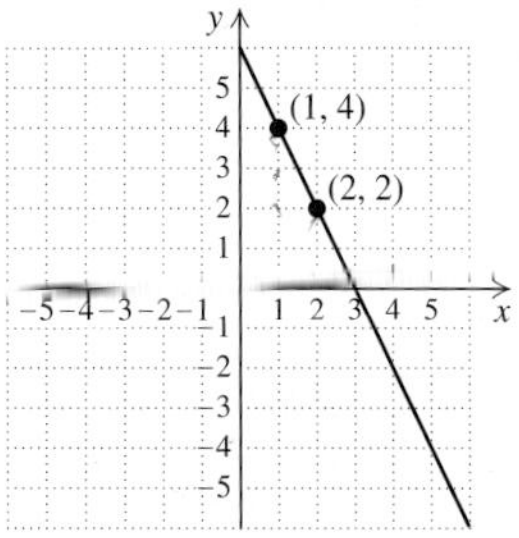

$y - 4 = -2(x - 1)$

13.

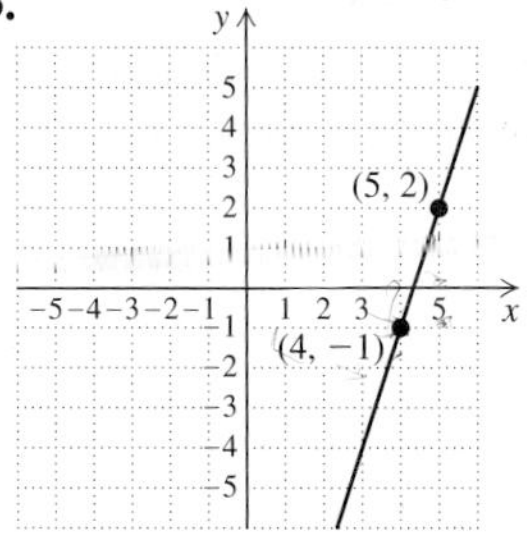

$y - 2 = 3(x - 5)$

15.

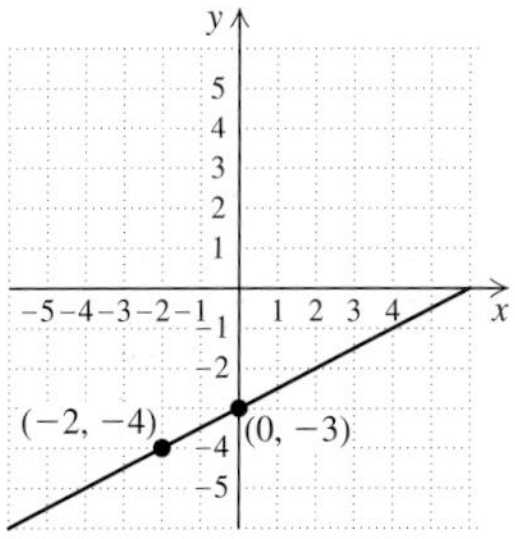

$y - (-4) = \frac{1}{2}(x - (-2))$, or

$y + 4 = \frac{1}{2}(x + 2)$

17.

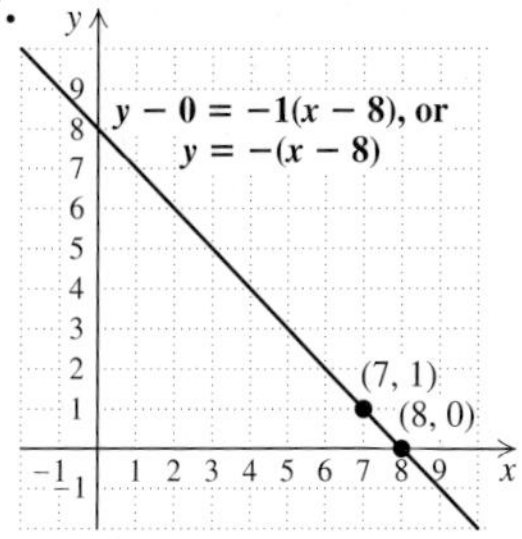

19. $\frac{2}{7}$; (8, 9) **21.** −5; (7, −2) **23.** $-\frac{5}{3}$; (−2, 4)
25. $\frac{4}{7}$; (0, 0)
27. $f(x) = 4x - 11$ **29.** $f(x) = -\frac{3}{5}x + \frac{28}{5}$

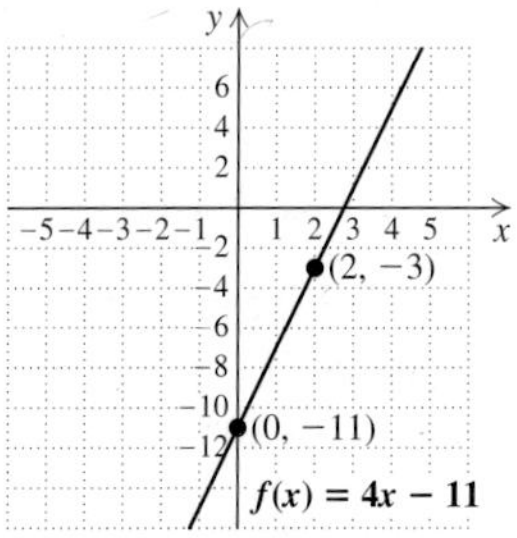

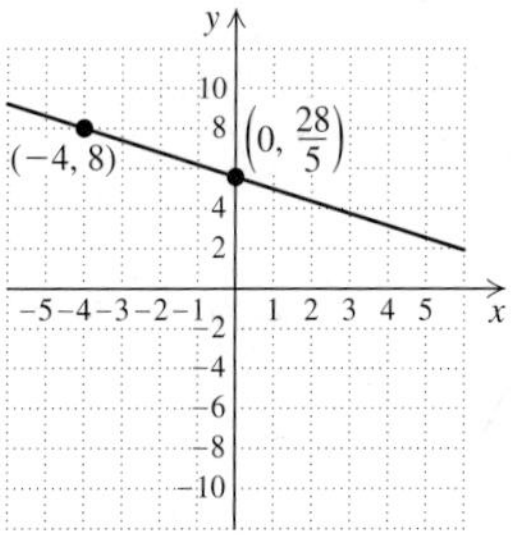

$f(x) = -\frac{3}{5}x + \frac{28}{5}$

31. $f(x) = -0.6x - 5.8$ **33.** $f(x) = \frac{2}{7}x - 6$

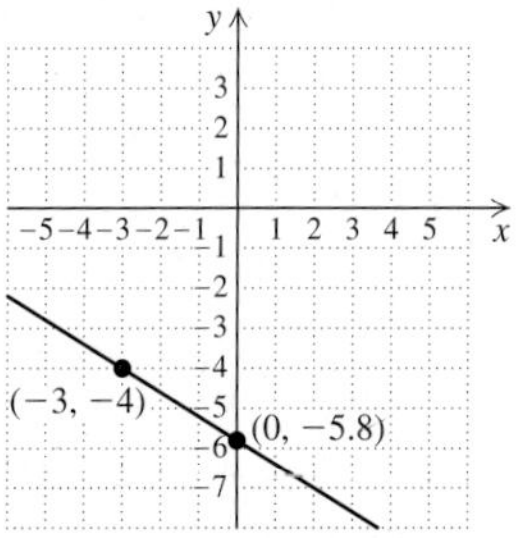

$f(x) = -0.6x - 5.8$

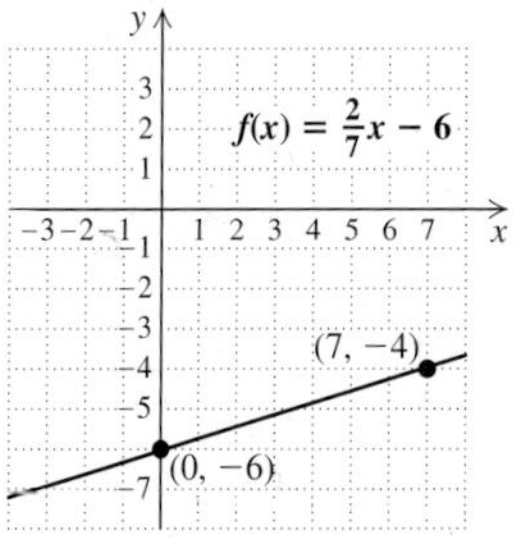

35. $f(x) = \frac{3}{5}x + \frac{42}{5}$

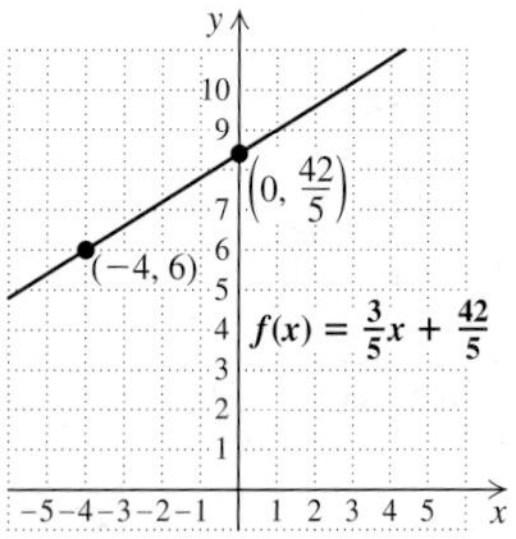

37. $f(x) = \frac{1}{2}x + \frac{7}{2}$ **39.** $f(x) = 1.5x - 6.75$
41. $f(x) = 5x - 2$ **43.** $f(x) = \frac{3}{2}x$

45.

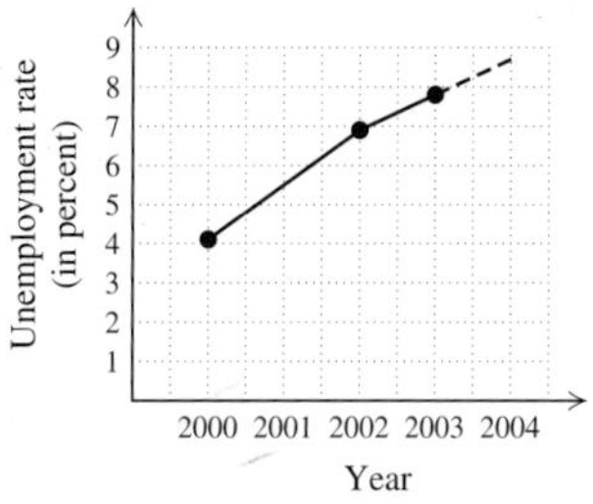

About 5.5%; about 8.7%

47.

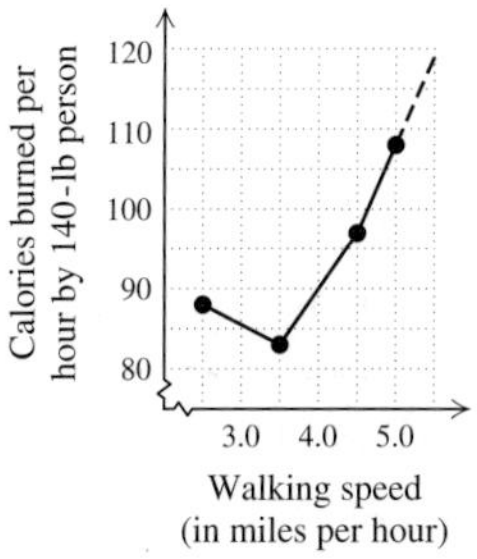

About 90 calories per hour; about 120 calories per hour

49. **(a)** $C(t) = \frac{335}{29}t + 1542$; **(b)** 1981 calories; **(c)** approximately 2011
51. **(a)** $E(t) = 0.26t + 71.42$; **(b)** 76.36
53. **(a)** $A(t) = 12.09t + 189.6$; **(b)** \$358.86 million
55. **(a)** $N(t) = \frac{15}{7}t + 57.3$; **(b)** 87.3 million tons
57. **(a)** $N(t) = \frac{31}{7}t + 48$; **(b)** approximately 92 million; **(c)** 2013
59. **(a)** $A(t) = 0.2t + 78.2$; **(b)** 80.2 million acres
61. Linear **63.** Not linear **65.** Linear
67. **(a)** $W(x) = 0.1673799884x + 63.2346443$; **(b)** 81.6 yr; this estimate is higher **69.** **(a)** $B(x) = 885.33x + 81{,}693$; **(b)** 93,202 financial institutions **71.** TW **73.** 3 **74.** −4 **75.** $-\frac{1}{3}$ **76.** 0 **77.** x-intercept: (10, 0); y-intercept: (0, −5) **78.** x-intercept: (1, 0); y-intercept: (0, 1) **79.** x-intercept: (5, 0); y-intercept: (0, 4) **80.** No x-intercept; y-intercept: (0, −4) **81.** TW **83.** 21.1°C **85.** \$30 **87.** \$8.33 per pound **89.** $y = 3x + 7$ **91.** $y = -3x + 12$ **93.** $\{p \mid p > 5.5\}$

Interactive Discovery, p. 163

1. $f(x) + g(x)$ **2.** Add the y-values of y_1 and y_2 for each x-value. **3.** $y_3 = y_4$

Interactive Discovery, p. 168

1. 0 **2.** −1 **3.** −1, 0 **4.** Yes **5.** Yes **6.** No; 3 is not in the domain of f/g.

Exercise Set 2.5, pp. 170–173

1. Domain **2.** Subtract **3.** Evaluate **4.** Domains **5.** Excluding **6.** Sum **7.** 1 **9.** -41 **11.** 12 **13.** $\frac{13}{18}$ **15.** 5 **17.** $x^2 - 3x + 3$ **19.** $x^2 - x + 3$ **21.** 23 **23.** 5 **25.** 56 **27.** $\frac{x^2 - 2}{5 - x}, x \neq 5$ **29.** $\frac{2}{7}$ **31.** 4% **33.** $1.3 + 2.2 = 3.5$ million **35.** About 50 million; the number of passengers using Newark Liberty and LaGuardia in 1998 **37.** About 81 million; the number of passengers using the three airports in 2002 **39.** About 51 million; the number of passengers using LaGuardia and Newark Liberty in 2002
41. **(a)** $\{x \mid x \text{ is a real number } and\ x \neq 3\}$; **(b)** $\{x \mid x \text{ is a real number } and\ x \neq 6\}$; **(c)** $\mathbb{R}$; **d)** $\mathbb{R}$; **(e)** $\{x \mid x \text{ is a real number } and\ x \neq \frac{5}{2}\}$; **(f)** $\mathbb{R}$ **43.** $\mathbb{R}$
45. $\{x \mid x \text{ is a real number } and\ x \neq 3\}$
47. $\{x \mid x \text{ is a real number } and\ x \neq 0\}$
49. $\{x \mid x \text{ is a real number } and\ x \neq 1\}$
51. $\{x \mid x \text{ is a real number } and\ x \neq 2\ and\ x \neq 4\}$
53. $\{x \mid x \text{ is a real number } and\ x \neq 3\}$
55. $\{x \mid x \text{ is a real number } and\ x \neq 4\}$
57. $\{x \mid x \text{ is a real number } and\ x \neq 4\ and\ x \neq 5\}$
59. 4; 3 **61.** 5; -1 **63.** $\{x \mid 0 \leq x \leq 9\}$; $\{x \mid 3 \leq x \leq 10\}$; $\{x \mid 3 \leq x \leq 9\}$; $\{x \mid 3 \leq x \leq 9\}$
65.

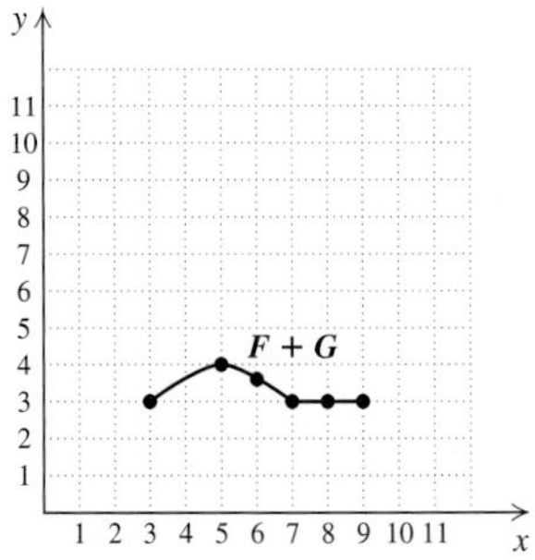

67. TW **69.** $x = \frac{7}{4}y + 2$ **70.** $y = \frac{3}{8}x - \frac{5}{8}$
71. $y = -\frac{5}{2}x - \frac{3}{2}$ **72.** $x = -\frac{5}{6}y - \frac{1}{3}$
73. Let n represent the number; $2n + 5 = 49$
74. Let x represent the number; $\frac{1}{2}x - 3 = 57$
75. Let x represent the first integer; $x + (x + 1) = 145$
76. Let n represent the number; $n - (-n) = 20$ **77.** TW
79. $\{x \mid x \text{ is a real number } and\ x \neq -\frac{5}{2}\ and\ x \neq -3\ and\ x \neq 1\ and\ x \neq -1\}$

81. Answers may vary.

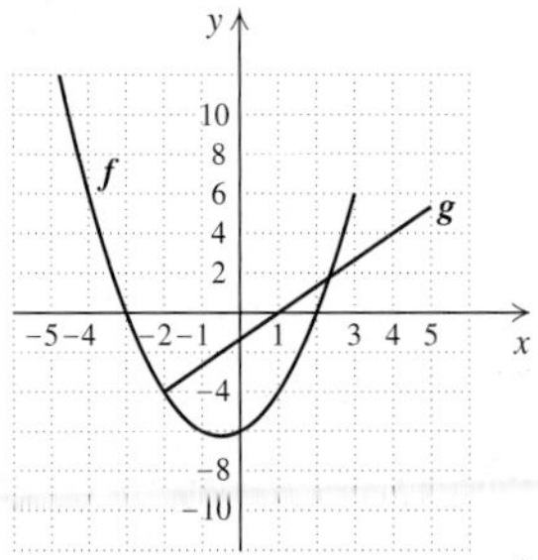

83. $\{x \mid x \text{ is a real number } and\ -1 < x < 5\ and\ x \neq \frac{3}{2}\}$
85. Answers may vary. $f(x) = \frac{1}{x + 2}$, $g(x) = \frac{1}{x - 5}$
87.

Review Exercises: Chapter 2, pp. 176–179

1. True **2.** False **3.** False **4.** False **5.** False **6.** True **7.** True **8.** True **9.** True **10.** True
11. **(a)** 3; **(b)** $\{x \mid -2 \leq x \leq 4\}$; **(c)** -1; **(d)** $\{y \mid 1 \leq y \leq 5\}$
12. 10.53 yr **13.** Yes **14.** No **15.** $3a + 2$ **16.** $3a$
17. -1 **18.** $\frac{11}{3}$ **19.** $\frac{1}{3}$
20. Slope: -4; y-intercept: $(0, -9)$

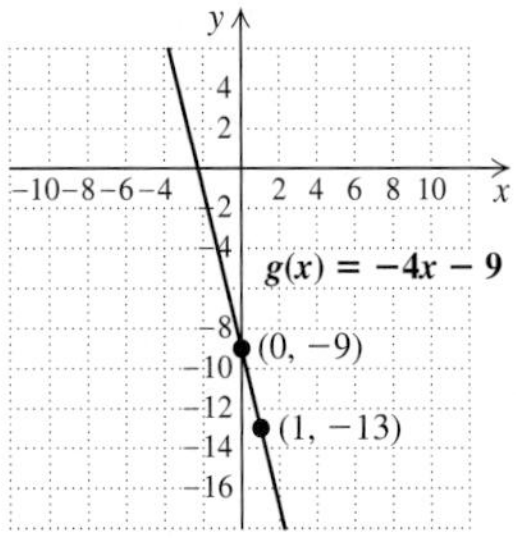

21. Slope: $\frac{1}{3}$; y-intercept: $\left(0, -\frac{7}{3}\right)$

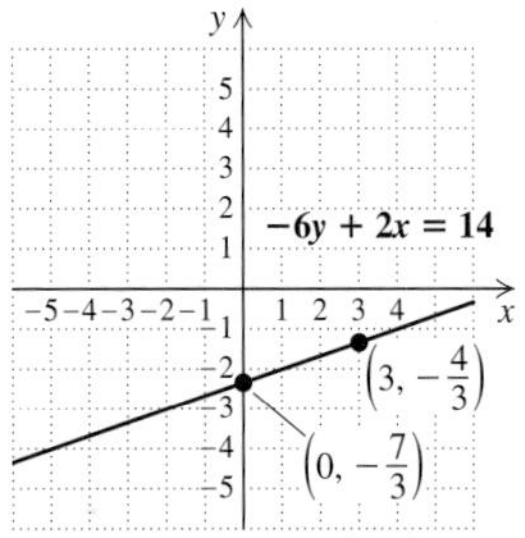

22. The value of the apartment is increasing at a rate of \$7500 per year. **23.** $\frac{4}{7}$ **24.** Undefined **25.** The number of Freecycle groups was increasing at a rate of 64.1 groups per month. **26.** 645 signifies that tuition is increasing at a rate of \$645 a year; 9800 represents tuition costs in 1997.
27. $f(x) = \frac{2}{7}x - 6$
28.

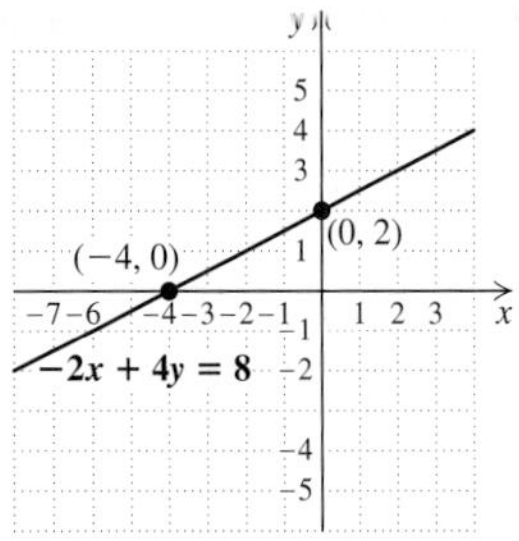

29.

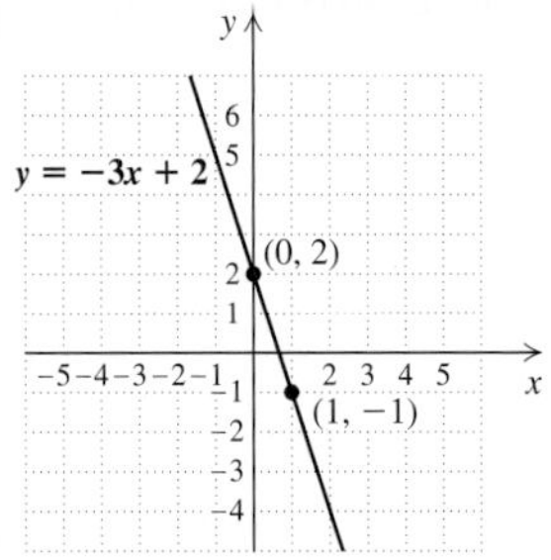

30.

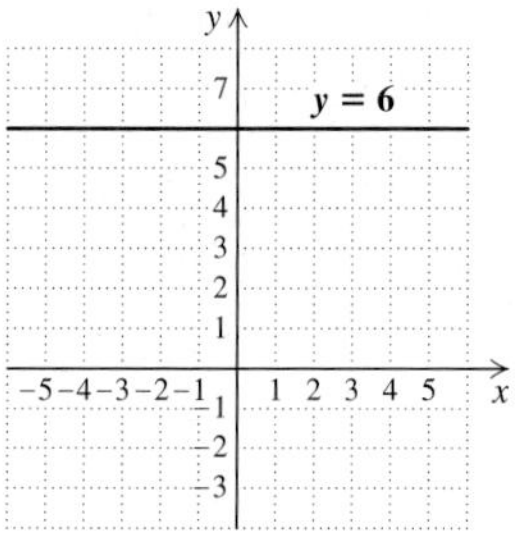

31.

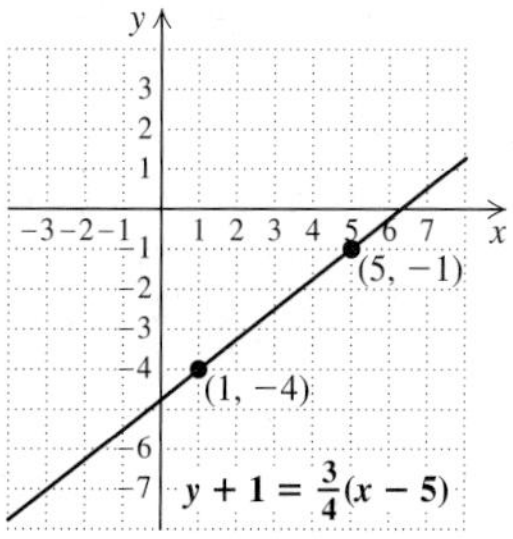

32.

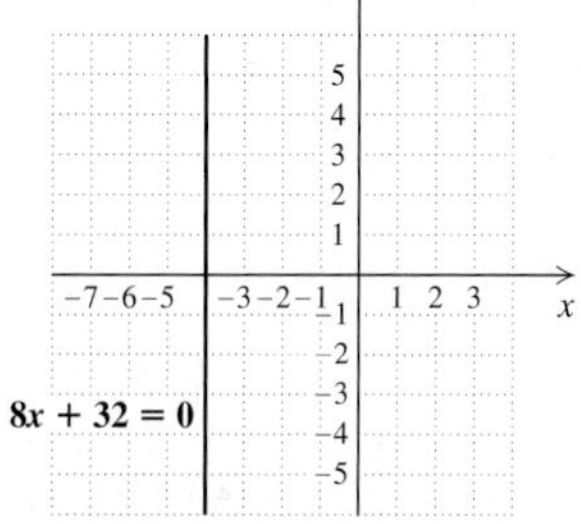

33.

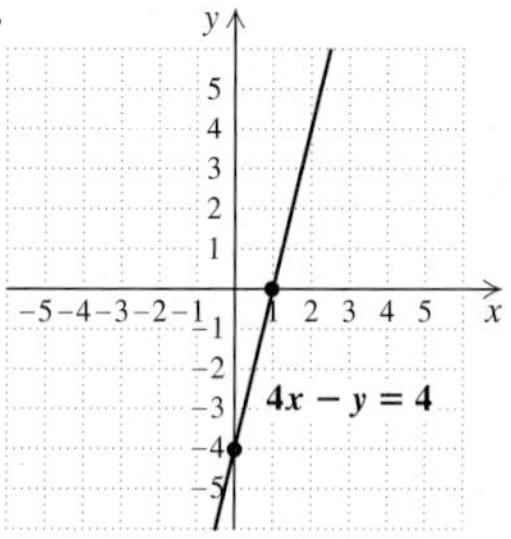

34. The window should show both intercepts: (98, 0) and (0, 14). One possibility is [−10, 120, −5, 15], with Xscl = 10. **35.** Perpendicular **36.** Parallel **37.** $y = -x - 3$ **38.** Yes **39.** Yes **40.** No **41.** No **42.** $y - 4 = -2(x-(-3))$, or $y - 4 = -2(x + 3)$ **43.** $f(x) = \frac{4}{3}x + \frac{7}{3}$ **44.** About \$3.50;

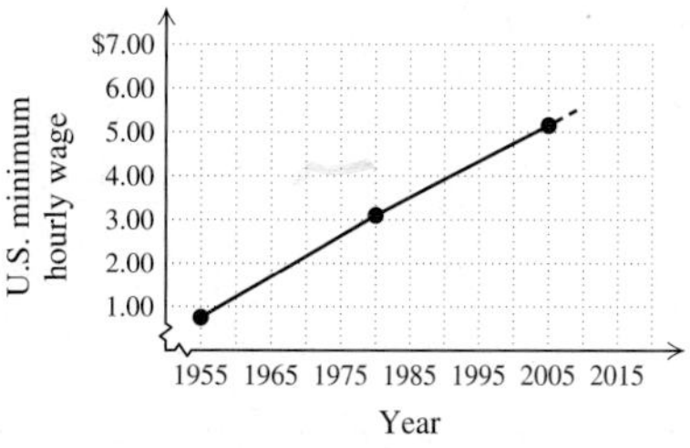

45. About \$5.50 **46.** **(a)** $R(t) = -0.0215t + 19.75$; **(b)** 19.21 sec; 19.11 sec

47. Linear

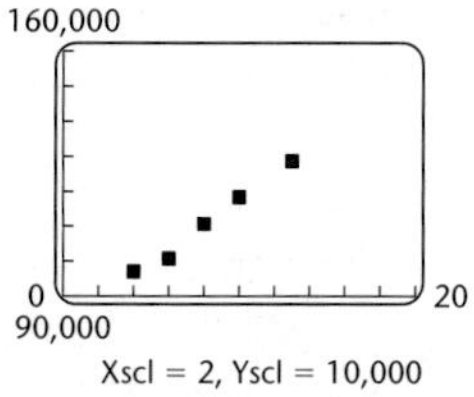

48. $B(t) = 3692.377049t + 80935.5082$ **49.** 151,091 twin births **50.** $\{x|x$ is a real number *and* $x \neq -3\}$ **51.** $\mathbb{R}$ **52.** 102 **53.** −17 **54.** $-\frac{9}{2}$ **55.** $x^2 + 3x - 5$ **56.** $\mathbb{R}$ **57.** $\{x|x$ is a real number *and* $x \neq 2\}$ **58.** TW For a function, every member of the domain corresponds to *exactly one* member of the range. Thus, for any function, each member of the domain corresponds to *at least one* member of the range. Therefore, a function is a relation. In a relation, every member of the domain corresponds to *at least one,* but not necessarily *exactly one,* member of the range. Therefore, a relation may or may not be a function. **59.** TW The slope of a line is the rise between two points on the line divided by the run between those points. For a vertical line, there is no run between any two points, and division by 0 is undefined. Therefore, the slope is undefined. For a horizontal line, there is no rise between any two points, so the slope is 0/run, or 0. **60.** −9 **61.** $-\frac{9}{2}$ **62.** $f(x) = 3.09x + 3.75$ **63.** **(a)** III; **(b)** IV; **(c)** I; **(d)** II

Test: Chapter 2, pp. 179–180

1. [2.1] **(a)** 1; **(b)** $\{x|-3 \le x \le 4\}$; **(c)** 3; **(d)** $\{y|-1 \le y \le 2\}$ **2.** **(a)** [2.1] \$45.66 billion; **(b)** [2.2] 1.24 signifies that the rate of increase in U.S. book sales was \$1.24 billion per year; 34.5 signifies that U.S. book sales were \$34.5 billion in 2000.

3. [2.4] About 47.1 million international visitors

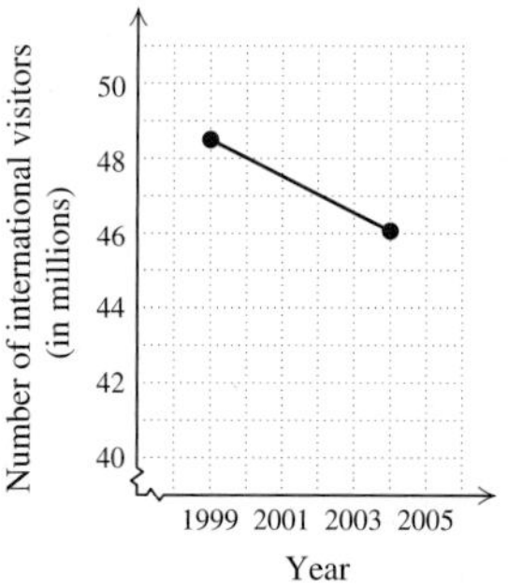

4. [2.2] Slope: $-\frac{3}{5}$; y-intercept: (0, 12)
5. [2.2] Slope: $-\frac{2}{5}$; y-intercept: $\left(0, -\frac{7}{5}\right)$ **6.** [2.2] $\frac{5}{8}$
7. [2.2] 0 **8.** [2.2] 75 calories per 30 minutes, or 2.5 calories per minute **9.** [2.2] $f(x) = -5x - 1$

10. [2.2]

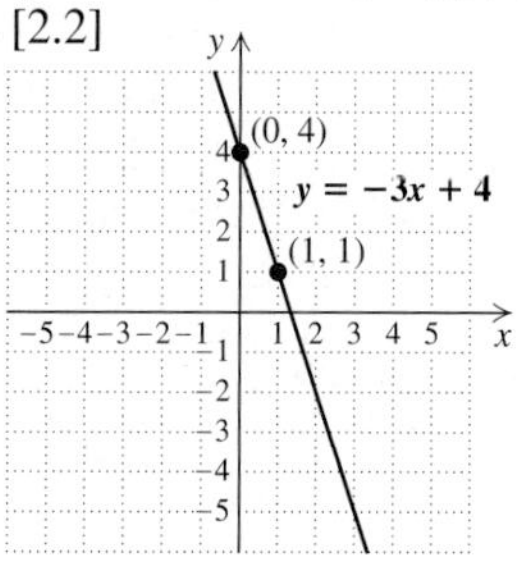

11. [2.3]

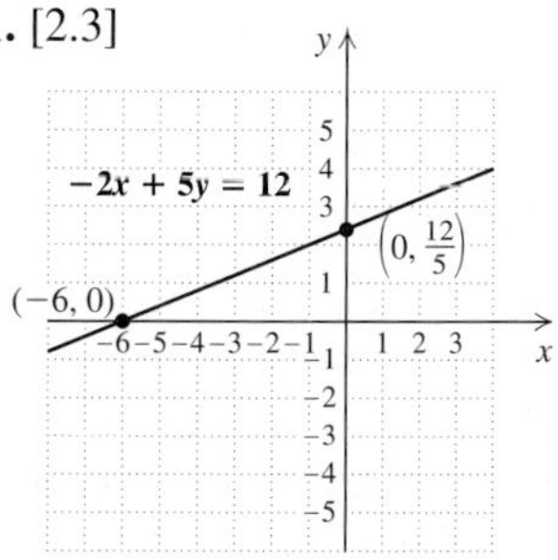

12. [2.4]

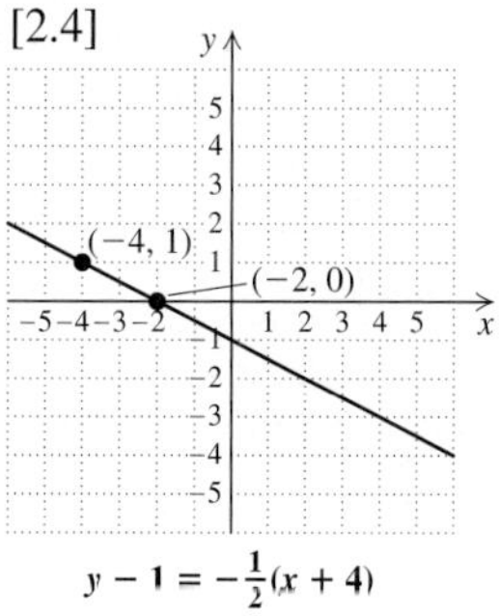

13. [2.3]

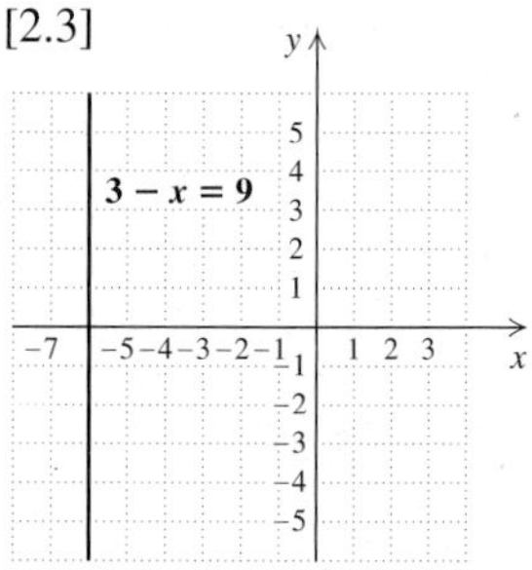

14. [2.3] Yes **15.** [2.3] Parallel **16.** [2.3] Perpendicular **17.** [2.3] (a), (c) **18.** [2.4] $y - (-4) = 4(x - (-2))$, or $y + 4 = 4(x + 2)$ **19.** [2.4] $f(x) = -x + 2$
20. [2.4] **(a)** $C(m) = 0.3m + 25$; **(b)** \$175
21. [2.4] **(a)** $A(x) = -0.5375687433x + 89.50755151$; **(b)** 30.4 accidental deaths
22. [2.5] $\{x \mid x$ is a real number *and* $x \neq -\frac{1}{2}\}$
23. **(a)** [2.1] 5; **(b)** [2.1] -4; **(c)** [2.1] $-3t + 1$; **(d)** [2.5] -130; **(e)** [2.1] $-\frac{4}{3}$; **(f)** [2.5] $\{x \mid x$ is a real number *and* $x \neq -\frac{4}{5}\}$
24. [2.1], [2.4] **(a)** 30 mi; **(b)** 15 mph
25. [2.3], [2.4] $y = \frac{2}{5}x + \frac{16}{5}$ **26.** [2.3], [2.4] $y = -\frac{5}{2}x - \frac{11}{2}$
27. [2.4] $s = -\dfrac{3}{2}r + \dfrac{27}{2}$, or $\dfrac{27 - 3r}{2}$
28. [2.5] $h(x) = 7x - 2$

Chapter 3

Visualizing the Graph, p. 192

1. C **2.** H **3.** J **4.** G **5.** D **6.** I **7.** A **8.** F **9.** E **10.** B

Exercise Set 3.1, pp. 193–197

1. True **2.** False **3.** True **4.** True **5.** True **6.** False **7.** False **8.** True **9.** Yes **11.** No **13.** Yes **15.** Yes **17.** (4, 1) **19.** (2, −1) **21.** (4, 3) **23.** (−3, −2) **25.** (−3, 2) **27.** (3, −7) **29.** (7, 2) **31.** (4, 0) **33.** No solution **35.** $\{(x, y) \mid y = 3 - x\}$ **37.** Approximately (1.53, 2.58) **39.** No solution **41.** Approximately (−6.37, −18.77) **43.** All except 33 and 39 **45.** 35 **47.** Let x represent the first number and y the second number; $x + y = 50$, $x = 0.25y$ **49.** Let m represent the number of ounces of mineral oil and v the number of ounces of vinegar; $m + v = 16$, $m = 2v + 4$ **51.** Let x and y represent the angles; $x + y = 180$, $x = 2y - 3$ **53.** Let x represent the number of two-point shots and y the number of foul shots; $x + y = 64$, $2x + y = 100$ **55.** Let x represent the number of \$8.50 brushes sold and y the number of \$9.75 brushes sold; $x + y = 45$, $8.50x + 9.75y = 398.75$
57. Let h represent the number of vials of Humalog sold and n the number of vials of Novolog sold; $h + n = 50$, $80.86h + 83.70n = 4125.36$
59. Let l represent the length, in feet, and w the width, in feet; $2l + 2w = 288$, $l = w + 44$ **61.** Full-time faculty: $y = 8.198x + 436.757$; part-time faculty: $y = 13.150x + 197.345$, where x is the number of years after 1980; approximately 2028 **63.** Waste generated: $y = 0.759x + 75.056$; waste recycled: $y = 1.521x + 28.965$, where x is the number of years after 1990; approximately 2050 **65.** TW **67.** 15 **68.** $\frac{19}{12}$ **69.** $\frac{9}{20}$ **70.** $\frac{13}{3}$ **71.** $y = -\frac{3}{4}x + \frac{7}{4}$ **72.** $y = \frac{2}{5}x - \frac{9}{5}$ **73.** TW
75. Answers may vary. **(a)** $x + y = 6$, $x - y = 4$; **(b)** $x + y = 1$, $2x + 2y = 3$; **(c)** $x + y = 1$, $2x + 2y = 2$

77. $A = -\frac{17}{4}$, $B = -\frac{12}{5}$ **79.** Let x and y represent the number of years that Lou and Juanita have taught at the university, respectively; $x + y = 46$, $x - 2 = 2.5(y - 2)$ **81.** Let s and v represent the number of ounces of baking soda and vinegar needed, respectively; $s = 4v$, $s + v = 16$
83. (0, 0), (1, 1) **85.** (c) **87.** (b)

Exercise Set 3.2, pp. 204–206

1. (d) **2.** (e) **3.** (a) **4.** (f) **5.** (c) **6.** (b) **7.** (2, −3) **9.** (−4, 3) **11.** (2, −2) **13.** $\{(x, y) \mid 2x - 3 = y\}$ **15.** (−2, 1) **17.** $\left(\frac{1}{2}, \frac{1}{2}\right)$ **19.** $\left(\frac{25}{23}, -\frac{11}{23}\right)$ **21.** No solution **23.** (1, 2) **25.** (2, 7) **27.** (−1, 2) **29.** $\left(\frac{128}{31}, -\frac{17}{31}\right)$ **31.** (6, 2) **33.** No solution **35.** $\left(\frac{110}{19}, -\frac{12}{19}\right)$ **37.** (3, −1) **39.** $\{(x, y) \mid -4x + 2y = 5\}$ **41.** $\left(\frac{140}{13}, -\frac{50}{13}\right)$ **43.** (−2, −9) **45.** (30, 6) **47.** $\{(x, y) \mid x = 2 + 3y\}$ **49.** No solution **51.** (140, 60) **53.** $\left(\frac{1}{3}, -\frac{2}{3}\right)$ **55.** (d) **57.** (b) **59.** TW **61.** 4 mi **62.** 86 **63.** $\$11\frac{2}{3}$ billion on bathrooms; $\$23\frac{1}{3}$ billion on kitchens **64.** 30 m, 90 m, 360 m **65.** 450.5 mi **66.** 460.5 mi **67.** TW **69.** $m = -\frac{1}{2}$, $b = \frac{5}{2}$ **71.** $a = 5$, $b = 2$ **73.** $\left(-\frac{32}{17}, \frac{38}{17}\right)$ **75.** $\left(-\frac{1}{5}, \frac{1}{10}\right)$

Exercise Set 3.3, pp. 218–222

1. 10, 40 **3.** Mineral oil: 12 oz; vinegar: 4 oz **5.** 119°, 61° **7.** Two-point shots: 36; foul shots: 28 **9.** \$8.50-brushes: 32; \$9.75-brushes: 13 **11.** Humalog vials: 21; Novolog vials: 29 **13.** Width: 50 ft; length: 94 ft **15.** Nonrecycled sheets: 38; recycled sheets: 112 **17.** General Electric bulbs: 60; SLi bulbs: 140 **19.** HP cartridges: 15; Epson cartridges: 35 **21.** Kenyan: 8 lb; Sumatran: 12 lb **23.** 10 lb of each **25.** Deep Thought: 12 lb; Oat Dream: 8 lb **27.** \$7500 at 6%; \$4500 at 9% **29.** Arctic Antifreeze: 12.5 L; Frost No-More: 7.5 L **31.** 87-octane: 4 gal; 93-octane: 8 gal **33.** Whole milk: $169\frac{3}{13}$ lb; cream: $30\frac{10}{13}$ lb **35.** 375 km **37.** 24 mph **39.** About 1489 mi **41.** Length: 265 ft; width: 165 ft **43.** Simon: 122 properties; DeBartolo: 61 properties **45.** 30-sec commercials: 4; 60-sec commercials: 8 **47.** Quarters: 17; fifty-cent pieces: 13 **49.** TW **51.** 16 **52.** 11 **53.** −28 **54.** −10 **55.** $\frac{49}{12}$ **56.** $\frac{13}{10}$ **57.** TW **59.** 0%: 20 reams; 30%: 40 reams **61.** 1.8 L **63.** 180 members **65.** Brown: 0.8 gal; neutral: 0.2 gal **67.** City: 261 mi; highway: 204 mi **69.** $P(x) = \dfrac{0.1 + x}{1.5}$ (This expresses the percent as a decimal quantity.)

Exercise Set 3.4, pp. 229–230

1. True **2.** False **3.** False **4.** True **5.** True **6.** False **7.** No **9.** (4, 0, 2) **11.** (2, −2, 2)

13. $(3, -2, 1)$ **15.** No solution **17.** $(2, 1, 3)$
19. $(2, -5, 6)$ **21.** The equations are dependent.
23. $\left(\frac{1}{2}, 4, -6\right)$ **25.** $\left(\frac{1}{2}, \frac{1}{3}, \frac{1}{6}\right)$ **27.** $\left(\frac{1}{2}, \frac{2}{3}, -\frac{5}{6}\right)$
29. $(15, 33, 9)$ **31.** $(3, 4, -1)$ **33.** $(10, 23, 50)$
35. No solution **37.** The equations are dependent.
39. TW **41.** $(9, -4)$ **42.** $(-12, -29)$ **43.** $\left(\frac{7}{4}, \frac{31}{8}\right)$
44. $\left(\frac{17}{7}, \frac{22}{7}\right)$ **45.** $(5, 1, 2)$ **46.** $(4, 5)$ **47.** TW
49. $(1, -1, 2)$ **51.** $(-3, -1, 0, 4)$ **53.** $\left(-\frac{1}{2}, -1, -\frac{1}{3}\right)$
55. 14 **57.** $z = 8 - 2x - 4y$

Exercise Set 3.5, pp. 235–238

1. 16, 19, 22 **3.** 8, 21, -3 **5.** 32°, 96°, 52°
7. Individual adult: \$64; spouse: \$57; child: \$43 **9.** Bran muffin: 1.5 g; banana: 3 g; 1 cup Wheaties: 3 g
11. \$27,415; 4WD: \$1970; sunroof: \$800 **13.** Elrod: 20 ft/hr; Dot: 24 ft/hr; Wendy: 30 ft/hr **15.** 12-oz cups: 17; 16-oz cups: 25; 20-oz cups: 13 **17.** Small: 10; medium: 25; large: 5 **19.** Roast beef: 2 servings; baked potato: 1 serving; broccoli: 2 servings **21.** Asia: 4.8 billion; Africa: 1.8 billion; rest of world: 2.5 billion **23.** Two-point field goals: 32; three-point field goals: 5; foul shots: 13 **25.** TW **27.** -8
28. 33 **29.** -55 **30.** -71 **31.** $-14x + 21y - 35z$
32. $-24a - 42b + 54c$ **33.** $-5a$ **34.** $11x$ **35.** TW
37. Applicant: \$102; spouse: \$58; first child: \$43; second child: \$40 **39.** 20 yr **41.** 35 tickets

Exercise Set 3.6, pp. 245–246

1. Horizontal; columns **2.** Equation **3.** Entry
4. Matrices **5.** Multiple **6.** First **7.** $\left(-\frac{1}{3}, -4\right)$
9. $(-4, 3)$ **11.** $\left(\frac{3}{2}, \frac{5}{2}\right)$ **13.** $\left(2, \frac{1}{2}, -2\right)$ **15.** $(2, -2, 1)$
17. $\left(4, \frac{1}{2}, -\frac{1}{2}\right)$ **19.** $(1, -3, -2, -1)$ **21.** Dimes: 18; nickels: 24 **23.** \$4.05-granola: 5 lb; \$2.70-granola: 10 lb
25. \$400 at 7%; \$500 at 8%; \$1600 at 9% **27.** TW
29. 13 **30.** -22 **31.** 37 **32.** 422 **33.** TW
35. 1324

Exercise Set 3.7, pp. 251–252

1. True **2.** True **3.** False **4.** False **5.** False
6. False **7.** 18 **9.** 36 **11.** 27 **13.** -3 **15.** -5
17. $(-3, 2)$ **19.** $\left(\frac{9}{19}, \frac{51}{38}\right)$ **21.** $\left(-1, -\frac{6}{7}, \frac{11}{7}\right)$
23. $(2, -1, 4)$ **25.** $(1, 2, 3)$ **27.** TW **29.** $(1, -3)$
30. Approximately (0.26, 1.65), or $\left(\frac{6}{23}, \frac{38}{23}\right)$ **31.** $\left(\frac{4}{3}, \frac{5}{3}\right)$
32. $(2, 2)$ **33.** $\left(\frac{31}{7}, -\frac{6}{7}\right)$ **34.** $\left(\frac{12}{7}, -\frac{17}{7}\right)$ **35.** TW
37. 12 **39.** 10

Exercise Set 3.8, pp. 257–260

1. (b) **2.** (a) **3.** (e) **4.** (f) **5.** (h) **6.** (c)
7. (g) **8.** (d) **9. (a)** $P(x) = 20x - 300{,}000$;
(b) (15,000 units, \$975,000)
11. (a) $P(x) = 50x - 120{,}000$; **(b)** (2400 units, \$144,000)
13. (a) $P(x) = 45x - 22{,}500$; **(b)** (500 units, \$42,500)
15. (a) $P(x) = 18x - 16{,}000$; **(b)** (889 units, \$35,560)
17. (a) $P(x) = 50x - 100{,}000$; **(b)** (2000 units, \$250,000)
19. (\$70, 300) **21.** (\$22, 474) **23.** (\$50, 6250)
25. (\$10, 1070) **27. (a)** $C(x) = 125{,}300 + 450x$;
(b) $R(x) = 800x$; **(c)** $P(x) = 350x - 125{,}300$; **(d)** \$90,300 loss, \$14,700 profit; **(e)** (358 computers, \$286,400)
29. (a) $C(x) = 16{,}404 + 6x$; **(b)** $R(x) = 18x$;
(c) $P(x) = 12x - 16{,}404$; **(d)** \$19,596 profit, \$4404 loss;
(e) (1367 dozen caps, \$24,606) **31. (a)** \$8.74; **(b)** 793 units
33. TW **35.** 12 **36.** 15 **37.** $\frac{8}{3}$ **38.** 4 **39.** $\frac{9}{2}$
40. $\frac{1}{3}$ **41.** TW **43.** (\$5, 300 yo-yo's)
45. (a) $S(p) = 15.97p - 1.05$;
(b) $D(p) = -11.26p + 41.16$; **(c)** (\$1.55, 23.7 million jars)

Review Exercises: Chapter 3, pp. 263–265

1. Substitution **2.** Elimination **3.** Approximate
4. Dependent **5.** Inconsistent **6.** Infinite **7.** Parallel
8. Square **9.** Determinant **10.** Zero **11.** $(-2, 1)$
12. $(2, 1)$ **13.** $\left(-\frac{11}{15}, -\frac{43}{30}\right)$ **14.** No solution
15. $\left(-\frac{4}{5}, \frac{2}{5}\right)$ **16.** $\left(\frac{37}{19}, \frac{53}{19}\right)$ **17.** $\left(\frac{76}{17}, -\frac{2}{119}\right)$ **18.** $(2, 2)$
19. $\{(x, y) \mid 3x + 4y = 6\}$ **20.** DVD: \$17; videocassette: \$14 **21.** 4 hr **22.** 8% juice: 10 L; 15% juice: 4 L **23.** $(4, -8, 10)$ **24.** The equations are dependent. **25.** $(2, 0, 4)$ **26.** No solution
27. $\left(\frac{8}{9}, -\frac{2}{3}, \frac{10}{9}\right)$ **28.** A: 90°; B: 67.5°; C: 22.5°
29. Oil: $21\frac{1}{3}$ oz; lemon juice: $10\frac{2}{3}$ oz
30. Lumber: 29 pallets; plywood: 13 pallets **31.** $\left(55, -\frac{89}{2}\right)$
32. $(-1, 1, 3)$ **33.** 2 **34.** 9 **35.** $(6, -2)$
36. $(-3, 0, 4)$ **37.** (\$3, 81)
38. (a) $C(x) = 0.75x + 9000$; **(b)** $R(x) = 5.25x$;
(c) $P(x) = 4.5x - 9000$; **(d)** \$2250 loss, \$13,500 profit;
(e) (2000 pints of honey, \$10,500)
39. TW To solve a problem involving four variables, go through the *Familiarize* and *Translate* steps as usual. The resulting system of equations can be solved using the elimination method just as for three variables but likely with more steps. **40.** TW A system of equations can be both dependent and inconsistent if it is equivalent to a system with fewer equations that has no solution. An example is a system of three equations in three unknowns in which two of the equations represent the same plane, and the third represents a parallel plane. **41.** 8000 pints **42.** (0, 2), (1, 3)
43. $a = -\frac{2}{3}, b = -\frac{4}{3}, c = 3; f(x) = -\frac{2}{3}x^2 - \frac{4}{3}x + 3$

Test: Chapter 3, pp. 265–266

1. [3.1] (2, 4) **2.** [3.2] $\left(3, -\frac{11}{3}\right)$ **3.** [3.2] $\left(\frac{15}{7}, -\frac{18}{7}\right)$
4. [3.2] $\left(-\frac{3}{2}, -\frac{3}{2}\right)$ **5.** [3.2] No solution **6.** [3.3] Length: 30 units; width: 18 units **7.** [3.3] Pepperidge Farm

Goldfish: 120 g; Rold Gold Pretzels: 500 g **8.** [3.4] The equations are dependent. **9.** [3.4] $\left(2, -\frac{1}{2}, -1\right)$
10. [3.4] No solution **11.** [3.4] (0, 1, 0)
12. [3.6] $\left(\frac{34}{107}, -\frac{104}{107}\right)$ **13.** [3.6] (3, 1, −2)
14. [3.7] 34 **15.** [3.7] 133 **16.** [3.7] $\left(\frac{13}{18}, \frac{7}{27}\right)$
17. [3.5] Electrician: 3.5 hr; carpenter: 8 hr; plumber: 10 hr
18. [3.8] ($3, 55) **19.** [3.8] **(a)** $C(x) = 25x + 40{,}000$; **(b)** $R(x) = 70x$; **(c)** $P(x) = 45x - 40{,}000$; **(d)** $26,500 loss, $500 profit; **(e)** (889 hammocks, $62,230) **20.** [2.3], [3.3] $m = 7, b = 10$ **21.** [3.5] Adults' tickets: 1346; senior citizens' tickets: 335; children's tickets: 1651

Cumulative Review: Chapters 1–3, pp. 266–268

1. 0.5 million bicycles per year
2. (a) $C(s) = \frac{85}{3}s + 70$, where s is the speed, in miles per hour, and C is the number of calories burned per hour; **(b)** approximately 353 calories per hour **3. (a)** $C(w) = 2.64w + 8.\overline{6}$, where w is weight, in pounds, and C is the number of calories burned per hour; **(b)** approximately 365 calories per hour
4. $210 billion **5.** 90 **6.** Length: 10 cm; width: 6 cm
7. Nickels: 19; dimes: 15 **8.** $120 **9.** Soakem: $48\frac{8}{9}$ oz; Rinsem: $71\frac{1}{9}$ oz **10.** Wins: 23; losses: 33; ties: 8
11. Cookie: 90 calories; banana: 80 calories; yogurt: 165 calories **12.** x^{11} **13.** $-\dfrac{40x}{y^5}$
14. $-288x^4y^{18}$ **15.** y^{10} **16.** $-\dfrac{2a^{11}}{5b^{33}}$
17. $\dfrac{81x^{36}}{256y^8}$ **18.** 1.12×10^6 **19.** 4.00×10^6
20. $b = \dfrac{2A}{h} - t$, or $\dfrac{2A - ht}{h}$ **21.** Yes
22.

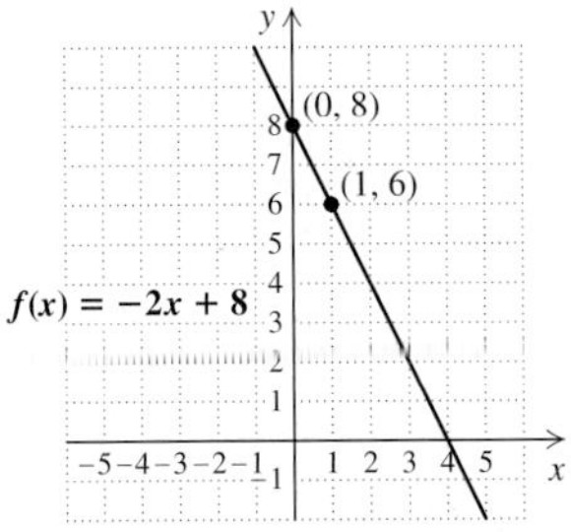

23.

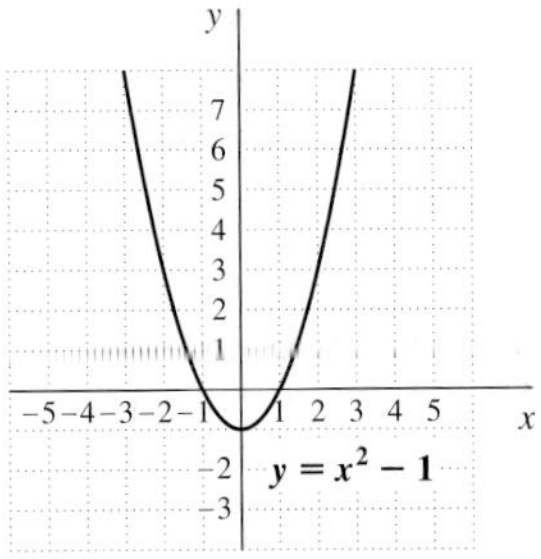

24.

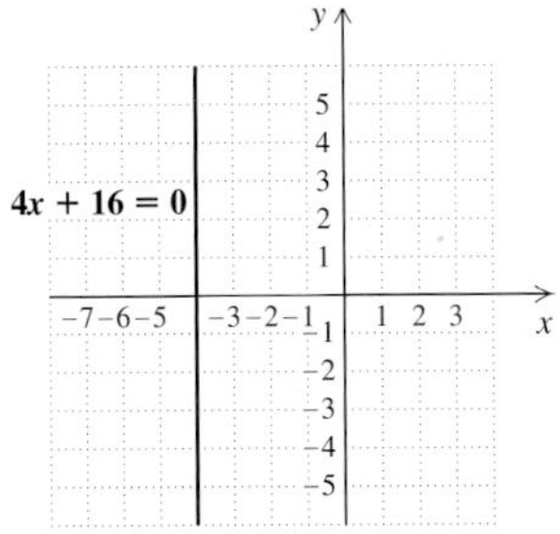

25.

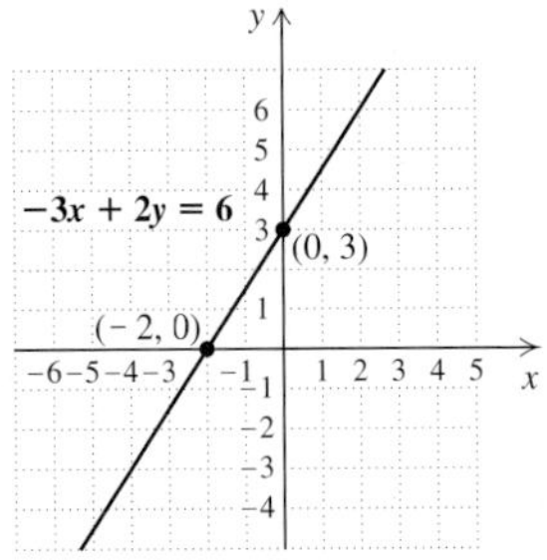

26. Slope: $\frac{9}{4}$; y-intercept: $(0, -3)$ **27.** $\frac{4}{3}$
28. $y = -3x - 5$ **29.** $y = -\frac{1}{10}x + \frac{12}{5}$ **30.** Parallel
31. $y = -2x + 5$
32. $\{-5, -3, -1, 1, 3\}$; $\{-3, -2, 1, 4, 5\}$; −2; 3
33. $\left\{x \mid x \text{ is a real number } and\ x \neq \frac{1}{2}\right\}$ **34.** −31 **35.** 3
36. 7 **37.** $8a^2 + 4a - 4$ **38.** −22 **39.** 20 **40.** −56
41. 6 **42.** −5 **43.** $\frac{10}{9}$ **44.** $-\frac{32}{5}$ **45.** $\frac{18}{17}$ **46.** (1, 1)
47. (−2, 3) **48.** $\left(-3, \frac{2}{5}\right)$ **49.** (−3, 2, −4)
50. (0, −1, 2) **51.** 14 **52.** 0 **53.** $-12x^{2a}y^{b+y+3}$
54. $151,000 **55.** $m = -\frac{5}{9}, b = -\frac{2}{9}$

Chapter 4

Exercise Set 4.1, pp. 278–281

1. Equivalent inequalities **2.** Not equivalent
3. Equivalent equations **4.** Equivalent expressions
5. Not equivalent **6.** Equivalent equations
7. Equivalent expressions **8.** Not equivalent
9. Not equivalent **10.** Equivalent inequalities
11. No, no, yes, yes **13.** No, yes, yes, no
15. 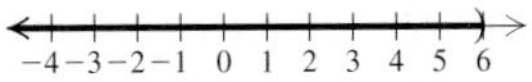 $(-\infty, 6)$, $\{y \mid y < 6\}$
17.  $[-4, \infty)$, $\{x \mid x \geq -4\}$
19.  $(-3, \infty)$, $\{t \mid t > -3\}$
21. 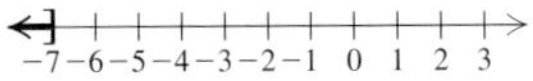 $(-\infty, -7]$, $\{x \mid x \leq -7\}$
23. $\{x \mid x > -6\}$, or $(-6, \infty)$ 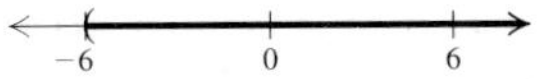
25. $\{a \mid a \leq -20\}$, or $(-\infty, -20]$ 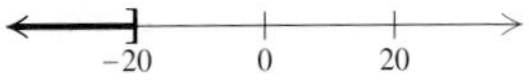
27. $\{x \mid x \leq 17\}$, or $(-\infty, 17]$ 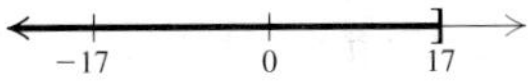
29. $\{y \mid y > -9\}$, or $(-9, \infty)$ 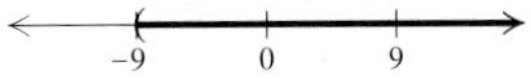
31. $\{y \mid y \leq 14\}$, or $(-\infty, 14]$ 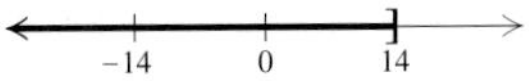
33. $\{t \mid t < -9\}$, or $(-\infty, -9)$ 
35. $\{x \mid x < -60\}$, or $(-\infty, -60)$ 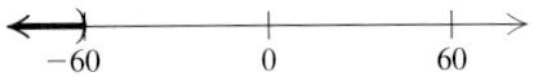
37. $\{x \mid x \leq 0.9\}$, or $(-\infty, 0.9]$ 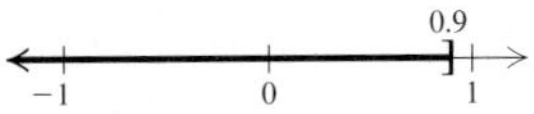
39. $\left\{x \mid x \leq \frac{5}{6}\right\}$, or $\left(-\infty, \frac{5}{6}\right]$ 
41. $\{x \mid x < -26\}$, or $(-\infty, -26)$ 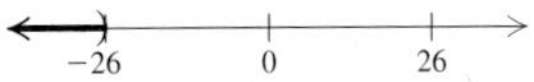
43. $\left\{t \mid t \geq -\frac{13}{3}\right\}$, or $\left[-\frac{13}{3}, \infty\right)$ 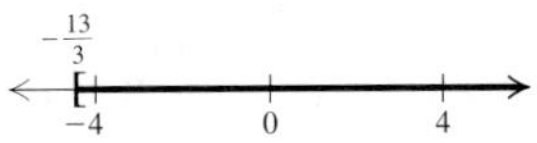
45. $\{x \mid x \geq 6\}$, or $[6, \infty)$ 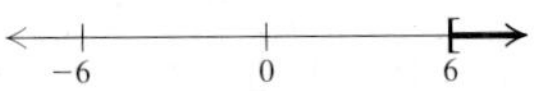

47. $\{x|x \geq 2\}$, or $[2, \infty)$ 

49. $\left\{x|x > \frac{2}{3}\right\}$, or $\left(\frac{2}{3}, \infty\right)$ 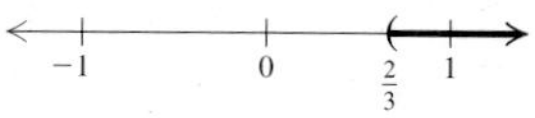

51. $\left\{x|x \geq \frac{1}{2}\right\}$, or $\left[\frac{1}{2}, \infty\right)$ 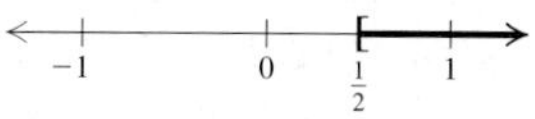

53. $\left\{y|y \leq -\frac{53}{6}\right\}$, or $\left(-\infty, -\frac{53}{6}\right]$ **55.** $\left\{t|t < \frac{29}{5}\right\}$, or $\left(-\infty, \frac{29}{5}\right)$ **57.** $\left\{m|m > \frac{7}{3}\right\}$, or $\left(\frac{7}{3}, \infty\right)$ **59.** $\{x|x \geq 2\}$, or $[2, \infty)$ **61.** $\{y|y < 5\}$, or $(-\infty, 5)$ **63.** $\left\{x|x \leq \frac{4}{7}\right\}$, or $\left(-\infty, \frac{4}{7}\right]$ **65.** Mileages less than or equal to 150 mi **67.** \$11,500 or more **69.** More than 25 checks **71.** Gross sales greater than \$7000 **73.** Parties of more than 80 guests **75.** About 6.8 gal or less **77.** TW **79.** $\{x|x$ is a real number *and* $x \neq 2\}$ **80.** $\{x|x$ is a real number *and* $x \neq -3\}$ **81.** $\left\{x|x \text{ is a real number and } x \neq \frac{7}{2}\right\}$ **82.** $\left\{x|x \text{ is a real number and } x \neq \frac{9}{4}\right\}$ **83.** $7x + 10$ **84.** $22x - 7$ **85.** TW **87.** $\left\{x|x \leq \frac{2}{a - 1}\right\}$ **89.** $\left\{y|y \geq \frac{2a + 5b}{b(a - 2)}\right\}$ **91.** $\left\{x|x > \frac{4m - 2c}{d - (5c + 2m)}\right\}$ **93.** False; $2 < 3$ and $4 < 5$, but $2 - 4 = 3 - 5$. **95.** TW **97.** $\mathbb{R}$ 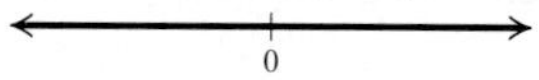

99. $\{x|x$ is a real number *and* $x \neq 0\}$
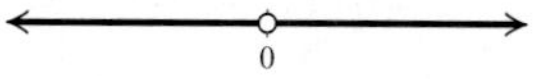

Exercise Set 4.2, pp. 292–297

1. (e) **2.** (d) **3.** (f) **4.** (b) **5.** (a) **6.** (c) **7.** (g) **8.** (h) **9.** −2 **11.** −4 **13.** 8 **15.** 0 **17.** 5 **19.** 7 **21.** 3 **23.** 9 **25.** 1 **27.** 6 **29.** $1\frac{1}{3}$ **31.** $\{x|x \geq 2\}$, or $[2, \infty)$ **33.** $\{x|x < 3\}$, or $(-\infty, 3)$ **35. (a)** $\{x|x \leq -2\}$, or $(-\infty, -2]$; **(b)** $\left\{x|x > \frac{8}{3}\right\}$ or $\left(\frac{8}{3}, \infty\right)$; **(c)** $\left\{x|x > \frac{4}{5}\right\}$, or $\left(\frac{4}{5}, \infty\right)$ **37.** $\{x|x < 7\}$, or $(-\infty, 7)$ **39.** $\{x|x \geq 2\}$, or $[2, \infty)$ **41.** $\{x|x < 8\}$, or $(-\infty, 8)$ **43.** $\{x|x \leq 2\}$, or $(-\infty, 2]$ **45.** $\left\{x|x \leq -\frac{4}{9}\right\}$, or $\left(-\infty, -\frac{4}{9}\right]$ **47.** \$10,500 over \$100 **49.** 5 months **51.** 2 hr 15 min **53.** 150 lb **55.** At least 625 people **57. (a)** $\left\{x|x < 8181\frac{9}{11}\right\}$, or $\{x|x \leq 8181\}$; **(b)** $\left\{x|x > 8181\frac{9}{11}\right\}$, or $\{x|x \geq 8182\}$ **59.** $n(x) = 1.274096386x + 58.45783133$; years after 2021 **61.** $n(x) = -2.05x + 178.6444444$, $v(x) = 4.566666667x + 55.4$; years after 2018 **63.** $f(x) = 0.0056862962x + 1.891866531$; years after 2006 **65.** TW **67.** $\frac{333}{245}$ **68.** −12 **69.** One piece: 20.8 ft; other piece: 12 ft **70.** Scientific calculators: 18; graphing calculators: 27 **71.** Mazzas: 28 rolls; Kranepools: 8 rolls **72.** Buckets: 17; dinners: 11 **73.** TW **75.** −2, 2 **77.** 1 **79.** −6, 2 **81.** −1, 2

83.
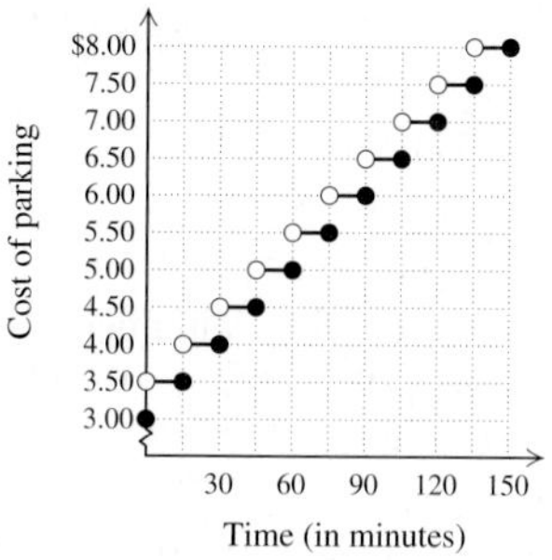

Interactive Discovery, p. 305

1. Domain of $f = (-\infty, 3]$; domain of $g = [-1, \infty)$ **2.** Domain of $f + g =$ domain of $f - g =$ domain of $f \cdot g = [-1, 3]$ **3.** By finding the intersection of the domains of f and g

Exercise Set 4.3, pp. 306–309

1. (h) **2.** (j) **3.** (f) **4.** (a) **5.** (e) **6.** (d) **7.** (b) **8.** (g) **9.** (c) **10.** (i) **11.** $\{9, 11\}$ **13.** $\{0, 5, 10, 15, 20\}$ **15.** $\{b, d, f\}$ **17.** $\{r, s, t, u, v\}$ **19.** $\varnothing$ **21.** $\{3, 5, 7\}$

23. 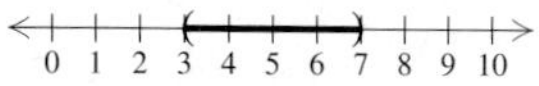 $(3, 7)$

25. 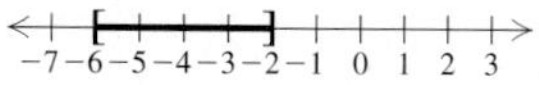 $[-6, -2]$

27. 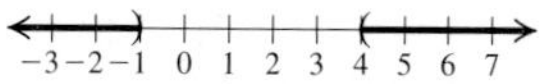 $(-\infty, -1) \cup (4, \infty)$

29. 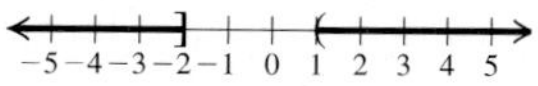 $(-\infty, -2] \cup (1, \infty)$

31. 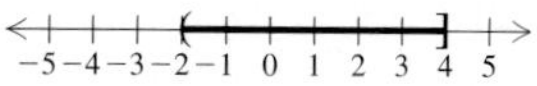 $(-2, 4]$

33. 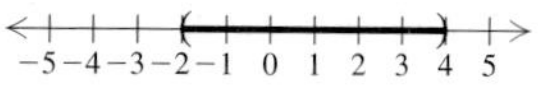 $(-2, 4)$

35. 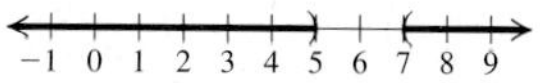 $(-\infty, 5) \cup (7, \infty)$

37. 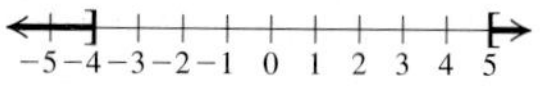 $(-\infty, -4] \cup [5, \infty)$

39. 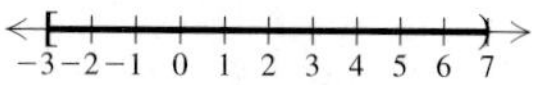 $[-3, 7)$

41. 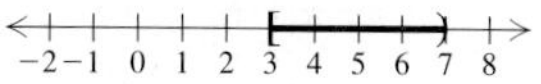 $[3, 7)$

43. 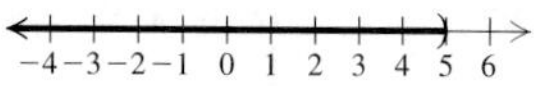 $(-\infty, 5)$

45. $\{t|-3 < t < 7\}$, or $(-3, 7)$ 

47. $\{x|-1 < x \leq 4\}$, or $(-1, 4]$ 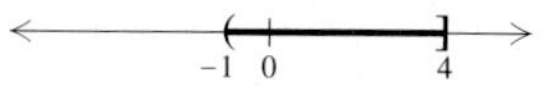

49. $\{a|-2 \leq a < 2\}$, or $[-2, 2)$ 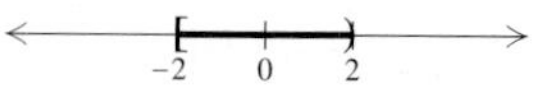

51. $\mathbb{R}$, or $(-\infty, \infty)$ 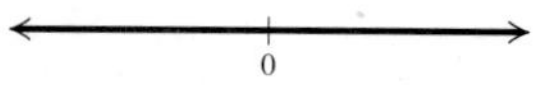

53. $\{x|7 < x < 23\}$, or $(7, 23)$ 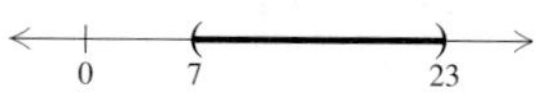

55. $\{x \mid -3 \leq x \leq 2\}$, or $[-3, 2]$ 

57. $\{x \mid 1 \leq x \leq 3\}$, or $[1, 3]$ 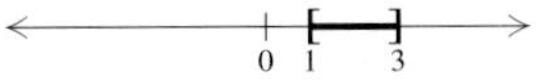

59. $\left\{x \mid -\frac{7}{2} < x \leq 7\right\}$, or $\left(-\frac{7}{2}, 7\right]$ 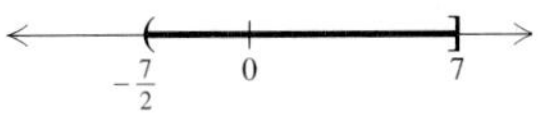

61. $\{x \mid x \leq 1 \text{ or } x \geq 3\}$, or $(-\infty, 1] \cup [3, \infty)$
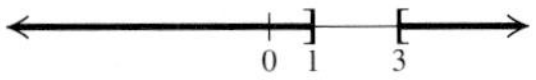

63. $\{x \mid x < 2 \text{ or } x > 6\}$, or $(-\infty, 2) \cup (6, \infty)$

65. $\left\{a \mid a < \frac{7}{2}\right\}$, or $\left(-\infty, \frac{7}{2}\right)$ 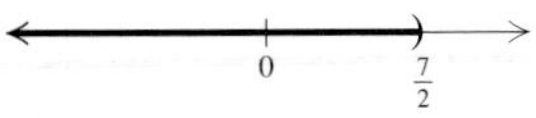

67. $\{a \mid a < -5\}$, or $(-\infty, -5)$ 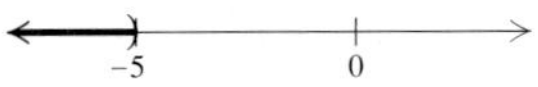

69. $\mathbb{R}$, or $(-\infty, \infty)$ 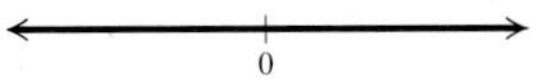

71. $\{t \mid t \leq 6\}$, or $(-\infty, 6]$ 

73. $(-1, 6)$ **75.** $(-\infty, -8) \cup (-8, \infty)$ **77.** $[6, \infty)$
79. $(-\infty, 4) \cup (4, \infty)$ **81.** $\left[-\frac{7}{2}, \infty\right)$ **83.** $(-\infty, 4]$
85. $[5, \infty)$ **87.** $\left[\frac{2}{3}, 3\right]$ **89.** TW

91.
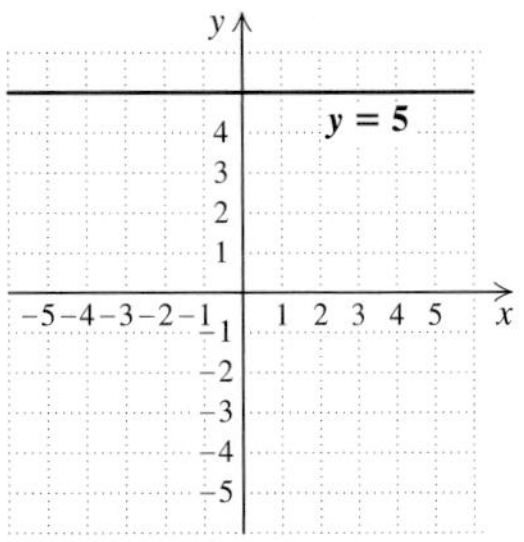

92.
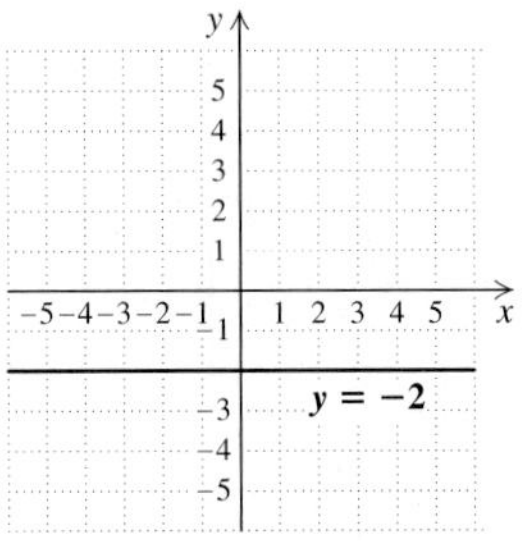

93.
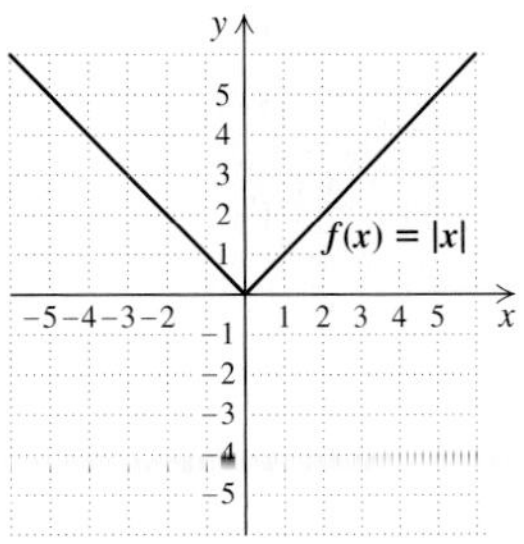

94.
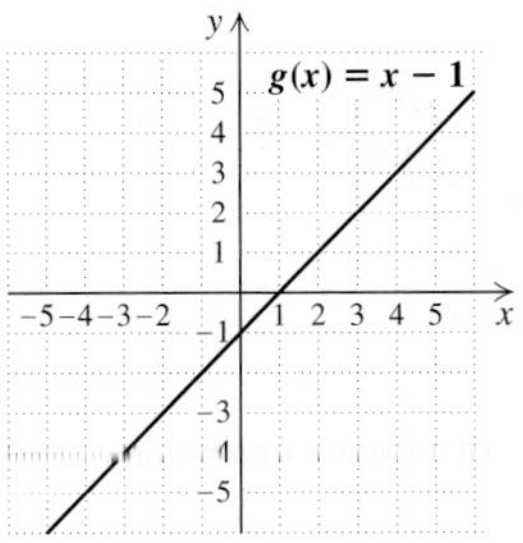

95. $(8, 5)$ **96.** $(-5, -3)$ **97.** TW **99.** From 2000 to 2025 **101.** Sizes between 6 and 13
103. $1965 \leq y \leq 1981$ **105.** Between 12 and 240 trips
107. $\left\{m \mid m < \frac{6}{5}\right\}$, or $\left(-\infty, \frac{6}{5}\right)$ 

109. $\left\{x \mid -\frac{1}{8} < x < \frac{1}{2}\right\}$, or $\left(-\frac{1}{8}, \frac{1}{2}\right)$ 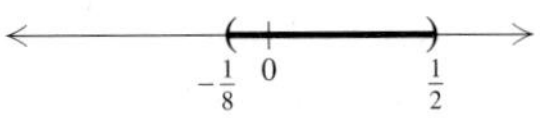

111. False **113.** True **115.** $(-\infty, -7) \cup \left(-7, \frac{3}{4}\right]$

Exercise Set 4.4, pp. 318–320

1. True **2.** True **3.** False **4.** True **5.** True
6. True **7.** False **8.** False **9.** $\{-5, 1\}$ **11.** $(-5, 1)$
13. $(-\infty, -5] \cup [1, \infty)$ **15.** $\{-7, 7\}$ **17.** $\varnothing$ **19.** $\{0\}$
21. $\{-5.5, 5.5\}$ **23.** $\left\{-\frac{1}{2}, \frac{7}{2}\right\}$ **25.** $\varnothing$ **27.** $\{-4, 8\}$
29. $\{2, 8\}$ **31.** $\{-2, 16\}$ **33.** $\{-8, 8\}$ **35.** $\left\{-\frac{11}{7}, \frac{11}{7}\right\}$
37. $\{-7, 8\}$ **39.** $\{-12, 2\}$ **41.** $\left\{-\frac{1}{3}, 3\right\}$ **43.** $\{-7, 1\}$
45. $\{-8.7, 8.7\}$ **47.** $\left\{-\frac{8}{3}, 4\right\}$ **49.** $\{1, 11\}$ **51.** $\left\{-\frac{1}{2}\right\}$
53. $\left\{-\frac{3}{5}, 5\right\}$ **55.** $\mathbb{R}$ **57.** $\{1\}$ **59.** $\left\{32, \frac{8}{3}\right\}$
61. $\{a \mid -9 \leq a \leq 9\}$, or $[-9, 9]$ 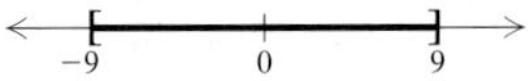

63. $\{x \mid x < -8 \text{ or } x > 8\}$, or $(-\infty, -8) \cup (8, \infty)$
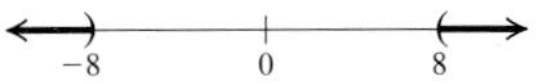

65. $\{t \mid t < 0 \text{ or } t > 0\}$, or $(-\infty, 0) \cup (0, \infty)$
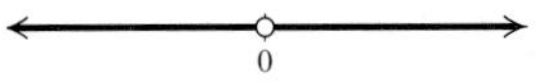

67. $\{x \mid -3 < x < 5\}$, or $(-3, 5)$ 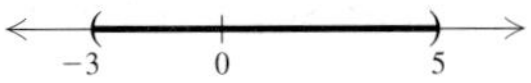

69. $\{x \mid -8 \leq x \leq 4\}$, or $[-8, 4]$ 

71. $\{x \mid x < -2 \text{ or } x > 8\}$, or $(-\infty, -2) \cup (8, \infty)$

73. $\mathbb{R}$, or $(-\infty, \infty)$ 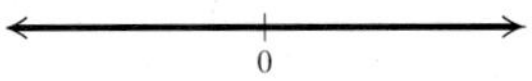

75. $\left\{a \mid a \leq -\frac{2}{3} \text{ or } a \geq \frac{10}{3}\right\}$, or $\left(-\infty, -\frac{2}{3}\right] \cup \left[\frac{10}{3}, \infty\right)$
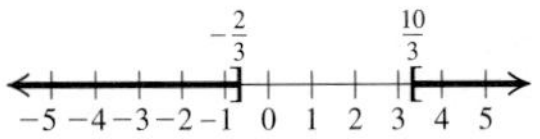

77. $\{y \mid -9 < y < 15\}$, or $(-9, 15)$
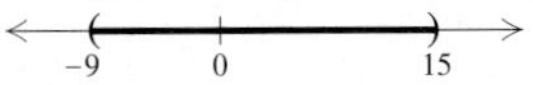

79. $\{x \mid x \leq -8 \text{ or } x \geq 0\}$, or $(-\infty, -8] \cup [0, \infty)$

81. $\left\{y \mid y < -\frac{4}{3} \text{ or } y > 4\right\}$, or $\left(-\infty, -\frac{4}{3}\right) \cup (4, \infty)$

83. $\varnothing$ **85.** $\left\{x \mid x \leq -\frac{2}{15} \text{ or } x \geq \frac{14}{15}\right\}$, or $\left(-\infty, -\frac{2}{15}\right] \cup \left[\frac{14}{15}, \infty\right)$

87. $\{m \mid -9 \leq m \leq 3\}$, or $[-9, 3]$ 

89. $\{a \mid -6 < a < 0\}$, or $(-6, 0)$ 

91. $\left\{x \mid -\frac{1}{2} \leq x \leq \frac{7}{2}\right\}$, or $\left[-\frac{1}{2}, \frac{7}{2}\right]$ 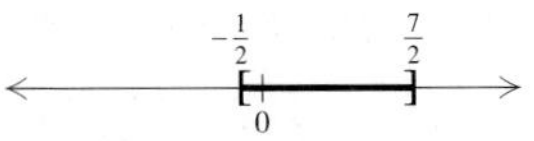

93. $\left\{x \mid x \leq -\frac{7}{3} \text{ or } x \geq 5\right\}$, or $\left(-\infty, -\frac{7}{3}\right] \cup [5, \infty)$
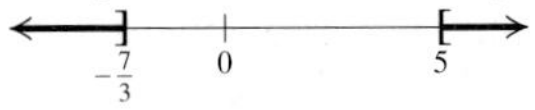

95. $\{x \mid -4 < x < 5\}$, or $(-4, 5)$ 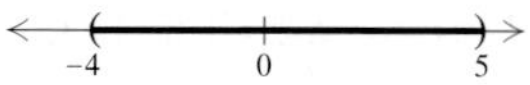

97. TW **99.** 6 **100.** $\{x \mid x \geq 6\}$, or $[6, \infty)$ **101.** (6, 21)
102. $\left\{x \mid x \leq -\frac{1}{3}\right\}$, or $\left(-\infty, -\frac{1}{3}\right]$
103. $\{x \mid -11 < x < 1\}$, or $(-11, 1)$
104. $\{-11, 1\}$ **105.** TW **107.** $\left\{t \mid t \geq \frac{5}{3}\right\}$, or $\left[\frac{5}{3}, \infty\right)$
109. $\mathbb{R}$, or $(-\infty, \infty)$ **111.** $\left\{-\frac{1}{7}, \frac{7}{3}\right\}$ **113.** $|x| < 3$
115. $|x| \geq 6$ **117.** $|x + 3| > 5$
119. $|x - 7| < 2$, or $|7 - x| < 2$ **121.** $|x - 3| \leq 4$
123. $|x + 4| < 3$ **125.** Between 80 ft and 100 ft

Visualizing the Graph, p. 330

1. B **2.** F **3.** J **4.** A **5.** E **6.** G **7.** C **8.** D
9. I **10.** H

Exercise Set 4.5, pp. 331–333

1. (e) **2.** (c) **3.** (d) **4.** (a) **5.** (b) **6.** (f)
7. Yes **9.** No

11.
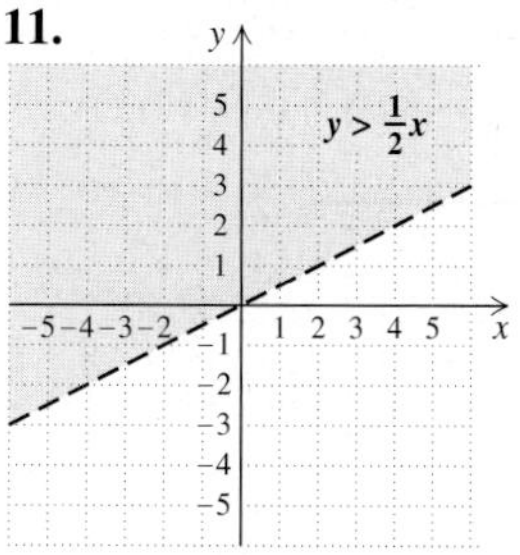

13.
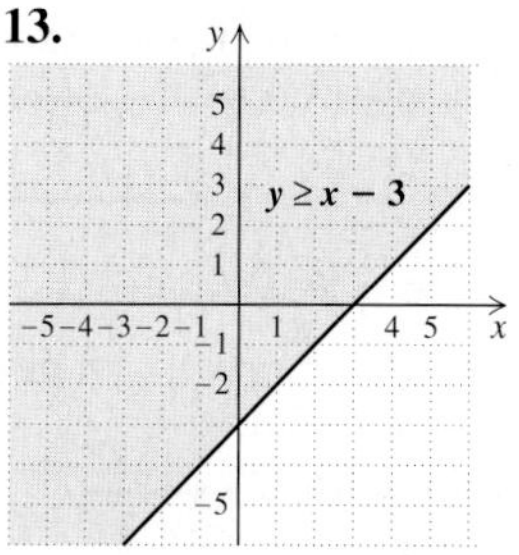

15.
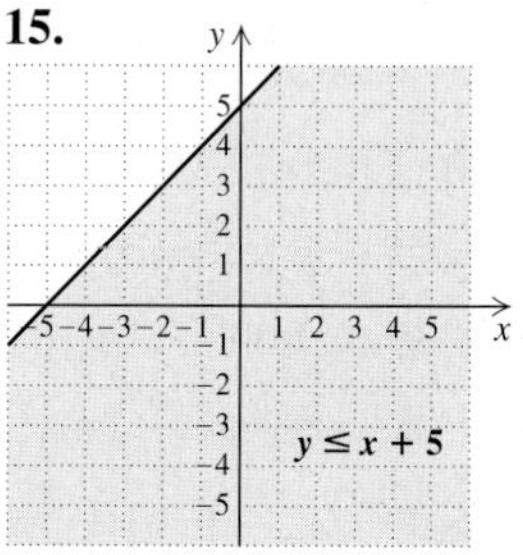

17.
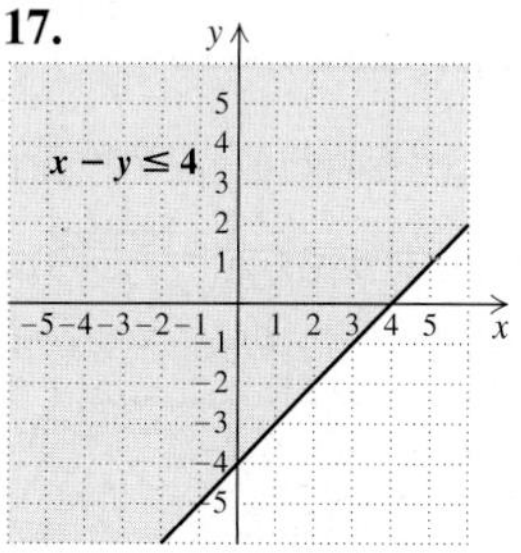

19.
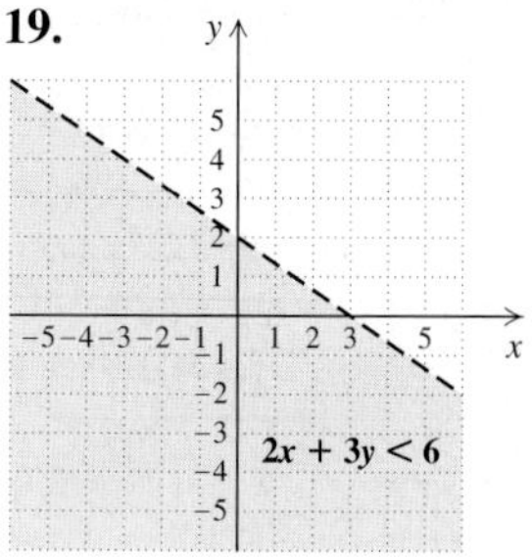

21.
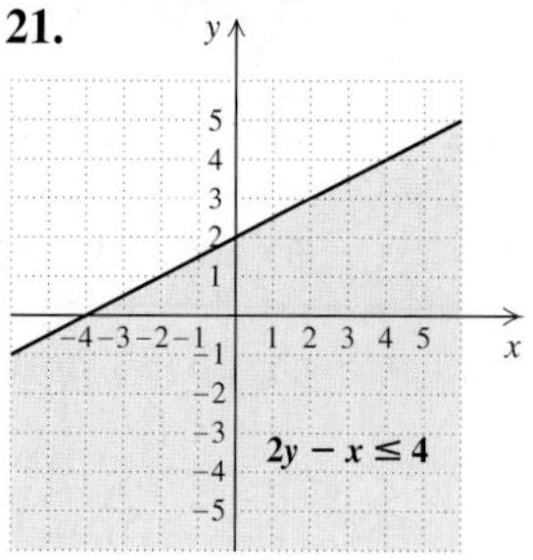

23.
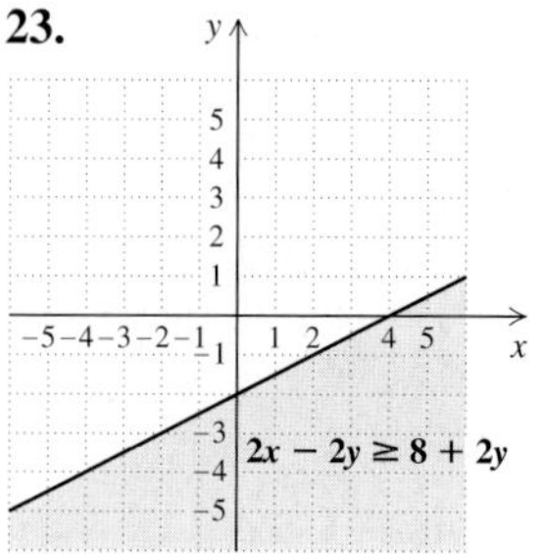

25.
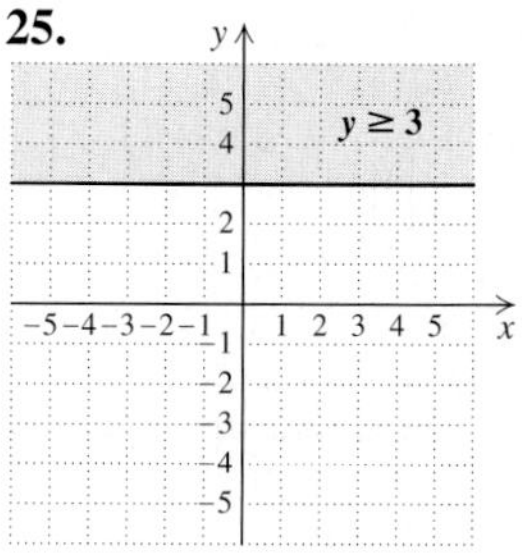

27.
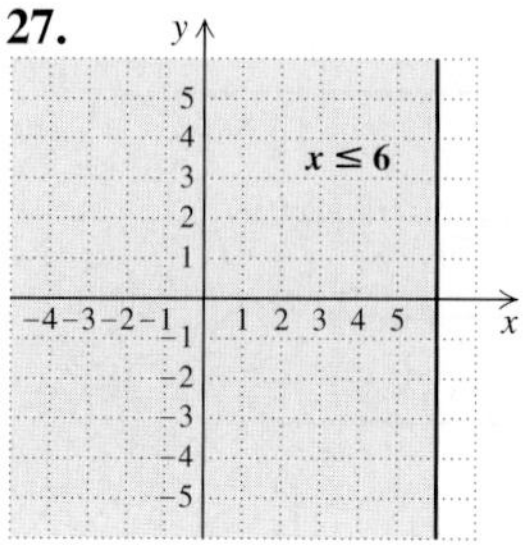

29.
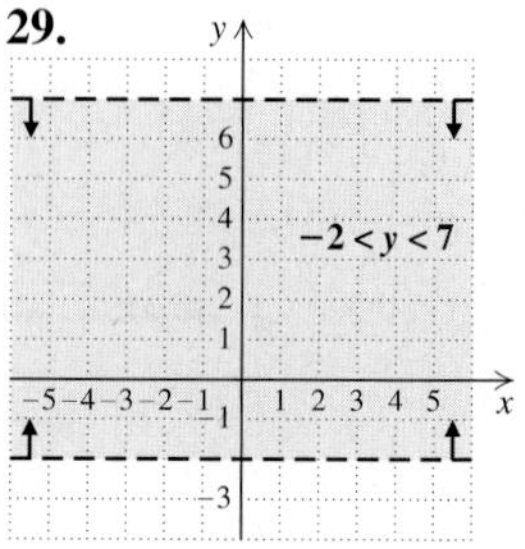

31.
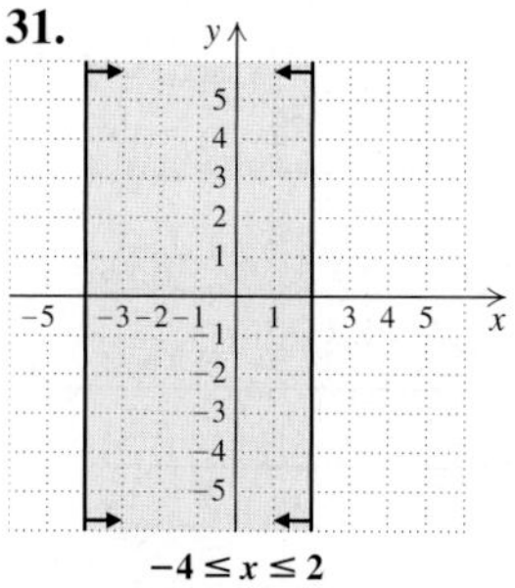

33.
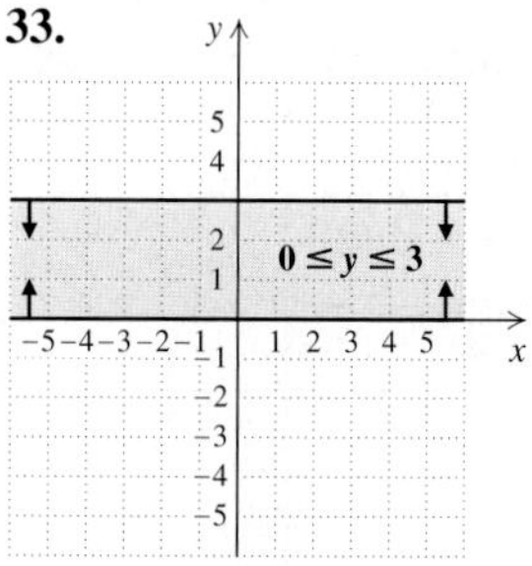

35. $y > x + 3.5$
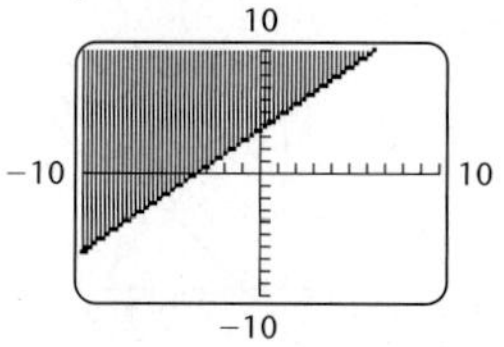

37. $8x - 2y < 11$
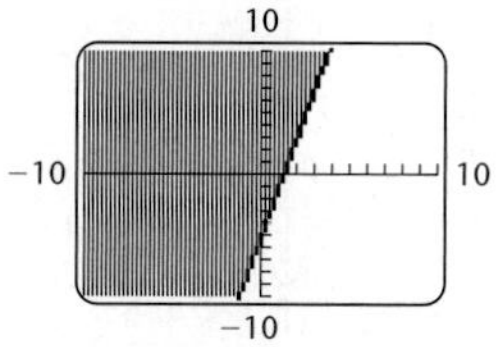

39.
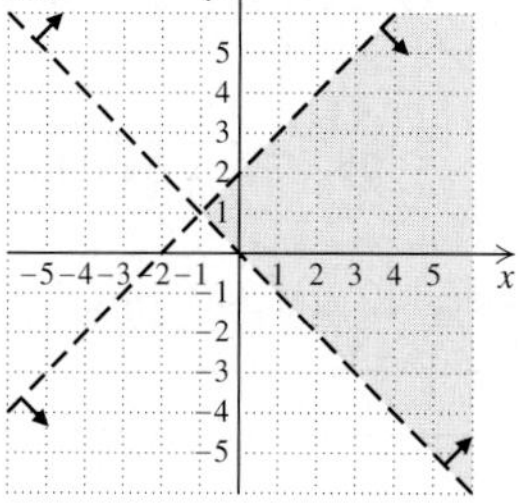

41.
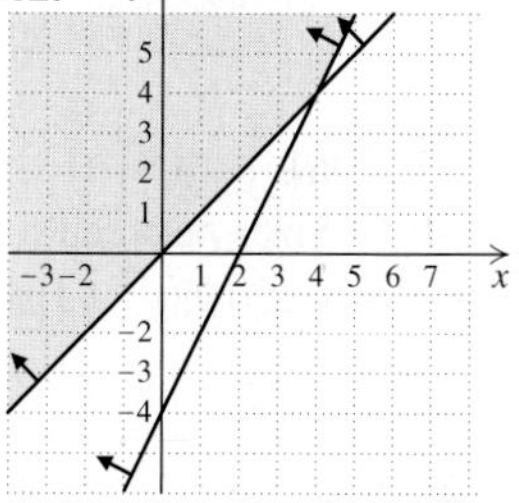

43.
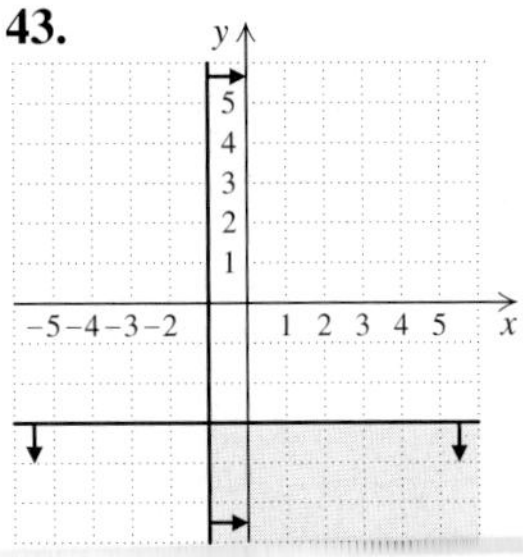

45.
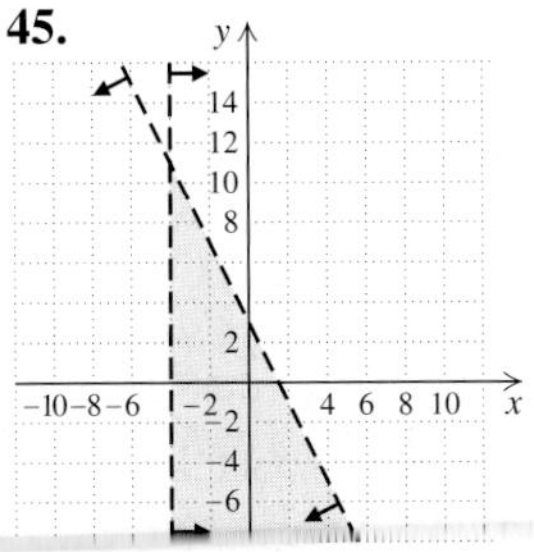

71.
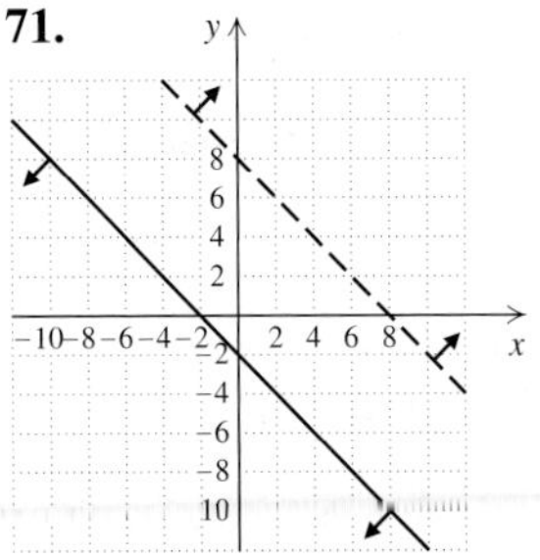

73.
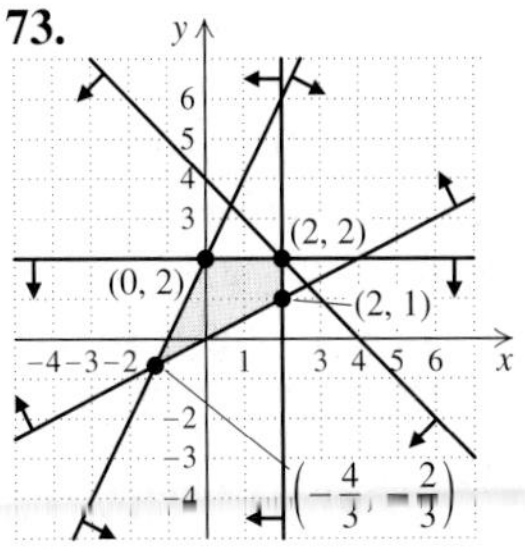

47.
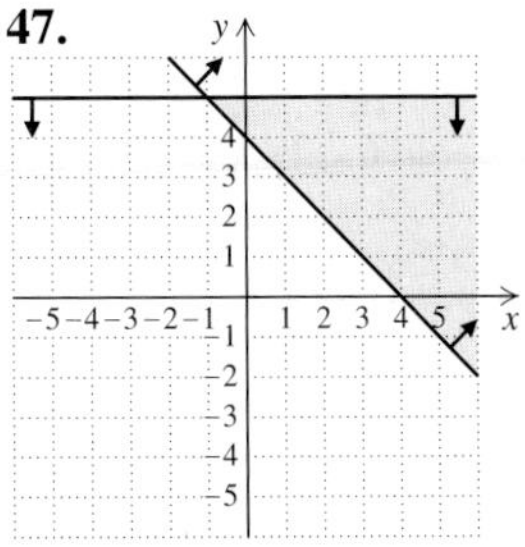

49.
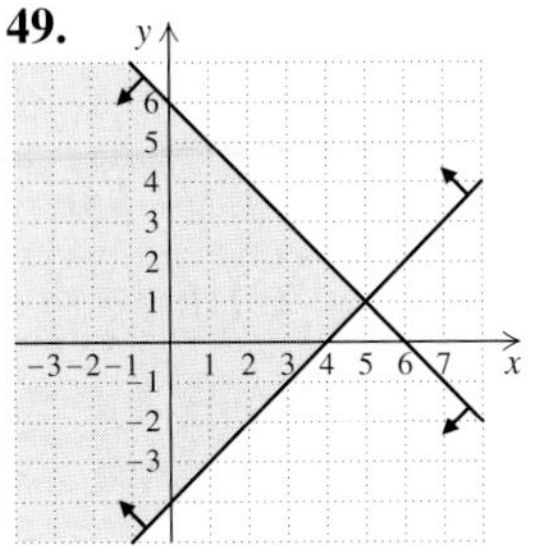

51.
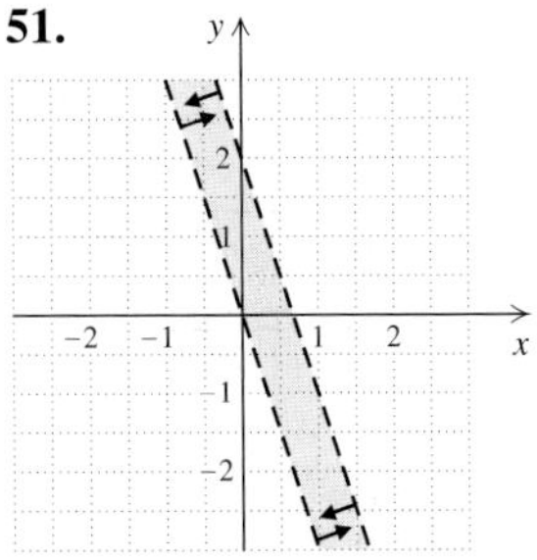

53.
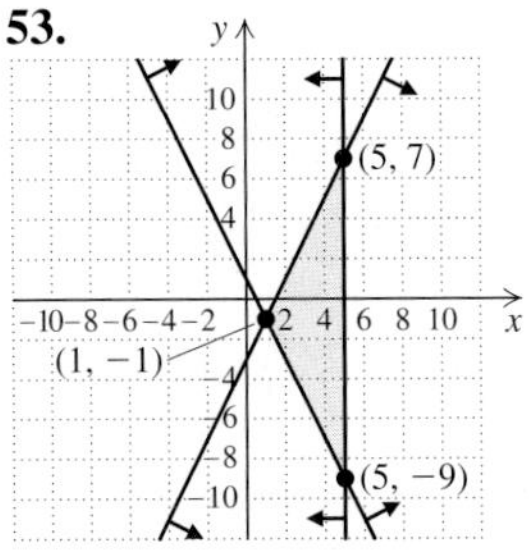

55.
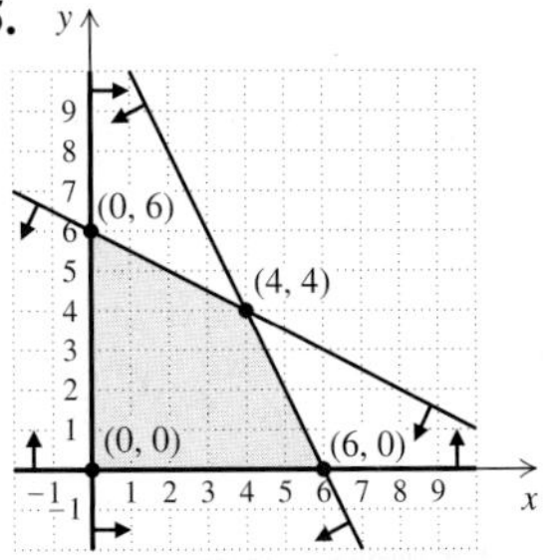

57.
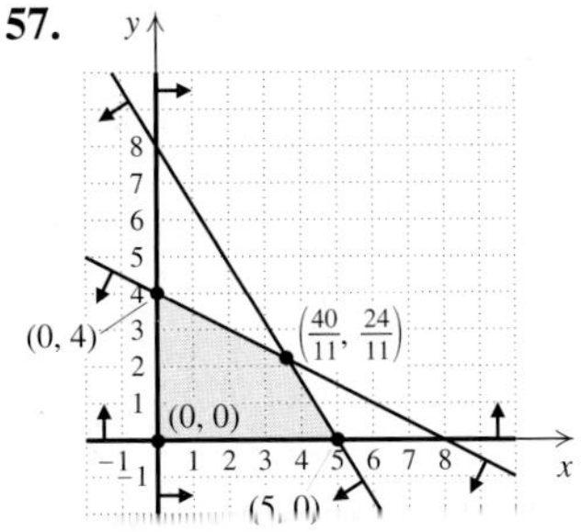

59.
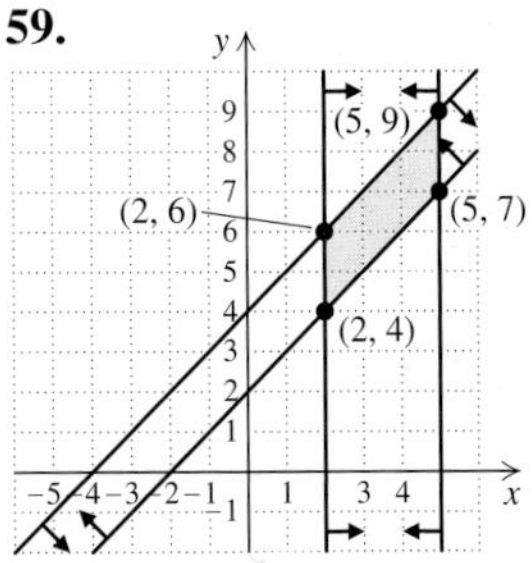

61. TW **63.** Peanuts: $6\frac{2}{3}$ lb; fancy nuts: $3\frac{1}{3}$ lb **64.** Hendersons: 10 bags; Savickis: 4 bags **65.** Activity-card holders: 128; noncard holders: 75 **66.** Students: 70; adults: 130 **67.** 25 ft **68.** 2.75% **69.** TW

75. $w > 0,$
$h > 0,$
$w + h + 30 \leq 62,$ or
$w + h \leq 32,$
$2w + 2h + 30 \leq 130,$ or
$w + h \leq 50$

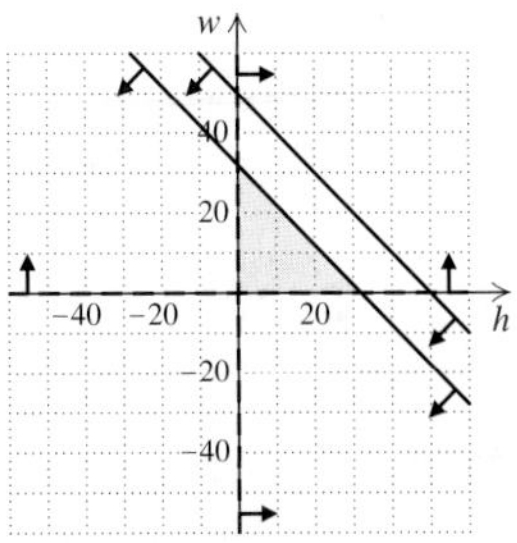

77. $35c + 75a > 1000,$
$c \geq 0,$
$a \geq 0$

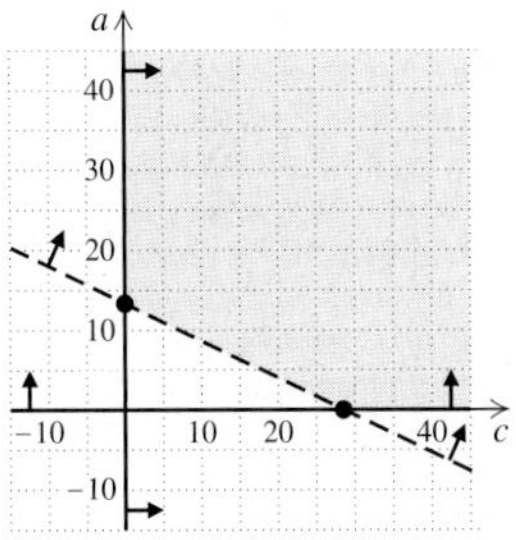

79. $y \leq x,$
$y \leq 2$

81. $y \leq x + 2,$
$y \leq -x + 4,$
$y \geq 0$

Review Exercises: Chapter 4, pp. 336–337

1. True **2.** False **3.** True **4.** True **5.** False
6. True **7.** True **8.** True **9.** False **10.** False

11. $\{x \mid x \leq -2\}$, or $(-\infty, -2]$;
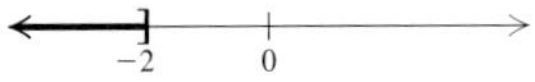

12. $\{u \mid u \leq -21\}$, or $(-\infty, -21]$;

13. $\{y \mid y \geq -7\}$, or $[-7, \infty)$;

14. $\left\{y \mid y > -\frac{15}{4}\right\}$, or $\left(-\frac{15}{4}, \infty\right)$;
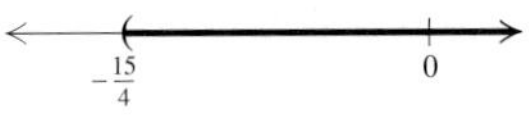

15. $\{y \mid y > -30\}$, or $(-30, \infty)$;
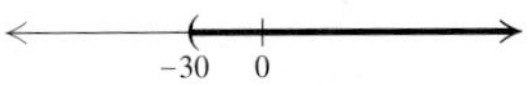

16. $\left\{x \mid x > -\frac{3}{2}\right\}$, or $\left(-\frac{3}{2}, \infty\right)$;
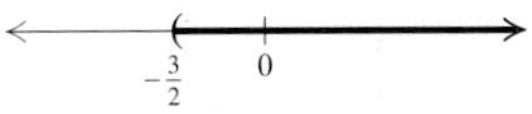

17. $\{x \mid x < -3\}$, or $(-\infty, -3)$;
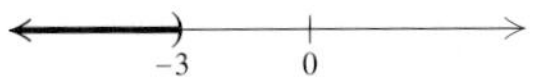

18. $\left\{y \mid y > -\frac{220}{23}\right\}$, or $\left(-\frac{220}{23}, \infty\right)$;
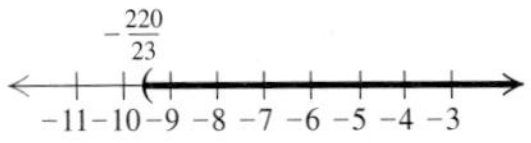

19. $\{x \mid x \le -\frac{5}{2}\}$, or $\left(-\infty, -\frac{5}{2}\right]$;
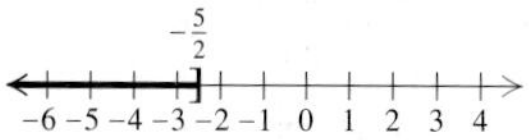

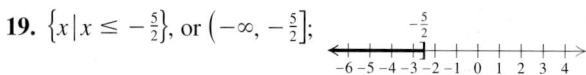

20. $\{x \mid x \le 4\}$, or $(-\infty, 4]$ **21.** More than 125 hr
22. \$3000 **23.** -4 **24.** $\{x \mid x \ge -6\}$, or $[-6, \infty)$
25. $\{1, 5, 9\}$ **26.** $\{1, 2, 3, 5, 6, 9\}$
27. 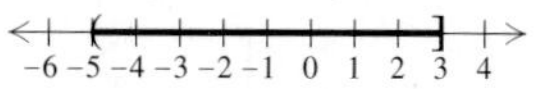 $(-5, 3]$
28. 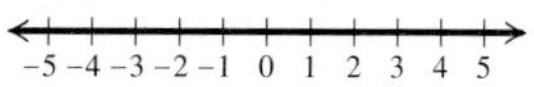 $(-\infty, \infty)$
29. $\{x \mid -12 < x \le -3\}$, or $(-12, -3]$
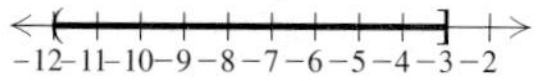
30. $\{x \mid -\frac{5}{4} < x < \frac{5}{2}\}$, or $\left(-\frac{5}{4}, \frac{5}{2}\right)$
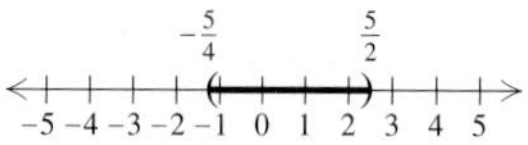
31. $\{x \mid x < -3 \text{ or } x > 1\}$, or $(-\infty, -3) \cup (1, \infty)$
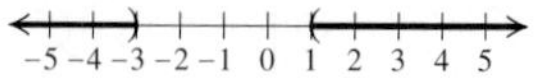
32. $\{x \mid x < -11 \text{ or } x \ge -6\}$, or $(-\infty, -11) \cup [-6, \infty)$
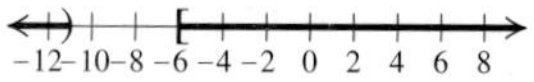
33. $\{x \mid x \le -6 \text{ or } x \ge 8\}$, or $(-\infty, -6] \cup [8, \infty)$
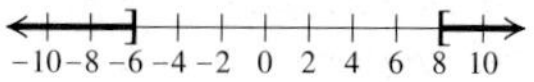
34. $\{x \mid x < -\frac{2}{5} \text{ or } x > \frac{8}{5}\}$, or $\left(-\infty, -\frac{2}{5}\right) \cup \left(\frac{8}{5}, \infty\right)$
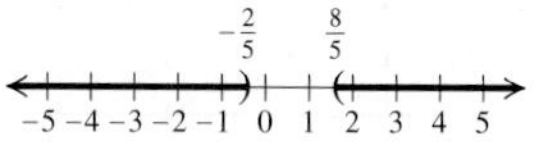

35. $(-\infty, 8) \cup (8, \infty)$ **36.** $[-5, \infty)$ **37.** $\left(-\infty, \frac{8}{3}\right]$
38. $\{-5, 5\}$ **39.** $\{t \mid t \le -3.5 \text{ or } t \ge 3.5\}$, or $(-\infty, -3.5] \cup [3.5, \infty)$ **40.** $\{-4, 10\}$
41. $\{x \mid -\frac{17}{2} < x < \frac{7}{2}\}$, or $\left(-\frac{17}{2}, \frac{7}{2}\right)$
42. $\{x \mid x \le -\frac{11}{3} \text{ or } x \ge \frac{19}{3}\}$, or $\left(-\infty, -\frac{11}{3}\right] \cup \left[\frac{19}{3}, \infty\right)$
43. $\{-14, \frac{4}{3}\}$ **44.** $\varnothing$ **45.** $\{x \mid -16 \le x \le 8\}$, or $[-16, 8]$
46. $\{x \mid x < 0 \text{ or } x > 10\}$, or $(-\infty, 0) \cup (10, \infty)$ **47.** $\varnothing$

48.
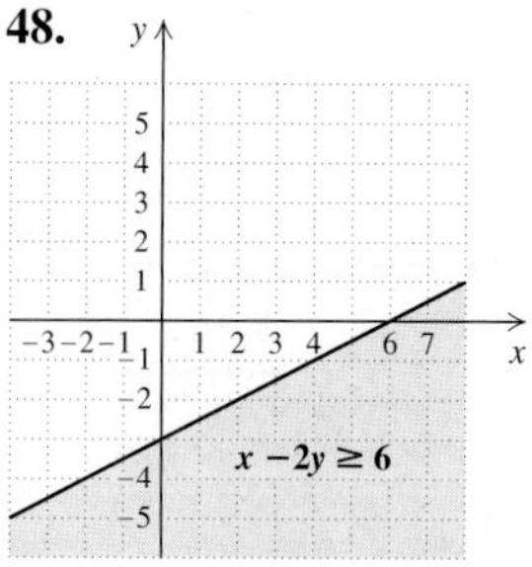

49.
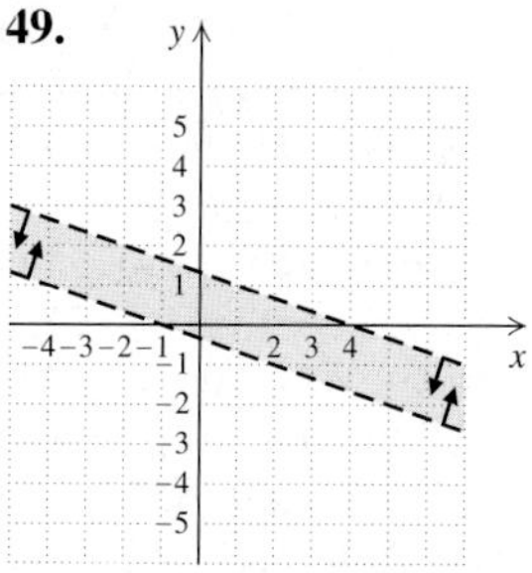

50.
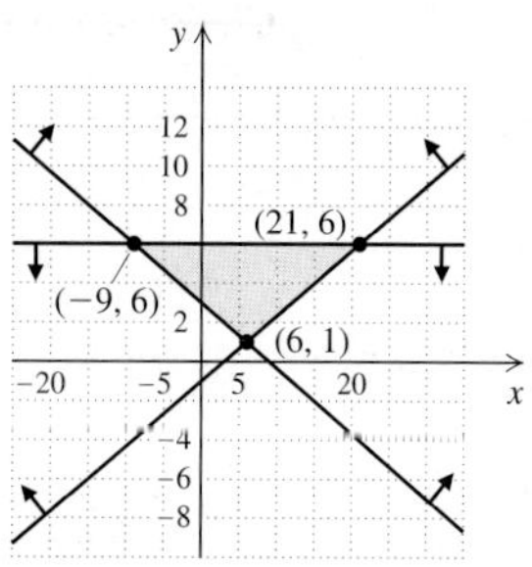

51. TW The equation $|X| = p$ has two solutions when p is positive because X can be either p or $-p$. The same equation has no solution when p is negative because no number has a negative absolute value. **52.** TW The solution set of a system of inequalities is all ordered pairs that make *all* the individual inequalities true. This consists of ordered pairs that are common to all the individual solution sets, or the intersection of the graphs.
53. $\{x \mid -\frac{8}{3} \le x \le -2\}$, or $\left[-\frac{8}{3}, -2\right]$ **54.** False; $-4 < 3$ is true, but $(-4)^2 < 9$ is false. **55.** $|d - 1.1| \le 0.03$
56. $|t - 21.5| \le 3.5$

Test: Chapter 4, p. 338

1. [4.1] $\{x \mid x < 12\}$, or $(-\infty, 12)$ 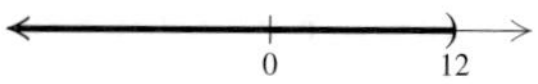
2. [4.1] $\{y \mid y > -50\}$, or $(-50, \infty)$

3. [4.1] $\{y \mid y \le -2\}$, or $(-\infty, -2]$ 
4. [4.1] $\{a \mid a \le \frac{11}{5}\}$, or $\left(-\infty, \frac{11}{5}\right]$ 
5. [4.1] $\{x \mid x > \frac{16}{5}\}$, or $\left(\frac{16}{5}, \infty\right)$ 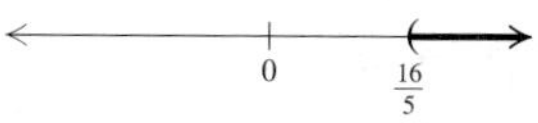
6. [4.1] $\{x \mid x \le \frac{7}{4}\}$, or $\left(-\infty, \frac{7}{4}\right]$ 

7. [4.1] $\{x \mid x > 1\}$, or $(1, \infty)$ **8.** [4.1] More than $166\frac{2}{3}$ mi
9. [4.1] Less than or equal to 2.5 hr **10.** [4.2] 3
11. [4.2] $\{x \mid x > -3\}$, or $(-3, \infty)$ **12.** [4.3] $\{3, 5\}$
13. [4.3] $\{1, 3, 5, 7, 9, 11, 13\}$ **14.** [4.3] $(-\infty, 4]$
15. [4.3] $\{x \mid 1 < x < 8\}$, or $(1, 8)$ 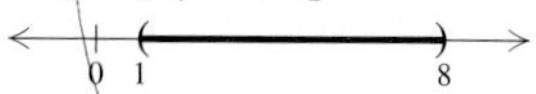
16. [4.3] $\{t \mid -\frac{2}{5} < t \le \frac{9}{5}\}$, or $\left(-\frac{2}{5}, \frac{9}{5}\right]$
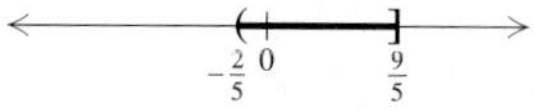
17. [4.3] $\{x \mid x < 3 \text{ or } x > 6\}$, or $(-\infty, 3) \cup (6, \infty)$

18. [4.3] $\{x \mid x < -4 \text{ or } x > -\frac{5}{2}\}$, or $(-\infty, -4) \cup \left(-\frac{5}{2}, \infty\right)$
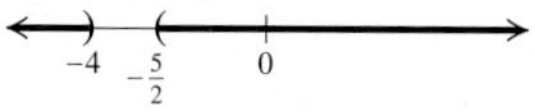
19. [4.3] $\{x \mid 4 \le x < \frac{15}{2}\}$, or $\left[4, \frac{15}{2}\right)$
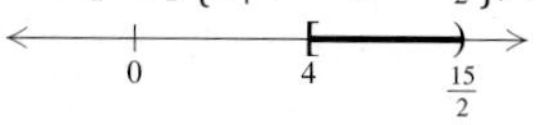
20. [4.4] $\{-13, 13\}$ 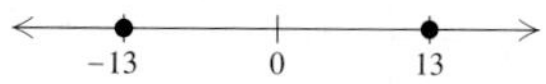
21. [4.4] $\{a \mid a < -7 \text{ or } a > 7\}$, or $(-\infty, -7) \cup (7, \infty)$
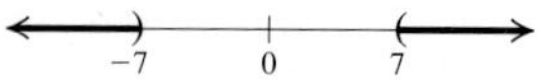
22. [4.4] $\{x \mid -2 < x < \frac{8}{3}\}$, or $\left(-2, \frac{8}{3}\right)$
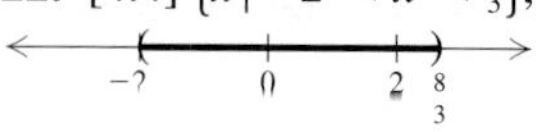

23. [4.4] $\{t \mid t \le -\frac{13}{5}\ or\ t \ge \frac{7}{5}\}$, or $\left(-\infty, -\frac{13}{5}\right] \cup \left[\frac{7}{5}, \infty\right)$

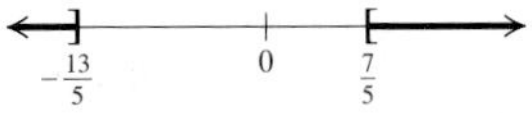

24. [4.4] $\varnothing$ **25.** [4.3] $\{x \mid x < \frac{1}{2}\ or\ x > \frac{7}{2}\}$, or $\left(-\infty, \frac{1}{2}\right) \cup \left(\frac{7}{2}, \infty\right)$

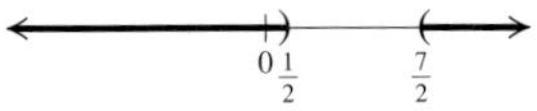

26. [4.4] $\{1\}$
27. [4.5]

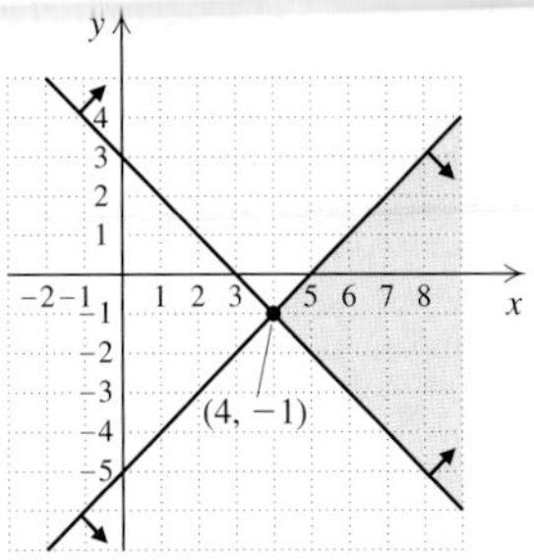

28. [4.5]

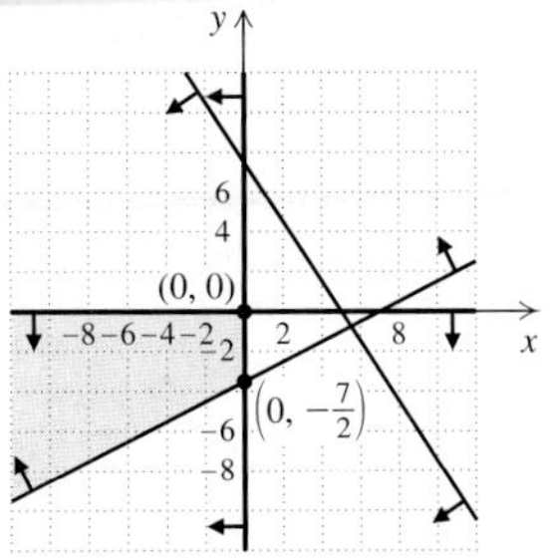

29. [4.4] $[-1, 0] \cup [4, 6]$ **30.** [4.3] $\left(\frac{1}{5}, \frac{4}{5}\right)$
31. [4.4] $|x + 3| \le 5$

Chapter 5

Interactive Discovery, p. 341

Answers may include: The graph of a polynomial function has no sharp corners; there are no holes or breaks; the domain is the set of all real numbers.

Interactive Discovery, pp. 348–349

1. The graphs of y_1 and y_2 should be the same.

$y_1 = (-3x^3 + 2x - 4) + (4x^3 + 3x^2 + 2)$,
$y_2 = x^3 + 3x^2 + 2x - 2$

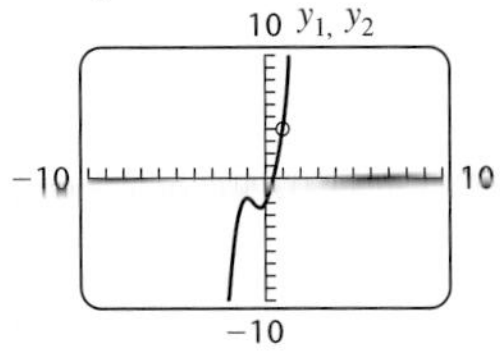

2. The graph of y_3 should be the x-axis.

$y_3 = y_2 - y_1$

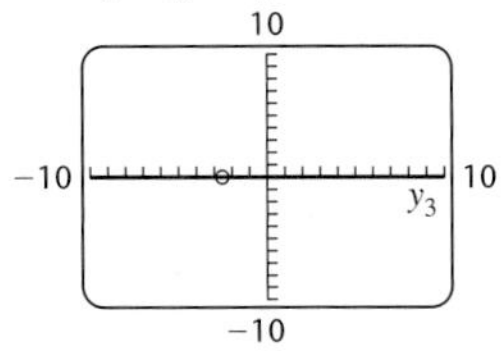

3.

X	Y1	Y2
−3	−8	−8
−2	−2	−2
−1	−2	−2
0	−2	−2
1	4	4
2	22	22
3	58	58
X = −3		

4. The graphs appear to coincide.

$y_1 = (-2x^4 + 3x^2 + 5) + (6x^4 - 2x - 8)$,
$y_2 = 4x^4 + 3x^2 - 2x$

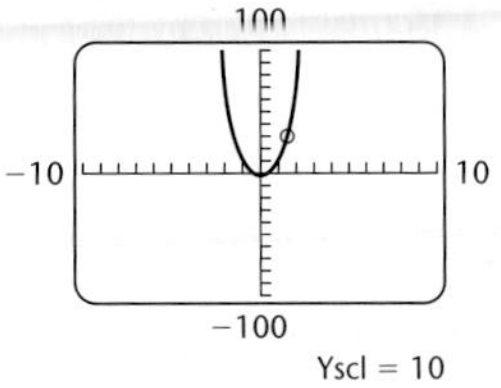

5. The table shows that the values of y_1 and y_2 are different for the same x-value.

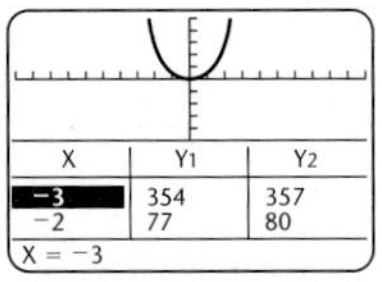

6. The graph of y_3 is not the x-axis. It may be easier to see that the difference is not zero than that the graphs do not coincide.

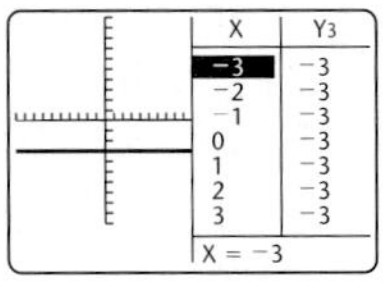

7. No

Exercise Set 5.1, pp. 351–357

1. (g) **2.** (d) **3.** (a) **4.** (h) **5.** (b) **6.** (c) **7.** (j) **8.** (e) **9.** (f) **10.** (i) **11.** 5, 3, 2, 1, 0; 5 **13.** 3, 7, 6, 0; 7 **15.** 5, 6, 2, 1, 0; 6 **17.** $-18y^4 + 11y^3 + 6y^2 - 5y + 3$; $-18y^4$; -18 **19.** $-a^7 + 8a^5 + 5a^3 - 19a^2 + a$; $-a^7$; -1 **21.** $-9 + 6x - 5x^2 + 3x^4$ **23.** $-9xy + 5x^2y^2 + 8x^3y^2 - 5x^4$ **25.** -23 **27.** -12 **29.** 9 **31.** -13; 11 **33.** 282; -9 **35.** About 25.5 mph **37.** About 29.6 ft **39.** 6840 **41.** 26.4 million **43.** 6.4 million **45.** 14; 55 oranges **47.** About 2.3 mcg/mL **49.** $[0, 10]$ **51.** 56.5 in^2 **53.** \$18,750 **55.** \$8375 **57.** $(-\infty, 3]$ **59.** $(-\infty, \infty)$ **61.** $[-4, \infty)$ **63.** $[-65, \infty)$ **65.** $[0, \infty)$ **67.** $(-\infty, 5]$ **69.** $(-\infty, \infty)$ **71.** $[-6.7, \infty)$ **73.** $2a^3 - a + 4$ **75.** $-6a^2b - 3b^2$ **77.** $10x^2 + 2xy + 15y^2$ **79.** $16x + 7y - 5z$ **81.** $-2x^2 + x - xy - 1$ **83.** $6x^2y - 4xy^2 + 5xy$ **85.** $9r^2 + 9r - 9$ **87.** $-\frac{5}{24}xy - \frac{27}{20}x^3y^2 + 1.4y^3$ **89.** $-(5x^3 - 7x^2 + 3x - 9)$, $-5x^3 + 7x^2 - 3x + 9$ **91.** $-(-12y^5 + 4ay^4 - 7by^2)$, $12y^5 - 4ay^4 + 7by^2$ **93.** $10x - 9$ **95.** $-4x^2 - 3x + 13$

97. $6a - 6b + 5c$ **99.** $-2a^2 + 12ab - 7b^2$
101. $8a^2b + 16ab + 3ab^2$ **103.** $x^4 - x^2 - 1$
105. $4x^2 + 2x + 6$ **107.** $13r^2 - 8r - 1$ **109.** $3x^2 - 9$
111. \$9700 **113.** (a) and (c) **115.** (b) and (d)
117. TW **119.** x^{10} **120.** t^{12} **121.** $25x^6$ **122.** $49t^8$
123. x^9 **124.** a^8 **125.** TW
127. $68x^5 - 81x^4 - 22x^3 + 52x^2 + 2x + 250$
129. $45x^5 - 8x^4 + 208x^3 - 176x^2 + 116x - 25$
131. 494.55 cm^3 **133.** $5x^2 - 8x$ **135.** $8x^{2a} + 7x^a + 7$
137. $x^{5b} + 4x^{4b} + x^{3b} - 6x^{2b} - 9x^b$

Interactive Discovery, p. 361

1. Not an identity **2.** Identity **3.** Not an identity
4. Not an identity **5.** Identity

Interactive Discovery, p. 362

1. Identity **2.** Not an identity **3.** Not an identity
4. Not an identity

Exercise Set 5.2, pp. 365–367

1. True **2.** False **3.** True **4.** True **5.** False
6. False **7.** True **8.** True **9.** $40a^3$ **11.** $-20x^3y$
13. $-20x^5y^6$ **15.** $21x - 7x^2$ **17.** $20c^3d^2 - 25c^2d^3$
19. $x^2 + 8x + 15$ **21.** $t^2 + 5t - 14$
23. $8a^2 + 14a + 3$ **25.** $10x^2 - 3xy - 4y^2$
27. $x^3 - x^2 - 5x + 2$ **29.** $t^3 - 3t^2 - 13t + 15$
31. $a^4 + 5a^3 - 2a^2 - 9a + 5$ **33.** $x^3 + 27$
35. $a^3 - b^3$ **37.** $t^2 - \frac{7}{12}t + \frac{1}{12}$ **39.** $r^3 + 4r^2 + r - 6$
41. $x^2 + 10x + 25$ **43.** $3t^2 + 1.5st - 15s^2$
45. $4y^2 - 28y + 49$ **47.** $25a^2 - 30ab + 9b^2$
49. $4a^6 - 12a^3b^2 + 9b^4$ **51.** $x^6y^8 + 10x^3y^4 + 25$
53. $12x^4 - 21x^3 - 17x^2 + 35x - 5$ **55.** $25x^2 - 20x + 4$
57. $4x^2 - \frac{4}{3}x + \frac{1}{9}$ **59.** $c^2 - 49$ **61.** $16x^2 - 1$
63. $9m^2 - 4n^2$ **65.** $x^6 - y^2z^2$ **67.** $-m^2n^2 + m^4$, or $m^4 - m^2n^2$ **69.** $14x + 58$ **71.** $3m^2 + 4mn - 5n^2$
73. $a^2 + 2ab + b^2 - 1$ **75.** $4x^2 + 12xy + 9y^2 - 16$
77. $A = P + 2Pi + Pi^2$ **79.** **(a)** $t^2 - 2t + 6$; **(b)** $2ah + h^2$; **(c)** $2ah - h^2$ **81.** TW **83.** $a = \dfrac{d}{b + c}$
84. $y = \dfrac{w}{x + z}$ **85.** $m = \dfrac{p}{n + 1}$ **86.** $s = \dfrac{t}{r + 1}$
87. Dimes: 5; nickels: 2; quarters: 6 **88.** 8 weekdays
89. TW **91.** $x^4 - y^{2n}$ **93.** $5x^{n+2}y^3 + 4x^2y^{n+3}$
95. $t^{3n} + t^{2n} - 5t^n + 3$
97. $a^2 + 2ac + c^2 - b^2 - 2bd - d^2$
99. $\frac{4}{9}x^2 - \frac{1}{9}y^2 - \frac{2}{3}y - 1$ **101.** $x^{4a} - y^{4b}$ **103.** $x^{a^2-b^2}$
105. 0 **107.** $2a + h$ **109.** (b) and (c) are identities.

Interactive Discovery, p. 371

1. 1 **2.** 0 **3.** 2 **4.** 3 **5.** 2 **6.** 1
7. The number of real-number zeros is less than or equal to the degree.

Exercise Set 5.3, pp. 378–382

1. False **2.** False **3.** True **4.** True **5.** True
6. True **7.** True **8.** False **9.** $-3, 5$ **11.** $-2, 0$
13. $-3, 1$ **15.** $-4, 2$ **17.** 0, 5 **19.** 1, 3 **21.** 10, 15
23. 0, 1, 2 **25.** $-15, 6, 12$ **27.** $-0.42857, 0.33333$
29. $-5, 9$ **31.** $-0.5, 7$ **33.** $-1, 0, 3$ **35.** III **37.** I
39. Expression **41.** Equation **43.** Equation
45. $2t(t + 4)$ **47.** $y^2(y + 9)$ **49.** $5x^2(3 - x^2)$
51. $4xy(x - 3y)$ **53.** $3(y^2 - y - 3)$
55. $2a(3b - 2d + 6c)$ **57.** $3x^2y^4z^2(3xy^2 - 4x^2z^2 + 5yz)$
59. $-5(x - 7)$ **61.** $-6(y + 12)$ **63.** $-2(x^2 - 2x + 6)$
65. $-3(-y + 8x)$, or $-3(8x - y)$ **67.** $-7(-s + 2t)$, or $-7(2t - s)$ **69.** $-(x^2 - 5x + 9)$
71. $-a(a^3 - 2a^2 + 13)$ **73.** $(b - 5)(a + c)$
75. $(x + 7)(2x - 3)$ **77.** $(x - y)(a^2 - 5)$
79. $(c + d)(a + b)$ **81.** $(b - 1)(b^2 + 2)$
83. $(a - 3)(a^2 - 2)$ **85.** $12x(6x^2 - 3x + 2)$
87. $x^3(x - 1)(x^2 + 1)$ **89.** $(y^2 + 3)(2y^2 + 5)$
91. **(a)** $h(t) = -8t(2t - 9)$; **(b)** $h(1) = 56$ ft
93. $R(n) = n(n - 1)$ **95.** $P(x) = x(x - 3)$
97. $R(x) = 0.4x(700 - x)$ **99.** $N(x) = \frac{1}{6}(x^3 + 3x^2 + 2x)$
101. $H(n) = \frac{1}{2}n(n - 1)$ **103.** $-1, 0$ **105.** 0, 3
107. $-3, 0$ **109.** $-\frac{1}{3}, 0$ **111.** TW **113.** -26
114. -1 **115.** -30 **116.** -19 **117.** 56, 58, 60
118. A: 75; B: 84; C: 63 **119.** TW
121. $(x + 4)(x - 2)$ **123.** $x^5y^4 + x^4y^6 = x^3y(x^2y^3 + xy^5)$
125. $(x^2 - x + 5)(r + s)$
127. $(x^4 + x^2 + 5)(a^4 + a^2 + 5)$ **129.** $x^{-9}(x^3 + 1 + x^6)$
131. $x^{1/3}(1 - 5x^{1/6} + 3x^{5/12})$ **133.** $x^{-5/2}(1 + x)$
135. $x^{-7/5}(x^{3/5} - 1 + x^{16/15})$ **137.** $3a^n(a + 2 - 5a^2)$
139. $y^{a+b}(7y^a - 5 + 3y^b)$

Exercise Set 5.4, pp. 390–392

1. True **2.** True **3.** False **4.** True **5.** False
6. False **7.** True **8.** True **9.** $(x + 2)(x + 6)$
11. $(t + 3)(t + 5)$ **13.** $(x - 9)(x + 3)$
15. $2(n - 5)(n - 5)$, or $2(n - 5)^2$ **17.** $a(a + 8)(a - 9)$
19. $(x + 9)(x + 5)$ **21.** $(x + 5)(x - 2)$
23. $3(x + 2)(x + 3)$ **25.** $-(x - 8)(x + 7)$, or $(-x + 8)(x + 7)$, or $(8 - x)(7 + x)$
27. $-y(y - 8)(y + 4)$, or $y(-y + 8)(y + 4)$, or $y(8 - y)(4 + y)$ **29.** $x^2(x + 16)(x - 5)$ **31.** Prime
33. $(p - 8q)(p + 3q)$ **35.** $(y + 4z)(y + 4z)$, or $(y + 4z)^2$ **37.** $p^2(p + 1)(p + 79)$ **39.** $-6, -2$
41. 5 **43.** $-8, 0, 9$ **45.** $-5, 1$ **47.** $-3, 2$

49. $-5, 9$ **51.** $-1, 0, 3$ **53.** $-9, 5$ **55.** $0, 9$
57. $-5, 0, 8$ **59.** $-4, 5$ **61.** $-7, 5$
63. $(x + 22)(x - 12)$ **65.** $(x + 24)(x + 16)$
67. $(x + 64)(x - 38)$ **69.** $f(x) = (x + 1)(x - 2)$, or $f(x) = x^2 - x - 2$ **71.** $f(x) = (x + 7)(x + 10)$, or $f(x) = x^2 + 17x + 70$ **73.** $f(x) = x(x - 1)(x - 2)$, or $f(x) = x^3 - 3x^2 + 2x$ **75.** TW **77.** $3x^2(3x^3 - 8x + 1)$
78. $x^2(x + 1)(x + 2)$ **79.** $(x + 2)(y + 3)$
80. $(x - 7)(2x - 3)$ **81.** $15x^2y^3(3xy + 2y^2 + 4x^2)$
82. $-2(t^3 + 2t - 4)$ **83.** $(w - z)(x - y)$
84. $(x + 5)(x + 10)$ **85.** TW **87.** $\{-1, 3\}$; $(-2, 4)$, or $\{x \mid -2 < x < 4\}$ **89.** $f(x) = 5x^3 - 20x^2 + 5x + 30$; answers may vary **91.** 6.90 **93.** 3.48
95. $ab^2(2a^3b^4 - 3ab - 20)$ **97.** $\left(x + \frac{4}{5}\right)\left(x - \frac{1}{5}\right)$
99. $(y - 0.1)(y + 0.5)$ **101.** $(x + a)(x + b)$
103. $a^2(p^a + 2)(p^a - 1)$
105. $76, -76, 28, -28, 20, -20$ **107.** $x - 365$

Exercise Set 5.5, pp. 399–401

1. (f) **2.** (c) **3.** (e) **4.** (a) **5.** (g) **6.** (d)
7. (h) **8.** (b) **9.** $(3x + 5)(2x - 5)$
11. $y(2y - 3)(5y + 4)$ **13.** $2(4a - 1)(3a - 1)$
15. $(7y + 4)(5y + 2)$ **17.** $2(5t - 3)(t + 1)$
19. $4(x - 4)(2x + 1)$ **21.** $x^2(2x - 3)(7x + 1)$
23. $4(3a - 4)(a + 1)$ **25.** $(3x + 4)(3x + 1)$
27. $(4x + 3)(x + 3)$ **29.** $-2(2t - 3)(2t + 5)$
31. $(4 + 3z)(2 - 3z)$ **33.** $xy(6y + 5)(3y - 2)$
35. $(x - 2)(24x + 1)$ **37.** $3x(7x + 3)(3x + 4)$
39. $2x^2(4x - 3)(6x + 5)$ **41.** $(4a - 3b)(3a - 2b)$
43. $(2x - 3y)(x + 2y)$ **45.** $(3x - 4y)(2x - 7y)$
47. $(3x - 5y)(3x - 5y)$, or $(3x - 5y)^2$
49. $(9xy - 4)(xy + 1)$ **51.** $-\frac{5}{3}, \frac{5}{2}$ **53.** $-\frac{4}{3}, \frac{2}{3}$
55. $-\frac{4}{3}, -\frac{3}{7}, 0$ **57.** $\frac{2}{3}, 2$ **59.** $-2, -\frac{3}{4}, 0$ **61.** $-\frac{1}{3}, \frac{5}{2}$
63. $-\frac{7}{4}, \frac{4}{3}$ **65.** $-\frac{1}{2}, 7$ **67.** $-8, -4$ **69.** $-4, \frac{3}{2}$
71. $-9, -3$ **73.** $\{x \mid x$ is a real number *and* $x \neq 5$ *and* $x \neq -1\}$, or $(-\infty, -1) \cup (-1, 5) \cup (5, \infty)$
75. $\{x \mid x$ is a real number *and* $x \neq 0$ *and* $x \neq \frac{1}{2}\}$, or $(-\infty, 0) \cup \left(0, \frac{1}{2}\right) \cup \left(\frac{1}{2}, \infty\right)$
77. $\{x \mid x$ is a real number *and* $x \neq 0$ *and* $x \neq 2$ *and* $x \neq 5\}$, or $(-\infty, 0) \cup (0, 2) \cup (2, 5) \cup (5, \infty)$
79. TW **81.** $(x + 1)(x + 3)$ **82.** $(2x - 5)(x + 1)$
83. $y(x + y)(x + 2)$ **84.** $x(3x - 1)(5x + 3)$
85. $(4x - 3)(3x + 1)$ **86.** $2(x - 7)(x + 2)$
87. $(a - b)(c - d)$ **88.** $4x^2y^3(3xy - 6y^2 + 1)$
89. $3a(a - 2)$ **90.** $-8(x^2 - x - 1)$ **91.** TW
93. $(2x + 15)(2x + 45)$ **95.** $3x(x + 68)(x - 18)$
97. $-\frac{11}{8}, -\frac{1}{4}, \frac{2}{3}$ **99.** 1 **101.** $(3ab + 2)(6ab - 5)$
103. Prime **105.** $(5t^5 - 1)^2$ **107.** $(10x^n + 3)(2x^n + 1)$
109. $[7(t - 3)^n - 2][(t - 3)^n + 1]$
111. $(2a^2b^3 + 5)(a^2b^3 - 4)$ **113.** $(2x^a - 3)(2x^a + 1)$
115. Since $ax^2 + bx + c = (mx + r)(nx + s)$, from FOIL we know that $a = mn$, $c = rs$, and $b = ms + rn$. If $P = ms$ and $Q = rn$, then $b = P + Q$. Since $ac = mnrs = msrn$, we have $ac = PQ$.

Exercise Set 5.6, pp. 408–409

1. Difference of two squares **2.** Perfect-square trinomial
3. Perfect-square trinomial **4.** Difference of two squares
5. None of these **6.** Polynomial having a common factor
7. Polynomial having a common factor **8.** None of these
9. Perfect-square trinomial **10.** Polynomial having a common factor **11.** $(t + 3)^2$ **13.** $(a - 7)^2$
15. $4(a - 2)^2$ **17.** $(y + 6)^2$ **19.** $y(y - 9)^2$
21. $2(x - 10)^2$ **23.** $(1 - 4d)^2$ **25.** $y(y + 4)^2$
27. $(0.5x + 0.3)^2$ **29.** $(p - q)^2$ **31.** $(5a + 3b)^2$
33. $5(a - b)^2$ **35.** $(y + 10)(y - 10)$
37. $(m + 8)(m - 8)$ **39.** $(pq + 5)(pq - 5)$
41. $8(x + y)(x - y)$ **43.** $7x(y^2 + z^2)(y + z)(y - z)$
45. $a(2a + 7)(2a - 7)$
47. $3(x^4 + y^4)(x^2 + y^2)(x + y)(x - y)$
49. $a^2(3a + 5b^2)(3a - 5b^2)$ **51.** $\left(\frac{1}{7} + x\right)\left(\frac{1}{7} - x\right)$
53. $(a + b + 3)(a + b - 3)$
55. $(x - 3 + y)(x - 3 - y)$ **57.** $(t + 8)(t + 1)(t - 1)$
59. $(r - 3)^2(r + 3)$ **61.** $(m - n + 5)(m - n - 5)$
63. $(6 + x + y)(6 - x - y)$
65. $(r - 1 + 2s)(r - 1 - 2s)$
67. $(4 + a + b)(4 - a - b)$
69. $(x + 5)(x + 2)(x - 2)$ **71.** $(a - 2)(a + b)(a - b)$
73. 1 **75.** 6 **77.** $-3, 3$
79. $-\frac{1}{5}, \frac{1}{5}$ **81.** $-\frac{1}{2}, \frac{1}{2}$ **83.** $-1, 1, 3$ **85.** $-1.541, 4.541$ **87.** $-3.871, -0.129$ **89.** $-2.414, -1, 0.414$
91. 6 **93.** $-4, 4$ **95.** -1 **97.** $-1, 1, 2$ **99.** TW
101. $(3x - 1)(2x - 5)$ **102.** $(4x + 1)(4x - 1)$
103. $(4x + 1)^2$ **104.** $3x(x + 1)(x - 2)$
105. $(x^2 - 2)(x + 3)$ **106.** $(2x + 5)(2x - 3)$
107. $\left(\frac{1}{8} + x\right)\left(\frac{1}{8} - x\right)$ **108.** $(10 + t + s)(10 - t - s)$
109. $12xy(x - y)^2$ **110.** $-3x(x^3 - 6x^2 - 1)$ **111.** TW
113. $-\frac{1}{54}(4r + 3s)^2$ **115.** $(0.3x^4 + 0.8)^2$, or $\frac{1}{100}(3x^4 + 8)^2$
117. $(r + s + 1)(r - s - 9)$ **119.** $(x^{2a} + y^b)(x^{2a} - y^b)$
121. $(5y^a + x^b - 1)(5y^a - x^b + 1)$ **123.** $3(x + 3)^2$
125. $(3x^n - 1)^2$ **127.** $(t + 1 - 3)(t + 1 - 1)$, or $t(t - 2)$
129. $(m + 2n)(m + 2n + 5)$ **131.** $h(2a + h)$
133. **(a)** $\pi h(R + r)(R - r)$; **(b)** $3{,}014{,}400\ \text{cm}^3$

Exercise Set 5.7, pp. 414–415

1. Difference of two cubes **2.** Sum of two cubes
3. Difference of two squares **4.** Prime polynomial
5. Sum of two cubes **6.** Difference of two cubes
7. None of these **8.** Both a difference of two cubes *and* a difference of two squares **9.** Both a difference of two cubes *and* a difference of two squares **10.** None of these
11. $(x + 4)(x^2 - 4x + 16)$ **13.** $(z - 1)(z^2 + z + 1)$

15. $(x - 3)(x^2 + 3x + 9)$ **17.** $(3x + 1)(9x^2 - 3x + 1)$
19. $(4 - 5x)(16 + 20x + 25x^2)$
21. $(3y + 4)(9y^2 - 12y + 16)$
23. $(x - y)(x^2 + xy + y^2)$
25. $\left(a + \frac{1}{2}\right)\left(a^2 - \frac{1}{2}a + \frac{1}{4}\right)$ **27.** $8(t - 1)(t^2 + t + 1)$
29. $2(3x + 1)(9x^2 - 3x + 1)$
31. $a(b + 5)(b^2 - 5b + 25)$
33. $5(x - 2z)(x^2 + 2xz + 4z^2)$
35. $(x + 0.1)(x^2 - 0.1x + 0.01)$
37. $8(2x^2 - t^2)(4x^4 + 2x^2t^2 + t^4)$
39. $2y(y - 4)(y^2 + 4y + 16)$
41. $(z + 1)(z^2 - z + 1)(z - 1)(z^2 + z + 1)$
43. $(t^2 + 4y^2)(t^4 - 4t^2y^2 + 16y^4)$
45. $(x^4 - yz^4)(x^8 + x^4yz^4 + y^2z^8)$ **47.** -1 **49.** $\frac{3}{2}$
51. 10 **53.** TW **55.** $5(x - z)(x^2 + xz + z^2)$
56. $(x - 3)(x + 7)$ **57.** $-8xz(x^2 - 3xz + 2z^2 + 1)$
58. $(1 + 10x)(1 - 10x + 100x^2)$
59. $3(x^2 + z^2)(x + z)(x - z)$ **60.** $(2x - 5)^2$
61. $(x + 1)(x^2 - x + 1)(x - 1)(x^2 + x + 1)$
62. $(x^2 + 1)(x + 1)$ **63.** $3x(x + 10)^2$
64. $x^2y^3(x + 2)(x + 6)$ **65.** $(4x + 3)(3x + 4)$
66. $(x + 1 + y)(x + 1 - y)$ **67.** TW
69. $(x^{2a} - y^b)(x^{4a} + x^{2a}y^b + y^{2b})$ **71.** $2x(x^2 + 75)$
73. $5\left(xy^2 - \frac{1}{2}\right)\left(x^2y^4 + \frac{1}{2}xy^2 + \frac{1}{4}\right)$ **75.** $-(3x^{4a} + 3x^{2a} + 1)$
77. $(t - 8)(t - 1)(t^2 + t + 1)$
79. $h(2a + h)(a^2 + ah + h^2)(3a^2 + 3ah + h^2)$

Visualizing the Graph, p. 423

1. D **2.** J **3.** A **4.** B **5.** E **6.** C **7.** I **8.** F **9.** G **10.** H

Exercise Set 5.8, pp. 424–428

1. $-12, 11$ **3.** Length: 12 cm; width: 7 cm **5.** 3 m **7.** 3 cm **9.** 10 ft **11.** 16, 18, 20 **13.** Height: 6 ft; base: 4 ft **15.** Height: 16 m; base: 7 m **17.** 41 ft **19.** Length: 100 m; width: 75 m **21.** 2 sets **23.** 2 sec **25.** 5 sec **27.** $1\frac{1}{2}$ yr, 6 yr
29. **(a)** $C(x) = 1.703571429x^2 - 11.75357143x + 32.2$; **(b)** \$47.2 million; **(c)** about 6.7 yr after 2000, or approximately 2007
31. **(a)** $L(x) = 0.0078989899x^3 - 1.040519481x^2 + 46.47265512x - 678.4848485$; **(b)** about \$58; **(c)** about 64
33. **(a)** $B(x) = 9.289417614x^4 - 263.4086174x^3 + 2056.113163x^2 - 939.209145x + 37459.76515$; **(b)** about 78,667 degrees; **(c)** approximately 1996, 2001, and 2004
35. TW **37.** $\frac{14}{3}$ **38.** $\left\{x \mid x > \frac{10}{3}\right\}$, or $\left(\frac{10}{3}, \infty\right)$ **39.** $(2, -1, 3)$
40. $\left\{x \mid x \le -\frac{4}{7} \text{ or } x \ge 2\right\}$, or $\left(-\infty, -\frac{4}{7}\right] \cup [2, \infty)$
41. $\left\{x \mid -\frac{4}{7} \le x \le 2\right\}$, or $\left[-\frac{4}{7}, 2\right]$
42. $\left\{x \mid x < \frac{14}{19}\right\}$, or $\left(-\infty, \frac{14}{19}\right)$ **43.** $\{-32, 26\}$ **44.** $\left(6, \frac{4}{3}\right)$

45. $-\frac{1}{3}, \frac{1}{3}$ **46.** $-1, 6$ **47.** TW **49.** Length: 28 cm; width: 14 cm **51.** About 5.7 sec

Review Exercises: Chapter 5, pp. 433–435

1. (g) **2.** (e) **3.** (j) **4.** (h) **5.** (c) **6.** (i) **7.** (b) **8.** (f) **9.** (a) **10.** (d) **11.** 7, 11, 3, 0; 11
12. $-5x^3 + 2x^2 + 3x + 9$; $-5x^3$; -5
13. $-3x^2 + 2x^3 + 8x^6y - 7x^8y^3$ **14.** 0; -6 **15.** 4
16. $-2a^3 + a^2 - 3a - 4$ **17.** $-x^2y - 2xy^2$
18. $-2x^3 + 2x^2 + 5x + 3$ **19.** $-3x^4 + 3x^3 - x + 16$
20. $-5xy^2 - 2xy + x^2y$ **21.** $14x - 7$
22. $-2a + 6b + 7c$ **23.** $6x^2 - 7xy + 3y^2$ **24.** $-18x^3y^4$
25. $x^8 - x^6 + 5x^2 - 3$ **26.** $8a^2b^2 + 2abc - 3c^2$
27. $4x^2 - 25y^2$ **28.** $9x^2 - 24xy + 16y^2$
29. $2x^2 + 5x - 3$ **30.** $x^4 + 8x^2y^3 + 16y^6$
31. $5t^2 - 42t + 16$ **32.** $x^2 - \frac{1}{2}x + \frac{1}{18}$ **33.** $x(7x + 6)$
34. $3y^2(3y^2 - 1)$ **35.** $3x(5x^3 - 6x^2 + 7x - 3)$
36. $(a - 9)(a - 3)$ **37.** $(3m + 2)(m + 4)$
38. $(5x + 2)^2$ **39.** $4(y + 2)(y - 2)$
40. $x(x - 2)(x + 7)$ **41.** $(a + 2b)(x - y)$
42. $(y + 2)(3y^2 - 5)$ **43.** $(a^2 + 9)(a + 3)(a - 3)$
44. $4(x^4 + x^2 + 5)$ **45.** $(3x - 2)(9x^2 + 6x + 4)$
46. $(0.4b - 0.5c)(0.16b^2 + 0.2bc + 0.25c^2)$
47. $y(y^4 + 1)$ **48.** $2z^6(z^2 - 8)$
49. $2y(3x^2 - 1)(9x^4 + 3x^2 + 1)$ **50.** $4(3x - 5)^2$
51. $(3t + p)(2t + 5p)$ **52.** $(x + 3)(x - 3)(x + 2)$
53. $(a - b + 2t)(a - b - 2t)$ **54.** $-2, 1, 5$ **55.** 4, 7
56. 10 **57.** $\frac{2}{3}, \frac{3}{2}$ **58.** $0, \frac{7}{4}$ **59.** $-4, 4$ **60.** $-3, 0, 7$
61. $-1, 6$ **62.** $-4, 4, 5$ **63.** 12, 15
64. $-1.646, 3.646$ **65.** $-4, 11$
66. $\left\{x \mid x \text{ is a real number } \textit{and } x \ne -7 \textit{ and } x \ne \frac{2}{3}\right\}$, or $(-\infty, -7) \cup \left(-7, \frac{2}{3}\right) \cup \left(\frac{2}{3}, \infty\right)$ **67.** 5 **68.** 3, 5, 7; $-7, -5, -3$ **69.** Length: 8 in.; width: 5 in. **70.** 17 ft
71. **(a)** $P(x) = 0.0256348234x^2 - 1.01295601x + 24.99905124$; **(b)** about 25.5; **(c)** approximately 1971 and 2009 **72.** TW The roots of a polynomial function are the x-coordinates of the points at which the graph of the function crosses the x-axis.
73. TW The principle of zero products states that if a product is equal to 0, at least one of the factors must be 0. If a product is nonzero, we cannot conclude that any one of the factors is a particular value.
74. $2(2x - y)(4x^2 + 2xy + y^2)(2x + y)(4x^2 - 2xy + y^2)$
75. $-2(3x^2 + 1)$ **76.** $a^3 - b^3 + 3b^2 - 3b + 1$
77. z^{5n^5} **78.** $-1, -\frac{1}{2}$

Test: Chapter 5, pp. 435–436

1. [5.1] 9 **2.** [5.1] $5x^5y^4 - 2x^4y - 4x^2y + 3xy^3$
3. [5.1] $-4a^3$ **4.** [5.1] 4; 2 **5.** [5.2] $2ah + h^2 - 5h$
6. [5.1] $3xy + 3xy^2$ **7.** [5.1] $-3x^3 + 3x^2 - 6y - 7y^2$
8. [5.1] $7m^3 + 2m^2n + 3mn^2 - 7n^3$ **9.** [5.1] $6a - 8b$

10. [5.1] $2y^2 + 5y + y^3$ **11.** [5.2] $64x^3y^3$
12. [5.2] $12a^2 - 4ab - 5b^2$ **13.** [5.2] $x^3 - 2x^2y + y^3$
14. [5.2] $4x^6 + 20x^3 + 25$ **15.** [5.2] $16y^2 - 72y + 81$
16. [5.2] $x^2 - 4y^2$ **17.** [5.3] $5x^2(3 - x^2)$
18. [5.6] $(y + 5)(y + 2)(y - 2)$
19. [5.4] $(p - 14)(p + 2)$ **20.** [5.5] $(6m + 1)(2m + 3)$
21. [5.6] $(3y + 5)(3y - 5)$
22. [5.7] $3(r - 1)(r^2 + r + 1)$ **23.** [5.6] $(3x - 5)^2$
24. [5.6] $(x^4 + y^4)(x^2 + y^2)(x + y)(x - y)$
25. [5.6] $(y + 4 + 10t)(y + 4 - 10t)$
26. [5.6] $5(2a - b)(2a + b)$ **27.** [5.5] $2(4x - 1)(3x - 5)$
28. [5.7] $2ab(2a^2 + 3b^2)(4a^4 - 6a^2b^2 + 9b^4)$
29. [5.3] $4xy(y^3 + 9x + 2x^2y - 4)$ **30.** [5.3] $-3, -1, 2, 4$
31. [5.5] $-\frac{5}{2}, 8$ **32.** [5.4] $-3, 6$ **33.** [5.6] $-5, 5$
34. [5.5] $-7, -\frac{3}{2}$ **35.** [5.3] $-\frac{1}{3}, 0$ **36.** [5.6] 9
37. [5.4] $-3.372, 0, 2.372$ **38.** [5.5] 0, 5
39. [5.5] $\{x \mid x \text{ is a real number } and\ x \neq -1\}$, or $(-\infty, -1) \cup (-1, \infty)$ **40.** [5.8] Length: 8 cm; width: 5 cm
41. [5.8] $4\frac{1}{2}$ sec **42.** [5.8] **(a)** $E(x) = 0.7x^2 - 0.6x + 33.28$; **(b)** about \$84.6 billion; **(c)** about 10.2 yr after 2000, or approximately 2010 **43.** **(a)** [5.2] $x^5 + x + 1$; **(b)** [5.2], [5.7] $(x^2 + x + 1)(x^3 - x^2 + 1)$
44. [5.5] $(3x^n + 4)(2x^n - 5)$

Chapter 6

Interactive Discovery, p. 447

1. Domain of f: $\{x \mid x \neq -\frac{1}{2}\ and\ x \neq 2\}$, or $(-\infty, -\frac{1}{2}) \cup (-\frac{1}{2}, 2) \cup (2, \infty)$; domain of g: $\{x \mid x \neq -\frac{1}{2}\}$, or $(-\infty, -\frac{1}{2}) \cup (-\frac{1}{2}, \infty)$
2. One vertical asymptote; $x = -\frac{1}{2}$ **3.** $x = -\frac{1}{2}$

Visualizing the Graph, p. 448

1. A **2.** D **3.** J **4.** H **5.** I **6.** B **7.** E **8.** F
9. C **10.** G

Exercise Set 6.1, pp. 449–453

1. (e) **2.** (j) **3.** (g) **4.** (c) **5.** (i) **6.** (h)
7. (a) **8.** (f) **9.** (d) **10.** (b) **11.** $\frac{40}{13}$ hr, or $3\frac{1}{13}$ hr
13. $\frac{2}{3}$; 28; $\frac{163}{10}$ **15.** $-\frac{9}{4}$; does not exist; $-\frac{11}{9}$
17. $\dfrac{4x(x - 3)}{4x(x + 2)}$ **19.** $\dfrac{(t - 2)(-1)}{(t + 3)(-1)}$ **21.** $\dfrac{3}{x}$ **23.** $\dfrac{2}{3t^4s^7}$
25. $a - 5$ **27.** $\dfrac{3}{5a - 6}$ **29.** $\dfrac{x - 4}{x + 5}$
31. $f(x) = \dfrac{3}{x}, x \neq -7, 0$ **33.** $g(x) = \dfrac{x - 3}{5}, x \neq -3$
35. $h(x) = -\frac{1}{5}, x \neq 4$ **37.** $f(t) = \dfrac{t + 4}{t - 4}, t \neq 4$
39. $g(t) = -\frac{7}{3}, t \neq 3$ **41.** $h(t) = \dfrac{t + 4}{t - 9}, t \neq -1, 9$
43. $f(x) = 3x + 2, x \neq \frac{2}{3}$ **45.** $g(t) = \dfrac{4 + t}{4 - t}, t \neq 4$
47. $\dfrac{7b^2}{6a^4}$ **49.** $\dfrac{8x^2}{25}$ **51.** $\dfrac{(x + 4)(x - 4)}{x(x + 3)}$ **53.** $-\dfrac{a + 1}{2 + a}$
55. 1 **57.** $c(c - 2)$ **59.** $\dfrac{a^2 + ab + b^2}{3(a + 2b)}$ **61.** $6x^4y^7$
63. $\dfrac{5}{x^5}$ **65.** $-\dfrac{5x + 2}{x - 3}$ **67.** $-\dfrac{1}{y^3}$ **69.** $\dfrac{(x + 4)(x + 2)}{3(x - 5)}$
71. $\dfrac{x^2 + 4x + 16}{(x + 4)^2}$ **73.** $f(t) = \dfrac{t + 4}{4}, t \neq -3, 4$
75. $g(x) = \dfrac{(x + 5)(2x + 3)}{7x}, x \neq 0, \dfrac{3}{2}, 7$
77. $f(x) = \dfrac{(x + 2)(x + 4)}{x^7}, x \neq -4, 0, 2$
79. $h(n) = \dfrac{n(n^2 + 3)}{(n + 3)(n - 2)}, n \neq -7, -3, 2, 3$
81. $\dfrac{3(x - 3y)}{2(2x - y)(2x - 3y)}$ **83.** $\dfrac{(2a - b)(a - 1)}{(a - b)(a + 1)}$
85. $x = -5$ **87.** No vertical asymptotes **89.** $x = -3$
91. $x = 2, x = 4$ **93.** (b) **95.** (f) **97.** (a) **99.** TW
101. $-\frac{7}{30}$ **102.** $-\frac{13}{40}$ **103.** $\frac{5}{14}$ **104.** $\frac{1}{35}$
105. $4x^3 - 7x^2 + 9x - 5$
106. $-2t^4 + 11t^3 - t^2 + 10t - 3$ **107.** TW
109. $2a + h$ **111.** **(a)** $\dfrac{2x + 2h + 3}{4x + 4h - 1}$; **(b)** $\dfrac{2x + 3}{8x - 9}$;
(c) $\dfrac{x + 5}{4x - 1}$ **113.** $\dfrac{4s^2}{(r + 2s)^2(r - 2s)}$
115. $\dfrac{6t^2 - 26t + 30}{8t^2 - 15t - 21}$ **117.** $\dfrac{a^2 + 2}{a^2 - 3}$
119. $\dfrac{(u^2 - uv + v^2)^2}{u - v}$ **121.** **(a)** $\dfrac{16(x + 1)}{(x - 1)^2(x^2 + x + 1)}$;
(b) $\dfrac{x^2 + x + 1}{(x + 1)^3}$; **(c)** $\dfrac{(x + 1)^3}{x^2 + x + 1}$
123. Domain: $(-\infty, -1) \cup (-1, 0) \cup (0, 1) \cup (1, \infty)$; range: $(-\infty, -3) \cup (-3, -1) \cup (-1, 0) \cup (0, \infty)$

Exercise Set 6.2, pp. 460–463

1. True **2.** True **3.** False **4.** False **5.** True
6. False **7.** True **8.** False **9.** $\dfrac{4}{y}$ **11.** $\dfrac{1}{3m^2n^2}$
13. 2 **15.** $\dfrac{2t + 4}{t - 4}$ **17.** $f(x) = \dfrac{45x^2 + 4x + 12}{5x(x + 3)}$, $x \neq -3, 0$ **19.** $f(x) = \dfrac{3}{x^2 - 9}, x \neq \pm 3$
21. $f(x) = \dfrac{13x^2 - 7}{3x^2(x - 1)(x + 1)}, x \neq 0, \pm 1$

23. $f(x) = \frac{1}{1-x}, x \neq 1$ **25.** $\frac{1}{x-7}$ **27.** $\frac{-1}{a+3}$
29. $-(s+r)$ **31.** $\frac{7}{a}$ **33.** $-\frac{1}{y+5}$ **35.** $\frac{1}{y^2+9}$
37. $\frac{1}{r^2+rs+s^2}$ **39.** $\frac{2a^2-a+14}{(a-4)(a+3)}$ **41.** $\frac{5x+1}{x+1}$
43. $\frac{x+y}{x-y}$ **45.** $\frac{3x^2+7x+14}{(2x-5)(x-1)(x+2)}$
47. $\frac{8x+1}{(x+1)(x-1)}$ **49.** $\frac{-x+34}{20(x+2)}$
51. $\frac{-a^2+7ab-b^2}{(a-b)(a+b)}$ **53.** $\frac{x-5}{(x+5)(x+3)}$
55. $\frac{y}{(y-2)(y-3)}$ **57.** $\frac{7x+1}{x-y}$
59. $\frac{3y^2-3y-29}{(y-3)(y+8)(y-4)}$ **61.** $\frac{-y}{(y+3)(y-1)}$
63. $-\frac{2y}{2y+1}$ **65.** $f(x) = \frac{3(x+4)}{x+3}, x \neq \pm 3$
67. $f(x) = \frac{(x-7)(2x-1)}{(x-4)(x-1)(x+3)}, x \neq -4, -3, 1, 4$
69. $f(x) = \frac{-2}{(x+1)(x+2)}, x \neq -3, -2, -1$ **71.** TW
73. $\frac{3y^6z^7}{7x^5}$ **74.** $\frac{7b^{11}c^7}{9a^2}$ **75.** $\frac{7}{9}$ **76.** $-\frac{5}{6}$
77. Dimes: 2 rolls; nickels: 5 rolls; quarters: 5 rolls
78. 30-min tapes: 4; 60-min tapes: 8 **79.** TW
81. 420 days **83.** 12 parts
85. $x^4(x^2+1)(x+1)(x-1)(x^2+x+1)(x^2-x+1)$
87. $8a^4, 8a^4b, 8a^4b^2, 8a^4b^3, 8a^4b^4, 8a^4b^5, 8a^4b^6, 8a^4b^7$
89. $\frac{x^4+6x^3+2x^2}{(x+2)(x-2)(x+5)}$ **91.** $\frac{x^5}{(x^2-4)(x^2+3x-10)}$
93. $\{x \mid x \neq -2 \textit{ and } x \neq 2 \textit{ and } x \neq -5\}$
95. $\frac{9x^2+28x+15}{(x-3)(x+3)^2}$ **97.** $\frac{1}{2x(x-5)}$ **99.** $-4t^4$
101. Domain: $(-\infty, -1) \cup (-1, \infty)$; range: $(-\infty, 3) \cup (3, \infty)$
103. Domain: $(-\infty, 0) \cup (0, 1) \cup (1, \infty)$; range: $(0, \infty)$

Exercise Set 6.3, pp. 470–473

1. (e) **2.** (d) **3.** (f) **4.** (c) **5.** (b) **6.** (a)
7. $\frac{(x+2)(x-3)}{(x-1)(x+4)}$ **9.** $\frac{5b-4a}{2b+3a}$ **11.** $\frac{2z+3y}{4y-z}$
13. $\frac{a+b}{a}$ **15.** $\frac{3}{3x+2}$ **17.** $\frac{1}{x-y}$ **19.** $\frac{1}{a(a-h)}$
21. $\frac{(a-2)(a-7)}{(a+1)(a-6)}$ **23.** $\frac{x+2}{x+3}$ **25.** $\frac{1+2y}{1-3y}$
27. $\frac{y^2+1}{y^2-1}$ **29.** $\frac{4x-7}{7x-9}$ **31.** $\frac{a^2-3a-6}{a^2-2a-3}$
33. $\frac{a+1}{2a+5}$ **35.** $\frac{-1-3x}{8-2x}$, or $\frac{3x+1}{2x-8}$ **37.** $\frac{1}{y+5}$
39. $-y$ **41.** $\frac{6a^2+30a+60}{3a^2+2a+4}$ **43.** $\frac{(2x+1)(x+2)}{2x(x-1)}$
45. $\frac{(2a-3)(a+5)}{2(a-3)(a+2)}$ **47.** -1 **49.** $\frac{2x^2-11x-27}{2x^2+21x+13}$
51. TW **53.** $\frac{15}{4}$ **54.** $\frac{4t-3}{4(t-3)}$
55. $\frac{a^2+9a-21}{(a-3)(a-1)(a+2)}$ **56.** $\frac{t-5}{t+5}$ **57.** 1
58. $\frac{x-1}{x-4}$ **59.** TW **61.** $\frac{5(y+x)}{3(y-x)}$ **63.** $\frac{8c}{17}$
65. $\frac{-3}{x(x+h)}$ **67.** $\frac{2}{(1+x+h)(1+x)}$
69. $\{x \mid x \text{ is a real number } \textit{and } x \neq \pm 1 \textit{ and } x \neq \pm 4 \textit{ and } x \neq \pm 5\}$ **71.** $\frac{2+a}{3+a}, a \neq -2, -3$
73. $\frac{x^4}{81}$; $\{x \mid x \text{ is a real number } \textit{and } x \neq 3\}$

Exercise Set 6.4, pp. 478–480

1. Equation **2.** Expression **3.** Expression
4. Equation **5.** Equation **6.** Equation **7.** Equation
8. Expression **9.** Expression **10.** Equation **11.** $\frac{42}{5}$
13. -8 **15.** $-\frac{11}{9}$ **17.** 5 **19.** -3 **21.** No solution
23. $-4, -1$ **25.** $-2, 6$ **27.** No solution **29.** -5
31. $-\frac{10}{3}$ **33.** -1 **35.** 2, 3 **37.** -145
39. No solution **41.** -1 **43.** 4 **45.** $-6, 5$
47. $-\frac{3}{2}, 5$ **49.** 14 **51.** $\frac{3}{4}$ **53.** $\frac{5}{14}$ **55.** $\frac{3}{5}$ **57.** TW
59. 38 **60.** $\{x \mid x > \frac{7}{10}\}$, or $(\frac{7}{10}, \infty)$ **61.** $\{x \mid -\frac{1}{2} \leq x \leq \frac{15}{2}\}$, or $[-\frac{1}{2}, \frac{15}{2}]$ **62.** $\{x \mid x < \frac{5}{3} \textit{ or } x > 10\}$, or $(-\infty, \frac{5}{3}) \cup (10, \infty)$
63. $(-3, -14)$ **64.** $(-1, 2, 5)$ **65.** $\{-\frac{7}{4}, \frac{13}{4}\}$
66. $-2, 3$ **67.** $-4, 5$ **68.** $-5, 0, \frac{1}{2}$ **69.** TW **71.** $\frac{1}{5}$
73. $-\frac{7}{2}$ **75.** 0.0854697 **77.** Yes

Exercise Set 6.5, pp. 488–491

1. 2 **3.** $-3, -2$ **5.** 6 and 7, -7 and -6 **7.** $3\frac{3}{14}$ hr
9. $8\frac{4}{7}$ hr **11.** $19\frac{4}{5}$ min **13.** Canon: 36 min; HP: 72 min
15. Blueair 402: 30 min; Panasonic F-P20HU1: 40 min
17. Erickson Air-Crane: 10 hr; S-58T: 40 hr
19. Zsuzanna: $\frac{4}{3}$ hr; Stan: 4 hr **21.** 8 hr **23.** 7 mph
25. 4.3 ft/sec **27.** Freight: 66 mph; passenger: 80 mph
29. Express: 45 mph; local: 38 mph **31.** 9 km/h
33. 2 km/h **35.** 25 mph **37.** 20 mph **39.** TW
41. $5a^4b^6$ **42.** $5x^6y^4$ **43.** $4s^{10}t^8$
44. $-2x^4 - 7x^2 + 11x$ **45.** $-8x^3 + 28x^2 - 25x + 19$
46. $11x^4 + 7x^3 - 2x^2 - 4x - 10$ **47.** TW **49.** $49\frac{1}{2}$ hr
51. 11.25 min **53.** 2250 people per hour **55.** $14\frac{7}{8}$ mi
57. Page 278 **59.** $8\frac{2}{11}$ min after 10:30 **61.** $51\frac{3}{7}$ mph

Exercise Set 6.6, pp. 496–497

1. True **2.** False **3.** True **4.** True **5.** False **6.** True **7.** $\frac{16}{3}x^4 + 3x^3 - \frac{9}{2}$ **9.** $3a^2 + a - \dfrac{3}{7} - \dfrac{2}{a}$ **11.** $-6t^3 + 5 - \dfrac{7}{t}$ **13.** $-4z + 2y^2z^3 - 3y^4z^2$ **15.** $8y - \dfrac{9}{2} - \dfrac{4}{y}$ **17.** $-5x^5 + 7x^2 + 1$ **19.** $1 - ab^2 - a^3b^4$ **21.** $x + 3$ **23.** $a - 12 + \dfrac{32}{a + 4}$ **25.** $x - 5 + \dfrac{1}{x - 4}$ **27.** $y - 5$ **29.** $y^2 - 2y - 1 + \dfrac{-8}{y - 2}$ **31.** $2x^2 - x + 1 + \dfrac{-5}{x + 2}$ **33.** $a^2 + 4a + 15 + \dfrac{70}{a - 4}$ **35.** $2y^2 + 2y - 1 + \dfrac{8}{5y - 2}$ **37.** $2x^2 - x - 9 + \dfrac{3x + 12}{x^2 + 2}$ **39.** $4x^2 + 6x + 9, x \neq \frac{3}{2}$ **41.** $2x - 5, x \neq -\frac{2}{3}$ **43.** $x^2 + 1, x \neq -5, x \neq 5$ **45.** $2x^3 - 3x^2 + 5, x \neq -1, x \neq 1$ **47.** TW **49.** $5x^3 + 3x^2 + 4x - 11$ **50.** $5x^3 - x^2 - 6x + 5$ **51.** $6x^2 - 11x - 35$ **52.** $\frac{1}{9}x^2 - 100$ **53.** $16x^2 - 64x + 64$ **54.** $-6x^2 + 2x - 1$ **55.** $x + 2 + \dfrac{-1}{x + 1}$ **56.** $-3x + 13$ **57.** TW **59.** $a^2 + ab$ **61.** $a^6 - a^5b + a^4b^2 - a^3b^3 + a^2b^4 - ab^5 + b^6$ **63.** $-\frac{3}{2}$ **65.**

Exercise Set 6.7, pp. 502–503

1. True **2.** True **3.** True **4.** False **5.** True **6.** False **7.** True **8.** False **9.** $x^2 - x + 1 + \dfrac{-6}{x - 1}$ **11.** $a + 5 + \dfrac{-4}{a + 3}$ **13.** $x^2 - 9x + 5 + \dfrac{-7}{x + 2}$ **15.** $3x^2 - 2x + 2 + \dfrac{-3}{x + 3}$ **17.** $y^2 + 2y + 1 + \dfrac{12}{y - 2}$ **19.** $x^4 + 2x^3 + 4x^2 + 8x + 16$ **21.** $3x^2 + 6x - 3 + \dfrac{2}{x + \frac{1}{3}}$ **23.** 6 **25.** 1 **27.** 54 **29.** TW **31.** $b = \dfrac{a - 9}{c + 1}$ **32.** $a = \dfrac{bd - 8}{c - b}$ **33.** $\{x \mid x \text{ is a real number } and\ x \neq -5\ and\ x \neq 5\}$ **34.** $\{x \mid x \text{ is a real number } and\ x \neq -\frac{9}{2}\ and\ x \neq 1\}$

35.

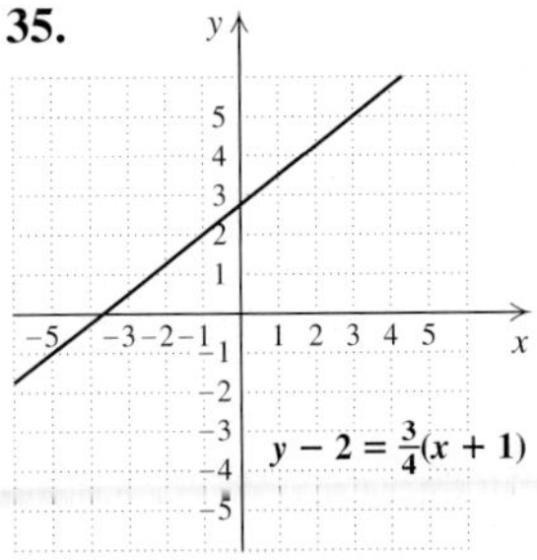

36.

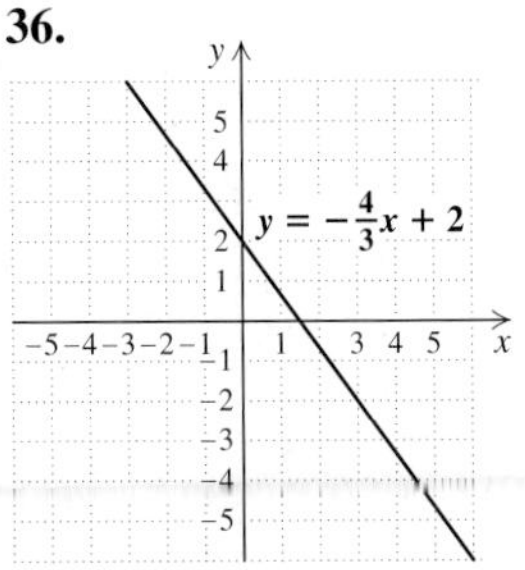

37. TW **39.** **(a)** The degree of R must be less than 1, the degree of $x - r$; **(b)** Let $x = r$. Then

$$\begin{aligned} P(r) &= (r - r) \cdot Q(r) + R \\ &= 0 \cdot Q(r) + R \\ &= R. \end{aligned}$$

41. $0; -3, -\frac{5}{2}, \frac{3}{2}$ **43.** **45.** 0

Exercise Set 6.8, pp. 513–518

1. LCD **2.** Same **3.** Factor **4.** Constant; proportionality **5.** Inverse **6.** Direct **7.** Direct **8.** Inverse **9.** Inverse **10.** Direct **11.** $d = \dfrac{L}{f}$ **13.** $v_1 = \dfrac{2s}{t} - v_2$, or $\dfrac{2s - tv_2}{t}$ **15.** $b = \dfrac{at}{a - t}$ **17.** $R = \dfrac{2V}{I} - 2r$, or $\dfrac{2V - 2Ir}{I}$ **19.** $g = \dfrac{Rs}{s - R}$ **21.** $n = \dfrac{IR}{E - Ir}$ **23.** $q = \dfrac{pf}{p - f}$ **25.** $t_1 = \dfrac{H}{Sm} + t_2$, or $\dfrac{H + Smt_2}{Sm}$ **27.** $r = \dfrac{Re}{E - e}$ **29.** $r = 1 - \dfrac{a}{S}$, or $\dfrac{S - a}{S}$ **31.** $a + b = \dfrac{f}{c^2}$ **33.** $r = \dfrac{A}{P} - 1$, or $\dfrac{A - P}{P}$ **35.** $t_2 = \dfrac{d_2 - d_1}{v} + t_1$, or $\dfrac{d_2 - d_1 + t_1v}{v}$ **37.** $b^2 = \dfrac{a^2y^2}{a^2 - x^2}$ **39.** $Q = \dfrac{2Tt - 2AT}{A - q}$ **41.** $k = 7; y = 7x$ **43.** $k = 1.7; y = 1.7x$ **45.** $k = 6; y = 6x$ **47.** $k = 60; y = \dfrac{60}{x}$ **49.** $k = 112; y = \dfrac{112}{x}$ **51.** $k = 9; y = \dfrac{9}{x}$ **53.** 241,920,000 cans **55.** 6 amperes **57.** 3.5 hr **59.** 32 kg **61.** 50 min **63.** 20 min **65.** 122,269,231 tons **67.** $y = \frac{2}{3}x^2$ **69.** $y = \dfrac{54}{x^2}$ **71.** $y = 0.3xz^2$ **73.** $y = \dfrac{4wx^2}{z}$ **75.** About 2.9 sec **77.** 308 cm^3 **79.** About 57.42 mph **81.** **(a)** Inverse; **(b)** $y = \dfrac{300}{x}$; **(c)** 100 min **83.** **(a)** Directly; **(b)** $y \approx 1.08x$; **(c)** 8.64 million people **85.** TW **87.** $\mathbb{R}$ **88.** $\mathbb{R}$

89.

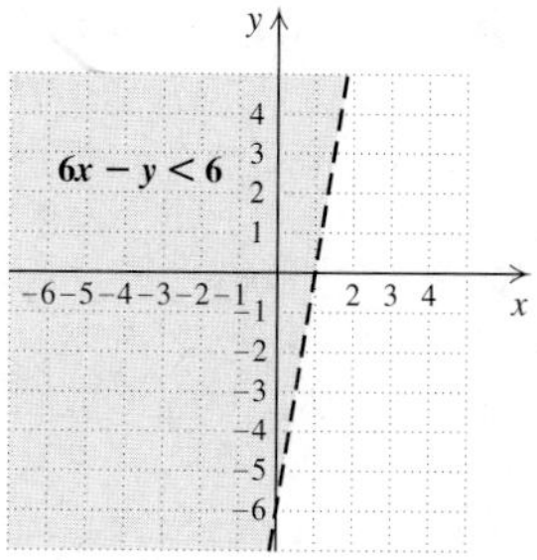

90. $8a^3 - 2a$ **91.** $(t + 2b)(t^2 - 2bt + 4b^2)$

92. $-\frac{5}{3}, \frac{7}{2}$ **93.** TW **95.** 567 mi **97.** Ratio is $\dfrac{a + 12}{a + 6}$; percent increase is $\dfrac{6}{a + 6} \cdot 100\%$, or $\dfrac{600}{a + 6}\%$

99. $t_1 = t_2 + \dfrac{(d_2 - d_1)(t_4 - t_3)}{a(t_4 - t_2)(t_4 - t_3) + d_3 - d_4}$

101. The intensity is halved. **103.** About 1.697 m

105. $d(s) = \dfrac{28}{s}$; 70 yd

Review Exercises: Chapter 6, pp. 521–523

1. True **2.** False **3.** False **4.** True **5.** False **6.** False **7.** True **8.** False **9.** True **10.** True **11. (a)** $-\frac{2}{9}$; **(b)** $-\frac{3}{4}$; **(c)** 0 **12.** $48x^3$

13. $(x + 5)(x - 2)(x - 4)$ **14.** $x + 3$ **15.** $\dfrac{b^2c^6d^2}{a^5}$

16. $\dfrac{15np + 14m}{18m^2n^4p^2}$ **17.** $\dfrac{(x - 2)(x + 5)}{x - 5}$

18. $\dfrac{(x^2 + 4x + 16)(x - 6)}{(x + 4)(x + 2)}$ **19.** $\dfrac{x - 3}{(x + 1)(x + 3)}$

20. $\dfrac{x - y}{x + y}$ **21.** $2(x + y)$ **22.** $\dfrac{-y}{(y + 4)(y - 1)}$

23. $f(x) = \dfrac{1}{x - 1}, x \neq 1, 4$

24. $f(x) = \dfrac{2}{x - 8}, x \neq -8, -5, 8$

25. $f(x) = \dfrac{3x - 1}{x - 3}, x \neq -3, -\frac{1}{3}, 3$ **26.** $\frac{4}{9}$

27. $\dfrac{a^2b^2}{2(b^2 - ba + a^2)}$ **28.** $\dfrac{(y + 11)(y + 5)}{(y - 5)(y + 2)}$

29. $\dfrac{(14 - 3x)(x + 3)}{2x^2 + 16x + 6}$ **30.** 2 **31.** 6 **32.** No solution

33. 0 **34.** $-1, 4$ **35.** $5\frac{1}{7}$ hr **36.** Celeron: 45 sec; Pentium 4: 30 sec **37.** 24 mph **38.** Motorcycle: 62 mph; car: 70 mph **39.** $6s^2 + 5s - 4rs^2$ **40.** $y^2 - 5y + 25$

41. $4x + 3 + \dfrac{-9x - 5}{x^2 + 1}$ **42.** $x^2 + 6x + 20 + \dfrac{54}{x - 3}$

43. 341 **44.** $s = \dfrac{Rg}{g - R}$ **45.** $m = \dfrac{H}{S(t_1 - t_2)}$

46. $c = \dfrac{b + 3a}{2}$ **47.** $t_1 = \dfrac{-A}{vT} + t_2$, or $\dfrac{-A + vTt_2}{vT}$

48. About 22 lb **49.** 64 L **50.** $y = \dfrac{\frac{3}{4}}{x}$

51. (a) Inverse; **(b)** $y = \dfrac{24}{x}$; **(c)** 3 oz

52. TW The least common denominator was used to add and subtract rational expressions, to simplify complex rational expressions, and to solve rational equations.
53. TW A rational *expression* is a quotient of two polynomials. Expressions can be simplified, multiplied, or added, but they cannot be solved for a variable. A rational *equation* is an equation containing rational expressions. In a rational equation, we often can solve for a variable.
54. All real numbers except 0 and 13 **55.** 45 **56.** $9\frac{9}{19}$ sec

Test: Chapter 6, pp. 523–524

1. [6.1] $\dfrac{3}{4(t - 1)}$ **2.** [6.1] $\dfrac{x^2 - 3x + 9}{x + 4}$

3. [6.2] $(x - 3)(x + 8)(x - 9)$ **4.** [6.2] $\dfrac{25x + x^3}{x + 5}$

5. [6.2] $3(a - b)$ **6.** [6.2] $\dfrac{a^3 - a^2b + 4ab + ab^2 - b^3}{(a - b)(a + b)}$

7. [6.2] $\dfrac{-2(2x^2 + 5x + 20)}{(x - 4)(x + 4)(x^2 + 4x + 16)}$

8. [6.2] $f(x) = \dfrac{x - 4}{(x + 3)(x - 2)}, x \neq -3, 2$

9. [6.1] $f(x) = \dfrac{(x - 1)^2(x + 1)}{x(x - 2)}, x \neq -2, 0, 1, 2$

10. [6.3] $\dfrac{a(2b + 3a)}{5a + b}$ **11.** [6.3] $\dfrac{(x - 9)(x - 6)}{(x + 6)(x - 3)}$

12. [6.3] $\dfrac{2x^2 - 7x + 1}{3x^2 + 10x - 17}$ **13.** [6.4] $-\frac{21}{4}$ **14.** [6.4] 15

15. [6.1] 5; 0 **16.** [6.4] $\frac{5}{3}$ **17.** [6.5] $1\frac{31}{32}$ hr

18. [6.6] $\dfrac{4b^2c}{a} - \dfrac{5bc^2}{2a} + 3bc$ **19.** [6.6] $y - 14 + \dfrac{-20}{y - 6}$

20. [6.6] $6x^2 - 9 + \dfrac{5x + 22}{x^2 + 2}$

21. [6.7] $x^2 + 9x + 40 + \dfrac{153}{x - 4}$ **22.** [6.7] 449

23. [6.8] $b_1 = \dfrac{2A}{h} - b_2$, or $\dfrac{2A - b_2h}{h}$ **24.** [6.5] 5 and 6; -6 and -5 **25.** [6.5] $3\frac{3}{11}$ mph **26.** [6.8] 30 workers
27. [6.8] 637 in^2 **28.** [6.4] $-\frac{19}{3}$
29. [6.4] $\{x \mid x \text{ is a real number } and\ x \neq 0\ and\ x \neq 15\}$
30. [6.3], [6.4] x-intercept: $(11, 0)$; y-intercept: $\left(0, -\frac{33}{5}\right)$
31. [6.5] Hans: 56 lawns; Franz: 42 lawns

Cumulative Review: Chapters 1–6, pp. 524–527

1. 1.8 min/yr **2. (a)** $F(t) = 1.45t + 5.5$, where t is the number of years since 1999 and F is in trillions of miles; **(b)** 17.1 trillion miles **3. (a)** $a(t) = 3.482704918t + 67.88319672$; **(b)** approximately 259 million automobiles
4. (a)

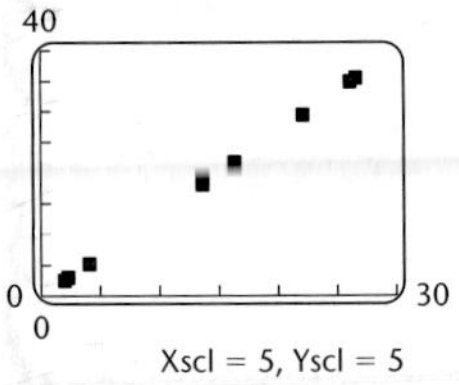

(b) $y \approx 1.33x$; **(c)** The fuel cost is approximately \$1.33/gal.
5. 3 teams received \$18.3 million, and 11 teams received \$750,000 **6.** (b) **7.** (a) **8.** (d) **9.** (c)
10.

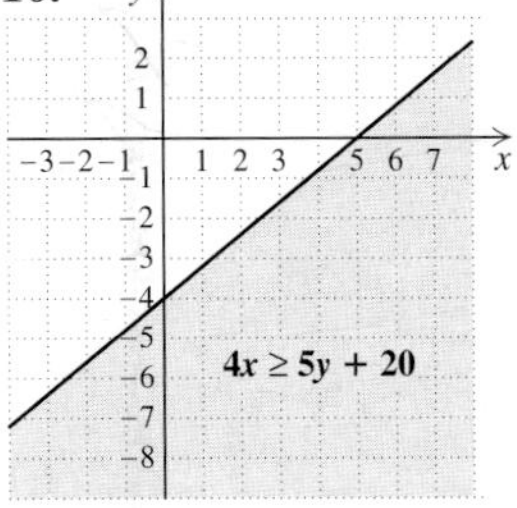

11.

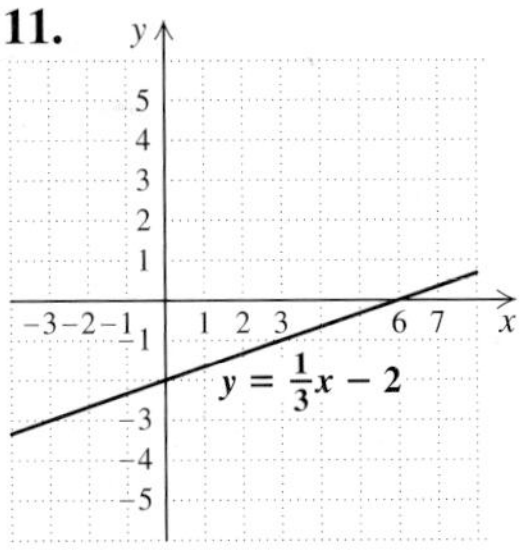

12. 10 **13.** 5.76×10^9 **14.** Slope: $\frac{7}{4}$; y-intercept: $(0, -3)$
15. $y = -\frac{10}{3}x + \frac{11}{3}$ **16.** $-3x^3 + 9x^2 + 3x - 3$
17. $-15x^4y^4$ **18.** $5a + 5b - 5c$
19. $15x^4 - x^3 - 9x^2 + 5x - 2$ **20.** $4x^4 - 4x^2y + y^2$
21. $4x^4 - y^2$ **22.** $-m^3n^2 - m^2n^2 - 5mn^3$
23. $\frac{y - 6}{2}$ **24.** $x - 1$ **25.** $\frac{a^2 + 7ab + b^2}{(a - b)(a + b)}$
26. $\frac{-m^2 + 5m - 6}{(m + 1)(m - 5)}$ **27.** $\frac{3y^2 - 2}{3y}$ **28.** $\frac{y - x}{xy(x + y)}$
29. $9x^2 - 13x + 26 + \frac{-50}{x + 2}$ **30.** $2x^2(2x + 9)$
31. $(x - 6)(x + 14)$ **32.** $(4y - 9)(4y + 9)$
33. $8(2x + 1)(4x^2 - 2x + 1)$ **34.** $(t - 8)^2$
35. $x^2(x - 1)(x + 1)(x^2 + 1)$
36. $(0.3b - 0.2c)(0.09b^2 + 0.06bc + 0.04c^2)$
37. $(4x - 1)(5x + 3)$ **38.** $(3x + 4)(x - 7)$
39. $(x^2 - y)(x^3 + y)$ **40. (a)** $-\frac{1}{2}$;
(b) $\{x \mid x \text{ is a real number } and\ x \neq 5\}$, or $(-\infty, 5) \cup (5, \infty)$
41. $[7, \infty)$ **42.** $\{x \mid x \text{ is a real number } and\ x \neq 2\ and\ x \neq 5\}$, or $(-\infty, 2) \cup (2, 5) \cup (5, \infty)$ **43.** $\frac{1}{4}$ **44.** $-\frac{25}{7}, \frac{25}{7}$
45. $\{x \mid x > -3\}$, or $(-3, \infty)$ **46.** $\{x \mid x \geq -1\}$, or $[-1, \infty)$
47. $\{x \mid -10 < x < 13\}$, or $(-10, 13)$
48. $\left\{x \mid x < -\frac{4}{3}\ or\ x > 6\right\}$, or $\left(-\infty, -\frac{4}{3}\right) \cup (6, \infty)$
49. $\{x \mid x < -6.4\ or\ x > 6.4\}$, or $(-\infty, -6.4) \cup (6.4, \infty)$
50. $\left\{x \mid -\frac{13}{4} \leq x \leq \frac{15}{4}\right\}$, or $\left[-\frac{13}{4}, \frac{15}{4}\right]$ **51.** $-\frac{5}{3}$ **52.** -1
53. No solution **54.** $\frac{1}{3}$ **55.** $(-3, 4)$ **56.** $(-2, -3, 1)$
57. $1, \frac{7}{3}$ **58.** $n = \frac{m - 12}{3}$ **59.** $a = \frac{Pb}{3 - P}$ **60.** 1
61. $\{x \mid x \geq 1\}$, or $[1, \infty)$ **62.** $(2, -3)$ **63.** 2 **64.** 1
65. $\{x \mid 1 \leq x \leq 5\}$, or $[1, 5]$ **66.** $\{1, 5\}$ **67.** $12, \frac{1}{2}, 7\frac{1}{2}$
68. $3\frac{1}{3}$ sec **69.** 30 ft **70.** All such sets of even integers satisfy this condition. **71.** 25 ft
72. $x^3 - 12x^2 + 48x - 64$ **73.** $-3, 3, -5, 5$
74. $\{x \mid -3 \leq x \leq -1\ or\ 7 \leq x \leq 9\}$, or $[-3, -1] \cup [7, 9]$
75. All real numbers except 9 and -5 **76.** $-\frac{1}{4}, 0, \frac{1}{4}$

Chapter 7

Interactive Discovery, p. 532

1. Not an identity **2.** Not an identity **3.** Identity
4. Not an identity **5.** Identity **6.** Identity

Visualizing the Graph, p. 541

1. B **2.** H **3.** C **4.** I **5.** D **6.** A **7.** F
8. J **9.** G **10.** E

Exercise Set 7.1, pp. 542–545

1. Two **2.** Negative **3.** Positive **4.** Negative
5. Irrational **6.** Real **7.** Nonnegative **8.** Negative
9. $7, -7$ **11.** $12, -12$ **13.** $20, -20$ **15.** $30, -30$
17. $-\frac{6}{7}$ **19.** 21 **21.** $-\frac{4}{9}$ **23.** 0.2 **25.** -0.05
27. p^2; 2 **29.** $\frac{x}{y + 4}$; 3 **31.** $\sqrt{20}$; 0; does not exist; does not exist **33.** -3; -1; does not exist; 0
35. 1; $\sqrt{2}$; $\sqrt{101}$ **37.** 1; does not exist; 6 **39.** $6|x|$
41. $6|b|$ **43.** $|8 - t|$ **45.** $|y + 8|$ **47.** $|2x + 7|$
49. 4 **51.** -1 **53.** $-\frac{2}{3}$ **55.** $|x|$ **57.** $6|a|$ **59.** 6
61. $|a + b|$ **63.** $|a^{11}|$ **65.** Cannot be simplified
67. $4x$ **69.** $3t$ **71.** $a + 1$ **73.** $2(x + 1)$, or $2x + 2$
75. $3t - 2$ **77.** 3 **79.** $2x$ **81.** -6 **83.** $5y$ **85.** t^9
87. $(x - 2)^4$ **89.** 2; 3; -2; -4 **91.** 2; does not exist; does not exist; 3 **93.** $\{x \mid x \geq 6\}$, or $[6, \infty)$
95. $\{t \mid t \geq -8\}$, or $[-8, \infty)$ **97.** $\{x \mid x \geq 5\}$, or $[5, \infty)$
99. $\mathbb{R}$ **101.** $\left\{z \mid z \geq -\frac{2}{5}\right\}$, or $\left[-\frac{2}{5}, \infty\right)$ **103.** $\mathbb{R}$
105. Domain: $\{x \mid x \leq 5\}$, or $(-\infty, 5]$; range: $\{y \mid y \geq 0\}$, or $[0, \infty)$
107. Domain: $\{x \mid x \geq -1\}$, or $[-1, \infty)$; range: $\{y \mid y \leq 1\}$, or $(-\infty, 1]$
109. Domain: $\mathbb{R}$; range: $\{y \mid y \geq 5\}$, or $[5, \infty)$
111. (c) **113.** (d) **115.** Yes **117.** Yes **119.** No
121. 20.5 cm; 36.0 cm **123.** TW **125.** $a^9b^6c^{15}$
126. $10a^{10}b^9$ **127.** $\frac{a^6c^{12}}{8b^9}$ **128.** $\frac{x^6y^2}{25z^4}$ **129.** $2x^4y^5z^2$
130. $\frac{5c^3}{a^4b^7}$ **131.** TW **133.** TW
135. (a) 13; **(b)** 15; **(c)** 18; **(d)** 20

137. $\{x \mid x \geq -5\}$, or $[-5, \infty)$ **139.** $\{x \mid x \geq 0\}$, or $[0, \infty)$

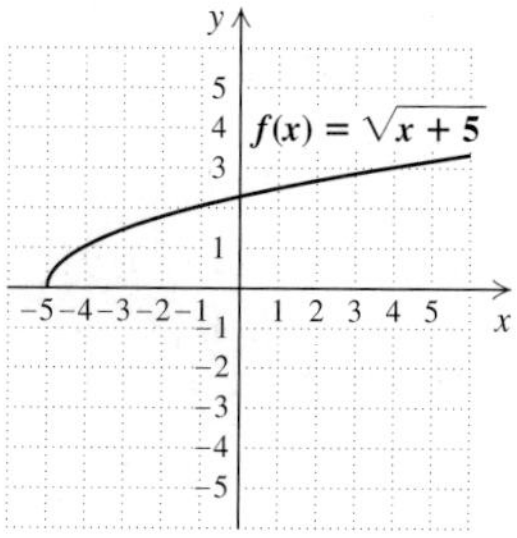

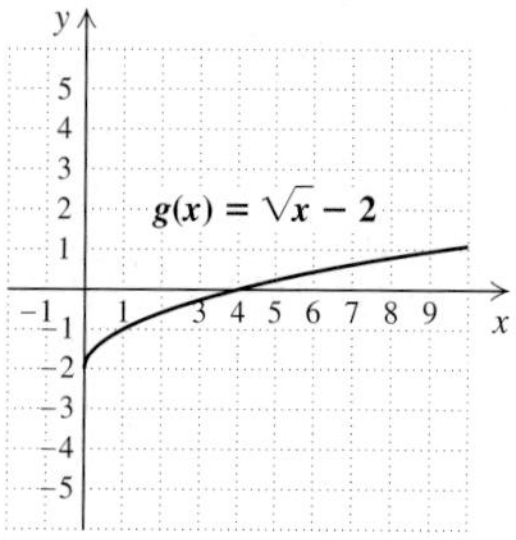

141. $\{x \mid -4 < x \leq 5\}$, or $(-4, 5]$ **143.** Cubic

Exercise Set 7.2, pp. 551–553

1. (g) **2.** (c) **3.** (e) **4.** (h) **5.** (a) **6.** (d) **7.** (b) **8.** (f) **9.** $\sqrt[6]{x}$ **11.** 4 **13.** 3 **15.** 3 **17.** $\sqrt[3]{xyz}$ **19.** $\sqrt[5]{a^2b^2}$ **21.** $\sqrt[5]{t^2}$ **23.** 8 **25.** 81 **27.** $27\sqrt[4]{x^3}$ **29.** $125x^6$ **31.** $20^{1/3}$ **33.** $17^{1/2}$ **35.** $x^{3/2}$ **37.** $m^{2/5}$ **39.** $(cd)^{1/4}$ **41.** $(xy^2z)^{1/5}$ **43.** $(3mn)^{3/2}$ **45.** $(8x^2y)^{5/7}$ **47.** $\frac{2x}{z^{2/3}}$ **49.** $\frac{1}{x^{1/3}}$ **51.** $\frac{1}{(2rs)^{3/4}}$ **53.** 8 **55.** $2a^{3/5}c$ **57.** $\frac{5y^{4/5}z}{x^{2/3}}$ **59.** $\frac{a^3}{3^{5/2}b^{7/3}}$ **61.** $\left(\frac{3c}{2ab}\right)^{5/6}$ **63.** $\frac{6a}{b^{1/4}}$

65.

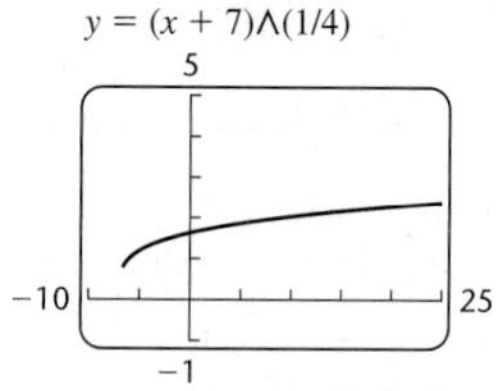

67.

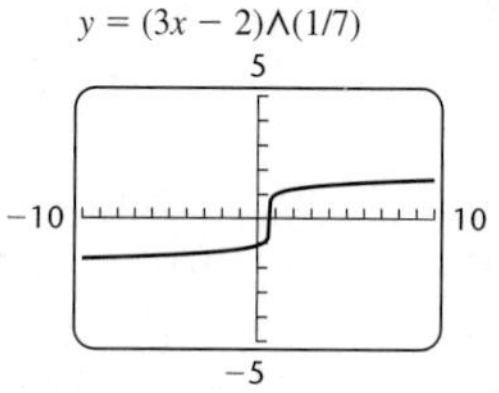

69.

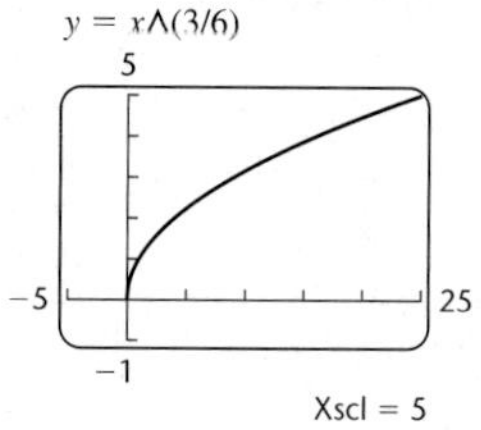

71. 1.552 **73.** 1.778 **75.** −6.240 **77.** $7^{7/8}$ **79.** $3^{3/4}$ **81.** $5.2^{1/2}$ **83.** $10^{6/25}$ **85.** $a^{23/12}$ **87.** 64 **89.** $\frac{m^{1/3}}{n^{1/8}}$ **91.** $\sqrt[3]{x^2}$ **93.** a^3 **95.** a^2 **97.** x^2y^2 **99.** $\sqrt{7a}$ **101.** $\sqrt[4]{8x^3}$ **103.** $\sqrt[18]{a}$ **105.** x^3y^3 **107.** a^6b^{12} **109.** $\sqrt[12]{xy}$ **111.** TW **113.** $11x^4 + 14x^3$ **114.** $-3t^6 + 28t^5 - 20t^4$ **115.** $15a^2 - 11ab - 12b^2$ **116.** $49x^2 - 14xy + y^2$ **117.** $93,500 **118.** 0, 1 **119.** TW **121.** $\sqrt[6]{x^5}$ **123.** $\sqrt[6]{p+q}$ **125.** $2^{7/12} \approx 1.498 \approx 1.5$ **127.** 53.0% **129.** 10 mg **131.**

Exercise Set 7.3, pp. 560–562

1. True **2.** False **3.** False **4.** False **5.** True **6.** True **7.** $\sqrt{35}$ **9.** $\sqrt[3]{14}$ **11.** $\sqrt[4]{18}$ **13.** $\sqrt{26xy}$ **15.** $\sqrt[5]{80y^4}$ **17.** $\sqrt{y^2 - b^2}$ **19.** $\sqrt[3]{0.21y^2}$ **21.** $\sqrt[5]{(x-2)^3}$ **23.** $\sqrt{\frac{7s}{11t}}$ **25.** $\sqrt[7]{\frac{5x-15}{4x+8}}$ **27.** $3\sqrt{2}$ **29.** $3\sqrt{3}$ **31.** $2\sqrt{2}$ **33.** $3\sqrt{22}$ **35.** $6a^2\sqrt{b}$ **37.** $2x\sqrt[3]{y^2}$ **39.** $-2x^2\sqrt[3]{2}$ **41.** $f(x) = 5x\sqrt[3]{x^2}$ **43.** $f(x) = |7(x-3)|$, or $7|x-3|$ **45.** $f(x) = |x-1|\sqrt{5}$ **47.** $a^3b^3\sqrt{b}$ **49.** $xy^2z^3\sqrt[3]{x^2z}$ **51.** $-2ab^2\sqrt[5]{a^2b}$ **53.** $x^2yz^3\sqrt[5]{x^3y^3z^2}$ **55.** $-2a^4\sqrt[3]{10a^2}$ **57.** $3\sqrt{2}$ **59.** $3\sqrt{35}$ **61.** 3 **63.** $18a^3$ **65.** $a\sqrt[3]{10}$ **67.** $2x^3\sqrt{5x}$ **69.** $s^2t^3\sqrt[3]{t}$ **71.** $(x+5)^2$ **73.** $2ab^3\sqrt[4]{5a}$ **75.** $x(y+z)^2\sqrt[5]{x}$ **77.** TW **79.** $\frac{12x^2 + 5y^2}{64xy}$ **80.** $\frac{2a + 6b^3}{a^4b^4}$ **81.** $\frac{-7x - 13}{2(x-3)(x+3)}$ **82.** $\frac{-3x + 1}{2(x-5)(x+5)}$ **83.** $3a^2b^2$ **84.** $3ab^5$ **85.** TW **87.** 175.6 mi **89.** **(a)** −3.3°C; **(b)** −16.6°C; **(c)** −25.5°C; **(d)** −54.0°C **91.** $25x^5\sqrt[3]{25x}$ **93.** $a^{10}b^{17}\sqrt{ab}$

95.

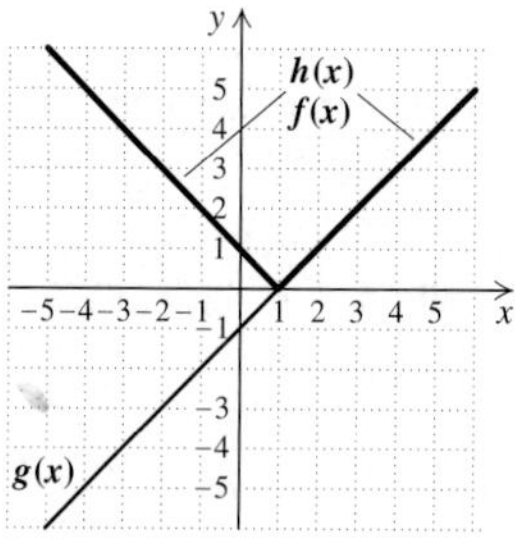

$f(x) = h(x); f(x) \neq g(x)$

97. $\{x \mid x \leq 2$ *or* $x \geq 4\}$, or $(-\infty, 2] \cup [4, \infty)$ **99.** 6 **101.** , TW

Exercise Set 7.4, pp. 566–568

1. (e) **2.** (b) **3.** (f) **4.** (c) **5.** (h) **6.** (d) **7.** (a) **8.** (g) **9.** $\frac{6}{5}$ **11.** $\frac{4}{3}$ **13.** $\frac{7}{y}$ **15.** $\frac{6y\sqrt{y}}{x^2}$ **17.** $\frac{3a\sqrt[3]{a}}{2b}$ **19.** $\frac{2a}{bc^2}$ **21.** $\frac{ab^2}{c^2}\sqrt[4]{\frac{a}{c^2}}$ **23.** $\frac{2x}{y^2}\sqrt[5]{\frac{x}{y}}$ **25.** $\frac{xy}{z^2}\sqrt[6]{\frac{y^2}{z^3}}$ **27.** $\sqrt{5}$ **29.** 3 **31.** $y\sqrt{5y}$ **33.** $2\sqrt[3]{a^2b}$ **35.** $\sqrt{2ab}$ **37.** $2x^2y^3\sqrt[4]{y^3}$ **39.** $\sqrt[3]{x^2 + xy + y^2}$ **41.** $\frac{\sqrt{6}}{2}$ **43.** $\frac{2\sqrt{15}}{21}$ **45.** $\frac{2\sqrt[3]{6}}{3}$ **47.** $\frac{\sqrt[3]{75ac^2}}{5c}$ **49.** $\frac{y\sqrt[3]{180x^2y}}{6x^2}$ **51.** $\frac{\sqrt[3]{2xy^2}}{xy}$ **53.** $\frac{\sqrt{14a}}{6}$ **55.** $\frac{3\sqrt{5y}}{10xy}$ **57.** $\frac{\sqrt{5b}}{6a}$ **59.** $\frac{5}{\sqrt{35x}}$ **61.** $\frac{2}{\sqrt{6}}$ **63.** $\frac{52}{3\sqrt{91}}$

65. $\frac{7}{\sqrt[3]{98}}$ **67.** $\frac{7x}{\sqrt{21xy}}$ **69.** $\frac{2a^2}{\sqrt[3]{20ab}}$ **71.** $\frac{x^2y}{\sqrt{2xy}}$
73. TW **75.** $\frac{3(x-1)}{(x-5)(x+5)}$ **76.** $\frac{7(x-2)}{(x+4)(x-4)}$
77. $\frac{a-1}{a+7}$ **78.** $\frac{t+11}{t+2}$ **79.** $125a^9b^{12}$ **80.** $225x^{10}y^6$
81. TW **83.** **(a)** 1.62 sec; **(b)** 1.99 sec; **(c)** 2.20 sec
85. $9\sqrt[3]{9n^2}$ **87.** $\frac{-3\sqrt{a^2-3}}{a^2-3}$, or $\frac{-3}{\sqrt{a^2-3}}$
89. Step 1: $\sqrt[n]{a} = a^{1/n}$, by definition; Step 2: $\left(\frac{a}{b}\right)^n = \frac{a^n}{b^n}$, raising a quotient to a power; Step 3: $a^{1/n} = \sqrt[n]{a}$, by definition
91. $(f/g)(x) = 3x$, where x is a real number and $x > 0$
93. $(f/g)(x) = \sqrt{x+3}$, where x is a real number and $x > 3$

Exercise Set 7.5, pp. 573–575

1. Radicands; indices **2.** Indices **3.** Bases
4. Denominators **5.** Numerator; conjugate **6.** Bases
7. $9\sqrt{5}$ **9.** $2\sqrt[3]{4}$ **11.** $10\sqrt[3]{y}$ **13.** $7\sqrt{2}$
15. $13\sqrt[3]{7} + \sqrt{3}$ **17.** $9\sqrt{3}$ **19.** $23\sqrt{5}$ **21.** $9\sqrt[3]{2}$
23. $(1 + 6a)\sqrt{5a}$ **25.** $(x + 2)\sqrt[3]{6x}$ **27.** $3\sqrt{a-1}$
29. $(x + 3)\sqrt{x-1}$ **31.** $4\sqrt{3} + 3$ **33.** $15 - 3\sqrt{10}$
35. $6\sqrt{5} - 4$ **37.** $3 - 4\sqrt[3]{63}$ **39.** $a + 2a\sqrt[3]{3}$
41. $4 + 3\sqrt{6}$ **43.** $\sqrt{6} - \sqrt{14} + \sqrt{21} - 7$ **45.** 4
47. -2 **49.** $2 - 8\sqrt{35}$ **51.** $7 + 4\sqrt{3}$ **53.** $5 - 2\sqrt{6}$
55. $2t + 5 + 2\sqrt{10t}$ **57.** $14 + x - 6\sqrt{x+5}$
59. $6\sqrt[4]{63} + 4\sqrt[4]{35} - 3\sqrt[4]{54} - 2\sqrt[4]{30}$ **61.** $\frac{20 + 5\sqrt{3}}{13}$
63. $\frac{12 - 2\sqrt{3} + 6\sqrt{5} - \sqrt{15}}{33}$ **65.** $\frac{a - \sqrt{ab}}{a - b}$ **67.** -1
69. $\frac{12 - 3\sqrt{10} - 2\sqrt{14} + \sqrt{35}}{6}$ **71.** $\frac{3}{5\sqrt{7} - 10}$
73. $\frac{2}{14 + 2\sqrt{3} + 3\sqrt{2} + 7\sqrt{6}}$ **75.** $\frac{x - y}{x + 2\sqrt{xy} + y}$
77. $\frac{1}{\sqrt{a+h} + \sqrt{a}}$ **79.** $a\sqrt[4]{a}$ **81.** $b\sqrt[10]{b^9}$
83. $xy\sqrt[6]{xy^5}$ **85.** $3a^2b\sqrt[4]{ab}$ **87.** $a^2b^2c^2\sqrt[6]{a^2bc^2}$
89. $\sqrt[12]{a^5}$ **91.** $\sqrt[12]{x^2y^5}$ **93.** $\sqrt[10]{ab^9}$ **95.** $\sqrt[20]{(3x-1)^3}$
97. $\sqrt[15]{(2x+1)^4}$ **99.** $x\sqrt[6]{xy^5} - \sqrt[15]{x^{13}y^{14}}$
101. $2m^2 + m\sqrt[4]{n} + 2m\sqrt[3]{n^2} + \sqrt[12]{n^{11}}$ **103.** $\sqrt[4]{2x^2} - x^3$
105. $x^2 - 7$ **107.** $27 + 10\sqrt{2}$ **109.** $8 - 2\sqrt{15}$
111. TW **113.** $7\sqrt{3}$ **114.** 30 **115.** $5\sqrt{x}$
116. $x^2y^2\sqrt[3]{z^2}$ **117.** $75 - x$ **118.** $2x\sqrt[4]{2x}$ **119.** TW
121. $f(x) = -6x\sqrt{5 + x}$
123. $f(x) = (x + 3x^2)\sqrt[4]{x - 1}$
125. $ac^2\left[(3a + 2c)\sqrt{ab} - 2\sqrt[3]{ab}\right]$
127. $9a^2(b + 1)\sqrt[6]{243a^5(b + 1)^5}$ **129.** $1 - \sqrt{w}$
131. $\left(\sqrt{x} + \sqrt{5}\right)\left(\sqrt{x} - \sqrt{5}\right)$
133. $\left(\sqrt{x} + \sqrt{a}\right)\left(\sqrt{x} - \sqrt{a}\right)$ **135.** $2x - 2\sqrt{x^2 - 4}$

Interactive Discovery, p. 576

1. $\{3\}; \{-3, 3\}$ **2.** $\{-2\}; \{-2, 2\}$ **3.** $\{25\}; \{25\}$
4. $\varnothing; \{9\}$

Exercise Set 7.6, pp. 582–584

1. False **2.** True **3.** True **4.** False **5.** True
6. True **7.** $\frac{51}{5}$ **9.** $\frac{[illegible]}{3}$ **11.** 168 **13.** 56 **15.** 3
17. 82 **19.** 0, 9 **21.** 64 **23.** -27 **25.** 125
27. No solution **29.** $\frac{80}{3}$ **31.** 57 **33.** $-\frac{5}{3}$ **35.** 1
37. $\frac{106}{27}$ **39.** 4 **41.** 3, 7 **43.** $\frac{80}{9}$ **45.** -1
47. No solution **49.** 2, 6 **51.** 2 **53.** 4 **55.** TW
57. $-3, 4$ **58.** $\frac{63}{4}$ **59.** $\{x \mid x < -1 \text{ or } x > 4\}$, or $(-\infty, -1) \cup (4, \infty)$ **60.** 28 **61.** $\left\{x \mid x \geq \frac{1}{3}\right\}$, or $\left[\frac{1}{3}, \infty\right)$
62. $-6, 6$ **63.** -7 **64.** $\frac{3}{5}$ **65.** $\left(\frac{31}{7}, \frac{2}{7}\right)$ **66.** $\left(\frac{1}{2}, 0, -1\right)$
67. TW **69.** 524.8°C **71.** $t = \frac{1}{9}\left(\frac{S^2 \cdot 2457}{1087.7^2} - 2617\right)$
73. 4480 rpm **75.** $r = \frac{v^2h}{2gh - v^2}$ **77.** 72.25 ft **79.** $-\frac{8}{9}$
81. $-8, 8$ **83.** 1, 8 **85.** $\left(\frac{1}{36}, 0\right), (36, 0)$ **87.** , TW

Exercise Set 7.7, pp. 590–593

1. Right; hypotenuse **2.** Legs **3.** Square roots
4. Isosceles **5.** 30°, 60°, 90°; leg **6.** $a\sqrt{2}$
7. $\sqrt{34}$; 5.831 **9.** $9\sqrt{2}$; 12.728 **11.** 5 **13.** 4 m
15. $\sqrt{19}$ in.; 4.359 in. **17.** 1 m **19.** 250 ft
21. $\sqrt{8450}$, or $65\sqrt{2}$ ft; 91.924 ft **23.** 12 in.
25. $\left(\sqrt{340} + 8\right)$ ft; 26.439 ft **27.** $\left(110 - \sqrt{6500}\right)$ paces; 29.377 paces **29.** Leg = 5; hypotenuse = $5\sqrt{2} \approx 7.071$
31. Shorter leg = 7; longer leg = $7\sqrt{3} \approx 12.124$
33. Leg = $5\sqrt{3} \approx 8.660$; hypotenuse = $10\sqrt{3} \approx 17.321$
35. Both legs = $\frac{13\sqrt{2}}{2} \approx 9.192$
37. Leg = $14\sqrt{3} \approx 24.249$; hypotenuse = 28
39. $3\sqrt{3} \approx 5.196$ **41.** $13\sqrt{2} \approx 18.385$
43. $\frac{19\sqrt{2}}{2} \approx 13.435$ **45.** $\sqrt{10{,}561}$ ft ≈ 102.767 ft
47. $h = 2\sqrt{3}$ ft ≈ 3.464 ft **49.** $(0, -4), (0, 4)$ **51.** TW
53.

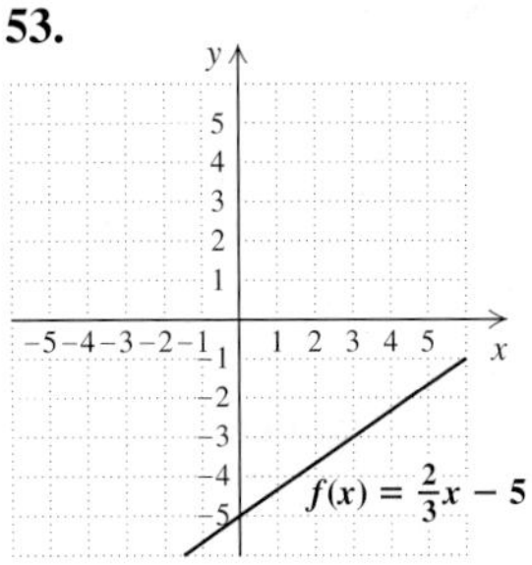

54.

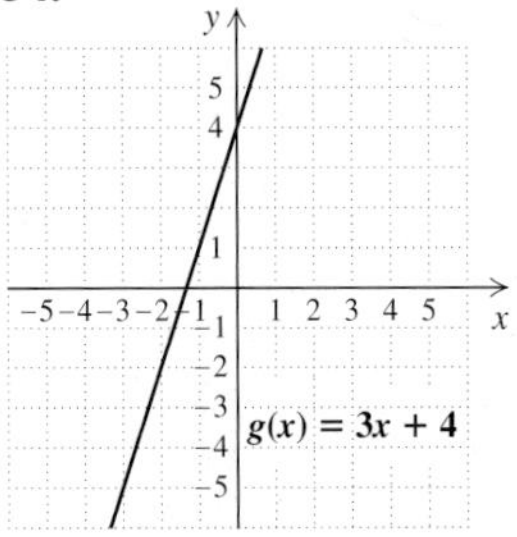

55.

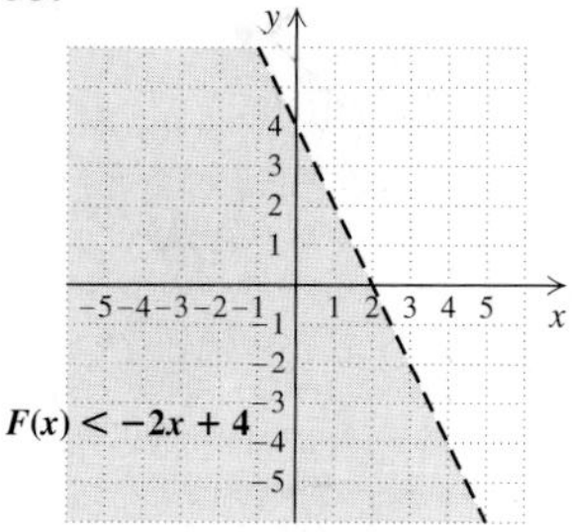

56.

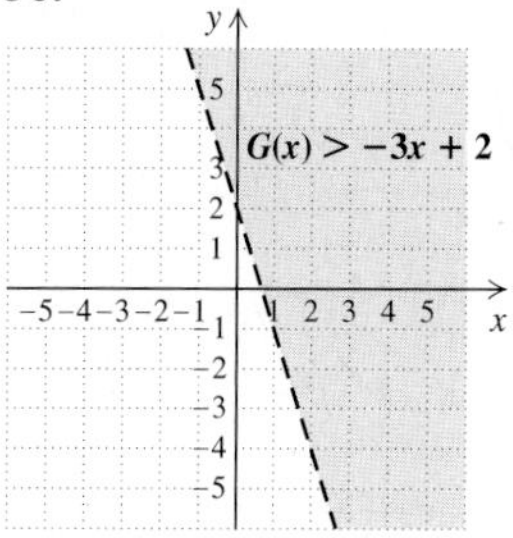

57.

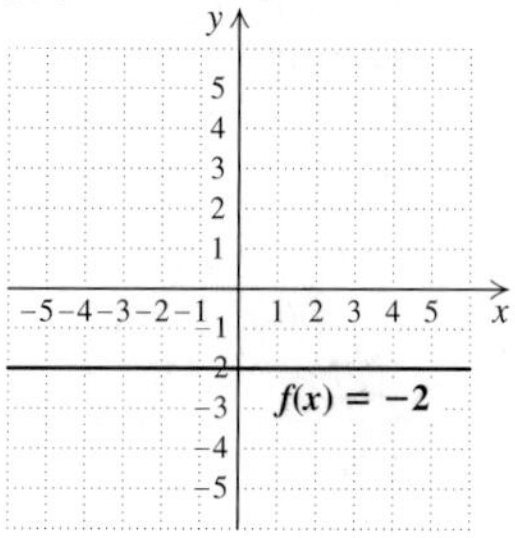

58.

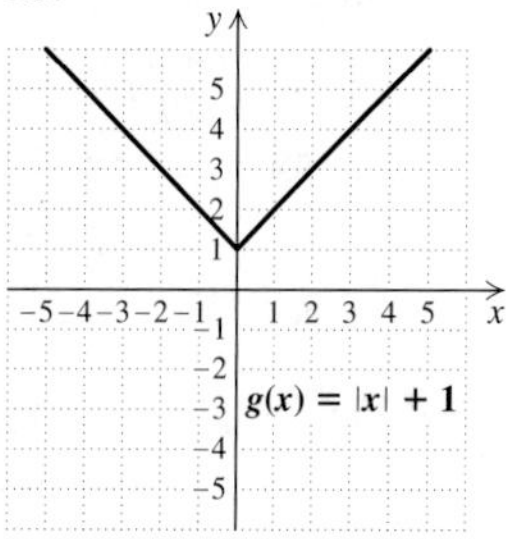

59. TW **61.** $36\sqrt{3}$ cm^2; 62.354 cm^2 **63.** $d = s + s\sqrt{2}$
65. 5 gal. The total area of the doors and windows is 134 ft^2 or more. **67.** 60.28 ft by 60.28 ft

Exercise Set 7.8, pp. 600–601

1. False **2.** False **3.** True **4.** True **5.** True **6.** True **7.** False **8.** True **9.** $6i$ **11.** $i\sqrt{13}$, or $\sqrt{13}i$ **13.** $3i\sqrt{2}$, or $3\sqrt{2}i$ **15.** $i\sqrt{3}$, or $\sqrt{3}i$ **17.** $9i$ **19.** $-10i\sqrt{3}$, or $-10\sqrt{3}i$ **21.** $6 - 2i\sqrt{21}$, or $6 - 2\sqrt{21}i$ **23.** $\left(-2\sqrt{19} + 5\sqrt{5}\right)i$ **25.** $\left(3\sqrt{2} - 10\right)i$ **27.** $11 + 10i$ **29.** $4 + 5i$ **31.** $2 - i$ **33.** $-12 - 5i$ **35.** -42 **37.** -24 **39.** -18 **41.** $-\sqrt{10}$ **43.** $-3\sqrt{14}$ **45.** $-30 + 10i$ **47.** $-28 - 21i$ **49.** $1 + 5i$ **51.** $38 + 9i$ **53.** $2 - 46i$ **55.** $-11 - 16i$ **57.** $13 - 47i$ **59.** $12 - 16i$ **61.** $-5 + 12i$ **63.** $-5 - 12i$ **65.** $\frac{28}{17} - \frac{7}{17}i$ **67.** $\frac{6}{13} + \frac{4}{13}i$ **69.** $\frac{3}{17} + \frac{5}{17}i$ **71.** $-\frac{5}{6}i$ **73.** $-\frac{3}{4} - \frac{5}{4}i$ **75.** $1 - 2i$ **77.** $-\frac{23}{58} + \frac{43}{58}i$ **79.** $\frac{19}{29} - \frac{4}{29}i$ **81.** $\frac{6}{25} - \frac{17}{25}i$ **83.** $-i$ **85.** 1 **87.** -1 **89.** i **91.** -1 **93.** $-125i$ **95.** 0 **97.** TW **99.** $(3x - 2)(4x - 5)$ **100.** $x(x + 2)(x - 5)$ **101.** $5x^2y^2(4y + 3x - 7x^2y)$ **102.** $(x^2 - 3)(y - x)$ **103.** $14y(x^2 + y^2)(x + y)(x - y)$ **104.** $(3x - 4y)(9x^2 + 12xy + 16y^2)$ **105.** TW **107.** $-9 - 27i$ **109.** $50 - 120i$ **111.** $\frac{250}{41} + \frac{200}{41}i$ **113.** 8 **115.** $\frac{3}{5} + \frac{9}{5}i$ **117.** 1

Review Exercises: Chapter 7, pp. 605–606

1. True **2.** False **3.** True **4.** True **5.** True **6.** True **7.** True **8.** False **9.** $\frac{7}{3}$ **10.** -0.5 **11.** 5 **12.** $\left\{x \middle| x \geq \frac{7}{2}\right\}$, or $\left[\frac{7}{2}, \infty\right)$ **13.** (a) 34.6 lb; (b) 10.0 lb; (c) 24.9 lb; (d) 42.3 lb **14.** $5|t|$ **15.** $|c + 8|$ **16.** $|x - 3|$ **17.** $|2x + 1|$ **18.** -2 **19.** $-\dfrac{4x^2}{3}$ **20.** $|x^3y^2|$, or $|x^3|y^2$ **21.** $2x^2$ **22.** $(5ab)^{4/3}$ **23.** $8a^4\sqrt{a}$ **24.** x^3y^5 **25.** $\sqrt[3]{x^2y}$ **26.** $\dfrac{1}{x^{2/5}}$ **27.** $7^{1/6}$ **28.** $f(x) = 5|x - 6|$ **29.** $\sqrt{6xy}$ **30.** $3a\sqrt[3]{a^2b^2}$ **31.** $-6x^5y^4\sqrt[3]{2x^2}$ **32.** $y\sqrt[3]{6}$ **33.** $\dfrac{5\sqrt{x}}{2}$ **34.** $\dfrac{2a^2\sqrt[4]{3a^3}}{c^2}$ **35.** $7\sqrt[3]{x}$ **36.** $\sqrt{3}$ **37.** $(2x + y^2)\sqrt[3]{x}$ **38.** $15\sqrt{2}$ **39.** $\sqrt{15} + 4\sqrt{6} - 6\sqrt{10} - 48$ **40.** $\sqrt[4]{x^3}$ **41.** $\sqrt[12]{x^5}$ **42.** $a^2 - 2a\sqrt{2} + 2$ **43.** $-4\sqrt{10} + 4\sqrt{15}$ **44.** $\dfrac{20}{\sqrt{10} + \sqrt{15}}$ **45.** 19 **46.** -126 **47.** 4 **48.** 14 **49.** $5\sqrt{2}$ cm; 7.071 cm **50.** $\sqrt{32}$ ft; 5.657 ft **51.** Short leg = 10; long leg = $10\sqrt{3} \approx 17.321$ **52.** $-2i\sqrt{2}$, or $-2\sqrt{2}i$ **53.** $-2 - 9i$ **54.** $6 + i$ **55.** 29 **56.** -1 **57.** $9 - 12i$ **58.** $\frac{13}{25} - \frac{34}{25}i$ **59.** TW A complex number $a + bi$ is real when $b = 0$. It is imaginary when $b \neq 0$.
60. TW An absolute-value sign must be used to simplify $\sqrt[n]{x^n}$ when n is even, since x may be negative. If x is negative while n is even, the radical expression cannot be simplified to x, since $\sqrt[n]{x^n}$ represents the principal, or positive, root. When n is odd, there is only one root, and it will be positive or negative depending on the sign of x. Thus there is no absolute-value sign when n is odd.

61. 3 **62.** $-\frac{2}{5} + \frac{9}{10}i$ **63.** $\dfrac{2i}{3i}$; answers may vary

Test: Chapter 7, p. 607

1. [7.3] $5\sqrt{2}$ **2.** [7.4] $-\dfrac{2}{x^2}$ **3.** [7.1] $9|a|$ **4.** [7.1] $|x - 4|$ **5.** [7.3] $x^2y\sqrt[5]{x^2y^3}$ **6.** [7.4] $\left|\dfrac{5x}{6y^2}\right|$, or $\dfrac{5|x|}{6y^2}$ **7.** [7.3] $\sqrt[3]{15y^2z}$ **8.** [7.4] $\sqrt[5]{x^2y^2}$ **9.** [7.5] $xy\sqrt[4]{x}$ **10.** [7.5] $\sqrt[20]{a^3}$ **11.** [7.5] $6\sqrt{2}$ **12.** [7.5] $(x^2 + 3y)\sqrt{y}$ **13.** [7.5] $14 - 19\sqrt{x} - 3x$ **14.** [7.2] $(7xy)^{1/2}$ **15.** [7.2] $\sqrt[6]{(4a^3b)^5}$ **16.** [7.1] $\{x | x \geq 5\}$, or $[5, \infty)$ **17.** [7.5] $27 + 10\sqrt{2}$ **18.** [7.5] $\dfrac{5\sqrt{3} - \sqrt{6}}{23}$ **19.** [7.6] 7 **20.** [7.6] No solution **21.** [7.1] 3.4 sec **22.** [7.7] $7\sqrt{3}$ cm or 12.124 cm (if the shorter leg is 7 cm); $\dfrac{7}{\sqrt{3}}$ cm or 4.041 cm (if the longer leg is 7 cm) **23.** [7.7] $\sqrt{10{,}600}$ ft; 102.956 ft **24.** [7.8] $5i\sqrt{2}$, or $5\sqrt{2}i$ **25.** [7.8] $12 + 2i$

26. [7.8] -24 **27.** [7.8] $15 - 8i$ **28.** [7.8] $-\frac{11}{34} - \frac{7}{34}i$
29. [7.8] i **30.** [7.6] 3 **31.** [7.8] $-\frac{17}{4}i$
32. [7.7] The isosceles right triangle is larger by 1.206 ft^2.

Chapter 8

Interactive Discovery, p. 610

1. 1 **2.** 1 **3.** 1 **4.** 0 **5.** 0 **6.** 2 **7.** A cup-shaped curve opening up or down

Exercise Set 8.1, pp. 620–622

1. $\sqrt{k}; -\sqrt{k}$ **2.** 7; -7 **3.** $t + 3; t + 3$ **4.** 16
5. 25; 5 **6.** 9; 3 **7.** 2 **9.** 1 **11.** 0 **13.** $\pm\sqrt{5}$
15. $\pm\frac{4}{3}i$ **17.** $\pm\sqrt{\frac{7}{5}}$, or $\pm\frac{\sqrt{35}}{5}$ **19.** $-6, 8$
21. $13 \pm 3\sqrt{2}$ **23.** $-1 \pm 3i$ **25.** $-\frac{3}{4} \pm \frac{\sqrt{17}}{4}$, or $\frac{-3 \pm \sqrt{17}}{4}$ **27.** $-3, 13$ **29.** 1, 9 **31.** $-4 \pm \sqrt{13}$
33. $-14, 0$ **35.** $x^2 + 16x + 64 = (x + 8)^2$
37. $t^2 - 10t + 25 = (t - 5)^2$ **39.** $x^2 + 3x + \frac{9}{4} = \left(x + \frac{3}{2}\right)^2$
41. $t^2 - 9t + \frac{81}{4} = \left(t - \frac{9}{2}\right)^2$ **43.** $x^2 + \frac{2}{5}x + \frac{1}{25} = \left(x + \frac{1}{5}\right)^2$
45. $t^2 - \frac{5}{6}t + \frac{25}{144} = \left(t - \frac{5}{12}\right)^2$ **47.** $-7, 1$ **49.** 4, 6
51. $-9, -1$ **53.** $-4 \pm \sqrt{19}$
55. $\left(-3 - \sqrt{2}, 0\right), \left(-3 + \sqrt{2}, 0\right)$
57. $\left(-6 - \sqrt{11}, 0\right), \left(-6 + \sqrt{11}, 0\right)$
59. $\left(5 - \sqrt{47}, 0\right), \left(5 + \sqrt{47}, 0\right)$ **61.** $-\frac{4}{3}, -\frac{2}{3}$
63. $-\frac{1}{3}, 2$ **65.** $-\frac{2}{5} \pm \frac{\sqrt{19}}{5}$, or $\frac{-2 \pm \sqrt{19}}{5}$
67. $\left(-\frac{1}{4} - \frac{\sqrt{13}}{4}, 0\right), \left(-\frac{1}{4} + \frac{\sqrt{13}}{4}, 0\right)$, or $\left(\frac{-1 - \sqrt{13}}{4}, 0\right), \left(\frac{-1 + \sqrt{13}}{4}, 0\right)$
69. $\left(\frac{3}{4} - \frac{\sqrt{17}}{4}, 0\right), \left(\frac{3}{4} + \frac{\sqrt{17}}{4}, 0\right)$, or $\left(\frac{3 - \sqrt{17}}{4}, 0\right), \left(\frac{3 + \sqrt{17}}{4}, 0\right)$ **71.** 10% **73.** 18.75%
75. 4% **77.** About 8.1 sec **79.** About 9.5 sec
81. TW **83.** 28 **84.** -92 **85.** $3\sqrt[3]{10}$ **86.** $4\sqrt{5}$
87. 5 **88.** 7 **89.** TW **91.** ± 18 **93.** $-\frac{7}{2}, -\sqrt{5}, 0, \sqrt{5}, 8$ **95.** Barge: 8 km/h; fishing boat: 15 km/h

Exercise Set 8.2, pp. 627–628

1. True **2.** False **3.** False **4.** True **5.** False
6. True **7.** $-\frac{7}{2} \pm \frac{\sqrt{61}}{2}$ **9.** $3 \pm \sqrt{7}$
11. $-\frac{1}{2} \pm \frac{\sqrt{3}}{2}i$ **13.** $2 \pm 3i$ **15.** $3 \pm \sqrt{5}$
17. $-\frac{4}{3} \pm \frac{\sqrt{19}}{3}$ **19.** $-\frac{1}{2} \pm \frac{\sqrt{17}}{2}$ **21.** $-\frac{3}{8} \pm \frac{\sqrt{129}}{24}$
23. $\frac{2}{5}$ **25.** $-\frac{11}{8} \pm \frac{\sqrt{41}}{8}$ **27.** 5, 10 **29.** $\frac{13}{10} \pm \frac{\sqrt{509}}{10}$
31. $2 \pm \sqrt{5}i$ **33.** $2, -1 \pm \sqrt{3}i$ **35.** $\frac{2}{3}, 1$
37. $5 \pm \sqrt{53}$ **39.** $\frac{7}{2} \pm \frac{\sqrt{85}}{2}$ **41.** $\frac{3}{2}, 6$
43. $-5.31662479, 1.31662479$ **45.** 0.7639320225, 5.236067978 **47.** $-1.265564437, 2.765564437$ **49.** TW
51. $\frac{x + y}{2}$ **52.** $\frac{a^2 - b^2}{b}$ **53.** $9a^2b^3\sqrt{2a}$
54. $4a^2b^3\sqrt{6}$ **55.** $\frac{3(x + 1)}{3x + 1}$ **56.** $\frac{4b}{3ab^2 - 4a^2}$
57. TW **59.** $(-2, 0), (1, 0)$ **61.** $4 - 2\sqrt{2}, 4 + 2\sqrt{2}$
63. $-1.179210116, 0.3392101158$ **65.** $\frac{-5\sqrt{2} \pm \sqrt{34}}{4}$
67. $\frac{1}{2}$ **69.** , TW

Exercise Set 8.3, pp. 633–636

1. First part: 60 mph; second part: 50 mph **3.** 40 mph
5. Cessna: 150 mph, Beechcraft: 200 mph; or Cessna: 200 mph, Beechcraft: 250 mph
7. To Hillsboro: 10 mph; return trip: 4 mph
9. About 14 mph **11.** 12 hr **13.** About 3.24 mph
15. $r = \frac{1}{2}\sqrt{\frac{A}{\pi}}$, or $\frac{\sqrt{A\pi}}{2\pi}$
17. $r = \frac{-\pi h + \sqrt{\pi^2h^2 + 2\pi A}}{2\pi}$
19. $s = \sqrt{\frac{kQ_1Q_2}{N}}$, or $\frac{\sqrt{kQ_1Q_2N}}{N}$ **21.** $g = \frac{4\pi^2 l}{T^2}$
23. $c = \sqrt{d^2 - a^2 - b^2}$ **25.** $t = \frac{-v_0 + \sqrt{v_0^2 + 2gs}}{g}$
27. $n = \frac{1 + \sqrt{1 + 8N}}{2}$ **29.** $h = \frac{V^2}{12.25}$
31. $t = \frac{-b \pm \sqrt{b^2 - 4ac}}{2a}$ **33.** **(a)** 10.1 sec; **(b)** 7.49 sec; **(c)** 272.5 m **35.** 2.9 sec **37.** 0.968 sec **39.** 2.5 m/sec
41. 7% **43.** TW **45.** $\{x \mid x > 3\}$, where x is the number of semesters **46.** Cream-filled: 46; glazed: 44 **47.** 1 in.
48. Jaime: 3 min; Cheri: 6 min **49.** 762 kilobytes
50. 45 mph **51.** TW
53. $t = \frac{-10.2 + 6\sqrt{-A^2 + 13A - 39.36}}{A - 6.5}$
55. $\pm\sqrt{2}$ **57.** $l = \frac{w + w\sqrt{5}}{2}$
59. $n = \pm\sqrt{\frac{r^2 \pm \sqrt{r^4 + 4m^4r^2p - 4mp}}{2m}}$ **61.** $A(S) = \frac{\pi S}{6}$

Exercise Set 8.4, pp. 640–642

1. Discriminant **2.** One **3.** Two **4.** Two **5.** Rational **6.** Imaginary **7.** Two irrational **9.** Two imaginary **11.** Two irrational **13.** Two rational **15.** Two imaginary **17.** One rational **19.** Two rational **21.** Two rational **23.** Two irrational **25.** Two imaginary **27.** Two irrational **29.** $x^2 + 4x - 21 = 0$ **31.** $x^2 - 6x + 9 = 0$ **33.** $x^2 + 4x + 3 = 0$ **35.** $4x^2 - 23x + 15 = 0$ **37.** $8x^2 + 6x + 1 = 0$ **39.** $x^2 - 2x - 0.96 = 0$ **41.** $x^2 - 3 = 0$ **43.** $x^2 - 20 = 0$ **45.** $x^2 + 16 = 0$ **47.** $x^2 - 4x + 53 = 0$ **49.** $x^2 - 6x - 5 = 0$ **51.** $3x^2 - 6x - 4 = 0$ **53.** $x^3 - 4x^2 - 7x + 10 = 0$ **55.** $x^3 - 2x^2 - 3x = 0$ **57.** TW **59.** $81a^8$ **60.** $16x^6$ **61.** $(-1, 0), (8, 0)$ **62.** $(2, 0), (4, 0)$ **63.** 6 commercials
64.

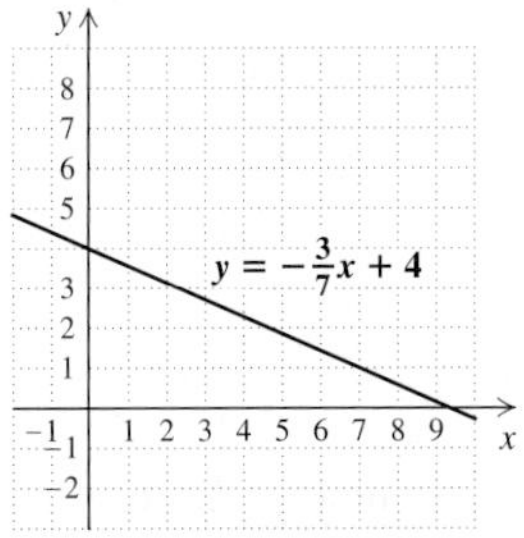

65. TW **67.** $a = 1, b = 2, c = -3$ **69.** **(a)** $-\frac{3}{5}$; **(b)** $-\frac{1}{3}$
71. **(a)** $9 + 9i$; **(b)** $3 + 3i$
73. The solutions of $ax^2 + bx + c = 0$ are $x = \dfrac{-b \pm \sqrt{b^2 - 4ac}}{2a}$. When there is just one solution, $b^2 - 4ac$ must be 0, so $x = \dfrac{-b \pm 0}{2a} = \dfrac{-b}{2a}$.
75. $a = 8, b = 20, c = -12$
77. $x^4 - 8x^3 + 21x^2 - 2x - 52 = 0$

Exercise Set 8.5, pp. 647–648

1. (f) **2.** (d) **3.** (h) **4.** (b) **5.** (g) **6.** (a) **7.** (e) **8.** (c) **9.** $\pm 1, \pm 2$ **11.** $\pm\sqrt{5}, \pm 2$ **13.** $\pm\dfrac{\sqrt{3}}{2}, \pm 2$ **15.** $8 + 2\sqrt{7}$ **17.** $\pm 2\sqrt{2}, \pm 3$ **19.** No solution **21.** $-\frac{1}{2}, \frac{1}{3}$ **23.** $-\frac{4}{5}, 1$ **25.** $-27, 8$ **27.** 729 **29.** 1 **31.** No solution **33.** $\frac{12}{5}$ **35.** $\left(\frac{4}{25}, 0\right)$ **37.** $\left(\dfrac{3}{2} + \dfrac{\sqrt{33}}{2}, 0\right), \left(\dfrac{3}{2} - \dfrac{\sqrt{33}}{2}, 0\right), (4, 0), (-1, 0)$ **39.** $(-243, 0), (32, 0)$ **41.** No x-intercepts **43.** TW **45.** $-3, 2$ **46.** $\dfrac{3}{4} \pm \dfrac{\sqrt{23}}{4}i$ **47.** $7 \pm \sqrt{5}$ **48.** $-\dfrac{1}{8} \pm \dfrac{\sqrt{97}}{8}$, or $\dfrac{-1 \pm \sqrt{97}}{8}$ **49.** $\pm\sqrt{2}, \pm\sqrt{3}$ **50.** No solution **51.** TW **53.** $\pm\sqrt{\dfrac{7 \pm \sqrt{29}}{10}}$ **55.** $-2, -1, 5, 6$ **57.** $\frac{100}{99}$ **59.** $-5, -3, -2, 0, 2, 3, 5$ **61.** $1, 3, -\dfrac{1}{2} + \dfrac{\sqrt{3}}{2}i, -\dfrac{1}{2} - \dfrac{\sqrt{3}}{2}i, -\dfrac{3}{2} + \dfrac{3\sqrt{3}}{2}i, -\dfrac{3}{2} - \dfrac{3\sqrt{3}}{2}i$

Interactive Discovery, p. 650

1. **(a)** $(0, 0)$; **(b)** $x = 0$; **(c)** upward; **(d)** narrower
2. **(a)** $(0, 0)$; **(b)** $x = 0$; **(c)** upward; **(d)** wider
3. **(a)** $(0, 0)$; **(b)** $x = 0$; **(c)** upward; **(d)** wider
4. **(a)** $(0, 0)$; **(b)** $x = 0$; **(c)** downward; **(d)** neither narrower nor wider **5.** **(a)** $(0, 0)$; **(b)** $x = 0$; **(c)** downward; **(d)** narrower **6.** **(a)** $(0, 0)$; **(b)** $x = 0$; **(c)** downward; **(d)** wider **7.** When $a > 1$, the graph of $y = ax^2$ is narrower than the graph of $y = x^2$. When $0 < a < 1$, the graph of $y = ax^2$ is wider than the graph of $y = x^2$.
8. When $a < -1$, the graph of $y = ax^2$ is narrower than the graph of $y = x^2$ and the graph opens downward. When $-1 < a < 0$, the graph of $y = ax^2$ is wider than the graph of $y = x^2$ and the graph opens downward.

Interactive Discovery, p. 651

1. **(a)** $(3, 0)$; **(b)** $x = 3$; **(c)** same shape **2.** **(a)** $(-1, 0)$; **(b)** $x = -1$; **(c)** same shape **3.** **(a)** $\left(\frac{3}{2}, 0\right)$; **(b)** $x = \frac{3}{2}$; **(c)** same shape **4.** **(a)** $(-2, 0)$; **(b)** $x = -2$; **(c)** same shape
5. The graph of $g(x) = a(x - h)^2$ looks like the graph of $f(x) = ax^2$, except that it is moved left or right.

Exercise Set 8.6, pp. 655–657

1. (h) **2.** (g) **3.** (f) **4.** (d) **5.** (b) **6.** (c) **7.** (e) **8.** (a) **9.** **(a)** Positive; **(b)** $(3, 1)$; **(c)** $x = 3$; **(d)** $[1, \infty)$ **11.** **(a)** Negative; **(b)** $(-2, -3)$; **(c)** $x = -2$; **(d)** $(-\infty, -3]$ **13.** **(a)** Positive; **(b)** $(-3, 0)$; **(c)** $x = -3$; **(d)** $[0, \infty)$
15.

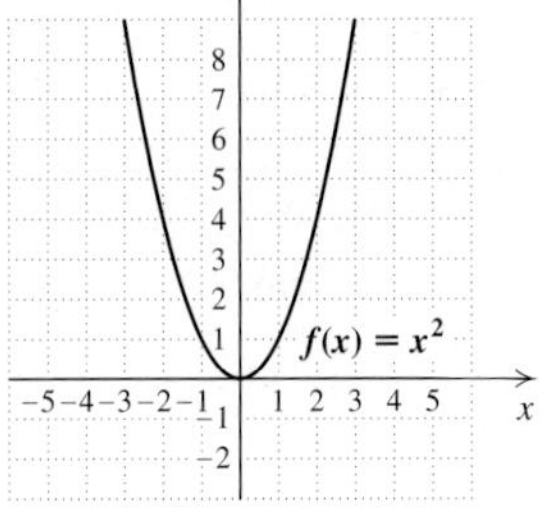

17.

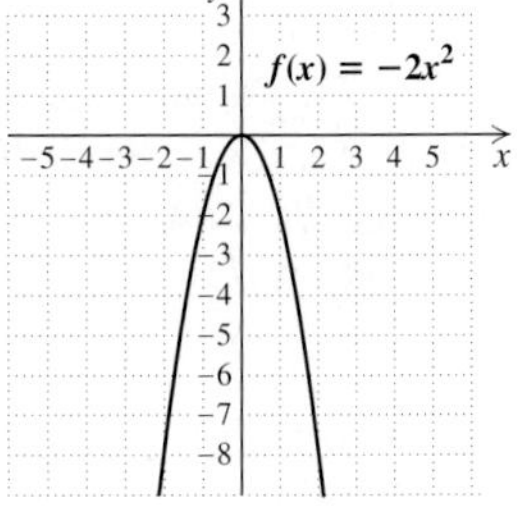

19.

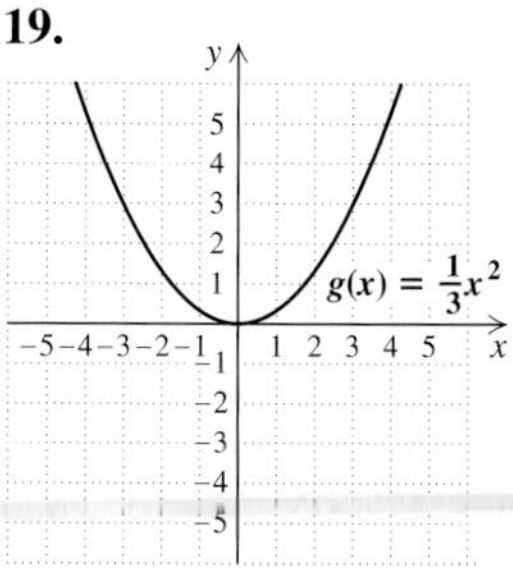

21.

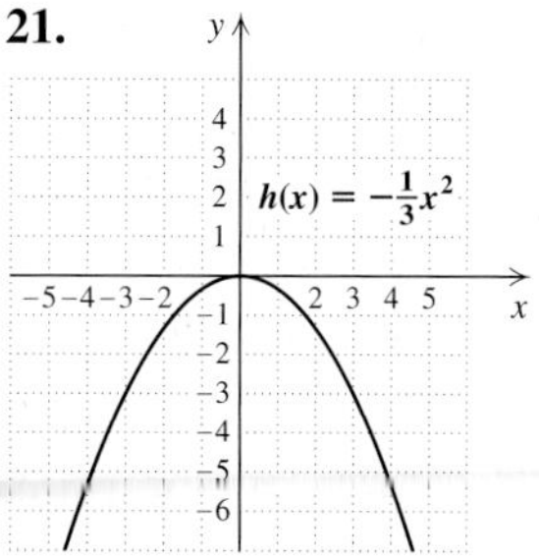

23.

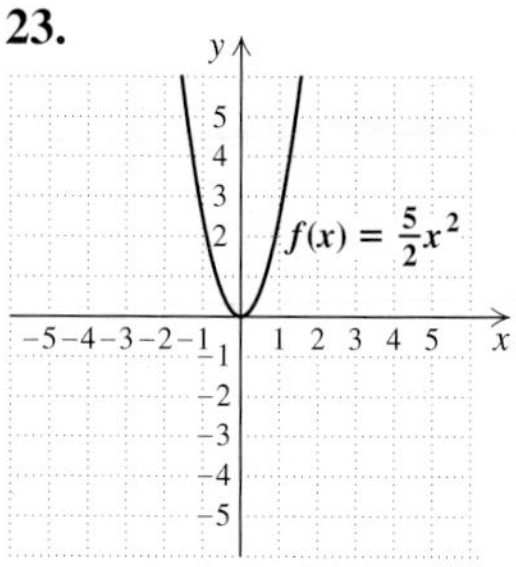

25. Vertex: $(-1, 0)$; axis of symmetry: $x = -1$

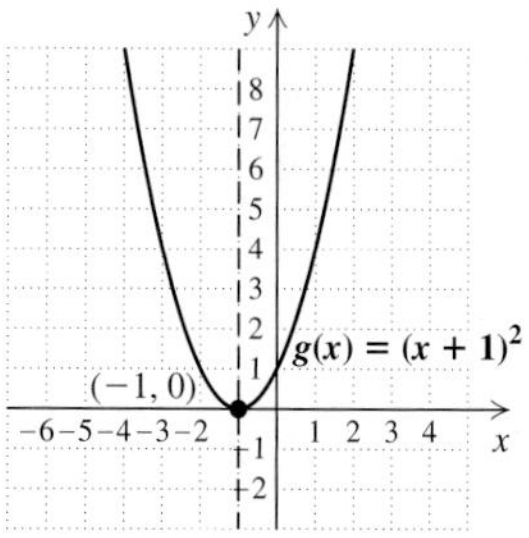

27. Vertex $(2, 0)$; axis of symmetry: $x = 2$

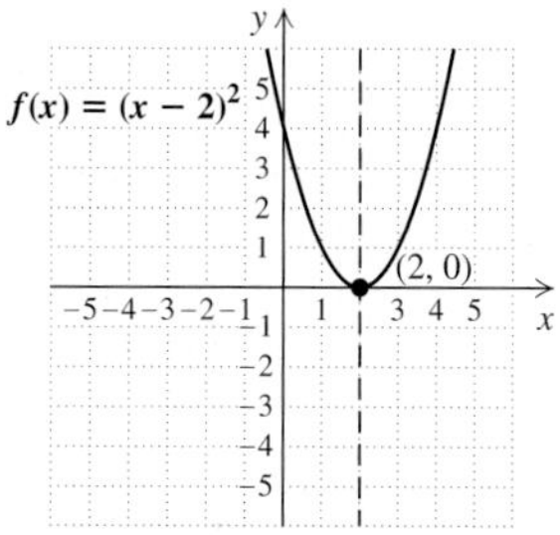

29. Vertex: $(3, 0)$; axis of symmetry: $x = 3$

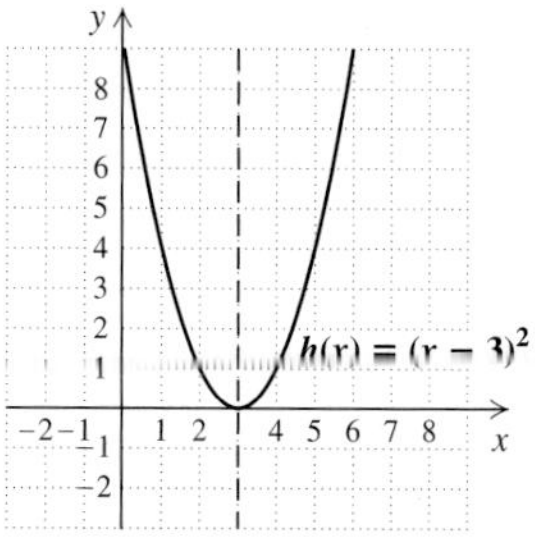

31. Vertex: $(-1, 0)$; axis of symmetry: $x = -1$

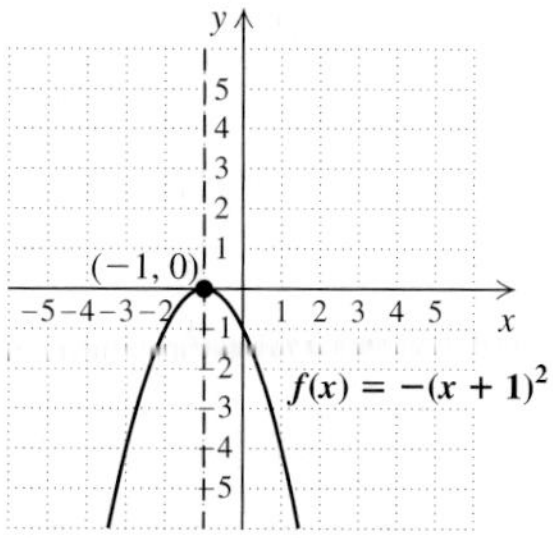

33. Vertex: $(2, 0)$; axis of symmetry: $x = 2$

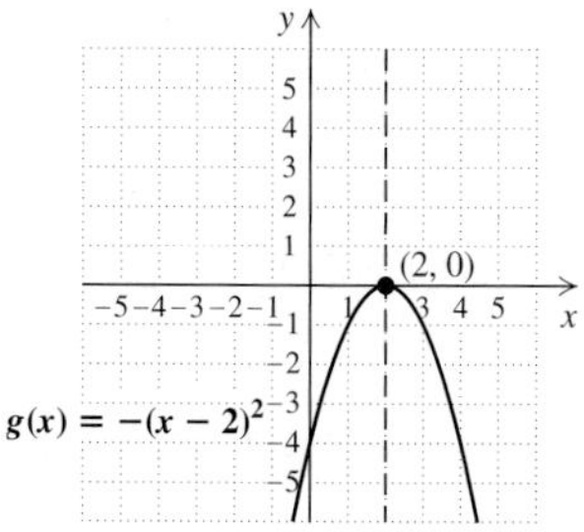

35. Vertex: $(-1, 0)$; axis of symmetry: $x = -1$

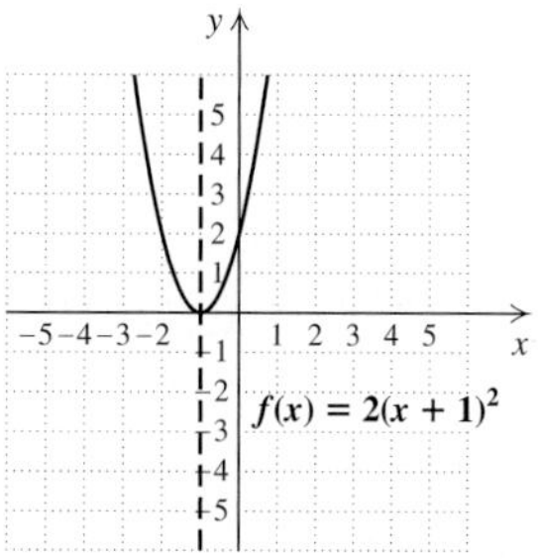

37. Vertex: $(4, 0)$; axis of symmetry: $x = 4$

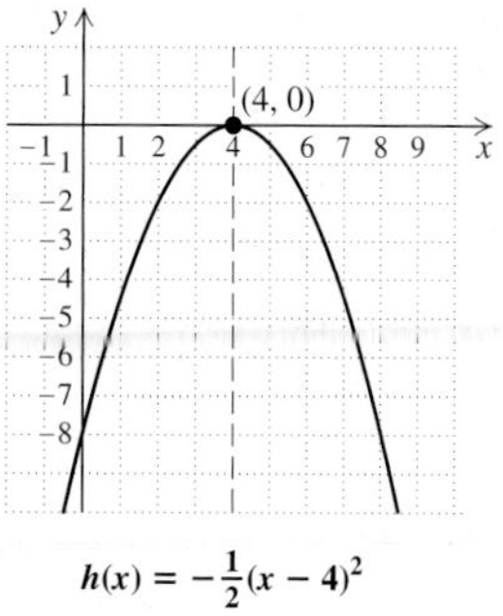

39. Vertex: $(1, 0)$; axis of symmetry: $x = 1$

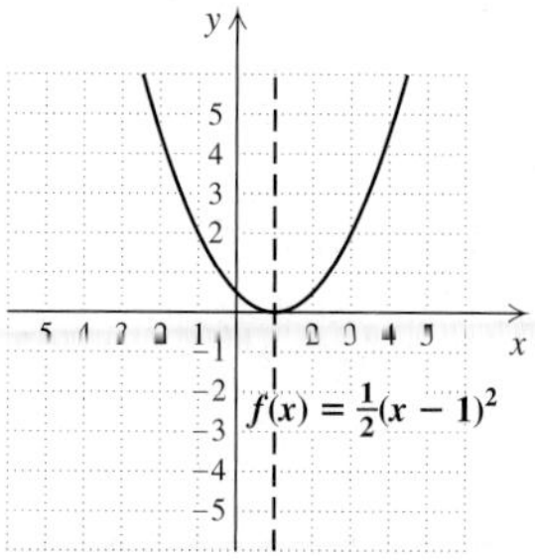

41. Vertex: $(-5, 0)$; axis of symmetry: $x = -5$

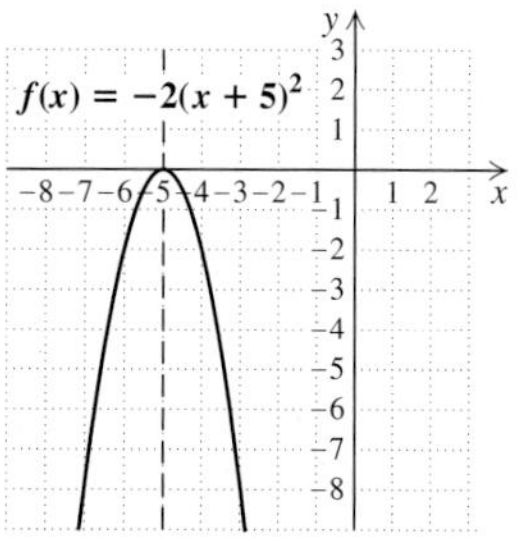

43. Vertex: $\left(\frac{1}{2}, 0\right)$; axis of symmetry: $x = \frac{1}{2}$

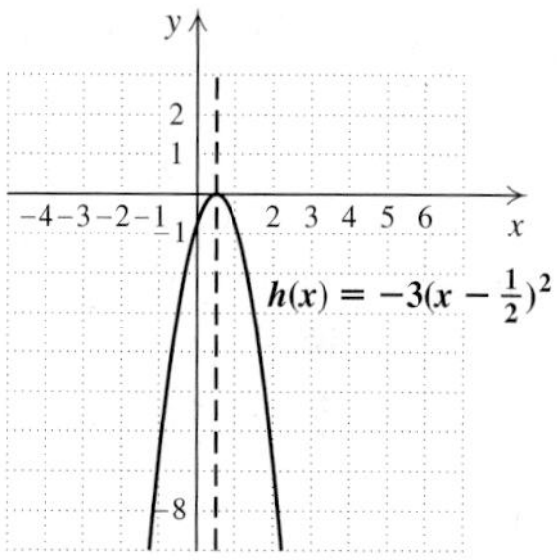

45. Vertex: $(5, 2)$; axis of symmetry: $x = 5$; minimum: 2

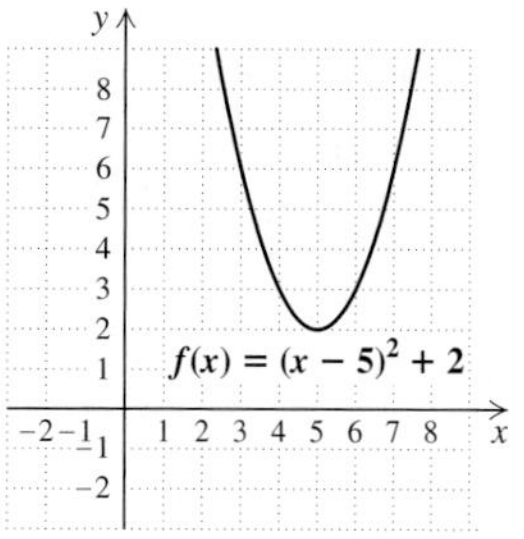

47. Vertex: $(-1, -3)$; axis of symmetry: $x = -1$; minimum: -3

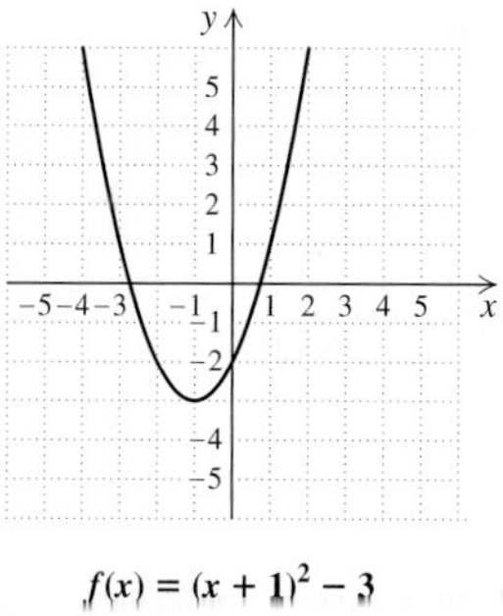

49. Vertex: $(-4, 1)$; axis of symmetry: $x = -4$; minimum: 1

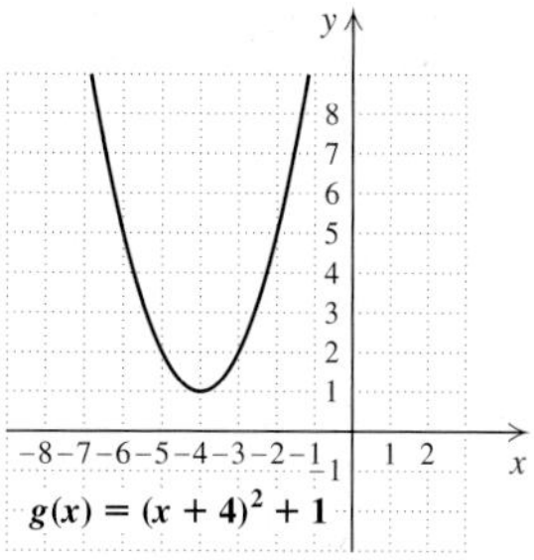

51. Vertex: $(1, -3)$; axis of symmetry: $x = 1$; maximum: -3

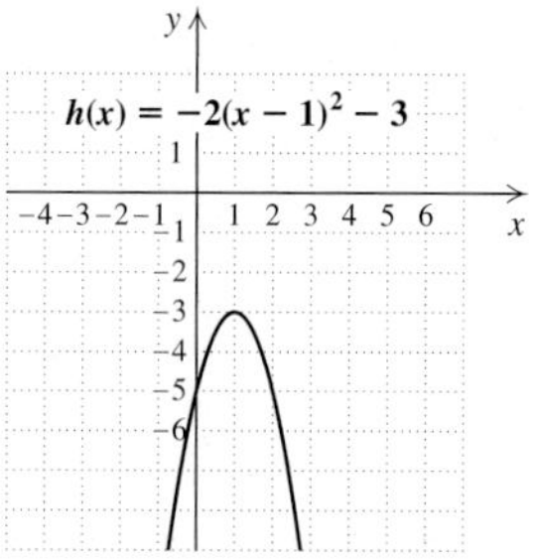

53. Vertex: $(-4, 1)$; axis of symmetry: $x = -4$; minimum: 1

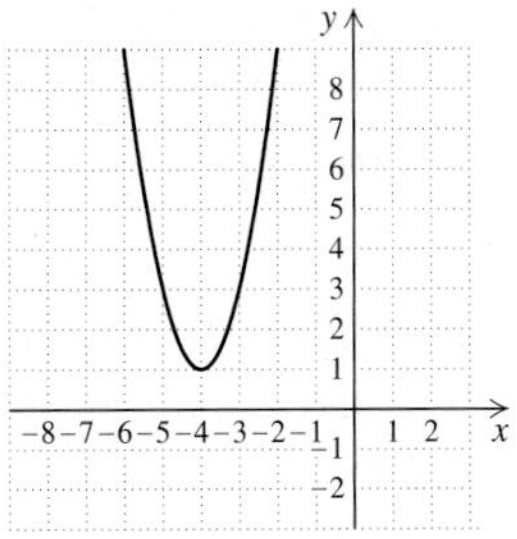

$f(x) = 2(x + 4)^2 + 1$

55. Vertex: $(1, 4)$; axis of symmetry: $x = 1$; maximum: 4

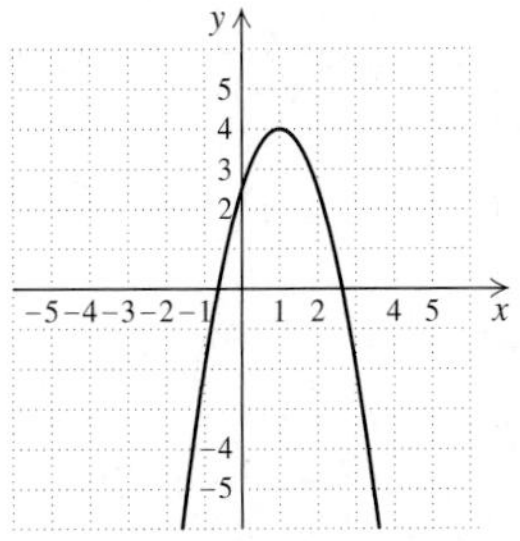

$g(x) = -\frac{3}{2}(x - 1)^2 + 4$

57. Vertex: $(8, 7)$; axis of symmetry: $x = 8$; minimum: 7
59. Vertex: $(-6, 11)$; axis of symmetry: $x = -6$; maximum: 11 **61.** Vertex: $\left(-\frac{1}{4}, -13\right)$; axis of symmetry: $x = -\frac{1}{4}$; minimum: -13
63. Vertex: $(-4.58, 65\pi)$; axis of symmetry: $x = -4.58$; minimum: 65π **65.** TW

67.

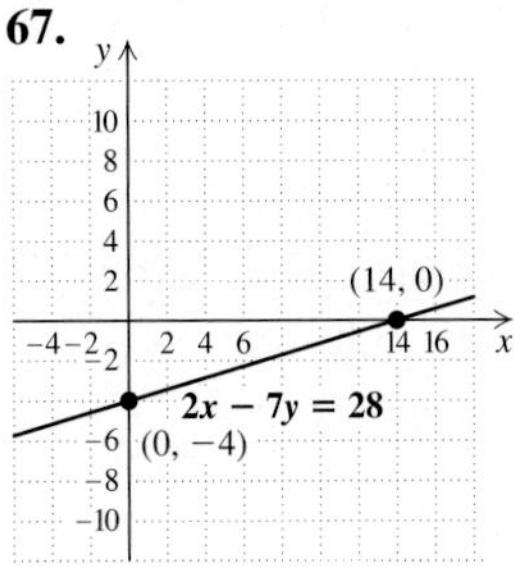

68.

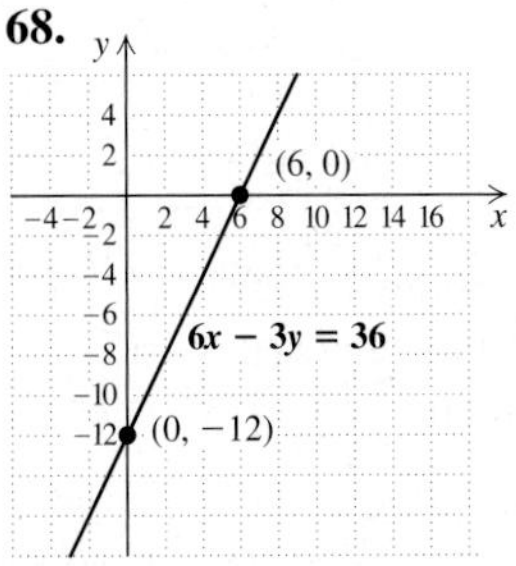

69. $(-5, -1)$ **70.** $(-1, 2)$ **71.** $x^2 + 5x + \frac{25}{4} = \left(x + \frac{5}{2}\right)^2$
72. $x^2 - 9x + \frac{81}{4} = \left(x - \frac{9}{2}\right)^2$ **73.** TW
75. $f(x) = \frac{3}{5}(x - 4)^2 + 1$ **77.** $f(x) = \frac{3}{5}(x - 3)^2 - 1$
79. $f(x) = \frac{3}{5}(x + 2)^2 - 5$ **81.** $f(x) = 2(x - 2)^2$
83. $g(x) = -2x^2 + 3$ **85.** $F(x) = 3(x - 5)^2 + 1$

87.

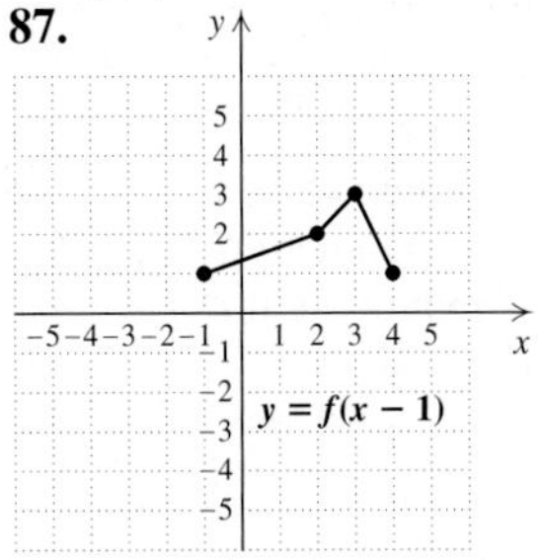

89.

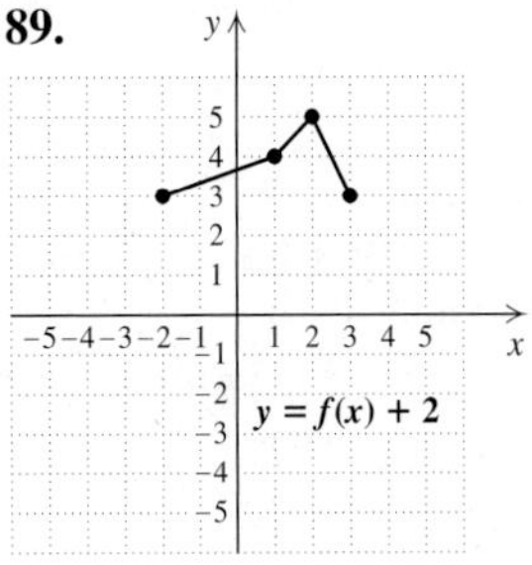

91.

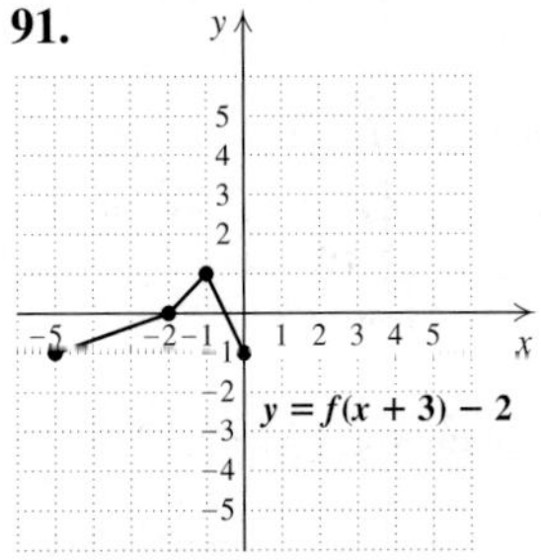

Visualizing the Graph, p. 665

1. B **2.** E **3.** A **4.** H **5.** C **6.** J **7.** F **8.** G **9.** I **10.** D

Exercise Set 8.7, pp. 666–667

1. 9 **2.** 16 **3.** 9 **4.** 1 **5.** 3 **6.** 1 **7.** $\frac{5}{2}$; (-4) **8.** (-2); $\left(-2, \frac{7}{2}\right)$ **9. (a)** $f(x) = (x - 2)^2 - 1$; **(b)** vertex: $(2, -1)$; axis of symmetry: $x = 2$ **11. (a)** $f(x) = -\left(x - \frac{3}{2}\right)^2 - \frac{31}{4}$; **(b)** vertex: $\left(\frac{3}{2}, -\frac{31}{4}\right)$; axis of symmetry: $x = \frac{3}{2}$ **13. (a)** $f(x) = 2\left(x - \frac{7}{4}\right)^2 - \frac{41}{8}$; **(b)** vertex: $\left(\frac{7}{4}, -\frac{41}{8}\right)$; axis of symmetry: $x = \frac{7}{4}$
15. (a) Vertex: $(-2, 1)$; axis of symmetry: $x = -2$; **(b)**

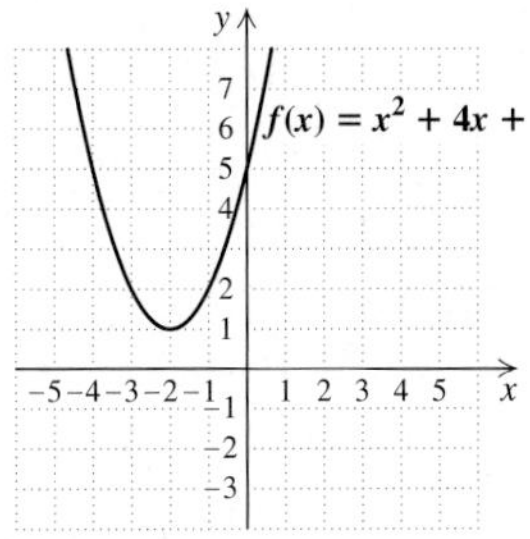

17. (a) Vertex: $(3, 4)$; axis of symmetry: $x = 3$; **(b)**

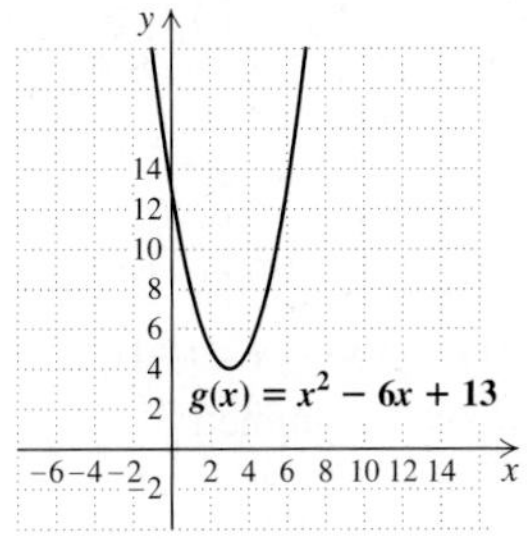

19. (a) Vertex: $(-4, 4)$; axis of symmetry: $x = -4$; **(b)**

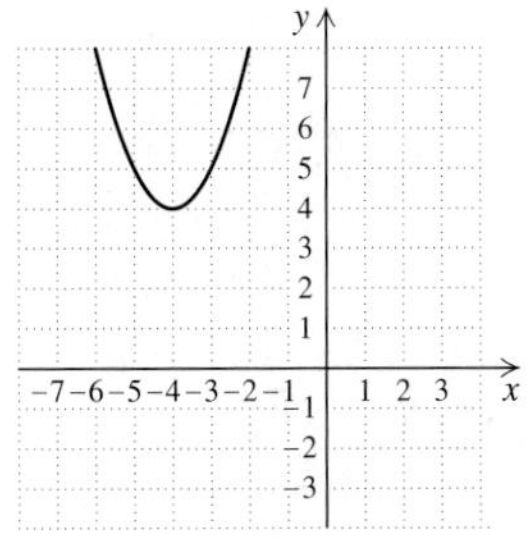

$f(x) = x^2 + 8x + 20$

21. **(a)** Vertex: $(4, -7)$; axis of symmetry: $x = 4$;
(b)
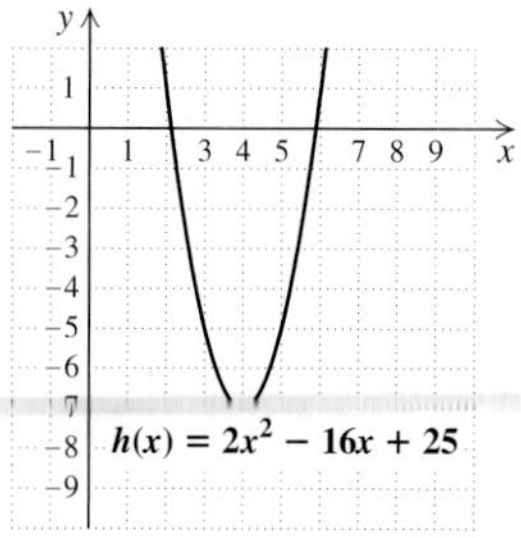

23. **(a)** Vertex: $(1, 6)$; axis of symmetry: $x = 1$;
(b)
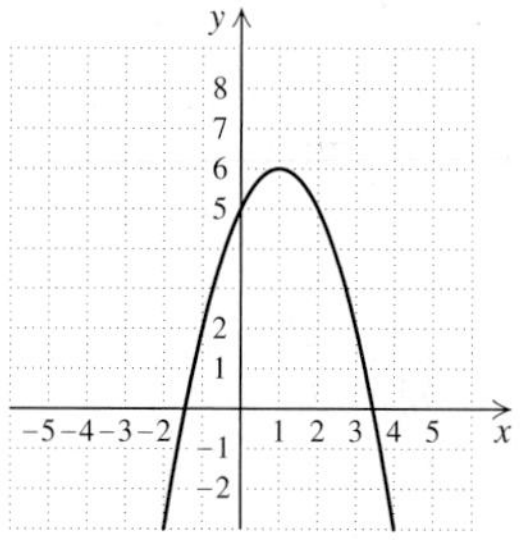

$f(x) = -x^2 + 2x + 5$

25. **(a)** Vertex: $\left(-\frac{3}{2}, -\frac{49}{4}\right)$; axis of symmetry: $x = -\frac{3}{2}$;
(b)
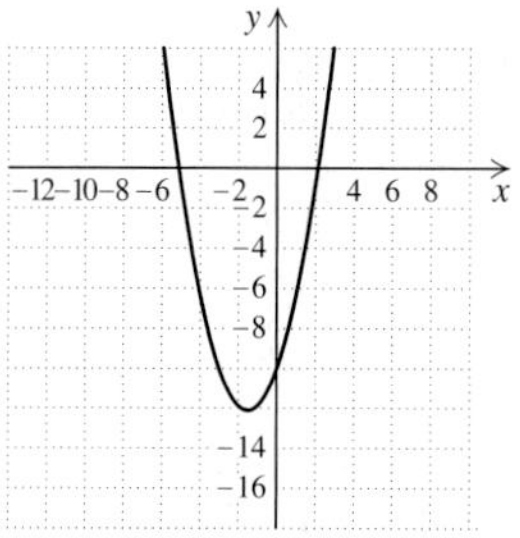

$g(x) = x^2 + 3x - 10$

27. **(a)** Vertex: $(4, 2)$; axis of symmetry: $x = 4$;
(b)
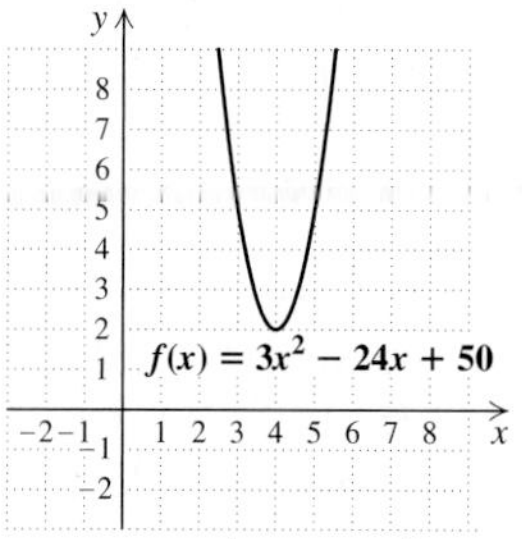

29. **(a)** Vertex: $\left(-\frac{7}{2}, -\frac{49}{4}\right)$; axis of symmetry: $x = -\frac{7}{2}$;
(b)
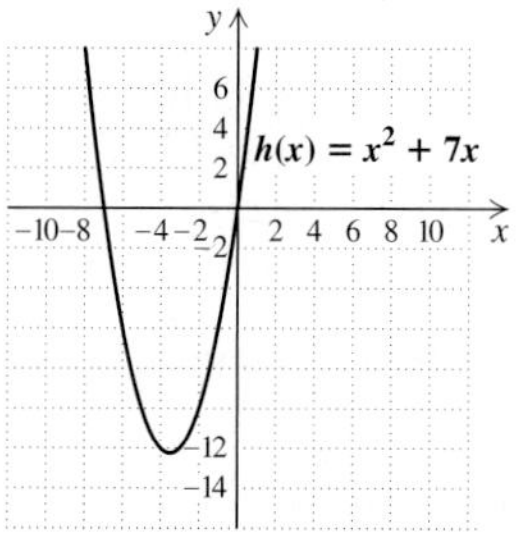

31. **(a)** Vertex: $(-1, -4)$; axis of symmetry: $x = -1$;
(b)
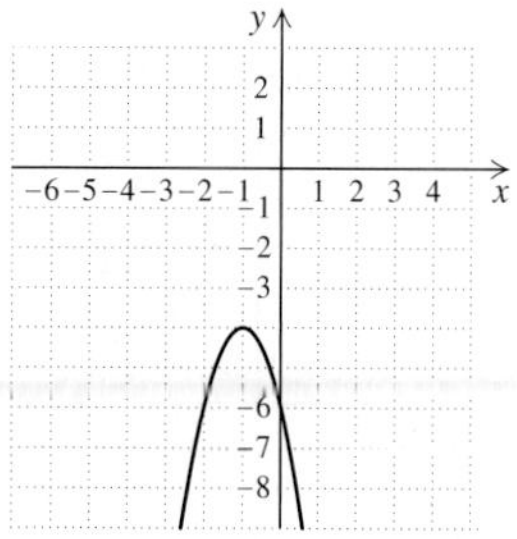

$f(x) = -2x^2 - 4x - 6$

33. **(a)** Vertex: $(2, -5)$; axis of symmetry: $x = 2$;
(b)
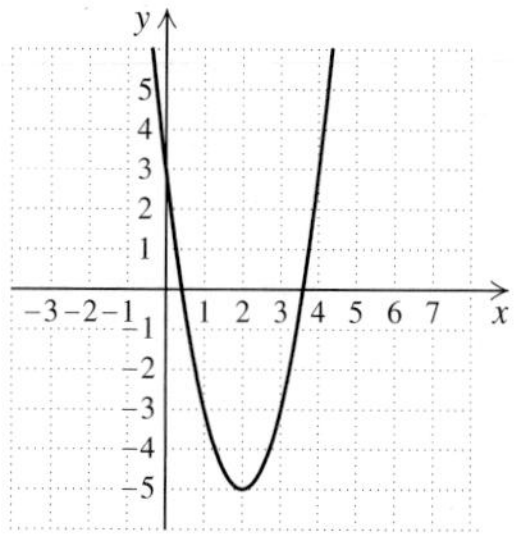

$g(x) = 2x^2 - 8x + 3$

35. **(a)** Vertex: $\left(\frac{5}{6}, \frac{1}{12}\right)$; axis of symmetry: $x = \frac{5}{6}$;
(b)
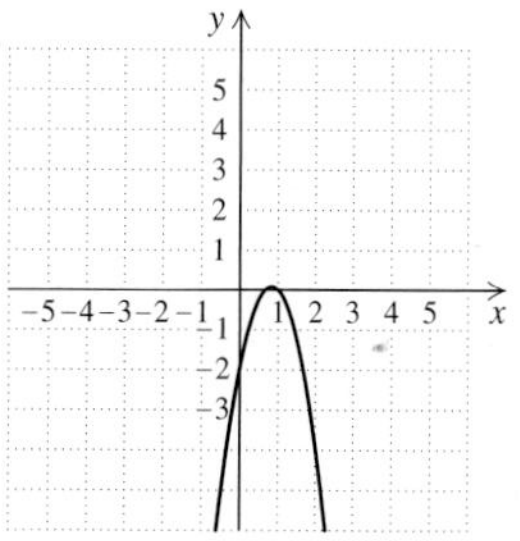

$f(x) = -3x^2 + 5x - 2$

37. **(a)** Vertex: $\left(-4, -\frac{5}{3}\right)$; axis of symmetry: $x = -4$;
(b)
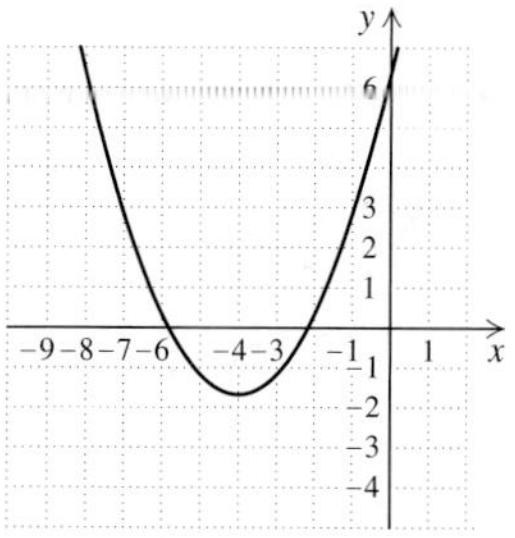

$h(x) = \frac{1}{2}x^2 + 4x + \frac{19}{3}$

39. $(-0.5, -6.25)$ **41.** $(0.1, 0.95)$ **43.** $(3.5, -4.25)$
45. $\left(3 - \sqrt{6}, 0\right), \left(3 + \sqrt{6}, 0\right)$; $(0, 3)$ **47.** $(-1, 0), (3, 0)$; $(0, 3)$ **49.** $(0, 0), (9, 0)$; $(0, 0)$ **51.** $(2, 0)$; $(0, -4)$
53. No x-intercept; $(0, 6)$ **55.** **(a)** Minimum: -6.95; **(b)** $(-1.06, 0), (2.41, 0)$; $(0, -5.89)$
57. **(a)** Maximum: -0.45; **(b)** no x-intercept; $(0, -2.79)$
59. TW

61.
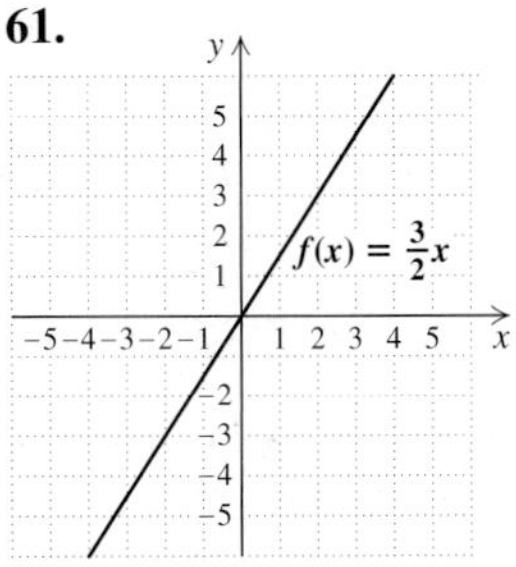

62.
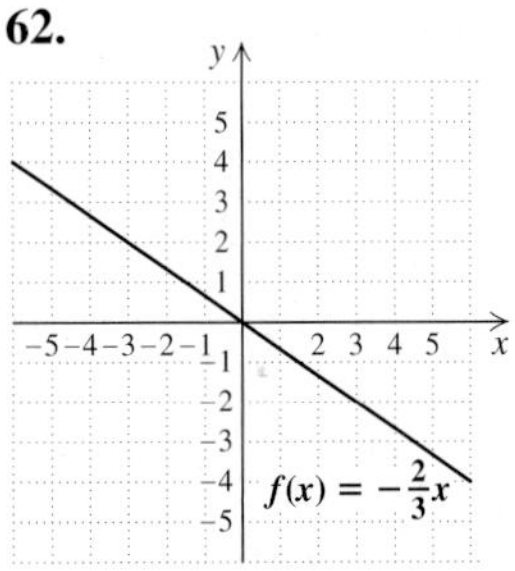

63.
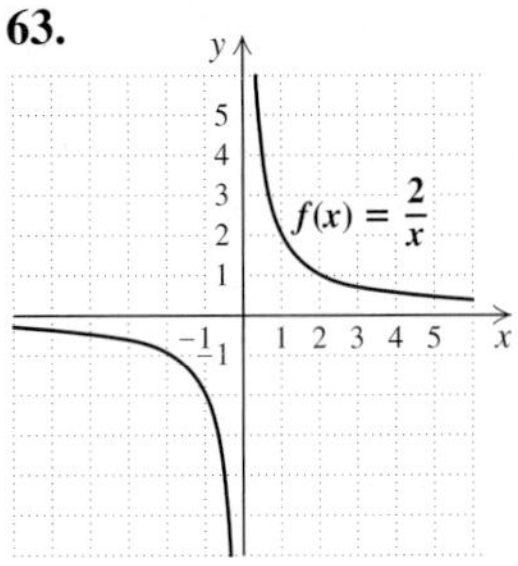

64.
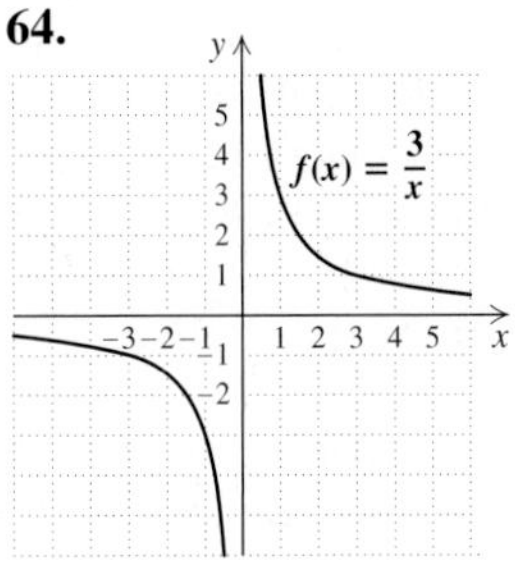

65.
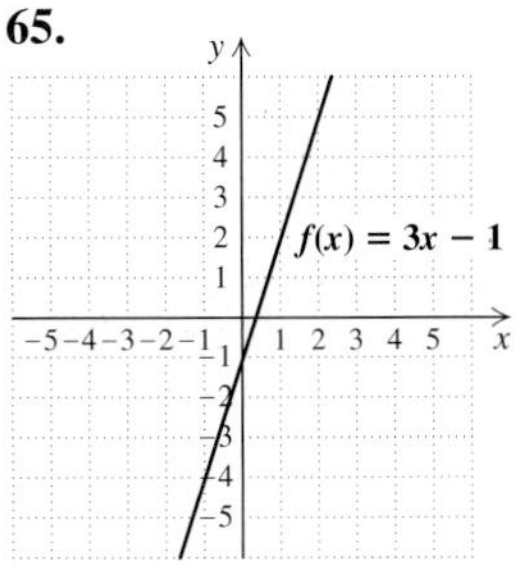

66.
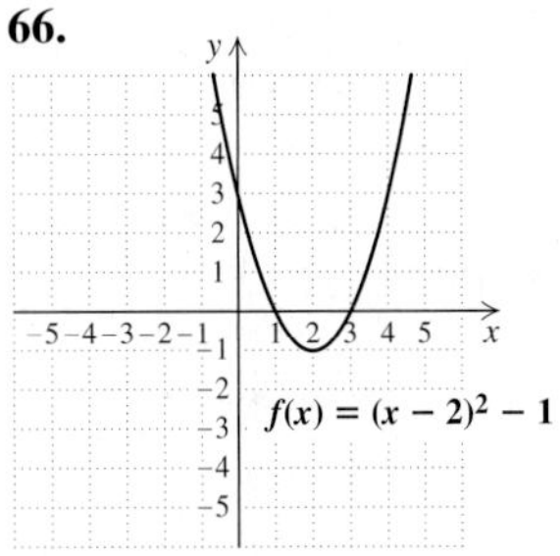

67.
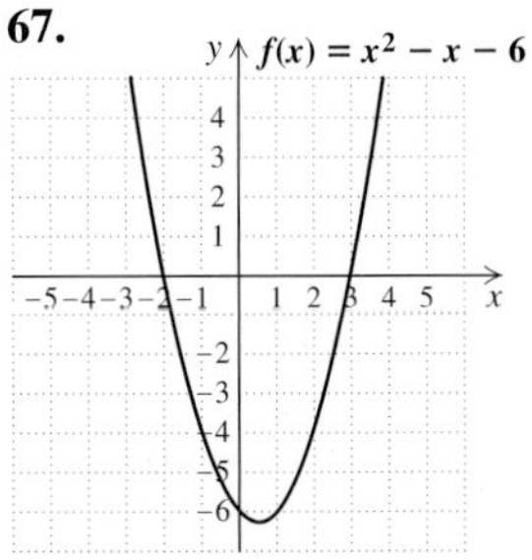

68.
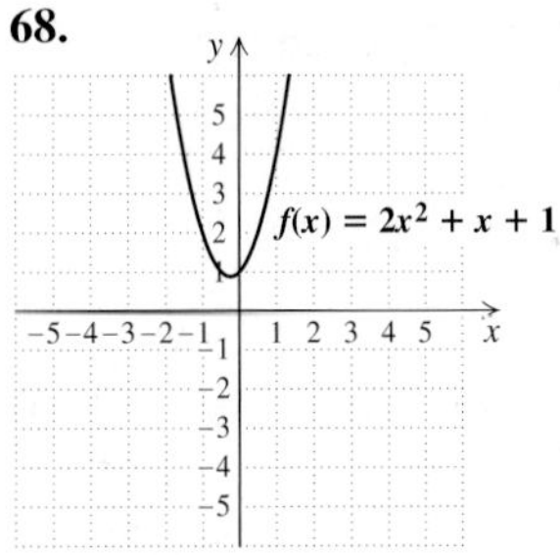

69. TW **71.** **(a)** $-2.4, 3.4$; **(b)** $-1.3, 2.3$

73. $f(x) = m\left(x - \frac{n}{2m}\right)^2 + \frac{4mp - n^2}{4m}$

75. $f(x) = \frac{5}{16}x^2 - \frac{15}{8}x - \frac{35}{16}$, or $f(x) = \frac{5}{16}(x - 3)^2 - 5$

77.
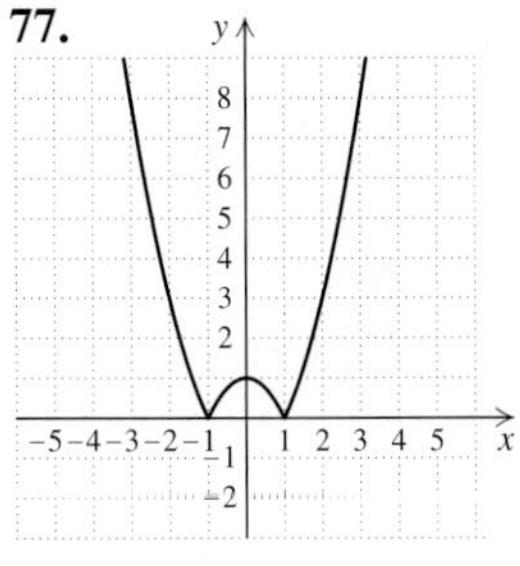

$f(x) = |x^2 - 1|$

79.
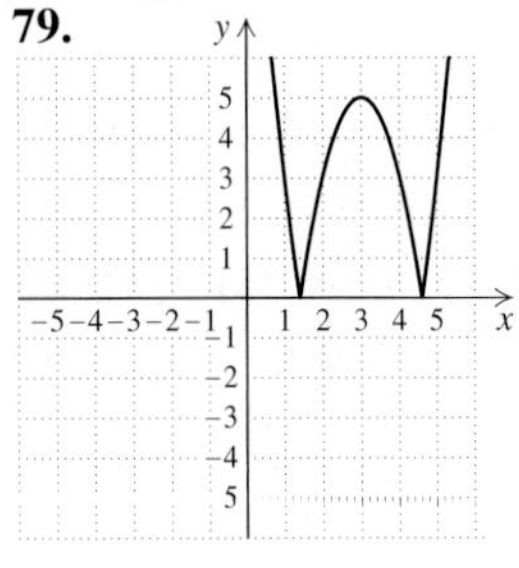

$f(x) = |2(x - 3)^2 - 5|$

Exercise Set 8.8, pp. 676–682

1. (e) **2.** (b) **3.** (c) **4.** (a) **5.** (d) **6.** (f)
7. 11 days after the concert was announced; about 62 tickets
9. \$120/Dobro; 350 Dobros **11.** 32 in. by 32 in.
13. 450 ft^2; 15 ft by 30 ft (The house serves as a 30-ft side.)
15. 3.5 in. **17.** 81; 9 and 9 **19.** -16; 4 and -4
21. 25; -5 and -5 **23.** $f(x) = mx + b$
25. $f(x) = ax^2 + bx + c, a < 0$
27. $f(x) = ax^2 + bx + c, a > 0$
29. $f(x) = ax^2 + bx + c, a > 0$
31. Neither quadratic nor linear
33. $f(x) = 2x^2 + 3x - 1$ **35.** $f(x) = -\frac{1}{4}x^2 + 3x - 5$
37. **(a)** $A(s) = \frac{3}{16}s^2 - \frac{135}{4}s + 1750$; **(b)** about 531 accidents
39. $h(d) = -0.0068d^2 + 0.8571d$
41. **(a)** $D(x) = -0.0082833093x^2 + 0.8242996891x + 0.2121786608$; **(b)** 17.325 ft
43. **(a)** $v(x) = 9982.696429x^2 - 18{,}401.85357x + 18{,}864.75$; **(b)** 510,542 vehicles **45.** TW
47. $\frac{x - 9}{(x + 9)(x + 7)}$ **48.** $\frac{(x - 3)(x + 1)}{(x - 7)(x + 3)}$
49. $\{x | x < 8\}$, or $(-\infty, 8)$ **50.** $\{x | x \geq 10\}$, or $[10, \infty)$
51. $r(x) = 1454.8125x + 17{,}685.75$
52. $r(x) = 1473.553571x + 17{,}775.5$ **53.** TW
55. 158 ft **57.** The radius of the circular portion of the window and the height of the rectangular portion should each be $\frac{24}{\pi + 4}$ ft. **59.** \$15

Exercise Set 8.9, pp. 688–691

1. True **2.** False **3.** True **4.** True **5.** True
6. True **7.** False **8.** False **9.** $\left[-4, \frac{3}{2}\right]$
11. $(-\infty, -2) \cup (0, 2) \cup (3, \infty)$ **13.** $\left(-\infty, -\frac{7}{2}\right) \cup (-2, \infty)$
15. $(-4, 3)$, or $\{x | -4 < x < 3\}$
17. $(-\infty, -7] \cup [2, \infty)$, or $\{x | x \leq -7 \text{ or } x \geq 2\}$
19. $(-\infty, -1) \cup (2, \infty)$, or $\{x | x < -1 \text{ or } x > 2\}$ **21.** $\varnothing$
23. $(-2, 6)$, or $\{x | -2 < x < 6\}$
25. $(-\infty, -2) \cup (0, 2)$, or $\{x | x < -2 \text{ or } 0 < x < 2\}$
27. $[-2, 1] \cup [4, \infty)$, or $\{x | -2 \leq x \leq 1 \text{ or } x \geq 4\}$
29. $[-0.78, 1.59]$, or $\{x | -0.78 \leq x \leq 1.59\}$
31. $(-\infty, -2) \cup (1, 3)$, or $\{x | x < -2 \text{ or } 1 < x < 3\}$
33. $[-2, 2]$, or $\{x | -2 \leq x \leq 2\}$ **35.** $(-1, 2) \cup (3, \infty)$, or $\{x | -1 < x < 2 \text{ or } x > 3\}$ **37.** $(-\infty, 0] \cup [2, 5]$, or $\{x | x \leq 0 \text{ or } 2 \leq x \leq 5\}$ **39.** $(-\infty, -5)$, or $\{x | x < -5\}$
41. $(-\infty, -1] \cup (3, \infty)$, or $\{x | x \leq -1 \text{ or } x > 3\}$
43. $(-\infty, -6)$, or $\{x | x < -6\}$ **45.** $(-\infty, -1] \cup [2, 5)$, or $\{x | x \leq -1 \text{ or } 2 \leq x < 5\}$ **47.** $(-\infty, -3) \cup [0, \infty)$, or $\{x | x < -3 \text{ or } x \geq 0\}$ **49.** $(0, \infty)$, or $\{x | x > 0\}$
51. $(-\infty, -4) \cup [1, 3)$, or $\{x | x < -4 \text{ or } 1 \leq x < 3\}$
53. $\left(-\frac{3}{4}, \frac{5}{2}\right]$, or $\left\{x \middle| -\frac{3}{4} < x \leq \frac{5}{2}\right\}$ **55.** $(-\infty, 2) \cup [3, \infty)$, or $\{x | x < 2 \text{ or } x \geq 3\}$ **57.** TW **59.** $[3, \infty)$, or $\{x | x \geq 3\}$
60. $\left(-\infty, -\frac{11}{2}\right) \cup \left(\frac{5}{2}, \infty\right)$, or $\left\{x \middle| x < -\frac{11}{2} \text{ or } x > \frac{5}{2}\right\}$
61. $\left[-\frac{8}{5}, \frac{12}{5}\right]$, or $\left\{x \middle| -\frac{8}{5} \leq x \leq \frac{12}{5}\right\}$ **62.** $\varnothing$

63. $(-1, 7)$, or $\{x \mid -1 < x < 7\}$
64. $(-\infty, 0) \cup [1, \infty)$, or $\{x \mid x < 0 \text{ or } x \geq 1\}$ **65.** TW
67. $(-1 - \sqrt{6}, -1 + \sqrt{6})$, or $\{x \mid -1 - \sqrt{6} < x < -1 + \sqrt{6}\}$ **69.** $\{0\}$
71. **(a)** $(10, 200)$, or $\{x \mid 10 < x < 200\}$; **(b)** $[0, 10) \cup (200, \infty)$, or $\{x \mid 0 \leq x < 10 \text{ or } x > 200\}$
73. $\{n \mid n \text{ is an integer and } 12 \leq n \leq 25\}$ **75.** $f(x)$ has no zeros; $f(x) < 0$ for $(-\infty, 0)$, or $\{x \mid x < 0\}$; $f(x) > 0$ for $(0, \infty)$, or $\{x \mid x > 0\}$ **77.** $f(x) = 0$ for $x = -1, 0$; $f(x) < 0$ for $(-\infty, -3) \cup (-1, 0)$, or $\{x \mid x < -3 \text{ or } -1 < x < 0\}$; $f(x) > 0$ for $(-3, -1) \cup (0, 2) \cup (2, \infty)$, or $\{x \mid -3 < x < -1 \text{ or } 0 < x < 2 \text{ or } x > 2\}$
79. **(a)** $h(x) = -0.3553571429x^2 + 4.185x + 16.77142857$; **(b)** more than 11 yr after 1994, or after 2005

Review Exercises: Chapter 8, pp. 694–697

1. False **2.** True **3.** True **4.** True **5.** True
6. False **7.** True **8.** True **9.** True **10.** True
11. **(a)** 2; **(b)** positive; **(c)** -3 **12.** $\pm\frac{3}{2}$ **13.** $0, -\frac{3}{4}$
14. 3, 9 **15.** $2 \pm 2i$ **16.** 3, 5 **17.** $-\frac{9}{2} \pm \frac{\sqrt{85}}{2}$
18. -0.3722813233, 5.3722813233 **19.** $-\frac{1}{4}, 1$
20. $x^2 - 12x + 36 = (x - 6)^2$
21. $x^2 + \frac{3}{5}x + \frac{9}{100} = \left(x + \frac{3}{10}\right)^2$ **22.** $3 \pm 2\sqrt{2}$
23. 10% **24.** 6.7 sec **25.** About 153 mph **26.** 6 hr
27. Two irrational **28.** Two imaginary **29.** $x^2 - 5 = 0$
30. $x^2 + 8x + 16 = 0$ **31.** $(-3, 0), (-2, 0), (2, 0), (3, 0)$
32. $-5, 3$ **33.** $\pm\sqrt{2}, \pm\sqrt{7}$
34.

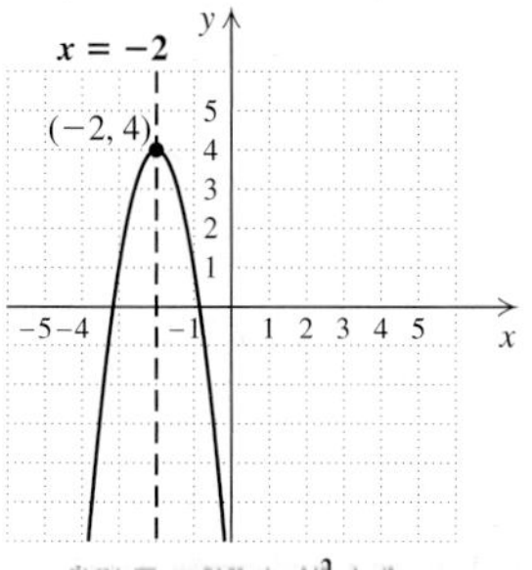

$f(x) = -3(x + 2)^2 + 4$
Maximum: 4

35. **(a)** Vertex: $(3, 5)$; axis of symmetry: $x = 3$;
(b)

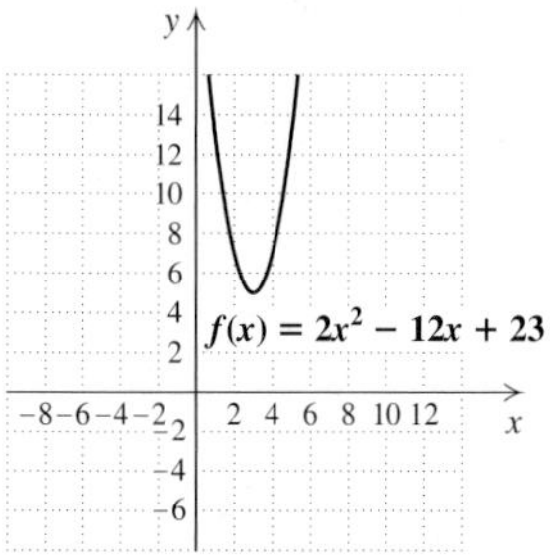

36. $(2, 0), (7, 0); (0, 14)$ **37.** $p = \frac{9\pi^2}{N^2}$
38. $T = \frac{1 \pm \sqrt{1 + 24A}}{6}$ **39.** Neither quadratic nor linear
40. $f(x) = ax^2 + bx + c, a > 0$ **41.** $f(x) = mx + b$
42. 225 ft²; 15 ft by 15 ft
43. **(a)** $M(x) = \frac{4063}{360}x^2 - \frac{31{,}639}{180}x + 1$; **(b)** 32,487 restaurants
44. **(a)** $M(x) = 14.14783742x^2 - 298.8611333x + 995.3006199$; **(b)** 36,850 restaurants
45. $(-1, 0) \cup (3, \infty)$, or $\{x \mid -1 < x < 0 \text{ or } x > 3\}$
46. $(-3, 5]$, or $\{x \mid -3 < x \leq 5\}$
47. TW Completing the square was used to solve quadratic equations and to graph quadratic functions by rewriting the function in the form $f(x) = a(x - h)^2 + k$.
48. TW The model found in Exercise 44 predicts 4363 more restaurants in 2010 than the model from Exercise 43. The higher prediction seems to fit the pattern better. Since the function in Exercise 44 considers all the data, we would expect it to be a better model.
49. $f(x) = \frac{7}{15}x^2 - \frac{14}{15}x - 7$ **50.** $h = 60, k = 60$
51. 18, 324 **52.** **(a)** $S(x) = 148.1428571x^2 - 15{,}211.71429x + 393{,}615.8571$; **(b)** more than 68 yr after 1948, or for years after 2016

Test: Chapter 8, pp. 697–698

1. **(a)** [8.1] 0; **(b)** [8.6] negative; **(c)** [8.6] -1
2. [8.1] $\pm\frac{4\sqrt{3}}{3}$ **3.** [8.2] 2, 9 **4.** [8.2] $\frac{-1 \pm i\sqrt{3}}{2}$
5. [8.2] $1 \pm \sqrt{6}$ **6.** [8.5] $-2, \frac{2}{3}$ **7.** [8.2] -4.192582404, 1.192582404 **8.** [8.2] $-\frac{3}{4}, \frac{7}{3}$
9. [8.1] $x^2 - 16x + 64 = (x - 8)^2$
10. [8.1] $x^2 + \frac{2}{7}x + \frac{1}{49} = \left(x + \frac{1}{7}\right)^2$ **11.** [8.1] $-5 \pm \sqrt{10}$
12. [8.3] 16 km/h **13.** [8.3] 2 hr
14. [8.4] Two imaginary **15.** [8.4] $3x^2 + 5x - 2 = 0$
16. [8.5] $(-3, 0), (-1, 0), (-2 - \sqrt{5}, 0), (-2 + \sqrt{5}, 0)$
17. [8.6]

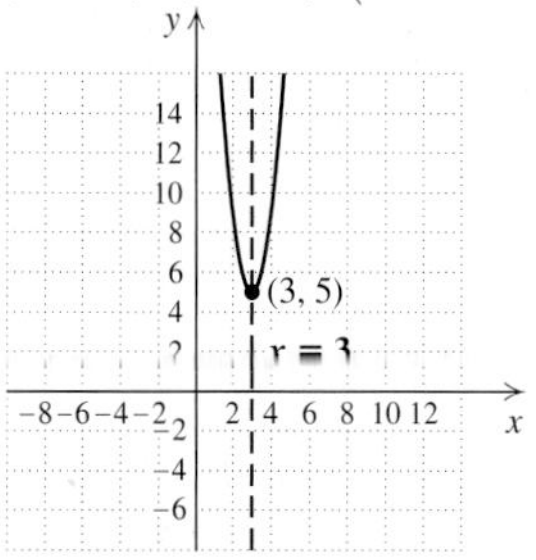

$f(x) = 4(x - 3)^2 + 5$
Minimum: 5

18. [8.7] **(a)** $(-1, -8), x = -1$;
(b)

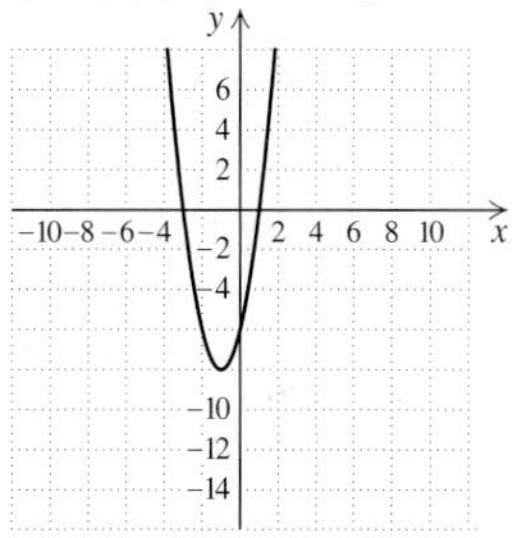

$f(x) = 2x^2 + 4x - 6$

19. [8.7] $(-2, 0), (3, 0); (0, -6)$
20. [8.3] $r = \sqrt{\frac{3V}{\pi} - R^2}$ **21.** [8.8] Quadratic; the data approximate a parabola opening downward.
22. [8.8] Minimum \$129/cap when 325 caps are built
23. [8.8] $p(x) = -\frac{165}{8}x^2 + \frac{605}{4}x + 35$
24. [8.8] $p(x) = -23.54166667x^2 + 162.2583333x + 57.61666667$ **25.** [8.9] $[-6, 1]$, or $\{x \mid -6 \le x \le 1\}$
26. [8.9] $(-1, 0) \cup (1, \infty)$, or $\{x \mid -1 < x < 0 \text{ or } x > 1\}$
27. [8.4] $\frac{1}{2}$ **28.** [8.4] $x^4 - 14x^3 + 67x^2 - 114x + 26 = 0$; answers may vary. **29.** [8.4] $x^6 - 10x^5 + 20x^4 + 50x^3 - 119x^2 - 60x + 150 = 0$; answers may vary

Chapter 9

Exercise Set 9.1, pp. 711–715

1. True **2.** True **3.** False **4.** True **5.** False **6.** False **7.** True **8.** True
9. $(f \circ g)(1) = 2$; $(g \circ f)(1) = 1$; $(f \circ g)(x) = 4x^2 - 12x + 10$; $(g \circ f)(x) = 2x^2 - 1$
11. $(f \circ g)(1) = -8$; $(g \circ f)(1) = 1$; $(f \circ g)(x) = 2x^2 - 10$; $(g \circ f)(x) = 2x^2 - 12x + 11$
13. $(f \circ g)(1) = 8$; $(g \circ f)(1) = \frac{1}{64}$; $(f \circ g)(x) = \frac{1}{x^2} + 7$; $(g \circ f)(x) = \frac{1}{(x+7)^2}$
15. $(f \circ g)(1) = 2$; $(g \circ f)(1) = 4$; $(f \circ g)(x) = \sqrt{x+3}$; $(g \circ f)(x) = \sqrt{x} + 3$
17. $(f \circ g)(1) = 2$; $(g \circ f)(1) = \frac{1}{2}$; $(f \circ g)(x) = \sqrt{\frac{4}{x}}$; $(g \circ f)(x) = \frac{1}{\sqrt{4x}}$ **19.** $(f \circ g)(1) = 4$; $(g \circ f)(1) = 2$; $(f \circ g)(x) = x + 3$; $(g \circ f)(x) = \sqrt{x^2 + 3}$ **21.** 8
23. -4 **25.** Not defined **27.** 4 **29.** Not defined
31. $f(x) = x^2$; $g(x) = 7 + 5x$ **33.** $f(x) = \sqrt{x}$; $g(x) = 2x + 7$ **35.** $f(x) = \frac{2}{x}$; $g(x) = x - 3$
37. $f(x) = \frac{1}{\sqrt{x}}$; $g(x) = 7x + 2$ **39.** $f(x) = \frac{1}{x} + x$; $g(x) = \sqrt{3x}$ **41.** Yes **43.** No **45.** Yes **47.** No
49. **(a)** Yes; **(b)** $f^{-1}(x) = x - 4$ **51.** **(a)** Yes; **(b)** $f^{-1}(x) = \frac{x}{2}$ **53.** **(a)** Yes; **(b)** $g^{-1}(x) = \frac{x+1}{3}$
55. **(a)** Yes; **(b)** $f^{-1}(x) = 2x - 2$ **57.** **(a)** No
59. **(a)** Yes; **(b)** $h^{-1}(x) = \frac{x-4}{-2}$
61. **(a)** Yes; **(b)** $f^{-1}(x) = \frac{1}{x}$ **63.** **(a)** No
65. **(a)** Yes; **(b)** $f^{-1}(x) = \frac{3x-1}{2}$
67. **(a)** Yes; **(b)** $f^{-1}(x) = \sqrt[3]{x+5}$ **69.** **(a)** Yes; **(b)** $g^{-1}(x) = \sqrt[3]{x} + 2$ **71.** **(a)** Yes; **(b)** $f^{-1}(x) = x^2, x \ge 0$

73.

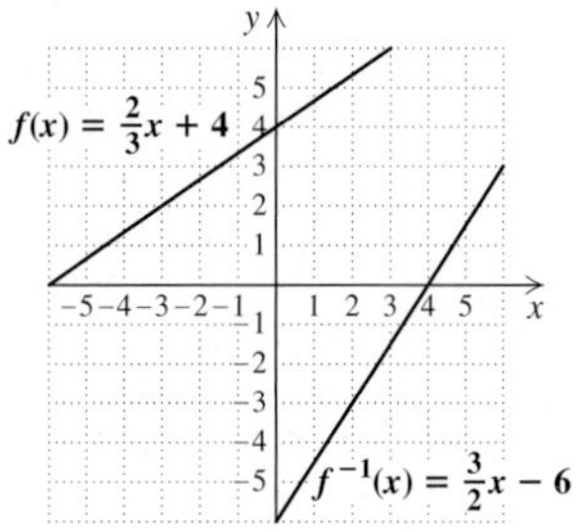

75.

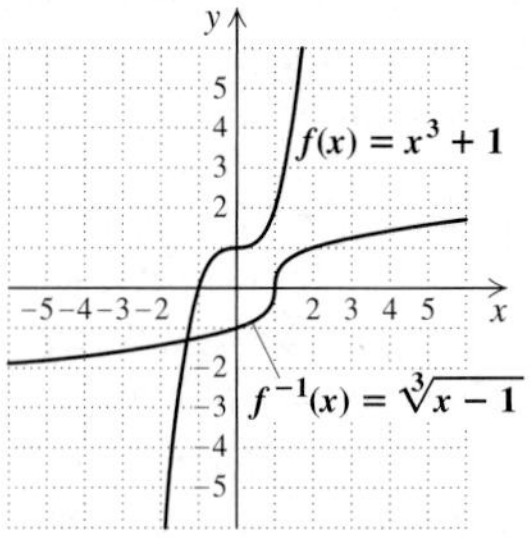

77.

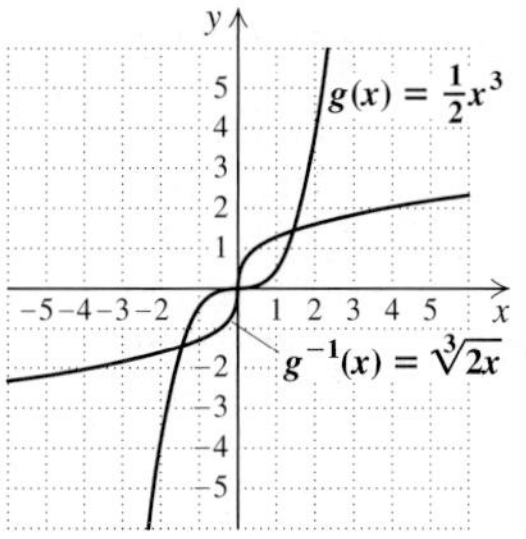

79.

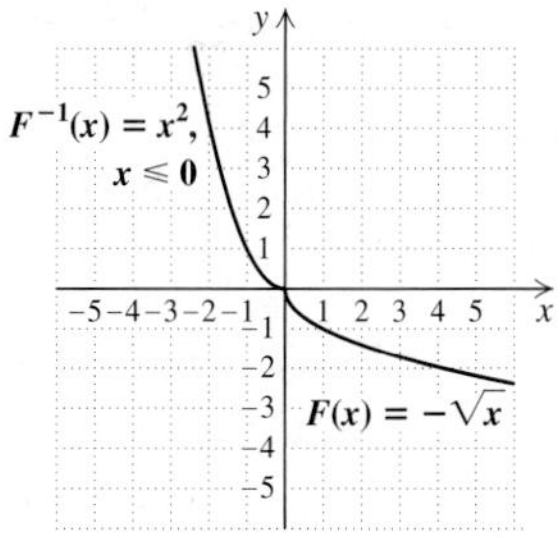

81.

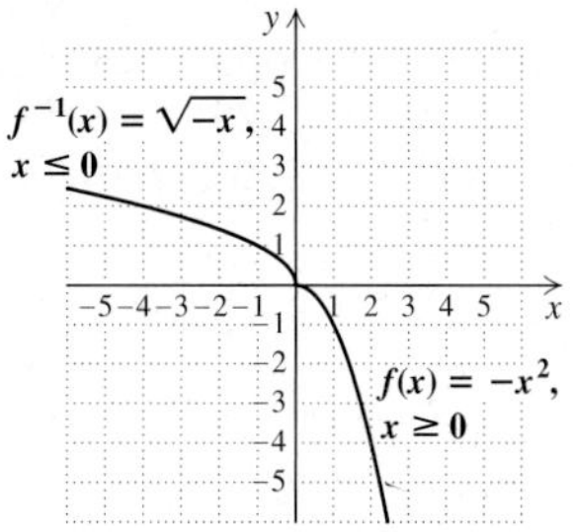

83. **(1)** $(f^{-1} \circ f)(x) = f^{-1}(f(x))$
$$= f^{-1}(\sqrt[3]{x-4}) = (\sqrt[3]{x-4})^3 + 4$$
$$= x - 4 + 4 = x;$$
(2) $(f \circ f^{-1})(x) = f(f^{-1}(x))$
$$= f(x^3 + 4) = \sqrt[3]{x^3 + 4 - 4}$$
$$= \sqrt[3]{x^3} = x$$

85. **(1)** $(f^{-1} \circ f)(x) = f^{-1}(f(x)) = f^{-1}\left(\frac{1-x}{x}\right)$
$$= \frac{1}{\left(\frac{1-x}{x}\right) + 1}$$
$$= \frac{1}{\frac{1-x+x}{x}}$$
$$= x;$$

(2) $(f \circ f^{-1})(x) = f(f^{-1}(x)) = f\left(\frac{1}{x+1}\right)$

$$= \frac{1 - \left(\frac{1}{x+1}\right)}{\left(\frac{1}{x+1}\right)}$$

$$= \frac{\frac{x+1-1}{x+1}}{\frac{1}{x+1}} = x$$

87. No **89.** Yes **91.** (1) C; (2) D; (3) B; (4) A
93. **(a)** 40, 42, 46, 50; **(b)** yes; $f^{-1}(x) = x - 32$;
(c) 8, 10, 14, 18 **95.** TW **97.** $6 + \sqrt{3}$
98. Not defined **99.** $x^2 + 2x + \sqrt{x}$ **100.** $\frac{2x + \sqrt{x}}{x^2}$
101. $\{x|x \geq 0\}$, or $[0, \infty)$ **102.** $\mathbb{R}$ **103.** $\{x|x \geq 0\}$, or $[0, \infty)$ **104.** $\{x|x > 0\}$, or $(0, \infty)$ **105.** $2x^2 + |x|$
106. $4x^2 + 4x\sqrt{x} + x$ **107.** TW

109.

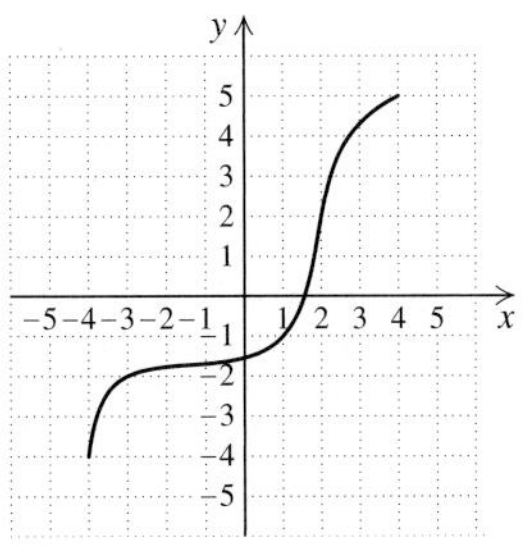

111. $g(x) = \frac{x}{2} + 20$

113. TW
115. Suppose that $h(x) = (f \circ g)(x)$. First, note that for $I(x) = x$, $(f \circ I)(x) = f(I(x)) = f(x)$ for any function f.
(i) $((g^{-1} \circ f^{-1}) \circ h)(x) = ((g^{-1} \circ f^{-1}) \circ (f \circ g))(x)$
$= ((g^{-1} \circ (f^{-1} \circ f)) \circ g)(x)$
$= ((g^{-1} \circ I) \circ g)(x)$
$= (g^{-1} \circ g)(x) = x$
(ii) $(h \circ (g^{-1} \circ f^{-1}))(x) = ((f \circ g) \circ (g^{-1} \circ f^{-1}))(x)$
$= ((f \circ (g \circ g^{-1})) \circ f^{-1})(x)$
$= ((f \circ I) \circ f^{-1})(x)$
$= (f \circ f^{-1})(x) = x.$
Therefore, $(g^{-1} \circ f^{-1})(x) = h^{-1}(x)$.
117. TW **119.** The cost of mailing n copies of the book
121. 22 mm **123.** 15 L/min
125. **(a)** $h(t) = -136t + 6322$; **(b)** $r(t) = 0.172t + 3.626$; **(c)** $H(r) = -776.5100671r + 9132.744966$; **(d)** $y_1 \approx y_4$; $h(t) = (H \circ r)(t)$

Interactive Discovery, p. 719

1. **(a)** Increases; **(b)** increases; **(c)** increases; **(d)** decreases; **(e)** decreases **2.** If $a > 1$, the graph increases; if $0 < a < 1$, the graph decreases. **3.** g **4.** r

Exercise Set 9.2, pp. 723–726

1. True **2.** True **3.** True **4.** False **5.** False
6. True **7.** $a > 1$ **9.** $0 < a < 1$

11.

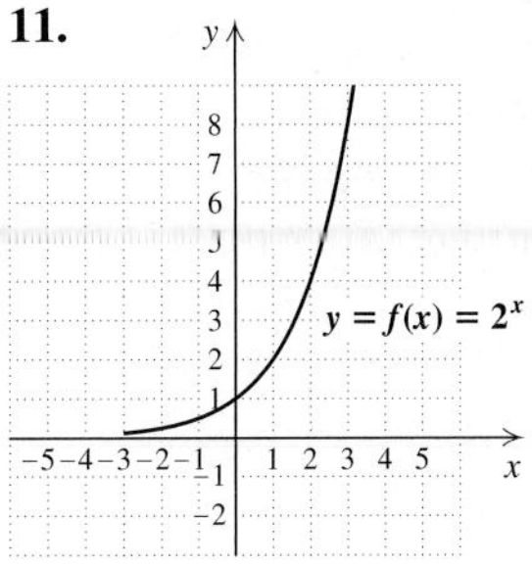

13.

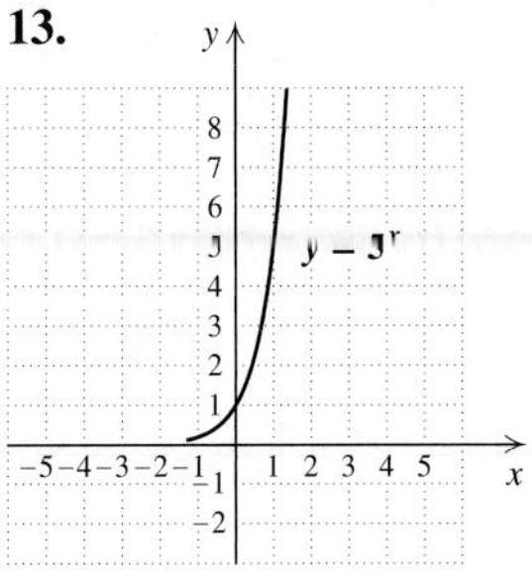

15.

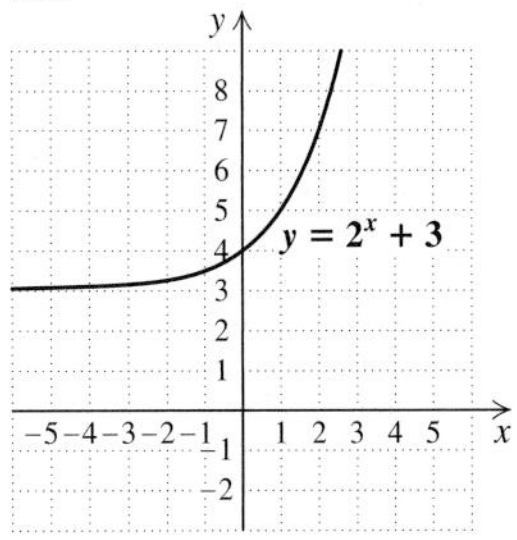

17.

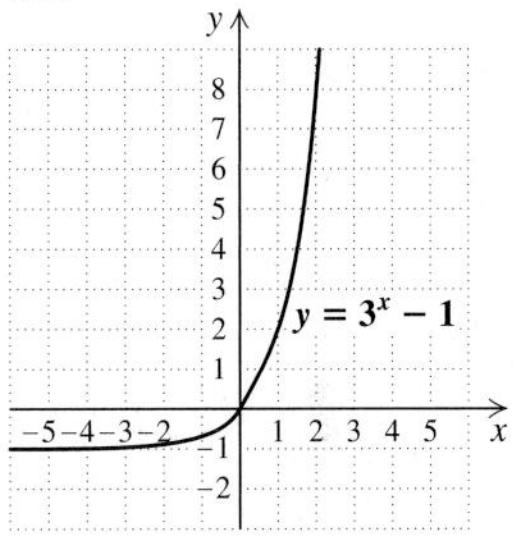

19.

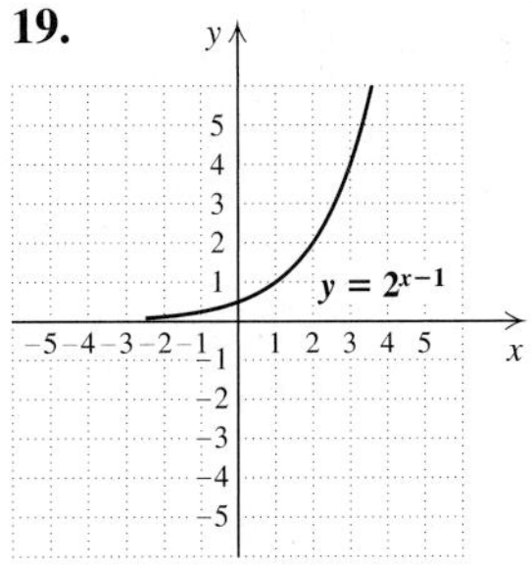

21.

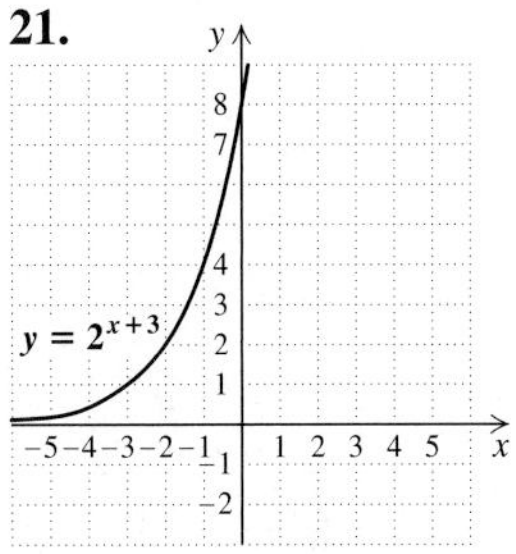

23.

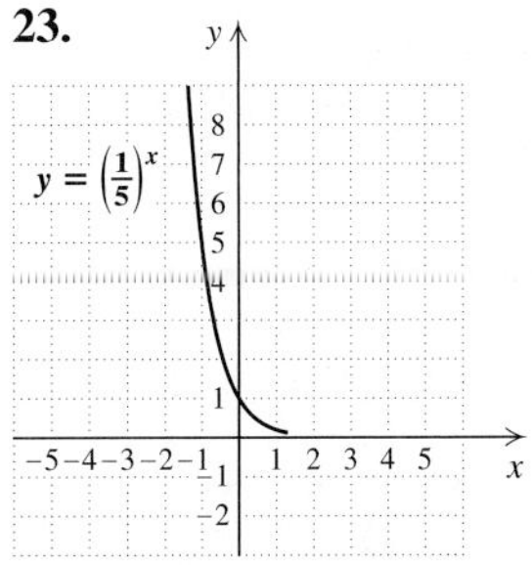

25.

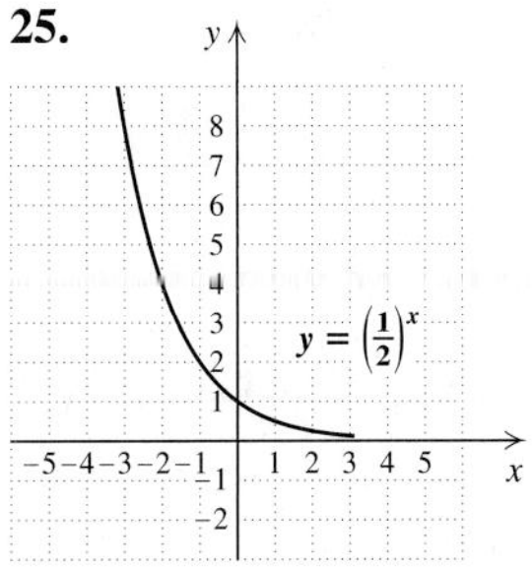

27.

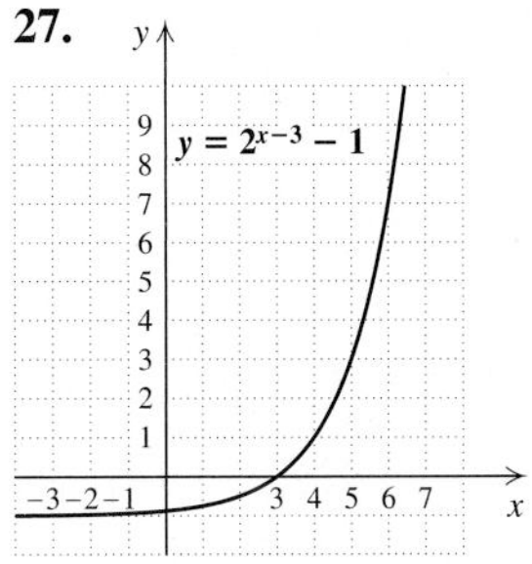

29.

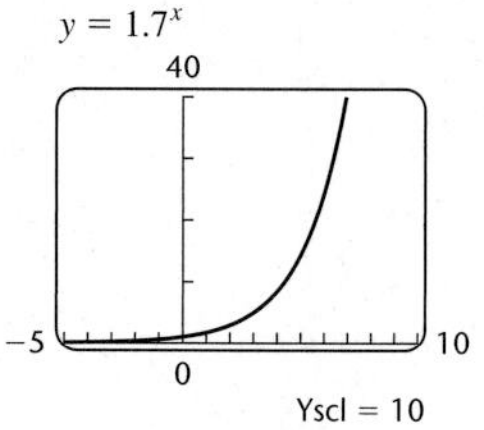

31.

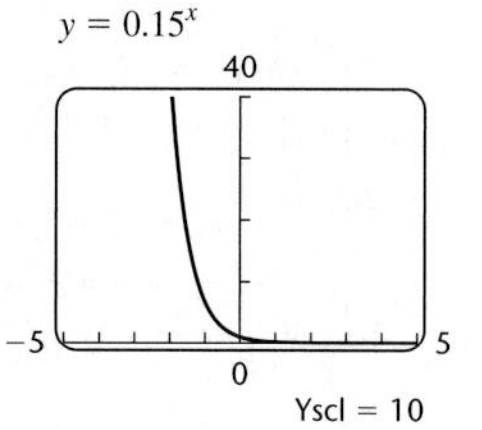

33.

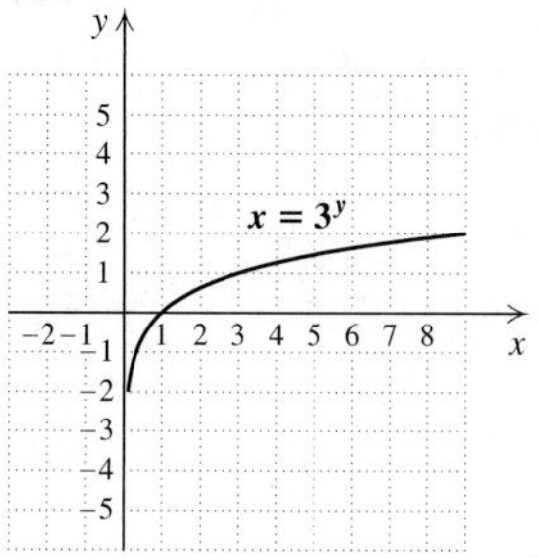

35.

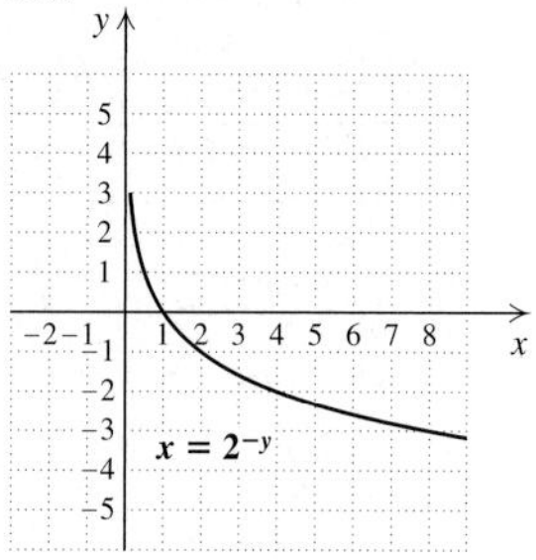

37.

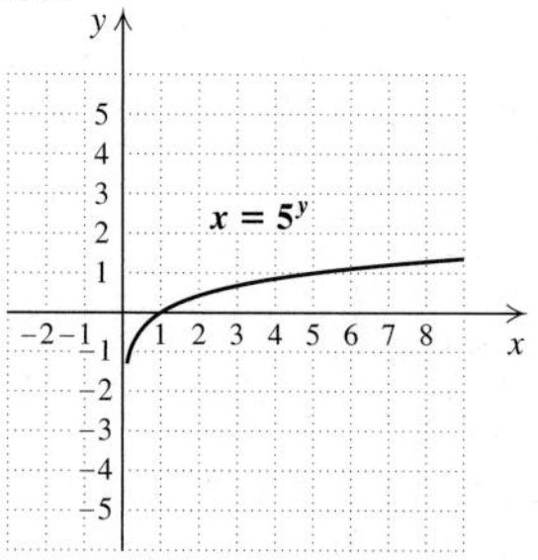

39.

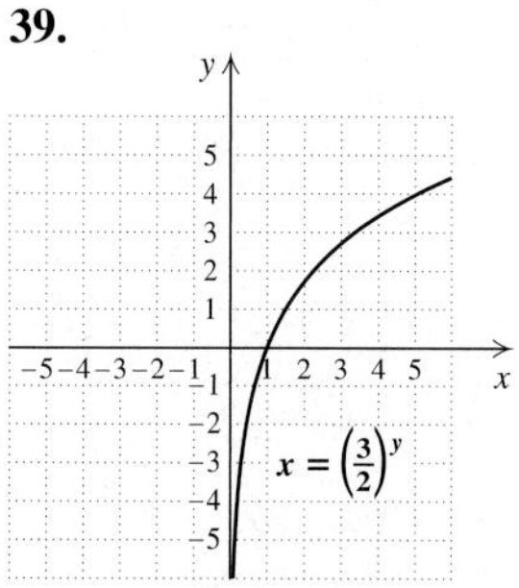

41.

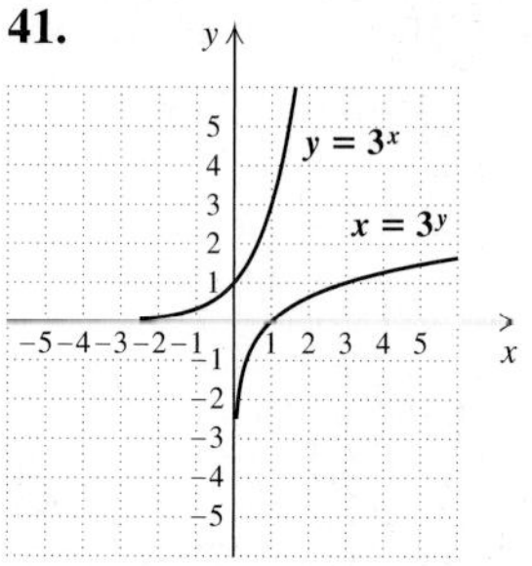

43.

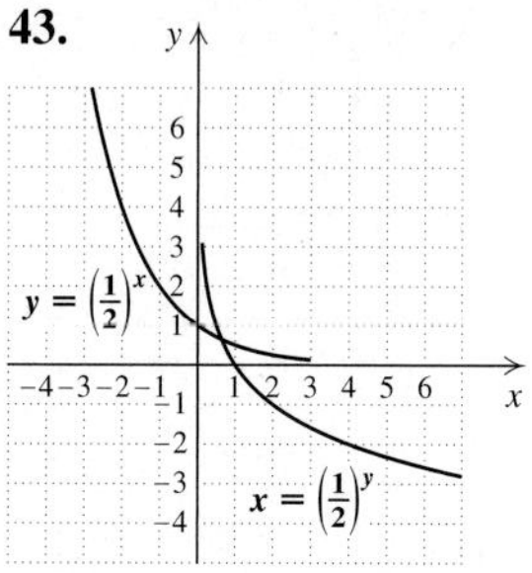

45. (d) **47.** (f) **49.** (c)

51. **(a)** About 6.8 billion; about 7.2 billion; about 7.7 billion;
(b)

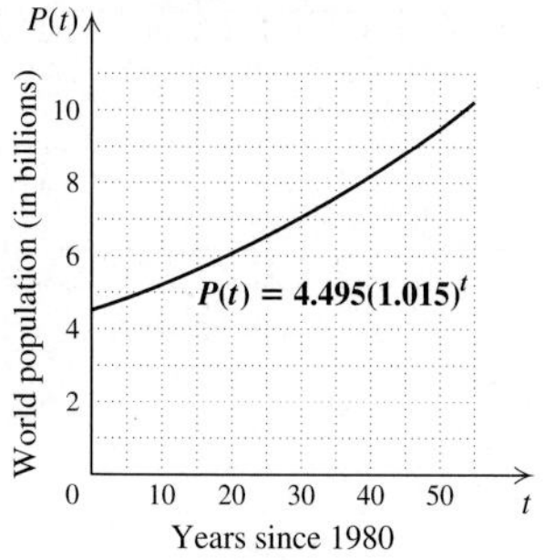

53. **(a)** 19.6%; 16.3%; 7.3%;
(b)

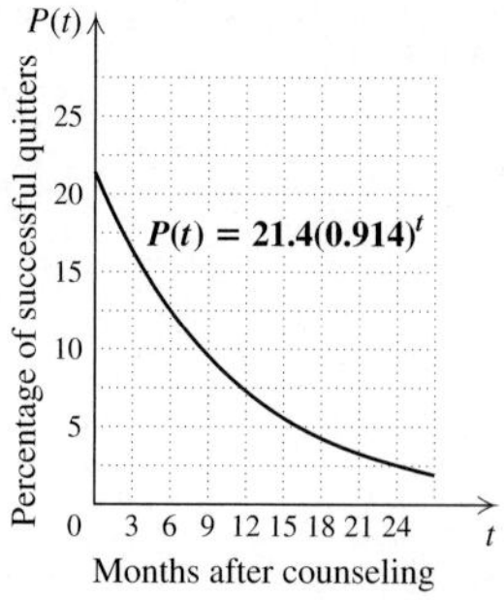

55. **(a)** About 44,079 whales; about 12,953 whales;
(b)

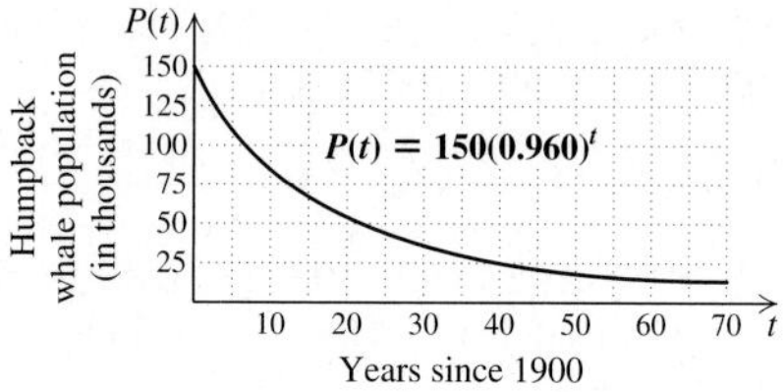

57. **(a)** About 11,900 whales; about 35,000 whales;
(b)

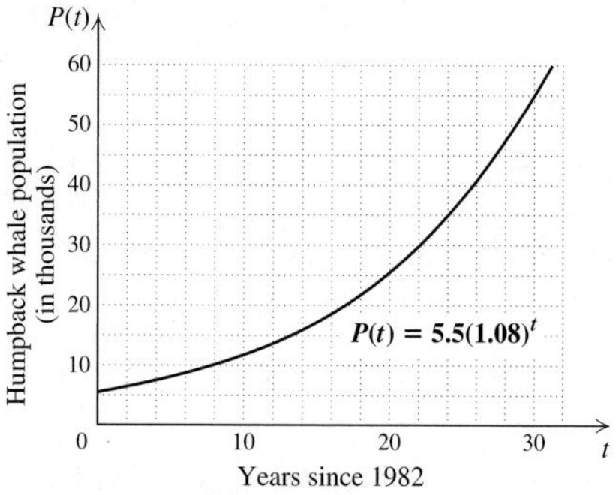

59. **(a)** 454,354,240 cm^2; 525,233,501,440 cm^2;
(b)

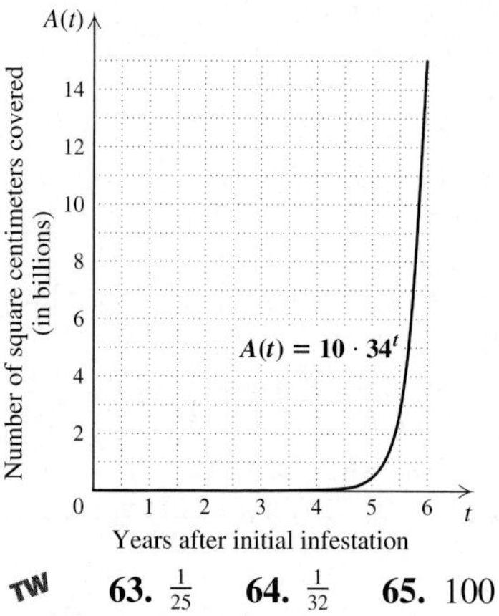

61. TW **63.** $\frac{1}{25}$ **64.** $\frac{1}{32}$ **65.** 100 **66.** $\frac{1}{125}$
67. $5a^6b^3$ **68.** $6x^4y$ **69.** TW **71.** $\pi^{2.4}$

73.

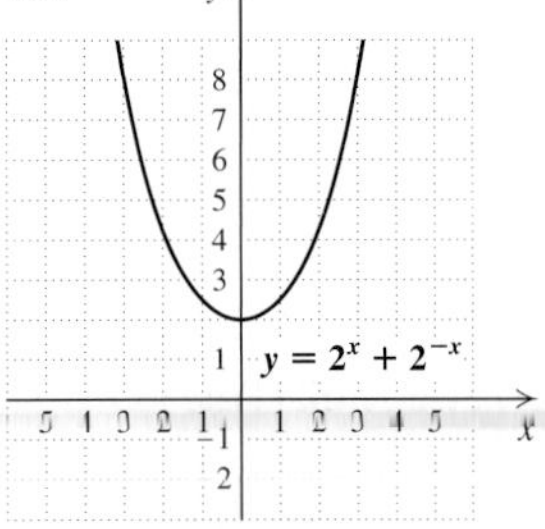

75.

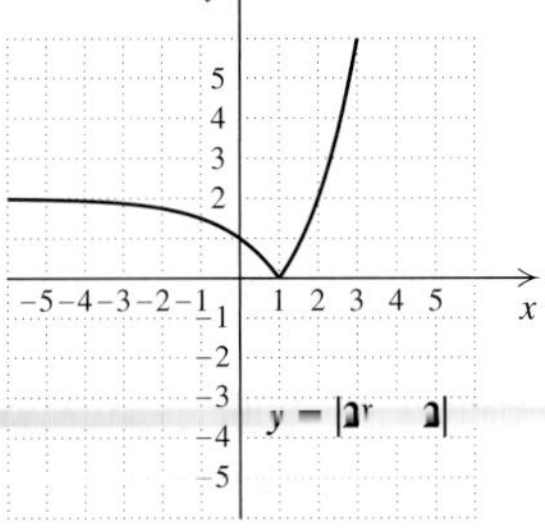

77.

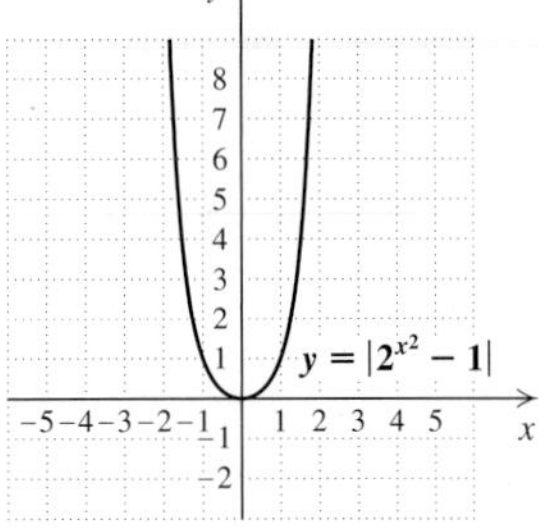

79.

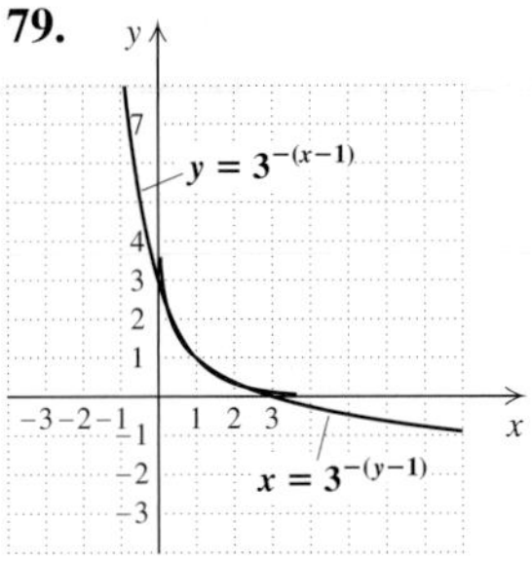

81. $m(x) = 69.67150549(1.969627617)^x$; \$15,781 million

Exercise Set 9.3, pp. 734–735

1. (g) **2.** (d) **3.** (a) **4.** (h) **5.** (b) **6.** (c) **7.** (e) **8.** (f) **9.** 3 **11.** 4 **13.** 4 **15.** -2 **17.** -1 **19.** 4 **21.** 1 **23.** 0 **25.** 5 **27.** -2 **29.** $\frac{1}{2}$ **31.** $\frac{3}{2}$ **33.** $\frac{2}{3}$ **35.** 7

37.

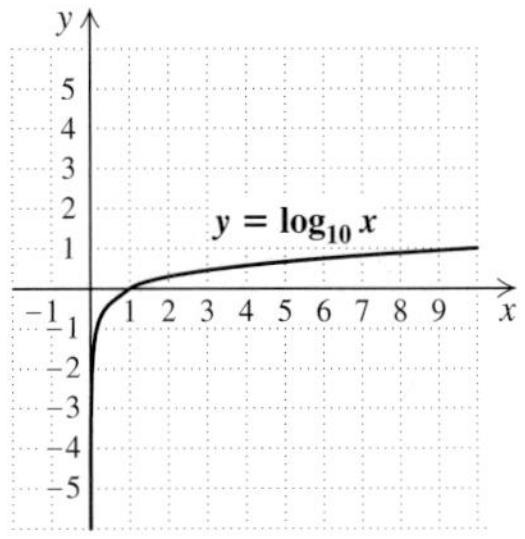

39.

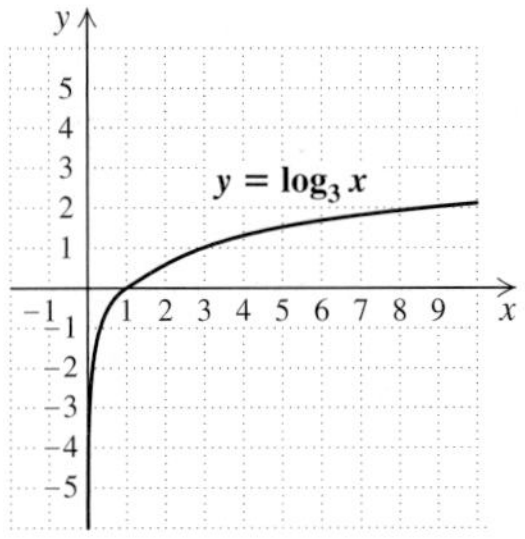

41.

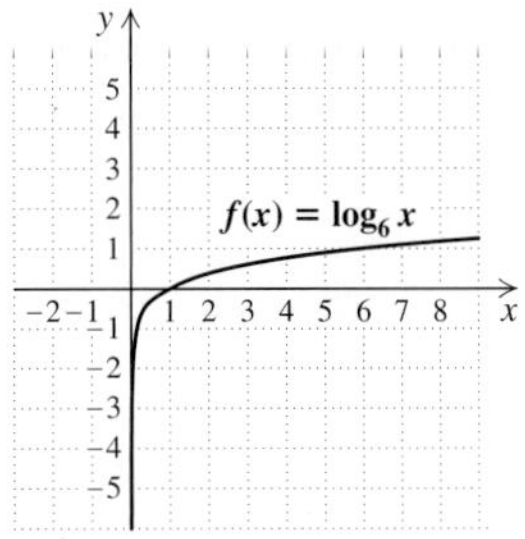

43.

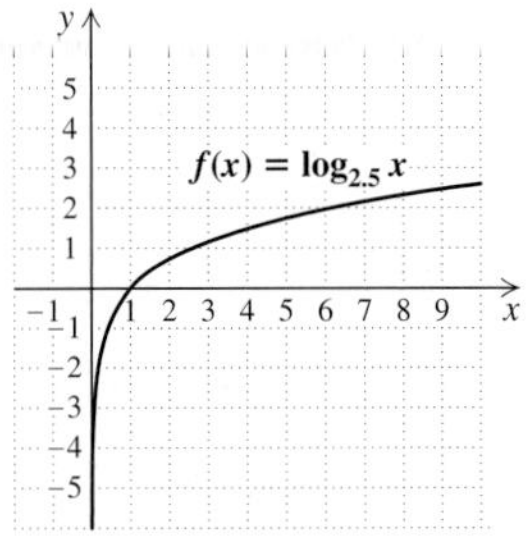

45.

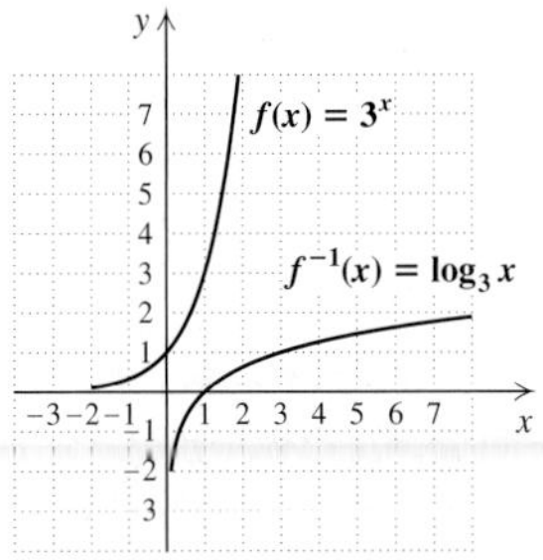

47. 0.6021 **49.** 4.1271

51. -0.2782 **53.** 199.5262 **55.** 0.0011 **57.** 1.0028

59. $y = \log(x + 2)$

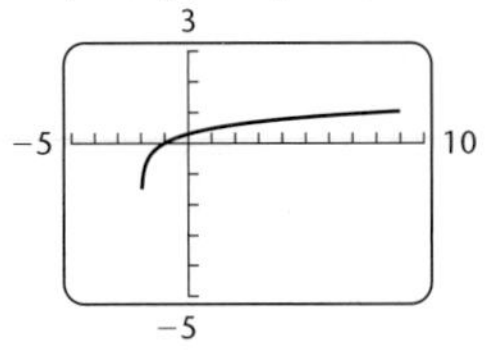

61. $y = \log(1 - 2x)$

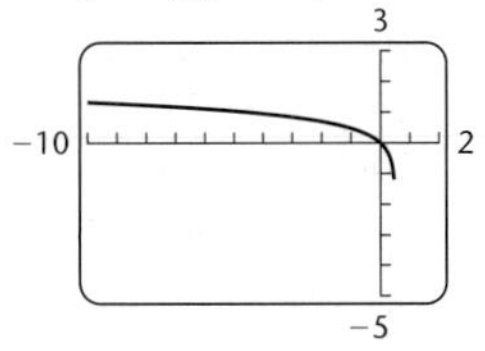

63. $y = \log(x^2)$

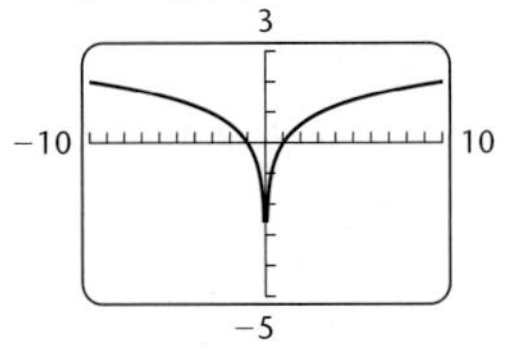

65. $5^t = 9$ **67.** $5^2 = 25$

69. $10^{-1} = 0.1$ **71.** $10^{0.845} = 7$ **73.** $c^8 = m$ **75.** $t^r = Q$ **77.** $e^{-1.3863} = 0.25$ **79.** $r^{-x} = T$ **81.** $2 = \log_{10} 100$ **83.** $-5 = \log_4 \frac{1}{1024}$ **85.** $\frac{3}{4} = \log_{16} 8$ **87.** $0.4771 = \log_{10} 3$ **89.** $m = \log_z 6$ **91.** $m = \log_p V$ **93.** $3 = \log_e 20.0855$ **95.** $-4 = \log_e 0.0183$ **97.** 9 **99.** 3 **101.** 4 **103.** 7 **105.** $\frac{1}{9}$ **107.** 4 **109.** TW **111.** x^8 **112.** a^{12} **113.** a^7b^8 **114.** x^5y^{12} **115.** $\frac{x(3y - 2)}{2y + x}$ **116.** $\frac{x + 2}{x + 1}$ **117.** TW

119.

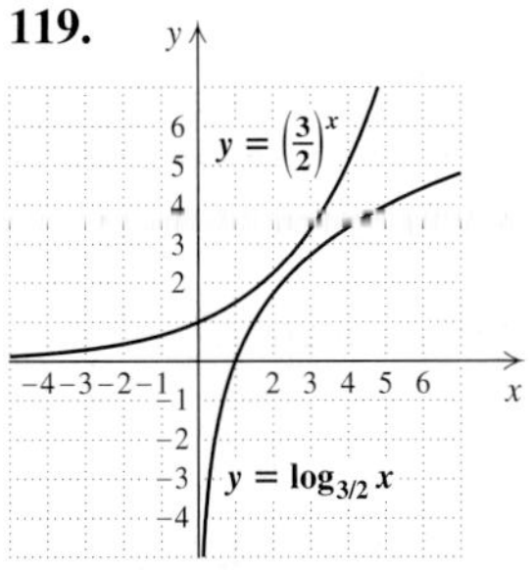

121.

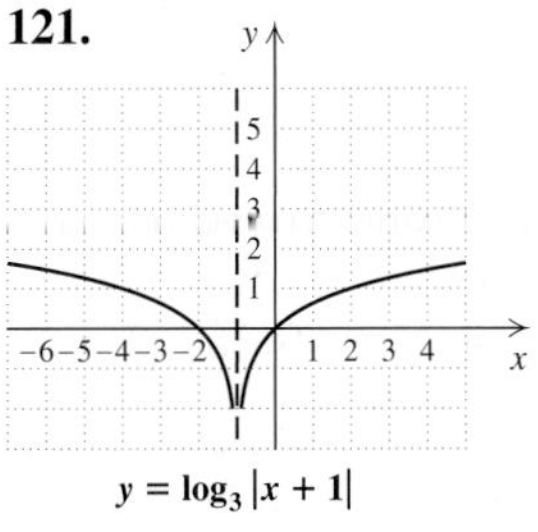

123. 6 **125.** $-25, 4$ **127.** -2 **129.** 0 **131.** Let $b = 0$, and suppose that $x_1 = 1$ and $x_2 = 2$. Then $0^1 = 0^2$, but $1 \neq 2$. Then let $b = 1$, and suppose that $x_1 = 1$ and $x_2 = 2$. Then $1^1 = 1^2$, but $1 \neq 2$.

Interactive Discovery, p. 736

1. (c) **2.** (b) **3.** (c) **4.** (a)

Exercise Set 9.4, pp. 741–743

1. (e) **2.** (f) **3.** (a) **4.** (b) **5.** (c) **6.** (d)
7. $\log_3 81 + \log_3 27$ **9.** $\log_4 64 + \log_4 16$
11. $\log_c r + \log_c s + \log_c t$ **13.** $\log_a (5 \cdot 14)$, or $\log_a 70$
15. $\log_c (t \cdot y)$ **17.** $8 \log_a r$ **19.** $6 \log_c y$
21. $-3 \log_b C$ **23.** $\log_2 25 - \log_2 13$
25. $\log_b m - \log_b n$ **27.** $\log_a \frac{17}{6}$ **29.** $\log_b \frac{36}{4}$, or $\log_b 9$
31. $\log_a \frac{7}{18}$ **33.** $\log_a x + \log_a y + \log_a z$
35. $3 \log_a x + 4 \log_a z$ **37.** $2 \log_a x - 2 \log_a y + \log_a z$
39. $4 \log_a x - 3 \log_a y - \log_a z$
41. $\log_b x + 2 \log_b y - \log_b w - 3 \log_b z$
43. $\frac{1}{2}(7 \log_a x - 5 \log_a y - 8 \log_a z)$
45. $\frac{1}{3}(6 \log_a x + 3 \log_a y - 2 - 7 \log_a z)$ **47.** $\log_a (x^8 z^3)$
49. $\log_a x$ **51.** $\log_a \dfrac{y^5}{x^{3/2}}$ **53.** $\log_a (x - 2)$ **55.** 1.953
57. -0.369 **59.** -1.161 **61.** $\frac{3}{2}$ **63.** Cannot be found
65. 7 **67.** m **69.** 7 **71.** 3 **73.** TW
75.

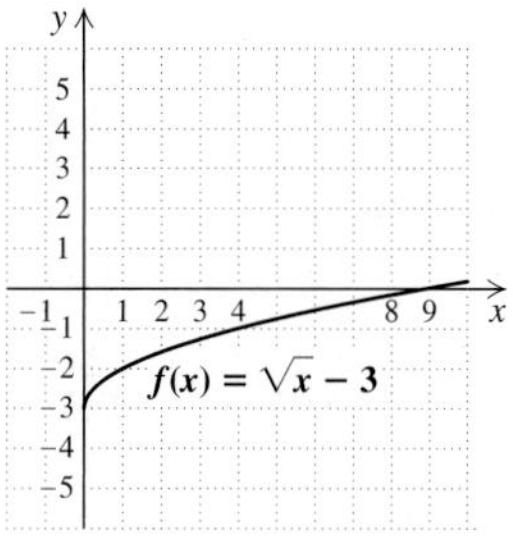

76.

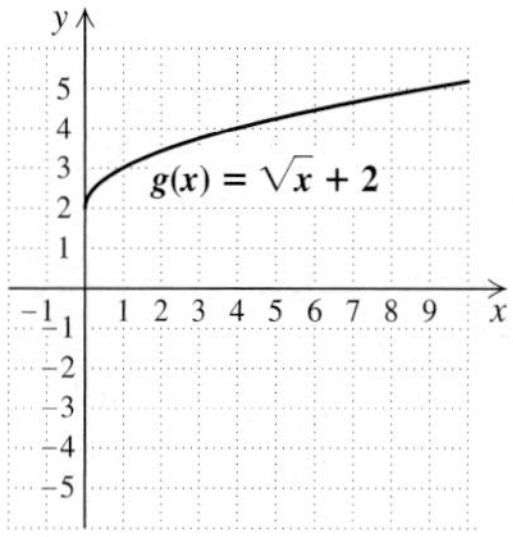

77.

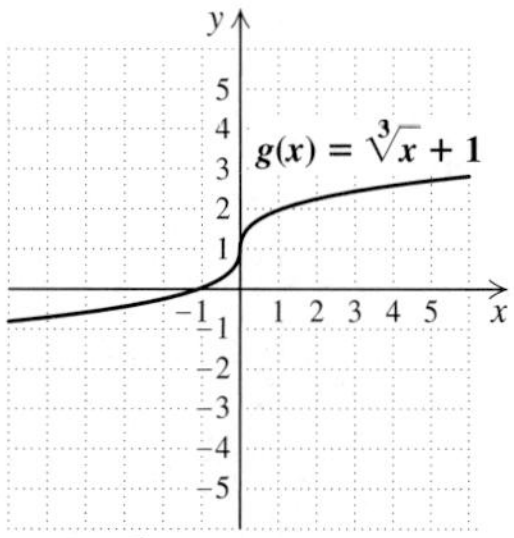

78.

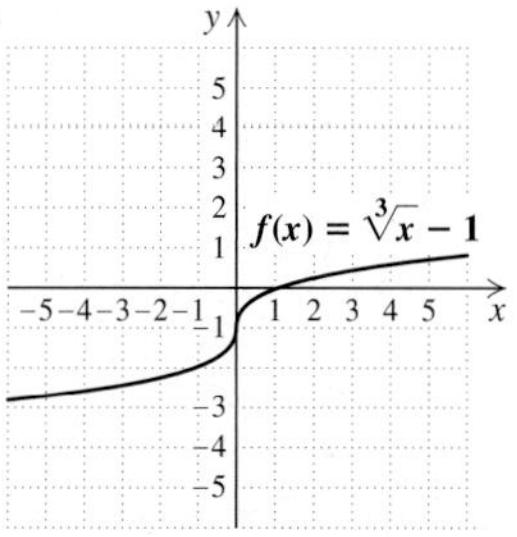

79. $a^{17}b^{17}$ **80.** $x^{11}y^6z^8$ **81.** TW
83. $\log_a (x^6 - x^4y^2 + x^2y^4 - y^6)$
85. $\frac{1}{2} \log_a (1 - s) + \frac{1}{2} \log_a (1 + s)$ **87.** $\frac{10}{3}$ **89.** -2
91. True

Interactive Discovery, p. 744

1. \$2.25; \$2.370370; \$2.441406; \$2.613035; \$2.692597; \$2.714567; \$2.718127 **2.** (c)

Visualizing the Graph, p. 749

1. J **2.** D **3.** B **4.** G **5.** H **6.** C **7.** F
8. I **9.** E **10.** A

Exercise Set 9.5, pp. 750–751

1. True **2.** True **3.** True **4.** False **5.** True
6. True **7.** True **8.** True **9.** 0.7782 **11.** 1.8621
13. 3 **15.** -0.2782 **17.** 1.7986 **19.** 199.5262
21. 1.4894 **23.** 0.0011 **25.** 1.6094 **27.** 4.0431
29. -5.0832 **31.** 96.7583 **33.** 15.0293 **35.** 0.0305
37. 109.9472 **39.** 2.5237 **41.** 6.6439 **43.** 2.1452
45. -2.3219 **47.** -2.3219 **49.** 3.5471
51. Domain: $\mathbb{R}$; range: $(0, \infty)$

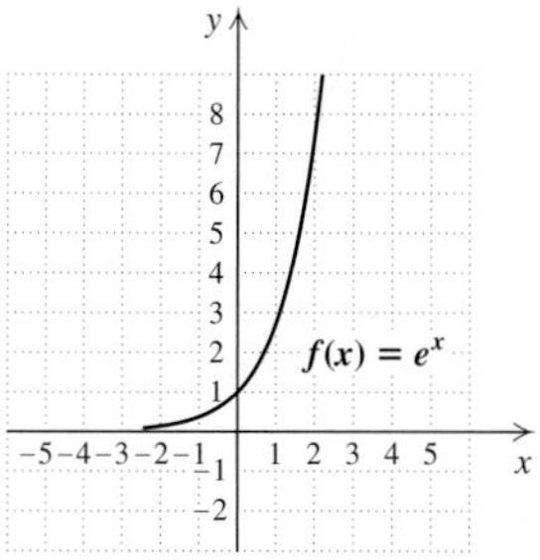

53. Domain: $\mathbb{R}$; range: $(3, \infty)$

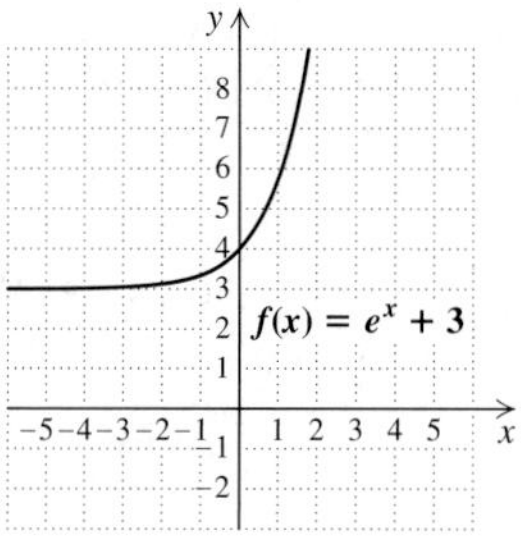

55. Domain: $\mathbb{R}$; range: $(-2, \infty)$

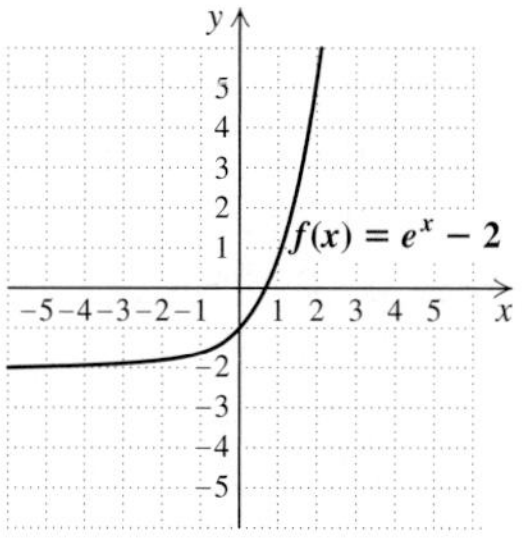

57. Domain: $\mathbb{R}$; range: $(0, \infty)$

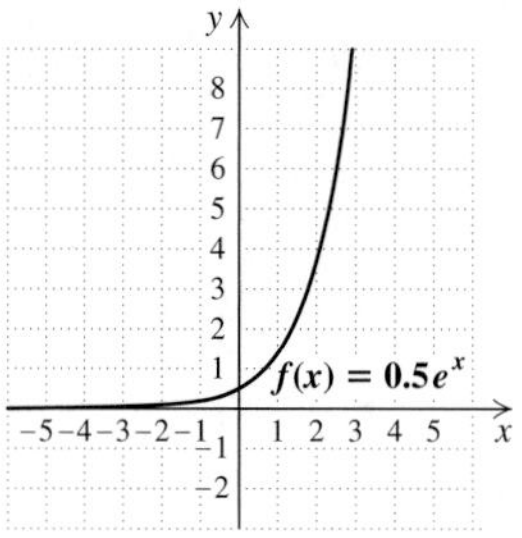

59.

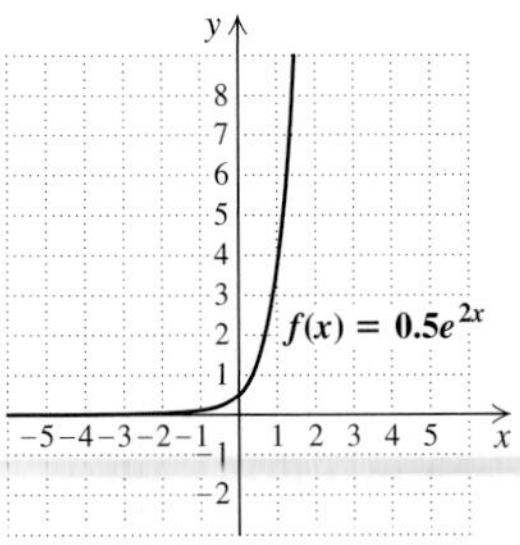

Domain: $\mathbb{R}$; range: $(0, \infty)$

61.

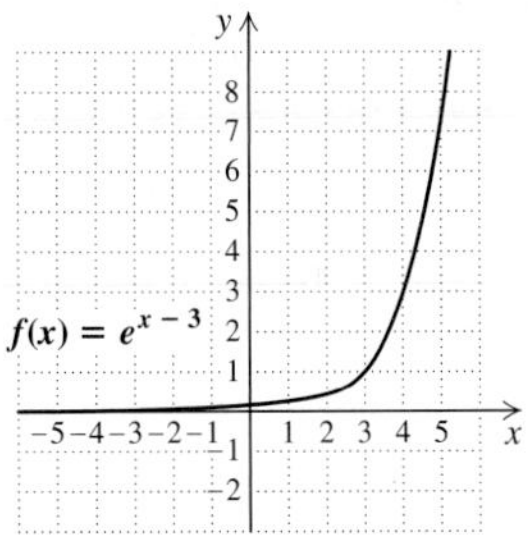

Domain: $\mathbb{R}$; range: $(0, \infty)$

63.

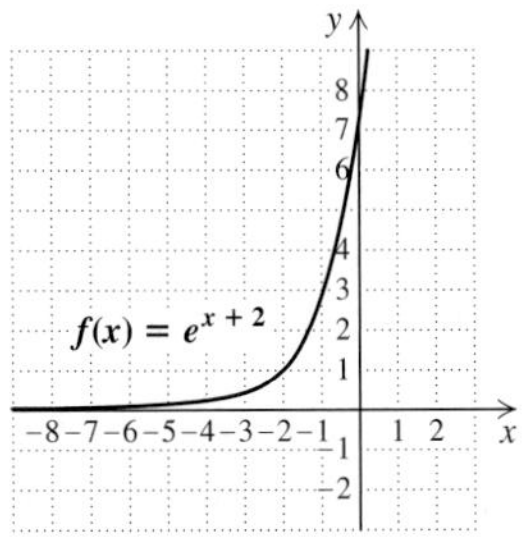

Domain: $\mathbb{R}$; range: $(0, \infty)$

65.

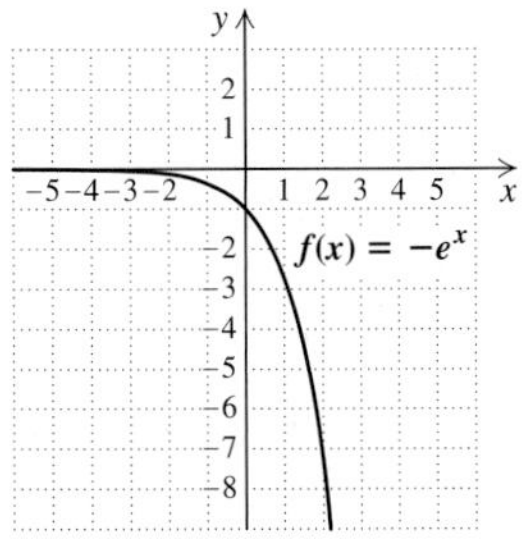

Domain: $\mathbb{R}$; range: $(-\infty, 0)$

67.

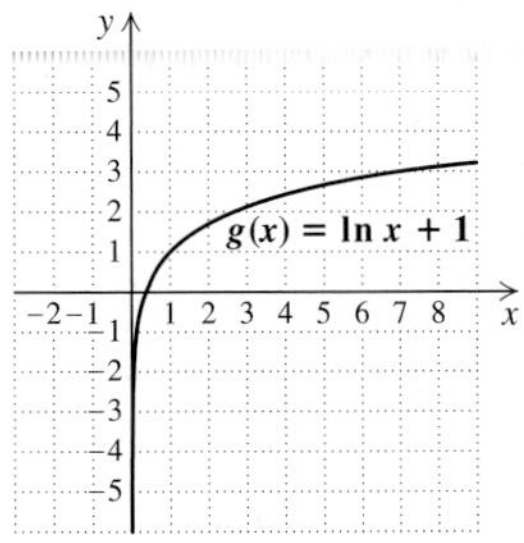

Domain: $(0, \infty)$; range: $\mathbb{R}$

69.

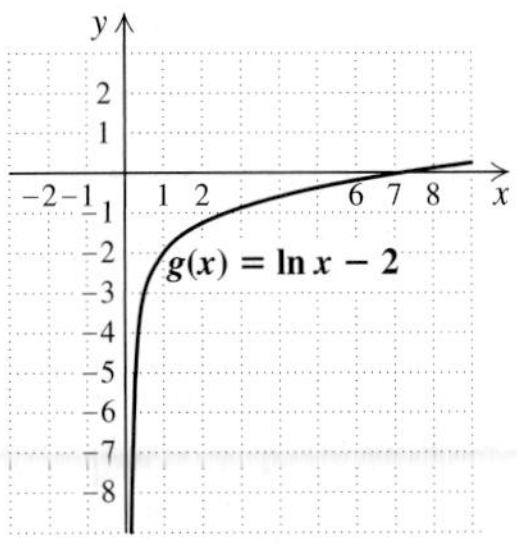

Domain: $(0, \infty)$; range: $\mathbb{R}$

71.

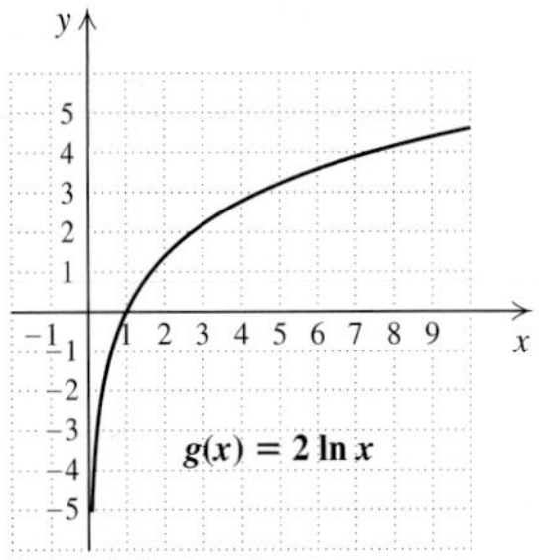

Domain: $(0, \infty)$; range: $\mathbb{R}$

73.

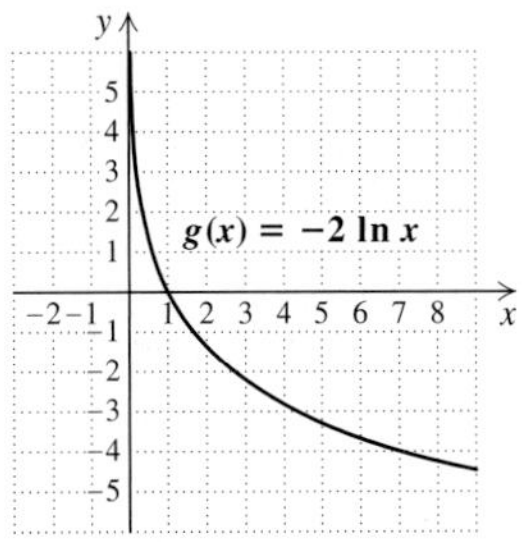

Domain: $(0, \infty)$; range: $\mathbb{R}$

75.

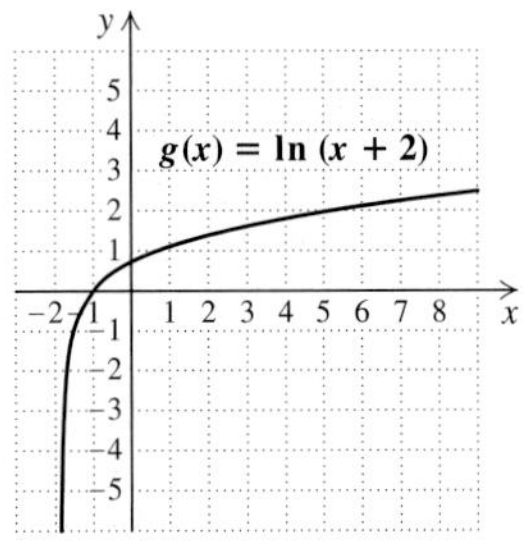

Domain: $(-2, \infty)$; range: $\mathbb{R}$

77.

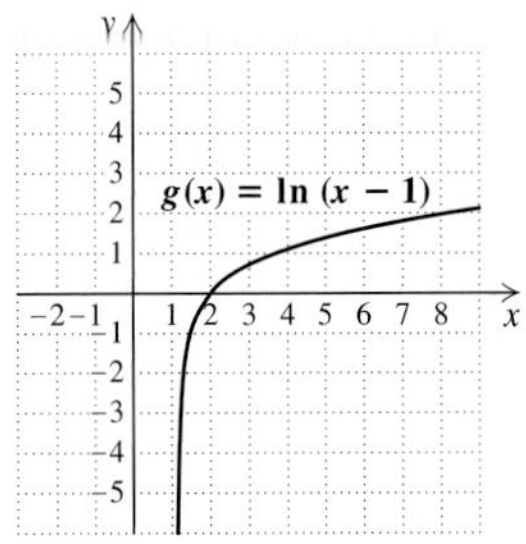

Domain: $(1, \infty)$; range: $\mathbb{R}$

79. $f(x) = \log(x)/\log(5)$, or $f(x) = \ln(x)/\ln(5)$

$y = \log(x)/\log(5)$, or $y = \ln(x)/\ln(5)$

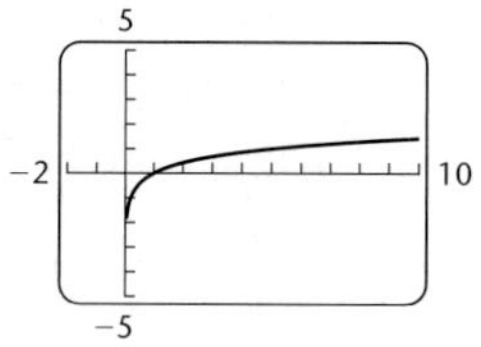

81. $f(x) = \log(x - 5)/\log(2)$, or $f(x) = \ln(x - 5)/\ln(2)$

$y = \log(x - 5)/\log(2)$, or $y = \ln(x - 5)/\ln(2)$

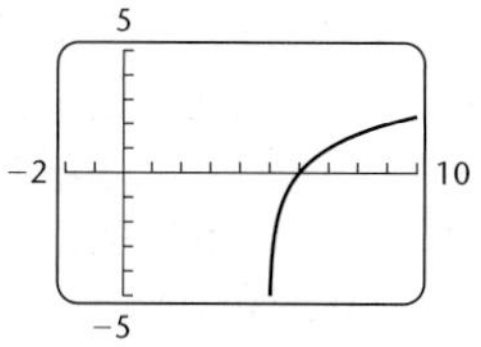

83. $f(x) = \log(x)/\log(3) + x$, or $f(x) = \ln(x)/\ln(3) + x$

$y = \log(x)/\log(3) + x$, or $y = \ln(x)/\ln(3) + x$

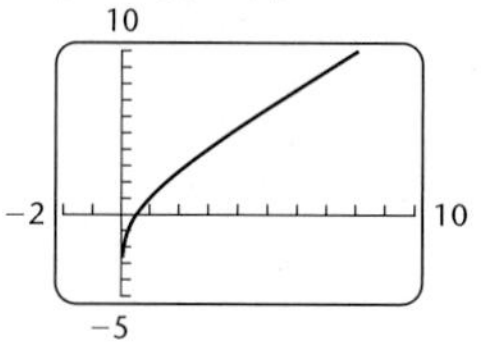

85. TW **87.** $y = 9x$ **88.** $y = \dfrac{21.35}{x}$ **89.** $L = \dfrac{8T^2}{\pi^2}$

90. $c = \sqrt{\dfrac{E}{m}}$ **91.** $1\frac{1}{5}$ hr

92. $9\frac{3}{8}$ min **93.** TW

95. 2.452 **97.** 1.442 **99.** $\log M = \dfrac{\ln M}{\ln 10}$

101. 1086.5129 **103.** 4.9855

105. **(a)** Domain: $\{x \mid x > 0\}$, or $(0, \infty)$; range: $\{y \mid y < 0.5135\}$, or $(-\infty, 0.5135)$; (b) $[-1, 5, -10, 5]$;

(c) $y = 3.4 \ln x - 0.25e^x$

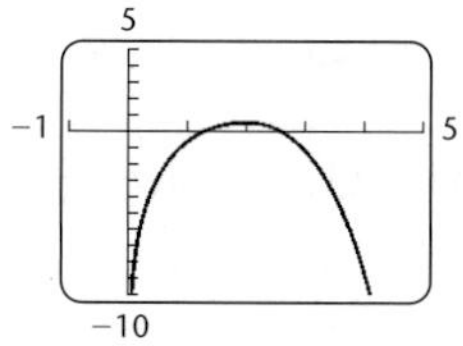

107. **(a)** Domain: $\{x \mid x > 0\}$, or $(0, \infty)$; range: $\{y \mid y > -0.2453\}$, or $(-0.2453, \infty)$; (b) $[-1, 5, -1, 10]$;

(c) $y = 2x^3 \ln x$

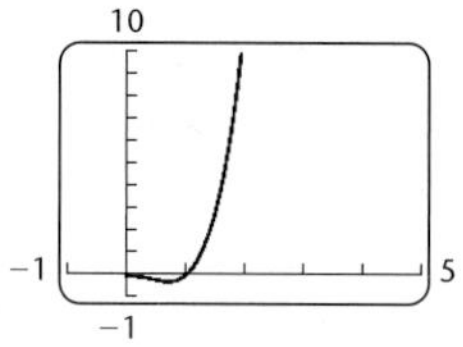

Exercise Set 9.6, pp. 758–760

1. (e) **2.** (a) **3.** (f) **4.** (h) **5.** (b) **6.** (d) **7.** (g) **8.** (c) **9.** $\dfrac{\log 19}{\log 2} \approx 4.248$

11. $\dfrac{\log 17}{\log 8} + 1 \approx 2.362$ **13.** $\ln 1000 \approx 6.908$

15. $\dfrac{\ln 5}{0.03} \approx 53.648$ **17.** $\dfrac{\log 5}{\log 3} - 1 \approx 0.465$ **19.** 1

21. $\dfrac{\log 87}{\log 4.9} \approx 2.810$ **23.** $\dfrac{\ln\left(\frac{19}{2}\right)}{4} \approx 0.563$

25. $\dfrac{\ln 2}{5} \approx 0.139$ **27.** 81 **29.** $\frac{1}{8}$ **31.** $e^5 \approx 148.413$

33. 2 **35.** $\dfrac{e^3}{4} \approx 5.021$ **37.** $10^{2.5} \approx 316.228$

39. $\dfrac{e^4 - 1}{2} \approx 26.799$ **41.** $e \approx 2.718$ **43.** $e^{-3} \approx 0.050$

45. -4 **47.** 10 **49.** No solution **51.** $\frac{83}{15}$ **53.** 1 **55.** 6 **57.** 1 **59.** 5 **61.** $\frac{17}{2}$ **63.** 4 **65.** $-6.480, 6.519$ **67.** 0.000112, 3.445 **69.** 1 **71.** TW **73.** $-\frac{5}{2}, \frac{5}{2}$ **74.** $0, \frac{7}{5}$ **75.** $\frac{4}{3}, 4$ **76.** 0

77. 16, 256 **78.** 5 **79.** $\dfrac{-3}{2} \pm \dfrac{\sqrt{11}}{2}i$ **80.** $\frac{5}{11}$

81. $\ln 1.5 \approx 0.405$ **82.** 5 **83.** $-4, 1$ **84.** No solution **85.** $(2, -2)$ **86.** $(3, 2, 1)$ **87.** TW **89.** $\frac{12}{5}$ **91.** $\sqrt[3]{3}$ **93.** -1 **95.** $-3, -1$ **97.** $-625, 625$ **99.** $\frac{1}{2}, 5000$ **101.** $-3, -1$ **103.** $\frac{1}{100{,}000}$, 100,000 **105.** $-\frac{1}{3}$ **107.** 38

Exercise Set 9.7, pp. 772–779

1. **(a)** About 2004; **(b)** 1.6 yr **3.** **(a)** 119,157 skateboarders; **(b)** 49 **5.** **(a)** 6.4 yr; **(b)** 23.4 yr **7.** **(a)** About 2005; **(b)** about 2018 **9.** **(a)** About 2008; **(b)** 9.0 yr **11.** 4.9 **13.** 10^{-7} moles per liter **15.** 65 dB **17.** $10^{-1.5}$ W/m² **19.** **(a)** -26.9; **(b)** 1.58×10^{-17} W/m² **21.** **(a)** $P(t) = P_0e^{0.025t}$; **(b)** \$5126.58; \$5256.36; **(c)** 27.7 yr **23.** **(a)** $P(t) = 300e^{0.009t}$, where t is the number of years after 2006 and $P(t)$ is in millions; **(b)** 311 million; **(c)** about 2015 **25.** 6.2 months **27.** **(a)** About 2010; **(b)** about 2019;

(c)

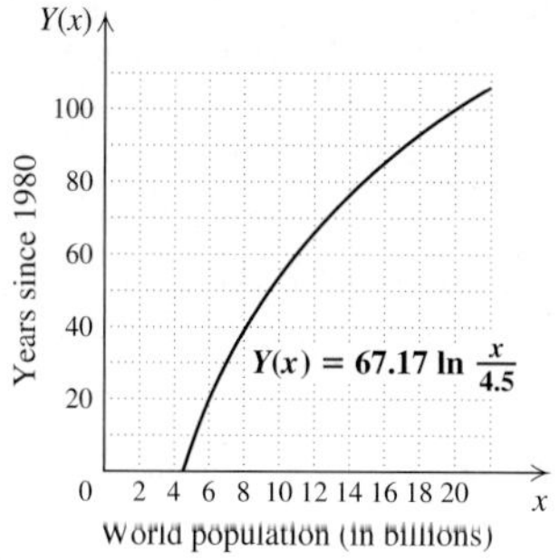

29. **(a)** 68%; **(b)** 54%, 40%
(c)

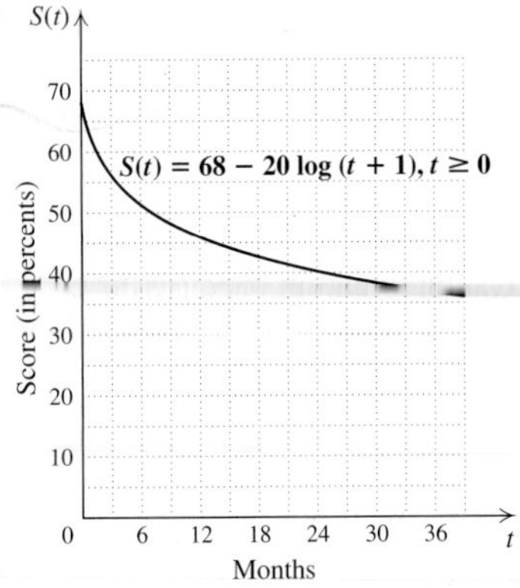

(d) 6.9 months

31. **(a)** $P(t) = 1.7e^{0.622t}$, where t is the number of years since 2002 and $P(t)$ is in millions; **(b)** 38.1 million units
33. **(a)** $k \approx 0.004$; $P(t) = 987e^{-0.004t}$, where t is the number of years after 1990 and $P(t)$ is in millions; **(b)** 918 million acres; **(c)** about 2043 **35.** About 2103 yr
37. About 7.2 days **39.** 69.3% per year
41. **(a)** $k \approx 0.099$; $V(t) = 451{,}000e^{0.099t}$, where t is the number of years after 1991; **(b)** about \$1.99 million; **(c)** about 7.0 yr; **(d)** about 2010 **43.** No **45.** Yes
47. **(a)** $p(x) = 7.019404696(1.099514502)^x$; **(b)** 0.095, or 9.5%; **(c)** about \$121
49. **(a)** $f(x) = 2097152(0.8705505633)^x$; **(b)** 4 hr **51.** TW

53.

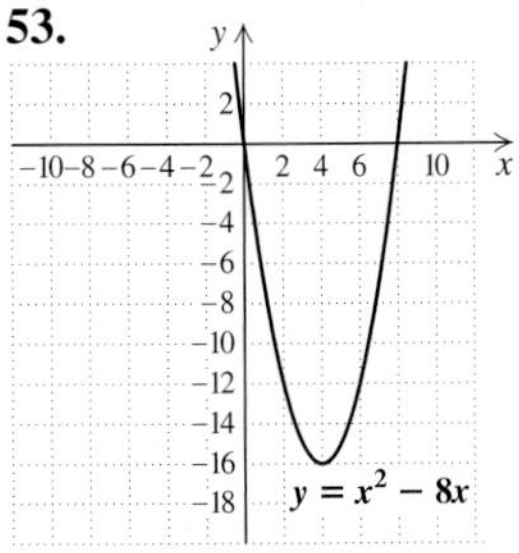

54.

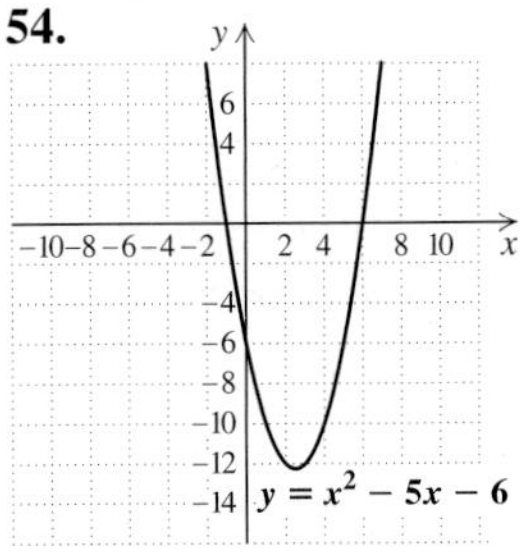

55.

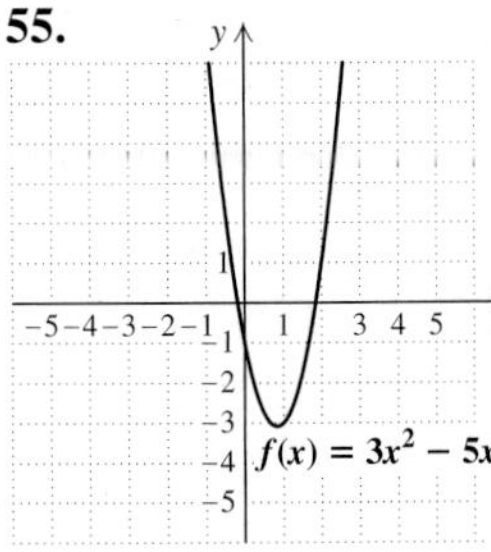

56.

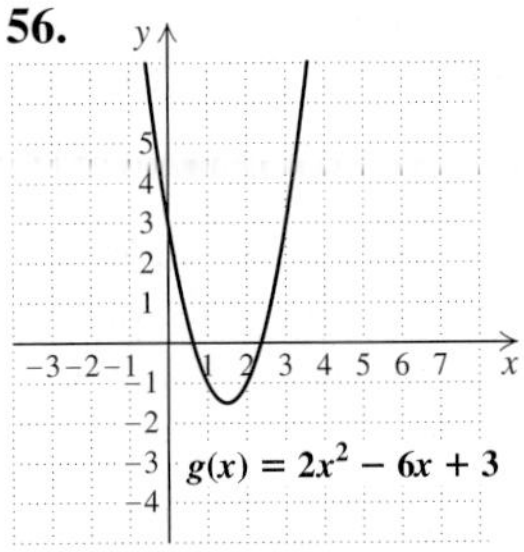

57. $4 \pm \sqrt{23}$ **58.** $-5 \pm \sqrt{31}$ **59.** TW
61. \$18.9 million **63.** $P(t) = 100 - 63.03(0.95)^t$
65. About 80,922 yr, or with rounding of k, about 80,792 yr
67. Consider an exponential growth function $P(t) = P_0e^{kt}$. At time T, $P(T) = 2P_0$.

Solve for T:

$$2P_0 = P_0e^{kT}$$
$$2 = e^{kT}$$
$$\ln 2 = kT$$
$$\frac{\ln 2}{k} = T.$$

69. **(a)**

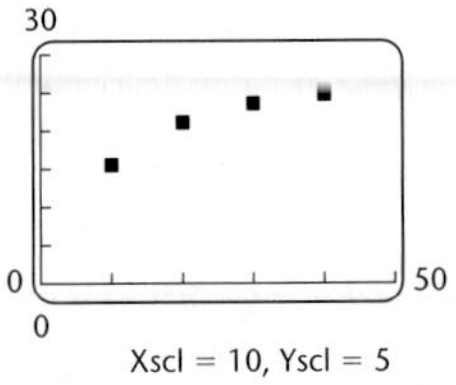

(b) $s(t) = -0.0300167067 + 6.879024467 \ln t$;
(c) about 55.5 min
71. **(a)**

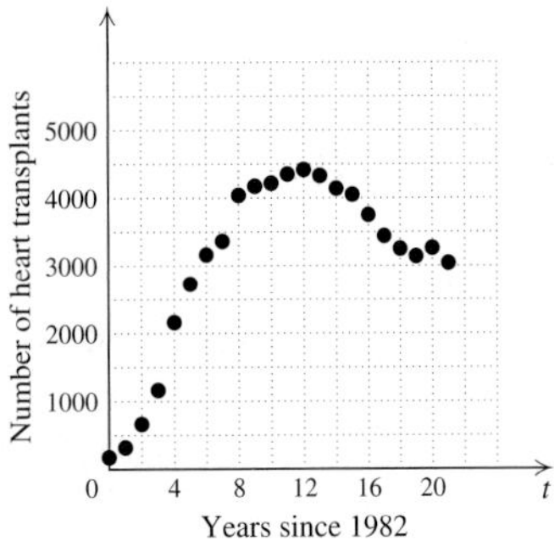

Review Exercises: Chapter 9, pp. 783–785

1. True **2.** True **3.** True **4.** False **5.** False
6. True **7.** False **8.** False **9.** True **10.** False
11. $(f \circ g)(x) = 4x^2 - 12x + 10$; $(g \circ f)(x) = 2x^2 - 1$
12. $f(x) = \sqrt{x}$; $g(x) = 3 - x$ **13.** No
14. $f^{-1}(x) = x + 8$ **15.** $g^{-1}(x) = \dfrac{2x - 1}{3}$
16. $f^{-1}(x) = \dfrac{\sqrt[3]{x}}{3}$

17.

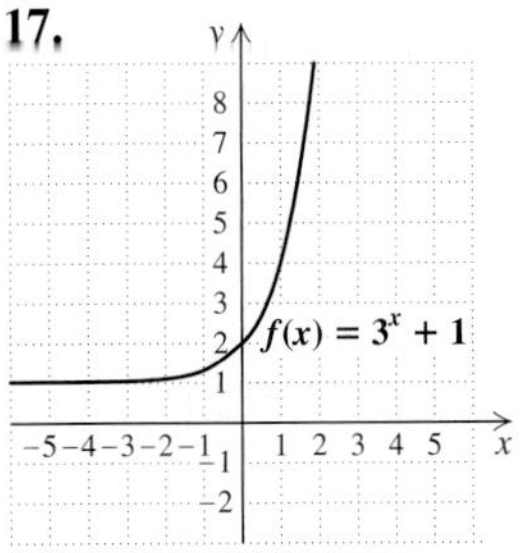

18.

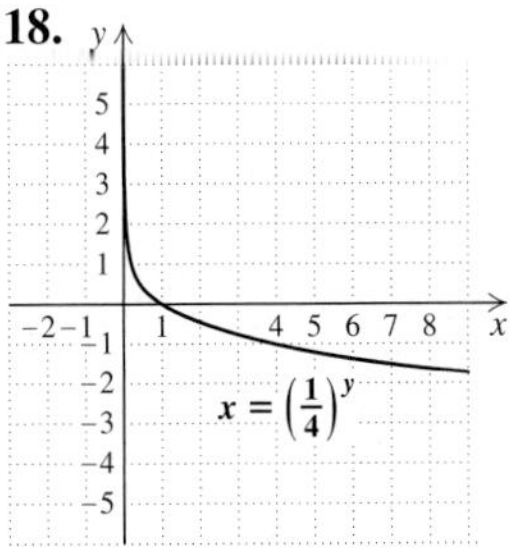

19. 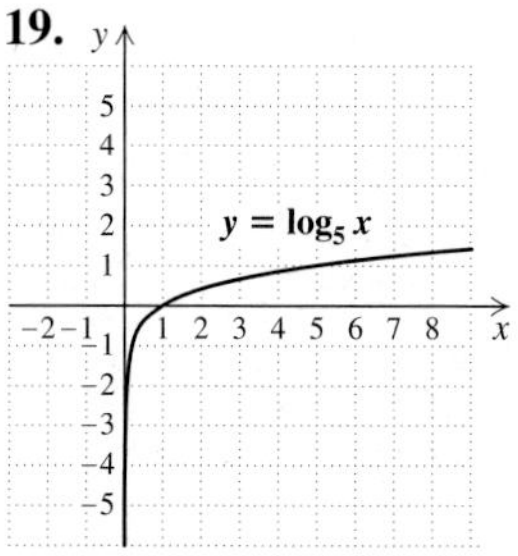

20. 2 **21.** -2 **22.** 7 **23.** $\frac{1}{2}$ **24.** $\log_{10} \frac{1}{100} = -2$
25. $\log_{25} 5 = \frac{1}{2}$ **26.** $16 = 4^x$ **27.** $1 = 8^0$
28. $4 \log_a x + 2 \log_a y + 3 \log_a z$
29. $5 \log_a x - (\log_a y + 2 \log_a z)$, or $5 \log_a x - \log_a y - 2 \log_a z$
30. $\frac{1}{4}(2 \log z - 3 \log x - \log y)$
31. $\log_a (7 \cdot 8)$, or $\log_a 56$ **32.** $\log_a \frac{72}{12}$, or $\log_a 6$
33. $\log \frac{a^{1/2}}{bc^2}$ **34.** $\log_a \sqrt[3]{\frac{x}{y^2}}$ **35.** 1 **36.** 0
37. 17 **38.** 6.93 **39.** -3.2698 **40.** 8.7601
41. 3.2698 **42.** 2.54995 **43.** -3.6602 **44.** 1.8751
45. 61.5177 **46.** -2.9957 **47.** 0.3753 **48.** 0.4307
49. 1.7097
50. 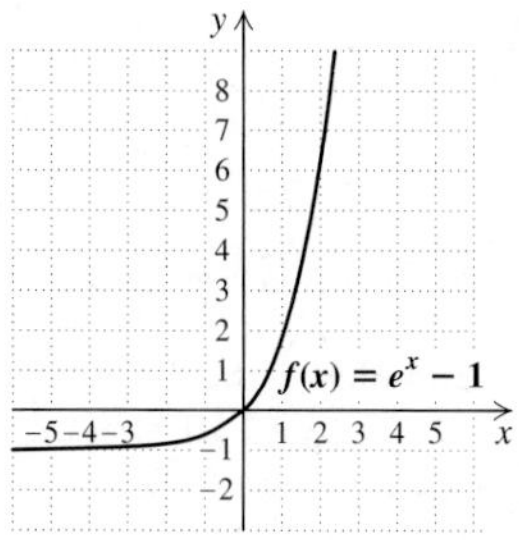

Domain: $\mathbb{R}$; range: $(-1, \infty)$

51. 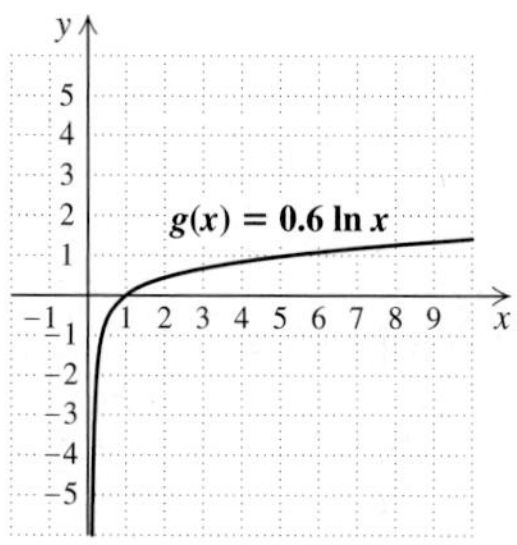

Domain: $(0, \infty)$; range: $\mathbb{R}$

52. 5 **53.** -2 **54.** $\frac{1}{81}$ **55.** 2 **56.** $\frac{1}{1000}$
57. $e^{-2} \approx 0.1353$ **58.** $\frac{1}{2}\left(\frac{\log 19}{\log 4} + 5\right) \approx 3.5620$
59. $-5, 1$ **60.** $\frac{\log 8.3}{\log 4} \approx 1.5266$ **61.** $\frac{\ln 0.03}{-0.1} \approx 35.0656$
62. $\frac{\ln 2}{2} \approx 0.3466$ **63.** 4 **64.** 8 **65.** 20 **66.** $\sqrt{43}$
67. **(a)** 82; **(b)** 66.8; **(c)** 35 months **68.** **(a)** 6.6 yr; **(b)** 3.1 yr **69.** **(a)** $k \approx 0.253$, $P(t) = 107.8e^{0.253t}$; **(b)** about 2.2 billion phones; **(c)** about 2007
70. **(a)** $f(x) = 35380.82167(1.139052882)^x$; **(b)** about \$3,840,147; **(c)** about 34% **71.** 23.105% per year
72. 16.5 yr **73.** 3463 yr **74.** 6.6 **75.** 90 dB
76. TW Negative numbers do not have logarithms because logarithm bases are positive, and there is no exponent to which a positive number can be raised to yield a negative number. **77.** TW Taking the logarithm on each side of an equation produces an equivalent equation because the logarithm function is one-to-one. If two quantities are equal, their logarithms must be equal, and if the logarithms of two quantities are equal, the quantities must be the same.
78. e^{e^3} **79.** $-3, -1$ **80.** $\left(\frac{8}{3}, -\frac{2}{3}\right)$ **81.** $P(t) = 0.6e^{0.116t}$, where P is the number of blogs, in millions, t months after January 2003

Test: Chapter 9, pp. 785–786

1. [9.1] $(f \circ g)(x) = 2 + 6x + 4x^2$; $(g \circ f)(x) = 2x^2 + 2x + 1$ **2.** [9.1] $f(x) = \frac{1}{x}$; $g(x) = 2x^2 + 1$ **3.** [9.1] No **4.** [9.1] $f^{-1}(x) = \frac{x - 4}{3}$
5. [9.1] $g^{-1}(x) = \sqrt[3]{x} - 1$
6. [9.2] 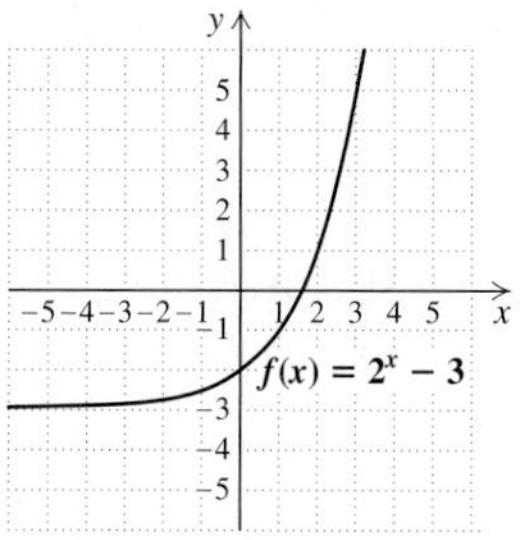

7. [9.3] 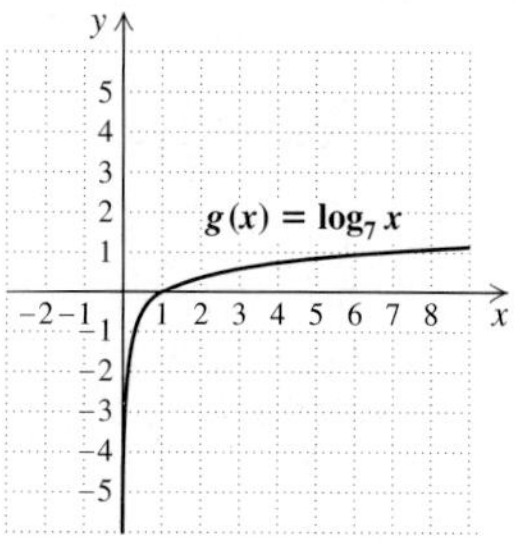

8. [9.3] 3 **9.** [9.3] $\frac{1}{2}$ **10.** [9.3] 18 **11.** [9.3] $\log_4 \frac{1}{64} = -3$
12. [9.3] $\log_{256} 16 = \frac{1}{2}$ **13.** [9.3] $49 = 7^m$
14. [9.3] $81 = 3^4$ **15.** [9.4] $3 \log a + \frac{1}{2} \log b - 2 \log c$
16. [9.4] $\log_a (z^2 \sqrt[3]{x})$ **17.** [9.4] 1 **18.** [9.4] 23
19. [9.4] 0 **20.** [9.4] 1.146 **21.** [9.4] 0.477
22. [9.4] 1.204 **23.** [9.3] 1.0899 **24.** [9.3] 0.1585
25. [9.5] -3.3524 **26.** [9.5] 121.5104 **27.** [9.5] 2.4022
28. [9.5] 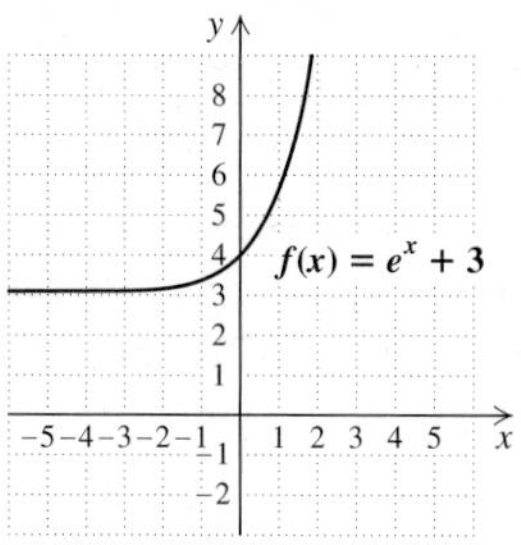

Domain: $\mathbb{R}$; range: $(3, \infty)$

29. [9.5]

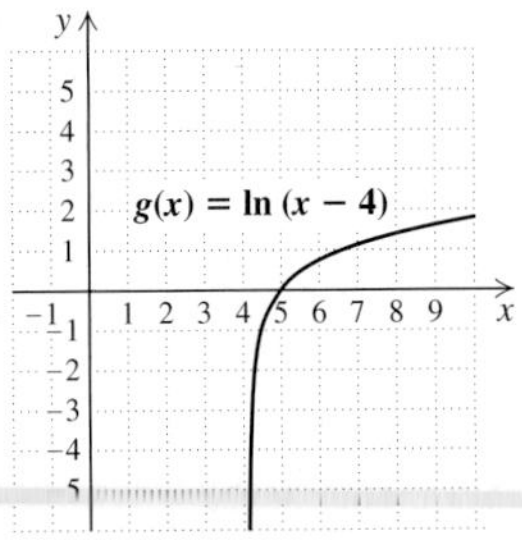

Domain: $(4, \infty)$; range: $\mathbb{R}$

30. [9.6] -5 **31.** [9.6] 5 **32.** [9.6] 2
33. [9.6] 10,000 **34.** [9.6] $-\frac{1}{3}\left(\frac{\log 87}{\log 5} - 4\right) \approx 0.4084$
35. [9.6] $\frac{\log 1.2}{\log 7} \approx 0.0937$ **36.** [9.6] $e^{1/4} \approx 1.2840$
37. [9.6] 4 **38.** [9.7] **(a)** 2.36 ft/sec; **(b)** about 1,517,000
39. [9.7] **(a)** $P(t) = 128.8e^{0.024t}$, where t is the number of years after 2005 and $P(t)$ is in millions; **(b)** 135.1 million; 152.4 million; **(c)** 2018; **(d)** 28.9 yr
40. [9.7] **(a)** $k \approx 0.044$; $C(t) = 18{,}039e^{0.044t}$, where t is the number of years after 1997; **(b)** \$31,962; **(c)** 2020
41. [9.7] **(a)** $a(t) = 4.975711955(1.20499705)^t$; **(b)** about \$22.1 billion **42.** [9.7] 4.6% **43.** [9.7] About 4684 yr **44.** [9.7] $10^{-4.5}$ W/m^2 **45.** [9.7] 7.0
46. [9.6] $-309, 316$ **47.** [9.4] 2

Cumulative Review: Chapters 1–9 , pp. 787–790

1. Linear **2.** \$1/min **3.** $f(x) = x + 3$ **4.** \$13
5. $m = 1$ signifies the cost per minute; $b = 3$ signifies the startup cost of each massage **6.** Exponential, or perhaps quadratic **7.** $m(x) = 1.893702275(1.059578007)^x$
8. \$25.60 **9.** Quadratic **10.** $c(t) = -0.5035714286x^2 + 3.043571429x + 64.30714286$ **11.** 64.4 million texts
12. Length: 36 m; width: 20 m **13.** A: 15°; B: 45°; C: 120°
14. $5\frac{5}{11}$ min **15.** Thick and Tasty: 6 oz; Light and Lean: 9 oz
16. 7.25 mph **17.** -49; -7 and 7 **18.** (c) **19.** (b)
20. (a) **21.** (d)
22. $\{x \mid x \text{ is a real number } and\ x \neq -\frac{1}{3}\ and\ x \neq 2\}$
23. -12 **24.** $f^{-1}(x) = \frac{x-9}{-2}$, or $f^{-1}(x) = \frac{9-x}{2}$
25. $f(x) = -10x - 8$ **26.** $y = \frac{1}{2}x + 7$

27.

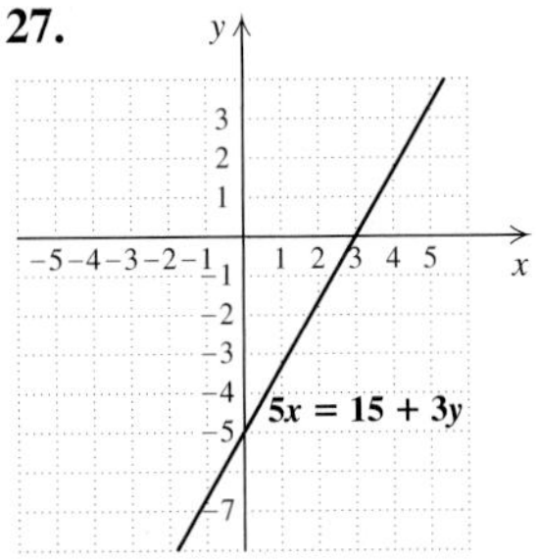

28.

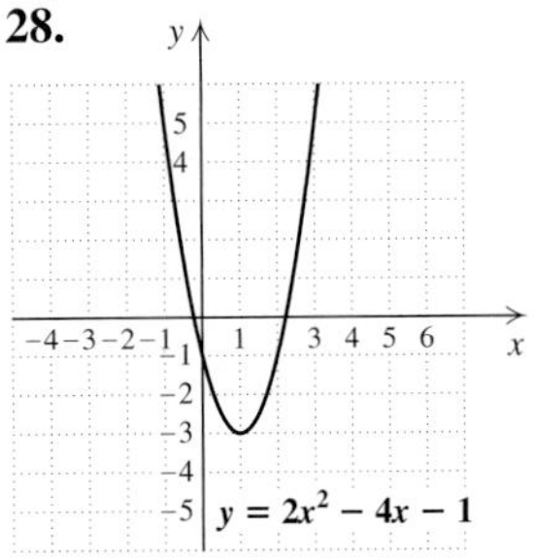

29.

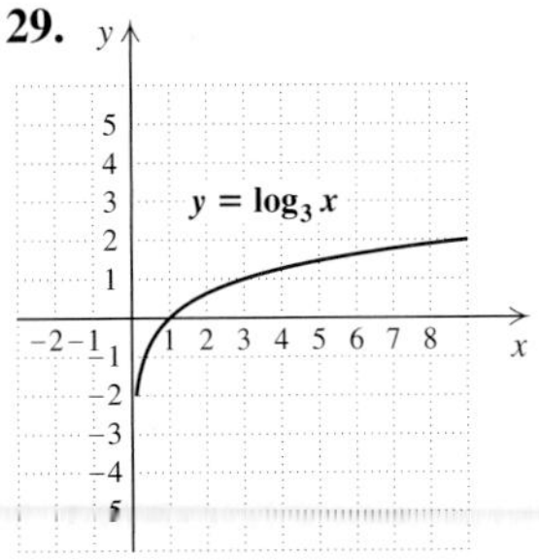

30.

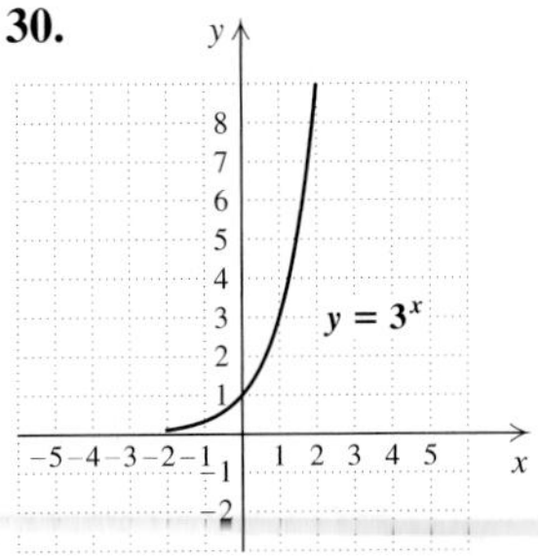

31.

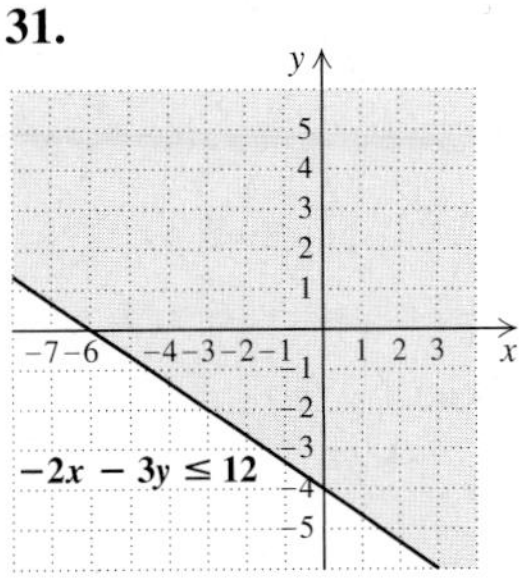

32.

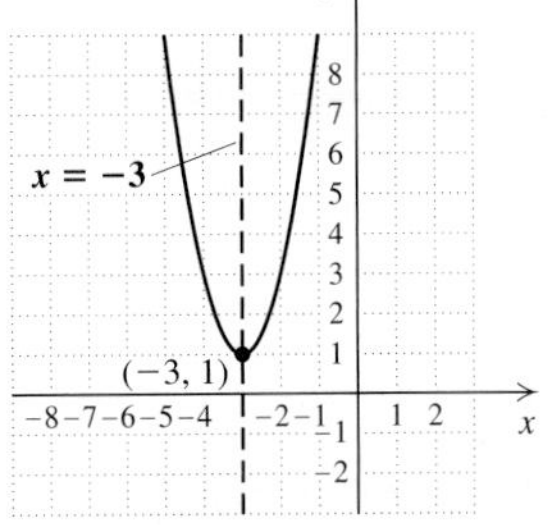

$f(x) = 2(x + 3)^2 + 1$
Minimum: 1

33.

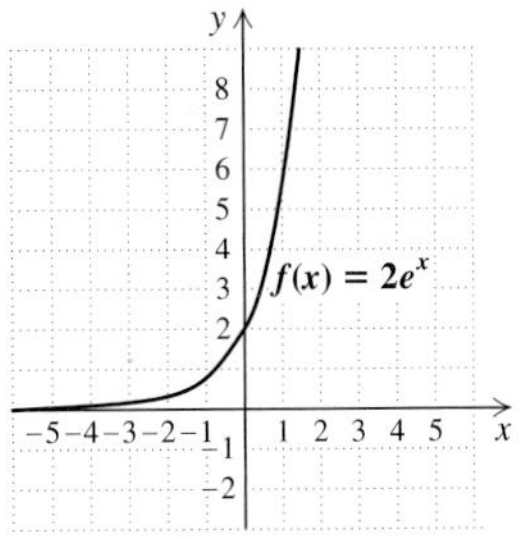

Domain: $\mathbb{R}$; range: $(0, \infty)$

34. 2 **35.** 6 **36.** $\frac{y^{12}}{16x^8}$ **37.** $\frac{20x^6z^2}{y}$ **38.** $\frac{-y^4}{3z^5}$
39. $-2x - 1$ **40.** 25 **41.** $7p^2q^3 + pq + p - 9$
42. $8x^2 - 11x - 1$ **43.** $9x^4 - 12x^2y + 4y^2$
44. $10a^2 - 9ab - 9b^2$ **45.** $\frac{(x+4)(x-3)}{2(x-1)}$ **46.** $\frac{1}{x-4}$
47. $\frac{a+2}{6}$ **48.** $\frac{7x+4}{(x+6)(x-6)}$ **49.** $x(y + 2z - w)$
50. $(2 - 5x)(4 + 10x + 25x^2)$ **51.** $2(3x - 2y)(x + 2y)$
52. $(x^3 + 7)(x - 4)$ **53.** $2(m + 3n)^2$
54. $(x - 2y)(x + 2y)(x^2 + 4y^2)$
55. $x^3 - 2x^2 - 4x - 12 + \frac{-42}{x-3}$ **56.** 1.8×10^{-1}
57. $2y^2\sqrt[3]{y}$ **58.** $14xy^2\sqrt{x}$ **59.** $81a^8b\sqrt[3]{b}$
60. $\frac{6 + \sqrt{y} - y}{4 - y}$ **61.** $\sqrt[10]{(x+5)^3}$ **62.** $18 - 2\sqrt{3}i$
63. $13 - i$ **64.** $2 \log a + 3 \log c - \log b$
65. $\log\left(\frac{x^3}{y^{1/2}z^2}\right)$ **66.** $a^x = 5$ **67.** $\log_x t = 3$
68. -1.2545 **69.** 776.2471 **70.** 2.5479 **71.** 0.2466
72. $\frac{14}{5}$ **73.** $(3, -1)$ **74.** $(1, -2, 0)$ **75.** $-2, 5$

76. $\frac{9}{2}$ **77.** $\frac{5}{8}$ **78.** $\frac{3}{4}$ **79.** $\frac{1}{2}$ **80.** $\pm 5i$ **81.** 9, 25
82. $\pm 2, \pm 3$ **83.** 7 **84.** 6 **85.** $\frac{3}{2}$
86. $\dfrac{\log 7}{5 \log 3} \approx 0.3542$ **87.** $\dfrac{8e}{e - 1} \approx 12.6558$
88. $(-\infty, -5) \cup (1, \infty)$, or $\{x \mid x < -5 \text{ or } x > 1\}$
89. $-3 \pm 2\sqrt{5}$ **90.** $\{x \mid x \leq -2 \text{ or } x \geq 5\}$, or $(-\infty, -2] \cup [5, \infty)$ **91.** $a = \dfrac{Db}{b - D}$ **92.** $q = \dfrac{pf}{p - f}$
93. $B = \dfrac{3M - 2A}{2}$, or $B = \frac{3}{2}M - A$ **94.** 78 **95.** 67.5
96. $P(t) = 33.8e^{0.026t}$, where t is the number of years after 2005 **97.** 36.5 million; 42.7 million **98.** 26.7 yr
99. 18 **100.** All real numbers except 1 and -2
101. $\frac{1}{3}, \frac{10{,}000}{3}$ **102.** 35 mph

Chapter 10

Exercise Set 10.1, pp. 800–804

1. (f) **2.** (e) **3.** (g) **4.** (h) **5.** (c) **6.** (b)
7. (d) **8.** (a)

9.

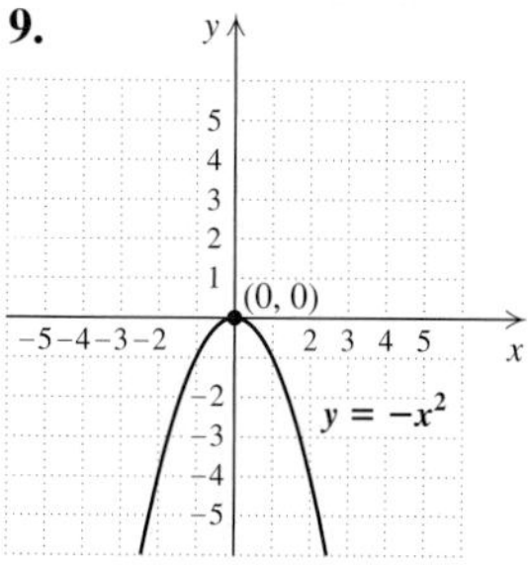

11.

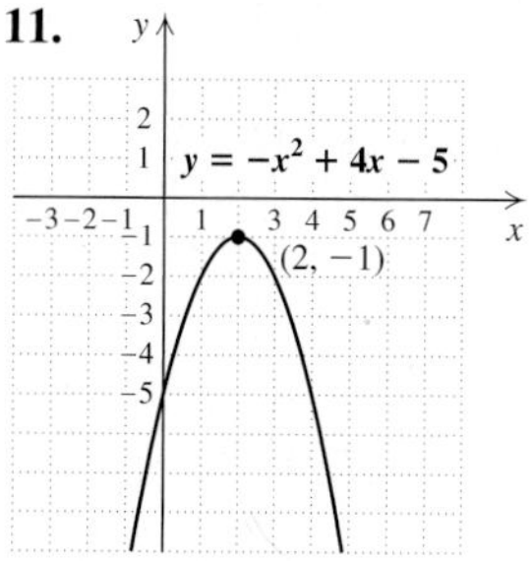

13.

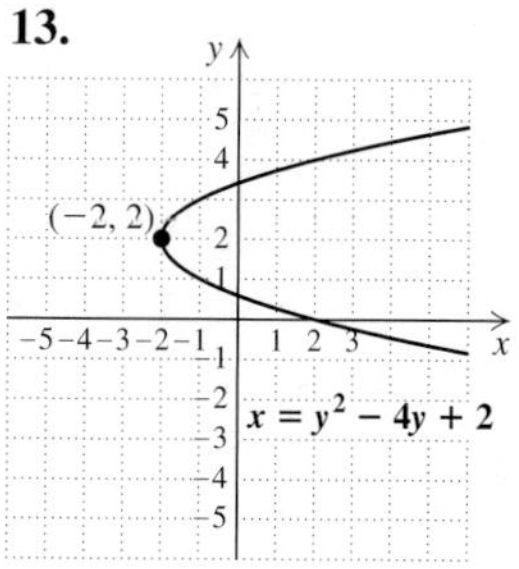

15.

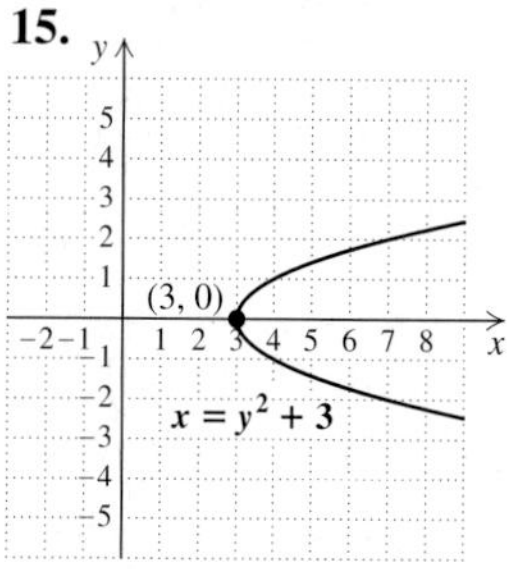

17.

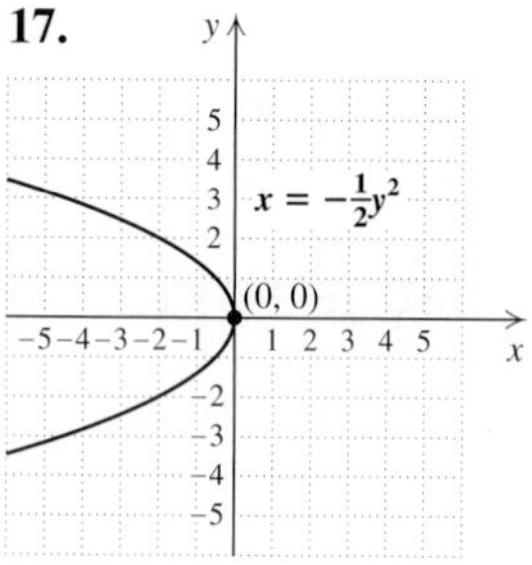

19.

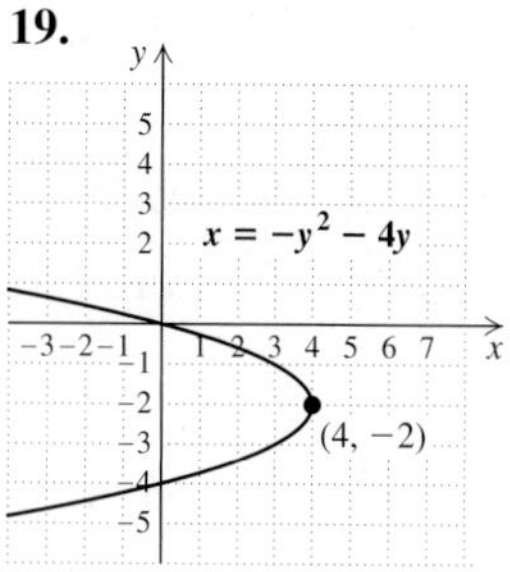

21.

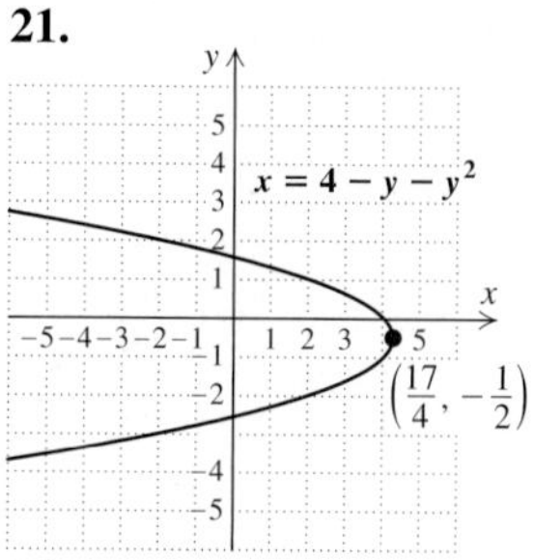

23.

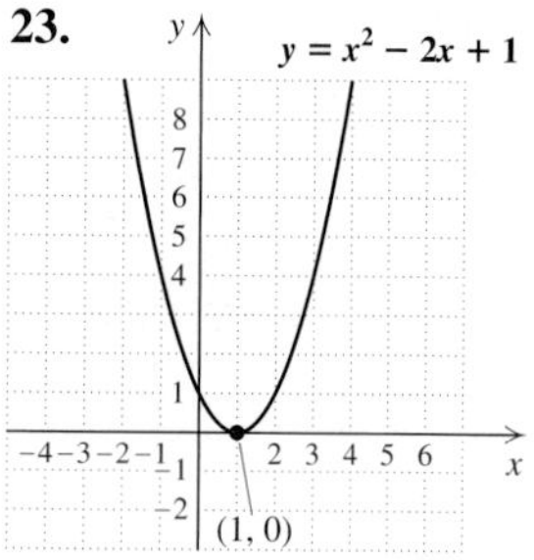

25.

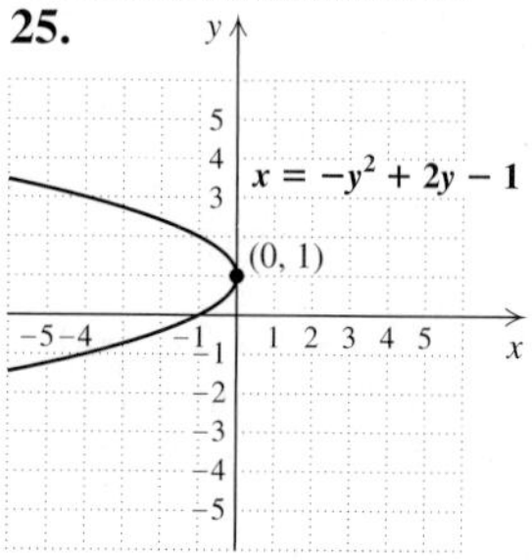

27.

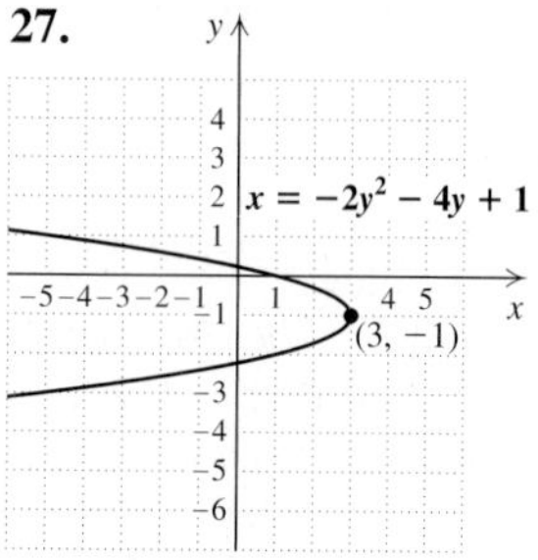

29. 5 **31.** $\sqrt{18} \approx 4.243$ **33.** $\sqrt{200} \approx 14.142$
35. 17.8 **37.** $\dfrac{\sqrt{41}}{7} \approx 0.915$ **39.** $\sqrt{8} \approx 2.828$
41. $\sqrt{90} \approx 9.487$ **43.** (1, 4) **45.** $\left(\frac{7}{2}, \frac{7}{2}\right)$
47. $(-1, -3)$ **49.** $(-0.25, -0.3)$ **51.** $\left(-\frac{1}{12}, \frac{1}{24}\right)$
53. $\left(\dfrac{\sqrt{2} + \sqrt{3}}{2}, \dfrac{3}{2}\right)$ **55.** $x^2 + y^2 = 36$
57. $(x - 7)^2 + (y - 3)^2 = 5$
59. $(x + 4)^2 + (y - 3)^2 = 48$
61. $(x + 7)^2 + (y + 2)^2 = 50$ **63.** $x^2 + y^2 = 25$
65. $(x + 4)^2 + (y - 1)^2 = 20$
67. (0, 0); 8 **69.** $(-1, -3)$; 6

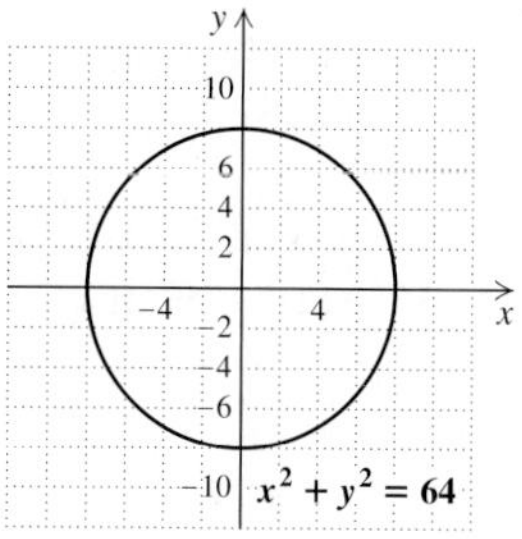

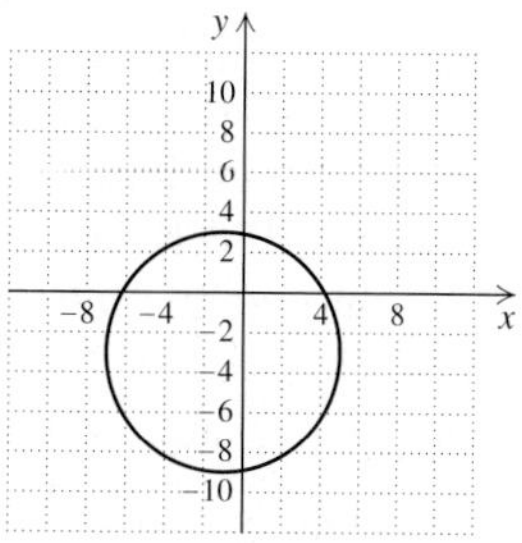

$(x + 1)^2 + (y + 3)^2 = 36$

71. $(4, -3)$; $\sqrt{10}$ **73.** (0, 0); $\sqrt{10}$

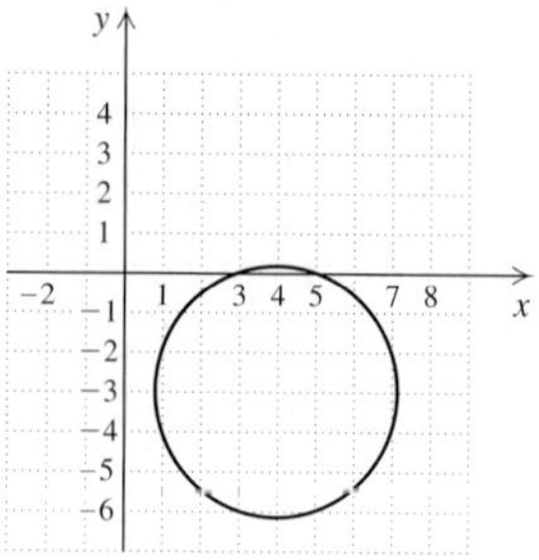

$(x - 4)^2 + (y + 3)^2 = 10$

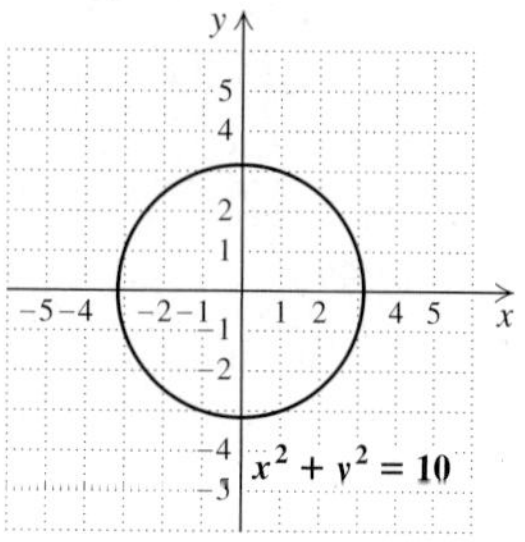

75. $(5, 0); \frac{1}{2}$

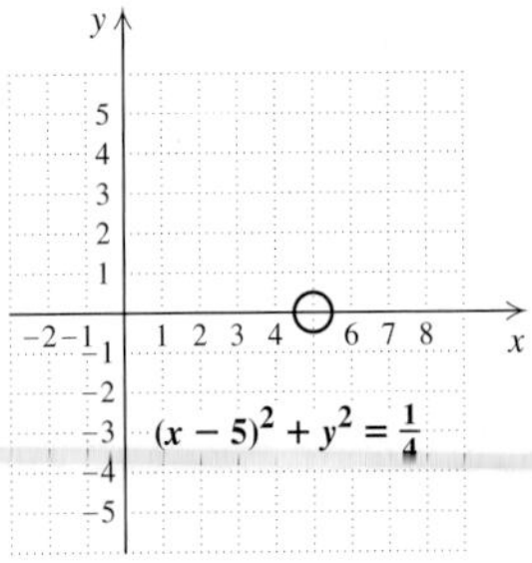

77. $(-4, 3); \sqrt{40}$, or $2\sqrt{10}$

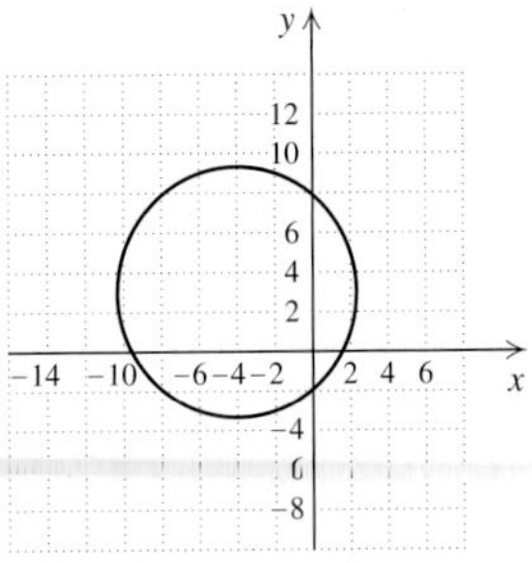

$x^2 + y^2 + 8x - 6y - 15 = 0$

79. $(4, -1); 2$

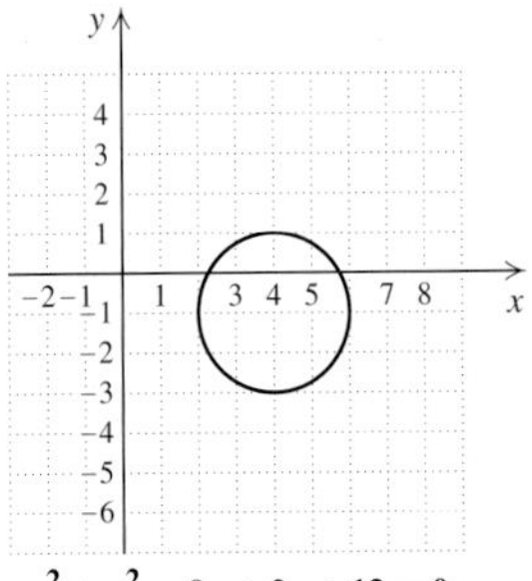

$x^2 + y^2 - 8x + 2y + 13 = 0$

81. $(0, -5); 10$

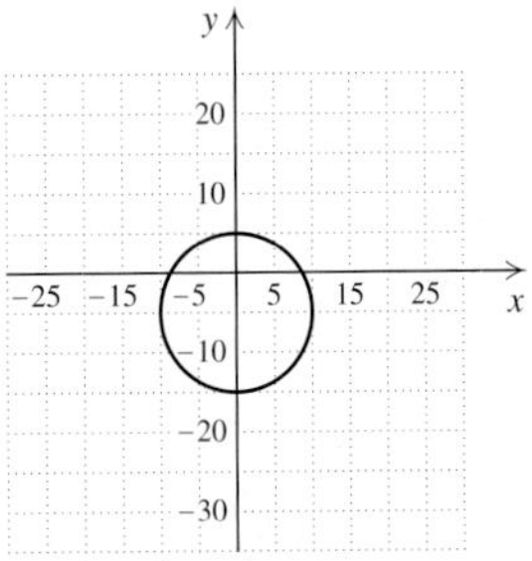

$x^2 + y^2 + 10y - 75 = 0$

83. $\left(-\frac{7}{2}, \frac{3}{2}\right); \sqrt{\frac{98}{4}}$, or $\frac{7\sqrt{2}}{2}$

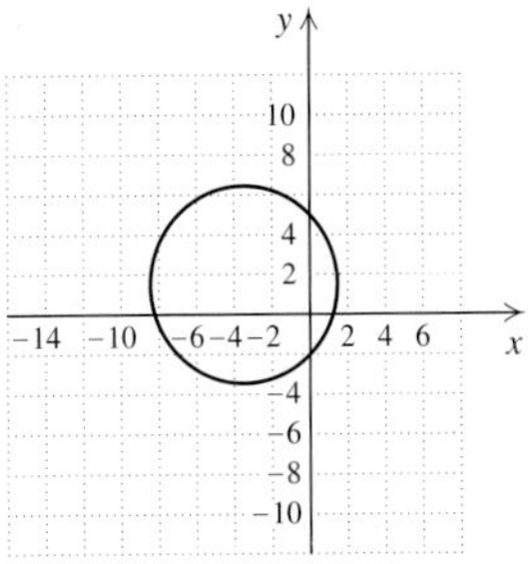

$x^2 + y^2 + 7x - 3y - 10 = 0$

85. $(0, 0); \frac{1}{6}$

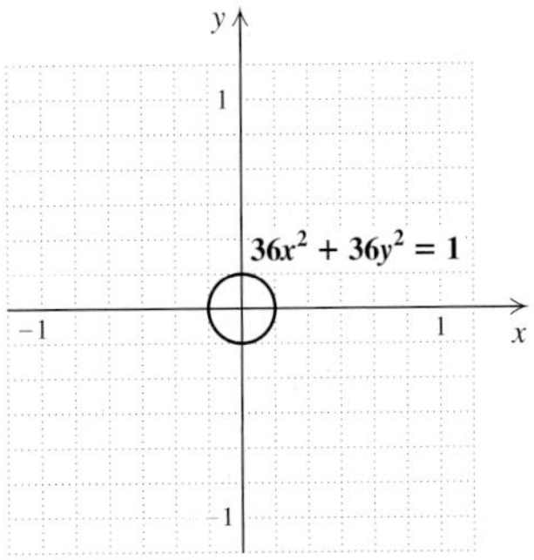

87. $x^2 + y^2 - 16 = 0$

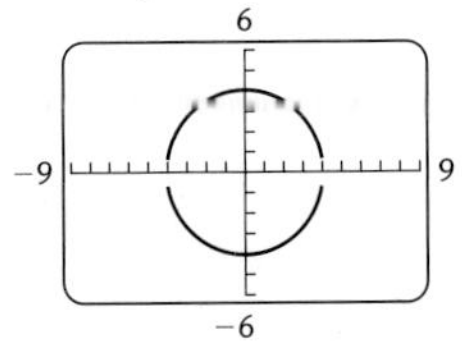

89. $x^2 + y^2 + 14x - 16y + 54 = 0$

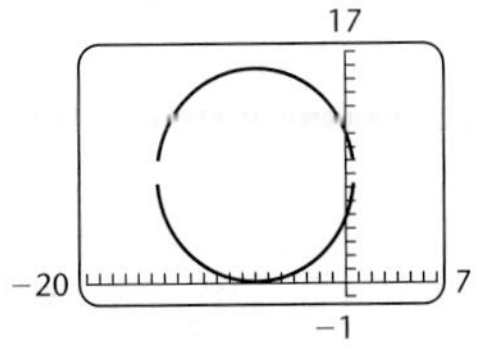

91. TW **93.** $-\frac{2}{3}$ **94.** $\frac{25}{6}$ **95.** 4 in. **96.** 2640 mi
97. $\left(\frac{35}{17}, \frac{5}{34}\right)$ **98.** $\left(0, -\frac{9}{5}\right)$ **99.** TW
101. $(x - 3)^2 + (y + 5)^2 = 9$ **103.** $(x - 3)^2 + y^2 = 25$
105. $(0, 4)$ **107.** $\frac{17}{4}\pi$ m^2, or approximately 13.4 m^2
109. 7700 mm **111. (a)** $(0, -3)$; **(b)** 5 ft
113. $x^2 + (y - 30.6)^2 = 590.49$

115.

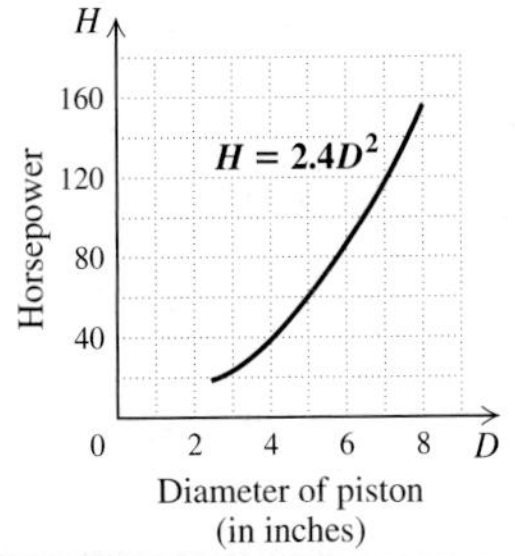

117. (a) $y = -1 \pm \sqrt{-x^2 + 6x + 7}$; **(b)**
119. TW

Exercise Set 10.2, pp. 808–811

1. True **2.** True **3.** True **4.** False **5.** False
6. False **7.** True **8.** True

9.

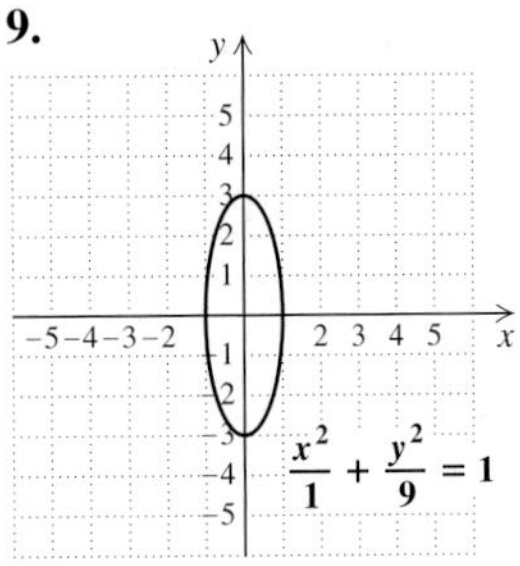

11.

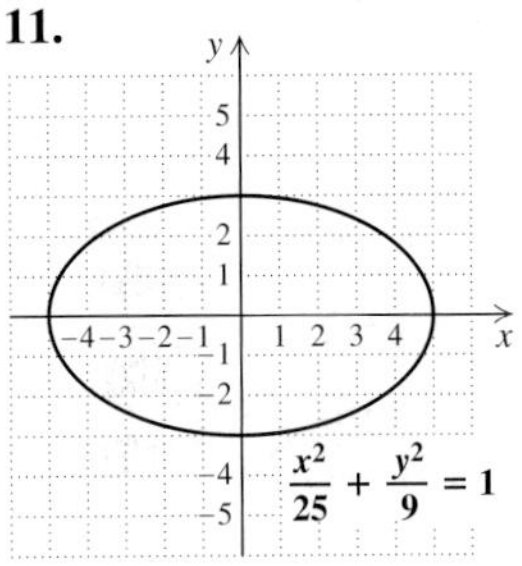

13.

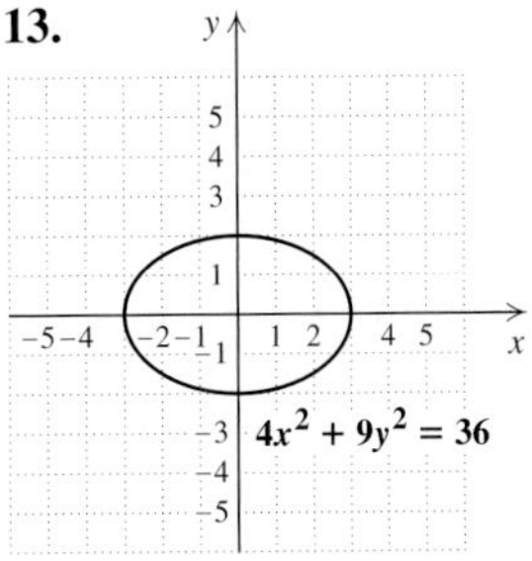

15.

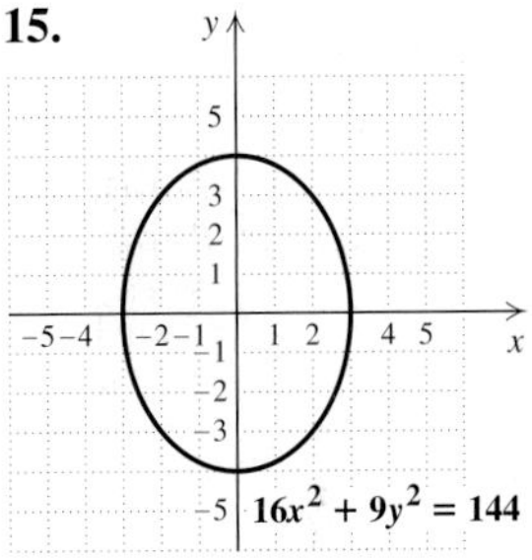

17.

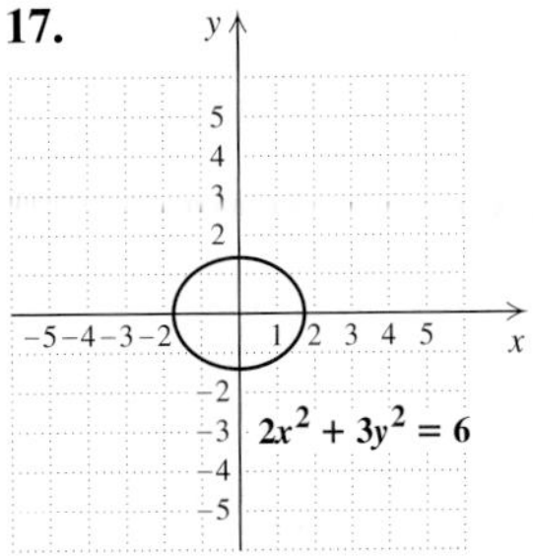

19.

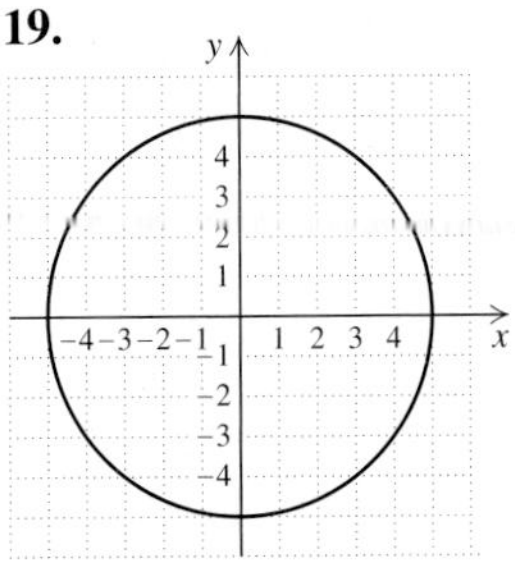

$5x^2 + 5y^2 = 125$

21.

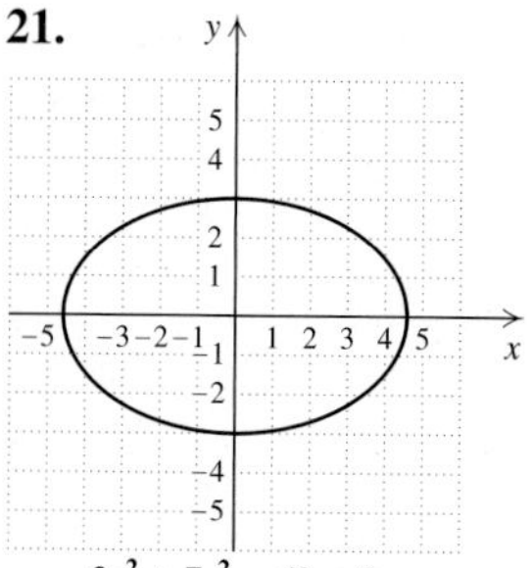

$3x^2 + 7y^2 - 63 = 0$

23.

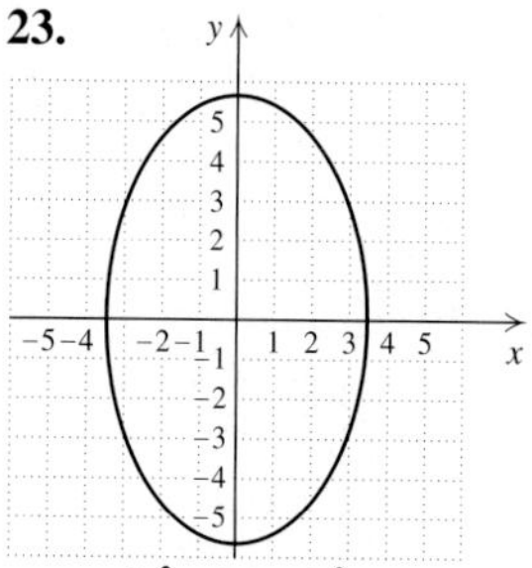

$8x^2 = 96 - 3y^2$

25.

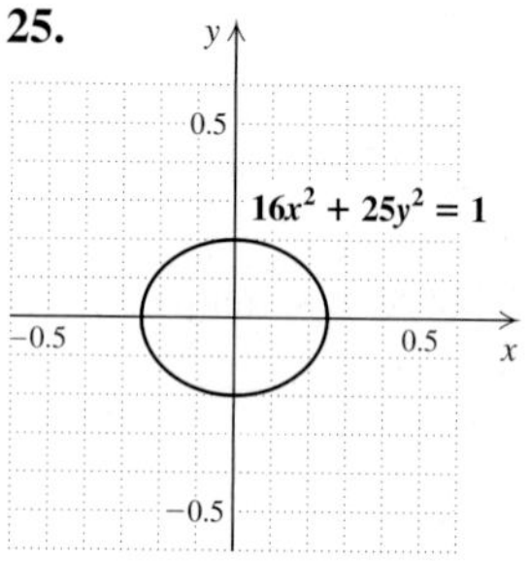

27.

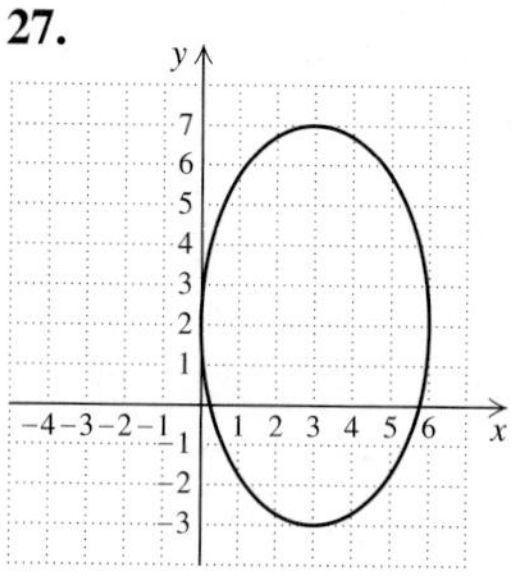

$\frac{(x-3)^2}{9} + \frac{(y-2)^2}{25} = 1$

29.

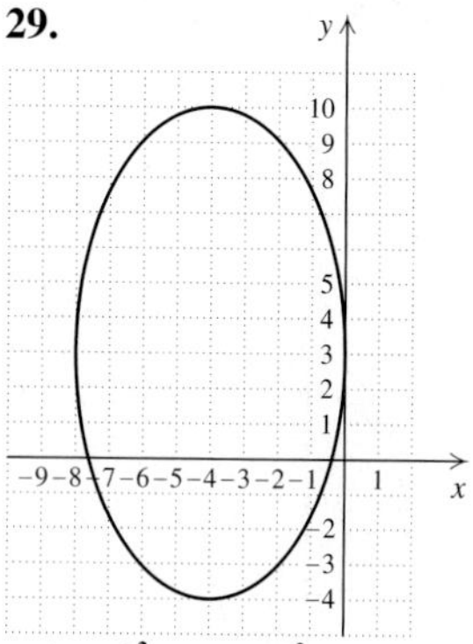

$\frac{(x+4)^2}{16} + \frac{(y-3)^2}{49} = 1$

31.

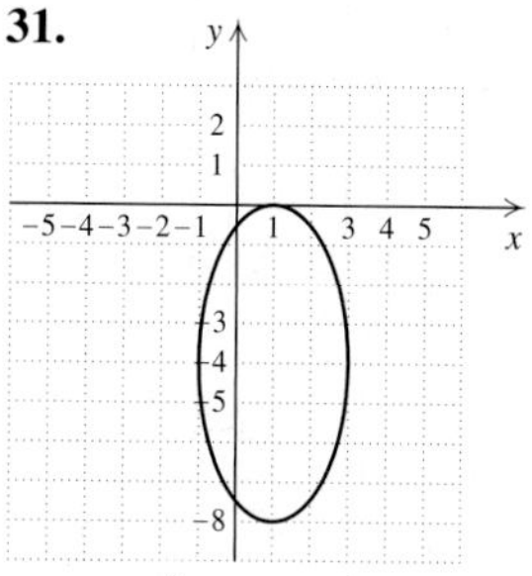

$12(x-1)^2 + 3(y+4)^2 = 48$

33.

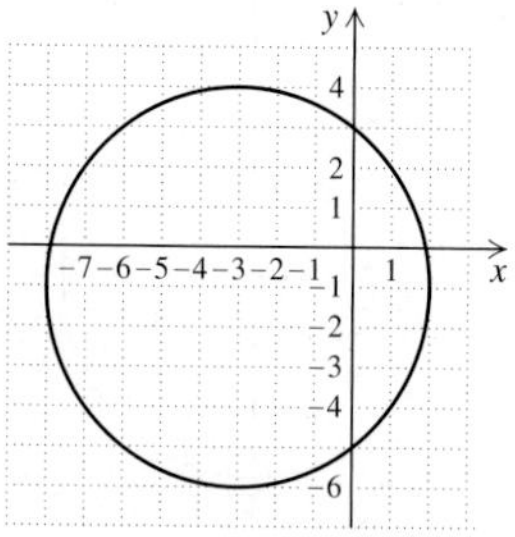

$4(x+3)^2 + 4(y+1)^2 - 10 = 90$

35. TW **37.** $\frac{16}{9}$ **38.** $-\frac{19}{8}$ **39.** $5 \pm \sqrt{3}$ **40.** $3 \pm \sqrt{7}$

41. $\frac{3}{2}$ **42.** No solution **43.** TW **45.** $\frac{x^2}{81} + \frac{y^2}{121} = 1$

47. $\frac{(x-2)^2}{16} + \frac{(y+1)^2}{9} = 1$ **49.** $\frac{x^2}{9} + \frac{y^2}{25} = 1$

51. (a) Let $F_1 = (-c, 0)$ and $F_2 = (c, 0)$. Then the sum of the distances from the foci to P is $2a$. By the distance formula,

$$\sqrt{(x+c)^2 + y^2} + \sqrt{(x-c)^2 + y^2} = 2a, \text{ or}$$
$$\sqrt{(x+c)^2 + y^2} = 2a - \sqrt{(x-c)^2 + y^2}.$$

Squaring, we get

$$(x+c)^2 + y^2 = 4a^2 - 4a\sqrt{(x-c)^2 + y^2} + (x-c)^2 + y^2,$$

or

$$x^2 + 2cx + c^2 + y^2 = 4a^2 - 4a\sqrt{(x-c)^2 + y^2} + x^2 - 2cx + c^2 + y^2.$$

Thus

$$-4a^2 + 4cx = -4a\sqrt{(x-c)^2 + y^2}$$
$$a^2 - cx = a\sqrt{(x-c)^2 + y^2}.$$

Squaring again, we get

$$a^4 - 2a^2cx + c^2x^2 = a^2(x^2 - 2cx + c^2 + y^2)$$
$$a^4 - 2a^2cx + c^2x^2 = a^2x^2 - 2a^2cx + a^2c^2 + a^2y^2,$$

or

$$x^2(a^2 - c^2) + a^2y^2 = a^2(a^2 - c^2)$$
$$\frac{x^2}{a^2} + \frac{y^2}{a^2 - c^2} = 1.$$

(b) When P is at $(0, b)$, it follows that $b^2 = a^2 - c^2$. Substituting, we have

$$\frac{x^2}{a^2} + \frac{y^2}{b^2} = 1.$$

53. 5.66 ft

55.

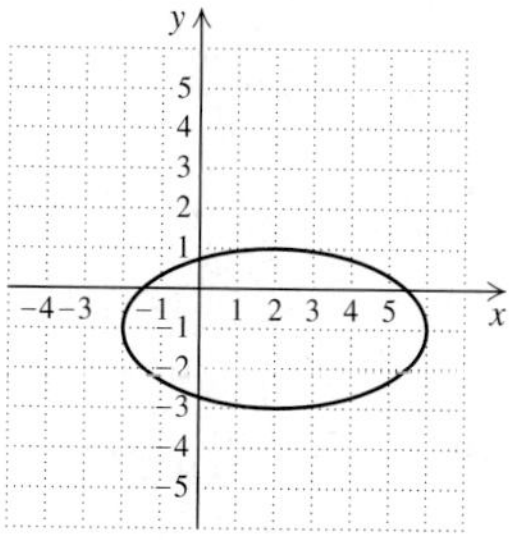

$\frac{(x-2)^2}{16} + \frac{(y+1)^2}{4} = 1$

57. 152,100,000 km

Visualizing the Graph, p. 821

1. C **2.** A **3.** F **4.** B **5.** J **6.** D **7.** H **8.** I **9.** G **10.** E

Exercise Set 10.3, pp. 822–823

1. (d) **2.** (f) **3.** (h) **4.** (a) **5.** (g) **6.** (b) **7.** (c) **8.** (e)

9.

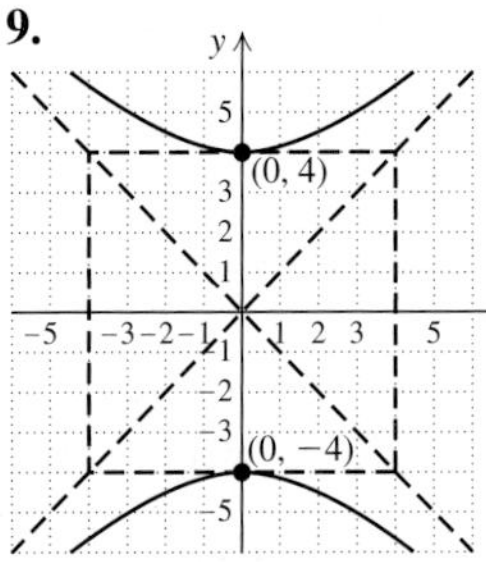

$\frac{y^2}{16} - \frac{x^2}{16} = 1$

11.

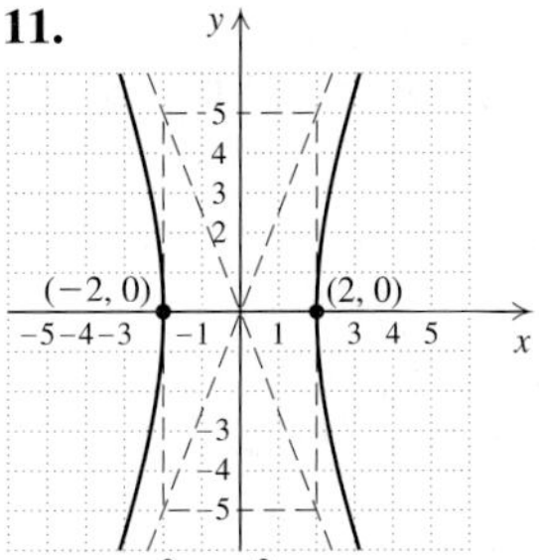

$\frac{x^2}{4} - \frac{y^2}{25} = 1$

13.

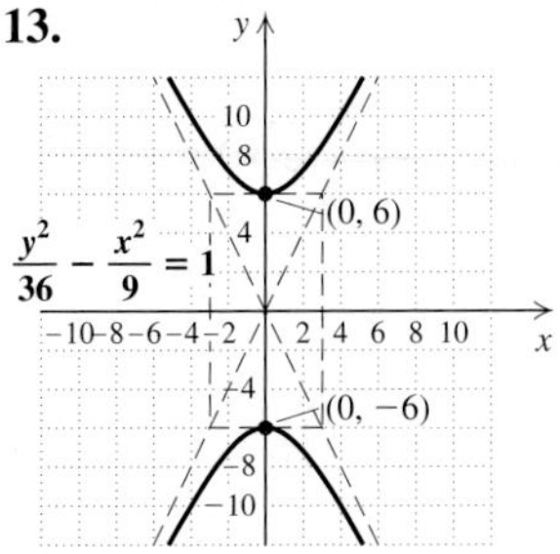

15.

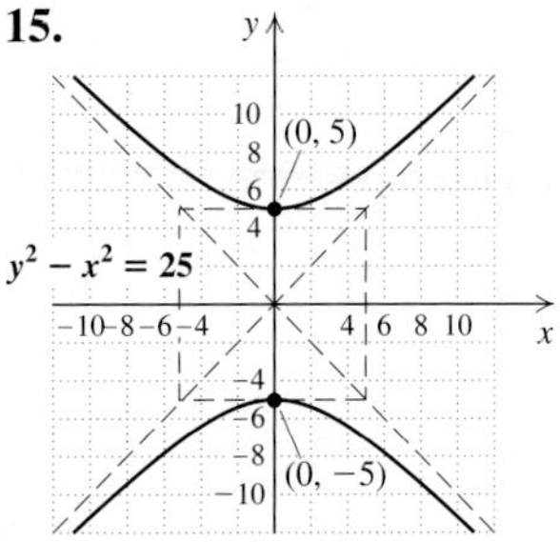

17.

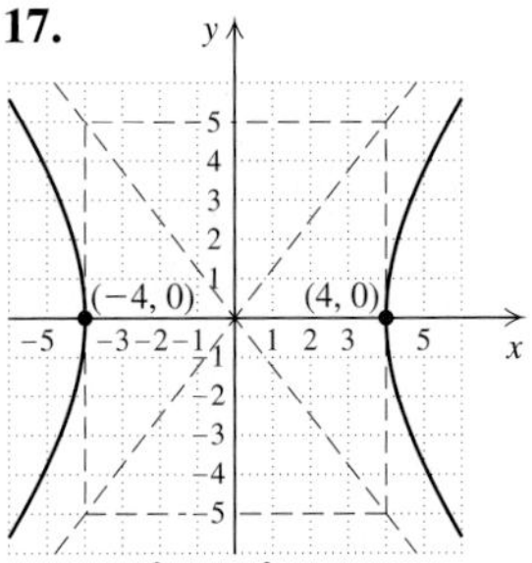

$25x^2 - 16y^2 = 400$

19.

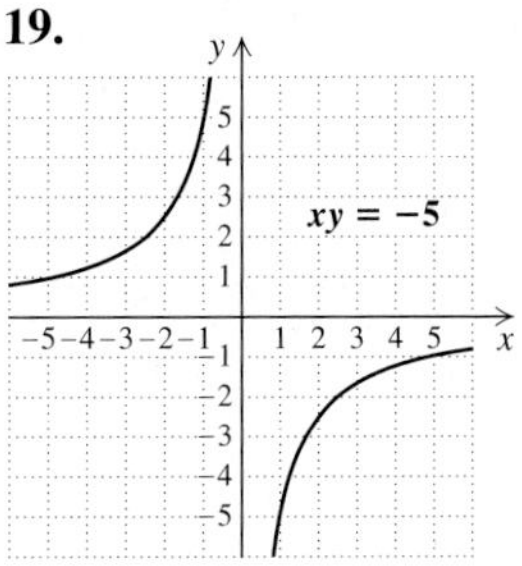

21.

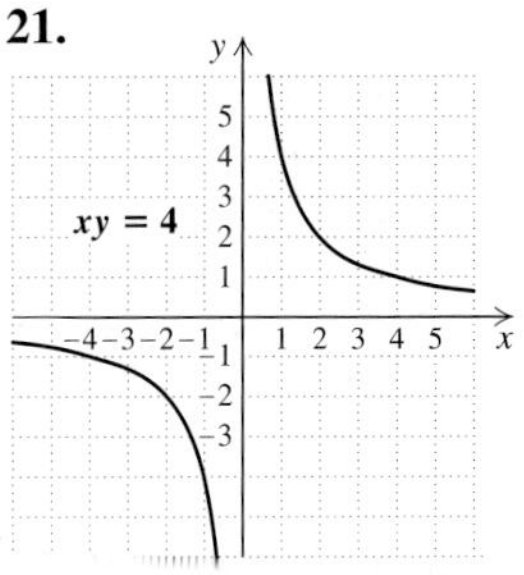

23.

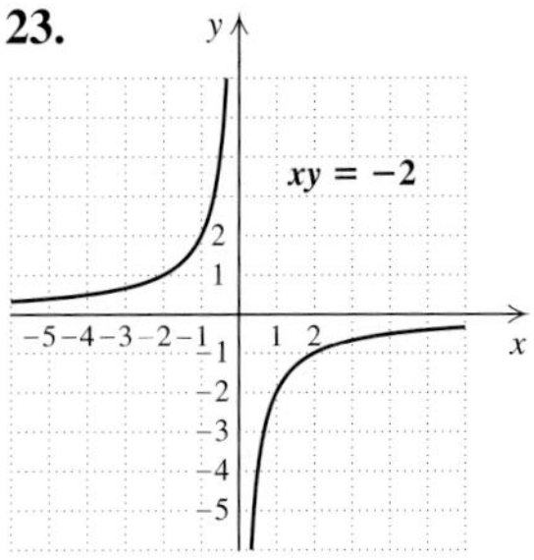

25.

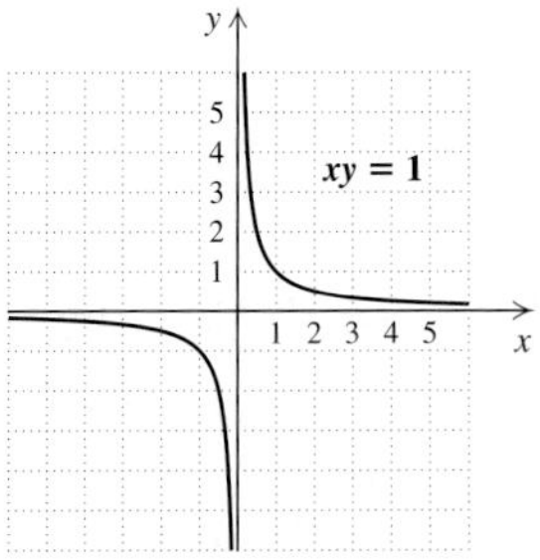

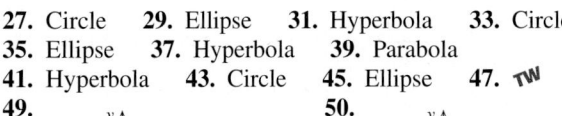

27. Circle **29.** Ellipse **31.** Hyperbola **33.** Circle **35.** Ellipse **37.** Hyperbola **39.** Parabola **41.** Hyperbola **43.** Circle **45.** Ellipse **47.** TW

49.

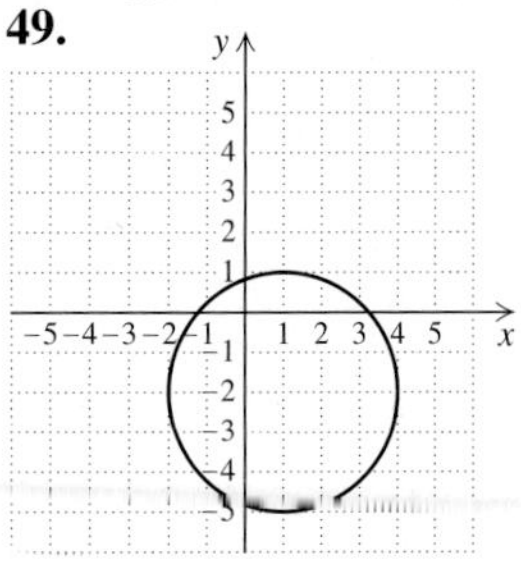

$(x - 1)^2 + (y + 2)^2 = 9$

50.

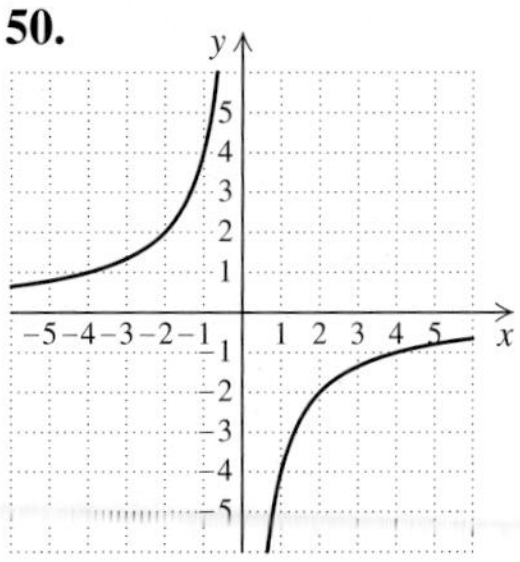

$xy = -4$

51.

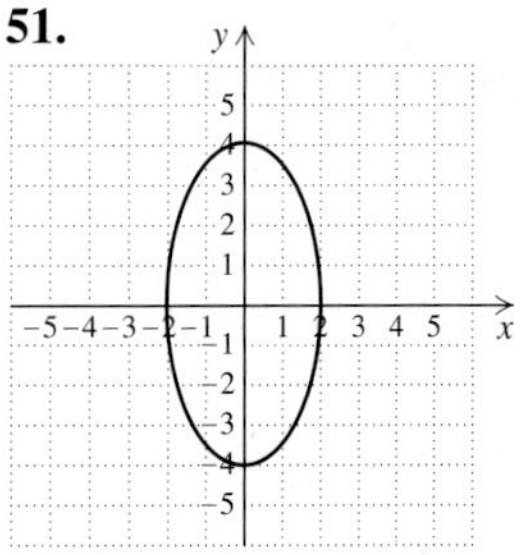

$4x^2 + y^2 = 16$

52.

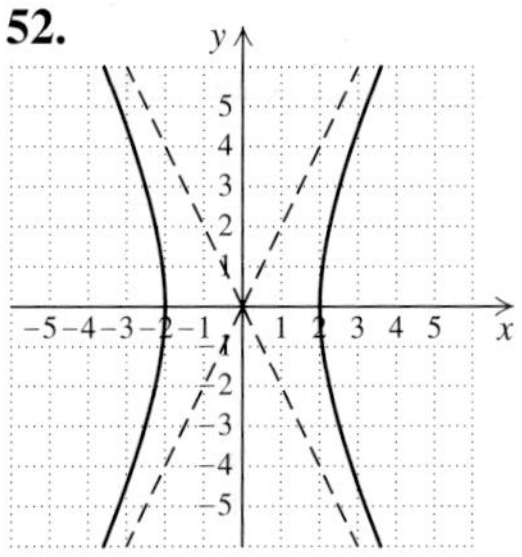

$4x^2 - y^2 = 16$

53.

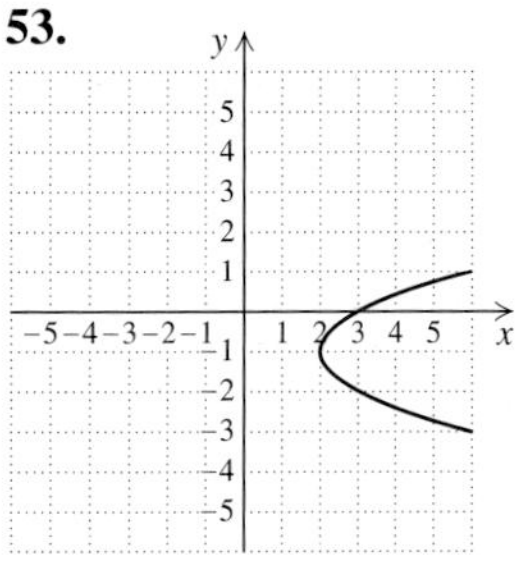

$x = y^2 + 2y + 3$

54.

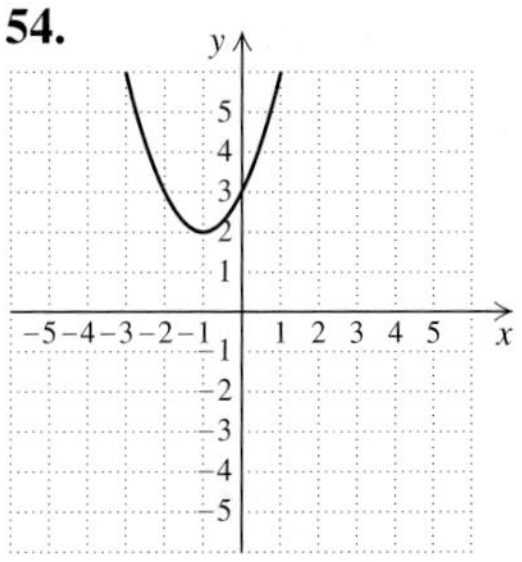

$y = x^2 + 2x + 3$

55. TW **57.** $\frac{y^2}{36} - \frac{x^2}{4} = 1$ **59.** C: (5, 2); V: (−1, 2), (11, 2); asymptotes: $y - 2 = \frac{5}{6}(x - 5)$, $y - 2 = -\frac{5}{6}(x - 5)$

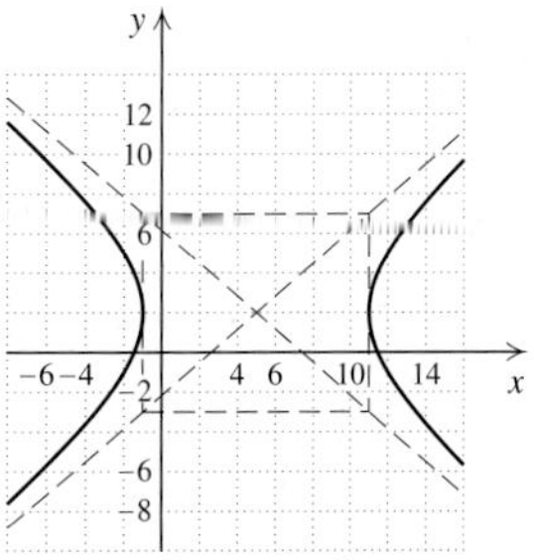

$\frac{(x - 5)^2}{36} - \frac{(y - 2)^2}{25} = 1$

61. $\frac{(y+3)^2}{4} - \frac{(x-4)^2}{16} = 1$; C: $(4, -3)$; V: $(4, -5)$, $(4, -1)$; asymptotes: $y + 3 = \frac{1}{2}(x - 4)$, $y + 3 = -\frac{1}{2}(x - 4)$

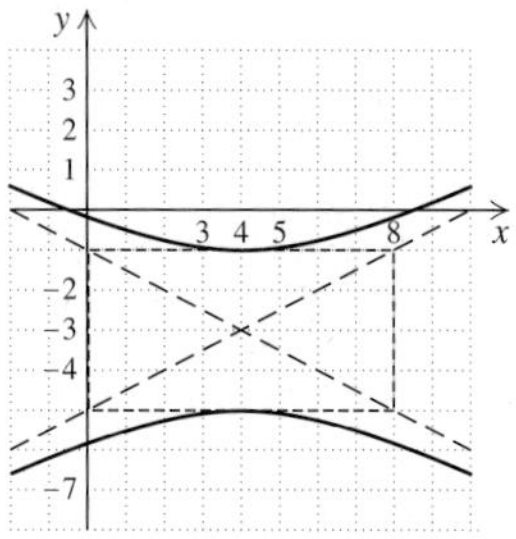

$8(y + 3)^2 - 2(x - 4)^2 = 32$

63. $\frac{(x+3)^2}{1} - \frac{(y-2)^2}{4} = 1$; C: $(-3, 2)$; V: $(-4, 2)$, $(-2, 2)$; asymptotes: $y - 2 = 2(x + 3)$, $y - 2 = -2(x + 3)$

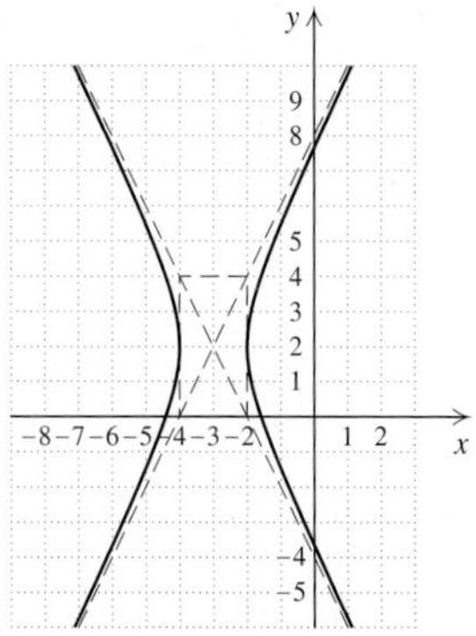

$4x^2 - y^2 + 24x + 4y + 28 = 0$

Exercise Set 10.4, pp. 830–833

1. True **2.** True **3.** False **4.** False **5.** True **6.** True **7.** $(-4, -3)$, $(3, 4)$ **9.** $(0, 2)$, $(3, 0)$ **11.** $(-2, 1)$ **13.** $\left(\frac{5+\sqrt{70}}{3}, \frac{-1+\sqrt{70}}{3}\right)$, $\left(\frac{5-\sqrt{70}}{3}, \frac{-1-\sqrt{70}}{3}\right)$ **15.** $\left(4, \frac{3}{2}\right)$, $(3, 2)$ **17.** $\left(\frac{7}{3}, \frac{1}{3}\right)$, $(1, -1)$ **19.** $\left(\frac{11}{4}, -\frac{5}{4}\right)$, $(1, 4)$ **21.** $\left(\frac{7-\sqrt{33}}{2}, \frac{7+\sqrt{33}}{2}\right)$, $\left(\frac{7+\sqrt{33}}{2}, \frac{7-\sqrt{33}}{2}\right)$ **23.** $(3, -5)$, $(-1, 3)$ **25.** $(-5, -8)$, $(8, 5)$ **27.** $(0, 0)$, $(1, 1)$, $\left(-\frac{1}{2} + \frac{\sqrt{3}}{2}i, -\frac{1}{2} - \frac{\sqrt{3}}{2}i\right)$, $\left(-\frac{1}{2} - \frac{\sqrt{3}}{2}i, -\frac{1}{2} + \frac{\sqrt{3}}{2}i\right)$ **29.** $(-3, 0)$, $(3, 0)$ **31.** $(-4, -3)$, $(-3, -4)$, $(3, 4)$, $(4, 3)$ **33.** $\left(\frac{16}{3}, \frac{5\sqrt{7}}{3}i\right)$, $\left(\frac{16}{3}, -\frac{5\sqrt{7}}{3}i\right)$, $\left(-\frac{16}{3}, \frac{5\sqrt{7}}{3}i\right)$, $\left(-\frac{16}{3}, -\frac{5\sqrt{7}}{3}i\right)$ **35.** $(-3, -\sqrt{5})$, $(-3, \sqrt{5})$, $(3, -\sqrt{5})$, $(3, \sqrt{5})$ **37.** $(4, 2)$, $(-4, -2)$, $(2, 4)$, $(-2, -4)$ **39.** $(4, 1)$, $(-4, -1)$, $(2, 2)$, $(-2, -2)$ **41.** $(2, 1)$, $(-2, -1)$ **43.** $\left(2, -\frac{4}{5}\right)$, $\left(-2, -\frac{4}{5}\right)$, $(5, 2)$, $(-5, 2)$ **45.** $(-\sqrt{2}, \sqrt{2})$, $(\sqrt{2}, -\sqrt{2})$ **47.** Length: 8 cm; width: 6 cm **49.** Length: 5 in.; width: 4 in. **51.** Length: 12 ft; width: 5 ft **53.** 6 and 10; -6 and -10 **55.** 24 ft, 16 ft **57.** 13 and 12 **59.** TW **61.** -16 **62.** -32 **63.** 1 **64.** $-\frac{1}{4}$ **65.** 44 **66.** 28 **67.** TW **69.** $(x + 2)^2 + (y - 1)^2 = 4$ **71.** $(-2, 3)$, $(2, -3)$, $(-3, 2)$, $(3, -2)$ **73.** Length: 55 ft; width: 45 ft **75.** 10 in. by 7 in. by 5 in. **77.** Length: 61.02 in.; height: 34.32 in. **79.** $(-1.50, -1.17)$; $(3.50, 0.50)$

Review Exercises: Chapter 10, pp. 836–837

1. True **2.** False **3.** False **4.** False **5.** True **6.** True **7.** False **8.** True **9.** 4 **10.** 5 **11.** $\sqrt{90.1} \approx 9.492$ **12.** $\sqrt{9 + 4a^2}$ **13.** $\left(\frac{9}{2}, -1\right)$ **14.** $(-3, 7)$ **15.** $\left(\frac{3}{4}, \frac{\sqrt{3}-\sqrt{2}}{2}\right)$ **16.** $\left(\frac{1}{2}, 2a\right)$ **17.** $(-3, 2)$, $\sqrt{7}$ **18.** $(5, 0)$, 7 **19.** $(3, 1)$, 3 **20.** $(-4, 3)$, $\sqrt{35}$ **21.** $(x + 4)^2 + (y - 3)^2 = 48$ **22.** $(x - 7)^2 + (y + 2)^2 = 20$

23. Circle

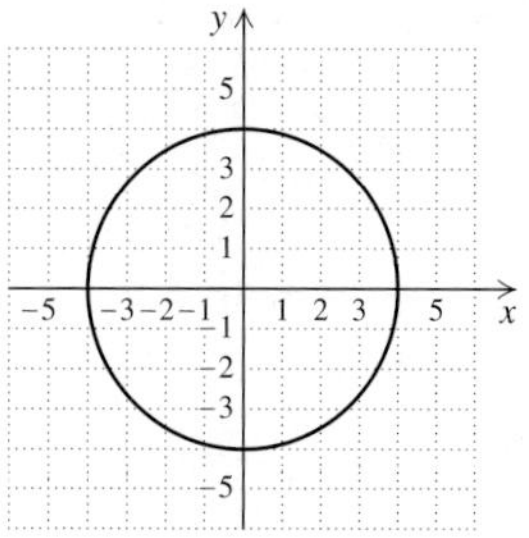

$5x^2 + 5y^2 = 80$

24. Ellipse

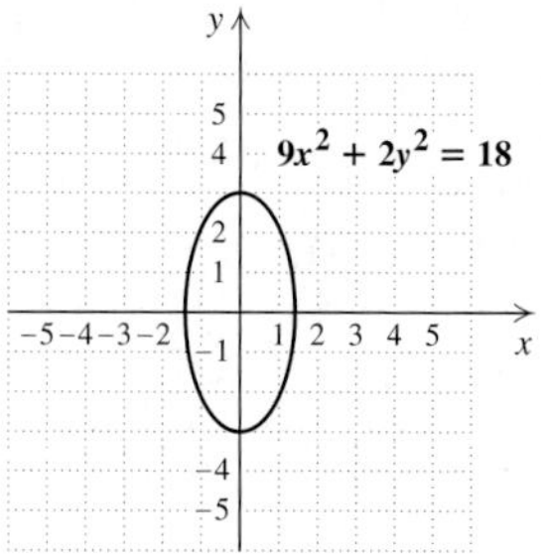

25. Parabola

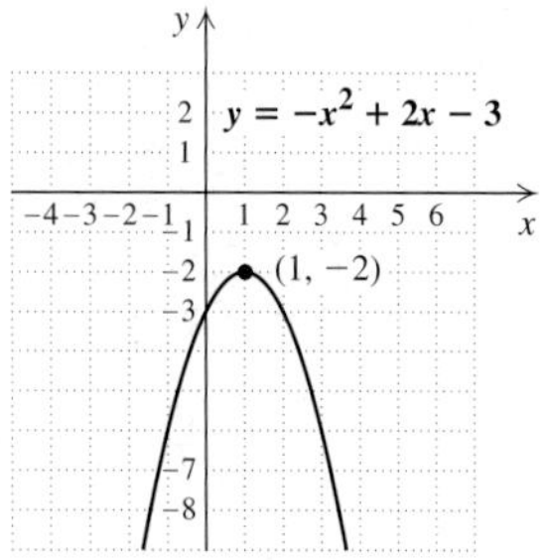

26. Hyperbola

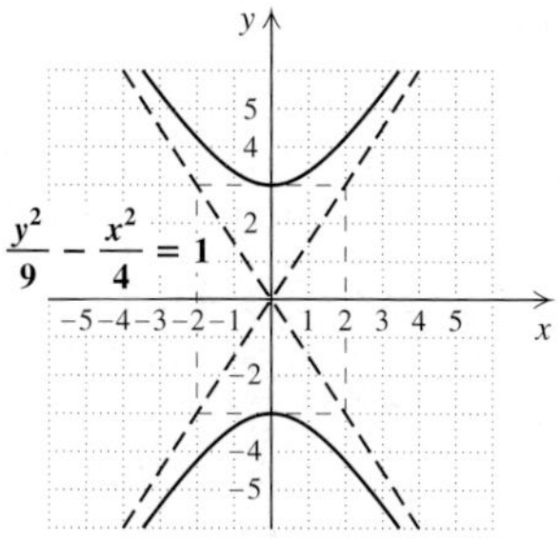

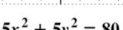

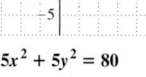

27. Hyperbola

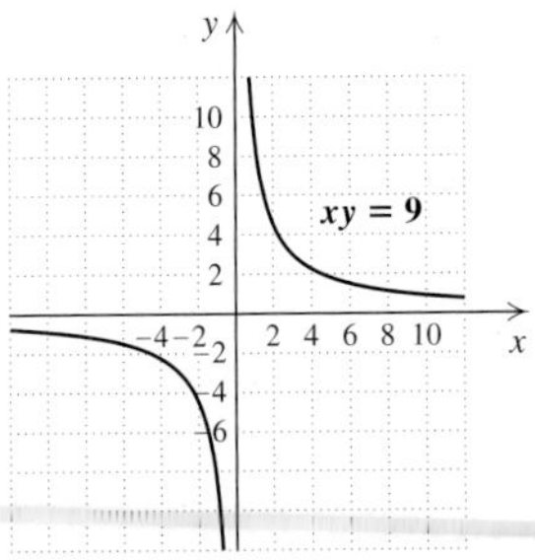

28. Parabola

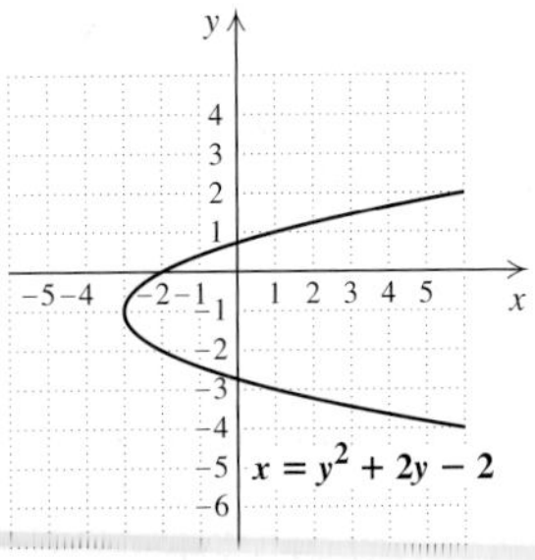

29. Ellipse

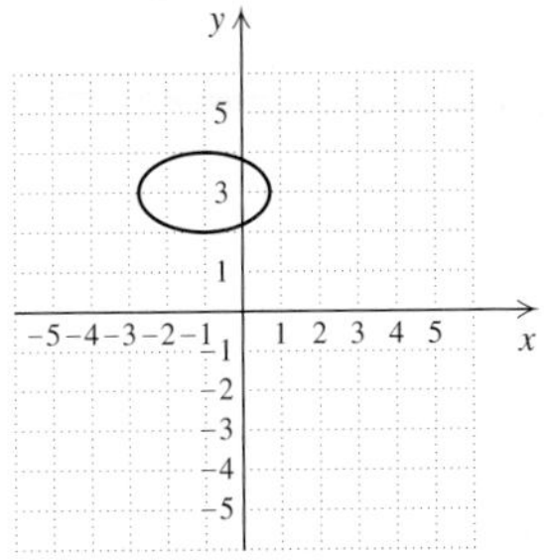

$\frac{(x + 1)^2}{3} + (y - 3)^2 = 1$

30. Circle

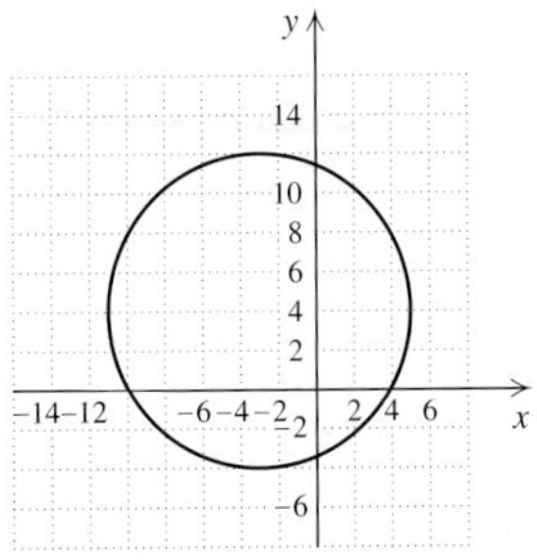

$x^2 + y^2 + 6x - 8y - 39 = 0$

31. (7, 4) **32.** (2, 2), $\left(\frac{32}{9}, -\frac{10}{9}\right)$ **33.** (0, −3), (2, 1)
34. (4, 3), (4, −3), (−4, 3), (−4, −3)
35. (2, 1), $(\sqrt{3}, 0)$, (−2, 1), $(-\sqrt{3}, 0)$
36. (3, −3), $\left(-\frac{3}{5}, \frac{21}{5}\right)$
37. (6, 8), (6, −8), (−6, 8), (−6, −8)
38. (2, 2), (−2, −2), $(2\sqrt{2}, \sqrt{2})$, $(-2\sqrt{2}, -\sqrt{2})$
39. Length: 12 m; width: 7 m **40.** Length: 12 in.; width: 9 in. **41.** 32 cm, 20 cm **42.** 3 ft, 11 ft
43. TW The graph of a parabola has one branch whereas the graph of a hyperbola has two branches. A hyperbola has asymptotes, but a parabola does not.
44. TW Function notation rarely appears in this chapter because many of the relations are not functions. Function notation could be used for vertical parabolas and for hyperbolas that have the axes as asymptotes.
45. $(-5, -4\sqrt{2})$, $(-5, 4\sqrt{2})$, $(3, -2\sqrt{2})$, $(3, 2\sqrt{2})$
46. (0, 6), (0, −6) **47.** $(x - 2)^2 + (y + 1)^2 = 25$
48. $\frac{x^2}{81} + \frac{y^2}{25} = 1$ **49.** $\left(\frac{9}{4}, 0\right)$

Test: Chapter 10, pp. 837–838

1. [10.1] $9\sqrt{2} \approx 12.728$ **2.** [10.1] $2\sqrt{9 + a^2}$
3. [10.1] $\left(-\frac{1}{2}, \frac{7}{2}\right)$ **4.** [10.1] (0, 0) **5.** [10.1] (−5, 1), 9
6. [10.1] (−2, 3), 3

7. [10.1], [10.3] Parabola

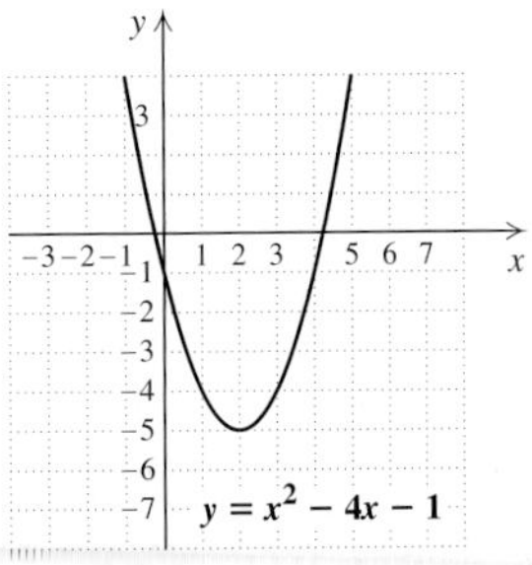

8. [10.1], [10.3] Circle

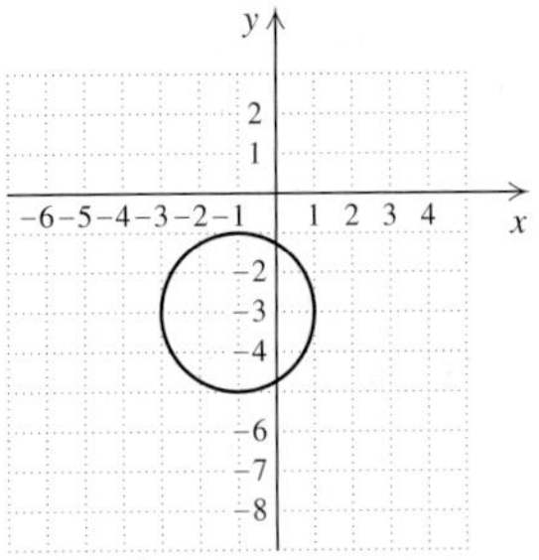

$x^2 + y^2 + 2x + 6y + 6 = 0$

9. [10.3] Hyperbola

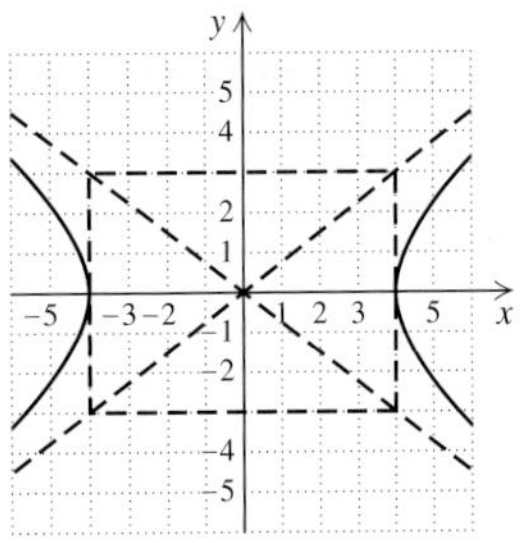

$\frac{x^2}{16} - \frac{y^2}{9} = 1$

10. [10.2], [10.3] Ellipse

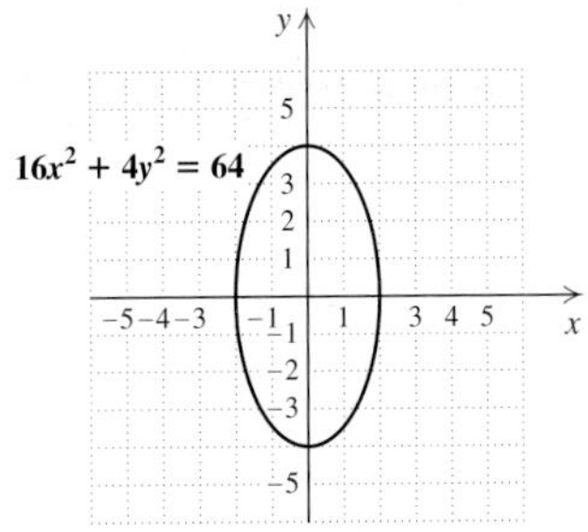

11. [10.3] Hyperbola

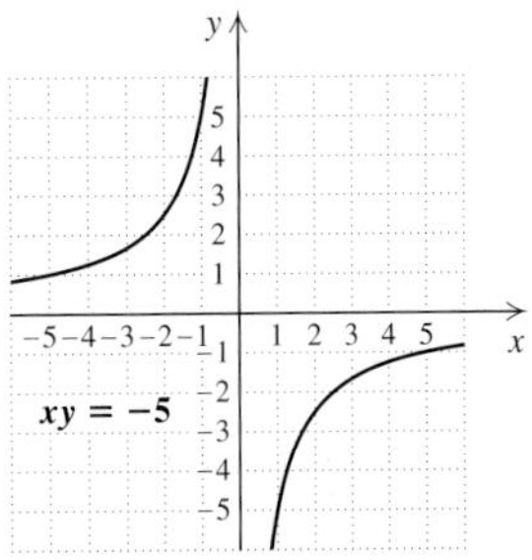

12. [10.1], [10.3] Parabola

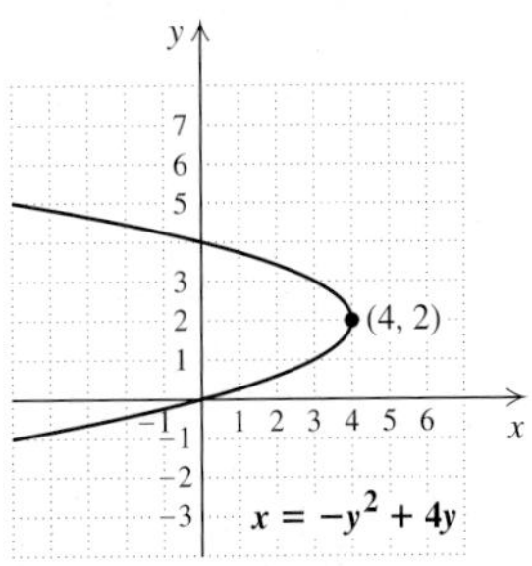

13. [10.4] (0, 3), $\left(\frac{8}{5}, \frac{9}{5}\right)$ **14.** [10.4] (4, 0), (−4, 0)
15. [10.4] (3, 2), (−3, −2), $\left(2\sqrt{2}i, -\frac{3\sqrt{2}}{2}i\right)$, $\left(-2\sqrt{2}i, \frac{3\sqrt{2}}{2}i\right)$ **16.** [10.4] $(\sqrt{6}, 2)$, $(\sqrt{6}, -2)$, $(-\sqrt{6}, 2)$, $(-\sqrt{6}, -2)$ **17.** [10.4] 2 by 11
18. [10.4] $\sqrt{5}$ m, $\sqrt{3}$ m **19.** [10.4] Length: 32 ft; width: 24 ft **20.** [10.4] \$1200, 6%
21. [10.2] $\frac{(x - 6)^2}{25} + \frac{(y - 3)^2}{9} = 1$ **22.** [10.1] $\left(0, -\frac{31}{4}\right)$
23. [10.4] 9 **24.** [10.2] $\frac{x^2}{16} + \frac{y^2}{49} = 1$

Chapter 11

Exercise Set 11.1, pp. 847–849

1. (f) **2.** (a) **3.** (d) **4.** (b) **5.** (c) **6.** (e)
7. 5, 7, 9, 11; 23; 33 **9.** 3, 6, 11, 18; 102; 227
11. $0, \frac{3}{5}, \frac{4}{5}, \frac{15}{17}; \frac{99}{101}; \frac{112}{113}$ **13.** $1, -\frac{1}{2}, \frac{1}{4}, -\frac{1}{8}; -\frac{1}{512}; \frac{1}{16{,}384}$
15. $-4, 5, -6, 7; 13; -18$ **17.** $0, 7, -26, 63; 999; -3374$
19. 13 **21.** 364 **23.** -23.5 **25.** -363 **27.** $\frac{441}{400}$
29. $2n$ **31.** $(-1)^{n+1}$ **33.** $(-1)^n \cdot n$ **35.** $2n + 1$
37. $(-1)^n \cdot 2 \cdot (3)^{n-1}$ **39.** $\frac{n}{n+1}$ **41.** 5^n **43.** $(-1)^n \cdot n^2$
45. 4 **47.** 30 **49.** $\frac{1}{2} + \frac{1}{4} + \frac{1}{6} + \frac{1}{8} + \frac{1}{10} = \frac{137}{120}$
51. $3^0 + 3^1 + 3^2 + 3^3 + 3^4 = 121$
53. $2 + \frac{3}{2} + \frac{4}{3} + \frac{5}{4} + \frac{6}{5} + \frac{7}{6} + \frac{8}{7} = \frac{1343}{140}$
55. $(-1)^2 2^1 + (-1)^3 2^2 + (-1)^4 2^3 + (-1)^5 2^4 + (-1)^6 2^5 + (-1)^7 2^6 + (-1)^8 2^7 + (-1)^9 2^8 = -170$
57. $(0^2 - 2 \cdot 0 + 3) + (1^2 - 2 \cdot 1 + 3) + (2^2 - 2 \cdot 2 + 3) + (3^2 - 2 \cdot 3 + 3) + (4^2 - 2 \cdot 4 + 3) + (5^2 - 2 \cdot 5 + 3) = 43$
59. $\frac{(-1)^3}{3 \cdot 4} + \frac{(-1)^4}{4 \cdot 5} + \frac{(-1)^5}{5 \cdot 6} = -\frac{1}{15}$ **61.** $\sum_{k=1}^{5} \frac{k+1}{k+2}$
63. $\sum_{k=1}^{6} k^2$ **65.** $\sum_{k=2}^{n} (-1)^k k^2$ **67.** $\sum_{k=1}^{\infty} 5k$
69. $\sum_{k=1}^{\infty} \frac{1}{k(k+1)}$ **71.** TW **73.** 77 **74.** 23
75. $x^3 + 3x^2y + 3xy^2 + y^3$ **76.** $a^3 - 3a^2b + 3ab^2 - b^3$
77. $8a^3 - 12a^2b + 6ab^2 - b^3$
78. $8x^3 + 12x^2y + 6xy^2 + y^3$ **79.** TW
81. 1, 3, 13, 63, 313, 1563 **83.** \$5200, \$3900, \$2925, \$2193.75, \$1645.31, \$1233.98, \$925.49, \$694.12, \$520.59, \$390.44 **85.** $S_{100} = 0; S_{101} = -1$ **87.** $i, -1, -i, 1, i; i$
89. 11th term

Interactive Discovery, p. 851

1. The points lie on a straight line with positive slope.
2. The points lie on a straight line with positive slope.
3. The points lie on a straight line with negative slope.
4. The points lie on a straight line with negative slope.

Exercise Set 11.2, pp. 856–859

1. True **2.** True **3.** False **4.** False **5.** True
6. True **7.** False **8.** False **9.** $a_1 = 2, d = 4$
11. $a_1 = 7, d = -4$ **13.** $a_1 = \frac{3}{2}, d = \frac{3}{4}$
15. $a_1 = \$5.12, d = \$0.12$ **17.** 49 **19.** -94
21. $-\$1628.16$ **23.** 26th **25.** 57th **27.** 82 **29.** 5
31. 28 **33.** $a_1 = 8; d = -3; 8, 5, 2, -1, -4$
35. $a_1 = 1; d = 1$ **37.** 780 **39.** 31,375 **41.** 2550
43. 918 **45.** 1030 **47.** 35 marchers; 315 marchers
49. 180 stones **51.** \$49.60 **53.** 722 seats **55.** TW
57. $\frac{13}{30x}$ **58.** $\frac{23}{36t}$ **59.** $a^k = P$ **60.** $e^a = t$
61. $x^2 + y^2 = 81$ **62.** $(x + 2)^2 + (y - 5)^2 = 18$
63. TW **65.** 33 jumps **67.** \$8760, \$7961.77, \$7163.54, \$6365.31, \$5567.08; \$4768.85, \$3970.62, \$3172.39, \$2374.16, \$1575.93 **69.** Let d = the common difference. Since p, m, and q form an arithmetic sequence, $m = p + d$ and $q = p + 2d$. Then $\frac{p+q}{2} = \frac{p + (p + 2d)}{2} = p + d = m$.
71. 156,375 **73.** Arithmetic; $a_n = -0.75n + 150.75$, where $n = 1$ corresponds to age 20, $n = 2$ to age 21, and so on **75.** Not arithmetic

Interactive Discovery, p. 861

1. The points lie on an exponential curve with $a > 1$.
2. The points lie on an exponential curve with $a > 1$.
3. The points lie on an exponential curve with $0 < a < 1$.
4. The points lie on an exponential curve with $0 < a < 1$.

Interactive Discovery, p. 864

1. **(a)** $\frac{1}{3}$; **(b)** 1, 1.33333, 1.44444, 1.48148, 1.49383, 1.49794, 1.49931, 1.49977, 1.49992, 1.49997 (sums are rounded to 5 decimal places); **(c)** 1.5
2. **(a)** $-\frac{7}{2}$; **(b)** 1, −2.5, 9.75, −33.125, 116.9375, −408.28125, 1429.984375, −5003.945313, 17514.80859, −61300.83008; **(c)** does not exist
3. **(a)** 2; **(b)** 4, 12, 28, 60, 124, 252, 508, 1020, 2044, 4092; **(c)** does not exist
4. **(a)** −0.4; **(b)** 1.3, 0.78, 0.988, 0.9048, 0.93808, 0.92477, 0.93009, 0.92796, 0.92881, 0.92847 (sums are rounded to 5 decimal places); **(c)** 0.928 **5.** S_∞ does not exist if $|r| > 1$.

Visualizing the Graph, p. 868

1. J **2.** G **3.** A **4.** H **5.** I **6.** B **7.** E
8. D **9.** F **10.** C

Exercise Set 11.3, pp. 869–872

1. Geometric sequence **2.** Arithmetic sequence
3. Arithmetic sequence **4.** Geometric sequence
5. Geometric series **6.** Arithmetic series
7. Geometric series **8.** None of these **9.** 2
11. -0.1 **13.** $-\frac{1}{2}$ **15.** $\frac{1}{5}$ **17.** $\frac{6}{m}$ **19.** 192
21. $112\sqrt{2}$ **23.** 52,488 **25.** \$2331.64 **27.** $a_n = 5^{n-1}$
29. $a_n = (-1)^{n-1}$, or $a_n = (-1)^{n+1}$

31. $a_n = \frac{1}{x^n}$, or $a_n = x^{-n}$ **33.** 3066 **35.** $\frac{547}{18}$ **37.** $\frac{1 - x^8}{1 - x}$, or $(1 + x)(1 + x^2)(1 + x^4)$ **39.** \$5134.51 **41.** $\frac{64}{3}$
43. $\frac{49}{4}$ **45.** No **47.** No **49.** $\frac{43}{99}$ **51.** \$25,000
53. $\frac{7}{9}$ **55.** $\frac{830}{99}$ **57.** $\frac{5}{33}$ **59.** $\frac{5}{1024}$ ft **61.** 155,797
63. 2710 flies **65.** 10,723,491 apartments and houses
67. 3100.35 ft **69.** 20.48 in. **71.** Arithmetic
73. Geometric **75.** Geometric **77.** TW
79. $a_1 = 2, d = 2$ **80.** $a_1 = 2, r = 2$ **81.** -19
82. $-\frac{1}{16}$ **83.** 210 **84.** 3,392,405 **85.** TW **87.** 54
89. $\frac{x^2[1 - (-x)^n]}{1 + x}$ **91.** 512 cm^2

Exercise Set 11.4, pp. 880–881

1. 2^5, or 32 **2.** 8 **3.** 9 **4.** 4! **5.** $\binom{8}{5}$ **6.** $\binom{10}{2}$, or 45 **7.** x^7y^2 **8.** 10 choose 4 **9.** 362,880
11. 39,916,800 **13.** 56 **15.** 3024 **17.** 35 **19.** 126
21. 4060 **23.** 780 **25.** $a^4 - 4a^3b + 6a^2b^2 - 4ab^3 + b^4$
27. $p^7 + 7p^6q + 21p^5q^2 + 35p^4q^3 + 35p^3q^4 + 21p^2q^5 + 7pq^6 + q^7$
29. $2187c^7 - 5103c^6d + 5103c^5d^2 - 2835c^4d^3 + 945c^3d^4 - 189c^2d^5 + 21cd^6 - d^7$
31. $t^{-12} + 12t^{-10} + 60t^{-8} + 160t^{-6} + 240t^{-4} + 192t^{-2} + 64$
33. $x^5 - 5x^4y + 10x^3y^2 - 10x^2y^3 + 5xy^4 - y^5$
35. $19{,}683s^9 + \frac{59{,}049s^8}{t} + \frac{78{,}732s^7}{t^2} + \frac{61{,}236s^6}{t^3} + \frac{30{,}618s^5}{t^4} + \frac{10{,}206s^4}{t^5} + \frac{2268s^3}{t^6} + \frac{324s^2}{t^7} + \frac{27s}{t^8} + \frac{1}{t^9}$
37. $x^{15} - 10x^{12}y + 40x^9y^2 - 80x^6y^3 + 80x^3y^4 - 32y^5$
39. $125 + 150\sqrt{5}t + 375t^2 + 100\sqrt{5}t^3 + 75t^4 + 6\sqrt{5}t^5 + t^6$
41. $x^{-3} - 6x^{-2} + 15x^{-1} - 20 + 15x - 6x^2 + x^3$
43. $15a^4b^2$ **45.** $-64{,}481{,}508a^3$ **47.** $1120x^{12}y^2$
49. $1{,}959{,}552u^5v^{10}$ **51.** y^8 **53.** TW **55.** 4 **56.** $\frac{5}{2}$
57. 5.6348 **58.** ± 5 **59.** TW
61. List all the subsets of size 3: $\{a, b, c\}, \{a, b, d\}, \{a, b, e\}, \{a, c, d\}, \{a, c, e\}, \{a, d, e\}, \{b, c, d\}, \{b, c, e\}, \{b, d, e\}, \{c, d, e\}$.
There are exactly 10 subsets of size 3 and $\binom{5}{3} = 10$, so there are exactly $\binom{5}{3}$ ways of forming a subset of size 3 from $\{a, b, c, d, e\}$.
63. $\binom{8}{5}(0.15)^3(0.85)^5 \approx 0.084$
65. $\binom{8}{6}(0.15)^2(0.85)^6 + \binom{8}{7}(0.15)(0.85)^7 + \binom{8}{8}(0.85)^8 \approx 0.89$
67. $\binom{n}{n-r} = \frac{n!}{[n-(n-r)]!\,(n-r)!} = \frac{n!}{r!\,(n-r)!} = \binom{n}{r}$
69. $\frac{-\sqrt[3]{q}}{2p}$ **71.** $x^7 + 7x^6y + 21x^5y^2 + 35x^4y^3 + 35x^3y^4 + 21x^2y^5 + 7xy^6 + y^7$

Review Exercises: Chapter 11, pp. 884–885

1. False **2.** True **3.** True **4.** False **5.** True
6. True **7.** False **8.** False **9.** 1, 5, 9, 13; 29; 45
10. $0, \frac{1}{5}, \frac{1}{5}, \frac{3}{17}; \frac{7}{65}; \frac{11}{145}$ **11.** $a_n = 7n$
12. $a_n = (-1)^n(2n - 1)$
13. $-2 + 4 + (-8) + 16 + (-32) = -22$
14. $-3 + (-5) + (-7) + (-9) + (-11) + (-13) = -48$
15. $\sum_{k=1}^{5} 4k$ **16.** $\sum_{k=1}^{5} \frac{1}{(-2)^k}$ **17.** 85 **18.** $\frac{8}{3}$
19. $a_1 = \frac{45}{4}, d = \frac{5}{4}$ **20.** -544 **21.** 8580 **22.** $1024\sqrt{2}$
23. $\frac{2}{3}$ **24.** $a_n = 2(-1)^n$ **25.** $a_n = 3\left(\frac{x}{4}\right)^{n-1}$ **26.** 4095
27. $-4095x$ **28.** 12 **29.** $\frac{49}{11}$ **30.** No **31.** No
32. \$40,000 **33.** $\frac{5}{9}$ **34.** $\frac{46}{33}$ **35.** \$24.30 **36.** 903 poles
37. \$15,791.18 **38.** 6 m **39.** 5040 **40.** 56
41. $190a^{18}b^2$ **42.** $x^4 - 8x^3y + 24x^2y^2 - 32xy^3 + 16y^4$
43. TW For a geometric sequence with $|r| < 1$, as n gets larger, the absolute value of the terms gets smaller, since $|r^n|$ gets smaller.
44. TW The first form of the binomial theorem draws the coefficients from Pascal's triangle; the second form uses factorial notation. The second form avoids the need to compute all preceding rows of Pascal's triangle, and is generally easier to use when only one term of an expansion is needed. When several terms of an expansion are needed and n is not large (say, $n \leq 8$), it is often easier to use Pascal's triangle.
45. $\frac{1 - (-x)^n}{x + 1}$
46. $x^{-15} + 5x^{-9} + 10x^{-3} + 10x^3 + 5x^9 + x^{15}$

Test: Chapter 11, pp. 885–886

1. [11.1] 1, 7, 13, 19, 25; 67 **2.** [11.1] $a_n = 4(\frac{1}{3})^n$
3. [11.1] $-1 + (-5) + (-13) + (-29) + (-61) = -109$
4. [11.1] $\sum_{k=1}^{5}(-1)^{k+1}k^3$ **5.** [11.2] -51 **6.** [11.2] $\frac{3}{8}$
7. [11.2] $a_1 = 31.2; d = -3.8$ **8.** [11.2] 2508
9. [11.3] $\frac{9}{128}$ **10.** [11.3] $\frac{2}{3}$ **11.** [11.3] 3^n
12. [11.3] $511 + 511x$ **13.** [11.3] 1 **14.** [11.3] No
15. [11.3] $\frac{\$25{,}000}{23} \approx \$1086.96$ **16.** [11.3] $\frac{85}{99}$
17. [11.2] 63 seats **18.** [11.2] \$17,100
19. [11.3] \$8981.05 **20.** [11.3] 36 m **21.** [11.4] 220

22. [11.4] $x^{10} - 15x^8y + 90x^6y^2 - 270x^4y^3 + 405x^2y^4 - 243y^5$
23. [11.4] $220a^9x^3$ **24.** [11.2] $n(n + 1)$

25. [11.3] $\dfrac{1 - \left(\dfrac{1}{x}\right)^n}{1 - \dfrac{1}{x}}$, or $\dfrac{x^n - 1}{x^{n-1}(x - 1)}$

Cumulative Review: Chapters 1–11, pp. 886–890

1. $1\frac{1}{4}$ in. **2.** $8\frac{2}{5}$ hr, or 8 hr 24 min **3.** 69
4. More than 4 purchases **5.** $11\frac{3}{7}$ **6.** \$2.68 herb: 10 oz; \$4.60 herb: 14 oz **7.** 350 mph **8.** 20 **9.** 5000 ft^2
10. 5 ft by 12 ft **11.** \$652.39 **12.** **(a)** $f(t) = 0.48t + 21.3$, where $f(t)$ is in millions; **(b)** 27.06 million households
13. **(a)** $k \approx 0.242$; $P(t) = 0.12e^{0.242t}$, where $P(t)$ is in millions; **(b)** about 50.9 million computers **14.** $y = 3x - 8$

15.
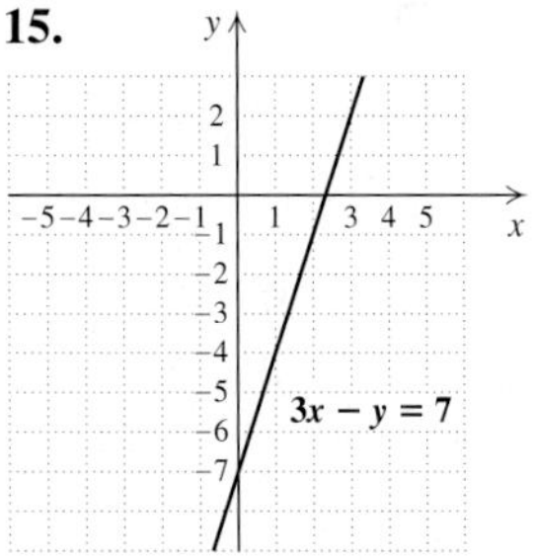

16.
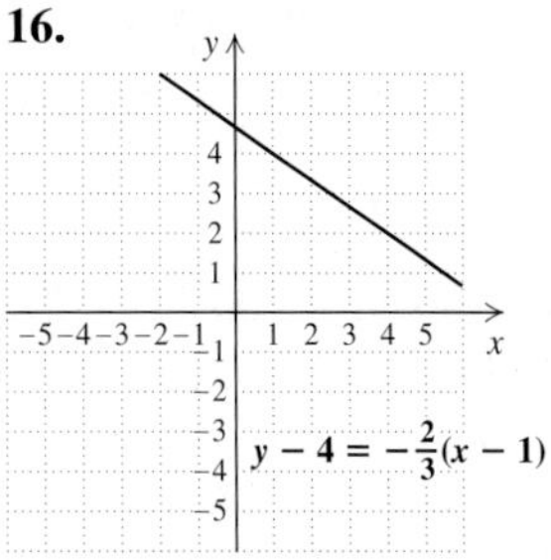

17.
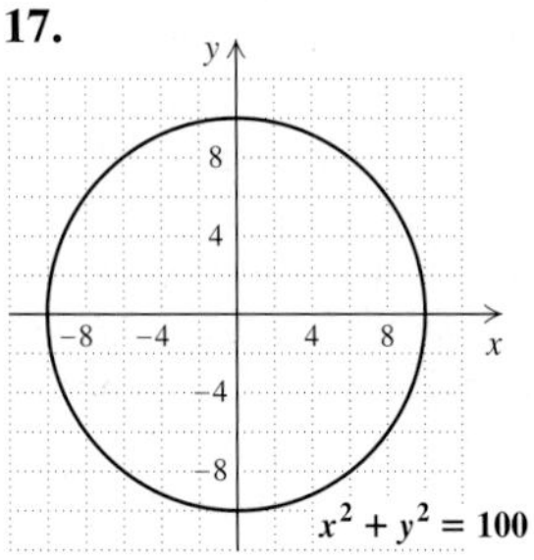

18.
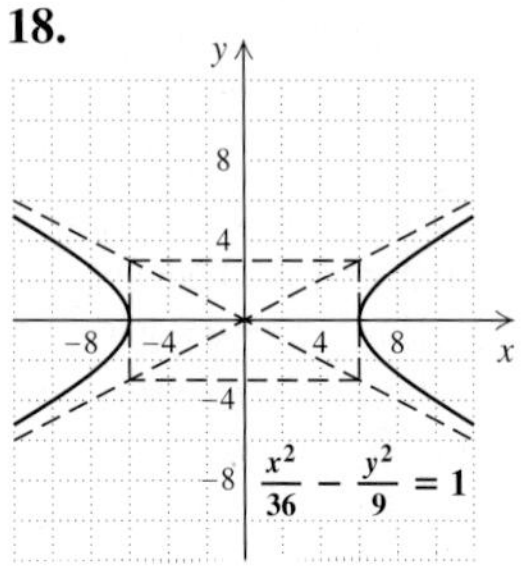

19.
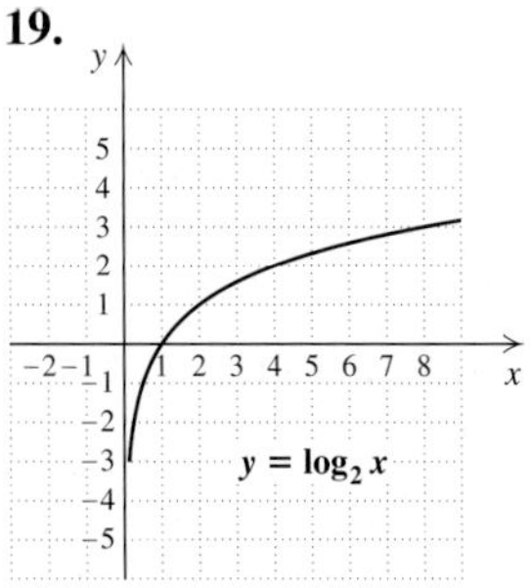

20.
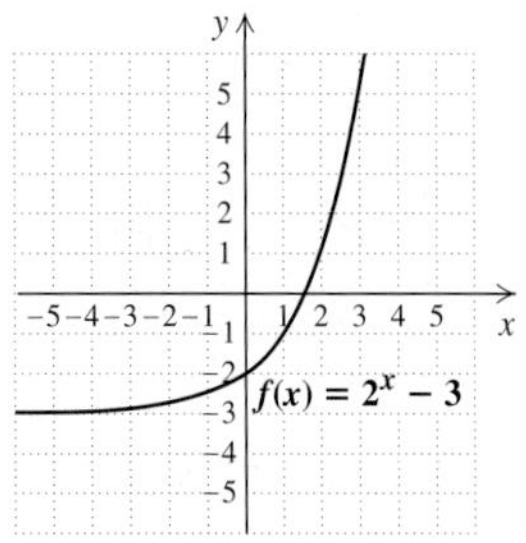

21.
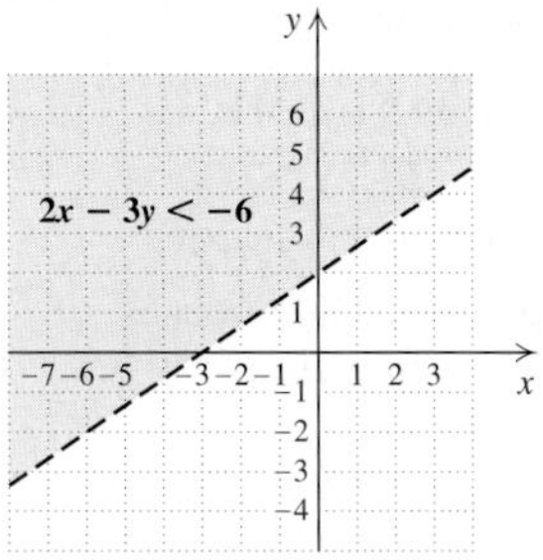

22.
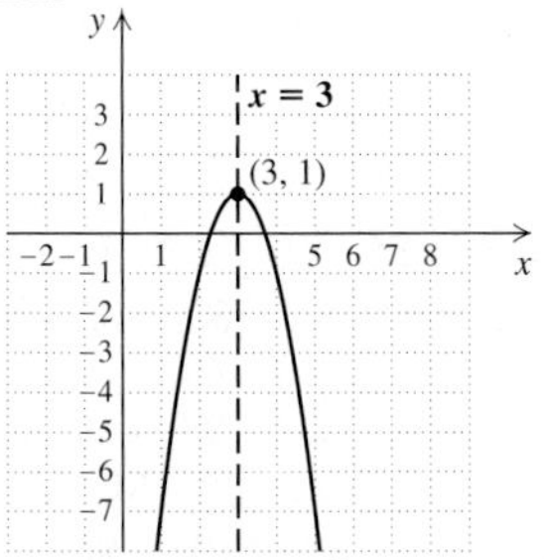

$f(x) = -2(x - 3)^2 + 1$
Maximum: 1

23.
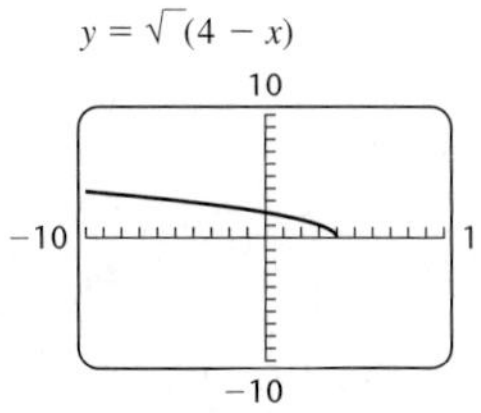

Domain: $(-\infty, 4]$;
range: $[0, \infty)$

24. 20 **25.** $\left\{x \mid x \le \frac{5}{3}\right\}$, or $\left(-\infty, \frac{5}{3}\right]$
26. $\{x \mid x$ is a real number *and* $x \ne 1\}$, or $(-\infty, 1) \cup (1, \infty)$
27. **(a)** Linear; **(b)** 2 **28.** **(a)** Logarithmic; **(b)** -1
29. **(a)** Quadratic; **(b)** $-1, 4$ **30.** **(a)** Exponential; **(b)** no real zeros **31.** $\frac{3}{5}$ **32.** $-\frac{6}{5}, 4$ **33.** $\mathbb{R}$, or $(-\infty, \infty)$
34. $(-1, 1)$ **35.** $(2, -1, 1)$ **36.** 2 **37.** $\pm 2, \pm 5$
38. $(\sqrt{5}, \sqrt{3}), (\sqrt{5}, -\sqrt{3}), (-\sqrt{5}, \sqrt{3}), (-\sqrt{5}, -\sqrt{3})$
39. 1.4037 **40.** 1005 **41.** $\frac{1}{25}$ **42.** $-\frac{1}{2}$
43. $\{x \mid -2 \le x \le 3\}$ or $[-2, 3]$ **44.** $\pm i\sqrt{2}$
45. $-2 \pm \sqrt{7}$ **46.** $\{y \mid y < -5$ *or* $y > 2\}$, or $(-\infty, -5) \cup (2, \infty)$ **47.** $-6, 8$ **48.** 3

49. $r = \dfrac{V - P}{-Pt}$, or $\dfrac{P - V}{Pt}$ **50.** $R = \dfrac{Ir}{1 - I}$

51. $-35x^6y^{-4}$, or $\dfrac{-35x^6}{y^4}$ **52.** 6.3 **53.** $-4y + 17$

54. 280 **55.** $\frac{7}{6}$ **56.** $3a^2 - 8ab - 15b^2$
57. $13x^3 - 7x^2 - 6x + 6$ **58.** $6a^2 + 7a - 5$

59. $9a^4 - 30a^2y + 25y^2$ **60.** $\dfrac{4}{x + 2}$ **61.** $\dfrac{x - 4}{4(x + 2)}$

62. $\dfrac{(x + y)(x^2 + xy + y^2)}{x^2 + y^2}$ **63.** $x - a$ **64.** $(2x - 3)^2$

65. $(3a - 2)(9a^2 + 6a + 4)$ **66.** $(a + 3)(a^2 - b)$
67. $3(y^2 + 3)(5y^2 - 4)$

68. $7x^3 + 9x^2 + 19x + 38 + \dfrac{72}{x - 2}$ **69.** 6.8×10^{-12}

70. $8x^2\sqrt{y}$ **71.** $125x^2y^{3/4}$ **72.** $\dfrac{\sqrt[3]{5xy}}{y}$

73. $\frac{1 - 2\sqrt{x} + x}{1 - x}$ **74.** $26 - 13i$ **75.** $x^2 - 50 = 0$
76. $(2, -3)$; 6 **77.** $\log_a \frac{\sqrt[3]{x^2} \cdot z^5}{\sqrt{y}}$ **78.** $a^5 = c$
79. 3.7541 **80.** 0.0003 **81.** 8.6442 **82.** 0.0277
83. 5 **84.** −121 **85.** 875 **86.** $16\left(\frac{1}{4}\right)^{n-1}$
87. $13{,}440a^4b^6$ **88.** $74.88671875x$ **89.** **(a)** Quadratic; **(b)** $A(t) = 0.1930762006t^2 - 3.682999186t + 39.4379563$
90. **(a)** Exponential; **(b)** $I(t) = 15.58543135(1.082414612)^t$
91. **(a)** Linear; **(b)** $S(t) = 1.889081456t + 14.48700173$
92. More than 23 yr after 1985, or years after 2008
93. All real numbers except 0 and −12 **94.** 81
95. y gets divided by 8 **96.** $-\frac{7}{13} + \frac{2\sqrt{30}}{13}i$ **97.** 84 yr

Appendix

1. 20 ft **3.** 11 g **5.** 24 ft^3 **7.** 12 cm^2 **9.** 12 yd
11. 4 ft-lb/sec **13.** 20 km/hr **15.** 4.4 cm/day
17. 48 cm^2 **19.** 25.1 in. **21.** 250 m
23. Approximately 15.9 cm **25.** 5625 kg-m^2/sec^2, or 5625 Joules **27.** 6.2 km = 6200 m
29. 35 mi/hr = 3080 ft/min **31.** 1 billion sec ≈ 31.7 yr
33. 216 in^2 = 1.5 ft^2 **35.** 5,865,696,000,000 mi/yr
37. 8 ft^2 **39.** Approximately 1337 calories **41.** A
43. B

Photo Credits

1, © Reuters/CORBIS **70, 78,** © Royalty-Free/Corbis **80,** © Reuters/CORBIS **86,** Delta Queen **88,** © Royalty-Free/Corbis **101,** PhotoDisc **129, 130, 149,** © Royalty-Free/Corbis **154,** PhotoDisc **160,** © Royalty-Free/Corbis **178,** Stockbyte **181,** © Royalty-Free/Corbis **182,** © Darrell Gulin/CORBIS **190,** © Royalty-Free/Corbis **218,** Digital Vision **219,** FoodCollection/Getty **235,** © Royalty-Free/Corbis **237,** © Kapoor Baldev/Sygma/Corbis **266,** © Royalty-Free/Corbis **267,** PhotoDisc **269,** Digital Vision **276,** © Royalty-Free/Corbis **291,** Digital Vision **295,** © Elizabeth Kreutz/NewSport/Corbis **308,** © Atlantide Phototravel/Corbis **339,** © Will & Deni McIntyre/CORBIS **344,** © Royalty-Free/Corbis **352,** © Matt Inden/Corbis **421,** © Will & Deni McIntyre/CORBIS **425,** PhotoDisc **437,** © Royalty-Free/Corbis **463,** © Ted Streshinsky/CORBIS **481,** PhotoDisc **483,** © Royalty-Free/Corbis **490,** © George Shelley/CORBIS **504,** © Royalty-Free/Corbis **508,** David Ellenbogen **509, 511,** © Royalty-Free/Corbis **518,** © Lester Lefkowitz/CORBIS **529, 539,** © Gunter Marx Photography/CORBIS **544,** © Craig Lovell/CORBIS **553,** © Royalty-Free/Corbis **568,** © Loren Winters, Visuals Unlimited, Getty Images **584,** © Michael S. Yamashita/CORBIS **609,** © Darrell Gulin/CORBIS **619,** © Royalty-Free/Corbis **621,** Bob Thomason, Stone, Getty Images **635,** Nathaniel S. Butler/NBAE via Getty Images **674,** © Darrell Gulin/CORBIS **699,** © Royalty-Free/Corbis **725,** © Marianna Day Massey/ZUMA/Corbis **760,** AFP/Getty Images **763,** Photo by Andrew Sacks/Time Life Pictures/Getty Images **766, 770, 772, 773 (left),** © Royalty-Free/Corbis **773 (right),** © Darren Staples/Reuters/Corbis **774,** © Sion Touhig/Corbis **776 (left),** © Kit Kittle/CORBIS **776 (right),** AFP/Getty Images **791,** © NASA/Roger Ressmeyer/CORBIS **810,** Patrick Wright, Photographic Services, Clemson University **811,** © NASA/Roger Ressmeyer/CORBIS **839, 859,** © Jose Luis Pelaez, Inc./CORBIS **870,** © Royalty-Free/Corbis

Glossary

Absolute value [1.2] The number of units that a number is from zero on the number line.

***ac*-method** [5.5] A method for factoring a trinomial that uses factoring by grouping.

Additive identity [1.2] The number 0.

Additive inverse [1.2] A number's opposite. Two numbers are additive inverses of each other if their sum is zero.

Algebraic expression [1.1] An expression consisting of a number or variable or a collection of numbers, variables, and operation signs.

Arithmetic sequence [11.2] A sequence in which the difference between any two successive terms is constant.

Arithmetic series [11.2] A series for which the associated sequence is arithmetic.

Ascending order [5.1] A polynomial in one variable written so that the exponents *increase* from left to right.

Associative law for addition [1.3] The statement that when three numbers are added, regrouping the addends gives the same sum.

Associative law for multiplication [1.3] The statement that when three numbers are multiplied, regrouping the factors gives the same product.

Asymptote [10.3] A line that a graph approaches more and more closely as x increases or as x decreases.

Average Most commonly, the mean of a set of numbers.

Axes [1.5] Two perpendicular number lines used to identify points in a plane.

Axis of symmetry [8.6] A line that can be drawn through a graph such that the part of the graph on one side of the line is an exact reflection of the part on the opposite side.

Bar graph A graphic display of data using bars proportional in length to the numbers represented.

Base [1.1] In exponential notation, the number being raised to an exponent.

Binomial [5.1] A polynomial with two terms.

Branches [11.3] The two curves that comprise a hyperbola.

Break-even point [3.8] In business, the point of intersection of the revenue function and the cost function.

Cartesian coordinate system [1.5] A coordinate system in which two perpendicular number lines are used to identify points in a plane.

Circle [10.1] A set of points in a plane that are a fixed distance r, called the radius, from a fixed point (h, k), called the center.

Circle graph A graphic display of data using sectors of a circle to represent percents.

Circumference The distance around a circle.

Closed interval [*a*, *b*] [4.1] The set of all numbers x for which $a \le x \le b$. Thus, $[a, b] = \{x \mid a \le x \le b\}$.

Coefficient [5.1] The numerical multiplier of a variable.

Columns of a matrix [3.6] Vertically aligned elements of a matrix.

Combined variation [6.8] A mathematical relationship in which a variable varies directly and/or inversely, at the same time, with more than one other variable.

Common logarithm [9.3] A logarithm with base 10.

Commutative law for addition [1.3] The statement that when two real numbers are added, the order in which the numbers are written does not affect the result.

Commutative law for multiplication [1.3] The statement that when two real numbers are multiplied, the order in which the numbers are written does not affect the result.

Completing the square [8.1] Adding a particular constant to an expression so that the resulting sum is a perfect square.

Complex number [7.8] Any number that can be written $a + bi$, where a and b are real numbers.

Complex rational expression [6.3] A rational expression that contains rational expressions within its numerator and/or denominator.

Complex-number system [7.8] A number system that contains the real-number system and is designed so that negative numbers do have square roots.

Composite function [9.1] A function in which a quantity depends on a variable that, in turn, depends on another variable.

Composite number A natural number, other than 1, that is not prime.

Compound inequality [4.3] A statement in which two or more inequalities are joined by the word *and* or the word *or*.

Compound interest [8.1] Interest computed on the sum of an original principal and the interest previously accrued by that principal.

Conditional equation [1.6] An equation that is true for some replacements and false for others.

Conic section [10.1] A curve formed by the intersection of a plane and a cone.

Conjugate of a complex number [7.8] The conjugate of a complex number $a + bi$ is $a - bi$.

Conjugates [7.5] Pairs of radical terms, like $\sqrt{a} + \sqrt{b}$ and $\sqrt{a} - \sqrt{b}$, for which the product does not have a radical term.

Conjunction [4.3] A sentence in which two statements are joined by the word *and*.

Consecutive numbers Integers that are one unit apart.

Consistent system of equations [3.1] A system of equations that has at least one solution.

Constant [1.1] A specific number that never changes.

Constant function [2.1] A function given by an equation of the form $f(x) = b$, where b is a real number.

Constant of proportionality [6.8] The constant, k, in an equation of direct or inverse variation.

Contradiction [1.6] An equation that is never true.

Coordinates [1.5] The numbers in an ordered pair.

Counting numbers [1.1] The set of numbers used for counting: $\{1, 2, 3, 4, 5, \ldots\}$.

Cube root [7.1] The number c is called the cube root of a if $c^3 = a$.

Cubic polynomial [5.1] A polynomial in one variable of degree 3.

Curve fitting [2.4] The process of analyzing data to determine if it has a recognized pattern and if it does, fitting an equation to the data.

Data point A given ordered pair of a function, usually found experimentally.

Degree of a monomial [5.1] The sum of the exponents of the variables.

Degree of a polynomial [5.1] The degree of the term of highest degree in a polynomial.

Demand function [3.8] A function modeling the relationship between the price of a good and the quantity of that good demanded.

Denominator The number below the fraction bar in a fraction.

Dependent equations [3.1], [3.4] The equations in a system are dependent if one equation can be removed without changing the solution set.

Descending order [5.1] A polynomial in one variable written so that the exponents *decrease* from left to right.

Determinant [3.7] The determinant of a two-by-two matrix $\begin{bmatrix} a & c \\ b & d \end{bmatrix}$ is denoted by $\begin{vmatrix} a & c \\ b & d \end{vmatrix}$ and represents $ad - bc$.

Direct variation [6.8] A situation that translates to an equation of the form $y = kx$, where k is a nonzero constant.

Discriminant [8.4] The expression $b^2 - 4ac$ from the quadratic formula.

Disjunction [4.3] A sentence in which two statements are joined by the word *or*.

Distributive law [1.3] The statement that multiplying a factor by the sum of two numbers gives the same result as multiplying the factor by each of the two numbers and then adding.

Domain [2.1] The set of all first coordinates of the ordered pairs in a function.

Double root [5.6] A repeated root that appears twice.

Doubling time [9.7] The time necessary for a population to double in size.

Element [1.1] Each object belonging to a set.

Elements of a matrix [3.6] The individual numbers in a matrix.

Elimination method [3.2] An algebraic method that uses the addition principle to solve a system of equations.

Ellipse [10.2] The set of all points in a plane for which the sum of the distances from two fixed points F_1 and F_2 is constant.

Empty set [1.6] The set containing no elements, denoted $\varnothing$ or $\{\ \}$.

Equation [1.1] A number sentence formed by placing an equals sign between two expressions.

Equation of variation [7.5] A equation used to represent direct, inverse, or combined variation.

Equilibrium point [3.8] The point of intersection between the demand function and the supply function.

Equivalent equations [1.6] Equations that have the same solutions.

Equivalent expressions [1.3] Expressions that have the same value for all allowable replacements.

Equivalent inequalities [4.1] Inequalities that have the same solution set.

Evaluate [1.1] To substitute a value for each occurrence of a variable in an expression and carry out the operations.
Exponent [1.1] In expressions of the form a^n, the number n is an exponent. For n a natural number, a^n represents n factors of a.
Exponential decay [9.7] A decrease in quantity over time that can be modeled by an exponential function of the form $P(t) = P_0e^{-kt}$, $k > 0$.
Exponential equation [9.6] An equation in which a variable appears as an exponent.
Exponential function [9.2] A function that can be described by an exponential equation.
Exponential growth [9.7] An increase in quantity over time that can be modeled by an exponential function of the form $P(t) = P_0e^{kt}$, $k > 0$.
Exponential notation [1.1] A representation of a number using a base raised to an exponent.
Extrapolation [2.4] The process of predicting a future value on the basis of given data.

Factor [1.3] *Verb*: to write an equivalent expression that is a product. *Noun*: a multiplier.
Factoring [1.3] The process of rewriting a sum or a difference as a product.
Finite sequence [11.1] A function having for its domain a set of natural numbers: $\{1, 2, 3, 4, 5, \ldots, n\}$, for some natural number n.
Fixed costs [3.8] In business, costs that must be paid whether or not a product is produced.
Focus [10.2] One of two fixed points that determine the points of an ellipse.
FOIL method [5.2] To multiply two binomials by multiplying the First terms, the Outside terms, the Inside terms, and then the Last terms.
Formula [1.6] An equation that uses letters to represent a relationship between two or more quantities.
Fraction notation A number written using a numerator and a denominator.
Function [2.1] A correspondence between a first set, called the *domain*, and a second set, called the *range*, such that each member of the domain corresponds to *exactly one* member of the range.

General term of a sequence [11.1] The nth term, denoted a_n.
Geometric sequence [11.3] A sequence in which the ratio of every pair of successive terms is constant.
Geometric series [11.3] A series for which the associated sequence is geometric.
Graph [1.5] A picture or drawing of data. A line, curve, or collection of points that represents all the solutions of an equation.

Half-life [9.7] The amount of time necessary for half of a quantity to decay.
Half-open interval [4.1] An interval that includes exactly one of two endpoints.
Half-plane [4.5] The graph of a linear inequality.
Horizontal-line test [9.1] If it is impossible to draw a horizontal line that intersects the graph of a function more than once, then that function is one-to-one.
Hyperbola [10.3] The set of all points P in the plane such that the difference of the distance from P to two fixed points is constant.
Hypotenuse [5.8] In a right triangle, the longest side that is opposite the right angle.

Identity [1.6] An equation that is true for all replacements.
Identity property of 0 The statement that the sum of a number and 0 is always the original number.
Identity property of 1 The statement that the product of a number and 1 is always the original number.
Imaginary number [7.8] A number that can be written in the form $a + bi$, where a and b are real numbers and $b \neq 0$.
Imaginary number i [7.8] The square root of -1. That is, $i = \sqrt{-1}$ and $i^2 = -1$.
Inconsistent system of equations [3.1] A system of equations for which there is no solution.
Independent equations [3.1] Equations that are not dependent.
Index [7.1] In the radical $\sqrt[n]{a}$, the number n is called the index.
Inequality [1.1] A mathematical sentence formed when an inequality symbol, $<$, $>$, $\leq$, $\geq$, or $\neq$, is placed between two algebraic expressions.
Infinite geometric series [11.3] The sum of the terms of an infinite geometric sequence.
Infinite sequence [11.1] A function having for its domain the set of natural numbers: $\{1, 2, 3, 4, 5, \ldots\}$.
Input [2.1] An element of the domain of a function.
Integers [1.1] The set of all whole numbers and their opposites.
Interpolation [2.4] The process of estimating a value between given values.
Intersection of A and B [4.3] The set of all elements that are common to *both* A and B.
Interval notation [4.1] The use of a pair of numbers inside parentheses and brackets to represent the set of all

numbers between those two numbers. *See also Closed interval* and *Open interval.*

Inverse relation [9.1] The relation formed by interchanging the members of the domain and the range of a relation.

Inverse variation [6.8] A situation that translates to an equation of the form $y = k/x$, where k is a nonzero constant.

Irrational number [1.1] A real number that when written as a decimal, neither terminates nor repeats.

Isosceles right triangle [7.7] A right triangle in which both legs have the same length.

Joint variation [6.8] A situation that translates to an equation of the form $y = kxz$, where k is a nonzero constant.

Leading coefficient [5.1] The coefficient of the term of highest degree in a polynomial.

Leading term [5.1] The term of highest degree in a polynomial.

Least common denominator [6.2] The least common multiple of the denominators.

Legs [5.8] In a right triangle, the two sides that form the right angle.

Like radicals [7.5] Two radical expressions that have the same indices and radicand.

Like terms [1.3] Terms that have exactly the same variable factors.

Line graph A graph in which quantities are represented as points connected by straight-line segments.

Linear equation [1.5] Any equation whose graph is a straight line and can be written in the form $y = mx + b$, or $Ax + By = C$, where x and y are variables.

Linear equation in three variables [3.4] An equation in the form $Ax + By + Cz = D$, where A, B, C, and D are real numbers.

Linear function [2.2] A function whose graph is a straight line; $f(x) = mx + b$.

Linear inequality [4.5] An inequality whose related equation is a linear equation.

Linear polynomial [5.1] A polynomial in one variable of degree 0 or 1.

Linear regression [2.4] A method of finding a "best" line to fit a set of data.

Logarithmic equation [9.6] An equation containing a logarithmic expression.

Logarithmic function, base *a* [9.3] The inverse of an exponential function, $f(x) = a^x$.

Matrix [3.6] A rectangular array of numbers.

Maximum value [8.6] The largest function value (output) achieved by a function.

Member [1.1] Each object belonging to a set.

Minimum value [8.6] The smallest function value (output) achieved by a function.

Monomial [5.1] A constant, a variable, or a product of a constant and one or more variables.

Motion problem [6.5] A problem that deals with distance, rate, and time.

Multiplicative identity [1.2] The number 1.

Multiplicative inverses [1.2] Reciprocals; two numbers whose product is 1.

Multiplicative property of zero The statement that the product of 0 and any real number is 0.

Natural logarithm [9.5] A logarithm with base e.

Natural numbers [1.1] The set of counting numbers: $\{1, 2, 3, 4, 5, \ldots\}$.

Nonlinear equation [1.5] An equation whose graph is not a straight line.

***n*th root** [7.1] A number c is called the nth root of a if $c^n = a$. In the case of n an even number, c is called *the* nth root of a (or the *principal square root of* a) if c is an nth root and c is nonnegative.

Numerator The number above the fraction bar in a fraction.

One-to-one function [9.1] A function for which different inputs have different outputs.

Open interval (*a*, *b*) [4.1] The set of all numbers x for which $a < x < b$. Thus, $(a, b) = \{x | a < x < b\}$.

Opposite [1.2] The opposite, or additive inverse, of a number a is written $-a$. Opposites are the same distance from 0 on the number line but on different sides of 0.

Opposite of a polynomial [5.1] The *opposite* of a polynomial P can be written $-P$ or, equivalently, by replacing each term in P with its opposite.

Ordered pair [1.5] A pair of numbers of the form (x, y) for which the order in which the numbers are listed is important.

Origin [1.5] The point $(0, 0)$ on a graph where the two axes intersect.

Output [2.1] An element of the range of a function.

Parabola [8.6] A graph of a quadratic function.

Parallel lines [2.3] Lines that extend indefinitely without intersecting.

Pascal's triangle [11.4] A triangular array of coefficients of the expansion $(a + b)^n$ for $n = 0, 1, 2, \ldots$.

Perfect-square trinomial [5.6] A trinomial that is the square of a binomial.
Perpendicular lines [2.3] Lines that form a right angle.
Point–slope form of an equation [2.4] An equation of the type $y - y_1 = m(x - x_1)$, where x and y are variables, m is the slope, and (x_1, y_1) is a point through which the line passes.
Polynomial [5.1] A monomial or a sum of monomials.
Polynomial equation [5.3] An equation in which two polynomials are set equal to each other.
Polynomial function [5.1] A function in which ordered pairs are determined by evaluating a polynomial.
Polynomial inequality [8.9] An inequality that is equivalent to an inequality with a polynomial as one side and 0 as the other.
Prime factorization The factorization of a whole number into a product of its prime factors.
Prime number A natural number that has exactly two different factors: the number itself and 1.
Principal square root [7.1] The nonnegative square root of a number.
Pure imaginary number [7.8] A complex number of the form $a + bi$, in which $a = 0$ and $b \neq 0$.
Pythagorean theorem [5.8] In any right triangle, if a and b are the lengths of the legs and c is the length of the hypotenuse, then $a^2 + b^2 = c^2$.

Quadrants [1.5] The four regions into which the axes divide a plane.
Quadratic equation [8.1] An equation equivalent to one of the form $ax^2 + bx + c = 0$, where $a \neq 0$.
Quadratic formula [8.2] The solutions of $ax^2 + bx + c = 0$, $a \neq 0$, are given by the equation $x = (-b \pm \sqrt{b^2 - 4ac})/(2a)$.
Quadratic function [8.1] A second-degree polynomial function in one variable.
Quadratic inequality [8.9] A second-degree polynomial inequality in one variable.
Quadratic polynomial [5.1] A polynomial in one variable of degree 2.
Quartic polynomial [5.1] A polynomial in one variable of degree 4.

Radical equation [7.6] An equation in which the variable appears in a radicand.
Radical expression [7.1] An algebraic expression in which a radical sign appears.
Radical function [7.1] A function that is described by a radical expression.
Radical sign [7.1] The symbol $\sqrt{\ }$.
Radical term [7.5] A term in which a radical sign appears.
Radicand [7.1] The expression under the radical sign.
Radius [10.1] The distance from the center of a circle to a point on the circle. Also, a segment connecting the center to a point on the circle.
Range [2.1] The set of all second coordinates of the ordered pairs in a function.
Ratio The ratio of a to b is a/b, also written $a:b$.
Rational equation [6.4] An equation containing one or more rational expressions.
Rational expression [6.1] An expression that consists of a polynomial divided by a nonzero polynomial.
Rational inequality [8.9] An inequality containing a rational expression.
Rational number [1.1] A number that can be written in the form p/q, where p and q are integers and $q \neq 0$.
Rationalizing the denominator [7.4] A procedure for finding an equivalent expression without a radical in the denominator.
Rationalizing the numerator [7.4] A procedure for finding an equivalent expression without a radical in the numerator.
Real number [1.1] A number that is either rational or irrational.
Reciprocal [1.2] A multiplicative inverse. Two numbers are reciprocals if their product is 1.
Reflection [8.6] The mirror image of a graph.
Relation [2.1] A correspondence between the domain and the range of a function such that each member of the domain corresponds to *at least one* member of the range.
Remainder theorem [6.7] The remainder obtained by dividing the polynomial $P(x)$ by $x - r$ is $P(r)$.
Repeated root [5.6] The number c is a repeated root of an equation of the form $P(x) = 0$ if $P(x)$ has $(x - c)$ as a factor more than once.
Repeating decimal [1.1] A decimal in which a number pattern repeats indefinitely.
Right triangle [5.8] A triangle that has a 90°, or right angle.
Root of an equation [5.3] Any solution of an equation in one variable.
Root of multiplicity two [5.6] A repeated root that appears twice.
Roster notation [1.1] Set notation in which the elements of the set are listed within { }.
Row-equivalent operations [3.6] Operations used to produce equivalent systems of equations.
Rows of a matrix [3.6] Horizontally aligned elements of a matrix.

Scientific notation [1.4] A number written in the form $N \times 10^m$, where m is an integer, $1 \leq N < 10$, and N is expressed in decimal notation.

Sequence [11.1] A function for which the domain is a set of consecutive positive integers beginning with 1.

Series [11.1] The sum of specified terms in a sequence.

Set [1.1] A collection of objects.

Set-builder notation [1.1] The naming of a set by describing specific conditions under which a number is in the set.

Sigma notation [11.1] The naming of a sum using the Greek letter Σ (capital sigma) as part of an abbreviated form.

Significant digits [1.4] When working with decimals in science, significant digits are used to determine how accurate a measurement is.

Simplify To rewrite an expression in an equivalent, abbreviated form.

Slope [2.2] The ratio of the rise to the run for any two points on a line.

Slope–intercept form of an equation [2.2] An equation of the form $y = mx + b$, where x and y are variables, m is the slope, and $(0, b)$ is the y-intercept.

Solution [1.1] A replacement or substitution that makes an equation or inequality true.

Solution set [1.6] The set of all solutions of an equation, an inequality, or a system of equations or inequalities.

Solve [1.1] To find all solutions of an equation, an inequality, or a system of equations or inequalities; to find the solution(s) of a problem.

Square matrix [3.7] A matrix with the same number of rows and columns.

Square root [7.1] The number c is a square root of a if $c^2 = a$.

Standard form of a linear equation [2.3] An equation of the form $Ax + By = C$, where A, B, and C are real numbers and A and B are not both 0.

Subset [1.1] The set A is a subset of B if every element in A is also in B.

Substitute [1.1] To replace a variable with a number.

Substitution method [3.2] An algebraic method for solving systems of equations.

Supply function [3.8] A function modeling the relationship between the price of a good and the quantity of that good supplied.

Synthetic division [6.7] A method used to divide a polynomial by a binomial of the form $x - a$.

System of equations [3.1] A set of two or more equations, in two or more variables, that are to be solved simultaneously.

Term [1.3] A number, a variable, or a product or a quotient of numbers and/or variables.

Terminating decimal [1.1] A decimal that can be written using a finite number of decimal places.

Total cost [3.8] The money spent to produce a product.

Total profit [3.8] The money taken in less the money spent, or total revenue minus total cost.

Total revenue [3.8] The money taken in from the sale of a product.

Trinomial [5.1] A polynomial with three terms.

Union of A and B [4.3] The collection of all elements belonging to A and/or B.

Variable [1.1] A letter that represents an unknown number.

Variable costs [3.8] In business, costs that vary according to the amount of products produced.

Vertex [8.6] The point at which the graph of a quadratic equation crosses its axis of symmetry.

Vertical asymptote [6.1] When graphing a rational function, one or more lines that the graph approaches, but never touches.

Vertical-line test [2.1] The statement that a graph represents a function if it is impossible to draw a vertical line that intersects the graph more than once.

Whole numbers [1.1] The set of natural numbers and 0: $\{0, 1, 2, 3, 4, 5, \ldots\}$.

x-intercept [2.3] The point at which a graph crosses the x-axis.

x, y-coordinate system [1.5] A coordinate system in which two perpendicular number lines are used to identify points in a plane. The variable x represents the horizontal axis and the variable y represents the vertical axis.

y-intercept [2.3] The point at which a graph crosses the y-axis.

Zeros [5.3] The x-values for which a function $f(x)$ is 0.

Index

F

I

J

L

S

T

Index of Applications

Construction

Consumer

Economics

Education

Engineering

Environment

Miscellaneous

Physics

Social Sciences

Sports/Entertainment

Statistics/Demographics

Technology

Transportation

Frequently Used Symbols and Formulas

SYMBOLS

$=$	Is equal to
$\approx$	Is approximately equal to
$>$	Is greater than
$<$	Is less than
$\geq$	Is greater than or equal to
$\leq$	Is less than or equal to
$\in$	Is an element of
$\subseteq$	Is a subset of
$\lvert x \rvert$	The absolute value of x
$\{x \mid x \ldots\}$	The set of all x such that $x \ldots$
∞	Infinity
$-x$	The opposite of x
$\sqrt{x}$	The square root of x
$\sqrt[n]{x}$	The nth root of x
LCM	Least Common Multiple
LCD	Least Common Denominator
π	Pi, approximately 3.14
i	$\sqrt{-1}$
$f(x)$	f of x, or f at x
$f^{-1}(x)$	f inverse of x
$(f \circ g)(x)$	$f(g(x))$
e	Approximately 2.7
Σ	Summation
$n!$	Factorial notation

FORMULAS

$m = \dfrac{y_2 - y_1}{x_2 - x_1}$	Slope of a line
$y = mx + b$	Slope–intercept form of a linear equation
$y - y_1 = m(x - x_1)$	Point–slope form of a linear equation
$(A + B)(A - B) = A^2 - B^2$	Product of the sum and difference of the same two terms
$\left.\begin{array}{l}(A + B)^2 = A^2 + 2AB + B^2, \\ (A - B)^2 = A^2 - 2AB + B^2\end{array}\right\}$	Square of a binomial
$d = rt$	Formula for distance traveled
$\dfrac{1}{a} \cdot t + \dfrac{1}{b} \cdot t = 1$	Work principle
$s = 16t^2$	Free-fall distance
$y = kx$	Direct variation
$y = \dfrac{k}{x}$	Inverse variation
$x = \dfrac{-b \pm \sqrt{b^2 - 4ac}}{2a}$	Quadratic formula
$P(t) = P_0 e^{kt}, k > 0$	Exponential growth
$P(t) = P_0 e^{-kt}, k > 0$	Exponential decay
$d = \sqrt{(x_2 - x_1)^2 + (y_2 - y_1)^2}$	Distance formula
$\dbinom{n}{r} = \dfrac{n!}{(n - r)!\,r!}$	$\dbinom{n}{r}$ notation

Geometric Formulas

PLANE GEOMETRY

Rectangle
Area: $A = lw$
Perimeter: $P = 2l + 2w$

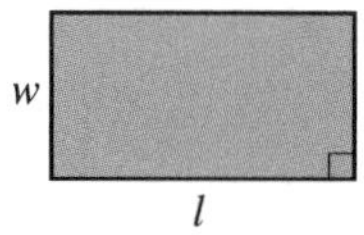

Square
Area: $A = s^2$
Perimeter: $P = 4s$

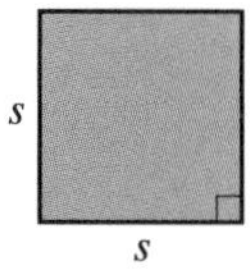

Triangle
Area: $A = \frac{1}{2}bh$

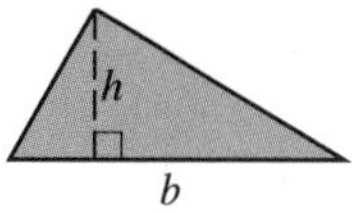

Triangle
Sum of Angle Measures:
$A + B + C = 180°$

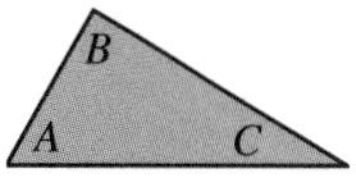

Right Triangle
Pythagorean Theorem (Equation):
$a^2 + b^2 = c^2$

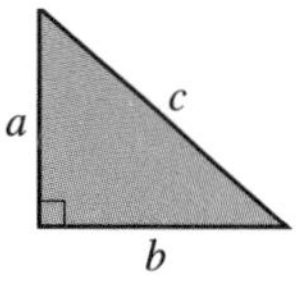

Parallelogram
Area: $A = bh$

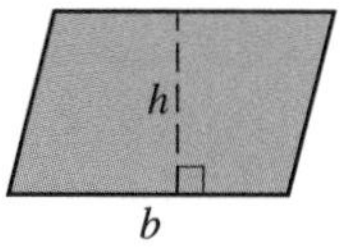

Trapezoid
Area: $A = \frac{1}{2}h(b_1 + b_2)$

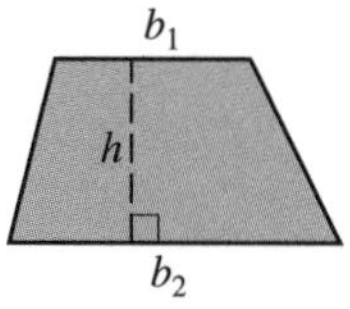

Circle
Area: $A = \pi r^2$
Circumference:
$C = \pi d = 2\pi r$
($\frac{22}{7}$ and 3.14 are different approximations for π)

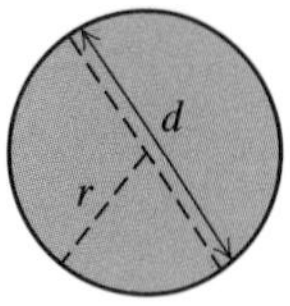

SOLID GEOMETRY

Rectangular Solid
Volume: $V = lwh$

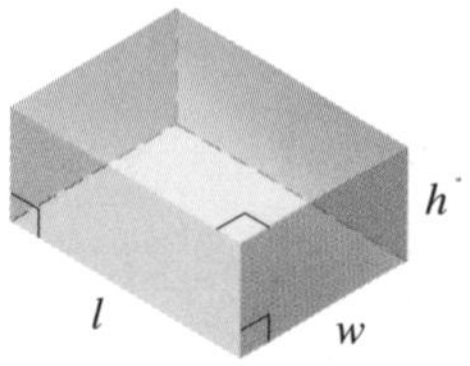

Cube
Volume: $V = s^3$

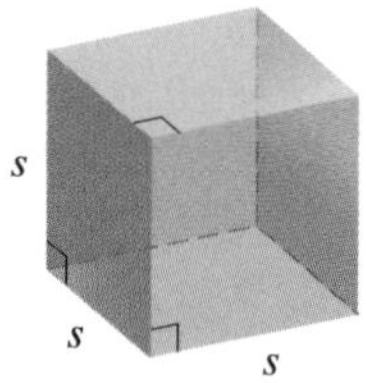

Right Circular Cylinder
Volume: $V = \pi r^2 h$
Total Surface Area:
$S = 2\pi rh + 2\pi r^2$

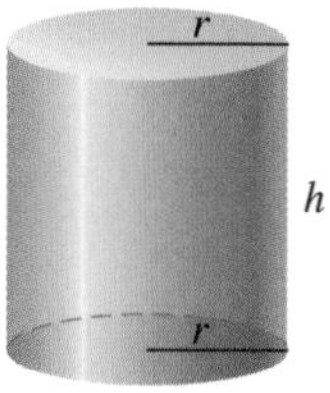

Right Circular Cone
Volume: $V = \frac{1}{3}\pi r^2 h$
Total Surface Area:
$S = \pi r^2 + \pi rs$
Slant Height:
$s = \sqrt{r^2 + h^2}$

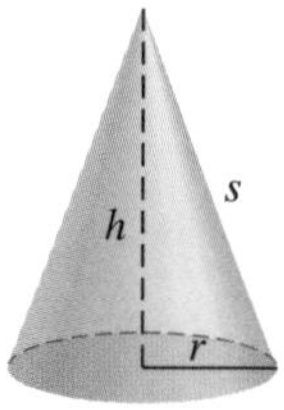

Sphere
Volume: $V = \frac{4}{3}\pi r^3$
Surface Area: $S = 4\pi r^2$

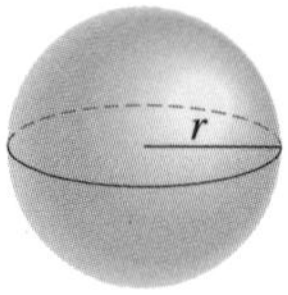